AF823179

Principles of Abstract Interpretation

Principles of Abstract Interpretation

Patrick Cousot

The MIT Press
Cambridge, Massachusetts
London, England

© 2021 Massachusetts Institute of Technology

All rights reserved. No part of this book may be reproduced in any form by any electronic or mechanical means (including photocopying, recording, or information storage and retrieval) without permission in writing from the publisher.

This book was set in New York City by Patrick Cousot. Printed and bound in the United States of America.

Library of Congress Cataloging-in-Publication Data

Names: Cousot, Patrick, 1948– author.
Title: Principles of abstract interpretation / Patrick Cousot.
Description: Cambridge, Massachusetts: The MIT Press, [2021] | Includes bibliographical references and index.
Identifiers: LCCN 2020041256 | ISBN 9780262044905
Subjects: LCSH: Semantic computing. | Abstract data types (Computer science)
Classification: LCC QA76.5913 .C578 2021 | DDC 006—dc23
LC record available at https://lccn.loc.gov/2020041256

10 9 8 7 6 5 4 3 2 1

To Radhia Cousot, my daughters Camille and Rebecca, my sons Laurent and Thibault, and my grandchildren Margot, Sacha, Maxime, and Maël.

Contents

I INTRODUCTION

After a short introduction to abstract interpretation (chapter 1), we provide a very minimal presentation of the core concepts of the theory and applications of abstract interpretation (chapter 3).

These core concepts of abstract interpretation can be presented as one or two classes of programming languages, compilation, program analysis, formal methods, or software engineering courses.

Before introducing abstract interpretation, we recall the basic concepts of set theory which are used almost everywhere in computer science and can be self-studied (chapter 2).

1 Abstract Interpretation and Its Main Applications

Contents

1.1 How to Write Correct Programs?

Since the beginning of computer science, in the 1940s, programmers have been writing programs that are increasingly powerful but still filled with errors. The most common method for tracking errors in programs is debugging. Run the program on as many data as possible and track for results that look incorrect. The obvious limitation is that only a finite number of cases—indeed very few—can be considered among all possible cases. Moreover, some properties cannot be checked in this way—for example, that a program will never stop (unless interrupted).

Therefore, since the 1940s, great programmers have thought about writing programs without errors. Alan Turing [933] was the first to propose a method "to check a routine in the sense of making sure that it is right." He applied it to a large addition routine/procedure of about 10 lines. This method was rediscovered in the 1960s, by Peter Naur [742], Bob Floyd [376], and Tony Hoare [501] and exemplified by programs of about 30 lines.

Things have changed a lot since then. Programs are much larger—millions of lines are not uncommon. The program properties to be checked are much more complicated: for example, security and privacy properties that no one has yet been able to define precisely enough to anticipate all possible attacks and leaks. There are many more programmers, not all of them scientists as great as Alan Turing. Furthermore even computer hardware has bugs. Computer science applications are no longer created for a few scientists but are offered commercially to an extremely large public. Moreover, program bugs can and do kill people.[1] Bugs are becoming very expensive. They can ruin the reputation of companies that are legally ruled against for software defects.[2]

Still, debugging is going on. It is inexpensive, because it is performed mainly by end users who have no alternative to software that needs debugging. At root, programmers are not responsible for their errors. In addition, the no-warranty clause in software license agreements typically states that the licensor makes no guarantees regarding the utility, usability, accuracy, or performance of the software and related documentation or information. The software is provided "as is," with faults, defects, errors and viruses and without warranty of any kind. The licensor does not warrant that the software will be free of bugs, errors, viruses, or other defects, and the licensor shall have no liability of any kind for the use of or inability to use the software, the software content, or any associated service.

This disgraceful Dantean situation is specific to computer science, in particular because of the lack of usable tools for engineers. Progress is needed. The final goal is to check programs for correctness without ever omitting a case that might go wrong. This is the ultimate, perhaps unfeasible, and surely not fully automatizable (see chapter 9) objective of formal methods.

1.2 Abstract Interpretation Theory

Abstract interpretation [235, 241] is a mathematical theory to reason on computer programs, more precisely on the executions of computer programs running on a computer (see [261] for an informal introduction). It formalizes formal methods and allows discussing the guarantees they provide, such as soundness (the conclusions about programs are always correct under suitable explicitly stated hypotheses), completeness (all true facts are provable), or incompleteness (showing the limits of applicability of the formal method). For example, abstract interpretation explains the power and limits of debugging, deductive/proof/verification methods, model checking, static analysis, and so forth.

1.3 Main Applications of Abstract Interpretation

Abstract interpretation is applied mainly to design semantics, proof methods, and static analysis of programs. The semantics of a programming language formally defines all possible executions of its programs at various levels of abstraction. Proof methods can be used to prove (manually or using theorem provers) that the semantics of a program satisfies some specification, which is a property of executions defining what the program is supposed to do. Static analyzers are programs that automatically extract properties of program's semantics (i.e., properties of their executions) using only the program text, without running the programs on a computer.

1. *Boeing's 737 Max: '60s Design Meets '90s Computing Power*, New York Times, April 9, 2019, page A1 of the New York edition.

2. *Toyota Seeks a Settlement for Sudden Acceleration Cases*, New York Times, December 14, 2013, page B3.

1.4 Objective

The objective of this book is to introduce the theory of abstract interpretation and discuss its application to the semantics, specification, verification, and static program analysis of safety and security properties of programs. We focus on principles and breadth of applicability. We do not consider in great depth the algorithmic aspects (data structures and complexity); instead we refer to the literature for the improvement of the reader's knowledge. The book is intended for readers interested in the theory of abstract interpretation, the understanding of formal methods, and the design of verifiers and static analyzers. For those interested only in the use of such static analyzers, *Introduction to Static Analysis* by Xavier Rival and Kwangkeun Yi [837] is a good reference. Over time, the fundamental ideas of abstract interpretation have found a wide range of applications, often unexpected, such as, the formalization of Boolean satisfiability (SAT) and satisfiability modulo theories (SMT) solvers [274, 284]; fuzz testing [851]; robustness and fairness of neural networks [893, 941]; biological networks [140]; and so on. This book provides all the foundations to understand these applications in depth.

1.5 Prerequisites

Readers have varying mathematical and computer science backgrounds. Therefore, all mathematical prerequisites are provided immediately before the chapters in which they are needed. This material can be skipped by knowledgeable readers. For course instruction, it can be given for self-study to prepare for classes in semantics, verification, and static analysis in which it is used. Alternatively, it can be used as a basis for preliminary courses on "mathematics for computer science."[3]

The only prerequisite in computer science is the mastery of any high-level programming language (our examples are given in a subset of C [569], and we use OCaml [628] for implementations). For readers not familiar with compilation, we have included chapter 5 on context-free grammars, parsing, and attributes.

For the necessary background in computability theory, we have introduced undecidability and Rice's theorem in chapter 9.

1.6 Reading Material

The book is organized in parts that may be read independently. In particular, readers interested in static analysis, but not as much in semantics and verification, can start with the reachability semantics of chapter 19 (and skip its calculational design

3. The mathematical chapters are chapter 2, "Basic Set Theory"; facultatively chapter 9, "Undecidability and Rice Theorem"; chapter 10, "Posets, Lattices, and Complete Lattices"; chapter 11, "Galois Connections and Abstraction"; chapter 15, "Fixpoints"; chapter 16, "Fixpoint, Deductive, Inductive, Structural, Coinductive, and Bi-inductive Definitions"; chapter 18, "Fixpoint Abstraction"; chapter 22, "Chaotic Iterations"; chapter 24, "Fixpoint Induction"; chapter 30, "Basic Number Theory"; facultatively chapter 32, "Dynamic Interval Analysis"; chapter 37, "Basic Linear Algebra"; chapter 39, "Graphs"; and chapter 48, "The Herbrand Abstract Domain of Symbolic Terms."

in chapter 20).[4] The various static analyses (and their mathematical prerequisites) can be studied independently of one another.[5]

1.7 Layout and Numbering

Parts, chapters, and sections are numbered using Dewey notation (1, 1.1, 1.2, 2, 2.1, etc.). Equations, definitions, lemmas, theorems, corollaries, remarks, examples, exercises, solutions, and so forth, are all numbered sequentially, restarting from 1 at the beginning of each chapter. This is unconventional, but together with the chapter number, this makes searching in the book much easier. Bibliographic references are numbered sequentially in alphabetic order by the first author's last name and, if necessary, next by the succeeding authors' names and then by the year of publication.

[The main definitions and results are enclosed in boxes. Intermediate definitions and results are indicated with a bracket on their left.

1.8 Notations

Semantics $\boldsymbol{\mathcal{S}}$, abstractions α, concretizations γ, and so forth are basic concepts of abstract interpretation, of which there are numerous instances in the book. Instead of using exactly the same notation for all these instances, we use a different index for each different instance. On one hand, this increases the number of notations. For example, we have over seventy similar but different semantics $\boldsymbol{\mathcal{S}}$ at various levels of abstraction in the book, which means seventy different symbols! For example in chapter 6, we have $\boldsymbol{\mathcal{S}}^{*}$, $\boldsymbol{\mathcal{S}}^{+}$, $\boldsymbol{\mathcal{S}}^{\widehat{*}}$, and $\boldsymbol{\mathcal{S}}^{\dot{*}}$. There are only a few per chapter, so locally, little memorization is necessary. We understand that using the same symbol for all the different instances of the same notion would definitely simplify the notations, but it would require a much more consequent parsing effort to determine which notion we were referring to. Occasionally, we refer to a notion defined several chapters previously, for which the specific denotation has certainly been forgotten. In that case, the symbol index is definitely essential.

4. More precisely read chapter 2, "Basic Set Theory"; chapter 3, "Syntax, Semantics, Properties, and Static Analysis of Expressions"; chapter 4, "Syntax"; facultatively chapter 9, "Undecidability and Rice Theorem"; chapter 10, "Posets, Lattices, and Complete Lattices"; chapter 11, "Galois Connections and Abstraction"; chapter 15, "Fixpoints"; chapter 16, "Fixpoint, Deductive, Inductive, Structural, Coinductive, and Bi-inductive Definitions"; chapter 18, "Fixpoint Abstraction"; chapter 19, "Structural Forward Reachability Semantics," and proceed from chapter 21 through the remaining chapters, skipping chapter 25, "Abstract Reachability / Invariance / Safety Verification Semantics" and chapter 26, "Hoare Logic" on verification.

5. Specifically, chapter 30, "Basic Number Theory," and chapter 31, "Cartesian Congruence Analysis"; chapter 32, "Dynamic Interval Analysis," and chapter 33, "Static Interval Analysis"; chapter 37, "Basic Linear Algebra," and chapter 38, "Linear Equality Analysis"; chapter 39, "Graphs," and chapter 40, "Zone and Octagon Analysis"; chapter 41, "Dataflow Analysis"; chapter 42, "Stateful Prefix Trace Semantics," and chapter 44, "Software Model Checking"; chapter 45, "Flow-Insensitive Static Analysis," and chapter 46, "Points-To Analysis"; chapter 47, "Dependency Analysis"; chapter 48, "The Herbrand Abstract Domain of Symbolic Terms" and chapter 49, "Typing"; chapter 50, "Backward Accessibility Semantics," and chapter 51, "Reduced Forward—Backward Analysis" are independent of one another and can be studied separately, in any order, or can be skipped depending on the specific interests of the reader.

1.9 Exercises and Project

We have provided exercises; with some, we provide solutions, particularly for exercises from which the results are subsequently used in proofs. We have provided a series of exercises, marked "Project," that lead to the design and implementation of an academic static analyzer with facultative parts. The list of these exercises is provided in the project index.

1.10 Repository

To limit the size of the book, some solutions of exercises, some mathematical proofs, the index of names, and a corrigendum are provided in electronic form on https://github.com/PrAbsInt/. The repository also contains slides and videos for each chapter of the book.

1.11 Errors

If you find an error in this book, please send an email to PrAbsInt@gmail.com, and an accompanying emendation is welcome.

1.12 Library of Congress Subject Headings

The LCSH chosen by the Library of Congress are inappropriate. A better choice would be

Subjects: LCSH: Programming languages (Electronic computers) | Programming languages (Electronic computers)–Semantics | Computer programs–Verification

1.13 Acknowledgments

I thank the institutions that have supported my research on abstract interpretation with Radhia over several decades: the University of Grenoble, Centre national de la recherche scientifique (CNRS), University of Lorraine (Metz), Centre de Recherche en Informatique de Nancy (CRIN), École polytechnique, École normale supérieure, MIT Department of Aeronautics and Astronautics, New York University, IMDEA Software, Agence nationale de la recherche (ANR), European Research Council (ERC), and National Science Foundation (NSF).

I thank Alex Hoopes, Marie Lufkin Lee, Elizabeth Swayze, and Roger Woods at MIT Press, the copyeditors Theresa Carcaldi, Virginia Schaefer, and Sherry Gerstein at Westchester Publishing Services, and the readers of parts of preliminary versions: Jade Alglave, Sandrine Blazy, Marc Chevalier, David Delmas, Chaoqiang Deng, Pietro Ferrara, Yu Gai, Pierre Ganty, Roberto Giacobazzi, Saatcioglu Goktug, Klaus Havelund, Manuel Hermenegildo, Xavier Leroy, Dragoş Cristian Lizan, Luc Maranget, Dominique Méry, Jan Midtgaard, Antoine Miné, Andrea Podelski, Francesco Ranzato, Xavier Rival, Axel Simon, Martin Sulzmann, Caterina Urban, Weiyi Wang, Jing Xia, and Enea Zaffanella.

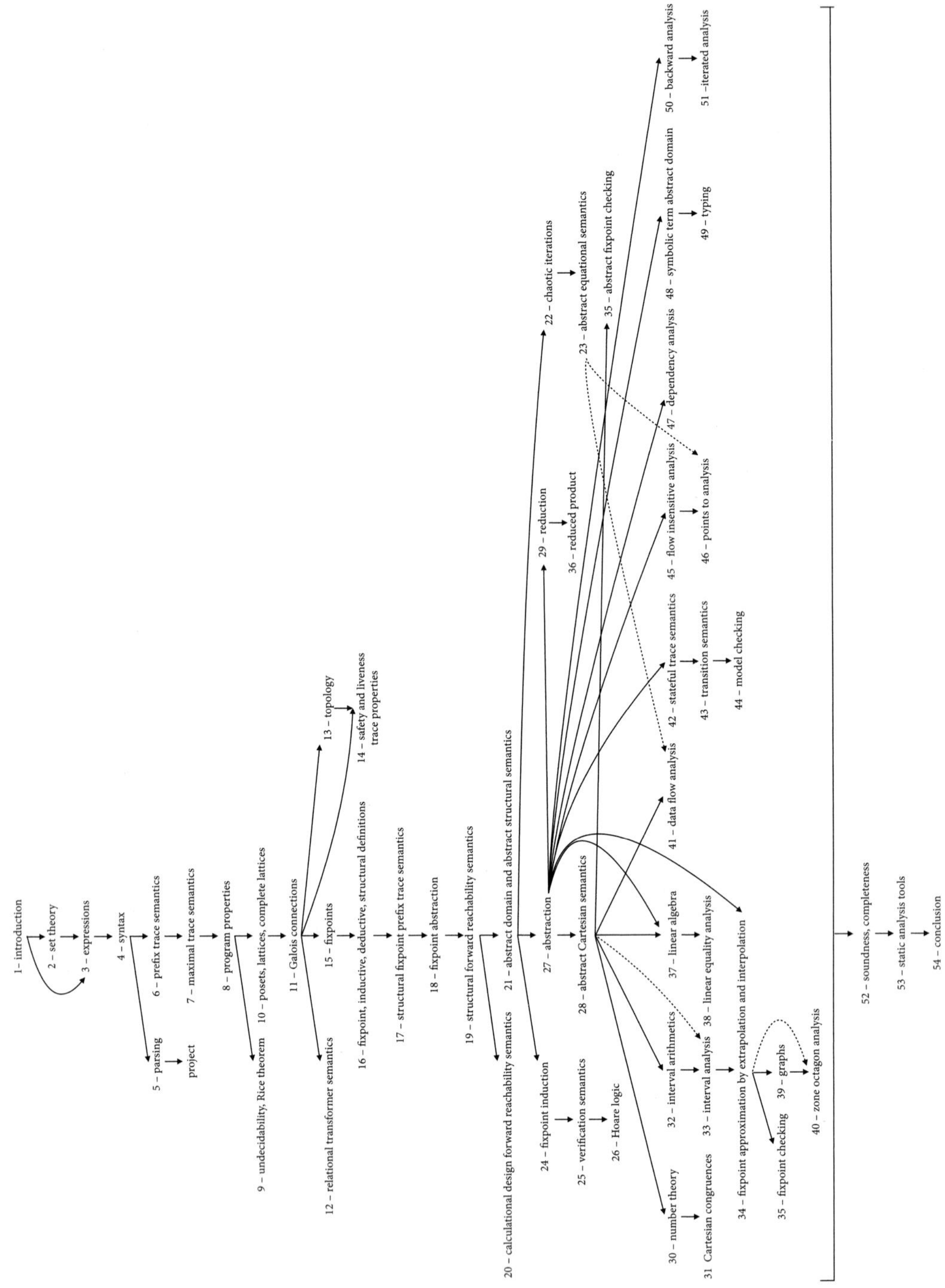
1- introduction
2 – set theory
3 – expressions
4 – syntax
5 – parsing
project
6 – prefix trace semantics
7 – maximal trace semantics
8 – program properties
9 – undecidability, Rice theorem
10 – posets, lattices, complete lattices
11 – Galois connections
12 – relational transformer semantics
13 – topology
14 – safety and liveness trace properties
15 – fixpoints
16 – fixpoint, inductive, deductive, structural definitions
17 – structural fixpoint prefix trace semantics
18 – fixpoint abstraction
19 – structural forward reachability semantics
20 – calculational design forward reachability semantics
21 – abstract domain and abstract structural semantics
22 – chaotic iterations
23 – abstract equational semantics
35 – abstract fixpoint checking
24 – fixpoint induction
25 – verification semantics
26 – Hoare logic
27 – abstraction
28 – abstract Cartesian semantics
29 – reduction
36 – reduced product
30 – number theory
31 Cartesian congruences
32 – interval arithmetics
33 – interval analysis
34 – fixpoint approximation by extrapolation and interpolation
35 – fixpoint checking
39 – graphs
40 – zone octagon analysis
37 – linear algebra
38 – linear equality analysis
41 – data flow analysis
42 – stateful trace semantics
43 – transition semantics
44 – model checking
45 – flow insensitive analysis
46 – points to analysis
47 – dependency analysis
48 – symbolic term abstract domain
49 – typing
50 – backward analysis
51 -iterated analysis
52 – soundness, completeness
53 – static analysis tools
54 – conclusion

2 Basic Set Theory

We assume a basic knowledge of mathematics as introduced in [218, chapters I and II]. We briefly summarize the mathematical prerequisites in set theory [558, ch. 1, Basic set theory] and logic [903, 917, 926].

Contents

2.1 Notations

2.1.1 Term Notations

Let $\mathbb{N}$, $\mathbb{N}^+$, $\mathbb{Z}$, and $\mathbb{R}$ be the set of all natural, strictly positive naturals,[1] integer, and real numbers.[2] Terms are symbolic expressions built out of constants (such as 0 and −42), mathematical variables such as x, y, and so on (denoting immutable but unknown entities such as x in the expression $x + 1$), and operations such as + (sum), −, (difference), × (product), $|\_|$ (absolute value), and so forth. We write $p \triangleq P$ to define the mathematical object p (the *definiendum*) to be equal to the mathematical expression P (the *definiens*), for example, $2 \triangleq 1 + 1$. We use mathematical variables x, p, Q, and so forth to denote mathematical entities that are fixed over time and possibly unknown (see [926, ch. I] "On the use of variables"). We use program variables x, p, Q, and so forth to store values and change them over time by assignments.

2.1.2 Predicate Notations

Let $\mathbb{B} \triangleq \{\mathtt{tt}, \mathtt{ff}\}$ be the set of Boolean truth values $\mathtt{tt}$ (true) and $\mathtt{ff}$ (false).[3] We can state properties of numbers (also called assertions, conditions, constraints, facts, predicates, propositions, requirements, statements, etc.) by logical predicates for which the value

1. Other notations for positive naturals are $\mathbb{N}^*$, $\mathbb{N}_1$, $\mathbb{N}_{>0}$, $\mathbb{Z}^+$, and so forth. We use $\mathbb{N}^+$ because a superscript or subscript * is often interpreted as "zero or more" whereas $^+$ is often interpreted as "one or more." Originally, Richard Dedekind [302] and Giuseppe Peano [787] used $\mathbb{N}$ for positive naturals and $\mathbb{N}_0$ to include 0 [788].

2. The series of historical footnotes aims at showing that good notations may take decades if not centuries to emerge.

3. This notation is not classic. For example George Boole [131] uses 0 and 1, Giuseppe Peano uses V and Λ, David Hilbert and Wilhelm Friedrich Ackermann [497] use $\curlyvee$ and $\curlywedge$, Stephen Cole Kleene [582] uses $\mathfrak{t}$ and $\mathfrak{f}$, Alfred Tarski [926] uses **T** and **F**, and so forth.

is either tt or ff. Relations such as $(x-1)+1=x$ or $x<x+1$ are basic logical predicates relating terms by relation operations = (equal), < (less than), $\leqslant$ (less than or equal), and so on. We can combine predicates using logical connectives $\vee$ (disjunction[4]), $\wedge$ (conjunction[5]), $\neg$ (negation[6]), $\Rightarrow$ (implication,[7] with inverse $\Leftarrow$), and $\Leftrightarrow$ (if and only if),[8] as in $((x-1)+1=x)\wedge\neg(x>x+1)$. This is true for all possible values x so $\forall x\in\mathbb{Z}\,.\,(x-1)+1=x$ where $\forall$ is the universal quantification.[9] This property also holds on reals, but not on naturals because $\exists x\in\mathbb{N}\,.\,(x-1)\notin\mathbb{N}$ where $\exists$ is existential quantification[10] and $\notin$ means "does not belongs to" (by choosing $x=0$). Of course, $\forall x\in\mathbb{N}\,.\,(x+1)-1=x$ holds for naturals but not $\forall x\in\mathbb{N}\,.\,(x-1)+1=x$ (because 0 is an exception). We use $\exists!$ for unique existential quantification (i.e., $\exists!x\,.\,P(x)$ if and only if $\exists x\,.\,P(x)\wedge\forall x,y\,.\,(P(x)\wedge P(y))\Rightarrow(x=y)$).[11]

If $P\Rightarrow Q$ then P is called the *premise* and Q the *conclusion*. We say that P is a *sufficient condition* for Q to hold (i.e., to be true). If $Q\Rightarrow P$ then we say that P is a *necessary condition* for Q to hold. If $P\Leftrightarrow Q$ then we say that P is a *necessary and sufficient condition* for Q. We say that $P\Rightarrow Q$ *holds vacuously* when P is ff or Q is tt.

Exercise 2.1 Provide the truth tables of disjunction $\vee$, negation $\neg$, and implication $\Rightarrow$ (given all possible arguments $x,y\in\mathbb{B}$, such a truth table specifies the value $x\vee y\in\mathbb{B}$).

Exercise 2.2 Using the truth tables of exercise 2.1, prove that for all $x,y\in\mathbb{B}$, $x\Rightarrow y$ if and only if $\neg x\vee y$ if and only if $\neg(x\wedge\neg y)$.

4. George Boole [131] uses + (by analogy with addition).

5. George Boole [131] uses juxtaposition, so $x\wedge y$ is written xy (by analogy with a multiplication); Alfred Whitehead and Bertrand Russell use $\cdot$, which confusingly is also used as a delimiter [978, v. I, ch. I, 9–10].

6. George Boole [131, prop. III] writes $1-P$ for $\neg P$, Eliakim Hastings Moore writes $-P$ [713], Alfred Whitehead and Bertrand Russell [978] use $\sim$, Jacques Herbrand [487] refers to Alfred Whitehead and Bertrand Russell but uses ∞ instead. Following David Hilbert and Wilhelm Friedrich Ackermann [497], $\neg P$ is sometimes denoted $\overline{P}$.

7. Charles Sanders Peirce [790] uses $\prec$, Giuseppe Peano [787] uses Ͻ, Alfred Whitehead and Bertrand Russell [978], and Stephen Cole Kleene [582] uses $\supset$. This is unfortunate because if $P\Rightarrow Q$ then $\{x\mid P(x)\}\subseteq\{x\mid Q(x)\}$ (see section 2.3.2). David Hilbert and Wilhelm Friedrich Ackermann [497] and Alfred Tarski [926] use $\rightarrow$.

8. Alfred Whitehead and Bertrand Russell [978] use $\equiv$.

9. Introduced by Charles Sanders Peirce [790] as Any or $\prod$. David Hilbert and Wilhelm Friedrich Ackermann [497] as well as Alfred Whitehead and Bertrand Russell [978] write $(x)P$ for $\forall x\,.\,P$. Kazimierz Kuratowski and Andrzej Mostowski [606] use $\bigwedge_x P$. Alfred Tarski in [926] writes *for any*.

10. Introduced by Charles Sanders Peirce [790] as Some or Σ; for example $\forall P\,.\,\exists x\in P\ldots$. is written $\prod_P\sum_{x_p}\ldots$. David Hilbert and Wilhelm Friedrich Ackermann [497] write $(Ex)P$ whereas Alfred Whitehead and Bertrand Russell [978] write $(\exists x)P$ for $\exists x\,.\,P$. Kazimierz Kuratowski and Andrzej Mostowski [606] use $\bigvee_x P$. Alfred Tarski [926] writes *there exists*.

11. i.e. stands for latin id est meaning "that is to say."

2.1.3 Set Notations

Naive set theory understands sets S as collections of elements x that belong to that set S, written $x \in S$.[12] The empty set $\varnothing$[13] has no elements so $\forall x \,.\, x \notin \varnothing$. A singleton is a set $\{x\}$ with only one element x, so $y \in \{x\} \Leftrightarrow y = x$. If $p(x)$ is a predicate on x, then the *set-builder notation* $\{x \mid p(x)\}$ denotes the set of all elements x satisfying the predicate p. Therefore, if $S \triangleq \{x \mid p(x)\}$ (i.e., S is defined in intension to be the set $\{x \mid p(x)\}$), then $x \in S$ if and only if $p(x)$ holds, that is, p is true of x. $S = \varnothing$ is empty if and only if $\forall x \,.\, \neg p(x)$.[14] For example, $2\mathbb{N} \triangleq \{x \in \mathbb{N} \mid \exists k \,.\, x = 2k\}$ is the set of all even natural numbers, and $2\mathbb{Z} + 1 \triangleq \{x \in \mathbb{Z} \mid \exists k \,.\, x = 2k + 1\}$ is the set of all odd integers. More generally, $d(a) \triangleq \{f(a, b, \ldots) \mid p(a, b, \ldots, x, \ldots)\}$ is a shorthand for the definition of function d as $\forall a \,.\, d(a) \triangleq \{f(a', b, \ldots) \mid \exists x \,.\, p(a', b, \ldots, x, \ldots) \wedge a' = a\}$. The free variables $a', b, \ldots, x, \ldots, a$ of the term $p(a', b, \ldots, x, \ldots) \wedge a' = a$ are existentially quantified unless they are free in $f(a', b, \ldots)$ or are a global variable a.[15] Contradictions such as the circular definition $S \triangleq \{x \mid x \notin S\}$ should be avoided. The cardinality $|S|$ of a finite set is the number of its elements (so, $|\varnothing| = 0$ and $|\{x\}| = 1$). For infinite sets, the cardinality is (informally) a measure of their size. For example the set $\mathbb{Z}$ of integers and the set $2\mathbb{Z}$ of even integers have the same cardinality (because there is a bijection $n \mapsto 2n$ between the two). We write $\{x \in X \mid p(x)\}$ for $\{x \mid x \in X \wedge p(x)\}$ and $\{x \in S\}$ for $\{x \mid x \in S\}$, which is S. We write $S \subseteq S'$ to mean that S is a subset of S', that is, $\forall x \in S \,.\, x \in S'$.[16] Therefore $\varnothing \subseteq S'$ is always true because there is no $x \in \varnothing$ so that all of them (and there is none) belong to any set. The union or join of sets is $\cup$ (defined as $x \in S \cup S' \Leftrightarrow x \in S \vee x \in S'$). The intersection or meet of sets is $\cap$. The difference of sets is $S \setminus S' \triangleq \{x \in S \mid x \notin S'\}$. The complement of a set S with respect to a set U (generally understood from the context) is $\neg S \triangleq U \setminus S$. Augustus De Morgan's laws state that $\neg(S \cup S') = (\neg S) \cap (\neg S')$ and $\neg(S \cap S') = (\neg S) \cup (\neg S')$. The powerset $\wp(S)$ of a set S is the set of all its subsets so $\wp(S) \triangleq \{S' \mid S' \subseteq S\}$.[17] $\wp_f(S)$ is the set of all finite subsets of S, so $\wp_f(S) \triangleq \{S' \mid S' \subseteq S \wedge |S| \in \mathbb{N}\}$. For example, $\wp(\{0, 1\}) = \wp_f(\{0, 1\}) = \{\varnothing, \{0\}, \{1\}, \{0, 1\}\}$.

Exercise 2.3 Formally define the property that a set S is a singleton.

12. $\in$ can be remembered as the "e" in "is an <u>e</u>lement of." The symbol $\in$ was introduced by Giuseppe Peano [787]. Ernst Schröder uses the same notation $\not\leqslant$ both for $\in$ and $\subseteq$, [866] so formulas must be parsed to sort between elements and sets! In fact $x \in X$ is $\{x\} \subseteq X$ so $\in$ is abstracted away in favor of $\subseteq$ in order theory; see chapter 10.

13. The denotation of the empty set by the letter $\varnothing$ from the Norwegian alphabet (similar to the German Ö) was introduced by the French mathematician André Weil.

14. Formally, [917, p. 34] defines $\{x \mid p(x)\} = y \Leftrightarrow (\forall x \,.\, (x \in y) \Leftrightarrow p(x) \wedge y \text{ is a set}) \vee (y = \varnothing \wedge \neg(\exists S \,.\, \forall x \,.\, x \in S \Leftrightarrow p(x)))$. "The point of the second member of the disjunction of the definiens is to put $\{x \mid p(x)\}$ equal to the empty set if there is no nonempty set having as members just those entities with property p."

15. A definition of the form $d(\vec{y}) \triangleq \{f(\vec{x}', \vec{y}) \mid P(\vec{x}', \vec{x}, \vec{y})\}$ has the global variables $\vec{y}$ of $f(\vec{x}', \vec{y})$ and $P(\vec{x}', \vec{x}, \vec{y})$ bound to the variables $\vec{y}$ in the definition of $d(\vec{y})$. The variables $\vec{x}'$ in $P(\vec{x}', \vec{x}, \vec{y})$ are bound to those of $f(\vec{x}', \vec{y})$ whereas $\vec{x}$ is free in $P(\vec{x}', \vec{x}, \vec{y})$ since it appears neither in $f(\vec{x}', \vec{y})$ nor in $d(\vec{y})$, nor (by assumption) under quantifiers in $P(\vec{x}', \vec{x}, \vec{y})$ so is implicitly existentially quantified. This is made more explicit by writing $d(\vec{y}) \triangleq \{f(\vec{x}', \vec{y}') \mid \exists \vec{x} \,.\, P(\vec{x}', \vec{x}, \vec{y}') \wedge y = y'\}$.

16. If $\forall x \,.\, p(x) \Rightarrow q(x)$ then $\{x \mid p(x)\} \subseteq \{x \mid a(x)\}$ so using $\supset$ for implication as [582] is confusing.

17. $\wp(S)$ is sometimes denoted 2^S. This is because a subset X of S can be represented by a Boolean *characteristic function* $\mathbb{c}_(x) \triangleq x \in X$ characterizing which elements $x \in S$ belong to the subset X of S, the booleans are isomorphic to $2 = \{0, 1\}$ (in John Von Neumann theory of ordinals [749]), and the set of functions from A to B which we denote as $A \to B$ is also denoted B^A.

Exercise 2.4 Formally define set intersection $\cap$.

Exercise 2.5 Prove De Morgan law $\neg(S \cup S') = (\neg S) \cap (\neg S')$.

More generally, by complement duality, if two predicates P and Q involving only variables X, $\vee/\exists$ (disjunction) or $\cup$ (union), $\wedge/\forall$ (conjunction) or $\cap$ (intersection), and $\neg$ (negation or complement) are logically equal $P \Leftrightarrow Q$ then their dual complements (replacing variables X by their negation $\neg X$, $\vee/\exists/\cup$ by $\wedge/\forall/\cap$, and $\wedge/\forall/\cap$ by $\vee/\exists/\cup$, and eliminating double negation $\neg\neg$) are also logically equal [497, ch I, § 6, and ch III, § 8]. This follows immediately from the fact that the dual complement of P is $\neg P$.

Exercise 2.6 Assume the set $\mathbb{R}$ of reals has already been defined (for example, by Tarski's axiomatization of the reals [926, section 61 or 63]). Provide a definition of $\mathbb{N}$ and then $\mathbb{Z}$ as subsets of $\mathbb{R}$.

Remark 2.7 (on finiteness) Because programs run on finite machines with a finite lifetime, one may argue that everything concerning the programs and their executions must be finite [508]. However, infinite sets can be considered as the limit of finite sets for which cardinality is bounded but the bound is unknown. It is often easier to reason on infinite sets than on finite sets which cardinality has an unknown bound. A classical example is fairness or unbounded nondeterminism in parallelism [328, 625, 778], in which verification methods not explicitly considering the unknown bound are simpler. Another argument is that computations are currently performed on networks, in which a dying machine is replaced by another one without affecting the well-behaved operation of the network, hence of program executions that can, in principle, last forever. Distributed memory is also finite, but in contrast to a machine, its size bound is not known. Considering that the network has a finite lifetime is impossible, because it is necessary to give a lower bound to be able to prove anything. This lower bound is itself completely unknown. Therefore, considering an infinite lifetime, hence infinite sets, is a reasonable simplification to get rid of these minimal and maximal unknown resource and lifetime bounds.

2.1.4 Interval Notation

Given a lower and an upper integer bound $\ell, u \in \mathbb{Z}$, we let the closed interval be $[\ell, u] \triangleq \{z \in \mathbb{Z} \mid \ell \leqslant z \leqslant u\}$ while the opened intervals are $]\ell, u] \triangleq \{z \in \mathbb{Z} \mid \ell < z \leqslant u\}$, $[\ell, u[\triangleq \{z \in \mathbb{Z} \mid \ell \leqslant z < u\}$, and $]\ell, u[\triangleq \{z \in \mathbb{Z} \mid \ell < z < u\}$. Hence $[\ell, u] = \varnothing$ when $u < \ell$.

2.2 Mathematical Constructions

Mathematicians (followed by computer scientists) construct new objects and operations on these objects from previously defined ones (sets for mathematicians, and bits for computer scientists). For example, Kazimierz Kuratowski [602, def. V, p. 171] defines a pair (sometimes called an ordered pair) as a set $\langle x, y\rangle \triangleq \{\{x\}, \{x, y\}\}$. The first coordinate is $\langle x, y\rangle_1$ defined by $z = \langle x, y\rangle_1 \Leftrightarrow \forall S \in \langle x, y\rangle : z \in S$ (so that $\langle x, y\rangle_1 = x$). The second coordinate is $\langle x, y\rangle_2$ defined by $z = \langle x, y\rangle_2 \Leftrightarrow (\exists S \in \langle x, y\rangle : z \in S) \wedge (\forall S_1, S_2 \in \langle x, y\rangle : S_1 \neq S_2 \rightarrow (z \notin S_1 \vee z \notin S_2))$ (so that $\langle x, y\rangle_2 = y$).

Exercise 2.8 Show that $\langle x, x\rangle_1 = \langle x, x\rangle_2 = x$.

2.2.1 Cartesian Product

The Cartesian product of two sets S_1 and S_2 is the set $S_1 \times S_2 \triangleq \{\langle x, y\rangle \mid x \in S_1 \wedge y \in S_2\}$ of pairs of elements of S_1 and S_2. The set $S \times S$ is sometimes written S^2. This generalizes to n-ary tuples, $n \in \mathbb{N}^+$, as $\langle x_1, x_2, \ldots, x_n\rangle \in S_1 \times S_2 \times \ldots \times S_n$ and S^n when $S_1 = S_2 \ldots = S_n = S$. The selection of the ith element of the tuple $\langle x_1, x_2, \ldots, x_n\rangle$ uses the index subscript notation $\langle x_1, x_2, \ldots, x_n\rangle_i \triangleq x_i$ or the function notation $\langle x_1, x_2, \ldots, x_n\rangle(i) \triangleq x_i$. When the index is not an integer, we write $\prod_{i\in\Delta} x_i$, which we understand as a function from the set Δ of indexes to S.

2.2.2 Relations

A binary relation r on sets S_1 and S_2 is a set of pairs $\langle x, y\rangle \in S_1 \times S_2$ of related elements $x \in S_1$ and $y \in S_2$ so $r \in \wp(S_1 \times S_2)$. y is said to be an image of x by r. We write $x \mathrel{r} y$, $x \xrightarrow{r} y$, or $r(x, y)$ for $\langle x, y\rangle \in r$. For example, equality on $\mathbb{Z}$ is $= \triangleq \{\langle x, x\rangle \mid x \in \mathbb{Z}\}$ whereas "less than or equal" is $\leqslant \triangleq \{\langle x, y\rangle \in \mathbb{Z} \times \mathbb{Z} \mid \exists z \in \mathbb{N} \,.\, x + z = y\}$, and the identity relation on a set S is $\mathbb{1}_S \triangleq \{\langle x, x\rangle \mid x \in S\}$. The domain of a relation r on sets S_1 and S_2 is $\mathsf{dom}(r) \triangleq \{x \in S_1 \mid \exists y \in S_2 \,.\, \langle x, y\rangle \in r\}$; the codomain is $\mathsf{cod}(r) \triangleq \{y \in S_2 \mid \exists x \in S_1 \,.\, \langle x, y\rangle \in r\}$; and the field is $\mathsf{fld}(r) \triangleq \mathsf{dom}(r) \cup \mathsf{cod}(r)$. The left (respectively right) restriction $r \rceil S$ (respectively $r \lceil S$) of a relation $r \in \wp(S_1 \times S_2)$ to a set S is $\{\langle x, y\rangle \in r \mid x \in S\}$ (respectively $\{\langle x, y\rangle \in r \mid y \in S\}$), which is equal to $\{\langle x, y\rangle \in r \mid x \in S_1 \cap S\}$ (respectively $\{\langle x, y\rangle \in r \mid y \in S_2 \cap S\}$). The composition of relations is $r_1 \mathbin{⨾} r_2 \triangleq \{\langle x, z\rangle \mid \exists y \,.\, \langle x, y\rangle \in r_1 \wedge \langle y, z\rangle \in r_2\}$. The inverse of a relation $r \in \wp(S_1 \times S_2)$ is $r^{-1} \triangleq \{\langle y, x\rangle \mid \langle x, y\rangle \in r\}$. $\langle \wp(S \times S), ⨾, \mathbb{1}_S\rangle$ is an example of *monoid* that is a mathematical structure $\langle \mathcal{S}, \oplus, 1\rangle$, where $\oplus$ is a binary relation on the set $\mathcal{S}$, which is associative (i.e., $(x \oplus y) \oplus z = x \oplus (y \oplus z)$ with neutral element 1 (i.e., $1 \oplus x = x \oplus 1 = x$).

Exercise 2.9 Show that $\langle \wp(S \times S), ⨾, \mathbb{1}_S, {}^{-1}\rangle$ is not a *group* that is a mathematical structure $\langle \mathcal{S}, \oplus, 1, {}^{-1}\rangle$ where $\oplus$ is a binary relation on the set $\mathcal{S}$, with neutral element 1, and inverse ${}^{-1}$ (i.e., $x \oplus x^{-1} = x^{-1} \oplus x = 1$).

2.2.3 Equivalence and Partial Order Relations

An *equivalence relation* $\equiv$ on a set S is *reflexive* $(\forall x \in S \,.\, x \equiv x)$, *symmetric* $(\forall x, y \in S \,.\, (x \equiv y) \Leftrightarrow (y \equiv x))$, and *transitive* $(\forall x, y, z \in S \,.\, (x \equiv y \wedge y \equiv z) \Rightarrow (x \equiv z))$. The equivalence class of an element $x \in S$ is the set $[x]_\equiv \triangleq \{y \in S \mid y \equiv x\}$ of all elements of S that are equivalent to x. The quotient $S|_\equiv \triangleq \{[x]_\equiv \mid x \in S\}$ is the set of all equivalence classes.

A *partial order* $\leqslant$ on a set S is reflexive, *antisymmetric* $(\forall x, y \in S \,.\, (x \leqslant y \wedge y \leqslant x) \Rightarrow (x = y))$ and transitive. The *strict* partial order is $x < y \triangleq (x \leqslant y) \wedge (x \neq y)$. An order is *total* if and only if any two elements of S are comparable $(\forall a, b \in S \,.\, (a \leqslant b) \vee (b \leqslant a))$. A set S equipped with a partial order $\leqslant$ is called a *poset* $\langle S, \leqslant\rangle$.

A *pointwise* definition of a relation is $\dot{r} \triangleq f, g \mapsto \forall x \,.\, r(f(x), g(x))$. The functional pointwise definition is $\ddot{r} \triangleq f, g \mapsto \forall X \,.\, \dot{r}(f(X), g(X)) = f, g \mapsto \forall X \,.\, \forall x \,.\, r(f(X)x, g(X)x)$, and so forth. For example, $f \mathrel{\dot{\sqsubseteq}} g$ is $\forall x \,.\, f(x) \sqsubseteq g(x)$. If $\langle\langle L_i, \sqsubseteq_i\rangle, i \in \Delta\rangle$ is a family of posets (see section 2.2.5), then the *componentwise* order $\dot{\sqsubseteq}$ on the Cartesian product $\prod_{i\in\Delta} L_i$ is $\prod_{i\in\Delta} x_i \mathrel{\dot{\sqsubseteq}} \prod_{i\in\Delta} y_i \triangleq \forall i \in \Delta \,.\, x_i \sqsubseteq_i y_i$. The componentwise order $\dot{\sqsubseteq}$ is sometimes denoted $\prod_{i\in\Delta} \sqsubseteq_i$ or $\sqsubseteq_1 \times \sqsubseteq_2$ when $\Delta = \{1, 2\}$.

Exercise 2.10 Prove that equality (=) on a nonempty set S is the only relation on that set S that is both an equivalence and a partial order.

Exercise 2.11 (lexicographic order) Let $\langle A, \leqslant\rangle$ and $\langle B, \sqsubseteq\rangle$ be two posets. Define $\langle a, b\rangle \leqslant\times\sqsubseteq \langle a', b'\rangle \triangleq (a < a') \vee (a = a' \wedge b \sqsubseteq b')$. Show that the Cartesian product $\langle A \times B, \leqslant\times\sqsubseteq\rangle$ is a poset. Show that if $\leqslant$ and $\sqsubseteq$ are total orders then $\leqslant\times\sqsubseteq$ is a total order.

More details on equivalence and partial order relations appear in chapter 10.

2.2.4 Partial and Total Functions

A relation $r \in \wp(S_1 \times S_2)$ is functional when any element of S_1 is in relation with at most one element of S_2 by r that is $\forall x \in S_1 . \forall y, y' \in S_2 . (\langle x, y\rangle \in r \wedge \langle x, y'\rangle \in r) \Rightarrow (y = y')$. In that case, we write $r \in \wp_{\mathsf{F}}(S_1 \times S_2)$. A relation $r \in \wp(S_1 \times S_2)$ is total when any element of S_1 has at least one image by r that is $\forall x \in S_1 . \exists y \in S_2 . \langle x, y\rangle \in r$.

A partial function $f \in S_1 \nrightarrow S_2$ of S_1 into S_2 is a functional relation r on sets S_1 and S_2 such that any element $x \in S_1$ has at most an image, written $f(x)$ when it exists (so $f(x)$ is the unique y such that $\langle x, y\rangle \in r$). We write $f \triangleq x \mapsto\!\!\!\!\!\nrightarrow e(x)$ when $\forall x \in \mathsf{dom}(f) . f(x) \triangleq e(x)$ and $f \triangleq x \in S \mapsto\!\!\!\!\!\nrightarrow e(x)$ when $S = \mathsf{dom}(f)$. We sometimes use the subscript notation f_x for $f(x)$. The composition of partial functions is $f \circ g = x \mapsto\!\!\!\!\!\nrightarrow f(g(x))$. Considered as relations, this is $g \fatsemi f$.

A total function $f \in S_1 \rightarrow S_2$ has $\mathsf{dom}(f) = S_1$, that is, is everywhere defined on S_1, which we write $x \in S_1 \mapsto f(x)$. If $S_1 = S_2 = S$ then $f \in S \rightarrow S$ is often called an *operator* on S or an *S-transformer*. A function $F \in (S_1 \rightarrow S_2) \rightarrow (S_1' \rightarrow S_2')$ taking functions as parameters is called a *functional*.

The right image of a relation $r \in \wp(S_1 \times S_2)$ is the function $x \in S_1 \mapsto \{y \in S_2 \mid \langle x, y\rangle \in r\} \in S_1 \rightarrow \wp(S_2)$.

A total function $f \in S_1 \rightarrow S_2$ is injective/one-to-one when $\forall x_1 \in S_1 . \forall x_2 \in S_2 . x_1 \neq x_2 \Rightarrow f(x_1) \neq f(x_2)$ (written $f \in S_1 \rightarrowtail S_2$). It is surjective/onto if and only if $\forall y \in S_2 . \exists x \in S_1 . f(x) = y$ (written $f \in S_1 \twoheadrightarrow S_2$). It is bijective if and only if both injective and surjective (written $f \in S_1 \rightarrowtail\!\!\!\!\twoheadrightarrow S_2$). Sets S_1 and S_2 are isomorphic when there exists a bijection of S_1 onto S_2.

It may be that the returned value of a function always belong to a set that depends upon one of its parameters (this is called a dependent type in computer science [166]). For example $f \in n \in \mathbb{N} \rightarrow \{k \in \mathbb{N} \mid k \geqslant n\}$ specifies a function $f \in \mathbb{N} \rightarrow \mathbb{N}$ such that $\forall n \in \mathbb{N} . f(n) \geqslant n$. More generally, $f \in x \in S_1 \rightarrow S_2(x)$ is the function $f \in x \in S_1 \rightarrow \bigcup_{x \in S_1} S_2(x)$ such that $\forall x \in S_1 . f(x) \in S_2(x)$ where S_2 maps each $x \in S_1$ to a set $S_2(x)$. Up to an isomorphism $f \in \prod_{x \in S_1} S_2(x)$.

Exercise 2.12 Show that $\mathbb{N}$, $2\mathbb{N}$, and $2\mathbb{N}+1$ are isomorphic.

A set S is *enumerable* if and only if there exists a bijection $\iota \in S \rightarrowtail\!\!\!\!\twoheadrightarrow \mathbb{N}$ between S and the naturals.

A function may return a function, and $f \in A \rightarrow B \rightarrow C$ stands for $f \in A \rightarrow (B \rightarrow C)$, the same for partial functions $\nrightarrow$.

A *pointwise* definition of a function is $\dot{f} \triangleq x \mapsto f(x)$, $\ddot{f} \triangleq X \mapsto \dot{f}(X) = X \mapsto x \mapsto f(X)x$ for functionals, and so forth.

2.2.5 Families

A family $F \in \Delta \to S$ of elements of S indexed by Δ is a map from a set Δ (called the domain or index set, which may be infinite) into a set S. Such a family defines a set $\{F(i) \mid i \in \Delta\}$ (where $F(i)$ is often denoted F_i with an index $i \in \Delta$). It also defines a Cartesian product $\prod_{i\in\Delta} F_i$ as well as a sequence $\langle F_i, i \in \Delta\rangle$ when Δ is totally ordered.

2.2.6 Recursive Definitions

An example of recursive definition of a function $f \in \mathbb{N} \to S$ where S is a set has the form $f(0) \triangleq c$ where $c \in S$ and $f(n) \triangleq F(n, f(n-1))$ where $F \in \mathbb{N} \times S \to S$. For example the factorial is $!0 \triangleq 1$ and $!n \triangleq n\times!(n-1)$. Recursive programming was promoted, among others, by [320, 673].

Recursive definitions may be ill defined, such as $f(0) \triangleq 0$ and $f(n) \triangleq f(n+1)$ when $n \neq 0$. We have $f(n) = 0$ for $n \leqslant 0$, whereas $f(n)$ is undefined when $n > 0$. For programs, *undefined* means "does not terminate" or "terminates with a runtime error" (such as `Stack overflow` or `Segmentation fault`, etc.). So recursive definitions must be proved to be well defined[18] (e.g., $! \in \mathbb{N} \to \mathbb{N}$).[19] Recursive definitions are generalized to inductive definitions in section 16.4.

2.3 Properties

2.3.1 Properties Are Sets

Properties (e.g., "to be an even integer" and "to be an odd natural") can be understood as the set of mathematical objects that have this property (e.g., $2\mathbb{Z} \triangleq \{x \in \mathbb{Z} \mid \exists k \in \mathbb{Z} \,.\, x = 2k\}$ and $2\mathbb{N} + 1 = \{x \in \mathbb{N} \mid \exists k \in \mathbb{N} \,.\, x = 2k + 1\}$). Hence if P is a property then $x \in P$ means "x has property P," and $x \notin P$ means "x does not have property P." For example, $42 \in 2\mathbb{Z}$[20] but $43 \notin 2\mathbb{Z}$ and $! \in \mathbb{N} \to \mathbb{N}$ and $! \notin \mathbb{Z} \to \mathbb{Z}$. The concept of property is fully developed in chapter 8, "Program Properties."

2.3.2 Implication, Weaker and Stronger Properties

In considering properties as sets, the logical implication is subset inclusion $\subseteq$. For example, "to be greater than 42 implies to be positive" is $\{x \in \mathbb{Z} \mid x > 42\} \subseteq \{x \in \mathbb{Z} \mid x \geqslant 0\}$. If $P \subseteq Q$ then P is said to be stronger/more precise than Q, and Q is said to be weaker/less precise than P. Stronger/more precise properties are satisfied by fewer elements, and weaker/less precise properties are satisfied by more elements. ff, that is, $\varnothing$, is the strongest property, and tt, that is, $\mathbb{Z}$, is the weakest property of integers.

18. We distinguish *mathematically well defined* (meaning that the definition of the object assigns it a unique interpretation or value in a mathematical space) from *computationally well defined* (meaning that the object is defined to be computable in finite time without error).

19. e.g. stands for Latin *exempli gratia* or example given.

20. At the end of the last chapter 25 of *The Hitchhiker's Guide to the Galaxy* by Douglas Adams, the supercomputer Deep Thought provides the answer to the question "Of Life, the Universe and Everything...," calculated over a period of seven and a half million years. " 'You're really not going to like it,' observed Deep Thought. ... 'Forty-two' said Deep Thought, with infinite majesty and calm." Moreover, at the end of chapter 33 of *The Restaurant at the End of the Universe*, Arthur Dent confirms that "Six by nine. Forty-two." "That's it. That's all there is."

2.4 Proofs

2.4.1 Proof by Contraposition

A proof of $P \Rightarrow Q$ by contraposition consists of proving the contrapositive $\neg Q \Rightarrow \neg P$. If P is true then $\neg P$ is false, so $\neg Q$ cannot be true because then $\neg P$ would be true and therefore Q is true.

2.4.2 Proof by Reductio Ad Absurdum or by Contradiction

A proof of P by *reductio ad absurdum* consists of finding a property Q that is known to be true and proving $\neg P \Rightarrow \neg Q$. By contraposition $Q \Rightarrow P$ that is $\mathtt{tt} \Rightarrow P$ and so P is true.

2.4.3 Proof by Recurrence

Theorem 2.13 (proof by recurrence) To prove that a property P holds for all natural numbers i.e. $\mathbb{N} \subseteq P$ equivalently $\forall n \in \mathbb{N} \,.\, n \in P$, the proof by recurrence consists of proving $0 \in P$ and $\forall n \in \mathbb{N} \,.\, (n \in P) \Rightarrow (n+1 \in P)$.

$n \in P$ is called the induction hypothesis (also called recurrence hypothesis) so $n+1 \in P$ must be proved assuming this induction hypothesis.

Proof (soundness of the proof by recurrence) Assume that we have made the proof by recurrence and $\mathbb{N} \not\subseteq P$. Then $\exists n \in \mathbb{N} \,.\, n \notin P$. The case $n = 0$ is impossible because we proved $0 \in P$. Therefore $n > 0$ hence $n = (n-1) + 1$. We proved that $\forall m \in \mathbb{N} \,.\, (m \in P) \Rightarrow (m+1 \in P)$ so $\neg(m+1 \in P) \Rightarrow \neg(m \in P)$. For $m = n-1$ we have $n-1 \notin P$. Going on this way, $n-2 \notin P, n-3 \notin P, \ldots, 0 \notin P$ in contradiction to the proof that $0 \in P$. By reductio ad absurdum $\neg(\exists n \in \mathbb{N} \,.\, n \notin P)$, that is, $\forall n \in \mathbb{N} \,.\, n \in P$. □

Proof (completeness of the proof by recurrence) If P holds (that is $\mathbb{N} \subseteq P$) then this can always be proved by recurrence. Let $Q \triangleq P \cap \mathbb{N}$, so that $\mathbb{N} \subseteq P$ implies $Q = \mathbb{N}$. So trivially, $0 \in Q$ and $\forall n \in Q \,.\, n+1 \in Q$. Therefore we have $\mathbb{N} \subseteq Q = P \cap \mathbb{N} \subseteq P$.

So $\mathbb{N} \subseteq P$ can be proved by recurrence (maybe with a stronger recurrence hypothesis Q and an additional implication $Q \subseteq P$). □

Proof by recurrence dates back to Pierre de Fermat infinite descent method [368] (originally formulated contrapositively as $\forall n \in \mathbb{N} \,.\, (n \notin P) \Rightarrow (\exists m < n \,.\, m \notin P)$ then $\forall n \in \mathbb{N} \,.\, n \in P$). The term mathematical induction is credited to Augustus De Morgan [720].

Exercise 2.14 Prove that factorial $! \in \mathbb{N} \rightarrow \mathbb{N}$ by recurrence.

2.5 Conclusion

Set theory is the logical basis for all mathematics and computer science. Additional topics in set theory will be covered subsequently, as needed. The reader may enjoy studying more advanced introductions to abstract set theory, such as [706].

2.6 Solutions to Selected Exercises

Solution to exercise 2.3 S is a singleton if and only if $\exists x \,.\, x \in S \wedge \forall x, y \in S \,.\, x = y$.

3 Syntax, Semantics, Properties, and Static Analysis of Expressions

We introduce basic concepts of abstract interpretation using arithmetic and Boolean expressions.

Contents

3.1 The Rule of Signs

The Indian mathematician and astronomer Brahmagupta (born c. 598, died after 665) was the first to give rules to compute with zero and invented the rule of signs [799, p. 151]. Verses 18.30–35 of his *Brāhma-sphuṭ-a-siddhānta* state

> [The sum] of two positives is positive, of two negatives negative; of a positive and a negative [the sum] is their difference; if they are equal it is zero. The sum of a negative and zero is negative, [that] of a positive and zero positive, [and that] of two zeros zero.
>
> ...
>
> A negative minus zero is negative, a positive [minus zero] positive; zero [minus zero] is zero. When a positive is to be subtracted from a negative or a negative from a positive, then it is to be added.
> The product of a negative and a positive is negative, of two negatives positive, and of positives positive; the product of zero and a negative, of zero and a positive, or of two zeros is zero.
> A positive divided by a positive or a negative divided by a negative is positive; a zero divided by a zero is zero; a positive divided by a negative is negative; a negative divided by a positive is [also] negative.
> A negative or a positive divided by zero has that [zero] as its divisor, or zero divided by a negative or a positive [has that negative or positive as its divisor]. The square of a negative or of a positive is positive; [the square] of zero is zero.

Exercise 3.1 What is the modern understanding of $\frac{0}{0}$?

Exercise 3.2 (erroneous sign analysis) Following the pseudo evaluation idea of Peter Naur in compilation [740, 741], Michel Sintzoff [899] postulates the sign analysis in the following way:

> "$a \times a + b \times b$ yields always the object "pos" when a and b are the objects "pos" or "neg," and when the valuation is defined as follows :

$$
\begin{array}{rcl@{\qquad}rcl}
\text{pos+pos} &=& \text{pos} & \text{pos} \times \text{pos} &=& \text{pos} \\
\text{pos+neg} &=& \text{pos,neg} & \text{pos} \times \text{neg} &=& \text{neg} \\
\text{neg+pos} &=& \text{pos,neg} & \text{neg} \times \text{pos} &=& \text{neg} \\
\text{neg+neg} &=& \text{neg} & \text{neg} \times \text{neg} &=& \text{pos} \\
\text{V(p+q)} &=& \text{V(p)+V(q)} & \text{V}(\text{p} \times \text{q}) &=& \text{V(p)} \times \text{V(q)} \\
\text{V(0)} &=& \text{V(1)} \;=\; \ldots \;=\; \text{pos} \\
\text{V(-1)} &=& \text{V(-2)} \;=\; \ldots \;=\; \text{neg}
\end{array}
$$

> The valuation of $a \times a + b \times b$ yields "pos" by the following computation :

$$
\begin{array}{rcl@{\quad}rcl}
\text{V}(a) &=& \text{pos,neg} & \text{V}(b) &=& \text{pos,neg} \\
\text{V}(a \times a) &=& \text{pos} \times \text{pos, neg} \times \text{neg} & \text{V}(b \times b) &=& \text{pos} \times \text{pos, neg} \times \text{neg} \\
&=& \text{pos,pos} \;=\; \text{pos} & &=& \text{pos,pos} \;=\; \text{pos}
\end{array}
$$

$$
\text{V}(a \times a + b \times b) = \text{V}(a \times a)\text{+V}(b \times b) = \text{pos+pos} = \text{pos"}
$$

What is wrong with this?

3.2 Sign Analysis of Iterative Programs

The rule of signs generalizes to programs. For example the sign of `x` in

```
x = 0; while (…) { x = x+1 }
```

where the iteration condition (…) is ignored, can be determined as follows:

- After zero iteration, when entering the loop, if ever, $\texttt{x} = 0$;
- After one iteration, the sign of `x` is zero, 1 is positive, so the sum `x+1` of zero and positive is positive;
- For the basis, we have shown that after zero or one iteration, the sign of `x` is zero (at iteration 0) or positive (at iteration 1), that is, positive after at most one iteration;

- For the induction step, if after at most $n \geqslant 0$ iterations, the sign of x is positive, then 1 is positive, so the sum x+1 of positive and positive is positive after the next iteration;
- After at most $n+1$ iterations, x is positive (at the previous $n \geqslant 0$ iterations) or positive (at the $n+1$th iteration), then x is positive after at most $n+1$ iterations; and
- By recurrence on the number of iterations in the loop, x is positive in the loop.

3.3 Sign Abstraction, Informally

The abstraction is that you do not (always) need to know the absolute value of the arguments to know the sign of the result of an operation. This is sometimes precise (for example, for the multiplication) but can be imprecise (for example, the sign of the sum of a positive and a negative is unknown when the absolute value of the arguments is ignored). This is nevertheless useful in practice if you know what to do when you don't know the sign. For example, a compiler will not suppress the lower bound check when accessing an array with an index not known to be positive. Moreover, it is always possible to refine the abstraction to get more precise results. For example, Brahmagupta states [799, page 151]

> [If] a smaller [positive] is to be subtracted from a larger positive, [the result] is positive; [if] a smaller negative from a larger negative, [the result] is negative; [if] a larger [negative or positive is to be subtracted] from a smaller [positive or negative, the algebraic sign of] their difference is reversed—negative [becomes] positive and positive negative...

Knowing an interval of the possible values is more precise than just knowing the sign. Interval analysis is considered in chapter 33, "Static Interval Analysis."

The objective of this chapter 3 is to formalize abstract interpretations of arithmetic expressions (such as the rule of signs) and to show how the abstraction can be formally calculated out of the semantics of arithmetic expressions.

3.4 Syntax of Expressions

Let us consider the language of expressions.

$\mathsf{x}, \mathsf{y}, \ldots \in \mathbb{V}$				variables ($\mathbb{V}$ not empty)
$\mathsf{A} \in \mathbb{A}$	::=	$1 \mid \mathsf{x} \mid \mathsf{A}_1 - \mathsf{A}_2$		arithmetic expressions[a]
$\mathsf{B} \in \mathbb{B}$	::=	$\mathsf{A}_1 < \mathsf{A}_2 \mid \mathsf{B}_1\ \mathsf{nand}\ \mathsf{B}_2$		Boolean expressions
$\mathsf{E} \in \mathbb{E}$	::=	$\mathsf{A} \mid \mathsf{B}$		expressions

a. All arithmetic operations can be defined from 1, −, and iteration. The same holds for Boolean expressions with < and nand.

This context-free grammar specifies sets of program syntactic entities; the sets $\mathbb{V}$ of variables, $\mathbb{A}$ of arithmetic expressions, $\mathbb{B}$ of Boolean expressions, and $\mathbb{E}$ of either arithmetic or Boolean expressions. The mathematical variables x, y, A, B, and E denote arbitrary elements of these sets.

Their syntax is defined by grammar rules such as $\mathsf{A} ::= \mathsf{1} \mid \mathsf{x} \mid \mathsf{A}_1 - \mathsf{A}_2$ specifying that an arithmetic expression A is either the constant $\mathsf{1}$, a variable $\mathsf{x} \in \mathbb{V}$, or the difference $\mathsf{A}_1 - \mathsf{A}_2$ of two arithmetic expressions A_1 and A_2. Hence ::= means "is defined as" and | denotes alternatives. The set $\mathbb{V}$ of variables is left unspecified (usually it is an identifier starting with a letter followed by 0 or more letters, digits, or special symbols such as _).

This grammar is ambiguous because $\mathsf{1} - \mathsf{1} - \mathsf{1}$ can be understood either as $(\mathsf{1} - \mathsf{1}) - \mathsf{1}$ or as $\mathsf{1} - (\mathsf{1} - \mathsf{1})$. We choose the first alternative so that the binary operator is left-associative. In Boolean expressions, nand is left-associative and the arithmetic operators have priority over Boolean operators (so $\mathsf{1} - \mathsf{1} < \mathsf{1} - \mathsf{1} - \mathsf{1}$ is $((\mathsf{1} - \mathsf{1}) < ((\mathsf{1} - \mathsf{1}) - \mathsf{1}))$ i.e. false ff). The description of syntax by grammar dates back to Noam Chomsky [180].

3.5 Structural Definitions

Structural definitions are generalizations of recursive definitions on naturals. Assume that we want to define a function $f \in \mathbb{E} \to S$ where S is a set. A structural definition is a recursive definition of the form

- $f(\mathsf{1})$ and $f(\mathsf{x})$ are defined to be constants (so $f(\mathsf{1}) \triangleq c_{\mathsf{1}}$ and $f(\mathsf{x}) \triangleq c_{\mathsf{x}}$ where $c_{\mathsf{1}}, c_{\mathsf{x}} \in S$);
- $f(\mathsf{A}_1 - \mathsf{A}_2)$ and $f(\mathsf{A}_1 < \mathsf{A}_2)$ are functions of $f(\mathsf{A}_1)$ and $f(\mathsf{A}_2)$ (so $f(\mathsf{A}_1 - \mathsf{A}_2) \triangleq F_{-}(f(\mathsf{A}_1), f(\mathsf{A}_2)))$, $f(\mathsf{A}_1 < \mathsf{A}_2) \triangleq F_{<}(f(\mathsf{A}_1), f(\mathsf{A}_2))$;[1]
- $f(\mathsf{B}_1 \ \mathsf{nand}\ \mathsf{B}_2) \triangleq F_{\mathsf{nand}}(f(\mathsf{B}_1), f(\mathsf{B}_2))$ where $F_{-}, F_{<}, F_{\mathsf{nand}} \in S \times S \to S$.

Exercise 3.3 Define $\mathsf{vars} \in \mathbb{E} \to \wp(\mathbb{V})$ such that $\mathsf{vars}[\![\mathsf{E}]\!]$ is the (possibly empty) set of variables occurring in expression E.

Structural definitions are the basis of denotational semantics, introduced by Dana Scott and Christopher Strachey [872] (and called compositional in this context).

3.6 Environments

In order to formally define the value of any expression, such as $\mathsf{1} - \mathsf{1} - \mathsf{1} = -1$, we need to know the value of variables occurring in expressions, for example $\mathsf{x} - \mathsf{1}$ is 2 when $\mathsf{x} = 3$, $\mathsf{x} - \mathsf{1}$ is 42 when $\mathsf{x} = 43$, and so forth. We cannot enumerate the infinitely many cases $\ldots, \mathsf{x} = -1, \mathsf{x} = 0, \mathsf{x} = 1, \ldots$. Therefore, we use an environment[2] $\rho \in \mathbb{E}\mathbb{v}$ where $\mathbb{E}\mathbb{v} \triangleq \mathbb{V} \to \mathbb{Z}$ that is a function ρ mapping a variable x to its value $\rho(\mathsf{x})$ in the set $\mathbb{Z}$ of all mathematical integers. By reasoning on the mathematical variable ρ denoting an unspecified function, we can handle infinitely many cases at once. For example, in environment ρ, the value of $\mathsf{x} - \mathsf{1}$ is $\rho(\mathsf{x}) - 1$ where $\rho(\mathsf{x})$ is the value of variable x, 1 is the mathematical integer one, and $-$ is the mathematical difference.

3.7 Structural Semantics of Expressions

Given an environment $\rho \in \mathbb{E}\mathbb{v} \triangleq \mathbb{V} \to \mathbb{Z}$ mapping variables $\mathsf{x} \in \mathbb{V}$ to their value $\rho(\mathsf{x}) \in \mathbb{Z}$, the value $\mathscr{A}[\![\mathsf{A}]\!]\rho \in \mathbb{Z}$ of an arithmetic expression $\mathsf{A} \in \mathbb{A}$ and $\mathscr{B}[\![\mathsf{B}]\!]\rho \in \mathbb{B}$ of a Boolean

1. $\triangleq$ is "is mathematically defined as" (whereas ::= is "is syntactically defined as").
2. Also called assignment in logic, valuation, and other names.

expression $\mathsf{B} \in \mathbb{B}$ is structurally defined as follows ($\mathbb{B} \triangleq \{\mathsf{tt}, \mathsf{ff}\}$ is the set of Booleans true tt and false ff).

$$
\begin{array}{rcl}
\mathscr{A}[\![\mathsf{1}]\!]\rho & \triangleq & 1 \\
\mathscr{A}[\![\mathsf{x}]\!]\rho & \triangleq & \rho(\mathsf{x}) \\
\mathscr{A}[\![\mathsf{A}_1 - \mathsf{A}_2]\!]\rho & \triangleq & \mathscr{A}[\![\mathsf{A}_1]\!]\rho - \mathscr{A}[\![\mathsf{A}_2]\!]\rho \\
\mathscr{B}[\![\mathsf{A}_1 < \mathsf{A}_2]\!]\rho & \triangleq & \mathscr{A}[\![\mathsf{A}_1]\!]\rho < \mathscr{A}[\![\mathsf{A}_2]\!]\rho \\
\mathscr{B}[\![\mathsf{B}_1 \ \mathsf{nand} \ \mathsf{B}_2]\!]\rho & \triangleq & \mathscr{B}[\![\mathsf{B}_1]\!]\rho \uparrow \mathscr{B}[\![\mathsf{B}_2]\!]\rho \\
\mathcal{S}[\![\mathsf{E}]\!] & \triangleq & \mathscr{A}[\![\mathsf{E}]\!] \qquad \text{when} \quad \mathsf{E} \in \mathbb{A} \\
\mathcal{S}[\![\mathsf{E}]\!] & \triangleq & \mathscr{B}[\![\mathsf{E}]\!] \qquad \text{when} \quad \mathsf{E} \in \mathbb{B}
\end{array}
\tag{3.4}
$$

$\mathsf{1}$, x, $-$, $<$, nand, A, and B are syntactic objects, for example, strings of characters. 1, ρ, $-$, $<$, and $\uparrow$ are mathematical objects. The recursive definition is structural, that is, by induction on the syntax of expressions E (either arithmetic A or Boolean B). The semantics of complex expressions $\mathscr{A}[\![\mathsf{A}]\!]$ or $\mathscr{B}[\![\mathsf{B}]\!]$ is defined in function of the semantics of simpler expressions until reaching basic cases $\mathscr{A}[\![\mathsf{1}]\!]\rho \triangleq 1$ and $\mathscr{A}[\![\mathsf{x}]\!]\rho \triangleq \rho(\mathsf{x})$, for which the value is constant. The "not and" or "nand" Boolean operator $\uparrow$ is defined by the following truth table

a	tt	tt	ff	ff
b	tt	ff	tt	ff
$a \uparrow b$	ff	tt	tt	tt

Observe that in (3.4), if a variable x has several occurrences in an expression E, the same value $\rho(\mathsf{x})$ should be taken by x for all its occurrences in E. The functions $\mathscr{A}$ and $\mathscr{B}$ are shown to be a total functions, that is, well defined for all their arguments in exercise 3.10. This shows that the recursion always terminates.

Remark 3.5 (variables $\mathbb{V}$) In section 3.4, we left the set $\mathbb{V}$ of variables undefined. So the type of $\mathscr{A}$ and $\mathscr{B}$ is also undefined in (3.4). There are essentially three possibilities.

- The first possibility is to define $\mathscr{A} \in \mathbb{A} \to (\mathbb{V} \to \mathbb{V}) \to \mathbb{V}$ where $\mathbb{V}$ is the infinite set of syntactically valid variables in the language (such as an underscore or a letter followed by zero or more underscores, letters, or numbers). The problem is that this is an infinite set hardly implementable in a compiler, interpreter, static analyzer, or the like.
- The second possibility is to define $\mathscr{A} \in \mathbb{A} \to (\mathbb{V} \to \mathbb{V}) \to \mathbb{V}$ where $\mathbb{V}$ is the set of variables of the program under consideration. This guarantees that $\rho(\mathsf{x})$ is well defined in the definition of $\mathscr{A}[\![\mathsf{x}]\!]\rho$ since $\mathsf{x} \in \mathbb{V}$ according to this definition of $\mathbb{V}$. But this is impractical because we want to understand the semantics of an arithmetic expression independently of the programs in which it occurs.
- The preferred third possibility is to consider $\mathbb{V}$ as an implicit formal parameter of the definition of the semantics $\mathscr{A}$ of an expression $\mathsf{A} \in \mathbb{A}$ satisfying the hypothesis than all variables occurring in the expression A do belong to $\mathbb{V}$. So, the dependent type of $\mathscr{A}$ is $\mathbb{V} \to (\mathsf{A} \in \mathbb{A}) \to (\mathbb{V} \to \mathbb{V}) \to \mathbb{V}$ where $\mathsf{vars}[\![\mathsf{A}]\!] \subseteq \mathbb{V}$. Then, if we want to use this semantic definition $\mathscr{A}$ in the definition of the semantics of a program P with variables $\mathsf{vars}[\![\mathsf{P}]\!]$,

we will pass $\mathtt{vars}[\![\mathsf{P}]\!]$ as an actual parameter and the semantics $\mathscr{A}(\mathtt{vars}[\![\mathsf{P}]\!])$ will be of type $(\mathsf{A} \in \mathbb{A}) \to (\mathtt{vars}[\![\mathsf{P}]\!] \to \mathbb{V}) \to \mathbb{V}$ where obviously $\mathtt{vars}[\![\mathsf{A}]\!] \subseteq \mathtt{vars}[\![\mathsf{P}]\!]$ because the variables of the program are defined as including all variables occurring in expressions (and assignments) of the program. To be precise in that third case, we should write (3.4) as follows

$$\begin{aligned}\mathscr{A}(\mathbb{V})[\![\mathtt{1}]\!]\rho &\triangleq 1\\ \mathscr{A}(\mathbb{V})[\![\mathsf{x}]\!]\rho &\triangleq \rho(\mathsf{x})\\ \mathscr{A}(\mathbb{V})[\![\mathsf{A}_1 - \mathsf{A}_2]\!]\rho &\triangleq \mathscr{A}(\mathbb{V})[\![\mathsf{A}_1]\!]\rho - \mathscr{A}(\mathbb{V})[\![\mathsf{A}_2]\!]\rho\end{aligned}$$

but this additional $\mathbb{V}$ is heavy, so we say that (3.4) should be understood as implicitly parameterized by $\mathbb{V}$ (or if you prefer, $\mathbb{V}$ is global in the recursive definition of $\mathscr{A}$).

Then no matter of the three possible definitions of $\mathbb{V}$ is chosen, (3.4) is a valid definition of $\mathscr{A}$ (and the same for $\mathscr{B}$).

Exercise 3.6 Define the logical operators (negation $\neg$, implication $\Rightarrow$, conjunction $\wedge$, disjunction $\vee$) in terms of the Sheffer stroke $\uparrow$.

Exercise 3.7 Write a program in the language of your choice that inputs (an encoding of) an expression (with no variables) and returns the value of this expression.

Exercise 3.8 Prove that any integer $z \in \mathbb{Z}$ has a finite denotation in the syntax $\mathbb{A}$ of arithmetic expressions.

Exercise 3.9 Let $\iota \in \mathbb{V} \to \mathbb{V}$ be an isomorphism (i.e., a renaming of variables). Define $\iota[\![\mathsf{E}]\!]$ to be the expression obtained by renaming of the variables x of E to $\iota(\mathsf{x})$. Prove that $\boldsymbol{S}[\![\mathsf{E}]\!]\rho = \boldsymbol{S}[\![\iota[\![\mathsf{E}]\!]]\!](\mathsf{x} \in \mathbb{V} \mapsto \rho(\iota(\mathsf{x})))$. This shows, intuitively, that the renaming of variables does not change the semantics of expressions [798].

3.8 Proofs by Structural Induction

Proofs by structural induction [157] generalize proofs by recurrence. They are well suited for proving properties of structural definitions. Rod Burstall [157] introduced them as follows “If for some set of structures a structure has a certain property whenever all its proper constituents have that property then all the structures in the set have the property.” So, to prove that a property P holds for all expressions $\mathsf{E} \in \mathbb{E}$, we prove that the property holds for the basic cases $\mathtt{1}$ and x. Then, assuming that the property holds for A_1 and A_2, we prove that it holds for $\mathsf{A}_1 - \mathsf{A}_2$ and $\mathsf{A}_1 < \mathsf{A}_2$. Moreover, assuming that the property holds for Boolean expressions B_1 and B_2, we prove that it also holds for B_1 $\mathtt{nand}$ B_2. We conclude that $\mathbb{E} \subseteq P$.

Exercise 3.10 Prove, by structural induction on the syntax of expressions, that $\mathscr{B}$ is a total function, that is, $\forall \mathsf{B} \in \mathbb{B}\,.\, \mathscr{B}[\![\mathsf{B}]\!] \in (\mathbb{V} \to \mathbb{Z}) \to \mathbb{B}$. (The property to be proved is therefore $P = \{\mathsf{B} \in \mathbb{B} \mid \mathscr{B}[\![\mathsf{B}]\!] \in (\mathbb{V} \to \mathbb{Z}) \to \mathbb{B}\}$.) (A similar property has to be stated and proved for arithmetic expressions.)

Exercise 3.11 Prove, by structural induction on the syntax of expressions, that if $\mathsf{x} \notin \mathtt{vars}[\![\mathsf{B}]\!]$ and $\forall \mathsf{y} \in \mathbb{V} \setminus \{\mathsf{x}\}\,.\, \rho'(\mathsf{y}) = \rho(\mathsf{y})$ then $\mathscr{B}[\![\mathsf{B}]\!]\rho = \mathscr{B}[\![\mathsf{B}]\!]\rho'$. (A similar property has to be stated and proved for arithmetic expressions.)

3.9 Semantic Properties of Expressions

By expression property we might mean a property of the syntax of the expression (such as A has 42 signs −, more precisely, A belongs to the set of expressions with 42 signs −). This is software metrics and metrology [523], of little interest to us.

Instead, an expression property will be understood as a semantic property that is a property of the semantics of expressions.

The semantics $\mathscr{A}[\![\mathsf{A}]\!]$ of an expression A maps environments $\rho \in \mathbb{V} \to \mathbb{Z}$ to a value in $\mathbb{Z}$, $\mathscr{A}[\![\mathsf{A}]\!] \in (\mathbb{V} \to \mathbb{Z}) \to \mathbb{Z}$. Following section 2.3, a semantic property of an expression is a set of possible semantics hence belongs to $\wp((\mathbb{V} \to \mathbb{Z}) \to \mathbb{Z})$. If $P \in \wp((\mathbb{V} \to \mathbb{Z}) \to \mathbb{Z})$ is a semantic property, then $\mathscr{A}[\![\mathsf{A}]\!] \in P$ means that "A has property P."

Example 3.12 $P = \{b \mid \forall \rho \in \mathbb{V} \to \mathbb{Z} \,.\, b(\rho) = \mathtt{tt}\} \cup \{b \mid \forall \rho \in \mathbb{V} \to \mathbb{Z} \,.\, b(\rho) = \mathtt{ff}\}$ is the semantic property of a Boolean expression "to always evaluate to tt" or "to always evaluate to ff." For example, $\mathsf{x} * \mathsf{x} + 1 > 0$ and $\mathsf{x} * \mathsf{x} < 0$ have this property in $\mathbb{Z}$ but not $\mathsf{x} * \mathsf{x} > 0$, because $\mathsf{x} * \mathsf{x} > 0$ is sometimes true (when $|\rho(\mathsf{x})| > 0$) and sometimes false (when $|\rho(\mathsf{x})| = 0$). So, $\mathscr{B}[\![\mathsf{x} * \mathsf{x} + 1 > 0]\!] \in P$ whereas $\mathscr{B}[\![\mathsf{x} * \mathsf{x} > 0]\!] \notin P$.

Notice that semantic properties P of expressions are just a particular case of a property of expressions; that is, the property $\{\mathsf{E} \in \mathbb{E} \mid \boldsymbol{S}[\![\mathsf{E}]\!] \in P\}$.

3.10 Collecting Semantics of Expressions

The collecting semantics of expressions is the strongest property of an expression.

$$\boldsymbol{S}^{\mathrm{C}}[\![\mathsf{A}]\!] \quad \triangleq \quad \{\mathscr{A}[\![\mathsf{A}]\!]\} \quad \in \quad \wp((\mathbb{V} \to \mathbb{Z}) \to \mathbb{Z}) \tag{3.13}$$

Arithmetic expression A is said to have semantic property $P \in \wp((\mathbb{V} \to \mathbb{Z}) \to \mathbb{Z})$ if and only if $\mathscr{A}[\![\mathsf{A}]\!] \in P$ or equivalently, $\boldsymbol{S}^{\mathrm{C}}[\![\mathsf{A}]\!] \subseteq P$ so that $\boldsymbol{S}^{\mathrm{C}}[\![\mathsf{A}]\!]$ is the strongest property of A. The idea of collecting semantics was introduced in [235] (under the qualifier *static semantics*) as a basis for proving the soundness of static analyses. The concept of collecting semantics is further developed in chapter 8.

The fact that $(\mathscr{A}[\![\mathsf{A}]\!] \in P) \Leftrightarrow (\boldsymbol{S}^{\mathrm{C}}[\![\mathsf{A}]\!] \subseteq P)$ may suggest that the concept of collecting semantics is of little interest. However, $x \in S \Leftrightarrow \{x\} \subseteq S$ is the basic idea for abstracting set theory into order/lattice theory [121]. This allows us subsequently to use order theory (which has the equivalent of $\subseteq$ but not of $\in$); see chapter 10, "Posets, Lattices, and Complete Lattices."

Similarly, the collecting semantics of Boolean expressions is

$$\boldsymbol{S}^{\mathrm{C}}[\![\mathsf{B}]\!] \quad \triangleq \quad \{\mathscr{B}[\![\mathsf{B}]\!]\} \quad \in \quad \wp((\mathbb{V} \to \mathbb{Z}) \to \mathbb{B})$$

Again, the collecting semantics $\boldsymbol{S}^{\mathrm{C}}[\![\mathsf{E}]\!]$ of expressions E is just a particular case of a property of expressions, that is, the property $\{\mathsf{E}' \in \mathbb{E} \mid \boldsymbol{S}[\![\mathsf{E}']\!] \in \boldsymbol{S}^{\mathrm{C}}[\![\mathsf{E}]\!]\}$, that is, all expressions E' that have the same semantics as E.

3.11 Proving Semantic Properties of Expressions by Structural Induction

Semantic properties can be proved by structural induction on expressions. For basic cases the proof is $\boldsymbol{S}^{\mathrm{C}}[\![\mathsf{1}]\!] \subseteq P$ and $\boldsymbol{S}^{\mathrm{C}}[\![\mathsf{x}]\!] \subseteq P$. Assuming $\boldsymbol{S}^{\mathrm{C}}[\![\mathsf{A}_1]\!] \subseteq P$ and $\boldsymbol{S}^{\mathrm{C}}[\![\mathsf{A}_2]\!] \subseteq P$, we prove $\boldsymbol{S}^{\mathrm{C}}[\![\mathsf{A}_1 - \mathsf{A}_2]\!] \subseteq P$ and $\boldsymbol{S}^{\mathrm{C}}[\![\mathsf{A}_1 < \mathsf{A}_2]\!] \subseteq P$. Assuming $\boldsymbol{S}^{\mathrm{C}}[\![\mathsf{B}_1]\!] \subseteq P$ and $\boldsymbol{S}^{\mathrm{C}}[\![\mathsf{B}_2]\!] \subseteq P$, we prove that for $\mathcal{S}^{c}[\![\mathsf{B}_1 \text{ nand } \mathsf{B}_2]\!] \subseteq P$. By structural induction, we conclude that $\mathbb{E} \subseteq \{\mathsf{E} \in \mathbb{E} \mid \mathcal{S}^{c}[\![\mathsf{E}]\!] \subseteq P\}$ i.e. $\forall \mathsf{E} \in \mathbb{E} \,.\, \mathcal{S}^{c}[\![\mathsf{E}]\!] \subseteq P$ or, equivalently, $\mathcal{S}[\![\mathsf{E}]\!] \in P$.

Exercise 3.14 Prove by structural induction on expressions that

$$\begin{array}{rcl}
\mathcal{S}^{c}[\![1]\!] &=& \{\rho \in (\mathbb{V} \to \mathbb{Z}) \mapsto 1\} \\
\mathcal{S}^{c}[\![\mathsf{x}]\!] &=& \{\rho \in (\mathbb{V} \to \mathbb{Z}) \mapsto \rho(\mathsf{x})\} \\
\mathcal{S}^{c}[\![\mathsf{A}_1 - \mathsf{A}_2]\!] &=& \{\rho \in (\mathbb{V} \to \mathbb{Z}) \mapsto f_1(\rho) - f_2(\rho) \mid f_1 \in \mathcal{S}^{c}[\![\mathsf{A}_1]\!] \wedge f_2 \in \mathcal{S}^{c}[\![\mathsf{A}_2]\!]\} \\
\mathcal{S}^{c}[\![\mathsf{A}_1 < \mathsf{A}_2]\!] &=& \{\rho \in (\mathbb{V} \to \mathbb{Z}) \mapsto f_1(\rho) < f_2(\rho) \mid f_1 \in \mathcal{S}^{c}[\![\mathsf{A}_1]\!] \wedge f_2 \in \mathcal{S}^{c}[\![\mathsf{A}_2]\!]\} \\
\mathcal{S}^{c}[\![\mathsf{B}_1 \text{ nand } \mathsf{B}_2]\!] &=& \{\rho \in (\mathbb{V} \to \mathbb{Z}) \mapsto f_1(\rho) \uparrow f_2(\rho) \mid f_1 \in \mathcal{S}^{c}[\![\mathsf{B}_1]\!] \wedge f_2 \in \mathcal{S}^{c}[\![\mathsf{B}_2]\!]\}
\end{array}$$

Exercise 3.15 Continuing exercise 3.14, prove that $\mathcal{S}^{c}[\![\mathsf{x} - \mathsf{x}]\!] = \{\rho \in (\mathbb{V} \to \mathbb{Z}) \mapsto 0\}$.

Program proof methods are further studied in chapter 24, "Fixpoint Induction," chapter 25, "Abstract Reachability / Invariance / Safety Verification Semantics," and chapter 26, "Hoare Logic."

3.12 Abstract Sign Properties

We let $\mathbb{P}^{\pm} \triangleq \{\bot_{\pm}, <0, =0, >0, \leqslant 0, \neq 0, \geqslant 0, \top_{\pm}\}$ be the abstract properties of the sign abstract domain $\mathbb{D}_{\pm}$. The sign property <0 is "strictly negative"; $\geqslant 0$ is "positive or zero" and so on; $=0$ is "equal to zero"; $\neq 0$ is "different from zero" (i.e., "strictly negative or strictly positive"); $\top_{\pm}$ (top) is "unknown sign" (i.e., tt that is "negative, zero, or positive"), $\bot_{\pm}$ (bottom) is "no sign" (i.e., ff that is "neither negative, zero, nor positive").

Exercise 3.16 Continuing section 3.2, provide an example of program in which the sign of a variable x is $\bot_{\pm}$.

The sign minus operation $-_{\pm} \in \mathbb{P}^{\pm} \times \mathbb{P}^{\pm} \to \mathbb{P}^{\pm}$ defines the sign $s_1 -_{\pm} s_2$ of $\mathsf{x} - \mathsf{y}$ when x has sign s_1 and y has sign s_2.

$s_1 -_{\pm} s_2$		s_2							
		$\bot_{\pm}$	<0	$=0$	>0	$\leqslant 0$	$\neq 0$	$\geqslant 0$	$\top_{\pm}$
s_1	$\bot_{\pm}$	$\bot_{\pm}$	$\bot_{\pm}$	$\bot_{\pm}$	$\bot_{\pm}$	$\bot_{\pm}$	$\bot_{\pm}$	$\bot_{\pm}$	$\bot_{\pm}$
	<0	$\bot_{\pm}$	$\top_{\pm}$	<0	<0	$\top_{\pm}$	$\top_{\pm}$	<0	$\top_{\pm}$
	$=0$	$\bot_{\pm}$	>0	$=0$	<0	$\geqslant 0$	$\neq 0$	$\leqslant 0$	$\top_{\pm}$
	>0	$\bot_{\pm}$	>0	>0	$\top_{\pm}$	>0	$\top_{\pm}$	$\top_{\pm}$	$\top_{\pm}$
	$\leqslant 0$	$\bot_{\pm}$	$\top_{\pm}$	$\leqslant 0$	<0	$\top_{\pm}$	$\top_{\pm}$	$\leqslant 0$	$\top_{\pm}$
	$\neq 0$	$\bot_{\pm}$	$\top_{\pm}$	$\neq 0$	$\top_{\pm}$	$\top_{\pm}$	$\top_{\pm}$	$\top_{\pm}$	$\top_{\pm}$
	$\geqslant 0$	$\bot_{\pm}$	>0	$\geqslant 0$	$\top_{\pm}$	$\geqslant 0$	$\top_{\pm}$	$\top_{\pm}$	$\top_{\pm}$
	$\top_{\pm}$	$\bot_{\pm}$	$\top_{\pm}$	$\top_{\pm}$	$\top_{\pm}$	$\top_{\pm}$	$\top_{\pm}$	$\top_{\pm}$	$\top_{\pm}$

Exercise 3.17 Show that sign operator $-_{\pm}$ is imprecise for difference ($-$).

Exercise 3.18 What is wrong with the `int abs(int x) { return (x<0) ? -x : x; }` method in Java?

Exercise 3.19 Prove that the sign minus operation $-_{\pm}$ is incorrect with machine integers.

Exercise 3.20 Design a sign operator for multiplication of mathematical integers (×). Prove that it is exact i.e. the sign of the result is exactly known from the sign of the parameters.

3.13 Structural Sign Semantics of Expressions

The sign of an expression depends upon the sign of its free variables. We represent the sign of variables by a sign environment $\overset{\pm}{\rho} \in \mathbb{V} \to \mathbb{P}^{\pm}$ such that $\overset{\pm}{\rho}(\mathsf{x})$ is the sign of variable x.

The sign semantics $\boldsymbol{\mathcal{S}}^{\pm}[\![\mathsf{A}]\!]\overset{\pm}{\rho}$ of an arithmetic expression A is the sign of the expression value when evaluated with variables for which the sign is given by the sign environment $\overset{\pm}{\rho}$. For example, if $\overset{\pm}{\rho}(\mathsf{x}) = {>}0$ and $\overset{\pm}{\rho}(\mathsf{y}) = {\leqslant}0$ then $\boldsymbol{\mathcal{S}}^{\pm}[\![\mathsf{x}-\mathsf{y}]\!]\overset{\pm}{\rho} = {>}0$.

The structural sign semantics $\boldsymbol{\mathcal{S}}^{\pm}[\![\mathsf{A}]\!] \in (\mathbb{V} \to \mathbb{P}^{\pm}) \to \mathbb{P}^{\pm}$ may be defined as follows

$$\begin{aligned}
\boldsymbol{\mathcal{S}}^{\pm}[\![1]\!]\overset{\pm}{\rho} &= {>}0 \\
\boldsymbol{\mathcal{S}}^{\pm}[\![\mathsf{x}]\!]\overset{\pm}{\rho} &= \overset{\pm}{\rho}(\mathsf{x}) \\
\boldsymbol{\mathcal{S}}^{\pm}[\![\mathsf{A}_1-\mathsf{A}_2]\!]\overset{\pm}{\rho} &= (\boldsymbol{\mathcal{S}}^{\pm}[\![\mathsf{A}_1]\!]\overset{\pm}{\rho}) -_{\pm} (\boldsymbol{\mathcal{S}}^{\pm}[\![\mathsf{A}_2]\!]\overset{\pm}{\rho})
\end{aligned} \tag{3.21}$$

Exercise 3.22 Implement the structural sign semantics (3.21) in the programming language of your choice.

To be more precise, if any of the variables has sign $\bot_{\pm}$, meaning "the expression is never evaluated," then the result is $\bot_{\pm}$, meaning "no result is ever returned." We say that signs are $\bot_{\pm}$ *strict* and define $\mathop{\downarrow}^{\pm}$ to enforce it.

$$\begin{aligned}
\mathop{\downarrow}^{\pm}[\overset{\pm}{\rho}]s &\triangleq (\!|\, \exists \mathsf{y} \in \mathbb{V}\,.\,\overset{\pm}{\rho}(\mathsf{y}) = \bot_{\pm} \mathrel{?} \bot_{\pm} \mathrel{\raisebox{0pt}{:}} s \,|\!) \\
\boldsymbol{\mathcal{S}}^{\pm}[\![1]\!]\overset{\pm}{\rho} &= \mathop{\downarrow}^{\pm}[\overset{\pm}{\rho}]({>}0) \\
\boldsymbol{\mathcal{S}}^{\pm}[\![\mathsf{x}]\!]\overset{\pm}{\rho} &= \mathop{\downarrow}^{\pm}[\overset{\pm}{\rho}](\overset{\pm}{\rho}(\mathsf{x})) \\
\boldsymbol{\mathcal{S}}^{\pm}[\![\mathsf{A}_1-\mathsf{A}_2]\!]\overset{\pm}{\rho} &= (\boldsymbol{\mathcal{S}}^{\pm}[\![\mathsf{A}_1]\!]\overset{\pm}{\rho}) -_{\pm} (\boldsymbol{\mathcal{S}}^{\pm}[\![\mathsf{A}_2]\!]\overset{\pm}{\rho})
\end{aligned} \tag{3.23}$$

The conditional expression is $(\!|\, b_1 \mathrel{?} x \parallel b_2 \mathrel{?} y : z \,|\!) = x$ when b_1 is tt; $(\!|\, b_1 \mathrel{?} x \parallel b_2 \mathrel{?} y : z \,|\!) = y$ when b_1 is ff, and b_2 is tt; otherwise $(\!|\, b_1 \mathrel{?} x \parallel b_2 \mathrel{?} y : z \,|\!) = z$ when both b_1 and b_2 are ff. There may be no or several alternatives $\parallel \ldots \mathrel{?} \ldots$.

Exercise 3.24 Prove by structural induction on A that if $\exists \mathsf{x} \in \mathbb{V}\,.\,\overset{\pm}{\rho}(\mathsf{x}) = \bot_{\pm}$ then $\boldsymbol{\mathcal{S}}^{\pm}[\![\mathsf{A}]\!]\overset{\pm}{\rho} = \bot_{\pm}$.

3.14 Soundness

We would like to prove that the sign semantics $\boldsymbol{\mathcal{S}}^{\pm}[\![\mathsf{A}]\!]$ of an arithmetic expression A is a weaker property than the collecting semantics $\boldsymbol{\mathcal{S}}^{\mathbb{C}}[\![\mathsf{A}]\!]$. But $\boldsymbol{\mathcal{S}}^{\pm}[\![\mathsf{A}]\!] \in (\mathbb{V} \to \mathbb{P}^{\pm}) \to \mathbb{P}^{\pm}$ whereas $\boldsymbol{\mathcal{S}}^{\mathbb{C}}[\![\mathsf{A}]\!] \in \wp((\mathbb{V} \to \mathbb{Z}) \to \mathbb{Z})$ and the concrete semantic properties in $\wp((\mathbb{V} \to \mathbb{Z}) \to \mathbb{Z})$ are hardly comparable to the abstract sign properties in $(\mathbb{V} \to \mathbb{P}^{\pm}) \to \mathbb{P}^{\pm}$.

The solution is to express abstract properties in $(\mathbb{V} \to \mathbb{P}^{\pm}) \to \mathbb{P}^{\pm}$ as a concrete property in $\wp((\mathbb{V} \to \mathbb{Z}) \to \mathbb{Z})$. For that purpose we define a concretization function $\ddot{\gamma}_{\pm} \in ((\mathbb{V} \to \mathbb{P}^{\pm}) \to \mathbb{P}^{\pm}) \to (\wp((\mathbb{V} \to \mathbb{Z}) \to \mathbb{Z}))$ mapping an abstract property to an "equivalent" concrete property.

Then the concrete semantics implies the abstract semantics up to concretization, in that for all arithmetic expressions A,

$$\boldsymbol{S}^{\mathrm{C}}[\![\mathsf{A}]\!] \quad \subseteq \quad \ddot{\gamma}_{\pm}(\boldsymbol{S}^{\pm}[\![\mathsf{A}]\!]).$$

3.15 Sign Concretization

We define the sign concretization function $\ddot{\gamma}_{\pm}$ in several steps.[3]

1. First we consider signs (in $\mathbb{P}^{\pm}$) as properties of integers (in $\wp(\mathbb{Z})$).

$$\begin{array}{lllllll} \gamma_{\pm}(\bot_{\pm}) & \triangleq & \varnothing & \quad & \gamma_{\pm}(\leqslant 0) & \triangleq & \{z \in \mathbb{Z} \mid z \leqslant 0\} \\ \gamma_{\pm}(<0) & \triangleq & \{z \in \mathbb{Z} \mid z < 0\} & & \gamma_{\pm}(\neq 0) & \triangleq & \{z \in \mathbb{Z} \mid z \neq 0\} \\ \gamma_{\pm}(=0) & \triangleq & \{0\} & & \gamma_{\pm}(\geqslant 0) & \triangleq & \{z \in \mathbb{Z} \mid z \geqslant 0\} \\ \gamma_{\pm}(>0) & \triangleq & \{z \in \mathbb{Z} \mid z > 0\} & & \gamma_{\pm}(\top_{\pm}) & \triangleq & \mathbb{Z} \end{array} \tag{3.25}$$

2. Then we consider sign environments $\overset{\pm}{\rho} \in \mathbb{V} \to \mathbb{P}^{\pm}$ as properties of environments (in $\wp(\mathbb{V} \to \mathbb{Z})$). $\overset{\pm}{\rho}$ is the abstract property of all concrete environments ρ such that for all variables x, the sign of $\rho(\mathsf{x})$ is $\overset{\pm}{\rho}(\mathsf{x})$.

$$\dot{\gamma}_{\pm}(\overset{\pm}{\rho}) \quad \triangleq \quad \{\rho \in \mathbb{V} \to \mathbb{Z} \mid \forall \mathsf{x} \in \mathbb{V} \,.\, \rho(\mathsf{x}) \in \gamma_{\pm}(\overset{\pm}{\rho}(\mathsf{x}))\} \tag{3.26}$$

Observe that if $\overset{\pm}{\rho}(\mathsf{x}) = \bot_{\pm}$ for some $\mathsf{x} \in \mathbb{V}$ then $\gamma_{\pm}(\overset{\pm}{\rho}(\mathsf{x})) = \varnothing$ so $\forall \mathsf{x} \in \mathbb{V} \,.\, \rho(\mathsf{x}) \in \gamma_{\pm}(\overset{\pm}{\rho}(\mathsf{x}))$ is false, proving that $\dot{\gamma}_{\pm}(\overset{\pm}{\rho}) = \varnothing$. Therefore the abstraction of false ($\varnothing \in \wp(\mathbb{V} \to \mathbb{Z})$) is any abstract environment $\overset{\pm}{\rho}$ with at least one variable x such that $\overset{\pm}{\rho}(\mathsf{x}) = \bot_{\pm}$.

3. Finally, the concretization of abstract properties $\overline{P} \in (\mathbb{V} \to \mathbb{P}^{\pm}) \to \mathbb{P}^{\pm}$ is the concrete property $\ddot{\gamma}_{\pm}(\overline{P}) \in \wp((\mathbb{V} \to \mathbb{Z}) \to \mathbb{Z})$ defined as

$$\ddot{\gamma}_{\pm}(\overline{P}) \quad \triangleq \quad \{\boldsymbol{S} \in (\mathbb{V} \to \mathbb{Z}) \to \mathbb{Z} \mid \forall \overset{\pm}{\rho} \in \mathbb{V} \to \mathbb{P}^{\pm} \,.\, \forall \rho \in \dot{\gamma}_{\pm}(\overset{\pm}{\rho}) \,.\, \boldsymbol{S}(\rho) \in \gamma_{\pm}(\overline{P}(\overset{\pm}{\rho}))\} \tag{3.27}$$

That is A has abstract property $\overline{P}$, which is $\mathscr{A}[\![\mathsf{A}]\!] \in \ddot{\gamma}_{\pm}(\overline{P})$, if and only if for all environments ρ with signs $\overset{\pm}{\rho}$, the value $\mathscr{A}[\![\mathsf{A}]\!]\rho$ of arithmetic expression A has sign $\overline{P}(\overset{\pm}{\rho})$.

Exercise 3.28 Assume that $\gamma(\text{pos}) = \{z \in \mathbb{Z} \mid z \geqslant 0\}$ and $\gamma(\text{neg}) = \{z \in \mathbb{Z} \mid z < 0\}$ as in exercise 3.2. Provide a sound definition of the rule of signs for multiplication $\times$ with this interpretation of the rule of sign.

Exercise 3.29 (a posteriori soundness) Prove that for all $\mathsf{A} \in \mathbb{A}$, $\boldsymbol{S}^{\mathrm{C}}[\![\mathsf{A}]\!] \subseteq \ddot{\gamma}_{\pm}(\boldsymbol{S}^{\pm}[\![\mathsf{A}]\!])$.

3.16 Sign Lattice

Sign properties $\mathscr{P}^{\pm} \triangleq \{\gamma_{\pm}(s) \mid s \in \mathbb{P}^{\pm}\}$ of integers can be partially ordered by $\subseteq$ (i.e., implication) as represented by the Hasse diagram that follows, in which the nodes are the elements of $\mathscr{P}^{\pm}$ and there is a bottom-up arrow from $P \in \mathscr{P}^{\pm}$ to $P' \in \mathscr{P}^{\pm}$ when $P \subsetneq P'$

3. γ is the third letter of the Greek alphabet with the same rank as the third letter c of the Latin alphabet in concretization. Phonetically, the Greek γ/Γ corresponds to the Latin g (English j; jä in phonetics), not to the Latin c. Another choice would have been the Greek for c, that is X/χ (chi), which might be confused with the Latin X/x (Ξ/ξ (xi) in Greek).

and no $Q \in \mathscr{P}^{\pm}$ such that $P \subsetneq Q \subsetneq P'$. So $P \subseteq Q$ if and only if there is a path from P to Q in the Hasse diagram.

The abstract signs $\mathbb{P}^{\pm}$ are an isomorphic representation of $\mathscr{P}^{\pm}$, as shown by the following righthand diagram, where the isomorphism is $\gamma_{\pm} \in \mathbb{P}^{\pm} \to \mathscr{P}^{\pm}$.

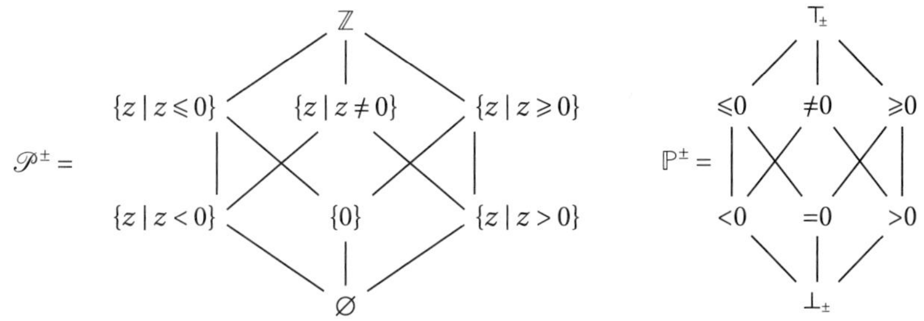

Therefore, the abstract sign properties are partially ordered by $\sqsubseteq_{\pm}$ defined by $s \sqsubseteq_{\pm} s'$ if and only if $\gamma_{\pm}(s) \subseteq \gamma_{\pm}(s')$.

Exercise 3.30 Specify algorithmically the inclusion $\sqsubseteq_{\pm}$ on $\mathbb{P}^{\pm}$ (mathematically defined above as $s \sqsubseteq_{\pm} s'$ if and only if $\gamma_{\pm}(s) \subseteq \gamma_{\pm}(s')$).

Exercise 3.31 Prove that $-_{\pm}$ is increasing in each of its parameters i.e. if $s_1 \sqsubseteq_{\pm} s_1'$ then $s_1 -_{\pm} s_2 \sqsubseteq_{\pm} s_1' -_{\pm} s_2$ and $s_2 \sqsubseteq_{\pm} s_2'$ then $s_1 -_{\pm} s_2 \sqsubseteq_{\pm} s_1 -_{\pm} s_2'$ so that if $s_1 \sqsubseteq_{\pm} s_1'$ and $s_2 \sqsubseteq_{\pm} s_2'$ then $s_1 -_{\pm} s_2 \sqsubseteq_{\pm} s_1' -_{\pm} s_2'$.

3.17 Sign Abstraction, Formally

3.17.1 Abstraction of Sign Properties

An integer property such as $2\mathbb{N}+1$ (odd naturals) can be overapproximated in $\mathbb{P}^{\pm}$ by sign properties $\{z \in \mathbb{Z} \mid z > 0\}$, $\{z \in \mathbb{Z} \mid z \geqslant 0\}$, and $\mathbb{Z}$. The best overapproximation of $2\mathbb{N}+1$ in $\mathbb{P}^{\pm}$ is $\{z \in \mathbb{Z} \mid z > 0\}$ since it is sound (in that $2\mathbb{N}+1 \subseteq \{z \in \mathbb{Z} \mid z > 0\}$) and the most precise/strongest (in that $\{z \in \mathbb{Z} \mid z > 0\} \subseteq \{z \in \mathbb{Z} \mid z \geqslant 0\} \subseteq \mathbb{Z}$). The term *approximate* does not imply any possibility of incorrectness/unsoundness but rather a loss of information. The *overapproximation* of a property P by a sign guarantees that no integer not of that sign may have property P whereas an integer of that sign may not have property P. The *underapproximation* of a property P by a sign guarantees that all integers of that sign do have property P while an integer not of that sign may also have property P.

More generally, the best overapproximation of any integer property $P \in \wp(\mathbb{Z})$ in $\mathbb{P}^{\pm}$ is given by the following abstraction function:[4]

4. α is the first letter of the Greek alphabet, corresponding to the Latin a that starts the word abstraction.

$$\begin{array}{rcl} \alpha_\pm(P) & \triangleq & (\!| P \subseteq \emptyset \,?\, \bot_\pm \\ & & \| P \subseteq \{z \mid z < 0\} \,?\, {<}0 \\ & & \| P \subseteq \{0\} \,?\, {=}0 \\ & & \| P \subseteq \{z \mid z > 0\} \,?\, {>}0 \\ & & \| P \subseteq \{z \mid z \leqslant 0\} \,?\, {\leqslant}0 \\ & & \| P \subseteq \{z \mid z \neq 0\} \,?\, {\neq}0 \\ & & \| P \subseteq \{z \mid z \geqslant 0\} \,?\, {\geqslant}0 \\ & & :\, \top_\pm \,|\!) \end{array} \tag{3.32}$$

$\alpha_\pm(P)$ is the best overapproximation of $P \in \wp(\mathbb{Z})$ in $\mathbb{P}^\pm$ since

- $P \subseteq \gamma_\pm(\alpha_\pm(P))$ i.e. $\alpha_\pm(P)$ is an overapproximation/sound abstraction of P;
- if $\overline{P} \in \mathbb{P}^\pm$ and $P \subseteq \gamma_\pm(\overline{P})$ then $\alpha_\pm(P) \sqsubseteq_\pm \overline{P}$ i.e. $\alpha_\pm(P)$ is more precise than any other overapproximation/sound abstraction of P.

Exercise 3.33 Prove that $s_1 -_\pm s_2 = \alpha_\pm(\{x - y \mid x \in \gamma_\pm(s_1) \wedge y \in \gamma_\pm(s_2)\})$.

Exercise 3.34 Represent elements of $\mathbb{P}^\pm$ by subsets of $\{<0, =0, >0\}$. Define the finite join $\sqcup_\pm$ on $\mathbb{P}^\pm$ such that $\sqcup_\pm\{s_i \mid i \in \Delta\} \triangleq \alpha_\pm(\bigcup\{\gamma_\pm(s_i) \mid i \in \Delta\})$. Prove that $s \sqcup_\pm s' = \{a \mid a \in \{<0, =0, >0\} \wedge (\{a\} \sqsubseteq_\pm s \vee \{a\} \sqsubseteq_\pm s')\}$. Specify the join $\sqcup_\pm$ on $\mathbb{P}^\pm$ algorithmically.

3.17.2 Abstraction of Environment Properties

The best abstraction of an environment property $P \in \wp(\mathbb{V} \to \mathbb{Z})$ is

$$\dot{\alpha}_\pm(P) \triangleq \mathsf{x} \in \mathbb{V} \mapsto \alpha_\pm(\{\rho(\mathsf{x}) \mid \rho \in P\}) \tag{3.35}$$

i.e. for each variable x it is the sign of the set of values $\rho(\mathsf{x})$ in all environments ρ satisfying P.

Observe that $\dot{\alpha}_\pm(\emptyset) = \dot{\bot}_\pm \triangleq \mathsf{x} \in \mathbb{V} \mapsto \bot_\pm$ whereas if $\overset{\pm}{\rho}(\mathsf{x}) = \bot_\pm$ then $\dot{\gamma}_\pm(\overset{\pm}{\rho}) = \emptyset$, so $\emptyset \in \wp(\mathbb{V} \to \mathbb{Z})$ has several possible abstractions in $\mathbb{P}^\pm$ but $\dot{\bot}_\pm$ is the $\dot{\sqsubseteq}_\pm$-smallest of them ($\dot{\sqsubseteq}_\pm$ is the pointwise partial order $\sqsubseteq_\pm$).

3.17.3 Abstraction of Semantic Properties

The best abstraction of a semantic property $P \in \wp((\mathbb{V} \to \mathbb{Z}) \to \mathbb{Z})$ is

$$\ddot{\alpha}_\pm(P) \triangleq \overset{\pm}{\rho} \in \mathbb{V} \to \mathbb{P}^\pm \mapsto \alpha_\pm(\{\boldsymbol{S}(\rho) \mid \boldsymbol{S} \in P \wedge \rho \in \dot{\gamma}_\pm(\overset{\pm}{\rho})\}) \tag{3.36}$$

that is, given a sign environment $\overset{\pm}{\rho}$, $\ddot{\alpha}_\pm(P)\overset{\pm}{\rho}$ is the sign of the possible results $\boldsymbol{S}(\rho)$ of the semantics $\boldsymbol{S} \in P$ with property P for all environments ρ with sign $\overset{\pm}{\rho}$.

3.18 Characteristic Property of Abstraction/Concretization

The abstraction/concretization functions $\langle\alpha_\pm, \gamma_\pm\rangle$ are closely related in that for all $P \in \wp(\mathbb{Z})$ and $\overline{P} \in \mathbb{P}^\pm$, they satisfy

$$\alpha_\pm(P) \sqsubseteq_\pm \overline{P} \iff P \subseteq \gamma_\pm(\overline{P})$$

Proof By definition (3.25) of $\gamma_\pm$ and (3.32) of $\alpha_\pm$, we observe that

- $\gamma_{\pm}$ is increasing, that is, if $s \sqsubseteq_{\pm} s'$ then $\gamma_{\pm}(s) \subseteq \gamma_{\pm}(s')$;
- $\alpha_{\pm}$ is increasing, that is, if $P \subseteq P'$ then $\alpha_{\pm}(P) \sqsubseteq_{\pm} \alpha_{\pm}(P')$; (3.37)
- if $\alpha_{\pm}(P) = s$ then $P \subseteq \gamma_{\pm}(s)$ so $\gamma_{\pm} \circ \alpha_{\pm}$ is extensive, that is, $P \subseteq \gamma_{\pm} \circ \alpha_{\pm}(P)$; (3.38)
- by case analysis, if $P = \gamma_{\pm}(s)$ then $\alpha_{\pm}(P) = s$ so $\alpha_{\pm} \circ \gamma_{\pm}$ is the identity hence reductive, that is, $\alpha_{\pm} \circ \gamma_{\pm}(s) \sqsubseteq_{\pm} s$ because $\sqsubseteq_{\pm}$ is reflexive. (3.39)

It follows that

$$\alpha_{\pm}(P) \sqsubseteq_{\pm} \overline{P}$$

$$\Rightarrow \gamma_{\pm} \circ \alpha_{\pm}(P) \subseteq \gamma_{\pm}(\overline{P}) \qquad \{\gamma_{\pm} \text{ is increasing and def. function composition } \circ\}$$

$$\Rightarrow P \subseteq \gamma_{\pm}(\overline{P}) \qquad \{\gamma_{\pm} \circ \alpha_{\pm} \text{ is extensive and } \subseteq \text{ transitive}\}$$

$$\Rightarrow \alpha_{\pm}(P) \sqsubseteq_{\pm} \alpha_{\pm} \circ \gamma_{\pm}(\overline{P}) \qquad \{\alpha_{\pm} \text{ is increasing and definition of function composition } \circ\}$$

$$\Rightarrow \alpha_{\pm}(P) \sqsubseteq_{\pm} \overline{P} \qquad \{\alpha_{\pm} \circ \gamma_{\pm} \text{ is reductive and definition of function composition } \circ\} \quad \square$$

Similar results hold for $\langle \dot{\alpha}_{\pm}, \dot{\gamma}_{\pm} \rangle$, and $\langle \ddot{\alpha}_{\pm}, \ddot{\gamma}_{\pm} \rangle$; see exercise 3.41.

3.19 Galois Connection

The abstraction/concretization functions $\langle \alpha_{\pm}, \gamma_{\pm} \rangle$ satisfy $\forall P \in \wp(\mathbb{Z}) \,.\, \forall \overline{P} \in \mathbb{P}^{\pm} \,.\, \alpha_{\pm}(P) \sqsubseteq_{\pm} \overline{P} \Leftrightarrow P \subseteq \gamma_{\pm}(\overline{P})$, which is the definition of a Galois connection, which we write $\langle \wp(\mathbb{Z}), \subseteq \rangle \stackrel{\gamma_{\pm}}{\underset{\alpha_{\pm}}{\leftrightarrows}} \langle \mathbb{P}^{\pm}, \sqsubseteq_{\pm} \rangle$.

More generally,

Definition 3.40 (Galois connection) A Galois connection $\langle \mathbb{P}, \sqsubseteq \rangle \stackrel{\gamma}{\underset{\alpha}{\leftrightarrows}} \langle \overline{\mathbb{P}}, \overline{\sqsubseteq} \rangle$ is such that the concrete domain $\langle \mathbb{P}, \sqsubseteq \rangle$ and the abstract domain $\langle \overline{\mathbb{P}}, \overline{\sqsubseteq} \rangle$ are partial orders; $\alpha \in \mathbb{P} \rightarrow \overline{\mathbb{P}}$ is the abstraction function; $\gamma \in \overline{\mathbb{P}} \rightarrow \mathbb{P}$ is the concretization function; and $\forall P \in \mathbb{P} \,.\, \forall \overline{P} \in \overline{\mathbb{P}} \,.\, \alpha(P) \,\overline{\sqsubseteq}\, \overline{P} \Leftrightarrow P \sqsubseteq \gamma(\overline{P})$.

Exercise 3.41 Prove that $\langle \wp(\mathbb{V} \rightarrow \mathbb{Z}), \subseteq \rangle \stackrel{\dot{\gamma}_{\pm}}{\underset{\dot{\alpha}_{\pm}}{\leftrightarrows}} \langle \mathbb{V} \rightarrow \mathbb{P}^{\pm}, \dot{\sqsubseteq}_{\pm} \rangle$, and $\langle \wp((\mathbb{V} \rightarrow \mathbb{Z}) \rightarrow \mathbb{Z}), \subseteq \rangle \stackrel{\ddot{\gamma}_{\pm}}{\underset{\ddot{\alpha}_{\pm}}{\leftrightarrows}} \langle (\mathbb{V} \rightarrow \mathbb{P}^{\pm}) \rightarrow \mathbb{P}^{\pm}, \dot{\sqsubseteq}_{\pm} \rangle$.

Exercise 3.42 Consider the following lattice of signs

$\langle \mathbb{P}^{\pm}, \sqsubseteq_{\pm} \rangle =$

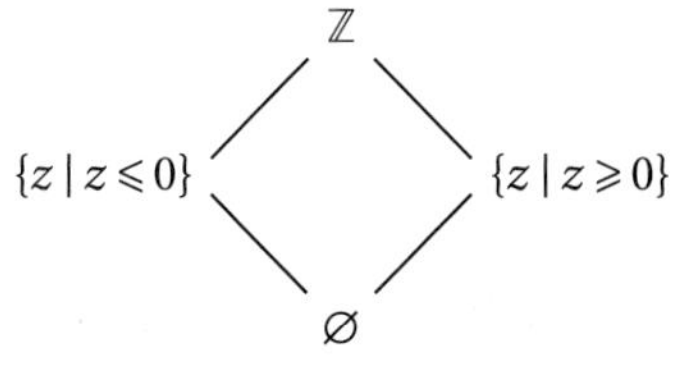

Prove that there can exist no Galois connection $\langle \wp(\mathbb{Z}), \subseteq \rangle \stackrel{\gamma_{\pm}}{\underset{\alpha_{\pm}}{\leftrightarrows}} \langle \mathbb{P}^{\pm}, \sqsubseteq_{\pm} \rangle$ with $\forall \overline{P} \in \mathbb{P}^{\pm} \,.\, \gamma_{\pm}(\overline{P}) = \overline{P} \in \wp(\mathbb{Z})$.

3.20 Calculational Design of the Sign Semantics of Expressions

The soundness requirement in section 3.15 is that $\forall \mathsf{A} \in \mathbb{A} \,.\, \boldsymbol{\mathcal{S}}^{\mathbb{C}}\llbracket \mathsf{A} \rrbracket \subseteq \ddot{\gamma}_{\pm}(\boldsymbol{\mathcal{S}}^{\pm}\llbracket \mathsf{A} \rrbracket)$. By the Galois connection of exercise 3.41, this is equivalent to $\ddot{\alpha}_{\pm}(\boldsymbol{\mathcal{S}}^{\mathbb{C}}\llbracket \mathsf{A} \rrbracket) \mathrel{\dot{\sqsubseteq}_{\pm}} \boldsymbol{\mathcal{S}}^{\pm}\llbracket \mathsf{A} \rrbracket$. Therefore the sign semantics is a sign abstraction of the collecting semantics. It follows that we can design $\boldsymbol{\mathcal{S}}^{\pm}\llbracket \mathsf{A} \rrbracket$ by calculus, calculating $\ddot{\alpha}_{\pm}(\boldsymbol{\mathcal{S}}^{\mathbb{C}}\llbracket \mathsf{A} \rrbracket)$ using $\dot{\sqsubseteq}_{\pm}$-overapproximation to avoid all computations made in the concrete domain.

- We first consider the case when $\exists \mathsf{x} \in \mathbb{V} \,.\, \overset{\pm}{\rho}(\mathsf{x}) = \bot_{\pm}$ so that $\dot{\gamma}_{\pm}(\overset{\pm}{\rho}) = \emptyset$.

$$
\begin{aligned}
- \;& \ddot{\alpha}_{\pm}(\boldsymbol{\mathcal{S}}^{\mathbb{C}}\llbracket \mathsf{A} \rrbracket)\overset{\pm}{\rho} \\
= \;& \alpha_{\pm}(\{\boldsymbol{\mathcal{S}}(\rho) \mid \boldsymbol{\mathcal{S}} \in \boldsymbol{\mathcal{S}}^{\mathbb{C}}\llbracket \mathsf{A} \rrbracket \wedge \rho \in \dot{\gamma}_{\pm}(\overset{\pm}{\rho})\}) && \langle \text{definition (3.36) of } \ddot{\alpha}_{\pm} \rangle \\
= \;& \alpha_{\pm}(\{\boldsymbol{\mathcal{A}}\llbracket \mathsf{A} \rrbracket(\rho) \mid \rho \in \dot{\gamma}_{\pm}(\overset{\pm}{\rho})\}) && \langle \text{definition (3.13) of } \boldsymbol{\mathcal{S}}^{\mathbb{C}}\llbracket \mathsf{A} \rrbracket \rangle \\
= \;& \alpha_{\pm}(\emptyset) && \langle \exists \mathsf{x} \in \mathbb{V} \,.\, \overset{\pm}{\rho}(\mathsf{x}) = \bot_{\pm} \text{ so that } \dot{\gamma}_{\pm}(\overset{\pm}{\rho}) = \emptyset \rangle \\
= \;& \bot_{\pm} && \langle \text{definition (3.32) of } \alpha_{\pm} \rangle \\
\triangleq \;& \boldsymbol{\mathcal{S}}^{\pm}\llbracket \mathsf{A} \rrbracket\overset{\pm}{\rho}
\end{aligned}
$$

$\langle$in accordance with (3.23) such that, by exercise 3.24, $\exists \mathsf{x} \in \mathbb{V} \,.\, \overset{\pm}{\rho}(\mathsf{x}) = \bot_{\pm}$ implies $\boldsymbol{\mathcal{S}}^{\pm}\llbracket \mathsf{A} \rrbracket\overset{\pm}{\rho} = \bot_{\pm}$.$\rangle$

- Then we consider the case when $\forall \mathsf{x} \in \mathbb{V} \,.\, \overset{\pm}{\rho}(\mathsf{x}) \neq \bot_{\pm}$ so that $\dot{\gamma}_{\pm}(\overset{\pm}{\rho}) \neq \emptyset$. We proceed by structural induction on A.

— For the basic case of a constant `1`, we just apply the definitions.

$$
\begin{aligned}
& \ddot{\alpha}_{\pm}(\boldsymbol{\mathcal{S}}^{\mathbb{C}}\llbracket \mathtt{1} \rrbracket)\overset{\pm}{\rho} \\
= \;& \alpha_{\pm}(\{\boldsymbol{\mathcal{S}}(\rho) \mid \boldsymbol{\mathcal{S}} \in \boldsymbol{\mathcal{S}}^{\mathbb{C}}\llbracket \mathtt{1} \rrbracket \wedge \rho \in \dot{\gamma}_{\pm}(\overset{\pm}{\rho})\}) && \langle \text{definition (3.36) of } \ddot{\alpha}_{\pm} \rangle \\
= \;& \alpha_{\pm}(\{\boldsymbol{\mathcal{A}}\llbracket \mathtt{1} \rrbracket(\rho) \mid \rho \in \dot{\gamma}_{\pm}(\overset{\pm}{\rho})\}) && \langle \text{definition (3.13) of } \boldsymbol{\mathcal{S}}^{\mathbb{C}}\llbracket \mathtt{1} \rrbracket \rangle \\
= \;& \alpha_{\pm}(\{1\}) && \langle \dot{\gamma}_{\pm}(\overset{\pm}{\rho}) \text{ is not empty and definition (3.4) of } \boldsymbol{\mathcal{A}}\llbracket \mathtt{1} \rrbracket \rangle \\
= \;& {>}0 && \langle \text{definition (3.32) of } \alpha_{\pm} \rangle \\
\triangleq \;& \boldsymbol{\mathcal{S}}^{\pm}\llbracket \mathtt{1} \rrbracket\overset{\pm}{\rho} && \langle \text{in accordance with (3.23) when } \forall \mathsf{y} \in \mathbb{V} \,.\, \overset{\pm}{\rho}(\mathsf{y}) \neq \bot_{\pm} \rangle
\end{aligned}
$$

— For the basic case of a variable x, we apply the definitions and then simplify.

$$
\begin{aligned}
& \ddot{\alpha}_{\pm}(\boldsymbol{\mathcal{S}}^{\mathbb{C}}\llbracket \mathsf{x} \rrbracket)\overset{\pm}{\rho} \\
= \;& \alpha_{\pm}(\{\boldsymbol{\mathcal{S}}(\rho) \mid \boldsymbol{\mathcal{S}} \in \boldsymbol{\mathcal{S}}^{\mathbb{C}}\llbracket \mathsf{x} \rrbracket \wedge \rho \in \dot{\gamma}_{\pm}(\overset{\pm}{\rho})\}) && \langle \text{definition of (3.36) of } \ddot{\alpha}_{\pm} \rangle \\
= \;& \alpha_{\pm}(\{\boldsymbol{\mathcal{A}}\llbracket \mathsf{x} \rrbracket(\rho) \mid \rho \in \dot{\gamma}_{\pm}(\overset{\pm}{\rho})\}) && \langle \text{definition of (3.13) of } \boldsymbol{\mathcal{S}}^{\mathbb{C}}\llbracket \mathsf{x} \rrbracket \rangle \\
= \;& \alpha_{\pm}(\{\rho(\mathsf{x}) \mid \rho \in \dot{\gamma}_{\pm}(\overset{\pm}{\rho})\}) && \langle \text{definition of (3.4) of } \boldsymbol{\mathcal{A}}\llbracket \mathsf{x} \rrbracket \rangle \\
= \;& \alpha_{\pm}(\{\rho(\mathsf{x}) \mid \forall \mathsf{y} \in \mathbb{V} \,.\, \rho(\mathsf{y}) \in \gamma_{\pm}(\overset{\pm}{\rho}(\mathsf{y}))\}) && \langle \text{definition of (3.26) of } \dot{\gamma}_{\pm} \rangle \\
= \;& \alpha_{\pm}(\{\rho(\mathsf{x}) \mid \rho(\mathsf{x}) \in \gamma_{\pm}(\overset{\pm}{\rho}(\mathsf{x}))\})
\end{aligned}
$$

$\langle$because $\gamma_{\pm}(\overset{\pm}{\rho}(\mathsf{y}))$ is not empty so for $\mathsf{y} \neq \mathsf{x}$, $\rho(\mathsf{y})$ can be chosen arbitrarily to satisfy $\rho(\mathsf{y}) \in \gamma_{\pm}(\overset{\pm}{\rho}(\mathsf{y}))$$\rangle$

$$= \alpha_\pm(\{x \mid x \in \gamma_\pm(\overset{\pm}{\rho}(\mathsf{x}))\}) \qquad \text{⦃letting } x = \rho(\mathsf{x})\text{⦄}$$

$$= \alpha_\pm(\gamma_\pm(\overset{\pm}{\rho}(\mathsf{x}))) \qquad \text{⦃because } S = \{x \mid z \in S\} \text{ for any set } S\text{⦄}$$

$$= \overset{\pm}{\rho}(\mathsf{x}) \qquad \text{⦃by (3.39), } \alpha_\pm \circ \gamma_\pm \text{ is the identity⦄}$$

$$\triangleq \boldsymbol{S}^\pm[\![\mathsf{x}]\!]\overset{\pm}{\rho} \qquad \text{⦃in accordance with (3.23) when } \forall \mathsf{y} \in \mathbb{V} \,.\, \overset{\pm}{\rho}(\mathsf{y}) \neq \bot_\pm\text{⦄}$$

— For the inductive case of $\mathsf{A}_1 - \mathsf{A}_2$, we assume, by structural induction hypothesis, that $\ddot{\alpha}_\pm(\boldsymbol{S}^{\mathrm{C}}[\![\mathsf{A}_1]\!]) \,\dot{\sqsubseteq}_\pm\, \boldsymbol{S}^\pm[\![\mathsf{A}_1]\!]$ and $\ddot{\alpha}_\pm(\boldsymbol{S}^{\mathrm{C}}[\![\mathsf{A}_2]\!]) \,\dot{\sqsubseteq}_\pm\, \boldsymbol{S}^\pm[\![\mathsf{A}_2]\!]$

$$\ddot{\alpha}_\pm(\boldsymbol{S}^{\mathrm{C}}[\![\mathsf{A}_1 - \mathsf{A}_2]\!])\overset{\pm}{\rho}$$

$$= \alpha_\pm(\{\boldsymbol{S}(\rho) \mid \boldsymbol{S} \in \boldsymbol{S}^{\mathrm{C}}[\![\mathsf{A}_1 - \mathsf{A}_2]\!] \wedge \rho \in \dot{\gamma}_\pm(\overset{\pm}{\rho})\}) \qquad \text{⦃definition of (3.36) of } \ddot{\alpha}_\pm\text{⦄}$$

$$= \alpha_\pm(\{\mathscr{A}[\![\mathsf{A}_1 - \mathsf{A}_2]\!](\rho) \mid \rho \in \dot{\gamma}_\pm(\overset{\pm}{\rho})\}) \qquad \text{⦃definition of (3.13) of } \boldsymbol{S}^{\mathrm{C}}[\![\mathsf{A}_1 - \mathsf{A}_2]\!]\text{⦄}$$

$$= \alpha_\pm(\{\mathscr{A}[\![\mathsf{A}_1]\!](\rho) - \mathscr{A}[\![\mathsf{A}_2]\!](\rho) \mid \rho \in \dot{\gamma}_\pm(\overset{\pm}{\rho})\}) \qquad \text{⦃definition of (3.4) of } \mathscr{A}\text{⦄}$$

$$\sqsubseteq_\pm \alpha_\pm(\{x - y \mid x \in \{\mathscr{A}[\![\mathsf{A}_1]\!](\rho') \mid \rho' \in \dot{\gamma}_\pm(\overset{\pm}{\rho})\} \wedge y \in \{\mathscr{A}[\![\mathsf{A}_2]\!](\rho'') \mid \rho'' \in \dot{\gamma}_\pm(\overset{\pm}{\rho})\}\}$$

⦃$\{f(\rho) - g(\rho) \mid \rho \in R\} \subseteq \{x - y \mid x \in \{f(\rho') \mid \rho' \in R\} \wedge y \in \{g(\rho'') \mid \rho'' \in R\}\}$ and $\alpha_\pm$ is increasing by (3.37).

This overapproximation allows for A_1 and A_2 to be evaluated in the concrete with different environments ρ' and ρ'' with the same sign of variables but possibly different values of variables. This accounts for the fact that the rule of signs does not take relationships between values of variables into account. For example, the sign of $\mathsf{x} - \mathsf{x}$ is not =0 in general; see exercise 3.17.⦄

$$\sqsubseteq_\pm \alpha_\pm(\{x - y \mid x \in \gamma_\pm(\alpha_\pm(\{\mathscr{A}[\![\mathsf{A}_1]\!](\rho) \mid \rho \in \dot{\gamma}_\pm(\overset{\pm}{\rho})\}) \wedge y \in \gamma_\pm(\alpha_\pm(\{\mathscr{A}[\![\mathsf{A}_2]\!](\rho) \mid \rho \in \dot{\gamma}_\pm(\overset{\pm}{\rho})\})\})$$

⦃$\{x - y \mid x \in P \wedge y \in Q\} \subseteq \{x - y \mid x \in \gamma_\pm(\alpha_\pm(P)) \wedge y \in \gamma_\pm(\alpha_\pm(Q))\}$ because $\gamma_\pm \circ \alpha_\pm$ is extensive by (3.38) and $\alpha_\pm$ is increasing by (3.37).

This overapproximation allows for the evaluation of the sign to be performed in the abstract with $-_\pm$ instead of the concrete.⦄

$$= \alpha_\pm(\{\mathscr{A}[\![\mathsf{A}_1]\!](\rho) \mid \rho \in \dot{\gamma}_\pm(\overset{\pm}{\rho})\}) -_\pm \alpha_\pm(\{\mathscr{A}[\![\mathsf{A}_2]\!](\rho) \mid \rho \in \dot{\gamma}_\pm(\overset{\pm}{\rho})\})$$

$$\text{⦃}s_1 -_\pm s_2 = \alpha_\pm(\{x - y \mid x \in \gamma_\pm(s_1) \wedge y \in \gamma_\pm(s_2)\}) \text{ by exercise 3.33⦄}$$

$$= \alpha_\pm(\{\boldsymbol{S}(\rho) \mid \boldsymbol{S} \in \boldsymbol{S}^{\mathrm{C}}[\![\mathsf{A}_1]\!] \wedge \rho \in \dot{\gamma}_\pm(\overset{\pm}{\rho})\}) -_\pm \alpha_\pm(\{\boldsymbol{S}(\rho) \mid \boldsymbol{S} \in \boldsymbol{S}^{\mathrm{C}}[\![\mathsf{A}_2]\!] \wedge \rho \in \dot{\gamma}_\pm(\overset{\pm}{\rho})\}) \text{⦃definition of (3.13) of } \boldsymbol{S}^{\mathrm{C}}\text{⦄}$$

$$= \ddot{\alpha}_\pm(\boldsymbol{S}^{\mathrm{C}}[\![\mathsf{A}_1]\!])\overset{\pm}{\rho} -_\pm \ddot{\alpha}_\pm(\boldsymbol{S}^{\mathrm{C}}[\![\mathsf{A}_2]\!])\overset{\pm}{\rho} \qquad \text{⦃definition of (3.36) of } \ddot{\alpha}_\pm\text{⦄}$$

$$\sqsubseteq_\pm (\boldsymbol{S}^\pm[\![\mathsf{A}_1]\!]\overset{\pm}{\rho}) -_\pm (\boldsymbol{S}^\pm[\![\mathsf{A}_2]\!]\overset{\pm}{\rho})$$

⦃induction hypothesis and $-_\pm$ is increasing in both parameters by exercise 3.31⦄

$$\triangleq \boldsymbol{S}^\pm[\![\mathsf{A}_1 - \mathsf{A}_2]\!]\overset{\pm}{\rho} \qquad \text{⦃in accordance with (3.23) when } \forall \mathsf{y} \in \mathbb{V} \,.\, \overset{\pm}{\rho}(\mathsf{y}) \neq \bot_\pm\text{⦄}$$

Remark 3.43 (Computable Abstractions Lose Precision) Notice that in general the inequality $\ddot{\alpha}_\pm(\boldsymbol{S}^{\mathrm{C}}[\![\mathsf{A}]\!]) \,\dot{\sqsubsetneq}_\pm\, \boldsymbol{S}^\pm[\![\mathsf{A}]\!]$ is strict because some mathematical properties of arithmetic expressions are lost in the abstraction. For example, $\boldsymbol{S}^\pm[\![1 - 1]\!] = \overset{\pm}{\rho} \mapsto \top_\pm$ whereas $\ddot{\alpha}_\pm(\boldsymbol{S}^{\mathrm{C}}[\![1 - 1]\!]) = \ddot{\alpha}_\pm(\{\rho \mapsto 0 \mid \mathbb{E}\mathrm{v}\}) = \overset{\pm}{\rho} \mapsto (=0)$. This can be extended to infinitely many examples, $1 - 1 - 1 \ldots - 1$. This is the price to be paid for having $\boldsymbol{S}^\pm[\![\mathsf{A}]\!]$ to be computable whereas $\boldsymbol{S}^{\mathrm{C}}[\![\mathsf{A}]\!]$ is not; see corollary 9.6.

3.21 Calculational Design of Abstract Interpretations

This concludes our formal design of the rule of signs for arithmetic expressions.

- We first define the semantics $\mathscr{A}[\![\mathsf{A}]\!]$ of arithmetic expressions A in (3.4);
- The strongest property of the semantics of arithmetic expressions A is their collecting semantics $\mathcal{S}^{\mathbb{C}}[\![\mathsf{A}]\!]$ in (3.13);
- Among the semantic properties $\wp((\mathbb{V} \to \mathbb{Z}) \to \mathbb{Z})$ of arithmetic expressions, we select a subset of properties of interest i.e. the sign properties and choose a computer representation, as defined by the abstraction function $\ddot{\alpha}_{\pm}$ in (3.36), which is the lower adjoint of the Galois connection; see exercise 3.41;
- The rule of sign $\mathcal{S}^{\pm}[\![\mathsf{A}]\!]$ is then formally derived by calculational design in section 3.20 by overapproximating the best abstraction $\ddot{\alpha}_{\pm}(\mathcal{S}^{\mathbb{C}}[\![\mathsf{A}]\!])$ of the collecting semantics $\mathcal{S}^{\mathbb{C}}[\![\mathsf{A}]\!]$.

It follows that $\mathcal{S}^{\pm}[\![\mathsf{A}]\!]$ is sound by construction.

3.22 Conclusion

We have defined the semantics of expressions, their properties, a proof method, and a sign analysis.

Instead of designing the rule of sign empirically and then proving its soundness (as proposed in exercise 3.29), we used the soundness requirement as a guideline for designing the abstract sign semantics by calculus.

This sign analysis discovers an abstract property of an arithmetic expression by computing in the abstract only. This may involve some loss of precision to eliminate computations in the concrete, which was the case for the sign analysis.

The sign semantics is finite, so it is an easily implementable static analysis (see exercise 5.11) whereas exhaustive case analysis would not scale up (see exercise 3.48). Other inventive examples of finite abstractions are liquid types [842] and cash-flow analysis of smart contracts [878]. Section 34.11 shows that it is always possible to use infinite abstractions to guarantee more precise results.

3.23 Exercises

Exercise 3.44 (parity) Design the parity analysis of expressions $\mathsf{E} \in \mathbb{E}$. Show that the parity analysis is correct with machine integers.

Exercise 3.45 (constancy analysis)
Design the analysis of expressions $\mathsf{E} \in \mathbb{E}$ using the opposite constancy properties such that $\gamma(\bot) = \emptyset$, $\gamma(i) = \{i\}$, $i \in \mathbb{Z}$, and $\gamma(\top) = \mathbb{Z}$ [572].

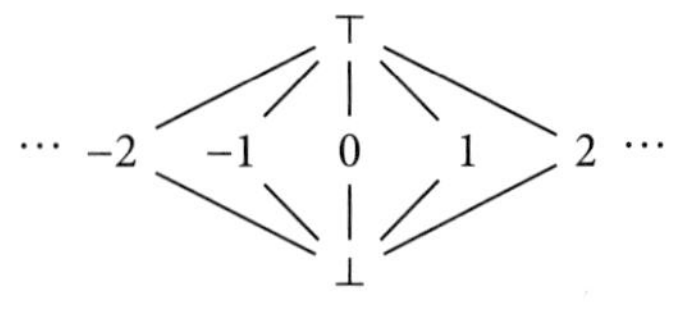

Exercise 3.46 (type analysis) Extend the semantics of expressions with the real constant `1.0` and a generic minus operation with implicit conversion from `nat` to `int` to

`real` where $\gamma(\texttt{nat}) \triangleq \mathbb{N}$, $\gamma(\texttt{int}) \triangleq \mathbb{Z}$, and $\gamma(\texttt{real}) \triangleq \mathbb{R}$. Design a type analysis for these extended expressions [739].

Exercise 3.47 Define the Cartesian set transformer $\mathscr{P}[\![\mathsf{A}]\!]$ of an arithmetic expression A to be

$$\begin{array}{rcl} \mathscr{P}[\![\mathsf{A}]\!] & \in & (\mathbb{V} \rightarrow \wp(\mathbb{Z})) \rightarrow \wp(\mathbb{Z}) \\ \mathscr{P}[\![\mathsf{A}]\!] & \triangleq & P \mapsto \{\mathscr{A}[\![\mathsf{A}]\!]\rho \mid \rho \in \gamma_{\times}(P)\} . \end{array}$$

For example, if $P(\mathsf{x}) = \{0, 1\}$ and $P(\mathsf{y}) = \{-1, 0\}$ then $\mathscr{P}[\![\mathsf{x} - \mathsf{y}]\!]P = \{0, 1, 2\}$ and $\mathscr{P}[\![\mathsf{x} - \mathsf{x}]\!]P = \{-1, 0, 1\}$.

Prove that

$$\begin{array}{rcl} \mathscr{P}[\![\mathsf{1}]\!]P & \triangleq & \{1\} \\ \mathscr{P}[\![\mathsf{x}]\!]P & \triangleq & P(\mathsf{x}) \\ \mathscr{P}[\![\mathsf{A}_1 - \mathsf{A}_2]\!]P & \triangleq & \{x - y \mid x \in \mathscr{P}[\![\mathsf{A}_1]\!]P \wedge y \in \mathscr{P}[\![\mathsf{A}_2]\!]P\} \\ \mathscr{P}[\![\mathsf{A}_1 < \mathsf{A}_2]\!]P & \triangleq & \{x < y \mid x \in \mathscr{P}[\![\mathsf{A}_1]\!]P \wedge y \in \mathscr{P}[\![\mathsf{A}_2]\!]P\} \\ \mathscr{P}[\![\mathsf{B}_1 \ \mathsf{nand}\ \mathsf{B}_2]\!]P & \triangleq & \{x \uparrow y \mid x \in \mathscr{P}[\![\mathsf{B}_1]\!]P \wedge y \in \mathscr{P}[\![\mathsf{B}_2]\!]P\} \end{array}$$

Exercise 3.48 (model checking) Implement the Cartesian property transformer of exercise 3.47 in the language of your choice. In order to prove that $\mathsf{x} - \mathsf{y} = 0$ when $\mathsf{x} = \mathsf{y}$, check that $\mathscr{P}[\![\mathsf{x} - \mathsf{y}]\!]P = \{0\}$ for all P such that $P(\mathsf{x}) = \{i\}$ and $P(\mathsf{y}) = \{i\}$, $i \in 2^p - 1$ where $p = 2, 3, \ldots, 32$ is the bit size of integers. Observe the state explosion problem [192].

3.24 Solutions to Selected Exercises

Solution to exercise 3.1 $\frac{0}{0}$ and more generally $\frac{z}{0}$, $z \in \mathbb{Z}$ is undefined (computer scientists would say the computation has a "runtime error").

Solution to exercise 3.2 0 is pos, −1 is neg, and the sign of $0 \times -1 = 0$ is pos, in contradiction to the rule pos × neg = neg.

Solution to exercise 3.7 In OCaml [628], we have

```
# type aexpr = Num of int | Minus of aexpr * aexpr;;
# type bexpr = Lt of aexpr * aexpr | Nand of bexpr * bexpr;;

# let rec eval_aexpr a = match a with
    | Num i          -> i
    | Minus (a1,a2) -> (eval_aexpr a1) - (eval_aexpr a2)
  let rec eval_bexpr b = match b with
    | Lt (a1,a2)     ->  (eval_aexpr a1) < (eval_aexpr a2)
    | Nand (b1,b2)   -> not((eval_bexpr b1) && (eval_bexpr b2));;

# eval_bexpr (Lt ((Minus (Num 1, Num 2)), (Minus (Num 0, Num 2))));;
- : bool = false
```

Solution to exercise 3.8 For all environments ρ, $\mathscr{A}[\![\mathsf{1} - \mathsf{1}]\!]\rho = 0$. If $\mathscr{A}[\![\mathsf{A}]\!]\rho = n$ then $\mathscr{A}[\![\mathsf{A} - ((\mathsf{1} - \mathsf{1}) - \mathsf{1})]\!]\rho = n + 1$ and $\mathscr{A}[\![(\mathsf{1} - \mathsf{1}) - \mathsf{A}]\!] = -n$.

Solution to exercise 3.16 The loop body in `x = 0; while (0) { x = x+1 }` is never executed so the set of possible values of `x` in this loop body is empty and its sign is $\perp_\pm$.

Solution to exercise 3.17 Consider $42 - 42 = 0$. The sign is $(>0) -_\pm (>0) = \top_\pm$ not $=0$.

Solution to exercise 3.22 In the OCaml [628] programming language, `List.assoc a l` returns the value associated with key `a` in the list of pairs `l`. That is, `List.assoc a [ ...; (a,b); ...] = b` if `(a,b)` is the leftmost binding of `a` in list `l`. Raises `Not_found` if there is no value associated with a in the list `l`.

```
type aexpr = One |Var of string | Minus of aexpr * aexpr;;

let bot = 0 and neg = 1 and is0 = 2 and pos = 3 and
    neg0 = 4 and not0 = 5 and pos0 = 6 and top = 7;;

let print s = match s with
     0 -> "bot" | 1 -> "neg" | 2 -> "is0" | 3 -> "pos" |
     4 -> "neg0" | 5 ->  "not0" | 6 -> "pos0" | 7 -> "top" |
     _ -> failwith "incorrect sign";;

let minus= [|[|bot;  bot;  bot;  bot;  bot;  bot;  bot;  bot|];
             [|bot;  top;  neg;  neg;  top;  top;  neg;  top|];
             [|bot;  pos;  is0;  neg;  pos0; not0; neg0; top|];
             [|bot;  pos;  pos;  top;  pos;  top;  top;  top|];
             [|bot;  pos;  neg0; top;  top;  top;  neg0; top|];
             [|bot;  top;  not0; top;  top;  top;  top;  top|];
             [|bot;  pos;  pos0; top;  pos0; top;  top;  top|];
             [|bot;  top;  top;  top;  top;  top;  top;  top|];
           |];;

type environment = (string * int) list;;

let rec sign a r = match a with
  | One -> pos
  | Var x -> List.assoc x r
  | Minus (a1, a2) -> minus.(sign a1 r).(sign a2 r);;

let r = [("x",pos); ("y",neg)];;
print (sign (Minus ((Var "x"),(Var "y"))) r);;
```

Solution to exercise 3.24 This is true for $\mathsf{A} = \mathsf{1}$ and $\mathsf{A} = \mathsf{x}$ by definition of $\dot{\mathbb{I}}^{\pm}[\overset{\pm}{\rho}]$. Assume, by structural induction ,that this is true for A_1 and A_2. Then $\boldsymbol{\mathcal{S}}^{\pm}[\![\mathsf{A}_1 - \mathsf{A}_2]\!]\overset{\pm}{\rho} = (\boldsymbol{\mathcal{S}}^{\pm}[\![\mathsf{A}_1]\!]\overset{\pm}{\rho}) -_\pm (\boldsymbol{\mathcal{S}}^{\pm}[\![\mathsf{A}_2]\!]\overset{\pm}{\rho}) = \perp_\pm -_\pm \perp_\pm = \perp_\pm$ by induction hypothesis and definition of $-_\pm$.

Solution to exercise 3.33 We can use the soundness requirement as a definition of $s_1 -_\pm s_2 \triangleq \alpha_\pm(\{x - y \mid x \in \gamma_\pm(s_1) \wedge y \in \gamma_\pm(s_2)\})$ to design $-_\pm$ by calculus. We have to consider all possible cases for s_1 and s_2. We show three cases $\top_\pm -_\pm \perp_\pm = \perp_\pm$, $<0 -_\pm \geqslant 0 = <0$, and $\geqslant 0 -_\pm \geqslant 0 = \top_\pm$.

— $\alpha_\pm(\{x - y \mid x \in \gamma_\pm(\top_\pm) \wedge y \in \gamma_\pm(\perp_\pm)\})$

$= \alpha_\pm(\{x - y \mid x \in \mathbb{Z} \wedge y \in \varnothing\})$ ⟨definition of $\gamma_\pm$⟩

$= \alpha_\pm(\varnothing) \quad = \quad \perp_\pm$ ⟨definition of $\alpha_\pm$⟩

— $\alpha_\pm(\{x - y \mid x \in \gamma_\pm(<0) \wedge y \in \gamma_\pm(\geqslant 0)\})$

$= \alpha_\pm(\{x - y \mid x < 0 \wedge y \geqslant 0\})$ $\quad$ ⦃definition of $\gamma_\pm$⦄

$= \alpha_\pm(\{x \mid x < 0\}) \quad = \quad <0$ $\quad$ ⦃definition of $\alpha_\pm$⦄

$-\ \alpha_\pm(\{x - y \mid x \in \gamma_\pm(\geqslant 0) \wedge y \in \gamma_\pm(\geqslant 0)\})$

$= \alpha_\pm(\{x - y \mid x \geqslant 0 \wedge y \geqslant 0\})$ $\quad$ ⦃definition of $\gamma_\pm$⦄

$= \alpha_\pm(\mathbb{Z}) \quad = \quad \top_\pm$ $\quad$ ⦃definition of $\alpha_\pm$⦄

The calculations can be formally certified by a proof verifier [161, 546].

One can also consider all cases $s \in \mathbb{P}^\pm$ for $s_1 -_\pm s_2$ for given s_1, s_2 when needed, using a theorem prover to make the proof that $\{x - y \mid x \in \gamma_\pm(s_1) \wedge y \in \gamma_\pm(s_2)\} \subseteq \gamma_\pm(s)$, and returning $\top_\pm$ when the proof fails (e.g., times out). Among all possible answers s for which the theorem prover could make the proof, the $\sqsubseteq_\pm$-minimal one is chosen, if any. Otherwise, an arbitrary $\sqsubseteq_\pm$-minimal one has to be selected. This is called predicate abstraction [436].

Solution to exercise 3.41 For all $P \in \wp(\mathbb{V} \to \mathbb{Z})$ and $\overset{\pm}{\rho} \in \mathbb{V} \to \mathbb{P}^\pm$, we have

$\dot{\alpha}_\pm(P) \mathrel{\dot{\sqsubseteq}_\pm} \overset{\pm}{\rho}$

$\Leftrightarrow \forall \mathtt{x} \in \mathbb{V} \,.\, \dot{\alpha}_\pm(P)\mathtt{x} \sqsubseteq_\pm \overset{\pm}{\rho}(\mathtt{x})$ $\quad$ ⦃pointwise definition of $\dot{\sqsubseteq}_\pm$⦄

$\Leftrightarrow \forall \mathtt{x} \in \mathbb{V} \,.\, \alpha_\pm(\{\rho(\mathtt{x}) \mid \rho \in P\}) \sqsubseteq_\pm \overset{\pm}{\rho}(\mathtt{x})$ $\quad$ ⦃definition (3.35) of $\dot{\alpha}_\pm$⦄

$\Leftrightarrow \forall \mathtt{x} \in \mathbb{V} \,.\, \{\rho(\mathtt{x}) \mid \rho \in P\} \subseteq \gamma_\pm(\overset{\pm}{\rho}(\mathtt{x}))$ $\quad$ ⦃$\langle \wp(\mathbb{Z}), \subseteq\rangle \underset{\alpha_\pm}{\overset{\gamma_\pm}{\leftrightarrows}} \langle \mathbb{P}^\pm, \sqsubseteq_\pm\rangle$⦄

$\Leftrightarrow \forall \mathtt{x} \in \mathbb{V} \,.\, \forall \rho \in P \,.\, \rho(\mathtt{x}) \in \gamma_\pm(\overset{\pm}{\rho}(\mathtt{x}))$ $\quad$ ⦃definition of $\in$⦄

$\Leftrightarrow \forall \rho \in P \,.\, \forall \mathtt{x} \in \mathbb{V} \,.\, \rho(\mathtt{x}) \in \gamma_\pm(\overset{\pm}{\rho}(\mathtt{x}))$ $\quad$ ⦃definition of $\forall$⦄

$\Leftrightarrow P \subseteq \{\rho \in \mathbb{V} \to \mathbb{Z} \mid \forall \mathtt{x} \in \mathbb{V} \,.\, \rho(\mathtt{x}) \in \gamma_\pm(\overset{\pm}{\rho}(\mathtt{x}))\}$ $\quad$ ⦃definition of $\subseteq$⦄

$\Leftrightarrow P \subseteq \dot{\gamma}_\pm(\overset{\pm}{\rho})$ $\quad$ ⦃definition (3.26) of $\dot{\gamma}_\pm$, proving $\langle \wp(\mathbb{V} \to \mathbb{Z}), \subseteq\rangle \underset{\dot{\alpha}_\pm}{\overset{\dot{\gamma}_\pm}{\leftrightarrows}} \langle \mathbb{V} \to \mathbb{P}^\pm, \dot{\sqsubseteq}_\pm\rangle$⦄

For all $P \in \wp((\mathbb{V} \to \mathbb{Z}) \to \mathbb{Z})$ and $\overline{P} \in (\mathbb{V} \to \mathbb{P}^\pm) \to \mathbb{P}^\pm$, we have

$\ddot{\alpha}_\pm(P) \mathrel{\dot{\sqsubseteq}_\pm} \overline{P}$

$\Leftrightarrow \forall \overset{\pm}{\rho} \in \mathbb{V} \to \mathbb{P}^\pm \,.\, \ddot{\alpha}_\pm(P)\overset{\pm}{\rho} \sqsubseteq_\pm \overline{P}(\overset{\pm}{\rho})$ $\quad$ ⦃pointwise definition of $\dot{\sqsubseteq}_\pm$⦄

$\Leftrightarrow \forall \overset{\pm}{\rho} \in \mathbb{V} \to \mathbb{P}^\pm \,.\, \alpha_\pm(\{\boldsymbol{S}(\rho) \mid \boldsymbol{S} \in P \wedge \rho \in \dot{\gamma}_\pm(\overset{\pm}{\rho})\}) \sqsubseteq_\pm \overline{P}(\overset{\pm}{\rho})$ $\quad$ ⦃definition (3.36) of $\ddot{\alpha}_\pm$⦄

$\Leftrightarrow \forall \overset{\pm}{\rho} \in \mathbb{V} \to \mathbb{P}^\pm \,.\, \{\boldsymbol{S}(\rho) \mid \boldsymbol{S} \in P \wedge \rho \in \dot{\gamma}_\pm(\overset{\pm}{\rho})\} \subseteq \gamma_\pm(\overline{P}(\overset{\pm}{\rho}))$ $\quad$ ⦃$\langle \wp(\mathbb{Z}), \subseteq\rangle \underset{\alpha_\pm}{\overset{\gamma_\pm}{\leftrightarrows}} \langle \mathbb{P}^\pm, \sqsubseteq_\pm\rangle$⦄

$\Leftrightarrow \forall \overset{\pm}{\rho} \in \mathbb{V} \to \mathbb{P}^\pm \,.\, \forall \boldsymbol{S} \in P \,.\, \forall \rho \in \dot{\gamma}_\pm(\overset{\pm}{\rho}) \,.\, \boldsymbol{S}(\rho) \in \gamma_\pm(\overline{P}(\overset{\pm}{\rho}))$ $\quad$ ⦃definition of $\subseteq$⦄

$\Leftrightarrow \forall \boldsymbol{S} \in P \,.\, \forall \overset{\pm}{\rho} \in \mathbb{V} \to \mathbb{P}^\pm \,.\, \forall \rho \in \dot{\gamma}_\pm(\overset{\pm}{\rho}) \,.\, \boldsymbol{S}(\rho) \in \gamma_\pm(\overline{P}(\overset{\pm}{\rho}))$ $\quad$ ⦃definition of $\forall$⦄

$\Leftrightarrow \forall \boldsymbol{S} \in P \,.\, \boldsymbol{S} \in \ddot{\gamma}_\pm(\overline{P})$ $\quad$ ⦃definition of $\in$ and definition (3.27) of $\ddot{\gamma}_\pm$⦄

$\Leftrightarrow P \subseteq \ddot{\gamma}_\pm(\overline{P})$

⦃definition of $\subseteq$, proving $\langle \wp((\mathbb{V} \to \mathbb{Z}) \to \mathbb{Z}), \subseteq\rangle \underset{\ddot{\alpha}_\pm}{\overset{\ddot{\gamma}_\pm}{\leftrightarrows}} \langle (\mathbb{V} \to \mathbb{P}^\pm) \to \mathbb{P}^\pm, \dot{\sqsubseteq}_\pm\rangle$.⦄

Solution to exercise 3.42 Consider $P = \{0\}$. We have $P \subseteq \{z \mid z \leqslant 0\} = \gamma_\pm(\{z \mid z \leqslant 0\})$ and $P \subseteq \{z \mid z \geqslant 0\} = \gamma_\pm(\{z \mid z \geqslant 0\})$. Therefore, by definition of a Galois connection, we have $\alpha_\pm(P) \sqsubseteq_\pm \{z \mid z \leqslant\}$ and $\alpha_\pm(P) \sqsubseteq_\pm \{z \mid z \geqslant 0\}$. The only element with this property in $\langle \mathbb{P}^\pm$,

$\sqsubseteq_{\pm}\rangle$ is $\varnothing$. Therefore, we must have $\alpha_{\pm}(P) = \varnothing$ so $\alpha_{\pm}(\{0\}) \sqsubseteq_{\pm} \varnothing$ by reflexivity. By definition of a Galois connection, it follows that $\{0\} \subseteq \gamma_{\pm}(\varnothing) = \varnothing$, a contradiction.

Solution to exercise 3.48 The model checking time is exponential in the bit size of integers, as shown by the following proof by exhaustive enumeration.

```
$ cat CartesianTransformer.py
import time
import matplotlib.pyplot as pyplot
from sets import *

# i   integer
# 'x' variables
# e ::= ('INT', i) | ('VAR', 'x') |
#        ('MINUS', e1, e2)

maxp = 32
natural = Set(range(0, 2 ** maxp))

def eval (e, P):
    if e[0] == 'INT':
        return Set([e[1]])
    if e[0] == 'VAR':
        return P(e[1])
    if e[0] == 'MINUS':
        P1 = eval(e[1], P)
        P2 = eval(e[2], P)
        res = []
        for x in P1:
            for y in P2:
                res = res + [x - y]
        return Set(res)
    handle_error()

expr = ('MINUS', ('VAR', 'x'), ('VAR'
    , 'y'))

def makeP(i):
   def P(x):
      if x == "x":
         return Set([i])
      if x == "y":
         return Set([i])
      return Set([0])
   return P

x = []
y = []
result = Set([])
for p in range (2, maxp):
    start_time = time.time()
    for i in range (0, ((2 ** p) - 1)
        ):
        P = makeP(i)
        result = result | eval (expr,
            P)
    finish_time = time.time()
    x = x + [p]
    t = (finish_time - start_time)
    y = y + [t]
    print p, result, t, "seconds"
pyplot.plot(x,y)
pyplot.xlabel('p')
pyplot.ylabel('time(2**p-1)')
pyplot.savefig('CartesianTransformer.
    png')
$ /usr/bin/python
    CartesianTransformer.py
2 Set([0]) 0.000694036483765 seconds
3 Set([0]) 0.000127077102661 seconds...

25 Set([0]) 262.113693953 seconds
26 Set([0]) 537.459581137 seconds
27 Set([0]) 1062.62982702 seconds
$
```

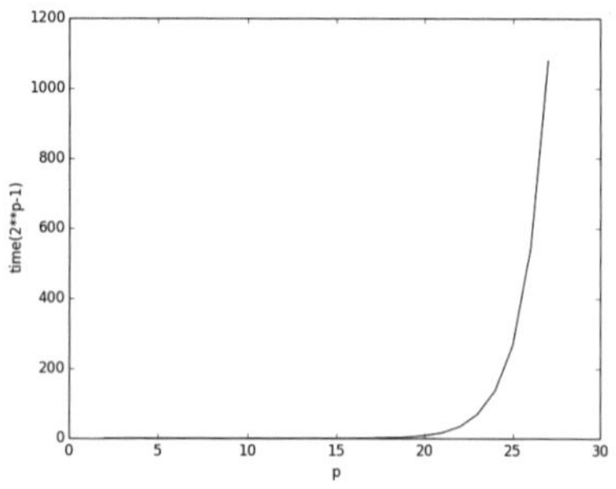

II SYNTAX

The syntax of a programming language specifies the rules to write well-formed programs in the language and how to describe and refer to the syntactic structure of programs (chapter 4).

For those students who would like to undertake the projects suggested in this book, we introduce bottom-up parsing to automatically build the syntactic structure of programs (chapter 5). Otherwise parsing is not required.

4 Syntax

In order to verify and analyze programs, it is necessary first to define their syntax and then their semantics. The syntax specifies how programs must be written. The semantics specifies the runtime behavior of programs, that is, what and how they compute when executed. This chapter defines the syntax of the programming language used throughout the book.

Contents

4.1 Syntax of Programs

Program statements[1] have the following context-free syntax [180, 983] (ϵ denotes the empty string).

$$
\begin{array}{rcll}
\mathtt{S} & ::= & & \text{statement } \mathtt{S} \in \mathbb{S} \\
 & & \mathtt{x = A\,;} & \quad\text{assignment} \\
 & | & \mathtt{;} & \quad\text{skip} \\
 & | & \mathtt{\textbf{if}\,(B)\,S} & \quad\text{conditional} \\
 & | & \mathtt{\textbf{if}\,(B)\,S\ \textbf{else}\ S} & \\
 & | & \mathtt{\textbf{while}\,(B)\,S} & \quad\text{iteration} \\
 & | & \mathtt{\textbf{break}\,;} & \quad\text{iteration break} \\
 & | & \mathtt{\{\,Sl\,\}} & \quad\text{compound statement} \\
\mathtt{Sl} & ::= & \mathtt{Sl\ S} \mid \epsilon & \text{statement list } \mathtt{Sl} \in \mathbb{Sl} \\
\mathtt{P} & ::= & \mathtt{Sl} & \text{program } \mathtt{P} \in \mathbb{P} \\
\mathbb{Pc} & \triangleq & \mathbb{S} \cup \mathbb{Sl} \cup \mathbb{P} & \text{program component } \mathtt{S} \in \mathbb{Pc}
\end{array}
$$

For example, the program

$$\mathtt{x = x + 1\,;\textbf{while}\,(tt)\,\{\,x = x + 1\,;\textbf{if}\,(x > 2)\,\textbf{break}\,;\};} \tag{4.1}$$

has the syntax tree (with explicit labels; see section 4.2.3) given in figure 4.2.

1. If P is a program then `int main () {P}` is a valid C program (after global variable declarations have been added).

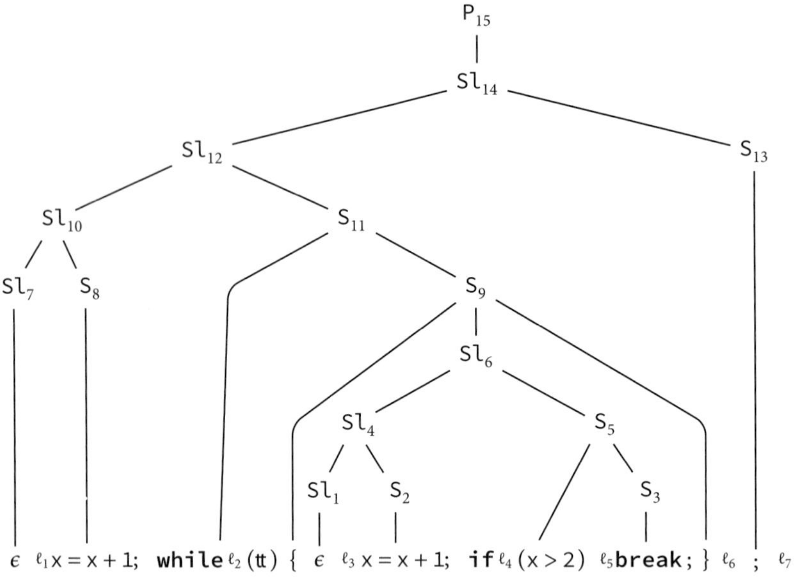

Figure 4.2
Syntax tree of the labeled program (4.1)

Exercise 4.3 Write, compile, and execute the program (4.1) in the C programming language.

Exercise 4.4 Show that the conditional statements can be eliminated from any program $\mathsf{P} \in \mathbb{P}$.

4.2 Labels of Programs

We use labels to designate program points of statements. They are not part of the program syntax; their syntax is therefore free. So we do not provide a syntax/model for labels but, instead, specify axiomatically the properties that they should satisfy, for example, that labels are unique. For example, program (4.1) can be explicitly labeled as follows (the labels ℓ_i would better be written as program comments)

$$
\begin{aligned}
&\ell_1\ \mathtt{x = x + 1;} \\
&\quad \mathtt{while}\ \ell_2\ (\mathtt{tt})\ \{ \\
&\qquad \ell_3\ \mathtt{x = x + 1;} \\
&\qquad \mathtt{if}\ \ell_4\ (\mathtt{x > 2})\ \ell_5\ \mathtt{break;}\}\ell_6;\ell_7
\end{aligned} \tag{4.5}
$$

If we change the labels, or even how the labels are encoded, the program semantics is the same. Hence we want to describe the program semantics independently of how labels are encoded. This is accomplished by an axiomatization (specifying which properties the labeling should satisfy) without specifying which interpretations of labels should be chosen among multiple interpretations.

We next define functions of program components (i.e., programs, statement lists, or statements), providing sets of labels to designate specific program points of these program components. This allows us to reason on program points without knowing how they are represented. Section 4.2.3 shows how programs can be labeled explicitly if such a representation is needed for clarity.

at⟦S⟧	the program point at which execution of program component S starts;
after⟦S⟧	the program exit point after program component S, at which execution of S is supposed to normally terminate, if ever;
escape⟦S⟧	a Boolean indicating whether or not the program component S contains a **break** ; statement escaping out of that component S;
break-to⟦S⟧	the program point to which execution of the program component S goes when a **break** ; statement escapes out of that component S;
breaks-of⟦S⟧	the set of labels of all **break** ; statements that can escape out of component S;
in⟦S⟧	the set of program points inside program component S (including at⟦S⟧ but excluding after⟦S⟧ and break-to⟦S⟧);
labs⟦S⟧	the potentially reachable program points while executing S either in or after the statement (excluding reachability by a break); and
labx⟦S⟧	the potentially reachable program points while executing program component S at, in, or after the program component, or resulting from a break.

The function definitions are reminiscent of attribute grammars [589–591] in that they are associated with the grammar rules and applied at each node of the syntax tree.

4.2.1 Label at a Statement

For each program component $\mathtt{S} \in \mathbb{P}c$ (program, statement list, and statement), we define

- at⟦S⟧: the program point at which execution of S starts should satisfy the following postulates.

$$
\begin{array}{rcll}
\mathtt{P} & ::= & \mathtt{Sl} & \mathsf{at}[\![\mathtt{P}]\!] \triangleq \mathsf{at}[\![\mathtt{Sl}]\!] \\
\mathtt{Sl} & ::= & \mathtt{Sl}'\ \mathtt{S} & \mathsf{at}[\![\mathtt{Sl}]\!] \triangleq \mathsf{at}[\![\mathtt{Sl}']\!] \\
\mathtt{Sl} & ::= & \epsilon & \mathsf{at}[\![\mathtt{Sl}]\!] \triangleq \mathsf{after}[\![\mathtt{Sl}]\!] \\
\mathtt{S} & ::= & \{\mathtt{Sl}\} & \mathsf{at}[\![\mathtt{S}]\!] \triangleq \mathsf{at}[\![\mathtt{Sl}]\!]
\end{array}
$$

4.2.2 Label after a Statement

- after⟦S⟧: the program point at which execution of S is supposed to terminate, if ever. After executing S, execution will continue at that program point (unless there is a **break** ; or the statement does not terminate). after⟦S⟧ should satisfy the following postulates.

$$
\begin{array}{lcll}
\mathsf{P} & ::= & \mathsf{Sl} & \mathsf{after}[\![\mathsf{P}]\!] \triangleq \mathsf{after}[\![\mathsf{Sl}]\!] \\
\mathsf{Sl} & ::= & \mathsf{Sl}'\ \mathsf{S} & \mathsf{after}[\![\mathsf{Sl}']\!] \triangleq \mathsf{at}[\![\mathsf{S}]\!], \quad \mathsf{after}[\![\mathsf{S}]\!] \triangleq \mathsf{after}[\![\mathsf{Sl}]\!] \\
\mathsf{S} & ::= & \mathtt{if}\,(\mathsf{B})\,\mathsf{S}_t & \mathsf{after}[\![\mathsf{S}_t]\!] \triangleq \mathsf{after}[\![\mathsf{S}]\!] \\
\mathsf{S} & ::= & \mathtt{if}\,(\mathsf{B})\,\mathsf{S}_t\ \mathtt{else}\ \mathsf{S}_f & \mathsf{after}[\![\mathsf{S}_t]\!] \triangleq \mathsf{after}[\![\mathsf{S}_f]\!] \triangleq \mathsf{after}[\![\mathsf{S}]\!] \\
\mathsf{S} & ::= & \mathtt{while}\,(\mathsf{B})\,\mathsf{S}_b & \mathsf{after}[\![\mathsf{S}_b]\!] \triangleq \mathsf{at}[\![\mathsf{S}]\!] \\
\mathsf{S} & ::= & \{\,\mathsf{Sl}\,\} & \mathsf{after}[\![\mathsf{Sl}]\!] \triangleq \mathsf{after}[\![\mathsf{S}]\!]
\end{array}
$$

4.2.3 Interpretations of Labels

The axiomatic definition of labels leaves open different possible interpretations. The most common is *explicit labeling*. When explicitly decorating programs with labels, we should have

$$
\begin{array}{lcll}
\mathsf{S} & ::= & \ell\ \mathsf{x} = \mathsf{E}\,; & \mathsf{at}[\![\mathsf{S}]\!] \triangleq \ell \\
\mathsf{S} & ::= & \ell; & \mathsf{at}[\![\mathsf{S}]\!] \triangleq \ell \\
\mathsf{S} & ::= & \mathtt{if}\,\ell\,(\mathsf{B})\,\mathsf{S}_t & \mathsf{at}[\![\mathsf{S}]\!] \triangleq \ell \\
\mathsf{S} & ::= & \mathtt{if}\,\ell\,(\mathsf{B})\,\mathsf{S}_t\ \mathtt{else}\ \mathsf{S}_f & \mathsf{at}[\![\mathsf{S}]\!] \triangleq \ell \\
\mathsf{S} & ::= & \mathtt{while}\,\ell\,(\mathsf{B})\,\mathsf{S}_b & \mathsf{at}[\![\mathsf{S}]\!] \triangleq \ell \\
\mathsf{S} & ::= & \ell\ \mathtt{break}\,; & \mathsf{at}[\![\mathsf{S}]\!] \triangleq \ell \\
\mathsf{P} & ::= & \mathsf{Sl}\,\ell & \mathsf{after}[\![\mathsf{P}]\!] \triangleq \mathsf{after}[\![\mathsf{Sl}]\!] \triangleq \ell
\end{array}
$$

(so that the program labeling can be generated automatically).

Exercise 4.6 Calculate $\mathsf{at}[\![\mathsf{P}]\!]$ for the program (4.5).

Exercise 4.7 Calculate $\mathsf{after}[\![\mathsf{S}]\!]$ for all statements S in the (syntax tree of) the program (4.5).

Exercise 4.8 Compare the labeling of $\mathsf{P}_1 ::= \ell_1;\ell_2$ (with a no-operation skip/NOP computation most commonly used for timing purposes) and of $\mathsf{P}_2 ::= \{\ \epsilon\ \}\ell$ (involving no computation).

Exercises 4.9, 4.10, and 4.11 propose three other possible representations of labels.

Exercise 4.9 Show that labels can be represented by a path in the program syntax tree. For example, for program (4.1), we would have $\ell_5 = \mathsf{P}_{15}\mathsf{Sl}_{14}\mathsf{Sl}_{12}\mathsf{Sl}_{11}\mathsf{S}_9\mathsf{Sl}_6\mathsf{S}_5\mathsf{S}_3$.

Exercise 4.10 Show that labels can be represented by statement lists; that is, a label is the statement list that remains to be executed when control is at that label. For example, $\ell_1\ \mathsf{x} = \mathsf{x} + 1\,;\ell_2\ \mathsf{x} = \mathsf{x} + 2\,;\ell_3$ would have $\ell_1 \triangleq \mathsf{x} = \mathsf{x} + 1\,;\mathsf{x} = \mathsf{x} + 2\,;$, $\ell_2 \triangleq \mathsf{x} = \mathsf{x} + 2\,;$, and $\ell_3 \triangleq$, that is the empty statement list ϵ. For an iteration $\mathsf{S} ::= \mathtt{while}\,(\mathsf{B})\,\mathsf{S}_b$, we would have $\mathsf{at}[\![\mathsf{S}]\!] = \mathtt{if}\,(\mathsf{B})\,\{\,\mathsf{S}_b\ \mathtt{while}\,(\mathsf{B})\,\mathsf{S}_b\,\}$ and $\mathsf{at}[\![\mathsf{S}_b]\!] = \mathsf{S}_b\ \mathtt{while}\,(\mathsf{B})\,\mathsf{S}_b$.

Exercise 4.11 Assume that the compiler of the program constructs a program control graph (or flowchart). Show that program labels can be chosen as the nodes of the program control graph [333].

4.2.4 Escape Label of a Statement

- $\mathsf{escape}[\![\mathsf{S}]\!]$: tt if and only if the program component $\mathsf{S} \in \mathbb{P}\mathbb{c}$ contains a **break**; statement escaping out of that component S (so that this **break**; statement is not inside an iteration within S). The definition checks that S (hence, in particular, the **break**; statements escaping S) is inside a loop.

P	::=	Sl	$\text{escape}[\![\text{P}]\!] \triangleq \text{escape}[\![\text{Sl}]\!]$, $\text{escape}[\![\text{P}]\!] = \text{ff}$
Sl	::=	Sl′ S	$\text{escape}[\![\text{Sl}]\!] \triangleq \text{escape}[\![\text{Sl}']\!] \vee \text{escape}[\![\text{S}]\!]$
Sl	::=	ϵ	$\text{escape}[\![\text{Sl}]\!] \triangleq \text{ff}$
S	::=	x = E ;	$\text{escape}[\![\text{S}]\!] \triangleq \text{ff}$
S	::=	;	$\text{escape}[\![\text{S}]\!] \triangleq \text{ff}$
S	::=	**if** (B) S_t	$\text{escape}[\![\text{S}]\!] \triangleq \text{escape}[\![\text{S}_t]\!]$
S	::=	**if** (B) S_t **else** S_f	$\text{escape}[\![\text{S}]\!] \triangleq \text{escape}[\![\text{S}_t]\!] \vee \text{escape}[\![\text{S}_f]\!]$
S	::=	**while** (B) S_b	$\text{escape}[\![\text{S}]\!] \triangleq \text{ff}$
S	::=	**break** ;	$\text{escape}[\![\text{S}]\!] \triangleq \text{tt}$
S	::=	{ Sl }	$\text{escape}[\![\text{S}]\!] \triangleq \text{escape}[\![\text{Sl}]\!]$

- $\text{break-to}[\![\text{S}]\!]$: the program point at which execution of the program component S ∈ $\mathbb{P}\mathbb{c}$ goes to when a **break**; statement escapes out of that component S. It is well defined only when $\text{escape}[\![\text{S}]\!] = \text{tt}$.

Sl	::=	Sl′ S	$\text{break-to}[\![\text{Sl}']\!] \triangleq \text{break-to}[\![\text{S}]\!] \triangleq \text{break-to}[\![\text{Sl}]\!]$
S	::=	**if** (B) S_t	$\text{break-to}[\![\text{S}_t]\!] \triangleq \text{break-to}[\![\text{S}]\!]$
S	::=	**if** (B) S_t **else** S_f	$\text{break-to}[\![\text{S}_t]\!] \triangleq \text{break-to}[\![\text{S}_f]\!] \triangleq \text{break-to}[\![\text{S}]\!]$
S	::=	**while** (B) S_b	$\text{break-to}[\![\text{S}_b]\!] \triangleq \text{after}[\![\text{S}]\!]$
S	::=	{ Sl }	$\text{break-to}[\![\text{Sl}]\!] \triangleq \text{break-to}[\![\text{S}]\!]$

Exercise 4.12 Calculate $\text{escape}[\![\text{S}]\!]$ and $\text{break-to}[\![\text{S}]\!]$ for all statements S in the (syntax tree of) the program (4.5).

4.2.5 Labels of break; Statements

- $\text{breaks-of}[\![\text{S}]\!]$: collects the labels of all **break**; program components that can escape out of S (so excluding **break**; statements inside an iteration statement within S). The definition checks that **break**; statements can appear only within loops;

P	::=	Sl	$\text{breaks-of}[\![\text{P}]\!] \triangleq \text{breaks-of}[\![\text{Sl}]\!]$	$\text{breaks-of}[\![\text{P}]\!] = \varnothing$
Sl	::=	Sl′ S	$\text{breaks-of}[\![\text{Sl}]\!] \triangleq \text{breaks-of}[\![\text{Sl}']\!] \cup \text{breaks-of}[\![\text{S}]\!]$	
Sl	::=	ϵ	$\text{breaks-of}[\![\text{Sl}]\!] \triangleq \varnothing$	
S	::=	x = E ;	$\text{breaks-of}[\![\text{S}]\!] \triangleq \varnothing$	
S	::=	;	$\text{breaks-of}[\![\text{S}]\!] \triangleq \varnothing$	
S	::=	**if** (B) S_t	$\text{breaks-of}[\![\text{S}]\!] \triangleq \text{breaks-of}[\![\text{S}_t]\!]$	
S	::=	**if** (B) S_t **else** S_f	$\text{breaks-of}[\![\text{S}]\!] \triangleq \text{breaks-of}[\![\text{S}_t]\!] \cup \text{breaks-of}[\![\text{S}_f]\!]$	

S	::=	**while** (B) S_b	breaks-of⟦S⟧ ≜ ∅
S	::=	ℓ **break**;	breaks-of⟦S⟧ ≜ {at⟦S⟧}
S	::=	{ Sl }	breaks-of⟦S⟧ ≜ breaks-of⟦Sl⟧

Exercise 4.13 Calculate breaks-of⟦S⟧ for all statements S in the (syntax tree of) the program (4.5).

4.2.6 Labels in a Statement

- in⟦S⟧: the program points inside S (including at⟦S⟧ but excluding after⟦S⟧ and break-to⟦S⟧). The definition checks that program labels are unique.

P	::=	Sl	in⟦P⟧ ≜ in⟦Sl⟧	after⟦Sl⟧ ∉ in⟦Sl⟧
Sl	::=	Sl′ S	in⟦Sl⟧ ≜ in⟦Sl′⟧ ∪ in⟦S⟧	in⟦Sl′⟧ ∩ in⟦S⟧ = ∅ when Sl′ ≠ {…{ ϵ }…}
Sl	::=	ϵ	in⟦Sl⟧ ≜ {at⟦Sl⟧}	
S	::=	x = E ;	in⟦S⟧ ≜ {at⟦S⟧}	
S	::=	;	in⟦S⟧ ≜ {at⟦S⟧}	
S	::=	**if** (B) S_t	in⟦S⟧ ≜ {at⟦S⟧} ∪ in⟦S_t⟧	at⟦S⟧ ∉ in⟦S_t⟧
S	::=	**if** (B) S_t **else** S_f	in⟦S⟧ ≜ {at⟦S⟧} ∪ in⟦S_t⟧ ∪ in⟦S_f⟧	at⟦S⟧ ∉ in⟦S_t⟧ ∪ in⟦S_f⟧ in⟦S_t⟧ ∩ in⟦S_f⟧ = ∅
S	::=	**while** (B) S_b	in⟦S⟧ ≜ {at⟦S⟧} ∪ in⟦S_b⟧	at⟦S⟧ ∉ in⟦S_b⟧
S	::=	**break**;	in⟦S⟧ ≜ {at⟦S⟧}	
S	::=	{ Sl }	in⟦S⟧ ≜ in⟦Sl⟧	

Exercise 4.14 Calculate in⟦P⟧ for the program (4.5).

Lemma 4.15 For all program components S ∈ $\mathbb{P}\mathbb{c}$ of a program P, at⟦S⟧ ∈ in⟦S⟧.

Proof of lemma 4.15 By structural induction on S.

In the base case, for example, if Sl ::= ϵ then in⟦Sl⟧ ≜ {at⟦Sl⟧} so at⟦Sl⟧ ∈ in⟦Sl⟧ and for S ::= ℓ x = E ; then in⟦S⟧ ≜ {ℓ} where at⟦S⟧ ≜ ℓ. Similarly for the other base cases, S ::= ;, S ::= **if** (B) S_t, S ::= **if** (B) S_t **else** S_f, S ::= **while** (B) S_b, and S ::= **break** ;.

For the induction cases, if S ::= { Sl } then at⟦S⟧ = at⟦Sl⟧ ∈ in⟦Sl⟧ = in⟦S⟧ by def. at, in, and induction hypothesis. If Sl ::= Sl′ S then at⟦Sl⟧ = at⟦Sl′⟧ ∈ in⟦Sl′⟧ ⊆ in⟦Sl⟧ by def. at, in, and induction hypothesis. Otherwise, P ::= Sl ℓ and at⟦P⟧ = at⟦Sl⟧ ∈ in⟦Sl⟧ ⊆ in⟦P⟧ by def. at, in, and induction hypothesis. □

Lemma 4.16 For all program nonempty components S ≠ {…{ ϵ }…} of a program P, after⟦S⟧ ∉ in⟦S⟧.

Proof of lemma 4.16 The proof is by induction on the distance $\delta(\mathsf{S})$ of S to the root of the abstract syntax tree of P.

- For the basis $\mathsf{P} ::= \mathtt{Sl}\ \ell$, where $\delta(\mathsf{P}) = 0$, we have $\mathsf{after}[\![\mathsf{P}]\!] \triangleq \mathsf{after}[\![\mathtt{Sl}]\!] \triangleq \ell$ and $\mathsf{in}[\![\mathsf{P}]\!] \triangleq \mathsf{in}[\![\mathtt{Sl}]\!]$ with $\ell \notin \mathsf{in}[\![\mathtt{Sl}]\!]$ so $\mathsf{after}[\![\mathsf{P}]\!] \notin \mathsf{in}[\![\mathsf{P}]\!]$ and $\mathsf{after}[\![\mathtt{Sl}]\!] \notin \mathsf{in}[\![\mathtt{Sl}]\!]$.
- For $\mathtt{Sl} ::= \mathtt{Sl}'\ \mathsf{S}$ where $\delta(\mathtt{Sl}') = \delta(\mathsf{S}) = \delta(\mathtt{Sl}) + 1$, we have $\mathsf{after}[\![\mathtt{Sl}']\!] \triangleq \mathsf{at}[\![\mathsf{S}]\!]$, $\mathsf{after}[\![\mathsf{S}]\!] \triangleq \mathsf{after}[\![\mathtt{Sl}]\!]$, $\mathsf{in}[\![\mathtt{Sl}]\!] \triangleq \mathsf{in}[\![\mathtt{Sl}']\!] \cup \mathsf{in}[\![\mathsf{S}]\!]$, $\mathsf{in}[\![\mathtt{Sl}']\!] \cap \mathsf{in}[\![\mathsf{S}]\!] = \emptyset$ because $\mathtt{Sl}' \neq \epsilon$ and, by lemma 4.15, $\mathsf{at}[\![\mathsf{S}]\!] \in \mathsf{in}[\![\mathsf{S}]\!]$ so $\mathsf{after}[\![\mathtt{Sl}']\!] = \mathsf{at}[\![\mathsf{S}]\!] \notin \mathsf{in}[\![\mathtt{Sl}']\!]$. Moreover, $\mathsf{after}[\![\mathsf{S}]\!] = \mathsf{after}[\![\mathtt{Sl}]\!] \notin \mathsf{in}[\![\mathtt{Sl}]\!]$ by induction hypothesis hence $\mathsf{after}[\![\mathsf{S}]\!] \notin \mathsf{in}[\![\mathsf{S}]\!]$.
- If $\mathsf{S} ::= \mathtt{if}\ \ell\ (\mathsf{B})\ \mathsf{S}_t$ then $\mathsf{after}[\![\mathsf{S}_t]\!] \triangleq \mathsf{after}[\![\mathsf{S}]\!]$, $\mathsf{after}[\![\mathsf{S}]\!] \notin \mathsf{in}[\![\mathsf{S}]\!]$ by induction hypothesis because $\delta(\mathsf{S}_t) = \delta(\mathsf{S}) + 1$, so $\mathsf{after}[\![\mathsf{S}_t]\!] \notin \mathsf{in}[\![\mathsf{S}_t]\!]$ because $\mathsf{in}[\![\mathsf{S}_t]\!] \subseteq \mathsf{in}[\![\mathsf{S}]\!]$.
- By a similar argument, $\mathsf{after}[\![\mathsf{S}_t]\!] \notin \mathsf{in}[\![\mathsf{S}_t]\!]$ and $\mathsf{after}[\![\mathsf{S}_f]\!] \notin \mathsf{in}[\![\mathsf{S}_f]\!]$ when $\mathsf{S} ::= \mathtt{if}\ \ell\ (\mathsf{B})\ \mathsf{S}_t\ \mathtt{else}\ \mathsf{S}_f$.
- If $\mathsf{S} ::= \mathtt{while}\ \ell\ (\mathsf{B})\ \mathsf{S}_b$ then $\mathsf{after}[\![\mathsf{S}_b]\!] \triangleq \ell$ and $\ell \notin \mathsf{in}[\![\mathsf{S}_b]\!]$ by def. $\mathsf{in}[\![\mathsf{S}]\!]$.
- If $\mathsf{S} ::= \{\ \mathtt{Sl}\ \}$ and $\mathtt{Sl} \neq \{\ \ldots \{\ \epsilon\ \} \ldots\ \}$ then $\mathsf{after}[\![\mathtt{Sl}]\!] \triangleq \mathsf{after}[\![\mathsf{S}]\!]$, $\mathsf{in}[\![\mathsf{S}]\!] \triangleq \mathsf{in}[\![\mathtt{Sl}]\!]$, and $\delta(\mathtt{Sl}) = \delta(\mathsf{S}) + 1$ so $\mathsf{after}[\![\mathsf{S}]\!] \notin \mathsf{in}[\![\mathsf{S}]\!]$ by induction hypothesis because $\mathsf{S} \neq \epsilon$, proving $\mathsf{after}[\![\mathtt{Sl}]\!] \notin \mathsf{in}[\![\mathtt{Sl}]\!]$. □

Lemma 4.17 For all program components $\mathsf{S} \in \mathbb{P}\mathbb{c}$ of a program P, $\mathsf{escape}[\![\mathsf{S}]\!] \Rightarrow (\mathsf{break\text{-}to}[\![\mathsf{S}]\!] \notin \mathsf{in}[\![\mathsf{S}]\!])$.

Proof of lemma 4.17 The lemma trivially holds if $\mathsf{escape}[\![\mathsf{S}]\!] = \mathrm{ff}$. Otherwise $\mathsf{escape}[\![\mathsf{S}]\!] = \mathrm{tt}$ and the proof that $\mathsf{break\text{-}to}[\![\mathsf{S}]\!] \notin \mathsf{in}[\![\mathsf{S}]\!]$ is by induction on the distance $\delta(\mathsf{S})$ of S to the root of the abstract syntax tree of P (where $\delta(\mathsf{P}) = 0$).

- If $\mathtt{Sl} ::= \mathtt{Sl}'\ \mathsf{S}$ then $\mathsf{break\text{-}to}[\![\mathtt{Sl}']\!] \triangleq \mathsf{break\text{-}to}[\![\mathsf{S}]\!] \triangleq \mathsf{break\text{-}to}[\![\mathtt{Sl}]\!]$ and $\mathsf{break\text{-}to}[\![\mathtt{Sl}]\!] \notin \mathsf{in}[\![\mathtt{Sl}]\!]$ by induction hypothesis because $\delta(\mathtt{Sl}') = \delta(\mathsf{S}) = \delta(\mathtt{Sl}) + 1$ so $\mathsf{break\text{-}to}[\![\mathtt{Sl}']\!] \notin \mathsf{in}[\![\mathtt{Sl}']\!]$ and $\mathsf{break\text{-}to}[\![\mathsf{S}]\!] \notin \mathsf{in}[\![\mathsf{S}]\!]$ because $\mathsf{in}[\![\mathtt{Sl}]\!] \triangleq \mathsf{in}[\![\mathtt{Sl}']\!] \cup \mathsf{in}[\![\mathsf{S}]\!]$.
- If $\mathsf{S} ::= \mathtt{if}\ \ell\ (\mathsf{B})\ \mathsf{S}_t$ then $\mathsf{break\text{-}to}[\![\mathsf{S}_t]\!] \triangleq \mathsf{break\text{-}to}[\![\mathsf{S}]\!]$ and $\mathsf{break\text{-}to}[\![\mathsf{S}]\!] \notin \mathsf{in}[\![\mathsf{S}]\!]$ by induction hypothesis so $\mathsf{break\text{-}to}[\![\mathsf{S}_t]\!] \notin \mathsf{in}[\![\mathsf{S}_t]\!]$ because $\mathsf{in}[\![\mathsf{S}_t]\!] \subseteq \mathsf{in}[\![\mathsf{S}]\!]$.
- The proof is similar for $\mathsf{S} ::= \mathtt{if}\ \ell\ (\mathsf{B})\ \mathsf{S}_t\ \mathtt{else}\ \mathsf{S}_f$.
- If $\mathsf{S} ::= \mathtt{while}\ \ell\ (\mathsf{B})\ \mathsf{S}_b$ then $\mathsf{break\text{-}to}[\![\mathsf{S}_b]\!] \triangleq \mathsf{after}[\![\mathsf{S}]\!]$, $\mathsf{after}[\![\mathsf{S}]\!] \notin \mathsf{in}[\![\mathsf{S}]\!]$ by lemma 4.16 so $\mathsf{break\text{-}to}[\![\mathsf{S}_b]\!] \notin \mathsf{in}[\![\mathsf{S}_b]\!]$ because $\mathsf{in}[\![\mathsf{S}_b]\!] \subseteq \mathsf{in}[\![\mathsf{S}]\!]$.
- If $\mathsf{S} ::= \{\ \mathtt{Sl}\ \}$ then $\mathsf{break\text{-}to}[\![\mathtt{Sl}]\!] \triangleq \mathsf{break\text{-}to}[\![\mathsf{S}]\!]$, $\mathsf{break\text{-}to}[\![\mathsf{S}]\!] \notin \mathsf{in}[\![\mathsf{S}]\!]$ by induction hypothesis, so $\mathsf{break\text{-}to}[\![\mathtt{Sl}]\!] \notin \mathsf{in}[\![\mathtt{Sl}]\!]$ because $\mathsf{in}[\![\mathtt{Sl}]\!] = \mathsf{in}[\![\mathsf{S}]\!]$. □

Lemma 4.18 For all program components $\mathsf{S} \in \mathbb{P}\mathbb{c}$ of a program P, $\mathsf{escape}[\![\mathsf{S}]\!] \Rightarrow (\mathsf{break\text{-}to}[\![\mathsf{S}]\!] \neq \mathsf{after}[\![\mathsf{S}]\!])$.

The proof of lemma 4.18 is available at https://github.com/PrAbsInt/.

4.2.7 Labels of a Statement

- $\mathsf{labs}[\![\mathsf{S}]\!]$: the potentially reachable program points while executing the program component S either in or after S (excluding reachability by a break) and

- labx⟦S⟧: the potentially reachable program points while executing the program component S either in or after S, or resulting from a break.

$$\mathsf{labs}[\![S]\!] \triangleq \mathsf{in}[\![S]\!] \cup \{\mathsf{after}[\![S]\!]\}$$
$$\mathsf{labx}[\![S]\!] \triangleq \mathsf{labs}[\![S]\!] \cup (\mathsf{escape}[\![S]\!] \mathbin{?} \{\mathsf{break\text{-}to}[\![S]\!]\} \mathbin{:} \varnothing)$$

as shown in figure 4.19.

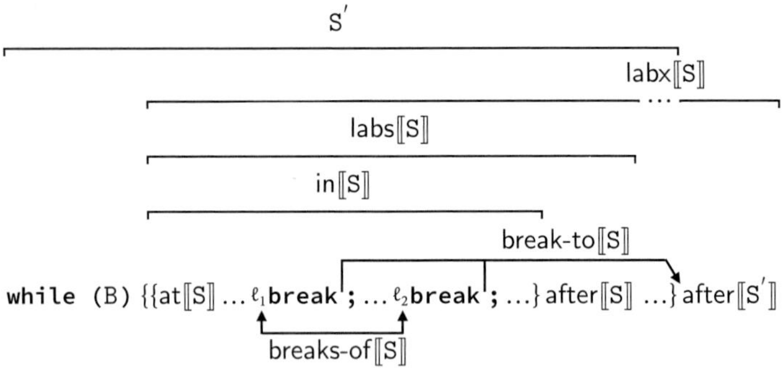

Figure 4.19
Labels of a statement

For an iteration S = **while** (B) S_b, labx⟦S⟧ = labs⟦S⟧ because escape⟦S⟧ = ff.

Example 4.20 Consider the program P = **while** ℓ_1 (tt) ℓ_2 **break** ; ℓ_3 x = 7 ; ℓ_4 whose grammatical structure is

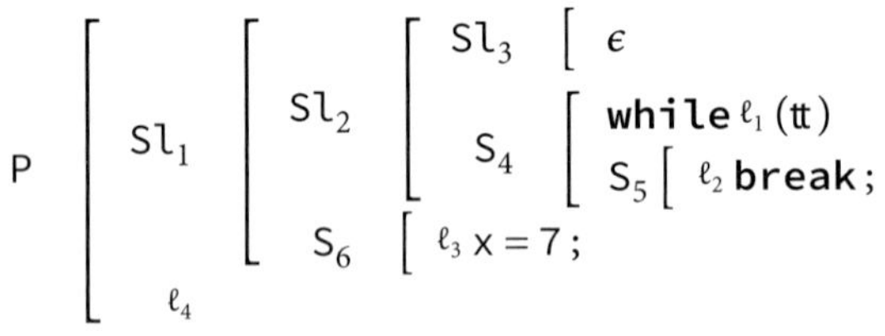

We have

S	at⟦S⟧	in⟦S⟧	escape⟦S⟧	break-to⟦S⟧	after⟦S⟧	labs⟦S⟧	labx⟦S⟧
P	ℓ_1	ℓ_1, ℓ_2, ℓ_3	ff	—	ℓ_4	$\ell_1, \ell_2, \ell_3, \ell_4$	$\ell_1, \ell_2, \ell_3, \ell_4$
Sl_1	ℓ_1	ℓ_1, ℓ_2, ℓ_3	ff	—	ℓ_4	$\ell_1, \ell_2, \ell_3, \ell_4$	$\ell_1, \ell_2, \ell_3, \ell_4$
Sl_2	ℓ_1	ℓ_1, ℓ_2	ff	—	ℓ_3	ℓ_1, ℓ_2, ℓ_3	ℓ_1, ℓ_2, ℓ_3
Sl_3	ℓ_1	ℓ_1	ff	—	ℓ_1	ℓ_1	ℓ_1
S_4	ℓ_1	ℓ_1, ℓ_2	ff	—	ℓ_3	ℓ_1, ℓ_2, ℓ_3	ℓ_1, ℓ_2, ℓ_3
S_5	ℓ_2	ℓ_2	tt	ℓ_3	ℓ_1	ℓ_1, ℓ_2	ℓ_1, ℓ_2, ℓ_3
S_6	ℓ_3	ℓ_3	ff	—	ℓ_4	ℓ_3, ℓ_4	ℓ_3, ℓ_4

Exercise 4.21 Calculate labx⟦S⟧ for all statements S in the (syntax tree of) the program (4.5).

Exercise 4.22 For each program, statement list, and statement S, define

- $\mathsf{dest}[\![\mathtt{S}]\!]$: the labels that are the destination of a break statement in $\mathtt{S}$ including the inner loops (empty if there is no break in statement $\mathtt{S}$ or its inner loops).

4.3 Conclusion

We have defined the syntax of the language by a context-free grammar and various auxiliary syntactic functions.

This syntax definition is used to define the semantics of the language in chapter 6.

This syntax definition also allows us to automatically generate a parser in chapter 5 to create a computer representation of the program for a verification or static analysis tool.

4.4 Solutions to Selected Exercises

Solution to exercise 4.4 An $\mathtt{if}\,(\mathtt{B})\;\mathtt{S}_t\;\mathtt{else}\;\mathtt{S}_f$ can be replaced with ($(\mathtt{B}) \uparrow (\mathtt{B})$ is $\neg(\mathtt{B})$)

$$\mathtt{while}\,(\mathtt{B})\,\{\,\mathtt{S}_t\;\mathtt{break}\,;\}\;\mathtt{while}\,((\mathtt{B}) \uparrow (\mathtt{B}))\,\{\,\mathtt{S}_f\;\mathtt{break}\,;\}.$$

Solution to exercise 4.6

$\mathsf{at}[\![\mathtt{Sl}_1]\!] = \mathsf{after}[\![\mathtt{Sl}_1]\!] = \ell_3$ (see exercise 4.7)
$\mathsf{at}[\![\mathtt{S}_2]\!] = \ell_3$
$\mathsf{at}[\![\mathtt{S}_3]\!] = \ell_5$
$\mathsf{at}[\![\mathtt{Sl}_4]\!] = \mathsf{at}[\![\mathtt{Sl}_1]\!] = \ell_3$
$\mathsf{at}[\![\mathtt{S}_5]\!] = \ell_4$
$\mathsf{at}[\![\mathtt{Sl}_6]\!] = \mathsf{at}[\![\mathtt{Sl}_4]\!] = \ell_3$
$\mathsf{at}[\![\mathtt{Sl}_7]\!] = \mathsf{after}[\![\mathtt{Sl}_7]\!] = \ell_1$ (see exercise 4.7)
$\mathsf{at}[\![\mathtt{S}_8]\!] = \ell_1$
$\mathsf{at}[\![\mathtt{Sl}_9]\!] = \mathsf{at}[\![\mathtt{Sl}_6]\!] = \ell_3$
$\mathsf{at}[\![\mathtt{Sl}_{10}]\!] = \mathsf{at}[\![\mathtt{Sl}_7]\!] = \ell_1$
$\mathsf{at}[\![\mathtt{S}_{11}]\!] = \ell_2$
$\mathsf{at}[\![\mathtt{Sl}_{12}]\!] = \mathsf{at}[\![\mathtt{Sl}_{10}]\!] = \ell_1$
$\mathsf{at}[\![\mathtt{S}_{13}]\!] = \ell_6$
$\mathsf{at}[\![\mathtt{Sl}_{14}]\!] = \mathsf{at}[\![\mathtt{Sl}_{12}]\!] = \ell_1$
$\mathsf{at}[\![\mathtt{P}_{15}]\!] = \mathsf{at}[\![\mathtt{Sl}_{14}]\!] = \ell_1$

Solution to exercise 4.7

$\mathsf{after}[\![\mathtt{P}_{15}]\!] = \mathsf{after}[\![\mathtt{Sl}_{14}]\!] = \ell_7$
$\mathsf{after}[\![\mathtt{Sl}_{12}]\!] = \mathsf{at}[\![\mathtt{S}_{13}]\!] = \ell_6$
$\mathsf{after}[\![\mathtt{S}_{13}]\!] = \mathsf{after}[\![\mathtt{Sl}_{14}]\!] = \ell_7$
$\mathsf{after}[\![\mathtt{Sl}_{10}]\!] = \mathsf{at}[\![\mathtt{S}_{11}]\!] = \ell_2$
$\mathsf{after}[\![\mathtt{S}_{11}]\!] = \mathsf{after}[\![\mathtt{Sl}_{12}]\!] = \ell_6$
$\mathsf{after}[\![\mathtt{Sl}_7]\!] = \mathsf{at}[\![\mathtt{S}_8]\!] = \ell_1$
$\mathsf{after}[\![\mathtt{S}_8]\!] = \mathsf{after}[\![\mathtt{Sl}_{10}]\!] = \ell_2$
$\mathsf{after}[\![\mathtt{S}_9]\!] = \ell_2$
$\mathsf{after}[\![\mathtt{Sl}_6]\!] = \mathsf{after}[\![\mathtt{S}_9]\!] = \ell_2$
$\mathsf{after}[\![\mathtt{Sl}_4]\!] = \mathsf{at}[\![\mathtt{S}_5]\!] = \ell_4$
$\mathsf{after}[\![\mathtt{S}_5]\!] = \mathsf{after}[\![\mathtt{Sl}_6]\!] = \ell_2$
$\mathsf{after}[\![\mathtt{Sl}_1]\!] = \mathsf{at}[\![\mathtt{S}_2]\!] = \ell_3$
$\mathsf{after}[\![\mathtt{S}_2]\!] = \mathsf{after}[\![\mathtt{Sl}_4]\!] = \ell_4$
$\mathsf{after}[\![\mathtt{S}_3]\!] = \mathsf{after}[\![\mathtt{S}_5]\!] = \ell_2$

Hence $\mathsf{after}[\![\mathtt{S}]\!]$ is the label at which execution goes on when $\mathtt{S}$ terminates without a **break**;.

Solution to exercise 4.8

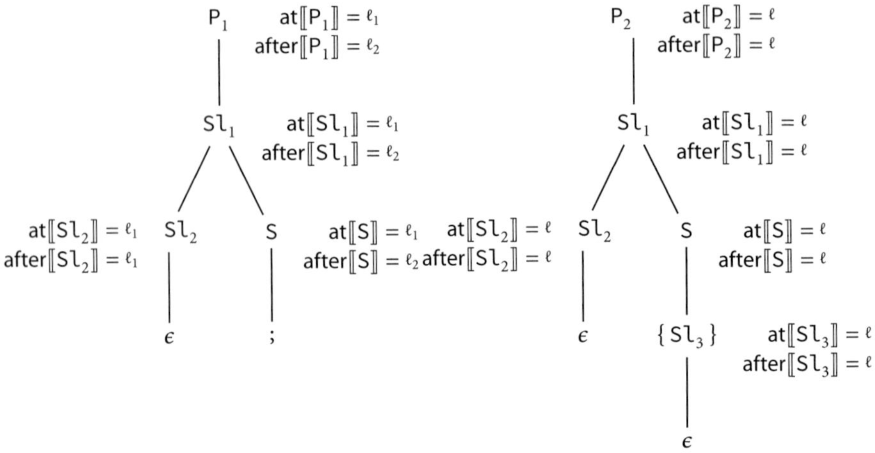

Solution to exercise 4.12

$$
\begin{aligned}
&\text{break-to}[\![\mathtt{P}_{15}]\!] = \text{break-to}[\![\mathtt{Sl}_{14}]\!] = \ell_7\\
&\text{break-to}[\![\mathtt{Sl}_{12}]\!] = \text{break-to}[\![\mathtt{S}_{13}]\!] = \text{break-to}[\![\mathtt{Sl}_{14}]\!] = \ell_7\\
&\text{break-to}[\![\mathtt{Sl}_{10}]\!] = \text{break-to}[\![\mathtt{S}_{11}]\!] = \text{break-to}[\![\mathtt{Sl}_{12}]\!] = \ell_7\\
&\text{break-to}[\![\mathtt{Sl}_{7}]\!] = \text{break-to}[\![\mathtt{S}_{8}]\!] = \text{break-to}[\![\mathtt{Sl}_{10}]\!] = \ell_7\\
&\text{break-to}[\![\mathtt{S}_{9}]\!] = \text{after}[\![\mathtt{S}_{11}]\!] = \ell_6\\
&\text{break-to}[\![\mathtt{Sl}_{6}]\!] = \text{break-to}[\![\mathtt{S}_{9}]\!] = \ell_6\\
&\text{break-to}[\![\mathtt{Sl}_{4}]\!] = \text{break-to}[\![\mathtt{S}_{5}]\!] = \text{break-to}[\![\mathtt{Sl}_{6}]\!] = \ell_6\\
&\text{break-to}[\![\mathtt{Sl}_{1}]\!] = \text{break-to}[\![\mathtt{S}_{2}]\!] = \text{break-to}[\![\mathtt{Sl}_{4}]\!] = \ell_6\\
&\text{break-to}[\![\mathtt{S}_{3}]\!] = \text{break-to}[\![\mathtt{S}_{5}]\!] = \ell_6
\end{aligned}
$$

so a `break` before the `while` loop would terminate the program at ℓ_7 whereas a `break` inside the `while` loop (like ℓ_5**`break`** ;) terminates this loop at ℓ_6.

Solution to exercise 4.22

$$
\begin{array}{rcll}
\mathtt{P} & ::= & \mathtt{Sl}\,\ell & \text{dest}[\![\mathtt{P}]\!] = \text{dest}[\![\mathtt{Sl}]\!]\\
\mathtt{Sl} & ::= & \mathtt{Sl'}\ \mathtt{S} & \text{dest}[\![\mathtt{Sl}]\!] = \text{dest}[\![\mathtt{Sl'}]\!] \cup \text{dest}[\![\mathtt{S}]\!]\\
\mathtt{Sl} & ::= & \epsilon & \text{dest}[\![\mathtt{Sl}]\!] = \emptyset\\
\mathtt{S} & ::= & \ell\ \mathtt{x = E;} & \text{dest}[\![\mathtt{S}]\!] = \emptyset\\
\mathtt{S} & ::= & \ell\mathtt{;} & \text{dest}[\![\mathtt{S}]\!] = \emptyset\\
\mathtt{S} & ::= & \mathbf{if}\,\ell\,\mathtt{(B)}\ \mathtt{S}_t & \text{dest}[\![\mathtt{S}]\!] = \text{dest}[\![\mathtt{S}_t]\!]\\
\mathtt{S} & ::= & \mathbf{if}\,\ell\,\mathtt{(B)}\ \mathtt{S}_t\ \mathbf{else}\ \mathtt{S}_f & \text{dest}[\![\mathtt{S}]\!] = \text{dest}[\![\mathtt{S}_t]\!] \cup \text{dest}[\![\mathtt{S}_f]\!]\\
\mathtt{S} & ::= & \mathbf{while}\,\ell\,\mathtt{(B)}\ \mathtt{S}_b & \text{dest}[\![\mathtt{S}]\!] = \text{dest}[\![\mathtt{S}_b]\!]\\
\mathtt{S} & ::= & \ell\ \mathbf{break}\mathtt{;} & \text{dest}[\![\mathtt{S}]\!] = \{\text{break-to}[\![\mathtt{S}]\!]\}\\
\mathtt{S} & ::= & \mathtt{\{\ Sl\ \}} & \text{dest}[\![\mathtt{S}]\!] = \text{dest}[\![\mathtt{Sl}]\!]
\end{array}
\tag{4.23}
$$

5 Parsing

Parsing consists of building an abstract syntax tree of a sentence recognized by a context-free grammar otherwise rejecting that sentence because of a syntax error.

Contents

5.1 Lexer

Consider the program `if (x < 0) x = - x ;`. In a first phase, a program (called a lexer) abstracts the terminals in this input sentence by lexemes, as follows:

```
IF LPARENT IDENT LT NUM RPARENT IDENT ASSIGN MINUS IDENT SEMICOLON END
if    (       x    <  0       )       x       =     -      x        ;
```

The `END` lexeme represents the end of the program, for example, the end of the input file.

5.2 Lexer Specification

The lexer can be generated automatically from a formal description of the lexemes by regular expressions as follows:

```
(* File lexer.mll *)
{
open Parser
exception Error of string
}
rule token = parse
    [' ' '\t' '\n'] { token lexbuf } (* skip blanks, tabs and newlines *)
  | "nand"      { NAND }
  | "if"        { IF }
  | "else"      { ELSE }
```

```
| "while"      { WHILE }
| "break"      { BREAK }
| (['a'-'z'] | ['A'-'Z'])(['a'-'z'] | ['A'-'Z'] | ['0'-'9'])* as idt
               { IDENT idt }
| ['0'-'9']+ as num
               { NUM (int_of_string num) }
| '-'          { MINUS }
| '<'          { LT }
| '('          { LPAREN }
| ')'          { RPAREN }
| '='          { ASSIGN }
| ';'          { SEMICOLON }
| '{'          { LBRACKET }
| '}'          { RBRACKET }
| eof          { END }
| _
  { raise (Error (Printf.sprintf "At offset %d: unexpected character.\n"
                                      (Lexing.lexeme_start lexbuf))) }
```

The `lexbuf` contains the input sentence, which is a sequence of characters. Each regular expression is considered in order. For the first that applies a lexeme is returned, and lexing goes on with the rest of the input sentence `lexbuf`.

`token lexbuf` moves to the next character in `lexbuf` so that space, tabulation, and end of line characters are skipped.

Then, reserved keywords are checked.

Otherwise, an attempt is made to recognize an identifier that is the longest possible string, if any, starting witty a lowercase or uppercase letter followed by zero or more letters or digits. If an identifier is recognized, this string of characters is returned as part of the lexeme.

Otherwise, an attempt is made to recognize a natural number as a nonempty sequence of digits. The value returned with the lexeme is the value of that sequence of digits considered with the decimal encoding.

Otherwise, an attempt is made to recognize symbols, where the end of file `eof` is considered to be the end of the program marked by the lexeme `END`.

If none of the above cases applies, an error exception is raised. For example, for the input `x=10; while (x>0) x=x-1;` we get the error `At offset 14: unexpected character.` because > is not a Boolean operator in our language.

Exercise 5.1 Provide the syntax and semantics of regular expressions on a given finite alphabet $\mathscr{A}$ (where the semantics is the language on the alphabet $\mathscr{A}$ specified by a regular expression).

Exercise 5.2 (Project: arithmetic and Boolean operators lexer) Extend the lexer for the usual arithmetic and Boolean operators.

5.3 Lexer Generation

The lexer generator ocamllex takes this specification and automatically produces a lexer, which is an OCaml program that whenever called, returns the next lexeme recognized in `lexbuf`. The `lexbuf` buffer can be chosen to be a file, the input keyboard, or a string. The lexemes will be OCaml values of the following type `token`:

```
type token =
```

```
  | WHILE
  | SEMICOLON
  | RPAREN
  | RBRACKET
  | NUM of (int)
  | NAND
  | MINUS
  | LT
  | LPAREN
  | LBRACKET
  | IF
  | IDENT of (string)
  | END
  | ELSE
  | BREAK
  | ASSIGN
```

(which is automatically generated from the parser specification in the file `parser.mly` below).

5.4 Grammar

The syntax of the language is specified by the following grammar. Terminals in the input sentence have been replaced by lexemes so that the grammar input sentence is a sequence of lexemes.

```
prog:
  | stmtlist END

stmt:
  | IDENT ASSIGN aexpr SEMICOLON
  | SEMICOLON
  | IF LPAREN bexpr RPAREN stmt
  | IF LPAREN bexpr RPAREN stmt ELSE stmt
  | WHILE LPAREN bexpr RPAREN stmt
  | BREAK SEMICOLON
  | LBRACKET stmtlist RBRACKET

stmtlist:
  | stmtlist stmt
  |

aexpr:
  | NUM
  | IDENT
  | aexpr MINUS aexpr
  | MINUS aexpr
  | LPAREN aexpr RPAREN

bexpr:
  | aexpr LT aexpr
  | bexpr NAND bexpr
  | LPAREN bexpr RPAREN
```

Exercise 5.3 (Project: arithmetic and boolean expressions parser) Continuing exercise 5.2, extend the grammar for the usual arithmetic and Boolean operators.

The objective of a parser is to built the parse/syntax tree of a program. For the example program `if (x < 0) x = - x ;`, the parse tree is the following.

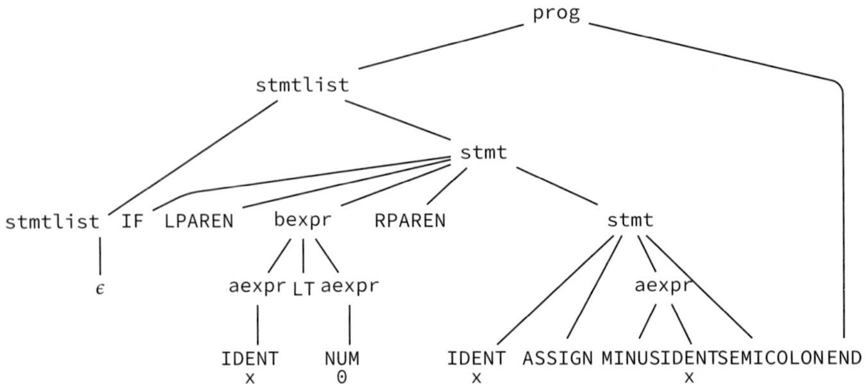

5.5 Parser

We consider left to right, bottom up parsing so that the syntax tree is built in prefix order from the input sentence of lexemes. Initially the syntax tree is empty and the input sentence is the sequence of lexemes to be recognized. An intermediate state of the parser would be the following:

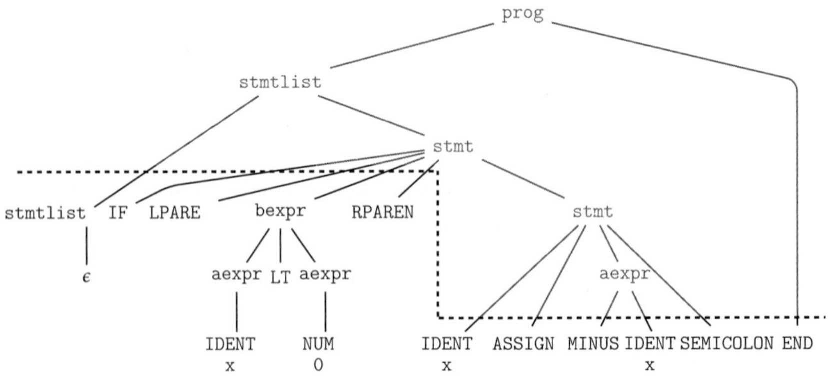

Only the top of the prefix of the parsing tree that has been recognized until this point is used so that it can be stored in a stack. The input is the rest of the sequence of lexemes to be recognized.

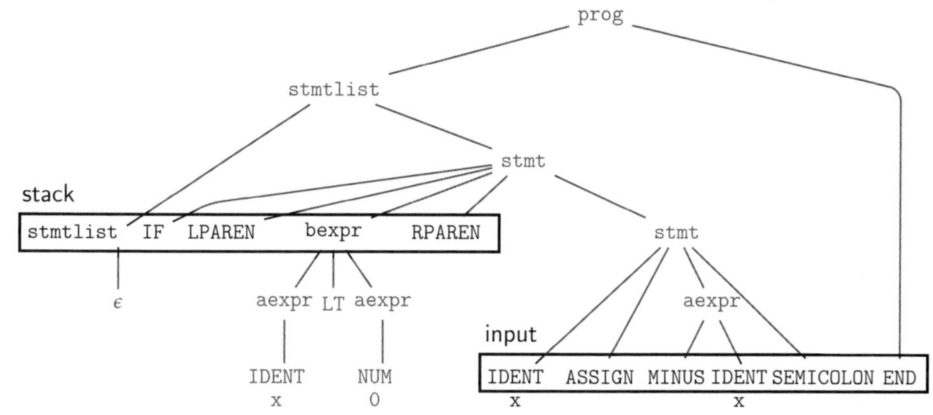

Depending on the state of the stack and the input, the next action of the parser is either a shift, a reduce, an acceptance, or a rejection of the sentence.

A *reduce* consists of replacing the right hand side of a grammar rule recognized on top of the stack by the left hand side of the rule.

For example, if the stack is `stmtlist IF LPARENT aexpr LT aexpr` and the input is `RPARENT IDENT ASSIGN MINUS IDENT SEMICOLON END` then a reduction replaces the right hand side `aexpr LT aexpr` of the grammar rule by its left hand side `bexpr` so that the stack becomes `stmtlist IF LPARENT bexpr` and the input is unchanged.

A *shift* is for the case when the stack contains part of the right hand side of a rule in construction and a lexeme is pushed from the input to the stack.

For example, if the stack is `stmtlist IF LPARENT bexpr` and the input is `RPARENT IDENT ASSIGN MINUS IDENT SEMICOLON END` then the next lexeme `RPARENT` is shifted to the stack, which becomes `stmtlist IF LPARENT bexpr RPARENT` while the input is at this point `IDENT ASSIGN MINUS IDENT SEMICOLON END`.

An *acceptance* is when the stack is reduced to the grammar axiom and the input is empty. Otherwise it is a rejection, that is, the stack cannot be extended by the input to a meaningful sentence of the language defined by the grammar.

The parsing of the input sentence

`IF LPARENT IDENT LT NUM RPARENT IDENT ASSIGN MINUS IDENT SEMICOLON END`

proceeds as follows (a state has the form stack | input, $\xrightarrow{S}$ is a shift and $\xrightarrow{R}$ is a reduction)

| `IF LPARENT IDENT LT NUM RPARENT IDENT ASSIGN MINUS IDENT SEMICOLON END`

$\xrightarrow{R}$ `stmtlist` | `IF LPARENT IDENT LT NUM RPARENT IDENT ASSIGN MINUS IDENT SEMICOLON END`

$\xrightarrow{S}$ `stmtlist IF` | `LPARENT IDENT LT NUM RPARENT IDENT ASSIGN MINUS IDENT SEMICOLON END`

$\xrightarrow{S}$ `stmtlist IF LPARENT` | `IDENT LT NUM RPARENT IDENT ASSIGN MINUS IDENT SEMICOLON END`

$\xrightarrow{S}$ `stmtlist IF LPARENT IDENT` | `LT NUM RPARENT IDENT ASSIGN MINUS IDENT SEMICOLON END`

$\xrightarrow{R}$ `stmtlist IF LPARENT aexpr` | `LT NUM RPARENT IDENT ASSIGN MINUS IDENT SEMICOLON END`

$\xrightarrow{S}$ `stmtlist IF LPARENT aexpr LT` | `NUM RPARENT IDENT ASSIGN MINUS IDENT SEMICOLON END`

$\xrightarrow{S}$ `stmtlist IF LPARENT aexpr LT NUM | RPARENT IDENT ASSIGN MINUS IDENT SEMICOLON END`

$\xrightarrow{R}$ `stmtlist IF LPARENT aexpr LT aexpr | RPARENT IDENT ASSIGN MINUS IDENT SEMICOLON END`

$\xrightarrow{R}$ `stmtlist IF LPARENT bexpr | RPARENT IDENT ASSIGN MINUS IDENT SEMICOLON END`

$\xrightarrow{S}$ `stmtlist IF bexpr | IDENT ASSIGN MINUS IDENT SEMICOLON END`

$\xrightarrow{R}$ `stmtlist IF bexpr | IDENT ASSIGN MINUS IDENT SEMICOLON END`

$\xrightarrow{S}$ `stmtlist IF bexpr IDENT | ASSIGN MINUS IDENT SEMICOLON END`

$\xrightarrow{S}$ `stmtlist IF bexpr IDENT ASSIGN | MINUS IDENT SEMICOLON END`

$\xrightarrow{S}$ `stmtlist IF bexpr IDENT ASSIGN MINUS | IDENT SEMICOLON END`

$\xrightarrow{S}$ `stmtlist IF bexpr IDENT ASSIGN MINUS IDENT | SEMICOLON END`

$\xrightarrow{R}$ `stmtlist IF bexpr IDENT ASSIGN aexpr | SEMICOLON END`

$\xrightarrow{S}$ `stmtlist IF bexpr IDENT ASSIGN aexpr SEMICOLON | END`

$\xrightarrow{R}$ `stmtlist IF bexpr stmt | END`

$\xrightarrow{R}$ `stmtlist stmt | END`

$\xrightarrow{R}$ `stmtlist | END`

$\xrightarrow{S}$ `stmtlist END |`

$\xrightarrow{R}$ `prog |`

A syntactically valid program has been recognized.

5.6 Ambiguities, Shift-Reduce Conflicts

The grammar of section 5.4 is ambiguous in that a sentence can be parsed in different ways. For example, `1-2-3` can be parsed as `((1-2)-3)` or `(1-(2-3))` which does not return the same result. In mathematics, `1-2-3` means `((1-2)-3)`, that is, the binary minus operation − is left associative and evaluated left to right. The ambiguity is therefore solved by the following declaration in the grammar specification

```
%left MINUS
```

that `MINUS` is left associative. Another way would be to change the grammar (see exercise 5.4).

Another ambiguity is for expressions like `1 - -2`, which should be evaluated as `(1 - (-2))`, that is, the unary operator − in `-2` should be evaluated before the binary operator − in `1 - ␣`. Another difficulty is that the unary and binary symbols are the same −. These two problems are solved by declaring

```
%left MINUS             /* evaluated second */
%nonassoc UMINUS        /* evaluated first */
```

that is the unary minus is evaluated before the binary minus and

```
aexpr:
  | NUM
```

```
  | IDENT
  | aexpr MINUS aexpr
  | MINUS aexpr %prec UMINUS
  | LPAREN aexpr RPAREN
```

the unary minus has the syntax of the binary minus but has the precedence of the unary minus, that is, should be evaluated first.

Another ambiguity is for expressions such as `1-2<3`, which should be parsed as `((1-2)<3)` and not `(1-(2<3))`, that is, the binary `-` should be evaluated before the logical comparison `<`, as indicated by the declaration

```
%left NAND              /* evaluated fourth */
%left LT                /* evaluated third  */
%left MINUS             /* evaluated second */
%nonassoc UMINUS        /* evaluated first  */
```

The `NAND` is left associative and should be evaluated last.

All these situations yield a shift-reduce conflict. For example, during the analysis of `1-2-x`, we reach the state `aexpr MINUS aexpr` | `MINUS IDENT`. We can either perform a reduce $\xrightarrow{R}$ `aexpr` | `MINUS IDENT` in the case of left associativity or a shift $\xrightarrow{S}$ `aexpr MINUS aexpr MINUS` | `IDENT` in the case of right associativity. The declaration indicates that the `MINUS` operator is left associative so the reduce should be performed and not the shift.

Exercise 5.4 Solve the expression ambiguity problem by modifying the grammar.

The grammar for the conditional is also ambiguous. For example `if (bexpr`$_1$`) if (bexpr`$_2$`) stmt`$_1$ `else stmt`$_2$ can be understood either as (`if (bexpr`$_1$`)` (`if (bexpr`$_2$`) stmt`$_1$) `else stmt`$_2$) or as (`if (bexpr`$_1$`)` (`if (bexpr`$_2$`) stmt`$_1$ `else stmt`$_2$)). In most programming languages, the `else` refers to the closest enclosing `if`, that is, the ambiguity is solved according to the second parsing. Again, the grammar can be modified or precedences can be used as follows

```
%nonassoc NO_ELSE       /* evaluated sixth  */
%nonassoc ELSE          /* evaluated fifth  */
%left NAND              /* evaluated fourth */
%left LT                /* evaluated third  */
%left MINUS             /* evaluated second */
%nonassoc UMINUS        /* evaluated first  */
```

The grammar is then

```
stmt:
  | ...
  | IF LPAREN b = bexpr RPAREN s = stmt %prec NO_ELSE
  | IF LPAREN b = bexpr RPAREN s1 = stmt ELSE s2 = stmt
  | ...
```

The `%prec NO_ELSE` states that the lexeme `NO_ELSE` is fake (an empty string in the input). Moreover it does not associate with itself and is evaluated last. Therefore, an input

$$\texttt{if (bexpr}_1\texttt{) if (bexpr}_2\texttt{) stmt}_1\texttt{ else if (bexpr}_3\texttt{) stmt}_2\texttt{ else stmt}_3$$

is parsed as

$$(\texttt{if (bexpr}_1\texttt{) if (bexpr}_2\texttt{) stmt}_1\texttt{ else if (bexpr}_3\texttt{) stmt}_2\texttt{ else stmt}_3\texttt{ NO\_ELSE})$$

Moreover, the `ELSE` does not associate with itself (i.e., interpreting ␣ `ELSE` ␣ `ELSE` ␣ as ((␣ `ELSE` ␣) `ELSE` ␣) or (␣ `ELSE` (␣ `ELSE` ␣)) is meaningless) so the parsing is

$$(\texttt{if (bexpr}_1\texttt{) (if (bexpr}_2\texttt{) stmt}_1\texttt{ else (if (bexpr}_3\texttt{) stmt}_2\texttt{ else stmt}_3\texttt{)) NO\_ELSE})$$

Exercise 5.5 Solve the dangling else problem by modifying the grammar.

5.7 Ambiguities, Reduce-Reduce Conflicts

A reduce-reduce conflict occurs during parsing when a reduction can be applied with two or more different grammar rule right hand sides, for the same input token. Consider, for example, the grammar

```
/* File parser.mly */
%{ %}
%token NUM
%start s /* the grammar entry point */
%type  <unit> s
%type  <unit> t

%% /* rules */
s:
  | t         { () }
  | NUM       { () }

t:
  | NUM       { () }

%%
```

The menhir parser generator produces the following error message

```
Warning: one state has reduce/reduce conflicts.
Warning: one reduce/reduce conflict was arbitrarily resolved.
File "parser.mly", line 14, characters 4-7:
Warning: production t -> NUM is never reduced.
```

menhir explains the problem in the `parser.conflicts` file. A `NUM` can be reduced to an `s` or to a `t` and then to an `s`. menhir arbitrarily chooses the first alternative so the second one is never used, which is most certainly an error in the design of the context free-grammar.

Exercise 5.6 Change the grammar to recognize the same language while avoiding the reduce-reduce conflict.

5.8 Abstract Syntax

A program can be represented by its abstract syntax which is a tree of the following type (parameterized by the type 'a of the attributes of the nodes of the abstract syntax tree, unit () for parsing.

```
(* file abstractSyntax.ml *)

type variable = string
type aexpr = Num of int | Var of string | Minus of aexpr * aexpr
type bexpr = Lt of aexpr * aexpr | Nand of bexpr * bexpr
type 'a tree =
  | Prog of 'a tree list * 'a
  | Assign of variable * aexpr * 'a
  | Emptystmt of 'a
  | If of bexpr * 'a tree * 'a
  | Ifelse of bexpr * 'a tree * 'a tree * 'a
  | While of bexpr * 'a tree * 'a
  | Break of 'a
  | Stmtlist of 'a tree list * 'a (* trees in inverse order *)
```

in the programming language OCaml [628].

The abstract syntax tree can be constructed during parsing by maintaining its abstract syntax tree with each terminal or non-terminal on the stack. Each shift extends the abstract syntax tree left to right while each reduction extends it bottom up.

Continuing the example of section 5.5, we have (⟨`nt`, `ast`⟩ is the pair of a nonterminal `nt` and its abstract syntax tree `ast` on the stack)

```
...
 S
──→ ⟨stmtlist, Emptystmt⟩ IF LPARENT ⟨bexpr, LT (Var "x",Num 0)⟩ RPARENT
     IDENT  ASSIGN MINUS ⟨IDENT, Var "x"⟩ | SEMICOLON END
      "x"
 R
──→ ⟨stmtlist, Emptystmt⟩ IF LPARENT ⟨bexpr, LT (Var "x",Num 0)⟩ RPARENT
     IDENT  ASSIGN ⟨aexpr, Minus (Num 0, Var "x")⟩ | SEMICOLON END
      "x"
 S
──→ ⟨stmtlist, Emptystmt⟩ IF LPARENT ⟨bexpr, LT (Var "x",Num 0)⟩ RPARENT
     IDENT  ASSIGN ⟨aexpr, Minus (Num 0, Var "x")⟩ SEMICOLON | END
      "x"
 R
──→ ⟨stmtlist, Emptystmt⟩ IF LPARENT ⟨bexpr, LT (Var "x",Num 0)⟩ RPARENT
     ⟨stmt, Assign ("x",Minus (Num 0, Var "x"))⟩ | END
 R
──→ ⟨stmtlist, Emptystmt⟩
     ⟨stmt, If (LT (Var "x",Num 0), Assign ("x",Minus (Num 0, Var "x")))⟩ | END
...
```

Exercise 5.7 (Project: program labeling) Continuing exercise 5.3, modify the abstract syntax to account for program labeling.

5.9 Parser Specification

The following parser specification lists the lexemes (`%token`), possibly with an associated OCaml type; the precedences (`%left`, `%right`, `%nonassoc`) in increasing priority order; the grammar axiom (`%start`); the OCaml type of the abstract syntax tree associated to the grammar nonterminals (`%type`); and the grammar rules.

The types are defined in the module `AbstractSyntax` (in file `abstractSyntax.ml`), which is opened in the header to avoid explicit qualification (such as `AbstractSyntax.aexpr`).

```
/* File parser.mly */

%{ (* header *)
open AbstractSyntax

%} /* declarations */

%token MINUS LT NAND LPAREN RPAREN ASSIGN /* lexer tokens */
%token SEMICOLON IF ELSE WHILE BREAK LBRACKET RBRACKET END
%token <string> IDENT
%token <int> NUM

%nonassoc NO_ELSE     /* evaluated fifth */
%nonassoc ELSE        /* evaluated fourth */
%left NAND            /* evaluated third */
%left MINUS           /* evaluated second */
%nonassoc UMINUS      /* evaluated first */

%start prog           /* the grammar entry point */
%type  <unit AbstractSyntax.tree> prog
%type  <aexpr> aexpr
%type  <bexpr> bexpr
%type  <unit AbstractSyntax.tree> stmt
%type  <unit AbstractSyntax.tree list> stmtlist

%% /* rules */

prog:
  | l = stmtlist END                          { Prog (l, ()) }
stmt:
  | x = IDENT ASSIGN a = aexpr SEMICOLON      { Assign (x, a, ()) }
  | SEMICOLON                                 { Emptystmt () }
  | IF LPAREN b = bexpr RPAREN s = stmt %prec NO_ELSE    { If (b, s, ()) }
  | IF LPAREN b = bexpr RPAREN s1 = stmt
    ELSE s2 = stmt                            { Ifelse (b, s1, s2, ()) }
  | WHILE LPAREN b = bexpr RPAREN s = stmt    { While (b, s, ()) }
  | BREAK SEMICOLON                           { Break () }
  | LBRACKET l = stmtlist RBRACKET            { Stmtlist (l, ()) }
stmtlist:
  | l = stmtlist s = stmt                     { s :: l (* nodes in inverse order
      *) }
  |                                           { [] }
aexpr:
  | n = NUM                                   { Num n }
  | x = IDENT                                 { Var x }
  | a1 = aexpr MINUS a2 = aexpr               { Minus (a1, a2) }
  | MINUS a = aexpr                           { Minus ((Num 0), a) } %prec UMINUS
  | LPAREN a = aexpr RPAREN                   { a }
bexpr:
  | a1 = aexpr LT a2 = aexpr                  { Lt (a1, a2) }
  | b1 = bexpr NAND b2 = bexpr                { Nand (b1, b2) }
```

```
  | LPAREN b = bexpr RPAREN                 { b }

%% (* trailer *)
```

The grammar rules give a name to the abstract syntax trees already constructed (e.g., `l` in `l = stmtlist`) and use these named abstract syntax trees to construct the abstract syntax tree of the rule left hand side nonterminal upon reduction (e.g., `{ Prog l }`).

5.10 Parser Generation

For the parser specification of section 5.9, parser generators such as ocamlyacc or menhir can automatically generate parsers in the OCaml programming language. The parser can be called in a main program (called `main.ml` in our example).

```
(* File main.ml *)

open AbstractSyntax

let rec print_aexpr a = ...;;
let rec print_bexpr a = ...;;

let rec print_program p = ...
and print_stmtlist sl = ...
and print_stmt s = ...;;

let lexbuf = Lexing.from_channel stdin in
  try
    print_program (Parser.prog Lexer.token lexbuf)
  with
  | Lexer.Error msg ->
      Printf.fprintf stderr "%s%!" msg
  | Parser.Error ->
      Printf.fprintf stderr "At offset %d: syntax error.\n%!"
                                          (Lexing.lexeme_start lexbuf);;
```

The lexing buffer is created by `lexbuf = Lexing.from_channel stdin` from the keyboard standard input (`stdin`) and is passed as a parameter to the parser `Parser.prog` of programs. The `Lexer` module is created by ocamllex from the lexer specification of section 5.2 and `Lexer.token` is the set of lexemes to be recognized. An exception is raised and captured in case of lexing or parsing error. Else, the parser returns an abstract syntax tree that is printed (in parenthesized form).

The compilation through `make` [363]

```
# file "makefile"

OCAMLBUILD := ocamlbuild -use-menhir -menhir "menhir --infer --
    explain"

all: delete built examples
built:
        $(OCAMLBUILD) main.native
```

```
examples:
        @echo "# using the parser:"
        @echo ";"| ./main.native
        @echo "x = 42;"| ./main.native
        @echo "break;"| ./main.native
        @echo "break; x = 7;"| ./main.native
        @echo "x = 7; ; break; "| ./main.native
        @echo "{}" | ./main.native
        @echo "x=-10-20--30;" | ./main.native
        @echo "x=1; y=2;" | ./main.native
        @echo "{x=10; ; y=20;}" | ./main.native
        @echo "if (1-2<3-4-5) x=-x;" | ./main.native
        @echo "if (0<x) if (x<0) x=1; else if (x<0) { x=2; x=3; }
            else { x=4; x=5; x=6; }" | ./main.native
        @echo "while (0<1){x = x - 1;} x = 42;" | ./main.native
        @echo "while (0<1){}" | ./main.native
        @echo "x=x-1;while (0<1){x=x-1;if(x<2)break;};" | ./main.
            native
        @echo "x=-10; while (x<0) if (x<0) if (0<x) x=-x;" | ./main.
            native
        @echo "x=-10; while (x<0) { x=x-1; break; }; x= 10;" | ./main
            .native
        @echo "x=0; while (x<0) { while (x<0) x=x-1; x= 10; }; x=
            100;" | ./main.native
        @echo "x=0; while (x<0) { while (x<0) x=x-1; break; }; x=
            100;" | ./main.native
        @echo "x=x-1; while (0<1) { x=x-1; if (2 < x) break; };" | ./
            main.native
        -@echo "x=10; while (x>0) x=x-1;" | ./main.native
        @echo "# end"

delete:
        rm -f *~ .*~ main.native
        ocamlbuild -clean
```

yields the following result (syntax trees are printed in parenthesized form)

```
# using the parser:
...
x = (x - 1); (while (0 < 1) { x = (x - 1); (if (x < (0 - 2)) break; ) } ) ;
At offset 14: unexpected character.
# end
```

ocamlbuild [343] automatically determines the sequence of calls to the OCaml compiler to build a native OCaml-centric software project. This can be done manually by the following `makefile`

```
all:
     ocamllex -q lexer.mll
     menhir --infer --explain parser.mly
     ocamlc -c parser.mli
     ocamlc -c lexer.ml
     ocamlc -c parser.ml
```

```
ocamlc -c main.ml
ocamlc parser.cmo lexer.cmo main.cmo -o main
```

Exercise 5.8 (Project: label generation) Continuing exercise 5.7, automatically generate the program labels in OCaml.

Exercise 5.9 (Project: labeling functions) Continuing exercise 5.8, write the syntactic labeling functions of section 4.2.1 to section 4.2.7 in OCaml. They can be memorized in the syntax tree. For example, the program (4.1) would be encoded as follows

```
<l1, l7, ff, l7>      Prog:
<l1, l2, ff, l7>         l1: x = (x - 1);
<l2, l6, ff, l6>         (while l2: (0 < 1)
<l3, l2, tt, l6>            Stmtlist: {
<l3, l4, ff, l6>               l3: x = (x - 1);
<l4, l2, tt, l6>               (if l4: (2 < x)
<l5, l2, tt, l6>                  l5: break; )
                            } )
<l6, l7, ff, l7>         l6: ;
                      l7:
```

where the quadruple attached to a statement `S` is at⟦`S`⟧, after⟦`S`⟧, escape⟦`S`⟧, and break-to⟦`S`⟧. For example, the statement list starting at `l3` may escape to `l6`.

Exercise 5.10 Consider programs that are sequences of assignments. Change `main.ml` in order to record the values of the variables after assignments and print the value last assigned.

Exercise 5.11 Consider programs that are sequences of assignments. Change `main.ml` in order to record the sign of the variables after assignments and print the sign last assigned.

5.11 Conclusion

Algol 60 was the first programming language for which syntax was formally defined by a grammar [62, 588]. This boosted research in parsing for programming languages in the 1960s and 1970s, a subject still very active for natural languages [812]. Although parsers can be generated for ambiguous grammars [345], it is in practice preferable to avoid ambiguity. This is ensured when considering LR(k) grammars [587], which lead to left to right, bottom up parsers using the first k lexemes of the input (as described in section 5.5). These parsers are simpler for the subclass of LALR(k) grammars [313, 779]. In practice $k = 1$ (or 0).

Popular lexer and parser generators are lex [629] and yacc [533], that is, ocamllex and ocamlyacc [628, chapter 12] for the OCaml functional programming language [628]. We prefer menhir [806, 807] to ocamlyacc because it reports errors with respect to the grammar (and not with respect to the more obscure parser code). `Jacc` [13] can be used for a development in Java.

We have introduced parsing from an empirical practical user point of view. For the theoretic background, see [900, 901]. Parsers are abstract interpretations of the semantics of grammars [258, 262].

5.12 Solutions to Selected Exercises

Solution to exercise 5.1 Given $\mathsf{a} \in \mathscr{A}$ and the empty string ϵ, a regular expression has syntax $R ::= \mathsf{a} \mid \epsilon \mid R? \mid R_1 \mid R_2 \mid R^+ \mid R^*$, and semantics $\boldsymbol{\mathcal{S}}[\![\mathsf{a}]\!] \triangleq \{\mathsf{a}\}$, $\boldsymbol{\mathcal{S}}[\![\epsilon]\!] \triangleq \{\epsilon\}$, $\boldsymbol{\mathcal{S}}[\![R?]\!] \triangleq \boldsymbol{\mathcal{S}}[\![R]\!] \cup \{\epsilon\}$, $\boldsymbol{\mathcal{S}}[\![R_1 \mid R_2]\!] \triangleq \boldsymbol{\mathcal{S}}[\![R_1]\!] \cup \boldsymbol{\mathcal{S}}[\![R_2]\!]$, $\boldsymbol{\mathcal{S}}[\![R^0]\!] \triangleq \{\epsilon\}$, $\boldsymbol{\mathcal{S}}[\![R^1]\!] \triangleq \boldsymbol{\mathcal{S}}[\![R]\!]$, $\boldsymbol{\mathcal{S}}[\![R^{n+1}]\!] \triangleq \boldsymbol{\mathcal{S}}[\![R]\!]\boldsymbol{\mathcal{S}}[\![R^n]\!]$, $\boldsymbol{\mathcal{S}}[\![R^+]\!] \triangleq \bigcup_{n>0} \boldsymbol{\mathcal{S}}[\![R^n]\!]$, $\boldsymbol{\mathcal{S}}[\![R^*]\!] \triangleq \bigcup_{n\geqslant 0} \boldsymbol{\mathcal{S}}[\![R^n]\!]$ and language concatenation $\mathscr{L}_1\mathscr{L}_2 \triangleq \{\sigma_1\sigma_2 \mid \sigma_1 \in \mathscr{L}_1 \wedge \sigma_2 \in \mathscr{L}_2\}$ is the concatenation of strings in the languages.

Solution to exercise 5.4

```
aterm:
  | NUM
  | IDENT
  | MINUS aterm
  | LPAREN aexpr RPAREN

aexpr:
  | aterm MINUS aexpr
  | aterm

bterm:
  | aexpr LT aexpr
  | LPAREN bexpr RPAREN

bexpr:
  | bterm NAND bexpr
  | bterm
```

Solution to exercise 5.5

```
stmt:
  | IDENT ASSIGN aexpr SEMICOLON
  | SEMICOLON
  | IF LPAREN bexpr RPAREN stmt
  | IF LPAREN bexpr RPAREN thenstmt ELSE stmt
  | WHILE LPAREN bexpr RPAREN stmt
  | BREAK SEMICOLON
  | LBRACKET stmtlist RBRACKET

thenstmt:
  | IDENT ASSIGN aexpr SEMICOLON
  | SEMICOLON
  | IF LPAREN bexpr RPAREN thenstmt ELSE thenstmt
  | WHILE LPAREN bexpr RPAREN thenstmt
  | BREAK SEMICOLON
  | LBRACKET stmtlist RBRACKET
```

III TRACE SEMANTICS

The trace semantics of a programming language formally defines the execution of any syntactically correct program of the language by describing the successive computation steps performed during this execution.

These traces may be partial (chapter 6) when the observation of the execution can stop at any time, or maximal (chapter 7) when the observation of the execution stops only on program termination and could last forever.

6 Structural Deductive Stateless Prefix Trace Semantics

In order to verify and analyze programs, it is necessary to define their runtime behavior, that is, what and how they compute when executed.

We first consider partial observations of executions, which are finite and can stop at any time. This partial observation of an execution is represented by a finite trace, which is a finite sequence of events describing the actions executed by the program. The prefix trace semantics of a program is the set of all possible partial executions traces of that program.

Maximal finite traces of program components are prefix executions that reach the point after the program component (e.g., through normal termination or, for iteration statements, a `break`).

The semantics is stateless because states record what to do next but do not record the values of variables. When a value of a variable is needed, the semantics specifies from which assignment to that variable it can come from. By abstraction of this stateless semantics, we can derive a stateful trace semantics formalizing the computer memory (see chapter 42). The semantics is structural, that is, by induction on the program syntax. It is deductive because the semantics of each program component is specified by a deductive system (or Hilbert system [497]) of axioms and rules of inference, a program execution being a proof in such a system (a formal study of deductive systems is given in section 16.2 of chapter 16).

Contents

6.1 Traces

A finite trace is a finite sequence of configurations separated by actions. Configurations are program labels designating the next action to be executed in the program. The actions record the effect of this execution: the value assigned to a variable, a test B that is true (marked B) or false (marked ¬(B)), a break exiting from a loop, or a skip.

$$\begin{array}{llll} v & \in & \mathbb{V} & & \text{values} \\ \mathsf{a} & \in & \mathbb{A} & ::= \mathsf{x} = \mathsf{A} = v \mid \mathsf{B} \mid \neg(\mathsf{B}) \mid \mathsf{break} \mid \mathsf{skip} & \text{actions} \end{array} \tag{6.1}$$

Here are a prefix trace and a finite maximal trace of program (4.5) (the second trace is maximal in that the computation it describes cannot be extended).

- $$\ell_1 \xrightarrow{\mathsf{x}=\mathsf{x}+1=1} \ell_2 \xrightarrow{\mathsf{tt}} \ell_3 \xrightarrow{\mathsf{x}=\mathsf{x}+1=2} \ell_4 \xrightarrow{\neg(\mathsf{x}>2)} \ell_2 \xrightarrow{\mathsf{tt}} \ell_3 \tag{6.2}$$
- $$\ell_1 \xrightarrow{\mathsf{x}=\mathsf{x}+1=1} \ell_2 \xrightarrow{\mathsf{tt}} \ell_3 \xrightarrow{\mathsf{x}=\mathsf{x}+1=2} \ell_4 \xrightarrow{\neg(\mathsf{x}>2)} \ell_2 \xrightarrow{\mathsf{tt}} \ell_3 \xrightarrow{\mathsf{x}=\mathsf{x}+1=3} \ell_4 \xrightarrow{\mathsf{x}>2} \ell_5 \xrightarrow{\mathsf{break}} \ell_6 \xrightarrow{\mathsf{skip}} \ell_7$$

A trace can be reduced to a single label ℓ_1.
A trace can also be infinite, as for

$$\ell_1\ \mathsf{x} = 0\ ;\mathbf{while}\ \ell_2\ (\mathsf{tt})\ \{\ \ell_3\ \mathsf{x} = \mathsf{x} + 1\ ;\ \}\ell_4$$

which has an infinite trace of the form

$$\ell_1 \xrightarrow{\mathsf{x}=0=0} \ell_2 \xrightarrow{\mathsf{tt}} \ell_3 \xrightarrow{\mathsf{x}=\mathsf{x}+1=1} \ell_2 \xrightarrow{\mathsf{tt}} \ell_3 \xrightarrow{\mathsf{x}=\mathsf{x}+1=2} \ell_2 \ldots \ell_2 \xrightarrow{\mathsf{tt}} \ell_3 \xrightarrow{\mathsf{x}=\mathsf{x}+1=n} \ell_2 \xrightarrow{\mathsf{tt}} \ell_3 \xrightarrow{\mathsf{x}=\mathsf{x}+1=n+1} \ell_2 \ldots$$

We let $\mathbb{T}^+$ be the set of all finite traces, $\mathbb{T}^\infty$ be the set of all infinite traces, and $\mathbb{T}^{+\infty} \triangleq \mathbb{T}^+ \cup \mathbb{T}^\infty$ be the set of all finite or infinite traces. Let $\mathbb{T}^* = \{\ni\} \cup \mathbb{T}^+$ and $\mathbb{T}^{*\infty} = \{\ni\} \cup \mathbb{T}^{+\infty}$ include the empty trace $\ni$.

If $\pi \in \mathbb{T}^{+\infty}$, we write $\pi = \ell\pi'$ to make clear that the trace π is assumed to start with the program label ℓ (π' is a pattern and is not in itself a properly formed trace). For finite traces $\pi \in \mathbb{T}^+$, we write $\pi = \pi'\ell$ when assuming that the trace π ends with label ℓ (although, again, π' is not in itself a properly formed trace).

Example 6.3 We use the notation π both for traces and for patterns matching parts of traces. For example, the pattern $\ell_1\pi\ell_2$ matches

$$\overbrace{\ell_1}^{\ell_1} \overbrace{\xrightarrow{\mathsf{x}=0=0}}^{\pi} \overbrace{\ell_2}^{\ell_2} \quad \text{and} \quad \overbrace{\ell_1}^{\ell_1} \overbrace{\xrightarrow{\mathsf{x}=0=0} \ell_2 \xrightarrow{\mathsf{tt}} \ell_3 \xrightarrow{\mathsf{x}=\mathsf{x}+1=1}}^{\pi} \overbrace{\ell_2}^{\ell_2}$$

but does not match $\ell_1 \xrightarrow{\mathsf{x}=0=0} \ell_3$, assuming that $\ell_2 \neq \ell_3$. By convention, $\ell_1\pi\ell_2$ also matches the trace ℓ by taking $\ell_1 = \ell_2 = \ell$ and $\pi = \ni$, where $\ni$ is the empty trace.

Trace concatenation $\frown$ is defined as follows

$$\begin{array}{rcll} \pi_1\ell_1 \frown \ell_2\pi_2 & & & \text{undefined if } \pi_1\ell_1 \text{ is finite and } \ell_1 \neq \ell_2 \\ \pi_1\ell_1 \frown \ell_1\pi_2 & \triangleq & \pi_1\ell_1\pi_2 & \text{if } \pi_1\ell_1 \text{ is finite} \\ \pi_1 \frown \pi_2 & \triangleq & \pi_1 & \text{if } \pi_1 \text{ is infinite} \end{array} \tag{6.4}$$

The first state $\langle \ell, \rho\rangle$ in a trace $\langle \ell, \rho\rangle\pi$ is called the *initial state* and $\rho(\mathsf{x})$ the *initial value* of variable $\mathsf{x} \in \mathbb{V}$. Similarly, in a finite trace $\pi\langle \ell, \rho\rangle$, $\langle \ell, \rho\rangle$ is the *final state* and $\rho(\mathsf{x})$ is the *final value* of $\mathsf{x} \in \mathbb{V}$.

Remark 6.5 (on the definition of the set of traces $\mathbb{T}^{+\infty}$) We have not defined very precisely what are traces in $\mathbb{T}^{+\infty}$. They depend on actions that depend on variables $\mathbb{V}$ for which there are several possibilities, as discussed in remark 3.5. The same holds for the set $\mathbb{L}$ of labels, which can be (1) an arbitrary set (e.g., in which the labels of the explicit labeling are chosen), (2) the labels of the program of interest, and (3) a formal parameter of the definition of traces (for which the actual parameter will be the labels of the program). In addition, the arbitrary set of labels can be given different interpretations (as shown in section 4.2.3 and exercises 4.9, 4.10, and 4.11). We prefer the third solution. Traces are parameterized by labels $\mathbb{L}$ and actions $\mathbb{A}(\mathbb{V})$, themselves parameterized by variables $\mathbb{V}$. Hence $\mathbb{T}^{+\infty} = \mathbb{V} \to \mathbb{L} \to (\mathbb{L}\,\mathbb{A}(\mathbb{V}))^*\mathbb{L} \cup (\mathbb{L}\,\mathbb{A}(\mathbb{V}))^\omega$ where $(\mathbb{L}\,\mathbb{A}(\mathbb{V}))^*\mathbb{L}$ is the set of finite sequences of labels separated by action in $\mathbb{A}(\mathbb{V})$ (possibly reduced to a single label) and $(\mathbb{L}\,\mathbb{A}(\mathbb{V}))^\omega$ is the infinite sequences of labels separated by actions. As in remark 3.5 for the semantics of expressions, we leave this parametrization by $\mathbb{V}$ and $\mathbb{L}$ implicit in the definition of the prefix trace semantics of programs.

6.2 Value of a Variable at the End of a Finite Trace

The prefix traces are stateless because they do not record the values of variables $\mathsf{x} \in \mathbb{V}$ during computation (see remark 3.5 for the choice of the definition of $\mathbb{V}$). These values can be recovered from the past computations. We define $\varrho(\pi)\mathsf{x} \in \mathbb{V}$ to be the value of variable $\mathsf{x} \in \mathbb{V}$ at the end of trace π, that is, for sequential computations the last value assigned to x (or 0 at the origin of the trace).

$$
\begin{aligned}
\varrho(\pi\ell \xrightarrow{\mathsf{x}\,=\,\mathsf{A}\,=\,v} \ell')\mathsf{x} &\triangleq v \qquad (6.6)\\
\varrho(\pi\ell \xrightarrow{\;\cdots\;} \ell')\mathsf{x} &\triangleq \varrho(\pi\ell)\mathsf{x} \quad \text{otherwise}\\
\varrho(\ell)\mathsf{x} &\triangleq 0
\end{aligned}
$$

Exercise 6.7 Prove that $\varrho(\pi)\mathsf{x}$ is the value assigned to x on the last assignment to x on π or 0 in absence of such an assignment.

Exercise 6.8 Assuming that all values $v \in \mathbb{V}$ have a denotation by an expression, prove that $\mathbb{E}\mathrm{v} \triangleq \mathbb{V} \to \mathbb{V} = \{\varrho(\pi\ell) \mid \pi\ell \in \mathbb{T}^+\}$.

6.3 Prefix Trace Semantics

Let $\pi_1\mathsf{at}[\![\mathsf{S}]\!]$ be an initialization trace ending on entry $\mathsf{at}[\![\mathsf{S}]\!]$ of statement S. $\boldsymbol{\mathcal{S}}^*[\![\mathsf{S}]\!](\pi_1\mathsf{at}[\![\mathsf{S}]\!])$ is the set of prefix traces $\mathsf{at}[\![\mathsf{S}]\!]\pi_2\ell$ of S continuing the trace $\pi_1\mathsf{at}[\![\mathsf{S}]\!]$ and reaching some program label $\ell \in \mathsf{labx}[\![\mathsf{S}]\!]$. Schematically,

$$
\xrightarrow{\quad\pi_1\quad} \underbrace{\mathsf{at}[\![\mathsf{S}]\!] \xrightarrow{\quad\pi_2\quad} \ell}_{\in\ \boldsymbol{\mathcal{S}}^*[\![\mathsf{S}]\!](\pi_1\mathsf{at}[\![\mathsf{S}]\!])}
$$

We call $\mathsf{at}[\![\mathsf{S}]\!]\pi_2\ell$ the continuation[1] of the initialization $\pi_1\mathsf{at}[\![\mathsf{S}]\!]$. Although our language is determinist, we consider a set of possible continuations to cope, for example, with inputs and random number generation. By convention, $\boldsymbol{\mathcal{S}}^*[\![\mathsf{S}]\!](\pi_1\ell) = \varnothing$ when $\ell \neq \mathsf{at}[\![\mathsf{S}]\!]$.

A pair $\langle\pi_0, \pi\rangle$ such that $\pi \in \boldsymbol{\mathcal{S}}^*[\![\mathsf{S}]\!](\pi_0)$ is called a *prefix execution* of program component S.

6.4 Maximal Finite Trace Semantics

Let $\pi_1\mathsf{at}[\![\mathsf{S}]\!]$ be an initialization trace ending on entry $\mathsf{at}[\![\mathsf{S}]\!]$ of statement S. $\boldsymbol{\mathcal{S}}^+[\![\mathsf{S}]\!](\pi_1\mathsf{at}[\![\mathsf{S}]\!])$ is the set of maximal finite traces $\mathsf{at}[\![\mathsf{S}]\!]\pi_2\mathsf{after}[\![\mathsf{S}]\!]$ of S continuing the trace $\pi_1\mathsf{at}[\![\mathsf{S}]\!]$ and reaching $\mathsf{after}[\![\mathsf{S}]\!]$. Schematically,

$$\xrightarrow{\quad\pi_1\quad} \underbrace{\mathsf{at}[\![\mathsf{S}]\!] \xrightarrow{\quad\pi_2\quad} \mathsf{after}[\![\mathsf{S}]\!]}_{\in\ \boldsymbol{\mathcal{S}}^+[\![\mathsf{S}]\!](\pi_1\mathsf{at}[\![\mathsf{S}]\!])}$$

Formally,

$$\boldsymbol{\mathcal{S}}^+[\![\mathsf{S}]\!](\pi_1\mathsf{at}[\![\mathsf{S}]\!]) \quad\triangleq\quad \{\pi_2\ell \in \boldsymbol{\mathcal{S}}^*[\![\mathsf{S}]\!](\pi_1\mathsf{at}[\![\mathsf{S}]\!]) \mid \ell = \mathsf{after}[\![\mathsf{S}]\!]\} \tag{6.9}$$

A pair $\langle\pi_0, \pi\rangle$ such that $\pi \in \boldsymbol{\mathcal{S}}^+[\![\mathsf{S}]\!](\pi_0)$ is called a *finite maximal execution* of program component S.

6.5 Introduction to Structural Definitions

In order to specify a set $F[\![S]\!]$ associated with a grammar nonterminal S, we use structural definitions (generalizing section 3.5 for expressions).

For the base case, there are axioms of the form $\dfrac{}{x \in F[\![S]\!]}$ meaning that x is an element of $F[\![S]\!]$.

For the inductive case $S ::= S_1 \ldots S_n$, there are inductive rules of the form

$$\frac{x_1 \in F[\![S_1]\!], \ldots, x_n \in F[\![S_n]\!]}{f(x_1, \ldots, x_n) \in F[\![S]\!]}$$

meaning that if $x_i \in F[\![S_i]\!], i = 1, \ldots, n$ then $f(x_1, \ldots, x_n)$ is an element of $F[\![S]\!]$ where the function f is well defined for $(x_1, \ldots, x_n)$.

Example 6.10 The positive integers $\mathbb{N}^+$ can be denoted by a collection of sticks $\mathbb{N}^+ ::= \mathbf{I} \mid \mathbb{N}^+\mathbf{I}$. For example $\mathbf{IIIIII}$ is six. The set $\mathbb{O}$ of odd positive integers is defined by the axiom $\dfrac{}{\mathbf{I} \in \mathbb{O}}$ and the inference rule $\dfrac{n \in \mathbb{O}}{n\mathbf{II} \in \mathbb{O}}$.

To define the set $s(n)$ of numbers smaller than or equal to n, we can specify recursively that $\dfrac{}{n \in s(n)}$ and $\dfrac{m\mathbf{I} \in s(n)}{m \in s(n)}$. For example $\mathbf{III} \in s(\mathbf{III})$ by the axiom so $\mathbf{II} \in s(\mathbf{III})$ by the inference rule so $\mathbf{I} \in s(\mathbf{III})$ by the inference rule proving that $s(\mathbf{III}) = \{\mathbf{III}, \mathbf{II}, \mathbf{I}\}$.

A more rigorous definition of recursive structural definitions is given in chapter 16.

1. Continuation has a specific connotation in denotational semantics [915]. We use the term to mean 'a partial trace that is attached to and is an extension of an initialization trace.' Extension would also have been adequate, but it is already overloaded in mathematics and software.

6.6 Structural Prefix Trace Semantics

We provide a structural definition $\widehat{\boldsymbol{\mathcal{S}}}^*[\![\mathtt{S}]\!]$ of the prefix trace semantics $\boldsymbol{\mathcal{S}}^*[\![\mathtt{S}]\!]$ of statements, statement lists, and programs $\mathtt{S}$ (and so $\boldsymbol{\mathcal{S}}^*[\![\mathtt{S}]\!] \triangleq \widehat{\boldsymbol{\mathcal{S}}}^*[\![\mathtt{S}]\!]$).

- A prefix continuation of the traces $\pi_1\mathsf{at}[\![\mathtt{S}]\!]$ arriving at a program, statement, or statement list $\mathtt{S}$ can be reduced to the program point $\mathsf{at}[\![\mathtt{S}]\!]$ at this program, statement, or statement list $\mathtt{S}$.

Prefix trace at a statement $\mathtt{S}$

- $$\frac{}{\mathsf{at}[\![\mathtt{S}]\!] \in \widehat{\boldsymbol{\mathcal{S}}}^*[\![\mathtt{S}]\!](\pi_1\mathsf{at}[\![\mathtt{S}]\!])} \tag{6.11}$$

- If $\mathtt{P} ::= \mathtt{Sl}\ \ell$ then the prefix continuations of the traces $\pi_1\mathsf{at}[\![\mathtt{Sl}]\!]$ arriving at program entry $\mathsf{at}[\![\mathtt{P}]\!] = \mathsf{at}[\![\mathtt{Sl}]\!]$ are the continuations of the statement list $\mathtt{Sl}$.

Prefix traces of a program $\mathtt{P} ::= \mathtt{Sl}\ \ell$

- $$\frac{\pi_2 \in \widehat{\boldsymbol{\mathcal{S}}}^*[\![\mathtt{Sl}]\!](\pi_1\mathsf{at}[\![\mathtt{Sl}]\!])}{\pi_2 \in \widehat{\boldsymbol{\mathcal{S}}}^*[\![\mathtt{P}]\!](\pi_1\mathsf{at}[\![\mathtt{P}]\!])} \tag{6.12}$$

- A prefix trace of $\mathtt{Sl}'\ \mathtt{S}$ continuing an initial trace π_1 can be a prefix trace of $\mathtt{Sl}'$ or a finite maximal trace of $\mathtt{Sl}'$ followed by a prefix trace of $\mathtt{S}$.

Prefix traces of a statement list $\mathtt{Sl} ::= \mathtt{Sl}'\ \mathtt{S}$

- $$\frac{\pi_2 \in \widehat{\boldsymbol{\mathcal{S}}}^*[\![\mathtt{Sl}']\!](\pi_1)}{\pi_2 \in \widehat{\boldsymbol{\mathcal{S}}}^*[\![\mathtt{Sl}]\!](\pi_1)} \tag{6.13}$$

- $$\frac{\pi_2 \in \widehat{\boldsymbol{\mathcal{S}}}^+[\![\mathtt{Sl}']\!](\pi_1), \quad \pi_3 \in \widehat{\boldsymbol{\mathcal{S}}}^*[\![\mathtt{S}]\!](\pi_1 \mathbin{\widehat{}} \pi_2)}{\pi_2 \mathbin{\widehat{}} \pi_3 \in \widehat{\boldsymbol{\mathcal{S}}}^*[\![\mathtt{Sl}]\!](\pi_1)} \tag{6.14}$$

Schematically in case (6.14),

$$\xrightarrow{\pi_1} \underbrace{\underbrace{\begin{matrix}\mathsf{at}[\![\mathtt{Sl}]\!]\\ \mathsf{at}[\![\mathtt{Sl}']\!]\end{matrix} \xrightarrow{\pi_2} \begin{matrix}\mathsf{after}[\![\mathtt{Sl}']\!]\\ \mathsf{at}[\![\mathtt{S}]\!]\end{matrix}}_{\in\ \widehat{\boldsymbol{\mathcal{S}}}^+[\![\mathtt{Sl}']\!](\pi_1\mathsf{at}[\![\mathtt{Sl}']\!])} \underbrace{\xrightarrow{\pi_3} \ell}_{\in\ \widehat{\boldsymbol{\mathcal{S}}}^*[\![\mathtt{S}]\!](\pi_1\mathsf{at}[\![\mathtt{Sl}']\!]\pi_2\mathsf{at}[\![\mathtt{S}]\!])}}_{\in\ \widehat{\boldsymbol{\mathcal{S}}}^*[\![\mathtt{Sl}]\!](\pi_1\mathsf{at}[\![\mathtt{Sl}]\!])}$$

Notice that by definition (6.6) of $\boldsymbol{\varrho}$, the values of variables on π_1, π_2, and π_3 are necessarily coherent. For example, $\underbrace{\ldots}_{\pi_1} \ell_1 \underbrace{\xrightarrow{\mathtt{x = 0 = 0}}}_{\pi_2} \ell_2 \underbrace{\xrightarrow{\mathtt{x = x - 1 = 42}}}_{\pi_3} \ell_3$ is impossible because the value of $\mathtt{x}$ at ℓ_2 is computed on $\pi_1 \mathbin{\widehat{}} \pi_2$.

- A prefix (and indeed maximal) trace π of the empty statement list ϵ continuing some trace is reduced to the program label at that empty statement (a redundant case already covered by (6.11)).

Prefix traces of an empty statement list $\texttt{Sl} ::= \epsilon$

$$\bullet \quad \frac{}{\text{at}[\![\texttt{Sl}]\!] \in \widehat{\boldsymbol{\mathcal{S}}}^*[\![\texttt{Sl}]\!](\pi\text{at}[\![\texttt{Sl}]\!])} \tag{6.15}$$

• A prefix (and indeed maximal) finite trace of an assignment $\ell\ \texttt{x} = \texttt{A}\,;$ continuing some trace $\pi\ell$ is ℓ followed by the event $\texttt{x} = \texttt{A} = \nu$ where ν is value assigned to $\texttt{x}$ that is the value of the arithmetic expression $\texttt{A}$ evaluated on $\pi\ell$ and finishing at the label $\text{after}[\![\texttt{S}]\!]$ after the assignment.

Prefix traces of an assignment statement $\texttt{S} ::= \ell\ \texttt{x} = \texttt{A}\,;$

$$\bullet \quad \frac{\nu = \boldsymbol{\mathscr{A}}[\![\texttt{A}]\!]\varrho(\pi\ell)}{\ell \xrightarrow{\texttt{x} = \texttt{A} = \nu} \text{after}[\![\texttt{S}]\!] \in \widehat{\boldsymbol{\mathcal{S}}}^*[\![\texttt{S}]\!](\pi\ell)} \tag{6.16}$$

We often omit the assignment action writing $\ell \xrightarrow{\texttt{x} = \upsilon} \ell'$ for $\ell \xrightarrow{\texttt{x} = \texttt{A} = \upsilon} \ell'$ (because the assignment $\ell\ \texttt{x} = \texttt{A}\,;$ can be recovered from the program text and the program label ℓ).

• A prefix (and indeed maximal) finite trace of a skip statement $\ell;$ continuing an initial trace $\pi\ell$ arriving at ℓ is just continuing after the skip statement.

Prefix traces of a skip statement $\texttt{S} ::= \ell;$

$$\bullet \quad \frac{}{\ell \xrightarrow{\texttt{skip}} \text{after}[\![\texttt{S}]\!] \in \widehat{\boldsymbol{\mathcal{S}}}^*[\![\texttt{S}]\!](\pi\ell)} \tag{6.17}$$

• A prefix trace of a conditional statement $\texttt{if}\ \ell\ (\texttt{B})\ \texttt{S}_t$ continuing some initial trace $\pi_1\ell$ is

- either ℓ (a case already covered by (6.11));
- or, in case (6.18), ℓ followed by the event $\neg(\texttt{B})$ when the value of this Boolean expression on $\pi_1\ell$ is ff and finishing at the label $\text{after}[\![\texttt{S}]\!]$ after the conditional statement;
- or, in case (6.19), when the value of the Boolean expression $\texttt{B}$ on $\pi_1\ell$ is tt, ℓ followed by the test event $\texttt{B}$ followed by a prefix trace of $\texttt{S}_t$ continuing $\pi_1\ell \xrightarrow{\texttt{B}} \text{at}[\![\texttt{S}_t]\!]$.

Prefix traces of a conditional statement $\texttt{S} ::= \texttt{if}\ \ell\ (\texttt{B})\ \texttt{S}_t$

$$\bullet \quad \frac{\boldsymbol{\mathscr{B}}[\![\texttt{B}]\!]\varrho(\pi_1\ell) = \text{ff}}{\ell \xrightarrow{\neg(\texttt{B})} \text{after}[\![\texttt{S}]\!] \in \widehat{\boldsymbol{\mathcal{S}}}^*[\![\texttt{S}]\!](\pi_1\ell)} \tag{6.18}$$

$$\bullet \quad \frac{\boldsymbol{\mathscr{B}}[\![\texttt{B}]\!]\varrho(\pi_1\ell) = \text{tt}, \quad \pi_2 \in \widehat{\boldsymbol{\mathcal{S}}}^*[\![\texttt{S}_t]\!](\pi_1\ell \xrightarrow{\texttt{B}} \text{at}[\![\texttt{S}_t]\!])}{\ell \xrightarrow{\texttt{B}} \text{at}[\![\texttt{S}_t]\!] \frown \pi_2 \in \widehat{\boldsymbol{\mathcal{S}}}^*[\![\texttt{S}]\!](\pi_1\ell)} \tag{6.19}$$

Remark 6.20 (break statements) The inference rule (6.19) includes the case of conditional within an iteration containing a break statement in the true branch $\texttt{S}_t$. That instance of the rule would be of the form

$$\frac{\mathscr{B}[\![\mathsf{B}]\!]\varrho(\pi_1\ell) = \mathsf{tt}, \quad \pi_2 \xrightarrow{\mathsf{break}} \mathsf{break\text{-}to}[\![\mathsf{S}_t]\!] \in \widehat{\mathcal{S}}^*[\![\mathsf{S}_t]\!](\pi_1\ell \xrightarrow{\mathsf{B}} \mathsf{at}[\![\mathsf{S}_t]\!])}{\ell \xrightarrow{\mathsf{B}} \mathsf{at}[\![\mathsf{S}_t]\!] \frown \pi_2 \xrightarrow{\mathsf{break}} \mathsf{break\text{-}to}[\![\mathsf{S}]\!] \in \widehat{\mathcal{S}}^*[\![\mathsf{S}]\!](\pi_1\ell)} \tag{6.21}$$

because $\mathsf{break\text{-}to}[\![\mathsf{S}]\!] = \mathsf{break\text{-}to}[\![\mathsf{S}_t]\!]$.

- A prefix finite trace of a conditional statement **if** ℓ (B) S_t **else** S_f continuing an initial trace $\pi_1\ell$ is the test event B (respectively $\neg(\mathsf{B})$) at ℓ followed by a prefix trace of S_t (respectively S_f) when Boolean expression B is tt (respectively ff) on $\pi_1\ell$ in case (6.22) (respectively (6.23)).

Prefix traces of a conditional statement S ::= **if** ℓ (B) S_t **else** S_f

- $$\frac{\mathscr{B}[\![\mathsf{B}]\!]\varrho(\pi_1\ell) = \mathsf{tt}, \quad \pi_2 \in \widehat{\mathcal{S}}^*[\![\mathsf{S}_t]\!](\pi_1\ell \xrightarrow{\mathsf{B}} \mathsf{at}[\![\mathsf{S}_t]\!])}{\ell \xrightarrow{\mathsf{B}} \mathsf{at}[\![\mathsf{S}_t]\!] \frown \pi_2 \in \widehat{\mathcal{S}}^*[\![\mathsf{S}]\!](\pi_1\ell)} \tag{6.22}$$
- $$\frac{\mathscr{B}[\![\mathsf{B}]\!]\varrho(\pi_1\ell) = \mathsf{ff}, \quad \pi_2 \in \widehat{\mathcal{S}}^*[\![\mathsf{S}_f]\!](\pi_1\ell \xrightarrow{\neg(\mathsf{B})} \mathsf{at}[\![\mathsf{S}_f]\!])}{\ell \xrightarrow{\neg(\mathsf{B})} \mathsf{at}[\![\mathsf{S}_f]\!] \frown \pi_2 \in \widehat{\mathcal{S}}^*[\![\mathsf{S}]\!](\pi_1\ell)} \tag{6.23}$$

Definitions (6.22) and (6.23) include the cases of breaks respectively from S_t and S_f to $\mathsf{break\text{-}to}[\![\mathsf{S}]\!] = \mathsf{break\text{-}to}[\![\mathsf{S}_t]\!] = \mathsf{break\text{-}to}[\![\mathsf{S}_f]\!]$.

- A prefix finite trace of an iteration statement **while** ℓ (B) S_b continuing some initial trace $\pi_1\ell$ is one and only one of the following:
- either ℓ (case (6.24), already covered by (6.11));
- or, in case (6.25), the trace starting at ℓ followed by the event $\neg(\mathsf{B})$ when the value of this Boolean expression on $\pi_1\ell$ is ff and finishing at the label $\mathsf{after}[\![\mathsf{S}]\!]$ after the iteration statement;
- or, in case (6.28), the trace starting at ℓ followed by the event B when the value of this Boolean expression on $\pi_1\ell$ is tt and finishing at the label $\mathsf{at}[\![\mathsf{S}_b]\!]$ followed by a prefix (indeed maximal) trace of the loop body S_b ending up in a break;
- or, in case (6.26), the trace starting at ℓ, followed by a prefix trace of the iteration statement **while** ℓ (B) S_b representing 0 or more of iterations ending at ℓ, followed by the test event B (where the expression B is tt), followed by a prefix finite trace of the body S_b.

Prefix traces of an iteration statement $\mathsf{S} ::= \mathtt{while}\ \ell\ (\mathsf{B})\ \mathsf{S}_b$

- $$\frac{}{\ell \in \widehat{\mathcal{S}}^*[\![\mathsf{S}]\!](\pi_1\ell)} \tag{6.24}$$

- $$\frac{\ell\pi_2\ell \in \widehat{\mathcal{S}}^*[\![\mathsf{S}]\!](\pi_1\ell), \quad \mathcal{B}[\![\mathsf{B}]\!]\varrho(\pi_1\ell\pi_2\ell) = \mathrm{ff}}{\ell\pi_2\ell \xrightarrow{\neg(\mathsf{B})} \mathsf{after}[\![\mathsf{S}]\!] \in \widehat{\mathcal{S}}^*[\![\mathsf{S}]\!](\pi_1\ell)} \tag{6.25}$$

- $$\frac{\ell\pi_2\ell \in \widehat{\mathcal{S}}^*[\![\mathsf{S}]\!](\pi_1\ell), \quad \mathcal{B}[\![\mathsf{B}]\!]\varrho(\pi_1\ell\pi_2\ell) = \mathrm{tt}, \quad \pi_3 \in \widehat{\mathcal{S}}^*[\![\mathsf{S}_b]\!](\pi_1\ell\pi_2\ell \xrightarrow{\mathsf{B}} \mathsf{at}[\![\mathsf{S}_b]\!])}{\ell\pi_2\ell \xrightarrow{\mathsf{B}} \mathsf{at}[\![\mathsf{S}_b]\!] \frown \pi_3 \in \widehat{\mathcal{S}}^*[\![\mathsf{S}]\!](\pi_1\ell)} \tag{6.26}$$

This is a forward, left-recursive definition where $n + 1$ iterations are n iterations followed by one more iteration.

Remark 6.27 (break statements) The inference rule (6.26) includes the case of an iteration ending by a break statement that would have the form

- $$\frac{\ell\pi_2\ell \in \widehat{\mathcal{S}}^*[\![\mathsf{S}]\!](\pi_1\ell), \quad \mathcal{B}[\![\mathsf{B}]\!]\varrho(\pi_1\ell\pi_2\ell) = \mathrm{tt}, \quad \pi_3 \xrightarrow{\mathsf{break}} \mathsf{break\text{-}to}[\![\mathsf{S}]\!] \in \widehat{\mathcal{S}}^*[\![\mathsf{S}_b]\!](\pi_1\ell\pi_2\ell \xrightarrow{\mathsf{B}} \mathsf{at}[\![\mathsf{S}_b]\!])}{\ell\pi_2\ell \xrightarrow{\mathsf{B}} \mathsf{at}[\![\mathsf{S}_b]\!] \frown \pi_3 \xrightarrow{\mathsf{break}} \mathsf{break\text{-}to}[\![\mathsf{S}]\!] \in \widehat{\mathcal{S}}^*[\![\mathsf{S}]\!](\pi_1\ell)} \tag{6.28}$$

- A prefix (indeed maximal) finite trace of a break ℓ **`break;`** continuing some initial trace $\pi\ell$ is the trace ℓ followed by the **`break;`** event and ending at the break label $\mathsf{break\text{-}to}[\![\mathsf{S}]\!]$ (which is the exit label of the closest enclosing iteration loop).

Prefix traces of a break statement $\mathsf{S} ::= \ell\ \mathtt{break;}$

- $$\frac{}{\ell \xrightarrow{\mathsf{break}} \mathsf{break\text{-}to}[\![\mathsf{S}]\!] \in \widehat{\mathcal{S}}^*[\![\mathsf{S}]\!](\pi\ell)} \tag{6.29}$$

- A prefix trace of a compound statement { `Sl` } is that of its statement list `Sl`.

Prefix traces of a compound statement $\mathsf{S} ::= \{\ \mathtt{Sl}\ \}$

- $$\frac{\pi_2 \in \widehat{\mathcal{S}}^*[\![\mathtt{Sl}]\!](\pi_1)}{\pi_2 \in \widehat{\mathcal{S}}^*[\![\mathsf{S}]\!](\pi_1)} \tag{6.30}$$

Example 6.31 Consider the program $\mathsf{P}_6 \triangleq \ell_1\ \mathtt{x = x + 1\ ;}\ell_1\ \mathtt{x = x + 1\ ;}\ell_3$ with abstract syntax

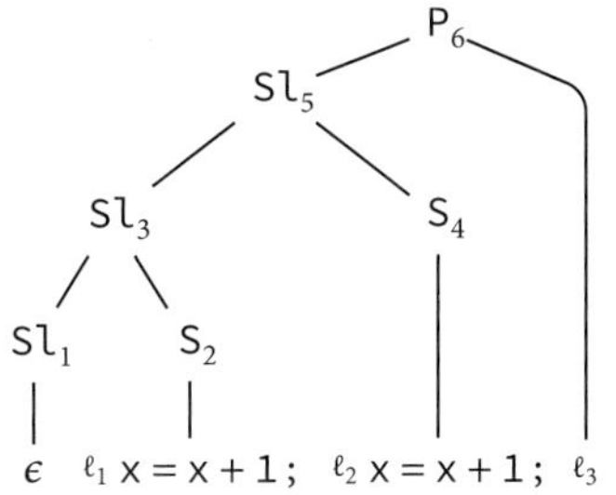

The derivation of the proof that (a) ℓ_1, (b) $\ell_1 \xrightarrow{\texttt{x}=1} \ell_2$, and (c) $\ell_1 \xrightarrow{\texttt{x}=1} \ell_2 \xrightarrow{\texttt{x}=2} \ell_3 \in \widehat{\mathcal{S}}^*[\![\texttt{P}_6]\!]\ell_1$ is the following:

- $\ell_1 \in \widehat{\mathcal{S}}^*[\![\texttt{P}_6]\!](\ell_1)$ $\quad$ (by $\mathsf{at}[\![\texttt{P}_6]\!] = \ell_1$, $\pi_1 = \ni$, and (6.11), proving (a))
- $\ell_1 \in \widehat{\mathcal{S}}^+[\![\texttt{Sl}_1]\!](\ell_1)$ $\quad$ ($\texttt{Sl}_1 = \epsilon$, (6.15), and (6.9))
- $\ell_1 \xrightarrow{\texttt{x}=1} \ell_2 \in \widehat{\mathcal{S}}^*[\![\texttt{S}_2]\!](\ell_1)$

  ($\texttt{S}_2 = \ell_1\ \texttt{x = x + 1 ;}$, $\mathcal{A}[\![\texttt{x + 1}]\!]\varrho(\ell_1) = 1$, $\mathsf{after}[\![\texttt{S}_2]\!] = \ell_2$, and (6.16))
- $\ell_1 \xrightarrow{\texttt{x}=1} \ell_2 \in \widehat{\mathcal{S}}^*[\![\texttt{Sl}_3]\!](\ell_1)$

  ($\texttt{Sl}_3 = \texttt{Sl}_1\ \texttt{S}_2$, $\ell_1 \in \widehat{\mathcal{S}}^+[\![\texttt{Sl}_1]\!](\ell_1)$, $\ell_1 \xrightarrow{\texttt{x}=1} \ell_2 \in \widehat{\mathcal{S}}^*[\![\texttt{S}_2]\!](\ell_1)$, and (6.14))
- $\ell_1 \xrightarrow{\texttt{x}=1} \ell_2 \in \widehat{\mathcal{S}}^*[\![\texttt{P}_6]\!](\ell_1)$ $\quad$ (by (6.13) for $\texttt{Sl}_5$ and $\texttt{P}_6$, proving (b))
- $\ell_1 \xrightarrow{\texttt{x}=1} \ell_2 \in \widehat{\mathcal{S}}^+[\![\texttt{Sl}_3]\!](\ell_1)$ $\quad$ (by (6.9))
- $\ell_2 \xrightarrow{\texttt{x}=2} \ell_3 \in \widehat{\mathcal{S}}^*[\![\texttt{S}_4]\!](\ell_1 \xrightarrow{\texttt{x}=1} \ell_2)$

  ($\texttt{S}_4 = \ell_2\ \texttt{x = x + 1 ;}$, $\mathcal{A}[\![\texttt{x + 1}]\!]\varrho(\ell_1 \xrightarrow{\texttt{x}=1} \ell_2) = 2$, $\mathsf{after}[\![\texttt{S}_4]\!] = \ell_3$, and (6.16))
- $\ell_1 \xrightarrow{\texttt{x}=1} \ell_2 \xrightarrow{\texttt{x}=2} \ell_3 \in \widehat{\mathcal{S}}^*[\![\texttt{Sl}_5]\!](\ell_1)$

  ($\texttt{Sl}_5 = \texttt{Sl}_3\ \ \texttt{S}_4$, $\ell_1 \xrightarrow{\texttt{x}=1} \ell_2 \in \widehat{\mathcal{S}}^*[\![\texttt{Sl}_3]\!](\ell_1)$, $\ell_1 \xrightarrow{\texttt{x}=1} \ell_2 \xrightarrow{\texttt{x}=2} \ell_3 \in \widehat{\mathcal{S}}^*[\![\texttt{S}_4]\!](\ell_1 \xrightarrow{\texttt{x}=1} \ell_2)$, and (6.14))
- $\ell_1 \xrightarrow{\texttt{x}=1} \ell_2 \xrightarrow{\texttt{x}=2} \ell_3 \in \widehat{\mathcal{S}}^*[\![\texttt{P}_6]\!]$

  ($\texttt{P}_6 = \texttt{Sl}_5\ \ell_3$, $\mathsf{at}[\![\texttt{Sl}_5]\!] = \ell_1$, $\ell_1 \xrightarrow{\texttt{x}=1} \ell_2 \xrightarrow{\texttt{x}=2} \ell_3 \in \widehat{\mathcal{S}}^*[\![\texttt{Sl}_5]\!](\ell_1)$, and (6.12), proving (c))

Example 6.32 Let us calculate the prefix semantics of the program

$$\texttt{S} = \texttt{while}\ \ell_1\ (\texttt{tt})\ \ell_2\ \texttt{x = x + 1 ;}\ell_3.$$

- $\ell_2 \xrightarrow{\texttt{x} = \mathcal{A}[\![\texttt{x + 1}]\!]\varrho(\pi\ell_2)} \ell_1 \in \widehat{\mathcal{S}}^*[\![\ell_2\ \texttt{x = x + 1 ;}]\!](\pi\ell_2)$ $\qquad$ (6.33)

  (by (6.16))
- $\ell_1 \in \widehat{\mathcal{S}}^*[\![\texttt{S}]\!](\ell_1)$ $\quad$ (by (6.24))
- $\ell_1 \xrightarrow{\texttt{tt}} \ell_2 \in \widehat{\mathcal{S}}^*[\![\texttt{S}]\!](\ell_1)$ $\quad$ (by (6.26) and $\ell_2 \in \widehat{\mathcal{S}}^*[\![\texttt{S}]\!](\ell_1 \xrightarrow{\texttt{tt}} \ell_2)$)
- $\ell_1 \xrightarrow{\texttt{tt}} \ell_2 \xrightarrow{\texttt{x}=1} \ell_1 \in \widehat{\mathcal{S}}^*[\![\texttt{S}]\!](\ell_1)$

$\{$by $\ell_1 \in \widehat{\mathcal{S}}^*\llbracket \mathsf{S} \rrbracket(\ell_1)$, $\mathcal{B}\llbracket \mathtt{tt} \rrbracket \varrho(\ell_1) = \mathtt{tt}$, $\ell_2 \xrightarrow{\mathsf{x} = 1} \ell_1 \in \widehat{\mathcal{S}}^*\llbracket \ell_2\ \mathsf{x} = \mathsf{x} + 1\ ; \rrbracket(\ell_1 \xrightarrow{\mathtt{tt}} \ell_2)$ by (6.33), and (6.26) $\}$

$$- \left(\ell_1 \xrightarrow{\mathtt{tt}} \ell_2 \xrightarrow{\mathsf{x} = i} \ell_1\right)_{i=1}^{n} \in \widehat{\mathcal{S}}^*\llbracket \mathsf{S} \rrbracket(\ell_1) \qquad \{\text{induction hypothesis}\}$$

$$\approx \left(\ell_1 \xrightarrow{\mathtt{tt}} \ell_2 \xrightarrow{\mathsf{x} = i} \ell_1\right)_{i=1}^{n} \xrightarrow{\mathtt{tt}} \ell_2 \xrightarrow{\mathsf{x} = n+1} \ell_1 \in \widehat{\mathcal{S}}^*\llbracket \mathsf{S} \rrbracket(\ell_1)$$

$\{$by $\left(\ell_1 \xrightarrow{\mathtt{tt}} \ell_2 \xrightarrow{\mathsf{x} = i} \ell_1\right)_{i=1}^{n} \in \widehat{\mathcal{S}}^*\llbracket \mathsf{S} \rrbracket(\ell_1)$,

$\mathcal{B}\llbracket \mathtt{tt} \rrbracket \varrho(\left(\ell_1 \xrightarrow{\mathtt{tt}} \ell_2 \xrightarrow{\mathsf{x} = i} \ell_1\right)_{i=1}^{n}) = \mathtt{tt}$,

$\mathcal{A}\llbracket \mathsf{x} + 1 \rrbracket \varrho(\left(\ell_1 \xrightarrow{\mathtt{tt}} \ell_2 \xrightarrow{\mathsf{x} = i} \ell_1\right)_{i=1}^{n} \xrightarrow{\mathtt{tt}} \ell_2) = n + 1$,

$\ell_2 \xrightarrow{\mathsf{x} = n+1} \ell_1 \in \widehat{\mathcal{S}}^*\llbracket \ell_2\ \mathsf{x} = \mathsf{x} + 1\ ; \rrbracket(\left(\ell_1 \xrightarrow{\mathtt{tt}} \ell_2 \xrightarrow{\mathsf{x} = i} \ell_1\right)_{i=1}^{n} \xrightarrow{\mathtt{tt}} \ell_2)$ by (6.33), and (6.26)$\}$

$$- \left(\ell_1 \xrightarrow{\mathtt{tt}} \ell_2 \xrightarrow{\mathsf{x} = i}\right)_{i=1}^{n+1} \ell_1 \in \widehat{\mathcal{S}}^*\llbracket \mathsf{S} \rrbracket(\ell_1) \qquad \{\text{by } (\ell_1 \pi(i) \ell_1)_{i=1}^{n}\ \pi(n+1)\ell_1 = (\ell_1 \pi(i))_{i=1}^{n+1}\ \ell_1\}$$

$$- \left(\ell_1 \xrightarrow{\mathtt{tt}} \ell_2 \xrightarrow{\mathsf{x} = i}\right)_{i=1}^{n+1} \ell_1 \xrightarrow{\mathtt{tt}} \ell_2 \in \widehat{\mathcal{S}}^*\llbracket \mathsf{S} \rrbracket(\ell_1)$$

$\{$by $\left(\ell_1 \xrightarrow{\mathtt{tt}} \ell_2 \xrightarrow{\mathsf{x} = i}\right)_{i=1}^{n+1} \ell_1 \in \widehat{\mathcal{S}}^*\llbracket \mathsf{S} \rrbracket(\ell_1)$, $\mathcal{B}\llbracket \mathtt{tt} \rrbracket \varrho(\ell_1) = \mathtt{tt}$, and (6.26) where $\pi_3 = \ell_2\}$

proving, by recurrence, that $\widehat{\mathcal{S}}^*\llbracket \mathsf{S} \rrbracket(\ell_1) = \left\{\left(\ell_1 \xrightarrow{\mathtt{tt}} \ell_2 \xrightarrow{\mathsf{x} = i} \ell_1\right)_{i=1}^{n}, \left(\ell_1 \xrightarrow{\mathtt{tt}} \ell_2 \xrightarrow{\mathsf{x} = i} \ell_1\right)_{i=1}^{n} \xrightarrow{\mathtt{tt}} \ell_2 \;\middle|\; n \in \mathbb{N}\right\}$ (reduced to ℓ_1 for $n = 0$).

Exercise 6.34 Prove that $\widehat{\mathcal{S}}^*\llbracket \mathsf{S} \rrbracket(\pi_1 \ell) = \varnothing$ when $\ell \neq \mathsf{at}\llbracket \mathsf{S} \rrbracket$.

Exercise 6.35 Prove that if $\pi_2 \in \widehat{\mathcal{S}}^*\llbracket \mathsf{S} \rrbracket(\pi_1 \mathsf{at}\llbracket \mathsf{S} \rrbracket)$ then π_2 starts with $\mathsf{at}\llbracket \mathsf{S} \rrbracket$ i.e. $\exists \overline{\pi}_2 \,.\, \pi_2 = \overline{\pi}_2 \mathsf{at}\llbracket \mathsf{S} \rrbracket$.

Exercise 6.36 Show that if $\forall \pi, \pi' \,.\, \forall \mathsf{x} \in \mathbb{V} \,.\, \varrho(\pi)\mathsf{x} = \varrho(\pi')\mathsf{x}$ then for any program, statement or statement list S, $\widehat{\mathcal{S}}^*\llbracket \mathsf{S} \rrbracket(\pi) = \widehat{\mathcal{S}}^*\llbracket \mathsf{S} \rrbracket(\pi')$.

Exercise 6.37 Given a trace π define the labels $\alpha_\ell(\pi)$ on π to be $\alpha_\ell(\ni) \triangleq \varnothing$ and $\alpha_\ell(\pi_1 \ell_1 \pi_2) \triangleq \alpha_\ell(\pi_1) \cup \{\ell_1\} \cup \alpha_\ell(\pi_2)$.[2] Prove that $\forall \pi \in \widehat{\mathcal{S}}^*\llbracket \mathsf{S} \rrbracket(\pi_1 \ell) \,.\, \alpha_\ell(\pi) \subseteq \mathsf{labx}\llbracket \mathsf{S} \rrbracket$.

Exercise 6.38 Add a `continue ;` statement to the syntax and prefix trace semantics of language of section 4.1, which forces the next iteration of the inner enclosing loop to take place immediately.

2. Technically, by coinduction for infinite traces; see section 16.6.

6.7 Characterization of the Left-Recursive Definition of Iteration

Given finite traces of the form $\ell\pi(i)\ell$, we write $\left(\ell\pi(i)\ell\right)_{i=1}^{n}$ for the finite trace $\ell\pi(1)\ell\pi(2)\ell \ldots \pi(n)\ell$. This is the trace ℓ for $n = 0$. The infinite trace $\left(\ell\pi(i)\ell\right)_{i=1}^{\infty}$ denotes $\ell\pi(1)\ell\pi(2)\ell \ldots \pi(n)\ell\pi(n+1)\ell \ldots$. The assignments and conditions along these traces are assumed to be consistent with the past trace (i.e., in $\pi\ell \xrightarrow{\mathsf{X} = \mathsf{A} = v} \ell' \cdots$ we have $v = \mathscr{A}[\![\mathsf{A}]\!]\varrho(\pi\ell)$ and in $\pi\ell \xrightarrow{\mathsf{B}} \ell' \cdots$, we have $\mathscr{B}[\![\mathsf{B}]\!]\varrho(\pi\ell) = \mathrm{tt}$, and similarly for negation $\neg(\mathsf{B})$).

Lemma 6.39 The prefix traces of an iteration statement $\mathsf{S} ::= \texttt{while}\,\ell\,(\mathsf{B})\,\mathsf{S}_b$ with entry label $\ell = \mathsf{at}[\![\mathsf{S}]\!] = \mathsf{after}[\![\mathsf{S}_b]\!]$ are

$$\widehat{\mathcal{S}}^*[\![\mathsf{S}]\!](\pi_1\ell) \quad \triangleq \quad \{\underline{\pi}(k-1) \mathbin{\widehat{\ }} \ell\pi'(k) \mid \pi_1\ell \in \mathbb{T}^+ \wedge k \in \mathbb{N}^+ \wedge \beta(k-1)\} \qquad (6.39)$$

where

- $\underline{\pi}(k) \triangleq \left(\ell \xrightarrow{B} \pi(j)\ell\right)_{j=1}^{k}$ is the trace of the first $k \geqslant 0$ iterations where $\pi(j)\ell$ is the maximal trace of the jth terminating execution of the loop body S_b starting from the initialization $\pi_1\ell$ of S. By convention, $\underline{\pi}(0) \triangleq \ell$. These traces are feasible provided that the tests along those traces are all true, as defined subsequently by β;
- The maximal trace $\pi(j)\ell$ of the jth terminating execution of that loop body S_b starting from the initialization $\pi_1\ell$ of S is such that
  - $\pi(0)\ell = \ell$;
  - $\pi(j)\ell$ is such that $\pi(j)\ell \in \widehat{\mathcal{S}}^{+\infty}[\![\mathsf{S}_b]\!](\pi_1\ell \mathbin{\widehat{\ }} \underline{\pi}(j-1) \mathbin{\widehat{\ }} \ell \xrightarrow{B} \mathsf{at}[\![\mathsf{S}_b]\!])$, for $j > 0$;
- The kth iteration of the loop is possible only if the iteration tests have all been true (tt) before this kth iteration, which we write $\beta(k)$.
  - $\beta(k) = \bigwedge_{j=1}^{k}\left(\mathscr{B}[\![\mathsf{B}]\!]\varrho(\pi_1\ell \mathbin{\widehat{\ }} \underline{\pi}(j-1)) = \mathrm{tt}\right)$ so that $\beta(1) = (\mathscr{B}[\![\mathsf{B}]\!]\varrho(\pi_1\ell) = \mathrm{tt})$. By convention, $\beta(0) \triangleq \mathrm{tt}$ is true;
- $\ell\pi'(n)$ traces the prefix execution of the nth iteration in the loop body S_b or a loop exit, so that, for $n \in \mathbb{N}^+$,

$$\ell\pi'(n) \quad \in \quad \{\ell\} \cup \qquad \text{(a)}$$

$$\{\ell \xrightarrow{\neg(\mathsf{B})} \mathsf{after}[\![\mathsf{S}]\!] \mid \mathscr{B}[\![\mathsf{B}]\!]\varrho(\pi_1\ell \mathbin{\widehat{\ }} \underline{\pi}(n-1)) = \mathrm{ff}\} \cup \qquad \text{(b)}$$

$$\{\ell \xrightarrow{B} \pi'' \mid \mathscr{B}[\![\mathsf{B}]\!]\varrho(\pi_1\ell \mathbin{\widehat{\ }} \underline{\pi}(n-1)) = \mathrm{tt} \wedge \qquad \text{(c)}$$

$$\pi'' \in \widehat{\mathcal{S}}^*[\![\mathsf{S}_b]\!](\pi_1\ell \mathbin{\widehat{\ }} \underline{\pi}(n-1) \xrightarrow{\mathsf{B}} \mathsf{at}[\![\mathsf{S}_b]\!])\}$$

(where the third term includes $\{\ell \xrightarrow{B} \mathsf{at}[\![\mathsf{S}_b]\!] \mid \mathscr{B}[\![\mathsf{B}]\!]\varrho(\pi_1\ell) = \mathsf{tt}\}$ for $n = 1$ and $\pi'' = \mathsf{at}[\![\mathsf{S}_b]\!]$ as well as $\{\ell \xrightarrow{\mathsf{B}} \pi_2 \xrightarrow{\mathsf{break}} \mathsf{break\text{-}to}[\![\mathsf{S}]\!] \mid \mathscr{B}[\![\mathsf{B}]\!]\varrho(\pi_1\ell \frown \underline{\pi}(n-1)) = \mathsf{tt} \wedge \pi_2 \xrightarrow{\mathsf{break}} \mathsf{break\text{-}to}[\![\mathsf{S}]\!] \in \widehat{\boldsymbol{\mathcal{S}}}^*[\![\mathsf{S}_b]\!](\pi_1\ell \frown \underline{\pi}(n-1) \xrightarrow{\mathsf{B}} \mathsf{at}[\![\mathsf{S}_b]\!])\}$ for $\pi'' = \pi_2 \xrightarrow{\mathsf{break}} \mathsf{break\text{-}to}[\![\mathsf{S}]\!]$ where $\mathsf{break\text{-}to}[\![\mathsf{S}]\!] = \mathsf{after}[\![\mathsf{S}]\!]$).

Proof

- In the basic case (6.24), $n = 1$ and $\underline{\pi}(1) = \ell$ so that, by taking $\ell\pi'(1) = \ell$ this reduces to the trace ℓ at the entry of the iteration $\mathsf{S} ::= \texttt{while}\,\ell\,(\mathsf{B})\,\mathsf{S}_b$.
- Assume by induction hypothesis that $\ell\pi_2\ell = \underline{\pi}(n-1) = \left(\ell \xrightarrow{B} \mathsf{at}[\![\mathsf{S}_b]\!]\pi(i)\ell\right)_{i=1}^{n-1}$ in the premises of rules (6.25), (6.28), and (6.26). Consider one more rule derivation; only three cases do apply.
  - In case (6.25), $\mathscr{B}[\![\mathsf{B}]\!]\varrho(\pi_1\ell\pi_2\ell) = \mathsf{ff}$ so the loop is exited and $\ell\pi_2\ell \xrightarrow{\neg(\mathsf{B})} \mathsf{after}[\![\mathsf{S}]\!] \in \widehat{\boldsymbol{\mathcal{S}}}^*[\![\mathsf{S}]\!](\pi_1\ell)$, which is of the form (6.39) requested by lemma 6.39;
  - In case (6.28), $\mathscr{B}[\![\mathsf{B}]\!]\varrho(\pi_1\ell\pi_2\ell) = \mathsf{tt}$, $\pi_3 \xrightarrow{\mathsf{break}} \mathsf{break\text{-}to}[\![\mathsf{S}]\!] \in \widehat{\boldsymbol{\mathcal{S}}}^*[\![\mathsf{S}_b]\!](\pi_1\ell\pi_2\ell \xrightarrow{\mathsf{B}} \mathsf{at}[\![\mathsf{S}_b]\!])$ so the loop is exited by a **break**; and $\ell\pi_2\ell \xrightarrow{\mathsf{B}} \mathsf{at}[\![\mathsf{S}_b]\!] \frown \pi_3 \xrightarrow{\mathsf{break}} \mathsf{break\text{-}to}[\![\mathsf{S}]\!] \in \widehat{\boldsymbol{\mathcal{S}}}^*[\![\mathsf{S}]\!](\pi_1\ell)$, which is of the form (6.39) requested by lemma 6.39;
  - Finally in case (6.26), $\mathscr{B}[\![\mathsf{B}]\!]\varrho(\pi_1\ell\pi_2\ell) = \mathsf{tt}$, $\pi_3 \in \widehat{\boldsymbol{\mathcal{S}}}^*[\![\mathsf{S}_b]\!](\pi_1\ell\pi_2\ell \xrightarrow{\mathsf{B}} \mathsf{at}[\![\mathsf{S}_b]\!])$. We must show that $\ell\pi_2\ell \xrightarrow{\mathsf{B}} \mathsf{at}[\![\mathsf{S}_b]\!] \frown \pi_3 \in \widehat{\boldsymbol{\mathcal{S}}}^*[\![\mathsf{S}]\!](\pi_1\ell)$ has the form (6.39) requested by lemma 6.39. By structural induction on S_b included in S, there are several cases.
    - if $\pi_3 = \mathsf{at}[\![\mathsf{S}_b]\!]$ then this is $\ell\pi_2\ell \xrightarrow{\mathsf{B}} \mathsf{at}[\![\mathsf{S}_b]\!] \frown \pi_3 = \ell\pi_2\ell \frown \ell \xrightarrow{\mathsf{B}} \mathsf{at}[\![\mathsf{S}_b]\!]$, which is of the requested form with $\ell\pi'(n) \in \{\ell = \ell \xrightarrow{B} \mathsf{at}[\![\mathsf{S}_b]\!]$;
    - the case when π_3 ends in a **break**; has already been handled;
    - else, if π_3 ends in $\mathsf{in}[\![\mathsf{S}_b]\!]$ (which does not contain $\mathsf{after}[\![\mathsf{S}_b]\!] = \mathsf{at}[\![\mathsf{S}]\!]$), which is of the requested form with $\ell\pi'(n) = \ell \xrightarrow{B} \mathsf{at}[\![\mathsf{S}_b]\!]\pi''(n)$ since $\mathsf{at}[\![\mathsf{S}_b]\!]\pi''(n)$ belongs to $\widehat{\boldsymbol{\mathcal{S}}}^*[\![\mathsf{S}_b]\!](\pi_1\ell\pi_2\ell \xrightarrow{\mathsf{B}} \mathsf{at}[\![\mathsf{S}_b]\!])\}$;
    - else π_3 ends at $\mathsf{after}[\![\mathsf{S}_b]\!]$, which is of the requested form with $\ell\pi'(n) = \ell \xrightarrow{\neg(\mathsf{B})} \mathsf{after}[\![\mathsf{S}]\!]$.

We have shown that the traces in $\widehat{\boldsymbol{\mathcal{S}}}^*[\![\mathsf{S}]\!](\pi_1\ell)$ are all generated by the rules (6.24), (6.25), and (6.26).

Conversely, we must show that all traces generated by the rules (6.24), (6.25), and (6.26) belong to $\widehat{\boldsymbol{\mathcal{S}}}^*[\![\mathsf{S}]\!](\pi_1\ell)$. This is by induction on the length of such a trace. This is obviously true for the basis ℓ. For this induction step, the premise is assumed to belong to $\widehat{\boldsymbol{\mathcal{S}}}^*[\![\mathsf{S}]\!](\pi_1\ell)$ by induction hypothesis and it can be easily checked that the conclusion of the rule also does. □

In (6.25), (6.28), and (6.26), iterations are defined by left recursion, that is, iterations are previous iterations followed by one more iteration i.e. (see lemma 6.39 when $\ell\pi'(n) = \ell$),

$$\left(\ell \xrightarrow{B} \mathsf{at}[\![\mathsf{S}_b]\!]\pi(i)\ell\right)_{i=1}^{n+1} = \left(\ell \xrightarrow{B} \mathsf{at}[\![\mathsf{S}_b]\!]\pi(i)\ell\right)_{i=1}^{n} \frown \ell \xrightarrow{B} \mathsf{at}[\![\mathsf{S}_b]\!]\pi(n+1)\ell.$$

6.8 Equivalent Right-Recursive Definition of Iteration

Iterations can also be defined backward by a right recursion; that is, $n+1$ iterations are one iteration followed by n more iterations i.e. $\left(\ell \xrightarrow{B} \mathsf{at}[\![\mathsf{S}_b]\!]\pi(i)\ell\right)_{i=1}^{n+1} = \ell \xrightarrow{B} \mathsf{at}[\![\mathsf{S}_b]\!]\pi(1)\ell \frown \left(\ell \xrightarrow{B} \mathsf{at}[\![\mathsf{S}_b]\!]\pi(i)\ell\right)_{i=2}^{n+1}$, by replacing (6.25), (6.28), and (6.26) by (6.40), (6.44), (6.41), and (6.42), as follows:

Prefix traces of an iteration statement $\mathsf{S} ::= \texttt{while}\ \ell\ (\mathsf{B})\ \mathsf{S}_b$

- $$\frac{\mathscr{B}[\![\mathsf{B}]\!]\varrho(\pi_1\ell) = \mathsf{ff}}{\ell \xrightarrow{\neg(\mathsf{B})} \mathsf{after}[\![\mathsf{S}]\!] \in \widehat{\boldsymbol{\mathcal{S}}}^*[\![\mathsf{S}]\!](\pi_1\ell)} \tag{6.40}$$

- $$\frac{\mathscr{B}[\![\mathsf{B}]\!]\varrho(\pi_1\ell) = \mathsf{tt},\quad \pi_2\ell' \in \widehat{\boldsymbol{\mathcal{S}}}^*[\![\mathsf{S}_b]\!](\pi_1\ell \xrightarrow{\mathsf{B}} \mathsf{at}[\![\mathsf{S}_b]\!]),\quad \ell' \neq \ell}{\ell \xrightarrow{\mathsf{B}} \mathsf{at}[\![\mathsf{S}_b]\!] \frown \pi_2\ell' \in \widehat{\boldsymbol{\mathcal{S}}}^*[\![\mathsf{S}]\!](\pi_1\ell)} \tag{6.41}$$

- $$\frac{\mathscr{B}[\![\mathsf{B}]\!]\varrho(\pi_1\ell) = \mathsf{tt},\quad \pi_2' \in \widehat{\boldsymbol{\mathcal{S}}}^*[\![\mathsf{S}_b]\!](\pi_1\ell \xrightarrow{\mathsf{B}} \mathsf{at}[\![\mathsf{S}_b]\!]),\quad \pi_3' \in \widehat{\boldsymbol{\mathcal{S}}}^*[\![\mathsf{S}]\!](\pi_1\ell \xrightarrow{\mathsf{B}} \mathsf{at}[\![\mathsf{S}_b]\!] \frown \pi_2')}{\ell \xrightarrow{\mathsf{B}} \mathsf{at}[\![\mathsf{S}_b]\!] \frown \pi_2' \frown \pi_3' \in \widehat{\boldsymbol{\mathcal{S}}}^*[\![\mathsf{S}]\!](\pi_1\ell)} \tag{6.42}$$

Remark 6.43 (on the break statement) Rule (6.41) includes the case of an iteration exited by a break statement when $\pi_2\ell' = \pi_2' \xrightarrow{\mathsf{break}} \mathsf{break\text{-}to}[\![\mathsf{S}]\!]$, that is,

- $$\frac{\mathscr{B}[\![\mathsf{B}]\!]\varrho(\pi_1\ell) = \mathsf{tt},\quad \pi_2 \xrightarrow{\mathsf{break}} \mathsf{break\text{-}to}[\![\mathsf{S}]\!] \in \widehat{\boldsymbol{\mathcal{S}}}^*[\![\mathsf{S}_b]\!](\pi_1\ell \xrightarrow{\mathsf{B}} \mathsf{at}[\![\mathsf{S}_b]\!])}{\ell \xrightarrow{\mathsf{B}} \mathsf{at}[\![\mathsf{S}_b]\!] \frown \pi_2 \xrightarrow{\mathsf{break}} \mathsf{break\text{-}to}[\![\mathsf{S}]\!] \in \widehat{\boldsymbol{\mathcal{S}}}^*[\![\mathsf{S}]\!](\pi_1\ell)} \tag{6.44}$$

Replacing the rules (6.25), (6.28), and (6.26) by rules (6.40), (6.44), (6.41), and (6.42) does not change the deductive definition of $\widehat{\boldsymbol{\mathcal{S}}}^*[\![\mathsf{S}]\!]$.

Lemma 6.45 The rules (6.40), (6.44), (6.41), and (6.42) are equivalent to (6.25), (6.28), and (6.26).

Proof The proof is by structural induction. The only nontrivial case is the iteration for which the rules differ. By lemma 6.39, we know that rules (6.25), (6.28), and (6.26) generate all and only those traces of the form (6.39). We prove that rules (6.40), (6.44), (6.41), and (6.42) allow us to derive exactly the same traces.

The rules (6.24), (6.40), (6.44), and (6.41) allow us to derive $\ell\pi'(n) \;\in\; \widehat{\boldsymbol{S}}^*[\![\mathsf{S}]\!](\pi_1\ell \frown \big(\ell \xrightarrow{B} \mathsf{at}[\![\mathsf{S}_b]\!]\pi(i)\ell\big)_{i=1}^{n-1})$ where $\ell\pi'(n)$ has the form requested by lemma 6.39.

Then rule (6.24) for the basis and (6.42) for the induction allow us to derive (6.39) by recurrence. □

Exercise 6.46 Show that $\widehat{\boldsymbol{S}}^*[\![\mathsf{S}]\!](\pi_1\ell)$ is *closed by prefix*, that is, if $\pi_1' \frown \pi_2 \in \widehat{\boldsymbol{S}}^*[\![\mathsf{S}]\!](\pi_1\ell)$ then $\pi_1' \in \widehat{\boldsymbol{S}}^*[\![\mathsf{S}]\!](\pi_1\ell)$.

Exercise 6.47 Continuing exercise 3.9 define what it means and show that renaming program variables and labels does not change the semantics of programs [798].

6.9 Side Effects

The evaluation of an assignment statement (6.16) has a *side effect*: the values of variables before and after the assignment may be different. By contrast, the evaluation of Boolean expressions in (6.18), (6.19), (6.22), (6.23), (6.25), and (6.26) has no side effect because it does not change the values of variables. Formally, an action a has a side effect in a stateless trace semantics $\boldsymbol{S}$ if and only if $\exists \mathsf{S} \in \mathbb{P}\mathrm{c} \,.\, \exists \pi \xrightarrow{\mathsf{a}} \ell\pi' \in \boldsymbol{S}[\![\mathsf{S}]\!] \,.\, \varrho(\pi) \neq \varrho(\pi \xrightarrow{\mathsf{a}} \ell)$.

6.10 The Prefix Trace Semantics as a Relation Between an Initialization and a Continuation Trace

Because $\mathbb{T}^+ \to \wp(\mathbb{T}^+)$ and $\wp(\mathbb{T}^+ \times \mathbb{T}^+)$ are isomorphic (exercise 11.52), we can, for any program component S understand the prefix trace semantics $\boldsymbol{S}^*[\![\mathsf{S}]\!]$ as a relation $\boldsymbol{S}^{\bar{*}}[\![\mathsf{S}]\!] \triangleq \{\langle\pi, \pi'\rangle \mid \pi \in \mathbb{T}^+ \wedge \pi' \in \boldsymbol{S}^*[\![\mathsf{S}]\!]\pi\}$ so that $\boldsymbol{S}^*[\![\mathsf{S}]\!]$ is the right image of $\boldsymbol{S}^{\bar{*}}[\![\mathsf{S}]\!]$. By abuse of notation, we usually write $\boldsymbol{S}^*[\![\mathsf{S}]\!]$ for $\boldsymbol{S}^{\bar{*}}[\![\mathsf{S}]\!]$. Therefore $\pi' \in \boldsymbol{S}^*[\![\mathsf{S}]\!](\pi)$ will be equivalently written $\langle\pi, \pi'\rangle \in \boldsymbol{S}^*[\![\mathsf{S}]\!]$ when implicitly referring to the right image $\boldsymbol{S}^{\bar{*}}[\![\mathsf{S}]\!]$ of $\boldsymbol{S}^*[\![\mathsf{S}]\!]$

6.11 Prefix Trace Semantics from Initial Traces

Given a set $\mathscr{P}_0$ of initial traces, the prefix trace semantics $\boldsymbol{S}^*[\![\mathsf{S}]\!]\,\mathscr{P}_0$ is the set of the prefix traces starting from $\mathscr{P}_0$.

$$\boldsymbol{S}^*[\![\mathsf{S}]\!]\,\mathscr{P}_0 \;\triangleq\; \bigcup\{\boldsymbol{S}^*[\![\mathsf{S}]\!](\pi\ell) \mid \pi\ell \in \mathscr{P}_0\}\,.$$

6.12 Pointwise Prefix Trace Semantics

Given a set $\mathscr{P}_0$ of initial traces, the pointwise prefix trace semantics $\boldsymbol{S}^{\dot{*}}[\![\mathsf{S}]\!]\,\mathscr{P}_0\,\ell$ is the set of the prefix traces starting from $\mathscr{P}_0$ and arriving at program label ℓ

$$\begin{aligned} \boldsymbol{S}^{\dot{*}}[\![\mathsf{S}]\!] &\in \wp(\mathbb{T}^+) \stackrel{\nearrow}{\longrightarrow} (\mathbb{L} \to \wp(\mathbb{T}^+)) \\ \boldsymbol{S}^{\dot{*}}[\![\mathsf{S}]\!]\,\mathscr{P}_0\,\ell &\triangleq \{\pi_0\ell_0\pi_1\ell_1 \mid \pi_0\ell_0 \in \mathscr{P}_0 \wedge \ell_0\pi_1\ell_1 \in \boldsymbol{S}^*[\![\mathsf{S}]\!](\pi_0\ell_0) \wedge \ell_1 = \ell\} \end{aligned} \tag{6.48}$$

which we can define by the inference rule

$$\frac{\pi_0\ell_0 \in \mathcal{P}_0, \quad \ell_0\pi_1\ell \in \boldsymbol{S}^*[\![\mathsf{S}]\!](\pi_0\ell_0)}{\pi_0\ell_0\pi_1\ell \in \boldsymbol{S}^{\hat{*}}[\![\mathsf{S}]\!]\,\mathcal{P}_0\,\ell}$$

which itself can be instantiated for all cases of S considered in (6.11) to (6.42).

6.13 Conclusion

We have defined the stateless prefix and finite maximal trace semantics of programs. The prefix trace semantics describes safety properties of programs [263]. The finite maximal trace semantics of programs abstracts to partial correctness program properties (where termination is assumed so infinite computations are abstracted away) [223]. Conventional trace semantics are stateful in that each configuration in the trace records the current value of variables. Peter Wegner [970] was an early advocate for modeling the semantics of programming languages by stateful traces generated by a transition system. Such stateful prefix trace semantics and transition systems are abstractions of a stateless prefix trace semantics, as shown in chapter 42. Stateful trace semantics do not generalize to more complex languages (such as parallel programs with weak memory models allowing to read from past and future writes; see, for example, [21]) as easily as stateless trace semantics do (e.g., [23]).

Our semantics maps an initialization trace describing how a computation ended at a program component to traces continuing this initial computation by executing this program component. A similar idea is used in [42] in which the initialization trace is abstracted into a store transformer, called a prelude, mapping the initial state of the whole program computation to the state before execution of a command reached from that initial state.

Our semantics does not describe the intricacies of the C programming language. A more realistic semantic definition of implementations of C is proposed in [178, section I.4]. Our semantics does not describe no more the intricacies of modern machine implementations. Hence we cannot discuss, for example, vulnerabilities such as Meltdown [637], Spectre [594], and ForeShadow [156]. To do that, we would have to use a different semantics taking speculative execution into account as proposed by, for example, [991]. For timing attacks, the timing of action, test, and branching executions may also have to be taken into account. So the semantics would be different, and therefore the static analyses would also be different. However, the principles of abstract interpretation to formally construct sound verification and analysis methods would be applicable in the same way for this alternative semantics.

6.14 Solutions to Selected Exercises

Solution to exercise 6.37 The proof is by structural induction on the program component $\mathsf{S} \in \mathbb{P}\mathbb{c}$.

— For the basic cases, this follows from $\mathsf{at}[\![\mathsf{S}]\!] \in \mathsf{labx}[\![\mathsf{S}]\!]$ for (6.11), (6.15), (6.29), and (6.24) and, in addition, $\mathsf{after}[\![\mathsf{S}]\!] \in \mathsf{labx}[\![\mathsf{S}]\!]$ for (6.16), (6.17) while $\mathsf{break\text{-}to}[\![\mathsf{S}]\!] \in \mathsf{labx}[\![\mathsf{S}]\!]$ for (6.29).

— For the induction step, this follows from the induction hypothesis that the labels in traces of subcomponents are labels of these subcomponents, $\alpha_\ell(\pi_1 \frown \pi_2) = \alpha_\ell(\pi_1) \cup \alpha_\ell(\pi_2)$, and

— from labx⟦P⟧ = labx⟦Sl⟧ (containing no escape label) in (6.12);

— from labx⟦Sl⟧ = labx⟦Sl′⟧ ∪ labx⟦S⟧ in (6.13) and (6.14);

— from labx⟦**if** ℓ (B) S_t⟧ = {ℓ} ∪ labx⟦S_t⟧ and after⟦S_t⟧ = after⟦**if** ℓ (B) S_t⟧ in (6.18) and (6.19) and similarly for (6.22) and (6.23);

— from labx⟦**while** ℓ (B) S_b⟧ = {ℓ} ∪ labx⟦S_b⟧ with break-to⟦S_b⟧ = after⟦**while** ℓ (B) S_b⟧ for (6.25) and (6.26).

7 Maximal Trace Semantics

A trace is maximal when it is finite and represents an execution that terminates or it is infinite and represents an execution that does not terminate. Maximal finite traces of statements are prefix executions that exit the statement (e.g., through normal termination after the statement or a `break`). Statements containing iteration are the only statements that may not terminate. If an iteration has infinitely many different prefix traces, this means that the iteration does not terminate. Passing to the limit, we get the infinite behavior of nonterminating iterations.

Contents

7.1 Maximal Finite Trace Semantics

Let $\pi_1\mathsf{at}[\![\mathsf{S}]\!]$ be an initialization trace ending on entry $\mathsf{at}[\![\mathsf{S}]\!]$ of statement S. We have defined the abstraction

$$\begin{aligned}\boldsymbol{\mathcal{S}}^+[\![\mathsf{S}]\!](\pi_1\mathsf{at}[\![\mathsf{S}]\!]) &\triangleq \alpha^+[\![\mathsf{S}]\!](\boldsymbol{\mathcal{S}}^*[\![\mathsf{S}]\!])(\pi_1\mathsf{at}[\![\mathsf{S}]\!]) \\ \alpha^+[\![\mathsf{S}]\!](\boldsymbol{\mathcal{S}})\pi &\triangleq \{\pi_2\ell \in \boldsymbol{\mathcal{S}}(\pi) \mid (\ell = \mathsf{after}[\![\mathsf{S}]\!]) \vee (\mathsf{escape}[\![\mathsf{S}]\!] \wedge \ell = \mathsf{break\text{-}to}[\![\mathsf{S}]\!])\}\end{aligned} \tag{6.9}$$

as the set of maximal finite traces $\mathsf{at}[\![\mathsf{S}]\!]\pi_2\mathsf{after}[\![\mathsf{S}]\!]$ of S, continuing the trace $\pi_1\mathsf{at}[\![\mathsf{S}]\!]$, and reaching $\mathsf{after}[\![\mathsf{S}]\!]$. Schematically,

$$\xrightarrow{\quad\pi_1\quad} \underbrace{\mathsf{at}[\![\mathsf{S}]\!] \xrightarrow{\quad\pi_2\quad} \left|\begin{array}{l}\mathsf{break\text{-}to}[\![\mathsf{S}]\!] \quad (\text{when } \mathsf{escape}[\![\mathsf{S}]\!]) \\ \mathsf{after}[\![\mathsf{S}]\!]\end{array}\right.}_{\in\ \boldsymbol{\mathcal{S}}^+[\![\mathsf{S}]\!](\pi_1\mathsf{at}[\![\mathsf{S}]\!])}$$

By exercise 6.34, $\boldsymbol{\mathcal{S}}^+[\![\mathsf{S}]\!](\pi_1\ell) = \varnothing$ when $\ell \neq \mathsf{at}[\![\mathsf{S}]\!]$.

Example 7.1 The program (4.5) has only one maximal trace as follows

$$\begin{aligned}\boldsymbol{\mathcal{S}}^+[\![\mathsf{S}]\!](\ell_1) = \{&\ell_1 \xrightarrow{\mathsf{x}=1} \ell_2 \xrightarrow{\mathsf{tt}} \ell_3 \xrightarrow{\mathsf{x}=2} \ell_4 \xrightarrow{\neg(\mathsf{x}>2)} \ell_2 \xrightarrow{\mathsf{tt}} \ell_3 \xrightarrow{\mathsf{x}=3} \\ &\ell_4 \xrightarrow{\mathsf{x}>2} \ell_5 \xrightarrow{\mathsf{break}} \ell_6 \xrightarrow{\mathsf{skip}} \ell_7\}\end{aligned}$$

7.2 Infinite Trace Semantics

If $\pi = \ell_0 \xrightarrow{\mathsf{a}_0} \ldots \ell_i \xrightarrow{\mathsf{a}_i} \ldots \ell_n$ is a finite trace then its prefix $\pi[0..p]$ at p is π when $p \geqslant n$ and $\ell_0 \xrightarrow{\mathsf{a}_0} \ldots \ell_j \xrightarrow{\mathsf{a}_j} \ldots \ell_p$ when $0 \leqslant p < n$. If $\pi = \ell_0 \xrightarrow{\mathsf{a}_0} \ldots \ell_i \xrightarrow{\mathsf{a}_i} \ldots$ is an infinite trace then its prefix $\pi[0..p]$ at p is $\ell_0 \xrightarrow{\mathsf{a}_0} \ldots \ell_j \xrightarrow{\mathsf{a}_j} \ldots \ell_p$.

Given a set $\mathcal{T} \in \wp(\mathbb{T}^+)$ of finite traces, its limit $\lim \mathcal{T}$ is the set of infinite traces the prefixes of which are traces in $\mathcal{T}$.

$$\lim \mathcal{T} \ \triangleq\ \{\pi \in \mathbb{T}^\infty \mid \forall n \in \mathbb{N} \,.\, \pi[0..n] \in \mathcal{T}\}\,. \tag{7.2}$$

(With this definition $\lim \varnothing = \varnothing$.)

Example 7.3 Continuing example 6.32, the prefix semantics of the program

$$\mathsf{S} = \mathbf{while}\ \ell_1\ (\mathsf{tt})\ \ell_2\ \mathsf{x} = \mathsf{x} + 1\,;\ell_3$$

is $\boldsymbol{\mathcal{S}}^*[\![\mathsf{S}]\!](\ell_1) = \Big\{\Big(\ell_1 \xrightarrow{\mathsf{tt}} \ell_2 \xrightarrow{\mathsf{x}=i} \ell_1\Big)_{i=1}^{n}, \Big(\ell_1 \xrightarrow{\mathsf{tt}} \ell_2 \xrightarrow{\mathsf{x}=i} \ell_1\Big)_{i=1}^{n} \xrightarrow{\mathsf{tt}} \ell_2 \,\Big|\, n \in \mathbb{N}\Big\}$. Its limit is $\lim(\boldsymbol{\mathcal{S}}^*[\![\mathsf{S}]\!](\ell_1)) = \{\pi\}$ where the infinite trace is $\pi = \Big(\ell_1 \xrightarrow{\mathsf{tt}} \ell_2 \xrightarrow{\mathsf{x}=i}\Big)_{i=1}^{\infty}$. All prefixes of π belong to $\boldsymbol{\mathcal{S}}^*[\![\mathsf{S}]\!](\ell_1)$.

A more general definition of the limit not requiring the set $\mathcal{T} \in \wp(\mathbb{T}^+)$ of finite traces to be closed by prefix, consists of defining the limit $\lim \mathcal{T}$ as the set of infinite traces the prefixes of which can be extended to a trace in $\mathcal{T}$.

$$\lim \mathcal{T} \ \triangleq\ \{\pi \in \mathbb{T}^\infty \mid \forall n \in \mathbb{N} \,.\, \exists p \geqslant n \,.\, \pi[0..p] \in \mathcal{T}\}\,. \tag{7.4}$$

Example 7.5 $\lim\Big\{\Big(\ell_1 \xrightarrow{\mathsf{tt}} \ell_2 \xrightarrow{\mathsf{x}=i} \ell_1\Big)_{i=1}^{n} \,\Big|\, n \in \mathbb{N}\Big\} = \{\pi\}$ where $\pi = \Big(\ell_1 \xrightarrow{\mathsf{tt}} \ell_2 \xrightarrow{\mathsf{x}=i}\Big)_{i=1}^{\infty}$. All prefixes of π are of the form $\Big(\ell_1 \xrightarrow{\mathsf{tt}} \ell_2 \xrightarrow{\mathsf{x}=i} \ell_1\Big)_{i=1}^{n}$ or $\Big(\ell_1 \xrightarrow{\mathsf{tt}} \ell_2 \xrightarrow{\mathsf{x}=i} \ell_1\Big)_{i=1}^{n} \xrightarrow{\mathsf{tt}} \ell_2$ and this last one can be extended to a finite trace $\Big(\ell_1 \xrightarrow{\mathsf{tt}} \ell_2 \xrightarrow{\mathsf{x}=i} \ell_1\Big)_{i=1}^{n+1}$.

We define the infinite trace semantics

$$\boldsymbol{\mathcal{S}}^\infty[\![\mathsf{S}]\!](\pi^\ell) \ \triangleq\ \lim(\boldsymbol{\mathcal{S}}^*[\![\mathsf{S}]\!](\pi^\ell)). \tag{7.6}$$

7.3 Maximal Finite and Infinite Trace Semantics

The maximal trace semantics is the set of traces that are either finite

$$\boldsymbol{\mathcal{S}}^+[\![\mathsf{S}]\!](\pi_1\mathsf{at}[\![\mathsf{S}]\!]) \ \triangleq\ \{\pi_2{}^\ell \in \boldsymbol{\mathcal{S}}^*[\![\mathsf{S}]\!](\pi_1\mathsf{at}[\![\mathsf{S}]\!]) \mid (\ell = \mathsf{after}[\![\mathsf{S}]\!]) \vee (\mathsf{escape}[\![\mathsf{S}]\!] \wedge \ell = \mathsf{break\text{-}to}[\![\mathsf{S}]\!])\} \tag{6.9}$$

or infinite defined as limits of finite prefix traces.

$$\begin{aligned}
\mathcal{S}^{+\infty}[\![\mathsf{S}]\!](\pi\ell) &\triangleq \mathcal{S}^{+}[\![\mathsf{S}]\!](\pi\ell) \cup \mathcal{S}^{\infty}[\![\mathsf{S}]\!](\pi\ell) \\
\mathcal{S}^{+\infty}[\![\mathsf{S}]\!]\Pi &\triangleq \bigcup\{\mathcal{S}^{+\infty}[\![\mathsf{S}]\!](\pi\ell) \mid \pi\ell \in \Pi\} \\
\mathcal{S}^{+\infty}[\![\mathsf{S}]\!] &\triangleq \mathcal{S}^{+\infty}[\![\mathsf{S}]\!](\mathbb{T}^{+}) \\
\mathcal{S}^{+\infty}[\![\mathsf{P}]\!] &\triangleq \mathcal{S}^{+\infty}[\![\mathsf{P}]\!](\{\mathsf{at}[\![\mathsf{P}]\!]\}).
\end{aligned} \tag{7.7}$$

A pair $\langle\pi_0, \pi\rangle$ such that $\pi \in \mathcal{S}^{+\infty}[\![\mathsf{S}]\!](\pi_0)$ is called a *maximal execution* of program component S.

Example 7.8 Continuing Examples 6.32, 7.3 and 7.5, the maximal trace semantics of the program $\mathsf{S} = \texttt{while}\,\ell_1\,(\mathsf{tt})\,\ell_2\,\mathsf{x} = \mathsf{x} + 1\,;\ell_3$ is $\mathcal{S}^{+\infty}[\![\mathsf{S}]\!](\ell_1) = \left\{\left(\ell_1 \xrightarrow{\mathsf{tt}} \ell_2 \xrightarrow{\mathsf{x}=i}\right)_{i=1}^{\infty}\right\}$.

Exercise 7.9 Continuing exercise 4.3, modify the program written in exercise 4.3 so that it prints its own trace, as follows

```
l1 -[x = 1]-> l2 -[tt]-> l3 -[x = 2]-> l4 -[¬(x > 2)]-> l2 -[tt]-> l3
-[x = 3]-> l4 -[(x > 2)]-> l5 -[break]-> l6 -[skip]-> l7 .
```

Exercise 7.10 Prove that $\mathcal{S}^{+}[\![\mathsf{S}]\!](\Pi) = \bigcup\{\mathcal{S}^{+}[\![\mathsf{S}]\!](\pi\ell) \mid \pi\ell \in \Pi\}$

Exercise 7.11 What is $\lim\{a^n \mid n \in \mathbb{N}\}$ where $a^0 = \epsilon$ is the empty string and a^n is the string made of n letters a? What is $\lim\{a^n b \mid n \in \mathbb{N}\}$?

Exercise 7.12 Show that $\alpha^{*\to}(X) \triangleq X \cup \lim X$ and $\gamma^{*\to}(Y) \triangleq \{\pi \in \mathbb{T}^{+} \mid \exists \pi' \mid \pi \mathbin{\widehat{}} \pi' \in Y\}$ is a Galois connection and therefore, by pointwise extension $\dot{\alpha}^{*\to}$ and $\dot{\gamma}^{*\to}$, that $\mathcal{S}^{*}[\![\mathsf{S}]\!] = \dot{\gamma}^{*\to}(\mathcal{S}^{+\infty}[\![\mathsf{P}]\!])$ and $\mathcal{S}^{+\infty}[\![\mathsf{P}]\!] = \dot{\alpha}^{*\to}(\mathcal{S}^{*}[\![\mathsf{S}]\!])$.

Exercise 7.13 Continuing exercise 5.1, let the syntax of regular expressions be

$e \in \mathbb{E}$			alphabet
$\mathsf{R} \in \mathbb{R}$	$::=$	$e \mid \varepsilon \mid \mathsf{R}_1\mathsf{R}_2 \mid \mathsf{R}_1 \parallel \mathsf{R}_2 \mid \mathsf{R}^{+}$	finite regular expression
		$\mid \mathsf{R}^{\infty}$	infinite regular expression

The semantics of a regular expression R is the set of all finite and infinite strings recognized by that regular expression R. It is informally defined as follows

- A regular expression reduced to a letter e of the alphabet $\mathbb{E}$ recognizes this letter and only this letter.
- The empty regular expression ε recognizes the empty string ϵ only.
- The concatenation $\mathsf{R}_1\mathsf{R}_2$ recognizes the strings $\sigma_1\sigma_2$ where σ_1 is recognized by R_1 and followed by a string recognized σ_2 by R_2.
- The alternative $\mathsf{R}_1 \parallel \mathsf{R}_2$ recognizes a string either recognized by R_1 or recognized by R_2.
- The finite repetition R^{+} recognizes a string made of one or more of the strings recognized by R.
- The infinite repetition R^{∞} denotes an infinite repetition of strings recognized by R.

Formally define the semantics of regular expressions.

Exercise 7.14 (Project: maximal trace interpreter) Continuing exercise 5.9, implement an interpreter in the programming language of your choice that reads a program with syntax given in section 4.1 and prints the maximal trace formalizing its execution (this does not terminate for infinite traces). For example,

```
$ echo "x=x+1;while (0<1){x=x+1;if(2<x)break;};"| ./interpreter
<l1, l7, ff, l7>      Prog:
<l1, l2, ff, l7>          l1: x = (0 - ((0 - x) - 1));
<l2, l6, ff, l6>          (while l2: (0 < 1)
<l3, l2, tt, l6>              Stmtlist: {
<l3, l4, ff, l6>                  l3: x = (0 - ((0 - x) - 1));
<l4, l2, tt, l6>                  (if l4: (2 < x)
<l5, l2, tt, l6>                      l5: break; )
                              } )
<l6, l7, ff, l7>          l6: ;
                      l7:
l1 --[x = (0 - ((0 - x) - 1)) = 1]-> l2 --[(0 < 1)]-> l3
--[x = (0 - ((0 - x) - 1)) = 2]-> l4 --[¬((2 < x))]-> l2
--[(0 < 1)]-> l3 --[x = (0 - ((0 - x) - 1)) = 3]-> l4
--[(2 < x)]-> l5 --[break]-> l6 --[skip]-> l7
$
```

7.4 Iteration

The finite prefix traces of an iteration have been characterized by lemma 6.39. The infinite traces describing nonterminating executions can be characterized as follows (where the infinite trace $\left(\ell\pi(i)\ell\right)_{i=1}^{\infty}$ denotes $\ell\pi(1)\ell\pi(2)\ell \ldots \pi(n)\ell\pi(n+1)\ell \ldots$):

Lemma 7.15 The infinite traces of $\widehat{\boldsymbol{\mathcal{S}}}^{\infty}[\![\mathsf{S}]\!](\pi_1\ell)$ for an iteration statement of the form $\mathsf{S} ::= \texttt{while}\,\ell\,(\mathsf{B})\,\mathsf{S}_b$ with entry label $\ell = \mathsf{at}[\![\mathsf{S}]\!] = \mathsf{after}[\![\mathsf{S}_b]\!]$ initialized by $\pi_1\ell$ have the form

$$\left(\ell \xrightarrow{B} \mathsf{at}[\![\mathsf{S}_b]\!]\pi(i)\ell\right)_{i=1}^{\infty} \tag{a}$$

$$\text{or} \quad \left(\ell \xrightarrow{B} \mathsf{at}[\![\mathsf{S}_b]\!]\pi(i)\ell\right)_{i=1}^{n-1} \mathbin{\hat{\frown}} \ell\pi'(n) \tag{b}$$

where

- for all $i \in [1, \infty]$ in case (a) and $i \in [1, n-1]$ in case (b), $\mathsf{at}[\![\mathsf{S}_b]\!]\pi(i)\ell \in \widehat{\boldsymbol{\mathcal{S}}}^{*}[\![\mathsf{S}_b]\!](\pi_1\ell \mathbin{\hat{\frown}} \left(\ell \xrightarrow{B} \mathsf{at}[\![\mathsf{S}_b]\!]\pi(j)\ell\right)_{j=1}^{i-1} \mathbin{\hat{\frown}} \ell \xrightarrow{B} \mathsf{at}[\![\mathsf{S}_b]\!])$ is the maximal finite trace of the ith iteration in the loop body S_b;
- $\ell\pi'(n)$ traces the prefix execution of the nth iteration in the loop body S_b or a loop exit, so

$$\ell\pi'(n) \in \{\ell \xrightarrow{B} \mathsf{at}[\![\mathsf{S}_b]\!]\pi''(n) \mid$$
$$\mathsf{at}[\![\mathsf{S}_b]\!]\pi''(n) \in \widehat{\mathcal{S}}^{\infty}[\![\mathsf{S}_b]\!](\pi_1\ell \mathbin{\widehat{\ }} \left(\ell \xrightarrow{B} \mathsf{at}[\![\mathsf{S}_b]\!]\pi(i)\ell\right)_{i=1}^{n-1} \xrightarrow{\mathsf{B}} \mathsf{at}[\![\mathsf{S}_b]\!])\}\ .$$

Proof Case (a) corresponds to infinitely many iterations of terminating executions of the loop body and follows directly form the definition (7.4) of lim and lemma 6.39. Case (b) corresponds to an infinite execution of the loop body S_b after zero or more previous terminating iterations of this loop body as described by lemma 6.39. □

7.5 Conclusion

We have defined the maximal trace semantics of programs. Infinite traces were obtained as limits of prefix traces. An alternative way of defining infinite traces (by coinductive definitions of section 16.6) or simultaneously finite and infinite traces (by bi-inductive definitions of section 16.7) is considered in [227, 247, 260].

7.6 Solutions to Selected Exercises

Solution to exercise 7.11 $\lim \{a^n \mid n \in \mathbb{N}\} = a^{\infty}$ where a^{∞} is the infinite string of letters a. $\lim \{a^n b \mid n \in \mathbb{N}\} = \emptyset$.

Solution to exercise 7.13

$$
\begin{aligned}
L_1 \cdot L_2 &\triangleq \{w_1 w_2 \mid w_1 \in L_1 \wedge w_2 \in L_2\} && \text{language concatenation}\\
L^1 &\triangleq L && \text{language powers}\\
L^{n+1} &\triangleq L^n \cdot L, \quad n \in \mathbb{N}\\
\mathcal{S}[\![e]\!] &\triangleq \{e\}\\
\mathcal{S}[\![\varepsilon]\!] &\triangleq \ni\\
\mathcal{S}[\![\mathrm{R}_1\mathrm{R}_2]\!] &\triangleq \mathcal{S}[\![\mathrm{R}_1]\!] \cdot \mathcal{S}[\![\mathrm{R}_2]\!]\\
\mathcal{S}[\![\mathrm{R}_2 \mid \mathrm{R}_1]\!] &\triangleq \mathcal{S}[\![\mathrm{R}_1]\!] \cup \mathcal{S}[\![\mathrm{R}_2]\!]\\
\mathcal{S}[\![\mathrm{R}^+]\!] &\triangleq \bigcup_{n \in \mathbb{N}} \mathcal{S}[\![\mathrm{R}]\!]^n\\
\mathcal{S}[\![\mathrm{R}^{\infty}]\!] &\triangleq \lim \mathcal{S}[\![\mathrm{R}^+]\!]
\end{aligned}
$$

IV PROPERTIES AND THEIR ABSTRACTIONS

By program's concrete properties, we understand the properties of their trace semantics. Because the trace semantics is a set of traces, its properties are sets of sets of traces (chapter 8).

The properties of a program cannot be calculated automatically (chapter 9).

This calls for the determination of more abstract properties that are sound (that is, satisfied by the trace semantics) but less expressive. The mathematical structure of such abstract program properties is that of complete lattices (chapter 10).

The formalization of abstraction (chapter 11) ensures that abstract properties will always be valid in the concrete (i.e. for the trace semantics).

Relational and transformer semantics (chapter 12) are a first example of abstraction of trace properties where intermediate computations are abstracted away. Although more abstract than trace semantics properties they are still not automatically computable. However they are the basis of many verification (part VIII) and static analysis methods (from chapter 27 on).

8 Program Properties

We specify program properties by their extension, that is, the set of elements that have such a property, and more precisely, properties of the trace semantics of programs. We introduce the program collecting semantics as the strongest property of a program. We consider abstractions of the collecting semantics such as trace and reachability properties.

Contents

8.1 What is a Program Property?

By the term *program property* we could mean a property of the program text P, such as a ten-line program or a syntax error at line 12. In that case, we instead use the term of a *program syntax property*.

Here, however, by *program property*, we mean a *program semantic property* that is a property of its maximal finite or infinite trace semantics $\boldsymbol{\mathcal{S}}^{+\infty}[\![\mathsf{P}]\!] \in \mathbb{T}^{+} \rightarrow \wp(\mathbb{T}^{+\infty})$.

8.2 What is a Formal Property?

When speaking of *properties* we have to say "Properties of what?" This *what* is a given set $\mathscr{E}$ of entities (individuals, things, elements, objects, etc.) of interest. Logicians would say *elements of a universe* (or universal set or universe of discourse $\mathbb{U}$).

By *formal property* we mean the set of entities that have this property. So, if $\mathscr{E}$ is a set of entities then the properties of these entities are sets belonging to $\wp(\mathscr{E})$.

An entity has the formal property if and only if it belongs to the formal property, that is, to the set of entities which have this property. Formally, an entity $e \in \mathscr{E}$ has property $P \in \wp(\mathscr{E})$ if and only if $e \in P$, or equivalently, $\{e\} \subseteq P$ where $\{e\}$ is the strongest property of element e and $\subseteq$ is implication.

A property P is stronger than a property P' (or P' is weaker than P) if and only if P implies P' that is $P \subseteq P'$. The strongest property is $\varnothing$ (false) and the weakest property is $\mathscr{E}$ (true). The strongest property of an entity $e \in \mathscr{E}$ is $\{e\}$, that is to be that entity (and nothing else).

Example 8.1 (even/odd) Take $\mathscr{E} = \mathbb{Z}$. The properties of integers are elements of $\wp(\mathbb{Z})$. For example, $\varnothing$ is false (no integer has the false property). $\mathbb{Z}$ is true (all integers have the true property). $2\mathbb{Z} \triangleq \{z \in \mathbb{Z} \mid \exists k \in \mathbb{Z} \,.\, z = 2k\}$ is the even property, and $2\mathbb{Z}{+}1 \triangleq \{z \in \mathbb{Z} \mid \exists k \in \mathbb{Z} \,.\, z = 2k + 1\}$ is the odd property. An integer z is either odd $z \in 2\mathbb{Z}$ or even $z \in 2\mathbb{Z} + 1$ (i.e., $\mathbb{Z} = 2\mathbb{Z} \cup 2\mathbb{Z}{+}1$).

The stronger a property P of entities is, the more information/knowledge we have about the entities that have this property. The abstraction of a property P into a weaker property P' loses information about these entities in that there are more entities satisfying P' than P. For example, abstracting "even positive integers" into "positive integers" loses the evenness information.

Remark 8.2 (properties as sets) Representing properties by sets, we do not have to consider a specific logic to describe these properties. Moreover, we avoid the confusion between a program variable x and the denotation of its value by the same name x (unfortunately, also often denoted x). So, "the variable x has an even value" (in environment ρ) is $\rho(\mathsf{x}) \in 2\mathbb{Z}$ instead of the predicate $\exists k \in \mathbb{Z} \,.\, \mathsf{x} = 2k$ (in which x implicitly denotes $\rho(\mathsf{x})$). Of course, the sets representing properties of x can be defined by predicates $p(x)$ (such as $x \in 2\mathbb{Z}$ or $\exists k \in \mathbb{Z} \,.\, x = 2k$) considered to be a shorthand for $\{\rho(\mathsf{x}) \mid p(\rho(\mathsf{x}))\}$.

Exercise 8.3 (constancy) What is the property that means "to be an integer constant."

Remark 8.4 (on the interpretation of logics) Consider a first-order language with variables $x \in \mathbb{V}$, functions f to build terms $t ::= x \mid f(t_1, \ldots, t_n)$ (e.g., the arithmetic expressions defined in section 3.4); and predicates p (e.g., the Boolean expressions of section 3.4), logical connectives $\Rightarrow$, $\neg$, and so forth, and quantifiers $\forall$ and $\exists$ to build logic formulas $\varphi ::= p(t_1, \ldots, t_m) \mid \varphi_1 \Rightarrow \varphi_2 \mid \neg\varphi_1 \mid \ldots \mid \forall x \,.\, \varphi_1 \mid \exists x \,.\, \varphi_1$.

Let $\mathscr{E}$ be a universe of discourse and $\mathscr{I}$ be an interpretation of the functions and predicates such that $\mathscr{I}(f) \in \mathscr{E}^n \mapsto \mathscr{E}$ and $\mathscr{I}(p) \in \wp(\mathscr{E}^m)$. Let $\rho \in \mathbb{Ev} \triangleq \mathbb{V} \rightarrow \mathscr{E}$ be environments assigning values to variables. The semantics $\boldsymbol{S}[\![\varphi]\!]$ of a formula φ is defined structurally as $\boldsymbol{S}[\![x]\!]\rho \triangleq \rho(x)$, $\boldsymbol{S}[\![f(t_1, \ldots, t_n)]\!]\rho \triangleq \mathscr{I}(f)(\boldsymbol{S}[\![t_1]\!]\rho, \ldots, \boldsymbol{S}[\![t_n]\!]\rho)$, $\boldsymbol{S}[\![p(t_1, \ldots, t_m)]\!] \triangleq \{\rho \mid \langle \boldsymbol{S}[\![t_1]\!]\rho, \ldots, \boldsymbol{S}[\![t_n]\!]\rho\rangle \in \mathscr{I}(p)\}$, $\boldsymbol{S}[\![\varphi_1 \Rightarrow \varphi_2]\!] \triangleq (\!|\, \boldsymbol{S}[\![\varphi_1]\!] \subseteq \boldsymbol{S}[\![\varphi_2]\!] \mathrel{?} \mathbb{Ev} \mathrel{\text{:}} \varnothing \,|\!)$, $\boldsymbol{S}[\![\neg\varphi_1]\!] \triangleq \mathbb{Ev} \setminus \boldsymbol{S}[\![\varphi_1]\!]$, ..., $\boldsymbol{S}[\![\forall x \,.\, \varphi_1]\!] \triangleq \{\rho \mid \{\rho[\mathsf{x} \leftarrow v] \mid v \in \mathscr{E}\} \subseteq \boldsymbol{S}[\![\varphi_1]\!]\rho\}$, and $\boldsymbol{S}[\![\exists x \,.\, \varphi_1]\!] \triangleq \{\rho \mid \{\rho[\mathsf{x} \leftarrow v] \mid v \in \mathscr{E}\} \cap \boldsymbol{S}[\![\varphi_1]\!]\rho \neq \varnothing\}$.

Observe that $\boldsymbol{S}[\![\varphi]\!] \in \wp(\mathbb{Ev})$ is what we called a property of environments in $\mathbb{Ev}$. So instead of reasoning on formulas φ, we reason on their semantics $\boldsymbol{S}[\![\varphi]\!]$.

Logicians would object that such reasonings are then dependent upon a particular interpretation $\mathscr{I}$. This is true because this interpretation $\mathscr{I}$ is determined by the semantics of the programming language of chapters 3 and 6 so $\mathscr{E} = \mathbb{Z}$ in discussing expressions (or $\mathscr{E} = \mathbb{T}^{+\infty}$ in discussing programs).

For full generality, we could have left the expressions in section 3.4 uninterpreted with very few hypotheses on their possible interpretations (such as $\mathscr{A}[\![\mathsf{A}]\!] \in \mathbb{E}\mathbb{v} \to \mathscr{E}$ and $\mathscr{B}[\![\mathsf{B}]\!] \in \wp(\mathbb{E}\mathbb{v})$), as logicians do. However, with regards to verification as discussed in part VIII, most results depend on which specific interpretation is chosen. Nevertheless, we can still write φ to be understood as $\boldsymbol{S}[\![\varphi]\!]$ for this interpretation. However, it is more difficult in static analysis to abstract formulas φ rather than their semantics $\boldsymbol{S}[\![\varphi]\!]$.

8.3 What is a Formal Semantic Property?

In the case of a formal program property, the entity is $\boldsymbol{S}^{+\infty}[\![\mathsf{P}]\!]$. $\boldsymbol{S}^{+\infty}[\![\mathsf{P}]\!]$ belongs to $\mathbb{T}^{+} \to \wp(\mathbb{T}^{+\infty})$, so a program property is an element of $\wp(\mathbb{T}^{+} \to \wp(\mathbb{T}^{+\infty}))$. This is the traditional notion of property in abstract interpretation. Because a relation in $\wp(\mathbb{T}^{+} \times \mathbb{T}^{+\infty})$ is isomorphic to its right image in $\mathbb{T}^{+} \to \wp(\mathbb{T}^{+\infty})$, a program property can also be understood as an element of $\wp(\wp(\mathbb{T}^{+} \times \mathbb{T}^{+\infty}))$. $\wp(\wp(\mathbb{T}^{+\infty}))$. It was later called hyperproperty [194] but we stick to the original set theoretic definition of property in abstract interpretation.[1]

Example 8.5 Assume that π_0 is a given initialization and that we are interested in the properties of $\boldsymbol{S}^{+\infty}[\![\mathsf{P}]\!]\pi_0 \in \wp(\mathbb{T}^{+\infty})$. The formal program property "all executions of P from π_0 always terminate with $\mathsf{x} = 0$ or all executions of P from π_0 always terminate with $\mathsf{x} = 1$" is the formally defined as $P \triangleq \wp(\{\pi \in \mathbb{T}^{+} \mid \varrho(\pi)\mathsf{x} = 0\}) \cup \wp(\{\pi \in \mathbb{T}^{+} \mid \varrho(\pi)\mathsf{x} = 1\}) \in \wp(\wp(\mathbb{T}^{+\infty}))$.

$P \ =$

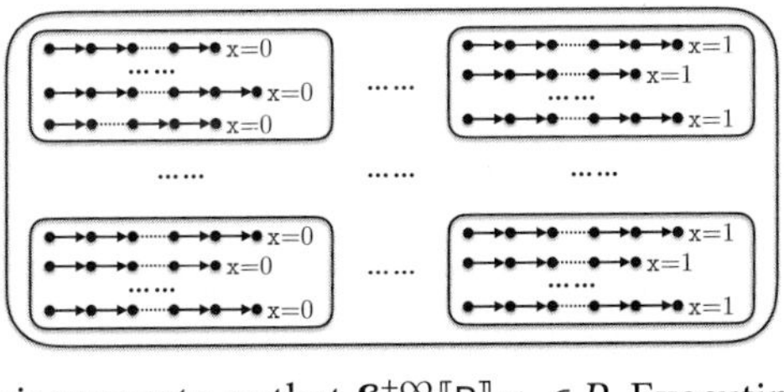

Assume program P has this property so that $\boldsymbol{S}^{+\infty}[\![\mathsf{P}]\!]\pi_0 \in P$. Executing program P from π_0 once, we know the result of all other executions possible from π_0. If the execution terminates with $\mathsf{x} = 0$ (respectively $\mathsf{x} = 1$) the property P implies that all other possible executions will always terminate with $\mathsf{x} = 0$ (respectively $\mathsf{x} = 1$).

8.4 Collecting Semantics

By definition of properties in section 8.1, the strongest semantic property of program P is therefore $\{\boldsymbol{S}^{+\infty}[\![\mathsf{P}]\!]\}$. It is traditionally called the *static* or *collecting semantics* of program P.

$$\boldsymbol{S}^{\mathrm{c}}[\![\mathsf{P}]\!] \ \triangleq \ \{\boldsymbol{S}^{+\infty}[\![\mathsf{P}]\!]\} \,. \tag{8.6}$$

1. Hence a property of a semantic property is a property in $\wp(\wp(\mathbb{T}^{+} \to \wp(\mathbb{T}^{+\infty})))$, not a hyper-hyperproperty.

Therefore program P has property $P \in \wp(\mathbb{T}^+ \rightarrow \wp(\mathbb{T}^{+\infty}))$ is $\boldsymbol{\mathcal{S}}^{+\infty}[\![\mathsf{P}]\!] \in P$, or equivalently, $\{\boldsymbol{\mathcal{S}}^{+\infty}[\![\mathsf{P}]\!]\} \subseteq P$, that is, $\boldsymbol{\mathcal{S}}^{\mathbb{C}}[\![\mathsf{P}]\!] \subseteq P$ meaning that P is implied by the collecting semantics of program P. This trivial remark allows us to use implication $\subseteq$ ($\Rightarrow$) instead of membership $\in$. This is useful in static analysis in which $\subseteq$ is generally easy to abstract but not $\in$.

8.5 What is a Formal Trace Property?

A trace property is a property of the continuation traces for a given initialization trace hence an element of $\mathbb{T}^+ \rightarrow \wp(\mathbb{T}^{+\infty})$. The program semantics is a trace property because $\boldsymbol{\mathcal{S}}^{+\infty}[\![\mathsf{P}]\!] \in \mathbb{T}^+ \rightarrow \wp(\mathbb{T}^{+\infty})$. Isomorphically, a trace property is an element of $\wp(\mathbb{T}^+ \times \mathbb{T}^{+\infty})$, inappropriately called property in [194] with the consequence of confusing program semantics and program properties.

Example 8.7 Continuing example 8.5, $\{\pi \in \mathbb{T}^+ \mid \varrho(\pi)\mathsf{x} = 0\} \in \wp(\mathbb{T}^{+\infty})$ is the trace property of "terminating with x=0." Program component S terminates with a Boolean value for initialization π_0 is $\boldsymbol{\mathcal{S}}^{+\infty}[\![\mathsf{S}]\!]\pi_0 \in \{\pi \in \mathbb{T}^+ \mid \varrho(\pi)\mathsf{x} = 0\} \cup \{\pi \in \mathbb{T}^+ \mid \varrho(\pi)\mathsf{x} = 1\}$. Contrary to example 8.5, knowing that one execution π of S from π_0 returns 0 (/1) does not guarantee that no other execution π' from π_0 returns 1 (/0). So the semantic property of example 8.5 is more expressive than a trace property.

Examples 8.5 and 8.7 show that trace properties are less expressive than semantic properties. This is made clear by showing that a trace property is an abstraction of a semantic property.

8.6 Abstraction of a Semantic Property into a Trace Property

Any semantic property P can be abstracted into a less precise trace property $\alpha^{\mathbb{T}}(P)$ defined as follows

$$\begin{array}{rcl} \alpha^{\mathbb{T}} & \in & \wp(\mathbb{T}^+ \rightarrow \wp(\mathbb{T}^{+\infty})) \rightarrow (\mathbb{T}^+ \rightarrow \wp(\mathbb{T}^{+\infty})) \\ \alpha^{\mathbb{T}}(P) & \triangleq & \pi \mapsto \bigcup\{\boldsymbol{\mathcal{S}}(\pi) \mid \boldsymbol{\mathcal{S}} \in P\} \\ \gamma^{\mathbb{T}} & \in & (\mathbb{T}^+ \rightarrow \wp(\mathbb{T}^{+\infty})) \rightarrow \wp(\mathbb{T}^+ \rightarrow \wp(\mathbb{T}^{+\infty})) \\ \gamma^{\mathbb{T}}(\overline{P}) & \triangleq & \{\boldsymbol{\mathcal{S}} \mid \boldsymbol{\mathcal{S}} \dot{\subseteq} \overline{P}\} \end{array} \tag{8.8}$$

Exercise 8.9 Prove that $\langle\wp(\mathbb{T}^+ \rightarrow \wp(\mathbb{T}^{+\infty})), \subseteq\rangle \underset{\alpha^{\mathbb{T}}}{\overset{\gamma^{\mathbb{T}}}{\leftrightarrows}} \langle\mathbb{T}^+ \rightarrow \wp(\mathbb{T}^{+\infty}), \dot{\subseteq}\rangle$ and, more generally, $\langle\wp(S \rightarrow \wp(S')), \subseteq\rangle \underset{\alpha^{\mathbb{T}}}{\overset{\gamma^{\mathbb{T}}}{\leftrightarrows}} \langle S \rightarrow \wp(S'), \dot{\subseteq}\rangle$ for any sets S and S'.

Example 8.10 Continuing example 8.5, let $P \triangleq \wp(\{\pi \in \mathbb{T}^+ \mid \varrho(\pi)\mathsf{x} = 0\}) \cup \wp(\{\pi \in \mathbb{T}^+ \mid \varrho(\pi)\mathsf{x} = 1\}) \in \wp(\wp(\mathbb{T}^{+\infty}))$. We have $\overline{P} = \alpha^{\mathbb{T}}(P) = \{\pi \in \mathbb{T}^+ \mid \varrho(\pi)\mathsf{x} \in \{0, 1\}\}$.

$\overline{P}$ =

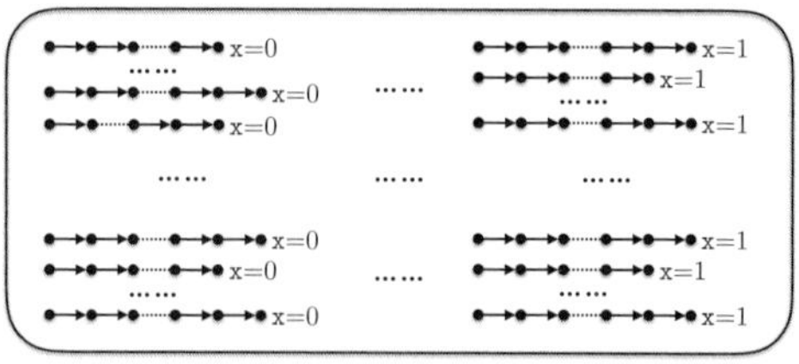

Hence P and $\overline{P}$ both express that program executions always terminate with a Boolean value for x. However, P is stronger because it expresses that the result is always the same, whereas $\overline{P}$ doesn't. No abstract trace property in $\mathbb{T}^+ \rightarrow \wp(\mathbb{T}^{+\infty})$ can express this (unless the result is constant).

Exercise 8.11 Assume that program P has property $P = \{\{\langle\pi_0, \pi_1\rangle\}, \{\langle\pi_0, \pi_2\rangle\}\}$. What is $\alpha^{\mathbb{T}}(P)$?, $\gamma^{\mathbb{T}}(\alpha^{\mathbb{T}}(P))$? Why is this an information loss?

8.7 Trace Property Semantics

The trace property semantics of a program is the strongest trace property of the trace semantics of that program, that is, the trace semantics itself.

$$\mathcal{S}^{\mathbb{T}}[\![\mathsf{P}]\!] \triangleq \alpha^{\mathbb{T}}(\mathcal{S}^{\mathbb{C}}[\![\mathsf{P}]\!]) = \mathcal{S}^{+\infty}[\![\mathsf{P}]\!] . \tag{8.12}$$

8.8 Invariance/Reachability Property

An invariance or reachability property is a relational between values of variables attached to each program point that does hold whenever execution reaches that program point.

Example 8.13 An invariant for the program (4.1) is the following (x denotes the value of variable x. The local invariant attached to a program point ℓ is given as a comment /* ... */ at that program point. For iteration and conditional statements the local invariant holds at that statement, just before the test. We assume all variables are initialized to 0.)

```
ℓ1  /* x=0 */
    x = x + 1;
    while ℓ2 (tt) /* 1≤x≤2 */ {
ℓ3      /* 1≤x≤2 */
        x = x + 1;
        if ℓ4 (x > 2) /* 2≤x≤3 */
ℓ5          /* x=3 */
            break;
    }
ℓ6  /* x=3 */
    ;
ℓ7  /* x=3 */
```

Formally, the strongest invariant $\mathcal{I}$ is

$$\begin{aligned}
&&&& \mathcal{I}(\ell_1) &\triangleq \{\rho \in \mathbb{E}\mathbb{v} \mid \forall \mathsf{y} \in \mathbb{V} \,.\, \rho(\mathsf{y}) = 0\} \\
&& \mathcal{I}(\ell_2) &\triangleq & \mathcal{I}(\ell_3) &\triangleq \{\rho \in \mathbb{E}\mathbb{v} \mid 1 \leqslant \rho(\mathsf{x}) \leqslant 2 \wedge \forall \mathsf{y} \in \mathbb{V} \setminus \{\mathsf{x}\} \,.\, \rho(\mathsf{y}) = 0\} \\
&&&& \mathcal{I}(\ell_4) &\triangleq \{\rho \in \mathbb{E}\mathbb{v} \mid 2 \leqslant \rho(\mathsf{x}) \leqslant 3 \wedge \forall \mathsf{y} \in \mathbb{V} \setminus \{\mathsf{x}\} \,.\, \rho(\mathsf{y}) = 0\} \\
\mathcal{I}(\ell_5) &\triangleq & \mathcal{I}(\ell_6) &\triangleq & \mathcal{I}(\ell_7) &\triangleq \{\rho \in \mathbb{E}\mathbb{v} \mid \rho(\mathsf{x}) = 3 \wedge \forall \mathsf{y} \in \mathbb{V} \setminus \{\mathsf{x}\} \,.\, \rho(\mathsf{y}) = 0\}
\end{aligned}$$

An invariant is a property of the program variables that remains unchanged under a program step computation. Methods for proving that a property is invariant were introduced by A. M. Turing [933], P. Naur [742], R. W. Floyd [376], and C. A. R. Hoare [501, 504]. They are developed by abstract interpretation in part VIII, "Specification and Verification."

8.9 Abstraction of a Trace Property into an Invariance/Reachability Property

The trace to invariance/reachability property abstraction $\alpha^{\mathbb{I}}$ abstracts a trace semantics $\boldsymbol{\mathcal{S}}$ to an invariant collecting at each program point ℓ on the trace the possible values of the variables at that point for executions satisfying a precondition P.

$$\begin{aligned} \alpha^{\mathbb{I}} &\in (\mathbb{T}^{+} \to \wp(\mathbb{T}^{+\infty})) \to (\wp(\mathbb{E}\mathbb{v}) \to \mathbb{L} \to \wp(\mathbb{E}\mathbb{v})) && (8.14)\\ \alpha^{\mathbb{I}}(\boldsymbol{\mathcal{S}}) &\triangleq P \mapsto \ell \mapsto \{\varrho(\pi_0 \smallfrown \pi\ell) \mid \exists \pi' \,.\, \varrho(\pi_0) \in P \wedge \pi\ell\pi' \in \boldsymbol{\mathcal{S}}(\pi_0)\} \end{aligned}$$

Exercise 8.15 Prove that $\alpha^{\mathbb{I}}$ is the lower adjoint of a Galois connection.

8.10 Invariance/Reachability Semantics

The invariance/reachability semantics of a program P is the trace to invariant abstraction of its trace semantics. This is the strongest invariant in that it implies all other invariants. In example 8.13, I is the invariance semantics of program (4.1).

$$\boldsymbol{\mathcal{S}}^{\mathbb{I}}[\![\mathsf{P}]\!] \triangleq \alpha^{\mathbb{I}}(\boldsymbol{\mathcal{S}}^{\mathbb{T}}[\![\mathsf{P}]\!])\,. \qquad (8.16)$$

8.11 Hierarchy of Program Properties

The abstraction of program properties into less expressive ones, as formalized by Galois connections, gives raise to a hierarchy of program properties. This induces a corresponding hierarchy of program semantics.

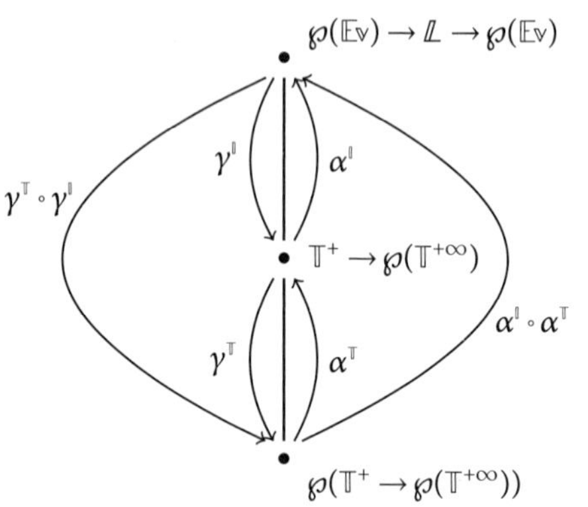

$$\begin{aligned} \boldsymbol{\mathcal{S}}^{\mathbb{I}}[\![\mathsf{P}]\!] &= \alpha^{\mathbb{I}}(\boldsymbol{\mathcal{S}}^{\mathbb{T}}[\![\mathsf{P}]\!]) \\ &= \alpha^{\mathbb{I}} \circ \alpha^{\mathbb{T}}(\boldsymbol{\mathcal{S}}^{\mathbb{C}}[\![\mathsf{P}]\!]) \end{aligned}$$

invariance/ reachability semantics

$$\begin{aligned} \boldsymbol{\mathcal{S}}^{\mathbb{T}}[\![\mathsf{P}]\!] &= \boldsymbol{\mathcal{S}}^{+\infty}[\![\mathsf{P}]\!] \\ &= \alpha^{\mathbb{T}}(\boldsymbol{\mathcal{S}}^{\mathbb{C}}[\![\mathsf{P}]\!]) \end{aligned}$$

trace semantics

$$\boldsymbol{\mathcal{S}}^{\mathbb{C}}[\![\mathsf{P}]\!] \triangleq \{\boldsymbol{\mathcal{S}}^{+\infty}[\![\mathsf{P}]\!]\}$$

collecting semantics

The idea that the specification of the properties of programs defines their semantics goes back to Robert Floyd "Assigning meaning to programs" [376]. The organization of various semantics in a hierarchy appeared in [49, 505] and in [227] as part of the lattice of abstract domains [241, section 8].

8.12 Conclusion

We have defined properties (of programs) as sets (of semantics), which is not unusual in mathematics (e.g., [880, ch. 2, § 1]). We have defined a hierarchy of more or less expressive semantic program properties related by abstractions. In verifying a program semantic property, it is important to do so at an appropriate level of abstraction, that is, precise enough to verify the semantic property and abstract enough not to consider irrelevant details of the semantics. This is the objective of, for example, the design of program proof methods given in chapters 25 and 26. Properties as sets is also a fruitful point of view in static analysis, in which sets are easier to abstract than logical formulas (see , for example, chapters 21 and 27).

8.13 Solutions to Selected Exercises

Solution to exercise 8.3 $\{\{n\} \mid n \in \mathbb{Z}\}$.

Solution to exercise 8.9

$$\begin{aligned}
& \alpha^{\intercal}(P) \mathrel{\dot{\subseteq}} \overline{P} \\
\Leftrightarrow\ & \pi \mapsto \bigcup\{\boldsymbol{S}(\pi) \mid \boldsymbol{S} \in P\} \mathrel{\dot{\subseteq}} \overline{P} && \wr\text{definition of } \alpha^{\intercal}\wr \\
\Leftrightarrow\ & \forall \pi \,.\, \bigcup\{\boldsymbol{S}(\pi) \mid \boldsymbol{S} \in P\} \subseteq \overline{P}(\pi) && \wr\text{pointwise definition of } \dot{\subseteq}\wr \\
\Leftrightarrow\ & \forall \pi \,.\, \{\pi' \mid \exists \boldsymbol{S} \in P \,.\, \pi' \in \boldsymbol{S}(\pi)\} \subseteq \overline{P}(\pi) && \wr\text{definition of } \textstyle\bigcup\wr \\
\Leftrightarrow\ & \forall \pi \,.\, \forall \boldsymbol{S} \in P \,.\, \forall \pi' \in \boldsymbol{S}(\pi) \,.\, \pi' \in \overline{P}(\pi) && \wr\text{definition of } \subseteq\wr \\
\Leftrightarrow\ & \forall \boldsymbol{S} \in P \,.\, \forall \pi \,.\, \forall \pi' \in \boldsymbol{S}(\pi) \,.\, \pi' \in \overline{P}(\pi) && \wr\text{definition of } \forall\wr \\
\Leftrightarrow\ & P \subseteq \{\boldsymbol{S} \mid \forall \pi \,.\, \forall \pi' \in \boldsymbol{S}(\pi) \,.\, \pi' \in \overline{P}(\pi)\} && \wr\text{definition of } \subseteq\wr \\
\Leftrightarrow\ & P \subseteq \{\boldsymbol{S} \mid \forall \pi \,.\, \boldsymbol{S}(\pi) \subseteq \overline{P}(\pi)\} && \wr\text{definition of } \subseteq\wr \\
\Leftrightarrow\ & P \subseteq \{\boldsymbol{S} \mid \boldsymbol{S} \mathrel{\dot{\subseteq}} \overline{P}\} && \wr\text{pointwise definition of } \dot{\subseteq}\wr \\
\Leftrightarrow\ & P \subseteq \gamma^{\intercal}(\overline{P}) && \wr\text{definition of } \gamma^{\intercal}(\overline{P}) \triangleq \{\boldsymbol{S} \mid \boldsymbol{S} \mathrel{\dot{\subseteq}} \overline{P}\}\wr
\end{aligned}$$

$\alpha^{\intercal}$ is surjective (because $\alpha^{\intercal}(\{\overline{P}\}) = \overline{P}$).

Solution to exercise 8.11 $\alpha^{\intercal}(P) = \{\langle\pi_0, \pi_1\rangle, \langle\pi_0, \pi_2\rangle\}$, $\gamma^{\intercal}(\alpha^{\intercal}(P)) = \{\varnothing, \{\langle\pi_0, \pi_1\rangle\}, \{\langle\pi_0, \pi_2\rangle\}, \{\langle\pi_0, \pi_1\rangle, \langle\pi_0, \pi_2\rangle\}\}$. The information that the program P has one and only one execution (either π_1 or π_2) from π_0 is lost.

Solution to exercise 8.15 Let $\dot{\subseteq}$ be the pointwise inclusion $\subseteq$. We have

$$\begin{aligned}
& \alpha^{\Vert}(\boldsymbol{S}) \mathrel{\ddot{\subseteq}} \mathcal{I} \\
\Leftrightarrow\ & P \mapsto \ell \mapsto \{\varrho(\pi_0 \mathbin{\frown} \pi\ell) \mid \exists \pi' \,.\, \varrho(\pi_0) \in P \wedge \pi\ell\pi' \in \boldsymbol{S}(\pi_0)\} \mathrel{\ddot{\subseteq}} \mathcal{I} && \wr\text{definition of } \alpha^{\Vert}\wr \\
\Leftrightarrow\ & \forall P \,.\, \forall \ell \,.\, \{\varrho(\pi_0 \mathbin{\frown} \pi\ell) \mid \exists \pi' \,.\, \varrho(\pi_0) \in P \wedge \pi\ell\pi' \in \boldsymbol{S}(\pi_0)\} \subseteq \mathcal{I}(P)\ell \\
& && \wr\text{pointwise definition of } \ddot{\subseteq}\wr \\
\Leftrightarrow\ & \forall P \,.\, \forall \ell \,.\, \{\varrho(\pi_0 \mathbin{\frown} \pi\ell) \mid \exists \overline{\pi} \in \boldsymbol{S}(\pi_0) \,.\, \exists \pi' \,.\, \varrho(\pi_0) \in P \wedge \overline{\pi} = \pi\ell\pi'\} \subseteq \mathcal{I}(P)\ell \\
& && \wr\text{definition of } \in\wr \\
\Leftrightarrow\ & \forall P \,.\, \forall \ell \,.\, \forall \pi_0, \overline{\pi}, \pi\ell\pi' \,.\, \varrho(\pi_0) \in P \wedge \overline{\pi} \in \boldsymbol{S}(\pi_0) \wedge \overline{\pi} = \pi\ell\pi' \Rightarrow \varrho(\pi_0 \mathbin{\frown} \pi\ell) \in \mathcal{I}(P)\ell
\end{aligned}$$

$$\text{\{definition of } \subseteq \text{\}}$$

$$\Leftrightarrow \forall \pi_0 \,.\, \forall \overline{\pi} \in \boldsymbol{\mathcal{S}}(\pi_0) \,.\, \forall P \,.\, \forall \ell \,.\, \forall \pi \ell \pi' \,.\, (\boldsymbol{\varrho}(\pi_0) \in P \wedge \overline{\pi} = \pi \ell \pi') \Rightarrow \boldsymbol{\varrho}(\pi_0 \mathbin{\widehat{\cdot}} \pi \ell) \in \mathcal{I}(P)\ell$$

$$\text{\{definition of } \subseteq \text{\}}$$

$$\Leftrightarrow \forall \pi_0 \,.\, \boldsymbol{\mathcal{S}}(\pi_0) \subseteq \{\overline{\pi} \mid \forall P \,.\, \forall \ell \,.\, \forall \pi \ell \pi' \,.\, (\boldsymbol{\varrho}(\pi_0) \in P \wedge \overline{\pi} = \pi \ell \pi') \Rightarrow \boldsymbol{\varrho}(\pi_0 \mathbin{\widehat{\cdot}} \pi \ell) \in \mathcal{I}(P)\ell\}$$

$$\text{\{definition of } \subseteq \text{\}}$$

$$\Leftrightarrow \forall \pi_0 \,.\, \boldsymbol{\mathcal{S}}(\pi_0) \subseteq \gamma^{\shortmid}(\mathcal{I})\pi_0$$

$$\text{\{by defining } \gamma^{\shortmid}(\mathcal{I}) \triangleq \pi_0 \mapsto \{\overline{\pi} \mid \forall P \,.\, \forall \ell \,.\, \forall \pi \ell \pi' \,.\, (\boldsymbol{\varrho}(\pi_0) \in P \wedge \overline{\pi} = \pi \ell \pi') \Rightarrow \boldsymbol{\varrho}(\pi_0 \mathbin{\widehat{\cdot}} \pi \ell) \in \mathcal{I}(P)\ell\}\text{\}}$$

$$\Leftrightarrow \boldsymbol{\mathcal{S}} \mathrel{\dot{\subseteq}} \gamma^{\shortmid}(\mathcal{I}) \qquad \text{\{pointwise definition of } \dot{\subseteq} \text{\}}$$

9 Undecidability and Rice Theorem

The problem of proving the termination of a program or, more generally, of verifying its correctness is shown to be undecidable. This means that it is impossible to construct always-terminating algorithms to solve these problems correctly.

Contents

9.1 Decidability, Semidecidability, and Undecidability

A question (for instance, on the semantics of a program) with Boolean answer is

- *Decidable* if and only if there exists an algorithm that takes the program as input and effectively computes the answer to the question in finite time;
- *Semidecidable* if and only if there exists an algorithm that takes the program as input and effectively computes the answer to the question in finite time when the answer is true and may be not terminating when it is false;
- *Undecidable* if and only if there exists no algorithm that takes the program as input and effectively computes the answer to the question in finite time (any algorithm providing correct answers will therefore answer *true*, *false*, *I don't know*, or will not terminate).

Turing's theorem 9.1 states that the halting/termination problem is algorithmically undecidable. Rice's theorem 9.12 states that the problem of whether a program has a given nontrivial semantic property is algorithmically undecidable. Our proofs are informal sketches in the style of Neil Jones [537] and Bernhard Reus [825]; see [841, 902] for rigorous proofs.

9.2 Turing Completeness

A programming language is *Turing complete* if and only if it is possible to write an interpreter for this language in this language. For example, a language without iteration or recursion is not Turing complete since it cannot read arbitrarily long programs.

A programming language is *deterministic* if any input (i.e., given initial values of variables) has a unique outcome (including the possibility of nontermination). The language of chapters 4 and 7 is deterministic.

The language of chapters 4 and 7 is Turing complete (because the initial values of variables are not sequences of characters but integers, an interpreter written in this language would have to encode programs as naturals, for example, the natural obtained by concatenating the fixed-size character codes of the sequence of characters constituting the program).

9.3 Undecidability of the Termination Problem

The *halting/termination problem* is to write a program `termination` that takes as input an arbitrary program `P` (represented as a string or file or an integer encoding) and its input data `D` (also represented as a string or file or an integer encoding) such that `termination(P,D)` always terminates for all programs `P` and input data `D` and returns tt if the execution of the program `P` on the data `D` does terminate and ff if the execution of the program `P` on the data `D` does not terminate.

The halting/termination problem is undecidable because no such algorithm does exist. The proof of the halting/termination problem undecidability is credited to Alan Turing [931, 932]. Similar arguments had been used to prove undecidability of other problems (e.g., in number theory by Alonzo Church [184] or the incompleteness theorems in logic by Kurt Gödel [420]), but the originality of Turing is that his theorem explicitly refers to the halting problem of a (Turing) machine modeling computers.

Theorem 9.1 (Turing's theorem) The halting/termination problem is undecidable for a deterministic Turing complete language.

Proof Sketch The proof that such a program `termination` cannot exist is by reductio ad absurdum. Let `interpret` be an interpreter for the language such that `interpret(P,D)` is the result of running the program `P` on the input data `D`.

Assume `termination` does exist. Define the deterministic program

```
contradiction  ≜
     function P = if (termination(P,P)) { while (tt) ; }
```

and run the program `contradiction` on its own text, that is,

```
contradiction(contradiction).
```

More precisely, that is computed by running the interpreter

```
interpret(contradiction,contradiction).
```

Then `termination(contradiction, contradiction)` is computed by running the interpreter

```
interpret(termination, (contradiction,contradiction))
```

and always terminates.

- If `contradiction(contradiction)` does terminate then

```
termination(contradiction, contradiction)
```

terminates and returns true, which is a contradiction because the iteration `while (tt) ;` does not terminate or

- `contradiction(contradiction)` does not terminate and then

  `termination(contradiction, contradiction)`

  terminates and returns false, which is a contradiction because the iteration `while (tt) ;` is not entered and the program terminates. □

A more rigorous proof must precisely define what is a program and what is the execution of this program, for which Alonzo Church introduced the lambda calculus and Alan Turing introduced the notion of Turing machine. Equivalently, we can formally define the semantics of programs, which is the objective of chapter 7, "Maximal Trace Semantics."

9.4 Examples of Algorithmically Undecidable Problems

Other decision problems for programs can be proved undecidable by reduction to the halting/termination problem.

Example 9.2 (constancy) Constancy analysis (see exercise 3.45) is the problem of deciding whether a program `P` assigns a value different from the initial value 0 to a variable `x`. It is undecidable because otherwise, given a program `P`, consider a fresh variable `x` not in `P` and the witness program `P′ = P; x = 1;`. `P′` assigns a value different from the initial value 0 to the variable `x` if and only if `P` terminates. Hence if the constancy problem were decidable, termination would also be decidable, in contradiction to Turing's theorem 9.1.

Exercise 9.3 (absence of runtime error) Prove that the problem of proving the absence of runtime error in an execution of a program `P` is undecidable.

Exercise 9.4 (sign) Prove that the problem of deciding whether a numerical variable has a given sign is undecidable.

Exercise 9.5 (type) Prove that the problem of deciding whether a variable has a given type is undecidable (it follows that the more difficult typing problem of inferring the type of a variable is also undecidable).

Corollary 9.6 (failures of algorithms "solving" undecidable problems) Let `A(P, d)` be an algorithm to solve an undecidable problem on program `P` for input data `d`. Assume that the algorithm is sound (otherwise, giving wrong answers is easy). Soundness means that `A(P,d)` correctly answers *true*, *false*, or *fails*, that is, answers *I don't know* or does not terminate.

Then `A(P,d)` must fail on infinitely many input data `d`.

Proof By reductio ad absurdum, assume that `A(P,d)` fails on finitely many input data $\mathtt{d}_i$, $i \in \Delta$ only. Then ask mathematicians to solve these finitely many problems and provide an $\mathtt{Answer(P,d}_i\mathtt{)} \in \{\mathit{true}, \mathit{false}\}$. Formally check the correctness of their answers for each of $\mathtt{d}_i$, $i \in \Delta$, with a proof assistant or theorem prover.
Consider the witness program:

```
A'(P,d) = if d∈{d_i | i∈Δ} then Answer(P,d_i)
                            else A(P,d)
```

`A'(P,d)` always terminates and never fails to solve the undecidable problem, in contradiction to Turing's theorem 9.1. □

An example was given in remark 3.43.

9.5 Semidecidability

A problem with Boolean answer is *semidecidable* if and only if there exists an algorithm that terminate with *true* when the answer is true and might not terminate when the answer is false.

Example 9.7 (termination) The termination problem is semidecidable. Simply run the program and answer *true* upon termination. If the program does not terminate then there is no answer that is allowed by semidecidability.

Theorem 9.8 (Emil L. Post, [805]) Decidable(P) ⇔ (semidecidable(P) and Semidecidable(not P)).

Proof For (⇒), we have either Decidable(P) ⇒ Semidecidable(P), or not Decidable(P) ⇒ Decidable(not P) ⇒ Semidecidable(not P). For (⇐), alternatively execute one step of the algorithms for Semidecidable(P) and Semidecidable(not P) and stop when one has an answer □

Theorem 9.9 (nontermination is undecidable) The non-termination question: "Will execution of program `P` on data `d` never terminate?" is not semidecidable (hence not decidable) for a deterministic Turing complete language.

Proof Sketch By reductio ad absurdum, Semidecidable (Termination) and Semidecidable (not Termination) would imply that Decidable (Termination). □

9.6 Rice's Theorem

As shown by example 9.2 and exercises 9.3, 9.4, and 9.5 the proof that a specific program semantic property is undecidable is always by the same reduction to the nontermination problem that was proved to be undecidable in section 9.3. Rice's theorem [830] is a generalization from these particular semantics properties to all nontrivial semantic properties: Rice's theorem states that all nontrivial semantic properties of programs are undecidable.

9.6.1 Semantics

The functional semantics $\mathcal{S}^{\text{pf}}[\![\mathsf{S}]\!]$ of program components $\mathsf{S} \in \mathbb{S}$ (as defined in chapter 4) considered in Rice's theorem is an integer partial function $\mathcal{S}^{\text{pf}}[\![\mathsf{S}]\!] \in \mathbb{N} \nrightarrow \mathbb{N}$.[1]

This is an abstraction $\mathcal{S}^{\text{pf}}[\![\mathsf{S}]\!] \triangleq \alpha^{\text{pf}}(\mathcal{S}^{+\infty}[\![\mathsf{S}]\!])$ of the maximal trace semantics $\mathcal{S}^{+\infty}[\![\mathsf{S}]\!] \in \mathbb{T}^{+} \to \mathbb{T}^{+\infty}$ (as defined in definition (7.7) of chapter 7 and later refined in section 17.2).

Assume that the set $\mathbb{S}$ of states is enumerable so that, by definition, there exists a bijective encoding $\iota \in \mathbb{S} \rightarrowtail\!\!\!\!\twoheadrightarrow \mathbb{N}$ of states to naturals.

The abstraction of $\mathcal{S} \in \mathbb{T}^{+} \to \mathbb{T}^{+\infty}$ is $\alpha^{\text{pf}}(\mathcal{S})n \triangleq m$ if and only if for any finite initialization trace $\pi\sigma$ ending in an initial state σ for which encoding is $\iota(\sigma) = n$, the execution is continued by a finite maximal trace $\sigma\pi'\sigma'$ ending in a final state σ' for which encoding $\iota(\sigma')$ is m. Otherwise, if a different final state σ'' is reachable from a different initialization trace $\pi'\sigma$ or $\pi\sigma$ is continued by a infinite maximal trace in $\mathbb{T}^{\infty}$ then $\alpha^{\text{pf}}(\mathcal{S})n$ is undefined.

Formally, $\alpha^{\text{pf}}(\mathcal{S})n = m$ if and only if $\forall \pi\sigma \in \mathbb{T}^{+} . \forall \sigma\pi' \in \mathcal{S}(\pi\sigma) . (\iota(\sigma) = n) \Rightarrow (\exists \sigma\pi'\sigma' \in \mathbb{T}^{+} . \sigma\pi' = \sigma\pi'\sigma' \wedge \iota(\sigma') = m)$.

9.6.2 Program Properties

In conformity with chapter 8, a program property is the set of all programs that have this property. Therefore, program properties belong to $\wp(\mathbb{S})$. The trivial properties are either the empty set $\varnothing$ of programs or the set $\mathbb{S}$ of all programs.

9.6.3 Semantic Program Properties

A program property $P \in \wp(\mathbb{S})$ is a *semantic property* if and only if $\mathsf{S} \in P \Leftrightarrow \mathcal{S}^{\text{pf}}[\![\mathsf{S}]\!] \in \gamma_{\mathcal{S}^{\text{pf}}}(P)$ where the semantic meaning $\gamma_{\mathcal{S}^{\text{pf}}}(P)$ of a property $P \in \wp(\mathbb{S})$ is $\gamma_{\mathcal{S}^{\text{pf}}}(P) \triangleq \{\mathcal{S}^{\text{pf}}[\![\mathsf{S}']\!] \mid \mathsf{S}' \in P\} \in \wp(\mathbb{N} \nrightarrow \mathbb{N})$.

Only the $\Leftarrow$ direction is not trivial. For example, "S has 42 minus operations –" is not a functional semantic property because given a statement S having 42 minus operations –, we can certainly find another program with a different number of minus operations – but with the same functional semantics.

9.6.4 Extensional Program Properties

Rice's theorem considers only program properties $P \in \wp(\mathbb{S})$ that are extensional.

A property $P \in \wp(\mathbb{S})$ is extensional if and only if $\forall \mathsf{S}_1, \mathsf{S}_2 \in \mathbb{S} . \mathcal{S}^{\text{pf}}[\![\mathsf{S}_1]\!] = \mathcal{S}^{\text{pf}}[\![\mathsf{S}_2]\!] \Rightarrow (\mathsf{S}_1 \in P \Leftrightarrow \mathsf{S}_2 \in P)$.

9.6.5 Extensional and Functional Semantic Program Properties are Equivalent

The extensional properties are exactly the functional semantics properties:

Lemma 9.10 A property $P \in \wp(\mathbb{S})$ is extensional if and only if P is a semantic property.

Proof ($\Rightarrow$) Assume that property $P \in \wp(\mathbb{S})$ is extensional.

1. Denotational semantics is similar but encodes $\nu \notin \text{dom}(\mathcal{S}^{\text{pf}}[\![\mathsf{S}]\!])$ by $\mathcal{S}^{\text{d}}[\![\mathsf{S}]\!]\nu = \bot$ where $\bot$ denotes non-termination. It also considers the case of parameters for which evaluation may not terminate so $\mathcal{S}^{\text{d}}[\![\mathsf{S}]\!] \in \mathbb{N} \cup \{\bot\} \nrightarrow \mathbb{N} \cup \{\bot\}$. Lazy evaluation of parameters allows $\mathcal{S}^{\text{d}}[\![\mathsf{S}]\!]\bot \neq \bot$

If $\mathcal{S}^{\text{pf}}[\![\mathsf{S}]\!] \in \gamma_{\mathcal{S}\text{pf}}(P) = \{\mathcal{S}^{\text{pf}}[\![\mathsf{S}']\!] \mid \mathsf{S}' \in P\}$ then there exists $\mathsf{S}' \in P$ such that $\mathcal{S}^{\text{pf}}[\![\mathsf{S}]\!] = \mathcal{S}^{\text{pf}}[\![\mathsf{S}']\!]$ and so, by extensionality, $\mathsf{S} \in P$.

Conversely, if $\mathsf{S} \in P$ then $\mathcal{S}^{\text{pf}}[\![\mathsf{S}]\!] \in \{\mathcal{S}^{\text{pf}}[\![\mathsf{S}']\!] \mid \mathsf{S}' \in P\} \triangleq \gamma_{\mathcal{S}\text{pf}}(P)$.

We conclude that $\mathsf{S} \in P \Leftrightarrow \mathcal{S}^{\text{pf}}[\![\mathsf{S}]\!] \in \gamma_{\mathcal{S}\text{pf}}(P)$ so that, by definition, P is a functional semantic property.

($\Leftarrow$) Assume that P is a functional semantic property that is $\forall \mathsf{S} \in \mathbb{S}\,.\,\mathcal{S}^{\text{pf}}[\![\mathsf{S}]\!] \in \gamma_{\mathcal{S}\text{pf}}(P) \Leftrightarrow \mathsf{S} \in P$. Let $\mathsf{S}_1, \mathsf{S}_2 \in \mathbb{S}$ such that $\mathcal{S}^{\text{pf}}[\![\mathsf{S}_1]\!] = \mathcal{S}^{\text{pf}}[\![\mathsf{S}_2]\!]$. Then $(\mathsf{S}_1 \in P \Leftrightarrow \mathsf{S}_2 \in P)$ is equivalent to $(\mathcal{S}^{\text{pf}}[\![\mathsf{S}_1]\!] \in \gamma_{\mathcal{S}\text{pf}}(P) \Leftrightarrow \mathcal{S}^{\text{pf}}[\![\mathsf{S}_2]\!] \in \gamma_{\mathcal{S}\text{pf}}(P)$ which is true because $\mathcal{S}^{\text{pf}}[\![\mathsf{S}_1]\!] = \mathcal{S}^{\text{pf}}[\![\mathsf{S}_2]\!]$. It follows that P is extensional. □

Lemma 9.11 The termination and nontermination properties are nontrivial and extensional.

Proof Expressed with $\mathcal{S}^{+\infty}[\![\mathsf{S}]\!]$, termination is $\mathcal{S}^{+\infty}[\![\mathsf{S}]\!]\pi\sigma \in \mathbb{T}^{+}$. For deterministic programs this is equivalent to $\mathcal{S}^{\text{pf}}[\![\mathsf{S}]\!]n \in \mathbb{N}$ where $n = \iota(\sigma)$. This is the semantic property $\{f \in \mathbb{N} \nrightarrow \mathbb{N} \mid n \in \mathsf{dom}(f)\}$, which by lemma 9.10 is extensional.

Similarly for nontermination, we consider $\mathcal{S}^{+\infty}[\![\mathsf{S}]\!]\pi\sigma \in \mathbb{T}^{\infty}$ abstracted in $\{f \in \mathbb{N} \nrightarrow \mathbb{N} \mid \iota(\sigma) \notin \mathsf{dom}(f)\}$.

There are programs that terminates on given data and others that don't so the termination and nontermination properties are nontrivial. □

9.6.6 Undecidability of Semantic Property Verification

Theorem 9.12 (Rice's theorem) Let P be a nontrivial and extensional/functional semantics property of a program S in a deterministic Turing complete language. Then the Boolean question $\mathsf{S} \in P$? is undecidable.

In short, an extensional/functional semantic property P is decidable if and only if it is trivial.

Proof Sketch of Theorem 9.12 We let $\mathtt{i}_\mathsf{S}$ be the variable in which statement S finds its input data ν and let $\mathtt{r}_\mathsf{S}$ be the variable in which statement S writes its final result ξ in the case of termination. The functional semantics of statement S is therefore $\nu \notin \mathsf{dom}(\mathcal{S}^{\text{pf}}[\![\mathsf{S}]\!])$ if S does not terminate and $\mathcal{S}^{\text{pf}}[\![\mathsf{S}]\!]\nu = \xi$ in the case of termination.

To define this abstraction more formally, let $\pi_0 = \ell \xrightarrow{\mathtt{i}_\mathsf{S} = \mathsf{A}_\nu = \nu} \mathsf{at}[\![\mathsf{S}]\!]$ be the initialization trace such that $\varrho(\pi_0)\mathtt{i}_\mathsf{S} = \nu$ (the existence of A_ν such that $\forall \rho\,.\,\mathcal{A}[\![\mathsf{A}_\nu]\!]\rho = \nu$ is proved in exercise 3.8). Let $\pi = \mathcal{S}^{+\infty}[\![\mathsf{S}]\!]\pi_0$ be the trace continuing π_0 when executing S so that the execution is $\pi_0 \frown \pi$. If $\pi_0 \frown \pi \in \mathbb{T}^{\infty}$ then $\mathcal{S}^{\text{pf}}[\![\mathsf{S}]\!]$ is undefined for ν i.e. $\nu \notin \mathsf{dom}(\mathcal{S}^{\text{pf}}[\![\mathsf{S}]\!])$. Otherwise $\pi_0 \frown \pi \in \mathbb{T}^{+}$ and $\mathcal{S}^{\text{pf}}[\![\mathsf{S}]\!]\nu = \varrho(\pi_0 \frown \pi)\mathtt{r}_\mathsf{S}$.

We want to determine whether statement S terminates for the value ν of its input variable $\mathtt{i}_\mathsf{S}$. Formally, the termination question is $\mathcal{S}^{+\infty}[\![\mathsf{S}]\!]\pi_0 \in \mathbb{T}^{+}$? By the preceding definition of the abstraction of the maximal trace into functional semantics, this question is equivalent to $\nu \in \mathsf{dom}(\mathcal{S}^{\text{pf}}[\![\mathsf{S}]\!])$?

The proof is by reduction ad absurdum, assuming that $\forall \mathtt{S} \in \mathbb{S}\,.\, \mathtt{S} \in P$? is decidable. Under this hypothesis, there would be an effective algorithm $\mathtt{in}_P(\mathtt{S})$ returning tt if $\mathtt{S} \in P$ and false otherwise.

Because P is not trivial, there are programs $\mathtt{S}_t \in P$ and $\mathtt{S}_f \notin P$. Moreover $\mathtt{S}_d \triangleq$ **while** (tt) ; never terminates on any input $\mathtt{x}_{\mathtt{S}_d}$ and so $\mathrm{dom}(\mathcal{S}^{\mathrm{pf}}[\![\mathtt{S}_d]\!]) = \varnothing$.

Consider the witness program (the value of expression $\mathtt{A}_v$ is v)

$$\begin{aligned}\mathtt{S}_w \;\triangleq\;& \mathtt{if}(\mathtt{in}_P(\mathtt{S}_d))\; \mathtt{i}_{\mathtt{S}} = \mathtt{A}_v\,;\;\; \mathtt{S}\;\; \mathtt{i}_{\mathtt{S}_f} = \mathtt{i}_{\mathtt{S}_w}\,;\;\; \mathtt{S}_f\;\; \mathtt{r}_{\mathtt{S}_w} = \mathtt{r}_{\mathtt{S}_f}\,;\\ & \mathtt{else}\; \mathtt{i}_{\mathtt{S}} = \mathtt{A}_v\,;\;\; \mathtt{S}\;\; \mathtt{i}_{\mathtt{S}_t} = \mathtt{i}_{\mathtt{S}_w}\,;\;\; \mathtt{S}_t\;\; \mathtt{r}_{\mathtt{S}_w} = \mathtt{r}_{\mathtt{S}_t}\,;\end{aligned}$$

The termination program is

$$\mathtt{termination}(\mathtt{S}, v) \;\triangleq\; \mathtt{if}(\mathtt{in}_P(\mathtt{S}_d))\, \neg\mathtt{in}_P(\mathtt{S}_w)\;\mathtt{else}\;\mathtt{in}_P(\mathtt{S}_w)$$

This program always terminates by hypothesis that P is decided by $\mathtt{in}_P$ which therefore always terminates.

It remains to prove that `termination` is correct. There are two cases.

- Assume that the divergent program $\mathtt{S}_d \in P$ has property P so $\mathtt{in}_P(\mathtt{S}_d) = \mathrm{tt}$. In that first case, $\mathtt{S}_w$ behaves as $\mathtt{i}_{\mathtt{S}} = \mathtt{A}_v\,;\;\; \mathtt{S}\;\; \mathtt{i}_{\mathtt{S}_f} = \mathtt{i}_{\mathtt{S}_w}\,;\;\; \mathtt{S}_f\;\; \mathtt{r}_{\mathtt{S}_w} = \mathtt{r}_{\mathtt{S}_f}\,;$. There are two subcases.
  - $v \notin \mathrm{dom}(\mathcal{S}^{\mathrm{pf}}[\![\mathtt{S}]\!])$ if and only if $\mathtt{S}$ does not terminate on input v.
    In that case $\mathtt{S}_w$ does not terminate on any input (in $\mathtt{i}_{\mathtt{S}_w}$ which is not used). Therefore $\mathtt{S}_w$ behaves as $\mathtt{S}_d \in P$ i.e. $\mathcal{S}^{\mathrm{pf}}[\![\mathtt{S}_w]\!] = \mathcal{S}^{\mathrm{pf}}[\![\mathtt{S}_d]\!]$ and $\mathtt{S}_d \in P$ so $\mathtt{S}_w \in P$ by extensionality.
  - $v \in \mathrm{dom}(\mathcal{S}^{\mathrm{pf}}[\![\mathtt{S}]\!])$ if and only if $\mathtt{S}$ does terminate on input v.
    In that case $\mathtt{S}_w$ behaves as $\mathtt{S}_f \notin P$ i.e. $\mathcal{S}^{\mathrm{pf}}[\![\mathtt{S}_w]\!] = \mathcal{S}^{\mathrm{pf}}[\![\mathtt{S}_f]\!]$ so that $\mathtt{S}_w \notin P$ by extensionality.

  In this first case $\mathtt{S}_d \in P$, $\mathtt{S}$ terminates on input v if and only if $\mathtt{S}_w \notin P$, so $\mathtt{termination}(\mathtt{S}, v) = \neg\mathtt{in}_P(\mathtt{S}_w)$ is correct.
- Otherwise, the divergent program $\mathtt{S}_d \notin P$ does not have property P. In that second case $\mathtt{S}_w$ behaves as $\mathtt{i}_{\mathtt{S}} = \mathtt{A}_v\,;\;\; \mathtt{S}\;\; \mathtt{i}_{\mathtt{S}_t} = \mathtt{i}_{\mathtt{S}_w}\,;\;\; \mathtt{S}_t\;\; \mathtt{r}_{\mathtt{S}_w} = \mathtt{r}_{\mathtt{S}_t}\,;$. Again, there are two subcases.
  - $v \notin \mathrm{dom}(\mathcal{S}^{\mathrm{pf}}[\![\mathtt{S}]\!])$ if and only if $\mathtt{S}$ does not terminate on input v.
    In that case $\mathtt{S}_w$ does not terminate on any input (in $\mathtt{i}_{\mathtt{S}_w}$ which is not used). Therefore $\mathtt{S}_w$ behaves as $\mathtt{S}_d$ for all its inputs i.e. $\mathcal{S}^{\mathrm{pf}}[\![\mathtt{S}_w]\!] = \mathcal{S}^{\mathrm{pf}}[\![\mathtt{S}_d]\!]$. Then$\mathtt{S}_d \notin P$ so $\mathtt{S}_w \notin P$ by extensionality.
  - $v \in \mathrm{dom}(\mathcal{S}^{\mathrm{pf}}[\![\mathtt{S}]\!])$ if and only if $\mathtt{S}$ does terminate on input v.
    In that case $\mathtt{S}_w$ behaves as $\mathtt{S}_t$ for all its inputs i.e. $\mathcal{S}^{\mathrm{pf}}[\![\mathtt{S}_w]\!] = \mathcal{S}^{\mathrm{pf}}[\![\mathtt{S}_t]\!]$. Then$\mathtt{S}_t \in P$ implies $\mathtt{S}_w \in P$ by extensionality.

  In this second case $\mathtt{S}_d \notin P$, $\mathtt{S}$ terminates on input v if and only if $\mathtt{S}_w \in P$, so $\mathtt{termination}(\mathtt{S}, v) = \mathtt{in}_P(\mathtt{S}_w)$ is correct. □

We have shown that $\mathtt{termination}(\mathtt{S}, v)$ solves the termination problem in contradiction to Turing's theorem 9.1. □

The following exercise shows, using Rice's theorem, that precise static analysis is infeasible.

Exercise 9.13 Prove that there is no algorithm for examining a program P and determining infallibly whether P is an implementation of a partial function f, which takes an integer n and returns $f(n)$ (whenever f is well defined for n).

Rice's theorem 9.12 remains valid when restricting the considered programs (e.g., programs of a given complexity class [55]).

9.7 Conclusion

By Rice's theorem 9.12, checking that a program semantics has any nontrivial semantic property is undecidable. This means that when a sound algorithm automatically checks for a program's properties, the algorithm may not terminate; or if it always terminates, then either there are restrictions on the considered programs (such as finite states only, e.g., model checking), a human interaction is necessary (e.g., theorem proving), or the result may be *true*, *false*, or undetermined, that is, *I don't know*. This is an indication that the verification and static program analysis problems are very difficult, even with approximate solutions [613], but not impossible, although without limit on the perfection of verification and static program analysis.

Rice's theorem 9.12 was extended to program verification and analysis by [277] to show that, from a computability perspective, and for infinite abstract properties, static program analysis is harder than checking/verification. Another extension in [74] is to abstract semantics more general than the extensional/functional semantics property of Rice's theorem 9.12.

9.8 Solutions to Selected Exercises

Solution to exercise 9.3 By reductio ad absurdum, let P be a program and X be a variable not in P. Define P'

```
var X : int = 0;
P;
X := 1 / X;
```

P terminates if and only if P' has a runtime error (division by 0, because P does not use or modify X). So if the absence of runtime error were decidable then termination would be decidable, which is a contradiction.

Solution to exercise 9.4 Sign analysis in section 3.2 is undecidable because otherwise, given a program P, consider a fresh variable x not in P and the derived program P′ = P;x = 1;. P′ assigns a strictly positive value to the variable x different from the initial value 0 of x if and only if P terminates. Therefore, if the sign problem were decidable, termination would also be decidable, which is a contradiction.

10 Posets, Lattices, and Complete Lattices

Partial order theory emerged from the work of George Boole, Ernst Schröder, Charles Peirce, and it was mainly developed by Garrett Birkhoff [120]. Partial order theory is an abstraction of set theory in which $\in$ is expressed in terms of $\subseteq$ and $\subseteq$ (e.g., property implication) is abstracted as a partial order $\sqsubseteq$ (e.g., abstract property implication). The abstract partial order $\sqsubseteq$ retains the essential properties of inclusion (reflexive, antisymmetric, and transitive). Many theorems of set theory remain valid, but not all, and this widely broadens the scope of applicability of order theory (see introductions in [121, 295, 445]). The poset representation $\langle \mathbb{P}, \sqsubseteq \rangle$ of program properties $\mathbb{P}$ with an abstract implication $\sqsubseteq$ provides a unified theory of program properties and their abstraction while giving us much more freedom and diversity in the choice of the possible encodings of abstract properties.

Contents

10.1 Posets

A *poset* $\langle \mathbb{P}, \sqsubseteq \rangle$ is a set $\mathbb{P}$ equipped with a *partial order* $\sqsubseteq$ that is

Reflexive: $\forall x \in \mathbb{P} \,.\, x \sqsubseteq x$;
Antisymmetric: $\forall x, y \in \mathbb{P} \,.\, ((x \sqsubseteq y) \wedge (y \sqsubseteq x)) \Rightarrow (x = y)$ and
Transitive: $\forall x, y, z \in \mathbb{P} \,.\, ((x \sqsubseteq y) \wedge (y \sqsubseteq z)) \Rightarrow (x \sqsubseteq z)$.

Two elements x and y are *comparable* when either $x \sqsubseteq y$ or $y \sqsubseteq x$ and *incomparable* when neither $x \sqsubseteq y$ nor $y \sqsubseteq x$. A partial order $\sqsubseteq$ is total whenever any two elements are comparable.

Total: $\forall x, y \in \mathbb{P} \,.\, (x \sqsubseteq y) \vee (y \sqsubseteq x)$.

A *strict partial order* $\sqsubset$ is irreflexive ($\forall x \in \mathbb{P} . x \not\sqsubset x$) and transitive. If $\sqsubseteq$ is a partial order then $x \sqsubset y \triangleq x \sqsubseteq y \wedge x \neq y$ is strict. If $\sqsubset$ is a strict partial order then $x \sqsubseteq y \triangleq x \sqsubset y \vee x = y$ is a partial order.

A *preorder* $\preceq$ is reflexive and transitive. Then $x \equiv y \triangleq x \preceq y \wedge y \preceq x$ is an *equivalence relation* (reflexive, symmetric ($\forall x, y \in \mathbb{P} . x \equiv y \Rightarrow y \equiv x$), and transitive). The *equivalence class* of $x \in \mathbb{P}$ for the equivalence relation $\equiv$ is $[x]_{\equiv} \triangleq \{y \in \mathbb{P} \mid x \equiv y\}$. The *quotient* of $\mathbb{P}$ by $\equiv$ is $\mathbb{P}|_{\equiv} \triangleq \{[x]_{\equiv} \mid x \in \mathbb{P}\}$. The *extension* of the preorder $\preceq$ to the quotient $\mathbb{P}|_{\equiv}$ is $[x]_{\equiv} \preceq_{\equiv} [y]_{\equiv} \Leftrightarrow \exists x' \in [x]_{\equiv}, y' \in [y]_{\equiv} . x' \preceq y'$. If $\preceq$ is a preorder on $\mathbb{P}$ then $\preceq_{\equiv}$ is a partial order on $\mathbb{P}|_{\equiv}$.

Exercise 10.1 Prove that the equality $=$ on a nonempty set is the only nonempty relation that is both a partial order and an equivalence relation.

10.2 Hasse Diagrams

Finite posets $\langle \mathbb{P}, \sqsubseteq \rangle$ can be represented by a Hasse diagram, which is a set of points $\{p_x \mid x \in \mathbb{P}\}$ in the plane, different two by two, and such that

- if $x \sqsubset y$ then p_x is strictly below p_y and
- p_x and p_y are linked by a segment when $x \lessdot y$ (y covers x) where $x \lessdot y \triangleq x \sqsubset y \wedge \nexists z \in \mathbb{P} . x \sqsubset z \wedge z \sqsubset y$.

$\sqsubseteq$ is derived from $\lessdot$ by reflexivity and transitivity. Two unlinked elements are incomparable. Examples are given in section 3.12, exercise 3.45, and section 8.11.

10.3 Least Upper Bound (lub), Greatest Lower Bound (glb), Minimum, Maximum, Infimum, and Supremum

Let $\langle \mathbb{P}, \sqsubseteq \rangle$ be a poset and $S \in \wp(\mathbb{P})$ be a subset. This subset S has

An upper bound u: if and only if $u \in \mathbb{P}$ and $\forall x \in S . x \sqsubseteq u$;

A least upper bound (lub/join) $\sqcup S$*:* if and only if $\sqcup S$ is an upper bound of S smaller that other upper bound of S (i.e., $\sqcup S \in \mathbb{P}$ and $\forall x \in S . x \sqsubseteq \sqcup S$ and $\forall u \in \mathbb{P} . (\forall x \in S . x \sqsubseteq u) \Rightarrow (\sqcup S \sqsubseteq u)$). $\sqcup\{x, y\}$ is denoted with the infix notation $x \sqcup y$;

A maximum M: if and only if $M = \sqcup S \in S$;

A supremum $\top$*:* if and only if $\top = \sqcup \mathbb{P} \in \mathbb{P}$.

10.4 Duality Principle

The *order dual* of an order-theoretic definition or statement is obtained by replacing $\sqsubseteq$ by its inverse $\sqsupseteq$, upper by lower, least by greatest, $\sqcup$ by $\sqcap$, $\sqcap$ by $\sqcup$, join by meet, meet by join, maximum by minimum, and so forth.

The *duality principle* states that if a definition or statement is valid for all partially ordered sets then the dual definition or dual statement is also valid for all partially ordered sets [121].

Example 10.2 Let $f \in \langle A, \leqslant \rangle \xrightarrow{\nearrow} \langle B, \sqsubseteq \rangle$ be increasing that is, $\forall x, y \in A . (x \leqslant y) \Rightarrow (f(x) \sqsubseteq f(y))$. The dual of "$f$ is increasing" is "f is increasing." Note that if duality is applied to $\langle A, \leqslant \rangle$ or $\langle B, \sqsubseteq \rangle$ only, then the semidual of "f is increasing" would be "f is decreasing," that is, $\forall x, y \in A . (x \leqslant y) \Rightarrow (f(x) \sqsupseteq f(y))$.

10.5 Lattices

Lattices are posets with the following properties.

Join semilattice: $\forall x, y \in \mathbb{P} \,.\, x \sqcup y$ exists in $\mathbb{P}$ (hence any nonempty finite subset of $\mathbb{P}$ has a lub);

Meet semilattice: $\forall x, y \in \mathbb{P} \,.\, x \sqcap y$ exists in $\mathbb{P}$ (hence any nonempty finite subset of $\mathbb{P}$ has a glb);

Lattice: $\forall x, y \in \mathbb{P} \,.\, x \sqcup y$ and $x \sqcap y$ exist in $\mathbb{P}$ (hence any nonempty finite subset of $\mathbb{P}$has a lub/join and a glb/meet).

Example 10.3

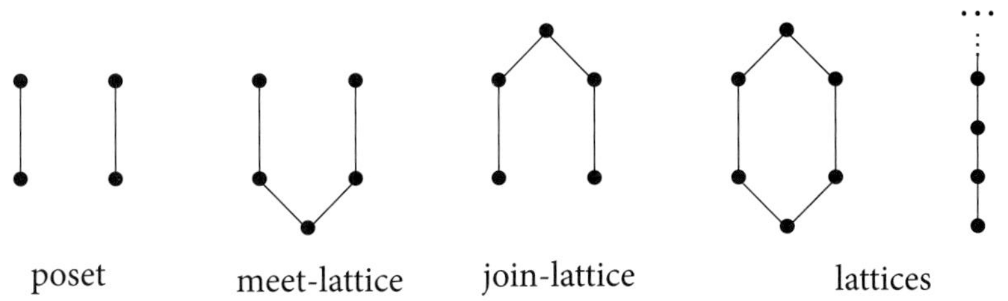

Exercise 10.4 Consider the following diagram $\bullet^{a}$ $\bullet^{b}$. Is it the Hasse diagram of a poset?

Lattices have unique joins/meets of two-elements hence, by associativity of the union of finite sets. lattices have unique joins/meets of finite subsets. For example, $\langle \mathbb{N}, \leqslant \rangle$ is a lattice where the glb/meet is min and the lub/join is max taken on nonempty finite subsets of $\mathbb{N}$.

Exercise 10.5 Show that the lub $\sqcup$ and glb $\sqcap$ of a lattice $\langle \mathbb{P}, \sqsubseteq \rangle$ have the following properties:

$x \sqcap x = x$	$x \sqcup x = x$	idempotency
$x \sqcap y = y \sqcap x$	$x \sqcup y = y \sqcup x$	commutativity
$x \sqcap (y \sqcap z) = (x \sqcap y) \sqcap z$	$x \sqcup (y \sqcup z) = (x \sqcup y) \sqcup z$	associativity
$(x \sqcap y) \sqcup x = x$	$(x \sqcup y) \sqcap x = x$	absorpsion

Conversely, show that a set equipped with binary operations $\sqcup$ and $\sqcap$ satisfying these properties is a lattice by defining $x \sqsubseteq y \triangleq x \sqcap y = x$ (or equivalently, $x \sqcup y = y$).

Let S be a subset of a lattice $\langle L, \sqsubseteq, \sqcup, \sqcap \rangle$. If $\forall x, y \in S \,.\, x \sqcup y \in S \wedge x \sqcap y \in S$, then $\langle S, \sqsubseteq, \sqcup, \sqcap \rangle$ is a *sublattice* of L. It is also a lattice.

Exercise 10.6 Show that a subset S of a lattice $\langle L, \sqsubseteq, \sqcup, \sqcap \rangle$ may be a lattice $\langle S, \sqsubseteq \rangle$ although not a sublattice of L.

10.6 Complete Lattices

A *complete lattice* is a poset $\langle \mathbb{P}, \sqsubseteq \rangle$ in which any subset $S \in \wp(\mathbb{P})$ has a lub/join $\sqcup S$ (not only the finite ones). Therefore a complete lattice has a supremum $\top = \sqcup \mathbb{P}$ and an infinum $\bot = \sqcup \varnothing$. (Any element x of $\mathbb{P}$ is an upper bound of $\varnothing$ since $\forall y \in \varnothing \,.\, y \sqsubseteq x$. Hence the lub of $\varnothing$ is the least element of $\mathbb{P}$.)

For example, $\langle \mathbb{N}, \leqslant \rangle$ is not a complete lattice because $\mathbb{N}$ has no lub. $\langle \mathbb{N} \cup \{\infty\}, \leqslant \rangle$ where $\forall n \in \mathbb{N} . n < \infty \leqslant \infty$ is a complete lattice with supremum ∞ and infimum 0. The powerset of a set S is a complete lattice $\langle \wp(S), \subseteq, \emptyset, S, \cup, \cap \rangle$.

Exercise 10.7 Prove that a complete lattice $\langle \mathbb{P}, \sqsubseteq, \bot, \top, \sqcup \rangle$ has a glb $\sqcap$ for arbitrary subsets.

Exercise 10.8 Prove that in a complete lattice $\langle \mathbb{P}, \sqsubseteq, \bot, \top, \sqcup \rangle$, if $X, Y \in \wp(\mathbb{P})$ and $X \subseteq Y$ then $\sqcup X \sqsubseteq \sqcup Y$.

Exercise 10.9 Prove that a pair $\langle P, \leqslant \rangle$ and $\langle Q, \sqsubseteq \rangle$ of posets (respectively semilattices, lattices, and complete lattices) is a poset $\langle P \times Q, \leqslant\times\sqsubseteq \rangle$ (respectively semilattice, lattice, complete lattice) for the componentwise ordering $\langle x, y \rangle \leqslant\times\sqsubseteq \langle x', y' \rangle$ if and only if $x \leqslant x' \wedge y \sqsubseteq y'$. Generalize to Cartesian products.

10.7 Pointwise Extension

Let $\langle \mathbb{P}, \sqsubseteq \rangle$ be a poset and S be a set. The pointwise extension $\dot{\sqsubseteq}$ of $\sqsubseteq$ to S is $\langle S \to \mathbb{P}, \dot{\sqsubseteq} \rangle$ where $f \mathrel{\dot{\sqsubseteq}} g$ if and only if $\forall x \in S . f(x) \sqsubseteq g(x)$. The pointwise join is $f \mathbin{\dot{\sqcup}} g \triangleq x \in S \mapsto f(x) \sqcup g(x)$, the meet is $f \mathbin{\dot{\sqcap}} g \triangleq x \in S \mapsto f(x) \sqcap g(x)$, and so on. The pointwise extension of $\dot{\sqsubseteq}$ is denoted $\ddot{\sqsubseteq}$, that of $\ddot{\sqsubseteq}$ is $\dddot{\sqsubseteq}$, and so forth.

Exercise 10.10 Show that the pointwise extension of a poset (respectively semilattice, lattice, complete lattice, etc.) is a poset (respectively semilattice, lattice, complete lattice, etc.).

Exercise 10.11 Let $\langle \mathbb{P}, \sqsubseteq, \bot, \top, \sqcup, \sqcap \rangle$ be a complete lattice. Prove that the functions $\mathbb{P} \to \mathbb{P}$, increasing functions $\mathbb{P} \xrightarrow{\nearrow} \mathbb{P} \triangleq \{f \in \mathbb{P} \to \mathbb{P} \mid \forall x, y \in \mathbb{P} . (x \sqsubseteq y) \Rightarrow f(x) \sqsubseteq f(y))\}$, arbitrary join preserving functions $\mathbb{P} \xrightarrow{\sqcup} \mathbb{P} \triangleq \{f \in \mathbb{P} \to \mathbb{P} \mid \forall S \in \wp(\mathbb{P}) . (\sqcup S \in \mathbb{P}) \Rightarrow (\sqcup f(S) \in \mathbb{P} \wedge f(\sqcup S) = \sqcup f(S))\}$ where $f(S) \triangleq \{f(x) \mid x \in S\}$, and dually arbitrary meet preserving functions $\mathbb{P} \xrightarrow{\sqcap} \mathbb{P}$ are a complete lattice for the pointwise ordering $f \mathrel{\dot{\sqsubseteq}} g \Leftrightarrow \forall x \in \mathbb{P} . f(x) \sqsubseteq g(x)$.

10.8 Chain

A *chain* C of a poset $\langle \mathbb{P}, \sqsubseteq \rangle$ is a subset of the poset $\mathbb{P}$ such that any two elements of the chain are comparable, that is, $C \subseteq \mathbb{P} \wedge \forall x, y \in C . x \sqsubseteq y \vee y \sqsubseteq x$. A *denumerable ascending/increasing chain* is a sequence $\langle x_i, i \in \mathbb{N} \rangle$ such that $x_0 \sqsubseteq x_1 \sqsubseteq \ldots x_n \sqsubseteq x_{n+1} \ldots$, that is, $\forall i < j \in \mathbb{N} . x_i \sqsubseteq x_j$ (so $x \in \mathbb{N} \xrightarrow{\nearrow} \mathbb{P}$). A denumerable sequence $\langle x_i, i \in \mathbb{N} \rangle$ is *ultimately stationary* if and only if $\exists \ell \in \mathbb{N} . \forall i \geqslant \ell . x_i = x_\ell$. A poset $\langle \mathbb{P}, \sqsubseteq \rangle$ is *Noetherian* (or satisfies the *increasing chain condition* (also called ascending chain condition (ACC)[1])) if and only if any increasing chain is ultimately stationary (so that any strictly ascending chain is finite). The *descending chain condition* (*DCC*) is dual.

Exercise 10.12 Is any nonempty chain a lattice? Is it a complete lattice?

Exercise 10.13 Show that any Noetherian lattice is a complete lattice.

1. *Voraussetzung des Teilerkettensatz* [755, Satz I], [756, Satz II].

10.9 CPO

A *complete partial order (CPO)* (or *countably chain-complete poset*) is a poset $\langle \mathbb{P}, \sqsubseteq, \bot, \sqcup \rangle$ with infimum $\bot$ such that any denumerable ascending chain $\langle x_i, i \in \mathbb{N} \rangle$ has a least upper bound $\bigsqcup_{i \in \mathbb{N}} x_i \in \mathbb{P}$. A dual CPO is defined dually.

Exercise 10.14 (flat domain) Let $\mathbb{P}$ be a set. Let $\bot \notin \mathbb{P}$. Define $\bot \sqsubseteq \bot \sqsubseteq x \sqsubseteq x$ for all $x \in \mathbb{P}$. Show that $\langle \mathbb{P} \cup \{\bot\}, \sqsubseteq \rangle$ is a CPO.

Exercise 10.15 (prefix order) Consider the prefix relation $\leqslant$ on traces $\forall \pi, \pi' \in \mathbb{T}^{+\infty}$. $(\pi \lessdot \pi') \triangleq (\exists \pi'' \in \mathbb{T}^{+\infty} . \pi \frown \pi'' = \pi')$ where trace concatenation $\frown$ is $\pi \frown \pi' = \pi$ when $\pi \in \mathbb{T}^{\infty}$. For example, $\ni \lessdot a \lessdot aa \lessdot aaa \lessdot \ldots \lessdot a^n \lessdot \ldots$. Define the limit $\dot{\curlyvee}_{i \in \mathbb{N}} \pi_i$ of increasing chains $\langle \pi_i, i \in \mathbb{N} \rangle$. For example, $\dot{\curlyvee}\{a^n \mid n \in \mathbb{N}\} = a^{\omega}$. Prove that $\langle \mathbb{T}^{+\infty} \cup \{\ni\}, \leqslant, \ni, \curlyvee \rangle$ is a CPO.

Exercise 10.16 Let X^k, $k \in \mathbb{N}$ by an infinite increasing chain in a CPO $\langle L, \sqsubseteq, \sqcup \rangle$. Let $f \in \mathbb{N} \to \mathbb{N}$ be a strictly increasing function on naturals. Prove that $\bigsqcup_{k \in \mathbb{N}} X^k = \bigsqcup_{k \in \mathbb{N}} X^{f(k)}$.

Exercise 10.17 Let $f \in P \to P$ be an extensive operator on a CPO $\langle P, \sqsubseteq, \bot, \sqcup \rangle$, that is, $\forall x \in P . x \sqsubseteq f(x)$. Prove that f may have a fixpoint but (contrary to the claim [295, 8.23 Fixpoint Theorem III]) may not have a minimal one.

10.10 Conclusion

Abstract program properties can be represented by posets. Finite Boolean lattices appeared in dataflow analysis [27, 29]. Gary Kildall [572] (and independently, Ben Wegbreit [967]) introduced the use of non-Boolean finite lattices in dataflow analysis. Gary Kildall [572] required the lattice to be distributive, which is not true of his main example: constant propagation. In many examples of dataflow analysis, lattices are used dually, the infimum is tt(true) and the supremum is ff(∅) [472]. Hassan Aït-Kaci et al. [15] provides an example of efficient implementation of finite lattices.

The use of complete lattices not satisfying the increasing chain condition (every strictly increasing chain is finite) in program static analysis was introduced in [234, 235].

10.11 Solutions to Selected Exercises

Solution to exercise 10.1 Let $r \in \wp(S \times S)$ be a nonempty relation on a nonempty set S which is both a partial order and an equivalence relation. By reflexivity, $\forall x \in S . \langle x, x \rangle \in r$, proving that $\mathbb{1}_S \subseteq r$. Assume by reductio ad absurdum that $r \neq \mathbb{1}_S$. Then $\exists \langle x, y \rangle \in r . x \neq y$. By symmetry $\langle y, x \rangle \in r$ so by antisymmetry $x = y$, a contradiction.

Solution to exercise 10.4 Yes, the poset is $\langle \{a, b\}, \{\langle a, a \rangle, \langle b, b \rangle\} \rangle$.

Solution to exercise 10.6

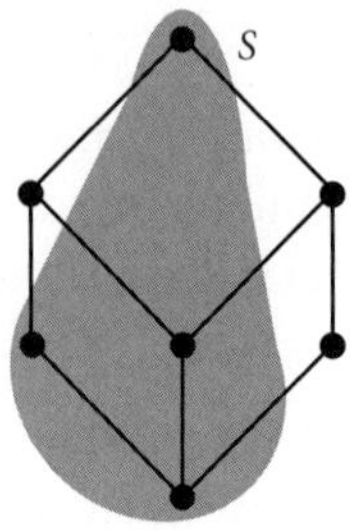

Solution to exercise 10.7 Let $L = \{x \in \mathbb{P} \mid \forall y \in S \,.\, x \sqsubseteq y\}$ be the set of lower bounds of S. $\forall y \in S \,.\, \forall x \in L \,.\, x \sqsubseteq y$, i.e. y is an upper bound of L. So, by existence and definition of $\bigsqcup L$

- $\bigsqcup L$ is an upper bound of L, that is, $\forall x \in L \,.\, x \sqsubseteq \bigsqcup L$
- it is the least one, so $\forall y \in S \,.\, \bigsqcup L \sqsubseteq y$

It follows that $\bigsqcup L$ is

- a lower bound of S
- the greatest one (since any lower bound x of S belongs to L so $x \subseteq \bigsqcup L$

Hence, by definition of a glb, $\bigsqcap S = \bigsqcup L$.

Solution to exercise 10.8 By definition of $\subseteq$, $\forall x \in X \,.\, x \in Y$, so $x \sqsubseteq \sqcup Y$ by definition of the lub $\sqcup$. It follows that $\sqcup Y$ is an upper bound of X, so $\sqcup X \sqsubseteq \sqcup Y$, by definition of the lub $\sqcup$.

Solution to exercise 10.12 Any finite nonempty subset S of a chain has a lub (max S) and a glb (min S) so is a lattice. However it may not be a complete lattice, as shown by the chain of naturals $\langle \mathbb{N}, \leqslant \rangle$ with their natural order $\leqslant$ such that the subset $\mathbb{N}$ has no lub.

Solution to exercise 10.15 Denote the limit of the $\dot{\leqslant}$-increasing denumerable chain $\langle \pi_i,\ i \in \mathbb{N} \rangle$ by $\pi' = \dot{\curlyvee}_{i \in \mathbb{N}} \pi_i$. If $\langle \pi_i,\ i \in \mathbb{N} \rangle$ is ultimately stationary at rank n, then $\pi' = \pi_n$. This includes the case of an infinite trace π_n since then $\forall m > n \,.\, \pi_m = \pi_n$. Otherwise the chain is infinite such that $\forall n \in \mathbb{N} \,.\, \exists m > n \,.\, \pi_n \lessdot \pi_m$ so all π_i, $i \in \mathbb{N}$ are finite. In that case, $\pi' \in \mathbb{T}^{\infty}$ is such that $\forall n \in \mathbb{N} \,.\, \pi'[0..|\pi_i| - 1] = \pi_i$ where $|\pi|$ is the length of the finite trace π.

Solution to exercise 10.16 We have $\{X^{f(k)} \mid k \in \mathbb{N}\} \subseteq \{X^k \mid k \in \mathbb{N}\}$ so $\bigsqcup_{k \in \mathbb{N}} X^{f(k)} \sqsubseteq \bigsqcup_{k \in \mathbb{N}} X^k$. Assume the inequality is strict. Then there exists an $\ell \in \mathbb{N}$ such that $\bigsqcup_{k \in \mathbb{N}} X^{f(k)} \sqsubset X^{\ell}$. However $X^{\ell} \sqsubseteq X^{f(\ell)}$, a contradiction.

Solution to exercise 10.17 Consider the negative integers ordered naturally with infimum $-\infty$. This is a CPO. Let $f(-\infty) = 0$ and $f(n) = n$ otherwise. f has infinitely many fixpoints but no minimal one.

11 Galois Connections and Abstraction

Galois connections (or Galois adjunctions) formalize the correspondence between concrete properties (e.g., sets of traces) and abstract properties (e.g., sets of reachable states) in case there is always a most precise abstract property overapproximating any concrete property.

Contents

11.1 Definitions of Galois Connections

The Galois connection between powersets of section 3.19 generalizes to posets.

Definition 11.1 (Galois connection) Given posets $\langle \mathscr{C}, \sqsubseteq \rangle$ (the *concrete domain*) and $\langle \mathscr{A}, \preccurlyeq \rangle$ (the *abstract domain*), the pair $\langle \alpha, \gamma \rangle$ of functions $\alpha \in \mathscr{C} \to \mathscr{A}$ (the *lower adjoint* or *abstraction*) and $\gamma \in \mathscr{A} \to \mathscr{C}$ (the *upper adjoint* or *concretization*) is a *Galois connection* (GC) if and only if

$$\forall P \in \mathscr{C} \,.\, \forall \overline{P} \in \mathscr{A} \,.\, \alpha(P) \preccurlyeq \overline{P} \Leftrightarrow P \sqsubseteq \gamma(\overline{P}) \tag{11.1}$$

which we write

$$\langle \mathscr{C}, \sqsubseteq \rangle \underset{\alpha}{\overset{\gamma}{\leftrightarrows}} \langle \mathscr{A}, \preccurlyeq \rangle \,.$$

Such Galois connections, introduced by Jürgen Schmidt [861, § 8], are sometimes called increasing, because α is increasing; see lemma 11.33 (as opposed to Galois correspondences, discussed in section 11.2, also called decreasing Galois connections, which were introduced by Oystein Ore [762] after Evarist Galois [388] and Garrett Birkhoff's polarities [120] of exercise 11.11). Definition 11.1 also applies to preorders (and most, but not all, properties considered below are preserved).

The intuition is that the concrete properties in $\mathscr{C}$ are approximated by abstract properties in $\mathscr{A}$. The concretization γ provides the concrete meaning/semantics/representation $\gamma(\overline{P}) \in \mathscr{C}$ of abstract properties $\overline{P} \in \mathscr{A}$. $\gamma(\overline{P})$ is the least precise element of $\mathscr{C}$ that can be overapproximated by $\overline{P}$. Therefore any element of $\mathscr{C}$, $\sqsubseteq$-less than $\gamma(\overline{P})$ can also be overapproximated by $\overline{P}$.

An abstract property $\overline{P} \in \mathscr{A}$ is a sound overapproximation of a concrete property $P \in \mathscr{C}$ whenever $P \sqsubseteq \gamma(\overline{P})$. $\sqsubseteq$ is understood as implication, so P is a stronger property than $\gamma(\overline{P})$. We choose $\alpha(P) \in \mathscr{A}$ to be the abstraction of the concrete property P. The stronger is the abstract property, the more precise is the abstraction.

- The implication $\alpha(P) \preccurlyeq \overline{P} \Rightarrow P \sqsubseteq \gamma(\overline{P})$ holds when $\overline{P} = \alpha(P)$ so $P \sqsubseteq \gamma(\alpha(P))$), meaning that $\alpha(P)$ is a sound overapproximation of the concrete property P and
- The converse implication $P \sqsubseteq \gamma(\overline{P}) \Rightarrow \alpha(P) \preccurlyeq \overline{P}$ means that for any sound overapproximation $\overline{P}$ of the concrete property P, $\alpha(P)$ is stronger, that is, more precise than $\overline{P}$.

It follows that $\alpha(P)$ is the best (most precise) sound overapproximation of P in the abstract domain $\mathscr{A}$. Moreover γ is increasing and so preserves the abstract implication $\preccurlyeq$ in the concrete. This implies that a proof in the abstract is valid in the concrete: if $\overline{P}_1 \preccurlyeq \overline{P}_2 \preccurlyeq \ldots \preccurlyeq \overline{P}_n$ then $\gamma(\overline{P}_1) \sqsubseteq \gamma(\overline{P}_2) \sqsubseteq \ldots \sqsubseteq \gamma(\overline{P}_n)$.

We say that the abstraction of a concrete property P is *exact* whenever $\gamma(\alpha(P)) = P$. This means that the abstraction $\alpha(P)$ of the property P loses no information at all, perhaps merely changing the abstract encoding of the concrete property P. For example, the sign abstraction of the property $\{0\}$, $\mathbb{N}$, $\mathbb{N}^+$, or $\mathbb{Z}$ mentioned in section 3.15 is exact whereas that of $\{0, 2\}$ is not.

We write $\langle \mathscr{C}, \sqsubseteq \rangle \underset{\alpha}{\overset{\gamma}{\longleftarrow\!\!\!\twoheadrightarrow}} \langle \mathscr{A}, \preccurlyeq \rangle$ when α is surjective (see section 11.6, Galois retraction), $\langle \mathscr{C}, \sqsubseteq \rangle \underset{\alpha}{\overset{\gamma}{\twoheadleftarrow\!\!\!\longrightarrow}} \langle \mathscr{A}, \preccurlyeq \rangle$ when γ is surjective, and $\langle \mathscr{C}, \sqsubseteq \rangle \underset{\alpha}{\overset{\gamma}{\twoheadleftarrow\!\!\!\twoheadrightarrow}} \langle \mathscr{A}, \preccurlyeq \rangle$ for a bijection (Galois isomorphism).

Exercise 11.1 Given $n \in \mathbb{Z}$, define $\alpha_n \in \mathbb{Z} \to \mathbb{Z}$ as $\alpha_n(x) = x - n$. Prove that this is a Galois isomorphism on $\langle \mathbb{Z}, \leqslant \rangle$.

Exercise 11.2 What is the Galois connection following from $n \leqslant \lfloor x \rfloor \Leftrightarrow n \leqslant x$?

Exercise 11.3 ([722, 889]) Define $x \div y = \max\{z \mid z \times y \leqslant x\}$ to be the largest natural number that, when multiplied by y is at most x. Obviously, $\forall x, z, y > 0 \,.\, z \times y \leqslant x \Leftrightarrow x \div y$. Show that this is a Galois connection.

Exercise 11.4 Let $\mathbb{K}$ be a field, $\mathbb{K}[x_1, \ldots, x_n]$ be the ring of n variate polynomials, and $I = (f_1, \ldots, f_s)$ be an ideal defining a variety $\mathscr{V}$ (see [616]). Show that $\alpha(I) = \{x \in \mathbb{K}^n \mid \forall f \in I \,.\, f(x) = 0\}$ and $\gamma(\mathscr{V}) = \{f \in \mathbb{K}[x_1, \ldots, x_n] \mid \forall x \in \mathscr{V} \,.\, f(x) = 0\}$ defines a Galois connection between $\langle \mathbb{K}[x_1, \ldots, x_n], \subseteq \rangle$ and $\langle \mathbb{K}^n, \subseteq \rangle$.

Exercise 11.5 (exclusion abstraction) Let C be a set with subset $S \subsetneq C$. Define $\alpha^S_\cap(X) = X \cap S$. Prove that $\langle \wp(C), \subseteq\rangle \xleftrightarrow[\alpha^S_\cap]{\gamma^S_\cap} \langle \wp(X), \subseteq\rangle$.

Exercise 11.6 (homomorphic/elementwise/right image abstraction) Given $f \in C \rightarrow A$, define $\forall X \in \wp(C) . \alpha_f(X) \triangleq f(X) \triangleq \{f(x) \mid x \in X\}$ and $\forall Y \in \wp(A) . \gamma_f(Y) \triangleq f^{-1}(Y) \triangleq \{x \mid f(x) \in Y\}$. Prove that $\langle \wp(C), \subseteq\rangle \xleftrightarrow[\alpha_f]{\gamma_f} \langle \wp(A), \subseteq\rangle$.[1]

Exercise 11.7 Continuing exercise 11.6, prove that $\langle \wp(C), \subseteq\rangle \xleftrightarrow[\alpha_f]{\gamma_f} \langle \wp(A), \subseteq\rangle$ is equivalent to the existence of $\mu \in A \rightarrow \wp(C)$ such that $\forall x \in C . \forall y \in A . x \in \mu(y) \Leftrightarrow f(x) = y$ [292].

Exercise 11.8 (join homomorphic/partitioning abstraction)[2] Let C be a set, $\langle A, \sqsubseteq, \sqcup\rangle$ be a complete lattice, and $h \in C \rightarrow A$. Define $\alpha(X) \triangleq \bigsqcup\{h(x) \mid x \in X\}$. Prove that α is the lower adjoint of a Galois connection. Which choice of A and h allows to recover exercise 11.6?

Exercise 11.9 (inclusion checking) Let S be a set, $R \in \wp(S) \setminus \{S\}$, and $\alpha_\subseteq(X) \triangleq (X \subseteq R)$. Prove that $\langle \wp(S), \subseteq\rangle \xleftrightarrow[\alpha_\subseteq]{\gamma_\subseteq} \langle \mathbb{B}, \Leftarrow\rangle$.

Exercise 11.10 (cover abstraction) Let C be a set and $A \in \wp(\wp(C))$ be a cover of C that is $C = \bigcup A$. For all $X \in C$ define $\alpha(X) \triangleq \{Y \in A \mid X \cap Y \neq \varnothing\}$. Prove that α is the lower adjoint of a Galois connection. Prove that if A is a partition of C that is $\forall Y, Y' \in A . Y \cap Y' \neq \varnothing \Rightarrow Y = Y'$ then γ is injective [249, section 5].

Exercise 11.11 (polarities) Given a relation $R \in \wp(\mathbb{P} \times \mathbb{Q})$, define the polarities $R^*(P) \triangleq \{y \in \mathbb{Q} \mid \forall x \in P . \langle x, y\rangle \in R\}$ and $R^\dagger(Q) \triangleq \{x \in \mathbb{P} \mid \forall y \in Q . \langle x, y\rangle \in R\}$ [120, p. V.7]. Prove that $\langle \wp(\mathbb{P}), \subseteq\rangle \xleftrightarrow[R^*]{R^\dagger} \langle \wp(\mathbb{Q}), \supseteq\rangle$.

Exercise 11.12 Let A and B be sets. Define $\alpha_{\mathsf{fr}}(F) \triangleq \{\langle a, b\rangle \mid b \in F(a)\}$ and $\gamma_{\mathsf{fr}}(R) \triangleq a \mapsto \{b \mid \langle a, b\rangle \in R\}$. Prove that $\langle A \rightarrow \wp(B), \dot{\subseteq}\rangle \xleftrightarrow[\alpha_{\mathsf{fr}}]{\gamma_{\mathsf{fr}}} \langle \wp(A \times B), \subseteq\rangle$.

Exercise 11.13 (partition abstraction) A partition of a set S is a family $P = \{B_i \mid i \in \Delta\}$ of disjoint nonempty blocks $(\forall i, j \in \Delta . (i \neq j) \Rightarrow (B_i \cap B_j = \varnothing))$ covering S $(\bigcup P = \bigcup\{B_i \mid i \in \Delta\} = S)$. If $x \in S$, we let $P(x)$ be the unique block of P to which x belongs to. Define $\alpha^\pi_P(X) \triangleq \prod_{i\in\Delta} X \cap B_i$. Prove that $\langle \wp(S), \subseteq\rangle \xleftrightarrow[\alpha^\pi_P]{\gamma^\pi_P} \langle \prod_{i\in\Delta} \wp(B_i), \dot{\subseteq}\rangle$, pointwise. Define $\alpha^\Delta_P(X) \triangleq \{i \in \Delta \mid X \cap B_i \neq \varnothing\}$. Prove that $\langle \wp(S), \subseteq\rangle \xleftrightarrow[\alpha^\Delta_P]{\gamma^\Delta_P} \langle \wp(\Delta), \subseteq\rangle$. Which abstraction is the most abstract?

Exercise 11.14 (language quotient) Given a finite alphabet Σ, let Σ^* be the possibly empty finite strings of letters of alphabet Σ called sentences (or words), and $L_1, L_2 \in$

1. We call this abstraction "homomorphic" because it is a structure-preserving map (homomorphic/partitioning abstractions of sets are sets, of joins are joins, of meets are meets, of functions are functions, of relations are relations, etc.).

2. Exercise 11.8 is the basis for *formal concept analysis* [335, 393, 394] to classify data symbolically. C is a set of studied objects, A is a set of descriptions $h(x) \in A$ of objects $x \in C$, and $d \sqsubseteq d'$ means that d' is more general than d. The pairs $\langle c, d\rangle$ s.t. $c = \gamma(d)$ and $d = \alpha(c)$ are called *conceptual classes* and, ordered componentwise, form a complete lattice.

$\wp(\Sigma^*)$ be languages, that is, sets of sentences. Sentence and language concatenation is denoted by juxtaposition. Given a sentence $w \in \Sigma^*$ and a language $L \in \wp(\Sigma^*)$, define the left quotient $Lw^{-1} \triangleq \{x \in \Sigma^* \mid xw \in L\}$ and the right $w^{-1}L \triangleq \{x \in \Sigma^* \mid wx \in L\}$. Show that concatenation and the quotients give rise to Galois connections.

Exercise 11.15 (big $\mathcal{O}$ notation) Show that the Bachmann-Landau notation $\mathcal{O}(f)$ [463, § 1.6, "Some Notations"] (in which one writes $f(x) = \mathcal{O}(g(x))$ as $x \to \infty$ if and only if for all sufficiently large values of x, the absolute value of $f(x)$ is at most a positive constant multiple of $g(x)$) is an abstraction of properties of real functions $\mathbb{R} \to \mathbb{R}$ by the positive real function $f \in \mathbb{R} \to \mathbb{R}_*$ where $\mathbb{R}_*$ is the set of all positive reals.

Exercise 11.16 (expected value) Let X be a variable taking its values in a set $\mathbb{V}$. A *set property* P_X of X is an element of $\wp(\mathbb{V})$. For example, $P_X = \{0\}$ states that X is constant equal to 0. A *probabilistic property* P_X of X is an element of $\mathbb{V} \to [0, 1]$. $P_X(x)$ (often written $\mathsf{P}[X = x]$) is the probability that variable X takes value $x \in \mathbb{V}$. For example, $P_X(0) = 1$ whereas $P_X(x) = 0$ whenever $x \neq 0$ states that X is constant equal to 0. Statistics deal with large data sets and probabilistic distributions that are too large to consider individually each possible values. Statistical measures are used to represent meaningful properties of the distribution, with some loss of information, as a basis for best-guess predictions. Thus statistical measures are abstract interpretations of properties of distributions. An example is the *expected value* or *mean* $\mathsf{E}(X) \triangleq \sum_{x \in \mathbb{V}} x.P_X(x)$. Show that this is a Galois connection.

Exercise 11.17 Let $\langle L, \sqsubseteq\rangle$ be a poset. Given $A \in \wp(L)$, define $\alpha(A) \triangleq \{x \in L \mid \forall y \in A \,.\, y \sqsubseteq x\}$ (the upper bounds of A) and $\gamma(A) \triangleq \{x \in L \mid \forall y \in A \,.\, x \sqsubseteq y\}$ (the lower bounds of A). Show that $\langle \wp(L), \subseteq\rangle \underset{\alpha}{\overset{\gamma}{\leftrightarrows}} \langle \wp(L), \supseteq\rangle$.

Exercise 11.18 (lub abstraction) Let $\langle \mathcal{C}, \leqslant\rangle$ be a poset and $\langle \mathcal{A}, \sqsubseteq, \sqcup, \sqcap\rangle$ be a complete lattice. Assume that $\forall i \in [1, n] \,.\, \langle \mathcal{C}, \leqslant\rangle \underset{\alpha_i}{\overset{\gamma_i}{\leftrightarrows}} \langle \mathcal{A}, \sqsubseteq\rangle$. Define $\alpha \triangleq \dot{\bigsqcup}_{i=1}^{n} \alpha_i$ and $\gamma = \dot{\bigsqcap}_{i=1}^{n} \gamma_i$, pointwise. Prove that $\langle \mathcal{C}, \leqslant\rangle \underset{\alpha}{\overset{\gamma}{\leftrightarrows}} \langle \mathcal{A}, \sqsubseteq\rangle$.

Exercise 11.19 (projection abstraction) Let $\langle \mathcal{C}_1, \leqslant_1, \top_1\rangle$ and $\langle \mathcal{C}_2, \leqslant_2, \top_2\rangle$ be posets with respective suprema $\top_1$ and $\top_2$. Prove that $\langle \mathcal{C}_1 \times \mathcal{C}_2, \leqslant_1 \times \leqslant_2\rangle \underset{\alpha_1}{\overset{\gamma_1}{\leftrightarrows}} \langle \mathcal{C}_1, \leqslant_1\rangle$ where the partial order $\langle x, y\rangle \leqslant_1 \times \leqslant_2 \langle x', y'\rangle \triangleq (x \leqslant_1 x') \wedge (y \leqslant_2 y')$ on pairs is componentwise, $\alpha_1(\langle x, y\rangle) = x$ is the first projection, and $\gamma_1(x) = \langle x, \top_2\rangle$. Generalize to the projection of any component of a Cartesian product.

Exercise 11.20 Considering a function $f \in \mathcal{X} \mapsto \mathcal{Y}$ as the relation $\{\langle x, f(x)\rangle \mid x \in \mathcal{X}\}$, show that $\langle \mathcal{C}, \leqslant\rangle \underset{\alpha}{\overset{\gamma}{\leftrightarrows}} \langle \mathcal{A}, \sqsubseteq\rangle$ if and only if $\alpha \mathbin{⨾} \sqsubseteq \,=\, \leqslant \mathbin{⨾} \gamma^{-1}$.

Exercise 11.21 (pointwise extension of a Galois connection) Show that the pointwise extension $\langle S \to \mathcal{C}, \dot{\sqsubseteq}\rangle \underset{\dot{\alpha}}{\overset{\dot{\gamma}}{\leftrightarrows}} \langle S \to \mathcal{A}, \dot{\preccurlyeq}\rangle$ of a Galois connection $\langle \mathcal{C}, \sqsubseteq\rangle \underset{\alpha}{\overset{\gamma}{\leftrightarrows}} \langle \mathcal{A}, \preccurlyeq\rangle$ where $\dot{\alpha}(f) \triangleq x \mapsto \alpha(f(x))$ is a Galois connection.

Exercise 11.22 (abstraction of an increasing function at a point) Let $\langle \mathcal{D}, \sqsubseteq, \top\rangle$ be a poset with supremum $\top$ and $p \in \mathcal{D}$. Given an increasing function f on $\mathcal{D}$, define

$\alpha_p(f) = f(p)$. Prove that $\langle \mathscr{D} \stackrel{\nearrow}{\longrightarrow} \mathscr{D}, \dot{\sqsubseteq}\rangle \underset{\alpha_p}{\overset{\gamma_p}{\leftrightarrows}} \langle \mathscr{D}, \sqsubseteq\rangle$ where $\dot{\sqsubseteq}$ is the pointwise extension of $\sqsubseteq$.

Exercise 11.23 (pointwise homomorphic abstraction) Let A, B, and C be sets. Assume $f \in A \to B \to C$. Define $\alpha_h(X)\, a \triangleq \{f(a)x \mid x \in X\}$. Prove that $\langle \wp(B), \subseteq\rangle \underset{\alpha_h}{\overset{\gamma_h}{\leftrightarrows}} \langle A \to \wp(C), \dot{\subseteq}\rangle$.

Exercise 11.24 (projection abstraction) Prove that $\langle \wp(\mathscr{C}_1) \times \ldots \times \wp(\mathscr{C}_n) \times \ldots \times \wp(\mathscr{C}_m), \dot{\subseteq}_{n+m}\rangle \underset{\alpha_\exists^{n,m}}{\overset{\gamma_\exists^{n,m}}{\leftrightarrows}} \langle \wp(\mathscr{C}_1) \times \ldots \times \wp(\mathscr{C}_n), \dot{\subseteq}_n\rangle$ for the projection abstraction $\alpha_\exists^{n,m}(\langle X_1, \ldots, X_n, \ldots, X_m\rangle) = \langle X_1, \ldots, X_n\rangle$ where $m \geqslant n > 0$ and $\dot{\subseteq}_k$ is the componentwise inclusion for k-tuples, $k > 0$.

11.2 Galois Correspondences

A Galois correspondence is a Galois connection $\langle \mathscr{C}, \sqsubseteq\rangle \underset{\alpha}{\overset{\gamma}{\leftrightarrows}} \langle \mathscr{A}, \succcurlyeq\rangle$, that is, the adjoint functions are decreasing with $\mathbb{1}_{\mathscr{C}} \sqsubseteq \alpha \circ \gamma$ and $\mathbb{1}_{\mathscr{A}} \succcurlyeq \gamma \circ \alpha$. So a Galois correspondence is the semidual of a Galois connection. Their composition is not a Galois correspondence but is a Galois connection. It is the original notion inherited from Evarist Galois' work.

We prefer Galois connections because they compose.

Galois correspondences are sometimes called decreasing Galois connections. Many other alternative names can be found in the literature, for example, residuated mappings for Galois connections [128, section 1.3] and Galois connections for Galois correspondences [128, section 1.6].

11.3 Duality of Galois Connections

Lemma 11.25 (duality) The order dual of an order-theoretic definition or statement is obtained by replacing $\sqsubseteq$ by its inverse $\sqsupseteq$, upper by lower, least by greatest, $\sqcup$ by $\sqcap$, join by meet, maximum by minimum, $\langle \mathscr{C}, \sqsubseteq\rangle \underset{\alpha}{\overset{\gamma}{\leftrightarrows}} \langle \mathscr{A}, \preccurlyeq\rangle$ by $\langle \mathscr{C}, \sqsupseteq\rangle \underset{\alpha}{\overset{\gamma}{\leftrightarrows}} \langle \mathscr{A}, \succcurlyeq\rangle$ or equivalently by $\langle \mathscr{A}, \preccurlyeq\rangle \underset{\gamma}{\overset{\alpha}{\leftrightarrows}} \langle \mathscr{C}, \sqsubseteq\rangle$, and so forth.

Therefore, dualization of a definition/statement/formula involving Galois connections consists of exchanging the adjoints.

Proof of lemma 11.25

$$\langle \mathscr{C}, \sqsubseteq\rangle \underset{\alpha}{\overset{\gamma}{\leftrightarrows}} \langle \mathscr{A}, \preccurlyeq\rangle$$
$$\Leftrightarrow \alpha(x) \preccurlyeq y \Leftrightarrow x \sqsubseteq \gamma(y) \qquad \text{(definition of Galois connections)}$$

The dual statement is

$$\langle \mathscr{C}, \sqsupseteq\rangle \underset{\alpha}{\overset{\gamma}{\leftrightarrows}} \langle \mathscr{A}, \succcurlyeq\rangle$$
$$\Leftrightarrow \alpha(x) \succcurlyeq y \Leftrightarrow x \sqsupseteq \gamma(y) \qquad \text{(definition of Galois connections)}$$
$$\Leftrightarrow \gamma(y) \sqsubseteq x \Leftrightarrow y \preccurlyeq \alpha(x) \qquad \text{(inverse order } x \sqsupseteq y \Leftrightarrow y \sqsubseteq x\text{)}$$
$$\Leftrightarrow \gamma(x) \sqsubseteq y \Leftrightarrow x \preccurlyeq \alpha(y) \qquad \text{(dummy variable renaming)}$$

$\Leftrightarrow \langle \mathscr{A}, \preccurlyeq \rangle \underset{\gamma}{\overset{\alpha}{\leftrightarrows}} \langle \mathscr{C}, \sqsubseteq \rangle$ ⦃definition of Galois connections⦄ □

Exercise 11.26 Prove that that if $\langle \mathscr{C}, \sqsubseteq \rangle \underset{\alpha}{\overset{\gamma}{\leftrightarrows}} \langle \mathscr{A}, \preccurlyeq \rangle$ and $\langle \mathscr{C}, \sqsubseteq \rangle$ has an infimum $\bot$ then $\langle \mathscr{A}, \preccurlyeq \rangle$ has an infimum. What is the dual?

Example 11.27 An example of duality is [190]. The first part recalls results of overapproximation by abstract interpretation. The second part on underapproximation is the dual of the first part.

11.4 Order, Join, and Meet Preservation

Definition 11.28 Let $\langle P_1, \sqsubseteq_1 \rangle$ and $\langle P_2, \sqsubseteq_2 \rangle$ be posets and $f \in P_1 \to P_2$ be a total function. A standard example is $\langle P_1, \sqsubseteq_1 \rangle = \langle \wp(S_1), \subseteq \rangle$, $\langle P_2, \sqsubseteq_2 \rangle = \langle \wp(S_2), \subseteq \rangle$, and $f \in \wp(S_1) \mapsto \wp(S_2)$ is a set transformer.

- f is *order preserving* or *increasing* when $\forall x, y \in P_1 \,.\, x \sqsubseteq_1 y$ implies $f(s) \sqsubseteq_2 f(y)$. This is written $f \in P_1 \xrightarrow{\nearrow} P_2$. The dual of increasing is increasing.
- *f preserves finite joins* (respectively *finite meets*) if and only if for all $x, y \in P_1$ such that the lub $x \sqcup_1 y$ exists, the lub $f(x) \sqcup_2 f(y)$ exists, and $f(x \sqcup_1 y) = f(x) \sqcup_2 f(y)$ (respectively for the glb, $f(x \sqcap_1 y) = f(x) \sqcap_2 f(y)$ when $x \sqcap_1 y$ exists).
- *f preserves the infimum* or is *infimum strict* whenever if P_1 has an infimum $\bot_1$ then P_2 has an infimum $\bot_2$ such that $f(\bot_1) = \bot_2$. *Supremum preservation* or *supremum strictness* is dual.
- *f preserves nonempty joins* if and only if given any nonempty indexing set Δ and index $X \in \Delta \mapsto P_1$ such that the lub $\bigsqcup_1 \{X(i) \mid i \in \Delta\}$ exists in P_1, we have $\bigsqcup_2 \{f(X(i)) \mid i \in \Delta\}$ exists in P_2 and $f(\bigsqcup_1 \{X(i) \mid i \in \Delta\}) = \bigsqcup_2 \{f(X(i)) \mid i \in \Delta\}$ (which is often written $f(\bigsqcup_{1_{i \in \Delta}} X_i) = \bigsqcup_{2_{i \in \Delta}} f(X_i)$ where X_i denotes the family $X(i), i \in \Delta$). *Nonempty meet preservation* is dual.
- *f preserves (arbitrary) joins* or is *strictly join preserving* if and only if it preserves both the infimum $\bot = \bigsqcup_1 \varnothing$ and nonempty joins. This is written $f \in P_1 \xrightarrow{\sqcup} P_2$ Dually, *meet preservation* is denoted $P_1 \xrightarrow{\sqcap} P_2$.

Exercise 11.29 Let f be an increasing function on a poset $\langle P, \sqsubseteq, \sqcup \rangle$. Assume that $X \in \wp(P)$ and the lubs $\sqcup X$ and $\sqcup f(X)$ do exist (with $f(X) \triangleq \{f(x) \mid x \in X\}$. Prove that $\sqcup f(X) \sqsubseteq f(\sqcup X)$.

Exercise 11.30 Prove that a function f on posets preserving joins or meets is increasing.

Exercise 11.31 Prove that if $\langle P_1, \sqsubseteq_1 \rangle$ and $\langle P_2, \sqsubseteq_2 \rangle$ be posets, $f \in P_1 \xrightarrow{\nearrow} P_2$ is increasing, and $X \subseteq P_1$ then $\sqcup_2 \{f(x) \mid x \in X\} \sqsubseteq_2 f(\sqcup_1 X)$ whenever the lubs do exist.

Exercise 11.32 Let $\langle D, \sqsubseteq, \bot, \top, \sqcup, \sqcap \rangle$ be a complete lattice. Prove that the set $D \xrightarrow{\sqcup} D$ of (arbitrary) joins functions ordered pointwise ($f \mathrel{\dot{\sqsubseteq}} g$ if and only if $\forall x \in D \,.\, f(x) \sqsubseteq g(x)$) is a complete lattice $\langle D \xrightarrow{\sqcup} D, \dot{\sqsubseteq}, \dot{\bot}, \dot{\top}, \dot{\sqcup} \rangle$ where $\dot{\bot}$ and $\dot{\top}$ are respectively $\bot$ and $\top$, pointwise.

11.5 Order Properties of Galois Connections

Lemma 11.33 (increasingness) If $\langle \mathscr{C}, \sqsubseteq\rangle \xrightleftharpoons[\alpha]{\gamma} \langle \mathscr{A}, \preccurlyeq\rangle$ then α and γ are increasing.

Proof Assume $P \sqsubseteq P'$. By $\alpha(P') \preccurlyeq \alpha(P')$ we have $P' \sqsubseteq \gamma(\alpha(P'))$ so that $P \sqsubseteq \gamma(\alpha(P'))$ by transitivity, hence $\alpha(P) \preccurlyeq \alpha(P')$ by definition of a GC, proving that α is increasing.

Increasing is self-dual, so the dual of "In $\langle \mathscr{C}, \sqsubseteq\rangle \xrightleftharpoons[\alpha]{\gamma} \langle \mathscr{A}, \preccurlyeq\rangle$, the lower adjoint α is increasing" is "In $\langle \mathscr{A}, \succcurlyeq\rangle \xrightleftharpoons[\gamma]{\alpha} \langle \mathscr{C}, \sqsupseteq\rangle$, the lower adjoint γ is increasing," proving that γ is increasing in $\langle \mathscr{C}, \sqsubseteq\rangle \xrightleftharpoons[\alpha]{\gamma} \langle \mathscr{A}, \preccurlyeq\rangle$. □

Exercise 11.34 Prove that in a Galois connection $\langle \mathscr{C}, \sqsubseteq\rangle \xrightleftharpoons[\alpha]{\gamma} \langle \mathscr{A}, \preccurlyeq\rangle$ we have $\alpha \circ \gamma \circ \alpha = \alpha$. Express the dual.

Exercise 11.35 (equivalent definition of Galois connections) Prove that $\langle \mathscr{C}, \sqsubseteq\rangle \xrightleftharpoons[\alpha]{\gamma} \langle \mathscr{A}, \preccurlyeq\rangle$ if and only if $\alpha \in \mathscr{C} \to \mathscr{A}$ and $\gamma \in \mathscr{A} \to \mathscr{C}$ satisfy

(1) α is increasing;
(2) γ is increasing;
(3) $\forall x \in \mathscr{C} \,.\, x \sqsubseteq \gamma \circ \alpha(x)$ (i.e., $\gamma \circ \alpha$ is extensive);
(4) $\forall y \in \mathscr{A} \,.\, \alpha \circ \gamma(y) \preccurlyeq y$ (i.e., $\alpha \circ \gamma$ is reductive).

Exercise 11.36 (equivalent definition of a Galois connection [61]) Show that $\langle \mathscr{C}, \sqsubseteq\rangle \xrightleftharpoons[\alpha]{\gamma} \langle \mathscr{A}, \preccurlyeq\rangle$ if and only if α is increasing, $\forall x \in \mathscr{A} \,.\, \alpha \circ \gamma(x) \sqsubseteq x$ and $\forall x, y \,.\, \alpha(x) \preccurlyeq y \Rightarrow x \sqsubseteq \gamma(y)$.

Exercise 11.37 Prove that there is no Galois connection $\langle \mathbb{N}, \leqslant\rangle \xrightleftharpoons[\alpha]{\gamma} \langle \{\bullet\}, =\rangle$ where $\bullet$ is the only element of the singleton domain $\{\bullet\}$.

Lemma 11.38 (existing join preservation) If $\langle \mathscr{C}, \sqsubseteq\rangle \xrightleftharpoons[\alpha]{\gamma} \langle \mathscr{A}, \preccurlyeq\rangle$ then α preserves lubs that may exist in $\mathscr{C}$; that is, let $\sqcup$ be the partially defined lub for $\sqsubseteq$ in $\mathscr{C}$ and $\curlyvee$ be the partially defined lub for $\preccurlyeq$ in $\mathscr{A}$. Let $S \in \wp(\mathscr{C})$ be any subset of $\mathscr{C}$. If $\bigsqcup S$ exists in $\mathscr{C}$ then the upper bound $\bigcurlyvee\{\alpha(e) \mid e \in S\}$ exists in $\mathscr{C}$ and is equal to $\alpha(\bigsqcup S)$.

Proof By existence and definition of the lub $\bigsqcup S$, we have $\forall e \in S \,.\, e \sqsubseteq \bigsqcup S$ so $\alpha(e) \preccurlyeq \alpha(\bigsqcup S)$ because α is increasing. It follows that $\alpha(\bigsqcup S)$ is an upper bound of $\{\alpha(e) \mid e \in S\}$. Let u be any upper bound of this set $\{\alpha(e) \mid e \in S\}$ so that $\forall e \in S \,.\, \alpha(e) \preccurlyeq u$. By definition of a GC, $\forall e \in S \,.\, e \sqsubseteq \gamma(u)$. Hence $\gamma(u)$ is an upper bound of S. By existence and definition of the lub $\bigsqcup S$, $\bigsqcup S \sqsubseteq \gamma(u)$ so $\alpha(\bigsqcup S) \preccurlyeq u$, proving that $\alpha(\bigsqcup S)$, which exists because α is a total function, is the lub of $\{\alpha(e) \mid e \in S\}$ denoted $\bigcurlyvee\{\alpha(e) \mid e \in S\}$. □

By duality, γ preserves existing meets.

Exercise 11.39 ([795]) Prove that if α preserves existing lubs and $\gamma(y) \triangleq \bigsqcup\{x \in \mathscr{C} \mid \alpha(x) \preccurlyeq y\}$ is well defined then $\langle \mathscr{C}, \sqsubseteq\rangle \xrightleftharpoons[\alpha]{\gamma} \langle \mathscr{A}, \preccurlyeq\rangle$.

Exercise 11.40 Consider the following concretization function. Prove that there is no corresponding Galois connection.

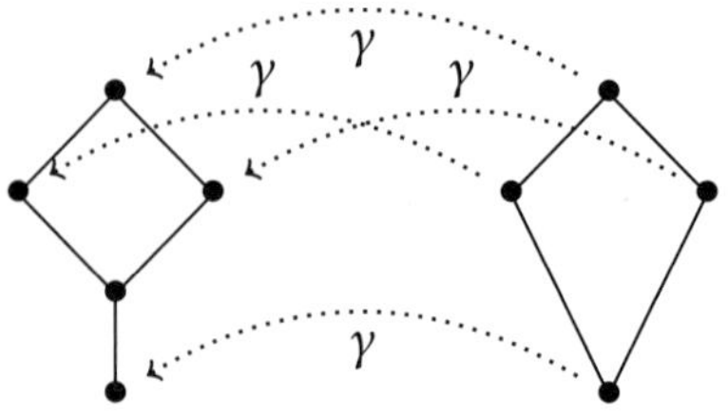

Exercise 11.41 Consider the following concretization function. Prove that there is no corresponding Galois connection.

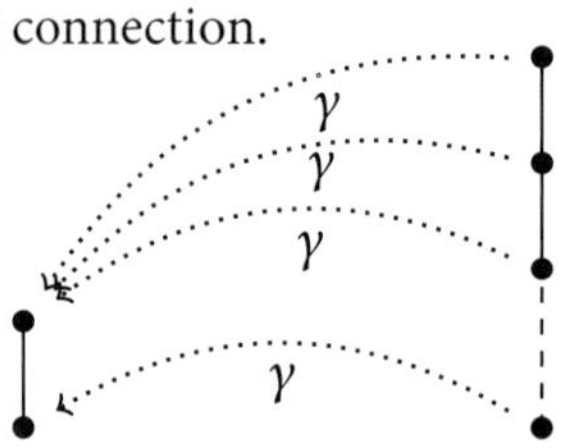

Lemma 11.42 In a Galois connection one adjoint uniquely determines the other i.e. $\alpha(P) = \sqcap\{\overline{P} \mid P \sqsubseteq \gamma(\overline{P})\}$ and $\gamma(\overline{P}) = \sqcup\{P \mid \alpha(P) \preccurlyeq \overline{P}\}$.

Proof of Lemma 11.42 Observe that $\forall P \in \mathscr{C} \,.\, \alpha(P) = \sqcap\{\overline{P} \mid \alpha(P) \preccurlyeq \overline{P}\}$ so, by definition of a GC, $\alpha(P) = \sqcap\{\overline{P} \mid P \sqsubseteq \gamma(\overline{P})\}$, that is, γ uniquely determines α. Dually α uniquely determines γ because $\forall \overline{P} \in \mathscr{A} \,.\, \gamma(\overline{P}) = \sqcup\{P \mid \alpha(P) \preccurlyeq \overline{P}\}$. □

It immediately follows that

Corollary 11.43 If $\langle \mathscr{C}, \sqsubseteq\rangle \underset{\alpha}{\overset{\gamma}{\leftrightarrows}} \langle \mathscr{A}, \preccurlyeq\rangle$ and $\langle \mathscr{C}, \sqsubseteq\rangle \underset{\alpha}{\overset{\gamma'}{\leftrightarrows}} \langle \mathscr{A}, \preccurlyeq\rangle$ then $\gamma = \gamma'$. If $\langle \mathscr{C}, \sqsubseteq\rangle \underset{\alpha}{\overset{\gamma}{\leftrightarrows}} \langle \mathscr{A}, \preccurlyeq\rangle$ and $\langle \mathscr{C}, \sqsubseteq\rangle \underset{\alpha'}{\overset{\gamma}{\leftrightarrows}} \langle \mathscr{A}, \preccurlyeq\rangle$ then $\alpha = \alpha'$.

An alternative proof is the following.

Proof of Corollary 11.43 By exercise 11.35.3, $\forall x \in \mathscr{C} \,.\, x \sqsubseteq \gamma' \circ \alpha(x)$ so in particular $\gamma(y) \sqsubseteq \gamma' \circ \alpha \circ \gamma(y)$. By exercise 11.35.4, $\alpha \circ \gamma(y) \preccurlyeq y$ so, by exercise 11.35.2, $\gamma' \circ \alpha \circ \gamma(y) \sqsubseteq \gamma'(y)$. By transitivity, $\gamma(y) \sqsubseteq \gamma'(y)$. By exchanging the rôles of γ and γ', a similar proof yields $\gamma'(y) \sqsubseteq \gamma(y)$. By antisymmetry $\gamma = \gamma'$. The proof for α and α' is dual. □

Lemma 11.42 and corollary 11.43 are useful in situations where only one adjoint is defined explicitly because then the other is also uniquely determined. It is important to note that the two Galois connections must have the same concrete and abstract partial orders. If these orders differ, this may no longer be true.

Exercise 11.44 Which Galois connection can be derived from the arithmetic property $\forall x, y, z \in \mathbb{N} \,.\, z \times y \leqslant x \Leftrightarrow z \leqslant x \div y$? Prove that this implies that $x \div y = \max\{z \mid x \times y \leqslant x\}$. Perform the same exercise with $z - y \leqslant x \Leftrightarrow z \leqslant x + y$ on $\mathbb{Z}$.

Exercise 11.45 Let $\langle \mathscr{C}, \sqsubseteq\rangle \underset{\alpha}{\overset{\gamma}{\leftrightarrows}} \langle \mathscr{A}, \preccurlyeq\rangle$. Prove that $\gamma \circ \alpha \circ \gamma = \gamma$. What is the dual?

Exercise 11.46 Let $\langle \mathscr{C}, \sqsubseteq\rangle \underset{\alpha}{\overset{\gamma}{\leftrightarrows}} \langle \mathscr{A}, \preccurlyeq\rangle$. Following exercise 11.45, prove that $\langle \gamma(\mathscr{A}), \sqsubseteq\rangle \underset{\overline{\alpha}}{\overset{\overline{\gamma}}{\stackrel{\twoheadleftarrow}{\twoheadrightarrow}}} \langle \alpha(\mathscr{C}), \preccurlyeq\rangle$ (where $f(X) = \{f(x) \mid x \in X\}$, $\overline{\alpha} \triangleq \alpha \rceil \gamma(\mathscr{A})$ is the restriction of α to $\gamma(\mathscr{X})$ and $\overline{\gamma} \triangleq \gamma \rceil \alpha(\mathscr{A})$).

Exercise 11.47 The *corresponding fixpoints* of the Galois connection $\langle \mathscr{C}, \sqsubseteq\rangle \underset{\alpha}{\overset{\gamma}{\leftrightarrows}} \langle \mathscr{A}, \preccurlyeq\rangle$ are $\langle x, y\rangle \in \mathscr{C} \times \mathscr{A}$ such that $\alpha(x) = y$ and $\gamma(y) = x$. Following exercise 11.45, show that for all $x \in \mathscr{C}$ and $y \in \mathscr{A}$, the pairs $\langle \gamma \circ \alpha(x), \alpha(x)\rangle$ and $\langle \gamma(y), \alpha \circ \gamma(y)\rangle$ are corresponding fixpoints.

Exercise 11.48 Let $\langle \mathscr{C}, \sqsubseteq\rangle \underset{\alpha}{\overset{\gamma}{\leftrightarrows}} \langle \mathscr{A}, \preccurlyeq\rangle$. Prove that $\gamma(a) = \max \alpha^{-1}(\downarrow a) = \bigsqcup \alpha^{-1}(\downarrow a)$ where $\max C$ is the maximum of the set $C \in \wp(\mathscr{C})$, $\alpha^{-1}(A) \triangleq \{c \in \mathscr{C} \mid \alpha(c) \in A\}$, and $\downarrow a \triangleq \{x \in \mathscr{A} \mid x \preccurlyeq a\}$. State the dual property (using $\uparrow c \triangleq \{x \in \mathscr{C} \mid c \sqsubseteq x\}$).

11.6 Galois Retraction

A Galois connection $\langle \mathscr{C}, \sqsubseteq\rangle \underset{\alpha}{\overset{\gamma}{\leftrightarrows}} \langle \mathscr{A}, \preccurlyeq\rangle$ such that α is surjective is called a Galois retraction.[3] (or Galois surjection, Galois insertion, Galois reflexion [353] or "perfect Galois connection" [762, 831]) denoted $\langle \mathscr{C}, \sqsubseteq\rangle \underset{\alpha}{\overset{\gamma}{\stackrel{\longleftarrow}{\twoheadrightarrow}}} \langle \mathscr{A}, \preccurlyeq\rangle$.

Example 11.49 A simple example of Galois connection that is not a retraction is

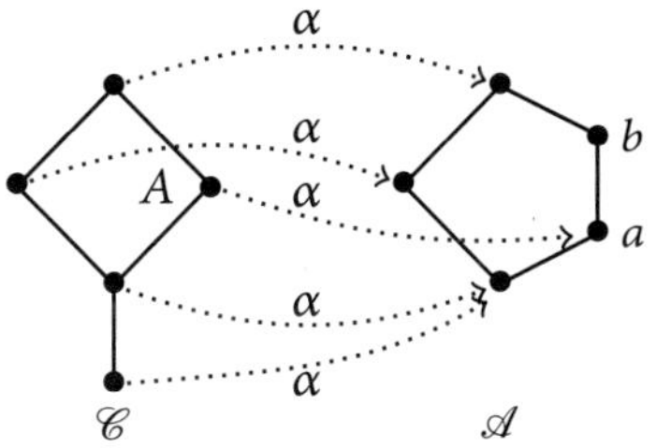

The abstract property b in $\mathscr{A}$ is useless because any concrete property P in $\mathscr{C}$ overapproximated by b in $\mathscr{A}$ (i.e., $\alpha(P) \preccurlyeq b$) can be overapproximated more precisely by a (because $\alpha(P) \preccurlyeq b$ implies $\alpha(P) \preccurlyeq a$) and a and b have the same concretization A. Hence the abstract domain $\mathscr{A}$ can be reduced by eliminating b; see section 36.3.3.

Exercise 11.50 (Galois retraction [762, Theorem 3]) Show that if $\langle \mathscr{C}, \sqsubseteq\rangle \underset{\alpha}{\overset{\gamma}{\leftrightarrows}} \langle \mathscr{A}, \preccurlyeq\rangle$ then α is surjective if and only if γ is injective if and only if $\forall \overline{P} \in \mathscr{A} \,.\, \alpha \circ \gamma(\overline{P}) = \overline{P}$ if and only if $\gamma(Q) = \max\{P \in \mathscr{C} \mid \alpha(P) = Q\}$ for all $Q \in A$. Express the dual $\langle \mathscr{C}, \sqsubseteq\rangle \underset{\alpha}{\overset{\gamma}{\stackrel{\twoheadleftarrow}{\longrightarrow}}} \langle \mathscr{A}, \preccurlyeq\rangle$. Why is any isomorphism between $\mathscr{C}$ and $\mathscr{A}$ not a Galois isomorphism (denoted $\langle \mathscr{C}, \sqsubseteq\rangle \underset{\alpha}{\overset{\gamma}{\stackrel{\twoheadleftarrow}{\twoheadrightarrow}}} \langle \mathscr{A}, \preccurlyeq\rangle$)?

3. We use this terminology by reference to a retraction in topology, which is a continuous mapping from a topological space into a subspace that preserves the position of all points in that subspace. In particular, given any idempotent continuous map, a retraction onto the image of the map is obtained by restricting the codomain; see exercise 11.51

A similar terminology is used in category theory. If $\alpha \in \mathscr{C} \to \mathscr{A}$, $\gamma \in \mathscr{A} \to \mathscr{C}$, and $\gamma \circ \alpha \in \mathscr{C} \to \mathscr{C}$ then α is called a section of γ and γ is a retraction of α [614, p. 19].

Exercise 11.51 Let ρ be an upper closure on $\langle \mathscr{C}, \sqsubseteq\rangle$ (that is, increasing, extensive, and idempotent; see next section 11.7). Prove that $\langle \mathscr{C}, \sqsubseteq\rangle \xrightleftharpoons[\rho]{\mathbb{1}_{\rho(\mathscr{C})}} \langle \rho(\mathscr{C}), \sqsubseteq\rangle$ where $\rho(\mathscr{C})$ is the image of $\mathscr{C}$ by ρ (called a *retract* of $\mathscr{C}$).

Exercise 11.52 Define α and γ such that $\langle \wp(\mathscr{D} \times \mathscr{C}), \subseteq\rangle \xrightleftharpoons[\alpha]{\gamma} \langle \mathscr{D} \rightarrow \wp(\mathscr{C}), \dot{\subseteq}\rangle$.

Exercise 11.53 Let $\langle L, \sqsubseteq, \bot, \top, \sqcup, \sqcap\rangle$ be a complete lattice, let $\langle L \xrightarrow{\nearrow} L, \dot{\sqsubseteq}, \dot{\bot}, \dot{\top}, \dot{\sqcup}, \dot{\sqcap}\rangle$ be the complete lattice of increasing maps on L ordered pointwise. Define $\alpha(f) \triangleq \{x \in L \mid f(x) \sqsubseteq x\}$. Prove that $\langle L \xrightarrow{\nearrow} L, \dot{\sqsubseteq}\rangle \xrightleftharpoons[\alpha]{\gamma} \langle \wp(L), \supseteq\rangle$.

Exercise 11.54 A subset $\mathscr{A}$ of $\mathscr{X}$ is said to be a *retract* of $\mathscr{X}$ if there is a so-called *retraction* $f \in \mathscr{X} \rightarrow \mathscr{A}$ such that $\forall a \in \mathscr{A} \,.\, f(a) = a$. Show that in a Galois retraction $\langle \mathscr{C}, \sqsubseteq\rangle \xrightleftharpoons[\alpha]{\gamma} \langle \mathscr{A}, \preccurlyeq\rangle$, $\mathscr{A}$ is a retract of $\alpha(\mathscr{C})$.

Exercise 11.55 Let $d \in \mathscr{C} \times \mathscr{C} \rightarrow \mathbb{R}_*$ be a distance on a set $\mathscr{C}$, meaning that for all $x, y, z \in \mathscr{C}$, we have $d(x, y) = 0 \Leftrightarrow x = y$, $d(x, y) \leqslant d(x, z) + d(z, y)$, and $d(x, y) = d(y, x)$. Assume that $\langle \mathscr{C}, \sqsubseteq\rangle \xrightleftharpoons[\alpha]{\gamma} \langle \mathscr{A}, \preccurlyeq\rangle$ is a Galois retraction. Prove that $D(\overline{x}, \overline{y}) \triangleq d(\gamma(\overline{x}), \gamma(\overline{y}))$ is a distance on $\mathscr{A}$.

11.7 Closure Operators

An *upper closure operator* $\rho \in L \rightarrow L$ on a poset $\langle L, \sqsubseteq\rangle$ is increasing, *idempotent* ($\rho \circ \rho = \rho$), and *extensive* $\forall x \in L \,.\, x \sqsubseteq \rho(x)$. The dual *lower closure operator* is increasing, idempotent, and *reductive* $\forall x \in L \,.\, \rho(x) \sqsubseteq x$. The *closed elements* $x \in L$ of a closure operator $\rho \in L \rightarrow L$ are such that $\rho(x) = x$. The *closed subsets* of a closure operator $\rho \in L \rightarrow L$ are such that $\rho(S) \triangleq \{\rho(x) \mid x \in S\} = S$. For example,

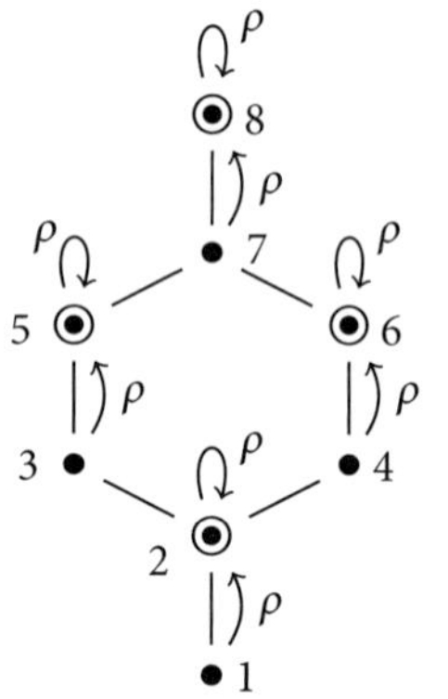

Figure 11.56
Upper closure operator

Exercise 11.57 Prove that in a Galois connection $\langle \mathscr{C}, \sqsubseteq\rangle \xrightleftharpoons[\alpha]{\gamma} \langle \mathscr{A}, \preccurlyeq\rangle$, $\gamma \circ \alpha$ is an upper closure operator on $\mathscr{C}$ and $\alpha \circ \gamma$ is a lower closure operator on $\mathscr{A}$.

Exercise 11.58 Continuing exercise 11.57, let the ceiling $\lceil x \rceil$ of $x \in \mathbb{R}$ be the least integer greater than or equal to x. Show that $\langle \mathbb{R}, \leqslant \rangle \underset{\alpha}{\overset{\gamma}{\substack{\longleftarrow \\ \twoheadrightarrow}}} \langle \mathbb{Z}, \leqslant \rangle$ where $\alpha(x) \triangleq \lceil x \rceil$ and $\gamma(y) = y$. Prove that $\lceil x \rceil$ is an upper closure on $\mathbb{R}$.

Exercise 11.59 Let $\langle \mathscr{C}, \sqsubseteq \rangle$ and $\langle \mathscr{A}, \preccurlyeq \rangle$ be posets, $\alpha \in \mathscr{C} \to \mathscr{A}$ and $\gamma \in \mathscr{A} \to \mathscr{C}$ be such that $\gamma \circ \alpha$ is an upper closure operator on $\mathscr{C}$ and $\alpha \circ \gamma$ is a lower closure operator on $\mathscr{A}$. Is $\langle \alpha, \gamma \rangle$ a Galois connection?

Exercise 11.60 Prove that if $\langle \mathscr{C}, \sqsubseteq \rangle \underset{\alpha}{\overset{\gamma}{\leftrightarrows}} \langle \mathscr{A}, \preccurlyeq \rangle$ is a Galois connection, then $\langle \mathscr{C}, \sqsubseteq \rangle \underset{\gamma \circ \alpha}{\overset{\mathbb{1}}{\leftrightarrows}} \langle \gamma \circ \alpha(\mathscr{C}), \sqsubseteq \rangle$ is a Galois connection where $\gamma \circ \alpha(\mathscr{C}) = \{\gamma(\alpha(x)) \mid x \in \mathscr{C}\}$ is the image of $\mathscr{C}$ by the upper closure $\gamma \circ \alpha$. What is the dual?

Exercise 11.61 (upper closure operator) Prove that ρ is an upper closure operator on $\langle L, \sqsubseteq \rangle$ if and only if $\langle L, \sqsubseteq \rangle \underset{\rho}{\overset{\mathbb{1}_L}{\substack{\longleftarrow \\ \twoheadrightarrow}}} \langle \rho(L), \sqsubseteq \rangle$ where $\mathbb{1}_L$ is the identity function on L and $\rho(L) = \{\rho(x) \mid x \in L\}$ is the right image of L by ρ. What is the dual?

Exercise 11.62 ([719]) Using exercise 11.61, prove that ρ is an upper closure operator on $\langle L, \sqsubseteq \rangle$ if and only if $x \sqsubseteq \rho(y) \Leftrightarrow \rho(x) \sqsubseteq \rho(y)$.

Exercise 11.63 Prove that $\langle \mathscr{C}, \sqsubseteq \rangle \underset{\alpha}{\overset{\gamma}{\leftrightarrows}} \langle \mathscr{A}, \preccurlyeq \rangle$ if and only if $\{\gamma(y) \mid y \in \mathscr{A}\}$ is the image of $\mathscr{C}$ by an upper closure operator. What is the dual?

Exercise 11.64 ([964, Theorem 4.1]) Prove that the image of a complete lattice $\langle \mathscr{L}, \sqsubseteq \rangle$ by a closure operator ρ is a complete lattice. Extend this result to Galois retractions.

Let $\langle \mathscr{P}, \sqsubseteq \rangle$ and $\langle \mathcal{Q}, \preceq \rangle$ be posets. An *order embedding* of $\langle \mathscr{P}, \sqsubseteq \rangle$ into $\langle \mathcal{Q}, \preceq \rangle$ is map $f \in \mathscr{P} \to \mathcal{Q}$ such that $\forall x, y \in \mathscr{P} \,.\, x \sqsubseteq y \Leftrightarrow f(x) \preceq f(y)$.

The Dedekind–MacNeille completion order embed any poset $\langle \mathscr{P}, \sqsubseteq \rangle$ into a complete lattice $\langle \mathscr{L}, \preceq \rangle$ so that distinct elements of $\mathscr{P}$ are be mapped to distinct elements of $\mathscr{L}$, and each pair of elements in $\langle \mathscr{P}, \sqsubseteq \rangle$ has the same ordering in $\langle \mathscr{L}, \preceq \rangle$ as they do in $\langle \mathscr{P}, \sqsubseteq \rangle$. This construction is sometimes used to construct abstract domains, for example, [530].

Exercise 11.65 (Dedekind–MacNeille completion) Let $\langle \mathscr{P}, \sqsubseteq \rangle$ be a poset. Define $\downarrow(X) \triangleq \{y \in \mathscr{P} \mid \forall x \in X \,.\, y \sqsubseteq x\}$ and $\uparrow(X) \triangleq \{x \in \mathscr{P} \mid \forall y \in X \,.\, y \sqsubseteq x\}$. Prove that $\langle \wp(\mathscr{P}), \subseteq \rangle \underset{\downarrow}{\overset{\uparrow}{\leftrightarrows}} \langle \wp(\mathscr{P}), \supseteq \rangle$. Prove that $\uparrow \circ \downarrow$ is an upper closure operator on $\wp(\mathscr{P})$. Its fixpoints in $\uparrow \circ \downarrow(\wp(\mathscr{P}))$ are called *stable* elements. Prove that $\langle \uparrow \circ \downarrow(\wp(\mathscr{P})), \subseteq \rangle$ is a complete lattice (called the *Dedekind–MacNeille completion* of $\langle \mathscr{P}, \sqsubseteq \rangle$). Prove that the map $\dot{\uparrow}(x) \triangleq \uparrow(\{x\})$ is an embedding of $\langle \mathscr{P}, \sqsubseteq \rangle$ into $\langle \uparrow \circ \downarrow(\wp(\mathscr{P})), \supseteq \rangle$. Prove that distinct elements of $\mathscr{P}$ are be mapped to distinct elements of $\uparrow \circ \downarrow(\wp(\mathscr{P}))$. Prove that each pair of elements in $\langle \mathscr{P}, \sqsubseteq \rangle$ has the same ordering in $\langle \uparrow \circ \downarrow(\wp(\mathscr{P})), \supseteq \rangle$ as they do in $\langle \mathscr{P}, \sqsubseteq \rangle$. Prove that $\langle \mathscr{P}, \sqsubseteq \rangle \underset{\dot{\uparrow}}{\overset{\dot{\uparrow}^{-1}}{\substack{\twoheadleftarrow \\ \twoheadrightarrow}}} \langle \uparrow \circ \downarrow(\wp(\mathscr{P})), \supseteq \rangle$.

The ideas of exercise 11.65 also apply to the completion of posets into chain-complete posets (where increasing chains have lubs) [656, 657].

11.8 Composition of Galois Connections

Exercise 11.66 (composition of Galois connections) Prove that the composition of Galois connections $\langle \mathscr{P}_1, \sqsubseteq \rangle \overset{\gamma_1}{\underset{\alpha_1}{\leftrightarrows}} \langle \mathscr{P}_2, \preccurlyeq \rangle$ and $\langle \mathscr{P}_2, \preccurlyeq \rangle \overset{\gamma^2}{\underset{\alpha^2}{\leftrightarrows}} \langle \mathscr{P}_3, \trianglelefteq \rangle$ is the Galois connection $\langle \mathscr{P}_1, \sqsubseteq \rangle \overset{\gamma_1 \circ \gamma^2}{\underset{\alpha^2 \circ \alpha_1}{\leftrightarrows}} \langle \mathscr{P}_3, \trianglelefteq \rangle$.

11.9 Galois Connection in a Complete Boolean Algebra

11.9.1 Complete Boolean Algebra

Let $\langle L, \sqsubseteq, \bot, \sqcap, \neg \rangle$ be a meet semilattice with infimum $\bot$ such that every element $y \in L$ has a *pseudocomplement* $\neg y$ such that, by definition, $x \sqcap y = \bot$ if and only if $x \sqsubseteq \neg y$, that is, for any $x \in L$ the set of elements y disjoint from x has a largest element $\neg y$.

It follows that $y \sqcap (\neg y) = \bot$.

Assume that $\neg\!\!\!\neg\, y$ is another pseudocomplement of y. $(\neg\!\!\!\neg\, y) \sqcap y = \bot$ implies $(\neg\!\!\!\neg\, y) \sqsubseteq \neg y$ and $(\neg y) \sqcap y = \bot$ implies $(\neg y) \sqsubseteq \neg\!\!\!\neg\, y$ proving $\neg\!\!\!\neg\, y = \neg y$ by antisymmetry so that the pseudocomplement is unique.

Moreover $x \sqcap \bot = \bot$ so that $x \sqsubseteq \neg\bot$ proving that $\neg\bot = \top$ is the supremum of L.

A *complete Boolean algebra* is a complete lattice such that any element x has a unique *complement* such that $x \sqcap \neg x = \bot$ and $x \sqcup \neg x = \top$. For example, the powerset is a complete Boolean algebra $\langle \wp(S), \subseteq, \varnothing, \bigcap, \neg \rangle$ where $\neg X \triangleq S \setminus X$.

We have $x \sqsubseteq y$ implies $x \sqcap \neg y \sqsubseteq y \sqcap \neg y = \bot$, which implies $x \sqcap \neg y = \bot$ so $\neg y \sqsubseteq \neg x$.

Because $y \sqcap (\neg y) = \bot$, $y \sqsubseteq (\neg\neg y)$.

By exercise 11.35, it follows that $\langle L, \sqsubseteq \rangle \overset{\neg}{\underset{\neg}{\leftrightarrows}} \langle L, \sqsupseteq \rangle$.

11.9.2 Conjugate Galois Connection

Let $\langle L, \sqsubseteq, \bot, \top, \sqcap, \sqcup, \neg \rangle$ be a complete Boolean algebra and $\langle L, \sqsubseteq \rangle \overset{\gamma}{\underset{\alpha}{\leftrightarrows}} \langle L, \sqsubseteq \rangle$. Then, by the composition of Galois connections defined in section 11.8, we have the *conjugate Galois connection* $\langle L, \sqsubseteq \rangle \overset{\neg \circ \gamma \circ \neg}{\underset{\neg \circ \alpha \circ \neg}{\leftrightarrows}} \langle L, \sqsubseteq \rangle$.

Similarly, $\langle L, \sqsubseteq \rangle \overset{\gamma \circ \neg}{\underset{\neg \circ \alpha}{\leftrightarrows}} \langle L, \sqsupseteq \rangle$, and by the duality lemma 11.25, $\langle L, \sqsubseteq \rangle \overset{\alpha \circ \neg}{\underset{\gamma \circ \neg}{\leftrightarrows}} \langle L, \sqsupseteq \rangle$.

We write $\tilde{\alpha}$ and $\tilde{\gamma}$ for the conjugates $\tilde{\alpha} \triangleq \neg \circ \alpha \circ \neg$ and $\tilde{\gamma} \triangleq \neg \circ \gamma \circ \neg$.

11.9.3 Characterization of Galois Connections in a Complete Boolean Algebra

Lemma 11.67 Let $\langle S \to \mathbb{B}, \dot{\Rightarrow}, \dot{\mathrm{ff}}, \dot{\mathrm{tt}}, \dot{\wedge}, \dot{\vee}, \dot{\neg} \rangle$ be a pointwise extension of the complete Boolean algebra $\langle \mathbb{B}, \Rightarrow, \mathrm{ff}, \mathrm{tt}, \wedge, \vee, \neg \rangle$. Then, $\langle S \to \mathbb{B}, \dot{\Rightarrow} \rangle \overset{\gamma}{\underset{\alpha}{\leftrightarrows}} \langle S \to \mathbb{B}, \dot{\Rightarrow} \rangle$ if and only if $\forall P, Q \in S \to \mathbb{B} \,.\, \tilde{\alpha}(P) \,\dot{\vee}\, Q = P \,\dot{\vee}\, \gamma(Q)$.

Proof of lemma 11.67

$$\langle S \to \mathbb{B}, \dot{\Rightarrow} \rangle \overset{\gamma}{\underset{\alpha}{\leftrightarrows}} \langle S \to \mathbb{B}, \dot{\Rightarrow} \rangle$$

$$\Leftrightarrow \forall P, Q \in S \to \mathbb{B} \,.\, (\alpha(P) \dot{\Rightarrow} Q) \Leftrightarrow (P \dot{\Rightarrow} \gamma(Q)) \qquad \text{(definition of a Galois connection)}$$

$$\Leftrightarrow \forall P, Q \in S \to \mathbb{B} \,.\, (\alpha(P) \dot{\Rightarrow} Q) = (P \dot{\Rightarrow} \gamma(Q)) \qquad \text{(}\Leftrightarrow \text{ is Boolean equality in } \mathbb{B}\text{)}$$

$$\Leftrightarrow\ \forall P, Q \in S \to \mathbb{B}\,.\,(\dot{\neg}\alpha(P)\,\dot{\vee}\,Q) = (\dot{\neg}P\,\dot{\vee}\,\gamma(Q)) \qquad \text{(because } P \Rightarrow Q = \neg P \vee Q \text{ in } \mathbb{B}\text{)}$$
$$\Leftrightarrow\ \forall P', Q \in S \to \mathbb{B}\,.\,(\dot{\neg}\alpha(\dot{\neg}P')\,\dot{\vee}\,Q) = (\dot{\neg}\dot{\neg}P'\,\dot{\vee}\,\gamma(Q)) \qquad \text{(letting } P' \triangleq \dot{\neg}P\text{)}$$
$$\Leftrightarrow\ \forall P', Q \in S \to \mathbb{B}\,.\,(\widetilde{\alpha}(P')\,\dot{\vee}\,Q) = (P'\,\dot{\vee}\,\gamma(Q)) \qquad \text{(definition of } \widetilde{\alpha}\text{)} \quad \square$$

11.10 Complete Heyting Algebra

In propositional calculus, we have $(a \wedge x) \Rightarrow y$ if and only if $x \Rightarrow (a \Rightarrow y)$, that is, $(\alpha_a(x) \Rightarrow y) \Leftrightarrow (x \Rightarrow \gamma_a(y))$ where $\alpha_a(x) \triangleq a \wedge x$ and $\gamma_a(y) \triangleq a \Rightarrow y$. Considering $\Rightarrow$ has a partial order, this extends to posets.

A *complete Heyting algebra* is a complete lattice $\langle L, \sqsubseteq, \sqcap\rangle$ such that for all elements $a \in L$, $\alpha_a(y) \triangleq a \sqcap y$ is the lower adjoint of a Galois connection $\langle L, \sqsubseteq\rangle \underset{\alpha_a}{\overset{\gamma_a}{\leftrightarrows}} \langle L, \sqsubseteq\rangle$. Arend Heyting used such algebras to formalize intuitionist logic [495, CH VII].

Exercise 11.68 Show that the definition of a complete Heyting algebra $\langle L, \sqsubseteq, \sqcap, \sqcup\rangle$ is equivalent to $\forall x \in L\,.\,\forall S \in \wp(L)\,.\,x \sqcap \bigsqcup_{s \in S} s = \bigsqcup_{s \in S}(x \sqcap s)$.

11.11 Sound Abstraction

Assume that $\langle \mathscr{C}, \sqsubseteq\rangle \underset{\alpha}{\overset{\gamma}{\leftrightarrows}} \langle \mathscr{A}, \preccurlyeq\rangle$ is a Galois connection between concrete properties in $\mathscr{C}$ with implication $\sqsubseteq$ and abstract properties in $\mathscr{A}$ with abstract implication $\preccurlyeq$.

> **Definition 11.69 (sound abstraction)** We say that $\overline{P} \in \mathscr{A}$ is a *sound abstraction* of $P \in \mathscr{C}$ if and only if $P \sqsubseteq \gamma(\overline{P})$.

The intuition is that the meaning $\gamma(\overline{P})$ of the abstract property $\overline{P}$ is weaker than the concrete property P hence is an overapproximation. The dual is underapproximation.

Example 11.70 Consider the lattice of abstract properties

$\langle \mathbb{P}^{\pm}, \sqsubseteq^{\pm}\rangle =$ 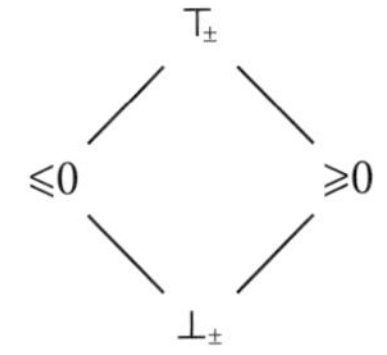

$$\begin{aligned} \gamma_{\pm}(\bot_{\pm}) &\triangleq \varnothing \\ \gamma_{\pm}(\leqslant 0) &\triangleq \{z \mid z \leqslant 0\} \\ \gamma_{\pm}(\geqslant 0) &\triangleq \{z \mid z \geqslant 0\} \\ \gamma_{\pm}(\top_{\pm}) &\triangleq \mathbb{Z} \end{aligned}$$

The sound abstractions of $\{1, 42\}$ are $\geqslant 0$ and $\top_{\pm}$ because $\{1, 42\} \subseteq \{z \mid z \geqslant 0\} = \gamma_{\pm}(\geqslant 0)$ and $\{1, 42\} \subseteq \mathbb{Z} = \gamma_{\pm}(\top_{\pm})$.

11.12 Best Abstraction

> **Definition 11.71 (better abstraction)** Let $\overline{P}_1, \overline{P}_2 \in \mathscr{A}$ be sound abstractions of the concrete property $P \in \mathscr{C}$. We say that $\overline{P}_1$ is better/more precise/stronger/less abstract than $\overline{P}_2$ if and only if $\overline{P}_1 \preccurlyeq \overline{P}_2$.

Theorem 11.72 (best abstraction) If $\langle \mathscr{C}, \sqsubseteq\rangle \xleftrightarrow[\alpha]{\gamma} \langle \mathscr{A}, \preccurlyeq\rangle$ then $\alpha(P)$ is the best/most precise/strongest/least abstract property which is a sound abstraction of the concrete property P.

Proof By definition 11.69, $\alpha(P)$ is a sound abstraction of P because $P \sqsubseteq \gamma(\alpha(P))$.

Let $\overline{P}$ be any sound abstraction of P so that, by definition 11.69, $P \sqsubseteq \gamma(\overline{P})$. Then by definition of a Galois connection, $\alpha(P) \preccurlyeq \overline{P}$ proving that $\alpha(P)$ is the least sound abstraction of P, that is, $\alpha(P) = \bigsqcap\{\overline{P} \mid P \sqsubseteq \gamma(\overline{P})\}$, as shown in lemma 11.42. □

Exercise 11.73 Prove a converse of theorem 11.72.

Example 11.74 Continuing example 11.70, the best abstraction of $\{1, 42\}$ is $\geqslant 0$ because $\{1, 42\} \subseteq \gamma_\pm(\geqslant 0)$, $\{1, 42\} \subseteq \gamma_\pm(\top_\pm)$, and $\geqslant 0 \sqsubseteq^\pm \top_\pm$.

The sound abstractions of $\{0\}$ are $\leqslant 0$, $\geqslant 0$, and $\top_\pm$ because $\{0\} \subseteq \{z \mid z \leqslant 0\} = \gamma_\pm(\leqslant 0)$, $\{0\} \subseteq \{z \mid z \geqslant 0\} = \gamma_\pm(\geqslant 0)$, and $\{0\} \subseteq \mathbb{Z} = \gamma_\pm(\top_\pm)$. However, there is no best approximation because $\leqslant 0$ and $\geqslant 0$ are not comparable by the partial order $\sqsubseteq^\pm$.

By the contraposition of theorem 11.72, it follows that there is no Galois connection between $\langle \wp(\mathbb{Z}), \subseteq\rangle$ and $\langle \mathbb{P}^\pm, \sqsubseteq^\pm\rangle$.

11.13 Combinations of Galois Connections

11.13.1 Galois Connections Pairing

The following theorem shows that Galois connections on domains can be extended to a Galois connection between their Cartesian product.

Theorem 11.75 If $\langle \mathscr{C}_1, \sqsubseteq_1\rangle \xleftrightarrow[\alpha_1]{\gamma_1} \langle \mathscr{A}_1, \preccurlyeq_1\rangle$ and $\langle \mathscr{C}_2, \sqsubseteq_2\rangle \xleftrightarrow[\alpha_2]{\gamma_2} \langle \mathscr{A}_2, \preccurlyeq_2\rangle$ then $\langle \mathscr{C}_1 \times \mathscr{C}_2, \dot{\sqsubseteq}\rangle \xleftrightarrow[\alpha]{\gamma} \langle \mathscr{A}_1 \times \mathscr{A}_2, \dot{\preccurlyeq}\rangle$ where $\dot{\sqsubseteq}$ and $\dot{\preccurlyeq}$ are componentwise, and $\alpha(\langle x, y\rangle) = \langle \alpha_1(x), \alpha_2(y)\rangle$, $\gamma(\langle \overline{x}, \overline{y}\rangle) = \langle \gamma_1(\overline{x}), \gamma_2(\overline{y})\rangle$.

Proof of theorem 11.75

$$
\begin{aligned}
& \alpha(\langle x, y\rangle) \dot{\preccurlyeq} \langle \overline{x}, \overline{y}\rangle \\
\Leftrightarrow\ & \langle \alpha_1(x), \alpha_2(y)\rangle \dot{\preccurlyeq} \langle \overline{x}, \overline{y}\rangle && \text{(definition of } \alpha\text{)} \\
\Leftrightarrow\ & \alpha_1(x) \preccurlyeq_1 \overline{x} \wedge \alpha_2(y) \preccurlyeq_2 \overline{y} && \text{(componentwise definition of } \dot{\preccurlyeq}\text{)} \\
\Leftrightarrow\ & x \sqsubseteq_1 \gamma_1(\overline{x}) \wedge y \sqsubseteq_2 \gamma_2(\overline{y}) && \text{(Galois connections)} \\
\Leftrightarrow\ & \langle x, y\rangle \dot{\sqsubseteq} \langle \gamma_1(\overline{x}), \gamma_2(\overline{y})\rangle && \text{(componentwise definition of } \dot{\sqsubseteq}\text{)} \\
\Leftrightarrow\ & \langle x, y\rangle \dot{\sqsubseteq} \gamma(\langle \overline{x}, \overline{y}\rangle) && \text{(definition of } \gamma\text{)}
\end{aligned}
$$
□

Exercise 11.76 Let $\Vdash_1 \in \langle \mathscr{C}_1, \sqsubseteq_1\rangle \otimes \langle \mathscr{A}_1, \preccurlyeq_1\rangle$ and $\Vdash_2 \in \langle \mathscr{C}_2, \sqsubseteq_2\rangle \otimes \langle \mathscr{A}_2, \preccurlyeq_2\rangle$. Define $\langle x, y\rangle \Vdash \langle \overline{x}, \overline{y}\rangle$ if and only if $x \Vdash_1 \overline{x} \wedge y \Vdash_1 \overline{y}$. Prove that $\Vdash$ is a logical relation.

11.13.2 Higher-order Galois Connections

The following theorem shows that Galois connections on the domain and codomain of increasing functions can be extended to a Galois connection between these functions, as shown in figure 11.77.

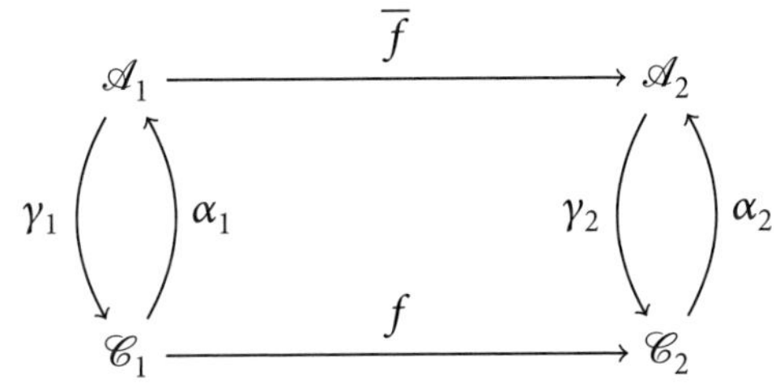

Figure 11.77
Higher-order Galois connection

Theorem 11.78 If $\langle \mathscr{C}_1, \sqsubseteq_1\rangle \xleftrightarrows[\alpha_1]{\gamma_1} \langle \mathscr{A}_1, \preccurlyeq_1\rangle$ and $\langle \mathscr{C}_2, \sqsubseteq_2\rangle \xleftrightarrows[\alpha_2]{\gamma_2} \langle \mathscr{A}_2, \preccurlyeq_2\rangle$ then $\langle \mathscr{C}_1 \xrightarrow{\nearrow} \mathscr{C}_2, \dot{\sqsubseteq}_2\rangle \xleftrightarrows[\vec{\alpha}]{\vec{\gamma}} \langle \mathscr{A}_1 \xrightarrow{\nearrow} \mathscr{A}_2, \dot{\preccurlyeq}_2\rangle$ where $\vec{\alpha} = f \mapsto \alpha_2 \circ f \circ \gamma_1$ and $\vec{\gamma} = \overline{f} \mapsto \gamma_2 \circ \overline{f} \circ \alpha_1$.

Proof of theorem 11.78

$\vec{\alpha}(f) \dot{\preccurlyeq}_2 \overline{f}$

$\Leftrightarrow \forall y \in \mathscr{A}_1 . \vec{\alpha}(f)y \preccurlyeq_2 \overline{f}(y)$ ⟨pointwise definition of $\dot{\preccurlyeq}_2$⟩

$\Leftrightarrow \forall y \in \mathscr{A}_1 . \alpha_2(f(\gamma_1(y))) \preccurlyeq_2 \overline{f}(y)$ ⟨definition of $\vec{\alpha}$ and function composition $\circ$⟩

$\Leftrightarrow \forall y \in \mathscr{A}_1 . f(\gamma_1(y))) \sqsubseteq_2 \gamma_2(\overline{f}(y))$

⟨$(\Rightarrow)$ for $y = \alpha_1(x)$, $f(x) \sqsubseteq_2 f(\gamma_1(\alpha_1(x))) \sqsubseteq_2 \gamma_2(\overline{f}(\alpha_1(x))$ because f is increasing and $x \sqsubseteq_1 \gamma_1(\alpha_1(x))$ by $\langle \mathscr{C}_1, \sqsubseteq_1\rangle \xleftrightarrows[\alpha_1]{\gamma_1} \langle \mathscr{A}_1, \preccurlyeq_1\rangle$;

$(\Leftarrow)$ for $x = \gamma_1(y)$, $f(\gamma_1(y)) \sqsubseteq_2 \gamma_2(\overline{f}(\alpha_1(\gamma_1(y)))) \sqsubseteq_2 \gamma_2(\overline{f}(y))$ because $\alpha_1(\gamma_1(y)) \dot{\preccurlyeq}_1 y$ by $\langle \mathscr{C}_1, \sqsubseteq_1\rangle \xleftrightarrows[\alpha_1]{\gamma_1} \langle \mathscr{A}_1, \preccurlyeq_1\rangle$, $\overline{f}$ is increasing, and γ_2 is increasing by $\langle \mathscr{C}_2, \sqsubseteq_2\rangle \xleftrightarrows[\alpha_2]{\gamma_2} \langle \mathscr{A}, \preccurlyeq_2\rangle$.⟩

$\Leftrightarrow \forall x \in \mathscr{C}_1 . f(x) \sqsubseteq_2 \gamma_2(\overline{f}(\alpha_1(x)))$

$\Leftrightarrow \forall x \in \mathscr{C}_2 . f(x) \sqsubseteq_2 \vec{\gamma}(\overline{f})x$ ⟨definition of $\vec{\gamma}$⟩

$\Leftrightarrow f \dot{\sqsubseteq}_2 \vec{\gamma}(\overline{f})$ ⟨pointwise definition of $\dot{\sqsubseteq}_2$⟩ □

Exercise 11.79 Let $\Vdash_1 \in \langle \mathscr{C}_1, \sqsubseteq_1\rangle \otimes \langle \mathscr{A}_1, \preccurlyeq_1\rangle$ and $\Vdash_2 \in \langle \mathscr{C}_2, \sqsubseteq_2\rangle \otimes \langle \mathscr{A}_2, \preccurlyeq_2\rangle$. Define $f \Vdash \overline{f}$ if and only if $\forall x \in \mathscr{C}_1, \overline{x} \in \mathscr{A}_1 . x \Vdash_1 \overline{x} \Rightarrow f(x) \Vdash_2 \overline{f}(\overline{x})$. Prove that $\Vdash$ is a logical relation.

11.14 Galois Connection, Logical Relation, Tensor Product, Soundness Relation

We formalize best abstractions by Galois connections. Soundness relation are mathematically equivalent formalizations useful to prove soundness (what is true in the abstract holds in the concrete), but, by experience, not as convenient to prove completeness (i.e., what is true in the abstract is provable in the abstract; see chapter 52, "Semantic Soundness, Completeness, and Definedness").

Definition 11.80 (logical relation) A relation $\Vdash \in \wp(\mathscr{C} \times \mathscr{A})$ between complete lattices $\langle \mathscr{C}, \sqsubseteq, \bigsqcup\rangle$ and $\langle \mathscr{A}, \preccurlyeq, \curlywedge\rangle$ is a *logical relation* if and only if

(1) $(P \sqsubseteq P' \wedge P' \Vdash \overline{P}' \wedge \overline{P}' \preccurlyeq \overline{P}) \Rightarrow (P \Vdash \overline{P})$;

(2) $(\forall i \in \Delta \,.\, P_i \Vdash \overline{P}) \Rightarrow \bigsqcup\limits_{i \in \Delta} P_i \Vdash \overline{P}$;

(3) $(\forall i \in \Delta \,.\, P \Vdash \overline{P}_i) \Rightarrow P \Vdash \underset{i \in \Delta}{\curlywedge} \overline{P}_i$.

Definition 11.81 (tensor product) [887] The tensor product $\langle \mathscr{C}, \sqsubseteq\rangle \otimes \langle \mathscr{A}, \preccurlyeq\rangle$ of two complete lattices $\langle \mathscr{C}, \sqsubseteq\rangle$ and $\langle \mathscr{A}, \preccurlyeq\rangle$ is

$$\langle \mathscr{C}, \sqsubseteq\rangle \otimes \langle \mathscr{A}, \preccurlyeq\rangle \triangleq \{\Vdash \in \wp(\mathscr{C} \times \mathscr{A}) \mid \Vdash \text{ is a logical relation}\}$$

Definition 11.82 (soundness relation) Assume that $\langle \mathscr{C}, \sqsubseteq\rangle \underset{\alpha}{\overset{\gamma}{\leftrightarrows}} \langle \mathscr{A}, \preccurlyeq\rangle$ then the relation $P \Vdash \overline{P} \triangleq (P \sqsubseteq \gamma(\overline{P}))$, or equivalently, $P \Vdash \overline{P} \triangleq (\alpha(P) \preccurlyeq \overline{P})$ is called the *soundness relation*.

The following theorem shows that the use of Galois connections or logical relations is mathematically equivalent in abstract interpretation.

Theorem 11.83 Let $\langle \mathscr{C}, \sqsubseteq\rangle$ and $\langle \mathscr{A}, \preccurlyeq\rangle$ be complete lattices. $\langle \mathscr{C}, \sqsubseteq\rangle \underset{\alpha}{\overset{\gamma}{\leftrightarrows}} \langle \mathscr{A}, \preccurlyeq\rangle$ if and only if $\Vdash \in \langle \mathscr{C}, \sqsubseteq\rangle \otimes \langle \mathscr{A}, \preccurlyeq\rangle$.

Proof of theorem 11.83 Assume $\langle \mathscr{C}, \sqsubseteq\rangle \underset{\alpha}{\overset{\gamma}{\leftrightarrows}} \langle \mathscr{A}, \preccurlyeq\rangle$.

– $P \sqsubseteq P' \wedge P' \Vdash \overline{P}' \wedge \overline{P}' \preccurlyeq \overline{P}$

$\Rightarrow$ $P \sqsubseteq P' \wedge P' \sqsubseteq \gamma(\overline{P}') \wedge \overline{P}' \preccurlyeq \overline{P}$ $\quad\wr$definition of $\Vdash\wr$

$\Rightarrow$ $P \sqsubseteq \gamma(\overline{P}') \wedge \overline{P}' \preccurlyeq \overline{P}$ $\quad\wr$transitivity$\wr$

$\Rightarrow$ $\alpha(P) \preccurlyeq \overline{P}' \wedge \overline{P}' \preccurlyeq \overline{P}$ $\quad\wr$Galois connection$\wr$

$\Rightarrow$ $\alpha(P) \preccurlyeq \overline{P}$ $\quad\wr$transitivity$\wr$

$\Rightarrow$ $P \Vdash \overline{P}$ $\quad\wr$transitivity$\wr$

– $\forall i \in \Delta \,.\, P_i \Vdash \overline{P} \wedge \bigsqcup\limits_{i \in \Delta} P_i$ exists

$\Rightarrow$ $\forall i \in \Delta \,.\, P_i \sqsubseteq \gamma(\overline{P}) \wedge \bigsqcup\limits_{i \in \Delta} P_i$ exists $\quad\wr$definition of $\Vdash\wr$

$\Rightarrow \bigsqcup_{i \in \Delta} P_i \sqsubseteq \gamma(\overline{P})$ ⦃definition of the lub⦄

$\Rightarrow \alpha(\bigsqcup_{i \in \Delta} P_i) \preccurlyeq \overline{P}$ ⦃Galois connection⦄

$\Rightarrow (\bigsqcup_{i \in \Delta} P_i) \Vdash \overline{P}$ ⦃definition of $\Vdash$⦄

– $(\forall i \in \Delta \,.\, P \Vdash \overline{P}_i \wedge \curlywedge_{i \in \Delta} \overline{P}_i \text{ exists}) \Rightarrow P \Vdash \curlywedge_{i \in \Delta} \overline{P}_i$ ⦃by duality⦄

Conversely, assume that $\Vdash \in \langle \mathscr{C}, \sqsubseteq \rangle \otimes \langle \mathscr{A}, \preccurlyeq \rangle$. Define

$$\alpha(P) \triangleq \curlywedge\{\overline{P} \mid P \Vdash \overline{P}\} \qquad \gamma(\overline{P}) \triangleq \bigsqcup\{P \mid P \Vdash \overline{P}\} \tag{11.84}$$

so that

$\alpha(P) \preccurlyeq \overline{P}$

$\Leftrightarrow \curlywedge\{\overline{P}' \mid P \Vdash \overline{P}'\} \preccurlyeq \overline{P}$ ⦃definition of α⦄

$\Leftrightarrow P \Vdash \overline{P}$

⦃($\Leftarrow$) $P \Vdash \overline{P}$ implies $\overline{P} \in \{\overline{P}' \mid P \Vdash \overline{P}'\}$ so $\curlywedge\{\overline{P}' \mid P \Vdash \overline{P}'\} \preccurlyeq \overline{P}$ by def. glb
($\Rightarrow$) $P \Vdash \curlywedge\{\overline{P}' \mid P \Vdash \overline{P}'\}$ by definition 11.80.(3), so $P \sqsubseteq P$ by reflexivity and $\curlywedge\{\overline{P}' \mid P \Vdash \overline{P}'\} \preccurlyeq \overline{P}$ implies $P \Vdash \overline{P}$ by definition 11.80.(1). ⦄

$\Leftrightarrow P \preccurlyeq \bigsqcup\{P \mid P \Vdash \overline{P}\}$ ⦃by duality⦄

$\Leftrightarrow P \preccurlyeq \gamma(\overline{P})$ ⦃definition of γ⦄ □

11.15 Hierarchy of Abstractions

In a Galois connection $\langle \mathscr{C}, \sqsubseteq \rangle \underset{\alpha}{\overset{\gamma}{\leftrightarrows}} \langle \mathscr{A}, \preccurlyeq \rangle$, $\gamma \circ \alpha$ is, by exercise 11.57, an upper closure operator on $\langle \mathscr{C}, \sqsubseteq \rangle$. Moreover, $\langle \mathscr{C}, \sqsubseteq \rangle \underset{\alpha}{\overset{\gamma}{\leftrightarrows\!\!\!\!\rightarrow}} \langle \alpha(\mathscr{C}), \preccurlyeq \rangle$ where $\alpha(\mathscr{C}) \triangleq \{\alpha(x) \mid x \in \mathscr{C}\}$, in which case $\gamma \circ \alpha(\mathscr{C})$ and $\alpha(\mathscr{C})$ are isomorphic by α according to exercise 11.45. It is therefore possible, at least in theory, to eliminate the redundant elements from the abstract domain $\langle \mathscr{A}, \preccurlyeq \rangle$ to get $\langle \mathscr{C}, \sqsubseteq \rangle \underset{\alpha}{\overset{\gamma}{\leftrightarrows}} \langle \alpha(\mathscr{C}), \preccurlyeq \rangle$ and then to reason in the concrete on $\langle \gamma \circ \alpha(\mathscr{C}), \sqsubseteq \rangle$ which is the isomorphic image of the abstract domain $\langle \alpha(\mathscr{C}), \preccurlyeq \rangle$ in the concrete. The mathematical advantage is to get rid of the specific representation of abstract properties in $\langle \mathscr{A}, \preccurlyeq \rangle$ in order to reason uniquely in the concrete $\langle \mathscr{C}, \sqsubseteq \rangle$ and the concrete representation $\langle \gamma \circ \alpha(\mathscr{C}), \sqsubseteq \rangle$ of the abstract properties. Of course, in practice this problem cannot be escaped for static analysis.
This motivates mathematically the study of upper closure operators (and lower closure operators by duality).

A closure operator is fully determined by its closed elements. This shows that if the abstract domain is known then the abstraction is uniquely known.

Theorem 11.86 ([712, th. 5.2]) Let $\langle L, \sqsubseteq, \bot, \top, \sqcup, \sqcap \rangle$ be a complete lattice and $\rho, \rho' \in L \to L$ be closure operators on L. If $\rho(L) = \rho'(L)$, then $\rho = \rho'$.

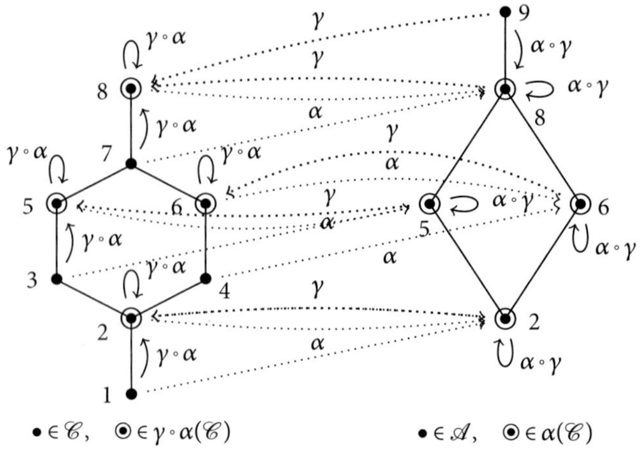

Figure 11.85
Galois connection $\langle \mathscr{C}, \sqsubseteq \rangle \xrightleftharpoons[\alpha]{\gamma} \langle \mathscr{A}, \preccurlyeq \rangle$, upper closure operator $\gamma \circ \alpha$, and lower closure operator $\alpha \circ \gamma$

Proof of theorem 11.86 $\forall z \in L \,.\, z \sqsubseteq \rho(z)$ so $\rho'(z) \sqsubseteq \rho'(\rho(z))$. By $\rho(L) = \rho'(L)$ and idempotence, ρ and ρ' have the same fixpoints. So $\rho(\rho(z)) = \rho(z)$ implies $\rho'(\rho(z)) = \rho(z)$ proving $\rho'(z) \sqsubseteq \rho(z)$. Exchanging ρ' and ρ, we get $\rho(z) \sqsubseteq \rho'(z)$, that is, $\rho(z) = \rho'(z)$ by antisymmetry. □

Theorem 11.87 Let $\langle L, \sqsubseteq, \bot, \top, \sqcup, \sqcap \rangle$ be a complete lattice and $f \in L \to L$ be an operator on L. Define $\mathsf{uco}(f) \triangleq x \mapsto \bigsqcap\{y \in L \mid x \sqcup f(y) \sqsubseteq y\}$. $\mathsf{uco}(f)$ is an upper closure operator that is smaller than any upper closure operator on L and $\dot\sqsubseteq$ greater than or equal to f.

If $f \in L \xrightarrow{\nearrow} L$ is increasing then $\mathsf{uco}(f)x = \mathsf{lfp}^{\sqsubseteq}\, y \mapsto x \sqcup f(y)$ is the $\dot\sqsubseteq$-least upper closure operator on L which is $\dot\sqsubseteq$ greater than or equal to f.

A dual result for $\mathsf{lco}(f) \triangleq x \mapsto \bigsqcup\{y \in L \mid y \sqsubseteq (x \sqcap f(y))\}$.

Proof of theorem 11.87

– $\forall y \in \{y \in L \mid x \sqcup f(y) \sqsubseteq y\} \,.\, x \sqsubseteq y$ so $x \sqsubseteq \bigsqcap\{y \in L \mid x \sqcup f(y) \sqsubseteq y\} = \mathsf{uco}(f)x$ so $\mathsf{uco}(f)$ is extensive.

– If $x \sqsubseteq x'$ then $x' \sqcup f(y) \sqsubseteq y$ implies $x \sqcup f(y) \sqsubseteq y$ so $\{y \mid x' \sqcup f(y) \sqsubseteq y\} \subseteq \{y \mid x \sqcup f(y) \sqsubseteq y\}$, proving that $\mathsf{uco}(f)x = \bigsqcap\{x \sqcup f(y) \sqsubseteq y\} \sqsubseteq \bigsqcap\{y \mid x' \sqcup f(y) \sqsubseteq y\} = \mathsf{uco}(f)x'$ so $\mathsf{uco}(f)$ is increasing.

– By extensivity and increasingness, $x \sqsubseteq \mathsf{uco}(f)(x) \sqsubseteq \mathsf{uco}(f)(\mathsf{uco}(f)x)$.

– If $f(y) \sqsubseteq y$ then $y \sqcup f(y) \sqsubseteq y$ so $y \in \{z \in L \mid y \sqcup f(z) \sqsubseteq z\}$, in which case $\mathsf{uco}(f)y = \bigsqcap\{z \in L \mid y \sqcup f(z) \sqsubseteq z\} \sqsubseteq y$. It follows that $x \sqcup f(y) \sqsubseteq y$ implies $x \sqsubseteq y$ and $\mathsf{uco}(f)y \sqsubseteq y$ so $x \sqcup \mathsf{uco}(f)y \sqsubseteq y$ and therefore $\{y \mid x \sqcup f(y) \sqsubseteq y\} \subseteq \{y \mid x \sqcup \mathsf{uco}(f)y \sqsubseteq y\}$, proving that $\mathsf{uco}(f)(\mathsf{uco}(f)x) = \bigsqcap\{y \mid x \sqcup \mathsf{uco}(f)y \sqsubseteq y\} \sqsubseteq \bigsqcap\{y \mid x \sqcup f(y) \sqsubseteq y\} = \mathsf{uco}(f)x$.

– By antisymmetry $\mathsf{uco}(f)(\mathsf{uco}(f)x) = \mathsf{uco}(f)x$ so $\mathsf{uco}(f)x$ is idempotent, hence an upper closure operator.

– Let ρ be a closure operator on L such that $f \mathrel{\dot{\sqsubseteq}} \rho$. We have $x \sqsubseteq \rho(x)$ and $f(\rho(x)) \mathrel{\dot{\sqsubseteq}} \rho(\rho(x)) = \rho(x)$ so $x \sqcup f(\rho(x)) \sqsubseteq \rho(x)$ which implies $\rho(x) \in \{y \mid x \sqcup f(y) \sqsubseteq y\}$ and therefore $\mathsf{uco}(f)x = \bigsqcap\{y \mid x \sqcup f(y) \sqsubseteq y\} \sqsubseteq \rho(x)$ proving that $\mathsf{uco}(f)$ is smaller than any upper closure operator on L $\dot{\sqsubseteq}$ greater than or equal to f.
– If f is increasing then the map $y \mapsto (x \sqcup f(y))$ is increasing so $\mathsf{uco}(f)x \triangleq \mathsf{lfp}^{\sqsubseteq}\, y \mapsto (x \sqcup f(y))$ is well defined by theorem 15.6 and equal to $\bigsqcap\{y \mid x \sqcup f(y) \sqsubseteq y\}$.
– If f is increasing and $y \in \{y \mid x \sqcup f(y) \sqsubseteq y\}$ then $x \sqsubseteq y$ so $f(x) \sqsubseteq f(y)$ and $f(y) \sqsubseteq y$ so $f(x) \sqsubseteq y$ and therefore $f(x) \sqsubseteq \bigsqcap\{y \mid x \sqcup f(y) \sqsubseteq y\} = \mathsf{uco}(f)x$, proving that $\mathsf{uco}(f)$ is the $\dot{\sqsubseteq}$-least upper closure operator on L which is $\dot{\sqsubseteq}$ greater than or equal to f. □

Theorem 11.88 The image of a complete lattice $\langle L, \sqsubseteq, \bot, \top, \sqcup, \sqcap\rangle$ by a closure operator $\rho \in L \to L$ is a complete lattice $\langle \rho(L), \sqsubseteq, \rho(\bot), \top, X \mapsto \rho(\sqcup X), \sqcap\rangle$.

For example, in figure 11.56, $\rho(L) = \{2, 5, 6, 8\}$ is the set of circled bullets. The join of 5 and 6 in L is $5 \sqcup 6 = 7$. The join of 5 and 6 in $\rho(L)$ is $\rho(5 \sqcup 6) = \rho(7) = 8$. The meet $5 \sqcap 6 = 2$ is the same in L and $\rho(L)$.

Theorem 11.88 explains why abstract domains discussed in chapter 3 are complete lattices. The collecting semantics $\langle \wp((\mathbb{V} \to \mathbb{Z}) \to \mathbb{Z}), \subseteq\rangle$ of arithmetic expressions is a complete lattice, so all its abstractions $\langle \rho(\wp((\mathbb{V} \to \mathbb{Z}) \to \mathbb{Z})), \subseteq\rangle$ by upper closure operators (or Galois retractions) ρ on expression properties $\wp((\mathbb{V} \to \mathbb{Z}) \to \mathbb{Z})$ are complete lattices.[4]

Proof of theorem 11.88 We have $\rho(L) = \{\rho(x) \mid x \in L\} \subseteq L$ so $\sqsubseteq$ is a partial order on $\rho(L)$. $\forall y \in \rho(L) \,.\, \exists x \in L \,.\, \rho(x) = y$ so $\bot \sqsubseteq x \sqsubseteq \top$ and increasingness imply $\rho(\bot) \sqsubseteq \rho(x) = y \sqsubseteq \rho(\top) = \top$ because ρ is extensive and $\top$ is the supremum, proving that $\rho(\bot)$ is the infimum and $\top$ is the supremum of $\rho(L)$. Let $\langle x_i,\ i \in \Delta\rangle$ be a family of elements of $\rho(L)$. Let us prove that $\bigsqcap_{i\in\Delta} x_i \in \rho(L)$. We have $\bigsqcap_{i\in\Delta} x_i \sqsubseteq x_j$ so $\rho(\bigsqcap_{i\in\Delta} x_i) \sqsubseteq \rho(x_j) = x_j$ because ρ is increasing, $x_j \in \rho(L)$, and idempotence. By definition of glb, $\rho(\bigsqcap_{i\in\Delta} x_i) \sqsubseteq \bigsqcap_{j\in\Delta} x_j$. Moreover, $\bigsqcap_{j\in\Delta} x_j \sqsubseteq \rho(\bigsqcap_{i\in\Delta} x_i)$ because ρ is extensive and therefore $\rho(\bigsqcap_{i\in\Delta} x_i) = \bigsqcap_{i\in\Delta} x_i$ by antisymmetry, proving that $\bigsqcap_{i\in\Delta} x_i \in \rho(L)$ so that $\bigsqcap$ is the glb in $\langle L, \sqsubseteq\rangle$. For the lub, let $\langle x_i,\ i \in \Delta\rangle$ be a family of elements of $\rho(L)$. We have $x_i = \rho(x_i) \sqsubseteq \rho(\bigsqcup_{i\in\Delta} x_i)$ by $x_i \in \rho(L)$, definition of the lub $\bigsqcup$, and ρ increasing, proving that $\rho(\bigsqcup_{i\in\Delta} x_i) \in \rho(L)$ is an upper bound of the $\langle x_i,\ i \in \Delta\rangle$. Let $u \in \rho(L)$ be another one. We have $x_i \sqsubseteq u$ and so $\bigsqcup_{i\in\Delta} x_i \sqsubseteq u$ that implies $\rho(\bigsqcup_{i\in\Delta} x_i) \sqsubseteq \rho(u) = u$, proving that $\rho(\bigsqcup_{i\in\Delta} x_i)$ is smaller than any other upper bound of the $\langle x_i,\ i \in \Delta\rangle$, hence their lub. □

11.16 Moore Families

By theorem 11.88, the image of a complete lattice by a closure operator is closed by arbitrary meets. A subset $\mathcal{M} \in \wp(\mathcal{P})$ of a poset $\langle \mathcal{P}, \sqsubseteq\rangle$ that is closed by arbitrary meets (i.e., $\forall S \in \mathcal{M} \,.\, \bigsqcap S$ exists in $\langle \mathcal{P}, \sqsubseteq\rangle$ and $\bigsqcap S \in \mathcal{M}$) is called a *Moore family* (after [713]).

4. More generally in chapter 21, the collecting semantics $\langle \wp(\wp(\mathbb{T}^+ \times \mathbb{T}^{+\infty})), \subseteq\rangle$ is a complete lattice, so its abstractions $\langle \rho(\wp(\wp(\mathbb{T}^+ \times \mathbb{T}^{+\infty}))), \subseteq\rangle$ by upper closure operators (or Galois retractions) ρ on program properties $\wp(\wp(\mathbb{T}^+ \times \mathbb{T}^{+\infty}))$ are also complete lattices.

So, by the dual of exercise 10.7, a Moore family $\langle \mathcal{M}, \sqsubseteq \rangle$ is a complete lattice (which includes the supremum $\top = \bigsqcap \varnothing$ of the poset $\langle \mathcal{P}, \sqsubseteq \rangle$ for the meet of the empty set).

Reciprocally, a Moore family of a complete lattice uniquely defines a closure operator (exercise 11.89), that is to say, an abstraction. Hence an abstraction can be uniquely defined by choosing an arbitrary subset of the concrete properties and then completing it to a Moore family by adding missing meets (including the supremum).

Exercise 11.89 (Moore family [712, Th. 5.3]) Let $\langle L, \sqsubseteq, \bot, \top, \sqcup, \sqcap \rangle$ be a complete lattice and $M \in \wp(L)$. Then there exists an upper closure operator on L such that $\rho(L) = M$ if and only if M is a Moore family. Then $\rho(x) = \bigsqcap \{y \in M \mid x \sqsubseteq M\}$.

The following theorem 11.90 shows that all possible abstractions by an upper closure operator (or Galois retractions) of a given collecting semantics (that is a complete lattice) form a complete lattice called "the hierarchy of abstractions." Part of this complete lattice of abstractions is illustrated in section 8.11.

Theorem 11.90 (Ward's theorem on upper closure operators [964]) Let $\langle L, \sqsubseteq, \bot, \top, \sqcup, \sqcap \rangle$ be a complete lattice. Let $\mathsf{Uco}(L)$ be the set of all upper closure operators on L. $\langle \mathsf{Uco}(L), \dot{\sqsubseteq}, x \mapsto x, x \mapsto \top, R \mapsto \mathsf{uco}(\dot{\bigsqcup} R), \dot{\sqcap} \rangle$ is a complete lattice (where $\mathsf{uco}(f)x = \mathsf{lfp}^{\sqsubseteq}\, y \mapsto x \sqcup f(y)$ for $f \in L \overset{\nearrow}{\longrightarrow} L$ increasing). A dual theorem holds for the lower closure operators $\mathsf{Lco}(L)$ on L.

Proof of theorem 11.90 By exercise 10.11, the set $\langle L \overset{\nearrow}{\longrightarrow} L, \dot{\sqsubseteq}, x \mapsto \bot, x \mapsto \top, \dot{\sqcup}, \dot{\sqcap} \rangle$ of increasing functions on L ordered pointwise is a complete lattice. By theorem 11.87, uco maps any $f \in L \overset{\nearrow}{\longrightarrow} L$ to an upper closure operator $\mathsf{uco}(f) \in \mathsf{Uco}(L)$. Any $\rho \in \mathsf{Uco}(L)$ is increasing so $\rho \in L \overset{\nearrow}{\longrightarrow} L$ with $\mathsf{uco}(\rho) = \rho$ by theorem 11.87. It follows that $\mathsf{Uco}(L)$ is the image of $L \overset{\nearrow}{\longrightarrow} L$ by the closure operator uco. It follows by theorem 11.88 that $\langle \mathsf{Uco}(L) = \mathsf{uco}(L \overset{\nearrow}{\longrightarrow} L), \dot{\sqsubseteq}, \mathsf{uco}(x \mapsto \bot), x \mapsto \top, X \mapsto \mathsf{uco}(\dot{\bigsqcup} X), \dot{\sqcap} \rangle$ is a complete lattice with $\mathsf{uco}(x \mapsto \bot) = x \mapsto \bigsqcap \{y \in L \mid x \sqcup \bot \sqsubseteq y\} = x \mapsto x$. □

More results on closure operators, Moore families, and equivalent formalizations of abstraction are given in [221, ch. 4] and [241].

11.17 Conclusion

We can represent abstract program properties by posets and, in case of existence of a best abstraction, establish the correspondence with the concrete properties using a Galois connection. The concrete order structure is preserved in the abstract and inversely. Otherwise stated, concrete and abstract implications coincide up to the Galois connection. Hence proofs in the abstract domain $\langle \mathcal{A}, \preccurlyeq \rangle$ using the abstract implication/order $\preccurlyeq$ are valid in the concrete $\langle \mathcal{C}, \sqsubseteq \rangle$ for $\sqsubseteq$, up to this Galois connection.

The original definition of Galois correspondences originating from the work of Evarist Galois [388] were between powersets and decreasing $\langle \wp(C), \subseteq \rangle \underset{\alpha}{\overset{\gamma}{\leftrightarrows}} \langle \wp(A), \supseteq \rangle$ so they do not compose as their semidual Galois connections. The order-theoretical Galois connection is credited to Oystein Ore [762] (spelled *connexion*). Galois connections have numerous applications in mathematics [309], including number theory

[609], modal logic [344], fuzzy logic [159, 949], fuzzy sets [993], μ-calculus [132], topology [19], metric spaces [154], formal concept analysis [394, 600], computability [146], constraint satisfaction problems [382, 383], entropy in thermodynamics [607], and others. Galois connections [492, 554] and closure operators [167] have been generalized in category theory. One can also study pairs of functions $\langle\alpha, \gamma\rangle$ with α increasing/decreasing, γ increasing/decreasing, $\gamma \circ \alpha$ which is extensive/reductive, and $\alpha \circ \gamma$ which is reductive/extensive; Galois connections being for the first choice [106, 281, 341, 364, 680].

The results on Galois connections apply to preorders (reflexive and transitive) when partially ordered up to the equivalence $x \equiv_{\sqsubseteq} y \triangleq (x \sqsubseteq y \wedge y \sqsubseteq x)$, in which case equalities must be replaced by the equivalence $\equiv_{\sqsubseteq}$.

Abstract interpretation with best abstraction has many equivalent formalizations equivalent to Galois connections [241]. For example, Moore families (exercise 11.89), closure operators (exercise 11.61), covers (exercise 11.10), logical relations [828], and soundness relations [738]. The advantage of Galois connections is that the abstraction α is explicit (and useful in discussing completeness) and abstract properties may be different from the concrete properties.

Logical relations are mathematically equivalent to Galois connections (section 11.14) and so can be used to establish the soundness of verification methods, static analyses, and types [738, 828].

Completeness involves considering the best abstraction given in section 11.12. By (11.84), this best abstraction involves considering *all* possible overapproximations specified by the logical relation, which is impractical (a point made in [249] and misinterpreted in [60], in which completeness is not considered). This is also the problem with completeness whenever considering only a concretization function, using lemma 11.42 to determine the best abstraction (e.g., [546] provides no completeness proof of Hoare logic). This situation is further studied in [245].

The combinations of Galois connections in section 11.13 can be extended to a Galois connection calculus [264] to build complex Galois connections out of basic ones. As shown in the following chapters, a proof method or a static analysis is uniquely defined by a program semantics and a Galois connection.

11.18 Solutions to Selected Exercises

Solution to exercise 11.1 Define $\gamma_n(x) = x + n$ so that $\alpha_n(x) \leqslant y \Leftrightarrow x - n \leqslant y \Leftrightarrow x \leqslant y + n \Leftrightarrow \gamma_n(x) \leqslant y$. Moreover, α_n is bijective.

Solution to exercise 11.5

$$\begin{aligned}
& \alpha(X) \subseteq Y \\
\Leftrightarrow\ & (X \cap S) \subseteq Y && \text{(definition of } \alpha\text{)} \\
\Leftrightarrow\ & X \subseteq (Y \cup \neg S) && \text{(definition of } \subseteq\text{)} \\
\Leftrightarrow\ & X \subseteq \gamma(Y) && \text{(by defining } \gamma(Y) \triangleq Y \cup \neg S\text{)}
\end{aligned}$$

Solution to exercise 11.6

$$
\begin{array}{ll}
\alpha_f(X) \subseteq Y & \\
\Leftrightarrow \{f(x) \mid x \in X\} \subseteq Y & \text{(definition of } \alpha_f\text{)} \\
\Leftrightarrow \forall x \in X \,.\, f(x) \in Y & \text{(definition of } \subseteq\text{)} \\
\Leftrightarrow X \subseteq \{x \mid f(x) \in Y\} & \text{(definition of } \subseteq\text{)} \\
\Leftrightarrow X \subseteq \gamma_f(Y) & \text{(definition of } \gamma\text{)}
\end{array}
$$

Solution to exercise 11.7 ($\Rightarrow$) Define $\mu(y) \triangleq \gamma_f(\{y\})$ so that $x \in \mu(y) \Leftrightarrow x \in \gamma_f(\{y\}) \Leftrightarrow x \in \{x' \mid f(x') \in \{x\}\} \Leftrightarrow x = f(y)$. Conversely ($\Leftarrow$), define $\gamma_f(Y) \triangleq \bigcup_{y \in Y} \mu(y)$ so that $X \subseteq \gamma_f(Y) \Leftrightarrow X \subseteq \bigcup_{y \in Y} \mu(y) \Leftrightarrow \forall x \in X \,.\, \exists y \in Y : x \in \mu(y) \Leftrightarrow \forall x \in X \,.\, \exists y \in Y : f(x) = y \Leftrightarrow \forall x \in X \,.\, f(x) \in Y \Leftrightarrow \alpha_f(X) \subseteq Y$.

Solution to exercise 11.9 Define $\gamma_{\subseteq}(b) \triangleq [\![b \,?\, R \,:\, S]\!]$.

Solution to exercise 11.13 We can define the Galois retraction $\langle \prod_{i \in \Delta} \wp(B_i), \dot{\subseteq} \rangle \xleftrightarrows[\alpha_P^{\pi\Delta}]{\gamma_P^{\pi\Delta}} \langle \wp(\Delta), \subseteq \rangle$ by $\alpha_P^{\pi\Delta}(\prod_{i \in \Delta} X_i) = \{i \in \Delta \mid X_i \neq \varnothing\}$ such that $\alpha^{\Delta} = \alpha^{\pi\Delta} \circ \alpha^{\pi}$.

Solution to exercise 11.15 We have $\langle \wp(\mathbb{R} \to \mathbb{R}), \subseteq \rangle \xleftrightarrows[\alpha_{\mathcal{O}}]{\gamma_{\mathcal{O}}} \langle \mathbb{R} \to \mathbb{R}_*, \dot{\leqslant} \rangle$ where

$$
\begin{array}{rcl}
\alpha_{\mathcal{O}}(P) & \triangleq & \dot{min}\{|g| \in \mathbb{R} \to \mathbb{R} \mid \forall f \in P \,.\, \exists x_0, c \in \mathbb{R} \,.\, \forall x \geqslant x_0 \,.\, |f(x)| \leqslant c|g(x)|\} \\
\gamma_{\mathcal{O}}(g) & \triangleq & \{f \in \mathbb{R} \to \mathbb{R} \mid \exists x_0, c \in \mathbb{R} \,.\, \forall x \geqslant x_0 \,.\, |f(x)| \leqslant c|g(x)|\}
\end{array}
$$

and $|g|$, $\dot{\leqslant}$, and $\dot{min}$ are pointwise.

$$
\begin{array}{ll}
\alpha_{\mathcal{O}}(P) \dot{\subseteq} |h| & \\
\Leftrightarrow \dot{min}\{g \in \mathbb{R} \to \mathbb{R} \mid \forall f \in P \,.\, \exists x_0, c \in \mathbb{R} \,.\, \forall x \geqslant x_0 \,.\, |f(x)| \leqslant c|g(x)|\} \dot{\leqslant} |h| & \\
& \text{(definition of } \alpha\text{)} \\
\Leftrightarrow \exists g \in \mathbb{R} \to \mathbb{R} \,.\, \forall f \in P \,.\, \exists x_0, c \in \mathbb{R} \,.\, \forall x \geqslant x_0 \,.\, |f(x)| \leqslant c|g(x)| \wedge |g(x)| \leqslant |h(x)| & \\
& \text{(definition of } \dot{min} \text{ and } \dot{\leqslant}\text{)} \\
\Rightarrow \exists g \in \mathbb{R} \to \mathbb{R} \,.\, \forall f \in P \,.\, \exists x_0, c \in \mathbb{R} \,.\, \forall x \geqslant x_0 \,.\, |f(x)| \leqslant c|h(x)| & \text{(} \leqslant \text{ transitive)} \\
\Leftrightarrow \forall f \in P \,.\, \exists x_0, c \in \mathbb{R} \,.\, \forall x \geqslant x_0 \,.\, |f(x)| \leqslant c|h(x)| & \text{(definition of } \exists\text{)} \\
\Leftrightarrow P \subseteq \{f \mid \exists x_0, c \in \mathbb{R} \,.\, \forall x \geqslant x_0 \,.\, |f(x)| \leqslant c|h(x)|\} & \text{(definition of } \subseteq\text{)} \\
\Leftrightarrow P \subseteq \gamma_{\mathcal{O}}(h) & \text{(definition of } \gamma_{\mathcal{O}}.\text{)}
\end{array}
$$

Conversely,

$$
\begin{array}{ll}
\forall f \in P \,.\, \exists x_0, c \in \mathbb{R} \,.\, \forall x \geqslant x_0 \,.\, |f(x)| \leqslant c|h(x)| & \\
\Rightarrow \exists g \in \mathbb{R} \to \mathbb{R} \,.\, \forall f \in P \,.\, \exists x_0, c \in \mathbb{R} \,.\, \forall x \geqslant x_0 \,.\, |f(x)| \leqslant c|g(x)| \wedge |g(x)| \leqslant |h(x)| & \\
& \text{(taking } g = h\text{)}
\end{array}
$$

Solution to exercise 11.32 That $\dot{\sqsubseteq}$ is a partial order on $D \xrightarrow{\sqcup} D$ follows directly from the hypothesis that $\langle D, \sqsubseteq\rangle$ is a poset. Define $\dot{\bigsqcup\limits_{i\in\Delta}} f_i \triangleq x \mapsto \bigsqcup_{i\in\Delta} f_i(x)$. $\dot{\bigsqcup\limits_{i\in\Delta}} f_i \in D \xrightarrow{\sqcup} D$ because

$$
\begin{aligned}
&(\dot{\bigsqcup_i} f_i)(\bigsqcup_j x_j)\\
=\ &\bigsqcup_i (f_i(\bigsqcup_j x_j)) && \text{(pointwise definition of } \dot{\textstyle\bigsqcup}\text{)}\\
=\ &\bigsqcup_i \bigsqcup_j (f_i(x_j)) && \text{(} f_i \in D \xrightarrow{\sqcup} D \text{ preserves joins)}\\
=\ &\bigsqcup_j \bigsqcup_i (f_i(x_j)) && \text{(definition of } \textstyle\bigsqcup\text{)}\\
=\ &\bigsqcup_j (\dot{\bigsqcup_i} f_i)(x_j) && \text{(pointwise definition of } \dot{\textstyle\bigsqcup}\text{, proving that } \dot{\textstyle\bigsqcup_{i\in\Delta}} f_i \text{ preserves arbitrary joins)}
\end{aligned}
$$

That $\dot{\bigsqcup}$ is a pointwise lub in $D \xrightarrow{\sqcup} D$ for $\dot{\sqsubseteq}$ follows directly from the fact that $\bigsqcup$ is a lub in D for $\sqsubseteq$. By exercise 10.7, $\langle D \xrightarrow{\sqcup} D, \dot{\sqsubseteq}, \dot{\sqcup}\rangle$ is a complete lattice. $\dot{\bot} = x \mapsto \bot \in D \xrightarrow{\sqcup} D$ and $\dot{\top} = x \mapsto \top \in D \xrightarrow{\sqcup} D$ are the infimum and supremum for $\dot{\sqsubseteq}$.

Solution to exercise 11.34 In a Galois connection $\langle \mathscr{C}, \sqsubseteq\rangle \$ \underset{\alpha}{\overset{\gamma}{\leftrightarrows}} \langle \mathscr{A}, \preccurlyeq\rangle$, we have

- $\forall x \in \mathscr{C} . \alpha(x) \preccurlyeq \alpha(x)$ (reflexivity)

$\Rightarrow \forall x \in \mathscr{C} . x \sqsubseteq \gamma \circ \alpha(x)$ (def. Galois connection)

$\Rightarrow \forall x \in \mathscr{C} . \alpha(x) \sqsubseteq \alpha \circ \gamma \circ \alpha(x)$ (α increasing)

- $\forall y \in \mathscr{A} . \gamma(y) \sqsubseteq \gamma(y)$ (reflexivity)

$\Rightarrow \forall y \in \mathscr{A} . \alpha \circ \gamma(y) \sqsubseteq y$ (definition of a Galois connection)

$\Rightarrow \forall x \in \mathscr{C} . \alpha \circ \gamma \circ \alpha(x) \sqsubseteq \alpha(x)$ (for $y = \alpha(x)$)

- $\alpha \circ \gamma \circ \alpha = \alpha$ (by antisymmetry)
- $\gamma \circ \alpha \circ \gamma = \gamma$ (by duality)

Solution to exercise 11.37 By $\alpha \in \mathbb{N} \to \{\bullet\}$, we have $\forall n \in \mathbb{N} . \alpha(n) = \bullet$. By $\gamma \in \{\bullet\} \to \mathbb{N}$, we have $\gamma(\bullet) = n$ for some $n \in \mathbb{N}$. Then $\gamma(\alpha(n+1)) = n \not\leq n+1$, in contradiction to $\gamma \circ \alpha$ is extensive in exercise 11.35. A fix is to consider $\mathbb{N} \cup \{\infty\}$ with $\gamma(\bullet) = \infty$.

Solution to exercise 11.39

$$
\begin{aligned}
&x \sqsubseteq \gamma(y)\\
\Rightarrow\ & x \sqsubseteq \bigsqcup\{x' \in \mathscr{C} \mid \alpha(x') \preccurlyeq y\} && \text{(definition of } \gamma\text{)}\\
\Rightarrow\ & \alpha(x) \preccurlyeq \alpha(\bigsqcup\{x' \in \mathscr{C} \mid \alpha(x') \preccurlyeq y\}) && \text{(}\alpha \text{ preserves existing lubs so is increasing)}\\
\Rightarrow\ & \alpha(x) \preccurlyeq \curlyvee\{\alpha(x') \mid x' \in \mathscr{C} \wedge \alpha(x') \preccurlyeq y\}) && \text{(}\alpha \text{ preserves existing lubs)}\\
\Rightarrow\ & \alpha(x) \preccurlyeq y
\end{aligned}
$$

(because y is an upper bound of $\{\alpha(x') \mid \alpha(x') \preccurlyeq y\}$ greater than or equal to the lub $\curlyvee\{\alpha(x') \mid \alpha(x') \preccurlyeq y\}$)

$\Rightarrow\ x \preccurlyeq \bigsqcup\{x' \in \mathscr{C} \mid \alpha(x') \preccurlyeq y\}$ $\qquad$ ⟅because $x \in \{x' \in \mathscr{C} \mid \alpha(x') \preccurlyeq y\}$⟆

$\Rightarrow\ x \preccurlyeq \gamma(y)$ $\qquad$ ⟅definition of γ⟆

Solution to exercise 11.40 γ does not preserve meets.

Solution to exercise 11.41 γ preserves finite meets but not infinite ones.

Solution to exercise 11.52 For $R \in \wp(\mathscr{D} \times \mathscr{C})$, we let $\alpha(R) \triangleq x \in \mathscr{D} \mapsto \{y \in \mathscr{C} \mid \langle x, y\rangle \in R\}$, and for $f \in \mathscr{D} \to \wp(\mathscr{C})$, we let $\gamma(f) \triangleq \{\langle x, y\rangle \mid x \in \mathscr{D} \wedge y \in f(x)\}$.

Solution to exercise 11.53 $\gamma(P)x \triangleq \bigsqcap\{y \in P \mid x \sqsubseteq y\}$.

Solution to exercise 11.57 By exercise 11.35 (1) and (2), α and γ are increasing so their composition $\gamma \circ \alpha$ is increasing. By exercise 11.45, $\gamma \circ \alpha \circ \gamma = \gamma$ so $\gamma \circ \alpha \circ \gamma \circ \alpha = \gamma \circ \alpha$ proving idempotence. By reflexivity $\alpha(x) \preccurlyeq \alpha(x)$ so $x \sqsubseteq \gamma \circ \alpha(x)$ by definition 11.1, proving extensivity. By duality, $\alpha \circ \gamma$ is a lower closure operator.

Solution to exercise 11.64 By exercise 11.61 and and lemma 11.38, $\rho(\mathscr{L})$ has arbitrary joins and a supremum, hence by exercise 10.7, it has arbitrary meets, so is a complete lattice.

Solution to exercise 11.89 We have $\forall y \in \{y \in M \mid x \sqsubseteq y\} \,.\, x \sqsubseteq y$ so x is a lower bound of $\{y \in M \mid x \sqsubseteq y\}$ hence less than or equal to the greatest lower bound $\rho(x)$, proving that ρ is extensive because $x \sqsubseteq \rho(x)$.

If $x \sqsubseteq x'$ then $x' \sqsubseteq y$ implies $x \sqsubseteq y$ so $\{y \in M \mid x' \sqsubseteq y\} \subseteq \{y \in M \mid x \sqsubseteq y\}$ proving $\rho(x) = \bigsqcap\{y \in M \mid x \sqsubseteq y\} \sqsubseteq \bigsqcap\{y \in M \mid x' \sqsubseteq y\} = \rho(x')$ by definition of the glb, proving ρ to be increasing.

Because $x \sqsubseteq \rho(x)$, $\rho(x) \sqsubseteq \rho(\rho(x))$. Inversely, by reflexivity, $\rho(x) \in \{y \in M \mid \rho(x) \sqsubseteq y\}$ so by definition of the glb, $\rho(\rho(x)) = \bigsqcap\{y \in M \mid \rho(x) \sqsubseteq y\} \sqsubseteq \rho(x)$. By antisymmetry, $\rho(x) = \rho(\rho(x))$ so ρ is idempotent.

ρ being extensive, increasing and idempotent is an upper closure operator on L.

For all $x \in L$, we have $\rho(x) = \bigsqcap\{y \in M \mid x \sqsubseteq y\} \in M$ because the $y \in M$ and M is a Moore family closed by $\bigsqcap$. So $\rho(L) \subseteq M$. If $x \in M$ then $x \in L$ and $x \in \{y \in M \mid x \sqsubseteq y\}$ so $x = \bigsqcap\{y \in M \mid x \sqsubseteq y\} = \rho(x)$ proving $M \subseteq \rho(L)$. By antisymmetry, $\rho(L) = M$.

Notice that M is closed by arbitrary meets so $\bigsqcap\varnothing \in M \subseteq L$ and $\bigsqcap\varnothing$ is a supremum for $\sqsubseteq$ so L has a supremum (which also belongs to M). Moreover, $\bigsqcap M$ is the infimum of M. So, as shown in section 10.6, $\langle M, \sqsubseteq, \bigsqcap M, \bigsqcap\varnothing, \sqcap\rangle$ is a dual complete lattice, which by exercise 10.7 has a lub $\sqcup$ and so is a complete lattice. Notice that the join in M is in general not the join in L. If the join $\bigvee X$ exists in L, then by exercise 11.61 and join preservation in Galois connections, $\bigsqcup(X) = \rho(\bigvee X)$ in M. Here is an example.

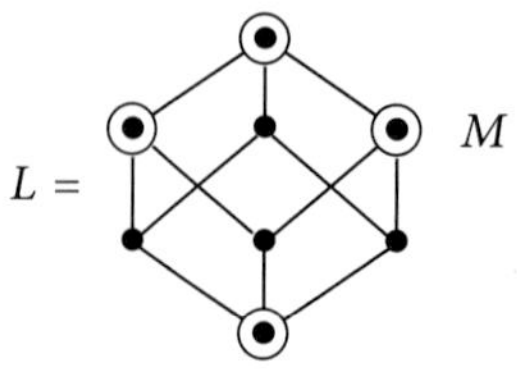

12 Relational and Transformer Semantics

We abstract the finite and infinite trace semantics into relational semantics and then into transformer semantics, generalizing the weakest predicate transformers. This is an illustration of Galois connections.

Contents

12.1 Relational Semantics

Let us abstract the maximal trace semantics defined in chapter 7 by abstracting a finite trace by its initial and final environments; an infinite trace by its initial environment and $\perp$ (designating nontermination after Dana Scott [872]); and a semantics by the abstraction of all its traces.

$$\begin{aligned}
\alpha_{\mathsf{R}}\langle\pi_1, \pi_2\rangle &\triangleq \langle\varrho(\pi_1), \varrho(\pi_1 \frown \pi_2)\rangle && \pi_2 \in \mathbb{T}^+ \qquad (12.1)\\
&\triangleq \langle\varrho(\pi_1), \perp\rangle && \pi_2 \in \mathbb{T}^\infty\\
\alpha_{\mathsf{R}}(\boldsymbol{\mathcal{S}}) &\triangleq \{\alpha_{\mathsf{R}}\langle\pi_1, \pi_2\rangle \mid \pi_1 \in \mathbb{T}^+ \wedge \pi_2 \in \boldsymbol{\mathcal{S}}(\pi_1)\}\\
\boldsymbol{\mathcal{S}}^{\mathsf{R}}[\![\mathsf{S}]\!] &\triangleq \alpha_{\mathsf{R}}(\boldsymbol{\mathcal{S}}^{+\infty}[\![\mathsf{S}]\!]).
\end{aligned}$$

This is a homomorphic/partitioning abstraction so that by exercise 11.6, $\langle\mathbb{T}^+ \rightarrow \wp(\mathbb{T}^{+\infty}), \dot{\subseteq}\rangle \underset{\alpha_{\mathsf{R}}}{\overset{\gamma_{\mathsf{R}}}{\leftrightarrows}} \langle\wp(\mathbb{E}\mathbb{v} \times (\mathbb{E}\mathbb{v} \cup \{\perp\}), \subseteq\rangle$. Therefore, the *infinitary relational semantics* $\boldsymbol{\mathcal{S}}^{\mathsf{R}}[\![\mathsf{S}]\!]$ of a statement S is an abstraction of the trace semantics $\boldsymbol{\mathcal{S}}^{+\infty}[\![\mathsf{S}]\!]$. By the further exclusion abstraction $\langle\wp(\mathbb{E}\mathbb{v} \times (\mathbb{E}\mathbb{v} \cup \{\perp\}), \subseteq\rangle \leftrightarrows \langle\wp(\mathbb{E}\mathbb{v} \times \mathbb{E}\mathbb{v}), \subseteq\rangle$ of exercise 11.5, it is also possible to ignore nontermination by defining a *finitary relational semantics* (also called *natural semantics*)

$$\boldsymbol{\mathcal{S}}^{\mathsf{R}^+}[\![\mathsf{S}]\!] \triangleq \{\langle\varrho(\pi_1), \varrho(\pi_1 \frown \pi_2)\rangle \mid \pi_1 \in \mathbb{T}^+ \wedge \pi_2 \in \boldsymbol{\mathcal{S}}^+[\![\mathsf{S}]\!](\pi_1)\}.$$

12.2 Property Transformers

Let us abstract relational semantics $R \in \wp(\mathbb{P} \times \mathbb{Q})$ into *transformers*, mapping a *preproperty* (belonging to $\wp(\mathbb{P})$) to a *postproperty* (belonging to $\wp(\mathbb{Q})$), or inversely. For example, $\mathbb{P} = \mathbb{E}\mathbb{v}$ and $\mathbb{Q} = \mathbb{E}\mathbb{v} \cup \{\perp\}$.

12.2.1 Forward Property Transformers

In *forward transformers* (post, $\widetilde{\text{post}}$), the preproperty is a hypothesis on the environments the execution is started with, and the postproperty is satisfied when execution terminates or is $\bot$ for nonterminating executions.

> *Postimage property transformer*
> - The *postimage* of a preproperty/precondition $P \in \wp(\mathbb{P})$ by relation $R \in \wp(\mathbb{P} \times \mathbb{Q})$ is $\text{post}[R]P \in \wp(\mathbb{Q})$ such that
>
> $$\text{post}[R] \;\triangleq\; P \mapsto \{y \in \mathbb{Q} \mid \exists x \in P\,.\, \langle x,\, y\rangle \in R\} \tag{12.2}$$
>
> (This is also called the *right image* of P by R, also written $R[P]$ or even $R(P)$.)

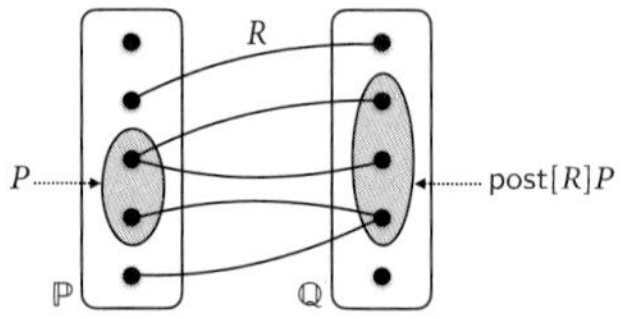

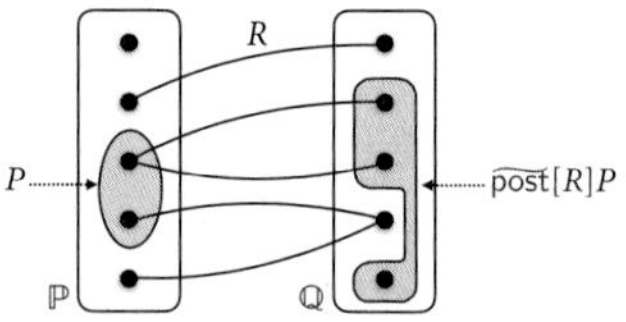

> *Dual postimage property transformer*
> - The *dual postimage* of a preproperty $P \in \wp(\mathbb{P})$ by relation R is $\widetilde{\text{post}}[R]P \in \wp(\mathbb{Q})$ such that
>
> $$\begin{aligned}\widetilde{\text{post}}[R] \;&\triangleq\; \neg \circ \text{post}[R] \circ \neg^{\text{a}} \\ &=\; P \mapsto \{y \in \mathbb{Q} \mid \forall x \in \mathbb{P}\,.\, \langle x,\, y\rangle \in R \Rightarrow x \in P\}\end{aligned} \tag{12.3}$$
>
> a. We write $\neg P$ for $\mathbb{P} \setminus P$ and $\neg Q$ for $\mathbb{Q} \setminus Q$, the ambiguity is solved by considering the powerset to which the negated property belongs.

Proof of (12.3)

$\widetilde{\text{post}}[R]P$

$\triangleq \neg(\text{post}[R](\neg P))$ $\qquad$ (definition (12.3) of $\widetilde{\text{post}}$ and function composition $\circ$)

$= \neg\{y \in \mathbb{Q} \mid \exists x\,.\, x \in \neg P \wedge \langle x,\, y\rangle \in R\}$ $\qquad$ (definition (12.2) of post)

$= \{y \in \mathbb{Q} \mid \forall x\,.\, \neg(\langle x,\, y\rangle \in R) \wedge x \in P\}$

$\qquad$ (definition of the complement $\neg$ and commutativity)

$= \{y \in \mathbb{Q} \mid \forall x\,.\, \langle x,\, y\rangle \in R \Rightarrow x \in P\}$ $\qquad$ (definition of the implication $\Rightarrow$) $\square$

Example 12.4 Consider the semantics $\mathscr{A}[\![\mathsf{A}]\!] \in \mathbb{E}\mathrm{v} \to \mathbb{Z}$ of arithmetic expressions A in (3.4). It can be regarded as a relation, $R \triangleq \{\langle \rho,\, \mathscr{A}[\![\mathsf{A}]\!]\rho\rangle \mid \rho \in \mathbb{E}\mathrm{v}\}$.

By property of variables x, one understands a property of their values $\rho(\mathsf{x})$ as given by an environment ρ. Therefore, a set $P \in \wp(\mathbb{E}\mathrm{v})$ of environments defines a property of variables that is a relation between their possible values, stating that the values $\rho(\mathsf{x})$ of variables x should be such that $\rho \in P$.

For example, the statement $\mathsf{x} > \mathsf{y} + 1$ is the property $\{\rho \in \mathbb{V} \to \mathbb{Z} \mid \rho(\mathsf{x}) > \rho(\mathsf{y}) + 1\}$. This makes explicit that in a program, a variable x denotes in fact its value $\rho(\mathsf{x})$ given by the state ρ of the computer memory. Such properties are called *relational* because they can express relations between the values of variables.

The postproperty $\mathsf{post}[\mathscr{A}[\![\mathsf{A}]\!]]P$ (more precisely $\mathsf{post}[R]P$) is the property of the result of the expression A when knowing the preproperty P of its variable.

For example, $\mathsf{post}[\mathscr{A}[\![\mathsf{x}-1]\!]]\{\rho \in \mathbb{V} \to \mathbb{Z} \mid \rho(\mathsf{x}) > \rho(\mathsf{y})+1\} = \{v \in \mathbb{Z} \mid v = \rho(\mathsf{x})-1 \geqslant \rho(\mathsf{y})+1\}$. Informally, if $\mathsf{x} > \mathsf{y}+1$ and $v = \mathsf{x}-1$ then $v = \mathsf{x}-1 \geqslant \mathsf{y}+1$.

The rule of signs is an abstraction of this transformer $\mathsf{post}[\mathscr{A}[\![\mathsf{A}]\!]]$.

Exercise 12.5 Prove that $\mathsf{post}[R]$ preserves arbitrary joins and $\widetilde{\mathsf{post}}[r]$ preserves arbitrary meets.

12.2.2 Isomorphism Between Relations and Property Transformers

Reasoning on the relation R or on the transformers $\mathsf{post}[R]$ or $\widetilde{\mathsf{post}}[R]$ is equivalent because one uniquely determines the other. Formally, define

$$\begin{aligned}
\mathsf{post}^{-1}[T] &\triangleq \{\langle x,\, y\rangle \in \mathbb{P}\times\mathbb{Q} \mid y \in T(\{x\})\} && (12.6)\\
\neg &\triangleq P \mapsto (\mathbb{S} \setminus P) && \text{(where } \mathbb{S} \text{ is } \mathbb{P} \text{ or } \mathbb{Q})\\
\widetilde{\mathsf{post}}^{-1}[T] &\triangleq \mathsf{post}^{-1}[\neg \circ T \circ \neg]\\
&= \{\langle x,\, y\rangle \in \mathbb{P}\times\mathbb{Q} \mid y \notin T(\mathbb{P} \setminus \{x\})\}
\end{aligned}$$

so that we have the Galois bijections $\langle \wp(\mathbb{P}\times\mathbb{Q}),\, \subseteq\rangle \underset{\mathsf{post}}{\overset{\mathsf{post}^{-1}}{\leftrightarrows}} \langle \wp(\mathbb{P}) \xrightarrow{\sqcup} \wp(\mathbb{Q}),\, \dot{\subseteq}\rangle$ and its dual $\langle \wp(\mathbb{P}\times\mathbb{Q}),\, \subseteq\rangle \underset{\widetilde{\mathsf{post}}}{\overset{\widetilde{\mathsf{post}}^{-1}}{\leftrightarrows}} \langle \wp(\mathbb{P}) \xrightarrow{\sqcap} \wp(\mathbb{Q}),\, \dot{\supseteq}\rangle$ (where $\dot{\supseteq}$ is pointwise inverse inclusion)

Proof

$$\begin{aligned}
&-\ \mathsf{post}[R] \mathrel{\dot{\subseteq}} T\\
&\Leftrightarrow \forall P \in \wp(\mathbb{P}) \,.\, \mathsf{post}[R]P \subseteq T(P) && \text{(pointwise definition of } \dot{\subseteq}\text{)}\\
&\Leftrightarrow \forall P \in \wp(\mathbb{P}) \,.\, \{y \in \mathbb{Q} \mid \exists x \in P \,.\, \langle x,\, y\rangle \in R\} \subseteq T(P) && \text{(definition (12.2) of post)}\\
&\Leftrightarrow \forall P \in \wp(\mathbb{P}) \,.\, \forall y \in \mathbb{Q} \,.\, \forall x \in P \,.\, \forall \langle x,\, y\rangle \in R \,.\, y \in T(P) && \text{(definition of } \subseteq\text{)}\\
&\Leftrightarrow \forall \langle x,\, y\rangle \in R \,.\, y \in T(\{x\})\\
&\qquad \text{((}\Rightarrow\text{) choosing } P = \{x\}\\
&\qquad \ (\Leftarrow)\ P = \textstyle\bigcup_{x\in P}\{x\} \text{ and } T \in \wp(\mathbb{P}) \xrightarrow{\sqcup} \wp(\mathbb{Q}) \text{ preserves joins)}\\
&\Leftrightarrow R \subseteq \{\langle x,\, y\rangle \in \mathbb{P}\times\mathbb{Q} \mid y \in T(\{x\})\} && \text{(definition of } \subseteq\text{)}\\
&\Leftrightarrow R \subseteq \mathsf{post}^{-1}[T] && \text{(definition (12.6) of } \mathsf{post}^{-1}[T]\text{)}
\end{aligned}$$

$$\begin{aligned}
&-\ \mathsf{post}^{-1}[\mathsf{post}[R]]\\
&= \{\langle x,\, y\rangle \in \mathbb{P}\times\mathbb{Q} \mid y \in \{y' \in \mathbb{Q} \mid \exists x' \in \{x\} \,.\, \langle x',\, y'\rangle \in R\}\}\\
& && \text{(definition of } \mathsf{post}^{-1} \text{ and post)}\\
&= R && \text{(definition of } \in\text{)}
\end{aligned}$$

$$\begin{aligned}
&-\ \mathsf{post}[\mathsf{post}^{-1}[T]]\\
&= P \mapsto \{y \in \mathbb{Q} \mid \exists x \in P \,.\, \langle x,\, y\rangle \in \{\langle x,\, y\rangle \in \mathbb{P}\times\mathbb{Q} \mid y \in T(\{x\})\}\}
\end{aligned}$$

$$
\begin{aligned}
& & &\text{\{ definition of post and post}^{-1}\text{\}}\\
&= P \mapsto \bigcup_{x\in P} T(\{x\}) & &\text{\{ definition of } \in \text{\}}\\
&= P \mapsto T(P) \;=\; T & &\text{\{ } T \in \wp(\mathbb{P}) \xrightarrow{\sqcup} \wp(\mathbb{Q}) \text{ preserves joins \}}
\end{aligned}
$$

$$
\begin{aligned}
&- \;\; \mathsf{post}[R] \,\dot{\subseteq}\, T \Leftrightarrow R \subseteq \mathsf{post}^{-1}[T]\\
&\Leftrightarrow \forall P \in \mathbb{P}\,.\,\mathsf{post}[R]P \subseteq T(P) \Leftrightarrow R \subseteq \mathsf{post}^{-1}[T] & &\text{\{ pointwise definition of } \dot{\subseteq} \text{\}}\\
&\Leftrightarrow \forall P' \in \mathbb{P}\,.\,\mathsf{post}[R]\neg P' \subseteq \neg\circ T'\circ\neg(\neg P') \Leftrightarrow R \subseteq \mathsf{post}^{-1}[\neg\circ T'\circ\neg]\\
& & &\text{\{ letting } T = \neg\circ T'\circ\neg \text{ and } P' = \neg P \text{\}}\\
&\Leftrightarrow \forall P \in \mathbb{P}\,.\,\mathsf{post}[R]\neg P \subseteq \neg\circ T(P) \Leftrightarrow R \subseteq \widetilde{\mathsf{post}}^{-1}[T] & &\text{\{ definition (12.6) of } \widetilde{\mathsf{post}}^{-1} \text{\}}\\
&\Leftrightarrow \mathsf{post}[R] \,\dot{\supseteq}\, T \Leftrightarrow R \subseteq \widetilde{\mathsf{post}}^{-1}[T] & &\text{\{ definition of } \neg \text{ and pointwise definition of } \dot{\supseteq} \text{\}}\\
&\Leftrightarrow \langle \wp(\mathbb{P}\times\mathbb{Q}),\, \subseteq\rangle \underset{\widetilde{\mathsf{post}}}{\overset{\widetilde{\mathsf{post}}^{-1}}{\leftrightarrows}} \langle \wp(\mathbb{P}) \xrightarrow{\sqcap} \wp(\mathbb{Q}),\, \dot{\supseteq}\rangle & &\text{\{ definition of a Galois connection \}}
\end{aligned}
$$

The bijection $\langle \widetilde{\mathsf{post}}, \widetilde{\mathsf{post}}^{-1}\rangle$ follows from the composition of bijections $\langle \mathsf{post}, \mathsf{post}^{-1}\rangle$ and $\langle\neg, \neg\rangle$. □

12.2.3 Abstraction of Functional/Relational Properties by Property Transformers

Let us show that property transformers are abstraction of properties of relations.

By section 8.2, properties of relations in $\wp(\mathbb{P}\times\mathbb{Q})$ are elements of $\wp(\wp(\mathbb{P}\times\mathbb{Q}))$. These properties of relations can be abstracted to property transformers by the Galois connection

$$
\langle \wp(\wp(\mathbb{P}\times\mathbb{Q})),\, \subseteq\rangle \underset{\alpha_{\mathscr{R}}}{\overset{\gamma_{\mathscr{R}}}{\leftrightarrows}} \langle \wp(\mathbb{P}) \xrightarrow{\sqcup} \wp(\mathbb{Q}),\, \dot{\subseteq}\rangle \tag{12.7}
$$

$$
\begin{aligned}
\alpha_{\mathscr{R}}(\mathscr{P}) &\triangleq P \mapsto \{y \mid \exists x \in P\,.\,\exists R \in \mathscr{P}\,.\,\langle x,\, y\rangle \in R\}\\
\gamma_{\mathscr{R}}(T) &\triangleq \wp(\mathsf{post}^{-1}[T])
\end{aligned}
$$

Proof of (12.7)

$$
\begin{aligned}
&\alpha_{\mathscr{R}}(\mathscr{P}) \,\dot{\subseteq}\, T\\
&\Leftrightarrow \forall P\,.\,\{y \mid \exists x \in P\,.\,\exists R \in \mathscr{P}\,.\,\langle x,\, y\rangle \in R\} \subseteq T(P) & &\text{\{ definition of } \dot{\subseteq} \text{ and } \alpha \text{\}}\\
&\Leftrightarrow \forall P\,.\,\forall y\,.\,(\exists x \in P\,.\,\exists R \in \mathscr{P}\,.\,\langle x,\, y\rangle \in R \Rightarrow y \in T(P) & &\text{\{ definition of } \subseteq \text{\}}\\
&\Leftrightarrow \forall P\,.\,\forall y\,.\,\forall x \in P\,.\,\forall R \in \mathscr{P}\,.\,\langle x,\, y\rangle \in R \Rightarrow y \in T(P) & &\text{\{ definition of } \Rightarrow \text{\}}\\
&\Leftrightarrow \forall R \in \mathscr{P}\,.\,R \in \{R \mid \forall P\,.\,\forall y\,.\,\forall x \in P\,.\,\langle x,\, y\rangle \in R \Rightarrow y \in T(P)\} & &\text{\{ definition of } \forall \text{ and } \in \text{\}}\\
&\Leftrightarrow \forall R \in \mathscr{P}\,.\,R \in \{R \mid \forall x, y\,.\,\langle x,\, y\rangle \in R \Rightarrow y \in T(\{x\})\}\\
&\qquad \text{\{ } (\Rightarrow) \text{ choosing } P = \{x\};\\
&\qquad\;\; (\Leftarrow) \;\; T \in \wp(\mathbb{P}) \xrightarrow{\sqcup} \wp(\mathbb{Q}) \text{ preserves joins and } P = \bigcup\{\{x\} \mid x \in P\} \text{\}}\\
&\Leftrightarrow \mathscr{P} \subseteq \{R \mid R \subseteq \{\langle x,\, y\rangle \mid y \in T(\{x\})\}\} & &\text{\{ definition of } \in \text{\}}\\
&\Leftrightarrow \mathscr{P} \subseteq \{R \mid R \subseteq \mathsf{post}^{-1}[T]\} & &\text{\{ definition (12.6) of } \mathsf{post}^{-1}[T] \text{\}}\\
&\Leftrightarrow \mathscr{P} \subseteq \wp(\mathsf{post}^{-1}[T]) & &\text{\{ definition of the powerset } \wp \text{\}}
\end{aligned}
$$

$\Leftrightarrow \ \mathscr{P} \subseteq \gamma_{\mathscr{R}}(T)$ $\qquad$ ⁅definition (12.7) of $\gamma_{\mathscr{R}}$⁆ □

As defined in section 2.2.4, functions in $\mathbb{P} \to \mathbb{Q}$ are relations in $\wp(\mathbb{P} \times \mathbb{Q})$, so we can abstract their properties in $\wp(\mathbb{P} \to \mathbb{Q})$ as we did for relation properties in $\wp(\wp(\mathbb{P} \to \mathbb{Q}))$ by transformers in $\wp(\mathbb{P}) \to \wp(\mathbb{Q})$.

$$\langle \wp(\mathbb{P} \to \mathbb{Q}), \subseteq\rangle \xleftrightarrows[\alpha_{\mathscr{F}}]{\gamma_{\mathscr{F}}} \langle \wp(\mathbb{P}) \xrightarrow{\sqcup} \wp(\mathbb{Q}), \dot{\subseteq}\rangle \tag{12.8}$$

$$\begin{array}{rcl} \alpha_{\mathscr{F}}(\mathscr{P}) & \triangleq & P \mapsto \{f(x) \mid \exists f \in \mathscr{P} \wedge x \in P\} \\ \gamma_{\mathscr{F}}(T) & \triangleq & \{f \in \mathbb{P} \to \mathbb{Q} \mid \forall x \in \mathbb{P} \,.\, f(x) \in T(\{x\})\} \end{array}$$

Exercise 12.9 Prove (12.8).

Exercise 12.10 Prove that the abstractions (12.7) and (12.8) lose information (i.e., are not isomorphisms).

12.2.4 Backward Property Transformers

In backward transformers (pre, $\widetilde{\text{pre}}$), the postproperty/postcondition is a hypothesis on the environments with which the execution may finish, or $\perp$ for nonterminating executions, and the preproperty is the set of initial environments for which this behavior is possible or guaranteed.

Preimage property transformer

- The *preimage* of a postproperty $Q \in \wp(\mathbb{Q})$ by relation $R \in \wp(\mathbb{P} \times \mathbb{Q})$ is $\text{pre}[R]Q \in \wp(\mathbb{P})$ such that

$$\text{pre}[R] \ \triangleq \ \text{post}[R^{-1}] \ = \ Q \mapsto \{x \in \mathbb{P} \mid \exists y \in Q \,.\, \langle x,\, y\rangle \in R\} \tag{12.11}$$

(This is also called the *left image* of P by R, also written $R^{-1}[P]$ or $R^{-1}(P)$.)

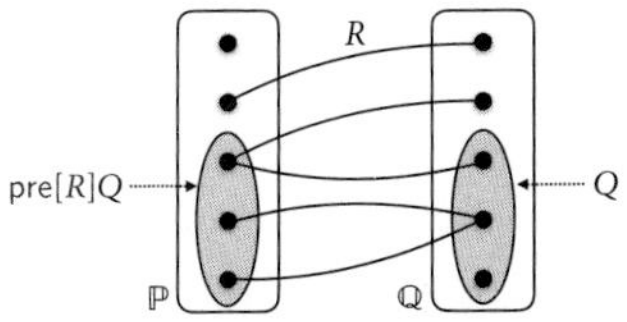

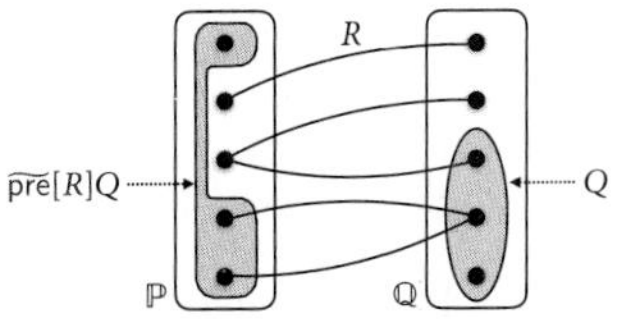

Dual preimage property transformer

- The *dual preimage* of a set $Q \in \wp(\mathbb{Q})$ by relation $R \in \wp(\mathbb{P} \times \mathbb{Q})$ is $\widetilde{\text{pre}}[R]Q \in \wp(\mathbb{P})$ such that

$$\begin{array}{rcl} \widetilde{\text{pre}}[R] & \triangleq & \neg \circ \text{pre}[R] \circ \neg \ = \ \widetilde{\text{post}}[R^{-1}] \\ & = & Q \mapsto \{x \in \mathbb{P} \mid \forall y \in \mathbb{Q} \,.\, \langle x,\, y\rangle \in R \Rightarrow y \in Q\} \end{array} \tag{12.12}$$

Proof

— $\text{pre}[R]Q$

$\triangleq \text{post}[R^{-1}]Q$ $\qquad$ ⁅definition (12.11) of pre⁆

$\triangleq \{y \in \mathbb{Q} \mid \exists x \in Q \,.\, \langle x,\, y\rangle \in R^{-1}\}$ $\qquad$ ⁅definition (12.2) of post⁆

$= \{x \in \mathbb{Q} \mid \exists y \in Q \,.\, \langle x,\, y\rangle \in R\}$ $\qquad$ ⁅renaming and definition of the inverse ${}^{-1}$⁆

— $\widetilde{\text{pre}}[R]Q$

$$\begin{aligned}
&\triangleq \neg \circ \mathsf{pre}[R] \circ \neg(Q) && \text{\{definition (12.12) of } \widetilde{\mathsf{pre}}\text{\}}\\
&= \neg \circ \mathsf{post}[R^{-1}] \circ \neg(Q) && \text{\{definition (12.11) of pre\}}\\
&= \{y \in \mathbb{Q} \mid \forall x \in \mathbb{P} \,.\, \langle x, y\rangle \in R^{-1} \Rightarrow x \in Q\} && \text{\{definition (12.3) of } \widetilde{\mathsf{post}}\text{\}}\\
&= \{x \in \mathbb{Q} \mid \forall y \,.\, \langle x, y\rangle \in R \Rightarrow x \in Q\} && \text{\{renaming and definition of the inverse } {}^{-1}\text{\}} \quad \square
\end{aligned}$$

Example 12.13 $\mathsf{pre}[\boldsymbol{\mathcal{S}}^{\mathsf{R}}[\![\mathsf{S}]\!]](\{\bot\}) \in \wp(\mathbb{E}\mathrm{v})$ is the weakest preproperty on environments for the existence of a nonterminating execution of statement S starting in any of these environments.

$\widetilde{\mathsf{pre}}[\boldsymbol{\mathcal{S}}^{\mathsf{R}}[\![\mathsf{S}]\!]](\{\bot\})$ is the weakest preproperty on environments for all executions of statement S starting in any of these environments to loop forever (which is trivially true for $\boldsymbol{\mathcal{S}}^{\mathsf{R}}[\![\mathsf{S}]\!] = \varnothing$).

Let $Q \in \wp(\mathbb{E}\mathrm{v})$ be a preproperty (excluding nontermination because $\bot \notin Q$). Then the weakest precondition on environments for an execution of statement S starting in any of these environments to terminate in a final environment satisfying property Q is $\mathsf{pre}[\boldsymbol{\mathcal{S}}^{\mathsf{R}}[\![\mathsf{S}]\!]](Q) \in \wp(\mathbb{E}\mathrm{v})$.

$\widetilde{\mathsf{pre}}[\boldsymbol{\mathcal{S}}^{\mathsf{R}}[\![\mathsf{S}]\!]](Q)$ is the weakest precondition on environments for any execution of statement S starting in any of these environments to terminate in a final environment satisfying property Q (which is trivially true for $\boldsymbol{\mathcal{S}}^{\mathsf{R}}[\![\mathsf{S}]\!] = \varnothing$).

If $Q \in \wp(\mathbb{E}\mathrm{v} \cup \{\bot\})$ is a postproperty then $\mathsf{pre}[\boldsymbol{\mathcal{S}}^{\mathsf{R}}[\![\mathsf{S}]\!]] \in \wp(\mathbb{E}\mathrm{v}))$ is the weakest precondition on environments for the existence of an execution of statement S starting in any of these environments that either terminates in a final environment satisfying property Q or does not terminate.

Exercise 12.14 Prove that $\mathsf{pre}[R]$ preserve infinite joins and $\widetilde{\mathsf{pre}}[R]$ preserve infinite meets.

Exercise 12.15 Prove that $\mathsf{pre}[R]Q \cap \widetilde{\mathsf{pre}}[R]Q = \mathsf{pre}[R]\mathbb{Q} \cap \widetilde{\mathsf{pre}}[R]Q$

Exercise 12.16 Prove that if R is total (i.e., for all $x \in \mathbb{P}$ there exists a $y \in \mathbb{Q}$ such that $\langle x, y\rangle \in R$) then $\mathsf{pre}[R]Q \cap \widetilde{\mathsf{pre}}[R]Q = \widetilde{\mathsf{pre}}[R]Q$.

Exercise 12.17 Prove that if R is deterministic (that is, if $\langle x, y\rangle \in R$ and $\langle x, y'\rangle \in R$ then $y = y'$) then $\mathsf{post}[R] = \widetilde{\mathsf{post}}[R]$ and $\mathsf{pre}[R] = \widetilde{\mathsf{pre}}[R]$.

Exercise 12.18 Prove that in the special case in which R is a partial function $f \in \mathbb{P} \nrightarrow \mathbb{Q}$, $\mathsf{post}[f] = P \mapsto \{f(x) \mid x \in \mathsf{dom}(f)\}$, $\mathsf{post}[f]\mathbb{P} = \mathsf{cod}(f)$, $\mathsf{pre}[f] = Q \mapsto \{x \in \mathsf{dom}(f) \mid f(x) \in Q\}$, $\mathsf{pre}[f]\mathbb{Q} = \mathsf{dom}(f)$, $\widetilde{\mathsf{post}}[f]\mathbb{P} = \mathbb{Q} \setminus \mathsf{cod}(f)$, and $\widetilde{\mathsf{pre}}[f]\mathbb{Q} = \mathbb{P} \setminus \mathsf{dom}(f)$. $\mathsf{post}[f]P$ is sometimes written $f[P]$ or $f(P)$. $\mathsf{pre}[f]Q$ is sometimes written $f^{-1}(Q)$.

Exercise 12.19 Is any Galois connection $\langle \wp(C), \subseteq\rangle \underset{\alpha}{\overset{\gamma}{\leftrightarrows}} \langle \wp(A), \subseteq\rangle$ between two powersets always of the form $\alpha = \mathsf{post}[r]$ and $\gamma = \widetilde{\mathsf{pre}}[r]$ for an appropriate $r \in \wp(C \times A)$?

Again, reasoning on the relation R or on the transformers $\mathsf{pre}[R]$ or $\widetilde{\mathsf{pre}}[R]$ is equivalent because one uniquely determines the other.

Exercise 12.20 Define

$$\mathsf{pre}^{-1}[T] \;\triangleq\; (\mathsf{post}^{-1}[T])^{-1} \tag{12.21}$$

Prove that $\text{pre}^{-1}[T] = \{\langle x, y\rangle \in \mathbb{P} \times \mathbb{Q} \mid x \in T(\{y\})\}$. Prove that $\langle \wp(\mathbb{P} \times \mathbb{Q}), \subseteq\rangle \underset{\text{pre}}{\overset{\text{pre}^{-1}}{\leftrightarrows}} \langle \wp(\mathbb{Q}) \xrightarrow{\sqcap} \wp(\mathbb{P}), \dot{\subseteq}\rangle$. Define $\widetilde{\text{pre}}^{-1}$ such that $\langle \wp(\mathbb{P} \times \mathbb{Q}), \subseteq\rangle \underset{\widetilde{\text{pre}}}{\overset{\widetilde{\text{pre}}^{-1}}{\leftrightarrows}} \langle \wp(\mathbb{Q}) \xrightarrow{\sqcup} \wp(\mathbb{P}), \dot{\supseteq}\rangle$.

12.3 Galois Connections between Property Transformers

If $R \in \wp(\mathbb{P} \times \mathbb{Q})$, we have the following Galois connections

$$\langle \wp(\mathbb{P}), \subseteq\rangle \underset{\text{post}[R]}{\overset{\widetilde{\text{pre}}[R]}{\leftrightarrows}} \langle \wp(\mathbb{Q}), \subseteq\rangle \tag{12.22}$$

$$\langle \wp(\mathbb{Q}), \subseteq\rangle \underset{\text{pre}[R]}{\overset{\widetilde{\text{post}}[R]}{\leftrightarrows}} \langle \wp(\mathbb{P}), \subseteq\rangle \tag{12.23}$$

showing that forward abstraction is essentially equivalent to backward abstraction.

Exercise 12.24 Prove (12.22) and (12.23).

Exercise 12.25 Given a trace semantics $\boldsymbol{S} \in \mathbb{T}^+ \to \wp(\mathbb{T}^{+\infty})$, define $\overrightarrow{\text{post}}[\boldsymbol{S}] \in \wp(\mathbb{T}^+) \to \wp(\mathbb{T}^{+\infty})$ by $\overrightarrow{\text{post}}[\boldsymbol{S}]P \triangleq \{\pi_2 \in \mathbb{T}^{+\infty} \mid \exists \pi_0, \pi_1 \,.\, \pi_0 \in P \wedge \pi_1 \mathbin{\hat{}} \pi_2 \in \boldsymbol{S}(\pi_0)\}$, $\widetilde{\overrightarrow{\text{post}}}[\boldsymbol{S}] \triangleq \neg \circ \overrightarrow{\text{post}}[\boldsymbol{S}] \circ \neg$, $\overrightarrow{\text{pre}}[\boldsymbol{S}] \in \wp(\mathbb{T}^{+\infty}) \to \wp(\mathbb{T}^+)$ by $\overrightarrow{\text{pre}}[\boldsymbol{S}]Q \triangleq \{\pi_0 \mid \exists \pi_1, \pi_2 \,.\, (\pi_1 \mathbin{\hat{}} \pi_2 \in \boldsymbol{S}(\pi_0)) \wedge (\pi_2 \in Q)\}$, and $\widetilde{\overrightarrow{\text{pre}}}[\boldsymbol{S}] \triangleq \neg \circ \overrightarrow{\text{pre}}[\boldsymbol{S}] \circ \neg$.

Show that $\langle \wp(\mathbb{T}^+), \subseteq\rangle \underset{\overrightarrow{\text{post}}[\boldsymbol{S}]}{\overset{\widetilde{\overrightarrow{\text{pre}}}[\boldsymbol{S}]}{\leftrightarrows}} \langle \wp(\mathbb{T}^{+\infty}), \subseteq\rangle$ and $\langle \wp(\mathbb{T}^{+\infty}), \subseteq\rangle \underset{\overrightarrow{\text{pre}}[\boldsymbol{S}]}{\overset{\widetilde{\overrightarrow{\text{post}}}[\boldsymbol{S}]}{\leftrightarrows}} \langle \wp(\mathbb{T}^+), \subseteq\rangle$.

If the traces in Q represent errors for a trace semantics $\boldsymbol{S}$, then the sufficient precondition of [222, 255] to guarantee the absence of errors is $P = \overrightarrow{\text{pre}}[\boldsymbol{S}](\neg Q) \cap \widetilde{\overrightarrow{\text{pre}}}[\boldsymbol{S}](\neg Q)$, that is, from an initial trace in P it is not possible to reach an error in Q. The necessary precondition to guarantee the presence of errors [265, 271] is $P' = \overrightarrow{\text{pre}}[\boldsymbol{S}]Q \cap \widetilde{\overrightarrow{\text{pre}}}[\boldsymbol{S}]Q$, that is, from an initial trace in P' it is not possible to escape an error in Q. For example,

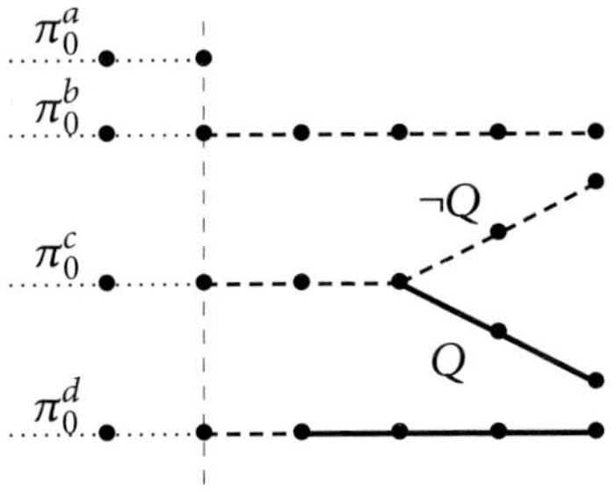

$P = \overrightarrow{\text{pre}}[\boldsymbol{S}](\neg Q) \cap \widetilde{\overrightarrow{\text{pre}}}[\boldsymbol{S}](\neg Q) = \{\pi_0^b, \pi_0^c\} \cap \{\pi_0^a, \pi_0^b\} = \{\pi_0^b\}$ precludes potential errors but excludes a correct trace and $P' = \overrightarrow{\text{pre}}[\boldsymbol{S}]Q \cap \widetilde{\overrightarrow{\text{pre}}}[\boldsymbol{S}]Q = \{\pi_0^c, \pi_0^d\} \cap \{\pi_0^a, \pi_0^d\} = \{\pi_0^d\}$, so its negation $\{\pi_0^a, \pi_0^b, \pi_0^c\}$ precludes definite errors without excluding any correct trace (which nevertheless may have an erroneous alternative).

12.4 Weakest Preconditions

The weakest preconditions introduced by Edger W. Dijkstra [321, 327] use logical predicates and so cannot refer to nontermination $\perp$ directly. It is however possible to handle non/termination indirectly by postulating the existence of a weakest precondition

($\mathsf{wp}[\![\mathsf{S}]\!]Q$ excluding the possibility of nontermination) and weakest liberal precondition ($\mathsf{wlp}[\![\mathsf{S}]\!]Q$ including the possibility of nontermination) for a statement to satisfy a postproperty Q upon termination. They can be defined as follows ($Q_\perp \in \mathbb{E}\mathrm{v} \cup \{\perp\}$ and $Q \in \mathbb{E}\mathrm{v}$).

$$\begin{aligned} \mathsf{gwp}[\![R]\!]Q_\perp &\triangleq \mathsf{pre}[R]Q_\perp \cap \widetilde{\mathsf{pre}}[R]Q_\perp \\ \mathsf{wp}[\![R]\!]Q &\triangleq \mathsf{gwp}[\![R]\!]Q \\ \mathsf{wlp}[\![R]\!]Q &\triangleq \mathsf{gwp}[\![R]\!](Q \cup \{\perp\}) \end{aligned} \tag{12.26}$$

For brevity, we write $\mathsf{post}[\mathsf{S}]$ for $\mathsf{post}[\boldsymbol{\mathcal{S}}^{\mathsf{R}}[\![\mathsf{S}]\!]]$, etc. We also write $\mathsf{gwp}[\![\mathsf{S}]\!]Q$ (respectively $\mathsf{wp}[\![\mathsf{S}]\!]Q$, $\mathsf{wlp}[\![\mathsf{S}]\!]$) for $\mathsf{gwp}[\![\boldsymbol{\mathcal{S}}^{\mathsf{R}}[\![\mathsf{S}]\!]]\!]$ (respectively $\mathsf{wp}[\![\boldsymbol{\mathcal{S}}^{\mathsf{R}}[\![\mathsf{S}]\!]]\!]$, $\mathsf{wlp}[\![\boldsymbol{\mathcal{S}}^{\mathsf{R}}[\![\mathsf{S}]\!]]\!]$).

By definition $\mathsf{wlp}[\![\mathsf{S}]\!]\mathsf{tt} = \mathsf{tt}$ (truth tt is $\mathbb{E}\mathrm{v}$) and $\mathsf{wlp}[\![\mathsf{S}]\!]\mathsf{ff} \wedge \mathsf{wp}[\![\mathsf{S}]\!]\mathsf{tt} = \mathsf{ff}$ (falsity ff is $\emptyset$). $\mathsf{wp}[\![\mathsf{S}]\!]\mathsf{tt}$ is a termination requirement of S and $\mathsf{wp}[\![\mathsf{S}]\!]Q = \mathsf{wlp}[\![\mathsf{S}]\!]Q \wedge \mathsf{wp}[\![\mathsf{S}]\!]\mathsf{tt}$.

Edger W. Dijkstra observed [322] that a deterministic statement either (a) terminates with a final environment satisfying a given postproperty Q, or (b) terminates with a final environment not satisfying the given postproperty Q, or (c) does not terminate. For a nondeterministic system, Edger W. Dijkstra [321] characterizes the possible behaviors as the following:

(a) Execution of S will terminate in a final environment satisfying Q ($\mathsf{gwp}[\![\mathsf{S}]\!]Q$).

$$\mathsf{wp}[\![\mathsf{S}]\!]Q = \mathsf{wlp}[\![\mathsf{S}]\!]Q \wedge \mathsf{wp}[\![\mathsf{S}]\!]\mathsf{tt}$$

(b) Execution of S will terminate in a final environment satisfying $\neg Q$ (i.e., $\mathbb{E}\mathrm{v} \setminus Q$, $\mathsf{gwp}[\![\mathsf{S}]\!]\neg Q$).

$$\mathsf{wp}[\![\mathsf{S}]\!](\neg Q) = \mathsf{wlp}[\![\mathsf{S}]\!](\neg Q) \wedge \mathsf{wp}[\![\mathsf{S}]\!]\mathsf{tt}$$

(c) Execution of S will fail to terminate properly ($\mathsf{gwp}[\![R]\!]\{\perp\}$).

$$\mathsf{wlp}[\![\mathsf{S}]\!]\mathsf{ff} = \mathsf{wlp}[\![\mathsf{S}]\!]Q \wedge \mathsf{wlp}[\![\mathsf{S}]\!](\neg Q)$$

(ab) Execution of S will terminate properly but the initial environment does not determine whether the final environment will satisfy Q or not ($\mathsf{gwp}[\![\mathsf{S}]\!]\mathsf{tt}$).

$$\mathsf{wp}[\![\mathsf{S}]\!]\mathsf{tt} \wedge \neg\mathsf{wlp}[\![\mathsf{S}]\!]Q \wedge \neg\mathsf{wp}[\![\mathsf{S}]\!](\neg Q)$$

(ac) Execution of S may terminate, in which case the final environment will satisfy Q but the initial environment does not determine whether S will terminate or not ($\mathsf{gwp}[\![\mathsf{S}]\!](Q \cup \{\perp\})$).

$$\mathsf{wlp}[\![\mathsf{S}]\!]Q \wedge \neg\mathsf{wp}[\![\mathsf{S}]\!]\mathsf{tt}$$

(bc) Execution of S may terminate, in which case the final environment will not satisfy Q but the initial environment does not determine whether S will terminate or not ($\mathsf{gwp}[\![\mathsf{S}]\!](\neg Q \cup \{\perp\})$).

$$\mathsf{wlp}[\![\mathsf{S}]\!](\neg Q) \wedge \neg\mathsf{wp}[\![\mathsf{S}]\!]\mathsf{tt}$$

(abc) The initial environment does not determine whether S will terminate or not and upon termination, if ever, whether the final environment will satisfy Q or not ($\mathsf{gwp}[\![\mathsf{S}]\!](\mathsf{tt} \cup \{\perp\})$).

$$\mathsf{pre}[\boldsymbol{\mathcal{S}}^{\mathsf{R}}[\![\mathsf{S}]\!]](\mathbb{E}\mathrm{v} \cup \{\perp\}) = \neg(\mathsf{wlp}[\![\mathsf{S}]\!]Q \vee \mathsf{wlp}[\![\mathsf{S}]\!](\neg Q) \vee \mathsf{wp}[\![\mathsf{S}]\!]\mathsf{tt})$$

Proof

— Assume $\mathsf{wp}[\![\mathsf{S}]\!]Q$ is true in case (a). By $\mathsf{pre}[\boldsymbol{\mathcal{S}}^{\mathsf{R}}[\![\mathsf{S}]\!]]Q$, there exists a pair $\langle \rho, \rho' \rangle \in \boldsymbol{\mathcal{S}}^{\mathsf{R}}[\![\mathsf{S}]\!]$ such that $\rho' \in Q$ so that, by definition (12.1) of α_{R} and $\boldsymbol{\mathcal{S}}^{\mathsf{R}}[\![\mathsf{S}]\!] \triangleq \alpha_{\mathsf{R}}(\boldsymbol{\mathcal{S}}^{+\infty}[\![\mathsf{S}]\!])$, there exists $\pi_1 \in \mathbb{T}^+$, $\pi_2 \in \mathbb{T}^{+\infty}$ such that $\pi_2 \in \boldsymbol{\mathcal{S}}^{+\infty}[\![\mathsf{S}]\!]\pi_1$ so $\boldsymbol{\mathcal{S}}^{+\infty}[\![\mathsf{S}]\!]$ is not empty. Let $\pi_1 \in \mathbb{T}^+$ and π_2 be any traces such that $\pi_2 \in \boldsymbol{\mathcal{S}}^{+\infty}[\![\mathsf{S}]\!]\pi_1$. If π_2 were infinite then, by (12.1), $\alpha_{\mathsf{R}}\langle \pi_1, \pi_2 \rangle = \langle \varrho(\pi_1), \bot \rangle$ so, by $\widetilde{\mathsf{pre}}[\boldsymbol{\mathcal{S}}^{\mathsf{R}}[\![\mathsf{S}]\!]]Q$, $\bot \in Q$ which is impossible since $Q \in \wp(\mathbb{E}\mathbb{v})$. So π_2 must be finite and, by (12.1) and (12.26), $\varrho(\pi_1 \mathbin{\widehat{\,}} \pi_2) \in Q$ and $\varrho(\pi_1) \in \mathsf{wp}[\![\mathsf{S}]\!]Q$. So execution of statement S starting with any environment in $\mathsf{wp}[\![\mathsf{S}]\!]Q$ (which by (12.1) and (7.7) is $\varrho(\pi_1)$ for some finite trace π_1) will always terminate in a final environment $\varrho(\pi_1 \mathbin{\widehat{\,}} \pi_2)$ satisfying Q which is the informal definition of $\mathsf{wp}[\![\mathsf{S}]\!]Q$.

— The proof of (b) is similar, replacing Q by $\neg Q$.

— In case (c), assume that $\rho \in \mathsf{wlp}[\![\mathsf{S}]\!]\mathsf{ff} = \mathsf{pre}[\boldsymbol{\mathcal{S}}^{\mathsf{R}}[\![\mathsf{S}]\!]](\{\bot\}) \cap \widetilde{\mathsf{pre}}[\boldsymbol{\mathcal{S}}^{\mathsf{R}}[\![\mathsf{S}]\!]](\{\bot\})$. By definition of $\boldsymbol{\mathcal{S}}^{\mathsf{R}}[\![\mathsf{S}]\!]$, α_{R}, and $\boldsymbol{\mathcal{S}}^{+\infty}[\![\mathsf{S}]\!]$, if $\pi_2 \in \boldsymbol{\mathcal{S}}^{+\infty}[\![\mathsf{S}]\!]\pi_1$ with $\varrho(\pi_1) = \rho$ then π_2 must be infinite. So any execution of statement S will fail to terminate, which is the informal definition of $\mathsf{wlp}[\![\mathsf{S}]\!]\mathsf{ff}$.

— The proof of all other cases is similar. □

12.5 Strongest Postconditions

If $R \in \wp(\mathbb{P} \times \mathbb{Q})$ is total (i.e., $\forall x \in \mathbb{P} \,.\, \exists y \in \mathbb{Q} \,.\, \langle x, y \rangle \in R$) then

$$\langle \wp(\mathbb{P}), \subseteq \rangle \underset{\mathsf{post}[R]}{\overset{\mathsf{wp}[\![\mathsf{S}]\!]}{\leftrightarrows}} \langle \wp(\mathbb{Q}), \subseteq \rangle$$

Proof

$$
\begin{aligned}
& \mathsf{post}[R]P \subseteq Q \\
\Leftrightarrow\ & P \subseteq \widetilde{\mathsf{pre}}[R]Q && \text{(12.22)} \\
\Leftrightarrow\ & P \subseteq \mathsf{pre}[R]Q \cap \widetilde{\mathsf{pre}}[R]Q && \text{exercise 12.16} \\
\Leftrightarrow\ & P \subseteq \mathsf{wp}[\![\mathsf{S}]\!]Q && \text{definition (12.26) of wp}
\end{aligned}
$$
□

The results on converse predicate transformers in [327, ch. 11] are then a consequence of lemma 11.67 and classic properties of Galois connections. Moreover, problems with handling nontermination with strongest postcondition transformers mentioned in [327, ch. 12] can be solved by considering $\boldsymbol{\mathcal{S}}^{\mathsf{R}}$ (not $\boldsymbol{\mathcal{S}}^{\mathsf{R}^+}$ ignoring infinite behaviors).

Exercise 12.27 Express cases (a), …, (abc) introduced in section 12.4 using strongest postconditions for a total relational semantics (if R is not total, let $\Omega \notin \mathbb{Q}$ and define $R_\Omega \triangleq R \cup \{\langle x, \Omega \rangle \mid \forall y \in \mathbb{Q} \,.\, \langle x, y \rangle \notin R\}$, which is similar to the solution adopted by Dana Scott for nontermination [872] which we have adopted for $\boldsymbol{\mathcal{S}}^{\mathsf{R}}$).

12.6 Conclusion

The relational semantics $\boldsymbol{\mathcal{S}}^{\mathsf{R}}$ is an abstraction of the trace semantics $\boldsymbol{\mathcal{S}}^{+\infty}$. The abstract semantics $\boldsymbol{\mathcal{S}}^{\mathsf{R}^+}$ abstracts away termination (and could be called the *natural semantics* after [551]). The trace semantics $\boldsymbol{\mathcal{S}}^{+\infty}$ itself is essentially a relation between an initialization trace and a finite or infinite continuation semantics in the spirit of [686].

Predicate transformers where introduced in mathematics by Jacques Riguet [831, p. 118]. See [918] for a generalization.

They were rediscovered in computer science. The weakest/backward/abductive preconditions are from [321], and the strongest/forward/deductice postconditions are found in [235, 241, 328]. Their definition as abstractions are from [222, 502] and [227] in which it is suggested to add nontermination $\bot$ to predicates for better expressivity. The special case of total and deterministic relations (exercise 12.17) is considered in [222].

Abstract interpretation was originally introduced [235, 241] using predicate transformers (called deductive semantics in [235]). The advantage is that the formulation is language independent. For example, it directly applies to Prolog [205]. The disadvantage is that it is sometimes too abstract (see, for example, chapter 44, "Software Model Checking") and the instantiation of the concepts for a specific language may not look simple.

It remains to provide a structural definition of these semantics. This is carried out in chapter 25, "Abstract Reachability / Invariance / Safety Verification Semantics," and chapter 26, "Hoare Logic," for partial correctness, namely case (ac) of section 12.4. Structural backward transformers are considered in chapter 50, "Backward Accessibility Semantics."

12.7 Solutions to Selected Exercises

Solution to exercise 12.9
Proof of (12.8)

$$\begin{aligned}
&\alpha_{\mathscr{F}}(\mathscr{P}) \dot{\subseteq} T \\
\Leftrightarrow\ & \forall P \in \wp(\mathbb{P}) \,.\, \{f(x) \mid \exists f \in \mathscr{P} \wedge x \in P\} \subseteq T(P) && \text{(definition of } \dot{\subseteq} \text{ and } \alpha\text{)} \\
\Leftrightarrow\ & \forall P \in \wp(\mathbb{P}) \,.\, \forall x \in P \,.\, \forall f \in \mathscr{P} \,.\, f(x) \in T(P) && \text{(definition of } \subseteq\text{)} \\
\Leftrightarrow\ & \forall f \in \mathscr{P} \,.\, \forall x \in \mathbb{P} \,.\, f(x) \in T(\{x\}) && \text{(for } P = \{x\} \text{ and conversely } T \text{ preserves joins)} \\
\Leftrightarrow\ & \mathscr{P} \subseteq \{f \mid \forall x \in \mathbb{P} \,.\, f(x) \in T(\{x\})\} && \text{(definition of } \subseteq\text{)} \\
\Leftrightarrow\ & \mathscr{P} \subseteq \gamma_{\mathscr{F}}(T) && \text{(definition (12.8) of } \gamma_{\mathscr{F}}\text{)} \quad \square
\end{aligned}$$

Solution to exercise 12.10 Take $\mathscr{P} = \{f, g\}$ where $f(x) = 0$ and $g(x) = 1$. Then $T = \alpha_{\mathscr{F}}(\mathscr{P})$ is $T(P) = (\!| P = \varnothing \ ?\ \varnothing \ ⸗\ \{0, 1\} |\!)$ so $\gamma_{\mathscr{F}}(T) = \{f \mid \forall x \,.\, f(x) \in \{0, 1\}\}$ which is different from $\{f, g\}$.

Solution to exercise 12.15

$$\begin{aligned}
&\mathsf{pre}[R]Q \cap \widetilde{\mathsf{pre}}[R]Q \\
=\ & \{x \in \mathbb{P} \mid \exists y \,.\, y \in Q \wedge \langle x,\ y\rangle \in R\} \cap \{x \in \mathbb{P} \mid \forall y \in \mathbb{Q} \,.\, \langle x,\ y\rangle \in R \Rightarrow y \in Q\} \\
& \hfill \text{(definition (12.11) of pre and } \widetilde{\mathsf{pre}}\text{)} \\
=\ & \{x \in \mathbb{P} \mid \exists y \,.\, y \in \mathbb{Q} \wedge \langle x,\ y\rangle \in R \wedge \forall y \in \mathbb{Q} \,.\, \langle x,\ y\rangle \in R \Rightarrow y \in Q\}
\end{aligned}$$

⦇definition of $\cap$ and ($\subseteq$). $y \in Q \wedge \langle x,\ y\rangle \in R$ implies $y \in \mathbb{Q}\,.\,\langle x,\ y\rangle \in R$ since $Q \subseteq \mathbb{Q}$, ($\supseteq$) if $y \in \mathbb{Q}\,.\,\langle x,\ y\rangle \in R \wedge \forall y' \in \mathbb{Q}\,.\,\langle x',\ y'\rangle \in R \Rightarrow y' \in Q$ then $y \in Q$ when $y' = y$, and antisymmetry⦈

$= \mathsf{pre}[R]\mathbb{Q} \cap \widetilde{\mathsf{pre}}[R]Q$ ⦇definition of $\cap$ and definition (12.11) of pre and $\widetilde{\mathsf{pre}}$⦈

Solution to exercise 12.16

$\mathsf{pre}[R]Q \cap \widetilde{\mathsf{pre}}[R]Q$

$= \mathsf{pre}[R]\mathbb{Q} \cap \widetilde{\mathsf{pre}}[R]Q$ ⦇exercise 12.15⦈

$= \{x \in \mathbb{P} \mid \exists y \in \mathbb{Q}\,.\,\langle x,\ y\rangle \in R\} \cap \widetilde{\mathsf{pre}}[R]Q$ ⦇definition (12.11) of pre⦈

$= \mathbb{P} \cap \widetilde{\mathsf{pre}}[R]Q$ ⦇$R \in \wp(\mathbb{P} \times \mathbb{Q})$ and R is total⦈

$= \widetilde{\mathsf{pre}}[R]Q$ ⦇definition (12.12) of $\widetilde{\mathsf{pre}}$ so that $\widetilde{\mathsf{pre}}[R]Q \subseteq \mathbb{P}$⦈

Solution to exercise 12.17

— $\widetilde{\mathsf{pre}}[R]Q$

$= \mathsf{pre}[R]Q \cap \widetilde{\mathsf{pre}}[R]Q$ ⦇exercise 12.16⦈

$= \{x \in \mathbb{P} \mid \exists y \in Q\,.\,\langle x,\ y\rangle \in R \wedge \forall y' \in \mathbb{Q}\,.\,\langle x,\ y'\rangle \in R \Rightarrow y' \in Q\}$

⦇definition (12.11) of pre and $\widetilde{\mathsf{pre}}$, definition of $\cap$⦈

$= \{x \in \mathbb{P} \mid \exists y \in Q\,.\,\langle x,\ y\rangle \in R\}$

⦇($\subseteq$) $A \wedge B \Rightarrow A$, ($\supseteq$) if $\langle x,\ y'\rangle \in R$ then $y' = y$ since R is deterministic so $y \in Q$, and antisymmetry⦈

$= \widetilde{\mathsf{pre}}[R]Q$ ⦇definition (12.11) of pre ⦈

— $\mathsf{pre}[R] = \mathsf{post}[R^{-1}] = \widetilde{\mathsf{post}}[R^{-1}] = \widetilde{\mathsf{pre}}[R]$ by (12.11), (12.2), (12.3), and (12.12).

Solution to exercise 12.24

— $\mathsf{post}[R]P \subseteq Q$

$\Leftrightarrow \{y \in \mathbb{Q} \mid \exists x \in P\,.\,\langle x,\ y\rangle \in R\} \subseteq Q$ ⦇definition (12.2) of post⦈

$\Leftrightarrow \forall y \in \mathbb{Q}\,.\,(\exists x \in P\,.\,\langle x,\ y\rangle \in R) \Rightarrow y \in Q$ ⦇definition of $\subseteq$⦈

$\Leftrightarrow \forall y \in \mathbb{Q}\,.\,\forall x \in P\,.\,\langle x,\ y\rangle \in R \Rightarrow y \in Q$ ⦇definition of $\subseteq$⦈

$\Leftrightarrow \forall x \in P\,.\,\forall y \in \mathbb{Q}\,.\,\langle x,\ y\rangle \in R \Rightarrow y \in Q$ ⦇definition of $\forall$⦈

$\Leftrightarrow P \subseteq \{x \in \mathbb{P} \mid \forall y \in \mathbb{Q}\,.\,\langle x,\ y\rangle \in R \Rightarrow y \in Q\}$ ⦇definition of $\subseteq$⦈

$\Leftrightarrow P \subseteq \widetilde{\mathsf{pre}}[R]Q$ ⦇definition (12.12) of $\widetilde{\mathsf{pre}}$⦈

— By lemma 11.38, it follows that $\mathsf{post}[R] \in \wp(\mathbb{P}) \xrightarrow{\sqcup} \wp(\mathbb{Q})$.

— $\mathsf{post}[R] \mathrel{\dot{\subseteq}} T$

$\Leftrightarrow \{y \in \mathbb{Q} \mid \exists x \in P\,.\,\langle x,\ y\rangle \in R\} \subseteq T(P)$

⦇pointwise definition of $\dot{\subseteq}$ and (12.2) of post⦈

$\Leftrightarrow \forall y \in \mathbb{Q} . (\exists x \in P . \langle x, y \rangle \in R) \Rightarrow y \in T(P)$ $\wr$definition of $\subseteq\int$

$\Leftrightarrow \forall x \in \mathbb{P} . \forall y \in \mathbb{Q} . (x \in P \Rightarrow (\langle x, y \rangle \in R) \Rightarrow y \in T(P))$ $\wr$definition of $\Rightarrow$ and $\forall\int$

$\Leftrightarrow \forall \langle x, y \rangle \in \mathbb{P} \times \mathbb{Q} . (\langle x, y \rangle \in R) \Rightarrow y \in T(\{x\})$

$\wr$($\Rightarrow$) for $P = \{x\}$ so $x \in P$ is true;
($\Leftarrow$) if $x \in P$ and $\langle x, y \rangle \in R$ then $y \in T(\{x\}) \subseteq T(P)$ since T preserves joins so is increasing hence $y \in T(P)$. $\int$

$\Leftrightarrow R \subseteq \{\langle x, y \rangle \in \mathbb{P} \times \mathbb{Q} \mid y \in T(\{x\})\}$ $\wr$definition of $\subseteq\int$

$\Leftrightarrow R \subseteq \mathsf{post}^{-1}[T]$ $\wr$definition (12.6) of $\mathsf{post}^{-1}\int$

— We have $\langle \wp(\mathbb{P}), \subseteq \rangle \underset{\mathsf{post}[R]}{\overset{\widetilde{\mathsf{pre}}[R]}{\leftrightarrows}} \langle \wp(\mathbb{Q}), \subseteq \rangle$, in particular $\langle \wp(\mathbb{P}), \subseteq \rangle \underset{\mathsf{post}[R^{-1}]}{\overset{\widetilde{\mathsf{pre}}[R^{-1}]}{\leftrightarrows}} \langle \wp(\mathbb{Q}), \subseteq \rangle$ when R is R^{-1} so $\langle \wp(\mathbb{P}), \subseteq \rangle \underset{\mathsf{pre}[R]}{\overset{\widetilde{\mathsf{post}}[R]}{\leftrightarrows}} \langle \wp(\mathbb{Q}), \subseteq \rangle$ by (12.11) and (12.12). Another proof would use the conjugate as in section 11.9.2.

— $\mathsf{pre}[R] \dot{\subseteq} T$

$\Leftrightarrow \mathsf{post}[R^{-1}] \dot{\subseteq} T$ $\wr$definition (12.11) of pre$\int$

$\Leftrightarrow R^{-1} \subseteq \mathsf{post}^{-1}[T]$ $\wr$by $\langle \wp(\mathbb{Q} \times \mathbb{P}), \subseteq \rangle \underset{\mathsf{post}}{\overset{\mathsf{post}^{-1}}{\leftrightarrows}} \langle \wp(\mathbb{Q}) \xrightarrow{\sqcup} \wp(\mathbb{P}), \dot{\subseteq} \rangle\int$

$\Leftrightarrow R \subseteq (\mathsf{post}^{-1}[T])^{-1}$ $\wr^{-1}$ is increasing and its own inverse$\int$

$R \subseteq \mathsf{pre}^{-1}[T]$ $\wr$definition (12.21) of $\mathsf{pre}^{-1}\int$

Solution to exercise 12.25 Let $P \in \wp(\mathbb{T}^{+})$ and $Q \in \wp(\mathbb{T}^{+\infty})$. We have

$\overrightarrow{\mathsf{post}}[\boldsymbol{S}]P \subseteq Q$

$\Leftrightarrow \{\pi_2 \in \mathbb{T}^{+\infty} \mid \exists \pi_0, \pi_1 . \pi_0 \in P \wedge \pi_1 \mathbin{\hat{\cdot}} \pi_2 \in \boldsymbol{S}(\pi_0)\} \subseteq Q$ $\wr$definition of $\overrightarrow{\mathsf{post}}[\boldsymbol{S}]\int$

$\Leftrightarrow \forall \pi_0, \pi_1, \pi_2 . (\pi_0 \in P \wedge \pi_1 \mathbin{\hat{\cdot}} \pi_2 \in \boldsymbol{S}(\pi_0)) \Rightarrow (\pi_2 \in Q)$ $\wr$definition of $\subseteq\int$

$\Leftrightarrow P \subseteq \{\pi_0 \mid \forall \pi_1, \pi_2 . (\pi_1 \mathbin{\hat{\cdot}} \pi_2 \in \boldsymbol{S}(\pi_0)) \Rightarrow (\pi_2 \in Q)\}$ $\wr$definition of $\subseteq\int$

$\Leftrightarrow P \subseteq \neg\{\pi_0 \mid \exists \pi_1, \pi_2 . (\pi_1 \mathbin{\hat{\cdot}} \pi_2 \in \boldsymbol{S}(\pi_0)) \wedge (\pi_2 \in \neg Q)\}$ $\wr$definition of the complement$\int$

$\Leftrightarrow P \subseteq \neg\overrightarrow{\mathsf{pre}}[\boldsymbol{S}](\neg Q)$ $\wr$definition of $\overrightarrow{\mathsf{pre}}[\boldsymbol{S}]\int$

$\Leftrightarrow P \subseteq \widetilde{\overrightarrow{\mathsf{pre}}}[\boldsymbol{S}]Q$ $\wr$definition of $\widetilde{\overrightarrow{\mathsf{pre}}}[\boldsymbol{S}]\int$

The second Galois connection is the conjugate of the first one; see section 11.9.2.

13 Topology

A topology can be thought of as abstracting away all properties irrelevant to the notion of limit (of functions, sequences, etc.). Topology is related to program trace semantics because infinite traces have been defined as the limit of prefix closed semantics (7.2). This limit could have been defined in topological terms (see example 13.11). We will use topology in chapter 14 to discuss program properties that do (liveness) or do not (safety) depend on this passage to the limit in the trace semantics. This chapter is a very lightweight introduction to topology; [53, 137, 890] are much more detailed introductions to topology.

Contents

13.1 Topology

Let $\mathcal{X}$ be a nonempty set. A *topology* $\mathcal{T}$ on $\mathcal{X}$ is a family $\mathcal{T} \in \wp(\mathcal{X})$ of subsets of $\mathcal{X}$, called the *open sets*, such that

(a) The union of open sets is an open set

$$\forall P \in \wp(\mathcal{T}) . \bigcup P \in \mathcal{T}$$

(b) The *finite* intersection of open sets is an open set

$$\forall n \in \mathbb{N} . \forall P_1, \ldots, P_n \in \mathcal{T} . \bigcap_{i=1}^{n} P_i \in \mathcal{T}$$

(so, in particular, $\bigcap \varnothing = \mathcal{X} \in \mathcal{T}$ and $\bigcup \varnothing = \varnothing \in \mathcal{T}$). The pair $\langle \mathcal{X}, \mathcal{T} \rangle$ is called a *topological space*.

Example 13.1

set $\mathcal{X}$:

topologies on $\mathcal{X}$:

$\mathcal{T}_A$ = 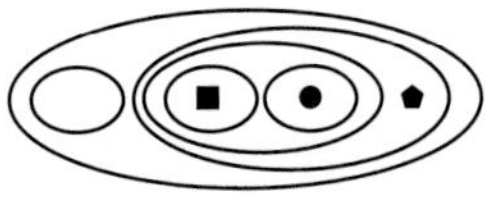 $\mathcal{T}_B$ = 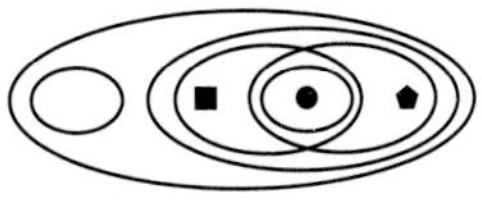

$\mathscr{T}_E$ = 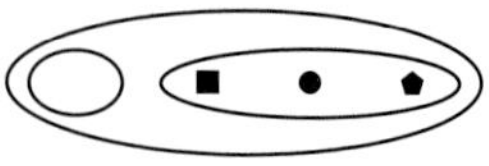 $\mathscr{T}_F$ = 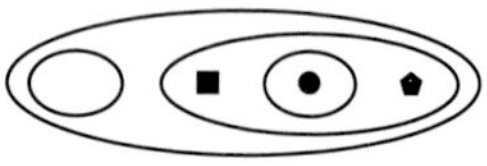

nontopologies on $\mathscr{X}$:

$\mathscr{S}_C$ = 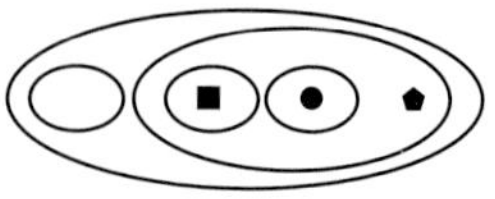 $\mathscr{S}_D$ = 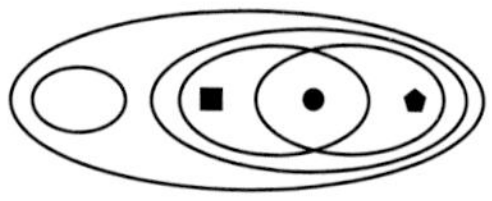

Exercise 13.2 What are the topologies of smallest and largest cardinality on a set $\mathscr{X}$?
The *neighborhoods* of an element $x \in \mathscr{X}$ are the open sets of the topology $\mathscr{T}$ containing x. The *interior* $\iota(X)$ of a subset $P \subseteq \mathscr{X}$ is the largest open set contained in P and therefore it is the union of all open sets contained in P, that is $\iota(X) = \bigcup\{U \mid U \in \mathscr{T} \wedge U \subseteq P\}$.

The *closed sets* $\overline{\mathscr{T}} \in \wp(\mathscr{X})$ of a topology $\mathscr{T}$ are the complements of the open sets. For $P \in \wp(\mathscr{X})$, define $\neg P \triangleq \mathscr{X} \setminus P$.

$$\overline{\mathscr{T}} \quad \triangleq \quad \{\neg U \mid U \in \mathscr{T}\}$$

Exercise 13.3 Provide an example of topology $\mathscr{T}$ on a set $\mathscr{X}$, which has some sets that are both open and closed (called *clopen* sets).

$\mathscr{T}$ is closed by union, so by De Morgan's laws, $\overline{\mathscr{T}}$ is closed by intersection, hence $\overline{\mathscr{T}}$ is a Moore family (see exercise 11.89).

Let ρ be associated upper closure (see exercise 11.89)

$$\rho(P) \quad \triangleq \quad \bigcap\{C \in \overline{\mathscr{T}} \mid P \subseteq C\} \tag{13.4}$$

The closure of $P \subseteq \mathscr{X}$ is therefore the smallest closed set contains P.

Exercise 13.5 Show that $\langle \wp(\mathscr{X}), \subseteq\rangle \xrightleftharpoons[\rho]{\mathbb{1}} \langle \overline{\mathscr{T}}, \subseteq\rangle$.

Exercise 13.6 Show that $\iota(P) \subseteq P \subseteq \rho(P)$.

Exercise 13.7 Show that $\iota(P) = \neg\rho(\neg P)$ and $\rho(P) = \neg\iota(\neg P)$.

13.2 Topological Closure

Lemma 13.8 $\rho \in \wp(\mathscr{X}) \mapsto \overline{\mathscr{T}}$ in (13.4) is a *topological closure* satisfying the Kazimierz Kuratowski closure axioms [604, 605]

(c) ρ is expansive, that is, $P \subseteq \rho(P)$
(d) ρ is idempotent, that is, $\rho \circ \rho = \rho$
(e) ρ is strict, that is, $\rho(\emptyset) = \emptyset$
(f) ρ preserves finite joins, that is, $\rho(P \cup Q) = \rho(P) \cup \rho(Q)$[a]

a. By recurrence, this holds for any finite numbers of joins.

Note that a topological closure ρ is an upper closure (because in lemma 13.8, (f) implies that ρ is increasing (if $P \subseteq Q$ then $P \cup Q = Q$ so $\rho(Q) = \rho(P \cup Q) = \rho(P) \cup \rho(Q)$, proving that $\rho(P) \subseteq \rho(Q)$)).

Proof of lemma 13.8

— Because the intersection of closed sets is closed, (13.4) implies that $\rho \in \wp(\mathcal{X}) \mapsto \overline{\mathcal{T}}$.

— By exercise 11.89, ρ is expansive, idempotent, and increasing.

— By exercise 11.51, $\langle \wp(\mathcal{X}), \subseteq\rangle \underset{\rho}{\overset{\mathbb{1}}{\leftrightarrows}} \langle \rho(\wp(\mathcal{X})), \subseteq\rangle$. Because $\varnothing \subseteq \mathbb{1}(\varnothing)$, we have $\rho(\varnothing) \subseteq \varnothing$ and so $\rho(\varnothing) = \varnothing$ by antisymmetry.

— We have $\rho(P) \cup \rho(Q) \subseteq \rho(P \cup Q)$ because ρ is $\subseteq$-increasing and $\cup$ is a lub. By expansivity, $P \subseteq \rho(P)$ and $Q \subseteq \rho(Q)$ so $P \cup Q \subseteq \rho(P) \cup \rho(Q)$ proving $\rho(P \cup Q) \subseteq \rho(\rho(P) \cup \rho(Q))$ because ρ is increasing. Because the finite intersection of open sets is an open set, the finite union of closed sets is a closed set so $\rho(\rho(P) \cup \rho(Q)) = \rho(P) \cup \rho(Q)$ and hence $\rho(P \cup Q) \subseteq \rho(P) \cup \rho(Q)$, proving $\rho(P \cup Q) = \rho(P) \cup \rho(Q)$ by antisymmetry. □

13.3 Topology Defined by a Topological Closure

Lemma 13.9 Any topological closure ρ on $\wp(\mathcal{X})$ defines a topology $\mathcal{T} \triangleq \{\neg\rho(P) \mid P \in \wp(\mathcal{X}\}$.

Proof of lemma 13.9 Let ρ be a topological closure operator (hence an upper closure) on $\wp(\mathcal{X}$. Let $\mathcal{T} \triangleq \{\neg\rho(P) \mid P \in \wp(\mathcal{X})\}$ be the complements of the closed sets $\{\rho(P) \mid P \in \wp(\mathcal{X})\}$.

— Because $\langle \wp(\mathcal{X}), \subseteq\rangle$ is a complete lattice, $\langle \rho(\wp(\mathcal{X})), \subseteq\rangle$ is a complete lattice by theorem 11.88 and hence closed by intersection. By De Morgan's laws, the complement $\mathcal{T}$ is therefore closed by union.

— Moreover by lemma 13.8, (f), the finite union of closed sets is closed, so by De Morgan's laws, the complement $\mathcal{T}$ is therefore closed by finite intersection. This proves that $\mathcal{T}$ is a topology on $\mathcal{X}$. □

13.4 Limits of Sequences

An element x of a set $\mathcal{X}$ with topology $\mathcal{T}$ is a *limit* of a sequence $\langle x_n \in \mathcal{X},\ n \in \mathbb{N}\rangle$ if and only if for every neighborhood U of x (i.e., $x \in U \in \mathcal{T}$), there exists $N \in \mathbb{N}$, such that for all $n \geqslant N$, $x_n \in U$.

Example 13.10 Let $\mathcal{X} = \{A_n \mid n \in \mathbb{N}\} \cup \{A_\omega\}$ be a family of sets A_n, $n \in \mathbb{N}$ and A_ω.

Let $\{\{A_0, A_1, \dots, A_{n-1}\} \mid n \in \mathbb{N}\}$ (including $\varnothing$ for $n = 0$) be the closed sets and their complements $\{\{A_n, A_{n+1}, A_{n+2}, \dots, A_\omega\} \mid n \in \mathbb{N}\}$ be the open sets (including $\mathcal{X}$ for $n = 0$) of the topology $\mathcal{T}$ of $\mathcal{X}$.

A_ω is the limit of the sequence $\langle A_k,\ k \in \mathbb{N}\rangle$ because the neighborhoods of A_ω all have the form $U_n = \{A_n, A_{n+1}, A_{n+2}, \dots, A_\omega\}$ where $n \in \mathbb{N}$ so that for any such U_n there exists $N = n \in \mathbb{N}$ such that $\forall k \geqslant N \,.\, A_k \in U_n$.

Example 13.11 Let $\mathcal{S} = \{a^n \mid n \in \mathbb{N}^+\} \cup \{a^\omega\}$ be the set of nonempty finite or infinite sequences of letters a (very like the semantics domain $\mathbb{T}^{+\infty}$).

Consider the transformer $F(X) = \{a\} \cup \{\pi a \mid \pi \in X\}$ (very like the prefix trace transformer of an iteration statement).

Its iterates are $A_0 = \varnothing$, $A_1 = F(A_0) = \{a\} = \{a^1\}, \ldots, A_{n+1} = F(A_n) = F(\{a^1, \ldots, a^{n-1}\}) = \{a\} \cup \{a^1 a, \ldots, a^{n-1} a\} = \{a^1, a^2, \ldots, a^{n-1}, a^n\}$. Define $A_\omega = \{a^\omega\}$.

By example 13.10, the limit of the iterates $\langle A_k,\ k \in \mathbb{N}\rangle$ of the transformer F from $\varnothing$ is $A_\omega = \{a^\omega\}$.

13.5 Dense Sets

Definition 13.12 The *dense sets* P of the topological space $\langle \mathcal{X}, \mathcal{T}\rangle$ are such that $\rho(P) = \mathcal{X}$, that is, belong to $\rho^{-1}(\mathcal{X})$.

Example 13.13 A classic example is the rationals $\mathbb{Q}$ are dense in the reals $\mathbb{R}$. Let $\rho(X)$, $X \in \wp(\mathbb{Q})$ be the operation that adds the limits of the sequences of elements of X. For example, the limit of the sequence of rationals 3, 3.1, 3.14, 3.141, 3.1415, 3.14159, … is π. Then $\rho(\mathbb{Q}) = \mathbb{R}$.

Lemma 13.14 $P \in \wp(\mathcal{X})$ is dense if and only if $P = \neg\rho(P) \cup P$.

Proof of lemma 13.14

— We first prove that for all $P \in \wp(\mathcal{X})$, $\rho(\neg\rho(P)) = \neg\rho(P)$. Indeed $\neg\rho(P) \subseteq \rho(\neg\rho(P))$ because ρ is expansive. Moreover $P \subseteq \rho(P)$ so $\neg\rho(P) \subseteq \neg P$ hence $\rho(\neg\rho(P)) \subseteq \rho(\neg P)$ proving $\neg\rho(P) = \rho(\neg\rho(P))$ by antisymmetry.

- If P is dense then $\rho(P) = \mathcal{X}$ so $\neg\rho(P) = \varnothing$, proving $P = \neg\rho(P) \cup P$.
- Reciprocally, if $P = \neg\rho(P) \cup P$ then $\rho(P) = \rho(\neg\rho(P) \cup P) = \rho(\neg\rho(P)) \cup \rho(P) = \neg\rho(P) \cup \rho(P) = \mathcal{X}$, proving that P is dense. □

Lemma 13.15 Any subset $P \in \wp(\mathcal{X})$ is the intersection $P = \rho(P) \cap (P \cup \neg\rho(P))$ of a closed set $\rho(P)$ and a dense set $(P \cup \neg\rho(P))$.

Proof of lemma 13.15 Let $P \in \wp(\mathcal{X})$. Because $P \subseteq \rho(P)$, we have $P = \rho(P) \cap P = (\rho(P) \cap P) \cup (\rho(P) \cap \neg\rho(P)) = \rho(P) \cap (P \cup \neg\rho(P))$. By lemma 13.9, $\rho(P)$ is a closed set. By lemma 13.14 and commutativity, $P \cup \neg\rho(P)$ is dense. □

13.6 Conclusion

The definitions of this intermezzo on topology will be used in the following chapter 14, "Safety and Liveness Trace Properties," to characterize safety and liveness trace properties.

14 Safety and Liveness Trace Properties

Safety properties are trace properties that can be checked at runtime. Examples are absence of runtime error or that the successive values taken by a variable are increasing. Liveness properties are trace properties that cannot be checked at runtime. Examples are termination or guarantee that the value of an integer variable eventually reach 0. Any trace property is the intersection of a safety and a liveness property [32]. Some program properties (like dependency studied in chapter 47) are not trace properties so are neither a safety nor a liveness nor the intersection of a safety and liveness property.

Contents

14.1 Safety Trace Properties

We explain and formally define safety trace properties as (topological) abstractions. In this chapter and the following ones, we use the Galois isomorphism $\langle \mathbb{T}^+ \rightarrow \wp(\mathbb{T}^{+\infty}), \dot{\subseteq}\rangle \xleftarrow[\alpha_{\text{fr}}]{\gamma_{\text{fr}}} \langle \wp(\mathbb{T}^+ \times \mathbb{T}^{+\infty}), \subseteq\rangle$ of exercise 11.12 between the maximal trace semantics $\boldsymbol{\mathcal{S}}^{+\infty}[\![\mathsf{S}]\!] \in \mathbb{T}^+ \rightarrow \wp(\mathbb{T}^{+\infty})$ and its isomorphic representation as a relation $\alpha_{\text{fr}}(\boldsymbol{\mathcal{S}}^{+\infty}[\![\mathsf{S}]\!])$, which we also denote $\boldsymbol{\mathcal{S}}^{+\infty}[\![\mathsf{S}]\!]$. Same for the prefix trace semantics $\widehat{\boldsymbol{\mathcal{S}}}^*[\![\mathsf{S}]\!] \in \mathbb{T}^+ \rightarrow \wp(\mathbb{T}^+)$.

14.1.1 Intuition for Safety

Safety properties S of programs are trace properties so that $S \in \wp(\mathbb{T}^+ \times \mathbb{T}^{+\infty})$ (and, by section 8.6, are the abstraction $\alpha^{\mathbb{T}}$ (8.8) of semantic properties in $\wp(\wp(\mathbb{T}^+ \times \mathbb{T}^{+\infty}))$).

The characteristics of a safety property S is that any program execution $\langle \pi_0, \pi\rangle \in \boldsymbol{\mathcal{S}}^{+\infty}[\![\mathsf{P}]\!]$ (where $\pi_0 \in \mathbb{T}^+$ and $\pi \in \mathbb{T}^{+\infty}$) that violates S has a *finite* prefix $\langle \pi_0, \pi'\rangle$ that violates S. Therefore if $\langle \pi_0, \pi\rangle \notin S$ then there exists $\pi' \in \mathbb{T}^+$ and $\pi'' \in \mathbb{T}^{*\infty}$ such that $\pi = \pi' \mathbin{\widehat{\cdot}} \pi''$ and $\langle \pi_0, \pi'\rangle \notin S$. By definition (7.6) of $\boldsymbol{\mathcal{S}}^{+\infty}[\![\mathsf{P}]\!]$, $\langle \pi_0, \pi'\rangle \in \boldsymbol{\mathcal{S}}^*[\![\mathsf{P}]\!]$. So the violation of a safety property can be observed using the prefix trace semantics only. This means that during execution π, the safety property can be checked on all prefixes to prove violation.

Example 14.1 The property "the program never divides by zero" is a safety property because the violation "the program eventually divides by zero" can be proved at runtime by testing for a division by zero and waiting long enough. How long is unknown, but definitely finite in case of violation.

The program property "all executions do terminate" (that is, $\boldsymbol{S}^{+\infty}[\![\mathsf{P}]\!] \subseteq (\mathbb{T}^+ \times \mathbb{T}^+)$) is not a safety property because if the program has not yet terminated in a prefix computation, it does not mean that it will terminate later or will never terminate. The violation of the termination property is "some execution never terminates," which is not observable in finite time, so this is not a safety property.

The violation of a safety property by a given trace is checkable at runtime, but its verification is another story. There might be infinitely many different finite traces and infinite ones, so an exhaustive search of possible violations is hopeless, as shown by Rice's theorem 9.12.

14.1.2 Definition and Examples of Prefix Closure

Define the *prefix closure* $\alpha_{\mathsf{pref}}(\Pi)$ of a set of executions (that is of trace properties $\Pi \in \wp(\mathbb{T}^+ \times \mathbb{T}^{+\infty})$) as taking all (finite and infinite) prefixes of traces in Π.

$$\begin{array}{rcll} \alpha_{\mathsf{pref}} & \in & \wp(\mathbb{T}^+ \times \mathbb{T}^{+\infty}) \mapsto \wp(\mathbb{T}^+ \times \mathbb{T}^{+\infty}) & \text{prefix closure} \\ \alpha_{\mathsf{pref}}(\Pi) & \triangleq & \{\langle \pi_0, \pi\rangle \in \mathbb{T}^+ \times \mathbb{T}^{+\infty} \mid \exists \pi' \in \mathbb{T}^{*\infty} . \langle \pi_0, \pi \mathbin{\widehat{}} \pi'\rangle \in \Pi\} & \end{array}$$

where, by (6.4) $\pi \mathbin{\widehat{}} \pi' = \pi$ when $\langle \pi_0, \pi\rangle \in \mathbb{T}^+ \times \mathbb{T}^{\infty}$ so $\Pi \subseteq \alpha_{\mathsf{pref}}(\Pi)$.

Example 14.2 Ignoring actions, let us write σ^n for the finite sequences of n consecutive states σ and σ^ω for the infinite sequence of states σ. The property $E = \{\langle \sigma, \sigma^{2n}\rangle \mid n \in \mathbb{N}^+\}$ is "once started in state σ, execution remains in that state, and terminates in an even number of steps." The prefix closure of E is $T = \alpha_{\mathsf{pref}}(E) = \{\langle \sigma, \sigma^n\rangle \mid n \in \mathbb{N}^+\}$. Observe that T is the weaker property, "once started in state σ, execution remains in that state, and terminates."

Observe that we can define the prefix ordering on traces such that

$$\begin{array}{rcll} \pi \leq \pi' & \triangleq & \exists \pi'' \in \mathbb{T}^{*\infty} . \pi \mathbin{\widehat{}} \pi'' = \pi' & \text{prefix ordering} \quad (14.3) \\ \pi < \pi' & \triangleq & \pi \leq \pi' \wedge \pi \neq \pi' & \text{strict prefix ordering} \end{array}$$

extended to executions as $\langle \pi_0, \pi\rangle \leq \langle \pi_0', \pi'\rangle \triangleq \pi_0 = \pi_0' \wedge \pi \leq \pi'$, so that we could equally have defined

$$\begin{array}{rcll} \alpha_{\mathsf{pref}} & \in & \wp(\mathbb{T}^+ \times \mathbb{T}^{+\infty}) \mapsto \wp(\mathbb{T}^+ \times \mathbb{T}^{+\infty}) & \text{prefix closure} \\ \alpha_{\mathsf{pref}}(\Pi) & \triangleq & \{\langle \pi_0, \pi\rangle \in \mathbb{T}^+ \times \mathbb{T}^{+\infty} \mid \exists \pi' \in \mathbb{T}^{+\infty} . \langle \pi_0, \pi'\rangle \in \Pi . \pi \leq \pi'\} & (14.4) \end{array}$$

Lemma 14.5 α_{pref} is a topological closure on $\wp(\mathbb{T}^+ \times \mathbb{T}^{+\infty})$.

Proof

— By reflexivity of $\leq$ in (14.4), α_{pref} is extensive.

— For idempotence, we have

$$\begin{array}{ll} \alpha_{\mathsf{pref}}(\alpha_{\mathsf{pref}}(\Pi)) & \\ = \alpha_{\mathsf{pref}}(\{\langle \pi_0, \pi\rangle \in \mathbb{T}^+ \times \mathbb{T}^{+\infty} \mid \exists \langle \pi_0, \pi'\rangle \in \Pi . \pi \leq \pi'\}) & \text{(definition (14.4) of } \alpha_{\mathsf{pref}}\text{)} \end{array}$$

$$
\begin{aligned}
&= \{\langle\pi_0', \pi''\rangle \in \mathbb{T}^+ \times \mathbb{T}^{+\infty} \mid \exists\langle\pi_0', \pi'''\rangle \in \{\langle\pi_0, \pi\rangle \in \mathbb{T}^+ \times \mathbb{T}^{+\infty} \mid \exists\langle\pi_0, \pi'\rangle \in \Pi \,.\, \pi \preceq \pi'\} \,.\\
&\qquad \pi'' \preceq \pi'''\} && \wr\text{definition (14.4) of } \alpha_{\text{pref}}\wr\\
&= \{\langle\pi_0, \pi''\rangle \in \mathbb{T}^+ \times \mathbb{T}^{+\infty} \mid \exists\pi \in \mathbb{T}^{+\infty} \,.\, \exists\langle\pi_0, \pi'\rangle \in \Pi \,.\, \pi \preceq \pi' \wedge \pi'' \preceq \pi\} && \wr\text{definition of } \in\wr\\
&= \{\langle\pi_0, \pi''\rangle \in \mathbb{T}^+ \times \mathbb{T}^{+\infty} \mid \exists\langle\pi_0, \pi'\rangle \in \Pi \,.\, \pi'' \preceq \pi'\}\\
&\qquad \wr(\subseteq) \preceq \text{ is transitive;}\\
&\qquad\ (\supseteq) \text{ take } \pi = \pi' \text{ and } \preceq \text{ is reflexive}\wr\\
&= \alpha_{\text{pref}}(\Pi) && \wr\text{definition (14.4) of } \alpha_{\text{pref}}\wr
\end{aligned}
$$

— $\alpha_{\text{pref}}(\varnothing) = \{\langle\pi_0, \pi\rangle \in \mathbb{T}^+ \times \mathbb{T}^{+\infty} \mid \exists\langle\pi_0, \pi'\rangle \in \varnothing \,.\, \pi \preceq \pi'\} = \{\langle\pi_0, \pi\rangle \in \mathbb{T}^+ \times \mathbb{T}^{+\infty} \mid \mathit{ff}\} = \varnothing$ proving α_{pref} to be strict.

— For the preservation of finite joins, we have

$$
\begin{aligned}
&\alpha_{\text{pref}}(\Pi_1 \cup \Pi_2)\\
&= \{\langle\pi_0, \pi\rangle \in \mathbb{T}^+ \times \mathbb{T}^{+\infty} \mid \exists\pi' \in \mathbb{T}^{+\infty} \,.\, \langle\pi_0, \pi \frown \pi'\rangle \in \Pi_1 \cup \Pi_2\} && \wr\text{definition of } \alpha_{\text{pref}}\wr\\
&= \{\langle\pi_0, \pi\rangle \in \mathbb{T}^+ \times \mathbb{T}^{+\infty} \mid (\exists\pi' \in \mathbb{T}^{+\infty} \,.\, \langle\pi_0, \pi \frown \pi'\rangle \in \Pi_1) \vee (\exists\pi' \in \mathbb{T}^{+\infty} \,.\, \langle\pi_0,\\
&\qquad \pi \frown \pi'\rangle \in \Pi_2)\} && \wr\text{definition of } \cup\wr\\
&= \{\langle\pi_0, \pi\rangle \in \mathbb{T}^+ \times \mathbb{T}^{+\infty} \mid \exists\pi' \in \mathbb{T}^{+\infty} \,.\, \langle\pi_0, \pi \frown \pi'\rangle \in \Pi_1\} \cup \{\pi \in \mathbb{T}^{+\infty} \mid \exists\pi' \in \mathbb{T}^{+\infty} \,.\, \langle\pi_0,\\
&\qquad \pi \frown \pi'\rangle \in \Pi_2\} && \wr\text{definition of } \cup\wr\\
&= \alpha_{\text{pref}}(\Pi_1) \cup \alpha_{\text{pref}}(\Pi_2) && \wr\text{definition of } \alpha_{\text{pref}}\wr \quad \square
\end{aligned}
$$

14.1.3 Definition and Examples of Limit Closure

Define the *limit closure* $\alpha_{\text{limit}}(\Pi)$ of a set of traces (that is on trace properties Π) as taking all infinite traces for which prefixes are in Π.

$$
\begin{aligned}
\alpha_{\text{limit}} &\in \wp(\mathbb{T}^+ \times \mathbb{T}^{+\infty}) \mapsto \wp(\mathbb{T}^+ \times \mathbb{T}^{+\infty}) && \text{limit closure}\\
\alpha_{\text{limit}}(\Pi) &\triangleq \Pi \cup \{\langle\pi_0, \pi\rangle \in \mathbb{T}^+ \times \mathbb{T}^{\infty} \mid \forall\pi' < \pi \,.\, \langle\pi_0, \pi'\rangle \in \Pi\}
\end{aligned}
$$

Example 14.6 Continuing example 14.2, with the property $T = \{\langle\sigma, \sigma^n\rangle \mid n \in \mathbb{N}^+\}$ stating that "once started in state σ, execution remains in that state, and terminates." Its limit closure is $\alpha_{\text{limit}}(T) = \{\langle\sigma, \sigma^n\rangle \mid n \in \mathbb{N}^+\} \cup \{\langle\sigma, \sigma^\omega\rangle\}$, which is the weaker property "once started in state σ, execution remains in that state" (and may or may not terminate).

[**Lemma 14.7** α_{limit} is a topological closure on $\wp(\mathbb{T}^+ \times \mathbb{T}^{+\infty})$.

Proof

— $\Pi \subseteq \Pi \cup \{\pi \in \mathbb{T}^+ \times \mathbb{T}^{\infty} \mid \forall\pi' < \pi \,.\, \pi' \in \Pi\} = \alpha_{\text{limit}}(\Pi)$ proving α_{limit} to be extensive.

— $\alpha_{\text{limit}}(\varnothing) = \varnothing \cup \{\langle\pi_0, \pi\rangle \in \mathbb{T}^+ \times \mathbb{T}^{\infty} \mid \forall\pi' < \pi \,.\, \pi' \in \varnothing\} = \varnothing \cup \{\langle\pi_0, \pi\rangle \in \mathbb{T}^+ \times \mathbb{T}^{\infty} \mid \mathit{ff}\} = \varnothing$ proving α_{limit} to be strict.

— For idempotence, we have

$$
\begin{aligned}
&\alpha_{\text{limit}}(\alpha_{\text{limit}}(\Pi))\\
&= \Pi \cup \{\langle\pi_0, \pi\rangle \in \mathbb{T}^+ \times \mathbb{T}^{\infty} \mid \forall\pi' < \pi \,.\, \langle\pi_0, \pi'\rangle \in \Pi \cup \{\langle\pi_0', \pi''\rangle \in \mathbb{T}^{\infty} \mid \forall\pi''' < \pi'' \,.\, \langle\pi_0',\\
&\qquad \pi'''\rangle \in \Pi\}\} && \wr\text{definition of } \alpha_{\text{limit}}\wr
\end{aligned}
$$

$= \Pi \cup \{\langle\pi_0, \pi\rangle \in \mathbb{T}^+ \times \mathbb{T}^\infty \mid \forall \pi' < \pi \,.\, \langle\pi_0, \pi'\rangle \in \Pi\}$

$\langle$because $\pi' < \pi$ implies $\langle\pi_0, \pi'\rangle \in \mathbb{T}^+ \times \mathbb{T}^+$ and so $\langle\pi_0', \pi'\rangle \notin \{\langle\pi_0', \pi''\rangle \in \mathbb{T}^+ \times \mathbb{T}^\infty \mid \forall \pi''' < \pi'' \,.\, \langle\pi_0', \pi'''\rangle \in \Pi\}\}\rangle$

$= \alpha_{\text{limit}}(\Pi)$ $\langle$definition of $\alpha_{\text{limit}}\rangle$

— Trivially, $\{\langle\pi_0, \pi\rangle \in \mathbb{T}^+ \times \mathbb{T}^\infty \mid \forall \pi' < \pi \,.\, \langle\pi_0, \pi'\rangle \in \Pi_1\} \cup \{\langle\pi_0, \pi\rangle \in \mathbb{T}^+ \times \mathbb{T}^\infty \mid \forall \pi' < \pi \,.\, \langle\pi_0, \pi'\rangle \in \Pi_2\} \subseteq \{\langle\pi_0, \pi\rangle \in \mathbb{T}^+ \times \mathbb{T}^\infty \mid \forall \pi' < \pi \,.\, \langle\pi_0, \pi'\rangle \in \Pi_1 \cup \Pi_2\}$.

For the inverse inclusion, we must prove that $\forall \pi' < \pi \,.\, \langle\pi_0, \pi'\rangle \in \Pi_1 \cup \Pi_2$ implies $\forall \pi' < \pi \,.\, \langle\pi_0, \pi'\rangle \in \Pi_1$ or $\forall \pi' < \pi \,.\, \langle\pi_0, \pi'\rangle \in \Pi_2$. This is trivial when $\langle\pi_0, \pi\rangle \in \Pi_1$ or $\langle\pi_0, \pi\rangle \in \Pi_2$. Otherwise $\langle\pi_0, \pi\rangle$ has infinitely many prefixes in $\Pi_1 \cup \Pi_2$. If there are finitely many in Π_1 then there are infinitely many in Π_2 so the limit trace $\langle\pi_0, \pi\rangle$ is in $\{\langle\pi_0, \pi\rangle \in \mathbb{T}^+ \times \mathbb{T}^\infty \mid \forall \pi' < \pi \,.\, \langle\pi_0, \pi'\rangle \in \Pi_2\}$. The case is the same if there are finitely many in Π_2 so that there are infinitely many in Π_1 hence the limit trace π is in $\{\langle\pi_0, \pi\rangle \in \mathbb{T}^+ \times \mathbb{T}^\infty \mid \forall \pi' < \pi \,.\, \langle\pi_0, \pi'\rangle \in \Pi_1\}$. Finally, if π has infinitely many prefixes in both Π_1 and Π_2 then the limits π are identical (assuming they are different, then there is a prefix π' in one not appearing in the other, which is impossible because the other has infinitely many prefixes, hence one longer than π'). Therefore $\langle\pi_0, \pi\rangle$ appears in both $\{\langle\pi_0, \pi\rangle \in \mathbb{T}^+ \times \mathbb{T}^\infty \mid \forall \pi' < \pi \,.\, \langle\pi_0, \pi'\rangle \in \Pi_1\}$ and $\{\langle\pi_0, \pi\rangle \in \mathbb{T}^+ \times \mathbb{T}^\infty \mid \forall \pi' < \pi \,.\, \langle\pi_0, \pi'\rangle \in \Pi_2\}$. By antisymmetry, we have equality.

— The preservation of finite joins follows, because

$\alpha_{\text{limit}}(\Pi_1 \cup \Pi_2)$

$= \Pi_1 \cup \Pi_2 \cup \{\langle\pi_0, \pi\rangle \in \mathbb{T}^+ \times \mathbb{T}^\infty \mid \forall \pi' < \pi \,.\, \langle\pi_0, \pi'\rangle \in \Pi_1 \cup \Pi_2\}$ $\langle$definition of $\alpha_{\text{limit}}\rangle$

$= \Pi_1 \cup \Pi_2 \cup \{\langle\pi_0, \pi\rangle \in \mathbb{T}^+ \times \mathbb{T}^\infty \mid \forall \pi' < \pi \,.\, \pi' \in \Pi_1\} \cup \{\langle\pi_0, \pi\rangle \in \mathbb{T}^+ \times \mathbb{T}^\infty \mid \forall \pi' < \pi \,.\, \langle\pi_0, \pi'\rangle \in \Pi_2\}$ $\langle$as shown above$\rangle$

$= \alpha_{\text{limit}}(\Pi_1) \cup \alpha_{\text{limit}}(\Pi_2)$ $\langle$definition of $\alpha_{\text{limit}}\rangle$ □

14.1.4 Definition and examples of Safety Closure

Define the *safety closure* α_{safety} on sets of traces (that is, on trace properties $\Pi \in \wp(\mathbb{T}^+ \times \mathbb{T}^{+\infty})$) such that $\alpha_{\text{safety}}(\Pi)$ is the set of limits of prefixes of Π.

$$\begin{aligned} \alpha_{\text{safety}} &\in \wp(\mathbb{T}^+ \times \mathbb{T}^{+\infty}) \mapsto \wp(\mathbb{T}^+ \times \mathbb{T}^{+\infty}) \\ \alpha_{\text{safety}} &\triangleq \alpha_{\text{limit}} \circ \alpha_{\text{pref}} \end{aligned} \tag{14.8}$$

Example 14.9 For example 14.2 with $E = \{\langle\sigma, \sigma^{2n}\rangle \mid n \in \mathbb{N}^+\}$ ("once started in state σ, execution remains in that state, and terminates in an even number of steps"), we have $\alpha_{\text{limit}} \circ \alpha_{\text{pref}}(E) = \{\langle\sigma, \sigma^n\rangle \mid n \in \mathbb{N}^+\} \cup \{\langle\sigma, \sigma^\omega\rangle\}$ ("once started in state σ, execution remains in that state").

Theorem 14.10 α_{safety} is a topological closure on $\wp(\mathbb{T}^+ \times \mathbb{T}^{+\infty})$.

Proof

— We first prove that the prefix closure of a set closed by limits and prefixes cannot bring in new prefixes that is $\alpha_{\text{pref}} \circ \alpha_{\text{limit}} \circ \alpha_{\text{pref}} = \alpha_{\text{limit}} \circ \alpha_{\text{pref}}$. To do so, we prove that any prefix of a trace in $\alpha_{\text{limit}} \circ \alpha_{\text{pref}}$ is in $\alpha_{\text{limit}} \circ \alpha_{\text{pref}}$ i.e. $\forall \pi_0, \pi, \pi' . (\langle \pi_0, \pi \rangle \in \alpha_{\text{limit}} \circ \alpha_{\text{pref}}(X) \wedge \pi' \preceq \pi) \Rightarrow (\langle \pi_0, \pi' \rangle \in \alpha_{\text{limit}} \circ \alpha_{\text{pref}}(X))$. By definition of α_{limit} we have

$$\alpha_{\text{limit}} \circ \alpha_{\text{pref}}(X) \quad = \quad \alpha_{\text{pref}}(X) \cup \{\langle \pi_0, \pi \rangle \in \mathbb{T}^+ \times \mathbb{T}^\infty \mid \forall \pi' \prec \pi . \langle \pi_0, \pi' \rangle \in \alpha_{\text{pref}}(X)\}$$

— If $\pi' = \pi$ then $\langle \pi_0, \pi \rangle \in \alpha_{\text{limit}} \circ \alpha_{\text{pref}}(X)$ obviously implies $\langle \pi_0, \pi' \rangle \in \alpha_{\text{limit}} \circ \alpha_{\text{pref}}(X)$.
— If $\langle \pi_0, \pi \rangle \in \alpha_{\text{pref}}(X)$ then, by definition of α_{pref}, $\pi' \preceq \pi$ implies $\langle \pi_0, \pi' \rangle \in \alpha_{\text{pref}}(X) \subseteq \alpha_{\text{limit}} \circ \alpha_{\text{pref}}(X)$.
— If $\langle \pi_0, \pi \rangle \in \{\langle \pi_0', \pi \rangle \in \mathbb{T}^+ \times \mathbb{T}^\infty \mid \forall \pi' \prec \pi . \langle \pi_0', \pi' \rangle \in \alpha_{\text{pref}}(X)\}$ and $\pi' \prec \pi$ then $\langle \pi_0, \pi' \rangle \in \alpha_{\text{limit}} \circ \alpha_{\text{pref}}(X)$.

So we have proved $\alpha_{\text{pref}} \circ \alpha_{\text{limit}} \circ \alpha_{\text{pref}} \mathrel{\dot{\subseteq}} \alpha_{\text{limit}} \circ \alpha_{\text{pref}}$. We have $\alpha_{\text{pref}} \circ \alpha_{\text{limit}} \circ \alpha_{\text{pref}} \mathrel{\dot{\supseteq}} \alpha_{\text{limit}} \circ \alpha_{\text{pref}}$ because α_{pref} is expansive by lemma 14.5. By antisymmetry, $\alpha_{\text{pref}} \circ \alpha_{\text{limit}} \circ \alpha_{\text{pref}} = \alpha_{\text{limit}} \circ \alpha_{\text{pref}}$.

— It follows that α_{safety} is idempotent.

$$\begin{aligned}
& \alpha_{\text{safety}} \circ \alpha_{\text{safety}} \\
= \; & \alpha_{\text{limit}} \circ \alpha_{\text{pref}} \circ \alpha_{\text{limit}} \circ \alpha_{\text{pref}} && \text{(definition of } \alpha_{\text{safety}}\text{)} \\
= \; & \alpha_{\text{limit}} \circ \alpha_{\text{limit}} \circ \alpha_{\text{pref}} && \text{(as shown previously)} \\
= \; & \alpha_{\text{limit}} \circ \alpha_{\text{pref}} && \text{(lemma 14.7 so that } \alpha_{\text{limit}} \text{ is idempotent)}
\end{aligned}$$

— By lemmas 14.5 and 14.7 both α_{pref} and α_{limit} are expansive, strict and preserve finite joins and so does their composition α_{safety}. □

14.1.5 Topological Definition of Safety

The safety properties are closed by prefix and limits.

Definition 14.11 (safety) The *safety properties* are the trace properties $P \in \wp(\mathbb{T}^+ \times \mathbb{T}^{+\infty})$ such that $\alpha_{\text{safety}}(P) = P$.

Example 14.12 We have seen in the example 14.9 that the safety closure of the property $E = \{\langle \sigma, \sigma^{2n} \rangle \mid n \in \mathbb{N}^+\}$, that is, "once started in state σ, execution remains in that state, and terminates in an even number of steps" is $S = \alpha_{\text{safety}}(E) = \{\langle \sigma, \sigma^n \rangle \mid n \in \mathbb{N}^+\} \cup \{\langle \sigma, \sigma^\omega \rangle\}$ which is the strictly weaker property "once started in state σ, execution remains in that state." Hence E is not a safety property because $\alpha_{\text{safety}}(E) = S \neq E$.

Example 14.13 The property $S = \{\langle \sigma, \sigma^n \rangle \mid n \in \mathbb{N}^+\} \cup \{\langle \sigma, \sigma^\omega \rangle\}$ i.e. "once started in state σ, execution remains in that state" is a safety property because $\alpha_{\text{safety}}(S) = S$. An execution satisfying this property (which may be infinite) will never reach state $\sigma' \neq \sigma$.

Theorem 14.14 The safety properties are the closed sets of the topology defined by α_{safety} on $\mathbb{T}^+ \times \mathbb{T}^{+\infty}$.

Proof α_{safety} is topological closure (theorem 14.10) so it defines a topology (lemma 13.9) in which closed sets are the $\Pi \in \wp(\mathbb{T}^+ \times \mathbb{T}^{+\infty})$ such that $\alpha_{\text{safety}}(\Pi) = \Pi$ (lemma 13.8). □

Theorem 14.15 The poset $\langle\alpha_{\mathsf{safety}}(\wp(\mathbb{T}^{+}\times\mathbb{T}^{+\infty})),\ \subseteq\rangle$ (i.e., the postimage of $\wp(\mathbb{T}^{+}\times\mathbb{T}^{+\infty})$ by α_{safety}) of safety properties is a complete lattice.

Proof $\langle\wp(\mathbb{T}^{+}\times\mathbb{T}^{+\infty}),\ \subseteq\rangle$ is a complete lattice, α_{safety} is a topological, hence upper closure, operator, and theorem 11.88. □

Example 14.16 Consider the trace properties $P_1=\{\langle\sigma,\ \sigma\rangle\}$, $P_2=\{\langle\sigma,\ \sigma\rangle,\langle\sigma,\ \sigma\sigma\rangle\}$, $P_3=\{\langle\sigma,\ \sigma\rangle,\langle\sigma,\ \sigma\sigma\rangle,\langle\sigma,\ \sigma\sigma\sigma\rangle\}$, …, $P_n=\{\langle\sigma,\ \sigma^k\rangle\mid 1\leqslant k\leqslant n\}$, …. They are closed by prefixes and limits so are all safety property. Their union $\bigcup_{n\geqslant 1}P_n$ is not a safety property because it does not contain the limit $\langle\sigma,\ \sigma^{\infty}\rangle$. By theorem 11.88, the union in the complete lattice of safety properties is precisely $\alpha_{\mathsf{safety}}(\bigcup_{n\geqslant 1}P_n)=\alpha_{\mathsf{limit}}(\bigcup_{n\geqslant 1}P_n)$ which does introduce the limit $\langle\sigma,\ \sigma^{\infty}\rangle$ and is therefore a safety property. Again by theorem 11.88, notice that the (infinite) intersection of safety properties is a safety property.

14.1.6 Examples of Safety Properties

Example 14.17 The property "after reaching program point ℓ, the value of variable x is always positive" that is $\{\langle\overline{\pi},\ \pi\rangle\in\mathbb{T}^{+}\times\mathbb{T}^{+\infty}\mid\forall i<|\pi|\,.\,(\pi_i=\ell)\Rightarrow(\forall j\,.\,i<j<|\pi|\,.\,\varrho(\overline{\pi}\smallfrown\pi[0..j])\mathsf{x}>0)\}$ is a safety property (where if $\pi=\ell_0\xrightarrow{\mathsf{a}_0}\ldots\ell_i\xrightarrow{\mathsf{a}_i}\ldots$ is a trace and $p<|\pi|$ then its prefix $\pi[0..p]$ at p is $\ell_0\xrightarrow{\mathsf{a}_0}\ldots\ell_j\xrightarrow{\mathsf{a}_j}\ldots\ell_p$ as defined in section 7.2).

Example 14.18 The property "the value of variable x is increasing," that is, $\{\langle\overline{\pi},\ \pi\rangle\in\mathbb{T}^{+}\times\mathbb{T}^{+\infty}\mid\forall i\leqslant j<|\pi|\,.\,\varrho(\overline{\pi}\smallfrown\pi[0..i])\mathsf{x}\leqslant\varrho(\overline{\pi}\smallfrown\pi[0..j])\mathsf{x})\}$ is a safety property.

Exercise 14.19 Prove that termination is not a safety property.

14.1.7 Runtime Checks of Safety Violation

The violation of a safety property by a computation trace can be verified in finite time,[1] on a finite prefix of that trace (theorem 14.20 is the definition of safety in [32, p. 182]). This explains the common explanation of safety as "nothing bad can happen."

Theorem 14.20 If $\alpha_{\mathsf{safety}}(\Pi)=\Pi$ then $\forall\langle\pi_0,\ \pi\rangle\notin\Pi\,.\,\exists\pi'\in\mathbb{T}^{+}\,.\,\langle\pi_0,\ \pi'\rangle\preceq\langle\pi_0,\ \pi\rangle\wedge\langle\pi_0,\ \pi'\rangle\notin\Pi$

Proof

— If $\pi\in\mathbb{T}^{+}$ then choosing $\pi'=\pi$, we have $\langle\pi_0,\ \pi'\rangle\preceq\langle\pi_0,\ \pi\rangle$ by reflexivity of $\preceq$ and $\langle\pi_0,\ \pi'\rangle\notin\Pi$ by hypothesis.

— Otherwise, $\pi\in\mathbb{T}^{\infty}$ and so for all $\langle\pi_0,\ \pi'\rangle\prec\langle\pi_0,\ \pi\rangle$, $\pi'\in\mathbb{T}^{+}$ and $\langle\pi_0,\ \pi'\rangle\in\Pi$ by prefix closure, proving $\langle\pi_0,\ \pi\rangle\in\{\langle\pi_0,\ \pi\rangle\in\mathbb{T}^{+}\times\mathbb{T}^{\infty}\mid\forall\langle\pi_0,\ \pi'\rangle\prec\langle\pi_0,\ \pi\rangle\,.\,\langle\pi_0,\ \pi'\rangle\in\Pi\}\subseteq\alpha_{\mathsf{limit}}(\Pi)=\alpha_{\mathsf{limit}}(\alpha_{\mathsf{safety}}(\Pi))=\alpha_{\mathsf{limit}}(\alpha_{\mathsf{limit}}\circ\alpha_{\mathsf{pref}}(\Pi))=\alpha_{\mathsf{limit}}\circ\alpha_{\mathsf{pref}}(\Pi)=\alpha_{\mathsf{safety}}(\Pi)=\Pi$ because α_{limit} is idempotent.

1. At least in theory because one may have to keep the full computation history in memory to check for the property.

We proved $\forall \pi_0 \in \mathbb{T}^+, \pi \in \mathbb{T}^{+\infty} . ((\forall \pi' \in \mathbb{T}^{+\infty} . \langle \pi_0, \pi' \rangle \leq \langle \pi_0, \pi \rangle) \Rightarrow (\langle \pi_0, \pi' \rangle \in \Pi))$ implies $\langle \pi_0, \pi \rangle \in \Pi$, and so by contraposition, $\langle \pi_0, \pi \rangle \notin \Pi$ implies $\exists \pi' \in \mathbb{T}^{+\infty} . (\langle \pi_0, \pi' \rangle \leq \langle \pi_0, \pi \rangle) \wedge (\langle \pi_0, \pi' \rangle \notin \Pi)$. □

Example 14.21 Continuing example 14.13, a runtime check at any step $i < |\pi|$ that an execution $\langle \sigma, \pi \rangle$ does satisfy the safety property $S \triangleq \{\langle \sigma, \sigma^n \rangle \mid n \in \mathbb{N}^+\} \cup \{\langle \sigma, \sigma^\omega \rangle\}$ is $\pi_i = \sigma$. If the safety property does not hold, the check will fail at some step $j < |\pi|$.

However, as shown by Rice's theorem 9.12, the satisfaction of such a property cannot be proved by runtime checks (there might be infinitely many finite traces and infinite traces to check in which case the tests never terminate).

Exercise 14.22 Let $\Pi = \{\langle a, a^n b^m \rangle \mid n, m \in \mathbb{N}^+\}$ be the executions starting with a and consisting of a finite number of a's followed by a finite number of b's. Is it a safety property?

14.1.8 Safety Specifications

The *safety specification domain* is $\langle \alpha_{\text{safety}}(\wp(\mathbb{T}^+ \times \mathbb{T}^{+\infty})), \subseteq \rangle$. A *safety specification* is a α_{safety}-closed set of computations. A program P with semantics $\boldsymbol{\mathcal{S}}^{+\infty}[\![\mathsf{P}]\!] \in \wp(\mathbb{T}^+ \times \mathbb{T}^{+\infty})$ *satisfies* the safety specification $P \in \alpha_{\text{safety}}(\wp(\mathbb{T}^+ \times \mathbb{T}^{+\infty}))$ whenever $\boldsymbol{\mathcal{S}}^{+\infty}[\![\mathsf{P}]\!] \in \gamma^{\mathbb{T}} \circ \mathbb{1}_{\mathbb{T}^{+\infty}}(P)$ (where $\gamma^{\mathbb{T}}$ is defined in (8.8) and $\mathbb{1}_{\mathbb{T}^{+\infty}}$ is th identity on $\mathbb{T}^{+\infty}$). Safety properties are an abstraction of semantic properties.

Lemma 14.23 $\langle \wp(\wp(\mathbb{T}^+ \times \mathbb{T}^{+\infty})), \subseteq \rangle \xrightleftharpoons[\alpha_{\text{safety}} \circ \alpha^{\mathbb{T}}]{\gamma^{\mathbb{T}} \circ \mathbb{1}_{\mathbb{T}^{+\infty}}} \langle \alpha_{\text{safety}}(\wp(\mathbb{T}^+ \times \mathbb{T}^{+\infty})), \subseteq \rangle$.

Proof By composing the Galois retractions $\langle \wp(\wp(\mathbb{T}^+ \times \mathbb{T}^{+\infty})), \subseteq \rangle \xrightleftharpoons[\alpha^{\mathbb{T}}]{\gamma^{\mathbb{T}}} \langle \wp(\mathbb{T}^+ \times \mathbb{T}^{+\infty}), \subseteq \rangle$ (see exercise 8.9 isomorphically for $\wp(\mathbb{T}^+ \to \wp(\mathbb{T}^+ \times \mathbb{T}^{+\infty}))$) and $\langle \wp(\mathbb{T}^+ \times \mathbb{T}^{+\infty}), \subseteq \rangle \xrightleftharpoons[\alpha_{\text{safety}}]{\mathbb{1}_{\mathbb{T}^{+\infty}}} \langle \alpha_{\text{safety}}(\mathbb{T}^+ \times \mathbb{T}^{+\infty}), \subseteq \rangle$ of theorem 14.10 and exercise 11.61. □

Exercise 14.24 Is the infinite union of safety properties a safety property? What is the lub in the complete lattice of safety properies of theorem 14.15?

14.1.9 Safety Satisfaction

Although safety is relative to both finite and infinite traces in $\mathbb{T}^{+\infty}$, the infinite traces need not be taken intp account in the proof that a program P with a prefix closed semantics $\boldsymbol{\mathcal{S}}^{+\infty}[\![\mathsf{P}]\!] \in \wp(\mathbb{T}^+ \times \mathbb{T}^{+\infty})$ satisfies the safety specification $P \in \alpha_{\text{safety}}(\wp(\mathbb{T}^+ \times \mathbb{T}^{+\infty}))$.

Theorem 14.25 $\forall S \in \alpha_{\text{pref}}(\wp(\mathbb{T}^+ \times \mathbb{T}^{+\infty})) . \forall P \in \alpha_{\text{safety}}(\wp(\mathbb{T}^+ \times \mathbb{T}^{+\infty})) . (S \subseteq P) \Leftrightarrow (S \cap (\mathbb{T}^+ \times \mathbb{T}^+)) \subseteq (P \cap (\mathbb{T}^+ \times \mathbb{T}^+))$.

Proof

— The implication $\Rightarrow$, is trivial by definition of the glb.

— Reciprocally $\Leftarrow$, if $\langle \pi_0, \pi \rangle$ is an infinite computation in $(S \cap (\mathbb{T}^+ \times \mathbb{T}^\infty))$ then $S = \alpha_{\text{pref}}(S)$ is closed by prefix, so all finite prefixes π' of π have $\langle \pi_0, \pi' \rangle$ in S, hence in $S \cap (\mathbb{T}^+ \times \mathbb{T}^+)$, which implies that they belong to $P \cap (\mathbb{T}^+ \times \mathbb{T}^+)$. Because $P = \alpha_{\text{safety}}(P)$ is a safety property, it is closed by limits and hence, containing all prefixes of π, it also contains π proving $(S \cap (\mathbb{T}^+ \times \mathbb{T}^\infty)) \subseteq P$. It follows that $S = (S \cap (\mathbb{T}^+ \times \mathbb{T}^+)) \cup (S \cap (\mathbb{T}^+ \times \mathbb{T}^\infty)) \subseteq (P \cap (\mathbb{T}^+ \times \mathbb{T}^+)) \cup P = P$. □

14.2 Liveness Trace Property

We formally define liveness trace properties as (topological) dense sets. Informally, liveness is explained as "something good must happen," that is, an eventuality/guarantee property. We show that this reduction of liveness to guarantee is improper.

14.2.1 Topological Definition of Liveness

Definition 14.26 (liveness) The *liveness properties* are the dense sets of the topology defined by α_{safety} (hence, by lemma 13.14, such that $\mathsf{live}(P) = P$ where $\mathsf{live}(P) \triangleq \neg\alpha_{\mathsf{safety}}(P) \cup P$.

Observe that live is extensive and idempotent but not increasing to that $\mathsf{live}(P)$ need not be the least liveness property implied by P.

Theorem 14.27 $P \in \wp(\mathbb{T}^+ \times \mathbb{T}^{+\infty})$ is a liveness property if and only if $\neg P \subseteq \alpha_{\mathsf{safety}}(P)$.

Proof of theorem 14.27

$$
\begin{aligned}
& \mathsf{live}(P) = P \\
\Leftrightarrow\ & \neg\alpha_{\mathsf{safety}}(P) \cup P = P && \text{(definition 14.26 of live}(P)\text{)} \\
\Leftrightarrow\ & \neg(\neg\alpha_{\mathsf{safety}}(P) \cup P) = \neg P && \text{(definition of the complement } \neg\text{)} \\
\Leftrightarrow\ & \alpha_{\mathsf{safety}}(P) \cap \neg P = \neg P && \text{(De Morgan's laws)} \\
\Leftrightarrow\ & \neg P \subseteq \alpha_{\mathsf{safety}}(P) && \text{(definition of the glb)}
\end{aligned}
$$
□

14.2.2 Examples of Liveness

Example 14.28 As discussed in exercise 14.19, given a nonempty set Π of executions, the property "all executions terminate" is $\Pi \in \wp(\wp(\mathbb{T}^+ \times \mathbb{T}^+))$ or equivalently $\Pi \subseteq \wp(\mathbb{T}^+ \times \mathbb{T}^+)$. Exercise 14.19 shows that $\alpha_{\mathsf{safety}}(\mathbb{T}^+ \times \mathbb{T}^+) = \mathbb{T}^+ \times \mathbb{T}^{+\infty}$.

It follows that $\mathsf{live}(\mathbb{T}^+ \times \mathbb{T}^+) \triangleq \neg\alpha_{\mathsf{safety}}(\mathbb{T}^+ \times \mathbb{T}^+) \cup (\mathbb{T}^+ \times \mathbb{T}^+) = \emptyset \cup (\mathbb{T}^+ \times \mathbb{T}^+) = (\mathbb{T}^+ \times \mathbb{T}^+)$. which is the definition of a liveness property. Hence termination $(\mathbb{T}^+ \times \mathbb{T}^+)$ is a liveness property.

Exercise 14.29 Consider infinite traces made out of two actions a_0 and a_1. A trace $\pi \in \mathbb{T}^\infty(\{\mathsf{a}_0, \mathsf{a}_1\})$ is *fair* if and only if both a_0 and a_1 do appear infinitely often in the infinite trace π. Formalize this fairness property and prove that it is a liveness property.

14.2.3 Impossible Runtime Checks of Liveness Violation

Liveness properties cannot be checked at runtime (because if the property is not satisfied after a finite time, there is always the possibility that it will be satisfied later). More precisely, if a prefix of a trace does not satisfy a liveness property then there is an extension that does satisfy this liveness property (theorem 14.30 is the definition of liveness in [32, p. 182]).

Theorem 14.30 For all $P \in \wp(\mathbb{T}^+ \times \mathbb{T}^{+\infty})$, we have $\mathsf{live}(P) = P$ if and only if $\forall \pi_0 \in \mathbb{T}^+ . \forall \pi \in \mathbb{T}^{+\infty} . \exists \pi' \in \mathbb{T}^{+\infty} . \langle \pi_0, \pi \frown \pi' \rangle \in P$.

Proof of theorem 14.30

$$
\begin{aligned}
& \mathsf{live}(P) = P \\
\Leftrightarrow\ & P \text{ is a dense set} && \text{(definition 14.26 of live and lemma 13.14)} \\
\Leftrightarrow\ & \alpha_{\mathsf{safety}}(P) = \mathbb{T}^+ \times \mathbb{T}^{+\infty} && \text{(definition 13.12 of dense sets)} \\
\Leftrightarrow\ & \mathbb{T}^+ \times \mathbb{T}^{+\infty} \subseteq \alpha_{\mathsf{safety}}(P) && \text{(because } \alpha_{\mathsf{safety}} \in \wp(\mathbb{T}^+ \times \mathbb{T}^{+\infty}) \mapsto \wp(\mathbb{T}^+ \times \mathbb{T}^{+\infty})\text{)} \\
\Leftrightarrow\ & \forall \langle \pi_0, \pi \rangle \in \mathbb{T}^+ \times \mathbb{T}^{+\infty} . \langle \pi_0, \pi \rangle \in \alpha_{\mathsf{limit}} \circ \alpha_{\mathsf{pref}}(P) && \text{(definition of } \subseteq \text{ and } \alpha_{\mathsf{safety}}\text{)} \\
\Leftrightarrow\ & \forall \langle \pi_0, \pi \rangle \in \mathbb{T}^+ \times \mathbb{T}^{+\infty} . \langle \pi_0, \pi \rangle \in (\alpha_{\mathsf{pref}}(P) \cup \{\langle \pi_0', \pi \rangle \in \mathbb{T}^+ \times \mathbb{T}^{\infty} \mid \forall \langle \pi_0', \pi' \rangle \lessdot \langle \pi_0', \\
& \quad \pi \rangle . \langle \pi_0', \pi' \rangle \in \alpha_{\mathsf{pref}}(P)\}) && \text{(definition of } \alpha_{\mathsf{limit}}\text{)} \\
\Leftrightarrow\ & \forall \langle \pi_0, \pi \rangle \in \mathbb{T}^+ \times \mathbb{T}^{+\infty} . \langle \pi_0, \pi \rangle \in \alpha_{\mathsf{pref}}(P) \vee (\langle \pi_0, \pi \rangle \in \mathbb{T}^+ \times \mathbb{T}^{\infty} \wedge \forall \langle \pi_0, \pi' \rangle \lessdot \langle \pi_0, \\
& \quad \pi \rangle . \langle \pi_0, \pi' \rangle \in \alpha_{\mathsf{pref}}(P)\}) && \text{(definition of } \cup\text{)} \\
\Leftrightarrow\ & \forall \langle \pi_0, \pi \rangle \in \mathbb{T}^+ \times \mathbb{T}^+ . \langle \pi_0, \pi \rangle \in \alpha_{\mathsf{pref}}(P) \\
& \quad \text{(}(\Rightarrow)\ \mathbb{T}^+ \cap \mathbb{T}^{\infty} = \varnothing \\
& \quad (\Leftarrow)\ \langle \pi_0, \pi \rangle \in \mathbb{T}^+ \times \mathbb{T}^{\infty} \wedge \langle \pi_0, \pi' \rangle \lessdot \langle \pi_0, \pi \rangle \text{ implies } \langle \pi_0, \pi' \rangle \in \mathbb{T}^+ \text{ and so } \langle \pi_0, \\
& \quad \pi' \rangle \in \alpha_{\mathsf{pref}}(P)\text{)} \\
\Leftrightarrow\ & \forall \langle \pi_0, \pi \rangle \in \mathbb{T}^+ \times \mathbb{T}^+ . \langle \pi_0, \pi \rangle \in \{\langle \pi_0', \pi \rangle \in \mathbb{T}^+ \times \mathbb{T}^{+\infty} \mid \exists \pi' \in \mathbb{T}^{+\infty} . \langle \pi_0', \pi \frown \pi' \rangle \in P\} \\
& && \text{(definition of } \alpha_{\mathsf{pref}}\text{)} \\
\Leftrightarrow\ & \forall \pi_0 . \pi \in \mathbb{T}^+ . \exists \pi' \in \mathbb{T}^{+\infty} . \langle \pi_0, \pi \frown \pi' \rangle \in P && \text{(definition of } \in\text{)} \quad \square
\end{aligned}
$$

14.2.4 Liveness Specifications

Given a set $\mathbb{T}^{+\infty}$ of finite and infinite traces, the *liveness specification domain* is $\langle \mathsf{live}(\wp(\mathbb{T}^+ \times \mathbb{T}^{+\infty})), \subseteq \rangle$. A *liveness specification* is a live-closed set of traces. A program P with semantics $\boldsymbol{\mathcal{S}}^{+\infty}[\![\mathsf{P}]\!] \in \wp(\wp(\mathbb{T}^+ \times \mathbb{T}^{+\infty}))$ *satisfies* the liveness specification $P \in \mathsf{live}(\wp(\mathbb{T}^+ \times \mathbb{T}^{+\infty}))$ whenever $\boldsymbol{\mathcal{S}}^{+\infty}[\![\mathsf{P}]\!] \in \gamma^{\mathbb{T}}(P)$. This time, we do not have a Galois connection.

Exercise 14.31 Characterize the properties which are both a safety and a liveness property.

14.3 Safety/Liveness Decomposition of Trace Properties

Finally, any trace property is the intersection of a safety (closed) and a liveness (dense) property.

Theorem 14.32 ([32, Th. 1]) $\forall P \in \wp(\mathbb{T}^+ \times \mathbb{T}^{+\infty}) \,.\, P = \alpha_{\mathsf{safety}}(P) \cap \mathsf{live}(P)$.

Proof This is a corollary of lemma 13.15. □

Exercise 14.33 Show that the program property "to be deterministic" (i.e., to have only one possible execution trace, either finite or infinite) is neither a safety nor a liveness property.

14.4 Guarantee

Because there is no strongest liveness property implied by a given property, it is interesting to look for weaker concepts of liveness having this property. For example, guarantee properties [651] are defined by the *guarantee closure*

$$\begin{array}{rcll} \alpha_{\mathsf{guarantee}} & \in & \wp(\mathbb{T}^+ \times \mathbb{T}^{+\infty}) \mapsto \wp(\mathbb{T}^+ \times \mathbb{T}^{+\infty}) & \text{guarantee closure} \\ \alpha_{\mathsf{guarantee}}(X) & \triangleq & (\mathbb{T}^+ \times \mathbb{T}^*) \circledcirc X \qquad \text{where} & \\ X \circledcirc Y & \triangleq & \{\langle \pi_0,\, \pi \frown \pi' \rangle \mid \langle \pi_0,\, \pi \rangle \in X \wedge \langle \pi_0 \frown \pi,\, \pi' \rangle \in Y\} & \text{concatenation} \end{array}$$

Theorem 14.34 $\alpha_{\mathsf{guarantee}}$ is an upper closure operator on $\wp(\mathbb{T}^+ \times \mathbb{T}^{+\infty})$.

Proof
— We have $X \subseteq \alpha_{\mathsf{guarantee}}(X)$ because $\langle \pi_0, \ni \rangle \in \mathbb{T}^*$ so $(\mathbb{T}^+ \times \mathbb{T}^*) \circledcirc X$ includes $\{\langle \pi_0, \ni \frown \pi' \rangle \mid \langle \pi_0 \frown \ni, \pi' \rangle \in X\} = X$.
— $\alpha_{\mathsf{guarantee}}$ is idempotent because $\alpha_{\mathsf{guarantee}}(\alpha_{\mathsf{guarantee}}(X)) = (\mathbb{T}^+ \times \mathbb{T}^*) \circledcirc ((\mathbb{T}^+ \times \mathbb{T}^*) \circledcirc X)$ and $\circledcirc$ is associative such that $(\mathbb{T}^+ \times \mathbb{T}^*) \circledcirc (\mathbb{T}^+ \times \mathbb{T}^*) = (\mathbb{T}^+ \times \mathbb{T}^*)$.
— If $X \subseteq Y$ then $\alpha_{\mathsf{guarantee}}(X) = \{\langle \pi_0, \pi \frown \pi' \rangle \mid \langle \pi_0, \pi \rangle \in \mathbb{T}^+ \times \mathbb{T}^* \wedge \langle \pi_0 \frown \pi, \pi' \rangle \in X\} \subseteq \{\langle \pi_0, \pi \frown \pi' \rangle \mid \langle \pi_0, \pi \rangle \in \mathbb{T}^+ \times \mathbb{T}^* \wedge \langle \pi_0 \frown \pi, \pi' \rangle \in Y\} = \alpha_{\mathsf{guarantee}}(Y)$. □

Definition 14.35 (guarantee) The *guarantee properties* are the trace properties $P \in \wp(\mathbb{T}^+ \times \mathbb{T}^{+\infty})$ such that $\alpha_{\mathsf{guarantee}}(P) = P$.

A guarantee property ensures that some event eventually happens at least once (and so it will still happen when delayed for a finite amount of time).

Example 14.36 Termination is a guarantee property because $\alpha_{\mathsf{guarantee}}(\mathbb{T}^+ \times \mathbb{T}^+) = ((\mathbb{T}^+ \times \mathbb{T}^*) \circledcirc (\mathbb{T}^+ \times \mathbb{T}^+)) = \mathbb{T}^+ \times \mathbb{T}^+$.

Exercise 14.37 Prove that fairness in exercise 14.29 is a guarantee property.

Theorem 14.38 Any guarantee property is a liveness property.

Proof We have $\alpha_{\mathsf{safety}}(\alpha_{\mathsf{guarantee}}(X)) = \alpha_{\mathsf{safety}}((\mathbb{T}^+ \times \mathbb{T}^*) \circledcirc X) = \mathbb{T}^+ \times \mathbb{T}^{+\infty}$ so $\mathsf{live}(X) = \neg\alpha_{\mathsf{safety}}(X) \cup X = \neg\alpha_{\mathsf{safety}}(\alpha_{\mathsf{guarantee}}(X)) \cup X = \neg(\mathbb{T}^+ \times \mathbb{T}^{+\infty}) \cup X = \varnothing \cup X = X$ □

Note that this shows that guarantee properties are a strict subset of the liveness properties because liveness properties do not have this complete lattice structure.

Example 14.39 The liveness property that "an interactive system can continue to provide services while its usage of resources cannot grow indefinitely" is not a guarantee property.

Exercise 14.40 Formalize example 14.39 and prove that the stated property is a liveness property but not a guarantee property.

Theorem 14.41 The poset $\langle \alpha_{\text{guarantee}}(\wp(\mathbb{T}^+ \times \mathbb{T}^{+\infty})), \subseteq\rangle$ of guarantee properties is a complete lattice $\langle \alpha_{\text{guarantee}}(\wp(\mathbb{T}^+ \times \mathbb{T}^{+\infty})), \subseteq, \emptyset, \mathbb{T}^{+\infty}, \cup, \cap\rangle$.

Proof A direct consequence of exercise 11.61 and theorem 11.88, $\alpha_{\text{guarantee}}(\emptyset) = \emptyset$ and if $\forall i \,.\, X_i \in \alpha_{\text{guarantee}}(\wp(\mathbb{T}^+ \times \mathbb{T}^{+\infty}))$ then $\alpha_{\text{guarantee}}(\bigcup_i X_i) = (\mathbb{T}^+ \times \mathbb{T}^*) \circledcirc (\bigcup_i X_i) = \bigcup_i ((\mathbb{T}^+ \times \mathbb{T}^*) \circledcirc X) = \bigcup_i \alpha_{\text{guarantee}}(X_i) = \bigcup_i X_i$. □

But for the true property, a property cannot be both a safety and a guarantee property.

Theorem 14.42 $\alpha_{\text{safety}}(\wp(\mathbb{T}^+ \times \mathbb{T}^{+\infty})) \cap \alpha_{\text{guarantee}}(\wp(\mathbb{T}^+ \times \mathbb{T}^{+\infty})) = \{\mathbb{T}^+ \times \mathbb{T}^{+\infty}\}$

Proof Assume $P \in \wp(\mathbb{T}^+ \times \mathbb{T}^{+\infty})$ has $\alpha_{\text{safety}}(P) = P$ and $\alpha_{\text{guarantee}}(P) = P$. Then $P = \alpha_{\text{safety}}(P) = \alpha_{\text{safety}}(\alpha_{\text{guarantee}}(P)) = \alpha_{\text{safety}}((\mathbb{T}^+ \times \mathbb{T}^*) \circledcirc P) = \mathbb{T}^+ \times \mathbb{T}^{+\infty}$. □

Exercise 14.43 (reduction of liveness to safety for finite systems) It is sometimes argued that actual programs executions can go through only finitely many states because machines are finite (see the discussion in remark 2.7). This means, informally, that if $\langle \pi_0, \pi\rangle \in \boldsymbol{\mathcal{S}}^+[\![\mathsf{P}]\!] \cap (\mathbb{T}^+ \times \mathbb{T}^{+\infty})$ then π has the form $\pi'(\pi'')^\infty$ where $\pi', \pi'' \in \mathbb{T}^+$. Prove that, in that case, the guarantee properties are safety properties. Does this hold, in that case, for all liveness properties?

14.5 Conclusion

This book concentrates on the verification and static analysis of safety properties of programs.

The safety/liveness classification of trace properties is credited to [32, 33] (see also [1, 611, 863]) with a topological presentation in [133, 654, 960]. The guarantee properties are part of Zohar Manna and Amir Pnueli's [170, 650, 651] hierarchy of weak liveness/progress properties. Reduction of some liveness properties to safety (exercise 14.4) has been studied, for example, by [118, 772, 867]. Topological presentations of abstract interpretation can be found in [238, 858, 859].

14.6 Solutions to Selected Exercises

Solution to exercise 14.19 The termination property is $\mathbb{T}^+ \times \mathbb{T}^+$. Its safety closure is

$$
\begin{aligned}
& \alpha_{\text{safety}}(\mathbb{T}^+ \times \mathbb{T}^+) \\
=\ & \alpha_{\text{pref}}(\mathbb{T}^+ \times \mathbb{T}^+) \cup \alpha_{\text{limit}}(\alpha_{\text{pref}}(\mathbb{T}^+ \times \mathbb{T}^+)) && \text{(definition of } \alpha_{\text{safety}}) \\
=\ & (\mathbb{T}^+ \times \mathbb{T}^+) \cup \alpha_{\text{limit}}(\mathbb{T}^+ \times \mathbb{T}^+) && \text{(definition of } \alpha_{\text{pref}}) \\
=\ & (\mathbb{T}^+ \times \mathbb{T}^+) \cup (\mathbb{T}^+ \times \mathbb{T}^\infty) && \text{(definition of } \alpha_{\text{limit}})
\end{aligned}
$$

$= (\mathbb{T}^+ \times \mathbb{T}^{+\infty})$ $\wr$definition of $\mathbb{T}^{+\infty}$$\int$

so $\alpha_{\text{safety}}(\mathbb{T}^+ \times \mathbb{T}^+) = (\mathbb{T}^+ \times \mathbb{T}^{+\infty}) \neq (\mathbb{T}^+ \times \mathbb{T}^+)$, proving that termination is not a safety property.

Solution to exercise 14.22 Because n is not bounded the observation of finite prefixes cannot distinguish between prefixes of correct traces and the incorrect ones $a^\infty = aaaa\ldots$ or $a^n b^\infty$ so the property is not a safty property.

Solution to exercise 14.24 The infinite union of safety properties is not a safety property. For example, terminating in exactly n steps is a safety property. To check it at runtime, simply count the number of steps and produce an alarm after n steps. Their infinite union is termination, which is not a safety property. By theorem 14.15, the safety properties form a complete lattice $\langle \alpha_{\text{safety}}(\wp(\mathbb{T}^+ \times \mathbb{T}^{+\infty})), \subseteq, \varnothing, \mathbb{T}^+ \times \mathbb{T}^{+\infty}, S \mapsto \alpha_{\text{limit}}(S), \cap \rangle$ (because the join of prefix closed traces is prefix closed). Therefore the intersection of safety properties is a safety property and the limit of their unions, not their union, is the lub. In the example, the union is termination but the limit of termination introduces the possibility of nontermination.

Solution to exercise 14.31 By definition 14.11, $P \in \wp(\mathbb{T}^+ \times \mathbb{T}^{+\infty})$ is a safety property if and only if $\alpha_{\text{safety}}(P) = P$ or, equivalently $\alpha_{\text{safety}}(P) \subseteq P$ since, by theorem 14.10, α_{safety} is extensive. If P is also a liveness property then, by definition 14.26, $\text{live}(P) = P$ so $\neg\alpha_{\text{safety}}(P) \cup P = P$ and therefore, $\neg\alpha_{\text{safety}}(P) \subseteq P$. It follows that $\mathbb{T}^+ \times \mathbb{T}^{+\infty} = \alpha_{\text{safety}}(P) \cup \neg\alpha_{\text{safety}}(P) \subseteq P$, proving that $P = \mathbb{T}^+ \times \mathbb{T}^{+\infty}$ is the only property which is both a safety and a liveness property.

Solution to exercise 14.37 Adding any finite prefix trace to a fair trace yields a fair trace.

Solution to exercise 14.40 Let r be the action of using a resource, s be the action of providing a service, and o be some other action. A server $\pi \in \mathbb{T}^\infty(\{r, s, o\})$ is an infinite sequence of actions (i.e., once the server is launched, its actions are recorded, and it is assumed to never terminate).

Consider the properties $P(N)$, $N \in \mathbb{N}$ of servers such that

$$P(N) \triangleq \{\pi \in \mathbb{T}^\infty(\{r, s, o\}) \mid (\forall i \,.\, \exists j \geqslant i \,.\, \pi_i = s) \Rightarrow (|\{i \in \mathbb{N} \mid \pi_i = r\}| \leqslant N)\}.$$

A fair server $\pi \in P(N)$ satisfies $\forall i \,.\, \exists j \geqslant i \,.\, \pi_i = s$, that is, at any point i in time, the service s will later be provided at time $j \geqslant i$. Under these circumstances, the trace π must use at most N resources. Otherwise, the trace is unfair, this is, at some point in time no service will ever be provided. Nothing is required on unfair traces, in particular an unbounded number of resources may be used (as in r^∞).

For all $N \in \mathbb{N}$, $P(N)$ is not a guarantee property, since if a fair trace π is prefixed by any finite trace using more that N resources, the trace is still fair, but the property is no longer true.

The safety closure of P contains the prefixes of all unfair traces, that is, $\mathbb{T}^+(\{r, s, o\})$ so the limit is $\mathbb{T}^{+\infty}(\{r, s, o\})$ with empty complement. It follows that $\text{live}(P) = P$, proving P to be a liveness property.

V FIXPOINTS AND THEIR ABSTRACTIONS

Fixpoints are the mathematical formalization of iteration and recursion (chapter 15).

They provide the rigorous foundations of the deductive systems used to describe semantics (chapter 16) and can replace them in the descriptions of semantics (chapter 17).

The abstraction of fixpoints (chapter 18) is the basis for the design of more abstract semantics and static analyzers.

15 Fixpoints

Iterating an *operator* f on a set S (i.e., $f \in S \to S$) from an element $a \in S$, that is, $f^0 \triangleq a$ and $f^{n+1} \triangleq f(f^n)$, can

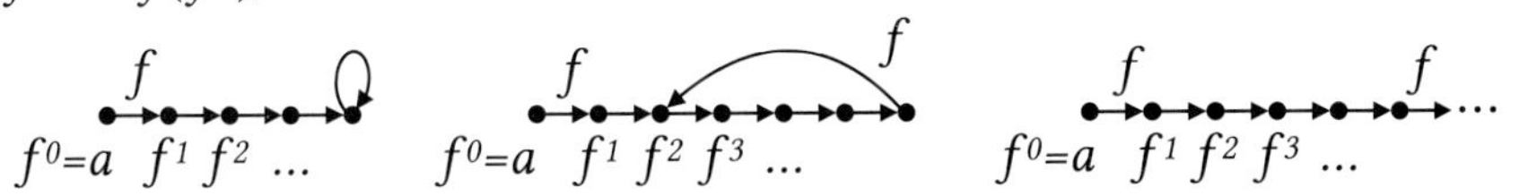

1. reach a *fixpoint* $f^p = f(f^p)$;
2. or reach a *cycle* and iterate forever along that cycle $f^{p+\ell} = f^p$, $\ell > 0$;
3. or *iterate forever* along infinitely many different elements $\forall p \neq q\,.\, f^p \neq f^q$. This is impossible if the set S is finite.

We study sufficient conditions on S, f, and a, ensuring that the iteration reaches a fixpoint.

Contents

15.1 Fixpoint Theorems

A *fixpoint* of a function $f \in S \to S$ is an element x of S such that $f(x) = x$. In general, a function may have no, one, or many fixpoints.

Exercise 15.1 Find functions $f \in \mathbb{Z} \to \mathbb{Z}$ that have no, one, and many fixpoints.

Let us recall definition 11.28 of *increasing*. A function $f \in S \to S$ on a poset $\langle S, \sqsubseteq\rangle$ is increasing (denoted $f \in S \xrightarrow{\nearrow} S$) if and only if $\forall x, y \in S\,.\,(x \sqsubseteq y) \Rightarrow (f(x) \sqsubseteq f(y))$.

Exercise 15.2 (alternative characterization of increasing functions) Prove that a function f is increasing if and only if, for all $X \in \wp(S)$, when the lubs do exist, $\bigsqcup\{f(x) \mid x \in X\} \sqsubseteq f(\bigsqcup\{x \mid x \in X\})$. What is the dual?

Exercise 15.3 (composition of increasing functions) Prove that the composition of increasing functions is increasing.

As shown by Alfred Tarski [925], an increasing function on a complete lattice has at least one fixpoint and, moreover, has a least one. This is illustrated in figure 15.4.

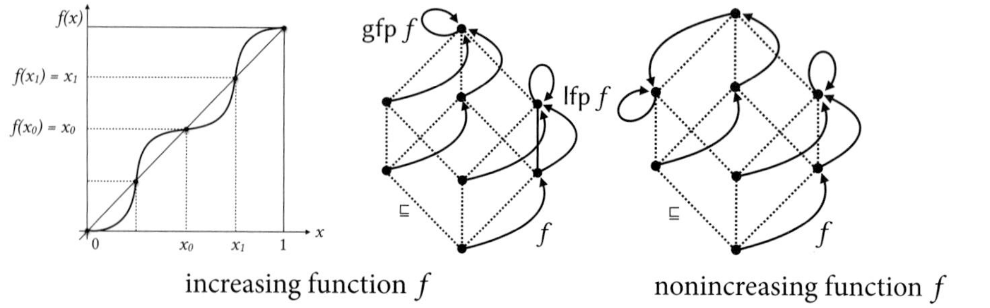

Figure 15.4
Fixpoints of increasing and nonincreasing functions

Exercise 15.5 What is the complete lattice considered in the left-hand example in figure 15.4. What is the least fixpoint?

$\mathsf{lfp}^{\sqsubseteq} f$ denotes the least fixpoint of $f \in L \to L$ for the partial order $\langle L, \sqsubseteq\rangle$, if any. If the partial order $\sqsubseteq$ is understood from the context, we write $\mathsf{lfp}\ f$. $\mathsf{lfp}_a^{\sqsubseteq} f$ is the least fixpoint of f for $\sqsubseteq$, which is $\sqsubseteq$-greater than or equal to $a \in L$, if any, that is, $a \sqsubseteq \mathsf{lfp}_a^{\sqsubseteq} f = f(\mathsf{lfp}_a^{\sqsubseteq} f)$ and if $a \sqsubseteq x = f(x)$ then $\mathsf{lfp}_a^{\sqsubseteq} f \sqsubseteq x$. Dually, for $\mathsf{gfp}_a^{\sqsubseteq} f$, $\mathsf{gfp}^{\sqsubseteq} f$, and $\mathsf{gfp}\ f$

> **Theorem 15.6 (Tarski's fixpoint theorem)** An increasing function $f \in L \xrightarrow{\nearrow} L$ on a complete lattice $\langle L, \sqsubseteq, \bot, \top, \sqcap, \sqcup\rangle$ has a least fixpoint $\mathsf{lfp}^{\sqsubseteq} f = \bigsqcap\{x \in L \mid f(x) \sqsubseteq x\}$.

Proof Let $a \triangleq \bigsqcap\{x \in L \mid f(x) \sqsubseteq x\}$ where the greatest lower bound $\bigsqcap$ is well defined in a complete lattice. For all $x \in L$,

$f(x) \sqsubseteq x$

$\Rightarrow a \sqsubseteq x$

$\wr$because $x \in \{x \in L \mid f(x) \sqsubseteq x\}$ and the greatest lower bound a is smaller than or equal to any element of the set$\wr$

$\Rightarrow f(a) \sqsubseteq f(x) \sqsubseteq x$ $\quad\wr f$ is increasing and $f(x) \sqsubseteq x\wr$

$\Rightarrow f(a) \sqsubseteq \bigsqcap\{x \in L \mid f(x) \sqsubseteq x\} = a$

$\wr$because $f(a)$ is a lower bound of $\{x \in L \mid f(x) \sqsubseteq x\}$ hence smaller than or equal to its greatest lower bound $a\wr$

$\Rightarrow f(f(a)) \sqsubseteq f(a)$ $\quad\wr$because f is increasing$\wr$

$\Rightarrow f(a) \in \{x \in L \mid f(x) \sqsubseteq x\}$ $\quad\wr$definition of $\in\wr$

$\Rightarrow a = \bigsqcap\{x \in L \mid f(x) \sqsubseteq x\} \sqsubseteq f(a)$

$\wr$definition of the greatest lower bound smaller than or equal to any element of $\{x \in L \mid f(x) \sqsubseteq x\}\wr$

$\Rightarrow a = f(a)$ $\quad\wr$by antisymmetry$\wr$

Let $x = f(x)$ be a fixpoint of f. Then $x \in \{x' \in L \mid f(x') \sqsubseteq x'\}$ by reflexivity. It follows that $a = \bigsqcap\{x' \in L \mid f(x') \sqsubseteq x'\} \sqsubseteq x$ by definition of the greatest lower bound. Therefore $a = \mathsf{lfp}^{\sqsubseteq} f$ is the least fixpoint of f. □

Tarski's theorem 15.6 was generalized to CPOs [656, theorem 9] and [780] as reported by [354, corollary 2.1.].

Remark 15.7 Tarski's theorem 15.6 shows that computing the least fixpoint $\mathsf{lfp}^{\sqsubseteq} f$ can be understood as solving the optimization problem [143]

$$\begin{aligned} &\underset{x}{\text{minimize}} \quad f(x) \\ &\text{subject to} \quad f(x) \sqsubseteq x \end{aligned}$$

where f is the objective function and $f(x) \sqsubseteq x$ is an inequality constraint. The optimization problem may be solvable for some complete lattices L [230, 635] (e.g., finite ones) but is, in general, uncomputable [634].

Exercise 15.8 Given an increasing function $f \in L \xrightarrow{\nearrow} L$ on a complete lattice $\langle L, \sqsubseteq, \bot, \top, \sqcap, \sqcup \rangle$, show that $\mathsf{lfp}^{\sqsubseteq} f = \mathsf{lfp}^{\sqsubseteq}\, x \mapsto x \sqcup f(x)$.

Exercise 15.9 Given an increasing function $f \in L \xrightarrow{\nearrow} L$ on a complete lattice $\langle L, \sqsubseteq, \bot, \top, \sqcap, \sqcup \rangle$, show that for all $a \in L$ such that $a \sqsubseteq f(a)$, $\mathsf{lfp}^{\sqsubseteq}\, x \mapsto a \sqcup f(x) = \mathsf{lfp}_a^{\sqsubseteq} f$ where $\mathsf{lfp}_a^{\sqsubseteq} f$ is the least fixpoint of f greater than or equal to $a \in L$.

Exercise 15.10 Given an increasing function $f \in L \xrightarrow{\nearrow} L$ on a complete lattice $\langle L, \sqsubseteq, \bot, \top, \sqcap, \sqcup \rangle$ show that $\mathsf{lfp}^{\sqsubseteq} f = \bigsqcap \{x \in L \mid f(x) = x\}$.

Exercise 15.11 Given an increasing function $f \in L \xrightarrow{\nearrow} L$ on a complete lattice $\langle L, \sqsubseteq, \bot, \top, \sqcap, \sqcup \rangle$ and a prefixpoint $a \in L$ such that $a \sqsubseteq f(a)$, show that $\mathsf{lfp}_a^{\sqsubseteq} f = \bigsqcap \{x \in L \mid a \sqsubseteq x \wedge f(x) \sqsubseteq x\}$ is the least fixpoint of f greater than or equal to a.

Exercise 15.12 (fixpoint induction) Let $f \in L \xrightarrow{\nearrow} L$ be an increasing function on the complete lattice $\langle L, \sqsubseteq, \bot, \top, \sqcap, \sqcup \rangle$ and $x \in L$. Prove that $\mathsf{lfp}^{\sqsubseteq} f \sqsubseteq x$ if and only if $\exists y \in L\,.\, f(y) \sqsubseteq y \wedge y \sqsubseteq x$.

Exercise 15.13 ([777, (2.2)]) Let $f \in L \xrightarrow{\nearrow} L$ be an increasing function on the complete lattice $\langle L, \sqsubseteq, \bot, \top, \sqcap, \sqcup \rangle$ and $x \in L$. Prove that $\mathsf{lfp}^{\sqsubseteq} f \sqsubseteq x$ if and only if $\mathsf{lfp}^{\sqsubseteq} f \sqcap f(x) \sqsubseteq x$.

Exercise 15.14 Let $f \in L \xrightarrow{\nearrow} L$ be an increasing function on the complete lattice $\langle L, \sqsubseteq, \bot, \top, \sqcap, \sqcup \rangle$. Prove that f has a greatest fixpoint $\mathsf{gfp}^{\sqsubseteq} f$.

Exercise 15.15 Let $f \in L \xrightarrow{\nearrow} L$ be an increasing function on the complete lattice $\langle L, \sqsubseteq, \bot, \top, \sqcap, \sqcup \rangle$. Prove that the set of fixpoints of f ordered by $\sqsubseteq$ form a complete lattice (but not necessarily a sublattice of L, meaning that the join and meet in this complete lattice of fixpoints may not be $\sqcup$ and $\sqcap$).

Exercise 15.16 (language recognized by an automaton) Consider the following automaton 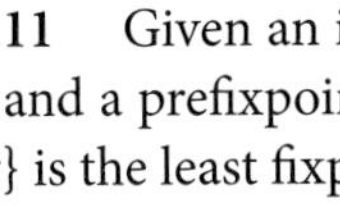

. Characterize the language recognized by this automaton [850, section 2.4.3] as a fixpoint. Express the solution as a regular expression.

Exercise 15.17 (Knaster-Tarski's fixpoint theorem [584]) Let S be a set and $f \in \wp(S) \xrightarrow{\nearrow} \wp(S)$ be increasing for $\subseteq$. If $A \subseteq S$ is such that $f(A) \subseteq A$ then there exists a subset $D \subseteq A$ such that $f(D) = D$.

Exercise 15.18 (Bekić–Leszczyłowski's fixpoint theorem [104, 630]) Let $\langle L, \sqsubseteq, \bot, \top, \sqcup, \sqcap\rangle$ be a complete lattice. Assume that $f \in L^{n+m} \xrightarrow{\nearrow} L^n$ and $g \in L^{n+m} \xrightarrow{\nearrow} L^m$ are increasing componentwise. We write $\langle X, Y\rangle = \langle X_1, \ldots, X_n, Y_1, \ldots, Y_m\rangle$ when $X \in L^n$ and $Y \in L^m$. Let us consider $F_1\langle X, Y\rangle = \langle f\langle X, Y\rangle, g\langle X, Y\rangle\rangle$. Let us first solve the fixpoint equation in X with the resolvant $r = Y \mapsto \mathsf{lfp}\, X \mapsto f\langle X, Y\rangle$. Let us next solve in Y with $F_2\langle X, Y\rangle = \langle r(Y), g(r(Y), Y)\rangle$. Prove that $\mathscr{F}_1 = \{\langle X, Y\rangle \mid \langle X, Y\rangle = F_1\langle X, Y\rangle\} \subseteq \mathscr{F}_2 = \{\langle X, Y\rangle \mid \langle X, Y\rangle = F_2\langle X, Y\rangle\}$ and that $\mathsf{lfp}^{\sqsubseteq} F_1 = \mathsf{lfp}^{\sqsubseteq} F_2$.

15.2 Iterative Fixpoint Theorems

In general, the iteration $f^0 = a$ and $f^{n+1} = f(f^n)$ for $n \in \mathbb{N}$ of a function $f \in S \to S$ on a set S from an element $a \in S$ either reaches a fixpoint, a cycle, or loops forever.

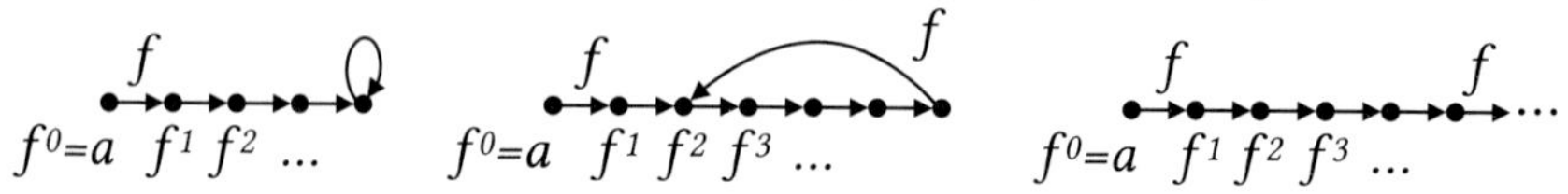

In the case of an increasing function on a poset $\langle P, \sqsubseteq\rangle$, the iterates from $a \in S$ such that $a \sqsubseteq f(a)$ are increasing, so reaching a cycle is impossible by antisymmetry. If limits exist and have a continuity property then a fixpoint is always reached, as illustrated in figure 15.19.

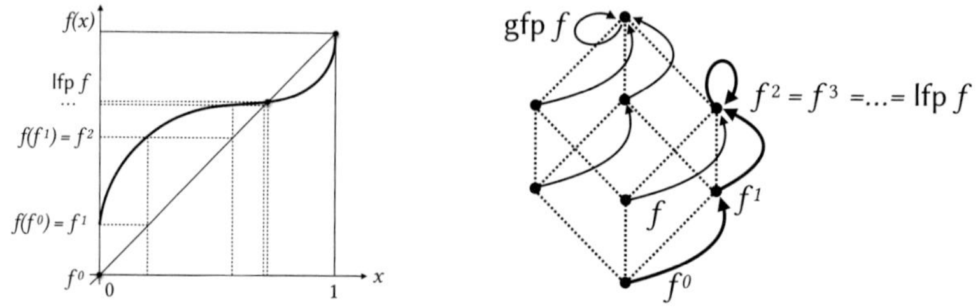

Figure 15.19
Iteration

Definition 15.20 (iterates) Let $f \in P \to P$ be a function on a poset $\langle P, \sqsubseteq, \bot, \sqcup\rangle$ with infimum $\bot$ and lub $\sqcup$. The (partially defined) *iterates* of f from $a \in P$ are $\langle f^n, n \in \mathbb{N} \cup \{\omega\}\rangle$ such that $f^0 \triangleq a$, $f^{n+1} \triangleq f(f^n)$ for $n \in \mathbb{N}$ and $f^\omega \triangleq \bigsqcup\{f^n \mid n \in \mathbb{N}\}$ (where, to denote the limit of infinite iterations, the naturals are extended by the infinity ω such that $\forall n \in \mathbb{N}\,.\, n < \omega \leqslant \omega$). By default, $a = \bot$. We write $\langle f^n(a), n \in \mathbb{N} \cup \{\omega\}\rangle$ to make clear that the iterates start from $a \neq \bot$.

The iterates *converge* in finitely many steps or are *ultimately stationary* at $\ell \in \mathbb{N}$ if and only if $\forall n \geqslant \ell\,.\, f^n = f^\ell$.

Theorem 15.21 (Tarski–Kantorovich's iterative fixpoint) Let $f \in P \xrightarrow{\nearrow} P$ be an increasing function on a poset $\langle P, \sqsubseteq, \bot, \sqcup\rangle$ with infimum $\bot$ and lub $\sqcup$. Let $\langle f^n, n \in \mathbb{N} \cup \{\omega\}\rangle$ be the iterates of f. Assuming that the least upper bound $f^\omega = \bigsqcup\{f^n \mid$

$n \in \mathbb{N}\}$ exists and $f(\bigsqcup\{f^n \mid n \in \mathbb{N}\}) = \bigsqcup\{f(f^n) \mid n \in \mathbb{N}\}$ then $\langle f^n,\ n \in \mathbb{N} \cup \{\omega\}\rangle$ is an increasing chain and f has a least fixpoint $\mathsf{lfp}^{\sqsubseteq} f = \bigsqcup\{f^n \mid n \in \mathbb{N}\}$.

Proof Observe that $\bigsqcup\{f^n \mid n \in \mathbb{N}\} = \bigsqcup(\{f^{n+1} \mid n \in \mathbb{N}\} \cup \{f^0\}) = (\bigsqcup\{f^{n+1} \mid n \in \mathbb{N}\}) \sqcup \bot = \bigsqcup\{f(f^n) \mid n \in \mathbb{N}\}$ because $f^0 = \bot$ is the infimum and $f^{n+1} = f(f^n)$ for $n \in \mathbb{N}$. Then by hypothesis, $f(\bigsqcup\{f^n \mid n \in \mathbb{N}\}) = \bigsqcup\{f(f^n) \mid n \in \mathbb{N}\} = \bigsqcup\{f^n \mid n \in \mathbb{N}\}$, proving that $\bigsqcup\{f^n \mid n \in \mathbb{N}\}$ is a fixpoint of f.

Let $x = f(x)$ be a fixpoint of f. $f^0 = \bot \sqsubseteq x$ because $\bot$ is the infimum (smallest element) of P. Assume that $f^n \sqsubseteq x$ by induction hypothesis. Then $f^{n+1} = f(f^n) \sqsubseteq f(x) = x$ by definition of the iterates, f increasing, and fixpoint property. By recurrence $\forall n \in \mathbb{N}\,.\, f^n \sqsubseteq x$. So x is an upper bound of the $\{f^n \mid n \in \mathbb{N}\}$, hence larger than or equal to the least upper bound $\bigsqcup\{f^n \mid n \in \mathbb{N}\}$, proving that $\bigsqcup\{f^n \mid n \in \mathbb{N}\} = \mathsf{lfp}^{\sqsubseteq} f$ is the least fixpoint of f.

We have $f^0 = \bot \sqsubseteq f^1$ by def. infimum. Assume $f^n \sqsubseteq f^{n+1}$. By definition 15.20 of the iterates and f increasing, we have $f^{n+1} = f(f^n) \sqsubseteq f(f^{n+1}) = f^{n+2}$, proving $\langle f^n,\ n \in \mathbb{N}\rangle$ to be increasing. The lub $f^\omega = \bigsqcup\{f^n \mid n \in \mathbb{N}\}$ is assumed to exist, so $\langle f^n,\ n \in \mathbb{N} \cup \{\omega\}\rangle$ is an increasing chain. □

The iterative fixpoint theorem is often proved with stronger hypotheses, and this is left as an exercise exercise 15.24 below.

Let us recall the definition 11.28 of (strict/arbitrary) join preservation. A function $f \in P \to P$ on a poset $\langle P, \sqsubseteq\rangle$ is nonempty join preserving if and and only if for any nonempty subset $\{x_i \mid i \in \Delta\} \subseteq P$ of P that has a least upper bound $\bigsqcup\{x_i \mid i \in \Delta\} \in P$ then $\bigsqcup\{f(x_i) \mid i \in \Delta\}$ exists in P and $f(\bigsqcup\{x_i \mid i \in \Delta\}) = \bigsqcup\{f(x_i) \mid i \in \Delta\}$.

It is strictly or arbitrary join preserving if whenever $\bot = \bigsqcup\emptyset \in P$ then $f(\bot) = \bot$.

Let us also recall the definition of increasing chain from section 10.8. A subset $\{x_i \mid i \in \mathbb{N}\} \subseteq P$ of a poset $\langle P, \sqsubseteq\rangle$ is an (enumerable) increasing chain if and only if $\forall i, j \in \mathbb{N}\,.\,(i \leqslant j) \Rightarrow (x_i \sqsubseteq x_j)$.

The family x_i, $i \in \mathbb{N}$ considered as a function of i is therefore increasing.

Definition 15.22 ((upper) continuous) A function $f \in P \to P$ on a poset $\langle P, \sqsubseteq\rangle$ is upper continuous (denoted $f \in P \xrightarrow{uc} P$) if and and only if for any enumerable increasing chain $\{x_i \mid i \in \mathbb{N}\} \subseteq P$ of P which has a least upper bound $\bigsqcup\{x_i \mid i \in \mathbb{N}\} \in P$ then $\bigsqcup\{f(x_i) \mid i \in \mathbb{N}\}$ exists in P and $f(\bigsqcup\{x_i \mid i \in \mathbb{N}\}) = \bigsqcup\{f(x_i) \mid i \in \mathbb{N}\}$. The dual is lower continuity (denoted $f \in P \xrightarrow{lc} P$).

Observe that join preservation implies upper continuity.

Exercise 15.24 Prove that an upper continuous function on a poset is increasing (but not all increasing functions are upper continuous).

Exercise 15.25 (composition of continuous functions) Prove that the composition of continuous functions is continuous.

Let us recall the definition of a complete partial order in section 10.9. A complete partial order (CPO) is a poset $\langle P, \sqsubseteq, \bot\rangle$ with infimum $\bot$ such that any increasing chain $\{x_i \mid i \in \mathbb{N}\}$ has a least upper bound $\bigsqcup\{x_i \mid i \in \mathbb{N}\} \in P$.

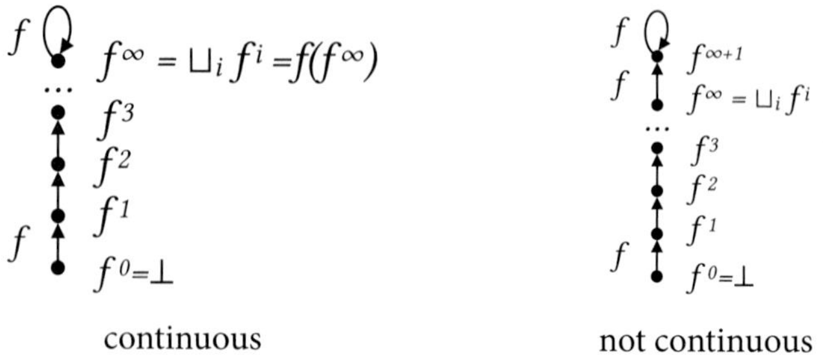

Figure 15.23
Continuity

> **Theorem 15.26 (Scott–Kleene's iterative fixpoint theorem)** If $f \in L \xrightarrow{uc} L$ be an upper continuous function on a CPO $\langle L, \sqsubseteq, \bot, \sqcup\rangle$ then f has a least fixpoint $\mathsf{lfp}^{\sqsubseteq} f = \bigsqcup_{n \in \mathbb{N}} f^n(\bot)$.

Exercise 15.27 Prove theorem 15.26 (taking inspiration from theorem 15.21).

Exercise 15.28 (Tarski's fixpoint theorem on a CPO) Let $f \in L \xrightarrow{uc} L$ be an upper continuous function on a CPO $\langle L, \sqsubseteq, \bot, \sqcup\rangle$. Assume that $a \in L$ is a prefixpoint, that is $a \sqsubseteq f(a)$. Prove that $\mathsf{lfp}^{\sqsubseteq}_{a} f = \bigsqcap\{x \in L \mid a \sqsubseteq x \wedge f(x) \sqsubseteq x\}$.

Exercise 15.29 Provide a fixpoint definition of the structural definition of the positive integers $\mathbb{N}^+$ in section 6.5. Define an isomorphic abstraction of sets of sticks into numbers. Prove that, up to this abstraction, $s(n)$ is the set of numbers smaller than or equal to n.

Exercise 15.30 (reflexive transitive closure) Let $r \in \wp(S \times S)$ be a relation on a set S. Define $r^0 \triangleq \mathbb{1}_S = \{\langle e, e\rangle \mid e \in S\}$, $r^{n+1} = r^n \circ r$ where $r \circ r' \triangleq \{\langle e, e''\rangle \mid \exists e' \in S \,.\, \langle e, e'\rangle \in r \wedge \langle e', e''\rangle \in r'\}$, and $r^* \triangleq \bigcup_{r \in \mathbb{N}} r^n$. Prove that $r^* = \mathsf{lfp}\, X \mapsto \mathbb{1}_S \cup X \circ r = \mathsf{lfp}\, X \mapsto \mathbb{1}_S \cup r \circ X$.

Exercise 15.31 (Arden's lemma [52]) Prove that $A^* \cdot B$ is the smallest language that is a solution for X in the equation $X = A \cdot X \cup B$ where X, A, B are sets of strings. Moreover, prove that if the set A does not contain the empty string, then this solution is unique.

> **Theorem 15.32** If $F \in L \xrightarrow{uc} L \xrightarrow{uc} L$ is an upper continuous function on a CPO $\langle L, \sqsubseteq, \bot\rangle$ then $R \in L \mapsto \mathsf{lfp}^{\sqsubseteq} F(R) \in L \xrightarrow{uc} L$.

Proof of theorem 15.32 $F(R) \in L \xrightarrow{uc} L$ and so $\mathsf{lfp}^{\sqsubseteq} F(R) = \bigsqcup_{n \in \mathbb{N}} F(R)^n(\bot)$ exists by theorem 15.26. By recurrence $R \in L \mapsto F(R)^n(\bot) \in L \xrightarrow{uc} L$. For the basis, $R \in L \mapsto F(R)^0(\bot) = R \in L \mapsto \bot$ is constant so continuous. For the induction, $R \in L \mapsto F(R)^n(\bot)$ is continuous by induction hypothesis so $R \in L \mapsto F(R)^{n+1}(\bot) = F(R) \circ R \in L \mapsto F(R)^n(\bot)$ is continuous because $F(R) \in L \xrightarrow{uc} L$ and, as by exercise 15.25, the composition of continuous functions is continuous.

To prove continuity $R \in L \mapsto \mathsf{lfp}^{\sqsubseteq} F(R) \in L \xrightarrow{uc} L$, let $\langle R_i, i \in \mathbb{N}\rangle$ be a $\sqsubseteq$-increasing chain.

$$
\begin{aligned}
&(R \in L \mapsto \mathsf{lfp}^{\sqsubseteq} F(R))(\bigsqcup_{i\in\mathbb{N}} R_i)\\
=\ &(R \in L \mapsto \bigsqcup_{n\in\mathbb{N}} F(R)^n(\bot))(\bigsqcup_{i\in\mathbb{N}} R_i) && \text{(theorem 15.26)}\\
=\ &\bigsqcup_{n\in\mathbb{N}} F(\bigsqcup_{i\in\mathbb{N}} R_i)^n(\bot) && \text{(function application)}\\
=\ &\bigsqcup_{n\in\mathbb{N}}\bigsqcup_{i\in\mathbb{N}} F(R_i)^n(\bot) && \text{(}R \in L \mapsto F(R)^n(\bot) \in L \xrightarrow{uc} L \text{ is continuous)}\\
=\ &\bigsqcup_{i\in\mathbb{N}}\bigsqcup_{n\in\mathbb{N}} F(R_i)^n(\bot) && \text{(definition of } \sqcup\text{)}\\
=\ &\bigsqcup_{i\in\mathbb{N}}(R \in L \mapsto \bigsqcup_{n\in\mathbb{N}} F(R)^n(\bot))(R_i) && \text{(definition of function application)}\\
=\ &\bigsqcup_{i\in\mathbb{N}}(R \in L \mapsto \mathsf{lfp}^{\sqsubseteq} F(R))(R_i) && \text{(theorem 15.26)} \quad \square
\end{aligned}
$$

15.3 Conjugate Fixpoint Theorem

Theorem 15.33 (Park's conjugate fixpoint theorem [777], [778, (4.1.2)]**)** Let S be a set and $f \in \wp(S) \xrightarrow{\nearrow} \wp(S)$ be $\subseteq$-increasing on the complete lattice $\langle \wp(S), \subseteq, \varnothing, S, \cap, \cup \rangle$. Let $\neg X \triangleq S \setminus X$ be the set complement. Define $\tilde{f} \triangleq X \mapsto \neg f(\neg X)$. Then $\mathsf{gfp}^{\subseteq} f = \neg \mathsf{lfp}^{\subseteq} \tilde{f}$.

Proof of theorem 15.33

$$
\begin{aligned}
&\neg \mathsf{lfp}^{\subseteq} \tilde{f}\\
=\ &\neg \mathsf{lfp}^{\subseteq} X \mapsto \neg f(\neg X) && \text{(definition of } \tilde{f}\text{)}\\
=\ &\neg \bigcap \{X \mid \neg f(\neg X) \subseteq X\}\\
& && \text{(}f \text{ is increasing so } x \mapsto \neg f(\neg X) \text{ is increasing and theorem 15.6)}\\
=\ &\bigcup \{\neg X \mid f(\neg X) \supseteq \neg X\} && \text{(De Morgan's laws)}\\
=\ &\bigcup \{Y \mid Y \subseteq f(Y)\} && \text{(letting } Y \triangleq \neg X\text{)}\\
=\ &\mathsf{gfp}^{\subseteq} f && \text{(dual of theorem 15.6)} \quad \square
\end{aligned}
$$

Exercise 15.34 ([777, (2.2.3)]) Prove that $\mathsf{lfp}^{\subseteq} f \cap \mathsf{lfp}^{\subseteq} \tilde{f} = \varnothing$.

Exercise 15.35 ([777, (2.2.4)]) Prove that $\mathsf{lfp}^{\subseteq} f \cup \mathsf{lfp}^{\subseteq} \tilde{f} = S$ if and only if $\mathsf{lfp}^{\subseteq} f$ is the unique fixpoint of f.

Exercise 15.36 Generalize theorem 15.33, exercise 15.34, and exercise 15.35 to a complete Boolean lattice [707].

15.4 Conclusion

The idea of solving equations iteratively dates back to the Isaac Newton – Joseph Raphson method [751, 820]. The study of fixpoints originates from the principle of mathematical induction for chains by Richard Dedekind [302], Ernst Zermelo set theory [1001], and Gerhard Hessenberg [494] study for reductive or, equivalently, extensive functions $\forall x \in L \,.\, x \sqsubseteq f(x)$.[1] This idea of iteration has produced numerous fixpoint theorems such as that of Casimir Kuratowski [603, Théorème II], Stefan Banach [83, Théorème 6], and Leonid Kantorovitch [555, theorem I] for increasing functions on bounded posets, and others.

Tarski's theorem is often called Knaster-Tarski's theorem because Bronisław Knaster (mentioning that the results were obtained in common with Tarski) previously proved a weaker version for powersets of a set [584]; see exercise 15.17. The iterative version of theorem 15.26 is in [925, p. 305] for complete join morphisms (i.e., $f(\bigsqcup_{i\in\Delta} x_i) = \bigsqcup_{i\in\Delta} f(x_i)$), indeed countable ones. It was extended by Dana Scott to upper continuity [870] and to increasing functions using transfinite iterations, [136, ch. III, example 6b, p. 48] and [5, 240, 656]. This result is often attributed to Stephen Cole Kleene, after its first recursion theorem [582, p. 348].[2]

15.5 Solutions to Selected Exercises

Solution to exercise 15.1 $f(x) = x + 1$ has no fixpoint on $\mathbb{Z}$, $f(x) = -x$ has one fixpoint (0), and $f(x) = x$ has infinitely many fixpoints.

Solution to exercise 15.5 The complete lattice is $\langle [0, 1], \leqslant \rangle$. The least fixpoint is 0.

Solution to exercise 15.8 Let $a = \mathsf{lfp}^{\sqsubseteq} f$ and $b = \mathsf{lfp}^{\sqsubseteq} x \mapsto x \sqcup f(x)$. $b = b \sqcup f(b)$ so $f(b) \sqsubseteq b \sqcup f(b) = b$ is a postfixpoint of f so that by Tarski's fixpoint theorem [925], $a = \mathsf{lfp}^{\sqsubseteq} f = \bigsqcap\{x \in L \mid f(x) \sqsubseteq x\} \sqsubseteq b$. $a = f(a)$, so $a = a \sqcup f(a)$ is a fixpoint of $x \mapsto x \sqcup f(x)$, proving $b \sqsubseteq a$. By antisymmetry, $a = b$

Solution to exercise 15.9 By Tarski's fixpoint theorem [925] and definition of the lub $\sqcup$, we have $\mathsf{lfp}^{\sqsubseteq}_a f \triangleq \bigsqcap\{x \in L \mid f(x) \sqsubseteq x \wedge a \sqsubseteq x\} = \bigsqcap\{x \in L \mid a \sqcup f(x) \sqsubseteq x\} = \mathsf{lfp}^{\sqsubseteq} x \mapsto a \sqcup f(x)$.

Solution to exercise 15.12 If $f(y) \sqsubseteq y \wedge y \sqsubseteq x$ then $y \in \{x \in L \mid f(x) \sqsubseteq x\}$ so by definition of $\bigsqcap$ and theorem 15.6, $\mathsf{lfp}^{\sqsubseteq} f = \bigsqcap\{x \in L \mid f(x) \sqsubseteq x\} \sqsubseteq y \sqsubseteq x$. Conversely, if $y = \mathsf{lfp}^{\sqsubseteq} f \sqsubseteq x$ then $f(y) = y$ so $f(y) \sqsubseteq y$ by reflexivity and $y \sqsubseteq x$ proving $\exists y \in L \,.\, f(y) \sqsubseteq y \wedge y \sqsubseteq x$.

Solution to exercise 15.13 If $\mathsf{lfp}^{\sqsubseteq} f \sqcap f(x) \sqsubseteq x$ then $\mathsf{lfp}^{\sqsubseteq} f \sqsubseteq x$ by definition of the greatest lower bound $\sqcap$ and transitivity.

Conversely, let $y = \mathsf{lfp}^{\sqsubseteq} f \sqcap f(x) \sqsubseteq x$. $y \sqsubseteq \mathsf{lfp}^{\sqsubseteq} f$ implies $f(y) \sqsubseteq f(\mathsf{lfp}^{\sqsubseteq} f) = \mathsf{lfp}^{\sqsubseteq} f$ by f increasing and fixpoint property. Moreover, $y \sqsubseteq x$ implies $f(y) \sqsubseteq f(x)$ so $f(y) \sqsubseteq \mathsf{lfp}^{\sqsubseteq} f \sqcap f(x) = y$. By exercise 15.12 and reflexivity, we conclude that $\mathsf{lfp}^{\sqsubseteq} f \sqsubseteq y$.

Solution to exercise 15.14 By duality, the greatest fixpoint is $\mathsf{gfp}^{\sqsubseteq} f = \bigsqcup\{x \in L \mid x \sqsubseteq f(x)\}$.

1. The iterates of f from any $a \in L$ form an increasing chain so have a maximal element by Teichmüller–Tukey maximal principle, which is a fixpoint because f is extensive. theorem 15.26 is similar because f is increasing, so extensive on its iterates from the infimum.

2. Given a partial recursive functional $F(f, x_1, \ldots, x_n)$ where f is a partial recursive (i.e., computable by an algorithm) function of n variable, there is a partial recursive function such that $f(x_1, \ldots, x_n) = F(f, x_1, \ldots, x_n)$, which is minimal, that is, terminates less often than any other possible solution [583].

Solution to exercise 15.16 Let X_i, $i = 1, 2$ be the language recognized from node (i) of the automaton. X_1 contains the empty word (which is indeed accepted starting from state 1) and also words starting with b and then a word recognized from X_2 (which corresponds to the transition from 1 to 2). X_2 contains the words starting with a and then a word recognized from X_2 (which corresponds to the transition from 2 to 2) and the words starting with a and then a word recognized from X_1 (which corresponds to the transition from 2 to 1).

$$\mathsf{lfp}\left(\begin{pmatrix} X_1 \\ X_2 \end{pmatrix} \mapsto \begin{pmatrix} \{\epsilon\} \cup bX_2 \\ aX_1 \cup aX_2 \end{pmatrix}\right)$$

The language recognized by the automaton is $X_1 = (ba^+)^*$ and $X_2 = a^+(ba^+)^*$.

Solution to exercise 15.18 • The Cartesian product of complete lattices is a complete lattice ordered componentwise. The composition of increasing functions is increasing. Hence by Tarski fixpoint theorem 15.6, the least fixpoints do exist.

- If $Y, Z \in L^m$ then $Y \sqsubseteq Z$ implies $\langle X, Y\rangle \dot{\sqsubseteq} \langle X, Z\rangle$ so $X \mapsto f(X, Y) \dot{\sqsubseteq} X \mapsto f(X, Z)$, so $\mathsf{lfp}\, X \mapsto f(X, Y) \dot{\sqsubseteq} \mathsf{lfp}\, X \mapsto f(X, Z)$ by theorem 18.7 whence $r(Y) \dot{\sqsubseteq} r(Z)$ pointwise, proving that $r \in L^m \xrightarrow{\nearrow} L^n$ whence the set $\mathcal{F}_2$ of fixpoints of F_2 is not empty.
- Let $\langle A_2, B_2\rangle \in \mathcal{F}_2$ be a fixpoint of F_2. Then $A_2 = r(B_2)$ so $\mathsf{lfp}\, X \mapsto f(X, B_2) = A_2$ whence $A_2 = f(A_2, B_2)$ and $B_2 = g(r(B_2), B_2)$ that is $B_2 = g(A_2, B_2)$ proving that $\langle A_2, B_2\rangle \in \mathcal{F}_1$ so $\mathcal{F}_2 \subseteq \mathcal{F}_1$.
- In general $\mathcal{F}_2 \neq \mathcal{F}_1$. A counterexample is provided by $L = \{\bot, \top\}$ with $\bot \sqsubseteq \bot \sqsubseteq \top \sqsubseteq \top$, $f(X, Y) = X \sqcap Y$, and $g(X, Y) = X \sqcup Y$ so that the resolvant is $r = Y \mapsto \mathsf{lfp}\, X \mapsto f(X, Y) = Y \mapsto \mathsf{lfp}\, X \mapsto X \sqcap Y = Y \mapsto \bot$. The system of equation F_1 has the solution $\langle \top, \top\rangle$ that is not a solution of F_2 in that particular case.
- Because $\mathsf{lfp}^{\dot{\sqsubseteq}}\, F_2 \in \mathcal{F}_1$ we have $\mathsf{lfp}^{\dot{\sqsubseteq}}\, F_1 \dot{\sqsubseteq} \mathsf{lfp}^{\dot{\sqsubseteq}}\, F_2$ by theorem 15.6 and reflexivity.
- Let $\langle A_1, B_1\rangle \in \mathcal{F}_1$ be a fixpoint of F_1. We have $F(A_1, B_1) = A_1$ whence $F(A_1, B_1) \dot{\sqsubseteq} A_1$ whence A_1 is a postfixpoint of $X \mapsto F(X, B_1)$ which implies by Tarski fixpoint theorem 15.6 that $\mathsf{lfp}\, X \mapsto f(X, B_1 \dot{\sqsubseteq} A_1$ that is $r(B_1) \dot{\sqsubseteq} A_1$. Because $\langle r(B_1), B_1\rangle \dot{\sqsubseteq} \langle A_1, B_1\rangle$ and g is increasing, $g(R(B_1), B_1) \dot{\sqsubseteq} g(A_1, B_1) \dot{\sqsubseteq} B_1$ because $\langle A_1, B_1\rangle$ is a postfixpoint of F_1. It follows that $\langle A_1, B_1\rangle$ is a postfixpoint of F_2, which implies $\mathsf{lfp}^{\dot{\sqsubseteq}}\, F_2 \dot{\sqsubseteq} \langle A_1, B_1\rangle$, in particular $\mathsf{lfp}^{\dot{\sqsubseteq}}\, F_2 \dot{\sqsubseteq} \mathsf{lfp}^{\dot{\sqsubseteq}}\, F_1$.
- By antisymmetry, $\mathsf{lfp}^{\dot{\sqsubseteq}}\, F_1 = \mathsf{lfp}^{\dot{\sqsubseteq}}\, F_2$.

Solution to exercise 15.24 If $x \sqsubseteq y \in P$ then $x \sqcup y = y$ so for the chain $x, y, y, \ldots$, we have by continuity $f(y) = f(x \sqcup y) = f(x \sqcup y \sqcup y \ldots) = f(x) \sqcup f(y) \sqcup f(y) \ldots) = f(x) \sqcup f(y)$ so $f(x) \sqsubseteq f(x) \sqcup f(y) = f(y)$, proving f to be increasing. The noncontinuous function in figure 15.23 is increasing.

Solution to exercise 15.28 Let $P = \{x \in L \mid a \sqsubseteq x \wedge f(x) \sqsubseteq x\}$. By theorem 15.26 applied to the CPO $\langle\{x \in L \mid a \sqsubseteq x\}, \sqsubseteq\rangle$, $\mathsf{lfp}_a^{\sqsubseteq} f$ does exist so $f(\mathsf{lfp}_a^{\sqsubseteq} f) = \mathsf{lfp}_a^{\sqsubseteq} f \in P$ by reflexivity. Let $\langle f^n, n \in \mathbb{N} \cup \{\omega\}\rangle$ be the iterates of f from a (with $f^\omega = \mathsf{lfp}_a^{\sqsubseteq} f$ by continuity of f). For any $x \in P$, we have $f^0 = a \sqsubseteq x$. Assume that $f^n \sqsubseteq x$ by induction hypothesis. Then $f^{n+1} = f(f^n) \sqsubseteq f(x) \sqsubseteq x$ since f is continuous hence increasing, proving that $\forall n \in \mathbb{N}$. $f^n \sqsubseteq x$ by recurrence. We have proved that $\forall x \in P$. $\mathsf{lfp}_a^{\sqsubseteq} f \sqsubseteq x$ and therefore $\mathsf{lfp}_a^{\sqsubseteq} f \sqsubseteq \bigsqcap P$ (since for all other lower bounds y of P, we have $\forall x \in P$. $y \sqsubseteq x$ so $y \sqsubseteq \mathsf{lfp}_a^{\sqsubseteq} f$ proving

$\mathsf{lfp}^{\sqsubseteq}_a\, f$ to be the greatest lower bound). Moreover $\mathsf{lfp}^{\sqsubseteq}_a\, f \in P$ so $\mathsf{lfp}^{\sqsubseteq}_a\, f = \bigsqcap . P$. (Notice that $\bigsqcap P$ is shown to exist despite the fact that $\bigsqcap X$ may not exist for an arbitrary subset $X \subseteq L$ of a CPO.)

Solution to exercise 15.34 We have $\mathsf{lfp}^{\subseteq}\, f \subseteq \mathsf{gfp}^{\subseteq}\, f$ so $\neg\mathsf{gfp}^{\subseteq}\, f \subseteq \neg\mathsf{lfp}^{\subseteq}\, f$ and, by theorem 15.33, $\neg\mathsf{gfp}^{\subseteq}\, f = \mathsf{lfp}^{\subseteq}\, \tilde{f}$ so $\mathsf{lfp}^{\subseteq}\, \tilde{f} \subseteq \neg\mathsf{lfp}^{\subseteq}\, f$. It follows that $\mathsf{lfp}^{\subseteq}\, f \cap \mathsf{lfp}^{\subseteq}\, \tilde{f} \subseteq \mathsf{lfp}^{\subseteq}\, f \cap \neg\mathsf{lfp}^{\subseteq}\, f = \varnothing$.

Solution to exercise 15.35 If $\mathsf{lfp}^{\subseteq}\, f$ is the unique fixpoint of f then, by theorem 15.33, $\mathsf{lfp}^{\subseteq}\, f = \mathsf{gfp}^{\subseteq}\, f = \neg\mathsf{lfp}^{\subseteq}\, \tilde{f}$ so $\mathsf{lfp}^{\subseteq}\, f \cup \mathsf{lfp}^{\subseteq}\, \tilde{f} = \mathsf{lfp}^{\subseteq}\, f \cup \neg\mathsf{lfp}^{\subseteq}\, f = S$.

Conversely, assume that $\mathsf{lfp}^{\subseteq}\, f \cup \mathsf{lfp}^{\subseteq}\, \tilde{f} = S$ and f has at least two distinct fixpoints so that by Tarski fixpoint theorem 15.6, $\mathsf{lfp}^{\subseteq}\, f \subsetneq \mathsf{gfp}^{\subseteq}\, f$. By theorem 15.33, $\mathsf{lfp}^{\subseteq}\, f \subsetneq \neg\mathsf{lfp}^{\subseteq}\, \tilde{f}$, in contradiction to $\mathsf{lfp}^{\subseteq}\, f \cup \mathsf{lfp}^{\subseteq}\, \tilde{f} = S$.

16 Fixpoint, Deductive, Inductive, Structural, Coinductive, and Bi-inductive Definitions

The problem is to formally define a subset $D \in \wp(\mathbb{U})$ of a set $\mathbb{U}$ (called the universe).

Example 16.1 Define the odd numbers $\mathbb{O}\mathbb{d}$ as a subset of the natural numbers $\mathbb{N}$. Do the same for the even numbers $\mathbb{E}\mathbb{n}$.

Example 16.2 The semantics of a program is a subset of its semantic domain. For example, in chapter 6, the prefix trace semantics $\widehat{\mathcal{S}}^*[\![\mathsf{P}]\!]$ of a program P is a map of initial finite traces to their continuations $\widehat{\mathcal{S}}^*[\![\mathsf{P}]\!] \in \mathbb{T}^+ \to \wp(\mathbb{T}^+)$ i.e., understanding this functions as the right image of a relation (see section 6.10), a subset of the semantic domain $\mathbb{T}^+ \times \mathbb{T}^+$.

Contents

16.1 Fixpoint Definitions

Because $\langle \wp(\mathbb{U}), \subseteq, \varnothing, \mathbb{U}, \cup, \cap \rangle$ is a complete lattice, D can be defined as the least fixpoint $D \triangleq \mathsf{lfp}^{\subseteq} F$ of an increasing function $F \in \wp(\mathbb{U}) \xrightarrow{\nearrow} \wp(\mathbb{U})$, that is, as the $\subseteq$-least solution of the equation $X = F(X)$. D is well defined (i.e., exists and is unique) by theorem 15.6.

Example 16.3 Continuing example 16.1, in the universe $\mathbb{N}$, the odd numbers are $\mathbb{O}\mathbb{d} \triangleq \mathsf{lfp}^{\subseteq} F$ where $F(X) \triangleq \{1\} \cup \{n+2 \mid n \in X\}$. Applying theorem 15.21, we get $\mathbb{O}\mathbb{d} = \varnothing \cup \{1\} \cup \{1,3\} \cup \ldots \cup \{1,3,\ldots,2k+1\} \cup \ldots$.

> **Definition 16.4 (fixpoint definition)** The fixpoint definition of $D \in \wp(\mathbb{U})$ by a $\subseteq$-increasing function $F \in \wp(\mathbb{U}) \xrightarrow{\nearrow} \wp(\mathbb{U})$ is $D \triangleq \mathsf{lfp}^{\subseteq} F$.

Theorem 16.5 D in definition 16.4 is well defined.

Proof By theorem 15.6. □

16.2 Deductive Definitions

A *deductive definition* of $D \in \wp(\mathbb{U})$ is given by a set of *inference rules* $R = \{\frac{P_i}{c_i} \mid i \in \Delta\}$ where $P_i \in \wp_f(\mathbb{U})$ is the finite premise and $c_i \in \mathbb{U}$ is the conclusion of the rule. A rule $\frac{P_i}{c_i} \in R$ states that if $P_i \subseteq D$ then $c_i \in D$. If $P_i = \varnothing$, the rule is called an *axiom* and states that $c_i \in D$. The set of rules is called a *deductive system* (or *Hilbert system*).

Example 16.6 Continuing example 16.3, in the universe $\mathbb{N}$, the odd numbers are $\{\frac{\varnothing}{1}\} \cup \{\frac{\{n\}}{n+2} \mid n \in \mathbb{N}\}$. 1 is an axiom and from n is odd, we infer that $n+2$ is odd. As a shorthand, this can be written in the form of an axiom $1 \in \mathbb{O}\mathrm{d}$ and an inference rule schema $\frac{n \in \mathbb{O}\mathrm{d}}{n+2 \in \mathbb{O}\mathrm{d}}$ which instantiation for all $n \in \mathbb{N}$ yields the rules $\frac{\varnothing}{1}, \frac{\{0\}}{2}, \frac{\{1\}}{3}, \frac{\{2\}}{4}, \frac{\{3\}}{5}, \frac{\{4\}}{6}, \ldots, \frac{\{n\}}{n+2}, \ldots$. Notice that the rules $\frac{\{0\}}{2}, \frac{\{2\}}{4}, \ldots$ are useless because their premises cannot be derived form the deductive definition (0 is not an axiom).

A proof of p by rules R is a finite sequence $t_0 \ldots t_n$ of elements of $\mathbb{U}$ such that each t_i, $i \in [0, n]$ is deduced from $t_0 \ldots t_{i-1}$ by application of a rule of R and $t_n = p$. Formally

> **Definition 16.7 (proof)** $\textsf{is-provable}(p, R) \triangleq \exists t_0 \ldots t_n \in \mathbb{U} \,.\, (\forall i \in [0, n] \,.\, \exists \frac{P}{c} \in R \,.\, P \subseteq \{t_0, \ldots, t_{i-1}\} \wedge t_i = c) \wedge t_n = p$.

Example 16.8 The proof that 5 in example 16.6 is odd is 1, 3, 5. To prove that 4 is not odd, observe that $\frac{\{2\}}{4}$ is the only rule allowing us proving that 4 would be odd, which requires to prove that 2 is odd, the only applicable rule being $\frac{\{0\}}{2}$. It remains to prove that 0 is odd, which is impossible because there is no rule with 0 as the conclusion.

The set D defined by a set of rules R is $D \triangleq \{p \in \mathbb{U} \mid \textsf{is-provable}(p, R)\}$.

Example 16.9 To prove that $\mathbb{O}\mathrm{d} = \{2k+1 \mid k \in \mathbb{N}\}$ in example 16.6, we must prove that $2k+1$ is provable for all $k \in \mathbb{N}$. 1 is provable by rule $\frac{\varnothing}{1}$. Assume, by recurrence hypothesis, that we have a proof $1, 3, 5, \ldots, 2k+1$ of $2k+1$. A proof of $2(k+1)+1 = 2k+3$ is by the rule $\frac{\{2k+1\}}{2k+3}$ such that $\{2k+1\} \subseteq \{1, 3, 5, \ldots, 2k+1\}$. By recurrence, all $2k+1$, $k \in \mathbb{N}$ are provable so that $\{2k+1 \mid k \in \mathbb{N}\} \subseteq \mathbb{O}\mathrm{d}$. For the inverse inclusion, we can use a reasoning by reductio ad absurdum as illustrated in example 16.8.

> **Definition 16.10 (deductive definition)** The deductive definition of $D \in \wp(\mathbb{U})$ by a deductive system of rules $\frac{P}{c} \in R$ is $D \triangleq \{p \in \mathbb{U} \mid \textsf{is-provable}(p, R)\}$.

Exercise 16.11 Show that a context-free grammar is a deductive definition.

16.3 Equivalence of the Least Fixpoint and Deductive Definition Methods

A deductive definition can be expressed as a fixpoint definition, and conversely.

16.3.1 Deductive Definition as a Fixpoint Definition

For a deductive definition by rules $R = \{\frac{P_i}{c_i} \mid i \in \Delta\}$, define the consequence operator $F_R(X) \triangleq \{c \mid \exists \frac{P}{c} \in R \,.\, P \subseteq X\}$. $F_R(X)$ is the set of consequences provable by R when X has already been proved.

> **Theorem 16.12** We have $D = \{p \in \mathbb{U} \mid \text{is-provable}(p, R)\} = \mathsf{lfp}^{\subseteq} F_R$ where $F_R(X) \triangleq \{c \mid \exists \frac{P}{c} \in R \,.\, P \subseteq X\}$ is the *consequence operator* of R.

Proof Let F_R^n be the iterates of F_R. We observe that F_R^n contains all elements with a proof of length less than or equal to n. This holds for $n = 0$ because $F_R^0 = \emptyset$ and there is no proof of length 0 or less. F_R^1 contains all elements with a proof of length 1 obtained by applying an axiom. Assume that F_R^n contains all elements with a proof of length less than or equal to n. If c has a proof of length less than or equal to $n + 1$ then it is deduced by a rule $\frac{P}{c} \in R$ where the elements of P are proved before c hence have proofs of length less than or equal to n. It follows that $c \in \{c \mid \exists \frac{P}{c} \in R \,.\, P \subseteq F_R^n\} = F_R(F_R^n) = F_R^{n+1}$. By recurrence, for all $n \in \mathbb{N}$, all elements c with a proof of length less than or equal to n belong to F_R^n. (F_R^n may contains proofs of longer length.) Now all elements in $\{p \in \mathbb{U} \mid \text{is-provable}(p, R)\}$ have a proof of some length $n \in \mathbb{N}$ so belong to $\bigcup\{F_R^n \mid n \in \mathbb{N}\}$, proving $D \subseteq \bigcup\{F_R^n \mid n \in \mathbb{N}\}$.

Conversely, assume that $\bigcup\{F_R^n \mid n \in \mathbb{N}\}$ contains an element c not in D. Because the $\langle F_R^n,\ n \in \mathbb{N}\rangle$ form a $\subseteq$-increasing chain, there exists a smallest n such that c belongs to F_R^n but does not belong to any of the F_R^m, $m < n$. Among the pairs $\langle c,\ n\rangle$ with this property, chose one such that n is minimal. By minimality, all elements in F_R^m, $m < n$ have a proof so belong to D. By definition of the iterates $F_R^n = F_R(F_R^{n-1})$ and of the consequence operator F_R, c has a proof, a contradiction. So $\bigcup\{F_R^n \mid n \in \mathbb{N}\} \subseteq D$.

By antisymmetry, we conclude that $D = \bigcup\{F_R^n \mid n \in \mathbb{N}\}$.

Let us prove that $\mathsf{lfp}^{\subseteq} F_R = \bigcup\{F_R^n \mid n \in \mathbb{N}\}$ using theorem 15.21. F_R is increasing because if $X \subseteq X'$ then $P \subseteq X$ implies $P \subseteq X'$ and so $c \in F_R(X)$ implies $c \in F_R(X')$, proving $F_R(X) \subseteq F_R(X')$. Because $\langle \wp(\mathbb{U}),\ \subseteq\rangle$ is a complete lattice, the lub $\cup$ exists. It remains to prove that $F_R(\bigcup\{F_R^n \mid n \in \mathbb{N}\}) = \bigcup\{F_R(F_R^n) \mid n \in \mathbb{N}\}$ i.e.. $F_R(D) = D$.

Let us first prove the $\supseteq$ inclusion.

$$
\begin{array}{ll}
\forall n \in \mathbb{N} \,.\, F_R^n \subseteq \bigcup\{F_R^n \mid n \in \mathbb{N}\} & \text{(definition of the lub } \cup\text{)} \\
\Rightarrow\ \forall n \in \mathbb{N} \,.\, F_R(F_R^n) \subseteq F_R(\bigcup\{F_R^n \mid n \in \mathbb{N}\}) & (F_R \text{ is } \subseteq\text{-increasing}) \\
\Rightarrow\ \forall n \in \mathbb{N} \,.\, F_R^{n+1} \subseteq F_R(\bigcup\{F_R^n \mid n \in \mathbb{N}\}) & \text{(definition of the iterates)} \\
\Rightarrow\ \forall n \in \mathbb{N} \,.\, F_R^n \subseteq F_R(\bigcup\{F_R^n \mid n \in \mathbb{N}\}) & \text{(because } F_R^0(X) = \emptyset\text{)} \\
\Rightarrow\ \bigcup\{F_R^n \mid n \in \mathbb{N}\} \subseteq F_R(\bigcup\{F_R^n \mid n \in \mathbb{N}\}) & \text{(definition of the lub)} \\
\Rightarrow\ D \subseteq F_R(D) & (\text{by } D = \bigcup\{F_R^n \mid n \in \mathbb{N}\})
\end{array}
$$

Conversely, assume by reductio ad absurdum that $F_R(D) \not\subseteq D$ so that $\exists c \in F_R(D) \,.\, c \notin D$. Because $c \notin D$ so there exists no finite proof of c. By definition of F_R, $\exists \frac{P}{c} \in R \,.\, P \subseteq D$.

Because $P \subseteq D$, all elements p of P have a proof. Because the premise P must be finite, P has a finite proof (which is the finite sequence of the proofs of the elements of P), and therefore, using the rule $\frac{P}{c}$, c has also a finite proof, a contradiction.

By antisymmetry, $F_R(D) = D$ so $D = \mathsf{lfp}^{\subseteq} F_R$ by theorem 15.21. □

Exercise 16.13 Show that the consequence operator F_R in theorem 16.12 does not necessarily preserve joins but is increasing.

Exercise 16.14 Show that theorem 16.12 may not hold when considering rules on which premises can be infinite sets.

Exercise 16.15 Show for the example $X ::= X\ X \mid a$ on the alphabet of terminals $\{a, b\}$ that a context-free grammar is a deductive definition of a language, and express this language as a fixpoint [419, 868].

16.3.2 Fixpoint Definition as a Deductive Definition

Theorem 16.16 For a fixpoint definition $\mathsf{lfp}^{\subseteq} F$ define $R = \{\frac{P}{c} \mid P \subseteq \mathbb{U} \wedge c \in F(P)\}$. Then $F = F_R$ so $\mathsf{lfp}^{\subseteq} F_R = \mathsf{lfp}^{\subseteq} F$.

Note that if R has finite premises only, then $\{p \in \mathbb{U} \mid \text{is-provable}(p, R)\} = \mathsf{lfp}^{\subseteq} F_R = \mathsf{lfp}^{\subseteq} F$.

Proof

$$
\begin{aligned}
& F_R(X) \\
\triangleq\ & \{c \mid \exists \frac{P}{c} \in R \,.\, P \subseteq X\} && \text{(definition of } F_R\text{)} \\
=\ & \{c \mid \exists P \subseteq \mathbb{U} \,.\, c \in F(P) \wedge P \subseteq X\} && \text{(definition of } R\text{)} \\
=\ & F(X)
\end{aligned}
$$

$\big($($\subseteq$) if $c \in F(P)$ and $P \subseteq X$ then $c \in F(X)$ because F is $\subseteq$-increasing.
($\supseteq$) if $c \in F(X)$ then $\exists P \subseteq \mathbb{U} \,.\, c \in F(P) \wedge P \subseteq X$ by choosing $P = X$.$\big)$ □

16.3.3 Well Definedness of Deductive Definitions

Theorem 16.17 D in definition 16.10 is well defined.

Proof The deductive definition of D by rules R is equivalent to $D = \mathsf{lfp}^{\subseteq} F_R$ where F_R is $\subseteq$-increasing so is well defined by theorem 15.6. □

16.4 Inductive Definitions

Inductive definitions are mathematical generalizations of recursive programs such as the factorial $\mathtt{f}(0) = 1$ and $\mathtt{f}(n) = n * \mathtt{f}(n-1)$ for $n \in \mathbb{Z}$.

```
$ cat factorial.c
#include <stdio.h>
int f(int n) {
    if (n==0) return 1;
    else return n * f(n - 1);
}
```

```
int main () {
    int n;
    scanf("%d", &n);
    printf("%d! = %d\n", n, f(n));
}
$ gcc factorial.c
$ echo "7" | ./a.out
7! = 5040
$
```

The difference is that at each recursive call a different function is called which parameters are all previously computed values.

$$
\begin{array}{lcllcl}
\texttt{f}(0) & = & 1 & D(0) & = & 1 \\
\texttt{f}(1) & = & 1 * \texttt{f}(0) & D(1) & = & F_1(\langle D(0)\rangle) \\
\texttt{f}(2) & = & 2 * \texttt{f}(1) & D(2) & = & F_2(\langle D(0),\ D(1)\rangle) \\
\texttt{f}(3) & = & 3 * \texttt{f}(2) & D(3) & = & F_3(\langle D(0),\ D(1),\ D(2)\rangle) \\
\dots & & & \dots & & \\
\texttt{f}(n) & = & n * \texttt{f}(n-1) & D(n) & = & F_n(\langle D(i),\ i \in [0, n-1]\rangle) \\
\dots & & & \dots & &
\end{array}
$$

Of course programmers would implement $F_n(\langle D(i),\ i \in [0, n-1]\rangle)$ by a function F taking n as a parameter of F and $\langle D(i),\ i \in [0, n-1]\rangle$ represented, for example, as a linear list of n elements.

The program might not terminate for negative values of the parameter, in which case the corresponding mathematical definition is not well defined.

```
$ echo "-7" | ./a.out
Segmentation fault: 11
$
```

Termination can be proved by recurrence. For $n = 0$, the function `f` returns the evaluation of expression 1, which terminates. Assume, by recurrence hypothesis that $\texttt{f}(n)$ terminates. For the parameter $n + 1$, the function call returns the evaluation of $(n + 1) \times \texttt{f}((n + 1) - 1)$. Because $\texttt{f}(n)$ terminates by induction hypothesis, the evaluation of the expression terminates. By recurrence, all calls $\texttt{f}(n), n \in \mathbb{N}$ do terminate. The corresponding reasoning on the mathematical inductive definition is by induction on a given well founded relation $\preccurlyeq$ ($\preccurlyeq$ is $\leqslant$ on $\mathbb{N}$ for the factorial example).

Of course, computer integers are limited in size which leads to errors.

```
$ echo "20" | ./a.out
20! = -2102132736
$ echo "40" | ./a.out
40! = 0
$ echo "10000000" | ./a.out
Segmentation fault: 11
$
```

The corresponding mathematical reasoning must consider the universe $\mathbb{U} = $ [INT_MIN, INT_MAX] from C directive `#include <limits.h>`, not $\mathbb{U} = \mathbb{Z}$.

Definition 16.18 (well founded relation) A relation $\preccurlyeq \in \wp(S \times S)$ on a set S is well founded if and only if there is no infinite (strictly decreasing chain if $\preccurlyeq$ is a partial order) sequence $x_0 \succ x_1 \succ x_2 \succ \ldots \succ x_n \succ x_{n+1} \succ \ldots$ of elements of S.

Theorem 16.19 (inductive proof) Let $\preccurlyeq$ be a well founded relation on S and $P \subseteq S$ be a property of the elements of S. We write $P(x)$ for $x \in P$. If

$$\forall x \in S\,.\,(\forall y \in S\,.\,(y \prec x) \Rightarrow P(y)) \Rightarrow P(x)$$

then $\forall x \in S\,.\,P(x)$.

Proof By reductio ad absurdum, assuming $\exists x_0 \in S\,.\,\neg P(x_0)$, we construct an infinite sequence $x_0 \succ x_1 \succ x_2 \succ \ldots \succ x_n \succ x_{n+1} \succ \ldots$ of elements of S such that $\forall n \in \mathbb{N}\,.\,\neg P(x_n)$.

Assume we have constructed the sequence up to x_n (i.e., x_0 to start with). Then, by contraposition, $\neg P(x_n)$ implies $\exists x_{n+1} \prec x_n\,.\,\neg P(x_{n+1})$. We get an infinite sequence of elements of S in contradiction to $\preccurlyeq$ is a well founded relation on S. □

Definition 16.20 (inductive definition) The inductive definition of $D \in S \to \mathbb{U}$ where $\langle S, \preccurlyeq\rangle$ is well founded has the form

(1) $D(m) \triangleq D_m$ where $D_m \in \mathbb{U}$ is a constant for minimal elements $m \in S$ (i.e., $\nexists s \in S\,.\,s \prec m$);
(2) otherwise, $D(s) \triangleq F_s(\langle D(s'),\ s' \prec s\rangle)$ where $F_s \in (\{s' \in S \mid s' \prec s\} \to \mathbb{U}) \to \mathbb{U}$.

Most often, we use definition 16.20 for $\mathbb{U} = \wp(\mathbb{S})$ where $\mathbb{S}$ is a set.

[**Theorem 16.21** *D* in definition 16.20 is well defined.

Proof We first observe that the first case is a special case of the second case by defining $F_m(\langle\rangle) = D_m$ for all $m \in S$ such that $\nexists s \in S\,.\,s \prec m$.

The proof is by the induction proof theorem 16.19 on the well-found set $\langle S, \preccurlyeq\rangle$.

Assume by induction hypothesis that $D(s')$ is well defined for all $s' \prec s$. Then $\langle D(s'),\ s' \prec s\rangle \in \{s' \in S \mid s' \prec s\} \to \mathbb{U}$ so $F_s(\langle D(s'),\ s' \prec s\rangle) \in \mathbb{U}$ is well defined, proving that $D(s)$ is well defined. By induction, $\forall s \in S\,.\,D(s) \in \mathbb{U}$ is well defined. So $D \in S \to \mathbb{U}$ is well defined. □

Example 16.22 The definition of a label at a statement in section 4.2.1 is an example of inductive definition, by induction on the syntax of programs (called *structural induction*, see section 16.5). at⟦S⟧ can also be understood as an induction on the number of characters in the S (but for programs that do not have more characters than their constituent statement list, but can be considered larger than this statement list).

16.4.1 Inductive Definition as a Fixpoint Definition

Exercise 16.23 Prove that an inductive definition can be expressed as an equivalent fixpoint definition. Representing functions $f \in A \to B$ as a relation $\{\langle a,\ \mathtt{f}(a)\rangle \mid a \in A\}$, show that the inductive definition 16.20 is $D = \mathsf{lfp}^{\subseteq}\, \mathscr{F}$ where

$$\begin{array}{rcl}\mathscr{F}(X) & \triangleq & \{\langle m,\ D_m\rangle \mid \forall s' \in S\,.\,s' \nprec m\}\\ & \cup & \{\langle s,\ F_s(\langle X(s'),\ s' \prec s\rangle)\rangle \mid \forall s' \prec s\,.\,s' \in \mathsf{dom}(X)\}\end{array}$$

16.5 Structural Definitions

A structural definition is a special case of inductive definition for the syntactic order $\trianglelefteq$ on programs.

Definition 16.24 (structural definition) A structural definition is an inductive definition of the form

$$\begin{cases} D[\![\mathtt{S}]\!] \triangleq f[\![\mathtt{S}]\!]\Big(\prod_{\mathtt{S}' \lhd \mathtt{S}} D[\![\mathtt{S}']\!]\Big) \\ \mathtt{S} \in \mathbb{P}\mathbb{c} \end{cases} \tag{16.24}$$

where the well founded order $\preccurlyeq$ is the syntactic order $\trianglelefteq$ on programs, that is, $\mathtt{S} \lhd \mathtt{S}'$ if and only if $\mathtt{S}$ is a strict syntactic component of $\mathtt{S}'$.

Example 16.25 (strict syntactic order) The *strict syntactic order* $\lhd$ for the programs of section 4.1 is defined as follows. This is the order used in the structural definitions of the trace semantics introduced in chapter 6.

$$\begin{array}{lcll}
\mathtt{P} & ::= & \mathtt{Sl} & \mathtt{Sl} \lhd \mathtt{P} \\
\mathtt{S} & ::= & & \\
 & & \mathtt{x = E\,;} & \mathtt{x} \lhd \mathtt{S},\quad \mathtt{E} \lhd \mathtt{S} \\
 & | & \mathtt{;} & \\
 & | & \mathtt{if\,(B)\,S}_t & \mathtt{B} \lhd \mathtt{S},\quad \mathtt{S}_t \lhd \mathtt{S} \\
 & | & \mathtt{if\,(B)\,S}_t\ \mathtt{else}\ \mathtt{S}_f & \mathtt{B} \lhd \mathtt{S},\quad \mathtt{S}_t \lhd \mathtt{S},\quad \mathtt{S}_f \lhd \mathtt{S} \\
 & | & \mathtt{while\,(B)\,S}_b & \mathtt{B} \lhd \mathtt{S},\quad \mathtt{S}_b \lhd \mathtt{S} \\
 & | & \mathtt{break\,;} & \\
 & | & \mathtt{\{Sl\}} & \mathtt{Sl} \lhd \mathtt{S} \\
\mathtt{Sl} & ::= & \mathtt{Sl'\ S} \mid \epsilon & \mathtt{Sl'} \lhd \mathtt{Sl},\quad \mathtt{S} \lhd \mathtt{Sl},\quad \epsilon \lhd \mathtt{Sl}
\end{array} \tag{16.26}$$

The *syntactic order* $\trianglelefteq$ is $\mathtt{S} \trianglelefteq \mathtt{S}' \triangleq \mathtt{S} \lhd \mathtt{S}' \vee \mathtt{S} = \mathtt{S}'$. The *recursive syntactic order* $\lhd^+$ is the transitive closure of $\lhd$. The *recursive subcomponent partial order* $\lhd^*$ is the reflexive transitive closure of $\lhd$, that is, the transitive closure of $\trianglelefteq$.

Exercise 16.27 Prove that $\lhd$ is well founded.

Exercise 16.28 Provide a fixpoint definition of $\lhd^+$ and $\lhd^*$ in terms of $\lhd$.

Example 16.29 The prefix trace semantics $\widehat{\boldsymbol{\mathcal{S}}}^*$ of section 6.6 is a structural definition based on the syntactic order $\trianglelefteq$ defined in example 16.25.

The base case definition 16.20.(1) is given by axioms (6.15) for empty statement lists, (6.16) for assignment, (6.17) for skip, and (6.29) for break statements.

For the inductive case definition 16.20.(2), the function F_s is specified by induction rules (6.12) for programs, (6.13) and (6.14) for statement lists, (6.19), (6.18), (6.22), (6.23) for conditional, (6.24), (6.25), (6.26) for iteration, and (6.30) for compound statements.

Exercise 16.30 Define the syntactic order for expressions in chapter 3.

Corollary 16.31 (structural proof) If

$$\forall \mathsf{S} \lhd^* \mathsf{P} \,.\, (\forall \mathsf{S}' \lhd^* \mathsf{P} \,.\, (\mathsf{S}' \lhd \mathsf{S}) \Rightarrow P(\mathsf{S}')) \Rightarrow P(\mathsf{S})$$

then $\forall \mathsf{S} \lhd^* \mathsf{P} \,.\, P(\mathsf{S})$.

The structural induction hypothesis $P(\mathsf{S}')$ is assumed to hold for all $\mathsf{S}' \lhd \mathsf{S}$ when proving $P(\mathsf{S})$.

Proof By theorem 16.19 for the syntactic order $\langle\{\mathsf{S} \mid \mathsf{S} \lhd^* \mathsf{P}\}, \unlhd\rangle$ of example 16.25 which is well founded by exercise 16.27. □

Corollary 16.32 A structural definition 16.20 for the syntactic order $\langle\{\mathsf{S} \mid \mathsf{S} \lhd^* \mathsf{P}\}, \unlhd\rangle$ is well defined.

Proof By structural induction and corollary 16.31. □

Structural proofs were originally introduced by Rod Burstall [157] for recursively defined structures such as data types.

Exercise 16.33 The label $\mathsf{after}[\![\mathsf{S}]\!]$ after a program component S in section 4.2.2 is *neither* by structural induction *nor* by induction on the size of S (e.g., measured in number of characters). Which well founded set is used for this inductive definition?

16.6 Coinductive Definitions

The same way that deductive definitions are equivalent to least fixpoint definitions by theorems 16.12 and 16.16 in section 16.3, coinductive definitions are equivalent to greatest fixpoint definitions.

Definition 16.34 (coinductive definition) The coinductive definition of $D \in \wp(\mathbb{U})$ by a deductive system of rules $\frac{P}{c} \in R$ is $\mathsf{gfp}^{\subseteq} F_R$ where $F_R(X) \triangleq \{c \mid \exists \frac{P}{c} \in R \,.\, P \subseteq X\}$ is the consequence operator of R.

Example 16.35 Let $\mathbb{U}$ be the set of infinite strings on the alphabet $\{\mathsf{a}, \mathsf{b}\}$. Let $R = \{\frac{\{\sigma\}}{\mathsf{a}\sigma} \mid \sigma \in \mathbb{U}\}$ stating that if $\sigma \in \mathbb{U}$ is an infinite string on the alphabet $\{\mathsf{a}, \mathsf{b}\}$ in D then $a\sigma$ is also an infinite string in D. This coinductive definition is equivalent to $\mathsf{gfp}^{\subseteq} F_R$ where $F_R(X) \triangleq \{\mathsf{a}\sigma \mid \sigma \in X\}$. F_R preserves arbitrary meets, so by the dual of theorem 15.21, the greatest fixpoint is the limit of the following $\subseteq$-decreasing chain of iterates of F_R

$$\begin{array}{rcl} F_R^0 & = & \mathbb{U} \\ F_R^1 & = & \{\mathsf{a}\sigma \mid \sigma \in \mathbb{U}\} \\ \dots & & \\ F_R^n & = & \{\mathsf{a}^n\sigma \mid \sigma \in \mathbb{U}\} \\ \dots & & \\ D & = & \mathsf{gfp}^{\subseteq} F_R \;=\; \bigcap_{n\in\mathbb{N}} F_R^n \;=\; \{\mathsf{aaaaa}\dots\} \;=\; \{\mathsf{a}^\omega\} \end{array}$$

For the limit, observe that all iterates contains a^ω and if an infinite string contains a b, at given rank n in the string, then it does not belong to F_R^n hence not to the limit $\bigcap_{n\in\mathbb{N}} F_R^n$.

Exercise 16.36 Reconsider example 16.35 when $\mathbb{U}$ be the set of finite and infinite strings on the alphabet $\{\mathsf{a}, \mathsf{b}\}$ and $R = \{\frac{\varnothing}{\mathsf{a}}\} \cup \{\frac{\{\sigma\}}{\mathsf{a}\sigma} \mid \sigma \in \mathbb{U}\}$. Prove that $D = \{\mathsf{a}^n \mid n \in \mathbb{N}^+\} \cup \{\mathsf{a}^\omega\}$.

16.7 Bi-inductive Definitions

Although finite and infinite traces of program semantics can be defined as coinductive definition/greatest fixpoints (see exercise 16.36), it is often more practical (in particular for fixpoint abstraction in chapter 18, "Fixpoint Abstraction") to define finite traces as least fixpoint (thus constructing longer and longer ones) and infinite traces as greatest fixpoint (thus eliminating the ones with longer and longer unfitting prefixes instead of passing to the limit as in chapter 7, "Maximal Trace Semantics"). This can be done with bi-inductive definitions thus avoiding having to consider finite and infinite traces separately.

Lemma 16.37 Let $\mathbb{U} = \mathbb{U}^+ \cup \mathbb{U}^-$ be a partitioned bi-universe such that $\mathbb{U}^+ \cap \mathbb{U}^- = \varnothing$. Define $X^+ \triangleq X \cap \mathbb{U}^+$, $X^- \triangleq X \cap \mathbb{U}^-$, $X \sqsubseteq^{\mp} Y \triangleq X^+ \subseteq Y^+ \wedge X^- \supseteq Y^-$, $\bigsqcup^{\mp}_{i\in\Delta} X_i \triangleq \bigcup_{i\in\Delta} X_i^+ \cup \bigcap_{i\in\Delta} X_i^-$, and $\bigsqcap^{\mp}_{i\in\Delta} X_i \triangleq \bigcap_{i\in\Delta} X_i^+ \cup \bigcup_{i\in\Delta} X_i^-$. Then $\langle \mathbb{U}, \sqsubseteq^{\mp}, \mathbb{U}^-, \mathbb{U}^+, \sqcup^{\mp}, \sqcap^{\mp}\rangle$ is a complete lattice.

Proof By the Galois isomorphism $\langle \wp(\mathbb{U}^+ \times \mathbb{U}^-, \subseteq\times\supseteq\rangle \xleftrightarrow[\langle X, Y\rangle \mapsto X \cup Y]{Z \mapsto \langle Z^+, Z^-\rangle} \langle \wp(\mathbb{U}), \sqsubseteq^{\mp}\rangle$, existing join preservation lemma 11.38, and exercise 10.7. □

Definition 16.38 (bi-inductive definition) The bi-inductive definition of $D \in \wp(\mathbb{U})$ on a bi-universe $\mathbb{U}$ by a deductive system $R = R^+ \cup R^-$ of inductive rules $+\frac{P}{c} \in R^+$ (where $P \subseteq \mathbb{U}^+$, $c \in \mathbb{U}^+$) and coinductive rules $-\frac{P}{c} \in R^-$ (where $P \subseteq \mathbb{U}^-$, $c \in \mathbb{U}^-$) is $\mathsf{lfp}^{\sqsubseteq^{\mp}} F_R$ where $F_R(X) \triangleq \{c \mid \exists \frac{P}{c} \in R \,.\, P \subseteq X\}$ is the consequence operator of D, and $\sqsubseteq^{\mp}$ is the bi-order of lemma 16.37.

Example 16.39 Continuing exercise 16.36 let $\mathbb{U}$ be the set of finite $\mathbb{U}^+$ and infinite $\mathbb{U}^-$ strings on the alphabet $\{\mathsf{a}, \mathsf{b}\}$, $R^+ = \{+\frac{\varnothing}{\mathsf{a}}\} \cup \{+\frac{\{\sigma\}}{\mathsf{a}\sigma} \mid \sigma \in \mathbb{U}^+\}$, $R^- = \{-\frac{\{\sigma\}}{\mathsf{a}\sigma} \mid \sigma \in \mathbb{U}^-\}$ so that $F_R(X) \triangleq \{\mathsf{a}\} \cup \{\mathsf{a}\sigma \mid \sigma \in X\}$. The iterates of F_R are

$$
\begin{aligned}
F_R^0 &= \mathbb{U}^- \\
F_R^1 &= \{a\} \cup \{\mathsf{a}\sigma \mid \sigma \in \mathbb{U}^-\} \\
&\ldots \\
F_R^n &= \{a, a^2, \ldots, a^n\} \cup \{\mathsf{a}^n\sigma \mid \sigma \in \mathbb{U}^-\} \\
&\ldots \\
D &= \mathsf{lfp}^{\sqsubseteq^\mp} F_R = \bigsqcup^{\mp}_{n\in\mathbb{N}} F_R^n \\
&= \bigcup_{n\in\mathbb{N}} \{a, a^2, \ldots, a^n\} \cup \bigcap_{n\in\mathbb{N}} \{\mathsf{a}^n\sigma \mid \sigma \in \mathbb{U}^-\} = \{\mathsf{a}^n \mid n \in \mathbb{N}^+\} \cup \{\mathsf{a}^\omega\}
\end{aligned}
$$

16.8 Conclusion

Fixpoint, deductive, and inductive definitions are used to provide well defined specifications of the structural semantics of programs and their abstractions. Context-free grammars are a particular case (see exercise 16.15 and [181, 419, 868]). An overview of set-theoretic formal definitions is given in [6]. A generalization from $\wp(\mathbb{U})$ to complete partial orders is considered in [250].

Coinduction originates from [7], [686], and category theory [8]. Bi-inductive definitions are from [247, 260]. Another interpretation between induction and coinduction considers infinite proof trees [285] that are regular [219], that is, containing only finitely many different subtrees.

16.9 Solutions to Selected Exercises

Solution to exercise 16.13 Consider $R = \left\{\frac{\varnothing}{a}, \frac{\varnothing}{b}, \frac{\{a\}}{b}, \frac{\{b\}}{a}, \frac{\{a,b\}}{c}\right\}$. Then $F_R(\{a\}) \cup F_R(\{b\}) = \{a, b\} \neq \{a, b, c\} = F_R(\{a, b\})$.

If $X \subseteq Y$ then $(P \subseteq X) \Rightarrow (P \subseteq Y)$ by transitivity so $\{c \mid \exists \frac{P}{c} \in R \,.\, P \subseteq X\} \subseteq \{c \mid \exists \frac{P}{c} \in R \,.\, P \subseteq Y\}$ proving that $F_R(X) \subseteq F_R(Y)$.

Solution to exercise 16.27 $\lhd$ is defined on program components that have finite tree representations and $\mathtt{S}' \lhd \mathtt{S}$ implies that finite tree representation of $\mathtt{S}'$ has at least one node less than the tree representation of $\mathtt{S}$.

Solution to exercise 16.33 The definition of function $\mathsf{after}[\![\bullet]\!]$ attaches information to the nodes $\bullet$ of a given program syntax tree. The information on the descendants depends only on the information on their ascendants. So the definition is by induction on the distance to the root of the tree, which is well founded by going up. For $\mathtt{Sl} ::= \mathtt{Sl}'\ \mathtt{S}$, we have $\mathsf{after}[\![\mathtt{S}]\!] \triangleq \mathsf{after}[\![\mathtt{Sl}]\!]$ and the distance of $\mathtt{S}$ to the tree root is 1 plus the distance of $\mathtt{Sl}$ to the tree root. For $\mathsf{after}[\![\mathtt{Sl}']\!] \triangleq \mathsf{at}[\![\mathtt{S}]\!]$, there is no induction because $\mathsf{at}[\![\bullet]\!]$ was completely defined before defining $\mathsf{after}[\![\bullet]\!]$.

17 Structural Fixpoint Prefix and Maximal Trace Semantics

As shown in chapter 16, structural inductive definitions can be given an equivalent fixpoint form. It follows that structural inductive semantics such as the prefix trace semantics of section 6.3 and the maximal trace semantics of section 7.3 have equivalent structural fixpoint presentations that we study in this chapter 17.

Contents

17.1 Structural Fixpoint Prefix Trace Semantics

The prefix trace semantics $\boldsymbol{\mathcal{S}}^*$ of section 6.3 can be given an equivalent structural definition

$$\begin{cases} \widehat{\boldsymbol{\mathcal{S}}}^*[\![\mathsf{S}]\!] = f^*[\![\mathsf{S}]\!]\Big(\prod_{\mathsf{S}' \lhd \mathsf{S}} \widehat{\boldsymbol{\mathcal{S}}}^*[\![\mathsf{S}']\!]\Big) \\ \mathsf{S} \in \mathbb{P}\mathrm{c} \end{cases} \tag{17.1}$$

based on the syntactic order $\trianglelefteq$ defined in example 16.25.

The base case in definition 16.20.(1) is given by axiom (6.11), as well as axioms (6.15) for empty statement lists, (6.16) for assignment, (6.17) for skip, (6.29) for break, and (6.24) for while statements.

For example, the prefix trace semantics of the assignment is derived from (6.11) and (6.16) as follows:

Prefix traces of an assignment statement $\mathsf{S} ::= {}^{\ell}\, \mathsf{x} = \mathsf{A}\,;$

$$\widehat{\boldsymbol{\mathcal{S}}}^*[\![\mathsf{S}]\!](\pi\ell) = \{\ell\} \cup \{\ell \xrightarrow{\mathsf{x}\,=\,\mathsf{A}\,=\,v} \mathsf{after}[\![\mathsf{S}]\!] \mid v = \boldsymbol{\mathcal{A}}[\![\mathsf{A}]\!]\varrho(\pi\ell)\} \tag{17.2}$$

$$\widehat{\boldsymbol{\mathcal{S}}}^*[\![\mathsf{S}]\!](\pi\ell') = \varnothing \quad \text{when} \quad \ell' \neq \ell$$

The inductive case definition 16.20.(2) is given by induction rules (6.12) for programs, (6.13), (6.14) for statement lists, (6.19), (6.18), (6.22), (6.23) for conditional, (6.25), (6.28), (6.26) for while, and (6.30) for compound statements.

For example, the prefix trace semantics of the statement list is

Prefix traces of a statement list $\texttt{Sl} ::= \texttt{Sl}'\ \texttt{S}$

$$\widehat{\boldsymbol{\mathcal{S}}}^*[\![\texttt{Sl}]\!](\pi_1) \;=\; \widehat{\boldsymbol{\mathcal{S}}}^*[\![\texttt{Sl}']\!](\pi_1) \cup \{\pi_2 \mathbin{\frown} \pi_3 \mid \pi_2 \in \widehat{\boldsymbol{\mathcal{S}}}^+[\![\texttt{Sl}']\!](\pi_1) \wedge \pi_3 \in \widehat{\boldsymbol{\mathcal{S}}}^*[\![\texttt{S}]\!](\pi_1 \mathbin{\frown} \pi_2)\} \tag{17.3}$$

The deductive definition of the prefix trace semantics $\boldsymbol{\mathcal{S}}^*[\![\texttt{while}\,\ell\,(\texttt{B})\,\texttt{S}_b]\!]$ of the while statement $\texttt{S} ::= \texttt{while}\,\ell\,(\texttt{B})\,\texttt{S}_b$ is by rules (6.24), (6.25), (6.28), and (6.26), including, by remark 6.27, breaks after the loop. Following section 16.3.1, this deductive semantics can be expressed in fixpoint form $\mathsf{lfp}^{\dot\subseteq}\,\boldsymbol{\mathcal{F}}^*[\![\texttt{S}]\!]$ where $\dot\subseteq$ is $\subseteq$ pointwise and

Prefix traces of an iteration statement $\texttt{S} ::= \texttt{while}\,\ell\,(\texttt{B})\,\texttt{S}_b$

$$\widehat{\boldsymbol{\mathcal{S}}}^*[\![\texttt{while}\,\ell\,(\texttt{B})\,\texttt{S}_b]\!] \;=\; \mathsf{lfp}^{\dot\subseteq}\,\boldsymbol{\mathcal{F}}^*[\![\texttt{while}\,\ell\,(\texttt{B})\,\texttt{S}_b]\!] \tag{17.4}$$

$$\boldsymbol{\mathcal{F}}^*[\![\texttt{while}\,\ell\,(\texttt{B})\,\texttt{S}_b]\!] \in (\mathbb{T}^+ \rightarrow \wp(\mathbb{T}^+)) \xrightarrow{\nearrow} (\mathbb{T}^+ \rightarrow \wp(\mathbb{T}^+))$$

$$\boldsymbol{\mathcal{F}}^*[\![\texttt{while}\,\ell\,(\texttt{B})\,\texttt{S}_b]\!](X)(\pi_1\ell') \triangleq \varnothing \qquad \text{when} \quad \ell' \neq \ell$$

$$\boldsymbol{\mathcal{F}}^*[\![\texttt{while}\,\ell\,(\texttt{B})\,\texttt{S}_b]\!](X)(\pi_1\ell) \triangleq \{\ell\} \tag{a}$$

$$\cup \{\ell'\pi_2\ell' \xrightarrow{\neg(\texttt{B})} \mathsf{after}[\![\texttt{S}]\!] \mid \ell'\pi_2\ell' \in X(\pi_1\ell') \wedge \boldsymbol{\mathcal{B}}[\![\texttt{B}]\!]\boldsymbol{\varrho}(\pi_1\ell'\pi_2\ell') = \mathsf{ff} \wedge \ell' = \ell\} \tag{b}$$

$$\cup \{\ell'\pi_2\ell' \xrightarrow{\texttt{B}} \mathsf{at}[\![\texttt{S}_b]\!] \mathbin{\frown} \pi_3 \mid \ell'\pi_2\ell' \in X(\pi_1\ell') \wedge \boldsymbol{\mathcal{B}}[\![\texttt{B}]\!]\boldsymbol{\varrho}(\pi_1\ell'\pi_2\ell') = \mathsf{tt} \wedge \pi_3 \in \boldsymbol{\mathcal{S}}^*[\![\texttt{S}_b]\!](\pi_1\ell'\pi_2\ell' \xrightarrow{\texttt{B}} \mathsf{at}[\![\texttt{S}_b]\!]) \wedge \ell' = \ell\} \tag{c}$$

Remark 17.5 (increasing transformer) By exercise 16.13, $\boldsymbol{\mathcal{F}}^*[\![\texttt{while}\,\ell\,(\texttt{B})\,\texttt{S}_b]\!]$ is $\dot\subseteq$ increasing and $\langle \mathbb{T}^+ \rightarrow \wp(\mathbb{T}^+),\ \dot\subseteq\rangle$ is a complete lattice so the least fixpoint (17.4) is well defined by Tarski fixpoint theorem 15.6.

Remark 17.6 (on break statements) The last case (17.4.c) covers the case of an iteration terminated by a break statement when π_3 is of the form $\pi_3' \xrightarrow{\texttt{break}} \ell''$ and $\ell'' \in \mathsf{dest}[\![\texttt{S}]\!]$.

Theorem 17.7 For any program component $\texttt{S} \in \mathbb{P}\mathbb{c}$, $\widehat{\boldsymbol{\mathcal{S}}}^*[\![\texttt{S}]\!] = \boldsymbol{\mathcal{S}}^*[\![\texttt{S}]\!]$.

Proof of theorem 17.7 By structural induction of S, using the equivalence of structural deductive definitions and structural fixpoint definitions given in chapter 16. □

Remark 17.8 (on free and bound/dummy variables) In a definition of the form $d \triangleq \{f(\vec{x}) \mid P(\vec{x})\}$, where $\vec{x}$ is a tuple of variables, the dummy variables of $\vec{x}$ occurring in $P(\vec{x})$ are bound to those $\vec{x}$ occurring in $f(\vec{x})$. The *definiendum*[1] d does not depend upon $\vec{x}$, which are dummy variables in the *definiens*[2] $\{f(\vec{x}) \mid P(\vec{x})\}$.

1. The *definiendum* (plural *definienda*) is the subject of a definition, what is being defined.
2. The *definiens* (plural *definientia*) is the term/symbolic expression that defines the *definiendum*.

On the contrary, a *definiens* of the form $d(\vec{x}) \triangleq \{f(\vec{x}') \mid P(\vec{x}', \vec{x})\}$ has the variables $\vec{x}'$ in $P(\vec{x}', \vec{x})$ bound to those of $f(\vec{x}')$ whereas $\vec{x}$ is free in $P(\vec{x}', \vec{x})$ because it appears neither in $f(\vec{x}')$ nor (by assumption) under quantifiers in $P(\vec{x}', \vec{x})$. The $\vec{x}$ of $P(\vec{x}', \vec{x})$ is therefore bound to the $\vec{x}$ of $d(\vec{x})$.

For example, the case $\mathscr{F}^*[\![\texttt{while}\,\ell\,(\mathsf{B})\,\mathsf{S}_b]\!](X)(\pi_1\ell) \triangleq \{\ell\}$ (which is also $\{\ell \mid \mathsf{tt}\}$) has ℓ free in $\{\ell\}$ so is the ℓ of $\mathscr{F}^*[\![\texttt{while}\,\ell\,(\mathsf{B})\,\mathsf{S}_b]\!](X)(\pi_1\ell)$. Notice that ℓ is a constant that could have been replaced by $\mathsf{at}[\![\texttt{while}\,\ell\,(\mathsf{B})\,\mathsf{S}_b]\!]$.

In case $\mathscr{F}^*[\![\texttt{while}\,\ell\,(\mathsf{B})\,\mathsf{S}_b]\!](X)(\pi_1\ell) \triangleq \{\ell'\pi_2\ell' \xrightarrow{\neg(\mathsf{B})} \mathsf{after}[\![\mathsf{S}]\!] \mid \ell'\pi_2\ell' \in X(\pi_1\ell') \wedge \mathscr{B}[\![\mathsf{B}]\!]\varrho(\pi_1\ell'\pi_2\ell') = \mathsf{ff} \wedge \ell' = \ell\}$, ℓ is free so is the ℓ of $\mathscr{F}^*[\![\texttt{while}\,\ell\,(\mathsf{B})\,\mathsf{S}_b]\!](X)(\pi_1\ell)$ whereas ℓ' is bound to $\ell'\pi_2\ell'$.

Because $\ell' = \ell$, we could have written $\mathscr{F}^*[\![\texttt{while}\,\ell\,(\mathsf{B})\,\mathsf{S}_b]\!](X)(\pi_1\ell) \triangleq \{\ell\pi_2\ell \xrightarrow{\neg(\mathsf{B})} \mathsf{after}[\![\mathsf{S}]\!] \mid \ell\pi_2\ell \in X(\pi_1\ell) \wedge \mathscr{B}[\![\mathsf{B}]\!]\varrho(\pi_1\ell\pi_2\ell) = \mathsf{ff}\}$ but then ℓ would be bound to $\ell\pi_2\ell$ hence equivalent to $\mathscr{F}^*[\![\texttt{while}\,\ell\,(\mathsf{B})\,\mathsf{S}_b]\!](X)(\pi_1\ell) \triangleq \{\ell'\pi_2\ell' \xrightarrow{\neg(\mathsf{B})} \mathsf{after}[\![\mathsf{S}]\!] \mid \ell'\pi_2\ell' \in X(\pi_1\ell') \wedge \mathscr{B}[\![\mathsf{B}]\!]\varrho(\pi_1\ell'\pi_2\ell') = \mathsf{ff}\}$ (because dummy variables can be renamed) which does not require that $\ell' = \ell$. Nevertheless, we sometimes use this notation where ℓ is global.

Example 17.9 Consider $\mathsf{S} = \texttt{while}\,\ell\,(\mathsf{tt})\,\ell'\,\mathsf{x} = \mathsf{x} + 1\,;$ so that $\mathsf{S}_b = \mathsf{x} = \mathsf{x} + 1\,;$. We have

$$\begin{aligned}\mathscr{F}^*[\![\mathsf{S}]\!](X)(\pi_1\ell) \triangleq\ & \{\ell\} \cup \{\ell\pi_2\ell \xrightarrow{\mathsf{tt}} \ell' \mid \ell\pi_2\ell \in X(\pi_1\ell)\} \cup \\ & \{\ell\pi_2\ell \xrightarrow{\mathsf{tt}} \ell' \xrightarrow{\mathsf{x} = \mathsf{x} + 1 = v} \ell \mid \ell\pi_2\ell \in X(\pi_1\ell) \wedge v = \varrho(\pi_1\ell\pi_2\ell) + 1\}\end{aligned}$$

The iterates $\langle \mathscr{F}^{*n}, n \in \mathbb{N}\rangle$ of $\mathscr{F}^*[\![\mathsf{S}]\!]$ from $\varnothing$ are (for all initial traces $(\pi_1\ell)$)

$$\begin{aligned}
\mathscr{F}^{*0}(\pi_1\ell) &= \varnothing \\
\mathscr{F}^{*1}(\pi_1\ell) &= \{\ell\} \\
\mathscr{F}^{*2}(\pi_1\ell) &= \{\ell, \ell \xrightarrow{\mathsf{tt}} \ell', \ell \xrightarrow{\mathsf{tt}} \ell' \xrightarrow{\mathsf{x} = \mathsf{x} + 1 = v} \ell \mid v = \varrho(\pi_1\ell) + 1\} \\
\mathscr{F}^{*3}(\pi_1\ell) &= \{\ell, \ell \xrightarrow{\mathsf{tt}} \ell', \ell \xrightarrow{\mathsf{tt}} \ell' \xrightarrow{\mathsf{x} = \mathsf{x} + 1 = v(1)} \ell, \ell \xrightarrow{\mathsf{tt}} \ell' \xrightarrow{\mathsf{x} = \mathsf{x} + 1 = v(1)} \\
&\quad \ell \xrightarrow{\mathsf{tt}} \ell', \ell \xrightarrow{\mathsf{tt}} \ell' \xrightarrow{\mathsf{x} = \mathsf{x} + 1 = v(1)} \ell \xrightarrow{\mathsf{tt}} \ell' \xrightarrow{\mathsf{x} = \mathsf{x} + 1 = v(2)} \ell \mid \forall i \in \\
&\quad [1, 2]\,.\,v(i) = \varrho(\pi_1\ell) + i\} \\
\dots \\
\mathscr{F}^{*n}(\pi_1\ell) &= \Big\{\Big(\ell \xrightarrow{\mathsf{tt}} \ell' \xrightarrow{\mathsf{x} = \mathsf{x} + 1 = v(i)} \ell\Big)_{i=0}^{k}, \Big(\ell \xrightarrow{\mathsf{tt}} \ell' \xrightarrow{\mathsf{x} = \mathsf{x} + 1 = v(i)} \ell\Big)_{i=0}^{k'} \,\widehat{}\, \\
&\quad \ell \xrightarrow{\mathsf{tt}} \ell' \Bigm| k \in [0, n[\wedge k' \in [0, n-2] \wedge \forall i \in [1, n-1]\,.\,v(i) = \varrho(\pi_1\ell) + i\Big\} \\
&\qquad \text{(induction hypothesis with } \Big(\ell \dots \ell\Big)^{0} = \ell\text{)} \\
\mathscr{F}^{*n+1}(\pi_1\ell) &= \mathscr{F}^*[\![\mathsf{S}]\!](\mathscr{F}^{*n})(\pi_1\ell) \\
&= \{\ell\} \cup \{\ell\pi_2\ell \xrightarrow{\mathsf{tt}} \ell' \mid \ell\pi_2\ell \in \mathscr{F}^{*n}(\pi_1\ell)\} \cup \{\ell\pi_2\ell \xrightarrow{\mathsf{tt}} \ell' \xrightarrow{\mathsf{x} = \mathsf{x} + 1 = v} \ell \mid \\
&\quad \ell\pi_2\ell \in \mathscr{F}^{*n}(\pi_1\ell) \wedge v = \varrho(\pi_1\ell\pi_2\ell) + 1\}
\end{aligned}$$

$$= \{\ell\} \cup \{\ell\pi_2\ell \xrightarrow{\mathsf{tt}} \ell' \mid \ell\pi_2\ell \in \{(\ell \xrightarrow{\mathsf{tt}} \ell' \xrightarrow{\mathsf{x}=\mathsf{x}+1=v(i)} \ell)_{i=0}^{k} \mid k \in [0, n[\wedge \forall i \in [1, n-1] \,.\, v(i) = \varrho(\pi_1\ell) + i\}\} \cup \{\ell\pi_2\ell \xrightarrow{\mathsf{tt}} \ell' \xrightarrow{\mathsf{x}=\mathsf{x}+1=v} \ell \mid \ell\pi_2\ell \in \{(\ell \xrightarrow{\mathsf{tt}} \ell' \xrightarrow{\mathsf{x}=\mathsf{x}+1=v(i)} \ell)_{i=0}^{k} \mid k \in [0, n[\wedge \forall i \in [1, n-1] \,.\, v(i) = \varrho(\pi_1\ell) + i\} \wedge v = \varrho(\pi_1\ell\pi_2\ell) + 1\}$$

$$= \{(\ell \xrightarrow{\mathsf{tt}} \ell' \xrightarrow{\mathsf{x}=\mathsf{x}+1=v(i)} \ell)_{i=0}^{k}, (\ell \xrightarrow{\mathsf{tt}} \ell' \xrightarrow{\mathsf{x}=\mathsf{x}+1=v(i)} \ell)_{i=0}^{k'} \frown \ell \xrightarrow{\mathsf{tt}} \ell' \mid k \in [0, n] \wedge k' \in [0, n-1] \wedge \forall i \in [1, n] \,.\, v(i) = \varrho(\pi_1\ell) + i\}$$

...

$$\begin{aligned}
\widehat{\mathcal{S}}^*[\![\mathsf{S}]\!](\pi_1\ell) &= (\mathsf{lfp}^{\dot{\subseteq}}\, \mathcal{F}^*[\![\mathsf{S}]\!])(\pi_1\ell) \\
&= \mathcal{F}^{*\omega} \\
&= \bigcup_{n \in \mathbb{N}} \mathcal{F}^{*n}(\pi_1\ell) \\
&= \{(\ell \xrightarrow{\mathsf{tt}} \ell' \xrightarrow{\mathsf{x}=\mathsf{x}+1=v(i)} \ell)_{i=0}^{k}, (\ell \xrightarrow{\mathsf{tt}} \ell' \xrightarrow{\mathsf{x}=\mathsf{x}+1=v(i)} \ell)_{i=0}^{k} \frown \ell \xrightarrow{\mathsf{tt}} \ell' \mid k \in \mathbb{N} \wedge \forall i \in \mathbb{N} \,.\, v(i) = \varrho(\pi_1\ell) + i\}
\end{aligned}$$

Exercise 17.10 Prove, by structural induction, that for each program component S, $\widehat{\mathcal{S}}^*[\![\mathsf{S}]\!] \in \wp(\mathbb{T}^+)$ is an $\preceq$-increasing denumerable chain for the prefix ordering $\preceq$ of exercise 10.15 defined as $\pi' \preceq \pi \triangleq \exists \pi'' \,.\, \pi = \pi'\pi''$ (with $\ni \not\preceq \pi$ to exclude empty traces). Prove that $\widehat{\mathcal{S}}^*[\![\mathsf{S}]\!]$ is prefix closed i.e., $\forall \pi \in \widehat{\mathcal{S}}^*[\![\mathsf{S}]\!] \,.\, \forall \pi' \preceq \pi \,.\, \pi' \in \widehat{\mathcal{S}}^*[\![\mathsf{S}]\!]$ (so that $\widehat{\mathcal{S}}^*[\![\mathsf{S}]\!]$ is equal to its $\preceq$-downward closure $\downarrow \widehat{\mathcal{S}}^*[\![\mathsf{S}]\!]$ where $\downarrow X \triangleq \{\pi' \mid \exists \pi \in X \,.\, \pi' \preceq \pi\}$). It follows that $\langle \widehat{\mathcal{S}}^*[\![\mathsf{S}]\!] \cup \{\ni\}, \preceq, \ni, \dot{\curlyvee} \rangle$ is a CPO where the lub $\dot{\curlyvee}$ of a finite chain is its maximal element whereas for an infinite chain $\mathcal{T}$ it is the unique chain of $\lim \mathcal{T}$ defined in (7.4).

In case of an iteration statement S ::= **while** ℓ (B) S_b, prove, by induction on the rank of the iterates, that this holds for the $\subseteq$-increasing chain of the iterates $\langle \mathcal{F}^{*n}, n \in \mathbb{N}^+ \cup \{\omega\} \rangle$ of the transformer $\mathcal{F}^*[\![\mathsf{S}]\!]$ in (17.4) from $\mathcal{F}^{*0} = \pi \mapsto \varnothing$.

Exercise 17.11 Prove that if X is an iterate of $\mathcal{F}^*[\![\mathbf{while}\ \ell\ (\mathsf{B})\ \mathsf{S}_b]\!]$ and $\ell'\pi \in X(\pi'\ell'')$ then $\ell' = \ell'' = \ell$.

Exercise 17.12 Prove that if X is an iterate of $\mathcal{F}^*[\![\mathbf{while}\ \ell\ (\mathsf{B})\ \mathsf{S}_b]\!]$, $\ell\pi_1\ell\pi_2 \in X(\pi_0\ell)$, and $\ell \notin \pi_1$ then $\ell\pi_2 \in X(\pi_0\ell\pi_1\ell)$.

Exercise 17.13 Prove that for all program components S, $\widehat{\mathcal{S}}^*[\![\mathsf{S}]\!]$ is a nonempty, prefix closed, $\preceq$-increasing chain of traces (where the trace prefix order $\preceq$ was defined in exercise 10.15 and a set Π of traces is prefix-closed if and only if $\forall \pi \in \Pi \,.\, \forall \pi' \in \mathbb{T}^+ \,.\, (\pi' \preceq \pi) \Rightarrow (\pi' \in \Pi)$).

17.2 Structural Fixpoint Maximal Trace Semantics

The maximal trace semantics $\boldsymbol{\mathcal{S}}^{+\infty}[\![\mathsf{S}]\!] \in \mathbb{T}^+ \to \wp(\mathbb{T}^{+\infty})$ of chapter 7 can also be easily expressed structurally. For the iteration, one can use (6.9) for the finite traces and (7.6) for the infinite traces. Since, by exercise 17.10, the language is deterministic, we get a singleton so $\boldsymbol{\mathcal{S}}^{+\infty}[\![\mathsf{S}]\!] \in \mathbb{T}^+ \to \{\{\pi\} \mid \pi \in \mathbb{T}^{+\infty}\}$.

Another solution is to define $\widehat{\boldsymbol{\mathcal{S}}}^{+\infty}[\![\mathsf{S}]\!]$ as the maximal trace only, as shown in exercise 17.14 (inspired by [278, section 5.3]). Then $\widehat{\boldsymbol{\mathcal{S}}}^{+\infty}[\![\mathsf{S}]\!] \in \mathbb{T}^+ \to \mathbb{T}^{+\infty}$ with $\forall \pi \in \mathbb{T}^+ .\ \boldsymbol{\mathcal{S}}^{+\infty}[\![\mathsf{S}]\!]\pi = \{\widehat{\boldsymbol{\mathcal{S}}}^{+\infty}[\![\mathsf{S}]\!]\pi\}$ (so the superfluous singleton is abstracted away).

Exercise 17.14 Following exercise 17.10, define the domain of the maximal trace semantics as

$$\mathcal{D}^{+\infty} \triangleq \{C \in \wp(\mathbb{T}^{+\infty}) \mid C \text{ is a } \lesssim\text{-prefix closed and a } \lesssim\text{-increasing chain}\}$$

Consider the abstraction $\alpha^{+\infty}(C) \triangleq \dot{\curlyvee} C$ and $\gamma^{+\infty}(C) \triangleq {\downarrow} C$ (which is well defined because $C \in \mathcal{D}^{+\infty}$ is a CPO) and its pointwise extension to $\mathbb{T}^+$ so that $\langle \mathbb{T}^+ \to \mathcal{D}^{+\infty}, \dot{\subseteq} \rangle \xleftarrow[\dot{\alpha}^{+\infty}]{\dot{\gamma}^{+\infty}} \langle \mathbb{T}^+ \to \mathbb{T}^{+\infty}, \dot{\lesssim} \rangle$. Define $\widehat{\boldsymbol{\mathcal{S}}}^{+\infty}[\![\mathsf{S}]\!] \triangleq \dot{\alpha}^{+\infty}(\widehat{\boldsymbol{\mathcal{S}}}^{*}[\![\mathsf{S}]\!])$ so that by calculational design, we obtain

- *Empty traces*

$$\widehat{\boldsymbol{\mathcal{S}}}^{+\infty}[\![\mathsf{S}]\!](\ni) = \ni \tag{17.15}$$

- *Maximal traces of an assignment statement* $\mathsf{S} ::= \ell\ \mathsf{x} = \mathsf{A}\ ;$

$$\widehat{\boldsymbol{\mathcal{S}}}^{+\infty}[\![\mathsf{S}]\!](\pi\ell) = \mathsf{let}\ v = \mathcal{A}[\![\mathsf{A}]\!]\varrho(\pi\ell)\ \mathsf{in}\ \ell \xrightarrow{\mathsf{x} = \mathsf{A} = v} \mathsf{after}[\![\mathsf{S}]\!] \tag{17.16}$$

$$\widehat{\boldsymbol{\mathcal{S}}}^{+\infty}[\![\mathsf{S}]\!](\pi\ell') = \ni \quad \text{when} \quad \ell' \neq \ell$$

- *Maximal traces of a statement list* $\mathsf{Sl} ::= \mathsf{Sl}'\ \mathsf{S}$

$$\widehat{\boldsymbol{\mathcal{S}}}^{+\infty}[\![\mathsf{Sl}]\!](\pi_1) = \widehat{\boldsymbol{\mathcal{S}}}^{+\infty}[\![\mathsf{Sl}']\!](\pi_1) \frown \widehat{\boldsymbol{\mathcal{S}}}^{+\infty}[\![\mathsf{S}]\!](\pi_1 \frown \widehat{\boldsymbol{\mathcal{S}}}^{+\infty}[\![\mathsf{Sl}']\!](\pi_1)) \tag{17.17}$$

(which is $\widehat{\boldsymbol{\mathcal{S}}}^{+\infty}[\![\mathsf{Sl}']\!](\pi_1)$ when $\widehat{\boldsymbol{\mathcal{S}}}^{+\infty}[\![\mathsf{Sl}']\!](\pi_1) \in \mathbb{T}^{\infty}$)

- *Maximal traces of an iteration statement* $\mathsf{S} ::= \mathtt{while}\ \ell\ (\mathsf{B})\ \mathsf{S}_b$

$$\widehat{\boldsymbol{\mathcal{S}}}^{+\infty}[\![\mathtt{while}\ \ell\ (\mathsf{B})\ \mathsf{S}_b]\!] = \mathsf{lfp}^{\dot{\lesssim}}\ \mathcal{F}^{+\infty}[\![\mathtt{while}\ \ell\ (\mathsf{B})\ \mathsf{S}_b]\!] \tag{17.18}$$

$$\mathcal{F}^{+\infty}[\![\mathtt{while}\ \ell\ (\mathsf{B})\ \mathsf{S}_b]\!] \in (\mathbb{T}^+ \to \mathbb{T}^{+\infty}) \xrightarrow{\nearrow} (\mathbb{T}^+ \to \mathbb{T}^{+\infty})$$

$$\mathcal{F}^{+\infty}[\![\mathtt{while}\ \ell\ (\mathsf{B})\ \mathsf{S}_b]\!](X)(\pi_1) \triangleq X(\pi_1) \quad \text{when} \quad X(\pi_1) \in \mathbb{T}^{\infty}$$

$$\mathcal{F}^{+\infty}[\![\mathtt{while}\ \ell\ (\mathsf{B})\ \mathsf{S}_b]\!](X)(\pi_1\ell') \triangleq \ni \quad \text{when} \quad \pi_1\ell' \in \mathbb{T}^+ \wedge \ell' \neq \ell$$

$$\mathcal{F}^{+\infty}[\![\mathtt{while}\ \ell\ (\mathsf{B})\ \mathsf{S}_b]\!](X)(\pi_1\ell) \triangleq \quad \text{when} \quad \pi_1\ell \in \mathbb{T}^+$$

$$(\!|\ \mathcal{B}[\![\mathsf{B}]\!]\varrho(\pi_1\ell \frown X(\pi_1\ell)) = \mathrm{ff}\ \text{?}\ X(\pi_1\ell) \frown \ell \xrightarrow{\neg(\mathsf{B})} \mathsf{after}[\![\mathsf{S}]\!]$$

$$\text{:}\ X(\pi_1\ell) \frown \ell \xrightarrow{\mathsf{B}} \mathsf{at}[\![\mathsf{S}_b]\!] \frown \boldsymbol{\mathcal{S}}^{+\infty}[\![\mathsf{S}_b]\!](\pi_1\ell \frown X(\pi_1\ell) \frown \xrightarrow{\mathsf{B}} \mathsf{at}[\![\mathsf{S}_b]\!])\ |\!)$$

(We locally define x to be E_x and y to be E_y in E by $\mathsf{let}\ x = E_x\ \mathsf{and}\ y = E_y\ \mathsf{in}\ E$. There may be no or several additional definitions $\mathsf{and}\ \ldots = \ldots$.) Notice that in (17.16) and (17.18), ℓ stands for $\mathsf{at}[\![\mathsf{S}]\!]$; see remark 17.8. The iterates of $\mathsf{lfp}^{\dot{\lesssim}}\ \mathcal{F}^{+\infty}[\![\mathtt{while}\ \ell\ (\mathsf{B})\ \mathsf{S}_b]\!]$

start form ∍. By remark 17.5, $\mathscr{F}^*[\![\mathtt{while}\,^\ell\,(\mathsf{B})\,\mathsf{S}_b]\!]$ is $\dot{\subseteq}$-increasing and this is preserved by the abstraction, so $\mathscr{F}^{+\infty}[\![\mathtt{while}\,^\ell\,(\mathsf{B})\,\mathsf{S}_b]\!] = \dot{\alpha}^{+\infty} \circ \mathscr{F}^*[\![\mathtt{while}\,^\ell\,(\mathsf{B})\,\mathsf{S}_b]\!] \circ \dot{\gamma}^{+\infty}$ is $\dot{\preceq}$-increasing.

Exercise 17.19 Calculate $\widehat{\mathcal{S}}^{+\infty}[\![\mathtt{while}\,^\ell\,(\mathsf{tt})\,^{\ell'}\mathsf{x} = \mathsf{x} + 1\,;]\!]$ where $\widehat{\mathcal{S}}^{+\infty}$ is defined by exercise 17.14.

Exercise 17.20 Prove that for any program component $\mathsf{S} \in \mathbb{P}\mathbb{c}$, $\mathcal{S}^{+\infty}[\![\mathsf{S}]\!] = \{\widehat{\mathcal{S}}^{+\infty}[\![\mathsf{S}]\!]\}$.

Exercise 17.21 Continuing exercises 10.15 and 17.13, prove that for all program components S, the maximal trace semantics $\mathcal{S}^{+\infty}[\![\mathsf{S}]\!]$ is *deterministic* i.e., $\forall \pi, \pi' \,.\, (\varrho(\pi) = \varrho(\pi')) \Rightarrow (\mathcal{S}^{+\infty}[\![\mathsf{S}]\!]\pi = \mathcal{S}^{+\infty}[\![\mathsf{S}]\!]\pi' \in \{\{\pi''\} \mid \pi'' \in \mathbb{T}^{+\infty}\})$.

Exercise 17.22 (sequential consistency) Let us provide a relaxed memory model [21, 22, 25, 886] of the language $\mathbb{P}$.

We first modify the definition (6.6) of ϱ to record where in the trace the value of a variable is read from.

$$\begin{aligned}
\overline{\varrho}\,(\pi\ell \xrightarrow{\;\mathsf{x}=\mathsf{A}=\nu\;} \ell')\mathsf{x} &\triangleq \langle \pi\ell \xrightarrow{\;\mathsf{x}=\mathsf{A}=\nu\;} \ell',\ \nu\rangle \\
\overline{\varrho}\,(\pi\ell \xrightarrow{\;\cdots\;} \ell')\mathsf{x} &\triangleq \overline{\varrho}\,(\pi\ell)\mathsf{x} \qquad\qquad \text{otherwise} \\
\overline{\varrho}\,(\ell)\mathsf{x} &\triangleq \langle \ell,\ 0\rangle
\end{aligned}$$

Observe that $\overline{\varrho}$ generalizes ϱ in that if $\overline{\varrho}\,(\pi)\mathsf{x} = \langle \pi',\ \nu\rangle$ then $\varrho(\pi)\mathsf{x} = \nu$.

Let us define a set $\mathbb{E}$ of events that contains

- the reads $\langle i,\ \ell,\ \mathsf{R}\,\mathsf{x}\,\nu\rangle$ of the value ν of a variable x executed at program point ℓ during the ith step of computation;
- the writes $\langle i,\ \ell,\ \mathsf{W}\,\mathsf{x}\,\nu\rangle$ of the value ν to a variable y executed at program point ℓ during the ith step of computation;
- the no-operations $\langle i,\ \ell,\ \mathsf{skip}\rangle$ or $\langle i,\ \ell,\ \mathsf{break}\rangle$ executed at program point ℓ during the ith step of computation.

Notice that a label is not enough to uniquely identify events when they appear in a loop whereas the position in the trace is sufficient. Using both is simply redundant.

Assume that all (arithmetic or Boolean) expressions have at most one free variable (as defined in exercise 3.3). This is always possible by rewriting the program and decomposing the evaluation of expressions using auxiliary variables.

Let us define $\mathsf{evt}(i, \ell, \mathsf{a}, \langle\overline{\pi},\ \overline{\nu}\rangle)$ as a pair of events such that

- if a is B or ¬(B) with $\mathbb{vars}[\![\mathsf{B}]\!] = \{\mathsf{x}\}$ then $\mathsf{evt}(i, \ell, \mathsf{a}, \langle\overline{\pi},\ \overline{\nu}\rangle)_1 = \mathsf{evt}(i, \ell, \mathsf{a}, \langle\overline{\pi},\ \overline{\nu}\rangle)_2 = \langle i,\ \ell,\ \mathsf{R}\,\mathsf{x}\,\overline{\nu}\rangle$;
- if a is $\mathsf{x} = \mathsf{A} = \nu$ with $\mathbb{vars}[\![\mathsf{A}]\!] = \varnothing$ then $\mathsf{evt}(i, \ell, \mathsf{a}, \langle\overline{\pi},\ \overline{\nu}\rangle)_1 = \mathsf{evt}(i, \ell, \mathsf{a}, \langle\overline{\pi},\ \overline{\nu}\rangle)_2 = \langle i,\ \ell,\ \mathsf{W}\,\mathsf{y}\,\nu\rangle$;
- if a is $\mathsf{x} = \mathsf{A} = \nu$ with $\mathbb{vars}[\![\mathsf{A}]\!] = \{\mathsf{y}\}$ then $\mathsf{evt}(i, \ell, \mathsf{a}, \langle\overline{\pi},\ \overline{\nu}\rangle)_1 = \langle i,\ \ell,\ \mathsf{R}\,\mathsf{x}\,(\overline{\nu})\rangle$ and $\mathsf{evt}(i, \ell, \mathsf{a}, \langle\overline{\pi},\ \overline{\nu}\rangle)_2 = \langle i,\ \ell,\ \mathsf{W}\,\mathsf{y}\,\nu\rangle$;
- otherwise $\mathsf{evt}(i, \ell, \mathsf{a}, \langle\overline{\pi},\ \overline{\nu}\rangle)_1 = \mathsf{evt}(i, \ell, \mathsf{a}, \langle\overline{\pi},\ \overline{\nu}\rangle)_2 = \langle i,\ \ell,\ \mathsf{a}\rangle$ for no-operations skip or break.

Let us abstract a trace $\pi = \ell_0 \xrightarrow{\;\mathsf{a}_0\;} \ell_1 \xrightarrow{\;\mathsf{a}_1\;} \ldots\ \ell_i \xrightarrow{\;\mathsf{a}_i\;} \ell_{i+1} \ldots$ into $\alpha_{\mathsf{w}}(\pi) = \langle E,\ \mathsf{po},\ \mathsf{co},\ \mathsf{rf}\rangle$ where

- $E = \{\mathsf{evt}(i, \ell_i, \mathsf{a}_i, \bar{\varrho}\,(\ell_0 \xrightarrow{\mathsf{a}_0} \ell_1 \xrightarrow{\mathsf{a}_1} \dots \ell_i))_1, \mathsf{evt}(i, \ell_i, \mathsf{a}_i, \bar{\varrho}\,(\ell_0 \xrightarrow{\mathsf{a}_0} \ell_1 \xrightarrow{\mathsf{a}_1} \dots \ell_i)_2 \mid i = 0, 1, \dots\} \cup \{\langle \bullet, \bullet, \mathsf{W\,y\,0}\rangle \mid \mathsf{y} \in \mathbb{V}\}$ is the set of unique *events* on the trace augmented by initial events $\langle \bullet, \bullet, \mathsf{W\,y\,0}\rangle$ abstracting the initialization all variables y to 0 at ℓ_0;
- $\mathsf{po} = \{\langle\langle \bullet, \bullet, \mathsf{W\,y\,0}\rangle, \mathsf{evt}(0, \ell_0, \mathsf{a}_0, \bar{\varrho}\,(\ell_0))\rangle \mid \mathsf{y} \in \mathbb{V}\} \cup \{\langle \mathsf{evt}(i, \ell_i, \mathsf{a}_i, \bar{\varrho}\,(\ell_0 \xrightarrow{\mathsf{a}_0} \ell_1 \xrightarrow{\mathsf{a}_1} \dots \ell_i))_2, \mathsf{evt}(i+1, \ell_{i+1}, \mathsf{a}_{i+1}, \bar{\varrho}\,(\ell_0 \xrightarrow{\mathsf{a}_0} \ell_1 \xrightarrow{\mathsf{a}_1} \dots \ell_i \xrightarrow{\mathsf{a}_i} \ell_{i+1}))_1\rangle \mid i = 0, 1, \dots\} \cup \{\langle e, e'\rangle \mid i = 0, 1, \dots \wedge e = \mathsf{evt}(i, \ell_i, \mathsf{a}_i, \bar{\varrho}\,(\ell_0 \xrightarrow{\mathsf{a}_0} \ell_1 \xrightarrow{\mathsf{a}_1} \dots \ell_i))_1 \neq e' = \mathsf{evt}(i, \ell_i, \mathsf{a}_i, \bar{\varrho}\,(\ell_0 \xrightarrow{\mathsf{a}_0} \ell_1 \xrightarrow{\mathsf{a}_1} \dots \ell_i))_2\}$ is the *program order* between two consecutive events during execution (in which initialization of variables to 0 is performed first and $\mathsf{a}_i = (\mathsf{y} = \mathsf{A} = v)$ with $\mathsf{vars}[\![\mathsf{A}]\!] = \{\mathsf{x}\}$ gives rise to a transition from the read to the write events for that read-modify-write action a_i);
- $\mathsf{co} = \{\langle\langle \bullet, \bullet, \mathsf{W\,x\,0}\rangle, \langle i, \ell_i, \mathsf{W\,x}\,v_i\rangle\rangle \mid \mathsf{x} \in \mathbb{V} \wedge \langle i, \ell_i, \mathsf{W\,x}\,v_i\rangle = \mathsf{evt}(i, \ell_i, \mathsf{a}_i, \bar{\varrho}\,(\ell_0 \xrightarrow{\mathsf{a}_0} \ell_1 \xrightarrow{\mathsf{a}_1} \dots \ell_i))_2\} \cup \{\langle\langle i, \ell_i, \mathsf{W\,x}\,v_i\rangle, \langle j, \ell_j, \mathsf{W\,x}\,v_j\rangle\rangle \mid \mathsf{x} \in \mathbb{V} \wedge \langle i, \ell_i, \mathsf{W\,x}\,v_i\rangle = \mathsf{evt}(i, \ell_i, \mathsf{a}_i, \bar{\varrho}\,(\ell_0 \xrightarrow{\mathsf{a}_0} \ell_1 \xrightarrow{\mathsf{a}_1} \dots \ell_i))_2 \wedge \langle j, \ell_j, \mathsf{W\,x}\,v_j\rangle = \mathsf{evt}(j, \ell_j, \mathsf{a}_{ji}, \bar{\varrho}\,(\ell_0 \xrightarrow{\mathsf{a}_0} \ell_1 \xrightarrow{\mathsf{a}_1} \dots \ell_j))_2 \wedge i < j\}$ is the *coherence order* between a write event to a variable and all consecutive write events to that same variable. Hence co totally orders all the write events to the same variable;
- The *read-from relation* rf is the set of pairs $\langle\langle i, \ell_i, \mathsf{W\,y}\,v_i\rangle, \langle j, \ell_j, \mathsf{R\,x}\,\bar{v}_j\rangle\rangle$ of a write event $\langle i, \ell_i, \mathsf{W\,y}\,v_i\rangle = \mathsf{evt}(i, \ell_i, \mathsf{a}_i, \bar{\varrho}\,(\ell_0 \xrightarrow{\mathsf{a}_0} \ell_1 \xrightarrow{\mathsf{a}_1} \dots \ell_i))_2$ and a read event $\langle j, \ell_j, \mathsf{R\,x}\,\bar{v}_j\rangle = \mathsf{evt}(j, \ell_j, \mathsf{a}_j, \bar{\varrho}\,(\ell_0 \xrightarrow{\mathsf{a}_0} \ell_1 \xrightarrow{\mathsf{a}_1} \dots \ell_j))_1$ of the trace $\pi = \ell_0 \dots \ell_i \xrightarrow{\mathsf{a}_i} \ell_{i+1} \dots \ell_j \xrightarrow{\mathsf{a}_j} \ell_{j+1} \dots$ such that $\bar{\varrho}\,(\ell_0 \xrightarrow{\mathsf{a}_0} \ell_1 \xrightarrow{\mathsf{a}_1} \dots \ell_j) = \langle \ell_0 \xrightarrow{\mathsf{a}_0} \ell_1 \xrightarrow{\mathsf{a}_1} \dots \ell_i, \bar{v}_j\rangle$ so that the value $\bar{v}_i$ assigned to y in action a_j is computed using the value v_i written to x in action a_i.
  In addition rf contains the pairs $\langle\langle \bullet, \bullet, \mathsf{W\,y\,0}\rangle, \langle j, \ell_j, \mathsf{R\,y\,0}\rangle\rangle$ when $\bar{\varrho}\,(\ell_0 \dots \ell_j)(\mathsf{y}) = \langle \ell_0, 0\rangle$ when reading the initial value of y;

Define the *from-read relation* $\mathsf{fr} \triangleq \mathsf{rf}^{-1} \mathbin{⨾} \mathsf{co}$ (where the composition $⨾$ of relations is defined in section 2.2.2). Intuitively, a from-read arrow starts from a read of a given variable and points to all the writes that overwrite the value this read has taken.

Continuing the example program $\ell_0\ \mathsf{y} = \mathsf{x}\ ;\ell_1\ \mathsf{y} = \mathsf{y}\ ;\ell_2$ where x and y are initialized to 0. Its execution trace is $\pi = \ell_0 \xrightarrow{\mathsf{y}=\mathsf{x}=0} \ell_1 \xrightarrow{\mathsf{y}=\mathsf{y}=0} \ell_2$. The abstraction $\alpha_{\mathsf{w}}(\pi)$ and fr are as follows:

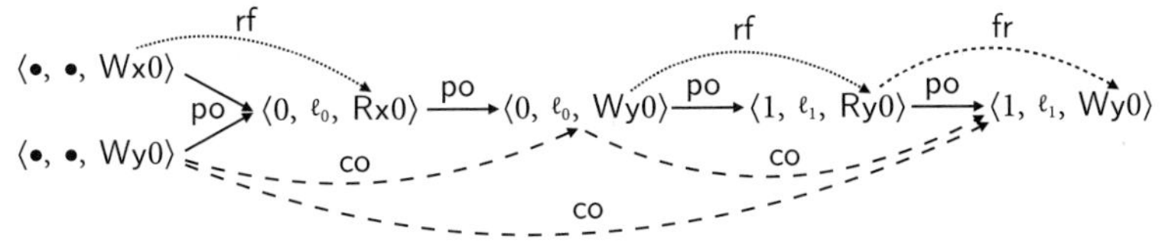

— Prove that the relation $(\mathsf{po} \cup \mathsf{co} \cup \mathsf{rf} \cup \mathsf{fr})^+$ is acyclic [22]. This specifies *sequential consistency* [20, section 4.2.1] that is writes hit the memory in their order of execution, a read is from a past write, and this write is the last one [610].

— Define $\alpha_{\mathsf{W}}(\mathcal{S}) = \{\alpha_{\mathsf{w}}(\pi) \mid \pi \in \mathcal{S}\}$. Prove that $\langle \wp(\mathbb{T}^{+\infty}), \subseteq\rangle \xrightleftharpoons[\alpha_{\mathsf{W}}]{\gamma_{\mathsf{W}}} \langle \wp(\wp(\mathbb{E}) \times \wp(\mathbb{E} \times \mathbb{E}) \times \wp(\mathbb{E} \times \mathbb{E}) \times \wp(\mathbb{E} \times \mathbb{E})), \subseteq\rangle$ is a Galois connection.

— Observe that a maximal trace semantics of programs $\mathsf{P} \in \mathbb{P}$ can be specified as $\gamma_{\mathsf{W}}(\{\langle E, \mathsf{po}, \mathsf{co}, \mathsf{rf}\rangle \mid (\mathsf{po} \cup \mathsf{co} \cup \mathsf{rf} \cup \mathsf{fr})^+$ is acyclic$\})$ where the set of events E, the program order po, and the coherence order co are defined from the syntax of program P while rf is chosen arbitrarily in order to satisfy $(\mathsf{po} \cup \mathsf{co} \cup \mathsf{rf} \cup \mathsf{fr})^+$ is acyclic.

- Prove that if $\pi, \pi' \in \gamma_{\mathsf{W}}(\{\langle E, \mathsf{po}, \mathsf{co}, \mathsf{rf}\rangle\}) \cap \mathbb{T}^+$ where $(\mathsf{po} \cup \mathsf{co} \cup \mathsf{rf} \cup \mathsf{fr})^+$ is acyclic then $\forall \mathsf{x} \in \mathbb{V} \,.\, \varrho(\pi)\mathsf{x} = \varrho(\pi')\mathsf{x}$ i.e., the program finite computations always produce the same result.
- Prove that if $\pi \in \mathcal{S}^+[\![\mathsf{P}]\!](\mathsf{at}[\![\mathsf{P}]\!])$ and $\pi' \in \gamma_{\mathsf{W}}(\alpha_{\mathsf{W}}(\{\pi\}))$ then $\forall \mathsf{x} \in \mathbb{V} \,.\, \varrho(\pi)\mathsf{x} = \varrho(\pi')\mathsf{x}$ i.e., the abstraction loses no information on the final values of maximal finite computations.
- What can be said about infinite computations?

— It is also possible to use a read-modify-write event so that if a is $\mathsf{y} = \mathsf{A} = v$ with $\mathsf{vars}[\![\mathsf{A}]\!] = \{\mathsf{y}\}$ then $\mathsf{evt}(i, \ell, \mathsf{a}, \langle \overline{\pi}, \overline{v}\rangle)_1 = \mathsf{evt}(i, \ell, \mathsf{a}, \langle \overline{\pi}, \overline{v}\rangle)_2 = \langle i, \ell, \mathsf{RMW}\ \mathsf{y}\ \overline{v}\ v\rangle$ (which replaces $\langle i, \ell, \mathsf{R}\ \mathsf{x}\ (\overline{v})\rangle \xrightarrow{\mathsf{po}} \langle i, \ell, \mathsf{W}\ \mathsf{y}\ v\rangle$). In that case define rmw to be the identity relation between such read-modify-write events. Show that these results remain valid when defining $\mathsf{fr} = (\mathsf{rf}^{-1} \mathbin{;} \mathsf{co}) \setminus \mathsf{rmw}$.

17.3 Conclusion

The first semantics using fixpoint definitions were the denotational semantics of Dana Scott and Christopher Strachey [871, 872]. In fact, all program semantics can be given fixpoint definitions (by abstract interpretations of fixpoint trace semantics [227]).

The correctness/soundness proof of any static analysis must relate the results of the analysis to a formal definition of the executions of any program of a given programming language. Maximal traces are a rather precise descriptions of these executions (although e.g. they abstract away execution times) so are a good basis for abstracting into less precise descriptions of program executions (like the reachability semantics of chapter 19) with respect to which static analyses can be proved sound (and so, by composition, are also sound with respect to traces). But some static analyses such at [200, 263, 804, 906] can be proved only with respect to traces (because any abstractions of traces (e.g., as a reachability semantics of chapter 19) would not be precise enough).

18 Fixpoint Abstraction

Given an abstraction $\langle \mathscr{C}, \sqsubseteq \rangle \xrightleftharpoons[\alpha]{\gamma} \langle \mathscr{A}, \preccurlyeq \rangle$ of a concrete domain $\mathscr{C}$ into an abstract domain $\mathscr{A}$ and an increasing concrete transformer $f \in \mathscr{C} \xrightarrow{\nearrow} \mathscr{C}$, we would like to abstract $\mathsf{lfp}^{\sqsubseteq} f$. We first abstract the concrete transformer f into an abstract transformer $\overline{f} \in \mathscr{A} \xrightarrow{\nearrow} \mathscr{A}$ and then $\alpha(\mathsf{lfp}^{\sqsubseteq} f)$ into $\mathsf{lfp}^{\preccurlyeq} \overline{f}$. This abstraction may be exact i.e., $\alpha(\mathsf{lfp}^{\sqsubseteq} f) = \mathsf{lfp}^{\preccurlyeq} \overline{f}$ (as in figures 18.1a and 18.1b) or sound but imprecise, in which case we get an overapproximation $\alpha(\mathsf{lfp}^{\sqsubseteq} f) \preccurlyeq \mathsf{lfp}^{\preccurlyeq} \overline{f}$ (as in figure 18.1c).

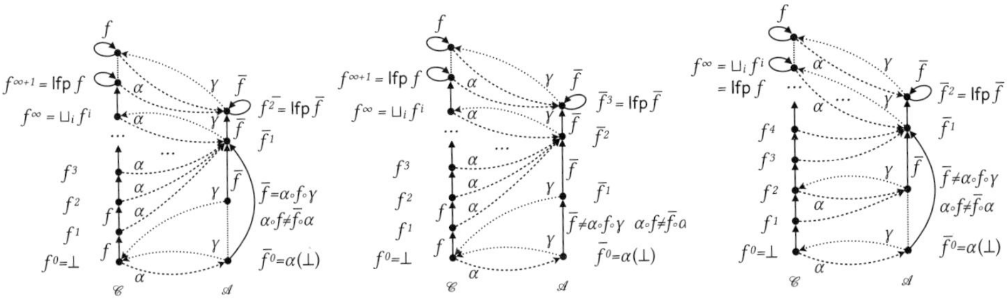

(a) (b) (c)

exact fixpoint abstractions — imprecise fixpoint abstraction

Figure 18.1
Fixpoint abstraction

Contents

18.1 Sound Transformer Abstraction

To abstract a least fixpoint $\alpha(\mathsf{lfp}^{\sqsubseteq} f)$, we first abstract its transformer $f \in \mathscr{C} \rightarrow \mathscr{C}$ into an abstract transformer $\alpha \circ f \circ \gamma \in \mathscr{A} \rightarrow \mathscr{A}$. Given an abstract argument $y \in \mathscr{A}$, the abstract transformer concretize it in $\gamma(y) \in \mathcal{C}$, applies the concrete transformer $f(\gamma(y)) \in \mathcal{C}$, and finally abstracts the result $\alpha(f(\gamma(y))) \in \mathcal{C}$; see figure 18.2 and note that $\overline{f} = \alpha \circ f \circ \gamma$ in figure 18.1a. Of course, in practice we would like to calculate the abstract transformer $\overline{f}$ using only abstract operations on $\mathcal{A}$, without resorting to concretization, but this may or may not be possible exactly, depending on α, f, and γ.

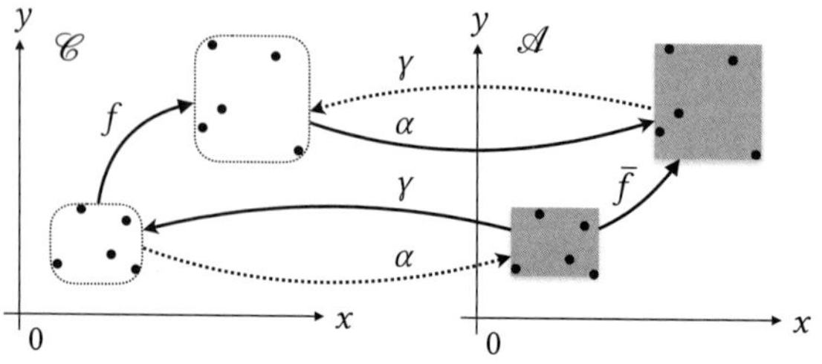

Figure 18.2
Transformer abstraction

> **Theorem 18.3 (sound transformer abstraction)** If $\langle \mathcal{C}, \sqsubseteq \rangle \underset{\alpha}{\overset{\gamma}{\leftrightarrows}} \langle \mathcal{A}, \preccurlyeq \rangle$ then $\langle \mathcal{C} \xrightarrow{\nearrow} \mathcal{C}, \dot{\sqsubseteq} \rangle \underset{\vec{\alpha}}{\overset{\vec{\gamma}}{\leftrightarrows}} \langle \mathcal{A} \xrightarrow{\nearrow} \mathcal{A}, \dot{\preccurlyeq} \rangle$ where $\dot{\sqsubseteq}$ and $\dot{\preccurlyeq}$ are pointwise (i.e., $f \mathrel{\dot{\sqsubseteq}} g$ if and only if $\forall x \in \mathcal{C} \,.\, f(x) \sqsubseteq g(x)$), $\vec{\alpha}(f) \triangleq \alpha \circ f \circ \gamma$, and $\vec{\gamma}(\overline{f}) \triangleq \gamma \circ \overline{f} \circ \alpha$.

Proof Let $f \in \mathcal{C} \xrightarrow{\nearrow} \mathcal{C}$ and $\overline{f} \in \mathcal{A} \xrightarrow{\nearrow} \mathcal{A}$ be increasing.

$\quad \vec{\alpha}(f) \mathrel{\dot{\preccurlyeq}} \overline{f}$

$\Leftrightarrow \forall \overline{x} \in \mathcal{A} \,.\, \vec{\alpha}(f)\overline{x} \preccurlyeq \overline{f}(\overline{x})$ $\quad$ ⟅pointwise definition of $\dot{\preccurlyeq}$⟆

$\Leftrightarrow \forall \overline{x} \in \mathcal{A} \,.\, \alpha \circ f \circ \gamma(\overline{x}) \preccurlyeq \overline{f}(\overline{x})$ $\quad$ ⟅definition of $\vec{\alpha}$⟆

$\Leftrightarrow \forall \overline{x} \in \mathcal{A} \,.\, f \circ \gamma(\overline{x}) \sqsubseteq \gamma \circ \overline{f}(\overline{x})$ $\quad$ ⟅$\langle \mathcal{C}, \sqsubseteq \rangle \underset{\alpha}{\overset{\gamma}{\leftrightarrows}} \langle \mathcal{A}, \preccurlyeq \rangle$⟆

$\Rightarrow \forall x \in \mathcal{C} \,.\, f \circ \gamma \circ \alpha(x) \sqsubseteq \gamma \circ \overline{f} \circ \alpha(x)$ $\quad$ ⟅for $\overline{x} = \alpha(x)$⟆

$\quad \forall x \in \mathcal{C} \,.\, x \sqsubseteq \gamma \circ \alpha(x)$ $\quad$ ⟅exercise 11.35.3⟆

$\Leftrightarrow \forall x \in \mathcal{C} \,.\, f(x) \sqsubseteq \gamma \circ \overline{f} \circ \alpha(x)$ $\quad$ ⟅f is increasing and $\sqsubseteq$ is transitive⟆

$\Leftrightarrow \forall x \in \mathcal{C} \,.\, f(x) \sqsubseteq \vec{\gamma}(\overline{f})(x)$ $\quad$ ⟅definition of $\vec{\gamma}$⟆

$\Leftrightarrow f \mathrel{\dot{\sqsubseteq}} \vec{\gamma}(\overline{f})$ $\quad$ ⟅pointwise definition of $\dot{\sqsubseteq}$⟆

Conversely,

$\quad f \mathrel{\dot{\sqsubseteq}} \vec{\gamma}(\overline{f})$

$\Leftrightarrow \forall x \in \mathcal{C} \,.\, f(x) \sqsubseteq \gamma \circ \overline{f} \circ \alpha(x)$ $\quad$ ⟅pointwise definition of $\dot{\sqsubseteq}$ and definition of $\vec{\gamma}$⟆

$\Leftrightarrow \forall x \in \mathcal{C} \,.\, \alpha \circ f(x) \preccurlyeq \overline{f} \circ \alpha(x)$ $\quad$ ⟅$\langle \mathcal{C}, \sqsubseteq \rangle \underset{\alpha}{\overset{\gamma}{\leftrightarrows}} \langle \mathcal{A}, \preccurlyeq \rangle$⟆

$\Rightarrow \forall \overline{x} \in \mathcal{A} \,.\, \alpha \circ f \circ \gamma(\overline{x}) \preccurlyeq \overline{f} \circ \alpha \circ \gamma(\overline{x})$ $\quad$ ⟅for $x = \gamma(\overline{x})$⟆

$\quad \forall \overline{x} \in \mathcal{A} \,.\, \alpha \circ \gamma(\overline{x}) \preccurlyeq \overline{x}$ $\quad$ ⟅exercise 11.35.4⟆

$\Rightarrow \forall \overline{x} \in \mathscr{A} \,.\, \alpha \circ f \circ \gamma(\overline{x}) \preccurlyeq \overline{f}(\overline{x})$ ⦇$\overline{f}$ is increasing and $\preccurlyeq$ is transitive⦈

$\Rightarrow \forall \overline{x} \in \mathscr{A} \,.\, \vec{\alpha}(f)(\overline{x}) \preccurlyeq \overline{f}(\overline{x})$ ⦇definition of $\vec{\alpha}(f)$⦈

$\Rightarrow \vec{\alpha}(f) \mathrel{\dot{\preccurlyeq}} \overline{f}$ ⦇pointwise definition of $\dot{\preccurlyeq}$⦈ □

Example 18.4 In figure 18.1a, we have $\overline{f} = \vec{\alpha}(f) \triangleq \alpha \circ f \circ \gamma$. In figure 18.1c, we have $\vec{\alpha}(f) \triangleq \alpha \circ f \circ \gamma \mathrel{\dot{\prec}} \overline{f}$ (where $\dot{\prec}$ is the strict version of $\dot{\preccurlyeq}$, $\dot{\preccurlyeq}$ is the pointwise extension of $\preccurlyeq$, and $\preccurlyeq$ is the partial order on the abstract domain $\langle \mathscr{A}, \preccurlyeq \rangle$).

Exercise 18.5 (transformer refinement) Let $\langle \mathscr{C}, \sqsubseteq \rangle \underset{\alpha}{\overset{\gamma}{\leftrightarrows}} \langle \mathscr{A}, \preceq \rangle$, $f \in \mathscr{C} \xrightarrow{\sqcup} \mathscr{C}$ preserves existing arbitrary joins, and $\alpha \circ f \circ \gamma \mathrel{\dot{\preceq}} \overline{f}$ where $\dot{\preceq}$ is $\preceq$ pointwise. For elements $\overline{x} \in \mathscr{A}$ which concretization can be decomposed into a disjunction $\gamma(\overline{x}) = \bigsqcup_{i=1}^{n} x^i$, the transformer $\overline{f}$ can be refined as $\overline{f}'(\overline{x}) \triangleq \bigvee_{i=1}^{n} \alpha(f(x^i))$. Prove that $\alpha \circ f \circ \gamma \mathrel{\dot{\preceq}} \overline{f}' \mathrel{\dot{\preceq}} \overline{f}$. One can also do the refinement with respect to an intermediate abstract domain, an operation called *materialization* in [996] and *focus* in [632].

18.2 Sound Concrete Least Fixpoint Overapproximation

In general abstracting the fixpoint transformer by a larger one yields a least fixpoint overapproximation.

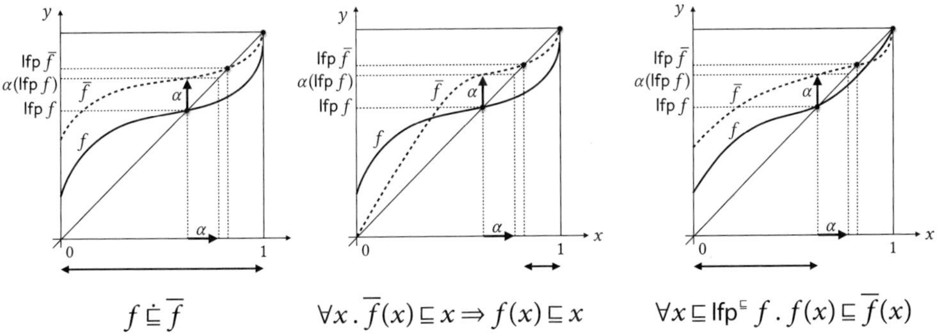

Figure 18.6
Fixpoint overapproximation

> **Theorem 18.7 (pointwise least fixpoint overapproximation for an increasing function)** Assume that $\langle \mathscr{C}, \sqsubseteq, \bot, \top, \sqcup, \sqcap \rangle$ is a complete lattice, $f, g \in \mathscr{C} \xrightarrow{\nearrow} \mathscr{C}$ are increasing, and $f \mathrel{\dot{\sqsubseteq}} g$ then $\mathsf{lfp}^{\sqsubseteq} f \sqsubseteq \mathsf{lfp}^{\sqsubseteq} g$.

Proof By $f \mathrel{\dot{\sqsubseteq}} g$, for all $x \in \mathscr{C}$, $g(x) \sqsubseteq x$ implies $f(x) \sqsubseteq x$ so $\{x \in \mathscr{C} \mid g(x) \sqsubseteq x\} \subseteq \{x \in \mathscr{C} \mid f(x) \sqsubseteq x\}$ and so, by Tarski's fixpoint theorem 15.6 and definition of glbs, $\mathsf{lfp}^{\sqsubseteq} f = \bigsqcap\{x \in \mathscr{C} \mid f(x) \sqsubseteq x\} \sqsubseteq \bigsqcap\{x \in \mathscr{C} \mid g(x) \sqsubseteq x\} = \mathsf{lfp}^{\sqsubseteq} g$. □

As shown by figure 18.6, the condition $\forall x \in \mathcal{C} \,.\, f(x) \sqsubseteq g(x)$ is not necessary when $g(x) \not\sqsubseteq x$.

Corollary 18.8 (least fixpoint overapproximation for an increasing function) Assume that $\langle \mathcal{C}, \sqsubseteq, \bot, \top, \sqcup, \sqcap \rangle$ is a complete lattice, $f, g \in \mathcal{C} \xrightarrow{\nearrow} \mathcal{C}$ are increasing, and $\forall x \,.\, g(x) \sqsubseteq x \Rightarrow f(x) \sqsubseteq x$ then $\mathsf{lfp}^{\sqsubseteq} f \sqsubseteq \mathsf{lfp}^{\sqsubseteq} g$.

Proof In the proof of theorem 18.7, the condition $f \mathrel{\dot{\sqsubseteq}} g$ is only used to prove that $\forall x \in \mathcal{C} \,.\, g(x) \sqsubseteq x \Rightarrow f(x) \sqsubseteq x$. □

Theorem 18.9 (least fixpoint overapproximation for a upper continuous function) Let $\langle \mathcal{C}, \sqsubseteq, \bot, \top, \sqcup, \sqcap \rangle$ be a complete lattice, $f \in \mathcal{C} \xrightarrow{uc} \mathcal{C}$ be upper continuous, $g \in \mathcal{C} \xrightarrow{\nearrow} \mathcal{C}$ be increasing. If $\forall x \sqsubseteq \mathsf{lfp}^{\sqsubseteq} f \,.\, f(x) \sqsubseteq g(x)$ then $\mathsf{lfp}^{\sqsubseteq} f \sqsubseteq \mathsf{lfp}^{\sqsubseteq} g$.

Proof Let $\langle f^n, n \in \mathbb{N} \rangle$ be the iterates of f from $f^0 = \bot$. We have $f^0 = \bot \sqsubseteq \mathsf{lfp}^{\sqsubseteq} g$ by definition of the infimum. Assume by induction hypothesis that $f^n \sqsubseteq \mathsf{lfp}^{\sqsubseteq} g$. Then

$$
\begin{aligned}
& f^{n+1} \\
= \;& f(f^n) && \text{(definition of the iterates)} \\
\sqsubseteq \;& g(f^n) && \text{(because } f^n \sqsubseteq \mathsf{lfp}^{\sqsubseteq} f \text{ so } f(f^n) \sqsubseteq g(f^n) \text{ by hypothesis)} \\
\sqsubseteq \;& g(\mathsf{lfp}^{\sqsubseteq} g) && \text{(induction hypothesis and } g \text{ increasing)} \\
= \;& \mathsf{lfp}^{\sqsubseteq} g && \text{(definition of a fixpoint)}
\end{aligned}
$$

By recurrence $\forall n \in \mathbb{N} \,.\, f^n \sqsubseteq \mathsf{lfp}^{\sqsubseteq} g$ so $\bigsqcup_{n \in \mathbb{N}} f^n \sqsubseteq \mathsf{lfp}^{\sqsubseteq} g$ by definition of the lub, and therefore $\mathsf{lfp}^{\sqsubseteq} f \sqsubseteq \mathsf{lfp}^{\sqsubseteq} g$ by Kleene/Scott iterative fixpoint theorem 15.26. □

18.3 Sound Abstract Fixpoint Overapproximation

An abstract fixpoint $\mathsf{lfp}^{\preccurlyeq} \overline{f}$ is a sound fixpoint abstraction or fixpoint overapproximation of a concrete fixpoint $\mathsf{lfp}^{\sqsubseteq} f$ whenever $\alpha(\mathsf{lfp}^{\sqsubseteq} f) \preccurlyeq \mathsf{lfp}^{\preccurlyeq} \overline{f}$.

Theorem 18.10 (least fixpoint overapproximation in a complete lattice) Assume that $\langle \mathcal{C}, \sqsubseteq, \bot, \top, \sqcup, \sqcap \rangle$ and $\langle \mathcal{A}, \preccurlyeq, 0, 1, \curlyvee, \curlywedge \rangle$ are complete lattices, $\langle \mathcal{C}, \sqsubseteq \rangle \underset{\alpha}{\overset{\gamma}{\leftrightarrows}} \langle \mathcal{A}, \preccurlyeq \rangle$, and $f \in \mathcal{C} \xrightarrow{\nearrow} \mathcal{C}$ is increasing. Then $\mathsf{lfp}^{\sqsubseteq} f \sqsubseteq \gamma(\mathsf{lfp}^{\preccurlyeq} \alpha \circ f \circ \gamma)$.

Example 18.11 In figure 18.13.a, we have $\mathsf{lfp}^{\sqsubseteq} f \sqsubset \gamma(\mathsf{lfp}^{\preccurlyeq} \overline{f})$ with $\overline{f} = \alpha \circ f \circ \gamma$.

Example 18.12 The two examples in figure 18.13.a and figure 18.13.b show that the inequality $\sqsubseteq$ may be strict or not.

Proof of theorem 18.10

$$
\begin{aligned}
& \mathsf{lfp}^{\sqsubseteq} f \\
= \;& \bigsqcap \{x \in \mathcal{C} \mid f(x) \sqsubseteq x\} && \text{(Tarski's fixpoint theorem 15.6)} \\
\sqsubseteq \;& \bigsqcap \{\gamma(\overline{x}) \mid f(\gamma(\overline{x})) \sqsubseteq \gamma(\overline{x})\}
\end{aligned}
$$

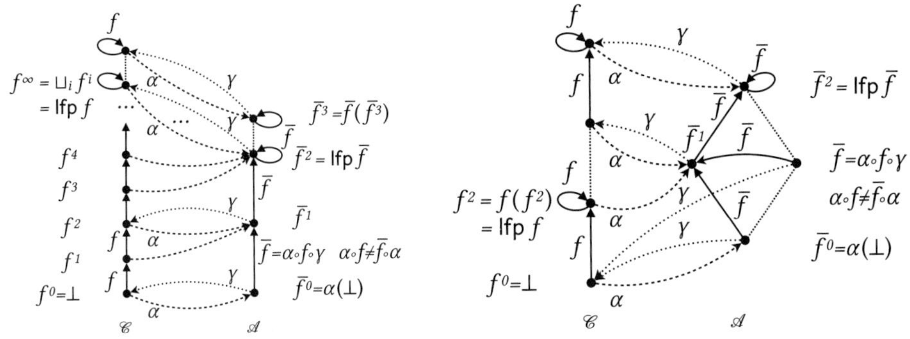

$\mathsf{lfp}^{\sqsubseteq} f = \gamma(\mathsf{lfp}^{\preccurlyeq} \alpha \circ f \circ \gamma)$
(a) exact fixpoint abstraction

$\mathsf{lfp}^{\sqsubseteq} f \sqsubsetneq \gamma(\mathsf{lfp}^{\preccurlyeq} \alpha \circ f \circ \gamma)$
(b) imprecise fixpoint abstraction

Figure 18.13
Fixpoint abstraction

⦃because $\{\gamma(\overline{x}) \mid f(\gamma(\overline{x})) \sqsubseteq \gamma(\overline{x})\} \subseteq \{x \in \mathscr{C} \mid f(x) \sqsubseteq x\}$ and definition of the glb $\bigsqcap$⦄

$= \gamma(\bigwedge\{\overline{x} \mid f(\gamma(\overline{x})) \sqsubseteq \gamma(\overline{x})\})$ ⦃γ preserves existing glbs, by dual of lemma 11.38⦄

$= \gamma(\bigwedge\{\overline{x} \mid \alpha \circ f \circ \gamma(\overline{x}) \preccurlyeq \overline{x}\})$ ⦃$\langle \mathscr{C}, \sqsubseteq \rangle \underset{\alpha}{\overset{\gamma}{\leftrightarrows}} \langle \mathscr{A}, \preccurlyeq \rangle$⦄

$= \gamma(\mathsf{lfp}^{\preccurlyeq} \alpha \circ f \circ \gamma)$

⦃because the composition of increasing functions is increasing and Tarski's fixpoint theorem 15.6⦄ □

Corollary 18.14 (least fixpoint approximation by transformer overapproximation) Assume that $\langle \mathscr{C}, \sqsubseteq, \bot, \top, \sqcup, \sqcap \rangle$ and $\langle \mathscr{A}, \preccurlyeq, 0, 1, \curlyvee, \curlywedge \rangle$ are complete lattices, $\langle \mathscr{C}, \sqsubseteq \rangle \underset{\alpha}{\overset{\gamma}{\leftrightarrows}} \langle \mathscr{A}, \preccurlyeq \rangle$, $f \in \mathscr{C} \xrightarrow{\nearrow} \mathscr{C}$ and $\overline{f} \in \mathscr{A} \xrightarrow{\nearrow} \mathscr{A}$ are increasing, and $\alpha \circ f \circ \gamma \mathrel{\dot{\preccurlyeq}} \overline{f}$. Then $\mathsf{lfp}^{\sqsubseteq} f \sqsubseteq \gamma(\mathsf{lfp}^{\preccurlyeq} \overline{f})$.

Proof By theorem 18.10 and theorem 18.7. □

Exercise 18.15 Prove that under the hypotheses of corollary 18.14, $\alpha \circ f \circ \gamma \mathrel{\dot{\preccurlyeq}} \overline{f} \Leftrightarrow f \circ \gamma \mathrel{\dot{\sqsubseteq}} \gamma \circ \overline{f} \Leftrightarrow \alpha \circ f \mathrel{\dot{\preccurlyeq}} \overline{f} \circ \alpha$. Show that the first equivalence does not require $\overline{f}$ to be increasing whereas the second, called pointwise soundness, may not hold if $\overline{f}$ is not increasing or $\alpha \circ \gamma \neq \mathbb{1}_{\mathscr{A}}$.

Corollary 18.16 (least fixpoint approximation by semicommuting transformer) Under the hypotheses of corollary 18.14 assume instead that $\alpha \circ f \mathrel{\dot{\preccurlyeq}} \overline{f} \circ \alpha$ (*semicommutation* or *pointwise soundness*). Then $\mathsf{lfp}^{\sqsubseteq} f \sqsubseteq \gamma(\mathsf{lfp}^{\preccurlyeq} \overline{f})$.

Proof If $\alpha \circ f \mathrel{\dot{\preccurlyeq}} \overline{f} \circ \alpha$ then, in particular, $\alpha \circ f \circ \gamma \mathrel{\dot{\preccurlyeq}} \overline{f} \circ \alpha \circ \gamma \mathrel{\dot{\preccurlyeq}} \overline{f}$ because $\alpha \circ \gamma$ is reductive by exercise 11.35.4 and $\overline{f}$ increasing by hypothesis. We conclude by corollary 18.14. □

Exercise 18.17 Assume that $\langle \mathscr{C}, \sqsubseteq, \bot, \top, \sqcup, \sqcap \rangle$ is a complete lattice, $f \in \mathscr{C} \xrightarrow{\nearrow} \mathscr{C}$ is increasing, $\langle \mathscr{A}, \preccurlyeq \rangle$ is a poset, $\overline{f} \in \mathscr{A} \to \mathscr{A}$, $f \circ \gamma \mathrel{\dot{\sqsubseteq}} \gamma \circ \overline{f}$, and $\gamma \in \mathscr{A} \xrightarrow{\nearrow} \mathscr{C}$ is increasing. Then $\forall y \in \mathscr{A}, \overline{f}(y) \preccurlyeq y$ implies $\mathsf{lfp}^{\sqsubseteq} f \sqsubseteq \gamma(y)$ (in particular when $y = \mathsf{lfp}^{\preccurlyeq} \overline{f}$).

Theorem 18.18 (least fixpoint overapproximation in a CPO) Assume that $\langle \mathscr{C}, \sqsubseteq, \bot, \sqcup \rangle$ and $\langle \mathscr{A}, \preccurlyeq, 0, \curlyvee \rangle$ are CPOs, $\langle \mathscr{C}, \sqsubseteq \rangle \underset{\alpha}{\overset{\gamma}{\leftrightarrows}} \langle \mathscr{A}, \preccurlyeq \rangle$, and $f \in \mathscr{C} \xrightarrow{uc} \mathscr{C}$ is upper continuous. Then $\mathsf{lfp}^{\sqsubseteq} f \sqsubseteq \gamma(\mathsf{lfp}^{\preccurlyeq} \alpha \circ f \circ \gamma)$.

Proof Let $f^n(\bot)$, $n \in \mathbb{N}$ and $(\alpha \circ f \circ \gamma)^n(\alpha(\bot))$, $n \in \mathbb{N}$ be the respective iterates of f and $\alpha \circ f \circ \gamma$ from $\bot$ and $\alpha(\bot)$.

By definition of the infimum $\bot \sqsubseteq f(\bot)$ and by exercise 15.24, f is increasing so that by recurrence, $\langle f^n(\bot), n \in \mathbb{N} \rangle$ is an increasing chain so has a lub $\bigsqcup\{f^n(\bot) \mid n \in \mathbb{N}\}$ in a CPO.

By exercise 11.35.1, exercise 15.24 and exercise 11.35.2, the composition $\alpha \circ f \circ \gamma$ of increasing functions is increasing. By definition of the infimum $\forall \overline{x} \in \mathscr{A} . \bot \sqsubseteq \gamma(\overline{x})$ so $\alpha(\bot) \preccurlyeq \overline{x}$ by $\langle \mathscr{C}, \sqsubseteq \rangle \underset{\alpha}{\overset{\gamma}{\leftrightarrows}} \langle \mathscr{A}, \preccurlyeq \rangle$, proving that $\alpha(\bot)$ is the infimum of $\langle \mathscr{A}, \preccurlyeq \rangle$. It follows, by recurrence, that the iterates of $\alpha \circ f \circ \gamma$ from $\alpha(\bot)$ is increasing so has a lub $\curlyvee\{(\alpha \circ f \circ \gamma)^n(\alpha(\bot)) \mid n \in \mathbb{N}\}$ in a CPO.

- $\alpha(f^0(\bot)) = \alpha(\bot) \preccurlyeq \alpha(\bot) = (\alpha \circ f \circ \gamma)^0(\alpha(\bot))$

  ⟅definition of the iterates and reflexivity⟆
- Assume by recurrence hypothesis that $\alpha(f^n(\bot)) \preccurlyeq (\alpha \circ f \circ \gamma)^n(\alpha(\bot))$.
- $\alpha(f^n(\bot)) \preccurlyeq (\alpha \circ f \circ \gamma)^n(\alpha(\bot))$

$\Rightarrow f^n(\bot) \sqsubseteq \gamma((\alpha \circ f \circ \gamma)^n(\alpha(\bot))$ ⟅$\langle \mathscr{C}, \sqsubseteq \rangle \underset{\alpha}{\overset{\gamma}{\leftrightarrows}} \langle \mathscr{A}, \preccurlyeq \rangle$⟆

$\Rightarrow \alpha(f(f^n(\bot))) \preccurlyeq \alpha \circ f \circ (\gamma(\alpha \circ f \circ \gamma)^n(\alpha(\bot))$

⟅f increasing by exercise 15.24, α increasing by exercise 11.35.1⟆

- by recurrence, $\forall n \in \mathbb{N} . \alpha(f^n(\bot)) \preccurlyeq (\alpha \circ f \circ \gamma)^n(\alpha(\bot))$.

By definition of lubs, it follows that $\forall n \in \mathbb{N} . \alpha(f^n(\bot)) \preccurlyeq \curlyvee\{(\alpha \circ f \circ \gamma)^n(\alpha(\bot)) \mid n \in \mathbb{N}\}$ so $\curlyvee\{\alpha(f^n(\bot)) \mid n \in \mathbb{N}\} \preccurlyeq \curlyvee\{(\alpha \circ f \circ \gamma)^n(\alpha(\bot)) \mid n \in \mathbb{N}\}$. By $\langle \mathscr{C}, \sqsubseteq \rangle \underset{\alpha}{\overset{\gamma}{\leftrightarrows}} \langle \mathscr{A}, \preccurlyeq \rangle$ and lemma 11.38, α preserves existing lubs, so $\alpha(\bigsqcup\{f^n(\bot) \mid n \in \mathbb{N}\}) \preccurlyeq \curlyvee\{(\alpha \circ f \circ \gamma)^n(\alpha(\bot)) \mid n \in \mathbb{N}\}$. By Scott–Kleene's iterative fixpoint theorem 15.26, we conclude that $\alpha(\mathsf{lfp}^{\sqsubseteq} f) \preccurlyeq \mathsf{lfp}^{\preccurlyeq} \alpha \circ f \circ \gamma$ so $\mathsf{lfp}^{\sqsubseteq} f \sqsubseteq \gamma(\mathsf{lfp}^{\preccurlyeq} \alpha \circ f \circ \gamma)$ □

Exercise 18.19 Assume that $\langle \mathscr{C}, \sqsubseteq, \bot, \sqcup \rangle$ and $\langle \mathscr{A}, \preccurlyeq, 0, \curlyvee \rangle$ are CPOs, $\langle \mathscr{C}, \sqsubseteq \rangle \underset{\alpha}{\overset{\gamma}{\leftrightarrows}} \langle \mathscr{A}, \preccurlyeq \rangle$, $f \in \mathscr{C} \xrightarrow{uc} \mathscr{C}$ is upper continuous, $\overline{f} \in \mathscr{A} \xrightarrow{\nearrow} \mathscr{A}$ is increasing, $\mathscr{X} \in \wp(\mathscr{C})$ contains the iterates of f (i.e., $\bot \in \mathscr{X} \wedge \forall x \in \mathscr{X} . f(x) \in \mathscr{X}$), and $\forall x \in \mathscr{X} . \alpha(f(x)) \preccurlyeq \overline{f}(\alpha(X))$. Then $\alpha(\mathsf{lfp}^{\sqsubseteq} f) \preccurlyeq \curlyvee \overline{f}^n(0)$ (which is $\mathsf{lfp}^{\preccurlyeq \overline{f}}$) when $\overline{f}$ is upper continuous).

Exercise 18.20 Assume that $\langle \mathscr{C}, \sqsubseteq, \bot, \sqcup \rangle$ is a CPO, $f \in \mathscr{C} \xrightarrow{\nearrow} \mathscr{C}$ is upper continuous, $\langle \mathscr{A}, \preccurlyeq \rangle$ is a poset, $\overline{f} \in \mathscr{A} \to \mathscr{A}$, $f \circ \gamma \mathrel{\dot{\sqsubseteq}} \gamma \circ \overline{f}$, and $\gamma \in \mathscr{A} \xrightarrow{\nearrow} \mathscr{C}$ is increasing. Then $\forall y \in \mathscr{A}$, $\overline{f}(y) \preccurlyeq y$ implies $\mathsf{lfp}^{\sqsubseteq} f \sqsubseteq \gamma(y)$ (in particular when $y = \mathsf{lfp}^{\preccurlyeq} \overline{f}$).

All previous fixpoint overapproximation theorems rely on the fact that the overapproximation is from the same partial order (also called the approximation ordering) as the fixpoint order (also called the calculational ordering). When the approximation and calculational orderings differ, one can use the following abstraction theorem (adapted from the more general [248, Prop. 2]).

Theorem 18.21 (least fixpoint overapproximation in a CPO) Assume that $\langle \mathscr{C}, \sqsubseteq, \bot, \sqcup \rangle$ is a CPO, $f \in \mathscr{C} \xrightarrow{uc} \mathscr{C}$ is $\sqcup$-upper continuous, $\mathscr{I} \in \wp(\mathscr{C})$ contains the iterates $f^0(\bot) = \bot$ and $f^{n+1}(\bot) = f(f^n(\bot))$ of f from $\bot$, $\langle \mathscr{A}, \preccurlyeq, 0, \curlyvee \rangle$ is a CPO, $\overline{f} \in \mathscr{A} \xrightarrow{uc} \mathscr{A}$ is $\curlyvee$-upper continuous, $\trianglelefteq$ is a preorder on $\mathscr{A}$, $\alpha \in \mathscr{C} \to \mathscr{A}$, $\alpha(\bot) \trianglelefteq 0$, $\forall x \in \mathscr{I}, y \in \mathscr{A} \,.\, (\alpha(x) \trianglelefteq y) \Rightarrow (\alpha(f(x)) \trianglelefteq \overline{f}(y))$, and for all $\sqsubseteq$-increasing chains $\langle x^i, i \in \mathbb{N} \rangle$ of $\mathscr{I}$ and $\preccurlyeq$-increasing chains $\langle y^i, i \in \mathbb{N} \rangle$ of $\mathscr{A}$, $\alpha(\bigsqcup_{i \in \mathbb{N}} x^i) \trianglelefteq \bigcurlyvee_{i \in \mathbb{N}} y^i$. Then

$\alpha(\mathsf{lfp}^{\sqsubseteq} f) \trianglelefteq \mathsf{lfp}^{\preccurlyeq} \overline{f}$.

Remark 18.22 If $\overline{f}$ is $\trianglelefteq$-increasing, and $\forall x \in \mathscr{I} \,.\, \alpha \circ f(x) \trianglelefteq \overline{f} \circ \alpha(x)$ (semicommutativity) then $\forall x \in \mathscr{I} \,.\, \alpha(x) \trianglelefteq y \Rightarrow \overline{f}(\alpha(x)) \trianglelefteq \overline{f}(y) \Rightarrow \alpha(f(x)) \trianglelefteq \overline{f}(y)$. Because, in general, this property does not hold for all $x \in \mathscr{C}$ and it is used for the iterates of f only, $\mathscr{I} \in \wp(\mathscr{C})$ can be used to restrict the elements of $\mathscr{C}$ for which the property is required to hold.

Proof of theorem 18.21 By Scott–Kleene's iterative fixpoint theorem theorem 15.26, $f \in \mathscr{C} \xrightarrow{uc} \mathscr{C}$ is $\sqcup$-continuous function on a CPO $\langle \mathscr{C}, \sqsubseteq, \bot, \sqcup \rangle$ so f has a least fixpoint $\mathsf{lfp}^{\sqsubseteq} f = \bigsqcup_{n \in \mathbb{N}} f^n(\bot)$. Similarly, $\mathsf{lfp}^{\preccurlyeq} \overline{f} = \bigcurlyvee_{n \in \mathbb{N}} \overline{f}^n(0)$. We have $\alpha(f^0(\bot)) = \alpha(\bot) \trianglelefteq 0 = \overline{f}^0(0)$. Then $f^n(\bot) \in \mathscr{I}$ and $\alpha(f^n(\bot)) \trianglelefteq \overline{f}^n(0)$ by induction hypothesis so that $\alpha(f^{n+1}(\bot)) = \alpha(f(f^n(\bot))) \trianglelefteq \overline{f}(\overline{f}^n(0)) = \overline{f}^{n+1}(0)$. By recurrence $\forall n \in \mathbb{N} \,.\, \alpha(f^n(\bot)) \trianglelefteq \overline{f}^n(0)$. Because the fixpoint iterates are increasing, it follows, by hypothesis, that $\alpha(\mathsf{lfp}^{\sqsubseteq} f) = \alpha(\bigsqcup_{n \in \mathbb{N}} f^n(\bot)) = \bigcurlyvee_{n \in \mathbb{N}} \overline{f}^n(0) = \mathsf{lfp}^{\preccurlyeq} \overline{f}$. □

18.4 Exact/Complete Least Fixpoint Abstraction

A sound fixpoint abstraction $\alpha(\mathsf{lfp}^{\sqsubseteq} f) \preccurlyeq \mathsf{lfp}^{\preccurlyeq} \overline{f}$ is exact (or complete) when $\alpha(\mathsf{lfp}^{\sqsubseteq} f) = \mathsf{lfp}^{\preccurlyeq} \overline{f}$. It is sound but approximate (or imprecise) when $\alpha(\mathsf{lfp}^{\sqsubseteq} f) \prec \mathsf{lfp}^{\preccurlyeq} \overline{f}$.

Theorem 18.23 (exact least fixpoint abstraction in a complete lattice) Assume that $\langle \mathscr{C}, \sqsubseteq, \bot, \top, \sqcup, \sqcap \rangle$ and $\langle \mathscr{A}, \preccurlyeq, 0, 1, \curlyvee, \curlywedge \rangle$ are complete lattices, $f \in \mathscr{C} \xrightarrow{\nearrow} \mathscr{C}$ is increasing, $\langle \mathscr{C}, \sqsubseteq \rangle \underset{\alpha}{\overset{\gamma}{\leftrightarrows}} \langle \mathscr{A}, \preccurlyeq \rangle$, $\overline{f} \in \mathscr{A} \xrightarrow{\nearrow} \mathscr{A}$ is increasing, and $\alpha \circ f = \overline{f} \circ \alpha$ (*commutation property* or *pointwise completeness*). Then $\alpha(\mathsf{lfp}^{\sqsubseteq} f) = \mathsf{lfp}^{\preccurlyeq} \overline{f}$.

Example 18.25 The commutation condition is not necessary as shown by figure 18.1a, but it is sufficient. It may hold whether $\overline{f} = \alpha \circ f \circ \gamma$ or not.

Proof $\mathsf{lfp}^{\sqsubseteq} f$ and $\mathsf{lfp}^{\preccurlyeq} \overline{f}$ do exist by Tarski's fixpoint theorem 15.6. We have

$\alpha(\mathsf{lfp}^{\sqsubseteq} f)$

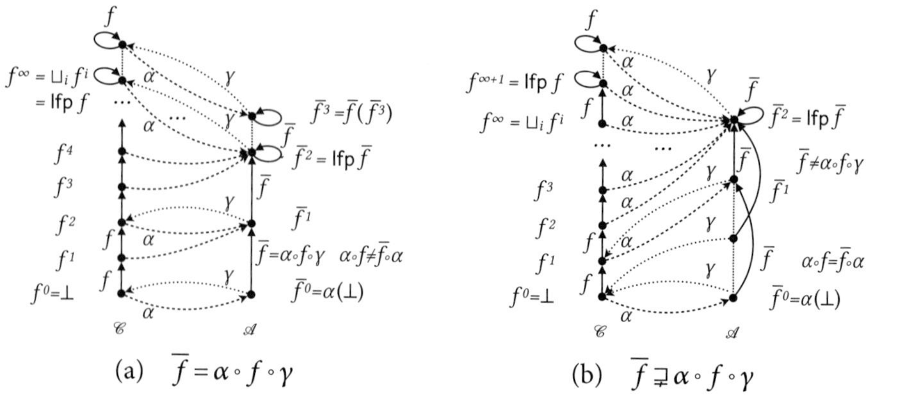

Figure 18.24
On commutation

$= \alpha \circ f(\mathsf{lfp}^{\sqsubseteq} f)$ ⦃fixpoint property⦄

$= \overline{f} \circ \alpha(\mathsf{lfp}^{\sqsubseteq} f)$ ⦃commutation property⦄

so $\alpha(\mathsf{lfp}^{\sqsubseteq} f)$ is a fixpoint of $\overline{f}$, proving that $\mathsf{lfp}^{\preccurlyeq} \overline{f} \preccurlyeq \alpha(\mathsf{lfp}^{\sqsubseteq} f)$ for the least fixpoint.

Conversely, $\alpha \circ f = \overline{f} \circ \alpha$ implies $\alpha \circ f \circ \gamma = \overline{f} \circ \alpha \circ \gamma \mathrel{\dot{\preccurlyeq}} \overline{f}$ because $\alpha \circ \gamma$ is reductive by exercise 11.35.4 and $\overline{f}$ is increasing. By corollary 18.14 and $\langle \mathscr{C}, \sqsubseteq \rangle \underset{\alpha}{\overset{\gamma}{\leftrightarrows}} \langle \mathscr{A}, \preccurlyeq \rangle$, $\alpha(\mathsf{lfp}^{\sqsubseteq} f) \preccurlyeq \mathsf{lfp}^{\preccurlyeq} \overline{f}$.

By antisymmetry, $\alpha(\mathsf{lfp}^{\sqsubseteq} f) = \mathsf{lfp}^{\preccurlyeq} \overline{f}$. □

It the image of the iterates of f by α are the iterates of $\overline{f}$ then $\alpha(\mathsf{lfp}^{\sqsubseteq} f) = \mathsf{lfp}^{\preccurlyeq} \overline{f}$.

Theorem 18.26 (exact least fixpoint abstraction in a CPO) Let $\langle \mathscr{C}, \sqsubseteq, \bot, \bigsqcup \rangle$ and $\langle \mathscr{A}, \preccurlyeq, 0, \curlyvee \rangle$ be CPOs, $f \in \mathscr{C} \xrightarrow{uc} \mathscr{C}$ and $\overline{f} \in \mathscr{A} \xrightarrow{uc} \mathscr{A}$ be upper continuous with respective iterates $\langle f^n(\bot), n \in \mathbb{N} \cup \{\omega\} \rangle$ and $\langle \overline{f}^n(0), n \in \mathbb{N} \cup \{\omega\} \rangle$. Let $\alpha \in \mathscr{C} \to \mathscr{A}$.

If $\forall n \in \mathbb{N} \cup \{\omega\} \,.\, \alpha(f^n(\bot)) = \overline{f}^n(0)$ then $\alpha(\mathsf{lfp}^{\sqsubseteq} f) = \mathsf{lfp}^{\preccurlyeq} \overline{f}$.

A sufficient condition is $\alpha(\bot) = 0$, $\forall x \in \mathscr{C} \,.\, (x \sqsubseteq \mathsf{lfp}^{\sqsubseteq} f) \Rightarrow (\alpha \circ f(x) = \overline{f} \circ \alpha(x))$ (*commutation property*), and α is upper continuous.

Proof of theorem 18.26 By hypothesis and the Tarski–Kantorovich's iterative fixpoint theorem 15.21, we have $\alpha(\mathsf{lfp}^{\sqsubseteq} f) = \alpha(f^\omega(\bot)) = \overline{f}^\omega(0) = \mathsf{lfp}^{\preccurlyeq} \overline{f}$.

For the sufficient condition, $\alpha(f^0(\bot)) = \alpha(\bot) = 0 = \overline{f}^0(0)$. Assume that $\alpha(f^n(\bot)) = \overline{f}^n(0)$. By definition 15.20 of the iterates, the commutation condition, and induction hypothesis, $\alpha(f^{n+1}(\bot)) = \alpha(f(f^n(\bot))) = \overline{f}(\alpha(f^n(\bot))) = \overline{f}(\overline{f}^n(0)) = \overline{f}^{n+1}(0)$. By recurrence $\langle f^n(\bot), n \in \mathbb{N} \rangle = \langle \overline{f}^n(0), n \in \mathbb{N} \rangle$. Theorem 15.21 implies that $\langle f^n(\bot), n \in \mathbb{N} \rangle$ and $\langle \overline{f}^n(0), n \in \mathbb{N} \rangle$ are increasing chains. By upper continuity $\alpha(f^\omega(\bot)) \triangleq \alpha(\bigsqcup\{f^n(\bot) \mid n \in \mathbb{N}\}) = \curlyvee\{\alpha(f^n(\bot)) \mid n \in \mathbb{N}\} = \curlyvee\{\overline{f}^n(0) \mid n \in \mathbb{N}\} \triangleq \overline{f}^\omega(0)$, proving $\forall n \in \mathbb{N} \cup \{\omega\} \,.\, \alpha(f^n(\bot)) = \overline{f}^n(0)$. □

Sometimes, the abstract function $\overline{f} \in \mathscr{A} \to \mathscr{A}$ is not upper continuous and not even increasing. In that case, we can apply the theorem 18.27.

> **Theorem 18.27 (exact least fixpoint abstraction in a CPO)** Assume that $\langle \mathscr{C}, \sqsubseteq, \bot, \sqcup \rangle$ is a CPO, $f \in \mathscr{C} \xrightarrow{uc} \mathscr{C}$ is upper continuous, $\langle \mathscr{C}, \sqsubseteq \rangle \underset{\alpha}{\overset{\gamma}{\leftrightarrows}} \langle \mathscr{A}, \preccurlyeq \rangle$ is a Galois retraction, and $\overline{f} \in \mathscr{A} \to \mathscr{A}$ satisfies the commutation property $\forall x \in \mathscr{C} . (x \sqsubseteq \mathsf{lfp}^{\sqsubseteq} f) \Rightarrow (\alpha \circ f(x) = \overline{f} \circ \alpha(x))$. Then $\overline{f} = \alpha \circ f \circ \gamma$ is increasing and $\alpha(\mathsf{lfp}^{\sqsubseteq} f) = \mathsf{lfp}^{\preccurlyeq} \overline{f} = \bigvee_{n \in \mathbb{N}} \overline{f}^{\,n}(\alpha(\bot))$.

Example 18.28 An example for theorem 18.27 is figure 18.24a of example 18.25.

Proof • $\alpha \circ f = \overline{f} \circ \alpha$ implies $\alpha \circ f \circ \gamma = \overline{f} \circ \alpha \circ \gamma = \overline{f}$ because, by exercise 11.50, $\alpha \circ \gamma$ is the identity in a Galois retraction. It follows that $\overline{f} = \alpha \circ f \circ \gamma$ is the composition of increasing functions hence increasing.

- $\alpha(\mathsf{lfp}^{\sqsubseteq} f) = \alpha \circ f(\mathsf{lfp}^{\sqsubseteq} f) = \overline{f} \circ \alpha(\mathsf{lfp}^{\sqsubseteq} f)$ by the commutation property, proving that $\alpha(\mathsf{lfp}^{\sqsubseteq} f)$ is a fixpoint of $\overline{f}$.
- f is is upper continuous so by Scott–Kleene's iterative fixpoint theorem 15.26, $\mathsf{lfp}^{\sqsubseteq} f = \bigsqcup_{n \in \mathbb{N}} f^{\,n}(\bot)$.
  - By the commutation property and definition of the iterates, $\overline{f}^{\,0}(\alpha(\bot)) = \alpha(\bot) = \alpha(f^{0}(\bot))$ where $\alpha(\bot)$ is the infimum of $\langle \mathscr{A}, \preccurlyeq \rangle$ because $\forall \overline{x} \in \mathscr{A} . \bot \sqsubseteq \gamma(\overline{x})$ implies $\forall \overline{x} \in \mathscr{A} . \alpha(\bot) \preccurlyeq \overline{x}$;
  - Assume, by induction hypothesis, that $\overline{f}^{\,n}(\alpha(\bot)) = \alpha(f^{\,n}(\bot))$;
  - Then $\overline{f}^{\,n+1}(\alpha(\bot)) = \overline{f}(\overline{f}^{\,n}(\alpha(\bot))) = \overline{f}((\alpha(f^{\,n}(\bot))) = \alpha(f(f^{\,n}(\bot))) = \alpha(f^{\,n+1}(\bot))$ by the commutation property $\alpha \circ f = \overline{f} \circ \alpha$;
  - By recurrence, $\forall n \in \mathbb{N} . \overline{f}^{\,n}(\alpha(\bot)) = \alpha(f^{\,n}(\bot))$.

If follows that $\alpha(\mathsf{lfp}^{\sqsubseteq} f) = \alpha(\bigsqcup_{n \in \mathbb{N}} f^{\,n}(\bot)) = \bigvee_{n \in \mathbb{N}} \alpha(f^{\,n}(\bot)) = \bigvee_{n \in \mathbb{N}} \overline{f}^{\,n}(\alpha(\bot))$ because, by lemma 11.38, α preserves existing lubs. So $\bigvee_{n \in \mathbb{N}} \overline{f}^{\,n}(\alpha(\bot))$ is a fixpoint of $\overline{f}$.

- Let us prove that $\mathsf{lfp}^{\preccurlyeq} \overline{f} = \bigvee_{n \in \mathbb{N}} \overline{f}^{\,n}(\alpha(\bot))$. So consider a fixpoint $\overline{x} \in \mathscr{A}$ of $\overline{f}$.
  - By definition of the iterates, $\overline{f}^{\,0}(\alpha(\bot)) = \alpha(\bot) \preccurlyeq \overline{x}$ because $\alpha(\bot)$ is the infimum of $\langle \mathscr{A}, \preccurlyeq \rangle$;
  - Assume by induction hypothesis that $\overline{f}^{\,n}(\alpha(\bot)) \preccurlyeq \overline{x}$;
  - By definition of the iterates, $\overline{f}$ increasing, and fixpoint property, $\overline{f}^{\,n+1}(\alpha(\bot)) = \overline{f}(\overline{f}^{\,n}(\alpha(\bot))) \preccurlyeq \overline{f}(\overline{x}) = \overline{x}$; □

If follows, by recurrence, that $\forall n \in \mathbb{N} . \overline{f}^{\,n}(\alpha(\bot)) \preccurlyeq \overline{x}$. By definition of the lub (previously proved to exist), $\alpha(\mathsf{lfp}^{\sqsubseteq})f = \bigvee_{n \in \mathbb{N}} \overline{f}^{\,n}(\alpha(\bot)) \preccurlyeq \overline{x}$, proving that $\alpha(\mathsf{lfp}^{\sqsubseteq} f) = \mathsf{lfp}^{\preccurlyeq} \overline{f}$ is the least fixpoint of $\overline{f}$. □

Exercise 18.29 Consider $F(X) \triangleq \{1\} \cup \{z+2 \mid z \in X\}$ on $\wp(\mathbb{Z})$. Prove that F preserves lubs $\cup$. Consider the sign abstraction $\alpha_{\pm}$ defined in section 3.12. Calculate $\overline{F} = \alpha_{\pm} \circ F \circ \gamma_{\pm}$ and $\mathsf{lfp}\,\overline{F}$.

Exercise 18.30 Let $r \in \wp(S \times S)$ be a binary relation on a set S. Define the reflexive transitive closure $r^* \triangleq \mathsf{lfp}^{\subseteq} F_r$ where $F_r(X) \triangleq \mathbb{1}_S \cup X \circ r$ and $\mathbb{1}_S \triangleq \{\langle e, e\rangle \mid e \in S\}$ is the identity relation. For example

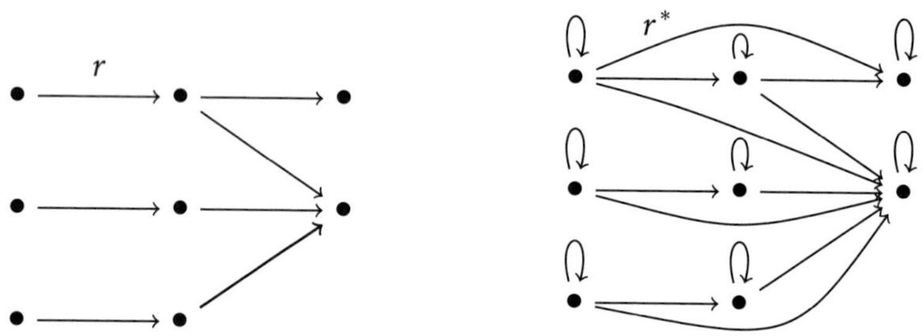

Given $P \in \wp(S)$, consider the abstraction $\alpha_P(r) \triangleq \mathsf{post}[r]P$ where $\mathsf{post}[r] \triangleq P \mapsto \{y \mid \exists x \in P \,.\, \langle x, y\rangle \in r\}$. Provide a fixpoint characterization of $\alpha_P(r^*)$. For example,

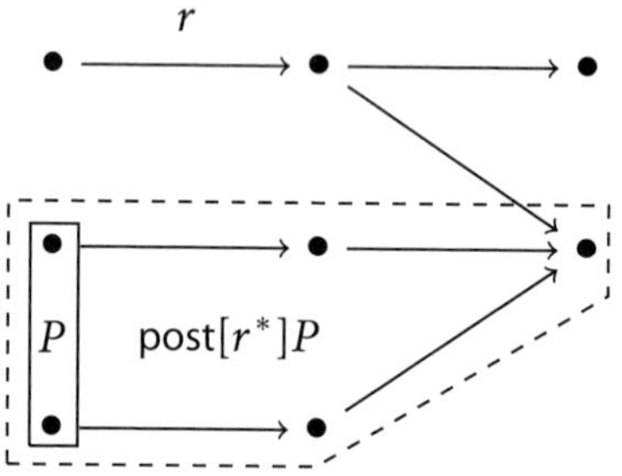

Exercise 18.31 Assume that $\langle \mathcal{C}, \sqsubseteq, \bot, \sqcup\rangle$ and $\langle \mathcal{A}, \preccurlyeq, 0, \curlyvee\rangle$ are posets, $f \in \mathcal{C} \xrightarrow{\nearrow} \mathcal{C}$ is increasing, $\langle \mathcal{C}, \sqsubseteq\rangle \overset{\gamma}{\underset{\alpha}{\leftrightarrows}} \langle \mathcal{A}, \preccurlyeq\rangle$ is a Galois retraction, $\overline{f} = \alpha \circ f \circ \gamma$. Consider

$$\forall x, y \in \mathcal{C} \,.\, (\alpha(x) = \alpha(y)) \Rightarrow (\alpha(f(x)) = \alpha(f(y))) \tag{18.32}$$

(a) Prove that (18.32) holds if and only if $\alpha \circ f = \overline{f} \circ \alpha$ (and so (18.32) implies, by theorem 18.23, that $\alpha(\mathsf{lfp}^{\sqsubseteq} f) = \mathsf{lfp}^{\preccurlyeq} \overline{f}$ for complete lattices). (b) Prove that f preserves existing lubs and (18.32) imply that $\overline{f}$ preserves existing lubs (and so (18.32) implies, by theorem 15.26, that $\mathsf{lfp}^{\preccurlyeq} \overline{f} = \curlyvee_{n \in \mathbb{N}} \overline{f}^{\,n}(0)$ for CPOs). (c) Prove that the hypotheses do not imply (18.32).

Exercise 18.33 Let $\langle D, \sqsubseteq, \bot, \top, \sqcup, \sqcap\rangle$ be a complete lattice. By exercise 11.32, the set $D \xrightarrow{\sqcup} D$ of (arbitrary) joins preserving functions ordered pointwise is a complete lattice $\langle D \xrightarrow{\sqcup} D, \dot\sqsubseteq, \dot\bot, \dot\top, \dot\sqcup\rangle$. Let $\mathcal{F} \in (D \xrightarrow{\sqcup} D) \xrightarrow{\sqcup} (D \xrightarrow{\sqcup} D)$ be an (arbitrary) join preserving operator on $D \xrightarrow{\sqcup} D$. Define $\forall x \in D \,.\, \alpha_x(f) \triangleq \mathcal{F}(f)x$. Prove that α_x is the lower adjoint of a Galois connection (with adjoint γ_x). Assume $F \in D \to (D \xrightarrow{\sqcup} D)$ satisfies $\alpha_x \circ \mathcal{F} = F(x) \circ \alpha_x$ for all $x \in D$. Prove that $\mathsf{lfp}^{\dot\sqsubseteq}\mathcal{F} = x \mapsto \mathsf{lfp}^{\sqsubseteq} F(x)$. The idea is that instead of calculating a fixpoint $\mathsf{lfp}^{\dot\sqsubseteq}\,\mathscr{F}$ on operators on D and then applying this function $(\mathsf{lfp}^{\dot\sqsubseteq}\,\mathscr{F})(x)$ at $x \in D$, we can calculate a fixpoint $\mathsf{lfp}^{\sqsubseteq}\,F(x)$ of a operator $F(x)$ on D, which is in general much simpler.

The hypotheses of theorem 18.27 on the exact fixpoint abstraction in a CPO can be weakened as was the case for Tarski–Kantorovich's iterative fixpoint theorem 15.21 for Scott–Kleene's iterative fixpoint theorem 15.26 by considering only the concrete iterates.

Corollary 18.34 (exact least fixpoint iterates abstraction) Assume $\langle \mathscr{C}, \sqsubseteq\rangle$ and $\langle \mathscr{A}, \preccurlyeq\rangle$ are posets, $\bot$ is the infimum of $\langle \mathscr{C}, \sqsubseteq\rangle$, $f \in \mathscr{C} \to \mathscr{C}$, the lub $\bigsqcup_{n\in\mathbb{N}} f^n(\bot)$ exists in $\langle \mathscr{C}, \sqsubseteq\rangle$ such that $\mathsf{lfp}^{\sqsubseteq} f = \bigsqcup_{n\in\mathbb{N}} f^n(\bot)$, $\langle \mathscr{C}, \sqsubseteq\rangle \underset{\alpha}{\overset{\gamma}{\leftrightarrows}} \langle \mathscr{A}, \preccurlyeq\rangle$, $\overline{f} \in \mathscr{A} \to \mathscr{A}$, and $\forall n \in \mathbb{N}\,.\,\alpha(f^{n+1}(\bot)) = \overline{f}(\alpha(f^n(\bot)))$. Then the lub $\curlyvee_{n\in\mathbb{N}} \overline{f}^{\,n}(\alpha(\bot))$ exists in $\langle \mathscr{A}, \preccurlyeq\rangle$ such that $\alpha(\mathsf{lfp}^{\sqsubseteq} f) = \curlyvee_{n\in\mathbb{N}} \overline{f}^{\,n}(\alpha(\bot))$.

Proof We have $\alpha(f^0(\bot)) = \alpha(\bot) = \overline{f}^{\,0}(\alpha(\bot))$. Assume that $\alpha(f^n(\bot)) = \overline{f}^{\,n}(\alpha(\bot))$ by induction hypothesis. Then $\alpha(f^{n+1}(\bot)) = \overline{f}(\alpha(f^n(\bot))) = \overline{f}(\overline{f}^{\,n}(\alpha(\bot))) = \overline{f}^{\,n+1}(\alpha(\bot))$, proving $\forall n \in \mathbb{N}\,.\,\alpha(f^n(\bot)) = \overline{f}^{\,n}(\alpha(\bot))$ and so, α preserving existing lubs, $\alpha(\mathsf{lfp}^{\sqsubseteq} f) = \alpha(\bigsqcup_{n\in\mathbb{N}} f^n(\bot)) = \curlyvee_{n\in\mathbb{N}} \alpha(f^n(\bot)) = \curlyvee_{n\in\mathbb{N}} \overline{f}^{\,n}(\alpha(\bot))$. □

Corollary 18.35 (exact iterates abstraction) Corollary 18.34 holds with $\forall n \in \mathbb{N}\,.\,\alpha(f^{n+1}(\bot)) = \overline{f}(\alpha(f^n(\bot)))$ is replaced by $\forall x \sqsubseteq \mathsf{lfp}^{\sqsubseteq} f\,.\,\alpha(f(x)) = \overline{f}(\alpha(x))$.

Proof For all $n \in \mathbb{N}$, $f^n(\bot) \sqsubseteq \mathsf{lfp}^{\sqsubseteq} f$ so $\alpha(f^{n+1}(\bot)) = \alpha(f(f^n(\bot))) = \overline{f}(\alpha(f^n(\bot)))$. □

18.5 Iterates Multiabstraction

It is sometimes necessary to abstract the iterates and their limit differently. Moreover, a different abstract transformer may have to be applied at each abstract iterate. An example is provided in section 39.14. In that case, we can use the following theorem 18.36 and its corollaries. The theorem essentially describes the following picture.

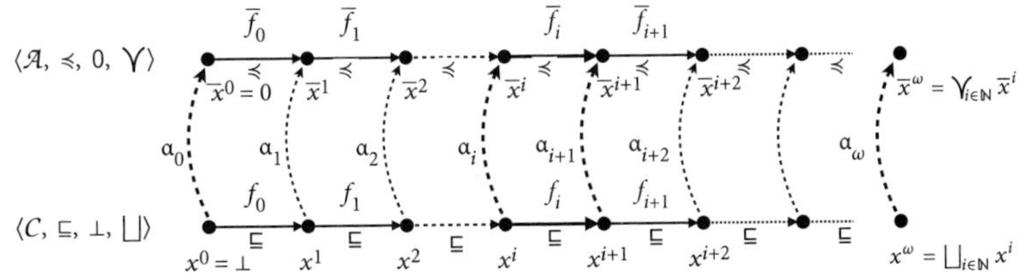

Two concrete $\langle x^i, i \in \mathbb{N} \cup \{\omega\}\rangle$ and abstract $\langle \overline{x}^i, i \in \mathbb{N} \cup \{\omega\}\rangle$ iterations involving different concrete $\langle f_i, i \in \mathbb{N}\rangle$ and abstract $\langle \overline{f}_i, i \in \mathbb{N}\rangle$ functions are related by different abstractions $\langle \alpha_i, i \in \mathbb{N} \cup \{\omega\}\rangle$ at each stage. A condition on the functions generalizing increasingness ensures that the iterates are increasing. The infimum and the limit are assumed to be preserved. The concrete and abstract functions are assumed to commute via the abstractions. The theorem states that under continuity hypotheses the abstraction of the least fixpoint of the limit $f \triangleq \dot{\bigsqcup}_{i\in\mathbb{N}} f_i$ of the concrete functions is the least fixpoint of the limit $\overline{f} \triangleq \dot{\curlyvee}_{i\in\mathbb{N}} \overline{f}_i$ of the abstract functions.

Theorem 18.36 (exact abstraction of iterates) Let $\langle \mathscr{C}, \sqsubseteq, \bot, \bigsqcup\rangle$ be a CPO, $\forall i \in \mathbb{N} \,.\, f_i \in \mathscr{C} \to \mathscr{C}$ be such that $\forall x, y \in \mathscr{C} \,.\, x \sqsubseteq y \Rightarrow f_i(x) \sqsubseteq f_{i+1}(y)$ with iterates $\langle x^i, i \in \mathbb{N} \cup \{\omega\}\rangle$ defined by $x^0 = \bot$, $x^{i+1} = f_i(x^i)$, $x^\omega = \bigsqcup_{i\in\mathbb{N}} x^i$. Then these concrete iterates form a well defined $\sqsubseteq$-increasing chain and $f \triangleq \dot{\bigsqcup}_{i\in\mathbb{N}} f_i$ is well defined.

Let $\langle \mathscr{A}, \preccurlyeq, 0, \curlyvee\rangle$ be a CPO, $\forall i \in \mathbb{N} \,.\, \overline{f}_i \in \mathscr{A} \to \mathscr{A}$ be such that $\forall \overline{x}, \overline{y} \in \mathscr{A} \,.\, \overline{x} \preccurlyeq \overline{y} \Rightarrow \overline{f}_i(\overline{x}) \preccurlyeq \overline{f}_{i+1}(\overline{y})$ with iterates $\langle \overline{x}^i, i \in \mathbb{N} \cup \{\omega\}\rangle$ defined by $\overline{x}^0 = 0$, $\overline{x}^{i+1} = \overline{f}_i(\overline{x}^i)$, $\overline{x}^\omega = \curlyvee_{i\in\mathbb{N}} \overline{x}^i$. Then these abstract iterates and $\overline{f} \triangleq \dot{\curlyvee}_{i\in\mathbb{N}} \overline{f}_i$ are well defined.

For all $i \in \mathbb{N} \cup \{\omega\}$, let $\alpha_i \in \mathscr{C} \to \mathscr{A}$ be such that $\alpha_0(x^0) = \overline{x}^0$, $\alpha_{i+1} \circ f_i(x^i) = \overline{f}_i \circ \alpha_i(x^i)$, and $\alpha_\omega(\bigsqcup_{i\in\mathbb{N}} x^i) = \curlyvee_{i\in\mathbb{N}} \alpha_i(x^i)$ for the iterates $\langle x^i, i \in \mathbb{N}\rangle$. It follows that $\alpha_\omega(x^\omega) = \overline{x}^\omega$.

If, moreover, $\forall i \in \mathbb{N} \,.\, f_i \in \mathscr{C} \xrightarrow{uc} \mathscr{C}$ is upper continuous then $x^\omega = \mathsf{lfp}^{\sqsubseteq} f$. Similarly $\overline{x}^\omega = \mathsf{lfp}^{\preccurlyeq} \overline{f}$ when the $\overline{f}_i$ are upper continuous. If both the f_i and $\overline{f}_i$ are upper continuous then $\alpha_\omega(\mathsf{lfp}^{\sqsubseteq} f) = \alpha_\omega(x^\omega) = \overline{x}^\omega = \mathsf{lfp}^{\preccurlyeq} \overline{f}$.

A trivial generalization is to have a different (concrete and) abstract domain at each iteration and the limit (as, for example, in *cofibered domains* [953]).

Proof of theorem 18.36 $x^0 \triangleq \bot \sqsubseteq x^1$ because $\bot$ is the infimum and if $x^i \sqsubseteq x^{i+1}$ then, by hypothesis, $x^{i+1} \triangleq f_i(x^i) \sqsubseteq f_{i+1}(x^{i+1}) \triangleq x^{i+2}$. Its follows that $\langle x^i, i \in \mathbb{N}\rangle$ is a $\sqsubseteq$-increasing chain so that its lub $x^\omega \triangleq \bigsqcup_{i\in\mathbb{N}} x^i$ is well defined in the CPO $\langle \mathscr{C}, \sqsubseteq\rangle$. The concrete iterates $\langle x^i, i \in \mathbb{N} \cup \{\omega\}\rangle$ are therefore well defined.

For $x \in \mathscr{C}$, reflexivity $x \sqsubseteq x$ implies $f_i(x) \sqsubseteq f_{i+1}(x)$ so $\langle f_i(x), i \in \mathbb{N}\rangle$ is an increasing chain which limit $f(x) \triangleq \bigsqcup_{i\in\mathbb{N}} f_i(x)$ is well defined in the CPO $\langle \mathscr{C}, \sqsubseteq\rangle$.

Similarly, the abstract iterates $\langle \overline{x}^i, i \in \mathbb{N} \cup \{\omega\}\rangle$ form a well defined $\preccurlyeq$-increasing chain and and $\overline{f}$ is well defined.

Let us prove by recurrence on i that $\forall i \in \mathbb{N} \,.\, \alpha_i(x^i) = \overline{x}^i$.

- For the basis, $\alpha_0(x^0) = \alpha_0(\bot) = 0 = \overline{x}^0$.
- Assume, by induction hypothesis, that $\alpha_i(x^i) = \overline{x}^i$. For the induction step,

$$\begin{aligned}
& \alpha_{i+1}(x^{i+1}) \\
= {} & \alpha_{i+1}(f_i(x^i)) && \text{(definition of the concrete iterates of the } f_i\text{)} \\
= {} & \overline{f}_i(\alpha_i(x^i)) && \text{(commutation } \alpha_{i+1} \circ f(x^i) = \overline{f}_i \circ \alpha_i(x^i)\text{)} \\
= {} & \overline{f}_i(\overline{x}^i) && \text{(induction hypothesis)} \\
= {} & \overline{x}^{i+1} && \text{(definition of the abstract iterates of the } \overline{f}_i\text{)}
\end{aligned}$$

It follows that $\alpha_\omega(x^\omega) = \alpha_\omega(\bigsqcup_{i\in\mathbb{N}} x^i) = \curlyvee_{i\in\mathbb{N}} \alpha_i(x^i) = \curlyvee_{i\in\mathbb{N}} \overline{x}^i = \overline{x}^\omega$.

If, moreover, $\forall i \in \mathbb{N} \,.\, f_i \in \mathscr{C} \xrightarrow{uc} \mathscr{C}$ is upper continuous, then we have

$$
\begin{aligned}
& f(x^\omega) \\
= {} & \bigsqcup_{j\in\mathbb{N}} f_j(\bigsqcup_{i\in\mathbb{N}} x^i) && \text{(definition of } f \text{ and } x^\omega\text{)} \\
= {} & \bigsqcup_{j\in\mathbb{N}} \bigsqcup_{i\in\mathbb{N}} f_j(x^i) && \text{(}\langle x^i,\ i\in\mathbb{N}\rangle \text{ is an } \sqsubseteq\text{-increasing chain and } f_i \text{ is upper continuous)} \\
= {} & \bigsqcup_{j\in\mathbb{N}} (\bigsqcup_{i<j} f_j(x^i) \sqcup \bigsqcup_{j=i} f_j(x^i) \sqcup \bigsqcup_{i>j} f_j(x^i)) && \text{(case analysis)} \\
= {} & \bigsqcup_{j\in\mathbb{N}} (f_j(x^j) \sqcup \bigsqcup_{i>j} f_j(x^i))
\end{aligned}
$$

(because, by recurrence using $x^i \sqsubseteq x^{i+1} \Rightarrow f_i(x^i) \sqsubseteq f_{i+1}(x^{i+1})$, we have $i<j \Rightarrow x^i \sqsubseteq x^j \Rightarrow f_i(x^i) \sqsubseteq f_j(x^j) \Rightarrow \bigsqcup_{i<j} f_j(x^i) \sqsubseteq f_j(x^j)$ and so, by definition of the lub $\bigsqcup$, $\bigsqcup_{i<j} f_j(x^i) \sqcup \bigsqcup_{j=i} f_j(x^i) = f_j(x^j)$)

$$
\begin{aligned}
= {} & (\bigsqcup_{j\in\mathbb{N}} f_j(x^j)) \sqcup (\bigsqcup_{j\in\mathbb{N}} \bigsqcup_{i>j} f_j(x^i)) && \text{(definition of the lub } \textstyle\bigsqcup\text{)} \\
= {} & (\bigsqcup_{j\in\mathbb{N}} f_j(x^j)) \sqcup \bigsqcup_{j\in\mathbb{N}} \bigsqcup_{i\geqslant j} f_i(x^i))
\end{aligned}
$$

(because, $j<i \Rightarrow x_j \sqsubseteq x^i \Rightarrow f_j(x^j) \sqsubseteq f_i(x^i)$ so $(\bigsqcup_{j\in\mathbb{N}} \bigsqcup_{i>j} f_j(x^i)) \sqsubseteq (\bigsqcup_{j\in\mathbb{N}} \bigsqcup_{i>j} f_i(x^i)) = (\bigsqcup_{j\in\mathbb{N}} \bigsqcup_{i\geqslant j} f_i(x^i))$)

$$
= (\bigsqcup_{j\in\mathbb{N}} f_j(x^j))
$$

($(\bigsqcup_{j\in\mathbb{N}} \bigsqcup_{i>j} f_j(x^i)) = (\bigsqcup_{j\in\mathbb{N}} f_j(x^j))$ by $\bigsqcup$ associative, commutative, and idempotent)

$$
\begin{aligned}
= {} & \bigsqcup_{i\in\mathbb{N}} x^{i+1} \quad = \quad x^0 \sqcup \bigsqcup_{j\in\mathbb{N}^+} x^j && \text{(definition of } x^{i+1} \text{ and } j=i+1 \text{ is positive)} \\
= {} & \bigsqcup_{i\in\mathbb{N}} x^i \quad = \quad x^\omega && \text{(}x^0=\bot \text{ is the infimum and definition of } x^\omega\text{)}
\end{aligned}
$$

Therefore x^ω is a fixpoint of f. Assume that $y \in \mathscr{C}$ is a fixpoint of f. Let us prove by recurrence that $\forall i \in \mathbb{N} \,.\, x^i \sqsubseteq y$. For the basis $x^0 = \bot \sqsubseteq y$, by definition of the infimum $\bot$. Assume that $x^i \sqsubseteq y$ by induction hypothesis. Then

$$
\begin{aligned}
& x^{i+1} \\
= {} & f_i(x^i) && \text{(definition of the abstract iterates)} \\
\sqsubseteq {} & f_i(y) && \text{(induction hypothesis } x^i \sqsubseteq y \text{ and } f_i \text{ upper continuous hence increasing)} \\
\sqsubseteq {} & \bigsqcup_{i\in\mathbb{N}} f_i(y) && \text{(definition of the lub, if it exists)} \\
= {} & f(y) && \text{(}\langle f_i(y),\ i\in\mathbb{N}\rangle \text{ is increasing with well defined limit } f(y) \triangleq \textstyle\bigsqcup_{i\in\mathbb{N}} f_i(y)\text{)} \\
= {} & y && \text{(fixpoint hypothesis)}
\end{aligned}
$$

It follows that $x^\omega \triangleq \bigsqcup_{i\in\mathbb{N}} x^i \sqsubseteq y$ proving that $x^\omega = \mathsf{lfp}^{\sqsubseteq} f$ is the $\sqsubseteq$-least fixpoint of f. □

Corollary 18.37 (exact iterates multiabstraction) Let $\langle \mathscr{C}, \sqsubseteq, \bot, \bigsqcup \rangle$ and $\langle \mathscr{A}, \preccurlyeq, 0, \curlyvee \rangle$ be CPOs, $f \in \mathscr{C} \xrightarrow{uc} \mathscr{C}$ be upper continuous, $\forall i \in \mathbb{N} . \overline{f}_i \in \mathscr{A} \rightarrow \mathscr{A}$ such that $\forall \overline{x}, \overline{y} \in \mathscr{A} . \overline{x} \preccurlyeq \overline{y} \Rightarrow \overline{f}_i(\overline{x}) \preccurlyeq \overline{f}_{i+1}(\overline{y})$, $\forall i \in \mathbb{N} . \alpha_i \in \mathscr{C} \rightarrow \mathscr{A}$ be such that $\alpha_0(\bot) = 0$, $\alpha_{i+1} \circ f(x^i) = \overline{f}_i \circ \alpha_i(x^i)$, and $\alpha \in \mathscr{C} \rightarrow \mathscr{A}$ satisfy $\alpha(\bigsqcup_{i \in \mathbb{N}} x^i) = \bigcurlyvee_{i \in \mathbb{N}} \alpha_i(x^i)$ for the iterates $x^0 = \bot$, $x^{i+1} = f(x^i)$, $i \in \mathbb{N}$. Let $\langle \overline{x}^i, i \in \mathbb{N} \cup \{\omega\} \rangle$ be the abstract iterates defined as $\overline{x}^0 \triangleq 0$, $\overline{x}^{i+1} \triangleq \overline{f}_i(\overline{x}^i)$, and $\overline{x}^\omega \triangleq \bigcurlyvee_{i \in \mathbb{N}} \overline{x}^i$.

Then $\alpha(\mathsf{lfp}^{\sqsubseteq} f) = \overline{x}^\omega$. Moreover, if the $\overline{f}_i \in \mathscr{A} \xrightarrow{uc} \mathscr{A}$, $i \in \mathbb{N}$, are upper continuous and $\overline{f}(\overline{x}) \triangleq \bigcurlyvee_{i \in \mathbb{N}} \overline{f}_i(\overline{x})$ is well defined then $\overline{x}^\omega = \mathsf{lfp}^{\preccurlyeq} \overline{f}$.

Proof of corollary 18.37 The f_i in theorem 18.36 can be chosen to be all identical equal to $f \in \mathscr{C} \xrightarrow{uc} \mathscr{C}$ in which case $x \sqsubseteq y \Rightarrow f_i(x) \sqsubseteq f_{i+1}(y)$ follows from f being upper continuous hence monotonically increasing. Then $\alpha_\omega(\mathsf{lfp}^{\sqsubseteq} f) = \alpha_\omega(x^\omega) = \overline{x}^\omega$. □

Similarly, the choice $\overline{f}_i = \overline{f} \in \mathscr{A} \xrightarrow{uc} \mathscr{A}$ yields $\alpha_\omega(x^\omega) = \overline{x}^\omega = \mathsf{lfp}^{\preccurlyeq} \overline{f}$.

Exercise 18.38 (approximate iterates multiabstraction) Show that if the hypotheses of theorem 18.36 or corollary 18.37 are weakened to $\alpha_0(\bot) \preccurlyeq 0$, $\alpha_{i+1} \circ f \preccurlyeq \overline{f}_i \circ \alpha_i$, and $\alpha(\bigsqcup_{i \in \mathbb{N}} x^i) \preccurlyeq \bigcurlyvee_{i \in \mathbb{N}} \alpha_i(x^i)$ for all increasing chains $\langle x^i \in \mathscr{C}, i \in \mathbb{N} \rangle$ then $\alpha(\mathsf{lfp}^{\sqsubseteq} f) \preccurlyeq \mathsf{lfp}^{\preccurlyeq} \overline{f}$.

A simpler case is when all concrete and abstract transformers are the same.

Corollary 18.39 (exact iterates multiabstraction) Let $\langle \mathscr{C}, \sqsubseteq, \bot, \bigsqcup \rangle$ and $\langle \mathscr{A}, \preccurlyeq, 0, \curlyvee \rangle$ be CPOs, $f \in \mathscr{C} \xrightarrow{uc} \mathscr{C}$ and $\overline{f} \in \mathscr{A} \xrightarrow{uc} \mathscr{A}$ be upper continuous, $\forall i \in \mathbb{N} . \alpha_i \in \mathscr{C} \rightarrow \mathscr{A}$ be such that $\alpha_0(\bot) = 0$ and $\alpha_{i+1} \circ f(x^i) = \overline{f} \circ \alpha_i(x^i)$, and $\alpha \in \mathscr{C} \rightarrow \mathscr{A}$ satisfy $\alpha(\bigsqcup_{i \in \mathbb{N}} x^i) = \bigcurlyvee_{i \in \mathbb{N}} \alpha_i(x^i)$ for the iterates $x^0 = \bot$, $x^{i+1} = f(x^i)$, $i \in \mathbb{N}$. Then $\alpha(\mathsf{lfp}^{\sqsubseteq} f) = \mathsf{lfp}^{\preccurlyeq} \overline{f}$.

Proof of corollary 18.39 Taking $\overline{f}_i = \overline{f}$, the hypotheses of theorem 18.36 are satisfied and $\bigcurlyvee_{i \in \mathbb{N}} \overline{f}_i(x) = \bigcurlyvee_{i \in \mathbb{N}} \overline{f} = \overline{f}(x)$ so $\alpha(\mathsf{lfp}^{\sqsubseteq} f) = \mathsf{lfp}^{\preccurlyeq} \overline{f}$. □

Theorem 18.36 and corollaries 18.37 and 18.39 have numerous variants (e.g., by restricting the commutation condition to iterates as in corollary 18.34, using Galois connections as in theorem 18.23, or considering overapproximations as in section 18.3) that are left as exercises. Obviously, the previous results follow from the choice $\alpha_i = \alpha$ in corollary 18.39.

18.6 Exact and Approximate Greatest Fixpoint Abstraction

By duality, including for Galois connections in lemma 11.25, the exact and approximate least fixpoint abstract theorems of sections 18.2, 18.3, and 18.4 apply dually to greatest fixpoint underapproximation for the concretization γ. However, as shown in chapter 50, it is sometimes necessary to overapproximate greatest fixpoints for the abstraction α. In that case, the commutation condition of corollary 18.35 is sufficient to ensure an exact/complete least fixpoint abstraction, but not for the greatest fixpoint, as shown by the following counterexample.

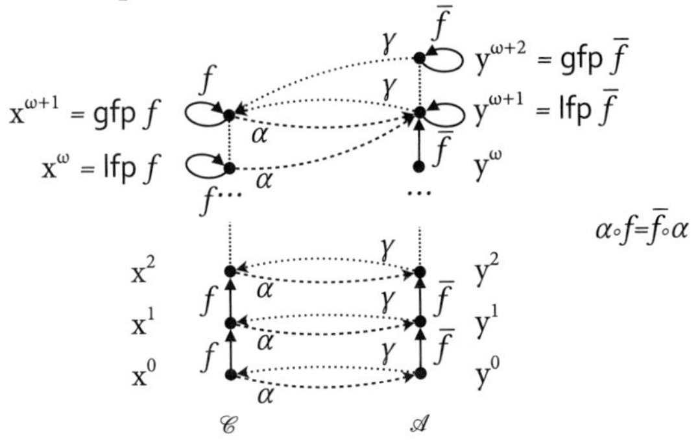

Nevertheless, the following results can be applied.

> **Theorem 18.40 (exact greatest fixpoint abstraction)** Let $\langle \mathscr{C}, \sqsubseteq, \top, \sqcap \rangle$ and $\langle \mathscr{A}, \preccurlyeq, \mathbf{1}, \curlywedge \rangle$ be dual CPOs, $f \in \mathscr{C} \xrightarrow{lc} \mathscr{C}$ and $\overline{f} \in \mathscr{A} \xrightarrow{lc} \mathscr{A}$ be lower continuous, $\langle \mathscr{C}, \sqsubseteq \rangle \underset{\alpha}{\overset{\gamma}{\leftrightarrows}} \langle \mathscr{A}, \preccurlyeq \rangle$, $\alpha \circ f = \overline{f} \circ \alpha$, and $\gamma \circ \alpha$ be lower continuous then $\alpha(\mathsf{gfp}^{\sqsubseteq} f) = \mathsf{gfp}^{\preccurlyeq} \overline{f}$.

(This result generalizes to increasingness using transfinite iterations [240].)

Proof Let $f^i(\top)$ and $\overline{f}^{\,i}(\mathbf{1})$ be the iterates of f and $\overline{f}$ from the suprema.

– α is surjective (exercise 11.50) so $\exists x \in \mathscr{C} \,.\, \alpha(x) = \mathbf{1}$. Because $x \sqsubseteq \top$ and α is increasing, we have $\mathbf{1} = \alpha(x) \preccurlyeq \alpha(\top) \preccurlyeq \mathbf{1}$, proving $\alpha(\top) = \mathbf{1}$ by antisymmetry.

– Assume $\alpha(f^i(\top)) = \overline{f}^{\,i}(\mathbf{1})$ by induction hypothesis. Then

$$\begin{array}{ll} \overline{f} \circ \alpha \circ f^i(\top) = \overline{f} \circ \overline{f}^{\,i}(\mathbf{1}) & \text{(function application)} \\ \Rightarrow \alpha \circ f \circ (f^i(\top)) = \overline{f} \circ \overline{f}^{\,i}(\mathbf{1}) & \text{(commutation } \alpha \circ f = \overline{f} \circ \alpha\text{)} \\ \Rightarrow \alpha(f^{i+1}(\top)) = \overline{f}^{\,i+1}(\mathbf{1}) & \text{(definition of the iterates)} \end{array}$$

proving $\forall i \in \mathbb{N} \,.\, \alpha(f^i(\top)) = \overline{f}^{\,i}(\mathbf{1})$ by recurrence.

– It follows that

$$\begin{array}{ll} \gamma(\mathsf{gfp}^{\preccurlyeq} \overline{f}) & \\ = \gamma(\displaystyle\curlywedge_{i\in\mathbb{N}} \overline{f}^{\,i}(\mathbf{1})) & \text{(dual of Scott–Kleene's iterative fixpoint theorem 15.26)} \\ = \displaystyle\bigsqcap_{i\in\mathbb{N}} \gamma(\overline{f}^{\,i}(\mathbf{1})) & \text{(}\gamma \text{ preserves existing meets by the dual of lemma 11.38)} \\ = \displaystyle\bigsqcap_{i\in\mathbb{N}} \gamma \circ \alpha(f^i(\top)) & \text{(}\forall i \in \mathbb{N} \,.\, \alpha(f^i(\top)) = \overline{f}^{\,i}(\mathbf{1})\text{)} \\ = \gamma \circ \alpha(\displaystyle\bigsqcap_{i\in\mathbb{N}} f^i(\top)) & \text{(the iterates are decreasing and } \gamma \circ \alpha \text{ is lower continuous)} \end{array}$$

and therefore $\mathsf{gfp}^{\preccurlyeq} \overline{f} = \alpha \circ \gamma(\mathsf{gfp}^{\preccurlyeq} \overline{f}) = \alpha \circ \gamma \circ \alpha(\bigsqcap_{i\in\mathbb{N}} f^i(\top)) = \alpha(\bigsqcap_{i\in\mathbb{N}} f^i(\top)) = \alpha(\mathsf{gfp}^{\sqsubseteq} f)$ because $\alpha \circ \gamma = \mathbb{1}_{\mathscr{A}}$ in the Galois retraction (exercise 11.50), $\alpha \circ \gamma \circ \alpha = \alpha$ in a Galois connection (exercise 11.34), and by the dual of Scott–Kleene's iterative fixpoint theorem 15.26. □

For an overapproximation, we have

Theorem 18.41 (greatest fixpoint overapproximation) Assume $\langle \mathscr{C}, \sqsubseteq, \top, \bigsqcap \rangle$ and $\langle \mathscr{A}, \preccurlyeq, \mathbf{1}, \curlywedge \rangle$ are dual CPOs, $f \in \mathscr{C} \xrightarrow{lc} \mathscr{C}$ is lower continuous, and $\langle \mathscr{C}, \sqsubseteq \rangle \underset{\alpha}{\overset{\gamma}{\leftrightarrows}} \langle \mathscr{A}, \preccurlyeq \rangle$ is a Galois connection. Then $\alpha(\mathsf{gfp}^{\sqsubseteq} f) \preccurlyeq \curlywedge_{n\in\mathbb{N}} \overline{f}^{\,i}(\mathbf{1})$ where $\overline{f} = \alpha \circ f \circ \gamma$ (which is $\mathsf{gfp}^{\preccurlyeq} \overline{f}$ when $\overline{f}$ is lower continuous).

(This result generalizes to increasingness using transfinite iterations [240].)

Proof Let $f^i(\top)$ and $\overline{f}^{\,i}(\mathbf{1})$ be the iterates of f and $\overline{f}$ from the suprema.

- $\alpha(f^0(\top)) = \alpha(\top) \preccurlyeq \mathbf{1} = \overline{f}^{\,0}(\mathbf{1})$ by definition of the iterates and of the supremum.
- Assume $\alpha(f^i(\top)) \preccurlyeq \overline{f}^{\,i}(\mathbf{1})$ by induction hypothesis. Then

$$
\begin{array}{ll}
f^i(\top) \sqsubseteq \gamma \circ \overline{f}^{\,i}(\mathbf{1}) & \text{(induction hypothesis and definition of Galois connections)} \\
\Rightarrow \alpha \circ f \circ f^i(\top) \preccurlyeq \alpha \circ f \circ \gamma \circ \overline{f}^{\,i}(\mathbf{1}) & \text{(f and α increasing)} \\
\Rightarrow \alpha(f^{i+1}(\top)) \preccurlyeq \overline{f}^{\,i+1}(\mathbf{1}) & \text{(definition of $\overline{f}$ and of the iterates)}
\end{array}
$$

proving $\forall i \in \mathbb{N} \,.\, \alpha(f^i(\top)) \preccurlyeq \overline{f}^{\,i}(\mathbf{1})$ by recurrence.

- It follows that

$$
\begin{array}{ll}
\alpha(\mathsf{gfp}^{\sqsubseteq} f) & \\
= (\curlywedge_{n\in\mathbb{N}} \alpha(f^i(\top)) & \text{(dual of Scott–Kleene's iterative fixpoint theorem 15.26)} \\
\preccurlyeq \curlywedge_{n\in\mathbb{N}} \alpha(f^i(\top)) & \text{(α increasing so $\alpha(\bigsqcap_{i\in\Delta} x^i) \preccurlyeq \curlywedge_{i\in\Delta} \alpha(x^i)$)} \\
\preccurlyeq \curlywedge_{n\in\mathbb{N}} \overline{f}^{\,i}(\mathbf{1}) & \text{(because $\forall i \in \mathbb{N} \,.\, \alpha(f^i(\top)) \preccurlyeq \overline{f}^{\,i}(\mathbf{1})$ and definition of the glb)}
\end{array}
$$

which, by the dual of Scott–Kleene's iterative fixpoint theorem 15.26, is $\mathsf{gfp}^{\preccurlyeq} \overline{f}$ when $\overline{f}$ is lower continuous. □

An alternative is to consider the concretization function.

Theorem 18.42 (exact greatest fixpoint concretization) Let $\langle \mathscr{C}, \sqsubseteq, \top, \bigsqcap \rangle$ and $\langle \mathscr{A}, \preccurlyeq, \mathbf{1}, \curlywedge \rangle$ be dual CPOs, $f \in \mathscr{C} \xrightarrow{lc} \mathscr{C}$ and $\overline{f} \in \mathscr{A} \xrightarrow{lc} \mathscr{A}$ be lower continuous, $\langle \mathscr{C}, \sqsubseteq \rangle \underset{\alpha}{\overset{\gamma}{\leftrightarrows}} \langle \mathscr{A}, \preccurlyeq \rangle$, $f \circ \gamma = \gamma \circ \overline{f}$ then $\mathsf{gfp}^{\sqsubseteq} f = \gamma(\mathsf{gfp}^{\preccurlyeq} \overline{f})$.

(This result generalizes to increasingness using transfinite iterations [240].)

Proof Let $f^i(\top)$ and $\overline{f}^{\,i}(\mathbf{1})$ be the iterates of f and $\overline{f}$ from $\top$ and $\mathbf{1}$.

- By definition of the supremum, $\alpha(\top) \preccurlyeq \mathbf{1}$ so $\top \sqsubseteq \gamma(\mathbf{1})$ by the Galois connection and therefore $f^0(\top) = \top = \gamma(\mathbf{1}) = \gamma(\overline{f}^{\,0}(\mathbf{1}))$ by definition of the supremum and of the iterates.
- Assume that $f^i(\top) = \gamma(\overline{f}^{\,i}(\mathbf{1}))$ by induction hypothesis. Then

$f^{i+1}(\top) = f(f^i(\top)) = f(\gamma(\overline{f}^{\,i}(\mathbf{1})))$ (definition of the iterates and induction hypothesis)

$= \gamma(\overline{f}(\overline{f}^{\,i}(\mathbf{1}))) = \overline{f}^{\,i+1}(\mathbf{1})$ (γ-commutation hypothesis and definition of the iterates)

– By recurrence, $\forall i \in \mathbb{N} \,.\, f^i(\top) = \gamma(\overline{f}^i(\mathbf{1}))$ and so

$$
\begin{array}{lll}
\gamma(\mathsf{gfp}^{\preccurlyeq}\, \overline{f}) & = \gamma(\bigcurlywedge_{i\in\mathbb{N}} \overline{f}^i(\mathbf{1})) & \text{⦇dual of Scott–Kleene's iterative fixpoint theorem 15.26⦈} \\
& = \bigsqcap_{i\in\mathbb{N}} \gamma(\overline{f}^i(\mathbf{1})) & \text{⦇the upper adjoint of a Galois connection preserves existing meets⦈} \\
& = \bigsqcap_{i\in\mathbb{N}} f^i(\top) & \text{⦇}f^i(\top) = \gamma(\overline{f}^i(\mathbf{1}))\text{⦈} \\
& = \mathsf{gfp}^{\sqsubseteq}\, f & \text{⦇dual of Scott–Kleene's iterative fixpoint theorem 15.26⦈} \quad \square
\end{array}
$$

In general, theorem 18.42 cannot be extended to least fixpoints, the problem being that γ does not preserve joins. However, if the abstract lattice $\mathscr{A}$ satisfies the ascending chain condition then the join of the abstract iterates is merely some of these iterates, so we never have to take a least upper bound in the abstract that would not be preserved by concretization.

Exercise 18.43 Let $\langle \mathscr{C}, \sqsubseteq, \bot, \bigsqcup \rangle$ be a CPO, $\langle \mathscr{A}, \preccurlyeq, \mathbf{0} \rangle$ be a Noetherian partial order with infimum $\mathbf{0}$ (see section 10.8). Let $f \in \mathscr{C} \xrightarrow{uc} \mathscr{C}$ be upper continuous and $\overline{f} \in \mathscr{A} \xrightarrow{\nearrow} \mathscr{A}$ be increasing. Let the increasing concretization $\gamma \in \mathscr{A} \xrightarrow{\nearrow} \mathscr{C}$ be such that $\gamma(\mathbf{0}) = \bot$ and $f \circ \gamma = \gamma \circ \overline{f}$. Prove that $\mathsf{lfp}^{\sqsubseteq} f = \gamma(\mathsf{lfp}^{\preccurlyeq} \overline{f})$.

18.7 Closure Operator Based Fixpoint Abstraction and Overapproximation

As noticed in section 11.15, abstract interpretation can be formalized using closure operators ρ [221, 241], for which all results in this chapter 18 can be applied using $\langle \mathscr{C}, \sqsubseteq \rangle \underset{\rho}{\overset{\mathbb{1}_{\mathscr{C}}}{\leftrightarrows}} \langle \mathscr{C}, \sqsubseteq \rangle$ and $\langle \mathscr{C}, \sqsubseteq \rangle \underset{\rho}{\overset{\mathbb{1}_{\mathscr{C}}}{\leftrightarrows}} \langle \rho(\mathscr{C}), \sqsubseteq \rangle$, and dually. In that case, the α-commutation condition $\alpha \circ f = \overline{f} \circ \alpha$ in theorem 18.27 where $\overline{f} = \alpha \circ f \circ \gamma$, that is $\alpha \circ f = \alpha \circ f \circ \gamma \circ \alpha$, becomes $\rho \circ f = \rho \circ f \circ \rho$ called backward completeness in [417]. In theorem 18.42, the γ-commutation condition $f \circ \gamma = \gamma \circ \overline{f}$ becomes $f \circ \rho = \rho \circ f \circ \rho$, called forward completeness in [414].

18.8 Exact and Approximate Deductive Definition Abstraction

Because least fixpoint and deductive definitions are equivalent (section 16.3), the fixpoint sound approximation and complete abstraction theorems of sections 18.2, 18.3, and 18.4 generalize to deductive definition abstraction.

Theorem 18.44 (deductive definition abstraction) Let $R = \{\frac{P_i}{c_i} \mid i \in \Delta\}$ be the inference rules of the deductive definition of $D \in \wp(\mathscr{C})$. Assume that $\langle \wp(\mathscr{C}), \subseteq \rangle \underset{\alpha}{\overset{\gamma}{\leftrightarrows}} \langle \wp(\mathscr{A}), \subseteq \rangle$. Let $\overline{R} = \{\frac{\alpha(P)}{\overline{c}} \mid \frac{P}{c} \in R \wedge \overline{c} \in \alpha(\{c\})\}$ and $\overline{D} \in \wp(\mathscr{A})$ be defined by $\overline{R}$. Then

- $\alpha(D) \subseteq \overline{D}$;
- if $\forall X \subseteq D \,.\, \gamma \circ \alpha(X) \subseteq X$ then $\alpha(D) = \overline{D}$ (this hypothesis on X is necessary only for the iterates $X = F_R^n(\varnothing)$ of the consequence operator F_R of rules R).

Proof By theorem 16.12, $D = \mathsf{lfp}^{\subseteq} F_R$ and $\overline{D} = \mathsf{lfp}^{\subseteq} \overline{F}_{\overline{R}}$ where $F_R(X) \triangleq \{c \mid \exists \frac{P}{c} \in R \,.\, P \subseteq X\}$ and $\overline{F}_{\overline{R}}(Y) \triangleq \{\overline{c} \mid \exists \frac{\alpha(P)}{\overline{c}} \,.\, \frac{P}{c} \in R \wedge \overline{c} \in \alpha(\{c\}) \wedge \alpha(P) \subseteq Y\}$. We have

$$\begin{aligned}
&\alpha(F_R(X))\\
=\;& \alpha(\bigcup\{\{c\} \mid \exists \frac{P}{c} \in R \,.\, P \subseteq X\}) && \text{(definition of } F_R \text{ and } S = \bigcup\{\{x\} \mid x \in S\}\text{)}\\
=\;& \bigcup\{\alpha(\{c\}) \mid \exists \frac{P}{c} \in R \,.\, P \subseteq X\} && \text{(}\alpha \text{ preserves existing joins)}\\
=\;& \{\overline{c} \mid \exists \frac{P}{c} \in R \,.\, \overline{c} \in \alpha(\{c\}) \wedge P \subseteq X\} && \text{(}S = \bigcup\{\{x\} \mid x \in S\}\text{)}\\
=\;& \{\overline{c} \mid \exists \frac{\alpha(P)}{\overline{c}} \in \overline{R} \,.\, P \subseteq X\} && \text{(definition of } \overline{R}\text{)}\\
\subseteq\;& \{\overline{c} \mid \exists \frac{\alpha(P)}{\overline{c}} \in \overline{R} \,.\, \alpha(P) \subseteq \alpha(X)\} && \text{(}\alpha \text{ increasing)}\\
=\;& \overline{F}_{\overline{R}}(\alpha(X)) && \text{(definition of } \overline{F}_{\overline{R}}\text{)}
\end{aligned}$$

proving, by corollary 18.16, that $\alpha(D) = \alpha(\mathsf{lfp}^{\subseteq} F_R) \subseteq \mathsf{lfp}^{\subseteq} \overline{F}_{\overline{R}} = \overline{D}$.

Assume that $\forall X \subseteq D \,.\, \gamma \circ \alpha(X) \subseteq X$. F_R preserves nonempty joins so by Tarski–Kantorovich's fixpoint theorem 15.21, $\mathsf{lfp}^{\subseteq} F_R = \bigcup_{n \in \mathbb{N}} F_R^n(\varnothing)$. We have $\forall n \in \mathbb{N} \,.\, F_R^n(\varnothing) \subseteq \mathsf{lfp}^{\subseteq} F_R = D$ so $\gamma \circ \alpha(F_R^n(\varnothing)) \subseteq F_R^n(\varnothing)$ by hypothesis.

In the preceding proof for $X = F_R^n(\varnothing)$ we have $\alpha(P) \subseteq \alpha(X)$ implies $P \subseteq \gamma \circ \alpha(X) \subseteq X$ so $\alpha(F_R(X)) = \overline{F}_{\overline{R}}(\alpha(X))$ that is $\alpha(F_R^{n+1}(\varnothing)) = \overline{F}_{\overline{R}}(\alpha(F_R^n(\varnothing)))$. By corollary 18.34, we conclude that $\alpha(D) = \alpha(\mathsf{lfp}^{\subseteq} F_R) = \bigcup_{n \in \mathbb{N}} \overline{F}_{\overline{R}}^{\,n}(\alpha(\varnothing)) = \bigcup_{n \in \mathbb{N}} \overline{F}_{\overline{R}}^{\,n}(\varnothing) = \mathsf{lfp}^{\subseteq} \overline{F}_{\overline{R}} = \overline{D}$ (because $\varnothing \subseteq \gamma(\varnothing)$ so $\alpha(\varnothing) \subseteq \varnothing$ and $\alpha(\varnothing) = \varnothing$ by antisymmetry). □

18.9 Inductive and Structural Abstraction

The abstraction of an inductive definition of $D \in S \to \wp(\mathbb{U})$ (where $\langle S, \preccurlyeq\rangle$ ($\lhd$ for structural definition) is well founded) by Galois connection $\langle \wp(\mathbb{U}), \subseteq\rangle \underset{\alpha}{\overset{\gamma}{\leftrightarrows}} \langle \mathscr{A}, \sqsubseteq\rangle$ is $\overline{D} \in S \to \mathscr{A}$ such that $\forall s \in S \,.\, \overline{D}(s) \triangleq \alpha(D(s))$. Each $D(s)$ is defined as a function of the $\langle D(s'), s' \prec s\rangle$ using in general a fixpoint definition or a deductive definition and so the abstraction is obtained by induction using the fixpoint and deductive definition abstraction theorems introduced in this chapter 18.

18.10 Conclusion

Fixpoints are used to define the semantics of iteration (and recursion). The fixpoint abstraction and approximation theorems [235, 241] provide methods for constructing exact or else sound abstractions of the semantics of iteration. The abstraction of fixpoint was introduced in [235, 241] and that of deductive definitions in [250].

18.11 Solutions to Selected Exercises

Solution to exercise 18.5

$\overline{f}'(\overline{x})$

$= \bigvee_{i=1}^{n} \alpha(f(x^i))$ $\wr$decomposition hypothesis $\gamma(\overline{x}) = \bigsqcup_{i=1}^{n} x^i$$\wr$

$= \alpha(\bigsqcup_{i=1}^{n} f(x^i))$ $\wr$the lower adjoint of a Galois connection preserves existing joins$\wr$

$= \alpha(f(\bigsqcup_{i=1}^{n} x^i))$ $\wr f \in \mathscr{C}$ *joinmorphismto*$\mathscr{C}$ preserves existing arbitrary joins$\wr$

$= \alpha \circ f \circ \gamma(\overline{x})$ $\wr$decomposition hypothesis $\gamma(\overline{x}) = \bigsqcup_{i=1}^{n} x^i$ and definition of function composition $\circ$$\wr$

$\preceq \overline{f}(\overline{x})$ $\wr$hypothesis $\alpha \circ f \circ \gamma \dot{\preceq} \overline{f}$.$\wr$

Solution to exercise 18.15 The proof that $\alpha \circ f \circ \gamma \dot{\preccurlyeq} \overline{f} \Leftrightarrow f \circ \gamma \sqsubseteq \gamma \circ \overline{f}$ does not use the fact that f and $\overline{f}$ are increasing.

$\alpha \circ f \circ \gamma \dot{\preccurlyeq} \overline{f}$

$\Rightarrow f \circ \gamma \sqsubseteq \gamma \circ \overline{f}$ $\wr$definition of Galois connections$\wr$

$\Rightarrow \alpha \circ f \circ \gamma \dot{\preccurlyeq} \alpha \circ \gamma \circ \overline{f}$ $\wr$Galois connection so α is increasing$\wr$

$\Rightarrow \alpha \circ f \circ \gamma \dot{\preccurlyeq} \overline{f}$ $\wr$Galois connection so $\alpha \circ \gamma \dot{\preccurlyeq} \mathbb{1}_{\mathscr{A}}$$\wr$

$\alpha \circ f \circ \gamma \dot{\preccurlyeq} \overline{f} \Leftrightarrow \alpha \circ f \dot{\preccurlyeq} \overline{f} \circ \alpha$ holds if $\overline{f}$ is increasing or the Galois connection is a retraction so $\alpha \circ \gamma = \mathbb{1}_{\mathscr{A}}$.

$\alpha \circ f \dot{\preccurlyeq} \overline{f} \circ \alpha$

$\Rightarrow \alpha \circ f \circ \gamma \dot{\preccurlyeq} \overline{f} \circ \alpha \circ \gamma$ $\wr$function application$\wr$

$\Rightarrow \alpha \circ f \circ \gamma \dot{\preccurlyeq} \overline{f}$

$\wr$if $\alpha \circ \gamma = \mathbb{1}_{\mathscr{A}}$ or $\overline{f}$ increasing because $\alpha \circ \gamma \dot{\preccurlyeq} \mathbb{1}_{\mathscr{A}}$ by the Galois connection$\wr$

$\Rightarrow \alpha \circ f \circ \gamma \circ \alpha \dot{\preccurlyeq} \overline{f} \circ \alpha$ $\wr$function application$\wr$

$\Rightarrow \alpha \circ f \dot{\preccurlyeq} \overline{f} \circ \alpha$

$\wr$by $\mathbb{1}_{\mathscr{C}} \sqsubseteq \gamma \circ \alpha$ and α increasing in a Galois connection and f increasing$\wr$

This may not hold when $\overline{f}$ is not increasing, as shown by the following counterexample where $\langle \mathscr{C}, \sqsubseteq \rangle \underset{\alpha}{\overset{\gamma}{\leftrightarrows}} \langle \mathscr{A}, \preccurlyeq \rangle$, $f \in \mathscr{C} \xrightarrow{\nearrow} \mathscr{C}$, $\alpha \circ f \dot{\preccurlyeq} \overline{f} \circ \alpha$ but $\alpha \circ f \circ \gamma(a) \not\preccurlyeq \overline{f}(a)$.

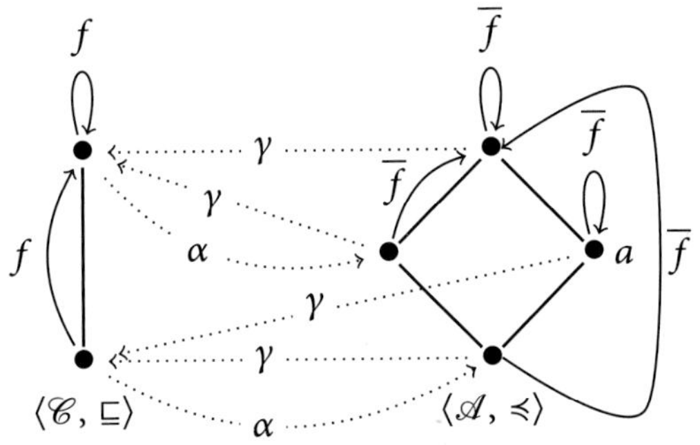

Solution to exercise 18.17 If $\overline{f}(y) \preccurlyeq y$ then $f(\gamma(y)) \sqsubseteq \gamma(\overline{f}(y)) \sqsubseteq \gamma(y)$ by semicommutation $f \circ \gamma \mathrel{\dot{\sqsubseteq}} \gamma \circ \overline{f}$ and γ increasing. So $\gamma(y) \in \{x \mid f(x) \sqsubseteq x\}$ proving, by Tarski's fixpoint theorem 15.6 and definition of the lub, that $\mathsf{lfp}^{\sqsubseteq} f = \sqcup\{x \mid f(x) \sqsubseteq x\} \sqsubseteq \gamma(y)$. This hold for any fixpoint y of $\overline{f}$, if any, by reflexivity.

Solution to exercise 18.19 Let $\langle f^n, n \in \mathbb{N}\rangle$ and $\langle \overline{f}^n, n \in \mathbb{N}\rangle$ be the iterates of f and $\overline{f}$. By recurrence, $\langle f^n, n \in \mathbb{N}\rangle \subseteq \mathcal{X}$. $\bot = f^0(\bot) \sqsubseteq \gamma(\overline{f}^0(0))$. Assume, by induction hypothesis that $\alpha(f^n(\bot)) \preccurlyeq \overline{f}^n(0)$. Then $\alpha(f^{n+1}(\bot)) = \alpha(f(f^n(\bot))) \preccurlyeq \overline{f}(\alpha(f^n(\bot))) \preccurlyeq \overline{f}(\overline{f}^n(0)) = \overline{f}^{n+1}(0)$ by definition of the iterates, semicommutation for elements of $\mathcal{X}$, $\overline{f}$ increasing, and definition of the iterates. Passing to the limit for increasing chains $\alpha(\mathsf{lfp}^{\sqsubseteq} f) = \alpha(\bigsqcup f^n(\bot)) = \curlyvee \alpha(f^n(\bot)) \preccurlyeq \curlyvee \overline{f}^n(0)$.

Solution to exercise 18.30 $\langle \wp(S \times S, \subseteq\rangle$ is a complete lattice. F_r preserves nonempty lubs so is continuous so, by Scott–Kleene's iterative fixpoint theorem 15.26, $r^* \triangleq \mathsf{lfp}^{\subseteq} F_r = \bigcup_{n \geqslant 0} r^n$ where $r^0 = \triangleq \mathbb{1}_S$ and $r^{n+1} \triangleq r^n \circ r$.

Then $\alpha_P(F_r(X)) = \alpha_P(\mathbb{1}_S \cup X \circ r) = \alpha_P(\mathbb{1}_S) \cup \alpha_P(X \circ r) = P \cup \mathsf{post}[X \circ r]P = P \cup \{y \mid \exists x \in P \,.\, \langle x, y\rangle \in X \circ r\} = P \cup \{y \mid \exists x \in P \,.\, \exists z \,.\, \langle x, z\rangle \in X \wedge \langle z, y\rangle \in r\} = P \cup \{y \mid \exists z \in \{z \mid \exists x \in P \,.\, \langle x, z\rangle \in X\} \,.\, \langle z, y\rangle \in r\} = P \cup \{y \mid \exists z \in \mathsf{post}[X]P \,.\, \langle z, y\rangle \in r\} = P \cup \mathsf{post}[r](\alpha_P(X))$ which, by theorem 18.23, provides a fixpoint characterization of $\alpha_P(r^*) = \mathsf{post}[r^*]P = \mathsf{lfp}^{\subseteq} X \mapsto P \cup \mathsf{post}[r]X$.

Solution to exercise 18.43 By $\gamma(\mathbf{0}) = \bot$, we have $\gamma(\overline{f}^0(\mathbf{0})) = f^0(\bot)$. By recurrence using $f \circ \gamma = \gamma \circ \overline{f}$, we have $\forall n \in \mathbb{N} \,.\, \gamma(\overline{f}^n(\mathbf{0})) = f^n(\bot)$. Because $\mathbf{0}$ is the infimum and $\overline{f}$ is increasing, the abstract iterates $\langle \overline{f}^n(\mathbf{0}), n \in \mathbb{N}\rangle$ form an increasing chain. By the ascending chain condition $\exists \ell \in \mathbb{N} \,.\, \forall n \geqslant \ell \,.\, \overline{f}^n(\mathbf{0}) = \overline{f}^\ell(\mathbf{0}) = \mathsf{lfp}^{\preccurlyeq} \overline{f}$. It follows, by theorem 15.26 and γ increasing, that $\mathsf{lfp}^{\sqsubseteq} f = \bigsqcup_{n \in \mathbb{N}} f^n(\bot) = \bigsqcup_{n \in \mathbb{N}} \gamma(\overline{f}^n(\mathbf{0})) = \bigsqcup_{n \in \mathbb{N}} \gamma(\overline{f}^\ell(\mathbf{0})) = \gamma(\overline{f}^\ell(\mathbf{0})) = \gamma(\mathsf{lfp}^{\preccurlyeq} \overline{f}))$.

VI REACHABILITY SEMANTICS

The forward/deductive *reachability semantics* of a syntactically correct program is an abstraction of its trace semantics specifying the set of states reachable at any program point for all executions of the program (chapter 19). It is an *invariant* property of the program executions.

Because it is an abstraction of the trace semantics, it can be formally derived from this semantics by *calculational design* (chapter 20).

19 Structural Forward Reachability Semantics

The (forward/deductive) reachability semantics of a program is an abstraction of the prefix trace semantics. The assertional reachability semantics $\vec{\mathcal{S}}^{r}[\![\mathsf{S}]\!]$ attaches to each program point $\ell \in \mathsf{labs}[\![\mathsf{S}]\!]$ of a statement S a set $\vec{\mathcal{S}}^{r}[\![\mathsf{S}]\!]\,\ell$ of environments that may be encountered during any execution of the statement S reaching that program point. The environments record the values of the program variables whenever reaching that point. The relational reachability semantics $\vec{\mathcal{S}}^{R}[\![\mathsf{S}]\!]$ is a variant where the values of variables at a program point are related to the initial values of variables.

Contents

19.1 Assertional Versus Relational Forward Reachability Semantics

We illustrate the difference between assertional and relational forward reachability semantics on the very simple program $\mathsf{P} = \mathtt{while}\,\ell_1\,(\mathsf{x} < 10)\,\ell_2\;\mathsf{x} = \mathsf{x} + 1\,;$.

19.1.1 Assertional Forward Reachability Semantics

Given a set $\mathcal{P}_0$ of initial environments, the assertional forward reachability semantics $\vec{\mathcal{S}}^{r}[\![\mathsf{S}]\!]\,\mathcal{P}_0\,\ell$ maps the set $\mathcal{P}_0$ of initial environments to the set of reachable environments at each program point ℓ of program component $\mathsf{S} \in \mathbb{P}\mathbb{c}$ (i.e., program, statement, or statement list). Therefore $\vec{\mathcal{S}}^{r}[\![\mathsf{S}]\!] \in \wp(\mathbb{E}\mathbb{v}) \rightarrow (\mathbb{L} \rightarrow \wp(\mathbb{E}\mathbb{v}))$.

Example 19.1 Consider the program $\mathsf{P} = \mathtt{while}\,\ell_1\,(\mathsf{x} < 10)\,\ell_2\;\mathsf{x} = \mathsf{x} + 1\,;\,\ell_3$ where all variables are initially 0. At ℓ_1, $0 \leqslant \mathsf{x} \leqslant 10$, at ℓ_2, $0 \leqslant \mathsf{x} < 10$, and at ℓ_3, $\mathsf{x} = 10$.

This is usually specified by inserting the logical propositions (called *local invariants*) at the corresponding program points.

```
/* x=0 (initialization hypothesis) */
while ℓ1 (x < 10) /* 0⩽x⩽10 (loop invariant) */
  ℓ2 /* 0⩽x<10 */
    x = x + 1;
ℓ3 /* x=10 */
```

Representing such logical propositions by sets of environments, we have

ℓ	$\boldsymbol{\mathcal{S}}^{\vec{r}}[\![\mathtt{S}]\!]\,\mathcal{P}_0\,\ell$ where $\mathcal{P}_0 = \{\rho \in \mathbb{E}\mathrm{v} \mid \forall \mathtt{y} \in \mathbb{V}\,.\,\rho(\mathtt{y}) = 0\}$
ℓ_1	$\{\rho \in \mathbb{E}\mathrm{v} \mid 0 \leqslant \rho(\mathtt{x}) \leqslant 10 \wedge \forall \mathtt{y} \in \mathbb{V} \setminus \{\mathtt{x}\}\,.\,\rho(\mathtt{y}) = 0\}$
ℓ_2	$\{\rho \in \mathbb{E}\mathrm{v} \mid 0 \leqslant \rho(\mathtt{x}) < 10 \wedge \forall \mathtt{y} \in \mathbb{V} \setminus \{\mathtt{x}\}\,.\,\rho(\mathtt{y}) = 0\}$
ℓ_3	$\{\rho \in \mathbb{E}\mathrm{v} \mid \rho(\mathtt{x}) = 10 \wedge \forall \mathtt{y} \in \mathbb{V} \setminus \{\mathtt{x}\}\,.\,\rho(\mathtt{y}) = 0\}$

This forward reachability semantics is considered by Alan Turing [933], Peter Naur [742], Bob Floyd [376], and Tony Hoare [501] proof methods (see part VIII, "Specification and Verification").

Exercise 19.2 Provide the assertional local invariants for the program (4.5).

19.1.2 Relational Forward Reachability Semantics

The forward relational reachability semantics $\boldsymbol{\mathcal{S}}^{\vec{R}}[\![\mathtt{S}]\!]$ relates the current environment at a program point to the initial environment. So $\boldsymbol{\mathcal{S}}^{\vec{R}}[\![\mathtt{S}]\!] \in \wp(\mathbb{E}\mathrm{v} \times \mathbb{E}\mathrm{v}) \to (\mathbb{L} \to \wp(\mathbb{E}\mathrm{v} \times \mathbb{E}\mathrm{v}))$.

Example 19.3 Consider again the program $\mathtt{P} = \mathtt{while}\,\ell_1\,(\mathtt{x} < 10)\,\ell_2\ \mathtt{x} = \mathtt{x} + 1\,;\ell_3$, assuming that all variables y are initialized to some unknown value symbolically denoted $\mathtt{y}_0$. We have at ℓ_1, $(10 \leqslant \mathtt{x}_0 = \mathtt{x}) \vee (\mathtt{x}_0 \leqslant \mathtt{x} \leqslant 10)$, at ℓ_2, $\mathtt{x}_0 \leqslant \mathtt{x} < 10$, and at ℓ_3, $(10 \leqslant \mathtt{x}_0 = \mathtt{x}) \vee (\mathtt{x}_0 < 10 \wedge \mathtt{x} = 10)$.

Again, this is usually specified by inserting local invariants in the program.

```
/* x=x0 (initialization hypothesis) */
while ℓ1 (x < 10) /* (10⩽x0=x) ∨ (x0⩽x⩽10) (loop invariant) */
  ℓ2 /* x0⩽x<10 */
    x = x + 1;
ℓ3 /* (10⩽x0=x) ∨ (x0<10 ∧ x=10) */
```

Representing such logical propositions by a binary relation between environments, we have

ℓ	$\boldsymbol{\mathcal{S}}^{\vec{R}}[\![\mathtt{S}]\!]\,\mathcal{P}_0\,\ell$ where $\mathcal{P}_0 = \{\langle\rho_0, \rho\rangle \in \mathbb{E}\mathrm{v} \times \mathbb{E}\mathrm{v} \mid \rho = \rho_0\}$
ℓ_1	$\{\langle\rho_0, \rho\rangle \in \mathbb{E}\mathrm{v} \times \mathbb{E}\mathrm{v} \mid (10 \leqslant \rho_0(\mathtt{x}) = \rho(\mathtt{x})) \vee (\rho_0(\mathtt{x}) \leqslant \rho(\mathtt{x}) \leqslant 10) \wedge \forall \mathtt{y} \in \mathbb{V} \setminus \{\mathtt{x}\}\,.\,\rho(\mathtt{y}) = \rho_0(\mathtt{y})\}$
ℓ_2	$\{\langle\rho_0, \rho\rangle \in \mathbb{E}\mathrm{v} \times \mathbb{E}\mathrm{v} \mid \rho_0(\mathtt{x}) \leqslant \rho(\mathtt{x}) < 10 \wedge \forall \mathtt{y} \in \mathbb{V} \setminus \{\mathtt{x}\}\,.\,\rho(\mathtt{y}) = \rho_0(\mathtt{y})\}$
ℓ_3	$\{\langle\rho_0, \rho\rangle \in \mathbb{E}\mathrm{v} \times \mathbb{E}\mathrm{v} \mid (10 \leqslant \rho_0(\mathtt{x}) = \rho(\mathtt{x})) \vee (\rho_0(\mathtt{x}) \leqslant \rho(\mathtt{x}) \leqslant 10) \wedge \forall \mathtt{y} \in \mathbb{V} \setminus \{\mathtt{x}\}\,.\,\rho(\mathtt{y}) = \rho_0(\mathtt{y})\}$

This is the forward reachability semantics considered by Zohar Manna [647].

Remark 19.4 The forward relational reachability semantics $\boldsymbol{\mathcal{S}}^{\vec{\mathsf{R}}}[\![\mathsf{S}]\!]$ can be simulated by the forward assertional reachability semantics $\boldsymbol{\mathcal{S}}^{\vec{\mathsf{r}}}[\![\mathsf{S}]\!]$ by introducing initial auxiliary variables. An equivalence proof is given in [242, theorem 10.0.4]. In our example, we would rewrite `x0 = x ; while (x < 10) x = x + 1;`.

Exercise 19.5 Provide the relational local invariants for the program (4.5).

Exercise 19.6 Provide the relational local invariants for the program

```
while ℓ1 (x > 0) {                    (19.7)
   ℓ2 x = x + 1;
   if ℓ3 (x > 9)
       ℓ4 break;
} ℓ5
```

19.1.3 Forward Reachability Semantics

In order to reason simultaneously on both assertional and relational reachability semantics in a wording, we write $\boldsymbol{\mathcal{S}}^{\vec{\varrho}}[\![\mathsf{S}]\!]$, $\mathbb{E}\mathbb{v}^{\vec{\varrho}}$, ... to mean either $\boldsymbol{\mathcal{S}}^{\vec{\mathsf{r}}}[\![\mathsf{S}]\!]$, $\mathbb{E}\mathbb{v}$, ... for $\vec{\varrho} = \vec{\mathsf{r}}$ or $\boldsymbol{\mathcal{S}}^{\vec{\mathsf{R}}}[\![\mathsf{S}]\!]$, $\mathbb{E}\mathbb{v} \times \mathbb{E}\mathbb{v}$, ... for $\vec{\varrho} = \vec{\mathsf{R}}$, consistently in this wording. For a given program component, we have

$$\boldsymbol{\mathcal{S}}^{\vec{\varrho}}[\![\mathsf{S}]\!] \quad \in \quad \wp(\mathbb{E}\mathbb{v}^{\vec{\varrho}}) \to (\mathbb{L} \to \wp(\mathbb{E}\mathbb{v}^{\vec{\varrho}})).$$

Therefore if $\mathscr{P}_0 \in \wp(\mathbb{E}\mathbb{v}^{\vec{\varrho}})$ is a precondition and $\ell \in \mathbb{L}$ is the program label then $\boldsymbol{\mathcal{S}}^{\vec{\varrho}}[\![\mathsf{S}]\!]\,\mathscr{P}_0\,\ell$ is an invariant at ℓ that holds if and when execution of the program component S started with an initial state satisfying the precondition $\mathscr{P}_0$ reaches program point ℓ. Formally,

$$\begin{aligned}
\boldsymbol{\mathcal{S}}^{\vec{\mathsf{r}}}[\![\mathsf{S}]\!]\,\mathscr{P}_0\,\ell &\triangleq \{\boldsymbol{\varrho}(\pi_0\ell_0\pi_1\ell') \mid \boldsymbol{\varrho}(\pi_0\ell_0) \in \mathscr{P}_0 \wedge \exists \pi_2\,.\,\ell_0\pi_1\ell'\pi_2 \in \boldsymbol{\mathcal{S}}^*[\![\mathsf{S}]\!](\pi_0\ell_0) \wedge \ell' = \ell\}\\
\boldsymbol{\mathcal{S}}^{\vec{\mathsf{R}}}[\![\mathsf{S}]\!]\,\mathscr{P}_0\,\ell &\triangleq \{\langle\rho_0, \boldsymbol{\varrho}(\pi_0\ell_0\pi_1\ell')\rangle \mid \langle\rho_0, \boldsymbol{\varrho}(\pi_0\ell_0)\rangle \in \mathscr{P}_0 \wedge\\
&\qquad\qquad \exists \pi_2\,.\,\ell_0\pi_1\ell'\pi_2 \in \boldsymbol{\mathcal{S}}^*[\![\mathsf{S}]\!](\pi_0\ell_0) \wedge \ell' = \ell\}
\end{aligned}$$

This formal definition is difficult to work with for reasoning about program semantics, so we look for an equivalent structural definition $\widehat{\boldsymbol{\mathcal{S}}}^{\vec{\varrho}}[\![\mathsf{S}]\!] = \boldsymbol{\mathcal{S}}^{\vec{\varrho}}[\![\mathsf{S}]\!]$.

Exercise 19.8 Show that in the preceding definition of $\boldsymbol{\mathcal{S}}^{\vec{\varrho}}[\![\mathsf{S}]\!]$, $\boldsymbol{\mathcal{S}}^*[\![\mathsf{S}]\!]$ can be replaced by the equivalent $\boldsymbol{\mathcal{S}}^{+\infty}[\![\mathsf{S}]\!]$.

Exercise 19.9 Show that $\boldsymbol{\mathcal{S}}^{\vec{\mathsf{r}}}[\![\mathsf{S}]\!]$ is an abstraction of $\boldsymbol{\mathcal{S}}^{\vec{\mathsf{R}}}[\![\mathsf{S}]\!]$.

19.2 Example for Assignment and Iteration

We first consider the forward reachability semantics of assignments and iterations that we introduce by means of very simple examples.

19.2.1 Forward Reachability Semantics of an Assignment

19.2.1.1 Environment assignment. Let us define the assignment $\rho[\mathsf{x} \leftarrow v]$ of a value $v \in \mathbb{V}$ to a variable $\mathsf{x} \in \mathbb{V}$ in an environment $\rho \in \mathbb{E}\mathbb{v}$.

$$\begin{aligned}
\rho[\mathsf{x} \leftarrow v](\mathsf{x}) &\triangleq v && (19.10)\\
\rho[\mathsf{x} \leftarrow v](\mathsf{y}) &\triangleq \rho(\mathsf{y}) \quad \text{when} \quad \mathsf{x} \neq \mathsf{y}
\end{aligned}$$

Example 19.11 If ρ encodes the values of variables before the assignment $\mathsf{x} = \mathtt{0}$; then $\rho[\mathsf{x} \leftarrow 0](\mathsf{x})$ is the value of x after the assignment while the value of the other variables is unchanged.

The value of x after the assignment $\mathsf{x} = \mathsf{x} + 1$; is the value $\rho(\mathsf{x})$ of x before the assignment incremented by 1, that is, $\rho(\mathsf{x}) + 1$. Hence the environment after the assignment is $\rho[\mathsf{x} \leftarrow \rho(\mathsf{x}) + 1](\mathsf{x})$ so that $\rho[\mathsf{x} \leftarrow \rho(\mathsf{x}) + 1](\mathsf{x}) = \rho(\mathsf{x}) + 1$ while the value of the other variables $\rho[\mathsf{x} \leftarrow \rho(\mathsf{x}) + 1](\mathsf{y}) = \rho(\mathsf{y})$ is unchanged.

19.2.1.2 Forward reachability semantics of an assignment statement. The forward reachability semantics of an assignment statement $\mathsf{S} ::= \mathsf{x} = \mathsf{A}$; maps the initial environments to themselves on entry and to themselves as modified by the assignment on exit. The set of reachable environments of a statement S not at a label in that statement is empty.

Reachability of an assignment statement $\mathsf{S} ::= \mathsf{x} = \mathsf{A}$;

$$\begin{aligned}
\widehat{\boldsymbol{\mathcal{S}}}^{\vec{\varrho}}[\![\mathsf{S}]\!]\,\mathcal{P}_0\,\ell \;&=\; (\!|\, \ell = \mathsf{at}[\![\mathsf{S}]\!] \;?\; \mathcal{P}_0 \\
&\qquad [\!]\; \ell = \mathsf{after}[\![\mathsf{S}]\!] \;?\; \mathsf{assign}^{\vec{\varrho}}[\![\mathsf{x}, \mathsf{A}]\!]\,\mathcal{P}_0 \\
&\qquad :\, \varnothing \,|\!) \qquad\qquad (19.12)\\
\mathsf{assign}^{\vec{r}}[\![\mathsf{x}, \mathsf{A}]\!]\,\mathcal{P}_0 \;&\triangleq\; \{\rho[\mathsf{x} \leftarrow \boldsymbol{\mathcal{A}}[\![\mathsf{A}]\!]\rho] \mid \rho \in \mathcal{P}_0\} \\
\mathsf{assign}^{\vec{R}}[\![\mathsf{x}, \mathsf{A}]\!]\,\mathcal{P}_0 \;&\triangleq\; \{\langle \rho_0, \rho[\mathsf{x} \leftarrow \boldsymbol{\mathcal{A}}[\![\mathsf{A}[\rho]\!]\rangle \mid \langle \rho_0, \rho\rangle \in \mathcal{P}_0\}
\end{aligned}$$

Example 19.13

$$\begin{aligned}
&\widehat{\boldsymbol{\mathcal{S}}}^{\vec{r}}[\![\ell_1\ \mathsf{x} = \mathsf{x} + 1\ ;\ell_2]\!]\ \{\rho \mid \rho(\mathsf{x}) = 0\}\ \ell_2 \\
\triangleq\;& \mathsf{assign}^{\vec{r}}[\![\mathsf{x}, \mathsf{x} + 1]\!]\ \{\rho \mid \rho(\mathsf{x}) = 0\} && \text{(definition (19.12) of } \widehat{\boldsymbol{\mathcal{S}}}^{\vec{r}}\text{)} \\
\triangleq\;& \{\rho[\mathsf{x} \leftarrow \boldsymbol{\mathcal{A}}[\![\mathsf{x} + 1]\!]\rho] \mid \rho \in \{\rho \mid \rho(\mathsf{x}) = 0\}\} && \text{(definition (19.12) of } \mathsf{assign}^{\vec{r}}\text{)} \\
\triangleq\;& \{\rho[\mathsf{x} \leftarrow \rho(\mathsf{x}) + 1] \mid \rho(\mathsf{x}) = 0\} \\
& && \text{(definition of } \in \text{ and semantics of arithmetic expressions in section 3.6)} \\
=\;& \{\rho[\mathsf{x} \leftarrow 1] \mid \rho \in \mathbb{E}\mathbb{v}\} && \text{(mathematical definition of } + \text{)}
\end{aligned}$$

Exercise 19.14 Calculate $\widehat{\boldsymbol{\mathcal{S}}}^{\vec{R}}[\![\ell_1\ \mathsf{x} = \mathsf{x} + 1\ ;\ell_2]\!]\ \{\langle \rho_0, \rho_0\rangle \mid \rho_0 \in \mathbb{E}\mathbb{v}\}\ \ell_2$.

19.2.2 Forward Reachability Semantics of an Iteration

The forward reachability semantics of an iteration is obtained by reasoning on the reachable environments after at most n iterations. This is the entry environments $\mathcal{P}_0$ for $n = 0$. For $n + 1$ this is either the entry environments $\mathcal{P}_0$ or the reachable environments after at most n iterations which pass the iteration condition and are modified by the loop body.

Example 19.15 Consider again the program $\mathsf{P} = \mathtt{while}\ \ell_1\ (\mathsf{x} < \mathtt{10})\ \ell_2\ \mathsf{x} = \mathsf{x} + \mathtt{1}\ ;\ \ell_3$ where all variables are initially 0. We let $R^n_{\ell_1}$ be the reachable values of x at ℓ_1 after at most $n \geqslant 0$ iterations.

- The initial value $\mathsf{x}=0$ is reachable at ℓ_1 on iteration entry, that is at iteration 0. So $R^0_{\ell_1} = \mathscr{P}_0 = \{0\}$
- After at most 1 iteration, the reachable values $R^1_{\ell_1}$ of x at ℓ_1 are those $\mathscr{P}_0$ reachable at iteration 0 plus those of iteration 0 which pass the test and have been incremented in the loop body. So $R^1_{\ell_1} = \mathscr{P}^0 \cup \{x+1 \mid x \in R^0_{\ell_1} \wedge x < 10\} = \{0, 1\}$;
- ...
- Similarly, $R^9_{\ell_1} = \{0, 1, \dots, 9\}$. Then, after at most 10 iterations, the reachable values $R^{10}_{\ell_1}$ of x at ℓ_1 are those $\mathscr{P}_0$ reachable at iteration 0 plus those $R^9_{\ell_1}$ of previous iterations that pass the test and have been incremented in the loop body. Hence
$$\begin{aligned} R^{10}_{\ell_1} &= \mathscr{P}^0 \cup \{x+1 \mid x \in R^9_{\ell_1} \wedge x < 10\} \\ &= \{0\} \cup \{1, \dots, 10\} = \{0, 1, \dots, 10\}; \end{aligned}$$
- After at most 11 iterations, the reachable values $R^{11}_{\ell_1}$ of x at ℓ_1 are those $\mathscr{P}_0$ reachable at iteration 0 plus those $R^{10}_{\ell_1}$ of previous iterations which pass the test and have been incremented in the loop body. So
$$\begin{aligned} R^{11}_{\ell_1} &= \mathscr{P}^0_{\ell_1} \cup \{x+1 \mid x \in R^{10}_{\ell_1} \wedge x < 10\} \\ &= \{0\} \cup \{1, \dots, 10\} = \{0, 1, \dots, 10\} = R^{10}_{\ell_1}; \end{aligned}$$
- Similarly, after at most $n > 10$ iterations, $R^n_{\ell_1} = R^{10}_{\ell_1}$.

Therefore we have
$$\begin{aligned} R^0_{\ell_1} &= \mathscr{P}^0 \\ R^{n+1}_{\ell_1} &= \mathscr{P}^0 \cup \{x+1 \mid x \in R^n_{\ell_1} \wedge x < 10\} \end{aligned}$$
with $R^0_{\ell_1} \subseteq R^1_{\ell_1} \subseteq \dots R^n_{\ell_1} \subseteq R^{n+1}_{\ell_1} \subseteq \dots$. Letting $R'^0_{\ell_1} = \varnothing$ and $R'^{n+1}_{\ell_1} = R^n_{\ell_1}$, this is the same as
$$\begin{aligned} R'^0_{\ell_1} &= \varnothing \\ R'^{n+1}_{\ell_1} &= \mathscr{P}^0 \cup \{x+1 \mid x \in R'^n_{\ell_1} \wedge x < 10\} \end{aligned}$$
with $R'^0_{\ell_1} \subseteq R'^1_{\ell_1} \subseteq \dots R'^n_{\ell_1} \subseteq R'^{n+1}_{\ell_1} \subseteq \dots$, which limit is the set of reachable values $\bigcup R'^n_{\ell_1} = \{0, 1, \dots, 10\}$ of x at ℓ_1. Obviously, the function $\mathsf{F}(R') \triangleq \mathscr{P}^0_{\ell_1} \cup \{x+1 \mid x \in R' \wedge x < 10\}$ preserves arbitrary joins $\bigcup$, so that, by theorem 15.26, the reachable values of x at ℓ_1 are $\hat{\boldsymbol{\mathcal{S}}}^{\vec{r}}[\![\mathsf{P}]\!]\, \mathscr{P}^0\, \ell_1 = \mathsf{lfp}^{\subseteq}\, \mathsf{F}$.

- The reachable values of x at ℓ_3 on loop exit are those reachable at ℓ_1 that do not pass the test, that is
$$\hat{\boldsymbol{\mathcal{S}}}^{\vec{r}}[\![\mathsf{P}]\!]\, \mathscr{P}^0\, \ell_3 = \{x \in \hat{\boldsymbol{\mathcal{S}}}^{\vec{r}}[\![\mathsf{P}]\!]\, \mathscr{P}^0\, \ell_1 \mid x \geqslant 10\} = \{10\}.$$

Generalizing from example 19.15, the forward reachability semantics of an iteration statement $\mathsf{S} ::= \mathtt{while}\,\ell\,(\mathsf{B})\,\mathsf{S}_b$, is a fixpoint for which successive iterates provide the reachable environments after at most 0, 1, ... or more iterations in the loop.

If the loop terminates, the fixpoint iterates beyond the maximal number of iterations add no new reachable environments.

If the loop does not terminate, the reachable environments after any number of iterations are obtained by passing to the limit in the fixpoint iterates.

Reachability of an iteration statement $\mathsf{S} ::= \texttt{while}\ \ell\ (\mathsf{B})\ \mathsf{S}_b$

$$\widehat{\mathcal{S}}^{\vec{e}}[\![\mathsf{S}]\!]\,\mathcal{P}_0\,\ell' \;=\; (\mathsf{lfp}^{\subseteq}\,\mathcal{F}^{\vec{e}}[\![\texttt{while}\ \ell\ (\mathsf{B})\ \mathsf{S}_b]\!]\,\mathcal{P}_0)\,\ell' \tag{19.16}$$

$$\mathcal{F}^{\vec{e}}[\![\texttt{while}\ \ell\ (\mathsf{B})\ \mathsf{S}_b]\!]\,\mathcal{P}_0 \;\in\; (\mathbb{L} \to \wp(\mathbb{E}\mathrm{v}^{\vec{e}})) \xrightarrow{\nearrow} (\mathbb{L} \to \wp(\mathbb{E}\mathrm{v}^{\vec{e}}))$$

$$\mathcal{F}^{\vec{e}}[\![\texttt{while}\ \ell\ (\mathsf{B})\ \mathsf{S}_b]\!]\,\mathcal{P}_0\,X\,\ell' \;=$$

$$\begin{aligned}
&(\!\!|\ \ell' = \ell \ ?\ \mathcal{P}_0 \cup \widehat{\mathcal{S}}^{\vec{e}}[\![\mathsf{S}_b]\!]\,(\mathsf{test}^{\vec{e}}[\![\mathsf{B}]\!]X(\ell))\,\ell \\
&\|\ \ell' \in \mathsf{in}[\![\mathsf{S}_b]\!] \setminus \{\ell\}\ ?\ \widehat{\mathcal{S}}^{\vec{e}}[\![\mathsf{S}_b]\!]\,(\mathsf{test}^{\vec{e}}[\![\mathsf{B}]\!]X(\ell))\,\ell' \\
&\|\ \ell' = \mathsf{after}[\![\mathsf{S}]\!]\ ?\ \overline{\mathsf{test}}^{\vec{e}}[\![\mathsf{B}]\!](X(\ell)) \cup \bigcup_{\ell'' \in \mathsf{breaks\text{-}of}[\![\mathsf{S}_b]\!]} \widehat{\mathcal{S}}^{\vec{e}}[\![\mathsf{S}_b]\!]\,(\mathsf{test}^{\vec{e}}[\![\mathsf{B}]\!]X(\ell))\,\ell'' \\
&\ :\ \emptyset\ |\!\!)
\end{aligned}$$

where

$$\begin{aligned}
\mathsf{test}^{r}[\![\mathsf{B}]\!]\mathcal{P} &\triangleq \{\rho \in \mathcal{P} \mid \mathcal{B}[\![\mathsf{B}]\!]\rho = \mathsf{tt}\} \\
\mathsf{test}^{R}[\![\mathsf{B}]\!]\mathcal{P} &\triangleq \{\langle \rho_0, \rho\rangle \in \mathcal{P} \mid \mathcal{B}[\![\mathsf{B}]\!]\rho = \mathsf{tt}\} \\
\overline{\mathsf{test}}^{r}[\![\mathsf{B}]\!]\mathcal{P} &\triangleq \{\rho \in \mathcal{P} \mid \mathcal{B}[\![\mathsf{B}]\!]\rho = \mathsf{ff}\} \\
\overline{\mathsf{test}}^{R}[\![\mathsf{B}]\!]\mathcal{P} &\triangleq \{\langle \rho_0, \rho\rangle \in \mathcal{P} \mid \mathcal{B}[\![\mathsf{B}]\!]\rho = \mathsf{ff}\}
\end{aligned}$$

The reachable environments are the entry environments or the reachable environments after one more iteration from reachable environments with test true, or the final environments which are reachable either from environments with test false or from breaking environments reachable from reachable environments with test true.

Remark 19.17 Following remark 17.8, notice that in (19.16), ℓ is a mathematical variable standing for $\mathsf{at}[\![\mathsf{S}]\!]$ (computer scientists would say that it is a constant). ℓ' is a mathematical variable standing for any program point in $\ell' \in \mathbb{L}$ (computer scientists would say that is is a variable or a formal parameter). ℓ'' is a mathematical variable standing for any program point before a break $\ell'' \in \mathsf{breaks\text{-}of}[\![\mathsf{S}_b]\!]$ (computer scientists would say that is is a local or dummy variable).

Exercise 19.18 (loop invariant) In (19.16), $\mathcal{F}^{\vec{e}}[\![\texttt{while}\ \ell\ (\mathsf{B})\ \mathsf{S}_b]\!]\,\mathcal{P}_0 \in (\mathbb{L} \to \wp(\mathbb{E}\mathrm{v}^{\vec{e}})) \xrightarrow{\nearrow} (\mathbb{L} \to \wp(\mathbb{E}\mathrm{v}^{\vec{e}}))$ but uses only $X(\ell)$ where ℓ is the loop entry point.

Show that $\widehat{\mathcal{S}}^{\vec{e}}[\![\texttt{while}\ \ell\ (\mathsf{B})\ \mathsf{S}_b]\!]\,\mathcal{P}_0\,\ell'$ can be defined using a least fixpoint of a transformer $\overline{\mathcal{F}}^{\vec{e}}[\![\texttt{while}\ \ell\ (\mathsf{B})\ \mathsf{S}_b]\!]\,\mathcal{P}_0 \in \wp(\mathbb{E}\mathrm{v}^{\vec{e}}) \xrightarrow{\nearrow} \wp(\mathbb{E}\mathrm{v}^{\vec{e}})$.

19.3 Structural Forward Reachability Semantics

We define the forward reachability semantics of a program P by structural induction on the syntax of program P. The definition is postulated, that is, claimed to be true without justification. This justification will be presented in chapter 20, "Calculational Design of the Forward Reachability Semantics."

The forward reachability semantics of assignment has already been postulated at (19.12) and that of iteration at (19.16).

The forward reachability semantics of a program $\mathsf{P} ::= \mathtt{Sl}\ \ell'$ is that of its statement list `Sl`.

Reachability of a program $\mathsf{P} ::= \mathtt{Sl}\ \ell'$

$$\hat{\mathcal{S}}^{\vec{e}}[\![\mathsf{P}]\!] \triangleq \hat{\mathcal{S}}^{\vec{e}}[\![\mathtt{Sl}]\!] \tag{19.19}$$

The empty statement list $\mathtt{Sl} ::= \epsilon$ does not involve any computation so that the reachable environments are the initial environments.

Reachability of an empty statement list $\mathtt{Sl} ::= \epsilon$

$$\hat{\mathcal{S}}^{\vec{e}}[\![\mathtt{Sl}]\!]\mathcal{P}_0\,\ell = (\!|\, \ell = \mathsf{at}[\![\mathtt{Sl}]\!] \,?\, \mathcal{P}_0 \,\mathbin{\text{:}}\, \varnothing \,|\!) \tag{19.20}$$

The forward reachability semantics of a skip statement $\mathsf{S} ::= ;$ maps initial environments to themselves because they are not modified.

Reachability of a skip statement $\mathsf{S} ::= ;$

$$\hat{\mathcal{S}}^{\vec{e}}[\![\mathsf{S}]\!]\,\mathcal{P}_0\,\ell = (\!|\, \ell \in \{\mathsf{at}[\![\mathsf{S}]\!], \mathsf{after}[\![\mathsf{S}]\!]\} \,?\, \mathcal{P}_0 \,\mathbin{\text{:}}\, \varnothing \,|\!) \tag{19.21}$$

The forward reachability semantics of a conditional statement $\mathsf{S} ::= \mathtt{if}\,(\mathsf{B})\,\mathsf{S}_t$ is the set of reachable environments in S_t under the hypothesis that the test B is true and, after the conditional statement, the union of the environments reaching the end of the true alternative S_t and those for which the conditional statement is skipped when the test B is false.

Reachability of a conditional statement $\mathsf{S} ::= \mathtt{if}\,(\mathsf{B})\,\mathsf{S}_t$

$$\begin{aligned}\hat{\mathcal{S}}^{\vec{e}}[\![\mathsf{S}]\!]\,\mathcal{P}_0\,\ell \;=\;& (\!|\, \ell = \mathsf{at}[\![\mathsf{S}]\!] \,?\, \mathcal{P}_0 \\ & [\!]\, \ell \in \mathsf{in}[\![\mathsf{S}_t]\!] \,?\, \hat{\mathcal{S}}^{\vec{e}}[\![\mathsf{S}_t]\!]\,(\mathsf{test}^{\vec{e}}[\![\mathsf{B}]\!]\mathcal{P}_0)\,\ell \\ & [\!]\, \ell = \mathsf{after}[\![\mathsf{S}]\!] \,?\, \hat{\mathcal{S}}^{\vec{e}}[\![\mathsf{S}_t]\!]\,(\mathsf{test}^{\vec{e}}[\![\mathsf{B}]\!]\mathcal{P}_0)\,\ell \cup (\overline{\mathsf{test}}^{\vec{e}}[\![\mathsf{B}]\!]\mathcal{P}_0) \\ & \mathbin{\text{:}}\, \varnothing \,|\!)\end{aligned} \tag{19.22}$$

The forward reachability semantics of a conditional statement $\mathsf{S} ::= \mathtt{if}\,(\mathsf{B})\,\mathsf{S}_t$ $\mathtt{else}$ S_f is the set of reachable environments in S_t under the hypothesis that the test B is true, the set of reachable environments in S_f under the hypothesis that the test B is false, and, after the conditional statement, the union of the environments reaching the end of the true alternative S_t or that of the false alternative S_f.

Reachability of a conditional statement $\mathsf{S} ::= \mathtt{if}\,(\mathsf{B})\,\mathsf{S}_t\ \mathtt{else}\ \mathsf{S}_f$

$$\begin{aligned}\hat{\mathcal{S}}^{\vec{e}}[\![\mathsf{S}]\!]\,\mathcal{P}_0\,\ell \;=\;& (\!|\, \ell = \mathsf{at}[\![\mathsf{S}]\!] \,?\, \mathcal{P}_0 \\ & [\!]\, \ell \in \mathsf{in}[\![\mathsf{S}_t]\!] \,?\, \hat{\mathcal{S}}^{\vec{e}}[\![\mathsf{S}_t]\!]\,(\mathsf{test}^{\vec{e}}[\![\mathsf{B}]\!]\mathcal{P}_0)\,\ell \\ & [\!]\, \ell \in \mathsf{in}[\![\mathsf{S}_f]\!] \,?\, \hat{\mathcal{S}}^{\vec{e}}[\![\mathsf{S}_f]\!]\,(\overline{\mathsf{test}}^{\vec{e}}[\![\mathsf{B}]\!]\mathcal{P}_0)\,\ell \\ & [\!]\, \ell = \mathsf{after}[\![\mathsf{S}]\!] \,? \\ & \qquad \hat{\mathcal{S}}^{\vec{e}}[\![\mathsf{S}_t]\!]\,(\mathsf{test}^{\vec{e}}[\![\mathsf{B}]\!]\mathcal{P}_0)\,\ell \cup \hat{\mathcal{S}}^{\vec{e}}[\![\mathsf{S}_f]\!]\,(\overline{\mathsf{test}}^{\vec{e}}[\![\mathsf{B}]\!]\mathcal{P}_0)\,\ell \\ & \mathbin{\text{:}}\, \varnothing \,|\!)\end{aligned} \tag{19.23}$$

The forward reachability semantics of a statement list $\texttt{Sl} ::= \texttt{Sl}'\ \texttt{S}$ are the environments reachable in $\texttt{Sl}'$ and those reachable in $\texttt{S}$ after executing $\texttt{Sl}'$. (As shown by (19.16) in the case that $\texttt{S}$ is an iteration, the environments reachable $\mathsf{at}[\![\texttt{S}]\!]$ are either those after $\texttt{Sl}'$ or those of the iterations of $\texttt{S}$.)

Reachability of a statement list $\texttt{Sl} ::= \texttt{Sl}'\ \texttt{S}$

$$\widehat{\boldsymbol{\mathcal{S}}}^{\vec{e}}[\![\texttt{Sl}]\!]\mathcal{P}_0\,\ell \;=\; (\!\!\mid \ell \in \mathsf{labs}[\![\texttt{Sl}']\!] \setminus \{\mathsf{at}[\![\texttt{S}]\!]\} \;?\; \widehat{\boldsymbol{\mathcal{S}}}^{\vec{e}}[\![\texttt{Sl}']\!]\mathcal{P}_0\,\ell \;\|\; \ell \in \mathsf{labs}[\![\texttt{S}]\!] \;?\; \widehat{\boldsymbol{\mathcal{S}}}^{\vec{e}}[\![\texttt{S}]\!](\widehat{\boldsymbol{\mathcal{S}}}^{\vec{e}}[\![\texttt{Sl}']\!]\mathcal{P}_0\,\mathsf{at}[\![\texttt{S}]\!])\,\ell \;:\; \varnothing \mid\!\!) \tag{19.24}$$

The forward reachability semantics of a break statement $\texttt{S} ::= \texttt{break};$ goes to the break label $\mathsf{break\text{-}to}[\![\texttt{S}]\!]$, so no environment is reachable after that statement.

Reachability of a break statement $\texttt{S} ::= \ell\ \texttt{break};$

$$\widehat{\boldsymbol{\mathcal{S}}}^{\vec{e}}[\![\texttt{S}]\!]\,\mathcal{P}_0\,\ell \;=\; (\!\!\mid \ell = \mathsf{at}[\![\texttt{S}]\!] \;?\; \mathcal{P}_0 \;:\; \varnothing \mid\!\!) \tag{19.25}$$

The forward reachability semantics of a compound statement $\texttt{S} ::= \{\ \texttt{Sl}\ \}$ is that of the statement list $\texttt{Sl}$.

Reachability of a compound statement $\texttt{S} ::= \{\ \texttt{Sl}\ \}$

$$\widehat{\boldsymbol{\mathcal{S}}}^{\vec{e}}[\![\texttt{S}]\!] \;=\; \widehat{\boldsymbol{\mathcal{S}}}^{\vec{e}}[\![\texttt{Sl}]\!] \tag{19.26}$$

Exercise 19.27 Is it true that the reachable environments of a program component $\texttt{S} \in \mathbb{P}\mathrm{c}$ (statement, statement list, or program) at that program component are the initial environments, that is, $\boldsymbol{\mathcal{S}}^{\vec{e}}[\![\texttt{S}]\!](\mathcal{P}_0)\mathsf{at}[\![\texttt{S}]\!] \;\triangleq\; \mathcal{P}_0$?

Exercise 19.28 To finish example 19.15, calculate, by structural induction on the program syntax, the forward reachability semantics $\widehat{\boldsymbol{\mathcal{S}}}^{\vec{e}}[\![\texttt{P}]\!]$ of the program

```
P [ Sl₁ [ Sl₂ [ ϵ
          S₃  [ S₄ [ while ℓ₁ (x < 10)
                       ℓ₂ x = x + 1;
    ℓ₅
```

Exercise 19.29 Prove that the set of reachable environments of a statement $\texttt{S}$ not at a label of that statement is empty.

$$\ell \notin \mathsf{labs}[\![\texttt{S}]\!] \;\Rightarrow\; \widehat{\boldsymbol{\mathcal{S}}}^{\vec{e}}[\![\texttt{S}]\!](\mathcal{P}_0)\ell = \varnothing \tag{19.30}$$

Exercise 19.31 Calculate $\widehat{\boldsymbol{\mathcal{S}}}^{\vec{r}}[\![\texttt{P}]\!]\ \mathbb{E}\mathrm{v}\ \ell_7$ for the following program $\texttt{P}$.

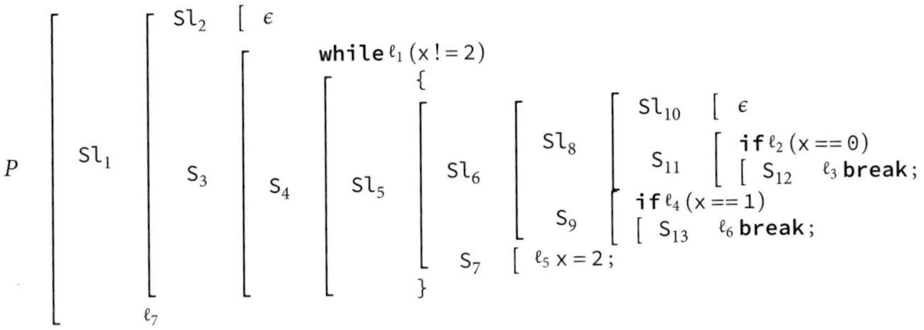

(19.32)

Exercise 19.33 Calculate the relational forward reachability semantics of the program P (19.7) of exercise 19.6.

Exercise 19.34 (Project: forward reachability interpreter) Continuing exercise 5.9, write a reachability interpreter that reads a program P, computes and prints $\widehat{\mathcal{S}}^{\vec{r}}[\![\mathsf{P}]\!]\,\mathscr{P}_0\,\ell$ for all $\ell \in \mathsf{labs}[\![\mathsf{P}]\!]$ where $\mathscr{P}_0 \triangleq \{\rho \in \mathbb{E}\mathbb{v} \mid \forall \mathsf{x} \in \mathbb{V}\,.\,\rho(\mathsf{x}) = 0\}\,\ell$. For example, each program component $\mathsf{S} \in \mathbb{P}\mathbb{c}$ in the abstract syntax tree is decorated with attributes $\langle \mathsf{at}[\![\mathsf{S}]\!],\ \widehat{\mathcal{S}}^{\vec{e}}[\![\mathsf{S}]\!](\mathscr{P}_0)\mathsf{at}[\![\mathsf{S}]\!],\ \mathsf{after}[\![\mathsf{S}]\!],\ \widehat{\mathcal{S}}^{\vec{e}}[\![\mathsf{S}]\!](\mathscr{P}_0)\mathsf{after}[\![\mathsf{S}]\!],\ \mathsf{escape}[\![\mathsf{S}]\!],\ \mathsf{break\text{-}to}[\![\mathsf{S}]\!],$ $\bigcup_{\ell'' \in \mathsf{breaks\text{-}of}[\![\mathsf{S}]\!]} \widehat{\mathcal{S}}^{\vec{e}}[\![\mathsf{S}]\!]\,(\mathscr{P}_0)\,\ell''\rangle$.

```
$ echo ”while (0<1) { x=x+1; if (x>1) break; }” | ./main
<l1:{(x=0),(x=1)}; l5:{(x=2)}; ff; l0:{}>
<l1:{(x=0),(x=1)}; l5:{(x=2)}; ff; l0:{}>
<l2:{(x=0),(x=1)}; l1:{(x=1)}; tt; l5:{(x=2)}>
<l2:{(x=0),(x=1)}; l3:{(x=1),(x=2)}; ff; l0:{}>
<l3:{(x=1),(x=2)}; l1:{(x=1)}; tt; l5:{(x=2)}
<l4:{(x=2)}; l1:{}; tt; l5:{(x=2)}>
```

```
Prog:
  (while l1: (0 < 1)
      Stmtlist: {
     l2: x = (x + 1);
      (if l3: (x > 1)
         l4: break; )
      } )
l5:
```

Exercise 19.35 (reachability compiler) Continuing exercise 19.34, write a program that reads a program P and outputs an object program $\mathsf{compile}[\![\mathsf{P}]\!]$ that, when executed, computes and prints $\widehat{\mathcal{S}}^{\vec{r}}[\![\mathsf{P}]\!]\,\{\rho \in \mathbb{E}\mathbb{v} \mid \forall \mathsf{x} \in \mathbb{V}\,.\,\rho(\mathsf{x}) = 0\}\,\ell$ for all $\ell \in \mathsf{labs}[\![\mathsf{P}]\!]$.

19.4 Reachability Transformers Preserve Joins

Theorem 19.36 (the forward reachability semantics preserves arbitrary joins) For all program components $\mathsf{S} \in \mathbb{P}\mathbb{c}$, $\widehat{\mathcal{S}}^{\vec{e}}[\![\mathsf{S}]\!]$ preserves arbitrary joins, that is,

$$\widehat{\mathcal{S}}^{\vec{e}}[\![\mathsf{S}]\!]\,(\bigcup_i P_i)\,\ell \quad = \quad \bigcup_i \widehat{\mathcal{S}}^{\vec{e}}[\![\mathsf{S}]\!]\,(P_i)\,\ell.$$

In particular $\widehat{\mathcal{S}}^{\vec{e}}[\![\mathsf{S}]\!]\,(\varnothing) = \varnothing$ and the loop transformer $\mathscr{F}^{\vec{e}}[\![\mathtt{while}\,\ell\,(\mathsf{B})\,\mathsf{S}]\!]$ preserves arbitrary joins $\bigcup$.

Proof The proof is by structural induction on S. We consider typical cases.

- For the break S ::= ℓ **break** ; in (19.25), either $\ell = \mathsf{at}[\![S]\!]$ in which case $\widehat{\mathcal{S}}^{\vec{\varrho}}[\![S]\!]\,(\bigcup_i P_i)\,\ell = \bigcup_i P_i = \bigcup_i(\widehat{\mathcal{S}}^{\vec{\varrho}}[\![S]\!]\,P_i\,\ell)$ because $\widehat{\mathcal{S}}^{\vec{\varrho}}[\![S]\!]\,P\,\ell = P$, or $\widehat{\mathcal{S}}^{\vec{\varrho}}[\![S]\!]\,(\bigcup_i P_i)\,\ell = \bigcup_i(\widehat{\mathcal{S}}^{\vec{\varrho}}[\![S]\!]\,P_i\,\ell) = \varnothing$ because $\widehat{\mathcal{S}}^{\vec{\varrho}}[\![S]\!]\,P\,\ell = \varnothing$.
- For the assignment S ::= x = A ; in (19.12), there are three cases:
  - If $\ell = \mathsf{at}[\![S]\!]$ then $\widehat{\mathcal{S}}^{\vec{\varrho}}[\![S]\!]\,(\bigcup_i P_i)\,\ell = \bigcup_i P_i = \bigcup_i(\widehat{\mathcal{S}}^{\vec{\varrho}}[\![S]\!]\,P_i\,\ell)$ because $\widehat{\mathcal{S}}^{\vec{\varrho}}[\![S]\!]\,P\,\ell = P$.
  - Else, if $\ell = \mathsf{after}[\![S]\!]$ then $\widehat{\mathcal{S}}^{\vec{\varrho}}[\![S]\!]\,P\,\ell = \mathsf{assign}^{\vec{\varrho}}[\![x, A]\!]\,P$ which is of the form $\{f(x) \mid x \in P\}$ so $\bigcup_{i\in\Delta} \mathsf{assign}^{\vec{\varrho}}[\![x, A]\!]\,P_i = \bigcup_{i\in\Delta}\{f(x) \mid x \in P_i\} = \{f(x) \mid x \in \bigcup_{i\in\Delta} P_i\} = \mathsf{assign}^{\vec{\varrho}}[\![x, A]\!]\,(\bigcup_{i\in\Delta} P_i)$.
  - Otherwise, $\widehat{\mathcal{S}}^{\vec{\varrho}}[\![S]\!]\,(\bigcup_i P_i)\,\ell = \bigcup_i(\widehat{\mathcal{S}}^{\vec{\varrho}}[\![S]\!]\,P_i\,\ell) = \varnothing$ because $\widehat{\mathcal{S}}^{\vec{\varrho}}[\![S]\!]\,P\,\ell = \varnothing$.
- For the conditional S ::= **if** (B) S_t in (19.22), we have $\mathsf{test}^{\vec{\varrho}}[\![B]\!](\bigcup_{i\in\Delta} P_i) \triangleq \{\rho \in (\bigcup_{i\in\Delta} P_i) \mid \mathcal{B}[\![B]\!]\rho = \mathsf{tt}\} = \bigcup_{i\in\Delta}\{\rho \in P_i \mid \mathcal{B}[\![B]\!]\rho = \mathsf{tt}\}$ and similarly $\overline{\mathsf{test}}^{\vec{r}}[\![B]\!]$ preserves arbitrary joins.
  - In case $\ell \in \mathsf{in}[\![S_t]\!]$, we have $\widehat{\mathcal{S}}^{\vec{\varrho}}[\![S]\!]\,(\bigcup_{i\in\Delta} P_i)\,\ell = \widehat{\mathcal{S}}^{\vec{\varrho}}[\![S_t]\!]\,(\mathsf{test}^{\vec{\varrho}}[\![B]\!](\bigcup_{i\in\Delta} P_i))\,\ell = \widehat{\mathcal{S}}^{\vec{\varrho}}[\![S_t]\!]\,(\bigcup_{i\in\Delta} \mathsf{test}^{\vec{\varrho}}[\![B]\!]P_i)\,\ell = \bigcup_{i\in\Delta} \widehat{\mathcal{S}}^{\vec{\varrho}}[\![S_t]\!]\,(\mathsf{test}^{\vec{\varrho}}[\![B]\!]P_i)\,\ell$, by structural induction hypothesis.
  - In case $\ell = \mathsf{after}[\![S]\!]$, we have

$$
\begin{aligned}
&\widehat{\mathcal{S}}^{\vec{\varrho}}[\![S]\!]\,(\bigcup_{i\in\Delta} P_i)\,\ell \\
=\ &\widehat{\mathcal{S}}^{\vec{\varrho}}[\![S_t]\!]\,(\mathsf{test}^{\vec{\varrho}}[\![B]\!](\bigcup_{i\in\Delta} P_i))\,\ell \cup (\overline{\mathsf{test}}^{\vec{\varrho}}[\![B]\!](\bigcup_{i\in\Delta} P_i)) && \text{(19.22)} \\
=\ &\widehat{\mathcal{S}}^{\vec{\varrho}}[\![S_t]\!]\,(\bigcup_{i\in\Delta} \mathsf{test}^{\vec{\varrho}}[\![B]\!]P_i)\,\ell \cup (\bigcup_{i\in\Delta} \overline{\mathsf{test}}^{\vec{\varrho}}[\![B]\!]P_i) && \text{test}^{\vec{r}}[\![B]\!] \text{ and } \overline{\text{test}}^{\vec{r}}[\![B]\!] \text{ preserve } \cup \\
=\ &\bigcup_{i\in\Delta} \widehat{\mathcal{S}}^{\vec{\varrho}}[\![S_t]\!]\,(\mathsf{test}^{\vec{\varrho}}[\![B]\!]P_i)\,\ell \cup (\bigcup_{i\in\Delta} \overline{\mathsf{test}}^{\vec{\varrho}}[\![B]\!]P_i) && \text{structural induction hypothesis} \\
=\ &\bigcup_{i\in\Delta} (\widehat{\mathcal{S}}^{\vec{\varrho}}[\![S_t]\!]\,(\mathsf{test}^{\vec{\varrho}}[\![B]\!]P_i)\,\ell \cup (\overline{\mathsf{test}}^{\vec{\varrho}}[\![B]\!]P_i)) && \text{definition of } \cup \\
=\ &\bigcup_{i\in\Delta} \widehat{\mathcal{S}}^{\vec{\varrho}}[\![S]\!]\,P_i\,\ell && \text{(19.22).}
\end{aligned}
$$

  - the other cases $\ell \notin \mathsf{labs}[\![S]\!]$ are handled as before.
- For the iteration S ::= **while** ℓ (B) S_b in (19.16), the proof that

$$\mathcal{F}^{\vec{\varrho}}[\![\mathtt{while}\ \ell\ (B)\ S_b]\!]\,\mathcal{P}_0\,(\dot{\bigcup_i} P_i)\,\ell' = \bigcup_i \mathcal{F}^{\vec{\varrho}}[\![\mathtt{while}\ \ell\ (B)\ S_b]\!]\,\mathcal{P}_0\,P_i\,\ell'$$

  that $\mathcal{F}^{\vec{\varrho}}[\![\mathtt{while}\ \ell\ (B)\ S_b]\!]\,\mathcal{P}_0$ preserves arbitrary pointwise joins, is similar to that for the conditional S ::= **if** (B) S_t.
  Trivially, $\mathcal{F}^{\vec{\varrho}}[\![\mathtt{while}\ \ell\ (B)\ S_b]\!]\,(\bigcup_i R_i)\,X\,\ell' = \bigcup_i \mathcal{F}^{\vec{\varrho}}[\![\mathtt{while}\ \ell\ (B)\ S_b]\!]\,R_i\,X\,\ell'$. It follows that $\mathcal{F}^{\vec{\varrho}}[\![\mathtt{while}\ \ell\ (B)\ S_b]\!] \in \wp(\mathbb{E}\mathrm{v}^{\varrho}) \xrightarrow{\sqcup} (\mathbb{L} \to \wp(\mathbb{E}\mathrm{v}^{\varrho})) \xrightarrow{\sqcup} (\mathbb{L} \to \wp(\mathbb{E}\mathrm{v}^{\varrho}))$ where $\mathbb{E}\mathrm{v}^{\varrho} = \mathbb{E}\mathrm{v}$ when $\vec{\varrho} = \vec{r}$ and $\mathbb{E}\mathrm{v}^{\varrho} = \mathbb{E}\mathrm{v} \times \mathbb{E}\mathrm{v}$ when $\vec{\varrho} = \vec{R}$.
  Let F^n, $n \in \mathbb{N}$, be the iterates of $\mathcal{F}^{\vec{\varrho}}[\![\mathtt{while}\ \ell\ (B)\ S_b]\!](\bigcup_i R_i)$.
  Let G_i^n be the iterates of $\mathcal{F}^{\vec{\varrho}}[\![\mathtt{while}\ \ell\ (B)\ S_b]\!]R_i$.
  Let us prove, by recurrence, that $\forall n \in \mathbb{N}\,.\,F^n = \dot{\bigcup_i} G_i^n$.
  - For $n = 0$, $F^0 = G^0 = \ell \mapsto \varnothing$ by definition of the iterates.

- Assume, by induction hypothesis, that $F^n = \dot{\bigcup_i} G_i^n$. Then
$F^{n+1} = \mathscr{F}^{\vec{\varrho}}[\![\texttt{while}\,\ell\,(\texttt{B})\,\texttt{S}_b]\!](\bigcup_i R_i)\, F^n = \mathscr{F}^{\vec{\varrho}}[\![\texttt{while}\,\ell\,(\texttt{B})\,\texttt{S}_b]\!](\bigcup_i R_i)\, (\dot{\bigcup_i} G_i^n) =$
$\dot{\bigcup_i} \mathscr{F}^{\vec{\varrho}}[\![\texttt{while}\,\ell\,(\texttt{B})\,\texttt{S}_b]\!](\bigcup_i R_i)\, G_i^n = \dot{\bigcup_i} G_i^{n+1}$.

We conclude by Scott–Kleene's iterative fixpoint theorem 15.26 that
$\widehat{\boldsymbol{\mathcal{S}}}^{\vec{\varrho}}[\![\texttt{S}]\!]\,(\bigcup_i R_i) = \mathsf{lfp}^{\dot{\subseteq}}\, \mathscr{F}^{\vec{\varrho}}[\![\texttt{while}\,\ell\,(\texttt{B})\,\texttt{S}_b]\!]\,(\bigcup_i R_i) = \dot{\bigcup_{n\in\mathbb{N}}} F^n = \dot{\bigcup_{n\in\mathbb{N}}} \dot{\bigcup_i} G_i^{n+1} = \dot{\bigcup_i} \dot{\bigcup_{n\in\mathbb{N}}} G_i^{n+1}$
$=$
$\dot{\bigcup_i} \mathsf{lfp}^{\dot{\subseteq}}\, \mathscr{F}^{\vec{\varrho}}[\![\texttt{while}\,\ell\,(\texttt{B})\,\texttt{S}_b]\!]\, R_i = \dot{\bigcup_i} \widehat{\boldsymbol{\mathcal{S}}}^{\vec{\varrho}}[\![\texttt{S}]\!]\, R_i$. □

Exercise 19.37 Generalize the proof of theorem 19.36 for the iteration statement and prove that, with appropriate hypotheses, $\mathsf{lfp}^{\sqsubseteq} f(\bigsqcup_i X_i) = \bigsqcup_i \mathsf{lfp}^{\sqsubseteq} f(X_i)$.

19.5 Uncomputability of the Reachability Semantics

It is possible, but in practice difficult, to calculate the reachable environments by hand for iteration. This computation would be better fully automated. By Rice's theorem 9.12, this is impossible. The following lesser-evil solutions have been considered:

- The set $\mathbb{E}\mathbb{v}$ is assumed to be finite (e.g., all variables are Boolean or $\mathbb{V}$ is finite). So elements of $\wp(\mathbb{E}\mathbb{v}^{\varrho})$ have finite computer representations. The reachability semantics $\widehat{\boldsymbol{\mathcal{S}}}^{\vec{\varrho}}[\![\texttt{while}\,\ell\,(\texttt{B})\,\texttt{S}_b]\!]$ can be encoded and computed with finite time and memory resources. This is, depending on the way elements of $\wp(\mathbb{E}\mathbb{v}^{\varrho})$ are represented, enumerative or symbolic model checking [193]. The finiteness hypothesis is restrictive and the unsolved problem is state explosion as shown by examples as trivial as exercise 3.48.
- Ask the programmer to provide the semantics $\widehat{\boldsymbol{\mathcal{S}}}^{\vec{\varrho}}[\![\texttt{while}\,\ell\,(\texttt{B})\,\texttt{S}_b]\!]\,\mathscr{P}_0$ or an inductive approximation of this semantics and check that it is correct by applying a proof method of chapter 25 using a theorem prover, for example, [562] or a proof assistant, for example, [210, 754]).
- Use an abstract interpretation to simplify the semantics $\widehat{\boldsymbol{\mathcal{S}}}^{\vec{\varrho}}[\![\texttt{while}\,\ell\,(\texttt{B})\,\texttt{S}_b]\!]\,\mathscr{P}_0$ and then use one of the previous techniques (e.g., abstract model checking) or effectively compute the simplified semantics (e.g., static analysis).

By corollary 9.6, all these methods will fail on infinitely many programs. Their practical effectiveness measured as scalability and precision can only, up to now, be determined experimentally. [] is a first formal attempt in measuring and comparing the precision of different analyses.

19.6 Sound, Complete, and Exact Structural Abstract Semantics

Let $\langle \boldsymbol{\mathcal{S}}[\![\texttt{S}]\!] \in \mathscr{D}[\![\texttt{S}]\!],\ \texttt{S} \in \mathbb{P}\mathbb{c} \rangle$ be a structural semantics defined as follows:

$$\begin{cases} \boldsymbol{\mathcal{S}}[\![\texttt{S}]\!] \;\triangleq\; \mathscr{F}[\![\texttt{S}]\!](\prod_{\texttt{S}' \lhd \texttt{S}} \boldsymbol{\mathcal{S}}[\![\texttt{S}']\!]) \\ \texttt{S} \in \mathbb{P}\mathbb{c} \end{cases} \tag{19.38}$$

where $\langle \mathtt{S}', \mathtt{S}' \lhd \mathtt{S} \rangle$ is the finite vector of immediate subcomponents of program components $\mathtt{S} \in \mathbb{P}\mathbb{c}$. The map $\mathscr{F}[\![\mathtt{S}]\!] \in \prod_{\mathtt{S}' \lhd \mathtt{S}} \mathscr{D}[\![\mathtt{S}']\!] \to \mathscr{D}[\![\mathtt{S}]\!]$ has no parameters in the basic cases (assignment, skip, etc.). It is defined as a fixpoint for iteration statements.

Let $\alpha[\![\mathtt{S}]\!] \in \wp(\mathscr{D}[\![\mathtt{S}]\!]) \to \langle \mathbb{D}[\![\mathtt{S}]\!], \sqsubseteq \rangle$ be an abstraction of the properties of the semantics $\boldsymbol{\mathcal{S}}[\![\mathtt{S}]\!] \in \mathscr{D}[\![\mathtt{S}]\!]$. The abstract semantics of interest is the abstraction of the collecting semantics.

$$\boldsymbol{\mathcal{S}}^{¤}[\![\mathtt{S}]\!] \triangleq \alpha[\![\mathtt{S}]\!](\{\boldsymbol{\mathcal{S}}[\![\mathtt{S}]\!]\}) \tag{19.39}$$

The definition of a structural abstract semantics has the form

$$\begin{cases} \widehat{\boldsymbol{\mathcal{S}}}^{¤}[\![\mathtt{S}]\!] \triangleq \mathscr{F}^{¤}[\![\mathtt{S}]\!](\prod_{\mathtt{S}' \lhd \mathtt{S}} \widehat{\boldsymbol{\mathcal{S}}}^{¤}[\![\mathtt{S}']\!]) \\ \mathtt{S} \in \mathbb{P}\mathbb{c} \end{cases} \tag{19.40}$$

where $\mathscr{F}^{¤}[\![\mathtt{S}]\!] \in \prod_{\mathtt{S}' \lhd \mathtt{S}} \mathbb{D}[\![\mathtt{S}']\!] \to \mathbb{D}[\![\mathtt{S}]\!]$. Therefore the calculation of the structural abstract semantics $\widehat{\boldsymbol{\mathcal{S}}}^{¤}[\![\mathtt{S}]\!]$ is purely in the abstract domains $\langle \mathbb{D}[\![\mathtt{S}]\!], \mathtt{S} \in \mathbb{P}\mathbb{c} \rangle$ as opposed to abstract semantics $\boldsymbol{\mathcal{S}}^{¤}[\![\mathtt{S}]\!]$ involving calculations in the concrete domains $\langle \mathscr{D}[\![\mathtt{S}]\!], \mathtt{S} \in \mathbb{P}\mathbb{c} \rangle$, which, in general, are more complicated.

The structural abstract semantics is *sound* when $\forall \mathtt{S} \in \mathbb{P}\mathbb{c} \,.\, \boldsymbol{\mathcal{S}}^{¤}[\![\mathtt{S}]\!] \sqsubseteq \widehat{\boldsymbol{\mathcal{S}}}^{¤}[\![\mathtt{S}]\!]$, *complete* when $\forall \mathtt{S} \in \mathbb{P}\mathbb{c} \,.\, \boldsymbol{\mathcal{S}}^{¤}[\![\mathtt{S}]\!] \sqsupseteq \widehat{\boldsymbol{\mathcal{S}}}^{¤}[\![\mathtt{S}]\!]$, and *sound and complete* or *exact* when $\forall \mathtt{S} \in \mathbb{P}\mathbb{c} \,.\, \boldsymbol{\mathcal{S}}^{¤}[\![\mathtt{S}]\!] = \widehat{\boldsymbol{\mathcal{S}}}^{¤}[\![\mathtt{S}]\!]$.

The structural reachability semantics $\widehat{\boldsymbol{\mathcal{S}}}^{\vec{\varrho}}$ is exact (as shown by theorem 20.16 in chapter 20). The structural sign semantics $\widehat{\boldsymbol{\mathcal{S}}}^{\pm}$ of section 3.13 is sound but not exact. For example, $\boldsymbol{\mathcal{S}}^{\pm}[\![2-1]\!] = \alpha_{\pm}(\{\boldsymbol{\mathcal{S}}[\![2-1]\!]\}) = \alpha_{\pm}(\{1\}) = (>0)$ whereas $\widehat{\boldsymbol{\mathcal{S}}}^{\pm}[\![2-1]\!] = \widehat{\boldsymbol{\mathcal{S}}}^{\pm}[\![2]\!] -_{\pm} \widehat{\boldsymbol{\mathcal{S}}}^{\pm}[\![1]\!] = (>0) -_{\pm} (>0) = \top_{\pm}$.

Exercise 19.41 Propose a mathematical method to prove the exactness of a structural abstract semantics.

19.7 Conclusion

The forward reachability semantics ignores all aspects of trace properties irrelevant to reachability. It is therefore a *collecting semantics* for reachability/invariance properties. This means that it is the most abstract of the abstract semantics that are precise enough to express reachability/invariance properties. Therefore it is the basis for designing invariance proof methods and can be further abstracted to get automatic static reachability program analysis methods.

The forward reachability semantics has been postulated in this chapter 19. It will be proved correct by calculational design in the following chapter 20. It can also be specified as a solution to a system of equations or constraints, as shown in chapter 23.

19.8 Solutions to Selected Exercises

Solution to exercise 19.2 The assertional local invariants for the program (4.5) are as follows:

ℓ_1 `/* x=0 */`	$\mathscr{P}_0 \;=\; \{\rho \in \mathbb{E}\mathrm{v} \mid \forall \mathrm{y} \in \mathbb{V} \,.\, \rho(\mathrm{y}) = 0\}$
`  x = x + 1;`	
`  while` ℓ_2 `(tt) /* 1⩽x⩽2 */ {`	$\{\rho \in \mathscr{P}_0 \mid \rho(\mathrm{x}) \in [1,2]\}$
`    ` ℓ_3 `/* 1⩽x⩽2 */`	$\{\rho \in \mathscr{P}_0 \mid \rho(\mathrm{x}) \in [1,2]\}$
`      x = x + 1;`	
`      /* 2⩽x⩽3 */`	$\{\rho \in \mathscr{P}_0 \mid \rho(\mathrm{x}) \in [2,3]\}$
`      if` ℓ_4 `(x > 2)`	
`        ` ℓ_5 `/* x=3 */`	$\{\rho[\mathrm{x} \leftarrow 3] \mid \rho \in \mathscr{P}_0\}$
`        break;`	
`  }`	
ℓ_6 `/* x=3 */`	$\{\rho[\mathrm{x} \leftarrow 3] \mid \rho \in \mathscr{P}_0\}$
`  ;`	
ℓ_7 `/* x=3 */`	$\{\rho[\mathrm{x} \leftarrow 3] \mid \rho \in \mathscr{P}_0\}$

Solution to exercise 19.5 The relational local invariants for the program (4.5) are as follows:

ℓ	$\mathsf{post}^{\vec{\mathrm{R}}}(\boldsymbol{\mathcal{S}})\,\overline{\mathscr{P}}_0\,\ell$
ℓ_1	$\overline{\mathscr{P}}_0 \;=\; \{\langle\rho_0,\,\rho\rangle \mid \forall \mathrm{y} \in \mathbb{V} \,.\, \rho_0(\mathrm{y}) = 0 \wedge \rho = \rho_0\}$
ℓ_2, ℓ_3	$\{\langle\rho_0,\,\rho[\mathrm{x} \leftarrow i]\rangle \mid \langle\rho_0,\,\rho\rangle \in \overline{\mathscr{P}}_0 \wedge i \in [1,2]\}$
ℓ_4	$\{\langle\rho_0,\,\rho[\mathrm{x} \leftarrow i]\rangle \mid \langle\rho_0,\,\rho\rangle \in \overline{\mathscr{P}}_0 \wedge i \in [2,3]\}$
ℓ_5, ℓ_6, ℓ_7	$\{\langle\rho_0,\,\rho[\mathrm{x} \leftarrow 3]\rangle \mid \langle\rho_0,\,\rho\rangle \in \overline{\mathscr{P}}_0\}$

Solution to exercise 19.18

Invariant of an iteration statement $\mathsf{S} ::= \mathtt{while}\,\ell\,(\mathsf{B})\,\mathsf{S}_b$

$$\widehat{\boldsymbol{\mathcal{S}}}^{\vec{e}}[\![\mathsf{S}]\!]\,\mathscr{P}_0\,\ell' \;=\; \mathsf{let}\,\vec{\boldsymbol{\mathscr{F}}}^{e}[\![\mathtt{while}\,\ell\,(\mathsf{B})\,\mathsf{S}_b]\!]\,\mathscr{P}_0 \;\in\; \wp(\mathbb{E}\mathrm{v}^{\vec{e}}) \xrightarrow{\nearrow} \wp(\mathbb{E}\mathrm{v}^{\vec{e}}) \tag{19.42}$$

$$\vec{\boldsymbol{\mathscr{F}}}^{e}[\![\mathtt{while}\,\ell\,(\mathsf{B})\,\mathsf{S}_b]\!]\,\mathscr{P}_0\,X \;=\; \mathscr{P}_0 \cup \widehat{\boldsymbol{\mathcal{S}}}^{\vec{e}}[\![\mathsf{S}_b]\!]\,(\mathsf{test}^{\vec{e}}[\![\mathsf{B}]\!]X)\,\ell$$

$$\mathsf{and}\ I \;=\; \mathsf{lfp}^{\subseteq}\,\vec{\boldsymbol{\mathscr{F}}}^{e}[\![\mathtt{while}\,\ell\,(\mathsf{B})\,\mathsf{S}_b]\!]\,\mathscr{P}_0\ \mathsf{in}$$

$$\big(\!\!\big|\ \ell' = \ell \ ?\ I$$

$$\big\|\ \ell' \in \mathsf{in}[\![\mathsf{S}_b]\!] \setminus \{\ell\} \ ?\ \widehat{\boldsymbol{\mathcal{S}}}^{\vec{e}}[\![\mathsf{S}_b]\!]\,(\mathsf{test}^{\vec{e}}[\![\mathsf{B}]\!]\,I)\,\ell'$$

$$\big\|\ \ell' = \mathsf{after}[\![\mathsf{S}]\!] \ ?\ \overline{\mathsf{test}}^{\vec{e}}[\![\mathsf{B}]\!]\,I \cup \bigcup_{\ell'' \in \mathsf{breaks\text{-}of}[\![\mathsf{S}_b]\!]} \widehat{\boldsymbol{\mathcal{S}}}^{\vec{e}}[\![\mathsf{S}_b]\!]\,(\mathsf{test}^{\vec{e}}[\![\mathsf{B}]\!]\,I)\,\ell''$$

$$:\ \varnothing\ \big|\!\!\big)$$

Proof of (19.42) Let $X = \mathsf{lfp}^{\subseteq}\,\boldsymbol{\mathscr{F}}^{\vec{e}}[\![\mathsf{S}]\!]\,\mathscr{P}_0$ in (19.16). By definition of $\boldsymbol{\mathscr{F}}^{\vec{e}}[\![\mathsf{S}]\!]\,\mathscr{P}_0$ in (19.16)and $\vec{\boldsymbol{\mathscr{F}}}^{e}[\![\mathsf{S}]\!]\,\mathscr{P}_0$ in (19.42), we have $X(\ell) = \mathscr{P}_0 \cup \widehat{\boldsymbol{\mathcal{S}}}^{\vec{e}}[\![\mathsf{S}_b]\!]\,(\mathsf{test}^{\vec{e}}[\![\mathsf{B}]\!]X(\ell))\,\ell =$

$\overrightarrow{\mathcal{F}}^{\vec{\varrho}}[\![\mathsf{S}]\!]\,\mathcal{P}_0\,X(\ell)$ so $X(\ell)$ is a fixpoint of $\overrightarrow{\mathcal{F}}^{\vec{\varrho}}[\![\mathsf{S}]\!]\,\mathcal{P}_0$. Let $\overline{Y}$ be a fixpoint of $\overrightarrow{\mathcal{F}}^{\vec{\varrho}}[\![\mathsf{S}]\!]\,\mathcal{P}_0$. Define $Y(\ell') = (\!|\,\ell' = \ell \mathrel{?} \overline{Y} \mathrel{\text{\textsf{8}}} \mathbb{E}\mathsf{v}^{\vec{\varrho}}\,|\!)$. Then $\mathcal{F}^{\vec{\varrho}}[\![\mathsf{S}]\!]\,\mathcal{P}_0\,Y \mathrel{\dot{\subseteq}} Y$ so, by Tarski's fixpoint theorem 15.6, $X = \mathsf{lfp}^{\dot{\subseteq}}\,\mathcal{F}^{\vec{\varrho}}[\![\mathsf{S}]\!]\,\mathcal{R}_0 \mathrel{\dot{\subseteq}} Y$ pointwise, so $X(\ell) \subseteq Y(\ell) = \overline{Y}$, proving that $X(\ell)$ is the least fixpoint I of $\overrightarrow{\mathcal{F}}^{\vec{\varrho}}[\![\mathsf{S}]\!]\,\mathcal{P}_0$. It follows that $\widehat{\mathcal{S}}^{\vec{\varrho}}_{(19.42)}[\![\mathsf{S}]\!]\,\mathcal{P}_0\,\ell = \widehat{\mathcal{S}}^{\vec{\varrho}}_{(19.16)}[\![\mathsf{S}]\!]\,\mathcal{P}_0\,\ell$. It immediately follows that for $\ell' \in \mathsf{in}[\![\mathsf{S}_b]\!] \setminus \{\ell\}$, we have $\widehat{\mathcal{S}}^{\vec{\varrho}}[\![\mathsf{S}]\!]\,\mathcal{P}_0\,\ell' = (\mathsf{lfp}^{\dot{\subseteq}}\,\mathcal{F}^{\vec{\varrho}}[\![\mathsf{S}]\!]\,\mathcal{R}_0)\,\ell' = \mathcal{F}^{\vec{\varrho}}[\![\mathsf{S}]\!]\,\mathcal{P}_0)\,(\mathsf{lfp}^{\dot{\subseteq}}\,\mathcal{F}^{\vec{\varrho}}[\![\mathsf{S}]\!]\,\mathcal{R}_0)\,\ell' = \widehat{\mathcal{S}}^{\vec{\varrho}}[\![\mathsf{S}_b]\!]\,(\mathsf{test}^{\vec{\varrho}}[\![\mathsf{B}]\!]((\mathsf{lfp}^{\dot{\subseteq}}\,\mathcal{F}^{\vec{\varrho}}[\![\mathsf{S}]\!]\,\mathcal{R}_0)(\ell)))\,\ell' = \widehat{\mathcal{S}}^{\vec{\varrho}}[\![\mathsf{S}_b]\!]\,(\mathsf{test}^{\vec{\varrho}}[\![\mathsf{B}]\!](I))\,\ell' = \widehat{\mathcal{S}}^{\vec{\varrho}}_{(19.42)}[\![\mathsf{S}]\!]\,\mathcal{P}_0\,\ell'$. The same reasoning is used for $\ell' = \mathsf{after}[\![\mathsf{S}]\!]$.

□

Solution to exercise 19.41 We first prove the commutation property

$$\forall \mathsf{S} \in \mathbb{P}\mathbb{c}\,.\,\alpha[\![\mathsf{S}]\!](\{\mathcal{F}[\![\mathsf{S}]\!](\prod_{\mathsf{S}' \lhd \mathsf{S}} X_{\mathsf{S}'})\}) \quad = \quad \mathcal{F}^{¤}[\![\mathsf{S}]\!](\prod_{\mathsf{S}' \lhd \mathsf{S}} \alpha[\![\mathsf{S}']\!](\{X_{\mathsf{S}'}\})) \tag{19.43}$$

for all $\mathsf{S} \in \mathbb{P}\mathbb{c}$ and $X_{\mathsf{S}'} \in \mathcal{D}[\![\mathsf{S}']\!]$, $\mathsf{S}' \lhd \mathsf{S}$.

For iteration statements, $\mathcal{F}[\![\mathsf{S}]\!](\prod_{\mathsf{S}' \lhd \mathsf{S}} X_{\mathsf{S}'})$ is a fixpoint, and this proof involves, for example, theorems 18.23 and 18.27, corollaries 18.34 and 18.35, or similar results.

The commutation condition (19.43) allows us to derive the abstract transformer $\mathcal{F}^{¤}[\![\mathsf{S}]\!]$, knowing the concrete transformer $\mathcal{F}[\![\mathsf{S}]\!]$ and the abstraction $\alpha[\![\mathsf{S}]\!]$.

Then the proof proceeds by structural induction on $\langle \mathbb{P}\mathbb{c}, \lhd \rangle$. Assuming, by structural induction hypothesis, that $\forall \mathsf{S}' \lhd \mathsf{S}\,.\,\mathcal{S}^{¤}[\![\mathsf{S}']\!] = \widehat{\mathcal{S}}^{¤}[\![\mathsf{S}']\!]$, we have

$$
\begin{aligned}
& \mathcal{S}^{¤}[\![\mathsf{S}]\!] \\
= {} & \alpha[\![\mathsf{S}]\!](\{\mathcal{S}[\![\mathsf{S}]\!]\}) && \text{(19.39)} \\
= {} & \alpha[\![\mathsf{S}]\!](\{\mathcal{F}[\![\mathsf{S}]\!](\prod_{\mathsf{S}' \lhd \mathsf{S}} \mathcal{S}[\![\mathsf{S}']\!])\}) && \text{(19.38)} \\
= {} & \mathcal{F}^{¤}[\![\mathsf{S}]\!](\prod_{\mathsf{S}' \lhd \mathsf{S}} \alpha[\![\mathsf{S}']\!](\{\mathcal{S}[\![\mathsf{S}']\!]\})) && \text{commutation property (19.43)} \\
= {} & \mathcal{F}^{¤}[\![\mathsf{S}]\!](\prod_{\mathsf{S}' \lhd \mathsf{S}} \mathcal{S}^{¤}[\![\mathsf{S}']\!]) && \text{(19.39)} \\
= {} & \mathcal{F}^{¤}[\![\mathsf{S}]\!](\prod_{\mathsf{S}' \lhd \mathsf{S}} \widehat{\mathcal{S}}^{¤}[\![\mathsf{S}']\!]) && \text{structural induction hypothesis} \\
= {} & \widehat{\mathcal{S}}^{¤}[\![\mathsf{S}]\!] && \text{(19.40)}
\end{aligned}
$$

20 Calculational Design of the Forward Reachability Semantics

The forward reachability semantics $\boldsymbol{\mathcal{S}}^{\vec{\varrho}}[\![\mathsf{S}]\!]$ of program components S was introduced in chapter 19 without a formal definition. The structural forward reachability semantics $\widehat{\boldsymbol{\mathcal{S}}}^{\vec{\varrho}}[\![\mathsf{S}]\!]$ was postulated, that is, claimed to be true without justification. No correspondence was established between the two semantics $\boldsymbol{\mathcal{S}}^{\vec{\varrho}}[\![\mathsf{S}]\!]$ and $\widehat{\boldsymbol{\mathcal{S}}}^{\vec{\varrho}}[\![\mathsf{S}]\!]$. This chapter fills the gaps in the informal introduction of chapter 19.

First, we formally define the forward reachability semantics $\boldsymbol{\mathcal{S}}^{\vec{\varrho}}[\![\mathsf{S}]\!]$ as an abstraction of the prefix trace semantics $\boldsymbol{\mathcal{S}}^{*}[\![\mathsf{S}]\!]$ of chapter 6.

Then we use the methodology presented in section 19.6 to prove that $\widehat{\boldsymbol{\mathcal{S}}}^{\vec{\varrho}}[\![\mathsf{S}]\!] = \boldsymbol{\mathcal{S}}^{\vec{\varrho}}[\![\mathsf{S}]\!]$ (theorem 20.16). More precisely, we show that we can design the structural abstract semantics $\widehat{\boldsymbol{\mathcal{S}}}^{\vec{\varrho}}[\![\mathsf{S}]\!]$ as an abstraction of the concrete semantics $\boldsymbol{\mathcal{S}}^{*}[\![\mathsf{S}]\!]$ by calculus, to ensure that $\widehat{\boldsymbol{\mathcal{S}}}^{\vec{\varrho}}[\![\mathsf{S}]\!] = \boldsymbol{\mathcal{S}}^{\vec{\varrho}}[\![\mathsf{S}]\!]$.

The motivation for presenting the results of the calculational design in chapter 19 before developing the calculational design in this chapter, was to facilitate understanding.

Contents

20.1 Assertional Forward Reachability Abstraction of the Prefix-Closed Trace Semantics

20.1.1 Assertional Forward Reachability Abstraction

The assertional reachable abstraction of a prefix closed trace semantics $\boldsymbol{\mathcal{S}}$ for initial environments $\mathcal{P}_0$ is

$$\mathsf{post}^{\vec{r}}(\boldsymbol{\mathcal{S}})\,\mathcal{P}_0\,\ell \;\triangleq\; \{\varrho(\pi_0\ell_0\pi_1\ell_1) \mid \varrho(\pi_0\ell_0) \in \mathcal{P}_0 \wedge \ell_0\pi_1\ell_1 \in \boldsymbol{\mathcal{S}}(\pi_0\ell_0) \wedge \ell_1 = \ell\} \tag{20.1}$$

(See remark 17.8 on free and bound variables.)

Assume that $\ell_0\pi_1\ell$ is the continuation of an initial trace $\pi_0\ell_0$ of the prefix trace semantics $\boldsymbol{\mathcal{S}}$ (so $\pi_0\ell_0\pi_1\ell$ is possibly continued by a trace $\ell\pi_2$ in the maximal semantics). Assume that the initial environment $\varrho(\pi_0\ell_0)$ satisfies the initial specification $\mathcal{P}_0$. Then the environment $\varrho(\pi_0\ell_0\pi_1\ell)$ is reachable at ℓ so belongs to $\mathsf{post}^{\vec{r}}(\boldsymbol{\mathcal{S}})\,\mathcal{P}_0\,\ell$.

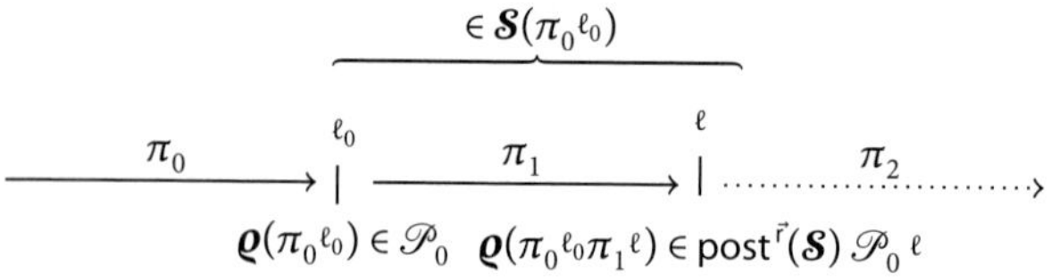

See example 19.1.

Exercise 20.2 Prove that $\langle \mathbb{T}^+ \to \wp(\mathbb{T}^+),\ \dot{\subseteq}\rangle \xleftrightarrow[\mathsf{post}^{\vec{r}}]{\gamma^{\vec{r}}} \langle \wp(\mathbb{E}\mathbb{v}) \to \mathbb{L} \mapsto \wp(\mathbb{E}\mathbb{v}),\ \ddot{\subseteq}\rangle$.

Exercise 20.3 Instead of the prefix trace semantics, we could have considered maximal finite or infinite traces. Assume that π is the (maximal) continuation of an initial trace $\pi_0\ell_0$. The reachable environments on π initialized by π_0 is the set of environments encountered on π.

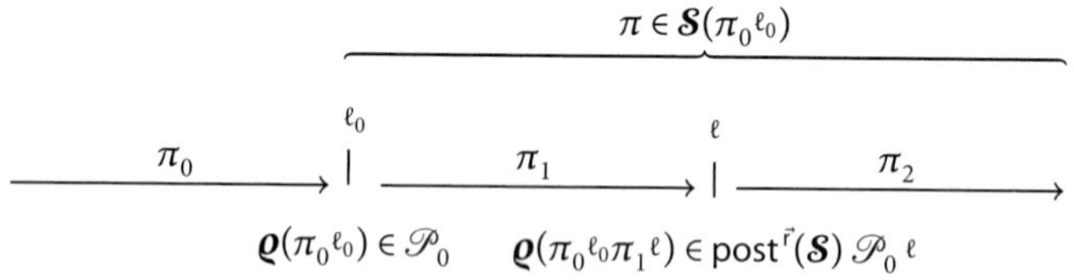

$$\begin{aligned}\mathsf{post}^{\vec{r}*}(\boldsymbol{S})\,\mathcal{P}_0\,\ell \;\triangleq\;& \{\varrho(\pi_0\ell_0\pi_1\ell') \mid \varrho(\pi_0\ell_0) \in \mathcal{P}_0 \wedge \\ & \exists \pi_2 \,.\, \ell_0\pi_1\ell'\pi_2 \in \boldsymbol{S}(\pi_0\ell_0) \wedge \ell' = \ell\}\end{aligned} \tag{20.4}$$

Prove that if $\boldsymbol{S}$ is closed by nonempty prefixes then $\mathsf{post}^{\vec{r}*}(\boldsymbol{S})\,\mathcal{P}_0 = \mathsf{post}^{\vec{r}}(\boldsymbol{S})\,\mathcal{P}_0$.

Exercise 20.5 Following exercise 20.3, prove that the reachable environments can be collected on the maximal trace semantics, that is, $\mathsf{post}^{\vec{R}*}(\boldsymbol{S}^{+\infty}[\![\mathsf{S}]\!]) = \mathsf{post}^{\vec{R}}(\boldsymbol{S}^{*}[\![\mathsf{S}]\!])$.

20.1.2 Assertional Forward Reachability Semantics

Given a set $\mathcal{P}_0$ of initial environments, the assertional forward reachability semantics $\boldsymbol{S}^{\vec{r}}[\![\mathsf{S}]\!]\,\mathcal{P}_0\,\ell$ maps the set $\mathcal{P}_0$ of initial environments to the set of reachable environments at each program point ℓ of program component $\mathsf{S} \in \mathbb{P}\mathbb{c}$ (i.e., program, statement, or statement list). $\boldsymbol{S}^{\vec{r}}[\![\mathsf{S}]\!]$ can be defined by abstraction of the prefix closed semantics $\boldsymbol{S}^{*}[\![\mathsf{S}]\!]$ or, by exercise 20.3, by abstraction of the maximal $\boldsymbol{S}^{+\infty}[\![\mathsf{S}]\!]$.

Definition of the assertional forward reachability semantics

$$\begin{aligned}\boldsymbol{S}^{\vec{r}}[\![\mathsf{S}]\!] &\in \wp(\mathbb{E}\mathbb{v}) \xrightarrow{\nearrow} (\mathbb{L} \to \wp(\mathbb{E}\mathbb{v})) \\ \boldsymbol{S}^{\vec{r}}[\![\mathsf{S}]\!] &\triangleq \mathsf{post}^{\vec{r}}(\boldsymbol{S}^{*}[\![\mathsf{S}]\!] \rceil \mathsf{labs}[\![\mathsf{S}]\!]) \;=\; \mathsf{post}^{\vec{r}*}(\boldsymbol{S}^{+\infty}[\![\mathsf{S}]\!] \rceil \mathsf{labs}[\![\mathsf{S}]\!]).\end{aligned} \tag{20.6}$$

In this formula, if $f \in A \to B$ and $X \subseteq A$, $f \rceil X \in X \nrightarrow B$ is the restriction of f to X, equal to f on X and undefined elsewhere.[1] It follows that $\boldsymbol{S}^{\vec{r}}[\![\mathsf{S}]\!] \in \wp(\mathbb{E}\mathbb{v}) \xrightarrow{\nearrow} (\mathsf{labs}[\![\mathsf{S}]\!] \to \wp(\mathbb{E}\mathbb{v}))$. Alternatively, we could have considered $\boldsymbol{S}^{\vec{r}}[\![\mathsf{S}]\!] \in \wp(\mathbb{E}\mathbb{v}) \xrightarrow{\nearrow} (\mathsf{labx}[\![\mathsf{S}]\!] \to \wp(\mathbb{E}\mathbb{v}))$. The restriction to $\mathsf{labs}[\![\mathsf{S}]\!]$ follows from the fact that we are not interested in the reachable

1. For a relation $r \in \wp(A \times B)$, $X \rceil r \lceil Y$ is $r \cap (X \times Y)$.

environments at the escape point of a statement because they will be available at that break point as the reachable environments of the program components containing that break point.

Exercise 20.7 Show that $\boldsymbol{S}^{\vec{r}}[\![\mathtt{S}]\!]$ can be equivalently defined by abstracting the prefix trace semantics to an input-output relational semantics as in section 12.1 and then abstracting by the postimage property transformer (12.2).

Remark 20.8 Define the finite trace to final environment abstraction $\alpha_\varrho(P) \triangleq \{\varrho(\pi) \mid \pi \in P\}$ which, by exercise 11.6 is the homomorphic/partitioning abstraction $\langle \wp(\mathbb{T}^{+\infty}), \subseteq\rangle \underset{\alpha_\varrho}{\overset{\gamma_\varrho}{\leftrightarrows}} \langle \wp(\mathbb{E}\mathrm{v}), \subseteq\rangle$ (the surjection follows form exercise 6.8). By exercise 11.21, the pointwise extension to the set $\mathbb{L}$ of all program labels is the abstraction $\langle \mathbb{L} \rightarrow \wp(\mathbb{T}^{+\infty}), \dot{\subseteq}\rangle \underset{\dot{\alpha}_\varrho}{\overset{\dot{\gamma}_\varrho}{\leftrightarrows}} \langle \mathbb{L} \rightarrow \wp(\mathbb{E}\mathrm{v}), \dot{\subseteq}\rangle$. By theorem 11.78, defining $\ddot{\alpha}_\varrho(\boldsymbol{S}) \triangleq \alpha_\varrho \circ \boldsymbol{S} \circ \dot{\alpha}_\varrho$, we have the abstraction $\langle \wp(\mathbb{T}^{+\infty}) \xrightarrow{\nearrow} (\mathbb{L} \rightarrow \wp(\mathbb{T}^{+\infty})), \ddot{\subseteq}\rangle \underset{\ddot{\alpha}_\varrho}{\overset{\ddot{\gamma}_\varrho}{\leftrightarrows}} \langle \wp(\mathbb{E}\mathrm{v}) \xrightarrow{\nearrow} (\mathbb{L} \rightarrow \wp(\mathbb{E}\mathrm{v})), \ddot{\subseteq}\rangle$ where $\ddot{\subseteq}$ is functional pointwise inclusion. Then $\boldsymbol{S}^{\vec{r}}[\![\mathtt{S}]\!] = \ddot{\alpha}_\varrho(\boldsymbol{S}^*[\![\mathtt{S}]\!])$ for the pointwise prefix trace semantics (6.48).

20.2 Relational ReachabilityAbstraction of the Prefix-Closed Trace Semantics

20.2.1 Relational Reachability Abstraction

The relational reachability abstraction $\mathsf{post}^{\vec{R}}(\boldsymbol{S})\,\mathscr{P}_0\,\ell$ of a prefix closed trace semantics $\boldsymbol{S}$ attaches at each program point ℓ a relation between the initial and current environments, assuming that the initial environments satisfy $\mathscr{P}_0$.

$$\mathsf{post}^{\vec{R}}(\boldsymbol{S})\,\mathscr{P}_0\,\ell \;\triangleq\; \{\langle \rho_0,\, \varrho(\pi_0\ell_0\pi_1\ell')\rangle \mid \langle \rho_0,\, \varrho(\pi_0\ell_0)\rangle \in \mathscr{P}_0 \land \ell_0\pi_1\ell' \in \boldsymbol{S}(\pi_0\ell_0) \land \ell' = \ell\} \tag{20.9}$$

(See remark 17.8 on free and bound variables.)

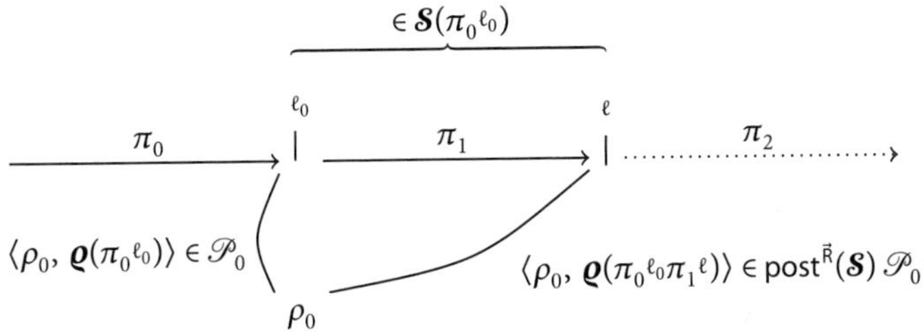

See example 19.3.

Exercise 20.10 Prove that for all $\langle \mathbb{T}^+ \rightarrow \wp(\mathbb{T}^+), \dot{\subseteq}\rangle \underset{\mathsf{post}^{\vec{R}}}{\overset{\gamma^{\vec{R}}}{\leftrightarrows}} \langle \wp(\mathbb{E}\mathrm{v} \times \mathbb{E}\mathrm{v}) \rightarrow \mathbb{L} \mapsto \wp(\mathbb{E}\mathrm{v} \times \mathbb{E}\mathrm{v}), \ddot{\subseteq}\rangle$.

Exercise 20.11 Define the abstraction

$$\alpha^2(\mathscr{P}) \;\triangleq\; \{\rho \mid \exists \rho_0 \,.\, \langle \rho_0,\, \rho\rangle \in \mathscr{P}\} \tag{20.12}$$

that gets rid of the initial environments. Prove that $\langle \wp(\mathbb{E}\mathrm{v} \times \mathbb{E}\mathrm{v}), \subseteq\rangle \underset{\alpha^2}{\overset{\gamma^2}{\leftrightarrows}} \langle \wp(\mathbb{E}\mathrm{v}), \subseteq\rangle$.

Exercise 20.13 Continuing exercise 20.11, prove that

$$\mathsf{post}^{\vec{r}}(\boldsymbol{\mathcal{S}})\, \mathscr{P}_0\, \ell = \alpha^2(\mathsf{post}^{\vec{R}}(\boldsymbol{\mathcal{S}})\, \gamma^2(\mathscr{P}_0)\, \ell).$$

20.2.2 Relational Forward Reachability Semantics

The forward relational reachability semantics $\boldsymbol{\mathcal{S}}^{\vec{R}}[\![\mathsf{S}]\!]$ relates the current environment at a program point to the initial environments.

Definition of the relational forward reachability semantics

$$\begin{array}{lcl} \boldsymbol{\mathcal{S}}^{\vec{R}}[\![\mathsf{S}]\!] & \in & \wp(\mathbb{E}\mathrm{v} \times \mathbb{E}\mathrm{v}) \rightarrow (\mathbb{L} \rightarrow \wp(\mathbb{E}\mathrm{v} \times \mathbb{E}\mathrm{v})) \\ \boldsymbol{\mathcal{S}}^{\vec{R}}[\![\mathsf{S}]\!] & \triangleq & \mathsf{post}^{\vec{R}}(\boldsymbol{\mathcal{S}}^*[\![\mathsf{S}]\!] \rceil \mathsf{labs}[\![\mathsf{S}]\!]) \quad = \quad \mathsf{post}^{\vec{R}*}(\boldsymbol{\mathcal{S}}^{+\infty}[\![\mathsf{S}]\!] \rceil \mathsf{labs}[\![\mathsf{S}]\!]). \end{array} \tag{20.14}$$

The forward assertional reachability semantics $\boldsymbol{\mathcal{S}}^{\vec{r}}[\![\mathsf{S}]\!]$ is an abstraction (20.12) of the forward relational forward reachability semantics $\boldsymbol{\mathcal{S}}^{\vec{R}}[\![\mathsf{S}]\!]$ because, by exercise 20.13, we have that $\mathsf{post}^{\vec{r}}(\boldsymbol{\mathcal{S}})\, \alpha^2(\mathscr{P})\, \ell = \alpha^2(\mathsf{post}^{\vec{R}}(\boldsymbol{\mathcal{S}})\, \mathscr{P})\, \ell$.

20.3 Reachability Abstraction of the Prefix-Closed Trace Semantics

Recall that in section 19.1.3, we defined $\boldsymbol{\mathcal{S}}^{\vec{\varrho}}[\![\mathsf{P}]\!]$, $\widehat{\boldsymbol{\mathcal{S}}}^{\vec{\varrho}}[\![\mathsf{P}]\!]$, $\mathsf{post}^{\vec{\varrho}}$, $\mathsf{post}^{\vec{\varrho}*}$, $\mathbb{E}\mathrm{v}^{\vec{\varrho}}$, ... to be consistently replaced either by $\boldsymbol{\mathcal{S}}^{\vec{r}}[\![\mathsf{P}]\!]$, $\widehat{\boldsymbol{\mathcal{S}}}^{\vec{r}}[\![\mathsf{P}]\!]$, $\mathsf{post}^{\vec{r}}$, $\mathsf{post}^{\vec{r}*}$, $\mathbb{E}\mathrm{v}$, ... for $\vec{\varrho} = \vec{r}$ or $\boldsymbol{\mathcal{S}}^{\vec{R}}[\![\mathsf{P}]\!]$, $\widehat{\boldsymbol{\mathcal{S}}}^{\vec{R}}[\![\mathsf{P}]\!]$, $\mathsf{post}^{\vec{R}}$, $\mathsf{post}^{\vec{R}*}$, $\mathbb{E}\mathrm{v} \times \mathbb{E}\mathrm{v}$, ... for $\vec{\varrho} = \vec{R}$. Hence we define

Definition of the relational forward reachability semantics

$$\begin{array}{lcl} \boldsymbol{\mathcal{S}}^{\vec{\varrho}}[\![\mathsf{P}]\!] & \in & \wp(\mathbb{E}\mathrm{v}^{\vec{\varrho}}) \rightarrow (\mathbb{L} \rightarrow \wp(\mathbb{E}\mathrm{v}^{\vec{\varrho}})) \\ \boldsymbol{\mathcal{S}}^{\vec{\varrho}}[\![\mathsf{P}]\!] & \triangleq & \mathsf{post}^{\vec{\varrho}}(\boldsymbol{\mathcal{S}}^*[\![\mathsf{S}]\!] \rceil \mathsf{labs}[\![\mathsf{S}]\!]) \quad = \quad \mathsf{post}^{\vec{\varrho}*}(\boldsymbol{\mathcal{S}}^{+\infty}[\![\mathsf{S}]\!] \rceil \mathsf{labs}[\![\mathsf{S}]\!]) \end{array} \tag{20.15}$$

20.4 Design of the Structural Forward Reachability Semantics

We design the structural reachability semantics $\widehat{\boldsymbol{\mathcal{S}}}^{\vec{\varrho}}[\![\mathsf{P}]\!]$ so that

Theorem 20.16 $\forall \mathsf{S} \in \mathbb{P}\mathrm{c}\,.\, \widehat{\boldsymbol{\mathcal{S}}}^{\vec{\varrho}}[\![\mathsf{S}]\!] = \boldsymbol{\mathcal{S}}^{\vec{\varrho}}[\![\mathsf{S}]\!]$.

Proof of theorem 20.16 Following exercise 19.41, $\widehat{\boldsymbol{\mathcal{S}}}^{\vec{\varrho}}[\![\mathsf{S}]\!]$ is given a structural definition and designed by calculus, as follows (the case $\ell \notin \mathsf{labs}[\![\mathsf{S}]\!]$ handled in (19.30) is trivial):

$$\begin{array}{ll} \widehat{\boldsymbol{\mathcal{S}}}^{\vec{\varrho}}[\![\mathsf{S}]\!]\, \mathscr{P}_0\, \ell & \text{(case } \ell \in \mathsf{labs}[\![\mathsf{S}]\!]\text{)} \\ = \mathsf{post}^{\vec{\varrho}}(\boldsymbol{\mathcal{S}}^*[\![\mathsf{S}]\!] \rceil \mathsf{labs}[\![\mathsf{S}]\!])\, \mathscr{P}_0\, \ell & \text{(ensuring that } \widehat{\boldsymbol{\mathcal{S}}}^{\vec{\varrho}}[\![\mathsf{S}]\!] = \boldsymbol{\mathcal{S}}^{\vec{\varrho}}[\![\mathsf{S}]\!] \text{ by (20.15))} \\ = \mathsf{post}^{\vec{\varrho}}(\widehat{\boldsymbol{\mathcal{S}}}^*[\![\mathsf{S}]\!] \rceil \mathsf{labs}[\![\mathsf{S}]\!])\, \mathscr{P}_0\, \ell & \text{(}\boldsymbol{\mathcal{S}}^*[\![\mathsf{S}]\!] \triangleq \widehat{\boldsymbol{\mathcal{S}}}^*[\![\mathsf{S}]\!] \text{ in section 6.6)} \\ = \mathsf{post}^{\vec{\varrho}}(f^*[\![\mathsf{S}]\!](\prod_{\mathsf{S}' \lhd \mathsf{S}} \widehat{\boldsymbol{\mathcal{S}}}^*[\![\mathsf{S}']\!]) \rceil \mathsf{labs}[\![\mathsf{S}]\!])\, \mathscr{P}_0\, \ell & \end{array}$$

$\wr$fixpoint definition $\widehat{\boldsymbol{\mathcal{S}}}^*[\![\mathsf{S}]\!] = f^*[\![\mathsf{S}]\!](\prod_{\mathsf{S}' \lhd \mathsf{S}} \widehat{\boldsymbol{\mathcal{S}}}^*[\![\mathsf{S}']\!])$ from chapter 17 equivalent to the inductive definition of chapter 6$\wr$

$= \ldots$ $\wr$calculational design$\wr$

$= f^{\vec{\varrho}}[\![\mathsf{S}]\!](\prod_{\mathsf{S}' \lhd \mathsf{S}} \mathsf{post}^{\vec{\varrho}}(\boldsymbol{\mathcal{S}}^*[\![\mathsf{S}']\!] \rceil \mathsf{labs}[\![\mathsf{S}']\!]))\, \mathscr{P}_0\, \ell$ $\wr$exhibiting the commutation property (19.43)$\wr$

$= f^{\vec{\varrho}}[\![\mathsf{S}]\!](\prod_{\mathsf{S}' \lhd \mathsf{S}} \boldsymbol{\mathcal{S}}^{\vec{\varrho}}[\![\mathsf{S}']\!])\, \mathscr{P}_0\, \ell$ $\wr$definition (20.15)$\wr$

$= f^{\vec{\varrho}}[\![\mathsf{S}]\!](\prod_{\mathsf{S}' \lhd \mathsf{S}} \widehat{\boldsymbol{\mathcal{S}}}^{\vec{\varrho}}[\![\mathsf{S}']\!])\, \mathscr{P}_0\, \ell$ $\wr$induction hypothesis of theorem 20.16$\wr$

The proof has to consider all possible program components S, which is shown for each component below. □

> *Reachability outside a statement* S
>
> $$\ell \notin \mathsf{labs}[\![\mathsf{S}]\!] \quad\Rightarrow\quad \widehat{\boldsymbol{\mathcal{S}}}^{\vec{\varrho}}[\![\mathsf{S}]\!](\mathscr{P}_0)\ell = \varnothing \tag{19.30}$$

Proof of (19.30) Assume that $\ell \notin \mathsf{labs}[\![\mathsf{S}]\!]$.

$\widehat{\boldsymbol{\mathcal{S}}}^{\vec{\varrho}}[\![\mathsf{S}]\!]\, \mathscr{P}_0\, \ell$

$= \mathsf{post}^{\vec{\varrho}}(\widehat{\boldsymbol{\mathcal{S}}}^*[\![\mathsf{S}]\!] \rceil \mathsf{labs}[\![\mathsf{S}]\!])\, \mathscr{P}_0\, \ell$ $\wr$(20.15)$\wr$

$= \{\varrho(\pi_0\ell_0\pi_1\ell') \mid \varrho(\pi_0\ell_0) \in \mathscr{P}_0 \wedge \ell_0\pi_1\ell' \in \boldsymbol{\mathcal{S}}^*[\![\mathsf{S}]\!](\pi_0\ell_0) \wedge \ell' = \ell \in \mathsf{labs}[\![\mathsf{S}]\!]\}$

$\wr$(20.1), similar for (20.9) and definition of $\rceil$$\wr$

$= \varnothing$ $\wr$because $\ell \notin \mathsf{labs}[\![\mathsf{S}]\!] \wedge \ell \in \mathsf{labs}[\![\mathsf{S}]\!]$ is false$\wr$ □

> *Reachability of a program* $\mathsf{P} ::= \mathsf{Sl}\ \ell'$
>
> $$\widehat{\boldsymbol{\mathcal{S}}}^{\vec{\varrho}}[\![\mathsf{P}]\!] \quad\triangleq\quad \widehat{\boldsymbol{\mathcal{S}}}^{\vec{\varrho}}[\![\mathsf{Sl}]\!] \tag{19.19}$$

Proof of (19.19)

$\widehat{\boldsymbol{\mathcal{S}}}^{\vec{\varrho}}[\![\mathsf{P}]\!]\, \mathscr{P}_0\, \ell$

$= \mathsf{post}^{\vec{\varrho}}(\widehat{\boldsymbol{\mathcal{S}}}^*[\![\mathsf{P}]\!] \rceil \mathsf{labs}[\![\mathsf{P}]\!])\, \mathscr{P}_0\, \ell$ $\wr$by definition (20.15)$\wr$

$= \mathsf{post}^{\vec{\varrho}}(\widehat{\boldsymbol{\mathcal{S}}}^*[\![\mathsf{Sl}]\!] \rceil \mathsf{labs}[\![\mathsf{Sl}]\!])\, \mathscr{P}_0\, \ell$ $\wr$by definition(6.12) and definition of labs$\wr$

$= \widehat{\boldsymbol{\mathcal{S}}}^{\vec{\varrho}}[\![\mathsf{Sl}]\!]\, \mathscr{P}_0\, \ell$ $\wr$by definition (20.15)$\wr$ □

> *Reachability of a statement list* $\mathsf{Sl} ::= \mathsf{Sl}'\ \mathsf{S}$
>
> $$\widehat{\boldsymbol{\mathcal{S}}}^{\vec{\varrho}}[\![\mathsf{Sl}]\!]\mathscr{P}_0\,\ell = (\!|\ \ell \in \mathsf{labs}[\![\mathsf{Sl}']\!] \setminus \{\mathsf{at}[\![\mathsf{S}]\!]\}\ ?\ \widehat{\boldsymbol{\mathcal{S}}}^{\vec{\varrho}}[\![\mathsf{Sl}']\!]\mathscr{P}_0\,\ell \ \|\ \ell \in \mathsf{labs}[\![\mathsf{S}]\!]\ ?\ \widehat{\boldsymbol{\mathcal{S}}}^{\vec{\varrho}}[\![\mathsf{S}]\!](\widehat{\boldsymbol{\mathcal{S}}}^{\vec{\varrho}}[\![\mathsf{Sl}']\!]\mathscr{P}_0\,\mathsf{at}[\![\mathsf{S}]\!])\,\ell \ :\ \varnothing\ |\!) \tag{19.24}$$

Proof of (19.24)

$\widehat{\boldsymbol{S}}^{\vec{e}}[\![\mathtt{Sl}]\!]\,\mathcal{P}_0\,\ell$

$= \mathsf{post}^{\vec{e}}(\widehat{\boldsymbol{S}}^*[\![\mathtt{Sl}]\!] \rceil \mathsf{labs}[\![\mathtt{Sl}]\!])\,\mathcal{P}_0\,\ell$ (by definition (20.15))

$= \{\varrho(\pi_0\ell_0\pi_1\ell') \mid \varrho(\pi_0\ell_0) \in \mathcal{P}_0 \wedge \ell_0\pi_1\ell' \in \widehat{\boldsymbol{S}}^*[\![\mathtt{Sl}'\ \mathtt{S}]\!](\pi_0\ell_0) \wedge \ell' = \ell \in \mathsf{labs}[\![\mathtt{Sl}]\!]\}$

(by definition (20.4) and $\mathtt{Sl} ::= \mathtt{Sl}'\ \mathtt{S}$, similar for (20.9))

$= \{\varrho(\pi_0\ell_0\pi_1\ell') \mid \varrho(\pi_0\ell_0) \in \mathcal{P}_0 \wedge \exists \pi_2 \,.\, \ell_0\pi_1\ell'\pi_2 \in \boldsymbol{S}^*[\![\mathtt{Sl}']\!](\pi_0\ell_0) \wedge \ell = \ell' \in \mathsf{labs}[\![\mathtt{Sl}]\!]\} \cup$
$\{\varrho(\pi_0\ell_0\pi_1\ell') \mid \varrho(\pi_0\ell_0) \in \mathcal{P}_0 \wedge \exists \pi_2' \in \boldsymbol{S}^+[\![\mathtt{Sl}']\!](\pi_0\ell_0), \pi_3' \in \boldsymbol{S}^*[\![\mathtt{S}]\!](\pi_0\ell_0 \mathbin{\frown} \pi_2'), \pi_2 \,.$
$\ell_0\pi_1\ell'\pi_2 = \pi_2' \mathbin{\frown} \pi_3' \wedge \ell = \ell' \in \mathsf{labs}[\![\mathtt{Sl}]\!]\}$ ((6.13) and (6.14))

$= \{\varrho(\pi_0\ell_0\pi_1\ell') \mid \varrho(\pi_0\ell_0) \in \mathcal{P}_0 \wedge \exists \pi_2 \,.\, \ell_0\pi_1\ell'\pi_2 \in \boldsymbol{S}^*[\![\mathtt{Sl}']\!](\pi_0\ell_0) \wedge \ell = \ell' \in \mathsf{labs}[\![\mathtt{Sl}']\!] \setminus$
$\{\mathsf{at}[\![\mathtt{S}]\!]\}\} \cup \{\varrho(\pi_0\ell_0\pi_1\ell') \mid \varrho(\pi_0\ell_0) \in \mathcal{P}_0 \wedge \exists \pi_2' \in \boldsymbol{S}^+[\![\mathtt{Sl}']\!](\pi_0\ell_0), \pi_3' \in \boldsymbol{S}^*[\![\mathtt{S}]\!](\pi_0\ell_0 \mathbin{\frown}$
$\pi_2'), \pi_2 \,.\, \ell_0\pi_1\ell'\pi_2 = \pi_2' \mathbin{\frown} \pi_3' \wedge \ell = \ell' \in \mathsf{labs}[\![\mathtt{S}]\!]\}$

(we have $\mathsf{labs}[\![\mathtt{Sl}]\!] = \mathsf{labs}[\![\mathtt{Sl}']\!] \cup \mathsf{labs}[\![\mathtt{S}]\!]$, $\mathsf{labs}[\![\mathtt{Sl}']\!] \cap \mathsf{labs}[\![\mathtt{S}]\!] = \{\mathsf{after}[\![\mathtt{Sl}']\!]\} = \{\mathsf{at}[\![\mathtt{S}]\!]\}$ so that $\mathsf{labs}[\![\mathtt{Sl}]\!] = (\mathsf{labs}[\![\mathtt{Sl}']\!] \setminus \{\mathsf{at}[\![\mathtt{S}]\!]\}) \cup \mathsf{labs}[\![\mathtt{S}]\!]$. Moreover, by exercise 6.37, the traces of a program component $\mathtt{S} \in \mathbb{P}\mathbb{c}$ belongs to $\mathsf{labx}[\![\mathtt{S}]\!]$, and the escape label in $\mathsf{labx}[\![\mathtt{S}]\!]$, if any does not belong to $\mathsf{labs}[\![\mathtt{S}]\!]$.

It follows, for the first term, that if $\ell_0\pi_1\ell'\pi_2 \in \boldsymbol{S}^*[\![\mathtt{Sl}']\!](\pi_0\ell_0)$ then $\ell' \notin \mathsf{labs}[\![\mathtt{S}]\!]$ or $\ell' = \mathsf{after}[\![\mathtt{Sl}']\!] = \mathsf{at}[\![\mathtt{S}]\!] \wedge \pi_2 = \ni$ and this last case is taken into account in the second term when π_2' is $\ell_0\pi_1\ell'$. Therefore in this first case $\ell = \ell' \in \mathsf{labs}[\![\mathtt{Sl}]\!]$ implies $\ell = \ell' \in \mathsf{labs}[\![\mathtt{Sl}']\!] \setminus \{\mathsf{at}[\![\mathtt{S}]\!]\}$ because the case $\ell = \ell' = \mathsf{at}[\![\mathtt{S}]\!]$ is handled by the second term;

And for the second term, π_2' has either the form $\ell_0\pi_4\mathsf{break\text{-}to}[\![\mathtt{Sl}']\!]$ or the form $\ell_0\pi_4\mathsf{after}[\![\mathtt{Sl}']\!] = \ell_0\pi_4\mathsf{at}[\![\mathtt{S}]\!]$ where $\ell \in \mathsf{in}[\![\mathtt{S}]\!] \cup \{\mathsf{after}[\![\mathtt{S}]\!]\}$ which in both cases implies that ℓ does not appear in $\ell_0\pi_4$ so can only appear in π_3' which is therefore of the form $\mathsf{at}[\![\mathtt{S}]\!]\pi_5\ell\pi_6$. If follows that $\ell = \ell' \in \mathsf{labs}[\![\mathtt{Sl}]\!]$ implies that $\ell = \ell' \in \mathsf{labs}[\![\mathtt{S}]\!]$)

$= (\!|\, \ell \in \mathsf{labs}[\![\mathtt{Sl}']\!] \setminus \{\mathsf{at}[\![\mathtt{S}]\!]\} \,?\, \{\varrho(\pi_0\ell_0\pi_1\ell') \mid \varrho(\pi_0\ell_0) \in \mathcal{P}_0 \wedge \exists \pi_2 \,.\, \ell_0\pi_1\ell'\pi_2 \in \boldsymbol{S}^*[\![\mathtt{Sl}']\!](\pi_0\ell_0) \wedge$
$\ell = \ell'\}\}$
$\|\, \ell \in \mathsf{labs}[\![\mathtt{S}]\!] \,?\, \{\varrho(\pi_0\ell_0\pi_1\ell') \mid \varrho(\pi_0\ell_0) \in \mathcal{P}_0 \wedge \exists \pi_2' \in \boldsymbol{S}^+[\![\mathtt{Sl}']\!](\pi_0\ell_0), \pi_3' \in \boldsymbol{S}^*[\![\mathtt{S}]\!](\pi_0\ell_0 \mathbin{\frown}$
$\pi_2'), \pi_2 \,.\, \ell_0\pi_1\ell'\pi_2 = \pi_2' \mathbin{\frown} \pi_3' \wedge \ell = \ell' \in \mathsf{labs}[\![\mathtt{S}]\!]\}$
$:\, \varnothing \,|\!)$ (because $(\mathsf{labs}[\![\mathtt{Sl}']\!] \setminus \{\mathsf{at}[\![\mathtt{S}]\!]\}) \cap \mathsf{labs}[\![\mathtt{S}]\!] = \varnothing$)

$= (\!|\, \ell \in \mathsf{labs}[\![\mathtt{Sl}']\!] \setminus \{\mathsf{at}[\![\mathtt{S}]\!]\} \,?\, \widehat{\boldsymbol{S}}^{\vec{e}}[\![\mathtt{Sl}']\!]\,\mathcal{P}_0\,\ell$
$\|\, \ell \in \mathsf{labs}[\![\mathtt{S}]\!] \,?\, \{\varrho(\pi_0\ell_0\pi_4\mathsf{at}[\![\mathtt{S}]\!]\pi_5\ell') \mid \varrho(\pi_0\ell_0) \in \mathcal{P}_0 \wedge \ell_0\pi_4\mathsf{at}[\![\mathtt{S}]\!] \in \boldsymbol{S}^+[\![\mathtt{Sl}']\!](\pi_0\ell_0) \wedge \exists \pi_2 \,.$
$\mathsf{at}[\![\mathtt{S}]\!]\pi_5\ell'\pi_2 \in \boldsymbol{S}^*[\![\mathtt{S}]\!](\pi_0\ell_0\pi_4\mathsf{at}[\![\mathtt{S}]\!]) \wedge \ell' = \ell\} :\, \varnothing \,|\!)$

(By (20.4) (and similarly (20.9)) for the first case.

In the second case, we have seen that π_2' has the form $\ell_0\pi_4\mathsf{at}[\![\mathtt{S}]\!]$ where ℓ can only appear in π_3' which is therefore of the form $\mathsf{at}[\![\mathtt{S}]\!]\pi_5\ell\pi_6$. It follows that $\ell_0\pi_1\ell\pi_2 = \pi_2' \mathbin{\frown} \pi_3'$ implies $\ell_0\pi_1\ell\pi_2 = \ell_0\pi_4\mathsf{at}[\![\mathtt{S}]\!]\pi_5\ell\pi_6$ where $\pi_1 = \pi_4\mathsf{at}[\![\mathtt{S}]\!]\pi_5$ and $\pi_6 = \pi_2$)

$= (\!|\, \ell \in \mathsf{labs}[\![\mathtt{Sl}']\!] \setminus \{\mathsf{at}[\![\mathtt{S}]\!]\} \,?\, \widehat{\boldsymbol{S}}^{\vec{e}}[\![\mathtt{Sl}']\!]\,\mathcal{P}_0\,\ell$
$\|\, \ell \in \mathsf{labs}[\![\mathtt{S}]\!] \,?\, \{\varrho(\pi_0\ell_0\pi_4\mathsf{at}[\![\mathtt{S}]\!]\pi_5\ell') \mid \varrho(\pi_0\ell_0\pi_4\mathsf{at}[\![\mathtt{S}]\!]) \in \{\varrho(\pi_0\ell_0\pi_4\mathsf{at}[\![\mathtt{S}]\!]) \mid \varrho(\pi_0\ell_0) \in \mathcal{P}_0 \wedge$
$\ell_0\pi_4\mathsf{at}[\![\mathtt{S}]\!] \in \boldsymbol{S}^+[\![\mathtt{Sl}']\!](\pi_0\ell_0)\} \wedge \exists \pi_2 \,.\, \mathsf{at}[\![\mathtt{S}]\!]\pi_5\ell'\pi_2 \in \boldsymbol{S}^*[\![\mathtt{S}]\!](\pi_0\ell_0\pi_4\mathsf{at}[\![\mathtt{S}]\!]) \wedge \ell' = \ell\} :\, \varnothing \,|\!)$

(because $x \in \{x' \mid P(x')\}$ if and only if $P(x)$)

$= (\!| \ell \in \mathsf{labs}[\![\mathtt{Sl}']\!] \setminus \{\mathsf{at}[\![\mathtt{S}]\!]\} \,?\, \widehat{\mathcal{S}}^{\vec{\varrho}}[\![\mathtt{Sl}']\!]\, \mathscr{P}_0\, \ell$
$\;\|\; \ell \in \mathsf{labs}[\![\mathtt{S}]\!] \,?\, \{\varrho(\pi_0\ell_0\pi_4\mathsf{at}[\![\mathtt{S}]\!]\pi_5\ell') \mid \varrho(\pi_0\ell_0\pi_4\mathsf{at}[\![\mathtt{S}]\!]) \in \{\varrho(\pi_0\ell_0\pi_4\mathsf{at}[\![\mathtt{S}]\!]) \mid \varrho(\pi_0\ell_0) \in \mathscr{P}_0 \wedge$
$\ell_0\pi_4\mathsf{at}[\![\mathtt{S}]\!] \in \mathcal{S}^*[\![\mathtt{Sl}']\!](\pi_0\ell_0)\} \wedge \exists \pi_2 \,.\, \mathsf{at}[\![\mathtt{S}]\!]\pi_5\ell'\pi_2 \in \mathcal{S}^*[\![\mathtt{S}]\!](\pi_0\ell_0\pi_4\mathsf{at}[\![\mathtt{S}]\!]) \wedge \ell' = \ell\} : \varnothing |\!)$

(because $\pi\mathsf{after}[\![\mathtt{Sl}']\!] \in \mathcal{S}^+[\![\mathtt{Sl}']\!](\pi_0\ell_0)$ if and only if $\pi\mathsf{after}[\![\mathtt{Sl}']\!] \in \mathcal{S}^*[\![\mathtt{Sl}']\!](\pi_0\ell_0)$ and $\mathsf{after}[\![\mathtt{Sl}']\!] = \mathsf{at}[\![\mathtt{S}]\!]$)

$= (\!| \ell \in \mathsf{labs}[\![\mathtt{Sl}']\!] \setminus \{\mathsf{at}[\![\mathtt{S}]\!]\} \,?\, \widehat{\mathcal{S}}^{\vec{\varrho}}[\![\mathtt{Sl}']\!]\, \mathscr{P}_0\, \ell$
$\;\|\; \ell \in \mathsf{labs}[\![\mathtt{S}]\!] \,?\, \widehat{\mathcal{S}}^{\vec{\varrho}}[\![\mathtt{Sl}']\!]\, (\{\varrho(\pi_0\ell_0\pi_4\mathsf{at}[\![\mathtt{S}]\!]) \mid \varrho(\pi_0\ell_0) \in \mathscr{P}_0 \wedge \ell_0\pi_4\mathsf{at}[\![\mathtt{S}]\!] \in \mathcal{S}^*[\![\mathtt{Sl}']\!](\pi_0\ell_0)\})\, \ell :$
$\varnothing |\!)$

(because $\widehat{\mathcal{S}}^{\vec{\varrho}}[\![\mathtt{Sl}']\!]\, \Sigma\, \ell \;=\; \{\varrho(\pi_0\ell_0\pi_4\mathsf{at}[\![\mathtt{S}]\!]\pi_5\ell') \mid \varrho(\pi_0\ell_0\pi_4\mathsf{at}[\![\mathtt{S}]\!]) \in \Sigma \wedge \exists \pi_2 \,.$ $\mathsf{at}[\![\mathtt{S}]\!]\pi_5\ell'\pi_2 \in \mathcal{S}^*[\![\mathtt{Sl}']\!](\pi_0\ell_0\pi_4\mathsf{at}[\![\mathtt{S}]\!]) \wedge \ell' = \ell\}$)

$= (\!| \ell \in \mathsf{labs}[\![\mathtt{Sl}']\!] \setminus \{\mathsf{at}[\![\mathtt{S}]\!]\} \,?\, \widehat{\mathcal{S}}^{\vec{\varrho}}[\![\mathtt{Sl}']\!]\, \mathscr{P}_0\, \ell$
$\;\|\; \ell \in \mathsf{labs}[\![\mathtt{S}]\!] \,?\, \widehat{\mathcal{S}}^{\vec{\varrho}}[\![\mathtt{Sl}']\!]\, (\widehat{\mathcal{S}}^{\vec{\varrho}}[\![\mathtt{S}]\!]\, \mathscr{P}_0\, \mathsf{at}[\![\mathtt{S}]\!])\, \ell : \varnothing |\!)$

(because $\widehat{\mathcal{S}}^{\vec{\varrho}}[\![\mathtt{S}]\!]\, \mathscr{P}_0\, \mathsf{at}[\![\mathtt{S}]\!] \;=\; \{\varrho(\pi_0\ell_0\pi_4\mathsf{at}[\![\mathtt{S}]\!]) \mid \varrho(\pi_0\ell_0) \in \mathscr{P}_0 \wedge \exists \pi_2 \,.\, \ell_0\pi_4\mathsf{at}[\![\mathtt{S}]\!]\pi_2 \in$ $\mathcal{S}^*[\![\mathtt{S}]\!](\pi_0\ell_0)\}$ and π_2 must be empty) □

Reachability of an empty statement list $\mathtt{Sl} ::= \epsilon$

$$\widehat{\mathcal{S}}^{\vec{\varrho}}[\![\mathtt{Sl}]\!]\mathscr{P}_0\,\ell \;=\; (\!| \ell = \mathsf{at}[\![\mathtt{Sl}]\!] \,?\, \mathscr{P}_0 : \varnothing |\!) \tag{19.20}$$

Proof of (19.20)

$\widehat{\mathcal{S}}^{\vec{\varrho}}[\![\mathtt{S}]\!]\, \mathscr{P}_0\, \ell$

$= \{\varrho(\pi_0\ell_0\pi_1\ell') \mid \varrho(\pi_0\ell_0) \in \mathscr{P}_0 \wedge \ell_0\pi_1\ell' \in \widehat{\mathcal{S}}^*[\![\mathtt{S}]\!](\pi_0\ell_0) \wedge \ell' = \ell \in \mathsf{labs}[\![\mathtt{S}]\!]\}$

(by definitions (20.15) and (20.1), similar for (20.9))

$= \{\varrho(\pi_0\ell_0\pi_1\ell') \mid \varrho(\pi_0\ell_0) \in \mathscr{P}_0 \wedge \ell_0\pi_1\ell' \in \{\mathsf{at}[\![\mathtt{Sl}]\!]\}) \wedge \ell' = \ell\})$

(definition of the prefix semantics same as (6.15) and $\mathsf{labs}[\![\mathtt{S}]\!] = \{\mathsf{at}[\![\mathtt{Sl}]\!]\}$ in section 4.2.7)

$= \{\varrho(\pi_0\mathsf{at}[\![\mathtt{Sl}]\!]) \mid \varrho(\pi_0\mathsf{at}[\![\mathtt{Sl}]\!]) \in \mathscr{P}_0 \wedge \ell = \mathsf{at}[\![\mathtt{Sl}]\!]\})$

(because $\ell_0 = \ell' = \ell = \mathsf{at}[\![\mathtt{Sl}]\!]$ and so $\ell_0\pi_1\ell' = \mathsf{at}[\![\mathtt{Sl}]\!]$)

$= (\!| \ell = \mathsf{at}[\![\mathtt{Sl}]\!] \,?\, \mathscr{P}_0 : \varnothing |\!)$ (by exercise 6.8) □

Reachability of an assignment statement $\mathtt{S} ::= \mathtt{x} = \mathtt{E}\,;$

$$\begin{aligned}\widehat{\mathcal{S}}^{\vec{\varrho}}[\![\mathtt{S}]\!]\, \mathscr{P}_0\, \ell \;&=\; (\!| \ell = \mathsf{at}[\![\mathtt{S}]\!] \,?\, \mathscr{P}_0 \\ &\quad\;\|\; \ell = \mathsf{after}[\![\mathtt{S}]\!] \,?\, \mathsf{assign}^{\vec{\varrho}}[\![\mathtt{x}, \mathtt{A}]\!]\, \mathscr{P}_0 \\ &\quad\; : \varnothing |\!)\end{aligned} \tag{19.12}$$

$$\mathsf{assign}^{\vec{\mathrm{r}}}[\![\mathtt{x}, \mathtt{A}]\!]\, \mathscr{P}_0 \;\triangleq\; \{\rho[\mathtt{x} \leftarrow \mathscr{A}[\![\mathtt{A}]\!]\rho] \mid \rho \in \mathscr{P}_0\}$$

$$\mathsf{assign}^{\vec{\mathrm{R}}}[\![\mathtt{x}, \mathtt{A}]\!]\, \mathscr{P}_0 \;\triangleq\; \{\langle \rho_0, \rho[\mathtt{x} \leftarrow \mathscr{A}[\![\mathtt{A}]\!]\rho]\rangle \mid \langle \rho_0, \rho\rangle \in \mathscr{P}_0\}$$

Proof of (19.12)

$\widehat{\mathcal{S}}^{\vec{\mathrm{r}}}[\![\mathtt{S}]\!]\, \mathscr{P}_0\, \ell$ (similar for $\widehat{\mathcal{S}}^{\vec{\mathrm{R}}}$)

$= \{\varrho(\pi_0\ell_0\pi_1\ell') \mid \varrho(\pi_0\ell_0) \in \mathcal{P}_0 \wedge \ell_0\pi_1\ell' \in \widehat{\mathcal{S}}^*[\![\mathsf{S}]\!](\pi_0\ell_0) \wedge \ell' = \ell \in \mathsf{labs}[\![\mathsf{S}]\!]\}$

(by definitions (20.15) and (20.1), similar for (20.9))

$= \{\varrho(\pi_0\ell_0\pi_1\ell') \mid \varrho(\pi_0\ell_0) \in \mathcal{P}_0 \wedge \ell_0\pi_1\ell' \in \{\mathsf{at}[\![\mathsf{S}]\!]\} \cup \{\mathsf{at}[\![\mathsf{S}]\!] \xrightarrow{\mathsf{x}=v} \mathsf{after}[\![\mathsf{S}]\!] \mid v = \mathcal{A}[\![\mathsf{A}]\!]\varrho(\pi_0\ell_0)\} \wedge \ell' = \ell \in \mathsf{labs}[\![\mathsf{S}]\!]\}$

(definition of the prefix semantics same as (6.11) and (6.16))

$= (\!| \ell = \mathsf{at}[\![\mathsf{S}]\!] \;?\; \mathcal{P}_0 \;|\!|\; \ell = \mathsf{after}[\![\mathsf{S}]\!] \;?\; \{\varrho(\pi_0\ell_0\pi_1\ell') \mid \varrho(\pi_0\ell_0) \in \mathcal{P}_0 \wedge \ell_0\pi_1\ell' \in \{\mathsf{at}[\![\mathsf{S}]\!] \xrightarrow{\mathsf{x}=\mathcal{A}[\![\mathsf{A}]\!]\varrho(\pi_0\ell_0)} \mathsf{after}[\![\mathsf{S}]\!]\} \wedge \ell' = \ell \in \mathsf{labs}[\![\mathsf{S}]\!]\} \;:\; \varnothing |\!)$

($\mathsf{labs}[\![\mathsf{S}]\!] = \{\mathsf{at}[\![\mathsf{S}]\!], \mathsf{after}[\![\mathsf{S}]\!]\}$ in section 4.2.7)

$= (\!| \ell = \mathsf{at}[\![\mathsf{S}]\!] \;?\; \mathcal{P}_0 \;|\!|\; \ell = \mathsf{after}[\![\mathsf{S}]\!] \;?\; \{\rho[\mathsf{x} \leftarrow \mathcal{A}[\![\mathsf{A}]\!]\rho] \mid \rho \in \mathcal{P}_0\} \;:\; \varnothing |\!)$

(letting $\rho = \varrho(\pi_0\ell_0)$ by exercise 6.8 and definition (6.6) of ϱ)

$= (\!| \ell = \mathsf{at}[\![\mathsf{S}]\!] \;?\; \mathcal{P}_0 \;|\!|\; \ell = \mathsf{after}[\![\mathsf{S}]\!] \;?\; \mathsf{assign}^{\vec{r}}[\![\mathsf{x}, \mathsf{A}]\!]\, \mathcal{P}_0 \;:\; \varnothing |\!)$

(definition of $\mathsf{assign}^{\vec{r}}[\![\mathsf{x}, \mathsf{A}]\!]\, \mathcal{P}_0$, similar for (20.9)) □

> *Reachability of a skip statement* $\mathsf{S} ::= ;$
>
> $$\widehat{\mathcal{S}}^{\vec{e}}[\![\mathsf{S}]\!]\, \mathcal{P}_0\, \ell \quad = \quad (\!| \ell \in \{\mathsf{at}[\![\mathsf{S}]\!], \mathsf{after}[\![\mathsf{S}]\!]\} \;?\; \mathcal{P}_0 \;:\; \varnothing |\!) \qquad (19.21)$$

Proof of (19.21)

$\widehat{\mathcal{S}}^{\vec{e}}[\![\mathsf{S}]\!]\, \mathcal{P}_0\, \ell$

$= \{\varrho(\pi_0\ell_0\pi_1\ell') \mid \varrho(\pi_0\ell_0) \in \mathcal{P}_0 \wedge \ell_0\pi_1\ell' \in \widehat{\mathcal{S}}^*[\![\mathsf{S}]\!](\pi_0\ell_0) \wedge \ell' = \ell \in \mathsf{labs}[\![\mathsf{S}]\!]\}$

(by definitions (20.15) and (20.1), similar for (20.9))

$= \{\varrho(\pi_0\ell_0\pi_1\ell') \mid \varrho(\pi_0\ell_0) \in \mathcal{P}_0 \wedge \ell_0\pi_1\ell' \in \{\mathsf{at}[\![\mathsf{S}]\!]\} \cup \{\mathsf{at}[\![\mathsf{S}]\!] \xrightarrow{\mathsf{skip}} \mathsf{after}[\![\mathsf{S}]\!]\} \wedge \ell' = \ell \in \{\mathsf{at}[\![\mathsf{S}]\!], \mathsf{after}[\![\mathsf{S}]\!]\}\}$

(definition of the prefix semantics same as (6.11) and (6.17) and $\mathsf{labs}[\![\mathsf{S}]\!] = \{\mathsf{at}[\![\mathsf{S}]\!], \mathsf{after}[\![\mathsf{S}]\!]\}$ in section 4.2.7)

$= (\!| \ell = \mathsf{at}[\![\mathsf{S}]\!] \;?\; \mathcal{P}_0 \;|\!|\; \ell = \mathsf{after}[\![\mathsf{S}]\!] \;?\; \{\varrho(\pi_0\ell_0\pi_1\ell') \mid \varrho(\pi_0\ell_0) \in \mathcal{P}_0 \wedge \ell_0\pi_1\ell' \in \{\mathsf{at}[\![\mathsf{S}]\!] \xrightarrow{\mathsf{skip}} \mathsf{after}[\![\mathsf{S}]\!]\} \wedge \ell' = \ell \in \mathsf{labs}[\![\mathsf{S}]\!]\} \;:\; \varnothing |\!)$ (cases $\ell = \ell_0 = \mathsf{at}[\![\mathsf{S}]\!]$ or $\ell_0 = \mathsf{at}[\![\mathsf{S}]\!]$ and $\ell = \ell' = \mathsf{after}[\![\mathsf{S}]\!]$)

$= (\!| \ell = \mathsf{at}[\![\mathsf{S}]\!] \;?\; \mathcal{P}_0 \;|\!|\; \ell = \mathsf{after}[\![\mathsf{S}]\!] \;?\; \{\varrho(\pi_0\ell_0) \mid \varrho(\pi_0\ell_0) \in \mathcal{P}_0\} \;:\; \varnothing |\!)$

(because $\varrho(\pi_0\ell_0 \xrightarrow{\mathsf{skip}} \ell) = \varrho(\pi_0\ell_0)$)

$= (\!| \ell \in \{\mathsf{at}[\![\mathsf{S}]\!], \mathsf{after}[\![\mathsf{S}]\!]\} \;?\; \mathcal{P}_0 \;:\; \varnothing |\!)$ (grouping cases) □

Reachability of a conditional statement $\mathsf{S} ::= \mathtt{if}\,(\mathsf{B})\,\mathsf{S}_t$

$$\begin{aligned}
\widehat{\mathcal{S}}^{\vec{\varrho}}[\![\mathsf{S}]\!]\,\mathcal{P}_0\,\ell \;=\; & (\!\!|\; \ell = \mathsf{at}[\![\mathsf{S}]\!] \;?\; \mathcal{P}_0 \\
& \|\; \ell \in \mathsf{in}[\![\mathsf{S}_t]\!] \;?\; \widehat{\mathcal{S}}^{\vec{\varrho}}[\![\mathsf{S}_t]\!]\,(\mathsf{test}^{\vec{\varrho}}[\![\mathsf{B}]\!]\mathcal{P}_0)\,\ell \\
& \|\; \ell = \mathsf{after}[\![\mathsf{S}]\!] \;? \\
& \qquad \widehat{\mathcal{S}}^{\vec{\varrho}}[\![\mathsf{S}_t]\!]\,(\mathsf{test}^{\vec{\varrho}}[\![\mathsf{B}]\!]\mathcal{P}_0)\,\ell \cup (\overline{\mathsf{test}}^{\vec{\varrho}}[\![\mathsf{B}]\!]\mathcal{P}_0) \\
& :\; \emptyset \;|\!\!)
\end{aligned} \tag{19.22}$$

$$\begin{aligned}
\mathsf{test}^{\vec{r}}[\![\mathsf{B}]\!]\mathcal{P}_0 &\triangleq \{\rho \in \mathcal{P}_0 \mid \mathcal{B}[\![\mathsf{B}]\!]\rho = \mathsf{tt}\} &\quad \mathsf{test}^{\vec{R}}[\![\mathsf{B}]\!]\mathcal{P}_0 &\triangleq \{\langle \rho_0, \rho\rangle \in \mathcal{P}_0 \mid \mathcal{B}[\![\mathsf{B}]\!]\rho = \mathsf{tt}\} \\
\overline{\mathsf{test}}^{\vec{r}}[\![\mathsf{B}]\!]\mathcal{P}_0 &\triangleq \{\rho \in \mathcal{P}_0 \mid \mathcal{B}[\![\mathsf{B}]\!]\rho = \mathsf{ff}\} &\quad \overline{\mathsf{test}}^{\vec{R}}[\![\mathsf{B}]\!]\mathcal{P}_0 &\triangleq \{\langle \rho_0, \rho\rangle \in \mathcal{P}_0 \mid \mathcal{B}[\![\mathsf{B}]\!]\rho = \mathsf{ff}\}
\end{aligned}$$

Proof of (19.22)

$$\widehat{\mathcal{S}}^{\vec{\varrho}}[\![\mathsf{S}]\!]\,\mathcal{P}_0\,\ell$$

$$= \{\varrho(\pi_0\ell_0\pi_1\ell') \mid \varrho(\pi_0\ell_0) \in \mathcal{P}_0 \wedge \ell_0\pi_1\ell' \in \widehat{\mathcal{S}}^*[\![\mathsf{S}]\!](\pi_0\ell_0) \wedge \ell' = \ell \in \mathsf{labs}[\![\mathsf{S}]\!]\}$$

(by definitions (20.15) and (20.1), similar for (20.9))

$$\begin{aligned}
= & \{\varrho(\pi_0\ell_0\pi_1\ell') \mid \varrho(\pi_0\ell_0) \in \mathcal{P}_0 \wedge \ell_0\pi_1\ell' \in \{\mathsf{at}[\![\mathsf{S}]\!]\} \wedge \ell' = \ell \in \mathsf{labs}[\![\mathsf{S}]\!]\} \cup \\
& \{\varrho(\pi_0\ell_0\pi_1\ell') \mid \varrho(\pi_0\ell_0) \in \mathcal{P}_0 \wedge \ell_0\pi_1\ell' \in \{\ell_0 \xrightarrow{\mathsf{B}} \mathsf{at}[\![\mathsf{S}_t]\!] \frown \pi_3 \mid \mathcal{B}[\![\mathsf{B}]\!]\varrho(\pi_0\ell_0) = \mathsf{tt} \wedge \pi_3 \in \\
& \mathcal{S}^*[\![\mathsf{S}_t]\!](\pi_0\ell_0 \xrightarrow{\mathsf{B}} \mathsf{at}[\![\mathsf{S}_t]\!])\} \wedge \ell' = \ell \in \mathsf{labs}[\![\mathsf{S}]\!]\} \cup \\
& \{\varrho(\pi_0\ell_0\pi_1\ell') \mid \varrho(\pi_0\ell_0) \in \mathcal{P}_0 \wedge \ell_0\pi_1\ell' \in \{\ell_0 \xrightarrow{\neg(\mathsf{B})} \mathsf{after}[\![\mathsf{S}]\!] \mid \mathcal{B}[\![\mathsf{B}]\!]\varrho(\pi_0\ell_0) = \mathsf{ff}\} \wedge \ell' = \ell \in \\
& \mathsf{labs}[\![\mathsf{S}]\!]\}
\end{aligned}$$

(definition of the prefix semantics same as (6.11), (6.18) and (6.19))

$$\begin{aligned}
= & (\!\!|\; \ell = \mathsf{at}[\![\mathsf{S}]\!] \;?\; \mathcal{P}_0 \\
& \|\; \ell \in \mathsf{in}[\![\mathsf{S}_t]\!] \;?\; \{\varrho(\pi_0\ell_0\pi_1\ell') \mid \varrho(\pi_0\ell_0) \in \mathcal{P}_0 \wedge \ell_0\pi_1\ell' \in \{\ell_0 \xrightarrow{\mathsf{B}} \mathsf{at}[\![\mathsf{S}_t]\!] \frown \pi_3 \mid \mathcal{B}[\![\mathsf{B}]\!]\varrho(\pi_0\ell_0) = \\
& \mathsf{tt} \wedge \pi_3 \in \mathcal{S}^*[\![\mathsf{S}_t]\!](\pi_0\ell_0 \xrightarrow{\mathsf{B}} \mathsf{at}[\![\mathsf{S}_t]\!])\} \wedge \ell' = \ell\} \\
& \|\; \ell = \mathsf{after}[\![\mathsf{S}]\!] \;?\; \{\varrho(\pi_0\ell_0\pi_1\ell') \mid \varrho(\pi_0\ell_0) \in \mathcal{P}_0 \wedge \ell_0\pi_1\ell' \in \{\ell_0 \xrightarrow{\mathsf{B}} \mathsf{at}[\![\mathsf{S}_t]\!] \frown \pi_3 \mid \\
& \mathcal{B}[\![\mathsf{B}]\!]\varrho(\pi_0\ell_0) = \mathsf{tt} \wedge \pi_3 \in \mathcal{S}^*[\![\mathsf{S}_t]\!](\pi_0\ell_0 \xrightarrow{\mathsf{B}} \mathsf{at}[\![\mathsf{S}_t]\!])\} \wedge \ell' = \ell\} \cup \\
& \{\varrho(\pi_0\ell_0\pi_1\ell') \mid \varrho(\pi_0\ell_0) \in \mathcal{P}_0 \wedge \ell_0\pi_1\ell' \in \{\ell_0 \xrightarrow{\neg(\mathsf{B})} \mathsf{after}[\![\mathsf{S}]\!] \mid \mathcal{B}[\![\mathsf{B}]\!]\varrho(\pi_0\ell_0) = \mathsf{ff}\} \wedge \ell' = \ell\} \\
& :\; \emptyset \;|\!\!)
\end{aligned}$$

(case analysis with $\mathsf{in}[\![\mathsf{S}]\!] = \{\mathsf{at}[\![\mathsf{S}]\!]\} \cup \mathsf{in}[\![\mathsf{S}_t]\!]$, $\mathsf{labs}[\![\mathsf{S}]\!] = \mathsf{in}[\![\mathsf{S}]\!] \cup \{\mathsf{after}[\![\mathsf{S}]\!]\}$ and $\mathsf{after}[\![\mathsf{S}_t]\!] = \mathsf{after}[\![\mathsf{S}]\!] \notin \mathsf{in}[\![\mathsf{S}]\!]$)

$$\begin{aligned}
= & (\!\!|\; \ell = \mathsf{at}[\![\mathsf{S}]\!] \;?\; \mathcal{P}_0 \\
& \|\; \ell \in \mathsf{in}[\![\mathsf{S}_t]\!] \;?\; \{\varrho(\pi_0\ell_0\pi_1\ell') \mid \varrho(\pi_0\ell_0) \in (\mathsf{test}^{\vec{\varrho}}[\![\mathsf{B}]\!]\mathcal{P}_0) \wedge \ell_0\pi_1\ell' \in \{\ell_0 \xrightarrow{\mathsf{B}} \mathsf{at}[\![\mathsf{S}_t]\!] \frown \pi_3 \mid \pi_3 \in \\
& \mathcal{S}^*[\![\mathsf{S}_t]\!](\pi_0\ell_0 \xrightarrow{\mathsf{B}} \mathsf{at}[\![\mathsf{S}_t]\!])\} \wedge \ell' = \ell\} \\
& \|\; \ell = \mathsf{after}[\![\mathsf{S}]\!] \;?\; \{\varrho(\pi_0\ell_0\pi_1\ell') \mid \varrho(\pi_0\ell_0) \in (\mathsf{test}^{\vec{\varrho}}[\![\mathsf{B}]\!]\mathcal{P}_0) \wedge \ell_0\pi_1\ell' \in \{\ell_0 \xrightarrow{\mathsf{B}} \mathsf{at}[\![\mathsf{S}_t]\!] \frown \pi_3 \mid \pi_3 \in \\
& \mathcal{S}^*[\![\mathsf{S}_t]\!](\pi_0\ell_0 \xrightarrow{\mathsf{B}} \mathsf{at}[\![\mathsf{S}_t]\!])\} \wedge \ell' = \ell\} \cup \{\rho \in \mathcal{P}_0 \mid \mathcal{B}[\![\mathsf{B}]\!]\rho = \mathsf{ff}\} \\
& :\; \emptyset \;|\!\!)
\end{aligned}$$

⦃grouping the conditions on $\boldsymbol{\varrho}(\pi_0\ell_0)$. For the false branch, $\ell_0\pi_1\ell' = \ell_0 \xrightarrow{\neg(\mathsf{B})} \mathsf{after}[\![\mathsf{S}]\!]$ where $\ell_0 = \mathsf{at}[\![\mathsf{S}]\!]$ and $\ell' = \ell = \mathsf{after}[\![\mathsf{S}]\!]$ so $\boldsymbol{\varrho}(\pi_0\ell_0\pi_1\ell') = \boldsymbol{\varrho}(\pi_0\ell_0 \xrightarrow{\neg(\mathsf{B})} \mathsf{after}[\![\mathsf{S}]\!]) = \boldsymbol{\varrho}(\pi_0\ell_0)$ by definition (6.6) of $\boldsymbol{\varrho}$. So the term $\{\boldsymbol{\varrho}(\pi_0\ell_0\pi_1\ell') \mid \boldsymbol{\varrho}(\pi_0\ell_0) \in \mathscr{P}_0 \wedge \ell_0\pi_1\ell' \in \{\ell_0 \xrightarrow{\neg(\mathsf{B})} \mathsf{after}[\![\mathsf{S}]\!] \mid \mathscr{B}[\![\mathsf{B}]\!]\boldsymbol{\varrho}(\pi_0\ell_0) = \mathsf{ff}\} \wedge \ell' = \ell\}$ reduces to $\{\rho \in \mathscr{P}_0 \mid \mathscr{B}[\![\mathsf{B}]\!]\rho = \mathsf{ff}\} = (\mathscr{P}_0 \cap \overline{\mathsf{test}}^{\vec{e}}[\![\mathsf{B}]\!])$.⦄

$= ⦇\, \ell = \mathsf{at}[\![\mathsf{S}]\!] \,?\, \mathscr{P}_0$

$\;\|\; \ell \in \mathsf{in}[\![\mathsf{S}_t]\!] \,?\, \widehat{\boldsymbol{\mathcal{S}}}^{\vec{e}}[\![\mathsf{S}_t]\!]\, \{\rho \in \mathscr{P}_0 \mid \mathscr{B}[\![\mathsf{B}]\!]\rho = \mathsf{tt}\}\, \ell$

$\;\|\; \ell = \mathsf{after}[\![\mathsf{S}]\!] \,?\, \widehat{\boldsymbol{\mathcal{S}}}^{\vec{e}}[\![\mathsf{S}_t]\!]\, (\mathsf{test}^{\vec{e}}[\![\mathsf{B}]\!]\mathscr{P}_0)\, \ell \cup \{\rho \in \mathscr{P}_0 \mid \mathscr{B}[\![\mathsf{B}]\!]\rho = \mathsf{ff}\} \,⨾\, \varnothing\, ⦈$

⦃— For the true branch in S_t, $\ell_0\pi_1\ell' = \ell_0 \xrightarrow{\mathsf{B}} \mathsf{at}[\![\mathsf{S}_t]\!] \frown \pi_3$ where $\ell_0 = \mathsf{at}[\![\mathsf{S}]\!]$, π_3 has the form $\pi_3 = \mathsf{at}[\![\mathsf{S}_t]\!]\pi_3'$, and $\ell' = \ell \in \mathsf{in}[\![\mathsf{S}_t]\!]$ so π_3 has the form $\mathsf{at}[\![\mathsf{S}_t]\!]\pi_4\ell'\pi_5$. Hence $\ell_0\pi_1\ell'$ has the form $\ell_0 \xrightarrow{\mathsf{B}} \mathsf{at}[\![\mathsf{S}_t]\!]\pi_4\ell'\pi_5$ so $\ell_0\pi_1\ell' = \ell_0 \xrightarrow{\mathsf{B}} \mathsf{at}[\![\mathsf{S}_t]\!]\pi_4\ell'$ and $\pi_5 = \ni$ is empty that is $\pi_3 = \mathsf{at}[\![\mathsf{S}_t]\!]\pi_4\ell'$. It follows that

$\{\boldsymbol{\varrho}(\pi_0\ell_0\pi_1\ell') \mid \boldsymbol{\varrho}(\pi_0\ell_0) \in (\mathsf{test}^{\vec{e}}[\![\mathsf{B}]\!]\mathscr{P}_0) \wedge \ell_0\pi_1\ell' \in \{\ell_0 \xrightarrow{\mathsf{B}} \mathsf{at}[\![\mathsf{S}_t]\!] \frown \pi_3 \mid \pi_3 \in \boldsymbol{\mathcal{S}}^*[\![\mathsf{S}_t]\!](\pi_0\ell_0 \xrightarrow{\mathsf{B}} \mathsf{at}[\![\mathsf{S}_t]\!])\} \wedge \ell' = \ell\}$

$=$

$\{\boldsymbol{\varrho}(\pi_0\ell_0 \xrightarrow{\mathsf{B}} \mathsf{at}[\![\mathsf{S}_t]\!]\pi_4\ell') \mid \boldsymbol{\varrho}(\pi_0\ell_0) \in \{\rho \in \mathscr{P}_0 \mid \mathscr{B}[\![\mathsf{B}]\!]\rho = \mathsf{tt}\} \wedge \mathsf{at}[\![\mathsf{S}_t]\!]\pi_4\ell' \in \boldsymbol{\mathcal{S}}^*[\![\mathsf{S}_t]\!](\pi_0\ell_0 \xrightarrow{\mathsf{B}} \mathsf{at}[\![\mathsf{S}_t]\!]) \wedge \ell' = \ell\}$

$=$

$\{\boldsymbol{\varrho}(\pi_0\ell_0 \xrightarrow{\mathsf{B}} \mathsf{at}[\![\mathsf{S}_t]\!]\pi_4\ell') \mid \boldsymbol{\varrho}(\pi_0\ell_0) \in (\mathsf{test}^{\vec{e}}[\![\mathsf{B}]\!]\mathscr{P}_0) \wedge \mathsf{at}[\![\mathsf{S}_t]\!]\pi_4\ell' \in \boldsymbol{\mathcal{S}}^*[\![\mathsf{S}_t]\!](\pi_0\ell_0 \xrightarrow{\mathsf{B}} \mathsf{at}[\![\mathsf{S}_t]\!]) \wedge \ell' = \ell\}$
⦃$\boldsymbol{\mathcal{S}}^*[\![\mathsf{S}_t]\!]$ is prefix closed⦄

$=$

$\{\boldsymbol{\varrho}(\pi_0\ell_0 \xrightarrow{\mathsf{B}} \mathsf{at}[\![\mathsf{S}_t]\!]\pi_4\ell') \mid \boldsymbol{\varrho}(\pi_0\ell_0 \xrightarrow{\mathsf{B}} \mathsf{at}[\![\mathsf{S}_t]\!]) \in (\mathsf{test}^{\vec{e}}[\![\mathsf{B}]\!]\mathscr{P}_0) \wedge \mathsf{at}[\![\mathsf{S}_t]\!]\pi_4\ell' \in \boldsymbol{\mathcal{S}}^*[\![\mathsf{S}_t]\!](\pi_0\ell_0 \xrightarrow{\mathsf{B}} \mathsf{at}[\![\mathsf{S}_t]\!]) \wedge \ell' = \ell\}$ ⦃because $\boldsymbol{\varrho}(\pi_0\ell_0) = \boldsymbol{\varrho}(\pi_0\ell_0 \xrightarrow{\mathsf{B}} \mathsf{at}[\![\mathsf{S}_t]\!])$⦄

$=$

$\{\boldsymbol{\varrho}(\pi_0'\ell_0'\pi_1\ell') \mid \boldsymbol{\varrho}(\pi_0'\ell_0') \in (\mathsf{test}^{\vec{e}}[\![\mathsf{B}]\!]\mathscr{P}_0) \wedge \ell_0'\pi_1\ell' \in \widehat{\boldsymbol{\mathcal{S}}}^*[\![\mathsf{S}_t]\!](\pi_0'\ell_0') \wedge \ell' = \ell\}$
⦃letting $\pi_0'\ell_0' = \pi_0\ell_0 \xrightarrow{\mathsf{B}} \mathsf{at}[\![\mathsf{S}_t]\!]$ and $\ell_0'\pi_1\ell' = \mathsf{at}[\![\mathsf{S}_t]\!]\pi_4\ell'$⦄

$=$

$\widehat{\boldsymbol{\mathcal{S}}}^{\vec{e}}[\![\mathsf{S}_t]\!]\, (\mathsf{test}^{\vec{e}}[\![\mathsf{B}]\!]\mathscr{P}_0)\, \ell$ ⦃by definitions (20.15) and (20.1), similar for (20.9)⦄

— For the exit of the true branch this is similar to the previous case except that $\ell = \ell' = \mathsf{after}[\![\mathsf{S}_t]\!] = \mathsf{after}[\![\mathsf{S}]\!]$.

$= ⦇\, \ell = \mathsf{at}[\![\mathsf{S}]\!] \,?\, \mathscr{P}_0$

$\;\|\; \ell \in \mathsf{in}[\![\mathsf{S}_t]\!] \,?\, \widehat{\boldsymbol{\mathcal{S}}}^{\vec{e}}[\![\mathsf{S}_t]\!]\, (\mathsf{test}^{\vec{e}}[\![\mathsf{B}]\!]\mathscr{P}_0)\, \ell$

$\;\|\; \ell = \mathsf{after}[\![\mathsf{S}]\!] \,?\, \widehat{\boldsymbol{\mathcal{S}}}^{\vec{e}}[\![\mathsf{S}_t]\!]\, (\mathsf{test}^{\vec{e}}[\![\mathsf{B}]\!]\mathscr{P}_0)\, \ell \cup (\overline{\mathsf{test}}^{\vec{e}}[\![\mathsf{B}]\!]\mathscr{P}_0) \,⨾\, \varnothing\, ⦈$

⦃definitions (19.16) of $\mathsf{test}^{\vec{e}}[\![\mathsf{B}]\!]$ and $\overline{\mathsf{test}}^{\vec{e}}[\![\mathsf{B}]\!]$⦄⦄ □

Reachability of a conditional statement $\mathsf{S} ::= \mathtt{if}\,(\mathsf{B})\,\mathsf{S}_t\ \mathtt{else}\ \mathsf{S}_f$

$$\widehat{\boldsymbol{\mathcal{S}}}^{\vec{e}}[\![\mathsf{S}]\!]\,\mathcal{P}_0\,\ell \;=\; \big(\!\!\big[\ \ell = \mathsf{at}[\![\mathsf{S}]\!]\ ?\ \mathcal{P}_0 \quad (19.23)$$

$$\big[\!\!\big]\ \ell \in \mathsf{in}[\![\mathsf{S}_t]\!]\ ?\ \widehat{\boldsymbol{\mathcal{S}}}^{\vec{e}}[\![\mathsf{S}_t]\!]\,(\mathsf{test}^{\vec{e}}[\![\mathsf{B}]\!]\mathcal{P}_0)\,\ell$$

$$\big[\!\!\big]\ \ell \in \mathsf{in}[\![\mathsf{S}_f]\!]\ ?\ \widehat{\boldsymbol{\mathcal{S}}}^{\vec{e}}[\![\mathsf{S}_f]\!]\,(\widetilde{\mathsf{test}}^{\vec{e}}[\![\mathsf{B}]\!]\mathcal{P}_0)\,\ell$$

$$\big[\!\!\big]\ \ell = \mathsf{after}[\![\mathsf{S}]\!]\ ?$$

$$\qquad \widehat{\boldsymbol{\mathcal{S}}}^{\vec{e}}[\![\mathsf{S}_t]\!]\,(\mathsf{test}^{\vec{e}}[\![\mathsf{B}]\!]\mathcal{P}_0)\,\ell \cup \widehat{\boldsymbol{\mathcal{S}}}^{\vec{e}}[\![\mathsf{S}_f]\!]\,(\widetilde{\mathsf{test}}^{\vec{e}}[\![\mathsf{B}]\!]\mathcal{P}_0)\,\ell$$

$$:\ \emptyset\ \big]\!\!\big)$$

Exercise 20.17 Prove (19.23) by calculational design.

Exercise 20.18 Define

$$\begin{aligned}
\overleftarrow{\mathsf{test}}^{\vec{r}}[\![\mathsf{B}]\!]\mathcal{P} &\triangleq \{\rho \mid (\mathcal{B}[\![\mathsf{B}]\!]\rho = \mathsf{tt}) \Rightarrow (\rho \in \mathcal{P})\}\\
\overleftarrow{\mathsf{test}}^{\vec{R}}[\![\mathsf{B}]\!]\mathcal{P} &\triangleq \{\langle\rho,\,\rho\rangle \mid (\mathcal{B}[\![\mathsf{B}]\!]\rho = \mathsf{tt}) \Rightarrow (\langle\rho,\,\rho\rangle \in \mathcal{P})\}\\
\overleftarrow{\widetilde{\mathsf{test}}}^{\vec{r}}[\![\mathsf{B}]\!]\mathcal{P} &\triangleq \{\rho \mid (\mathcal{B}[\![\mathsf{B}]\!]\rho = \mathsf{ff}) \Rightarrow (\rho \in \mathcal{P})\}\\
\overleftarrow{\widetilde{\mathsf{test}}}^{\vec{R}}[\![\mathsf{B}]\!]\mathcal{P} &\triangleq \{\langle\rho,\,\rho\rangle \mid (\mathcal{B}[\![\mathsf{B}]\!]\rho = \mathsf{ff}) \Rightarrow (\langle\rho,\,\rho\rangle \in \mathcal{P})\}
\end{aligned}$$

Prove that $\langle \mathbb{E}\mathrm{v}^{\vec{e}},\,\subseteq\rangle \underset{\mathsf{test}^{\vec{e}}[\![\mathsf{B}]\!]}{\overset{\overleftarrow{\mathsf{test}}^{\vec{e}}[\![\mathsf{B}]\!]}{\leftrightarrows}} \langle \mathbb{E}\mathrm{v}^{\vec{e}},\,\subseteq\rangle$ and $\langle \mathbb{E}\mathrm{v}^{\vec{e}},\,\subseteq\rangle \underset{\widetilde{\mathsf{test}}^{\vec{e}}[\![\mathsf{B}]\!]}{\overset{\overleftarrow{\widetilde{\mathsf{test}}}^{\vec{e}}[\![\mathsf{B}]\!]}{\leftrightarrows}} \langle \mathbb{E}\mathrm{v}^{\vec{e}},\,\subseteq\rangle$.

To calculate the forward reachability semantics of an iteration statement $\mathsf{S} ::= \mathtt{while}\,\ell\,(\mathsf{B})\,\mathsf{S}_b$, we consider the prefix trace semantics in fixpoint form of section 17.1 and apply the exact iterates abstraction corollary 18.34.

Reachability of an iteration statement $\mathsf{S} ::= \mathtt{while}\,\ell\,(\mathsf{B})\,\mathsf{S}_b$

$$\widehat{\boldsymbol{\mathcal{S}}}^{\vec{e}}[\![\mathsf{S}]\!]\,\mathcal{P}_0\,\ell' \;=\; (\mathsf{lfp}^{\dot{\subseteq}}\,\mathcal{F}^{\vec{e}}[\![\mathtt{while}\,\ell\,(\mathsf{B})\,\mathsf{S}_b]\!]\,\mathcal{P}_0)\,\ell' \quad (19.16)$$

$$\mathcal{F}^{\vec{e}}[\![\mathtt{while}\,\ell\,(\mathsf{B})\,\mathsf{S}_b]\!]\,\mathcal{P}_0\,X\,\ell' \;=$$

$$\big(\!\!\big[\ \ell' = \ell\ ?\ \mathcal{P}_0 \cup \widehat{\boldsymbol{\mathcal{S}}}^{\vec{e}}[\![\mathsf{S}_b]\!]\,(\mathsf{test}^{\vec{e}}[\![\mathsf{B}]\!]X(\ell))\,\ell \quad \text{(a)}$$

$$\big[\!\!\big]\ \ell' \in \mathsf{in}[\![\mathsf{S}_b]\!] \setminus \{\ell\}\ ?\ \widehat{\boldsymbol{\mathcal{S}}}^{\vec{e}}[\![\mathsf{S}_b]\!]\,(\mathsf{test}^{\vec{e}}[\![\mathsf{B}]\!]X(\ell))\,\ell' \quad \text{(b)}$$

$$\big[\!\!\big]\ \ell' = \mathsf{after}[\![\mathsf{S}]\!]\ ?\ \widetilde{\mathsf{test}}^{\vec{e}}[\![\mathsf{B}]\!](X(\ell)) \cup \bigcup_{\ell'' \in \mathsf{breaks\text{-}of}[\![\mathsf{S}_b]\!]} \widehat{\boldsymbol{\mathcal{S}}}^{\vec{e}}[\![\mathsf{S}_b]\!]\,(\mathsf{test}^{\vec{e}}[\![\mathsf{B}]\!]X(\ell))\,\ell'' \quad \text{(c)}$$

$$:\ \emptyset\ \big]\!\!\big)$$

The reachable states are (a) the entry states at ℓ or the states at ℓ reached after zero or more iterates followed by one more iteration ending at ℓ, or (b) the states at ℓ' in the loop body reached from a state at ℓ for which the test is true, followed by one more execution reaching ℓ', or (c) the final states at $\mathsf{after}[\![\mathsf{S}]\!]$ reached from a state at ℓ for which the test is false, or reached from a state at ℓ for which the test is true, followed by one more execution reaching a break statement to $\mathsf{after}[\![\mathsf{S}]\!]$.

Proof of (19.16) Let X be an iterate of $\mathcal{F}^{\vec{e}}[\![\mathsf{S}]\!]\,\mathcal{P}_0$ where $\mathsf{S} \triangleq \mathtt{while}\,\ell\,(\mathsf{B})\,\mathsf{S}_b$.

$$\mathsf{post}^{\vec{e}}(\mathcal{F}^{*}[\![\mathtt{while}\,\ell\,(\mathsf{B})\,\mathsf{S}_b]\!](X))\,\mathcal{P}_0\,\ell'$$

$= \{\varrho(\pi_0\ell_0\pi_1\ell'') \mid \varrho(\pi_0\ell_0) \in \mathcal{P}_0 \wedge \ell_0 = \ell \wedge \ell_0\pi_1\ell'' \in \mathcal{F}^*[\![\texttt{while}\,\ell\,(\mathsf{B})\,\mathsf{S}_b]\!](X)(\pi_0\ell_0) \wedge \ell'' = \ell' \in \mathsf{labs}[\![\mathsf{S}]\!]\}$

⟨definition (20.1) of $\mathsf{post}^{\vec{r}}$, similar with (20.9) for $\mathsf{post}^{\vec{R}}$, and definition $\mathcal{F}^*[\![\texttt{while}\,\ell\,(\mathsf{B})\,\mathsf{S}_b]\!] = \varnothing$ when $\ell_0 \neq \ell$⟩

$= \{\varrho(\pi_0\ell\pi_1\ell'') \mid \varrho(\pi_0\ell) \in \mathcal{P}_0 \wedge \ell\pi_1\ell'' \in \mathcal{F}^*[\![\texttt{while}\,\ell\,(\mathsf{B})\,\mathsf{S}_b]\!](X)(\pi_0\ell) \wedge \ell'' = \ell' \in \mathsf{labs}[\![\mathsf{S}]\!]\}$

⟨because $\ell_0 = \ell$⟩

By definition (17.4) of $\mathcal{F}^*[\![\texttt{while}\,\ell\,(\mathsf{B})\,\mathsf{S}_b]\!](X)(\pi_0\ell)$, we have to consider the union of three cases.

(1) $\{\varrho(\pi_0\ell\pi_1\ell'') \mid \varrho(\pi_0\ell) \in \mathcal{P}_0 \wedge \ell\pi_1\ell'' \in \{\ell\} \wedge \ell'' = \ell' \in \mathsf{labs}[\![\mathsf{S}]\!]\}$

$= \{\varrho(\pi_0\ell) \mid \varrho(\pi_0\ell) \in \mathcal{P}_0 \wedge \ell = \ell'\}$ ⟨because $\ell\pi_1\ell'' = \ell = \ell'' = \ell' \in \mathsf{labs}[\![\mathsf{S}]\!]$⟩

$= \llparenthesis\, \ell' = \ell \;?\; \mathcal{P}_0 \;¿\; \varnothing \,\rrparenthesis$ ⟨by exercise 6.8⟩

(2) $\{\varrho(\pi_0\ell\pi_1\ell'') \mid \varrho(\pi_0\ell) \in \mathcal{P}_0 \wedge \ell\pi_1\ell'' \in \{\ell\pi_2\ell \xrightarrow{\neg(\mathsf{B})} \mathsf{after}[\![\mathsf{S}]\!] \mid \ell\pi_2\ell \in X(\pi_0\ell) \wedge \mathcal{B}[\![\mathsf{B}]\!]\varrho(\pi_0\ell\pi_2\ell) = \mathsf{ff}\} \wedge \ell'' = \ell' \in \mathsf{labs}[\![\mathsf{S}]\!]\}$

$= \{\varrho(\pi_0\ell\pi_1\ell'') \mid \varrho(\pi_0\ell) \in \mathcal{P}_0 \wedge \exists \pi_2 \,.\, \ell\pi_1\ell'' = \ell\pi_2\ell \xrightarrow{\neg(\mathsf{B})} \mathsf{after}[\![\mathsf{S}]\!] \wedge \ell\pi_2\ell \in X(\pi_0\ell) \wedge \mathcal{B}[\![\mathsf{B}]\!]\varrho(\pi_0\ell\pi_2\ell) = \mathsf{ff} \wedge \ell'' = \ell' \in \mathsf{labs}[\![\mathsf{S}]\!]\}$ ⟨definition of $\in$⟩

$= \{\varrho(\pi_0\ell\pi_2\ell \xrightarrow{\neg(\mathsf{B})} \mathsf{after}[\![\mathsf{S}]\!]) \mid \varrho(\pi_0\ell) \in \mathcal{P}_0 \wedge \ell\pi_2\ell \in X(\pi_0\ell) \wedge \mathcal{B}[\![\mathsf{B}]\!]\varrho(\pi_0\ell\pi_2\ell) = \mathsf{ff} \wedge \mathsf{after}[\![\mathsf{S}]\!] = \ell'\}$ ⟨because $\ell\pi_1\ell'' = \ell\pi_2\ell \xrightarrow{\neg(\mathsf{B})} \mathsf{after}[\![\mathsf{S}]\!]$ so $\ell'' = \mathsf{after}[\![\mathsf{S}]\!] \in \mathsf{labs}[\![\mathsf{S}]\!]$⟩

$= \{\varrho(\pi_0\ell_0\pi_2\ell'') \mid \varrho(\pi_0\ell_0) \in \mathcal{P}_0 \wedge \ell_0\pi_2\ell'' \in X(\pi_0\ell_0) \wedge \mathcal{B}[\![\mathsf{B}]\!]\varrho(\pi_0\ell_0\pi_2\ell'') = \mathsf{ff} \wedge \mathsf{after}[\![\mathsf{S}]\!] = \ell' \wedge \ell_0 = \ell'' = \ell\}$

⟨because $\varrho(\pi_0\ell\pi_2\ell \xrightarrow{\neg(\mathsf{B})} \mathsf{after}[\![\mathsf{S}]\!]) = \varrho(\pi_0\ell\pi_2\ell)$ by (6.6) and renaming $\ell_0 = \ell'' = \ell$⟩

$= \llparenthesis\, \ell' = \mathsf{after}[\![\mathsf{S}]\!] \;?\; \overline{\mathsf{test}}^{\vec{e}}[\![\mathsf{B}]\!](\{\varrho(\pi_0\ell_0\pi_2\ell'') \mid \varrho(\pi_0\ell_0) \in \mathcal{P}_0 \wedge \ell_0\pi_2\ell'' \in X(\pi_0\ell_0) \wedge \ell = \ell_0 = \ell''\}) \;¿\; \varnothing \,\rrparenthesis$ ⟨by definition of $\overline{\mathsf{test}}^{\vec{e}}[\![\mathsf{B}]\!]$⟩

$= \llparenthesis\, \ell' = \mathsf{after}[\![\mathsf{S}]\!] \;?\; \overline{\mathsf{test}}^{\vec{e}}[\![\mathsf{B}]\!](\{\varrho(\pi_0\ell_0\pi_2\ell'') \mid \varrho(\pi_0\ell_0) \in \mathcal{P}_0 \wedge \ell_0\pi_2\ell'' \in X(\pi_0\ell_0) \wedge \ell'' = \ell\}) \;¿\; \varnothing \,\rrparenthesis$

⟨by exercise 17.11 because X is an iterate of $\mathcal{F}^{\vec{e}}[\![\texttt{while}\,\ell\,(\mathsf{B})\,\mathsf{S}_b]\!]\,\mathcal{P}_0$ so $\ell_0 = \ell$⟩

$= \llparenthesis\, \ell' = \mathsf{after}[\![\mathsf{S}]\!] \;?\; \overline{\mathsf{test}}^{\vec{e}}[\![\mathsf{B}]\!](\mathsf{post}^{\vec{e}}(X)\,\mathcal{P}_0\,\ell) \;¿\; \varnothing \,\rrparenthesis$

⟨definition (20.1) of $\mathsf{post}^{\vec{r}}$ or (20.9) of $\mathsf{post}^{\vec{R}}$⟩

(3) $\{\varrho(\pi_0\ell\pi_1\ell'') \mid \varrho(\pi_0\ell) \in \mathcal{P}_0 \wedge \ell\pi_1\ell'' \in \{\ell\pi_2\ell \xrightarrow{\mathsf{B}} \mathsf{at}[\![\mathsf{S}_b]\!] \frown \pi_3\ell'' \mid \ell\pi_2\ell \in X(\pi_0\ell) \wedge \mathcal{B}[\![\mathsf{B}]\!]\varrho(\pi_0\ell\pi_2\ell) = \mathsf{tt} \wedge \pi_3\ell'' \in \mathcal{S}^*[\![\mathsf{S}_b]\!](\pi_0\ell\pi_2\ell \xrightarrow{\mathsf{B}} \mathsf{at}[\![\mathsf{S}_b]\!])\} \wedge \ell'' = \ell' \in \mathsf{labs}[\![\mathsf{S}]\!]\}$ (where ℓ of $\texttt{while}\,\ell\,(\mathsf{B})\,\mathsf{S}_b$ is global)

$= \{\varrho(\pi_0\ell\pi_1\ell'') \mid \varrho(\pi_0\ell) \in \mathcal{P}_0 \wedge \exists \pi_2, \pi_2 \,.\, \ell\pi_1\ell'' = \ell\pi_2\ell \xrightarrow{\mathsf{B}} \mathsf{at}[\![\mathsf{S}_b]\!] \frown \pi_3\ell'' \wedge \ell\pi_2\ell \in X(\pi_0\ell) \wedge \mathcal{B}[\![\mathsf{B}]\!]\varrho(\pi_0\ell\pi_2\ell) = \mathsf{tt} \wedge \pi_3\ell'' \in \mathcal{S}^*[\![\mathsf{S}_b]\!](\pi_0\ell\pi_2\ell \xrightarrow{\mathsf{B}} \mathsf{at}[\![\mathsf{S}_b]\!]) \wedge \ell'' = \ell' \in \mathsf{labs}[\![\mathsf{S}]\!]\}$

⟨definition of $\in$⟩

$= \{\varrho(\pi_0\ell\pi_1\ell') \mid \varrho(\pi_0\ell) \in \mathcal{P}_0 \wedge \exists \pi_2, \pi_3 \,.\, \ell\pi_1\ell' = \ell\pi_2\ell \xrightarrow{\mathsf{B}} \mathsf{at}[\![\mathsf{S}_b]\!] \frown \pi_3\ell' \wedge \ell\pi_2\ell \in X(\pi_0\ell) \wedge \mathcal{B}[\![\mathsf{B}]\!]\varrho(\pi_0\ell\pi_2\ell) = \mathsf{tt} \wedge \pi_3\ell' \in \mathcal{S}^*[\![\mathsf{S}_b]\!](\pi_0\ell\pi_2\ell \xrightarrow{\mathsf{B}} \mathsf{at}[\![\mathsf{S}_b]\!])\}$

$\wr$because $\ell'' = \ell' \in \mathsf{labs}[\![\mathsf{S}]\!]$ where ℓ and ℓ' are global$\int$

$= \quad \{\boldsymbol{\varrho}(\pi_0{}^{\ell}\pi_2{}^{\ell} \xrightarrow{\mathsf{B}} \mathsf{at}[\![\mathsf{S}_b]\!] \frown \pi_3{}^{\ell'}) \mid \boldsymbol{\varrho}(\pi_0{}^{\ell}) \in \mathscr{P}_0 \wedge {}^{\ell}\pi_2{}^{\ell} \in X(\pi_0{}^{\ell}) \wedge \mathscr{B}[\![\mathsf{B}]\!]\boldsymbol{\varrho}(\pi_0{}^{\ell}\pi_2{}^{\ell}) = \mathsf{tt} \wedge \pi_3{}^{\ell'} \in \boldsymbol{\mathcal{S}}^*[\![\mathsf{S}_b]\!](\pi_0{}^{\ell}\pi_2{}^{\ell} \xrightarrow{\mathsf{B}} \mathsf{at}[\![\mathsf{S}_b]\!])\}$ $\wr$because ${}^{\ell}\pi_1{}^{\ell'} = {}^{\ell}\pi_2{}^{\ell} \xrightarrow{\mathsf{B}} \mathsf{at}[\![\mathsf{S}_b]\!] \frown \pi_3{}^{\ell'}$$\int$

$= \quad \{\boldsymbol{\varrho}(\pi_0{}^{\ell_0}\pi_2{}^{\ell''} \xrightarrow{\mathsf{B}} \mathsf{at}[\![\mathsf{S}_b]\!] \frown \pi_3{}^{\ell'}) \mid \boldsymbol{\varrho}(\pi_0{}^{\ell_0}) \in \mathscr{P}_0 \wedge {}^{\ell}\pi_2{}^{\ell''} \in X(\pi_0{}^{\ell_0}) \wedge \ell'' = \ell \wedge \mathscr{B}[\![\mathsf{B}]\!]\boldsymbol{\varrho}(\pi_0{}^{\ell_0}\pi_2{}^{\ell''}) = \mathsf{tt} \wedge \pi_3{}^{\ell'} \in \boldsymbol{\mathcal{S}}^*[\![\mathsf{S}_b]\!](\pi_0{}^{\ell_0}\pi_2{}^{\ell''} \xrightarrow{\mathsf{B}} \mathsf{at}[\![\mathsf{S}_b]\!])\}$

$\wr$by exercise 17.11 because X is an iterate of $\mathscr{F}^{\vec{\varrho}}[\![$`while` ℓ $(\mathsf{B})\ \mathsf{S}_b]\!]\ \mathscr{P}_0$ so $\ell_0 = \ell$ and $\ell'' = \ell$$\int$

$= \quad \{\boldsymbol{\varrho}(\pi_0{}^{\ell_0}\pi_2{}^{\ell} \xrightarrow{\mathsf{B}} \mathsf{at}[\![\mathsf{S}_b]\!] \frown \pi_3{}^{\ell'}) \mid \boldsymbol{\varrho}(\pi_0{}^{\ell_0}\pi_2{}^{\ell}) \in \mathsf{test}^{\vec{\varrho}}[\![\mathsf{B}]\!](\mathsf{post}^{\vec{\varrho}}(X)\ \mathscr{P}_0\ \ell) \wedge \pi_3{}^{\ell'} \in \boldsymbol{\mathcal{S}}^*[\![\mathsf{S}_b]\!](\pi_0{}^{\ell_0}\pi_2{}^{\ell} \xrightarrow{\mathsf{B}} \mathsf{at}[\![\mathsf{S}_b]\!])\}$

$\wr$by definition of $\mathsf{test}^{\vec{\varrho}}[\![\mathsf{B}]\!]$ and definition (20.1) of $\mathsf{post}^{\vec{r}}(X)\ \mathscr{P}_0\ \ell \triangleq \{\boldsymbol{\varrho}(\pi_0{}^{\ell_0}\pi_2{}^{\ell''}) \mid \boldsymbol{\varrho}(\pi_0{}^{\ell_0}) \in \mathscr{P}_0 \wedge {}^{\ell_0}\pi_2{}^{\ell''} \in X(\pi_0{}^{\ell_0}) \wedge \ell'' = \ell\}$ or definition (20.9) of $\mathsf{post}^{\vec{\mathsf{R}}}$$\int$

$= \quad \{\boldsymbol{\varrho}(\pi_1{}^{\ell} \xrightarrow{\mathsf{B}} \mathsf{at}[\![\mathsf{S}_b]\!] \frown \pi_3{}^{\ell'}) \mid \boldsymbol{\varrho}(\pi_1{}^{\ell}) \in (\mathsf{test}^{\vec{\varrho}}[\![\mathsf{B}]\!](\mathsf{post}^{\vec{\varrho}}(X)\ \mathscr{P}_0\ \ell)) \wedge \pi_3{}^{\ell'} \in \boldsymbol{\mathcal{S}}^*[\![\mathsf{S}_b]\!](\pi_1{}^{\ell} \xrightarrow{\mathsf{B}} \mathsf{at}[\![\mathsf{S}_b]\!])\}$ $\wr$letting $\pi_1{}^{\ell} = \pi_0{}^{\ell_0}\pi_2{}^{\ell}$$\int$

$= \quad \{\boldsymbol{\varrho}(\pi_1{}^{\ell} \xrightarrow{\mathsf{B}} \mathsf{at}[\![\mathsf{S}_b]\!] \frown \pi_3{}^{\ell'}) \mid \boldsymbol{\varrho}(\pi_1{}^{\ell} \xrightarrow{\mathsf{B}} \mathsf{at}[\![\mathsf{S}_b]\!]) \in \mathsf{test}^{\vec{\varrho}}[\![\mathsf{B}]\!](\mathsf{post}^{\vec{\varrho}}(X)\ \mathscr{P}_0\ \ell) \wedge \pi_3{}^{\ell'} \in \boldsymbol{\mathcal{S}}^*[\![\mathsf{S}_b]\!](\pi_1{}^{\ell} \xrightarrow{\mathsf{B}} \mathsf{at}[\![\mathsf{S}_b]\!])\}$ $\wr$definition (6.6) of $\boldsymbol{\varrho}$$\int$

$= \quad \{\boldsymbol{\varrho}(\pi_0{}^{\ell_0}\pi_1{}^{\ell''}) \mid \boldsymbol{\varrho}(\pi_0{}^{\ell_0}) \in \mathsf{test}^{\vec{\varrho}}[\![\mathsf{B}]\!](\mathsf{post}^{\vec{\varrho}}(X)\ \mathscr{P}_0\ \ell) \wedge {}^{\ell_0}\pi_1{}^{\ell''} \in \widehat{\boldsymbol{\mathcal{S}}}^*[\![\mathsf{S}_b]\!](\pi_0{}^{\ell_0}) \wedge \ell'' = \ell' \in \mathsf{labs}[\![\mathsf{S}]\!]\}$

$\wr(\subseteq)$ letting $\pi_0{}^{\ell_0} = \pi_1{}^{\ell} \xrightarrow{\mathsf{B}} \mathsf{at}[\![\mathsf{S}_b]\!]$ and ${}^{\ell_0}\pi_1{}^{\ell''} = \pi_3{}^{\ell'}$.$\int$

$\wr(\supseteq)$ ${}^{\ell_0}\pi_1{}^{\ell''} \in \widehat{\boldsymbol{\mathcal{S}}}^*[\![\mathsf{S}_b]\!](\pi_0{}^{\ell_0})$ implies that $\ell_0 = \mathsf{at}[\![\mathsf{S}_b]\!]$.

By definition of dest in exercise 4.22, $\ell' \in \mathsf{dest}[\![\mathsf{S}]\!] = \mathsf{dest}[\![\mathsf{S}_b]\!]$ and ${}^{\ell_0}\pi_1{}^{\ell''} = \mathsf{at}[\![\mathsf{S}_b]\!]\pi_1{}^{\ell''} \in \widehat{\boldsymbol{\mathcal{S}}}^*[\![\mathsf{S}_b]\!](\pi_0{}^{\ell_0})$ imply, by definition of $\widehat{\boldsymbol{\mathcal{S}}}^*$ in section 6.6, and together with $\ell'' = \ell' \in \mathsf{labs}[\![\mathsf{S}]\!]$, that ${}^{\ell_0}\pi_1{}^{\ell''}$ has the form $\pi_3{}^{\ell'}$. Hence $\pi_0{}^{\ell_0}\pi_1{}^{\ell''} = \pi_0\mathsf{at}[\![\mathsf{S}_b]\!] \frown \pi_3{}^{\ell'}$.

For any $\rho \in \mathsf{test}^{\vec{\varrho}}[\![\mathsf{B}]\!](\mathsf{post}^{\vec{\varrho}}(X)\ \mathscr{P}_0\ \ell)$ there exists, by exercise 6.8, traces of the form $\pi_0{}^{\ell_0}$ such that $\boldsymbol{\varrho}(\pi_0{}^{\ell_0}) = \rho$. By exercise 6.8 and definition (6.6) of $\boldsymbol{\varrho}$ (which implies $\boldsymbol{\varrho}(\pi^{\ell} \xrightarrow{\mathsf{B}} \ell')\mathsf{x} = \boldsymbol{\varrho}(\pi^{\ell})$), there exists traces of the form $\pi_1{}^{\ell} \xrightarrow{\mathsf{B}} \mathsf{at}[\![\mathsf{S}_b]\!]$ such that $\boldsymbol{\varrho}(\pi_1{}^{\ell} \xrightarrow{\mathsf{B}} \mathsf{at}[\![\mathsf{S}_b]\!]) = \rho$. By exercise 6.36, we conclude that $\widehat{\boldsymbol{\mathcal{S}}}^*[\![\mathsf{S}_b]\!](\pi_0{}^{\ell_0}) = \boldsymbol{\mathcal{S}}^*[\![\mathsf{S}_b]\!](\pi_1{}^{\ell} \xrightarrow{\mathsf{B}} \mathsf{at}[\![\mathsf{S}_b]\!])$, proving the reverse inclusion.$\int$

$= \quad \mathsf{post}^{\vec{\varrho}}(\widehat{\boldsymbol{\mathcal{S}}}^*[\![\mathsf{S}_b]\!])\ (\mathsf{test}^{\vec{\varrho}}[\![\mathsf{B}]\!](\mathsf{post}^{\vec{\varrho}}(X)\ \mathscr{P}_0\ \ell))\ \ell', \quad \ell' \in \mathsf{labs}[\![\mathsf{S}]\!]$

$\wr$by definition (20.1) or (20.9)$\int$

$= \quad \widehat{\boldsymbol{\mathcal{S}}}^{\vec{\varrho}}[\![\mathsf{S}_b]\!]\ (\mathsf{test}^{\vec{\varrho}}[\![\mathsf{B}]\!](\mathsf{post}^{\vec{\varrho}}(X)\ \mathscr{P}_0\ \ell))\ \ell', \quad \ell' \in \mathsf{labs}[\![\mathsf{S}]\!]$ $\wr$by definition (20.15)$\int$

By section 4.2.7, $\mathsf{labs}[\![\mathsf{S}]\!] = \{\ell\} \cup \mathsf{in}[\![\mathsf{S}_b]\!] \cup \{\mathsf{after}[\![\mathsf{S}]\!]\}$ and by section 4.2.4, $\mathsf{break\text{-}to}[\![\mathsf{S}_b]\!] \triangleq \mathsf{after}[\![\mathsf{S}]\!]$ so we can group the various terms as follows

(a) for $\ell' = \ell$ we have the term (1) and the case $\ell' = \ell$ of (3);
(b) for $\ell' \in \mathsf{in}[\![\mathsf{S}_b]\!] \setminus \{\ell\}$ we have the terms (3);

(c) for $\ell' = \mathsf{after}[\![\mathsf{S}]\!]$, we have the term (2) and the terms (3) such that $\ell' = \mathsf{after}[\![\mathsf{S}]\!] = \mathsf{break\text{-}to}[\![\mathsf{S}_b]\!]$ that is, by section 4.2.5, belong to $\mathsf{breaks\text{-}of}[\![\mathsf{S}_b]\!]$. □

This regrouping of cases yields (19.16). □

Reachability of a break statement $\mathsf{S} ::= \ell\ \mathtt{break}\,;$

$$\widehat{\boldsymbol{\mathcal{S}}}^{\vec{e}}[\![\mathsf{S}]\!]\,\mathscr{P}_0\,\ell \quad = \quad (\!|\, \ell = \mathsf{at}[\![\mathsf{S}]\!]\ ?\ \mathscr{P}_0 \ :\ \varnothing\,|\!) \tag{19.25}$$

Proof of (19.25)

$$\widehat{\boldsymbol{\mathcal{S}}}^{\vec{e}}[\![\mathsf{S}]\!]\,\mathscr{P}_0\,\ell$$

$$= \{\varrho(\pi_0\ell_0\pi_1\ell') \mid \varrho(\pi_0\ell_0) \in \mathscr{P}_0 \wedge \ell_0\pi_1\ell' \in \widehat{\boldsymbol{\mathcal{S}}}^*[\![\ell\ \mathtt{break}\,;]\!](\pi_0\ell_0) \wedge \ell' = \ell \in \mathsf{labs}[\![\mathsf{S}]\!]\}$$

$\langle$by definition (20.15), (20.1) or (20.9), and $\mathsf{S} ::= \ell\ \mathtt{break}\,;\rangle$

$$= \{\varrho(\pi_0\ell_0\pi_1\ell') \mid \varrho(\pi_0\ell_0) \in \mathscr{P}_0 \wedge (\ell_0 = \ell' \vee \ell_0\pi_1\ell' = \ell_0 \xrightarrow{\mathsf{break}} \mathsf{break\text{-}to}[\![\mathsf{S}]\!]) \wedge \ell' = \ell \in \{\mathsf{at}[\![\mathsf{S}]\!]\}\}$$

$\langle$(6.11), definition (6.29) of $\widehat{\boldsymbol{\mathcal{S}}}^*[\![\mathsf{S}]\!](\pi_0\ell_0)$ and definition of $\mathsf{labs}[\![\mathsf{S}]\!]$ in section 4.2.7$\rangle$

$$= (\!|\, \ell = \mathsf{at}[\![\mathsf{S}]\!]\ ?\ \{\varrho(\pi_0\ell_0) \mid \varrho(\pi_0\ell_0) \in \mathscr{P}_0\} \ :\ \varnothing\,|\!)$$

$\langle$because $\mathsf{break\text{-}to}[\![\mathsf{S}]\!] \neq \mathsf{at}[\![\mathsf{S}]\!]$ in section 4.2.1 and section 4.2.4$\rangle$

$$= (\!|\, \ell = \mathsf{at}[\![\mathsf{S}]\!]\ ?\ \mathscr{P}_0 \ :\ \varnothing\,|\!)$$

$\langle$by exercise 6.8$\rangle$ □

Reachability of a compound statement $\mathsf{S} ::= \{\,\mathsf{Sl}\,\}$

$$\widehat{\boldsymbol{\mathcal{S}}}^{\vec{e}}[\![\mathsf{S}]\!] \quad = \quad \widehat{\boldsymbol{\mathcal{S}}}^{\vec{e}}[\![\mathsf{Sl}]\!] \tag{19.26}$$

Proof of (19.26)

$$\widehat{\boldsymbol{\mathcal{S}}}^{\vec{e}}[\![\mathsf{S}]\!]\,\mathscr{P}_0\,\ell$$

$$= \mathsf{post}^{\vec{e}}(\widehat{\boldsymbol{\mathcal{S}}}^*[\![\mathsf{S}]\!] \rceil \mathsf{labs}[\![\mathsf{S}]\!])\,\mathscr{P}_0\,\ell$$

$\langle$by definition (20.15)$\rangle$

$$= \mathsf{post}^{\vec{e}}(\widehat{\boldsymbol{\mathcal{S}}}^*[\![\mathsf{Sl}]\!] \rceil \mathsf{labs}[\![\mathsf{Sl}]\!])\,\mathscr{P}_0\,\ell$$

$\langle$by definition (6.30) and $\mathsf{labs}[\![\mathsf{S}]\!] = \mathsf{labs}[\![\mathsf{Sl}]\!]$ in section 4.2.7$\rangle$

$$= \widehat{\boldsymbol{\mathcal{S}}}^{\vec{e}}[\![\mathsf{Sl}]\!]\,\mathscr{P}_0\,\ell$$

$\langle$by definition (20.15)$\rangle$ □

Exercise 20.19 Design by calculus the complement dual forward reachability semantics defined by $\widehat{\boldsymbol{\mathcal{S}}}^{\tilde{\vec{e}}}[\![\mathsf{S}]\!] \triangleq \dot{\neg}\,\widehat{\boldsymbol{\mathcal{S}}}^{\vec{e}}[\![\mathsf{S}]\!]\neg$ using David Park conjugate fixpoint theorem 15.33.

20.5 Conclusion

Abstract interpretation can be used to relate semantics at various levels of abstraction describing executions with more or fewer details [227]. As illustrated by this chapter 20, the formal definition of the concrete semantics and the abstraction can be used to design the abstract semantics by calculus, using fixpoint abstraction given in chapter 18.

The forward reachability semantics is the basis for invariance proof methods and can be further abstracted to get automatic static reachability program analysis methods.

VII ABSTRACT DOMAIN AND ABSTRACT INTERPRETER

The program semantics can be abstracted away in a mathematical structure given by an abstract domain (formalizing the program properties defined by the semantics) and an abstract interpreter (formalizing how the semantics can be calculated), chapter 21.

This abstract interpreter is a formal specification of a generic static analyzer parameterized by an abstract domain.

This abstract interpreter can be understood as providing the solutions to equations (chapter 23) computed by iteration and by passing to the limit (chapter 22).

21 Abstract Domain and Abstract Structural Semantics

The maximal trace semantics $\boldsymbol{\mathcal{S}}^{+\infty}[\![\mathsf{P}]\!]$ of chapter 7, the relational reachability semantics $\widehat{\boldsymbol{\mathcal{S}}}^{\vec{\mathsf{R}}}[\![\mathsf{P}]\!]$ of chapter 19, the assertional reachability semantics $\boldsymbol{\mathcal{S}}^{\vec{\mathsf{r}}}[\![\mathsf{P}]\!]$ of chapter 19 all have the same structure. They are instances of a generic abstract semantics that can be implemented as an abstract interpreter parameterized by an abstract domain representing the information collected about program executions and the basic operations involved in this collection.

Contents

21.1 Abstract Domain

The domain of properties, inclusion (that is, logical implication), and the structural definitions of the semantics have the following common structure.

semantics	prefix trace $\widehat{\boldsymbol{\mathcal{S}}}^{\dot{*}}$	reachability $\widehat{\boldsymbol{\mathcal{S}}}^{\vec{\mathsf{r}}}$	abstract $\widehat{\boldsymbol{\mathcal{S}}}^{¤}$
	$\wp(\mathbb{T}^+) \xrightarrow{\nearrow} (\mathbb{L} \longrightarrow \wp(\mathbb{T}^+))$	$\wp(\mathbb{E}\mathbb{v}) \xrightarrow{\nearrow} (\mathbb{L} \longrightarrow \wp(\mathbb{E}\mathbb{v}))$	$\mathbb{P}^{¤} \xrightarrow{\nearrow} (\mathbb{L} \longrightarrow \mathbb{P}^{¤})$
	chapters 6 and 17	chapter 19	chapter 21
domain	$\wp(\mathbb{T}^+)$	$\wp(\mathbb{E}\mathbb{v})$	$\mathbb{P}^{¤}$
inclusion	$\subseteq$	$\subseteq$	$\sqsubseteq^{¤}$
abstraction	$\mathbb{1}_{\wp(\mathbb{T}^+)}$ [1]	$\ddot{\alpha}_{\varrho}$	$\alpha_{¤}$
infimum	$\varnothing$	$\varnothing$	$\bot^{¤}$
join	$\cup$	$\cup$	$\sqcup^{¤}$
assignment	$\mathsf{assign}^{\dot{*}}[\![\mathsf{x}, \mathsf{A}]\!]$	$\mathsf{assign}^{\vec{\mathsf{r}}}[\![\mathsf{x}, \mathsf{A}]\!]$	$\mathsf{assign}^{¤}[\![\mathsf{x}, \mathsf{A}]\!]$
test	$\mathsf{test}^{\dot{*}}[\![\mathsf{B}]\!]$	$\mathsf{test}^{\vec{\mathsf{r}}}[\![\mathsf{B}]\!]$	$\mathsf{test}^{¤}[\![\mathsf{B}]\!]$
	$\overline{\mathsf{test}}^{\dot{*}}[\![\mathsf{B}]\!]$	$\overline{\mathsf{test}}^{\vec{\mathsf{r}}}[\![\mathsf{B}]\!]$	$\overline{\mathsf{test}}^{¤}[\![\mathsf{B}]\!]$

1. $\mathbb{1}_S \triangleq x \in S \mapsto x$ is the identity function on the set S.

Definition 21.1 (domain well definedness) We say that a domain

$$\mathbb{D}^{¤} \triangleq \langle \mathbb{P}^{¤}, \sqsubseteq^{¤}, \bot^{¤}, \sqcup^{¤}, \mathsf{assign}^{¤}[\![\mathsf{x}, \mathsf{A}]\!], \mathsf{test}^{¤}[\![\mathsf{B}]\!], \overline{\mathsf{test}}^{¤}[\![\mathsf{B}]\!]\rangle$$

is *well defined* when $\langle \mathbb{P}^{¤}, \sqsubseteq^{¤}\rangle$ is a poset of properties with infimum $\bot^{¤}$; the lub $\sqcup^{¤}$ is well defined both for pairs of properties and for $\sqsubseteq^{¤}$-increasing chains (so that $\langle \mathbb{P}^{¤}, \sqsubseteq^{¤}\rangle$ is a join-lattice and a CPO); the assignment $\mathsf{assign}^{¤}$ is well defined in $(\mathbb{V} \times \mathbb{E}) \to \mathbb{P}^{¤} \xrightarrow{\nearrow} \mathbb{P}^{¤}$; and the tests $\mathsf{test}^{¤}[\![\mathsf{B}]\!]$ and $\overline{\mathsf{test}}^{¤}[\![\mathsf{B}]\!]$ are well defined in $\mathbb{B} \to \mathbb{P}^{¤} \xrightarrow{\nearrow} \mathbb{P}^{¤}$.

The abstract domain $\mathbb{D}^{¤}$ is an algebra whereas the domain of abstract properties $\mathbb{P}^{¤}$ is a set. So the mathematical structures are different. However, following mathematicians that call $\mathbb{Z}$ the *ring of integers* where a ring is an algebraic structure and $\mathbb{Z}$ is a set, we often say, by abuse of language, that $\mathbb{P}^{¤}$ an abstract domain or even less accurately, call $\mathbb{D}^{¤}$ the set of abstract properties (while meaning $\mathbb{P}^{¤}$). For expressivity, the set of $\mathbb{P}^{¤}$ may have to include a supremum $\top^{¤}$ (encoding tt).

Exercise 21.2 What would be the signature of a module implementing an abstract domain?

21.2 Forward/Deductive Abstract Interpreter

The semantics $\widehat{\mathcal{S}}^{¤}$ can be implemented as instances of a generic abstract interpreter, defined as follows.

- *Abstract semantics outside a statement* S

$$\ell \notin \mathsf{labs}[\![\mathsf{S}]\!] \quad\Rightarrow\quad \widehat{\mathcal{S}}^{¤}[\![\mathsf{S}]\!]\,\mathcal{P}_0\,\ell \triangleq \bot^{¤} \tag{21.3}$$

- *Abstract semantics of a program* P ::= Sl ℓ'

$$\widehat{\mathcal{S}}^{¤}[\![\mathsf{P}]\!] \;\triangleq\; \widehat{\mathcal{S}}^{¤}[\![\mathtt{Sl}]\!] \tag{21.4}$$

- *Abstract semantics of a statement list* Sl ::= Sl$'$ S

$$\begin{aligned}\widehat{\mathcal{S}}^{¤}[\![\mathtt{Sl}]\!]\,\mathcal{P}_0\,\ell \;\triangleq\; & (\!\!\mid \ell \in \mathsf{labs}[\![\mathtt{Sl}']\!] \setminus \{\mathsf{at}[\![\mathsf{S}]\!]\} \;?\; \widehat{\mathcal{S}}^{¤}[\![\mathtt{Sl}']\!]\,\mathcal{P}_0\,\ell\\ & \;\|\; \ell \in \mathsf{labs}[\![\mathsf{S}]\!] \;?\; \widehat{\mathcal{S}}^{¤}[\![\mathsf{S}]\!](\widehat{\mathcal{S}}^{¤}[\![\mathtt{Sl}']\!]\,\mathcal{P}_0\,\mathsf{at}[\![\mathsf{S}]\!])\,\ell\\ & \;{:}\; \bot^{¤} \,\mid\!\!)\end{aligned} \tag{21.5}$$

- *Abstract semantics of an empty statement list* Sl ::= ϵ

$$\widehat{\mathcal{S}}^{¤}[\![\mathtt{Sl}]\!]\,\mathcal{P}_0\,\ell \;\triangleq\; (\!\!\mid \ell = \mathsf{at}[\![\mathtt{Sl}]\!] \;?\; \mathcal{P}_0 \;{:}\; \bot^{¤} \,\mid\!\!) \tag{21.6}$$

- *Abstract semantics of an assignment statement* S ::= x = A ;

$$\begin{aligned}\widehat{\mathcal{S}}^{¤}[\![\mathsf{S}]\!]\,\mathcal{P}_0\,\ell \;=\; & (\!\!\mid \ell = \mathsf{at}[\![\mathsf{S}]\!] \;?\; \mathcal{P}_0\\ & \;\|\; \ell = \mathsf{after}[\![\mathsf{S}]\!] \;?\; \mathsf{assign}^{¤}[\![\mathsf{x}, \mathsf{A}]\!]\,\mathcal{P}_0\\ & \;{:}\; \bot^{¤} \,\mid\!\!)\end{aligned} \tag{21.7}$$

  where $\mathsf{assign}[\![\mathsf{x}, \mathsf{A}]\!] \circ \gamma \sqsubseteq \gamma \circ \mathsf{assign}^{¤}[\![\mathsf{x}, \mathsf{A}]\!]$.

- *Abstract semantics of a skip statement* S ::= ;

$$\widehat{\mathcal{S}}^{¤}[\![\mathsf{S}]\!]\,\mathcal{P}_0\,\ell \;=\; (\!\!\mid \ell \in \{\mathsf{at}[\![\mathsf{S}]\!], \mathsf{after}[\![\mathsf{S}]\!]\} \;?\; \mathcal{P}_0 \;{:}\; \bot^{¤} \,\mid\!\!) \tag{21.8}$$

- *Abstract semantics of a conditional statement* $\mathsf{S} ::= \mathtt{if}\,(\mathsf{B})\,\mathsf{S}_t$

$$\begin{aligned}\hat{\mathcal{S}}^{¤}[\![\mathsf{S}]\!]\,\mathcal{P}_0\,\ell \;=\; & (\!\!|\; \ell = \mathsf{at}[\![\mathsf{S}]\!] \;?\; \mathcal{P}_0\\ & \|\; \ell \in \mathsf{in}[\![\mathsf{S}_t]\!] \;?\; \hat{\mathcal{S}}^{¤}[\![\mathsf{S}_t]\!]\,(\mathsf{test}^{¤}[\![\mathsf{B}]\!]\,\mathcal{P}_0)\,\ell\\ & \|\; \ell = \mathsf{after}[\![\mathsf{S}]\!] \;?\\ & \qquad \hat{\mathcal{S}}^{¤}[\![\mathsf{S}_t]\!]\,(\mathsf{test}^{¤}[\![\mathsf{B}]\!]\,\mathcal{P}_0)\,\ell \sqcup^{¤} \overline{\mathsf{test}}^{¤}[\![\mathsf{B}]\!]\,\mathcal{P}_0\\ & \;\text{⸗}\; \bot^{¤} \;|\!\!)\end{aligned} \tag{21.9}$$

where $\mathsf{test}[\![\mathsf{B}]\!] \circ \gamma \sqsubseteq \gamma \circ \mathsf{test}^{¤}[\![\mathsf{B}]\!]$ and $\overline{\mathsf{test}}[\![\mathsf{B}]\!] \circ \gamma \sqsubseteq \gamma \circ \overline{\mathsf{test}}^{¤}[\![\mathsf{B}]\!]$.

- *Abstract semantics of a conditional statement* $\mathsf{S} ::= \mathtt{if}\,(\mathsf{B})\,\mathsf{S}_t\;\mathtt{else}\;\mathsf{S}_f$

$$\begin{aligned}\hat{\mathcal{S}}^{¤}[\![\mathsf{S}]\!]\,\mathcal{P}_0\,\ell \;=\; & (\!\!|\; \ell = \mathsf{at}[\![\mathsf{S}]\!] \;?\; \mathcal{P}_0\\ & \|\; \ell \in \mathsf{in}[\![\mathsf{S}_t]\!] \;?\; \hat{\mathcal{S}}^{¤}[\![\mathsf{S}_t]\!]\,(\mathsf{test}^{¤}[\![\mathsf{B}]\!]\,\mathcal{P}_0)\,\ell\\ & \|\; \ell \in \mathsf{in}[\![\mathsf{S}_f]\!] \;?\; \hat{\mathcal{S}}^{¤}[\![\mathsf{S}_f]\!]\,(\overline{\mathsf{test}}^{¤}[\![\mathsf{B}]\!]\,\mathcal{P}_0)\,\ell\\ & \|\; \ell = \mathsf{after}[\![\mathsf{S}]\!] \;?\\ & \qquad \hat{\mathcal{S}}^{¤}[\![\mathsf{S}_t]\!]\,(\mathsf{test}^{¤}[\![\mathsf{B}]\!]\,\mathcal{P}_0)\,\ell \sqcup^{¤} \hat{\mathcal{S}}^{¤}[\![\mathsf{S}_f]\!]\,(\overline{\mathsf{test}}^{¤}[\![\mathsf{B}]\!]\,\mathcal{P}_0)\,\ell\\ & \;\text{⸗}\; \bot^{¤} \;|\!\!)\end{aligned} \tag{21.10}$$

- *Abstract semantics of an iteration statement* $\mathsf{S} ::= \mathtt{while}\,\ell\,(\mathsf{B})\,\mathsf{S}_b$

$$\hat{\mathcal{S}}^{¤}[\![\mathsf{S}]\!]\,\mathcal{P}_0\,\ell' \;=\; \mathsf{lfp}^{\dot{\sqsubseteq}^{¤}}\,(\mathcal{F}^{¤}[\![\mathtt{while}\,\ell\,(\mathsf{B})\,\mathsf{S}_b]\!]\,\mathcal{P}_0)\,\ell' \tag{21.11}$$

$$\mathcal{F}^{¤}[\![\mathtt{while}\,\ell\,(\mathsf{B})\,\mathsf{S}_b]\!] \;\in\; \mathbb{P}^{¤} \to ((\mathbb{L} \to \mathbb{P}^{¤}) \to (\mathbb{L} \to \mathbb{P}^{¤}))$$

$$\begin{aligned}&\mathcal{F}^{¤}[\![\mathtt{while}\,\ell\,(\mathsf{B})\,\mathsf{S}_b]\!]\,\mathcal{P}_0\,X\,\ell' \;=\\ &(\!\!|\; \ell' = \ell \;?\; \mathcal{P}_0 \sqcup^{¤} \hat{\mathcal{S}}^{¤}[\![\mathsf{S}_b]\!]\,(\mathsf{test}^{¤}[\![\mathsf{B}]\!]X(\ell))\,\ell\\ &\|\; \ell' \in \mathsf{in}[\![\mathsf{S}_b]\!] \setminus \{\ell\} \;?\; \hat{\mathcal{S}}^{¤}[\![\mathsf{S}_b]\!]\,(\mathsf{test}^{¤}[\![\mathsf{B}]\!]X(\ell))\,\ell'\\ &\|\; \ell' = \mathsf{after}[\![\mathsf{S}]\!] \;?\; \overline{\mathsf{test}}^{¤}[\![\mathsf{B}]\!]X(\ell) \sqcup^{¤} \bigsqcup^{¤}_{\ell'' \in \mathsf{breaks\text{-}of}[\![\mathsf{S}_b]\!]} \hat{\mathcal{S}}^{¤}[\![\mathsf{S}_b]\!]\,(\mathsf{test}^{¤}[\![\mathsf{B}]\!]X(\ell))\,\ell''\\ &\;\text{⸗}\; \bot^{¤} \;|\!\!)\end{aligned}$$

- *Abstract semantics of a break statement* $\mathsf{S} ::= \ell\;\mathtt{break}\,;$

$$\hat{\mathcal{S}}^{¤}[\![\mathsf{S}]\!]\,\mathcal{P}_0\,\ell \;=\; (\!\!|\; \ell = \mathsf{at}[\![\mathsf{S}]\!] \;?\; \mathcal{P}_0 \;\text{⸗}\; \bot^{¤} \;|\!\!) \tag{21.12}$$

- *Abstract semantics of a compound statement* $\mathsf{S} ::= \{\,\mathsf{Sl}\,\}$

$$\hat{\mathcal{S}}^{¤}[\![\mathsf{S}]\!] \;=\; \hat{\mathcal{S}}^{¤}[\![\mathsf{Sl}]\!] \tag{21.13}$$

See remark 19.17 for the rôles of ℓ, ℓ', and ℓ'' in (21.11).

Remark 21.14 Following remarks 3.5 and 6.5, the abstract domain $\mathbb{D}^{¤}$ and the abstract interpreter $\hat{\mathcal{S}}^{¤}$ are implicitly parameterized by a set $\mathbb{L}$ of program labels and a set $\mathbb{V}$ of program variables that can be defined in different ways, which we want to leave implicit and unspecified. The only constraint is that all program variables belong to $\mathbb{V}$ and all program labels belong to $\mathbb{L}$. It follows that the definitions and hypotheses on the abstract interpreter and the abstract domains will be the same whichever specific definition of $\mathbb{L}$ and $\mathbb{V}$ is chosen in remarks 3.5 and 6.5.

Example 21.15 For the program

$$\mathsf{P} \;=\; \mathtt{while}\,\ell_1\,(\mathsf{x}\,!=2)\;\ell_2\;\mathtt{break}\,;\ell_3$$

with grammatical structure

$$\mathsf{P}\left[\begin{array}{l}\mathsf{Sl}_1\left[\begin{array}{l}\epsilon\,\big[\\ \mathsf{S}_2\left[\begin{array}{l}\mathtt{while}\,\ell_1\,(\mathsf{x}\,!=2)\\ \mathsf{S}_3\,\big[\;\ell_2\;\mathtt{break}\,;\end{array}\right.\end{array}\right.\\ \ell_3\end{array}\right.$$

we have

$$\begin{aligned}\widehat{\mathcal{S}}^{\sharp}[\![\mathsf{P}]\!]\,\mathcal{P}_0\,\ell \;=\;& (\!\!\mid \ell=\ell_1\;?\;\mathcal{P}_0 \;[\!]\; \ell=\ell_2\;?\;\mathsf{test}^{\sharp}[\![\mathsf{x}\,!=2]\!]\mathcal{P}_0\\ & [\!]\; \ell=\ell_3\;?\;\overline{\mathsf{test}}^{\sharp}[\![\mathsf{x}\,!=2]\!]\mathcal{P}_0 \sqcup^{\sharp} \mathsf{test}^{\sharp}[\![\mathsf{x}\,!=2]\!]\mathcal{P}_0 \;⨾\; \bot^{\sharp} \mid\!\!)\end{aligned}$$

as shown by the following formal derivation

$$\begin{aligned}
&\widehat{\mathcal{S}}^{\sharp}[\![\mathsf{P}]\!]\,\mathcal{P}_0\,\ell_3\\
=\;&\widehat{\mathcal{S}}^{\sharp}[\![\mathsf{Sl}_1]\!]\,\mathcal{P}_0\,\ell_3 && \text{(21.4)}\\
=\;&\widehat{\mathcal{S}}^{\sharp}[\![\mathsf{S}_2]\!]\,\mathcal{P}_0\,\ell_3 && \text{(21.5), } \mathsf{labs}[\![\epsilon]\!]=\mathsf{at}[\![\mathsf{S}_2]\!]\ell_1\text{, and (21.6)}\\
=\;&\mathsf{lfp}^{\sqsubseteq}\,(\mathcal{F}^{\sharp}[\![\mathtt{while}\,\ell_1\,(\mathsf{x}\,!=2)\;\mathsf{S}_3]\!]\,\mathcal{P}_0)\,\ell_3 && \text{(21.11)}
\end{aligned}$$

where $\mathcal{F}^{\sharp}[\![\mathtt{while}\,\ell_1\,(\mathsf{x}\,!=2)\;\ell_2\;\mathtt{break}\,;\,\ell_3]\!]\,\mathcal{P}_0\,X\,\ell'$

$$\begin{aligned}
=\;&(\!\!\mid \ell'=\ell_1\;?\;\mathcal{P}_0\\
&[\!]\; \ell'=\ell_2\;?\;\mathsf{test}^{\sharp}[\![\mathsf{x}\,!=2]\!]X(\ell_1)\\
&[\!]\; \ell'=\ell_3\;?\;\overline{\mathsf{test}}^{\sharp}[\![\mathsf{x}\,!=2]\!]X(\ell_1)\sqcup^{\sharp}\mathsf{test}^{\sharp}[\![\mathsf{x}\,!=2]\!]X(\ell_1)\\
&⨾\;\bot^{\sharp}\mid\!\!)
\end{aligned}$$

(21.11), $\mathsf{breaks\text{-}of}[\![\mathsf{S}_3]\!]=\{\ell_2\}$, $\mathsf{in}[\![\mathsf{S}_3]\!]=\{\ell_1,\ell_2\}$, $\widehat{\mathcal{S}}^{\sharp}[\![\mathsf{S}_3]\!]\,(\mathsf{test}^{\sharp}[\![\mathsf{x}\,!=2]\!]X(\ell_1))\,\ell_1=\bot^{\sharp}$ by (21.12), $\mathcal{P}_0\sqcup^{\sharp}\bot^{\sharp}=\mathcal{P}_0$, $\widehat{\mathcal{S}}^{\sharp}[\![\mathsf{S}_3]\!]\,(\mathsf{test}^{\sharp}[\![\mathsf{x}\,!=2]\!]X(\ell_1))\,\ell_2=\mathsf{test}^{\sharp}[\![\mathsf{x}\,!=2]\!]X(\ell_1)$ by (21.12), and $\mathsf{after}[\![\mathsf{S}]\!]=\ell_3$

$$\begin{aligned}
=\;&(\!\!\mid \ell'=\ell_1\;?\;\mathcal{P}_0 \;[\!]\; \ell'=\ell_2\;?\;\mathsf{test}^{\sharp}[\![\mathsf{x}\,!=2]\!]\mathcal{P}_0\\
&[\!]\; \ell'=\ell_3\;?\;\overline{\mathsf{test}}^{\sharp}[\![\mathsf{x}\,!=2]\!]\mathcal{P}_0\sqcup^{\sharp}\mathsf{test}^{\sharp}[\![\mathsf{x}\,!=2]\!]\mathcal{P}_0\;⨾\;\bot^{\sharp}\mid\!\!) && \text{because } X(\ell_1)=\mathcal{P}_0
\end{aligned}$$

$\mathcal{F}^{\sharp}[\![\mathsf{S}_2]\!]\,\mathcal{P}_0\,X\,\ell'$ does not depend upon X, so it is the unique hence least fixpoint. In general, $\overline{\mathsf{test}}^{\sharp}[\![\mathsf{x}\,!=2]\!]\mathcal{P}_0\sqcup^{\sharp}\mathsf{test}^{\sharp}[\![\mathsf{x}\,!=2]\!]\mathcal{P}_0=\mathcal{P}_0$ but this might not be the case for imprecise abstract domains (e.g. $\mathsf{test}^{\sharp}[\![\mathsf{B}]\!]\mathcal{P}_0=\top^{\sharp}$ is sound).

Theorem 21.16 (well definedness of the abstract interpreter) The abstract interpreter $\widehat{\mathcal{S}}^{\sharp}[\![\mathsf{S}]\!]$ for well defined abstract domain by $\mathbb{D}^{\sharp}$ of definition 21.1 (assuming $\mathsf{assign}^{\sharp}[\![\mathsf{x},\mathsf{A}]\!]$, $\mathsf{test}^{\sharp}[\![\mathsf{B}]\!]$, and $\overline{\mathsf{test}}^{\sharp}[\![\mathsf{B}]\!]$ to be continuous) is well defined and continuous for any program component $\mathsf{S}\in\mathbb{P}\mathbb{c}$.

Proof By structural induction, using definition 21.1 for basic cases and Scott–Kleene's iterative fixpoint theorem 15.26 for (19.16). The proof shows simultaneously that $\widehat{\mathcal{S}}^{\sharp}[\![\mathsf{S}]\!]\,\mathcal{P}_0$ is continuous in $\mathcal{P}_0$ for any program component S. This follows from the fact that the composition of continuous functions is continuous by exercise 15.25 and theorem 15.32 for the iteration. □

Once again upper bounds of increasing chains need only to exist for the fixpoint iterations (theorem 15.21) and the continuity hypothesis can be dispensed by considering transfinite fixpoint iterations [240].

Corollary 21.17 (well definedness of the reachability semantics) The structural forward reachability semantics of section 19.3 is well defined.

Proof The reachability semantics $\boldsymbol{S}^{\vec{r}}[\![\mathsf{S}]\!]$ and $\boldsymbol{S}^{\vec{R}}[\![\mathsf{S}]\!]$ and their common definition $\boldsymbol{S}^{\vec{\varrho}}[\![\mathsf{S}]\!]$ are easily checked to be an instance of the abstract interpreter and so by theorem 21.16 are well defined. □

Exercise 21.18 Express the prefix trace semantics $\widehat{\boldsymbol{S}}^*$ of chapter 6 as an instance of the abstract semantics.

Exercise 21.19 Express the relational reachability semantics $\widehat{\boldsymbol{S}}^{\vec{R}}$ of chapter 19 as an instance of the abstract semantics.

Exercise 21.20 (loop invariant) Continuing exercise 19.18, propose an alternative formulation of (21.11) using a loop invariant.

Exercise 21.21 Show that the reachability interpreter of exercise 19.34 is an instance of the abstract interpreter of exercise 21.22 for an appropriate choice of the abstract domain.

Exercise 21.22 (Project: forward abstract interpreter) Continuing exercise 19.34, transform the reachability interpreter into an abstract interpreter $\boldsymbol{S}^{\alpha}[\![\mathsf{P}]\!]$ of programs P parameterized by an abstract domain $\mathbb{D}^{\alpha}$ (as represented in exercise 21.2). The static analyzer reads a program P, computes its abstract syntax, calls the abstract interpreter for an initial abstract property $\mathscr{P}_0$, and prints the result of the static analysis. The abstract interpreter expects an abstract syntax tree of the following type.

```
(* File abstractTree.mli  © P. Cousot 2021 *)

open AbstractSyntaxExpressions
open AbstractDomain

type 'a tree =
  | Prog of 'a tree list * 'a
  | Assign of variable * aexpr * 'a
  | Emptystmt of 'a
  | If of bexpr * 'a tree * 'a
  | Ifelse of bexpr * 'a tree * 'a tree * 'a
  | While of bexpr * 'a tree * 'a
  | Break of 'a
  | Stmtlist of 'a tree list * 'a (* trees in inverse order *)
type program_label = int
type labelling =
    program_label * AbstractDomain.t * program_label * AbstractDomain.t *
    bool * program_label * AbstractDomain.t
    (* escape, where to break, abstract property when breaking *)
type bare_tree = unit tree (* without labeling attributes *)
type labelled_tree = labelling tree (* with labeling attributes *)
val vars : 'a tree -> AbstractSyntaxExpressions.variable list
```

In the abstract syntax tree, each attribute `(at,atP,af,afP,es,br,brP)` of a program component $\mathsf{S} \in \mathbb{P}\mathrm{c}$ is such that `at` = at$[\![\mathsf{S}]\!]$, `af` = after$[\![\mathsf{S}]\!]$, `es` = escape$[\![\mathsf{S}]\!]$, `br` =

break-to⟦S⟧, and initially `atP` = `afP` = `brP` = $\bot^{¤}$. The abstract interpreter computes `atP` = $\widehat{\mathcal{S}}^{¤}$⟦S⟧ $\mathscr{P}_0$ at⟦S⟧, `afP` = $\widehat{\mathcal{S}}^{¤}$⟦S⟧ $\mathscr{P}_0$ after⟦S⟧, and `brP` = $\bigcup_{\ell \in \text{breaks-of}[\![S]\!]} \widehat{\mathcal{S}}^{¤}[\![S]\!]\ \mathscr{P}_0\ \ell$.

Exercise 21.23 (Project: simple abstract domains) Provide a few simple abstract domains to run the abstract interpreter (e.g., the abstract domain with only one top $\top^{¤}$ value, or with only $\bot^{¤}$ (no environment is reachable) and $\top^{¤}$ (all environments are reachable)).

21.3 Finitary Abstract Domains

An abstract domain $\mathbb{P}^{¤}$ is *finitary* when it satisfies the *increasing chain condition*, that is, any strictly $\sqsubset^{¤}$-increasing chain is finite. A finite abstract domain is finitary.

Theorem 21.24 The abstract interpreter (i.e., a program implementing the specification $\widehat{\mathcal{S}}^{¤}$) is computationally well defined (i.e., terminating) for finitary abstract domains.

Proof By theorem 21.16, $\widehat{\mathcal{S}}^{¤}$ is mathematically well defined. By hypothesis the elements of $\mathbb{P}^{¤}$ are computer representable. But fixpoint iterations might be infinite (e.g., for the reachability semantics of a nonterminating program). However, the abstract domain being finitary the least fixpoint can be computed by iteration from the infimum. □

Note that only the fixpoint iteration strictly increasing chains need to be finite (so the definition of finitary can be weakened to the iterates of fixpoints).

21.4 Conclusion

The idea of an abstract interpreter is from [572, 741, 899, 967] and [234]. Infinitary abstract domains and the notion of soundness with respect to an operational semantics were first introduced in [234]. Abstract domains and abstract interpreter are available online such as APRON/Interproc [528, 529], PPL (Parma Polyhedra Library) [69] and EL/NA [897] (see section 53.2.8). Abstract domain can be integrated in constraint solvers [791, 921]. [972] shows how to transform an abstract interpreter into an abstract compiler [134] by partial evaluation [538], [257, sect. 7] using the first Futamura projection [384].

21.5 Solutions to Selected Exercises

Solution to exercise 21.2 In Ocaml (where we have added an initialization `initialize vl` (depending on the set `vl` of program variables) called after building the abstract syntax tree, an initial abstract property `initialP` specifying $\mathscr{P}_0$ on program entry, and a conversion `stringofaP` of an abstract value into a representation as a string).

```
(* file abstractDomain.mli  © P. Cousot 2021 *)

open AbstractSyntaxExpressions

type t
```

```
val initialize : variable list -> unit
val leq : t -> t -> bool
val bot : unit -> t
val join :  t -> t -> t
val initialP : unit -> t
val assign : variable -> aexpr -> t -> t
val test : bexpr -> t -> t
val nottest : bexpr -> t -> t
val stringofaP : t -> string
```

Solution to exercise 21.20

- *Abstract invariant of an iteration statement* $\mathsf{S} ::= \mathtt{while}\ \ell\ (\mathsf{B})\ \mathsf{S}_b$
[leftmargin=*]

$$\begin{aligned}
\hat{\mathcal{S}}^{¤}[\![\mathsf{S}]\!]\,\mathscr{P}_0\,\ell' \;=\;& \text{let}\ \overrightarrow{\mathscr{F}}^{¤}[\![\mathtt{while}\ \ell\ (\mathsf{B})\ \mathsf{S}_b]\!] \in \mathbb{P}^{¤} \xrightarrow{\nearrow} (\mathbb{P}^{¤} \xrightarrow{\nearrow} \mathbb{P}^{¤})\\
& \quad \overrightarrow{\mathscr{F}}^{¤}[\![\mathtt{while}\ \ell\ (\mathsf{B})\ \mathsf{S}_b]\!]\,\mathscr{P}_0\,X = \mathscr{P}_0 \sqcup^{¤} \hat{\mathcal{S}}^{¤}[\![\mathsf{S}_b]\!]\,(\mathsf{test}^{¤}[\![\mathsf{B}]\!]X)\,\ell\\
& \text{and}\ I = \mathsf{lfp}^{\sqsubseteq^{¤}}\,\overrightarrow{\mathscr{F}}^{¤}[\![\mathtt{while}\ \ell\ (\mathsf{B})\ \mathsf{S}_b]\!]\,\mathscr{P}_0\ \text{in}\\
& \quad (\!\!|\ \ell' = \ell\ ?\ I\\
& \quad \|\ \ell' \in \mathsf{in}[\![\mathsf{S}_b]\!] \setminus \{\ell\}\ ?\ \hat{\mathcal{S}}^{¤}[\![\mathsf{S}_b]\!]\,(\mathsf{test}^{¤}[\![\mathsf{B}]\!]\,I)\,\ell'\\
& \quad \|\ \ell' = \mathsf{after}[\![\mathsf{S}]\!]\ ?\ \overline{\mathsf{test}}^{¤}[\![\mathsf{B}]\!]\,I \sqcup^{¤} \bigsqcup^{¤}_{\ell'' \in \mathsf{breaks\text{-}of}[\![\mathsf{S}_b]\!]} \hat{\mathcal{S}}^{¤}[\![\mathsf{S}_b]\!]\,(\mathsf{test}^{¤}[\![\mathsf{B}]\!]\,I)\,\ell''\\
& \quad :\ \bot^{¤}\ |\!\!)
\end{aligned}$$

The proof is similar to that of exercise 19.18 with $I = (\mathsf{lfp}^{\sqsubseteq^{¤}}\,\mathscr{F}^{¤}[\![\mathtt{while}\ \ell\ (\mathsf{B})\ \mathsf{S}_b]\!]\,\mathscr{R}_0)\,\ell$.

Solution to exercise 21.22 An attribute `(at,atP,af,afP,es,br,brP)` decorates each program component $\mathsf{S} \in \mathbb{P}c$ of the program abstract syntax tree. The components `at`, `af`, `br` are the labels at S, after S, and where S breaks to when escape `es` is true (`l0` if none). `atP` and `afP` are the abstract property attached to `at` and `af` while `brP` is the join of all the abstract properties attached to the program points of S that break to `br` (when `es` is true). When the interpreter is initially called for the program, the labels are set to $\mathsf{at}[\![\mathsf{S}]\!]$, $\mathsf{after}[\![\mathsf{S}]\!]$, $\mathsf{escape}[\![\mathsf{S}]\!]$, and $\mathsf{break\text{-}to}[\![\mathsf{S}]\!]$, the properties are all set to the infimum `_|_`, and `r0` is $\mathscr{P}_0$ (defined by `initialP` ($\mathsf{vars}[\![\mathsf{P}]\!]$)). For the program component $\mathsf{S} \in \mathbb{P}c$ of the program abstract syntax tree, these properties are accessed by `property_at S` = `at`, `property_after S` = `afP`, and `property_break S` = `brP`, respectively. If the interpreter does terminate, they are equal to $\mathcal{S}^{¤}[\![\mathsf{S}]\!]\,\mathscr{P}_0\,\ell$ where ℓ is `at` or `af`, and `brP` = $\bigsqcup_{\ell \in \mathsf{breaks\text{-}of}[\![\mathsf{S}]\!]} \hat{\mathcal{S}}^{¤}[\![\mathsf{S}]\!]\,\mathscr{P}_0\,\ell$, respectively.

```
(* file abstractInterpreter.ml  © P. Cousot 2021 *)

open AbstractDomain
open AbstractTree
open AbstractSyntax

let rec lfp a f leq = let fa = f a in
                      if leq fa a then fa
                      else lfp fa f leq
```

```
let leq x y = match x, y with
  | (While (b, sb, (at,atP,af,afP,es,br,brP))),
      (While (b', sb', (at',atP',af',afP',es',br',brP'))) ->
        (AbstractDomain.leq atP atP') && (AbstractDomain.leq afP afP') && (
            AbstractDomain.leq brP brP')
  | _,_ -> failwith "abstractInterpreter: incorrect leq argument"

let rec fWhile r0 x = match x with
  | (While (b, sb, (at,xl,af,afP,es,br,brP))) ->
      let atP = join r0 xl in
      let sb' = abstractInterpreter sb (test b atP) in
        let afP' = property_after sb' and brP' = property_break sb' in
          let atP'' = join atP afP' in
            let afP'' = join (nottest b atP) brP' in
              While (b, sb', (at,atP'',af,afP'',es,br,bot ()))
  | _ -> failwith "abstractInterpreter: incorrect fWhile argument"
and abstractInterpreter s r0 =
  match s with
  | Prog (sl, (at,atP,af,afP,es,br,brP)) ->
      let sl',atP',afP',brP' = abstractInterpreterStmtlist sl r0 in
          Prog (sl', (at,atP',af,afP',es,br,bot ()))
  | Assign (v, a, (at,atP,af,afP,es,br,brP)) ->
      let afP' = assign v a r0 in
        Assign (v, a, (at,r0,af,afP',es,br,bot ()))
  | Emptystmt (at,atP,af,afP,es,br,brP) ->
      Emptystmt (at,r0,af,r0,es,br,bot ())
  | If (b, st, (at,atP,af,afP,es,br,brP)) ->
      let st' = abstractInterpreter st (test b r0) in
        let afP' = property_after st' and brP' = property_break st' in
          let afP'' = join afP' (nottest b r0) in
            If (b, st', (at,r0,af,afP'',es,br,brP'))
  | Ifelse (b, st, se, (at,atP,af,afP,es,br,brP)) ->
      let st' = abstractInterpreter st (test b r0)
      and se' = abstractInterpreter se (nottest b r0) in
        let afP' = property_after st' and brP' = property_break st'
        and afP'' = property_after se' and brP'' = property_break se' in
          let afP''' = join afP' afP'' and brP''' = join brP' brP'' in
            Ifelse (b, st', se', (at,r0,af,afP''',es,br,brP'''))
  | Break (at,atP,af,afP,es,br,brP) ->
      Break (at,r0,af,bot (),es,br,r0)
  | While (b, sb, (at,atP,af,afP,es,br,brP)) ->
      let wbot = (While (b, sb, (at,bot (),af,bot (),es,br,bot ()))) in
          let fp = lfp wbot (fWhile r0) leq in
            (match fp with
            | (While (b, sb, (at,atP,af,afP,es,br,brP))) ->
                (* update loop body after convergence of loop invariant *)
                let sb' = abstractInterpreter sb (test b atP) in
                    While (b, sb', (at,atP,af,afP,es,br,brP))
            | _ -> failwith "abstractInterpreter: incorrect while
                transformer")
  | Stmtlist (sl, (at,atP,af,afP,es,br,brP)) ->
      let sl',atP',afP',brP' = abstractInterpreterStmtlist sl r0 in
          Stmtlist (sl', (at,atP',af,afP',es,br,brP'))
and abstractInterpreterStmtlist sl r0 = match sl with
  | [] -> [],r0,r0,(bot ())
  | [s] ->
      let s' = abstractInterpreter s r0 in
        [s'],property_at s',property_after s',property_break s'
  | s :: sl' ->
      let sl'',atP',afP'',brP'' = abstractInterpreterStmtlist sl' r0 in
        let s' = abstractInterpreter s afP'' in
          let brP = join (property_break s') brP'' in
            s' :: sl'',atP',(property_after s'),brP
```

22 Chaotic Iterations

An alternative to structural definitions of semantics is to use solutions to systems of equations. Chaotic iterations for solving systems of equations allow to choose at each iteration which components evolve while the others are unchanged (but none can be forgotten forever). In full generality, different iteration strategies lead to different results. On the contrary, for continuous equations on complete partial orders, the limit of the iterates is always the least fixpoint.

Contents

22.1 Systems of Equations

Let $\vec{D} = \prod_{i=1}^{n} D_i$ be the Cartesian product of $n \geqslant 1$ sets D_i. Let $\vec{F} \in \vec{D} \to \vec{D}$. When we write the vectorial equation

$$\vec{X} \quad = \quad \vec{F}(\vec{X})$$

we mean the system of equations

$$\begin{cases} X_i & = & F_i(X_1, \ldots, X_n) \\ i = 1, \ldots, n \end{cases}$$

where $\vec{X} = \langle X_1, \ldots, X_n \rangle \in \vec{D}$ is a vector of variables and

$$F_i(X_1, \ldots, X_n) \quad \triangleq \quad \text{let } \langle X'_1, \ldots, X'_n \rangle = \vec{F}(\langle X_1, \ldots, X_n \rangle) \text{ in } X'_i$$

is the ith component of $\vec{F}(\vec{X})$.

The variables $X_i \in \mathbb{V}$, $i = 1, \ldots, n$ are identifiers with values in D_i whereas $\vec{F} \in \vec{D} \to \vec{D}$ so $\vec{F}(\vec{X})$ is an abuse of notation because $\vec{D} \neq \prod_{i=1}^{n} \mathbb{V}$. In this notation, it is implicitly understood that a solution to this system of equations is a map ρ of variables $\{X_i \in \mathbb{V} \mid i = 1, \ldots, n\}$ to their value such that $\rho(X_i) \in D_i$ and

$$\begin{cases} \rho(X_i) & = & F_i(\rho(X_1), \ldots, \rho(X_n)) \\ i = 1, \ldots, n \end{cases}$$

where now $F_i \in \vec{D} \to D_i$. To be fully rigorous, we cannot write $\vec{X} = \vec{F}(\vec{X})$ but we must consider a denotation $\vec{\mathsf{F}}$ of the function $\vec{F}$ written in some formally defined language, of which the semantics is $\mathcal{S}[\![\vec{\mathsf{F}}]\!] = \vec{F}$. So, a solution to $\vec{X} = \vec{\mathsf{F}}(\vec{X})$ is $\rho(\vec{X}) = \mathcal{S}[\![\vec{\mathsf{F}}]\!](\rho(\vec{X}))$. Following the well-established mathematical tradition (outside of mathematical logic) we identify function notations $\vec{\mathsf{F}}$ with the function $\vec{F} = \mathcal{S}[\![\vec{\mathsf{F}}]\!]$ that they denote and variables X_i with their value $\rho(X_i)$ and, ignoring the incoherence, write $\vec{X} = \vec{F}(\vec{X})$ for brevity.

22.2 Historical Iterative Methods

Iterative methods (e.g., in numerical linear algebra [524, section 4.1]) solve equations $\vec{X} = \vec{F}(\vec{X})$ iteratively by Jacobi (or simultaneous) iteration(s) method from a given $\vec{X}^0$ as the limit of iterates $\vec{X}^{k+1} = \vec{F}(\vec{X}^k)$. All components evolve simultaneously at all iterations.

$$\begin{cases} X_i^{k+1} = F_i(X_1^k, \ldots, X_n^k) \\ i = 1, \ldots, n, \quad k = 1, \ldots, +\infty \end{cases}$$

of which programming requires two arrays to record both $\vec{X}^k$ and $\vec{X}^{k+1}$. This is the iteration method considered in chapter 15, "Fixpoints."

One array is enough to program Gauss-Seidel (or successive) iteration(s) method [524, section 4.2]

$$\begin{cases} X_i^{k+1} = F_i(X_1^{k+1}, \ldots, X_{i-1}^{k+1}, X_i^k, \ldots, X_n^k) \\ i = 1, \ldots, n, \quad k = 1, \ldots, +\infty \end{cases}$$

where the components evolve one after another.

In general the two iteration strategies yield different results.

Example 22.1 Consider the Jacobi and Gauss-Seidel iterations of $\vec{F}$ defined as follows: In this example, the iterations lead to two different fixpoints.

$\langle x, y\rangle$	$F_1(x, y)$	$F_2(x, y)$
$\langle a, a\rangle$	b	b
$\langle a, b\rangle$	b	b
$\langle b, a\rangle$	b	a
$\langle b, b\rangle$	b	b

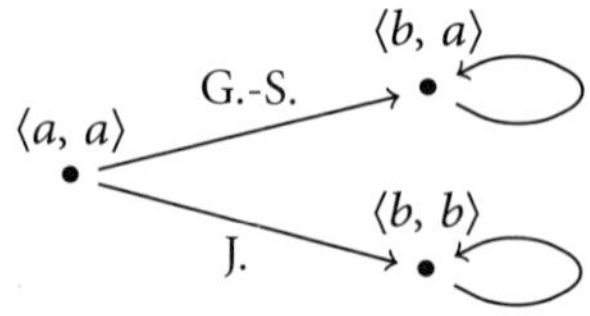

Jacobi iterations

$$\begin{aligned} \vec{X}^0 &= \langle a, a\rangle \\ \vec{X}^1 &= \langle F_1(a,a), F_2(a,a)\rangle = \langle b, b\rangle \\ \vec{X}^2, \vec{X}^3, \ldots &= \langle F_1(b,b), F_2(b,b)\rangle = \langle b, b\rangle \end{aligned}$$

Gauss-Seidel iterations

$$\begin{aligned} \vec{X}^0 &= \langle a, a\rangle \\ \vec{X}^1 &= \langle F_1(a,a), a\rangle = \langle b, a\rangle \\ \vec{X}^2, \vec{X}^4, \ldots &= \langle b, F_2(b,a)\rangle = \langle b, a\rangle \\ \vec{X}^3, \vec{X}^5, \ldots &= \langle F_1(b,a), a\rangle = \langle b, a\rangle \end{aligned}$$

22.3 Chaotic Iterations

For solving systems of linear equations, Daniel Chazan and Willard Miranker proposed a generalization to chaotic relaxation. A simple instance for general systems of equations is *chaotic iterations*. At each step, one can arbitrarily choose which components do evolve with the fairness condition that no component is omitted forever. If all components evolve at each iteration, that is Jacobi iterations. If the components evolve one after another, that is Gauss-Seidel.

A chaotic iteration is defined by an infinite sequence $\mathfrak{I}$ of subsets of $\{1,\ldots,n\}$ specifying that all components in $\mathfrak{I}(k)$ should evolved at iterate k (while those in $\{1,\ldots,n\}\setminus\mathfrak{I}(k)$ remain unchanged). This choice must be fair (i.e., no component is forgotten forever).

Definition 22.2 (chaotic iterations) Let $\vec{D}=\prod_{i=1}^{n} D_i$ be the Cartesian product of $n\geqslant 1$ sets D_i. Let $\vec{F}\in\vec{D}\to\vec{D}$. Let $\mathfrak{I}\in\mathbb{N}^+\to\wp(\{1,\ldots,n\})\setminus\{\varnothing\}$ satisfying the fairness condition

$$\forall i\in\{1,\ldots,n\}\,.\,\forall k\in\mathbb{N}\,.\,\exists k'>k\,.\,i\in\mathfrak{I}(k')\;.$$

The chaotic iterations $\vec{X}^k$, $k\in\mathbb{N}$ from $\vec{X}_0$ defined by $\mathfrak{I}$ for the system of equations $\vec{X}=\vec{F}(\vec{X})$ is

$$\begin{cases} \vec{X}_i^{k+1} \triangleq F_i(\vec{X}^k) & \text{when } i\in\mathfrak{I}(k)\\ \vec{X}_i^{k+1} \triangleq \vec{X}_i^k & \text{when } i\notin\mathfrak{I}(k)\;. \end{cases}$$

Exercise 22.3 Define $\mathfrak{I}$ to obtain Jacobi and Gauss-Seidel iterations.

22.4 Convergence of Chaotic Iterations of Continuous Operators on Complete Partial Orders

As shown by example 22.1, different chaotic iteration strategies lead to different results. However, as shown by [236], for continuous operators on a CPO, chaotic iterations starting from the infimum always converge to the least fixpoint.

Theorem 22.4 (convergence of chaotic iterations) The chaotic iterations of componentwise continuous operator $\vec{F}\in L^n \xrightarrow{uc} L^n$ on a CPO $\langle L^n,\,\dot{\sqsubseteq},\,\dot{\bot},\,\dot{\sqcup}\rangle$ from $\dot{\bot}$ converge to the $\mathsf{lfp}^{\dot{\sqsubseteq}}\,\vec{F}$.

Proof of theorem 22.4

— We first prove that the iterates form an increasing chain less than $\mathsf{lfp}^{\dot{\sqsubseteq}}\,\vec{F}$ (which exists by theorem 15.26).

By recurrence. For the basis, $\vec{X}^0=\dot{\bot}\,\dot{\sqsubseteq}\,\mathsf{lfp}^{\dot{\sqsubseteq}}\,\vec{F}$ by definition of the infimum $\dot{\bot}$. Assume, for the induction step, that for $k\in\mathbb{N}$, $\forall k'\leqslant k\,.\,\vec{X}^{k'}\,\dot{\sqsubseteq}\,\vec{X}^k\,\dot{\sqsubseteq}\,\mathsf{lfp}^{\dot{\sqsubseteq}}\,\vec{F}$ so that $\forall i\in\{1,\ldots,n\}\,.$ $\forall k'\leqslant k\,.\,\vec{X}_i^{k'}\sqsubseteq\vec{X}_i^k\sqsubseteq(\mathsf{lfp}^{\dot{\sqsubseteq}}\,\vec{F})_i$ componentwise.

Let $i\in\{1,\ldots,n\}$ be any component.

- If $i\notin\mathfrak{I}(k+1)$ then $\vec{X}_i^{k+1}=\vec{X}_i^k$ by definition of the iterates so $\vec{X}_i^k\sqsubseteq\vec{X}_i^{k+1}\sqsubseteq(\mathsf{lfp}^{\dot{\sqsubseteq}}\,\vec{F})_i$ by reflexivity and induction hypothesis so $\forall k'\leqslant k\,.\,\vec{X}_i^{k'}\sqsubseteq\vec{X}_i^{k+1}\sqsubseteq(\mathsf{lfp}^{\dot{\sqsubseteq}}\,\vec{F})_i$ by transitivity.

- Else $i \in \mathfrak{I}(k+1)$.
  - If there is no $k' \leqslant k$ such that $i \in \mathfrak{I}(k')$ then $\bot = \vec{X}_i^0 = \ldots = \vec{X}_i^k$ so $\vec{X}_i^k \sqsubseteq \vec{X}_i^{k+1} = F_i(\vec{X}^k) \sqsubseteq (\mathsf{lfp}^{\dot\sqsubseteq}\, \vec{F})_i$ by definition of the infimum $\bot$ and for all $\vec{X} \dot\sqsubseteq \mathsf{lfp}^{\dot\sqsubseteq}\, \vec{F}$, we have $\vec{F}(\vec{X}) \dot\sqsubseteq \vec{F}(\mathsf{lfp}^{\dot\sqsubseteq}\, \vec{F}) = \mathsf{lfp}^{\dot\sqsubseteq}\, \vec{F}$ because, by exercise 15.24, $\vec{F}$ is continuous hence increasing and by definition of fixpoints so $F_i(\vec{X}) \triangleq \vec{F}(\vec{X})_i \sqsubseteq (\mathsf{lfp}^{\dot\sqsubseteq}\, \vec{F})_i$ by componentwise definition of $\dot\sqsubseteq$;
  - Otherwise let $k' \leqslant k$ be the largest such that $i \in \mathfrak{I}(k')$ so that $\vec{X}_i^{k'-1} \sqsubseteq \vec{X}_i^{k'} = \vec{X}_i^{k'+1} = \ldots = \vec{X}_i^k \sqsubseteq (\mathsf{lfp}^{\dot\sqsubseteq}\, \vec{F})_i$ by definition of the iterates and induction hypothesis. By definition of the iterates, it follows that $F_i(\vec{X}^{k'-1}) = \vec{X}_i^{k'} = \vec{X}_i^{k'+1} = \ldots = \vec{X}_i^k \sqsubseteq F_i(\vec{X}^k) = \vec{X}_i^{k+1} \sqsubseteq (\mathsf{lfp}^{\dot\sqsubseteq}\, \vec{F})_i$ because, by exercise 15.24, F_i is continuous hence increasing. By transitivity, $\forall k' \leqslant k \,.\, \vec{X}_i^{k'} \sqsubseteq \vec{X}_i^{k+1} \sqsubseteq (\mathsf{lfp}^{\dot\sqsubseteq}\, \vec{F})_i$.

By componentwise definition of $\dot\sqsubseteq$, we conclude that $\forall k' \leqslant k \,.\, \vec{X}^{k'} \dot\sqsubseteq \vec{X}^k \dot\sqsubseteq \mathsf{lfp}^{\dot\sqsubseteq}\, \vec{F}$ proving, by recurrence, that the iterates form an increasing chain bounded by $\mathsf{lfp}^{\dot\sqsubseteq}\, \vec{F}$. By definition of a CPO in section 10.9 and that of a lub, they have a limit $\dot{\bigsqcup}_{k\in\mathbb{N}} \vec{X}^k \dot\sqsubseteq \mathsf{lfp}^{\dot\sqsubseteq}\, \vec{F}$.

— Let $\vec{J}^0 = \dot\bot$, $\vec{J}^{k+1} = \vec{F}(\vec{J}^k)$ be the Jacobi iterates, such that, by theorem 15.26, $\vec{J}^\omega = \dot{\bigsqcup}_{k\in\mathbb{N}} \vec{J}^k = \mathsf{lfp}^{\dot\sqsubseteq}\, \vec{F}$. We prove that any Jacobi iterate is ultimately bounded by a chaotic iterate, that is, $\forall k \in \mathbb{N} \,.\, \exists k' \geqslant k \,.\, \vec{J}^k \dot\sqsubseteq \vec{X}^{k'}$.

By the fairness hypothesis of definition 22.2, $\eta(k) = \max\{\min\{k' \geqslant k \mid i \in \mathfrak{I}(k')\} \mid i \in \{1, \ldots, n\}\}$ is well defined. In the chaotic iterations, all components have evolved at least once between k and $\eta(k)$. Let us extract the subsequence $\lambda(0) = 0$ and $\lambda(k+1) = \eta(\lambda(k))$. We have $\forall k \in \mathbb{N} \,.\, \vec{J}^k \dot\sqsubseteq \vec{X}^{\lambda(k)}$, that is, by waiting long enough, any Jacobi iterate will be overapproximated by some chaotic iterate. The proof is by recurrence.

For $k = 0$, $\vec{J}^0 = \vec{X}^{\lambda(0)} = \vec{X}^0 = \dot\bot$ by definition of λ so we conclude by reflexivity.

Assume by induction hypothesis that $\forall k' \leqslant k \in \mathbb{N} \,.\, \vec{J}^{k'} \dot\sqsubseteq \vec{X}^{\lambda(k')}$ where the $\vec{J}^\ell$ and $\vec{X}^\ell$, $\ell \in \mathbb{N}$, are $\dot\sqsubseteq$-increasing. $\vec{F}$ is continuous hence increasing so $\vec{J}^{k+1} = \vec{F}(\vec{J}^k) \dot\sqsubseteq \vec{F}(\vec{X}^{\lambda(k)})$.

For all $i \in \{1, \ldots, n\}$, the ith component has evolved at least once at k_i' between $\lambda(k)$ and $\eta(\lambda(k)) = \lambda(k+1)$ so the increasing chain has the form $\vec{X}^{\lambda(k)} \sqsubseteq \vec{X}^{k_i'-1} \sqsubseteq \vec{X}^{k_i'} \sqsubseteq \vec{X}^{\eta(\lambda(k))} = \vec{X}^{\lambda(k+1)}$ where $\vec{X}_i^{k_i'} = \vec{F}_i(\vec{X}^{k_i'-1})$. $\vec{F}$ hence $\vec{F}_i$ continuous hence increasing so by definition of the iterates, $\vec{F}_i(\vec{X}^{\lambda(k)}) \sqsubseteq \vec{F}_i(\vec{X}^{k_i'-1}) = \vec{X}^{k_i'} \sqsubseteq \vec{X}_i^{\lambda(k+1)}$. By componentwise definition of $\vec{F}$ and $\dot\sqsubseteq$, we have $\vec{F}(\vec{X}^{\lambda(k)}) \dot\sqsubseteq \vec{X}^{\lambda(k+1)}$ hence, by transitivity $\vec{J}^{k+1} \dot\sqsubseteq \vec{X}^{\lambda(k+1)}$.

By recurrence $\forall k \in \mathbb{N} \,.\, \vec{J}^k \dot\sqsubseteq \vec{X}^{\lambda(k)}$.

— Finally, $\{\vec{X}^{\lambda(k)} \mid k \in \mathbb{N}\} \subseteq \{\vec{X}^k \mid k \in \mathbb{N}\}$ so that by definition of lubs, $\mathsf{lfp}^{\dot\sqsubseteq \vec{F}} = \dot{\bigsqcup}_{k\in\mathbb{N}} \vec{J}^k \dot\sqsubseteq \dot{\bigsqcup}_{k\in\mathbb{N}} \vec{X}^{\lambda(k)} \dot\sqsubseteq \dot{\bigsqcup}_{k\in\mathbb{N}} \vec{X}^k \dot\sqsubseteq \mathsf{lfp}^{\dot\sqsubseteq}\, \vec{F}$ so that $\dot{\bigsqcup}_{k\in\mathbb{N}} \vec{X}^k = \mathsf{lfp}^{\dot\sqsubseteq}\, \vec{F}$ by antisymmetry. □

Exercise 22.5 Show that under the hypotheses of theorem 22.4, if $\vec{F}(\vec{X}_0) \dot\sqsubseteq \vec{X}_0$ then the chaotic iterates converge to $\mathsf{lfp}^{\dot\sqsubseteq}_{\vec{X}_0}\, \vec{F}$ which is the $\dot\sqsubseteq$-least fixpoint of $\vec{F}$ greater than or equal to $\vec{X}_0$.

Chaotic iterations are also applicable in the context of an iteration operator depending on the rank of the iteration.

Corollary 22.6 (convergence of finite chaotic multi-iterations) Let $\langle L^n, \dot{\sqsubseteq}, \dot{\bot}, \dot{\sqcup}\rangle$ be a CPO. Let $\vec{F}^k \in L^n \rightarrow L^n$, $k \in \mathbb{N}$ be functions such that $\vec{X} \dot{\sqsubseteq} \vec{Y} \Rightarrow \vec{F}^k(\vec{X}) \dot{\sqsubseteq} \vec{F}^{k+1}(\vec{Y})$. Assume the iteration $\vec{X}^0 = \dot{\bot}$, $\vec{X}^{k+1} = \vec{F}^k(\vec{X}^k)$ is converging is finitely many steps to $\vec{X}^\ell$ (i.e., $\forall k \geqslant \ell . \vec{X}^k = \vec{X}^\ell$). Then any chaotic iteration for the $\langle \vec{F}^k, k \in \mathbb{N}\rangle$ (i.e., $\vec{X}_i^{k+1} \triangleq F_i^k(\vec{X}^k)$ when $i \in \mathfrak{J}(k)$) also converges in finitely many steps to $\vec{X}^\ell$.

Proof of corollary 22.6 The proof is essentially the same as that of theorem 22.4, replacing the increasingness hypothesis by $\vec{X} \dot{\sqsubseteq} \vec{Y} \Rightarrow \vec{F}^k(\vec{X}) \dot{\sqsubseteq} \vec{F}^{k+1}(\vec{Y})$ and $\mathsf{lfp}^{\dot{\sqsubseteq}} \vec{F}$ by $\vec{X}^\ell$. □

22.5 Conclusion

Chaotic iterations cover all iteration algorithms used to solve systems of equations in static analysis, such as the *work list* [572]. An efficient chaotic iterations strategy was proposed by François Bourdoncle for static analysis [138] (the iteration strategy is also explained in [574]; see [574, 575] for memory allocation). Static analysis by constraint resolution amounts to a fixpoint computation by chaotic iteration [47, 251].

The chaotic iterations generalize from continuous to increasing operators on CPOs using transfinite iterations and to asynchronous iterations where the components evolve in parallel [220, 221][1] in order to minimize idle times on parallel or distributed computers. In this last case the iterates may not be increasing so without a limit in a CPO but this limit exists and is the least fixpoint in complete lattices. Dijkstra-Scholten's algorithm can be used to detect convergence of asynchronous iterations [325, 326]. Parallel fixpoint computations have been experimented successfully in Astrée [708] and Pikos [574], in both cases with good performance with four parallel threads and possibly degraded performance beyond that [447, 574].

22.6 Solutions to Selected Exercises

Solution to exercise 22.3 For Jacobi, $\mathfrak{J}(k) = \{1, \ldots, n\}$. For Gauss-Seidel, $\mathfrak{J}(1) = \{1\}$ and $\mathfrak{J}(k+1) = \{1 + (\mathfrak{J}(k) \mod n)\}$ for $k \in \mathbb{N}$.

1. Also [973] (with stronger hypotheses).

23 Abstract Equational Semantics

Traditional data flow analyses are expressed as solutions to systems of Boolean equations attached to program graphs [195, pp. 429–461], not as an abstract structural semantics [472]. These equations are abstractions of the structural semantics essentially losing the structural information on programs. For example, circular dependencies appearing in equations because of iterations must be rebuilt [195, pp. 435–437] (nowadays using Tarjan's strongly connected components algorithm [923]), whereas, knowing the program syntax, iterations are straightforward to locate.

A variable $\mathcal{X}_\ell$ is attached to each label ℓ of the program P. The equations must be designed so that their solutions satisfy $\mathcal{X}_\ell = \widehat{\boldsymbol{S}}^{\text{¤}}[\![\mathsf{P}]\!]\,\mathcal{P}_0\,\ell$ for all program labels ℓ. Hence the relations between the variables $\mathcal{X}_\ell$ of the equations directly follow from the structural definition of the semantics $\widehat{\boldsymbol{S}}^{\text{¤}}[\![\mathsf{P}]\!]\,\mathcal{P}_0\,\ell$ by expressing the relationships between the abstract properties of the subcomponents of each $\mathsf{S} \in \mathbb{P}c$. Then $\widehat{\boldsymbol{S}}^{\text{¤}}[\![\mathsf{P}]\!]\,\mathcal{P}_0\,\ell$ corresponds to a particular iteration strategy for solving the equations by structural induction, all such strategies being equivalent by the chaotic iteration theorem 22.4.

Contents

23.1 Example of Equational Semantics of a Program

23.1.1 The Program

We consider the program (19.7) of exercise 19.6, of which the grammatical structure is the following.

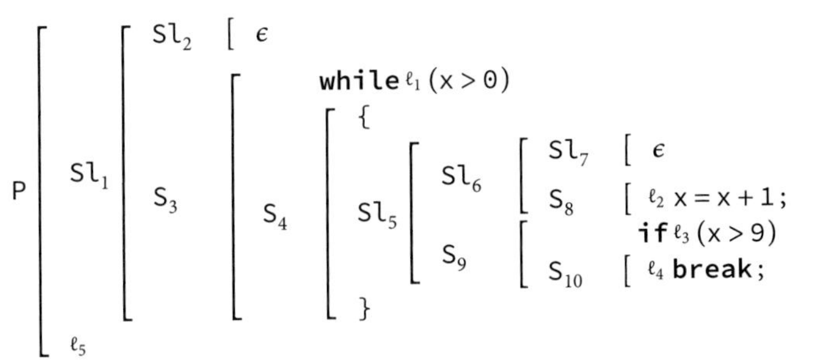

(23.1)

The labeling of program (23.1) is as follows.

S	P, Sl$_1$	Sl$_2$	S$_3$	S$_4$, Sl$_5$	Sl$_6$, S$_8$	Sl$_7$	S$_9$	S$_{10}$
at⟦S⟧	ℓ_1	ℓ_1	ℓ_1	ℓ_2	ℓ_2	ℓ_2	ℓ_3	ℓ_4
after⟦S⟧	ℓ_5	ℓ_1	ℓ_5	ℓ_1	ℓ_3	ℓ_2	ℓ_1	ℓ_1

and breaks-of⟦S$_4$⟧ $= \{\ell_4\}$, breaks-of⟦Sl$_1$⟧ $= \varnothing$.

23.1.2 The Relational Invariants

As shown in chapter 19, the structural relational forward reachability semantics $\widehat{\mathcal{S}}^{\vec{R}}[\![\mathsf{P}]\!]\,\mathcal{P}_0\,\ell_i$ is a relation between environments that can be encountered during an execution of program P starting from any initial environment statisfying $\mathcal{P}_0$ and reaching program label ℓ_i. We denote by x_0 the initial value of the variable x and by x the value of this variable x during execution, if and when reaching program label ℓ_i. The relational invariants are the following:

P	ℓ_i	$\widehat{\mathcal{S}}^{\vec{R}}[\![\mathsf{P}]\!]\mathcal{P}_0\,\ell_i,\quad i=1,\ldots,5$
`/*` $x=x_0$ `*/`		$\mathcal{P}_0 \;=\; \{\langle\rho_0,\rho\rangle \mid \rho=\rho_0\}$
`/*` $x=x_0\leqslant 0 \vee x_0>0$ `*/`		
while ℓ_1 (x > 0)	ℓ_1	$\{\langle\rho_0,\rho\rangle \mid \rho(\mathsf{x})=\rho_0(\mathsf{x})\leqslant 0 \vee \rho_0(\mathsf{x})>0\}$
S$_4$ {		
ℓ_2 `/*` $x_0>0\wedge x>0$ `*/`	ℓ_2	$\{\langle\rho_0,\rho\rangle \mid \rho_0(\mathsf{x})>0 \wedge \rho(\mathsf{x})>0\}$
x = x + 1;		
`/*` $x_0>0\wedge x>1$ `*/`		
if ℓ_3 (x > 9)	ℓ_3	$\{\langle\rho_0,\rho\rangle \mid \rho_0(\mathsf{x})>0 \wedge \rho(\mathsf{x})>1\}$
ℓ_4 `/*` $x_0>0\wedge x>9$ `*/`	ℓ_4	$\{\langle\rho_0,\rho\rangle \mid \rho_0(\mathsf{x})>0 \wedge \rho(\mathsf{x})>9\}$
break;		
}		
ℓ_5 `/*` $(x=x_0\leqslant 0)\vee$ $(x_0>0\wedge x>9)$ `*/`	ℓ_5	$\{(\langle\rho_0,\rho\rangle \mid \rho(\mathsf{x})=\rho_0(\mathsf{x})\leqslant 0)\vee$ $(\rho_0(\mathsf{x})>0\wedge\rho(\mathsf{x})>9)\}$

23.1.3 The System of Equations

The system of equations $\widehat{\mathcal{E}}^{\vec{R}}[\![\mathsf{P}]\!]\,\mathcal{P}_0$ for the program P (23.1) is

$$\begin{cases}\mathcal{X}_{\ell_1} &=& \mathcal{P}_0 \cup \overline{\mathsf{test}}^{\vec{R}}[\![\mathsf{x}>9]\!]\mathcal{X}_{\ell_3}\\ \mathcal{X}_{\ell_2} &=& \mathsf{test}^{\vec{R}}[\![\mathsf{x}>0]\!]\mathcal{X}_{\ell_1}\\ \mathcal{X}_{\ell_3} &=& \mathsf{assign}^{\vec{R}}[\![\mathsf{x},\mathsf{x}+1]\!]\,\mathcal{X}_{\ell_2}\\ \mathcal{X}_{\ell_4} &=& \mathsf{test}^{\vec{R}}[\![\mathsf{x}>9]\!]\mathcal{X}_{\ell_3}\\ \mathcal{X}_{\ell_5} &=& \overline{\mathsf{test}}^{\vec{R}}[\![\mathsf{x}>0]\!]\mathcal{X}_{\ell_1}\cup\mathcal{X}_{\ell_4}\end{cases} \tag{23.2}$$

The equations express that

- The values of x at ℓ_1 are the initial values of $\mathcal{P}_0$ or those after the loop body, that is, those at ℓ_3 that do not pass the test `(x > 9)`;
- The values of x at ℓ_2 are those at ℓ_1 that pass the test `(x > 0)`;
- The values of x at ℓ_3 are those at ℓ_2 after incrementation;

- The values of x at ℓ_4 are those at ℓ_3 that pass the test `(x > 9)`; and
- The values of x at ℓ_5 are are those at ℓ_1 that do not pass the test `(x > 0)` or those at ℓ_4 when the **break**; statement exits the loop.

Exercise 23.15 shows how the preceding specific equations for program (23.1) are derived from the rules (23.5)–(23.14) for constructing the equations $\vec{\mathscr{E}}^{\mathsf{R}}[\![\mathsf{S}]\!]$ of a program component $\mathsf{S} \in \mathbb{P}\mathbb{c}$.

23.1.4 Solving the System of Equations Iteratively

In the example (23.2), $\mathcal{X}_{\ell_1}$ recursively depends upon itself whereas all other $\mathcal{X}_{\ell_i}$, $i = 2, \ldots, 5$ depend only of this $\mathcal{X}_{\ell_1}$. So, we first have to solve the fixpoint equation

$$\mathcal{X}_{\ell_1} \;=\; \mathcal{P}_0 \cup \overline{\mathsf{test}}^{\vec{\mathsf{R}}}[\![\mathsf{x} > 9]\!](\mathsf{assign}^{\vec{\mathsf{R}}}[\![\mathsf{x}, \mathsf{x} + 1]\!]\,(\mathsf{test}^{\vec{\mathsf{R}}}[\![\mathsf{x} > 0]\!]\mathcal{X}_{\ell_1}))$$

that is (by definitions (19.12) and (19.16))

$$\mathcal{X}_{\ell_1} \;=\; \mathcal{P}_0 \cup \{\langle \rho_0, \rho[\mathsf{x} \leftarrow \rho(\mathsf{x}) + 1]\rangle \mid \langle \rho_0, \rho\rangle \in \mathcal{X}_{\ell_1} \wedge 0 < \rho(\mathsf{x}) \leqslant 9\} \tag{23.3}$$

and then carry forward the result to all other $\mathcal{X}_{\ell_i}$, $i = 2, \ldots, 5$. The equation is solved iteratively using Tarski's iterative fixpoint theorem 15.21. The iterates $\mathcal{X}^n_{\ell_1}$, $n \geqslant 0$ are as follows:

$$\begin{aligned}
\mathcal{X}^0_{\ell_1} &= \varnothing\\
\mathcal{X}^1_{\ell_1} &= \mathcal{P}_0 \cup \{\langle \rho_0, \rho[\mathsf{x} \leftarrow \rho(\mathsf{x}) + 1]\rangle \mid \langle \rho_0, \rho\rangle \in \mathcal{X}^0_{\ell_1} \wedge 0 < \rho(\mathsf{x}) \leqslant 9\}\\
&= \mathcal{P}_0\\
&= \{\langle \rho_0, \rho\rangle \mid \rho = \rho_0\}\\
\mathcal{X}^2_{\ell_1} &= \mathcal{P}_0 \cup \{\langle \rho_0, \rho[\mathsf{x} \leftarrow \rho(\mathsf{x}) + 1]\rangle \mid \langle \rho_0, \rho\rangle \in \mathcal{X}^1_{\ell_1} \wedge 0 < \rho(\mathsf{x}) \leqslant 9\}\\
&= \{\langle \rho_0, \rho\rangle \mid \rho = \rho_0\} \cup \{\langle \rho_0, \rho_0[x \leftarrow \rho_0(\mathsf{x}) + 1]\rangle \mid 0 < \rho_0(\mathsf{x}) \leqslant 9\}\\
\mathcal{X}^3_{\ell_1} &= \mathcal{P}_0 \cup \{\langle \rho_0, \rho[\mathsf{x} \leftarrow \rho(\mathsf{x}) + 1]\rangle \mid \langle \rho_0, \rho\rangle \in \mathcal{X}^2_{\ell_1} \wedge 0 < \rho(\mathsf{x}) \leqslant 9\}\\
&= \{\langle \rho_0, \rho\rangle \mid \rho = \rho_0\} \cup \{\langle \rho_0, \rho[\mathsf{x} \leftarrow \rho(\mathsf{x}) + 1]\rangle \mid \langle \rho_0, \rho\rangle \in (\{\langle \rho_0, \rho\rangle \mid \rho = \rho_0\} \cup \{\langle \rho_0,\\
&\qquad \rho_0[x \leftarrow \rho_0(\mathsf{x}) + 1]\rangle \mid 0 < \rho_0(\mathsf{x}) \leqslant 9\}) \wedge 0 < \rho(\mathsf{x}) \leqslant 9\}\\
&= \{\langle \rho_0, \rho\rangle \mid \rho = \rho_0\} \cup \{\langle \rho_0, \rho[\mathsf{x} \leftarrow \rho(\mathsf{x}) + 1]\rangle \mid \rho = \rho_0 \wedge 0 < \rho_0(\mathsf{x}) \leqslant 9\} \cup \{\langle \rho_0,\\
&\qquad \rho[\mathsf{x} \leftarrow \rho(\mathsf{x}) + 1]\rangle \mid \rho = \rho_0[x \leftarrow \rho_0(\mathsf{x}) + 1] \wedge 0 < \rho_0(\mathsf{x}) \leqslant 9 \wedge 0 < \rho(\mathsf{x}) \leqslant 9\}\\
&= \{\langle \rho_0, \rho\rangle \mid \rho = \rho_0\} \cup \{\langle \rho_0, \rho_0[x \leftarrow \rho_0(\mathsf{x}) + 1]\rangle \mid 0 < \rho_0(\mathsf{x}) \leqslant 9\} \cup \{\langle \rho_0, \rho_0[x \leftarrow\\
&\qquad \rho_0(\mathsf{x}) + 2]\rangle \mid 0 < \rho_0(\mathsf{x}) \leqslant 8\}
\end{aligned}$$

Therefore we make the following induction hypothesis (which holds for $1 \leqslant n \leqslant 3$ where $\bigcup \varnothing = \varnothing$).

$$\mathcal{X}^n_{\ell_1} \;=\; \{\langle \rho_0, \rho\rangle \mid \rho = \rho_0\} \cup \bigcup_{i=1}^{n-1} \{\langle \rho_0, \rho_0[x \leftarrow \rho_0(\mathsf{x}) + i]\rangle \mid 0 < \rho_0(\mathsf{x}) \leqslant (9 - i)\}$$

We check that the hypothesis is inductive.

$$\mathcal{X}^{n+1}_{\ell_1} = \mathcal{P}_0 \cup \{\langle \rho_0, \rho[\mathsf{x} \leftarrow \rho(\mathsf{x}) + 1]\rangle \mid \langle \rho_0, \rho\rangle \in \mathcal{X}^n_{\ell_1} \wedge 0 < \rho(\mathsf{x}) \leqslant 9\}$$

$$= \{\langle\rho_0, \rho\rangle \mid \rho = \rho_0\} \cup \{\langle\rho_0, \rho[x \leftarrow \rho(x)+1]\rangle \mid \langle\rho_0, \rho\rangle \in \{\langle\rho_0, \rho\rangle \mid \rho = \rho_0\} \cup \bigcup_{i=1}^{n-1}\{\langle\rho_0, \rho_0[x \leftarrow \rho_0(x)+i]\rangle \mid 0 < \rho_0(x) \leqslant (9-i)\}) \wedge 0 < \rho(x) \leqslant 9\}$$

$$= \{\langle\rho_0, \rho\rangle \mid \rho = \rho_0\} \cup \{\langle\rho_0, \rho_0[x \leftarrow \rho_0(x)+1]\rangle \mid \rho_0(x) \leqslant 9\} \cup \{\langle\rho_0, \rho[x \leftarrow \rho(x)+1]\rangle \mid \langle\rho_0, \rho\rangle \in \bigcup_{i=1}^{n-1}\{\langle\rho_0, \rho_0[x \leftarrow \rho_0(x)+i]\rangle \mid 0 < \rho_0(x) \leqslant (9-i)\}) \wedge 0 < \rho(x) \leqslant 9\}$$

$$= \{\langle\rho_0, \rho\rangle \mid \rho = \rho_0\} \cup \{\langle\rho_0, \rho_0[x \leftarrow \rho_0(x)+1]\rangle \mid \rho_0(x) \leqslant 9\} \cup \bigcup_{i=1}^{n-1}\{\langle\rho_0, \rho_0[x \leftarrow \rho_0(x)+i][x \leftarrow \rho_0[x \leftarrow \rho_0(x)+i](x)+1]\rangle \mid 0 < \rho_0(x) \leqslant (9-i) \wedge 0 < \rho_0[x \leftarrow \rho_0(x)+i](x) \leqslant 9\}$$

$$= \{\langle\rho_0, \rho\rangle \mid \rho = \rho_0\} \cup \{\langle\rho_0, \rho_0[x \leftarrow \rho_0(x)+1]\rangle \mid \rho_0(x) \leqslant 9\} \cup \bigcup_{i=1}^{n-1}\{\langle\rho_0, \rho_0[x \leftarrow \rho_0(x)+i+1]\rangle \mid i+1 < \rho_0(x)+i+1 \leqslant 10 \wedge 1 < \rho_0(x)+i+1 \leqslant 9\}$$

$$= \{\langle\rho_0, \rho\rangle \mid \rho = \rho_0\} \cup \bigcup_{j=1}^{(n+1)-1} \{\langle\rho_0, \rho_0[x \leftarrow \rho_0(x)+j]\rangle \mid 0 < \rho_0(x) \leqslant (9-j)\}$$

$\wr$letting $j = i+1$$\int$

Passing to the limit, the solution to the fixpoint equation (23.3) is

$$\mathcal{X}_{\ell_1} = \bigcup_{n\geqslant 0} \mathcal{X}^n_{\ell_1}$$
$$= \{\langle\rho_0, \rho\rangle \mid \rho = \rho_0\} \cup \{\langle\rho_0, \rho_0[x \leftarrow \rho_0(x)+i]\rangle \mid i \in [1,9] \wedge 0 < \rho_0(x) \leqslant (9-i)\}$$

$\wr$because the join is empty when $9 - i \leqslant 0$$\int$

We derive the other components of the equations (23.2).

$$\mathcal{X}_{\ell_2} = \{\langle\rho_0, \rho\rangle \in \mathcal{X}_{\ell_1} \mid \rho(x) > 0\}$$
$$= \{\langle\rho_0, \rho_0\rangle \mid \rho_0(x) > 0\} \cup \{\langle\rho_0, \rho_0[x \leftarrow \rho_0(x)+i]\rangle \mid i \in [1,9] \wedge 0 < \rho_0(x) \leqslant (9-i)\}$$

$$\mathcal{X}_{\ell_3} = \{\langle\rho_0, \rho[x \leftarrow \rho(x)+1]\rangle \mid \langle\rho_0, \rho\rangle \in \mathcal{X}_{\ell_1} \wedge \rho(x) > 0\}$$
$$= \{\langle\rho_0, \rho[x \leftarrow \rho(x)+1]\rangle \mid \langle\rho_0, \rho\rangle \in (\{\langle\rho_0, \rho\rangle \mid \rho = \rho_0\} \cup \{\langle\rho_0, \rho_0[x \leftarrow \rho_0(x)+i]\rangle \mid i \in [1,9] \wedge 0 < \rho_0(x) \leqslant (9-i)\}) \wedge \rho(x) > 0\}$$
$$= \{\langle\rho_0, \rho[x \leftarrow \rho(x)+1]\rangle \mid \rho = \rho_0 \wedge \rho(x) > 0\} \cup \{\langle\rho_0, \rho[x \leftarrow \rho(x)+1]\rangle \mid \langle\rho_0, \rho\rangle \in \{\langle\rho_0, \rho_0[x \leftarrow \rho_0(x)+i]\rangle \mid i \in [1,9] \wedge 0 < \rho_0(x) \leqslant (9-i)\}) \wedge \rho(x) > 0\}$$
$$= \{\langle\rho_0, \rho[x \leftarrow \rho(x)+1]\rangle \mid \rho = \rho_0 \wedge \rho(x) > 0\} \cup \{\langle\rho_0, \rho[x \leftarrow \rho(x)+1]\rangle \mid \exists i \in [1,9] \,.\, \rho = \rho_0[x \leftarrow \rho_0(x)+i] \wedge 0 < \rho_0(x) \leqslant (9-i) \wedge \rho(x) > 0\}$$
$$= \{\langle\rho_0, \rho[x \leftarrow \rho(x)+1]\rangle \mid \rho = \rho_0 \wedge \rho(x) > 0\} \cup \{\langle\rho_0, \rho_0[x \leftarrow \rho_0(x)+i][x \leftarrow \rho_0[x \leftarrow \rho_0(x)+i](x)+1]\rangle \mid i \in [1,9] \wedge 0 < \rho_0(x) \leqslant (9-i) \wedge \rho_0[x \leftarrow \rho_0(x)+i](x) > 0\}$$
$$= \{\langle\rho_0, \rho_0[x \leftarrow \rho_0(x)+1]\rangle \mid \rho_0(x) > 0\} \cup \{\langle\rho_0, \rho_0[x \leftarrow \rho_0(x)+i]\rangle \mid i \in [1,9] \wedge 0 < \rho_0(x) \leqslant (9-i)\}$$

$$\begin{aligned}\mathcal{X}_{\ell_4} &= \{\langle\rho_0, \rho[\mathsf{x} \leftarrow \rho(\mathsf{x})+1]\rangle \mid \langle\rho_0, \rho\rangle \in \mathcal{X}_{\ell_1} \wedge \rho(\mathsf{x}) > 9\}\\
&= \{\langle\rho_0, \rho[\mathsf{x} \leftarrow \rho(\mathsf{x})+1]\rangle \mid \langle\rho_0, \rho\rangle \in (\{\langle\rho_0, \rho\rangle \mid \rho = \rho_0\} \cup \{\langle\rho_0, \rho_0[x \leftarrow \rho_0(\mathsf{x})+i]\rangle \mid\\
&\quad i \in [1,9] \wedge 0 < \rho_0(\mathsf{x}) \leqslant (9-i)\}) \wedge \rho(\mathsf{x}) > 9\}\\
&= \{\langle\rho_0, \rho_0[x \leftarrow \rho_0(\mathsf{x})+1]\rangle \mid \rho_0(\mathsf{x}) > 9\} \cup \{\langle\rho_0, \rho_0[x \leftarrow \rho_0(\mathsf{x})+i+1]\rangle \mid i \in\\
&\quad [1,9] \wedge \rho_0(\mathsf{x}) = (9-i)\}\\
&= \{\langle\rho_0, \rho_0[x \leftarrow \rho_0(\mathsf{x})+1]\rangle \mid \rho_0(\mathsf{x}) > 9\} \cup \{\langle\rho_0, \rho_0[x \leftarrow 10]\rangle \mid 1 \leqslant \rho_0(\mathsf{x}) \leqslant 9\}\end{aligned}$$

$$\begin{aligned}\mathcal{X}_{\ell_5} &= \{\langle\rho_0, \rho[\mathsf{x} \leftarrow \rho(\mathsf{x})+1]\rangle \mid \langle\rho_0, \rho\rangle \in \mathcal{X}_{\ell_1} \wedge \rho(\mathsf{x}) > 9\} \cup \{\langle\rho_0, \rho\rangle \in \mathcal{P}_0 \mid \rho(\mathsf{x}) \leqslant 0\}\\
&= \{\langle\rho_0, \rho_0\rangle \mid \rho_0(\mathsf{x}) \leqslant 0\} \cup \{\langle\rho_0, \rho_0[x \leftarrow \rho_0(\mathsf{x})+1]\rangle \mid \rho_0(\mathsf{x}) > 9\} \cup \{\langle\rho_0, \rho_0[x \leftarrow 10]\rangle \mid\\
&\quad 1 \leqslant \rho_0(\mathsf{x}) \leqslant 9\}\end{aligned}$$

So the least solution $\boldsymbol{S}$ to the system of equations (23.2) is the following.

$$\begin{cases}\boldsymbol{S}_{\ell_1} &= \{\langle\rho_0, \rho\rangle \mid \rho = \rho_0\} \cup \{\langle\rho_0, \rho_0[x \leftarrow \rho_0(\mathsf{x})+i]\rangle \mid i \in [1,9] \wedge 0 < \rho_0(\mathsf{x}) \leqslant (9-i)\}\\
\boldsymbol{S}_{\ell_2} &= \{\langle\rho_0, \rho_0\rangle \mid \rho_0(\mathsf{x}) > 0\} \cup \{\langle\rho_0, \rho_0[x \leftarrow \rho_0(\mathsf{x})+i]\rangle \mid i \in [1,9] \wedge 0 < \rho_0(\mathsf{x}) \leqslant\\
&\quad (9-i)\}\\
\boldsymbol{S}_{\ell_3} &= \{\langle\rho_0, \rho_0[x \leftarrow \rho_0(\mathsf{x})+1]\rangle \mid \rho_0(\mathsf{x}) > 0\} \cup \{\langle\rho_0, \rho_0[x \leftarrow \rho_0(\mathsf{x})+i]\rangle \mid i \in\\
&\quad [1,9] \wedge 0 < \rho_0(\mathsf{x}) \leqslant (9-i)\}\\
\boldsymbol{S}_{\ell_4} &= \{\langle\rho_0, \rho_0[x \leftarrow \rho_0(\mathsf{x})+1]\rangle \mid \rho_0(\mathsf{x}) > 9\} \cup \{\langle\rho_0, \rho_0[x \leftarrow \rho_0(\mathsf{x})+i+1]\rangle \mid i \in\\
&\quad [1,9] \wedge \rho_0(\mathsf{x}) = (9-i)\}\\
\boldsymbol{S}_{\ell_5} &= \{\langle\rho_0, \rho_0\rangle \mid \rho_0(\mathsf{x}) \leqslant 0\} \cup \{\langle\rho_0, \rho_0[x \leftarrow \rho_0(\mathsf{x})+1]\rangle \mid \rho_0(\mathsf{x}) > 9\} \cup \{\langle\rho_0,\\
&\quad \rho_0[x \leftarrow 10]\rangle \mid 1 \leqslant \rho_0(\mathsf{x}) \leqslant 9\}\end{cases}$$

On program exit at ℓ_5, $\boldsymbol{S}_{\ell_5}$ specifies that the value x of x is

- the initial value x_0 of x when x_0 is negative or zero (because the loop is never entered);
- or, the initial value x_0 of x plus 1 when x_0 is greater than 9;
- or 10, when the initial value x_0 of x is between 1 and 9 because in that case x is incremented by successive loop iterations until reaching 10, in which case the loop exits by the break statement.

23.2 Structural Equational Semantics

The equational semantics of a program P is designed so that its least solution is the abstract interpreter of chapter 21. We define the equational semantics of a program P by structural induction on the syntax of program P. Hence the equational semantics (specifying how the equations are derived for a program) is structural but the equations are not (i.e., the program structure is not apparent in the equations).

The equations $\widehat{\mathscr{E}}^{\alpha}[\![\mathsf{S}]\!]$ for a program component $\mathsf{S} \in \mathbb{P}\mathrm{c}$ have the form

$$\mathcal{X} \quad = \quad \mathsf{E}[\![\mathsf{S}]\!]\, \mathcal{P}_0\, \mathcal{X} \tag{23.4}$$

which can be detailled as

$$\begin{cases}\mathcal{X}_\ell &= \mathsf{E}[\![\mathsf{S}]\!]\, \mathcal{P}_0 \,(\prod_{\ell' \in \mathsf{labs}[\![\mathsf{S}]\!]} \mathcal{X}_{\ell'})\, \ell\\
\ell \in \mathsf{labs}[\![\mathsf{S}]\!]\end{cases}$$

so that the equational semantics is the set of equations

$$\widehat{\mathscr{E}}^{\texttt{¤}}[\![\mathsf{S}]\!] \;=\; \{\mathscr{X}_{\ell} \;=\; d\mathsf{E}[\![\mathsf{S}]\!]\,\mathscr{P}_0\,(\prod_{\ell' \in \mathsf{labs}[\![\mathsf{S}]\!]} \mathscr{X}_{\ell'})\,\ell \mid \ell \in \mathsf{labs}[\![\mathsf{S}]\!]\}.$$

By (19.30), the set of reachable environments of a statement S not at a label of that statement is empty, abstracted by $\bot^{\texttt{¤}}$. So $\mathscr{X}^{\ell} = \bot^{\texttt{¤}}$ for all $\ell \notin \mathsf{labs}[\![\mathsf{S}]\!]$ and to determine this no specific equation need be considered.

The equational semantics of a program $\mathsf{P} ::= \mathsf{Sl}\ \ell'$ with variables $\mathscr{X}^{\ell}$, $\ell \in \mathsf{labs}[\![\mathsf{S}]\!]$ is that of its statement list Sl.

Equational semantics of a program $\mathsf{P} ::= \mathsf{Sl}\ \ell'$

$$\widehat{\mathscr{E}}^{\texttt{¤}}[\![\mathsf{P}]\!] \;\triangleq\; \widehat{\mathscr{E}}^{\texttt{¤}}[\![\mathsf{Sl}]\!] \tag{23.5}$$

The equational semantics of a skip statement $\mathsf{S} ::= ;$ maps the (abstraction of the) initial environments to themselves because they are not modified.

Equational semantics of a skip statement $\mathsf{S} ::= ;$

$$\widehat{\mathscr{E}}^{\texttt{¤}}[\![\mathsf{S}]\!]\,\mathscr{P}_0 \;=\; \{\mathscr{X}_{\mathsf{at}[\![\mathsf{S}]\!]} = \mathscr{P}_0, \mathscr{X}_{\mathsf{after}[\![\mathsf{S}]\!]} = \mathscr{X}_{\mathsf{at}[\![\mathsf{S}]\!]}\} \tag{23.6}$$

The equational semantics of an assignment statement $\mathsf{S} ::= \mathsf{x} = \mathsf{A}\,;$ maps the (abstraction of the) initial environments to themselves on entry and to themselves as modified by the assignment on exit.

Equational semantics of an assignment statement $\mathsf{S} ::= \mathsf{x} = \mathsf{A}\,;$

$$\begin{aligned}\widehat{\mathscr{E}}^{\texttt{¤}}[\![\mathsf{S}]\!]\,\mathscr{P}_0 \;=\; &\{\mathscr{X}_{\mathsf{at}[\![\mathsf{S}]\!]} = \mathscr{P}_0,\\ &\mathscr{X}_{\mathsf{after}[\![\mathsf{S}]\!]} = \mathsf{assign}^{\texttt{¤}}[\![\mathsf{x}, \mathsf{A}]\!]\,\mathscr{X}_{\mathsf{at}[\![\mathsf{S}]\!]}\}\end{aligned} \tag{23.7}$$

The equational semantics of a conditional statement $\mathsf{S} ::= \mathtt{if}\,(\mathsf{B})\,\mathsf{S}_t$ is the (abstraction of the) set of reachable environments in S_t under the hypothesis that the test B is true and, after the conditional statement, the union of the (abstraction of the) environments reaching the end of the true alternative S_t (as given by the equation for $\mathscr{X}_{\mathsf{after}[\![\mathsf{S}_t]\!]} = \mathscr{X}_{\mathsf{after}[\![\mathsf{S}]\!]}$ in the alternative S_t) and those for which the conditional statement is skipped when the test B is false. RHS stands for the right-hand side of an equation $\mathscr{X}_{\ell} = \text{RHS}$ defining $\mathscr{X}_{\ell}$.

Equational semantics of a conditional statement $\mathsf{S} ::= \mathtt{if}\,(\mathsf{B})\,\mathsf{S}_t$

$$\begin{aligned}\widehat{\mathscr{E}}^{\texttt{¤}}[\![\mathsf{S}]\!]\,\mathscr{P}_0 \;=\; &\mathsf{let}\,\{\mathscr{X}_{\mathsf{after}[\![\mathsf{S}]\!]} = \text{RHS}\} \uplus \mathscr{E} = \widehat{\mathscr{E}}^{\texttt{¤}}[\![\mathsf{S}_t]\!](\mathsf{test}^{\texttt{¤}}[\![\mathsf{B}]\!]\mathscr{X}_{\mathsf{at}[\![\mathsf{S}]\!]})\ \mathsf{in}\\ &\{\mathscr{X}_{\mathsf{at}[\![\mathsf{S}]\!]} = \mathscr{P}_0\} \cup \mathscr{E} \cup \{\mathscr{X}_{\mathsf{after}[\![\mathsf{S}]\!]} = \text{RHS} \sqcup^{\texttt{¤}} \overline{\mathsf{test}}^{\texttt{¤}}[\![\mathsf{B}]\!]\mathscr{X}_{\mathsf{at}[\![\mathsf{S}]\!]}\}\end{aligned} \tag{23.8}$$

(We write $X \uplus Y = Z$ to mean that X, Y is a partition of Z, that is, $X \cap Y = \varnothing$ and $X \cup Y = Z$. This is essential to ensure that there is only one equation per variable $\mathscr{X}_{\ell}$, $\ell \in \mathsf{labs}[\![\mathsf{S}]\!]$, see exercise 23.17).

The equational semantics of a conditional statement $\mathsf{S} ::= \mathtt{if}\,(\mathsf{B})\,\mathsf{S}_t\ \mathtt{else}\ \mathsf{S}_f$ is the (abstraction of the) set of reachable environments in S_t under the hypothesis that the test $\mathtt{B}$ is true, the set of reachable environments in $\mathtt{S}_f$ under the hypothesis that the test $\mathtt{B}$ is false, and, after the conditional statement, the union of the sets of environments reaching the end of the true and false alternatives $\mathtt{S}_t$ and $\mathtt{S}_f$.

Equational semantics of a conditional statement $\mathtt{S} ::= \mathtt{if}\,(\mathtt{B})\,\mathtt{S}_t\ \mathtt{else}\ \mathtt{S}_f$

$$\begin{array}{rcl}\widehat{\mathscr{E}}^{\natural}[\![\mathtt{S}]\!]\,\mathscr{P}_0 & = & \mathsf{let}\,(\{\mathscr{X}_{\mathsf{after}[\![\mathtt{S}]\!]} = \mathrm{RHS}_t\} \cup \mathscr{E}_t) = \widehat{\mathscr{E}}^{\natural}[\![\mathtt{S}_t]\!](\mathsf{test}^{\natural}[\![\mathtt{B}]\!]\mathscr{X}_{\mathsf{at}[\![\mathtt{S}]\!]}) \\ & & \mathsf{and}\ (\{\mathscr{X}_{\mathsf{after}[\![\mathtt{S}]\!]} = \mathrm{RHS}_f\} \cup \mathscr{E}_f) = \widehat{\mathscr{E}}^{\natural}[\![\mathtt{S}_f]\!](\overline{\mathsf{test}}^{\natural}[\![\mathtt{B}]\!]\mathscr{X}_{\mathsf{at}[\![\mathtt{S}]\!]})\ \mathsf{in} \\ & & \quad \{\mathscr{X}_{\mathsf{at}[\![\mathtt{S}]\!]} = \mathscr{P}_0\} \cup \\ & & \quad \mathscr{E}_t \cup \mathscr{E}_f \cup \\ & & \quad \{\mathscr{X}_{\mathsf{after}[\![\mathtt{S}]\!]} = \mathrm{RHS}_t \sqcup^{\natural} \mathrm{RHS}_f\}\end{array} \tag{23.9}$$

The equational semantics of a statement list $\mathtt{Sl} ::= \mathtt{Sl}'\ \mathtt{S}$ is the set of equations of $\mathtt{Sl}'$ and those of $\mathtt{S}$ after executing $\mathtt{Sl}'$ where the equation for $\mathscr{X}_{\mathsf{after}[\![\mathtt{Sl}']\!]} = \mathscr{X}_{\mathsf{at}[\![\mathtt{S}]\!]}$ is shared.

Equational semantics of a statement list $\mathtt{Sl} ::= \mathtt{Sl}'\ \mathtt{S}$

$$\begin{array}{rcll}\widehat{\mathscr{E}}^{\natural}[\![\mathtt{Sl}]\!]\mathscr{P}_0 & = & \widehat{\mathscr{E}}^{\natural}[\![\mathtt{S}]\!]\mathscr{P}_0 & \text{when}\quad \mathtt{Sl}' ::= \epsilon \\ & = & \mathsf{let}\,(\{\mathscr{X}_{\mathsf{after}[\![\mathtt{Sl}']\!]} = \mathrm{RHS}'\} \cup \mathscr{E}') = \widehat{\mathscr{E}}^{\natural}[\![\mathtt{Sl}']\!]\mathscr{P}_0 & \text{otherwise} \\ & & \mathsf{and}\,(\{\mathscr{X}_{\mathsf{at}[\![\mathtt{S}]\!]} = \mathrm{RHS}\} \cup \mathscr{E}) = \widehat{\mathscr{E}}^{\natural}[\![\mathtt{S}]\!]\mathscr{X}_{\mathsf{at}[\![\mathtt{S}]\!]}\ \mathsf{in} & \\ & & \quad \{\mathscr{X}_{\mathsf{at}[\![\mathtt{S}]\!]} = \mathrm{RHS}' \sqcup^{\natural} \mathrm{RHS}\} \cup \mathscr{E}' \cup \mathscr{E} & \end{array} \tag{23.10}$$

The empty statement list $\mathtt{Sl} ::= \epsilon$ does not involve any computation so that the (abstraction of the) reachable environments are the initial environments.

Equational semantics of an empty statement list $\mathtt{Sl} ::= \epsilon$

$$\widehat{\mathscr{E}}^{\natural}[\![\mathtt{Sl}]\!]\mathscr{P}_0 \quad = \quad \{\mathscr{X}_{\mathsf{at}[\![\mathtt{Sl}]\!]} = \mathscr{P}_0\} \tag{23.11}$$

The equational semantics of an iteration statement $\mathtt{S} ::= \mathtt{while}\,\ell\,(\mathtt{B})\,\mathtt{S}_b$, is a fixpoint at which successive iterates provide the (abstraction of the) reachable environments after at most 0, 1, …or more iterations in the loop. If the loop terminates, the fixpoint iterates beyond the maximal number of iterations add no new reachable environment. If the loop does not terminate, the (abstraction of the) reachable environments after any number of iterations (is) are obtained by passing to the limit in the fixpoint iterates.

Equational semantics of an iteration statement $\mathtt{S} ::= \mathtt{while}\,\ell\,(\mathtt{B})\,\mathtt{S}_b$

$$\begin{array}{rcl}\widehat{\mathscr{E}}^{\natural}[\![\mathtt{S}]\!]\,\mathscr{P}_0 & = & \mathsf{let}\,(\{\mathscr{X}_{\mathsf{at}[\![\mathtt{S}]\!]} = \mathrm{RHS}\} \cup \mathscr{E}) = \widehat{\mathscr{E}}^{\natural}[\![\mathtt{S}_b]\!](\mathsf{test}^{\natural}[\![\mathtt{B}]\!]\mathscr{X}_{\mathsf{at}[\![\mathtt{S}]\!]})\ \mathsf{in} \\ & & \quad \{\mathscr{X}_{\mathsf{at}[\![\mathtt{S}]\!]} = \mathscr{P}_0 \sqcup^{\natural} \mathrm{RHS}\} \cup \\ & & \quad \mathscr{E} \cup \\ & & \quad \{\mathscr{X}_{\mathsf{after}[\![\mathtt{S}]\!]} = \overline{\mathsf{test}}^{\natural}[\![\mathtt{B}]\!]\mathscr{X}_{\mathsf{at}[\![\mathtt{S}]\!]} \sqcup^{\natural} \bigsqcup^{\natural}_{\ell \in \mathsf{breaks\text{-}of}[\![\mathtt{S}_b]\!]} \mathscr{X}_{\ell}\}\end{array} \tag{23.12}$$

The reachable environments $\mathscr{X}_{\mathsf{at}[\![\mathsf{S}]\!]}$ are the entry environments or the reachable environments after one more iteration from reachable states with test true. The reachable environments $\mathscr{X}_{\mathsf{after}[\![\mathsf{S}]\!]}$ on loop exit, are those $\mathsf{at}[\![\mathsf{S}]\!]$ for which the test is false or the breaking environments reachable from reachable environments in an iteration with test true.

The equational semantics of a break statement $\mathsf{S} ::= \mathtt{break};$ states that execution never goes after the break at $\mathsf{after}[\![\mathsf{S}]\!]$.

Equational semantics of a break statement $\mathsf{S} ::= \ell\ \mathtt{break};$

$$\widehat{\mathscr{E}}^{\texttt{¤}}[\![\mathsf{S}]\!]\,\mathscr{P}_0 \quad = \quad \{\mathscr{X}_{\mathsf{at}[\![\mathsf{S}]\!]} = \mathscr{P}_0,\ \mathscr{X}_{\mathsf{after}[\![\mathsf{S}]\!]} = \bot^{\texttt{¤}}\} \tag{23.13}$$

The equational semantics of a compound statement $\mathsf{S} ::= \{\ \mathtt{Sl}\ \}$ is that of the statement list $\mathtt{Sl}$.

Equational semantics of a compound statement $\mathsf{S} ::= \{\ \mathtt{Sl}\ \}$

$$\widehat{\mathscr{E}}^{\texttt{¤}}[\![\mathsf{S}]\!] \quad = \quad \widehat{\mathscr{E}}^{\texttt{¤}}[\![\mathtt{Sl}]\!] \tag{23.14}$$

Exercise 23.15 Calculate the equations $\widehat{\mathscr{E}}^{\texttt{¤}}[\![\mathsf{P}]\!]$ of program (19.7) with parsing (23.1).

Exercise 23.16 Encode the equations of exercise 23.15 in the programming language of your choice and calculate their solution e.g. for constancy analysis.

Exercise 23.17 Show that for all program components S, $\widehat{\mathscr{E}}^{\texttt{¤}}[\![\mathsf{S}]\!]\,\mathscr{P}_0$ has only one equation per variable $\mathscr{X}_\ell$, $\ell \in \mathsf{labs}[\![\mathsf{S}]\!]$.

Exercise 23.18 Show that the equations (23.5)–(23.14) have the form (23.4).

Exercise 23.19 Modify the abstract interpreter of exercise 21.22 to output the system of equations and add a chaotic fixpoint iterator to solve them.

Theorem 23.20 shows that the equational definition of the semantics is exactly the same as the structural functional one of chapter 21. The idea is that the abstract interpreter is a function f that can be abstracted by an equation $\mathscr{X} = F(\mathscr{X})$ which least solution is $f = \mathsf{lfp}\, F$, and conversely.

Theorem 23.20 The equations $\widehat{\mathscr{E}}^{\texttt{¤}}[\![\mathsf{P}]\!]\,\mathscr{P}_0$ have the form

$$\left\{\begin{array}{lcl} \mathscr{X}_\ell & = & \mathsf{E}[\![\mathsf{P}]\!]\,\mathscr{P}_0\Big(\prod\limits_{\ell' \in \mathsf{labs}[\![\mathsf{P}]\!]} \mathscr{X}_{\ell'}\Big)\ell \\ \ell \in \mathsf{labs}[\![\mathsf{P}]\!] & & \end{array}\right.$$

such that

- There is exactly one equation $\mathscr{X}_\ell = \mathsf{E}[\![\mathsf{P}]\!]\,\mathscr{P}_0\,(\prod_{\ell' \in \mathsf{labs}[\![\mathsf{P}]\!]} \mathscr{X}_{\ell'})\,\ell$ for each program point $\ell \in \mathsf{labs}[\![\mathsf{P}]\!]$;
- $\mathsf{E}[\![\mathsf{P}]\!]\,\mathscr{P}_0 \in \prod_{\ell \in \mathsf{labs}[\![\mathsf{P}]\!]} \mathbb{P}^{\texttt{¤}} \xrightarrow{uc} \prod_{\ell \in \mathsf{labs}[\![\mathsf{P}]\!]} \mathbb{P}^{\texttt{¤}}$ is well defined and continuous; and
- The equations have a least pointwise fixpoint $\mathsf{lfp}^{\dot{\sqsubseteq}^{\texttt{¤}}}\ \mathsf{E}[\![\mathsf{P}]\!]\,\mathscr{P}_0 = \widehat{\mathcal{S}}^{\texttt{¤}}[\![\mathsf{P}]\!]\,\mathscr{P}_0$, as defined in section 21.2.

Proof of theorem 23.20 We prove the theorem for all program components $\mathsf{S} \in \mathbb{P}\mathbb{c}$, by structural induction of S. The equations $\widehat{\mathscr{E}}^{¤}[\![\mathsf{S}]\!]\,\mathcal{P}_0$ have the form

$$\left\{\begin{array}{l} \mathcal{X}_\ell = \mathsf{E}[\![\mathsf{S}]\!]\,\mathcal{P}_0\,(\prod\limits_{\ell' \in \mathsf{labs}[\![\mathsf{S}]\!]} \mathcal{X}_{\ell'})\,\ell \\ \ell \in \mathsf{labs}[\![\mathsf{S}]\!] \end{array}\right.$$

— The upper continuity of these equations follows pointwise from the continuity of each equation $\mathcal{X} \mapsto \mathsf{E}[\![\mathsf{P}]\!]\,\mathcal{P}_0\,\mathcal{X}\,\ell$, $\ell \in \mathsf{labs}[\![\mathsf{P}]\!]$, which itself follows from the continuity of the identity, of the primitives of the abstract domain (including the join) in definition 21.1, of the conditional with continuous alternatives, and the fact that the composition of continuous functions is continuous.

— The fact that there is one equation per program label is proved by structural induction.

For the basis, this is true for the assignment, empty statement list, skip, and break statement.

For the induction step, for a program component made of several subcomponents, this remains true, by structural induction, for disjoint labels. Moreover, when a label is shared between subcomponents their equations are regrouped into one single equation.

For example, for the statement list $\mathtt{Sl} ::= \mathtt{Sl}'\ \mathtt{S}$, $\mathtt{Sl}' \neq \epsilon$, the equations $\widehat{\mathscr{E}}^{¤}[\![\mathtt{Sl}']\!]\mathcal{P}_0$ and $\widehat{\mathscr{E}}^{¤}[\![\mathsf{S}]\!]\mathcal{P}_0$ share, by definition of labs in section 4.2.7, only one common variable $\mathcal{X}_{\mathsf{after}[\![\mathtt{Sl}']\!]} = \mathcal{X}_{\mathsf{at}[\![\mathsf{S}]\!]}$ so, by (23.10), the system of equations for $\mathtt{Sl} ::= \mathtt{Sl}'\ \mathsf{S}$ has exactly one equation for each program point of $\mathtt{Sl}$ because the equation $\mathcal{X}_{\mathsf{after}[\![\mathtt{Sl}']\!]} = \textsc{rhs}'$ for $\mathtt{Sl}'$ and the equation $\mathcal{X}_{\mathsf{at}[\![\mathsf{S}]\!]} = \textsc{rhs}$ for S are grouped into a single equation $\mathcal{X}_{\mathsf{at}[\![\mathsf{S}]\!]} = \textsc{rhs}' \sqcup^{¤} \textsc{rhs}$ for $\mathtt{Sl}$.

— It remains to prove, by structural induction, that the equations $\widehat{\mathscr{E}}^{¤}[\![\mathsf{S}]\!]\,\mathcal{P}_0$ for a statement S, of the form $\mathcal{X} = E(\mathcal{X})$ have a least pointwise fixpoint $\mathsf{lfp}^{\dot{\sqsubseteq}^{¤}}\,\mathsf{E}[\![\mathsf{P}]\!]\,\mathcal{P}_0 = \widehat{\boldsymbol{\mathcal{S}}}^{¤}[\![\mathsf{P}]\!]\,\mathcal{P}_0$, as defined in section 21.2.

— For the assignment, (23.7) the equations are $\{\mathcal{X}_{\mathsf{at}[\![\mathsf{S}]\!]} = \mathcal{P}_0, \mathcal{X}_{\mathsf{after}[\![\mathsf{S}]\!]} = \mathsf{assign}^{¤}[\![\mathsf{x}, \mathsf{E}]\!]\,\mathcal{X}_{\mathsf{at}[\![\mathsf{S}]\!]}\}$. The unique solution (hence $\dot{\sqsubseteq}^{¤}$-least solution) is $\boldsymbol{\mathcal{S}}_{\mathsf{at}[\![\mathsf{S}]\!]} = \mathcal{P}_0$, $\boldsymbol{\mathcal{S}}_{\mathsf{after}[\![\mathsf{S}]\!]} = \mathsf{assign}^{¤}[\![\mathsf{x}, \mathsf{E}]\!]\,\mathcal{P}_0$, and, by convention, $\boldsymbol{\mathcal{S}}_\ell = \bot^{¤}$ when $\ell \notin \{\mathsf{at}[\![\mathsf{S}]\!], \mathsf{after}[\![\mathsf{S}]\!]\}$. By (21.7), this solution is $\boldsymbol{\mathcal{S}} = \widehat{\boldsymbol{\mathcal{S}}}^{¤}[\![\mathsf{S}]\!]\,\mathcal{P}_0$.

The proof is similar for an empty statement list (23.11), a skip statement (23.6), a break statement (23.13), trivially by structural induction, for a compound statement (23.14).

— The equations for a program $\mathsf{P} ::= \mathtt{Sl}\ \ell'$ are the same as those for the statement list $\mathtt{Sl}$ because $\mathsf{labs}[\![\mathsf{P}]\!] = \mathsf{in}[\![\mathtt{Sl}]\!] \cup \{\mathsf{after}[\![\mathtt{Sl}]\!]\}$. By structural induction the equations $\widehat{\mathscr{E}}^{¤}[\![\mathtt{Sl}]\!]$ have solution $\boldsymbol{\mathcal{S}} = \widehat{\boldsymbol{\mathcal{S}}}^{¤}[\![\mathtt{Sl}]\!]\,\mathcal{P}_0$ so the solution is the same for the program in accordance with (21.4).

— For the statement list $\mathtt{Sl} ::= \epsilon\ \mathsf{S}$, reduced to a statement S, the result directly follows from $\mathsf{in}[\![\mathtt{Sl}]\!] = \mathsf{in}[\![\mathsf{S}]\!]$ and $\mathsf{after}[\![\mathtt{Sl}]\!] = \mathsf{after}[\![\mathsf{S}]\!]$, (21.5), (23.10), and the induction hypothesis.

— For the statement list $\mathtt{Sl} ::= \mathtt{Sl}'\ \mathsf{S}$ and by induction hypothesis, the equations $\{\mathcal{X}_{\mathsf{after}[\![\mathtt{Sl}']\!]} = \textsc{rhs}'\} \cup \mathscr{E}' = \widehat{\mathscr{E}}^{¤}[\![\mathtt{Sl}']\!]\mathcal{P}_0$ have the solution $\boldsymbol{\mathcal{S}}' = \widehat{\boldsymbol{\mathcal{S}}}^{¤}[\![\mathtt{Sl}']\!]\,\mathcal{P}_0$. Hence the value of $\mathcal{X}_{\mathsf{at}[\![\mathsf{S}]\!]} = \mathcal{X}_{\mathsf{after}[\![\mathtt{Sl}']\!]}$ is $\boldsymbol{\mathcal{S}}'_{\mathsf{at}[\![\mathsf{S}]\!]} = \widehat{\boldsymbol{\mathcal{S}}}^{¤}[\![\mathtt{Sl}']\!]\,\mathcal{P}_0\,\mathsf{at}[\![\mathsf{S}]\!]$.

By induction hypothesis, the equations $\{\mathscr{X}_{\mathsf{at}[\![\mathtt{S}]\!]} = \text{RHS}\} \cup \mathscr{E} = \widehat{\mathscr{E}}^{¤}[\![\mathtt{S}]\!]\mathscr{X}_{\mathsf{at}[\![\mathtt{S}]\!]}$ have therefore the solution $\widehat{\boldsymbol{\mathcal{S}}}^{¤}[\![\mathtt{S}]\!]\, \boldsymbol{\mathcal{S}}'_{\mathsf{at}[\![\mathtt{S}]\!]} = \widehat{\boldsymbol{\mathcal{S}}}^{¤}[\![\mathtt{S}]\!]\, (\widehat{\boldsymbol{\mathcal{S}}}^{¤}[\![\mathtt{Sl}']\!]\, \mathscr{P}_0\, \mathsf{at}[\![\mathtt{S}]\!]) = \widehat{\boldsymbol{\mathcal{S}}}^{¤}[\![\mathtt{Sl}]\!]\, \mathscr{P}_0$ by (21.5).

It remains to show that the equations $\{\mathscr{X}_{\mathsf{at}[\![\mathtt{S}]\!]} = \text{RHS}' \sqcup^{¤} \text{RHS}\} \cup \mathscr{E}' \cup \mathscr{E} = \widehat{\mathscr{E}}^{¤}[\![\mathtt{Sl}]\!]\mathscr{P}_0$ for $\mathtt{Sl}$ have the same solution.

First we observe, by structural induction, that no system of equations for a program component $\mathtt{S} \in \mathbb{P}\mathbb{c}$ has any equations depending on $\mathscr{X}_{\mathsf{after}[\![\mathtt{S}]\!]}$.

This means that a chaotic iteration strategy can solve $\mathscr{E}'$ first to get the solution $\boldsymbol{\mathcal{S}}'_\ell = \widehat{\mathscr{E}}^{¤}[\![\mathtt{Sl}]\!]\mathscr{P}_0\, \ell$ for $\mathscr{X}_\ell$ when $\ell \in \mathsf{labs}[\![\mathtt{Sl}']\!] \setminus \{\mathsf{after}[\![\mathtt{Sl}']\!]\}$.

Then, we can calculate the value of RHS' which depends only on $\boldsymbol{\mathcal{S}}'_\ell$, $\ell \in \mathsf{labs}[\![\mathtt{Sl}']\!] \setminus \{\mathsf{after}[\![\mathtt{Sl}']\!]\}$ and is $\widehat{\boldsymbol{\mathcal{S}}}^{¤}[\![\mathtt{Sl}']\!]\, \mathscr{P}_0\, \mathsf{after}[\![\mathtt{Sl}']\!] = \widehat{\boldsymbol{\mathcal{S}}}^{¤}[\![\mathtt{Sl}']\!]\, \mathscr{P}_0\, \mathsf{at}[\![\mathtt{S}]\!]$.

So the chaotic iteration strategy is left with solving $\{\mathscr{X}_{\mathsf{at}[\![\mathtt{S}]\!]} = \widehat{\boldsymbol{\mathcal{S}}}^{¤}[\![\mathtt{Sl}']\!]\, \mathscr{P}_0\, \mathsf{at}[\![\mathtt{S}]\!] \sqcup^{¤} \text{RHS}\} \cup \mathscr{E}$. There are two subcases.

– For all program components $\mathtt{S} \in \mathbb{P}\mathbb{c}$ but the iteration, the equations $\widehat{\mathscr{E}}^{¤}[\![\mathtt{S}]\!]\, \mathscr{P}_0$ have an entry equation of the form $\mathscr{X}_{\mathsf{at}[\![\mathtt{S}]\!]} = \mathscr{P}_0$. Therefore, $\mathscr{X}_{\mathsf{at}[\![\mathtt{S}]\!]} = \text{RHS}$ is $\mathscr{X}_{\mathsf{at}[\![\mathtt{S}]\!]} = \mathscr{X}_{\mathsf{after}[\![\mathtt{Sl}']\!]}$. Then the equations $\{\mathscr{X}_{\mathsf{at}[\![\mathtt{S}]\!]} = \widehat{\boldsymbol{\mathcal{S}}}^{¤}[\![\mathtt{Sl}']\!]\, \mathscr{P}_0\, \mathsf{at}[\![\mathtt{S}]\!] \sqcup^{¤} \text{RHS}\} \cup \mathscr{E}$ are $\{\mathscr{X}_{\mathsf{at}[\![\mathtt{S}]\!]} = \widehat{\boldsymbol{\mathcal{S}}}^{¤}[\![\mathtt{Sl}']\!]\, \mathscr{P}_0\, \mathsf{at}[\![\mathtt{S}]\!] \sqcup^{¤} \mathscr{X}_{\mathsf{after}[\![\mathtt{Sl}']\!]}\} \cup \mathscr{E}$ that is $\{\mathscr{X}_{\mathsf{at}[\![\mathtt{S}]\!]} = \widehat{\boldsymbol{\mathcal{S}}}^{¤}[\![\mathtt{Sl}']\!]\, \mathscr{P}_0\, \mathsf{at}[\![\mathtt{S}]\!] \sqcup^{¤} \mathscr{X}_{\mathsf{at}[\![\mathtt{S}]\!]}\} \cup \mathscr{E}$, which least solution is, by definition of the lub, the same as $\{\mathscr{X}_{\mathsf{at}[\![\mathtt{S}]\!]} = \widehat{\boldsymbol{\mathcal{S}}}^{¤}[\![\mathtt{Sl}']\!]\, \mathscr{P}_0\, \mathsf{at}[\![\mathtt{S}]\!]\} \cup \mathscr{E}$ which we have shown to have solution $\widehat{\boldsymbol{\mathcal{S}}}^{¤}[\![\mathtt{S}]\!]\, (\widehat{\boldsymbol{\mathcal{S}}}^{¤}[\![\mathtt{Sl}']\!]\, \mathscr{P}_0\, \mathsf{at}[\![\mathtt{S}]\!]) = \widehat{\boldsymbol{\mathcal{S}}}^{¤}[\![\mathtt{Sl}]\!]\, \mathscr{P}_0$.

– Otherwise, $\mathtt{S} ::= \mathtt{while}\, \ell\, (\mathtt{B})\, \mathtt{S}_b$ is an iteration, and by (23.12), the entry equation is $\mathscr{X}_{\mathsf{at}[\![\mathtt{S}]\!]} = \mathscr{X}_{\mathsf{after}[\![\mathtt{Sl}']\!]} \sqcup^{¤} \text{RHS}''$ where $\{\mathscr{X}_{\mathsf{at}[\![\mathtt{S}]\!]} = \text{RHS}''\} \cup \mathscr{E}'' = \widehat{\mathscr{E}}^{¤}[\![\mathtt{S}_b]\!](\mathsf{test}^{¤}[\![\mathtt{B}]\!]\mathscr{X}_{\mathsf{at}[\![\mathtt{S}]\!]})$ (where $\mathsf{after}[\![\mathtt{Sl}']\!] = \mathsf{at}[\![\mathtt{S}]\!] = \mathsf{after}[\![\mathtt{S}_b]\!]$ is the iteration label). So $\text{RHS} = \mathscr{X}_{\mathsf{after}[\![\mathtt{Sl}']\!]} \sqcup^{¤} \text{RHS}''$ and the chaotic iterations are left to solve $\{\mathscr{X}_{\mathsf{at}[\![\mathtt{S}]\!]} = \widehat{\boldsymbol{\mathcal{S}}}^{¤}[\![\mathtt{Sl}']\!]\, \mathscr{P}_0\, \mathsf{at}[\![\mathtt{S}]\!] \sqcup^{¤} \text{RHS}\} \cup \mathscr{E}$, that is, $\{\mathscr{X}_{\mathsf{at}[\![\mathtt{S}]\!]} = \widehat{\boldsymbol{\mathcal{S}}}^{¤}[\![\mathtt{Sl}']\!]\, \mathscr{P}_0\, \mathsf{at}[\![\mathtt{S}]\!] \sqcup^{¤} \mathscr{X}_{\mathsf{after}[\![\mathtt{Sl}']\!]} \sqcup^{¤} \text{RHS}''\} \cup \mathscr{E}$ which simplifies to $\{\mathscr{X}_{\mathsf{at}[\![\mathtt{S}]\!]} = \widehat{\boldsymbol{\mathcal{S}}}^{¤}[\![\mathtt{Sl}']\!]\, \mathscr{P}_0\, \mathsf{at}[\![\mathtt{S}]\!] \sqcup^{¤} \text{RHS}''\} \cup \mathscr{E}$ because the solution to $\mathscr{X}_{\mathsf{after}[\![\mathtt{Sl}']\!]}$ is $\widehat{\boldsymbol{\mathcal{S}}}^{¤}[\![\mathtt{Sl}']\!]\, \mathscr{P}_0\, \mathsf{at}[\![\mathtt{S}]\!]$ and $\sqcup^{¤}$ is idempotent. This is exactly the equations $\widehat{\mathscr{E}}^{¤}[\![\mathtt{S}]\!]\mathscr{X}_{\mathsf{at}[\![\mathtt{S}]\!]}$ that we have shown to have solution $\widehat{\boldsymbol{\mathcal{S}}}^{¤}[\![\mathtt{S}]\!]\, (\widehat{\boldsymbol{\mathcal{S}}}^{¤}[\![\mathtt{Sl}']\!]\, \mathscr{P}_0\, \mathsf{at}[\![\mathtt{S}]\!]) = \widehat{\boldsymbol{\mathcal{S}}}^{¤}[\![\mathtt{Sl}]\!]\, \mathscr{P}_0$.

By theorem 22.4, all chaotic iterations are equivalent and yields the least fixpoint proving that, in case of a statement list, $\mathsf{lfp}^{\dot{\sqsubseteq}^{¤}}\, \widehat{\mathscr{E}}^{¤}[\![\mathtt{Sl}]\!]\, \mathscr{P}_0 = \widehat{\boldsymbol{\mathcal{S}}}^{¤}[\![\mathtt{Sl}]\!]\, \mathscr{P}_0$.

— The proof for the conditionals (23.8) and (23.9) is very similar and left as an exercise.

— For the iteration $\mathtt{S} ::= \mathtt{while}\, \ell\, (\mathtt{B})\, \mathtt{S}_b$ (23.12), the Jacobi calculation of the least fixpoint $\widehat{\boldsymbol{\mathcal{S}}}^{¤}[\![\mathtt{S}]\!]\, \mathscr{P}_0 = \mathsf{lfp}^{\dot{\sqsubseteq}}\, (\mathscr{F}^{¤}[\![\mathtt{while}\, \ell\, (\mathtt{B})\, \mathtt{S}_b]\!]\, \mathscr{P}_0)$ corresponds to a specific chaotic iteration strategy for the loop equations (23.12) where the equations $\widehat{\mathscr{E}}^{¤}[\![\mathtt{S}_b]\!](\mathsf{test}^{¤}[\![\mathtt{B}]\!]\mathscr{X}_{\mathsf{at}[\![\mathtt{S}]\!]})$ for the loop body are solved by iteration before considering another iteration of the loop fixpoint $\widehat{\mathscr{E}}^{¤}[\![\mathtt{S}]\!]\, \mathscr{P}_0$.

More precisely, by (21.11), $\widehat{\boldsymbol{\mathcal{S}}}^{¤}[\![\mathtt{S}]\!]\, \mathscr{P}_0$ is the least solution to the system of equations

$$\begin{cases} X(\ell') &= \mathscr{F}^{¤}[\![\mathtt{while}\, \ell\, (\mathtt{B})\, \mathtt{S}_b]\!]\, \mathscr{P}_0\, X\, \ell' \\ \ell' \in \mathsf{labs}[\![\mathtt{S}]\!] \end{cases}$$

(where $\mathsf{labs}[\![\mathtt{S}]\!] = \mathsf{labs}[\![\mathtt{S}_b]\!] \cup \{\mathsf{after}[\![\mathtt{S}]\!]\}$ for iteration), that is,

$$\left\{\begin{array}{lcl} X(\ell') & = & (\!|\, \ell' = \ell \;?\; \mathscr{P}_0 \sqcup^{\natural} \widehat{\boldsymbol{\mathcal{S}}}^{\natural}[\![\mathsf{S}_b]\!]\,(\mathsf{test}^{\natural}[\![\mathsf{B}]\!]X(\ell))\,\ell \\ & & [\!]\ \ell' \in \mathsf{in}[\![\mathsf{S}_b]\!] \setminus \{\ell\} \;?\; \widehat{\boldsymbol{\mathcal{S}}}^{\natural}[\![\mathsf{S}_b]\!]\,(\mathsf{test}^{\natural}[\![\mathsf{B}]\!]X(\ell))\,\ell' \\ & & [\!]\ \ell' = \mathsf{after}[\![\mathsf{S}]\!] \;?\; \overline{\mathsf{test}}^{\natural}[\![\mathsf{B}]\!]X(\ell) \sqcup^{\natural} \bigsqcup^{\natural}_{\ell'' \in \mathsf{breaks\text{-}of}[\![\mathsf{S}_b]\!]} \widehat{\boldsymbol{\mathcal{S}}}^{\natural}[\![\mathsf{S}_b]\!]\,(\mathsf{test}^{\natural}[\![\mathsf{B}]\!]X(\ell))\,\ell'' \\ & & \;\text{:}\; \bot^{\natural}\,|\!) \\ \ell' \in \mathsf{labs}[\![\mathsf{S}]\!] & & \end{array}\right. \qquad \text{(a)}$$

We let X^k, $k \in \mathbb{N}$ be the Jacobi iterations for this system of equations (a).

Let $\{\mathscr{X}_{\mathsf{at}[\![\mathsf{S}]\!]} = \text{RHS}\} \cup \mathscr{E} = \widehat{\mathscr{E}}^{\natural}[\![\mathsf{S}_b]\!](\mathsf{test}^{\natural}[\![\mathsf{B}]\!]\mathscr{X}_{\mathsf{at}[\![\mathsf{S}]\!]})$ be the equations of the loop body S_b. By (23.12), the equations $\widehat{\mathscr{E}}^{\natural}[\![\mathsf{S}]\!]\,\mathscr{P}_0$ for the iteration $\mathsf{S} ::= \mathbf{while}\,\ell\,(\mathsf{B})\,\mathsf{S}_b$ are

$$\{\mathscr{X}_{\mathsf{at}[\![\mathsf{S}]\!]} = \mathscr{P}_0 \sqcup^{\natural} \text{RHS}\} \cup \mathscr{E} \cup \{\mathscr{X}_{\mathsf{after}[\![\mathsf{S}]\!]} = \overline{\mathsf{test}}^{\natural}[\![\mathsf{B}]\!]\mathscr{X}_{\mathsf{at}[\![\mathsf{S}]\!]} \sqcup^{\natural} \bigsqcup^{\natural}_{\ell \in \mathsf{breaks\text{-}of}[\![\mathsf{S}_b]\!]} \mathscr{X}_{\ell}\} \qquad \text{(b)}$$

We consider a chaotic iteration (generalized to transfinite iteration [220, 221]) for these equations (b) with a subsequence $\mathscr{X}^k$, $k \in \mathbb{N}$ defined as follows:

- The initialization is the infimum $\mathscr{X}^0 = X^0 = \dot{\bot}$;
- Assume, by recurrence hypothesis, that $\mathscr{X}^k = X^k$;
- We next iterate the subequations $\{\mathscr{X}_{\mathsf{at}[\![\mathsf{S}]\!]} = \mathscr{P}_0 \sqcup^{\natural} \text{RHS}\} \cup \mathscr{E}$ in (b) and pass to the limit. By $\mathscr{P}_0 \sqsubseteq^{\natural} \mathscr{X}^k_{\mathsf{at}[\![\mathsf{S}]\!]}$ the limit is the least solution to $\widehat{\mathscr{E}}^{\natural}[\![\mathsf{S}_b]\!](\mathsf{test}^{\natural}[\![\mathsf{B}]\!]\mathscr{X}^k_{\mathsf{at}[\![\mathsf{S}]\!]})$, which by structural induction is $\widehat{\boldsymbol{\mathcal{S}}}^{\natural}[\![\mathsf{S}_b]\!]\,(\mathsf{test}^{\natural}[\![\mathsf{B}]\!]\mathscr{X}^k(\mathsf{at}[\![\mathsf{S}]\!])) = \widehat{\boldsymbol{\mathcal{S}}}^{\natural}[\![\mathsf{S}_b]\!]\,(\mathsf{test}^{\natural}[\![\mathsf{B}]\!]X^k(\mathsf{at}[\![\mathsf{S}]\!]))$ because $\mathscr{X}^k = X^k$ by recurrence hypothesis. Then we do a Jacobi iteration on the remaining equations in (b), which is then exactly the same as the Jacobi iteration in (a), so we get $\mathscr{X}^{k+1} = X^{k+1}$. □

By recurrence, $\forall k \in \mathbb{N}\,.\,\mathscr{X}^k = X^k$ so the infinite increasing chaotic iterations for (b) contains as a subsequence the infinite increasing Jacobi iteration for (a) so the limits are the same (see exercise 10.16) that is, by theorem 22.4 and (21.11), $\widehat{\boldsymbol{\mathcal{S}}}^{\natural}[\![\mathsf{S}]\!]\,\mathscr{P}_0$. □

23.3 Constraints

Many static program analysis methods solve inequations instead of equations (e.g., [475] and the related [251], or chapter 46, among many others). This follows from the fact that the least solution $\mathsf{lfp}^{\dot{\sqsubseteq}}\,\vec{F}$ to the system of equations $\vec{X} = \vec{F}(\vec{X})$, that is

$$\left\{\begin{array}{lcl} X_i & = & F_i(X_1, \ldots, X_n) \\ \text{i=1,...,n} & & \end{array}\right.$$

on the Cartesian product $\vec{D} = \prod_{i=1}^{n} D_i$ of complete lattices $\langle D_i, \sqsubseteq_i\rangle, i = 1, \ldots, n$, $n \geqslant 1$ ordered pointwise is also, by Tarski's fixpoint theorem 15.6, the least solution to the inequations $\vec{F}(\vec{X}) \mathrel{\dot{\sqsubseteq}} \vec{X}$, that is

$$\left\{\begin{array}{lcl} F_i(X_1, \ldots, X_n) & \sqsubseteq_i & X_i. \\ \text{i=1,...,n} & & \end{array}\right.$$

It is trivial to modify section 23.2 to generate such inequations from the program text.

Moreover, when the complete lattices $\langle D_i, \sqsubseteq_i\rangle, i = 1, \ldots, n$ are powersets, constraints such as $\{x\} \subseteq X$ can be rewritten $x \in X$, $X \cup Y \subseteq Z$ as $X \subseteq Z \wedge Y \subseteq Z$, $X \subseteq Y$ as $\forall x \in$

$X.x \in Y$, and so on, which may give the false impression that solving such inequations is a problem quite different from solving equations iteratively. Krzysztof Apt [47] proves the contrary. Therefore, static analysis by constraint resolution fits perfectly in the abstract interpretation framework; see chapter 25, "Abstract Reachability / Invariance / Safety Verification Semantics" for an example of constraint generation.

23.4 Conclusion

The understanding of static program analysis as solving fixpoint equations associated to the program (graph) dates back from the origin of dataflow analysis [27, 195]. This point of view is much broader and applies equally well to semantics in part III, verification in part VIII, and static analysis in parts IX and XI. We give more importance to the functional semantics $\widehat{\boldsymbol{\mathcal{S}}}^{\sharp}[\![\mathtt{S}]\!]$ over the equational semantics $\widehat{\boldsymbol{\mathcal{E}}}^{\sharp}[\![\mathtt{S}]\!]$ because we prefer to reason on the solution $\widehat{\boldsymbol{\mathcal{S}}}^{\sharp}[\![\mathtt{S}]\!]$ of the equations rather than on the equations $\widehat{\boldsymbol{\mathcal{E}}}^{\sharp}[\![\mathtt{S}]\!]$. The abstract interpreter $\widehat{\boldsymbol{\mathcal{S}}}^{\sharp}[\![\mathtt{S}]\!]$ is simpler, can be reasoned upon by structural induction, and does not introduce an intermediate step of building equations before solving them. However, $\mathcal{X} = \boldsymbol{\mathcal{F}}^{\sharp}(\mathcal{X})$, $\boldsymbol{\mathcal{F}}^{\sharp}(\mathcal{X}) \sqsubseteq^{\sharp} \mathcal{X}$, and $\mathsf{lfp}^{\sqsubseteq^{\sharp}}\, \boldsymbol{\mathcal{F}}^{\sharp}$ are perfectly interchangeable because they depend solely on the choice of the abstract domain of definition 21.1 and its soundness (and completeness) of definition 27.1.

23.5 Solutions to Selected Exercises

Solution to exercise 23.15

— $\widehat{\boldsymbol{\mathcal{E}}}^{\sharp}[\![\mathtt{S}_8]\!]\mathcal{P}_0 \;=\; \{\mathcal{X}_{\ell_2} = \mathcal{P}_0, \mathcal{X}_{\ell_3} = \mathsf{assign}^{\sharp}[\![\mathtt{x}, \mathtt{x}+\mathtt{1}]\!]\,\mathcal{X}_{\ell_2}\}$ ⦃by (23.7)⦄

— $\widehat{\boldsymbol{\mathcal{E}}}^{\sharp}[\![\mathtt{Sl}_6]\!]\mathcal{P}_0 \;=\; \widehat{\boldsymbol{\mathcal{E}}}^{\sharp}[\![\mathtt{S}_8]\!]$ ⦃by (23.10) because $\mathtt{Sl}_6 = \mathtt{Sl}_7\ \mathtt{S}_8$ and $\mathtt{Sl}_7 = \epsilon$⦄

— $\widehat{\boldsymbol{\mathcal{E}}}^{\sharp}[\![\mathtt{S}_{10}]\!](\mathsf{test}^{\sharp}[\![\mathtt{x} > \mathtt{9}]\!]\mathcal{X}_{\ell_3}) =$
$\{\mathcal{X}_{\ell_4} = \mathsf{test}^{\sharp}[\![\mathtt{x} > \mathtt{9}]\!]\mathcal{X}_{\ell_3},\quad \mathcal{X}_{\ell_1} = \bot^{\sharp},\quad \mathcal{X}_{\ell_5} = \mathcal{X}_{\ell_4}\}$

⦃by (23.13) because $\mathsf{break\text{-}to}[\![\mathtt{S}_{10}]\!] = \ell_5$⦄

— $\widehat{\boldsymbol{\mathcal{E}}}^{\sharp}[\![\mathtt{S}_9]\!]\mathcal{P}_0 \;=$ ⦃by (23.8)⦄
$\mathsf{let}\ (\{\mathcal{X}_{\ell_1} = \mathrm{RHS}\} \cup \mathcal{E}) = \widehat{\boldsymbol{\mathcal{E}}}^{\sharp}[\![\mathtt{S}_{10}]\!](\mathsf{test}^{\sharp}[\![\mathtt{x} > \mathtt{9}]\!]\mathcal{X}_{\ell_3})\ \mathsf{in}$
$\{\mathcal{X}_{\ell_3} = \mathcal{P}_0\} \cup\ \ \mathcal{E} \cup\ \ \{\mathcal{X}_{\ell_1} = \mathrm{RHS} \sqcup^{\sharp} \overline{\mathsf{test}}^{\sharp}[\![\mathtt{x} > \mathtt{9}]\!]\mathcal{X}_{\ell_3}\}$
$= \{\mathcal{X}_{\ell_3} = \mathcal{P}_0,\quad \mathcal{X}_{\ell_4} = \mathsf{test}^{\sharp}[\![\mathtt{x} > \mathtt{9}]\!]\mathcal{X}_{\ell_3},\quad \mathcal{X}_{\ell_5} = \mathcal{X}_{\ell_4},$
$\mathcal{X}_{\ell_1} = \overline{\mathsf{test}}^{\sharp}[\![\mathtt{x} > \mathtt{9}]\!]\mathcal{X}_{\ell_3}\}$

— $\widehat{\boldsymbol{\mathcal{E}}}^{\sharp}[\![\mathtt{S}_4]\!]\,\mathcal{P}_0 = \widehat{\boldsymbol{\mathcal{E}}}^{\sharp}[\![\mathtt{Sl}_5]\!]\,\mathcal{P}_0$ ⦃by (23.14)⦄
$= \mathsf{let}\ (\{\mathcal{X}_{\mathsf{after}[\![\mathtt{Sl}_6]\!]} = \mathrm{RHS}_6\} \cup \mathcal{E}_6) = \widehat{\boldsymbol{\mathcal{E}}}^{\sharp}[\![\mathtt{Sl}_6]\!]\mathcal{P}_0$ ⦃by (23.10)⦄
$\mathsf{and}\ (\{\mathcal{X}_{\mathsf{at}[\![\mathtt{S}_9]\!]} = \mathrm{RHS}_9\} \cup \mathcal{E}_9) = \widehat{\boldsymbol{\mathcal{E}}}^{\sharp}[\![\mathtt{S}_9]\!]\mathcal{X}_{\mathsf{at}[\![\mathtt{S}_9]\!]}\ \mathsf{in}$
$\{\mathcal{X}_{\mathsf{at}[\![\mathtt{S}_9]\!]} = \mathrm{RHS}_6 \sqcup^{\sharp} \mathrm{RHS}_9\} \cup \mathcal{E}_6 \cup \mathcal{E}_9$
$= \{\mathcal{X}_{\ell_2} = \mathcal{P}_0,\quad \mathcal{X}_{\ell_3} = \mathsf{assign}^{\sharp}[\![\mathtt{x}, \mathtt{x}+\mathtt{1}]\!]\,\mathcal{X}_{\ell_2} \sqcup^{\sharp} \mathcal{X}_{\ell_3},\quad \mathcal{X}_{\ell_4} =$
$\mathsf{test}^{\sharp}[\![\mathtt{x} > \mathtt{9}]\!]\mathcal{X}_{\ell_3},\qquad \mathcal{X}_{\ell_1} = \overline{\mathsf{test}}^{\sharp}[\![\mathtt{x} > \mathtt{9}]\!]\mathcal{X}_{\ell_3}\}$ ⦃$\mathsf{after}[\![\mathtt{Sl}_6]\!] = \mathsf{at}[\![\mathtt{S}_9]\!] = \ell_3$⦄

so $\widehat{\boldsymbol{\mathcal{E}}}^{\sharp}[\![\mathtt{S}_4]\!]\,(\mathsf{test}^{\sharp}[\![\mathtt{x} > \mathtt{0}]\!]\mathcal{X}_{\ell_1}) = \{\mathcal{X}_{\ell_2} = \mathsf{test}^{\sharp}[\![\mathtt{x} > \mathtt{0}]\!]\mathcal{X}_{\ell_1},$
$\mathcal{X}_{\ell_3} = \mathsf{assign}^{\sharp}[\![\mathtt{x}, \mathtt{x}+\mathtt{1}]\!]\,\mathcal{X}_{\ell_2} \sqcup^{\sharp} \mathcal{X}_{\ell_3},\quad \mathcal{X}_{\ell_4} = \mathsf{test}^{\sharp}[\![\mathtt{x} > \mathtt{9}]\!]\mathcal{X}_{\ell_3},$
$\mathcal{X}_{\ell_1} = \overline{\mathsf{test}}^{\sharp}[\![\mathtt{x} > \mathtt{9}]\!]\mathcal{X}_{\ell_3}\}$

Remark: The equational semantics (23.10) of a statement list can be simplified because the term $\mathscr{X}_{\ell_3}$ in $\mathscr{X}_{\ell_3} = \mathsf{assign}^{¤}[\![\mathtt{x}, \mathtt{x}+1]\!]\, \mathscr{X}_{\ell_2} \sqcup^{¤} \mathscr{X}_{\ell_3}$ is superfluous and can be eliminated but not in the case when $\mathtt{S}_9$ would be an iteration because initial and iterate environments are mixed at iteration entry.

— $\widehat{\mathscr{E}}^{¤}[\![\mathtt{P}]\!]\,\mathscr{P}_0 = \widehat{\mathscr{E}}^{¤}[\![\mathtt{Sl}_1]\!]\,\mathscr{P}_0 = \widehat{\mathscr{E}}^{¤}[\![\mathtt{S}_3]\!]\,\mathscr{P}_0 =$ ⟨(23.5), (23.10),(23.12)⟩

$\mathsf{let}\ (\{\mathscr{X}_{\ell_1} = \text{RHS}\} \cup \mathscr{E}) = \widehat{\mathscr{E}}^{¤}[\![\mathtt{S}_4]\!](\mathsf{test}^{¤}[\![\mathtt{x} > 0]\!]\mathscr{X}_{\ell_1})\ \mathsf{in}$ ⟨(23.12)⟩

$\{\mathscr{X}_{\ell_1} = \mathscr{P}_0 \sqcup^{¤} \text{RHS}\} \cup$

$\mathscr{E} \cup$

$\{\mathscr{X}_{\ell_5} = \overline{\mathsf{test}}^{¤}[\![\mathtt{x} > 0]\!]\mathscr{X}_{\ell_1} \sqcup^{¤} \bigsqcup^{¤}_{\ell \in \mathsf{breaks\text{-}of}[\![\mathtt{S}_4]\!]} \mathscr{X}_{\ell}\}$

$= \{\mathscr{X}_{\ell_1} = \mathscr{P}_0 \sqcup^{¤} \overline{\mathsf{test}}^{¤}[\![\mathtt{x} > 9]\!]\mathscr{X}_{\ell_3},\quad \mathscr{X}_{\ell_2} = \mathsf{test}^{¤}[\![\mathtt{x} > 0]\!]\mathscr{X}_{\ell_1},$
$\mathscr{X}_{\ell_3} = \mathsf{assign}^{¤}[\![\mathtt{x}, \mathtt{x}+1]\!]\, \mathscr{X}_{\ell_2} \sqcup^{¤} \mathscr{X}_{\ell_3},\quad \mathscr{X}_{\ell_4} = \mathsf{test}^{¤}[\![\mathtt{x} > 9]\!]\mathscr{X}_{\ell_3}$
$\mathscr{X}_{\ell_5} = \overline{\mathsf{test}}^{¤}[\![\mathtt{x} > 0]\!]\mathscr{X}_{\ell_1} \sqcup^{¤} \mathscr{X}_{\ell_4}\}$

⟨This instantiates to equations (23.2) for relational reachability. For the program $\mathsf{breaks\text{-}of}[\![\mathtt{Sl}_1]\!] = \varnothing$ so $\text{RHS} \sqcup^{¤} \bigsqcup^{¤}_{\ell \in \mathsf{breaks\text{-}of}[\![\mathtt{Sl}_1]\!]} \mathscr{X}_{\ell} = \text{RHS} \sqcup^{¤} \bot^{¤} = \text{RHS}$⟩

Solution to exercise 23.16 In OCaml,

```
type abstract_property = BOT | INT of int | TOP

let leq a1 a2 = match (a1,a2) with
| (BOT,_) -> true
| (_,BOT) -> false
| (_,TOP) -> true
| (TOP,_) -> false
| (INT v1, INT v2) -> (v1=v2)

let join a1 a2 = match (a1,a2) with
| (BOT,a2) -> a2
| (a1,BOT) -> a1
| (_,TOP) -> TOP
| (TOP,_) -> TOP
| (INT v1, INT v2) -> if (v1=v2) then INT v1 else TOP

let test_x_gt i a = match a with
| BOT -> BOT
| INT v -> if (v>i) then a else BOT
| TOP -> TOP

let negtest_x_gt i a = match a with
| BOT -> BOT
| INT v -> if (v<=i) then a else BOT
| TOP -> TOP

let assign_incr_x i a = match a with
| BOT -> BOT
| INT v -> INT (v+i)
| TOP -> TOP
```

```
let eqns r0  (xl1, xl2, xl3, xl4, xl5) =
  (join r0 (negtest_x_gt 9 xl3),
   test_x_gt 0 xl1,
   assign_incr_x 1 xl2,
   test_x_gt 9 xl3,
   join (test_x_gt 0 xl1) xl4)

let pbot = (BOT, BOT, BOT, BOT, BOT)

let pleq (a1, a2, a3, a4, a5) (a'1, a'2, a'3, a'4, a'5) = (leq a1 a
    '1)
   && (leq a2 a'2) && (leq a3 a'3) && (leq a4 a'4) && (leq a5 a'5)

let rec lfp a f leq = if leq (f a) a then a else lfp (f a) f leq

lfp pbot (eqns (INT 0)) pleq;; (* = (INT 0, BOT, BOT, BOT, BOT) *)
lfp pbot (eqns (INT 1)) pleq;; (* = (TOP, TOP, TOP, TOP, TOP) *)
```

Solution to exercise 23.18 The proof is by structural induction. For a base case like the assignment (23.7), we have

$$\begin{cases} \mathcal{X}_{\mathsf{at}\llbracket \mathsf{S}\rrbracket} = \mathsf{E}\llbracket \mathsf{S}\rrbracket\, \mathcal{P}_0\, \mathsf{at}\llbracket \mathsf{S}\rrbracket\, \langle \mathcal{X}_{\mathsf{at}\llbracket \mathsf{S}\rrbracket}, \mathcal{X}_{\mathsf{after}\llbracket \mathsf{S}\rrbracket}\rangle = \mathcal{P}_0 \\ \mathcal{X}_{\mathsf{after}\llbracket \mathsf{S}\rrbracket} = \mathsf{E}\llbracket \mathsf{S}\rrbracket\, \mathcal{P}_0\, \mathsf{after}\llbracket \mathsf{S}\rrbracket\, \langle \mathcal{X}_{\mathsf{at}\llbracket \mathsf{S}\rrbracket}, \mathcal{X}_{\mathsf{after}\llbracket \mathsf{S}\rrbracket}\rangle = \mathcal{X}_{\mathsf{at}\llbracket \mathsf{S}\rrbracket} \end{cases}$$

The other base cases (23.6), (23.11), and (23.13) are similar.

For an inductive case consider the iteration (23.12). The equations of the body are assumed to be of the required form so, by (23.12), the equations for the iteration are ($\mathsf{labs}\llbracket \mathsf{S}_b\rrbracket \subseteq \mathsf{labs}\llbracket \mathsf{S}\rrbracket$ and $\mathsf{at}\llbracket \mathsf{S}\rrbracket = \mathsf{after}\llbracket \mathsf{S}_b\rrbracket$)

$$\begin{cases} \mathcal{X}_{\mathsf{at}\llbracket \mathsf{S}\rrbracket} = \mathsf{E}\llbracket \mathsf{S}\rrbracket\, \mathcal{P}_0\, \mathsf{at}\llbracket \mathsf{S}\rrbracket\, (\prod_{\ell' \in \mathsf{labs}\llbracket \mathsf{S}\rrbracket} \mathcal{X}_{\ell'}) \\ \quad = \mathcal{P}_0 \sqcup^{¤} \mathsf{E}\llbracket \mathsf{S}_b\rrbracket\, (\mathsf{test}^{¤}\llbracket \mathsf{B}\rrbracket \mathcal{X}_{\mathsf{at}\llbracket \mathsf{S}\rrbracket})\, \mathsf{after}\llbracket \mathsf{S}_b\rrbracket\, (\prod_{\ell' \in \mathsf{labs}\llbracket \mathsf{S}_b\rrbracket} \mathcal{X}_{\ell'}) \\ \mathcal{X}_{\ell} = \mathsf{E}\llbracket \mathsf{S}\rrbracket\, \mathcal{P}_0\, \ell\, (\prod_{\ell' \in \mathsf{labs}\llbracket \mathsf{S}\rrbracket} \mathcal{X}_{\ell'}) \qquad \text{for } \ell \in \mathsf{labs}\llbracket \mathsf{S}_b\rrbracket \setminus \{\mathsf{at}\llbracket \mathsf{S}\rrbracket\} \\ \quad = \mathsf{E}\llbracket \mathsf{S}_b\rrbracket\, (\mathsf{test}^{¤}\llbracket \mathsf{B}\rrbracket \mathcal{X}_{\mathsf{at}\llbracket \mathsf{S}\rrbracket})\, \ell\, (\prod_{\ell' \in \mathsf{labs}\llbracket \mathsf{S}_b\rrbracket} \mathcal{X}_{\ell'}) \\ \mathcal{X}_{\mathsf{after}\llbracket \mathsf{S}\rrbracket} = \mathsf{E}\llbracket \mathsf{S}\rrbracket\, \mathcal{P}_0\, \mathsf{after}\llbracket \mathsf{S}\rrbracket\, (\prod_{\ell' \in \mathsf{labs}\llbracket \mathsf{S}\rrbracket} \mathcal{X}_{\ell'}) \\ \quad = \overline{\mathsf{test}}^{¤}\llbracket \mathsf{B}\rrbracket \mathcal{X}_{\mathsf{at}\llbracket \mathsf{S}\rrbracket} \sqcup^{¤} \bigsqcup^{¤}_{\ell \in \mathsf{breaks\text{-}of}\llbracket \mathsf{S}_b\rrbracket} \mathcal{X}_{\ell} \end{cases}$$

The other inductive cases (23.5), (23.8), (23.9), and (23.14) are similar.

VIII SPECIFICATION AND VERIFICATION

Program verification consists in proving that the semantics of a program $\mathsf{P} \in \mathbb{P}$ in an abstract domain $\mathbb{D}^{\sharp}$ satisfies a given *specification* $\mathcal{S}^{\sharp}[\![\mathsf{P}]\!] \in \mathbb{D}^{\sharp}$.

A *verifier* needs a stronger inductive invariant $\mathcal{I} \in \mathbb{D}^{\sharp}$ satisfying verification conditions (given in chapter 25) and implying the specification $\mathcal{S}^{\sharp}[\![\mathsf{P}]\!]$.

These verification conditions for the invariant $\mathcal{I}$ to be true rely on *fixpoint induction* (chapter 24) and are therefore *sound* (i.e. what is proved is true) and *complete* (i.e. what is true can be proved).

Specification and verification can be presented equivalently as a *program logic* (chapter 26).

24 Fixpoint Induction

Tarski's fixpoint theorems provide a method for proving properties of fixpoints called *fixpoint induction*. Alternatively, *iteration induction* is a fixpoint proof method based on fixpoint iteration. Both proof methods are sound and complete. They provide the underlying principles of formal methods for proving program correctness.

Contents

24.1 Fixpoint Induction

Theorem 24.1 (fixpoint induction I) Let $f \in \mathscr{L} \xrightarrow{\nearrow} \mathscr{L}$ be an increasing function on a complete lattice $\langle \mathscr{L}, \sqsubseteq, \bot, \top, \sqcap, \sqcup \rangle$ and $P \in \mathscr{L}$. We have

$$\mathsf{lfp}^{\sqsubseteq} f \sqsubseteq P \quad \Leftrightarrow \quad \exists I \in \mathscr{L} . \quad \begin{array}{l} f(I) \sqsubseteq I \\ \wedge \; I \sqsubseteq P \end{array} \qquad \begin{array}{r} \text{(a)} \\ \text{(b)} \end{array}$$

I is called an *invariant* of f when $\mathsf{lfp}^{\sqsubseteq} f \sqsubseteq I$ and an *inductive invariant* when satisfying $f(I) \sqsubseteq I$. Exercise 24.5 shows that an invariant is not necessarily inductive.

Proof of theorem 24.1 By Tarski's fixpoint theorem 15.6, $\mathsf{lfp}^{\sqsubseteq} f = \bigsqcap\{x \in \mathscr{L} \mid f(x) \sqsubseteq x\}$.

Soundness ($\Leftarrow$): If $I \in \mathscr{L}$ satisfies $f(I) \sqsubseteq I$ then $I \in \{x \in \mathscr{L} \mid f(x) \sqsubseteq x\}$ so by definition of the glb $\bigsqcap$, $\mathsf{lfp}^{\sqsubseteq} f = \bigsqcap\{x \in \mathscr{L} \mid f(x) \sqsubseteq x\} \sqsubseteq I \sqsubseteq P$.

Completeness ($\Rightarrow$): If $\mathsf{lfp}^{\sqsubseteq} f \sqsubseteq P$ then take $I = \mathsf{lfp}^{\sqsubseteq} f$ then $I = f(I)$ so $f(I) \sqsubseteq I$ by reflexivity and $I \sqsubseteq P$ by hypothesis. □

Soundness states that if a statement is proved by the proof method then that statement is true. Completeness states that the proof method is always applicable to prove a true statement.

Remark 24.2 By exercise 15.28, theorem 24.1 applies to upper continuous functions on CPO.

Example 24.3 Consider $f \in \wp(\mathbb{Z}) \to \wp(\mathbb{Z})$ such that $f(X) = \{0\} \cup \{z + 4 \mid z \in X\}$. Let us prove that $\mathsf{lfp}^{\subseteq} f \subseteq P \triangleq \{2k \mid k \in \mathbb{N}\}$. $\langle \wp(\mathbb{Z}), \subseteq \rangle$ is a complete lattice and f is $\subseteq$ increasing. Take $I = \{4k \mid k \in \mathbb{N}\}$ so that $I \subseteq P$. We have

$f(I)$

$$= \{0\} \cup \{z+4 \mid z \in \{4k \mid k \in \mathbb{N}\}\} \qquad \text{(definition of } f \text{ and } I\text{)}$$
$$= \{4k \mid k \in \{0\}\} \cup \{4k \mid k \in \mathbb{N}^+\} \qquad \text{(definition of } \mathbb{N}^+\text{)}$$
$$= \{4k \mid k \in \mathbb{N}\} \qquad (\mathbb{N} = \{0\} \cup \mathbb{N}^+)$$
$$\subseteq I \qquad (\subseteq \text{ reflexive and definition of } I)$$

We conclude by theorem 24.1.

Example 24.4 As a simple answer to exercise 15.17, $f(A) \subseteq A$ and $A \subseteq A$ imply $\mathsf{lfp}^{\subseteq} f \subseteq A$ by theorem 24.1 so we can take $D = \mathsf{lfp}^{\subseteq} f$.

Exercise 24.5 Consider the increasing function $f \in \mathscr{L} \xrightarrow{\nearrow} \mathscr{L}$ on the complete lattice L represented by the following Hasse diagram.

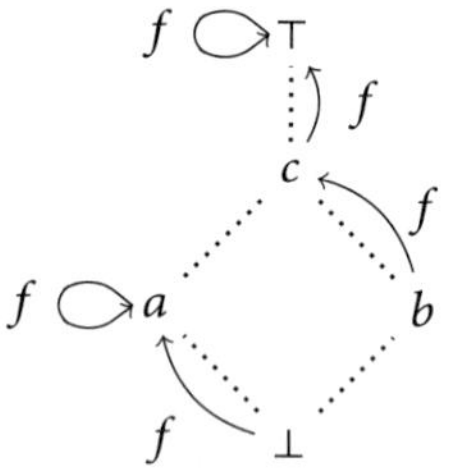

Prove that c is not an inductive invariant of f. Prove that $\mathsf{lfp}^{\sqsubseteq} f \sqsubseteq c$ by theorem 24.1.

Exercise 24.6 Let $f \in \mathscr{L} \xrightarrow{\nearrow} \mathscr{L}$ be an increasing function on a complete lattice $\langle \mathscr{L}, \sqsubseteq, \bot, \top, \sqcap, \sqcup \rangle$ and $a \in L$. Let $\mathsf{lfp}_a^{\sqsubseteq} f$ be the $\sqsubseteq$-smallest element x of L, if any, such that $a \sqsubseteq x = f(x)$. Prove that if $a \sqsubseteq f(a)$ then $\mathsf{lfp}_a^{\sqsubseteq} f$ exists.

Exercise 24.7 Continuing exercise 24.6, assuming $a \sqsubseteq f(a)$, prove that $\mathsf{lfp}_a^{\sqsubseteq} f \sqsubseteq P \Leftrightarrow \exists I \in \mathscr{L} \,.\, a \sqsubseteq I \wedge f(I) \sqsubseteq I \wedge I \sqsubseteq P$.

Exercise 24.8 (k-induction [336]) Let f be continuous on a CPO $\langle L, \bot, \sqsubseteq, \sqcup \rangle$. Let $k \in \mathbb{N}^+$ and $P \in L$. Prove that $f^k(P) \sqsubseteq P$ implies $\mathsf{lfp}^{\sqsubseteq} f \sqsubseteq P$.

Exercise 24.9 (functional fixpoint induction) Let $\langle \mathscr{L}, \sqsubseteq, \bot, \top, \sqcap, \sqcup \rangle$ be a complete lattice, $\mathscr{X}$ be a set, $F \in (\mathscr{X} \to \mathscr{L}) \xrightarrow{\nearrow} (\mathscr{X} \to \mathscr{L})$ be increasing for the pointwise order $\dot{\sqsubseteq}$, and $P \in \mathscr{X} \to \mathscr{L}$.

Prove that $\forall x \in \mathscr{X} \,.\, (\mathsf{lfp}^{\dot{\subseteq}} F)x \sqsubseteq P(x)$ if and only if $\exists I \in \mathscr{X} \to \mathscr{L} \,.\, F(I)x \subseteq I(x) \wedge I(x) \subseteq P(x)$.

Exercise 24.10 (factorial) Define the factorial relation $f_! \triangleq \{\langle n, !n \rangle \mid n \in \mathbb{N}\} \subseteq \mathbb{Z} \times \mathbb{Z}$ where $!0 = 1$ and $!n = 1 \times 2 \times \ldots \times n$ for $n > 0$. Define $F_!(r) \triangleq \{\langle 0, 1 \rangle\} \cup \{\langle n, n \times y \rangle \mid \langle n-1, y \rangle \in r\}$. Prove that $\mathsf{lfp}^{\subseteq} F_! \subseteq f_!$.[1]

1. For readers who know denotational semantics [337, 928], let $f_!'(n) = (\!| n \geqslant 0 \,?\, n! \,:\, \bot |\!) \in \mathbb{Z} \to \mathbb{Z}_\bot$ where $\bot$ denotes nontermination and $F_!'(f)n = (\!| n = 0 \,?\, 1 \,:\, n \times f(n-1) |\!)$ with $\mathsf{lfp}^{\dot{\sqsubseteq}} F_!' = f_!'$ where $\dot{\sqsubseteq}$ is the pointwise extension of Scott's flat CPO $\langle \mathbb{Z}_\bot, \sqsubseteq \rangle$ and $\bot \sqsubseteq \bot \sqsubsetneq z \sqsubseteq z$ for all $z \in \mathbb{Z}$. The correspondence is $f_! = \alpha_\partial(f_!')$ and $F_!(\alpha_\partial(f)) = \alpha_\partial(F_!'(f))$ with $\langle \mathbb{Z} \to \mathbb{Z}_\bot, \dot{\sqsubseteq} \rangle \underset{\alpha_\partial}{\overset{\gamma_\partial}{\leftrightarrows}} \langle \wp_{\mathsf{F}}(\mathbb{Z} \times \mathbb{Z}_\bot), \subseteq \rangle$ where $\alpha_\partial(f) = \{\langle z, f(z) \rangle \mid z \in \mathbb{Z} \wedge f(z) \neq \bot\}$, $\wp_{\mathsf{F}}(S \times S') \triangleq \{r \in \wp(S \times S') \mid \forall x, y \,.\, (\langle x, y \rangle \in r \wedge \langle x, y' \rangle \in r) \Rightarrow (y = y')\}$ are the functional relations over $S \times S'$ and $\gamma_\partial(r) \triangleq x \mapsto (\!| \exists! y \,.\, \langle x, y \rangle \in r \,?\, y \,:\, \bot |\!)$ where y is unique because r is a functional relation.

Exercise 24.11 The proof of theorem 24.1 shows that fixpoint induction follows from Tarski's fixpoint theorem 15.6. Conversely, show that Tarski's fixpoint theorem 15.6 follows from the fixpoint induction theorem 24.1.

Exercise 24.12 (characterization of fixpoint completeness [815]) As in theorem 18.23, assume that $\langle \mathscr{C}, \sqsubseteq, \bot, \top, \sqcup, \sqcap \rangle$ and $\langle \mathscr{A}, \preccurlyeq, 0, 1, \curlyvee, \curlywedge \rangle$ are complete lattices, $f \in \mathscr{C} \xrightarrow{\nearrow} \mathscr{C}$ is increasing, and $\langle \mathscr{C}, \sqsubseteq \rangle \underset{\alpha}{\overset{\gamma}{\leftrightarrows}} \langle \mathscr{A}, \preccurlyeq \rangle$. Prove that $\alpha(\mathsf{lfp}^{\sqsubseteq} f) = \mathsf{lfp}^{\preccurlyeq} \alpha \circ f \circ \gamma$ if and only if $\forall P \in \mathscr{A} \,.\, (\alpha(\mathsf{lfp}^{\sqsubseteq} f) \preccurlyeq P) \Leftrightarrow (\exists I \in \mathscr{A} \,.\, \alpha \circ f \circ \gamma(I) \preccurlyeq I \wedge I \preccurlyeq P)$.

24.2 Iteration Induction

Let $F \in S \to S$ and $\langle x_i, i \in \Delta \rangle$ be a family of elements of S. The family is *nonempty* if and only if $\Delta \neq \varnothing$. It is *infinite* when the cardinality of Δ is greater than or equal to that of $\mathbb{N}$. It is *denumerable* if and only if $\Delta \subseteq \mathbb{N}$ (up to an isomorphism). It is a $\sqsubseteq$-*increasing chain* if and only if $\forall i, j \in \Delta \,.\, (i \leqslant j) \Rightarrow (x_i \sqsubseteq x_j)$. It is a *strictly increasing chain* if and only if $\forall i, j \in \Delta \,.\, (i < j) \Rightarrow (x_i \sqsubsetneq x_j)$. It is *in* $S' \subseteq S$ if and only if $\forall i \in \Delta \,.\, x_i \in S'$. The sequence $\langle x_i, i \in \mathbb{N} \rangle$ is F-*maximally increasing* when it is infinite (hence nonempty), denumerable, iterating F (i.e., $\forall i \in \mathbb{N} \,.\, x_{i+1} = F(x_i)$), and either strictly increasing (i.e., $\forall i, j \in \mathbb{N} \,.\, (i < j) \Rightarrow (x_i \sqsubsetneq x_j)$) or is first strictly increasing and then stationary (i.e., $\exists k \in \mathbb{N} \,.\, \forall i, j \in \mathbb{N} \,.\, (i < j \leqslant k) \Rightarrow (x_i \sqsubsetneq x_j) \wedge (k \leqslant i) \Rightarrow (x_k = x_i)$).

Theorem 24.13 (iteration induction) Let $F \in \mathscr{L} \xrightarrow{uc} \mathscr{L}$ be a continuous function on a CPO $\langle \mathscr{L}, \sqsubseteq, \bot, \sqcup \rangle$ and $\mathscr{P} \in \wp(\mathscr{L})$.

$$
\begin{array}{lll}
 & \mathsf{lfp}^{\sqsubseteq} F \in \mathscr{P} & \\
\Leftrightarrow & \exists \mathscr{Q} \in \wp(\mathscr{L}) \,.\, \quad \bot \in \mathscr{Q} & (24.13.a)\\
 & \quad \wedge \quad \forall x \in \mathscr{Q} \,.\, F(x) \in \mathscr{Q} & (24.13.b)\\
 & \quad \wedge \quad \text{for any } F\text{-maximally } \sqsubseteq\text{-increasing chain } \langle x_i, i \in \mathbb{N} \rangle \text{ in } \mathscr{Q}, & \\
 & \qquad \displaystyle\bigsqcup_{i \in \mathbb{N}} x_i \in \mathscr{P} & (24.13.c)
\end{array}
$$

The proof shows that the hypotheses (a), (b), and (c) are necessary only for the iterates of F.

Proof of theorem 24.13
Soundness ($\Leftarrow$): Let $F^{i+1}(\bot) = F(F^i(\bot))$ be the iterates of F from $F^0(\bot) = \bot$. $F^0(\bot) \in \mathscr{Q}$ by (24.13.a). By recurrence, $\forall i \in \mathbb{N} \,.\, F^i(\bot) \in \mathscr{Q}$ by (24.13.b). F is continuous hence monotonically increasing so $\langle F^i(\bot) \in \mathscr{Q}, i \in \mathbb{N} \rangle$ is a $\sqsubseteq$-increasing enumerable chain iterating F. If it is finite then $F^0(\bot) \sqsubset F^1(\bot) \sqsubset \ldots \sqsubset F^{n-1}(\bot) = F^n(\bot) = \ldots = \ldots$ for some $n \in \mathbb{N}$ proving that the chain is F-maximally increasing in $\mathscr{Q}$. So, by (24.13.c), $\mathsf{lfp}^{\sqsubseteq} F = F^{n-1}(\bot) = \bigsqcup_{i \in \mathbb{N}} F^i(\bot) \in \mathscr{P}$. Otherwise, the chain is infinite and strictly increasing so F-maximally increasing in $\mathscr{Q}$. By theorem 15.26 and (24.13.c), we conclude that $\mathsf{lfp}^{\sqsubseteq} F = \bigsqcup_{i \in \mathbb{N}} F^i(\bot) \in \mathscr{P}$.

Completeness ($\Rightarrow$): Let $F^{i+1}(\bot) = F(F^i(\bot))$ be the iterates of F from $F^0(\bot) = \bot$. Choosing $\mathscr{Q} = \{F^i(\bot) \mid i \in \mathbb{N}\}$, we have (24.13.a) and (24.13.b). By theorem 15.26, $\langle F^i(\bot) \in \mathscr{Q}, i \in \mathbb{N} \rangle$ is a $\sqsubseteq$-increasing chain in $\mathscr{Q}$. It is enumerable and the only F-maximally increasing one so $\{x_i \in \mathscr{Q} \mid i \in \mathbb{N}\} = \{F^i(\bot) \in \mathscr{Q} \mid i \in \mathbb{N}\}$. By theorem 15.26,

$\mathsf{lfp}^{\sqsubseteq} F = \bigsqcup_{i\in\mathbb{N}} F^i(\bot)$. By hypothesis, $\mathsf{lfp}^{\sqsubseteq} F \in \mathscr{P}$, and so $\bigsqcup_{i\in\mathbb{N}} F^i(\bot) = \bigsqcup_{i\in\mathbb{N}} x_i \in \mathscr{P}$, proving (24.13.c). □

Example 24.14 Continuing example 24.3, let $f \in \wp(\mathbb{Z}) \to \wp(\mathbb{Z})$ such that $f(X) = \{0\} \cup \{z+4 \mid z \in X\}$. We have $X \subseteq P \triangleq \{2k \mid k \in \mathbb{N}\}$ if and only if $X \in \mathscr{P} \triangleq \wp(\{2k \mid k \in \mathbb{N}\})$. Let us prove that $\mathsf{lfp}^{\subseteq} f \in \mathscr{P}$. f preserves nonempty joins so is continuous and $\langle \mathscr{L}, \sqsubseteq\rangle$ is a complete lattice so a CPO.

Define $\mathcal{Q} \triangleq \{x_i \mid i \in \mathbb{N}\}$ such that $x_i \triangleq \{4k \mid 0 \leqslant k < i\}$.

- We have $\varnothing \in \mathcal{Q}$ for $i = 0$ because $x_0 = \varnothing$
- If $x \in \mathcal{Q}$ then $\exists i \in \mathbb{N} \,.\, x = x_i = \{4k \mid 0 \leqslant k < i\}$ and so

$$
\begin{aligned}
& f(x) \\
= {} & \{0\} \cup \{z+4 \mid z \in \{4k \mid 0 \leqslant k < i\}\} && \text{(definition of } x\text{)} \\
= {} & \{4k \mid k = 0\} \cup \{4k \mid 0 < k \leqslant i\}\} \\
= {} & \{4k \mid 0 \leqslant k < i+1\} \quad = \quad x_{i+1} \quad \in \quad \mathcal{Q}
\end{aligned}
$$

- For any increasing chain $\{y_i \mid i \in \mathbb{N}\}$, $\forall i \in \mathbb{N} \,.\, y_i \in \mathcal{Q}$ implies $\bigcup\{y_i \mid i \in \mathbb{N}\} \subseteq \bigcup \mathcal{Q} = \bigcup\{x_i \mid i \in \mathbb{N}\} = \bigcup_{i\in\mathbb{N}}\{4k \mid 0 \leqslant k < i\} = \{4k \mid k \in \mathbb{N}\} \subseteq \{2k \mid k \in \mathbb{N}\} = P$ and so $\bigcup_{i\in\mathbb{N}} y_i \in \mathscr{P}$

By theorem 24.13, we conclude that $\mathsf{lfp}^{\subseteq} f \in \mathscr{P}$ or equivalently $\mathsf{lfp}^{\subseteq} f \subseteq P$.

Exercise 24.15 (factorial termination) Continuing exercise 24.10, let $F_!(r) \triangleq \{\langle 0, 1\rangle\} \cup \{\langle n, n \times y\rangle \mid \langle n-1, y\rangle \in r\}$. Prove termination that is $\forall n \in \mathbb{N} \,.\, \exists y \in \mathbb{Z} \,.\, \langle n, y\rangle \in \mathsf{lfp}^{\subseteq} F_!$ (which, together with exercise 24.10, shows that $\mathsf{lfp}^{\subseteq} F_! = f_! \triangleq \{\langle n, !n\rangle \mid n \in \mathbb{N}\}$).

Exercise 24.16 (similar to exercise 24.11) Theorem 24.13 follows from theorem 15.26. Conversely, prove that theorem 15.26 follows from theorem 24.13.

Exercise 24.17 Assume that $\langle \mathscr{L}, \sqsubseteq, \bot, \sqcup\rangle$ is a complete lattice and $\langle (\mathscr{L} \to \mathscr{L}), \dot\sqsubseteq\rangle \underset{\vec{F}}{\overset{\overleftarrow{F}}{\leftrightarrows}} \langle (\mathscr{L} \to \mathscr{L}), \dot\sqsubseteq\rangle$. Prove that $\langle \mathscr{L}, \sqsubseteq\rangle \underset{\mathsf{lfp}^{\sqsubseteq} \vec{F}}{\overset{\mathsf{gfp}^{\sqsubseteq} \overleftarrow{F}}{\leftrightarrows}} \langle \mathscr{L}, \sqsubseteq\rangle$.

Exercise 24.18 Assume that $\langle \mathscr{L}, \sqsubseteq, \bot, \sqcup\rangle$ is a CPO, $f \in \mathscr{L} \xrightarrow{uc} \mathscr{L}$ is upper continuous, and $P \in \mathscr{L}$. Design a sound and complete induction principle to prove that $P \sqsubseteq \mathsf{lfp}^{\sqsubseteq} f$.

24.3 Conclusion

Fixpoint induction in theorem 24.1 directly follows from Tarski's fixpoint theorem 15.6, as observed by [777, (2.0.2)], [778, (2.3)], and [221, p. 3.4.1]. It relies on properties of f above its least fixpoint. In contrast, iteration induction directly follows from Scott–Kleene's fixpoint theorem 15.26, as observed by [648, 652, 869]. It relies on properties of f below its least fixpoint. Iteration induction generalizes to transfinite iterations [240]. Dual coinduction theorems can be easily stated (see section 10.4). As a direct application, proof methods for verifying semantic properties of programs are by induction on the fixpoint definition of these semantics [232, 243].

24.4 Solutions to Selected Exercises

Solution to exercise 24.5 $f(c) \sqsubseteq \top \not\sqsubseteq c$ so c is not inductive. We have $f(a) \sqsubseteq a$ and $a \sqsubseteq c$ so, by theorem 24.1, $\mathsf{lfp}^{\sqsubseteq} f \sqsubseteq c$.

Solution to exercise 24.6 Apply Tarski's theorem to the complete lattice $\langle \{x \in L \mid a \sqsubseteq x\}, \sqsubseteq\rangle$.

Solution to exercise 24.8 By the iterative fixpoint theorem 15.26, $\mathsf{lfp}^{\sqsubseteq} f = \bigsqcup_{i\in\mathbb{N}} f^i(\bot)$ where the iterates $\langle f^i(\bot),\ i \in \mathbb{N}\rangle$ of f from $\bot$ form an increasing chain. f is continuous hence increasing and so f^k is increasing. It follows, by recurrence and hypothesis $f^k(P) \sqsubseteq P$, that $\forall n \in \mathbb{N} \,.\, f^{nk}(\bot) \sqsubseteq P$. Because $\forall i \in \mathbb{N} \,.\, \exists n \in \mathbb{N} \,.\, f^i(\bot) \sqsubseteq f^{nk}(\bot)$, we have $f^i(\bot) \sqsubseteq P$ by transitivity. We conclude that $\mathsf{lfp}^{\sqsubseteq} f = \bigsqcup_{i\in\mathbb{N}} f^i(\bot) \sqsubseteq P$.

Solution to exercise 24.9 $\langle \mathcal{X} \to \mathcal{L},\ \dot{\sqsubseteq},\ \dot{\bot},\ \dot{\top},\ \dot{\sqcap},\ \dot{\sqcup}\rangle$ is a complete lattice for the pointwise partial order $\dot{\sqsubseteq}$. By Tarski fixpoint theorem 15.6, $\mathsf{lfp}^{\dot{\sqsubseteq}} F = \dot{\bigsqcap}\{f \mid F(f) \mathrel{\dot{\sqsubseteq}} f\}$ so for all $x \in \mathcal{X}$, $(\mathsf{lfp}^{\dot{\sqsubseteq}} F)x = \bigsqcap\{f(x) \mid F(f)x \sqsubseteq f(x)\}$. Following the proof of theorem 24.1, $(\mathsf{lfp}^{\dot{\sqsubseteq}} F)x \sqsubseteq P(x)$ if and only if $\exists I \in \mathcal{X} \to \mathcal{L} \,.\, F(I)x \subseteq I(x) \wedge I(x) \subseteq P(x)$.

Solution to exercise 24.10 We apply theorem 24.1 with $P = f_!$. $F_!$ is increasing on the complete lattice $\langle \wp(\mathbb{Z} \times \mathbb{Z}), \subseteq\rangle$. Let $I = f_!$ so that (b) holds. Moreover, $F_!(I) = F_!(f_!) = \{\langle 0,\ 1\rangle\} \cup \{\langle n,\ n \times y\rangle \mid \langle n-1,\ y\rangle \in f_!\} = \{\langle 0,\ 1\rangle\} \cup \{\langle n,\ n \times (n-1)!\rangle \mid n-1 \geqslant \mathbb{N}\} = f_! = I$, proving (b) by reflexivity of $\subseteq$. We conclude that $\mathsf{lfp}^{\sqsubseteq} F \subseteq P = f_!$.

Solution to exercise 24.11 Assume that theorem 24.1 holds. In a complete lattice $\langle L, \sqsubseteq, \bigsqcap\rangle$, if $\forall y \in S \,.\, x \sqsubseteq y$ and $x \in S$ then $x = \bigsqcup S$. Take $x = \mathsf{lfp}^{\sqsubseteq} f$ and $S = \{y \in L \mid f(y) \sqsubseteq y\}$. Assume that $y \in S$ then $f(y) \sqsubseteq y$ so $f(y) \sqsubseteq y \wedge y \sqsubseteq y$ proving $x = \mathsf{lfp}^{\sqsubseteq} f \sqsubseteq y$ by theorem 24.1. Moreover, $f(\mathsf{lfp}^{\sqsubseteq} f) = \mathsf{lfp}^{\sqsubseteq} f \sqsubseteq \mathsf{lfp}^{\sqsubseteq} f$ so $x = \mathsf{lfp}^{\sqsubseteq} f \in S$. We conclude that $\mathsf{lfp}^{\sqsubseteq} f = x = \bigsqcup S = \bigsqcup\{x \in L \mid f(x) \sqsubseteq x\}$ which is theorem 15.6.

Solution to exercise 24.12 $(\Leftarrow)$ Take $P = I = \mathsf{lfp}^{\preccurlyeq} \alpha \circ f \circ \gamma$. Then, by fixpoint property and reflexivity, $\alpha \circ f \circ \gamma(I) = I$ and $I = P$ implies $\alpha(\mathsf{lfp}^{\sqsubseteq} f) \preccurlyeq P = \mathsf{lfp}^{\preccurlyeq} \alpha \circ f \circ \gamma$.

Take $P = \alpha(\mathsf{lfp}^{\sqsubseteq} f)$ so that $\alpha(\mathsf{lfp}^{\sqsubseteq} f) \preccurlyeq P$ implies $\exists I \in \mathcal{A} \,.\, \alpha \circ f \circ \gamma(I) \preccurlyeq I \wedge I \preccurlyeq \alpha(\mathsf{lfp}^{\sqsubseteq} f)$. By fixpoint induction in theorem 24.1, $\mathsf{lfp}^{\preccurlyeq} \alpha \circ f \circ \gamma \preccurlyeq P = \alpha(\mathsf{lfp}^{\sqsubseteq} f)$.

By antisymmetry, we conclude that $\alpha(\mathsf{lfp}^{\sqsubseteq} f) = \mathsf{lfp}^{\preccurlyeq} \alpha \circ f \circ \gamma$.

$(\Rightarrow)$ If $\alpha(\mathsf{lfp}^{\sqsubseteq} f) = \mathsf{lfp}^{\preccurlyeq} \alpha \circ f \circ \gamma$ then for all $P \in \mathcal{A}$, $\alpha(\mathsf{lfp}^{\sqsubseteq} f) \preccurlyeq P \Leftrightarrow \mathsf{lfp}^{\preccurlyeq} \alpha \circ f \circ \gamma \preccurlyeq P$, which, by fixpoint induction in theorem 24.1, is equivalent to $\exists I \in \mathcal{A} \,.\, \alpha \circ f \circ \gamma(I) \preccurlyeq I \wedge I \preccurlyeq P$.

Solution to exercise 24.15 Define $\mathcal{P} = \{r \in \wp(\mathbb{Z} \times \mathbb{Z}) \mid \forall n \in \mathbb{N} \,.\, \exists y \in \mathbb{Z} \,.\, \langle n,\ y\rangle \in r\}$. We use theorem 24.13 to prove that $\mathsf{lfp}^{\subseteq} F_! \in \mathcal{P}$. Let us define $\forall i \in \mathbb{N} \,.\, \mathcal{Q}_i \triangleq \{r \in \wp(\mathbb{Z} \times \mathbb{Z}) \mid \forall n \in [0, i[\,.\, \exists y \in \mathbb{Z} \,.\, \langle n,\ y\rangle \in r \wedge \forall n \geqslant i \,.\, \forall y \in \mathbb{Z} \,.\, \langle n,\ y\rangle \notin r\}$ and $\mathcal{Q} \triangleq \bigcup_{i\in\mathbb{N}} \mathcal{Q}_i$.

- We have $\varnothing \in \{\varnothing\} = \mathcal{Q}_0 \subseteq \mathcal{Q}$, proving (24.13.a).
- Assume that $i \in \mathbb{N}$ and $r \in \mathcal{Q}_i$. We have

$$
\begin{aligned}
& F_!(r) \\
= \ & \{\langle 0,\ 1\rangle\} \cup \{\langle n,\ n \times y\rangle \mid \langle n-1,\ y\rangle \in r\} && \text{(definition of } F_! \text{ in exercise 2.14)} \\
\subseteq \ & \{\langle 0,\ 1\rangle\} \cup \{\langle n,\ n \times y\rangle \mid \langle n-1,\ y\rangle \in r \wedge \forall k \in [0, i[\,.\, \exists z \in \mathbb{Z} \,.\, \langle k,\ z\rangle \in r\} && (r \in \mathcal{Q}_i)
\end{aligned}
$$

$\subsetneq \{\langle 0, 1\rangle\} \cup \{\langle k, z\rangle \mid k \in [1, i] \wedge z \in \mathbb{Z}\} \in \mathcal{Q}_{i+1} \subseteq \mathcal{Q}$ $\quad$ (definition of $\mathcal{Q}_{i+1}$)
proving $F_!(r) \in \mathcal{Q}_{i+1}$.

- If $r \in \mathcal{Q}$ then $r \in \mathcal{Q}_i$ for some $i \in \mathbb{N}$. Then $F_!(r) \in \mathcal{Q}_{i+1} \subseteq \mathcal{Q}$ proving (24.13.b).
- Let $r_0 \subseteq r_1 \subseteq \ldots \subseteq r_i \subseteq \ldots$ be a $F_!$-maximal increasing chain of elements of Q. There are two cases.

Either the chain is stationary at some rank i such that $r_0 \subsetneq r_1 \subsetneq \ldots \subsetneq r_i = r_{i+1} = r_{i+2} = \ldots$. Then $r_i \in \mathcal{Q}_j$ for some $j \in \mathbb{N}$. So $r_{i+1} = r_i \in \mathcal{Q}_j$ and $r_{i+1} = F_!(r_i) \in \mathcal{Q}_{j+1}$, which is a contradiction because $\mathcal{Q}_j \cap \mathcal{Q}_{j+1} = \emptyset$.

Because the chain is assumed to be a $F_!$-maximal increasing chain, the only other possibility is that it is strictly increasing $r_0 \subsetneq r_1 \subsetneq \ldots \subsetneq r_n \subsetneq \ldots$. Then, by definition of $\mathcal{Q}$, we have $r_0 \in \mathcal{Q}_{j_0}, r_1 \in \mathcal{Q}_{j_1}, \ldots, r_n \in \mathcal{Q}_{j_n}, r_{n+1} \in \mathcal{Q}_{j_{n+1}}, \ldots$. Because the chain $r_0 \subsetneq r_1 \subsetneq \ldots \subsetneq r_n \subsetneq \ldots$ is strictly increasing, we have $j_0 < j_1 < \ldots < j_n < j_{n+1} < \ldots$ so $j_{n+1} > n+1$. Because $r_{n+1} \in \mathcal{Q}_{j_{n+1}}$, $\exists y \in \mathbb{Z} \,.\, \langle n, y\rangle \in r_{n+1}$.

To prove that $\bigcup_{i\in\mathbb{N}} r_i \in \mathcal{P}$, assume by reductio ad absurdum, that $\exists n \in \mathbb{N} \,.\, \forall y \in \mathbb{Z} \,.\, \langle n, y\rangle \notin \bigcup_{i\in\mathbb{N}} r_i$ so $\exists n \in \mathbb{N} \,.\, \forall y \in \mathbb{Z} \,.\, \forall i \in \mathbb{N} \,.\, \langle n, y\rangle \notin r_i$. In particular $\forall y \in \mathbb{Z} \,.\, \langle n, y\rangle \notin r_{n+1}$, which is a contradiction.

We have proved (24.13.c) hence $\mathsf{lfp}^{\subseteq} F_! \in \mathcal{P}$, that is, $\forall n \in \mathbb{N} \,.\, \exists y \in \mathbb{Z} \,.\, \langle n, y\rangle \in \mathsf{lfp}^{\subseteq} F_!$.

Another way of proving termination for a positive parameter is to observe that it strictly decreases on recursive call and remains positive, which can be done only a finite number of times because $\langle \mathbb{N}, <\rangle$ is well founded [263].

Solution to exercise 24.16 Take $\mathcal{Q} = \{f^n(\bot) \mid n \in \mathbb{N}\}$ and $\mathcal{P} = \{\bigsqcup_{i\in\mathbb{N}} f^n(\bot)\}$.

Solution to exercise 24.18 We have $P \sqsubseteq \mathsf{lfp}^{\sqsubseteq} f$ if and only if $\exists I \in \mathbb{N} \to \mathcal{L} \,.\, I(0) = \bot, \forall i \in \mathbb{N} \,.\, I(i) \sqsubseteq I(i+1) \sqsubseteq f(I(i))$, and $P \sqsubseteq \bigsqcup_{i\in\mathbb{N}} I(i)$.

For soundness, let $\langle f^i, i \in \mathbb{N}\rangle$ be the iterates of f from $f^0 = \bot$. We have $I(0) \sqsubseteq f^0 = \bot \sqsubseteq I(1)$. Assume, by induction hypothesis, that $I(i) \sqsubseteq f^i$. Then f is upper continuous hence increasing, so $I(i) \sqsubseteq I(i+1) \sqsubseteq f(I(i)) \sqsubseteq f(f^i) = f^{i+1}$. By recurrence $\forall i \in \mathbb{N} \,.\, I(i) \sqsubseteq f^i$. Because $\langle I(i), i \in \mathbb{N}\rangle$ is an increasing chain, its limit $\bigsqcup_{i\in\mathbb{N}} I(i)$ exists in the CPO $\mathcal{L}$ such that, by theorem 15.26, $P \sqsubseteq \bigsqcup_{i\in\mathbb{N}} I(i) \sqsubseteq \bigsqcup_{i\in\mathbb{N}} f^i = \mathsf{lfp}^{\sqsubseteq} f$.

For completeness, take $I(i) = f^i, i \in \mathbb{N}$.

25 Abstract Reachability / Invariance / Safety Verification Semantics

The invariance proof method of Alan Turing [933], Peter Naur [742] and Bob Floyd [376] essentially consist in asking the programmer to provide an assertional invariant (or relational, for Zohar Manna [647]) and checking verification conditions that ensure that this invariant is implied by the assertional (or relational) reachability semantics.

This invariance proof method is an application of the fixpoint induction theorem 24.1 to the fixpoint equations of chapter 23 for the forward reachability semantics of chapter 19. By fixpoint induction theorem 24.1, the proof method is sound (correct) and complete (always applicable) although, by Rice's theorem 9.12, undecidable (i.e., not fully automatizable).

We present a structural definition of the verification conditions that proceeds by induction on the program syntax in the spirit of Tony Hoare's logic [501] or Edger W. Dijkstra's [321] predicate transformers. Moreover, we do not consider the particular case of reachable states but instead consider any abstract domain for the equational semantics of chapter 23, or equivalently the abstract semantics of chapter 21. It follows that the proof method is sound for all possible abstractions of the abstract semantics of chapter 21. In particular, when instantiating with the prefix trace semantics, we get a structural proof method [263, section 6] for safety specification section 14.1.

Contents

25.1 Reachability Specification and Proof Method

A reachability proof of a statement S for preconditions $\mathscr{P}_0$ is a proof that the reachability semantics $\widehat{\boldsymbol{\mathcal{S}}}^{\vec{e}}[\![\mathsf{S}]\!]\,\mathscr{P}_0$ of that statement satisfies a reachability specification $\mathcal{S}^{\vec{e}}[\![\mathsf{S}]\!]$. Any environment reachable at any program point ℓ of the statement S should satisfy the property $\mathcal{S}^{\vec{e}}[\![\mathsf{S}]\!]\,\ell$ attached to that program point, that is, $\widehat{\boldsymbol{\mathcal{S}}}^{\vec{e}}[\![\mathsf{S}]\!]\,\mathscr{P}_0 \mathrel{\dot{\subseteq}} \mathcal{S}^{\vec{e}}[\![\mathsf{S}]\!]$.

A proof method is a recipe explaining how to make the proof. This recipe should ensure that the proof is sound (if the recipe is applied, then this provides a reachability proof) and complete (if a reachability proof exists then the recipe can be applied to find a proof). The proof method consists of checking local verification conditions relative to a single program computation step.

25.1.1 Reachability Specification

Definition 25.1 (reachability specification) A reachability specification $\mathcal{S}^{\vec{\varrho}}[\![\mathsf{S}]\!] \in \mathsf{labs}[\![\mathsf{S}]\!] \rightarrow \wp(\mathbb{E}\mathrm{v}^{\vec{\varrho}})$ attaches an environment property $\mathcal{S}^{\vec{\varrho}}[\![\mathsf{S}]\!]\ell$ at each point ℓ of a program component $\mathsf{S} \in \mathbb{P}\mathrm{c}$.

Following section 20.4, the reachability specification is assertional when $\vec{\varrho} = \vec{r}$ and $\mathbb{E}\mathrm{v}^{\vec{\varrho}} = \mathbb{E}\mathrm{v}$ and relational when $\vec{\varrho} = \vec{\mathsf{R}}$ and $\mathbb{E}\mathrm{v}^{\vec{\varrho}} = \mathbb{E}\mathrm{v} \times \mathbb{E}\mathrm{v}$.

In general, a subset of program points is considered, for example, $\{\mathsf{at}[\![\mathsf{S}]\!], \mathsf{after}[\![\mathsf{S}]\!]\}$ for a precondition $\mathcal{S}^{\vec{\varrho}}[\![\mathsf{S}]\!]\,\mathsf{at}[\![\mathsf{S}]\!]$ and a postcondition $\mathcal{S}^{\vec{\varrho}}[\![\mathsf{S}]\!]\,\mathsf{after}[\![\mathsf{S}]\!]$, and it is left implicit that one does not care about other program points, that is, $\mathcal{S}^{\vec{\varrho}}[\![\mathsf{S}]\!]\ell = \mathbb{E}\mathrm{v}^{\vec{\varrho}}$ for all $\ell \in \mathsf{labs}[\![\mathsf{S}]\!] \setminus \{\mathsf{at}[\![\mathsf{S}]\!], \mathsf{after}[\![\mathsf{S}]\!]\}$.

Example 25.2 In the example below, either the specification or the program or both are wrong.

P	ℓ_i	$\mathcal{S}^{\vec{\varrho}}[\![\mathsf{P}]\!]\ell_i,\quad i=1,2$
ℓ_1 `/* x⩾0 */`	ℓ_1	$\{\rho \in \mathbb{E}\mathrm{v} \mid \rho(\mathsf{x}) \geqslant 0)\}$
`x = x - 1;`		
ℓ_2 `/* x<0 */`	ℓ_2	$\{\rho \in \mathbb{E}\mathrm{v} \mid \rho_0(\mathsf{x}) < 0\}$

Exercise 25.3 Provide a fix of example 25.2 assuming (1) that the specification is wrong and (2) that the program is wrong.

Exercise 25.4 Provide the specification of a statement S to compute the factorial $\mathsf{f} = !\mathsf{n}$.

25.1.2 Invariant Specification

Definition 25.5 (invariant reachability specification) A reachability specification $\mathcal{S}^{\vec{\varrho}}[\![\mathsf{S}]\!] \in \mathsf{labs}[\![\mathsf{S}]\!] \rightarrow \wp(\mathbb{E}\mathrm{v}^{\vec{\varrho}})$ is invariant for a program component $\mathsf{S} \in \mathbb{P}\mathrm{c}$ and a precondition $\mathscr{P}_0 \in \wp(\mathbb{E}\mathrm{v}^{\vec{\varrho}})$ if and only if $\widehat{\boldsymbol{\mathcal{S}}}^{\vec{\varrho}}[\![\mathsf{S}]\!]\,\mathscr{P}_0 \,\dot{\subseteq}\, \mathcal{S}^{\vec{\varrho}}[\![\mathsf{S}]\!]$.

The intuition is that wherever an execution of the program component $\mathsf{S} \in \mathbb{P}\mathrm{c}$ starts $\mathsf{at}[\![\mathsf{S}]\!]$ with an environment (initial values of variables) in $\mathscr{P}_0$ and reaches a point $\ell \in \mathsf{labs}[\![\mathsf{S}]\!]$ then the environment (value of the variables) at that point ℓ should belong to $\mathcal{S}^{\vec{\varrho}}[\![\mathsf{S}]\!]\ell$ (i.e., should have the specified property $\mathcal{S}^{\vec{\varrho}}[\![\mathsf{S}]\!]\ell$).

Example 25.6 The specification of example 25.2 is not invariant because if execution starts at ℓ_1 with a value $x = 1$ of the variable x that satisfies $x \geqslant 0$, then execution will reach ℓ_2 with a value $x' = 0$ of the variable x which does not satisfy $x < 0$.

Example 25.7 Consider the program P of (19.32). A possible specification is the following.

P	ℓ_i	$\mathcal{S}^{\vec{e}}[\![\mathsf{P}]\!]\ell_i, \quad i=1,\ldots,76$
`/*` $x=x_0$ `*/`		$\mathcal{P}_0 \;=\; \{\langle \rho_0, \rho\rangle \mid \rho_0=\rho\in\mathbb{E}\mathbb{v}\}$
`while` ℓ_1 `(x != 2) {`	ℓ_1	$\{\langle \rho_0, \rho\rangle \mid (\rho=\rho_0)\vee(\rho_0(\mathsf{x})\notin\{0,1,2\}\wedge\rho(\mathsf{x})=2)\}$
`/*` $\ell_2 : x=x_0\neq 2$ `*/`		
`if` ℓ_2 `(x == 0)`	ℓ_2	$\{\langle \rho, \rho\rangle \mid \rho(\mathsf{x})\neq 2\}$
ℓ_3 `/*` $x=x_0=0$ `*/`	ℓ_3	$\{\langle \rho, \rho\rangle \mid \rho(\mathsf{x})=0\}$
`break``;`		
`/*` $\ell_4 : x=x_0\neq 0$ `*/`	ℓ_4	$\{\langle \rho, \rho\rangle \mid \rho(\mathsf{x})\notin\{0,2\}\}$
`if` ℓ_4 `(x == 1)`		
ℓ_5 `/*` $x=x_0=1$ `*/`	ℓ_5	$\{\langle \rho, \rho\rangle \mid \rho(\mathsf{x})=1\}$
`break``;`		
ℓ_6 `/*` $x=x_0\notin\{0,1,2\}$ `*/`	ℓ_6	$\{\langle \rho, \rho\rangle \mid \rho(\mathsf{x})\notin\{0,1,2\}\}$
`x = 2;`		
`}`		
ℓ_7 `/*` $(x_0\in\{0,1\}\wedge x=x_0)$ $\vee\, x=2$ `*/`	ℓ_7	$\{\langle \rho_0, \rho\rangle \mid (\rho_0(\mathsf{x})\in\{0,1\}\wedge\rho(\mathsf{x})=\rho_0(\mathsf{x}))\vee\rho(\mathsf{x})=2\}$

This specification is (the strongest) invariant.

25.1.3 Structural Invariance Proof

The invariance proof $\widehat{\boldsymbol{\mathcal{S}}}^{\vec{e}}[\![\mathsf{S}]\!]\,\mathcal{P}_0 \mathrel{\dot{\subseteq}} \mathcal{S}^{\vec{e}}[\![\mathsf{S}]\!]$ can be made by structural induction on the program component $\mathsf{S}\in\mathbb{P}\mathbb{c}$.

Example 25.8 To prove

$$\ell_1\texttt{/*}\; x=0 \;\texttt{*/ x = x - 1;}\,\ell_2\texttt{/*}\; x=-1 \;\texttt{*/ x = x + 1;}\,\ell_3\texttt{/*}\; x=0 \;\texttt{*/}$$

we can prove separately ℓ_1`/*` $x=0$ `*/ x = x - 1;` ℓ_2`/*` $x=-1$ `*/` and ℓ_2`/*` $x=-1$ `*/ x = x + 1;` ℓ_3`/*` $x=0$ `*/`.

However, for the following reachability specification

$$\ell_1\;\texttt{/*}\; x=0 \;\texttt{*/ x = x - 1;}\,\ell_2\;\texttt{/*}\; \mathsf{tt} \;\texttt{*/ x = x + 1;}\,\ell_3\;\texttt{/*}\; x=0 \;\texttt{*/}$$

we can prove ℓ_1 `/*` $x=0$ `*/ x = x - 1;` ℓ_2 `/*` tt `*/` but not ℓ_2 `/*` tt `*/ x = x + 1;` ℓ_3 `/*` $x=0$ `*/` since the precondition `/*` tt `*/` is to weak for a purely local reasoning.

25.1.4 Inductive Invariant

Definition 25.9 (inductive invariant) An inductive reachability specification $\mathcal{I}^{\vec{e}}[\![\mathsf{S}]\!]\in\mathsf{labs}[\![\mathsf{S}]\!]\to\wp(\mathbb{E}\mathbb{v}^{\vec{e}})$ is stronger than the reachability specification $\mathcal{S}^{\vec{e}}[\![\mathsf{S}]\!]$ (i.e. $\mathcal{I}^{\vec{e}}[\![\mathsf{S}]\!]\mathrel{\dot{\subseteq}}\mathcal{S}^{\vec{e}}[\![\mathsf{S}]\!]$) and can be proved to be invariant by induction on S's computation steps:

- The invariant holds on program entry; and
- If the invariant holds at any program point and a computation step is executed, then the invariant holds at the next program point.

Example 25.10 Consider the specification and inductive invariant the following program.

P	ℓ_i	noninductive specification $\mathcal{S}^{\vec{e}}[\![\mathtt{S}]\!](\ell_i)$	inductive invariant $\mathcal{I}^{\vec{e}}[\![\mathtt{S}]\!](\ell_i)$
`/* tt */`		$\mathcal{P}_0 \;=\; \mathbb{E}\mathrm{v}$	$\mathcal{P}_0 \;=\; \mathbb{E}\mathrm{v}$
ℓ_1 `x = 1;`	ℓ_1	$\mathbb{E}\mathrm{v}$	$\mathbb{E}\mathrm{v}$
`while` ℓ_2 `(x > 0)`	ℓ_2	$\mathbb{E}\mathrm{v}$	$\{\rho \in \mathbb{E}\mathrm{v} \mid \rho(\mathtt{x}) > 0\}$
ℓ_3 `x = x + 1;`	ℓ_3	$\mathbb{E}\mathrm{v}$	$\{\rho \in \mathbb{E}\mathrm{v} \mid \rho(\mathtt{x}) > 0\}$
ℓ_4 `/* ff */`	ℓ_4	$\varnothing$	$\varnothing$

The specification is not inductive for the iteration statement S because $\overline{\mathsf{test}}^{\vec{e}}[\![\mathtt{x} > 0]\!](\mathcal{S}^{\vec{e}}[\![\mathtt{S}]\!](\ell_2)) = \{\rho \in \mathbb{E}\mathrm{v} \mid \rho(\mathtt{x}) \leqslant 0\} \not\subseteq \varnothing = \mathcal{S}^{\vec{e}}[\![\mathtt{S}]\!](\ell_4)$ where $\ell_4 = \mathsf{after}[\![\mathtt{S}]\!]$.

Not all invariant specifications are inductive invariants, as shown by counterexample 25.10. It is sometimes necessary and always possible to strengthen a noninductive specification into an inductive invariant. This is similar to the proof of a theorem that requires proving a stronger theorem. This may not be true when specifications and invariants are expressed in a logic, because the logic may not be expressive enough to express the inductive invariant. This is the case, for example, for the logic $\{\mathtt{tt}, \mathtt{ff}\}$, which can express the noninductive specification of example 25.10 that the program never terminates but cannot express the inductive invariant necessary to prove that fact.

25.1.5 Proof Method

This discussion shows that to prove that a reachability specification $\mathcal{S}^{\vec{e}}[\![\mathtt{S}]\!]$ is invariant, it is necessary and sufficient to find an invariant specification $\mathcal{I}^{\vec{e}}[\![\mathtt{S}]\!]$ (there always exists one) such that

1. the invariant specification $\mathcal{I}^{\vec{e}}[\![\mathtt{S}]\!]$ is inductive (which implies that $\widehat{\boldsymbol{\mathcal{S}}}^{\vec{e}}[\![\mathtt{S}]\!]\,\mathcal{P}_0 \mathrel{\dot{\subseteq}} \mathcal{I}^{\vec{e}}[\![\mathtt{S}]\!]$);
2. the invariant specification is stronger than the reachability specification $\mathcal{I}^{\vec{e}}[\![\mathtt{S}]\!] \mathrel{\dot{\subseteq}} \mathcal{S}^{\vec{e}}[\![\mathtt{S}]\!]$

proving, by transitivity that, $\widehat{\boldsymbol{\mathcal{S}}}^{\vec{e}}[\![\mathtt{S}]\!]\,\mathcal{P}_0 \mathrel{\dot{\subseteq}} \mathcal{S}^{\vec{e}}[\![\mathtt{S}]\!]$.

25.2 Abstract Specification, Invariant, Inductive Invariant, and Structural Proof Method

Section 25.1 was specific to the reachability semantics of chapter 19, "Structural Forward Reachability Semantics," but can be made more general by considering the abstract domain and abstract semantics of chapter 21 and the equivalent abstract equational semantics of chapter 23. The proof method is based on structural induction corollary 16.31 and fixpoint induction in chapter 24.

Theorem 25.11 (sound and complete abstract invariance proof method) Let $\mathbb{D}^{¤}$ be an abstract domain in definition 21.1, $\mathcal{S}^{¤}[\![\mathsf{S}]\!] \in \mathsf{labs}[\![\mathsf{S}]\!] \rightarrow \overline{\mathbb{P}}^{¤}$ be an abstract specification and $\mathcal{P}_0 \in \overline{\mathbb{P}}^{¤}$ be an abstract precondition of program component $\mathsf{S} \in \mathbb{P}\mathbb{c}$. The *invariance proof method* is

$$\widehat{\boldsymbol{\mathcal{S}}}^{¤}[\![\mathsf{S}]\!]\,\mathcal{P}_0 \mathrel{\dot{\sqsubseteq}^{¤}} \mathcal{S}^{¤}[\![\mathsf{S}]\!] \quad\Leftrightarrow\quad \exists \mathcal{I} \in \mathsf{labs}[\![\mathsf{S}]\!] \rightarrow \overline{\mathbb{P}}^{¤} \,.\, \widehat{\mathcal{V}}^{¤}[\![\mathsf{S}]\!]\,\mathcal{P}_0\,\mathcal{I} \wedge \mathcal{I} \mathrel{\dot{\sqsubseteq}^{¤}} \mathcal{S}^{¤}[\![\mathsf{S}]\!]$$

($\Leftarrow$ is soundness and $\Rightarrow$ is completeness)

Informally, $\mathcal{S}^{¤}[\![\mathsf{S}]\!]$ is *invariant* for S (i.e. $\widehat{\boldsymbol{\mathcal{S}}}^{¤}[\![\mathsf{S}]\!]\,\mathcal{P}_0 \mathrel{\dot{\sqsubseteq}^{¤}} \mathcal{S}^{¤}[\![\mathsf{S}]\!]$) if and only if there exists an *inductive invariant* $\mathcal{I} \in \mathsf{labs}[\![\mathsf{S}]\!] \rightarrow \overline{\mathbb{P}}^{¤}$ that is *stronger* than the specification (i.e. $\mathcal{I} \mathrel{\dot{\sqsubseteq}^{¤}} \mathcal{S}^{¤}[\![\mathsf{S}]\!]$) and is *inductive* that is, by definition, satisfies the *verification conditions* (25.12) to (25.21) below.

- Verification conditions for a program $\mathsf{P} ::= \mathsf{Sl}\ \ell'$

$$\widehat{\mathcal{V}}^{¤}[\![\mathsf{P}]\!] \;=\; \widehat{\mathcal{V}}^{¤}[\![\mathsf{Sl}]\!] \tag{25.12}$$

- Verification conditions for a skip statement $\mathsf{S} ::= \mathsf{;}$

$$\begin{aligned}\widehat{\mathcal{V}}^{¤}[\![\mathsf{S}]\!]\,\mathcal{P}_0\,\mathcal{I} \;=\;& \mathcal{P}_0 \sqsubseteq^{¤} \mathcal{I}_{\mathsf{at}[\![\mathsf{S}]\!]} && \text{(a)} \quad (25.13)\\ \wedge\;& \mathcal{I}_{\mathsf{at}[\![\mathsf{S}]\!]} \sqsubseteq^{¤} \mathcal{I}_{\mathsf{after}[\![\mathsf{S}]\!]} && \text{(b)}\end{aligned}$$

- Verification conditions for an assignment statement $\mathsf{S} ::= \mathsf{x} = \mathsf{E}\,;$

$$\begin{aligned}\widehat{\mathcal{V}}^{¤}[\![\mathsf{S}]\!]\,\mathcal{P}_0\,\mathcal{I} \;=\;& \mathcal{P}_0 \sqsubseteq^{¤} \mathcal{I}_{\mathsf{at}[\![\mathsf{S}]\!]} && \text{(a)} \quad (25.14)\\ \wedge\;& \mathsf{assign}^{¤}[\![\mathsf{x}, \mathsf{E}]\!]\,\mathcal{I}_{\mathsf{at}[\![\mathsf{S}]\!]} \sqsubseteq^{¤} \mathcal{I}_{\mathsf{after}[\![\mathsf{S}]\!]} && \text{(b)}\end{aligned}$$

- Verification conditions for a conditional statement $\mathsf{S} ::= \mathtt{if}\,(\mathsf{B})\,\mathsf{S}_t$

$$\begin{aligned}\widehat{\mathcal{V}}^{¤}[\![\mathsf{S}]\!]\,\mathcal{P}_0\,\mathcal{I} \;=\;& \mathcal{P}_0 \sqsubseteq^{¤} \mathcal{I}_{\mathsf{at}[\![\mathsf{S}]\!]} && \text{(a)} \quad (25.15)\\ \wedge\;& \widehat{\mathcal{V}}^{¤}[\![\mathsf{S}_t]\!]\,(\mathsf{test}^{¤}[\![\mathsf{B}]\!]\mathcal{I}_{\mathsf{at}[\![\mathsf{S}]\!]})\,\mathcal{I} && \text{(b)}\\ \wedge\;& \overline{\mathsf{test}}^{¤}[\![\mathsf{B}]\!]\mathcal{I}_{\mathsf{at}[\![\mathsf{S}]\!]} \sqsubseteq^{¤} \mathcal{I}_{\mathsf{after}[\![\mathsf{S}]\!]} && \text{(c)}\end{aligned}$$

- Verification conditions for a conditional statement $\mathsf{S} ::= \mathtt{if}\,(\mathsf{B})\,\mathsf{S}_t\ \mathtt{else}\ \mathsf{S}_f$

$$\begin{aligned}\widehat{\mathcal{V}}^{¤}[\![\mathsf{S}]\!]\,\mathcal{P}_0\,\mathcal{I} \;=\;& \mathcal{P}_0 \sqsubseteq^{¤} \mathcal{I}_{\mathsf{at}[\![\mathsf{S}]\!]} && \text{(a)} \quad (25.16)\\ \wedge\;& \widehat{\mathcal{V}}^{¤}[\![\mathsf{S}_t]\!]\,(\mathsf{test}^{¤}[\![\mathsf{B}]\!]\mathcal{I}_{\mathsf{at}[\![\mathsf{S}]\!]})\,\mathcal{I} && \text{(b)}\\ \wedge\;& \widehat{\mathcal{V}}^{¤}[\![\mathsf{S}_f]\!]\,(\overline{\mathsf{test}}^{¤}[\![\mathsf{B}]\!]\mathcal{I}_{\mathsf{at}[\![\mathsf{S}]\!]})\,\mathcal{I} && \text{(c)}\end{aligned}$$

- Verification conditions for an empty statement list $\mathsf{Sl} ::= \epsilon$

$$\widehat{\mathcal{V}}^{¤}[\![\mathsf{Sl}]\!]\,\mathcal{P}_0\,\mathcal{I} \;=\; \mathcal{P}_0 \sqsubseteq^{¤} \mathcal{I}_{\mathsf{at}[\![\mathsf{Sl}]\!]} \tag{25.17}$$

- Verification conditions for a statement list $\mathsf{Sl} ::= \mathsf{Sl}'\ \mathsf{S}$

$$\begin{aligned}\widehat{\mathcal{V}}^{¤}[\![\mathsf{Sl}]\!]\,\mathcal{P}_0\,\mathcal{I} \;&=\; \widehat{\mathcal{V}}^{¤}[\![\mathsf{S}]\!]\,\mathcal{P}_0\,\mathcal{I} && \text{when } \mathsf{Sl}' ::= \epsilon \quad (25.18)\\ &=\; \widehat{\mathcal{V}}^{¤}[\![\mathsf{Sl}']\!]\,\mathcal{P}_0\,\mathcal{I} \wedge \widehat{\mathcal{V}}^{¤}[\![\mathsf{S}]\!]\,\mathcal{I}_{\mathsf{at}[\![\mathsf{S}]\!]}\,\mathcal{I} && \text{otherwise}\end{aligned}$$

- Verification conditions for an iteration statement $\mathsf{S} ::= \mathtt{while}\ \ell\,(\mathsf{B})\,\mathsf{S}_b$

$$\widehat{\mathscr{V}}^{¤}[\![\mathsf{S}]\!]\,\mathscr{P}_0\,\mathscr{I} \;=\; \mathscr{P}_0 \sqsubseteq^{¤} \mathscr{I}_{\mathsf{at}[\![\mathsf{S}]\!]} \quad \text{(a)} \quad (25.19)$$
$$\wedge\; \widehat{\mathscr{V}}^{¤}[\![\mathsf{S}_b]\!](\mathsf{test}^{¤}[\![\mathsf{B}]\!]\mathscr{I}_{\mathsf{at}[\![\mathsf{S}]\!]})\,\mathscr{I} \quad \text{(b)}$$
$$\wedge\; \overline{\mathsf{test}}^{¤}[\![\mathsf{B}]\!]\mathscr{I}_{\mathsf{at}[\![\mathsf{S}]\!]} \sqsubseteq^{¤} \mathscr{I}_{\mathsf{after}[\![\mathsf{S}]\!]} \quad \text{(c)}$$
$$\wedge\; \forall \ell \in \mathsf{breaks\text{-}of}[\![\mathsf{S}_b]\!]\,.\,\mathscr{I}_\ell \sqsubseteq^{¤} \mathscr{I}_{\mathsf{after}[\![\mathsf{S}]\!]} \quad \text{(d)}$$

- Verification conditions for a break statement $\mathsf{S} ::= \ell\ \mathtt{break}\,;$

$$\widehat{\mathscr{V}}^{¤}[\![\mathsf{S}]\!]\,\mathscr{P}_0\,\mathscr{I} \;=\; \mathscr{P}_0 \sqsubseteq^{¤} \mathscr{I}_{\mathsf{at}[\![\mathsf{S}]\!]} \quad (25.20)$$

- Verification conditions for a compound statement $\mathsf{S} ::= \{\,\mathsf{Sl}\,\}$

$$\widehat{\mathscr{V}}^{¤}[\![\mathsf{S}]\!] \;=\; \widehat{\mathscr{V}}^{¤}[\![\mathsf{Sl}]\!] \quad (25.21)$$

Proof of theorem 25.11 By theorem 23.20, the solution of the system of equations $\widehat{\mathscr{E}}^{¤}[\![\mathsf{P}]\!]\,\mathscr{P}_0$ of the form $\mathscr{X} = \mathsf{E}[\![\mathsf{P}]\!]\,\mathscr{P}_0\,(\mathscr{X})$ is $\mathsf{lfp}^{\dot\sqsubseteq^{¤}}\,\mathsf{E}[\![\mathsf{P}]\!]\,\mathscr{P}_0 = \widehat{\boldsymbol{\mathcal{S}}}^{¤}[\![\mathsf{P}]\!]\,\mathscr{P}_0$ where $\mathsf{E}[\![\mathsf{P}]\!]\,\mathscr{P}_0$ is pointwise $\sqsubseteq^{¤}$-continuous and hence increasing.

Therefore, the proof that $\widehat{\boldsymbol{\mathcal{S}}}^{¤}[\![\mathsf{S}]\!]\,\mathscr{P}_0 \,\dot\sqsubseteq^{¤}\, \mathcal{S}^{¤}[\![\mathsf{S}]\!]$ is equivalent to $\mathsf{lfp}^{\dot\sqsubseteq^{¤}}\,\mathsf{E}[\![\mathsf{P}]\!]\,\mathscr{P}_0 \,\dot\sqsubseteq^{¤}\, \mathcal{S}^{¤}[\![\mathsf{S}]\!]$.

Applying the fixpoint induction theorem 24.1, this is equivalent to $\exists \mathscr{I} \in \mathsf{labs}[\![\mathsf{S}]\!] \to \overline{\mathbb{P}}^{¤}\,.\,\mathsf{E}[\![\mathsf{P}]\!]\,\mathscr{P}_0(\mathscr{I}) \,\dot\sqsubseteq^{¤}\, \mathscr{I} \wedge \mathscr{I} \,\dot\sqsubseteq^{¤}\, \mathcal{S}^{¤}[\![\mathsf{S}]\!]$.

It remains to state the Boolean verification conditions

$$\widehat{\mathscr{V}}^{¤}[\![\mathsf{S}]\!]\,\mathscr{P}_0\,\mathscr{I} \;\triangleq\; \mathsf{E}[\![\mathsf{S}]\!]\,\mathscr{P}_0(\mathscr{I}) \,\dot\sqsubseteq^{¤}\, \mathscr{I} \quad (25.22)$$

which are an abstraction $\widehat{\mathscr{V}}^{¤}[\![\mathsf{S}]\!] = \vec{\alpha}\,(\mathsf{E}[\![\mathsf{P}]\!])$ of the equations $\mathsf{E}[\![\mathsf{P}]\!]$ where $\vec{\alpha}\,(F) \triangleq \mathscr{P}_0 \mapsto \mathscr{I} \mapsto F(\mathscr{I}) \,\dot\sqsubseteq^{¤}\, \mathscr{I}$.

We calculate $\widehat{\mathscr{V}}^{¤}[\![\mathsf{S}]\!]$ by applying $\vec{\alpha}$ to $\mathsf{E}[\![\mathsf{P}]\!])$ and simplifying. We proceed by structural induction on S.

- For an assignment statement $\mathsf{S} ::= \mathsf{x} = \mathsf{E}\,;$,

$$\widehat{\mathscr{V}}^{¤}[\![\mathsf{S}]\!]\,\mathscr{P}_0\,\mathscr{I}$$
$$= \mathsf{E}[\![\mathsf{S}]\!]\,\mathscr{P}_0(\mathscr{I}) \,\dot\sqsubseteq^{¤}\, \mathscr{I} \qquad \text{(25.22)}$$
$$= \mathsf{E}[\![\mathsf{S}]\!]\,\mathscr{P}_0(\mathscr{I})\,\mathsf{at}[\![\mathsf{S}]\!] \sqsubseteq^{¤} \mathscr{I}_{\mathsf{at}[\![\mathsf{S}]\!]} \wedge \mathsf{E}[\![\mathsf{S}]\!]\,\mathscr{P}_0(\mathscr{I})\,\mathsf{after}[\![\mathsf{S}]\!] \sqsubseteq^{¤} \mathscr{I}_{\mathsf{after}[\![\mathsf{S}]\!]}$$
$$\text{(pointwise definition of } \dot\sqsubseteq^{¤} \text{ and definition of } \mathsf{labs}[\![\mathsf{S}]\!] = \{\mathsf{at}[\![\mathsf{S}]\!], \mathsf{after}[\![\mathsf{S}]\!]\}\text{)}$$
$$= \mathscr{P}_0 \sqsubseteq^{¤} \mathscr{I}_{\mathsf{at}[\![\mathsf{S}]\!]} \wedge \mathsf{assign}^{¤}[\![\mathsf{x}, \mathsf{E}]\!]\,\mathscr{I}_{\mathsf{at}[\![\mathsf{S}]\!]} \sqsubseteq^{¤} \mathscr{I}_{\mathsf{after}[\![\mathsf{S}]\!]}$$

(by (23.7) because the equations have the form

$$\begin{cases} \mathscr{I}_{\mathsf{at}[\![\mathsf{S}]\!]} = \mathsf{E}[\![\mathsf{S}]\!]\,\mathscr{P}_0(\mathscr{I})\,\mathsf{at}[\![\mathsf{S}]\!] = \mathscr{P}_0 \\ \mathscr{I}_{\mathsf{after}[\![\mathsf{S}]\!]} = \mathsf{E}[\![\mathsf{S}]\!]\,\mathscr{P}_0(\mathscr{I})\,\mathsf{after}[\![\mathsf{S}]\!] = \mathsf{assign}^{¤}[\![\mathsf{x}, \mathsf{E}]\!]\,\mathscr{I}_{\mathsf{at}[\![\mathsf{S}]\!]} \end{cases}$$)

- The base cases (25.13), (25.17), and (25.20) are similar. For (25.20), we get $\widehat{\mathscr{V}}^{¤}[\![\mathsf{S}]\!]\,\mathscr{P}_0\,\mathscr{I} = \mathscr{P}_0 \sqsubseteq^{¤} \mathscr{I}_{\mathsf{at}[\![\mathsf{S}]\!]} \wedge \bot^{¤} \sqsubseteq^{¤} \mathscr{I}_{\mathsf{after}[\![\mathsf{S}]\!]}$ and the term $\bot^{¤} \sqsubseteq^{¤} \mathscr{I}_{\mathsf{after}[\![\mathsf{S}]\!]}$ is always true, by definition of the infimum $\bot^{¤}$.
- For the iteration statement $\mathsf{S} ::= \mathtt{while}\,\ell\,(\mathsf{B})\,\mathsf{S}_b$, the equations (23.12) have the form

$$\begin{cases} \mathscr{I}_{\mathsf{at}[\![\mathsf{S}]\!]} = \mathscr{P}_0 \sqcup^{¤} \mathsf{E}[\![\mathsf{S}_b]\!]\,(\mathsf{test}^{¤}[\![\mathsf{B}]\!]\mathscr{I}_{\mathsf{at}[\![\mathsf{S}]\!]})\,(\mathscr{I})\,\mathsf{at}[\![\mathsf{S}]\!] \\ \mathscr{I}_\ell = \mathsf{E}[\![\mathsf{S}_b]\!]\,(\mathsf{test}^{¤}[\![\mathsf{B}]\!]\mathscr{I}_\ell)\,(\mathscr{I})\,\ell, \quad \ell \in \mathsf{labs}[\![\mathsf{S}_b]\!] \setminus \{\mathsf{at}[\![\mathsf{S}]\!]\} \\ \mathscr{I}_{\mathsf{after}[\![\mathsf{S}]\!]} = \overline{\mathsf{test}}^{¤}[\![\mathsf{B}]\!]\mathscr{I}_{\mathsf{at}[\![\mathsf{S}]\!]} \sqcup^{¤} \bigsqcup^{¤}_{\ell \in \mathsf{breaks\text{-}of}[\![\mathsf{S}_b]\!]} \mathscr{I}_\ell \end{cases}$$

Noting that $A \sqcup^{¤} B \sqsubseteq^{¤} C \Leftrightarrow A \sqsubseteq^{¤} C \wedge B \sqsubseteq^{¤} C$, we have

$$\begin{aligned} &\widehat{\mathscr{V}}^{¤}[\![\mathsf{S}]\!]\,\mathscr{P}_0\,\mathscr{I} \\ = \;& \mathsf{E}[\![\mathsf{S}]\!]\,\mathscr{P}_0(\mathscr{I}) \,\dot{\sqsubseteq}^{¤}\, \mathscr{I} && \text{(25.22)} \\ = \;& \mathscr{P}_0 \sqsubseteq^{¤} \mathscr{I}_{\mathsf{at}[\![\mathsf{S}]\!]} \wedge \\ & \forall \ell \in \mathsf{labs}[\![\mathsf{S}_b]\!]\,.\,\mathsf{E}[\![\mathsf{S}_b]\!]\,(\mathsf{test}^{¤}[\![\mathsf{B}]\!]\mathscr{I}_\ell)\,(\mathscr{I})\,\ell \sqsubseteq^{¤} \mathscr{I}_\ell \wedge \\ & \overline{\mathsf{test}}^{¤}[\![\mathsf{B}]\!]\mathscr{I}_{\mathsf{at}[\![\mathsf{S}]\!]} \sqsubseteq^{¤} \mathscr{I}_{\mathsf{after}[\![\mathsf{S}]\!]} \wedge \\ & \forall \ell \in \mathsf{breaks\text{-}of}[\![\mathsf{S}_b]\!]\,.\,\mathscr{I}_\ell \sqsubseteq^{¤} \mathscr{I}_{\mathsf{after}[\![\mathsf{S}]\!]} \end{aligned}$$

which, by structural induction hypothesis, is (25.19).

- The remaining cases (25.12), (25.15), (25.16), (25.18), and (25.21) are handled similarly, by structural induction. □

Remark 25.23 In (25.19.d), $\ell \in \mathsf{breaks\text{-}of}[\![\mathsf{S}_b]\!]$ if and only if there is a break statement $\mathsf{S} ::= \ell$ **`break`**; such that $\ell = \mathsf{at}[\![\mathsf{S}]\!]$ and $\mathsf{break\text{-}to}[\![\mathsf{S}]\!] = \mathsf{after}[\![\mathsf{S}]\!]$ so that the verification condition (25.19.d) that is $\forall \ell \in \mathsf{breaks\text{-}of}[\![\mathsf{S}_b]\!]\,.\,\mathscr{I}_\ell \sqsubseteq^{¤} \mathscr{I}_{\mathsf{after}[\![\mathsf{S}]\!]}$ can be distributed as $\mathscr{I}_{\mathsf{at}[\![\mathsf{S}]\!]} \sqsubseteq^{¤} \mathscr{I}_{\mathsf{break\text{-}to}[\![\mathsf{S}]\!]}$ to all such break statements. We obtain equivalently

- Verification conditions for an iteration statement $\mathsf{S} ::= \mathtt{while}\ \ell\ (\mathsf{B})\ \mathsf{S}_b$

$$\begin{aligned} \widehat{\mathscr{V}}^{¤}[\![\mathsf{S}]\!]\,\mathscr{P}_0\,\mathscr{I} \;=\;& \mathscr{P}_0 \sqsubseteq^{¤} \mathscr{I}_{\mathsf{at}[\![\mathsf{S}]\!]} && \text{(a)} \quad (25.19') \\ \wedge\;& \widehat{\mathscr{V}}^{¤}[\![\mathsf{S}_b]\!](\mathsf{test}^{¤}[\![\mathsf{B}]\!]\mathscr{I}_{\mathsf{at}[\![\mathsf{S}]\!]})\,\mathscr{I} && \text{(b)} \\ \wedge\;& \overline{\mathsf{test}}^{¤}[\![\mathsf{B}]\!]\mathscr{I}_{\mathsf{at}[\![\mathsf{S}]\!]} \sqsubseteq^{¤} \mathscr{I}_{\mathsf{after}[\![\mathsf{S}]\!]} && \text{(c)} \end{aligned}$$

- Verification conditions for a break statement $\mathsf{S} ::= \ell$ **`break`**;

$$\widehat{\mathscr{V}}^{¤}[\![\mathsf{S}]\!]\,\mathscr{P}_0\,\mathscr{I} \;=\; \mathscr{P}_0 \sqsubseteq^{¤} \mathscr{I}_{\mathsf{at}[\![\mathsf{S}]\!]} \wedge \mathscr{I}_{\mathsf{at}[\![\mathsf{S}]\!]} \sqsubseteq^{¤} \mathscr{I}_{\mathsf{break\text{-}to}[\![\mathsf{S}]\!]} \qquad (25.20')$$

Example 25.24 The local invariants of example 25.7 satisfy the following verification conditions stating that if the invariant holds at any program point then it holds at the next reachable program program point after one more computation step (the numbering corresponds to forthcoming formal verification conditions).

$\mathscr{P}_0 \subseteq \mathscr{I}_{\ell_1}$ (25.12) (25.18) (25.19.a)

$\mathsf{test}^{\vec{R}}[\![\mathsf{x} \mathrel{!}= 2]\!]\mathscr{I}_{\ell_1} \subseteq \mathscr{I}_{\ell_2}$ (25.12) (25.18) (25.19.b) (25.18) (25.15.a)

$\overline{\mathsf{test}}^{\vec{R}}[\![\mathsf{x} \mathrel{!}= 2]\!]\mathscr{I}_{\ell_1} \subseteq \mathscr{I}_{\ell_7}$ (25.12) (25.18) (25.19.b) (25.18) (25.15.a)

$\mathsf{test}^{\vec{R}}[\![\mathsf{x} == 0]\!]\mathscr{I}_{\ell_2} \subseteq \mathscr{I}_{\ell_3}$ (25.12) (25.18) (25.19.b) (25.18) (25.15.b) (25.20)

$\overline{\mathsf{test}}^{\vec{R}}[\![\mathsf{x} == 0]\!]\mathscr{I}_{\ell_2} \subseteq \mathscr{I}_{\ell_4}$ (25.12) (25.18) (25.19.b) (25.18) (25.15.c)

$\mathsf{test}^{\vec{\mathsf{R}}}[\![\mathsf{x} == 1]\!]\mathcal{J}_{\ell_4} \subseteq \mathcal{J}_{\ell_5}$ (25.12) (25.18) (25.19.b) (25.15.b) (25.20)

$\overline{\mathsf{test}}^{\vec{\mathsf{R}}}[\![\mathsf{x} == 1]\!]\mathcal{J}_{\ell_4} \subseteq \mathcal{J}_{\ell_6}$ (25.12) (25.18) (25.19.b) (25.15.c)

$\mathsf{assign}^{\vec{\varrho}}[\![\mathsf{x}, 2]\!]\, \mathcal{J}_{\ell_6} \subseteq \mathcal{J}_{\ell_1}$ (25.12) (25.18) (25.19.b) (25.14.b)

that is, by (19.12), $\{\langle \rho_0, \rho[\mathsf{x} \leftarrow 2]\rangle \mid \langle \rho_0, \rho\rangle \in \mathcal{J}_{\ell_6}\} \subseteq \mathcal{J}_{\ell_1}$

$\mathcal{J}_{\ell_3} \subseteq \mathcal{J}_{\ell_7}$ (25.12) (25.18) (25.19.d)

$\mathcal{J}_{\ell_5} \subseteq \mathcal{J}_{\ell_7}$ (25.12) (25.18) (25.19.d)

Exercise 25.25 Prove that the intersection of structurally inductive invariants is structurally inductive, that is, if $\forall i \in \Delta\,.\,\widehat{\mathcal{J}}^{\vec{\varrho}}[\![\mathsf{S}]\!]\,\mathcal{J}_i$ then $\widehat{\mathcal{J}}^{\vec{\varrho}}[\![\mathsf{S}]\!] \bigcap_{i\in\Delta}\mathcal{J}_i$.

Exercise 25.26 Show that $\widehat{\mathcal{J}}^{\vec{\varrho}}[\![\mathsf{S}]\!]$ is the adjoint of a Galois connection.

Exercise 25.27 Prove that the program invariants given in example 19.1 are inductive.

Exercise 25.28 Design an algorithm for the Euclidean division of integers and prove it correct.

Exercise 25.29 (Project: verification conditions) Modify the abstract interpreter of exercise 21.23 in order to print the verification conditions of a program. For example, `while` ℓ_1 `(x!=2) { if` ℓ_2 `(x==0)` ℓ_3 `break; if` ℓ_4 `(x==1)` ℓ_5 `break;` ℓ_6 `x=2; }` ℓ_7 yields

```
R0 => l1                     (23.1) (23.6) (23.8.a)
test[(x != 2)]l1 => l2       (23.1) (23.6) (23.8.b) (23.6) (23.4.a)
test[(x == 0)]l2 => l3      (23.1) (23.6) (23.8.b) (23.6) (23.4.b) (23.9)
ntest[(x == 0)]l2 => l4      (23.1) (23.6) (23.8.b) (23.6) (23.4.c)
test[(x == 1)]l4 => l5       (23.1) (23.6) (23.8.b) (23.4.b) (23.9)
ntest[(x == 1)]l4 => l6      (23.1) (23.6) (23.8.b) (23.4.c)
assign[x=2]l6 => l1          (23.1) (23.6) (23.8.b) (23.3.b)
ntest[(x != 2)]l1 => l7      (23.1) (23.6) (23.8.c)
l3 => l7                     (23.1) (23.6) (23.8.d)
l5 => l7                     (23.1) (23.6) (23.8.d)
```

25.3 Verifying that an Abstract Invariant is Inductive

Invariance proof methods are a special case with $\overline{\mathbb{P}}^{\texttt{¤}} = \wp(\mathbb{Ev}^{\vec{\varrho}})$ where $\vec{\varrho} \in \{\vec{\mathsf{r}}, \vec{\mathsf{R}}\}$, $\mathbb{P}^{\texttt{¤}} = \mathbb{Ev}^{\vec{\mathsf{r}}} = \mathbb{Ev}$ for assertional invariance, and $\mathbb{P}^{\texttt{¤}} = \mathbb{Ev}^{\vec{\mathsf{R}}} = \mathbb{Ev} \times \mathbb{Ev}$ for relational invariance.

An abstract invariant $\mathcal{J} \in \mathbb{L} \to \mathbb{P}^{\texttt{¤}}$ is defined independently of a precondition $\mathcal{P}_0$ which is equivalent to defining $\mathcal{P}_0 \triangleq \mathcal{J}(\mathsf{at}[\![\mathsf{S}]\!])$ and assuming that $\mathcal{J}(\mathsf{at}[\![\mathsf{S}]\!])$ is a hypothesis. An invariant can be checked to be inductive by

$$\begin{aligned} \widehat{\mathcal{J}}^{\vec{\varrho}}[\![\mathsf{S}]\!] &\in (\mathbb{L} \to \mathbb{P}^{\vec{\varrho}}) \to \mathbb{B} \\ \widehat{\mathcal{J}}^{\vec{\varrho}}[\![\mathsf{S}]\!]\,\mathcal{J} &\triangleq \widehat{\mathcal{V}}^{\vec{\varrho}}[\![\mathsf{S}]\!]\,\mathcal{J}(\mathsf{at}[\![\mathsf{S}]\!])\,\mathcal{J} \end{aligned} \tag{25.30}$$

We can also express $\widehat{\mathscr{V}}^{\vec{\varrho}}[\![\mathsf{S}]\!]$ in terms of $\widehat{\mathscr{F}}^{\vec{\varrho}}[\![\mathsf{S}]\!]$.

Lemma 25.31 $\widehat{\mathscr{V}}^{\vec{\varrho}}[\![\mathsf{S}]\!]\,\mathscr{P}_0\,\mathscr{I} = \mathscr{P}_0 \sqsubseteq^{¤} \mathscr{I}(\mathsf{at}[\![\mathsf{S}]\!]) \wedge \widehat{\mathscr{F}}^{\vec{\varrho}}[\![\mathsf{S}]\!]\,\mathscr{I}$.

Proof of lemma 25.31

$$\widehat{\mathscr{V}}^{\vec{\varrho}}[\![\mathsf{S}]\!]\,\mathscr{P}_0\,\mathscr{I}$$

$$= \mathscr{P}_0 \sqsubseteq^{¤} \mathscr{I}_{\mathsf{at}[\![\mathsf{S}]\!]} \wedge \widehat{\mathscr{V}}^{\vec{\varrho}}[\![\mathsf{S}]\!]\,\mathscr{I}(\mathsf{at}[\![\mathsf{S}]\!])\,\mathscr{I}$$

≀by structural induction of S checking that this is valid for all verification conditions (25.12) to (25.21)≀

$$= \mathscr{P}_0 \sqsubseteq^{¤} \mathscr{I}_{\mathsf{at}[\![\mathsf{S}]\!]} \wedge \widehat{\mathscr{F}}^{\vec{\varrho}}[\![\mathsf{S}]\!]\,\mathscr{I}$$

≀definition (25.30) of $\widehat{\mathscr{F}}^{\vec{\varrho}}[\![\mathsf{S}]\!]$≀ □

By lemma 25.31, the verification conditions (25.12) to (25.21) do apply (with the simplification that $\mathscr{P}_0 \triangleq \mathscr{I}(\mathsf{at}[\![\mathsf{S}]\!])$ so $\mathscr{P}_0 \sqsubseteq^{¤} \mathscr{I}_{\mathsf{at}[\![\mathsf{S}]\!]}$, that is, $\mathscr{P}_0 \subseteq \mathscr{I}_{\mathsf{at}[\![\mathsf{S}]\!]}$ is always tt). For example (25.14.a) and (25.14.b) become

$$\widehat{\mathscr{F}}^{\vec{\varrho}}[\![\mathsf{S}]\!]\,\mathscr{I} \quad = \quad \mathsf{assign}^{\vec{\varrho}}[\![\mathsf{x}, \mathsf{E}]\!]\,\mathscr{I}_{\mathsf{at}[\![\mathsf{S}]\!]} \subseteq \mathscr{I}_{\mathsf{after}[\![\mathsf{S}]\!]}$$

25.4 Automation of Program Verification

Automated program verification of a program component S for initial precondition $\mathscr{P}_0$ requires the end user to provide an inductive invariant $\mathscr{I}$ (at least for $\mathsf{at}[\![\mathsf{S}]\!]$ and loops because the other ones can be formally derived using the verification conditions). The abstract domain $\mathbb{D}^{¤}$ is most often chosen to be a logic (that is a set of logical formulas closed by $\vee$, $\wedge$, etc.). Then the checking of the verification conditions $\widehat{\mathscr{V}}^{¤}[\![\mathsf{S}]\!]\,\mathscr{P}_0\,\mathscr{I}$ essentially consist in proving implications between logical formulas of $\mathbb{D}^{¤}$, as specified by (25.12)–(25.21). These implications are proved by a theorem prover specialized in this logic $\mathbb{D}^{¤}$, hopefully fully automatically. There are several theoretical and practical difficulties.

- The theorem prover might not be able to prove valid implications. This is why the abstract domain $\mathbb{D}^{¤}$ is often chosen to be a reduced product of decidable theories (see example 36.32). By restricting expressivity, implication becomes decidable, as for satisfiability modulo theories (SMT)-solvers [91], although not necessarily in guaranteed reasonable time.
- However, the inductive invariant $\mathscr{I}$ may not be strong enough to do the proof.
  - For decidable theories, the theorem prover/SMT-solver will provide a counterexample. This can help provide a stronger invariant.
  - For undecidable theories, the theorem prover will fail to prove the implications. The end user must understand that the inductive invariant is too weak and must provide a stronger one.
- However, the abstract domain $\mathbb{D}^{¤}$ might not be expressive enough to express this stronger invariant. Then there is no way out, except to use a more powerful theorem prover for a more expressive logic, which may end up to be undecidable.
- If $\mathbb{D}^{¤}$ is very general such as a first-order logic with arithmetics used in Isabelle genetic proof assistant, the Prototype Verification System (PVS) [771], or the Computational

Logic for Applicative Common Lisp (ACL2) [561, 714], then the implication $\sqsubseteq^{\alpha}$ is undecidable (by Gödel's incompleteness theorems [420]; see theorem 9.1). Therefore the theorem prover must be helped to do the implication proofs, a nontrivial task that often requires understanding the algorithms used by the theorem prover and why they fail.

- An alternative consists of providing the proof by hand (or at least the essential arguments) and have this proof automatically checked by a proof verifier such as COQ [520].
- A final difficulty is that the inductive invariant is generally as complex as the program and the proof an order of magnitude more complex. The program, the inductive invariant, and the proof must be maintained for modification which may be a fantastic task.

Example 25.32 (Presburger arithmetics) Presburger arithmetics is a first-order logic with constants 0 and 1, addition +, equality =, logical operation ↑ (nand), and quantifiers ∃ and ∀ over variables. The axiomatization $\neg(0 = x + 1)$, $x + 1 = y + 1 \Rightarrow x = y$, $x + 0 = x$, and $x + (y + 1) = (x + y) + 1$ includes an axiom schema providing an induction rule for proofs by recurrence $(P(0) \wedge \forall x . (P(x) \Rightarrow P(x + 1))) \Rightarrow \forall y . P(y)$ (where $P(x)$ is a first-order formula in the language of Presburger arithmetic with a free variable x (and possibly other free variables)). Presburger arithmetics is interpreted over the naturals $\mathbb{N}$. Mojżesz Presburger showed that there exists an algorithm that decides whether any given formula in the Presburger arithmetic is a theorem or is a not a theorem (and so the theory is sound [a theorem and its negation cannot both be derived from the axioms] and complete [any formula or its negation can be derived from the axioms]). The decision algorithm is doubly exponential in the size n of the formula for time and space [373, 760]. It is implemented in COQ [520], Z3 [299], Omega[811], and so forth.

It is easy to write a program to compute the multiplication of two numbers x and y (expressed as $0 + 1 + 1 + \ldots + 1$) using only Presburger arithmetics primitives `p = 0; i = 0; while (i < y) { i = i + 1; p = p + x; }`. The verification conditions of theorem 25.11 can all be expressed in Presburger arithmetics. However Presburger arithmetics does not contain multiplication. It is therefore impossible to prove the correctness of this program with that abstract domain because neither the specification ($p = x \times y$) nor the loop invariant ($p = x \times i$) can be expressed in Presburger arithmetics. Of course, one can include multiplication in the logic, but this is Peano arithmetics [787], which is undecidable [420].

Exercise 25.33 Tarski showed that the first-order theory of reals with 0, 1, +, ×, =, < is decidable (by elimination of the quantifiers ∀ and ∃) [927]. The decision algorithm by cylindrical algebraic decomposition [201] is implemented in Mathematica [987] and doubly exponential. Are the verification conditions for the program of example 25.32 (for integers encoded as reals) provable in Tarski's theory of reals?

25.5 Conclusion

We have designed by calculus the verification conditions for proving that the reachable states imply an abstract invariance specification. This involves finding an abstract invariant and proving that is is inductive. For reachability, this proof method is a structural version of the Alan Turing [933], Peter Naur [742], Bob Floyd [376], and Zohar Manna [647] invariance proof method. The proof method has many equivalent variants [243] (including James Morris and Ben Wegbreit subgoal induction [323, 721] for backward reasoning).

As noticed by Bob Floyd [376], a proof method "assigns meaning to programs," that is, define their semantics. The difference between the semantics of chapter 21, "Abstract Domain and Abstract Structural Semantics," chapter 23, "Abstract Equational Semantics," and this chapter is that between increasing functions F, equations $X = F(X)$, and inequations $F(X) \sqsubseteq X$, which by Tarski fixpoint theorem 15.6 and fixpoint induction theorem 24.1 are essentially the same. Each one gives rise to a static analysis method that looks different but is essentially equivalent, that is, computing or approximating fixpoints by iteration, solving equations, or solving constraints [47].

One difficulty for automation is that implication is undecidable in the proof that the inductive invariant satisfies the verification conditions. The main difficulty however is to automatically infer an inductive invariant. Tools like Why3 [129, 371] automatically generate the verification conditions, use abstract interpretation to help the user to infer part of an inductive invariant, and resort to external provers to prove that the user-provided inductive invariant satisfies the verification conditions.

25.6 Solutions to Selected Exercises

Solution to exercise 25.3

(1)	ℓ_1 /* $x \geqslant 0$ */	$\{\rho \in \mathbb{E}\mathrm{v} \mid \rho(\mathrm{x}) \geqslant 0)\}$	(2)	ℓ_1 /* $x \geqslant 0$ */	$\{\rho \in \mathbb{E}\mathrm{v} \mid \rho(\mathrm{x}) \geqslant 0)\}$
	x = x - 1;			x = - x - 1;	
	ℓ_2 /* $x \geqslant -1$ */	$\{\rho \in \mathbb{E}\mathrm{v} \mid \rho_0(\mathrm{x}) \geqslant -1\}$		ℓ_2 /* $x < 0$ */	$\{\rho \in \mathbb{E}\mathrm{v} \mid \rho_0(\mathrm{x}) < 0\}$

Solution to exercise 25.4 $\mathcal{S}^{\vec{\varrho}}[\![\mathrm{S}]\!]\,\mathrm{at}[\![\mathrm{S}]\!]\,\mathrm{n} \in \mathbb{N} \Rightarrow (\mathcal{S}^{\vec{\varrho}}[\![\mathrm{S}]\!]\,\mathrm{after}[\![\mathrm{S}]\!]\,\mathrm{f} = !(\mathcal{S}^{\vec{\varrho}}[\![\mathrm{S}]\!]\,\mathrm{at}[\![\mathrm{S}]\!]\,\mathrm{n}))$. This does not take possible arithmetic overflows into account [319].

Solution to exercise 25.25 For a noniterative program component in (25.30), $\forall i \in \Delta\,.\,\widehat{\boldsymbol{\mathcal{F}}}^{\vec{\varrho}}[\![\mathrm{S}]\!]\,\mathcal{I}_i$ implies $\widehat{\boldsymbol{\mathcal{S}}}^{\vec{\varrho}}[\![\mathrm{S}]\!]\,\dot{\bigcap\limits_{i\in\Delta}}\,\mathcal{I}_i(\mathrm{at}[\![\mathrm{S}]\!]) \,\dot{\subseteq}\, \widehat{\boldsymbol{\mathcal{S}}}^{\vec{\varrho}}[\![\mathrm{S}]\!]\,\mathcal{I}_i(\mathrm{at}[\![\mathrm{S}]\!]) \,\dot{\subseteq}\, \mathcal{I}_i$ by definition of $\cap$ and $\widehat{\boldsymbol{\mathcal{S}}}^{\vec{\varrho}}[\![\mathrm{S}]\!]$ is $\subseteq$-increasing (by (20.6) and (20.1) or (20.14) and (20.9)). By definition of $\cap$, it follows that $\widehat{\boldsymbol{\mathcal{S}}}^{\vec{\varrho}}[\![\mathrm{S}]\!]\,\dot{\bigcap\limits_{i\in\Delta}}\,\mathcal{I}_i(\mathrm{at}[\![\mathrm{S}]\!]) \,\dot{\subseteq}\, \dot{\bigcap\limits_{i\in\Delta}}\,\mathcal{I}_i$ which implies $\widehat{\boldsymbol{\mathcal{F}}}^{\vec{\varrho}}[\![\mathrm{S}]\!]\,\dot{\bigcap\limits_{i\in\Delta}}\,\mathcal{I}_i$. The proof for the iteration is similar and we conclude by structural induction.

Solution to exercise 25.28 The solution proposed by Jean-Christophe Filliâtre in Why3 [129] (http://toccata.lri.fr/gallery/division.en.html).

```
(* Euclidean division *)

module Division
```

```
  use import int.Int
  use import ref.Refint

  let division (a b: int) : int
    requires { $0$ <= a && $0$ < b }
    ensures  { exists r: int. result * b + r = a && $0$ <= r < b }
  =
    let q = ref $0$ in
    let r = ref a in
    while !r >= b do
      invariant { !q * b + !r = a && $0$ <= !r }
      variant   { !r }
      incr q;
      r -= b
    done;
    !q

end
```

26 Hoare Logic

Contents

26.1 Introduction to Hoare Logic

Hoare logic [48, 501] can be used to prove a relation between initial and final states of a program component $\mathtt{S} \in \mathbb{P}\mathbb{c}$. Traditionally, the specification $\{\!|Q|\!\}\ \mathtt{S}\ \{\!|R|\!\}$ (originally written $Q\{\mathtt{S}\}R$) means that if execution of S starts in an initial state satisfying a precondition predicate Q and this execution terminates, then upon termination after⟦S⟧, if ever occurs, the final state satisfies the postcondition predicate R.

For example, $\{\!|x=0|\!\}$ `x = 1 ;` $\{\!|x=1|\!\}$, $\{\!|\mathsf{tt}|\!\}$ `if (x != 0) x = 0 ;` $\{\!|x=0|\!\}$, $\{\!|\mathsf{tt}|\!\}\ \mathtt{S}\ \{\!|\mathsf{ff}|\!\}$ holds if and only if statement S never terminates so $\{\!|\mathsf{tt}|\!\}$ `x = 0 ; while (x == 0) ;` $\{\!|\mathsf{ff}|\!\}$ and $\mathsf{ff} \Rightarrow (x=42)$ so that $\{\!|\mathsf{tt}|\!\}$ `x = 0 ; while (x == 0) ;` $\{\!|x=42|\!\}$ because anything holds on loop exit that is not reachable.

As shown by example $\{\!|\mathsf{tt}|\!\}$ `x = 0 ; while (x == 0) ;` $\{\!|x=42|\!\}$, termination is abstracted away, so Hoare logic is for partial correctness (as opposed to total correctness that requires termination). The problem of proving $\{\!|Q|\!\}\ \mathtt{S}\ \{\!|R|\!\}$ is undecidable because nontermination is expressible by $\{\!|\mathsf{tt}|\!\}\ \mathtt{S}\ \{\!|\mathsf{ff}|\!\}$.

26.2 Hoare Triples with Exceptions, Informally

We have break statements, so we use the notation $\{\!|Q|\!\}\ \mathtt{S}\ \{\!|R \mid T|\!\}$ to state that precondition R holds upon normal termination of S at after⟦S⟧ whereas break condition T holds upon termination of program component $\mathtt{S} \in \mathbb{P}\mathbb{c}$ by a break outside this program component S.

For example $\{\!|\mathsf{tt}|\!\}$`if (x == 0) break; else x = 1 ;`$\{\!|x=1 \mid x=0|\!\}$, $\{\!|\mathsf{tt}|\!\}$`x = 1 ;`$\{\!|x=1 \mid \mathsf{ff}|\!\}$, and $\{\!|\mathsf{tt}|\!\}$`x = 1 ;`$\{\!|x=1 \mid x=42|\!\}$ because $\mathsf{ff} \Rightarrow (x=42)$ so anything holds on an escape that is impossible.

In $\{\!|Q|\!\}$ S $\{\!|R \mid T|\!\}$, $T = \text{ff}$ means that there is no possible escape from S by a break. This is the case if S has no break statement or if all its break statements are captured by an iteration statement.

In classic Hoare logic there are no `break`; i.e. there is no possible termination by a break statement outside S, so the classic notation $\{\!|Q|\!\}$ S $\{\!|R|\!\}$ is an abbreviation for $\{\!|Q|\!\}$ S $\{\!|R \mid \text{ff}|\!\}$.

We also need to express relations between initial and final values, as in $\{\!|x = x_0|\!\}$ x = x + 1 ; $\{\!|x = x_0 + 1|\!\}$. Therefore, we represent predicates by relations Q, R, T between environments. Predicate $\langle \rho_0, \rho \rangle \in R$ means that if ρ_0 is an initial environment (that is, $\rho_0(\mathsf{x}) = x_0$) then ρ is the current environment at program point to which R is attached (that is, $\rho(\mathsf{x}) = x$).

Therefore we consider Hoare triples $\{\!|Q|\!\}$ S $\{\!|R \mid T|\!\}$ where $Q, R, T \in \wp(\mathbb{E}\mathrm{v} \times \mathbb{E}\mathrm{v})$ to be relations between environments. Typically, $Q = \{\langle \rho, \rho \rangle \mid \rho \in \mathbb{E}\mathrm{v}\}$ so that R is a relation between the initial and final environments upon normal termination of S, whereas T holds upon termination of S by a break in S, if any. So ff is $\emptyset$, and tt is $\mathbb{E}\mathrm{v} \times \mathbb{E}\mathrm{v}$. By this convention, a Hoare triple $\{\!|Q|\!\}$ S $\{\!|R \mid T|\!\}$ denotes a triple of $\wp(\mathbb{E}\mathrm{v}^{\vec{\varrho}}) \times \wp(\mathbb{E}\mathrm{v}^{\vec{\varrho}}) \times \wp(\mathbb{E}\mathrm{v}^{\vec{\varrho}})$ where $\vec{\varrho} \in \{\vec{r}, \vec{R}\}$, $\mathbb{E}\mathrm{v}^{\vec{r}} = \mathbb{E}\mathrm{v}$, and $\mathbb{E}\mathrm{v}^{\vec{R}} = \mathbb{E}\mathrm{v} \times \mathbb{E}\mathrm{v}$.

26.3 Hoare Triples, Formally

The information on the semantics of a program component $\mathsf{S} \in \mathbb{P}\mathrm{c}$ provided by a Hoare triple $\{\!|Q|\!\}$ S $\{\!|R \mid T|\!\}$ is

- Q attached to at⟦S⟧;
- R attached to after⟦S⟧; and
- In case S has a break which escapes outside of S (so escape⟦S⟧ = tt and break-to⟦S⟧ ≠ after⟦S⟧ by lemma 4.18), T is attached to its break label break-to⟦S⟧. Otherwise escape⟦S⟧ = ff and T is meaningless and so can be chosen arbitrarily.

We formalize Hoare logic for an arbitrary abstract domain $\mathbb{D}^{¤} = \langle \mathbb{P}^{¤}, \sqsubseteq^{¤}, \bot^{¤}, \sqcup^{¤}, \mathsf{assign}^{¤}[\![\mathsf{x}, \mathsf{A}]\!], \mathsf{test}^{¤}[\![\mathsf{B}]\!], \overline{\mathsf{test}}^{¤}[\![\mathsf{B}]\!] \rangle$ satisfying definition 21.1, typically $\mathbb{P}^{¤} = \wp(\mathbb{E}\mathrm{v}^{\vec{\varrho}})$. Therefore the extended Hoare triple notation is defined to be

$$\{\!|Q|\!\}\ \mathsf{S}\ \{\!|R \mid T|\!\} \quad \triangleq \quad \{\langle Q,\ R,\ (\!|\ \mathsf{escape}[\![\mathsf{S}]\!]\ ?\ T : U\ |\!) \rangle \mid U \in \mathbb{P}^{¤}\} \tag{26.1}$$

The domain $\mathbb{H}\mathfrak{t}^{¤}[\![\mathsf{S}]\!]$ of all Hoare triples is

$$\mathbb{H}\mathfrak{t}^{¤}[\![\mathsf{S}]\!] \quad \triangleq \quad \{\{\!|Q|\!\}\ \mathsf{S}\ \{\!|R \mid T|\!\} \mid Q, R, T \in \mathbb{P}^{¤}\}.$$

The extended Hoare triple notation differs from the tradition [501] in that

- The T component accounts for possible `break`; statements in the statement S. Notice that its presence (when escape⟦S⟧ = tt) or absence (when escape⟦S⟧ = ff) is determined purely syntactically, without any reference to the semantics of statement S;
- $\mathbb{P}^{¤}$ is an abstract domain, not only first-order logic;
- In the particular case $\mathbb{P}^{¤} = \wp(\mathbb{E}\mathrm{v}^{\varrho})$, the properties $Q, R, T \in \wp(\mathbb{E}\mathrm{v}^{\varrho})$ are represented by sets and not expressed in a first-order logic.
  - One reason for inexpressivity is the iteration. For example, a logic with arithmetic addition operation + but missing the multiplication × will be inexpressive,

because a program can compute a multiplication with an iteration using addition only. Therefore $\times$ must be added to the logic for expressivity. Then, however, exponentiation will be computable but not expressible in the logic, etc.

- Another source of inexpressivity comes from the fact that properties of entities can be expressed in first-order logic only by giving a name to these entities using an identifier. For example, one can express *x is odd* as a first-order logic formula using 1, +, and $\times$, as $\exists k \,.\, x = 2 \times k + 1$. However, one cannot express "to be odd" in this logic. Some semantics can manipulate entities that are not named by an identifier or obtained by applying selection operations from an identifier. This is the case of [539], which directly leads to an inexpressivity result [187], which can be gotten around by reasoning directly on the semantic entities (the environment, in our case) [287, 288, 822].

26.4 Abstraction of an Abstract Invariant into a Hoare Triple

A Hoare triple is the abstraction $\alpha^{\mathsf{H}}[\![\mathsf{S}]\!](\mathcal{J})$ of an abstract invariant $\mathcal{J} \in \mathsf{labx}[\![\mathsf{S}]\!] \to \mathbb{P}^{¤}$ by

$$\alpha^{\mathsf{H}}[\![\mathsf{S}]\!] \in (\mathsf{labx}[\![\mathsf{S}]\!] \to \mathbb{P}^{¤}) \to \mathbb{H}\mathbb{t}^{¤}[\![\mathsf{S}]\!]$$

where the abstraction of an invariant $\mathcal{J}$ into a Hoare triple $\alpha^{\mathsf{H}}[\![\mathsf{S}]\!](\mathcal{J})$ is

$$\alpha^{\mathsf{H}}[\![\mathsf{S}]\!](\mathcal{J}) \triangleq \{\!|\mathcal{J}(\mathsf{at}[\![\mathsf{S}]\!])|\!\}\ \mathsf{S}\ \{\!|\mathcal{J}(\mathsf{after}[\![\mathsf{S}]\!]) \mathrel{¦} \mathcal{J}(\mathsf{break\text{-}to}[\![\mathsf{S}]\!])|\!\} \tag{26.2}$$

26.5 Valid Hoare Triples

A Hoare triple is a specification that may be valid or invalid. For example, $\{\!|\{\rho \mid \rho(\mathsf{x}) = 0\}|\!\}\ \mathsf{x} = 1\,;\ \{\!|\{\rho \mid \rho(\mathsf{x}) = 2\} \mathrel{¦} \varnothing|\!\}$ is invalid because the value of variable x cannot be 2 after being assigned 1.

A Hoare triple is valid if and only if it is the abstraction of an inductive invariant as defined in section 25.3. The intuition is that if initial states satisfy $\mathcal{J}(\mathsf{at}[\![\mathsf{S}]\!])$ then the abstract invariant holds during execution, in particular when reaching $\mathsf{after}[\![\mathsf{S}]\!]$, if ever.

The Hoare logic of a statement S is the set of all valid Hoare triples of this statement, so

$$\widehat{\mathsf{H}}^{¤}[\![\mathsf{S}]\!] \triangleq \{\alpha^{\mathsf{H}}[\![\mathsf{S}]\!](\mathcal{J}) \mid \widehat{\boldsymbol{\mathcal{J}}}^{¤}[\![\mathsf{S}]\!]\,\mathcal{J}\} \tag{26.3}$$

where $\widehat{\boldsymbol{\mathcal{J}}}^{¤}[\![\mathsf{S}]\!]\,\mathcal{J}$ checks that an invariant $\mathcal{J} \in \mathbb{L} \to \mathbb{P}^{¤}$ is inductive.

$$\widehat{\boldsymbol{\mathcal{J}}}^{¤}[\![\mathsf{S}]\!] \in (\mathbb{L} \to \mathbb{P}^{¤}) \to \mathbb{B}$$

has been defined in (25.30) and (25.12) to (25.21).

26.6 Hoare Logic is an Abstract Interpretation of the Invariance Semantics

$\widehat{\mathsf{H}}^{¤}[\![\mathsf{S}]\!]$ is an abstract interpretation of the invariance semantics $\widehat{\boldsymbol{\mathcal{J}}}^{¤}[\![\mathsf{S}]\!]$ of a program component $\mathsf{S} \in \mathbb{P}\mathbb{c}$

$$\widehat{\mathsf{H}}^{¤}[\![\mathsf{S}]\!] \triangleq \alpha^{\mathbb{H}\mathbb{t}}[\![\mathsf{S}]\!](\widehat{\boldsymbol{\mathcal{J}}}^{¤}[\![\mathsf{S}]\!]) \tag{26.3}$$

by the abstraction

$$\alpha^{\mathbb{H}\mathbb{t}}[\![\mathsf{S}]\!](\mathsf{I}) \triangleq \{\alpha^{\mathsf{H}}[\![\mathsf{S}]\!](\mathcal{J}) \mid \mathsf{I}(\mathcal{J})\}$$

such that

$$\langle \mathbb{P}^{¤} \to (\mathbb{L} \to \mathbb{P}^{¤}) \to \mathbb{B},\ \dot{\Leftarrow}\rangle \xrightleftharpoons[\alpha^{\mathbb{H}t}[\![S]\!]]{\gamma^{\mathbb{H}t}[\![S]\!]} \langle \wp(\mathbb{H}t^{¤}[\![S]\!]),\ \supseteq\rangle.$$

where $\dot{\Leftarrow}$ is functional inverse implication.

Remark 26.4 There is a Galois isomorphism

$$\langle \mathbb{P}^{¤} \to (\mathbb{L} \to \mathbb{P}^{¤}) \to \mathbb{B},\ \dot{\Leftarrow}\rangle \xrightleftharpoons[\alpha_{\mathbb{1}}]{\gamma_{\mathbb{1}}} \langle \wp(\mathbb{P}^{¤} \to (\mathbb{L} \to \mathbb{P}^{¤})),\ \supseteq\rangle$$

where $\gamma_{\mathbb{1}}(S) \triangleq x \mapsto (x \in S)$ is the characteristic function (or indicator function) and $\alpha_{\mathbb{1}}(I) \triangleq \{x \mid I(x)\}$. Therefore defining the homomorphic/partitioning abstraction

$$\alpha^{\mathbb{H}t}[\![S]\!](I) \quad \triangleq \quad \{\alpha^{H}[\![S]\!](\mathcal{I}) \mid \mathcal{I} \in I\}$$

of a set I of inductive invariants into the set of all valid Hoare triples $\alpha^{\mathbb{H}t}[\![S]\!](I)$, such that

$$\langle \wp(\mathbb{P}^{¤} \to (\mathbb{L} \to \mathbb{P}^{¤})),\ \supseteq\rangle \xrightleftharpoons[\alpha^{\mathbb{H}t}[\![S]\!]]{\gamma^{\mathbb{H}t}[\![S]\!]} \langle \wp(\mathbb{H}t^{¤}[\![S]\!]),\ \supseteq\rangle$$

we have

$$\begin{aligned}
\widehat{H}^{¤}[\![S]\!] &= \alpha^{\mathbb{H}t}[\![S]\!](I)(\alpha_{\mathbb{1}}(\widehat{\mathcal{I}}^{¤}[\![S]\!]))\\
&= \{\alpha^{H}[\![S]\!](\mathcal{I}) \mid \mathcal{I} \in \alpha_{\mathbb{1}}(\widehat{\mathcal{I}}^{¤}[\![S]\!])\}\\
&= \{\alpha^{H}[\![S]\!](\mathcal{I}) \mid \widehat{\mathcal{I}}^{¤}[\![S]\!]\ \mathcal{I}\}
\end{aligned}$$

26.7 Calculational Design of Hoare Logic Rules

By (26.3), we have $\widehat{H}^{¤}[\![S]\!] \triangleq \alpha^{\mathbb{H}t}[\![S]\!](\widehat{\mathcal{I}}^{¤}[\![S]\!])$ so we can infer Hoare logic structural rules by calculational design. This consists of

1. Calculating the set of all valid Hoare triples $\widehat{H}^{¤}[\![S]\!] \triangleq \alpha^{\mathbb{H}t}[\![S]\!](\widehat{\mathcal{I}}^{¤}[\![S]\!])$ from the structural definition of $\widehat{\mathcal{I}}^{¤}[\![S]\!]$ of a program component $S \in \mathbb{P}c$.
   We proceed by structural induction on this program component $S \in \mathbb{P}c$.
   We get a definition of $\widehat{H}^{¤}[\![S]\!]$ as a function of the subcomponents $\widehat{H}^{¤}[\![S']\!]$, $S' \triangleleft S$.
2. Expressing this definition of $\widehat{H}^{¤}[\![S]\!]$ as a structural rule-based definition 16.24. (There is no need for fixpoint/inductive definitions that have already been eliminated in chapter 25 for $\widehat{\mathcal{I}}^{¤}[\![S]\!]$ using the fixpoint induction theorem 24.1.)

Example 26.5 (structural Hoare triple derivation) An example is given in figure 26.6

Rule for a program P ::= Sl

$$\frac{\{\!|Q|\!\}\ \mathtt{Sl}\ \{\!|R \mathrel{\vdots} T|\!\}}{\{\!|Q|\!\}\ \mathtt{P}\ \{\!|R \mathrel{\vdots} T|\!\}} \tag{26.7}$$

Proof of (26.7)

$$\begin{aligned}
&\widehat{H}^{¤}[\![P]\!]\\
=\ &\{\alpha^{H}[\![P]\!](\mathcal{I}) \mid \widehat{\mathcal{I}}^{¤}[\![P]\!]\ \mathcal{I}\} && \text{(26.3)}\\
=\ &\{\{\!|\mathcal{I}(\text{at}[\![P]\!])|\!\}\ P\ \{\!|\mathcal{I}(\text{after}[\![P]\!]) \mathrel{\vdots} \mathcal{I}(\text{break-to}[\![P]\!])|\!\} \mid \widehat{\mathcal{I}}^{¤}[\![P]\!]\ \mathcal{I}\} && \text{(26.2)}
\end{aligned}$$

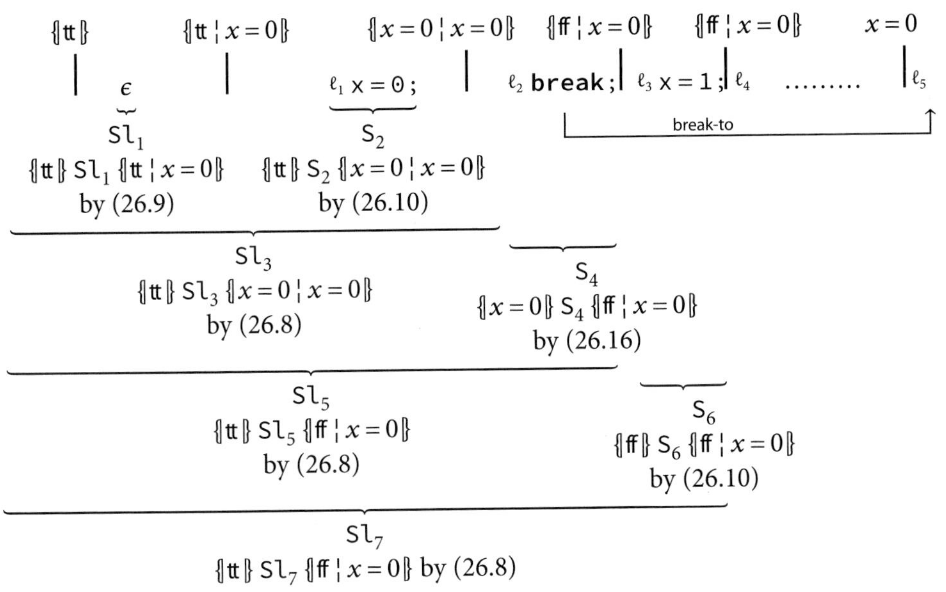

Figure 26.6
Example of structural Hoare triple derivation

$= \{\{\langle \mathscr{I}(\mathsf{at}[\![\mathtt{P}]\!]),\ \mathscr{I}(\mathsf{after}[\![\mathtt{P}]\!]),\ (\!|\ \mathsf{escape}[\![\mathtt{P}]\!]\ ?\ \mathscr{I}(\mathsf{break\text{-}to}[\![\mathtt{P}]\!]) : U\ |\!)\rangle \mid U \in \mathbb{P}^{¤}\} \mid \widehat{\mathscr{I}}^{¤}[\![\mathtt{P}]\!]\ \mathscr{I}\}$ ≀(26.1)ʃ

$= \{\{\langle \mathscr{I}(\mathsf{at}[\![\mathtt{Sl}]\!]),\ \mathscr{I}(\mathsf{after}[\![\mathtt{Sl}]\!]),\ (\!|\ \mathsf{escape}[\![\mathtt{Sl}]\!]\ ?\ \mathscr{I}(\mathsf{break\text{-}to}[\![\mathtt{Sl}]\!]) : U\ |\!)\rangle \mid U \in \mathbb{P}^{¤}\} \mid \widehat{\mathscr{I}}^{¤}[\![\mathtt{P}]\!]\ \mathscr{I}\}$

≀$\mathsf{escape}[\![\mathtt{P}]\!] = \mathsf{escape}[\![\mathtt{Sl}]\!] = \mathsf{ff}$ in section 4.2.4, (25.30) and (25.12) so $\widehat{\mathscr{I}}^{¤}[\![\mathtt{P}]\!] = \widehat{\mathscr{I}}^{¤}[\![\mathtt{Sl}]\!]$ʃ

$= \{\{\!|\mathscr{I}(\mathsf{at}[\![\mathtt{Sl}]\!])|\!\}\ \mathtt{Sl}\ \{\!|\mathscr{I}(\mathsf{after}[\![\mathtt{Sl}]\!]) \mathbin{¦} \mathscr{I}(\mathsf{break\text{-}to}[\![\mathtt{Sl}]\!])|\!\} \mid \widehat{\mathscr{I}}^{¤}[\![\mathtt{Sl}]\!]\ \mathscr{I}\}$ ≀(26.1)ʃ

$= \{\alpha^{\mathrm{H}}[\![\mathtt{Sl}]\!](\mathscr{I}) \mid \widehat{\mathscr{I}}^{¤}[\![\mathtt{Sl}]\!]\ \mathscr{I}\}$ ≀(26.2)ʃ

$= \widehat{\mathrm{H}}^{¤}[\![\mathtt{Sl}]\!]$ ≀(26.3)ʃ

By definition 16.24, we get the symbolic rule $\dfrac{\{\!|Q|\!\}\ \mathtt{Sl}\ \{\!|R \mathbin{¦} T|\!\}}{\{\!|Q|\!\}\ \mathtt{P}\ \{\!|R \mathbin{¦} T|\!\}}$ (defining $\widehat{\mathrm{H}}^{¤}[\![\mathtt{P}]\!]$ by the inference rules $\left\{\dfrac{\varnothing}{\{\!|Q|\!\}\ \mathtt{P}\ \{\!|R \mathbin{¦} T|\!\}} \;\middle|\; \{\!|Q|\!\}\ \mathtt{Sl}\ \{\!|R \mathbin{¦} T|\!\} \in \widehat{\mathrm{H}}^{¤}[\![\mathtt{Sl}]\!]\right\}$).

□

Rule for a statement list $\mathtt{Sl} ::= \mathtt{Sl}'\ \mathtt{S}$

$$\frac{\{\!|Q|\!\}\ \mathtt{Sl}'\ \{\!|R \mathbin{¦} T|\!\},\quad \{\!|R|\!\}\ \mathtt{S}\ \{\!|S \mathbin{¦} T|\!\}}{\{\!|Q|\!\}\ \mathtt{Sl}\ \{\!|S \mathbin{¦} T|\!\}} \tag{26.8}$$

(defining $\widehat{\mathrm{H}}^{¤}[\![\mathtt{Sl}]\!]$ by the inference rules $\left\{\dfrac{\varnothing}{\{\!|Q|\!\}\ \mathtt{Sl}\ \{\!|S \mathbin{¦} T|\!\}} \;\middle|\; \{\!|Q|\!\}\ \mathtt{Sl}'\ \{\!|R \mathbin{¦} T|\!\} \in \widehat{\mathrm{H}}^{¤}[\![\mathtt{Sl}']\!] \wedge \{\!|R|\!\}\ \mathtt{S}\ \{\!|S \mathbin{¦} T|\!\} \in \widehat{\mathrm{H}}^{¤}[\![\mathtt{S}]\!]\right\}$).

Proof of (26.8)

$$\widehat{\mathsf{H}}^{¤}\llbracket \mathtt{Sl} \rrbracket$$

$$= \{\alpha^{\mathsf{H}}\llbracket \mathtt{Sl}\rrbracket(\mathscr{I}) \mid \widehat{\mathscr{I}}^{¤}\llbracket \mathtt{Sl}\rrbracket\mathscr{I}\} \qquad \wr(26.3)\int$$

$$= \{\alpha^{\mathsf{H}}\llbracket \mathtt{Sl}\rrbracket(\mathscr{I}) \mid \widehat{\mathscr{I}}^{¤}\llbracket \mathtt{Sl}'\rrbracket\mathscr{I} \wedge \widehat{\mathscr{I}}^{¤}\llbracket \mathtt{S}\rrbracket\mathscr{I}\} \qquad \wr(25.30) \text{ and } (25.18)\int$$

$$= \{\alpha^{\mathsf{H}}\llbracket \mathtt{Sl}\rrbracket(\mathscr{I}) \mid \alpha^{\mathsf{H}}\llbracket \mathtt{Sl}'\rrbracket(\mathscr{I}) \in \widehat{\mathsf{H}}^{¤}\llbracket \mathtt{Sl}'\rrbracket \wedge \alpha^{\mathsf{H}}\llbracket \mathtt{S}\rrbracket(\mathscr{I}) \in \widehat{\mathsf{H}}^{¤}\llbracket \mathtt{S}\rrbracket\} \qquad \wr(26.3)\int$$

$$= \{\{\!|\mathscr{I}(\mathsf{at}\llbracket \mathtt{Sl}\rrbracket)|\!\}\ \mathtt{Sl}\ \{\!|\mathscr{I}(\mathsf{after}\llbracket \mathtt{Sl}\rrbracket) \mathbin{\vdots} \mathscr{I}(\mathsf{break\text{-}to}\llbracket \mathtt{Sl}\rrbracket)|\!\} \mid \{\!|\mathscr{I}(\mathsf{at}\llbracket \mathtt{Sl}'\rrbracket)|\!\}\ \mathtt{Sl}'$$
$$\{\!|\mathscr{I}(\mathsf{after}\llbracket \mathtt{Sl}'\rrbracket) \mathbin{\vdots} \mathscr{I}(\mathsf{break\text{-}to}\llbracket \mathtt{Sl}'\rrbracket)|\!\} \in \widehat{\mathsf{H}}^{¤}\llbracket \mathtt{Sl}'\rrbracket \wedge \{\!|\mathscr{I}(\mathsf{at}\llbracket \mathtt{S}\rrbracket)|\!\}\ \mathtt{S}\ \{\!|\mathscr{I}(\mathsf{after}\llbracket \mathtt{S}\rrbracket) \mathbin{\vdots}$$
$$\mathscr{I}(\mathsf{break\text{-}to}\llbracket \mathtt{S}\rrbracket)|\!\} \in \widehat{\mathsf{H}}^{¤}\llbracket \mathtt{S}\rrbracket\} \qquad \wr(26.2)\int$$

$$= \{\{\!|Q|\!\}\ \mathtt{Sl}\ \{\!|S \mathbin{\vdots} T|\!\} \mid \{\!|Q|\!\}\ \mathtt{Sl}'\ \{\!|R \mathbin{\vdots} T|\!\} \in \widehat{\mathsf{H}}^{¤}\llbracket \mathtt{Sl}'\rrbracket \wedge \{\!|R|\!\}\ \mathtt{S}\ \{\!|S \mathbin{\vdots} T|\!\} \in \widehat{\mathsf{H}}^{¤}\llbracket \mathtt{S}\rrbracket\}$$

$\wr$Renaming the local invariants as
$Q = \mathscr{I}(\mathsf{at}\llbracket \mathtt{Sl}\rrbracket) = \mathscr{I}(\mathsf{at}\llbracket \mathtt{Sl}'\rrbracket)$,
$R = \mathscr{I}(\mathsf{after}\llbracket \mathtt{Sl}'\rrbracket) = \mathscr{I}(\mathsf{at}\llbracket \mathtt{S}\rrbracket)$,
$S = \mathscr{I}(\mathsf{after}\llbracket \mathtt{Sl}\rrbracket) = \mathscr{I}(\mathsf{after}\llbracket \mathtt{S}\rrbracket)$,
$T = \mathscr{I}(\mathsf{break\text{-}to}\llbracket \mathtt{Sl}\rrbracket) = \mathscr{I}(\mathsf{break\text{-}to}\llbracket \mathtt{Sl}'\rrbracket) = \mathscr{I}(\mathsf{break\text{-}to}\llbracket \mathtt{S}\rrbracket)\int$ □

Rule for an empty statement list $\mathtt{Sl} ::= \epsilon$	
$\{\!|Q|\!\}\ \epsilon\ \{\!|Q \mathbin{\vdots} T|\!\}$	(26.9)

$\left(\text{defining } \widehat{\mathsf{H}}^{¤}\llbracket \mathtt{Sl}\rrbracket \text{ by the inference rules } \left\{\dfrac{\varnothing}{\{\!|Q|\!\}\ \epsilon\ \{\!|Q \mathbin{\vdots} T|\!\}} \;\middle|\; Q, T \in \mathbb{P}^{¤}\right\}\right).$

Proof of (26.9)

$$\widehat{\mathsf{H}}^{¤}\llbracket \mathtt{Sl} \rrbracket$$

$$= \{\alpha^{\mathsf{H}}\llbracket \mathtt{Sl}\rrbracket(\mathscr{I}) \mid \widehat{\mathscr{I}}^{¤}\llbracket \mathtt{Sl}\rrbracket\mathscr{I}\} \qquad \wr(26.3)\int$$

$$= \{\{\!|\mathscr{I}(\mathsf{at}\llbracket \mathtt{Sl}\rrbracket)|\!\}\ \mathtt{Sl}\ \{\!|\mathscr{I}(\mathsf{after}\llbracket \mathtt{Sl}\rrbracket) \mathbin{\vdots} \mathscr{I}(\mathsf{break\text{-}to}\llbracket \mathtt{Sl}\rrbracket)|\!\} \mid \widehat{\mathscr{I}}^{¤}\llbracket \mathtt{Sl}\rrbracket\mathscr{I}\} \qquad \wr(26.2)\int$$

$$= \{\{\langle \mathscr{I}(\mathsf{at}\llbracket \mathtt{Sl}\rrbracket), \mathscr{I}(\mathsf{after}\llbracket \mathtt{Sl}\rrbracket), (\!\!| \mathsf{escape}\llbracket \mathtt{Sl}\rrbracket \mathbin{?} \mathscr{I}(\mathsf{break\text{-}to}\llbracket \mathtt{Sl}\rrbracket) \mathbin{:} T |\!\!) \rangle \mid T \in \mathbb{P}^{¤}\} \mid$$
$$\widehat{\mathscr{I}}^{¤}\llbracket \mathtt{Sl}\rrbracket\mathscr{I}\} \qquad \wr(26.1)\int$$

$$= \{\{\langle Q, Q, T\rangle\} \mid Q, T \in \mathbb{P}^{¤}\}$$

$\wr$letting $Q = \mathscr{I}(\mathsf{at}\llbracket \mathtt{Sl}\rrbracket)$, $\mathsf{at}\llbracket \mathtt{Sl}\rrbracket) = \mathsf{after}\llbracket \mathtt{Sl}\rrbracket$ in section 4.2.1, $\mathsf{escape}\llbracket \mathtt{Sl}\rrbracket = \mathsf{ff}$, and $\widehat{\mathscr{I}}^{¤}\llbracket \mathtt{Sl}\rrbracket\mathscr{I} = \mathsf{tt}$ by (25.30) and (25.17)$\int$

$$= \{\langle Q, Q, (\!\!| \mathsf{escape}\llbracket \mathtt{S}\rrbracket \mathbin{?} T \mathbin{:} U |\!\!)\rangle \mid U \in \mathbb{P}^{¤}\} \qquad \wr\mathsf{escape}\llbracket \mathtt{Sl}\rrbracket = \mathsf{ff}\int$$

$$= \{\{\!|Q|\!\}\ \mathtt{S}\ \{\!|Q \mathbin{\vdots} T|\!\} \mid Q, T \in \mathbb{P}^{¤}\} \qquad \wr(26.1)\int \quad \square$$

Rule for an assignment statement $\mathtt{S} ::= \mathtt{x} = \mathtt{E}\,;$	
$\dfrac{\mathsf{assign}^{¤}\llbracket x, \mathtt{E}\rrbracket\, Q \sqsubseteq^{¤} R}{\{\!|Q|\!\}\ \mathtt{x} = \mathtt{E}\,;\ \{\!|R \mathbin{\vdots} T|\!\}}$	(26.10)

$\left(\text{defining } \widehat{\mathsf{H}}^{¤}\llbracket \mathtt{S}\rrbracket \text{ by the inference rules } \left\{\dfrac{\varnothing}{\{\!|Q|\!\}\ \mathtt{x} = \mathtt{E}\,;\ \{\!|R \mathbin{\vdots} T|\!\}} \;\middle|\; \mathsf{assign}^{¤}\llbracket x, \mathtt{E}\rrbracket\, Q \sqsubseteq^{¤} R\right\}\right).$

Proof of (26.10)

$$\widehat{\mathsf{H}}^{¤}[\![\mathsf{S}]\!]$$

$$= \{\{\langle \mathcal{I}(\mathsf{at}[\![\mathsf{S}]\!]),\ \mathcal{I}(\mathsf{after}[\![\mathsf{S}]\!]),\ (\!|\ \mathsf{escape}[\![\mathsf{S}]\!]\ ?\ \mathcal{I}(\mathsf{break\text{-}to}[\![\mathsf{S}]\!]) : U\ |\!) \rangle \mid U \in \mathbb{P}^{¤}\} \mid \widehat{\mathcal{I}}^{¤}[\![\mathsf{S}]\!]\ \mathcal{I}\}$$
$$\text{(26.3), (26.2), and (26.1)}$$

$$= \{\{\langle \mathcal{I}(\mathsf{at}[\![\mathsf{S}]\!]),\ \mathcal{I}(\mathsf{after}[\![\mathsf{S}]\!]),\ (\!|\ \mathsf{escape}[\![\mathsf{S}]\!]\ ?\ \mathcal{I}(\mathsf{break\text{-}to}[\![\mathsf{S}]\!]) : U\ |\!) \rangle \mid U \in \mathbb{P}^{¤}\} \mid \mathsf{assign}^{¤}[\![x, \mathsf{E}]\!]\ \mathcal{I}(\mathsf{at}[\![\mathsf{S}]\!]) \sqsubseteq^{¤} \mathcal{I}(\mathsf{after}[\![\mathsf{S}]\!])\}$$
$$\text{(25.30) and (25.14)}$$

$$= \{\{\langle Q, R, (\!|\ \mathsf{escape}[\![\mathsf{S}]\!]\ ?\ T : U\ |\!) \rangle \mid U \in \mathbb{P}^{¤}\} \mid \mathsf{assign}^{¤}[\![x, \mathsf{E}]\!]\ Q \sqsubseteq^{¤} R \wedge T \in \mathbb{P}^{¤}\}$$
$$\text{letting } Q = \mathcal{I}(\mathsf{at}[\![\mathsf{S}]\!]),\ R = \mathcal{I}(\mathsf{after}[\![\mathsf{S}]\!]),\ T = \mathcal{I}(\mathsf{break\text{-}to}[\![\mathsf{S}]\!]), \text{ and } \mathsf{escape}[\![\mathsf{S}]\!] = \mathit{ff}$$

$$= \{\{\!|Q|\!\}\ \mathsf{S}\ \{\!|R \mid T|\!\} \mid \mathsf{assign}^{¤}[\![x, \mathsf{E}]\!]\ Q \sqsubseteq^{¤} R \wedge T \in \mathbb{P}^{¤}\} \qquad \text{(26.1)} \quad \square$$

Rule for a skip statement S ::= ;

$$\frac{Q \sqsubseteq^{¤} R}{\{\!|Q|\!\}\ ;\ \{\!|R \mid T|\!\}} \qquad (26.11)$$

$\Big($defining $\widehat{\mathsf{H}}^{¤}[\![\mathsf{S}]\!]$ by the inference rules $\left\{\frac{\varnothing}{\{\!|Q|\!\}\ \mathsf{S}\ \{\!|R \mid T|\!\}} \;\middle|\; Q \sqsubseteq^{¤} R\right\}\Big)$.

Proof of (26.11)

$$\widehat{\mathsf{H}}^{¤}[\![\mathsf{S}]\!]$$

$$= \{\{\langle \mathcal{I}(\mathsf{at}[\![\mathsf{S}]\!]),\ \mathcal{I}(\mathsf{after}[\![\mathsf{S}]\!]),\ (\!|\ \mathsf{escape}[\![\mathsf{S}]\!]\ ?\ \mathcal{I}(\mathsf{break\text{-}to}[\![\mathsf{S}]\!]) : U\ |\!) \rangle \mid U \in \mathbb{P}^{¤}\} \mid \widehat{\mathcal{I}}^{¤}[\![\mathsf{S}]\!]\ \mathcal{I}\}$$
$$\text{(26.3), (26.2), and (26.1)}$$

$$= \{\{\langle \mathcal{I}(\mathsf{at}[\![\mathsf{S}]\!]),\ \mathcal{I}(\mathsf{after}[\![\mathsf{S}]\!]),\ (\!|\ \mathsf{escape}[\![\mathsf{S}]\!]\ ?\ \mathcal{I}(\mathsf{break\text{-}to}[\![\mathsf{S}]\!]) : U\ |\!) \rangle \mid U \in \mathbb{P}^{¤}\} \mid \mathcal{I}(\mathsf{at}[\![\mathsf{S}]\!]) \sqsubseteq^{¤} \mathcal{I}(\mathsf{after}[\![\mathsf{S}]\!])\}$$
$$\text{(25.30) and (25.13)}$$

$$= \{\{\langle Q, R, (\!|\ \mathsf{escape}[\![\mathsf{S}]\!]\ ?\ T : U\ |\!) \rangle \mid U \in \mathbb{P}^{¤}\} \mid Q \sqsubseteq^{¤} R\}$$
$$\text{letting } Q = \mathcal{I}(\mathsf{at}[\![\mathsf{S}]\!]),\ R = \mathcal{I}(\mathsf{after}[\![\mathsf{S}]\!]), \text{ and } T = \mathcal{I}(\mathsf{break\text{-}to}[\![\mathsf{S}]\!])$$

$$= \{\{\!|Q|\!\}\ \mathsf{S}\ \{\!|R \mid T|\!\} \mid Q \sqsubseteq^{¤} R\} \qquad \text{(26.1)} \quad \square$$

Rule for a conditional statement S ::= `if` (B) S_t

$$\frac{\mathsf{test}^{¤}[\![\mathsf{B}]\!]\ Q \sqsubseteq^{¤} Q', \quad \{\!|Q'|\!\}\ \mathsf{S}_t\ \{\!|R \mid T|\!\}, \quad \overline{\mathsf{test}}^{¤}[\![\mathsf{B}]\!]\ Q \sqsubseteq^{¤} R}{\{\!|Q|\!\}\ \mathtt{if}\ (\mathsf{B})\ \mathsf{S}_t\ \{\!|R \mid T|\!\}} \qquad (26.12)$$

$\Big($defining $\widehat{\mathsf{H}}^{¤}[\![\mathsf{S}]\!]$ by the inference rules $\left\{\frac{\varnothing}{\{\!|Q|\!\}\ \mathtt{if}\ (\mathsf{B})\ \mathsf{S}_t\ \{\!|R \mid T|\!\}} \;\middle|\; \mathsf{test}^{¤}[\![\mathsf{B}]\!]\ Q \sqsubseteq^{¤} Q' \wedge \{\!|Q'|\!\}\ \mathsf{S}_t\ \{\!|R \mid T|\!\} \in \widehat{\mathsf{H}}^{¤}[\![\mathsf{S}_t]\!] \wedge \overline{\mathsf{test}}^{¤}[\![\mathsf{B}]\!]\ Q \sqsubseteq^{¤} R\right\}\Big)$.

Proof of (26.12)

$$\widehat{\mathsf{H}}^{¤}[\![\mathsf{S}]\!]$$

$$= \{\{\langle \mathcal{I}(\mathsf{at}[\![\mathsf{S}]\!]),\ \mathcal{I}(\mathsf{after}[\![\mathsf{S}]\!]),\ (\!|\ \mathsf{escape}[\![\mathsf{S}]\!]\ ?\ \mathcal{I}(\mathsf{break\text{-}to}[\![\mathsf{S}]\!]) : U\ |\!) \rangle \mid U \in \mathbb{P}^{¤}\} \mid \widehat{\mathcal{I}}^{¤}[\![\mathsf{S}]\!]\ \mathcal{I}\}$$

$$\wr (26.3), (26.2), \text{ and } (26.1) \int$$

$$= \{\{\langle \mathscr{I}(\mathsf{at}[\![\mathsf{S}]\!]), \mathscr{I}(\mathsf{after}[\![\mathsf{S}]\!]), (\!| \mathsf{escape}[\![\mathsf{S}]\!] \,?\, \mathscr{I}(\mathsf{break\text{-}to}[\![\mathsf{S}]\!]) \,\text{:}\, U |\!) \rangle \mid U \in \mathbb{P}^{\text{¤}}\} \mid \mathsf{test}^{\text{¤}}[\![\mathsf{B}]\!]\mathscr{I}(\mathsf{at}[\![\mathsf{S}]\!]) \sqsubseteq^{\text{¤}} \mathscr{I}(\mathsf{at}[\![\mathsf{S}_t]\!]) \wedge \widehat{\mathscr{F}}^{\text{¤}}[\![\mathsf{S}_t]\!]\,\mathscr{I} \wedge \overline{\mathsf{test}}^{\text{¤}}[\![\mathsf{B}]\!]\mathscr{I}(\mathsf{at}[\![\mathsf{S}]\!]) \sqsubseteq^{\text{¤}} \mathscr{I}(\mathsf{after}[\![\mathsf{S}]\!])\}$$

$$\wr (25.30) \text{ and } (25.15) \int$$

$$= \{\{\langle \mathscr{I}(\mathsf{at}[\![\mathsf{S}]\!]), \mathscr{I}(\mathsf{after}[\![\mathsf{S}]\!]), (\!| \mathsf{escape}[\![\mathsf{S}]\!] \,?\, \mathscr{I}(\mathsf{break\text{-}to}[\![\mathsf{S}]\!]) \,\text{:}\, U |\!) \rangle \mid U \in \mathbb{P}^{\text{¤}}\} \mid \mathsf{test}^{\text{¤}}[\![\mathsf{B}]\!]\mathscr{I}(\mathsf{at}[\![\mathsf{S}]\!]) \sqsubseteq^{\text{¤}} \mathscr{I}(\mathsf{at}[\![\mathsf{S}_t]\!]) \wedge \alpha^{\natural}[\![\mathsf{S}_t]\!](\mathscr{I}) \in \widehat{\mathsf{H}}^{\text{¤}}[\![\mathsf{S}_t]\!] \wedge \overline{\mathsf{test}}^{\text{¤}}[\![\mathsf{B}]\!]\mathscr{I}(\mathsf{at}[\![\mathsf{S}]\!]) \sqsubseteq^{\text{¤}} \mathscr{I}(\mathsf{after}[\![\mathsf{S}]\!])\}$$

$$\wr (26.3) \text{ so that } \widehat{\mathscr{F}}^{\text{¤}}[\![\mathsf{S}_t]\!]\,\mathscr{I} \text{ if and only if } \alpha^{\natural}[\![\mathsf{S}_t]\!](\mathscr{I}) \in \widehat{\mathsf{H}}^{\text{¤}}[\![\mathsf{S}_t]\!] \int$$

$$= \{\{\langle \mathscr{I}(\mathsf{at}[\![\mathsf{S}]\!]), \mathscr{I}(\mathsf{after}[\![\mathsf{S}]\!]), (\!| \mathsf{escape}[\![\mathsf{S}]\!] \,?\, \mathscr{I}(\mathsf{break\text{-}to}[\![\mathsf{S}]\!]) \,\text{:}\, U |\!) \rangle \mid U \in \mathbb{P}^{\text{¤}}\} \mid \mathsf{test}^{\text{¤}}[\![\mathsf{B}]\!]\mathscr{I}(\mathsf{at}[\![\mathsf{S}]\!]) \sqsubseteq^{\text{¤}} \mathscr{I}(\mathsf{at}[\![\mathsf{S}_t]\!]) \wedge \{\!|\mathscr{I}(\mathsf{at}[\![\mathsf{S}_t]\!])|\!\}\, \mathsf{S}_t\, \{\!|\mathscr{I}(\mathsf{after}[\![\mathsf{S}_t]\!]) \mid \mathscr{I}(\mathsf{break\text{-}to}[\![\mathsf{S}_t]\!])|\!\} \in \widehat{\mathsf{H}}^{\text{¤}}[\![\mathsf{S}_t]\!] \wedge \overline{\mathsf{test}}^{\text{¤}}[\![\mathsf{B}]\!]\mathscr{I}(\mathsf{at}[\![\mathsf{S}]\!]) \sqsubseteq^{\text{¤}} \mathscr{I}(\mathsf{after}[\![\mathsf{S}]\!])\}$$

$$\wr (26.2) \int$$

$$= \{\{\langle Q, R, (\!| \mathsf{escape}[\![\mathsf{S}]\!] \,?\, T \,\text{:}\, U |\!) \rangle \mid U \in \mathbb{P}^{\text{¤}}\} \mid \mathsf{test}^{\text{¤}}[\![\mathsf{B}]\!]\, Q \sqsubseteq^{\text{¤}} Q' \wedge \{\!|Q'|\!\}\, \mathsf{S}_t\, \{\!|R \mid T|\!\} \in \widehat{\mathsf{H}}^{\text{¤}}[\![\mathsf{S}_t]\!] \wedge \overline{\mathsf{test}}^{\text{¤}}[\![\mathsf{B}]\!]\, Q \sqsubseteq^{\text{¤}} R\}$$

$\wr$letting $Q = \mathscr{I}(\mathsf{at}[\![\mathsf{S}]\!])$, $R = \mathscr{I}(\mathsf{after}[\![\mathsf{S}]\!]) = \mathscr{I}(\mathsf{after}[\![\mathsf{S}_t]\!])$, $Q' = \mathscr{I}(\mathsf{at}[\![\mathsf{S}_t]\!])$, and $T = \mathscr{I}(\mathsf{break\text{-}to}[\![\mathsf{S}]\!]) = \mathscr{I}(\mathsf{break\text{-}to}[\![\mathsf{S}_t]\!])$ because $\mathsf{escape}[\![\mathsf{S}]\!] = \mathsf{escape}[\![\mathsf{S}_t]\!]$ and $\mathsf{break\text{-}to}[\![\mathsf{S}]\!] = \mathsf{break\text{-}to}[\![\mathsf{S}_t]\!]$ in section 4.2.4 and $\mathsf{after}[\![\mathsf{S}]\!] = \mathsf{after}[\![\mathsf{S}_t]\!]$ in section 4.2.2$\int$

$$= \{\{\!|Q|\!\}\, \mathsf{S}\, \{\!|R \mid T|\!\} \mid \mathsf{test}^{\text{¤}}[\![\mathsf{B}]\!]\, Q \sqsubseteq^{\text{¤}} Q' \wedge \{\!|Q'|\!\}\, \mathsf{S}_t\, \{\!|R \mid T|\!\} \in \widehat{\mathsf{H}}^{\text{¤}}[\![\mathsf{S}_t]\!] \wedge \overline{\mathsf{test}}^{\text{¤}}[\![\mathsf{B}]\!]\, Q \sqsubseteq^{\text{¤}} R\}$$

$$\wr (26.1) \int \quad \square$$

Rule for a conditional statement $\mathsf{S} ::= \texttt{if}\,(\mathsf{B})\; \mathsf{S}_t\; \texttt{else}\; \mathsf{S}_f$

$$\frac{\mathsf{test}^{\text{¤}}[\![\mathsf{B}]\!]Q \sqsubseteq^{\text{¤}} Q_t \wedge \{\!|Q_t|\!\}\, \mathsf{S}_t\, \{\!|R \mid T|\!\} \wedge \overline{\mathsf{test}}^{\text{¤}}[\![\mathsf{B}]\!]Q \sqsubseteq^{\text{¤}} Q_f \wedge \{\!|Q_f|\!\}\, \mathsf{S}_f\, \{\!|R \mid T|\!\}}{\{\!|Q|\!\}\, \mathsf{S}\, \{\!|R \mid T|\!\}} \qquad (26.13)$$

Proof of (26.13)

$$\widehat{\mathsf{H}}^{\text{¤}}[\![\mathsf{S}]\!]$$

$$= \{\{\langle \mathscr{I}(\mathsf{at}[\![\mathsf{S}]\!]), \mathscr{I}(\mathsf{after}[\![\mathsf{S}]\!]), (\!| \mathsf{escape}[\![\mathsf{S}]\!] \,?\, \mathscr{I}(\mathsf{break\text{-}to}[\![\mathsf{S}]\!]) \,\text{:}\, U |\!) \rangle \mid U \in \mathbb{P}^{\text{¤}}\} \mid \widehat{\mathscr{F}}^{\text{¤}}[\![\mathsf{S}]\!]\,\mathscr{I}\}$$

$$\wr (26.3), (26.2), \text{ and } (26.1) \int$$

$$= \{\{\langle \mathscr{I}(\mathsf{at}[\![\mathsf{S}]\!]), \mathscr{I}(\mathsf{after}[\![\mathsf{S}]\!]), (\!| \mathsf{escape}[\![\mathsf{S}]\!] \,?\, \mathscr{I}(\mathsf{break\text{-}to}[\![\mathsf{S}]\!]) \,\text{:}\, U |\!) \rangle \mid U \in \mathbb{P}^{\text{¤}}\} \mid \mathsf{test}^{\text{¤}}[\![\mathsf{B}]\!]\mathscr{I}(\mathsf{at}[\![\mathsf{S}]\!]) \sqsubseteq^{\text{¤}} \mathscr{I}(\mathsf{at}[\![\mathsf{S}_t]\!]) \wedge \widehat{\mathscr{F}}^{\text{¤}}[\![\mathsf{S}_t]\!]\,\mathscr{I} \wedge \overline{\mathsf{test}}^{\text{¤}}[\![\mathsf{B}]\!]\mathscr{I}(\mathsf{at}[\![\mathsf{S}]\!]) \sqsubseteq^{\text{¤}} \mathscr{I}(\mathsf{at}[\![\mathsf{S}_f]\!]) \wedge \widehat{\mathscr{F}}^{\text{¤}}[\![\mathsf{S}_f]\!]\,\mathscr{I}\}$$

$$\wr (25.30) \text{ and } (25.16) \int$$

$$= \{\{\langle \mathscr{I}(\mathsf{at}[\![\mathsf{S}]\!]), \mathscr{I}(\mathsf{after}[\![\mathsf{S}]\!]), (\!| \mathsf{escape}[\![\mathsf{S}]\!] \,?\, \mathscr{I}(\mathsf{break\text{-}to}[\![\mathsf{S}]\!]) \,\text{:}\, U |\!) \rangle \mid U \in \mathbb{P}^{\text{¤}}\} \mid \mathsf{test}^{\text{¤}}[\![\mathsf{B}]\!]\mathscr{I}(\mathsf{at}[\![\mathsf{S}]\!]) \sqsubseteq^{\text{¤}} \mathscr{I}(\mathsf{at}[\![\mathsf{S}_t]\!]) \wedge \alpha^{\natural}[\![\mathsf{S}_t]\!](\mathscr{I}) \in \widehat{\mathsf{H}}^{\text{¤}}[\![\mathsf{S}_t]\!] \wedge \overline{\mathsf{test}}^{\text{¤}}[\![\mathsf{B}]\!]\mathscr{I}(\mathsf{at}[\![\mathsf{S}]\!]) \sqsubseteq^{\text{¤}} \mathscr{I}(\mathsf{at}[\![\mathsf{S}_f]\!]) \wedge \alpha^{\natural}[\![\mathsf{S}_f]\!](\mathscr{I}) \in \widehat{\mathsf{H}}^{\text{¤}}[\![\mathsf{S}_f]\!]\}$$

$$\wr (26.3) \text{ so that } \widehat{\mathscr{F}}^{\text{¤}}[\![\mathsf{S}]\!]\,\mathscr{I} \text{ if and only if } \alpha^{\natural}[\![\mathsf{S}]\!](\mathscr{I}) \in \widehat{\mathsf{H}}^{\text{¤}}[\![\mathsf{S}]\!] \int$$

$= \{\{\langle \mathcal{I}(\mathsf{at}[\![\mathsf{S}]\!]),\ \mathcal{I}(\mathsf{after}[\![\mathsf{S}]\!]),\ (\!|\,\mathsf{escape}[\![\mathsf{S}]\!]\ ?\ \mathcal{I}(\mathsf{break\text{-}to}[\![\mathsf{S}]\!]) : U\,|\!)\rangle \mid U \in \mathbb{P}^{¤}\} \mid$
$\mathsf{test}^{¤}[\![\mathsf{B}]\!]\mathcal{I}(\mathsf{at}[\![\mathsf{S}]\!]) \sqsubseteq^{¤} \mathcal{I}(\mathsf{at}[\![\mathsf{S}_t]\!]) \wedge \{\!|\mathcal{I}(\mathsf{at}[\![\mathsf{S}_t]\!])|\!\}\ \mathsf{S}_t\ \{\!|\mathcal{I}(\mathsf{after}[\![\mathsf{S}_t]\!]) \,\vert\, \mathcal{I}(\mathsf{break\text{-}to}[\![\mathsf{S}_t]\!])|\!\} \in$
$\widehat{\mathsf{H}}^{¤}[\![\mathsf{S}_t]\!] \wedge \overline{\mathsf{test}}^{¤}[\![\mathsf{B}]\!]\mathcal{I}(\mathsf{at}[\![\mathsf{S}]\!]) \sqsubseteq^{¤} \mathcal{I}(\mathsf{at}[\![\mathsf{S}_f]\!]) \wedge \{\!|\mathcal{I}(\mathsf{at}[\![\mathsf{S}_f]\!])|\!\}\ \mathsf{S}_f\ \{\!|\mathcal{I}(\mathsf{after}[\![\mathsf{S}_f]\!]) \,\vert$
$\mathcal{I}(\mathsf{break\text{-}to}[\![\mathsf{S}_f]\!])|\!\} \in \widehat{\mathsf{H}}^{¤}[\![\mathsf{S}_f]\!]\}$ ⟨(26.2)⟩

$= \{\{\langle Q,\ R,\ (\!|\,\mathsf{escape}[\![\mathsf{S}]\!]\ ?\ T : U\,|\!)\rangle \mid U \in \mathbb{P}^{¤}\} \mid \mathsf{test}^{¤}[\![\mathsf{B}]\!]Q \sqsubseteq^{¤} Q_t \wedge \{\!|Q_t|\!\}\ \mathsf{S}_t\ \{\!|R \,\vert\, T|\!\} \in$
$\widehat{\mathsf{H}}^{¤}[\![\mathsf{S}_t]\!] \wedge \overline{\mathsf{test}}^{¤}[\![\mathsf{B}]\!]Q \sqsubseteq^{¤} Q_f \wedge \{\!|Q_f|\!\}\ \mathsf{S}_f\ \{\!|R \,\vert\, T|\!\} \in \widehat{\mathsf{H}}^{¤}[\![\mathsf{S}_f]\!]\}$

⟨letting $Q = \mathcal{I}(\mathsf{at}[\![\mathsf{S}]\!])$, $Q_t = \mathcal{I}(\mathsf{at}[\![\mathsf{S}_t]\!])$, $Q_f = \mathcal{I}(\mathsf{at}[\![\mathsf{S}_f]\!])$, $R = \mathcal{I}(\mathsf{after}[\![\mathsf{S}]\!])$ with $\mathsf{after}[\![\mathsf{S}_t]\!] = \mathsf{after}[\![\mathsf{S}_f]\!] = \mathsf{after}[\![\mathsf{S}]\!]$ in section 4.2.2, and $T = \mathcal{I}(\mathsf{break\text{-}to}[\![\mathsf{S}]\!])$ with $\mathsf{break\text{-}to}[\![\mathsf{S}_t]\!] = \mathsf{break\text{-}to}[\![\mathsf{S}_f]\!] = \mathsf{break\text{-}to}[\![\mathsf{S}]\!]$ in section 4.2.4⟩

$= \{\{\!|Q|\!\}\ \mathsf{S}\ \{\!|R \,\vert\, T|\!\} \mid \mathsf{test}^{¤}[\![\mathsf{B}]\!]Q \sqsubseteq^{¤} Q_t \wedge \{\!|Q_t|\!\}\ \mathsf{S}_t\ \{\!|R \,\vert\, T|\!\} \in \widehat{\mathsf{H}}^{¤}[\![\mathsf{S}_t]\!] \wedge \overline{\mathsf{test}}^{¤}[\![\mathsf{B}]\!]Q \sqsubseteq^{¤}$
$Q_f \wedge \{\!|Q_f|\!\}\ \mathsf{S}_f\ \{\!|R \,\vert\, T|\!\} \in \widehat{\mathsf{H}}^{¤}[\![\mathsf{S}_f]\!]\}$ ⟨(26.1)⟩ □

Rule for an iteration statement S ::= **while** ℓ (B) S_b

$$\frac{\mathsf{test}^{¤}[\![\mathsf{B}]\!]\ Q \sqsubseteq^{¤} Q_b,\ \{\!|Q_b|\!\}\ \mathsf{S}_b\ \{\!|Q \,\vert\, R|\!\},\ \overline{\mathsf{test}}^{¤}[\![\mathsf{B}]\!]\ Q \sqsubseteq^{¤} R}{\{\!|Q|\!\}\ \mathtt{while}\ (\mathsf{B})\ \mathsf{S}_b\ \{\!|R \,\vert\, T|\!\}} \qquad (26.14)$$

(defining $\widehat{\mathsf{H}}^{¤}[\![\mathsf{S}]\!]$ by the inference rules $\left\{\dfrac{\varnothing}{\{\!|Q|\!\}\ \mathsf{S}\ \{\!|R \,\vert\, T|\!\}}\ \middle|\ \mathsf{test}^{¤}[\![\mathsf{B}]\!]\ Q \sqsubseteq^{¤} Q_b \wedge \{\!|Q_b|\!\}\ \mathsf{S}_b\ \{\!|Q|\!\} \in \widehat{\mathsf{H}}^{¤}[\![\mathsf{S}_b]\!] \wedge \overline{\mathsf{test}}^{¤}[\![\mathsf{B}]\!]\ Q \sqsubseteq^{¤} R\right\}$).

In case the loop body cannot be escaped by a **break**; statement, this is Hoare classic rule [501] where Q is the loop invariant and $T = \mathrm{ff}$. In case the loop body may escape to after the iteration statement, the postcondition R must be established both through an escape of the loop body and through a normal termination, if ever. [911] extends the rule to nonstandard control flow including break, return, throw exception, and continue statements.

Proof of (26.14)

$\widehat{\mathsf{H}}^{¤}[\![\mathsf{S}]\!]$

$= \{\{\langle \mathcal{I}(\mathsf{at}[\![\mathsf{S}]\!]),\ \mathcal{I}(\mathsf{after}[\![\mathsf{S}]\!]),\ (\!|\,\mathsf{escape}[\![\mathsf{S}]\!]\ ?\ \mathcal{I}(\mathsf{break\text{-}to}[\![\mathsf{S}]\!]) : U\,|\!)\rangle \mid U \in \mathbb{P}^{¤}\} \mid \widehat{\boldsymbol{\mathcal{F}}}^{¤}[\![\mathsf{S}]\!]\,\mathcal{I}\}$

⟨(26.3), (26.2), and (26.1)⟩

$= \{\{\langle \mathcal{I}(\mathsf{at}[\![\mathsf{S}]\!]),\ \mathcal{I}(\mathsf{after}[\![\mathsf{S}]\!]),\ (\!|\,\mathsf{escape}[\![\mathsf{S}]\!]\ ?\ \mathcal{I}(\mathsf{break\text{-}to}[\![\mathsf{S}]\!]) : U\,|\!)\rangle \mid U \in \mathbb{P}^{¤}\} \mid$
$\mathsf{test}^{¤}[\![\mathsf{B}]\!]\mathcal{I}(\mathsf{at}[\![\mathsf{S}]\!]) \sqsubseteq^{¤} \mathcal{I}(\mathsf{at}[\![\mathsf{S}_b]\!]) \wedge \alpha^{\mathsf{H}}[\![\mathsf{S}_b]\!](\mathcal{I}) \in \widehat{\mathsf{H}}^{¤}[\![\mathsf{S}_b]\!] \wedge \overline{\mathsf{test}}^{¤}[\![\mathsf{B}]\!]\mathcal{I}(\mathsf{at}[\![\mathsf{S}]\!]) \sqsubseteq^{¤}$
$\mathcal{I}(\mathsf{after}[\![\mathsf{S}]\!])\}$ ⟨(25.30) and (25.19′) in remark 25.23⟩

$= \{\{\langle \mathcal{I}(\mathsf{at}[\![\mathsf{S}]\!]),\ \mathcal{I}(\mathsf{after}[\![\mathsf{S}]\!]),\ (\!|\,\mathsf{escape}[\![\mathsf{S}]\!]\ ?\ \mathcal{I}(\mathsf{break\text{-}to}[\![\mathsf{S}]\!]) : U\,|\!)\rangle \mid U \in \mathbb{P}^{¤}\} \mid$
$\mathsf{test}^{¤}[\![\mathsf{B}]\!]\mathcal{I}(\mathsf{at}[\![\mathsf{S}]\!]) \sqsubseteq^{¤} \mathcal{I}(\mathsf{at}[\![\mathsf{S}_b]\!]) \wedge \{\!|\mathcal{I}(\mathsf{at}[\![\mathsf{S}_b]\!])|\!\}\ \mathsf{S}_b\ \{\!|\mathcal{I}(\mathsf{after}[\![\mathsf{S}_b]\!]) \,\vert\, \mathcal{I}(\mathsf{break\text{-}to}[\![\mathsf{S}_b]\!])|\!\} \in$
$\widehat{\mathsf{H}}^{¤}[\![\mathsf{S}_b]\!] \wedge \overline{\mathsf{test}}^{¤}[\![\mathsf{B}]\!]\mathcal{I}(\mathsf{at}[\![\mathsf{S}]\!]) \sqsubseteq^{¤} \mathcal{I}(\mathsf{after}[\![\mathsf{S}]\!])\}$ ⟨(26.2)⟩

$= \{\{\langle Q,\ R,\ (\!|\,\mathsf{escape}[\![\mathsf{S}]\!]\ ?\ T : U\,|\!)\rangle \mid U \in \mathbb{P}^{\text{¤}}\} \mid \mathsf{test}^{\text{¤}}[\![\mathsf{B}]\!]Q \sqsubseteq^{\text{¤}} Q_b \wedge \{\!|Q_b|\!\}\ \mathsf{S}_b\ \{\!|Q \mid R|\!\} \in \widehat{\mathsf{H}}^{\text{¤}}[\![\mathsf{S}_b]\!] \wedge \overline{\mathsf{test}}^{\text{¤}}[\![\mathsf{B}]\!]Q \sqsubseteq^{\text{¤}} R\}$

  (letting $Q = \mathscr{I}(\mathsf{at}[\![\mathsf{S}]\!]) = \mathscr{I}(\mathsf{after}[\![\mathsf{S}_b]\!])$, $T = \mathscr{I}(\mathsf{break\text{-}to}[\![\mathsf{S}]\!])$, $R = \mathscr{I}(\mathsf{after}[\![\mathsf{S}]\!]) = \mathscr{I}(\mathsf{break\text{-}to}[\![\mathsf{S}_b]\!])$, $Q_b = \mathscr{I}(\mathsf{at}[\![\mathsf{S}_b]\!])$)

$= \{\{\!|Q|\!\}\ \mathsf{S}\ \{\!|R \mid T|\!\} \mid \mathsf{test}^{\text{¤}}[\![\mathsf{B}]\!]\,Q \sqsubseteq^{\text{¤}} Q_b \wedge \{\!|Q_b|\!\}\ \mathsf{S}_b\ \{\!|Q \mid R|\!\} \in \widehat{\mathsf{H}}^{\text{¤}}[\![\mathsf{S}_b]\!] \wedge \overline{\mathsf{test}}^{\text{¤}}[\![\mathsf{B}]\!]\,Q \sqsubseteq^{\text{¤}} R\}$

(26.1) □

Exercise 26.15 (nontermination) Prove $\{\!|\mathsf{tt}|\!\}$ `while (0 < 1);` $\{\!|\mathsf{ff}|\!\}$.

Rule for a break statement S ::= `break;`

$$\frac{Q \sqsubseteq^{\text{¤}} T}{\{\!|Q|\!\}\ \texttt{break;}\ \{\!|R \mid T|\!\}} \qquad (26.16)$$

(defining $\widehat{\mathsf{H}}^{\text{¤}}[\![\mathsf{S}]\!]$ by the inference rules $\left\{\dfrac{\varnothing}{\{\!|Q|\!\}\ \texttt{break;}\ \{\!|R \mid T|\!\}}\ \middle|\ Q \sqsubseteq^{\text{¤}} T\right\}$).

Proof of (26.16)

$\widehat{\mathsf{H}}^{\text{¤}}[\![\mathsf{S}]\!]$

$= \{\{\langle \mathscr{I}(\mathsf{at}[\![\mathsf{S}]\!]),\ \mathscr{I}(\mathsf{after}[\![\mathsf{S}]\!]),\ (\!|\,\mathsf{escape}[\![\mathsf{S}]\!]\ ?\ \mathscr{I}(\mathsf{break\text{-}to}[\![\mathsf{S}]\!]) : U\,|\!)\rangle \mid U \in \mathbb{P}^{\text{¤}}\} \mid \widehat{\boldsymbol{\mathscr{I}}}^{\text{¤}}[\![\mathsf{S}]\!]\,\mathscr{I}\}$

(26.3), (26.2), and (26.1)

$= \{\{\langle \mathscr{I}(\mathsf{at}[\![\mathsf{S}]\!]),\ \mathscr{I}(\mathsf{after}[\![\mathsf{S}]\!]),\ (\!|\,\mathsf{escape}[\![\mathsf{S}]\!]\ ?\ \mathscr{I}(\mathsf{break\text{-}to}[\![\mathsf{S}]\!]) : U\,|\!)\rangle \mid U \in \mathbb{P}^{\text{¤}}\} \mid \mathscr{I}(\mathsf{at}[\![\mathsf{S}]\!]) \sqsubseteq^{\text{¤}} \mathscr{I}(\mathsf{break\text{-}to}[\![\mathsf{S}]\!])\}$

(25.30) and (25.20′) in remark 25.23

$= \{\langle Q,\ R,\ (\!|\,\mathsf{escape}[\![\mathsf{S}]\!]\ ?\ T : \varnothing\,|\!)\rangle \mid Q \sqsubseteq^{\text{¤}} T\}$

(letting $Q = \mathscr{I}(\mathsf{at}[\![\mathsf{S}]\!])$, $R = \mathscr{I}(\mathsf{after}[\![\mathsf{S}]\!])$, and $T = \mathscr{I}(\mathsf{break\text{-}to}[\![\mathsf{S}]\!])$)

$= \{\{\!|Q|\!\}\ \texttt{break;}\ \{\!|R \mid T|\!\} \mid Q \sqsubseteq^{\text{¤}} T\}$

(26.1) □

Rule for a compound statement S ::= `{ Sl }`

$$\frac{\{\!|Q|\!\}\ \mathsf{Sl}\ \{\!|R \mid T|\!\}}{\{\!|Q|\!\}\ \{\ \mathsf{Sl}\ \}\ \{\!|R \mid T|\!\}} \qquad (26.17)$$

(defining $\widehat{\mathsf{H}}^{\text{¤}}[\![\mathsf{S}]\!]$ by the inference rules $\left\{\dfrac{\varnothing}{\{\!|Q|\!\}\ \mathsf{S}\ \{\!|R \mid T|\!\}}\ \middle|\ \{\!|Q|\!\}\ \mathsf{Sl}\ \{\!|R \mid T|\!\} \in \widehat{\mathsf{H}}^{\text{¤}}[\![\mathsf{Sl}]\!]\right\}$)

Proof of (26.17)

$\widehat{\mathsf{H}}^{\text{¤}}[\![\mathsf{S}]\!]$

$= \{\{\langle \mathscr{I}(\mathsf{at}[\![\mathsf{S}]\!]),\ \mathscr{I}(\mathsf{after}[\![\mathsf{S}]\!]),\ (\!|\,\mathsf{escape}[\![\mathsf{S}]\!]\ ?\ \mathscr{I}(\mathsf{break\text{-}to}[\![\mathsf{S}]\!]) : U\,|\!)\rangle \mid U \in \mathbb{P}^{\text{¤}}\} \mid \widehat{\boldsymbol{\mathscr{I}}}^{\text{¤}}[\![\mathsf{S}]\!]\,\mathscr{I}\}$

(26.3), (26.2), and (26.1)

$= \{\{\langle \mathscr{I}(\mathsf{at}[\![\mathsf{Sl}]\!]),\ \mathscr{I}(\mathsf{after}[\![\mathsf{Sl}]\!]),\ (\!|\,\mathsf{escape}[\![\mathsf{Sl}]\!]\ ?\ \mathscr{I}(\mathsf{break\text{-}to}[\![\mathsf{Sl}]\!]) : U\,|\!)\rangle \mid U \in \mathbb{P}^{\text{¤}}\} \mid \widehat{\boldsymbol{\mathscr{I}}}^{\text{¤}}[\![\mathsf{Sl}]\!]\,\mathscr{I}\}$

$\wr$at⟦Sl⟧ = at⟦S⟧, break-to⟦Sl⟧ = break-to⟦S⟧, after⟦Sl⟧ = after⟦S⟧, so escape⟦Sl⟧ = escape⟦S⟧, (25.30) and (25.21)$\wr$

= $\widehat{\mathsf{H}}^{\texttt{¤}}$⟦Sl⟧ $\wr$(26.3), (26.2), and (26.1)$\wr$ □

Exercise 26.18 Prove the partial correctness of the following factorial program by Hoare logic ($n! \triangleq \prod_{i=1}^{n} i = 1 \times 2 \times \ldots \times (n-1) \times n$ where $\prod_{i=1}^{0} i \triangleq 1$ so $0! \triangleq 1$)

```
/* n = n0 > 0 */
ℓ1 f = n;
while ℓ2 (n > 1) {
    ℓ3 n = n - 1;
    ℓ4 f = f * n;
}
ℓ5 /* f = n0! */
```

Exercise 26.19 Prove the consequence rule

$$\frac{Q' \sqsubseteq^{\texttt{¤}} Q, \quad \{\!|Q|\!\}\ \mathsf{S}\ \{\!|R \,\mathop{\mid}\, T|\!\}, \quad R \sqsubseteq^{\texttt{¤}} R', \quad T \sqsubseteq^{\texttt{¤}} T'}{\{\!|Q'|\!\}\ \mathsf{S}\ \{\!|R' \,\mathop{\mid}\, T'|\!\}}.$$

In classic Hoare logic, i.e. $\mathbb{P}^{\texttt{¤}} = \wp(\mathbb{E}\mathbb{v}^{\vec{\ell}})$, the disjunction $\dfrac{\forall i \in \Delta\,.\,\{\!|Q_i|\!\}\ \mathsf{S}\ \{\!|R_i|\!\}}{\{\!|\bigcup_{i\in\Delta} Q_i|\!\}\ \mathsf{S}\ \{\!|\bigcup_{i\in\Delta} R_i|\!\}}$ and conjunction rules $\dfrac{\forall i \in \Delta\,.\,\{\!|Q_i|\!\}\ \mathsf{S}\ \{\!|R_i|\!\}}{\{\!|\bigcap_{i\in\Delta} Q_i|\!\}\ \mathsf{S}\ \{\!|\bigcap_{i\in\Delta} R_i|\!\}}$ are sound.

Exercise 26.20 Derive the disjunction and conjunction rules from the other rules by structural induction on the program component $\mathsf{S} \in \mathbb{P}\mathbb{c}$.

As observed by [272], this is not, in general, valid in the abstract.

26.8 Conclusion

The success of the Turing/Floyd/Naur invariance verification method of chapter 25 is that it is the most abstract, sound, and complete invariance proof method. Being the most abstract, it cannot be further simplified. Being sound, it is infallible. Being complete, it is always applicable and never fails. The success of Hoare logic is that it is a logical reformulation and, compared to previous formulations of invariance proofs, that it works by structural induction on programs. We have used sets to encode predicates. In practice, they are encoded using a (first-order) logic so that the necessary inductive invariant for iteration might not be expressible in that logic, and the logic inexpressibility is a source of incompleteness, not Hoare logic in itself.

The original definition of Hoare logic [501] postulated rules to prove that a Hoare triple $\{\!|Q|\!\}\ \mathsf{S}\ \{\!|R|\!\}$ is valid, by structural induction on the statement S. These rules were later proved sound and relatively complete [208, 209] where the “relatively” means that the considered first-order logic can express the inductive invariant.

The Hoare logic that we calculated is different from the classic one [501, 503, 504] because of the **break**; (which has led to inconsistencies [758]).

Postulating rules to derive inductive invariants [376] or valid Hoare triples [501] is a way to "assign meaning to programs" [376], that is, to define their semantics (called the *axiomatic semantics*). However, this axiomatic semantics is too abstract for many applications such as static analysis. For example, [46, 546] use Hoare logic as collecting semantics, so cannot consider execution traces, and so this work cannot fully account for the trace-based soundness of Astrée [113].

We have inferred structural Hoare logic rules defining the set $\widehat{\mathsf{H}}^{¤}[\![\mathsf{S}]\!]$ of all valid Hoare triples for a statement S by calculational design using an abstraction $\alpha^{\mathsf{Ht}}[\![\mathsf{S}]\!](\widehat{\mathscr{I}}^{¤}[\![\mathsf{S}]\!])$ of the structural reachability semantics $\widehat{\mathscr{I}}^{¤}[\![\mathsf{S}]\!]$.

Although this calculational design methodology might appear superfluous for an already known logic, it is effective in designing invariance proof methods for new languages or semantic models [24].

Program verification methods have been implemented in interactive tools for actual programming languages. These tools often incorporate abstract interpretation-based static analysis to help inferring inductive invariants. This is the case of Frama-C for C [579] and of Key for Java [10, chapter 6].

26.9 Solutions to Selected Exercises

Solution to exercise 26.15

$$\frac{\mathsf{test}^{\vec{e}}[\![\mathsf{tt}]\!]\,\mathsf{tt} \subseteq \mathsf{tt},\quad \dfrac{\mathsf{tt} \subseteq \mathsf{tt}}{\{\!|\mathsf{tt}|\!\}\,;\,\{\!|\mathsf{tt}|\!\}},\ (26.11),\ \overline{\mathsf{test}}^{\vec{e}}[\![\mathsf{tt}]\!]\,\mathsf{tt} \subseteq \mathsf{ff}}{\{\!|\mathsf{tt}|\!\}\ \mathtt{while}\ (\mathsf{tt})\,;\ \{\!|\mathsf{ff}|\!\}},\ (26.14)$$

Solution to exercise 26.18

```
/* n = n_0 > 0 */
ℓ1 f = n;
while ℓ2 /* n > 0 ∧ f = ∏_{i=n}^{n_0} i */ (n > 1) {
    ℓ3 /* n > 1 ∧ f = ∏_{i=n}^{n_0} i */
    n = n - 1;
    ℓ4 /* n > 0 ∧ f = ∏_{i=n+1}^{n_0} i */
    f = f * n;
}
ℓ5 /* f = ∏_{i=1}^{n_0} i = n_0! */
```

IX CARTESIAN STATIC ANALYSIS

Abstractions consider program properties that are less expressive than their semantic properties (chapter 27).

An abstraction is *Cartesian* when it does not relate the values of variables (e.g. signs, parity, and constancy in chapter 3). This abstraction yields abstract Cartesian reachability semantics, which associate with each program point a characterization of the values of the variables, but not of their interrelationships (chapter 28). However, reduction (chapter 29) is able to capture some of these interrelationships by iterating Cartesian analyses.

As examples we provide Cartesian congruence analysis (chapter 31) based upon elementary notions of number theory (chapter 30) and interval/range analysis that can be performed dynamically, at runtime (chapter 32) or statically, at compile time (chapter 33).

Because the Cartesian interval domain has infinite strictly increasing chains, fixpoint iterations may diverge. We use interval analysis to introduce convergence acceleration to overapproximate these fixpoints. This study is further refined in part X.

27 Abstraction

The assertional reachability semantics $\mathcal{S}^{\vec{r}}[\![\mathsf{P}]\!]$ of chapter 19 is an abstraction of the relational reachability semantics $\widehat{\mathcal{S}}^{\vec{\mathsf{R}}}[\![\mathsf{P}]\!]$ of chapter 19 by α^2 (exercise 20.13).

This relational reachability semantics $\widehat{\mathcal{S}}^{\vec{\mathsf{R}}}[\![\mathsf{P}]\!]$ of chapter 19 is an abstraction of the maximal trace semantics $\mathcal{S}^{+\infty}[\![\mathsf{P}]\!]$ of chapter 7 by $\mathsf{post}^{\vec{\mathsf{R}}*}$ (20.14).

This maximal trace semantics $\mathcal{S}^{+\infty}[\![\mathsf{P}]\!]$ of chapter 7 is, by $\dot{\alpha}^{*\to}$ (exercise 7.12), an abstraction of the prefix trace semantics $\mathcal{S}^{*}[\![\mathsf{P}]\!]$ of chapter 6.

We have the following hierarchy of semantics.

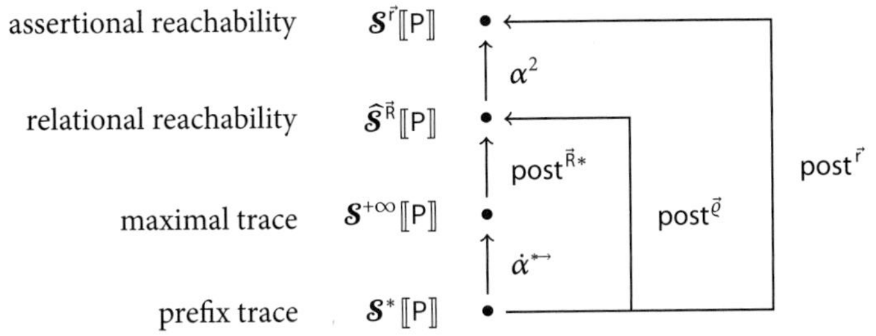

These semantics have the same structure and are all instances of the generic abstract semantics of chapter 21 implemented as an abstract interpreter parameterized by an abstract domain representing the information collected about program executions and the basic operations involved in this collection. This preservation of structure follows from the fact that the abstraction of an instance of the abstract interpreter is an instance of the abstract interpreter, as shown in this chapter.

Contents

27.1 Domain Abstraction

A concrete domain is an algebra that encodes program properties and operations on those properties. The abstraction of a concrete domain is an abstract domain that encodes less expressive program properties and operations on these abstract properties. The relation between the concrete and abstract domains is formalized by a concretization function and, in the case of best abstractions, by a Galois connection. Of course, the notion of concrete and abstract domains is not absolute, but relative to the expressiveness of one compared to the other. For example, assertional reachability is more abstract than relational reachability, which itself is more abstract than the maximal trace semantics.

Definition 27.1 (I — approximate domain abstraction) We say that an abstract domain

$$\mathbb{D}^{\sharp} \triangleq \langle \mathbb{P}^{\sharp}, \sqsubseteq^{\sharp}, \bot^{\sharp}, \sqcup^{\sharp}, \mathsf{assign}^{\sharp}[\![\mathsf{x}, \mathsf{A}]\!], \mathsf{test}^{\sharp}[\![\mathsf{B}]\!], \overline{\mathsf{test}}^{\sharp}[\![\mathsf{B}]\!]\rangle$$

is a sound *approximate abstraction* of a concrete domain

$$\mathbb{D}^{\text{¤}} \triangleq \langle \mathbb{P}^{\text{¤}}, \sqsubseteq^{\text{¤}}, \bot^{\text{¤}}, \sqcup^{\text{¤}}, \mathsf{assign}^{\text{¤}}[\![\mathsf{x}, \mathsf{A}]\!], \mathsf{test}^{\text{¤}}[\![\mathsf{B}]\!], \overline{\mathsf{test}}^{\text{¤}}[\![\mathsf{B}]\!]\rangle$$

for concretization γ whenever

1. $\gamma \in \mathbb{P}^{\sharp} \xrightarrow{\nearrow} \mathbb{P}^{\text{¤}}$. (concretization)
2. $\mathsf{assign}^{\text{¤}}[\![\mathsf{x}, \mathsf{A}]\!] \circ \gamma \mathrel{\dot{\sqsubseteq}^{\text{¤}}} \gamma \circ \mathsf{assign}^{\sharp}[\![\mathsf{x}, \mathsf{A}]\!]$. ($\gamma$-semicommutation)
3. $\mathsf{test}^{\text{¤}}[\![\mathsf{B}]\!] \circ \gamma \mathrel{\dot{\sqsubseteq}^{\text{¤}}} \gamma \circ \mathsf{test}^{\sharp}[\![\mathsf{B}]\!]$ and $\overline{\mathsf{test}}^{\text{¤}}[\![\mathsf{B}]\!] \circ \gamma \mathrel{\dot{\sqsubseteq}^{\text{¤}}} \gamma \circ \overline{\mathsf{test}}^{\sharp}[\![\mathsf{B}]\!]$.

(II — exact domain abstraction) We say that an abstract domain $\mathbb{D}^{\sharp}$ is an *exact abstraction* of a concrete domain $\mathbb{D}^{\text{¤}}$ for an abstraction $\langle \alpha, \gamma\rangle$ whenever

1. $\langle \mathbb{P}^{\text{¤}}, \sqsubseteq^{\text{¤}}\rangle \underset{\alpha}{\overset{\gamma}{\leftrightarrows\!\!\!\!\rightarrow}} \langle \mathbb{P}^{\sharp}, \sqsubseteq^{\sharp}\rangle$. (exact abstraction)
2. $\alpha \circ \mathsf{assign}^{\text{¤}}[\![\mathsf{x}, \mathsf{A}]\!] = \mathsf{assign}^{\sharp}[\![\mathsf{x}, \mathsf{A}]\!] \circ \alpha$. ($\alpha$-commutation)
3. $\alpha \circ \mathsf{test}^{\text{¤}}[\![\mathsf{B}]\!] = \mathsf{test}^{\sharp}[\![\mathsf{B}]\!] \circ \alpha$ and $\alpha \circ \overline{\mathsf{test}}^{\text{¤}}[\![\mathsf{B}]\!] = \overline{\mathsf{test}}^{\sharp}[\![\mathsf{B}]\!] \circ \alpha$.

The hypothesis that γ is increasing in definition 27.1.I.1 is sufficient to ensure soundness, as shown in theorem 27.8. It states that a logical implication in the abstract is valid in the concrete, that is, reasonings in the abstract are preserved in the concrete. We have $\bot^{\text{¤}} \sqsubseteq^{\text{¤}} \gamma(\bot^{\sharp})$ because $\bot^{\text{¤}}$ is the infimum of $\mathbb{P}^{\text{¤}}$ and, for all $A, A' \in \mathbb{P}^{\sharp}$, $\gamma(A) \sqcup^{\text{¤}} \gamma(A') \sqsubseteq^{\text{¤}} \gamma(A \sqcup^{\sharp} A')$ because γ is increasing.

The stronger hypotheses in definition 27.1.II yield more precise results because there is a best abstraction, a Galois retraction, and commutation (instead of semicommutation). This ensures completeness (see theorem 27.8).

A stronger alternative to definition 27.1.I would be

1. $\langle \mathbb{P}^{\text{¤}}, \sqsubseteq^{\text{¤}}\rangle \underset{\alpha}{\overset{\gamma}{\leftrightarrows}} \langle \mathbb{P}^{\sharp}, \sqsubseteq^{\sharp}\rangle$. (best abstraction)
2. $\alpha \circ \mathsf{assign}^{\text{¤}}[\![\mathsf{x}, \mathsf{A}]\!] \mathrel{\dot{\sqsubseteq}^{\sharp}} \mathsf{assign}^{\sharp}[\![\mathsf{x}, \mathsf{A}]\!] \circ \alpha$. ($\alpha$-semicommutation)
3. $\alpha \circ \mathsf{test}^{\text{¤}}[\![\mathsf{B}]\!] \mathrel{\dot{\sqsubseteq}^{\sharp}} \mathsf{test}^{\sharp}[\![\mathsf{B}]\!] \circ \alpha$ and $\alpha \circ \overline{\mathsf{test}}^{\text{¤}}[\![\mathsf{B}]\!] \mathrel{\dot{\sqsubseteq}^{\sharp}} \overline{\mathsf{test}}^{\sharp}[\![\mathsf{B}]\!] \circ \alpha$.

Then $\gamma \in \mathbb{P}^{\sharp} \xrightarrow{\nearrow} \mathbb{P}^{\text{¤}}$ and for all $A \in \mathbb{P}^{\text{¤}}$,

$$\alpha \circ \mathsf{assign}^{\text{¤}}[\![\mathsf{x}, \mathsf{A}]\!]\, \gamma(A) \sqsubseteq^{\sharp} \mathsf{assign}^{\sharp}[\![\mathsf{x}, \mathsf{A}]\!] \circ \alpha(\gamma(A))$$

$\Leftrightarrow \alpha \circ \mathsf{assign}^{\natural}[\![\mathsf{x}, \mathsf{A}]\!]\, \gamma(A) \sqsubseteq^{\sharp} \mathsf{assign}^{\sharp}[\![\mathsf{x}, \mathsf{A}]\!]\, A$

$\wr(\Rightarrow)$ $\alpha \circ \gamma$ reductive and $\mathsf{assign}^{\sharp}[\![\mathsf{x}, \mathsf{A}]\!]$ increasing
$(\Leftarrow)$ for $A' = \alpha(\gamma(A))$ and $\gamma(\alpha(\gamma(A'))) = \gamma(A')$ by the dual of exercise 11.34$\wr$

$\Leftrightarrow \mathsf{assign}^{\natural}[\![\mathsf{x}, \mathsf{A}]\!] \circ \gamma(A) \sqsubseteq^{\natural} \gamma \circ \mathsf{assign}^{\sharp}[\![\mathsf{x}, \mathsf{A}]\!]\,(A)$ $\quad\wr$definition of a Galois connection$\wr$

$\Leftrightarrow \mathsf{assign}^{\natural}[\![\mathsf{x}, \mathsf{A}]\!] \circ \gamma \,\dot{\sqsubseteq}^{\natural}\, \gamma \circ \mathsf{assign}^{\sharp}[\![\mathsf{x}, \mathsf{A}]\!]$ $\quad\wr$pointwise, that is, definition 27.1.I.2$\wr$

and similarly for tests. So semicommutation hypotheses for α are equivalent to definition 27.1.I.

Remark 20.8 and the design of transformers in chapter 20 show that the reachability semantics $\widehat{\boldsymbol{S}}^{\vec{r}}$ is the best abstraction of the pointwise prefix trace semantics $\widehat{\boldsymbol{S}}^{*}$ by $\langle \ddot{\alpha}_{\varrho}, \ddot{\gamma}_{\varrho} \rangle$.

Exercise 27.2 Prove that the assertional reachability domain is an abstraction of the relational reachability domain.

27.2 Concrete and Abstract Semantics

The semantics all have the same structure because this structure is preserved by structural abstraction. Given an concrete semantics

$$\widehat{\boldsymbol{S}}^{\natural}[\![\mathsf{S}]\!] \quad \in \quad \mathbb{P}^{\natural} \to \mathbb{L} \to \mathbb{P}^{\natural}$$

on a concrete domain $\mathbb{D}^{\natural} = \langle \mathbb{P}^{\natural}, \sqsubseteq^{\natural}, \bot^{\natural}, \sqcup^{\natural}, \mathsf{assign}^{\natural}[\![\mathsf{x}, \mathsf{A}]\!], \mathsf{test}^{\natural}[\![\mathsf{B}]\!], \overline{\mathsf{test}}^{\natural}[\![\mathsf{B}]\!] \rangle$, the objective is to design an abstract semantics

$$\widehat{\boldsymbol{S}}^{\sharp}[\![\mathsf{S}]\!] \quad \in \quad \mathbb{P}^{\sharp} \to \mathbb{L} \to \mathbb{P}^{\sharp}$$

on an abstract domain $\mathbb{D}^{\sharp} = \langle \mathbb{P}^{\sharp}, \sqsubseteq^{\sharp}, \bot^{\sharp}, \sqcup^{\sharp}, \mathsf{assign}^{\sharp}[\![\mathsf{x}, \mathsf{A}]\!], \mathsf{test}^{\sharp}[\![\mathsf{B}]\!], \overline{\mathsf{test}}^{\sharp}[\![\mathsf{B}]\!] \rangle$ that is sound (and sometimes complete) for a concretization $\gamma \in \mathbb{P}^{\sharp} \xrightarrow{\nearrow} \mathbb{P}^{\natural}$ (or abstraction $\langle \mathbb{P}^{\natural}, \sqsubseteq \rangle \underset{\alpha}{\overset{\gamma}{\leftrightarrows}} \langle \mathbb{P}^{\sharp}, \sqsubseteq^{\sharp} \rangle$) expressing the meaning of abstract properties in terms of concrete properties. Let us extend γ pointwise to $\dot{\gamma} \in (\mathbb{L} \to \mathbb{P}^{\sharp}) \xrightarrow{\nearrow} (\mathbb{L} \to \mathbb{P}^{\natural})$ by $\dot{\gamma}(\overline{\mathscr{F}})\, \ell \triangleq \gamma(\overline{\mathscr{F}}(\ell))$. Then, given $\mathscr{P}_0 \in \mathbb{P}^{\sharp}$, theorem 27.4 shows that the global soundness condition

$$\boldsymbol{S}^{\natural}[\![\mathsf{S}]\!](\gamma(\mathscr{P}_0)) \quad \dot{\sqsubseteq}^{\natural} \quad \dot{\gamma}(\boldsymbol{S}^{\sharp}[\![\mathsf{S}]\!](\mathscr{P}_0))$$

on the semantics directly follows from the local soundness conditions on abstract domains stated in definition 27.1.I. Notice that we have chosen $\mathscr{P}_0 \in \mathbb{P}^{\sharp}$ because if instead we had $\mathscr{P}_0' \in \mathbb{P}^{\natural}$ we would have no way to abstract $\mathscr{P}_0'$ into a parameter of $\boldsymbol{S}^{\sharp}[\![\mathsf{S}]\!]$ with a concretization only (this would be possible in case of a Galois connection; see exercise 27.3).

In case of an abstraction with a Galois retraction $\langle \mathbb{P}^{\natural}, \sqsubseteq^{\natural} \rangle \underset{\alpha}{\overset{\gamma}{\leftrightarrows}} \langle \mathbb{P}^{\sharp}, \sqsubseteq^{\sharp} \rangle$, we define the pointwise extension $\ddot{\alpha} \in (\mathbb{P}^{\natural} \to \mathbb{L} \to \mathbb{P}^{\natural}) \xrightarrow{\nearrow} (\mathbb{P}^{\sharp} \to \mathbb{L} \to \mathbb{P}^{\sharp})$ by $\ddot{\alpha}(\boldsymbol{S}^{\natural})\, \mathscr{P}_0\, \ell \triangleq \alpha(\boldsymbol{S}^{\natural}\, \gamma(\mathscr{P}_0)\, \ell)$. Then, by theorem 27.8, the exact abstractions hypotheses of definition 27.1.II ensure both soundness and completeness.

$$\ddot{\alpha}(\widehat{\boldsymbol{S}}^{\natural}[\![\mathsf{S}]\!]) \quad = \quad \widehat{\boldsymbol{S}}^{\sharp}[\![\mathsf{S}]\!].$$

Exercise 27.3 Assume $\langle \mathbb{P}^{\natural}, \sqsubseteq^{\natural} \rangle \underset{\alpha}{\overset{\gamma}{\leftrightarrows}} \langle \mathbb{P}^{\sharp}, \sqsubseteq^{\sharp} \rangle$. Define $\ddot{\gamma}(\boldsymbol{S}^{\sharp})\, \mathscr{P}_0\, \ell \triangleq \gamma(\boldsymbol{S}^{\sharp}\, (\alpha(\mathscr{P}_0))\, \ell)$. Prove that $\langle \mathbb{P}^{\natural} \to \mathbb{L} \to \mathbb{P}^{\natural}, \ddot{\sqsubseteq}^{\natural} \rangle \underset{\ddot{\alpha}}{\overset{\ddot{\gamma}}{\leftrightarrows}} \langle \mathbb{P}^{\sharp} \to \mathbb{L} \to \mathbb{P}^{\sharp}, \ddot{\sqsubseteq}^{\sharp} \rangle$.

27.3 Approximate Abstraction of the Abstract Interpreter

Theorem 27.4 (soundness of the abstract interpreter) Let $\widehat{\boldsymbol{\mathcal{S}}}^{\texttt{¤}}$ and $\widehat{\boldsymbol{\mathcal{S}}}^{\sharp}$ be structural abstract interpreters for well defined concrete $\mathbb{D}^{\texttt{¤}}$ and abstract domains $\mathbb{D}^{\sharp}$ by definition 21.1 such that $\mathbb{D}^{\sharp}$ is an approximate abstraction of $\mathbb{D}^{\texttt{¤}}$ by definition 27.1.I). Then for all $\mathscr{P}_0 \in \mathbb{P}^{\sharp}$,

$$\boldsymbol{\mathcal{S}}^{\texttt{¤}}[\![\mathsf{S}]\!](\gamma(\mathscr{P}_0)) \;\;\dot{\sqsubseteq}^{\texttt{¤}}\;\; \dot{\gamma}(\boldsymbol{\mathcal{S}}^{\sharp}[\![\mathsf{S}]\!](\mathscr{P}_0))$$

Proof of theorem 27.4 The proof is by structural induction, using definition 27.1.I for basic cases and exercise 18.20 for fixpoint abstraction.

- The abstract semantics $\widehat{\boldsymbol{\mathcal{S}}}^{\sharp}[\![\mathsf{S}]\!]\,\mathscr{P}_0\,\ell = \bot^{\sharp}$ of a statement S at a label $\ell \notin \mathsf{labs}[\![\mathsf{S}]\!]$ not of that statement is empty.

Proof Assume that $\ell \notin \mathsf{labs}[\![\mathsf{S}]\!]$.

$$\begin{aligned}
& \widehat{\boldsymbol{\mathcal{S}}}^{\texttt{¤}}[\![\mathsf{S}]\!](\gamma(\mathscr{P}_0))\ell && \text{≀hypothesis of theorem 27.4≀} \\
=\;& \bot^{\texttt{¤}} && \text{≀(21.3)≀} \\
\sqsubseteq^{\texttt{¤}}\;& \gamma(\bot^{\sharp}) && \text{≀definition of the infimum in definition 21.1≀}
\end{aligned}$$

proving theorem 27.4 by defining $\widehat{\boldsymbol{\mathcal{S}}}^{\sharp}[\![\mathsf{S}]\!](\mathscr{P}_0)\ell \triangleq \bot^{\sharp}$, which is (21.3) for $\mathbb{D}^{\sharp}$. □

- The abstract semantics $\widehat{\boldsymbol{\mathcal{S}}}^{\sharp}[\![\mathsf{P}]\!] \triangleq \widehat{\boldsymbol{\mathcal{S}}}^{\sharp}[\![\mathsf{Sl}]\!]$ of a program $\mathsf{P} ::= \mathsf{Sl}\;\ell'$ is that of its statement list Sl.

Proof

$$\begin{aligned}
& \widehat{\boldsymbol{\mathcal{S}}}^{\texttt{¤}}[\![\mathsf{P}]\!](\gamma(\mathscr{P}_0))\ell && \text{≀hypothesis of theorem 27.4≀} \\
=\;& \widehat{\boldsymbol{\mathcal{S}}}^{\texttt{¤}}[\![\mathsf{Sl}]\!](\gamma(\mathscr{P}_0))\ell && \text{≀(21.4)≀} \\
\sqsubseteq^{\texttt{¤}}\;& \gamma(\widehat{\boldsymbol{\mathcal{S}}}^{\sharp}[\![\mathsf{Sl}]\!](\mathscr{P}_0)\ell) && \text{≀structural induction hypothesis for theorem 27.4≀} \\
=\;& \gamma(\widehat{\boldsymbol{\mathcal{S}}}^{\sharp}[\![\mathsf{P}]\!](\mathscr{P}_0)\ell) && \text{≀by defining } \widehat{\boldsymbol{\mathcal{S}}}^{\sharp}[\![\mathsf{P}]\!] \triangleq \widehat{\boldsymbol{\mathcal{S}}}^{\sharp}[\![\mathsf{Sl}]\!] \text{ of the form (21.4)≀} \\
=\;& \dot{\gamma}(\widehat{\boldsymbol{\mathcal{S}}}^{\sharp}[\![\mathsf{P}]\!](\mathscr{P}_0))\ell && \text{≀pointwise≀} \quad \square
\end{aligned}$$

- The abstract semantics of a statement list $\mathsf{Sl} ::= \mathsf{Sl}'\;\mathsf{S}$ collects the values of variables reachable in Sl' and those reachable in S after executing Sl'.

$$\begin{aligned}
\widehat{\boldsymbol{\mathcal{S}}}^{\sharp}[\![\mathsf{Sl}]\!]\mathscr{P}_0\,\ell \;\triangleq\;& (\!\![\, \ell \in \mathsf{labs}[\![\mathsf{Sl}']\!] \setminus \{\mathsf{at}[\![\mathsf{S}]\!]\} \;?\; \widehat{\boldsymbol{\mathcal{S}}}^{\sharp}[\![\mathsf{Sl}']\!]\,\mathscr{P}_0\,\ell \\
& \;[\!]\; \ell \in \mathsf{labs}[\![\mathsf{S}]\!] \;?\; \widehat{\boldsymbol{\mathcal{S}}}^{\sharp}[\![\mathsf{S}]\!](\widehat{\boldsymbol{\mathcal{S}}}^{\sharp}[\![\mathsf{Sl}']\!]\,\mathscr{P}_0\;\mathsf{at}[\![\mathsf{S}]\!])\,\ell \\
& \;:\; \bot^{\sharp} \,]\!\!)
\end{aligned}$$

Proof

$$\begin{aligned}
& \widehat{\boldsymbol{\mathcal{S}}}^{\texttt{¤}}[\![\mathsf{Sl}]\!](\gamma(\mathscr{P}_0))\ell && \text{≀hypothesis of theorem 27.4≀} \\
=\;& (\!\![\, \ell \in \mathsf{labs}[\![\mathsf{Sl}']\!] \setminus \{\mathsf{at}[\![\mathsf{S}]\!]\} \;?\; \widehat{\boldsymbol{\mathcal{S}}}^{\texttt{¤}}[\![\mathsf{Sl}']\!]\,(\gamma(\mathscr{P}_0))\,\ell && \text{≀(21.5)≀} \\
& \;[\!]\; \ell \in \mathsf{labs}[\![\mathsf{S}]\!] \;?\; \widehat{\boldsymbol{\mathcal{S}}}^{\texttt{¤}}[\![\mathsf{S}]\!](\widehat{\boldsymbol{\mathcal{S}}}^{\texttt{¤}}[\![\mathsf{Sl}']\!]\,(\gamma(\mathscr{P}_0))\;\mathsf{at}[\![\mathsf{S}]\!])\,\ell \\
& \;:\; \bot^{\texttt{¤}} \,]\!\!)
\end{aligned}$$

$$
\begin{aligned}
&\mathrel{\dot\sqsubseteq^{\natural}} (\!| \ell \in \mathsf{labs}[\![\mathtt{Sl'}]\!] \setminus \{\mathsf{at}[\![\mathtt{S}]\!]\} \mathrel{?} \gamma(\widehat{\boldsymbol{\mathcal{S}}}^{\sharp}[\![\mathtt{Sl'}]\!]\,(\gamma(\mathscr{P}_0))\,\ell) \\
&\quad [\!] \; \ell \in \mathsf{labs}[\![\mathtt{S}]\!] \mathrel{?} \widehat{\boldsymbol{\mathcal{S}}}^{\natural}[\![\mathtt{S}]\!](\gamma(\widehat{\boldsymbol{\mathcal{S}}}^{\sharp}[\![\mathtt{Sl'}]\!]\,(\mathscr{P}_0)\,\mathsf{at}[\![\mathtt{S}]\!]))\,\ell \\
&\quad : \bot^{\natural} |\!) \\
&\qquad \text{\{structural induction hypothesis of theorem 28.44, } \widehat{\boldsymbol{\mathcal{S}}}^{\natural}[\![\mathtt{S}]\!] \text{ and } \gamma \text{ is increasing,} \\
&\qquad \text{and } \bot^{\natural} \mathrel{\dot\sqsubseteq^{\natural}} \gamma(\bot^{\sharp})\text{\}} \\
&\mathrel{\dot\sqsubseteq^{\natural}} (\!| \ell \in \mathsf{labs}[\![\mathtt{Sl'}]\!] \setminus \{\mathsf{at}[\![\mathtt{S}]\!]\} \mathrel{?} \gamma(\widehat{\boldsymbol{\mathcal{S}}}^{\sharp}[\![\mathtt{Sl'}]\!]\,\mathscr{P}_0\,\ell) \\
&\quad [\!] \; \ell \in \mathsf{labs}[\![\mathtt{S}]\!] \mathrel{?} \gamma(\widehat{\boldsymbol{\mathcal{S}}}^{\sharp}[\![\mathtt{S}]\!](\widehat{\boldsymbol{\mathcal{S}}}^{\sharp}[\![\mathtt{Sl'}]\!]\,\mathscr{P}_0\,\mathsf{at}[\![\mathtt{S}]\!])\,\ell) \\
&\quad : \gamma(\bot^{\sharp}) |\!) \qquad \text{\{structural induction hypothesis of theorem 28.44\}} \\
&= \gamma((\!| \ell \in \mathsf{labs}[\![\mathtt{Sl'}]\!] \setminus \{\mathsf{at}[\![\mathtt{S}]\!]\} \mathrel{?} \widehat{\boldsymbol{\mathcal{S}}}^{\sharp}[\![\mathtt{Sl'}]\!]\,\mathscr{P}_0\,\ell \\
&\quad [\!] \; \ell \in \mathsf{labs}[\![\mathtt{S}]\!] \mathrel{?} \widehat{\boldsymbol{\mathcal{S}}}^{\sharp}[\![\mathtt{S}]\!](\widehat{\boldsymbol{\mathcal{S}}}^{\sharp}[\![\mathtt{Sl'}]\!]\,\mathscr{P}_0\,\mathsf{at}[\![\mathtt{S}]\!])\,\ell \\
&\quad : \bot^{\sharp} |\!)) \qquad \text{\{factoring the } \gamma \text{ over the conditional\}} \\
&= \gamma(\widehat{\boldsymbol{\mathcal{S}}}^{\sharp}[\![\mathtt{Sl}]\!]\mathscr{P}_0\,\ell) \qquad \text{\{by definition of } \widehat{\boldsymbol{\mathcal{S}}}^{\sharp}[\![\mathtt{Sl}]\!]\mathscr{P}_0\,\ell \text{ above, which is of the form (21.5)\}} \\
&= \dot\gamma(\widehat{\boldsymbol{\mathcal{S}}}^{\sharp}[\![\mathtt{Sl}]\!]\mathscr{P}_0)\,\ell \qquad \text{\{pointwise.\}} \quad \square
\end{aligned}
$$

- For the empty statement list $\mathtt{Sl} ::= \epsilon$, $\widehat{\boldsymbol{\mathcal{S}}}^{\sharp}[\![\mathtt{Sl}]\!]\,\mathscr{P}_0\,\ell \triangleq (\!| \ell = \mathsf{at}[\![\mathtt{Sl}]\!] \mathrel{?} \mathscr{P}_0 : \bot^{\sharp} |\!)$

Proof

$$
\begin{aligned}
&\widehat{\boldsymbol{\mathcal{S}}}^{\natural}[\![\mathtt{Sl}]\!]\,(\gamma(\mathscr{P}_0)\ell) \qquad \text{\{hypothesis of theorem 27.4\}} \\
&\mathrel{\dot\sqsubseteq^{\natural}} \gamma((\!| \ell = \mathsf{at}[\![\mathtt{Sl}]\!] \mathrel{?} \mathscr{P}_0 : \bot^{\sharp} |\!)) \qquad \text{\{(21.6), } \bot^{\natural} \mathrel{\dot\sqsubseteq^{\natural}} \gamma(\bot^{\sharp})\text{, and factoring } \gamma\text{\}} \\
&= \dot\gamma(\widehat{\boldsymbol{\mathcal{S}}}^{\sharp}[\![\mathtt{Sl}]\!](\mathscr{P}_0))\ell \qquad \text{\{by definition of } \widehat{\boldsymbol{\mathcal{S}}}^{\sharp}[\![\mathtt{S}]\!]\mathscr{P}_0\,\ell \text{ above, of the form (21.6)\}} \quad \square
\end{aligned}
$$

- The abstract semantics of an assignment statement $\mathtt{S} ::= \mathtt{x} = \mathtt{A}\,;$ is

$$
\widehat{\boldsymbol{\mathcal{S}}}^{\sharp}[\![\mathtt{S}]\!]\,\mathscr{P}_0\,\ell \quad = \quad (\!| \ell = \mathsf{at}[\![\mathtt{S}]\!] \mathrel{?} \mathscr{P}_0 \;[\!]\; \ell = \mathsf{after}[\![\mathtt{S}]\!] \mathrel{?} \mathsf{assign}^{\sharp}[\![\mathtt{x}, \mathtt{A}]\!]\,\mathscr{P}_0 : \bot^{\sharp} |\!)
$$

Proof

$$
\begin{aligned}
&\widehat{\boldsymbol{\mathcal{S}}}^{\natural}[\![\mathtt{S}]\!]\,(\gamma(\mathscr{P}_0))\ell \qquad \text{\{hypothesis of theorem 27.4\}} \\
&= (\!| \ell = \mathsf{at}[\![\mathtt{S}]\!] \mathrel{?} \gamma(\mathscr{P}_0) \;[\!]\; \ell = \mathsf{after}[\![\mathtt{S}]\!] \mathrel{?} \mathsf{assign}^{\natural}[\![\mathtt{x}, \mathtt{A}]\!]\,\gamma(\mathscr{P}_0) : \bar\bot^{\natural} |\!) \qquad \text{\{(21.7)\}} \\
&\mathrel{\dot\sqsubseteq^{\natural}} \gamma((\!| \ell = \mathsf{at}[\![\mathtt{S}]\!] \mathrel{?} \mathscr{P}_0 \;[\!]\; \ell = \mathsf{after}[\![\mathtt{S}]\!] \mathrel{?} \mathsf{assign}^{\sharp}[\![\mathtt{x}, \mathtt{A}]\!]\,\mathscr{P}_0 : \bot^{\sharp} |\!)) \\
&\qquad \text{\{}\bot^{\natural} \mathrel{\dot\sqsubseteq^{\natural}} \gamma(\bot^{\sharp})\text{, definition 27.1.2, and factoring } \gamma \text{ over the conditional\}} \\
&= \dot\gamma(\widehat{\boldsymbol{\mathcal{S}}}^{\sharp}[\![\mathtt{S}]\!](\mathscr{P}_0))\ell \qquad \text{\{by definition of } \widehat{\boldsymbol{\mathcal{S}}}^{\sharp}[\![\mathtt{S}]\!]\mathscr{P}_0\,\ell \text{ above, of the form (21.7)\}} \quad \square
\end{aligned}
$$

- The abstract semantics of a skip statement $\mathtt{S} ::= ;$ is $\widehat{\boldsymbol{\mathcal{S}}}^{\sharp}[\![\mathtt{S}]\!]\,\mathscr{P}_0\,\ell = (\!| \ell \in \{\mathsf{at}[\![\mathtt{S}]\!], \mathsf{after}[\![\mathtt{S}]\!]\} \mathrel{?} \mathscr{P}_0 : \bot^{\sharp} |\!)$.

Proof similar to that of the empty statement. □

- The abstract semantics of a conditional statement $\mathtt{S} ::= \mathtt{if}\,(\mathtt{B})\,\mathtt{S}_t$ is

$$
\begin{aligned}
\widehat{\boldsymbol{\mathcal{S}}}^{\sharp}[\![\mathtt{S}]\!]\,\mathscr{P}_0\,\ell \quad = \quad & (\!| \ell = \mathsf{at}[\![\mathtt{S}]\!] \mathrel{?} \mathscr{P}_0 \\
& [\!] \; \ell \in \mathsf{in}[\![\mathtt{S}_t]\!] \mathrel{?} \widehat{\boldsymbol{\mathcal{S}}}^{\sharp}[\![\mathtt{S}_t]\!]\,(\mathsf{test}^{\sharp}[\![\mathtt{B}]\!]\,\mathscr{P}_0)\,\ell \\
& [\!] \; \ell = \mathsf{after}[\![\mathtt{S}]\!] \mathrel{?} \widehat{\boldsymbol{\mathcal{S}}}^{\sharp}[\![\mathtt{S}_t]\!]\,(\mathsf{test}^{\sharp}[\![\mathtt{B}]\!]\,\mathscr{P}_0)\,\ell \sqcup^{\natural} \overline{\mathsf{test}}^{\sharp}[\![\mathtt{B}]\!]\,\mathscr{P}_0 \\
& : \bot^{\natural} |\!)
\end{aligned}
$$

Proof

$$\begin{aligned}
& \widehat{\mathcal{S}}^{\natural}[\![\mathsf{S}]\!]\,(\gamma(\mathscr{P}_0))\ell && \text{(hypothesis of theorem 27.4)}\\
= \ & (\!|\,\ell = \mathsf{at}[\![\mathsf{S}]\!] \ ?\ \gamma(\mathscr{P}_0)\\
& \ \|\ \ell \in \mathsf{in}[\![\mathsf{S}_t]\!] \ ?\ \widehat{\mathcal{S}}^{\natural}[\![\mathsf{S}_t]\!]\,(\mathsf{test}^{\natural}[\![\mathsf{B}]\!]\,\gamma(\mathscr{P}_0))\,\ell\\
& \ \|\ \ell = \mathsf{after}[\![\mathsf{S}]\!] \ ?\ \widehat{\mathcal{S}}^{\natural}[\![\mathsf{S}_t]\!]\,(\mathsf{test}^{\natural}[\![\mathsf{B}]\!]\,\gamma(\mathscr{P}_0))\,\ell \sqcup^{\natural} \overline{\mathsf{test}}^{\natural}[\![\mathsf{B}]\!]\,\gamma(\mathscr{P}_0)\\
& \ \text{\textdivorce}\ \bot^{\natural}\,|\!) && \text{((21.9))}\\
\sqsubseteq^{\natural} \ & (\!|\,\ell = \mathsf{at}[\![\mathsf{S}]\!] \ ?\ \gamma(\mathscr{P}_0)\\
& \ \|\ \ell \in \mathsf{in}[\![\mathsf{S}_t]\!] \ ?\ \widehat{\mathcal{S}}^{\natural}[\![\mathsf{S}_t]\!]\,(\gamma(\mathsf{test}^{\sharp}[\![\mathsf{B}]\!]\,\mathscr{P}_0))\,\ell\\
& \ \|\ \ell = \mathsf{after}[\![\mathsf{S}]\!] \ ?\ \widehat{\mathcal{S}}^{\natural}[\![\mathsf{S}_t]\!]\,(\gamma(\mathsf{test}^{\sharp}[\![\mathsf{B}]\!]\,\mathscr{P}_0))\,\ell \sqcup^{\natural} \gamma(\overline{\mathsf{test}}^{\sharp}[\![\mathsf{B}]\!]\,\mathscr{P}_0)\\
& \ \text{\textdivorce}\ \gamma(\bot^{\sharp})\,|\!)
\end{aligned}$$

(definition 27.1.I.3, $\widehat{\mathcal{S}}^{\natural}[\![\mathsf{S}_t]\!]$ and $\sqcup^{\natural}$ are increasing, and $\bot^{\natural} \sqsubseteq^{\natural} \gamma(\bot^{\sharp})$)

$$\begin{aligned}
\sqsubseteq^{\natural} \ & (\!|\,\ell = \mathsf{at}[\![\mathsf{S}]\!] \ ?\ \gamma(\mathscr{P}_0)\\
& \ \|\ \ell \in \mathsf{in}[\![\mathsf{S}_t]\!] \ ?\ \gamma(\widehat{\mathcal{S}}^{\sharp}[\![\mathsf{S}_t]\!]\,(\mathsf{test}^{\sharp}[\![\mathsf{B}]\!]\,\mathscr{P}_0)\,\ell)\\
& \ \|\ \ell = \mathsf{after}[\![\mathsf{S}]\!] \ ?\ \gamma(\widehat{\mathcal{S}}^{\sharp}[\![\mathsf{S}_t]\!]\,(\mathsf{test}^{\sharp}[\![\mathsf{B}]\!]\,\mathscr{P}_0)\,\ell) \sqcup^{\natural} \gamma(\overline{\mathsf{test}}^{\sharp}[\![\mathsf{B}]\!]\,\mathscr{P}_0)\\
& \ \text{\textdivorce}\ \gamma(\bot^{\sharp})\,|\!)
\end{aligned}$$

(structural induction hypothesis for the hypothesis of theorem 27.4 and $\sqcup^{\natural}$ is increasing)

$$\begin{aligned}
\sqsubseteq^{\natural} \ & \gamma((\!|\,\ell = \mathsf{at}[\![\mathsf{S}]\!] \ ?\ \mathscr{P}_0\\
& \ \|\ \ell \in \mathsf{in}[\![\mathsf{S}_t]\!] \ ?\ \widehat{\mathcal{S}}^{\sharp}[\![\mathsf{S}_t]\!]\,(\mathsf{test}^{\sharp}[\![\mathsf{B}]\!]\,\mathscr{P}_0)\,\ell\\
& \ \|\ \ell = \mathsf{after}[\![\mathsf{S}]\!] \ ?\ \widehat{\mathcal{S}}^{\sharp}[\![\mathsf{S}_t]\!]\,(\mathsf{test}^{\sharp}[\![\mathsf{B}]\!]\,\mathscr{P}_0)\,\ell \sqcup^{\sharp} \gamma(\overline{\mathsf{test}}^{\sharp}[\![\mathsf{B}]\!]\,\mathscr{P}_0)\\
& \ \text{\textdivorce}\ \bot^{\sharp}\,|\!))
\end{aligned}$$

(γ is increasing so $\gamma(A) \sqcup^{\natural} \gamma(A') \sqsubseteq^{\natural} \gamma(A \sqcup^{\sharp} A')$ and factoring γ out of the conditional)

$$= \ \dot{\gamma}(\widehat{\mathcal{S}}^{\sharp}[\![\mathsf{S}]\!](\gamma(\mathscr{P}_0)))\ell \qquad \text{(by definition of } \widehat{\mathcal{S}}^{\sharp}[\![\mathsf{S}]\!]\mathscr{P}_0\,\ell\text{, of the form (21.9))} \quad \square$$

Exercise 27.5 Prove theorem 27.4 for the conditional statement $\mathsf{S} ::= \mathtt{if}\,(\mathsf{B})\,\mathsf{S}_t$ $\mathtt{else}\ \mathsf{S}_f$ by calculational design.

- For the iteration $\mathsf{S} ::= \mathtt{while}\,\ell\,(\mathsf{B})\,\mathsf{S}_b$, we have

$$\widehat{\mathcal{S}}^{\sharp}[\![\mathsf{S}]\!]\,\mathscr{P}_0 \quad = \quad \mathsf{lfp}^{\dot{\sqsubseteq}^{\sharp}}\,(\mathscr{F}^{\sharp}[\![\mathtt{while}\,\ell\,(\mathsf{B})\,\mathsf{S}_b]\!]\,\mathscr{P}_0)$$

where the transformer $\mathscr{F}^{\sharp}[\![\mathtt{while}\,\ell\,(\mathsf{B})\,\mathsf{S}_b]\!]$ is given by (21.11).

Proof We want to prove that $\widehat{\mathcal{S}}^{\natural}[\![\mathsf{S}]\!]\,\gamma(\mathscr{P}_0) \mathrel{\dot{\sqsubseteq}^{\natural}} \gamma(\widehat{\mathcal{S}}^{\sharp}[\![\mathsf{S}]\!]\,\mathscr{P}_0)$ that is $\forall \ell' \in \mathsf{labs}[\![\mathsf{S}]\!]$, we have $\widehat{\mathcal{S}}^{\natural}[\![\mathsf{S}]\!]\,\gamma(\mathscr{P}_0)\,\ell' \sqsubseteq^{\natural} \gamma(\widehat{\mathcal{S}}^{\sharp}[\![\mathsf{S}]\!]\,\mathscr{P}_0)\,\ell'$ pointwise using the concretization $\gamma \in \mathbb{P}^{\sharp} \xrightarrow{\nearrow} \mathbb{P}^{\natural}$.

We look, by calculational design, for an iteration transformer $\mathscr{F}^{\sharp}[\![\mathtt{while}\,\ell\,(\mathsf{B})\,\mathsf{S}_b]\!]$ such that for all $\ell' \in \mathsf{labs}[\![\mathsf{S}]\!]$, iterates X of $\mathscr{F}^{\natural}[\![\mathtt{while}\,\ell\,(\mathsf{B})\,\mathsf{S}_b]\!]\,\gamma(\mathscr{P}_0)$, we have

$$\mathscr{F}^{\natural}[\![\mathtt{while}\,\ell\,(\mathsf{B})\,\mathsf{S}_b]\!]\,\gamma(\mathscr{P}_0)\,\dot{\gamma}(X) \quad \dot{\sqsubseteq}^{\natural} \quad \dot{\gamma}(\mathscr{F}^{\sharp}[\![\mathtt{while}\,\ell\,(\mathsf{B})\,\mathsf{S}_b]\!]\,\mathscr{P}_0\,X)$$

or equivalently for all $\ell' \in \mathsf{labs}[\![\mathsf{S}]\!]$, we have

$$\mathscr{F}^{\natural}[\![\mathtt{while}\,\ell\,(\mathsf{B})\,\mathsf{S}_b]\!]\,\gamma(\mathscr{P}_0)\,X\,\ell' \quad \sqsubseteq^{\natural} \quad \gamma(\mathscr{F}^{\sharp}[\![\mathtt{while}\,\ell\,(\mathsf{B})\,\mathsf{S}_b]\!]\,\mathscr{P}_0\,\dot{\alpha}(X)\,\ell')$$

We conclude by exercise 18.20 for $\ddot{\sqsubseteq}^{\sharp}$.

The commutation calculational design is as follows:

$$\mathscr{F}^{\natural}[\![\mathtt{while}\,\ell\,(\mathsf{B})\,\mathsf{S}_b]\!]\,\gamma(\mathscr{P}_0)\,\dot{\gamma}(X)\,\ell'$$

$$
\begin{aligned}
&= \llparenthesis\, \ell' = \ell \mathrel{?} \gamma(\mathscr{P}_0) \sqcup^{¤} \widehat{\boldsymbol{\mathcal{S}}}^{¤}[\![\mathsf{S}_b]\!]\,(\mathsf{test}^{¤}[\![\mathsf{B}]\!]\dot{\gamma}(X)(\ell))\,\ell \\
&\quad \| \; \ell' \in \mathsf{in}[\![\mathsf{S}_b]\!] \setminus \{\ell\} \mathrel{?} \widehat{\boldsymbol{\mathcal{S}}}^{¤}[\![\mathsf{S}_b]\!]\,(\mathsf{test}^{¤}[\![\mathsf{B}]\!]\dot{\gamma}(X)(\ell))\,\ell' \\
&\quad \| \; \ell' = \mathsf{after}[\![\mathsf{S}]\!] \mathrel{?} \overline{\mathsf{test}}^{¤}[\![\mathsf{B}]\!]\dot{\gamma}(X)(\ell) \sqcup^{¤} \bigsqcup^{¤}_{\ell'' \in \mathsf{breaks\text{-}of}[\![\mathsf{S}_b]\!]} \widehat{\boldsymbol{\mathcal{S}}}^{¤}[\![\mathsf{S}_b]\!]\,(\mathsf{test}^{¤}[\![\mathsf{B}]\!]\dot{\gamma}(X)(\ell))\,\ell'' \\
&\quad : \bot^{¤} \,\rrparenthesis \qquad \text{\textlbrackdbl(21.11)\textrbrackdbl} \\
&= \llparenthesis\, \ell' = \ell \mathrel{?} \gamma(\mathscr{P}_0) \sqcup^{¤} \widehat{\boldsymbol{\mathcal{S}}}^{¤}[\![\mathsf{S}_b]\!]\,(\mathsf{test}^{¤}[\![\mathsf{B}]\!]\gamma(X(\ell)))\,\ell \\
&\quad \| \; \ell' \in \mathsf{in}[\![\mathsf{S}_b]\!] \setminus \{\ell\} \mathrel{?} \widehat{\boldsymbol{\mathcal{S}}}^{¤}[\![\mathsf{S}_b]\!]\,(\mathsf{test}^{¤}[\![\mathsf{B}]\!]\gamma(X(\ell)))\,\ell' \\
&\quad \| \; \ell' = \mathsf{after}[\![\mathsf{S}]\!] \mathrel{?} \overline{\mathsf{test}}^{¤}[\![\mathsf{B}]\!]\gamma(X(\ell)) \sqcup^{¤} \bigsqcup^{¤}_{\ell'' \in \mathsf{breaks\text{-}of}[\![\mathsf{S}_b]\!]} \widehat{\boldsymbol{\mathcal{S}}}^{¤}[\![\mathsf{S}_b]\!]\,(\mathsf{test}^{¤}[\![\mathsf{B}]\!]\gamma(X(\ell)))\,\ell'' \\
&\quad : \bot^{¤} \,\rrparenthesis \qquad \text{(pointwise definition of } \dot{\gamma}\text{)} \\
&\sqsubseteq^{¤} \llparenthesis\, \ell' = \ell \mathrel{?} \gamma(\mathscr{P}_0) \sqcup^{¤} \widehat{\boldsymbol{\mathcal{S}}}^{¤}[\![\mathsf{S}_b]\!]\,\gamma(\mathsf{test}^{\sharp}[\![\mathsf{B}]\!]X(\ell))\,\ell \\
&\quad \| \; \ell' \in \mathsf{in}[\![\mathsf{S}_b]\!] \setminus \{\ell\} \mathrel{?} \widehat{\boldsymbol{\mathcal{S}}}^{¤}[\![\mathsf{S}_b]\!]\,\gamma(\mathsf{test}^{\sharp}[\![\mathsf{B}]\!]X(\ell))\,\ell' \\
&\quad \| \; \ell' = \mathsf{after}[\![\mathsf{S}]\!] \mathrel{?} \gamma(\overline{\mathsf{test}}^{\sharp}[\![\mathsf{B}]\!]X(\ell)) \sqcup^{¤} \bigsqcup^{¤}_{\ell'' \in \mathsf{breaks\text{-}of}[\![\mathsf{S}_b]\!]} \widehat{\boldsymbol{\mathcal{S}}}^{¤}[\![\mathsf{S}_b]\!]\,\gamma(\mathsf{test}^{\sharp}[\![\mathsf{B}]\!]X(\ell))\,\ell'' \\
&\quad : \gamma(\bot^{\sharp}) \,\rrparenthesis \\
&\qquad \text{(definition 27.1.I.3, } \widehat{\boldsymbol{\mathcal{S}}}^{¤}[\![\mathsf{S}_t]\!] \text{ and } \sqcup^{¤} \text{ are increasing, and } \bot^{¤} \sqsubseteq^{¤} \gamma(\bot^{\sharp})\text{)} \\
&\sqsubseteq^{¤} \llparenthesis\, \ell' = \ell \mathrel{?} \gamma(\mathscr{P}_0) \sqcup^{¤} \gamma(\widehat{\boldsymbol{\mathcal{S}}}^{\sharp}[\![\mathsf{S}_b]\!]\,\mathsf{test}^{\sharp}[\![\mathsf{B}]\!]X(\ell)\,\ell) \\
&\quad \| \; \ell' \in \mathsf{in}[\![\mathsf{S}_b]\!] \setminus \{\ell\} \mathrel{?} \gamma(\widehat{\boldsymbol{\mathcal{S}}}^{\sharp}[\![\mathsf{S}_b]\!]\,\mathsf{test}^{\sharp}[\![\mathsf{B}]\!]X(\ell)\,\ell') \\
&\quad \| \; \ell' = \mathsf{after}[\![\mathsf{S}]\!] \mathrel{?} \gamma(\overline{\mathsf{test}}^{\sharp}[\![\mathsf{B}]\!]X(\ell)) \sqcup^{¤} \bigsqcup^{¤}_{\ell'' \in \mathsf{breaks\text{-}of}[\![\mathsf{S}_b]\!]} \gamma(\widehat{\boldsymbol{\mathcal{S}}}^{\sharp}[\![\mathsf{S}_b]\!]\,\mathsf{test}^{\sharp}[\![\mathsf{B}]\!]X(\ell)\,\ell'') \\
&\quad : \gamma(\bot^{\sharp}) \,\rrparenthesis \\
&\qquad \text{(structural induction hypothesis of theorem 28.44)} \\
&\sqsubseteq^{¤} \llparenthesis\, \ell' = \ell \mathrel{?} \gamma(\mathscr{P}_0 \sqcup^{\sharp} \widehat{\boldsymbol{\mathcal{S}}}^{\sharp}[\![\mathsf{S}_b]\!]\,\mathsf{test}^{\sharp}[\![\mathsf{B}]\!]X(\ell)\,\ell) \\
&\quad \| \; \ell' \in \mathsf{in}[\![\mathsf{S}_b]\!] \setminus \{\ell\} \mathrel{?} \gamma(\widehat{\boldsymbol{\mathcal{S}}}^{\sharp}[\![\mathsf{S}_b]\!]\,\mathsf{test}^{\sharp}[\![\mathsf{B}]\!]X(\ell)\,\ell') \\
&\quad \| \; \ell' = \mathsf{after}[\![\mathsf{S}]\!] \mathrel{?} \gamma(\overline{\mathsf{test}}^{\sharp}[\![\mathsf{B}]\!]X(\ell) \sqcup^{\sharp} \bigsqcup^{\sharp}_{\ell'' \in \mathsf{breaks\text{-}of}[\![\mathsf{S}_b]\!]} \widehat{\boldsymbol{\mathcal{S}}}^{\sharp}[\![\mathsf{S}_b]\!]\,\mathsf{test}^{\sharp}[\![\mathsf{B}]\!]X(\ell)\,\ell'') \\
&\quad : \gamma(\bot^{\sharp}) \,\rrparenthesis \\
&\qquad \text{(}\gamma \text{ is increasing so } \bigsqcup^{¤}_{i\in\Delta} \gamma(x_i) \sqsubseteq^{¤} \gamma(\bigsqcup^{\sharp}_{i\in\Delta} x_i)\text{)} \\
&= \gamma(\llparenthesis\, \ell' = \ell \mathrel{?} \mathscr{P}_0 \sqcup^{\sharp} \widehat{\boldsymbol{\mathcal{S}}}^{\sharp}[\![\mathsf{S}_b]\!]\,\mathsf{test}^{\sharp}[\![\mathsf{B}]\!]X(\ell)\,\ell \\
&\quad \| \; \ell' \in \mathsf{in}[\![\mathsf{S}_b]\!] \setminus \{\ell\} \mathrel{?} \widehat{\boldsymbol{\mathcal{S}}}^{\sharp}[\![\mathsf{S}_b]\!]\,\mathsf{test}^{\sharp}[\![\mathsf{B}]\!]X(\ell)\,\ell' \\
&\quad \| \; \ell' = \mathsf{after}[\![\mathsf{S}]\!] \mathrel{?} \overline{\mathsf{test}}^{\sharp}[\![\mathsf{B}]\!]X(\ell) \sqcup^{\sharp} \bigsqcup^{\sharp}_{\ell'' \in \mathsf{breaks\text{-}of}[\![\mathsf{S}_b]\!]} \widehat{\boldsymbol{\mathcal{S}}}^{\sharp}[\![\mathsf{S}_b]\!]\,\mathsf{test}^{\sharp}[\![\mathsf{B}]\!]X(\ell)\,\ell'' \\
&\quad : \bot^{\sharp} \,\rrparenthesis) \\
&\qquad \text{(factoring } \gamma \text{ over the conditional)} \\
&= \gamma(\boldsymbol{\mathscr{F}}^{\sharp}[\![\mathtt{while}\,\ell\,(\mathsf{B})\,\mathsf{S}_b]\!]\,\mathscr{P}_0\,X\,\ell') \quad \text{(by definition (21.11) of } \boldsymbol{\mathscr{F}}^{\sharp}[\![\mathtt{while}\,\ell\,(\mathsf{B})\,\mathsf{S}_b]\!]\text{)} \qquad \square
\end{aligned}
$$

- The abstract semantics $\widehat{\boldsymbol{\mathcal{S}}}^{\sharp}[\![\mathsf{S}]\!]\,\mathscr{P}_0\,\ell = \llparenthesis\, \ell = \mathsf{at}[\![\mathsf{S}]\!] \mathrel{?} \mathscr{P}_0 : \bot^{\sharp} \,\rrparenthesis$ of a break statement $\mathsf{S} ::= \ell\ \mathtt{break}\,;$ is that of its entry point.

Proof similar to that of an empty statement. □

- The abstract semantics $\widehat{\boldsymbol{\mathcal{S}}}^{\sharp}[\![\mathsf{S}]\!] = \widehat{\boldsymbol{\mathcal{S}}}^{\sharp}[\![\mathsf{Sl}]\!]$ of a compound statement $\mathsf{S} ::= \{\,\mathsf{Sl}\,\}$ is that of the statement list Sl.

Proof similar to that of a program. □

Remark 27.6 (soundness of reachability analyses) The soundness of reachability analyses $\widehat{\mathcal{S}}^\sharp$ follows from theorem 27.4 where the concrete semantics is $\mathcal{S}^\natural[\![\mathsf{S}]\!] = \mathcal{S}^{\vec{e}}[\![\mathsf{S}]\!]$. To prove this it is sufficient to check the soundness conditions of definition 27.1 for the abstract domain.

Remark 27.7 (increasingness of the abstract semantics) The proof of theorem 27.4 relies on the hypothesis that $\mathcal{S}^\natural[\![\mathsf{S}]\!]$ is increasing but this hypothesis is never used for $\mathcal{S}^\sharp[\![\mathsf{S}]\!]$. So, the proof remains valid when replacing $\mathcal{S}^\sharp[\![\mathsf{S}]\!]$ by any possibly nonincreasing overapproximation. This is an essential remark for proving the soundness of widening in chapter 34.

27.4 Exact Abstraction of the Abstract Interpreter

Theorem 27.8 (soundness and completeness of the abstract interpreter) Let $\widehat{\mathcal{S}}^\natural$ and $\widehat{\mathcal{S}}^\sharp$ be structural abstract interpreters for well defined concrete $\mathbb{D}^\natural$ and abstract domains $\mathbb{D}^\sharp$ by definition 21.1 such that $\mathbb{P}^\sharp$ is an abstraction of $\mathbb{D}^\natural$ by definition 27.1.II. Then for all $\mathcal{P}_0 \in \mathbb{D}^\sharp$,

$$\ddot{\alpha}(\mathcal{S}^\natural[\![\mathsf{S}]\!])\mathcal{P}_0 \quad = \quad \dot{\alpha}(\mathcal{S}^\natural[\![\mathsf{S}]\!]\,\gamma(\mathcal{P}_0)) \quad = \quad \mathcal{S}^\sharp[\![\mathsf{S}]\!]\,\mathcal{P}_0$$

Proof of theorem 27.8 The proof is by structural induction, using definition 27.1.I for basic cases and the fixpoint abstraction results of theorem 18.23 for (19.16). We consider a few basic cases only, because the proof is quite similar to that of theorem 27.4.

- $\widehat{\mathcal{S}}^\sharp[\![\mathsf{S}]\!]\,\mathcal{P}_0\,\ell = \bot^\sharp$.

Proof Assume that $\ell \notin \mathsf{labs}[\![\mathsf{S}]\!]$.

$$\begin{aligned}
&\dot{\alpha}(\widehat{\mathcal{S}}^\natural[\![\mathsf{S}]\!](\gamma(\mathcal{P}_0)))\ell && \text{(hypothesis of theorem 27.4)}\\
=\;&\alpha(\widehat{\mathcal{S}}^\natural[\![\mathsf{S}]\!](\gamma(\mathcal{P}_0))\ell) && \text{(pointwise definition of } \dot{\alpha}\text{)}\\
=\;&\alpha(\bot^\natural) && \text{((21.3))}\\
=\;&\bot^\sharp && \text{(definition 27.1.II.1 so } \alpha \text{ preserves infimum)}\\
=\;&\widehat{\mathcal{S}}^\sharp[\![\mathsf{S}]\!](\mathcal{P}_0)\ell && \text{((21.3) for } \mathbb{D}^\sharp\text{)} \quad \square
\end{aligned}$$

- The abstract semantics of an assignment statement $\mathsf{S} ::= \mathsf{x} = \mathsf{A}\,;$ is

$$\widehat{\mathcal{S}}^\sharp[\![\mathsf{S}]\!]\,\mathcal{P}_0\,\ell \quad = \quad \alpha(\widehat{\mathcal{S}}^\natural[\![\mathsf{S}]\!]\,\gamma(\mathcal{P}_0)\,\ell)$$

Proof

$$\begin{aligned}
&\alpha(\widehat{\mathcal{S}}^\natural[\![\mathsf{S}]\!]\,(\gamma(\mathcal{P}_0))\ell) && \text{(hypothesis of theorem 27.4)}\\
=\;&\alpha((\!| \ell = \mathsf{at}[\![\mathsf{S}]\!] \;?\; \gamma(\mathcal{P}_0) \;\|\; \ell = \mathsf{after}[\![\mathsf{S}]\!] \;?\; \mathsf{assign}^\natural[\![\mathsf{x}, \mathsf{A}]\!]\,\gamma(\mathcal{P}_0) : \bot^\natural |\!)) && \text{((21.7))}\\
=\;&(\!| \ell = \mathsf{at}[\![\mathsf{S}]\!] \;?\; \mathcal{P}_0 \;\|\; \ell = \mathsf{after}[\![\mathsf{S}]\!] \;?\; \mathsf{assign}^\sharp[\![\mathsf{x}, \mathsf{A}]\!]\,\mathcal{P}_0 : \bot^\sharp |\!)
\end{aligned}$$

(distributing α over the conditional, $\alpha \circ \gamma$ is the identity in a Galois retraction, commutation hypothesis in definition 27.1.II.2, and $\alpha(\bot^\natural) = \bot^\sharp$ in a Galois connection)

$$= \widehat{\mathcal{S}}^\sharp[\![\mathsf{S}]\!]\,\mathcal{P}_0\,\ell \qquad \text{(by definition of } \widehat{\mathcal{S}}^\sharp[\![\mathsf{S}]\!]\mathcal{P}_0\,\ell \text{ in (21.7))} \quad \square$$

- For the iteration $\mathsf{S} ::= \mathtt{while}\ \ell\ (\mathsf{B})\ \mathsf{S}_b$, we have

$$\widehat{\boldsymbol{\mathcal{S}}}^{\sharp}\llbracket \mathsf{S} \rrbracket\, \mathscr{P}_0 \quad = \quad \mathsf{lfp}^{\dot{\sqsubseteq}^{\text{¤}}}\,(\boldsymbol{\mathscr{F}}^{\sharp}\llbracket \mathtt{while}\ \ell\ (\mathsf{B})\ \mathsf{S}_b \rrbracket\, \mathscr{P}_0)$$

where the transformer $\boldsymbol{\mathscr{F}}^{\sharp}\llbracket \mathtt{while}\ \ell\ (\mathsf{B})\ \mathsf{S}_b \rrbracket$ is given by (21.11).

Proof We want to prove that $\dot{\alpha}(\widehat{\boldsymbol{\mathcal{S}}}^{\text{¤}}\llbracket \mathsf{S} \rrbracket\, \gamma(\mathscr{P}_0)) = \widehat{\boldsymbol{\mathcal{S}}}^{\sharp}\llbracket \mathsf{S} \rrbracket\, \mathscr{P}_0$ that is $\forall \ell' \in \mathsf{labs}\llbracket \mathsf{S} \rrbracket$, we have $\alpha(\widehat{\boldsymbol{\mathcal{S}}}^{\text{¤}}\llbracket \mathsf{S} \rrbracket\, \gamma(\mathscr{P}_0)\, \ell') = \widehat{\boldsymbol{\mathcal{S}}}^{\sharp}\llbracket \mathsf{S} \rrbracket\, \mathscr{P}_0\, \ell'$ pointwise using the Galois retraction $\langle \mathbb{P}^{\text{¤}}, \sqsubseteq^{\text{¤}} \rangle \underset{\alpha}{\overset{\gamma}{\leftrightarrows}} \langle \mathbb{P}^{\sharp}, \sqsubseteq^{\sharp} \rangle$.

We look, by calculational design, for an iteration transformer $\boldsymbol{\mathscr{F}}^{\sharp}\llbracket \mathtt{while}\ \ell\ (\mathsf{B})\ \mathsf{S}_b \rrbracket$ such that for all $\ell' \in \mathsf{labs}\llbracket \mathsf{S} \rrbracket$, iterates X of $\boldsymbol{\mathscr{F}}^{\text{¤}}\llbracket \mathtt{while}\ \ell\ (\mathsf{B})\ \mathsf{S}_b \rrbracket\, \gamma(\mathscr{P}_0)$, we have

$$\dot{\alpha}(\boldsymbol{\mathscr{F}}^{\text{¤}}\llbracket \mathtt{while}\ \ell\ (\mathsf{B})\ \mathsf{S}_b \rrbracket\, \gamma(\mathscr{P}_0)\, X) \quad = \quad \boldsymbol{\mathscr{F}}^{\sharp}\llbracket \mathtt{while}\ \ell\ (\mathsf{B})\ \mathsf{S}_b \rrbracket\, \mathscr{P}_0\, \dot{\alpha}(X)$$

or equivalently for all $\ell' \in \mathsf{labs}\llbracket \mathsf{S} \rrbracket$, we have

$$\alpha(\boldsymbol{\mathscr{F}}^{\text{¤}}\llbracket \mathtt{while}\ \ell\ (\mathsf{B})\ \mathsf{S}_b \rrbracket\, \gamma(\mathscr{P}_0)\, X\, \ell') \quad = \quad \boldsymbol{\mathscr{F}}^{\sharp}\llbracket \mathtt{while}\ \ell\ (\mathsf{B})\ \mathsf{S}_b \rrbracket\, \mathscr{P}_0\, \dot{\alpha}(X)\, \ell'$$

We conclude by theorem 18.23. The commutation calculational design is as follows:

$$\begin{aligned}
& \alpha(\boldsymbol{\mathscr{F}}^{\text{¤}}\llbracket \mathtt{while}\ \ell\ (\mathsf{B})\ \mathsf{S}_b \rrbracket\, \gamma(\mathscr{P}_0)\, X\, \ell') \\
= {} & \alpha(\llparenthesis\, \ell' = \ell \mathrel{?} \gamma(\mathscr{P}_0) \sqcup^{\text{¤}} \widehat{\boldsymbol{\mathcal{S}}}^{\text{¤}}\llbracket \mathsf{S}_b \rrbracket\, (\mathsf{test}^{\text{¤}}\llbracket \mathsf{B} \rrbracket X(\ell))\, \ell \\
& \quad \| \ \ell' \in \mathsf{in}\llbracket \mathsf{S}_b \rrbracket \setminus \{\ell\} \mathrel{?} \widehat{\boldsymbol{\mathcal{S}}}^{\text{¤}}\llbracket \mathsf{S}_b \rrbracket\, (\mathsf{test}^{\text{¤}}\llbracket \mathsf{B} \rrbracket X(\ell))\, \ell' \\
& \quad \| \ \ell' = \mathsf{after}\llbracket \mathsf{S} \rrbracket \mathrel{?} \overline{\mathsf{test}}^{\text{¤}}\llbracket \mathsf{B} \rrbracket X(\ell) \sqcup^{\text{¤}} \bigsqcup^{\text{¤}}_{\ell'' \in \mathsf{breaks\text{-}of}\llbracket \mathsf{S}_b \rrbracket} \widehat{\boldsymbol{\mathcal{S}}}^{\text{¤}}\llbracket \mathsf{S}_b \rrbracket\, (\mathsf{test}^{\text{¤}}\llbracket \mathsf{B} \rrbracket X(\ell))\, \ell'' \\
& \quad \mathrel{:} \bot^{\text{¤}}\, \rrparenthesis) \qquad \wr (21.11) \int \\
= {} & \llparenthesis\, \ell' = \ell \mathrel{?} \mathscr{P}_0 \sqcup^{\sharp} \dot{\alpha}(\widehat{\boldsymbol{\mathcal{S}}}^{\text{¤}}\llbracket \mathsf{S}_b \rrbracket\, (\mathsf{test}^{\text{¤}}\llbracket \mathsf{B} \rrbracket X(\ell)))\, \ell \\
& \quad \| \ \ell' \in \mathsf{in}\llbracket \mathsf{S}_b \rrbracket \setminus \{\ell\} \mathrel{?} \dot{\alpha}(\widehat{\boldsymbol{\mathcal{S}}}^{\text{¤}}\llbracket \mathsf{S}_b \rrbracket\, (\mathsf{test}^{\text{¤}}\llbracket \mathsf{B} \rrbracket X(\ell)))\, \ell' \\
& \quad \| \ \ell' = \mathsf{after}\llbracket \mathsf{S} \rrbracket \mathrel{?} \alpha(\overline{\mathsf{test}}^{\text{¤}}\llbracket \mathsf{B} \rrbracket X(\ell)) \sqcup^{\sharp} \bigsqcup^{\sharp}_{\ell'' \in \mathsf{breaks\text{-}of}\llbracket \mathsf{S}_b \rrbracket} \dot{\alpha}(\widehat{\boldsymbol{\mathcal{S}}}^{\text{¤}}\llbracket \mathsf{S}_b \rrbracket\, (\mathsf{test}^{\text{¤}}\llbracket \mathsf{B} \rrbracket X(\ell)))\, \ell'' \\
& \quad \mathrel{:} \bot^{\sharp}\, \rrparenthesis
\end{aligned}$$

$\wr$distributing α over the conditionals, $\alpha \circ \gamma$ is the identity by definition 27.1.II.1 and exercise 11.50, α preserves joins in definition 27.1.II.1 and the infimum, pointwise definition of $\dot{\alpha}$.$\int$

$$\begin{aligned}
= {} & \llparenthesis\, \ell' = \ell \mathrel{?} \mathscr{P}_0 \sqcup^{\sharp} \widehat{\boldsymbol{\mathcal{S}}}^{\sharp}\llbracket \mathsf{S}_b \rrbracket\, \alpha(\mathsf{test}^{\text{¤}}\llbracket \mathsf{B} \rrbracket X(\ell))\, \ell \\
& \quad \| \ \ell' \in \mathsf{in}\llbracket \mathsf{S}_b \rrbracket \setminus \{\ell\} \mathrel{?} \widehat{\boldsymbol{\mathcal{S}}}^{\sharp}\llbracket \mathsf{S}_b \rrbracket\, \alpha(\mathsf{test}^{\text{¤}}\llbracket \mathsf{B} \rrbracket X(\ell))\, \ell' \\
& \quad \| \ \ell' = \mathsf{after}\llbracket \mathsf{S} \rrbracket \mathrel{?} \alpha(\overline{\mathsf{test}}^{\text{¤}}\llbracket \mathsf{B} \rrbracket X(\ell)) \sqcup^{\sharp} \bigsqcup^{\sharp}_{\ell'' \in \mathsf{breaks\text{-}of}\llbracket \mathsf{S}_b \rrbracket} \widehat{\boldsymbol{\mathcal{S}}}^{\sharp}\llbracket \mathsf{S}_b \rrbracket\, \alpha(\mathsf{test}^{\text{¤}}\llbracket \mathsf{B} \rrbracket X(\ell))\, \ell'' \\
& \quad \mathrel{:} \bot^{\sharp}\, \rrparenthesis \qquad \wr\text{by structural induction hypothesis}\int \\
= {} & \llparenthesis\, \ell' = \ell \mathrel{?} \mathscr{P}_0 \sqcup^{\sharp} \widehat{\boldsymbol{\mathcal{S}}}^{\sharp}\llbracket \mathsf{S}_b \rrbracket\, \mathsf{test}^{\sharp}\llbracket \mathsf{B} \rrbracket(\dot{\alpha}(X)(\ell))\, \ell \\
& \quad \| \ \ell' \in \mathsf{in}\llbracket \mathsf{S}_b \rrbracket \setminus \{\ell\} \mathrel{?} \widehat{\boldsymbol{\mathcal{S}}}^{\sharp}\llbracket \mathsf{S}_b \rrbracket\, \mathsf{test}^{\sharp}\llbracket \mathsf{B} \rrbracket(\dot{\alpha}(X)(\ell))\, \ell' \\
& \quad \| \ \ell' = \mathsf{after}\llbracket \mathsf{S} \rrbracket \mathrel{?} \overline{\mathsf{test}}^{\sharp}\llbracket \mathsf{B} \rrbracket(\dot{\alpha}(X)(\ell)) \sqcup^{\sharp} \bigsqcup^{\sharp}_{\ell'' \in \mathsf{breaks\text{-}of}\llbracket \mathsf{S}_b \rrbracket} \widehat{\boldsymbol{\mathcal{S}}}^{\sharp}\llbracket \mathsf{S}_b \rrbracket\, \mathsf{test}^{\sharp}\llbracket \mathsf{B} \rrbracket(\dot{\alpha}(X)(\ell))\, \ell'' \\
& \quad \mathrel{:} \bot^{\sharp}\, \rrparenthesis \qquad \wr\text{by commutation, definition 27.1.II.3, and pointwise definition of } \dot{\alpha}\int \\
= {} & \boldsymbol{\mathscr{F}}^{\sharp}\llbracket \mathtt{while}\ \ell\ (\mathsf{B})\ \mathsf{S}_b \rrbracket\, \mathscr{P}_0\, \dot{\alpha}(X)\, \ell' \qquad \wr\text{by definition (21.11) of } \boldsymbol{\mathscr{F}}^{\sharp}\llbracket \mathtt{while}\ \ell\ (\mathsf{B})\ \mathsf{S}_b \rrbracket\int \qquad \square
\end{aligned}$$

Exercise 27.9 Complete the missing cases in the proof of theorem 27.8.

27.5 Best Abstraction (Continuing Section 11.12)

Given a domain abstraction of definition 27.1, a concrete property $P \in \mathbb{P}^{¤}$ can be overapproximated by any abstract property $\overline{P} \in \mathbb{P}^{\sharp}$ such that $P \sqsubseteq^{¤} \gamma(\overline{P})$. If the set of abstract overapproximations $\{\overline{P} \in \mathbb{P}^{\sharp} \mid P \sqsubseteq^{¤} \gamma(\overline{P})\}$ of concrete property P has a greatest lower bound then we call it "the best abstraction of the property P in the abstract domain $\mathbb{P}^{\sharp}$."

Example 27.10 For signs, the property $P = \{0\}$ (to be zero) can either be overapproximated by 0 (to be zero), by $\geqslant 0$ (to be positive or zero), by $\leqslant 0$ (to be negative or zero), or by $\top_{\pm}$ (to be integer). <0 would be unsound. 0 (to be zero) is the most precise abstraction (and $\top_{\pm}$ (to be integer), the less precise one).

Example 27.11 Consider the poset of signs

```
  ⊤
 / \
neg pos
```

of exercise 3.2 from [899] where we have added $\top$ to express that the rule of signs is inconclusive like "neg − neg = $\top$." [899] informally defined $\gamma(\text{pos}) \triangleq \{z \in \mathbb{Z} \mid z \geqslant 0\}$ and $\gamma(\text{neq}) \triangleq \{z \in \mathbb{Z} \mid z < 0\}$. Then $P = \emptyset$ (false) has no best abstraction and the classic rule of signs is incorrect. It is correct with $\gamma(\text{neq}) \triangleq \{z \in \mathbb{Z} \mid z \leqslant 0\}$ but then $P = \{0\}$ (to be zero) has no best abstraction. In $(0 - 1) - 0$ is should be neg for the first instance and pos for the second to get (neg−pos)-pos=neg-pos=neg. Trying all possibilities would yield a combinatorial explosion, which is avoided by adding zero to the lattice

```
   ⊤
  / \
neg  pos
  \ /
  zero
```

. Then the best abstraction of $P = \emptyset$ (false) and $P = \{0\}$ (to be zero), both have zero as best abstraction. Of course adding $\bot$ in

```
   ⊤
  / \
neg  pos
  \ /
  zero
   .
   ⊥
```

such that $\gamma(\bot) = \emptyset$ is more precise because one can express that an expression is never evaluated, for example, in dead code.

Best abstractions correspond to Galois connections between the concrete and abstract domains.

Theorem 27.12 Let $\langle \mathbb{P}^{¤}, \sqsubseteq^{¤} \rangle$ be a concrete domain, $\langle \mathbb{P}^{\sharp}, \sqsubseteq^{\sharp} \rangle$ be an abstract domain, and $\langle \mathbb{P}^{¤}, \sqsubseteq^{¤} \rangle \underset{\alpha}{\overset{\gamma}{\leftrightarrows}} \langle \mathbb{P}^{\sharp}, \sqsubseteq^{\sharp} \rangle$. Then $\alpha(P)$ is the best abstraction of P in $\mathbb{P}^{\sharp}$.

Proof By lemma 11.42 so that $\alpha(P) = \sqcap^{\sharp}\{\overline{P} \in \mathbb{P}^{\sharp} \mid P \sqsubseteq^{¤} \gamma(\overline{P})\}$. ☐

Theorem 27.13 Let $\langle \mathbb{P}^{¤}, \sqsubseteq^{¤} \rangle$ be a concrete complete lattice domain and $\langle \mathbb{P}^{\sharp}, \sqsubseteq^{\sharp} \rangle$ be an abstract complete lattice domain with meet-preserving concretization $\gamma \in \mathbb{P}^{\sharp} \xrightarrow{\sqcap} \mathbb{P}^{¤}$. Assume that any concrete property $P \in \mathbb{P}^{¤}$ has a best abstraction in $\mathbb{P}^{\sharp}$ (i.e., $\sqcap^{\sharp}\{\overline{P} \in \mathbb{P}^{\sharp} \mid P \sqsubseteq^{¤} \gamma(\overline{P})\}$ exists in $\mathbb{P}^{\sharp}$). Define $\alpha(P) \triangleq \sqcap^{\sharp}\{\overline{P} \in \mathbb{P}^{\sharp} \mid P \sqsubseteq^{¤} \gamma(\overline{P})\}$. Then $\langle \mathbb{P}^{¤}, \sqsubseteq^{¤} \rangle \underset{\alpha}{\overset{\gamma}{\leftrightarrows}} \langle \mathbb{P}^{\sharp}, \sqsubseteq^{\sharp} \rangle$.

Proof

$$
\begin{aligned}
& P \sqsubseteq^{\natural} \gamma(\overline{P}) \\
\Rightarrow\ & \overline{P} \in \{\overline{P}' \in \mathbb{P}^{\sharp} \mid P \sqsubseteq^{\natural} \gamma(\overline{P}')\} && \text{(definition of } \in\text{)} \\
\Rightarrow\ & \sqcap^{\sharp}\{\overline{P}' \in \mathbb{P}^{\sharp} \mid P \sqsubseteq^{\natural} \gamma(\overline{P}')\} \sqsubseteq^{\sharp} \overline{P} && \text{(definition of the glb } \sqcap^{\sharp}\text{)} \\
\Rightarrow\ & \alpha(P) \sqsubseteq^{\sharp} \overline{P} && \text{(definition of } \alpha\text{)}
\end{aligned}
$$

$$
\begin{aligned}
& \alpha(P) \sqsubseteq^{\sharp} \overline{P} \\
\Rightarrow\ & \gamma(\alpha(P)) \sqsubseteq^{\natural} \gamma(\overline{P}) && \text{(}\gamma\text{ preserves meets, so is increasing)} \\
\Rightarrow\ & \gamma(\sqcap^{\sharp}\{\overline{P}' \in \mathbb{P}^{\sharp} \mid P \sqsubseteq^{\natural} \gamma(\overline{P}')\}) \sqsubseteq^{\natural} \gamma(\overline{P}) && \text{(definition of } \alpha\text{)} \\
\Rightarrow\ & \sqcap^{\natural}\{\gamma(\overline{P}') \in \mathbb{P}^{\sharp} \mid P \sqsubseteq^{\natural} \gamma(\overline{P}')\}) \sqsubseteq^{\natural} \gamma(\overline{P}) && \text{(}\gamma\text{ preserves meets)} \\
\Rightarrow\ & P \sqsubseteq^{\natural} \gamma(\overline{P}) && \text{(definition of the glb)} \quad \square
\end{aligned}
$$

Exercise 27.14 Provide a simple example showing that theorem 27.12 does not hold when some meets are missing in the concrete domain $\mathbb{P}^{\natural}$.

Exercise 27.15 With the hypotheses of theorem 27.13, prove that $\mathbb{P}^{\sharp}$ is closed by glb (including for the empty set so $\mathbb{P}^{\sharp}$ is a so-called *Moore family*).

Exercise 27.16 With the hypotheses of theorem 27.13, prove that $\alpha \circ \gamma$ is an upper closure operator (increasing, extensive, and idempotent).

Exercise 27.17 With the hypotheses of theorem 27.13, prove that $\gamma \circ \alpha$ is a lower closure operator (increasing, reductive, and idempotent).

27.6 Best Sound (and Complete) Abstract Interpreter

In the case of existence of best abstraction, there is a best way of choosing the approximate transformers of an abstract domain. Recall that we say that f *is better than* g if and only if $f \mathrel{\dot{\sqsubseteq}} g$. If P is a property of functions then f *is the best satisfying* P if and only if $f \in P$ and $\forall g \in P\,.\, f \mathrel{\dot{\sqsubseteq}} g$. Therefore f is the minimum of P

Theorem 27.18 If $\langle \mathbb{P}^{\natural}, \sqsubseteq^{\natural} \rangle \underset{\alpha}{\overset{\gamma}{\leftrightarrows}} \langle \mathbb{P}^{\sharp}, \sqsubseteq^{\sharp} \rangle$ then $F^{\sharp} \triangleq \alpha \circ F \circ \gamma$ is the best abstraction of $F \in \mathbb{P}^{\natural} \xrightarrow{\nearrow} \mathbb{P}^{\natural}$ satisfying the *semicommutation condition* $F \circ \gamma \mathrel{\dot{\sqsubseteq}^{\natural}} \gamma \circ F^{\sharp}$.

Proof of theorem 27.18 By the Galois connection, $\gamma \circ \alpha$ is extensive so $F \circ \gamma \mathrel{\dot{\sqsubseteq}^{\natural}} \gamma \circ \alpha \circ F \circ \gamma \mathrel{\dot{\sqsubseteq}^{\natural}} \gamma \circ \alpha \circ \gamma \circ F^{\sharp} = \gamma \circ F^{\sharp}$, proving that $F^{\sharp} \triangleq \alpha \circ F \circ \gamma$ satisfies the semicommutation condition.

If $F^{\sharp}$ satisfies the semicommutation condition then $F \circ \gamma \mathrel{\dot{\sqsubseteq}^{\natural}} \gamma \circ F^{\sharp}$ so, by definition of a Galois connection, $\alpha \circ F \circ \gamma \mathrel{\dot{\sqsubseteq}^{\sharp}} F^{\sharp}$ proving that $\alpha \circ F \circ \gamma$ is the best abstract transformer satisfying the semicommutation condition. □

Moreover, in the case of existence of a best abstraction, the choice of the best abstraction of an abstract domain is unique.

Theorem 27.19 If $\langle \mathbb{P}^{\natural}, \sqsubseteq^{\natural} \rangle \underset{\alpha}{\overset{\gamma}{\leftrightarrows\!\!\!\!\twoheadrightarrow}} \langle \mathbb{P}^{\sharp}, \sqsubseteq^{\sharp} \rangle$ and $\alpha \circ F = F^{\sharp} \circ \alpha$ then $F^{\sharp} \triangleq \alpha \circ F \circ \gamma$.

Proof of theorem 27.19 If $\alpha \circ F = F^\sharp \circ \alpha$ then $\alpha \circ F \circ \gamma = F^\sharp \circ \alpha \circ \gamma = F^\sharp$ because $\alpha \circ \gamma$ is the identity for Galois retractions. □

Corollary 27.20 (best sound abstract interpreter) Let $\widehat{\mathcal{S}}^{¤}$ and $\widehat{\mathcal{S}}^{\sharp}$ be structural abstract interpreters for well defined concrete $\mathbb{D}^{¤}$ and abstract domains $\mathbb{D}^\sharp$ by definition 21.1 such that $\mathbb{D}^\sharp$ is an abstraction of $\mathbb{D}^{¤}$ by definition 27.1.I where definition 27.1.I.1 is strengthened to $\langle \mathbb{P}^{¤}, \sqsubseteq^{¤} \rangle \underset{\alpha}{\overset{\gamma}{\leftrightarrows}} \langle \mathbb{P}^\sharp, \sqsubseteq^\sharp \rangle$.

If $\alpha \circ \mathsf{assign}^{¤}[\![\mathsf{x}, \mathsf{A}]\!] \circ \gamma$ satisfies definition 27.1.I.2 and $\alpha \circ \mathsf{test}^{¤}[\![\mathsf{B}]\!] \circ \gamma$ as well as $\alpha \circ \overline{\mathsf{test}}^{¤}[\![\mathsf{B}]\!] \circ \gamma$ satisfy definition 27.1.I.3 then $\ddot{\alpha} \circ \mathcal{S}^{¤}[\![\mathsf{S}]\!] \circ \gamma$ is the best sound abstract interpreter, in that,

$$\ddot{\alpha} \circ \mathcal{S}^{¤}[\![\mathsf{S}]\!] \circ \gamma \quad \ddot{\sqsubseteq}^{¤} \quad \mathcal{S}^\sharp[\![\mathsf{S}]\!].$$

Proof of corollary 27.20 By theorem 27.4 and theorem 27.18. □

Corollary 27.21 (best sound and complete abstract interpreter) Let $\widehat{\mathcal{S}}^{¤}$ and $\widehat{\mathcal{S}}^{\sharp}$ be structural abstract interpreters for well defined concrete $\mathbb{D}^{¤}$ and abstract domains $\mathbb{D}^\sharp$ by definition 21.1 such that $\mathbb{D}^\sharp$ is an exact abstraction of $\mathbb{D}^{¤}$ by definition 27.1.II.

If $\alpha \circ \mathsf{assign}^{¤}[\![\mathsf{x}, \mathsf{A}]\!] \circ \gamma$ satisfies definition 27.1.II.2 and $\alpha \circ \mathsf{test}^{¤}[\![\mathsf{B}]\!] \circ \gamma$ as well as $\alpha \circ \overline{\mathsf{test}}^{¤}[\![\mathsf{B}]\!] \circ \gamma$ satisfy definition 27.1.II.3 then

$$\mathcal{S}^\sharp[\![\mathsf{S}]\!] \quad = \quad \ddot{\alpha} \circ \mathcal{S}^{¤}[\![\mathsf{S}]\!] \circ \gamma$$

is the best sound and complete abstract interpreter.

Proof of corollary 27.21 By theorem 27.8 and theorem 27.19 □

27.7 Example: Predicate Abstraction

Predicate abstraction [437] is a simple form of abstract interpretation [80, 228]. It consists of selecting finitely many predicates $\{\varphi_i \mid i \in \Delta\}$ in some logic with an associated automatic theorem prover to form the atomic predicate of the abstract domain and closing by Boolean operations (we choose conjunction $\wedge$ only for this simple presentation[1]). For the sign lattice of section 3.16 that would be $\mathsf{x} \leqslant 0$, $\mathsf{x} \neq 0$, and $\mathsf{x} \geqslant 0$ for all program variables x. All other abstract properties in the lattice are obtained by conjunction, for example, $\mathsf{x} < 0 = \mathsf{x} \leqslant 0 \wedge \mathsf{x} \neq 0$ and $\varnothing = \mathsf{x} \leqslant 0 \wedge \mathsf{x} \neq 0 \wedge \mathsf{x} \geqslant 0$ (i.e., false in the logic). The transformers are designed by querying an automatic theorem prover. $\mathsf{assign}^{¤}[\![\mathsf{x}, \mathsf{A}]\!] \bigwedge_{i \in \Theta} \varphi_i$, $\Theta \subseteq \Delta$, is the conjunction the subset of all φ_j, $j \in \Delta$, for which the prover can prove that $\mathsf{assign}^{¤}[\![\mathsf{x}, \mathsf{A}]\!] \bigwedge_{i \in \Theta} \varphi_i \Rightarrow \varphi_j$ in a bounded amount of time. Otherwise the result is $\bigwedge \varnothing = \mathbb{Z}$ (i.e., true in the logic). The method is similar for the Boolean transformers $\mathsf{test}^{¤}[\![\mathsf{B}]\!]$ and $\overline{\mathsf{test}}^{¤}[\![\mathsf{B}]\!]$. Abstract properties can be encoded as a bit vector so that the fixpoint engine of the model checker can be reused in the abstract interpreter. Convergence of the static analysis is no problem because the lattice is finite (but can be

1. That is, predicates in conjunctive normal form because the $\{\varphi_i \mid i \in \Delta\}$ may be disjunctions.

very large). An example is SLAM [78, 80] and its successors [683] presented as a great success but were ultimately abandoned because of the high failure rate by timeout in an industrial environment. The whole difficulty is to select the atomic predicates of the abstract domain. Theoretically, this amounts to doing a proof of the program and selecting a basis for the invariants [226]. In practice, heuristics are used; most often the basic predicates are selected from the program text. This is very limited because interesting invariants are not usually made explicit in the program text (as shown, for example, by Cartesian arithmetic congruences in chapter 31 or affine equalities in chapter 38), and many predicates extracted for the program may be useless for the program verification. Predicates can be selected from the failures of a SMT solvers, but they are in the abstract domains considered by the SMT solver, without guarantee of being inductive. Another way to select the useful basic predicates is to extract them by monitoring program executions [117, 448] (but the problem of defining what to observe is the same as choosing an abstract domain, and is limited by this arbitrary choice).

27.8 Conclusion

An abstraction of an abstract structural semantics (specified as an instance of the abstract interpreter) is an abstract structural semantics (specified as another instance of the abstract interpreter), obtained by abstraction of the abstract domain. Hence the abstract interpretation problem is reduced to the abstract interpretation of the primitives of an abstract domain, and the abstract interpreter can be reused without needing any other change.

The program verification methods of chapters 25 and 26 for a concrete structural semantics $\widehat{\boldsymbol{\mathcal{S}}}^{\texttt{¤}}[\![\mathsf{S}]\!]$ on a concrete domain $\mathbb{D}^{\texttt{¤}}$ (e.g., the reachability semantics $\widehat{\boldsymbol{\mathcal{S}}}^{\vec{\varrho}}[\![\mathsf{S}]\!]$ on $\wp(\mathbb{E}\mathbb{v}^{\vec{\varrho}})$ of chapter 19) are exactly the same for an abstract structural semantics $\widehat{\boldsymbol{\mathcal{S}}}^{\sharp}[\![\mathsf{S}]\!]$ on an abstract domain $\mathbb{D}^{\sharp}$. But this abstraction may not simplify the program verification because the existence of an invariant in the abstract domain $\mathbb{D}^{\sharp}$ may still be undecidable [711]. Of course one can always simplify the abstract domain but then the program property of interest can no longer be expressed in the simplified abstract domain. So, the remaining problem with which we cope in the following chapters is that proofs for iteration statements require some form of mechanized induction.

The idea of a hierarchy of semantics applies in other scientific domains, for example, systems biology [304, 358].

27.9 Solutions to Selected Exercises

Solution to exercise 27.14 In the following example, c and d have no greatest lower bound.

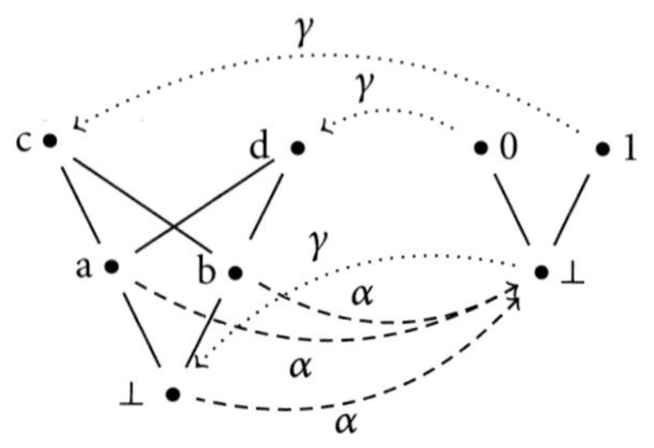

We have $a \sqsubseteq c = \gamma(1)$, $a \sqsubseteq d = \gamma(0)$, $a \not\sqsubseteq \bot = \gamma(\bot)$ so $\alpha(a) = 0 \sqcap 1 = \bot$ but $a \not\sqsubseteq b = \gamma(\bot)$ so $a \not\sqsubseteq \gamma \circ \alpha(a)$, proving by exercise 11.35.3 that α is not the adjoint of γ.

28 Abstract Cartesian Semantics

The view that abstraction consists of replacing concrete values by abstract values in [572, 741, 899, 967] is restrictive and does not allow, for example, relational analyses.

Such abstract value-based static analyses can be formalized as a Cartesian abstraction in which relationships between values of variables are lost.

Contents

28.1 Cartesian Semantics

A Cartesian semantics captures the possible values of variables at each program point but not their relationships. It is named after René Descartes, a French philosopher, mathematician, and scientist, who developed analytic geometry, using algebra to describe geometry by means of a coordinate system [315]. Cartesian analyses are called "attribute independent" in [540, section 12.2].

28.1.1 Cartesian Abstraction

The Cartesian abstraction abstracts a relational property expressing a relation between values of variables as a conjunction of properties on the values of each variable that are unrelated which are independent of the values of the other variables.

For example, the Cartesian abstraction of property $P \triangleq (x = y + 1 \wedge y \in [1, 4])$ is $\dot{\alpha}_{\times}(P) \triangleq (x \in [2, 5] \wedge y \in [1, 4])$ so that the relation $x = y + 1$ is lost. This is René Descartes' projection on the axes of a coordinate system.

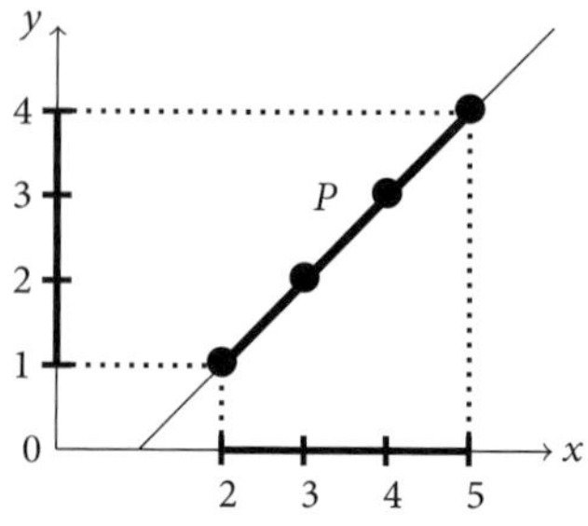

Because relational information is lost, the Cartesian abstraction is mainly applied to assertional invariants (rather than relational invariants, because the relation to the initial values of variables would be lost by the abstraction).

Formally, let the set of environments be $\mathbb{E}\mathbb{v} \triangleq \mathbb{V} \to \mathbb{V}$ where $\mathbb{V}$ are program variables and $\mathbb{V}$ are values. Let $P \in \wp(\mathbb{E}\mathbb{v}) = \wp(\mathbb{V} \to \mathbb{V})$ be a property of the variables. The Cartesian variable property abstraction $\dot{\alpha}_{\vec{\times}}(P)$ collects the possible values of each variable in environments satisfying P.

$$
\begin{aligned}
\dot{\alpha}_{\vec{\times}}(P) &\triangleq \mathsf{x} \in \mathbb{V} \mapsto \{\rho(\mathsf{x}) \mid \rho \in P\} \qquad (28.1)\\
\dot{\gamma}_{\vec{\times}}(\overline{P}) &\triangleq \{\rho \in \mathbb{V} \to \mathbb{V} \mid \forall \mathsf{x} \in \mathbb{V} \,.\, \rho(\mathsf{x}) \in \overline{P}(\mathsf{x})\}
\end{aligned}
$$

So the abstraction is componentwise (or pointwise when encoding the Cartesian product $\prod_{i \in \Delta} x_i$ as a function $i \in \Delta \mapsto x_i$), from the complete lattice $\langle \wp(\mathbb{V} \to \mathbb{V}), \subseteq, \emptyset, \mathbb{V}, \cup, \cap \rangle$ into the Cartesian product $\langle \mathbb{V} \to \wp(\mathbb{V}), \dot{\subseteq}, \dot{\emptyset}, \mathsf{x} \in \mathbb{V} \mapsto \mathbb{V}, \dot{\cup}, \dot{\cap} \rangle$.

Exercise 28.2 (Cartesian abstraction) Prove that $\langle \wp(\mathbb{V} \to \mathbb{V}), \subseteq \rangle \xleftarrow[\dot{\alpha}_{\vec{\times}}]{\dot{\gamma}_{\vec{\times}}} \langle \mathbb{V} \to \wp(\mathbb{V}), \dot{\subseteq} \rangle$.

Remark 28.3 By remark 21.14, $\mathbb{V}$ is a formal parameter of the semantics so the abstract domain $\langle \wp(\mathbb{V} \to \mathbb{V}), \subseteq, \emptyset, \mathbb{V}, \cup, \cap \rangle$ is in fact $\mathbb{V} \mapsto \langle \wp(\mathbb{V} \to \mathbb{V}), \subseteq, \emptyset, \mathbb{V}, \cup, \cap \rangle$ where we have left the formal parameter $\mathbb{V}$ implicit. Similarly, the Cartesian product is $\mathbb{V} \mapsto \langle \mathbb{V} \to \wp(\mathbb{V}), \dot{\subseteq}(\mathbb{V}), \dot{\emptyset}(\mathbb{V}), \mathsf{x} \in \mathbb{V} \mapsto \mathbb{V}, \dot{\cup}(\mathbb{V}), \dot{\cap}(\mathbb{V}) \rangle$ where we leave $\mathbb{V}$ implicit (although it is explicit in the project implementation of exercises 28.46 and 28.47 and following). For each program to be analyzed. the actual parameter bounded to $\mathbb{V}$ will be the set variables of this program. Similarly, because the program properties are attached to program points, the corresponding abstract domains in this chapter 28 and subsequent chapters are implicitly parameterized by $\mathbb{L}$.

Remark 28.4 The Cartesian abstraction is a projection of relational program properties along variables. It provides properties of variables independently of those of the other variables. Other projections are possible, such as, for example, along expressions, including expressions appearing in tests, nonlinear subexpressions appearing multiple times in the program, or symbolic addresses in memory models [691, 693] and [178, ch. 21]. The Cartesian abstraction will then provide independent properties of the selected expressions.

By exercise 11.21, the variable property abstraction can be extended pointwise to variable properties attached program points $\ddot{\alpha}_{\vec{\times}}(P)\ell \triangleq \dot{\alpha}_{\vec{\times}}(P(\ell))$

$$\langle \mathbb{L} \rightarrow \wp(\mathbb{V} \rightarrow \mathbb{V}),\ \dot{\subseteq}\rangle \underset{\ddot{\alpha}_{\vec{\times}}}{\overset{\ddot{\gamma}_{\vec{\times}}}{\leftrightarrows}} \langle \mathbb{L} \rightarrow (\mathbb{V} \rightarrow \wp(\mathbb{V})),\ \ddot{\subseteq}\rangle$$

Then, by theorem 11.78, a further abstraction allows us to abstract transformers of initial properties of variables to variable properties attached to program points into Cartesian transformers of initial Cartesian properties of variables to Cartesian variable properties attached to program points $\vec{\alpha}_{\vec{\times}}(T)\overline{\rho} \triangleq \ddot{\alpha}_{\vec{\times}}(T(\dot{\gamma}_{\vec{\times}}(\overline{\rho})))$.

$$\langle \wp(\mathbb{V} \rightarrow \mathbb{V}) \rightarrow (\mathbb{L} \rightarrow \wp(\mathbb{V} \rightarrow \mathbb{V})),\ \ddot{\subseteq}\rangle \underset{\vec{\alpha}_{\vec{\times}}}{\overset{\vec{\gamma}_{\vec{\times}}}{\leftrightarrows}} \langle (\mathbb{V} \rightarrow \wp(\mathbb{V})) \rightarrow (\mathbb{L} \rightarrow (\mathbb{V} \rightarrow \wp(\mathbb{V}))),\ \dddot{\subseteq}\rangle$$

28.1.2 Cartesian Reachability Semantics

The Cartesian reachability semantics is then

$$\vec{\alpha}_{\vec{\times}}(\boldsymbol{\mathcal{S}}^{\vec{r}}[\![\mathsf{P}]\!]) \quad \dddot{\subseteq} \quad \boldsymbol{\mathcal{S}}^{\vec{\times}}[\![\mathsf{P}]\!] \tag{28.5}$$

(or equality for an exact abstraction in the case of completeness).

28.1.3 Abstract Cartesian Semantics

Because the set $\mathbb{V}$ of values is usually very large, if not infinite, the Cartesian reachability domain $(\mathbb{V} \rightarrow \wp(\mathbb{V})) \rightarrow (\mathbb{L} \rightarrow (\mathbb{V} \rightarrow \wp(\mathbb{V})))$ is hardly machine representable. It is therefore necessary to introduce a further value property abstraction

$$\langle \wp(\mathbb{V}),\ \subseteq\rangle \underset{\alpha_{\times}}{\overset{\gamma_{\times}}{\leftrightarrows}} \langle \mathbb{P}^{\times},\ \sqsubseteq^{\times}\rangle$$

Examples include parity in section 28.8.1, sign in sections 3.20 and 28.8.2, constancy in section 28.8.3, congruences in chapter 31, intervals in chapter 33, and others. This value property abstraction is extended pointwise to Cartesian environment properties $\dot{\alpha}_{\times}(\overline{\rho})\mathsf{x} \triangleq \alpha_{\times}(\overline{\rho}(\mathsf{x}))$

$$\langle \mathbb{V} \rightarrow \wp(\mathbb{V}),\ \dot{\subseteq}\rangle \underset{\dot{\alpha}_{\times}}{\overset{\dot{\gamma}_{\times}}{\leftrightarrows}} \langle \mathbb{V} \rightarrow \mathbb{P}^{\times},\ \dot{\sqsubseteq}^{\times}\rangle$$

and pointwise again to local Cartesian properties $\ddot{\alpha}_{\times}(\dot{\overline{P}})\ell \triangleq \dot{\alpha}_{\times}(\dot{\overline{P}}(\ell))$ attached to program points ℓ

$$\langle \mathbb{L} \rightarrow (\mathbb{V} \rightarrow \wp(\mathbb{V})),\ \ddot{\subseteq}\rangle \underset{\ddot{\alpha}_{\times}}{\overset{\ddot{\gamma}_{\times}}{\leftrightarrows}} \langle \mathbb{L} \rightarrow (\mathbb{V} \rightarrow \mathbb{P}^{\times}),\ \ddot{\sqsubseteq}^{\times}\rangle$$

and functionally to transformers of Cartesian initial conditions $\vec{\alpha}_{\times}(\overline{T}) \triangleq \ddot{\alpha}_{\times} \circ \overline{T} \circ \dot{\gamma}_{\times}$

$$\langle (\mathbb{V} \rightarrow \wp(\mathbb{V})) \rightarrow (\mathbb{L} \rightarrow (\mathbb{V} \rightarrow \wp(\mathbb{V}))),\ \dddot{\subseteq}\rangle \underset{\vec{\alpha}_{\times}}{\overset{\vec{\gamma}_{\times}}{\leftrightarrows}} \langle (\mathbb{V} \rightarrow \mathbb{P}^{\times}) \rightarrow (\mathbb{L} \rightarrow (\mathbb{V} \rightarrow \mathbb{P}^{\times})),\ \vec{\sqsubseteq}^{\times}\rangle$$

By composition, we have

$$\langle \wp(\mathbb{V} \rightarrow \mathbb{V}) \rightarrow (\mathbb{L} \rightarrow \wp(\mathbb{V} \rightarrow \mathbb{V})),\ \dddot{\subseteq}\rangle \underset{\vec{\alpha}_{\times} \circ \vec{\alpha}_{\vec{\times}}}{\overset{\vec{\gamma}_{\vec{\times}} \circ \vec{\gamma}_{\times}}{\leftrightarrows}} \langle (\mathbb{V} \rightarrow \mathbb{P}^{\times}) \rightarrow (\mathbb{L} \rightarrow (\mathbb{V} \rightarrow \mathbb{P}^{\times})),\ \vec{\sqsubseteq}^{\times}\rangle$$

so that the abstract Cartesian reachability semantics $\boldsymbol{\mathcal{S}}^{\times}[\![\mathsf{P}]\!]$ must then be such that

$$\vec{\alpha}_{\times}(\boldsymbol{\mathcal{S}}^{\vec{\times}}[\![\mathsf{P}]\!]) \quad = \quad \vec{\alpha}_{\times}(\vec{\alpha}_{\vec{\times}}(\boldsymbol{\mathcal{S}}^{\vec{r}}[\![\mathsf{P}]\!])) \quad \vec{\sqsubseteq}^{\times} \quad \boldsymbol{\mathcal{S}}^{\times}[\![\mathsf{P}]\!] \tag{28.6}$$

(or equality for an exact abstraction in case of completeness).

It might be that there is no best abstraction (as for example in exercise 3.42). In that case, we would use increasing concretizations only, with $\boldsymbol{\mathcal{S}}^{\vec{r}}[\![\mathsf{P}]\!] \dddot{\subseteq} \vec{\gamma}_{\vec{\times}} \circ \vec{\gamma}_{\times}(\boldsymbol{\mathcal{S}}^{\times}[\![\mathsf{P}]\!])$.

The abstraction $\dot{\alpha}_{\times}$ of $\langle \mathbb{V} \to \wp(\mathbb{V})), \dot{\subseteq}\rangle$ into $\langle \mathbb{V} \to \mathbb{P}^{\times}, \dot{\sqsubseteq}^{\times}\rangle$ will in general use a different abstract domain $\mathbb{P}^{\times}$ for different variables.

Moreover, the Cartesian abstraction may not be pointwise for each variable,[1] but into an abstract domain $\langle \mathbb{P}^{¤}, \dot{\sqsubseteq}^{¤}\rangle$. By exercises 28.2 and 11.57, $\dot{\gamma}_{\times} \circ \dot{\alpha}_{\times}$ is an upper closure operator. This can be used to define Cartesian abstractions as those for which $\dot{\gamma}_{\times} \circ \dot{\alpha}_{\times}$ causes no additional loss of information.

Definition 28.7 (Cartesian abstraction) An abstraction $\gamma \in \mathbb{D}^{¤} \xrightarrow{\nearrow} \wp(\mathbb{V} \mapsto \mathbb{V})$ is *Cartesian* if and only if $\gamma = \dot{\gamma}_{\times} \circ \dot{\alpha}_{\times} \circ \gamma$.

28.1.4 The Structural Cartesian Semantics Is Incomplete

The definition (28.5) of the Cartesian semantics involves the calculation of the concrete assertional semantics $\boldsymbol{\mathcal{S}}^{\vec{r}}[\![\mathsf{P}]\!]$ in the concrete domain $\wp(\mathbb{V} \to \mathbb{V}) \to (\mathbb{L} \to \wp(\mathbb{V} \to \mathbb{V}))$ and then its abstraction $\vec{\alpha}_{\ddot{\times}}(\boldsymbol{\mathcal{S}}^{\vec{r}}[\![\mathsf{P}]\!])$ into the Cartesian domain $(\mathbb{V} \to \wp(\mathbb{V})) \to (\mathbb{L} \to (\mathbb{V} \to \wp(\mathbb{V})))$ by the Cartesian abstraction $\vec{\alpha}_{\ddot{\times}}$.

We look for a structural Cartesian abstract semantics $\widehat{\boldsymbol{\mathcal{S}}}^{\vec{\times}}[\![\mathsf{P}]\!]$ calculated by structural induction on the program in the Cartesian abstract domain $(\mathbb{V} \to \wp(\mathbb{V})) \to (\mathbb{L} \to (\mathbb{V} \to \wp(\mathbb{V})))$ only, without any reference to the concrete semantics. This semantics would be an instance of the abstract interpreter of chapter 21.

The fact that such a sound semantics $\widehat{\boldsymbol{\mathcal{S}}}^{\vec{\times}}[\![\mathsf{P}]\!]$ exists is trivial (take the semantics that always returns the top element of the abstract domain, that is, *I don't know*). The following example shows that this structural Cartesian semantics $\widehat{\boldsymbol{\mathcal{S}}}^{\vec{\times}}[\![\mathsf{P}]\!]$ must be incomplete.

Example 28.8 Consider the reachability semantics of the following program and its Cartesian abstraction.

$\boldsymbol{\mathcal{S}}^{\vec{r}}[\![\mathsf{P}]\!](\dot{\gamma}_{\ddot{\times}}(\overline{\rho}))\ell$	$\dot{\alpha}_{\ddot{\times}}(\boldsymbol{\mathcal{S}}^{\vec{r}}[\![\mathsf{P}]\!](\dot{\gamma}_{\ddot{\times}}(\overline{\rho}))\ell')$
ℓ_1 /* $\dot{\gamma}_{\ddot{\times}}(\overline{\rho}) = 0 \leqslant x \leqslant 2$ */	ℓ'_1 /* $\dot{\gamma}_{\ddot{\times}}(\overline{\rho}) = 0 \leqslant x \leqslant 2$ */
y = x ;	y = x ;
ℓ_2 /* $0 \leqslant x = y \leqslant 2$ */	ℓ'_2 /* $0 \leqslant x \leqslant 2 \wedge 0 \leqslant y \leqslant 2$ */
z = y − x ;	z = y − x ;
ℓ_3 /* $0 \leqslant x = y \leqslant 2 \wedge z = 0$ */	ℓ'_3 /* $0 \leqslant x \leqslant 2 \wedge 0 \leqslant y \leqslant 2 \wedge z = 0$ */

The structural reachability semantics and its Cartesian abstraction both derive that $z = 0$ on program exit. However the structural Cartesian reachability semantics that follows cannot infer this information at ℓ'_3 from the Cartesian invariant at ℓ'_2. It is unknown that $z = x - y$ at ℓ'_2 so that, without this information, it is impossible to know that $z = 0$ after the assignment z = y − x ;. Therefore the structural Cartesian reachability semantics will be the overapproximation

1. For example, one might be interested in the union of the sets of possible values of a group of variables.

$\widehat{\boldsymbol{\mathcal{S}}}^{\times}[\![\mathsf{P}]\!](\overline{\rho})^{\ell}$

$/\star\ \dot{\gamma}_{\vec{\times}}(\overline{\rho}) = 0 \leqslant x \leqslant 2\ \star/$

$y = \mathsf{x}\,;$

$/\star\ 0 \leqslant x \leqslant 2 \wedge 0 \leqslant y \leqslant 2\ \star/$

$z = \mathsf{y} - \mathsf{x}\,;$

$/\star\ 0 \leqslant x \leqslant 2 \wedge 0 \leqslant y \leqslant 2 \wedge -2 \leqslant z \leqslant 2\ \star/$

28.2 Inductive Hierarchy of Abstract Cartesian Semantics

In general, the value abstraction $\mathbb{P}^{\times}$ of value properties $\wp(\mathbb{V})$ is designed by successive abstractions of the form $\dot{\gamma}_{\vec{\times}} \circ \dot{\gamma}_{\times}^{1} \circ \ldots \circ \dot{\gamma}_{\times}^{n} \in (\mathbb{V} \to \mathbb{P}^{\times n}) \xrightarrow{\nearrow} \wp(\mathbb{V} \to \mathbb{V})$ where $\dot{\gamma}_{\times}^{i+1} \in (\mathbb{V} \to \mathbb{P}^{\times i+1}) \xrightarrow{\nearrow} (\mathbb{V} \to \mathbb{P}^{\times i})$. This induces an equivalent hierarchy for the Cartesian abstractions $\vec{\gamma}_{\vec{\times}} \circ \vec{\gamma}_{\times} \circ \vec{\gamma}_{\times}^{1} \circ \cdots \circ \vec{\gamma}_{\times}^{n}$.

In this way, we get a hierarchy of semantics $\widehat{\boldsymbol{\mathcal{S}}}^{\vec{r}}[\![\mathsf{P}]\!] \ddot{\subseteq} \vec{\gamma}_{\vec{\times}}(\widehat{\boldsymbol{\mathcal{S}}}^{\vec{\times}}[\![\mathsf{P}]\!]) \ddot{\subseteq} \vec{\gamma}_{\vec{\times}} \circ \vec{\gamma}_{\times}(\widehat{\boldsymbol{\mathcal{S}}}^{\times}[\![\mathsf{P}]\!]) \ddot{\subseteq} \vec{\gamma}_{\vec{\times}} \circ \vec{\gamma}_{\times} \circ \dot{\gamma}_{\times}^{1}(\widehat{\boldsymbol{\mathcal{S}}}_{1}^{\times}[\![\mathsf{P}]\!]) \ddot{\subseteq} \ldots \ddot{\subseteq} \vec{\gamma}_{\vec{\times}} \circ \vec{\gamma}_{\times} \circ \dot{\gamma}_{\times}^{1} \circ \cdots \circ \vec{\gamma}_{\times}^{n}(\widehat{\boldsymbol{\mathcal{S}}}_{n}^{\times}[\![\mathsf{P}]\!])$.

Because the concretization functions are increasing, this follows from $\widehat{\boldsymbol{\mathcal{S}}}^{\vec{r}}[\![\mathsf{P}]\!] \ddot{\subseteq} \vec{\gamma}_{\vec{\times}}(\widehat{\boldsymbol{\mathcal{S}}}^{\vec{\times}}[\![\mathsf{P}]\!])$, $\widehat{\boldsymbol{\mathcal{S}}}^{\vec{\times}}[\![\mathsf{P}]\!] \vec{\sqsubseteq}^{\vec{\times}} \vec{\gamma}_{\times}(\widehat{\boldsymbol{\mathcal{S}}}^{\times}[\![\mathsf{P}]\!])$, $\widehat{\boldsymbol{\mathcal{S}}}^{\times}[\![\mathsf{P}]\!] \vec{\sqsubseteq}_{1}^{\times} \dot{\gamma}_{\times}^{1}(\widehat{\boldsymbol{\mathcal{S}}}_{1}^{\times}[\![\mathsf{P}]\!])$, …, $\widehat{\boldsymbol{\mathcal{S}}}_{n-1}^{\times}[\![\mathsf{P}]\!] \vec{\sqsubseteq}_{n}^{\times} \vec{\gamma}_{\times}^{n}(\widehat{\boldsymbol{\mathcal{S}}}_{n}^{\times}[\![\mathsf{P}]\!])$.

This hierarchy is designed by induction. For the basis, we must design $\widehat{\boldsymbol{\mathcal{S}}}^{\vec{\times}}[\![\mathsf{P}]\!]$ such that $\widehat{\boldsymbol{\mathcal{S}}}^{\vec{r}}[\![\mathsf{P}]\!] \ddot{\subseteq} \vec{\gamma}_{\vec{\times}}(\widehat{\boldsymbol{\mathcal{S}}}^{\vec{\times}}[\![\mathsf{P}]\!])$. For the inductive abstraction in the inductive hierarchy of abstractions, given a Cartesian semantics $\widehat{\boldsymbol{\mathcal{S}}}^{\times}[\![\mathsf{P}]\!]$ (which is $\widehat{\boldsymbol{\mathcal{S}}}^{\vec{\times}}[\![\mathsf{P}]\!]$ for the basis), we must design an abstract Cartesian semantics $\widehat{\boldsymbol{\mathcal{S}}}^{\mathtt{¤}}[\![\mathsf{P}]\!]$ such that $\widehat{\boldsymbol{\mathcal{S}}}^{\times}[\![\mathsf{P}]\!] \dddot{\sqsubseteq}^{\times} \vec{\gamma}_{\mathtt{¤}}(\widehat{\boldsymbol{\mathcal{S}}}^{\mathtt{¤}})$.

For both the basis and the induction step, we apply theorem 27.4 which shows that it is sufficient to abstract the corresponding abstract domains. Therefore, for the basis, we must design a Cartesian reachability abstract domain $\dddot{\mathbb{D}}^{\vec{\times}}$ abstracting the assertional reachability concrete domain $\ddot{\mathbb{D}}^{\vec{r}}$. For the inductive abstraction in the inductive hierarchy of abstractions, given a Cartesian abstract domain $\dddot{\mathbb{D}}^{\times}$ (which is $\dddot{\mathbb{D}}^{\vec{\times}}$ for the basis), we must design an abstract Cartesian domain $\dddot{\mathbb{D}}^{\mathtt{¤}}$ abstracting the Cartesian abstract domain $\dddot{\mathbb{D}}^{\times}$. In all cases, the domain abstraction should satisfy definition 27.1. Most of the work to prove the soundness of the abstract interpreter has been done in chapter 27, "Abstraction." By definition 27.1, it remains to check the soundness of assignments (definition 27.1.I.2 or definition 27.1.II.2) and tests (definition 27.1.I.3 or definition 27.1.II.3), which constitutes the substance of this chapter 28, "Abstract Cartesian Semantics."

The structure of the abstraction hierarchy is then as follows:

$$
\begin{array}{lll}
\dddot{\mathbb{D}}^{¤} \triangleq \langle((\mathbb{V} \to \mathbb{P}^{¤}) \to (\mathbb{L} \to (\mathbb{V} \to \mathbb{P}^{¤})), \dddot{\sqsubseteq}^{¤}, \dddot{\bot}^{¤}, \dddot{\sqcup}^{¤}, & \widehat{\boldsymbol{\mathcal{S}}}^{¤}[\![\mathsf{P}]\!] & \text{abstract Cartesian semantics} \qquad (28.9)\\
\quad \mathsf{assign}^{¤}[\![\mathsf{x}, \mathsf{A}]\!], \mathsf{test}^{¤}[\![\mathsf{B}]\!], \overline{\mathsf{test}}^{¤}[\![\mathsf{B}]\!]\rangle & \big\downarrow \vec{\gamma}_{¤} & \\
\qquad \vec{\alpha}_{¤} \uparrow\downarrow \vec{\gamma}_{¤} & \overset{\times}{\dddot{\sqsubseteq}} & \\
\dddot{\mathbb{D}}^{\times} \triangleq \langle((\mathbb{V} \to \mathbb{P}^{\times}) \to (\mathbb{L} \to (\mathbb{V} \to \mathbb{P}^{\times})), \dddot{\sqsubseteq}^{\times}, \dddot{\bot}^{\times}, \dddot{\sqcup}^{\times}, & \widehat{\boldsymbol{\mathcal{S}}}^{\times}[\![\mathsf{P}]\!] & \text{Cartesian semantics}\\
\quad \mathsf{assign}^{\times}[\![\mathsf{x}, \mathsf{A}]\!], \mathsf{test}^{\times}[\![\mathsf{B}]\!], \overline{\mathsf{test}}^{\times}[\![\mathsf{B}]\!]\rangle & & \\
\qquad \vec{\alpha}_{\times} \uparrow\downarrow \vec{\gamma}_{\times} & \big\downarrow \vec{\gamma}_{\times} & \\
\qquad \ldots & \ldots & \\
\qquad \uparrow\downarrow & \big\downarrow & \\
 & \dot{\sqsubseteq} & \\
\dddot{\mathbb{D}}^{\vec{\times}} \triangleq \langle((\mathbb{V} \to \wp(\mathbb{V})) \to (\mathbb{L} \to (\mathbb{V} \to \wp(\mathbb{V}))), \ddot{\subseteq}, \ddot{\varnothing}, \ddot{\cup}, & \widehat{\boldsymbol{\mathcal{S}}}^{\vec{\times}}[\![\mathsf{P}]\!] & \text{Cartesian reachability semantics}\\
\quad \mathsf{assign}^{\vec{\times}}[\![\mathsf{x}, \mathsf{A}]\!], \mathsf{test}^{\vec{\times}}[\![\mathsf{B}]\!], \overline{\mathsf{test}}^{\vec{\times}}[\![\mathsf{B}]\!]\rangle & \big\downarrow \vec{\gamma}_{\vec{\times}} & \\
\qquad \vec{\alpha}_{\vec{\times}} \uparrow\downarrow \vec{\gamma}_{\vec{\times}} & \dot{\sqsubseteq} & \\
\dddot{\mathbb{D}}^{\vec{r}} \triangleq \langle\wp(\mathbb{E}\mathbb{v}) \to (\mathbb{L} \to \wp(\mathbb{E}\mathbb{v})), \ddot{\subseteq}, \ddot{\varnothing}, \ddot{\cup}, \mathsf{assign}^{\vec{r}}[\![\mathsf{x}, \mathsf{A}]\!], & \widehat{\boldsymbol{\mathcal{S}}}^{\vec{r}}[\![\mathsf{P}]\!] & \text{assertional reachability semantics}\\
\quad \mathsf{test}^{\vec{r}}[\![\mathsf{B}]\!], \overline{\mathsf{test}}^{\vec{r}}[\![\mathsf{B}]\!]\rangle & &
\end{array}
$$

It is clear that, except for the assertional reachability semantics $\widehat{\boldsymbol{\mathcal{S}}}^{\vec{r}}$, the abstraction of $(\mathbb{V} \to \wp(\mathbb{V})) \to (\mathbb{L} \to (\mathbb{V} \to \wp(\mathbb{V})))$ into $(\mathbb{V} \to \mathbb{P}^{\times}) \to (\mathbb{L} \to (\mathbb{V} \to \mathbb{P}^{\times}))$ and then into $(\mathbb{V} \to \mathbb{P}^{¤}) \to (\mathbb{L} \to (\mathbb{V} \to \mathbb{P}^{¤}))$ can be done by pointwise and functional extension of an abstraction of $\wp(\mathbb{V})$ into $\mathbb{P}^{\times}$ and next to $\mathbb{P}^{¤}$). This is confirmed by the following calculational design showing that all Cartesian semantics in the hierarchy (that is, $\widehat{\boldsymbol{\mathcal{S}}}^{\vec{\times}}$ for the basis and the abstraction of $\widehat{\boldsymbol{\mathcal{S}}}^{\times}$ into $\widehat{\boldsymbol{\mathcal{S}}}^{¤}$ for the induction step) have exactly the same structure and can be defined out of an abstraction of properties of values to their abstract properties.

$$
\begin{array}{ll}
\mathbb{D}^{¤} \triangleq \langle\mathbb{P}^{¤}, \sqsubseteq^{¤}, \bot^{¤}, \top^{¤}, \sqcup^{¤}, \sqcap^{¤}, 1^{¤}, \ominus^{¤}, \ominus^{¤}_{-1}, \oslash^{¤}, \overline{\oslash}^{¤}\rangle & \text{abstract Cartesian value domain} \qquad (28.10)\\
\qquad \alpha_{¤} \uparrow\downarrow \gamma_{¤} & \\
\mathbb{D}^{\times} \triangleq \langle\mathbb{P}^{\times}, \sqsubseteq^{\times}, \bot^{\times}, \top^{\times}, \sqcup^{\times}, \sqcap^{\times}, 1^{\times}, \ominus^{\times}, \ominus^{\times}_{-1}, \oslash^{\times}, \overline{\oslash}^{\times}\rangle & \text{Cartesian value domain}\\
\qquad \alpha_{\times} \uparrow\downarrow \gamma_{\times} & \\
\qquad \ldots & \\
\qquad \uparrow\downarrow & \\
\mathbb{D}^{\vec{\times}} \triangleq \langle\wp(\mathbb{V}), \subseteq, \varnothing, \mathbb{V}, \cup, \cap, \{1\}, \ominus^{\vec{\times}}, \ominus^{\vec{\times}}_{-1}, \oslash^{\vec{\times}}, \overline{\oslash}^{\vec{\times}}\rangle & \text{value collecting domain}
\end{array}
$$

So, for the basis, $\dddot{\mathbb{D}}^{\vec{\times}}$ derives from $\mathbb{D}^{\vec{\times}}$ by pointwise and functional extension, $\dddot{\mathbb{D}}^{\times}$ from $\mathbb{D}^{\times}$, and $\dddot{\mathbb{D}}^{¤}$ from $\mathbb{D}^{¤}$ for the induction step in the hierarchy.

It follows that the generic abstract interpreter of section 21.2 can take as a parameter a generic Cartesian abstract domain $\dot{\mathbb{D}}^{\times} = \langle\mathbb{V} \to \mathbb{P}^{\times}, \dot{\sqsubseteq}^{\times}\rangle$, which is itself parameterized by a Cartesian value domain $\mathbb{D}^{\times} = \langle\mathbb{P}^{\times}, \sqsubseteq^{\times}\rangle$.

We illustrate a small part of this hierarchy (28.10) as follows:

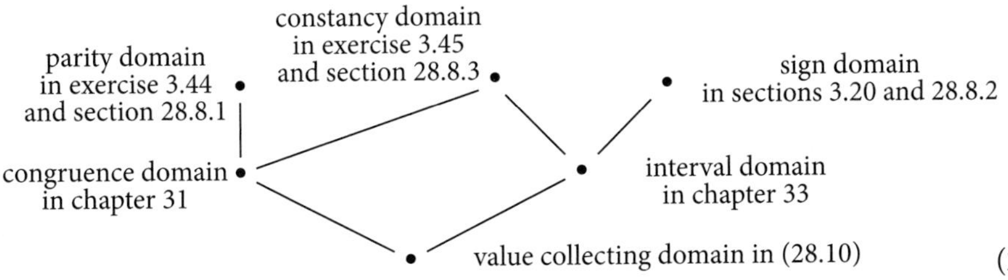

(28.11)

These posets (28.9), (28.10), and (28.11) express the same idea of hierarchy of abstractions as illustrated in section 8.11. However, whereas in section 8.11 the vertices of the hierarchy are semantics (and arcs abstractions), the vertices of (28.11) are abstract domains (and arcs are still abstractions). The reason is that an abstract domain together with the abstract interpreter of section 21.2 completely and uniquely specify a well defined abstract semantics.

28.3 Principle of the Calculational Design of the Cartesian Semantics

Theorem 27.4 shows that to design a structural abstract semantics by abstraction of a concrete semantics, it is sufficient to abstract its abstract domains as in definition 27.1. This consists of designing an abstraction of $\mathsf{assign}[\![\mathsf{x}, \mathsf{A}]\!]$ for all arithmetic expressions A, $\mathsf{test}[\![\mathsf{B}]\!]$ and $\overline{\mathsf{test}}[\![\mathsf{B}]\!]$ for all Boolean expressions B.

28.3.1 Exact Domain Abstraction

In the case definition 27.1.II of an exact abstraction $\langle \mathbb{P}^{\times}, \sqsubseteq^{\times}\rangle \xrightleftharpoons[\alpha]{\gamma} \langle \mathbb{P}^{¤}, \sqsubseteq^{¤}\rangle$, we have to calculate the best abstraction $f^{¤}$ (e.g., $\mathsf{assign}^{¤}[\![\mathsf{x}, \mathsf{A}]\!]$) of a transformer $f^{\times}$ (e.g., $\mathsf{assign}^{\times}[\![\mathsf{x}, \mathsf{A}]\!]$). As shown in corollary 27.21, $f^{¤} = \alpha \circ f^{\times} \circ \gamma$, provided that the α-commutation condition $\alpha \circ f^{\times} = f^{¤} \circ \alpha$ holds. There are essentially two alternatives for the calculational design of $f^{¤}$ from $f^{\times}$.

1. Expand the definitions of α, γ, and the concrete transformer $f^{\times}$ in the term $\alpha \circ f^{\times} \circ \gamma$ and then simplify this term until no reference is made to operations on the concrete domain $\mathbb{P}^{\times}$. The simplified term provides a definition of $f^{¤} = \alpha \circ f^{\times} \circ \gamma$. $f^{¤}$ is therefore increasing when $f^{\times}$ is increasing.
   Then check the commutation condition $\alpha \circ f^{\times} = f^{¤} \circ \alpha$.
2. Expand the definition of α and $f^{\times}$ in the term $\alpha \circ f^{\times}(x)$ while pushing α toward the parameters x, which yields a term of the form $t(\alpha(x))$ satisfying $\alpha \circ f^{\times}(x) = t(\alpha(x))$ so that we can define $f^{¤}(y) = t(y)$ by letting $y = \alpha(x)$.

We prefer the second strategy, which, from experience, requires fewer calculations. The reason is that we have $f^{¤} = \alpha \circ f^{\times} \circ \gamma$ by the commutation condition and α is surjective (see theorem 27.19). Hence we have a guarantee that $f^{¤}$ is the best possible abstraction, thereby avoiding the calculation of $\alpha \circ f^{\times} \circ \gamma$ and needing to calculate only $\alpha \circ f^{\times}$.

An example is the calculational design of the structural forward reachability semantics in section 20.4.

28.3.2 Approximate Domain Abstraction with Best Abstraction

In the case definition 27.1.I of an abstraction $\langle \mathbb{P}^\times, \sqsubseteq^\times \rangle \underset{\alpha}{\overset{\gamma}{\leftrightarrows}} \langle \mathbb{P}^\sharp, \sqsubseteq^\sharp \rangle$ when it is impossible to calculate the term $\alpha \circ f^\times \circ \gamma$ exactly in the abstract only, we have to overapproximate according to corollary 27.20. Again, there are essentially two alternatives.

1. Expand the definitions of α, γ, and the concrete transformer $f^\times$ in the term $\alpha \circ f^\times \circ \gamma$, and then simplify this term by $\sqsubseteq^\sharp$-overapproximations whenever a reference is made to operations on the concrete domain $\mathbb{P}^\times$. The $\sqsubseteq^\sharp$-overapproximated term provides a definition of $f^\sharp$. Then check the semicommutation condition $\alpha \circ f^\times \sqsubseteq^\sharp f^\sharp \circ \alpha$ (which implies that $f^\times \circ \gamma \sqsubseteq^\times \gamma \circ f^\sharp$ when $f^\sharp$ is increasing).
2. Expand the definition of α and $f^\times$ in the term $\alpha \circ f^\times(x)$, using $\sqsubseteq^\times$-overapproximations to push α toward the parameters x, which yields a term of the form $t(\alpha(x))$ satisfying $\alpha \circ f^\times(x) \sqsubseteq^\times t(\alpha(x))$ so that we can define $f^\sharp(y) = t(y)$ by letting $y = \alpha(x)$ (which implies that $f^\times \circ \gamma \sqsubseteq^\times \gamma \circ f^\sharp$ when $f^\sharp$ is increasing).

Again, we prefer the second strategy, which from experience, requires fewer calculations. Of course this strategy also applies to the exact domain abstraction of section 28.3.1 when the term $\alpha \circ f^\times(x)$ cannot be expressed in the abstract only with equality, which requires an overapproximation.

An example is the calculational design of the Cartesian assignment $\mathsf{assign}^\times[\![\mathsf{x}, \mathsf{A}]\!]$ in theorem 28.15 below.

28.3.3 Approximate Domain Abstraction in Absence of Best Abstraction

Otherwise there is no best abstraction and so we are in the case definition 27.1.I with an increasing concretization function $\gamma \in \mathbb{P}^\sharp \xrightarrow{\nearrow} \mathbb{P}^\times$. Knowing $f^\times$ and γ, we must calculate $f^\sharp$ satisfying $f^\times \circ \gamma \sqsubseteq^\sharp \gamma \circ f^\sharp$. We conclude by theorem 27.4.

The calculation strategy is to expand the definition of γ and $f^\times$ in the term $f^\times \circ \gamma(x)$, using $\sqsubseteq^\times$-overapproximations to push $\gamma(x)$ toward the result of the function $f^\times$ which yields a term of the form $\gamma(t(x))$ satisfying $f^\times(x) \circ \gamma \sqsubseteq^\times \gamma(t(x))$ so that we can define $f^\sharp(x) = t(x)$.

An example is the following proof of theorems 28.18 and 28.19.

28.4 Hierarchy of Cartesian Reachability Semantics of Assignment

We first consider the assignment $\mathsf{assign}[\![\mathsf{x}, \mathsf{A}]\!]$ in the hierarchy (28.9). First the basis in section 28.4.2 and then the induction in section 28.4.3.

28.4.1 The Collecting Domain Assignment in the Inductive Hierarchy of Abstractions is the Assertional Reachability Assignment of the Concrete Domain $\ddot{\mathbb{D}}^{\vec{r}}$

For the assertional reachability concrete domain $\ddot{\mathbb{D}}^{\vec{r}}$, we have defined for all $\overline{\rho} \in \wp(\mathbb{V} \to \mathbb{V})$,

$$\mathsf{assign}^{\vec{r}}[\![\mathsf{x}, \mathsf{A}]\!]\,\overline{\rho} \;\triangleq\; \{\rho[\mathsf{x} \leftarrow \mathscr{A}[\![\mathsf{A}]\!]\rho] \mid \rho \in \overline{\rho}\} \tag{19.12}$$

For arithmetic expressions A, we define

$$\mathscr{A}^{\vec{r}}[\![\mathsf{A}]\!]\,\overline{\rho} \;\triangleq\; \mathsf{post}[\mathscr{A}[\![\mathsf{A}]\!]]\,\overline{\rho} \;=\; \{\mathscr{A}[\![\mathsf{A}]\!]\,\rho \mid \rho \in \overline{\rho}\} \tag{28.12}$$

28.4.2 For the Basic Abstraction in the Inductive Hierarchy of Abstractions, the Cartesian Reachability Assignment of the Abstract Domain $\ddot{\mathbb{D}}^{\vec{\times}}$ Abstracts the Assertional Reachability Assignment of the Concrete Domain $\ddot{\mathbb{D}}^{\vec{r}}$

We first consider the Cartesian abstraction of concrete arithmetic expressions in section 28.4.2.1 and then the Cartesian abstraction of concrete assignments in section 28.4.2.2.

28.4.2.1 Calculational design of the Cartesian reachability semantics of arithmetic expressions. Let us define the Cartesian evaluation of arithmetic expressions for $\overline{\rho} \in \mathbb{V} \to \mathbb{P}^{\vec{\times}}$, $\mathbb{P}^{\vec{\times}} = \wp(\mathbb{V})$,

Structural Cartesian reachability semantics of an arithmetic expression A

$$\begin{array}{rcll}
\circledcirc^{\vec{\times}}(\overline{P})\,\overline{\rho} & \triangleq & (\!|\,\overline{\rho} = \dot{\varnothing}\ ?\ \varnothing\ \text{:}\ \overline{P}\,|\!) & \text{smash} \\
\mathscr{A}^{\vec{\times}}[\![1]\!]\,\overline{\rho} & \triangleq & \circledcirc^{\vec{\times}}(1^{\vec{\times}})\,\overline{\rho} \quad \text{where} \quad 1^{\vec{\times}} \triangleq \{1\} & (28.13) \\
\mathscr{A}^{\vec{\times}}[\![\mathsf{x}]\!]\,\overline{\rho} & \triangleq & \circledcirc^{\vec{\times}}(\overline{\rho}(\mathsf{x}))\,\overline{\rho} & \\
\mathscr{A}^{\vec{\times}}[\![\mathsf{A}_1 - \mathsf{A}_2]\!]\,\overline{\rho} & \triangleq & \circledcirc^{\vec{\times}}(\mathscr{A}^{\vec{\times}}[\![\mathsf{A}_1]\!]\,\overline{\rho} \ominus^{\vec{\times}} \mathscr{A}^{\vec{\times}}[\![\mathsf{A}_2]\!]\,\overline{\rho})\,\overline{\rho} & \\
X \ominus^{\vec{\times}} Y & \triangleq & \{x - y \mid x \in X \wedge y \in Y\} &
\end{array}$$

so that

Lemma 28.14 (soundness of the structural Cartesian reachability semantics of an arithmetic expression A)

$$\mathscr{A}^{\vec{r}}[\![\mathsf{A}]\!] \mathrel{\dot{\subseteq}} \mathscr{A}^{\vec{\times}}[\![\mathsf{A}]\!] \circ \dot{\alpha}_{\vec{\times}}$$

Proof of lemma 28.14 The inclusion is trivial when $\overline{\rho} = \varnothing$ (which is equivalent to $\dot{\alpha}_{\vec{\times}}(\overline{\rho}) = \dot{\varnothing}$). Otherwise, by structural induction on arithmetic expressions, for all $\overline{\rho} \in \wp(\mathbb{V} \to \mathbb{V})$.

$$\begin{array}{lll}
- & \mathscr{A}^{\vec{r}}[\![1]\!]\,\overline{\rho} & \\
= & \{\mathscr{A}[\![1]\!]\,\rho \mid \rho \in \overline{\rho}\} & \text{(28.12)} \\
= & \{1\} & \text{definition (3.4) of } \mathscr{A} \text{ and } \overline{\rho} \neq \varnothing \\
= & \mathscr{A}^{\vec{\times}}[\![1]\!]\dot{\alpha}_{\vec{\times}}(\overline{\rho}) & \text{because } \overline{\rho} \neq \varnothing \text{ so } \dot{\alpha}_{\vec{\times}}(\overline{\rho}) \neq \dot{\varnothing} \text{ and (28.13)}
\end{array}$$

$$\begin{array}{lll}
- & \mathscr{A}^{\vec{r}}[\![\mathsf{x}]\!]\,\overline{\rho} & \\
= & \{\mathscr{A}[\![\mathsf{x}]\!]\,\rho \mid \rho \in \overline{\rho}\} & \text{(28.12)} \\
= & \{\rho(\mathsf{x}) \mid \rho \in \overline{\rho}\} & \text{definition (3.4) of } \mathscr{A} \text{ and } \overline{\rho} \neq \varnothing \\
= & \dot{\alpha}_{\vec{\times}}(\overline{\rho})\mathsf{x} & \text{definition (28.1) of } \dot{\alpha}_{\vec{\times}} \\
= & \mathscr{A}^{\vec{\times}}[\![\mathsf{x}]\!]\dot{\alpha}_{\vec{\times}}(\overline{\rho}) & \text{because } \overline{\rho} \neq \varnothing \text{ so } \dot{\alpha}_{\vec{\times}}(\overline{\rho}) \neq \dot{\varnothing} \text{ and (28.13)}
\end{array}$$

$$\begin{array}{lll}
- & \mathscr{A}^{\vec{r}}[\![\mathsf{A}_1 - \mathsf{A}_2]\!]\,\overline{\rho} & \\
= & \{\mathscr{A}[\![\mathsf{A}_1 - \mathsf{A}_2]\!]\,\rho \mid \rho \in \overline{\rho}\} & \text{(28.12)} \\
= & \{\mathscr{A}[\![\mathsf{A}_1]\!]\,\rho - \mathscr{A}[\![\mathsf{A}_2]\!]\,\rho \mid \rho \in \overline{\rho}\} & \text{definition (3.4) of } \mathscr{A}
\end{array}$$

$$\subseteq \{x - y \mid x \in \{\mathscr{A}[\![\mathsf{A}_1]\!]\,\rho \mid \rho \in \overline{\rho}\} \wedge y \in \{\mathscr{A}[\![\mathsf{A}_2]\!]\,\rho \mid \rho \in \overline{\rho}\}\}$$

(losing relationships between variables)

$$\subseteq \{x - y \mid x \in \mathscr{A}^{\vec{\times}}[\![\mathsf{A}_1]\!]\dot{\alpha}_{\vec{\times}}(\overline{\rho}) \wedge y \in \mathscr{A}^{\vec{\times}}[\![\mathsf{A}_2]\!]\dot{\alpha}_{\vec{\times}}(\overline{\rho})\}$$

(structural induction hypothesis and $\overline{\rho} \neq \varnothing$)

$$= \mathscr{A}^{\vec{\times}}[\![\mathsf{A}_1]\!]\dot{\alpha}_{\vec{\times}}(\overline{\rho}) \ominus^{\vec{\times}} \mathscr{A}^{\vec{\times}}[\![\mathsf{A}_2]\!]\dot{\alpha}_{\vec{\times}}(\overline{\rho})$$ (by definition of $\ominus^{\vec{\times}}$ in (28.13))

$$= \mathscr{A}^{\vec{\times}}[\![\mathsf{A}_1 - \mathsf{A}_2]\!]\dot{\alpha}_{\vec{\times}}(\overline{\rho})$$ (because $\overline{\rho} \neq \varnothing$ so $\dot{\alpha}_{\vec{\times}}(\overline{\rho}) \neq \dot{\varnothing}$ and definition (28.13) of $\mathscr{A}^{\vec{\times}}[\![\mathsf{A}_1 - \mathsf{A}_2]\!]$)

□

28.4.2.2 Calculational design of the structural Cartesian reachability semantics of assignment. We can now define the Cartesian abstract semantics assignment based upon the Cartesian evaluation of arithmetic expressions.

Theorem 28.15 (soundness of the Cartesian reachability assignment)

$$\dot{\alpha}_{\vec{\times}} \circ \mathsf{assign}^{\vec{r}}[\![\mathsf{x}, \mathsf{A}]\!] \;\;\dot{\subseteq}\;\; \mathsf{assign}^{\vec{\times}}[\![\mathsf{x}, \mathsf{A}]\!] \circ \dot{\alpha}_{\vec{\times}}$$

where

$$\mathsf{assign}^{\vec{\times}}[\![\mathsf{x}, \mathsf{A}]\!]\,\overline{\rho} \;\triangleq\; \overline{\rho}[\mathsf{x} \leftarrow \mathscr{A}^{\vec{\times}}[\![\mathsf{A}]\!]\overline{\rho}] \tag{28.16}$$

Proof of theorem 28.15

$$\dot{\alpha}_{\vec{\times}} \circ \mathsf{assign}^{\vec{r}}[\![\mathsf{x}, \mathsf{A}]\!]\overline{\rho}$$

$$= \dot{\alpha}_{\vec{\times}}(\{\rho[\mathsf{x} \leftarrow \mathscr{A}[\![\mathsf{A}]\!]\,\rho] \mid \rho \in \overline{\rho}\})$$ (definition (19.12) of $\mathsf{assign}^{\vec{r}}[\![\_, \_]\!]$)

$$= \mathsf{y} \in \mathbb{V} \mapsto \{\rho[\mathsf{x} \leftarrow \mathscr{A}[\![\mathsf{A}]\!]\,\rho](\mathsf{y}) \mid \rho \in \overline{\rho}\}$$ (definition (28.1) of $\dot{\alpha}_{\vec{\times}}$)

$$= \mathsf{y} \in \mathbb{V} \mapsto (\!\![\, \overline{\rho} = \varnothing \;?\; \varnothing \;[\!]\; \mathsf{y} = \mathsf{x} \;?\; \{\mathscr{A}[\![\mathsf{A}]\!]\,\rho \mid \rho \in \overline{\rho}\} \;:\; \{\rho(\mathsf{y}) \mid \rho \in \overline{\rho}\} \,]\!\!)$$

(definition (19.10) of environment assignment)

$$= \mathsf{y} \in \mathbb{V} \mapsto (\!\![\, \overline{\rho} = \varnothing \;?\; \varnothing \;[\!]\; \mathsf{y} = \mathsf{x} \;?\; \{\mathscr{A}[\![\mathsf{A}]\!]\,\rho \mid \rho \in \overline{\rho}\} \;:\; \dot{\alpha}_{\vec{\times}}(\overline{\rho})(\mathsf{y}) \,]\!\!)$$ (definition (28.1) of $\dot{\alpha}_{\vec{\times}}$)

$$\subseteq \mathsf{y} \in \mathbb{V} \mapsto (\!\![\, \overline{\rho} = \varnothing \;?\; \varnothing \;[\!]\; \mathsf{y} = \mathsf{x} \;?\; \mathscr{A}^{\vec{\times}}[\![\mathsf{A}]\!](\dot{\alpha}_{\vec{\times}}(\overline{\rho})) \;:\; \dot{\alpha}_{\vec{\times}}(\overline{\rho})(\mathsf{y}) \,]\!\!)$$ (by lemma 28.14)

$$= (\!\![\, \dot{\alpha}_{\vec{\times}}(\overline{\rho}) = \dot{\varnothing} \;?\; \dot{\varnothing} \;:\; \dot{\alpha}_{\vec{\times}}(\overline{\rho})[\mathsf{x} \leftarrow \mathscr{A}^{\vec{\times}}[\![\mathsf{A}]\!](\dot{\alpha}_{\vec{\times}}(\overline{\rho}))] \,]\!\!)$$

(definition (19.10) of environment assignment and $\overline{\rho} = \varnothing \Leftrightarrow \dot{\alpha}_{\vec{\times}}(\overline{\rho}) = \dot{\varnothing}$)

$$= \dot{\alpha}_{\vec{\times}}(\overline{\rho})[\mathsf{x} \leftarrow \mathscr{A}^{\vec{\times}}[\![\mathsf{A}]\!](\dot{\alpha}_{\vec{\times}}(\overline{\rho}))]$$ (because $\mathscr{A}^{\vec{\times}}[\![\mathsf{A}]\!](\dot{\varnothing}) = \varnothing$ implies $\dot{\varnothing}[\mathsf{x} \leftarrow \mathscr{A}^{\vec{\times}}[\![\mathsf{A}]\!](\dot{\varnothing})] = \dot{\varnothing}$)

$$= \mathsf{assign}^{\vec{\times}}[\![\mathsf{x}, \mathsf{A}]\!](\dot{\alpha}_{\vec{\times}}(\overline{\rho}))$$ (definition of $\mathsf{assign}^{\vec{\times}}[\![\mathsf{x}, \mathsf{A}]\!]\overline{\rho} \triangleq \overline{\rho}[\mathsf{x} \leftarrow \mathscr{A}^{\vec{\times}}[\![\mathsf{A}]\!]\overline{\rho}]$) □

28.4.3 For the Inductive Abstraction in the Inductive Hierarchy of Abstractions, the Assignment of the Abstract Cartesian Domain $\ddot{\mathbb{D}}^{¤}$ Abstracts the Assignment of the Cartesian Abstract Domain $\ddot{\mathbb{D}}^{\times}$

We next consider the induction step for the arithmetic expressions and assignment in the hierarchy $\mathbb{D}^{\vec{\times}} = \langle \wp(\mathbb{V}), \subseteq \rangle, \ldots, \mathbb{D}^{\times} = \langle \mathbb{P}^{\times}, \sqsubseteq^{\times} \rangle, \mathbb{D}^{¤} = \langle \mathbb{P}^{¤}, \sqsubseteq^{¤} \rangle, \ldots$ of abstract domains in (28.9).

An essential remark is that the Cartesian abstract forward semantics $\mathscr{A}^{\vec{\times}}[\![\mathsf{A}]\!], \ldots, \mathscr{A}^{\times}[\![\mathsf{A}]\!], \mathscr{A}^{¤}[\![\mathsf{A}]\!], \ldots$ of an arithmetic expression A and $\mathsf{assign}^{\vec{\times}}[\![\mathsf{x}, \mathsf{A}]\!], \ldots, \mathsf{assign}^{\times}[\![\mathsf{x}, \mathsf{A}]\!]$,

$\mathsf{assign}^{\text{¤}}[\![\mathsf{x}, \mathsf{A}]\!]$, ... of an assignment statement $\mathsf{x} = \mathsf{A}\,;$ all have exactly the same structure given subsequently by (28.17) and (28.20).

The basis to prove this is that (28.13) is an instance of (28.17) and (28.16) is an instance of (28.20).

The rest of this paragraph proves the induction step. We consider an instance $\mathscr{A}^{\times}[\![\mathsf{A}]\!]$ of (28.17) and prove in theorem 28.18 that its abstraction $\mathscr{A}^{\text{¤}}[\![\mathsf{A}]\!]$ is an also instance of (28.17). This requires hypotheses stated in theorem 28.18 that are satisfied for the basis $\mathscr{A}^{\vec{\times}}[\![\mathsf{A}]\!]$ and all its abstractions. Similarly, theorem 28.19 shows that an instance $\mathsf{assign}^{\times}[\![\mathsf{x}, \mathsf{A}]\!]$ of (28.20) (including the basis $\mathsf{assign}^{\vec{\times}}[\![\mathsf{x}, \mathsf{A}]\!]$ in (28.16)) is abstracted in an instance $\mathsf{assign}^{\text{¤}}[\![\mathsf{x}, \mathsf{A}]\!]$ of (28.20).

Therefore the proofs have to consider only the abstraction $\gamma_{\text{¤}} \in \langle \mathbb{P}^{\text{¤}}, \sqsubseteq^{\text{¤}}\rangle \xrightarrow{\nearrow} \langle \mathbb{P}^{\times}, \sqsubseteq^{\times}\rangle$ and its pointwise extension to $\dot{\gamma}_{\text{¤}} \in \langle \mathbb{V} \to \mathbb{P}^{\text{¤}}, \dot{\sqsubseteq}^{\text{¤}}\rangle \xrightarrow{\nearrow} \langle \mathbb{V} \to \mathbb{P}^{\times}, \dot{\sqsubseteq}^{\times}\rangle$.

Structural Cartesian abstract forward semantics of an arithmetic expression A

Define the abstract Cartesian semantics of an expression A and an assignment statement $\mathsf{x} = \mathsf{A}\,;$ as follows ($\overline{P}, \overline{\rho} \in \mathbb{V} \to \mathbb{P}^{\text{¤}}$)

$$
\begin{array}{rcll}
\oslash^{\text{¤}}(\overline{P})\,\overline{\rho} & \triangleq & \llparenthesis\, \overline{\rho} = \dot{\bot}^{\text{¤}} \mathrel{?} \dot{\bot}^{\text{¤}} ⨾ \overline{P}\,\rrparenthesis & \text{if } \gamma_{\text{¤}}(\bot^{\text{¤}}) = \bot^{\times}, \ldots, \gamma_{\text{¤}}(\bot^{\times}) = \varnothing \quad \text{smash} \\
& \triangleq & \overline{P} & \text{otherwise} \\
\mathscr{A}^{\text{¤}}[\![1]\!]\,\overline{\rho} & \triangleq & \oslash^{\text{¤}}(1^{\text{¤}})\,\overline{\rho} & \\
\mathscr{A}^{\text{¤}}[\![\mathsf{x}]\!]\,\overline{\rho} & \triangleq & \oslash^{\text{¤}}(\overline{\rho}(\mathsf{x}))\,\overline{\rho} & \\
\mathscr{A}^{\text{¤}}[\![\mathsf{A}_1 - \mathsf{A}_2]\!]\,\overline{\rho} & \triangleq & \oslash^{\text{¤}}(\mathscr{A}^{\text{¤}}[\![\mathsf{A}_1]\!]\,\overline{\rho} \ominus^{\text{¤}} \mathscr{A}^{\text{¤}}[\![\mathsf{A}_2]\!]\,\overline{\rho})\,\overline{\rho} &
\end{array}
\tag{28.17}
$$

Notice that the smash operator $\oslash^{\text{¤}}$ is defined with respect to the concrete domain $\mathbb{D}^{\vec{\times}} = \langle \wp(\mathbb{V}), \subseteq\rangle$ and that choosing the first alternative is always a sound overapproximation.

Theorem 28.18 (soundness of the structural Cartesian abstract forward semantics of an arithmetic expression A**)**

If $\mathscr{A}^{\times}[\![\mathsf{A}]\!]$ is an instance of (28.17) for $\langle \mathbb{P}^{\times}, \sqsubseteq^{\times}\rangle$ (including (28.13)), $\mathscr{A}^{\text{¤}}[\![\mathsf{A}]\!]$ is an instance of (28.17) for $\langle \mathbb{P}^{\text{¤}}, \sqsubseteq^{\text{¤}}\rangle$, $\gamma_{\text{¤}} \in \langle \mathbb{P}^{\text{¤}}, \sqsubseteq^{\text{¤}}\rangle \xrightarrow{\nearrow} \langle \mathbb{P}^{\times}, \sqsubseteq^{\times}\rangle$ is increasing, $1^{\times} \sqsubseteq^{\times} \gamma_{\text{¤}}(1^{\text{¤}})$, $\ominus^{\times}$ is $\sqsubseteq^{\times}$-increasing argumentwise, and $\forall X, Y \in \mathbb{P}^{\text{¤}} \,.\, \gamma_{\text{¤}}(X) \ominus^{\times} \gamma_{\text{¤}}(Y) \sqsubseteq^{\times} \gamma_{\text{¤}}(X \ominus^{\text{¤}} Y)$ then

$$\mathscr{A}^{\times}[\![\mathsf{A}]\!] \circ \dot{\gamma}_{\text{¤}} \quad \sqsubseteq^{\times} \quad \gamma_{\text{¤}} \circ \mathscr{A}^{\text{¤}}[\![\mathsf{A}]\!]$$

Proof of theorem 28.18 Let $\overline{\rho} \in \mathbb{V} \to \mathbb{P}^{\text{¤}}$. Le us first consider the case when $\gamma_{\text{¤}}(\bot^{\text{¤}}) = \bot^{\times}, \ldots, \gamma_{\text{¤}}(\bot^{\times}) = \varnothing$ so that, necessarily, $\gamma_{\times}(\ldots \gamma_{\text{¤}}(\bot^{\text{¤}}) \ldots) = \varnothing$.

— $\mathscr{A}^{\times}[\![1]\!](\dot{\gamma}_{\text{¤}}(\overline{\rho}))$

$= \llparenthesis\, \dot{\gamma}_{\text{¤}}(\overline{\rho}) = \dot{\bot}^{\times} \mathrel{?} \bot^{\times} ⨾ 1^{\times}\,\rrparenthesis$

(definition (28.17) of $\mathscr{A}^{\times}[\![1]\!]$ (or (28.13) when $\dot{\bot}^{\text{¤}} = \dot{\varnothing}$ and $\bot^{\times} = \varnothing$))

$\sqsubseteq^{\times} \llparenthesis\, \dot{\gamma}_{\text{¤}}(\overline{\rho}) = \dot{\bot}^{\times} \mathrel{?} \gamma_{\text{¤}}(\bot^{\text{¤}}) ⨾ \gamma_{\text{¤}}(1^{\text{¤}})\,\rrparenthesis$

($\gamma_{\text{¤}}(\bot^{\text{¤}}) = \bot^{\times}$ and $1^{\times} \sqsubseteq^{\times} \gamma_{\text{¤}}(1^{\text{¤}})$ by hypothesis (or $\varnothing \subseteq \gamma_{\text{¤}}(\bot^{\text{¤}})$ because $\sqsubseteq^{\times} = \subseteq$ for (28.13)))

$= \gamma_{\text{¤}}(\llparenthesis\, \dot{\gamma}_{\text{¤}}(\overline{\rho}) = \dot{\bot}^{\times} \mathrel{?} \bot^{\text{¤}} ⨾ 1^{\text{¤}}\,\rrparenthesis)$ (factoring out of the conditional)

$= \gamma_{¤}(\llparenthesis\, \dot{\gamma}_{¤}(\overline{\rho}) = \dot{\gamma}_{¤}(\dot{\bot}^{¤}) \mathrel{?} \bot^{¤} \mathrel{:} 1^{¤} \,\rrparenthesis)$ ⟨because $\gamma_{¤}(\bot^{¤}) = \bot^{\times}$ implies $\dot{\gamma}_{¤}(\dot{\bot}^{¤}) = \dot{\bot}^{\times}$ pointwise⟩

$\sqsubseteq^{\times} \gamma_{¤}(\llparenthesis\, \overline{\rho} = \dot{\bot}^{¤} \mathrel{?} \bot^{¤} \mathrel{:} 1^{¤} \,\rrparenthesis)$

⟨because either $\overline{\rho} = \dot{\bot}^{¤}$, which implies $\dot{\gamma}_{¤}(\overline{\rho}) = \dot{\gamma}_{¤}(\dot{\bot}^{¤})$ and $\dot{\bot}^{¤} \sqsubseteq^{\times} \dot{\bot}^{¤}$ or $\overline{\rho} \neq \dot{\bot}^{¤}$ and $\dot{\gamma}_{¤}(\overline{\rho}) = \dot{\gamma}_{¤}(\dot{\bot}^{¤})$ and $\bot^{¤} \sqsubseteq^{\times} 1^{¤}$ or else $\overline{\rho} \neq \dot{\bot}^{¤}$ and $\dot{\gamma}_{¤}(\overline{\rho}) \neq \dot{\gamma}_{¤}(\dot{\bot}^{¤})$ and $1^{¤} \sqsubseteq^{\times} 1^{¤}$⟩

$= \gamma_{¤}(\mathscr{A}^{¤}[\![1]\!](\overline{\rho}))$ ⟨definition (28.17) of $\mathscr{A}^{¤}[\![1]\!]\,\overline{\rho}$⟩

— The proof that $\mathscr{A}^{\times}[\![x]\!] \circ \dot{\gamma}_{¤} \sqsubseteq^{\times} \gamma_{¤} \circ \mathscr{A}^{¤}[\![x]\!]$ is similar.

— $\mathscr{A}^{\times}[\![A_1 - A_2]\!](\dot{\gamma}_{¤}(\overline{\rho}))$

$= \llparenthesis\, \dot{\gamma}_{¤}(\overline{\rho}) = \dot{\bot}^{¤} \mathrel{?} \bot^{\times} \mathrel{:} \mathscr{A}^{¤}[\![A_1]\!]\,(\dot{\gamma}_{¤}(\overline{\rho})) \ominus^{\times} \mathscr{A}^{¤}[\![A_2]\!]\,(\dot{\gamma}_{¤}(\overline{\rho})) \,\rrparenthesis$

⟨definition (28.17) of $\mathscr{A}^{\times}[\![A_1 - A_2]\!]$ (or (28.13) when $\dot{\bot}^{¤} = \dot{\varnothing}$ and $\bot^{\times} = \varnothing$)⟩

$\sqsubseteq^{\times} \llparenthesis\, \dot{\gamma}_{¤}(\overline{\rho}) = \dot{\bot}^{¤} \mathrel{?} \bot^{\times} \mathrel{:} \gamma_{¤}(\mathscr{A}^{¤}[\![A_1]\!]\,\overline{\rho}) \ominus^{\times} \gamma_{¤}(\mathscr{A}^{¤}[\![A_2]\!]\,\overline{\rho}) \,\rrparenthesis$

⟨structural induction hypothesis of theorem 28.18 and $\ominus^{\times}$ is $\sqsubseteq^{\times}$-increasing argumentwise⟩

$\sqsubseteq^{\times} \llparenthesis\, \dot{\gamma}_{¤}(\overline{\rho}) = \dot{\bot}^{¤} \mathrel{?} \bot^{\times} \mathrel{:} \gamma_{¤}(\mathscr{A}^{¤}[\![A_1]\!]\,\overline{\rho} \ominus^{¤} \mathscr{A}^{¤}[\![A_2]\!]\,\overline{\rho}) \,\rrparenthesis$ ⟨hypothesis of theorem 28.18⟩

$\sqsubseteq^{\times} \gamma_{¤}(\llparenthesis\, \overline{\rho} = \dot{\bot}^{¤} \mathrel{?} \bot^{¤} \mathrel{:} \mathscr{A}^{¤}[\![A_1]\!]\,\overline{\rho} \ominus^{¤} \mathscr{A}^{¤}[\![A_2]\!]\,\overline{\rho} \,\rrparenthesis)$ ⟨as above⟩

$= \gamma_{¤}(\mathscr{A}^{¤}[\![A_1 - A_2]\!](\overline{\rho}))$ ⟨definition (28.17) of $\mathscr{A}^{¤}[\![A_1 - A_2]\!]\,\overline{\rho}$⟩

— Soundness in case $\gamma_{\times}(\ldots \gamma_{¤}(\bot^{¤}) \ldots) \neq \varnothing$ follows immediately because we get an overapproximation of $\bot^{¤}$. □

Theorem 28.19 (soundness of the Cartesian abstract forward semantics of an assignment $x = A$ **;)** With the hypotheses of theorem 28.18,

$$\text{assign}^{\times}[\![x, A]\!] \circ \dot{\gamma}_{¤} \;\;\dot{\sqsubseteq}^{\times}\;\; \dot{\gamma}_{¤} \circ \text{assign}^{¤}[\![x, A]\!]$$

where

$$\text{assign}^{¤}[\![x, A]\!]\,\overline{\rho} \;\triangleq\; \overline{\rho}[x \leftarrow \mathscr{A}^{¤}[\![A]\!]\overline{\rho}] \tag{28.20}$$

Proof of theorem 28.19

— $\text{assign}^{\times}[\![x, A]\!](\dot{\gamma}_{¤}(\overline{\rho}))$

$= \dot{\gamma}_{¤}(\overline{\rho})[x \leftarrow \mathscr{A}^{\times}[\![A]\!](\dot{\gamma}_{¤}(\overline{\rho}))]$ ⟨definition (28.20) of $\text{assign}^{\times}[\![x, A]\!]$ (or (28.16) for the basis)⟩

$\dot{\sqsubseteq}^{\times} \dot{\gamma}_{¤}(\overline{\rho})[x \leftarrow \gamma_{¤}(\mathscr{A}^{¤}[\![A]\!](\overline{\rho}))]$

⟨$\mathscr{A}^{\times}[\![A]\!] \circ \dot{\gamma}_{¤} \sqsubseteq^{\times} \gamma_{¤} \circ \mathscr{A}^{¤}[\![A]\!]$ in theorem 28.18 and pointwise definition of $\dot{\sqsubseteq}^{\times}$⟩

$= \dot{\gamma}_{¤}(\overline{\rho}[x \leftarrow \mathscr{A}^{¤}[\![A]\!](\overline{\rho})])$ ⟨definition (19.10) of environment assignment⟩

$= \dot{\gamma}_{¤}(\text{assign}^{¤}[\![x, A]\!]\overline{\rho})$ ⟨definition (28.20) of $\text{assign}^{¤}[\![x, A]\!]\overline{\rho}$⟩ □

Remark 28.21 As shown by the proofs of theorems 28.18 and 28.19, they are valid for $\mathbb{P}^{\times} = \wp(\mathbb{V})$ because (28.13) is an instance of (28.17). Moreover, the hypothesis that $\ominus$ is increasing argumentwise is necessary only in the concrete, that is, when composing abstractions, for $\ominus^{\times}$, which follows from (28.13).

28.5 Hierarchy of Cartesian Accessibility Semantics of Assignment

Before handling tests test and their negations $\overline{\text{test}}$ of the hierarchy (28.9) in the next section 28.6, we address the problem of deducing a condition on variables when knowing a condition on an arithmetic expression over these variables. Section 28.5.1 is for the collecting semantics $\ddot{\mathbb{D}}^{\vec{r}}$ in the (28.9). Section 28.5.2 is for the abstraction basis from $\ddot{\mathbb{D}}^{\vec{r}}$ to $\ddot{\mathbb{D}}^{\vec{\times}}$ in the hierarchy. Section 28.5.3 is for the induction step of abstraction from $\ddot{\mathbb{D}}^{\times}$ (which initially is $\ddot{\mathbb{D}}^{\vec{\times}}$) to a more abstract Cartesian domain $\ddot{\mathbb{D}}^{\texttt{¤}}$.

28.5.1 The Collecting Domain Assignment in the Inductive Hierarchy of Abstractions is the Accessibility Assignment of the Concrete Domain $\ddot{\mathbb{D}}^{\vec{r}}$

From a condition on an expression for a test to be true, we need to infer a condition on the variables of this expression for this test to be true. For example from $1 \leqslant x+1 \leqslant 2$, we want to infer that $0 \leqslant x \leqslant 1$. For that purpose, we define the accessibility semantics of arithmetic expressions.

Accessibility semantics of an arithmetic expression A

$$\mathscr{A}^{-1}[\![\mathsf{A}]\!] \in \wp(\mathbb{V}) \to \wp(\mathcal{V} \to \mathbb{V}) \to \wp(\mathcal{V} \to \mathbb{V}) \tag{28.22}$$
$$\mathscr{A}^{-1}[\![\mathsf{A}]\!]\,\chi\,P \triangleq P \cap \mathsf{pre}[\mathscr{A}[\![\mathsf{A}]\!]]\chi = \{\rho \in P \mid \mathscr{A}[\![\mathsf{A}]\!]\rho \in \chi\}$$

$\mathscr{A}^{-1}[\![\mathsf{A}]\!]\,\chi\,P$ restricts the precondition P to those environments in which the arithmetic expression A may return a value in the set of values χ (considered as a postcondition). $\mathscr{A}^{-1}[\![\mathsf{A}]\!]\,\chi$ is a lower closure operator on $\mathbb{E}\mathrm{v}$.

Example 28.23 Assume the precondition $P = \{\langle x,\, y\rangle \mid 0 \leqslant x \leqslant 6 \wedge 2 \leqslant y \leqslant 7\}$. After the test $0 \leqslant \mathsf{x}+\mathsf{y} \leqslant 5$, the set of possible values of $x+y$ is $\chi = \{z \mid 0 \leqslant z \leqslant 5\}$. We have $\mathscr{A}^{-1}[\![\mathsf{A}]\!]\,\chi\,P = \{\langle x,\, y\rangle \mid 0 \leqslant x \leqslant 3 \wedge 2 \leqslant y \leqslant 5\}$ because otherwise $\max x + \min y > 5$ or $\min x + \max y > 5$.

The accessibility semantics of arithmetic expressions is expressed structurally by the following:

Theorem 28.24 (structural accessibility semantics of an arithmetic expression A)

$$\mathscr{A}^{-1}[\![1]\!]\,\chi\,P = (\!|\, 1 \in \chi \,?\, P \,\mathrel{\raisebox{0.5ex}{$\scriptstyle\circ$}}\, \varnothing \,|\!)$$
$$\mathscr{A}^{-1}[\![\mathsf{x}]\!]\,\chi\,P = \{\rho \in P \mid \rho(\mathsf{x}) \in \chi\}$$
$$\mathscr{A}^{-1}[\![\mathsf{A}_1 - \mathsf{A}_2]\!]\,\chi\,P = \{\rho \in P \mid (\mathscr{A}[\![\mathsf{A}_1]\!]\,\rho - \mathscr{A}[\![\mathsf{A}_2]\!]\,\rho) \in \chi\}$$

Exercise 28.25 Prove theorem 28.24.

28.5.2 For the Basic Abstraction in the Inductive Hierarchy of Abstractions, Calculational Design of the Structural Cartesian Accessibility Backward Semantics of Arithmetic Expressions

We look for a Cartesian overapproximation

$$\mathscr{A}^{\vec{\times}_1}[\![\mathsf{A}]\!] \in \wp(\mathbb{V}) \to (\mathcal{V} \to \wp(\mathbb{V})) \to (\mathcal{V} \to \wp(\mathbb{V}))$$

of $\mathscr{A}^{-1}[\![\mathsf{A}]\!]$.

Given a binary operation $b \in \mathbb{V} \times \mathbb{V} \to \mathbb{V}$ define

$$\begin{aligned} b^{\vec{\times}_1} &\in \wp(\mathbb{V}) \to (\wp(\mathbb{V}) \times \wp(\mathbb{V})) \to (\wp(\mathbb{V}) \times \wp(\mathbb{V})) \qquad (28.26)\\ b^{\vec{\times}_1}\,\chi\,\langle\chi_1,\,\chi_2\rangle &\triangleq \langle\{v_1 \in \chi_1 \mid \exists v_2 \in \chi_2\,.\,(v_1\,b\,v_2) \in \chi\},\\ &\qquad \{v_2 \in \chi_2 \mid \exists v_1 \in \chi_1\,.\,(v_1\,b\,v_2) \in \chi\}\rangle \end{aligned}$$

$b^{\vec{\times}_1}\,\chi$ is a lower closure operator on $\wp(\mathbb{V}) \times \wp(\mathbb{V})$.

Example 28.27 Continuing example 28.23, $+^{\vec{\times}_1}\,[0,5]\,\langle[0,6],\,[2,7]\rangle = \langle[0,3],\,[2,5]\rangle$.

Then, by calculational design where $b^{\vec{\times}_1}$ is denoted $\ominus^{\vec{\times}_1}$ when $b = -$, we get [225]

Cartesian accessibility semantics of an arithmetic expression A

$$\begin{aligned} \mathscr{A}^{\vec{\times}_1}[\![1]\!]\,\chi\,\overline{\rho} &\triangleq (\![\,1 \in \chi \,?\, \overline{\rho} \,:\, \dot{\varnothing}\,]\!) \qquad (\dot{\varnothing} \triangleq \mathsf{x} \in \mathcal{V} \mapsto \varnothing) \qquad (28.28)\\ \mathscr{A}^{\vec{\times}_1}[\![\mathsf{x}]\!]\,\chi\,\overline{\rho} &\triangleq (\![\,\overline{\rho}(\mathsf{x}) \cap \chi \neq \varnothing \,?\, \overline{\rho}[\mathsf{x} \leftarrow \overline{\rho}(\mathsf{x}) \cap \chi] \,:\, \dot{\varnothing}\,]\!)\\ \mathscr{A}^{\vec{\times}_1}[\![\mathsf{A}_1 - \mathsf{A}_2]\!]\,\chi\,\overline{\rho} &\triangleq \mathsf{let}\ \langle\chi_1,\,\chi_2\rangle = \ominus^{\vec{\times}_1}\,\chi\,\langle\mathscr{A}^{\times}[\![\mathsf{A}_1]\!]\overline{\rho},\,\mathscr{A}^{\times}[\![\mathsf{A}_2]\!]\overline{\rho}\rangle\ \mathsf{in}\\ &\qquad \mathscr{A}^{\vec{\times}_1}[\![\mathsf{A}_1]\!]\,\chi_1\,\overline{\rho}\,\dot{\cap}\,\mathscr{A}^{\vec{\times}_1}[\![\mathsf{A}_2]\!]\,\chi_2\,\overline{\rho} \end{aligned}$$

$\mathscr{A}^{\vec{\times}_1}[\![\mathsf{A}]\!]\,\chi$ is a lower closure operator on $\mathcal{V} \to \wp(\mathbb{V})$ and $\mathscr{A}^{\vec{\times}_1}[\![\mathsf{A}]\!]\,\chi$ is $\dot{\subseteq}$-increasing in χ.

Theorem 28.29 (soundness of the Cartesian accessibility semantics of an arithmetic expression A)

$\dot{\alpha}_{\vec{\times}} \circ \mathscr{A}^{-1}[\![\mathsf{A}]\!]\,\chi \mathrel{\dot{\subseteq}} \mathscr{A}^{\vec{\times}_1}[\![\mathsf{A}]\!]\,\chi \circ \dot{\alpha}_{\vec{\times}}$.

Proof of theorem 28.29 By structural induction on A for $\mathscr{P}_0 \in \wp(\mathcal{V} \to \mathbb{V})$.

— $\dot{\alpha}_{\vec{\times}} \circ \mathscr{A}^{-1}[\![1]\!]\,\chi\,(\mathscr{P}_0)$

$= \dot{\alpha}_{\vec{\times}}((\![\,1 \in \chi \,?\, \mathscr{P}_0 \,:\, \varnothing\,]\!))$ ⟨theorem 28.24⟩

$= (\![\,1 \in \chi \,?\, \dot{\alpha}_{\vec{\times}}(\mathscr{P}_0) \,:\, \dot{\alpha}_{\vec{\times}}(\varnothing)\,]\!)$ ⟨distributing $\dot{\alpha}_{\vec{\times}}$ over the conditional⟩

$= (\![\,1 \in \chi \,?\, \dot{\alpha}_{\vec{\times}}(\mathscr{P}_0) \,:\, \mathsf{x} \in \mathcal{V} \mapsto \varnothing)\,]\!)$

⟨$\dot{\alpha}_{\vec{\times}}$ preserves infimum in Galois connection of exercise 28.2⟩

$= \mathscr{A}^{\vec{\times}_1}[\![1]\!]\,\chi\,\dot{\alpha}_{\vec{\times}}(\mathscr{P}_0)$ ⟨definition (28.28) of $\mathscr{A}^{\vec{\times}_1}[\![1]\!]\,\chi$⟩

— $\dot{\alpha}_{\vec{\times}} \circ \mathscr{A}^{-1}[\![\mathsf{x}]\!]\,\chi\,(\mathscr{P}_0)$

$= \dot{\alpha}_{\vec{\times}}(\{\rho \in \mathscr{P}_0 \mid \rho(\mathsf{x}) \in \chi\})$ ⟨theorem 28.24⟩

$= \mathsf{z} \in \mathcal{V} \mapsto \{\rho(\mathsf{z}) \mid \rho \in \{\rho \in \mathscr{P}_0 \mid \rho(\mathsf{x}) \in \chi\}\}$ ⟨definition of $\dot{\alpha}_{\vec{\times}}$⟩

$= \mathsf{z} \in \mathcal{V} \mapsto \{\rho(\mathsf{z}) \mid \rho \in \mathscr{P}_0 \wedge \rho(\mathsf{x}) \in \chi\}$ ⟨definition of $\in$⟩

$\dot{\subseteq}\ \mathsf{z} \in \mathcal{V} \mapsto \{\rho(\mathsf{z}) \in \dot{\alpha}_{\vec{\times}}(\mathscr{P}_0)(\mathsf{z}) \mid \rho(\mathsf{x}) \in \chi \cap \dot{\alpha}_{\vec{\times}}(\mathscr{P}_0)(\mathsf{x})\}$

⟨definition of $\dot{\alpha}_{\vec{\times}}$ so that $\rho \in \mathscr{P}_0$ implies $\rho(\mathsf{y}) \in \dot{\alpha}_{\vec{\times}}(\mathscr{P}_0)(\mathsf{y})$⟩

$= \llparenthesis\, \chi \cap \dot{\alpha}_{\vec{\times}}(\mathscr{P}_0)(\mathsf{x}) \neq \emptyset \,?\, \mathsf{z} \in \mathbb{V} \mapsto \llparenthesis\, \mathsf{z} = \mathsf{x} \,?\, \{\rho(\mathsf{x}) \in \dot{\alpha}_{\vec{\times}}(\mathscr{P}_0)(\mathsf{x}) \mid \rho(\mathsf{x}) \in \chi \cap \dot{\alpha}_{\vec{\times}}(\mathscr{P}_0)(\mathsf{x})\} \mathbin{:}$
$\quad \{\rho(\mathsf{z}) \in \dot{\alpha}_{\vec{\times}}(\mathscr{P}_0)(\mathsf{z}) \mid \rho(\mathsf{x}) \in \chi \cap \dot{\alpha}_{\vec{\times}}(\mathscr{P}_0)(\mathsf{x})\} \,\rrparenthesis \mathbin{:} \mathsf{z} \in \mathbb{V} \mapsto \emptyset \,\rrparenthesis$

≀definition of the conditionalʃ

$\dot{\subseteq}\ \llparenthesis\, \chi \cap \dot{\alpha}_{\vec{\times}}(\mathscr{P}_0)(\mathsf{x}) \neq \emptyset \,?\, \mathsf{z} \in \mathbb{V} \mapsto \llparenthesis\, \mathsf{z} = \mathsf{x} \,?\, \chi \cap \dot{\alpha}_{\vec{\times}}(\mathscr{P}_0)(\mathsf{x}) \mathbin{:} \dot{\alpha}_{\vec{\times}}(\mathscr{P}_0)(\mathsf{z}) \,\rrparenthesis \mathbin{:} \mathsf{z} \in \mathbb{V} \mapsto \emptyset \,\rrparenthesis$

≀$\{\rho(\mathsf{z}) \in \dot{\alpha}_{\vec{\times}}(\mathscr{P}_0)(\mathsf{z}) \mid \rho(\mathsf{x}) \in \chi \cap \dot{\alpha}_{\vec{\times}}(\mathscr{P}_0)(\mathsf{x})\} \subseteq \dot{\alpha}_{\vec{\times}}(\mathscr{P}_0)(\mathsf{z})$ʃ

$= \llparenthesis\, \dot{\alpha}_{\vec{\times}}(\mathscr{P}_0)(\mathsf{x}) \cap \chi \neq \emptyset \,?\, \dot{\alpha}_{\vec{\times}}(\mathscr{P}_0)[\mathsf{x} \leftarrow \dot{\alpha}_{\vec{\times}}(\mathscr{P}_0)(\mathsf{x}) \cap \chi] \mathbin{:} \mathsf{z} \in \mathbb{V} \mapsto \emptyset \,\rrparenthesis$ ≀definition (19.10)ʃ

$= \boldsymbol{\mathscr{A}}^{\overset{\vec{\times}}{-1}} [\![\mathsf{x}]\!]\, \chi\, \dot{\alpha}_{\vec{\times}}(\mathscr{P}_0)$ ≀definition (28.28)ʃ

— $\dot{\alpha}_{\vec{\times}} \circ \boldsymbol{\mathscr{A}}^{-1} [\![\mathsf{A}_1 - \mathsf{A}_2]\!]\, \chi\, (\mathscr{P}_0)$

$= \dot{\alpha}_{\vec{\times}}(\{\rho \in \mathscr{P}_0 \mid (\boldsymbol{\mathscr{A}} [\![\mathsf{A}_1]\!]\, \rho - \boldsymbol{\mathscr{A}} [\![\mathsf{A}_2]\!]\, \rho) \in \chi\})$ ≀theorem 28.24ʃ

$\dot{\subseteq}\ \dot{\alpha}_{\vec{\times}}(\{\rho \in \mathscr{P}_0 \mid \exists v_2 \in \{\boldsymbol{\mathscr{A}} [\![\mathsf{A}_2]\!]\, \rho \mid \rho \in \mathscr{P}_0\} \,.\, (\boldsymbol{\mathscr{A}} [\![\mathsf{A}_1]\!] \rho\, b\, v_2) \in \chi\} \cap \{\rho \in \mathscr{P}_0 \mid \exists v_1 \in$
$\quad \{\boldsymbol{\mathscr{A}} [\![\mathsf{A}_1]\!]\, \rho \mid \rho \in \mathscr{P}_0\} \,.\, (v_1\, b\, \boldsymbol{\mathscr{A}} [\![\mathsf{A}_2]\!] \rho) \in \chi\})$

≀because $\{\rho \in S \mid f(g(\rho), h(\rho)) \in X\} \subseteq \{\rho \in S \mid \exists v_1 \in \{g(\rho) \mid \rho \in S\} \,.\, f(v_1, h(\rho)) \in X\}$ and similarly for hʃ

$= \dot{\alpha}_{\vec{\times}}(\{\rho \in \mathscr{P}_0 \mid \boldsymbol{\mathscr{A}} [\![\mathsf{A}_1]\!] \rho \in \{v_1 \in \{\boldsymbol{\mathscr{A}} [\![\mathsf{A}_1]\!]\, \rho \mid \rho \in \mathscr{P}_0\} \mid \exists v_2 \in \{\boldsymbol{\mathscr{A}} [\![\mathsf{A}_2]\!]\, \rho \mid \rho \in \mathscr{P}_0\} \,.\, (v_1\, b\, v_2) \in$
$\quad \chi\}\} \cap \{\rho \in \mathscr{P}_0 \mid \boldsymbol{\mathscr{A}} [\![\mathsf{A}_2]\!] \rho \in \{v_2 \in \{\boldsymbol{\mathscr{A}} [\![\mathsf{A}_2]\!]\, \rho \mid \rho \in \mathscr{P}_0\} \mid \exists v_1 \in \{\boldsymbol{\mathscr{A}} [\![\mathsf{A}_1]\!]\, \rho \mid \rho \in \mathscr{P}_0\} \,.$
$\quad (v_1\, b\, v_2) \in \chi\}\})$ ≀definition of $\in$, letting $v1 = \boldsymbol{\mathscr{A}} [\![\mathsf{A}_1]\!]\, \rho$ and $v_2 = \boldsymbol{\mathscr{A}} [\![\mathsf{A}_2]\!] \rho$ʃ

$= \dot{\alpha}_{\vec{\times}}(\boldsymbol{\mathscr{A}}^{\overset{\vec{\times}}{-1}} [\![\mathsf{A}_1]\!]\, \{v_1 \in \{\boldsymbol{\mathscr{A}} [\![\mathsf{A}_1]\!]\, \rho \mid \rho \in \mathscr{P}_0\} \mid \exists v_2 \in \{\boldsymbol{\mathscr{A}} [\![\mathsf{A}_2]\!]\, \rho \mid \rho \in \mathscr{P}_0\} \,.\, (v_1\, b\, v_2) \in \chi\}\, \mathscr{P}_0 \cap$
$\quad \boldsymbol{\mathscr{A}}^{\overset{\vec{\times}}{-1}} [\![\mathsf{A}_2]\!]\, \{v_2 \in \{\boldsymbol{\mathscr{A}} [\![\mathsf{A}_2]\!]\, \rho \mid \rho \in \mathscr{P}_0\} \mid \exists v_1 \in \{\boldsymbol{\mathscr{A}} [\![\mathsf{A}_1]\!]\, \rho \mid \rho \in \mathscr{P}_0\} \,.\, (v_1\, b\, v_2) \in \chi\}\, \mathscr{P}_0)$

≀definition (28.22) of $\boldsymbol{\mathscr{A}}^{\overset{\vec{\times}}{-1}} [\![\mathsf{A}]\!]$ʃ

$\dot{\subseteq}\ \dot{\alpha}_{\vec{\times}}(\boldsymbol{\mathscr{A}}^{\overset{\vec{\times}}{-1}} [\![\mathsf{A}_1]\!]\, \{v_1 \in \{\boldsymbol{\mathscr{A}} [\![\mathsf{A}_1]\!]\, \rho \mid \rho \in \mathscr{P}_0\} \mid \exists v_2 \in \{\boldsymbol{\mathscr{A}} [\![\mathsf{A}_2]\!]\, \rho \mid \rho \in \mathscr{P}_0\} \,.\, (v_1\, b\, v_2) \in \chi\}\, \mathscr{P}_0)\, \dot{\cap}$
$\quad \dot{\alpha}_{\vec{\times}}(\boldsymbol{\mathscr{A}}^{\overset{\vec{\times}}{-1}} [\![\mathsf{A}_2]\!]\, \{v_2 \in \{\boldsymbol{\mathscr{A}} [\![\mathsf{A}_2]\!]\, \rho \mid \rho \in \mathscr{P}_0\} \mid \exists v_1 \in \{\boldsymbol{\mathscr{A}} [\![\mathsf{A}_1]\!]\, \rho \mid \rho \in \mathscr{P}_0\} \,.\, (v_1\, b\, v_2) \in \chi\}\, \mathscr{P}_0)$

≀$\dot{\alpha}_{\vec{\times}}$ is $\dot{\subseteq}$-increasing by Galois connection of exercise 28.2, and $\dot{\cap}$ is $\dot{\subseteq}$ increasing.ʃ

$= \mathsf{let}\ \langle \chi_1, \chi_2 \rangle = \langle \{v_1 \in \{\boldsymbol{\mathscr{A}} [\![\mathsf{A}_1]\!]\, \rho \mid \rho \in \mathscr{P}_0\} \mid \exists v_2 \in \{\boldsymbol{\mathscr{A}} [\![\mathsf{A}_2]\!]\, \rho \mid \rho \in \mathscr{P}_0\} \,.\, (v_1\, b\, v_2) \in \chi\},$
$\quad \{v_2 \in \{\boldsymbol{\mathscr{A}} [\![\mathsf{A}_2]\!]\, \rho \mid \rho \in \mathscr{P}_0\} \mid \exists v_1 \in \{\boldsymbol{\mathscr{A}} [\![\mathsf{A}_1]\!]\, \rho \mid \rho \in \mathscr{P}_0\} \,.\, (v_1\, b\, v_2) \in \chi\} \rangle\ \mathsf{in}$
$\quad \dot{\alpha}_{\vec{\times}}(\boldsymbol{\mathscr{A}}^{\overset{\vec{\times}}{-1}} [\![\mathsf{A}_1]\!]\, \chi_1\, \mathscr{P}_0) \,\dot{\cap}\, \dot{\alpha}_{\vec{\times}}(\boldsymbol{\mathscr{A}}^{\overset{\vec{\times}}{-1}} [\![\mathsf{A}_2]\!]\, \chi_2\, \mathscr{P}_0)$ ≀definition of letʃ

$= \mathsf{let}\ \langle \chi_1, \chi_2 \rangle = b^{\overset{\vec{\times}}{-1}}\, \chi\, \langle \{\boldsymbol{\mathscr{A}} [\![\mathsf{A}_1]\!]\, \rho \mid \rho \in \mathscr{P}_0\},\ \{\boldsymbol{\mathscr{A}} [\![\mathsf{A}_2]\!]\, \rho \mid \rho \in \mathscr{P}_0\} \rangle\ \mathsf{in}$
$\quad \dot{\alpha}_{\vec{\times}}(\boldsymbol{\mathscr{A}}^{\overset{\vec{\times}}{-1}} [\![\mathsf{A}_1]\!]\, \chi_1\, \mathscr{P}_0) \,\dot{\cap}\, \dot{\alpha}_{\vec{\times}}(\boldsymbol{\mathscr{A}}^{\overset{\vec{\times}}{-1}} [\![\mathsf{A}_2]\!]\, \chi_2\, \mathscr{P}_0)$ ≀definition of $b^{\overset{\vec{\times}}{-1}}\, \chi$ʃ

$\dot{\subseteq}\ \mathsf{let}\ \langle \chi_1, \chi_2 \rangle = b^{\overset{\vec{\times}}{-1}}\, \chi\, \langle \boldsymbol{\mathscr{A}}^{\times} [\![\mathsf{A}_1]\!]\, \dot{\alpha}_{\vec{\times}}(\mathscr{P}_0),\ \boldsymbol{\mathscr{A}}^{\times} [\![\mathsf{A}_2]\!]\, \dot{\alpha}_{\vec{\times}}(\mathscr{P}_0) \rangle\ \mathsf{in}$
$\quad \dot{\alpha}_{\vec{\times}}(\boldsymbol{\mathscr{A}}^{\overset{\vec{\times}}{-1}} [\![\mathsf{A}_1]\!]\, \chi_1\, \mathscr{P}_0) \,\dot{\cap}\, \dot{\alpha}_{\vec{\times}}(\boldsymbol{\mathscr{A}}^{\overset{\vec{\times}}{-1}} [\![\mathsf{A}_2]\!]\, \chi_2\, \mathscr{P}_0)$

≀$\{\boldsymbol{\mathscr{A}} [\![\mathsf{A}]\!]\, \rho \mid \rho \in \mathscr{P}_0\} \subseteq \boldsymbol{\mathscr{A}}^{\times} [\![\mathsf{A}]\!] \dot{\alpha}_{\vec{\times}}(\mathscr{P}_0)$ by lemma 28.14, $b^{\overset{\vec{\times}}{-1}}\, \chi$ is a lower closure hence increasing, $\boldsymbol{\mathscr{A}}^{\overset{\vec{\times}}{-1}} [\![\mathsf{A}_i]\!]\, \chi_i$, $i = 1, 2$ is increasing in χ_i, $\dot{\cap}$ is $\dot{\subseteq}$-increasing, and $\dot{\alpha}_{\vec{\times}}$ increasing by the Galois connection of exercise 28.2ʃ

$\dot{\subseteq}\ \mathsf{let}\ \langle \chi_1, \chi_2 \rangle = b^{\overset{\vec{\times}}{-1}}\, \chi\, \langle \boldsymbol{\mathscr{A}}^{\times} [\![\mathsf{A}_1]\!] \dot{\alpha}_{\vec{\times}}(\mathscr{P}_0),\ \boldsymbol{\mathscr{A}}^{\times} [\![\mathsf{A}_2]\!] \dot{\alpha}_{\vec{\times}}(\mathscr{P}_0) \rangle\ \mathsf{in}$
$\quad \boldsymbol{\mathscr{A}}^{\overset{\vec{\times}}{-1}} [\![\mathsf{A}_1]\!]\, \chi_1\, \dot{\alpha}_{\vec{\times}}(\mathscr{P}_0) \,\dot{\cap}\, \boldsymbol{\mathscr{A}}^{\overset{\vec{\times}}{-1}} [\![\mathsf{A}_2]\!]\, \chi_2\, \dot{\alpha}_{\vec{\times}}(\mathscr{P}_0)$ ≀induction hypothesisʃ

$= \boldsymbol{\mathscr{A}}^{\overset{\vec{\times}}{-1}} [\![\mathsf{A}_1 - \mathsf{A}_2]\!]\, \chi\, \dot{\alpha}_{\vec{\times}}(\mathscr{P}_0)$ ≀definition (28.28) of $\boldsymbol{\mathscr{A}}^{\overset{\vec{\times}}{-1}} [\![\mathsf{A}_1 - \mathsf{A}_2]\!]\, \chi$ʃ □

28.5.3 For the Inductive Abstraction in the Inductive Hierarchy of Abstractions, Calculational Design of the Structural Cartesian Abstract Backward Semantics of Arithmetic Expressions

As noticed in section 28.4.3, if $\mathbb{D}^{\times} = \langle \mathbb{P}^{\times}, \sqsubseteq^{\times} \rangle$ is large or infinite (in particular when $\mathbb{D}^{\times} = \mathbb{D}^{\vec{\times}} = \langle \wp(\mathbb{V}), \subseteq \rangle$), a further abstraction $\gamma_{¤} \in \langle \mathbb{P}^{¤}, \sqsubseteq^{¤} \rangle \xrightarrow{\nearrow} \langle \mathbb{P}^{\times}, \sqsubseteq^{\times} \rangle$ may be needed. This abstraction can be extended pointwise to $\dot{\gamma}_{¤} \in \langle \mathbb{V} \to \mathbb{P}^{¤}, \dot{\sqsubseteq}^{¤} \rangle \xrightarrow{\nearrow} \langle \mathbb{V} \to \mathbb{P}^{\times}, \dot{\sqsubseteq}^{\times} \rangle$.

Structural Cartesian accessibility semantics of an arithmetic expression A

$$\mathscr{A}^{\overset{¤}{-1}}[\![1]\!]\,\chi\,\overline{\rho} \;\triangleq\; (\!|\, 1^{¤} \sqcap^{¤} \chi \neq \bot^{¤} \mathrel{?} \overline{\rho} \mathrel{:} \dot{\bot}^{¤} \,|\!) \tag{28.30}$$

$$\mathscr{A}^{\overset{¤}{-1}}[\![\mathsf{x}]\!]\,\chi\,\overline{\rho} \;\triangleq\; (\!|\, \overline{\rho}(\mathsf{x}) \sqcap^{¤} \chi \neq \bot^{¤} \mathrel{?} \overline{\rho}[\mathsf{x} \leftarrow \overline{\rho}(\mathsf{x}) \sqcap^{¤} \chi] \mathrel{:} \dot{\bot}^{¤} \,|\!)$$

$$\mathscr{A}^{\overset{¤}{-1}}[\![\mathsf{A}_1 - \mathsf{A}_2]\!]\,\chi\,\overline{\rho} \;\triangleq\; \mathsf{let}\ \langle \chi_1, \chi_2 \rangle = \ominus^{\overset{¤}{-1}}\,\chi\,\langle \mathscr{A}^{¤}[\![\mathsf{A}_1]\!]\overline{\rho}, \mathscr{A}^{¤}[\![\mathsf{A}_2]\!]\overline{\rho} \rangle\ \mathsf{in}$$
$$\mathscr{A}^{\overset{¤}{-1}}[\![\mathsf{A}_1]\!]\,\chi_1\,\overline{\rho}\,\dot{\sqcap}^{¤}\,\mathscr{A}^{\overset{¤}{-1}}[\![\mathsf{A}_2]\!]\,\chi_2\,\overline{\rho}$$

$\mathscr{A}^{\overset{¤}{-1}}[\![\mathsf{A}]\!]\,\chi$ is a lower closure operator on $\mathbb{V} \to \mathbb{P}^{¤}$ and $\mathscr{A}^{\overset{¤}{-1}}[\![\mathsf{A}]\!]\,\chi$ is $\sqsubseteq^{¤}$-increasing in χ.

Theorem 28.31 (soundness of the structural Cartesian accessibility semantics of an arithmetic expression A)

If $\mathscr{A}^{\overset{\times}{-1}}$ is an instance of (28.30) for $\langle \mathbb{P}^{\times}, \sqsubseteq^{\times} \rangle$, $\mathscr{A}^{\overset{¤}{-1}}$ is an instance of (28.30) for $\langle \mathbb{P}^{¤}, \sqsubseteq^{¤} \rangle$, $\gamma_{¤} \in \langle \mathbb{P}^{¤}, \sqsubseteq^{¤} \rangle \xrightarrow{\sqcap} \langle \mathbb{P}^{\times}, \sqsubseteq^{\times} \rangle$ preserves finite meets and satisfies $\forall P \in \mathbb{P}^{¤} . (P = \bot^{¤}) \Rightarrow (\gamma_{¤}(P) = \bot^{\times})$, $\gamma_{¤}(1^{¤}) = 1^{\times}$, the hypotheses of theorem 28.18 hold, $(\ominus^{\overset{\times}{-1}}\,\chi)$ is increasing argumentwise, and $\ominus^{\overset{\times}{-1}}\,\gamma_{¤}(\chi)\,\langle \gamma_{¤}(P_1), \gamma_{¤}(P_2) \rangle \sqsubseteq^{\times} \dot{\gamma}_{¤}(\ominus^{\overset{¤}{-1}}\,\chi\,\langle P_1, P_2 \rangle)$ then

$$\mathscr{A}^{\overset{\times}{-1}}[\![\mathsf{A}]\!]\,\gamma_{¤}(\chi) \circ \dot{\gamma}_{¤} \;\;\dot{\sqsubseteq}^{\times}\;\; \dot{\gamma}_{¤} \circ \mathscr{A}^{\overset{¤}{-1}}[\![\mathsf{A}]\!]\,\chi.$$

Proof of theorem 28.31 Assume $P, \chi \in \mathbb{P}^{¤}$ and $\overline{\rho} \in \mathbb{V} \to \mathbb{P}^{¤}$.

$$
\begin{array}{ll}
- & \gamma_{¤}(P) \sqcap^{\times} \gamma_{¤}(\chi) = \bot^{\times} \\
\Leftarrow & \gamma_{¤}(P \sqcap^{¤} \chi) = \bot^{\times} \hfill \text{\{}\gamma_{¤}\ \text{preserves meets\}} \\
\Leftarrow & P \sqcap^{¤} \chi = \bot^{¤} \hfill \text{\{}\forall P \in \mathbb{P}^{¤} . (P = \bot^{¤}) \Rightarrow (\gamma_{¤}(P) = \bot^{\times})\text{\}}
\end{array}
$$

and so, by contraposition, $(\gamma_{¤}(P) \sqcap^{\times} \gamma_{¤}(\chi) \neq \bot^{\times}) \Rightarrow (P \sqcap^{¤} \chi \neq \bot^{¤})$.

$$
\begin{array}{ll}
- & \mathscr{A}^{\overset{\times}{-1}}[\![1]\!]\,\gamma_{¤}(\chi)\,(\dot{\gamma}_{¤}(\overline{\rho})) \\
= & (\!|\, 1^{\times} \sqcap^{\times} \gamma_{¤}(\chi) \neq \bot^{\times} \mathrel{?} \dot{\gamma}_{¤}(\overline{\rho}) \mathrel{:} \dot{\bot}^{\times} \,|\!) \hfill \text{\{definition (28.30) of } \mathscr{A}^{\overset{\times}{-1}}[\![1]\!]\text{\}} \\
= & (\!|\, \gamma_{¤}(1^{¤}) \sqcap^{\times} \gamma_{¤}(\chi) \neq \bot^{\times} \mathrel{?} \dot{\gamma}_{¤}(\overline{\rho}) \mathrel{:} \dot{\bot}^{\times} \,|\!) \hfill \text{\{}\gamma_{¤}(1^{¤}) = 1^{\times}\text{\}} \\
\dot{\sqsubseteq}^{\times} & (\!|\, P \sqcap^{¤} \chi \neq \bot^{¤} \mathrel{?} \dot{\gamma}_{¤}(\overline{\rho}) \mathrel{:} \dot{\bot}^{\times} \,|\!)
\end{array}
$$

{because either $\gamma_{¤}(1^{¤}) \sqcap^{\times} \gamma_{¤}(\chi) \neq \bot^{\times}$ holds, which implies $P \sqcap^{¤} \chi \neq \bot^{¤}$, and we conclude by reflexivity or else $\dot{\bot}^{\times}$ is the infimum}

$$
\begin{array}{ll}
\dot{\sqsubseteq}^{\times} & (\!|\, P \sqcap^{¤} \chi \neq \bot^{¤} \mathrel{?} \dot{\gamma}_{¤}(\overline{\rho}) \mathrel{:} \dot{\gamma}_{¤}(\dot{\bot}^{¤}) \,|\!) \hfill \text{\{}\dot{\bot}^{\times}\ \text{is the infimum\}} \\
= & \dot{\gamma}_{¤}((\!|\, P \sqcap^{¤} \chi \neq \bot^{¤} \mathrel{?} \overline{\rho} \mathrel{:} \dot{\bot}^{¤} \,|\!)) \hfill \text{\{factoring } \dot{\gamma}_{¤} \text{ out of the conditional\}} \\
= & \dot{\gamma}_{¤}(\mathscr{A}^{\overset{¤}{-1}}[\![1]\!]\,\chi\,\overline{\rho}) \hfill \text{\{definition (28.30) of } \mathscr{A}^{\overset{¤}{-1}}[\![1]\!]\text{\}}
\end{array}
$$

$$
\begin{aligned}
&- \quad \mathcal{A}^{\overleftarrow{\times}_1}[\![\mathsf{x}]\!]\,\gamma_{\alpha}(\chi)\,(\dot{\gamma}_{\alpha}(\overline{\rho}))\\
&= \quad (\!|\,\dot{\gamma}_{\alpha}(\overline{\rho})(\mathsf{x}) \sqcap^{\times} \gamma_{\alpha}(\chi) \neq \bot^{\times} \,?\, \dot{\gamma}_{\alpha}(\overline{\rho})[\mathsf{x} \leftarrow \dot{\gamma}_{\alpha}(\overline{\rho})(\mathsf{x}) \sqcap^{\times} \gamma_{\alpha}(\chi)] \,\text{:}\, \dot{\bot}^{\times}\,|\!)\\
&\qquad\qquad \wr \text{definition (28.30) of } \mathcal{A}^{\overleftarrow{\times}_1}[\![\mathsf{x}]\!] \int\\
&= \quad (\!|\,\gamma_{\alpha}(\overline{\rho}(\mathsf{x})) \sqcap^{\times} \gamma_{\alpha}(\chi) \neq \bot^{\times} \,?\, \dot{\gamma}_{\alpha}(\overline{\rho})[\mathsf{x} \leftarrow \gamma_{\alpha}(\overline{\rho}(\mathsf{x})) \sqcap^{\times} \gamma_{\alpha}(\chi)] \,\text{:}\, \dot{\bot}^{\times}\,|\!)\\
&\qquad\qquad \wr \text{pointwise definition of } \dot{\gamma}_{\alpha} \int\\
&\dot{\sqsubseteq}^{\times} \quad (\!|\,\overline{\rho}(\mathsf{x}) \sqcap^{\alpha} \chi \neq \bot^{\alpha} \,?\, \dot{\gamma}_{\alpha}(\overline{\rho})[\mathsf{x} \leftarrow \gamma_{\alpha}(\overline{\rho}(\mathsf{x})) \sqcap^{\times} \gamma_{\alpha}(\chi)] \,\text{:}\, \dot{\bot}^{\times}\,|\!)\\
&\qquad \wr \text{because either } \gamma_{\alpha}(\overline{\rho}(\mathsf{x})) \sqcap^{\times} \gamma_{\alpha}(\chi) \neq \bot^{\times} \text{ holds, which implies } \overline{\rho}(\mathsf{x}) \sqcap^{\alpha} \chi \neq \bot^{\alpha}\text{, and}\\
&\qquad \text{we conclude by reflexivity or else } \dot{\bot}^{\times} \text{ is the infimum} \int\\
&= \quad (\!|\,\overline{\rho}(\mathsf{x}) \sqcap^{\alpha} \chi \neq \bot^{\alpha} \,?\, \dot{\gamma}_{\alpha}(\overline{\rho}[\mathsf{x} \leftarrow \overline{\rho}(\mathsf{x}) \sqcap^{\alpha} \chi]) \,\text{:}\, \dot{\bot}^{\times}\,|\!)\\
&\qquad \wr \text{pointwise definition of } \gamma_{\alpha} \text{ and definition (19.10) of the environment assignment} \int\\
&\dot{\sqsubseteq}^{\times} \quad (\!|\,\overline{\rho}(\mathsf{x}) \sqcap^{\alpha} \chi \neq \bot^{\alpha} \,?\, \dot{\gamma}_{\alpha}(\overline{\rho}[\mathsf{x} \leftarrow \overline{\rho}(\mathsf{x}) \sqcap^{\alpha} \chi]) \,\text{:}\, \dot{\gamma}_{\alpha}(\dot{\bot})\,|\!) \qquad \wr \dot{\bot}^{\times} \text{ is the infimum} \int\\
&= \quad \dot{\gamma}_{\alpha}((\!|\,\overline{\rho}(\mathsf{x}) \sqcap^{\alpha} \chi \neq \bot^{\alpha} \,?\, \overline{\rho}[\mathsf{x} \leftarrow \overline{\rho}(\mathsf{x}) \sqcap^{\alpha} \chi] \,\text{:}\, \dot{\bot}^{\alpha}\,|\!)) \qquad \wr \text{factoring } \dot{\gamma}_{\alpha} \text{ out of the conditional} \int\\
&= \quad \dot{\gamma}_{\alpha}(\mathcal{A}^{\overleftarrow{\alpha}_1}[\![\mathsf{x}]\!]\,\chi\,\overline{\rho}) \qquad \wr \text{definition (28.30) of } \mathcal{A}^{\overleftarrow{\alpha}_1}[\![\mathsf{x}]\!] \int
\end{aligned}
$$

$$
\begin{aligned}
&- \quad \mathcal{A}^{\overleftarrow{\times}_1}[\![\mathsf{A}_1 - \mathsf{A}_2]\!]\,\gamma_{\alpha}(\chi)\,(\dot{\gamma}_{\alpha}(\overline{\rho}))\\
&= \quad \mathsf{let}\ \langle \chi_1, \chi_2\rangle = \overleftarrow{\ominus}^{\times}_1\ \gamma_{\alpha}(\chi)\ \langle \mathcal{A}^{\times}[\![\mathsf{A}_1]\!]\dot{\gamma}_{\alpha}(\overline{\rho}), \mathcal{A}^{\times}[\![\mathsf{A}_2]\!]\dot{\gamma}_{\alpha}(\overline{\rho})\rangle\ \mathsf{in}\\
&\qquad \mathcal{A}^{\overleftarrow{\times}_1}[\![\mathsf{A}_1]\!]\,\chi_1\,\dot{\gamma}_{\alpha}(\overline{\rho}) \,\dot{\sqcap}^{\times}\, \mathcal{A}^{\overleftarrow{\times}_1}[\![\mathsf{A}_2]\!]\,\chi_2\,\dot{\gamma}_{\alpha}(\overline{\rho}) \qquad \wr \text{definition (28.30) of } \mathcal{A}^{\overleftarrow{\times}_1}[\![\mathsf{A}_1 - \mathsf{A}_2]\!] \int\\
&\sqsubseteq^{\times} \quad \mathsf{let}\ \langle \chi_1', \chi_2'\rangle = \overleftarrow{\ominus}^{\times}_1\ \gamma_{\alpha}(\chi)\ \langle \gamma_{\alpha}(\mathcal{A}^{\alpha}[\![\mathsf{A}_1]\!]\overline{\rho}), \gamma_{\alpha}(\mathcal{A}^{\alpha}[\![\mathsf{A}_2]\!]\overline{\rho})\rangle\ \mathsf{in}\\
&\qquad \mathcal{A}^{\overleftarrow{\times}_1}[\![\mathsf{A}_1]\!]\,\chi_1'\,\dot{\gamma}_{\alpha}(\overline{\rho}) \,\dot{\sqcap}^{\times}\, \mathcal{A}^{\overleftarrow{\times}_1}[\![\mathsf{A}_2]\!]\,\chi_2'\,\dot{\gamma}_{\alpha}(\overline{\rho})\\
&\qquad \wr \mathcal{A}^{\times}[\![\mathsf{A}_i]\!] \circ \dot{\gamma}_{\alpha} \,\dot{\sqsubseteq}^{\times}\, \gamma_{\alpha} \circ \mathcal{A}^{\alpha}[\![\mathsf{A}_i]\!] \text{ by theorem 28.18, } (\overleftarrow{\ominus}^{\times}_1\ \gamma_{\alpha}(\chi)) \text{ is increasing argu-}\\
&\qquad \text{mentwise so } \langle \chi_1, \chi_2\rangle \,\dot{\sqsubseteq}^{\times}\, \langle \chi_1', \chi_2'\rangle \text{ and } \mathcal{A}^{\overleftarrow{\times}_1}[\![\mathsf{A}_i]\!]\,\chi \text{ is } \dot{\sqsubseteq}^{\times}\text{-increasing in } \chi,\ i =\\
&\qquad 1, 2 \int\\
&\sqsubseteq^{\times} \quad \mathsf{let}\ \langle \chi_1'', \chi_2''\rangle = \overleftarrow{\ominus}^{\alpha}_1\ \chi\ \langle \mathcal{A}^{\alpha}[\![\mathsf{A}_1]\!]\overline{\rho}, \mathcal{A}^{\alpha}[\![\mathsf{A}_2]\!]\overline{\rho}\rangle\ \mathsf{in}\\
&\qquad \mathcal{A}^{\overleftarrow{\times}_1}[\![\mathsf{A}_1]\!]\,\gamma_{\alpha}(\chi_1'')\,\dot{\gamma}_{\alpha}(\overline{\rho}) \,\dot{\sqcap}^{\times}\, \mathcal{A}^{\overleftarrow{\times}_1}[\![\mathsf{A}_2]\!]\,\gamma_{\alpha}(\chi_2'')\,\dot{\gamma}_{\alpha}(\overline{\rho})\\
&\qquad \wr \overleftarrow{\ominus}^{\times}_1\ \gamma_{\alpha}(\chi)\ \langle \gamma_{\alpha}(P_1), \gamma_{\alpha}(P_2)\rangle \,\dot{\sqsubseteq}^{\times}\, \dot{\gamma}_{\alpha}(\overleftarrow{\ominus}^{\alpha}_1\ \chi\ \langle P_1, P_2\rangle) \text{ so } \chi_i' \subseteq \gamma_{\alpha}(\chi_i'') \text{ and } \mathcal{A}^{\overleftarrow{\times}_1}[\![\mathsf{A}_i]\!]\,\chi\\
&\qquad \text{is } \dot{\sqsubseteq}^{\times}\text{-increasing in } \chi,\ i = 1, 2 \int\\
&\sqsubseteq^{\times} \quad \mathsf{let}\ \langle \chi_1'', \chi_2''\rangle = \overleftarrow{\ominus}^{\alpha}_1\ \chi\ \langle \mathcal{A}^{\alpha}[\![\mathsf{A}_1]\!]\overline{\rho}, \mathcal{A}^{\alpha}[\![\mathsf{A}_2]\!]\overline{\rho}\rangle\ \mathsf{in}\\
&\qquad \dot{\gamma}_{\alpha}(\mathcal{A}^{\overleftarrow{\alpha}_1}[\![\mathsf{A}_1]\!]\,\chi_1''\,\overline{\rho}) \,\dot{\sqcap}^{\times}\, \dot{\gamma}_{\alpha}(\mathcal{A}^{\overleftarrow{\alpha}_1}[\![\mathsf{A}_2]\!]\,\chi_2''\,\overline{\rho})\\
&\qquad\qquad \wr \text{structural induction hypothesis and } \dot{\sqcap}^{\times} \text{ is } \dot{\sqsubseteq}^{\times}\text{-increasing} \int\\
&= \quad \dot{\gamma}_{\alpha}(\mathcal{A}^{\overleftarrow{\alpha}_1}[\![\mathsf{A}_1 - \mathsf{A}_2]\!]\,\chi\,\overline{\rho}) \quad \wr \text{definition of } \mathsf{let} \text{ and definition (28.30) of } \mathcal{A}^{\overleftarrow{\alpha}_1}[\![\mathsf{A}_1 - \mathsf{A}_2]\!] \int \quad \square
\end{aligned}
$$

Remark 28.32 Following remark 28.21, (28.28) is an instance of (28.30) so theorem 28.31 is valid for $\mathbb{P}^{\times} = \wp(\mathbb{V})$.

28.6 Hierarchy of Cartesian Reachability Semantics of Tests

We are now handling the tests test and their negations $\overline{\text{test}}$ in the hierarchy (28.9). Section 28.6.2 is for the basis, and section 28.6.3 for the induction.

28.6.1 The Collecting Domain Test in the Inductive Hierarchy of Abstractions, is the Assertional Reachability Test of the Concrete Domain $\ddot{\mathbb{D}}^{\vec{r}}$

For the assertional reachability concrete domain $\ddot{\mathbb{D}}^{\vec{r}}$, we have defined for all $\overline{\rho} \in \wp(\mathbb{V} \to \mathbb{V})$,

$$
\begin{aligned}
\mathsf{test}^{\vec{r}}[\![\mathsf{B}]\!]\overline{\rho} &\triangleq \{\rho \in \overline{\rho} \mid \mathscr{B}[\![\mathsf{B}]\!]\rho = \mathrm{tt}\} \qquad (19.16)\\
\overline{\mathsf{test}}^{\vec{r}}[\![\mathsf{B}]\!]\overline{\rho} &\triangleq \{\rho \in \overline{\rho} \mid \mathscr{B}[\![\mathsf{B}]\!]\rho = \mathrm{ff}\}
\end{aligned}
$$

28.6.2 For the Basic Abstraction in the Inductive Hierarchy of Abstractions, the Cartesian Reachability Test of the Abstract Domain $\ddot{\mathbb{D}}^{\vec{\times}}$ Abstracts the Assertional Reachability Test of the Concrete Domain $\ddot{\mathbb{D}}^{\vec{r}}$

The Cartesian reachability semantics of a conditional or an iteration statement involves the Cartesian reachability semantics of Boolean expressions defined as follows:

Given a relation operation $r \in \mathbb{V} \times \mathbb{V} \to \mathbb{B}$, $x \in \chi_1$. and $y \in \chi_2$, we define $\langle \chi_1', \chi_2' \rangle = r^{\vec{\times}} \langle \chi_1, \chi_2 \rangle$ to be the subset of values of $x \in \chi_1' \subseteq \chi_1$ and $y \in \chi_2' \subseteq \chi_2$ for which the relation $x\,r\,y$ may hold.

Example 28.33 If $x \in [1,4]$ and $y \in [0,3]$ then $x < y$ implies $x < \max\{y \mid y \in [0,3]\} = 3$ and $1 = \min\{x \mid x \in [1,4]\} < y$ so $x \in [1,2]$ and $y \in [2,3]$. Therefore $r^{\vec{\times}}\langle [1,4],\ [0,3] \rangle = \langle [1,2],\ [2,3] \rangle$ when $r = <$.

$$
\begin{aligned}
r^{\vec{\times}}, \overline{r}^{\vec{\times}} &\in (\wp(\mathbb{V}) \times \wp(\mathbb{V})) \to (\wp(\mathbb{V}) \times \wp(\mathbb{V}))\\
r^{\vec{\times}} \langle \chi_1, \chi_2 \rangle &\triangleq \langle \{v_1 \in \chi_1 \mid \exists v_2 \in \chi_2 \,.\, v_1\, r\, v_2\},\ \{v_2 \in \chi_2 \mid \exists v_1 \in \chi_1 \,.\, v_1\, r\, v_2\} \rangle\\
\overline{r}^{\vec{\times}} \langle \chi_1, \chi_2 \rangle &\triangleq \langle \{v_1 \in \chi_1 \mid \exists v_2 \in \chi_2 \,.\, \neg(v_1\, r\, v_2)\},\ \{v_2 \in \chi_2 \mid \exists v_1 \in \chi_1 \,.\, \neg(v_1\, r\, v_2)\} \rangle
\end{aligned}
$$

In (28.34), as follows, $r^{\vec{\times}}$ (respectively $\overline{r}^{\vec{\times}}$) is denoted $\otimes^{\vec{\times}}$ (respectively $\overline{\otimes}^{\vec{\times}}$) when $r = <$.

Structural Cartesian reachability semantics of tests B

$$
\begin{aligned}
\mathsf{test}^{\vec{\times}}[\![\mathsf{A}_1 < \mathsf{A}_2]\!]\overline{\rho} &\triangleq \mathsf{let}\ \langle A_1,\ A_2 \rangle = \otimes^{\vec{\times}} \langle \mathscr{A}^{\vec{\times}}[\![\mathsf{A}_1]\!]\,\overline{\rho},\ \mathscr{A}^{\vec{\times}}[\![\mathsf{A}_2]\!]\,\overline{\rho} \rangle\ \mathsf{in} \qquad (28.34)\\
&\qquad \mathscr{A}^{\vec{\times}_{-1}}[\![\mathsf{A}_1]\!]\, A_1\, \overline{\rho} \,\dot{\cap}\, \mathscr{A}^{\vec{\times}_{-1}}[\![\mathsf{A}_2]\!]\, A_2\, \overline{\rho}\\
\overline{\mathsf{test}}^{\vec{\times}}[\![\mathsf{A}_1 < \mathsf{A}_2]\!]\overline{\rho} &\triangleq \mathsf{let}\ \langle A_1,\ A_2 \rangle = \overline{\otimes}^{\vec{\times}} \langle \mathscr{A}^{\vec{\times}}[\![\mathsf{A}_1]\!]\,\overline{\rho},\ \mathscr{A}^{\vec{\times}}[\![\mathsf{A}_2]\!]\,\overline{\rho} \rangle\ \mathsf{in}\\
&\qquad \mathscr{A}^{\vec{\times}_{-1}}[\![\mathsf{A}_1]\!]\, A_1\, \overline{\rho} \,\dot{\cap}\, \mathscr{A}^{\vec{\times}_{-1}}[\![\mathsf{A}_0]\!]\, A_2\, \overline{\rho}\\
\mathsf{test}^{\vec{\times}}[\![\mathsf{B}_1\ \mathsf{nand}\ \mathsf{B}_1]\!]\overline{\rho} &\triangleq \quad (\mathsf{test}^{\vec{\times}}[\![\mathsf{B}_1]\!]\overline{\rho} \,\dot{\cap}\, \overline{\mathsf{test}}^{\vec{\times}}[\![\mathsf{B}_2]\!]\overline{\rho})\\
&\qquad \dot{\cup}\, (\overline{\mathsf{test}}^{\vec{\times}}[\![\mathsf{B}_1]\!]\overline{\rho} \,\dot{\cap}\, \mathsf{test}^{\vec{\times}}[\![\mathsf{B}_2]\!]\overline{\rho})\\
&\qquad \dot{\cup}\, (\overline{\mathsf{test}}^{\vec{\times}}[\![\mathsf{B}_1]\!]\overline{\rho} \,\dot{\cap}\, \overline{\mathsf{test}}^{\vec{\times}}[\![\mathsf{B}_2]\!]\overline{\rho})\\
\overline{\mathsf{test}}^{\vec{\times}}[\![\mathsf{B}_1\ \mathsf{nand}\ \mathsf{B}_2]\!]\overline{\rho} &\triangleq \mathsf{test}^{\vec{\times}}[\![\mathsf{B}_1]\!]\overline{\rho} \,\dot{\cap}\, \mathsf{test}^{\vec{\times}}[\![\mathsf{B}_2]\!]\overline{\rho}
\end{aligned}
$$

Theorem 28.35 (soundness of the structural Cartesian reachability semantics of tests B)

$$\dot{\alpha}_{\vec{\times}} \circ \mathsf{test}^{\vec{r}}\llbracket \mathsf{B} \rrbracket \mathrel{\dot{\subseteq}} \mathsf{test}^{\vec{\times}}\llbracket \mathsf{B} \rrbracket \circ \dot{\alpha}_{\vec{\times}} \text{ and } \dot{\alpha}_{\vec{\times}} \circ \overline{\mathsf{test}}^{\vec{r}}\llbracket \mathsf{B} \rrbracket \mathrel{\dot{\subseteq}} \overline{\mathsf{test}}^{\vec{\times}}\llbracket \mathsf{B} \rrbracket \circ \dot{\alpha}_{\vec{\times}}.$$

Proof of theorem 28.35

— The proof for $\mathsf{test}^{\vec{\times}}\llbracket \mathsf{A}_1 < \mathsf{A}_2 \rrbracket \overline{\rho}$ and $\overline{\mathsf{test}}^{\vec{\times}}\llbracket \mathsf{A}_1 < \mathsf{A}_2 \rrbracket \overline{\rho}$ are similar to that for $\mathscr{A}^{\vec{\times}_1}\llbracket \mathsf{A}_1 - \mathsf{A}_2 \rrbracket\, \chi\, \overline{\rho}$ where $X = \{\mathtt{tt}\}$ or $\{\mathtt{ff}\}$; see exercise 28.36.

— $\dot{\alpha}_{\vec{\times}}(\mathsf{test}^{\vec{r}}\llbracket \mathsf{B}_1 \text{ nand } \mathsf{B}_2 \rrbracket \mathscr{P}_0)$

$= \dot{\alpha}_{\vec{\times}}(\{\rho \in \mathscr{P}_0 \mid \mathscr{B}\llbracket \mathsf{B}_1 \text{ nand } \mathsf{B}_2 \rrbracket \rho = \mathtt{tt}\})$ ≀definition (19.16) of $\mathsf{test}^{\vec{r}}\llbracket \mathsf{B} \rrbracket \mathscr{P}_0 \triangleq \{\rho \in \mathscr{P}_0 \mid \mathscr{B}\llbracket \mathsf{B} \rrbracket \rho = \mathtt{tt}\}$ʃ

$= \dot{\alpha}_{\vec{\times}}(\{\rho \in \mathscr{P}_0 \mid (\mathscr{B}\llbracket \mathsf{B}_1 \rrbracket \rho \uparrow \mathscr{B}\llbracket \mathsf{B}_2 \rrbracket \rho) = \mathtt{tt}\})$

≀definition of $\mathscr{B}\llbracket \mathsf{B}_1 \text{ nand } \mathsf{B}_2 \rrbracket \rho \triangleq \mathscr{B}\llbracket \mathsf{B}_1 \rrbracket \rho \uparrow \mathscr{B}\llbracket \mathsf{B}_2 \rrbracket \rho$ in section 3.6ʃ

$= \dot{\alpha}_{\vec{\times}}(\{\rho \in \mathscr{P}_0 \mid (\mathscr{B}\llbracket \mathsf{B}_1 \rrbracket \rho = \mathtt{tt} \wedge \mathscr{B}\llbracket \mathsf{B}_2 \rrbracket \rho = \mathtt{ff}) \vee (\mathscr{B}\llbracket \mathsf{B}_1 \rrbracket \rho = \mathtt{ff} \wedge \mathscr{B}\llbracket \mathsf{B}_2 \rrbracket \rho = \mathtt{tt}) \vee (\mathscr{B}\llbracket \mathsf{B}_1 \rrbracket \rho = \mathtt{ff} \wedge \mathscr{B}\llbracket \mathsf{B}_2 \rrbracket \rho = \mathtt{ff})\})$ ≀truth table of $\uparrow$ in section 3.6ʃ

$= \dot{\alpha}_{\vec{\times}}(\{\rho \in \mathscr{P}_0 \mid \mathscr{B}\llbracket \mathsf{B}_1 \rrbracket \rho = \mathtt{tt} \wedge \mathscr{B}\llbracket \mathsf{B}_2 \rrbracket \rho = \mathtt{ff}\}) \mathbin{\dot{\cup}} \dot{\alpha}_{\vec{\times}}(\{\rho \in \mathscr{P}_0 \mid \mathscr{B}\llbracket \mathsf{B}_1 \rrbracket \rho = \mathtt{ff} \wedge \mathscr{B}\llbracket \mathsf{B}_2 \rrbracket \rho = \mathtt{tt}\}) \mathbin{\dot{\cup}} \dot{\alpha}_{\vec{\times}}(\{\rho \in \mathscr{P}_0 \mid \mathscr{B}\llbracket \mathsf{B}_1 \rrbracket \rho = \mathtt{ff} \wedge \mathscr{B}\llbracket \mathsf{B}_2 \rrbracket \rho = \mathtt{ff}\})$

≀ $\dot{\alpha}_{\vec{\times}}$ preserves lubs by the Galois connection of exercise 28.2ʃ

$\mathrel{\dot{\subseteq}} (\dot{\alpha}_{\vec{\times}}(\{\rho \in \mathscr{P}_0 \mid \mathscr{B}\llbracket \mathsf{B}_1 \rrbracket \rho = \mathtt{tt}\}) \mathbin{\dot{\cap}} \dot{\alpha}_{\vec{\times}}(\{\rho \in \mathscr{P}_0 \mid \mathscr{B}\llbracket \mathsf{B}_2 \rrbracket \rho = \mathtt{ff}\})) \mathbin{\dot{\cup}} (\dot{\alpha}_{\vec{\times}}(\{\rho \in \mathscr{P}_0 \mid \mathscr{B}\llbracket \mathsf{B}_1 \rrbracket \rho = \mathtt{ff}\}) \mathbin{\dot{\cap}} \dot{\alpha}_{\vec{\times}}(\{\rho \in \mathscr{P}_0 \mid \mathscr{B}\llbracket \mathsf{B}_2 \rrbracket \rho = \mathtt{tt}\})) \mathbin{\dot{\cup}} (\dot{\alpha}_{\vec{\times}}(\{\rho \in \mathscr{P}_0 \mid \mathscr{B}\llbracket \mathsf{B}_1 \rrbracket \rho = \mathtt{ff}\}) \mathbin{\dot{\cap}} \dot{\alpha}_{\vec{\times}}(\{\rho \in \mathscr{P}_0 \mid \mathscr{B}\llbracket \mathsf{B}_2 \rrbracket \rho = \mathtt{ff}\}))$

≀$\dot{\alpha}_{\vec{\times}}$ is increasing by the Galois connection of exercise 28.2 so $\dot{\alpha}_{\vec{\times}}(P_1 \cap P_2) \mathrel{\dot{\subseteq}} \dot{\alpha}_{\vec{\times}}(P_1) \mathbin{\dot{\cap}} \dot{\alpha}_{\vec{\times}}(P_2)$ and $\dot{\cup}$ is increasingʃ

$\mathrel{\dot{\subseteq}} (\mathsf{test}^{\vec{\times}}\llbracket \mathsf{B}_1 \rrbracket \dot{\alpha}_{\vec{\times}}(\mathscr{P}_0) \mathbin{\dot{\cap}} \overline{\mathsf{test}}^{\vec{\times}}\llbracket \mathsf{B}_2 \rrbracket \dot{\alpha}_{\vec{\times}}(\mathscr{P}_0)) \mathbin{\dot{\cup}} (\overline{\mathsf{test}}^{\vec{\times}}\llbracket \mathsf{B}_1 \rrbracket \dot{\alpha}_{\vec{\times}}(\mathscr{P}_0) \mathbin{\dot{\cap}} \mathsf{test}^{\vec{\times}}\llbracket \mathsf{B}_2 \rrbracket \dot{\alpha}_{\vec{\times}}(\mathscr{P}_0)) \mathbin{\dot{\cup}} (\overline{\mathsf{test}}^{\vec{\times}}\llbracket \mathsf{B}_1 \rrbracket \dot{\alpha}_{\vec{\times}}(\mathscr{P}_0) \mathbin{\dot{\cap}} \overline{\mathsf{test}}^{\vec{\times}}\llbracket \mathsf{B}_2 \rrbracket \dot{\alpha}_{\vec{\times}}(\mathscr{P}_0))$

≀induction hypothesis, $\dot{\cup}$ and $\dot{\cap}$ are $\dot{\subseteq}$-increasingʃ

$= \mathsf{test}^{\vec{\times}}\llbracket \mathsf{B}_1 \text{ nand } \mathsf{B}_1 \rrbracket \dot{\alpha}_{\vec{\times}}(\mathscr{P}_0)$ ≀definition (28.34)ʃ

— $\dot{\alpha}_{\vec{\times}}(\overline{\mathsf{test}}^{\vec{r}}\llbracket \mathsf{B}_1 \text{ nand } \mathsf{B}_2 \rrbracket \mathscr{P}_0)$

$= \dot{\alpha}_{\vec{\times}}(\{\rho \in \mathscr{P}_0 \mid (\mathscr{B}\llbracket \mathsf{B}_1 \rrbracket \rho \uparrow \mathscr{B}\llbracket \mathsf{B}_2 \rrbracket \rho) = \mathtt{ff}\})$

≀definition (19.16) of $\overline{\mathsf{test}}^{\vec{r}}\llbracket \mathsf{B} \rrbracket \mathscr{P}_0 \triangleq \{\rho \in \mathscr{P}_0 \mid \mathscr{B}\llbracket \mathsf{B} \rrbracket \rho = \mathtt{ff}\}$ and definition of $\mathscr{B}\llbracket \mathsf{B}_1 \text{ nand } \mathsf{B}_2 \rrbracket \rho \triangleq \mathscr{B}\llbracket \mathsf{B}_1 \rrbracket \rho \uparrow \mathscr{B}\llbracket \mathsf{B}_2 \rrbracket \rho$ in section 3.6ʃ

$= \dot{\alpha}_{\vec{\times}}(\{\rho \in \mathscr{P}_0 \mid \mathscr{B}\llbracket \mathsf{B}_1 \rrbracket \rho = \mathtt{tt} \wedge \mathscr{B}\llbracket \mathsf{B}_2 \rrbracket \rho = \mathtt{tt}\})$ ≀truth table of $\uparrow$ in section 3.6ʃ

$\mathrel{\dot{\subseteq}} (\dot{\alpha}_{\vec{\times}}(\{\rho \in \mathscr{P}_0 \mid \mathscr{B}\llbracket \mathsf{B}_1 \rrbracket \rho = \mathtt{tt}\}) \mathbin{\dot{\cap}} \dot{\alpha}_{\vec{\times}}(\{\rho \in \mathscr{P}_0 \mid \mathscr{B}\llbracket \mathsf{B}_2 \rrbracket \rho = \mathtt{tt}\}))$

≀$\dot{\alpha}_{\vec{\times}}$ is increasing by the Galois connection of exercise 28.2 so $\dot{\alpha}_{\vec{\times}}(P_1 \cap P_2) \mathrel{\dot{\subseteq}} \dot{\alpha}_{\vec{\times}}(P_1) \mathbin{\dot{\cap}} \dot{\alpha}_{\vec{\times}}(P_2)$ and $\dot{\cup}$ is increasingʃ

$= (\dot{\alpha}_{\vec{\times}}(\overline{\mathsf{test}}^{\vec{r}}\llbracket \mathsf{B}_1 \rrbracket \mathscr{P}_0) \mathbin{\dot{\cap}} \dot{\alpha}_{\vec{\times}}(\overline{\mathsf{test}}^{\vec{r}}\llbracket \mathsf{B}_2 \rrbracket \mathscr{P}_0))$ ≀definition (19.16) of $\overline{\mathsf{test}}^{\vec{r}}\llbracket \mathsf{B} \rrbracket \mathscr{P}_0 \triangleq \{\rho \in \mathscr{P}_0 \mid \mathscr{B}\llbracket \mathsf{B} \rrbracket \rho = \mathtt{ff}\}$ʃ

$\mathrel{\dot{\subseteq}} (\mathsf{test}^{\vec{\times}}\llbracket \mathsf{B}_1 \rrbracket \dot{\alpha}_{\vec{\times}}(\mathscr{P}_0) \mathbin{\dot{\cap}} \mathsf{test}^{\vec{\times}}\llbracket \mathsf{B}_2 \rrbracket \dot{\alpha}_{\vec{\times}}(\mathscr{P}_0))$ ≀induction hypothesis, $\dot{\cup}$ and $\dot{\cap}$ are $\dot{\subseteq}$-increasingʃ

$= \overline{\mathsf{test}}^{\vec{\times}}\llbracket \mathsf{B}_1 \text{ nand } \mathsf{B}_1 \rrbracket \dot{\alpha}_{\vec{\times}}(\mathscr{P}_0)$ ≀definition (28.34)ʃ □

Exercise 28.36 Prove that $\dot{\alpha}_{\ddot{\times}} \circ \mathsf{test}^{\vec{r}}[\![\mathsf{A}_1 < \mathsf{A}_2]\!] \mathrel{\dot{\subseteq}} \mathsf{test}^{\vec{\times}}[\![\mathsf{A}_1 < \mathsf{A}_2]\!] \circ \dot{\alpha}_{\ddot{\times}}$.

Exercise 28.37 Prove that $\mathsf{test}^{\vec{\times}}[\![\mathsf{B}]\!]$ is a lower closure operator.

28.6.3 For the Inductive Abstraction in the Inductive Hierarchy of Abstractions, the Test of the Abstract Cartesian Domain $\ddot{\mathbb{D}}^{\text{¤}}$ Abstracts the Test of the Cartesian AbstractDomain $\ddot{\mathbb{D}}^{\times}$

We now consider the induction step in the hierarchy (28.9), that is a further Cartesian abstraction $\gamma_{\text{¤}} \in \langle \mathbb{P}^{\text{¤}}, \sqsubseteq^{\text{¤}}\rangle \xrightarrow{\sqcap} \langle \mathbb{P}^{\times}, \sqsubseteq^{\times}\rangle$ of the Cartesian abstract domain and extend it pointwise to $\dot{\gamma}_{\text{¤}} \in \langle \mathbb{V} \to \mathbb{P}^{\text{¤}}, \dot{\sqsubseteq}^{\text{¤}}\rangle \xrightarrow{\sqcap} \langle \mathbb{V} \to \mathbb{P}^{\times}, \dot{\subseteq}\rangle$. The Cartesian abstract semantics of tests

Structural Cartesian abstract semantics of tests B

$$
\begin{aligned}
\mathsf{test}^{\text{¤}}[\![\mathsf{A}_1 < \mathsf{A}_2]\!]\overline{\rho} &\triangleq \mathsf{let}\ \langle A_1, A_2\rangle = \otimes^{\text{¤}} \langle \mathscr{A}^{\text{¤}}[\![\mathsf{A}_1]\!]\,\overline{\rho}, \mathscr{A}^{\text{¤}}[\![\mathsf{A}_2]\!]\,\overline{\rho}\rangle\ \mathsf{in} \qquad (28.38)\\
&\qquad \overset{\text{¤}}{\mathscr{A}}^{-1}[\![\mathsf{A}_1]\!]\, A_1\, \overline{\rho} \mathbin{\dot{\sqcap}^{\text{¤}}} \overset{\text{¤}}{\mathscr{A}}^{-1}[\![\mathsf{A}_2]\!]\, A_2\, \overline{\rho}\\
\overline{\mathsf{test}}^{\text{¤}}[\![\mathsf{A}_1 < \mathsf{A}_2]\!]\overline{\rho} &\triangleq \mathsf{let}\ \langle A_1, A_2\rangle = \overline{\otimes}^{\text{¤}} \langle \mathscr{A}^{\text{¤}}[\![\mathsf{A}_1]\!]\,\overline{\rho}, \mathscr{A}^{\text{¤}}[\![\mathsf{A}_2]\!]\,\overline{\rho}\rangle\ \mathsf{in}\\
&\qquad \overset{\text{¤}}{\mathscr{A}}^{-1}[\![\mathsf{A}_1]\!]\, A_1\, \overline{\rho} \mathbin{\dot{\sqcap}^{\text{¤}}} \overset{\text{¤}}{\mathscr{A}}^{-1}[\![\mathsf{A}_0]\!]\, A_2\, \overline{\rho}\\
\mathsf{test}^{\text{¤}}[\![\mathsf{B}_1\ \mathsf{nand}\ \mathsf{B}_1]\!]\overline{\rho} &\triangleq \quad (\mathsf{test}^{\text{¤}}[\![\mathsf{B}_1]\!]\overline{\rho} \mathbin{\dot{\sqcap}^{\text{¤}}} \overline{\mathsf{test}}^{\text{¤}}[\![\mathsf{B}_2]\!]\overline{\rho})\\
&\quad \dot{\sqcup}^{\text{¤}}\ (\overline{\mathsf{test}}^{\text{¤}}[\![\mathsf{B}_1]\!]\overline{\rho} \mathbin{\dot{\sqcap}^{\text{¤}}} \mathsf{test}^{\text{¤}}[\![\mathsf{B}_2]\!]\overline{\rho})\\
&\quad \dot{\sqcup}^{\text{¤}}\ (\overline{\mathsf{test}}^{\text{¤}}[\![\mathsf{B}_1]\!]\overline{\rho} \mathbin{\dot{\sqcap}^{\text{¤}}} \overline{\mathsf{test}}^{\text{¤}}[\![\mathsf{B}_2]\!]\overline{\rho})\\
\overline{\mathsf{test}}^{\text{¤}}[\![\mathsf{B}_1\ \mathsf{nand}\ \mathsf{B}_2]\!]\overline{\rho} &\triangleq \mathsf{test}^{\text{¤}}[\![\mathsf{B}_1]\!]\overline{\rho} \mathbin{\dot{\sqcap}^{\text{¤}}} \mathsf{test}^{\text{¤}}[\![\mathsf{B}_2]\!]\overline{\rho}
\end{aligned}
$$

is designed by calculus to be sound.

Theorem 28.39 (soundness of the structural Cartesian abstract semantics of tests B**)** If $\otimes^{\times}$ is increasing argumentwise, $\otimes^{\times} \langle \gamma_{\text{¤}}(X), \gamma_{\text{¤}}(Y)\rangle \mathrel{\dot{\sqsubseteq}^{\times}} \dot{\gamma}_{\text{¤}}(\otimes^{\text{¤}} \langle X, Y\rangle))$, and $\gamma_{\text{¤}}$ preserves finite glbs then $\mathsf{test}^{\times}[\![\mathsf{B}]\!] \circ \dot{\gamma}_{\text{¤}} \mathrel{\dot{\sqsubseteq}^{\times}} \dot{\gamma}_{\text{¤}} \circ \mathsf{test}^{\text{¤}}[\![\mathsf{B}]\!]$ and $\overline{\mathsf{test}}^{\times}[\![\mathsf{B}]\!] \circ \dot{\gamma}_{\text{¤}} \mathrel{\dot{\sqsubseteq}^{\times}} \dot{\gamma}_{\text{¤}} \circ \overline{\mathsf{test}}^{\text{¤}}[\![\mathsf{B}]\!]$.

Proof of theorem 28.39 By structural induction on the Boolean expression B.

$$
\begin{aligned}
&-\ \mathsf{test}^{\times}[\![\mathsf{A}_1 < \mathsf{A}_2]\!](\dot{\gamma}_{\text{¤}}(\overline{\rho}))\\
&=\ \mathsf{let}\ \langle A_1, A_2\rangle = \otimes^{\times} \langle \mathscr{A}^{\times}[\![\mathsf{A}_1]\!]\,(\dot{\gamma}_{\text{¤}}(\overline{\rho})), \mathscr{A}^{\times}[\![\mathsf{A}_2]\!]\,(\dot{\gamma}_{\text{¤}}(\overline{\rho}))\rangle\ \mathsf{in}\\
&\qquad \overset{\times}{\mathscr{A}}^{-1}[\![\mathsf{A}_1]\!]\, A_1\, (\dot{\gamma}_{\text{¤}}(\overline{\rho})) \mathbin{\dot{\sqcap}^{\times}} \overset{\times}{\mathscr{A}}^{-1}[\![\mathsf{A}_2]\!]\, A_2\, (\dot{\gamma}_{\text{¤}}(\overline{\rho}))\\
&\qquad\qquad \text{(definition (28.38), identical to (28.34) when } \dot{\sqcap}^{\times} = \dot{\cap}\text{)}\\
&\dot{\sqsubseteq}^{\times}\ \mathsf{let}\ \langle A_1, A_2\rangle = \otimes^{\times} \langle \gamma_{\text{¤}}(\mathscr{A}^{\times}[\![\mathsf{A}_1]\!]\,\overline{\rho}), \gamma_{\text{¤}}(\mathscr{A}^{\times}[\![\mathsf{A}_2]\!]\,\overline{\rho})\rangle\ \mathsf{in}\\
&\qquad \overset{\times}{\mathscr{A}}^{-1}[\![\mathsf{A}_1]\!]\, A_1\, (\dot{\gamma}_{\text{¤}}(\overline{\rho})) \mathbin{\dot{\sqcap}^{\times}} \overset{\times}{\mathscr{A}}^{-1}[\![\mathsf{A}_2]\!]\, A_2\, (\dot{\gamma}_{\text{¤}}(\overline{\rho}))
\end{aligned}
$$

(theorem 28.18 and remark 28.21 so that $\mathscr{A}^{\times}[\![\mathsf{A}]\!] \circ \dot{\gamma}_{\text{¤}} \sqsubseteq^{\times} \gamma_{\text{¤}} \circ \mathscr{A}^{\text{¤}}[\![\mathsf{A}]\!]$, $\otimes^{\times}$ is increasing argumentwise, $\overset{\times}{\mathscr{A}}^{-1}[\![\mathsf{A}_i]\!]\, A_i$ is increasing in argument A_i, $i = 1, 2$, and the glb $\dot{\sqcap}^{\times}$ is increasing argumentwise)

$$
\begin{aligned}
&\dot{\sqsubseteq}^{\times}\ \mathsf{let}\ \langle A_1, A_2\rangle = \otimes^{\text{¤}} \langle \mathscr{A}^{\times}[\![\mathsf{A}_1]\!]\,\overline{\rho}, \mathscr{A}^{\times}[\![\mathsf{A}_2]\!]\,\overline{\rho}\rangle\ \mathsf{in}\\
&\qquad \overset{\times}{\mathscr{A}}^{-1}[\![\mathsf{A}_1]\!]\, \gamma_{\text{¤}}(A_1)\, (\dot{\gamma}_{\text{¤}}(\overline{\rho})) \mathbin{\dot{\sqcap}^{\times}} \overset{\times}{\mathscr{A}}^{-1}[\![\mathsf{A}_2]\!]\, \gamma_{\text{¤}}(A_2)\, (\dot{\gamma}_{\text{¤}}(\overline{\rho}))
\end{aligned}
$$

⦇by hypothesis $\otimes^{\times}\langle\gamma_{\text{¤}}(X),\ \gamma_{\text{¤}}(Y)\rangle \mathrel{\dot{\sqsubseteq}^{\times}} \dot{\gamma}_{\text{¤}}(\otimes^{\text{¤}}\langle X,\ Y\rangle))$, $\otimes^{\times}$ is increasing argumentwise, $\mathscr{A}^{\overset{\times}{-1}}[\![\mathsf{A}_i]\!]\,A_i$ is increasing in argument A_i, $i = 1, 2$, and the glb $\dot{\sqcap}^{\times}$ is increasing argumentwise⦈

$\dot{\sqsubseteq}^{\times}$ $\mathsf{let}\ \langle A_1,\ A_2\rangle = \otimes^{\text{¤}}\langle\mathscr{A}^{\times}[\![\mathsf{A}_1]\!]\,\overline{\rho},\ \mathscr{A}^{\times}[\![\mathsf{A}_2]\!]\,\overline{\rho}\rangle\ \mathsf{in}$
$\quad\dot{\gamma}_{\text{¤}}(\mathscr{A}^{\overset{\text{¤}}{-1}}[\![\mathsf{A}_1]\!]\,A_1\,\overline{\rho}) \mathbin{\dot{\sqcap}^{\times}} \dot{\gamma}_{\text{¤}}(\mathscr{A}^{\overset{\text{¤}}{-1}}[\![\mathsf{A}_2]\!]\,A_2\,\overline{\rho})$

⦇theorem 28.31 and remark 28.32 so that $\mathscr{A}^{\overset{\times}{-1}}[\![\mathsf{A}]\!]\,\gamma_{\text{¤}}(\chi)\circ\dot{\gamma}_{\text{¤}} \mathrel{\dot{\subseteq}} \dot{\gamma}_{\text{¤}}\circ\mathscr{A}^{\overset{\text{¤}}{-1}}[\![\mathsf{A}]\!]\,\chi$ (where $\dot{\subseteq}$ is $\dot{\sqsubseteq}^{\times}$) and the lub $\dot{\sqcap}^{\text{¤}}$ is increasing) ⦈

$=$ $\dot{\gamma}_{\text{¤}}(\mathsf{let}\ \langle A_1,\ A_2\rangle = \otimes^{\text{¤}}\langle\mathscr{A}^{\times}[\![\mathsf{A}_1]\!]\,\overline{\rho},\ \mathscr{A}^{\times}[\![\mathsf{A}_2]\!]\,\overline{\rho}\rangle\ \mathsf{in}$
$\quad\mathscr{A}^{\overset{\text{¤}}{-1}}[\![\mathsf{A}_1]\!]\,A_1\,\overline{\rho} \mathbin{\dot{\sqcap}^{\text{¤}}} \mathscr{A}^{\overset{\text{¤}}{-1}}[\![\mathsf{A}_2]\!]\,A_2\,\overline{\rho})$

⦇$\gamma_{\text{¤}}$ hence $\dot{\gamma}_{\text{¤}}$ preserves finite glbs and factoring $\dot{\gamma}_{\text{¤}}$ out of the let⦈

$=$ $\dot{\gamma}_{\text{¤}}(\overline{\mathsf{test}}^{\text{¤}}[\![\mathsf{A}_1 < \mathsf{A}_2]\!]\overline{\rho})$ ⦇definition (28.38) of $\overline{\mathsf{test}}^{\text{¤}}[\![\mathsf{A}_1 < \mathsf{A}_2]\!]$⦈

– The case $\overline{\mathsf{test}}^{\text{¤}}[\![\mathsf{A}_1 < \mathsf{A}_2]\!]$ is similar.

– $\mathsf{test}^{\times}[\![\mathsf{B}_1\ \mathsf{nand}\ \mathsf{B}_1]\!](\dot{\gamma}_{\text{¤}}(\overline{\rho}))$

$=$ $(\mathsf{test}^{\times}[\![\mathsf{B}_1]\!](\dot{\gamma}_{\text{¤}}(\overline{\rho})) \mathbin{\dot{\sqcap}^{\times}} \overline{\mathsf{test}}^{\times}[\![\mathsf{B}_2]\!](\dot{\gamma}_{\text{¤}}(\overline{\rho}))) \mathbin{\dot{\sqcup}^{\times}} (\overline{\mathsf{test}}^{\times}[\![\mathsf{B}_1]\!](\dot{\gamma}_{\text{¤}}(\overline{\rho})) \mathbin{\dot{\sqcap}^{\times}} \mathsf{test}^{\times}[\![\mathsf{B}_2]\!](\dot{\gamma}_{\text{¤}}(\overline{\rho}))) \mathbin{\dot{\sqcup}^{\times}}$
$(\overline{\mathsf{test}}^{\times}[\![\mathsf{B}_1]\!](\dot{\gamma}_{\text{¤}}(\overline{\rho})) \mathbin{\dot{\sqcap}^{\times}} \overline{\mathsf{test}}^{\times}[\![\mathsf{B}_2]\!](\dot{\gamma}_{\text{¤}}(\overline{\rho})))$

⦇definition (28.38), identical to (28.34) when $\dot{\sqcap}^{\times} = \dot{\cap}$ and $\dot{\sqcup}^{\times} = \dot{\cup}$⦈

$\dot{\sqsubseteq}^{\times}$ $((\dot{\gamma}_{\text{¤}}(\mathsf{test}^{\text{¤}}[\![\mathsf{B}_1]\!]\overline{\rho})) \mathbin{\dot{\sqcap}^{\times}} (\dot{\gamma}_{\text{¤}}(\overline{\mathsf{test}}^{\text{¤}}[\![\mathsf{B}_2]\!]\overline{\rho}))) \mathbin{\dot{\sqcup}^{\times}} ((\dot{\gamma}_{\text{¤}}(\overline{\mathsf{test}}^{\text{¤}}[\![\mathsf{B}_1]\!]\overline{\rho})) \mathbin{\dot{\sqcap}^{\times}} (\dot{\gamma}_{\text{¤}}(\mathsf{test}^{\text{¤}}[\![\mathsf{B}_2]\!]\overline{\rho}))) \mathbin{\dot{\sqcup}^{\times}}$
$((\dot{\gamma}_{\text{¤}}(\overline{\mathsf{test}}^{\text{¤}}[\![\mathsf{B}_1]\!]\overline{\rho})) \mathbin{\dot{\sqcap}^{\times}} (\dot{\gamma}_{\text{¤}}(\overline{\mathsf{test}}^{\text{¤}}[\![\mathsf{B}_2]\!]\overline{\rho})))$

⦇structural induction hypothesis and $\dot{\sqcap}^{\times}$, $\dot{\sqcup}^{\times}$ are increasing in both arguments⦈

$=$ $(\dot{\gamma}_{\text{¤}}(\mathsf{test}^{\text{¤}}[\![\mathsf{B}_1]\!]\overline{\rho} \mathbin{\dot{\sqcap}^{\text{¤}}} \overline{\mathsf{test}}^{\text{¤}}[\![\mathsf{B}_2]\!]\overline{\rho})) \mathbin{\dot{\sqcup}^{\times}} (\dot{\gamma}_{\text{¤}}(\overline{\mathsf{test}}^{\text{¤}}[\![\mathsf{B}_1]\!]\overline{\rho} \mathbin{\dot{\sqcap}^{\text{¤}}} \mathsf{test}^{\text{¤}}[\![\mathsf{B}_2]\!]\overline{\rho})) \mathbin{\dot{\sqcup}^{\times}}$
$(\dot{\gamma}_{\text{¤}}(\overline{\mathsf{test}}^{\text{¤}}[\![\mathsf{B}_1]\!]\overline{\rho} \mathbin{\dot{\sqcap}^{\text{¤}}} \overline{\mathsf{test}}^{\text{¤}}[\![\mathsf{B}_2]\!]\overline{\rho}))$ ⦇$\gamma_{\text{¤}}$ hence $\dot{\gamma}_{\text{¤}}$ preserves finite meets⦈

$\dot{\sqsubseteq}^{\times}$ $\dot{\gamma}_{\text{¤}}((\mathsf{test}^{\text{¤}}[\![\mathsf{B}_1]\!]\overline{\rho} \mathbin{\dot{\sqcap}^{\text{¤}}} \overline{\mathsf{test}}^{\text{¤}}[\![\mathsf{B}_2]\!]\overline{\rho}) \mathbin{\dot{\sqcup}^{\text{¤}}} (\overline{\mathsf{test}}^{\text{¤}}[\![\mathsf{B}_1]\!]\overline{\rho} \mathbin{\dot{\sqcap}^{\text{¤}}} \mathsf{test}^{\text{¤}}[\![\mathsf{B}_2]\!]\overline{\rho}) \mathbin{\dot{\sqcup}^{\text{¤}}} (\overline{\mathsf{test}}^{\text{¤}}[\![\mathsf{B}_1]\!]\overline{\rho} \mathbin{\dot{\sqcap}^{\text{¤}}}$
$\overline{\mathsf{test}}^{\text{¤}}[\![\mathsf{B}_2]\!]\overline{\rho}))$

⦇$\gamma_{\text{¤}}$ hence $\dot{\gamma}_{\text{¤}}$ preserves finite meets hence is increasing and definition of the glb⦈

$=$ $\dot{\gamma}_{\text{¤}}(\mathsf{test}^{\text{¤}}[\![\mathsf{B}_1\ \mathsf{nand}\ \mathsf{B}_1]\!]\,\overline{\rho})$ ⦇definition (28.38) of $\mathsf{test}^{\text{¤}}[\![\mathsf{B}_1\ \mathsf{nand}\ \mathsf{B}_1]\!]$⦈

– The case $\overline{\mathsf{test}}^{\text{¤}}[\![\mathsf{B}_1\ \mathsf{nand}\ \mathsf{B}_2]\!]$ is similar. □

Remark 28.40 The proof of theorem 28.39 shows that the theorem holds for an abstraction of the Cartesian semantics as well as for abstractions of abstractions of this Cartesian semantics. The hypotheses are on the concrete semantics hence, by composition of abstractions, need only hold for the Cartesian semantics, which they do.

Exercise 28.41 Prove that $\mathsf{test}^{\text{¤}}[\![\mathsf{B}]\!]$ is increasing and reductive.

28.7 Cartesian Domain

28.7.1 Cartesian Abstract Domain of Abstract Value Properties

Theorems 28.15, 28.29, and 28.35 show that all Cartesian abstract domains have the same mathematical structure while Theorems 28.18, 28.19, 28.31, and 28.39 show that this mathematical structure is preserved by further Cartesian abstractions. They all have the same generic structure

$$\mathbb{D}^{¤} \triangleq \langle \mathbb{P}^{¤}, \sqsubseteq^{¤}, \bot^{¤}, \top^{¤}, \sqcup^{¤}, \sqcap^{¤}, 1^{¤}, \ominus^{¤}, \ominus^{\overset{¤}{-1}}, \oslash^{¤}, \overline{\oslash}^{¤} \rangle \tag{28.42}$$

They are all abstractions of the value collecting domain

$$\langle \wp(\mathbb{V}), \subseteq, \emptyset, \mathbb{V}, \cup, \cap, \{1\}, \ominus^{\times}, \ominus^{\overset{\times}{-1}}, \oslash^{\times}, \overline{\oslash}^{\times} \rangle$$

for a concretization $\gamma_{¤} \in \mathbb{P}^{¤} \xrightarrow{\nearrow} \wp(\mathbb{V})$ satisfying the hypotheses of Theorems 28.18, 28.19, 28.31, and 28.39 (which are satisfied by the Cartesian collecting semantics).

28.7.2 Cartesian Abstract Domain of Abstract Reachability Properties

The abstract interpreter of chapter 21 can be used with the Cartesian abstract domain of abstract reachability properties of the form

$$\mathbb{D}^{\dot{¤}} \triangleq \langle \mathcal{V} \to \mathbb{P}^{¤}, \dot{\sqsubseteq}^{¤}, \dot{\bot}^{¤}, \dot{\sqcup}^{¤}, \text{assign}^{¤}[\![\mathsf{x}, \mathsf{A}]\!], \text{test}^{¤}[\![\mathsf{B}]\!], \overline{\text{test}}^{¤}[\![\mathsf{B}]\!] \rangle \tag{28.43}$$

for the concretizations $\dot{\gamma}_{¤} \in (\mathcal{V} \to \mathbb{P}^{¤}) \xrightarrow{\nearrow} (\mathcal{V} \to \wp(\mathbb{V}))$ and $\dot{\gamma}_{\ddot{\times}} \in (\mathcal{V} \to \wp(\mathbb{V})) \xrightarrow{\nearrow} \wp(\mathcal{V} \to \mathbb{V})$.

Theorem 28.44 (well definedness and soundness of the Cartesian static analysis) The Cartesian abstract domain $\mathbb{D}^{\dot{¤}}$ (28.43) is well defined according to definition 21.1 so the abstract interpreter $\widehat{\mathcal{S}}^{\dot{¤}}[\![\mathsf{S}]\!]$ is well defined and a sound abstraction of the assertional forward reachability semantics $\mathcal{S}^{\vec{r}}[\![\mathsf{S}]\!]$ for any program component $\mathsf{S} \in \mathbb{P}\mathbb{c}$.

Proof of theorem 28.44 By theorems 28.18, 28.19, 28.31, and 28.39, the Cartesian abstract domain $\mathbb{D}^{\dot{¤}}$ is well defined according to definition 21.1 and an approximate domain abstraction according to definition 27.1.I. We conclude, by theorems 21.16 and 27.4, that the abstract interpreter of 21.3–21.13 of chapter 21, instantiated with the Cartesian abstract domain $\mathbb{D}^{\dot{¤}}$ of abstract reachability properties of (28.43) is a sound abstraction of the assertional forward reachability semantics $\mathcal{S}^{\vec{r}}[\![\mathsf{S}]\!]$ of chapter 19. □

Remark 28.45 Observe that the concrete/abstract domains are *universal algebræ* or, briefly *algebræ*, a set S equipped with finitary operations on S, that is, taking a finite number of parameters, which is 0 for constants [444].

All operations of the Cartesian algebræ $\mathbb{D}^{\dot{¤}}$ are defined in term of the operations of the Cartesian algebra $\mathbb{D}^{¤}$, so there is a functor $\dot{\mathfrak{F}}^{\times}[\![\mathcal{V}]\!]$ mapping the Cartesian algebræ $\mathbb{D}^{¤}$ to the Cartesian algebræ $\mathbb{D}^{\dot{¤}}$, for which the operations are defined in theorems 28.18, 28.19, 28.31, and 28.39. Some programming languages directly support such algebræ and functors such as OCaml modules and module functors [628]. The abstract domain $\mathbb{D}^{\dot{¤}}$ can be efficiently implemented as a functor taking as parameter the module implementing $\mathbb{D}^{¤}$. The abstract interpreter is then itself a functor taking as parameter the module implementing $\mathbb{D}^{\dot{¤}}$.

Exercise 28.46 (Project: Cartesian reachability interpreter) Continuing exercises 5.9 and 19.34, write a program that reads a program P, and computes and prints $\widehat{\mathcal{S}}^{\times}[\![\mathsf{P}]\!]\,(\mathsf{x} \in \mathcal{V} \mapsto \{0\})\,\ell$ for all $\ell \in \mathbb{L}$ (instantiated to $\mathcal{V} = \mathsf{vars}[\![\mathsf{P}]\!]$ and $\mathbb{L} = \mathsf{labs}[\![\mathsf{P}]\!]$).

Exercise 28.47 (Project: Cartesian reachability compiler) Continuing exercise 19.35 and exercise 28.46, write a program that reads a program P and outputs an object program compile⟦P⟧ that, when executed, computes and prints $\widehat{\mathcal{S}}^{\times}$⟦P⟧ $(\mathsf{x} \in \mathbb{V} \mapsto \{0\})\,\ell$ for all $\ell \in \mathbb{L}$ (instantiated to $\mathbb{V}$ = vars⟦P⟧ and $\mathbb{L}$ = labs⟦P⟧).

28.8 Examples of Cartesian Abstractions

28.8.1 Parity Abstraction

The Cartesian parity abstract domain $\mathbb{D}^{\dot{2}} \triangleq \langle \mathbb{V} \to \mathbb{P}^2, \sqsubseteq^{\dot{2}}, \bot^{\dot{2}}, \sqcup^{\dot{2}}, \mathsf{assign}^{\dot{2}}[\![\mathsf{x}, \mathsf{A}]\!], \mathsf{test}^{\dot{2}}[\![\mathsf{B}]\!], \overline{\mathsf{test}}^{\dot{2}}[\![\mathsf{B}]\!]\rangle$ is the pointwise extension of the parity value abstract domain $\langle \mathbb{P}^2, \sqsubseteq^2, \bot^2, \top^2, \sqcup^2, \sqcap^2\rangle$, as follows:

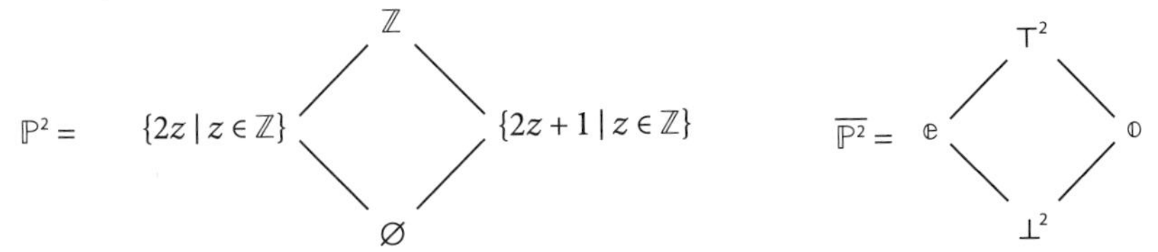

For the Cartesian reachability semantics of arithmetic expressions $\mathsf{assign}^2[\![\mathsf{x}, \mathsf{A}]\!]$, we use (28.17) with $\alpha_2(\{1\}) = \mathbb{o}$ and

x	e	e	o	o	$\_$	$\bot^2/\top^2$
y	e	o	e	o	$\bot^2/\top^2$	$\_$
$x \ominus^2 y$	e	o	o	e	$\bot^2/\top^2$	$\bot^2/\top^2$

For the Cartesian accessibility semantics $\mathscr{A}^{\vec{2}_1}[\![\mathsf{A}]\!]$ of an arithmetic expression A, we use (28.30) with

χ	$p1$	$p2$	$\ominus^{\vec{2}_1}\chi\,\langle p1, p2\rangle$
$\top^2$	$\_$	$\_$	$\langle p1, p2\rangle$
e	e	e	⟨e, e⟩
e	o	o	⟨o, o⟩
o	e	o	⟨e, o⟩
o	o	e	⟨o, e⟩
$\_$	$\_$	$\_$	$\langle \bot^2, \bot^2\rangle$

For tests $\mathsf{test}^2[\![\mathsf{B}]\!]$, the equality implies that the parities are the same. The other comparison tests provide no information on parity.

Exercise 28.48 Following exercise 21.22, implement the parity abstract domain (e.g., for simplicity, for programs with only two variables x and y). For example

```
<l1:x=e, y=e; l4:x=o, y=e; ff; l0:x=_|_, y=_|_>  Prog:
<l1:x=e, y=e; l2:x=o, y=e; ff; l0:x=_|_, y=_|_>     l1: x = 1;
<l2:x=o, y=e; l4:x=o, y=e; ff; l0:x=_|_, y=_|_>     (while l2: (0 < 1)
<l3:x=o, y=e; l2:x=o, y=e; ff; l0:x=_|_, y=_|_>        l3: x = (x + 2); )
                                                    l4:
```

28.8.2 Sign Abstraction

The Cartesian sign abstract domain $\mathbb{D}^{\dot{\pm}} \triangleq \langle \mathbb{V} \to \mathbb{P}^{\pm}, \sqsubseteq^{\dot{\pm}}, \bot^{\dot{\pm}}, \sqcup^{\dot{\pm}}, \mathsf{assign}^{\dot{\pm}}[\![\mathsf{x}, \mathsf{A}]\!], \mathsf{test}^{\dot{\pm}}[\![\mathsf{B}]\!], \overline{\mathsf{test}}^{\dot{\pm}}[\![\mathsf{B}]\!]\rangle$ is the pointwise extension of the sign value abstract domain $\langle \mathbb{P}^{\pm}, \sqsubseteq^{\pm}, \bot^{\pm}, \top^{\pm}, \sqcup^{\pm}, \sqcap^{\pm}\rangle$ considered in section 3.12. The assignment is $A[x \leftarrow \mathscr{A}^{\pm}[\![\mathsf{A}]\!]A]$ where the analysis $\mathscr{A}^{\pm}[\![\mathsf{A}]\!]A$ expression A applies the rule of signs as designed in section 3.20. A simple handling of Boolean expressions would be case analysis

$$
\begin{array}{rcl}
\mathsf{test}^{\dot{\pm}}[\![\mathsf{x} == \mathsf{0}]\!]\overline{\rho} & \triangleq & \llparenthesis\, \overline{\rho}(\mathsf{x}) \sqcap^{\pm} (=0) = \bot_{\pm} \mathrel{?} \mathsf{x} \mapsto \bot_{\pm} \mathrel{⁏} \overline{\rho}[x \leftarrow (=0)] \,\rrparenthesis \\
\mathsf{test}^{\dot{\pm}}[\![\mathsf{x} == \mathsf{y}]\!]\overline{\rho} & \triangleq & \llparenthesis\, \overline{\rho}(\mathsf{x}) \sqcap^{\pm} \overline{\rho}(\mathsf{y}) = \bot_{\pm} \mathrel{?} \mathsf{x} \mapsto \bot_{\pm} \\
 & & \mathrel{⁏} \overline{\rho}[x \leftarrow \overline{\rho}(\mathsf{x}) \sqcap^{\pm} \overline{\rho}(\mathsf{y})][y \leftarrow \overline{\rho}(\mathsf{x}) \sqcap^{\pm} \overline{\rho}(\mathsf{y})] \,\rrparenthesis \\
\mathsf{test}^{\dot{\pm}}[\![\mathsf{x} > \mathsf{0}]\!]\overline{\rho} & \triangleq & \llparenthesis\, \overline{\rho}(\mathsf{x}) \sqcap^{\pm} (>0) = \bot_{\pm} \mathrel{?} \mathsf{x} \mapsto \bot_{\pm} \mathrel{⁏} \overline{\rho}[x \leftarrow \overline{\rho}(\mathsf{x}) \sqcap^{\pm} (>0)] \,\rrparenthesis \\
\ldots & \ldots & \ldots \\
\mathsf{test}^{\dot{\pm}}[\![\mathsf{A}]\!]\overline{\rho} & \triangleq & \overline{\rho}
\end{array}
$$

The treatment of tests in theorem 28.39 is more precise. The operators $\otimes^{\pm}_{-1}$ and $\overline{\otimes}^{\pm}_{-1}$ are given in [225].

28.8.3 Constancy Abstraction

The Cartesian constancy abstract domain $\mathbb{D}^{\dot{c}} \triangleq \langle \mathbb{V} \to \mathbb{P}^{c}, \sqsubseteq^{\dot{c}}, \bot^{\dot{c}}, \top^{\dot{c}}, \sqcup^{\dot{c}}, \mathsf{assign}^{\dot{c}}[\![\mathsf{x}, \mathsf{A}]\!], \mathsf{test}^{\dot{c}}[\![\mathsf{B}]\!], \overline{\mathsf{test}}^{\dot{c}}[\![\mathsf{B}]\!]\rangle$ is the pointwise extension of the constant value abstract domain $\mathbb{D}^{c} \triangleq \langle \mathbb{P}^{c}, \sqsubseteq^{c}, \bot^{c}, \top^{c}, \sqcup^{c}, \mathsf{assign}^{c}[\![\mathsf{x}, \mathsf{A}]\!], \mathsf{test}^{c}[\![\mathsf{B}]\!], \overline{\mathsf{test}}^{c}[\![\mathsf{B}]\!]\rangle$ of exercise 3.45.

For assignments and tests, $\bot^{c}$ and $\top^{\dot{c}}$ are absorbent. Otherwise the assignments and tests involve only constants and can be evaluated in the concrete.

28.9 Conclusion

We have designed a collecting semantics for Cartesian abstract interpretation by abstraction of the reachability semantics. Although the given examples are for finitary abstract domains, this is not mandatory (contrary to the understanding of static analysis in which values are replaced by symbolic abstract values [572, 741, 899, 967]). Interval analysis [234] is an example of infinitary Cartesian abstraction. Cartesian abstraction is the basis of SLAM [80], a widely advertised static analyzer of device drivers lacking precision and scalability. It is also used in Astrée [113] for analyzing control-command software, but in conjunction with many complementary relational scalable abstract domains for precision. Cartesian abstraction need not be on all dimensions of the abstract domain; an example is [506]. Moreover, the Cartesian abstraction need not necessarily be on variables. It can be on any finite projection set, for example, on selected program expressions [919, 920].

Exercise 28.49 (Project: Cartesian abstract interpreter) Continuing exercise 21.22, implement the Cartesian abstract domain functor $\mathbb{D}^{\dot{\text{¤}}}$ of (28.43) as well as the Cartesian abstract domains $\mathbb{D}^{\text{¤}}$ for the parity, constancy, and/or sign static analyses.

28.10 Solutions to Selected Exercises

Solution to exercise 28.2

$$\dot{\alpha}_{\vec{\times}}(P) \dot{\subseteq} \overline{P}$$

$\Leftrightarrow \forall \mathsf{x} \in \mathbb{V} . \dot{\alpha}_{\vec{\times}}(P)\mathsf{x} \subseteq \overline{P}(\mathsf{x})$ ⟅pointwise definition of $\dot{\subseteq}$⟆

$\Leftrightarrow \forall \mathsf{x} \in \mathbb{V} . \{\rho(\mathsf{x}) \mid \rho \in P\} \subseteq \overline{P}(\mathsf{x})$ ⟅definition (28.1) of $\dot{\alpha}_{\vec{\times}}$⟆

$\Leftrightarrow \forall \mathsf{x} \in \mathbb{V} . \forall \rho \in P . \rho(\mathsf{x}) \in \overline{P}(\mathsf{x})$ ⟅definition of $\subseteq$⟆

$\Leftrightarrow \forall \rho \in P . \forall \mathsf{x} \in \mathbb{V} . \rho(\mathsf{x}) \in \overline{P}(\mathsf{x})$ ⟅definition of $\forall$⟆

$\Leftrightarrow P \subseteq \{\rho \in \mathbb{E}\mathbb{v} \mid \forall \mathsf{x} \in \mathbb{V} . \rho(\mathsf{x}) \in \overline{P}(\mathsf{x})\}$ ⟅definition of $\subseteq$⟆

$\Leftrightarrow P \subseteq \dot{\gamma}_{\vec{\times}}(\overline{P})$ ⟅definition (28.1) of $\dot{\gamma}_{\vec{\times}}$⟆

Solution to exercise 28.36

$-\ \dot{\alpha}_{\vec{\times}}(\mathsf{test}^{\vec{r}}[\![\mathsf{A}_1 < \mathsf{A}_2]\!](\mathscr{P}_0))$

$= \dot{\alpha}_{\vec{\times}}(\{\rho \in \mathscr{P}_0 \mid \mathscr{B}[\![\mathsf{A}_1 < \mathsf{A}_2]\!]\,\rho = \mathsf{tt}\})$ ⟅definition (19.16) of $\mathsf{test}^{\vec{r}}[\![\mathsf{B}]\!]$⟆

$= \dot{\alpha}_{\vec{\times}}(\{\rho \in \mathscr{P}_0 \mid \mathscr{A}[\![\mathsf{A}_1]\!]\,\rho < \mathscr{A}[\![\mathsf{A}_2]\!]\,\rho\})$ ⟅definition (3.4) of $\mathscr{B}[\![\mathsf{A}_1 < \mathsf{A}_2]\!]$⟆

$\dot{\subseteq}\ \dot{\alpha}_{\vec{\times}}(\{\rho \in \mathscr{P}_0 \mid \exists v_2 \in \{\mathscr{A}[\![\mathsf{A}_2]\!]\,\rho \mid \rho \in \mathscr{P}_0\} . \mathscr{A}[\![\mathsf{A}_1]\!]\,\rho < v_2\} \cap \{\rho \in \mathscr{P}_0 \mid \exists v_1 \in$
$\{\mathscr{A}[\![\mathsf{A}_1]\!]\,\rho \mid \rho \in \mathscr{P}_0\} . v_1 < \mathscr{A}[\![\mathsf{A}_2]\!]\,\rho\})$
⟅because $\{\rho \in S \mid f(\rho) < g(\rho)\} \subseteq \{\rho \in S \mid \exists v_1 \in \{f(\rho) \mid \rho \in S\} . v_1 < g(\rho)\}$ and similarly for g, and $\dot{\alpha}_{\vec{\times}}$ increasing⟆

$= \dot{\alpha}_{\vec{\times}}(\{\rho \in \mathscr{P}_0 \mid \mathscr{A}[\![\mathsf{A}_1]\!]\rho \in \{v_1 \in \{\mathscr{A}[\![\mathsf{A}_1]\!]\,\rho \mid \rho \in \mathscr{P}_0\} \mid \exists v_2 \in \{\mathscr{A}[\![\mathsf{A}_2]\!]\,\rho \mid \rho \in \mathscr{P}_0\} . v_1 < v_2\}\} \cap$
$\{\rho \in \mathscr{P}_0 \mid \mathscr{A}[\![\mathsf{A}_2]\!]\rho \in \{v_2 \in \{\mathscr{A}[\![\mathsf{A}_2]\!]\,\rho \mid \rho \in \mathscr{P}_0\} \mid \exists v_1 \in \{\mathscr{A}[\![\mathsf{A}_1]\!]\,\rho \mid \rho \in \mathscr{P}_0\} . v_1 < v_2\}\})$
⟅definition of $\in$, letting $v1 = \mathscr{A}[\![\mathsf{A}_1]\!]\,\rho$ and $v_2 = \mathscr{A}[\![\mathsf{A}_2]\!]\rho$⟆

$\dot{\subseteq}\ \dot{\alpha}_{\vec{\times}}(\mathscr{A}^{\times^{-1}}[\![\mathsf{A}_1]\!]\,\{v_1 \in \{\mathscr{A}[\![\mathsf{A}_1]\!]\,\rho \mid \rho \in \mathscr{P}_0\} \mid \exists v_2 \in \{\mathscr{A}[\![\mathsf{A}_2]\!]\,\rho \mid \rho \in \mathscr{P}_0\} . v_1 < v_2\}\,\mathscr{P}_0) \,\dot{\cap}$
$\dot{\alpha}_{\vec{\times}}(\mathscr{A}^{\times^{-1}}[\![\mathsf{A}_2]\!]\,\{v_2 \in \{\mathscr{A}[\![\mathsf{A}_2]\!]\,\rho \mid \rho \in \mathscr{P}_0\} \mid \exists v_1 \in \{\mathscr{A}[\![\mathsf{A}_1]\!]\,\rho \mid \rho \in \mathscr{P}_0\} . v_1 < v_2\}\,\mathscr{P}_0)$
⟅$\dot{\alpha}_{\vec{\times}}$ is $\dot{\subseteq}$-increasing by Galois connection of exercise 28.2, and $\dot{\cap}$ is $\dot{\subseteq}$ increasing.⟆

$= \mathsf{let}\ \langle \chi_1, \chi_2 \rangle = \langle \{v_1 \in \{\mathscr{A}[\![\mathsf{A}_1]\!]\,\rho \mid \rho \in \mathscr{P}_0\} \mid \exists v_2 \in \{\mathscr{A}[\![\mathsf{A}_2]\!]\,\rho \mid \rho \in \mathscr{P}_0\} . v_1 < v_2\},$
$\{v_2 \in \{\mathscr{A}[\![\mathsf{A}_2]\!]\,\rho \mid \rho \in \mathscr{P}_0\} \mid \exists v_1 \in \{\mathscr{A}[\![\mathsf{A}_1]\!]\,\rho \mid \rho \in \mathscr{P}_0\} . v_1 < v_2\}\rangle\ \mathsf{in}$
$\dot{\alpha}_{\vec{\times}}(\mathscr{A}^{\times^{-1}}[\![\mathsf{A}_1]\!]\,\chi_1\,\mathscr{P}_0) \,\dot{\cap}\, \dot{\alpha}_{\vec{\times}}(\mathscr{A}^{\times^{-1}}[\![\mathsf{A}_2]\!]\,\chi_2\,\mathscr{P}_0)$ ⟅definition of let⟆

$= \mathsf{let}\ \langle \chi_1, \chi_2 \rangle = \otimes^{\vec{\times}}\ \langle \{\mathscr{A}[\![\mathsf{A}_1]\!]\,\rho \mid \rho \in \mathscr{P}_0\}, \{\mathscr{A}[\![\mathsf{A}_2]\!]\,\rho \mid \rho \in \mathscr{P}_0\}\rangle\ \mathsf{in}$
$\dot{\alpha}_{\vec{\times}}(\mathscr{A}^{\times^{-1}}[\![\mathsf{A}_1]\!]\,\chi_1\,\mathscr{P}_0) \,\dot{\cap}\, \dot{\alpha}_{\vec{\times}}(\mathscr{A}^{\times^{-1}}[\![\mathsf{A}_2]\!]\,\chi_2\,\mathscr{P}_0)$ ⟅definition $\otimes^{\vec{\times}}$⟆

$\dot{\subseteq}\ \mathsf{let}\ \langle \chi_1, \chi_2 \rangle = \otimes^{\vec{\times}}\ \langle \mathscr{A}^{\times}[\![\mathsf{A}_1]\!]\dot{\alpha}_{\vec{\times}}(\mathscr{P}_0), \mathscr{A}^{\times}[\![\mathsf{A}_2]\!]\dot{\alpha}_{\vec{\times}}(\mathscr{P}_0)\rangle\ \mathsf{in}$
$\mathscr{A}^{\times^{-1}}[\![\mathsf{A}_1]\!]\,\chi_1\,\dot{\alpha}_{\vec{\times}}(\mathscr{P}_0) \,\dot{\cap}\, \mathscr{A}^{\times^{-1}}[\![\mathsf{A}_2]\!]\,\chi_2\,\dot{\alpha}_{\vec{\times}}(\mathscr{P}_0)$ ⟅induction hypothesis⟆

$= \mathsf{test}^{\vec{\times}}[\![\mathsf{A}_1 < \mathsf{A}_2]\!]\dot{\alpha}_{\vec{\times}}(\mathscr{P}_0)$ ⟅definition (28.34) of $\mathsf{test}^{\vec{\times}}[\![\mathsf{A}_1 < \mathsf{A}_2]\!]$⟆

29 Reduction

The idea of iterating an increasing and reductive abstract operator to get a more precise lower closure abstract operator was used to define the "reduced product" of [241]; in the "local decreasing iterations" [440] to handle backward assignments and conditionals (the same idea was later exploited in logic program analysis, as "reexecution" [621]); and in the iterated forward-backward program analysis of [221, section 5.7.5, pp. (5)-53–(5)-56]. We study this reduction idea and apply it to the analysis (28.38) of Boolean expressions.

Contents

29.1 Reduction

Lemma 29.1 ([239]) Let $g \in L \to L$ by an increasing and extensive operator on a complete lattice $\langle L, \sqsubseteq, \bot, \top, \sqcap, \sqcup\rangle$. Define $\hat{\rho}(x) \triangleq \mathsf{lfp}^{\sqsubseteq}_{x}\, g$ where $\mathsf{lfp}^{\sqsubseteq}_{a}\, f$ is the least fixpoint of f greater than or equal to a, if any.

Then the upper reduction $\hat{\rho}$ is the smallest upper closure operator pointwise greater than or equal to g (i.e., $g \dot{\sqsubseteq} \hat{\rho}$ and if ρ' is also increasing, extensive, and idempotent such that $g \dot{\sqsubseteq} \rho'$ then $g \dot{\sqsubseteq} \hat{\rho} \dot{\sqsubseteq} \rho'$).

Proof of lemma 29.1

— Let us first prove that $\hat{\rho}$ is an upper closure operator. By the variant exercise 15.11 of Tarski's fixpoint theorem 15.6 and g extensive so $x \sqsubseteq g(x)$, we have $\hat{\rho}(x) = \mathsf{lfp}^{\sqsubseteq}_{x}\, g = \bigsqcap\{y \mid g(y) \sqsubseteq y \wedge x \sqsubseteq y\}$.

$\hat{\rho}$ is obviously extensive because all y in $\{y \mid g(y) \sqsubseteq y \wedge x \sqsubseteq y\}$ are an upper bound of x hence $x \sqsubseteq \bigsqcap\{y \mid g(y) \sqsubseteq y \wedge x \sqsubseteq y\} = \mathsf{lfp}^{\sqsubseteq}_{x}\, g = \hat{\rho}(x)$ by definition of the lub $\bigsqcap$.

If $x \sqsubseteq x'$ then $\{y \mid g(y) \sqsubseteq y \wedge x' \sqsubseteq y\} \subseteq \{y \mid g(y) \sqsubseteq y \wedge x \sqsubseteq y\}$ so, by the order-dual of exercise 10.8, $\hat{\rho}(x) = \mathsf{lfp}^{\sqsubseteq}_{x}\, g = \bigsqcap\{y \mid g(y) \sqsubseteq y \wedge x \sqsubseteq y\} \sqsubseteq \bigsqcap\{y \mid g(y) \sqsubseteq y \wedge x' \sqsubseteq y\} = \mathsf{lfp}^{\sqsubseteq}_{x'}\, g = \hat{\rho}(x')$, proving that $\hat{\rho}$ is increasing.

By extension $x \sqsubseteq \hat{\rho}(x)$ so $\hat{\rho}(x) \sqsubseteq \hat{\rho}(\hat{\rho}(x))$ by increasingness. Moreover, $\hat{\rho}(x) = \mathsf{lfp}^{\sqsubseteq}_{x}\, g$ is a fixpoint of g so $g(\hat{\rho}(x)) \sqsubseteq \hat{\rho}(x)$ by reflexivity. It follows that $\hat{\rho}(x) \in \{y \mid g(y) \sqsubseteq y \wedge \hat{\rho}(x) \sqsubseteq y\}$ proving that $\hat{\rho}(\hat{\rho}(x)) = \mathsf{lfp}^{\sqsubseteq}_{\hat{\rho}(x)}\, g = \bigsqcap\{y \mid g(y) \sqsubseteq y \wedge \hat{\rho}(x) \sqsubseteq y\} \sqsubseteq \hat{\rho}(x)$. By antisymmetry, $\hat{\rho}(\hat{\rho}(x)) = \hat{\rho}(x)$ proving idempotency.

— To prove that $g \mathrel{\dot\sqsubseteq} \hat{\rho}$ that is $\forall x \in L \,.\, g(x) \sqsubseteq \hat{\rho}(x)$ pointwise, observe that, by definition of the least fixpoint, $x \sqsubseteq \mathsf{lfp}_x^{\sqsubseteq}\, g$ so, by increasingness, $g(x) \sqsubseteq g(\mathsf{lfp}_x^{\sqsubseteq}\, g) = \mathsf{lfp}_x^{\sqsubseteq}\, g \triangleq \hat{\rho}(x)$.

— Let ρ' be an upper closure operator greater than of equal to g, that is, $g \mathrel{\dot\sqsubseteq} \rho'$. We have $x \sqsubseteq \rho'(x) = \rho'(\rho'(x))$ so $x \sqsubseteq \mathsf{lfp}_x^{\sqsubseteq}\, \rho' \sqsubseteq \rho'(x)$ hence $\rho'(x) \sqsubseteq \rho'(\mathsf{lfp}_x^{\sqsubseteq}\, \rho') = \mathsf{lfp}_x^{\sqsubseteq}\, \rho' \sqsubseteq \rho'(\rho'(x)) = \rho'(x)$ by increasingness and fixpoint property, proving $\mathsf{lfp}_x^{\sqsubseteq}\, \rho' = \rho'(x)$ by antisymmetry.

By increasingness of g and ρ' and theorem 18.7, $\hat{\rho}(x) = \mathsf{lfp}_x^{\sqsubseteq}\, g \sqsubseteq \mathsf{lfp}_x^{\sqsubseteq}\, \rho' = \rho'(x)$, and so $\hat{\rho} \mathrel{\dot\sqsubseteq} \rho'$ pointwise, proving that $\hat{\rho}$ is the smallest upper closure operator pointwise greater than or equal to g. □

Theorem 29.2 Let $\langle C, \preceq, 0, 1, \curlywedge, \curlyvee\rangle$ and $\langle A, \sqsubseteq, \bot, \top, \sqcap, \sqcup\rangle$ be complete lattices such that $\langle C, \preceq\rangle \underset{\alpha}{\overset{\gamma}{\leftrightarrows}} \langle A, \sqsubseteq\rangle$, $f \in C \to C$ be a lower closure operator on C, and $g \in A \to A$ be an increasing and reductive operator on A such that $\alpha \circ f \circ \gamma \mathrel{\dot\sqsubseteq} g$. Then the lower reduction $\check{\rho}(x) \triangleq \mathsf{gfp}_x^{\sqsubseteq}\, g$ satisfies $\alpha \circ f \circ \gamma \mathrel{\dot\sqsubseteq} \check{\rho} \mathrel{\dot\sqsubseteq} g$.

Proof of theorem 29.2

— We have $\alpha \circ f \circ \gamma(x) \sqsubseteq \alpha \circ f \circ f \circ \gamma(x)$ because f is idempotent. Moreover $\alpha \circ f \circ f \circ \gamma(x) \sqsubseteq \alpha \circ f \circ \gamma \circ \alpha \circ f \circ \gamma(x)$ because $\gamma \circ \alpha$ is extensive, α and f are increasing. By transitivity, $\alpha \circ f \circ \gamma(x) \sqsubseteq \alpha \circ f \circ \gamma \circ \alpha \circ f \circ \gamma(x)$.

We have $f \circ \gamma(x) \preceq \gamma(x)$ because f is reductive so $\alpha \circ f \circ \gamma(x) \sqsubseteq x$ by definition of the Galois connection.

It follows that $\alpha \circ f \circ \gamma(x) \in \{y \mid y \sqsubseteq \alpha \circ f \circ \gamma(y) \wedge y \sqsubseteq x\}$ proving $\alpha \circ f \circ \gamma(x) \sqsubseteq \bigsqcup\{y \mid y \sqsubseteq \alpha \circ f \circ \gamma(y) \wedge y \sqsubseteq x\} = \mathsf{gfp}_x^{\sqsubseteq}\, \alpha \circ f \circ \gamma$ by definition of the lub $\sqcup$ and the order-dual of the variant exercise 15.11 of Tarski's fixpoint theorem 15.6.

By self-duality of $\mathrel{\dot\sqsubseteq}$ and the dual of theorem 18.7, we have $\alpha \circ f \circ \gamma \mathrel{\dot\sqsubseteq} g$ implies $\mathsf{gfp}_x^{\sqsubseteq}\, \alpha \circ f \circ \gamma \sqsubseteq \mathsf{gfp}_x^{\sqsubseteq}\, g = \check{\rho}(x)$.

By transitivity, we conclude that $\alpha \circ f \circ \gamma(x) \sqsubseteq \check{\rho}(x)$, that is, $\alpha \circ f \circ \gamma \mathrel{\dot\sqsubseteq} \check{\rho}$ pointwise.

— By the dual of lemma 29.1, $\check{\rho}$ is the largest lower closure operator pointwise less than or equal to g. Hence $\check{\rho} \mathrel{\dot\sqsubseteq} g$ and therefore $\alpha \circ f \circ \gamma \mathrel{\dot\sqsubseteq} \check{\rho} \mathrel{\dot\sqsubseteq} g$. □

Corollary 29.3 Let g^n be the iterates of g, that is, $g^1 \triangleq g$ and $g^{n+1} \triangleq g \circ g^n$. Then under the hypotheses of theorem 29.2, $\alpha \circ f \circ \gamma \mathrel{\dot\sqsubseteq} \check{\rho} \mathrel{\dot\sqsubseteq} g^n \mathrel{\dot\sqsubseteq} g$.

Proof of corollary 29.3 By recurrence on $n \in \mathbb{N}^+$. Follows from theorem 29.2 and reflexivity for $n = 1$. For the inductive step, g is increasing so $g \circ \check{\rho} \mathrel{\dot\sqsubseteq} g \circ g^n = g^{n+1} \mathrel{\dot\sqsubseteq} g \circ g \mathrel{\dot\sqsubseteq} g$ because g is reductive and increasing. Moreover the dual proof of lemma 29.1 shows that $\check{\rho}$ is a fixpoint of g so $g \circ \check{\rho} = \check{\rho}$, proving $\alpha \circ f \circ \gamma \mathrel{\dot\sqsubseteq} \check{\rho} \mathrel{\dot\sqsubseteq} g^{n+1} \mathrel{\dot\sqsubseteq} g$ by reflexivity and transitivity. □

29.2 Test Reduction

By exercise 28.37, $\mathsf{test}^{\times}[\![\mathsf{B}]\!]$ is a lower closure operator and, by exercise 28.41, $\mathsf{test}^{¤}[\![\mathsf{B}]\!]$ is increasing and reductive so we can apply corollary 29.3 and iterate $\mathsf{test}^{¤}[\![\mathsf{B}]\!]$. If the abstract domain has no infinite strictly decreasing chains, this iteration will stop at the greatest fixpoint. Otherwise, convergence can be enforced by a narrowing (introduced in definition 34.14, for example, by stopping after any number of iterations).

For example, the parity analysis of the program `y=1;if(z<0) z=x; else z=y; if ((x==z) nand (y==z)) x=1; else y=2;` with (28.38) is

```
l1: [x:e; y:e; z:e] y = 1;
if l2: (z < 0) [x:e; y:o; z:e]
    l3: [x:e; y:o; z:e] z = x;
 else
    l4: [x:e; y:o; z:e] z = y;
if l5: ((x == z) nand (y == z)) [x:e; y:o; z:T]
    l6: [x:e; y:o; z:T] x = 1;
 else
    l7: [x:e; y:o; z:_|_] y = 2;
l8: [x:T; y:T; z:T]
```

That is z cannot be both even and odd when `((x == z) nand (y == z))` is false, that is, $(x = z) \wedge (y = z)$. With local iterations for tests (28.38), this information is propagated to x and y

```
l1: [x:e; y:e; z:e] y = 1;
if l2: (z < 0) [x:e; y:o; z:e]
    l3: [x:e; y:o; z:e] z = x;
 else
    l4: [x:e; y:o; z:e] z = y;
if l5: ((x == z) nand (y == z)) [x:e; y:o; z:T]
    l6: [x:e; y:o; z:T] x = 1;
 else
    l7: [x:_|_; y:_|_; z:_|_] y = 2;
l8: [x:o; y:o; z:T]
```

See section 33.7 for further examples.

Exercise 29.4 (Project: local iterations) Implement local iterations for tests (28.38) for the Cartesian reachability interpreter of exercise 28.46.

29.3 Conclusion

Reductions can improve the precision of analyses. This was shown for local iterations for tests on Cartesian abstract domains. This is also useful for type inference. For example, Ada [521] allows for user-defined overloading of predefined arithmetic operators not only on the basis of argument types (as previously done in Algol68 [979]), but also on result types. So, Bernd Krieg-Brückner designed an iterated reduction algorithm of the top-down type inference of the result type from the argument types and a bottom-up type inference of the argument types from the type inference that was adopted in Ada [521, section 7.5.1, rule 5]. This extends to higher-order types, as in the CASL formal specification language [57].

However, the precision gain for relational domains may not be that spectacular when Boolean operators can be taken into account precisely. This is why, for example, the Astrée static analyzer [113] does not use local iterations for tests.

Another example of reduction is given in chapter 51, "Reduced Forward—Backward Analysis."

29.4 Solutions to Selected Exercises

Solution to exercise 29.4 Assuming that (28.38) is implemented by `basic_test b` and `notbasic_test b`, the local iterations are the following:

```
let rec iterate f a  =
    let x = (narrow a (f a)) in
        if (leq a x) then a else iterate f x
let test b r = iterate (basic_test b) r
let nottest b r = iterate (notbasic_test b) r
```

30 Basic Number Theory

We recall basic number theory results on congruences [462, 763, 888, 956].

Contents

30.1 Euclidean Division

The *Euclidean division* of an integer $n \in \mathbb{Z}$ by the *divisor* $d \in \mathbb{Z} \setminus \{0\}$ is the unique pair $\langle q, r\rangle$ of a *quotient* q and *rest* r such that $n = qd + r$ with $0 \leqslant r < |d|$. If $r = 0$ then d *divides* n (or n is a *multiple* of d) written $d \,/\, n$.

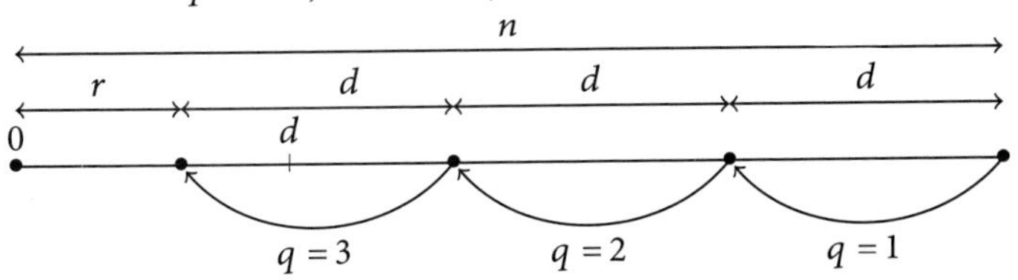

Exercise 30.1 Prove that if $n \in \mathbb{Z}$ and $d \in \mathbb{N}^+$ there is a unique pair $\langle q, r\rangle$ of integers such that $n = qd + r$ with $0 \leqslant r < |d|$.

Exercise 30.2 Prove the correctness of the following Euclidean division algorithm for $n \in \mathbb{N}$ and $d \in \mathbb{N}^+$ by repeated subtractions (`let x = ref e` declares a mutable variable `x` initialized to the value of expression `e` where `!x` denotes the value of `x` and `x := e'` is an assignment of the value of expression `e'` to the mutable variable `x`).

```
let divide n d = (* n>=0 && d>0 *)
   if n<0 || d <= 0 then failwith "divide: incorrect parameter";
   let q = ref 0 and r = ref n in
      while !r >= d do
         q := !q + 1; r := !r - d
      done;
      (!q,!r)
```

Exercise 30.3 Prove the correctness of the following Euclidean division algorithm for $n \in \mathbb{Z}$ by divisor $d \in \mathbb{Z} \setminus \{0\}$.

```
let rec Euclid n d =
   if d = 0 then if n=0 then (0,0)
                 else failwith "Euclid: division by 0"
   else if d < 0 then
           let (q,r) = Euclid n (-d) in (-q,r)
        else if n < 0 then
           let (q,r) = Euclid (-n) d in
              if r = 0 then (-q, 0)
              else (-q-1,d-r)
   else divide n d
# Euclid 17 7;;                  # Euclid 17 (-7);;
- : int * int = (2, 3)            - : int * int = (-2, 3)
# Euclid (-17) 7;;               # Euclid (-17) (-7);;
- : int * int = (-3, 4)           - : int * int = (3, 4)
```

Exercise 30.4 Prove that if d divides $x \in \mathbb{Z}$ and $y \in \mathbb{Z}$ then d divides any integral linear combination $ux + vy$ for all $u, v \in \mathbb{Z}$.

30.2 Congruences

We let $c + m\mathbb{Z} \triangleq \{z \in \mathbb{Z} \mid \exists k \in \mathbb{Z} \,.\, z = c + km\}$ be the *congruence class* of c modulo m. This is the set of all integers z equal to $c \in \mathbb{Z}$ modulo $m \in \mathbb{Z}$ so that m *divides* $z - c$, written $m \,/\, z - c$ or $z \equiv c \pmod{m}$. c is called the *residue* of z modulo m [398, p. 2].

Exercise 30.5 Show that for a given modulo m, the relation $x \equiv y \pmod{m}$ is an equivalence relation.

30.3 Canonical Congruences

The same set $c + m\mathbb{Z}$ can be defined by different residues and moduli. For example, $2 + 4\mathbb{Z} = 6 + -4\mathbb{Z}$ because if $z = 2 + 4k$ then $z = 6 + (-4)(1 - k)$. A canonical representation is obtained by requiring $m \geqslant 0$ and, if $m \neq 0$, by requiring c to be the least residue such that $0 \leqslant c < m$. For example, $c + 1\mathbb{Z} = \mathbb{Z}$ since $\forall z \in \mathbb{Z} \,.\, \forall c \in \mathbb{Z} \,.\, z = c + 1(z - c) \Rightarrow z \equiv c \pmod{1}$. The least residue is 0 with $0 + 1\mathbb{Z} = \mathbb{Z}$. We use $\varnothing$ for the empty set which cannot be represented by a congruence class. In the following we leave implicit that any congruence we consider has to be normalized into one of $\mathbb{P}^{\equiv} \triangleq \{c + m\mathbb{Z} \mid (m = 0) \vee (m > 0 \wedge 0 \leqslant c < m)\} \cup \{\varnothing\}$.

30.4 Greatest Common Divisor

The set of common divisors of two integers x and y, not both 0, contains 1 so is not empty and is bounded by $|x|$ by definition of divisibility so the divisors have a largest/highest element called the *greatest common divisor* of x and y. If $x = y = 0$ then all integers divide them so there is no greatest divisor. By convention, we define $0 \operatorname{gcd} 0 = 0$.

Define the *greatest common divisor* of two integers, not both 0, as the largest positive integer which divides both integers.

$$\begin{aligned} 0 \operatorname{gcd} 0 &\triangleq 0 \\ x \operatorname{gcd} y &\triangleq \max\{z \in \mathbb{N} \mid z \,/\, x \wedge z \,/\, y\} \qquad x \neq 0 \vee y \neq 0 \end{aligned}$$

Exercise 30.6 Prove that the following algorithm computes $x \gcd y$, $x, y \in \mathbb{Z}$.

```
let rec gcd x y = if (x=0) && (y=0) then 0
                  else if (x=0) then y
                  else if (y=0) then x
                  else if (x<0) || (y<0) then gcd (abs x) (abs y)
                  else if x > y then gcd (x-y) y
                  else if x = y then x
                  else gcd x (y-x)
```

30.5 Bachet–Bézout Identity

The following theorem is due to Claude-Gaspard Bachet de Méziriac [59, proposition XVIII] and was generalized by Étienne Bézout [116] to polynomials.

Theorem 30.7 (Bachet–Bézout's identity) let a and b be integers. There exist (nonunique) Bézout coefficients $x, y \in \mathbb{Z}$ such that $ax + by = a \gcd b$.

Moreover, $a \gcd b$ is the smallest positive integer that can be written as $ax + by$. Any integer of the form $ax + by$ is a multiple of the greatest common divisor $a \gcd b$.

Proof of theorem 30.7 Bézout's coefficients are not unique, for example $2 \times 1 + 4 \times 0 = 2 \times -1 + 4 \times 1 = 2 \times 3 + 4 \times -1 = 2 = 2 \gcd 4$.

If $a = b = 0$ then $ax + by = a \gcd b = 0 \gcd 0 = 0$ by choosing, for example, $x = y = 0$. If $a = 0$, $b \neq 0$ then $ax + by = by = 0 \gcd b = b$ by choosing $y = 1$. It is the same when $b = 0$ and $a \neq 0$.

Otherwise, $a \neq 0$ and $b \neq 0$. Let $S = \{ax + by \in \mathbb{N}^+ \mid x, y \in \mathbb{Z}\}$ be the set of integral positive linear combinations of a and b. $|a| \in S$ so S is not empty. Because $S \subseteq \mathbb{N}^+$, S is well founded, and therefore S has a minimal element $m \in S$. By definition of S, there exists $u, v \in \mathbb{Z}$ such that $m = au + bv$. By Euclidean division we have $a = mq + r$ where $0 \leqslant r < m$. Assume by contradiction that $r > 0$. Then $r = a - mq = a - (au + bv)q = a(1 - qu) + b(vq) > 0$ proving that $r \in S$ (for $x = 1 - qu$ and $y = vq$) in contradiction to $r < m$ and m is the minimum of S. So $r = 0$ proving that m divides a. Similarly m divides b.

By definition $a \gcd b$ is the greatest divisor of both a and b so $m \leqslant a \gcd b$. Because $a \gcd b$ divides both a and b, it divides $au + bv = m$ so $a \gcd b \leqslant m$. By antisymmetry, $m = a \gcd b$.

If $i = ax + by > 0$ then $i \in S$ so $a \gcd b = m \leqslant i$ because m is minimal in S, proving that $a \gcd b$ is the smallest positive integer that can be written as $ax + by$.

If $i = ax + by \in S$ then $a \gcd b$ divides a and b so divides $ax + by = i$ so i is a multiple of $a \gcd b$. □

30.6 Extended Euclidean Division

The extended Euclidean division computes the Bézout coefficients of any two integers a and b. Assume that $a \geqslant b \geqslant 0$. Using Euclidean division, define the following sequence of integers.

$$
\begin{array}{lll|l|l}
r_0 & = & a & x_0 = 1 & y_0 = 0 \\
r_1 & = & b & x_1 = 0 & y_1 = 1 \\
r_2 & s.t. & r_0 = r_1 q_1 + r_2,\ 0 < r_2 < r1 & x_2 = x_0 - x_1 q_1 & y_2 = y_0 - y_1 q_1 \\
\ldots & & & & \\
r_{k+1} & s.t. & r_{k-1} = r_k q_k + r_{k+1},\ 0 < r_{k+1} < r_k & x_{k+1} = x_{k-1} - x_k q_k & y_{k+1} = y_{k-1} - y_k q_k \\
\ldots & & & & \\
r_\ell & s.t. & r_{\ell-1} = r_\ell q_\ell + r_{\ell+1},\ r_{\ell+1} = 0 & x_\ell = x_{\ell-2} - x_{\ell-1} q_{\ell-1} & y_\ell = y_{\ell-2} - y_{\ell-1} q_{\ell-1} \\
r_{\ell+1} & = & 0 & x_{\ell+1} = x_{\ell-1} - x_\ell q_\ell & y_{\ell+1} = y_{\ell-1} - y_\ell q_\ell
\end{array}
$$

Theorem 30.8 $r_\ell = ax_\ell + by_\ell = a \operatorname{gcd} b$.

Proof of theorem 30.8

— The sequence $\langle r_k,\ k \in \mathbb{N}\rangle$ is well defined by the correctness of Euclid's algorithm in exercise 30.1. It is positive and strictly decreasing, so finite. The sequence continues while $r_k > 0$ so must end at some limit r_ℓ with $r_{\ell+1} = 0$.

— By recurrence, we have $a \operatorname{gcd} b = r_0 \operatorname{gcd} r_1 = \ldots = r_k \operatorname{gcd} r_{k+1} = \ldots = r_\ell \operatorname{gcd} r_\ell + 1 = r_\ell \operatorname{gcd} 0 = r_\ell$ because $r_0 = a$ and $r_1 = b$ and for the recurrence $r_k \operatorname{gcd} r_{k+1} = r_{k-1} \operatorname{gcd} r_k$ because $r_{k-1} = r_k q_k + r_{k+1}$, from which, by exercise 30.4, it follows that the common divisors of r_{k-1} and r_k are the same as the common divisors of r_k and r_{k+1}. For the limit, $r_{\ell+1} = 0$ so $r_\ell = a \operatorname{gcd} b$.

— Let us prove by recurrence that $\forall k \in [0, \ell+1] \,.\, ax_k + by_k = r_k$.
For the basis, we have $ax_0 + by_0 = a \times 1 + b \times 0 = a = r_0$.
For the induction step, we have

$$
\begin{aligned}
& ax_{k+1} + by_{k+1} \\
= \ & a(x_{k-1} - x_k q_k) + b(y_{k-1} - y_k q_k) && \wr \text{definitions of } x_{k+1} \text{ and } y_{k+1} \wr \\
= \ & ax_{k-1} + by_{k-1} - (ax_k + by_k) q_k \\
= \ & r_{k-1} - r_k q_k && \wr \text{induction hypothesis} \wr \\
= \ & r_{k+1} && \wr \text{definition of } r_{k+1} \wr
\end{aligned}
$$

— Combining the two previous results, we have $ax_\ell + by_\ell = a \operatorname{gcd} b$. □

Exercise 30.9 (extended Euclidean division algorithm) Prove that the following algorithm returns $a \operatorname{gcd} b = ax + by$.

```
let rec bezout a b =
  if a=0 || b=0 then (0,0,0)
  else if a<0 then let (r,x,y) = bezout (-a) b in (r,-x,y)
  else if b<0 then let (r,x,y) = bezout a (-b) in (r,x,-y)
  else let r = ref a and r' = ref b
  and x  = ref 1 and  y  = ref 0
  and x' = ref 0 and  y' = ref 1 in (* a,b > 0 *)
     while !r' <> 0 do
        (* r = ax+by et r' = ax'+by' *)
```

```
        let (q,_) = divide !r !r'
        and nr = !r and nx = !x and ny = !y in
           r := !r'; x := !x'; y := !y';
           r' := nr - q*(!r'); x' := nx - q*(!x'); y' := ny - q*(!y)
    done;
    (!r, !x, !y) (* r = (a gcd b) = ax+by *)
```

30.7 Least Common Multiple

Define the lowest or *least common multiple* of two integers, as the least natural number that is divisible by both integers.

$$x \operatorname{lcm} y \ \triangleq\ \min\{z \in \mathbb{N} \mid x \mathbin{/} z \wedge y \mathbin{/} z\} \qquad\qquad x, y \neq 0$$

Exercise 30.10 Show that a congruence holds for two different moduli if and only if it holds for their least common multiple.

Lemma 30.11 $(x \gcd y)(x \operatorname{lcm} y) = |xy|$

Proof of lemma 30.11 Let $g = x \gcd y$. $g \mathbin{/} a$ and $g \mathbin{/} b$ implies $g \mathbin{/} ab$ that is $\exists n \in \mathbb{N} \,.\, gn = xy$.

$g \mathbin{/} x$ and $g \mathbin{/} y$ imply $\exists u, v \in \mathbb{Z} \,.\, gu = x \wedge vg = y$ so $gn = xgv = guy$ hence $n = xv = uy$ so $x \mathbin{/} n$ and $y \mathbin{/} n$ and n is a multiplier of x and y.

To prove that n is the least multiplier, consider another one m such that $x \mathbin{/} m$ and $y \mathbin{/} m$ that is $\exists r, s \in \mathbb{Z} \,.\, rx = m \wedge sy = m$. By Bachet–Bézout identity theorem 30.7, we have $a, b \in \mathbb{Z}$ such that $xa + yb = x \gcd y = g$. It follows that $mg = mxa + myb = syxa + rxyb = xy(sa + rb) = gn(sa + rb)$ so $m = n(sa + rb)$ proving that $n \mathbin{/} m$ hence $n \leqslant |m|$ proving that $n = x \operatorname{lcm} y$.

To conclude, we have $xy = gn = x \gcd y x \operatorname{lcm} y$ so $|xy| = x \gcd y x \operatorname{lcm} y$ because $x \gcd y$ and $x \operatorname{lcm} y$ are positive. □

Exercise 30.12 Prove the correctness of the following `lcm` algorithm for $x, y \in \mathbb{Z}$.

```
let lcm x y = if (x=0) && (y=0) then 0
              else (abs(x) / (gcd x y)) * abs(y)
```

Exercise 30.13 Extend divisibility by $\forall x \,.\, x \mathbin{/} 0$, $x \operatorname{lcm} 0 = 0 \operatorname{lcm} x = 0$, and $x \gcd 0 = 0 \gcd x = x$. Prove that $\langle \mathbb{N}, \mathbin{/}, 1, 0, \operatorname{lcm}, \gcd\rangle$ is a distributive complete lattice (where a lattice $\langle L, \sqsubseteq, \sqcup, \sqcap\rangle$ is distributive if it satisfies the distributive law $\forall x, y, z \in L \,.\, x \sqcap (y \sqcup z) = (x \sqcap y) \sqcup (x \sqcap z)$).

Exercise 30.14 Carl Friedrich Gauss showed in his *Disquisitiones Arithmeticae* that casting out nines is an application of congruences for checking numerical computations [763, section 9.5]. Prove that it is an abstract interpretation.

30.8 Conclusion

The theory of congruences was initiated by Carl Friedrich Gauss [398] (and we still use his notations), [763, ch. 9]. It is used in the Cartesian integer congruence static analysis of following chapter 31.

30.9 Solutions to Selected Exercises

Solution to exercise 30.1 Let $S = \{n - qd \in \mathbb{N} \mid q \in \mathbb{Z}\}$. This set if not empty for $n \geqslant 0$ and $k = 0$ or $n < 0$ and $k = -a$. Because $S \subseteq \mathbb{N}$ it is well founded and has a minimum element r. We have $r = n - qd \geqslant 0$ for some $q \in \mathbb{Z}$. Assume by reduction ad absurdum that $r \geqslant d$. Then $r - d = n - (q + 1)d \in S$ in contradiction to the minimality of $r \in S$ proving $r < d$.

To prove unicity, assume that $n = qd + r$ and $n = q'd + r'$ with $0 \leqslant r, r' < d$. Then $|r - r'| < d$. Moreover $r' - r = n - q'd - (n - qd) = q(q - q')$ so $r' - r$ is a multiple of d which impossible unless $r' - r = 0$. So $r = r'$ and $n - qd = n - q'd$ proving $q = q'$.

Solution to exercise 30.4 By hypothesis, $\exists k, l \in \mathbb{Z} \,.\, x = kd \wedge y = ld$ so $ux + vx = (uk + vl)d$ proving that $ux + vy$ is a multiple of d.

Solution to exercise 30.14 The abstraction of n written in decimal notation as $n = d_k 10^k + \ldots + d_1 10^1 + d_0 10^0$ is $\alpha_9(n) \triangleq d_k + \ldots + d_1 + d_0 \pmod 9$. We have $10^i = 1 \pmod 9$ so if $\alpha_9(r) \neq \alpha_9(n) + \alpha_9(m) \pmod 9$ then $r \neq nm$ while the result may be incorrect when the congruence equality holds.

31 Cartesian Congruence Analysis

The Cartesian congruence analysis discovers congruence properties $x = a\,(\mathrm{mod}\ b)$ of values $x \in \mathbb{Z}$ of integer variables x where the integer coefficients $a, b \in \mathbb{N}$ are automatically inferred by the analysis. It generalizes the constancy analysis $x = c\,(\mathrm{mod}\ 0)$ of section 28.8.3 and the parity analysis $x = p\,(\mathrm{mod}\ 2)$, $p \in \{0, 1\}$ of section 28.8.1.

Contents

31.1 Congruence Abstract Properties

After normalization of section 30.3, the congruence abstract properties are

$$\mathbb{P}^{\equiv} \triangleq \{\emptyset\} \cup \{c + m\mathbb{Z} \mid c \in \mathbb{Z} \wedge m \in \mathbb{N} \wedge (\!| m > 0 \mathrel{?} 0 \leqslant c < m \mathrel{:} \mathrm{tt} |\!)\}$$

where $c + m\mathbb{Z} \triangleq c[m] \triangleq \{z \in \mathbb{Z} \mid \exists k \in \mathbb{Z} \,.\, z = c + km\}$.

Example 31.1

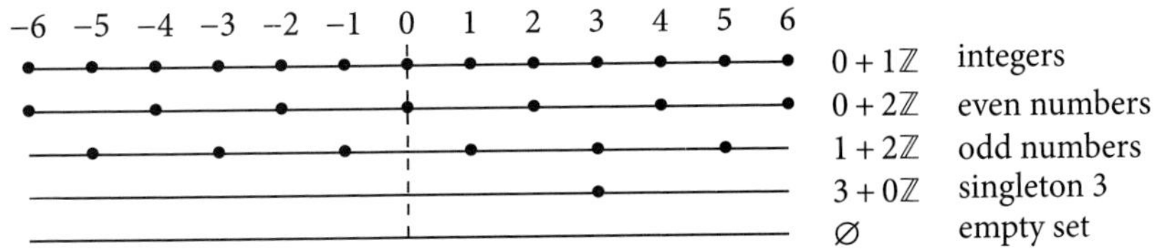

31.2 Example of Cartesian Congruence Analysis

Consider the program

$$\texttt{while}\,^{\ell_1}(\texttt{x} < 100)\{\texttt{if}\,^{\ell_2}(\texttt{odd(y)})\,^{\ell_3}\texttt{x} = \texttt{x} + 3;\ \texttt{else}\,^{\ell_4}\texttt{x} = \texttt{x} + 6;\,^{\ell_5}\texttt{y} = \texttt{y} + 1;\}^{\ell_6}$$

where variables are initialized to 0.

Initially $x = y = 0 + 0\mathbb{Z}$ at ℓ_1 and ℓ_2. $y = 0 + 0\mathbb{Z}$ is even so after execution of the iteration body $x = 0 + 0\mathbb{Z}$ or $x = 6 + 0\mathbb{Z}$ and $y = 0 + 0\mathbb{Z}$ or $y = 1 + 0\mathbb{Z}$ at ℓ_1. It follows that $x = 0 + 6\mathbb{Z}$ and $y = 0 + 1\mathbb{Z}$ at ℓ_1 hence at ℓ_2 because the test `x < 100` provides no congruence information. $y = 0 + 1\mathbb{Z}$ can be odd or even at ℓ_2 so at ℓ_5 we get either $x = 3 + 6\mathbb{Z}$ in the first case and $x = 6 + 6\mathbb{Z} = 0 + 6\mathbb{Z}$ in the second case. Therefore $x = 0 + 3\mathbb{Z}$ and $y = 1 + 1\mathbb{Z} = 0 + 1\mathbb{Z}$ at the end of the loop body. The join of $x = 0 + 6\mathbb{Z}$ and $x = 0 + 3\mathbb{Z}$ at ℓ_1 yields $x = 0 + 3\mathbb{Z}$ while $y = 0 + 1\mathbb{Z}$ is stable. One more iteration yields $x = 3 + 3\mathbb{Z}$ or $x = 6 + 3\mathbb{Z}$ that is $x = 0 + 3\mathbb{Z}$, which is stable. We conclude that x is congruent to 3 whereas y can be any integer value.

31.3 Congruence Abstraction

Let us define the *congruence abstraction*

$$
\begin{aligned}
\alpha^{\equiv}(\emptyset) &\triangleq \emptyset && (31.2)\\
\alpha^{\equiv}(\{c\}) &\triangleq c + 0\mathbb{Z} && \text{singleton}\\
\alpha^{\equiv}(Z) &\triangleq \text{let } m = \gcd\{|x - y| \mid x, y \in Z \wedge x \neq y\} && \text{otherwise}\\
&\qquad \text{and } c = \min\{c' \mid 0 \leqslant c' < m \wedge \exists k \in \mathbb{Z} \,.\, c' + km \in Z\} \text{ in}\\
&\qquad\quad c + m\mathbb{Z}
\end{aligned}
$$

The *congruence concretization* is

$$
\begin{aligned}
\gamma^{\equiv}(\emptyset) &\triangleq \emptyset && (31.3)\\
\gamma^{\equiv}(c + m\mathbb{Z}) &\triangleq \{z \in \mathbb{Z} \mid \exists k \in \mathbb{Z} \,.\, z = c + km\}
\end{aligned}
$$

Example 31.4

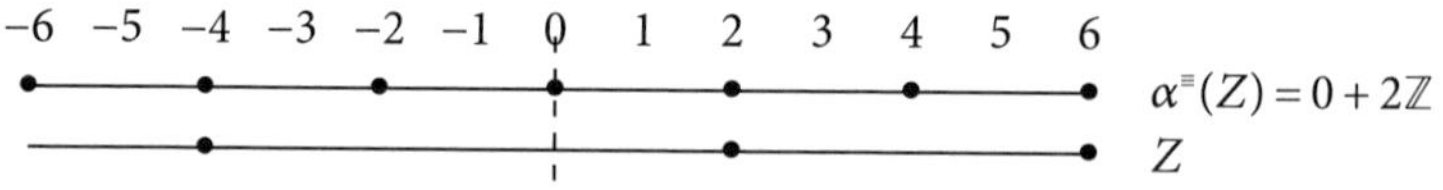

$Z = \{-4, 2, 6\}$. The gcd of 6, 4 and 10 is $m = 2$.

The abstract inclusion order $\sqsubseteq^{\equiv}$ is

$$P \sqsubseteq^{\equiv} Q \;\triangleq\; \gamma^{\equiv}(P) \subseteq \gamma^{\equiv}(Q)$$

Theorem 31.5 $\forall P \in \mathbb{P}^{\equiv} \,.\, \emptyset \sqsubseteq^{\equiv} P$ and

$$c + m\mathbb{Z} \sqsubseteq^{\equiv} c' + m'\mathbb{Z} \;\Leftrightarrow\; c \equiv c' \pmod{m'} \wedge m' \,/\, m$$

Proof of theorem 31.5

$$
\begin{aligned}
& c + m\mathbb{Z} \sqsubseteq^{\equiv} c' + m'\mathbb{Z}\\
\Leftrightarrow\;& \gamma^{\equiv}(c + m\mathbb{Z}) \subseteq \gamma^{\equiv}(c' + m'\mathbb{Z}) && \text{(definition of } \sqsubseteq^{\equiv}\text{)}\\
\Leftrightarrow\;& \{z \in \mathbb{Z} \mid \exists k \in \mathbb{Z} \,.\, z = c + km\} \subseteq \{z' \in \mathbb{Z} \mid \exists k' \in \mathbb{Z} \,.\, z' = c' + k'm'\} && \text{(definition of } \gamma^{\equiv}\text{)}\\
\Leftrightarrow\;& \forall k \in \mathbb{Z} \,.\, \exists k' \in \mathbb{Z} \,.\, c - c' = k'm' - km && \text{(definition of } \subseteq\text{)}
\end{aligned}
$$

$\Leftrightarrow\ c \equiv c' \pmod{m'} \wedge m' \,/\, m$

$\wr$($\Rightarrow$) In particular for $k = 0$, $\exists k' . c - c' = k'm'$ so $m' \,/\, c - c'$ hence $c \equiv c' \pmod{m'}$. Moreover, for $k = 1$, $\exists k'' . c - c' = k''m' - m$ hence $m = (k'' - k')m'$ proving that $m' \,/\, m$.

($\Leftarrow$) Conversely, $c \equiv c' \pmod{m'}$ implies $\exists k_1 \in \mathbb{Z} . c - c' = k_1 m'$ and $m' \,/\, m$ implies $\exists k_2 \in \mathbb{Z} . m = k_2 m'$ so given any $k \in \mathbb{Z}$, we have $c - c' = k_1 m' + km - km = (k_1 + kk_2)m' - km$ so the existence of k' is proved by choosing $k' = (k_1 + kk_2)$.$\smallint$ □

Theorem 31.6 We have the Galois connection $\langle \wp(\mathbb{Z}), \subseteq \rangle \xrightleftharpoons[\alpha^{\equiv}]{\gamma^{\equiv}} \langle \mathbb{P}^{\equiv}, \sqsubseteq^{\equiv} \rangle$.

Proof of theorem 31.6 The cases of the empty set $Z = \varnothing$ or a singleton $Z = \{c\}$ are easy. Otherwise, let $\alpha^{\equiv}(Z) = c + m\mathbb{Z}$, we have

$\alpha^{\equiv}(Z) \sqsubseteq^{\equiv} c' + m'\mathbb{Z}$

$\Leftrightarrow\ c + m\mathbb{Z} \sqsubseteq^{\equiv} c' + m'\mathbb{Z}$ $\wr$because we let $\alpha^{\equiv}(Z) = c + m\mathbb{Z}$$\smallint$

$\Leftrightarrow\ c \equiv c' \pmod{m'} \wedge m' \,/\, m$ $\wr$theorem 31.5$\smallint$

$\Leftrightarrow\ m = \gcd\{|x - y| \mid x, y \in Z \wedge x \neq y\} \wedge c = \min\{c'' \in \mathbb{N} \mid 0 \leqslant c'' < m \wedge \exists k \in \mathbb{Z} . c'' + km \in Z\} \wedge \exists k' \in \mathbb{Z} . c = c' + k'm' \wedge m' \,/\, m$ $\wr$definitions of $\alpha^{\equiv}$ and equivalence class$\smallint$

$\Leftrightarrow\ \forall z \in Z . \exists k' \in \mathbb{Z} . z = c' + k'm'$

$\wr$($\Rightarrow$) By definition of c, $0 \leqslant c < m \wedge \exists k_1 \in \mathbb{Z} . c + k_1 m \in Z$.

If $z = c + k_1 m$ then $m' \,/\, m$ so $m = k_2 m'$ and $z = c' + k'm' + k_1 k_2 m'$ of the form $z = c' + km'$ with $k = k' + k_1 k_2$.

Otherwise, for all $z \in Z$, $z = z - (c + k_1 m) + c + k_1 m$. If $z > c + k_1 m$ this is $z = |z - (c + k_1 m)| + c + k_1 m$ else $z < c + k_1 m$ and this is $z = -|z - (c + k_1 m)| + c + k_1 m$. Because $z \neq c + k_1 m \in Z$, $m \,/\, |z - (c + k_1 m)|$ by definition of m so $|z - (c + k_1 m)| = k_3 m$. So either $z = k_3 m + c + k_1 m$ or $z = -k_3 m + c + k_1 m$, so $\exists k_4 . z = k_4 m + c + k_1 m$. $m' \,/\, m$ so $m = k_5 m'$. It follows that $z = k_4 k_5 m' + c' + k'm' + k_1 k_5 m'$ or the form $z = c' + km'$ with $k = k_4 k_5 + k' + k_1 k_5$.

($\Leftarrow$) Conversely, assume that all $z \in Z$ have the form $z = c' + km'$ for some $k \in \mathbb{Z}$. Define $m = \gcd\{|x - y| \mid x, y \in Z \wedge x \neq y\}$ and $c = \min\{c'' \in \mathbb{N} \mid 0 \leqslant c'' < m \wedge \exists k \in \mathbb{Z} . c'' + km \in Z\}$, which depend only on Z. We must find $k' \in \mathbb{Z}$ such that $c = c' + k'm'$ and prove that $m' \,/\, m$.

We have $m = \gcd\{|x - y| \mid x, y \in Z \wedge x \neq y\} = \gcd\{|(c' + k_1'm') - (c' + k_2'm')| \mid k_1', k_2' \in \mathbb{Z} \wedge k_1' \neq k_2'\} = \gcd\{|km'| \mid k \in \mathbb{Z} \setminus \{0\}\}$ [letting $k = k_1' - k_2'$] $= k'm'$ [for some $k' \in \mathbb{Z} \setminus \{0\}$ by definition of the gcd]. Therefore m is a multiple of m' so $m' \,/\, m$.

By $c = \min\{c'' \in \mathbb{N} \mid 0 \leqslant c'' < m \wedge \exists k \in \mathbb{Z} . c'' + km \in Z\}$, we have $c \in \{c'' \in \mathbb{N} \mid 0 \leqslant c'' < m \wedge \exists k \in \mathbb{Z} . c'' + km \in Z\}$ and therefore $0 \leqslant c < m \wedge \exists k, k' = \in \mathbb{Z} . c + km = c' + k'm'$. m being a multiple of m', $\exists k, k', k'' \in \mathbb{Z} . c + kk''m' = c' + k'm'$ so $c = c' + km'$ by choosing $k = k' - kk''$. $\smallint$

$\Leftrightarrow\ Z \subseteq \{z \in \mathbb{Z} \mid \exists k' \in \mathbb{Z} . z = c' + k'm'\}$ $\wr$definition of $\subseteq$$\smallint$

$\Leftrightarrow Z \subseteq \gamma^{\equiv}(c' + m'\mathbb{Z})$ ⟅definition of $\gamma^{\equiv}$⟆

$\alpha^{\equiv}$ is surjective because $\mathbb{P}^{\equiv}$ uses the canonical representation. □

Exercise 31.7 Prove that $\alpha^{\equiv}(Z) = \bigsqcup^{\equiv}_{z \in Z} z + 0\mathbb{Z}$.

31.4 The Congruence Complete Lattice

The congruence abstract properties form a complete lattice, as illustrated in figure 31.9.

Corollary 31.8 The image of $\langle \wp(\mathbb{Z}), \subseteq, \emptyset, \mathbb{Z}, \cup, \cap \rangle$ by the lower adjoint $\alpha^{\equiv}$ is the complete lattice $\langle \mathbb{P}^{\equiv}, \sqsubseteq^{\equiv}, \emptyset, 0 + 1\mathbb{Z}, \sqcup^{\equiv}, \sqcap^{\equiv} \rangle$.

Proof of corollary 31.8 $\alpha^{\equiv}$ is surjective so any subset C of $\mathbb{P}^{\equiv}$ has an inverse image $\gamma^{\equiv}(Z)$ by the injective $\gamma^{\equiv}$. $\gamma^{\equiv}(C)$ has a lub $\cup\gamma^{\equiv}(C)$ in $\wp(\mathbb{Z})$ which, by lemma 11.38, is preserved by $\alpha^{\equiv}$, so $\alpha^{\equiv}(\cup\gamma^{\equiv}(C)) = \sqcup^{\equiv}\alpha^{\equiv}(\gamma^{\equiv}(C)) = \sqcup^{\equiv}C$ because $\alpha^{\equiv} \circ \gamma^{\equiv}$ is the identity by exercise 11.50. So any $C \subseteq \mathbb{P}^{\equiv}$ has a lub $\sqcup^{\equiv}C$ in $\mathbb{P}^{\equiv}$ which, by section 10.6, is a complete lattice, which, by exercise 10.7, has also a glb $\sqcap^{\equiv}$. □

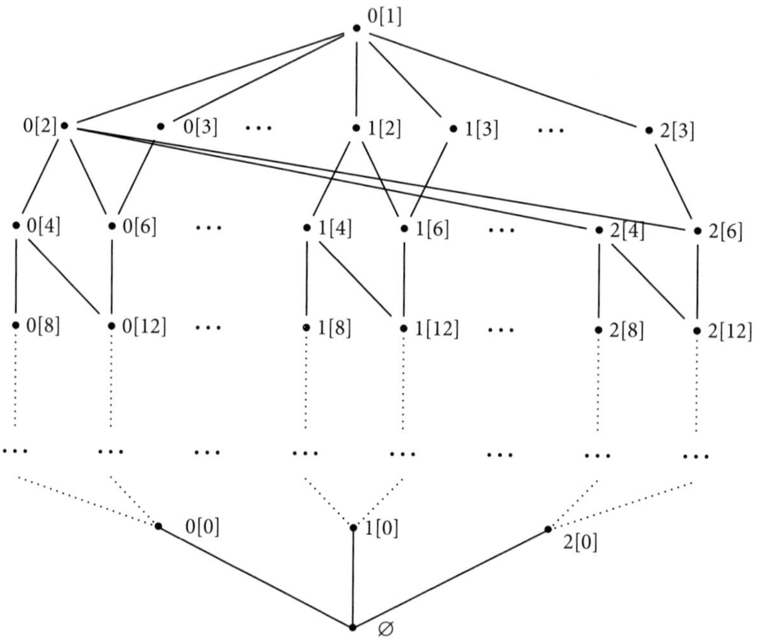

Figure 31.9
The congruence complete lattice

31.5 Congruence Least Upper Bound

Theorem 31.10 (congruence join/lub)

$$c + m\mathbb{Z} \sqcup^{\equiv} c' + m'\mathbb{Z} \quad = \quad c + (m \operatorname{gcd} m' \operatorname{gcd} |c - c'|)\mathbb{Z}$$

Example 31.11

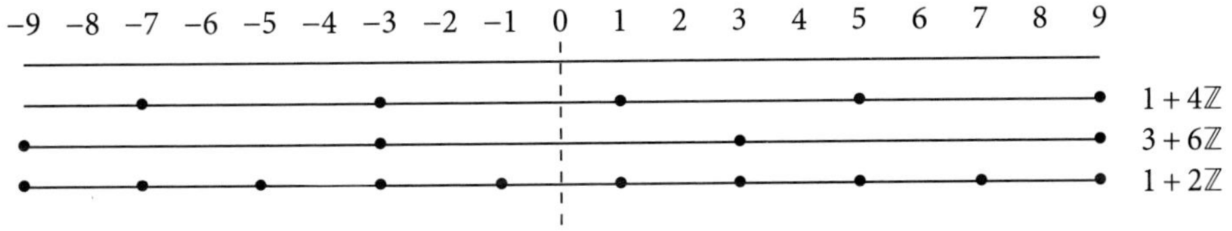

$$1 + 4\mathbb{Z} \sqcup^{\equiv} 3 + 6\mathbb{Z} = 1 + (4 \operatorname{gcd} 6 \operatorname{gcd} |1 - 3|)\mathbb{Z} = 1 + 2\mathbb{Z}$$

Proof of theorem 31.10 Let $g = m \operatorname{gcd} m' \operatorname{gcd} |c - c'|$ be the gcd of m, m', and $|c - c'|$. We have $c \equiv c \pmod{g}$ and $g \,/\, m$ so that by theorem 31.5, $c + m\mathbb{Z} \sqsubseteq^{\equiv} c + g\mathbb{Z}$. Moreover $c' \equiv c \pmod{g}$ and $g \,/\, m'$ so that by theorem 31.5, $c' + m'\mathbb{Z} \sqsubseteq^{\equiv} c + g\mathbb{Z}$. It follows that $c + g\mathbb{Z}$ is an upper bound of $c + m\mathbb{Z}$ and $c' + m'\mathbb{Z}$.

Let $c'' + m''\mathbb{Z}$ be any $\sqsubseteq^{\equiv}$-upper bound of $c + m\mathbb{Z}$ and $c' + m'\mathbb{Z}$. By theorem 31.5, we have $c \equiv c'' \pmod{m''} \wedge m'' \,/\, m$ and $c' \equiv c'' \pmod{m''} \wedge m'' \,/\, m'$. It follows that $c'' \equiv c \pmod{m''}$ and m'' divides m, m', and $c - c'$ so by definition of the gcd, $m'' \,/\, g$. By theorem 31.5, $c + g\mathbb{Z} \sqsubseteq^{\equiv} c'' + m''\mathbb{Z}$ proving that $c + g\mathbb{Z}$ is the lub of $c + m\mathbb{Z}$ and $c' + m'\mathbb{Z}$.

□

31.6 Congruence Disjointness

For the test x == y to be possibly true, the equivalence classes of x and y must not be disjoint.

Theorem 31.12 (disjointness) $c + m\mathbb{Z} \sqcap^{\equiv} c' + m'\mathbb{Z} \neq \emptyset$ if and only if $c \equiv c' \pmod{m \operatorname{gcd} m'}$.

Example 31.13 $1 + 4\mathbb{Z} \sqcap^{\equiv} 0 + 6\mathbb{Z} = \emptyset$ because $1 \not\equiv 0 \pmod{2}$. $1 + 4\mathbb{Z} \sqcap^{\equiv} 3 + 6\mathbb{Z} \neq \emptyset$ because $1 \equiv 3 \pmod{2}$.

Proof of theorem 31.12

$$\begin{aligned}
& c + m\mathbb{Z} \sqcap^{\equiv} c' + m'\mathbb{Z} \neq \emptyset \\
\Leftrightarrow\; & \exists k, k' \in \mathbb{Z} \,.\, c + km = c' + k'm' && \text{(definition of a congruence class)} \\
\Leftrightarrow\; & \exists k'', k' \in \mathbb{Z} \,.\, c - c' = k''m + k'm' && (k'' = -k) \\
\Leftrightarrow\; & \exists k'', k', \ell, \ell' \in \mathbb{Z} \,.\, c - c' = (k''\ell + k'\ell')(m \operatorname{gcd} m') \\
& && (m = \ell(m \operatorname{gcd} m'), \text{ and } m' = \ell'(m \operatorname{gcd} m')) \\
\Leftrightarrow\; & \exists k \in \mathbb{Z} \,.\, c - c' = k(m \operatorname{gcd} m') && (k = k''\ell + k'\ell') \\
\Leftrightarrow\; & c \equiv c' \pmod{m \operatorname{gcd} m'} && \text{(definition of a congruence class)}
\end{aligned}$$

□

31.7 Congruence Greatest Lower Bound

When the test x == y is true, both x and y belong to the intersection of their congruence classes.

Theorem 31.14 (congruence meet/glb)

$$
\begin{array}{rcll}
c+m\mathbb{Z} \sqcap^{\equiv} c'+m'\mathbb{Z} & = & c''+(m \operatorname{lcm} m')\mathbb{Z} & \text{if } c \equiv c' \pmod{(m \gcd m')} \\
 & = & \varnothing & \text{otherwise}
\end{array}
$$

with $c'' \equiv c + \frac{c'-c}{m \gcd m'} xm \equiv c' + \frac{c-c'}{m \gcd m'} ym' \pmod{m \operatorname{lcm} m'}$ where, by Bachet–Bézout's identity of theorem 30.7, $x, y \in \mathbb{Z}$ are such that $xm + ym' = m \gcd m'$. In case $m = m' = 0$, we have $c + 0\mathbb{Z} \sqcap^{\equiv} c' + 0\mathbb{Z} = \varnothing$ when $c \neq c'$ and otherwise $c + 0\mathbb{Z} \sqcap^{\equiv} c + 0\mathbb{Z} = c + 0\mathbb{Z}$.

Example 31.15

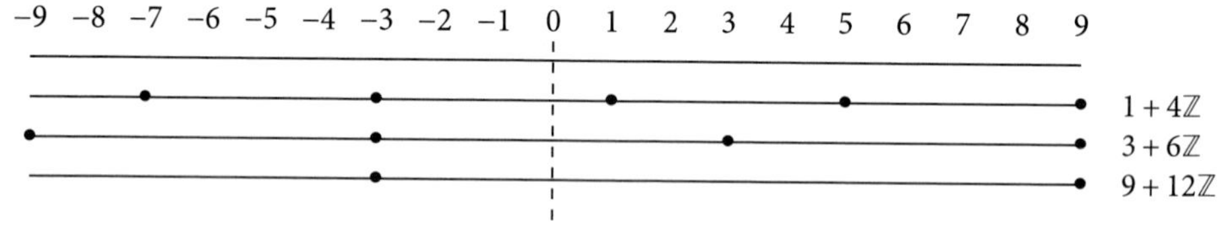

We have $(-1) \times 4 + 1 \times 6 = gcd(4, 6) = 2$ so the intersection is $c'' \equiv 1 + \frac{3-1}{4 \gcd 6} \times (-1) \times 4 \equiv -3 \equiv 3 + \frac{1-3}{4 \gcd 6} \times 1 \times 6 \equiv -3 \pmod{(4 \operatorname{lcm} 6)} \equiv 9 \pmod{12}$.

Exercise 31.16 What is $3 + 5\mathbb{Z} \sqcap^{\equiv} 1 + 7\mathbb{Z}$?

Proof of theorem 31.14 By the disjointness theorem 31.12, the intersection is not empty if and only if $c \equiv c' \pmod{(m \gcd m')}$.

By definition of lcm, we have $\exists k\,.\, m \operatorname{lcm} m' = km$ so $m \,/\, (m \operatorname{lcm} m')$. Moreover,

$$c'' \equiv c + \frac{c'-c}{m \gcd m'} xm \pmod{m \operatorname{lcm} m'}$$

so there exists k' such that $c'' = c + \frac{c'-c}{m \gcd m'} xm + k'(m \operatorname{lcm} m') = c + \frac{c'-c}{m \gcd m'} xm + k'(km)$, proving that $c'' \equiv c \pmod{m}$. By definition of $\sqcap^{\equiv}$ and theorem 31.5, we have $c + m\mathbb{Z} \sqcap^{\equiv} c' + m'\mathbb{Z} = c'' + (m \operatorname{lcm} m')\mathbb{Z} \sqsubseteq^{\equiv} c + m\mathbb{Z}$.

Moreover, we have

$$
\begin{array}{ll}
(m \gcd m')c + (m \gcd m')c' = (m \gcd m')c' + (m \gcd m')c & \wr + \text{ commutative} \wr \\
\Rightarrow (m \gcd m')c + c'xm + c'ym' = (m \gcd m')c' + cxm + cym' & \wr xm + ym' = m \gcd m' \wr \\
\Rightarrow c + \frac{c'-c}{m \gcd m'} xm = c' + \frac{c-c'}{m \gcd m'} ym' & \wr \text{dividing by } m \gcd m' \neq 0 \wr
\end{array}
$$

This shows that by a reasoning similar to the preceding proof of $c + m\mathbb{Z} \sqcap^{\equiv} c' + m'\mathbb{Z} \sqsubseteq^{\equiv} c + m\mathbb{Z}$, we have $c'' \equiv c' \pmod{m'}$ and therefore $c + m\mathbb{Z} \sqcap^{\equiv} c' + m'\mathbb{Z} \sqsubseteq^{\equiv} c' + m'\mathbb{Z}$. We conclude that $c + m\mathbb{Z} \sqcap^{\equiv} c' + m'\mathbb{Z}$ is a $\sqsubseteq^{\equiv}$-lower bound of $c + m\mathbb{Z}$ and $c' + m'\mathbb{Z}$.

Let $c + m\mathbb{Z}$ be any $\sqsubseteq^{\equiv}$-lower bound of $c + m\mathbb{Z}$ and $c' + m'\mathbb{Z}$. By theorem 31.5, we have $c \equiv c \pmod{m} \wedge m \,/\, m \wedge c \equiv c' \pmod{m'} \wedge m' \,/\, m$. It follows that $c \equiv c'' \pmod{m}$ and $c \equiv c'' \pmod{m'}$ so $c \equiv c'' \pmod{m \operatorname{lcm} m'}$ because $(m \operatorname{lcm} m')$ is a multiple of m and m'. Moreover $m \,/\, m$ and $m' \,/\, m$ so $(m \operatorname{lcm} m') \,/\, m$. We have $c \equiv c'' \pmod{(m \operatorname{lcm} m')} \wedge (m \operatorname{lcm} m') \,/\, m$ which, by theorem 31.5, implies $c + m\mathbb{Z} \sqsubseteq^{\equiv} c'' + m'\mathbb{Z}$.

We conclude that $\sqcap^{\equiv}$ is a lower bound greater than or equal to any lower bound that is the greatest lower bound. □

Exercise 31.17 Implement $\sqcap^{\equiv}$.

31.8 Abstract Congruence Operations

Theorem 31.18 $\ominus^{\equiv}(c + m\mathbb{Z}) = -c + m\mathbb{Z}$.

Proof of theorem 31.18

$$
\begin{aligned}
& \ominus^{\equiv} c + m\mathbb{Z} \\
= \;& (\alpha^{\equiv}(\ominus\gamma^{\equiv}(c + m\mathbb{Z})) \\
= \;& \alpha^{\equiv}(\ominus\{z \in \mathbb{Z} \mid \exists k \in \mathbb{Z} \,.\, z = c + km\}) && \text{(definition (31.3) of } \gamma^{\equiv}\text{)} \\
= \;& \alpha^{\equiv}(\{-z \in \mathbb{Z} \mid \exists k \in \mathbb{Z} \,.\, z = c + km\}) && \text{(definition of } \ominus\text{)} \\
= \;& \alpha^{\equiv}(\{z \in \mathbb{Z} \mid \exists k' \in \mathbb{Z} \,.\, z = -c + k'm\}) && (k' = -k) \\
= \;& \alpha^{\equiv}(\gamma^{\equiv}(-c + m\mathbb{Z})) && \text{(definition (31.3) of } \gamma^{\equiv}\text{)} \\
= \;& -c + m\mathbb{Z} && \text{(theorem 31.6 where } \alpha^{\equiv} \text{ is surjective and exercise 11.50)} \quad \square
\end{aligned}
$$

Theorem 31.19 $\varnothing \oplus^{\equiv} \varnothing = c + m\mathbb{Z} \oplus^{\equiv} \varnothing = \varnothing \oplus^{\equiv} c' + m'\mathbb{Z} = \varnothing$ and $c + m\mathbb{Z} \oplus^{\equiv} c' + m'\mathbb{Z} = (c + c') + (m \gcd m')\mathbb{Z}$.

Proof of theorem 31.19

$$
\begin{aligned}
& c + m\mathbb{Z} \oplus^{\equiv} c' + m'\mathbb{Z} \\
= \;& \alpha^{\equiv}(\gamma^{\equiv}(c + m\mathbb{Z}) \oplus \gamma^{\equiv}(c' + m'\mathbb{Z})) && \text{(best abstraction)} \\
= \;& \alpha^{\equiv}(\{x + y \mid x \in \gamma^{\equiv}(c + m\mathbb{Z}) \wedge y \in \gamma^{\equiv}(c' + m'\mathbb{Z})\} \\
& && \text{(definition (31.3) of } \gamma^{\equiv} \text{ and definition of } \oplus\text{)} \\
= \;& \alpha^{\equiv}(\{x + y \mid x \in \{z \in \mathbb{Z} \mid \exists k \in \mathbb{Z} \,.\, z = c + km\} \wedge y \in \{z' \in \mathbb{Z} \mid \exists k' \in \mathbb{Z} \,.\, z' = c' + k'm'\}\} \\
& && \text{(definition (31.3) of } \gamma^{\equiv}\text{)} \\
= \;& \alpha^{\equiv}(\{c + km + c' + k'm' \mid k, k' \in \mathbb{Z}\} \\
= \;& \alpha^{\equiv}(\{c + c' + k\ell(m \gcd m') + k'\ell'(m \gcd m') \mid k, k, \ell, \ell' \in \mathbb{Z}\} \\
& && (m = \ell(m \gcd m') \text{ and } m' = \ell'(m \gcd m')) \\
= \;& \alpha^{\equiv}(\{c + c' + k''(m \gcd m') \mid k'' \in \mathbb{Z}\} && (\text{letting } k'' = k\ell + k'\ell') \\
= \;& \alpha^{\equiv}(\gamma^{\equiv}((c + c') + (m \gcd m')\mathbb{Z})) && \text{(definition (31.3) of } \gamma^{\equiv}\text{)} \\
= \;& (c + c') + (m \gcd m')\mathbb{Z} && \text{(theorem 31.6 and exercise 11.50)} \quad \square
\end{aligned}
$$

Exercise 31.20 Prove that $c + m\mathbb{Z} \ominus^{\equiv} c' + m'\mathbb{Z} = (c - c') + (m \operatorname{gcd} m')\mathbb{Z}$.

Exercise 31.21 Prove that $c + m\mathbb{Z} \otimes^{\equiv} c' + m'\mathbb{Z} = (cc') + (mm' \operatorname{gcd} (mc' \operatorname{gcd} m'c))\mathbb{Z}$.

Exercise 31.22 Prove that

$$
\begin{aligned}
c + m\mathbb{Z} \oslash^{\equiv} c' + m'\mathbb{Z} &= \varnothing && \text{if } c' + m'\mathbb{Z} = 0 + 0\mathbb{Z} \\
&= (c/c') + m/|c'|\mathbb{Z} && \text{if } m' = 0,\ c' \neq 0,\ c' \,/\, m, \text{ and } c' \,/\, c \\
&= 0 + 1\mathbb{Z} && \text{otherwise}
\end{aligned}
$$

[438] proposes a more precise abstract division $\oslash^{\equiv}$ as well as an abstract modulo operation.

Exercise 31.23 Propose a definition of $\ominus^{\overline{\equiv}1}$.

31.9 The Congruence Abstract Domain

The value congruence domain (28.42) is

$$\mathbb{D}^{\equiv} \triangleq \langle \mathbb{P}^{\equiv}, \sqsubseteq^{\equiv}, \bot^{\equiv}, \top^{\equiv}, \sqcup^{\equiv}, \sqcap^{\equiv}, 1^{\equiv}, \ominus^{\equiv}, \ominus^{\overline{\equiv}1}, \circledless^{\equiv}, \overline{\circledless}^{\equiv} \rangle \tag{31.24}$$

from which the reachability congruence domain (28.43) is derived as in chapter 28. By corollary 31.8, $\langle \mathbb{P}^{\equiv}, \sqsubseteq^{\equiv}, \bot^{\equiv}, \top^{\equiv}, \sqcup^{\equiv}, \sqcap^{\equiv} \rangle$ is a complete lattice where $\sqsubseteq^{\equiv}$ is the order of theorem 31.5, $\bot^{\equiv} \triangleq \varnothing$, $\top^{\equiv} \triangleq 0 + 1\mathbb{Z}$, the lub $\sqcup^{\equiv}$ is defined in theorem 31.10, and the glb $\sqcap^{\equiv}$ by theorem 31.14. $1^{\equiv} \triangleq 1 + 0\mathbb{Z}$, $\ominus^{\equiv}$ is given in exercise 31.20, $\ominus^{\overline{\equiv}1}$ in exercise 31.23. The comparison operators $\circledless^{\equiv}$ and $\overline{\circledless}^{\equiv}$ are the identity but when one argument is $\bot^{\equiv}$, both are constants that can be compared, or the comparison is an equality.

31.10 Iteration

Increasing iterations always converge in finitely many steps.

Theorem 31.25 (no infinite strictly increasing chain) $\langle \mathbb{P}^{\equiv}, \sqsubseteq^{\equiv} \rangle$ has no infinite strictly increasing chain.

Proof of theorem 31.25 Assume the increasing chain starts with $c + m\mathbb{Z}$ with $m \neq 0$. Any strictly greater congruence class has the form $c + n\mathbb{Z}$ where n divides m and $n \neq m$. It follows that $2|n| \leqslant |m|$, so if the chain is strictly increasing it contains at most $1 + \log_2(m)$ elements ☐

$\langle \mathbb{P}^{\equiv}, \sqsubseteq^{\equiv} \rangle$ has infinite strictly decreasing chains.

Example 31.26 (infinite strictly decreasing chains)

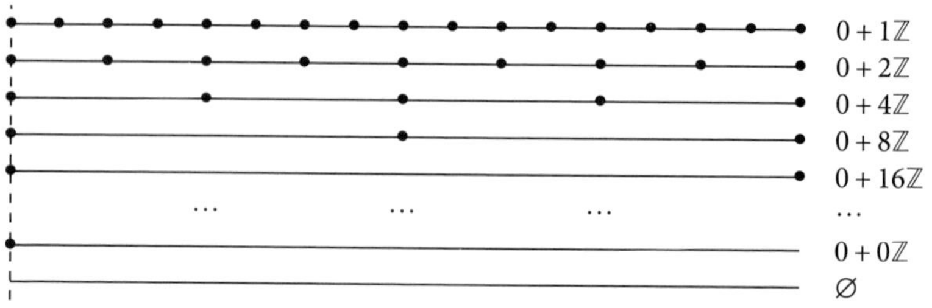

It follows that the convergence of the local iterations of section 29.2 must be enforced by a narrowing such as $\top^{\equiv} \mathbin{\Delta^{\equiv}} \overline{P} \triangleq \overline{P}$ and $\overline{P} \mathbin{\Delta^{\equiv}} \overline{Q} \triangleq \overline{P}$ otherwise. More details on narrowings in sections 33.6 and 34.8.

Exercise 31.27 (Project: Cartesian congruence domain) Continuing exercise 28.49, implement the Cartesian congruence domain $\mathbb{D}^{\equiv}$ and use it with the Cartesian abstract domain functor and the abstract interpreter for congruence analysis. For example (with implicit initialization to 0)

```
while l1: (x < 100) [x:0+3Z; y:0+6Z]
   {
      if l2: (y == 0) [x:0+3Z; y:0+6Z]
         l3: [x:0+3Z; y:0+6Z] x = (x + 3);
       else
         l4: [x:0+3Z; y:0+6Z] x = (x + 6);
      l5: [x:0+3Z; y:0+6Z] y = (y + 6);
   }
l6: [x:0+3Z; y:0+6Z]
```

31.11 Conclusion

The Cartesian integer congruence static analysis is due to Philippe Granger [438], who generalized to rationals [441] and linear congruence equalities [439]. François Masdupuy introduced interval congruences $[a, b] + m\mathbb{Z}$ [664] and trapezoidal congruences [663]. Antoine Miné designed zone congruences [689].

The Cartesian congruence analysis is used in Astrée [113], for example, to check that data structures are well aligned on word boundaries in computer memory, for example, to multiples of the word size.

31.12 Solutions to Selected Exercises

Solution to exercise 31.7 By theorem 31.6 and lemma 11.38, $\alpha^{\equiv}$ preserves existing joins so $\alpha^{\equiv}(Z) = \alpha^{\equiv}(\bigcup_{z \in Z} \{z\}) = \bigsqcup^{\equiv}_{z \in Z} \alpha^{\equiv}(\{z\}) = \bigsqcup^{\equiv}_{z \in Z} z + 0\mathbb{Z}$.

Solution to exercise 31.16 $3 + 5\mathbb{Z} \sqcap^{\equiv} 1 + 7\mathbb{Z} \equiv 8 + 35\mathbb{Z}$.

Solution to exercise 31.17

```
let abs n = if n=min_int then failwith "abs: overflow"
            else if n<0 then -n else n
type abstractProperty = BOT | CC of (int * int) (* c+mZ, $0$ <= c < m *)
let normalize c m = if m=0 then (c,m) else (c mod (abs m), abs(m))
let meet a1 a2 = match (a1,a2) with
    | (BOT,_) -> BOT
    | (_,BOT) -> BOT
    | (CC (c1,m1),CC (c2,m2)) ->
      let g = gcd m1 m2 in
         let (_,r) = divide (abs (c1-c2)) g in
            if r=0 then
               let m = lcm m1 m2 in
                  let (_,c) = divide c1 m in
                     CC (normalize c m)
```

```
          else BOT
# meet (CC (0,3)) (CC (3,7));;
- : abstractProperty = CC (0, 21)
```

```
#  meet (CC (0,3)) (CC (1,3));;
- : abstractProperty = BOT
```

Solution to exercise 31.23 Assume that $\langle X', Y'\rangle = \ominus^{\equiv -1} R\,\langle X, Y\rangle$. Let us handle X', Y' is symmetric. This means $x' \in X'$ if and only if $\exists y \in Y\,.\,x' \in X \wedge x' - y = r \in R$ so $x' = y + r$. Therefore $x' \in X \sqcap^{\equiv} (Y \oplus^{\equiv} R)$ and we define $\ominus^{\equiv -1} R\,\langle X, Y\rangle \triangleq \langle X \sqcap^{\equiv} (Y \oplus^{\equiv} R), Y \sqcap^{\equiv} (X \oplus^{\equiv} R)\rangle$.

32 Dynamic Interval Analysis

Interval arithmetics in scientific computing was developed by Ramon E. Moore [715–717] to put bounds on rounding errors in floating-point computations [522, 769]. Instead of computing with a floating-point number representing an uncertain real x, the computation is performed with the two ends of an interval $[\underline{x}, \overline{x}]$) where $\underline{x}$ and $\overline{x}$ are floating-point numbers such that $x \in [\underline{x}, \overline{x}]$. This is an abstraction of a real trace semantics in which each real trace is abstracted by one interval trace (which is a property of that real trace) where reals are abstracted by float intervals along the trace. So, the real computation is replaced by the computation of a float interval property. Whereas a test on a real takes one or the other alternative, tests on float intervals can take both true and false alternatives. So the real trace cannot be overapproximated by a single interval trace only. One must consider a set of interval traces (see section 32.5), each one being explored in turn, which may lead to combinatorial explosion.

Contents

32.1 Soundness of Interval Arithmetics

In what follows we let $\mathbb{I}$ be either the set $\mathbb{Z}$ of mathematical integers, `[min_int,max_int]` of machine integers, $\mathbb{Q}$ of rationals, $\mathbb{F}$ of floating-point numbers [423, 709], or $\mathbb{R}$ of reals. For machine integers, we consider overflow as an error. For modular integer arithmetics, see [389, 731, 891]. We write $\mathbb{P}^i_{\mathbb{I}}$ for the intervals on $\mathbb{I}$.

Given parameters $x \in [\underline{x}, \overline{x}]$, $y \in [\underline{y}, \overline{y}]$, ... in $\mathbb{P}^i_{\mathbb{I}}$, the interval computation of a function $f \in \mathbb{I}^n \to \mathbb{I}$ must return a sound interval $[\underline{f}, \overline{f}]$ that contains all possible results for all possible values of the parameters.

$$\{f(x, y, \ldots) \mid x \in [\underline{x}, \overline{x}] \wedge y \in [\underline{y}, \overline{y}] \wedge \ldots\} \quad \subseteq \quad [\underline{f}, \overline{f}]$$

32.2 Interval Abstraction

The interval abstraction abstracts a set of numerical values, possibly unbounded, by their minimum and maximal values. We define $\min \mathbb{I} \triangleq -\infty$, $\max \mathbb{I} \triangleq \infty$, $\min \varnothing \triangleq \infty$, $\max \varnothing \triangleq -\infty$ where $-\infty = \texttt{min\_int}$ (the smallest representable integer) and $\infty = \texttt{max\_int}$ (the greatest representable integer) for machine integers (where overflow is considered an error). The interval abstraction is

$$\begin{array}{rcl} \alpha_i(S) & \triangleq & [\min S, \max S] \\ \gamma_i([\underline{x}, \overline{x}]) & \triangleq & \{z \in \mathbb{I} \mid \underline{x} \leqslant z \leqslant \overline{x}\} \end{array}$$

Example 32.1 In interval arithmetics, a real is abstracted by the pair of enclosing floats, which is also the abstraction of the set of reals between these two floats, as shown by the concretization of a float interval as the set of reals within that interval.

We let the abstract domain of numerical intervals be

$$\begin{array}{rcl} \mathbb{P}^i_{\mathbb{I}} & \triangleq & \{\varnothing\} \cup \{[\underline{x}, \overline{x}] \mid \underline{x} \in \mathbb{I} \wedge \underline{x} \leqslant \overline{x} \wedge \overline{x} \in \mathbb{I}\} \\ & \cup & \{[-\infty, \overline{x}] \mid \overline{x} \in \mathbb{I}\} \cup \{[\underline{x}, \infty] \mid \underline{x} \in \mathbb{I}\} \cup \{[-\infty, \infty]\} \end{array}$$

where the empty interval $\perp^i = \varnothing$ can be encoded by any $[\underline{x}, \overline{x}]$ with $\overline{x} < \underline{x}$ (e.g., normalized to $[\infty, -\infty]$). The intervals $[-\infty, -\infty] \notin \mathbb{P}^i_{\mathbb{I}}$ and $[\infty, \infty] \notin \mathbb{P}^i_{\mathbb{I}}$ are excluded. For machine integers $\mathbb{I} = [\texttt{min\_int}, \texttt{max\_int}]$ where $-\infty = \texttt{min\_int}$ and $\infty = \texttt{max\_int}$, we have $\mathbb{P}^i_{\mathbb{I}} \triangleq \{\varnothing\} \cup \{[\underline{x}, \overline{x}] \mid \texttt{min\_int} \leqslant \underline{x} \leqslant \overline{x} \leqslant \texttt{max\_int}\}$.

The partial order $\sqsubseteq^i$ on $\mathbb{P}^i_{\mathbb{I}}$ is interval inclusion

$$\perp^i \sqsubseteq^i \perp^i \sqsubseteq^i [\underline{x}, \overline{x}] \sqsubseteq^i [\underline{y}, \overline{y}] \quad \text{if and only if} \quad \underline{y} \leqslant \underline{x} \leqslant \overline{x} \leqslant \overline{y}. \tag{32.2}$$

We have the Galois connection

$$\langle \wp(\mathbb{I}), \subseteq \rangle \underset{\alpha_i}{\overset{\gamma_i}{\leftrightarrows}} \langle \mathbb{P}^i_{\mathbb{I}}, \sqsubseteq^i \rangle \tag{32.3}$$

where $\langle \mathbb{P}^i, \sqsubseteq^i, \varnothing, [-\infty, +\infty], \sqcap^i, \sqcup^i \rangle$ is a complete lattice (more details will be given in section 33.1).

The soundness condition becomes

$$\alpha_i(\{f(x, y, \ldots) \mid x \in \gamma_i([\underline{x}, \overline{x}]) \wedge y \in \gamma_i([\underline{y}, \overline{y}]) \wedge \ldots\}) \quad \sqsubseteq^i \quad [\underline{f}, \overline{f}].$$

32.3 Interval Arithmetics

32.3.1 Constants

If the program contains a constant c, its interval is $[c, c]$. However, the compilation may introduce an error, that is, overflow with integers or rounding error for a float that must be taken into account. For example, the decimal 0.1 is $0.000(1100)^\infty$ in binary so has no exact binary representation on finitely many bits. It is approximated as 0.10000000149011611938476562500... in single precision.

32.3.2 Addition and Substraction

$$\begin{array}{rcl}[\underline{x},\overline{x}]\oplus^i\varnothing &=& \varnothing\oplus^i[\underline{x},\overline{x}] \;=\; [\underline{x},\overline{x}]\ominus^i\varnothing \;=\; \varnothing\ominus^i[\underline{x},\overline{x}] \;=\; \varnothing \\ [\underline{x},\overline{x}]\oplus^i[\underline{y},\overline{y}] &=& [\underline{x}+\underline{y},\overline{x}+\overline{y}] \\ [\underline{x},\overline{x}]\ominus^i[\underline{y},\overline{y}] &=& [\underline{x}-\overline{y},\overline{x}-\underline{y}] \\ \ominus^i[\underline{x},\overline{x}] &=& [-\overline{x},-\underline{x}]\end{array} \tag{32.4}$$

We assume that $-\infty+-\infty=-\infty$, $-\infty+z=-\infty$, $\infty+z=\infty$, and $\infty+\infty=\infty$ for any $z\in\mathbb{Z}$. For example, $[10,\infty]\ominus^i[-\infty,5]=[10-5,\infty-(-\infty)]=[5,\infty]$.

For floating-point numbers, the lower bound is rounded toward $-\infty$ and the upper bound toward $+\infty$. This implies that the computed value is always included in the concretization of the interval value.

Proof The proof is by calculational design. The case of $\varnothing$ is trivial. Otherwise

$\alpha_i(\{x+y \mid x\in\gamma_i([\underline{x},\overline{x}])\wedge y\in\gamma_i([\underline{y},\overline{y}])\})$

$\triangleq$ let $S=\{x+y \mid \underline{x}\leqslant x\leqslant\overline{x}\wedge\underline{y}\leqslant y\leqslant\overline{y}\}$ in $[\min S,\max S]$ ⦃definitions of γ_i and γ_i⦄

$=$ let $S=\{x+y \mid \underline{x}+\underline{y}\leqslant x+y\leqslant\overline{x}+\overline{y}\}$ in $[\min S,\max S]$

$=[\underline{x}+\underline{y},\overline{x}+\overline{y}]$

⦃by considering all cases when $\underline{x}$, $\overline{x}$, $\underline{y}$, and $\overline{y}$ are infinite or integer bounds.

- When $\underline{x}\leqslant\overline{x}$ and $\underline{y}\leqslant\overline{y}$ belong to $\mathbb{Z}$, we have $\{x+y \mid \underline{x}\leqslant x\leqslant\overline{x}\wedge\underline{y}\leqslant y\leqslant\overline{y}\}$ $=\{x+y \mid \underline{x}+\underline{y}\leqslant x+y\leqslant\overline{x}+\overline{y}\}$ since $(\underline{x}\leqslant x\leqslant\overline{x}\wedge\underline{y}\leqslant y\leqslant\overline{y})\Rightarrow(\underline{x}+\underline{y}\leqslant x+y\leqslant\overline{x}+\overline{y})$ and conversely, $(\underline{x}+\underline{y}\leqslant x+y\leqslant\overline{x}+\overline{y})\Rightarrow(\exists x',y'\in\mathbb{Z}\,.\,\underline{x}\leqslant x'\leqslant\overline{x}\wedge\underline{y}\leqslant y'\leqslant\overline{y}\wedge x'+y'=x+y)$. It follows that $[\min S,\max S]=[\underline{x}+\underline{y},\overline{x}+\overline{y}]$;
- When $\underline{x}=\underline{y}=-\infty$ and $\overline{x}=\overline{y}=\infty$, we have $\{x+y \mid \underline{x}+\underline{y}\leqslant x+y\leqslant\overline{x}+\overline{y}\}=\{x+y \mid x\in\mathbb{Z}\wedge y\in\mathbb{Z}\}=\mathbb{Z}$ with $[\min\mathbb{Z},\max\mathbb{Z}]=[-\infty,\infty];=[-\infty+-\infty,\infty+\infty]$;
- etc. ⦄

$=[\underline{x},\overline{x}]\oplus^i[\underline{y},\overline{y}]$ ⦃by defining $[\underline{x},\overline{x}]\oplus^i[\underline{y},\overline{y}]=[\underline{x}+\underline{y},\overline{x}+\overline{y}]$⦄

The proof for the binary minus $\ominus^i$ is similar and for the unary minus $\ominus^i[\underline{y},\overline{y}]=[0,0]\ominus^i[\underline{y},\overline{y}]$.

□

32.3.3 Multiplication

$$\begin{array}{rcl}[\underline{x},\overline{x}]\otimes^i\varnothing &=& \varnothing\otimes^i[\underline{x},\overline{x}] \;=\; \varnothing \\ [\underline{x},\overline{x}]\otimes^i[\underline{y},\overline{y}] &=& [\min(\underline{x}\underline{y},\underline{x}\overline{y},\overline{x}\underline{y},\overline{x}\,\overline{y}),\max(\underline{x}\underline{y},\underline{x}\overline{y},\overline{x}\underline{y},\overline{x}\,\overline{y})]\end{array}$$

which reduces to $[\underline{x}\underline{y},\overline{x}\,\overline{y}]$ when the lower bounds $\underline{x}$ and $\underline{y}$ are greater than zero.

Exercise 32.5 Prove the soundness of the interval multiplication $\otimes^i$.

32.3.4 Reciprocal and Division

The interval reciprocal is

$$
\begin{aligned}
[1,1]\oslash^i[\underline{x},\overline{x}] &\triangleq [1/\overline{x},1/\underline{x}] && \text{when } \overline{x}<0 \text{ or } 0<\underline{x}\\
[1,1]\oslash^i[\underline{x},\overline{x}] &\triangleq [-\infty,\infty] && \text{otherwise } (\tfrac{1}{0}=\pm\infty)
\end{aligned}
$$

The division can then be handled as $[\underline{x},\overline{x}]\oslash^i[\underline{y},\overline{y}] = [\underline{x},\overline{x}]\otimes([1,1]\oslash^i[\underline{y},\overline{y}])$. A more precise treatment of interval division is proposed in [496, section 4.7].

32.3.5 Algebraic Properties

The interval operations have the usual algebraic properties of arithmetic operations

$$
\begin{aligned}
(x\oplus^i y)\oplus^i z &= x\oplus^i(y\oplus^i kz) && \text{associativity}\\
(x\ominus^i y)\ominus^i z &= x\ominus^i(y\ominus^i z) &&\\
x\oplus^i y &= y\oplus^i x && \text{commutativity}\\
x\otimes y &= y\otimes^i x &&\\
x\oplus^i[0,0] &= x && \text{neutral element}\\
x\otimes^i[1,1] &= x &&
\end{aligned}
$$

However distributivity does not hold. We have

$$
x\otimes^i(y\oplus^i z) \sqsubseteq^i (x\otimes^i y)\oplus^i(x\otimes^i z) \qquad \text{subdistributivity}
$$

Exercise 32.6 Provide an example showing that subdistributivity can be strict.

32.3.6 Arithmetic Expressions

The interval semantics $\mathscr{A}^i[\![\mathsf{A}]\!]$ of an arithmetic expression A is similar to the cartesian evaluation of arithmetic expressions (28.13) which is the semantics of expressions (3.4) on $\mathbb{Z}$ where values in $\mathbb{Z}$ are replaced by intervals in $\mathbb{P}^i$ abstracting $\wp(\mathbb{Z})$.

The interval semantics $\mathscr{A}^i[\![\mathsf{A}]\!]$ does not preverve lubs in general. For example $\forall n\in\mathbb{Z}\,.\,\mathscr{A}^i[\![\mathsf{x}-\mathsf{x}]\!]\rho[\mathsf{x}\leftarrow[n,n]]=[0,0]$ whereas for $\rho[\mathsf{x}\leftarrow[-\infty,\infty]]=\dot{\bigsqcup}^i_{n\in\mathbb{Z}}\rho[\mathsf{x}\leftarrow[n,n]]$, $\mathscr{A}^i[\![\mathsf{x}-\mathsf{x}]\!]\rho[\mathsf{x}\leftarrow[-\infty,\infty]]=[-\infty,\infty]$. More generally, relations between variables are lost which is the cause of imprecision. This can be improved by coupling interval arithmetics with a relational analysis (through a reduced product of chapter 36), see, for example, [293, 296].

32.3.7 Conditions

Although in computing with $\mathbb{Z}$ only one branch of a conditional will be taken, interval computation with $\mathbb{P}^i_{\mathbb{Z}}$ may have to take both. This gives, in the worst case, an exponential number of cases to consider. The Boolean comparison operators $x \circledcirc y$ take two intervals for x and y and return two intervals for x and y such that the comparison may hold (and cannot hold outside these intervals).

$$
\begin{array}{rcll}
[\underline{x}, \overline{x}] \mathbin{\textcircled{=}}^i [\underline{y}, \overline{y}] & \triangleq & \langle \emptyset,\ \emptyset \rangle & \text{if } \overline{x} < \underline{y} \text{ or } \overline{y} < \underline{x} \\
& \triangleq & \langle [\max(\underline{x}, \underline{y}), \min(\overline{x}, \overline{y})],\ [\max(\underline{x}, \underline{y}), \min(\overline{x}, \overline{y})] \rangle & \text{otherwise} \\
[\underline{x}, \overline{x}] \mathbin{\textcircled{<}}^i [\underline{y}, \overline{y}] & \triangleq & \langle \emptyset,\ \emptyset \rangle & \text{if } \underline{x} \geqslant \overline{y} \\
& \triangleq & \langle [\underline{x}, \min(\overline{x}, \overline{y})],\ [\max(\underline{x}, \underline{y}), \overline{y}] \rangle & \text{otherwise, } \mathbb{I} \neq \mathbb{Z} \\
& \triangleq & \langle [\underline{x}, \min(\overline{x}, \overline{y} - 1)],\ [\max(\underline{x} + 1, \underline{y}), \overline{y}] \rangle & \text{otherwise, } \mathbb{I} = \mathbb{Z}
\end{array}
$$

Exercise 32.7 Provide the interval version of the other logical comparison operators.

Exercise 32.8 Define the interval arithmetic abstraction of the prefix trace semantics of section 6.6 from a semantics on reals $\mathbb{R}$ to a semantics on intervals $\mathbb{P}^i$ of floating-point numbers $\mathbb{F}$ where the abstraction of a real is the smallest float interval containing it and the abstraction of a set of traces on reals in a set of traces on float intervals.

Exercise 32.9 (Project: maximal trace float interpreter) Modify the maximal trace interpreter of exercise 7.14, in order to operate on integers or floats.

32.4 Backward Interval Arithmetics

For interval static analysis in chapter 33, we need the following backward interval arithmetics operators.

- $\overset{i}{\ominus}{}^{-1} [a, b]\langle [\underline{x}, \overline{x}],\ [\underline{y}, \overline{y}] \rangle \triangleq \langle [\max(\underline{x}, a + \underline{y}), \min(\overline{x}, b + \overline{y})],\ [\max(\underline{y}, \underline{x} - b), \min(\overline{y}, \overline{x} - a)] \rangle;$
- $\overset{i}{\textcircled{<}}{}^{-1} \langle [\underline{x}, \overline{x}],\ [\underline{y}, \overline{y}] \rangle \triangleq \langle [\underline{x}, \min(\overline{x}, \overline{y} - 1)],\ [\max(\underline{y}, \underline{x} + 1), \overline{y}] \rangle;$
- $\overset{i}{\overline{\textcircled{<}}}{}^{-1} \langle [\underline{x}, \overline{x}],\ [\underline{y}, \overline{y}] \rangle \triangleq \langle [\max(\underline{x}, \underline{y}), \overline{x}],\ [\underline{y}, \min(\overline{x}, \overline{y})] \rangle.$

For example $\overset{i}{\ominus}{}^{-1} [0, 0]\langle [0, 1],\ [-1, 0] \rangle = \langle [0, 0],\ [0, 0] \rangle$ because for the difference of $a \in [0, 1]$ and $b \in [-1, 0]$ to be 0, we must have $a = b = 0$. $\overset{i}{\ominus}{}^{-1} [0, 0]\langle [1, \infty],\ [-\infty, 0] \rangle = \langle [1, 0],\ [1, 0] \rangle = \langle \emptyset,\ \emptyset \rangle$ because the difference of a strictly positive number and a negative number cannot be null.

Exercise 32.10 (Project: interval arithmetic) Continuing exercise 29.4, implement the interval arithmetic for machine integers.

32.5 Dynamic Interval Analysis

We now show that the dynamic interval analysis of Ramon E. Moore [715–717] is an abstract interpretation. We first define the real and float semantics, then the interval abstraction, and finally we construct the interval semantics by calculus.

32.5.1 Syntax and Real or Float Semantics of the Programming Language

The syntax of the language is that of chapter 4, "Syntax," except that we add a constant `0.1` to arithmetic expressions in section 3.4.

$$\mathtt{A} \in \mathbb{A} \ ::= \ \mathtt{1} \mid \mathtt{0.1} \mid \mathtt{x} \mid \mathtt{A}_1 - \mathtt{A}_2 \qquad \text{arithmetic expressions}$$

The semantics of the language is the trace semantics $\boldsymbol{\mathcal{S}}^*[\![\mathtt{S}]\!]$ defined in chapters 6, "Structural Deductive Stateless Prefix Trace Semantics," and 7, "Maximal Trace Semantics" in structural deductive form and chapter 17 in fixpoint form, except that values are meant to be computed in the set $\mathbb{V}$ that was previously the set $\mathbb{Z}$ of integers and is now either the set $\mathbb{R}$ of reals (at least mathematically) or the set $\mathbb{F}$ of floats (and later the set $\mathbb{P}^i_{\mathbb{F}}$ of all float intervals). We write $\boldsymbol{\mathcal{S}}^*_{\mathbb{V}}[\![\mathtt{S}]\!]$ to make explicit which set of values $\mathbb{V}$ is considered in computations ($\widehat{\boldsymbol{\mathcal{S}}}^*_{\mathbb{V}}$ for the structural form and $\widehat{\boldsymbol{\mathcal{S}}}^+_{\mathbb{V}}$ for the maximal finite traces). The value (3.4) of an arithmetic expression $\mathtt{A}$ in environment $\rho \in \mathbb{E}\mathbb{v}_{\mathbb{V}} \triangleq \mathcal{V} \to \mathbb{V}$ is now $\boldsymbol{\mathcal{A}}_{\mathbb{V}}[\![\mathtt{A}]\!]\rho \in \mathbb{V}$ defined as

$$\begin{array}{rclrcl} \boldsymbol{\mathcal{A}}_{\mathbb{V}}[\![\mathtt{1}]\!]\rho & \triangleq & 1 & \boldsymbol{\mathcal{A}}_{\mathbb{V}}[\![\mathtt{0.1}]\!]\rho & \triangleq & 0.1_{\mathbb{V}} \\ \boldsymbol{\mathcal{A}}_{\mathbb{V}}[\![\mathtt{x}]\!]\rho & \triangleq & \rho(\mathtt{x}) & \boldsymbol{\mathcal{A}}_{\mathbb{V}}[\![\mathtt{A}_1 - \mathtt{A}_2]\!]\rho & \triangleq & \boldsymbol{\mathcal{A}}_{\mathbb{V}}[\![\mathtt{A}_1]\!]\rho -_{\mathbb{V}} \boldsymbol{\mathcal{A}}_{\mathbb{V}}[\![\mathtt{A}_2]\!]\rho \end{array} \tag{32.11}$$

where 1 denotes the real and float values one, $0.1_{\mathbb{V}}$ denotes either the real 0.1 or its float approximation and $-_{\mathbb{V}}$ the difference in $\mathbb{V}$. For example $-_{\mathbb{F}}$ is the difference found on IEEE-754 machines and must take rounding mode (and the machine specificities [709]) into account.

32.5.2 Float Interval Abstraction

Let ${}^{\leftarrow}\!\rrbracket x$ (which can be $-\infty$) be the largest float smaller than or equal to $x \in \mathbb{R}$ (or ${}^{\leftarrow}\!\rrbracket x = x$ for $x \in \mathbb{F}$) and $x\llbracket^{\rightarrow}$ (which can be $+\infty$) be the smallest float greater than or equal to $x \in \mathbb{R}$ (or $x\llbracket^{\rightarrow} = x$ for $x \in \mathbb{F}$). We let ${}^{\leftarrow}\!\rceil x$ be the largest floating-point number strictly less than $x \in \mathbb{F}$ (which can be $-\infty$) and $x\lceil^{\rightarrow}$ be the smallest floating-point number strictly larger than $x \in \mathbb{F}$ (which can be $+\infty$). We assume that

$$\begin{array}{rcll} {}^{\leftarrow}\!\rrbracket x -_{\mathbb{F}} y\llbracket^{\rightarrow} & \leqslant & {}^{\leftarrow}\!\rrbracket(x -_{\mathbb{V}} y) & (\mathbb{V} \text{ is } \mathbb{R} \text{ or } \mathbb{F}) \quad (32.12)\\ x\llbracket^{\rightarrow} -_{\mathbb{F}} {}^{\leftarrow}\!\rrbracket y & \geqslant & (x -_{\mathbb{V}} y)\llbracket^{\rightarrow} & \end{array}$$

$$(\mathtt{x} \in [\underline{x}, \overline{x}] \wedge \mathtt{y} \in [\underline{y}, \overline{y}] \wedge \mathtt{x} < \mathtt{y}) \Rightarrow (\mathtt{x} \in [\underline{x}, \min(\overline{x}, \overline{y})] \wedge \mathtt{y} \in [\max(\underline{x}, \underline{y}), \overline{y}]) \tag{32.13}$$

$$(\mathtt{x} \in [\underline{x}, \overline{x}] \wedge \mathtt{y} \in [\underline{y}, \overline{y}] \wedge \mathtt{x} < \mathtt{y}) \Rightarrow (\mathtt{x} \in [\underline{x}, \min(\overline{x}, \overline{y}\lceil^{\rightarrow})] \wedge \mathtt{y} \in [\max({}^{\leftarrow}\!\rceil\underline{x}, \underline{y}), \overline{y}]) \tag{32.13.bis}$$

Machine implementations of IEEE-754 floating-point arithmetics [522] are sometimes incorrect [423, 709]. Therefore the preceding hypotheses (32.12) and (32.13) on floats must be adjusted accordingly, for example, replacing (32.13) by (32.13.bis). In particular, (32.13.bis) follows the recommendation of [709, sect. 6.1.2]. If $\mathtt{x} < \mathtt{y}$ then the value of $\mathtt{x}$ is smaller than its maximal value $\overline{x}$ and the maximal $\overline{y}$ value of $\mathtt{y}$, by precaution, certainly smaller or equal to the next float greater than $\overline{y}$.

$$\begin{array}{rcll} \alpha^i_{\mathbb{F}}(x) & \triangleq & [{}^{\leftarrow}\!\rrbracket x, x\llbracket^{\rightarrow}] & \text{real/float abstraction by float interval} \quad (32.14)\\ \gamma^i_{\mathbb{F}}([\underline{x}, \overline{x}]) & \triangleq & \{x \in \mathbb{R} \mid \underline{x} \leqslant x \leqslant \overline{x}\} & \\ \dot{\alpha}^i_{\mathbb{F}}(\rho) & \triangleq & \mathtt{x} \in \mathcal{V} \mapsto \dot{\alpha}^i_{\mathbb{F}}(\rho(\mathtt{x})) & \text{environment abstraction}\\ \dot{\gamma}^i_{\mathbb{F}}(\overline{\rho}) & \triangleq & \{\rho \in \mathcal{V} \to \mathbb{R} \mid \forall \mathtt{x} \in \mathcal{V} \,.\, \rho(\mathtt{x}) \in \gamma^i_{\mathbb{F}}(\overline{\rho}(\mathtt{x}))\} & \end{array}$$

The Galois connection (32.3) for float intervals is extended pointwise $\rho \mathrel{\dot{\sqsubseteq}^i} \rho' \triangleq \forall \mathtt{x} \in \mathcal{V} \,.\, \rho(\mathtt{x}) \sqsubseteq^i \rho'(\mathtt{x})$ to interval environments (mapping variables to their interval of values).

$$\langle \wp(\mathbb{R}), \subseteq\rangle \underset{\alpha^i_{\mathbb{F}}}{\overset{\gamma^i_{\mathbb{F}}}{\leftrightarrows}} \langle \mathbb{P}^i_{\mathbb{F}}, \sqsubseteq^i\rangle \quad (32.3) \qquad \langle \wp(\mathbb{V} \to \mathbb{R}), \subseteq\rangle \underset{\dot\alpha^i_{\mathbb{F}}}{\overset{\dot\gamma^i_{\mathbb{F}}}{\leftrightarrows}} \langle \mathbb{V} \to \mathbb{P}^i_{\mathbb{F}}, \dot\sqsubseteq^i\rangle \quad (32.15)$$

32.5.3 Float Interval Abstraction of an Arithmetic Expression

Given $\rho \in \mathbb{E}\mathrm{v}_{\mathbb{V}}$ (where $\mathbb{V}$ is $\mathbb{R}$ or $\mathbb{F}$), let us evaluate $\alpha^i_{\mathbb{F}}(\mathscr{A}_{\mathbb{V}}\llbracket \mathsf{A}\rrbracket\rho)$ by structural induction on A and define $\mathscr{A}^i_{\mathbb{F}}\llbracket \mathsf{A}\rrbracket$ such that

$$\alpha^i_{\mathbb{F}}(\mathscr{A}_{\mathbb{V}}\llbracket \mathsf{A}\rrbracket\rho) \quad \sqsubseteq^i \quad \mathscr{A}^i_{\mathbb{F}}\llbracket \mathsf{A}\rrbracket\dot\alpha^i_{\mathbb{F}}(\rho). \tag{32.16}$$

By calculational design, we get

$$\left[\begin{array}{lcll}
\mathscr{A}^i_{\mathbb{F}}\llbracket 1\rrbracket\rho & \triangleq & 1_{\mathbb{F}} & \text{where } 1_{\mathbb{F}} = [1,1] \text{ and } 1 \in \mathbb{F} \qquad (32.17)\\
\mathscr{A}^i_{\mathbb{F}}\llbracket 0.1\rrbracket\rho & \triangleq & 0.1_{\mathbb{F}} & \text{where } 0.1_{\mathbb{F}} \triangleq [\lfloor 0.1_{\mathbb{V}}, 0.1_{\mathbb{V}}\rceil]\\
\mathscr{A}^i_{\mathbb{F}}\llbracket \mathsf{x}\rrbracket\rho & \triangleq & \rho(\mathsf{x}) & \\
\mathscr{A}^i_{\mathbb{F}}\llbracket \mathsf{A}_1 - \mathsf{A}_2\rrbracket\rho & \triangleq & \mathscr{A}^i_{\mathbb{F}}\llbracket \mathsf{A}_1\rrbracket\rho \ominus^i_{\mathbb{F}} \mathscr{A}^i_{\mathbb{F}}\llbracket \mathsf{A}_2\rrbracket\rho & \text{where } [\underline{x},\overline{x}] \ominus^i_{\mathbb{F}} [\underline{y},\overline{y}] \triangleq [\underline{x} -_{\mathbb{F}} \overline{y}, \overline{x} -_{\mathbb{F}} \underline{y}]
\end{array}\right.$$

as follows

$-\ \alpha^i_{\mathbb{F}}(\mathscr{A}_{\mathbb{V}}\llbracket 0.1\rrbracket\rho)$

$=\ \alpha^i_{\mathbb{F}}(0.1_{\mathbb{V}})$ ⦃definition of $\mathscr{A}_{\mathbb{V}}$ in (3.4)⦄

$=\ [\lfloor 0.1_{\mathbb{V}}, 0.1_{\mathbb{V}}\rceil]$ ⦃real abstraction by float interval in (32.14)⦄

$\triangleq\ \mathscr{A}^i_{\mathbb{F}}\llbracket 0.1\rrbracket(\dot\alpha^i_{\mathbb{F}}(\rho))$ ⦃by defining $\mathscr{A}^i_{\mathbb{F}}\llbracket 0.1\rrbracket\overline{\rho} \triangleq [\lfloor 0.1_{\mathbb{V}}, 0.1_{\mathbb{V}}\rceil]$⦄

$-\ \alpha^i_{\mathbb{F}}(\mathscr{A}_{\mathbb{V}}\llbracket \mathsf{x}\rrbracket\rho)$

$=\ \alpha^i_{\mathbb{F}}(\rho(\mathsf{x}))$ ⦃definition of $\mathscr{A}_{\mathbb{V}}$ in (3.4)⦄

$=\ \dot\alpha^i_{\mathbb{F}}(\rho)(\mathsf{x})$ ⦃definition of environment abstraction in (32.14)⦄

$\triangleq\ \mathscr{A}^i_{\mathbb{F}}\llbracket \mathsf{x}\rrbracket(\dot\alpha^i_{\mathbb{F}}(\rho))$ ⦃by defining $\mathscr{A}^i_{\mathbb{F}}\llbracket \mathsf{x}\rrbracket\overline{\rho} \triangleq \overline{\rho}(\mathsf{x})$⦄

$-\ \alpha^i_{\mathbb{F}}(\mathscr{A}_{\mathbb{V}}\llbracket \mathsf{A}_1 - \mathsf{A}_2\rrbracket\rho)$

$=\ \alpha^i_{\mathbb{F}}(\mathscr{A}_{\mathbb{V}}\llbracket \mathsf{A}_1\rrbracket\rho -_{\mathbb{V}} \mathscr{A}_{\mathbb{V}}\llbracket \mathsf{A}_2\rrbracket\rho)$ ⦃definition of $\mathscr{A}_{\mathbb{V}}$ in (3.4)⦄

$=\ [\lfloor(\mathscr{A}_{\mathbb{V}}\llbracket \mathsf{A}_1\rrbracket\rho -_{\mathbb{V}} \mathscr{A}_{\mathbb{V}}\llbracket \mathsf{A}_2\rrbracket\rho), (\mathscr{A}_{\mathbb{V}}\llbracket \mathsf{A}_1\rrbracket\rho -_{\mathbb{V}} \mathscr{A}_{\mathbb{V}}\llbracket \mathsf{A}_2\rrbracket\rho)\rceil]$

⦃value abstraction by float interval in (32.14)⦄

$\sqsubseteq^i\ [\lfloor(\mathscr{A}_{\mathbb{V}}\llbracket \mathsf{A}_1\rrbracket\rho) -_{\mathbb{F}} (\mathscr{A}_{\mathbb{V}}\llbracket \mathsf{A}_2\rrbracket\rho)\rceil), (\mathscr{A}_{\mathbb{V}}\llbracket \mathsf{A}_1\rrbracket\rho)\rceil -_{\mathbb{F}} \lfloor(\mathscr{A}_{\mathbb{V}}\llbracket \mathsf{A}_2\rrbracket\rho)]$

⦃(32.2) and hypothesis (32.12)⦄

$\sqsubseteq^i$ let $[\underline{x},\overline{x}] = \mathscr{A}^i_{\mathbb{F}}\llbracket \mathsf{A}_1\rrbracket\dot\alpha^i_{\mathbb{F}}(\rho)$ and $[\underline{y},\overline{y}] = \mathscr{A}^i_{\mathbb{F}}\llbracket \mathsf{A}_2\rrbracket\dot\alpha^i_{\mathbb{F}}(\rho)$ in $[\underline{x} -_{\mathbb{F}} \overline{y}, \overline{x} -_{\mathbb{F}} \underline{y}]$

⦃By induction hypothesis $[\lfloor\mathscr{A}_{\mathbb{V}}\llbracket \mathsf{A}_i\rrbracket\rho, \mathscr{A}_{\mathbb{V}}\llbracket \mathsf{A}_i\rrbracket\rho\rceil] = \alpha^i_{\mathbb{F}}(\mathscr{A}_{\mathbb{V}}\llbracket \mathsf{A}_i\rrbracket\rho) \sqsubseteq^i \mathscr{A}^i_{\mathbb{F}}\llbracket \mathsf{A}_i\rrbracket\dot\alpha^i_{\mathbb{F}}(\rho)$, $i = 1, 2$⦄

$=\ \mathscr{A}^i_{\mathbb{F}}\llbracket \mathsf{A}_1\rrbracket\dot\alpha^i_{\mathbb{F}}(\rho) \ominus^i_{\mathbb{F}} \mathscr{A}^i_{\mathbb{F}}\llbracket \mathsf{A}_2\rrbracket\dot\alpha^i_{\mathbb{F}}(\rho)$

⦃by defining $[\underline{x},\overline{x}] \ominus^i_{\mathbb{F}} [\underline{y},\overline{y}] \triangleq [\underline{x} -_{\mathbb{F}} \overline{y}, \overline{x} -_{\mathbb{F}} \underline{y}]$ as in (32.4)⦄

$\triangleq\ \mathscr{A}^i_{\mathbb{F}}\llbracket \mathsf{A}_1 - \mathsf{A}_2\rrbracket\dot\alpha^i_{\mathbb{F}}(\rho)$ ⦃by defining $\mathscr{A}^i_{\mathbb{F}}\llbracket \mathsf{A}_1 - \mathsf{A}_2\rrbracket\overline{\rho} \triangleq \mathscr{A}^i_{\mathbb{F}}\llbracket \mathsf{A}_1\rrbracket\overline{\rho} \ominus^i_{\mathbb{F}} \mathscr{A}^i_{\mathbb{F}}\llbracket \mathsf{A}_2\rrbracket\overline{\rho}$⦄

We observe that $\mathscr{A}^i_{\mathbb{F}}$ is $\dot{\sqsubseteq}^i$-increasing.

If we had a division, we would have to handle NaN (Not a Number) interpreted as a value that is undefined or unrepresentable. A simple way is to stop execution, by choosing $\mathscr{A}^i_{\mathbb{F}}[\![\mathtt{1}\ /\ \mathtt{0}]\!]\overline{\rho} \triangleq \varnothing$. Another way would be to include the NaN in the abstraction by considering $\mathsf{N}[\underline{x}, \overline{x}]$ meaning a float between the bounds while $\mathsf{NaN}[\underline{x}, \overline{x}]$ would mean a float between the bounds or NaN. We choose the first alternative, which is simpler.

32.5.4 Float Interval Abstraction of a Boolean Expression

The situation is more complicated for conditionals. Whereas a test is true or false for $\mathbb{V} = \mathbb{R}$ and $\mathbb{V} = \mathbb{F}$, it might be true for part of a float interval and false for another part of this interval when $\mathbb{V} = \mathbb{P}^i_{\mathbb{F}}$. Moreover in case of uncertainty (e.g., < is handled as ⩽) the two part may overlap.

Therefore we assume that the abstract interpretation $\mathscr{B}^i_{\mathbb{F}}[\![\mathsf{B}]\!]$ of a Boolean expression B is defined such that

$$
\left[\begin{array}{lll}
\mathsf{let}\ \langle\overline{\rho}_{\mathsf{tt}}, \overline{\rho}_{\mathsf{ff}}\rangle = \mathscr{B}^i_{\mathbb{F}}[\![\mathsf{B}]\!](\dot{\alpha}^i_{\mathbb{F}}(\rho))\ \mathsf{in} & & \\
\quad \dot{\alpha}^i_{\mathbb{F}}(\rho) \ \dot{\sqsubseteq}^i\ \overline{\rho}_{\mathsf{tt}} & \text{if} & \mathscr{B}_{\mathbb{V}}[\![\mathsf{B}]\!]\rho = \mathsf{tt} \\
\quad \dot{\alpha}^i_{\mathbb{F}}(\rho) \ \dot{\sqsubseteq}^i\ \overline{\rho}_{\mathsf{ff}} & \text{if} & \mathscr{B}_{\mathbb{V}}[\![\mathsf{B}]\!]\rho = \mathsf{ff} \\
\mathsf{and}\ (\langle\overline{\rho}_{\mathsf{tt}}, \overline{\rho}_{\mathsf{ff}}\rangle = \mathscr{B}^i_{\mathbb{F}}[\![\mathsf{B}]\!]\overline{\rho}) \Rightarrow (\overline{\rho}_{\mathsf{tt}} \dot{\sqsubseteq}^i \overline{\rho} \wedge \overline{\rho}_{\mathsf{ff}} \dot{\sqsubseteq}^i \overline{\rho}) & & \\
\mathsf{and}\ \mathscr{B}^i_{\mathbb{F}}[\![\mathsf{B}]\!]\ \text{is increasing} & &
\end{array}\right. \tag{32.18}
$$

stating that no concrete state passing the test is omitted in the abstract and that the postcondition $\overline{\rho}_{\mathsf{tt}}$ or $\overline{\rho}_{\mathsf{ff}}$ is stronger than the precondition $\overline{\rho}$ because, in absence of side effects, the test cannot change values of variables. If $\mathscr{B}_{\mathbb{V}}[\![\mathsf{B}]\!]\rho = \mathsf{tt}$ (respectively ff), there is no constraint on $\overline{\rho}_{\mathsf{ff}}$ (respectively $\overline{\rho}_{\mathsf{tt}}$), the best choice being the $\dot{\sqsubseteq}^i$-infimum empty interval environment $\dot{\varnothing}$. On the basis of section 32.3.7, and following [225], definitions of $\mathscr{B}^i_{\mathbb{F}}[\![\mathsf{B}]\!]$, will be discussed in sections 33.3 and 33.7.

32.5.5 Interval Boolean Expressions Have Side Effects

The evaluation of integer, float and real Boolean expressions has no side effect in that the value of a variable before and after a test is exactly the same, whether the condition is true or false. Formally,

$$\varrho(\pi\ell) \quad = \quad \varrho(\pi\ell \xrightarrow{\mathsf{B}/\neg(\mathsf{B})} \ell')$$

as shown by the definition (6.6) of ϱ.

This is no longer the case with interval arithmetics! For example, assume that $\varrho(\pi\ell)\mathsf{x} = [-0.1, 0.1]$. Then $\varrho(\pi\ell \xrightarrow{(\mathsf{x} > \mathsf{0})} \ell') = [0, 0.1]$ while $\varrho(\pi\ell \xrightarrow{\neg(\mathsf{x} > \mathsf{0})} \ell') = [-0.1, 0]$. Observe that > is handled as >= because intervals are closed and that the interval of value of x is changed by the test, which has a side effect.

To take such side effects into account in the stateless prefix and maximal traces semantics, we have

- to record the value of variables after the test in the Boolean actions, which become $\mathsf{B} = \rho$ and $\neg(\mathsf{B}) = \rho$ where $\rho \in \mathbb{E}\mathbb{v}$ is the environment passing the test; and
- to change the computation (6.6) of the value of variable $\mathsf{x} \in \mathbb{V}$ at the end of a trace π as follows:

$$\begin{aligned}
\varrho(\pi\ell \xrightarrow{\mathsf{x} = \mathsf{A} = \nu} \ell')\mathsf{x} &\triangleq \nu \\
\varrho(\pi\ell \xrightarrow{\mathsf{B} = \rho} \ell')\mathsf{x} &\triangleq \rho(\mathsf{x}) \\
\varrho(\pi\ell \xrightarrow{\neg(\mathsf{B}) = \rho} \ell')\mathsf{x} &\triangleq \rho(\mathsf{x}) \\
\varrho(\pi\ell \xrightarrow{\cdots} \ell')\mathsf{x} &\triangleq \varrho(\pi\ell)\mathsf{x} \quad \text{otherwise} \\
\varrho(\ell)\mathsf{x} &\triangleq 0
\end{aligned} \tag{32.19}$$

This can be taken into account in chapters 6, "Structural Deductive Stateless Prefix Trace Semantics," 7, "Maximal Trace Semantics," and 17, "Structural Fixpoint Prefix and Maximal Trace Semantics," by changing traces $\pi \xrightarrow{\mathsf{B}/\neg(\mathsf{B})} \pi'$ to $\pi \xrightarrow{\mathsf{B} = \overline{\rho}_{\mathtt{tt}}} \pi'$ and $\pi \xrightarrow{\neg(\mathsf{B}) = \overline{\rho}_{\mathtt{ff}}} \pi'$ where $\langle \overline{\rho}_{\mathtt{tt}}, \overline{\rho}_{\mathtt{ff}} \rangle = \mathscr{B}^i_{\mathbb{F}}[\![\mathsf{B}]\!]\, \varrho(\pi)$ in these semantics.

32.5.6 Abstraction of Real Traces by Float Interval Traces

We write $\mathbb{T}^+_{\mathbb{V}}$, $\mathbb{T}^\infty_{\mathbb{V}}$, and $\mathbb{T}^{+\infty}_{\mathbb{V}}$ for the nonempty finite, infinite, and nonempty finite or infinite traces considered in chapters 6 and 7 for $\mathbb{V} = \mathbb{Z}$ but now for $\mathbb{V} = \mathbb{R}$, $\mathbb{V} = \mathbb{F}$, and $\mathbb{V} = \mathbb{P}^i_{\mathbb{F}}$.

Given a real trace semantics, that is, a set $\Pi \in \wp(\mathbb{T}^{+\infty}_{\mathbb{R}})$, we define a float interval trace semantics $\mathring{\alpha}^i_{\mathbb{F}}(\Pi) \in \wp(\mathbb{T}^{+\infty}_{\mathbb{P}^i_{\mathbb{F}}})$ by abstracting the real $x \in \mathbb{R}$ values by an interval $\alpha^i_{\mathbb{F}}(x) = [\,\lceil\!\!\!\!\lceil x, x \rceil\!\!\!\!\rceil\,]$. More precisely, because abstract interpretation is about the abstraction of properties, the strongest property $\{x\} \in \wp(\mathbb{R})$ of this value is overapproximated by a weaker interval property, that is, $\{x\} \subseteq [\,\lceil\!\!\!\!\lceil x, x \rceil\!\!\!\!\rceil\,]$, or equivalently $x \in [\,\lceil\!\!\!\!\lceil x, x \rceil\!\!\!\!\rceil\,]$. Formally, the abstraction of a trace abstracts real values computed along the trace into a float interval.

$$\begin{aligned}
\bar{\alpha}^i_{\mathbb{F}}(\mathsf{x} = \mathsf{A} = \nu) &\triangleq \mathsf{x} = \mathsf{A} = \alpha^i_{\mathbb{F}}(\nu) && \text{action abstraction} \\
\bar{\alpha}^i_{\mathbb{F}}(\mathsf{B} = \rho) &\triangleq \mathsf{B} = \overline{\rho}_{\mathtt{tt}} && \text{if } \langle \overline{\rho}_{\mathtt{tt}}, \overline{\rho}_{\mathtt{ff}} \rangle = \mathscr{B}^i_{\mathbb{F}}[\![\mathsf{B}]\!](\dot{\alpha}^i_{\mathbb{F}}(\rho)) \\
\bar{\alpha}^i_{\mathbb{F}}(\neg(\mathsf{B}) = \rho) &\triangleq \neg(\mathsf{B}) = \overline{\rho}_{\mathtt{ff}} && \text{if } \langle \overline{\rho}_{\mathtt{tt}}, \overline{\rho}_{\mathtt{ff}} \rangle = \mathscr{B}^i_{\mathbb{F}}[\![\mathsf{B}]\!](\dot{\alpha}^i_{\mathbb{F}}(\rho)) \\
\bar{\alpha}^i_{\mathbb{F}}(\mathsf{a}) &\triangleq \mathsf{a} && \text{otherwise} \\
\bar{\gamma}^i_{\mathbb{F}}(\mathsf{x} = \mathsf{A} = \overline{\nu}) &\triangleq \{\mathsf{x} = \mathsf{A} = \nu \mid \nu \in \gamma^i_{\mathbb{F}}(\overline{\nu})\} \\
\bar{\gamma}^i_{\mathbb{F}}(\mathsf{B} = \overline{\rho}) &\triangleq \{\mathsf{B} = \rho \mid \rho \in \dot{\gamma}^i_{\mathbb{F}}(\overline{\rho})\} \\
\bar{\gamma}^i_{\mathbb{F}}(\neg(\mathsf{B}) = \overline{\rho}) &\triangleq \{\neg(\mathsf{B}) = \rho \mid \rho \in \dot{\gamma}^i_{\mathbb{F}}(\overline{\rho})\} \\
\bar{\gamma}^i_{\mathbb{F}}(\mathsf{a}) &\triangleq \{\mathsf{a}\} && \text{otherwise}
\end{aligned} \tag{32.20}$$

$$\vec{\alpha}^{i}_{\mathbb{F}}(\ell_1 \xrightarrow{\mathsf{a}_1} \ell_2 \dots \ell_n \xrightarrow{\mathsf{a}_n} \ell_{n+1} \dots) \triangleq \quad \text{[in]finite trace abstraction}$$
$$(\ell_1 \xrightarrow{\bar{\alpha}^{i}_{\mathbb{F}}(\mathsf{a}_1)} \ell_2 \dots \ell_n \xrightarrow{\bar{\alpha}^{i}_{\mathbb{F}}(\mathsf{a}_n)} \ell_{n+1} \dots) \tag{32.21}$$
$$\vec{\gamma}^{i}_{\mathbb{F}}(\overline{\pi}) \triangleq \mathsf{let}\ \ell_1 \xrightarrow{\overline{\mathsf{a}}_1} \ell_2 \dots \ell_n \xrightarrow{\overline{\mathsf{a}}_n} \ell_{n+1} \dots = \overline{\pi}\ \mathsf{in}$$
$$\{\pi \mid |\pi| = |\overline{\pi}| \wedge \pi = \ell_1 \xrightarrow{\mathsf{a}_1} \ell_2 \dots \ell_n \xrightarrow{\mathsf{a}_n} \ell_{n+1} \dots \wedge$$
$$\forall i = 1, \dots, n, \dots .\ \mathsf{a}_i \in \bar{\gamma}^{i}_{\mathbb{F}}(\overline{\mathsf{a}}_i)\}$$

such that $\dot{\alpha}^{i}_{\mathbb{F}}(\boldsymbol{\varrho}(\pi)) \mathrel{\dot{\sqsubseteq}^{i}} \boldsymbol{\varrho}(\vec{\alpha}^{i}_{\mathbb{F}}(\pi))$ (by definition (32.19) of $\boldsymbol{\varrho}$) (32.22)

This trace abstraction is extended componentwise to pairs $\langle\pi, \pi'\rangle$ of an initialization trace π and a continuation trace π'.

$$\ddot{\vec{\alpha}}^{i}_{\mathbb{F}}(\langle\pi, \pi'\rangle) \triangleq \langle\vec{\alpha}^{i}_{\mathbb{F}}(\pi), \vec{\alpha}^{i}_{\mathbb{F}}(\pi')\rangle \qquad \text{pairwise abstraction} \tag{32.23}$$
$$\ddot{\vec{\gamma}}^{i}_{\mathbb{F}}(\langle\overline{\pi}, \overline{\pi}'\rangle) \triangleq \{\langle\pi, \pi'\rangle \mid \pi \in \vec{\gamma}^{i}_{\mathbb{F}}(\overline{\pi}) \wedge \pi' \in \vec{\gamma}^{i}_{\mathbb{F}}(\overline{\pi}')\}$$

This pair of traces abstraction is extended elementwise to trace semantics, that is, sets of pairs of traces (see section 6.10)

$$\mathring{\alpha}^{i}_{\mathbb{F}}(\Pi) \triangleq \{\ddot{\vec{\alpha}}^{i}_{\mathbb{F}}(\langle\pi, \pi'\rangle) \mid \langle\pi, \pi'\rangle \in \Pi\} \qquad \text{set of traces abstraction} \tag{32.24}$$
$$\mathring{\gamma}^{i}_{\mathbb{F}}(\overline{\Pi}) \triangleq \{\langle\pi, \pi'\rangle \mid \ddot{\vec{\alpha}}^{i}_{\mathbb{F}}(\langle\pi, \pi'\rangle) \in \overline{\Pi}\} = \bigcup\{\ddot{\vec{\gamma}}^{i}_{\mathbb{F}}(\langle\overline{\pi}, \overline{\pi}'\rangle) \mid \langle\overline{\pi}, \overline{\pi}'\rangle \in \overline{\Pi}\}$$

so that, by exercise 11.6, we have the Galois connection

$$\langle\wp(\mathbb{T}^{+\infty}_{\mathbb{V}}), \subseteq\rangle \underset{\mathring{\alpha}^{i}_{\mathbb{F}}}{\overset{\mathring{\gamma}^{i}_{\mathbb{F}}}{\leftrightarrows}} \langle\wp(\mathbb{T}^{+\infty}_{\mathbb{P}^{i}_{\mathbb{F}}}), \subseteq\rangle \tag{32.25}$$

Because the floats are a subset of the reals, we can use $\mathring{\alpha}^{i}_{\mathbb{F}}$ to abstract sets of both real and float traces (i.e., whether $\mathbb{V}$ be $\mathbb{R}$ or $\mathbb{F}$).

32.5.7 Sound Overapproximation in the Concrete

Let $\Pi = \{\langle\ell_1, \ell_1 \xrightarrow{\mathtt{x = 1} = 1_{\mathbb{R}}} \ell_2 \xrightarrow{\mathtt{x = x + 0.1} = 1.1_{\mathbb{R}}} \ell_3\rangle\}$ be a semantics on $\mathbb{R}$.

Let $\overline{\Pi}_1 = \alpha^{i}_{\mathbb{F}}(\Pi) = \{\langle\ell_1, \ell_1 \xrightarrow{\mathtt{x = 1} = [0.99, 1.01]} \ell_2 \xrightarrow{\mathtt{x = x + 0.1} = [1.09, 1.11]} \ell_3\rangle\}$ where each trace π of Π is overapproximated by a trace $\vec{\alpha}^{i}_{\mathbb{F}}(\pi)$ of $\overline{\Pi}_1$ with a ± 0.01 rounding interval. We have $\Pi \subseteq \mathring{\gamma}^{i}_{\mathbb{F}}(\overline{\Pi}_1)$ so $\overline{\Pi}_1$ is a sound overapproximation of Π.

But

$$\overline{\Pi}_2 = \{\langle\ell_1, \ell_1 \xrightarrow{\mathtt{x = 1} = [0.99, 1.01]} \ell_2 \xrightarrow{\mathtt{x = x + 0.1} = [0, 1.1]} \ell_3\rangle,$$
$$\langle\ell_1, \ell_1 \xrightarrow{\mathtt{x = 1} = [0.99, 1.01]} \ell_2 \xrightarrow{\mathtt{x = x + 0.1} = [1.1, 2]} \ell_3\rangle\}$$

is also a sound overapproximation of Π because $\Pi \subseteq \mathring{\gamma}^{i}_{\mathbb{F}}(\overline{\Pi}_2)$.

Although $\overline{\Pi}_1 \in \wp(\mathbb{T}^{+\infty}_{\mathbb{P}^{i}_{\mathbb{F}}})$ is more precise than $\overline{\Pi}_2 \in \wp(\mathbb{T}^{+\infty}_{\mathbb{P}^{i}_{\mathbb{F}}})$, they are not comparable as abstract elements of $\langle\wp(\mathbb{T}^{+\infty}_{\mathbb{P}^{i}_{\mathbb{F}}}), \subseteq\rangle$ in (32.25) (because $[1.09, 1.11] \not\subseteq [0, 1.1]$, $[1.09, 1.11] \not\subseteq [1.1, 2]$, $[0, 1.1] \not\subseteq [1.09, 1.11]$, and $[1.1, 2] \not\subseteq [1.09, 1.11]$).

The intuition that $\overline{\Pi}_1$ is more precise than $\overline{\Pi}_2$ is by comparison in the concrete that is $\mathring{\gamma}^i_{\mathbb{F}}(\overline{\Pi}_1) \subseteq \mathring{\gamma}^i_{\mathbb{F}}(\overline{\Pi}_2)$. We now express this preorder relation $\mathring{\sqsubseteq}^i$ between $\overline{\Pi}_1$ and $\overline{\Pi}_2$ which will allow us to overapproximate intervals when needed.

$$\begin{aligned}
&\overline{\Pi} \mathring{\sqsubseteq}^i \overline{\Pi}' && (32.26)\\
&\quad \triangleq \mathring{\gamma}^i_{\mathbb{F}}(\overline{\Pi}) \subseteq \mathring{\gamma}^i_{\mathbb{F}}(\overline{\Pi}')\\
&\quad = \forall \langle \overline{\pi}_0, \overline{\pi}_1\rangle \in \overline{\Pi} \,.\, \forall \langle \pi_0, \pi_1\rangle \in \ddot{\vec{\gamma}}^i_{\mathbb{F}}(\langle \overline{\pi}_0, \overline{\pi}_1\rangle) \,.\, \exists \langle \overline{\pi}'_0, \overline{\pi}'_1\rangle \in \overline{\Pi}' \,. && (32.27)\\
&\qquad\qquad \langle \pi_0, \pi_1\rangle \in \ddot{\vec{\gamma}}^i_{\mathbb{F}}(\langle \overline{\pi}'_0, \overline{\pi}'_1\rangle)
\end{aligned}$$

Proof of (32.26)

$$\begin{aligned}
&\overline{\Pi} \mathring{\sqsubseteq}^i \overline{\Pi}'\\
\triangleq\ & \mathring{\gamma}^i_{\mathbb{F}}(\overline{\Pi}) \subseteq \mathring{\gamma}^i_{\mathbb{F}}(\overline{\Pi}') && \text{(32.26)}\\
=\ & \bigcup\{\ddot{\vec{\gamma}}^i_{\mathbb{F}}(\langle \overline{\pi}_0, \overline{\pi}_1\rangle) \mid \langle \overline{\pi}_0, \overline{\pi}_1\rangle \in \overline{\Pi}\} \subseteq \bigcup\{\ddot{\vec{\gamma}}^i_{\mathbb{F}}(\langle \overline{\pi}'_0, \overline{\pi}'_1\rangle) \mid \langle \overline{\pi}'_0, \overline{\pi}'_1\rangle \in \overline{\Pi}'\}\\
& && \text{(definition (32.24) of } \mathring{\gamma}^i_{\mathbb{F}}\text{)}\\
=\ & \forall \langle \overline{\pi}_0, \overline{\pi}_1\rangle \in \overline{\Pi} \,.\, \ddot{\vec{\gamma}}^i_{\mathbb{F}}(\langle \overline{\pi}_0, \overline{\pi}_1\rangle) \subseteq \bigcup\{\ddot{\vec{\gamma}}^i_{\mathbb{F}}(\langle \overline{\pi}'_0, \overline{\pi}'_1\rangle) \mid \langle \overline{\pi}'_0, \overline{\pi}'_1\rangle \in \overline{\Pi}'\} && \text{(definition of } \subseteq\text{)}\\
=\ & \forall \langle \overline{\pi}_0, \overline{\pi}_1\rangle \in \overline{\Pi} \,.\, \forall \langle \pi_0, \pi_1\rangle \in \ddot{\vec{\gamma}}^i_{\mathbb{F}}(\langle \overline{\pi}_0, \overline{\pi}_1\rangle) \,.\, \langle \pi_0, \pi_1\rangle \in \bigcup\{\ddot{\vec{\gamma}}^i_{\mathbb{F}}(\langle \overline{\pi}'_0, \overline{\pi}'_1\rangle) \mid \langle \overline{\pi}'_0,\\
& \overline{\pi}'_1\rangle \in \overline{\Pi}'\} && \text{(definition of } \subseteq\text{)}\\
=\ & \forall \langle \overline{\pi}_0, \overline{\pi}_1\rangle \in \overline{\Pi} \,.\, \forall \langle \pi_0, \pi_1\rangle \in \ddot{\vec{\gamma}}^i_{\mathbb{F}}(\langle \overline{\pi}_0, \overline{\pi}_1\rangle) \,.\, \exists \langle \overline{\pi}'_0, \overline{\pi}'_1\rangle \in \overline{\Pi}' \,.\, \langle \pi_0, \pi_1\rangle \in \ddot{\vec{\gamma}}^i_{\mathbb{F}}(\langle \overline{\pi}'_0, \overline{\pi}'_1\rangle)\\
& && \text{(definition of } \textstyle\bigcup\text{)} \quad \square
\end{aligned}$$

Exercise 32.28 Prove that we have a Galois connection (where the abstract preorder and the concretization to be determined are different from (32.25))

$$\langle \wp(\mathbb{S}^{+\infty}_{\mathbb{V}}), \subseteq\rangle \underset{\mathring{\alpha}^i_{\mathbb{F}}}{\overset{\mathring{\gamma}}{\leftrightarrows}} \langle \wp(\mathbb{S}^{+\infty}_{\mathbb{P}^i}), \mathring{\sqsubseteq}^i\rangle \tag{32.29}$$

Soundness is now $\mathring{\alpha}^i_{\mathbb{F}}(\boldsymbol{S}^*_{\mathbb{V}}[\![\mathsf{S}]\!]) \mathring{\sqsubseteq}^i \boldsymbol{S}^*_{\mathbb{P}^i_{\mathbb{F}}}[\![\mathsf{S}]\!]$. However $\mathring{\sqsubseteq}^i$ in (32.26) is defined by concretization to $\wp(\mathbb{T}^{+\infty}_{\mathbb{V}})$. We look for a definition $\mathring{\sqsubseteq}^i$ in the abstract $\wp(\mathbb{T}^{+\infty}_{\mathbb{P}^i_{\mathbb{F}}})$ only, that provides a sufficient soundness condition $(\overline{\Pi} \mathring{\sqsubseteq}^i \overline{\Pi}') \Rightarrow (\overline{\Pi} \mathring{\sqsubseteq}^i \overline{\Pi}')$.

32.5.8 Sound Overapproximation in the Abstract

We define $\overline{\Pi} \mathring{\sqsubseteq}^i \overline{\Pi}'$ so that the traces of $\overline{\Pi}'$ have the same control as the traces of $\overline{\Pi}$ but intervals are larger (and $\overline{\Pi}'$ may contain extra traces due to the imprecision of interval tests). $\mathring{\sqsubseteq}^i$ is Hoare preorder [985] on sets of traces.

$$\overline{\Pi} \mathring{\sqsubseteq}^i \overline{\Pi}' \triangleq \forall \langle \overline{\pi}_0, \overline{\pi}_1\rangle \in \overline{\Pi} \,.\, \exists \langle \overline{\pi}'_0, \overline{\pi}'_1\rangle \in \overline{\Pi}' \,.\, \langle \overline{\pi}_0, \overline{\pi}_1\rangle \ddot{\vec{\sqsubseteq}}^i \langle \overline{\pi}'_0, \overline{\pi}'_1\rangle \tag{32.30}$$

Observe that $\varnothing \mathring{\sqsubseteq}^i \overline{\Pi}'$ and if $\forall k \in \Delta \,.\, \overline{\Pi}_k \mathring{\sqsubseteq}^i \overline{\Pi}'_k$ then $\bigcup_{k\in\Delta} \overline{\Pi}_k \mathring{\sqsubseteq}^i \bigcup_{k\in\Delta} \overline{\Pi}'_k$.

In (32.30), $\ddot{\vec{\sqsubseteq}}^i$ is the pairwise extension of $\vec{\sqsubseteq}^i$

$$\langle \overline{\pi}_0, \overline{\pi}_1 \rangle \ddot{\vec{\sqsubseteq}}^i \langle \overline{\pi}'_0, \overline{\pi}'_1 \rangle \triangleq \overline{\pi}_0 \vec{\sqsubseteq}^i \overline{\pi}'_0 \wedge \overline{\pi}_1 \vec{\sqsubseteq}^i \overline{\pi}'_1$$

and $\vec{\sqsubseteq}^i$ is the tracewise extension of $\dot{\sqsubseteq}^i$, comparing traces of the same length, same control, but larger intervals.

$$\overline{\pi} \vec{\sqsubseteq}^i \overline{\pi}' \triangleq \text{let } \ell_1 \xrightarrow{\overline{\mathsf{a}}_1} \ell_2 \ldots \ell_n \xrightarrow{\overline{\mathsf{a}}_n} \ell_{n+1} \ldots = \overline{\pi} \text{ and } \ell'_1 \xrightarrow{\overline{\mathsf{a}}'_1} \ell'_2 \ldots \ell'_n \xrightarrow{\overline{\mathsf{a}}'_n} \ell'_{n+1} \ldots = \overline{\pi}' \text{ in}$$
$$\forall j \leqslant |\overline{\pi}| = |\overline{\pi}'| \,.\, \ell_j = \ell'_j \wedge \overline{\mathsf{a}}_j \dot{\sqsubseteq}^i \overline{\mathsf{a}}'_j \tag{32.31}$$

because the action intervals are larger

$$(\mathsf{x} = \mathsf{A} = \overline{\nu}) \dot{\sqsubseteq}^i (\mathsf{x} = \mathsf{A} = \overline{\nu}') \triangleq \overline{\nu} \sqsubseteq^i \overline{\nu}' \tag{32.32}$$
$$(\mathsf{B} = \overline{\rho}) \dot{\sqsubseteq}^i (\mathsf{B} = \overline{\rho}') \triangleq \overline{\rho} \dot{\sqsubseteq}^i \overline{\rho}'$$
$$(\neg(\mathsf{B}) = \overline{\rho}) \dot{\sqsubseteq}^i (\neg(\mathsf{B}) = \overline{\rho}') \triangleq \overline{\rho} \dot{\sqsubseteq}^i \overline{\rho}'$$
$$\mathsf{a} \dot{\sqsubseteq}^i \mathsf{a} \qquad \text{for all other actions a}$$

$$\overline{\pi} \vec{\sqsubseteq}^i \overline{\pi}' \text{ implies } \varrho(\overline{\pi}) \dot{\sqsubseteq}^i \varrho(\overline{\pi}') \text{ (by (32.19) and (32.20)) and } \vec{\gamma}^i_{\mathbb{F}}(\overline{\pi}) \subseteq \vec{\gamma}^i_{\mathbb{F}}(\overline{\pi}') \tag{32.33}$$

The following lemma 32.34 shows that $\mathring{\sqsubseteq}^i$ allows for more abstract traces than $\mathring{\sqsubseteq}^i$.

Lemma 32.34 $(\overline{\Pi} \mathring{\sqsubseteq}^i \overline{\Pi}') \Rightarrow (\overline{\Pi} \mathring{\sqsubseteq}^i \overline{\Pi}')$.

Proof of lemma 32.34

$\overline{\Pi} \mathring{\sqsubseteq}^i \overline{\Pi}'$

$\Leftrightarrow \forall \langle \overline{\pi}_0, \overline{\pi}_1 \rangle \in \overline{\Pi} \,.\, \exists \langle \overline{\pi}'_0, \overline{\pi}'_1 \rangle \in \overline{\Pi}' \,.\, \langle \overline{\pi}_0, \overline{\pi}_1 \rangle \ddot{\vec{\sqsubseteq}}^i \langle \overline{\pi}'_0, \overline{\pi}'_1 \rangle$ ⟨definition (32.30) of $\mathring{\sqsubseteq}^i$⟩

$\Rightarrow \forall \langle \overline{\pi}_0, \overline{\pi}_1 \rangle \in \overline{\Pi} \,.\, \exists \langle \overline{\pi}'_0, \overline{\pi}'_1 \rangle \in \overline{\Pi}' \,.\, \ddot{\vec{\gamma}}^i_{\mathbb{F}}(\langle \overline{\pi}_0, \overline{\pi}_1 \rangle) \subseteq \ddot{\vec{\gamma}}^i_{\mathbb{F}}(\langle \overline{\pi}'_0, \overline{\pi}'_1 \rangle)$

⟨because $\langle \overline{\pi}_0, \overline{\pi}_1 \rangle \ddot{\vec{\sqsubseteq}}^i \langle \overline{\pi}'_0, \overline{\pi}'_1 \rangle$ implies $\ddot{\vec{\gamma}}^i_{\mathbb{F}}(\langle \overline{\pi}_0, \overline{\pi}_1 \rangle) \subseteq \ddot{\vec{\gamma}}^i_{\mathbb{F}}(\langle \overline{\pi}'_0, \overline{\pi}'_1 \rangle)$⟩

$\Rightarrow \forall \langle \overline{\pi}_0, \overline{\pi}_1 \rangle \in \overline{\Pi} \,.\, \exists \langle \overline{\pi}'_0, \overline{\pi}'_1 \rangle \in \overline{\Pi}' \,.\, \forall \langle \pi_0, \pi_1 \rangle \in \ddot{\vec{\gamma}}^i_{\mathbb{F}}(\langle \overline{\pi}_0, \overline{\pi}_1 \rangle) \,.\, \langle \pi_0, \pi_1 \rangle \in \ddot{\vec{\gamma}}^i_{\mathbb{F}}(\langle \overline{\pi}'_0, \overline{\pi}'_1 \rangle)$

⟨definition of $\subseteq$⟩

$\Rightarrow \forall \langle \overline{\pi}_0, \overline{\pi}_1 \rangle \in \overline{\Pi} \,.\, \forall \langle \pi_0, \pi_1 \rangle \in \ddot{\vec{\gamma}}^i_{\mathbb{F}}(\langle \overline{\pi}_0, \overline{\pi}_1 \rangle) \,.\, \exists \langle \overline{\pi}'_0, \overline{\pi}'_1 \rangle \in \overline{\Pi}' \,.\, \langle \pi_0, \pi_1 \rangle \in \ddot{\vec{\gamma}}^i_{\mathbb{F}}(\langle \overline{\pi}'_0, \overline{\pi}'_1 \rangle)$

⟨choosing the same $\langle \overline{\pi}'_0, \overline{\pi}'_1 \rangle$ for all $\langle \overline{\pi}_0, \overline{\pi}_1 \rangle$⟩

$\Leftrightarrow \overline{\Pi} \mathring{\sqsubseteq}^i \overline{\Pi}'$ ⟨(32.26)⟩ □

Exercise 32.35 Prove that we have a Galois connection (where the abstract preorder and concretization are different from both (32.25) and (32.29))

$$\langle \wp(\mathbb{S}^{+\infty}_{\mathbb{V}}), \subseteq \rangle \underset{\mathring{\alpha}^i_{\mathbb{F}}}{\overset{\mathring{\overline{\overline{\gamma}}}}{\leftrightarrows}} \langle \wp(\mathbb{S}^{+\infty}_{\mathbb{P}^i}), \mathring{\sqsubseteq}^i \rangle \tag{32.36}$$

32.5.9 Calculational Design of the Float Interval Trace Semantics

The float interval prefix trace semantics $\widehat{\mathcal{S}}^{*}_{\mathbb{P}^i_{\mathbb{F}}}[\![\mathsf{S}]\!]$ of a program component S replaces concrete real or float traces (as defined by $\widehat{\mathcal{S}}^{*}_{\mathbb{V}}[\![\mathsf{S}]\!]$) by interval traces. It is sound if and only if the concrete traces are included in the abstract traces that is $\widehat{\mathcal{S}}^{*}_{\mathbb{V}}[\![\mathsf{S}]\!] \subseteq \mathring{\gamma}^i_{\mathbb{F}}(\widehat{\mathcal{S}}^{*}_{\mathbb{P}^i_{\mathbb{F}}}[\![\mathsf{S}]\!])$ or, equivalently, by (32.25), $\mathring{\alpha}^i_{\mathbb{F}}(\widehat{\mathcal{S}}^{*}_{\mathbb{V}}[\![\mathsf{S}]\!]) \subseteq \widehat{\mathcal{S}}^{*}_{\mathbb{P}^i_{\mathbb{F}}}[\![\mathsf{S}]\!]$.

Although the soundness condition $\mathring{\alpha}^i_{\mathbb{F}}(\widehat{\mathcal{S}}^{*}_{\mathbb{V}}[\![\mathsf{S}]\!]) \subseteq \widehat{\mathcal{S}}^{*}_{\mathbb{P}^i_{\mathbb{F}}}[\![\mathsf{S}]\!]$ allows the abstract semantics $\widehat{\mathcal{S}}^{*}_{\mathbb{P}^i_{\mathbb{F}}}[\![\mathsf{S}]\!]$ to contain more traces, including some with larger intervals, it requires the abstract traces in $\alpha^i_{\mathbb{F}}(\widehat{\mathcal{S}}^{*}_{\mathbb{V}}[\![\mathsf{S}]\!])$ (which are the best float interval abstractions of real computations) to all belong to the abstract semantics $\widehat{\mathcal{S}}^{*}_{\mathbb{P}^i_{\mathbb{F}}}[\![\mathsf{S}]\!]$.

We introduced $\mathring{\sqsubseteq}^i$ in (32.2) to relax this requirement about the presence of best interval trace abstractions of real computations in the abstract semantics. The weaker requirement $\mathring{\alpha}^i_{\mathbb{F}}(\widehat{\mathcal{S}}^{*}_{\mathbb{V}}[\![\mathsf{S}]\!]) \mathring{\sqsubseteq}^i \widehat{\mathcal{S}}^{*}_{\mathbb{P}^i_{\mathbb{F}}}[\![\mathsf{S}]\!]$ implies, by lemma 32.34, that $\mathring{\alpha}^i_{\mathbb{F}}(\widehat{\mathcal{S}}^{*}_{\mathbb{V}}[\![\mathsf{S}]\!]) \mathring{\sqsubseteq}^i \widehat{\mathcal{S}}^{*}_{\mathbb{P}^i_{\mathbb{F}}}[\![\mathsf{S}]\!]$ so that, by (32.26), $\mathring{\gamma}^i_{\mathbb{F}}(\mathring{\alpha}^i_{\mathbb{F}}(\widehat{\mathcal{S}}^{*}_{\mathbb{V}}[\![\mathsf{S}]\!])) \subseteq \mathring{\gamma}^i_{\mathbb{F}}(\widehat{\mathcal{S}}^{*}_{\mathbb{P}^i_{\mathbb{F}}}[\![\mathsf{S}]\!])$, which, together with $\widehat{\mathcal{S}}^{*}_{\mathbb{V}}[\![\mathsf{S}]\!] \subseteq \mathring{\gamma}^i_{\mathbb{F}}(\mathring{\alpha}^i_{\mathbb{F}}(\widehat{\mathcal{S}}^{*}_{\mathbb{V}}[\![\mathsf{S}]\!]))$ from the Galois connection (32.25) yields, by transitivity, that $\widehat{\mathcal{S}}^{*}_{\mathbb{V}}[\![\mathsf{S}]\!] \subseteq \mathring{\gamma}^i_{\mathbb{F}}(\widehat{\mathcal{S}}^{*}_{\mathbb{P}^i_{\mathbb{F}}}[\![\mathsf{S}]\!])$.

This weaker soundness requirement $\mathring{\alpha}^i_{\mathbb{F}}(\widehat{\mathcal{S}}^{*}_{\mathbb{V}}[\![\mathsf{S}]\!]) \mathring{\sqsubseteq}^i \widehat{\mathcal{S}}^{*}_{\mathbb{P}^i_{\mathbb{F}}}[\![\mathsf{S}]\!])$ yields a calculational design method where $\mathring{\alpha}^i_{\mathbb{F}}(\widehat{\mathcal{S}}^{*}_{\mathbb{V}}[\![\mathsf{S}]\!])$ is $\mathring{\sqsubseteq}^i$-overapproximated in order to eliminate any reference to the concrete semantics $\widehat{\mathcal{S}}^{*}_{\mathbb{V}}[\![\mathsf{S}]\!]$. We proceed by structural induction on $\triangleleft$, assuming $\mathring{\alpha}^i_{\mathbb{F}}(\widehat{\mathcal{S}}^{*}_{\mathbb{V}}[\![\mathsf{S}']\!]) \mathring{\sqsubseteq}^i \widehat{\mathcal{S}}^{*}_{\mathbb{P}^i_{\mathbb{F}}}[\![\mathsf{S}']\!]$ for all $\mathsf{S}' \triangleleft \mathsf{S}$.

However this is not a strong enough induction hypothesis to design $\widehat{\mathcal{S}}^{*}_{\mathbb{P}^i_{\mathbb{F}}}[\![\mathsf{S}]\!]$ such that $\mathring{\alpha}^i_{\mathbb{F}}(\mathcal{S}_{\mathbb{R}}[\![\mathsf{S}]\!]) \mathring{\sqsubseteq}^i \widehat{\mathcal{S}}^{*}_{\mathbb{P}^i_{\mathbb{F}}}[\![\mathsf{S}]\!]$ by structural induction.

In conjunction with $\mathring{\alpha}^i_{\mathbb{F}}(X) \mathring{\sqsubseteq}^i \overline{X}$ stating that any concrete execution $\langle \pi, \pi' \rangle \in X$ has an abstraction $\langle \overline{\pi}, \overline{\pi}' \rangle$ in $\overline{X}$, we will need a stronger induction hypothesis stating that any sound abstraction $\overline{\pi}$ of a concrete prelude π (i.e. $\vec{\alpha}^i_{\mathbb{F}}(\pi) \vec{\sqsubseteq}^i \overline{\pi}$) has an abstract continuation $\overline{\pi}'$ in $\overline{X}$ abstracting the concrete continuation π' (i.e. $\vec{\alpha}^i_{\mathbb{F}}(\pi') \vec{\sqsubseteq}^i \overline{\pi}'$). Formally,

$$\forall \langle \pi, \pi' \rangle \in X \,.\, \forall \overline{\pi} \,.\, (\vec{\alpha}^i_{\mathbb{F}}(\pi) \vec{\sqsubseteq}^i \overline{\pi}) \Rightarrow (\exists \overline{\pi}' \,.\, \langle \overline{\pi}, \overline{\pi}' \rangle \in \overline{X} \wedge \vec{\alpha}^i_{\mathbb{F}}(\pi') \vec{\sqsubseteq}^i \overline{\pi}') \qquad (32.37)$$

which we will use for $X = \widehat{\mathcal{S}}^{*}_{\mathbb{V}}[\![\mathsf{S}]\!]$ and $\overline{X} = \widehat{\mathcal{S}}^{*}_{\mathbb{P}^i_{\mathbb{F}}}[\![\mathsf{S}]\!]$ is well as for the concrete X and abstract $\overline{X}$ fixpoint iterates in (17.4) for iteration statements.

32.5.9.1 Interval trace semantics of an assignment statement. We can now abstract the semantics of real ($\mathbb{V} = \mathbb{R}$) or float ($\mathbb{V} = \mathbb{F}$) assignments by float intervals. By calculational design, we discover that the interval arithmetic semantics of an assignment $\mathsf{S} = \ell\ \mathsf{x} = \mathsf{A}\,;$ is

$$\widehat{\mathcal{S}}^{*}_{\mathbb{P}^i_{\mathbb{F}}}[\![\ell\ \mathsf{x} = \mathsf{A}\,;]\!] \ \triangleq\ \{\langle \overline{\pi}\ell, \ell \rangle\} \cup \{\langle \overline{\pi}\ell, \ell \xrightarrow{\mathsf{x} = \mathsf{A} = \overline{\nu}} \mathsf{after}[\![\mathsf{S}]\!] \rangle \mid \overline{\nu} = \mathcal{A}^i_{\mathbb{F}}[\![\mathsf{A}]\!]\varrho(\overline{\pi}\ell)\}$$

as follows

$$\begin{aligned}
&\mathring{\alpha}^i_{\mathbb{F}}(\widehat{\mathcal{S}}^*_{\mathbb{V}}[\![\ell\ \mathsf{x} = \mathsf{A}\ ;]\!])\\
=\ &\{\ddot{\vec{\alpha}}^i_{\mathbb{F}}(\langle\pi,\pi'\rangle) \mid \langle\pi,\pi'\rangle \in \widehat{\mathcal{S}}^*_{\mathbb{V}}[\![\ell\ \mathsf{x} = \mathsf{A}\ ;]\!]\} && \text{(set of traces abstraction (32.24))}\\
=\ &\{\langle\vec{\alpha}^i_{\mathbb{F}}(\pi), \vec{\alpha}^i_{\mathbb{F}}(\pi')\rangle \mid \langle\pi,\pi'\rangle \in \{\langle\pi^\ell, \ell\rangle\} \cup \{\langle\pi^\ell, \ell \xrightarrow{\mathsf{x} = \mathsf{A} = \nu} \mathsf{after}[\![\mathsf{S}]\!]\rangle \mid \nu = \\
&\mathcal{A}[\![\mathsf{A}]\!]\boldsymbol{\varrho}(\pi^\ell)\}\} && \text{(definition (32.23) of } \ddot{\vec{\alpha}}^i_{\mathbb{F}} \text{ and definition of } \widehat{\mathcal{S}}^*_{\mathbb{V}}[\![\ell\ \mathsf{x} = \mathsf{A}\ ;]\!] \text{ in (17.2))}\\
=\ &\{\langle\vec{\alpha}^i_{\mathbb{F}}(\pi^\ell), \ell\rangle\} \cup \{\langle\vec{\alpha}^i_{\mathbb{F}}(\pi^\ell), \ell \xrightarrow{\mathsf{x} = \mathsf{A} = \overline{\nu}} \mathsf{after}[\![\mathsf{S}]\!]\rangle \mid \overline{\nu} = \alpha^i_{\mathbb{F}}(\mathcal{A}[\![\mathsf{A}]\!]\boldsymbol{\varrho}(\pi^\ell))\}\\
& && \text{(definition of } \in \text{ and definition (32.20) and (32.21) of trace abstraction)}\\
\mathrel{\dot{\mathrel{\ddot{\sqsubseteq}}}}^i\ &\{\langle\vec{\alpha}^i_{\mathbb{F}}(\pi^\ell), \ell\rangle\} \cup \{\langle\vec{\alpha}^i_{\mathbb{F}}(\pi^\ell), \ell \xrightarrow{\mathsf{x} = \mathsf{A} = \overline{\nu}} \mathsf{after}[\![\mathsf{S}]\!]\rangle \mid \overline{\nu} = \mathcal{A}^i_{\mathbb{F}}[\![\mathsf{A}]\!]\dot{\alpha}^i_{\mathbb{F}}(\boldsymbol{\varrho}(\pi^\ell))\}\\
& && \text{((32.16) and definition (32.30) of } \mathrel{\dot{\mathrel{\ddot{\sqsubseteq}}}}^i\text{)}\\
\mathrel{\dot{\mathrel{\ddot{\sqsubseteq}}}}^i\ &\{\langle\overline{\pi}^\ell, \ell\rangle\} \cup \{\langle\overline{\pi}^\ell, \ell \xrightarrow{\mathsf{x} = \mathsf{A} = \overline{\nu}} \mathsf{after}[\![\mathsf{S}]\!]\rangle \mid \overline{\nu} = \mathcal{A}^i_{\mathbb{F}}[\![\mathsf{A}]\!]\boldsymbol{\varrho}(\overline{\pi}^\ell)\}\\
& && \text{(letting } \overline{\pi}^\ell = \vec{\alpha}^i_{\mathbb{F}}(\pi^\ell) \text{ so } \dot{\alpha}^i_{\mathbb{F}}(\boldsymbol{\varrho}(\pi^\ell)) = \boldsymbol{\varrho}(\overline{\pi}^\ell) \text{ and definition (32.30) of } \mathrel{\dot{\mathrel{\ddot{\sqsubseteq}}}}^i\text{)}\\
\triangleq\ &\widehat{\mathcal{S}}^*_{\mathbb{P}^i_{\mathbb{F}}}[\![\ell\ \mathsf{x} = \mathsf{A}\ ;]\!] && \text{(by defining } \widehat{\mathcal{S}}^*_{\mathbb{P}^i_{\mathbb{F}}}[\![\ell\ \mathsf{x} = \mathsf{A}\ ;]\!] \text{ as in (17.2) for } \mathbb{V} = \mathbb{P}^i_{\mathbb{F}}\text{)}
\end{aligned}$$

Assume that $\langle\pi,\pi'\rangle \in \widehat{\mathcal{S}}^*_{\mathbb{V}}[\![\ell\ \mathsf{x} = \mathsf{A}\ ;]\!]$ and $\vec{\alpha}^i_{\mathbb{F}}(\pi) \mathrel{\ddot{\sqsubseteq}}^i \overline{\pi}$. By definition of $\widehat{\mathcal{S}}^*_{\mathbb{V}}[\![\ell\ \mathsf{x} = \mathsf{A}\ ;]\!]$ in (17.2), it follows that the trace π ends at ℓ and either $\pi' = \ell$ (which is a trivial case) or else $\pi' = \ell \xrightarrow{\mathsf{x} = \mathsf{A} = \mathcal{A}[\![\mathsf{A}]\!]\boldsymbol{\varrho}(\pi)} \mathsf{after}[\![\mathsf{S}]\!]$. Because the trace π ends at ℓ, the hypothesis $\vec{\alpha}^i_{\mathbb{F}}(\pi) \mathrel{\ddot{\sqsubseteq}}^i \overline{\pi}$ implies that $\overline{\pi}$ ends at ℓ, and so there exists $\overline{\pi}'$ s.t. $\langle\overline{\pi},\overline{\pi}'\rangle \in \widehat{\mathcal{S}}^*_{\mathbb{P}^i_{\mathbb{F}}}[\![\ell\ \mathsf{x} = \mathsf{A}\ ;]\!]$ with $\overline{\pi}' = \ell \xrightarrow{\mathsf{x} = \mathsf{A} = \mathcal{A}^i_{\mathbb{F}}[\![\mathsf{A}]\!]\boldsymbol{\varrho}(\overline{\pi})} \mathsf{after}[\![\mathsf{S}]\!]$. Let us calculate

$$\begin{aligned}
&\vec{\alpha}^i_{\mathbb{F}}(\mathsf{x} = \mathsf{A} = \mathcal{A}[\![\mathsf{A}]\!]\boldsymbol{\varrho}(\pi))\\
=\ &\mathsf{x} = \mathsf{A} = \alpha^i_{\mathbb{F}}(\mathcal{A}[\![\mathsf{A}]\!]\boldsymbol{\varrho}(\pi)) && \text{(definition (32.20) of } \vec{\alpha}^i_{\mathbb{F}}\text{)}\\
\mathrel{\dot{\sqsubseteq}}^i\ &\mathsf{x} = \mathsf{A} = \mathcal{A}^i_{\mathbb{F}}[\![\mathsf{A}]\!]\dot{\alpha}^i_{\mathbb{F}}(\boldsymbol{\varrho}(\pi))\\
& && \text{(} \alpha^i_{\mathbb{F}}(\mathcal{A}_{\mathbb{V}}[\![\mathsf{A}]\!]\rho) \sqsubseteq^i \mathcal{A}^i_{\mathbb{F}}[\![\mathsf{A}]\!]\dot{\alpha}^i_{\mathbb{F}}(\rho) \text{ by (32.16) and definition of } \mathrel{\dot{\sqsubseteq}}^i\text{)}\\
\mathrel{\dot{\sqsubseteq}}^i\ &\mathsf{x} = \mathsf{A} = \mathcal{A}^i_{\mathbb{F}}[\![\mathsf{A}]\!]\boldsymbol{\varrho}(\vec{\alpha}^i_{\mathbb{F}}(\pi)) && \text{(} \dot{\alpha}^i_{\mathbb{F}}(\boldsymbol{\varrho}(\pi)) \mathrel{\dot{\sqsubseteq}}^i \boldsymbol{\varrho}(\vec{\alpha}^i_{\mathbb{F}}(\pi)) \text{ by (32.22) and } \mathcal{A}^i_{\mathbb{F}}[\![\mathsf{A}]\!] \text{ increasing)}\\
\mathrel{\dot{\sqsubseteq}}^i\ &\mathsf{x} = \mathsf{A} = \mathcal{A}^i_{\mathbb{F}}[\![\mathsf{A}]\!]\boldsymbol{\varrho}(\overline{\pi})\\
& && \text{(} \vec{\alpha}^i_{\mathbb{F}}(\pi) \mathrel{\ddot{\sqsubseteq}}^i \overline{\pi} \text{ so } \boldsymbol{\varrho}(\vec{\alpha}^i_{\mathbb{F}}(\pi)) \sqsubseteq^i \boldsymbol{\varrho}(\overline{\pi}) \text{ by (32.33) and } \mathcal{A}^i_{\mathbb{F}}[\![\mathsf{A}]\!] \text{ increasing)}
\end{aligned}$$

It follows that $\vec{\alpha}^i_{\mathbb{F}}(\pi') = \ell \xrightarrow{\vec{\alpha}^i_{\mathbb{F}}(\mathsf{x} = \mathsf{A} = \mathcal{A}[\![\mathsf{A}]\!]\boldsymbol{\varrho}(\pi))} \mathsf{after}[\![\mathsf{S}]\!] \mathrel{\ddot{\sqsubseteq}}^i \ell \xrightarrow{\mathsf{x} = \mathsf{A} = \mathcal{A}^i_{\mathbb{F}}[\![\mathsf{A}]\!]\boldsymbol{\varrho}(\overline{\pi})} \mathsf{after}[\![\mathsf{S}]\!] = \overline{\pi}'$, by definition (32.31) of $\mathrel{\ddot{\sqsubseteq}}^i$, proving (32.37).

32.5.9.2 Interval trace semantics of a break statement. For a break statement $\mathsf{S} = \ell\ \mathbf{break}\ ;$, we discover, by calculational design, that

$$\widehat{\mathcal{S}}^*_{\mathbb{P}^i_{\mathbb{F}}}[\![\mathsf{S}]\!] \ \triangleq\ \{\langle\overline{\pi}^\ell, \ell\rangle\} \cup \{\langle\overline{\pi}^\ell, \ell \xrightarrow{\mathsf{break}} \mathsf{break\text{-}to}[\![\mathsf{S}]\!]\rangle\}$$

as follows

$$\begin{aligned}
&\mathring{\alpha}^i_{\mathbb{F}}(\widehat{\mathcal{S}}^*_{\mathbb{V}}[\![\mathsf{S}]\!]) && \text{(where } \mathsf{S} = \ell\ \mathbf{break}\ ;\text{)}\\
\triangleq\ &\mathring{\alpha}^i_{\mathbb{F}}(\{\langle\pi^\ell, \ell\rangle\} \cup \{\langle\pi^\ell, \ell \xrightarrow{\mathsf{break}} \mathsf{break\text{-}to}[\![\mathsf{S}]\!]\rangle\}) && \text{((6.29))}
\end{aligned}$$

$= \mathring{\alpha}^i_{\mathbb{F}}(\{\langle \pi\ell,\ \ell\rangle\}) \cup \mathring{\alpha}^i_{\mathbb{F}}(\{\langle \pi\ell,\ \ell \xrightarrow{\mathsf{break}} \mathsf{break\text{-}to}[\![\mathsf{S}]\!]\rangle\})$

$\wr$the abstraction $\mathring{\alpha}^i_{\mathbb{F}}$ preserves joins in the Galois connection (32.25)$\int$

$= \{\langle \vec{\alpha}^i_{\mathbb{F}}(\pi\ell),\ \ell\rangle\} \cup \{\langle \vec{\alpha}^i_{\mathbb{F}}(\pi\ell),\ \ell \xrightarrow{\mathsf{break}} \mathsf{break\text{-}to}[\![\mathsf{S}]\!]\rangle\}$ $\wr$definition (32.14) of $\alpha^i_{\mathbb{F}}\int$

$\mathring{\sqsubseteq}^i\ \{\langle \overline{\pi}\ell,\ \ell\rangle\} \cup \{\langle \overline{\pi}\ell,\ \ell \xrightarrow{\mathsf{break}} \mathsf{break\text{-}to}[\![\mathsf{S}]\!]\rangle\}$

$\wr\{\langle \vec{\alpha}^i_{\mathbb{F}}(\pi),\ \vec{\alpha}^i_{\mathbb{F}}(\pi')\rangle \mid \pi, \pi' \in \mathbb{T}^{+\infty}_{\mathbb{V}}\} \subseteq \{\langle \overline{\pi},\ \overline{\pi}'\rangle \mid \overline{\pi}, \overline{\pi}' \in \mathbb{T}^{+\infty}_{\mathbb{P}^i_{\mathbb{F}}}\}$ and definition (32.30) of $\mathring{\sqsubseteq}^i\int$

$\triangleq\ \widehat{\boldsymbol{\mathcal{S}}}^*_{\mathbb{P}^i_{\mathbb{F}}}[\![\mathsf{S}]\!]$ $\wr$by defining $\widehat{\boldsymbol{\mathcal{S}}}^*_{\mathbb{P}^i_{\mathbb{F}}}[\![\mathsf{S}]\!]$ as in (6.29) for $\mathbb{V} = \mathbb{P}^i_{\mathbb{F}}\int$

(32.37) trivially follows from $\vec{\alpha}^i_{\mathbb{F}}(\ell \xrightarrow{\mathsf{break}} \mathsf{break\text{-}to}[\![\mathsf{S}]\!]) = \ell \xrightarrow{\mathsf{break}} \mathsf{break\text{-}to}[\![\mathsf{S}]\!]$.

32.5.9.3 Interval trace semantics of the statement list. The case of an empty statement list $\mathsf{Sl} ::= \epsilon$ is trivial. For a nonempty statement list $\mathsf{Sl} ::= \mathsf{Sl}'\ \mathsf{S}$, we discover, by calculational design, that

$\widehat{\boldsymbol{\mathcal{S}}}^*_{\mathbb{P}^i_{\mathbb{F}}}[\![\mathsf{Sl}]\!] \triangleq \{\langle \overline{\pi}\mathsf{at}[\![\mathsf{Sl}]\!],\ \mathsf{at}[\![\mathsf{Sl}]\!]\rangle \mid \overline{\pi}\mathsf{at}[\![\mathsf{Sl}]\!] \in \mathbb{T}^+_{\mathbb{P}^i_{\mathbb{F}}}\}$		$\mathsf{Sl} ::= \epsilon$
$\triangleq \widehat{\boldsymbol{\mathcal{S}}}^*_{\mathbb{P}^i_{\mathbb{F}}}[\![\mathsf{Sl}']\!] \cup \{\langle \overline{\pi}_1,\ \overline{\pi}_2 \mathbin{\frown} \overline{\pi}_3\rangle \mid \langle \overline{\pi}_1,\ \overline{\pi}_2\rangle \in \widehat{\boldsymbol{\mathcal{S}}}^*_{\mathbb{P}^i_{\mathbb{F}}}[\![\mathsf{Sl}']\!]) \wedge \langle \overline{\pi}_1 \mathbin{\frown} \overline{\pi}_2,\ \overline{\pi}_3\rangle \in \widehat{\boldsymbol{\mathcal{S}}}^*_{\mathbb{P}^i_{\mathbb{F}}}[\![\mathsf{S}]\!]\}$		$\mathsf{Sl} ::= \mathsf{Sl}'\ \mathsf{S}$

as follows

$\mathring{\alpha}^i_{\mathbb{F}}(\widehat{\boldsymbol{\mathcal{S}}}^*_{\mathbb{V}}[\![\mathsf{Sl}]\!])$

$\triangleq\ \mathring{\alpha}^i_{\mathbb{F}}(\widehat{\boldsymbol{\mathcal{S}}}^*_{\mathbb{V}}[\![\mathsf{Sl}']\!] \cup \{\langle \pi_1,\ \pi_2 \mathbin{\frown} \pi_3\rangle \mid \langle \pi_1,\ \pi_2\rangle \in \widehat{\boldsymbol{\mathcal{S}}}^+_{\mathbb{V}}[\![\mathsf{Sl}']\!] \wedge \langle \pi_1 \mathbin{\frown} \pi_2,\ \pi_3\rangle \in \widehat{\boldsymbol{\mathcal{S}}}^*_{\mathbb{V}}[\![\mathsf{S}]\!]\})$ $\wr$(17.3)$\int$

$=\ \mathring{\alpha}^i_{\mathbb{F}}(\widehat{\boldsymbol{\mathcal{S}}}^*_{\mathbb{V}}[\![\mathsf{Sl}']\!]) \cup \mathring{\alpha}^i_{\mathbb{F}}(\{\langle \pi_1,\ \pi_2 \mathbin{\frown} \pi_3\rangle \mid \langle \pi_1,\ \pi_2\rangle \in \widehat{\boldsymbol{\mathcal{S}}}^+_{\mathbb{V}}[\![\mathsf{Sl}']\!] \wedge \langle \pi_1 \mathbin{\frown} \pi_2,\ \pi_3\rangle \in \widehat{\boldsymbol{\mathcal{S}}}^*_{\mathbb{V}}[\![\mathsf{S}]\!]\})$

$\wr$the abstraction preserves joins in the Galois connection (32.25)$\int$

$=\ \mathring{\alpha}^i_{\mathbb{F}}(\widehat{\boldsymbol{\mathcal{S}}}^*_{\mathbb{V}}[\![\mathsf{Sl}']\!]) \cup \{\langle \vec{\alpha}^i_{\mathbb{F}}(\pi_1),\ \vec{\alpha}^i_{\mathbb{F}}(\pi_2 \mathbin{\frown} \pi_3)\rangle \mid \langle \pi_1,\ \pi_2\rangle \in \widehat{\boldsymbol{\mathcal{S}}}^+_{\mathbb{V}}[\![\mathsf{Sl}']\!] \wedge \langle \pi_1 \mathbin{\frown} \pi_2,\ \pi_3\rangle \in \widehat{\boldsymbol{\mathcal{S}}}^*_{\mathbb{V}}[\![\mathsf{S}]\!]\}$

$\wr$definition (32.23) of $\ddot{\alpha}^i_{\mathbb{F}}$ and (32.24) of $\mathring{\alpha}^i_{\mathbb{F}}\int$

$=\ \mathring{\alpha}^i_{\mathbb{F}}(\widehat{\boldsymbol{\mathcal{S}}}^*_{\mathbb{V}}[\![\mathsf{Sl}']\!]) \cup \{\langle \vec{\alpha}^i_{\mathbb{F}}(\pi_1),\ \vec{\alpha}^i_{\mathbb{F}}(\pi_2) \mathbin{\frown} \vec{\alpha}^i_{\mathbb{F}}(\pi_3)\rangle \mid \langle \pi_1,\ \pi_2\rangle \in \widehat{\boldsymbol{\mathcal{S}}}^+_{\mathbb{V}}[\![\mathsf{Sl}']\!] \wedge \langle \pi_1 \mathbin{\frown} \pi_2,\ \pi_3\rangle \in \widehat{\boldsymbol{\mathcal{S}}}^*_{\mathbb{V}}[\![\mathsf{S}]\!]\}$

$\wr\vec{\alpha}^i_{\mathbb{F}}(\pi_2 \mathbin{\frown} \pi_3) = \vec{\alpha}^i_{\mathbb{F}}(\pi_2) \mathbin{\frown} \vec{\alpha}^i_{\mathbb{F}}(\pi_3)$ by definition (32.21) of $\vec{\alpha}^i_{\mathbb{F}}$ and (6.4) of $\frown\int$

$\mathring{\sqsubseteq}^i\ \mathring{\alpha}^i_{\mathbb{F}}(\widehat{\boldsymbol{\mathcal{S}}}^*_{\mathbb{V}}[\![\mathsf{Sl}']\!]) \cup \{\langle \vec{\alpha}^i_{\mathbb{F}}(\pi_1),\ \vec{\alpha}^i_{\mathbb{F}}(\pi_2) \mathbin{\frown} \vec{\alpha}^i_{\mathbb{F}}(\pi_3)\rangle \mid \ddot{\alpha}^i_{\mathbb{F}}(\langle \pi_1,\ \pi_2\rangle) \in \mathring{\alpha}^i_{\mathbb{F}}(\widehat{\boldsymbol{\mathcal{S}}}^*_{\mathbb{V}}[\![\mathsf{Sl}']\!]) \wedge \ddot{\alpha}^i_{\mathbb{F}}(\langle \pi_1 \mathbin{\frown} \pi_2,\ \pi_3\rangle) \in \mathring{\alpha}^i_{\mathbb{F}}(\widehat{\boldsymbol{\mathcal{S}}}^*_{\mathbb{V}}[\![\mathsf{S}]\!])\}$

$\wr$definition (32.30) of $\mathring{\sqsubseteq}^i$ because if $\langle \vec{\alpha}^i_{\mathbb{F}}(\pi_1),\ \vec{\alpha}^i_{\mathbb{F}}(\pi_2) \mathbin{\frown} \vec{\alpha}^i_{\mathbb{F}}(\pi_3)\rangle$ belongs to the set $\{\langle \vec{\alpha}^i_{\mathbb{F}}(\pi_1),\ \vec{\alpha}^i_{\mathbb{F}}(\pi_2) \mathbin{\frown} \vec{\alpha}^i_{\mathbb{F}}(\pi_3)\rangle \mid \langle \pi_1,\ \pi_2\rangle \in \widehat{\boldsymbol{\mathcal{S}}}^+_{\mathbb{V}}[\![\mathsf{Sl}']\!] \wedge \langle \pi_1 \mathbin{\frown} \pi_2,\ \pi_3\rangle \in \widehat{\boldsymbol{\mathcal{S}}}^*_{\mathbb{V}}[\![\mathsf{S}]\!]\}$ then $\langle \pi_1,\ \pi_2\rangle \in \widehat{\boldsymbol{\mathcal{S}}}^+_{\mathbb{V}}[\![\mathsf{Sl}']\!] \wedge \langle \pi_1 \mathbin{\frown} \pi_2,\ \pi_3\rangle \in \widehat{\boldsymbol{\mathcal{S}}}^*_{\mathbb{V}}[\![\mathsf{S}]\!]$ and so, by definition (32.24) of $\mathring{\alpha}^i_{\mathbb{F}}$, $\ddot{\alpha}^i_{\mathbb{F}}(\langle \pi_1,\ \pi_2\rangle) \in \mathring{\alpha}^i_{\mathbb{F}}(\widehat{\boldsymbol{\mathcal{S}}}^*_{\mathbb{V}}[\![\mathsf{Sl}']\!]) \wedge \ddot{\alpha}^i_{\mathbb{F}}(\langle \pi_1 \mathbin{\frown} \pi_2,\ \pi_3\rangle) \in \mathring{\alpha}^i_{\mathbb{F}}(\widehat{\boldsymbol{\mathcal{S}}}^*_{\mathbb{V}}[\![\mathsf{S}]\!]$ which implies that there exists $\langle \vec{\alpha}^i_{\mathbb{F}}(\pi_1),\ \vec{\alpha}^i_{\mathbb{F}}(\pi_2) \mathbin{\frown} \vec{\alpha}^i_{\mathbb{F}}(\pi_3)\rangle$ which belongs to the set $\{\langle \vec{\alpha}^i_{\mathbb{F}}(\pi_1),\ \vec{\alpha}^i_{\mathbb{F}}(\pi_2) \mathbin{\frown} \vec{\alpha}^i_{\mathbb{F}}(\pi_3)\rangle \mid \ddot{\alpha}^i_{\mathbb{F}}(\langle \pi_1,\ \pi_2\rangle) \in \mathring{\alpha}^i_{\mathbb{F}}(\widehat{\boldsymbol{\mathcal{S}}}^*_{\mathbb{V}}[\![\mathsf{Sl}']\!]) \wedge \ddot{\alpha}^i_{\mathbb{F}}(\langle \pi_1 \mathbin{\frown} \pi_2,\ \pi_3\rangle) \in \mathring{\alpha}^i_{\mathbb{F}}(\widehat{\boldsymbol{\mathcal{S}}}^*_{\mathbb{V}}[\![\mathsf{S}]\!]\}$ $\int$

$\mathring{\sqsubseteq}^i\ \widehat{\boldsymbol{\mathcal{S}}}^*_{\mathbb{P}^i_{\mathbb{F}}}[\![\mathsf{Sl}']\!] \cup \{\langle \overline{\pi}_1,\ \overline{\pi}_2 \mathbin{\frown} \overline{\pi}_3\rangle \mid \langle \overline{\pi}_1,\ \overline{\pi}_2\rangle \in \widehat{\boldsymbol{\mathcal{S}}}^*_{\mathbb{P}^i_{\mathbb{F}}}[\![\mathsf{Sl}']\!]) \wedge \langle \overline{\pi}_1 \mathbin{\frown} \overline{\pi}_2,\ \overline{\pi}_3\rangle \in \widehat{\boldsymbol{\mathcal{S}}}^*_{\mathbb{P}^i_{\mathbb{F}}}[\![\mathsf{S}]\!]\}$

$\wr\mathring{\alpha}^i_{\mathbb{F}}(\widehat{\boldsymbol{\mathcal{S}}}^*_{\mathbb{V}}\llbracket \mathtt{Sl}'\rrbracket) \mathrel{\mathring{\sqsubseteq}^i} \widehat{\boldsymbol{\mathcal{S}}}^*_{\mathbb{P}^i_{\mathbb{F}}}\llbracket \mathtt{Sl}'\rrbracket$ and $\mathring{\alpha}^i_{\mathbb{F}}(\widehat{\boldsymbol{\mathcal{S}}}^*_{\mathbb{V}}\llbracket \mathtt{S}\rrbracket) \mathrel{\mathring{\sqsubseteq}^i} \widehat{\boldsymbol{\mathcal{S}}}^*_{\mathbb{P}^i_{\mathbb{F}}}\llbracket \mathtt{S}\rrbracket)$ by induction hypothesis, definition (32.23) of $\ddot{\alpha}^i_{\mathbb{F}}$, letting $\overline{\pi}_1 = \ddot{\alpha}^i_{\mathbb{F}}(\pi_1)$, $\overline{\pi}_2 = \ddot{\alpha}^i_{\mathbb{F}}(\pi_2)$, and $\overline{\pi}_3 = \ddot{\alpha}^i_{\mathbb{F}}(\pi_3)$, and $\mathring{\sqsubseteq}^i$ is preserved by $\cup\wr$

$\triangleq\ \widehat{\boldsymbol{\mathcal{S}}}^*_{\mathbb{P}^i_{\mathbb{F}}}\llbracket \mathtt{Sl}'\ \mathtt{S}\rrbracket$ $\qquad\qquad \wr$by defining $\widehat{\boldsymbol{\mathcal{S}}}^*_{\mathbb{P}^i_{\mathbb{F}}}\llbracket \mathtt{Sl}\rrbracket$ as in (17.3)$\wr$

(32.37) follows by induction hypothesis and definition of $\frown$.

32.5.9.4 Interval trace semantics of a conditional statement. We can now abstract the semantics of real tests using float intervals. By calculational design, we get

$$
\begin{aligned}
&\widehat{\boldsymbol{\mathcal{S}}}^*_{\mathbb{P}^i_{\mathbb{F}}}\llbracket \mathtt{if}\,\ell\,(\mathtt{B})\,\mathtt{S}_t\rrbracket \ \triangleq \qquad\qquad (32.38)\\
&\quad \{\langle\overline{\pi}\ell,\ \ell\rangle \mid \overline{\pi}\ell \in \mathbb{T}^+_{\mathbb{P}^i_{\mathbb{R}}}\} \cup\\
&\quad \{\langle\overline{\pi}\ell,\ \ell \xrightarrow{\neg(\mathtt{B}) = \overline{\rho}_{\mathrm{ff}}} \mathsf{after}\llbracket \mathtt{S}\rrbracket\rangle \mid \exists\overline{\rho}_{\mathrm{tt}}\,.\,\boldsymbol{\mathcal{B}}^i_{\mathbb{F}}\llbracket \mathtt{B}\rrbracket\boldsymbol{\varrho}(\overline{\pi}\ell) = \langle\overline{\rho}_{\mathrm{tt}},\ \overline{\rho}_{\mathrm{ff}}\rangle\}^{a} \cup\\
&\quad \{\langle\overline{\pi}\ell,\ \ell \xrightarrow{\mathtt{B} = \overline{\rho}_{\mathrm{tt}}} \mathsf{at}\llbracket \mathtt{S}_t\rrbracket \frown \overline{\pi}'\rangle \mid \exists\overline{\rho}_{\mathrm{ff}}\,.\,\boldsymbol{\mathcal{B}}^i_{\mathbb{F}}\llbracket \mathtt{B}\rrbracket\boldsymbol{\varrho}(\overline{\pi}\ell) = \langle\overline{\rho}_{\mathrm{tt}},\ \overline{\rho}_{\mathrm{ff}}\rangle \wedge\\
&\qquad\qquad \langle\overline{\pi}\ell \xrightarrow{\mathtt{B} = \overline{\rho}_{\mathrm{tt}}} \mathsf{at}\llbracket \mathtt{S}_t\rrbracket,\ \overline{\pi}'\rangle \in \widehat{\boldsymbol{\mathcal{S}}}^*_{\mathbb{P}^i_{\mathbb{F}}}\llbracket \mathtt{S}_t\rrbracket\}^{b}
\end{aligned}
$$

a. If $\dot{\gamma}^i_{\mathbb{F}}(\dot{\varnothing}) = \varnothing$ then this case is $\varnothing$ when $\overline{\rho}_{\mathrm{ff}} = \dot{\varnothing}$.
b. If $\dot{\gamma}^i_{\mathbb{F}}(\dot{\varnothing}) = \varnothing$ then this case is $\varnothing$ when $\overline{\rho}_{\mathrm{tt}} = \dot{\varnothing}$.

(and similarly for $\mathtt{if}\,\ell\,(\mathtt{B})\,\mathtt{S}_t\ \mathtt{else}\ \mathtt{S}_f$) as follows:

$$
\begin{aligned}
&\mathring{\alpha}^i_{\mathbb{F}}(\widehat{\boldsymbol{\mathcal{S}}}^*_{\mathbb{V}}\llbracket \mathtt{if}\,\ell\,(\mathtt{B})\,\mathtt{S}_t\rrbracket)\\
\triangleq\ &\mathring{\alpha}^i_{\mathbb{F}}(\{\langle\pi\ell,\ \ell\rangle \mid \pi\ell \in \mathbb{T}^+_{\mathbb{R}}\} \cup \{\langle\pi\ell,\ \ell \xrightarrow{\neg(\mathtt{B}) = \boldsymbol{\varrho}(\pi\ell)} \mathsf{after}\llbracket \mathtt{S}\rrbracket\rangle \mid \boldsymbol{\mathcal{B}}_{\mathbb{V}}\llbracket \mathtt{B}\rrbracket\boldsymbol{\varrho}(\pi\ell) = \mathrm{ff}\} \cup \{\langle\pi\ell,\\
&\ell \xrightarrow{\mathtt{B} = \boldsymbol{\varrho}(\pi\ell)} \mathsf{at}\llbracket \mathtt{S}_t\rrbracket \frown \pi'\rangle \mid \boldsymbol{\mathcal{B}}_{\mathbb{V}}\llbracket \mathtt{B}\rrbracket\boldsymbol{\varrho}(\pi\ell) = \mathrm{tt} \wedge \langle\pi\ell \xrightarrow{\mathtt{B} = \boldsymbol{\varrho}(\pi\ell)} \mathsf{at}\llbracket \mathtt{S}_t\rrbracket,\ \pi'\rangle \in\\
&\widehat{\boldsymbol{\mathcal{S}}}^*_{\mathbb{V}}\llbracket \mathtt{S}_t\rrbracket\})\\
&\qquad\qquad \wr\text{(6.11), (6.18), (6.19), section 16.3.1, and section 32.5.5}\wr\\
=\ &\mathring{\alpha}^i_{\mathbb{F}}(\{\langle\pi\ell,\ \ell\rangle \mid \pi\ell \in \mathbb{T}^+_{\mathbb{R}}\}) \cup \mathring{\alpha}^i_{\mathbb{F}}(\{\langle\pi\ell,\ \ell \xrightarrow{\neg(\mathtt{B}) = \boldsymbol{\varrho}(\pi\ell)} \mathsf{after}\llbracket \mathtt{S}\rrbracket\rangle \mid \boldsymbol{\mathcal{B}}_{\mathbb{V}}\llbracket \mathtt{B}\rrbracket\boldsymbol{\varrho}(\pi\ell) =\\
&\mathrm{ff}\}) \cup \mathring{\alpha}^i_{\mathbb{F}}(\{\langle\pi\ell,\ \ell \xrightarrow{\mathtt{B} = \boldsymbol{\varrho}(\pi\ell)} \mathsf{at}\llbracket \mathtt{S}_t\rrbracket \frown \pi'\rangle \mid \boldsymbol{\mathcal{B}}_{\mathbb{V}}\llbracket \mathtt{B}\rrbracket\boldsymbol{\varrho}(\pi\ell) = \mathrm{tt} \wedge \langle\pi\ell \xrightarrow{\mathtt{B} = \boldsymbol{\varrho}(\pi\ell)} \mathsf{at}\llbracket \mathtt{S}_t\rrbracket,\\
&\pi'\rangle \in \widehat{\boldsymbol{\mathcal{S}}}^*_{\mathbb{V}}\llbracket \mathtt{S}_t\rrbracket\}) \qquad\qquad \wr\text{join preservation in (32.25)}\wr\\
=\ &\{\ddot{\alpha}^i_{\mathbb{F}}(\langle\pi\ell,\ \ell\rangle) \mid \pi\ell \in \mathbb{T}^+_{\mathbb{R}}\} \cup \{\ddot{\alpha}^i_{\mathbb{F}}(\langle\pi\ell,\ \ell \xrightarrow{\neg(\mathtt{B}) = \boldsymbol{\varrho}(\pi\ell)} \mathsf{after}\llbracket \mathtt{S}\rrbracket\rangle) \mid \boldsymbol{\mathcal{B}}_{\mathbb{V}}\llbracket \mathtt{B}\rrbracket\boldsymbol{\varrho}(\pi\ell) =\\
&\mathrm{ff}\} \cup \{\ddot{\alpha}^i_{\mathbb{F}}(\langle\pi\ell,\ \ell \xrightarrow{\mathtt{B} = \boldsymbol{\varrho}(\pi\ell)} \mathsf{at}\llbracket \mathtt{S}_t\rrbracket \frown \pi'\rangle) \mid \boldsymbol{\mathcal{B}}_{\mathbb{V}}\llbracket \mathtt{B}\rrbracket\boldsymbol{\varrho}(\pi\ell) = \mathrm{tt} \wedge \langle\pi\ell \xrightarrow{\mathtt{B} = \boldsymbol{\varrho}(\pi\ell)} \mathsf{at}\llbracket \mathtt{S}_t\rrbracket,\\
&\pi'\rangle \in \widehat{\boldsymbol{\mathcal{S}}}^*_{\mathbb{V}}\llbracket \mathtt{S}_t\rrbracket\} \qquad\qquad \wr\text{definition (32.24) of } \mathring{\alpha}^i_{\mathbb{F}}\wr\\
\mathring{\sqsubseteq}^i\ &\{\langle\overline{\pi}\ell,\ \ell\rangle \mid \overline{\pi}\ell \in \mathbb{T}^+_{\mathbb{P}^i_{\mathbb{R}}}\} \cup \{\langle\overline{\pi}\ell,\ \ell \xrightarrow{\neg(\mathtt{B}) = \overline{\rho}_{\mathrm{ff}}} \mathsf{after}\llbracket \mathtt{S}\rrbracket\rangle \mid \exists\overline{\rho}_{\mathrm{tt}}\,.\,\boldsymbol{\mathcal{B}}^i_{\mathbb{F}}\llbracket \mathtt{B}\rrbracket\boldsymbol{\varrho}(\overline{\pi}\ell) = \langle\overline{\rho}_{\mathrm{tt}},\\
&\overline{\rho}_{\mathrm{ff}}\rangle\} \cup \{\langle\overline{\pi}\ell,\ \ell \xrightarrow{\mathtt{B} = \overline{\rho}_{\mathrm{tt}}} \mathsf{at}\llbracket \mathtt{S}_t\rrbracket \frown \overline{\pi}'\rangle \mid \exists\overline{\rho}_{\mathrm{ff}}\,.\,\boldsymbol{\mathcal{B}}^i_{\mathbb{F}}\llbracket \mathtt{B}\rrbracket\boldsymbol{\varrho}(\overline{\pi}\ell) = \langle\overline{\rho}_{\mathrm{tt}},\ \overline{\rho}_{\mathrm{ff}}\rangle \wedge \langle\overline{\pi}\ell \xrightarrow{\mathtt{B} = \overline{\rho}_{\mathrm{tt}}}\\
&\mathsf{at}\llbracket \mathtt{S}_t\rrbracket,\ \overline{\pi}'\rangle \in \widehat{\boldsymbol{\mathcal{S}}}^*_{\mathbb{P}^i_{\mathbb{F}}}\llbracket \mathtt{S}_t\rrbracket\}
\end{aligned}
$$

$\wr$— For the first term, by definition (32.30) of $\overset{\circ}{\sqsubseteq}^{i}$, we must prove that $\forall \ddot{\vec{\alpha}}^{i}_{\mathbb{F}}(\langle \pi\ell, \ell\rangle) . \exists \langle \overline{\pi}\ell, \ell\rangle . \ddot{\vec{\alpha}}^{i}_{\mathbb{F}}(\langle \pi\ell, \ell\rangle) \overset{\ddot{\rightarrow}}{\sqsubseteq}^{i} \langle \overline{\pi}\ell, \ell\rangle$. We can simply choose $\overline{\pi}\ell = \vec{\alpha}^{i}_{\mathbb{F}}(\pi\ell)$.

— The second term may be empty, in which case $\overline{\rho}_{\mathsf{ff}} = \dot{\varnothing}$. Otherwise, by definition (32.30) of $\overset{\circ}{\sqsubseteq}^{i}$, we must prove that $\forall \ddot{\vec{\alpha}}^{i}_{\mathbb{F}}(\langle \pi\ell, \ell \xrightarrow{\neg(\mathsf{B}) = \varrho(\pi\ell)} \mathsf{after}[\![\mathsf{S}]\!]\rangle) = \langle \vec{\alpha}^{i}_{\mathbb{F}}(\pi\ell), \ell \xrightarrow{\neg(\mathsf{B}) = \dot{\alpha}^{i}_{\mathbb{F}}(\varrho(\pi\ell))} \mathsf{after}[\![\mathsf{S}]\!]\rangle . \exists \langle \overline{\pi}\ell, \ell \xrightarrow{\neg(\mathsf{B}) = \overline{\rho}_{\mathsf{ff}}} \mathsf{after}[\![\mathsf{S}]\!]\rangle . \langle \vec{\alpha}^{i}_{\mathbb{F}}(\pi\ell), \ell \xrightarrow{\neg(\mathsf{B}) = \dot{\alpha}^{i}_{\mathbb{F}}(\varrho(\pi\ell))} \mathsf{after}[\![\mathsf{S}]\!]\rangle \overset{\ddot{\rightarrow}}{\sqsubseteq}^{i} \langle \overline{\pi}\ell, \ell \xrightarrow{\neg(\mathsf{B}) = \overline{\rho}_{\mathsf{ff}}} \mathsf{after}[\![\mathsf{S}]\!]\rangle$. The control abstraction is the same. We can choose $\overline{\pi}\ell = \vec{\alpha}^{i}_{\mathbb{F}}(\pi\ell)$ so that $\mathscr{B}_{\mathbb{V}}[\![\mathsf{B}]\!]\varrho(\pi\ell) = \mathsf{ff}$, (32.18), and $\mathscr{B}^{i}_{\mathbb{F}}[\![\mathsf{B}]\!]\varrho(\overline{\pi}\ell) = \langle \overline{\rho}_{\mathsf{tt}}, \overline{\rho}_{\mathsf{ff}}\rangle$ imply that $\dot{\alpha}^{i}_{\mathbb{F}}(\varrho(\pi\ell)) \sqsubseteq^{i} \overline{\rho}_{\mathsf{ff}}$.

— The third term may be empty, in which case $\overline{\rho}_{\mathsf{tt}} = \dot{\varnothing}$. Otherwise, by definition (32.30) of $\overset{\circ}{\sqsubseteq}^{i}$, we must prove that $\forall \ddot{\vec{\alpha}}^{i}_{\mathbb{F}}(\langle \pi\ell, \ell \xrightarrow{\mathsf{B} = \varrho(\pi\ell)} \mathsf{at}[\![\mathsf{S}_t]\!] \frown \pi'\rangle) = \langle \vec{\alpha}^{i}_{\mathbb{F}}(\pi\ell), \ell \xrightarrow{\mathsf{B} = \dot{\alpha}^{i}_{\mathbb{F}}(\varrho(\pi\ell))} \mathsf{at}[\![\mathsf{S}_t]\!] \frown \vec{\alpha}^{i}_{\mathbb{F}}(\pi')\rangle . \exists \langle \overline{\pi}\ell, \ell \xrightarrow{\mathsf{B} = \overline{\rho}_{\mathsf{tt}}} \mathsf{at}[\![\mathsf{S}_t]\!] \frown \overline{\pi}'\rangle . \ddot{\vec{\alpha}}^{i}_{\mathbb{F}}(\langle \pi\ell, \ell \xrightarrow{\mathsf{B} = \varrho(\pi\ell)} \mathsf{at}[\![\mathsf{S}_t]\!] \frown \pi'\rangle) \overset{\ddot{\rightarrow}}{\sqsubseteq}^{i} \langle \overline{\pi}\ell, \ell \xrightarrow{\mathsf{B} = \overline{\rho}_{\mathsf{tt}}} \mathsf{at}[\![\mathsf{S}_t]\!] \frown \overline{\pi}'\rangle$.

The control abstraction is the same. We can choose $\overline{\pi}\ell = \vec{\alpha}^{i}_{\mathbb{F}}(\pi\ell)$ so that $\mathscr{B}_{\mathbb{V}}[\![\mathsf{B}]\!]\varrho(\pi\ell) = \mathsf{tt}$, (32.18), and $\mathscr{B}^{i}_{\mathbb{F}}[\![\mathsf{B}]\!]\varrho(\overline{\pi}\ell) = \langle \overline{\rho}_{\mathsf{tt}}, \overline{\rho}_{\mathsf{ff}}\rangle$ imply that $\dot{\alpha}^{i}_{\mathbb{F}}(\varrho(\pi\ell)) \dot{\sqsubseteq}^{i} \overline{\rho}_{\mathsf{tt}}$.

It remains to find $\overline{\pi}'$ such that $\vec{\alpha}^{i}_{\mathbb{F}}(\pi') \overset{\rightarrow}{\sqsubseteq}^{i} \overline{\pi}'$ and $\langle \overline{\pi}\ell \xrightarrow{\mathsf{B} = \overline{\rho}_{\mathsf{tt}}} \mathsf{at}[\![\mathsf{S}_t]\!], \overline{\pi}'\rangle \in \widehat{\boldsymbol{\mathcal{S}}}^{*}_{\mathbb{P}^{i}_{\mathbb{F}}}[\![\mathsf{S}_t]\!]$.

It is given by the induction hypothesis (32.37) where $\langle \pi\ell \xrightarrow{\mathsf{B} = \varrho(\pi\ell)} \mathsf{at}[\![\mathsf{S}_t]\!], \pi'\rangle \in \widehat{\boldsymbol{\mathcal{S}}}^{*}_{\mathbb{V}}[\![\mathsf{S}_t]\!]$ and $\vec{\alpha}^{i}_{\mathbb{F}}(\pi\ell \xrightarrow{\mathsf{B} = \varrho(\pi\ell)} \mathsf{at}[\![\mathsf{S}_t]\!]) \overset{\rightarrow}{\sqsubseteq}^{i} \overline{\pi}\ell \xrightarrow{\mathsf{B} = \overline{\rho}_{\mathsf{tt}}} \mathsf{at}[\![\mathsf{S}_t]\!]$ imply that $\exists \overline{\pi}' . \langle \overline{\pi}\ell \xrightarrow{\mathsf{B} = \overline{\rho}_{\mathsf{tt}}} \mathsf{at}[\![\mathsf{S}_t]\!], \overline{\pi}'\rangle \in \widehat{\boldsymbol{\mathcal{S}}}^{*}_{\mathbb{P}^{i}_{\mathbb{F}}}[\![\mathsf{S}_t]\!] \wedge \vec{\alpha}^{i}_{\mathbb{F}}(\pi') \overset{\rightarrow}{\sqsubseteq}^{i} \overline{\pi}'$. $\int$

$\triangleq \widehat{\boldsymbol{\mathcal{S}}}^{*}_{\mathbb{P}^{i}_{\mathbb{F}}}[\![\texttt{if}\ \ell\ (\mathsf{B})\ \mathsf{S}_t]\!]$

$\wr$because the above term involves only computations in $\mathbb{S}_{\mathbb{P}^i}$ and none in $\mathbb{S}_{\mathbb{V}}$ $\int$

It remains to show that $\widehat{\boldsymbol{\mathcal{S}}}^{*}_{\mathbb{P}^{i}_{\mathbb{F}}}[\![\texttt{if}\ \ell\ (\mathsf{B})\ \mathsf{S}_t]\!]$ satisfies (32.37), which is trivial for the first two terms. For the third term, this follows from the induction hypothesis.

32.5.9.5 Interval trace semantics of the iteration statement. By calculational design, we get

$$\widehat{\boldsymbol{S}}^{*}_{\mathbb{P}^{i}_{\mathbb{F}}}[\![\texttt{while}\,\ell\,(\mathsf{B})\,\mathsf{S}_b]\!] \;=\; \mathsf{lfp}^{\dot\subseteq}\,\mathscr{F}^{*}_{\mathbb{P}^{i}_{\mathbb{F}}}[\![\texttt{while}\,\ell\,(\mathsf{B})\,\mathsf{S}_b]\!]$$

$$\mathscr{F}^{*}_{\mathbb{P}^{i}_{\mathbb{F}}}[\![\texttt{while}\,\ell\,(\mathsf{B})\,\mathsf{S}_b]\!]\,X \;\triangleq\; \{\langle\overline{\pi}\ell,\,\ell\rangle \mid \overline{\pi}\ell \in \mathbb{T}^{+}_{\mathbb{P}^{i}_{\mathbb{R}}}\} \tag{32.39}$$

$$\cup\,\{\langle\overline{\pi}_1\ell,\,\ell\overline{\pi}_2\ell \xrightarrow{\neg(\mathsf{B})=\overline{\rho}_{\mathsf{ff}}} \mathsf{after}[\![\mathsf{S}]\!]\rangle \mid \langle\overline{\pi}_1\ell,\,\ell\overline{\pi}_2\ell\rangle \in X \wedge \exists\overline{\rho}_{\mathsf{tt}}\,.\;\mathscr{B}^{i}_{\mathbb{F}}[\![\mathsf{B}]\!]\varrho(\overline{\pi}_1\ell\overline{\pi}_2\ell) = \langle\overline{\rho}_{\mathsf{tt}},\,\overline{\rho}_{\mathsf{ff}}\rangle\}^{\mathrm{a}}$$

$$\cup\,\{\langle\overline{\pi}_1\ell,\,\ell\overline{\pi}_2\ell \xrightarrow{\mathsf{B}=\overline{\rho}_{\mathsf{tt}}} \mathsf{at}[\![\mathsf{S}_b]\!] \mathbin{\widehat{}} \overline{\pi}_3\rangle \mid \langle\overline{\pi}_1\ell,\,\ell\overline{\pi}_2\ell\rangle \in X \wedge \exists\overline{\rho}_{\mathsf{ff}}\,.$$

$$\mathscr{B}^{i}_{\mathbb{F}}[\![\mathsf{B}]\!]\varrho(\overline{\pi}_1\ell\overline{\pi}_2\ell) = \langle\overline{\rho}_{\mathsf{tt}},\,\overline{\rho}_{\mathsf{ff}}\rangle \wedge \langle\overline{\pi}_1\ell\overline{\pi}_2\ell \xrightarrow{\mathsf{B}=[\rho}_{\mathsf{tt}} \mathsf{at}[\![\mathsf{S}_b]\!],\,\overline{\pi}_3\rangle \in \widehat{\boldsymbol{S}}^{*}_{\mathbb{P}^{i}_{\mathbb{F}}}[\![\mathsf{S}_b]\!]\}^{\mathrm{b}}$$

a. If $\dot\gamma^{i}_{\mathbb{F}}(\dot\varnothing) = \varnothing$ then this case is $\varnothing$ when $\overline{\rho}_{\mathsf{ff}} = \dot\varnothing$.
b. If $\dot\gamma^{i}_{\mathbb{F}}(\dot\varnothing) = \varnothing$ then this case is $\varnothing$ when $\overline{\rho}_{\mathsf{tt}} = \dot\varnothing$.

as follows:

Proof We let $\mathscr{F}^{*}_{\mathbb{V}}$ be $\mathscr{F}^{*}$ as defined in (17.4) for $\mathbb{V} = \mathbb{Z}$ generalized to $\mathbb{V} = \mathbb{R}$ and $\mathbb{V} = \mathbb{F}$. The proof is based on theorem 18.21. Assuming that $\mathring{\alpha}^{i}_{\mathbb{F}}(X) \mathrel{\mathring{\sqsubseteq}^{i}} \overline{X}$ and (32.37), we must prove that $\mathring{\alpha}^{i}_{\mathbb{F}}(\mathscr{F}^{*}_{\mathbb{V}}[\![\texttt{while}\,\ell\,(\mathsf{B})\,\mathsf{S}_b]\!]\,X) \mathrel{\mathring{\sqsubseteq}^{i}} \mathscr{F}^{*}_{\mathbb{P}^{i}_{\mathbb{F}}}[\![\texttt{while}\,\ell\,(\mathsf{B})\,\mathsf{S}_b]\!]\,\overline{X}$ to conclude that

$$\mathring{\alpha}^{i}_{\mathbb{F}}(\mathsf{lfp}^{\subseteq}\,\mathscr{F}^{*}_{\mathbb{R}}[\![\texttt{while}\,\ell\,(\mathsf{B})\,\mathsf{S}_b]\!]) \mathrel{\mathring{\sqsubseteq}^{i}} \mathsf{lfp}^{\dot\subseteq}\,\mathscr{F}^{*}_{\mathbb{P}^{i}_{\mathbb{F}}}[\![\texttt{while}\,\ell\,(\mathsf{B})\,\mathsf{S}_b]\!].$$

We have complete lattices hence CPOs. $\mathscr{F}^{*}_{\mathbb{R}}$ does not preserves joins but can be shown to be continuous. We define $\mathscr{I}$ in theorem 18.21 by assuming that iterates X and $\overline{X}$ satisfy the induction hypothesis (32.37), which is trivially satisfied by the first iterates $\varnothing$.

$$\begin{aligned}
&\mathring{\alpha}^{i}_{\mathbb{F}}(\mathscr{F}^{*}_{\mathbb{V}}[\![\texttt{while}\,\ell\,(\mathsf{B})\,\mathsf{S}_b]\!]\,X)\\
=\;& \mathring{\alpha}^{i}_{\mathbb{F}}(\{\langle\pi_1\ell,\,\ell\rangle\} \cup \{\langle\pi_1\ell',\,\ell'\pi_2\ell' \xrightarrow{\neg(\mathsf{B})=\varrho(\pi_1\ell'\pi_2\ell')} \mathsf{after}[\![\mathsf{S}]\!]\rangle \mid \langle\pi_1\ell',\,\ell'\pi_2\ell'\rangle \in\\
& X \wedge \quad \mathscr{B}_{\mathbb{V}}[\![\mathsf{B}]\!]\varrho(\pi_1\ell'\pi_2\ell') = \mathsf{ff} \wedge \ell' = \ell\} \cup \{\langle\pi_1\ell',\,\ell'\pi_2\ell' \xrightarrow{\mathsf{B}=\varrho(\pi_1\ell'\pi_2\ell')} \mathsf{at}[\![\mathsf{S}_b]\!] \mathbin{\widehat{}} \pi_3\rangle \mid\\
& \langle\pi_1\ell',\,\ell'\pi_2\ell'\rangle \in X \wedge \mathscr{B}_{\mathbb{V}}[\![\mathsf{B}]\!]\varrho(\pi_1\ell'\pi_2\ell') = \mathsf{tt} \wedge \langle\pi_1\ell'\pi_2\ell' \xrightarrow{\mathsf{B}=\varrho(\pi_1\ell'\pi_2\ell')} \mathsf{at}[\![\mathsf{S}_b]\!],\\
& \pi_3\rangle \in \boldsymbol{S}^{*}[\![\mathsf{S}_b]\!] \wedge \ell' = \ell\}) \qquad \text{(def (17.4) of } \mathscr{F}^{*}_{\mathbb{V}}[\![\texttt{while}\,\ell\,(\mathsf{B})\,\mathsf{S}_b]\!] \text{ and section 32.5.5)}\\
=\;& \mathring{\alpha}^{i}_{\mathbb{F}}(\{\langle\pi_1\ell,\,\ell\rangle\}) \cup \mathring{\alpha}^{i}_{\mathbb{F}}(\{\langle\pi_1\ell',\,\ell'\pi_2\ell' \xrightarrow{\neg(\mathsf{B})=\varrho(\pi_1\ell'\pi_2\ell')} \mathsf{after}[\![\mathsf{S}]\!]\rangle \mid \langle\pi_1\ell',\,\ell'\pi_2\ell'\rangle \in\\
& X \wedge \quad \mathscr{B}_{\mathbb{V}}[\![\mathsf{B}]\!]\varrho(\pi_1\ell'\pi_2\ell') = \mathsf{ff} \wedge \ell' = \ell\}) \cup \mathring{\alpha}^{i}_{\mathbb{F}}(\{\langle\pi_1\ell',\,\ell'\pi_2\ell' \xrightarrow{\mathsf{B}=\varrho(\pi_1\ell'\pi_2\ell')} \mathsf{at}[\![\mathsf{S}_b]\!] \mathbin{\widehat{}}\\
& \pi_3\rangle \mid \langle\pi_1\ell',\,\ell'\pi_2\ell'\rangle \in X \wedge \mathscr{B}_{\mathbb{V}}[\![\mathsf{B}]\!]\varrho(\pi_1\ell'\pi_2\ell') = \mathsf{tt} \wedge \langle\pi_1\ell'\pi_2\ell' \xrightarrow{\mathsf{B}=\varrho(\pi_1\ell'\pi_2\ell')} \mathsf{at}[\![\mathsf{S}_b]\!],\\
& \pi_3\rangle \in \boldsymbol{S}^{*}[\![\mathsf{S}_b]\!] \wedge \ell' = \ell\}) \qquad \text{(join preservation in the Galois connection (32.25))}
\end{aligned}$$

The first term has already been handled for $\texttt{if}\,\ell\,(\mathsf{B})\,\mathsf{S}_t$. It is the same for the second term, but for $\langle\pi_1\ell',\,\ell'\pi_2\ell'\rangle \in X$ handled as in the third term below. For this third term (simplified by abuse of notation with $\ell' = \ell$), we have

$$\mathring{\alpha}^i_{\mathbb{F}}(\{\langle\pi_1{}^\ell,\ {}^\ell\pi_2{}^\ell \xrightarrow{\mathsf{B}=\varrho(\pi_1{}^\ell\pi_2{}^\ell)} \mathsf{at}[\![\mathsf{S}_b]\!] \frown \pi_3\rangle \mid \langle\pi_1{}^\ell,\ {}^\ell\pi_2{}^\ell\rangle \in X \wedge \mathscr{B}_{\mathbb{V}}[\![\mathsf{B}]\!]\varrho(\pi_1{}^\ell\pi_2{}^\ell) = \mathsf{tt} \wedge \langle\pi_1{}^\ell\pi_2{}^\ell \xrightarrow{\mathsf{B}=\varrho(\pi_1{}^\ell\pi_2{}^\ell)} \mathsf{at}[\![\mathsf{S}_b]\!],\ \pi_3\rangle \in \boldsymbol{\mathcal{S}}^*[\![\mathsf{S}_b]\!]\})$$

$$= \{\langle\vec{\alpha}^i_{\mathbb{F}}(\pi_1{}^\ell),\ \vec{\alpha}^i_{\mathbb{F}}({}^\ell\pi_2{}^\ell) \xrightarrow{\mathsf{B}=\dot{\alpha}^i_{\mathbb{F}}(\varrho(\pi_1{}^\ell\pi_2{}^\ell))} \mathsf{at}[\![\mathsf{S}_b]\!] \frown \vec{\alpha}^i_{\mathbb{F}}(\pi_3)\rangle \mid \langle\pi_1{}^\ell,\ {}^\ell\pi_2{}^\ell\rangle \in X \wedge \mathscr{B}_{\mathbb{V}}[\![\mathsf{B}]\!]\varrho(\pi_1{}^\ell\pi_2{}^\ell) = \mathsf{tt} \wedge \langle\pi_1{}^\ell\pi_2{}^\ell \xrightarrow{\mathsf{B}=\varrho(\pi_1{}^\ell\pi_2{}^\ell)} \mathsf{at}[\![\mathsf{S}_b]\!],\ \pi_3\rangle \in \boldsymbol{\mathcal{S}}^*[\![\mathsf{S}_b]\!]\}$$

(definition (32.14) of $\mathring{\alpha}^i_{\mathbb{F}}$)

$$\mathring{\sqsubseteq}^i \{\langle\overline{\pi}_1{}^\ell,\ {}^\ell\overline{\pi}_2{}^\ell \xrightarrow{\mathsf{B}=\overline{\rho}_{\mathsf{tt}}} \mathsf{at}[\![\mathsf{S}_b]\!] \frown \overline{\pi}_3\rangle \mid \langle\overline{\pi}_1{}^\ell,\ {}^\ell\overline{\pi}_2{}^\ell\rangle \in \overline{X} \wedge \exists\overline{\rho}_{\mathsf{ff}}\ .\ \mathscr{B}^i_{\mathbb{F}}[\![\mathsf{B}]\!]\varrho(\overline{\pi}_1{}^\ell\overline{\pi}_2{}^\ell) = \langle\overline{\rho}_{\mathsf{tt}},\ \overline{\rho}_{\mathsf{ff}}\rangle \wedge \langle\overline{\pi}_1{}^\ell\overline{\pi}_2{}^\ell \xrightarrow{\mathsf{B}=\overline{\rho}_{\mathsf{tt}}} \mathsf{at}[\![\mathsf{S}_b]\!],\ \overline{\pi}_3\rangle \in \widehat{\boldsymbol{\mathcal{S}}}^*_{\mathbb{P}^i_{\mathbb{F}}}[\![\mathsf{S}_b]\!]\}$$

(By definition (32.30) of $\mathring{\sqsubseteq}^i$, we must find an interval execution s.t. $\langle\vec{\alpha}^i_{\mathbb{F}}(\pi_1{}^\ell),\ \vec{\alpha}^i_{\mathbb{F}}({}^\ell\pi_2{}^\ell) \xrightarrow{\mathsf{B}=\dot{\alpha}^i_{\mathbb{F}}(\varrho(\pi_1{}^\ell\pi_2{}^\ell))} \mathsf{at}[\![\mathsf{S}_b]\!] \frown \vec{\alpha}^i_{\mathbb{F}}(\pi_3)\rangle\ \ddot{\sqsubseteq}^i\ \langle\overline{\pi}_1{}^\ell,\ {}^\ell\overline{\pi}_2{}^\ell \xrightarrow{\mathsf{B}=\overline{\rho}_{\mathsf{tt}}} \mathsf{at}[\![\mathsf{S}_b]\!] \frown \overline{\pi}_3\rangle$.

The control abstraction is the same.

We can choose $\overline{\pi}_1{}^\ell = \vec{\alpha}^i_{\mathbb{F}}(\pi_1{}^\ell)$ so that, by induction hypothesis (32.37) and $\langle\pi_1{}^\ell,\ {}^\ell\pi_2{}^\ell\rangle \in X, \exists\overline{\pi}'\ .\ \langle\overline{\pi}_1{}^\ell,\ \overline{\pi}'\rangle \in \overline{X} \wedge \vec{\alpha}^i_{\mathbb{F}}({}^\ell\pi_2{}^\ell)\ \vec{\sqsubseteq}^i\ \overline{\pi}'$. Therefore $\overline{\pi}'$ has the form $\overline{\pi}' = {}^\ell\overline{\pi}_2{}^\ell$ and so $\langle\overline{\pi}_1{}^\ell,\ {}^\ell\overline{\pi}_2{}^\ell\rangle \in \overline{X}$ with $\vec{\alpha}^i_{\mathbb{F}}(\pi_1{}^\ell\pi_2{}^\ell)\ \vec{\sqsubseteq}^i\ \overline{\pi}_1{}^\ell\overline{\pi}_2{}^\ell$.

$\mathscr{B}_{\mathbb{V}}[\![\mathsf{B}]\!]\varrho(\pi_1{}^\ell\pi_2{}^\ell) = \mathsf{tt}$, $\langle\overline{\rho}^1_{\mathsf{tt}},\ \overline{\rho}^1_{\mathsf{ff}}\rangle = \mathscr{B}^i_{\mathbb{F}}[\![\mathsf{B}]\!]\varrho(\vec{\alpha}^i_{\mathbb{F}}(\pi_1{}^\ell\pi_2{}^\ell))$ and (32.18) imply that $\dot{\alpha}^i_{\mathbb{F}}(\varrho(\pi_1{}^\ell\pi_2{}^\ell))\ \dot{\sqsubseteq}^i\ \overline{\rho}^1_{\mathsf{tt}}$.

$\mathscr{B}^i_{\mathbb{F}}[\![\mathsf{B}]\!]$ increasing, $\vec{\alpha}^i_{\mathbb{F}}(\pi_1{}^\ell\pi_2{}^\ell)\ \vec{\sqsubseteq}^i\ \overline{\pi}_1{}^\ell\overline{\pi}_2{}^\ell$ and so $\varrho(\vec{\alpha}^i_{\mathbb{F}}(\pi_1{}^\ell\pi_2{}^\ell))\ \dot{\sqsubseteq}^i\ \varrho(\overline{\pi}_1{}^\ell\overline{\pi}_2{}^\ell)$ implies that $\langle\overline{\rho}_{\mathsf{tt}},\ \overline{\rho}_{\mathsf{ff}}\rangle = \mathscr{B}^i_{\mathbb{F}}[\![\mathsf{B}]\!]\varrho(\overline{\pi}_1{}^\ell\overline{\pi}_2{}^\ell)$ has $\overline{\rho}^1_{\mathsf{tt}}\ \dot{\sqsubseteq}^i\ \overline{\rho}_{\mathsf{tt}}$

By transitivity, $\dot{\alpha}^i_{\mathbb{F}}(\varrho(\pi_1{}^\ell\pi_2{}^\ell))\ \dot{\sqsubseteq}^i\ \overline{\rho}_{\mathsf{tt}}$.

It follows that $\ddot{\alpha}^i_{\mathbb{F}}(\langle\pi_1{}^\ell,\ {}^\ell\pi_2{}^\ell \xrightarrow{\mathsf{B}=\varrho(\pi_1{}^\ell\pi_2{}^\ell)} \mathsf{at}[\![\mathsf{S}_b]\!]\rangle)\ \ddot{\sqsubseteq}^i\ \langle\overline{\pi}_1{}^\ell,\ {}^\ell\overline{\pi}_2{}^\ell \xrightarrow{\mathsf{B}=\overline{\rho}_{\mathsf{tt}}} \mathsf{at}[\![\mathsf{S}_b]\!]\rangle$ and therefore $\vec{\alpha}^i_{\mathbb{F}}(\pi_1{}^\ell\pi_2{}^\ell \xrightarrow{\mathsf{B}=\varrho(\pi_1{}^\ell\pi_2{}^\ell)} \mathsf{at}[\![\mathsf{S}_b]\!])\ \vec{\sqsubseteq}^i\ \overline{\pi}_1{}^\ell\overline{\pi}_2{}^\ell \xrightarrow{\mathsf{B}=\overline{\rho}_{\mathsf{tt}}} \mathsf{at}[\![\mathsf{S}_b]\!]$

By $\langle\pi_1{}^\ell\pi_2{}^\ell \xrightarrow{\mathsf{B}=\varrho(\pi_1{}^\ell\pi_2{}^\ell)} \mathsf{at}[\![\mathsf{S}_b]\!],\ \pi_3\rangle \in \boldsymbol{\mathcal{S}}^*[\![\mathsf{S}_b]\!]$, $\vec{\alpha}^i_{\mathbb{F}}(\pi_1{}^\ell\pi_2{}^\ell \xrightarrow{\mathsf{B}=\varrho(\pi_1{}^\ell\pi_2{}^\ell)} \mathsf{at}[\![\mathsf{S}_b]\!])\ \vec{\sqsubseteq}^i\ \overline{\pi}_1{}^\ell\overline{\pi}_2{}^\ell \xrightarrow{\mathsf{B}=\overline{\rho}_{\mathsf{tt}}} \mathsf{at}[\![\mathsf{S}_b]\!]$ and induction hypothesis (32.37), there exists $\overline{\pi}_3$ such that $\vec{\alpha}^i_{\mathbb{F}}(\pi_3)\ \vec{\sqsubseteq}^i\ \overline{\pi}_3$ and $\langle\overline{\pi}_1{}^\ell\overline{\pi}_2{}^\ell \xrightarrow{\mathsf{B}=\overline{\rho}_{\mathsf{tt}}} \mathsf{at}[\![\mathsf{S}_b]\!],\ \overline{\pi}_3\rangle \in \widehat{\boldsymbol{\mathcal{S}}}^*_{\mathbb{P}^i_{\mathbb{F}}}[\![\mathsf{S}_b]\!]$.

As required to prove the inclusion $\mathring{\sqsubseteq}^i$, we conclude that $\ddot{\alpha}^i_{\mathbb{F}}(\langle\pi_1{}^\ell,\ {}^\ell\pi_2{}^\ell \xrightarrow{\mathsf{B}=\varrho(\pi_1{}^\ell\pi_2{}^\ell)} \mathsf{at}[\![\mathsf{S}_b]\!] \frown \pi_3\rangle)\ \ddot{\sqsubseteq}^i\ \langle\overline{\pi}_1{}^\ell,\ {}^\ell\overline{\pi}_2{}^\ell \xrightarrow{\mathsf{B}=\overline{\rho}_{\mathsf{tt}}} \mathsf{at}[\![\mathsf{S}_b]\!] \frown \overline{\pi}_3\rangle$.)

The above terms involves only computations in $\mathbb{S}_{\mathbb{P}^i}$ and none in $\mathbb{S}_{\mathbb{V}}$. Therefore, we can define $\mathscr{F}^*_{\mathbb{P}^i_{\mathbb{F}}}[\![\mathtt{while}\,\ell\,(\mathsf{B})\,\mathsf{S}_b]\!]$ as in (32.39), so that we proved that $\mathring{\alpha}^i_{\mathbb{F}}(X)\,\mathring{\sqsubseteq}^i\,\overline{X}$ implies

$$\mathring{\alpha}^i_{\mathbb{F}}(\mathscr{F}^*_{\mathbb{R}}[\![\mathtt{while}\,\ell\,(\mathsf{B})\,\mathsf{S}_b]\!]\,X)\,\mathring{\sqsubseteq}^i\,\mathscr{F}^*_{\mathbb{P}^i_{\mathbb{F}}}[\![\mathtt{while}\,\ell\,(\mathsf{B})\,\mathsf{S}_b]\!]\,\overline{X}.$$

We have to show that the next iterate $\mathscr{F}^*_{\mathbb{P}^i_{\mathbb{F}}}[\![\mathtt{while}\,\ell\,(\mathsf{B})\,\mathsf{S}_b]\!]\,X$ satisfies (32.37), which is trivial for the first term. For the second and third term this follows from the induction hypothesis. It follows that:

$$\begin{aligned}
& \mathring{\alpha}^i_{\mathbb{F}}(\widehat{\mathcal{S}}^*_{\mathbb{V}}[\![\mathtt{while}\,\ell\,(\mathsf{B})\,\mathsf{S}_b]\!]) && (32.40)\\
= \;& \mathring{\alpha}^i_{\mathbb{F}}(\mathsf{lfp}^{\subseteq}\,\mathscr{F}^*_{\mathbb{R}}[\![\mathtt{while}\,\ell\,(\mathsf{B})\,\mathsf{S}_b]\!]) && \text{by (17.4)}\\
\mathring{\sqsubseteq}^i \;& \mathsf{lfp}^{\mathring{\subseteq}}\,\mathscr{F}^*_{\mathbb{P}^i_{\mathbb{F}}}[\![\mathtt{while}\,\ell\,(\mathsf{B})\,\mathsf{S}_b]\!] && \text{by theorem 18.21}\\
\triangleq \;& \widehat{\mathcal{S}}^*_{\mathbb{P}^i_{\mathbb{F}}}[\![\mathtt{while}\,\ell\,(\mathsf{B})\,\mathsf{S}_b]\!] && \text{definition (32.39)}
\end{aligned}$$

It remains to show that $\widehat{\mathcal{S}}^*_{\mathbb{P}^i_{\mathbb{F}}}[\![\mathtt{while}\,\ell\,(\mathsf{B})\,\mathsf{S}_b]\!]$ satisfies (32.37). We have shown that it holds for all fixpoint iterates. Moreover, it is trivially preserved by trace set union. □

In conclusion of this section, $\widehat{\mathcal{S}}^*_{\mathbb{P}^i_{\mathbb{F}}}$ is similar to $\widehat{\mathcal{S}}^*_{\mathbb{V}}$ in (6.11)–(6.30), except for statements involving tests for which we have (32.38) for the conditional and (32.39) for the iteration statement.

Exercise 32.41 (Project: maximal trace float interval interpreter) Modify the maximal trace interpreter of exercises 7.14 and 32.9, in order to operate on float intervals.

32.6 On Floating Point Computations

Unfortunately, real computations are usually performed using floating-point arithmetics. One computes only one floating-point value hoping it is not too far from the real one. This problem has been deeply studied in static analysis [64, 289, 305, 411, 428–433, 659]. Another dynamic analysis solution is to check the precision with an interval analysis.

Consider the execution with reals (at least, their semantics), floats and float intervals, maybe with different possible execution traces for float intervals due to the nondeterminacy of tests. These interval executions abstract both the real and float executions.

If there is only one interval execution trace or we can prove that the real and float executions follow exactly the same control path then the real execution is in the join of the interval executions to which the float execution belongs, when projected on all program points.

Otherwise, the real and float executions may have followed different paths but both are guaranteed to belong to the union of all interval executions projected on all program points.

In both cases this provides an estimate of the rounding error of the float execution compared to the ideal real execution. Of course, the estimate might be rough because specific properties of the computation are not taken into account (e.g., [524, pp. 91–94]).

32.7 Affine Arithmetic

Interval arithmetic is imprecise. For example, if $x \in [1,4]$ then $x - x \in [1-4, 4-1] = [-3, 3]$ instead of $[0, 0]$. The problem as that the arguments of functions cannot be correlated by a Cartesian abstraction. So we have to independently take into consideration all possible values of variables within their interval of variation.

Several solutions have been proposed to solves this imprecision problem [746]. One of them, *affine arithmetics* [206, 913], represents an interval $x \in [\underline{x}, \overline{x}]$ by $x = a_0 + a_1\epsilon_x$ where $a_0 = \frac{\overline{x}+\underline{x}}{2}$, $a_1 = \frac{\overline{x}-\underline{x}}{2}$, and $\epsilon_x \in [-1, 1]$ is a fresh auxiliary variable. Then $x - x = (a_0 + a_1\epsilon_x) - (a_0 + a_1\epsilon_x) = 0 + 0\epsilon_x$, as required. In general a program involves several variables so that we have an affine form $x = a_0 + a_1\epsilon_1 + a_2\epsilon_2 + \cdots + a_n\epsilon_n$. This implies $x \in [a_0 - d, a_0 + d]$ where $d = \sum_{i=1}^{n} |a_i|$ is the total deviation of x. This is, by interval arithmetic, the smallest interval that contains all possible values of x, assuming that each ϵ_i ranges independently over the interval $[-1, +1]$. For m variables, the affine constraints determine a *zonotope* [679], a center-symmetric convex polytope in $\mathbb{R}^m$, whose faces are themselves center symmetric [100, ch. 9].

32.8 Conclusion

Ramon E. Moore [715–717] may have introduced the first dynamic program analysis ever to put bounds on rounding (or roundoff) errors in floating-point computations. As shown in section 32.5, it is an abstract interpretation of the trace semantics of chapters 6 and 7 so can be computed at runtime for one trace at a time (i.e., during one program execution). Tests may have to consider many executions, which can be quite inefficient.

Our formalization of the float interval semantics as an abstraction of the real semantics uses an approximation preorder $\mathring{\sqsubseteq}^{i}$ different from the fixpoint ordering $\subseteq$ (also called computational ordering). This is a rare example of application of theorem 18.21 in abstract interpretation with [248, 737].

Other dynamic analyses have been designed to cope with the imprecision of floating-point computations. For example, [107, 846] duplicate every floating-point computation with computations in higher precision to detect discrepancies at runtime. [952] embeds an abstract interpreter into a runtime execution for verifying numerical accuracy and robustness properties of floating point computations.

Dynamic interval analysis can be extended to ball analysis (also known as midpoint-radius interval arithmetic) [950, 951].

Dynamic interval analysis is different from other instrumented dynamic analyses (or runtime verification) [96, 361, 468, 469] in that it does not collect information upon executions and checks it to verify a specification but replaces that execution (on reals or floats) by another one (on float intervals).

We think that these other instrumented dynamic analyses can also be formalized by abstract interpretation of one execution trace. An example is regular model checking [233] in which executions are monitored by a regular expression specifying sequences of invariants. Other examples of runtime analyses are the Firefox's Javascript JIT compiler integer interval analysis [149] and the static analyzer of the Extended Berkeley Packet Filter in Linux Kernel [957]. A dynamic analysis can be directly constructed by considering a single execution at a time, as defined by abstraction to a transition system (see chapter 42, "Stateful Prefix Trace Semantics"), instead of all possible executions (as in static analysis). [809] formalizes the coverage and precision of dynamic analysis (on finitely many execution traces).

Instrumented dynamic analysis consists essentially in checking the abstraction of one execution trace (e.g., [631]). The program is instrumented to generate an event stream. This observed stream is an abstraction of the execution trace, which is checked by the runtime verifier. The runtime verifier can also be embedded into the observed program to check, for example, for temporal properties [233, 340].

Static interval analysis cannot be performed at runtime because it must deal with all possible program computation traces in finite time (it is nevertheless possible to consider only one execution and perform an analysis of this execution, for example, [759]). We study static interval analysis in the next chapter 33. It is a Cartesian abstraction of chapter 28 that is an abstraction of the reachability semantics of chapter 19 (itself an abstraction of the trace semantics of chapters 6 and 7).

32.9 Solutions to Selected Exercises

Solution to exercise 32.28
Proof of (32.29)

$$\mathring{\alpha}^i_{\mathbb{F}}(\Pi) \mathring{\sqsubseteq}^i \overline{\Pi}$$
$$\Leftrightarrow \{\vec{\alpha}^i_{\mathbb{F}}(\pi) \mid \pi \in \Pi\} \mathring{\sqsubseteq}^i \overline{\Pi} \qquad \wr \text{def. (32.14) of } \mathring{\alpha}^i_{\mathbb{F}} \int$$
$$\Leftrightarrow \forall \overline{\pi} \in \{\vec{\alpha}^i_{\mathbb{F}}(\pi') \mid \pi' \in \Pi\} \,.\, \forall \pi \in \vec{\gamma}^i_{\mathbb{F}}(\overline{\pi}) \,.\, \exists \overline{\pi}' \in \overline{\Pi} \,.\, \pi \in \vec{\gamma}^i_{\mathbb{F}}(\overline{\pi}') \qquad \wr \text{def. (32.26) of } \mathring{\sqsubseteq}^i \int$$
$$\Leftrightarrow \forall \pi' \in \Pi \,.\, \forall \pi \in \vec{\gamma}^i_{\mathbb{F}}(\vec{\alpha}^i_{\mathbb{F}}(\pi')) \,.\, \exists \overline{\pi}' \in \overline{\Pi} \,.\, \pi \in \vec{\gamma}^i_{\mathbb{F}}(\overline{\pi}') \qquad \wr \text{def. } \in \int$$
$$\Leftrightarrow \Pi \subseteq \{\pi' \mid \forall \pi \in \vec{\gamma}^i_{\mathbb{F}}(\vec{\alpha}^i_{\mathbb{F}}(\pi')) \,.\, \exists \overline{\pi}' \in \overline{\Pi} \,.\, \pi \in \vec{\gamma}^i_{\mathbb{F}}(\overline{\pi}')\} \qquad \wr \text{def. } \subseteq \int$$
$$\Leftrightarrow \Pi \subseteq \mathring{\overline{\gamma}}(\overline{\Pi})$$

by defining $\mathring{\overline{\gamma}}(\overline{\Pi}) \triangleq \{\pi' \mid \forall \pi \in \vec{\gamma}^i_{\mathbb{F}}(\vec{\alpha}^i_{\mathbb{F}}(\pi')) \,.\, \exists \overline{\pi}' \in \overline{\Pi} \,.\, \pi \in \vec{\gamma}^i_{\mathbb{F}}(\overline{\pi}')\}$. □

Solution to exercise 32.35

Proof of (32.36)

$$\mathring{\alpha}^i_{\mathbb{F}}(\Pi) \mathring{\sqsubseteq}^i \overline{\Pi}$$
$$\Leftrightarrow \{\vec{\alpha}^i_{\mathbb{F}}(\pi) \mid \pi \in \Pi\} \mathring{\sqsubseteq}^i \overline{\Pi} \qquad \wr \text{def. (32.14) of } \mathring{\alpha}^i_{\mathbb{F}} \int$$
$$\Leftrightarrow \forall \overline{\pi} \in \{\vec{\alpha}^i_{\mathbb{F}}(\pi) \mid \pi \in \Pi\} \,.\, \exists \overline{\pi}' \in \overline{\Pi} \,.\, \overline{\pi} \vec{\sqsubseteq}^i \overline{\pi}' \qquad \wr \text{def. (32.2) of } \mathring{\sqsubseteq}^i \int$$
$$\Leftrightarrow \forall \pi \in \Pi \,.\, \exists \overline{\pi}' \in \overline{\Pi} \,.\, \vec{\alpha}^i_{\mathbb{F}}(\pi) \vec{\sqsubseteq}^i \overline{\pi}' \qquad \wr \text{def. } \in \int$$
$$\Leftrightarrow \Pi \subseteq \{\pi \mid \exists \overline{\pi}' \in \overline{\Pi} \,.\, \vec{\alpha}^i_{\mathbb{F}}(\pi) \vec{\sqsubseteq}^i \overline{\pi}'\} \qquad \wr \text{def. } \subseteq \int$$
$$\Leftrightarrow \Pi \subseteq \mathring{\overline{\overline{\gamma}}}(\overline{\Pi})$$

by defining $\mathring{\overline{\overline{\gamma}}}(\overline{\Pi}) \triangleq \{\pi \mid \exists \overline{\pi}' \in \overline{\Pi} \,.\, \vec{\alpha}^i_{\mathbb{F}}(\pi) \vec{\sqsubseteq}^i \overline{\pi}'\}$. □

33 Static Interval Analysis

We generalize Cartesian static analyses like sign or constant analysis to interval analysis (also called range or box analysis) where the value of numerical variables is bounded by a minimal and a maximal value between which all reachable values of a variable must stand.

Contrary to interval arithmetics in scientific computing of chapter 32 which is an abstraction of *one* maximal execution trace of chapter 7, range analysis of this chapter 33 is a Cartesian abstraction (developed in chapter 28) of the reachability analysis of chapter 19 (hence of *all* possible execution traces of chapter 7).

Contents

33.1 Interval/Range Domain

The Cartesian interval abstract domain $\mathbb{D}_i \triangleq \langle \mathbb{P}^i, \sqsubseteq^i, \bot^i, \top^i, \sqcup^i, \sqcap^i \rangle$ is

- The set $\mathbb{P}^i \triangleq \{[\ell, h] \mid \ell \in \mathbb{I} \cup \{-\infty\} \wedge \ell \leqslant h \wedge h \in \mathbb{I} \cup \{\infty\}\} \cup \{\bot^i\}$ of numerical intervals (excluding $[-\infty, -\infty]$ and $[\infty, \infty]$) where the empty interval $\bot^i = \varnothing$ can be encoded by any $[\ell, h]$ with $h < \ell$ (e.g., normalized to $[\infty, -\infty]$);
- The partial order $\sqsubseteq^i$ on $\mathbb{P}^i$ is interval inclusion $\bot^i \sqsubseteq^i \bot^i \sqsubseteq^i \langle \ell, h \rangle \sqsubseteq^i \langle \ell', h' \rangle$ if and only if $\ell' \leqslant \ell \leqslant h \leqslant h'$. (The inverse is $\sqsupseteq^i$. The pointwise extensions are $\dot{\sqsubseteq}^i$ and $\dot{\sqsupseteq}^i$.);
- The interval abstract domain is a complete lattice:
  - The infimum is the empty interval $\bot^i \triangleq \varnothing$,
  - The supremum is $\top^i \triangleq [-\infty, \infty]$;
  - The join is the lub $\bigsqcup\limits_{i \in \Delta}^i [\ell_k, h_k] \triangleq [\min_{k \in \Delta}(\ell_k), \max_{k \in \Delta}(h_k)]$; and
  - The meet is the glb $\bigsqcap\limits_{k \in \Delta}^i [\ell_k, h_k] \triangleq [\max_{k \in \Delta}(\ell_k), \min_{i \in \Delta}(h_k)]$

where $\min \mathbb{I} \triangleq -\infty$, $\max \mathbb{I} \triangleq \infty$, $\min \emptyset \triangleq \infty$, $\max \emptyset \triangleq -\infty$, and $[\ell, h] = \emptyset = \bot^i$ when $h < \ell$.

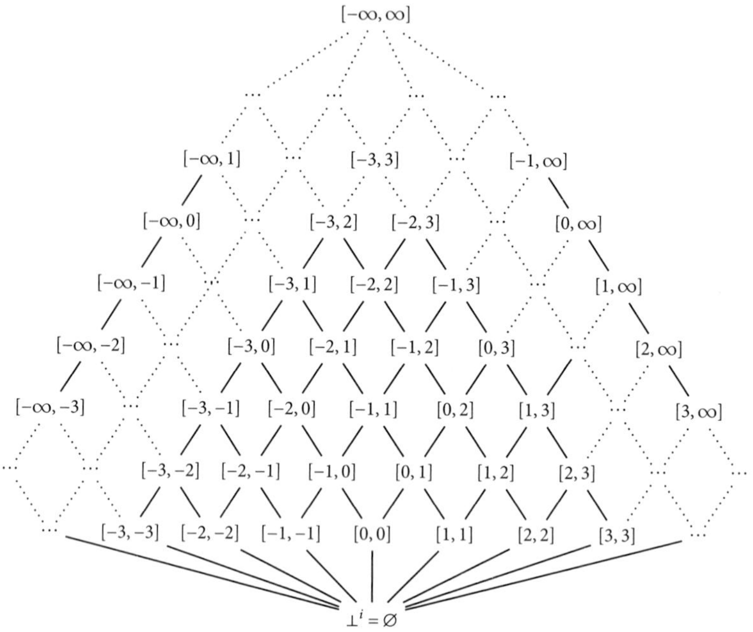

Figure 33.1
Interval complete lattice

33.2 Interval/Range Abstraction

The interval abstraction is

$$\langle \wp(\mathbb{I}), \subseteq \rangle \xleftarrow[\alpha_i]{\gamma_i} \langle \mathbb{P}^i, \sqsubseteq^i \rangle$$

such that

$$\begin{aligned} \alpha_i(S) &\triangleq [\min S, \max S] \\ \gamma_i([\ell, h]) &\triangleq \{z \in \mathbb{I} \mid \ell - 1 < z < h + 1\} \end{aligned}$$

where $-\infty - 1 = -\infty$, $\infty + 1 = \infty$, and $\alpha_i(\emptyset) \triangleq [\min \emptyset, \max \emptyset] = [\infty, -\infty]$.

Exercise 33.2 Generalize the interval abstraction to the abstraction of any powerset $\wp(\mathbb{V})$ where $\langle \mathbb{V}, \preceq, \bot, \top, \vee, \wedge \rangle$ is a complete lattice.

The abstraction is extended pointwise to $\langle \mathbb{V} \to \wp(\mathbb{I}), \dot{\subseteq} \rangle \xleftarrow[\dot{\alpha}_i]{\dot{\gamma}_i} \langle \mathbb{V} \to \mathbb{P}^i, \dot{\sqsubseteq}^i \rangle$ for an assertional reachability analysis. (The Cartesian abstraction is not expressive enough to be useful for a relational reachability analysis).

33.3 Interval/Range Cartesian Reachability Abstract Interpreter

To get an interval static analyzer, we can instantiate the Cartesian abstract interpreter of chapter 21 with the Cartesian abstract domain of (28.43) in chapter 28 where the value abstract domain of (28.42) is that $\mathbb{D}_i$ of intervals. In addition to the complete lattice operations $\langle \overline{\mathbb{P}}^i, \sqsubseteq^i, \bot^i, \top^i, \sqcup^i, \sqcap^i \rangle$, the basic abstract operations are

- $1^i \triangleq [1, 1]$;
- $\ominus^i \langle [\ell_1, h_1], [\ell_2, h_2] \rangle \triangleq [\ell_1 - h_2, h_1 - \ell_2]$.
  For example $\ominus^i \langle [0, 1], [-1, 0] \rangle = [0, 2]$ and $\ominus^i \langle [1, \infty], [-\infty, 0] \rangle = [1 - 0, \infty - (-\infty)] = [1, \infty]$ because $\infty + \infty = \infty$;
- $\overset{i}{\ominus^{-1}}[a, b] \langle [\ell_1, h_1], [\ell_2, h_2] \rangle \triangleq \langle [\max(\ell_1, a + \ell_2), \min(h_1, b + h_2)], [\max(\ell_2, \ell_1 - b), \min(h_2, h_1 - a)] \rangle$;
- $\overset{i}{\otimes^{-1}} \langle [\ell_1, h_1], [\ell_2, h_2] \rangle \triangleq \langle [\ell_1, \min(h_1, h_2 - 1)], [\max(\ell_2, \ell_1 + 1), h_2] \rangle$.
  For example $\overset{i}{\otimes^{-1}} \langle [0, 0], [1, 1] \rangle = \langle [0, 0], [1, 1] \rangle$ because $a \in [0, 0]$ is always less than $b \in [1, 1]$. $\overset{i}{\otimes^{-1}} \langle [0, 0], [0, 0] \rangle = \langle \varnothing, \varnothing \rangle$ because $a \in [0, 0]$ is never less than $b \in [0, 0]$. $\overset{i}{\otimes^{-1}} \langle [1, 4], [0, 3] \rangle = \langle [1, 2], [2, 3] \rangle$ because $a \in [1, 4]$ must be less than $\max b \in [2, 3] = 3$ and $b \in [2, 3]$ must be greater than $\min a \in [1, 4] = 1$; and
- $\overset{i}{\overline{\otimes}^{-1}} \langle [\ell_1, h_1], [\ell_2, h_2] \rangle \triangleq \langle [\max(\ell_1, \ell_2), h_1], [\ell_2, \min(h_1, h_2)] \rangle$.

Observe that the interval Cartesian reachability semantics $\mathscr{A}^i[\![\mathsf{A}]\!]$ of an arithmetic expression A in (28.17) does not preverve lubs in general. For example $\forall i \in \mathbb{I} \,.\, \mathscr{A}^i[\![\mathsf{x} - \mathsf{x}]\!]\overline{\rho}[x \leftarrow [i, i]] = [0, 0]$ whereas for $\overline{\rho}[x \leftarrow [-\infty, \infty]] = \dot{\bigsqcup}^i_{i \in \mathbb{I}} \overline{\rho}[x \leftarrow [i, i]]$, $\mathscr{A}^i[\![\mathsf{x} - \mathsf{x}]\!]\overline{\rho}[x \leftarrow [-\infty, \infty]] = [-\infty, \infty]$.

Exercise 33.3 (interval arithmetic) Define interval addition and multiplication.

Exercise 33.4 Define an operator neq restricting the interval of value of variables x and y when x ≠ y.

Contrary to interval arithmetic in chapter 32 to put bounds on rounding errors in floating-point arithmetic at runtime (i.e., during one program computation along an execution trace that tests may be split into several traces considered one at a time), interval analysis is static (and must deal with all possible program computations in finite time).

33.4 Divergence

Consider the simple diverging program

$$\mathsf{P}_1 \;=\; \mathtt{while}\ \ell_1\ (\mathsf{tt})\ \ \ell_2\ \mathsf{x} = \mathsf{x} + 1;\ \ \ell_3$$

By (21.11), the interval static analysis from an initial assignment $\overline{\rho}_0$ of intervals to variables is

$$\hat{\mathcal{S}}^i[\![\mathsf{P}_1]\!]\,\overline{\rho}_0 \;=\; \mathsf{lfp}^{\sqsubseteq^i}\,(\mathscr{F}^i[\![\mathtt{while}\ \ell_1\ (\mathsf{tt})\ \ell_2\ \mathsf{x} = \mathsf{x} + 1;]\!]\,\overline{\rho}_0)$$

where

$$
\begin{array}{ll}
& \mathscr{F}^i[\![\texttt{while}\,\ell_1\,(\mathrm{tt})\,\ell_2\;\mathsf{x}=\mathsf{x}+1\,;]\!]\,\overline{\rho}_0\,X\,\ell' \\
= & (\!|\,\ell'=\ell_1\;?\;\overline{\rho}_0\;⁏\;\mathsf{x}\in\mathbb{V}\mapsto\bot^i\,|\!)\sqcup^i \\
& (\!|\,\ell'\in\{\ell_1,\ell_2\}\;?\;\widehat{\boldsymbol{\mathcal{S}}}^i[\![\ell_2\;\mathsf{x}=\mathsf{x}+1\,;]\!]\,(\mathrm{test}^i[\![\mathrm{tt}]\!]X(\ell_1))\,\ell'\;⁏\;\mathsf{x}\in\mathbb{V}\mapsto\bot^i\,|\!)\sqcup^i \\
& (\!|\,\ell'=\ell_3\;?\;\overline{\mathrm{test}}^i[\![\mathrm{tt}]\!]X(\ell_1)\;⁏\;\mathsf{x}\in\mathbb{V}\mapsto\bot^i\,|\!) \\
= & (\!|\,\ell'=\ell_1\;?\;\overline{\rho}_0\;⁏\;\mathsf{x}\in\mathbb{V}\mapsto\bot^i\,|\!)\sqcup^i \\
& (\!|\,\ell'\in\{\ell_1,\ell_2\}\;?\;\widehat{\boldsymbol{\mathcal{S}}}^i[\![\ell_2\;\mathsf{x}=\mathsf{x}+1\,;]\!]\,X(\ell_1)\,\ell'\;⁏\;\mathsf{x}\in\mathbb{V}\mapsto\bot^i\,|\!)\sqcup^i \\
& \mathsf{x}\in\mathbb{V}\mapsto\bot^i \\
= & (\!|\,\ell'=\ell_1\;?\;\overline{\rho}_0\sqcup^i X(\ell_1)[x\leftarrow X(\ell_1)(\mathsf{x})\oplus^i[1,1]] \\
& [\!]\;\ell'=\ell_2\;?\;X(\ell_1) \\
& ⁏\;/\star\;\;\ell'=\ell_3\;\;\star/\;\mathsf{x}\in\mathbb{V}\mapsto\bot^i\,|\!)
\end{array}
$$

by $\mathrm{test}^i[\![\mathrm{tt}]\!]X(\ell_1) = X(\ell_1)$, and $\overline{\mathrm{test}}^i[\![\mathrm{tt}]\!]X(\ell_1) = \mathsf{x}\in\mathbb{V}\mapsto\bot^i$ which is the identity element (or neutral element) of $\sqcup^i$, and

$$
\begin{array}{lll}
\widehat{\boldsymbol{\mathcal{S}}}^i[\![\ell_2\;\mathsf{x}=\mathsf{x}+1\,;]\!]\,X(\ell_1)\,\ell & = & (\!|\,\ell=\ell_2\;?\;X(\ell_1) \\
& & [\!]\;\ell=\ell_1\;?\;X(\ell_1)[x\leftarrow X(\ell_1)(\mathsf{x})\oplus^i[1,1]] \\
& & ⁏\;\mathsf{x}\in\mathbb{V}\mapsto\varnothing\,|\!)
\end{array}
$$

in (21.7), $\mathscr{A}^i[\![\mathsf{x}+1]\!]X(\ell_1) = X(\ell_1)(\mathsf{x})\oplus^i[1,1]$, $[\ell_1,h_1])\oplus^i[\ell_2,h_2] = [\ell_1,h_1])\ominus^i([0,0]\ominus^i[\ell_2,h_2]) = [\ell_1+\ell_2,h_1+h_2]$.

Assume that initially $\overline{\rho}_0(\mathsf{x}) = [0,0]$ and let $x = X(\ell_1)(\mathsf{x})$. The fixpoint computation amounts to solving the fixpoint equation

$$
x \;=\; \mathscr{F}^i(x) \quad\text{where}\quad \mathscr{F}^i(x) \;=\; [0,0]\sqcup^i(x\oplus^i[1,1])
$$

iteratively (theorem 15.21).

$$
\begin{array}{lllllllll}
x^0 & = & \bot^i \\
x^1 & = & \mathscr{F}^i(x^0) & = & [0,0]\sqcup^i(x^0\oplus^i[1,1]) & = & [0,0] \\
x^2 & = & \mathscr{F}^i(x^1) & = & [0,0]\sqcup^i(x^1\oplus^i[1,1]) & = & [0,0]\sqcup^i[1,1] & = & [0,1] \\
x^3 & = & \mathscr{F}^i(x^2) & = & [0,0]\sqcup^i(x^2\oplus^i[1,1]) & = & [0,0]\sqcup^i[1,2] & = & [0,2] \\
\ldots
\end{array}
$$

$$
\begin{array}{lllllll}
\ldots \\
x^n & = & [0,n-1] & & & & \text{induction hypothesis} \\
x^{n+1} & = & \mathscr{F}^i(x^n) & = & [0,0]\sqcup^i(x^n\oplus^i[1,1]) \\
& & & = & [0,0]\sqcup^i[1,n] & = & [0,(n+1)-1] \\
\ldots \\
x^\omega & = & \bigsqcup^i_{n<\omega}[0,n-1] & = & [0,\infty] & & \text{limit} \\
x^{\omega+1} & = & \mathscr{F}^i(x^\omega) & = & [0,0]\sqcup^i(x^\omega\oplus^i[1,1]) & = & [0,0]\sqcup^i[1,\infty+1] \;=\; [0,\infty] \\
& = & \mathsf{lfp}^{\sqsubseteq^i}\,x\mapsto[0,0]\sqcup^i(x\oplus^i[1,1])
\end{array}
$$

Unfortunately computerized methods to infer induction hypotheses, to simplify the iteration terms, and to pass to the limit are not effective. The most popular cases are (a mixture of)

- Consider finitary abstractions only (of which limitations are studied in section 34.11);
- Ask for human help to get an inductive solution of the fixpoint equation (as considered in the verification methods of chapter 25), which can be difficult and costly, for example, when the inductive invariant is larger than the program and hard to change whenever the program is modified;
- To soundly automatize the induction and passage to the limit, possibly at the price of a loss of precision, to enforce rapid convergence. This is the purpose of widenings and narrowings.
- To unsoundly stop the iterates after a few iterations, for example, [182], which has no convincing scientific justification for verification objectives .

33.5 Extrapolation by Widening

The idea of the widening is to extrapolate from an iterate x^n and the next one x^{n+1} to an upper bound $x^n \nabla x^{n+1}$ in order to accelerate or enforce the convergence of the iterates in finitely many steps. The price to be paid is a loss of precision, a sound although imprecise answer being considered better than no answer at all by memory and time resources exhaustion. In the following example, the derivative is used to accelerate convergence by extrapolation (as in Newton–Raphson iteration method [524]).

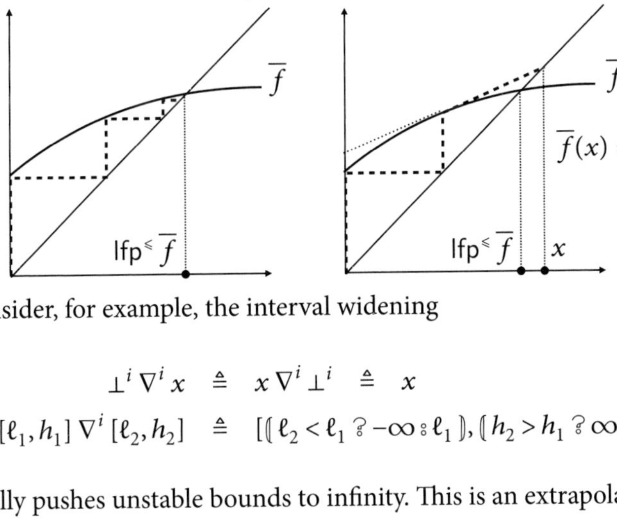

Let us consider, for example, the interval widening

$$\left[\begin{array}{rcl} \bot^i \nabla^i x & \triangleq & x \nabla^i \bot^i \;\triangleq\; x \\ [\ell_1, h_1] \nabla^i [\ell_2, h_2] & \triangleq & [\llparenthesis\, \ell_2 < \ell_1 \,?\, -\infty \,\text{°}\, \ell_1 \,\rrparenthesis, \llparenthesis\, h_2 > h_1 \,?\, \infty \,\text{°}\, h_1 \,\rrparenthesis] \end{array}\right. \quad (33.5)$$

that essentially pushes unstable bounds to infinity. This is an extrapolation

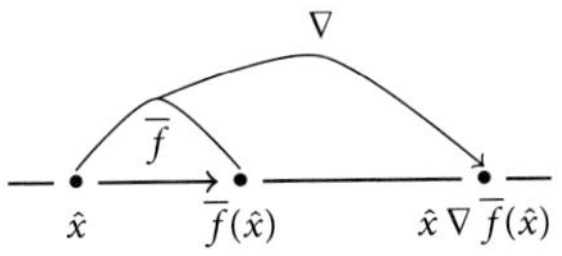

The upward iteration with widening is now

$$
\begin{array}{rcl}
\hat{x}^0 &=& \bot^i \\
\hat{x}^1 &=& \hat{x}^0 \,\nabla^i\, \mathscr{F}^i(\hat{x}^0) \;=\; \hat{x}^0 \,\nabla^i\, ([0,0] \sqcup^i (\hat{x}^0 \oplus^i [1,1])) \;=\; \bot^i \,\nabla^i\, [0,0] \\
&=& [0,0] \qquad \text{because } \mathscr{F}^i(\hat{x}^0) = [0,0] \not\sqsubseteq^i \bot^i = \hat{x}^0 \\
\hat{x}^2 &=& \hat{x}^1 \,\nabla^i\, \mathscr{F}^i(\hat{x}^1) \;=\; \hat{x}^1 \,\nabla^i\, ([0,0] \sqcup^i (\hat{x}^1 \oplus^i [1,1])) \\
&=& [0,0] \,\nabla^i\, ([0,0] \sqcup^i [1,1]) \;=\; [0,0] \,\nabla^i\, [0,1] \;=\; [0,\infty] \\
&& \qquad \text{because } \mathscr{F}^i(\hat{x}^1) = [0,1] \not\sqsubseteq^i \hat{x}^1 = [0,0] \\
\hat{x}^n &=& \hat{x}^2, \qquad n \geqslant 2 \\
&& \text{because } \mathscr{F}^i(\hat{x}^2) = ([0,0] \sqcup^i (\hat{x}^2 \oplus^i [1,1])) \;=\; ([0,0] \sqcup^i ([0,\infty] \oplus^i [1,1])) \\
&& = ([0,0] \sqcup^i [1,\infty+1]) \;=\; [0,\infty] \sqsubseteq^i \hat{x}^2 = [0,\infty]
\end{array}
$$

In general, the result with widening is less precise than the limit of the infinite iterations. An example is

```
P^1001  =  while ℓ1 (x < 1001)
              ℓ2 x = x + 1;
           ℓ3
```

Assuming initially $\overline{\rho}_0(\mathsf{x}) = [0,0]$, $x = X(\ell_1)(\mathsf{x})$, and $y = X(\ell_3)(\mathsf{x})$, the fixpoint computation amounts to solving the fixpoint system of equations

$$
\left\{\begin{array}{rcl} x &=& \mathscr{F}^i(x) \\ y &=& x \sqcap^i [1001,\infty] \end{array}\right. \qquad \text{where} \quad \mathscr{F}^i(x) \;=\; [0,0] \sqcup^i ((x \sqcap^i [-\infty,1000]) \oplus^i [1,1])
$$

Exercise 33.6 Formally derive these equations from (21.11) instantiated for intervals. Show that the least solution is $x = [0,1001]$ and $y = [1001,1001]$.

The upward iterates with widening are now

$$
\begin{array}{rcl}
\hat{x}^0 &=& \bot^i, \quad \hat{y}^0 \;=\; \bot^i \\
\hat{x}^1 &=& \hat{x}^0 \,\nabla^i\, \mathscr{F}^i(\hat{x}^0) \;=\; \hat{x}^0 \,\nabla^i\, ([0,0] \sqcup^i ((\hat{x}^0 \sqcap^i [-\infty,1000]) \oplus^i [1,1])) \\
&=& \bot^i \,\nabla^i\, [0,0] \;=\; [0,0] \qquad \text{because } \mathscr{F}^i(\hat{x}^0) \;=\; [0,0] \not\sqsubseteq^i \bot^i \;=\; \hat{x}^0 \\
\hat{x}^2 &=& \hat{x}^1 \,\nabla^i\, \mathscr{F}^i(\hat{x}^1) \;=\; \hat{x}^1 \,\nabla^i\, ([0,0] \sqcup^i ((\hat{x}^1 \sqcap^i [-\infty,1000]) \oplus^i [1,1])) \\
&=& [0,0] \,\nabla^i\, ([0,0] \sqcup^i [1,1]) \;=\; [0,0] \,\nabla^i\, [0,1] \\
&=& [0,\infty] \qquad \text{because } \mathscr{F}^i(\hat{x}^1) \;=\; [0,1] \not\sqsubseteq^i \hat{x}^1 \;=\; [0,0] \\
\hat{x}^n &=& \hat{x}^2, \qquad n \geqslant 2 \\
&& \text{because } \mathscr{F}^i(\hat{x}^2) \;=\; ([0,0] \sqcup^i ((\hat{x}^2 \sqcap^i [-\infty,1000]) \oplus^i [1,1])) \\
&& \;=\; ([0,0] \sqcup^i (([0,\infty] \sqcap^i [-\infty,1000]) \oplus^i [1,1])) \\
&& \;=\; ([0,0] \sqcup^i [1,1001]) \;=\; [0,1001] \;\sqsubseteq^i\; \hat{x}^2 \;=\; [0,\infty] \\
\hat{y} &=& \hat{x}^2 \sqcap^i [1001,\infty] \;=\; [0,\infty] \sqcap^i [1001,\infty] \;=\; [1001,\infty]
\end{array}
$$

33.6 Interpolation by Narrowing

The solution found is therefore $\hat{x} = [0, \infty]$ and $\hat{y} = [1001, \infty]$, which is frustrating because $\mathscr{F}^i(\hat{x}) = [0, 1001]$ provides a better solution. The reason is that the upward iteration with widening was stopped when $\mathscr{F}^i(\hat{x}) \sqsubseteq^i \hat{x}$ and $\mathscr{F}^i$ was increasing, so that the downward iterates $\check{x} = \hat{x} \sqsupseteq^i \mathscr{F}^i(\check{x}) \sqsupseteq^i \mathscr{F}^i(\mathscr{F}^i(\check{x})) \sqsupseteq^i \ldots$ were all improving the precision of the solution until reaching a fixpoint.

For intervals, the decreasing iterates can be very long and, in general, infinite (e.g., [441]). So we can stop at any point in the downward iterates or use a narrowing to enforce convergence on semantic grounds. An example of interval narrowing is

$$\left[\begin{array}{rcl} \bot^i \Delta^i x & \triangleq & x \,\Delta^i \bot^i \;\triangleq\; \bot^i \\ [\ell_1, h_1]\, \Delta^i\, [\ell_2, h_2] & \triangleq & [(\!| \ell_1 = -\infty \,?\, \ell_2 \,{:}\, \ell_1 |\!), (\!| h_1 = \infty \,?\, h_2 \,{:}\, h_1 |\!)] \end{array}\right. \tag{33.7}$$

which attempts to improve infinite bounds only. This is an interpolation

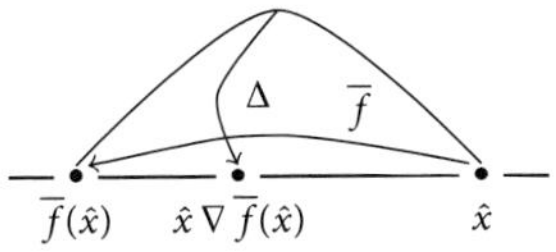

The downward iterates with narrowing are now

$$\begin{array}{rcl} \check{x}^0 & = & \hat{x} \;=\; [0, \infty], \quad \check{y} \;=\; \hat{y} \;=\; [1001, \infty] \\ \check{x}^1 & = & \check{x}^0 \,\Delta^i\, \mathscr{F}^i(\check{x}^0) \;=\; \check{x}^0 \,\Delta^i\, ([0,0] \sqcup^i ((\check{x}^0 \sqcap^i [-\infty, 1000]) \oplus^i [1,1])) \\ & = & [0, \infty]\, \Delta^i\, [0, 1001] \;=\; [0, 1001] \\ & & \qquad \text{because } \mathscr{F}^i(\check{x}^0) \;=\; [0, 1001] \;\neq\; [0, \infty] \;=\; \check{x}^0 \\ \check{x}^n & = & \check{x}^1, \qquad n \geqslant 1 \\ & & \text{because } \mathscr{F}^i(\check{x}^1) \;=\; ([0,0] \sqcup^i ((\check{x}^2 \sqcap^i [-\infty, 1000]) \oplus^i [1,1])) \\ & & \qquad = \; ([0,0] \sqcup^i (([0, 1001] \sqcap^i [-\infty, 1000]) \oplus^i [1,1])) \\ & & \qquad = \; ([0,0] \sqcup^i [1, 1001]) \;=\; [0, 1001] \;=\; \check{x}^1 \\ \check{y} & = & \check{x}^1 \sqcap^i [1001, \infty] \;=\; [0, 1001] \sqcap^i [1001, \infty] \;=\; [1001, 1001]. \end{array}$$

Of course, no narrowing can in general recover the information lost by the widening. An example is

```
P3  =  ℓ1 x = 0;
       while ℓ2 (x != 1001)
         ℓ3 x = x + 2;
       ℓ4
```

The widening can always be improved; see section 34.5.

The narrowing cannot recover more information than that lost by the interval abstraction of section 33.2. For example, the interval analysis of the program

$$P_4 \quad = \quad \texttt{while}^{\ell_1} (\texttt{x} < \texttt{1001}) \\ \quad {}^{\ell_2} \texttt{x} = \texttt{x} + \texttt{2}; \\ {}^{\ell_3}$$

will yield $y = [1001, 1002]$ because parity is lost by intervals (but can be recovered by a reduced product with the parity analysis or a arithmetic congruence analysis as shown in chapter 36).

Exercise 33.8 (Project: abstract interpreter for infinitary abstract domains) Continuing exercises 28.46, 28.49, 32.10, and 33.9, modify the abstract interpreter to include the widening and narrowing to enforce convergence (including for the local iterations of exercise 29.4). Experiment with the interval analysis. For example

```
$ echo "x=1;while (x<1000) x=x+2;" | ./main
l1: [x:[0, 0]] x = 1;
while l2: (x < 1000) [x:[1, 1001]]
   l3: [x:[1, 999]] x = (x + 2);
l4: [x:[1000, 1001]]
```

33.7 Interval Test Reduction

The interval abstract domain has no infinite strictly decreasing chains, but they can be very long. Therefore test reduction of section 29.2 can be time consuming without narrowing. For example the test `(x<y && y<x)` will reduce the bounds of `x` and `y` by 1 at each iteration until reaching $\bot^i$. Assuming that variables are initialized to $\top^i$ = [`min_int`, `max_int`], the analysis of the program `if ((x<y nand y<x) nand (x<y nand y<x)) r=1; else r=3;` has the following test reduction iteration:

```
[r:T; x:T; y:T]
[r:T; x:[-4611686018427387903, 4611686018427387902]; y:[-4611686018427387903,
                                                        4611686018427387902]]
[r:T; x:[-4611686018427387902, 4611686018427387901]; y:[-4611686018427387902,
                                                        4611686018427387901]]
[r:T; x:[-4611686018427387901, 4611686018427387900]; y:[-4611686018427387901,
                                                        4611686018427387900]]
[r:T; x:[-4611686018427387900, 4611686018427387899]; y:[-4611686018427387900,
                                                        4611686018427387899]]
...
[r:T; x:[-3, 3]; y:[-3, 3]]
[r:T; x:[-2, 2]; y:[-2, 2]]
[r:T; x:[-1, 1]; y:[-1, 1]]
[r:T; x:[0, 0]; y:[0, 0]]
[r:_|_; x:_|_; y:_|_]
```

and so the test is always false.

```
if l1: (((x < y) nand (y < x)) nand ((x < y) nand (y < x))) [r:T; x:T; y:T]
   l2: [r:_|_; x:_|_; y:_|_] r = 1;
 else
   l3: [r:T; x:T; y:T] r = 3;
l4: [r:[3, 3]; x:T; y:T]
```

With a narrowing, the iterations are much faster, but the result less precise.

```
r:T; x:T; y:T]
[r:T; x:[-4611686018427387903, 4611686018427387902]; y:[-4611686018427387903,
                                                        4611686018427387902]]
```

but the test is not found to be always false and so the analysis is less precise.

```
if l1: (((x < y) nand (y < x)) nand ((x < y) nand (y < x))) [r:T; x:T; y:T]
```

```
    l2: [r:T; x:[-4611686018427387903, 4611686018427387902]; y:[-4611686018427387903,
                                                          4611686018427387902]] r = 1;
  else
    l3: [r:T; x:T; y:T] r = 3;
l4: [r:[1, 3]; x:T; y:T]
```

Because the narrowing (33.7) only improves infinite bounds, a few iterations up to a threshold will be allowed without narrowing to improve finite bounds before narrowing the infinite ones.

Exercise 33.9 (Project: interval analysis) Continuing exercise 28.49, implement the Cartesian interval abstract domain.

Exercise 33.10 Using the interval abstract interpreter of exercise 33.8 with the interval abstract domain of exercise 33.9, prove that the following skeleton of an iterative factorial `i = n; while (i!=1) i = i-1;` does not terminate when initially $\mathsf{n} < 1$.

33.8 Modular Interval Analysis

In practice, machines have bounded integers between `min_int` and `max_int`. In case of overflow outside these bounds, the program semantics may specify that execution stops and goes on only when all operations provide results between these bounds. This is easy to take into account (provided the abstract operations do not themselves involve overflows) using $-\infty = \mathtt{min\_int}$ and $\infty = \mathtt{max\_int}$ for the widening/narrowing. This is also easy for floating-point numbers, rounding toward $-\infty$ for the lower bound and toward $+\infty$ for the upper bound calculation of intervals [124, 268]. A precise nonrelational abstract domain for floating-point operations is made by splitting into positive values, negative values, zeroes, infinities, and NaNs [387].

Another option is modular arithmetic, making it possible for computations to "wrap around" (e.g., $\mathtt{min\_int} - 1 = \mathtt{max\_int}$) which must be handled soundly in the interval abstract domain [389, 732, 884, 891].

33.9 Constancy Abstraction

Continuing section 28.8.3, observe that the constancy abstract domain of exercise 3.45 satisfies the ascending chain condition. Hence if $\mathbb{V}$ is finite in section 28.8.3, the Cartesian constancy abstract domain $\mathbb{V} \rightarrow \mathbb{P}^c$ also satisfies the ascending chain condition. This is not the case if $\mathbb{V} = \{\mathsf{x}_i \mid i \in \mathbb{N}\}$ is infinite, because we have the infinite strictly increasing chain $\overline{\rho}_0(\mathsf{x}_i) = \bot \mathrel{\dot{\sqsubsetneq}^c} \overline{\rho}_1 = \overline{\rho}_0[\mathsf{x}_1 \leftarrow 1] \mathrel{\dot{\sqsubsetneq}^c} \overline{\rho}_2 = \overline{\rho}_1[\mathsf{x}_2 \leftarrow 2] \mathrel{\dot{\sqsubsetneq}^c} \ldots$. This means that $\mathbb{V}$ must be chosen as the finite set of variables appearing in the program, in which case no extrapolation/interpolation is necessary.[1] Symbolic constancy analysis [452, 453, 969]) involves a non-Noetherian abstract domain and requires a widening.

1. This is the confusion made by [474] in claiming that the abstract domain is infinite but requires no widening thanks to a constraint resolution algorithm. In fact, the abstract domain is finite for a given program and it becomes infinite only when considering all possible programs. Moreover, for a given program, the constraint resolution algorithm is a least fixpoint iteration in an abstract domain satisfying the ascending condition [251], so without surprise, no widening is necessary [251].

33.10 Zonotopic Abstraction

In the same way that zonotopes have been used to improve the precision of dynamic interval arithmetics in scientific computing (see section 32.7), they have also been used to improve the precision of static interval reachability analysis of programs [411, 429, 433] and hybrid systems [432, 593].

33.11 Conclusion

Intervals were the first non-Noetherian abstract domain thus requiring iteration extrapolation operators for convergence acceleration [234, 235]. Some programming languages such as Pascal [986] and its followers have interval types. The compiler uses a simple type inference with an integer base type and the bounds are checked at runtime. Most of these checks can be eliminated at compile-time by a static interval analysis. The same observation applies to runtime array-bound checking.

Although the basic abstract operations are the same, static interval analysis should not be confused with the runtime interval analysis of chapter 32 used in numerical analysis to cope with rounding errors of floats [715]. Of course, this dynamic/runtime interval arithmetic abstract interpretation of program executions can be overapproximated by a static/compile-time one to statically analyze programs manipulating floating-point numbers [428, 525, 659].

An interval analysis is used in [482] to detect unseen inputs of neural networks as novel to avoid missclassification. The interval abstract domain is used in [58, 963] to show that for any function f, there exists a neural network whose abstract interpretation is an arbitrarily close approximation of the collecting semantics of f.

In addition to the zonotopes of section 33.10, various refinements of interval analysis have been studied, such as symbolic intervals [745, 853]; interval congruences [664]; anti-intervals (to close the domain by complement with $[-\infty, \infty]$, as used by the PolySpace Verifier [317]); by lists of intervals such as $\langle[-\infty, 1000], [1002, \infty]\rangle$ or boxes [451] (union of disjoint intervals encoded by a propositional formula over interval constraints and represented by Linear Decision Diagrams [169], an extension of Binary Decision Diagrams [150]); by symbolic bounds [992]; and more generally by relational domains. The relative precision of different numerical abstract domains with respect to the interval domain is studied in [37].

X EFFECTIVE FIXPOINT APPROXIMATION

A static analyzer is an instance of the abstract interpreter that automatically infers in inductive program property in a given abstract domain $\mathbb{D}^{\sharp}$. The main challenge is to compute the limit of iterates for fixpoints.

Convergence acceleration techniques ensures that can be done in finite time up to a loss of precision but not a loss of soundness (chapter 34). In checking a specification, this specification can be used to improve the precision of the fixpoint approximation (chapter 35).

The *reduced product* is a useful technique to decompose the design of the abstract domain $\mathbb{D}^{\sharp}$ into parts to get a modular and extensible abstract interpreter (chapter 36).

34 Fixpoint Approximation by Extrapolation and Interpolation

The fixpoint approximation by extrapolation with widening and interpolation by narrowing that we illustrated in chapter 33 for intervals can be generalized to any analysis in a non-Noetherian abstract domain (i.e., not satisfying the ascending (dually descending) chain condition (ACC/DCC) that every infinite increasing (decreasing) chain is finite).

Contents

34.1 Upward Iteration Convergence Acceleration by Overapproximating Extrapolation with Widening

Let us extend the naturals $\langle \mathbb{N}, \leqslant \rangle$ into $\langle \mathbb{N}_\omega, \leqslant \rangle$ with an infinite supremum ω such that $\mathbb{N}_\omega \triangleq \mathbb{N} \cup \{\omega\}$ and $\forall n \in \mathbb{N} \,.\, n < \omega \leqslant \omega$.

Definition 34.1 (iteration) The iterates of $f \in L \to L$ from $a \in L$ on a poset $\langle L, \sqsubseteq, \sqcup \rangle$ are $\langle f^n, n \in \mathbb{N}_\omega \rangle$ such that

1. $f^0 \triangleq a$
2. $f^{n+1} \triangleq f(f^n), n \in \mathbb{N}$
3. $f^\omega \triangleq \bigsqcup_{n<\omega} f^n$

(where the least upper bound $\sqcup$ in $f^\omega \triangleq \bigsqcup\{f^n \mid n < \omega\}$ is assumed to exist in L, for example, for CPOs when the iterates are increasing).

Definition 34.2 (convergence) The iterates $\langle f^n, n \in \mathbb{N}_\omega \rangle$ of $f \in L \to L$ from $a \in L$ *converge* to the limit $\ell \in L$ at $\epsilon \in \mathbb{N}$ if and only if $\forall n \geqslant \epsilon \,.\, f^n = \ell$ (the iterates are then said to be *ultimately stationary* at ϵ).

In general, the iterates do not converge and we must compute or overapproximate the limit f^ω. Observe that if L is a logic with quantifiers and the iterates f^n can be expressed as logical formulas depending upon $n \in \mathbb{N}$, then computing the limit $f^\omega \triangleq \exists n \,.\, f^n$ amounts to the elimination of the quantifier $\exists$ [710]. This may be feasible. For example, the Tarski-Seidenberg theorem [873, 924] states that for every first-order formula over the real field there exists an equivalent quantifier-free formula. However, the algorithm to compute this quantifier-free formula is very costly [201].

In the absence of an algorithm to compute the limit explicitly, we can enforce or speed up convergence using a widening in an abstract domain.

Definition 34.3 (widening) Let $\langle L, \sqsubseteq \rangle$ and $\langle \overline{L}, \overline{\sqsubseteq} \rangle$ be concrete and abstract posets with increasing concretization $\gamma \in \overline{L} \xrightarrow{\nearrow} L$.

- A *limit widening* $\boldsymbol{\nabla} \in \wp(\overline{L}) \to \overline{L}$ is *sound* if and only if it is an abstract upper bound, that is, $\forall X \in \wp(\overline{L}) \,.\, \forall \overline{x} \in X \,.\, \overline{x} \,\overline{\sqsubseteq}\, \boldsymbol{\nabla} X$.
- A *successor widening* $\triangledown \in \overline{L} \times \overline{L} \to \overline{L}$ is *sound* if and only if $\forall \overline{x}, \overline{y} \in \overline{L} \,.\, \overline{x} \,\overline{\sqsubseteq}\, \overline{x} \triangledown \overline{y}$ and $\overline{y} \,\overline{\sqsubseteq}\, \overline{x} \triangledown \overline{y}$.

Definition 34.4 (iteration with widening) The *upward iterates* of $\overline{f} \in \overline{L} \to \overline{L}$ from $a \in \overline{L}$ with widening $\boldsymbol{\nabla}$ on a poset $\langle \overline{L}, \overline{\sqsubseteq} \rangle$ are $\langle \hat{f}^n, n \in \mathbb{N}_\omega \rangle$ such that

1. $\hat{f}^0 \triangleq a$;
2. $\hat{f}^{n+1} \triangleq (\!| \overline{f}(\hat{f}^n) \,\overline{\sqsubseteq}\, \hat{f}^n \,?\, \hat{f}^n \,\text{\textcolon}\, \hat{f}^n \triangledown \overline{f}(\hat{f}^n) |\!)$ when $n \in \mathbb{N}$; and
3. $\hat{f}^\omega \triangleq \boldsymbol{\nabla}_{n<\omega} \hat{f}^n$.

The successor widening $\hat{f}^n \triangledown \overline{f}(\hat{f}^n)$ depends on the previous iterate $\hat{f}^n$ to extrapolate the next iterate $\overline{f}(\hat{f}^n)$. More generally, it might be made more precise by depending on all previous iterates, as in $\langle \hat{f}^\delta, \delta \leqslant n \rangle \triangledown \overline{f}(\hat{f}^n)$.

Definition 34.5 (terminating widening) A successor widening $\triangledown \in \overline{L} \times \overline{L} \to \overline{L}$ on a poset $\langle \overline{L}, \overline{\sqsubseteq} \rangle$ is *terminating* if and only if for any increasing sequence $\langle x^n, n \in \mathbb{N} \rangle$ of elements of $\overline{L}$ and any arbitrary sequence $\langle y^n, n \in \mathbb{N} \rangle$ and such that $\forall n \in \mathbb{N} \,.\, x^{n+1} = x^n \triangledown y^n$, the sequence $\langle x^n, n \in \mathbb{N} \rangle$ is converging, that is, $\exists \epsilon \in \mathbb{N} \,.\, x^{\epsilon+1} = x^\epsilon$.

In this definition $\langle x^n, n \in \mathbb{N} \rangle$ stands for the increasing iterates $\langle \hat{f}^n, n \in \mathbb{N} \rangle$ of $\overline{f}$ and $\langle y^n, n \in \mathbb{N} \rangle$ stands for their image $\langle \overline{f}(\hat{f}^n), n \in \mathbb{N} \rangle$ by $\overline{f}$, which is not assumed to be increasing, so that $\langle y^n, n \in \mathbb{N} \rangle$ can be an arbitrary overapproximation of $\langle x^n, n \in \mathbb{N} \rangle$. Of course, the hypotheses of the definition can be restricted to the case of an increasing chain.

Theorem 34.6 (upward iteration with [terminating] widening) Let $f \in L \xrightarrow{uc} L$ be an upper continuous function on a CPO $\langle L, \sqsubseteq, \bot \rangle$ with infimum $\bot$. Let $\langle f^n, n \in \mathbb{N}_\omega \rangle$ be the iterates of f from $\bot$.

Let $\langle \overline{L}, \overline{\sqsubseteq} \rangle$ be an abstract domain with increasing concretization $\gamma \in \overline{L} \xrightarrow{\nearrow} L$.

Let $\overline{f} \in \overline{L} \to \overline{L}$ be such that $f \circ \gamma \,\dot{\sqsubseteq}\, \gamma \circ \overline{f}$ (semicommutation).

Let $\triangledown$ be a widening on $\overline{L}$.

Then

- The iterates $\langle \hat{f}^n, n \in \mathbb{N}_\omega \rangle$ of $\overline{f}$ from any $a \in \overline{L}$ with widening $\triangledown$ are increasing.
- There exists $\Delta \in \mathbb{N}_\omega$ such that $\mathsf{lfp}^{\sqsubseteq} f \sqsubseteq \gamma(\hat{f}^\Delta)$.
- if $\triangledown$ is terminating then $\Delta \in \mathbb{N}$ and $\overline{f}(\hat{f}^\Delta) \,\overline{\sqsubseteq}\, \hat{f}^\Delta$.

Note that $\langle \overline{L}, \overline{\sqsubseteq} \rangle$ is *not* assumed to have lubs (even for increasing chains) and $\overline{f}$ is *not* assumed to be increasing.

Proof of theorem 34.6 Let $\langle f^n, n \in \mathbb{N}_\omega \rangle$ be the iterates of f from $\bot$ in definition 34.1 and $\langle \hat{f}^n, n \in \mathbb{N}_\omega \rangle$ be the iterates of $\overline{f}$ from $\overline{a} \in \overline{L}$ with widening $\triangledown$ in definition 34.4.

(1) The abstract iterates overapproximate the concrete iterates.

- $f^0 = \bot \sqsubseteq \gamma(\overline{a}) = \gamma(\hat{f}^0)$ by definition of the concrete and abstract iterates and $\bot$ is the infimum of L;
- For a positive natural $n+1$, assume, by induction hypothesis, that $f^n \sqsubseteq \gamma(\hat{f}^n)$.
  - If $\overline{f}(\hat{f}^n) \,\overline{\sqsubseteq}\, \hat{f}^n$ then, $f^{n+1} = f(f^n) \sqsubseteq f(\gamma(\hat{f}^n)) \sqsubseteq \gamma(\overline{f}(\hat{f}^n)) \sqsubseteq \gamma(\hat{f}^n) = \gamma(\hat{f}^{n+1})$ by definition of the iterates, f is upper continuous hence increasing, induction hypothesis, $\overline{f}$ and f semicommute, $\overline{f}(\hat{f}^n) \,\overline{\sqsubseteq}\, \hat{f}^n$, and γ is increasing.
  - Otherwise, $\overline{f}(\hat{f}^n) \,\overline{\not\sqsubseteq}\, \hat{f}^n$, in which case $f^{n+1} = f(f^n) \sqsubseteq f(\gamma(\hat{f}^n)) \sqsubseteq \gamma(\overline{f}(\hat{f}^n)) \sqsubseteq \gamma(\hat{f}^n \triangledown \overline{f}(\hat{f}^n)) = \gamma(\hat{f}^{n+1})$ by definition of the iterates, f is upper continuous hence increasing, induction hypothesis, $\overline{f}$ and f semicommute, soundness of the successor widening, and γ is increasing.
- For the limit ω, assume by induction hypothesis, that $\forall n < \omega \,.\, f^n \sqsubseteq \gamma(\hat{f}^n)$. By soundness of the limit widening and definition 34.3, $f^n \sqsubseteq \gamma(\hat{f}^n) \sqsubseteq \gamma(\underset{n<\omega}{\triangledown} \hat{f}^n)$. By definition of the iterates and of the least upper bound $\bigsqcup\{f^n \mid n < \omega\}$, which is assumed to exist in L, $f^\omega = \bigsqcup_{n<\omega} f^n \sqsubseteq \gamma(\underset{n<\omega}{\triangledown} \hat{f}^n) = \gamma(\hat{f}^\omega)$.

By induction, we conclude that $\forall n \in \mathbb{N}_\omega \,.\, f^n \sqsubseteq \gamma(\hat{f}^n)$.

(2) The abstract iterates are increasing.

The abstract iterates $\langle \hat{f}^n, n \in \mathbb{N}_\omega \rangle$ form an increasing chain. For positive naturals, either $\overline{f}(\hat{f}^n) \mathrel{\overline{\sqsubseteq}} \hat{f}^n$ and the following iterates are stationary or else $\hat{f}^{n+1} = \hat{f}^n \mathbin{\nabla} \overline{f}(\hat{f}^n)$ so $\hat{f}^n \mathrel{\overline{\sqsubseteq}} \hat{f}^{n+1}$ by soundness of the widening in definition 34.3. For the limit, $\hat{f}^\omega \triangleq \mathop{\nabla}\limits_{n<\omega} \hat{f}^n \mathrel{\overline{\sqsupseteq}} \hat{f}^n$ for all $n < \omega$.

(3) Fixpoint overapproximation.

By Scott–Kleene's iterative fixpoint theorem 15.26, $\langle f^n, n \in \mathbb{N}_\omega \rangle$ is an increasing chain that has a limit $f^\omega = \bigsqcup_{n \in \mathbb{N}_\omega} f^n = \mathsf{lfp}^{\sqsubseteq} f$. By (1), $f^\omega \sqsubseteq \gamma(\hat{f}^\omega)$ proving $\exists \Delta \in \mathbb{N}_\omega \,.\, \mathsf{lfp}^{\sqsubseteq} f \sqsubseteq \gamma(\hat{f}^\Delta)$ for $\Delta = \omega$.

(4) Terminating widenings enforce convergence.

By (2) in this theorem, the abstract iterates $\langle \hat{f}^n, n \in \mathbb{N} \rangle$ form an increasing chain. Assume, by reductio ad absurdum that $\forall n \in \mathbb{N} \,.\, \overline{f}(\hat{f}^n) \mathrel{\overline{\not\sqsubseteq}} \hat{f}^n$. Let $x^n = \hat{f}^n$ and $y^n = \overline{f}(\hat{f}^n)$ in definition 34.5 so that $\langle x^n, n \in \mathbb{N} \rangle$ is increasing and $\forall n \in \mathbb{N} \,.\, x^{n+1} = x^n \mathbin{\nabla} y^n$. By definition 34.5, $\exists \epsilon \in \mathbb{N} \,.\, x^{\epsilon+1} = x^\epsilon$, that is, $\hat{f}^{\epsilon+1} = \hat{f}^\epsilon$. By definition 34.4.2 of the iterates either $\overline{f}(\hat{f}^\epsilon) \mathrel{\overline{\sqsubseteq}} \hat{f}^\epsilon$ or else $\hat{f}^{\epsilon+1} = \hat{f}^\epsilon \mathbin{\nabla} \overline{f}(\hat{f}^\epsilon)$. By definition 34.3, $\overline{f}(\hat{f}^\epsilon) \mathrel{\overline{\sqsubseteq}} \hat{f}^\epsilon \mathbin{\nabla} \overline{f}(\hat{f}^\epsilon) = \hat{f}^{\epsilon+1} = \hat{f}^\epsilon$ proving that, in both cases, $\overline{f}(\hat{f}^\epsilon) \mathrel{\overline{\sqsubseteq}} \hat{f}^\epsilon$.

(5) The limit of the converging abstract iterates overapproximates all concrete iterates. The proof is by induction.

- $f^0 = \bot \sqsubseteq \gamma(\hat{f}^\epsilon)$ by definition of the iterates and of the infimum;
- If $f^n \sqsubseteq \gamma(\hat{f}^\epsilon)$ by induction hypothesis, then $f^{n+1} = f(f^n) \sqsubseteq f(\gamma(\hat{f}^\epsilon)) \sqsubseteq \gamma(\overline{f}(\hat{f}^\epsilon)) = \gamma(\hat{f}^\epsilon)$;
- For the limit ω, for all $n < \omega$, $f^n \sqsubseteq \gamma(\hat{f}^\epsilon)$ then $f^\omega = \bigsqcup_{n \in \mathbb{N}_\omega} f^n = \mathsf{lfp}^{\sqsubseteq} f \sqsubseteq \gamma(\hat{f}^\epsilon)$ by existence hypothesis and definition of the lub $\bigsqcup$ in L. It follows that we can choose $\Delta = \epsilon$.

□

Note that the proof of theorem 34.6 remains valid when weakening the hypothesis $\forall \overline{x}, \overline{y} \in \overline{L} \,.\, \overline{y} \mathrel{\overline{\sqsubseteq}} \overline{x} \mathbin{\nabla} \overline{y}$ to $\forall \overline{x}, \overline{y} \in \overline{L} \,.\, \gamma(\overline{y}) \subseteq \gamma(\overline{x} \mathbin{\nabla} \overline{y})$.

Exercise 34.7 Consider the interval widenings $x \mathbin{\nabla_1} y \triangleq [-\infty, \infty]$, $[\ell_1, h_1] \mathbin{\nabla_2} [\ell_2, h_2] \triangleq [(\!|\, \ell_2 \neq \ell_1 \;?\; -\infty \;\text{:}\; \ell_1 \,|\!), (\!|\, h_2 \neq h_1 \;?\; \infty \;\text{:}\; h_1 \,|\!)]$, and the widening ∇^i of (33.5). Prove that ∇_1, ∇_2, and ∇^i are terminating widenings. Prove that the upward iteration of theorem 34.6 with widening ∇_1 is less precise but shorter than the upward iteration, with widening ∇_2 itself less precise but shorter than the upward iteration with widening ∇^i.

34.2 Widening Increasing in Their First Parameter Cannot Enforce Termination

The widening for intervals is not increasing in its first parameter. For example, $[0, 0] \mathbin{\nabla^i} [0, 1] = [0, \infty]$, $[0, 0] \sqsubseteq^i [0, 1]$, $[0, 1] \mathbin{\nabla^i} [0, 1] = [0, 1]$, and $[0, \infty] \not\sqsubseteq^i [0, 1]$. This is a general property of terminating widenings.

Theorem 34.8 (terminating widenings are not increasing) Assume the hypotheses of theorem 34.6 in an abstract domain $\langle \overline{L}, \mathrel{\overline{\sqsubseteq}} \rangle$ where $\overline{f}$ has infinite non-converging increasing iterations $\langle \hat{f}^n, n \in \mathbb{N}_\omega \rangle$. Define $x \mathbin{\nabla'} y \triangleq (\!|\, y \mathrel{\overline{\sqsubseteq}} x \;?\; x \;\text{:}\; x \mathbin{\nabla} y \,|\!)$

so that definition 34.4.2 is $\hat{f}^{n+1} \triangleq \hat{f}^n \triangledown' \overline{f}(\hat{f}^n)$. Then $\triangledown'$ is a widening and $\triangledown$ cannot enforce convergence if $\triangledown'$ is $\overline{\sqsubseteq}$-increasing in its first parameter.

Proof $\triangledown'$ is a widening because
(a) $x \overline{\sqsubseteq} x \triangledown' y$ because either $y \overline{\sqsubseteq} x = x \triangledown' y$ or $x \overline{\sqsubseteq} x \triangledown y = x \triangledown' y$ by definition 34.3.
(b) $y \overline{\sqsubseteq} x \triangledown' y$ because either $y \overline{\sqsubseteq} x = x \triangledown' y$ or $y \overline{\sqsubseteq} x \triangledown y = x \triangledown' y$ by definition 34.3.
Moreover,
(c) $y \overline{\sqsubseteq} x \Rightarrow x \triangledown' y = x$ by definition of $\triangledown'$. In particular $y \overline{\sqsubseteq} y$ so $y \triangledown' y = y$.

Assume that $\triangledown'$ is increasing in its first parameter. Then $x \overline{\sqsubseteq} y$ implies that $y \overline{\sqsubseteq} x \triangledown' y \overline{\sqsubseteq} y \triangledown' y = y$ by (b) and (c) so $x \triangledown' y = y$ by antisymmetry.

The contradiction is that definition 34.4.2 is $\hat{f}^{n+1} \triangleq \hat{f}^n \triangledown' \overline{f}(\hat{f}^n) = \overline{f}(\hat{f}^n)$ and so if $\overline{f}$ has infinite nonconverging iterations, that is, such that $\forall \ell \in \mathbb{N} . \overline{f}(\hat{f}^\ell) \not\overline{\sqsubseteq} \hat{f}^\ell$, the widening $\triangledown'$ yields the same iterates and hence $\triangledown$ does not enforce convergence. □

Exercise 34.9 Find an example of transformer and terminating widening such that by making the transformer less precise, the iteration of theorem 34.6 with the same widening is more precise.

Exercise 34.10 Find an example of transformer and terminating widening such that by making the widening less precise or incomparable the iteration of theorem 34.6 with the same transformer is more precise.

34.3 Non-increasing Abstract Iteration Transformers

When considering nested loops, the transformer $\mathscr{F}^{\alpha}[\![\texttt{while}\ \ell\ (\texttt{B})\ \texttt{S}_b]\!]$ in (21.11) of the outer loop does depend on the approximation of the fixpoint semantics of the inner loop using theorem 34.6, which, because the widening is not increasing, may itself be non-increasing. Therefore in theorem 34.6, the abstract transformer $\overline{f}$ cannot be assumed to be increasing. Soundness relies only on the increasingness of the concrete transformer f, which is always the case because, intuitively, more possible interactions with the execution environments yield more possible program executions.

34.4 Comparing the Precision of Static Analyzes Using Widenings

More precise abstractions may yield less precise analyses. For example, consider the program

```
P5  =  ℓ1 x = 1;
       while ℓ2 (tt)
          ℓ3 x = 0;
       ℓ4
```

The sign analysis yields $\texttt{x} \geqslant 0$ whereas interval analysis yields $\texttt{x} \in [-\infty, 1]$. This is because of the widening $[1, 1] \triangledown^i [0, 1] = [-\infty, 1]$. Therefore the interval abstraction α_i in section 33.2 is more precise than the sign abstraction in section 3.12, but the interval analysis is not always more precise than the sign analysis.

To ensure that the interval analysis is more precise than the sign analysis we must ensure that the interval widening is more precise than the sign join, that is, $\forall x, y \in$

$\mathbb{P}^i . x \nabla^i y \sqsubseteq^i \alpha_i(\gamma_\pm(\alpha_\pm(\gamma_i(x)) \sqcup_\pm \alpha_\pm(\gamma_i(y))))$ where $\sqcup_\pm$ is the join of signs defined in section 3.12. (The interval lattice operations will all be more precise than the sign lattice operations by virtue of the Galois connections $\alpha_\pm$ and α_i, but all operations $1^{\mathtt{¤}}, \ominus^{\mathtt{¤}}, \ominus^{\mathtt{¤}}_{-1}$, $\oslash^{\mathtt{¤}}_{-1}, \overline{\oslash}^{\mathtt{¤}}_{-1}$ must also be more or equally precise).

34.5 Terminating Widenings Can Be Refined Forever

Fortunately, the widening can be improved to satisfy this requirement by introducing thresholds $\langle -\infty, -1, 0, 1, \infty \rangle$. The thresholds T can be any strictly increasing sequence of elements of $\mathbb{Z}$ bounded by $-\infty$ and ∞. The widening with thresholds T is then

$$\begin{aligned} \bot^i \nabla^i_T x &\triangleq x \nabla^i_T \bot^i \triangleq x \qquad (34.11) \\ [\ell_1, h_1] \nabla^i_T [\ell_2, h_2] &\triangleq [(\!| \ell_2 < \ell_1 \mathbin{?} \max\{t \in T \mid t \leqslant \ell_2\} \mathbin{:} \ell_1 |\!), \\ &\qquad (\!| h_2 > h_1 \mathbin{?} \min\{t \in T \mid t \geqslant h_2\} \mathbin{:} h_1 |\!)] \end{aligned}$$

(33.5) was for $\langle -\infty, \infty \rangle$. By adding more thresholds (and, more generally, more patterns of abstract properties to widen to), the precision of the analysis can be strictly improved as it will be strictly more precise for at least one program. [123, 135, 267, 268, 459, 573, 608, 682] propose methods to choose widening thresholds.

Exercise 34.12 Consider the analysis of program $\mathtt{P}^{1001}$ = `while (x < 1001) x = x + 1;` with widening in section 33.5 but this time with thresholds $\langle -\infty, 1001, \infty \rangle$. Show that the solution is $\hat{x} = [0, 1001]$ and $\hat{y} = [1001, 1001]$, as would give iterations without widening.

34.6 Delayed Widening

In theorem 34.6 the iteration with terminating widening applies the same widening at each iteration step. More generally, the widening ∇^i_n may depend on the iteration step n. For example, the widening may become coarser over time n. An example is delayed widening [124], which is the join $\sqcup$ for the k iterates after entering the loop, or after a new path has been explored in a loop [124, 268]. The further iterates use ∇ to enforce termination.

34.7 History Widening

The delayed widening is a special case of history widening in which the widening depends on the previous iterates. Therefore definition 34.4.2 is changed to

$$\hat{f}^{n+1} \triangleq (\!| \overline{f}(\hat{f}^n) \mathrel{\overline{\sqsubseteq}} \hat{f}^n \mathbin{?} \hat{f}^n \mathbin{:} \bar{\nabla}(\langle \hat{f}^\delta, \delta \leqslant n \rangle, \langle \overline{f}(\hat{f}^\delta), \delta \leqslant n \rangle) |\!) \text{ when } n \in \mathbb{N}.$$

where $\bar{\nabla}$ is a history widening. The successor widening is when $\bar{\nabla}(\langle x^\delta, \delta \leqslant n \rangle, \langle y^\delta, \delta \leqslant n \rangle) = x^n \nabla y^n$ so that a sequence of previous iterates is abstracted by its last element. A delayed widening would have $\bar{\nabla}(\langle x^\delta, \delta \leqslant n \rangle, \langle y^\delta, \delta \leqslant n \rangle) = (\!| n < n \mathbin{?} x^n \sqcup y^n \mathbin{:} x^n \nabla y^n |\!)$. More generally, the widening depends on an abstraction of the past iterates (which best can be computed incrementally at each iteration step). An example of history widening is given in section 40.1.8 of the form $\bar{\nabla}(\langle x^\delta, \delta \leqslant n \rangle, \langle y^\delta, \delta \leqslant n \rangle) =$

$(x^n \triangledown y^n) \sqcup \bigsqcup_{\delta<n}(x^\delta \triangledown y^\delta)$ where $\bigsqcup_{\delta<n}(x^\delta \triangledown y^\delta)$ records the effect of previous widenings to make sure the next one will not undo their effects. Other examples of history widening are [896, 897].

Exercise 34.13 (Project: history-based widening) Modify the abstract interpreter of exercise 33.8 to use a widening of the form $\triangledown(h, x, y) = (h', r)$ where h is an abstraction of the past history of widenings, x is the previous iterate, y is the transformation of x by one more iteration, r is the result of widening x by y, and h' is h updated by this last widening.

Instantiate the abstract interpreter for interval analysis with delayed widening/narrowing (so that h is a number of remaining iterations without widening/narrowing).

Provide an example in which the history widening is more precise than the successor widening.

34.8 Downward Iteration Convergence Acceleration by Overapproximating Interpolation with Narrowing

Definition 34.14 (narrowing) Let $\langle \overline{L}, \overline{\sqsubseteq}\rangle$ be a poset. A successor narrowing $\Delta \in \overline{L} \times \overline{L} \to \overline{L}$ is *sound* if and only if $\forall \overline{x}, \overline{y} \in \overline{L}\,.\,(\overline{y} \mathrel{\overline{\sqsubseteq}} \overline{x}) \Rightarrow (\overline{y} \mathrel{\overline{\sqsubseteq}} \overline{x} \,\Delta\, \overline{y} \mathrel{\overline{\sqsubseteq}} \overline{x})$.

Definition 34.15 (downward iteration with narrowing) The iterates of $\overline{f} \in \overline{L} \to \overline{L}$ from $a \in \overline{L}$ with narrowing Δ on a poset $\langle \overline{L}, \overline{\sqsubseteq}\rangle$ are $\langle \check{f}^{\,n}, n \in \mathbb{N}\rangle$ such that $(x \mathrel{\overline{\sqsubset}} y \triangleq x \mathrel{\overline{\sqsubseteq}} y \wedge x \neq y)$.

1. $\check{f}^{0} \triangleq a$; and
2. $\check{f}^{\,n+1} \triangleq (\!| \overline{f}(\check{f}^{\,n}) \mathrel{\overline{\sqsubset}} \check{f}^{\,n} \mathbin{?} \check{f}^{\,n} \,\Delta\, \overline{f}(\check{f}^{\,n}) \mathbin{⸰} \check{f}^{\,n} |\!)$.

Theorem 34.16 (downward iteration with narrowing) Let $f \in L \xrightarrow{uc} L$ be an upper continuous function on a CPO $\langle L, \sqsubseteq, \bot\rangle$ with infimum $\bot$. Let $\langle f^{\,n}, n \in \mathbb{N}_\omega\rangle$ be the iterates of f from $\bot$.

Let $\langle \overline{L}, \overline{\sqsubseteq}\rangle$ be an abstract domain with increasing concretization $\gamma \in \overline{L} \xrightarrow{\nearrow} L$.

Let $\overline{f} \in \overline{L} \to \overline{L}$ semicommute with f, that is, $\forall \overline{x} \in \overline{L}\,.\, f(\gamma(\overline{x})) \sqsubseteq \gamma(\overline{f}(\overline{x}))$.

Let Δ be a narrowing on $\overline{L}$.

Then the iterates $\langle \check{f}^{\,n}, n \in \mathbb{N}\rangle$ of $\overline{f}$ from any $b \in \overline{L}$ such that $\overline{f}(b) \mathrel{\overline{\sqsubseteq}} b$ with narrowing Δ are decreasing and $\forall n \in \mathbb{N}\,.\,\mathsf{lfp}^{\sqsubseteq} f \sqsubseteq \gamma(\check{f}^{\,n})$.

Moreover if $\overline{f}$ is increasing and the downward iteration with narrowing $\langle \check{f}^{\,n}, n \in \mathbb{N}\rangle$ converges to the limit $\check{f}^{\,\ell}$, then $\check{f}^{\,\ell}$ is a fixpoint of $\overline{f}$.

Proof *(1) Lemma.*

Let us first show that if $\overline{f}(x) \mathrel{\overline{\sqsubseteq}} x$ then $\forall n < \omega\,.\, f^{\,n} \sqsubseteq \gamma(x)$. The proof is by induction.

- We have $f^{\,0} = \bot \sqsubseteq \gamma(x)$ by definition of the iterates and of the infimum;
- If $f^{\,n} \sqsubseteq \gamma(x)$ then $f^{\,n+1} = f(f^{\,n}) \sqsubseteq f(\gamma(x)) \sqsubseteq \gamma(\overline{f}(\overline{x})) \sqsubseteq \gamma(x)$ by definition of the iterates, induction hypothesis and γ increasing, $\overline{f}$ and f semicommute, hypothesis $\overline{f}(x) \mathrel{\overline{\sqsubseteq}} x$ and γ increasing. By transitivity, $f^{\,n+1} \sqsubseteq \gamma(x)$;

- If for the limit $\omega \in \mathbb{N}_\omega$, $\forall n < \omega \,.\, f^{\,n} \sqsubseteq \gamma(x)$ by recurrence so $f^{\,\omega} = \bigsqcup_{n<\omega} f^{\,n} \sqsubseteq \gamma(x)$ by definition of the iterates, existence hypothesis of the lub $\bigsqcup$, smaller than or equal to any upper bound. □

By induction, $\overline{f}(x) \mathrel{\overline{\sqsubseteq}} x$ implies $\forall n < \omega \,.\, f^{\,n} \sqsubseteq \gamma(x)$.

(2) Concrete iterates overapproximation.

Let us show that $\forall n, m \in \mathbb{N} \,.\, f^{\,n} \sqsubseteq \gamma(\check{f}^{\,m})$ by recurrence on m.

For the basis $m = 0$, we have $\overline{f}(b) \mathrel{\overline{\sqsubseteq}} b$ that implies $\forall n \in \mathbb{N} \,.\, f^{\,n} \sqsubseteq \gamma(b) = \gamma(\check{f}^{0})$ by the lemma and definition 34.15.1 of the abstract iterates.

Assume by induction hypothesis that $\forall n \in \mathbb{N} \,.\, f^{\,n} \sqsubseteq \gamma(\check{f}^{\,m})$

By definition 34.15.2, either $\overline{f}(\check{f}^{\,m}) \mathrel{\overline{\not\sqsubseteq}} \check{f}^{\,m}$, so, by definition 34.15.2, $\check{f}^{\,m+1} = \check{f}^{\,m}$ and hence $\forall n \in \mathbb{N} \,.\, f^{\,n} \sqsubseteq \gamma(\check{f}^{\,m+1})$ by induction hypothesis.

Otherwise, $\overline{f}(\check{f}^{\,m}) \mathrel{\overline{\sqsubseteq}} \check{f}^{\,m}$, in which $\check{f}^{\,m+1} = \check{f}^{\,m} \mathbin{\Delta} \overline{f}(\check{f}^{\,m})$ with $\overline{f}(\check{f}^{\,m}) \mathrel{\overline{\sqsubseteq}} \check{f}^{\,m+1} \mathrel{\overline{\sqsubseteq}} \check{f}^{\,m}$ by soundness 34.14 of the narrowing Δ. Because f and γ are increasing, and $\overline{f}$ and f semicommute, we have $f(\gamma(\check{f}^{\,m+1})) \sqsubseteq f(\gamma(\check{f}^{\,m})) \sqsubseteq \gamma(\overline{f}(\check{f}^{\,m})) \sqsubseteq \gamma(\check{f}^{\,m+1})$. Therefore $\forall n \in \mathbb{N} \,.\, f^{\,n} \sqsubseteq f^{\,n+1} = f(f^{\,n}) \sqsubseteq f(\gamma(\check{f}^{\,m})) \sqsubseteq \gamma(\check{f}^{\,m+1})$.

(3) Concrete fixpoint overapproximation.

By $\forall n, m \in \mathbb{N} \,.\, f^{\,n} \sqsubseteq \gamma(\check{f}^{\,m})$ and Scott–Kleene's iterative fixpoint theorem 15.26, $\mathsf{lfp}^{\sqsubseteq} f = \bigsqcup_{n \in \mathbb{N}} f^{\,n} \sqsubseteq \gamma(\check{f}^{\,m})$ for all $m \in \mathbb{N}$.

(4) Decreasing abstract iterates.

By definition 34.15.2, either $\overline{f}(\check{f}^{\,n}) \mathrel{\overline{\not\sqsubseteq}} \check{f}^{\,n}$ and $\check{f}^{\,n+1} \triangleq \check{f}^{\,n}$ or $\overline{f}(\check{f}^{\,n}) \mathrel{\overline{\sqsubseteq}} \check{f}^{\,n}$ and $\check{f}^{\,n+1} \triangleq \check{f}^{\,n} \mathbin{\Delta} \overline{f}(\check{f}^{\,n}) \mathrel{\overline{\sqsubseteq}} \check{f}^{\,n}$ by definition 34.14.

(5) Fixpoint limit for increasing transformer $\overline{f}$.

By hypothesis $\overline{f}(b) \mathrel{\overline{\sqsubseteq}} b$ so $\overline{f}(\check{f}^{0}) \mathrel{\overline{\sqsubseteq}} \check{f}^{0}$.

Assume by induction hypothesis that $\overline{f}(\check{f}^{\,n}) \mathrel{\overline{\sqsubseteq}} \check{f}^{\,n}$.

If $\overline{f}(\check{f}^{n}) \mathrel{\overline{\not\sqsubseteq}} \check{f}^{n}$ then $\check{f}^{\,n+1} = \check{f}^{\,n}$ and so $\overline{f}(\check{f}^{\,n+1}) \mathrel{\overline{\sqsubseteq}} \check{f}^{\,n+1}$ by induction hypothesis.

Otherwise, $\overline{f}(\check{f}^{\,n}) \mathrel{\overline{\sqsubseteq}} \check{f}^{\,n}$ so $\check{f}^{\,n+1} = \check{f}^{\,n} \mathbin{\Delta} \overline{f}(\check{f}^{\,n})$ by definition 34.15 and therefore $\overline{f}(\check{f}^{\,n}) \mathrel{\overline{\sqsubseteq}} \check{f}^{\,n+1} \mathrel{\overline{\sqsubseteq}} \check{f}^{\,n}$ by definition 34.14. Because $\overline{f}$ is assumed to be increasing, $\overline{f}(\check{f}^{\,n+1}) \mathrel{\overline{\sqsubseteq}} \overline{f}(\check{f}^{\,n})$ and so by transitivity, we conclude that $\overline{f}(\check{f}^{\,n+1}) \mathrel{\overline{\sqsubseteq}} \check{f}^{\,n+1}$.

By recurrence, $\forall n \in \mathbb{N} \,.\, \overline{f}(\check{f}^{\,n}) \mathrel{\overline{\sqsubseteq}} \check{f}^{\,n}$.

Assuming the iterates do converge to $\check{f}^{\,\ell}$. We have $\overline{f}(\check{f}^{\ell}) \mathrel{\overline{\not\sqsubseteq}} \check{f}^{\ell}$ by definition 34.15 as well as $\overline{f}(\check{f}^{\,\ell}) \mathrel{\overline{\sqsubseteq}} \check{f}^{\,\ell}$, as shown previously, so that $\overline{f}(\check{f}^{\,\ell}) = \check{f}^{\,\ell}$. □

Remark 34.17 (trivial narrowing) The trivial narrowing $x \mathbin{\Delta} y = y$ satisfies definition 34.14. Then theorem 34.16 shows that it is sound to stop the iterates at any rank n to artificially enforce convergence. (Because this is not in general the limit of the iterates, the solution $\check{f}^{\,n}$ may not be a fixpoint of $\overline{f}$, even if $\overline{f}$ is increasing.)

34.9 Upward Iteration with Widening Refined by a Downward Iteration with Narrowing

In practice a concrete fixpoint is first abstracted in an abstract fixpoint, the upward iterates are accelerated by extrapolation with widening and next refined by downward iterates accelerated by interpolation with narrowing.

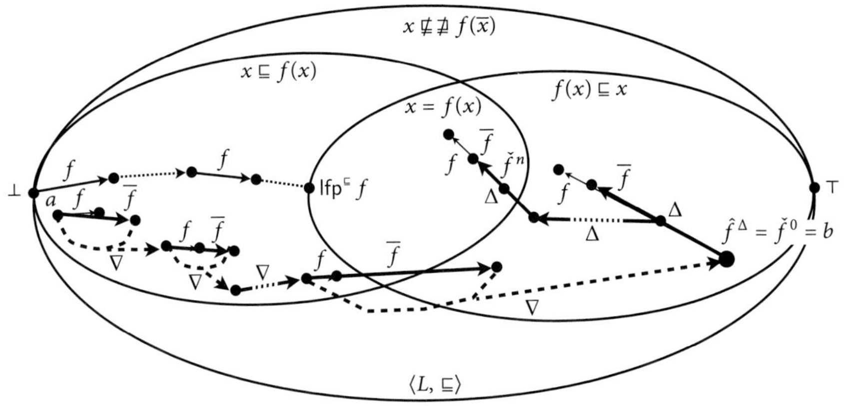

For an increasing abstract transformer $\overline{f}$, the downward iterates accelerated by interpolation with narrowing converge to a fixpoint without jumping over any of them. Therefore, the information lost by the widening may not always be recovered by the narrowing. If the upward widening iteration jumps over a fixpoint which is not the least one, then the downward narrowing is blocked by this fixpoint.

Further improvements are considered in [231, 455].

34.10 Chaotic and Asynchronous Iterations

Because the abstract interpreter of chapter 21 proceeds by structural induction, the widening needs to be done only once at the head of loops. With unstructured abstract interpreters based on the equational semantics of chapter 23, the static analysis is specified by a system of equations

$$\begin{cases} x_i &= \overline{f}_i(x_1, \ldots, x_n) \\ i = 1, \ldots, n \end{cases},$$

that can be solved by the chaotic iterations of chapter 22 (where at each iteration an arbitrary nonempty subset of the equations evolves) or asynchronous iterations (the equations evolve in parallel) with fairness hypotheses (no equation is delayed forever and the result of any equation evolution is eventually available to all equations) [220, 236]. A dependency analysis of the equations can be used to identify cycles that must be broken by a widening [221, ch. 4]. Good iterations strategies should ensure rapid convergence by avoiding useless repetitions of computations and loss of precision [138].

34.11 Finitary Versus Infinitary Abstractions

Following an argument of [246], an interval analysis with widening/narrowing will always terminate and so can explore only a finite part of the lattice of intervals. For example, for program P^{1001} = `while (x < n) x = x + 1 ;`, this is

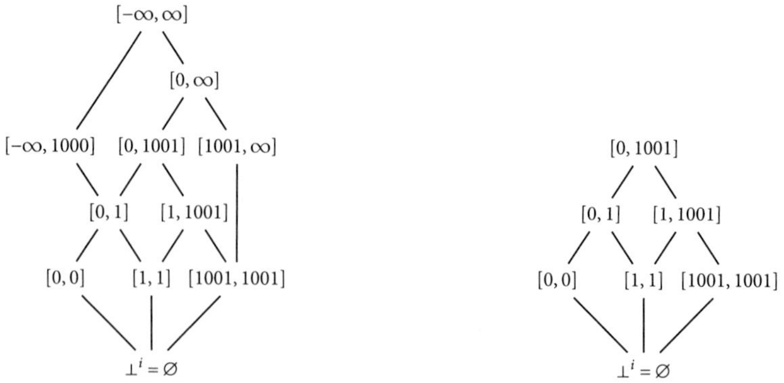

and even the smaller part on the right when it is using a widening with threshold 1001. A static analysis with these finite domains would find the same result for program P^{1001}. This is not an argument in favor of using only abstractions in finite domains (or in Noetherian domains, that is, domains with no infinite strictly increasing chains).

- Using a single Noetherian abstract domain for all programs will fail on infinitely many programs where the widening/narrowing analysis will succeed. Consider for example the family of programs

$$\mathsf{P}^n \quad = \quad \texttt{while}\,\ell_1\,(\texttt{x<n})\;\; \ell_2\,\texttt{x = x + 1;}\;\; \ell_3$$

  where $n \geqslant 0$ is an integer constant. The interval analysis examplified in sections 33.5 and 33.6 for $n = 1001$ will always yield the loop invariant $[0, n]$ for any particular $n \in \mathbb{N}$. These invariants form an infinite strictly increasing chain of intervals. Any Noetherian abstract domain will not be able to represent infinitely many of these invariants. Therefore an analysis using a Noetherian domain will fail on infinitely many programs where the analysis with widening/narrowing succeeds.

- Finding the adequate domain for each particular program amounts to doing a pre-analysis (globally or at each iteration step) of this particular program to discover the necessary invariants to be included in the abstract domain. However to determine these useful invariants (or a finite superset) the pre-analysis must know the result of the analysis, so this solution is circular.

34.12 Hidden Widenings and Narrowings

As noticed in [251, section 6.2], a solution to the convergence problem is to mask the explicit use of a widening ∇ in the iterations of $\overline{f}$ by iterating $\overline{g}(x) \triangleq (\!|\, \overline{f}(x) \sqsubseteq x \mathbin{?} x \mathbin{:} x \mathbin{\nabla} \overline{f}(x) \,|\!)$.

This is, for example, the case in the Robin Milner polymorphic typing algorithm [685], which involves a hidden widening to handle recursion and can be potentially improved [224].

Another example is [11] in the abstract domain of first-order formulas where, for example, $\forall i\,.\,(0 \leqslant i < n) \Rightarrow (x^{i+1} = x^i + 1)$ is widened to $\forall i\,.\,x^i = x^0 + i$.

A narrowing can be hidden by choosing $x \,\Delta\, y = y$, which is sound by theorem 34.16 and termination enforced by stopping the downwards iterations after an arbitrary number of iterates.

When restricted to specific classes of programs, such as non-nested simple loops, guarded by a linear condition on integer variables and consisting of incrementations/decrementations of these variables, the limit of the fixpoint iterations can be computed exactly as a Presburger formula (first-order theory of the natural numbers with addition and equality only, without multiplication). So the widening can deliver exact results, at a high complexity cost (e.g., doubly exponential). The restrictions and costs limit the applicability of the method [207, 989].

34.13 Duality

Widenings and narrowings are not dual but have duals that can be used to approximate fixpoints iteratively with convergence acceleration, as follows:

	widening	dual widening	narrowing	dual narrowing
	∇	$\widetilde{\nabla}$	Δ	$\widetilde{\Delta}$
Initialization of the iterates with respect to the fixpoint	below	above	above	below
Limit of the iterates with respect to the fixpoint	above	below	above	below

and shown on figure 34.18.

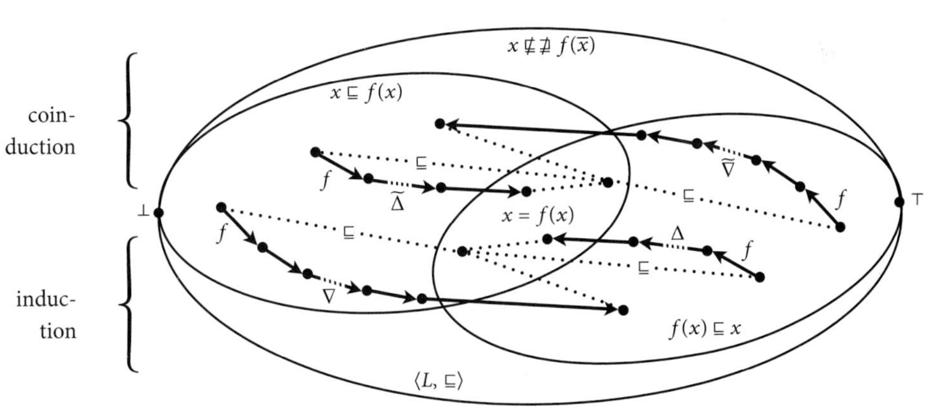

Figure 34.18
Inductive and coinductive fixpoint iterations

An example of dual narrowing is Graig interpolation [677] in the abstract domain of first-order formulas (this is not a widening [231, 678] and it does not enforce termination).

34.14 Conclusion

We have assumed upper continuous functions but the results extend to increasing functions using transfinite iterations [221, ch. 4].

Widening and narrowing were introduced in [234], and their duals in [235]. [221] distinguishes between extrapolating and terminating widening/narrows and duals because these hypotheses are independent; see also [392].

Terminating widenings are not increasing and cannot be increasing, as shown in section 34.2. Therefore fixpoints might not exist but soundness can be established with respect to an increasing concrete semantics with fixpoint existence guarantee.

The fact that widenings are not increasing and cannot be increasing if termination of infinite iterations is to be ensured is often misunderstood. An example, among others, is [705, ch. 5]. Interval analysis is considered as an instance of the so-called "monotone data flow analysis framework" [553] assuming that transformers are increasing, which is not the case of widening. The widening is therefore introduced in a framework assuming hypotheses that exclude it, so conclusions are simply wrong.

Widening and narrowing were the central idea for scalability of static analysis. They do not fit in previous data-flow analysis frameworks requiring finitary domains and monotonicity [27, 553, 572, 967], which severely limits their expressive power (section 34.11 and [246]).

Some fixpoint solvers such as TD [875] incorporate widenings and narrowings to cope with non-Noetherian abstract domains.

More general results on widenings, narrowings, and their duals are provided in [231]. Widenings and narrowing will not improve a fixpoint. This can be done by perturbation of the fixpoint, analysis of the iterations to identify losses of information, restarting increasing and decreasing sequences from well-chosen iterates, and other methods [141]. [39, 45] proposed to interwind widening and narrowing during iteration. Extrapolation and interpolation operators can be designed by analysis of the fixpoint iterations [278].

So-called *acceleration techniques* identify those fixpoint equations on those abstract domains for which an exact solution can be computed directly [425, 426]. *Policy iteration* applies to those fixpoints of increasing functions F which is a combination of "simpler" increasing functions g, for which there are fast algorithms to compute their least fixpoints [9, 216, 400, 401, 404–406, 874]. [559] combines iterations and, locally, policy iteration.

35 Fixpoint Checking

Many static analyses are performed to check properties of program components S. This frequently amounts to proving that $\mathsf{lfp}^{\sqsubseteq}\,\mathscr{F}[\![\mathsf{S}]\!] \sqsubseteq P$ where property P is a specification. A common solution based on the fixpoint induction theorem 24.1 consists of computing an inductive property I that is shown to be invariant $\mathscr{F}[\![\mathsf{S}]\!](I) \sqsubseteq I$ so that $\mathsf{lfp}^{\sqsubseteq}\,\mathscr{F}[\![\mathsf{S}]\!] \sqsubseteq I$ and stronger than P, that is, $I \sqsubseteq P$.

A better solution is to use the specification P to improve the precision of the invariant I. The invariant I is inferred assuming the specification P does hold. Then the invariant I is checked to imply that $\mathsf{lfp}^{\sqsubseteq}\,\mathscr{F}[\![\mathsf{S}]\!] \sqsubseteq P$.

Contents

35.1 Concrete Fixpoint Checking

The following theorem 35.1 can be used to check a specification for a semantics, using the specification to refine the verification condition.

> **Theorem 35.1 (concrete fixpoint checking)** Let $f \in \mathscr{L} \xrightarrow{\nearrow} \mathscr{L}$ be an increasing function on a complete lattice $\langle \mathscr{L}, \sqsubseteq, \bot, \top, \sqcap, \sqcup \rangle$ and $P \in \mathscr{L}$. Then $\mathsf{lfp}^{\sqsubseteq}\, f \sqsubseteq P$ if and only if there exists $I \in \mathscr{L}$ such that $(f(I) \sqcap P) \sqsubseteq I$ and $f(I) \sqsubseteq P$.

In static analysis, the advantage of theorem 35.1 over theorem 24.1 is that the invariant can be computed as an overapproximation of $\mathsf{lfp}^{\sqsubseteq}\, x \mapsto f(x) \sqcap P$ which is more precise than $\mathsf{lfp}^{\sqsubseteq}\, f$. For example, the intersection with P may limit and reduce the extrapolation performed by a widening and so improve the precision of the static analysis .

Proof of theorem 35.1 Because f is increasing, $x \mapsto f(x) \sqcap P$ is also increasing so $\mathsf{lfp}^{\sqsubseteq}\, f$ and $\mathsf{lfp}^{\sqsubseteq}\, x \mapsto f(x) \sqcap P$ do exist by Tarski's fixpoint theorem 15.6.

- $\Leftarrow$, soundness.

$$
\begin{array}{ll}
 & (f(I) \sqcap P) \sqsubseteq I \wedge f(I) \sqsubseteq P \\
\Rightarrow & (f(I) \sqcap P) \sqsubseteq I \wedge f(I) \sqcap P = f(I) \wedge f(I) \sqsubseteq P \qquad \text{(definition of } \sqsubseteq\text{)} \\
\Rightarrow & f(I) \sqsubseteq I \wedge f(I) \sqsubseteq P \qquad \text{(substitution)} \\
\Rightarrow & \mathsf{lfp}^{\sqsubseteq}\, f \sqsubseteq I \wedge f(I) \sqsubseteq P \qquad \text{(Tarski's fixpoint theorem 15.6)}
\end{array}
$$

$\Rightarrow \mathsf{lfp}^{\sqsubseteq} f \sqsubseteq P$ ⟨because f is increasing so $\mathsf{lfp}^{\sqsubseteq} f = f(\mathsf{lfp}^{\sqsubseteq} f) \sqsubseteq f(I) \sqsubseteq P$ and transitivity⟩

- $\Rightarrow$, completeness.

$\mathsf{lfp}^{\sqsubseteq} f \sqsubseteq P$

$\Rightarrow \mathsf{lfp}^{\sqsubseteq} f \sqcap P = \mathsf{lfp}^{\sqsubseteq} f \wedge \mathsf{lfp}^{\sqsubseteq} f \sqsubseteq P$ ⟨definition of $\sqsubseteq$ and $\sqcap$⟩

$\Rightarrow f(\mathsf{lfp}^{\sqsubseteq} f) \sqcap P \sqsubseteq \mathsf{lfp}^{\sqsubseteq} f \wedge \mathsf{lfp}^{\sqsubseteq} f \sqsubseteq P$ ⟨fixpoint property and $\sqsubseteq$ reflexive⟩

$\Rightarrow (\exists I \,.\, f(I) \sqcap P) \sqsubseteq I \wedge I \sqsubseteq P$ ⟨take $I = \mathsf{lfp}^{\sqsubseteq} f$⟩ □

Exercise 35.2 ([226]) Prove that under the hypotheses of theorem 35.1, the following algorithm returns the Boolean value $\mathsf{lfp}^{\sqsubseteq} f \sqsubseteq P$ if and when it terminates.

```
x := ⊥; Go := tt;
while Go do
    x' := f(x);
    Go := (x ≠ x') & (x' ⊑ P);
    x := x';
od;
return (x ⊑ P);
```

Exercise 35.3 ([226, theorem 4]) Let $f \in \mathscr{L} \xrightarrow{\nearrow} \mathscr{L}$ be an increasing function on a complete lattice $\langle \mathscr{L}, \sqsubseteq, \bot, \top, \sqcap, \sqcup \rangle$. We say that f has an adjoint $\tilde{f}$ whenever $\langle \mathscr{L}, \sqsubseteq \rangle \underset{f}{\overset{\tilde{f}}{\leftrightarrows}} \langle \mathscr{L}, \sqsubseteq \rangle$. Prove that $\langle \mathscr{L}, \sqsubseteq \rangle \underset{P \mapsto \mathsf{lfp}^{\sqsubseteq} x \mapsto P \sqcup f(x)}{\overset{Q \mapsto \mathsf{gfp}^{\sqsubseteq} y \mapsto Q \sqcap \tilde{f}(y)}{\leftrightarrows}} \langle \mathscr{L}, \sqsubseteq \rangle$.

Exercise 35.4 ([226, theorem 6]) Let $f \in \mathscr{L} \xrightarrow{\nearrow} \mathscr{L}$ be an increasing function on a complete lattice $\langle \mathscr{L}, \sqsubseteq, \bot, \top, \sqcap, \sqcup \rangle$ with adjoint $\tilde{f}$ such that $\langle \mathscr{L}, \sqsubseteq \rangle \underset{f}{\overset{\tilde{f}}{\leftrightarrows}} \langle \mathscr{L}, \sqsubseteq \rangle$. Define the set of invariants to be $\mathscr{I} \triangleq \{I \in \mathscr{L} \mid f(I) \sqsubseteq I \wedge I \sqsubseteq P\}$. Prove that $\langle \mathscr{I}, \sqsubseteq, \mathsf{lfp}^{\sqsubseteq} f, \mathsf{gfp}^{\sqsubseteq} y \mapsto P \sqcap \tilde{f}, \sqcap, \sqcup \rangle$ is a complete lattice.

35.2 Abstract Fixpoint Checking

In the abstract, theorem 35.1 becomes

> **Theorem 35.5 (abstract fixpoint checking)** Let $f \in L \xrightarrow{uc} L$ be an upper continuous function on a CPO $\langle L, \sqsubseteq, \bot \rangle$ with infimum $\bot$.
> Let $\langle \overline{L}, \overline{\sqsubseteq} \rangle$ be an abstract domain with increasing concretization $\gamma \in \overline{L} \xrightarrow{\nearrow} L$.
> Let $\overline{f} \in \overline{L} \to \overline{L}$ be such that $f \circ \gamma \,\dot{\sqsubseteq}\, \gamma \circ \overline{f}$ (semicommutation).
> Let $\overline{P} \in \overline{L}$ be an abstract specification.
> If, assuming the glb $\overline{\sqcap}$ exists, $\exists \overline{I} \in \overline{L} \,.\, \overline{f}(\overline{I}) \,\overline{\sqcap}\, \overline{P} \,\overline{\sqsubseteq}\, \overline{I} \wedge \overline{f}(\overline{I}) \,\overline{\sqsubseteq}\, \overline{P}$ then $\mathsf{lfp}^{\sqsubseteq} f \sqsubseteq \gamma(\overline{P})$.

This theorem is useful in the context, for example, of widening in theorem 34.6 because $\overline{f}$ is not assumed to be increasing.

Proof of theorem 35.5

$\overline{f}(\overline{I}) \,\overline{\sqcap}\, \overline{P} \,\overline{\sqsubseteq}\, \overline{I} \wedge \overline{f}(\overline{I}) \,\overline{\sqsubseteq}\, \overline{P}$

$$\Rightarrow\ \overline{f}(\overline{I}) \mathbin{\overline{\sqcap}} \overline{P} \mathrel{\overline{\sqsubseteq}} \overline{I} \wedge \overline{f}(\overline{I}) \mathbin{\overline{\sqcap}} \overline{P} = \overline{f}(\overline{I}) \wedge \overline{f}(\overline{I}) \mathrel{\overline{\sqsubseteq}} \overline{P} \qquad \text{(definition of } \overline{\sqsubseteq}\text{)}$$

$$\Rightarrow\ \overline{f}(\overline{I}) \mathrel{\overline{\sqsubseteq}} \overline{I} \wedge \overline{f}(\overline{I}) \mathrel{\overline{\sqsubseteq}} \overline{P} \qquad \text{(because } \overline{f}(\overline{I}) \mathbin{\overline{\sqcap}} \overline{P} = \overline{f}(\overline{I})\text{)}$$

$$\Rightarrow\ \gamma(\overline{f}(\overline{I})) \sqsubseteq \gamma(\overline{I}) \wedge \gamma(\overline{f}(\overline{I})) \sqsubseteq \gamma(\overline{P}) \qquad \text{(}\gamma\text{ increasing)}$$

$$\Rightarrow\ f(\gamma(\overline{I})) \sqsubseteq \gamma(\overline{I}) \wedge f(\gamma(\overline{I})) \sqsubseteq \gamma(\overline{P}) \qquad \text{(semicommutation and transitivity)}$$

$$\Rightarrow\ \mathsf{lfp}^{\sqsubseteq} f \sqsubseteq \gamma(\overline{P}) \qquad \text{(theorem 35.1 with } I = \gamma(\overline{I}) \text{ and } P = \gamma(\overline{P})\text{)} \quad \square$$

35.3 Abstract Invariants May Not Help Enough

Theorem 35.1 shows that adding information about the program behavior can only help the analysis. Therefore if a static analysis is not precise enough, for example, because of excessive extrapolations or interpolations, adding a specification of what the static analyzer should infer is sound and can only help.

However, this may be insufficient in general. The main reason is that if the specification P is given in the concrete then it will be abstracted in $\overline{P}$ in the abstract domain, where the concrete information may be lost. Moreover, if the specification $\overline{P}$ is given in the abstract then it might not be inductive.

35.4 Conclusion

Astrée [113] is based on theorem 35.5. In a first phase, the static analysis is done assuming the specification holds. Then, in a second phase, it is verified that the specification does hold.

[226] considers equivalent forms of theorems 35.1 and 35.5 based on duality. [232] introduces other induction principles for proving properties of fixpoints.

36 Reduced Product

The *direct product* of several static analyses is the conjunction of the separately computed analyses. The *reduced product* of several static analyses [241] computes in the conjunction of their abstract domains. In that case, information on program executions captured by one domain can be used to improve the precision of the information provided by other abstract domains. This is called a *meaning-preserving reduction.* As shown in chapter 29, reductions can be iterated to obtain more precise and still sound analyses.

Contents

36.1 Direct Product

The direct product of abstract domains can be used to perform *simultaneous independent analyses* of a program. The set of properties of the direct product is the Cartesian product of the component abstract properties, the concretization is the conjunction (meet) of the concretizations for each abstract domain, and the transformers are performed componentwise without interactions.

Definition 36.1 (direct product) Let $\overline{\mathbb{D}}^i \triangleq \langle\overline{\mathbb{P}}^i, \sqsubseteq^i, \bot^i, \sqcup^i, \mathsf{assign}^i[\![\mathsf{x}, \mathsf{A}]\!], \mathsf{test}^i[\![\mathsf{B}]\!], \overline{\mathsf{test}}^i[\![\mathsf{B}]\!]\rangle$, $i \in \Delta$, Δ finite, be several abstract domains, each abstracting a concrete domain $\mathbb{D} \triangleq \langle\mathbb{P}, \sqsubseteq, \bot, \sqcap, \sqcup, \mathsf{assign}[\![\mathsf{x}, \mathsf{A}]\!], \mathsf{test}[\![\mathsf{B}]\!], \overline{\mathsf{test}}[\![\mathsf{B}]\!]\rangle$ by a concretization $\gamma^i \in \overline{\mathbb{P}}^i \xrightarrow{\nearrow} \mathbb{P}$ (or a Galois connection).

Their *direct product* $\times$ is the abstract domain $\mathbb{D}^\times \triangleq \bigtimes_{i\in\Delta} \overline{\mathbb{D}}^i \triangleq \langle\mathbb{P}^\times, \sqsubseteq^\times, \bot^\times, \top^\times, \sqcap^\times, \sqcup^\times, \mathsf{assign}^\times[\![\mathsf{x}, \mathsf{A}]\!], \mathsf{test}^\times[\![\mathsf{B}]\!], \overline{\mathsf{test}}^\times[\![\mathsf{B}]\!]\rangle$ where the abstract operations are combined componentwise. For example $\vec{P} \in \mathbb{P}^\times \triangleq \{\prod_{i\in\Delta} P_i \mid \forall i \in \Delta \,.\, P_i \in \overline{\mathbb{P}}^i\}$, $\prod_{i\in\Delta} P_i \sqsubseteq^\times \prod_{i\in\Delta} P'_i \triangleq \bigwedge_{i\in\Delta} P_i \sqsubseteq^i P'_i$, $\mathsf{test}^\times[\![\mathsf{B}]\!] \prod_{i\in\Delta} P_i \triangleq \prod_{i\in\Delta} \mathsf{test}^i[\![\mathsf{B}]\!]P_i$, etc.

The concretization is $\gamma_\times \in \mathbb{P}^\times \xrightarrow{\nearrow} \mathbb{P}$ defined by $\gamma_\times(\prod_{i\in\Delta} P_i) \triangleq \bigsqcap_{i\in\Delta} \gamma^i(P_i)$.

The direct product can be used to perform the conjunction of several independent static analyses. Because abstract transformers are applied componentwise, there is no interaction between the abstract domains $\overline{\mathbb{D}}^i$, $i \in \Delta$, so the concretization of the static analysis of programs by the abstract interpreter of chapter 21 with the direct product $\mathbb{D}^{\times}$ yields exactly the same result as the conjunction of the results produced by the abstract interpreter of chapter 21 with each abstract domain $\overline{\mathbb{D}}^i$, $i \in \Delta$ separately.

Example 36.2 Consider the separate sign analysis (of section 28.8.2) and parity analysis (of section 28.8.1) of the following program.

program	sign of x	parity of x	conjunction of separate analyses
`x = -1977;`	<0	$\mathbb{o}$	$<0 \wedge \mathbb{o}$
`/* loop invariant */`	$\top_{\pm}$	$\mathbb{o}$	$\top_{\pm} \wedge \mathbb{o}$
`while (x < 0) {`	<0	$\mathbb{o}$	$<0 \wedge \mathbb{o}$
`  x = x + 2;`	$\top_{\pm}$	$\mathbb{o}$	$\top_{\pm} \wedge \mathbb{o}$
`}`	$\geqslant 0$	$\mathbb{o}$	$>0 \wedge \mathbb{o}$
`x = x - 1;`	$\top_{\pm}$	$\mathbb{e}$	$\top_{\pm} \wedge \mathbb{e}$

Their conjunction reduces the loop exit invariant from $(\geqslant 0 \wedge \mathbb{o})$ to $(>0 \wedge \mathbb{o})$ because x being positive and odd cannot be 0. However the program exit invariant $(\top_{\pm} \wedge \mathbb{e})$ is not improved.

36.2 Reduced Product, Informally

The reduced product of abstract domains can be used to perform *simultaneous dependent analyses* of a program in which the abstract domains share common information during the analysis.

Example 36.3 Continuing example 36.2, if the two analyses are performed simultaneously and the reduction is performed on the fly during these analyses, then the sign analysis gets $\geqslant 0$ on loop exit and the parity analysis gets $\mathbb{o}$. The reduction yields >0 for signs and $\mathbb{o}$ unchanged for parity. Then the assignment `x = x - 1;` yields $\geqslant 0$ for signs and $\mathbb{e}$ for parity which is more precise than the conjunction $(\top_{\pm} \wedge \mathbb{e})$ after the separate analyses.

The reduction of sign and parity goes as follows:

$$\begin{array}{lcl@{\qquad}lcl}
\langle \bot_{\pm}, _\rangle & \to & \langle \bot_{\pm}, \bot^2\rangle & \langle 0, \top^2\rangle & \to & \langle 0, \mathbb{e}\rangle \\
\langle _, \bot^2\rangle & \to & \langle \bot_{\pm}, \bot^2\rangle & \langle 0, \mathbb{o}\rangle & \to & \langle \bot_{\pm}, \bot^2\rangle \\
\langle \leqslant 0, \mathbb{o}\rangle & \to & \langle <0, \mathbb{o}\rangle & \langle \geqslant 0, \mathbb{o}\rangle & \to & \langle >0, \mathbb{o}\rangle
\end{array}$$

The reduced product of static analyses consist in computing them simultaneously with a reduction at each step. This is in general much easier to develop than designing a new analysis combining the two and is more precise than computing them separately.

The reduction of example 36.3 is strong in that it is the most precise possible. An example of weak reduction is the *smash product* that extends an infimum to all components of the product i.e. $\langle x_1, \ldots, x_{i-1}, \bot^i, x_{i+1}, \ldots, x_n\rangle \to \langle \bot^1, \ldots, \bot^{i-1}, \bot^i, \bot^{i+1}, \ldots, \bot^n\rangle$ when $\gamma_{\times}(\bot^i) = \bot$ is the empty set of reachable states. For example, the operation $\rfloor^{\pm}$ in (3.23) is a smash reduction.

Exercise 36.4 (reduced product of simple sign and parity) Draw the lattice of properties that is the reduction of the following simple sign analysis

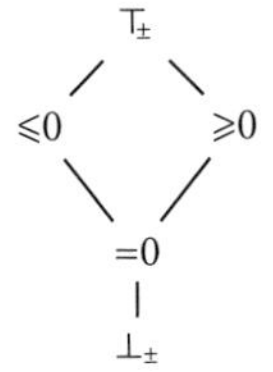

and parity analysis of section 28.8.1.

Exercise 36.5 (reduced product of simple sign and constant) Draw the lattice of properties that is the reduction of simple sign analysis of exercise 36.4 and the constant analysis of exercise 3.45.

Exercise 36.6 (reduced product of simple sign, parity, and constancy) Draw the lattice of properties that is the reduction of simple sign analysis of exercise 36.4, the parity analysis of section 28.8.1, and the constancy analysis of exercise 3.45.

The effect of the reduced product of two abstractions

$$\langle \mathbb{P}, \subseteq \rangle \xleftarrow[\alpha_{¤}]{\gamma_{¤}} \langle \mathbb{P}^{¤}, \overline{\sqsubseteq}^{¤} \rangle \otimes \langle \mathbb{P}, \subseteq \rangle \xleftarrow[\alpha_{\sharp}]{\gamma_{\sharp}} \langle \overline{\mathbb{P}}^{\sharp}, \overline{\sqsubseteq}^{\sharp} \rangle \triangleq$$

$$\langle \mathbb{P}, \subseteq \rangle \xleftarrow[\alpha_{\otimes}]{\gamma_{\otimes}} \langle \{ \langle \alpha_{¤}(\gamma_{¤}(\overline{P}^{¤}) \sqcap \gamma_{\sharp}(\overline{P}^{\sharp})), \alpha_{\sharp}(\gamma_{¤}(\overline{P}^{¤}) \sqcap \gamma_{\sharp}(\overline{P}^{\sharp})) \rangle \mid \overline{P}^{¤} \in \mathbb{P}^{¤} \wedge \overline{P}^{\sharp} \in \overline{\mathbb{P}}^{\sharp} \}, \overline{\sqsubseteq}^{¤} \times \overline{\sqsubseteq}^{\sharp} \rangle$$

(where $\overline{\sqsubseteq}^{¤} \times \overline{\sqsubseteq}^{\sharp}$ is componentwise) is to bring into each analysis the information of the other analysis which is expressible in this analysis. The property $\langle \overline{P}^{¤}, \overline{P}^{\sharp} \rangle$ meaning $(\gamma_{¤}(\overline{P}^{¤}) \sqcap \gamma_{\sharp}(\overline{P}^{\sharp}))$ is reduced to $\langle \alpha_{¤}(\gamma_{¤}(\overline{P}^{¤}) \sqcap \gamma_{\sharp}(\overline{P}^{\sharp})), \alpha_{\sharp}(\gamma_{¤}(\overline{P}^{¤}) \sqcap \gamma_{\sharp}(\overline{P}^{\sharp})) \rangle$. Therefore the information of one abstract domain pertinent to the other abstract domain is collected by this other abstract domain. For example the information $\overline{P}^{¤}$ of abstract domain $\mathbb{P}^{¤}$ pertinent to $\overline{\mathbb{P}}^{\sharp}$ is collected by $\overline{\mathbb{P}}^{\sharp}$ thanks to $\alpha_{\sharp}(\gamma_{¤}(\overline{P}^{¤}) \sqcap \gamma_{\sharp}(\overline{P}^{\sharp}))$.

36.3 Reduced Product, Formally

To improve the precision of the direct product, the reduced product was defined [241] so that each analysis benefits from the information brought by the other analyses. We will introduce three equivalent definition of the reduced product.

Definition 36.7 (reduced product (I)) Let $\mathbb{D} \triangleq \langle \mathbb{P}, \sqsubseteq, \perp, \sqcup, \sqcap, \mathsf{assign}[\![\mathsf{x}, \mathsf{A}]\!], \mathsf{test}[\![\mathsf{B}]\!], \overline{\mathsf{test}}[\![\mathsf{B}]\!] \rangle$ be a concrete domain. Let $\overline{\mathbb{D}}^{i} \triangleq \langle \overline{\mathbb{P}}^{i}, \overline{\sqsubseteq}^{i}, \overline{\perp}^{i}, \overline{\sqcap}^{i}, \overline{\sqcup}^{i}, \overline{\mathsf{assign}}^{i}[\![\mathsf{x}, \mathsf{A}]\!], \overline{\mathsf{test}}^{i}[\![\mathsf{B}]\!], \overline{\overline{\mathsf{test}}}^{i}[\![\mathsf{B}]\!] \rangle$, $i \in \Delta$, Δ finite, be abstract domains with increasing concretization $\gamma_i \in \overline{\mathbb{P}}_i \xrightarrow{\nearrow} \mathbb{P}$. Let $\mathbb{D}^{\times} = \bigtimes_{i \in \Delta} \overline{\mathbb{D}}^{i}$ be their direct product with concretization $\gamma_{\times}$.

Their *reduced product* is $\mathbb{D}^{\otimes} \triangleq \bigotimes_{i \in \Delta} \overline{\mathbb{D}}^{i} \triangleq \langle \mathbb{P}^{\otimes}, \sqsubseteq^{\otimes}, \perp^{\otimes}, \top^{\otimes}, \sqcup^{\otimes}, \sqcap^{\otimes}, \mathsf{assign}^{\otimes}[\![\mathsf{x}, \mathsf{A}]\!], \mathsf{test}^{\otimes}[\![\mathsf{B}]\!], \overline{\mathsf{test}}^{\otimes}[\![\mathsf{B}]\!] \rangle$ where the reduced properties $\mathbb{P}^{\otimes} \triangleq \mathbb{P}^{\times}/_{\equiv^{\otimes}}$ are the equivalence classes of the equivalence relation $(\vec{P} \equiv^{\otimes} \vec{Q}) \triangleq (\gamma_{\times}(\vec{P}) = \gamma_{\times}(\vec{Q}))$

on the direct product $\mathbb{P}^\times$ and all operations are extended to these equivalence classes $\mathbb{P}^\otimes = \{[\vec{P}]_{\equiv^\otimes} \mid \vec{P} \in \mathbb{P}^\times\}$ of $\equiv^\otimes$ as follows:

- $\forall \vec{P} \in \mathbb{P}^\times \,.\, \gamma_\otimes([\vec{P}]_{\equiv^\otimes}) \triangleq \gamma_\times(\vec{P})$;
- $[\vec{P}]_{\equiv^\otimes} \sqsubseteq^\otimes [\vec{Q}]_{\equiv^\otimes} \quad\triangleq\quad \exists \vec{P}' \in [\vec{P}]_{\equiv^\otimes} \,.\, \exists \vec{Q}' \in [\vec{Q}]_{\equiv^\otimes} \,.\, \vec{P}' \sqsubseteq^\times \vec{Q}'$;
- $\bot^\otimes \quad\triangleq\quad [\bot^\times]_{\equiv^\otimes}$;
- $\top^\otimes \quad\triangleq\quad [\top^\times]_{\equiv^\otimes}$;
- $\bigsqcup^\otimes_{k\in K} [\vec{P}_k]_{\equiv^\otimes} \quad\triangleq\quad [\bigsqcup^\times_{k\in K} \vec{P}_k]_{\equiv^\otimes}$;
- $\mathsf{assign}^\otimes[\![\mathsf{x}, \mathsf{A}]\!][\vec{P}]_{\equiv^\otimes} \quad\triangleq\quad [\mathsf{assign}^\times[\![\mathsf{x}, \mathsf{A}]\!]\vec{P}]_{\equiv^\otimes}$;
- $\mathsf{test}^\otimes[\![\mathsf{B}]\!][\vec{P}]_{\equiv^\otimes} \quad\triangleq\quad [\mathsf{test}^\times[\![\mathsf{B}]\!]\vec{P}]_{\equiv^\otimes}$;
- $\overline{\mathsf{test}}^\otimes[\![\mathsf{B}]\!][\vec{P}]_{\equiv^\otimes} \quad\triangleq\quad [\overline{\mathsf{test}}^\times[\![\mathsf{B}]\!]\vec{P}]_{\equiv^\otimes}$.

Notice that the definitions are independent of the choices of the representatives (e.g., $\vec{P}$) of the equivalence classes (e.g., $[\vec{P}]_{\equiv^\otimes}$). In practice the equivalence classes are implemented using a representant chosen by a meaning-preserving reduction operation (see section 36.3.4).

Example 36.8 (smash product) Assuming $\gamma(\bot) = \emptyset$, the smash product reduces the product of properties with a $\bot$ component to a tuple of $\bot$s.

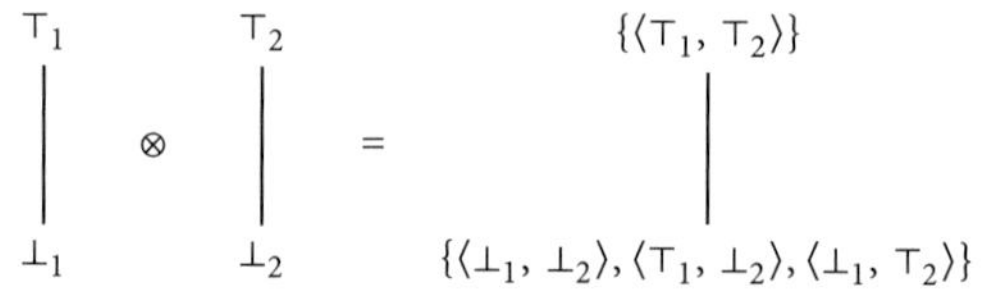

36.3.1 The Hierarchy (Poset/Complete Lattice) of Abstract Domains

We can compare the expressiveness of abstract properties by defining abstract properties $\overline{\mathbb{P}}_1$ to be more expressive/precise than $\overline{\mathbb{P}}_2$ whenever any property exactly expressible by $\overline{\mathbb{P}}_2$ is also expressible by $\overline{\mathbb{P}}_1$.[1] Two sets of abstract properties are equivalent when they are equally expressive.

Example 36.9 The lattice of signs (a) of section 3.12 is strictly more expressive than the lattice (b) of exercise 36.4 because all sign properties of (b) can be expressed by (a) whereas *strictly positive* in (a) must be overapproximated by *positive or zero* in (b). Assuming $\gamma(\leqslant 0') = \gamma(\leqslant 0) = \mathbb{N}$, (b) and (b') are equality expressive. Both lattices (a), (b) and (b') are incomparable with (c) because (c) lacks *positive or zero* found in (a) and (b), and (a), (b) and (b') all lack *strictly greater than one* found in (c).

1. This extends to abstract domains by requiring the transformers also to be more precise.

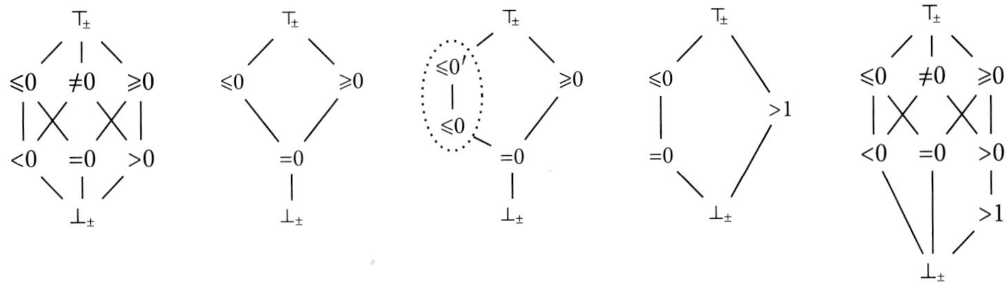

(a) (b) (b') (c) (d)

The two lattices in section 3.12 are equally expressive because they have the same concretization. Signs in section 3.12 or exercise 36.4 and parity in section 28.8.1 are incomparable.

These abstract domains form a lattice

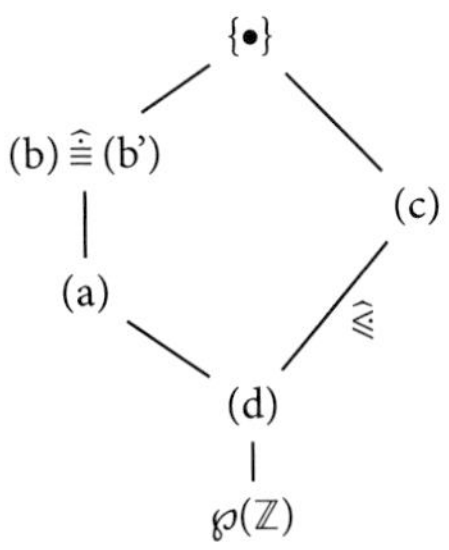

of which the top element is the less precise abstract domain $\{\bullet\}$ with a single abstract property $\bullet$, meaning $\mathbb{Z}$ (i.e., true). The most precise abstract domain is $\wp(\mathbb{Z})$, which can express any property of an integer. $\wp(\mathbb{Z})$ and (d) are both more precise than (a) and (c). The less precise of the two is the greatest lower bound (d) = (a) $\bigotimes$ (c), which is the reduced product of (a) and (c).

Definition 36.10 (precision of abstractions) Let $\langle \mathbb{P}, \sqsubseteq \rangle$ be a complete lattice of concrete properties.[a]

Let $\langle \overline{\mathbb{P}}_i, \sqsubseteq_i \rangle$, $i \in \{1, 2\}$, be abstract properties with concretization $\gamma_i \in \overline{\mathbb{P}}_i \to \mathbb{P}$.

We say that $\overline{\mathbb{P}}_2$ is *less precise* (also *less expressive, less refined*, etc.) *than* $\overline{\mathbb{P}}_1$ (written $\overline{\mathbb{P}}_1 \hat{\leqslant} \overline{\mathbb{P}}_2$[b]) whenever $\gamma_2(\overline{\mathbb{P}}_2) \subseteq \gamma_1(\overline{\mathbb{P}}_1)$.[c]

They are *equivalent* whenever $\gamma_1(\overline{\mathbb{P}}_1) = \gamma_2(\overline{\mathbb{P}}_2)$ (written $\overline{\mathbb{P}}_1 \hat{\equiv} \overline{\mathbb{P}}_2$).

a. For example, $\langle \mathbb{P}, \sqsubseteq \rangle = \langle \wp(\wp(\mathbb{T}^{+\infty})), \subseteq \rangle$ in chapter 8, "Program Properties," or $\langle \mathbb{P}, \sqsubseteq \rangle = \langle \wp(\mathbb{E}\mathbb{v}^{\varrho}) \to (\mathbb{L} \to \wp(\mathbb{E}\mathbb{v}^{\varrho})), \ddot{\subseteq} \rangle$ in chapter 19, "Structural Forward Reachability Semantics").

b. $\hat{\leqslant}$-smaller abstract domains are more precise/expressive while $\hat{\leqslant}$-larger abstract domains are less precise/expressive, like $\{\bot, \top\} \hat{\leqslant} \{\top\}$. $\hat{<}$ is strictly more expressive.

c. The right image of a set X by a function f is $f(X) = \{f(x) \mid x \in X\}$).

Any domain of abstract properties $\overline{\mathbb{P}}$ can express a subset $\gamma(\overline{\mathbb{P}})$ of the concrete properties $\mathbb{P}$. Two $\hat{\equiv}$-equivalent sets of abstract properties $\overline{\mathbb{P}}_1$ and $\overline{\mathbb{P}}_2$ are simply different encoding of some element of the complete lattice $\langle \wp(\mathbb{P}), \supseteq \rangle$. Therefore the set of all sets of abstract properties $\overline{\mathbb{P}}$ quotiented by $\hat{\equiv}$ and ordered by precision $\hat{\leqslant}$ is isomorphic to the complete lattice $\langle \wp(\mathbb{P}), \supseteq \rangle$ [241]. When the abstractions are defined by Galois connections $\langle \mathbb{P}, \sqsubseteq \rangle \xleftarrow[\alpha_1]{\gamma_1} \langle \overline{\mathbb{P}}_1, \sqsubseteq_1 \rangle$ and $\langle \mathbb{P}, \sqsubseteq \rangle \xleftarrow[\alpha_2]{\gamma_2} \langle \overline{\mathbb{P}}_2, \sqsubseteq_2 \rangle$, we have $\overline{\mathbb{P}}_1 \hat{\leqslant} \overline{\mathbb{P}}_2 \Leftrightarrow \gamma_1 \circ \alpha_1(\mathbb{P}) \supseteq \gamma_2 \circ \alpha_2(\mathbb{P})$ [241].

Exercise 36.11 Prove that $\overline{\mathbb{P}}_1 \hat{\leqslant} \overline{\mathbb{P}}_2 \Leftrightarrow \gamma_1 \circ \alpha_1(\mathbb{P}) \supseteq \gamma_2 \circ \alpha_2(\mathbb{P})$.

36.3.2 The Reduced Product is the Greatest Lower Bound in the Poset of Abstract Domains

We show that the reduced product is the greatest lower bound (glb) for $\hat{\leqslant}$. By definition of the glb, the reduced product is the most imprecise abstract domain that is more precise than all other domains in the product.

Definition 36.12 (closure by intersection) An abstract domain $\langle \overline{\mathbb{P}}, \overline{\sqsubseteq} \rangle$ with concretization $\gamma \in \overline{\mathbb{P}} \xrightarrow{\nearrow} \mathbb{P}$ into a meet semilattice (respectively complete lattice) $\langle \mathbb{P}, \sqsubseteq, \sqcap \rangle$ is *closed by finite (respectively infinite) intersection* if and only if $\forall P, Q \in \overline{\mathbb{P}} \,.\, \exists R \in \overline{\mathbb{P}} \,.\, \gamma(R) = \gamma(P) \sqcap \gamma(Q)$ (respectively $\forall \mathscr{P} \in \wp(\overline{\mathbb{P}}) \,.\, \exists R \in \overline{\mathbb{P}} \,.\, \gamma(R) = \bigsqcap \gamma(\mathscr{P})$).

Exercise 36.13 Provide an example of abstract domain and concretization function that does not satisfies definition definition 36.12 of closure by intersection.

When considering only abstract domains that are closed by finite intersection, the reduced product can equivalently be understood as the greatest lower bound in the complete lattice of abstract domains, up to the equivalence $\hat{\equiv}$.

Theorem 36.14 (equivalent definition of the reduced product (II)) Let the meet semilattice $\langle \mathbb{P}, \sqsubseteq, \top, \sqcap \rangle$ be concrete properties, $\langle \overline{\mathbb{P}}_i, \sqsubseteq_i, \top_i \rangle$, $i \in \Delta$, Δ finite, be abstract properties with increasing concretization $\gamma_i \in \overline{\mathbb{P}}_i \xrightarrow{\nearrow} \mathbb{P}$ and supremum $\top_i$ such that $\gamma_i(\top_i) = \top$. The reduced product $\langle \mathbb{P}^{\times}/_{\equiv^{\otimes}}, \sqsubseteq^{\otimes}, \top^{\otimes} \rangle$ is the greatest lower bound of the $\langle \overline{\mathbb{P}}_i, \sqsubseteq_i \rangle$, $i \in \Delta$ in that it is

- more precise than $\langle \overline{\mathbb{P}}_i, \overline{\sqsubseteq}_i \rangle$, $i \in \Delta$, and
- less precise than any other $\langle \overline{\mathbb{P}}, \overline{\sqsubseteq} \rangle$, which is closed by finite intersection and more precise than all the $\langle \overline{\mathbb{P}}_i, \overline{\sqsubseteq}_i \rangle$, $i \in \Delta$.

If $\langle \overline{\mathbb{P}}_i, \sqsubseteq_i, \top_i \rangle$, $i \in \Delta$ are closed by finite intersection then their reduced product $\langle \mathbb{P}^{\times}/_{\equiv^{\otimes}}, \sqsubseteq^{\otimes}, \top^{\otimes} \rangle$ is the unique such domain of abstract properties (up to the equivalence $\hat{\equiv}$).

Proof of theorem 36.14

— For all $i \in \Delta$ and $P_i \in \overline{\mathbb{P}}_i$, we have $\gamma_{\otimes}(\top^{\otimes}[i \leftarrow P_i]) \triangleq \bigsqcap_{j \in \Delta \setminus \{i\}} \gamma_i(\top_j) \sqcap \gamma_i(P_i) = \bigsqcap_{j \in \Delta \setminus \{i\}} \top \sqcap \gamma_i(P_i) = \top \sqcap \gamma_i(P_i) = \gamma_i(P_i)$ proving that $\gamma_i(\overline{\mathbb{P}}_i) \subseteq \gamma_{\otimes}(\mathbb{P}^{\times}/_{\equiv^{\otimes}})$ so that $\langle \mathbb{P}^{\times}/_{\equiv^{\otimes}}, \sqsubseteq^{\otimes} \rangle$ is more precise than the $\langle \overline{\mathbb{P}}_i, \overline{\sqsubseteq}_i \rangle$, $i \in \Delta$.

— If $\langle \overline{\mathbb{P}}, \overline{\sqsubseteq} \rangle$ with $\overline{\gamma} \in \overline{\mathbb{P}} \xrightarrow{\nearrow} \mathbb{P}$ is more precise than the $\langle \overline{\mathbb{P}}_i, \overline{\sqsubseteq}_i \rangle$ then $\gamma_i(\overline{\mathbb{P}}_i) \subseteq \overline{\gamma}(\overline{\mathbb{P}})$. Given $\vec{P} \in \mathbb{P}^{\times}/_{\equiv^{\otimes}}$, we have $\gamma_i(\vec{P}_i) \in \overline{\gamma}(\overline{\mathbb{P}})$ so there exists $\overline{P} \in \overline{\mathbb{P}}$ such that $\overline{\gamma}(\overline{P}) = \bigsqcap_{i \in \Delta} \gamma_i(\vec{P}_i) \triangleq \gamma_{\otimes}(\vec{P})$ because $\langle \overline{\mathbb{P}}, \overline{\sqsubseteq} \rangle$ is closed by finite intersection and Δ is finite. It follows that $\gamma_{\otimes}(\mathbb{P}^{\times}/_{\equiv^{\otimes}}) \subseteq \overline{\gamma}(\overline{\mathbb{P}})$.

— The property is characteristic because the $\langle \overline{\mathbb{P}}_i, \overline{\sqsubseteq}_i, \overline{\top}_i\rangle$, $i \in \Delta$ are closed by finite intersection so that their reduced product $\langle \mathbb{P}^{\times}/_{\equiv^{\otimes}}, \sqsubseteq^{\otimes}, \top^{\otimes}\rangle$ is also closed by intersection and therefore any other abstract domain $\langle \overline{\overline{\mathbb{P}}}, \overline{\overline{\sqsubseteq}}\rangle$ with the same property would have both $\gamma_{\otimes}(\mathbb{P}^{\times}/_{\equiv^{\otimes}}) \subseteq \overline{\overline{\gamma}}(\overline{\overline{\mathbb{P}}})$ and $\overline{\overline{\gamma}}(\overline{\overline{\mathbb{P}}}) \subseteq \gamma_{\otimes}(\mathbb{P}^{\times}/_{\equiv^{\otimes}})$ and hence would, by antisymmetry, be an equivalent domain of abstract properties. □

Remark 36.15 In the case that $\langle \mathbb{P}, \sqsubseteq\rangle$ is a complete lattice and the concretizations $\gamma_i \in \overline{\mathbb{P}}_i$ are upper adjoints of Galois connections, the abstraction amounts to an upper closure operator (see exercise 27.16). The possible abstract properties are the images $\{\rho(\mathbb{P}) \mid \rho \in \mathbb{P} \to \mathbb{P}$ is an upper closure operator$\}$ of the concrete properties $\mathbb{P}$ by these upper closure operators ρ. By Ward's theorem 11.90, it is a complete lattice (called the *lattice of abstractions* or the *lattice of abstract interpretations*). In this complete lattice the reduced product is the glb.

36.3.3 Abstract Domain Reduction

Different abstract domains $\overline{\mathbb{P}}_1 \mathrel{\hat{\equiv}} \overline{\mathbb{P}}_2$ with the same expressive power may yield different abstract properties that, after iteration, may yield sound but quite different results. In particular, this is the case when the best transformer is difficult to compute algorithmically, so that a strict overapproximation $\overline{\mathscr{F}}[\![\mathsf{P}]\!] \mathrel{\dot{\sqsupset}} \alpha \circ \mathscr{F}[\![\mathsf{P}]\!] \circ \gamma$ has to be used instead ($\dot{\sqsupset}$ is the pointwise strict inverse partial order $\sqsupset$). In such a case, reduction may be useful.

Example 36.16 Consider the abstraction of $\langle \wp(\mathbb{Z}), \subseteq\rangle$ by the complete lattice $\langle \overline{\mathbb{P}}, \overline{\sqsubseteq}\rangle$ such that

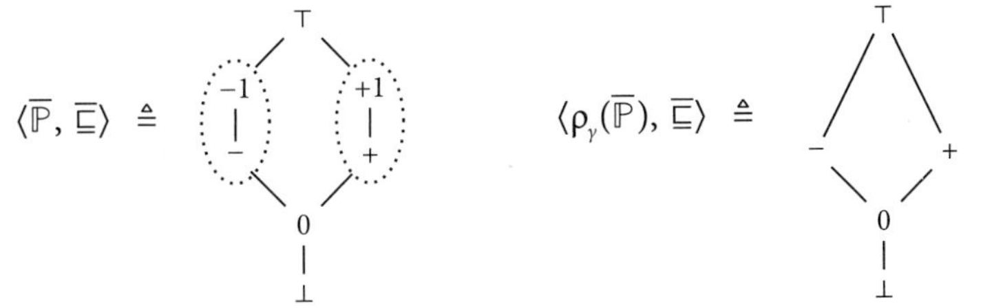

where $\overline{\mathbb{P}} \triangleq \{\bot, 0, +, +1, -, -1, \top\}$, $\bot \sqsubset 0 \sqsubset + \sqsubset +1 \sqsubset \top$ and $0 \sqsubset - \sqsubset -1 \sqsubset \top$ with $\gamma(+) = \gamma(+1) = \{z \in \mathbb{Z} \mid z \geqslant 0\}$ and $\gamma(-) = \gamma(-1) = \{z \in \mathbb{Z} \mid z \leqslant 0\}$. The positive and negative properties have distinct but equivalent encodings in the abstract. The two transformers $f_1(0) = f_1(+) = +$, $f_1(+1) = \top$ and $f_2(0) = +$, $f_2(+) = +1$, $f_2(+1) = \top$ are equivalent in the concrete in that $\forall \overline{P} \in \mathbb{P} \,.\, \gamma(f_1(\overline{P})) = \gamma(f_2(\overline{P}))$ but their composition is not because $\gamma(f_1(f_1(f_1(0)))) = \gamma(+) \neq \gamma(\top) = \gamma(f_2(f_2(f_2(0))))$.

Another example is the reduction of the direct product of abstract domains into their reduced product.

Example 36.17 Consider the direct product of sign and parity in section 36.2 with a reduction of the abstract value of x from *positive or zero* and *odd* to *strictly positive*. The forward transformer for x = x − 1; was the postcondition *unknown* and *even* for precondition *positive or zero* and *odd* of x but now yields *positive or zero* and *even*, a strict improvement of precision of signs.

The reduction of an abstract domain consists of eliminating the redundant abstract properties by putting them in normal/canonical form.

Example 36.18 Continuing example 36.16, let us define the reduction $\rho_\gamma(a) \triangleq \overline{\sqcap}\{a' \in \overline{\mathbb{P}} \mid \gamma(a) \sqsubseteq \gamma(a')\}$ such that $\rho_\gamma(+1) = +$, $\rho_\gamma(-1) = -$ and otherwise $\rho_\gamma(a) = a$. The reduced abstract domain is $\rho_\gamma(\overline{\mathbb{P}}) = \{\bot, 0, +, -, \top\}$ from which the redundant abstract properties $+1$ and -1 have been eliminated.

Theorem 36.19 (reduction operator) Let $\langle \mathbb{P}, \sqsubseteq\rangle \underset{\alpha}{\overset{\gamma}{\leftrightarrows}} \langle \overline{\mathbb{P}}, \overline{\sqsubseteq}\rangle$ where $\langle \overline{\mathbb{P}}, \overline{\sqsubseteq}, \overline{\bot}, \overline{\top}, \overline{\sqcup}, \overline{\sqcap}\rangle$ is a complete lattice. Define

$$\rho_\gamma(a) \triangleq \overline{\bigsqcap}\{a' \in \overline{\mathbb{P}} \mid \gamma(a) \sqsubseteq \gamma(a')\}$$

Then ρ_γ is a lower closure (reductive, increasing, and idempotent) on $\langle \overline{\mathbb{P}}, \overline{\sqsubseteq}\rangle$ and

$$\langle \mathbb{P}, \sqsubseteq\rangle \underset{\rho_\gamma \circ \alpha}{\overset{\gamma}{\leftrightarrows}} \langle \rho_\gamma(\overline{\mathbb{P}}), \overline{\sqsubseteq}\rangle .$$

Proof of theorem 36.19

— ρ_γ is reductive because $a \in \{a' \in \overline{\mathbb{P}} \mid \gamma(a) \sqsubseteq \gamma(a')\}$ by reflexivity and so $\rho_\gamma(a) \overline{\sqsubseteq} a$ by definition of the glb $\overline{\sqcap}$.

— If $a \overline{\sqsubseteq} b$ then $\gamma(a) \sqsubseteq \gamma(b)$ so $\gamma(b) \sqsubseteq \gamma(b')$ implies $\gamma(a) \sqsubseteq \gamma(b')$ hence $\{b' \mid \gamma(b) \sqsubseteq \gamma(b')\} \subseteq \{a' \mid \gamma(a) \sqsubseteq \gamma(a')\}$ so $\rho_\gamma(a) = \overline{\bigsqcap}\{a' \mid \gamma(a) \sqsubseteq \gamma(a')\} \overline{\sqsubseteq} \overline{\bigsqcap}\{b' \mid \gamma(b) \sqsubseteq \gamma(b')\} = \rho_\gamma(b)$.

— For idempotence, we have

$$
\begin{aligned}
& \rho_\gamma(\rho_\gamma(a)) \\
= {} & \overline{\bigsqcap}\{a' \in \overline{\mathbb{P}} \mid \gamma(\rho_\gamma(a)) \sqsubseteq \gamma(a')\} && \text{(definition of } \rho_\gamma\text{)} \\
= {} & \overline{\bigsqcap}\{a' \in \overline{\mathbb{P}} \mid \gamma(\overline{\bigsqcap}\{a'' \in \overline{\mathbb{P}} \mid \gamma(a) \sqsubseteq \gamma(a'')\}) \sqsubseteq \gamma(a')\} && \text{(definition of } \rho_\gamma\text{)} \\
= {} & \overline{\bigsqcap}\{a' \in \overline{\mathbb{P}} \mid \bigsqcap\{\gamma(a'') \in \overline{\mathbb{P}} \mid \gamma(a) \sqsubseteq \gamma(a'')\} \sqsubseteq \gamma(a')\} \\
& && \text{(in a Galois connection, } \gamma \text{ preserves preserves existing glbs)} \\
= {} & \overline{\bigsqcap}\{a' \in \overline{\mathbb{P}} \mid \gamma(a) \sqsubseteq \gamma(a')\} \\
& \text{(because } \gamma(a) = \bigsqcap\{\gamma(a'') \in \overline{\mathbb{P}} \mid \gamma(a) \sqsubseteq \gamma(a'')\} \text{ by reflexivity and definition of the glb)} \\
= {} & \rho_\gamma(a) && \text{(definition of } \rho_\gamma\text{)}
\end{aligned}
$$

— We conclude that ρ_γ is a lower closure (reductive, increasing, and idempotent) on $\overline{\mathbb{P}}$.

— If $x \in \mathbb{P}$ and $y \in \rho_\gamma(\overline{\mathbb{P}})$ then $y \in \overline{\mathbb{P}}$ so, by the Galois connection, $x \sqsubseteq \gamma(y)$ implies $\alpha(x) \overline{\sqsubseteq} y$ implies $\rho_\gamma \circ \alpha(x) \overline{\sqsubseteq} y$ because ρ_γ is reductive.

— Conversely, if $x \in \mathbb{P}$ and $y \in \rho_\gamma(\overline{\mathbb{P}})$ then

$$
\begin{aligned}
& \rho_\gamma \circ \alpha(x) \overline{\sqsubseteq} y \\
\Rightarrow {} & \overline{\bigsqcap}\{a' \in \overline{\mathbb{P}} \mid \gamma(\alpha(x)) \sqsubseteq \gamma(a')\} \overline{\sqsubseteq} y && \text{(definitions of } \circ \text{ and } \rho_\gamma\text{)} \\
\Rightarrow {} & \gamma(\overline{\bigsqcap}\{a' \in \overline{\mathbb{P}} \mid \gamma(\alpha(x)) \sqsubseteq \gamma(a')\}) \sqsubseteq \gamma(y) && \text{(in a Galois connection, } \gamma \text{ increasing)}
\end{aligned}
$$

$$\Rightarrow \quad (\overline{\bigsqcap}\{\gamma(a') \in \overline{\mathbb{P}} \mid \gamma(\alpha(x)) \sqsubseteq \gamma(a')\}) \sqsubseteq \gamma(y)$$

⦃in a Galois connection, γ preserves existing glbs⦄

$$\Rightarrow \quad \gamma \circ \alpha(x) \sqsubseteq \gamma(y)$$ ⦃reflexivity for $a' = \alpha(x)$ and definition of the glb⦄

$$\Rightarrow \quad x \sqsubseteq \gamma(y)$$ ⦃in a Galois connection, $\gamma \circ \alpha$ is extensive and transitivity⦄ □

Definition 36.20 (meaning-preserving map/reduction) Let the poset $\langle \overline{\mathbb{P}}, \overline{\sqsubseteq} \rangle$ be an abstract domain with concretization $\gamma \in \overline{\mathbb{P}} \xrightarrow{\nearrow} \mathbb{P}$ into the concrete poset $\langle \mathbb{P}, \sqsubseteq \rangle$. A *meaning-preserving map* is $\rho \in \overline{\mathbb{P}} \to \overline{\mathbb{P}}$ such that $\forall \overline{P} \in \overline{\mathbb{P}} \,.\, \gamma(\rho(\overline{P})) = \gamma(\overline{P})$. The map is a *reduction* if and only if $\forall \overline{P} \in \overline{\mathbb{P}} \,.\, \rho(\overline{P}) \overline{\sqsubseteq} \overline{P}$.

Lemma 36.21 The composition of finitely many meaning-preserving reductions is a meaning-preserving reduction.

Proof By recurrence. For the induction step, observe that $\gamma \circ (\rho_1 \circ \rho_2) = (\gamma \circ \rho_1) \circ \rho_2 = \gamma \circ \rho_2 = \gamma$ and $\forall \overline{P} \in \overline{\mathbb{P}} \,.\, (\rho_1 \circ \rho_2)(\overline{P}) = \rho_1(\rho_2(\overline{P})) \overline{\sqsubseteq} \rho_2(\overline{P}) \overline{\sqsubseteq} \overline{P}$. □

The following theorem shows that the reduction ρ_γ is meaning-preserving, that is, does not change the expressiveness of the reduced abstract properties.

Theorem 36.22 With the hypotheses of theorem 36.19, ρ_γ is meaning-preserving and $\gamma = \gamma \circ \rho_\gamma$.

Proof of theorem 36.22 For all $x \in \mathbb{P}$:

$$\gamma \circ \rho_\gamma \circ \alpha(x)$$
$$= \gamma(\overline{\bigsqcap}\{a \mid \gamma(\alpha(x)) \sqsubseteq \gamma(a)\})$$ ⦃definition of ρ_γ⦄
$$= \bigsqcap\{\gamma(a) \mid \gamma(\alpha(x)) \sqsubseteq \gamma(a)\})$$ ⦃in a Galois connection, γ preserves meets⦄
$$= \gamma(\alpha(x))$$ ⦃choosing $a = \alpha(x)$ and definition of the glb⦄

and so

$$\gamma$$
$$= \gamma \circ \alpha \circ \gamma$$ ⦃Galois connection⦄
$$= \gamma \circ \rho_\gamma \circ \alpha \circ \gamma$$ ⦃because $\gamma \circ \alpha = \gamma \circ \rho_\gamma \circ \alpha$⦄
$$\dot{\sqsubseteq} \; \gamma \circ \rho_\gamma$$ ⦃$\alpha \circ \gamma$ is reductive and γ and ρ_γ are increasing⦄

Moreover ρ_γ is a lower closure on $\langle \overline{\mathbb{P}}, \overline{\sqsubseteq} \rangle$ so ρ_γ is reductive ($\rho_\gamma \dot{\overline{\sqsubseteq}} \mathbb{1}$), hence $\gamma \circ \rho_\gamma \dot{\sqsubseteq} \gamma$ because γ is increasing. By antisymmetry, $\gamma \circ \rho_\gamma = \gamma$. □

As shown by example 36.16, although abstract domains and transformers may have equivalent concretizations, the equivalent but more precise abstract properties may propagate through several transformers with the resulting transformed properties and fixpoint computations that are more precise in the concrete.

Notice also that an implementation of the reduction of an abstract domain $\overline{\mathbb{P}}$ does not need considerable modifications of the implementation of $\overline{\mathbb{P}}$. The reduction operator ρ_γ may simply be applied before and after the operations of abstract domain $\overline{\mathbb{P}}$ (although maybe not after a widening/narrowing because the reduction might introduce divergences; see section 36.4.5).

Exercise 36.23 Prove that ρ is a lower closure on a poset $\langle P, \sqsubseteq\rangle$ (reductive, increasing, and idempotent) on the poset $\langle P, \sqsubseteq\rangle$ if and only if $\langle P, \sqsubseteq\rangle \xleftarrow[\rho]{\mathbb{1}_P} \langle P, \sqsubseteq\rangle$. What is the dual?

36.3.4 The Reduced Product is the Meaning-Preserving Reduction of the Direct Product

The reduced product of abstract domains $\overline{\mathbb{P}}_i$, $i \in \Delta$ can be encoded by a reduction of the direct product $\prod_{i\in\Delta} \overline{\mathbb{P}}_i$ using a meaning-preserving reduction operator $\vec{\rho}$ mapping any element of the direct product to the smallest representative of its equivalence class. This yields a more constructive definition of reduction and later leads to algorithms to perform or approximate this reduction.

Theorem 36.24 (equivalent definition of the reduced product (III)) Assume that $\langle \mathbb{P}, \sqsubseteq\rangle$ is a poset and $\forall i \in \Delta$, Δ finite, each $\langle \overline{\mathbb{P}}_i, \overline{\sqsubseteq}_i, \overline{\bot}_i, \overline{\top}_i, \overline{\sqcup}_i, \overline{\sqcap}_i\rangle$ is a complete lattice such that $\langle \mathbb{P}, \sqsubseteq\rangle \xleftarrow[\alpha_i]{\gamma_i} \langle \overline{\mathbb{P}}_i, \overline{\sqsubseteq}_i\rangle$. Let $\langle \mathbb{P}^\times, \sqsubseteq^\times\rangle$ be the direct product of the $\langle \overline{\mathbb{P}}_i, \overline{\sqsubseteq}_i\rangle$, $i \in \Delta$, with concretization $\gamma_\times$ as in definition 36.1. Define $\alpha_\times \triangleq x \mapsto \prod_{i\in\Delta} \alpha_i(x)$ such that $\langle \mathbb{P}, \sqsubseteq\rangle \xleftarrow[\alpha_\times]{\gamma_\times} \langle \mathbb{P}^\times, \sqsubseteq^\times\rangle$. Let $\vec{\rho} \triangleq \vec{P} \mapsto \vec{\bigsqcap}\{\vec{P}' \mid \gamma_\times(\vec{P}) \sqsubseteq \gamma_\times(\vec{P}')\}$ such that $\vec{\bigsqcap}$ is the componentwise join and $\langle \mathbb{P}, \sqsubseteq\rangle \xleftarrow[\vec{\rho}\circ\alpha_\times]{\gamma_\times} \langle \vec{\rho}(\mathbb{P}^\times), \sqsubseteq^\times\rangle$.

Then $\vec{\rho}$ is meaning-preserving and $\langle \vec{\rho}(\mathbb{P}^\times), \sqsubseteq^\times\rangle$ is the reduced product of the $\langle \overline{\mathbb{P}}_i, \overline{\sqsubseteq}_i\rangle$, $i \in \Delta$.

Proof of theorem 36.24

— $\langle \mathbb{P}, \sqsubseteq\rangle \xleftarrow[\alpha_\times]{\gamma_\times} \langle \mathbb{P}^\times, \sqsubseteq^\times\rangle$ and $\langle \mathbb{P}^\times, \sqsubseteq^\times\rangle$ is a complete lattice follow componentwise from the hypothesis $\forall i \in \Delta \,.\, \langle \mathbb{P}, \sqsubseteq\rangle \xleftarrow[\alpha_i]{\gamma_i} \langle \overline{\mathbb{P}}_i, \overline{\sqsubseteq}_i\rangle$ and $\langle \overline{\mathbb{P}}_i, \overline{\sqsubseteq}_i\rangle$ is a complete lattice.

— $\vec{\rho}$ is defined as ρ_γ in theorem 36.19 and so $\langle \mathbb{P}, \sqsubseteq\rangle \xleftarrow[\vec{\rho}\circ\alpha_\times]{\gamma_\times} \langle \vec{\rho}(\mathbb{P}^\times), \sqsubseteq^\times\rangle$, $\vec{\rho}$ is a lower closure operator. Therefore, by theorem 36.22, it is meaning preserving.

— By definition of $(\vec{P} \equiv^\otimes \vec{Q}) \triangleq (\gamma^\times(\vec{P}) = \gamma^\times(\vec{Q}))$ in definition 36.7, it follows that $[\vec{P}]_{\equiv^\otimes} = [\vec{\rho}(\vec{P})]_{\equiv^\otimes}$. Therefore $\langle \vec{\rho}(\mathbb{P}^\times), \sqsubseteq^\times\rangle$ is a complete lattice isomorphic to the reduced product $\langle \mathbb{P}^\otimes, \sqsubseteq^\otimes\rangle$ of the $\langle \overline{\mathbb{P}}_i, \overline{\sqsubseteq}_i\rangle$, $i \in \Delta$. □

36.4 Iterated Pairwise Reduction

36.4.1 Pairwise Reduction

The strong reduction operators ρ_γ of theorem 36.19 and $\vec{\rho}$ of theorem 36.24 may be difficult to compute algorithmically because they involve all abstract domains of the direct product. In adding a new abstract domain to increase precision of the analysis, the strong reduction between all domains must be completely redesigned.

For simplicity, weaker reductions are useful. Pairwise reduction is an example of such a weak meaning-preserving of the direct product reduction. It applies between two abstract domains only and leaves the other domains unchanged. When a new abstract domain is added to the product of abstract domains of an abstract interpreter, one need only to design a pairwise reductions between the new abstract domain and (some of) the existing ones.

Several weak reductions can be strengthen by iteration. This provides a general algorithm for constructing/approximating reduced products by iterated pairwise reductions, which can be implemented once for all in the static analyzer/verifier.

Definition 36.25 (pairwise reduction) Let $\langle \overline{\mathbb{P}}_i, \overline{\sqsubseteq}_i \rangle$ be abstract domains with increasing concretization $\gamma_i \in \overline{\mathbb{P}}_i \xrightarrow{\nearrow} \mathbb{P}$ into the concrete domain $\langle \mathbb{P}, \sqsubseteq, \sqcap \rangle$. Define $\gamma_\times(\vec{P}) \triangleq \bigwedge_{i \in \Delta} \gamma_i(\vec{P}_i)$ (as in definition 36.1).

For $i, j \in \Delta$, $i \neq j$, let $\rho_{ij} \in \langle \overline{\mathbb{P}}_i \times \overline{\mathbb{P}}_j, \overline{\sqsubseteq}_{ij} \rangle \mapsto \langle \overline{\mathbb{P}}_i \times \overline{\mathbb{P}}_j, \overline{\sqsubseteq}_{ij} \rangle$ be pairwise meaning-preserving reductions (so that $\forall \langle x, y \rangle \in \overline{\mathbb{P}}_i \times \overline{\mathbb{P}}_j \,.\, \rho_{ij}(\langle x, y \rangle) \overline{\sqsubseteq}_{ij} \langle x, y \rangle$ and $(\gamma_i \times \gamma_j) \circ \rho_{ij} = (\gamma_i \times \gamma_j)$[a]).

Define the pairwise reductions $\vec{\rho}_{ij} \in \langle \mathbb{P}^\times, \sqsubseteq^\times \rangle \mapsto \langle \mathbb{P}^\times, \sqsubseteq^\times \rangle$ of the direct product as

$$\vec{\rho}_{ij}(\vec{P}) \;\triangleq\; \text{let } \langle \vec{P}'_i, \vec{P}'_j \rangle \triangleq \rho_{ij}(\langle \vec{P}_i, \vec{P}_j \rangle) \text{ in } \vec{P}[i \leftarrow \vec{P}'_i][j \leftarrow \vec{P}'_j]$$

where $\vec{P}[i \leftarrow x]_i = x$ and $\vec{P}[i \leftarrow x]_j = \vec{P}_j$ when $i \neq j$.

a. We define $(f \times g)(\langle x, y \rangle) \triangleq \langle f(x), g(y) \rangle$.

Exercise 36.26 Design a meaning-preserving reduction between the interval abstract domain of chapter 33 and the Cartesian congruences of chapter 31, and inversely. See [147] for a generalization to linear congruences.

As shown in example 36.18, the reduction can be useful in a single domain to put abstract properties into so-called normal or canonical form.

Definition 36.27 (composition of pairwise reductions) Define the composition $\vec{\rho}$ of pairwise reductions $\{\vec{\rho}_{ij} \mid i, j \in \Delta\}$ of the direct product

$$\vec{\rho} \;\triangleq\; \bigcirc_{i,j \in \Delta} \vec{\rho}_{ij} \qquad (36.28)$$

where $\bigcirc_{i=1}^{n} f_i \triangleq f_{\pi_1} \circ \ldots \circ f_{\pi_n}$ is the function composition for some arbitrary permutation π of $[1, n]$.

Observe that $\vec{\rho}$ is the composition of finitely many meaning-preserving reductions so, by lemma 36.21, it is itself a meaning-preserving reduction.

36.4.2 Finitely Iterated Weak Reduction

The precision of weak reductions during a static analysis can be improved in the abstract by iteration without altering its soundness. This yields a weaker form of reduced product by iteration of weak reductions.

Definition 36.29 (iterated reduction) Let the poset $\langle \overline{\mathbb{P}}, \overline{\sqsubseteq} \rangle$ be an abstract domain with concretization $\gamma \in \overline{\mathbb{P}} \xrightarrow{\nearrow} \mathbb{P}$ where $\langle \mathbb{P}, \sqsubseteq \rangle$ is the concrete domain and let $\vec{\rho} \in \overline{\mathbb{P}} \to \overline{\mathbb{P}}$ be a meaning-preserving reduction ($\gamma \circ \vec{\rho} = \gamma$ and $\vec{\rho} \,\dot{\overline{\sqsubseteq}}\, \mathbb{1}_{\overline{\mathbb{P}}}$).

The *finite iterates* of the reduction are $\vec{\rho}^{\,0} \triangleq \overline{P} \mapsto \overline{P}$, $\vec{\rho}^{\,n+1} = \vec{\rho} \circ \vec{\rho}^{\,n}$ for $n \in \mathbb{N}$.

Theorem 36.30 (finite iterated reduction) The finite iterates $\vec{\rho}^{\,n}$, $n \in \mathbb{N}$ of a meaning-preserving reduction $\vec{\rho}$ on $\langle \mathbb{P}, \overline{\sqsubseteq} \rangle$ are more precise in the abstract (because $\forall \vec{P} \in \mathbb{P}^{\times} . \forall n \in \mathbb{N}^{+} . \vec{\rho}^{\,n}(\vec{P}) \sqsubseteq^{\otimes} \vec{\rho}(\vec{P}) \sqsubseteq^{\otimes} \vec{\rho}_{ij}(\vec{P}) \sqsubseteq^{\otimes} \vec{P}$, $i, j \in \Delta$) and meaning-preserving (because $\vec{\rho}^{\,n}(\vec{P}), \vec{\rho}(\vec{P}), \vec{\rho}_{ij}(\vec{P}), \vec{P} \in [\vec{P}]_{\equiv^{\otimes}}$).

Proof of theorem 36.30 Let $\langle \rho^n, n \in \mathbb{N} \rangle$ be the iterates of a meaning-preserving reduction ρ. Observe, by recurrence, that the iterates form a descending chain because ρ is reductive so $\forall n < m . \rho^n(\overline{P}) \sqsubseteq \rho^n(\overline{P}) \sqsubseteq \overline{P}$. Meaning-preservation follows by recurrence. For the basis, $\gamma(\rho^0(\overline{P})) = \gamma(\overline{P})$ by definition of ρ^0. For induction, $\gamma(\rho^{n+1}(\overline{P})) \triangleq \gamma(\rho(\rho^n(\overline{P}))) = \gamma(\rho^n(\overline{P})) = \gamma(\overline{P})$, because ρ is meaning-preserving and by induction hypothesis. □

36.4.3 Iterated Weak Reduction is Less Precise than the Reduced Product

The following examples prove that the iterated reduction may not be as precise as the reduced product. It is nevertheless easier to implement.

Example 36.31 (iterated pairwise reduction of the direct product is not minimal) Let us consider $\mathbb{P} = \wp(\{a, b, c\})$, $\overline{\mathbb{P}}_1 = \{\varnothing, \{a\}, \top\}$, $\overline{\mathbb{P}}_2 = \{\varnothing, \{a, b\}, \top\}$, $\overline{\mathbb{P}}_3 = \{\varnothing, \{a, c\}, \top\}$, and $\vec{P} = \langle \top, \{a, b\}, \{a, c\} \rangle$ where $\top = \{a, b, c\}$. We have $\langle \top, \{a, b\}, \{a, c\} \rangle /_{\equiv^{\otimes}} = \langle \{a\}, \{a, b\}, \{a, c\} \rangle$. However, $\vec{\rho}_{\vec{ij}}(\langle \top, \{a, b\}, \{a, c\} \rangle) = \langle \top, \{a, b\}, \{a, c\} \rangle$ for $\Delta = \{1, 2, 3\}$, $i, j \in \Delta, i \neq j$ and so $\vec{\rho}^{\,*}(\langle \top, \{a, b\}, \{a, c\} \rangle) = \langle \top, \{a, b\}, \{a, c\} \rangle$, proving, for that example, that $\vec{\rho}^{\,\star}(\langle \top, \{a, b\}, \{a, c\} \rangle)$ is not a minimal element of $[\langle \top, \{a, b\}, \{a, c\} \rangle]_{\equiv^{\otimes}}$.

Example 36.32 (Nelson-Oppen algorithm) The satisfiability modulo theories (SMT) problem is a decision problem for logical formulas with respect to combinations of background theories expressed in classic first-order logic with equality. The Nelson-Oppen algorithm [748] combines decision procedures for these background theories. It is an iterated pairwise reduction [275]. The reduction consists of propagating the equalities and inequalities between variables derived from one theory to all other theories. This is incomplete. For example, assume that a theory of signs derives $x \geqslant 0$ and a theory of parity deriving that x is odd. None derives $x \neq 0$ because this is not an inequality between variables. Hence $x \geqslant 0$ is not reduced to $x > 0$. Completeness can be obtained by putting restrictions on the background theories, for example, that they do not share symbols. Hence we cannot have both sign and parity background theories that share +, −, and so forth. Such restrictions to ensure completeness are meaningless in program verification using SMT solvers because program verification and analysis are, by Rice's theorem 9.12, undecidable anyway.

36.4.4 (Reduced) product transformers

The forward transformers $\mathsf{assign}^{\vec{\varrho}}[\![\mathsf{x}, \mathsf{A}]\!]$ (19.12) are

$$\begin{aligned}
\mathsf{assign}^{\vec{\mathsf{r}}}[\![\mathsf{x}, \mathsf{A}]\!]\, P &\triangleq \mathsf{post}[\{\langle\rho,\ \rho[\mathsf{x} \leftarrow \mathscr{A}[\![\mathsf{A}]\!]\rho]\rangle \mid \rho \in \mathbb{E}\mathbb{v}\}]P \\
&= \{\rho[\mathsf{x} \leftarrow \mathscr{A}[\![\mathsf{A}]\!]\rho] \mid \rho \in P\} \qquad (19.12)\\
\mathsf{assign}^{\vec{\mathsf{R}}}[\![\mathsf{x}, \mathsf{A}]\!]\, P &\triangleq \mathsf{post}[\{\langle\langle\rho_0,\ \rho\rangle,\ \langle\rho_0,\ \rho[\mathsf{x} \leftarrow \mathscr{A}[\![\mathsf{A}]\!]\rho]\rangle\rangle \mid \rho_0, \rho \in \mathbb{E}\mathbb{v}\}]P \\
&= \{\langle\rho_0,\ \rho[\mathsf{x} \leftarrow \mathscr{A}[\![\mathsf{A}]\!]\rho]\rangle \mid \langle\rho_0,\ \rho\rangle \in P\}
\end{aligned}$$

Their reduced product abstractions proceed componentwise and reduce the result.

Lemma 36.33 Let us consider a reduced product $\langle(\prod_{i\in\Delta}\overline{\mathbb{P}}_i)\,/_{\equiv^{\otimes}},\ \sqsubseteq^{\otimes}\rangle$ of abstract domains $\langle\overline{\mathbb{P}}_i,\ \sqsubseteq_i\rangle$, $i \in \Delta$ with concretizations $\gamma_i \in \overline{\mathbb{P}}_i \stackrel{\nearrow}{\longrightarrow} C$ and sound forward transformers $\mathsf{assign}^{\dot{\vec{\text{¤}}}_i}[\![\mathsf{x}, \mathsf{A}]\!]$ such that $\mathsf{assign}^{\vec{\varrho}}[\![\mathsf{x}, \mathsf{A}]\!]\gamma_i(\overline{P}) \subseteq \gamma_i(\mathsf{assign}^{\dot{\vec{\text{¤}}}_i}[\![\mathsf{x}, \mathsf{A}]\!]\overline{P})$ where $\mathsf{assign}^{\vec{\varrho}}[\![\mathsf{x}, \mathsf{A}]\!] \in C \stackrel{\nearrow}{\longrightarrow} C$ is the increasing concrete forward transformer. It is similar for backward transformers.

The corresponding transformer of a property $\vec{P} \in \prod_{i\in\Delta}\overline{\mathbb{P}}_i$ in the product is the reduction

$$\left(\prod_{i\in\Delta}\mathsf{assign}^{\dot{\vec{\text{¤}}}_i}[\![\mathsf{x}, \mathsf{A}]\!](\vec{P}_i)\right)/_{\equiv^{\otimes}}$$

of the componentwise transformation. (It may also be useful to reduce the argument $\vec{P}$ if this is not already the case.)

This is sound because $\mathsf{assign}^{\vec{\varrho}}[\![\mathsf{x}, \mathsf{A}]\!](\gamma_{\otimes}(\vec{P})) \subseteq \gamma_{\otimes}\left(\left(\prod_{i\in\Delta}\mathsf{assign}^{\dot{\vec{\text{¤}}}_i}[\![\mathsf{x}, \mathsf{A}]\!](\vec{P}_i)\right)/_{\equiv^{\otimes}}\right)$ and it is similar for other transformers.

Proof of lemma 36.33

$$\begin{aligned}
&\gamma_{\otimes}\left(\left(\prod_{i\in\Delta}\mathsf{assign}^{\dot{\vec{\text{¤}}}_i}[\![\mathsf{x}, \mathsf{A}]\!](\vec{P}_i)\right)/_{\equiv^{\otimes}}\right)\\
=\ &\gamma_{\otimes}\left(\prod_{i\in\Delta}\mathsf{assign}^{\dot{\vec{\text{¤}}}_i}[\![\mathsf{x}, \mathsf{A}]\!](\vec{P}_i)\right) && \text{(definition of the reduced product)}\\
=\ &\bigcap_{i\in\Delta}\gamma_i(\mathsf{assign}^{\dot{\vec{\text{¤}}}_i}[\![\mathsf{x}, \mathsf{A}]\!](\vec{P}_i)) && \text{(definition of } \gamma_{\otimes}\text{)}\\
\supseteq\ &\bigcap_{i\in\Delta}\mathsf{assign}^{\vec{\varrho}}[\![\mathsf{x}, \mathsf{A}]\!](\gamma_i(\vec{P}_i)) && \text{(soundness of the } \mathsf{assign}^{\dot{\vec{\text{¤}}}_i}[\![\mathsf{x}, \mathsf{A}]\!]\text{)}\\
\supseteq\ &\mathsf{assign}^{\vec{\varrho}}[\![\mathsf{x}, \mathsf{A}]\!]\left(\bigcap_{i\in\Delta}\gamma_i(\vec{P}_i)\right) && \text{(} \mathsf{assign}^{\vec{\varrho}}[\![\mathsf{x}, \mathsf{A}]\!] \text{ is increasing)}\\
=\ &\mathsf{assign}^{\vec{\varrho}}[\![\mathsf{x}, \mathsf{A}]\!](\gamma_{\otimes}(\vec{P})) && \text{(definition of } \gamma_{\otimes}\text{.)} \quad \square
\end{aligned}$$

Unfortunately, this definition of the product transformer is not modular because it must be entirely redesigned in adding a new abstract domain to the product. Notice, however, that abstract transformers themselves are elements of a reduced product, by defining their concretization as follows:

Lemma 36.34 $\gamma_{\otimes}\left(\prod_{i\in\Delta} \mathsf{assign}^{\dot{\vec{\alpha}}_i}[\![\mathsf{x}, \mathsf{A}]\!](\vec{P}_i)\right) = \vec{\gamma}\left(\prod_{i\in\Delta} \mathsf{assign}^{\dot{\vec{\alpha}}_i}[\![\mathsf{x}, \mathsf{A}]\!]\right)(\vec{P})$.

Proof of lemma 36.34

$$
\begin{array}{ll}
& \gamma_{\otimes}\left(\prod\limits_{i\in\Delta} \mathsf{assign}^{\dot{\vec{\alpha}}_i}[\![\mathsf{x}, \mathsf{A}]\!](\vec{P}_i)\right) \\
= & \bigcap\limits_{i\in\Delta} \gamma_i(\mathsf{assign}^{\dot{\vec{\alpha}}_i}[\![\mathsf{x}, \mathsf{A}]\!](\vec{P}_i)) \qquad \text{(definition of } \gamma_{\otimes}\text{)} \\
= & \bigcap\limits_{i\in\Delta} \dot{\gamma}_i(\mathsf{assign}^{\dot{\vec{\alpha}}_i}[\![\mathsf{x}, \mathsf{A}]\!])(\vec{P}) \qquad \text{(pointwise definition of } \dot{\gamma}_i(f)(\vec{x}) \triangleq \gamma_i(f(\vec{x}_i))\text{)} \\
= & \left(\dot{\bigcap\limits_{i\in\Delta}} \dot{\gamma}_i(\mathsf{assign}^{\dot{\vec{\alpha}}_i}[\![\mathsf{x}, \mathsf{A}]\!])\right)(\vec{P}) \qquad \text{(pointwise definition of } \left(\dot{\bigcap\limits_{i\in\Delta}} f_i\right)(x) \triangleq \bigcap\limits_{i\in\Delta} f_i(x)\text{)} \\
= & \vec{\gamma}\left(\prod\limits_{i\in\Delta} \mathsf{assign}^{\dot{\vec{\alpha}}_i}[\![\mathsf{x}, \mathsf{A}]\!]\right)(\vec{P}) \qquad \text{(definition of } \vec{\gamma}(\prod\limits_{i\in\Delta} f_i) \triangleq \dot{\bigcap\limits_{i\in\Delta}} \dot{\gamma}_i(f_i)) \text{ for products.)} \quad \square
\end{array}
$$

A direct consequence is that we can approximate the product transformer by iterated reduction of the componentwise transformers. Let us define the backward transformer $\mathsf{assign}^{\tilde{\bar{\varrho}}}[\![\mathsf{x}, \mathsf{A}]\!]$

$$
\begin{aligned}
\mathsf{assign}^{\tilde{\bar{\mathsf{r}}}}[\![\mathsf{x}, \mathsf{A}]\!]\, Q &\triangleq \widetilde{\mathsf{pre}}[\{\langle\rho,\ \rho[\mathsf{x} \leftarrow \mathscr{A}[\![\mathsf{A}]\!]\rho]\rangle \mid \rho \in \mathbb{E}\mathrm{v}\}]Q \\
&= \{\rho \mid \rho[\mathsf{x} \leftarrow \mathscr{A}[\![\mathsf{A}]\!]\rho] \in Q\} \qquad (36.35)\\
\mathsf{assign}^{\tilde{\bar{\mathsf{R}}}}[\![\mathsf{x}, \mathsf{A}]\!]\, Q &\triangleq \widetilde{\mathsf{pre}}[\{\langle\langle\rho_0,\ \rho\rangle,\ \langle\rho_0,\ \rho[\mathsf{x} \leftarrow \mathscr{A}[\![\mathsf{A}]\!]\rho]\rangle\rangle \mid \rho_0, \rho \in \mathbb{E}\mathrm{v}\}] \\
&= \{\langle\rho_0,\ \rho\rangle \mid \langle\rho_0,\ \rho[\mathsf{x} \leftarrow \mathscr{A}[\![\mathsf{A}]\!]\rho]\rangle \in Q\}
\end{aligned}
$$

Observe that, by (12.22), we have the following Galois connection

Lemma 36.36 $\mathsf{assign}^{\vec{\varrho}}[\![\mathsf{x}, \mathsf{A}]\!]\, P \subseteq Q \Leftrightarrow P \subseteq \mathsf{assign}^{\tilde{\bar{\varrho}}}[\![\mathsf{x}, \mathsf{A}]\!]\, Q$.

It follows that if $\mathsf{assign}^{\vec{\varrho}}[\![\mathsf{x}, \mathsf{A}]\!]\, P \subseteq Q$ then $\mathsf{assign}^{\vec{\varrho}}[\![\mathsf{x}, \mathsf{A}]\!]\ (\mathsf{assign}^{\tilde{\bar{\varrho}}}[\![\mathsf{x}, \mathsf{A}]\!]\, Q)$ is a more precise, sound overapproximation of $\mathsf{assign}^{\vec{\varrho}}[\![\mathsf{x}, \mathsf{A}]\!]\, P$ than Q, which suggests the following pairwise reduction $\dot{\rho}_{ij}$ of transformers (based on the pairwise reduction ρ_{ij} of abstract properties):

$$
\begin{aligned}
&\dot{\rho}_{ij}(\langle \mathsf{assign}^{\dot{\vec{\alpha}}_i}[\![\mathsf{x}, \mathsf{A}]\!],\ \mathsf{assign}^{\dot{\vec{\alpha}}_j}[\![\mathsf{x}, \mathsf{A}]\!]\rangle) \triangleq \\
&\quad \langle x,\ y\rangle \mapsto \text{let } \langle x',\ y'\rangle \triangleq \rho_{ij}(\langle \mathsf{assign}^{\dot{\vec{\alpha}}_i}[\![\mathsf{x}, \mathsf{A}]\!](x),\ \mathsf{assign}^{\dot{\vec{\alpha}}_j}[\![\mathsf{x}, \mathsf{A}]\!]\rangle(y)) \text{ in} \\
&\qquad \text{let } \langle\, \grave{x},\ \grave{y}\rangle \triangleq \rho_{ij}(\langle x' \sqcap_i \mathsf{assign}^{\dot{\overleftarrow{\alpha}}_i}[\![\mathsf{x}, \mathsf{A}]\!](x),\ y' \sqcap_j \mathsf{assign}^{\dot{\overleftarrow{\alpha}}_j}[\![\mathsf{x}, \mathsf{A}]\!](y)\rangle) \text{ in} \\
&\qquad\quad \rho_{ij}(\langle \mathsf{assign}^{\dot{\vec{\alpha}}_i}[\![\mathsf{x}, \mathsf{A}]\!](\grave{x}),\ \mathsf{assign}^{\dot{\vec{\alpha}}_j}[\![\mathsf{x}, \mathsf{A}]\!](\grave{y})\rangle)
\end{aligned}
$$

which defines a reduction $\dot{\vec{\rho}}$ of transformers by (36.28) lifting the reduction $\vec{\rho}$ to the product of higher-order abstract properties.

Example 36.37 Consider the product of equality and sign analysis. The componentwise forward propagation of $\langle \mathsf{a} = \mathsf{b},\ \top\rangle$ through the assignment $\mathsf{a} := \sqrt{\mathsf{b}} + \mathsf{a}$ is $\langle\top,\ \mathsf{b} \geqslant 0\rangle$ (with runtime error when $\mathsf{b} < 0$ in which case execution is assumed to stop). The backward propagation yields the precondition $\langle \mathsf{a} = \mathsf{b},\ \mathsf{b} \geqslant 0\rangle$ reduced to $\langle \mathsf{a} = \mathsf{b},$

$\mathsf{b} \geqslant 0 \wedge \mathsf{a} \geqslant 0\rangle$, whose forward propagation is now reduced to $\langle \top,\ \mathsf{a} \geqslant 0 \wedge \mathsf{b} \geqslant 0\rangle$. Hence the reduced componentwise forward propagation of $\langle \mathsf{a} = \mathsf{b},\ \top\rangle$ through the assignment $\mathsf{a} := \sqrt{\mathsf{b}} + \mathsf{a}$ is $\langle \top,\ \mathsf{a} \geqslant 0 \wedge \mathsf{b} \geqslant 0\rangle$, which is more precise than $\langle \top,\ \mathsf{b} \geqslant 0\rangle$.

Example 36.38 An iterated reduction of the product of linear equalities and sign analyses of $\mathsf{test}^{\vec{e}}[\![(\mathsf{x} = \mathsf{y}) \wedge ((\mathsf{z} + 1) = \mathsf{x}) \wedge (\mathsf{y} = \mathsf{z})]\!]$ with precondition $\mathsf{x} = 0$ yields the postcondition $\mathsf{x} = 0 \wedge \mathsf{y} = 0 \wedge \mathsf{z} < 0$ (See [225, Sect. 13.9]).

36.4.5 Widening

The widening/narrowing [235] of a reduced product is often defined componentwise using widenings/narrowings of the component abstract domains. This ensures convergence for the product. However, it must be proved that the reduction does not break down the termination of the product widening, in which case reduction must be weakened or the widening strengthened [215].

Example 36.39 The closure operation in the zone abstract domain of chapter 40 can be considered a reduction between separate domains, each considering only a pair of variables: if one applies the classic widening operation on zones followed by closure (reduction), then termination is no longer ensured as shown in section 40.1.8 (the same holds for the octagon abstract domain; see [692, fig. 25–26]).

36.4.6 Practical Implementation of the Iterated Pairwise Reduction

Abstract interpreters are often built incrementally, by successively adding new abstract domains and combining them with already existing ones by pairwise iterated reduction.

In order to avoid modifying existing domains each time a new one is introduced, one can use a common interface $\mathbb{C}\mathrm{c}$, called a communication channel [178, 267], between abstract domains. The communication channel represents the information shared between the abstract domains of the reduced product, as defined by a concretization $\gamma_{\mathbb{C}\mathrm{c}} \in \mathbb{C}\mathrm{c} \to \mathbb{P}$. Then abstract domains $\overline{\mathbb{P}}_i$ are enriched with two primitive operations.

$\mathsf{send} \in \overline{\mathbb{P}}_i \times \mathbb{C}\mathrm{c} \to \mathbb{C}\mathrm{c}$ is used to communicate information of abstract domain $\overline{\mathbb{P}}_i$ to the other abstract domains involved in the reduced product, through the communication channel. For soundness, it must satisfy $\forall \overline{P} \in \overline{\mathbb{P}}_i, \overline{Q} \in \mathbb{C}\mathrm{c} \,.\, \gamma_i(\overline{P}) \sqcap \gamma_{\mathbb{C}\mathrm{c}}(\overline{Q}) \sqsubseteq \gamma_{\mathbb{C}\mathrm{c}}(\mathsf{send}(\overline{P}, \overline{Q}))$.

$\mathsf{receive} \in \overline{\mathbb{P}}_i \times \mathbb{C}\mathrm{c} \to \overline{\mathbb{P}}_i$ is used to receive information from the other abstract domains, through the communication channel, to improve the information encoded by the abstract domain $\overline{\mathbb{P}}_i$. It must satisfy $\forall \overline{P} \in \overline{\mathbb{P}}_i, \overline{Q} \in \mathbb{C}\mathrm{c} \,.\, \gamma_i(\overline{P}) \sqcap \gamma_{\mathbb{C}\mathrm{c}}(\overline{Q}) \sqsubseteq \gamma_i(\mathsf{receive}(\overline{P}, \overline{Q}))$.

The abstract domain $\overline{\mathbb{P}}^i$ uses the received information to improve the precision of its operations $\sqsubseteq^i$, $\bot^i$, $\sqcap^i$, $\sqcup^i$, $\mathsf{assign}^i[\![\mathsf{x}, \mathsf{A}]\!]$, $\mathsf{test}^i[\![\mathsf{B}]\!]$, and $\overline{\mathsf{test}}^i[\![\mathsf{B}]\!]$. These operations may have to be adjusted when introducing a new abstract domain that may help improve their precision.

For example 36.32, the communication channel is equalities and disequalities between variables. For the reduced product of sign and parity in exercise 36.4, the communication channel can be equalities and disequalities between a variable and a constant. By asking the question $\mathsf{x} \neq 0$?, the sign abstract domain can improve $\mathsf{x} \leqslant 0$ to $\mathsf{x} < 0$, $\mathsf{x} \geqslant 0$ to $\mathsf{x} > 0$, $\mathsf{x} = 0$ to $\bot$, and so on. This question can be answered, for example, by the parity abstract domain (when the parity of the variable and the constant differ), the constant abstract domain (when the variable is equal to a different constant), and so forth.

Finally, a scheduler is in charge of orchestrating the reduction communications during the fixpoint computation. It must be extended, for each new abstract domain introduced in the reduced product, to decide which meaningful communications have to be performed, and in which order.

Only a number of the possible pairwise reductions of definition 36.25 may be performed, maybe in a given order specified by the scheduler [125, 214, 267, 546]. For cost saving, the reduction in definition 36.29 need not be iterated until reaching a fixpoint. It can be stopped at any time.

Many verification methods and static analyzers use auxiliary/intermediate/ghost variables. They are used to memoize the abstract value of a function of the program semantics. This can be done at some time in the computation, such as the auxiliary variables used to record the initial value of variables in the relational reachability semantics of section 19.1.2. More generally, this can be a value that changes during the computation (such as the value of an expression). Whereas the value of program variables are modified explicitly though an assignment (maybe indirectly through pointers), ghost variables are modified implicitly when the quantities upon which they depend are modified. This may be problematic in a reduced product because an abstract domain modifying a ghost variable must inform the other abstract domains. An elegant solution to this problem, extending the expressiveness of the reduced product and the way abstract interpreters are designed, is proposed by [179] and [178, part IV].

36.5 Conclusion

More properties of reduced products are studied in [274]. (It is also shown that the Nelson-Oppen combinations of theories [748] used in satisfiability modulo theories (SMT) solvers is a reduced product of abstract domains.) [1002] defines the reduced product of abstract domains for constraint-solving.

The reduced product and its variants are essential for extending static analyzers by introducing new abstractions without having to redo the whole static analyzer [267]. Some static analyzers with a monolithic design, such as Infer [162] and more generally type inference [224] lack this capacity of being simply extensible.

Because the full fixpoint iterated reduction of the reduced product may be costly, partial reductions are cheaper and often preferred in practice. There are many possible variants, with many different names, which are mostly variants for specific partial reductions [198, 211, 372, 449, 626].

The decision of where, when, and in which order to do reductions given in chapter 29 or this chapter is a question of balance between the admissible cost and desired precision of the analysis. This is often a matter of experimentation starting with minimal reductions and adding those indispensable for precision. See, for example, [267] which proposes an extensible implementation of reduction to facilitate the inclusion of new abstract domains through a common interface.

As mentioned in section 36.4.6, [179] extends the reduced product to allow for abstract domains to record information about the history of computation using ghost variables. Other compositions of abstract domains have been proposed in [36, 213, 241, 415, 418, 547]. In particular, disjunctive combinations of abstract domains allow for a case analysis based on program control [174] or values of variables [273] or for an arbitrary one, when considering all possible disjunctions [241, 370] with various widening strategies [66, 67, 330, 331, 451].

The inverse problem of composing abstract domains is that of decomposition into simpler ones; see for example, [213] for the reduced product inverse and [852] for tree decompositions to define an abstract domain.

Exercise 36.40 (Project: iterated pairwise reduction) Implement the (iterated) pairwise reduction of abstract domains parameterized by these abstract domains and the pairwise reductions.

- $\mathtt{cong(x}, v) = \langle \ell, h \rangle$ such that $\rho(\mathtt{x}) = v \Rightarrow (\ell \leqslant v < h)$
- $\mathtt{int(x)} = \langle \ell, h \rangle$ such that $\ell \leqslant \rho(\mathtt{x}) \leqslant h$
- $\mathtt{zone(x, y)} = (\rho(\mathtt{x}) \leqslant \rho(\mathtt{y}))$

36.6 Solutions to Selected Exercises

Solution to exercise 36.4

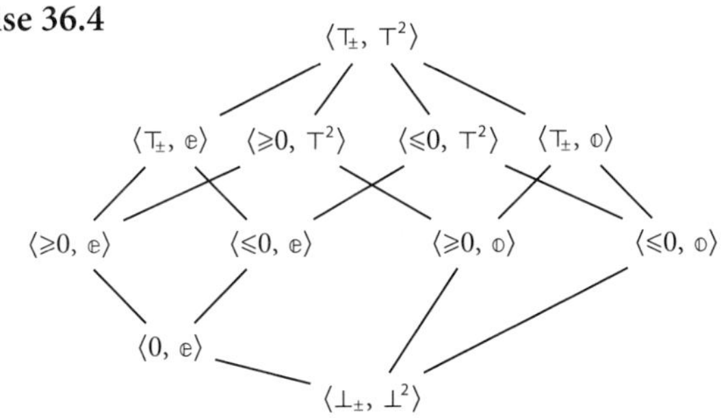

Solution to exercise 36.5

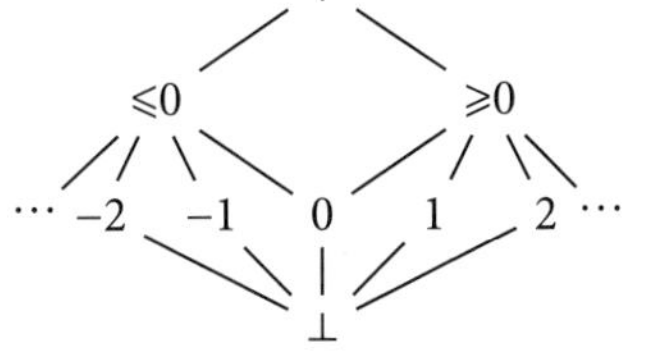

Solution to exercise 36.13 Consider the lattice of signs $\{\bot, -, +, \top\}$ with $\gamma(\bot) \triangleq \emptyset$, $\gamma(-) \triangleq \{z \in \mathbb{Z} \mid z \leqslant 0\}$, $\gamma(+) \triangleq \mathbb{N}$, and $\gamma(\top) = \mathbb{Z}$. We have $\gamma(-) \cap \gamma(+) = \{0\}$, but there no abstract property $R \in \{\bot, -, +, \top\}$ such that $\gamma(R) = \{0\}$. In Galois connections, γ preserves meets, so R can be chosen as the abstract meet.

Solution to exercise 36.26 Define $\min(c + m\mathbb{Z}, a) \triangleq a + ((c - a) \pmod{|m|})$ when $m \neq 0$ to be the least element of $c + m\mathbb{Z}$ greater than or equal to a. We have $a \leqslant \min(c + m\mathbb{Z}, a) \leqslant a + |m|$ and $\min(c + m\mathbb{Z}, a) = a + (c - a) - |m|((c - a) \pmod{|m|}) = c - |m|((c - a) \pmod{|m|}) \in c + m\mathbb{Z}$. Similarly, let $\max(c + m\mathbb{Z}, a) \triangleq a - ((a - c) \pmod{|m|})$ when

$m \neq 0$ be the greatest element of $c + m\mathbb{Z}$ less than of equal to a. By convention $\min(c + m\mathbb{Z}, -\infty) = -\infty$ and $\max(c + m\mathbb{Z}, \infty) \triangleq \infty$.

The reduction is as follows [438, proposition 6.1].

$$
\begin{aligned}
\rho(\varnothing, i) &\triangleq \langle \varnothing, \varnothing \rangle \\
\rho(c, \varnothing) &\triangleq \langle \varnothing, \varnothing \rangle \\
\rho(c + 0\mathbb{Z}, i) &\triangleq \langle \varnothing, \varnothing \rangle && \text{when } c \notin i \\
\rho(c + 0\mathbb{Z}, i) &\triangleq \langle c + 0\mathbb{Z},\ [c, c] \rangle && \text{when } c \in i \\
\rho(c + m\mathbb{Z}, [\ell, h]) &\triangleq \langle \varnothing, \varnothing \rangle && \text{when } \min(c + m\mathbb{Z}, \ell) > \max(c + m\mathbb{Z}, h) \\
\rho(c + m\mathbb{Z}, [\ell, h]) &\triangleq \langle \min(c + m\mathbb{Z}, \ell) + 0\mathbb{Z},\ [\min(c + m\mathbb{Z}, \ell), \min(c + m\mathbb{Z}, \ell)] \rangle \\
& && \text{when } \min(c + m\mathbb{Z}, \ell) = \max(c + m\mathbb{Z}, h) \\
\rho(c + m\mathbb{Z}, [\ell, h]) &\triangleq \langle c + m\mathbb{Z},\ [\min(c + m\mathbb{Z}, \ell), \max(c + m\mathbb{Z}, h)] \rangle \\
& && \text{when } \min(c + m\mathbb{Z}, \ell) < \max(c + m\mathbb{Z}, h)
\end{aligned}
$$

XI RELATIONAL STATIC ANALYSIS

A static analysis is relational when the inferred properties do relate the values of variables.

We provide two examples.

The linear equality analysis (chapter 38) is based on basic mathematical concepts of linear algebra (chapter 37).

The zone and octagon inequality analyses (chapter 40) are based on basic mathematical concepts of graphs (chapter 39).

37 Basic Linear Algebra

We introduce fields, vector spaces, matrices, and affine spaces. Linear algebra originates from the work of Hermann Grassmann [443]. There are numerous introductory books on vector spaces and linear algebra such as [467], [54, pp. 1–20], and [615].

Contents

37.1 Fields

A *field* $\langle \mathbb{F}, +, -, \times, / \rangle$ as formalized by Heinrich Weber [966] is a set of *scalars* $a, b, c, \ldots \in \mathbb{F}$ on which are defined addition $+$, subtraction $-$, multiplication $\times$ ($a \times b$ is often denoted ab), and division $/$ such that $+, -, \times, / \in \mathbb{F}^2 \to \mathbb{F}$ and $\forall a, b, c, \ldots \in \mathbb{F}$,

- $a + (b + c) = (a + b) + c$ — associativity of addition
- $a \times (b \times c) = (a \times b) \times c$ — associativity of multiplication
- $a + b = b + a$ — commutativity of addition
- $a \times b = b \times a$ — commutativity of multiplication
- $\exists 0 \in \mathbb{F} . \forall a \in \mathbb{F} . a + 0 = a$ — additive identity
- $\exists 1 \in \mathbb{F} . 1 \neq 0 \wedge \forall a \in \mathbb{F} . a \times 1 = a$ — multiplicative identity
- $\forall a \in \mathbb{F} . \exists -a \in \mathbb{F} . a + (-a) = 0$ — additive inverse
- $\forall a \in \mathbb{F} \setminus \{0\} . \exists a^{-1} \in \mathbb{F} . a \times a^{-1} = 1$ — multiplicative inverse (also denoted $1/a$ or $\frac{1}{a}$)
- $a \times (b + c) = (a \times b) + (a \times c)$ — distributivity of multiplication over addition
- $a - b \triangleq a + (-b)$ — definition of subtraction
- $\forall b \neq 0 . a/b \triangleq a \times b^{-1}$ — definition of division

The reals $\mathbb{R}$ and the rational numbers $\mathbb{Q}$ are examples of fields. A counterexample is the set $\mathbb{Z}$ of integers, by lack of multiplicative inverse.

37.2 Vector Spaces

The mathematical structure of vector spaces emerged with Richard Dedekind and Heinrich Weber [303], Giuseppe Peano [786], and others [446]. Vector spaces are sets of vectors $\vec{v} =$  that can be

- added: $\vec{v}_1 =$  , $\vec{v}_2 =$  , $\vec{v}_1 + \vec{v}_2 =$ 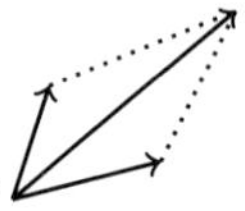
- scaled: $\vec{v}_1 =$  , $3\vec{v}_1 =$ 

Vectors represent, for example, a direction of displacement from one point to another or forces in physics.

37.2.1 Definition of a Vector Space

A *vector space* $\langle \vec{\mathbb{V}}, \langle \mathbb{F}, +, -, \times, / \rangle, +, -, \times, / \rangle$ (or *linear space*) on a field $\langle \mathbb{F}, +, -, \times, / \rangle$ of scalars $a, b, c, \ldots \in \mathbb{F}$ is a set of *vectors* $\vec{u}, \vec{v}, \vec{w}, \ldots \in \vec{\mathbb{V}}$ equipped with a vector addition $+$ and subtraction $-$ and a scalar multiplication $\times$ and division $/$ such that $+, - \in \vec{\mathbb{V}}^2 \to \vec{\mathbb{V}}$, $\times \in \mathbb{F} \times \vec{\mathbb{V}} \to \vec{\mathbb{V}}$, $/ \in \vec{\mathbb{V}} \times \mathbb{F} \to \vec{\mathbb{V}}$, and

- $\vec{u} + (\vec{v} + \vec{w}) = (\vec{u} + \vec{v}) + \vec{w}$ — associativity of vector addition
- $\vec{u} + \vec{v} = \vec{v} + \vec{u}$ — commutativity of vector addition
- $\exists \vec{0} \in \vec{\mathbb{V}} . \forall \vec{u} \in \vec{\mathbb{V}} . \vec{0} + \vec{u} = \vec{u}$ — zero vector, identity element of addition
- $\forall \vec{u} \in \mathbb{F} . \exists - \vec{u} \in \vec{\mathbb{V}} . \vec{u} + (-\vec{u}) = \vec{0}$ — additive inverse
- $a \times (b \times \vec{u}) = (a \times b) \times \vec{u}$ — mixed associativity
- $1 \times \vec{u} = \vec{u}$ — identity element of scalar multiplication
- $a \times (\vec{u} + \vec{v}) = (a \times \vec{u}) + (a \times \vec{v})$ — distributivity of scalar multiplication with respect to vector addition
- $(a + b) \times \vec{u} = (a \times \vec{u}) + (b \times \vec{u})$ — distributivity of scalar multiplication with respect to field addition
- $\vec{u} - \vec{v} \triangleq \vec{u} + (-\vec{v})$ — definition of vector subtraction
- $\forall a \neq 0 . \vec{u}/a \triangleq a^{-1} \times \vec{u}$ — definition of vector division

The null vector $\vec{0}$ plays the role of an origin of the vector space.[1]

Exercise 37.1 Prove the unicity of the additive identity $\vec{0}$, the unicity of the additive inverse of a vector, that $0\vec{v} = \vec{0}$, that $\forall a \in \mathbb{F} . a\vec{0} = \vec{0}$, that $(-1)\vec{v} = -\vec{v}$.

37.2.2 Vector Subspace

A *vector subspace* of a vector space $\vec{u}, \vec{v}, \vec{w}, \ldots \in \vec{\mathbb{V}}$ on a field $a, b, c, \ldots \in \mathbb{F}$ is a nonempty subset $\overrightarrow{W}$ of $\vec{\mathbb{V}}$ such that

- $\emptyset \neq \overrightarrow{W} \subseteq \vec{\mathbb{V}}$
- $\forall \vec{u}, \vec{v} \in \overrightarrow{W} . \vec{u} + \vec{v} \in \overrightarrow{W}$ — closure by vector addition
- $\forall \vec{u} \in \overrightarrow{W} . \forall a \in \mathbb{F} . a \times \vec{u} \in \overrightarrow{W}$ — closure by scalar multiplication

1. Affine spaces, described in section 37.6, will get rid of the origin.

It follows that the zero vector belongs to the vector subspace $\vec{0} \in \overrightarrow{W}$ as well as the additive inverses $\forall \vec{u} \in \overrightarrow{W} \,.\, -\vec{u} \in \overrightarrow{W}$ so that the subspace $\overrightarrow{W}$ is itself a vector space $\langle \overrightarrow{W}, \langle \mathbb{F}, +, -, \times, / \rangle, +, -, \times, / \rangle$.

Exercise 37.2 (intersection of vector subspaces is a vector subspace) Prove that if $\langle \overrightarrow{W}_i,\ i \in \Delta \rangle$ are subspaces of $\vec{\mathbb{V}}$ then so is $\bigcap_{i \in \Delta} \overrightarrow{W}_i$.

37.2.3 Span

Given a family $\overrightarrow{W} = \langle \vec{v}_i,\ i \in \Delta \rangle$ of vectors in the vector space $\vec{\mathbb{V}}$, the *span* or *linear hull* or set of all *linear combinations* of these vectors $\overrightarrow{W}$ is

$$\begin{aligned}\mathsf{Span}(\varnothing) &\triangleq \{\vec{0}\} \\ \mathsf{Span}(\langle \vec{v}_i,\ i \in \Delta \rangle) &\triangleq \Big\{ \sum_{i \in \Delta} a_i \times \vec{v}_i \ \Big|\ \forall i \in \Delta \,.\, a_i \in \mathbb{F} \Big\}\end{aligned}$$

37.2.4 Subspace Generated by a Family of Vectors

The vector addition and scalar multiplication of linear combinations is a linear combination so the linear combinations are a subspace of the vector space.

$$\mathsf{Span}(\langle \vec{v}_i,\ i \in \Delta \rangle) \text{ is a subspace of the vector space } \vec{\mathbb{V}}.$$

37.2.5 Linear Independence

A nonempty family $\langle \vec{v}_i,\ i \in \Delta \rangle$ of vectors is *free* (i.e., its vectors are *linearly independent*) if and only if the only linear combination equal to the zero vector has all scalar coefficients equal to zero, that is,

$$\forall \langle a_i,\ i \in \Delta \rangle \,.\, \Big(\sum_{i \in \Delta} a_i \times \vec{v}_i = \vec{0} \Big) \Rightarrow \Big(\forall i \in \Delta \,.\, a_i = 0 \Big)$$

Moreover, $\varnothing$ is linearly independent.

Otherwise, the vectors are *linearly dependent*, that is, one vector in the family can be defined as a linear combination of the others.

37.2.6 Basis of a Vector Space

A *basis* (or *linear basis*) of a vector space $\vec{\mathbb{V}}$ is a set $\langle \vec{b}_i,\ i \in \Delta \rangle$ of linearly independent vectors that generates this vector space $\mathsf{Span}(\langle \vec{b}_i,\ i \in \Delta \rangle) = \vec{\mathbb{V}}$.

Any vector $\vec{u} \in \vec{\mathbb{V}}$ of the vector space $\vec{\mathbb{V}}$ is then a unique linear combination $\sum_{i \in \Delta} a_i \times \vec{b}_i$ of the base vectors $\langle \vec{b}_i,\ i \in \Delta \rangle$ for coefficients $\prod_{i \in \Delta} a_i$ called the *coordinates* of the vector in this basis.

For example, for the standard basis $\vec{b}_1$ and $\vec{b}_2$, we would have

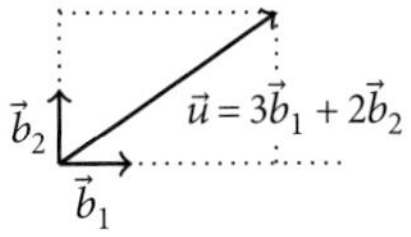

with coodinates $(3, 2)$.

37.2.7 Dimension of a Vector Space

All bases of the vector space have the same cardinality called the *dimension* $\mathsf{dim}(\vec{\mathbb{V}})$ of the vector space $\vec{\mathbb{V}}$. A vector line has dimension 1, a vector plane has dimension 2.

37.2.8 Example of the Coordinate Vector Space

For any strictly positive integer n, the space $\vec{\mathbb{F}}^n \triangleq \{\vec{x} \mid \vec{x} = \langle \vec{x}_1, \ldots, \vec{x}_n \rangle \wedge \forall i \in [0, n] . \vec{x}_i \in \mathbb{F}\}$ of all n-tuples of elements of the field $\mathbb{F}$ is a vector space over $\mathbb{F}$ of dimension n with operations

- $\vec{x} + \vec{y} \triangleq \langle \vec{x}_1 + \vec{y}_1, \ldots, \vec{x}_n + \vec{y}_n \rangle$ — vector addition
- $a \times \vec{x} \triangleq \langle a \times \vec{x}_1, \ldots, a \times \vec{x}_n \rangle$ — scalar multiplication
- $\vec{0} = \langle 0, \ldots, 0 \rangle$ — zero vector
- $-\vec{x} = \langle -\vec{x}_1, \ldots, -\vec{x}_n \rangle$ — additive inverse

The *standard (or canonical) basis* is $\{\vec{b}_i \mid i \in [1, n] \wedge \vec{b}_{i_i} = 1 \wedge \forall j \in [1, n] \setminus \{i\} . \vec{b}_{i_j} = 0\}$ where 0 is the additive identity and 1 is the multiplicative identity in $\mathbb{F}$. $\mathsf{dim}(\vec{\mathbb{F}}^n) = n$.

Typically, in program analysis, $\mathbb{F}$ is $\mathbb{Q}$ or $\mathbb{R}$ and an element $\langle \vec{x}_1, \ldots, \vec{x}_n \rangle$ of $\vec{\mathbb{F}}^n$ is the vector of values $\rho(\mathsf{x})$ of the n variables $\mathsf{x} \in \mathbb{V}$ of the program.

If a family of vectors $\langle \vec{v}_i, i \in \Delta \rangle$ of a vector space $\vec{\mathbb{V}}$ has a cardinality greater than the dimension $\mathsf{dim}(\vec{\mathbb{V}})$ of the vector space then these vectors are linearly dependent.

37.3 Systems of Linear Equations

The affine or linear equality analysis of programs in chapter 38, aims at automatically discovering a conjunction of linear equalities of the form $a_1 \times x_1 + \ldots + a_j \times x_j + \ldots + a_n \times x_n = b$ among the values $x_1, \ldots, x_n$ of the numerical variables $\mathsf{x}_1, \ldots, \mathsf{x}_n$ of the program where the coefficients $a_1, \ldots, a_j, \ldots, a_n, b$ are constants automatically inferred by the analysis. This motivates the study of system of linear equations, matrices, and affine spaces.

37.3.1 Definition of a System of Linear Equations

A system of m linear equations in n *unknowns* $x_1, \ldots, x_n$ has the form

$$\left\{\begin{array}{ccccccc} a_{11} \times x_1 + & \ldots & +a_{1j} \times x_j + & \ldots & +a_{1n} \times x_n & = & b_1 \\ a_{21} \times x_1 + & \ldots & +a_{2j} \times x_j + & \ldots & +a_{2n} \times x_n & = & b_2 \\ \ldots & \ldots & \ldots & \ldots & \ldots & & \\ a_{m1} \times x_1 + & \ldots & +a_{mj} \times x_j + & \ldots & +a_{mn} \times x_n & = & b_m \end{array}\right.$$

where the *coefficients* a_{ij}, $i \in [1, m]$, $j \in [1, n]$, b_j, $j \in [1, m]$, and the unknowns x_j, $j \in [1, n]$ all belong to a field $\mathbb{F}$. It is understood as a conjunction of the linear constraints on the unknowns $x_1, \ldots, x_n$.

For example, the system of linear equations

$$\left\{\begin{array}{ccccc} 2x & + & 3y & = & 0 \\ -x & + & y & = & -1 \end{array}\right.$$

has solution $x_0 = 3/5$ and $y_0 = -2/5$

$$\left\{\begin{array}{ccccccccccc} 2x_0 & + & 3y_0 & = & 2 \times 3/5 & + & 3 \times -2/5 & = & 6/5 - 6/5 & = & 0 \\ -x_0 & + & y_0 & = & -3/5 & + & -2/5 & = & -5/5 & = & -1 \end{array}\right.$$

37.3.2 Matrix Notation

Following Arthur Cayley [168], linear equations can be written as a single *matrix equation* $A \times \vec{x} = \vec{b}$ or $A \times \vec{x} - \vec{b} = \vec{0}$, by defining

- the matrix $A = (a_{ij})_{\substack{i=1,m \\ j=1,n}} \in \mathbb{F}^{m \times n}$ over a field $\mathbb{F}$ (also denoted $(a_{ij})^{m \times n}$) is the array of numbers $a_{ij} \in \mathbb{F}$ with m rows and n columns

$$A = \begin{bmatrix} a_{11} & \dots & a_{1j} & \dots & a_{1n} \\ a_{21} & \dots & a_{2j} & \dots & a_{2n} \\ \dots & \dots & \dots & \dots & \dots \\ a_{m1} & \dots & a_{mj} & \dots & a_{mn} \end{bmatrix}$$

- The $n \times 1$ *column vectors* in $\mathbb{F}^n$

$$\vec{x} = \begin{bmatrix} x_1 \\ x_2 \\ \dots \\ x_n \end{bmatrix} \quad \vec{b} = \begin{bmatrix} b_1 \\ b_2 \\ \dots \\ b_m \end{bmatrix} \quad \vec{0} = \begin{bmatrix} 0 \\ 0 \\ \dots \\ 0 \end{bmatrix}$$

- The *multiplication* of a matrix A by a vector $\vec{x}$, $A \times \vec{x} =$

$$\begin{bmatrix} a_{11} & \dots & a_{1j} & \dots & a_{1n} \\ a_{21} & \dots & a_{2j} & \dots & a_{2n} \\ \dots & \dots & \dots & \dots & \dots \\ a_{m1} & \dots & a_{mj} & \dots & a_{mn} \end{bmatrix} \times \begin{bmatrix} x_1 \\ x_2 \\ \dots \\ x_n \end{bmatrix} = \begin{bmatrix} a_{11}x_1 + & \dots & +a_{1j}x_j + & \dots & +a_{1n}x_n \\ a_{21}x_1 + & \dots & +a_{2j}x_j + & \dots & +a_{2n}x_n \\ \dots & \dots & \dots & \dots & \dots \\ a_{m1}x_1 + & \dots & +a_{mj}x_j + & \dots & +a_{mn}x_n \end{bmatrix}$$

(We omit the multiplication operator writing ab and $A\vec{b}$ for $a \times b$ and $A \times \vec{b}$.)
- The difference of column vectors

$$\begin{bmatrix} x_1 \\ x_2 \\ \dots \\ x_n \end{bmatrix} - \begin{bmatrix} y_1 \\ y_2 \\ \dots \\ y_n \end{bmatrix} = \begin{bmatrix} x_1 - y_1 \\ x_2 - y_2 \\ \dots \\ x_n - y_n \end{bmatrix}$$

- The system $A \times \vec{x} = \vec{b}$ of equations is *homogeneous* if $\vec{b} = \vec{0}$, *inhomogeneous* otherwise.

If $A = (a_{ij})_{\substack{i=1,m \\ j=1,n}} \in \mathbb{F}^{m \times n}$, $B = (b_{ij})_{\substack{i=1,m \\ j=1,n}} \in \mathbb{F}^{m \times n}$, and $c \in \mathbb{F}$, we also define

- the matrix addition $A + B = (a_{ij} + b_{ij})_{\substack{i=1,m \\ j=1,n}}$,
- the matrix difference $A - B = (a_{ij} - b_{ij})_{\substack{i=1,m \\ j=1,n}}$, and
- the matrix multiplication by the scalar $c \in \mathbb{F}$, $cA = (c \times a_{ij})_{\substack{i=1,m \\ j=1,n}}$.

Exercise 37.3 Show that the set $\mathbb{F}^{m \times n}$ of all $m \times n$ matrices with entries in $\mathbb{F}$ is a vector space over the scalars $\mathbb{F}$ with matrix addition and scalar multiplication.

37.4 Solving a System of Linear Equations

A system of linear equations $A\vec{x} = \vec{b}$ may have 0, 1, or many solutions. The Gauss–Jordan elimination algorithm can be used to effectively compute them.

37.4.1 Basic Row Transformations

The system $\mathbf{A}\vec{x} = \vec{b}$ is augmented to $(\mathbf{A}|\vec{b})$ by adding the extra column $\vec{b}$ and putting in reduced row echelon form by Gauss–Jordan elimination, with three *basic row transformations*:

1. the swap of two rows,
2. the multiplication of a row by a nonzero scalar $a \in \mathbb{F} \setminus \{0\}$, and
3. addition of a multiple of one row to another different row.

These basic row transformations of $(\mathbf{A}|\vec{b})$ do not change the solutions to $\mathbf{A}\vec{x} = \vec{b}$.

37.4.2 Row Echelon Form

Carl Friedrich Gauss [399, pp. 21–25] showed that by successive applications of the basic row transformations, any augmented matrix can be transformed in (nonunique) *row echelon form*. A matrix is in row echelon form if and only if

- all nonzero rows (rows with at least one nonzero element) are above any rows of all zeros (so all zero rows, if any, belong at the bottom of the matrix), and
- the leading coefficient (the first nonzero number from the left, also called the pivot) of a nonzero row is always equal to 1 and strictly to the right of the leading coefficient of the row above it.

These two conditions imply that all entries in a column below a pivot are zeros.

For example $\begin{bmatrix} 1 & \frac{3}{2} & 0 \\ 0 & 1 & -\frac{2}{5} \end{bmatrix}$ is in row echelon form. Backward substitution yields the solution $y = -\frac{2}{5}$ and replacing y by its value in the first equation $x + \frac{3}{2}(-\frac{2}{5}) = 0$, we get $x = \frac{3}{5}$.

Example 37.4 Assume that $\mathbf{A}\vec{x} = \vec{b}$ is

$$\begin{cases} 2x & + & 3y & = & 0 \\ -x & + & y & = & -1 \end{cases}$$

and must be put into row echelon form. The augmented matrix is

$$\begin{bmatrix} 2 & 3 & 0 \\ -1 & 1 & -1 \end{bmatrix}$$

Dividing the first row by 2.

$$\begin{bmatrix} 1 & \frac{3}{2} & 0 \\ -1 & 1 & -1 \end{bmatrix}$$

Adding the first row to the second.

$$\begin{bmatrix} 1 & \frac{3}{2} & 0 \\ 0 & \frac{5}{2} & -1 \end{bmatrix}$$

Dividing the second row by $\frac{5}{2}$.

$$\begin{bmatrix} 1 & \frac{3}{2} & 0 \\ 0 & 1 & -\frac{2}{5} \end{bmatrix}$$

which yields $y = -\frac{2}{5}$ and $x = \frac{3}{5}$.

37.4.3 Reduced Row Echelon Form

Wilhelm Jordan [543, pp. 62–83] showed that by successive applications of the basic row transformations, the augmented matrix can be transformed in a unique *reduced row echelon form*. A matrix is in reduced row echelon form if and only if it is in row echelon form and for each column vector containing a pivot, this pivot is the only nonzero element in the column. Example 37.4 is not in reduced form, but $\begin{bmatrix} 1 & 0 & \frac{3}{5} \\ 0 & 1 & -\frac{2}{5} \end{bmatrix}$ is. The solution is obtained directly without needing a final backward substitution. More importantly (and unlike the row echelon form), the reduced row echelon form of a matrix is unique so provides a canonical form.

37.4.4 Gauss–Jordan Elimination Algorithm

Assume that $A(n \times m)$ is the augmented matrix with n rows and m columns. After the $(i-1)$th-step ($i = 0, \ldots, n-1$), the Gauss–Jordan elimination algorithm [34] transforms the matrix A in the following form

$$
\begin{array}{r}
\\ \\ \\ \\ \text{i-1} \\ \text{i} \\ \\ \\
\end{array}
\left[
\begin{array}{cccccccccccc}
 & & & & & & & j & & & & \\
1 & * & 0 & * & \ldots & 0 & * & * & * & * & * \\
 & & 1 & * & \ldots & 0 & * & * & * & * & * \\
 & & & & \ldots & 0 & & \ldots & \ldots & \ldots & \\
 & \mathbf{0} & & & & 1 & * & * & * & * & * \\
 & & & & & & * & * & * & * & * \\
 & & & & & & * & * & * & * & * \\
 & & & & & & * & * & * & * & *
\end{array}
\right]
$$

where the pivots are 1 and unique nonzero elements in their column, and $*$ represents an arbitrary value. The ith-step is the following [524, p. 50]:

1. Find the first column j containing a nonzero element at row k, at or below row i. If no such j exists, the bottom right of the matrix is 0 and we are done.
2. If $k \neq i$, exchange rows k and i so that the pivot at $A(i, j)$ is nonzero.
3. Set the pivot to 1 by dividing line i by $A(i, j)$.
4. Set the column j above and below the pivot to 0 by updating each row $A(\ell, k)$, $\ell \in [0, n-1] \setminus \{i\}$, $k \in [0, m-1]$ to $A(\ell, k) - A(i, k)\frac{A(\ell, j)}{A(i, j)}$.

Contrary to the row echelon form, the reduced row echelon form is unique. The number of nonzero rows is less than or equal to the number of variables.

Example 37.5 For the augmented matrix

$$\begin{bmatrix} 2 & 3 & 0 \\ -1 & 1 & -1 \end{bmatrix}$$

the first step locates the pivot at $A(0, 0)$, divides the first row by 2 to set the pivot to 1

$$\begin{bmatrix} 1 & \frac{3}{2} & 0 \\ -1 & 1 & -1 \end{bmatrix}$$

and then nullifies the column below $A(0, 0)$ by adding the first to the second row.

$$\begin{bmatrix} 1 & \frac{3}{2} & 0 \\ 0 & \frac{5}{2} & -1 \end{bmatrix}$$

The second step locates the pivot at $A(1,1)$, divides the second row by $\frac{5}{2}$ to set the pivot to 1

$$\begin{bmatrix} 1 & \frac{3}{2} & 0 \\ 0 & 1 & -\frac{2}{5} \end{bmatrix}$$

and then nullifies the column above $A(1,1)$ by subtracting the second row multiplied by $\frac{3}{2}$ from the first row.

$$\begin{bmatrix} 1 & 0 & \frac{3}{5} \\ 0 & 1 & -\frac{2}{5} \end{bmatrix}$$

Exercise 37.6 (Project: Gauss–Jordan elimination algorithm) Program the Gauss–Jordan elimination algorithm using rationals to avoid rounding errors.[2]

```
# print_matrix a;;
2; 3; 0;
-1; 1; -1;
- : unit = ()
# print_matrix (gaussJordanElimination a);;
1; 0; 3/5;
0; 1; -2/5;
- : unit = ()
```

Exercise 37.7 (Project: linear equations resolution) Continuing exercise 37.6, write a program that returns a solution to $\mathbf{A}\vec{x} = \vec{b}$, if any.

Exercise 37.8 (Project: constant variables) Continuing exercise 37.6, write a program that given $\mathbf{A}\vec{x} = \vec{b}$ and a variable $\vec{x}_i$ returns the value of this variable $\vec{x}_i$ if and when it is constant for all solutions.

Theorem 37.9 After elimination of the zero rows, a system of linear equations $\mathbf{A}\vec{x} = \vec{b}$ in reduced row echelon form has solutions if and only if there is no row $\vec{0}\,\vec{b}_\ell$ with $\vec{b}_\ell \neq 0$.

Proof of theorem 37.9 Let $(\mathbf{A}|\vec{b})$ be the reduced row echelon form of a system of linear equations. Assume $\mathbf{A}$ has a zero row $\mathbf{A}_{i,j=0,\ldots,m}$. Then either $\vec{b}_i = 0$ and $\vec{0}\vec{x} = 0$ is satisfied by any $\vec{x}$ so this row can be eliminated without changing the solutions or $\vec{b}_i \neq 0$ and then there is no solution. Else $\mathbf{A}$ has no zero line. By definition of the reduced row echelon form, all rows start with a pivot 1, and these pivots are the only nonzero in a column and strictly to the right of the pivot of the row above. It follows that there are fewer lines than columns because otherwise a line would have two pivots, which is impossible. □

Example 37.10 The reduced row echelon form of $\begin{bmatrix} 1 & 1 & 1 \\ 1 & 1 & 2 \\ 1 & 1 & 3 \end{bmatrix}$ is $\begin{bmatrix} 1 & 1 & 0 \\ 0 & 0 & 1 \\ 0 & 0 & 0 \end{bmatrix}$ so that $\{x+y=1, x+y=2, x+y=3\}$ has obviously no solution.

2. The `OCaml/Zarith` library implements arithmetic and logical operations over rational numbers of arbitrary precision, this avoiding rounding errors and overflows. https://github.com/ocaml/Zarith.

37.5 Representation of the Solution Set of a System of Linear Equations

We show that the solutions $y_{\overrightarrow{=}}(\mathbf{A}|\vec{b})$ of $\mathbf{A}\vec{x} = \vec{b}$ can be represented by a frame or system of generators consisting of a point $\vec{x}_0$ (which is any solution to $\mathbf{A}\vec{x} = \vec{b}$) and a vector space $\mathsf{Ker}(\mathbf{A})$ (called the kernel of $\mathbf{A}$). For example, the solutions to $x - y = 2$ in $\mathbb{R}^2$ can be represented by the point $\langle 2, 0\rangle$ and the vector space generated by the vector $\langle 1, 1\rangle$.

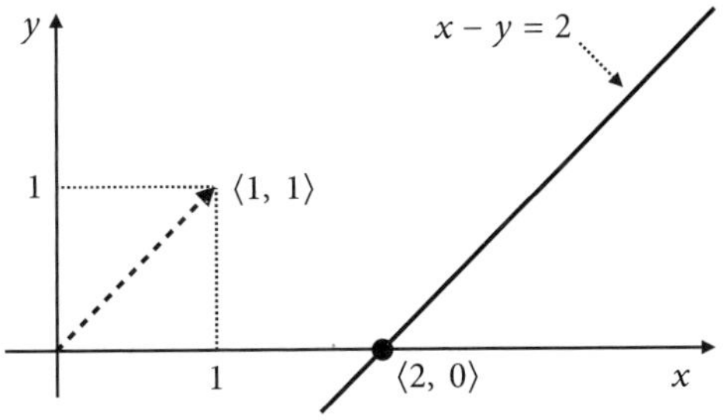

37.5.1 Kernel of a Matrix

The *kernel* or *nullspace* $\mathsf{Ker}(\mathbf{A})$ of a matrix $\mathbf{A}$ is the set of solutions of the system of equations $\mathbf{A}\vec{x} = \vec{0}$.

$$\mathsf{Ker}(\mathbf{A}) \quad \triangleq \quad \{\vec{x} \in \mathbb{F}^m \mid \mathbf{A}\vec{x} = \vec{0}\}$$

If $\vec{x}_1, \vec{x}_2 \in \mathsf{Ker}(\mathbf{A})$ and $a \in \mathbb{F}$ then $\vec{x}_1 + \vec{x}_2 \in \mathsf{Ker}(\mathbf{A})$ and $a\vec{x}_1 \in \mathsf{Ker}(\mathbf{A})$ and $\vec{0} \in \mathsf{Ker}(\mathbf{A})$ so that $\mathsf{Ker}(\mathbf{A})$ is a nonempty vector subspace of the coordinate vector space $\langle \vec{\mathbb{F}}^n, \langle \mathbb{F}, +, -, \times, /\rangle, +, -, \times, /\rangle$ of section 37.2.8.

37.5.2 Basis of the Kernel of a Matrix

To find a basis of $\mathsf{Ker}(\mathbf{A})$, we establish the equation $a_1\mathbf{A}_{i=1,n,1} + a_2\mathbf{A}_{i=1,n,2} + \ldots + a_m\mathbf{A}_{i=1,n,m} = 0$ to find which vectors are independent. So, we put $(\mathbf{A}|\vec{0})$ in reduced echelon form $\mathbf{A}'$ by the Gauss–Jordan algorithm. The pivot column vectors $\mathbf{A}'_{i=1,n,j}$ of $\mathbf{A}'$ are those with a 1 in the column with zeros below and to the left of this 1. The other non-pivot columns $\mathbf{A}'_{i=1,n,k}$ of $\mathbf{A}'$ are called free. The variables corresponding to the pivot columns depend upon the variables corresponding to the free columns. The variables corresponding to the free columns depend upon no other variable. So, the free column vectors form a basis.

Example 37.11 Consider the matrix $\mathbf{A} = \begin{bmatrix} 1 & 1 & -1 & 0 & 0 \\ 1 & 1 & 1 & -4 & 0 \\ 1 & 0 & 1 & -3 & -1 \\ 1 & 0 & -1 & 1 & -1 \end{bmatrix}$. Its kernel is the solution to $\mathbf{A}\vec{x} = \vec{0}$. The reduced row echelon form is $\mathbf{A}' = \begin{bmatrix} \boxed{1} & 0 & 0 & -1 & -1 \\ 0 & \boxed{1} & 0 & -1 & 1 \\ 0 & 0 & \boxed{1} & -2 & 0 \\ 0 & 0 & 0 & 0 & 0 \end{bmatrix}$ (the 1s in the pivot columns have been framed). The kernel of $\mathbf{A}$ consists of the solutions of $\mathbf{A}'\vec{x} = \vec{0}$, that is

$$\left\{\begin{array}{cccccc} x_1 & & & -x_4 & -x_5 & = 0 \\ & x_2 & & -x_4 & +x_5 & = 0 \\ & & x_3 & -2x_4 & & = 0 \end{array}\right.$$ This is the set of vectors $\vec{x} = \begin{bmatrix} x_1 \\ x_2 \\ x_3 \\ x_4 \\ x_5 \end{bmatrix} = \begin{bmatrix} a+b \\ a-b \\ 2a \\ a \\ b \end{bmatrix} = a \begin{bmatrix} 1 \\ 1 \\ 2 \\ 1 \\ 0 \end{bmatrix} + b \begin{bmatrix} 1 \\ -1 \\ 0 \\ 0 \\ 1 \end{bmatrix}$ where the values $a, b \in \mathbb{F}$ of x_4 and x_5 are arbitrary (there is no corresponding pivot in the reduced row echelon form). Hence $\mathsf{Ker}(\mathbf{A}) = \mathsf{Span}(\langle \vec{b}_4, \vec{b}_5 \rangle)$ for the basis $\vec{b}_4 = \begin{bmatrix} 1 \\ 1 \\ 2 \\ 1 \\ 0 \end{bmatrix}$ and $\vec{b}_5 = \begin{bmatrix} 1 \\ -1 \\ 0 \\ 0 \\ 1 \end{bmatrix}$.

Let $\mathbf{A}_{i=1,n,j}$ be the jth column of matrix $\mathbf{A}$ in reduced row echelon form. This column is *free* if and only if

- it is a zero column ($\forall i \in [1, n]\,.\,\mathbf{A}_{i,j} = 0$), or else
- let $k = max\{i \in [1, n] \mid \mathbf{A}_{i,j} \neq 0\}$ be the row of the last non-null element $\mathbf{A}_{k,j} \neq 0$ in this column j, and then there is a non-null element in this row k strictly before column j, that is, $\exists \ell \in [0, j[\,.\,\mathbf{A}_{k,\ell} \neq 0$.

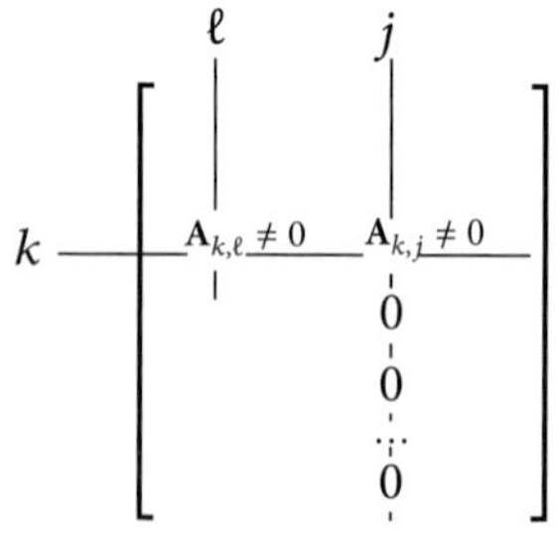

$\mathbf{A}_{i=1,n,j}$ is a free column

Lemma 37.12 Let $\{\mathbf{A}'_{i=1,n,j} \mid j \in F\}$ be the free columns of the reduced row echelon form $\mathbf{A}'$ of the matrix $\mathbf{A}$. Define the column vectors $\vec{b}^j$, $j \in F$ such that $\vec{b}^j_j = 1$, $\vec{b}^j_i = -\mathbf{A}'_{i,j}$, $i \in [1, n] \setminus \{j\}$, and $\vec{b}^j_i = 0$, $i \in]n, m] \setminus \{j\}$. Then $\{\vec{b}^j \mid j \in F\}$ is a basis of $\mathsf{Ker}(\mathbf{A})$. Each $\vec{b}^j$ is the solution of the linear equations when fixing $x_j = 1$ and $x_k = 0$ for $k \in F \setminus \{j\}$.

The variables corresponding to the free columns are called *free*, and the others, corresponding to pivot columns, are called *principal*.

Proof of lemma 37.12 Let $\mathbf{A}$ be a matrix and $\mathbf{A}'$ its reduced row echelon form. By the Gauss–Jordan elimination algorithm of section 37.4.4, the solutions to $\mathbf{A}\vec{x} = \vec{0}$ are those to $\mathbf{A}'\vec{x} = \vec{0}$ because all transformations of $(\mathbf{A}|\vec{0})$ into $(\mathbf{A}'|\vec{b})$ maintain $\vec{b} = \vec{0}$. Then by definition of the free columns $\mathbf{A}'_{i=1,n,j}$, $j \in F$ of $\mathbf{A}'$ we can fix arbitrarily the values v_j of the unknown $\vec{x}_j$. For any such fixed choice $v_{j,j\in F}$, there exists a solution of $\mathbf{A}'\vec{x} = \vec{0}$, hence of $\mathbf{A}\vec{x} = \vec{0}$ such that $\forall j \in F \,.\, \vec{x}_j = v_j$. So, $\vec{x} = \Sigma_{j\in F} v_j \vec{b}_j$ is a basis of $\mathsf{Ker}(\mathbf{A}') = \mathsf{Ker}(\mathbf{A})$ because the $\vec{b}_j$ are linearly independent. □

Exercise 37.13 (Project: system of generators of matrix) Continuing exercises 37.6 and 37.7, write a program that, from $(\mathbf{A}|\vec{b})$ determines whether a column of $\mathbf{A}$ is a pivot column (with a one with zeros below and to its left), computes a basis of the kernel of the matrix $\mathbf{A}$, and returns a system of generators $\langle \mathbf{B}, \vec{x_0}\rangle$ spanning the affine space defined by $(\mathbf{A}|\vec{b})$.

37.5.3 Solution Set of a System of Linear Equations

The concretization or *solution set* $\gamma_{\equiv}(\mathbf{A}|\vec{b})$ of a system of linear equations $\mathbf{A}\vec{x} = \vec{b}$ where $\mathbf{A} \in \mathbb{F}^{m\times n}$ and $\vec{b} \in \mathbb{F}^m$ is the set of all its possible solutions (which may be empty, a singleton, ..., infinite).

$$\gamma_{\equiv}(\mathbf{A}|\vec{b}) \quad \triangleq \quad \{\vec{x} \in \mathbb{F}^m \mid \mathbf{A}\vec{x} = \vec{b}\}$$

Lemma 37.14 If $\vec{x}_0 \in \mathbb{F}^m$ is any solution, that is, $\mathbf{A}\vec{x}_0 = \vec{b}$ then

$$\gamma_{\equiv}(\mathbf{A}|\vec{b}) \quad = \quad \{\vec{x}_0 + \vec{x} \mid \vec{x} \in \mathsf{Ker}(\mathbf{A})\}$$

Proof of lemma 37.14 If $\vec{x} \in \gamma_{\equiv}(\mathbf{A}|\vec{b})$ then $\mathbf{A}\vec{x} = \vec{b}$ and $\mathbf{A}\vec{x}_0 = \vec{b}$ so $\mathbf{A}(\vec{x} - \vec{x}_0) = (\vec{b} - \vec{b}) = \vec{0}$ so $(\vec{x} - \vec{x}_0) \in \mathsf{Ker}(\mathbf{A})$ so , by definition of $\in$, $\exists \vec{x}\,' \in \mathsf{Ker}(\mathbf{A}) \,.\, (\vec{x} - \vec{x}_0) = \vec{x}\,'$, in which case $\vec{x} = \vec{x}_0 + \vec{x}\,'$ so that $\vec{x} \in \{\vec{x}_0 + \vec{x}\,' \mid \vec{x}\,' \in \mathsf{Ker}(\mathbf{A})\}$.

Conversely, if $\vec{x} \in \{\vec{x}_0 + \vec{x}\,' \mid \vec{x}\,' \in \mathsf{Ker}(\mathbf{A})\}$ then $\exists \vec{x}\,' \,.\, \vec{x} = \vec{x}_0 + \vec{x}\,' \wedge \mathbf{A}\vec{x}\,' = \vec{0}$ so $\vec{x}\,' = \vec{x}_0 - \vec{x}$ implies $\mathbf{A}(\vec{x} - \vec{x}_0) = \vec{0}$, which implies $\mathbf{A}(\vec{x}) - \mathbf{A}(\vec{x}_0) = \vec{0}$, that is, $\mathbf{A}(\vec{x}) = \vec{b}$ because $\mathbf{A}(\vec{x}_0) = \vec{b}$, proving that $\vec{x} \in \gamma_{\equiv}(\mathbf{A}|\vec{b})$. □

The following picture illustrates lemma 37.14 on $x - y = 2$ (i.e., the 1×1 matrix $[x - y]$ and column vector $\vec{2}$).

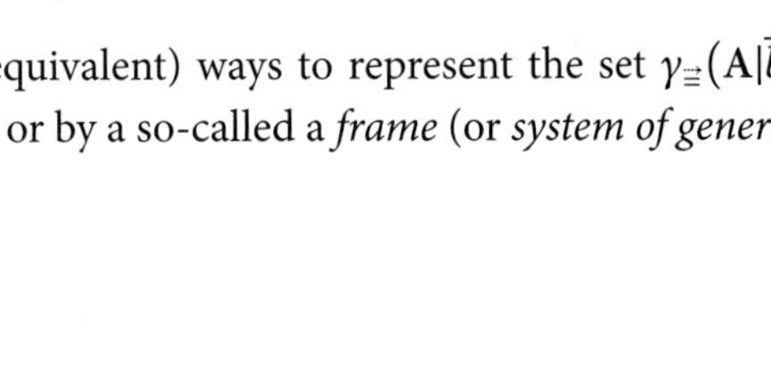

We now have two (equivalent) ways to represent the set $\gamma_{\equiv}(\mathbf{A}|\vec{b})$: either by the augmented matrix $(\mathbf{A}|\vec{b})$ or by a so-called a *frame* (or *system of generators*) $\langle \vec{x}_0, \mathsf{Ker}(\mathbf{A})\rangle$.

37.6 Affine Spaces

In order to describe lines, planes, hyperplanes, and so on, which do not necessarily pass through the origin, we can choose any point on that line, plane, hyperplane, etc., as a new origin so that we are reduced to describing a vector subspace through this origin. Therefore, an *affine space* $\mathbb{A}$ on a vector space $\vec{\mathbb{V}}$ with field $\mathbb{F}$ is a set of points equipped with a translation operation that given a point $A \in \mathbb{A}$ and a vector $\vec{u} \in \vec{\mathbb{V}}$, defines the point $A \vec{+} \vec{u}$ translated of point A by vector $\vec{u}$: 

. The translation aims at forgetting about the origin $\vec{0}$ of the vector space $\vec{\mathbb{V}}$ in that the translation of any point of $\mathbb{A}$ is a point of $\mathbb{A}$ so translating the origin does not matter—a way to forget about it! Without origin, the addition of vectors becomes meaningless in an affine space. However, the position of any point in the affine space can be uniquely expressed by a linear combination of vectors $\langle \vec{v}_1, \vec{v}_2, \ldots \rangle$ (of a vector space) relative to a fixed point $\vec{x}_0$.

The difference between a vector space and an affine space is illustrated as follows (the choice of the origin point O of coordinates $\vec{x}_0$ on the plane is arbitrary).

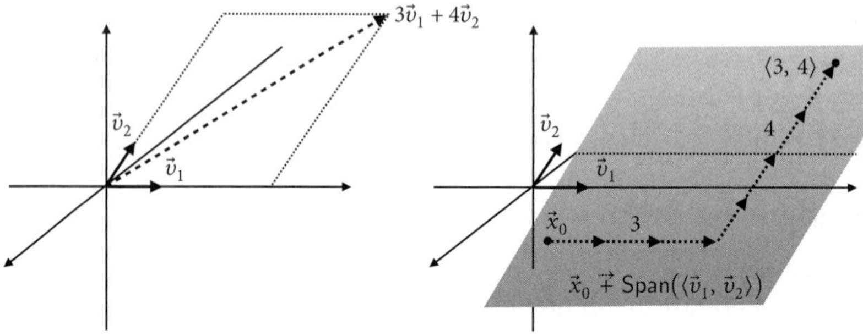

37.6.1 Definition of an Affine Space

An *affine space* $\langle \mathbb{A}, \langle \vec{\mathbb{V}}, \langle \mathbb{F}, +, -, \times, / \rangle, \vec{+}, \vec{-}, \vec{\times}, \vec{/} \rangle, \vec{+} \rangle$ on a vector space $\langle \vec{\mathbb{V}}, \langle \mathbb{F}, +, -, \times, / \rangle, \vec{+}, \vec{-}, \vec{\times}, \vec{/} \rangle$ with field $\langle \mathbb{F}, +, -, \times, / \rangle$ is a set of *points* $A, B, C, \ldots \in \mathbb{A}$ equipped with a *translation* operation $A \vec{+} \vec{u}$ of a point A by vector $\vec{u}$ such that for all $\vec{u}, \vec{v}, \vec{w}, \ldots \in \vec{\mathbb{V}}$ and $a, b, c, \ldots \in \mathbb{F}$,

- $\vec{+} \in \mathbb{A} \times \vec{\mathbb{V}} \to \mathbb{A}$,
- $\forall A \in \mathbb{A} \,.\, \forall \vec{u}, \vec{v} \in \vec{\mathbb{V}} \,.\, (A \vec{+} \vec{u}) \vec{+} \vec{v} = A \vec{+} (\vec{u} + \vec{v})$, and
- $\forall A, B \in \mathbb{A} \,.\, \exists! \vec{u} \in \vec{\mathbb{V}} \,.\, A \vec{+} \vec{u} = B$.

The zero vector $\vec{0}$ is the only vector such that $\forall A \in \mathbb{A} \,.\, A \vec{+} \vec{0} = A$.

37.6.2 Affine Coordinates

A *coordinate system* or *frame* or *system of generators* of an affine space $\mathbb{A}$ on a vector space $\vec{\mathbb{V}}$ with field $\mathbb{F}$ is a point $O \in \mathbb{A}$ (called the *origin*) and a basis $\langle \vec{b}_i, i \in \Delta \rangle$ of the vector space $\vec{\mathbb{V}}$. A point can be uniquely written in the form

$$A = O \vec{+} \sum_{i \in \Delta} a_i \vec{\times} \vec{b}_i$$

where the $\langle a_i, i \in \Delta \rangle$ are the *affine coordinates* of A.

37.6.3 Affine Subspace

An affine subspace of a vector space is a translation of a linear subspace. An *affine subspace* of an affine space $A, B, C \ldots \in \mathbb{A}$ on a vector space $\vec{u}, \vec{v}, \vec{w}, \ldots \in \vec{\mathbb{V}}$ with field $a, b, c, \ldots \in \mathbb{F}$ is a subset $\mathbb{B}$ of $\mathbb{A}$ of the form

- $A \vec{+} \vec{U} \triangleq \{A \vec{+} \vec{u} \mid \vec{u} \in \vec{U}\}$

where $A \in \mathbb{A}$ and $\vec{U}$ is a vector subspace of $\vec{\mathbb{V}}$.

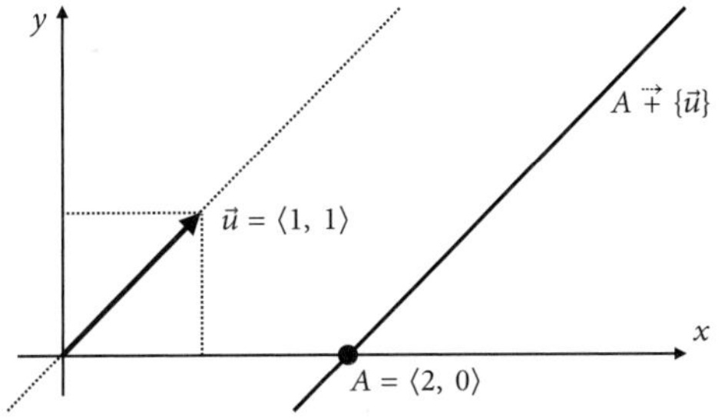

An affine subspace of an affine space is itself an affine space for the operations of this affine space.

Moreover the intersection $\bigcap_{i\in\Delta} \mathbb{B}_i$ of affine subspaces $\langle \mathbb{B}_i, i \in \Delta\rangle$ of an affine space $\mathbb{A}$ is an affine subspace of $\mathbb{A}$.

37.6.4 Example of the Coordinate Affine Space

Continuing section 37.2.8, consider the affine space $\langle \mathbb{F}^n, \langle \vec{\mathbb{F}}^n, \langle \mathbb{F}, +, -, \times, /\rangle, +, -, \times, /\rangle, \vec{+}\rangle$ of finite dimension $n > 0$ with

$$A \vec{+} \vec{u} \quad = \quad \langle A_1, \ldots, A_n\rangle \vec{+} \langle \vec{u}_1, \ldots, \vec{u}_n\rangle \quad \triangleq \quad \langle A_1 + \vec{u}_1, \ldots, A_n + \vec{u}_n\rangle$$

Lemma 37.15 If $\vec{x}_0 \in \mathbb{F}^m$ is any solution, that is, $\mathbf{A}\vec{x}_0 = \vec{b}$ then

$$\gamma_{\overrightarrow{\equiv}}(\mathbf{A}|\vec{b}) \quad = \quad \vec{x}_0 \vec{+} \mathsf{Ker}(\mathbf{A})$$

Proof of lemma 37.15

$$\begin{aligned} & \gamma_{\overrightarrow{\equiv}}(\mathbf{A}|\vec{b}) \\ = \; & \{\vec{x}_0 + \vec{x} \mid \vec{x} \in \mathsf{Ker}(\mathbf{A})\} && \text{(where } \mathbf{A}\vec{x}_0 = \vec{b}\text{, by lemma 37.14)} \\ = \; & \vec{x}_0 \vec{+} \mathsf{Ker}(\mathbf{A}) && \text{(}\mathsf{Ker}(\mathbf{A}) \text{ is a vector subspace of } \vec{\mathbb{F}}^n\text{)} \quad \square \end{aligned}$$

So an affine subspace is the solution set of an inhomogeneous linear system and inversely.

Because $\mathsf{Ker}(\mathbf{A})$ is a vector space of dimension n, it has a basis, and so we have

Lemma 37.16 If $\vec{x}_0 \in \mathbb{F}^m$ is any solution, that is, $\mathbf{A}\vec{x}_0 = \vec{b}$ and $\langle \vec{v}_i, i \in [1, n]\rangle$ is a basis of $\mathsf{Ker}(\mathbf{A})$ then

$$\gamma_{\overrightarrow{\equiv}}(\mathbf{A}|\vec{b}) \quad = \quad \vec{x}_0 \vec{+} \mathsf{Span}(\langle \vec{v}_i, i \in [1, n]\rangle)$$

37.6.5 Machine Representation of Subspaces of the Coordinate Affine Space

In the affine space $\langle \mathbb{F}^n, \langle \vec{\mathbb{F}}^n, \langle \mathbb{F}, +, -, \times, / \rangle, +, -, \times, / \rangle, \vec{+} \rangle$ of finite dimension $n > 0$, we can represent an affine subspace

- by a *system of equalities* represented by the pair $(\mathbf{A}|\vec{b})$ of a matrix $\mathbf{A} \in \mathbb{F}^{m \times n}$ and vector $\vec{b} \in \mathbb{F}^n$ encoding the affine subspace $\gamma_{\equiv}(\mathbf{A}|\vec{b})$. The matrix $\mathbf{A}$ is in reduced row-echelon form; or
- by a *frame* or *system of generators* $\langle \vec{x}_0, \langle \vec{v}_i, i \in [1, n] \rangle \rangle$ consisting of the coordinates $\vec{x}_0$ of a point O and a basis $\langle \vec{v}_i, i \in [1, n] \rangle$ of a vector space (precisely $\mathsf{Ker}(\mathbf{A})$) encoding the affine space $\gamma_{\rightleftharpoons}(\langle \vec{x}_0, \langle \vec{v}_i, i \in [1, n] \rangle \rangle) = \vec{x}_0 \vec{+} \mathsf{Span}(\langle \vec{v}_i, i \in [1, n] \rangle)$.

37.6.6 Matrix Representation of an Affine Coordinate Subspace

Let W be an affine subspace of the coordinate affine space given by a system of generators $W = \vec{x}_0 \vec{+} \mathsf{Span}(\mathbf{B})$ where $\mathbf{B} = \langle \mathbf{B}_i, i \in [1, n] \rangle$ is a basis. We look for a matrix $(\mathbf{A}|\vec{b})$ for which the solution set is W. We have $\vec{x} \in W$ if and only if $\exists \lambda_1, \ldots, \lambda_n \,.\, \vec{x} = \lambda_1 \mathbf{B}_1 + \ldots + \lambda_n \mathbf{B}_n + \vec{x}_0$ that is $\vec{x} = \lambda_1 \mathbf{B}_1 + \ldots + \lambda_n \mathbf{B}_n - \vec{x} + \vec{x}_0 = \vec{0}$ or in matrix form

$$(\mathbf{B}, \left\{ \begin{array}{cccc} -1 & 0 & \ldots & 0 \\ \ldots & \ldots & \ldots & \ldots \\ 0 & \ldots & 0 & -1 \end{array} \right\}, \vec{x}_0) \left\{ \begin{array}{c} \lambda_1 \\ \ldots \\ \lambda_n \\ \vec{x}_1 \\ \ldots \\ \vec{x}_n \\ 1 \end{array} \right\} = \vec{0}.$$

Transforming this matrix in reduced row echelon form yields the λ_i in function of the free $\vec{x}$. The matrix $(\mathbf{A}|\vec{b})$ is given by the rows with zero coefficients for the λ_i in the first n columns. They put constraints on $\vec{x}$ for the solution to exist, that is, for $\vec{x} \in W$.

Example 37.17 Consider the affine space

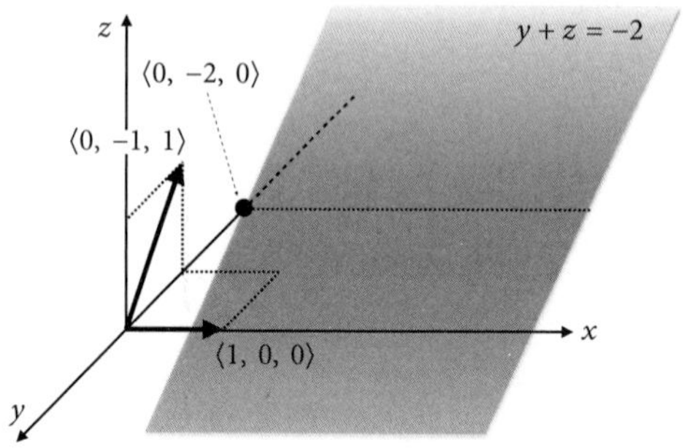

generated by $\mathbf{B} = \begin{bmatrix} 1 & 0 \\ 0 & -1 \\ 0 & 1 \end{bmatrix}$ and $\vec{x}_0 = \begin{bmatrix} 0 \\ -2 \\ 0 \end{bmatrix}$.

The equation matrix is $(\mathbf{B}, \overrightarrow{-1}, \vec{x}_0) = \begin{bmatrix} 1 & 0 & -1 & 0 & 0 & 0 \\ 0 & -1 & 0 & -1 & 0 & -2 \\ 0 & 1 & 0 & 0 & -1 & 0 \end{bmatrix}$ which is transformed in row echelon form as $\begin{bmatrix} 1 & 0 & -1 & 0 & 0 & 0 \\ 0 & 1 & 0 & 0 & -1 & 0 \\ 0 & 0 & 0 & 1 & 1 & 2 \end{bmatrix}$. The matrix $(\mathbf{A}|\vec{b})$ is given by the rows with zero coefficients for the λs in the first 2 columns, that is, the last row $0x + 1y + 1z + 2 = 0$.

Exercise 37.18 (Project: matrix of system of generators) Design an algorithm that given a system of generators $\langle \vec{x}_0, \langle \vec{v}_i, i \in [1,n] \rangle \rangle$, $n \geqslant 0$, of an affine subspace W of the coordinate affine space computes a matrix $(\mathbf{A}|\vec{b})$ for which the solution set $\{\vec{x} \mid \mathbf{A}\vec{x} = \vec{b}\}$ is W. For the affine subspace of section 37.6.3,

```
# print_matrix b_x0;;
1; 2;
1; 0;
# print_affine_space (mat_of_gen b_x0);;
1; -1; -2;
- : unit = ()
```

37.6.7 Projection / Elimination of a Variable

If $\vec{x} \in \mathbb{F}^n$ is a vector of scalars, $i \in [1,n]$, and $v \in \mathbb{F}$, we let $\vec{x}' = \vec{x}[i \leftarrow v]$ be the assignment of v to $\vec{x}$ at position i such that $\forall j \in [1,n] \setminus \{i\} . \vec{x}'_j = \vec{x}_j$ and $\vec{x}'_i = v$.

Given an affine space $W = \{\vec{x} \mid \mathbf{A}\vec{x} = \vec{b}\}$, we look for a characterization of the set $W' = \{\vec{x} \mid \exists v \in \mathbb{F} . \vec{x}[i \leftarrow v] \in W\}$. For example $\exists z . x + z = 0 \wedge y = z + 1$ is $x + y = 1$ so if $W = \left(\begin{bmatrix} 1 & 0 & 1 \\ 0 & 1 & -1 \end{bmatrix} \middle| \begin{bmatrix} 0 \\ 1 \end{bmatrix}\right)$ then $W' = \left(\begin{bmatrix} 1 & 1 & 0 \\ 0 & 0 & 0 \end{bmatrix} \middle| \begin{bmatrix} 1 \\ 0 \end{bmatrix}\right)$.

When considering the system of generators, the elimination of a variable consists of adding a vector to the basis in the dimension of this variable. In the example below, $\{\langle x, y, z \rangle \mid z = 0 \wedge x = 2\}$ is generated by $\langle \begin{bmatrix} 0 \\ 1 \\ 0 \end{bmatrix}, \begin{bmatrix} 2 \\ 0 \\ 0 \end{bmatrix} \rangle$ while $\{\langle x, y, z \rangle \mid \exists z . z = 0 \wedge x = 2\} = \{\langle x, y, z \rangle \mid x = 2\}$ is generated by $\langle \begin{bmatrix} 0 & 0 \\ 1 & 0 \\ 0 & 1 \end{bmatrix}, \begin{bmatrix} 2 \\ 0 \\ 0 \end{bmatrix} \rangle$.

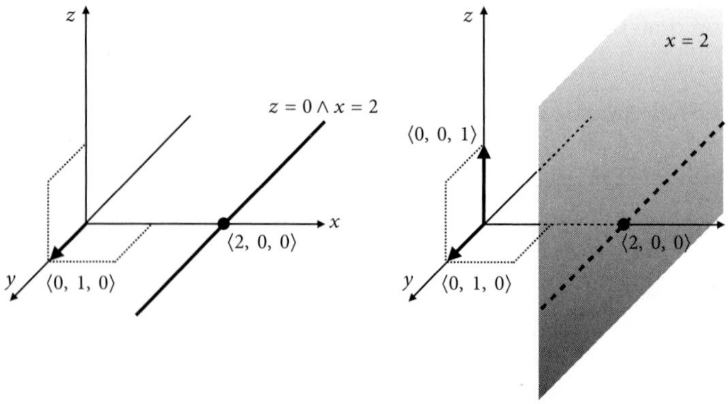

More generally,

Lemma 37.19 If W is generated by $\langle \mathbf{B}, \vec{x}_0 \rangle$ then $\{\vec{x} \mid \exists v \in \mathbb{F} \,.\, \vec{x}[i \leftarrow v] \in W\}$ is generated by $\langle (\mathbf{B}, \vec{0}[i \leftarrow 1]), \vec{x}_0 \rangle$ (where $\vec{x}[i \leftarrow v]_j = \vec{x}_j$ when $j \neq i$ and $\vec{x}[i \leftarrow v]_i = v$).

Exercise 37.20 Prove lemma 37.19.

Exercise 37.21 (Project: variable elimination) Design an algorithm that given a system of generators $\langle \vec{x}_0, \langle \vec{v}_i, i \in [1, n] \rangle \rangle$, $n \geqslant 0$, of an affine subspace W will eliminate the ith variable.

37.7 Conclusion

Finite dimensional linear algebra and matrix theory pervade mathematics and other sciences.

Considering variable properties, that is, sets of environments as sets of points, it is possible to abstract them to affine spaces, which yields the linear equality static analysis of programs of chapter 38.

37.8 Solutions to Selected Exercises

Solution to exercise 37.2 $\forall i \in \Delta \,.\, \vec{0} \in \overrightarrow{W}_i \subseteq \vec{\mathbb{V}}$ so $\vec{0} \in \bigcap_{i \in \Delta} \overrightarrow{W}_i \subseteq \vec{\mathbb{V}}$ proving $\bigcap_{i \in \Delta} \overrightarrow{W}_i$ to be a nonempty subset of $\vec{\mathbb{V}}$.

If $\vec{u}, \vec{v} \in \bigcap_{i \in \Delta} \overrightarrow{W}_i$ then, by definition of intersection, $\forall i \in \Delta \,.\, \vec{u}, \vec{v} \in \overrightarrow{W}_i$ so $\vec{u} + \vec{v} \in \overrightarrow{W}_i$ proving that $\vec{u} + \vec{v} \in \bigcap_{i \in \Delta} \overrightarrow{W}_i$.

If $\vec{u} \in \bigcap_{i \in \Delta} \overrightarrow{W}_i$ and $a \in \mathbb{F}$ then $\forall i \in \Delta \,.\, \vec{u} \in \overrightarrow{W}_i$ so $a\vec{u} \in \overrightarrow{W}_i$ proving that $a\vec{u} \in \bigcap_{i \in \Delta} \overrightarrow{W}_i$.

38 Linear Equality Analysis

Cartesian analyses cannot infer relations between variables, so the analysis of `x=0; y=0; while (x<10) {x=x+1; y=y+2;}` cannot infer an upper bound for y. The linear equality analysis aims at discovering linear equality relations $\mathrm{A} \times \vec{x} = \vec{b}$ between values $\vec{x}$ of the program variables where A and $\vec{b}$ are inferred by the static analysis. In this example, $2\mathrm{x} - \mathrm{y} = 0$, so $\mathrm{x} \leqslant 10$ implies $\mathrm{y} \leqslant 20$. Linear equality analysis was introduced by Michael Karr [560].

Contents

38.1 Affine Properties

A property of program variables $\mathbb{V}$ is a set of environments in $\mathbb{P} = \wp(\mathbb{V} \to \mathbb{F})$ (where $\mathbb{F}$ is the set $\mathbb{Q}$ of rationals (including integers and floats) or $\mathbb{R}$ of reals).

Let $m = |\mathbb{V}|$ be the cardinality of $\mathbb{V}$, that is, the finite number of program variables. If $\mathbb{V} = \{\mathrm{x}_1, \ldots, \mathrm{x}_m\}$, we let $\vec{x}$ be the column vector of values $\rho(\mathrm{x}_1), \ldots, \rho(\mathrm{x}_m)$ of these variables in environment $\rho \in \mathbb{V} \to \mathbb{F}$. So, up to the isomorphism $\rho \mapsto \vec{x}$, a program property $P \in \mathbb{P}$ is a set of points $\vec{x}$ in $P \in \mathbb{F}^m$ (and we write $\vec{x} \in P$ for $\rho \in P$ up to this isomorphism).

The affine space abstract domain consists of those subsets of $\wp(\mathbb{F}^m)$ (i.e., $\mathbb{P} = \wp(\mathbb{V} \to \mathbb{F})$), which are affine subspaces of the affine space $\langle \mathbb{F}^m, \langle \vec{\mathbb{F}}^m, \langle \mathbb{F}, +, -, \times, / \rangle, +, -, \times, / \rangle, \vec{+} \rangle$ of finite dimension $m > 0$:

$$\vec{\mathbb{P}} \triangleq \{\vec{\bot}\} \cup \{A + \vec{U} \mid A \in \mathbb{F}^m \wedge \vec{U} \text{ is a vector subspace of } \overrightarrow{\mathbb{F}^m}\}, \quad \vec{\bot} = \varnothing.$$

The infimum $\vec{\bot}$ encodes any system of linear equalities without a solution, and otherwise, following section 37.6.5, we have two representations of the affine subspaces $P \in \vec{\mathbb{P}}$:

- by a unique $n \times m+1$ matrix $(\mathrm{A}|\vec{b})$ in reduced row echelon form with no zero row encoding $\gamma_{\vec{=}}(\mathrm{A}|\vec{b}) = P$ (where the empty matrix with $n = 0$ encodes the supremum $(\mathbf{0}|\vec{0})$, and otherwise there are no zero rows so $n \leqslant m$ by theorem 37.9);

- by a frame or system of generators $\langle \vec{x}_0, \mathbf{B}\rangle$ where $\mathbf{A}\vec{x}_0 = \vec{b}, \mathbf{B} = \langle \vec{v}_i, i \in [1, m]\rangle$ is a basis such that $\mathsf{Span}(\mathbf{B}) = \mathsf{Ker}(\mathbf{A})$ so that $\vec{x}_0 \mathbin{\vec{+}} \mathsf{Ker}(\mathbf{A}) = \gamma_{\vec{=}}(\mathbf{A}|\vec{b}) = P$.

The system of generators is computed from $(\mathbf{A}|\vec{b})$ by finding a solution $\vec{x}_0$ using exercise 37.7 and the basis of the kernel of $\mathbf{A}$ as determined by lemma 37.12 and exercise 37.13.

The matrix $(\mathbf{A}|\vec{b})$ is computed from the system of generators by exercise 37.18 by transformation in reduced row echelon form and elimination of the zero rows.

Exercise 38.1 Express the Galois isomorphisms between the solution set of the system of linear equations, the matrices, and the systems of generators.

The implementation uses one of the two representations at a time and lazily switches to the other one if needed by the next operation to be performed.

38.2 Affine Abstraction

Up to the isomorphism between $\wp(\mathcal{V} \to \mathbb{F})$ and $\wp(\mathbb{F}^m)$, a property of the m variables is an element $P \in \wp(\mathbb{F}^m)$ is a variable property, its affine abstraction is

$$\alpha_{\mathbb{A}}(\emptyset) \triangleq \vec{\bot}$$
$$\alpha_{\mathbb{A}}(P) \triangleq \bigcap\{A + \vec{U} \mid A \in \mathbb{F}^m \wedge \vec{U} \text{ is a vector subspace of } \overrightarrow{\mathbb{F}^m} \wedge P \subseteq A + \vec{U}\}$$

That is, the least affine subspace of $\mathbb{F}^m$ that contains P. $\alpha_{\mathbb{A}}(P)$ is a well defined affine space because, by exercise 37.2, the intersection of affine subspaces is affine. This is a definition by an upper closure condition in a Moore family so that by exercise 11.89,

$$\langle \wp(\mathbb{F}^m), \subseteq\rangle \underset{\alpha_{\mathbb{A}}}{\overset{\mathbb{1}}{\leftrightarrows}} \langle \vec{\mathbb{P}}, \subseteq\rangle \tag{38.2}$$

where $\alpha_{\mathbb{A}}$ is an upper closure operator and $\vec{\mathbb{P}}$ is a complete lattice.

38.3 Affine Abstract Domain

The affine abstract domain is

$$\vec{\mathbb{D}} \triangleq \langle \vec{\mathbb{P}}, \vec{\sqsubseteq}, \vec{\bot}, \vec{\sqcup}, \vec{\sqcap}, \overrightarrow{\mathsf{assign}}[\![x, \mathsf{A}]\!], \overrightarrow{\mathsf{test}}[\![\mathsf{B}]\!], \overrightarrow{\overline{\mathsf{test}}}[\![\mathsf{B}]\!]\rangle.$$

- The supremum $\vec{\top} \triangleq \mathbb{F}^m$ is represented by the null matrix $(\mathbf{0}|\vec{0})$ with zero row.
- Equality $P \mathrel{\vec{=}} P'$ is the equality $(\mathbf{A}|\vec{b}) = (\mathbf{A}'|\vec{b}')$ of the matrices in reduced row echelon form without zero row encoding P and P' because $P = \gamma_{\vec{=}}(\mathbf{A}|\vec{b})$, $P' = \gamma_{\vec{=}}(\mathbf{A}'|\vec{b}')$, and unicity of the reduced row echelon form without zero row.
- The meet $P \mathbin{\vec{\sqcap}} P'$ of P and P' represented by $(\mathbf{A}|\vec{b})$ and $(\mathbf{A}'|\vec{b}')$ is the conjunction of the linear equalities that is $\left(\begin{array}{c|c} \mathbf{A} & \vec{b} \\ \mathbf{A}' & \vec{b}' \end{array}\right)$ normalized in reduced row echelon form without zero row.
- The inclusion $P \mathrel{\vec{\sqsubseteq}} P'$ is $P \mathbin{\vec{\sqcap}} P' \mathrel{\vec{=}} P$.
- The join $P \mathbin{\vec{\sqcup}} P'$ of P and P' represented by their systems of generators $\langle \vec{x_0}, \mathbf{B}\rangle$ and $\langle \vec{x_0}', \mathbf{B}'\rangle$ has system of generators $\langle \vec{x_0}, (\mathbf{B}, \mathbf{B}', \vec{x}_0' - \vec{x}_0)\rangle$ (or $\langle \vec{x_0}, (\mathbf{B}, \mathbf{B}')\rangle$ when $\vec{x}_0' = \vec{x}_0$).

Exercise 38.3 Prove the soundness of the linear equality join $\vec{\sqcup}$.

Exercise 38.4 Show that $\alpha_{\mathbb{A}}(P) = \vec{\sqcup}\{Y \in \vec{\mathbb{P}} \mid Y \vec{\sqsubseteq} P\}$.

38.4 Affine Abstract Assignment

An affine assignment $\mathsf{x} = \mathsf{A}\,;$ has the form

$$\mathsf{x}_i \;=\; v_1\mathsf{x}_1 + \ldots + v_i\mathsf{x}_i + \ldots + v_m\mathsf{x}_m + v_{m+1}$$

where $v_1, \ldots, v_i, \ldots, v_{m+1} \in \mathbb{F}$ are scalars and $\mathsf{x}_1, \ldots, \mathsf{x}_i, \ldots, \mathsf{x}_m$ are program variables. It is said to be *invertible* if and only if $v_i \neq 0$.

Let $(\mathbf{A}'|\vec{b}')$ be the affine abstraction of the reachable values of variables before the assignment (as defined by the forward reachability semantics of chapter 19). Let x_i' and x_i denote the value of the x_i before and after the assignment. So we have $\mathbf{A}'\vec{x}' = \vec{b}'$ and we must compute $(\mathbf{A}|\vec{b}) \triangleq \overrightarrow{\mathsf{assign}}[\![\mathsf{x}_i, \mathsf{A}]\!]\,(\mathbf{A}'|\vec{b}')$ such that $\mathbf{A}\vec{x} = \vec{b}$ after the affine assignment. We have $x_j = x_j'$ for $j \in [0, m] \setminus \{i\}$ because the values of all other variables but x_i are unchanged.

38.4.1 Invertible Affine Abstract Assignment

For invertible assignments $v_i \neq 0$, so we can express the old value x_i' of variable x_i before the assignment in terms of its new value x_i after the assignment. Therefore

$$x_i' \;=\; \frac{x_i}{v_i} - \frac{v_1}{v_i}x_1 - \ldots - \frac{v_{i-1}}{v_i}x_{i-1} - \frac{v_{i+1}}{v_i}x_{i+1} \ldots - \frac{v_m}{v_i}x_m - \frac{v_{m+1}}{v_i}.$$

Because $\forall j \in [0, m] \setminus \{i\}\,.\, x_j' = x_j$, the equality constraint before the assignment for each line ℓ of $\mathbf{A}'\vec{x}' = \vec{b}'$ was of the form

$$a_\ell^1 x_1 + \ldots + a_\ell^i x_i' + \ldots + a_\ell^m x_m = a_\ell^{m+1}.$$

Replacing x_i' by its value in terms of the values of the variables after the assignment, we get

$$a_\ell^1 x_1 + \ldots + a_\ell^{i-1} x_{i-1} + a_\ell^i\Big(\frac{x_i}{v_i} - \frac{v_1}{v_i}x_1 - \ldots - \frac{v_{i-1}}{v_i}x_{i-1} - \frac{v_{i+1}}{v_i}x_{i+1} \ldots - \frac{v_m}{v_i}x_m - \frac{v_{m+1}}{v_i}\Big)$$
$$+ a_\ell^{i+1} x_{i+1} + \ldots + a_\ell^m x_m = a_\ell^{m+1}.$$

Grouping the coefficients per variable, we get the line ℓ of $\mathbf{A}\vec{x} = \vec{b}$ after the affine assignment.

$$\Big(a_\ell^1 - \frac{a_\ell^i.v_1}{v_i}\Big)x_1 + \ldots + \Big(a_\ell^{i-1} - \frac{a_\ell^i.v_{i-1}}{v_i}\Big)x_{i-1} + \frac{a_\ell^i}{v_i}x_i + \Big(a_\ell^{i+1} - \frac{a_\ell^i.v_{i+1}}{v_i}\Big)x_{i+1} +$$
$$\ldots + \Big(a_\ell^m - \frac{a_\ell^i.v_m}{v_i}\Big)x_m = a_\ell^{m+1} + \frac{a_\ell^i.v_{m+1}}{v_i}.$$

38.4.2 Non-invertible Affine Abstract Assignment

If $v_i = 0$ in the assignment

$$\mathsf{x}_i \;=\; v_1\mathsf{x}_1 + \ldots + 0.\mathsf{x}_i + \ldots + v_m\mathsf{x}_m + v_{m+1}$$

the assignment is noninvertible. There is no relationship between the old value x_i' of variable x_i in $(\mathrm{A}'|\vec{b}')$ before the assignment and the new value x_i in $(\mathrm{A}|\vec{b})$ after the assignment. It follows that $\mathrm{A}\vec{x} = \vec{b}$ after the affine assignment is given by

$$\exists v \in \mathbb{F} \mathrel{.} \mathrm{A}'\vec{x}[i \leftarrow v] = \vec{b}' \wedge x_i = v_1 x_1 + \ldots + v_{i-1}x_{i-1} + v_{i+1}x_{i+1} + \ldots + v_m x_m + v_{m+1}.$$

So we first eliminate variable x_i by lemma 37.19 and then add the constraint $x_i = v_1 x_1 + \ldots + v_{i-1}x_{i-1} + v_{i+1}x_{i+1} + \ldots + v_m x_m + v_{m+1}$.

38.4.3 Abstraction of an Expression into an Affine Expression

The program assignments $\mathsf{x} = \mathsf{A}$; in section 4.1 are not necessarily in affine form. Therefore we use the following affine abstraction of the arithmetic expression A in section 3.4. The idea is that for nonlinear expressions like $\mathsf{x} * \mathsf{y}$, the static analysis may have determined that the value of x is a scalar c, so we can use the linear form $c.\mathsf{y}$.

$$\begin{array}{rclcll}
\alpha(\mathsf{1}) & \triangleq & 1 & & & \\
\alpha(\mathsf{x}) & \triangleq & 1.\mathsf{x} & & & \\
\alpha(\mathsf{A}_1 - \mathsf{A}_2) & \triangleq & \alpha(\mathsf{A}_1) - \alpha(\mathsf{A}_2) & & & \\
\alpha(\mathsf{c} * \mathsf{A}) & \triangleq & \alpha(\mathsf{A} * \mathsf{c}) & \triangleq & c.\alpha(\mathsf{A}) & c \text{ is the value of constant } \mathsf{c} \\
\alpha(\mathsf{x} * \mathsf{A}) & \triangleq & \alpha(\mathsf{A} * \mathsf{x}) & \triangleq & c.\alpha(\mathsf{A}) & \text{the analysis has determined that } \rho(\mathsf{x}) = c \\
\alpha(\mathsf{A}) & \triangleq & \top & & & \text{otherwise}
\end{array}$$

If the abstraction returns $\top$ then the assignment is handled by lemma 37.19 for eliminating the assigned variable. Otherwise the coefficients of the variables are summed up to get a linear assignment of section 38.4.

38.5 Affine Abstract Test

If a test B can be put into linear form $\vec{a}\vec{x} = b$ i.e. $a_1 x_1 + \ldots + a_{i-1}x_{i-1} + a_{i+1}x_{i+1} + \ldots + a_m x_m = b$ by transformation of B as in section 38.4.3, then the expression is conjuncted with P. $\overrightarrow{\mathsf{test}}[\![\mathsf{B}]\!](\mathrm{A} \mid \vec{b}) = \left(\begin{array}{c|c} \mathrm{A} & \vec{b} \\ \vec{a} & b \end{array}\right)$ and put into reduced row echelon form.

For linear equality tests, $\overrightarrow{\mathsf{test}}[\![\mathsf{B}]\!](P) = \overrightarrow{\overline{\mathsf{test}}}[\![\mathsf{B}]\!](P) = P$.

38.6 Fixpoint Computation

The abstract domain has no infinite ascending nor descending chain, so no widening/narrowing is needed.

38.7 Conclusion

Michael Karr [560] represents affine spaces as the solution set of linear equation systems $\mathrm{A}\vec{x} = \vec{b}$ represented by the matrix $(\mathrm{A} \mid \vec{b})$ (the number of affine relations between values of variables will be small hence the dimension of the affine space and the size of the system of generators will be large). He had to design an original algorithm for the join $\overrightarrow{\sqcup}$. A simpler and more efficient algorithm is proposed by [695, section 5.2.2].

Following what is traditionally done for polyhedral analysis [279], Markus Müller-Olm and Helmut Seidl [729] introduced the use of systems of generators [729, 730]. [348, 885] studied variations on matrix representations of affine domains. Integers can be analyzed using rationals with a concretization in the integers. A better solution is to generalize affine equalities to affine congruences [439]. Other generalizations include modular arithmetics [348] and interval [176] or polynomial equalities [160, 515, 531, 576, 642, 727, 840, 856, 998].

Exercise 38.5 (Project: linear equality abstract domain) Continuing the exercises of chapter 37, implement the linear equality abstract domain.

Exercise 38.6 Design the reduced product of the intervals and the linear equality abstract domains.

38.8 Solutions to Selected Exercises

Solution to exercise 38.3 If $\vec{x} \in P$ then $\vec{x} = \overrightarrow{x_0} + \mathbf{B}\vec{a}$ for some coefficients $\vec{a}$. So $\vec{x} = \overrightarrow{x_0} + (\mathbf{B}, \mathbf{B}', \vec{x}_0' - \vec{x}_0)\begin{bmatrix}\vec{a}\\ \vec{0}\\ 0\end{bmatrix}$ proving $\vec{x} \in P \mathbin{\vec{\sqcup}} P'$ (or $\vec{x} = \overrightarrow{x_0} + (\mathbf{B}, \mathbf{B}')\begin{bmatrix}\vec{a}\\ \vec{0}\end{bmatrix}$ when $\vec{x}_0' = \vec{x}_0$).

Similarly, if $\vec{x} \in P'$ then $\vec{x} = \overrightarrow{x_0}' + \mathbf{B}'\vec{a}'$ for some coefficients $\vec{a}'$. So $\vec{x} = \overrightarrow{x_0} + (\mathbf{B}, \mathbf{B}', \vec{x}_0' - \vec{x}_0)\begin{bmatrix}\vec{0}\\ \vec{a}'\\ 1\end{bmatrix}$ proving $\vec{x} \in P \mathbin{\vec{\sqcup}} P'$ (or $\vec{x} = \overrightarrow{x_0}' + (\mathbf{B}, \mathbf{B}')\begin{bmatrix}\vec{0}\\ \vec{a}'\end{bmatrix}$ when $\vec{x}_0' = \vec{x}_0$). So $P \cup P' \subseteq P \mathbin{\vec{\sqcup}} P'$.

Let Q with system of generators $\langle \overrightarrow{x_Q}, \mathbf{B}_Q\rangle$ be another upper bound of P and P'. Let us prove that $P \mathbin{\vec{\sqcup}} P' \mathrel{\vec{\sqsubseteq}} Q$, that is, $P \mathbin{\vec{\sqcup}} P' \subseteq Q$.

If $\vec{x} \in P \subseteq Q$ then $\exists \vec{a} \,.\, \vec{x} = \overrightarrow{x_0} + \mathbf{B}\vec{a}$ so $\exists \vec{b}$ such that $\vec{x} = \overrightarrow{x_Q} + \mathbf{B}_Q\vec{b}$. Similarly, if $\vec{x}' \in P' \subseteq Q$ then $\exists \vec{a}' \,.\, \vec{x}' = \overrightarrow{x_0}' + \mathbf{B}'\vec{a}'$ so $\exists \vec{b}'$ such that $\vec{x} = \overrightarrow{x_Q} + \mathbf{B}_Q\vec{b}'$. Therefore if $\vec{x} \in P \mathbin{\vec{\sqcup}} P'$ then $\exists \vec{a}, \vec{a}', c \,.\, \vec{x} = \overrightarrow{x_0} + (\mathbf{B}, \mathbf{B}', \vec{x}_0' - \vec{x}_0)\begin{bmatrix}\vec{a}\\ \vec{a}'\\ c\end{bmatrix} = \overrightarrow{x_0} + \mathbf{B}\vec{a} + \mathbf{B}'\vec{a}' + c(\vec{x}_0' - \vec{x}_0) = (\overrightarrow{x_Q} + \mathbf{B}_Q\vec{b}) + (\overrightarrow{x_Q} - \overrightarrow{x_0}' + \mathbf{B}_Q\vec{b}') + c(\vec{x}_0' - \vec{x}_0) = 2\overrightarrow{x_Q} + (c-1)\vec{x}_0' - c\vec{x}_0 + \mathbf{B}_Q(2\vec{b})$, which is a point of the affine space Q with the origin translated by $\overrightarrow{x_Q} + (c-1)\vec{x}_0' - c\vec{x}_0$, which, by definition of an affine space in section 37.6.1 is the affine space Q itself. The reasoning is the same when $\vec{x}_0' = \vec{x}_0$. So $P \mathbin{\vec{\sqcup}} P' \subseteq Q$. It follows, by the Galois connection (38.2) and lemma 11.38, that for all $P, Q \in \wp(\mathbb{F}^m) \,.\, \alpha_{\mathbb{A}}(P \cup Q) = \alpha_{\mathbb{A}}(P) \mathbin{\vec{\sqcup}} \alpha_{\mathbb{A}}(Q)$.

Solution to exercise 38.4 Because the concretization is the identity in (38.2), hence injective, we have $P = \bigcup\{X \in \wp(\mathbb{F}^m) \mid X \subseteq \mathbb{1}(P)\} = \bigcup\{X \in \wp(\mathbb{F}^m) \mid \alpha_{\mathbb{A}}(X) \mathrel{\vec{\sqsubseteq}} P\}$ so $\alpha_{\mathbb{A}}(P) = \vec{\bigsqcup}\{\alpha_{\mathbb{A}}(X) \mid \alpha_{\mathbb{A}}(X) \mathrel{\vec{\sqsubseteq}} P\} = \vec{\bigsqcup}\{Y \in \vec{\mathbb{P}} \mid Y \mathrel{\vec{\sqsubseteq}} P\}$ because $\alpha_{\mathbb{A}}$ preserves joins and is surjective.

39 Graphs

The Roy–Floyd–Warshall algorithm [375, 844, 965] finds shortest paths in a weighted graph (with no cycle with strictly negative weight). It is an abstract interpretation of a fixpoint path-finding algorithm. References [109, 165, 318, 860] are introductions to graph theory.

Contents

39.1 Graphs

A (directed) *graph* or *digraph* $G = \langle V, E\rangle$ is a pair of a set V of *vertices* (or *nodes* or *points*) and a set $E \in \wp(V \times V)$ of *edges* (or *arcs*). A (directed) edge $\langle x, y\rangle \in V$ has *origin* x and *end* y collectively called *extremities* (so the graphs that we consider are always directed). Therefore E is a binary relation of section 2.2.2 on V. Conversely, a binary relation can be understood as a graph on its field [860]. A graph is *finite* when the set of V of vertices (hence E) is finite.

Example 39.1

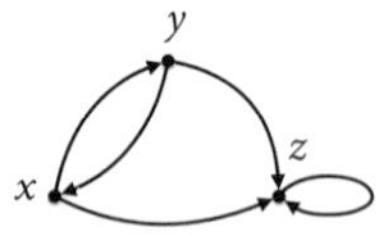

$$\mathrm{G} = \left[\begin{array}{ccl} V & = & \{x, y, z\} \\ E & = & \{\langle x,\ y\rangle, \langle x,\ z\rangle, \langle y,\ x\rangle, \\ & & \langle y,\ z\rangle, \langle z,\ z\rangle\} \end{array}\right] \qquad \mathrm{G} = \left[\begin{array}{c|ccc} & x & y & z \\ \hline x & 0 & 1 & 1 \\ y & 1 & 0 & 1 \\ z & 0 & 0 & 1 \end{array}\right]$$

39.2 Paths and Cycles of a Graph

A *path* π from y to z in a graph $\mathrm{G} = \langle V,\ E\rangle$ is a *finite* sequence of vertices $\pi = x_1 \dots x_n \in V^n$, $n > 1$, starting at *origin* $y = x_1$, finishing at *end* $z = x_n$, and linked by edges $\langle x_i, x_{i+1}\rangle \in E$, $i \in [1, n[$.[1] Let $V^{>1} \triangleq \bigcup_{n>1} V^n$ be the sequences of vertices of length at least 2. Formally, the set $\Pi(\mathrm{G}) \in \wp(V^{>1})$ of all paths of a graph $\mathrm{G} = \langle V,\ E\rangle$ is

$$\begin{array}{rcl} \Pi(\mathrm{G}) & \triangleq & \bigcup\limits_{n>1} \Pi^n(\mathrm{G}) \\ \Pi^n(\mathrm{G}) & \triangleq & \{x_1 \dots x_n \in V^n \mid \forall i \in [1, n[\,.\, \langle x_i,\ x_{i+1}\rangle \in E\} \quad (n > 1) \end{array} \tag{39.2}$$

The length $|\pi|$ of the path $\pi = x_1 \dots x_n \in V^n$ is the number of edges that is $n - 1 > 0$. We do not consider the case $n = 1$ of paths of length 0 with only one vertex because paths must have at least one edge. A *subpath* forms a strict part of another path (which, being strict, is not equal to that path).

The *vertices of a path* $\pi = x_1 \dots x_n \in \Pi^n(\mathrm{G})$ of a graph G is the set $\mathsf{V}(\pi) = \{x_1 \dots x_n\}$ of vertices appearing in that path π.

A *cycle* is a path $x_1 \dots x_n \in \Pi^n(\mathrm{G})$ with $x_n = x_1$, $n > 1$. Self-loops, that is, $\langle x,\ x\rangle \in E$ yield a cycle $x\,x$ of length 1.

39.3 Fixpoint Characterization of the Paths of a Graph

The concatenation of sets of finite paths is

$$P \circledcirc Q \ \triangleq\ \{x_1 \dots x_n y_2 \dots y_m \mid x_1 \dots x_n \in P \wedge y_1 y_2 \dots y_m \in Q \wedge x_n = y_1\}. \tag{39.3}$$

We have the following fixpoint characterization of the paths of a graph [259, Sect. 4], which is well defined, by Tarski–Kantorovich's iterative fixpoint theorem 15.21.

Theorem 39.4 (fixpoint characterization of the paths of a graph) The paths of a graph $\mathrm{G} = \langle V,\ E\rangle$ are

$$\begin{array}{rcllcll} \Pi(\mathrm{G}) & = & \mathsf{lfp}^{\subseteq}\, \overrightarrow{\mathscr{F}}_{\Pi}, & \overrightarrow{\mathscr{F}}_{\Pi}(X) & \triangleq & E \cup X \circledcirc E & (39.4.a) \\ & = & \mathsf{lfp}^{\subseteq}\, \overleftarrow{\mathscr{F}}_{\Pi}, & \overleftarrow{\mathscr{F}}_{\Pi}(X) & \triangleq & E \cup E \circledcirc X & (39.4.b) \\ & = & \mathsf{lfp}^{\subseteq}\, \overleftrightarrow{\mathscr{F}}_{\Pi}, & \overleftrightarrow{\mathscr{F}}_{\Pi}(X) & \triangleq & E \cup X \circledcirc X & (39.4.c) \\ & = & \mathsf{lfp}^{\subseteq}_{E}\, \widehat{\mathscr{F}}_{\Pi}, & \widehat{\mathscr{F}}_{\Pi}(X) & \triangleq & X \cup X \circledcirc X & (39.4.d) \end{array}$$

1. In order to consider infinite paths, we would have to introduce limits, as in section 7.2.

Proof of theorem 39.4 We observe that $\bigcup_{i\in\Delta}(X_i \circledcirc E) = \bigcup_{i\in\Delta}\{\pi x y \mid \pi x \in X_i \wedge \langle x,\ y\rangle \in E\} = \{\pi x y \mid \pi x \in \bigcup_{i\in\Delta} X_i \wedge \langle x,\ y\rangle \in E\} = (\bigcup_{i\in\Delta} X_i) \circledcirc E$ so that the transformer $\overrightarrow{\mathcal{F}}_{\Pi}$ preserves nonempty joins so is upper continuous. The same holds for $\overleftarrow{\mathcal{F}}_{\Pi}$.

Let $\langle X_i,\ i \in \mathbb{N}\rangle$ be a $\subseteq$-increasing chain of elements of $\wp(V^{>1})$. $\circledcirc$ is increasing componentwise so $\bigcup_{i\in\mathbb{N}}(X_i \circledcirc X_i) \subseteq (\bigcup_{i\in\mathbb{N}} X_i \circledcirc \bigcup_{i\in\mathbb{N}} X_i)$. Conversely, if $\pi \in (\bigcup_{i\in\mathbb{N}} X_i \circledcirc \bigcup_{i\in\mathbb{N}} X_i)$ then $\pi = \pi_i x \pi_j$ where $\pi_i x \in X_i$ and $x\pi_j \in X_j$. Assume $i \leqslant j$. Because $X_i \subseteq X_j$, $\pi_i x \in X_j$ so $\pi = \pi_i x \pi_j \in X_j \circledcirc X_j \subseteq \bigcup_{k\in\mathbb{N}} X_k \circledcirc X_k$ proving that $\bigcup_{i\in\mathbb{N}}(X_i \circledcirc X_i) \supseteq (\bigcup_{i\in\mathbb{N}} X_i \circledcirc \bigcup_{i\in\mathbb{N}} X_i)$. We conclude, by antisymmetry, that $\overrightarrow{\mathcal{F}}_{\Pi}$ and $\widehat{\mathcal{F}}_{\Pi}$ are upper continuous.

It follows, by Scott–Kleene's fixpoint theorem 15.26, that the least fixpoints do exist.

We consider case (39.4.c). By upper continuity, we can apply theorem 18.27. Let us calculate the iterates $\langle \overleftrightarrow{\mathcal{F}}_{\Pi}^{k},\ k \in \mathbb{N}\rangle$ of the fixpoint of transformer $\overleftrightarrow{\mathcal{F}}_{\Pi}$ from $\varnothing$.

$\overleftrightarrow{\mathcal{F}}_{\Pi}^{0} = \varnothing$, by definition of the iterates from $\varnothing$.

$\overleftrightarrow{\mathcal{F}}_{\Pi}^{1} = \overleftrightarrow{\mathcal{F}}_{\Pi}(\overleftrightarrow{\mathcal{F}}_{\Pi}^{0}) = E = \Pi^2(G)$ contains the paths of length 1 which are made of a single arc. If the graph has no paths longer than mere arcs, all paths are covered after 1 iteration.

Assume, by recurrence hypothesis on k, that $\overleftrightarrow{\mathcal{F}}_{\Pi}^{k} = \bigcup_{n=2}^{2^{k-1}} \Pi^n(G)$ contains exactly all paths of G of length less than or equal to 2^{k-1}. We have

$$
\begin{aligned}
\overleftrightarrow{\mathcal{F}}_{\Pi}^{k+1} &\triangleq \overleftrightarrow{\mathcal{F}}_{\Pi}(\overleftrightarrow{\mathcal{F}}_{\Pi}^{k}) && \text{(definition of the iterates)}\\
&= E \cup \overleftrightarrow{\mathcal{F}}_{\Pi}^{k} \circledcirc \overleftrightarrow{\mathcal{F}}_{\Pi}^{k} && \text{(definition of } \overleftrightarrow{\mathcal{F}}_{\Pi}\text{)}\\
&= E \cup \{x_1 \ldots x_n y_2 \ldots y_m \mid x_1 \ldots x_n \in \overleftrightarrow{\mathcal{F}}_{\Pi}^{k} \wedge x_n y_2 \ldots y_m \in \overleftrightarrow{\mathcal{F}}_{\Pi}^{k}\}\\
& && \text{(definition of } \circledcirc\text{)}\\
&= E \cup \{x_1 \ldots x_n y_2 \ldots y_m \mid x_1 \ldots x_n \in \bigcup_{n=2}^{2^{k-1}} \Pi^n(G) \wedge x_n y_2 \ldots y_m \in \bigcup_{n=2}^{2^{k-1}} \Pi^n(G)\}\\
& && \text{(induction hypothesis)}\\
&= E \cup \bigcup_{n=3}^{2^k} \Pi^n(G)
\end{aligned}
$$

($(\subseteq)$ the concatenation of two paths of length at least 1 and at most 2^{k-1} is at least of length 2 and at most of length $2 \times 2^{k-1} = 2^k$.
$(\supseteq)$ Conversely, any path of length at most 2^k has either length 1 in E or can be decomposed into two paths $\pi = x_1 \ldots x_n$ and $\pi' = x_n y_2 \ldots y_m$ of length at most 2^{k-1}. By induction hypothesis, $\pi, \pi' \in \bigcup_{n=2}^{2^{k-1}} \Pi^n(G)$)

By recurrence on k, for all $k \in \mathbb{N}^+$, $\overleftrightarrow{\mathcal{F}}_{\Pi}^{k} = \displaystyle\bigcup_{n=2}^{2^{k-1}} \Pi^n(G)$ contains exactly all paths from x to y of length less than or equal to 2^{k-1}.

Finally, we must prove that the limit $\mathsf{lfp}^{\subseteq}\, \overleftrightarrow{\mathcal{F}}_{\Pi} = \displaystyle\bigcup_{k\in\mathbb{N}} \overleftrightarrow{\mathcal{F}}_{\Pi}^{k}$ is $\Pi(G)$ that is contains exactly all paths of G.

Any path in $\Pi(G)$ has a length $n > 0$ such that $n \leqslant 2^{n-1}$ so belongs to $\overleftrightarrow{\mathcal{F}}_{\Pi}^{n}(\varnothing)$ hence to the limit, proving $\Pi(G) \subseteq \mathsf{lfp}^{\subseteq}\, \overleftrightarrow{\mathcal{F}}_{\Pi}$.

Conversely any path in $\mathsf{lfp}^{\subseteq}\, \overleftrightarrow{\mathcal{F}}_{\Pi} = \displaystyle\bigcup_{k\in\mathbb{N}} \overleftrightarrow{\mathcal{F}}_{\Pi}^{k}$ belongs to some iterate $\overleftrightarrow{\mathcal{F}}_{\Pi}^{k}$ which contains exactly all paths of length less than or equal to 2^k so belongs to $\Pi^{2^k}(G)$ hence to $\Pi(G)$, proving $\mathsf{lfp}^{\subseteq}\, \overleftrightarrow{\mathcal{F}}_{\Pi} \subseteq \Pi(G)$. By antisymmetry $\Pi(G) = \mathsf{lfp}^{\subseteq}\, \overleftrightarrow{\mathcal{F}}_{\Pi}$.

The equivalent form $\widehat{\mathscr{F}}_{\Pi}$ follows from exercises 15.8 and 15.9. The proofs for (39.4.a) and (39.4.b) are similar. □

Exercise 39.5 Show that in cases (39.4.a) and (39.4.b), the kth-iterate is $\bigcup_{n=2}^{k} \Pi^n(G)$.

39.4 Abstraction of the Paths of a Graph

A *path problem* in a graph $G = \langle V, E\rangle$ consists of specifying/computing an abstraction $\alpha(\Pi(G))$ of its paths $\Pi(G)$ defined by a Galois connection

$$\langle \wp(V^{>1}), \subseteq, \cup\rangle \xleftrightarrows[\alpha]{\gamma} \langle A, \sqsubseteq, \sqcup\rangle.$$

A path problem can be solved by a fixpoint definition/computation.

Theorem 39.6 (fixpoint characterization of a path problem) Let $G = \langle V, E\rangle$ be a graph with paths $\Pi(G)$ and $\langle \wp(V^{>1}), \subseteq, \cup\rangle \xleftrightarrows[\alpha]{\gamma} \langle A, \sqsubseteq, \sqcup\rangle$.

$$\begin{aligned}
\alpha(\Pi(G)) &= \mathsf{lfp}^{\sqsubseteq}\, \overrightarrow{\mathscr{F}}_{\Pi}^{\sharp}, & \overrightarrow{\mathscr{F}}_{\Pi}^{\sharp}(X) &\triangleq \alpha(E) \sqcup X \mathbin{\overline{\circledcirc}} \alpha(E) && (39.6.a)\\
&= \mathsf{lfp}^{\sqsubseteq}\, \overleftarrow{\mathscr{F}}_{\Pi}^{\sharp}, & \overleftarrow{\mathscr{F}}_{\Pi}^{\sharp}(X) &\triangleq \alpha(E) \sqcup \alpha(E) \mathbin{\overline{\circledcirc}} X && (39.6.b)\\
&= \mathsf{lfp}^{\sqsubseteq}\, \overleftrightarrow{\mathscr{F}}_{\Pi}^{\sharp}, & \overleftrightarrow{\mathscr{F}}_{\Pi}^{\sharp}(X) &\triangleq \alpha(E) \sqcup X \mathbin{\overline{\circledcirc}} X && (39.6.c)\\
&= \mathsf{lfp}^{\sqsubseteq}_{\alpha(E)}\, \widehat{\mathscr{F}}_{\Pi}^{\sharp}, & \widehat{\mathscr{F}}_{\Pi}^{\sharp}(X) &\triangleq X \sqcup X \mathbin{\overline{\circledcirc}} X && (39.6.d)
\end{aligned}$$

where $\alpha(X) \mathbin{\overline{\circledcirc}} \alpha(Y) = \alpha(X \circledcirc Y)$.

Proof of theorem 39.6 All cases are similar. Let us check the commutation for (39.6.c).

$\alpha(\overleftrightarrow{\mathscr{F}}_{\Pi}(X))$

$= \alpha(E \cup X \circledcirc X)$ ⟨definition (39.4.c) of $\overleftrightarrow{\mathscr{F}}_{\Pi}(X)$⟩

$= \alpha(E) \sqcup \alpha(X \circledcirc X)$
⟨by lemma 11.38, the abstraction of Galois connections preserves existing joins⟩

$= \alpha(E) \sqcup \alpha(X) \mathbin{\overline{\circledcirc}} \alpha(X)$ ⟨by hypothesis⟩

$= \overleftrightarrow{\mathscr{F}}_{\Pi}^{\sharp}(\alpha(X))$ ⟨definition (39.6.c) of $\overleftrightarrow{\mathscr{F}}_{\Pi}^{\sharp}$⟩

We conclude by theorem 39.4 and exact least fixpoint abstraction theorem 18.23. The equivalent form $\widehat{\mathscr{F}}_{\Pi}^{\sharp}$ follows from exercises 15.8 and 15.9. □

An essential remark is that the fixpoint definitions of the set of paths in $\wp(V^{>1})$ of a graph $G = \langle V, E\rangle$ in theorem 39.4 based on the primitives E, $\cup$, and $\circledcirc$ are preserved in theorem 39.6 by the abstraction $\langle \wp(V^{>1}), \subseteq, \cup\rangle \xleftrightarrows[\alpha]{\gamma} \langle A, \sqsubseteq, \sqcup\rangle$ for the primitives $\alpha(E)$, $\sqcup$, and $\overline{\circledcirc}$ on A, which explains the origin of the observation by [108, 314, 424, 774, 775] that path problems all have the same algebraic structure.

39.5 Calculational Design of the Paths Between Any Two Vertices

As a direct application of theorem 39.6, let us consider the abstraction of all paths $\Pi(G)$ into the paths between any two vertices. This is $\mathsf{p} \triangleq \alpha^{\circ\!\!-\!\!\circ}(\Pi(G))$ with the projection abstraction

$$\alpha^{\circ\!\!-\!\!\circ}(X) \quad \triangleq \quad (y, z) \mapsto \{x_1 \ldots x_n \in X \mid y = x_1 \wedge x_n = z\}$$

such that

$$\langle \wp(V^{>1}), \subseteq, \cup\rangle \underset{\alpha^{\circ\!\!-\!\!\circ}}{\overset{\gamma^{\circ\!\!-\!\!\circ}}{\leftrightarrows}} \langle V \times V \to \wp(V^{>1}), \dot{\subseteq}, \dot{\cup}\rangle \tag{39.7}$$

where $\mathsf{p} \mathrel{\dot{\subseteq}} \mathsf{p}' \Leftrightarrow \forall x, y \in V \,.\, \mathsf{p}(x, y) \subseteq \mathsf{p}'(x, y)$ and $(\dot{\bigcup}_{i\in\Delta} \mathsf{p}_i)(x, y) \triangleq \bigcup_{i\in\Delta}(\mathsf{p}_i(x, y))$ are defined pointwise.

By (39.2) and lemma 11.38 on existing join preservation, we have

$$\begin{aligned} \mathsf{p}(y, z) \quad &\triangleq \quad \bigcup_{n\in\mathbb{N}^+} \mathsf{p}^n(y, z) & (39.8)\\ \mathsf{p}^n(y, z) \quad &\triangleq \quad \{x_1 \ldots x_n \in \Pi^n(G) \mid y = x_1 \wedge x_n = z\}\\ &= \quad \{x_1 \ldots x_n \in V^n \mid y = x_1 \wedge x_n = z \wedge \forall i \in [1, n[\,.\, \langle x_i,\ x_{i+1}\rangle \in E\}. \end{aligned}$$

$\mathsf{p}(x, x)$ is empty if and only if there is no cycle from x to x (which requires, in particular, that the graph has no self-loops, that is, $\forall x \in V \,.\, \langle x,\ x\rangle \notin E$). We define the concatenation of finite paths

$$\begin{aligned} x_1 \ldots x_n \odot y_1 y_2 \ldots y_m \quad &\triangleq \quad x_1 \ldots x_n y_2 \ldots y_m && \text{if } x_n = y_1 & (39.9)\\ &\triangleq \quad \mathsf{undefined} && \text{otherwise} \end{aligned}$$

As a direct application of the path problem theorem 39.6, we have the following fixpoint characterization of the paths of a graph between any two vertices [259, Sect. 5], which, by Tarski–Kantorovich's iterative fixpoint theorem 15.21 and its variants yields an iterative algorithm (converging in finitely many iterations for finite graphs without infinite paths).

Theorem 39.10 (fixpoint characterization of the paths of a graph between any two vertices) Let $G = \langle V,\ E\rangle$ be a graph. The paths between any two vertices of G are $\mathsf{p} = \alpha^{\circ\!\!-\!\!\circ}(\Pi(G))$ such that

$$\begin{aligned} \mathsf{p} \quad &= \quad \mathsf{lfp}^{\dot{\subseteq}}\, \overrightarrow{\mathscr{F}}^{\circ\!\!-\!\!\circ}_{\Pi}, & \overrightarrow{\mathscr{F}}^{\circ\!\!-\!\!\circ}_{\Pi}(\mathsf{p}) \quad &\triangleq \quad \dot{E} \mathbin{\dot{\cup}} \mathsf{p} \mathbin{\overset{\circ\!\circ}{\circledcirc}} \dot{E} & (39.10.a)\\ &= \quad \mathsf{lfp}^{\dot{\subseteq}}\, \overleftarrow{\mathscr{F}}^{\circ\!\!-\!\!\circ}_{\Pi}, & \overleftarrow{\mathscr{F}}^{\circ\!\!-\!\!\circ}_{\Pi}(\mathsf{p}) \quad &\triangleq \quad \dot{E} \mathbin{\dot{\cup}} \dot{E} \mathbin{\overset{\circ\!\circ}{\circledcirc}} \mathsf{p} & (39.10.b)\\ &= \quad \mathsf{lfp}^{\dot{\subseteq}}\, \overleftrightarrow{\mathscr{F}}^{\circ\!\!-\!\!\circ}_{\Pi}, & \overleftrightarrow{\mathscr{F}}^{\circ\!\!-\!\!\circ}_{\Pi}(\mathsf{p}) \quad &\triangleq \quad \dot{E} \mathbin{\dot{\cup}} \mathsf{p} \mathbin{\overset{\circ\!\circ}{\circledcirc}} \mathsf{p} & (39.10.c)\\ &= \quad \mathsf{lfp}^{\dot{\subseteq}}_{\dot{E}}\, \widehat{\mathscr{F}}^{\circ\!\!-\!\!\circ}_{\Pi}, & \widehat{\mathscr{F}}^{\circ\!\!-\!\!\circ}_{\Pi}(\mathsf{p}) \quad &\triangleq \quad \mathsf{p} \mathbin{\dot{\cup}} \mathsf{p} \mathbin{\overset{\circ\!\circ}{\circledcirc}} \mathsf{p} & (39.10.d) \end{aligned}$$

where $\dot{E} \triangleq x, y \mapsto (E \cap \{\langle x,\ y\rangle\})$ and $\mathsf{p}_1 \mathbin{\overset{\circ\!\circ}{\circledcirc}} \mathsf{p}_2 \triangleq x, y \mapsto \bigcup_{z\in V} \mathsf{p}_1(x, z) \circledcirc \mathsf{p}_2(z, y)$.

Proof of theorem 39.10 We apply theorem 39.6 with $\alpha^{\circ\!\!-\!\!\circ}(E) = x, y \mapsto (E \cap \{\langle x,\ y\rangle\}) = \dot{E}$ and

$$\begin{aligned} &\alpha^{\circ\!\!-\!\!\circ}(X \circledcirc Y)\\ = \ &(x, y) \mapsto \{z_1 \ldots z_n \in X \circledcirc Y \mid x = z_1 \wedge z_n = y\} && \text{(definition (39.7) of } \alpha^{\circ\!\!-\!\!\circ}\text{)} \end{aligned}$$

$$= (x, y) \mapsto \{z_1 \ldots z_n \in \{x_1 \ldots x_k y_2 \ldots y_m \mid x_1 \ldots x_k \in X \wedge x_k y_2 \ldots y_m \in Y\} \mid x = z_1 \wedge z_n = y\}$$
⟅definition (39.3) of $\circledcirc$⟆

$$= (x, y) \mapsto \bigcup_{z \in V} \{x x_2 \ldots x_{k-1} z y_2 \ldots, y_{m-1} y \mid x x_2 \ldots x_{k-1} z \in X \wedge z y_2 \ldots y_{m-1} y \in Y\}$$
⟅definitions of $\in$ and $\cup$ with $x = x_1$, $y_m = y$, and $z = x_k$⟆

$$= (x, y) \mapsto \bigcup_{z \in V} \{x x_2 \ldots x_{k-1} z \odot z y_2 \ldots y_{m-1} y \mid x x_2 \ldots x_{k-1} z \in X \wedge z y_2 \ldots y_{m-1} y \in Y\}$$
⟅definition (39.9) of $\odot$⟆

$$= (x, y) \mapsto \bigcup_{z \in V} \{p \odot p' \mid p \in \alpha^{\circ\!-\!\circ}(X)(y, z) \wedge p' \in \alpha^{\circ\!-\!\circ}(Y)(z, y)\}$$
⟅definition of $\alpha^{\circ\!-\!\circ}(X)$ with $p = x x_2 \ldots x_{k-1} z$ and $p' = z y_2 \ldots y_{m-1} y$⟆

$$= \alpha^{\circ\!-\!\circ}(X) \mathbin{\overset{\circ\circ}{\circledcirc}} \alpha^{\circ\!-\!\circ}(Y)$$

by defining $X \mathbin{\overset{\circ\circ}{\circledcirc}} Y \triangleq (x, y) \mapsto \bigcup_{z \in V} \{p \odot p' \mid p \in X(y, z) \wedge p' \in Y(z, y)\} = (x, y) \mapsto \bigcup_{z \in V} X(y, z) \circledcirc Y(z, y)$ by (39.3) and (39.9). □

We now equip graphs with weights, for example, to measure the distance between vertices.

39.6 Groups

A *group* $\langle \mathbb{G}, 0, + \rangle$[2] is a set $\mathbb{G}$ with $0 \in \mathbb{G}$ and $+ \in \mathbb{G} \times \mathbb{G} \to \mathbb{G}$ such that

- $\forall a, b, c \in \mathbb{G} . (a + b) + c = a + (b + c)$ (associativity)
- $\forall a \in \mathbb{G} . 0 + a = a + 0 = a$ (identity denoted 0)
- $\forall a \in \mathbb{G} . \exists b \in \mathbb{G} . a + b = 0$ (inverse, b is denoted $-a$ or a^{-1}.)

For example, the scalars $\langle \mathbb{F}, 0, + \rangle$ are a group where 0 is the null scalar and + is scalar addition. Integers modulo n are a group under addition.

39.7 Weighted Graphs

Let $\langle \mathbb{G}, 0, + \rangle$ be a group. A (directed) graph $G = \langle V, E, \boldsymbol{\omega} \rangle$ weighted on the group $\mathbb{G}$ is a finite graph $\langle V, E \rangle$ equipped with a weight $\boldsymbol{\omega} \in E \to \mathbb{G}$ mapping arcs to their weight.

Example 39.11 Continuing example 39.1, with weights in the group $\langle \mathbb{Z} \cup \{\infty\}, 0, + \rangle$ of integers extended with a supremum ∞,

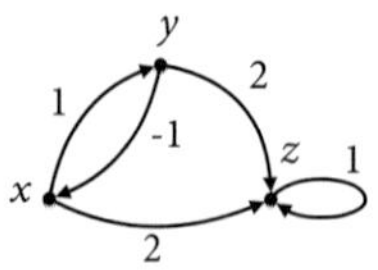

$$\left[\begin{array}{ll} \boldsymbol{\omega}(\langle x, y \rangle) = 1 & \boldsymbol{\omega}(\langle x, z \rangle) = 2 \\ \boldsymbol{\omega}(\langle x, z \rangle) = 2 & \boldsymbol{\omega}(\langle y, x \rangle) = -1 \\ \boldsymbol{\omega}(\langle y, z \rangle) = 2 & \boldsymbol{\omega}(\langle z, z \rangle) = 1 \end{array}\right] \qquad G = \left[\begin{array}{c|ccc} & x & y & z \\ \hline x & \infty & 1 & 2 \\ y & -1 & \infty & 2 \\ z & \infty & \infty & 1 \end{array}\right]$$

2. We call group, both the algebraic structure $\langle \mathbb{G}, 0, + \rangle$ and its support $\mathbb{G}$.

39.8 Totally Ordered Groups

A *totally (or linearly) ordered group* $\langle \mathbb{G}, \leqslant, 0, + \rangle$ is a group $\langle \mathbb{G}, 0, + \rangle$ with a total order $\leqslant$ on $\mathbb{G}$ such that

- $\forall a, b, c \in \mathbb{G} . (a \leqslant b) \Rightarrow (a + c \leqslant b + c)$ (translation-invariance.)

An element $x \in \mathbb{G}$ of a totally ordered group $\langle \mathbb{G}, \leqslant, 0, + \rangle$ is said to be strictly negative if and only if $x \leqslant 0 \wedge x \neq 0$.

Following section 10.3, if $S \subseteq \mathbb{G}$ then we define $\min S$ to be the greatest lower bound of S in $\mathbb{G}$ or $-\infty$:

$$
\begin{aligned}
\min S &= m &&\Leftrightarrow\quad m \in \mathbb{G} \wedge (\forall x \in S . m \leqslant x) \wedge (\forall y \in S . (\forall x \in S . y \leqslant x) \Rightarrow y \leqslant m) \\
&= -\infty &&\Leftrightarrow\quad \forall x \in S . \exists y \in S . y < x \qquad \text{(where } -\infty \notin \mathbb{G}) \\
&= \infty &&\Leftrightarrow\quad S = \varnothing \qquad \text{(where } \infty \notin \mathbb{G})
\end{aligned}
$$

Therefore if $\mathbb{G}$ has no infimum $\min \mathbb{G} = \max \varnothing = -\infty \notin \mathbb{G}$. Similarly, $\max S$ is the least upper bound of S in $\mathbb{G}$, if any, ∞ otherwise, with $\max \mathbb{G} = \min \varnothing = \infty \notin \mathbb{G}$ when $\mathbb{G}$ has no supremum.

Exercise 39.12 (min **of a sum**) Let $\langle \mathbb{G}, \leqslant, 0, + \rangle$ be a totally ordered group, $S_1, S_2 \in \wp(\mathbb{G})$. Extend $+$ with $x + \infty = \infty + x = \infty + \infty = \infty$. Prove that $\min\{x + y \mid x \in S_1 \wedge y \in S_2\} = \min S_1 + \min S_2$.

39.9 Minimal Weight of a Set of Paths

The weight of a path is

$$
\boldsymbol{\omega}(x_1 \dots x_n) \quad \triangleq \quad \sum_{i=1}^{n-1} \boldsymbol{\omega}(\langle x_i, x_{i+1} \rangle) \tag{39.13}
$$

which is 0 when $n \leqslant 1$ and $\boldsymbol{\omega}(\langle x_1, x_2 \rangle) + \sum_{i=2}^{n-1} \boldsymbol{\omega}(\langle x_i, x_{i+1} \rangle)$ when $n > 1$. The (minimal) weight of a set of paths is

$$
\boldsymbol{\omega}(P) \quad \triangleq \quad \min\{\boldsymbol{\omega}(\pi) \mid \pi \in P\}. \tag{39.14}
$$

We have $\boldsymbol{\omega}(\bigcup_{i \in \Delta} P_i) = \min_{i \in \Delta} \boldsymbol{\omega}(P_i)$ so a Galois connection

$$
\langle \wp(\bigcup_{n \in \mathbb{N}^+} V^n), \subseteq \rangle \underset{\boldsymbol{\omega}}{\overset{}{\leftrightarrows}} \langle \mathbb{G} \cup \{-\infty, \infty\}, \geqslant \rangle
$$

between path sets and the complete lattice $\langle \mathbb{G} \cup \{-\infty, \infty\}, \geqslant, \infty, -\infty, \min, \max \rangle$.

Extending pointwise to $V \times V \rightarrow \wp(\bigcup_{n \in \mathbb{N}^+} V^n)$ with $\dot{\boldsymbol{\omega}}(\mathsf{p})\langle x, y \rangle \triangleq \boldsymbol{\omega}(\mathsf{p}(x, y))$, $\mathsf{d} \mathrel{\dot{\leqslant}} \mathsf{d}' \triangleq \forall x, y . \mathsf{d}\langle x, y \rangle \leqslant \mathsf{d}'\langle x, y \rangle$, and $\dot{\geqslant}$ is the inverse of $\dot{\leqslant}$, we have

$$
\langle V \times V \rightarrow \wp(\bigcup_{n \in \mathbb{N}^+} V^n), \dot{\subseteq} \rangle \underset{\dot{\boldsymbol{\omega}}}{\overset{}{\leftrightarrows}} \langle V \times V \rightarrow \mathbb{G} \cup \{-\infty, \infty\}, \dot{\geqslant} \rangle. \tag{39.15}
$$

Example 39.16 (graph with cycle of strictly negative weight) The following graph

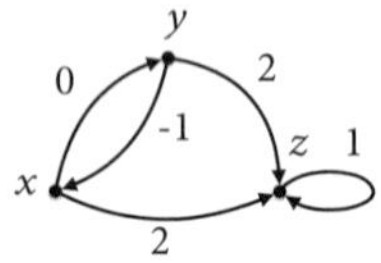

has a cycle x, y, x of weight -1, x, y, x, y, x of weight -2, x, y, x, y, x, y, x of weight -3, and so on, so that the minimum distance between x and y is $-\infty$.

39.10 Shortest Distance Along the Graph Paths

The distance $\mathsf{d}(x, y)$ between an origin $x \in V$ and an extremity $y \in V$ of a weighted finite graph $G = \langle V, E, \boldsymbol{\omega}\rangle$ on a totally ordered group $\langle \mathbb{G}, \leqslant, 0, +\rangle$ is defined as the length $\boldsymbol{\omega}(\mathsf{p}(x, y))$ of the shortest path between these vertices

$$\mathsf{d} \triangleq \dot{\boldsymbol{\omega}}(\mathsf{p})$$

where p has a fixpoint characterization given by theorem 39.10.

39.11 Calculation Design of the Shortest Distances Between Any Two Vertices

The shortest distance between vertices of a weighted graph is a path problem solved by theorem 39.6, the composition of the abstractions and (39.15) and (39.7), and the path abstraction theorem 39.6. Theorem 39.17 is based on (39.6.d) whereas (39.6.a—c) provide three other solutions.

Theorem 39.17 (fixpoint characterization of the shortest distances of a graph) Let $G = \langle V, E, \boldsymbol{\omega}\rangle$ be a graph weighted on the totally ordered group $\langle \mathbb{G}, \leqslant, 0, +\rangle$. Then the distances between any two vertices are

$$\mathsf{d} = \dot{\boldsymbol{\omega}}(\mathsf{p}) = \mathsf{gfp}^{\dot{\leqslant}}_{E^{\boldsymbol{\omega}}}\, \widehat{\mathscr{F}}^{\delta}_{G} \quad \text{where} \tag{39.17}$$

$$E^{\boldsymbol{\omega}} \triangleq (x, y) \mapsto (\!|\, \langle x, y\rangle \in E \,?\, \boldsymbol{\omega}(x, y) \,\text{\textthreeoldstyle}\, \infty \,|\!)$$

$$\widehat{\mathscr{F}}^{\delta}_{G}(X) \triangleq (x, y) \mapsto \min\{X(x, y), \min_{z \in V}\{X(x, z) + X(z, y)\}\}$$

Proof of theorem 39.17 We apply theorem 39.6 with abstraction $\dot{\boldsymbol{\omega}} \circ \alpha^{\multimap}$ so that we have to abstract the transformers in theorem 39.10 using an exact fixpoint abstraction of theorem 18.23. The initialization and commutation condition yield the transformers by calculational design.

$$
\begin{aligned}
&-\quad \dot{\boldsymbol{\omega}} \circ \alpha^{\multimap}(E)(x, y) \\
&= \boldsymbol{\omega}(E \cap \{\langle x, y\rangle\}) && \text{(as proved for theorem 39.10 and definition of } \dot{\boldsymbol{\omega}}\text{)} \\
&= (\!|\, \langle x, y\rangle \in E \,?\, \boldsymbol{\omega}(x, y) \,:\, \min \varnothing \,|\!) && \text{(definitions of } \cap\text{, of the conditional, and of } \boldsymbol{\omega}\text{)} \\
&= (\!|\, \langle x, y\rangle \in E \,?\, \boldsymbol{\omega}(x, y) \,:\, \infty \,|\!) && \text{(definition of min)}
\end{aligned}
$$

$$
\begin{aligned}
&-\quad \dot{\boldsymbol{\omega}} \circ \alpha^{\multimap}(X \circledcirc Y)(x, y) \\
&= \dot{\boldsymbol{\omega}}(\alpha^{\multimap}(X) \,\mathring{\circledcirc}\, \alpha^{\multimap}(Y))(x, y) && \text{(as proved for theorem 39.10)} \\
&= \boldsymbol{\omega}(\alpha^{\multimap}(X) \,\mathring{\circledcirc}\, \alpha^{\multimap}(Y))(x, y)) && \text{(pointwise definition (39.15) of } \dot{\boldsymbol{\omega}}\text{)}
\end{aligned}
$$

$$
\begin{aligned}
&= \boldsymbol{\omega}(\bigcup_{z \in V} \alpha^{\circ\!\!-\!\!\circ}(X)(x,z) \circledcirc \alpha^{\circ\!\!-\!\!\circ}(Y)(z,y))) && \text{(definition of } \overset{\circ\circ}{\circledcirc} \text{ in theorem 39.10)}\\
&= \min_{z \in V} \boldsymbol{\omega}(\alpha^{\circ\!\!-\!\!\circ}(X)(x,z) \circledcirc \alpha^{\circ\!\!-\!\!\circ}(Y)(z,y))) && \text{(Galois connection (39.14))}\\
&= \min_{z \in V} \boldsymbol{\omega}(\{x_1 \ldots x_n y_2 \ldots y_m \mid x_1 \ldots x_n \in \alpha^{\circ\!\!-\!\!\circ}(X)(x,z) \wedge x_n y_2 \ldots y_m \in \alpha^{\circ\!\!-\!\!\circ}(Y)(z,y)\})\\
& && \text{(definition (39.3) of } \circledcirc)\\
&= \min_{z \in V}\{\boldsymbol{\omega}(x_1 \ldots x_n y_2 \ldots y_m) \mid x_1 \ldots x_n \in \alpha^{\circ\!\!-\!\!\circ}(X)(x,z) \wedge x_n y_2 \ldots y_m \in \alpha^{\circ\!\!-\!\!\circ}(Y)(z,y)\})\\
& && \text{(definition (39.14) of } \boldsymbol{\omega})\\
&= \min_{z \in V}\{\boldsymbol{\omega}(x_1 \ldots x_n) + \boldsymbol{\omega}(x_n y_2 \ldots y_m) \mid x_1 \ldots x_n \in \alpha^{\circ\!\!-\!\!\circ}(X)(x,z) \wedge x_n y_2 \ldots y_m \in\\
&\quad \alpha^{\circ\!\!-\!\!\circ}(Y)(z,y)\}) && \text{(definition (39.13) of } \boldsymbol{\omega})\\
&= \min_{z \in V}\{\boldsymbol{\omega}(x_1 \ldots x_n) + \boldsymbol{\omega}(y_1 y_2 \ldots y_m) \mid x_1 \ldots x_n \in \alpha^{\circ\!\!-\!\!\circ}(X)(x,z) \wedge y_1 y_2 \ldots y_m \in\\
&\quad \alpha^{\circ\!\!-\!\!\circ}(Y)(z,y)\}) && \text{(definition of } \alpha^{\circ\!\!-\!\!\circ} \text{ so that } x_1 = x,\ x_n = y_1 = z, \text{ and } y_m = y)\\
&= \min_{z \in V} \min\{\boldsymbol{\omega}(x_1 \ldots x_n) \mid x_1 \ldots x_n \in \alpha^{\circ\!\!-\!\!\circ}(X)(x,z)\} + \min\{\boldsymbol{\omega}(y_1 y_2 \ldots y_m) \mid y_1 y_2 \ldots y_m \in\\
&\quad \alpha^{\circ\!\!-\!\!\circ}(Y)(z,y)\} && \text{(min of a sum in exercise 39.12)}\\
&= \min_{z \in V} \min\{\boldsymbol{\omega}(\pi_1) \mid \pi_1 \in \alpha^{\circ\!\!-\!\!\circ}(X)(x,z)\} + \min\{\boldsymbol{\omega}(\pi_2) \mid \pi_2 \in \alpha^{\circ\!\!-\!\!\circ}(Y)(z,y)\}\\
& && \text{(letting } \pi_1 = x_1 \ldots x_n \text{ and } \pi_2 = y_1 y_2 \ldots y_m)\\
&= \min_{z \in V} \boldsymbol{\omega}(\alpha^{\circ\!\!-\!\!\circ}(X)(x,z)) + \boldsymbol{\omega}(\alpha^{\circ\!\!-\!\!\circ}(Y)(z,y)) && \text{(definition (39.14) of } \boldsymbol{\omega})\\
&= \min_{z \in V} \dot{\boldsymbol{\omega}} \circ \alpha^{\circ\!\!-\!\!\circ}(X)(x,z) + \dot{\boldsymbol{\omega}} \circ \alpha^{\circ\!\!-\!\!\circ}(Y)(z,y) && \text{(pointwise definition (39.15) of } \dot{\boldsymbol{\omega}})
\end{aligned}
$$

By theorem 39.6 and (39.10.d), we get the transformer $\widehat{\mathscr{F}}_G^{\delta}$. □

Of course, the greatest fixpoint in theorem 39.17 is not computable for infinite graphs.

Exercise 39.18 Calculate the shortest distances between vertices of the infinite weighted graph $\langle \mathbb{N}, \{\langle n, n+1\rangle \mid n \in \mathbb{N}\}, \langle n, m\rangle \mapsto 1\rangle$.

For finite graphs, there is a problem with cycles with strictly negative weights. As shown by example 39.16, the minimal distance between the extremities x and y of the cycle xyx is $-\infty$. It is obtained as the limit of an infinite iteration for the greatest fixpoint in theorem 39.17. Following Roy-Floyd-Warshall, we will assume that the graph has no cycle with negative weight in which case the iterative computation of the greatest fixpoint in theorem 39.17 does converge to the shortest distance between any two vertices.

For a finite graph of n vertices, the computation of $\mathsf{gfp}^{\dot{\leqslant}}_{E^{\omega}} \widehat{\mathscr{F}}_G^{\delta}$ in (39.17) has to consider all pairs of vertices in n^2, for each such pair $\langle x, y\rangle$ the n vertices $z \in V$, and n iterations may be necessary to converge along a path going through all vertices, so would be in $\mathcal{O}(n^4)$.

However, the iteration in Roy-Floyd-Warshall algorithm is much more efficient in $\mathcal{O}(n^3)$, because it does not consider all paths in the graph but only simple paths that overapproximate paths with no cycles (called elementary paths). Let us design the Roy-Floyd-Warshall algorithm by calculus.

39.12 Elementary Paths and Cycles

A cycle is *elementary* if and only if it contains no internal subcycle (i.e., subpath which is a cycle). A path is *elementary* if and only if it contains no subpath that is an internal cycle (so an elementary cycle is an elementary path). The only vertices that can occur twice in an elementary path are its extremities, in which case it is an elementary cycle.

Example 39.19 In example 39.1, xyz and xz are elementary paths (zz is a path of length 1). The path xyx is an elementary cycle (zz is a cycle of length 1). The paths $xyzz$ and $xyxyxz$ and the cycles $yxyxy$ and zzz are not elementary. We do not consider infinite paths such as $xyxyx\ldots$.

Lemma 39.20 (elementary path) A path $x_1 \ldots x_n \in \Pi^n(G)$ is elementary if and only if

$$\begin{aligned}\mathsf{elem?}(x_1 \ldots x_n) \;\triangleq\;& (\forall i, j \in [1, n] \,.\, (i \neq j) \Rightarrow (x_i \neq x_j)) \vee \qquad (39.20)\\ & (x_1 = x_n \wedge \forall i, j \in [1, n[\,.\, (i \neq j) \Rightarrow (x_i \neq x_j)) \quad \text{(case of a cycle)}\end{aligned}$$

is true.

Proof of lemma 39.20

— The necessary condition ($x_1 \ldots x_n \in \Pi^n(G)$ is elementary implies that $\mathsf{elem?}(x_1 \ldots x_n)$) is proved contrapositively.

$$\begin{aligned}&\neg(\mathsf{elem?}(x_1 \ldots x_n))\\ =\;& \neg((\forall i, j \in [1, n] \,.\, (i \neq j) \Rightarrow (x_i \neq x_j)) \vee (x_1 = x_n \wedge \mathsf{elem?}(x_1 \ldots x_n)))\\ &\qquad \text{(definition of elem?)}\\ =\;& (\exists i, j \in [1, n] \,.\, i \neq j \wedge x_i = x_j) \wedge ((x_1 = x_n) \Rightarrow (\exists i, j \in [1, n[\,.\, i \neq j \wedge x_i \neq x_j))\\ &\qquad \text{(De Morgan's laws)}\end{aligned}$$

By $\exists i, j \in [1, n] \,.\, i \neq j \wedge x_i = x_j$ the path $x_1 \ldots x_n$ must have a cycle, but this is not forbidden if $x_1 = x_n$. In that case, the second condition $(x_1 = x_n) \Rightarrow (\exists i, j \in [1, n[\,.\, i \neq j \wedge x_i \neq x_j)$ implies that there is a subcycle within $x_1 \ldots x_{n-1}$, so the cycle $x_1 \ldots x_{n-1}x_1$ is not elementary.

— Conversely, the sufficient condition ($\mathsf{elem?}(x_1 \ldots x_n) \Rightarrow x_1 \ldots x_n \in \Pi^n(G)$ is elementary) is proved by reductio ad absurdum. Assume $\mathsf{elem?}(x_1 \ldots x_n)$ and $x_1 \ldots x_n \in \Pi^n(G)$ is not elementary so has an internal subcycle.

- If $x_1 = x_n$, the internal subcycle is $x_1 \ldots x_{n-1} = \pi_1 a \pi_2 a \pi_3$ so $\exists i, j \in [1, n[\,.\, i \neq j \wedge x_i \neq x_j$ in contradiction to $\mathsf{elem?}(x_1 \ldots x_n)$.
- Otherwise $x_1 \neq x_n$ and the internal subcycle has the form $x_1 \ldots x_n = \pi_1 a \pi_2 a \pi_3$ where, possibly $\pi_1 a = x_1$ or $a \pi_3 = x_n$, but not both, so $\exists i, j \in [1, n] \,.\, i \neq j \wedge x_i \neq x_j$ in contradiction to $\mathsf{elem?}(x_1 \ldots x_n)$. □

39.13 Calculational Design of the Elementary Paths Between Any Two Vertices

Restricting paths to elementary ones is the abstraction

$$\alpha^{\partial}(P) \;\triangleq\; \{\pi \in P \mid \mathsf{elem?}(\pi)\}.$$

Notice that, by (39.20), cycles (such as x, x for a self-loop $\langle x, x\rangle \in E$) are not excluded, provided it is through the path extremities. By exercise 11.5, this exclusion abstraction is a Galois connection.

$$\langle \wp(V^{>1}), \subseteq\rangle \xrightleftharpoons[\alpha^{\partial}]{\gamma^{\partial}} \langle \wp(V^{>1}), \subseteq\rangle$$

which extends pointwise between any two vertices

$$\langle V \times V \to \wp(V^{>1}), \dot{\subseteq}\rangle \xrightleftharpoons[\dot{\alpha}^{\partial}]{\dot{\gamma}^{\partial}} \langle V \times V \to \wp(V^{>1}), \dot{\subseteq}\rangle$$

The following lemma 39.21 provides a necessary and sufficient condition for the concatenation of two elementary paths to be elementary.

Lemma 39.21 (concatenation of elementary paths) If $x\pi_1 z$ and $z\pi_2 y$ are elementary paths then their concatenation $x\pi_1 z \odot z\pi_2 y = x\pi_1 z\pi_2 y$ is elementary if and only if

$$\begin{aligned}\mathsf{elem\text{-}conc?}(x\pi_1 z, z\pi_2 y) \ &\triangleq\ (x \neq z \wedge y \neq z \wedge \mathsf{V}(x\pi_1 z) \cap \mathsf{V}(\pi_2 y) = \varnothing) \\ &\quad \vee (x = y \neq z \wedge \mathsf{V}(\pi_1 z) \cap \mathsf{V}(\pi_2) = \varnothing)\end{aligned} \tag{39.21}$$

is true.

Example 39.22 Assume x, y, z, a, and b are all different vertices. $xaz \odot zbx = xazbx$ of the form $x\pi_1 z \odot z\pi_2 y$ with $x = y \neq z \wedge \mathsf{V}(az) \cap \mathsf{V}(b) = \varnothing$ is elementary. $xyz \odot zay = xyzay$ with $x \neq y \wedge \mathsf{V}(xyz) \cap \mathsf{V}(ay) = \{y\} \neq \varnothing$ is not elementary. $xx \odot xx = xxx$ and $xax \odot xbx = xaxbx$ of the form $x\pi_1 z \odot z\pi_2 y$ with $x = z = y$ are not elementary. $xz \odot zz = xzz$ of the form $x\pi_1 z \odot z\pi_2 y$ with $y = z$ is not.

Proof of lemma 39.21 Assume that $x\pi_1 z$ and $z\pi_2 y$ are elementary. We must prove that

$\mathsf{elem\text{-}conc?}(x\pi_1 z, z\pi_2 y)$ is true $\Leftrightarrow x\pi_1 z \odot z\pi_2 y = x\pi_1 z\pi_2 y$ is elementary.

— We prove the necessary condition ($\pi_1 \odot \pi_2$ is elementary $\Rightarrow \mathsf{elem\text{-}conc?}(\pi_1, \pi_2)$) by contraposition ($\neg\mathsf{elem\text{-}conc?}(\pi_1, \pi_2) \Rightarrow \pi_1 \odot \pi_2$ has an internal cycle). We have

$$\begin{aligned}&\neg((x \neq z \wedge y \neq z \wedge \mathsf{V}(x\pi_1 z) \cap \mathsf{V}(\pi_2 y) = \varnothing) \vee (x = y \wedge x \neq z \wedge y \neq z \wedge \mathsf{V}(\pi_1 z) \cap \mathsf{V}(\pi_2) = \\ &\quad \varnothing)) \\ =\ &(x = z \vee y = z \vee (\mathsf{V}(x\pi_1 z) \cap \mathsf{V}(\pi_2 y) \neq \varnothing) \wedge (x \neq y \vee x = z \vee y = z \vee \mathsf{V}(\pi_1 z) \cap \mathsf{V}(\pi_2) \neq \\ &\quad \varnothing)) \qquad \text{(de Morgan laws)}\end{aligned}$$

- If $x = z$ then $x\pi_1 x\pi_2 y$ has a cycle and is not elementary;
- else, if $y = z$ then $x\pi_1 y\pi_2 y$ has a cycle and is not elementary;
- else $x \neq z \wedge y \neq z$, and then
  - either $x \neq z \wedge y \neq z \wedge x = y$ so $\mathsf{V}(\pi_1 z) \cap \mathsf{V}(\pi_2) \neq \varnothing$. There are two cases
    - either $\mathsf{V}(\pi_1) \cap \mathsf{V}(\pi_2) \neq \varnothing$ so $\pi_1 = \pi_1' a \pi_1''$ and $\pi_2 = \pi_2' a \pi_2''$ and therefore $\pi_1 \odot \pi_2 = x\pi_1 z\pi_2 y = x\pi_1' a\pi_1'' z\pi_2' a\pi_2'' x$ has an internal cycle $a\pi_1'' z\pi_2' a$,
    - or $z \in \mathsf{V}(\pi_2)$ so $\pi_2 = \pi_2' z \pi_2''$ and therefore $\pi_1 \odot \pi_2 = x\pi_1 z\pi_2 y = x\pi_1 z\pi_2' z\pi_2'' x$ has an internal cycle $z\pi_2' z$;
  - otherwise $x \neq z \wedge y \neq z \wedge x \neq y$ and we have $\mathsf{V}(x\pi_1 z) \cap \mathsf{V}(\pi_2 y) \neq \varnothing$. By cases.
    - If x appears in $\pi_2 y$ that is in π_2 because $x \neq y$ we have $\pi_2 = \pi_2' x \pi_2''$ and then $x\pi_1 z \odot z\pi_2 y = x\pi_1 z\pi_2 y = x\pi_1 z\pi_2' x\pi_2'' y$ has an internal cycle $x\pi_1 z\pi_2' x$;

- Else, if $\mathsf{V}(\pi_1) \cap \mathsf{V}(\pi_2 y) \neq \emptyset$ then
  - Either $\mathsf{V}(\pi_1) \cap \mathsf{V}(\pi_2) \neq \emptyset$ so $\pi_1 = \pi_1' a \pi_1''$ and $\pi_2 = \pi_2' a \pi_2''$ and therefore $x\pi_1 z \odot z\pi_2 y = x\pi_1 z\pi_2 y = x\pi_1' a\pi_1'' z\pi_2' a\pi_2'' x$ has an internal cycle $a\pi_1'' z\pi_2' a$,
  - Or $y \in \mathsf{V}(\pi_1)$ so $\pi_1 = \pi_1' y \pi_1''$ and then $x\pi_1 z \odot z\pi_2 y = x\pi_1 z\pi_2 y = x\pi_1' y\pi_1'' z\pi_2 y$ has an internal cycle $y\pi_1'' z\pi_2 y$;
- Otherwise, $z \in \mathsf{V}(\pi_2 y) \neq \emptyset$ and then
  - Either $z \in \mathsf{V}(\pi_2)$ so $\pi_2 = \pi_2' z \pi_2''$ and $x\pi_1 z \odot z\pi_2 y = x\pi_1 z\pi_2 y = x\pi_1 z\pi_2' z\pi_2'' y$ has an internal cycle $z\pi_2' z$,
  - Or $z = y$ and $x\pi_1 z \odot z\pi_2 y = x\pi_1 z\pi_2 y = x\pi_1 z\pi_2 z$ has an internal cycle $z\pi_2 z$. □

— We prove that the condition is sufficient ($\mathsf{elem\text{-}conc?}(\pi_1, \pi_2) \Rightarrow \pi_1 \odot \pi_2$ is elementary) by reductio ad absurdum. Assume $x\pi_1 z$, and $z\pi_2 y$ are elementary, $\mathsf{elem\text{-}conc?}(x\pi_1 z, z\pi_2 y)$ holds, but that $x\pi_1 z \odot z\pi_2 y = x\pi_1 z\pi_2 y$ is not elementary.

- if the internal cycle is in $x\pi_1 z$ then, by hypothesis, $x = z$ so $\mathsf{elem\text{-}conc?}(x\pi_1 z, z\pi_2 y)$ does not hold, a contradiction;
- else, if the internal cycle is in $z\pi_2 y$ then, by hypothesis, $z = y$ so $\mathsf{elem\text{-}conc?}(x\pi_1 z, z\pi_2 y)$ does not hold, a contradiction;
- otherwise, the internal cycle is neither in $x\pi_1 z$ nor in $z\pi_2 y$ so $\mathsf{V}(x\pi_1 z) \cap \mathsf{V}(\pi_2 y) \neq \emptyset$. Because $\mathsf{elem\text{-}conc?}(\pi_1, \pi_2)$ holds, it follows that $x = y \neq z \wedge \mathsf{V}(\pi_1 z) \cap \mathsf{V}(\pi_2) = \emptyset$ in contradiction to the existence of an internal cycle $a\pi''\pi_2' a$ requiring $\pi_1 z = \pi' a \pi''$ and $\pi_2 = \pi_2' a \pi_2''$ and so $a \in \mathsf{V}(\pi' a \pi'') \cap \mathsf{V}(\pi_2' a \pi_2'') = \mathsf{V}(\pi_1 z) \cap \mathsf{V}(\pi_2) \neq \emptyset$. □

We have the following fixpoint characterization of the elementary paths of a graph (converging in finitely many iterations for graphs without infinite paths).

Theorem 39.23 (fixpoint characterization of the elementary paths of a graph) Let $G = \langle V, E \rangle$ be a graph. The elementary paths between any two vertices of G are $\mathsf{p}^{\partial} \triangleq \alpha^{\circ\!\!-\!\!\circ} \circ \alpha^{\partial}(\Pi(G))$ such that

$$
\begin{array}{rclrcll}
\mathsf{p}^{\partial} & = & \mathsf{lfp}^{\dot{\subseteq}}\, \overrightarrow{\mathscr{F}}^{\partial}_{\Pi}, & \overrightarrow{\mathscr{F}}^{\partial}_{\Pi}(\mathsf{p}) & \triangleq & \dot{E} \,\dot{\cup}\, \mathsf{p} \,\dot{\circledcirc}^{\partial}\, \dot{E} & (39.23.a)\\
& = & \mathsf{lfp}^{\dot{\subseteq}}\, \overleftarrow{\mathscr{F}}^{\partial}_{\Pi}, & \overleftarrow{\mathscr{F}}^{\partial}_{\Pi}(\mathsf{p}) & \triangleq & \dot{E} \,\dot{\cup}\, \dot{E} \,\dot{\circledcirc}^{\partial}\, \mathsf{p} & (39.23.b)\\
& = & \mathsf{lfp}^{\dot{\subseteq}}\, \overleftrightarrow{\mathscr{F}}^{\partial}_{\Pi}, & \overleftrightarrow{\mathscr{F}}^{\partial}_{\Pi}(\mathsf{p}) & \triangleq & \dot{E} \,\dot{\cup}\, \mathsf{p} \,\dot{\circledcirc}^{\partial}\, \mathsf{p} & (39.23.c)\\
& = & \mathsf{lfp}^{\dot{\subseteq}}_{\dot{E}}\, \widehat{\mathscr{F}}^{\partial}_{\Pi}, & \widehat{\mathscr{F}}^{\partial}_{\Pi}(\mathsf{p}) & \triangleq & \mathsf{p} \,\dot{\cup}\, \mathsf{p} \,\dot{\circledcirc}^{\partial}\, \mathsf{p} & (39.23.d)
\end{array}
$$

where

$$
\mathsf{p}_1 \,\dot{\circledcirc}^{\partial}\, \mathsf{p}_2 \triangleq x, y \mapsto \bigcup_{z \in V} \{\pi_1 \odot \pi_2 \mid \pi_1 \in \mathsf{p}_1(x, z) \wedge \pi_2 \in \mathsf{p}_2(z, y) \wedge \mathsf{elem\text{-}conc?}(\pi_1, \pi_2)\}
$$

and $\dot{E} \triangleq x, y \mapsto (E \cap \{\langle x, y \rangle\})$ in theorem 39.10.

Proof of theorem 39.23 We apply theorem 39.6 with abstraction $\dot{\alpha}^{\partial} \circ \alpha^{\circ\!\!-\!\!\circ}$ so that we have to abstract the transformers in theorem 39.10 using an exact fixpoint abstraction of theorem 18.23. The commutation condition yields the transformers by calculational design.

$\dot{\alpha}^{\partial}(\mathsf{p}_1 \overset{\circ\!\!-\!\!\circ}{\circledcirc} \mathsf{p}_2)$

$$= \dot{\alpha}^{\partial}(x, y \mapsto \bigcup_{z \in V} \mathsf{p}_1(x, z) \circledcirc \mathsf{p}_2(z, y)) \qquad \text{\{definition of } \ddot{\circledcirc} \text{ in theorem 39.10\}}$$

$$= x, y \mapsto \alpha^{\partial}(\bigcup_{z \in V} \mathsf{p}_1(x, z) \circledcirc \mathsf{p}_2(z, y)) \qquad \text{\{pointwise definition of } \dot{\alpha}^{\partial}\text{\}}$$

$$= x, y \mapsto \bigcup_{z \in V} \alpha^{\partial}(\mathsf{p}_1(x, z) \circledcirc \mathsf{p}_2(z, y))$$

$$\text{\{the abstraction preserves joins in a Galois connection\}}$$

$$= x, y \mapsto \bigcup_{z \in V} \alpha^{\partial}(\{\pi_1 \odot \pi_2 \mid \pi_1 \in \mathsf{p}_1(x, z) \wedge \pi_2 \in \mathsf{p}_2(z, y)\})$$

$$\text{\{definition (39.3) of } \circledcirc \text{ and (39.9) of } \odot\text{\}}$$

$$= x, y \mapsto \bigcup_{z \in V} \alpha^{\partial}(\{\pi_1 \odot \pi_2 \mid \pi_1 \in \alpha^{\partial}(\mathsf{p}_1(x, z)) \wedge \pi_2 \in \alpha^{\partial}(\mathsf{p}_2(z, y))\})$$

$$\text{\{because if } \pi_1 \text{ or } \pi_2 \text{ are not elementary so is their concatenation } \pi_1 \odot \pi_2\text{\}}$$

$$= x, y \mapsto \bigcup_{z \in V} \{\pi_1 \odot \pi_2 \mid \pi_1 \in \alpha^{\partial}(\mathsf{p}_1(x, z)) \wedge \pi_2 \in \alpha^{\partial}(\mathsf{p}_2(z, y)) \wedge \mathsf{elem\text{-}conc?}(\pi_1, \pi_2)\}$$

{because, by lemma 39.21, π_1 and π_2 being elementary, their concatenation $\pi_1 \odot \pi_2$ is elementary if and only if $\mathsf{elem\text{-}conc?}(\pi_1, \pi_2)$ is true}

$$= x, y \mapsto \bigcup_{z \in V} \{\pi_1 \odot \pi_2 \mid \pi_1 \in \dot{\alpha}^{\partial}(\mathsf{p}_1)(x, z) \wedge \pi_2 \in \dot{\alpha}^{\partial}(\mathsf{p}_2)(z, y) \wedge \mathsf{elem\text{-}conc?}(\pi_1, \pi_2)\}$$

$$\text{\{pointwise definition of } \dot{\alpha}^{\partial}\text{\}}$$

$$= \dot{\alpha}^{\partial}(\mathsf{p}_1) \dot{\circledcirc}^{\partial} \dot{\alpha}^{\partial}(\mathsf{p}_2) \qquad \text{\{definition of } \dot{\circledcirc}^{\partial} \text{ in theorem 39.23\}} \quad \square$$

39.14 Calculational Design of the Elementary Paths Between Vertices of Finite Graphs

In finite graphs $G = \langle V, E\rangle$ with $|V| = n > 0$ vertices, elementary paths in are of length at most $n + 1$ (for a cycle that would go through all vertices of the graph). This ensures that the fixpoint iterates in theorem 39.23 starting from $\dot{\varnothing}$ do converge in at most $n + 2$ iterates.

Moreover, if $V = \{z_1 \ldots z_n\}$ is finite, then the elementary paths of the $k + 2^{\text{nd}}$ iterate can be restricted to $\{z_1, \ldots, z_k\}$. This yields an iterative algorithm by application of the exact iterates multi-abstraction theorem 18.36 with[3]

$$\begin{aligned}
\alpha^{\partial}(\mathsf{p}) &\triangleq \mathsf{p} && (39.24)\\
\alpha^{\partial}_k(\mathsf{p}) &\triangleq x, y \mapsto \{\pi \in \mathsf{p}(x, y) \mid \mathsf{V}(\pi) \subseteq \{z_1, \ldots, z_k\} \cup \{x, y\}\}, && k \in [1, n]\\
\alpha^{\partial}_k(\mathsf{p}) &\triangleq \mathsf{p}, && k > n
\end{aligned}$$

By the exclusion abstraction of exercise 11.5 and pointwise extension of exercise 11.21, these are Galois connections

$$\langle V \times V \to \wp(V^{>1}), \dot{\subseteq}\rangle \underset{\alpha^{\partial}_k}{\overset{\gamma^{\partial}_k}{\leftrightarrows}} \langle V \times V \to \bigcup_{k=2}^{n+1} V^k, \dot{\subseteq}\rangle. \qquad (39.25)$$

Applying theorem 18.36, we get the following iterative characterization of the elementary paths of a finite graph. Notice that $\dot{\circledcirc}_z$ in (39.26.a) and (39.26.b) does not require to

3. This is for case (39.26.d). For cases (a–c), we also have $\alpha^{\partial}(\mathsf{p}) \triangleq \mathsf{p}$ whereas the second definition is for $k \in [2, n + 2]$.

test that the concatenation of two elementary paths is elementary while $\dot{\circledcirc}^{\partial}_{z}$ in (39.26.c) and (39.26.d) definitely does (because the concatenated elementary paths may have vertices in common). Notice also that the iteration $\langle \overrightarrow{\mathscr{F}}^{\partial\,k}_{\pi},\ k \in [0, n+2]\rangle$ in (39.26.a) is not the same as the iterates $\langle \overrightarrow{\mathscr{F}}^{\partial k}_{\Pi}(\dot{\varnothing}),\ k \in \mathbb{N}\rangle$ of $\overrightarrow{\mathscr{F}}^{\partial}_{\Pi}$ from $\dot{\varnothing}$, because using $\dot{\circledcirc}_{z}$ or $\dot{\circledcirc}^{\partial}_{z}$ instead of $\dot{\circledcirc}^{\partial}$ is the key to efficiency. This is also the case for (39.26.b—d).

Theorem 39.26 (iterative characterization of the elementary paths of a finite graph) Let $G = \langle V,\ E\rangle$ be a finite graph with $V = \{z_1, \ldots, z_n\}$, $n > 0$. Then

$$\mathsf{p}^{\partial} = \mathsf{lfp}^{\dot\subseteq}\, \overrightarrow{\mathscr{F}}^{\partial}_{\Pi} = \overrightarrow{\mathscr{F}}^{\partial n+2}_{\pi} \quad \text{where} \qquad (39.26.a)$$

$$\overrightarrow{\mathscr{F}}^{\partial}_{\Pi}(\mathsf{p}) \triangleq \dot{E} \mathbin{\dot\cup} \mathsf{p} \mathbin{\dot\circledcirc^{\partial}} \dot{E} \text{ in (39.23.a)} \quad \text{and} \quad \overrightarrow{\mathscr{F}}^{\partial 0}_{\pi} \triangleq \dot{\varnothing}, \quad \overrightarrow{\mathscr{F}}^{\partial 1}_{\pi} \triangleq \dot{E},$$

$$\overrightarrow{\mathscr{F}}^{\partial k+2}_{\pi} \triangleq \dot{E} \mathbin{\dot\cup} \overrightarrow{\mathscr{F}}^{\partial k+1}_{\pi} \mathbin{\dot\circledcirc_{z_{k+1}}} \dot{E},\quad k \in [0, n],\quad \overrightarrow{\mathscr{F}}^{\partial k+1}_{\pi} = \overrightarrow{\mathscr{F}}^{\partial k}_{\pi},\quad k \geqslant n+2$$

$$= \mathsf{lfp}^{\dot\subseteq}\, \overleftarrow{\mathscr{F}}^{\partial}_{\Pi} = \overleftarrow{\mathscr{F}}^{\partial}_{\pi}{}^{n+2} \quad \text{where} \qquad (39.26.b)$$

$$\overleftarrow{\mathscr{F}}^{\partial}_{\Pi}(\mathsf{p}) \triangleq \dot{E} \mathbin{\dot\cup} \dot{E} \mathbin{\dot\circledcirc^{\partial}} \mathsf{p} \text{ in (39.23.b)} \quad \text{and} \quad \overleftarrow{\mathscr{F}}^{\partial 0}_{\pi} \triangleq \dot{\varnothing}, \quad \overleftarrow{\mathscr{F}}^{\partial 0}_{\pi} \triangleq \dot{E},$$

$$\overleftarrow{\mathscr{F}}^{\partial k+2}_{\pi} \triangleq \dot{E} \mathbin{\dot\cup} \dot{E} \mathbin{\dot\circledcirc_{z_{k+1}}} \overleftarrow{\mathscr{F}}^{\partial k+1}_{\pi},\quad k \in [0, n],\quad \overleftarrow{\mathscr{F}}^{\partial k+1}_{\pi} = \overleftarrow{\mathscr{F}}^{\partial k}_{\pi},\quad k \geqslant n+2$$

$$= \mathsf{lfp}^{\dot\subseteq}\, \overleftrightarrow{\mathscr{F}}^{\partial}_{\Pi} = \overleftrightarrow{\mathscr{F}}^{\partial}_{\pi}{}^{n+2} \quad \text{where} \qquad (39.26.c)$$

$$\overleftrightarrow{\mathscr{F}}^{\partial}_{\Pi}(\mathsf{p}) \triangleq \dot{E} \mathbin{\dot\cup} \mathsf{p} \mathbin{\dot\circledcirc^{\partial}} \mathsf{p} \text{ in (39.23.c)} \quad \text{and} \quad \overleftrightarrow{\mathscr{F}}^{\partial 0}_{\pi} \triangleq \dot{\varnothing}, \quad \overleftrightarrow{\mathscr{F}}^{\partial 1}_{\pi} \triangleq \dot{E},$$

$$\overleftrightarrow{\mathscr{F}}^{\partial k+2}_{\pi} \triangleq \dot{E} \mathbin{\dot\cup} \overleftrightarrow{\mathscr{F}}^{\partial k+1}_{\pi} \mathbin{\dot\circledcirc^{\partial}_{z_{k+1}}} \overleftrightarrow{\mathscr{F}}^{\partial k+1}_{\pi},\quad k \in [0, n],\quad \overleftrightarrow{\mathscr{F}}^{\partial k+1}_{\pi} = \overleftrightarrow{\mathscr{F}}^{\partial k}_{\pi},\quad k \geqslant n+2$$

$$= \mathsf{lfp}^{\dot\subseteq}_{\dot{E}}\, \widehat{\mathscr{F}}^{\partial}_{\Pi} = \widehat{\mathscr{F}}^{\partial}_{\pi}{}^{n+1} \quad \text{where} \qquad (39.26.d)$$

$$\widehat{\mathscr{F}}^{\partial}_{\Pi}(\mathsf{p}) \triangleq \mathsf{p} \mathbin{\dot\cup} \mathsf{p} \mathbin{\dot\circledcirc^{\partial}} \mathsf{p} \text{ in (39.23.d)}, \quad \widehat{\mathscr{F}}^{\partial 0}_{\pi} \triangleq \dot{E},$$

$$\widehat{\mathscr{F}}^{\partial k+1}_{\pi} \triangleq \widehat{\mathscr{F}}^{\partial k}_{\pi} \mathbin{\dot\cup} \widehat{\mathscr{F}}^{\partial k}_{\pi} \mathbin{\dot\circledcirc^{\partial}_{z_{k+1}}} \widehat{\mathscr{F}}^{\partial k}_{\pi},\quad k \in [0, n],\quad \widehat{\mathscr{F}}^{\partial k+1}_{\pi} = \widehat{\mathscr{F}}^{\partial k}_{\pi},\quad k \geqslant n+2$$

$\mathsf{p}_1 \mathbin{\dot\circledcirc_{z}} \mathsf{p}_2 \triangleq x, y \mapsto \{\pi_1 \odot \pi_2 \mid \pi_1 \in \mathsf{p}_1(x, z) \wedge \pi_2 \in \mathsf{p}_2(z, y) \wedge z \notin \{x, y\}\}$, and
$\mathsf{p}_1 \mathbin{\dot\circledcirc^{\partial}_{z}} \mathsf{p}_2 \triangleq x, y \mapsto \{\pi_1 \odot \pi_2 \mid \pi_1 \in \mathsf{p}_1(x, z) \wedge \pi_2 \in \mathsf{p}_2(z, y) \wedge \mathsf{elem\text{-}conc?}(\pi_1, \pi_2)\}$.

Proof of theorem 39.26

— The proofs in cases (39.26.c) and (39.26.d) are similar. Let us consider (39.26.d). Assume $V = \{z_1 \ldots z_n\}$ and let $\widehat{\mathscr{F}}^{\partial k+1}_{\Pi} = \widehat{\mathscr{F}}^{\partial}_{\Pi}(\widehat{\mathscr{F}}^{\partial k}_{\Pi})$ be the iterates of $\widehat{\mathscr{F}}^{\partial}_{\Pi}$ from $\widehat{\mathscr{F}}^{\partial 0}_{\Pi} = \dot{E}$ in (39.23.d). To apply theorem 18.36, we consider the concrete CPO $\langle \mathscr{C}, \dot\subseteq\rangle$ and the abstract CPO $\langle \mathscr{A}, \dot\subseteq\rangle$ to be $\langle \mathscr{C}, \dot\subseteq, \dot{E}, \dot{\bigcup}\rangle$ with $\mathscr{C} \triangleq x \in V \times y \in V \to \{x\pi y \mid x\pi y \in E \cup \bigcup_{k=3}^{n+1} V^k \wedge x\pi y \text{ is elementary}\}$, and the functions $\widehat{\mathscr{F}}^{\partial}_{\pi k}(X) \triangleq X \mathbin{\dot\cup} X \mathbin{\dot\circledcirc^{\partial}_{z_{k+1}}} X$, $k \in [1, n]$, and $\widehat{\mathscr{F}}^{\partial}_{\pi k}(X) \triangleq X$, $k = 0$ or $k > n$ which iterates from the infimum $\dot{E}$ are precisely $\langle \widehat{\mathscr{F}}^{\partial k}_{\pi},\ i \in \mathbb{Z} \cup \{\omega\}\rangle$ where $\widehat{\mathscr{F}}^{\partial \omega}_{\pi} = \dot{\bigcup}_{i \in \mathbb{Z}} \widehat{\mathscr{F}}^{\partial i}_{\pi} = \widehat{\mathscr{F}}^{\partial n+1}_{\pi} = \widehat{\mathscr{F}}^{\partial k}_{\pi}$, $k > n$.

- For the infimum $\widehat{\mathscr{F}}^{\partial}_{\pi 0} = \dot{E}$ the paths $x\pi y \in \dot{E}(x, y)$ of which are elementary and have all intermediate states of π in $\varnothing = \{z_1, \ldots, z_0\}$ because π is empty.

- For the commutation, the case $k > n$ is trivial. Otherwise let $X \in \mathscr{A}$ so $x\pi y \in X(x, y)$ is elementary and has all states of π in $\{z_1, \ldots, z_k\}$

$$
\begin{aligned}
&\alpha^{\partial}(\widehat{\mathscr{F}}^{\partial}_{\Pi}(X)) \\
= \;& \alpha^{\partial}(X \mathbin{\dot\cup} X \mathbin{\dot\circledcirc}^{\partial} X) && \text{\wr definition (39.23.d) of } \widehat{\mathscr{F}}^{\partial}_{\Pi} \text{\wr} \\
= \;& \alpha^{\partial}(X) \mathbin{\dot\cup} \alpha^{\partial}(X \mathbin{\dot\circledcirc}^{\partial} X) && \text{\wr } \alpha^{\partial} \text{ preserves joins in (39.25)\wr} \\
= \;& \alpha^{\partial}_k(X) \mathbin{\dot\cup} \alpha^{\partial}(X \mathbin{\dot\circledcirc}^{\partial} X)
\end{aligned}
$$

(definition (39.24) of α^{∂} and hypothesis that all paths in X have all intermediate states in $\{z_1, \ldots, z_k\}$)

$$
= \alpha^{\partial}_k(X) \mathbin{\dot\cup} x, y \mapsto \{\pi \in X \mathbin{\dot\circledcirc}^{\partial} X(x, y) \mid \mathsf{V}(\pi) \subseteq \{z_1, \ldots, z_{k+1}\} \cup \{x, y\}\}
$$

(definition of α^{∂} in (39.24))

$$
= \alpha^{\partial}_k(X) \mathbin{\dot\cup} x, y \mapsto \{\pi \in \bigcup_{z \in V}\{\pi_1 \odot \pi_2 \mid \pi_1 \in X(x, z) \wedge \pi_2 \in X(z, y) \wedge \mathsf{elem\text{-}conc?}(\pi_1, \pi_2)\} \mid \mathsf{V}(\pi) \subseteq \{z_1, \ldots, z_{k+1}\} \cup \{x, y\}\}
$$

(definition of $\dot\circledcirc^{\partial}$ in theorem 39.23)

$$
= \alpha^{\partial}_k(X) \mathbin{\dot\cup} x, y \mapsto \bigcup_{z \in V}\{\pi_1 \odot \pi_2 \mid \pi_1 \in X(x, z) \wedge \pi_2 \in X(z, y) \wedge \mathsf{elem\text{-}conc?}(\pi_1, \pi_2) \wedge \mathsf{V}(\pi_1 \odot \pi_2) \subseteq \{z_1, \ldots, z_{k+1}\} \cup \{x, y\}\}
$$

(definition of $\in$)

$$
= \alpha^{\partial}_k(X) \mathbin{\dot\cup} x, y \mapsto \bigcup_{z \in V}\{x\pi_1 z\pi_2 y \mid x\pi_1 z \in X(x, z) \wedge z\pi_2 y \in X(z, y) \wedge \mathsf{elem\text{-}conc?}(x\pi_1 z, z\pi_2 y) \wedge \mathsf{V}(\pi_1) \cup \mathsf{V}(\pi_2) \cup \{z\} \subseteq \{z_1, \ldots, z_{k+1}\} \cup \{x, y\}\}
$$

(definitions of $\odot$ and V, and induction hypothesis)

$$
= \alpha^{\partial}_k(X) \mathbin{\dot\cup} x, y \mapsto \{x\pi_1 z_{k+1}\pi_2 y \mid x\pi_1 z_{k+1} \in \alpha^{\partial}_k(X)(x, z_{k+1}) \wedge z_{k+1}\pi_2 y \in \alpha^{\partial}_k(X)(z_{k+1}, y) \wedge \mathsf{elem\text{-}conc?}(x\pi_1 z_{k+1}, z_{k+1}\pi_2 y)\}
$$

((⊇) follows from taking $z = z_{k+1}$; (39.27)
(⊆) For $z \in \{z_1, \ldots, z_k\}$, the paths in $\alpha^{\partial}_k(X)$ are elementary through $\{z_1, \ldots, z_k\}$, so if there exist paths $x\pi_1 z \in X(x, z)$ and $z\pi_2 y \in X(z, y)$ then either $x\pi_1 z\pi_2 x$ is also elementary through $\{z_1, \ldots, z_k\}$ and already therefore belongs to $\alpha^{\partial}_k(X)$ or it is not elementary and then does not pass the test $\mathsf{elem\text{-}conc?}(x\pi_1 z_{k+1}, z_{k+1}\pi_2 y)$;
Otherwise, if $z \in \{z_{k+2}, \ldots, z_n\}$, then the path $x\pi_1 z_{k+1}\pi_2 y$ is eliminated by $\mathsf{V}(\pi_1) \cup \mathsf{V}(\pi_2) \cup \{z\} \subseteq \{z_1, \ldots, z_{k+1}\} \cup \{x, y\}$;
Finally, the only possibility is $z = z_{k+1}$, in which case all paths have the form $x\pi_1 z_{k+1}\pi_2 y$, are elementary, and with $\mathsf{V}(\pi) \subseteq \{z_1, \ldots, z_{k+1}\}$, as required by the definition of $\mathscr{A}$. It also holds for $\alpha^{\partial}_k(X)$ which is equal to $\alpha^{\partial}(X)$.)

$$
= \alpha^{\partial}_k(X) \mathbin{\dot\cup} x, y \mapsto \{x\pi_1 z_{k+1} \odot z_{k+1}\pi_2 y \mid x\pi_1 z_{k+1} \in \alpha^{\partial}_k(X)(x, z_{k+1}) \wedge z_{k+1}\pi_2 y \in \alpha^{\partial}_k(X)(z_{k+1}, y) \wedge \mathsf{elem\text{-}conc?}(x\pi_1 z_{k+1}, z_{k+1}\pi_2 y)\}
$$

(definition of $\odot$)

$$
= \alpha^{\partial}_k(X) \mathbin{\dot\cup} x, y \mapsto \{\pi_1 \odot \pi_2 \mid \pi_1 \in \alpha^{\partial}_k(X)(x, z_{k+1}) \wedge \pi_2 \in \alpha^{\partial}_k(X)(z_{k+1}, y) \wedge \mathsf{elem\text{-}conc?}(\pi_1, \pi_2)\}
$$

(by induction hypothesis all paths in $X(x, y)$ have the form $x\pi y$)

$$
\begin{aligned}
= \;& \alpha^{\partial}_k(X) \mathbin{\dot\cup} \alpha^{\partial}_k(X) \mathbin{\dot\circledcirc}^{\partial}_{z_{k+1}} \alpha^{\partial}_k(X) && \text{\wr definition of } \dot\circledcirc^{\partial}_{z_{k+1}} \text{ in (39.26.d)\wr} \\
= \;& \mathscr{F}^{\partial}_{\pi k}(\alpha^{\partial}_k(X)) && \text{\wr(39.26.d)\wr}
\end{aligned}
$$

We conclude by corollary 18.37.

— In cases (39.26.a) and (39.26.b), $\dot{\circledcirc}^{\partial}_{z_{k+1}}$ can be replaced by $\dot{\circledcirc}_{z_{k+1}}$ because in these cases the paths are elementary by construction. To see this, observe that for (39.26.a), the iterates $\langle \vec{\mathscr{F}}^{\partial k}_{\pi}(\dot{\varnothing}),\ k \in \mathbb{N} \cup \{\omega\}\rangle$ are those of the functions $\vec{\mathscr{F}}^{\partial}_{\pi 0}(X) \triangleq \dot{\varnothing}$, $\vec{\mathscr{F}}^{\partial}_{\pi 1}(X) \triangleq \dot{E}$, and $\vec{\mathscr{F}}^{\partial}_{\pi k}(X) \triangleq \dot{E} \mathbin{\dot{\cup}} X \mathbin{\dot{\circledcirc}^{\partial}_{z_{k-1}}} \dot{E}$, $k \in [2, n+2]$, and $\vec{\mathscr{F}}^{\partial}_{\pi k}(X) \triangleq X$, $k > n$, so that we can consider the iterates from 1 to apply theorem 18.36.

- By (39.23.a), the initialization is $\vec{\mathscr{F}}^{\partial}_{\Pi}(\dot{\varnothing}) \triangleq \dot{E} \mathbin{\dot{\cup}} \dot{\varnothing} \mathbin{\dot{\circledcirc}^{\partial}} \dot{E} = \dot{E}$ such that all paths $x\pi y$ in $\dot{E}(x, y)$ are elementary with π empty so $\mathsf{V}(\pi) \subseteq \varnothing = \{z_1, \ldots, z_0\}$.

- For the commutation, let $X \in \mathscr{A}$ such that all $x\pi y \in X(x, y)$ are elementary and have all states of π in $\{z_1, \ldots, z_k\}$. Then

$$
\begin{aligned}
& \alpha^{\partial}(\vec{\mathscr{F}}^{\partial}_{\Pi}(X)) && \text{\wr definition of the iterates\int}\\
= \ & \alpha^{\partial}(\dot{E} \mathbin{\dot{\cup}} X \mathbin{\dot{\circledcirc}^{\partial}} \dot{E}) && \text{\wr definition (39.23.a) of } \vec{\mathscr{F}}^{\partial}_{\Pi}\text{\int}\\
= \ & \alpha^{\partial}(\dot{E}) \mathbin{\dot{\cup}} \alpha^{\partial}(X \mathbin{\dot{\circledcirc}^{\partial}} \dot{E}) && \text{\wr} \alpha^{\partial} \text{ preserves joins in (39.25)\int}\\
= \ & \dot{E} \mathbin{\dot{\cup}} x, y \mapsto \{\pi \in X \mathbin{\dot{\circledcirc}^{\partial}} \dot{E} \mid \mathsf{V}(\pi) \subseteq \{z_1, \ldots, z_{k+2}\} \cup \{x, y\}\} && \text{\wr definition of } \alpha^{\partial} \text{ in (39.24)\int}\\
= \ & \dot{E} \mathbin{\dot{\cup}} x, y \mapsto \{\pi \in \bigcup_{z \in V} \{\pi_1 \odot \pi_2 \mid \pi_1 \in X(x, z) \wedge \pi_2 \in \dot{E}(z, y) \wedge \mathsf{elem\text{-}conc?}(\pi_1, \pi_2)\} \mid \\
& \mathsf{V}(\pi) \subseteq \{z_1, \ldots, z_{k+2}\} \cup \{x, y\}\} && \text{\wr definition of } \dot{\circledcirc}^{\partial} \text{ in theorem 39.23\int}\\
= \ & \dot{E} \mathbin{\dot{\cup}} x, y \mapsto \bigcup_{z \in V} \{\pi_1 \odot \pi_2 \mid \pi_1 \in X(x, z) \wedge \pi_2 \in \dot{E}(z, y) \wedge \mathsf{elem\text{-}conc?}(\pi_1, \pi_2) \wedge \mathsf{V}(\pi_1 \odot \\
& \pi_2) \subseteq \{z_1, \ldots, z_{k+2}\} \cup \{x, y\}\} && \text{\wr definition of } \in\text{\int}\\
= \ & \dot{E} \mathbin{\dot{\cup}} x, y \mapsto \bigcup_{z \in V} \{x\pi_1 z y \mid x\pi_1 z \in X(x, z) \wedge z y \in \dot{E}(z, y) \wedge \mathsf{elem\text{-}conc?}(x\pi_1 z, z y) \wedge \\
& \mathsf{V}(\pi_1) \cup \{z\} \subseteq \{z_1, \ldots, z_{k+2}\} \cup \{x, y\}\} \\
& \qquad \text{\wr definitions of } \odot, \mathsf{V}, \text{ and } \dot{E} \text{ in theorem 39.10, and induction hypothesis\int}\\
= \ & \dot{E} \mathbin{\dot{\cup}} x, y \mapsto \{x\pi_1 z_{k+2} \odot z_{k+2}\pi_2 y \mid x\pi_1 z_{k+2} \in \alpha^{\partial}(X)(x, z_{k+2}) \wedge z_{k+2}\pi_2 y \in \dot{E}(z_{k+2}, y) \wedge \\
& \mathsf{elem\text{-}conc?}(x\pi_1 z_{k+2}, z_{k+2}\pi_2 y)\} && \text{\wr by an argument similar to (39.27)\int}\\
= \ & \dot{E} \mathbin{\dot{\cup}} x, y \mapsto \{x\pi_1 z_{k+2} \odot z_{k+2} y \mid x\pi_1 z_{k+2} \in \alpha^{\partial}(X)(x, z_{k+2}) \wedge \langle z_{k+2}, y\rangle \in E \wedge \\
& \mathsf{elem\text{-}conc?}(x\pi_1 z_{k+2}, z_{k+2} y)\} && \text{\wr definition (39.24) of } \alpha^{\partial} \text{ and } \dot{E} \text{ in theorem 39.10\int}\\
= \ & \dot{E} \mathbin{\dot{\cup}} x, y \mapsto \{x\pi_1 z_{k+2} \odot z_{k+2} y \mid x\pi_1 z_{k+2} \in \alpha^{\partial}(X)(x, z_{k+2}) \wedge \langle z_{k+2}, y\rangle \in E\} \\
& \qquad \text{\wr because } z_{k+2} \notin \mathsf{V}(\pi_1) \text{ by induction hypothesis path so that the path } x\pi_1 z_{k+2} y \text{ is elementary\int}\\
= \ & \dot{E} \mathbin{\dot{\cup}} \alpha^{\partial}(X) \mathbin{\dot{\circledcirc}_{z_{k+2}}} \dot{E} && \text{\wr definition of } \dot{\circledcirc}^{\partial}_{z_{k+2}} \text{ in theorem 39.26\int}\\
= \ & \vec{\mathscr{F}}^{\partial}_{\pi k+2}(\alpha^{\partial}(X)) && \text{\wr(39.26.a)\int} \quad \square
\end{aligned}
$$

39.15 Calculational Design of an Overapproximation of the Elementary Paths Between Vertices of Finite Graphs

Because shortest paths are necessarily elementary, one could expect that Roy-Floyd-Warshall algorithm simply abstracts the elementary paths by their length. This is *not* the case, because the iterations in (39.26.c) and (39.26.d) for elementary paths are too expensive. They require checking elem-conc? in $\dot{\circledcirc}^{\partial}$ to make sure that the concatenation of elementary paths does yield an elementary path. But we can overapproximate by replacing $\dot{\circledcirc}^{\partial}$ by $\dot{\circledcirc}$ because the length of the shortest paths in the graph is the same as the length of the shortest paths in any subset of the graph paths provided this subset contains all elementary paths.

Corollary 39.28 (iterative characterization of an overapproximation of the elementary paths of a finite graph) Let $G = \langle V, E\rangle$ be a finite graph with $V = \{z_1, \ldots, z_n\}$, $n > 0$. Then

$$\mathsf{p}^{\partial} = \mathsf{lfp}^{\dot{\subseteq}}\, \overleftrightarrow{\mathscr{F}}^{\partial}_{\Pi} \mathrel{\dot{\subseteq}} \overleftrightarrow{\mathscr{F}}_{\pi}^{n+2} \quad \text{where} \tag{39.28.c}$$
$$\overleftrightarrow{\mathscr{F}}_{\pi}^{0} \triangleq \dot{\varnothing}, \quad \overleftrightarrow{\mathscr{F}}_{\pi}^{1} \triangleq \dot{E}, \quad \overleftrightarrow{\mathscr{F}}_{\pi}^{k+2} \triangleq \dot{E} \mathbin{\dot{\cup}} \overleftrightarrow{\mathscr{F}}_{\pi}^{k+1} \mathbin{\dot{\circledcirc}_{z_{k+1}}} \overleftrightarrow{\mathscr{F}}_{\pi}^{k+1}$$

$$= \mathsf{lfp}^{\dot{\subseteq}}_{\dot{E}}\, \widehat{\mathscr{F}}^{\partial}_{\Pi} \mathrel{\dot{\subseteq}} \widehat{\mathscr{F}}_{\pi}^{n+1} \quad \text{where} \tag{39.28.d}$$
$$\widehat{\mathscr{F}}_{\pi}^{0} \triangleq \dot{E}, \quad \widehat{\mathscr{F}}_{\pi}^{k+1} \triangleq \widehat{\mathscr{F}}_{\pi}^{k} \mathbin{\dot{\cup}} \widehat{\mathscr{F}}_{\pi}^{k} \mathbin{\dot{\circledcirc}_{z_k}} \widehat{\mathscr{F}}_{\pi}^{k}$$

Proof of corollary 39.28 Obviously $\dot{\circledcirc}^{\partial}_{z} \mathrel{\dot{\subseteq}} \dot{\circledcirc}_{z}$ so the iterates $\langle \overleftrightarrow{\mathscr{F}}_{\pi}^{k}, k \in [0, n+2]\rangle$ of (39.28.c) overapproximate those $\langle \overleftrightarrow{\mathscr{F}}_{\pi}^{\partial k}, k \in [0, n+2]\rangle$ of (39.26.c). Same for case (39.26.d). □

Exercise 39.29 Show that in (39.28.d), the subset inclusion can be strict.

39.16 The Roy-Floyd-Warshall Algorithm Overapproximating the Elementary Paths of a Finite Graph

The Roy-Floyd-Warshall algorithm shown in figure 39.31 does *not* compute elementary paths in (39.26.d) but the overapproximation of the set of elementary paths in (39.28.d), thus avoiding the potentially costly test in theorem 39.26 that the concatenation of elementary paths in $\dot{\circledcirc}^{\partial}_{z}$ is elementary.

Proof of algoritnm 39.30 The first for iteration computes $\widehat{\mathscr{F}}_{\pi}^{0} \triangleq \dot{E}$ in (39.28.d). Then, the second for iteration should compute $\widehat{\mathscr{F}}_{\pi}^{k+1} \triangleq \widehat{\mathscr{F}}_{\pi}^{k} \mathbin{\dot{\cup}} \widehat{\mathscr{F}}_{\pi}^{k} \mathbin{\dot{\circledcirc}_{z_k}} \widehat{\mathscr{F}}_{\pi}^{k}$ in (39.28.d) because $\mathsf{p}_1 \circledcirc_z \mathsf{p}_2 = \varnothing$ in (39.26.d) when $z \in \{x, y\}$, in which case, $\widehat{\mathscr{F}}_{\pi}^{k+1} = \widehat{\mathscr{F}}_{\pi}^{k}$, which is similar to the Jacobi iterative method in section 22.2. However, similar to the Gauss-Seidel iteration method in section 22.2, we reuse the last computed $\mathsf{p}(x, z)$ and $\mathsf{p}(z, y)$, not necessarily those of the previous iterate. For the convergence of the first n iterates of the second for iteration of the algorithm, the justification is similar to the convergence corollary 22.6 for chaotic iterations in definition 22.2. □

Exercise 39.32 Design an elementary path-finding algorithm based in (39.26.a).

39.17 Calculational Design of the Roy-Floyd-Warshall Shortest Path Algorithm

The shortest path algorithm of Bernard Roy [844, 845], Bob Floyd [375], and Steve Warshall [965] for finite graphs is based on the assumption that the graph has no cycles with strictly negative weights, that is, $\forall x \in V \,.\, \forall \pi \in \mathsf{p}(x, x) \,.\, \boldsymbol{\omega}(\pi) \geqslant 0$. In that case the

Algorithm 39.30 (Roy-Floyd-Warshall algorithm overapproximating the elementary paths of a finite graph) Let $G = \langle V, E\rangle$ be a graph with $|V| = n > 0$ vertices. The Roy-Floyd-Warshall algorithm

```
for x, y ∈ V do
    p(x, y) := E ∩ {⟨x, y⟩}
done;
for z ∈ V do
  for x, y ∈ V \ {z} do
    p(x, y) := p(x, y) ∪ p(x, z) ⊙ p(z, y)
  done
done
```

computes an overapproximation of all elementary paths p of G.

Figure 39.31
Roy-Floyd-Warshall shortest distance algorithm

shortest paths are all elementary because adding a cycle of weight 0 leaves the distance unchanged whereas a cycle of positive weight would strictly increase the distance on the path. Otherwise, as shown by example 39.16, if the graph has cycle with strictly negative weights, the convergence between two vertices containing a cycle with strictly negative weights is infinite to the limit $-\infty$.

The essential consequence is that we do not have to consider all paths as in theorem 39.10 but instead we can consider any subset provided that it contains all elementary paths. Therefore we can base the design of the shortest path algorithm on corollary 39.28. Observe that, although p may contain paths that are not elementary, d is precisely the minimal path lengths and not some strict overapproximation because

- p contains all elementary paths (so nonelementary paths are longer than the elementary path between their extremities), and
- no arc has a strictly negative weight (so path lengths are always positive and therefore the elementary paths are the shortest ones).

We derive the Roy-Floyd-Warshall algorithm by a calculation design applying theorem 18.36 for finite iterates to (39.28.d) with the abstraction $\dot{\boldsymbol{\omega}}$ (or a variant when considering (39.28.c)).

– for the infimum $\dot{E}$ in (39.28.d), we have

$$\begin{aligned}
&\dot{\boldsymbol{\omega}}(\dot{E})\langle x, y\rangle \\
=\ &\boldsymbol{\omega}(\dot{E}(x, y)) && \text{(pointwise definition of } \dot{\boldsymbol{\omega}}\text{)} \\
=\ &\boldsymbol{\omega}(\llbracket \langle x, y\rangle \in E \mathrel{?} \{\langle x, y\rangle\} \mathrel{:} \varnothing \rrbracket) && \text{(definition of } \dot{E} \text{ in theorem 39.10)} \\
=\ &\llbracket \langle x, y\rangle \in E \mathrel{?} \boldsymbol{\omega}(\{\langle x, y\rangle\}) \mathrel{:} \boldsymbol{\omega}(\varnothing) \rrbracket && \text{(definition of the conditional)} \\
=\ &\llbracket \langle x, y\rangle \in E \mathrel{?} \min\{\boldsymbol{\omega}(\pi) \mid \pi \in \{\langle x, y\rangle\}\} \mathrel{:} \infty \rrbracket && \text{((39.14))} \\
=\ &\llbracket \langle x, y\rangle \in E \mathrel{?} \boldsymbol{\omega}(x, y) \mathrel{:} \infty \rrbracket && \text{((39.13))}
\end{aligned}$$

– for the commutation with $\widehat{\mathscr{F}}_{\pi k+1}(X) \triangleq X \dot{\cup} X \dot{\circledcirc}_{z_k} X$, we have

$$
\begin{aligned}
& \dot{\boldsymbol{\omega}}(\widehat{\mathscr{F}}_{\pi k+1}(X))\langle x, y\rangle \\
= \; & \dot{\boldsymbol{\omega}}(X \dot{\cup} X \dot{\circledcirc}_{z_k} X)\langle x, y\rangle && \text{⦃(39.28.d)⦄} \\
= \; & \min(\dot{\boldsymbol{\omega}}(X)\langle x, y\rangle, \dot{\boldsymbol{\omega}}(X \dot{\circledcirc}_{z_k} X)\langle x, y\rangle) \\
& && \text{⦃the abstraction } \dot{\boldsymbol{\omega}} \text{ of Galois connection (39.15) preserves existing joins⦄}
\end{aligned}
$$

Let us evaluate

$$
\begin{aligned}
& \dot{\boldsymbol{\omega}}(X \dot{\circledcirc}_{z_k} X)\langle x, y\rangle \\
= \; & \boldsymbol{\omega}((X \dot{\circledcirc}_{z_k} X)(x, y)) && \text{⦃pointwise definition of } \dot{\boldsymbol{\omega}}\text{⦄} \\
= \; & \boldsymbol{\omega}(\{\pi_1 \odot \pi_2 \mid \pi_1 \in X(x, z_k) \wedge \pi_2 \in X(z_k, y) \wedge z_k \notin \{x, y\}\}) \\
& && \text{⦃definition of } \dot{\circledcirc}_{z_k} \text{ in theorem 39.26⦄} \\
= \; & \min\{\boldsymbol{\omega}(\pi_1 \odot \pi_2) \mid \pi_1 \in X(x, z_k) \wedge \pi_2 \in X(z_k, y) \wedge z_k \notin \{x, y\}\} && \text{⦃(39.14)⦄} \\
= \; & \min\{\boldsymbol{\omega}(\pi_1) + \boldsymbol{\omega}(\pi_2) \mid \pi_1 \in X(x, z_k) \wedge \pi_2 \in X(z_k, y) \wedge z_k \notin \{x, y\}\} \\
& && \text{⦃definition (39.13) of } \boldsymbol{\omega}\text{⦄} \\
= \; & ⟦ z_k \in \{x, y\} \; ? \; \infty \; ⁏ \; \min\{\boldsymbol{\omega}(\pi_1) \mid \pi_1 \in X(x, z_k)\} + \min\{\boldsymbol{\omega}(\pi_2) \mid \pi_1 \in X(x, z_k) \wedge \pi_2 \in \\
& X(z_k, y)\} ⟧ && \text{⦃definition of min⦄} \\
= \; & ⟦ z_k \in \{x, y\} \; ? \; \infty \; ⁏ \; \min(\dot{\boldsymbol{\omega}}(X)(x, z_k)) + \min(\dot{\boldsymbol{\omega}}(X)(z_k, y)) ⟧ \\
& && \text{⦃(39.14) and pointwise definition of } \dot{\boldsymbol{\omega}}\text{⦄}
\end{aligned}
$$

so that $\dot{\boldsymbol{\omega}}(\widehat{\mathscr{F}}_{\pi k+1}(X)) = \widehat{\mathscr{F}}_{\delta k}(\dot{\boldsymbol{\omega}}(X))$ with $\widehat{\mathscr{F}}_{\delta k}(X)(x, y) \triangleq ⟦ z_k \in \{x, y\} \; ? \; X(x, y) \; ⁏ \; \min(X(x, y), X(x, z_k) + X(z_k, y)) ⟧$.

We have proved

> **Theorem 39.33 (iterative characterization of the shortest path length of a graph)** Let $G = \langle V, E, \boldsymbol{\omega}\rangle$ be a finite graph with $V = \{z_1, \ldots, z_n\}$, $n > 0$ weighted on the totally ordered group $\langle \mathbb{G}, \leqslant, 0, +\rangle$ with no strictly negative weight. Then the distances between any two vertices are
>
> $$\mathsf{d} = \dot{\boldsymbol{\omega}}(\mathsf{p}) = \widehat{\mathscr{F}}_{\delta}^{\,n+1} \quad \text{where} \tag{39.33}$$
>
> $$\widehat{\mathscr{F}}_{\delta}^{\,0}(x, y) \triangleq ⟦ \langle x, y\rangle \in E \; ? \; \boldsymbol{\omega}(x, y) \; ⁏ \; \infty ⟧,$$
>
> $$\widehat{\mathscr{F}}_{\delta}^{\,k+1}(x, y) \triangleq ⟦ z_k \in \{x, y\} \; ? \; \widehat{\mathscr{F}}_{\delta}^{\,k}(x, y) \; ⁏ \; \min(\widehat{\mathscr{F}}_{\delta}^{\,k}(x, y), \widehat{\mathscr{F}}_{\delta}^{\,k}(x, z_k) + \widehat{\mathscr{F}}_{\delta}^{\,k}(z_k, y)) ⟧$$

and directly get the Roy-Floyd-Warshall distances algorithm in figure 39.35.

Proof of algorithm 39.34 Instead of calculating the next iterate $\widehat{\mathscr{F}}_{\delta}^{\,k+1}$ as a function of the previous one $\widehat{\mathscr{F}}_{\delta}^{\,k}$ (à la Jacobi), we reuse the latest assigned values (à la Gauss-Seidel), as authorized by the chaotic iteration corollary 22.6. □

Exercise 39.36 Continuing exercise 39.32, design an algorithm based in (39.26.a) to find the shortest distance between any two vertices of a graph without cycles with strictly negative weight.

Algorithm 39.34 (Roy-Floyd-Warshall shortest distances of a graph) $G = \langle V, E, \boldsymbol{\omega}\rangle$ be a finite graph with $|V| = n > 0$ vertices weighted on the totally ordered group $\langle \mathbb{G}, \leqslant, 0, +\rangle$. Let $\mathsf{d} \in V \times V \to \mathbb{G} \cup \{-\infty, \infty\}$ be computed by the Roy-Floyd-Warshall algorithm

```
for x, y ∈ V do
    d(x, y) := if ⟨x, y⟩ ∈ E then ω(x, y) else ∞
  done;
for z ∈ V do
  for x, y ∈ V do
    d(x, y) := min(d(x, y), d(x, z) + d(z, y))
  done
done.
```

The graph has no cycle with strictly negative weight if and only if $\forall x \in V . \mathsf{d}(x, x) \geqslant 0$, in which case $\mathsf{d}(x, y)$ is the length of the shortest path from x to y.

Figure 39.35
The Roy-Floyd-Warshall algorithm

Exercise 39.37 Show that the Roy-Floyd-Warshall algorithm would be incorrect for the longest elementary path length of a graph. Which elementary path fixpoint computation would you recommend for this problem? Design the algorithm by calculus.

Exercise 39.38 (irreflexive transitive closure) Consider the abstraction $\alpha_+(P) \triangleq \bigcup\{\langle x, y\rangle \in V^2 \mid \exists n \in \mathbb{N} . \exists z_1 \dots z_n \in V^n . x, z_1 \dots z_n, y \in P\}$. Define $E^+ \triangleq \alpha_+(\bigcup\{\mathsf{p}(x, y) \mid \langle x, y\rangle \in V\})$. Prove that $E^+ = \mathsf{lfp}^{\subseteq} X \mapsto E \cup X \mathbin{⨾} X$ using theorem 39.10 and theorem 18.23.

Exercise 39.39 (reflexive transitive closure) Continuing exercise 39.38 prove that the reflexive transitive closure is $E^* \triangleq \alpha_*(E^+) = \mathsf{lfp}^{\subseteq} X \mapsto \mathbb{1}_V \cup E \cup X \mathbin{⨾} X$ where $\alpha_*(R) \triangleq \mathbb{1}_V \cup R$ and $\mathbb{1}_V$ is the identity relation on V.

39.18 Adjacency Matrix

The Boolean adjacency matrix of a finite graph $G = \langle [1, n], E\rangle$, $n \in \mathbb{N}^+$ is $\mathbf{G} \triangleq (\llbracket \langle i, j\rangle \in E \mathbin{?} 1 \mathbin{⸗} 0 \rrbracket)_{\substack{i=1,n \\ j=1,n}} \in \{0, 1\}^{n \times n}$; see example 39.1.

For a graph $G = \langle V, E, \boldsymbol{\omega}\rangle$ weighted on the group $\mathbb{G}$, the adjacency matrix is $\mathbf{G} = (\llbracket \langle i, j\rangle \in E \mathbin{?} \boldsymbol{\omega}(i, j) \mathbin{⸗} \infty \rrbracket)_{\substack{i=1,n \\ j=1,n}} \in (\mathbb{G} \cup \{\infty\})^{n \times n}$; see example 39.11.

The infimum is the empty graph $\langle \varnothing, \varnothing\rangle$ encoded with the empty matrix $[\varnothing]$. The supremum is $(\infty)_{\substack{i=1,n \\ j=1,n}}$.

There is a Galois isomorphism between $\mathbf{G}$ and $\langle i, j\rangle \mapsto \boldsymbol{\omega}(E \cap \{\langle i, j\rangle\})$ and similarly the distance d can be encoded, up to an isomorphism into $\mathbf{D} = (\mathsf{d}(i, j))_{\substack{i=1,n \\ j=1,n}} \in$

$(\mathbb{G} \cup \{\infty\})^{n \times n}$, so that by an abstraction of (39.23.c) similar to theorem 39.17, it follows that

Corollary 39.40 (shortest distances in a weighted graph with adjacency matrix) Let $G = \langle V, E, \boldsymbol{\omega}\rangle$ be a finite graph weighted on the totally ordered group $\langle \mathbb{G}, \leqslant, 0, +\rangle$. Then the matrix of distances between any two vertices is $\mathbf{D} = \mathsf{lfp}^{\dot{\geqslant}}\, \mathscr{F}_\mu = \mathsf{gfp}^{\dot{\leqslant}}\, \mathscr{F}_\mu$ where $\mathscr{F}_\mu^{\mathrm{G}}(\mathrm{D})_{ij} \triangleq \min(\mathrm{G}_{ij}, \min\limits_{k \in [1,n]} (\mathrm{D}_{ik} + \mathrm{D}_{kj}))$.

Example 39.41 The iterates for example 39.11 starting from G are $\mathscr{F}_\mu^{\,1} = \begin{bmatrix} \infty & 1 & 2 \\ -1 & \infty & 2 \\ \infty & \infty & 1 \end{bmatrix}, \mathscr{F}_\mu^{\,2} = \begin{bmatrix} 0 & 1 & 2 \\ -1 & 0 & 1 \\ \infty & \infty & 1 \end{bmatrix} = \mathscr{F}_\mu^{\,3} = \mathrm{D}$.

Example 39.42 (graph with cycle of strictly negative weight) Continuing example 39.16, the iterates of corollary 39.40 starting from G are $\mathscr{F}_\mu^{\,0} = \begin{bmatrix} \infty & 0 & 2 \\ -1 & \infty & 2 \\ \infty & \infty & 1 \end{bmatrix}$, $\mathscr{F}_\mu^{\,1} = \begin{bmatrix} -1 & 0 & 2 \\ -1 & -1 & 1 \\ \infty & \infty & 1 \end{bmatrix}$, $\mathscr{F}_\mu^{\,2} = \begin{bmatrix} -2 & -1 & 1 \\ -2 & -2 & 0 \\ \infty & \infty & 1 \end{bmatrix}$, $\mathscr{F}_\mu^{\,3} = \begin{bmatrix} -3 & -3 & -1 \\ -4 & -4 & -2 \\ \infty & \infty & 1 \end{bmatrix}$, $\mathscr{F}_\mu^{\,4} = \begin{bmatrix} -7 & -7 & -5 \\ -8 & -8 & -6 \\ \infty & \infty & 1 \end{bmatrix}$, …, and passing to the limit, $\mathscr{F}_\mu^{\,\omega} = \begin{bmatrix} -\infty & -\infty & -\infty \\ -\infty & -\infty & -\infty \\ \infty & \infty & 1 \end{bmatrix}$.

An efficient iteration is obtained by abstracting (39.28.d).

Corollary 39.43 (shortest distances in a weighted graph with adjacency matrix iteration) Let $G = \langle V, E, \boldsymbol{\omega}\rangle$ be a finite graph with $V = \{z_1, \ldots, z_n\}$, $n > 0$ weighted on the totally ordered group $\langle \mathbb{G}, \leqslant, 0, +\rangle$ with no strictly negative weight. Then the matrix of distances between any two vertices is $\mathbf{D} = \mathscr{F}_\mu^{\,n+1}$ where $\mathscr{F}_\mu^{\,0} = \mathrm{G}$ and $\mathscr{F}_{\mu\,ij}^{\,\ell+1} = \min(\mathscr{F}_{\mu\,ij}^{\,\ell}, \min\limits_{k \in [1,n]} (\mathscr{F}_{\mu\,ik}^{\,\ell} + \mathscr{F}_{\mu\,kj}^{\,\ell}))$.

Exercise 39.44 Continue exercise 39.38 using the Boolean adjacency matrix of the finite graph (to get the original Roy-Warshall iterative algorithm [844, 965]).

Exercise 39.45 Program the Roy-Floyd-Warshall algorithm of exercise 39.44 in the programming language of your choice.

Corollaries 39.40 and 39.43 generalize by showing that the matrix algebraic structure is preserved by abstraction, as follows:

Exercise 39.46 ([165, 314]) Let $G = \langle V, E\rangle$ be a finite graph with $|V| = m$ and $\iota \in V \rightarrowtail\!\!\!\!\twoheadrightarrow [1, m]$ be a bijection between vertices and indexes. Let $\langle \wp(V^{>1}), \subseteq, \cup, ⊚\rangle \underset{\alpha}{\overset{\gamma}{\leftrightarrows}} \langle A, \sqsubseteq, \sqcup, \overline{⊚}\rangle$ define a path problem. Consider the representation of abstractions of sets of paths between extremities by a $m \times m$ matrix $\mathbf{M}$ of elements of A that is the isomorphism $\langle V \times V \to A, \dot{\sqsubseteq}\rangle \underset{\alpha^M}{\overset{\gamma^M}{\leftrightarrows}} \langle A^{m \times m}, \sqsubseteq\rangle$ such that $\gamma^M(\mathbf{M})(x, y) \triangleq \mathbf{M}_{\iota(x),\iota(y)}$. Define the operations

$$\mathbf{M} \sqsubseteq \mathbf{M}' \triangleq \forall i, j \in [1, m] \,.\, \mathbf{M}_{ij} \sqsubseteq \mathbf{M}'_{ij} \qquad \text{matrix partial order}$$

$$\overline{E}_{ij} \triangleq (\!| \langle \iota^{-1}(i), \iota^{-1}(j)\rangle \in E \,?\, \alpha(\{\langle \iota^{-1}(i), \iota^{-1}(j)\rangle\}) \,:\, \alpha(\varnothing) |\!)$$

$$(\mathbf{M} \sqcup \mathbf{M}')_{ij} \triangleq \mathbf{M}_{ij} \sqcup \mathbf{M}'_{ij} \qquad \text{matrix addition}$$

$$(\mathbf{M} \,\overline{\circledcirc}\, \mathbf{M}')_{ij} \triangleq \bigsqcup_{k=1}^{m} \mathbf{M}_{ik} \,\overline{\circledcirc}\, \mathbf{M}'_{kj} \qquad \text{matrix multiplication}$$

Show that $\alpha^{M} \circ \dot{\alpha} \circ \alpha^{\circ\!\!-\!\!\circ}(\mathsf{lfp}^{\subseteq}\, \vec{\mathscr{F}}_{\Pi}) = \mathsf{lfp}^{\sqsubseteq}\, \mathbf{M} \mapsto \overline{E} \sqcup \overline{E} \,\overline{\circledcirc}\, \mathbf{M}$ and similarly for (39.4.b–d). Show that the matrix algebraic structure in this exercise is preserved by a further abstraction of $\langle A, \sqsubseteq, \sqcup, \overline{\circledcirc}\rangle$.

39.19 Conclusion

Graph theory originates from Leonhard Euler's work on the so-called Seven Bridges of Königsberg problem [357]. Graphs are used to model many types of relations and processes in physical, biological, social, and information systems.

[108, 314, 424, 774, 775] observed that various path algorithms can be designed and proved correct on the basis of a common algebraic structure and then instantiated to various path problems up to homomorphisms. The explanation of this observation originates from the fixpoint characterization of graph paths in theorem 39.4 based on the primitives E, $\cup$ and $\circledcirc$ on sets of paths. Then theorem 39.6 shows that any path problem formalized by an abstraction α can be expressed in terms of abstract operations $\alpha(E)$, $\sqcup$, and $\overline{\circledcirc}$ derived from the primitives E, $\cup$ and $\circledcirc$ by calculational design. This is the case for the shortest path problem for which the composition of successive abstractions (which is not a homomorphism) allows us to formally derive the classical Roy-Floyd-Warshall algorithm (which therefore need not be postulated out of similarity observations, for example, [775, Sect. 1] and [165, ch. 3]). The derivation of Roy-Floyd-Warshall algorithm was tricky because it is based on the abstraction of an overapproximation of the elementary paths that is an underapproximation of all graph paths.

For comments on the Roy-Floyd-Warshall algorithm; see [460, p. 26–29], [743] and [865, p. 129]. The calculational design of the transitive closure by abstraction of a fixpoint path semantics is in [259]. See [877] for the calculational design of dominance and shortest path algorithms by abstract interpretation (based on exact abstractions of (39.4.a)). There are shortest path algorithms more efficient than the Roy-Floyd-Warshall algorithm, such as [532], and [14, 15] for acyclic graphs.

The semantics of a program can be understood as a graph (see chapter 43) so that program static analysis can be understood as graph abstractions.

40 Zone and Octagon Analysis

Inequalities between values of variables are necessary, for example, to bound indexes used in the traversal of arrays with a symbolic bound n, as in

```
  l1: {T} i = 0;
while l2: (i < n) {i>=0}
    l3: {i>=0, i<=n-1} i = (i + 1);
l4: {i=n}
```

Zones can express inequalities of the form $\pm x_i \leqslant c$ or $x_i - x_j \leqslant c$ where x_i and x_j are values of scalar variables x_i, x_j and c is a scalar constant determined by the analysis. The canonical encoding of zones by normalized difference bounded matrices with an extra zero variable was introduced by Vaughan Pratt [808]. Zones were first used to analyze timed systems [111, 329]. Octagons add constraints of the form $x_i + x_j \geqslant c$. Octagons were first used for program parallelization in [73] (where they are called "simple sections"). The missing operations to form an abstract domain, in particular assignment, test, join, widening, and narrowing were designed by Antoine Miné [687, 688, 692].

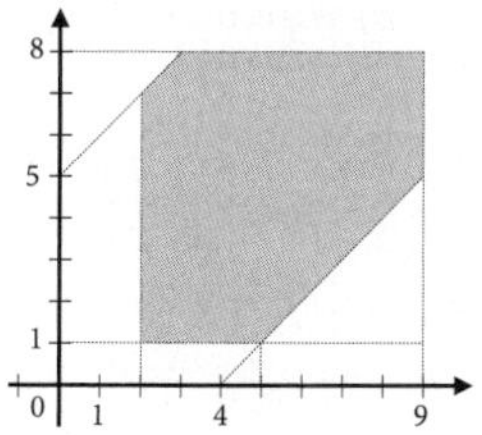

Zone

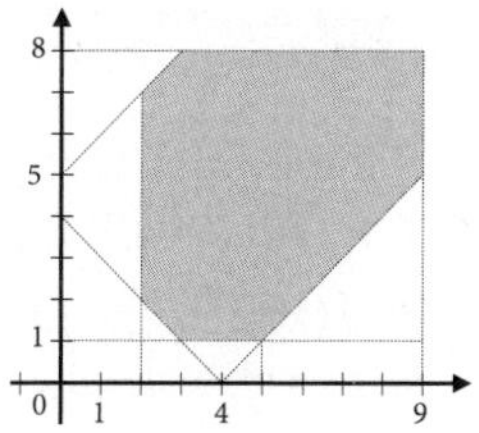

Octagon

Contents

40.1 Zone Analysis

We study the zone abstract domain

$$\mathbb{D}^{\Box} \triangleq \langle \mathbb{P}^{\Box}, \sqsubseteq^{\Box}, \bot^{\Box}, \sqcup^{\Box}, \sqcap^{\Box}, \mathsf{assign}^{\Box}[\![\mathsf{x}, \mathsf{A}]\!], \mathsf{test}^{\Box}[\![\mathsf{B}]\!], \overline{\mathsf{test}}^{\Box}[\![\mathsf{B}]\!], \nabla^{\Box}, \Delta^{\Box} \rangle.$$

40.1.1 Zone Abstract Properties

As in section 38.1, we let $\langle x_1, \ldots, x_m \rangle$ denote the values of the program variables $\mathbb{V} = \{\mathtt{x}_1, \ldots, \mathtt{x}_m\}$. More precisely, we consider the isomorphism between environment properties $P \in \mathbb{P} = \wp(\mathbb{V} \rightarrow \mathbb{F})$ and $\overline{P} \in \wp(\mathbb{F}^m)$ where $\langle \mathbb{F}, \leqslant, +, -, \times, / \rangle$ is a totally ordered field, $m = |\mathbb{V}|$ is the cardinality of the finite set $\mathbb{V}$ of variables and $\rho \in P$ if and only if $\langle \rho(\mathtt{x}_1), \ldots, \rho(\mathtt{x}_m) \rangle \in \overline{P}$. The field $\mathbb{F}$ can be the rationals $\mathbb{Q}$, or the reals $\mathbb{R}$. $\mathbb{F}$ can also be chosen to be the integers $\mathbb{Z}$, which is not a field, because division is not used by the zone analysis.[1]

To express $\pm x_i \leqslant c$, we consider an additional variable $\mathtt{x}_0$ which value is assumed to be always $0 \in \mathbb{F}$ so $x_i \leqslant c$ is $x_i - x_0 \leqslant c$ and $-x_i \leqslant c$ is $x_0 - x_i \leqslant c$. In this way all constraints are of the form $x_i - x_j \leqslant c_{ij}$, $i, j \in [0, m]$ and $i \neq j$.

40.1.2 Adjacency Matrix Representation of Zones

The constraints are represented by a weighted graph $G = \langle [0, m], E, \boldsymbol{\omega} \rangle$ (section 39.7) where the set of vertices $[0, m]$ is isomorphic to the set of program variables $\{\mathtt{x}_0, \mathtt{x}_1, \ldots, \mathtt{x}_m\}$ augmented by $\mathtt{x}_0$ and there is an edge $\langle i, j \rangle$ with weight $\boldsymbol{\omega}(\langle i, j \rangle) = c_{ij}$ for each constraint $x_i - x_j \leqslant c_{ij}$ between the values x_i, x_j of the program variables $\mathtt{x}_i, \mathtt{x}_j$, $i, j \in [0, m]$ and $i \neq j$. We let $\boldsymbol{\omega}(\langle i, i \rangle) = 0$, $i \in [0, m]$.

The concretization of $G = \langle [0, m], E, \boldsymbol{\omega} \rangle$ is

$$\gamma^{\square}(G) \triangleq \{\langle x_1, \ldots, x_m \rangle \mid \forall \langle i, j \rangle \in E \,.\, x_i - x_j \leqslant \boldsymbol{\omega}(\langle i, j \rangle) \wedge x_0 = 0\}$$

If the graph G has a cycle $x_{i_1}, x_{i_2}, x_{i_3}, \ldots, x_{i_n}, x_{i_1}$ with strictly negative weight then $x_{i_1} - x_{i_2} \leqslant c_{i_1 i_2}$, $x_{i_2} - x_{i_3} \leqslant c_{i_2 i_3}$, ..., $x_{i_n} - x_{i_1} \leqslant c_{i_n i_1}$ with $c_{i_1 i_2} + c_{i_2 i_3} + \ldots + c_{i_n i_1} < 0$ so that $0 = x_{i_1} - x_{i_2} + x_{i_2} - x_{i_3} + \ldots, + x_{i_n} - x_{i_1} \leqslant c_{i_1 i_2} + c_{i_2 i_3} + \ldots + c_{i_n i_1} < 0$, which is false, and therefore $\gamma^{\square}(G) = \varnothing$.

Typically, weighted graphs are represented by their isomorphic adjacency matrix (see section 39.18) $\mathbf{G} = (\llbracket \langle i, j \rangle \in E \,?\, \boldsymbol{\omega}(i, j) \,\mathbin{:}\, \infty \rrbracket)_{\substack{i=1,m \\ j=1,m}} \in (\mathbb{F} \cup \{\infty\})^{n \times n}$ so that

$$\gamma^{\square}(\mathbf{G}) \triangleq \{\langle x_1, \ldots, x_m \rangle \mid \forall i, j \in [0, m] \,.\, x_i - x_j \leqslant \mathbf{G}_{ij} \wedge x_0 = 0\}.$$

Example 40.1

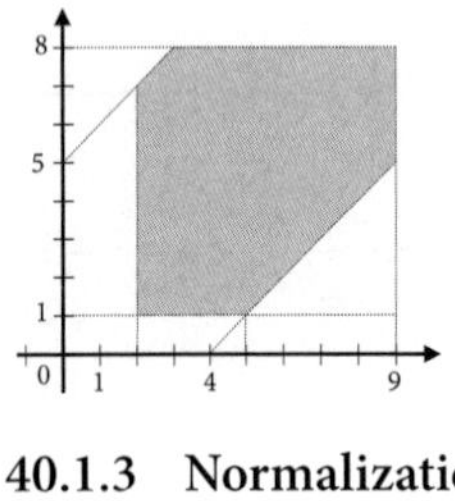


```
x>=2, y>=1,
x<=9, x-y<=4,
y<=8, y-x<=5
```

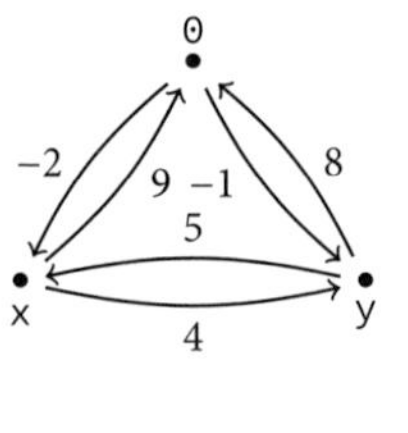

$$\begin{array}{c} \\ 0 \\ \mathtt{x} \\ \mathtt{y} \end{array} \begin{array}{c} \begin{array}{ccc} 0 & \mathtt{x} & \mathtt{y} \end{array} \\ \begin{bmatrix} 0 & -2 & -1 \\ 9 & 0 & 4 \\ 8 & 5 & 0 \end{bmatrix} \end{array}$$

40.1.3 Normalization

Even if the graph G has no cycle with strictly negative weight, several different graphs may have the same concretization. For example, $x - y \leqslant 0 \wedge y - z \leqslant 1$ and $x - y \leqslant 0 \wedge y - z \leqslant 1 \wedge x - z \leqslant 1$.

1. Octagon analysis in section 40.2 uses a division by 2 that does not need to be exact.

So, we need to normalize the abstract domain as in section 36.3.3. For that purpose all graphs with a cycle of strictly negative weight are normalized to the infimum $\perp^{\varnothing}$. All other graphs G with no cycle of strictly negative weight (represented by their adjacency matrix **G** such that $\gamma^{\varnothing}(G) = \gamma^{\varnothing}(\mathbf{G})$) are normalized to a unique matrix $\mathsf{norm}^{\varnothing}(\mathbf{G})$ which is

- meaning-preserving, that is, $\gamma^{\varnothing}(\mathbf{G}) = \gamma^{\varnothing}(\mathsf{norm}^{\varnothing}(\mathbf{G}))$, and
- normalizing, that is, $\gamma^{\varnothing}(\mathsf{norm}^{\varnothing}(\mathbf{G}_1)) = \gamma^{\varnothing}(\mathsf{norm}^{\varnothing}(\mathbf{G}_2))$ implies $\mathsf{norm}^{\varnothing}(\mathbf{G}_1) = \mathsf{norm}^{\varnothing}(\mathbf{G}_2)$ (and conversely).

(It follows that $\mathsf{norm}^{\varnothing}$ is idempotent).

The set of normalized zone abstract properties is

$$\overline{\mathbb{P}}^{\varnothing} \triangleq \{\perp^{\varnothing}\} \cup \{\mathbf{D} \in (\mathbb{F} \cup \{\infty\})^{m+1 \times m+1} \mid \forall i \in [0, m] \,.\, \mathbf{D}_{ii} = 0 \wedge \mathsf{norm}^{\varnothing}(\mathbf{D}) = \mathbf{D}\}.$$

with

$$\gamma^{\varnothing}(\mathbf{D}) \triangleq \{\langle x_1, \dots, x_m\rangle \mid \forall i, j \in [0, m] \,.\, x_i - x_j \leqslant \mathbf{D}_{ij} \wedge x_0 = 0\}$$

$$\gamma^{\varnothing}(\perp^{\varnothing}) \triangleq \varnothing$$

where, for all $v \in \mathbb{F}$, $v \leqslant \infty$ and $\infty \leqslant \infty$ are always true.

The normalization by saturation adds as much many constraints as possible by computing the distance matrix with Roy-Floyd-Warshall algorithm (theorem 39.26). Normalization by saturation provides a normal form. An alternative, suggested by [65], is to look for a minimal number of constraints. A total order must be imposed on variables for the minimization (of [617]) to provide a normal form.

Distance graph	Normalization by saturation	Normalization by minimization
$\begin{bmatrix} 0 & -1 & -2 \\ 1 & 0 & -1 \\ 2 & \infty & 0 \end{bmatrix}$	$\begin{bmatrix} 0 & -1 & -2 \\ 1 & 0 & -1 \\ 2 & 1 & 0 \end{bmatrix}$	$\begin{bmatrix} 0 & \infty & -2 \\ 1 & 0 & \infty \\ \infty & 1 & 0 \end{bmatrix}$
`{x=1, y=2, x<=y-1}`	`{x=1, y=2, x=y-1}`	`{y>=2, x<=1, y<=x+1}`

We prefer the normalization by saturation which introduces redundant constraints that may be propagated by incomplete assignment and test transformers and that would definitely be missed with the normalization by minimization (which is nevertheless useful for widening; see section 40.1.8).

40.1.4 Normalization by Saturation

In the normalization by saturation, the graph G (as represented by its adjacency matrix **G**) is normalized to its distance graph $G^* = \langle [0, m],\ [0, m] \times [0, m],\ \mathsf{d}\rangle$ (as represented by its matrix of distances[2] (section 39.18) $\mathbf{D} \in (\mathbb{F} \cup \{\infty\})^{m+1 \times m+1}$). The normal form is computed by the Roy-Floyd-Warshall algorithm $\mathbf{D} = \mathsf{RFW}(\mathbf{G})$ of exercise 39.45 such that

$$\mathsf{RFW}(\mathbf{D}) \triangleq \perp^{\varnothing} \qquad \text{if } \mathbf{D} \text{ has a cycle of strictly negative weight}$$

$$\mathsf{RFW}(\mathbf{D})_{ij} \triangleq \min(\mathbf{D}_{ij}, \min_{k \in [0,m]} (\mathbf{D}_{ik} + \mathbf{D}_{kj})) \qquad \text{otherwise, (corollary 39.40)}$$

where, for all $v \in \mathbb{F}$, $v \leqslant \infty$ and $\infty \leqslant \infty$ are always true.

2. Called the difference bounded matrix in [329].

Lemma 40.2 (meaning preservation) $\forall \mathbf{D} \in (\mathbb{F} \cup \{\infty\})^{m+1\times m+1}$. $\gamma^{\boxslash}(\mathsf{RFW}(\mathbf{D})) = \gamma^{\boxslash}(\mathbf{D})$.

Proof of lemma 40.2 $\mathsf{RFW}(\mathbf{D})$ returns $\bot^{\boxslash}$ if and only if the zone constrains are not satisfiable, so the concretization is $\varnothing$ in both cases.

Otherwise, there is no cycle with negative weight, and so if the graph G has a cycle $x_{i_1}, x_{i_2}, x_{i_3}, \ldots, x_{i_n}, x_{i_1}$ then $x_{i_1} - x_{i_2} + x_{i_2} - x_{i_3} + \ldots, +x_{i_n} - x_{i_1} \leqslant c_{i_1 i_2} + c_{i_2 i_3} + \ldots + c_{i_n i_1} \geqslant 0$ and therefore setting $\mathsf{RFW}(\mathbf{D})_{i_1 i_1} = 0$ does not change the concretization because $x_{i_1} - x_{i_1} = 0$. Therefore before the main loop on k of the Roy-Floyd-Warshall algorithm in exercise 39.45, $\gamma^{\boxslash}(\mathbf{D}) = \gamma^{\boxslash}(\mathbf{D}^0)$ where $\mathbf{D}^0$ is the initial matrix.

Let us show that a loop iteration does not change this invariant and so holds upon termination of the for loops. Assume by induction hypothesis that $\gamma^{\boxslash}(\mathbf{D}) = \gamma^{\boxslash}(\mathbf{D}^0)$. Then $\mathbf{D}$ is changed into $\mathbf{D}'_{ij} = \min(\mathbf{D}_{ij}, (\mathbf{D}_{ik} + \mathbf{D}_{kj}))$ so $\gamma^{\boxslash}(\mathbf{D}') = \gamma^{\boxslash}(\mathbf{D}) = \gamma^{\boxslash}(\mathbf{D}^0)$ because if $\mathbf{D}'_{ij} \neq \mathbf{D}_{ij}$ then $\mathbf{D}'_{ij} = \mathbf{D}_{ik} + \mathbf{D}_{kj}$ so by definition of $\gamma^{\boxslash}$ the term $x_i - x_j \leqslant \mathbf{D}'_{ij}$ in the concretization of $\mathbf{D}'$ is logically equivalent to the term $x_i - x_j \leqslant \mathbf{D}_{ij} \wedge x_i - x_k \leqslant \mathbf{D}_{ik} \wedge x_k - x_j \leqslant \mathbf{D}_{kj}$ $\Leftrightarrow x_i - x_k \leqslant \mathbf{D}_{ik} \wedge x_k - x_j \leqslant \mathbf{D}_{kj}$ appearing in the concretization of $\mathbf{D}$. □

Lemma 40.3 (normalization) $\gamma^{\boxslash}(\mathsf{RFW}(\mathbf{G}_1)) = \gamma^{\boxslash}(\mathsf{RFW}(\mathbf{G}_2)) \Leftrightarrow \mathsf{RFW}(\mathbf{G}_1) = \mathsf{RFW}(\mathbf{G}_2)$.

Proof of lemma 40.3 Let us first show that if the concretizations are the same then the graphs are the same.

For all i, j, there exists a point in $\gamma^{\boxslash}(\mathsf{RFW}(\mathbf{G}))$ satisfying the equality $x_i - x_j = \mathsf{RFW}(\mathbf{G})_{ij}$, that is, any constraint of the closure by shortest paths can be saturated.

If $\mathsf{RFW}(\mathbf{G}_1) \neq \mathsf{RFW}(\mathbf{G}_2)$ then they differ by at least one arc $\langle i, j\rangle$. By lemma 40.2, there exists a point that is in the zone with the arc $\langle i, j\rangle$ strictly greater. By construction, this point cannot be in the other zone , proving $\gamma^{\boxslash}(\mathsf{RFW}(\mathbf{G}_1)) \neq \gamma^{\boxslash}(\mathsf{RFW}(\mathbf{G}_2))$.

For the nontrivial implication $\Rightarrow$, there must exist i, j such that $\mathsf{RFW}(\mathbf{G}_1)_{ij} \neq \mathsf{RFW}(\mathbf{G}_2)_{ij}$, for example, $\mathsf{RFW}(\mathbf{G}_1)_{ij} < \mathsf{RFW}(\mathbf{G}_2)_{ij}$, because $\mathbb{F} \cup \{\infty\}$ is totally ordered. Because the concretizations are the same, the graphs are the same, so by theorem 39.33, the shortest distance between i and j should be the same in both of these cases, a contradiction. □

40.1.5 Lattice of Zones

Because an operation on normalized zones may return a non-normalized zone, the results on all operations on zones must be normalized, as necessary. The zone union $\sqcup^{\boxslash}$ is an example of operation which returns a normalized zone for normalized parameters. On the contrary, the intersection $\sqcap^{\boxslash}$ of normalized zones may not be normalized, which requires a renormalization. However, this normalization may prevent the convergence of the widening, as shown and solved in section 40.1.8.

Theorem 40.4 (lattice of zones) The domain of zone properties $\mathbb{D}^{\varpi} = \langle \mathbb{P}^{\varpi}, \sqsubseteq^{\varpi}, \bot^{\varpi}, \top^{\varpi}, \sqcup^{\varpi}, \sqcap^{\varpi} \rangle$ and the subdomain of normalized zone properties $\overline{\mathbb{D}}^{\varpi} = \langle \overline{\mathbb{P}}^{\varpi}, \sqsubseteq^{\varpi}, \bot^{\varpi}, \top^{\varpi}, \sqcup^{\varpi}, \overline{\sqcap}^{\varpi} \rangle$ are lattices where, for all $\mathbf{D}, \mathbf{D}' \in \mathbb{P}^{\varpi}$,

$$\bot^{\varpi} \sqsubseteq^{\varpi} \mathbf{D} \sqsubseteq^{\varpi} \top^{\varpi} \tag{40.4}$$
$$\mathbf{D} \sqsubseteq^{\varpi} \mathbf{D}' = \forall i, j \in [0, m] \,.\, \mathbf{D}_{ij} \leqslant \mathbf{D}'_{ij}$$
$$\mathbf{D} = \mathbf{D}' \triangleq \forall i, j \in [0, m] \,.\, \mathbf{D}_{ij} = \mathbf{D}'_{ij}$$
$$\top^{\varpi} \triangleq (\infty)^{m+1 \times m+1}$$
$$\mathbf{D} \sqcup^{\varpi} \mathbf{D}' \triangleq (\max(\mathbf{D}_{ij}, \mathbf{D}'_{ij}))^{m+1 \times m+1}$$
$$\mathbf{D} \sqcap^{\varpi} \mathbf{D}' \triangleq (\min(\mathbf{D}_{ij}, \mathbf{D}'_{ij}))^{m+1 \times m+1} \in \mathbb{P}^{\varpi}$$
$$\mathbf{D} \overline{\sqcap}^{\varpi} \mathbf{D}' = \mathsf{RFW}\left((\min(\mathbf{D}_{ij}, \mathbf{D}'_{ij}))^{m+1 \times m+1}\right) \in \overline{\mathbb{P}}^{\varpi}$$

Proof of (40.4)

— We define $P \sqsubseteq^{\varpi} Q$ if and only if $\gamma^{\varpi}(P) \subseteq \gamma^{\varpi}(Q)$. This is a preorder in zones $\mathbb{P}^{\varpi}$ but a partial order in normalized zones $\overline{\mathbb{P}}^{\varpi}$. We have $\gamma^{\varpi}(\bot^{\varpi}) = \varnothing$ and $\gamma^{\varpi}(\top^{\varpi}) = \mathbb{F}^m$ so $\bot^{\varpi} \sqsubseteq^{\varpi} \mathbf{D} \sqsubseteq^{\varpi} \top^{\varpi}$ because $\gamma^{\varpi}(\mathbf{D}) \in \wp(\mathbb{F}^m)$. Moreover

$$\begin{aligned}
& \mathbf{D} \sqsubseteq^{\varpi} \mathbf{D}' \\
\triangleq\ & \gamma^{\varpi}(\mathbf{D}) \subseteq \gamma^{\varpi}(\mathbf{D}') && \text{(definition of } \sqsubseteq^{\varpi}\text{)} \\
\Leftrightarrow\ & (\forall i, j \in [0, m] \,.\, x_i - x_j \leqslant \mathbf{D}_{ij}) \Rightarrow (\forall i, j \in [0, m] \,.\, x_i - x_j \leqslant \mathbf{D}'_{ij}) && \text{(definitions of } \gamma^{\varpi} \text{ and } \subseteq\text{)} \\
\Leftrightarrow\ & \forall i, j \in [0, m] \,.\, \mathbf{D}_{ij} \leqslant \mathbf{D}'_{ij} && \text{(definition of } \leqslant\text{)}
\end{aligned}$$

— Equality $P = Q$ is $P \sqsubseteq^{\varpi} Q \wedge Q \sqsubseteq^{\varpi} P$ and follows from the unicity of the matrix of distances returned by Roy-Floyd-Warshall algorithm.

— We have $\mathbf{D}_{ij} \leqslant \max(\mathbf{D}_{ij}, \mathbf{D}'_{ij})$ and $\mathbf{D}'_{ij} \leqslant \max(\mathbf{D}_{ij}, \mathbf{D}'_{ij})$ so that, by definition of γ^{ϖ}, $\mathbf{D} \sqsubseteq^{\varpi} \mathbf{D} \sqcup^{\varpi} \mathbf{D}'$ and $\mathbf{D}' \sqsubseteq^{\varpi} \mathbf{D} \sqcup^{\varpi} \mathbf{D}'$.

Given an upper bound $\mathbf{D}''$ of $\mathbf{D}$ and $\mathbf{D}'$, we have $\mathbf{D}_{ij} \leqslant \mathbf{D}''_{ij}$ and $\mathbf{D}'_{ij} \leqslant \mathbf{D}''_{ij}$ so that $\max(\mathbf{D}_{ij}, \mathbf{D}'_{ij}) \leqslant \mathbf{D}''_{ij}$ proving that $\mathbf{D} \sqcup^{\varpi} \mathbf{D}' \sqsubseteq^{\varpi} \mathbf{D}''$, and therefore that $\mathbf{D} \sqcup^{\varpi} \mathbf{D}'$ is the least upper bound.

Let $\mathbf{D}'' = \mathbf{D} \sqcup^{\varpi} \mathbf{D}'$. We must prove that $\mathbf{D}'' \in \overline{\mathbb{P}}^{\varpi}$. Obviously $\mathbf{D}''_{ii} = \max(\mathbf{D}_{ii}, \mathbf{D}'_{ii}) = \max(0, 0) = 0$. We must show that $\mathsf{RFW}(\mathbf{D}'') = \mathbf{D}''$. If not, we have some $\mathbf{D}''_{ij} \neq \min(\mathbf{D}''_{ij}, \min_{k \in [0,m]}(\mathbf{D}''_{ik} + \mathbf{D}''_{kj}))$ so $\min_{k \in [0,m]}(\mathbf{D}''_{ik} + \mathbf{D}''_{kj}) < \mathbf{D}''_{ij}$, hence $\mathbf{D}''_{ik} + \mathbf{D}''_{kj} < \mathbf{D}''_{ij}$ for some $k \in [0, m]$. By definition of $\sqcup^{\varpi}$, this implies that $\max(\mathbf{D}_{ik}, \mathbf{D}'_{ik}) + \max(\mathbf{D}_{kj}, \mathbf{D}'_{kj}) < \max(\mathbf{D}_{ij}, \mathbf{D}'_{ij})$. If $\max(\mathbf{D}_{ij}, \mathbf{D}'_{ij}) = \mathbf{D}_{ij}$ then $\mathbf{D}_{ik} + \mathbf{D}_{kj} < \mathbf{D}_{ij}$, in contradiction to $\mathsf{RFW}(\mathbf{D}) = \mathbf{D}$. Otherwise, $\max(\mathbf{D}_{ij}, \mathbf{D}'_{ij}) = \mathbf{D}'_{ij}$ and then $\mathbf{D}'_{ik} + \mathbf{D}'_{kj} < \mathbf{D}'_{ij}$, in contradiction to $\mathsf{RFW}(\mathbf{D}') = \mathbf{D}'$. By contradiction, $\mathsf{RFW}(\mathbf{D}'') = \mathbf{D}''$.

— By $\leqslant$-order duality, $\mathbf{D}'' = (\min(\mathbf{D}_{ij}, \mathbf{D}'_{ij}))^{m+1 \times m+1}$ is a $\sqsubseteq^{\varpi}$-glb of $\mathbf{D}$ and $\mathbf{D}'$ with 0-diagonal. However, it might not be in normal form. An example is $x - y \leqslant 1$ and $y - z \leqslant 2$ for which the meet yields $x - y \leqslant 1 \wedge y - z \leqslant 2 \wedge x - z \leqslant \min(\infty, \infty) = \infty$ and not $x - z \leqslant 3$. So the meet is $\mathsf{RFW}(\mathbf{D}'') \in \mathbb{P}^{\varpi}$ such that $\gamma^{\varpi}(\mathbf{D}'') = \gamma^{\varpi}(\mathsf{RFW}(\mathbf{D}''))$ by the normalization lemma 40.2. □

Exercise 40.5 Prove that if $\mathbb{F}$ is $\mathbb{Z}$ or $\mathbb{R}$ then $\gamma^{\varnothing}$ preserves infinite meets so $\langle\wp(\mathbb{F}^m), \subseteq\rangle \xrightleftharpoons[\alpha^{\varnothing}]{\gamma^{\varnothing}} \langle\mathbb{P}^{\varnothing}, \sqsubseteq^{\varnothing}\rangle$ (but not when $\mathbb{F}$ is $\mathbb{Q}$).

Exercise 40.6 Prove that if $\mathbb{F}$ is $\mathbb{Z}$ or $\mathbb{R}$ then $\langle\mathbb{P}^{\varnothing}, \sqsubseteq^{\varnothing}\rangle$ is a complete lattice (but not when $\mathbb{F}$ is $\mathbb{Q}$).

40.1.6 Calculational Design of the Zone Abstract Assignment Transformers

Following (21.7), we look for $\mathsf{assign}^{\varnothing}[\![\mathsf{x}, \mathsf{A}]\!]$ satisfying

$$\mathsf{assign}^{\vec{r}}[\![\mathsf{x}, \mathsf{A}]\!] \circ \gamma^{\varnothing} \subseteq \gamma^{\varnothing} \circ \mathsf{assign}^{\varnothing}[\![\mathsf{x}, \mathsf{A}]\!],$$

by calculational design. The assertional assignment transformer $\mathsf{assign}^{\vec{r}}[\![\mathsf{x}, \mathsf{A}]\!]$ in (19.12) is $\varnothing$-strict, so we get

$$\begin{array}{rcl}\mathsf{assign}^{\varnothing}[\![\mathsf{x}_k, \mathsf{A}]\!]\, \bot^{\varnothing} & \triangleq & \bot^{\varnothing}\\ \mathsf{test}^{\varnothing}[\![\mathsf{B}]\!]\bot^{\varnothing} & \triangleq & \bot^{\varnothing}\\ \overline{\mathsf{test}}^{\varnothing}[\![\mathsf{B}]\!]\bot^{\varnothing} & \triangleq & \bot^{\varnothing}.\end{array}$$

For the invertible assignment ($c \in \mathbb{F}$ is the value denoted by c and $\mathbf{D} \neq \bot^{\varnothing}$),

$$(\mathsf{assign}^{\varnothing}[\![\mathsf{x}_k, \mathsf{x}_k + \mathsf{c}]\!]\, \mathbf{D})_{ij} \triangleq \begin{cases}\mathbf{D}_{ij} + c & \text{if } i = k \wedge j \neq k\\ \mathbf{D}_{ij} - c & \text{if } i \neq k \wedge j = k\\ \mathbf{D}_{ij} & \text{otherwise}\end{cases} \tag{40.7}$$

we have $\mathsf{assign}^{\vec{r}}[\![\mathsf{x}_k, \mathsf{x}_k + \mathsf{c}]\!] \circ \gamma^{\varnothing} = \gamma^{\varnothing} \circ \mathsf{assign}^{\varnothing}[\![\mathsf{x}_k, \mathsf{x}_k + \mathsf{c}]\!]$.

Proof A short proof sketch is

— $x_k' - x_j \leqslant \mathbf{D}_{kj}$ ⟨precondition before the assignment⟩

$\Rightarrow (x_k - c) - x_j \leqslant \mathbf{D}_{kj}$ ⟨postcondition after the assignment so $x_k = x_k' + c$ so $x_k' = x_k - c$⟩

$\Rightarrow x_k - x_j \leqslant \mathbf{D}_{kj} + c$

— $x_i - x_k' \leqslant \mathbf{D}_{ik}$ ⟨precondition before the assignment⟩

$\Rightarrow x_i - (x_k - c) \leqslant \mathbf{D}_{ik}$ ⟨postcondition after the assignment so $x_k = x_k' + c$ so $x_k' = x_k - c$⟩

$\Rightarrow x_i - x_k + c \leqslant \mathbf{D}_{ik}$

$\Rightarrow x_i - x_k \leqslant \mathbf{D}_{ik} - c$

A formal proof is the following.

$\mathsf{assign}^{\vec{r}}[\![\mathsf{x}_k, \mathsf{x}_k + \mathsf{c}]\!](\gamma^{\varnothing}(\mathbf{D}))$

$= \{\rho[\mathsf{x}_k \leftarrow \mathscr{A}[\![\mathsf{x}_k + \mathsf{c}]\!]\rho] \mid \rho \in \gamma^{\varnothing}(\mathbf{D})\}$ ⟨definition (19.12) of $\mathsf{assign}^{\vec{r}}[\![\mathsf{x}, \mathsf{A}]\!]$⟩

$= \{\rho[\mathsf{x}_k \leftarrow \rho(\mathsf{x}_k) + c] \mid \rho \in \gamma^{\varnothing}(\mathbf{D})\}$ ⟨definition (3.4) of $\mathscr{A}$ and $\mathscr{A}[\![\mathsf{c}]\!]\rho = c$⟩

$= \{\langle x_1, \ldots, x_{k-1}, x_k + c, x_{k+1}, \ldots, x_m\rangle \mid \langle x_1, \ldots, x_k, \ldots, x_m\rangle \in \gamma^{\varnothing}(\mathbf{D})\}$

⟨encoding of variable values in section 40.1.1⟩

$= \{\langle x_1', \ldots, x_k', \ldots, x_m'\rangle \mid \langle x_1', \ldots, x_{k-1}', x_k' - c, x_{k+1}', \ldots, x_m'\rangle \in \gamma^{\varnothing}(\mathbf{D})\}$

⟨letting $x_i' = x_i$ for $k \neq i$ and $x_k' = x_k + c$⟩

$= \{\langle x_1, \ldots, x_k, \ldots, x_m\rangle \mid \forall i,j \in [0,m]\setminus\{k\} \,.\, x_i - x_j \leqslant \mathbf{D}_{ij} \wedge \forall i \in [0,m] \,.\, x_i - (x_k - c) \leqslant \mathbf{D}_{ij} \wedge \forall j \in [0,m] \,.\, (x_k - c) - x_j \leqslant \mathbf{D}_{ij} \wedge x_0 = 0\}$ ⟅definition of γ^{ϖ}⟆

$= \{\langle x_1, \ldots, x_k, \ldots, x_m\rangle \mid \forall i,j \in [0,m]\setminus\{k\} \,.\, x_i - x_j \leqslant \mathbf{D}_{ij} \wedge \forall i \in [0,m] \,.\, x_i - x_k \leqslant \mathbf{D}_{ij} - c \wedge \forall j \in [0,m] \,.\, x_k - x_j \leqslant \mathbf{D}_{ij} + c \wedge x_0 = 0\}$

⟅definition of the totally ordered field $\mathbb{F}$⟆

$= \{\langle x_1, \ldots, x_m\rangle \mid \forall i,j \in [0,m] \,.\, x_i - x_j \leqslant (\mathsf{assign}^{\varpi}[\![\mathsf{x}_k, \mathsf{x}_k + \mathsf{c}]\!]\,\mathbf{D})_{ij} \wedge x_0 = 0\}$

⟅by defining $(\mathsf{assign}^{\varpi}[\![\mathsf{x}_k, \mathsf{x}_k + \mathsf{c}]\!]\,\mathbf{D})_{ij}$ as in (40.7)⟆

$= \gamma^{\varpi}(\mathsf{assign}^{\varpi}[\![\mathsf{x}_k, \mathsf{x}_k + \mathsf{c}]\!]\,\mathbf{D})$ ⟅definition of γ^{ϖ}⟆

Observe that if $\mathbf{D} \in \overline{\mathbb{P}}^{\varpi}$ then $\mathsf{assign}^{\varpi}[\![\mathsf{x}_k, \mathsf{x}_k + \mathsf{c}]\!]\,\mathbf{D} \in \overline{\mathbb{P}}^{\varpi}$ so no normalization is necessary. □

For the elimination of a variable x_k, define $\mathsf{elim}^{\varpi}\ \mathsf{x}_k\ \bot^{\varpi} = \bot^{\varpi}$ and

$$(\mathsf{elim}^{\varpi}\ \mathsf{x}_k\ \mathbf{D})_{ij} \triangleq \begin{cases} 0 & \text{if } i = k \text{ and } j = k \\ \infty & \text{else, if } i = k \text{ or } j = k \\ \mathbf{D}_{ij} & \text{otherwise} \end{cases} \tag{40.8}$$

such that $\mathsf{elim}^{\vec{r}}\ \mathsf{x}_k \circ \gamma^{\varpi} \subseteq / = \gamma^{\varpi} \circ \mathsf{elim}^{\varpi}\ \mathsf{x}_k$ (with $=$ if $\mathbf{D}$ is normalized) where, up to the encoding of variable values in section 40.1.1, $\mathsf{elim}^{\vec{r}}\ \mathsf{x}_k\ P \triangleq \{\langle x_1, \ldots, x_k, v, x_{k+1}, \ldots, x_m\rangle \mid v \in \mathbb{F} \wedge \exists x_k \in \mathbb{F} \,.\, \langle x_1, \ldots, x_k, \ldots, x_m\rangle \in P\}$.

Proof

$\mathsf{elim}^{\vec{r}}\ \mathsf{x}_k\ (\gamma^{\varpi}(\mathbf{D}))$

$= \{\langle x_1, \ldots, x_{k-1}, v, x_{k+1}, \ldots, x_m\rangle \mid v \in \mathbb{F} \wedge \exists x_k \in \mathbb{F} \,.\, \langle x_1, \ldots, x_k, \ldots, x_m\rangle \in \gamma^{\varpi}(\mathbf{D})\}$
⟅definition of $\mathsf{elim}^{\vec{r}}$⟆

$= \{\langle x_1, \ldots, x_{k-1}, v, x_{k+1}, \ldots, x_m\rangle \mid v \in \mathbb{F} \wedge \exists x_k \in \mathbb{F} \,.\, \forall i,j \in [0,m] \,.\, x_i - x_j \leqslant \mathbf{D}_{ij} \wedge x_0 = 0\}$ ⟅definition of γ^{ϖ}⟆

$\subseteq \{\langle x_1, \ldots, x_{k-1}, v, x_{k+1}, \ldots, x_m\rangle \mid v \in \mathbb{F} \wedge \forall i,j \in [0,m]\setminus\{k\} \,.\, x_i - x_j \leqslant \mathbf{D}_{ij} \wedge x_0 = 0\}$

⟅$\mathbf{D} \neq \bot^{\varpi}$, $=$ if $\mathbf{D}$ is normalized⟆

$= \{\langle x_1, \ldots, x_{k-1}, v, x_{k+1}, \ldots, x_m\rangle \mid v \in \mathbb{F} \wedge \forall i,j \in [0,m]\setminus\{k\} \,.\, x_i - x_j \leqslant \mathbf{D}_{ij} \wedge x_0 = 0 \wedge v - v = 0 \wedge \forall i \in [0,m]\setminus\{k\} \,.\, x_i - v \leqslant \infty \wedge \forall j \in [0,m]\setminus\{k\} \,.\, v - x_j \leqslant \infty\}$

⟅definitions of 0 and ∞⟆

$= \{\langle x_1, \ldots, x_k, \ldots, x_m\rangle \mid v \in \mathbb{F} \wedge \forall i,j \in [0,m]\setminus\{k\} \,.\, x_i - x_j \leqslant \mathbf{D}_{ij} \wedge x_0 = 0 \wedge x_k - x_k = 0 \wedge \forall i \in [0,m]\setminus\{k\} \,.\, x_i - x_k \leqslant \infty \wedge \forall j \in [0,m]\setminus\{k\} \,.\, x_k - x_j \leqslant \infty\}$ ⟅letting $v = x_k$⟆

$= \{\langle x_1, \ldots, x_k, \ldots, x_m\rangle \mid \forall i,j \in [0,m] \,.\, x_i - x_j \leqslant \mathsf{elim}^{\varpi}\ x_k\ \mathbf{D}\}$

⟅definition (40.8) of elim^{ϖ}⟆

$= \gamma^{\varpi}(\mathsf{elim}^{\varpi}\ \mathsf{x}_k\ \mathbf{D})$ ⟅definition of γ^{ϖ}⟆

$\mathsf{elim}^{\varpi}\ \mathsf{x}_k\ \mathbf{D} \in \overline{\mathbb{P}}^{\varpi}$ because it satisfies the normal form condition of corollary 39.40 when $\mathbf{D} \in \overline{\mathbb{P}}^{\varpi}$ does. □

For the noninvertible assignment $\ell \neq k$, $\mathbf{D} \neq \bot^{\square}$, and c is the value of constant c,

$$(\mathsf{assign}^{\square}[\![\mathsf{x}_k, \mathsf{x}_\ell + \mathsf{c}]\!]\,\mathbf{D})_{ij} \quad \triangleq \quad \mathsf{let}\ \mathbf{D}' = \mathsf{elim}^{\square}(\mathsf{x}_k)\,\mathbf{D}\ \mathsf{in} \qquad (40.9)$$

$$\begin{cases} \mathbf{D}'_{ij} - c & \text{if } i \neq \ell \text{ and } j = \ell \\ \mathbf{D}'_{ij} + c & \text{if } i = \ell \text{ and } j \neq \ell \\ \mathbf{D}'_{ij} & \text{otherwise} \end{cases}$$

Proof

$\mathsf{assign}^{\vec{r}}[\![\mathsf{x}_k, \mathsf{x}_\ell + \mathsf{c}]\!](\gamma^{\square}(\mathbf{D}))$ $\quad ⦃k \neq \ell⦄$

$= \{\langle x_1, \ldots, x_{k-1}, x_\ell + c, x_{k+1}, \ldots, x_m\rangle \mid \exists x_k . \langle x_1, \ldots, x_{k-1}, x_k, x_{k+1}, \ldots, x_m\rangle \in \gamma^{\square}(\mathbf{D})\}$

⦃definition (19.12) of $\mathsf{assign}^{\vec{r}}[\![\mathsf{x}, \mathsf{A}]\!]$, definition (3.4) of $\mathscr{A}$ and $\mathscr{A}[\![\mathsf{c}]\!]\rho = c$, encoding of variable values in section 40.1.1, and letting v to be the value of variable x_k⦄

$= \{\langle x_1, \ldots, x_{k-1}, x_\ell + c, x_{k+1}, \ldots, x_m\rangle \mid \exists v . \langle x_1, \ldots, x_{k-1}, v, x_{k+1}, \ldots, x_m\rangle \in \{\langle x_1, \ldots, x_k, v, x_{k+1}, \ldots, x_m\rangle \mid v \in \mathbb{F} \wedge \exists x_k \in \mathbb{F} . \langle x_1, \ldots, x_k, \ldots, x_m\rangle \in \gamma^{\square}(\mathbf{D})\}\}$

⦃definition of $\in$ and letting $v = x_k$⦄

$= \{\langle x_1, \ldots, x_{k-1}, x_\ell + c, x_{k+1}, \ldots, x_m\rangle \mid \langle x_1, \ldots, x_m\rangle \in \mathsf{elim}^{\vec{r}}(\mathsf{x}_k)\,(\gamma^{\square}(\mathbf{D}))\}$

⦃definition (40.8) of $\mathsf{elim}^{\vec{r}}$⦄

$= \{\langle x_1, \ldots, x_{k-1}, x_\ell + c, x_{k+1}, \ldots, x_m\rangle \mid \langle x_1, \ldots, x_\ell, \ldots, x_m\rangle \in \gamma^{\square}(\mathsf{elim}^{\square}(\mathsf{x}_k)\,(\mathbf{D}))\}$

⦃definition (40.8) of $\mathsf{elim}^{\square}$⦄

$= \{\langle x'_1, \ldots, x'_{k-1}, x'_\ell, x'_{k+1}, \ldots, x'_m\rangle \mid \langle x'_1, \ldots, x'_{\ell-1}, x'_\ell - c, x'_{\ell+1}, \ldots, x'_m\rangle \in \gamma^{\square}(\mathsf{elim}^{\square}(\mathsf{x}_k)\,(\mathbf{D}))\}$ ⦃letting $x'_i = x_i$ for $i \neq \ell$ and $x'_\ell = x_\ell + c$⦄

$= \{\langle x_1, \ldots, x_\ell, \ldots, x_m\rangle \mid \forall i, j \in [0, m] \setminus \{\ell\} . x_i - x_j \leqslant \mathbf{D}'_{ij} \wedge \forall i \in [0, m] . x_i - (x_\ell - c) \leqslant \mathbf{D}'_{i\ell} \wedge \forall j \in [0, m] . (x_\ell - c) - x_j \leqslant \mathbf{D}'_{\ell j} \wedge x_0 = 0\}$

⦃definition of $\gamma^{\square}$ where $\mathbf{D}' = \mathsf{elim}^{\square}(\mathsf{x}_k)\,(\mathbf{D})$⦄

$= \{\langle x_1, \ldots, x_\ell, \ldots, x_m\rangle \mid \forall i, j \in [0, m] \setminus \{\ell\} . x_i - x_j \leqslant \mathbf{D}'_{ij} \wedge \forall i \in [0, m] . x_i - x_\ell \leqslant \mathbf{D}'_{i\ell} - c \wedge \forall j \in [0, m] . x_\ell - x_j \leqslant \mathbf{D}'_{\ell j} + c \wedge x_0 = 0\}$

⦃definition of + and − in totally ordered field $\mathbb{F}$⦄

$= \{\langle x_1, \ldots, x_m\rangle \mid \forall i, j \in [0, m] . x_i - x_j \leqslant (\mathsf{assign}^{\square}[\![\mathsf{x}_k, \mathsf{x}_\ell + \mathsf{c}]\!]\,\mathbf{D}')_{ij} \wedge x_0 = 0\}$

⦃by defining $(\mathsf{assign}^{\square}[\![\mathsf{x}_k, \mathsf{x}_\ell + \mathsf{c}]\!]\,\mathbf{D}')_{ij}$ as in (40.7)⦄

$= \gamma^{\square}(\mathsf{assign}^{\square}[\![\mathsf{x}_k, \mathsf{x}_k + \mathsf{c}]\!]\,\mathbf{D}')$ ⦃definition of $\gamma^{\square}$⦄ □

As a special case $\ell = 0$ such that $x_0 = 0$, we get

$$(\mathsf{assign}^{\square}[\![\mathsf{x}_k, \mathsf{c}]\!]\,\mathbf{D})_{ij} \quad \triangleq \quad \mathsf{let}\ \mathbf{D}' = \mathsf{elim}^{\square}(\mathsf{x}_k)\,\mathbf{D}\ \mathsf{in} \begin{cases} -c & \text{if } i \neq 0 \text{ and } j = 0 \\ +c & \text{if } i = 0 \text{ and } j \neq 0 \\ \mathbf{D}'_{ij} & \text{otherwise} \end{cases}$$

Otherwise,

$$\mathsf{assign}^{\varpi}[\![\mathsf{x}_k, \mathsf{A}]\!]\,\mathbf{D} \;\triangleq\; \mathsf{elim}^{\varpi}\ \mathsf{x}_k\ \mathbf{D} \tag{40.10}$$

Proof

$$\begin{aligned}
&\mathsf{assign}^{\vec{r}}[\![\mathsf{x}_k, \mathsf{A}]\!](\gamma^{\varpi}(\mathbf{D}))\\
=\ &\{\rho[\mathsf{x}_k \leftarrow \mathscr{A}[\![\mathsf{A}]\!]\rho] \mid \rho \in \gamma^{\varpi}(\mathbf{D})\} && \text{(definition (19.12) of } \mathsf{assign}^{\vec{r}}[\![\mathsf{x}, \mathsf{A}]\!]\text{)}\\
=\ &\{\langle x_1, \ldots, x_{k-1}, \mathscr{A}[\![\mathsf{A}]\!]\langle x_1, \ldots, x_m\rangle, x_{k+1}, \ldots, x_m\rangle \mid \langle x_1, \ldots, x_k, \ldots, x_m\rangle \in \gamma^{\varpi}(\mathbf{D})\}\\
& && \text{(encoding of variable values in section 40.1.1)}\\
\subseteq\ &\{\langle x_1, \ldots, x_{k-1}, v, x_{k+1}, \ldots, x_m\rangle \mid v \in \mathbb{F} \wedge \langle x_1, \ldots, x_k, \ldots, x_m\rangle \in \gamma^{\varpi}(\mathbf{D})\}\\
& && \text{(definition of } \subseteq \text{ and } \mathscr{A}[\![\mathsf{A}]\!]\langle x_1, \ldots, x_m\rangle \in \mathbb{F}\text{)}\\
=\ &\mathsf{elim}^{\vec{r}}\ \mathsf{x}_k\ (\gamma^{\varpi}(\mathbf{D})) && \text{(definition (40.8) of } \mathsf{elim}^{\vec{r}}\text{)}\\
=\ &\gamma^{\varpi}(\mathsf{elim}^{\varpi}\ \mathsf{x}_k\ \mathbf{D}) && \text{(definition (40.8) of } \mathsf{elim}^{\varpi}\text{)}\\
=\ &\gamma^{\varpi}(\mathsf{assign}^{\varpi}[\![\mathsf{x}_k, \mathsf{A}]\!]\,\mathbf{D}) && \text{(definition (40.10) of } \mathsf{assign}^{\varpi}[\![\mathsf{x}_k, \mathsf{A}]\!]\text{)} \quad \square
\end{aligned}$$

Remark 40.11 Following section 38.4.3, nonlinear assignments may be linearizable in one of the form (40.7) or (40.9), in particular when replacing constant variables by their value, as possibly determined by the zone analysis.

40.1.7 The Zone Abstract Test Transformers

For the test (>= is similar and = reduces to two inequalities),

$$(\mathsf{test}^{\varpi}[\![\mathsf{x}_k - \mathsf{x}_\ell <= \mathsf{c}]\!]\mathbf{D})_{ij} \;\triangleq\; \begin{cases} \min(\mathbf{D}_{ij}, c) & \text{if } i = k \text{ and } j = \ell\\ \mathbf{D}_{ij} & \text{otherwise}\end{cases}$$

$$(\mathsf{test}^{\varpi}[\![\mathsf{x}_k <= \mathsf{c}]\!]\mathbf{D})_{ij} \;\triangleq\; \begin{cases} \min(\mathbf{D}_{ij}, c) & \text{if } i = k \text{ and } j = 0\\ \mathbf{D}_{ij} & \text{otherwise}\end{cases}$$

which is $\perp^{\varpi}$ if this creates a cycle with strictly negative weight. In general, tests require renormalization (checking for strictly negative cycles in passing). Otherwise,

$$\mathsf{test}^{\varpi}[\![\mathsf{B}]\!]\mathbf{D} \;\triangleq\; \mathbf{D}$$

$$\overline{\mathsf{test}}^{\varpi}[\![\mathsf{B}]\!]\mathbf{D} \;\triangleq\; \mathbf{D}$$

40.1.8 Zone Widening and Narrowing

The widening and narrowing are inspired by those of intervals in chapter 32 where unstable constraints are eliminated.

$$\perp^{\varpi}\,\nabla^{\varpi}\,\mathbf{D}' \;\triangleq\; \mathbf{D}'$$

$$\mathbf{D}\,\nabla^{\varpi}\,\perp^{\varpi} \;\triangleq\; \mathbf{D}$$

$$(\mathbf{D}\,\nabla^{\varpi}\,\mathbf{D}')_{ij} \;\triangleq\; \begin{cases} \mathbf{D}_{ij} & \text{if } \mathbf{D}'_{ij} \leqslant \mathbf{D}_{ij}\\ \infty & \text{otherwise}\end{cases}$$

$$(\mathbf{D}\,\Delta^{\varpi}\,\mathbf{D}')_{ij} \;\triangleq\; \begin{cases} \mathbf{D}'_{ij} & \text{if } \mathbf{D}_{ij} = \infty\\ \mathbf{D}_{ij} & \text{otherwise}\end{cases}$$

Example 40.12 The widening on zones is similar to the widening on intervals.

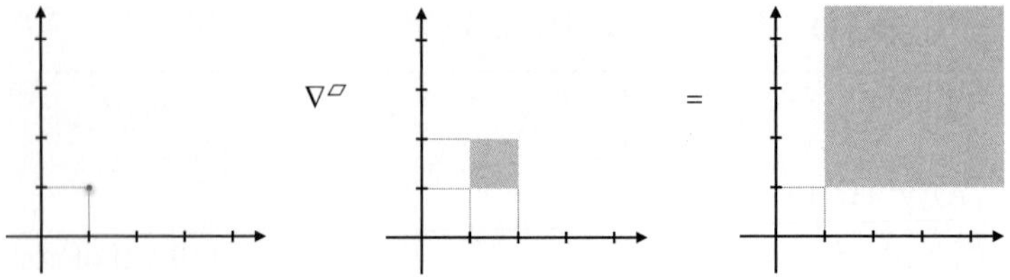

$$\begin{pmatrix} 0 & -1 & -1 \\ 1 & 0 & 0 \\ 1 & 0 & 0 \end{pmatrix} \;\nabla^{\square}\; \begin{pmatrix} 0 & -1 & -1 \\ 2 & 0 & 1 \\ 2 & 1 & 0 \end{pmatrix} \;=\; \begin{pmatrix} 0 & -1 & -1 \\ \infty & 0 & \infty \\ \infty & \infty & 0 \end{pmatrix}$$

`{y>=1, x<=1, y<=x}` `{x>=1, y>=1, x<=2, y<=2}` `{x>=1, y>=1}`

Normalization by saturation may prevent convergence of the iteration sequence with widening.

Example 40.13

$a = \mathsf{RFW}(a)$ $b = \mathsf{RFW}(b)$ $a \nabla^{\square} b$ $\mathsf{RFW}(a \nabla^{\square} b)$

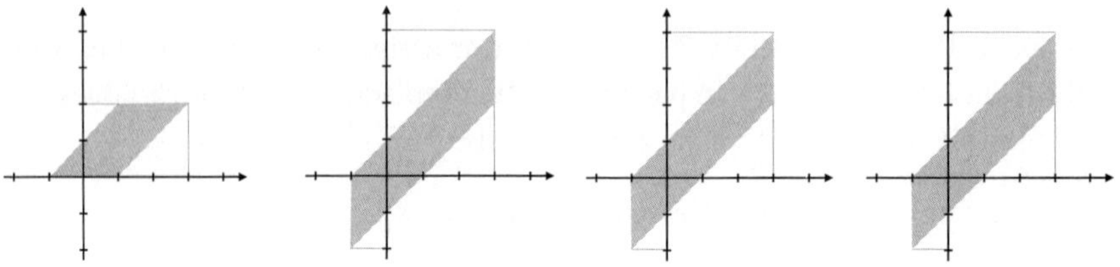

$$\begin{pmatrix} 0 & 1 & 0 \\ 3 & 0 & 1 \\ 2 & 1 & 0 \end{pmatrix} \quad \begin{pmatrix} 0 & 1 & 2 \\ 3 & 0 & 1 \\ 4 & 1 & 0 \end{pmatrix} \quad \begin{pmatrix} 0 & 1 & \infty \\ 3 & 0 & 1 \\ \infty & 1 & 0 \end{pmatrix} \quad \begin{pmatrix} 0 & 1 & 2 \\ 3 & 0 & 1 \\ 4 & 1 & 0 \end{pmatrix}$$

```
{x>=-1, y>=0,       {x>=-1, y>=-2,     {x>=-1, x<=3,          {x>=-1, y>=-2,
 x<=3, x<=y+1,       x<=3, x<=y+1,      x<=y+1, y<=x+1}        x<=3, x<=y+1,
 y<=2, y<=x+1}       y<=4, y<=x+1}                             y<=4, y<=x+1}
```

The problem is that normalization by saturation reintroduces the constraints on y eliminated by the widening to ∞.

A solution is to limit normalizations for widening (all other operations remained normalized to be precise) [692]. It is also possible to use normalization by minimization of constraints [65] but this potentially impairs the precision of incomplete transformers so normalization by minimization is used for the widening only while normalization by saturation is used for the other transformers [65]. Another solution is to use a history widening of section 34.7, where the constraints eliminated by the previous widening are recorded and cannot be reintroduction by the normalization.

The upward iteration with history widening of section 34.7 of $\overline{f} \in \overline{L} \to \overline{L}$ from $a \in \overline{L}$ with widening ∇ and normalization norm on a poset $\langle \overline{L}, \overline{\sqsubseteq} \rangle$ is modified as follows $\langle \hat{f}^{\beta}, \beta \in \mathbb{N}_{\omega} \rangle$ with an abstraction of the iteration history as follows:

1. $\hat{f}^{0} \triangleq \mathsf{norm}(a)$,
2. $\hat{f}^{\beta+1} \triangleq \llbracket \overline{f}(\hat{f}^{\beta}) \mathrel{\overline{\sqsubseteq}} \hat{f}^{\beta} \mathrel{?} \hat{f}^{\beta} \mathrel{:} \mathsf{norm}(\hat{f}^{\beta} \nabla \overline{f}(\hat{f}^{\beta})) \sqcup \bigsqcup_{0 \leqslant \delta < \beta} \hat{f}^{\delta} \nabla \overline{f}(\hat{f}^{\delta}) \rrbracket$ when $\beta \in \mathbb{N}$,

3. $\hat{f}^{\omega} \triangleq \mathsf{norm}(\mathop{\nabla}\limits_{\beta<\omega} \hat{f}^{\beta})$.

Theorem 40.14 The zone history iterates with widening $\nabla^{\square}$ and normalization RFW do converge in finitely many steps (so that case 3 is not used).

Proof of theorem 40.14 If the iteration is not stabilized, the widening extrapolate at least one element of the matrix to ∞. Even if it is eliminated by the normalization, joining with the term $\bigsqcup_{0\leqslant\delta<\beta} \hat{f}^{\delta} \nabla \overline{f}(\hat{f}^{\delta})$ ensures that from the next iterate on, the component will stay stabilized to ∞. Because the matrix is finite of dimension $(m+1)^2$, this ensures convergence in at most $(m+1)^2$ steps. □

Exercise 40.15 (Project: zone abstract domain) Implement the zone abstract domain with abstract interpreter of exercise 34.13.

40.2 Octagon Analysis

Octagons include constraints of the form $x_i + x_j \geqslant c$ (or $-x_i - x_j \leqslant c$) where x_i and x_j are values of variables x_i and x_j and $c \in \mathbb{V} \cup \{\infty\}$ is a constant inferred by the analysis. The encoding proposed by Antoine Miné [690, ch. 4.2] consists of using only constraints of the form $\pm x - \pm y \leqslant c$ where x and y are values of variables and $c \in \mathbb{V}$ is a constant. The translation is the following:

$$\begin{array}{rclcrcl} x \leqslant c & \rightarrow & x-(-x) \leqslant 2c & \quad & x \geqslant c & \rightarrow & (-x)-x \leqslant -2c \\ x-y \leqslant c & \rightarrow & x-y \leqslant c & & x-y \geqslant c & \rightarrow & y-x \leqslant -c \\ x+y \leqslant c & \rightarrow & x-(-y) \leqslant c & & x+y \geqslant c & \rightarrow & (-x)-y \leqslant -c \end{array}$$

This transformation results is distance constraints $x - y \leqslant c$ between values x_i of variables x_i and their opposite value $-x_i$. This can be encoded by a distance matrix $\mathbf{M} \in (\mathbb{V} \cup \{\infty\})^{2m}$ where m is the number of variables where x_i is at line/column position i and x_i is at line/column position $m+i$. The concretization is

$$\gamma^{\circ}(\mathbf{M}) \triangleq \{\langle x_1, \ldots, x_m\rangle \mid \forall i, j \in [1, 2m] \,.\, v(x,i) - v(x,j) \leqslant \mathbf{M}_{ij}\}$$

where $v(x,i) = (\!| i \leqslant m \mathbin{?} x_i \mathbin{:} -x_{i-m} |\!)$. Two different components of the matrix matrix may express the same constraint $x - y \leqslant c \Leftrightarrow (-y) - (-x) \leqslant c$. This is taken into account by requiring that $\forall i, j \in [1, m] \,.\, \mathbf{M}_{ij} = \mathbf{M}_{m+j,m+i}$.

Example 40.16

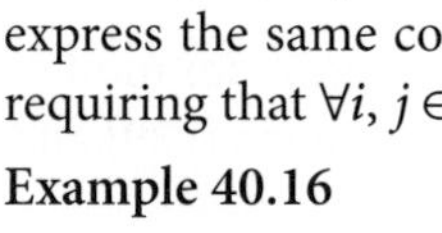


```
x>=2, y>=1,
x<=9, x-y<=4,
y<=8, y-x<=5,
x+y>=4
```

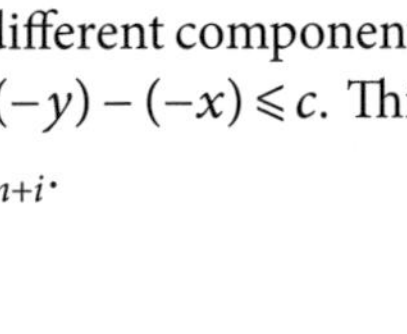

	x	y	−x	−y
x	0	4	−18	∞
y	5	0	∞	16
−x	−4	−4	0	5
−y	∞	16	4	0

The lattice structure is similar to that for zones. The normalization by saturation is performed by a variant of the Roy-Floyd-Warshall algorithm [692]. The assignment and test transformers are also similar to the zone case [692]. The normalization after widening of unstable elements to ∞ may impair convergence and so a history widening must be used.

All operations of the octagon abstract domain are reasonably efficient with quadratic or cubic complexity on the number of variables. However, this does not scale to tens of thousands of variables [268]. Scalability efforts have been done to increase the sparsity of the matrix representation [545, 895].

Exercise 40.17 (Project: octagon analysis) Implement the octagon analysis with the abstract interpreter of exercise 34.13 using an octagon library such as the APRON [529] (http://apron.cri.ensmp.fr/library/), PPL (Parma Polyhedra Library) [69] (http://www.bugseng.com/ppl), or ELINA (https://github.com/eth-sri/ELINA) numerical abstract domain libraries.

40.3 Relational Numerical Abstract Domains

Depending on the required precision of the linear relational analysis, several other linear abstract domains can be used beyond intervals, zones, octagons, and linear equalities such as pentagons [639], parallelotopes [38], octahedra [186], two variables per inequality [892], 4-octahedron $x \lesseqgtr \alpha$, $x - y \lesseqgtr \beta$, $(x - y) - (z - t) \lesseqgtr \lambda$ where $\lesseqgtr \in \{\leqslant, \geqslant\}$ [766, 767], Logahedra [512], gauges [954], and so forth; see [695] for a tutorial on inferring numerical properties of programs. They all consist in considering specific templates of polyhedra [279]. Generalizing predicate abstraction of section 27.7, the templates may be inferred by a pre-analysis or specified by the end user, for example [854, 855]. Unfortunately, the exactly appropriate templates can be found only from the program proof, which is the chicken or the egg causality dilemma. Examples of linear and nonlinear abstract domains are given here.

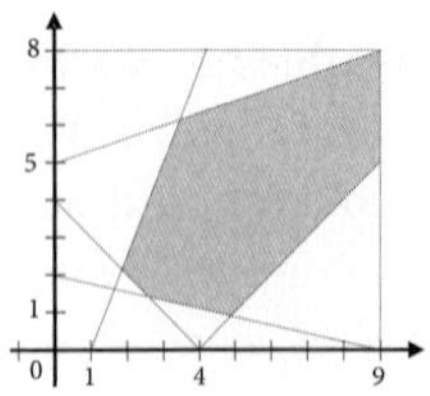

Polyhedra [279]

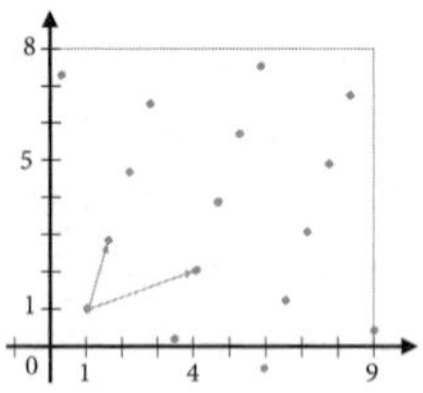

Linear congruences [439]

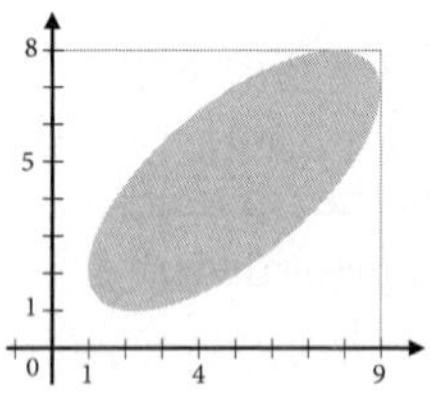

Ellipses [365]

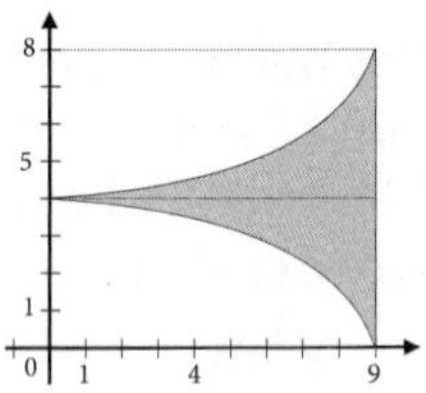

Exponentials [366]

Nonlinear numerical abstract domains must often be considered for program modeling or controlling physical systems, such as ellipses [365, 768, 843] for filters [789] and exponentials [366], introduced in Astrée [113] by Jérôme Feret.

40.4 Conclusion

More details on zones and octagons are provided in [173, 514, 687, 692], details of proofs are found in [690]. [687, 690] used normalization by saturation, and [65] suggested normalization by minimization. [690, definition 3.6.4] proposes an abstraction of linear assignments more precise than (40.10). An implementation of octagons is tested in [153] and another one certified in [544, 546]. Sparse [391, 545] and compact [172] representations have been proposed for the zone distance matrix. All results remain valid when replacing $\leqslant$ by strict equalities $<$, but not both. [794] adds disequality constraints.

Vector instructions may be used for fast implementations of the matrix operations [89]. [402, 403] introduces a policy iteration for zones. [68] consider integer-weighted octagonal graphs. [144] shows that the transitive closure of an octagonal relation is definable in Presburger arithmetic and hence decidable. The octagon domain has been extended to include non-convex properties of absolute value [175].

Zone and octagon abstract domains have many applications such as array index bounds checking of real Java programs [785], complexity analysis [898], program parallelization [937], the static analysis of software patches [306], and the analysis of cryptographic implementations [31]. [409] proposes an abstraction for Extended Berkeley Packet Filter (eBPF) programs that uses the zone abstract domain adapted to track the contents of low-level memory. It shows that the abstraction is both precise and efficient for real-world eBPF programs and is as precise in practice as more costly abstract domains like Octagons and Polyhedra.

Compared to non-relational analyses, zone/octagon-based static analyses decrease the performance but increase the precision, for example, [30]. Zone/octagon and more generally relational analyses hardly scale up for program with thousands of variables. It is therefore necessary to look for relations not between all variables but between some sets of variables (called "packs" or "variable-clusters") of the program. For large programs, these packs may be different at different program points. The packs can be determined statically (before the analysis), for example, on the basis of where relations between variables might be useful (e.g., between array indexes and the symbolic bounds of the array) [102, 268, 485]. The packs may also be determined dynamically during the analysis, as proposed by [484]. A bad packing may considerably degrade the precision of the analysis, for example in [362], it is less precise than intervals.

For a given cluster of variables, [456, 457, 895–897] consider a further decomposition of the relational domains into independent components, thus speeding up the analysis. This same idea applies to all relational abstract domains, the numerical as well as the symbolic ones [278]. Another idea to speed up the computation is to eliminate redundant constraints using machine learning techniques [471].

Relational numerical abstract domains are useful beyond static analysis, for example, in constraint-solving [791, 792]. They have been used, for example, for the Resource-Constrained Project Scheduling Problem (RCPSP) [308, ch. 6, 7] in project scheduling that aims at the minimization of the total project duration [922].

XII SYMBOLIC STATIC ANALYSIS

A static analysis is symbolic when the inferred properties do relate values of variables but the main interest is on discrete structures and not mainly on numerical values.

As examples, we consider data flow analysis (chapter 41), model checking (chapter 44) using a structural stateful semantics (chapter 42), as opposed to a more classic transition semantics (chapter 43); flow-insensitive (chapter 45) points-to analysis (chapter 46); dependency analysis (chapter 47); and typing (chapter 49) based on the symbolic term abstract domain (chapter 48).

41 Dataflow Analysis

Dataflow analysis [568] aims at statically gathering information on the flows of data between variables of a program by solving fixpoint equations over a Noetherian lattice (often Boolean). The classic algorithms of Frances Allen and John Cocke [26–29] were postulated without formal justification. Gary Kildall proposed to reason on paths [572]. The "merge over all paths" abstraction [552] is first postulated on a trace (i.e., a path in a program flowchart by composition of syntactic transfer functions along the path) and then merged (met or joined) over all traces (i.e., all paths in the program flowchart) reaching a program point. The dataflow fixpoint equations are justified by showing that their solution overapproximates the "merge over all paths" abstraction (or is equal for distributive equations). This justification is purely syntactic in that it relies only on occurrences of variables in assignments and expressions, ignoring the values of these variables and the program semantics. In an attempt to provide semantic justifications of the equations [857, section 7] reports that "Some Flow Analyses are Unsound." As observed by [254, section 15.3], the counterexample is based on a confusion between potential (join) and definite (meet) liveness.[1] This calls for an abstract interpretation-based approach to dataflow analysis.

We start by the classic syntactic "merge over all paths" (using a structural abstract semantics instead of equations derived from a flowchart, so that no fixpoint is necessary). We then study the soundness of the syntactic "merge over all paths" with respect to an abstraction of the program semantics. Finally we construct a liveness analysis algorithm by calculational design for which the soundness with respect to the program semantics is stated unambiguously. The result is quite different from classic dataflow analysis [472, 571] and might look puzzling and confusing for people used to this classic presentation of dataflow analysis. In particular, no flowchart, no equations, no fixpoint iteration is needed.

Contents

1. In this chapter, *liveness* is a synonym for "live variable analysis" and completely unrelated to the liveness trace properties of chapter 14, "Safety and Liveness Trace Properties."

41.1 Dead and Live Variables

A classic example of dataflow analysis is Ken Kennedy's live variable analysis [565–567] to overapproximate the set of variables the evaluation of a program component may depend on. It consists [567, Sect. 2.2] in determining whether the value of a variable may be used before being modified in a statement S (in which case it is said to be *live*) or else whether it is definitely modified before being used in the statement (in which case it is said to be *dead*). In the classic, purely syntactic, dataflow analysis, used variables appear in expressions whereas modified variables are assigned to.

Example 41.1 Consider the following statement S on integers (without overflows that could stop execution). Assume that variable x is live and y is dead on exit of the statement S. Liveness at each program point is as follows:

ℓ_1 $\longleftarrow$ x and y dead (modified in ℓ_1 and ℓ_2 before being used)

`x = 1;`

ℓ_2 $\longleftarrow$ x live (used in expression at ℓ_2) and y dead (modified in ℓ_2 before use)

`y = x;`

ℓ_3 $\longleftarrow$ x dead (modified in ℓ_3 before use) and y live (used in expression at ℓ_3)

`x = y + 1;`

ℓ_4 $\longleftarrow$ x and y live (used in assigned expression at ℓ_4)

`x = x - y;`

ℓ_5 $\longleftarrow$ x live and y dead (by hypothesis)

$L_e = \{\mathsf{x}\}$ $\longleftarrow$ hypothesis that x is live and y is dead on statement exit.

Variable x would be dead at program point ℓ_5 with $L_e = \varnothing$ (meaning that no variable is live on exit of the statement). Notice that by a simple constancy analysis of exercise 3.45, x = 1 at ℓ_5 so liveness at ℓ_1 to ℓ_5 does not hold after this program transformation.

This example is straight-line code involving only one execution trace. With conditionals and iterations there may be several execution paths. Then we must distinguish between *potential liveness* (on some of these paths) from *definitive liveness* (on all of these paths). Moreover, the execution of a statement can terminate via its exit or through a break, and both cases have to be considered.

41.2 The Semantic and Syntactic Liveness/Deadness Abstractions

We define the semantic and syntactic liveness/deadness abstractions for a trace and then definite/potential liveness/deadness for a set of traces. The definition is generic in that the notions of "use" and "modify" are left as parameters that will be later instantiated by semantic and syntactic definitions.

41.2.1 The Generic Liveness/Deadness Abstractions

Informally, "a variable is (potentially/definitely) live at some point during execution if it holds a value that is be used in the future before the next time the variable is modified." The liveness abstraction $\alpha^{l}_{use,mod}[\![\mathsf{S}]\!]\, L_b, L_e\, \langle \pi_0, \pi\rangle$ of a program trace π continuing an initial trace π_0 of a program component S is parameterized by

- *use* defining the set $use[\![\mathsf{a}]\!]\rho$ of variables which value is used when executing action a in environment ρ;
- *mod* defining the set $mod[\![\mathsf{a}]\!]\rho$ of variables which value is modified when executing action a in environment ρ.

Liveness depends on the set L_b of variables assumed to be live on exit of the program component S by a break statement and the set L_e on a normal exit after S.

Generic definition of live and dead variables

The set $\alpha^{l}_{use,mod}[\![\mathsf{S}]\!]\, L_b, L_e\, \langle \pi_0, \pi\rangle$ of live variables at the beginning of the execution of π is defined inductively as follows: [a]

$$\alpha^{l}_{use,mod}[\![\mathsf{S}]\!]\, L_b, L_e\, \langle \pi_0, \ell\rangle \triangleq \{\mathsf{x} \in \mathcal{V} \mid (\ell = \mathsf{after}[\![\mathsf{S}]\!] \wedge \mathsf{x} \in L_e) \vee (\mathsf{escape}[\![\mathsf{S}]\!] \wedge \ell = \mathsf{break\text{-}to}[\![\mathsf{S}]\!] \wedge \mathsf{x} \in L_b)\} \quad (41.2)\,(\mathrm{a})$$

$$\alpha^{l}_{use,mod}[\![\mathsf{S}]\!]\, L_b, L_e\, \langle \pi_0, \ell \xrightarrow{\mathsf{a}} \ell'\pi_1\rangle \triangleq \{\mathsf{x} \in \mathcal{V} \mid \mathsf{x} \in use[\![\mathsf{a}]\!]\varrho(\pi_0) \vee (\mathsf{x} \notin mod[\![\mathsf{a}]\!]\varrho(\pi_0) \wedge \mathsf{x} \in \alpha^{l}_{use,mod}[\![\mathsf{S}]\!]\, L_b, L_e\, \langle \pi_0 \mathbin{\widehat{\ }} \ell \xrightarrow{\mathsf{a}} \ell', \ell'\pi_1\rangle)\} \quad (\mathrm{b})$$

The potential and definite liveness at $\mathsf{at}[\![\mathsf{S}]\!]$ are abstractions of the maximal trace semantics $\boldsymbol{\mathcal{S}} = \boldsymbol{\mathcal{S}}^{+\infty}[\![\mathsf{S}]\!]$ is by merge (join or meet) over all traces.

$$\alpha^{\exists l}_{use,mod}[\![\mathsf{S}]\!]\, \boldsymbol{\mathcal{S}}\, L_b, L_e \triangleq \bigcup_{\langle \pi_0, \pi\rangle \in \boldsymbol{\mathcal{S}}} \alpha^{l}_{use,mod}[\![\mathsf{S}]\!]\, L_b, L_e\, \langle \pi_0, \pi\rangle \quad \text{potential liveness} \quad (41.3)$$

$$\alpha^{\forall l}_{use,mod}[\![\mathsf{S}]\!]\, \boldsymbol{\mathcal{S}}\, L_b, L_e \triangleq \bigcap_{\langle \pi_0, \pi\rangle \in \boldsymbol{\mathcal{S}}} \alpha^{l}_{use,mod}[\![\mathsf{S}]\!]\, L_b, L_e\, \langle \pi_0, \pi\rangle \quad \text{definite liveness} \quad (41.4)$$

Potential liveness means live on at least one execution whereas definite liveness means live on all executions. Potential and definite deadness $\mathsf{at}[\![\mathsf{S}]\!]$ are defined dually.

$$\alpha^{\exists d}_{use,mod}[\![\mathsf{S}]\!]\, \boldsymbol{\mathcal{S}}\, D_b, D_e \triangleq \neg\alpha^{\forall l}_{use,mod}[\![\mathsf{S}]\!]\, \boldsymbol{\mathcal{S}}\, \neg D_b, \neg D_e \quad \text{potential deadness} \quad (41.5)$$

$$\alpha^{\forall d}_{use,mod}[\![\mathsf{S}]\!]\, \boldsymbol{\mathcal{S}}\, D_b, D_e \triangleq \neg\alpha^{\exists l}_{use,mod}[\![\mathsf{S}]\!]\, \boldsymbol{\mathcal{S}}\, \neg D_b, \neg D_e \quad \text{definite deadness} \quad (41.6)$$

a. For infinite traces this definition must be understood by passing to the limit as in (7.4) or, equivalently, as a bi-inductive definition of section 16.7 [260].

Exercise 41.7 Prove that the potential liveness abstraction $\boldsymbol{\mathcal{S}} \mapsto \alpha^{\exists l}_{use,mod}[\![\mathsf{S}]\!]\, \boldsymbol{\mathcal{S}}\, L_b, L_e$ in (41.3) and definite deadness abstraction $\boldsymbol{\mathcal{S}} \mapsto \alpha^{\forall d}_{use,mod}[\![\mathsf{S}]\!]\, \boldsymbol{\mathcal{S}}\, D_b, D_e$ in (41.6) define Galois connections.

Unfolding the recursive definition (41.2), we get an iterative definition of the generic potential liveness abstraction.

Lemma 41.8 If $\pi_1 = \ell_1 \xrightarrow{a_1} \ell_2 \xrightarrow{a_2} \dots \xrightarrow{a_{n-1}} \ell_n$ and $\langle \pi_0, \pi_1 \rangle \in \mathcal{S}^*[\![S]\!]$ then

$$\alpha^l_{use,mod}[\![S]\!]\, L_b, L_e \langle \pi_0, \pi_1 \rangle = \{x \in \mathbb{V} \mid \exists i \in [1, n-1] \,.\, \forall j \in [1, i-1] \,.\, x \notin mod[\![a_j]\!]\varrho(\pi_0 \mathbin{\frown} \ell_1 \xrightarrow{a_1} \ell_2 \dots \xrightarrow{a_{j-1}} \ell_j) \wedge x \in use[\![a_i]\!]\varrho(\pi_0 \mathbin{\frown} \ell_1 \xrightarrow{a_1} \ell_2 \dots \xrightarrow{a_{i-1}} \ell_i)\}$$
$$\cup\, (\!|\, \ell_n = \mathsf{after}[\![S]\!] \;?\; L_e \;\text{:}\; \emptyset \,|\!) \cup (\!|\, \mathsf{escape}[\![S]\!] \wedge \ell_n = \mathsf{break\text{-}to}[\![S]\!] \;?\; L_b \;\text{:}\; \emptyset \,|\!).$$

Proof of lemma 41.8 For the basis $n = 1$, only the first clause (a) of (41.2) is applicable with $\pi_1 = \ell_1$, $[1, n-1]$ is empty, and $\alpha^l_{use,mod}[\![S]\!]\, L_b, L_e \langle \pi_0, \pi_1 \rangle = (\!|\, \ell_1 = \mathsf{after}[\![S]\!] \;?\; L_e \;\text{:}\; \emptyset \,|\!) \cup (\!|\, \mathsf{escape}[\![S]\!] \wedge \ell_1 = \mathsf{break\text{-}to}[\![S]\!] \;?\; L_b \;\text{:}\; \emptyset \,|\!)$ which is precisely what is given by lemma 41.8 because $[1, n-1] = \emptyset$, so the first term is empty.

For the induction step $n + 1 > 1$, we have $\pi_1 = \ell_1 \xrightarrow{a_1} \ell_2 \xrightarrow{a_2} \ell_3 \xrightarrow{a_3} \dots \xrightarrow{a_n} \ell_{n+1}$ so by the second clause (b) of (41.2), we get

$$\alpha^l_{use,mod}[\![S]\!]\, L_b, L_e \langle \pi_0, \pi_1 \rangle$$

$$= \{x \in \mathbb{V} \mid x \in use[\![a_1]\!]\varrho(\pi_0) \vee (x \notin mod[\![a_1]\!]\varrho(\pi_0)) \wedge x \in \alpha^l_{use,mod}[\![S]\!]\, L_b, L_e \langle \pi_0 \mathbin{\frown} \ell_1 \xrightarrow{a_1} \ell_2,\ \ell_2 \xrightarrow{a_2} \ell_3 \xrightarrow{a_3} \dots \xrightarrow{a_n} \ell_{n+1} \rangle)\}$$
$\lbag$(41.2) with $n \geqslant 1\rbag$

$$= \{(x \in \mathbb{V} \mid x \in use[\![a_1]\!]\varrho(\pi_0)) \vee (x \notin mod[\![a_1]\!]\varrho(\pi_0) \wedge \exists i \in [2, n] \,.\, \forall j \in [2, i-1] \,.\, x \notin mod[\![a_j]\!]\varrho(\pi_0 \mathbin{\frown} \ell_1 \xrightarrow{a_1} \ell_2 \dots \xrightarrow{a_{j-1}} \ell_j) \wedge x \in use[\![a_i]\!]\varrho(\pi_0 \mathbin{\frown} \ell_1 \xrightarrow{a_1} \ell_2 \dots \xrightarrow{a_{i-1}} \ell_i)) \vee$$
$$(\!|\, \ell_{n+1} = \mathsf{after}[\![S]\!] \;?\; x \in L_e \;\text{:}\; \mathsf{ff} \,|\!) \vee (\!|\, \mathsf{escape}[\![S]\!] \wedge \ell_{n+1} = \mathsf{break\text{-}to}[\![S]\!] \;?\; x \in L_b \;\text{:}\; \mathsf{ff} \,|\!)\}$$

$\lbag$because $\alpha^l_{use,mod}[\![S]\!]\, L_b, L_e \langle \pi_0 \mathbin{\frown} \ell_1 \xrightarrow{a_1}, \ell_2 \xrightarrow{a_2} \dots \xrightarrow{a_n} \ell_{n+1} \rangle = \{x \in \mathbb{V} \mid \exists i \in [2, n] \,.\, \forall j \in [2, i-1] \,.\, x \notin mod[\![a_j]\!]\varrho(\pi_0 \mathbin{\frown} \ell_1 \xrightarrow{a_1} \ell_2 \dots \xrightarrow{a_{j-1}} \ell_j) \wedge x \in use[\![a_i]\!]\varrho(\pi_0 \mathbin{\frown} \ell_1 \xrightarrow{a_1} \ell_2 \dots \xrightarrow{a_{i-1}} \ell_i)\} \cup (\!|\, \ell_{n+1} = \mathsf{after}[\![S]\!] \;?\; L_e \;\text{:}\; \emptyset \,|\!) \cup (\!|\, \mathsf{escape}[\![S]\!] \wedge \ell_{n+1} = \mathsf{break\text{-}to}[\![S]\!] \;?\; L_b \;\text{:}\; \emptyset \,|\!)$ by induction hypothesis for lemma 41.8$\rbag$

$$= \{x \in \mathbb{V} \mid \exists i \in [1, n] \,.\, \forall j \in [1, i-1] \,.\, x \notin mod[\![a_j]\!]\varrho(\pi_0 \mathbin{\frown} \ell_1 \xrightarrow{a_1} \ell_2 \dots \xrightarrow{a_{j-1}} \ell_j) \wedge x \in use[\![a_i]\!]\varrho(\pi_0 \mathbin{\frown} \ell_1 \xrightarrow{a_1} \ell_2 \dots \xrightarrow{a_{i-1}} \ell_i)\} \cup (\!|\, \ell_{n+1} = \mathsf{after}[\![S]\!] \;?\; L_e \;\text{:}\; \emptyset \,|\!) \cup (\!|\, \mathsf{escape}[\![S]\!] \wedge \ell_{n+1} = \mathsf{break\text{-}to}[\![S]\!] \;?\; L_b \;\text{:}\; \emptyset \,|\!)$$

$\lbag$incorporating a_1 in the case $i = 1$ that proves lemma 41.8 for the induction step$\rbag$

We conclude by recurrence on n. □

We also observe that potential liveness (hence dually definite deadness) can be equivalently defined using maximal or prefix traces.

Lemma 41.9 $\alpha^{\exists l}_{use,mod}[\![\mathsf{S}]\!]\,(\boldsymbol{\mathcal{S}}^{+\infty}[\![\mathsf{S}]\!]) = \alpha^{\exists l}_{use,mod}[\![\mathsf{S}]\!]\,(\boldsymbol{\mathcal{S}}^{*}[\![\mathsf{S}]\!])$.

Proof of lemma 41.9 To show that $\alpha^{\exists l}_{use,mod}[\![\mathsf{S}]\!]\,(\boldsymbol{\mathcal{S}}^{+\infty}[\![\mathsf{S}]\!]) = \alpha^{\exists l}_{use,mod}[\![\mathsf{S}]\!]\,(\boldsymbol{\mathcal{S}}^{*}[\![\mathsf{S}]\!])$ we must, by (41.3), prove that

$$A = \bigcup_{\langle\pi_0,\,\pi\rangle\,\in\,\boldsymbol{\mathcal{S}}^{+\infty}[\![\mathsf{S}]\!]} \alpha^{l}_{use,mod}[\![\mathsf{S}]\!]\,L_b, L_e\,\langle\pi_0,\,\pi\rangle = \bigcup_{\langle\pi_0,\,\pi'\rangle\,\in\,\boldsymbol{\mathcal{S}}^{*}[\![\mathsf{S}]\!]} \alpha^{l}_{use,mod}[\![\mathsf{S}]\!]\,L_b, L_e\,\langle\pi_0,\,\pi'\rangle = B.$$

- Assume $\mathsf{x} \in A$ because of some $\langle\pi_0,\,\pi\rangle \in \boldsymbol{\mathcal{S}}^{+\infty}[\![\mathsf{S}]\!]$. There are two cases.
  - Either $\mathsf{x} \in A$ follows from (41.2.a) and so the second alternative in (41.2.b) has always been chosen before reaching the end of the trace π with a label $\ell = \mathsf{after}[\![\mathsf{S}]\!]$ or $\mathsf{escape}[\![\mathsf{S}]\!] = \mathsf{tt}$ and $\ell = \mathsf{break\text{-}to}[\![\mathsf{S}]\!]$. In both cases, π is maximal, $\langle\pi_0,\,\pi\rangle \in \boldsymbol{\mathcal{S}}^{*}[\![\mathsf{S}]\!]$, and so $\mathsf{x} \in B$ by (41.2).
  - Otherwise, $\mathsf{x} \in A$ follows from (41.2.b) where the second alternative has been chosen finitely many times (so x is unmodified) until the first alternative is chosen because x is used. Consider the prefix of π up to that point of use. By (7.4), it is, or an extension of it, π' is in the prefix semantics $\pi' \in \boldsymbol{\mathcal{S}}^{*}[\![\mathsf{S}]\!]$ and so from this trace we derive from (41.2.b) that $\mathsf{x} \in B$.

  It follows that $A \subseteq B$.
- Conversely, assume $\mathsf{x} \in B$. Then there exists $\langle\pi_0,\,\pi'\rangle \in \boldsymbol{\mathcal{S}}^{*}[\![\mathsf{S}]\!]$ such that $\mathsf{x} \in \alpha^{l}_{use,mod}[\![\mathsf{S}]\!]\,L_b, L_e\,\langle\pi_0,\,\pi'\rangle$. Consider a maximal extension of π' so that there exists π'' with $\pi' \cdot \pi'' \in \boldsymbol{\mathcal{S}}^{+\infty}[\![\mathsf{S}]\!]$. There are two cases, depending of whether $\mathsf{x} \in B$ in (41.2.a) or (41.2.b).
  - If $\mathsf{x} \in B$ because of (41.2.a) then the π' ends at $\mathsf{after}[\![\mathsf{S}]\!]$ or at $\mathsf{break\text{-}to}[\![\mathsf{S}]\!]$ and so π' is maximal that is $\pi' \in \boldsymbol{\mathcal{S}}^{+\infty}[\![\mathsf{S}]\!]$ and so $\mathsf{x} \in A$.
  - If $\mathsf{x} \in B$ because of (41.2.b) then $\mathsf{x} \in B$ is used in π' without being modified before and so this is also the case in $\langle\pi_0,\,\pi' \cdot \pi''\rangle \in \boldsymbol{\mathcal{S}}^{+\infty}[\![\mathsf{S}]\!]$, $\pi'' = \ni$, and then $\mathsf{x} \in A$ by (41.2). □

  In both cases, $B \subseteq A$.
- By antisymmetry, $A = B$. □

41.2.2 Definite versus Potential Liveness/Deadness

[857, section 7] claims that some flow analyses are unsound. The program `y=1; while` ℓ `(x==2) x=y;` (where x is initialized to 0) is given as a counterexample to the classic syntactic potential liveness analysis (to be designed in section 41.3). The live variable analysis finds that y is live at ℓ whereas no execution of the program ever uses the value of y. The classic syntactic potential liveness analysis is, by theorem 41.24, an overapproximation, so it is sound to say that y is live at ℓ even though its value is never used later. Therefore, as noticed by [254], this counterexample is not at all a contradiction for potential liveness (but would be for definite liveness, hence the confusion about the counterexample).

[857, section 7] adds "Of course, dataflow practitioners are well aware of the above problem, and disaster does not arise in practice, because live variables analysis is used 'dually'—it is used to detect dead variables." Unfortunately, if potential live variables

(41.3) were unsound then the complement dual definite dead variables (41.6) would be unsound, too.

This leads us to define semantic and syntactic liveness/deadness abstractions and to study the correspondence between the two.

41.2.3 The Semantic Liveness/Deadness Abstractions

Semantically, an action a uses variable y in a given environment ρ if and only if it is possible to change the value of y in order to change the effect of action a on program execution. For an assignment, the assigned value will be changed. For example, $\mathsf{x} \notin \mathsf{use}[\![\mathsf{x} = \mathsf{x}]\!]\,\rho$ and $\mathsf{y} \notin \mathsf{use}[\![\mathsf{x} = \mathsf{y} - \mathsf{y}]\!]\,\rho$. For a test, which has no side effect on the values of variables, the branch taken will be different. Formally,

$$
\begin{aligned}
\mathsf{use}[\![\mathsf{skip}]\!]\,\rho &\triangleq \varnothing \qquad\qquad (41.10)\\
\mathsf{use}[\![\mathsf{break}]\!]\,\rho &\triangleq \varnothing\\
\mathsf{use}[\![\mathsf{x} = \mathsf{A}]\!]\,\rho &\triangleq \{\mathsf{y} \mid \exists \nu \in \mathbb{V} \,.\, \mathscr{A}[\![\mathsf{A}]\!]\,\rho \neq \mathscr{A}[\![\mathsf{A}]\!]\,\rho[\mathsf{y} \leftarrow \nu] \wedge \rho(\mathsf{x}) \neq \mathscr{A}[\![\mathsf{A}]\!]\,\rho\}\\
\mathsf{use}[\![\mathsf{a}]\!]\,\rho &\triangleq \{\mathsf{y} \mid \exists \nu \in \mathbb{V} \,.\, \mathscr{B}[\![\mathsf{a}]\!]\,\rho \neq \mathscr{B}[\![\mathsf{a}]\!]\,\rho[\mathsf{y} \leftarrow \nu]\} \qquad \mathsf{a} \in \{\mathsf{B}, \neg(\mathsf{B})\}
\end{aligned}
$$

Exercise 41.11 What is $\mathsf{use}[\![\mathsf{x} = \mathsf{A}]\!]\,\rho$ when $\mathsf{A} = \mathsf{y}$?

Notice that $\mathsf{x} \in \mathsf{use}[\![\mathsf{a}]\!]$ in (41.10) compares two executions of action a in different environments so that (41.2) is a dependency analysis involving a trace and the abstraction of another one by a different environment (see exercise 47.15).

An action a modifies variable x in an environment ρ if and only if the execution of action a in environment ρ changes the value of x. This corresponds to

$$\mathsf{mod}[\![\mathsf{a}]\!]\,\rho \triangleq \{\mathsf{x} \mid \mathsf{a} = (\mathsf{x} = \mathsf{A}) \wedge (\rho(\mathsf{x}) \neq \mathscr{A}[\![\mathsf{A}]\!]\,\rho)\}.$$

Let us define $\alpha^{l}_{\mathsf{use},\mathsf{mod}}$ by replacing $\langle use,\ mod\rangle$ in the definition (41.2) of $\alpha^{l}_{use,mod}$ by $\langle \mathsf{use},\ \mathsf{mod}\rangle$. The semantic potential liveness abstract semantics is

> *Semantic potential liveness*
>
> $$\boldsymbol{\mathcal{S}}^{\exists l}[\![\mathsf{S}]\!] \triangleq \alpha^{\exists l}_{\mathsf{use},\mathsf{mod}}[\![\mathsf{S}]\!]\,(\boldsymbol{\mathcal{S}}^{+\infty}[\![\mathsf{S}]\!]) \qquad (41.12)$$

(and similarly for the other cases (41.4), (41.5), and (41.6)).

41.2.4 The Classical Syntactic Liveness/Deadness Abstractions

Classical dataflow analysis is purely syntactic, that is, it approximates semantic properties by coarser syntactic ones on the basis of the program syntax only. The set $\mathbb{use}[\![\mathsf{a}]\!]$ of variables used and the set $\mathbb{mod}[\![\mathsf{a}]\!]$ of variables assigned to/modified in an action $\mathsf{a} \in \mathbb{A}$ are postulated to be as follows (the parameter ρ is useless but added for consistency with (41.2)).

$$
\begin{aligned}
\mathbb{use}[\![\mathsf{skip}]\!]\rho &\triangleq \varnothing & \mathbb{mod}[\![\mathsf{skip}]\!]\rho &\triangleq \varnothing \qquad (41.13)\\
\mathbb{use}[\![\mathsf{break}]\!]\rho &\triangleq \varnothing & \mathbb{mod}[\![\mathsf{break}]\!]\rho &\triangleq \varnothing\\
\mathbb{use}[\![\mathsf{x} = \mathsf{A}]\!]\rho &\triangleq \mathbb{vars}[\![\mathsf{A}]\!] & \mathbb{mod}[\![\mathsf{x} = \mathsf{A}]\!]\rho &\triangleq \{\mathsf{x}\}\\
\mathbb{use}[\![\mathsf{B}]\!]\rho &\triangleq \mathbb{vars}[\![\mathsf{B}]\!] & \mathbb{mod}[\![\mathsf{B}]\!]\rho &\triangleq \varnothing\\
\mathbb{use}[\![\neg(\mathsf{B})]\!]\rho &\triangleq \mathbb{vars}[\![\mathsf{B}]\!] & \mathbb{mod}[\![\neg(\mathsf{B})]\!]\rho &\triangleq \varnothing
\end{aligned}
$$

where $\mathbb{vars}[\![E]\!]$ is the set of program variables occurring in expression E, as defined in exercise 3.3.

Let us define $\alpha^{l}_{\mathbb{use},\mathbb{mod}}$ by replacing $\langle use,\ mod\rangle$ in the definition (41.2) of $\alpha^{l}_{use,mod}$ by $\langle \mathbb{use},\ \mathbb{mod}\rangle$. The syntactic potential liveness abstract semantics is

> *Syntactic potential liveness*
>
> $$\boldsymbol{\mathcal{S}}^{\exists \mathbb{l}}[\![\mathsf{S}]\!] \;\triangleq\; \alpha^{\exists l}_{\mathbb{use},\mathbb{mod}}[\![\mathsf{S}]\!]\,(\boldsymbol{\mathcal{S}}^{+\infty}[\![\mathsf{S}]\!]) \qquad (41.14)$$

instantiating (41.3) with *use* as $\mathbb{use}$ and *mod* as $\mathbb{mod}$ (and similarly for the other cases (41.4), (41.5), and (41.6)).

41.2.5 Syntactic versus Semantic Liveness

Because liveness is used by compilers, one would expect that the notion of liveness/deadness is the same for semantically equivalent programs and that syntactic liveness/deadness overapproximates/underapproximate semantic liveness/deadness, that is, $\alpha^{l}_{\mathsf{use,mod}} \mathrel{\dot{\subseteq}} \alpha^{l}_{\mathbb{use},\mathbb{mod}}$. The following two examples show that this is not the case. The problem comes from the fact that the semantic definition takes values into account that the abstract syntactic and hence the resulting dataflow analysis algorithm captures incorrectly.

Example 41.15 Consider $\mathsf{S} \triangleq \mathtt{if}\,\ell_1\,(\mathsf{x}==\mathsf{0})\,\ell_2\,\mathsf{x}=\mathsf{x}-\mathsf{x}\,;$ where x is dead on exit. The classic equational and path-based definitions of potential and definite liveness both yield x is live at ℓ_1 and ℓ_2.

Semantically, x is used and hence live at ℓ_1 (the values 0 and 1 of x yield different values of the Boolean expression $\mathsf{x}{=}{=}\mathsf{0}$). However, x is not used at ℓ_2. If program point ℓ_2 is reached during execution, by the test, x must be 0 at ℓ_1. This value is not modified while the program execution reaches ℓ_2. Therefore the assignment of $\mathsf{x}-\mathsf{x}=0$ does not modify the value of x. It follows that x is not semantically potentially or definitely live at ℓ_2, so it is dead at ℓ_2.

Notice that by constancy analysis of exercise 3.45 and section 33.9, a compiler might optimize $\mathtt{if}\,\ell_1\,(\mathsf{x}==\mathsf{0})\,\ell_2\,\mathsf{x}=\mathsf{x}-\mathsf{x}\,;$ into $\mathtt{if}\,\ell_1\,(\mathsf{x}==\mathsf{0})\,\ell_2\,\mathsf{x}=\mathsf{0}\,;$ and then $\mathtt{if}\,\ell_1\,(\mathsf{x}==\mathsf{0})\,\ell_2;$ that is $\ell_1;$ (skip) in which x is no longer live at ℓ_1. This shows that liveness is not preserved by program transformation.

Example 41.16 Consider the statement $\mathsf{S} \triangleq \ell_0\ \mathsf{x}=\mathsf{y}-\mathsf{y}\,;\ell_1$ where x is live on exit at ℓ_1. We have

$$\boldsymbol{\mathcal{S}}^{+\infty}[\![\mathsf{S}]\!] = \{\langle \pi_0\ell_0,\ \ell_0 \xrightarrow{\ \mathsf{x}=\mathsf{y}-\mathsf{y}=0\ } \ell_1\rangle \mid \pi_0\ell_0 \in \mathbb{T}^+\}$$

$$\mathsf{use}[\![\mathsf{x}=\mathsf{y}-\mathsf{y}]\!]\,\rho = \varnothing$$

$$\mathsf{mod}[\![\mathsf{x}=\mathsf{y}-\mathsf{y}]\!]\,\rho = \{\mathsf{x} \mid \rho(\mathsf{x}) \neq 0\}$$

$$\alpha^{l}_{\mathsf{use,mod}}[\![\mathsf{S}]\!]\,\varnothing,\{\mathsf{x}\}\,\langle \pi_0\ell_0,\ \ell_0 \xrightarrow{\ \mathsf{x}=\mathsf{y}-\mathsf{y}=0\ } \ell_1\rangle = \{\mathsf{x} \mid \boldsymbol{\varrho}(\pi_0\ell_0)(\mathsf{x}) = 0\}$$

$$\boldsymbol{\mathcal{S}}^{\exists \mathsf{l}}[\![\mathsf{S}]\!]\,\varnothing,\{\mathsf{x}\} = \alpha^{\exists l}_{\mathsf{use,mod}}[\![\mathsf{S}]\!]\,(\boldsymbol{\mathcal{S}}^{+\infty}[\![\mathsf{S}]\!])\,\varnothing,\{\mathsf{x}\} \;=\; \{\mathsf{x}\}$$

$$\mathbb{use}[\![x = y - y]\!]\rho = \{y\}$$
$$\mathbb{mod}[\![x = y - y]\!]\rho = \{x\}$$
$$\alpha^{l}_{\mathsf{use},\mathbb{mod}}[\![S]\!]\, \varnothing, \{x\}\, \langle \pi_0\ell_0,\ \ell_0 \xrightarrow{x = y - y = 0} \ell_1\rangle = \{y\}$$
$$\mathcal{S}^{\exists l}[\![S]\!]\, \varnothing, \{x\} = \alpha^{\exists l}_{\mathsf{use},\mathbb{mod}}[\![S]\!]\, (\mathcal{S}^{+\infty}[\![S]\!])\, \varnothing, \{x\} = \{y\}$$

Therefore $\mathcal{S}^{\exists l}[\![S]\!]\, \varnothing, \{x\} \not\subseteq \mathcal{S}^{\exists l}[\![S]\!]\, \varnothing, \{x\}$. This is contrary to soundness, expecting that the set $\mathcal{S}^{\exists l}[\![S]\!]$ of potentially live variables determined syntactically by (41.14) is a pointwise overapproximation of the set $\mathcal{S}^{\exists l}[\![S]\!]$ of potentially live variables determined semantically by (41.12).

41.2.6 Unsoundness of the Syntactic Liveness/Deadness Abstractions

The problem at the origin of unsoundness $\mathcal{S}^{\exists l}[\![S]\!] \dot{\not\subseteq} \mathcal{S}^{\exists l}[\![S]\!]$[2] is that

$$\exists \rho \in \mathbb{E}\mathrm{v} \,.\, y \in \mathsf{use}[\![a]\!]\, \rho \Rightarrow \forall \rho \in \mathbb{E}\mathrm{v} \,.\, y \in \mathbb{use}[\![a]\!]\, \rho \tag{41.17}$$

but in general not conversely, as shown by example 41.16, $\exists \rho \in \mathbb{E}\mathrm{v} \,.\, x \in \mathbb{mod}[\![a]\!]\, \rho \wedge x \notin \mathsf{mod}[\![a]\!]\, \rho$.

Proof of (41.17) Let us first remark that if $x \notin \mathbb{vars}[\![B]\!]$ and $\forall y \in \mathbb{V} \setminus \{x\} \,.\, \rho'(y) = \rho(y)$ then $\mathcal{B}[\![B]\!]\rho = \mathcal{B}[\![B]\!]\rho'$ and similarly for arithmetic expressions.

(41.17) is trivial for skip because $\mathsf{use}[\![\mathsf{skip}]\!]\, \rho\, y = \mathrm{ff}$ in (41.10) and similarly for break. Otherwise, by contraposition, assume that $y \notin \mathbb{use}[\![a]\!]\rho$.

— If $a = x = A$ then $y \notin \mathbb{vars}[\![A]\!]$ by (41.13) so $\forall \nu \in \mathbb{V} \,.\, \mathcal{A}[\![A]\!]\, \rho = \mathcal{A}[\![A]\!]\, \rho[y \leftarrow \nu]$, proving $\neg(\mathsf{use}[\![x = A]\!]\, \rho\, y)$ by (41.10).

— Similarly if $a = B$ or $a = \neg(B)$, then changing y does not change the value of the Boolean expression so y is not semantically used by (41.10). □

41.2.7 Soundness of the Syntactic Liveness/Deadness Abstractions with Respect to the Revised Syntactic/Semantic Liveness/Deadness Abstractions

To fix the problem $\mathcal{S}^{\exists l}[\![S]\!] \dot{\not\subseteq} \mathcal{S}^{\exists l}[\![S]\!]$, we can either find an alternative for $\alpha^{\exists l}_{\mathsf{use},\mathsf{mod}}$ or for $\alpha^{\exists l}_{\mathbb{use},\mathbb{mod}}$. An alternative to the syntactic liveness abstraction $\alpha^{\exists l}_{\mathbb{use},\mathbb{mod}}$ would not be reasonable, because this abstraction is the basis for the classic potential live variable algorithm and we want to justify this algorithm because all compilers use it. Therefore we have to find an alternative for the semantic liveness abstraction $\alpha^{\exists l}_{\mathsf{use},\mathsf{mod}}$ in order to explain exactly in what sense the unchanged classic potential live variable algorithm is sound for the program semantics (even if this is not the most semantically intuitive one (41.12)). The basis for that change is the following lemma 41.18 in which the use use is semantics and the modification $\mathbb{mod}$ is syntactic (meaning an assignment). $\alpha^{l}_{\mathsf{use},\mathbb{mod}}$ is defined by replacing $\langle use,\ mod\rangle$ in the definition (41.2) of $\alpha^{l}_{use,mod}$ by $\langle \mathsf{use},\ \mathbb{mod}\rangle$.

Lemma 41.18 $\alpha^{\exists l}_{\mathsf{use},\mathbb{mod}}\, (\mathcal{S}^{+\infty}[\![S]\!]) \,\dot{\subseteq}\, \alpha^{\exists l}_{\mathbb{use},\mathbb{mod}}[\![S]\!]\, (\mathcal{S}^{+\infty}[\![S]\!])$.

2. $f \dot{\not\subseteq} g \triangleq \neg(f \dot{\subseteq} g)$.

Proof of lemma 41.18 We first prove that $\alpha^{l}_{\mathsf{use,mod}}[\![\mathsf{S}]\!] \mathrel{\dot{\subseteq}} \alpha^{l}_{\mathsf{use,mod}}[\![\mathsf{S}]\!]$. We proceed by induction (more precisely bi-induction [260] to account for infinite traces).

— For the basis

$$\begin{aligned}
&\alpha^{l}_{\mathsf{use,mod}}[\![\mathsf{S}]\!]\, L_b, L_e\, \langle \pi_0, \ell\rangle \\
&= \{\mathsf{x} \in \mathbb{V} \mid (\ell = \mathsf{after}[\![\mathsf{S}]\!] \wedge \mathsf{x} \in L_e) \vee (\mathsf{escape}[\![\mathsf{S}]\!] \wedge \ell = \mathsf{break\text{-}to}[\![\mathsf{S}]\!] \wedge \mathsf{x} \in L_b)\} && \text{(41.2)} \\
&\subseteq \alpha^{l}_{\mathsf{use,mod}}[\![\mathsf{S}]\!]\, L_b, L_e\, \langle \pi_0, \ell\rangle && \text{(41.2) and } \subseteq \text{ reflexive}
\end{aligned}$$

— For the induction step

$$\begin{aligned}
&\alpha^{l}_{\mathsf{use,mod}}[\![\mathsf{S}]\!]\, L_b, L_e\, \langle \pi_0, \langle \pi_0, \ell \xrightarrow{\mathsf{a}} \ell'\pi_1\rangle\rangle \\
&= \{\mathsf{x} \in \mathbb{V} \mid \mathsf{x} \in \mathsf{use}[\![\mathsf{a}]\!]\varrho(\pi_0) \vee (\mathsf{x} \notin \mathsf{mod}[\![\mathsf{a}]\!]\varrho(\pi_0) \wedge \mathsf{x} \in \alpha^{l}_{\mathsf{use,mod}}[\![\mathsf{S}]\!]\, L_b, L_e\, \langle \pi_0 \mathbin{\frown} \ell \xrightarrow{\mathsf{a}} \ell', \\
&\qquad \ell'\pi_1\rangle)\} && \text{(41.2)} \\
&\subseteq \{\mathsf{x} \in \mathbb{V} \mid \mathsf{x} \in \mathsf{use}[\![\mathsf{a}]\!]\varrho(\pi_0) \vee (\mathsf{x} \notin \mathsf{mod}[\![\mathsf{a}]\!]\varrho(\pi_0) \wedge \mathsf{x} \in \alpha^{l}_{\mathsf{use,mod}}[\![\mathsf{S}]\!]\, L_b, L_e\, \langle \pi_0 \mathbin{\frown} \ell \xrightarrow{\mathsf{a}} \ell', \\
&\qquad \ell'\pi_1\rangle)\} && \text{(41.17)} \\
&\subseteq \{\mathsf{x} \in \mathbb{V} \mid \mathsf{x} \in \mathsf{use}[\![\mathsf{a}]\!]\varrho(\pi_0) \vee (\mathsf{x} \notin \mathsf{mod}[\![\mathsf{a}]\!]\varrho(\pi_0) \wedge \mathsf{x} \in \alpha^{l}_{\mathsf{use,mod}}[\![\mathsf{S}]\!]\, L_b, L_e\, \langle \pi_0 \mathbin{\frown} \ell \xrightarrow{\mathsf{a}} \ell', \\
&\qquad \ell'\pi_1\rangle)\} && \text{induction hypothesis} \\
&= \alpha^{l}_{\mathsf{use,mod}}[\![\mathsf{S}]\!]\, L_b, L_e\, \langle \pi_0, \langle \pi_0, \ell \xrightarrow{\mathsf{a}} \ell'\pi_1\rangle\rangle && \text{(41.2)}
\end{aligned}$$

It follows that $\alpha^{\exists l}_{\mathsf{use,mod}}[\![\mathsf{S}]\!] \mathrel{\dot{\subseteq}} \alpha^{\exists l}_{\mathsf{use,mod}}[\![\mathsf{S}]\!]$, pointwise.

$$\begin{aligned}
&\alpha^{\exists l}_{\mathsf{use,mod}}[\![\mathsf{S}]\!]\, \mathcal{S}\, L_b, L_e \\
&= \bigcup_{\langle \pi_0, \pi\rangle \in \mathcal{S}} \alpha^{l}_{\mathsf{use,mod}}[\![\mathsf{S}]\!]\, L_b, L_e\, \langle \pi_0, \pi\rangle && \text{(41.3)} \\
&\subseteq \bigcup_{\langle \pi_0, \pi\rangle \in \mathcal{S}} \alpha^{l}_{\mathsf{use,mod}}[\![\mathsf{S}]\!]\, L_b, L_e\, \langle \pi_0, \pi\rangle && \alpha^{l}_{\mathsf{use,mod}}[\![\mathsf{S}]\!] \mathrel{\dot{\subseteq}} \alpha^{l}_{\mathsf{use,mod}}[\![\mathsf{S}]\!] \\
&= \alpha^{\exists l}_{\mathsf{use,mod}}[\![\mathsf{S}]\!]\, \mathcal{S}\, L_b, L_e && \text{(41.3)} \quad \square
\end{aligned}$$

This leads us to define

Semantico-syntactic potential liveness

$$\mathcal{S}^{\exists l}[\![\mathsf{S}]\!] \;\triangleq\; \alpha^{\exists l}_{\mathsf{use,mod}}[\![\mathsf{S}]\!]\, (\mathcal{S}^{+\infty}[\![\mathsf{S}]\!]) \tag{41.19}$$

and to adopt a semantico-syntaxic definition of soundness (weaker than the semantic definition $\mathcal{S}^{\exists l}[\![\mathsf{S}]\!] \not\subseteq \mathcal{S}^{\exists l}[\![\mathsf{S}]\!]$) as follows

Definition 41.20 (soundness of syntactic liveness) A potential live variable algorithm $\hat{\mathcal{S}}^{\exists l}[\![\mathsf{S}]\!]$ is sound if and only if

$$\mathcal{S}^{\exists l}[\![\mathsf{S}]\!] \;\dot{\subseteq}\; \hat{\mathcal{S}}^{\exists l}[\![\mathsf{S}]\!].$$

so that the classic potential live variable algorithm is sound because it overapproximates $\alpha^{\exists l}_{\mathsf{use,mod}}[\![\mathsf{S}]\!]\, (\mathcal{S}^{+\infty}[\![\mathsf{S}]\!])$ (but in general, the algorithm is incomplete).

Theorem 41.21 If $\alpha^{\exists l}_{\mathsf{use},\mathsf{mod}}[\![\mathsf{S}]\!]\ (\mathcal{S}^{+\infty}[\![\mathsf{S}]\!]) \dot{\subseteq} \mathcal{S}^{\exists l}[\![\mathsf{S}]\!]$ then $\mathcal{S}^{\exists l}[\![\mathsf{S}]\!] \dot{\subseteq} \mathcal{S}^{\exists l}[\![\mathsf{S}]\!]$ so a potential live variable algorithm $\widehat{\mathcal{S}}^{\exists l}[\![\mathsf{S}]\!]$ is sound if $\alpha^{\exists l}_{\mathsf{use},\mathsf{mod}}[\![\mathsf{S}]\!]\ (\mathcal{S}^{*}[\![\mathsf{S}]\!]) \dot{\subseteq} \widehat{\mathcal{S}}^{\exists l}[\![\mathsf{S}]\!]$.

Proof of theorem 41.21 If $\alpha^{\exists l}_{\mathsf{use},\mathsf{mod}}[\![\mathsf{S}]\!]\ (\mathcal{S}^{+\infty}[\![\mathsf{S}]\!]) \dot{\subseteq} \mathcal{S}^{\exists l}[\![\mathsf{S}]\!]$ then $\alpha^{\exists l}_{\mathsf{use},\mathsf{mod}}[\![\mathsf{S}]\!]\ (\mathcal{S}^{+\infty}[\![\mathsf{S}]\!]) \dot{\subseteq} \mathcal{S}^{\exists l}[\![\mathsf{S}]\!]$ by lemma 41.18 and therefore, by (41.19), $\mathcal{S}^{\exists l}[\![\mathsf{S}]\!] \triangleq \alpha^{\exists l}_{\mathsf{use},\mathsf{mod}}\ (\mathcal{S}^{+\infty}[\![\mathsf{S}]\!]) \dot{\subseteq} \mathcal{S}^{\exists l}[\![\mathsf{S}]\!]$. □

However, the program transformations that preserve use and mod but not mod may not preserve liveness.

The other cases $\mathcal{S}^{\forall l}[\![\mathsf{S}]\!]$, $\mathcal{S}^{\exists d}[\![\mathsf{S}]\!]$, and $\mathcal{S}^{\forall d}[\![\mathsf{S}]\!]$ are similar.

41.3 Calculational Design of the Structural Syntactic Potential Liveness Static Analysis

By theorem 41.21, we must design a liveness inference algorithm $\widehat{\mathcal{S}}^{\exists l}[\![\mathsf{S}]\!]$ so that

$$\alpha^{\exists l}_{\mathsf{use},\mathsf{mod}}[\![\mathsf{S}]\!]\ (\mathcal{S}^{*}[\![\mathsf{S}]\!]) \dot{\subseteq} \widehat{\mathcal{S}}^{\exists l}[\![\mathsf{S}]\!].$$

Hence we can construct this algorithm $\widehat{\mathcal{S}}^{\exists l}[\![\mathsf{S}]\!]$ by a calculus aiming at simplifying the term $\alpha^{\exists l}_{\mathsf{use},\mathsf{mod}}[\![\mathsf{S}]\!]\ (\mathcal{S}^{*}[\![\mathsf{S}]\!])$. Because the semantics $\mathcal{S}^{*}[\![\mathsf{S}]\!]$ is structural, we get a structural algorithm that proceeds by elimination, without any fixpoint iteration. We first give the result in (41.22) and then show the systematic calculational design [225]. We omit the unused parameter of use and mod.

Notice that although the semantics is forward, the analysis is backward (see e.g. the statement list and iteration). Moreover, contrary to the classic algorithm, there is neither an iterative fixpoint computation nor two or more passes over the control flow graph as in [29, 130, 848].

Structural syntactic potential liveness analysis

$$\begin{aligned}
\widehat{\mathcal{S}}^{\exists l}[\![\mathtt{Sl}\ \ell]\!]\, L_e &\triangleq \widehat{\mathcal{S}}^{\exists l}[\![\mathtt{Sl}\ \ell]\!]\, \varnothing, L_e \qquad (41.22)\\
\widehat{\mathcal{S}}^{\exists l}[\![\mathsf{x} = \mathsf{E}\ ;]\!]\, L_b, L_e &\triangleq \mathsf{use}[\![\mathsf{x} = \mathsf{E}]\!] \cup (L_e \setminus \mathsf{mod}[\![\mathsf{x} = \mathsf{E}]\!])\\
\widehat{\mathcal{S}}^{\exists l}[\![;]\!]\, L_b, L_e &\triangleq L_e\\
\widehat{\mathcal{S}}^{\exists l}[\![\mathtt{Sl}'\ \mathsf{S}]\!]\, L_b, L_e &\triangleq \widehat{\mathcal{S}}^{\exists l}[\![\mathtt{Sl}']\!]\, L_b, (\widehat{\mathcal{S}}^{\exists l}[\![\mathsf{S}]\!]\, L_b, L_e)\\
\widehat{\mathcal{S}}^{\exists l}[\![\epsilon]\!]\, L_b, L_e &\triangleq L_e\\
\widehat{\mathcal{S}}^{\exists l}[\![\mathtt{if}\ (\mathsf{B})\ \mathsf{S}_t]\!]\, L_b, L_e &\triangleq \mathsf{use}[\![\mathsf{B}]\!] \cup L_e \cup \widehat{\mathcal{S}}^{\exists l}[\![\mathsf{S}_t]\!]\, L_b, L_e\\
\widehat{\mathcal{S}}^{\exists l}[\![\mathtt{if}\ (\mathsf{B})\ \mathsf{S}_t\ \mathtt{else}\ \mathsf{S}_f]\!]\, L_b, L_e &\triangleq \mathsf{use}[\![\mathsf{B}]\!] \cup \widehat{\mathcal{S}}^{\exists l}[\![\mathsf{S}_t]\!]\, L_b, L_e \cup \widehat{\mathcal{S}}^{\exists l}[\![\mathsf{S}_f]\!]\, L_b, L_e\\
\widehat{\mathcal{S}}^{\exists l}[\![\mathtt{while}\ (\mathsf{B})\ \mathsf{S}_b]\!]\, L_b, L_e &\triangleq \mathsf{use}[\![\mathsf{B}]\!] \cup L_e \cup \widehat{\mathcal{S}}^{\exists l}[\![\mathsf{S}_b]\!]\, L_b, L_e\\
\widehat{\mathcal{S}}^{\exists l}[\![\mathtt{break}\ ;]\!]\, L_b, L_e &\triangleq L_b\\
\widehat{\mathcal{S}}^{\exists l}[\![\{\ \mathtt{Sl}\ \}]\!]\, L_b, L_e &\triangleq \widehat{\mathcal{S}}^{\exists l}[\![\mathtt{Sl}]\!]\, L_b, L_e
\end{aligned}$$

Exercise 41.23 Postulate a structural definition of syntactic definite liveness.

Theorem 41.24 $\widehat{\boldsymbol{S}}^{\exists l}[\![\mathsf{S}]\!]$ defined by (41.22) is sound according to definition 41.20.

Proof of theorem 41.24 We design $\widehat{\boldsymbol{S}}^{\exists l}[\![\mathsf{S}]\!]$ by structural induction on S. The calculation design is such that $\alpha^{\exists l}_{\mathsf{use,mod}}[\![\mathsf{S}]\!]\,(\boldsymbol{S}^*[\![\mathsf{S}]\!]) \mathrel{\dot{\subseteq}} \widehat{\boldsymbol{S}}^{\exists l}[\![\mathsf{S}]\!]$. Then theorem 41.24 follows from theorem 41.21.

- For the ***assignment*** $\mathsf{S} ::= \ell\ \mathsf{x} = \mathsf{E}\,;$, by (17.2) (following (6.11) and (6.16)), we have $\boldsymbol{S}^*[\![\mathsf{S}]\!] = \{\langle\pi_0\mathsf{at}[\![\mathsf{S}]\!],\ \mathsf{at}[\![\mathsf{S}]\!]\rangle\} \cup \{\langle\pi_0\mathsf{at}[\![\mathsf{S}]\!],\ \mathsf{at}[\![\mathsf{S}]\!] \xrightarrow{\mathsf{x}\,=\,\mathsf{E}\,=\,\mathscr{E}[\![\mathsf{E}]\!]\varrho(\pi_0\mathsf{at}[\![\mathsf{S}]\!])} \mathsf{after}[\![\mathsf{S}]\!]\rangle\}$. Let us calculate

$$\alpha^{\exists l}_{\mathsf{use,mod}}[\![\mathsf{S}]\!](\boldsymbol{S}^*[\![\mathsf{S}]\!])\,L_b, L_e$$

$$= \bigcup\{\alpha^{l}_{\mathsf{use,mod}}[\![\mathsf{S}]\!]\,L_b, L_e\,\langle\pi_0,\ \pi_1\rangle \mid \langle\pi_0,\ \pi_1\rangle \in \widehat{\boldsymbol{S}}^*[\![\mathsf{S}]\!]\} \quad \text{(definition (41.3) of } \alpha^{\exists l}_{\mathsf{use,mod}}[\![\mathsf{S}]\!]\text{)}$$

$$= \bigcup\{\alpha^{l}_{\mathsf{use,mod}}[\![\mathsf{S}]\!]\,L_b, L_e\,\langle\pi_0\mathsf{at}[\![\mathsf{S}]\!],\ \mathsf{at}[\![\mathsf{S}]\!]\rangle\} \cup \bigcup\{\alpha^{l}_{\mathsf{use,mod}}[\![\mathsf{S}]\!]\,L_b, L_e\,\langle\pi_0\mathsf{at}[\![\mathsf{S}]\!],\ \mathsf{at}[\![\mathsf{S}]\!] \xrightarrow{\mathsf{x}\,=\,\mathsf{E}\,=\,\mathscr{E}[\![\mathsf{E}]\!]\varrho(\pi_0\mathsf{at}[\![\mathsf{S}]\!])} \mathsf{after}[\![\mathsf{S}]\!]\rangle\} \quad \text{(definition of } \boldsymbol{S}^*[\![\mathsf{S}]\!] \text{ for } \mathsf{S} ::= \ell\ \mathsf{x} = \mathsf{E}\,;\text{)}$$

$$= \bigcup\{\alpha^{l}_{\mathsf{use,mod}}[\![\mathsf{S}]\!]\,L_b, L_e\,\langle\pi_0\mathsf{at}[\![\mathsf{S}]\!],\ \mathsf{at}[\![\mathsf{S}]\!] \xrightarrow{\mathsf{x}\,=\,\mathsf{E}\,=\,\mathscr{E}[\![\mathsf{E}]\!]\varrho(\pi_0\mathsf{at}[\![\mathsf{S}]\!])} \mathsf{after}[\![\mathsf{S}]\!]\rangle\}$$
$$\text{(definition (41.2.a) of } \alpha^{l}_{\mathsf{use,mod}}[\![\mathsf{S}]\!]\,L_b, L_e\,\langle\pi_0\mathsf{at}[\![\mathsf{S}]\!],\ \mathsf{at}[\![\mathsf{S}]\!]\rangle = \varnothing\text{)}$$

$$= \bigcup\{\mathsf{y} \in \mathbb{V} \mid \mathsf{y} \in \mathsf{use}[\![\mathsf{x} = \mathsf{E}]\!]\varrho(\pi_0\mathsf{at}[\![\mathsf{S}]\!]) \vee (\mathsf{y} \notin \mathsf{mod}[\![\mathsf{x} = \mathsf{E}]\!]\varrho(\pi_0\mathsf{at}[\![\mathsf{S}]\!]) \wedge \mathsf{y} \in \alpha^{l}_{\mathsf{use,mod}}[\![\mathsf{S}]\!]\,L_b, L_e\,\langle\pi_0\mathsf{at}[\![\mathsf{S}]\!] \mathbin{\frown} \mathsf{at}[\![\mathsf{S}]\!] \xrightarrow{\mathsf{x}\,=\,\mathsf{E}\,=\,\mathscr{E}[\![\mathsf{E}]\!]\varrho(\pi_0\mathsf{at}[\![\mathsf{S}]\!])} \mathsf{after}[\![\mathsf{S}]\!],\ \mathsf{after}[\![\mathsf{S}]\!]\rangle)\}$$
$$\text{(definition (41.2.b) of } \alpha^{l}_{\mathsf{use,mod}}\,L_b, L_e\,\langle\pi_0\mathsf{at}[\![\mathsf{S}]\!],\ \mathsf{at}[\![\mathsf{S}]\!] \xrightarrow{\mathsf{x}\,=\,\mathsf{E}\,=\,\mathscr{E}[\![\mathsf{E}]\!]\varrho(\pi_0\mathsf{at}[\![\mathsf{S}]\!])} \mathsf{after}[\![\mathsf{S}]\!]\rangle\text{)}$$

$$= \{\mathsf{y} \in \mathbb{V} \mid \mathsf{y} \in \mathsf{use}[\![\mathsf{x} = \mathsf{E}]\!] \vee (\mathsf{y} \notin \mathsf{mod}[\![\mathsf{x} = \mathsf{E}]\!] \wedge \mathsf{y} \in L_e)\}$$
$$\text{(definition (41.2.a) of } \alpha^{l}_{\mathsf{use,mod}}[\![\mathsf{S}]\!]\,L_b, L_e\,\langle\pi_0,\ \mathsf{after}[\![\mathsf{S}]\!]\rangle \triangleq \{\mathsf{x} \in \mathbb{V} \mid \mathsf{x} \in L_e\} = L_e \text{ because } \mathsf{escape}[\![\mathsf{S}]\!] = \mathsf{ff} \text{ and omitting the useless parameters of } \mathsf{use} \text{ and } \mathsf{mod}\text{)}$$

$$= \mathsf{use}[\![\mathsf{x} = \mathsf{E}]\!] \cup (L_e \setminus \mathsf{mod}[\![\mathsf{x} = \mathsf{E}]\!]) \quad \text{(definition of } \in\text{)}$$

$$= \widehat{\boldsymbol{S}}^{\exists l}[\![\mathsf{S}]\!]\,L_b, L_e \quad \text{((41.22), Q.E.D.)}$$

- For the ***statement list*** $\mathsf{Sl} ::= \mathsf{Sl}'\ \mathsf{S}$, see the proof at https://github.com/PrAbsInt/.

- For the ***iteration*** $\mathsf{S} ::= \mathtt{while}\,(\mathsf{B})\,\mathsf{S}_b$, we apply the fixpoint approximation result of exercise 18.19 to the fixpoint definition (17.4) of the prefix trace semantics of the iteration. For the semicommutation, we have

$$\alpha^{\exists l}_{\mathsf{use,mod}}[\![\mathsf{S}]\!]\,(\mathscr{F}^*[\![\mathtt{while}\,\ell\,(\mathsf{B})\,\mathsf{S}_b]\!](X))\,L_b, L_e$$

$$= \bigcup\{\alpha^{l}_{\mathsf{use,mod}}[\![\mathsf{S}]\!]\,L_b, L_e\,\langle\pi_0,\ \pi_1\rangle \mid \langle\pi_0,\ \pi_1\rangle \in \mathscr{F}^*[\![\mathtt{while}\,\ell\,(\mathsf{B})\,\mathsf{S}_b]\!](X)\} \quad \text{((41.3))}$$

$$= \bigcup\{\alpha^{l}_{\text{use,mod}}[\![\mathsf{S}]\!]\, L_b, L_e\, \langle\pi_0,\ \pi_1\rangle \mid \langle\pi_0,\ \pi_1\rangle \in \{\langle\pi_1\ell,\ \ell\rangle\}\} \cup \qquad (a)$$

$$\bigcup\{\alpha^{l}_{\text{use,mod}}[\![\mathsf{S}]\!]\, L_b, L_e\, \langle\pi_0,\ \pi_1\rangle \mid \langle\pi_0,\ \pi_1\rangle \in \{\langle\pi_1\ell,\ \ell'\pi_2\ell' \xrightarrow{\neg(\mathsf{B})} \mathsf{after}[\![\mathsf{S}]\!]\rangle \mid \langle\pi_1\ell',\ \ell'\pi_2\ell'\rangle \in X \wedge \mathscr{B}[\![\mathsf{B}]\!]\varrho(\pi_1\ell'\pi_2\ell') = \mathsf{ff} \wedge \ell' = \ell\}\} \cup \qquad (b)$$

$$\bigcup\{\alpha^{l}_{\text{use,mod}}[\![\mathsf{S}]\!]\, L_b, L_e\, \langle\pi_0,\ \pi_1\rangle \mid \langle\pi_0,\ \pi_1\rangle \in \{\langle\pi_1\ell,\ \ell'\pi_2\ell' \xrightarrow{\mathsf{B}} \mathsf{at}[\![\mathsf{S}_b]\!] \mathbin{\frown} \pi_3\rangle \mid \langle\pi_1\ell',\ \ell'\pi_2\ell'\rangle \in X \wedge \mathscr{B}[\![\mathsf{B}]\!]\varrho(\pi_1\ell'\pi_2\ell') = \mathsf{tt} \wedge \pi_3 \in \boldsymbol{\mathcal{S}}^*[\![\mathsf{S}_b]\!](\pi_1\ell'\pi_2\ell' \xrightarrow{\mathsf{B}} \mathsf{at}[\![\mathsf{S}_b]\!]) \wedge \ell' = \ell\}\} \quad (c)$$

$$\text{\textbackslash(17.4)}$$

We go on by cases.

- For the case (a), we have

$$\bigcup\{\alpha^{l}_{\text{use,mod}}[\![\mathsf{S}]\!]\, L_b, L_e\, \langle\pi_0,\ \pi_1\rangle \mid \langle\pi_0,\ \pi_1\rangle \in \{\langle\pi_1\ell',\ \ell'\rangle \mid \pi_1\ell' \in \mathbb{T}^+ \wedge \ell' = \ell\}\} \quad \text{\textlangle(a)\textrangle}$$

$$= \bigcup\{\alpha^{l}_{\text{use,mod}}[\![\mathsf{S}]\!]\, L_b, L_e\, \langle\pi_1\ell,\ \ell\rangle \mid \pi_1\ell \in \mathbb{T}^+\} \quad \text{(where } \ell = \mathsf{at}[\![\texttt{while}\ \ell\ (\mathsf{B})\ \mathsf{S}_b]\!]\text{)}$$

$$= \{\mathsf{x} \in \mathbb{V} \mid (\ell = \mathsf{after}[\![\mathsf{S}]\!] \wedge \mathsf{x} \in L_e) \vee (\mathsf{escape}[\![\mathsf{S}]\!] \wedge \ell = \mathsf{break\text{-}to}[\![\mathsf{S}]\!] \wedge \mathsf{x} \in L_b)\} \quad \text{((41.2.a))}$$

$$= \emptyset \quad \text{(} \ell = \mathsf{at}[\![\mathsf{S}]\!] \neq \mathsf{after}[\![\mathsf{S}]\!] \text{ and } \ell = \mathsf{at}[\![\mathsf{S}]\!] \neq \mathsf{break\text{-}to}[\![\mathsf{S}]\!] \text{ for iteration in section 4.2)}$$

- For the case (b) where $X \subseteq \boldsymbol{\mathcal{S}}^*[\![\mathsf{S}]\!]$ is a subset of the iterates, we have

$$\bigcup\{\alpha^{l}_{\text{use,mod}}[\![\mathsf{S}]\!]\, L_b, L_e\, \langle\pi_1\ell,\ \ell\pi_2\ell \xrightarrow{\neg(\mathsf{B})} \mathsf{after}[\![\mathsf{S}]\!]\rangle \mid \langle\pi_1\ell,\ \ell\pi_2\ell\rangle \in X \wedge \mathscr{B}[\![\mathsf{B}]\!]\varrho(\pi_1\ell\pi_2\ell) = \mathsf{ff}\} \quad \text{(definition of } \in \text{ and } \ell' = \ell = \mathsf{at}[\![\mathsf{S}]\!]\text{)}$$

$$= \bigcup\{\{\mathsf{x} \in \mathbb{V} \mid \exists i \in [1, n-1] \,.\, \forall j \in [1, i-1] \,.\, \mathsf{x} \notin \mathsf{mod}[\![\mathsf{a}_j]\!] \wedge \mathsf{x} \in \mathsf{use}[\![\mathsf{a}_i]\!]\} \cup L_e \cup \mid \langle\pi_1\ell,\ \ell\pi_2\ell\rangle \in X \wedge \mathscr{B}[\![\mathsf{B}]\!]\varrho(\pi_1\ell\pi_2\ell) = \mathsf{ff}\}$$

(by lemma 41.8 where $\ell\pi_2\ell \xrightarrow{\neg(\mathsf{B})} \mathsf{after}[\![\mathsf{S}]\!] \triangleq \ell = \ell_1 \xrightarrow{\mathsf{a}_1} \ell_2 \xrightarrow{\mathsf{a}_2} \ldots \xrightarrow{\mathsf{a}_{n-2}} \ell_{n-1} = \ell \xrightarrow{\mathsf{a}_{n-1} = \neg(\mathsf{B})} \ell_n = \mathsf{after}[\![\mathsf{S}]\!]$, $\langle\pi_1\ell,\ \ell\pi_2\ell\rangle \in X \subseteq \boldsymbol{\mathcal{S}}^*[\![\mathsf{S}]\!]$ so $\langle\pi_1\ell,\ \ell\pi_2\ell \xrightarrow{\neg(\mathsf{B})} \mathsf{after}[\![\mathsf{S}]\!]\rangle \in \boldsymbol{\mathcal{S}}^*[\![\mathsf{S}]\!]$, and $\mathsf{escape}[\![\mathsf{S}]\!] = \mathsf{ff}$ by section 4.2.4)

$$= \bigcup\{\{\mathsf{x} \in \mathbb{V} \mid \exists i \in [1, n-2] \,.\, \forall j \in [1, i-1] \,.\, \mathsf{x} \notin \mathsf{mod}[\![\mathsf{a}_j]\!] \wedge \mathsf{x} \in \mathsf{use}[\![\mathsf{a}_i]\!]\} \cup \{\mathsf{x} \in \mathbb{V} \mid \forall j \in [1, n-2] \,.\, \mathsf{x} \notin \mathsf{mod}[\![\mathsf{a}_j]\!] \wedge \mathsf{x} \in \mathsf{use}[\![\mathsf{B}]\!]\} \cup L_e \mid \langle\pi_1\ell,\ \ell\pi_2\ell\rangle \in X \wedge \mathscr{B}[\![\mathsf{B}]\!]\varrho(\pi_1\ell\pi_2\ell) = \mathsf{ff}\}$$

($[1, n-1] = [1, n-2] \cup \{n-1\}$, $\mathsf{a}_{n-1} = \neg(\mathsf{B})$, and $\mathsf{use}[\![\neg(\mathsf{B})]\!] = \mathsf{use}[\![\mathsf{B}]\!]$)

$$\subseteq \bigcup\{\{\mathsf{x} \in \mathbb{V} \mid \exists i \in [1, n-2] \,.\, \forall j \in [1, i-1] \,.\, \mathsf{x} \notin \mathsf{mod}[\![\mathsf{a}_j]\!] \wedge \mathsf{x} \in \mathsf{use}[\![\mathsf{a}_i]\!] \mid \langle\pi_1\ell,\ \ell\pi_2\ell\rangle \in X\}\} \cup \mathsf{use}[\![\mathsf{B}]\!] \cup L_e$$

(ignoring the check $\forall j \in [1, n-2] \,.\, \mathsf{x} \notin \mathsf{mod}[\![\mathsf{a}_j]\!]$ that x has not been modified before its use in $\neg(\mathsf{B})$, that the test B is false, and $\ell\pi_2\ell \triangleq \ell = \ell_1 \xrightarrow{\mathsf{a}_1} \ell_2 \xrightarrow{\mathsf{a}_2} \ldots \xrightarrow{\mathsf{a}_{n-2}} \ell_{n-1} = \ell$)

$$\subseteq \bigcup\{\alpha^{l}_{\text{use,mod}}[\![\mathsf{S}]\!]\, L_b, L_e\, \langle\pi_0,\ \pi_1\rangle \mid \langle\pi_0,\ \pi_1\rangle \in X\} \cup \mathsf{use}[\![\mathsf{B}]\!] \cup L_e \quad \text{(lemma 41.8)}$$

$$= \alpha^{\exists l}_{\text{use,mod}}[\![\mathsf{S}]\!]\,(X)\, L_b, L_e \cup \mathsf{use}[\![\mathsf{B}]\!] \cup L_e \quad \text{((41.3))}$$

- For the case (c) where $X \subseteq \boldsymbol{\mathcal{S}}^*[\![\mathsf{S}]\!]$ is a subset of the iterates, we have

$\bigcup\{\alpha^{l}_{\mathsf{use,mod}}[\![\mathsf{S}]\!]\,L_b, L_e\,\langle\pi_1\ell,\ \ell\pi_2\ell \xrightarrow{\mathsf{B}} \mathsf{at}[\![\mathsf{S}_b]\!] \frown \pi_3\rangle \mid \langle\pi_1\ell,\ \ell\pi_2\ell\rangle \in X \wedge$
$\mathscr{B}[\![\mathsf{B}]\!]\varrho(\pi_1{}^\ell\pi_2{}^\ell) = \mathsf{tt} \wedge \pi_3 \in \mathcal{S}^*[\![\mathsf{S}_b]\!](\pi_1{}^\ell\pi_2{}^\ell \xrightarrow{\mathsf{B}} \mathsf{at}[\![\mathsf{S}_b]\!])\}$
(definition of $\in$ and $\ell' = \ell = \mathsf{at}[\![\mathsf{S}]\!]$)

$= \bigcup\{\{\mathsf{x} \in \mathbb{V} \mid \exists i \in [1, n-1]\,.\,\forall j \in [1, i-1]\,.\,\mathsf{x} \notin \mathsf{mod}[\![\mathsf{a}_j]\!] \wedge \mathsf{x} \in \mathsf{use}[\![\mathsf{a}_i]\!]\} \cup (\!\![\ \ell_n = \mathsf{after}[\![\mathsf{S}]\!]\ ?\ L_e : \varnothing\ ]\!\!) \cup (\!\![\ \mathsf{escape}[\![\mathsf{S}]\!] \wedge \ell_n = \mathsf{break\text{-}to}[\![\mathsf{S}]\!]\ ?\ L_b : \varnothing\ ]\!\!) \mid \langle\pi_1\ell,\ \ell\pi_2\ell\rangle \in X \wedge$
$\mathscr{B}[\![\mathsf{B}]\!]\varrho(\pi_1{}^\ell\pi_2{}^\ell) = \mathsf{tt} \wedge \pi_3 \in \mathcal{S}^*[\![\mathsf{S}_b]\!](\pi_1{}^\ell\pi_2{}^\ell \xrightarrow{\mathsf{B}} \mathsf{at}[\![\mathsf{S}_b]\!]) \wedge \ell\pi_2\ell \xrightarrow{\mathsf{B}} \mathsf{at}[\![\mathsf{S}_b]\!] \frown \pi_3 =$
$\ell_1 \xrightarrow{\mathsf{a}_1} \ell_2 \xrightarrow{\mathsf{a}_2} \ldots \xrightarrow{\mathsf{a}_{n-1}} \ell_n\}$

(by lemma 41.8 where $X \subseteq \mathcal{S}^*[\![\mathsf{S}]\!]$ implies that $\langle\pi_1\ell,\ \ell\pi_2\ell \xrightarrow{\mathsf{B}} \mathsf{at}[\![\mathsf{S}_b]\!] \frown \pi_3\rangle \in \mathcal{S}^*[\![\mathsf{S}]\!]$)

$= \bigcup\{\{\mathsf{x} \in \mathbb{V} \mid \exists i \in [1, n-1]\,.\,\forall j \in [1, i-1]\,.\,\mathsf{x} \notin \mathsf{mod}[\![\mathsf{a}_j]\!] \wedge \mathsf{x} \in \mathsf{use}[\![\mathsf{a}_i]\!]\} \mid \langle\pi_1\ell,$
$\ell\pi_2\ell\rangle \in X \wedge \mathscr{B}[\![\mathsf{B}]\!]\varrho(\pi_1{}^\ell\pi_2{}^\ell) = \mathsf{tt} \wedge \pi_3 \in \mathcal{S}^*[\![\mathsf{S}_b]\!](\pi_1{}^\ell\pi_2{}^\ell \xrightarrow{\mathsf{B}} \mathsf{at}[\![\mathsf{S}_b]\!]) \wedge \ell\pi_2\ell =$
$\ell_1 \xrightarrow{\mathsf{a}_1} \ell_2 \xrightarrow{\mathsf{a}_2} \ldots \xrightarrow{\mathsf{a}_{m-1}} \ell_m \wedge \ell \xrightarrow{\mathsf{B}} \mathsf{at}[\![\mathsf{S}_b]\!] = \ell_m \xrightarrow{\mathsf{a}_m = \mathsf{B}} \ell_{m+1} \wedge \pi_3 = \ell_{m+1} \xrightarrow{\mathsf{a}_{m+1}}$
$\ldots \xrightarrow{\mathsf{a}_{n-1}} \ell_n\}$

(by decomposing the trace according to its pattern, $\pi_3 \in \mathcal{S}^*[\![\mathsf{S}_b]\!](\pi_1{}^\ell\pi_2{}^\ell \xrightarrow{\mathsf{B}} \mathsf{at}[\![\mathsf{S}_b]\!])$ so $\ell_n \neq \mathsf{after}[\![\mathsf{S}]\!]$, and $\mathsf{escape}[\![\mathsf{S}]\!] = \mathsf{ff}$)

$= \bigcup\{\{\mathsf{x} \in \mathbb{V} \mid \exists i \in [1, m-1]\,.\,\forall j \in [1, i-1]\,.\,\mathsf{x} \notin \mathsf{mod}[\![\mathsf{a}_j]\!] \wedge \mathsf{x} \in \mathsf{use}[\![\mathsf{a}_i]\!]\} \cup \{\mathsf{x} \in \mathbb{V} \mid$
$\forall j \in [1, m-1]\,.\,\mathsf{x} \notin \mathsf{mod}[\![\mathsf{a}_j]\!] \wedge \mathsf{x} \in \mathsf{use}[\![\mathsf{a}_m]\!]\} \cup \{\mathsf{x} \in \mathbb{V} \mid \exists i \in [m+1, n-1]\,.\,\forall j \in$
$[1, i-1]\,.\,\mathsf{x} \notin \mathsf{mod}[\![\mathsf{a}_j]\!] \wedge \mathsf{x} \in \mathsf{use}[\![\mathsf{a}_i]\!]\} \mid \langle\pi_1\ell,\ \ell\pi_2\ell\rangle \in X \wedge \mathscr{B}[\![\mathsf{B}]\!]\varrho(\pi_1{}^\ell\pi_2{}^\ell) = \mathsf{tt} \wedge \pi_3 \in$
$\mathcal{S}^*[\![\mathsf{S}_b]\!](\pi_1{}^\ell\pi_2{}^\ell \xrightarrow{\mathsf{B}} \mathsf{at}[\![\mathsf{S}_b]\!]) \wedge \ell\pi_2\ell = \ell_1 \xrightarrow{\mathsf{a}_1} \ell_2 \xrightarrow{\mathsf{a}_2} \ldots \xrightarrow{\mathsf{a}_{m-1}} \ell_m \wedge \ell \xrightarrow{\mathsf{B}}$
$\mathsf{at}[\![\mathsf{S}_b]\!] = \ell_m \xrightarrow{\mathsf{a}_m = \mathsf{B}} \ell_{m+1} \wedge \pi_3 = \ell_{m+1} \xrightarrow{\mathsf{a}_{m+1}} \ldots \xrightarrow{\mathsf{a}_{n-1}} \ell_n\}$
(by decomposing $[1, n-1] = [1, m-1] \cup \{m\} \cup [m+1, n-1]$)

$\subseteq \bigcup\{\{\mathsf{x} \in \mathbb{V} \mid \exists i \in [1, m-1]\,.\,\forall j \in [1, i-1]\,.\,\mathsf{x} \notin \mathsf{mod}[\![\mathsf{a}_j]\!] \wedge \mathsf{x} \in \mathsf{use}[\![\mathsf{a}_i]\!]\} \mid \langle\pi_1\ell,$
$\ell\pi_2\ell\rangle \in X \wedge \ell\pi_2\ell = \ell_1 \xrightarrow{\mathsf{a}_1} \ell_2 \xrightarrow{\mathsf{a}_2} \ldots \xrightarrow{\mathsf{a}_{m-1}} \ell_m\} \cup \mathsf{use}[\![\mathsf{B}]\!] \cup \bigcup\{\{\mathsf{x} \in \mathbb{V} \mid \exists i \in$
$[m+1, n-1]\,.\,\forall j \in [1, i-1]\,.\,\mathsf{x} \notin \mathsf{mod}[\![\mathsf{a}_j]\!] \wedge \mathsf{x} \in \mathsf{use}[\![\mathsf{a}_i]\!]\} \mid \pi_3 \in \mathcal{S}^*[\![\mathsf{S}_b]\!](\pi_1{}^\ell\pi_2{}^\ell \xrightarrow{\mathsf{B}}$
$\mathsf{at}[\![\mathsf{S}_b]\!]) \wedge \pi_3 = \ell_{m+1} \xrightarrow{\mathsf{a}_{m+1}} \ldots \xrightarrow{\mathsf{a}_{n-1}} \ell_n\}$

(definition of $\cup$, ignoring the check $\forall j \in [1, m-1]\,.\,\mathsf{x} \notin \mathsf{mod}[\![\mathsf{a}_j]\!]$ that x has not been modified before its use in $\mathsf{a}_m = \mathsf{B}$, ignoring the value of $\mathscr{B}[\![\mathsf{B}]\!]\varrho(\pi_1{}^\ell\pi_2{}^\ell) = \mathsf{tt}$)

$\subseteq \bigcup\{\alpha^{l}_{\mathsf{use,mod}}[\![\mathsf{S}]\!]\,L_b, L_e\,\langle\pi_1\ell,\ \ell\pi_2\ell\rangle \mid \langle\pi_1\ell,\ \ell\pi_2\ell\rangle \in X\} \cup \mathsf{use}[\![\mathsf{B}]\!] \cup \bigcup\{\{\mathsf{x} \in \mathbb{V} \mid \exists i \in$
$[m+1, n-1]\,.\,\forall j \in [1, i-1]\,.\,\mathsf{x} \notin \mathsf{mod}[\![\mathsf{a}_j]\!] \wedge \mathsf{x} \in \mathsf{use}[\![\mathsf{a}_i]\!]\} \cup (\!\![\ \ell_n = \mathsf{after}[\![\mathsf{S}_b]\!]\ ?$
$L_e : \varnothing\ ]\!\!) \cup (\!\![\ \mathsf{escape}[\![\mathsf{S}_b]\!] \wedge \ell_n = \mathsf{break\text{-}to}[\![\mathsf{S}_b]\!]\ ?\ L_b : \varnothing\ ]\!\!) \mid \pi_3 \in \mathcal{S}^*[\![\mathsf{S}_b]\!](\pi_1{}^\ell\pi_2{}^\ell \xrightarrow{\mathsf{B}}$
$\mathsf{at}[\![\mathsf{S}_b]\!]) \wedge \pi_3 = \ell_{m+1} \xrightarrow{\mathsf{a}_{m+1}} \ldots \xrightarrow{\mathsf{a}_{n-1}} \ell_n\}$

(by lemma 41.8 for the first term because $\mathsf{after}[\![\mathsf{S}]\!] \neq \ell$ and $\mathsf{break\text{-}to}[\![\mathsf{S}]\!] \neq \ell$ and overapproximating the third term)

$\subseteq \bigcup\{\alpha^{l}_{\text{use,mod}}[\![\mathsf{S}]\!]\, L_b, L_e\, \langle \pi_1\ell,\ \ell\pi_2\ell\rangle \mid \langle \pi_1\ell,\ \ell\pi_2\ell\rangle \in X\} \cup \text{use}[\![\mathsf{B}]\!] \cup$
$\bigcup\{\alpha^{l}_{\text{use,mod}}[\![\mathsf{S}_b]\!]\, L_b, L_e\, \langle \pi_1\ell\pi_2\ell \xrightarrow{\mathsf{B}} \text{at}[\![\mathsf{S}_b]\!],\ \pi_3\rangle \mid \langle \pi_1\ell\pi_2\ell \xrightarrow{\mathsf{B}} \text{at}[\![\mathsf{S}_b]\!],\ \pi_3\rangle \in \boldsymbol{\mathcal{S}}^*[\![\mathsf{S}_b]\!]\}$
⟨by lemma 41.8⟩

$\subseteq \bigcup\{\alpha^{l}_{\text{use,mod}}[\![\mathsf{S}]\!]\, L_b, L_e\, \langle \pi_0,\ \pi_1\rangle \mid \langle \pi_0,\ \pi_1\rangle \in X\} \cup \text{use}[\![\mathsf{B}]\!] \cup \bigcup\{\alpha^{l}_{\text{use,mod}}[\![\mathsf{S}_b]\!]\, L_b, L_e\, \langle \pi_0,$
$\pi_1\rangle \mid \langle \pi_0,\ \pi_1\rangle \in \boldsymbol{\mathcal{S}}^*[\![\mathsf{S}_b]\!]\}$ ⟨overapproximating the semantics X and $\boldsymbol{\mathcal{S}}^*[\![\mathsf{S}_b]\!]$⟩

$\subseteq (\alpha^{\exists l}_{\text{use,mod}}[\![\mathsf{S}]\!]\,(X)\, L_b, L_e) \cup \text{use}[\![\mathsf{B}]\!] \cup (\alpha^{\exists l}_{\text{use,mod}}\mathsf{S}_b]\,(\boldsymbol{\mathcal{S}}^*[\![\mathsf{S}_b]\!])\, L_b, L_e)$ ⟨(41.3)⟩

$\subseteq (\alpha^{\exists l}_{\text{use,mod}}\mathsf{S}]\,(X)\, L_b, L_e) \cup \text{use}[\![\mathsf{B}]\!] \cup \widehat{\boldsymbol{\mathcal{S}}}^{\exists l}[\![\mathsf{S}_b]\!]\, L_b, L_e$
⟨structural induction hypothesis of theorem 41.24⟩

- Gathering the three cases (*a*), (*b*), and (*c*), we have proved the semicommutation condition

  $\alpha^{\exists l}_{\text{use,mod}}\mathsf{S}]\,(\boldsymbol{\mathcal{F}}^*[\![\texttt{while}\,\ell\,(\mathsf{B})\,\mathsf{S}_b]\!](X))\, L_b, L_e \subseteq$
  $L_e \cup (\alpha^{\exists l}_{\text{use,mod}}\mathsf{S}]\,(X)\, L_b, L_e \cup \text{use}[\![\mathsf{B}]\!] \cup L_e) \cup (\alpha^{\exists l}_{\text{use,mod}}\mathsf{S}]\,(X)\, L_b, L_e) \cup \text{use}[\![\mathsf{B}]\!] \cup$
  $\widehat{\boldsymbol{\mathcal{S}}}^{\exists l}[\![\mathsf{S}_b]\!]\, L_b, L_e$

  So we define

  $$\boldsymbol{\mathcal{B}}^{\exists l}[\![\texttt{while}\,(\mathsf{B})\,\mathsf{S}_b]\!]\, L_b, L_e\, X \ \triangleq\ L_e \cup \text{use}[\![\mathsf{B}]\!] \cup \widehat{\boldsymbol{\mathcal{S}}}^{\exists l}[\![\mathsf{S}_b]\!]\, L_b, L_e$$

  to get $\widehat{\boldsymbol{\mathcal{S}}}^{\exists l}[\![\texttt{while}\,(\mathsf{B})\,\mathsf{S}_b]\!]\, L_b, L_e \triangleq \mathsf{lfp}^{\subseteq}\, \boldsymbol{\mathcal{B}}^{\exists l}[\![\texttt{while}\,(\mathsf{B})\,\mathsf{S}_b]\!]\, L_b, L_e$ using exercise 18.19 and exercise 15.8 to get rid of the redundant first term X. Moreover, the transformer $\boldsymbol{\mathcal{B}}^{\exists l}[\![\texttt{while}\,(\mathsf{B})\,\mathsf{S}_b]\!]\, L_b, L_e\, X$ does not depend upon X, so the least fixpoint is that of a constant function

  $$\widehat{\boldsymbol{\mathcal{S}}}^{\exists l}[\![\texttt{while}\,(\mathsf{B})\,\mathsf{S}_b]\!]\, L_b, L_e \ =\ L_e \cup \text{use}[\![\mathsf{B}]\!] \cup \widehat{\boldsymbol{\mathcal{S}}}^{\exists l}[\![\mathsf{S}_b]\!]\, L_b, L_e$$

  as stated in (41.22), Q.E.D. □

- For the *empty statement list* $\mathtt{Sl} ::= \epsilon$, see the proof at https://github.com/PrAbsInt/.

- For the *conditional* $\mathsf{S} ::= \texttt{if}\,\ell\,(\mathsf{B})\,\mathsf{S}_t$, by (6.11), (6.18), and (6.19), we have $\boldsymbol{\mathcal{S}}^*[\![\mathsf{S}]\!](\pi_0\ell)$ $= \{\ell\} \cup \{\ell \xrightarrow{\neg(\mathsf{B})} \text{after}[\![\mathsf{S}]\!] \mid \boldsymbol{\mathcal{B}}[\![\mathsf{B}]\!]\boldsymbol{\varrho}(\pi_0\ell) = \text{ff}\} \cup \{\ell \xrightarrow{\mathsf{B}} \text{at}[\![\mathsf{S}_t]\!] \frown \pi_2 \mid \boldsymbol{\mathcal{B}}[\![\mathsf{B}]\!]\boldsymbol{\varrho}(\pi_0\ell) = \text{tt} \wedge$ $\pi_2 \in \boldsymbol{\mathcal{S}}^*[\![\mathsf{S}_t]\!](\pi_0\ell \xrightarrow{\mathsf{B}} \text{at}[\![\mathsf{S}_t]\!])\}$ where $\ell = \text{at}[\![\mathsf{S}]\!]$. Let us calculate

  $\alpha^{\exists l}_{\text{use,mod}}[\![\mathsf{S}]\!](\boldsymbol{\mathcal{S}}^*[\![\mathsf{S}]\!])\, L_b, L_e$

  $= \bigcup\{\alpha^{l}_{\text{use,mod}}[\![\mathsf{S}]\!]\, L_b, L_e\, \langle \pi_0,\ \pi_1\rangle \mid \langle \pi_0,\ \pi_1\rangle \in \boldsymbol{\mathcal{S}}^*[\![\mathsf{S}]\!]\}$ ⟨definition (41.3) of $\alpha^{\exists l}_{\text{use,mod}}[\![\mathsf{S}]\!]$⟩

  $= \bigcup\{\alpha^{l}_{\text{use,mod}}[\![\mathsf{S}]\!]\, L_b, L_e\, \langle \pi_0\ell,\ \ell\rangle\} \cup \bigcup\{\alpha^{l}_{\text{use,mod}}[\![\mathsf{S}]\!]\, L_b, L_e\, \langle \pi_0\text{at}[\![\mathsf{S}]\!],\ \ell \xrightarrow{\neg(\mathsf{B})}$
  $\text{after}[\![\mathsf{S}]\!]\rangle \mid \boldsymbol{\mathcal{B}}[\![\mathsf{B}]\!]\boldsymbol{\varrho}(\pi_0\ell) = \text{ff}\} \cup \bigcup\{\alpha^{l}_{\text{use,mod}}[\![\mathsf{S}]\!]\, L_b, L_e\, \langle \pi_0\text{at}[\![\mathsf{S}]\!],\ \ell \xrightarrow{\mathsf{B}} \text{at}[\![\mathsf{S}_t]\!] \frown \pi_2\rangle \mid$
  $\boldsymbol{\mathcal{B}}[\![\mathsf{B}]\!]\boldsymbol{\varrho}(\pi_0\ell) = \text{tt} \wedge \pi_2 \in \boldsymbol{\mathcal{S}}^*[\![\mathsf{S}_t]\!](\pi_0\ell \xrightarrow{\mathsf{B}} \text{at}[\![\mathsf{S}_t]\!])\}$ ⟨definition of $\boldsymbol{\mathcal{S}}^*[\![\mathsf{S}]\!](\pi_0\text{at}[\![\mathsf{S}]\!])$⟩

$$= \bigcup\{\alpha^l_{\mathsf{use,mod}}[\![\mathsf{S}]\!]\, L_b, L_e\, \langle \pi_0\mathsf{at}[\![\mathsf{S}]\!],\ \ell \xrightarrow{\neg(\mathsf{B})} \mathsf{after}[\![\mathsf{S}]\!]\rangle \mid \mathcal{B}[\![\mathsf{B}]\!]\varrho(\pi_0\ell) = \mathsf{ff}\} \cup$$
$$\bigcup\{\alpha^l_{\mathsf{use,mod}}[\![\mathsf{S}]\!]\, L_b, L_e\, \langle \pi_0\mathsf{at}[\![\mathsf{S}]\!],\ \ell \xrightarrow{\mathsf{B}} \mathsf{at}[\![\mathsf{S}_t]\!] \mathbin{\smallfrown} \pi_2\rangle \mid \mathcal{B}[\![\mathsf{B}]\!]\varrho(\pi_0\ell) = \mathsf{tt} \wedge \pi_2 \in \boldsymbol{\mathcal{S}}^*[\![\mathsf{S}_t]\!](\pi_0\ell \xrightarrow{\mathsf{B}} \mathsf{at}[\![\mathsf{S}_t]\!])\}$$

$\wr \ell = \mathsf{at}[\![\mathsf{S}]\!]$ and definition (41.3) of $\alpha^l_{\mathsf{use,mod}}[\![\mathsf{S}]\!]\, L_b, L_e\, \langle \pi_0\mathsf{at}[\![\mathsf{S}]\!],\ \mathsf{at}[\![\mathsf{S}]\!]\rangle = \varnothing \wr$

$$= \bigcup\{\mathsf{x} \in \mathbb{V} \mid \mathsf{x} \in \mathsf{use}[\![\neg(\mathsf{B})]\!] \vee (\mathsf{x} \notin \mathsf{mod}[\![\neg(\mathsf{B})]\!] \wedge \mathsf{x} \in \alpha^l_{\mathsf{use,mod}}[\![\mathsf{S}]\!]\, L_b, L_e\, \langle \pi_0\ell \xrightarrow{\neg(\mathsf{B})} \mathsf{after}[\![\mathsf{S}]\!],\ \mathsf{after}[\![\mathsf{S}]\!]\rangle) \mid \mathcal{B}[\![\mathsf{B}]\!]\varrho(\pi_0\ell) = \mathsf{ff}\} \cup \bigcup\{\alpha^l_{\mathsf{use,mod}}[\![\mathsf{S}]\!]\, L_b, L_e\, \langle \pi_0\ell,\ \ell \xrightarrow{\mathsf{B}} \mathsf{at}[\![\mathsf{S}_t]\!] \mathbin{\smallfrown} \pi_2\rangle \mid \mathcal{B}[\![\mathsf{B}]\!]\varrho(\pi_0\ell) = \mathsf{tt} \wedge \pi_2 \in \boldsymbol{\mathcal{S}}^*[\![\mathsf{S}_t]\!](\pi_0\ell \xrightarrow{\mathsf{B}} \mathsf{at}[\![\mathsf{S}_t]\!])\}$$

$\wr$(41.3)$\wr$

$$\subseteq (\mathsf{use}[\![\mathsf{B}]\!] \cup L_e) \cup \bigcup\{\alpha^l_{\mathsf{use,mod}}[\![\mathsf{S}]\!]\, L_b, L_e\, \langle \pi_0\ell,\ \ell \xrightarrow{\mathsf{B}} \mathsf{at}[\![\mathsf{S}_t]\!] \mathbin{\smallfrown} \pi_2\rangle \mid \mathcal{B}[\![\mathsf{B}]\!]\varrho(\pi_0\ell) = \mathsf{tt} \wedge \pi_2 \in \boldsymbol{\mathcal{S}}^*[\![\mathsf{S}_t]\!](\pi_0\ell \xrightarrow{\mathsf{B}} \mathsf{at}[\![\mathsf{S}_t]\!])\}$$

$\wr \mathsf{use}[\![\neg(\mathsf{B})]\!] = \mathsf{use}[\![\mathsf{B}]\!], \mathsf{mod}[\![\neg(\mathsf{B})]\!] = \varnothing$, ignoring the condition $\mathcal{B}[\![\mathsf{B}]\!]\varrho(\pi_0\ell) = \mathsf{ff}$, and $\alpha^l_{\mathsf{use,mod}}[\![\mathsf{S}]\!]\, L_b, L_e\, \langle \pi_0\ell \xrightarrow{\neg(\mathsf{B})} \mathsf{after}[\![\mathsf{S}]\!],\ \mathsf{after}[\![\mathsf{S}]\!]\rangle = L_e$ by (41.3)$\wr$

$$= \mathsf{use}[\![\mathsf{B}]\!] \cup L_e \cup \{\mathsf{x} \in \mathbb{V} \mid \mathsf{x} \in \mathsf{use}[\![\mathsf{B}]\!] \vee (\mathsf{x} \notin \mathsf{mod}[\![\mathsf{B}]\!] \wedge \mathsf{x} \in \alpha^l_{\mathsf{use,mod}}[\![\mathsf{S}]\!]\, L_b, L_e\, \langle \pi_0\ell \xrightarrow{\mathsf{B}} \mathsf{at}[\![\mathsf{S}_t]\!],\ \pi_2\rangle) \wedge \mathcal{B}[\![\mathsf{B}]\!]\varrho(\pi_0\ell) = \mathsf{tt} \wedge \pi_2 \in \boldsymbol{\mathcal{S}}^*[\![\mathsf{S}_t]\!](\pi_0\ell \xrightarrow{\mathsf{B}} \mathsf{at}[\![\mathsf{S}_t]\!])\}$$

$\wr$(41.3)$\wr$

$$\subseteq \mathsf{use}[\![\mathsf{B}]\!] \cup L_e \cup \{\mathsf{x} \in \mathbb{V} \mid \mathsf{x} \in \alpha^l_{\mathsf{use,mod}}[\![\mathsf{S}]\!]\, L_b, L_e\, \langle \pi_1,\ \pi_2\rangle \wedge \pi_2 \in \boldsymbol{\mathcal{S}}^*[\![\mathsf{S}_t]\!](\pi_0\ell \xrightarrow{\mathsf{B}} \mathsf{at}[\![\mathsf{S}_t]\!])\}$$

$\wr \mathsf{mod}[\![\mathsf{B}]\!] = \varnothing$, letting $\pi_1 = \pi_0\ell \xrightarrow{\mathsf{B}} \mathsf{at}[\![\mathsf{S}_t]\!]$, and ignoring the condition $\mathcal{B}[\![\mathsf{B}]\!]\varrho(\pi_0\ell) = \mathsf{tt}\wr$

$$= \mathsf{use}[\![\mathsf{B}]\!] \cup L_e \cup \bigcup\{\alpha^l_{\mathsf{use,mod}}[\![\mathsf{S}]\!]\, L_b, L_e\, \langle \pi_0,\ \pi_1\rangle \mid \langle \pi_0,\ \pi_1\rangle \in \boldsymbol{\mathcal{S}}^*[\![\mathsf{S}_t]\!]\}$$ $\wr$definition of $\cup\wr$

$$= \mathsf{use}[\![\mathsf{B}]\!] \cup L_e \cup \bigcup\{\alpha^l_{\mathsf{use,mod}}[\![\mathsf{S}_t]\!]\, L_b, L_e\, \langle \pi_0,\ \pi_1\rangle \mid \langle \pi_0,\ \pi_1\rangle \in \boldsymbol{\mathcal{S}}^*[\![\mathsf{S}_t]\!]\}$$

$\wr$(41.3) because $\mathsf{after}[\![\mathsf{S}_t]\!] = \mathsf{after}[\![\mathsf{S}]\!]$, $\mathsf{escape}[\![\mathsf{S}_t]\!] = \mathsf{escape}[\![\mathsf{S}]\!]$, and $\mathsf{break\text{-}to}[\![\mathsf{S}_t]\!] = \mathsf{break\text{-}to}[\![\mathsf{S}]\!]$ by sections 4.2.2 and 4.2.4$\wr$

$$= \mathsf{use}[\![\mathsf{B}]\!] \cup L_e \cup \alpha^{\exists l}_{\mathsf{use,mod}}[\![\mathsf{S}_t]\!](\boldsymbol{\mathcal{S}}^*[\![\mathsf{S}_t]\!])\, L_b, L_e$$ $\wr$definition (41.3) of $\alpha^{\exists l}_{\mathsf{use,mod}}[\![\mathsf{S}]\!]\wr$

$$\subseteq \mathsf{use}[\![\mathsf{B}]\!] \cup L_e \cup \widehat{\boldsymbol{\mathcal{S}}}^{\exists l}[\![\mathsf{S}_t]\!]\, L_b, L_e$$ $\wr$induction hypothesis of theorem 41.24$\wr$

- All *other cases* are similar and simple. □

We conclude that algorithm (41.22) is sound with respect to the revised syntactic/semantic definition of liveness in section 41.2.7.

Theorem 41.25 $\alpha^{\exists l}_{\mathsf{use,mod}}(\boldsymbol{\mathcal{S}}^{+\infty}[\![\mathsf{S}]\!]) \mathrel{\dot\subseteq} \boldsymbol{\mathcal{S}}^{\exists l}[\![\mathsf{S}]\!]$.

Proof of theorem 41.25

$$\alpha^{\exists l}_{\mathsf{use,mod}}(\boldsymbol{\mathcal{S}}^{+\infty}[\![\mathsf{S}]\!])$$
$$= \alpha^{\exists l}_{\mathsf{use,mod}}(\boldsymbol{\mathcal{S}}^{*}[\![\mathsf{S}]\!])$$ $\wr$lemma 41.9$\wr$
$$\dot\subseteq \boldsymbol{\mathcal{S}}^{\exists l}[\![\mathsf{S}]\!]$$ $\wr$theorem 41.24 and theorem 41.21$\wr$ □

41.4 Calculational Design of the Syntactic Structural Deadness Static Analysis

By duality we obtain the syntactic definite deadness analysis which is the information actually needed in compilers.

Structural syntactic definite deadness analysis

$$\begin{aligned}
\widehat{\mathcal{S}}^{\forall d}[\![\texttt{Sl}\ \ell]\!]\ D_e &= \mathcal{S}^{\forall d}[\![\texttt{Sl}\ \ell]\!]\ \mathbb{V}, D_e \qquad (41.26)\\
\mathcal{S}^{\forall d}[\![\texttt{x = E ;}]\!]\ D_b, D_e &= \neg\,\mathsf{use}[\![\texttt{x = E}]\!] \cap (D_e \cup \mathsf{mod}[\![\texttt{x = E}]\!])\\
\widehat{\mathcal{S}}^{\forall d}[\![\texttt{;}]\!]\ D_b, D_e &= D_e\\
\widehat{\mathcal{S}}^{\forall d}[\![\texttt{Sl}'\ \texttt{S}]\!]\ D_b, D_e &= \widehat{\mathcal{S}}^{\forall d}[\![\texttt{Sl}']\!]\ D_b, (\widehat{\mathcal{S}}^{\forall d}[\![\texttt{S}]\!]\ D_b, D_e)\\
\widehat{\mathcal{S}}^{\forall d}[\![\epsilon]\!]\ D_b, D_e &= D_e\\
\widehat{\mathcal{S}}^{\forall d}[\![\texttt{if (B) S}_t]\!]\ D_b, D_e &= \neg\,\mathsf{use}[\![\texttt{B}]\!] \cap D_e \cap \widehat{\mathcal{S}}^{\forall d}[\![\texttt{S}_t]\!]\ D_b, D_e\\
\widehat{\mathcal{S}}^{\forall d}[\![\texttt{if (B) S}_t\ \texttt{else S}_f]\!]\ D_b, D_e &= \neg\,\mathsf{use}[\![\texttt{B}]\!] \cap \widehat{\mathcal{S}}^{\forall d}[\![\texttt{S}_t]\!]\ D_b, D_e \cap \widehat{\mathcal{S}}^{\forall d}[\![\texttt{S}_f]\!]\ D_b, D_e\\
\widehat{\mathcal{S}}^{\forall d}[\![\texttt{while (B) S}_b]\!]\ D_b, D_e &= \neg\,\mathsf{use}[\![\texttt{B}]\!] \cap D_e \cap \widehat{\mathcal{S}}^{\forall d}[\![\texttt{S}_b]\!]\ D_b, D_e\\
\widehat{\mathcal{S}}^{\forall d}[\![\texttt{break ;}]\!]\ D_b, D_e &= D_b\\
\widehat{\mathcal{S}}^{\forall d}[\![\texttt{\{ Sl \}}]\!]\ D_b, D_e &= \widehat{\mathcal{S}}^{\forall d}[\![\texttt{Sl}]\!]\ D_b, D_e
\end{aligned}$$

Theorem 41.27 (structural syntactic definite deadness analysis) For all program components S, define $\widehat{\mathcal{S}}^{\forall d}[\![\texttt{S}]\!]\ D_b, D_e \triangleq \neg\widehat{\mathcal{S}}^{\exists l}[\![\texttt{S}]\!]\ \neg D_b, \neg D_e$. $\mathcal{S}^{\forall d}$ is equivalently defined by (41.6).

The proof of theorem 41.27 applies duality to (41.22) by structural induction.

The proof of theorem 41.27 is available at https://github.com/PrAbsInt/.

41.5 Is Liveness Analysis Correctly Used for Code Optimization?

41.5.1 Liveness Specification

Following [909, 910], [115, 857] proposed to specify liveness on a path by a temporal formula and to specify paths as an abstract interpretation of a trace semantics. Therefore liveness is specified syntactically as $\mathcal{S}^{\exists l}[\![\texttt{S}]\!]$ in (41.14), that is, with respect to a syntactic abstraction of the semantics, not the semantics itself. This is problematic, as noticed in section 41.2.6 for example 41.16, because syntactic liveness is not an overapproximation of semantic liveness. The problem is not so much the liveness analysis algorithm (41.22) which can be proved sound by theorem 41.24 with respect to a well defined notion of soundness, definition 41.20. The problem is in the use of liveness information to make program transformations that is provably wrong with respect to this definition 41.20 of soundness.

41.5.2 What Could Go Wrong When Optimizing Programs?

Consider a compiler that successively performs

(1) a (syntactic) liveness analysis;
(2) code optimization by removal
   (2.a) of assignments to variables that are dead after this assignment, and
   (2.b) of assignments to variables that do not change the value of this variable (using constancy analysis of exercise 3.45 and section 33.9 or better symbolic constancy analysis [453, 969]); and
(3) register allocation such that
   (3.a) simultaneously live variables are stored in different registers, and
   (3.b) when no register is left and one is needed, one of those containing the value of a dead variable is preferred (to avoid saving the value of the variable to its memory location as would be needed for live variables).

For the following program (where all variables are dead on exit)

		semantically		syntactically	
		live	dead	live	dead
`x=0; scanf(y);`					
`if (x==0){`					
ℓ_1 `...` `x` and `y` neither used nor modified `...`	ℓ_1	{x}	{y}	{y}	{x}
ℓ_2 `x = y - y; }`	ℓ_2	{x}	{y}	{y}	{x}
`else {`					
`x=42;`					
`}`					
ℓ_3 `print(x);`	ℓ_3	{x}	{y}	{x}	{y}

x is semantically live at ℓ_1, ℓ_2, and ℓ_3 because it is never modified (in particular not modified at ℓ_2) before being used at ℓ_3. However, it is syntactically dead at ℓ_1 and ℓ_2 because it is not used before being assigned at ℓ_2. Code elimination (2.b) will suppress the assignment at ℓ_2 because the value of x is unchanged. Assume that x is in a register at ℓ_1 and a fresh register is needed but none remains available. By (3.b) the register containing x may be selected because its value need not be saved to memory because x is syntactically dead at ℓ_1. Then the value of x is lost at ℓ_3, a compilation bug. The problem is the notion of modification assimilated to an assignment in (41.13), (41.14), and definition 41.20 when this assignment is redundant and may be eliminated.

This error does not occur with semantic liveness (41.12) which declares x live at ℓ_1 so the register containing its value will be saved to memory (and reloaded at ℓ_3).

41.5.3 Why It Does Not Go Wrong?

One solution is to prevent program transformations (such as (2.b) and (3.b) in the preceding explanation) that do not preserve the soundness of the liveness information (according to definition 41.20). Because (2.b) does not depend on the liveness analysis, it can be moved beforehand. Another solution is to redo the liveness analysis after any program transformation that does not preserve the information. A better solution is adopted in CompCert [627]: the liveness analysis and code elimination are performed simultaneously, and the liveness analysis is designed to be valid *after* code elimination. The soundness of the liveness analysis is stated and proved as "after code elimination, the program execution does not depend on the values of the variables declared dead by the analysis." More generally, a program transformation based on a sound static program analysis must be formally proved to be correct. This can be done in the framework of abstract interpretation [257].

41.6 Order Dual Abstract Interpretation

Classical data flow analysis [472] uses an order-dual abstract interpretation in which the lattice of abstract properties $\langle \overline{\mathbb{P}}, \sqsubseteq, \bot, \sqcup \rangle$ is represented upside-down $\langle \overline{\mathbb{P}}, \sqsupseteq, \top, \sqcap \rangle$. This duality can be formalized by a dualizing Galois isomorphism $\langle \mathbb{P}, \sqsubseteq \rangle \overset{\delta}{\underset{\delta}{\leftrightarrows}} \langle \mathbb{P}, \sqsupseteq \rangle$ that is decreasing ($x \sqsubseteq y$ implies $\delta(y) \sqsupseteq \delta(x)$ and of period two ($\delta(\delta(x)) = x$). The infimum $\bot$ becomes the supremum $\top = \delta(\bot)$. Joins become meets by the De Morgan identity $x \sqcap y = \delta(\delta(x) \sqcup \delta(y))$. Least fixpoints become greatest fixpoints, and so forth; see section 10.4.

41.7 Conclusion

Dataflow analysis [26, 29, 568, 810] was the first static program analysis method effectively used in compilers. [723] is comprehensive on dataflow analysis. [241, section 7.0.6.3] first showed that dataflow analysis is an (order dual) abstract interpretation.

Dataflow analysis is an abstract interpretation of a trace semantics restricted to abstract domains satisfying the ascending chain condition (often Boolean ones), most often does not take values of data into account, is often applied to program flow charts, and almost always lacks semantic justifications. [947] follows this line of thought as well as [380], using types, or [112] to provide formal proofs. Most dataflow analyses are still based on flowcharts and fixpoint equations although structural approaches [883] may be more efficient [703], in particular because, as we showed, fixpoint iterations are useless (41.22) for well-structured programs.

Our structural syntactic potential liveness analysis does not need to compute fixpoints by iteration. This is comparable to equation resolution by elimination for reducible flowcharts [848, 864, 883] but much simpler and more efficient.

The equational dataflow analysis (or equivalently the structural definitions in the style of (41.22) and (41.6)) might be less precise than the merge over all paths definitions (in the style of (41.3) and (41.4)). In that case, [241, section 9.2] showed that equational/structural definitions can achieve the same precision by using the powerdomain construction (which consists of adding missing joins/meets and was further studied in [370]).

Abstract interpretation makes formal and clear that traditional liveness is not an overapproximation of "not used before being modified" defined semantically. This shows the limits of the merge over all paths approach to dataflow analysis with postulated syntactic definitions on a path with no semantic justification. So, to prove the soundness of the classic algorithm, we had to change the soundness criterion to "not used before being assigned to," where "not used" is defined semantically and "assigned to" is defined syntactically. However then liveness is not preserved by elimination of useless assignments.

42 Stateful Prefix Trace Semantics

We abstract the stateless prefix trace semantics of chapter 6 (in which values of variables are retrieved from past computations) into a stateful trace semantics (in which values of variables are stored in memory modeled by environments).

Contents

42.1 Stateful Prefix Trace Semantics Abstraction

The prefix trace semantics of program components $\mathsf{S} \in \mathbb{P}\mathbb{c}$ of chapter 6 is the set $\boldsymbol{\mathcal{S}}^*[\![\mathsf{S}]\!]\pi$ of prefix traces continuing an initial trace π. This semantics is stateless in that the value of a variable is recovered from past computations as defined in (6.6). This approach has proved useful, for example, in defining the semantics of parallelism with weak memory models [23] without having to describe the hardware implementation. However, for sequential programs, stateful semantics are more traditional. The hardware memory is formalized by an environment $\rho \in \mathbb{E}\mathbb{v} \triangleq \mathcal{V} \to \mathbb{V}$ mapping variables $\mathsf{x} \in \mathcal{V}$ to their values $\rho(\mathsf{x}) \in \mathbb{V}$.

In this chapter, we are not interested in the actions along traces but only in sequences of states $\sigma = \langle \ell, \rho \rangle \in \mathbb{S} \triangleq (\mathbb{L} \times \mathbb{E}\mathbb{v})$ specifying the value ρ of variables when reaching a program point ℓ, if ever. This stateful abstraction is defined as follows:

$$\begin{aligned} \alpha^{\mathbb{S}}(\langle \pi_0\ell, \ell \rangle) &\triangleq \langle \ell, \varrho(\pi_0\ell) \rangle \\ \alpha^{\mathbb{S}}(\langle \pi_0, \pi \xrightarrow{\mathsf{a}} \ell \rangle) &\triangleq \alpha^{\mathbb{S}}(\langle \pi_0, \pi \rangle) \cdot \langle \ell, \varrho(\pi_0 \mathbin{\widehat{\cdot}} \pi \xrightarrow{\mathsf{a}} \ell) \rangle \end{aligned} \tag{42.1}$$

so that actions are abstracted away while values of variables are recorded everywhere in the trace (instead of being retrieved from past computations). Therefore

$$\alpha^{\mathbb{S}}(\langle \pi_0, \ell_1 \xrightarrow{\mathsf{a}_1} \ell_2 \xrightarrow{\mathsf{a}_2} \ell_3 \ldots \ell_{n-1} \xrightarrow{\mathsf{a}_{n-1}} \ell_n \rangle) = \langle \ell_1, \varrho(\pi_0 \mathbin{\widehat{\cdot}} \ell_1) \rangle \langle \ell_2, \varrho(\pi_0 \mathbin{\widehat{\cdot}} \ell_1 \xrightarrow{\mathsf{a}_1} \ell_2) \rangle \langle \ell_3, \varrho(\pi_0 \mathbin{\widehat{\cdot}} \ell_1 \xrightarrow{\mathsf{a}_1} \ell_2 \xrightarrow{\mathsf{a}_2} \ell_3) \rangle \ldots \langle \ell_{n-1}, \varrho(\pi_0 \mathbin{\widehat{\cdot}} \ell_1 \xrightarrow{\mathsf{a}_1} \ell_2 \xrightarrow{\mathsf{a}_2} \ell_3 \ldots \ell_{n-1}) \rangle \langle \ell_n, \varrho(\pi_0 \mathbin{\widehat{\cdot}} \ell_1 \xrightarrow{\mathsf{a}_1} \ell_2 \xrightarrow{\mathsf{a}_2} \ell_3 \ldots \ell_{n-1} \xrightarrow{\mathsf{a}_{n-1}} \ell_n) \rangle$$

The abstraction of a set of trace pairs is the homomorphic/partitioning abstraction of these trace pairs.

$$\alpha^{\mathbb{S}}(\Pi) \triangleq \{\alpha^{\mathbb{S}}(\langle \pi_0, \pi \rangle) \mid \langle \pi_0, \pi \rangle \in \Pi\}$$

so that, by exercise 11.6, $\langle \wp(\mathbb{T}^+ \times \mathbb{T}^+), \subseteq\rangle \overset{\gamma^{\mathbb{S}}}{\underset{\alpha^{\mathbb{S}}}{\leftrightarrows}} \langle \wp(\mathbb{S}^+), \subseteq\rangle$.

Moreover, we consider all possible initialization traces that is the abstraction $\langle \mathbb{T}^+ \to \wp(\mathbb{T}^+), \dot{\subseteq}\rangle \overset{\gamma^{\mathbb{S}}}{\underset{\alpha^{\mathbb{S}}}{\leftrightarrows}} \langle \wp(\mathbb{S}^+), \subseteq\rangle$ defined by $\alpha^{\mathbb{S}}(\boldsymbol{S}) \triangleq \alpha^{\mathbb{S}}(\{\langle \pi_0, \pi\rangle \mid \pi_0 \in \mathbb{T}^+ \wedge \pi \in \boldsymbol{S}(\pi_0)\}) = \{\alpha^{\mathbb{S}}(\langle \pi_0, \pi\rangle) \mid \pi_0 \in \mathbb{T}^+ \wedge \pi \in \boldsymbol{S}(\pi_0)\}$ where $\mathbb{S}^+$ is the set of all nonempty sequences of states in $\mathbb{S}$.

So we define the prefix state trace semantics $\boldsymbol{S}^*_{\mathbb{S}}[\![\mathsf{S}]\!]$ of a program component $\mathsf{S} \in \mathbb{P}\mathbb{c}$

$$\boldsymbol{S}^*_{\mathbb{S}}[\![\mathsf{S}]\!] \triangleq \alpha^{\mathbb{S}}(\boldsymbol{S}^*[\![\mathsf{S}]\!]) \qquad (42.2)$$

42.2 Stateful Prefix Trace Semantics

We now look for a structural specification $\widehat{\boldsymbol{S}}^*_{\mathbb{S}}[\![\mathsf{S}]\!]$ of the prefix state trace semantics

$$\widehat{\boldsymbol{S}}^*_{\mathbb{S}}[\![\mathsf{S}]\!] = f^*_{\mathbb{S}}[\![\mathsf{S}]\!](\prod_{\mathsf{S}' \triangleleft \mathsf{S}} \widehat{\boldsymbol{S}}^*_{\mathbb{S}}[\![\mathsf{S}']\!]) \qquad (42.3)$$

by calculational design. Let us start with the basic cases.

Prefix state traces of an assignment statement $\mathsf{S} ::= \ell\ \mathsf{x} = \mathsf{A}\,;$

$$\widehat{\boldsymbol{S}}^*_{\mathbb{S}}[\![\mathsf{S}]\!] = \{\langle \ell, \rho\rangle \mid \rho \in \mathbb{E}\mathbb{v}\} \cup \{\langle \ell, \rho\rangle\langle \mathsf{after}[\![\mathsf{S}]\!], \rho[\mathsf{x} \leftarrow \nu]\rangle \mid \rho \in \mathbb{E}\mathbb{v} \wedge \nu = \mathcal{A}[\![\mathsf{A}]\!]\rho\} \quad (42.4)$$

Proof of (42.4)

$$\begin{aligned}
&\widehat{\boldsymbol{S}}^*_{\mathbb{S}}[\![\mathsf{S}]\!]\\
\triangleq\ &\boldsymbol{S}^*_{\mathbb{S}}[\![\mathsf{S}]\!] && \text{(definition of } \widehat{\boldsymbol{S}}^*_{\mathbb{S}}\text{, structural version of } \boldsymbol{S}^*_{\mathbb{S}}\text{)}\\
\triangleq\ &\alpha^{\mathbb{S}}(\boldsymbol{S}^*[\![\mathsf{S}]\!]) && \text{(definition (42.2) of } \boldsymbol{S}^*_{\mathbb{S}}\text{)}\\
=\ &\alpha^{\mathbb{S}}(\widehat{\boldsymbol{S}}^*[\![\mathsf{S}]\!]) && \text{(theorem 17.7)}\\
=\ &\{\alpha^{\mathbb{S}}(\langle \pi, \pi'\rangle) \mid \pi \in \mathbb{T}^+ \wedge \pi' \in \widehat{\boldsymbol{S}}^*[\![\mathsf{S}]\!](\pi)\} && \text{(definition (42.1) of } \alpha^{\mathbb{S}}\text{)}\\
=\ &\{\alpha^{\mathbb{S}}(\langle \pi\ell', \pi'\rangle) \mid \pi\ell' \in \mathbb{T}^+ \wedge \pi' \in \{\ell'\} \cup \{\ell' \xrightarrow{\mathsf{x} = \mathsf{A} = \upsilon} \mathsf{after}[\![\mathsf{S}]\!] \mid \upsilon = \mathcal{A}[\![\mathsf{A}]\!]\boldsymbol{\varrho}(\pi\ell')\} \wedge \ell' = \ell\}\\
& && \text{(definition (17.2) of } \widehat{\boldsymbol{S}}^*[\![\mathsf{S}]\!] \text{ and remark 17.8)}\\
=\ &\{\alpha^{\mathbb{S}}(\langle \pi\ell', \ell'\rangle) \mid \pi\ell' \in \mathbb{T}^+ \wedge \ell' = \ell\} \cup \{\alpha^{\mathbb{S}}(\langle \pi\ell', \ell' \xrightarrow{\mathsf{x} = \mathsf{A} = \upsilon} \mathsf{after}[\![\mathsf{S}]\!]\rangle) \mid \pi\ell' \in \mathbb{T}^+ \wedge \upsilon =\\
&\mathcal{A}[\![\mathsf{A}]\!]\boldsymbol{\varrho}(\pi\ell') \wedge \ell' = \ell\} && \text{(definition of } \cup\text{)}\\
=\ &\{\langle \ell, \boldsymbol{\varrho}(\pi\ell)\rangle \mid \pi\ell' \in \mathbb{T}^+ \wedge \ell' = \ell\} \cup \{\langle \ell', \boldsymbol{\varrho}(\pi\ell')\rangle\langle \mathsf{after}[\![\mathsf{S}]\!], \boldsymbol{\varrho}(\pi\ell' \xrightarrow{\mathsf{x} = \mathsf{A} = \upsilon} \mathsf{after}[\![\mathsf{S}]\!])\rangle \mid\\
&\pi\ell' \in \mathbb{T}^+ \wedge \upsilon = \mathcal{A}[\![\mathsf{A}]\!]\boldsymbol{\varrho}(\pi\ell') \wedge \ell' = \ell\} && \text{(definition of } \alpha^{\mathbb{S}}\text{)}\\
=\ &\{\langle \ell, \rho\rangle \mid \rho \in \mathbb{E}\mathbb{v}\} \cup \{\langle \ell, \rho\rangle\langle \mathsf{after}[\![\mathsf{S}]\!], \rho[\mathsf{x} \leftarrow \upsilon]\rangle \mid \rho \in \mathbb{E}\mathbb{v} \wedge \upsilon = \mathcal{A}[\![\mathsf{A}]\!]\rho\}
\end{aligned}$$

(letting $\rho = \boldsymbol{\varrho}(\pi\ell')$ and conversely, by (6.6) and exercise 6.8, $\forall \rho \in \mathbb{E}\mathbb{v}\,.\,\exists \pi\ell' \in \mathbb{T}^+\,.\,\rho = \boldsymbol{\varrho}(\pi\ell')$, and $\ell' = \ell = \mathsf{at}[\![\mathsf{S}]\!]$) □

Prefix state traces of a statement list $\mathsf{Sl} ::= \mathsf{Sl}'\ \mathsf{S}$

$$\begin{aligned}
\widehat{\boldsymbol{S}}^*_{\mathbb{S}}[\![\mathsf{Sl}]\!] = \ &\widehat{\boldsymbol{S}}^*_{\mathbb{S}}[\![\mathsf{Sl}']\!] \cup && (42.5)\\
&\{\pi \cdot \langle \mathsf{at}[\![\mathsf{S}]\!], \rho\rangle \cdot \pi' \mid \pi \cdot \langle \mathsf{at}[\![\mathsf{S}]\!], \rho\rangle \in \widehat{\boldsymbol{S}}^*_{\mathbb{S}}[\![\mathsf{Sl}']\!] \wedge \langle \mathsf{at}[\![\mathsf{S}]\!], \rho\rangle \cdot \pi' \in \widehat{\boldsymbol{S}}^*_{\mathbb{S}}[\![\mathsf{S}]\!]\}
\end{aligned}$$

Proof of (42.5)

$$\widehat{\mathcal{S}}^*_{\mathbb{S}}[\![\mathtt{Sl}]\!]$$

$= \{\alpha^{\mathbb{S}}(\langle\pi, \pi'\rangle) \mid \pi \in \mathbb{T}^+ \wedge \pi' \in \widehat{\mathcal{S}}^*[\![\mathtt{Sl}]\!](\pi)\}$ ⦃as in the proof of (42.4)⦄

$= \{\alpha^{\mathbb{S}}(\langle\pi_1, \pi'\rangle) \mid \pi_1 \in \mathbb{T}^+ \wedge \pi' \in (\widehat{\mathcal{S}}^*[\![\mathtt{Sl}']\!](\pi_1) \cup \{\pi_2 \frown \pi_3 \mid \pi_2 \in \widehat{\mathcal{S}}^+[\![\mathtt{Sl}']\!](\pi_1) \wedge \pi_3 \in \widehat{\mathcal{S}}^*[\![\mathtt{S}]\!](\pi_1 \frown \pi_2))\}$ ⦃definition (17.3) of $\widehat{\mathcal{S}}^*[\![\mathtt{Sl}]\!]$⦄

$= \{\alpha^{\mathbb{S}}(\langle\pi_1, \pi'\rangle) \mid \pi_1 \in \mathbb{T}^+ \wedge \pi' \in \widehat{\mathcal{S}}^*[\![\mathtt{Sl}']\!](\pi_1)\} \cup \{\alpha^{\mathbb{S}}(\langle\pi_1, \pi_2 \frown \pi_3\rangle) \mid \pi_1 \in \mathbb{T}^+ \wedge \pi_2 \in \widehat{\mathcal{S}}^+[\![\mathtt{Sl}']\!](\pi_1) \wedge \pi_3 \in \widehat{\mathcal{S}}^*[\![\mathtt{S}]\!](\pi_1 \frown \pi_2)\}$ ⦃definitions of $\in$ and $\cup$⦄

$= \widehat{\mathcal{S}}^*_{\mathbb{S}}[\![\mathtt{Sl}']\!] \cup \{\alpha^{\mathbb{S}}(\langle\pi_1, \pi_2\rangle) \cdot \alpha^{\mathbb{S}}(\langle\pi_1 \frown \pi_2, \pi_3\rangle) \mid \pi_1 \in \mathbb{T}^+ \wedge \pi_2 \in \widehat{\mathcal{S}}^+[\![\mathtt{Sl}']\!](\pi_1) \wedge \pi_3 \in \widehat{\mathcal{S}}^*[\![\mathtt{S}]\!](\pi_1 \frown \pi_2)\}$ ⦃induction hypothesis and definition (42.1) of $\alpha^{\mathbb{S}}$⦄

$= \widehat{\mathcal{S}}^*_{\mathbb{S}}[\![\mathtt{Sl}']\!] \cup \{\pi \cdot \langle \mathsf{at}[\![\mathtt{S}]\!], \rho\rangle \cdot \pi' \mid \pi \cdot \langle \mathsf{at}[\![\mathtt{S}]\!], \rho\rangle \in \{\alpha^{\mathbb{S}}(\langle\pi_1, \pi_2\rangle) \mid \pi_1 \in \mathbb{T}^+ \wedge \pi_2 \in \widehat{\mathcal{S}}^*[\![\mathtt{Sl}']\!](\pi_1)\} \wedge \langle \mathsf{at}[\![\mathtt{S}]\!], \rho\rangle \cdot \pi' \in \{\alpha^{\mathbb{S}}(\langle\pi_4, \pi_3\rangle) \mid \pi_4 \in \mathbb{T}^+ \wedge \pi_3 \in \widehat{\mathcal{S}}^*[\![\mathtt{S}]\!](\pi_4)\}\}$

⦃by $\pi_3 \in \widehat{\mathcal{S}}^*[\![\mathtt{S}]\!](\pi_1 \frown \pi_2)$, we have $\pi_2 = \pi_2'\mathsf{at}[\![\mathtt{S}]\!] = \pi_2'\mathsf{after}[\![\mathtt{Sl}']\!]$. So, by (42.1), $\alpha^{\mathbb{S}}(\langle\pi_1, \pi_2\rangle) = \alpha^{\mathbb{S}}(\langle\pi_1, \pi_2'\mathsf{at}[\![\mathtt{S}]\!]\rangle) = \pi\langle\mathsf{after}[\![\mathtt{Sl}']\!], \rho\rangle = \pi\langle\mathsf{at}[\![\mathtt{S}]\!], \rho\rangle$ for some sequence of states π where $\rho = \varrho(\pi_1 \frown \pi_2) = \varrho(\pi_1 \frown \pi_2'\mathsf{at}[\![\mathtt{S}]\!])$. Again, by (42.1), $\alpha^{\mathbb{S}}(\langle\pi_1 \frown \pi_2, \pi_3\rangle) = \langle\mathsf{at}[\![\mathtt{S}]\!], \rho\rangle\pi'$ for some sequence of states π'. So we have proved inclusion $\subseteq$, by letting $\pi_4 = \pi_1 \frown \pi_2$.

Conversely, for any $\pi_4 \in \mathbb{T}^+$ such that $\pi_3 \in \widehat{\mathcal{S}}^*[\![\mathtt{S}]\!](\pi_4)$ and $\varrho(\pi_4) \neq \varrho(\pi_1 \frown \pi_2) = \rho$, we have $\langle\mathsf{at}[\![\mathtt{S}]\!], \rho\rangle \cdot \pi' \neq \alpha^{\mathbb{S}}(\langle\pi_4, \pi_3\rangle)$. Otherwise, for any $\pi_4 \in \mathbb{T}^+$ such that $\pi_3 \in \widehat{\mathcal{S}}^*[\![\mathtt{S}]\!](\pi_4)$ and $\varrho(\pi_4) = \varrho(\pi_1 \frown \pi_2) = \rho$, we have $\langle\mathsf{at}[\![\mathtt{S}]\!], \rho\rangle \cdot \pi' = \alpha^{\mathbb{S}}(\langle\pi_4, \pi_3\rangle) = \alpha^{\mathbb{S}}(\langle\pi_1 \frown \pi_2, \pi_3\rangle)$, proving the reverse inclusion $\supseteq$.⦄

$= \widehat{\mathcal{S}}^*_{\mathbb{S}}[\![\mathtt{Sl}']\!] \cup \{\pi \cdot \langle \mathsf{at}[\![\mathtt{S}]\!], \rho\rangle \cdot \pi' \mid \pi \cdot \langle \mathsf{at}[\![\mathtt{S}]\!], \rho\rangle \in \widehat{\mathcal{S}}^*_{\mathbb{S}}[\![\mathtt{Sl}']\!] \wedge \langle \mathsf{at}[\![\mathtt{S}]\!], \rho\rangle \cdot \pi' \in \widehat{\mathcal{S}}^*_{\mathbb{S}}[\![\mathtt{S}]\!]\}$

⦃induction hypothesis⦄

$= \widehat{\mathcal{S}}^*_{\mathbb{S}}[\![\mathtt{Sl}]\!]$ ⦃(42.5)⦄ □

Prefix state traces of an iteration statement $\mathtt{S} ::= \mathtt{while}\ \ell\ (\mathtt{B})\ \mathtt{S}_b$

$$\widehat{\mathcal{S}}^*_{\mathbb{S}}[\![\mathtt{while}\ \ell\ (\mathtt{B})\ \mathtt{S}_b]\!] = \mathsf{lfp}^{\subseteq}\, \mathcal{F}^*_{\mathbb{S}}[\![\mathtt{while}\ \ell\ (\mathtt{B})\ \mathtt{S}_b]\!] \quad (42.6)$$

$$\begin{aligned}\mathcal{F}^*_{\mathbb{S}}[\![\mathtt{while}\ \ell\ (\mathtt{B})\ \mathtt{S}_b]\!]X \triangleq\ & \{\langle\ell, \rho\rangle \mid \rho \in \mathbb{E}\mathbb{v}\} & \text{(a)}\\ & \cup \{\pi_2\langle\ell', \rho\rangle\langle\mathsf{after}[\![\mathtt{S}]\!], \rho\rangle \mid \pi_2\langle\ell', \rho\rangle \in X \wedge \mathcal{B}[\![\mathtt{B}]\!]\rho = \mathsf{ff} \wedge \ell' = \ell\} & \text{(b)}\\ & \cup \{\pi_2\langle\ell', \rho\rangle\langle\mathsf{at}[\![\mathtt{S}_b]\!], \rho\rangle \cdot \pi_3 \mid \pi_2\langle\ell', \rho\rangle \in X \wedge \mathcal{B}[\![\mathtt{B}]\!]\rho = \mathsf{tt} \wedge & \text{(c)}\\ & \qquad \langle\mathsf{at}[\![\mathtt{S}_b]\!], \rho\rangle \cdot \pi_3 \in \widehat{\mathcal{S}}^*_{\mathbb{S}}[\![\mathtt{S}_b]\!] \wedge \ell' = \ell\}\end{aligned}$$

Proof of (42.6) By corollary 18.34 and (17.4), we have $\mathcal{S}^*[\![\mathtt{S}]\!] = \widehat{\mathcal{S}}^*[\![\mathtt{S}]\!] = \mathsf{lfp}^{\subseteq}\, \mathcal{F}^*[\![\mathtt{S}]\!]$ so we apply theorem 18.23 and use the commutation condition to infer $\mathcal{F}^*_{\mathbb{S}}[\![\mathtt{S}]\!]$. For any iterate X of $\mathcal{F}^*[\![\mathtt{S}]\!]$, we have ($\ell = \mathsf{at}[\![\mathtt{S}]\!]$)

$$\alpha^{\mathbb{S}}(\mathcal{F}^*[\![\mathtt{S}]\!](X))$$

$$= \alpha^{\mathsf{S}}(\pi_1\ell'' \mapsto \{\ell \mid \ell'' = \ell\} \cup \{\ell'\pi_2\ell' \xrightarrow{\neg(\mathsf{B})} \mathsf{after}[\![\mathsf{S}]\!] \mid \ell'' = \ell \wedge \ell'\pi_2\ell' \in X(\pi_1\ell') \wedge$$
$$\mathscr{B}[\![\mathsf{B}]\!]\boldsymbol{\varrho}(\pi_1\ell'\pi_2\ell') = \mathsf{ff} \wedge \ell' = \ell\} \cup \{\ell'\pi_2\ell' \xrightarrow{\mathsf{B}} \mathsf{at}[\![\mathsf{S}_b]\!] \frown \pi_3 \mid \ell'' = \ell \wedge \ell'\pi_2\ell' \in X(\pi_1\ell') \wedge$$
$$\mathscr{B}[\![\mathsf{B}]\!]\boldsymbol{\varrho}(\pi_1\ell'\pi_2\ell') = \mathsf{tt} \wedge \pi_3 \in \boldsymbol{\mathcal{S}}^*[\![\mathsf{S}_b]\!](\pi_1\ell'\pi_2\ell' \xrightarrow{\mathsf{B}} \mathsf{at}[\![\mathsf{S}_b]\!]) \wedge \ell' = \ell\}$$

(definition (17.4) of $\mathscr{F}^*[\![\mathsf{S}]\!]$)

$$= \alpha^{\mathsf{S}}(\pi_1\ell'' \mapsto \{\ell \mid \ell'' = \ell\}) \cup \alpha^{\mathsf{S}}(\pi_1\ell'' \mapsto \{\ell'\pi_2\ell' \xrightarrow{\neg(\mathsf{B})} \mathsf{after}[\![\mathsf{S}]\!] \mid \ell'' = \ell \wedge \ell'\pi_2\ell' \in X(\pi_1\ell') \wedge$$
$$\mathscr{B}[\![\mathsf{B}]\!]\boldsymbol{\varrho}(\pi_1\ell'\pi_2\ell') = \mathsf{ff} \wedge \ell' = \ell\}) \cup \alpha^{\mathsf{S}}(\pi_1\ell'' \mapsto \{\ell'\pi_2\ell' \xrightarrow{\mathsf{B}} \mathsf{at}[\![\mathsf{S}_b]\!] \frown \pi_3 \mid \ell'' = \ell \wedge \ell'\pi_2\ell' \in$$
$$X(\pi_1\ell') \wedge \mathscr{B}[\![\mathsf{B}]\!]\boldsymbol{\varrho}(\pi_1\ell'\pi_2\ell') = \mathsf{tt} \wedge \pi_3 \in \boldsymbol{\mathcal{S}}^*[\![\mathsf{S}_b]\!](\pi_1\ell'\pi_2\ell' \xrightarrow{\mathsf{B}} \mathsf{at}[\![\mathsf{S}_b]\!]) \wedge \ell' = \ell\})$$

(join preservation in a Galois connection)

$$= \{\alpha^{\mathsf{S}}(\langle\pi\ell', \ell'\rangle) \mid \pi\ell' \in \mathbb{T}^+ \wedge \ell' = \ell\} \cup$$
$$\{\alpha^{\mathsf{S}}(\langle\pi_1\ell', \ell'\pi_2\ell' \xrightarrow{\neg(\mathsf{B})} \mathsf{after}[\![\mathsf{S}]\!]\rangle \mid \pi_1\ell' \in \mathbb{T}^+ \wedge \ell'\pi_2\ell' \in X(\pi_1\ell') \wedge \mathscr{B}[\![\mathsf{B}]\!]\boldsymbol{\varrho}(\pi_1\ell'\pi_2\ell') =$$
$$\mathsf{ff} \wedge \ell' = \ell\} \cup$$
$$\{\alpha^{\mathsf{S}}(\langle\pi_1\ell', \ell'\pi_2\ell' \xrightarrow{\mathsf{B}} \mathsf{at}[\![\mathsf{S}_b]\!] \frown \pi_3\rangle \mid \pi_1\ell' \in \mathbb{T}^+ \wedge \ell'\pi_2\ell' \in X(\pi_1\ell') \wedge \mathscr{B}[\![\mathsf{B}]\!]\boldsymbol{\varrho}(\pi_1\ell'\pi_2\ell') =$$
$$\mathsf{tt} \wedge \pi_3 \in \boldsymbol{\mathcal{S}}^*[\![\mathsf{S}_b]\!](\pi_1\ell'\pi_2\ell' \xrightarrow{\mathsf{B}} \mathsf{at}[\![\mathsf{S}_b]\!]) \wedge \ell' = \ell\}$$

(definition of α^{S})

$$= \{\langle\ell, \rho\rangle \mid \rho \in \mathbb{E}\mathbb{v}\} \cup$$
$$\{\alpha^{\mathsf{S}}(\langle\pi_1\ell', \ell'\pi_2\ell'\rangle\langle\mathsf{after}[\![\mathsf{S}]\!], \boldsymbol{\varrho}(\pi_1\ell'\pi_2\ell')\rangle \mid \pi_1\ell' \in \mathbb{T}^+ \wedge \ell'\pi_2\ell' \in X(\pi_1\ell') \wedge$$
$$\mathscr{B}[\![\mathsf{B}]\!]\boldsymbol{\varrho}(\pi_1\ell'\pi_2\ell') = \mathsf{ff} \wedge \ell' = \ell\} \cup$$
$$\{\alpha^{\mathsf{S}}(\langle\pi_1\ell', \ell'\pi_2\ell'\rangle) \cdot \alpha^{\mathsf{S}}(\langle\pi_1\ell'\pi_2\ell' \xrightarrow{\mathsf{B}} \mathsf{at}[\![\mathsf{S}_b]\!], \mathsf{at}[\![\mathsf{S}_b]\!]\pi_3'\rangle) \mid \pi_1\ell' \in \mathbb{T}^+ \wedge \ell'\pi_2\ell' \in$$
$$X(\pi_1\ell') \wedge \mathscr{B}[\![\mathsf{B}]\!]\boldsymbol{\varrho}(\pi_1\ell'\pi_2\ell') = \mathsf{tt} \wedge \mathsf{at}[\![\mathsf{S}_b]\!]\pi_3' \in \boldsymbol{\mathcal{S}}^*[\![\mathsf{S}_b]\!](\pi_1\ell'\pi_2\ell' \xrightarrow{\mathsf{B}} \mathsf{at}[\![\mathsf{S}_b]\!]) \wedge \ell' = \ell\}$$

(for the first term, definition (42.1) of α^{S}, letting $\rho = \boldsymbol{\varrho}(\pi\ell')$ and conversely, by (6.6) and exercise 6.8, $\forall \rho \in \mathbb{E}\mathbb{v} \,.\, \exists \pi\ell' \in \mathbb{T}^+ \,.\, \rho = \boldsymbol{\varrho}(\pi\ell')$, and $\ell' = \ell = \mathsf{at}[\![\mathsf{S}]\!]$; for the second and third terms, definitions (42.1) of α^{S}, and (6.6) of $\boldsymbol{\varrho}$, $\pi_3 = \mathsf{at}[\![\mathsf{S}_b]\!]\pi_3'$ for the third)

$$= \{\langle\ell, \rho\rangle \mid \rho \in \mathbb{E}\mathbb{v}\} \cup$$
$$\{\pi_1' \cdot \langle\ell', \boldsymbol{\varrho}(\pi_1\ell'\pi_2\ell')\rangle\langle\mathsf{after}[\![\mathsf{S}]\!], \boldsymbol{\varrho}(\pi_1\ell'\pi_2\ell')\rangle \mid \pi_1\ell'\pi_2\ell' \in \mathbb{T}^+ \wedge \pi_1'\langle\ell', \boldsymbol{\varrho}(\pi_1\ell'\pi_2\ell')\rangle \in$$
$$\{\alpha^{\mathsf{S}}(\langle\pi_1\ell', \ell'\pi_2\ell'\rangle) \mid \ell'\pi_2\ell' \in X(\pi_1\ell')\} \wedge \mathscr{B}[\![\mathsf{B}]\!]\boldsymbol{\varrho}(\pi_1\ell'\pi_2\ell') = \mathsf{ff} \wedge \ell' = \ell\} \cup$$
$$\{\pi_1' \cdot \langle\ell', \boldsymbol{\varrho}(\pi_1\ell'\pi_2\ell')\rangle \cdot \langle\mathsf{at}[\![\mathsf{S}_b]\!], \boldsymbol{\varrho}(\pi_1\ell'\pi_2\ell')\rangle \cdot \pi_2' \mid \pi_1\ell'\pi_2\ell' \in \mathbb{T}^+ \wedge \pi_1'\langle\ell', \boldsymbol{\varrho}(\pi_1\ell'\pi_2\ell')\rangle \in$$
$$\{\alpha^{\mathsf{S}}(\langle\pi_1\ell', \ell'\pi_2\ell'\rangle) \mid \ell'\pi_2\ell' \in X(\pi_1\ell')\} \wedge \mathscr{B}[\![\mathsf{B}]\!]\boldsymbol{\varrho}(\pi_1\ell'\pi_2\ell') = \mathsf{tt} \wedge \langle\mathsf{at}[\![\mathsf{S}_b]\!], \boldsymbol{\varrho}(\pi_1\ell'\pi_2\ell')\rangle \cdot$$
$$\pi_2' \in \{\alpha^{\mathsf{S}}(\langle\pi_1\ell'\pi_2\ell' \xrightarrow{\mathsf{B}} \mathsf{at}[\![\mathsf{S}_b]\!], \mathsf{at}[\![\mathsf{S}_b]\!]\pi_3'\rangle) \mid \mathsf{at}[\![\mathsf{S}_b]\!]\pi_3' \in \boldsymbol{\mathcal{S}}^*[\![\mathsf{S}_b]\!](\pi_1\ell'\pi_2\ell' \xrightarrow{\mathsf{B}}$$
$$\mathsf{at}[\![\mathsf{S}_b]\!])\} \wedge \ell' = \ell\}$$

(definition of α^{S} so that $\alpha^{\mathsf{S}}(\langle\pi_1\ell', \ell'\pi_2\ell'\rangle)$ has the form $\pi_1'\langle\ell', \boldsymbol{\varrho}(\pi_1\ell'\pi_2\ell')\rangle$, $\alpha^{\mathsf{S}}(\langle\pi_1\ell'\pi_2\ell' \xrightarrow{\mathsf{B}} \mathsf{at}[\![\mathsf{S}_b]\!], \mathsf{at}[\![\mathsf{S}_b]\!]\pi_3'\rangle)$ has the form $\langle\mathsf{at}[\![\mathsf{S}_b]\!], \boldsymbol{\varrho}(\pi_1\ell'\pi_2\ell')\rangle \cdot \pi_2'$ with $\boldsymbol{\varrho}(\pi_1\ell'\pi_2\ell' \xrightarrow{\mathsf{B}} \mathsf{at}[\![\mathsf{S}_b]\!]) = \boldsymbol{\varrho}(\pi_1\ell'\pi_2\ell')$ by definition (6.6) of $\boldsymbol{\varrho}$, and definition of $\in$)

$= \{\langle \ell, \rho\rangle \mid \rho \in \mathbb{E}\mathrm{v}\} \cup$
$\{\pi_1' \cdot \langle \ell', \rho\rangle\langle \mathsf{after}[\![\mathsf{S}]\!], \rho\rangle \mid \rho \in \mathbb{E}\mathrm{v} \wedge \pi_1'\langle \ell', \rho\rangle \in \{\alpha^{\mathsf{S}}(\langle \pi_1\ell', \ell'\pi_2\ell'\rangle) \mid \ell'\pi_2\ell' \in X(\pi_1\ell')\} \wedge$
$\mathcal{B}[\![\mathsf{B}]\!]\rho = \mathrm{ff} \wedge \ell' = \ell\} \cup$
$\{\pi_1' \cdot \langle \ell', \rho\rangle \cdot \langle \mathsf{at}[\![\mathsf{S}_b]\!], \rho\rangle \cdot \pi_2' \mid \rho \in \mathbb{E}\mathrm{v} \wedge \pi_1'\langle \ell', \rho\rangle \in \{\alpha^{\mathsf{S}}(\langle \pi_1\ell', \ell'\pi_2\ell'\rangle) \mid \ell'\pi_2\ell' \in$
$X(\pi_1\ell')\} \wedge \mathcal{B}[\![\mathsf{B}]\!]\rho = \mathrm{tt} \wedge \langle \mathsf{at}[\![\mathsf{S}_b]\!], \rho\rangle \cdot \pi_2' \in \{\alpha^{\mathsf{S}}(\langle \pi_1\ell'\pi_2\ell' \xrightarrow{\mathsf{B}} \mathsf{at}[\![\mathsf{S}_b]\!], \mathsf{at}[\![\mathsf{S}_b]\!]\pi_3'\rangle) \mid$
$\mathsf{at}[\![\mathsf{S}_b]\!]\pi_3' \in \mathcal{S}^*[\![\mathsf{S}_b]\!](\pi_1\ell'\pi_2\ell' \xrightarrow{\mathsf{B}} \mathsf{at}[\![\mathsf{S}_b]\!])\} \wedge \ell' = \ell\}$

(letting $\rho = \varrho(\pi_1\ell'\pi_2\ell')$ and conversely, by (6.6) and exercise 6.8, $\forall \rho \in \mathbb{E}\mathrm{v} . \exists \pi_1\ell' \in \mathbb{T}^+ . \rho = \varrho(\pi_1\ell')$ so that, X being an iterate of $\mathcal{F}^*[\![\mathsf{S}]\!]$, we can choose $\ell'\pi_2\ell' = \ell'$)

$= \{\langle \ell, \rho\rangle \mid \rho \in \mathbb{E}\mathrm{v}\} \cup \{\pi_1' \cdot \langle \ell', \rho\rangle\langle \mathsf{after}[\![\mathsf{S}]\!], \rho\rangle \mid \rho \in \mathbb{E}\mathrm{v} \wedge \pi_1'\langle \ell', \rho\rangle \in \alpha^{\mathsf{S}}(X) \wedge \mathcal{B}[\![\mathsf{B}]\!]\rho =$
$\mathrm{ff} \wedge \ell' = \ell\} \cup \{\pi_1' \cdot \langle \ell', \rho\rangle \cdot \langle \mathsf{at}[\![\mathsf{S}_b]\!], \rho\rangle \cdot \pi_2' \mid \rho \in \mathbb{E}\mathrm{v} \wedge \pi_1'\langle \ell', \rho\rangle \in \alpha^{\mathsf{S}}(X) \wedge \mathcal{B}[\![\mathsf{B}]\!]\rho =$
$\mathrm{tt} \wedge \langle \mathsf{at}[\![\mathsf{S}_b]\!], \rho\rangle \cdot \pi_2' \in \alpha^{\mathsf{S}}(\mathcal{S}^*[\![\mathsf{S}_b]\!]) \wedge \ell' = \ell\}$ (definition of α^{S})

$= \{\langle \ell, \rho\rangle \mid \rho \in \mathbb{E}\mathrm{v}\} \cup \{\pi_1' \cdot \langle \ell', \rho\rangle\langle \mathsf{after}[\![\mathsf{S}]\!], \rho\rangle \mid \rho \in \mathbb{E}\mathrm{v} \wedge \pi_1'\langle \ell', \rho\rangle \in \alpha^{\mathsf{S}}(X) \wedge \mathcal{B}[\![\mathsf{B}]\!]\rho =$
$\mathrm{ff} \wedge \ell' = \ell\} \cup \{\pi_1' \cdot \langle \ell', \rho\rangle \cdot \langle \mathsf{at}[\![\mathsf{S}_b]\!], \rho\rangle \cdot \pi_2' \mid \rho \in \mathbb{E}\mathrm{v} \wedge \pi_1'\langle \ell', \rho\rangle \in \alpha^{\mathsf{S}}(X) \wedge \mathcal{B}[\![\mathsf{B}]\!]\rho =$
$\mathrm{tt} \wedge \langle \mathsf{at}[\![\mathsf{S}_b]\!], \rho\rangle \cdot \pi_2' \in \widehat{\mathcal{S}}^*_{\mathsf{S}}[\![\mathsf{S}_b]\!] \wedge \ell' = \ell\}$ (induction hypothesis)

$= \mathcal{F}^*_{\mathsf{S}}[\![\mathsf{S}]\!](\alpha^{\mathsf{S}}(X))$ (definition (42.6) of $\mathcal{F}^*_{\mathsf{S}}[\![\mathsf{S}]\!]$) □

Exercise 42.7 Provide the prefix state trace semantics of a program $\mathsf{P} ::= \mathsf{Sl}\ \ell$, an empty statement list $\mathsf{Sl} ::= \epsilon$, a skip statement $\mathsf{S} ::= \ell;$, conditional statements $\mathsf{S} ::=$ **if** ℓ (B) S_t and $\mathsf{S} ::=$ **if** ℓ (B) S_t **else** S_f, a break statement $\mathsf{S} ::= \ell$ **break**;, a compound statement $\mathsf{S} ::= \{\ \mathsf{Sl}\ \}$.

Exercise 42.8 Provide a prefix trace semantics of the pointer language of chapter 46, in which states are memories $\mu \in \mathbb{M}_\lambda$ for a given memory allocation λ and the evaluation of expressions is given by (46.2).

42.3 Conclusion

In this chapter we made the link between the stateless prefix trace semantics of chapter 6 and the more traditional stateful trace as found, for example, in [970] stating "Implementation-dependent models may be referred to as *operational models* since they characterize functions constructively in terms of the 'observable' sequences of state transitions by which they may be evaluated."

42.4 Solutions to Selected Exercises

Solution to exercise 42.7
program $\mathsf{P} ::= \mathsf{Sl}\ \ell$

$$\mathcal{S}^*[\![\mathsf{P}]\!] = \mathcal{S}^*[\![\mathsf{Sl}]\!] \tag{42.9}$$

empty statement list $\mathsf{Sl} ::= \epsilon$

$$\mathcal{S}^*[\![\mathsf{Sl}]\!] = \{\langle \mathsf{at}[\![\mathsf{Sl}]\!], \rho\rangle \mid \rho \in \mathbb{E}\mathrm{v}\} \tag{42.10}$$

skip statement $\mathsf{S} ::= \ell;$

$$\mathcal{S}^*[\![\mathsf{S}]\!] = \{\langle \ell, \rho\rangle \mid \rho \in \mathbb{E}\mathrm{v}\} \tag{42.11}$$

conditional statement $\mathsf{S} ::= \mathtt{if}\ \ell\ (\mathsf{B})\ \mathsf{S}_t$

$$\begin{aligned} \boldsymbol{\mathcal{S}}^*[\![\mathsf{S}]\!] &= \{\langle \ell, \rho\rangle \mid \rho \in \mathbb{Ev}\} \cup \\ &\quad \{\langle \ell, \rho\rangle\langle \mathsf{after}[\![\mathsf{S}]\!], \rho\rangle \mid \boldsymbol{\mathcal{B}}[\![\mathsf{B}]\!]\,\rho = \mathsf{ff}\} \cup \\ &\quad \{\langle \ell, \rho\rangle\langle \mathsf{at}[\![\mathsf{S}_t]\!], \rho\rangle \frown \pi_2 \mid \boldsymbol{\mathcal{B}}[\![\mathsf{B}]\!]\,\rho = \mathsf{tt} \wedge \pi_2 \in \boldsymbol{\mathcal{S}}^*[\![\mathsf{S}_t]\!]\} \end{aligned} \tag{42.12}$$

conditional statement $\mathsf{S} ::= \mathtt{if}\ \ell\ (\mathsf{B})\ \mathsf{S}_t\ \mathtt{else}\ \mathsf{S}_f$

$$\begin{aligned} \boldsymbol{\mathcal{S}}^*[\![\mathsf{S}]\!] &= \{\langle \ell, \rho\rangle \mid \rho \in \mathbb{Ev}\} \cup \\ &\quad \{\langle \ell, \rho\rangle\langle \mathsf{at}[\![\mathsf{S}_t]\!], \rho\rangle \frown \pi_2 \mid \boldsymbol{\mathcal{B}}[\![\mathsf{B}]\!]\,\rho = \mathsf{tt} \wedge \pi_2 \in \boldsymbol{\mathcal{S}}^*[\![\mathsf{S}_t]\!]\,\rho\} \cup \\ &\quad \{\langle \ell, \rho\rangle\langle \mathsf{at}[\![\mathsf{S}_f]\!], \rho\rangle \frown \pi_2 \mid \boldsymbol{\mathcal{B}}[\![\mathsf{B}]\!]\,\rho = \mathsf{ff} \wedge \pi_2 \in \boldsymbol{\mathcal{S}}^*[\![\mathsf{S}_f]\!]\,\rho\} \end{aligned} \tag{42.13}$$

break statement $\mathsf{S} ::= \ell\ \mathtt{break}\ ;$

$$\boldsymbol{\mathcal{S}}^*[\![\mathsf{S}]\!] = \{\langle \ell, \rho\rangle \mid \rho \in \mathbb{Ev}\} \cup \{\langle \ell, \rho\rangle\langle \mathsf{break\text{-}to}[\![\mathsf{S}]\!], \rho\rangle \mid \rho \in \mathbb{Ev}\} \tag{42.14}$$

compound statement $\mathsf{S} ::= \{\ \mathtt{Sl}\ \}$

$$\boldsymbol{\mathcal{S}}^*[\![\mathsf{S}]\!] = \boldsymbol{\mathcal{S}}^*[\![\mathtt{Sl}]\!] \tag{42.15}$$

43 Transition Semantics

We abstract the stateful prefix trace semantics of chapter 42 (where values of variables are stored in memory modeled by environments) into a transition system that formally describes the individual steps of a program computation, from which the stateful trace semantics can be rebuilt.

Contents

43.1 Transition System

A transition system is a triple $\langle \Sigma, \mathbb{J}, \xrightarrow{\tau} \rangle$ where Σ is a nonempty set of states σ, $\mathbb{J} \subseteq \Sigma$ is a set of initial states, and $\xrightarrow{\tau} \in \wp(\Sigma \times \Sigma)$ is a transition relation between a state and its possible successors.

A transition system $\langle \Sigma, \mathbb{J}, \xrightarrow{\tau} \rangle$ can be used to define a state prefix trace semantics as follows:

$$\gamma^\tau(\langle \Sigma, \mathbb{J}, \xrightarrow{\tau} \rangle) \triangleq \{\pi_0 \cdot \cdots \cdot \pi_n \mid n \in \mathbb{N} \wedge \pi_0 \in \mathbb{J} \wedge \forall i \in [0, n[\,.\, \pi_i \xrightarrow{\tau} \pi_{i+1}\} \tag{43.1}$$

(where $\sigma \xrightarrow{\tau} \sigma'$ is a shorthand for $\langle \sigma, \sigma' \rangle \in \xrightarrow{\tau}$.)

Conversely a prefix trace semantics S can be abstracted in a transition system

$$\alpha^\tau(S) \triangleq \langle \Sigma, \mathbb{J}, \xrightarrow{\tau} \rangle \tag{43.2}$$

where

$$\begin{array}{rcll} \Sigma & \triangleq & \{\pi_i \mid \exists n \in \mathbb{N}, \pi_0, \ldots, \pi_{i-1}, \pi_{i+1}, \ldots, \pi_n \,.\, \pi_0 \cdots \pi_n \in S\} & (\text{or } \mathbb{S}) \\ \mathbb{J} & \triangleq & \{\pi_0 \mid \exists n \in \mathbb{N}, \pi_1, \ldots, \pi_n \,.\, \pi_0 \cdots \pi_n \in S\} & \\ \xrightarrow{\tau} & \triangleq & \{\pi_i \longrightarrow \pi_{i+1} \mid \exists n \in \mathbb{N}^+, \pi_0, \ldots, \pi_{i-1}, \pi_{i+2}, \ldots, \pi_n \,.\, \pi_0 \cdots \pi_n \in S\} & \end{array}$$

This is a Galois connection

$$\langle \wp(\mathbb{T}^+), \subseteq \rangle \xleftarrow[\alpha^\tau]{\gamma^\tau} \langle \{\langle \Sigma, \mathbb{J}, \xrightarrow{\tau} \rangle \mid \Sigma \in \wp(\mathbb{S}) \wedge \mathbb{J} \subseteq \Sigma \wedge \xrightarrow{\tau} \subseteq \Sigma \times \Sigma\}, \dot{\subseteq} \rangle$$

In general information is lost by the abstraction of a prefix trace semantics to a transition system (take for example $\Pi = \{a, aa\}$ so that $\gamma^\tau \circ \alpha^\tau(\Pi) = a^+$ is the set of all nonempty finite sequences of "a"s).

Notice that the abstraction of the prefix trace semantics of a program into a transition system will comprehend only reachable states. So, the transition semantics for a language is the join of all transition systems of the prefix trace semantics of all programs in the semantics. This may still be a strict overapproximation.

Exercise 43.3 Define $\boldsymbol{S} = \{a^n b \mid n \in \mathbb{N}\}$. Show that $\boldsymbol{S} \subsetneq \gamma^\tau \circ \alpha^\tau(\boldsymbol{S})$.

43.2 Transition Semantics

The transition semantics of the programming language $\mathbb{P}$ of chapters 4 and 6 is

$$\alpha^\tau(\widehat{\boldsymbol{S}}^*_{\mathrm{s}}[\![\mathsf{S}]\!]) \quad = \quad \langle \mathbb{S}, \{\langle \mathsf{at}[\![\mathsf{S}]\!], \rho\rangle \mid \mathsf{S} \in \mathbb{P}\mathbb{c} \wedge \rho \in \mathbb{E}\mathbb{v}\}, \widehat{\boldsymbol{S}}^\tau[\![\mathsf{S}]\!]\rangle$$

defined by structural induction on program components $\mathsf{S} \in \mathbb{P}\mathbb{c}$ as follows:

Transition semantics of an assignment statement $\mathsf{S} ::= \ell\ \mathsf{x} = \mathsf{A}\,;$

$$\widehat{\boldsymbol{S}}^\tau[\![\mathsf{S}]\!] \quad = \quad \{\langle \ell, \rho\rangle \longrightarrow \langle \mathsf{after}[\![\mathsf{S}]\!], \rho[\mathsf{x} \leftarrow \mathscr{A}[\![\mathsf{A}]\!]\rho]\rangle \mid \rho \in \mathbb{E}\mathbb{v}\} \tag{43.4}$$

Proof of (43.4)

$\widehat{\boldsymbol{S}}^\tau[\![\mathsf{S}]\!]$

$= \{\pi_i \longrightarrow \pi_{i+1} \mid \exists n \in \mathbb{N}^+, \pi_0, \ldots, \pi_{i-1}, \pi_{i+2}, \ldots, \pi_n \,.\, \pi_0 \cdots \pi_n \in \widehat{\boldsymbol{S}}^*_{\mathrm{s}}[\![\mathsf{S}]\!]\}$ $\quad\wr(43.2)\wr$

$= \{\langle \ell, \rho\rangle \longrightarrow \langle \mathsf{after}[\![\mathsf{S}]\!], \rho[\mathsf{x} \leftarrow \mathscr{A}[\![\mathsf{A}]\!]\rho]\rangle \mid \rho \in \mathbb{E}\mathbb{v}\}$ $\quad\wr(42.4)\wr$ □

Transition semantics of a statement list $\mathsf{Sl} ::= \mathsf{Sl}'\ \mathsf{S}$

$$\widehat{\boldsymbol{S}}^\tau[\![\mathsf{Sl}]\!] \quad = \quad \widehat{\boldsymbol{S}}^\tau[\![\mathsf{Sl}']\!] \cup \widehat{\boldsymbol{S}}^\tau[\![\mathsf{S}]\!] \tag{43.5}$$

Proof of (42.4)

$\widehat{\boldsymbol{S}}^\tau[\![\mathsf{Sl}]\!]$

$= \{\pi_i \longrightarrow \pi_{i+1} \mid \exists n \in \mathbb{N}^+, \pi_0, \ldots, \pi_{i-1}, \pi_{i+2}, \ldots, \pi_n \,.\, \pi_0 \cdots \pi_n \in \widehat{\boldsymbol{S}}^*_{\mathrm{s}}[\![\mathsf{Sl}]\!]\}$ $\quad\wr(43.2)\wr$

$= \{\pi_i \longrightarrow \pi_{i+1} \mid \exists n \in \mathbb{N}^+, \pi_0, \ldots, \pi_{i-1}, \pi_{i+2}, \ldots, \pi_n \,.\, \pi_0 \cdots \pi_n \in \widehat{\boldsymbol{S}}^*_{\mathrm{s}}[\![\mathsf{Sl}']\!] \cup \{\pi \cdot \langle \mathsf{at}[\![\mathsf{S}]\!], \rho\rangle \cdot \pi' \mid \pi \cdot \langle \mathsf{at}[\![\mathsf{S}]\!], \rho\rangle \in \widehat{\boldsymbol{S}}^*_{\mathrm{s}}[\![\mathsf{Sl}']\!] \wedge \langle \mathsf{at}[\![\mathsf{S}]\!], \rho\rangle \cdot \pi' \in \widehat{\boldsymbol{S}}^*_{\mathrm{s}}[\![\mathsf{S}]\!]\}\}$

$\wr$definition (42.5) of $\widehat{\boldsymbol{S}}^*_{\mathrm{s}}[\![\mathsf{Sl}]\!]\wr$

$= \{\pi_i \longrightarrow \pi_{i+1} \mid \exists n \in \mathbb{N}^+, \pi_0, \ldots, \pi_{i-1}, \pi_{i+2}, \ldots, \pi_n \,.\, \pi_0 \cdots \pi_n \in \widehat{\boldsymbol{S}}^*_{\mathrm{s}}[\![\mathsf{Sl}']\!]\} \cup \{\pi_i \longrightarrow \pi_{i+1} \mid \exists n \in \mathbb{N}^+, \pi_0, \ldots, \pi_{i-1}, \pi_{i+2}, \ldots, \pi_n \,.\, \pi_0 \cdots \pi_n \in \{\pi \cdot \langle \mathsf{at}[\![\mathsf{S}]\!], \rho\rangle \cdot \pi' \mid \pi \cdot \langle \mathsf{at}[\![\mathsf{S}]\!], \rho\rangle \in \widehat{\boldsymbol{S}}^*_{\mathrm{s}}[\![\mathsf{Sl}']\!] \wedge \langle \mathsf{at}[\![\mathsf{S}]\!], \rho\rangle \cdot \pi' \in \widehat{\boldsymbol{S}}^*_{\mathrm{s}}[\![\mathsf{S}]\!]\}\}$ $\quad\wr$definition of $\cup\wr$

$= \{\pi_i \longrightarrow \pi_{i+1} \mid \exists n \in \mathbb{N}^+, \pi_0, \ldots, \pi_{i-1}, \pi_{i+2}, \ldots, \pi_n \,.\, \pi_0 \cdots \pi_n \in \widehat{\boldsymbol{S}}^*_{\mathrm{s}}[\![\mathsf{Sl}']\!]\} \cup \{\pi_i \longrightarrow \pi_{i+1} \mid \exists n \in \mathbb{N}^+, \pi_0, \ldots, \pi_{i-1}, \pi_{i+2}, \ldots, \pi_n \,.\, \pi_0 \cdots \pi_n \in \{\langle \mathsf{at}[\![\mathsf{S}]\!], \rho\rangle \cdot \pi' \mid \langle \mathsf{at}[\![\mathsf{S}]\!], \rho\rangle \cdot \pi' \in \widehat{\boldsymbol{S}}^*_{\mathrm{s}}[\![\mathsf{S}]\!]\}\}$

$\wr$because all transitions originating from $\pi \cdot \langle \mathsf{at}[\![\mathsf{S}]\!], \rho\rangle = \pi \cdot \langle \mathsf{after}[\![\mathsf{Sl}']\!], \rho\rangle \in \widehat{\boldsymbol{S}}^*_{\mathrm{s}}[\![\mathsf{Sl}']\!]$ have already been collected in the first term of the $\cup\wr$

$= \{\pi_i \longrightarrow \pi_{i+1} \mid \exists n \in \mathbb{N}^+, \pi_0, \ldots, \pi_{i-1}, \pi_{i+2}, \ldots, \pi_n \,.\, \pi_0 \cdots \pi_n \in \widehat{\boldsymbol{S}}^*_{\mathbb{S}}[\![\mathtt{Sl'}]\!]\} \cup \{\pi_i \longrightarrow \pi_{i+1} \mid$
$\exists n \in \mathbb{N}^+, \pi_0, \ldots, \pi_{i-1}, \pi_{i+2}, \ldots, \pi_n \,.\, \pi_0 \cdots \pi_n \in \widehat{\boldsymbol{S}}^*_{\mathbb{S}}[\![\mathtt{S}]\!]\}$
(because all traces of $\widehat{\boldsymbol{S}}^*_{\mathbb{S}}[\![\mathtt{S}]\!]$ start with state $\langle \mathsf{at}[\![\mathtt{S}]\!], \rho\rangle$)

$= \widehat{\boldsymbol{S}}^\tau[\![\mathtt{Sl'}]\!] \cup \widehat{\boldsymbol{S}}^\tau[\![\mathtt{S}]\!]$ ((43.2) and induction hypothesis) □

Transition semantics of an iteration statement $\mathtt{S} ::= \mathtt{while}\ \ell\ (\mathtt{B})\ \mathtt{S}_b$

$$\widehat{\boldsymbol{S}}^\tau[\![\mathtt{while}\ \ell\ (\mathtt{B})\ \mathtt{S}_b]\!] = \{\langle \ell, \rho\rangle \longrightarrow \langle \mathsf{after}[\![\mathtt{S}]\!], \rho\rangle \mid \mathscr{B}[\![\mathtt{B}]\!]\rho = \mathsf{ff}\} \cup \{\langle \ell, \rho\rangle \longrightarrow \langle \mathsf{at}[\![\mathtt{S}_b]\!], \rho\rangle \mid \mathscr{B}[\![\mathtt{B}]\!]\rho = \mathsf{tt}\} \cup \widehat{\boldsymbol{S}}^\tau[\![\mathtt{S}_b]\!] \quad (43.6)$$

By definition of $\mathsf{after}[\![\mathtt{S}_b]\!] = \mathsf{at}[\![\mathtt{while}\ \ell\ (\mathtt{B})\ \mathtt{S}_b]\!] = \ell$, there is no need for a transition from after the loop body $\mathtt{S}_b$ to the start ℓ of the loop.

Transition semantics of a break statement $\mathtt{S} ::= \ell\ \mathtt{break}\,;$

$$\widehat{\boldsymbol{S}}^\tau[\![\mathtt{break}\,;]\!] = \{\langle \ell, \rho\rangle \longrightarrow \langle \mathsf{break\text{-}to}[\![\mathtt{S}]\!], \rho\rangle \mid \rho \in \mathbb{E}\mathrm{v}\} \quad (43.7)$$

Proof of (43.6) and (43.7)

$\widehat{\boldsymbol{S}}^\tau[\![\mathtt{S}]\!]$ (where $\mathtt{S} ::= \mathtt{while}\ \ell\ (\mathtt{B})\ \mathtt{S}_b$)

$= \{\pi_i \longrightarrow \pi_{i+1} \mid \exists n \in \mathbb{N}^+, \pi_0, \ldots, \pi_{i-1}, \pi_{i+2}, \ldots, \pi_n \,.\, \pi_0 \cdots \pi_n \in \widehat{\boldsymbol{S}}^*_{\mathbb{S}}[\![\mathtt{S}]\!]\}$ ((43.2))

$= \{\pi_i \longrightarrow \pi_{i+1} \mid \exists n \in \mathbb{N}^+, \pi_0, \ldots, \pi_{i-1}, \pi_{i+2}, \ldots, \pi_n \,.\, \pi_0 \cdots \pi_n \in \alpha^{\mathbb{S}}(\boldsymbol{S}^*[\![\mathtt{S}]\!])\}$ ((42.2))

$= \{\pi_i \longrightarrow \pi_{i+1} \mid \exists n \in \mathbb{N}^+, \pi_0, \ldots, \pi_{i-1}, \pi_{i+2}, \ldots, \pi_n \,.\, \pi_0 \cdots \pi_n \in \{\alpha^{\mathbb{S}}(\langle \pi', \pi\rangle) \mid \pi' \in \mathbb{T}^+ \wedge \pi \in \boldsymbol{S}^*[\![\mathtt{S}]\!](\pi')\}\}$ (definition of $\alpha^{\mathbb{S}}$)

$= \{\langle \ell_i, \varrho(\pi' \mathbin{\widehat{}} \pi[0,i])\rangle \longrightarrow \langle \ell_{i+1}, \varrho(\pi' \mathbin{\widehat{}} \pi[0,i+1])\rangle \mid \pi' \in \mathbb{T}^+ \wedge \pi \in \boldsymbol{S}^*[\![\mathtt{S}]\!](\pi') \wedge \pi = \ell_0 \xrightarrow{\mathsf{a}_0} \ell_1 \xrightarrow{\mathsf{a}_1} \cdots \ell_{n-1} \xrightarrow{\mathsf{a}_{n-1}} \ell_n\}$
(definition of $\in$, definitions (42.1) of $\alpha^{\mathbb{S}}$ and of $\boldsymbol{S}^*[\![\mathtt{S}]\!]$ in chapter 6 so that π has the form $\ell_0 \xrightarrow{\mathsf{a}_0} \ell_1 \xrightarrow{\mathsf{a}_1} \cdots \ell_{n-1} \xrightarrow{\mathsf{a}_{n-1}} \ell_n$, and $(\pi_0 \cdots \pi_n)[0,i] \triangleq \pi_0 \cdots \pi_i$)

Lemma 6.39 identifies all possible cases for $\widehat{\boldsymbol{S}}^*[\![\mathtt{S}]\!](\pi)$, as follows:

- $\{\ell\}$ yields no transition;
- $\{\ell \xrightarrow{B} \mathsf{at}[\![\mathtt{S}_b]\!] \mid \mathscr{B}[\![\mathtt{B}]\!]\varrho(\pi_1 \ell \mathbin{\widehat{}} \underline{\pi}) = \mathsf{tt}\}$ yields transitions $\{\langle \ell, \rho\rangle \longrightarrow \langle \mathsf{at}[\![\mathtt{S}_b]\!], \rho\rangle \mid \mathscr{B}[\![\mathtt{B}]\!]\rho = \mathsf{tt}\}$ where $\rho = \varrho(\pi_1 \ell \mathbin{\widehat{}} \underline{\pi})$ and, by (6.6) and exercise 6.8, ρ can be chosen arbitrarily on the first iteration for the converse inclusion;
- $\{\ell \xrightarrow{\neg(\mathtt{B})} \mathsf{after}[\![\mathtt{S}]\!] \mid \mathscr{B}[\![\mathtt{B}]\!]\varrho(\pi_1 \ell \mathbin{\widehat{}} \underline{\pi}) = \mathsf{ff}\}$ yields transitions $\{\langle \ell, \rho\rangle \longrightarrow \langle \mathsf{after}[\![\mathtt{S}]\!], \rho\rangle \mid \mathscr{B}[\![\mathtt{B}]\!]\rho = \mathsf{ff}\}$ where $\rho = \varrho(\pi_1 \ell \mathbin{\widehat{}} \underline{\pi})$;
- In addition to the iteration entry transition, $\{\ell \xrightarrow{B} \mathsf{at}[\![\mathtt{S}_b]\!]\pi''(n) \mid \mathscr{B}[\![\mathtt{B}]\!]\varrho(\pi_1 \ell \mathbin{\widehat{}} \underline{\pi}) = \mathsf{tt} \wedge \mathsf{at}[\![\mathtt{S}_b]\!]\pi''(n) \in \widehat{\boldsymbol{S}}^*[\![\mathtt{S}_b]\!](\pi_1 \ell \mathbin{\widehat{}} \underline{\pi} \xrightarrow{\mathtt{B}} \mathsf{at}[\![\mathtt{S}_b]\!])\}$ yields the transitions in $\alpha^\tau(\widehat{\boldsymbol{S}}^*_{\mathbb{S}}[\![\mathtt{S}_b]\!])$, which, by structural induction hypothesis, is $\widehat{\boldsymbol{S}}^\tau[\![\mathtt{S}_b]\!]$;
- In addition to the iteration entry and the $\widehat{\boldsymbol{S}}^\tau[\![\mathtt{S}_b]\!]$ transitions, the last term of (6.39), which is $\{\ell \xrightarrow{\mathtt{B}} \mathsf{at}[\![\mathtt{S}_b]\!] \mathbin{\widehat{}} \pi_3 \xrightarrow{\mathtt{break}} \mathsf{break\text{-}to}[\![\mathtt{S}]\!] \mid \mathscr{B}[\![\mathtt{B}]\!]\varrho(\pi_1 \ell \mathbin{\widehat{}} \underline{\pi}) = \mathsf{tt} \wedge$

$\pi_3 \xrightarrow{\text{break}}$ break-to⟦S⟧ $\in \widehat{\mathcal{S}}^*$⟦$\mathsf{S}_b$⟧$(\pi_1 \ell \cdot \underline{\pi} \xrightarrow{\mathsf{B}}$ at⟦S_b⟧$)\}$ yields transitions $\{\langle \ell, \rho\rangle \longrightarrow$ ⟨break-to⟦S⟧, $\rho\rangle \mid \rho \in \mathbb{E}\mathrm{v}\}$ which are those of the break statement in (43.7). These transitions need not be included for the iteration, because in general, by the labeling in chapter 4, the transitions of a program component include all transitions of its syntactic subcomponents, and by the syntax of the language this last term of (6.39) is not empty only if the break statement is a syntactic subcomponent of the iteration so that the transitions need not be counted twice. □

Exercise 43.8 Provide the transition semantics of a program P ::= Sl ℓ, an empty statement list Sl ::= ϵ, a skip statement S ::= ℓ;, conditional statements S ::= **if** ℓ (B) S_t and S ::= **if** ℓ (B) S_t **else** S_f, a break statement S ::= ℓ **break** ;, a compound statement S ::= { Sl }.

Let $\widehat{\mathcal{S}}^{+\infty}_{\mathrm{s}}$ be the structural stateful maximal finite and infinite trace semantics (obtained by limit of $\widehat{\mathcal{S}}^*_{\mathrm{s}}$, as in chapter 7).

Exercise 43.9 Prove that $\alpha^\tau(\widehat{\mathcal{S}}^{+\infty}_{\mathrm{s}}$⟦S⟧$) = \alpha^\tau(\widehat{\mathcal{S}}^*_{\mathrm{s}}$⟦S⟧$)$.

Exercise 43.10 Provide a structural deductive definition of $\widehat{\mathcal{S}}^\tau$⟦S⟧ in the style of [802].

> **Theorem 43.11** The stateful prefix trace semantics $\mathcal{S}^*_{\mathrm{s}}$ of section 42.2 for the programming language $\mathbb{P}$ of chapter 4 is generated by the transition semantics of chapter 43.

Proof of theorem 43.11 The proof is by structural induction on the program components S $\in \mathbb{P}\mathrm{c}$ of the language $\mathbb{P}$. Let ⟨S, $\widehat{\mathcal{S}}^\tau$⟦S⟧, $\mathfrak{I}$⟦S⟧⟩ where $\mathfrak{I}$⟦S⟧ $\triangleq \{\langle$at⟦S⟧, $\rho\rangle \mid \rho \in \mathbb{E}\mathrm{v}\}$ be the transition system of the program components S, as defined in this chapter. Let $\mathcal{S}^*$(⟨S, $\widehat{\mathcal{S}}^\tau$⟦S⟧, $\mathfrak{I}$⟦S⟧⟩) be the set of stateful prefix traces generated by this transition system, as defined in (43.1). We must show that $\mathcal{S}^*$(⟨S, $\widehat{\mathcal{S}}^\tau$⟦S⟧, $\mathfrak{I}$⟦S⟧⟩) $= \widehat{\mathcal{S}}^*_{\mathrm{s}}$⟦S⟧ for the structural stateful prefix semantics $\widehat{\mathcal{S}}^*_{\mathrm{s}}$⟦S⟧ defined in section 42.2.

– We observe that traces of length 1 in $\widehat{\mathcal{S}}^*_{\mathrm{s}}$⟦S⟧ are all of the form $\{\langle \ell, \rho\rangle \mid \rho \in \mathbb{E}\mathrm{v}\}$, which are exactly the same for $n = 0$ is (43.1). So, in the following, we just have to consider prefix traces of length strictly greater than 1.

– For the assignment statement S ::= ℓ x = A ;, we have

$$\{\pi_0 \cdot \cdots \cdot \pi_n \mid n \in \mathbb{N} \wedge \pi_0 \in \mathfrak{I}[\![\mathsf{S}]\!] \wedge \forall i \in [0, n[\,.\, \langle \pi_i,\ \pi_{i+1}\rangle \in \widehat{\mathcal{S}}^\tau[\![\mathsf{S}]\!]\} \qquad \text{(43.1)}$$

$$= \{\pi_0 \cdot \cdots \cdot \pi_n \mid n \in \mathbb{N} \wedge \pi_0 \in \mathfrak{I}[\![\mathsf{S}]\!] \wedge \forall i \in [0, n[\,.\, \langle \pi_i,\ \pi_{i+1}\rangle \in \{\langle \ell, \rho\rangle \longrightarrow \langle \mathsf{after}[\![\mathsf{S}]\!],\ \mathcal{A}[\![\mathsf{A}]\!]\rho\rangle \mid \rho \in \mathbb{E}\mathrm{v}\}\} \qquad \text{(43.4)}$$

$$= \{\pi_0\pi_1 \mid \langle \pi_0,\ \pi_1\rangle \in \{\langle \ell, \rho\rangle \longrightarrow \langle \mathsf{after}[\![\mathsf{S}]\!],\ \mathcal{A}[\![\mathsf{A}]\!]\rho\rangle \mid \rho \in \mathbb{E}\mathrm{v}\}\}$$

(because ℓ = at⟦S⟧ $\neq$ after⟦S⟧ and $\pi_0 = \langle \ell, \rho\rangle \in \mathfrak{I}$⟦S⟧ by definition of $\mathfrak{I}$⟦S⟧)

$$= \{\langle \ell, \rho\rangle\langle \mathsf{after}[\![\mathsf{S}]\!],\ \mathcal{A}[\![\mathsf{A}]\!]\rho\rangle \mid \rho \in \mathbb{E}\mathrm{v}\} \qquad \text{(definition of } \in\text{)}$$

$$= \widehat{\mathcal{S}}^*_{\mathrm{s}}[\![\mathsf{S}]\!] \setminus \{\langle \ell, \rho\rangle \mid \rho \in \mathbb{E}\mathrm{v}\} \qquad \text{(42.4)}$$

so we conclude by adding the traces of length 1.

– For the statement list Sl ::= Sl′ S, we have

$$\{\pi_0 \cdot \cdots \cdot \pi_n \mid n \in \mathbb{N} \wedge \pi_0 \in \mathfrak{I}[\![\texttt{Sl}]\!] \wedge \forall i \in [0, n[\,.\, \langle \pi_i,\ \pi_{i+1}\rangle \in \widehat{\boldsymbol{\mathcal{S}}}^{\tau}[\![\texttt{Sl}]\!]\} \qquad ⦃(43.1)⦄$$

$$= \{\pi_0 \cdot \cdots \cdot \pi_n \mid n \in \mathbb{N} \wedge \pi_0 \in \mathfrak{I}[\![\texttt{Sl}]\!] \wedge \forall i \in [0, n[\,.\, \langle \pi_i,\ \pi_{i+1}\rangle \in \widehat{\boldsymbol{\mathcal{S}}}^{\tau}[\![\texttt{Sl}']\!] \cup \widehat{\boldsymbol{\mathcal{S}}}^{\tau}[\![\texttt{S}]\!]\} \qquad ⦃(43.5)⦄$$

$$= \{\pi'_0 \cdot \cdots \cdot \pi'_{n'} \mid n' \in \mathbb{N} \wedge \pi'_0 \in \mathfrak{I}[\![\texttt{Sl}]\!] \wedge \forall i \in [0, n'[\,.\, \langle \pi'_i,\ \pi'_{i+1}\rangle \in \widehat{\boldsymbol{\mathcal{S}}}^{\tau}[\![\texttt{Sl}']\!]\} \cup \{\pi'_0 \cdot \cdots \cdot \pi'_{n'} \frown \pi''_0 \cdot \cdots \cdot \pi''_m \mid n' \in \mathbb{N} \wedge \pi'_0 \in \mathfrak{I}[\![\texttt{Sl}]\!] \wedge \forall i \in [0, n'[\,.\, \langle \pi'_i,\ \pi'_{i+1}\rangle \in \widehat{\boldsymbol{\mathcal{S}}}^{\tau}[\![\texttt{Sl}']\!] \wedge m \in \mathbb{N} \wedge \pi''_0 \in \mathfrak{I}[\![\texttt{S}]\!] \wedge \forall i \in [0, m[\,.\, \langle \pi''_i,\ \pi''_{i+1}\rangle \in \widehat{\boldsymbol{\mathcal{S}}}^{\tau}[\![\texttt{S}]\!]\}$$

⦃because all states involved in $\langle \ell, \rho\rangle \in \widehat{\boldsymbol{\mathcal{S}}}^{\tau}[\![\texttt{Sl}']\!]$ have $\ell \in \mathsf{labx}[\![\texttt{Sl}']\!]$ and similarly all states involved in $\langle \ell, \rho\rangle \in \widehat{\boldsymbol{\mathcal{S}}}^{\tau}[\![\texttt{S}]\!]$ have $\ell \in \mathsf{labx}[\![\texttt{S}]\!]$ and $\pi_0 = \pi'_0 \in \mathfrak{I}[\![\texttt{Sl}]\!] = \mathfrak{I}[\![\texttt{Sl}']\!]$ so that $\mathsf{labx}[\![\texttt{Sl}']\!] \cap \mathsf{labx}[\![\texttt{S}]\!] = \{\mathsf{after}[\![\texttt{Sl}']\!]\} = \{\mathsf{at}[\![\texttt{S}]\!]\}$ implies that the first transitions in $\pi'_0 \cdot \cdots \cdot \pi'_{n'}$ from $\pi_0 = \pi'_0$ must be transitions in $\widehat{\boldsymbol{\mathcal{S}}}^{\tau}[\![\texttt{Sl}']\!]$.

Then there are two cases. The first case is when $\pi'_0 \cdot \cdots \cdot \pi'_{n'}$ never reaches a final state of $\pi'_{n'} \neq \pi''_0$ of $\texttt{Sl}'$ so that $\pi'_0 \cdot \cdots \cdot \pi'_{n'} = \pi_0 \cdot \cdots \cdot \pi_n$.

Otherwise, a final state $\pi'_{n'} = \pi''_0$ is reached for which $\ell = \mathsf{after}[\![\texttt{Sl}']\!] = \mathsf{at}[\![\texttt{S}]\!]$ so $\pi'_0 \cdot \cdots \cdot \pi'_{n'} \frown \pi''_0 \cdot \cdots \cdot \pi''_m = \pi_0 \cdot \cdots \cdot \pi_n$ because after $\pi'_{n'} = \pi''_0$ the transitions in $\pi''_0 \cdot \cdots \cdot \pi''_m$, if any when $m > 0$, must belong to $\widehat{\boldsymbol{\mathcal{S}}}^{\tau}[\![\texttt{S}]\!]$. Both cases include the empty statement list $\texttt{Sl}' ::= \epsilon$ for which $n' = 0$⦄

$$= \widehat{\boldsymbol{\mathcal{S}}}^{*}_{\mathbb{S}}[\![\texttt{Sl}']\!] \cup \{\pi \cdot \langle \mathsf{at}[\![\texttt{S}]\!],\ \rho\rangle \cdot \pi' \mid \pi \cdot \langle \mathsf{at}[\![\texttt{S}]\!],\ \rho\rangle \in \widehat{\boldsymbol{\mathcal{S}}}^{*}_{\mathbb{S}}[\![\texttt{Sl}']\!] \wedge \langle \mathsf{at}[\![\texttt{S}]\!],\ \rho\rangle \cdot \pi' \in \widehat{\boldsymbol{\mathcal{S}}}^{*}_{\mathbb{S}}[\![\texttt{S}]\!]\}$$

⦃induction hypothesis⦄

$$= \widehat{\boldsymbol{\mathcal{S}}}^{*}_{\mathbb{S}}[\![\texttt{Sl}]\!] \qquad ⦃(42.5)⦄ \quad \square$$

– For the iteration statement $\texttt{S} ::= \textbf{while}\ \ell\ (\texttt{B})\ \texttt{S}_b$, we use the fact (42.2) that $\boldsymbol{\mathcal{S}}^{*}_{\mathbb{S}}[\![\texttt{S}]\!] \triangleq \alpha^{\mathbb{S}}(\boldsymbol{\mathcal{S}}^{*}[\![\texttt{S}]\!])$ as well as lemma 6.39 (a trivial rephrasing of this lemma for $\boldsymbol{\mathcal{S}}^{*}_{\mathbb{S}}[\![\texttt{S}]\!]$ would be useful). Obviously the traces in (6.39) can be generated by the transition system.

Conversely, let $\langle \ell_0,\ \rho_0\rangle \cdots \langle \ell_n,\ \rho_n\rangle$ be a trace generated by the transition system for an iteration $\texttt{S}$ so that $\ell_0 = \mathsf{at}[\![\texttt{S}]\!]$. By (6.6) and exercise 6.8, there exists $\pi_0\ell_0$ such that $\varrho(\pi_0\ell_0) = \rho_0$ and $\langle \ell_0,\ \rho_0\rangle$ in $\alpha^{\mathbb{S}}(\boldsymbol{\mathcal{S}}^{*}[\![\texttt{S}]\!])$. By reductio ad absurdum, let $k < n$ such that $\langle \ell_0,\ \rho_0\rangle \cdots \langle \ell_k,\ \rho_k\rangle$ is in $\alpha^{\mathbb{S}}(\boldsymbol{\mathcal{S}}^{*}[\![\texttt{S}]\!])$ but $\langle \ell_0,\ \rho_0\rangle \cdots \langle \ell_k,\ \rho_k\rangle\langle \ell_{k+1},\ \rho_{k+1}\rangle$ is not. By recurrence, there exists $\pi_k\ell_k$ such that $\rho_k = \varrho(\pi_0\ell_0 \frown \pi_k\ell_k)$. Because we have the transition $\langle \ell_k,\ \rho_k\rangle \longrightarrow \langle \ell_{k+1},\ \rho_{k+1}\rangle$ there is one statement $\texttt{S}'$ in section 43.1 from which this transition comes from such that $\ell_k \in \mathsf{labs}[\![\texttt{S}']\!]$. The contradiction is that a similar step is possible in $\alpha^{\mathbb{S}}(\boldsymbol{\mathcal{S}}^{*}[\![\texttt{S}]\!])$ (or $\boldsymbol{\mathcal{S}}^{*}[\![\texttt{S}]\!]$). We have to go on by considering all possible cases for $\texttt{S}'$. Because the reasoning is similar for all these cases, let us consider the typical cases (43.4) and (43.6).

In case (43.4), ℓ_k is at an assignment statement $\texttt{S}' ::= \ell\ \texttt{x} = \texttt{A}\,;$, we have $\rho_k = \varrho(\pi_0\ell_0 \frown \pi_k\ell_k)$. Because of unicity of the labeling, the transition $\langle \ell_k,\ \rho_k\rangle \longrightarrow \langle \ell_{k+1},\ \rho_{k+1}\rangle$ in (43.4) cannot come from any other statement. The contradiction is that (6.16) provides a transition in $\widehat{\boldsymbol{\mathcal{S}}}^{*}[\![\texttt{S}]\!](\pi_0\ell_0 \frown \pi_k\ell_k)$ that abstracts to $\langle \ell_{k+1},\ \rho_{k+1}\rangle$.

The reasoning is the same in case (43.6) for $\{\langle \ell,\ \rho\rangle \longrightarrow \langle \mathsf{after}[\![\texttt{S}]\!],\ \rho\rangle \mid \mathcal{B}[\![\texttt{B}]\!]\,\rho = \mathsf{ff}\}$ and $\{\langle \ell,\ \rho\rangle \longrightarrow \langle \mathsf{at}[\![\texttt{S}_b]\!],\ \rho\rangle \mid \mathcal{B}[\![\texttt{B}]\!]\,\rho = \mathsf{tt}\}$. Otherwise, $\ell_k \in \mathsf{labs}[\![\texttt{S}_b]\!]$ and we consider recursively the contradiction within $\texttt{S}' = \widehat{\boldsymbol{\mathcal{S}}}^{\tau}[\![\texttt{S}_b]\!]$. $\square$

It follows from theorem 43.11 that we could have followed the traditional way of defining a small-step operational semantics by first postulating the transition semantics of this chapter, then deriving the stateful prefix trace semantics by (43.1), and finally deriving the maximal trace semantics by taking limits as in chapter 7.

Exercise 43.12 Design, by calculus, the transition semantics $\widehat{\mathcal{S}}^{\tau}_{\mathbb{P}^i_{\mathbb{F}}}$ of the float interval prefix trace semantics $\widehat{\mathcal{S}}^{*}_{\mathbb{P}^i_{\mathbb{F}}}[\![\mathtt{S}]\!]$ of section 32.5.9.

43.3 Conclusion

In this chapter we made the link between the stateful prefix trace semantics of chapter 42 and the more traditional stateful small-step operational semantics in which execution traces or reachable states are defined by a transition semantics, for example, [970] stating "Each of the formalisms above involves the notion of a *state set* consisting of the set of all information configurations that can occur during computations and a *state transition relation* which defines for each state a next state or set of next states."

The transition semantics can be implemented in the language that it defines, which is John Reynolds' idea of definitional interpreters in [829] (following the definition of Lisp in Lisp [674, appendix B, p. 70–72]). [701] introduces monadic transition systems as a generalization of transition systems that unifies a wide range of models, including deterministic automata, nondeterministic automata, Markov chains, and probabilistic automata.

The operational semantics dates back to John McCarthy [673]. The Vienna Definition Language (VDL) [536, 971] was a formal specification language used essentially for giving operational semantics descriptions [480]. The aura of operational semantics faded away with the emergence of Dana Scott and Christopher Strachey's denotational semantics [684, 872, 914, 929], most people thinking that describing what is done instead of how it is done is more concise and elegant. Denotational semantics introduced the idea of structural definition and the use of fixpoints to handle iteration and recursion. It reached its limits in dealing with parallelism, for which no viable denotational solution ever emerged. The comeback of operational semantics is due to Gordon Plotkin [802], thanks to the use of structural rule-based deductive definitions (chapter 16), considered a gracefulness of style by many. Moreover, it applies to parallelism by interleaving of atomic actions. However, with weak memory models of modern machines [21], the use of states is somewhat heavy so stateless models might be an interesting alternative [23], hardly describable elegantly and concisely by state-based transition systems. This motivates our choice of starting from a stateless prefix trace semantics in chapter 6, instead of the more traditional stateful semantics of chapter 42, which has been easily recovered from chapter 6 by abstract interpretation.

43.4 Solutions to Selected Exercises

Solution to exercise 43.3 If $\mathcal{S} = \{a^n b \mid n \in \mathbb{N}\}$ then $\alpha^{\tau}(\mathcal{S}) = \{\langle a, a\rangle, \langle a, b\rangle\}$ so $\gamma^{\tau}(\{\langle a, a\rangle, \langle a, b\rangle\}) = \{a^n b \mid n \in \mathbb{N}\} \cup \{a^{\infty}\}$.

44 Software Model Checking

Model checking [188, 813] consists of proving that a model of a given program/computer system satisfies a temporal specification.[1] Traditionally, the model of the given program/computer system is a transition system and its semantics is the set of traces generated by the transition system. The temporal specification is usually one of the many variants of temporal logics such as the linear time temporal logic (LTL) [803], the computation tree logic (CTL) [188], the combination of the two (CTL*) [351], or the μ-calculus [597]. The semantics of the temporal specification is a set of traces. The problem is therefore to check that the set of traces of the semantics of the given program/computer system is included in the set of traces of the semantics of the temporal specification. This is an abstraction (formalized by the Galois connection of exercise 11.9) and so a model checking algorithm is an abstract interpretation of the program semantics that can be designed by calculus.

The main point of this chapter is to show that software model checking is an abstract interpretation of the program semantics, and not to provide an exhaustive introduction to model checking as found in [70, 189, 191].

To favor structural reasoning on programs, we use the stateful prefix trace semantics of section 42.2 as opposed to the transition system of chapter 43. For the specification, we use path conditions expressed by regular (or rational) expressions [581] as in [297, 876] but with an interpretation for prefix closed instead of maximal trace semantics. There are properties of traces that are not expressible in temporal logic but are easily expressible using regular expressions [988]. More expressive regular temporal specifications were proposed by [298, 355, 464, 988].

Contents

1. We define model checking as the verification of temporal properties and do not reduce it to the reachability analysis of chapter 20 (as done, for example, in [191, ch. 15, 16, 17, etc.]) because reachability analysis predates model checking [235] including for the use of transition systems [221].

44.1 Specifying Computations by Regular Expressions

In this chapter, the semantics properties that we consider are trace properties only, belonging to $\alpha^{\cup}(\wp(\wp(\mathbb{S}^+))) = \wp(\mathbb{S}^+)$ where $\alpha^{\cup}(P) \triangleq \bigcup P$ so for the collecting semantics $\alpha^{\cup}(\{\widehat{\boldsymbol{\mathcal{S}}}^*_{\mathsf{s}}[\![\mathsf{S}]\!]\}) = \widehat{\boldsymbol{\mathcal{S}}}^*_{\mathsf{s}}[\![\mathsf{S}]\!]$ (see section 8.6). Therefore we are in the simple case where we directly abstract the prefix trace semantics $\widehat{\boldsymbol{\mathcal{S}}}^*_{\mathsf{s}}[\![\mathsf{S}]\!]$.

We use regular expressions $\mathsf{R} \in \mathbb{R}$ to specify trace properties and define their syntax in section 44.1.1 and semantics $\boldsymbol{\mathcal{S}}^{\mathsf{r}}[\![\mathsf{R}]\!] \in \wp(\mathbb{S}^+)$ in section 44.1.2. Given a program component S and a specification R model checking is then the check that S satisfies R that is $\widehat{\boldsymbol{\mathcal{S}}}^*_{\mathsf{s}}[\![\mathsf{S}]\!] \subseteq \boldsymbol{\mathcal{S}}^{\mathsf{r}}[\![\mathsf{R}]\!]$.

The definition of regular model checking in section 44.2 will be slightly more complicated because we want the specification R to be able to relate the values taken at a program point to the initial values of program variables, as was the case for relational invariants in section 19.1.2 (as opposed to the assertional invariants of section 19.1.1).

Stephen Cole Kleene introduced regular expressions and finite automata to specify execution traces (called events) of automata (called nerve nets) [581]. Kleene proved in [581] that regular expressions and (deterministic or nondeterministic) finite automata can describe exactly the same classes of languages (see [850, ch. 1, sect. 4]). He noted that not all computable execution traces of nerve nets can be (exactly) represented by a regular expression. The situation is the same for programs for which regular expressions (or equivalently finite automata) can specify a superset of the prefix state trace semantics $\widehat{\boldsymbol{\mathcal{S}}}^*_{\mathsf{s}}[\![\mathsf{S}]\!]$ of program components $\mathsf{S} \in \mathbb{P}\mathbb{c}$. An early example is the testing with finite automata specifications [478, 479]. A similar example is security monitors.

Example 44.1 (security monitors) Fred Schneider's security monitors [644, 862] use a finite automata specification to state requirements of hardware or software systems. They have been used to check for safety program properties at runtime, that is, for any given execution trace in the semantics $\widehat{\boldsymbol{\mathcal{S}}}^*_{\mathsf{s}}[\![\mathsf{S}]\!]$. The safety property specified by the finite automaton or, equivalently, a regular expression is temporal, that is, links events occurring at different times in the computation (such as that a file must be opened before being accessed, must eventually be closed, and cannot be accessed after closing). Checking this safety temporal specification for all possible execution traces in the semantics $\widehat{\boldsymbol{\mathcal{S}}}^*_{\mathsf{s}}[\![\mathsf{S}]\!]$ is a static analysis problem, more precisely, a model checking problem [907], because model checking specializes on temporal properties.

The use of regular expressions is one distinctive feature of this chapter. Regular expressions are commonly used in text editing and understood by all programmers, which is not the case for temporal logics. This is the reason why we use regular expressions as specification language. This also allows us to construct a new model checking algorithm, which would not have been the case with classical temporal logic specifications (except that most model checking algorithms are not structural).

44.1.1 Syntax of Regular Expressions

Classical regular expressions consist of constants (empty string ε, literal characters a, b, etc.), which denote sets of strings, and operator symbols (concatenation $\bullet$, alternation $\mid$, repetition zero or more times * or one or more times $^+$), which denote operations over these sets. We replace the literal characters by invariant specifications $\mathsf{L} : \mathsf{B}$ stating that Boolean expression B should be true whenever control reaches any program point in the set L of program labels.[2] The Boolean expression B may depend on program variables $\mathsf{x}, \mathsf{y}, \ldots \in \mathbb{V}$ and their initial values denoted $\underline{\mathsf{x}}, \underline{\mathsf{y}}, \ldots \in \underline{\mathbb{V}}$ where $\underline{\mathbb{V}} \triangleq \{\underline{\mathsf{x}} \mid \mathsf{x} \in \mathbb{V}\}$.

$$
\begin{array}{rcll}
\mathsf{L} & \in & \wp(\mathbb{L}) & \text{sets of program labels} \\
\mathsf{x}, \mathsf{y}, \ldots & \in & \mathbb{V} & \text{program variables} \\
\underline{\mathsf{x}}, \underline{\mathsf{y}}, \ldots & \in & \underline{\mathbb{V}} & \text{initial values of variables} \\
\mathsf{B} & \in & \mathbb{B} & \text{Boolean expressions such that } \mathbb{vars}[\![\mathsf{B}]\!] \subseteq \mathbb{V} \cup \underline{\mathbb{V}}
\end{array}
$$

$$
\begin{array}{rcll}
\mathsf{R} & \in & \mathbb{R} & \text{regular expressions} \\
\mathsf{R} & ::= & \varepsilon & \text{empty} \\
& | & \mathsf{L} : \mathsf{B} & \text{invariant } \mathsf{B} \text{ at } \mathsf{L} \\
& | & \mathsf{R}_1 \mathsf{R}_2 \quad (\text{or } \mathsf{R}_1 \bullet \mathsf{R}_2) & \text{concatenation} \\
& | & \mathsf{R}_1 \mid \mathsf{R}_2 & \text{alternative} \\
& | & \mathsf{R}_1^* & \text{zero or more occurrences of } \mathsf{R} \\
& | & \mathsf{R}_1^+ & \text{one or more occurrences of } \mathsf{R} \\
& | & (\mathsf{R}_1) & \text{grouping}
\end{array}
\tag{44.2}
$$

We use abbreviations to designate sets of labels such as $? : \mathsf{B} \triangleq \mathbb{L} : \mathsf{B}$ so that B is invariant, $\ell : \mathsf{B} \triangleq \{\ell\} : \mathsf{B}$ so that B is invariant at program label ℓ, $\neg\ell : \mathsf{B} \triangleq \mathbb{L} \setminus \{\ell\} : \mathsf{B}$ when B holds everywhere but at program point ℓ, and so on.

Example 44.3 $(? : \mathtt{tt})^*$ holds for any program. $(? : \mathsf{x} >= 0)^*$ states that the value of x is always positive or zero during program execution. $(? : \mathsf{x} >= \underline{\mathsf{x}})^*$ states that the value of x is always greater than or equal to its initial value $\underline{\mathsf{x}}$ during execution. $(\neg\ell : \mathsf{x} >= 0)^* \bullet \ell : \mathsf{x} == 0 \bullet (? : \mathsf{x} < 0)^*$ states that the value of x should be positive or zero and if program point ℓ is ever reached then x should be 0, and if computations go on after program point ℓ then x should be negative afterward (including if and when execution comes back to ℓ).

Example 44.4 Continuing example 44.1 for security monitors, the basic regular expressions are names a of program actions. We can understand such an action a as designating the set L of labels of all its occurrences in the program. If necessary, the Boolean expression B can be used to specify the parameters of the action.

There are many regular expressions denoting the language $\{\ni\}$ containing only the empty sequence denoted $\ni$ (such that ε, $\varepsilon\varepsilon$, ε^*, etc.), as shown by the following grammar.

2. Alternatively, using the trace semantics of chapters 6 and 7, regular expressions could be used to specify requirements on the sequences of actions allowed in traces.

$$\mathsf{R} \in \mathbb{R}_\varepsilon \qquad \text{empty regular expressions}$$

$$\mathsf{R} ::= \varepsilon \mid \mathsf{R}_1\mathsf{R}_2 \mid \mathsf{R}_1 \mid \mathsf{R}_2 \mid \mathsf{R}_1^* \mid \mathsf{R}_1^+ \mid (\mathsf{R}_1) \tag{44.5}$$

For specification we use only nonempty regular expressions $\mathsf{R} \in \mathbb{R}^+$ because traces cannot be empty. The definition takes into account that, for example, $\mathsf{L} : \mathsf{B}$ can be written as $\varepsilon\varepsilon\varepsilon \ldots \varepsilon\mathsf{L} : \mathsf{B}\varepsilon\varepsilon \ldots \varepsilon$.

$$\mathsf{R} \in \mathbb{R}^+ \qquad \text{nonempty regular expressions}$$

$$\mathsf{R} ::= \mathsf{L} : \mathsf{B} \mid \varepsilon\mathsf{R}_2 \mid \mathsf{R}_1\varepsilon \mid \mathsf{R}_1\mathsf{R}_2 \mid \mathsf{R}_1 \mid \mathsf{R}_2 \mid \mathsf{R}_1^+ \mid (\mathsf{R}_1)$$

We will also have to consider regular expressions $\mathsf{R} \in \mathbb{R}^{\dagger}$ containing no alternative $\mid$.

$$\mathsf{R} \in \mathbb{R}^{\dagger} \qquad \mid\text{-free regular expressions}$$

$$\mathsf{R} ::= \varepsilon \mid \mathsf{L} : \mathsf{B} \mid \mathsf{R}_1\mathsf{R}_2 \mid \mathsf{R}_1^* \mid \mathsf{R}_1^+ \mid (\mathsf{R}_1)$$

44.1.2 Relational Semantics of Regular Expressions

The semantics (3.4) of expressions is changed as follows ($\underline{\varrho}$ denotes the initial values $\underline{\mathsf{x}}$ of variables x and ρ their current value, $\uparrow$ is the alternative denial logical operation defined in section 3.6)

$$\begin{array}{lll} \mathscr{A}[\![1]\!]\underline{\varrho}, \rho \triangleq 1 & \mathscr{A}[\![\mathsf{A}_1 - \mathsf{A}_2]\!]\underline{\varrho}, \rho \triangleq \mathscr{A}[\![\mathsf{A}_1]\!]\underline{\varrho}, \rho - \mathscr{A}[\![\mathsf{A}_2]\!]\underline{\varrho}, \rho & (44.6) \\ \mathscr{A}[\![\underline{\mathsf{x}}]\!]\underline{\varrho}, \rho \triangleq \underline{\varrho}(\mathsf{x}) & \mathscr{B}[\![\mathsf{A}_1 < \mathsf{A}_2]\!]\underline{\varrho}, \rho \triangleq \mathscr{A}[\![\mathsf{A}_1]\!]\underline{\varrho}, \rho < \mathscr{A}[\![\mathsf{A}_2]\!]\underline{\varrho}, \rho & \\ \mathscr{A}[\![\mathsf{x}]\!]\underline{\varrho}, \rho \triangleq \rho(\mathsf{x}) & \mathscr{B}[\![\mathsf{B}_1 \text{ nand } \mathsf{B}_2]\!]\underline{\varrho}, \rho \triangleq \mathscr{B}[\![\mathsf{B}_1]\!]\underline{\varrho}, \rho \uparrow \mathscr{B}[\![\mathsf{B}_2]\!]\underline{\varrho}, \rho & \end{array}$$

Let us represent a nonempty finite sequence $\sigma_1 \ldots \sigma_n \in \mathbb{S}^+ \triangleq \bigcup_{n \in \mathbb{N}^+} [1, n] \to \mathbb{S}$ of states $\sigma_i \in \mathbb{S} \triangleq (\mathbb{L} \times \mathbb{E}\mathrm{v})$ by a map $\sigma \in [1, n] \to \mathbb{S}$ (which is the empty sequence $\sigma = \ni$ when $n = 0$).

The relational semantics $\boldsymbol{\mathcal{S}}^{\mathsf{r}}[\![\mathsf{R}]\!] \in \wp(\mathbb{E}\mathrm{v} \times \mathbb{S}^*)$ of regular expressions R relates an arbitrary initial environment $\underline{\varrho} \in \mathbb{E}\mathrm{v}$ to a trace $\pi \in \mathbb{S}^*$ by defining how the states of the trace π are related to that initial environment $\underline{\varrho}$.

$$\begin{array}{ll} \boldsymbol{\mathcal{S}}^{\mathsf{r}}[\![\varepsilon]\!] \triangleq \{\langle\underline{\varrho}, \ni\rangle \mid \underline{\varrho} \in \mathbb{E}\mathrm{v}\} & \boldsymbol{\mathcal{S}}^{\mathsf{r}}[\![\mathsf{R}]\!]^0 \triangleq \boldsymbol{\mathcal{S}}^{\mathsf{r}}[\![\ni]\!] \qquad (44.7) \\ \boldsymbol{\mathcal{S}}^{\mathsf{r}}[\![\mathsf{L} : \mathsf{B}]\!] \triangleq \{\langle\underline{\varrho}, \langle\ell, \rho\rangle\rangle \mid \ell \in \mathsf{L} \wedge \mathscr{B}[\![\mathsf{B}]\!]\underline{\varrho}, \rho\} & \boldsymbol{\mathcal{S}}^{\mathsf{r}}[\![\mathsf{R}]\!]^{n+1} \triangleq \boldsymbol{\mathcal{S}}^{\mathsf{r}}[\![\mathsf{R}]\!]^n \circledcirc \boldsymbol{\mathcal{S}}^{\mathsf{r}}[\![\mathsf{R}]\!] \\ \boldsymbol{\mathcal{S}}^{\mathsf{r}}[\![\mathsf{R}_1\mathsf{R}_2]\!] \triangleq \boldsymbol{\mathcal{S}}^{\mathsf{r}}[\![\mathsf{R}_1]\!] \circledcirc \boldsymbol{\mathcal{S}}^{\mathsf{r}}[\![\mathsf{R}_2]\!] & \boldsymbol{\mathcal{S}}^{\mathsf{r}}[\![\mathsf{R}^*]\!] \triangleq \bigcup_{n \in \mathbb{N}} \boldsymbol{\mathcal{S}}^{\mathsf{r}}[\![\mathsf{R}]\!]^n \\ \boldsymbol{\mathcal{S}} \circledcirc \boldsymbol{\mathcal{S}}' \triangleq \{\langle\underline{\varrho}, \pi \cdot \pi'\rangle \mid \langle\underline{\varrho}, \pi\rangle \in \boldsymbol{\mathcal{S}} \wedge \langle\underline{\varrho}, \pi'\rangle \in \boldsymbol{\mathcal{S}}'\} & \boldsymbol{\mathcal{S}}^{\mathsf{r}}[\![\mathsf{R}^+]\!] \triangleq \bigcup_{n \in \mathbb{N}^+} \boldsymbol{\mathcal{S}}^{\mathsf{r}}[\![\mathsf{R}]\!]^n \\ \boldsymbol{\mathcal{S}}^{\mathsf{r}}[\![\mathsf{R}_1 \mid \mathsf{R}_2]\!] \triangleq \boldsymbol{\mathcal{S}}^{\mathsf{r}}[\![\mathsf{R}_1]\!] \cup \boldsymbol{\mathcal{S}}^{\mathsf{r}}[\![\mathsf{R}_2]\!] & \boldsymbol{\mathcal{S}}^{\mathsf{r}}[\![(\mathsf{R})]\!] \triangleq \boldsymbol{\mathcal{S}}^{\mathsf{r}}[\![\mathsf{R}]\!] \end{array}$$

Exercise 44.8 Prove that if $\mathsf{R} \in \mathbb{R}_\varepsilon$ then $\boldsymbol{\mathcal{S}}^{\mathsf{r}}[\![\mathsf{R}]\!] = \boldsymbol{\mathcal{S}}^{\mathsf{r}}[\![\varepsilon]\!]$ and if $\mathsf{R} \in \mathbb{R}^+$ then $\boldsymbol{\mathcal{S}}^{\mathsf{r}}[\![\mathsf{R}]\!] \in \wp(\mathbb{E}\mathrm{v} \times \mathbb{S}^+)$.

Example 44.9 In the following example, the initial environment $\underline{\varrho}$ is used to record the initial value of variables of a program, and the trace π describes the execution of some component of that program. The semantics $\boldsymbol{\mathcal{S}}^{\mathsf{r}}[\![\mathsf{R}]\!]$ of the regular expression $\mathsf{R} \triangleq \ell : \mathsf{x} = \underline{\mathsf{x}} \bullet \ell' : \mathsf{x} = \underline{\mathsf{x}} + 1$ is

$$\begin{aligned}
\boldsymbol{\mathcal{S}}^{\mathsf{r}}[\![\ell : \mathsf{x} = \underline{\mathsf{x}}]\!] &\triangleq \{\langle \underline{\varrho}, \langle \ell, \rho\rangle\rangle \mid \rho(\mathsf{x}) = \underline{\varrho}(\mathsf{x})\} \\
\boldsymbol{\mathcal{S}}^{\mathsf{r}}[\![\ell' : \mathsf{x} = \underline{\mathsf{x}} + 1]\!] &\triangleq \{\langle \underline{\varrho}', \langle \ell', \rho'\rangle\rangle \mid \rho'(\mathsf{x}) = \underline{\varrho}'(\mathsf{x}) + 1\} \\
\boldsymbol{\mathcal{S}}^{\mathsf{r}}[\![\ell : \mathsf{x} = \underline{\mathsf{x}} \bullet \ell' : \mathsf{x} = \underline{\mathsf{x}} + 1]\!] &\triangleq \{\langle \underline{\varrho}, \langle \ell, \rho\rangle\langle \ell', \rho'\rangle\rangle \mid \langle \underline{\varrho}, \langle \ell, \rho\rangle\rangle \in \boldsymbol{\mathcal{S}}^{\mathsf{r}}[\![\ell : \mathsf{x} = \underline{\mathsf{x}}]\!] \wedge \\
&\qquad \langle \underline{\varrho}, \langle \ell', \rho'\rangle\rangle \in \boldsymbol{\mathcal{S}}^{\mathsf{r}}[\![\ell' : \mathsf{x} = \underline{\mathsf{x}} + 1]\!]\} \\
&= \{\langle \underline{\varrho}, \langle \ell, \rho\rangle\langle \ell', \rho'\rangle\rangle \mid \rho(\mathsf{x}) = \underline{\varrho}(\mathsf{x}) \wedge \underline{\varrho}' = \underline{\varrho} \wedge \rho'(\mathsf{x}) = \underline{\varrho}'(\mathsf{x}) + 1\} \\
&= \{\langle \underline{\varrho}, \langle \ell, \rho\rangle\langle \ell', \rho'\rangle\rangle \mid \rho(\mathsf{x}) = \underline{\varrho}(\mathsf{x}) \wedge \rho'(\mathsf{x}) = \underline{\varrho}(\mathsf{x}) + 1\}
\end{aligned}$$

meaning that if the initial value of x is $\underline{\mathsf{x}}$ at ℓ then, after one step of computation, it is $\underline{\mathsf{x}} + 1$ at ℓ'

Exercise 44.10 Prove that $\forall n \in \mathbb{N} \,.\, \boldsymbol{\mathcal{S}}^{\mathsf{r}}[\![\mathsf{R}]\!]^{n+1} = \boldsymbol{\mathcal{S}}^{\mathsf{r}}[\![\mathsf{R}]\!] \circledcirc \boldsymbol{\mathcal{S}}^{\mathsf{r}}[\![\mathsf{R}]\!]^{n}$.

Exercise 44.11 (Arden's lemma [52]) Prove that $\forall \mathsf{R} \in \mathbb{R} \,.\, \boldsymbol{\mathcal{S}}^{\mathsf{r}}[\![\mathsf{R}^{+}]\!] = \mathsf{lfp}^{\subseteq}\, X \mapsto \boldsymbol{\mathcal{S}}^{\mathsf{r}}[\![\mathsf{R}]\!] \cup \boldsymbol{\mathcal{S}}^{\mathsf{r}}[\![\mathsf{R}]\!] \circledcirc X = \mathsf{lfp}^{\subseteq}\, X \mapsto \boldsymbol{\mathcal{S}}^{\mathsf{r}}[\![\mathsf{R}]\!] \cup X \circledcirc \boldsymbol{\mathcal{S}}^{\mathsf{r}}[\![\mathsf{R}]\!]$.

Arden lemma in exercise 44.1.2 is the basis for solving of regular equations, for example, to convert a deterministic finite automaton into a regular expression [916].

44.2 Definition of Regular Model Checking

Let the prefix closure $\alpha_{\mathsf{prefix}}(\Pi)$ of a set $\Pi \in \wp(\mathbb{E}\mathrm{v} \times \mathbb{S}^{+})$ of stateful traces be

$$\alpha_{\mathsf{prefix}}(\Pi) \triangleq \{\langle \underline{\varrho}, \pi\rangle \mid \pi \in \mathbb{S}^{+} \wedge \exists \pi' \in \mathbb{S}^{*} \,.\, \langle \underline{\varrho}, \pi \cdot \pi'\rangle \in \Pi\} \qquad \text{prefix closure.}$$

The following definition 44.12 defines the model checking problem $\mathsf{P}, \underline{\varrho} \models \mathsf{R}$ as checking that the semantics of the given program $\mathsf{P} \in \mathbb{P}$ meets the regular[3] specification $\mathsf{R} \in \mathbb{R}^{+}$ for the initial environment $\underline{\varrho}$.

Definition 44.12 (model checking)

$$\mathsf{P}, \underline{\varrho} \models \mathsf{R} \triangleq (\{\underline{\varrho}\} \times \widehat{\boldsymbol{\mathcal{S}}}^{*}_{\mathbb{S}}[\![\mathsf{P}]\!]) \subseteq \alpha_{\mathsf{prefix}}(\boldsymbol{\mathcal{S}}^{\mathsf{r}}[\![\mathsf{R} \bullet (? : \mathsf{tt})^{*}]\!]) \qquad (44.12)$$

Remark 44.13 The prefix closure α_{prefix} allows the regular specification R to specify traces satisfying a prefix of the specification only, as in $^{\ell}\,\mathsf{x} = \mathsf{x} + 1\,;^{\ell'}, \underline{\varrho} \models \mathsf{R}$ where $\mathsf{R} = \ell : \mathsf{x} = \underline{\mathsf{x}} \bullet \ell' : \mathsf{x} = \underline{\mathsf{x}} + 1 \bullet \ell'' : \mathsf{x} = \underline{\mathsf{x}} + 3$ and $\underline{\varrho}(\mathsf{x}) = \underline{\mathsf{x}}$.

The extension of the specification by $(? : \mathsf{tt})^{*}$ allows for the regular specification R to specify only a prefix of the traces, such as $^{\ell}\,\mathsf{x} = \mathsf{x} + 1\,;^{\ell'}\,\mathsf{x} = \mathsf{x} + 2\,;^{\ell''}, \underline{\varrho} \models \ell : \mathsf{x} = \underline{\mathsf{x}} \bullet \ell' : \mathsf{x} = \underline{\mathsf{x}} + 1$.

Model checking is a Boolean abstraction $\langle \wp(\mathbb{S}^{+}), \subseteq\rangle \underset{\alpha_{\underline{\varrho},\mathsf{R}}}{\overset{\gamma_{\underline{\varrho},\mathsf{R}}}{\leftrightarrows}} \langle \mathbb{B}, \Leftarrow\rangle$ where $\alpha_{\underline{\varrho},\mathsf{R}}(\Pi) \triangleq (\{\underline{\varrho}\} \times \Pi) \subseteq \alpha_{\mathsf{prefix}}(\boldsymbol{\mathcal{S}}^{\mathsf{r}}[\![\mathsf{R} \bullet (? : \mathsf{tt})^{*}]\!])$.

3. We understand "regular model checking" as checking temporal specifications given by a regular expression. This is different from [4, 374, 570] model checking transition systems of which the states are regular word or tree languages.

44.3 Properties of Regular Expressions

We state properties of regular expressions that we will use to design a structural model checking abstract semantics,

44.3.1 Equivalence of Regular Expressions

We say that regular expressions are equivalent when they have the same semantics.

$$\mathsf{R}_1 \Leftrightarrow \mathsf{R}_2 \quad \triangleq \quad \mathcal{S}^{\mathsf{r}}[\![\mathsf{R}_1]\!] = \mathcal{S}^{\mathsf{r}}[\![\mathsf{R}_2]\!]$$

44.3.2 Disjunctive Normal Form dnf of Regular Expressions

As noticed by Kleene [581, p. 14], regular expressions can be put into the equivalent disjunctive normal form of Hilbert–Ackermann [497, ch I, § 3 and ch III, § 8]. A regular expression is in disjunctive normal form if it is of the form $(\mathsf{R}_1 \mid \ldots \mid \mathsf{R}_n)$ for some $n \geqslant 1$, in which none of the R_i, for $1 \leqslant i \leqslant n$, contains an occurrence of $\mid$. Any regular expression R has a disjunctive normal form $\mathsf{dnf}(\mathsf{R})$ defined as follows:

$$\begin{aligned}
\mathsf{dnf}(\varepsilon) &\triangleq \varepsilon \\
\mathsf{dnf}(\mathsf{L} : \mathsf{B}) &\triangleq \mathsf{L} : \mathsf{B} \\
\mathsf{dnf}(\mathsf{R}_1\mathsf{R}_2) &\triangleq \mathsf{let}\ \mathsf{R}_1^1 \mid \ldots \mid \mathsf{R}_1^{n_1} = \mathsf{dnf}(\mathsf{R}_1)\ \mathsf{and}\ \mathsf{R}_2^1 \mid \ldots \mid \mathsf{R}_2^{n_2} = \mathsf{dnf}(\mathsf{R}_2)\ \mathsf{in} \mathop{\big|}_{i=1}^{n_1} \mathop{\big|}_{j=1}^{n_2} \mathsf{R}_1^i\mathsf{R}_2^j \\
\mathsf{dnf}(\mathsf{R}_1 \mid \mathsf{R}_2) &\triangleq \mathsf{dnf}(\mathsf{R}_1) \mid \mathsf{dnf}(\mathsf{R}_2) \\
\mathsf{dnf}(\mathsf{R}^*) &\triangleq \mathsf{let}\ \mathsf{R}^1 \mid \ldots \mid \mathsf{R}^n = \mathsf{dnf}(\mathsf{R})\ \mathsf{in}\ ((\mathsf{R}^1)^* \ldots (\mathsf{R}^n)^*)^* \\
\mathsf{dnf}(\mathsf{R}^+) &\triangleq \mathsf{dnf}(\mathsf{R}\mathsf{R}^*) \\
\mathsf{dnf}((\mathsf{R})) &\triangleq (\mathsf{dnf}(\mathsf{R}))
\end{aligned} \tag{44.14}$$

The proof that $\mathsf{dnf}(\mathsf{R})$ is in disjunctive normal form is by structural induction using the fact that the $\mathsf{R}^1, \ldots, \mathsf{R}^n$ do not contain any $\mid$. Lemma 44.16 shows that normalization leaves the semantics unchanged. It uses the following exercise.

Exercise 44.15 ([510, section 3.4.6, p. 118]) Prove that $(\mathsf{R}_1 \mid \mathsf{R}_2)^* \Leftrightarrow (\mathsf{R}_1{}^*\mathsf{R}_2{}^*)^*$ where the R_1 and R_2 do not contain any $\mid$.

[**Lemma 44.16** $\mathsf{dnf}(\mathsf{R}) \Leftrightarrow \mathsf{R}$.

Lemma 44.16 shows that normalization in (44.14) can be further simplified by $\varepsilon\mathsf{R} \Leftrightarrow \mathsf{R}\varepsilon \Leftrightarrow \mathsf{R}$ and $(\varepsilon)^* \Leftrightarrow \varepsilon$ which have equivalent semantics.

Exercise 44.17 Prove lemma 44.16.

44.3.3 fstnxt Derivative and Continuation of Regular Expressions

Janusz Brzozowski [151] introduced the notion of derivation for regular expressions (extended with arbitrary Boolean operations). The derivative of a regular expression R with respect to a symbol a, typically denoted as $D_a(\mathsf{R})$ or $a^{-1}\mathsf{R}$, is a regular expression given by a simple recursive definition on the syntactic structure of R. The crucial property of these derivatives is that a string of the form $a\sigma$ (starting with the symbol a) matches an expression R iff the suffix σ matches the derivative $D_a(\mathsf{R})$ [35, 151, 770].

For example, let $\mathsf{R} = a \bullet b^*$ over the alphabet $\{a, b, c\}$. The semantics of the derivative of R with respect to a is that of b^*, and the semantics of the derivative of R with respect to b or c is the empty set.

Following this idea, assume that a nonempty regular expression $\mathsf{R} \in \mathbb{R}^+$ has been decomposed into disjunctive normal form $(\mathsf{R}_1 \mid \ldots \mid \mathsf{R}_n)$ for some $n \geqslant 1$, in which none of the R_i, for $i \in [1, n]$, contains an occurrence of $\mid$. We can further decompose each $\mathsf{R}_i \in \mathbb{R}^+ \cap \mathbb{R}^{\dagger}$ into $\langle \mathsf{L} : \mathsf{B},\ \mathsf{R}_i' \rangle = \mathsf{fstnxt}(\mathsf{R}_i)$ such that

- the derivative $\mathsf{L} : \mathsf{B}$ recognizes the first state of sequences of states recognized by R_i; and
- the continuation R_i' is a regular expression that recognizes sequences of states after the first state of sequences of states recognized by R_i.

We define fstnxt for nonempty $\mid$-free regular expressions $\mathsf{R} \in \mathbb{R}^+ \cap \mathbb{R}^{\dagger}$ by structural induction, as follows:

$$\begin{aligned}
\mathsf{fstnxt}(\mathsf{L} : \mathsf{B}) &\triangleq \langle \mathsf{L} : \mathsf{B},\ \varepsilon \rangle && (44.18)\\
\mathsf{fstnxt}(\mathsf{R}_1\mathsf{R}_2) &\triangleq \mathsf{fstnxt}(\mathsf{R}_2) && \text{if } \mathsf{R}_1 \in \mathbb{R}_\varepsilon\\
\mathsf{fstnxt}(\mathsf{R}_1\mathsf{R}_2) &\triangleq \mathsf{let}\ \langle \mathsf{R}_1^f,\ \mathsf{R}_1^n \rangle = \mathsf{fstnxt}(\mathsf{R}_1)\ \mathsf{in}\ (\!\lvert\ \mathsf{R}_1^n \in \mathbb{R}_\varepsilon \mathbin{?} \langle \mathsf{R}_1^f,\ \mathsf{R}_2 \rangle \mathbin{⸰} \langle \mathsf{R}_1^f,\ \mathsf{R}_1^n \bullet \mathsf{R}_2 \rangle\ \rvert\!) && \text{if } \mathsf{R}_1 \notin \mathbb{R}_\varepsilon\\
\mathsf{fstnxt}(\mathsf{R}^+) &\triangleq \mathsf{let}\ \langle \mathsf{R}^f,\ \mathsf{R}^n \rangle = \mathsf{fstnxt}(\mathsf{R})\ \mathsf{in}\ (\!\lvert\ \mathsf{R}^n \in \mathbb{R}_\varepsilon \mathbin{?} \langle \mathsf{R}^f,\ \mathsf{R}^* \rangle \mathbin{⸰} \langle \mathsf{R}^f,\ \mathsf{R}^n \bullet \mathsf{R}^* \rangle\ \rvert\!)\\
\mathsf{fstnxt}((\mathsf{R})) &\triangleq \mathsf{fstnxt}(\mathsf{R})
\end{aligned}$$

For example, if $\mathsf{R} = a \bullet b^*$ where $a = \mathsf{L}_1 : \mathsf{B}_1$ and $b = \mathsf{L}_2 : \mathsf{B}_2$ then $\mathsf{fstnxt}(\mathsf{R}) = \langle a,\ b^* \rangle$.

The following lemma 44.19 shows the equivalence of an alternative-free regular expression and its first and next decomposition.

Lemma 44.19 Let $\mathsf{R} \in \mathbb{R}^+ \cap \mathbb{R}^{\dagger}$ be a nonempty $\mid$-free regular expression and $\langle \mathsf{L} : \mathsf{B},\ \mathsf{R}' \rangle = \mathsf{fstnxt}(\mathsf{R})$. Then $\mathsf{R}' \in \mathbb{R}^{\dagger}$ is $\mid$-free and $\mathsf{R} \simeq \mathsf{L} : \mathsf{B} \bullet \mathsf{R}'$.

Exercise 44.20 Prove lemma 44.19.

44.4 The Model Checking Abstraction

The model checking abstraction in section 44.2 is impractical for structural model checking. Assume, for example, that we check a trace concatenation $\pi_1 \frown \pi_2$ of a statement list $\mathsf{Sl} ::= \mathsf{Sl}'\ \mathsf{S}$ for a specification R where π_1 is a trace of Sl' and π_2 is a trace of S. We first check that π_1 satisfies R. We must then check π_2 for a continuation R_2 of R which should be derived from π_1 and R. This continuation is not provided by the Boolean abstraction $\alpha_{\underline{\varrho},\mathsf{R}}$, which needs to be refined as shown thereafter.

Example 44.21 Assume we want to check $\ell_1\ \mathsf{x} = \mathsf{x} + 1\ ;\ell_2\ \mathsf{x} = \mathsf{x} + 2\ ;\ell_3$ for the regular specification $? : \mathsf{x} = \underline{\mathsf{x}} \bullet ? : \mathsf{x} = \underline{\mathsf{x}} + 1 \bullet ? : \mathsf{x} = \underline{\mathsf{x}} + 3$ where $?$ can be any program point. We should first check the first statement and then the second. Knowing the Boolean information that $\ell_1\ \mathsf{x} = \mathsf{x} + 1\ ;\ell_2$ model checks for $? : \mathsf{x} = \underline{\mathsf{x}} \bullet ? : \mathsf{x} = \underline{\mathsf{x}} + 1$ is not enough. We must also know what to check the continuation $\ell_2\ \mathsf{x} = \mathsf{x} + 2\ ;\ell_3$ for. (This is $? : \mathsf{x} = \underline{\mathsf{x}} + 1 \bullet ? : \mathsf{x} = \underline{\mathsf{x}} + 3$,that is, if x is equal to the initial value plus 1 at ℓ_2, it is equal to this initial value plus 3 at ℓ_3.)

The model-checking $\mathcal{M}^t\langle \underline{\varrho}, \mathsf{R}\rangle\pi$ of a trace π with initial environment $\underline{\varrho}$ for a $\mid$-free specification $\mathsf{R} \in \mathbb{R}^{\dagger}$ is a pair $\langle b, \mathsf{R}'\rangle$ where the Boolean b states whether the specification R holds for the trace π and R' specifies the possible continuations of π according to R, ε if none.

Example 44.22 • For $\mathsf{S}_1 = \ell_1\ \mathsf{x} = \mathsf{x} + 1\ ;\ell_2$, we have

- $\widehat{\mathcal{S}}^*_{\mathsf{s}}[\![\ell_1\ \mathsf{x} = \mathsf{x} + 1\ ;\ell_2]\!] = \{\langle \ell_1, \rho\rangle\langle \ell_2, \rho[\mathsf{x} \leftarrow \rho(\mathsf{x}) + 1]\rangle \mid \rho \in \mathbb{E}\mathbb{v}\}$.
- $\mathcal{M}^t\langle \rho,\ ?:\mathsf{x} = \underline{\mathsf{x}} \bullet ?:\mathsf{x} = \underline{\mathsf{x}} + 1 \bullet ?:\mathsf{x} = \underline{\mathsf{x}} + 3\rangle(\langle \ell_1, \rho\rangle\langle \ell_2, \rho[\mathsf{x} \leftarrow \rho(\mathsf{x}) + 1]\rangle) = \langle \mathsf{tt},$ $?:\mathsf{x} = \underline{\mathsf{x}} + 1 \bullet ?:\mathsf{x} = \underline{\mathsf{x}} + 3\rangle$.

- For $\mathsf{S}_2 = \ell_2\ \mathsf{x} = \mathsf{x} + 2\ ;\ell_3$, we have
  - $\widehat{\mathcal{S}}^*_{\mathsf{s}}[\![\ell_2\ \mathsf{x} = \mathsf{x} + 2\ ;\ell_3]\!] = \{\langle \ell_2, \rho'\rangle\langle \ell_3, \rho'[\mathsf{x} \leftarrow \rho'(\mathsf{x}) + 2]\rangle \mid \rho' \in \mathbb{E}\mathbb{v}\}$.
  - $\mathcal{M}^t\langle \rho,\ ?:\mathsf{x} = \underline{\mathsf{x}} + 1 \bullet ?:\mathsf{x} = \underline{\mathsf{x}} + 3\rangle(\langle \ell_2, \rho[\mathsf{x} \leftarrow \rho(\mathsf{x}) + 1]\rangle\langle \ell_3, \rho[\mathsf{x} \leftarrow \rho(\mathsf{x}) + 3]\rangle) =$ $\langle \mathsf{tt},\ ?:\mathsf{x} = \underline{\mathsf{x}} + 3\rangle$.
- For $\mathsf{S}_3 = \ell_3$, we have
  - $\widehat{\mathcal{S}}^*_{\mathsf{s}}[\![\ell_3]\!] = \{\langle \ell_3, \rho''\rangle \mid \rho'' \in \mathbb{E}\mathbb{v}\}$.
  - $\mathcal{M}^t\langle \rho,\ ?:\mathsf{x} = \underline{\mathsf{x}} + 3\rangle\langle \ell_3, \rho[\mathsf{x} \leftarrow \rho(\mathsf{x}) + 3]\rangle = \langle \mathsf{tt}, \varepsilon\rangle$.
- For $\mathsf{S}_1\ \mathsf{S}_2\ \mathsf{S}_3 = \ell_1\ \mathsf{x} = \mathsf{x} + 1\ ;\ell_2\ \mathsf{x} = \mathsf{x} + 2\ ;\ell_3$, we have
  - $\widehat{\mathcal{S}}^*_{\mathsf{s}}[\![\ell_1\ \mathsf{x} = \mathsf{x} + 1\ ;\ell_2\ \mathsf{x} = \mathsf{x} + 2\ ;\ell_3]\!] = \{\langle \ell_1, \rho\rangle\langle \ell_2, \rho[\mathsf{x} \leftarrow \rho(\mathsf{x}) + 1]\rangle\langle \ell_3, \rho[\mathsf{x} \leftarrow \rho(\mathsf{x}) + 3]\rangle$ $\mid \rho \in \mathbb{E}\mathbb{v}\}$.
  - $\mathcal{M}^t\langle \rho,\ ?:\mathsf{x} = \underline{\mathsf{x}} \bullet ?:\mathsf{x} = \underline{\mathsf{x}} + 1 \bullet ?:\mathsf{x} = \underline{\mathsf{x}} + 3\rangle(\langle \ell_1, \rho\rangle\langle \ell_2, \rho[\mathsf{x} \leftarrow \rho(\mathsf{x}) + 1]\rangle\langle \ell_3,$ $\rho[\mathsf{x} \leftarrow \rho(\mathsf{x}) + 3]\rangle) = \langle \mathsf{tt}, \varepsilon\rangle$ (we have ignored the empty statement list in $\epsilon\ \mathsf{S}_1\ \mathsf{S}_2\ \mathsf{S}_3$ to simplify the specification).

The fact that $\mathcal{M}^t\langle \underline{\varrho}, \mathsf{R}\rangle\pi$ returns a pair $\langle b, \mathsf{R}'\rangle$ where R' is to be satisfied by continuations of π allows us to perform program model checking by structural induction on the program in section 44.7. The formal definition is the following.

Definition 44.23 (regular model checking)

- Trace model checking ($\underline{\varrho} \in \mathbb{E}\mathbb{v}$ is an initial environment and $\mathsf{R} \in \mathbb{R}^+ \cap \mathbb{R}^{\dagger}$ is a nonempty and $\mid$-free regular expression):

$$\mathcal{M}^t\langle \underline{\varrho}, \varepsilon\rangle\pi \triangleq \langle \mathsf{tt}, \varepsilon\rangle \tag{44.24}$$

$$\mathcal{M}^t\langle \underline{\varrho}, \mathsf{R}\rangle\ni \triangleq \langle \mathsf{tt}, \mathsf{R}\rangle$$

$$\mathcal{M}^t\langle \underline{\varrho}, \mathsf{R}\rangle\pi \triangleq \mathsf{let}\ \langle \ell_1, \rho_1\rangle\pi' = \pi\ \mathsf{and}\ \langle \mathsf{L} : \mathsf{B}, \mathsf{R}'\rangle = \mathsf{fstnxt}(\mathsf{R})\ \mathsf{in} \qquad \pi \neq \ni$$
$$(\!|\ \langle \underline{\varrho}, \langle \ell_1, \rho_1\rangle\rangle \in \mathcal{S}^{\mathrm{r}}[\![\mathsf{L} : \mathsf{B}]\!]\ ?\ \mathcal{M}^t\langle \underline{\varrho}, \mathsf{R}'\rangle\pi'\ \text{\textcolon}\ \langle \mathsf{ff}, \mathsf{R}'\rangle\ |\!)$$

- Set of traces model checking (for an $\mid$-free regular expression $\mathsf{R} \in \mathbb{R}^{\dagger}$):

$$\mathcal{M}^{\dagger}\langle \underline{\varrho}, \mathsf{R}\rangle\Pi \triangleq \{\langle \pi, \mathsf{R}'\rangle \mid \pi \in \Pi \wedge \langle \mathsf{tt}, \mathsf{R}'\rangle = \mathcal{M}^t\langle \underline{\varrho}, \mathsf{R}\rangle\pi\} \tag{44.25}$$

- Program component $\mathsf{S} \in \mathbb{P}\mathbb{c}$ model checking (for an $\mid$-free regular expression $\mathsf{R} \in \mathbb{R}^{\dagger}$):

$$\mathcal{M}^{\dagger}[\![\mathsf{S}]\!]\langle \underline{\varrho}, \mathsf{R}\rangle \triangleq \mathcal{M}^{\dagger}\langle \underline{\varrho}, \mathsf{R}\rangle(\widehat{\mathcal{S}}^*_{\mathsf{s}}[\![\mathsf{S}]\!]) \tag{44.26}$$

- Set of traces model checking (for regular expression $\mathsf{R} \in \mathbb{R}$):

$$\mathscr{M}\langle \underline{\varrho}, \mathsf{R}\rangle \Pi \triangleq \mathsf{let}\ (\mathsf{R}_1 \mid \ldots \mid \mathsf{R}_n) = \mathsf{dnf}(\mathsf{R})\ \mathsf{in} \quad \bigcup_{i=1}^{n} \{\pi \mid \exists \mathsf{R}' \in \mathbb{R}\,.\,\langle \pi, \mathsf{R}'\rangle \in \mathscr{M}^{+}\langle \underline{\varrho}, \mathsf{R}_i\rangle \Pi\} \tag{44.27}$$

- Model checking of a program component $\mathsf{S} \in \mathbb{P}\mathbb{c}$ (for regular expression $\mathsf{R} \in \mathbb{R}$):

$$\mathscr{M}[\![\mathsf{S}]\!]\langle \underline{\varrho}, \mathsf{R}\rangle \triangleq \mathscr{M}\langle \underline{\varrho}, \mathsf{R}\rangle(\widehat{\mathcal{S}}^{*}_{s}[\![\mathsf{S}]\!]) \tag{44.28}$$

The model checking $\mathscr{M}^{t}\langle \underline{\varrho}, \mathsf{R}\rangle\pi$ of a stateful trace π in (44.24) returns a pair $\langle b, \mathsf{R}'\rangle$ specifying whether π satisfies the specification R (when $b = \mathtt{tt}$) or not (when $b = \mathtt{ff}$). So, if $\mathscr{M}^{t}\langle \rho, \mathsf{R}\rangle(\pi) = \langle \mathtt{ff}, \mathsf{R}'\rangle$ in (44.25) then the trace π is a counterexample to the specification R. R' specifies what a continuation π' of π would have to satisfy for $\pi \frown \pi'$ to satisfy R (nothing specific when $\mathsf{R}' = \varepsilon$).

Notice that $\mathscr{M}^{t}\langle \underline{\varrho}, \mathsf{R}\rangle\pi$ checks whether the given trace π satisfies the regular specification R for initial environment $\underline{\varrho}$. Because only one trace is involved, this check can be done at runtime using a monitoring of the program execution. This is the case for Fred Schneider's security monitors [862] in example 44.1 (using an equivalent specification by finite automata).

The set of traces model checking $\mathscr{M}^{+}\langle \underline{\varrho}, \mathsf{R}\rangle\Pi$ returns the subset of traces of Π satisfying the specification R for the initial environment $\underline{\varrho}$. Because all program executions $\widehat{\mathcal{S}}^{*}_{s}[\![\mathsf{P}]\!]$ are involved, the model checking $\mathscr{M}^{+}[\![\mathsf{P}]\!]\langle \underline{\varrho}, \mathsf{R}\rangle$ of a program P becomes undecidable (by Rice's theorem 9.12).

The regular specification R is relational, as in section 19.1.2. If $\pi\langle \ell, \rho\rangle\pi' \in \widehat{\mathcal{S}}^{*}_{s}[\![\mathsf{S}]\!]$ is an execution trace satisfying the specification R then R in (44.28) determines a relationship between the initial environment $\underline{\varrho}$ and the current environment ρ. For example, $\mathsf{R} = \langle\{\mathsf{at}[\![\mathsf{S}]\!]\}, \mathsf{B}\rangle \bullet \mathsf{R}'$ with $\mathscr{B}[\![\mathsf{B}]\!]\underline{\varrho}, \rho = \forall \mathsf{x} \in \mathbb{V}\,.\,\underline{\varrho}(\underline{\mathsf{x}}) = \rho(\mathsf{x})$ expresses that the initial values of variables x are denoted $\underline{\mathsf{x}}$. $\mathscr{B}[\![\mathsf{B}]\!]\underline{\varrho}, \rho = \mathtt{tt}$ would state that there is no constraint on the initial value of variables. The difference with the invariant specifications of chapter 19 is that the order of computations is preserved. R can specify in which order program points may be reached, which is impossible with the invariants of chapter 19.[4] Moreover, the assertional case of section 19.1.1 occurs when the assertions in the regular specification R never refer to the initial environment $\underline{\varrho}$.

Exercise 44.29 Define the abstraction of a regular specification $\mathsf{R} \in \mathbb{R}$ into a local relation invariant in $\mathbb{L} \to \wp(\mathbb{E}\mathbb{v} \times \mathbb{E}\mathbb{v})$.

The model checking abstraction (44.25) that, given an initial environment $\underline{\varrho} \in \mathbb{E}\mathbb{v}$ and an $\mid$-free regular specification $\mathsf{R} \in \mathbb{R}^{+}$, returns the set of traces satisfying this specification is the lower adjoint of the Galois connection

4. By introduction of an auxiliary variable C incremented at each program step, one can simulate a trace with invariants. But then the reasoning is not on the original program P but on a transformed program $\overline{\mathsf{P}}$. Invariants in $\overline{\mathsf{P}}$ holding for a given value of c of C also hold at the position c of the traces in P. This kind of indirect reasoning is usually heavy and painful to maintain when programs are modified because values of counters are no longer the same. The use of temporal specifications has the advantage of avoiding the reasoning on explicit positions in the trace.

$$\langle\wp(\mathbb{S}^+), \subseteq\rangle \xleftrightarrow[\mathcal{M}^t\langle\underline{\varrho}, \mathsf{R}\rangle]{\gamma_{\mathcal{M}^t\langle\underline{\varrho}, \mathsf{R}\rangle}} \langle\wp(\mathbb{S}^+ \times \mathbb{R}^+), \subseteq\rangle \quad \text{for } \mathsf{R} \in \mathbb{R}^+ \text{ in (44.25)} \tag{44.30}$$

By exercise 11.18, this implies that

$$\langle\wp(\mathbb{S}^+), \subseteq\rangle \xleftrightarrow[\mathcal{M}\langle\underline{\varrho}, \mathsf{R}\rangle]{\gamma_{\mathcal{M}\langle\underline{\varrho}, \mathsf{R}\rangle}} \langle\wp(\mathbb{S}^+), \subseteq\rangle \quad \text{for } \mathsf{R} \in \mathbb{R} \text{ in (44.27)} \tag{44.31}$$

Boolean To follow the tradition that model checking returns a Boolean answer this abstraction can be composed with the Boolean abstraction

$$\langle\wp(\mathbb{S}^+), \subseteq\rangle \xleftrightarrow[\alpha_{\mathcal{M}\langle\underline{\varrho}, \mathsf{R}\rangle}]{\gamma_{\mathcal{M}\langle\underline{\varrho}, \mathsf{R}\rangle}} \langle\mathbb{B}, \Leftarrow\rangle \tag{44.32}$$

where $\alpha_{\mathcal{M}\langle\underline{\varrho}, \mathsf{R}\rangle}(X) \triangleq (\{\underline{\varrho}\} \times X) \subseteq \mathcal{M}\langle\underline{\varrho}, \mathsf{R}\rangle(X)$.

Exercise 44.33 Prove that (44.30) and (44.32) are Galois connections.

44.5 Soundness and Completeness of the Model Checking Abstraction

The following theorem 44.34 shows that the definition 44.12 of model checking a program semantics for a regular specification is a sound and complete abstraction of this semantics.

Theorem 44.34 (model checking soundness ($\Leftarrow$) and completeness ($\Rightarrow$))

$$\mathsf{P}, \underline{\varrho} \models \mathsf{R} \quad \Leftrightarrow \quad \alpha_{\mathcal{M}\langle\underline{\varrho}, \mathsf{R}\rangle}(\widehat{\boldsymbol{\mathcal{S}}}^*_{\mathsf{s}}[\![\mathsf{P}]\!]) \tag{44.34}$$

We decompose the proof of theorem 44.34 into soundness ($\Leftarrow$) and completeness ($\Rightarrow$). Soundness follows directly from the following

Lemma 44.35 (model checking soundness) $\mathcal{M}[\![\mathsf{P}]\!]\mathsf{R} \subseteq \alpha_{\text{prefix}}(\boldsymbol{\mathcal{S}}^{\mathsf{r}}[\![\mathsf{R} \bullet (? : \mathsf{tt})^*]\!])$.

Proof of lemma 44.35 Let $\mathcal{M}^t\langle\underline{\varrho}, \mathsf{R}\rangle\pi$ be the model checking of a prefix stateful trace $\pi = \sigma_1 \ldots \sigma_\ell = \langle\ell_1, \rho_1\rangle \ldots \langle\ell_\ell, \rho_\ell\rangle$ for an initial environment $\underline{\varrho}$ and a specification $\mathsf{R} \notin \mathbb{R}_\varepsilon$. Consider the repeated unrolling of R by fstnxt, which by lemma 44.19, is as follows:

$$\begin{array}{lll}
 & \mathsf{R} & \\
\Leftrightarrow & \mathsf{L}_1 : \mathsf{B}_1 \bullet \mathsf{R}_1 & \mathsf{fstnxt}(\mathsf{R}) \triangleq \langle\mathsf{L}_1 : \mathsf{B}_1, \mathsf{R}_1\rangle \\
\Leftrightarrow & \mathsf{L}_1 : \mathsf{B}_1 \bullet \mathsf{L}_2 : \mathsf{B}_2 \bullet \mathsf{R}_2 & \mathsf{fstnxt}(\mathsf{R}_1) \triangleq \langle\mathsf{L}_2 : \mathsf{B}_2, \mathsf{R}_2\rangle \\
 & \ldots & \\
\Leftrightarrow & \mathsf{L}_1 : \mathsf{B}_1 \bullet \mathsf{L}_2 : \mathsf{B}_2 \bullet \ldots \bullet \mathsf{L}_k : \mathsf{B}_k \bullet \mathsf{R}_k & \mathsf{fstnxt}(\mathsf{R}_{k-1}) \triangleq \langle\mathsf{L}_k : \mathsf{B}_k, \mathsf{R}_k\rangle \\
 & \ldots &
\end{array}$$

Either this development stops at some $k = n$ with $\mathsf{R}_n = \varepsilon$ or goes on forever, in which case $n = +\infty$. Let us prove that

(44.35.1) Either the specification R is satisfied, in which case let $m = \min(n, \ell)$ in $\forall i \in [1, m] . \langle\underline{\varrho}, \langle\ell_i, \rho_i\rangle\rangle \in \boldsymbol{\mathcal{S}}^{\mathsf{r}}[\![\mathsf{L}_i : \mathsf{B}_i]\!]$ and so $\mathcal{M}^t\langle\underline{\varrho}, \mathsf{R}\rangle\pi = \langle\mathsf{tt}, \mathsf{R}_m\rangle$;

(44.35.2) Or the specification R is not satisfied, in which case $\exists i \in [1, \min(n, \ell)]$. $\forall j \in [1, i[\,.\, \langle \underline{\varrho}, \langle \ell_j, \rho_j \rangle\rangle \in \boldsymbol{\mathcal{S}}^{\mathsf{r}}[\![\mathsf{L}_j : \mathsf{B}_j]\!] \wedge \langle \underline{\varrho}, \langle \ell_i, \rho_i \rangle\rangle \notin \boldsymbol{\mathcal{S}}^{\mathsf{r}}[\![\mathsf{L}_i : \mathsf{B}_i]\!]$ and so $\mathcal{M}^t \langle \underline{\varrho}, \mathsf{R}\rangle\pi = \langle \mathsf{ff}, \mathsf{R}_i \rangle$. □

The proof is by induction on $m = \min(n, \ell)$. We have $n \geqslant 1$ because $\mathsf{R} \notin \mathbb{R}_\varepsilon$ so $\mathsf{fstnxt}(\mathsf{R}) = \langle \mathsf{L}_1 : \mathsf{B}_1, \mathsf{R}_1 \rangle$ and $\pi = \langle \ell_1, \rho_1 \rangle \pi_1$ because stateful traces in chapter 42 are not empty.

— For the basis $n = 1$ or $\ell = 1$, (44.24) implies $\mathcal{M}^t \langle \underline{\varrho}, \mathsf{R}\rangle\pi = \langle b, \mathsf{R}_1 \rangle$ where $b = \langle \underline{\varrho}, \langle \ell_1, \rho_1 \rangle\rangle \in \boldsymbol{\mathcal{S}}^{\mathsf{r}}[\![\mathsf{L}_1 : \mathsf{B}_1]\!]$ which is case (44.35.1) when $b = \mathsf{tt}$ and (44.35.2) when $b = \mathsf{ff}$.

— For the induction case $m = \min(n, \ell) > 1$.

- Either $\langle \underline{\varrho}, \langle \ell_1, \rho_1 \rangle\rangle \notin \boldsymbol{\mathcal{S}}^{\mathsf{r}}[\![\mathsf{L}_1 : \mathsf{B}_1]\!]$ and, by (44.24), $\mathcal{M}^t \langle \underline{\varrho}, \mathsf{R}\rangle\pi \triangleq \langle \mathsf{ff}, \mathsf{R}_1 \rangle$, so we are in case (44.35.2) with $i = 1$;
- Or $\langle \underline{\varrho}, \langle \ell_1, \rho_1 \rangle\rangle \in \boldsymbol{\mathcal{S}}^{\mathsf{r}}[\![\mathsf{L}_1 : \mathsf{B}_1]\!]$ and there are two subcases.
  - If the specification R_1 is satisfied from 2 up to m, in which case $\forall i \in [2, m]$. $\langle \underline{\varrho}, \langle \ell_i, \rho_i \rangle\rangle \in \boldsymbol{\mathcal{S}}^{\mathsf{r}}[\![\mathsf{L}_i : \mathsf{B}_i]\!]$ and so $\mathcal{M}^t \langle \underline{\varrho}, \mathsf{R}_1\rangle\pi_1 = \langle \mathsf{tt}, \mathsf{R}_m \rangle$ by induction hypothesis (44.35.1). Then let $m = \min(n, \ell)$ in $\forall i \in [1, m]$. $\langle \underline{\varrho}, \langle \ell_i, \rho_i \rangle\rangle \in \boldsymbol{\mathcal{S}}^{\mathsf{r}}[\![\mathsf{L}_i : \mathsf{B}_i]\!]$. Moreover, (44.24) implies that $\mathcal{M}^t \langle \underline{\varrho}, \mathsf{R}\rangle\pi = \langle \mathsf{tt}, \mathsf{R}_m \rangle$
  - Otherwise, the specification R_1 is not satisfied from 2 up to m, so, by induction hypothesis (44.35.2), $\exists i \in [2, m]$. $\forall j \in [2, i[\,.\, \langle \underline{\varrho}, \langle \ell_j, \rho_j \rangle\rangle \in \boldsymbol{\mathcal{S}}^{\mathsf{r}}[\![\mathsf{L}_j : \mathsf{B}_j]\!] \wedge \langle \underline{\varrho}, \langle \ell_i, \rho_i \rangle\rangle \notin \boldsymbol{\mathcal{S}}^{\mathsf{r}}[\![\mathsf{L}_i : \mathsf{B}_i]\!]$ and so $\mathcal{M}^t \langle \underline{\varrho}, \mathsf{R}\rangle\pi_1 = \langle \mathsf{ff}, \mathsf{R}_i \rangle$. It follows that $\exists i \in [1, \min(n, \ell)]$. $\forall j \in [1, i[\,.\, \langle \underline{\varrho}, \langle \ell_j, \rho_j \rangle\rangle \in \boldsymbol{\mathcal{S}}^{\mathsf{r}}[\![\mathsf{L}_j : \mathsf{B}_j]\!] \wedge \langle \underline{\varrho}, \langle \ell_i, \rho_i \rangle\rangle \notin \boldsymbol{\mathcal{S}}^{\mathsf{r}}[\![\mathsf{L}_i : \mathsf{B}_i]\!]$. Moreover, by (44.24), $\mathcal{M}^t \langle \underline{\varrho}, \mathsf{R}\rangle\pi = \langle \mathsf{ff}, \mathsf{R}_i \rangle$. *Q.E.D.*

Let us now prove that $\mathcal{M}[\![\mathsf{P}]\!]\mathsf{R} \subseteq \alpha_{\mathsf{prefix}}(\boldsymbol{\mathcal{S}}^{\mathsf{r}}[\![\mathsf{R} \bullet (? : \mathsf{tt})^*]\!])$. By (44.28), $\mathsf{dnf}(\mathsf{R}) = (\mathsf{R}_1 \mid \ldots \mid \mathsf{R}_p)$ and, by definition of $\cup$, $\exists a \in [1, p]$. $\pi \in \mathcal{M}[\![\mathsf{P}]\!]\mathsf{R}_a$ where $\mathsf{R}_a \in \mathbb{R}^+$ is $\mid$-free by (44.14). So, by (44.25), $\exists \mathsf{R}'_a \in \mathbb{R}$. $\pi = \langle \ell_1, \rho_1 \rangle\pi' \in \widehat{\boldsymbol{\mathcal{S}}}^{\,*}_{\mathsf{s}}[\![\mathsf{P}]\!] \wedge \langle \mathsf{tt}, \mathsf{R}'_a \rangle = \mathcal{M}^t \langle \rho_1, \mathsf{R}_a \rangle(\langle \ell_1, \rho_1 \rangle\pi')$. So we are in case (44.35.1), and there are two subcases.

- If $n \leqslant \ell$ where $\ell \in \mathbb{N}$, and then $\mathsf{R}_a \simeq \mathsf{L}_1 : \mathsf{B}_1 \bullet \mathsf{L}_2 : \mathsf{B}_2 \bullet \ldots \bullet \mathsf{L}_n : \mathsf{B}_n \bullet \mathsf{R}'_a$ and $\forall i \in [1, n]$. $\langle \rho_1, \langle \ell_i, \rho_i \rangle\rangle \in \boldsymbol{\mathcal{S}}^{\mathsf{r}}[\![\mathsf{L}_i : \mathsf{B}_i]\!]$. Moreover, by (44.7), $\forall i \in [n+1, \ell]$. $\langle \rho_1, \langle \ell_i, \rho_i \rangle\rangle \in \boldsymbol{\mathcal{S}}^{\mathsf{r}}[\![? : \mathsf{tt}]\!]$. Therefore

$$
\begin{aligned}
& \langle \rho_1, \pi \rangle \in \boldsymbol{\mathcal{S}}^{\mathsf{r}}[\![\mathsf{L}_1 : \mathsf{B}_1]\!] \circledcirc \cdots \circledcirc \boldsymbol{\mathcal{S}}^{\mathsf{r}}[\![\mathsf{L}_n : \mathsf{B}_n]\!] \circledcirc \overbrace{\boldsymbol{\mathcal{S}}^{\mathsf{r}}[\![? : \mathsf{tt}]\!] \circledcirc \cdots \circledcirc \boldsymbol{\mathcal{S}}^{\mathsf{r}}[\![? : \mathsf{tt}]\!]}^{n-\ell \text{ times}} \\
\Rightarrow\ & \langle \rho_1, \pi \rangle \in \boldsymbol{\mathcal{S}}^{\mathsf{r}}[\![\mathsf{R}_a \bullet (? : \mathsf{tt})^{\ell-n}]\!] && \text{(44.7) for } \bullet \\
\Rightarrow\ & \langle \rho_1, \pi \rangle \in \boldsymbol{\mathcal{S}}^{\mathsf{r}}[\![\mathsf{R} \bullet (? : \mathsf{tt})^{\ell-n}]\!] && \text{(44.7) for } \mid \\
\Rightarrow\ & \langle \rho_1, \pi \rangle \in \alpha_{\mathsf{prefix}}(\boldsymbol{\mathcal{S}}^{\mathsf{r}}[\![\mathsf{R} \bullet (? : \mathsf{tt})^{\ell-n}]\!]) && \alpha_{\mathsf{prefix}} \text{ is } \subseteq\text{-extensive} \\
\Rightarrow\ & \langle \rho_1, \pi \rangle \in \alpha_{\mathsf{prefix}}(\boldsymbol{\mathcal{S}}^{\mathsf{r}}[\![\mathsf{R} \bullet (? : \mathsf{tt})^*]\!]) && \text{(44.7) for }^* \text{ and } \alpha_{\mathsf{prefix}} \text{ is } \subseteq\text{-increasing}
\end{aligned}
$$

- Otherwise, $n > \ell$, and then $\mathsf{R}_a \simeq \mathsf{L}_1 : \mathsf{B}_1 \bullet \mathsf{L}_2 : \mathsf{B}_2 \bullet \ldots \bullet \mathsf{L}_\ell : \mathsf{B}_\ell \bullet \mathsf{R}'_a$ and $\forall i \in [1, \ell]$. $\langle \rho_1, \langle \ell_i, \rho_i \rangle\rangle \in \boldsymbol{\mathcal{S}}^{\mathsf{r}}[\![\mathsf{L}_i : \mathsf{B}_i]\!]$. Therefore

$$
\begin{aligned}
& \langle \rho_1, \pi \rangle \in \boldsymbol{\mathcal{S}}^{\mathsf{r}}[\![\mathsf{L}_1 : \mathsf{B}_1]\!] \circledcirc \cdots \circledcirc \boldsymbol{\mathcal{S}}^{\mathsf{r}}[\![\mathsf{L}_\ell : \mathsf{B}_\ell]\!] \\
\Rightarrow\ & \langle \rho_1, \pi \rangle \in \boldsymbol{\mathcal{S}}^{\mathsf{r}}[\![\mathsf{L}_1 : \mathsf{B}_1 \bullet \cdots \bullet \mathsf{L}_\ell : \mathsf{B}_\ell]\!] && \text{(44.7) for } \bullet \\
\Rightarrow\ & \langle \rho_1, \pi \rangle \in \alpha_{\mathsf{prefix}}(\boldsymbol{\mathcal{S}}^{\mathsf{r}}[\![\mathsf{R} \bullet (? : \mathsf{tt})^*]\!]) && \text{(44.7) for } \bullet \text{ and }^* \text{ and definition of } \alpha_{\mathsf{prefix}}
\end{aligned}
$$
□

Lemma 44.36 (model checking completeness) $(\{\underline{\varrho}\} \times \widehat{\mathcal{S}}^{*}_{s}\llbracket \mathsf{P} \rrbracket) \subseteq \alpha_{\mathsf{prefix}}(\mathcal{S}^{r}\llbracket \mathsf{R} \bullet (? : \mathsf{tt})^{*} \rrbracket) \Rightarrow (\widehat{\mathcal{S}}^{*}_{s}\llbracket \mathsf{P} \rrbracket \subseteq \mathcal{M}\llbracket \mathsf{P} \rrbracket \mathsf{R})$.

Proof of lemma 44.36 By contradiction, assume that $\pi = \langle \ell_1, \rho_1 \rangle \pi' \in \widehat{\mathcal{S}}^{*}_{s}\llbracket \mathsf{P} \rrbracket$ and $\langle \underline{\varrho}, \pi \rangle \notin \alpha_{\mathsf{prefix}}(\mathcal{S}^{r}\llbracket \mathsf{R} \bullet (? : \mathsf{tt})^{*} \rrbracket)$. By decomposition of R by fstnxt as in the proof of lemma 44.35, we have $\exists i \in [1, \min(n, \ell)] \,.\, \forall j \in [1, i[\,.\, \langle \underline{\varrho}, \langle \ell_j, \rho_j \rangle \rangle \in \mathcal{S}^{r}\llbracket \mathsf{L}_j : \mathsf{B}_j \rrbracket \wedge \langle \underline{\varrho}, \langle \ell_i, \rho_i \rangle \rangle$. Therefore, we are in case (44.35.2) and so $\mathcal{M}^{t}\langle \underline{\varrho}, \mathsf{R} \rangle \pi = \langle \mathsf{ff}, \mathsf{R}_i \rangle$. By (44.25), $\pi \notin \mathcal{M}\llbracket \mathsf{P} \rrbracket \mathsf{R}$. □

Proof of theorem 44.34 Follows from lemmas 44.35 and 44.36.

$$\mathsf{P}, \underline{\varrho} \models \mathsf{R}$$

$$\Leftrightarrow (\{\underline{\varrho}\} \times \widehat{\mathcal{S}}^{*}_{s}\llbracket \mathsf{P} \rrbracket) \subseteq \alpha_{\mathsf{prefix}}(\mathcal{S}^{r}\llbracket \mathsf{R} \bullet (? : \mathsf{tt})^{*} \rrbracket) \quad \text{(definition 44.12)}$$

$$\Leftrightarrow (\{\underline{\varrho}\} \times \widehat{\mathcal{S}}^{*}_{s}\llbracket \mathsf{P} \rrbracket) \subseteq \mathcal{M}\langle \underline{\varrho}, \mathsf{R} \rangle (\widehat{\mathcal{S}}^{*}_{s}\llbracket \mathsf{P} \rrbracket) \quad \text{(lemma 44.35 for} \Leftarrow \text{and lemma 44.36 for} \Rightarrow\text{)}$$

$$\Leftrightarrow \alpha_{\mathcal{M}\langle \underline{\varrho}, \mathsf{R} \rangle}(\widehat{\mathcal{S}}^{*}_{s}\llbracket \mathsf{P} \rrbracket) \quad \text{(definition (44.32) of } \alpha_{\mathcal{M}\langle \underline{\varrho}, \mathsf{R} \rangle}(X) \triangleq (\{\underline{\varrho}\} \times X) \subseteq \mathcal{M}\langle \underline{\varrho}, \mathsf{R} \rangle (X)\text{)} \quad \square$$

At this point we know, by (44.28) and theorem 44.34 that a model checker $\mathcal{M}\llbracket \mathsf{S} \rrbracket \langle \underline{\varrho}, \mathsf{R} \rangle$ is a sound and complete abstraction $\mathcal{M}\langle \underline{\varrho}, \mathsf{R} \rangle (\widehat{\mathcal{S}}^{*}_{s}\llbracket \mathsf{S} \rrbracket)$ of the program component semantics $\widehat{\mathcal{S}}^{*}_{s}\llbracket \mathsf{S} \rrbracket$ which provides a counterexample in case of failure. This allows us to derive a structural model checker $\widehat{\mathcal{M}}\llbracket \mathsf{P} \rrbracket \langle \underline{\varrho}, \mathsf{R} \rangle$ in section 44.7 by calculational design.

44.6 Model Checking Trace Concatenation

To model check $\widehat{\mathcal{M}}\llbracket \mathsf{Sl} \rrbracket \langle \underline{\varrho}, \mathsf{R} \rangle$ statement lists $\mathsf{Sl} ::= \mathsf{Sl}'\ \mathsf{S}$, we will need to model check the concatenation of traces of Sl' and S using the following lemma.

Lemma 44.37 $\mathcal{M}^{t}\langle \underline{\varrho}, \mathsf{R} \rangle (\pi \frown \pi') = \langle \mathsf{tt}, \mathsf{R}' \rangle \Leftrightarrow (\exists \mathsf{R}'' \in \mathbb{R} \,.\, \mathcal{M}^{t}\langle \underline{\varrho}, \mathsf{R} \rangle (\pi) = \langle \mathsf{tt}, \mathsf{R}'' \rangle \wedge \mathcal{M}^{t}\langle \underline{\varrho}, \mathsf{R}'' \rangle (\pi') = \langle \mathsf{tt}, \mathsf{R}' \rangle)$.

Proof of lemma 44.37 Because the case of empty traces is trivial by $\mathcal{M}^{t}\langle \underline{\varrho}, \mathsf{R} \rangle \ni \triangleq \langle \mathsf{tt}, \mathsf{R} \rangle$ in (44.24), we can assume that $\pi = \langle \ell_1, \rho_1 \rangle \ldots \langle \ell_p, \rho_p \rangle$, $\pi' = \langle \ell_p, \rho_p \rangle \ldots \langle \ell_\ell, \rho_\ell \rangle$ with $1 \leqslant p \leqslant \ell$ so that, by definition (6.4) of trace concatenation, $\pi \frown \pi' = \langle \ell_1, \rho_1 \rangle \ldots \langle \ell_p, \rho_p \rangle \ldots \langle \ell_\ell, \rho_\ell \rangle$.

Let us consider the decomposition of $\mathsf{R} \approx \mathsf{L}_1 : \mathsf{B}_1 \bullet \mathsf{R}_1 \approx \mathsf{L}_1 : \mathsf{B}_1 \bullet \mathsf{L}_2 : \mathsf{B}_2 \bullet \mathsf{R}_2 \approx \ldots \approx \mathsf{L}_1 : \mathsf{B}_1 \bullet \mathsf{L}_2 : \mathsf{B}_2 \bullet \ldots \bullet \mathsf{L}_n : \mathsf{B}_n \bullet \mathsf{R}_n$ of lemma 44.35 where, by convention, $n \in \mathbb{N}^{+}$ and $\mathsf{R}_n = \varepsilon$ or $n = +\infty$ and R_n is undefined.

Because $\mathcal{M}^{t}\langle \underline{\varrho}, \mathsf{R} \rangle (\pi \frown \pi') = \langle \mathsf{tt}, \mathsf{R}' \rangle$, we are in case (44.35.1) because the specification R is satisfied, in which case

$$\mathsf{let}\ m = \min(n, \ell)\ \mathsf{in}\ \forall i \in [1, m] \,.\, \langle \underline{\varrho}, \langle \ell_i, \rho_i \rangle \rangle \in \mathcal{S}^{r}\llbracket \mathsf{L}_i : \mathsf{B}_i \rrbracket \text{ and so } \mathsf{R}_m = \mathsf{R}'. \quad \text{(a)}$$

There are three cases.

- If $n \leqslant p \leqslant \ell$ in (a), then $m = n$ so $\mathsf{R}_n = \mathsf{R}' = \varepsilon$. First (b), we have $\forall i \in [1, p] = [1, \min(p, \ell)] \,.\, \langle \underline{\varrho}, \langle \ell_i, \rho_i \rangle \rangle \in \mathcal{S}^{r}\llbracket \mathsf{L}_i : \mathsf{B}_i \rrbracket$ and so $\mathcal{M}^{t}\langle \underline{\varrho}, \mathsf{R} \rangle \pi = \langle \mathsf{tt}, \mathsf{R}_p \rangle$ with $\mathsf{R}'' = \mathsf{R}_p$, which in this case is ε. Second (c), by (44.24), $\mathcal{M}^{t}\langle \underline{\varrho}, \mathsf{R}'' \rangle (\pi') = \mathcal{M}^{t}\langle \underline{\varrho}, \varepsilon \rangle (\pi') = \langle \mathsf{tt},$

$\varepsilon\rangle = \langle \text{tt}, \text{R}'\rangle$, proving $\Rightarrow$. Conversely, $\mathcal{M}^t\langle \underline{\varrho}, \text{R}\rangle(\pi) = \langle \text{tt}, \text{R}''\rangle$ implies (b) and $\mathcal{M}^t\langle \underline{\varrho}, \text{R}''\rangle(\pi') = \langle \text{tt}, \text{R}'\rangle$ implies (c) which imply (a), hence $\mathcal{M}^t\langle \underline{\varrho}, \text{R}\rangle(\pi \frown \pi') = \langle \text{tt}, \text{R}'\rangle$, proving $\Leftarrow$.

- If $p \leqslant n \leqslant \ell$ in (a), then again $m = n$ so $\text{R}_n = \text{R}' = \varepsilon$ hence (b) holds. Moreover, we have (d) $\forall i \in [p, n] . \langle \underline{\varrho}, \langle \ell_i, \rho_i\rangle\rangle \in \mathcal{S}^r[\![\text{L}_i : \text{B}_i]\!]$ and so $\mathcal{M}^t\langle \underline{\varrho}, \text{R}_p\rangle(\pi') = \langle \text{tt}, \varepsilon\rangle$, proving $\Leftarrow$ for $\text{R}'' = \text{R}_p$. Conversely, $\mathcal{M}^t\langle \underline{\varrho}, \text{R}\rangle(\pi) = \langle \text{tt}, \text{R}''\rangle$ implies (b) and $\mathcal{M}^t\langle \underline{\varrho}, \text{R}''\rangle(\pi') = \langle \text{tt}, \text{R}'\rangle$ implies (d) which imply (a), hence $\mathcal{M}^t\langle \underline{\varrho}, \text{R}\rangle(\pi \frown \pi') = \langle \text{tt}, \text{R}'\rangle$, proving $\Leftarrow$.
- If $p \leqslant \ell \leqslant n$ in (a), then $m = \ell$ with (e) $\forall i \in [1, \ell] . \langle \underline{\varrho}, \langle \ell_i, \rho_i\rangle\rangle \in \mathcal{S}^r[\![\text{L}_i : \text{B}_i]\!]$ and so $\text{R}_\ell = \text{R}'$. This implies $\mathcal{M}^t\langle \underline{\varrho}, \text{R}\rangle(\pi) = \langle \text{tt}, \text{R}_p\rangle$ because (f) $\forall i \in [1, p] . \langle \underline{\varrho}, \langle \ell_i, \rho_i\rangle\rangle \in \mathcal{S}^r[\![\text{L}_i : \text{B}_i]\!]$ so $\text{R}'' = \text{R}_p$ and $\mathcal{M}^t\langle \underline{\varrho}, \text{R}_p\rangle(\pi') = \langle \text{tt}, \text{R}_\ell\rangle$ because (g) $\forall i \in [p, \ell] . \langle \underline{\varrho}, \langle \ell_i, \rho_i\rangle\rangle \in \mathcal{S}^r[\![\text{L}_i : \text{B}_i]\!]$ so $\text{R}' = \text{R}_\ell$, proving $\Rightarrow$. Conversely, (f) and (g) imply (a), hence $\Leftarrow$. □

44.7 Structural Model Checking

By definition 44.12 of the model checking of $\text{S}, \underline{\varrho} \models \text{R}$ of a program $\text{P} \in \mathbb{P}$ for a regular specification $\text{R} \in \mathbb{R}^+$ and initial environment $\underline{\varrho}$, theorem 44.34 shows that a model checker $\mathcal{M}[\![\text{P}]\!]\langle \underline{\varrho}, \text{R}\rangle$ is a sound and complete abstraction $\mathcal{M}\langle \underline{\varrho}, \text{R}\rangle(\widehat{\mathcal{S}}^*_{\text{s}}[\![\text{P}]\!])$ of the program semantics $\widehat{\mathcal{S}}^*_{\text{s}}[\![\text{P}]\!]$. This abstraction does not provide a model checking algorithm specification.

The standard model checking algorithms [70, 189, 191] use a transition system (or a Kripke structure variation [599]) for hardware and software modeling and proceed by induction on computation steps.

In contrast, we proceed by structural induction on programs, which will be shown in theorem 44.38 to be logically equivalent (but maybe more efficient because fixpoints are computed locally). The structural model checking $\widehat{\mathcal{M}}[\![\text{P}]\!]\langle \underline{\varrho}, \text{R}\rangle$ of the program P proceeds by structural induction on the program structure:

$$\begin{cases} \widehat{\mathcal{M}}[\![\text{S}]\!]\langle \underline{\varrho}, \text{R}\rangle \triangleq \widehat{\mathcal{F}}[\![\text{S}]\!](\prod_{\text{S}' \lhd \text{S}} \widehat{\mathcal{M}}[\![\text{S}']\!])\langle \underline{\varrho}, \text{R}\rangle \\ \text{S} \in \mathbb{P}\mathbb{c} \end{cases}$$

where the transformer $\widehat{\mathcal{F}}$ uses the results of model checking of the immediate components $\text{S}' \lhd \text{S}$ and involves a fixpoint computation for iteration statements.

The following theorem 44.38 shows that the algorithm specification is correct, that is, $\widehat{\mathcal{M}}[\![\text{S}]\!] = \mathcal{M}[\![\text{S}]\!]$ for all program components S. So together with theorem 44.34, the structural model checking is proved sound and complete.

Theorem 44.38 $\forall \text{S} \in \mathbb{P}\mathbb{c}, \text{R} \in \mathbb{R}, \underline{\varrho} \in \mathbb{E}\text{v} . \widehat{\mathcal{M}}^+[\![\text{S}]\!]\langle \underline{\varrho}, \text{R}\rangle = \mathcal{M}^+[\![\text{S}]\!]\langle \underline{\varrho}, \text{R}\rangle$ and $\widehat{\mathcal{M}}[\![\text{S}]\!]\langle \underline{\varrho}, \text{R}\rangle = \mathcal{M}[\![\text{S}]\!]\langle \underline{\varrho}, \text{R}\rangle$.

Following exercise 19.41, the proof of theorem 44.38 is by calculational design and proceeds by structural induction on the program component S. Assuming $\mathcal{M}[\![\text{S}']\!] = \widehat{\mathcal{M}}[\![\text{S}']\!]$ for all immediate components $\text{S}' \lhd \text{S}$ of statement, the proof for each program component S is as follows:

$$\begin{aligned}
&\mathcal{M}[\![\mathsf{S}]\!]\langle \underline{\varrho}, \mathsf{R}\rangle \\
\triangleq\;& \mathcal{M}\langle \underline{\varrho}, \mathsf{R}\rangle(\widehat{\mathcal{S}}^{*}_{s}[\![\mathsf{S}]\!]) && \text{(44.28)} \\
=\;& \mathcal{M}\langle \underline{\varrho}, \mathsf{R}\rangle(\mathcal{F}[\![\mathsf{S}]\!](\prod_{\mathsf{S}'\lhd\mathsf{S}} \widehat{\mathcal{S}}^{*}_{s}[\![\mathsf{S}']\!])\langle \underline{\varrho}, \mathsf{R}\rangle)
\end{aligned}$$

(by structural definition $\widehat{\mathcal{S}}^{*}_{s}[\![\mathsf{S}]\!] = \mathcal{F}[\![\mathsf{S}]\!](\prod_{\mathsf{S}'\lhd\mathsf{S}} \widehat{\mathcal{S}}^{*}_{s}[\![\mathsf{S}']\!])$ of the stateful prefix trace semantics in chapter 42)

$= \ldots$ (calculus to expand definitions, rewrite and simplify formulas by algebraic laws)

$$= \widehat{\mathcal{F}}[\![\mathsf{S}]\!](\prod_{\mathsf{S}'\lhd\mathsf{S}} \mathcal{M}[\![\mathsf{S}']\!])\langle \underline{\varrho}, \mathsf{R}\rangle$$

(by calculational design to commute the model checking abstraction on the result to the model checking of the arguments of $\widehat{\mathcal{S}}^{*}_{s}[\![\mathsf{S}]\!]$)

$$= \widehat{\mathcal{F}}[\![\mathsf{S}]\!](\prod_{\mathsf{S}'\lhd\mathsf{S}} \widehat{\mathcal{M}}[\![\mathsf{S}']\!])\langle \underline{\varrho}, \mathsf{R}\rangle \qquad \text{(induction hypothesis)}$$

$$\triangleq \widehat{\mathcal{M}}[\![\mathsf{S}]\!]\langle \underline{\varrho}, \mathsf{R}\rangle \qquad \text{(by defining } \widehat{\mathcal{M}}[\![\mathsf{S}]\!] \triangleq \widehat{\mathcal{F}}[\![\mathsf{S}]\!](\textstyle\prod_{\mathsf{S}'\lhd\mathsf{S}} \widehat{\mathcal{M}}[\![\mathsf{S}']\!])\text{)}$$

For iteration statements, $\mathcal{F}[\![\mathsf{S}]\!](\prod_{\mathsf{S}'\lhd\mathsf{S}} \widehat{\mathcal{S}}^{*}_{s}[\![\mathsf{S}']\!])\langle \underline{\varrho}, \mathsf{R}\rangle$ is a fixpoint, and this proof involves corollary 18.34.

We first provide the definition of structural model checking $\widehat{\mathcal{M}}[\![\mathsf{S}]\!]$ (although we got it from the proof given later). It checks all prefixes of the command for the specification $\langle \underline{\varrho}, \mathsf{R}\rangle$. Explanations of the formulas follow the definition.

Definition 44.39 (structural model checking)

- Model checking a program $\mathsf{P} ::= \mathsf{Sl}\;\ell$ for a temporal specification $\mathsf{R} \in \mathbb{R}$ with alternatives.

$$\widehat{\mathcal{M}}[\![\mathsf{P}]\!]\langle \underline{\varrho}, \mathsf{R}\rangle \triangleq \mathsf{let}\,(\mathsf{R}_1 \mid \ldots \mid \mathsf{R}_n) = \mathsf{dnf}(\mathsf{R})\;\mathsf{in} \quad \bigcup_{i=1}^{n} \{\pi \mid \exists \mathsf{R}' \in \mathbb{R} \,.\, \langle \pi, \mathsf{R}'\rangle \in \widehat{\mathcal{M}}^{+}[\![\mathsf{Sl}]\!]\langle \underline{\varrho}, \mathsf{R}_i\rangle\} \tag{44.40}$$

- Model checking an empty temporal specification ε.

$$\widehat{\mathcal{M}}^{+}[\![\mathsf{S}]\!]\langle \underline{\varrho}, \varepsilon\rangle \triangleq \{\langle \pi, \varepsilon\rangle \mid \pi \in \widehat{\mathcal{S}}^{*}_{s}[\![\mathsf{S}]\!]\} \tag{44.41}$$

- It is assumed below that $\mathsf{R} \in \mathbb{R}^{\dagger} \cap \mathbb{R}^{+}$ is a nonempty, alternative $\mid$-free regular expression.

- Model checking a statement list $\mathsf{Sl} ::= \mathsf{Sl}'\;\mathsf{S}$

$$\begin{aligned}
\widehat{\mathcal{M}}^{+}[\![\mathsf{Sl}]\!]\langle \underline{\varrho}, \mathsf{R}\rangle \triangleq\;& \widehat{\mathcal{M}}^{+}[\![\mathsf{Sl}']\!]\langle \underline{\varrho}, \mathsf{R}\rangle \\
&\cup \{\langle \pi \cdot \langle \mathsf{at}[\![\mathsf{S}]\!], \rho\rangle \cdot \pi', \mathsf{R}''\rangle \mid \langle \pi \cdot \langle \mathsf{at}[\![\mathsf{S}]\!], \rho\rangle, \mathsf{R}'\rangle \in \widehat{\mathcal{M}}^{+}[\![\mathsf{Sl}']\!]\langle \underline{\varrho}, \mathsf{R}\rangle \\
&\qquad \wedge \langle\langle \mathsf{at}[\![\mathsf{S}]\!], \rho\rangle \cdot \pi', \mathsf{R}''\rangle \in \widehat{\mathcal{M}}^{+}[\![\mathsf{S}]\!]\langle \underline{\varrho}, \mathsf{R}'\rangle\}
\end{aligned} \tag{44.42}$$

- Model checking an empty statement list $\mathsf{Sl} ::= \epsilon$

$$\widehat{\mathcal{M}}^{+}[\![\mathtt{Sl}]\!]\langle \underline{\varrho}, \mathsf{R}\rangle \triangleq \text{let } \langle \mathsf{L} : \mathsf{B}, \mathsf{R}'\rangle = \mathsf{fstnxt}(\mathsf{R}) \text{ in} \tag{44.43}$$
$$\{\langle\langle \mathsf{at}[\![\mathtt{Sl}]\!], \rho\rangle, \mathsf{R}'\rangle \mid \langle \underline{\varrho}, \langle \mathsf{at}[\![\mathtt{Sl}]\!], \rho\rangle\rangle \in \boldsymbol{\mathcal{S}}^{\mathsf{r}}[\![\mathsf{L} : \mathsf{B}]\!]\}$$

(In practice the empty statement list ϵ needs not be specified, so we could eliminate that need by ignoring ϵ in the specification R and defining $\widehat{\mathcal{M}}^{+}[\![\mathtt{Sl}]\!]\langle \underline{\varrho}, \mathsf{R}\rangle \triangleq \{\langle\langle \mathsf{at}[\![\mathtt{Sl}]\!], \rho\rangle, \mathsf{R}\rangle \mid \rho \in \mathbb{E}\mathrm{v}\}$.)

- Model checking a skip statement S ::= ;

$$\widehat{\mathcal{M}}^{+}[\![\mathtt{S}]\!]\langle \underline{\varrho}, \mathsf{R}\rangle \triangleq \text{let } \langle \mathsf{L} : \mathsf{B}, \mathsf{R}'\rangle = \mathsf{fstnxt}(\mathsf{R}) \text{ in} \tag{44.44}$$
$$\{\langle\langle \mathsf{at}[\![\mathtt{S}]\!], \rho\rangle, \mathsf{R}'\rangle \mid \langle \underline{\varrho}, \langle \mathsf{at}[\![\mathtt{S}]\!], \rho\rangle\rangle \in \boldsymbol{\mathcal{S}}^{\mathsf{r}}[\![\mathsf{L} : \mathsf{B}]\!]\}$$

(In practice the skip statement ; needs not be specified so we could eliminate that need by ignoring ; in the specification R and defining $\widehat{\mathcal{M}}^{+}[\![;]\!]\langle \underline{\varrho}, \mathsf{R}\rangle \triangleq \{\langle\langle \mathsf{at}[\![\mathtt{S}]\!], \rho\rangle, \mathsf{R}\rangle \mid \rho \in \mathbb{E}\mathrm{v}\}$.)

— For the assignment S ::= ℓ x = A ; in (44.45), case (a) checks the prefixes that stops at ℓ whereas (b) and (c) checks the maximal traces stopping after the assignment. In each trace checked for the specification R, the states are checked successively and the continuation specification is returned together with the checked trace, unless the check fails.

Definition 44.39 (Structural model checking, continued)

- Model checking an assignment statement S ::= ℓ x = A ;

$$\widehat{\mathcal{M}}^{+}[\![\mathtt{S}]\!]\langle \underline{\varrho}, \mathsf{R}\rangle \triangleq \text{let } \langle \mathsf{L} : \mathsf{B}, \mathsf{R}'\rangle = \mathsf{fstnxt}(\mathsf{R}) \text{ in} \tag{44.45}$$
$$\{\langle\langle \mathsf{at}[\![\mathtt{S}]\!], \rho\rangle, \mathsf{R}'\rangle \mid \langle \underline{\varrho}, \langle \mathsf{at}[\![\mathtt{S}]\!], \rho\rangle\rangle \in \boldsymbol{\mathcal{S}}^{\mathsf{r}}[\![\mathsf{L} : \mathsf{B}]\!]\} \qquad \text{(a)}$$
$$\cup \{\langle\langle \mathsf{at}[\![\mathtt{S}]\!], \rho\rangle\langle \mathsf{after}[\![\mathtt{S}]\!], \rho[\mathsf{x} \leftarrow \mathcal{A}[\![\mathsf{A}]\!]\rho]\rangle, \varepsilon\rangle \mid \mathsf{R}' \in \mathbb{R}_{\varepsilon} \wedge \qquad \text{(b)}$$
$$\langle \underline{\varrho}, \langle \mathsf{at}[\![\mathtt{S}]\!], \rho\rangle\rangle \in \boldsymbol{\mathcal{S}}^{\mathsf{r}}[\![\mathsf{L} : \mathsf{B}]\!]\}$$
$$\cup \{\langle\langle \mathsf{at}[\![\mathtt{S}]\!], \rho\rangle\langle \mathsf{after}[\![\mathtt{S}]\!], \rho[\mathsf{x} \leftarrow \mathcal{A}[\![\mathsf{A}]\!]\rho]\rangle, \mathsf{R}''\rangle \mid \mathsf{R}' \notin \mathbb{R}_{\varepsilon} \wedge \qquad \text{(c)}$$
$$\langle \underline{\varrho}, \langle \mathsf{at}[\![\mathtt{S}]\!], \rho\rangle\rangle \in \boldsymbol{\mathcal{S}}^{\mathsf{r}}[\![\mathsf{L} : \mathsf{B}]\!] \wedge \langle \mathsf{L}' : \mathsf{B}', \mathsf{R}''\rangle = \mathsf{fstnxt}(\mathsf{R}') \wedge$$
$$\langle \underline{\varrho}, \langle \mathsf{after}[\![\mathtt{S}]\!], \rho[\mathsf{x} \leftarrow \mathcal{A}[\![\mathsf{A}]\!]\rho]\rangle\rangle \in \boldsymbol{\mathcal{S}}^{\mathsf{r}}[\![\mathsf{L}' : \mathsf{B}']\!]\}$$

Example 44.46 Checking the assignment example S ::= ℓ x = A ; in (44.45) for $\langle \mathsf{L} : \mathsf{B}, \mathsf{R}'\rangle = \mathsf{fstnxt}(\mathsf{R})$ consists of first checking L : B at ℓ and then checking on R′ after the statement. In case (b), R′ is empty so is trivially satisfied. Otherwise, in case (c), $\langle \mathsf{L}' : \mathsf{B}', \mathsf{R}''\rangle = \mathsf{fstnxt}(\mathsf{R}')$ so L′ : B′ is checked after the statement while R″ is the continuation specification.

Definition 44.39 (Structural model checking, continued)

- Model checking a conditional statement S ::= **if** ℓ (B) $\mathtt{S}_t$

$$\widehat{\mathscr{M}}^{+}[\![\mathsf{S}]\!]\langle \underline{\varrho}, \mathsf{R}\rangle \triangleq \text{let } \langle \mathsf{L}' : \mathsf{B}', \mathsf{R}'\rangle = \text{fstnxt}(\mathsf{R}) \text{ in} \tag{44.47}$$
$$\{\langle\langle \text{at}[\![\mathsf{S}]\!], \rho\rangle, \mathsf{R}'\rangle \mid \langle \underline{\varrho}, \langle \text{at}[\![\mathsf{S}]\!], \rho\rangle\rangle \in \boldsymbol{\mathcal{S}}^{\mathsf{r}}[\![\mathsf{L}' : \mathsf{B}']\!]\}$$
$$\cup \{\langle\langle \text{at}[\![\mathsf{S}]\!], \rho\rangle\langle \text{at}[\![\mathsf{S}_t]\!], \rho\rangle\pi, \mathsf{R}''\rangle \mid \mathscr{B}[\![\mathsf{B}]\!]\rho = \text{tt} \wedge$$
$$\langle \underline{\varrho}, \langle \text{at}[\![\mathsf{S}]\!], \rho\rangle\rangle \in \boldsymbol{\mathcal{S}}^{\mathsf{r}}[\![\mathsf{L}' : \mathsf{B}']\!] \wedge$$
$$\langle\langle \text{at}[\![\mathsf{S}_t]\!], \rho\rangle\pi, \mathsf{R}''\rangle \in \widehat{\mathscr{M}}^{+}[\![\mathsf{S}_t]\!]\langle \underline{\varrho}, \mathsf{R}'\rangle\}$$
$$\cup \{\langle\langle \text{at}[\![\mathsf{S}]\!], \rho\rangle\langle \text{after}[\![\mathsf{S}]\!], \rho\rangle, \varepsilon\rangle \mid \mid \mathscr{B}[\![\mathsf{B}]\!]\rho = \text{ff} \wedge \mathsf{R}' \in \mathbb{R}_\varepsilon \wedge$$
$$\langle \underline{\varrho}, \langle \text{at}[\![\mathsf{S}]\!], \rho\rangle\rangle \in \boldsymbol{\mathcal{S}}^{\mathsf{r}}[\![\mathsf{L}' : \mathsf{B}']\!]\}$$
$$\cup \{\langle\langle \text{at}[\![\mathsf{S}]\!], \rho\rangle\langle \text{after}[\![\mathsf{S}]\!], \rho\rangle, \mathsf{R}''\rangle \mid \mathscr{B}[\![\mathsf{B}]\!]\rho = \text{ff} \wedge \mathsf{R}' \notin \mathbb{R}_\varepsilon \wedge$$
$$\langle \underline{\varrho}, \langle \text{at}[\![\mathsf{S}]\!], \rho\rangle\rangle \in \boldsymbol{\mathcal{S}}^{\mathsf{r}}[\![\mathsf{L}' : \mathsf{B}']\!] \wedge \langle \mathsf{L}'' : \mathsf{B}'', \mathsf{R}''\rangle = \text{fstnxt}(\mathsf{R}') \wedge$$
$$\langle \underline{\varrho}, \langle \text{after}[\![\mathsf{S}]\!], \rho\rangle\rangle \in \boldsymbol{\mathcal{S}}^{\mathsf{r}}[\![\mathsf{L}'' : \mathsf{B}'']\!]\}$$

- Model checking an alternative statement $\mathsf{S} ::= \texttt{if}\,^{\ell}\,(\mathsf{B})\;\mathsf{S}_t\;\texttt{else}\;\mathsf{S}_f$

$$\widehat{\mathscr{M}}^{+}[\![\mathsf{S}]\!]\langle \underline{\varrho}, \mathsf{R}\rangle \triangleq \text{let } \langle \mathsf{L}' : \mathsf{B}', \mathsf{R}'\rangle = \text{fstnxt}(\mathsf{R}) \text{ in} \tag{44.48}$$
$$\{\langle\langle \text{at}[\![\mathsf{S}]\!], \rho\rangle, \mathsf{R}'\rangle \mid \langle \underline{\varrho}, \langle \text{at}[\![\mathsf{S}]\!], \rho\rangle\rangle \in \boldsymbol{\mathcal{S}}^{\mathsf{r}}[\![\mathsf{L}' : \mathsf{B}']\!]\}$$
$$\cup \{\langle\langle \text{at}[\![\mathsf{S}]\!], \rho\rangle\langle \text{at}[\![\mathsf{S}_t]\!], \rho\rangle\pi, \mathsf{R}''\rangle \mid \mathscr{B}[\![\mathsf{B}]\!]\rho = \text{tt} \wedge$$
$$\langle \underline{\varrho}, \langle \text{at}[\![\mathsf{S}]\!], \rho\rangle\rangle \in \boldsymbol{\mathcal{S}}^{\mathsf{r}}[\![\mathsf{L}' : \mathsf{B}']\!] \wedge$$
$$\langle\langle \text{at}[\![\mathsf{S}_t]\!], \rho\rangle\pi, \mathsf{R}''\rangle \in \widehat{\mathscr{M}}^{+}[\![\mathsf{S}_t]\!]\langle \underline{\varrho}, \mathsf{R}'\rangle\}$$
$$\cup \{\langle\langle \text{at}[\![\mathsf{S}]\!], \rho\rangle\langle \text{at}[\![\mathsf{S}_f]\!], \rho\rangle\pi, \mathsf{R}''\rangle \mid \mathscr{B}[\![\mathsf{B}]\!]\rho = \text{ff} \wedge$$
$$\langle \underline{\varrho}, \langle \text{at}[\![\mathsf{S}]\!], \rho\rangle\rangle \in \boldsymbol{\mathcal{S}}^{\mathsf{r}}[\![\mathsf{L}' : \mathsf{B}']\!] \wedge$$
$$\langle\langle \text{at}[\![\mathsf{S}_f]\!], \rho\rangle\pi, \mathsf{R}''\rangle \in \widehat{\mathscr{M}}^{+}[\![\mathsf{S}_f]\!]\langle \underline{\varrho}, \mathsf{R}'\rangle\}$$

- Model checking a break statement $\mathsf{S} ::= {}^{\ell}\,\texttt{break}\,;$

$$\widehat{\mathscr{M}}^{+}[\![{}^{\ell}\,\texttt{break}\,;]\!]\langle \underline{\varrho}, \mathsf{R}\rangle \triangleq \text{let } \langle \mathsf{L} : \mathsf{B}, \mathsf{R}'\rangle = \text{fstnxt}(\mathsf{R}) \text{ in} \tag{44.49}$$
$$\{\langle\langle \text{at}[\![\mathsf{S}]\!], \rho\rangle, \mathsf{R}'\rangle \mid \langle \underline{\varrho}, \langle \text{at}[\![\mathsf{S}]\!], \rho\rangle\rangle \in \boldsymbol{\mathcal{S}}^{\mathsf{r}}[\![\mathsf{L} : \mathsf{B}]\!]\}$$
$$\cup \{\langle\langle \text{at}[\![\mathsf{S}]\!], \rho\rangle\langle \text{break-to}[\![\mathsf{S}]\!], \rho\rangle, \varepsilon\rangle \mid \mathsf{R}' \in \mathbb{R}_\varepsilon \wedge$$
$$\langle \underline{\varrho}, \langle \text{at}[\![\mathsf{S}]\!], \rho\rangle\rangle \in \boldsymbol{\mathcal{S}}^{\mathsf{r}}[\![\mathsf{L} : \mathsf{B}]\!]\}$$
$$\cup \{\langle\langle \text{at}[\![\mathsf{S}]\!], \rho\rangle\langle \text{break-to}[\![\mathsf{S}]\!], \rho\rangle, \mathsf{R}''\rangle \mid \mathsf{R}' \notin \mathbb{R}_\varepsilon \wedge$$
$$\langle \underline{\varrho}, \langle \text{at}[\![\mathsf{S}]\!], \rho\rangle\rangle \in \boldsymbol{\mathcal{S}}^{\mathsf{r}}[\![\mathsf{L} : \mathsf{B}]\!] \wedge \langle \mathsf{L}' : \mathsf{B}', \mathsf{R}''\rangle = \text{fstnxt}(\mathsf{R}')$$
$$\wedge \langle \underline{\varrho}, \langle \text{break-to}[\![\mathsf{S}]\!], \rho\rangle\rangle \in \boldsymbol{\mathcal{S}}^{\mathsf{r}}[\![\mathsf{L}' : \mathsf{B}']\!]\}$$

— The model checking of an iteration statement $\mathsf{S} ::= \texttt{while}\,^{\ell}\,(\mathsf{B})\;\mathsf{S}_b$ in (44.51) checks one more iteration (after checking the previous ones as recorded by X) while the fixpoint (44.50) repeats this check for all iterations. Case (a) checks the prefixes that stops at loop entry ℓ. (b) and (c) check the exit of an iteration when the iteration condition is false, (b) when the specification stops at loop entry ℓ before leaving and (c) when the specification goes further. (d), (e), and (f) check one more iteration when the iteration condition is true. In case (d), the continuation after the check of the iterates is empty so trivially satisfied by any continuation of the prefix trace. In case (e), the continuation after the check of the iterates just imposes to verify $\mathsf{L} : \mathsf{B}$ on iteration entry and nothing afterward. In case (f) the continuation after the check of the iterates requires verifying $\mathsf{L} : \mathsf{B}$ at the loop entry, $\mathsf{L}' : \mathsf{B}'$ at the body entry, and the rest R''' of the specification for the loop body (which returns the possibly empty continuation specification R'). The cases (b) to (f) are mutually exclusive.

Definition 44.39 (Structural model checking, continued)

- Model checking an iteration statement $\mathsf{S} ::= \mathtt{while}\,^{\ell}\,(\mathsf{B})\;\mathsf{S}_b$

$$\widehat{\mathcal{M}}^{+}[\![\mathsf{S}]\!]\langle\underline{\varrho},\,\mathsf{R}\rangle \;\triangleq\; \mathsf{lfp}^{\subseteq}\,(\widehat{\mathcal{F}}^{+}[\![\mathsf{S}]\!]\langle\underline{\varrho},\,\mathsf{R}\rangle) \tag{44.50}$$

$$\begin{aligned}
\widehat{\mathcal{F}}^{+}[\![\mathsf{S}]\!]\langle\underline{\varrho},\,\mathsf{R}\rangle\,X \;\triangleq\;& \mathsf{let}\,\langle\mathsf{L}' : \mathsf{B}',\,\mathsf{R}'\rangle = \mathsf{fstnxt}(\mathsf{R})\;\mathsf{in} && (44.51)\\
& \{\langle\langle\mathsf{at}[\![\mathsf{S}]\!],\,\rho\rangle,\,\mathsf{R}'\rangle \mid \rho\in\mathbb{E}\mathrm{v} \wedge \langle\underline{\varrho},\,\langle\mathsf{at}[\![\mathsf{S}]\!],\,\rho\rangle\rangle \in \boldsymbol{\mathcal{S}}^{\mathsf{r}}[\![\mathsf{L}' : \mathsf{B}']\!]\} && \text{(a)}\\
&\cup\{\langle\pi_2\langle\mathsf{at}[\![\mathsf{S}]\!],\,\rho\rangle\langle\mathsf{after}[\![\mathsf{S}]\!],\,\rho\rangle,\,\varepsilon\rangle \mid \langle\pi_2\langle\mathsf{at}[\![\mathsf{S}]\!],\,\rho\rangle,\,\varepsilon\rangle\in X \wedge\\
&\qquad \mathcal{B}[\![\mathsf{B}]\!]\rho = \mathsf{ff}\}\\
&\cup\{\langle\pi_2\langle\mathsf{at}[\![\mathsf{S}]\!],\,\rho\rangle\langle\mathsf{after}[\![\mathsf{S}]\!],\,\rho\rangle,\,\varepsilon\rangle \mid \langle\pi_2\langle\mathsf{at}[\![\mathsf{S}]\!],\,\rho\rangle,\,\mathsf{R}''\rangle\in X \wedge && \text{(b)}\\
&\qquad \mathcal{B}[\![\mathsf{B}]\!]\rho = \mathsf{ff} \wedge \mathsf{R}''\notin\mathbb{R}_{\varepsilon} \wedge \langle\mathsf{L}' : \mathsf{B}',\,\mathsf{R}'\rangle = \mathsf{fstnxt}(\mathsf{R}'') \wedge \mathsf{R}'\in\mathbb{R}_{\varepsilon}\wedge\\
&\qquad\quad \langle\underline{\varrho},\,\langle\mathsf{at}[\![\mathsf{S}]\!],\,\rho\rangle\rangle\in\boldsymbol{\mathcal{S}}^{\mathsf{r}}[\![\mathsf{L}' : \mathsf{B}']\!]\}\\
&\cup\{\langle\pi_2\langle\mathsf{at}[\![\mathsf{S}]\!],\,\rho\rangle\langle\mathsf{after}[\![\mathsf{S}]\!],\,\rho\rangle,\,\mathsf{R}'\rangle \mid \langle\pi_2\langle\mathsf{at}[\![\mathsf{S}]\!],\,\rho\rangle,\,\mathsf{R}''\rangle\in X \wedge && \text{(c)}\\
&\qquad \mathcal{B}[\![\mathsf{B}]\!]\rho = \mathsf{ff} \wedge \mathsf{R}''\notin\mathbb{R}_{\varepsilon} \wedge \langle\mathsf{L}' : \mathsf{B}',\,\mathsf{R}'''\rangle = \mathsf{fstnxt}(\mathsf{R}'') \wedge \mathsf{R}'''\notin\mathbb{R}_{\varepsilon}\wedge\\
&\qquad\quad \langle\underline{\varrho},\,\langle\mathsf{at}[\![\mathsf{S}]\!],\,\rho\rangle\rangle\in\boldsymbol{\mathcal{S}}^{\mathsf{r}}[\![\mathsf{L}' : \mathsf{B}']\!] \wedge \langle\mathsf{L}'' : \mathsf{B}'',\,\mathsf{R}'\rangle = \mathsf{fstnxt}(\mathsf{R}''')\wedge\\
&\qquad\qquad \langle\underline{\varrho},\,\langle\mathsf{after}[\![\mathsf{S}]\!],\,\rho\rangle\rangle\in\boldsymbol{\mathcal{S}}^{\mathsf{r}}[\![\mathsf{L}'' : \mathsf{B}'']\!]\}\\
&\cup\{\langle\pi_2\langle\mathsf{at}[\![\mathsf{S}]\!],\,\rho\rangle\langle\mathsf{at}[\![\mathsf{S}_b]\!],\,\rho\rangle\pi_3,\,\varepsilon\rangle \mid \langle\pi_2\langle\mathsf{at}[\![\mathsf{S}]\!],\,\rho\rangle,\,\varepsilon\rangle\in X \wedge && \text{(d)}\\
&\qquad \mathcal{B}[\![\mathsf{B}]\!]\rho = \mathsf{tt} \wedge \langle\mathsf{at}[\![\mathsf{S}_b]\!],\,\rho\rangle\pi_3 \in \widehat{\boldsymbol{\mathcal{S}}}^{*}_{\mathsf{s}}[\![\mathsf{S}_b]\!]\}\\
&\cup\{\langle\pi_2\langle\mathsf{at}[\![\mathsf{S}]\!],\,\rho\rangle\langle\mathsf{at}[\![\mathsf{S}_b]\!],\,\rho\rangle\pi_3,\,\varepsilon\rangle \mid \langle\pi_2\langle\mathsf{at}[\![\mathsf{S}]\!],\,\rho\rangle,\,\mathsf{R}''\rangle\in X \wedge && \text{(e)}\\
&\qquad \mathcal{B}[\![\mathsf{B}]\!]\rho = \mathsf{tt} \wedge \mathsf{R}''\notin\mathbb{R}_{\varepsilon} \wedge \langle\mathsf{L} : \mathsf{B},\,\varepsilon\rangle = \mathsf{fstnxt}(\mathsf{R}'')\wedge\\
&\qquad\quad \langle\underline{\varrho},\,\langle\mathsf{at}[\![\mathsf{S}]\!],\,\rho\rangle\rangle\in\boldsymbol{\mathcal{S}}^{\mathsf{r}}[\![\mathsf{L} : \mathsf{B}]\!] \wedge \langle\mathsf{at}[\![\mathsf{S}_b]\!],\,\rho\rangle\pi_3\in\widehat{\boldsymbol{\mathcal{S}}}^{*}_{\mathsf{s}}[\![\mathsf{S}_b]\!]\}\\
&\cup\{\langle\pi_2\langle\mathsf{at}[\![\mathsf{S}]\!],\,\rho\rangle\langle\mathsf{at}[\![\mathsf{S}_b]\!],\,\rho\rangle\pi_3,\,\mathsf{R}'\rangle \mid \langle\pi_2\langle\mathsf{at}[\![\mathsf{S}]\!],\,\rho\rangle,\,\mathsf{R}''\rangle\in X \wedge && \text{(f)}\\
&\qquad \mathcal{B}[\![\mathsf{B}]\!]\rho = \mathsf{tt} \wedge \mathsf{R}''\notin\mathbb{R}_{\varepsilon} \wedge \langle\mathsf{L} : \mathsf{B},\,\mathsf{R}''''\rangle = \mathsf{fstnxt}(\mathsf{R}'')\wedge\\
&\qquad\quad \langle\underline{\varrho},\,\langle\mathsf{at}[\![\mathsf{S}]\!],\,\rho\rangle\rangle\in\boldsymbol{\mathcal{S}}^{\mathsf{r}}[\![\mathsf{L} : \mathsf{B}]\!] \wedge \mathsf{R}''''\notin\mathbb{R}_{\varepsilon}\wedge\\
&\qquad\qquad \langle\mathsf{L}' : \mathsf{B}',\,\mathsf{R}'''\rangle = \mathsf{fstnxt}(\mathsf{R}'''') \wedge \langle\underline{\varrho},\,\langle\mathsf{at}[\![\mathsf{S}_b]\!],\,\rho\rangle\rangle\in\boldsymbol{\mathcal{S}}^{\mathsf{r}}[\![\mathsf{L}' : \mathsf{B}']\!]\\
&\qquad\qquad \wedge\langle\langle\mathsf{at}[\![\mathsf{S}_b]\!],\,\rho\rangle\pi_3,\,\mathsf{R}'\rangle\in\widehat{\mathcal{M}}^{+}[\![\mathsf{S}_b]\!]\langle\underline{\varrho},\,\mathsf{R}'''\rangle\}
\end{aligned}$$

- Model checking a compound statement $\mathsf{S} ::= \{\,\mathsf{Sl}\,\}$

$$\widehat{\mathcal{M}}^{+}[\![\{\,\mathsf{Sl}\,\}]\!] \;\triangleq\; \widehat{\mathcal{M}}^{+}[\![\mathsf{Sl}]\!] \tag{44.52}$$

Proof of theorem 44.38 As explained at the beginning of section 44.7, the proof is by calculational design and proceeds by structural induction on the program component S.

- In case (44.40) of a program $\mathsf{P} ::= \mathsf{Sl}\ \ell$, we have

$$
\begin{aligned}
& \mathcal{M}[\![\mathsf{P}]\!]\langle \underline{\varrho}, \mathsf{R}\rangle \\
\triangleq\ & \mathcal{M}\langle \underline{\varrho}, \mathsf{R}\rangle(\widehat{\mathcal{S}}^*_{\mathsf{s}}[\![\mathsf{P}]\!]) && \text{(44.28)} \\
\triangleq\ & \mathsf{let}\ (\mathsf{R}_1 \mid \ldots \mid \mathsf{R}_n) = \mathsf{dnf}(\mathsf{R})\ \mathsf{in} \bigcup_{i=1}^{n} \{\pi \mid \exists \mathsf{R}' \in \mathbb{R} \,.\, \langle \pi, \mathsf{R}'\rangle \in \mathcal{M}^{+}\langle \underline{\varrho}, \mathsf{R}_i\rangle(\widehat{\mathcal{S}}^*_{\mathsf{s}}[\![\mathsf{P}]\!])\} && \text{(44.27)} \\
=\ & \mathsf{let}\ (\mathsf{R}_1 \mid \ldots \mid \mathsf{R}_n) = \mathsf{dnf}(\mathsf{R})\ \mathsf{in} \bigcup_{i=1}^{n} \{\pi \mid \exists \mathsf{R}' \in \mathbb{R} \,.\, \langle \pi, \mathsf{R}'\rangle \in \mathcal{M}^{+}\langle \underline{\varrho}, \mathsf{R}_i\rangle(\widehat{\mathcal{S}}^*_{\mathsf{s}}[\![\mathsf{Sl}]\!])\} && \text{(42.9)} \\
=\ & \mathsf{let}\ (\mathsf{R}_1 \mid \ldots \mid \mathsf{R}_n) = \mathsf{dnf}(\mathsf{R})\ \mathsf{in} \bigcup_{i=1}^{n} \{\pi \mid \exists \mathsf{R}' \in \mathbb{R} \,.\, \langle \pi, \mathsf{R}'\rangle \in \widehat{\mathcal{M}}^{+}\langle \underline{\varrho}, \mathsf{R}_i\rangle(\widehat{\mathcal{S}}^*_{\mathsf{s}}[\![\mathsf{Sl}]\!])\} && \text{induction hypothesis} \\
=\ & \mathsf{let}\ (\mathsf{R}_1 \mid \ldots \mid \mathsf{R}_n) = \mathsf{dnf}(\mathsf{R})\ \mathsf{in} \bigcup_{i=1}^{n} \{\pi \mid \exists \mathsf{R}' \in \mathbb{R} \,.\, \langle \pi, \mathsf{R}'\rangle \in \widehat{\mathcal{M}}^{+}[\![\mathsf{Sl}]\!]\langle \underline{\varrho}, \mathsf{R}_i\rangle\} && \text{(44.26)} \\
=\ & \widehat{\mathcal{M}}[\![\mathsf{Sl}]\!]\langle \underline{\varrho}, \mathsf{R}\rangle && \text{(44.40)}
\end{aligned}
$$

- In case (44.41) of an empty temporal specification ε, see the proof at https://github.com/PrAbsInt/.

- In case (44.42) of a statement list $\mathsf{Sl} ::= \mathsf{Sl}'\ \mathsf{S}$

$$
\begin{aligned}
& \mathcal{M}^{+}[\![\mathsf{Sl}]\!]\, \langle \underline{\varrho}, \mathsf{R}\rangle \\
=\ & \mathcal{M}^{+}\langle \underline{\varrho}, \mathsf{R}\rangle(\widehat{\mathcal{S}}^*_{\mathsf{s}}[\![\mathsf{Sl}]\!]) && \text{(44.26)} \\
=\ & \{\langle \pi, \mathsf{R}'\rangle \mid \pi \in \widehat{\mathcal{S}}^*_{\mathsf{s}}[\![\mathsf{Sl}]\!] \wedge \langle \mathsf{tt}, \mathsf{R}'\rangle = \mathcal{M}^{t}\langle \underline{\varrho}, \mathsf{R}\rangle\pi\} && \text{(44.25)} \\
=\ & \{\langle \pi, \mathsf{R}'\rangle \mid \pi \in \widehat{\mathcal{S}}^*_{\mathsf{s}}[\![\mathsf{Sl}']\!] \cup \{\pi \cdot \langle \mathsf{at}[\![\mathsf{S}]\!], \rho\rangle \cdot \pi' \mid \pi \cdot \langle \mathsf{at}[\![\mathsf{S}]\!], \rho\rangle \in \widehat{\mathcal{S}}^*_{\mathsf{s}}[\![\mathsf{Sl}']\!] \wedge \langle \mathsf{at}[\![\mathsf{S}]\!], \\
& \rho\rangle \cdot \pi' \in \widehat{\mathcal{S}}^*_{\mathsf{s}}[\![\mathsf{S}]\!]\} \wedge \langle \mathsf{tt}, \mathsf{R}'\rangle = \mathcal{M}^{t}\langle \underline{\varrho}, \mathsf{R}\rangle\pi\} && \text{(42.5)} \\
=\ & \{\langle \pi, \mathsf{R}'\rangle \mid \pi \in \widehat{\mathcal{S}}^*_{\mathsf{s}}[\![\mathsf{Sl}']\!] \wedge \langle \mathsf{tt}, \mathsf{R}'\rangle = \mathcal{M}^{t}\langle \underline{\varrho}, \mathsf{R}\rangle\pi\} \cup \\
& \{\langle \pi, \mathsf{R}'\rangle \mid \pi \in \{\pi' \cdot \langle \mathsf{at}[\![\mathsf{S}]\!], \rho\rangle \cdot \pi'' \mid \pi' \cdot \langle \mathsf{at}[\![\mathsf{S}]\!], \rho\rangle \in \widehat{\mathcal{S}}^*_{\mathsf{s}}[\![\mathsf{Sl}']\!] \wedge \langle \mathsf{at}[\![\mathsf{S}]\!], \rho\rangle \cdot \pi'' \in \\
& \widehat{\mathcal{S}}^*_{\mathsf{s}}[\![\mathsf{S}]\!]\} \wedge \langle \mathsf{tt}, \mathsf{R}'\rangle = \mathcal{M}^{t}\langle \underline{\varrho}, \mathsf{R}\rangle\pi\} && \text{definitions of } \cup \text{ and } \in \\
=\ & \mathcal{M}^{+}[\![\mathsf{Sl}']\!]\, \langle \underline{\varrho}, \mathsf{R}\rangle \cup \\
& \{\langle \pi' \cdot \langle \mathsf{at}[\![\mathsf{S}]\!], \rho\rangle \cdot \pi'', \mathsf{R}'\rangle \mid \pi' \cdot \langle \mathsf{at}[\![\mathsf{S}]\!], \rho\rangle \in \widehat{\mathcal{S}}^*_{\mathsf{s}}[\![\mathsf{Sl}']\!] \wedge \langle \mathsf{at}[\![\mathsf{S}]\!], \rho\rangle \cdot \pi'' \in \widehat{\mathcal{S}}^*_{\mathsf{s}}[\![\mathsf{S}]\!] \wedge \langle \mathsf{tt}, \\
& \mathsf{R}'\rangle = \mathcal{M}^{t}\langle \underline{\varrho}, \mathsf{R}\rangle(\pi' \cdot \langle \mathsf{at}[\![\mathsf{S}]\!], \rho\rangle \cdot \pi'')\} && \text{(44.25) and (42.5), definition of } \in \\
=\ & \mathcal{M}^{+}[\![\mathsf{Sl}']\!]\, \langle \underline{\varrho}, \mathsf{R}\rangle \cup \\
& \{\langle \pi' \cdot \langle \mathsf{at}[\![\mathsf{S}]\!], \rho\rangle \cdot \pi'', \mathsf{R}'\rangle \mid \pi' \cdot \langle \mathsf{at}[\![\mathsf{S}]\!], \rho\rangle \in \widehat{\mathcal{S}}^*_{\mathsf{s}}[\![\mathsf{Sl}']\!] \wedge \langle \mathsf{at}[\![\mathsf{S}]\!], \rho\rangle \cdot \pi'' \in \\
& \widehat{\mathcal{S}}^*_{\mathsf{s}}[\![\mathsf{S}]\!] \wedge (\exists \mathsf{R}'' \in \mathbb{R} \,.\, \mathcal{M}^{t}\langle \underline{\varrho}, \mathsf{R}\rangle(\pi' \cdot \langle \mathsf{at}[\![\mathsf{S}]\!], \rho\rangle) = \langle \mathsf{tt}, \mathsf{R}''\rangle \wedge \mathcal{M}^{t}\langle \underline{\varrho}, \mathsf{R}''\rangle(\langle \mathsf{at}[\![\mathsf{S}]\!], \\
& \rho\rangle \cdot \pi'') = \langle \mathsf{tt}, \mathsf{R}'\rangle)\} && \text{lemma 44.37}
\end{aligned}
$$

$= \mathcal{M}^{+}[\![\mathtt{Sl}']\!]\,\langle\underline{\varrho},\ \mathsf{R}\rangle \cup$
$\{\langle\pi'\cdot\langle\mathsf{at}[\![\mathsf{S}]\!],\ \rho\rangle\cdot\pi'',\ \mathsf{R}'\rangle \mid \langle\pi'\cdot\langle\mathsf{at}[\![\mathsf{S}]\!],\ \rho\rangle,\ \mathsf{R}''\rangle \in \mathcal{M}^{+}\langle\underline{\varrho},\ \mathsf{R}\rangle(\widehat{\mathcal{S}}^{*}_{\mathsf{s}}[\![\mathtt{Sl}']\!]) \wedge \langle\langle\mathsf{at}[\![\mathsf{S}]\!],$
$\rho\rangle\cdot\pi'',\ \mathsf{R}'\rangle \in \mathcal{M}^{+}\langle\underline{\varrho},\ \mathsf{R}''\rangle(\widehat{\mathcal{S}}^{*}_{\mathsf{s}}[\![\mathsf{S}]\!])\}$ (44.25)

$= \mathcal{M}^{+}[\![\mathtt{Sl}']\!]\,\langle\underline{\varrho},\ \mathsf{R}\rangle \cup \{\langle\pi\cdot\langle\mathsf{at}[\![\mathsf{S}]\!],\ \rho\rangle\cdot\pi',\ \mathsf{R}'\rangle \mid \langle\pi\cdot\langle\mathsf{at}[\![\mathsf{S}]\!],\ \rho\rangle,\ \mathsf{R}''\rangle \in \mathcal{M}^{t}[\![\mathtt{Sl}']\!]\langle\underline{\varrho},$
$\mathsf{R}\rangle \wedge \langle\langle\mathsf{at}[\![\mathsf{S}]\!],\ \rho\rangle\cdot\pi',\ \mathsf{R}'\rangle \in \mathcal{M}^{t}[\![\mathsf{S}]\!]\langle\underline{\varrho},\ \mathsf{R}''\rangle\}$ (44.26)

$\mathcal{M}^{+}[\![\mathtt{Sl}']\!]\,\langle\underline{\varrho},\ \mathsf{R}\rangle \cup \{\langle\pi\cdot\langle\mathsf{at}[\![\mathsf{S}]\!],\ \rho\rangle\cdot\pi',\ \mathsf{R}'\rangle \mid \langle\pi\cdot\langle\mathsf{at}[\![\mathsf{S}]\!],\ \rho\rangle,\ \mathsf{R}''\rangle \in \widehat{\mathcal{M}}^{+}[\![\mathtt{Sl}']\!]\langle\underline{\varrho},$
$\mathsf{R}\rangle \wedge \langle\langle\mathsf{at}[\![\mathsf{S}]\!],\ \rho\rangle\cdot\pi',\ \mathsf{R}'\rangle \in \widehat{\mathcal{M}}^{+}[\![\mathsf{S}]\!]\langle\underline{\varrho},\ \mathsf{R}''\rangle\}$ (induction hypothesis)

$\triangleq \widehat{\mathcal{M}}^{+}[\![\mathtt{Sl}]\!]\langle\underline{\varrho},\ \mathsf{R}\rangle$ (44.42)

- In case (44.43) of an empty statement list $\mathtt{Sl} ::= \epsilon$, see the proof at https://github.com/PrAbsInt/.

- In case (44.45) of an assignment statement $\mathsf{S} ::= \ell\ \mathsf{x} = \mathsf{A}\ ;$

$\mathcal{M}^{+}[\![\mathsf{S}]\!]\,\langle\underline{\varrho},\ \mathsf{R}\rangle$

$= \{\langle\pi,\ \mathsf{R}'\rangle \mid \pi \in \widehat{\mathcal{S}}^{*}_{\mathsf{s}}[\![\mathsf{S}]\!] \wedge \langle\mathsf{tt},\ \mathsf{R}'\rangle = \mathcal{M}^{t}\langle\underline{\varrho},\ \mathsf{R}\rangle\pi\}$ (44.26) and (44.25)

$= \{\langle\pi,\ \mathsf{R}'\rangle \mid \pi \in \{\langle\ell,\ \rho\rangle \mid \rho \in \mathbb{E}\mathrm{v}\} \cup \{\langle\ell,\ \rho\rangle\langle\mathsf{after}[\![\mathsf{S}]\!],\ \rho[\mathsf{x}\leftarrow v]\rangle \mid \rho \in \mathbb{E}\mathrm{v} \wedge v =$
$\mathcal{A}[\![\mathsf{A}]\!]\rho \wedge \langle\mathsf{tt},\ \mathsf{R}'\rangle = \mathcal{M}^{t}\langle\underline{\varrho},\ \mathsf{R}\rangle\pi\}$ (42.4)

$= \{\langle\langle\ell,\ \rho\rangle,\ \mathsf{R}'\rangle \mid \rho \in \mathbb{E}\mathrm{v} \wedge \langle\mathsf{tt},\ \mathsf{R}'\rangle = \mathcal{M}^{t}\langle\underline{\varrho},\ \mathsf{R}\rangle\langle\ell,\ \rho\rangle\} \cup$
$\{\langle\langle\ell,\ \rho\rangle\langle\mathsf{after}[\![\mathsf{S}]\!],\ \rho[\mathsf{x}\leftarrow v]\rangle,\ \mathsf{R}'\rangle \mid \rho \in \mathbb{E}\mathrm{v} \wedge v = \mathcal{A}[\![\mathsf{A}]\!]\rho\langle\mathsf{tt},\ \mathsf{R}'\rangle = \mathcal{M}^{t}\langle\underline{\varrho},\ \mathsf{R}\rangle\langle\ell,$
$\rho\rangle\langle\mathsf{after}[\![\mathsf{S}]\!],\ \rho[\mathsf{x}\leftarrow v]\rangle\}$ (definitions of $\cup$ and $\in$)

$= \{\langle\langle\ell,\ \rho\rangle,\ \mathsf{R}'\rangle \mid \langle\mathsf{tt},\ \mathsf{R}'\rangle = (\mathsf{let}\ \langle\mathsf{L} : \mathsf{B},\ \mathsf{R}''\rangle = \mathsf{fstnxt}(\mathsf{R})\ \mathsf{in}\ (\!\!|\ \langle\underline{\varrho},\ \langle\ell,\ \rho\rangle\rangle \in \mathcal{S}^{\mathsf{r}}[\![\mathsf{L} : \mathsf{B}]\!]\ ?\ \langle\mathsf{tt},$
$\mathsf{R}''\rangle\ {:}\ \langle\mathsf{ff},\ \mathsf{R}'\rangle\ |\!\!))\} \cup$
$\{\langle\langle\ell,\ \rho\rangle\langle\mathsf{after}[\![\mathsf{S}]\!],\ \rho[\mathsf{x}\leftarrow v]\rangle,\ \mathsf{R}'\rangle \mid v = \mathcal{A}[\![\mathsf{A}]\!]\rho \wedge \langle\mathsf{tt},\ \mathsf{R}'\rangle = (\mathsf{let}\ \langle\mathsf{L} : \mathsf{B},\ \mathsf{R}''\rangle =$
$\mathsf{fstnxt}(\mathsf{R})\ \mathsf{in}\ (\!\!|\ \langle\underline{\varrho},\ \langle\ell,\ \rho\rangle\rangle \in \mathcal{S}^{\mathsf{r}}[\![\mathsf{L} : \mathsf{B}]\!]\ ?\ \mathcal{M}^{t}\langle\underline{\varrho},\ \mathsf{R}''\rangle\langle\mathsf{after}[\![\mathsf{S}]\!],\ \rho[\mathsf{x}\leftarrow v]\rangle\ {:}\ \langle\mathsf{ff},\ \mathsf{R}''\rangle\ |\!\!))\}$
(44.24)

$= \{\langle\langle\ell,\ \rho\rangle,\ \mathsf{R}'\rangle \mid \langle\mathsf{L} : \mathsf{B},\ \mathsf{R}'\rangle = \mathsf{fstnxt}(\mathsf{R}) \wedge \langle\underline{\varrho},\ \langle\ell,\ \rho\rangle\rangle \in \mathcal{S}^{\mathsf{r}}[\![\mathsf{L} : \mathsf{B}]\!]\} \cup$
$\{\langle\langle\ell,\ \rho\rangle\langle\mathsf{after}[\![\mathsf{S}]\!],\ \rho[\mathsf{x}\leftarrow v]\rangle,\ \mathsf{R}'\rangle \mid v = \mathcal{A}[\![\mathsf{A}]\!]\rho \wedge \exists\mathsf{R}'' \in \mathbb{R}\,.\,\langle\mathsf{L} : \mathsf{B},\ \mathsf{R}''\rangle = \mathsf{fstnxt}(\mathsf{R}) \wedge$
$\langle\underline{\varrho},\ \langle\ell,\ \rho\rangle\rangle \in \mathcal{S}^{\mathsf{r}}[\![\mathsf{L} : \mathsf{B}]\!] \wedge (\!\!|\ \mathsf{R}'' \in \mathbb{R}_{\varepsilon}\ ?\ \mathsf{tt}\ {:}\ \mathcal{M}^{t}\langle\underline{\varrho},\ \mathsf{R}''\rangle\langle\mathsf{after}[\![\mathsf{S}]\!],\ \rho[\mathsf{x}\leftarrow v]\rangle = \langle\mathsf{tt},\ \mathsf{R}'\rangle\ |\!\!)\}$
(definition of $=$ and $\mathcal{M}^{t}\langle\underline{\varrho},\ \varepsilon\rangle\pi \triangleq \langle\mathsf{tt},\ \varepsilon\rangle$ by (44.24))

$= \{\langle\langle\ell,\ \rho\rangle,\ \mathsf{R}'\rangle \mid \langle\mathsf{L} : \mathsf{B},\ \mathsf{R}'\rangle = \mathsf{fstnxt}(\mathsf{R}) \wedge \langle\underline{\varrho},\ \langle\ell,\ \rho\rangle\rangle \in \mathcal{S}^{\mathsf{r}}[\![\mathsf{L} : \mathsf{B}]\!]\} \cup$
$\{\langle\langle\ell,\ \rho\rangle\langle\mathsf{after}[\![\mathsf{S}]\!],\ \rho[\mathsf{x}\leftarrow v]\rangle,\ \mathsf{R}'\rangle \mid v = \mathcal{A}[\![\mathsf{A}]\!]\rho \wedge \exists\mathsf{R}'' \in \mathbb{R}\,.\,\langle\mathsf{L} : \mathsf{B},\ \mathsf{R}''\rangle = \mathsf{fstnxt}(\mathsf{R}) \wedge$
$\langle\underline{\varrho},\ \langle\ell,\ \rho\rangle\rangle \in \mathcal{S}^{\mathsf{r}}[\![\mathsf{L} : \mathsf{B}]\!] \wedge (\!\!|\ \mathsf{R}'' \in \mathbb{R}_{\varepsilon}\ ?\ \mathsf{tt}\ {:}\ \mathsf{let}\ \langle\mathsf{L}' : \mathsf{B}',\ \mathsf{R}'''\rangle = \mathsf{fstnxt}(\mathsf{R}'')\ \mathsf{in}\ \langle\underline{\varrho},$
$\langle\mathsf{after}[\![\mathsf{S}]\!],\ \rho[\mathsf{x}\leftarrow v]\rangle\rangle \in \mathcal{S}^{\mathsf{r}}[\![\mathsf{L}' : \mathsf{B}']\!]\ |\!\!)\}$ (44.24)

$= \mathsf{let}\ \langle\mathsf{L} : \mathsf{B},\ \mathsf{R}'\rangle = \mathsf{fstnxt}(\mathsf{R})\ \mathsf{in}$
$\{\langle\langle\ell,\ \rho\rangle,\ \mathsf{R}'\rangle \mid \langle\underline{\varrho},\ \langle\ell,\ \rho\rangle\rangle \in \mathcal{S}^{\mathsf{r}}[\![\mathsf{L} : \mathsf{B}]\!]\}$
$\cup \{\langle\langle\ell,\ \rho\rangle\langle\mathsf{after}[\![\mathsf{S}]\!],\ \rho[\mathsf{x}\leftarrow v]\rangle,\ \varepsilon\rangle \mid v = \mathcal{A}[\![\mathsf{A}]\!]\rho \wedge \langle\underline{\varrho},\ \langle\ell,\ \rho\rangle\rangle \in \mathcal{S}^{\mathsf{r}}[\![\mathsf{L} : \mathsf{B}]\!] \wedge \mathsf{R}' \in \mathbb{R}_{\varepsilon}\}$
$\cup \{\langle\langle\ell,\ \rho\rangle\langle\mathsf{after}[\![\mathsf{S}]\!],\ \rho[\mathsf{x}\leftarrow v]\rangle,\ \mathsf{R}''\rangle \mid v = \mathcal{A}[\![\mathsf{A}]\!]\rho \wedge \langle\underline{\varrho},\ \langle\ell,\ \rho\rangle\rangle \in \mathcal{S}^{\mathsf{r}}[\![\mathsf{L} : \mathsf{B}]\!] \wedge \mathsf{R}' \notin$
$\mathbb{R}_{\varepsilon} \wedge \mathsf{let}\ \langle\mathsf{L}' : \mathsf{B}',\ \mathsf{R}''\rangle = \mathsf{fstnxt}(\mathsf{R}')\ \mathsf{in}\ \langle\underline{\varrho},\ \langle\mathsf{after}[\![\mathsf{S}]\!],\ \rho[\mathsf{x}\leftarrow v]\rangle\rangle \in \mathcal{S}^{\mathsf{r}}[\![\mathsf{L}' : \mathsf{B}']\!]\}$
(definition of $\cup$)

$= \widehat{\mathcal{M}}^{+}[\![\mathsf{S}]\!]\,\langle \underline{\varrho}, \mathsf{R}\rangle$ $\wr(44.45)\wr$

- In case (44.44) of a skip statement $\mathsf{S} ::= ;$, see the proof at https://github.com/PrAbsInt/.

- In case (44.47) of a conditional statement $\mathsf{S} ::= \mathtt{if}\ \ell\ (\mathsf{B})\ \mathsf{S}_t$

$\mathcal{M}^{+}[\![\mathsf{S}]\!]\,\langle \underline{\varrho}, \mathsf{R}\rangle$

$= \mathcal{M}^{+}\langle \underline{\varrho}, \mathsf{R}\rangle(\widehat{\mathcal{S}}^{*}_{s}[\![\mathsf{S}]\!])$ $\wr(44.26)\wr$

$= \{\langle \pi, \mathsf{R}'\rangle \mid \pi \in \widehat{\mathcal{S}}^{*}_{s}[\![\mathsf{S}]\!] \wedge \langle \mathsf{tt}, \mathsf{R}'\rangle = \mathcal{M}^{t}\langle \underline{\varrho}, \mathsf{R}\rangle\pi\}$ $\wr(44.25)\wr$

$= \{\langle \pi, \mathsf{R}'\rangle \mid \pi \in \{\langle \mathsf{at}[\![\mathsf{S}]\!], \rho\rangle \mid \rho \in \mathbb{E}\mathbb{v}\} \wedge \langle \mathsf{tt}, \mathsf{R}'\rangle = \mathcal{M}^{t}\langle \underline{\varrho}, \mathsf{R}\rangle\pi\} \cup$
$\{\langle \pi, \mathsf{R}'\rangle \mid \pi \in \{\langle \mathsf{at}[\![\mathsf{S}]\!], \rho\rangle\langle \mathsf{after}[\![\mathsf{S}]\!], \rho\rangle \mid \mathcal{B}[\![\mathsf{B}]\!]\,\rho = \mathsf{ff}\} \wedge \langle \mathsf{tt}, \mathsf{R}'\rangle = \mathcal{M}^{t}\langle \underline{\varrho}, \mathsf{R}\rangle\pi\} \cup$
$\{\langle \pi, \mathsf{R}'\rangle \mid \pi \in \{\langle \mathsf{at}[\![\mathsf{S}]\!], \rho\rangle\langle \mathsf{at}[\![\mathsf{S}_t]\!], \rho\rangle \frown \pi_2 \mid \mathcal{B}[\![\mathsf{B}]\!]\,\rho = \mathsf{tt} \wedge \pi_2 \in \mathcal{S}^{*}[\![\mathsf{S}_t]\!]\} \wedge \langle \mathsf{tt}, \mathsf{R}'\rangle = \mathcal{M}^{t}\langle \underline{\varrho}, \mathsf{R}\rangle\pi\}$

$\wr$transition semantics of the conditional in exercise 43.12 and definition of $\cup\wr$

We go on separately for each of the previous three terms.

— The first term was already handled previously in the proof of (44.44) for the a skip statement $\mathsf{S} ::= ;$.

$\{\langle \pi, \mathsf{R}'\rangle \mid \pi \in \{\langle \mathsf{at}[\![\mathsf{S}]\!], \rho\rangle \mid \rho \in \mathbb{E}\mathbb{v}\} \wedge \langle \mathsf{tt}, \mathsf{R}'\rangle = \mathcal{M}^{t}\langle \underline{\varrho}, \mathsf{R}\rangle\pi\}$

$= \mathsf{let}\ \langle \mathsf{L} : \mathsf{B}, \mathsf{R}'\rangle = \mathsf{fstnxt}(\mathsf{R})\ \mathsf{in}\ \{\langle\langle \mathsf{at}[\![\mathsf{S}]\!], \rho\rangle, \mathsf{R}'\rangle \mid \langle \underline{\varrho}, \langle \mathsf{at}[\![\mathsf{S}]\!], \rho\rangle\rangle \in \mathcal{S}^{\mathsf{r}}[\![\mathsf{L} : \mathsf{B}]\!]\}$

— The second term is

$\{\langle \pi, \mathsf{R}'\rangle \mid \pi \in \{\langle \mathsf{at}[\![\mathsf{S}]\!], \rho\rangle\langle \mathsf{after}[\![\mathsf{S}]\!], \rho\rangle \mid \mathcal{B}[\![\mathsf{B}]\!]\,\rho = \mathsf{ff}\} \wedge \langle \mathsf{tt}, \mathsf{R}'\rangle = \mathcal{M}^{t}\langle \underline{\varrho}, \mathsf{R}\rangle\pi\}$

$= \{\langle\langle \mathsf{at}[\![\mathsf{S}]\!], \rho\rangle\langle \mathsf{after}[\![\mathsf{S}]\!], \rho\rangle, \mathsf{R}'\rangle \mid \mathcal{B}[\![\mathsf{B}]\!]\,\rho = \mathsf{ff} \wedge \langle \mathsf{tt}, \mathsf{R}'\rangle = \mathcal{M}^{t}\langle \underline{\varrho}, \mathsf{R}\rangle(\langle \mathsf{at}[\![\mathsf{S}]\!], \rho\rangle\langle \mathsf{after}[\![\mathsf{S}]\!], \rho\rangle)\}$ $\wr$definition of $\in\wr$

By (44.24), there are two subcases. If $\mathsf{R} = \varepsilon$, this is

$= \{\langle\langle \mathsf{at}[\![\mathsf{S}]\!], \rho\rangle\langle \mathsf{after}[\![\mathsf{S}]\!], \rho\rangle, \varepsilon\rangle \mid \mathsf{R} \in \mathbb{R}_{\varepsilon} \wedge \mathcal{B}[\![\mathsf{B}]\!]\,\rho = \mathsf{ff}\}$ $\wr(44.24)\wr$

This term appears as (44.41). Otherwise $\mathsf{R} \neq \varepsilon$, and this is

$= \{\langle\langle \mathsf{at}[\![\mathsf{S}]\!], \rho\rangle\langle \mathsf{after}[\![\mathsf{S}]\!], \rho\rangle, \mathsf{R}'\rangle \mid \mathsf{R} \notin \mathbb{R}_{\varepsilon} \wedge \mathcal{B}[\![\mathsf{B}]\!]\,\rho = \mathsf{ff} \wedge \langle \mathsf{tt}, \mathsf{R}'\rangle = \mathsf{let}\ \langle \mathsf{L} : \mathsf{B}, \mathsf{R}'\rangle = \mathsf{fstnxt}(\mathsf{R})\ \mathsf{in}\ (\!\!|\ \langle \underline{\varrho}, \langle \mathsf{at}[\![\mathsf{S}]\!], \rho\rangle\rangle \in \mathcal{S}^{\mathsf{r}}[\![\mathsf{L} : \mathsf{B}]\!]\ ?\ \mathcal{M}^{t}\langle \underline{\varrho}, \mathsf{R}'\rangle\langle \mathsf{after}[\![\mathsf{S}]\!], \rho\rangle : \langle \mathsf{ff}, \mathsf{R}'\rangle\ |\!\!)\}$

$\wr(44.24)\wr$

$= \{\langle\langle \mathsf{at}[\![\mathsf{S}]\!], \rho\rangle\langle \mathsf{after}[\![\mathsf{S}]\!], \rho\rangle, \mathsf{R}'\rangle \mid \mathsf{R} \notin \mathbb{R}_{\varepsilon} \wedge \mathcal{B}[\![\mathsf{B}]\!]\,\rho = \mathsf{ff} \wedge \mathsf{let}\ \langle \mathsf{L} : \mathsf{B}, \mathsf{R}'\rangle = \mathsf{fstnxt}(\mathsf{R})\ \mathsf{in}\ \langle \underline{\varrho}, \langle \mathsf{at}[\![\mathsf{S}]\!], \rho\rangle\rangle \in \mathcal{S}^{\mathsf{r}}[\![\mathsf{L} : \mathsf{B}]\!] \wedge \langle \mathsf{tt}, \mathsf{R}'\rangle = \mathcal{M}^{t}\langle \underline{\varrho}, \mathsf{R}'\rangle\langle \mathsf{after}[\![\mathsf{S}]\!], \rho\rangle\}$

$\wr$definition of the conditional$\wr$

By (44.24), there are two subsubcases. If $\mathsf{R}' = \varepsilon$, this is

$= \{\langle\langle \mathsf{at}[\![\mathsf{S}]\!], \rho\rangle\langle \mathsf{after}[\![\mathsf{S}]\!], \rho\rangle, \varepsilon\rangle \mid \mathsf{R} \notin \mathbb{R}_{\varepsilon} \wedge \mathcal{B}[\![\mathsf{B}]\!]\,\rho = \mathsf{ff} \wedge \langle \mathsf{L} : \mathsf{B}, \varepsilon\rangle = \mathsf{fstnxt}(\mathsf{R}) \wedge \langle \underline{\varrho}, \langle \mathsf{at}[\![\mathsf{S}]\!], \rho\rangle\rangle \wedge \mathcal{S}^{\mathsf{r}}[\![\mathsf{L} : \mathsf{B}]\!]\}$ $\wr(44.24)\wr$

$= \{\langle\langle \mathsf{at}[\![\mathsf{S}]\!], \rho\rangle\langle \mathsf{after}[\![\mathsf{S}]\!], \rho\rangle, \varepsilon\rangle \mid \mathsf{R} \notin \mathbb{R}_\varepsilon \wedge \mathscr{B}[\![\mathsf{B}]\!]\rho = \mathrm{ff} \wedge \langle \mathsf{L}' : \mathsf{B}', \mathsf{R}'\rangle = \mathsf{fstnxt}(\mathsf{R}) \wedge \mathsf{R}' \in \mathbb{R}_\varepsilon \wedge \langle \underline{\varrho}, \langle \mathsf{at}[\![\mathsf{S}]\!], \rho\rangle\rangle \in \boldsymbol{\mathcal{S}}^{\mathrm{r}}[\![\mathsf{L}' : \mathsf{B}']\!]\}$

Otherwise $\mathsf{R} \neq \varepsilon$, and this is

$= \{\langle\langle \mathsf{at}[\![\mathsf{S}]\!], \rho\rangle\langle \mathsf{after}[\![\mathsf{S}]\!], \rho\rangle, \mathsf{R}''\rangle \mid \mathsf{R} \notin \mathbb{R}_\varepsilon \wedge \mathscr{B}[\![\mathsf{B}]\!]\rho = \mathrm{ff} \wedge \langle \mathsf{L} : \mathsf{B}, \mathsf{R}'\rangle = \mathsf{fstnxt}(\mathsf{R}) \wedge \langle \underline{\varrho}, \langle \mathsf{at}[\![\mathsf{S}]\!], \rho\rangle\rangle \in \boldsymbol{\mathcal{S}}^{\mathrm{r}}[\![\mathsf{L} : \mathsf{B}]\!] \wedge \mathsf{R}' \notin \mathbb{R}_\varepsilon \wedge \langle \mathsf{L}'' : \mathsf{B}'', \mathsf{R}''\rangle = \mathsf{fstnxt}(\mathsf{R}') \wedge \langle \underline{\varrho}, \langle \mathsf{after}[\![\mathsf{S}]\!], \rho\rangle\rangle \in \boldsymbol{\mathcal{S}}^{\mathrm{r}}[\![\mathsf{L}'' : \mathsf{B}'']\!]\}$ ⟅(44.24)⟆

— The third term is

$\{\langle \pi, \mathsf{R}'\rangle \mid \pi \in \{\langle \mathsf{at}[\![\mathsf{S}]\!], \rho\rangle\langle \mathsf{at}[\![\mathsf{S}_t]\!], \rho\rangle \mathbin{\frown} \pi_2 \mid \mathscr{B}[\![\mathsf{B}]\!]\rho = \mathrm{tt} \wedge \pi_2 \in \boldsymbol{\mathcal{S}}^*[\![\mathsf{S}_t]\!]\} \wedge \langle \mathrm{tt}, \mathsf{R}'\rangle = \mathscr{M}^t\langle \underline{\varrho}, \mathsf{R}\rangle\pi\}$

$= \{\langle\langle \mathsf{at}[\![\mathsf{S}]\!], \rho\rangle\langle \mathsf{at}[\![\mathsf{S}_t]\!], \rho\rangle\pi_2, \mathsf{R}'\rangle \mid \mathscr{B}[\![\mathsf{B}]\!]\rho = \mathrm{tt} \wedge \langle \mathsf{at}[\![\mathsf{S}_t]\!], \rho\rangle\pi_2 \in \boldsymbol{\mathcal{S}}^*[\![\mathsf{S}_t]\!] \wedge \langle \mathrm{tt}, \mathsf{R}'\rangle = \mathscr{M}^t\langle \underline{\varrho}, \mathsf{R}\rangle(\langle \mathsf{at}[\![\mathsf{S}]\!], \rho\rangle\langle \mathsf{at}[\![\mathsf{S}_t]\!], \rho\rangle\pi_2)\}$ ⟅definitions of $\in$ and $\frown$⟆

By (44.24), there are two subases. If $\mathsf{R} = \varepsilon$, this is

$= \{\langle\langle \mathsf{at}[\![\mathsf{S}]\!], \rho\rangle\langle \mathsf{at}[\![\mathsf{S}_t]\!], \rho\rangle\pi_2, \varepsilon\rangle \mid \mathsf{R} \in \mathbb{R}_\varepsilon \wedge \mathscr{B}[\![\mathsf{B}]\!]\rho = \mathrm{tt} \wedge \langle \mathsf{at}[\![\mathsf{S}_t]\!], \rho\rangle\pi_2 \in \boldsymbol{\mathcal{S}}^*[\![\mathsf{S}_t]\!]\}$ ⟅(44.24)⟆

This term is incorporated in (44.41). Otherwise $\mathsf{R} \neq \varepsilon$, and this is

$= \{\langle\langle \mathsf{at}[\![\mathsf{S}]\!], \rho\rangle\langle \mathsf{at}[\![\mathsf{S}_t]\!], \rho\rangle\pi_2, \mathsf{R}''\rangle \mid \mathsf{R} \notin \mathbb{R}_\varepsilon \wedge \mathscr{B}[\![\mathsf{B}]\!]\rho = \mathrm{tt} \wedge \langle \mathsf{at}[\![\mathsf{S}_t]\!], \rho\rangle\pi_2 \in \boldsymbol{\mathcal{S}}^*[\![\mathsf{S}_t]\!] \wedge \langle \mathsf{L} : \mathsf{B}, \mathsf{R}'\rangle = \mathsf{fstnxt}(\mathsf{R}) \wedge \langle \underline{\varrho}, \langle \mathsf{at}[\![\mathsf{S}]\!], \rho\rangle\rangle \in \boldsymbol{\mathcal{S}}^{\mathrm{r}}[\![\mathsf{L} : \mathsf{B}]\!] \wedge \langle \mathrm{tt}, \mathsf{R}''\rangle = \mathscr{M}^t\langle \underline{\varrho}, \mathsf{R}'\rangle(\langle \mathsf{at}[\![\mathsf{S}_t]\!], \rho\rangle\pi_2)\}$ ⟅(44.24)⟆

$= \{\langle\langle \mathsf{at}[\![\mathsf{S}]\!], \rho\rangle\langle \mathsf{at}[\![\mathsf{S}_t]\!], \rho\rangle\pi_2, \mathsf{R}''\rangle \mid \mathsf{R} \notin \mathbb{R}_\varepsilon \wedge \mathscr{B}[\![\mathsf{B}]\!]\rho = \mathrm{tt} \wedge \langle \mathsf{L} : \mathsf{B}, \mathsf{R}'\rangle = \mathsf{fstnxt}(\mathsf{R}) \wedge \langle \underline{\varrho}, \langle \mathsf{at}[\![\mathsf{S}]\!], \rho\rangle\rangle \in \boldsymbol{\mathcal{S}}^{\mathrm{r}}[\![\mathsf{L} : \mathsf{B}]\!] \wedge \langle\langle \mathsf{at}[\![\mathsf{S}_t]\!], \rho\rangle\pi_2, \mathsf{R}''\rangle \in \{\langle \pi, \mathsf{R}''\rangle \mid \pi \in \widehat{\boldsymbol{\mathcal{S}}}^*_{\mathbb{S}}[\![\mathsf{S}_t]\!] \wedge \langle \mathrm{tt}, \mathsf{R}''\rangle = \mathscr{M}^t\langle \underline{\varrho}, \mathsf{R}'\rangle\pi\}\}$ ⟅definition of $\in$⟆

$= \{\langle\langle \mathsf{at}[\![\mathsf{S}]\!], \rho\rangle\langle \mathsf{at}[\![\mathsf{S}_t]\!], \rho\rangle\pi_2, \mathsf{R}''\rangle \mid \mathsf{R} \notin \mathbb{R}_\varepsilon \wedge \mathscr{B}[\![\mathsf{B}]\!]\rho = \mathrm{tt} \wedge \langle \mathsf{L} : \mathsf{B}, \mathsf{R}'\rangle = \mathsf{fstnxt}(\mathsf{R}) \wedge \langle \underline{\varrho}, \langle \mathsf{at}[\![\mathsf{S}]\!], \rho\rangle\rangle \in \boldsymbol{\mathcal{S}}^{\mathrm{r}}[\![\mathsf{L} : \mathsf{B}]\!] \wedge \langle\langle \mathsf{at}[\![\mathsf{S}_t]\!], \rho\rangle\pi_2, \mathsf{R}''\rangle \in \mathscr{M}^{\dagger}[\![\mathsf{S}_t]\!]\langle \underline{\varrho}, \mathsf{R}'\rangle\}$ ⟅(44.26) and (44.25)⟆

— Grouping all cases and subcases together, we get (44.47) and (44.41) when R is empty. The case (44.48) is left as exercise 44.53.

- In case (44.50) of an iteration statement $\mathsf{S} ::= \texttt{while}\,\ell\,(\mathsf{B})\,\mathsf{S}_b$, see the proof at https://github.com/PrAbsInt/. □

Exercise 44.53 Prove (44.48) for an alternative statement $\mathsf{S} ::= \texttt{if}\,\ell\,(\mathsf{B})\,\mathsf{S}_t\ \texttt{else}\ \mathsf{S}_f$ and (44.52) for a compound statement $\mathsf{S} ::= \{\,\mathsf{Sl}\,\}$.

44.8 Notes on Implementations and Expressivity

Of course, further hypotheses and refinements would be necessary to get an effective algorithm as specified by definition 44.39 of structural model checking. A common hypothesis in model checking is that the set of states $\mathbb{S}$ is finite. Traces may still be infinite, so the fixpoint computation (44.50) may not converge. However, as noticed in chapter 15, infinite traces on finite states must involve an initial finite prefix followed by a finite cycle (often called a lasso). It follows that the infinite set of prefix traces can be finitely represented by a finite set of maximal finite traces and finite lassos. Regular expressions `L : B` can be attached to the states as determined by the analysis, and there are finitely many of them in the specification. These finiteness properties can be taken into account to ensure the convergence of the fixpoint computation in (44.50).

A symbolic representation of the states in finite/lasso stateful traces may be useful as in symbolic execution [578] or using BDDs [150] for Boolean encodings of programs. By Kleene's theorem [850, theorem 2.1, p. 87], a convenient representation of regular expressions is by (deterministic) finite automata, for example, [70, 636], which may improve the performances but will not solve the state explosion problem. Symbolic automata-based algorithms can be used to implement a data structure for operations over sets of sequences [477].

Of course, the hypothesis that the state space is finite and small enough to scale up and limit the combinatorial blow up of the finite state-space [192] is unrealistic, as illustrated by exercise 3.48. In practice, the set of states $\mathbb{S}$ is very large, so abstraction of invariant properties at program points and a widening/dual narrowing are necessary. A typical trivial widening is bounded model checking (e.g., widen to all states after n fixpoint iterations) [119]. Those of [669–671] are more elaborated.

Exercise 44.54 (prefix trace semantics of a transition system) A transition system is a triple $\tau = \langle \mathbb{S}, \iota_0, \mathrm{\tau} \rangle$ where $\mathbb{S}$ is a set of states, $\iota_0 \in \mathbb{S}$ is the initial state and $\mathrm{\tau} \in \wp(\mathbb{S} \times \mathbb{S})$ is the transition relation. For example, $\mathbb{S} = \mathbb{Z}$, $\iota_0 = 0$, $\mathrm{\tau} = \{\langle 0, 2 \rangle\} \cup \{\langle z, z+4 \rangle \mid z \in \mathbb{Z}\}$ will have two infinite executions $0\,2\,6\,10\ \ldots$ and $0\,4\,8\,12\ \ldots$.

We let $\mathbb{S}^n$, $n \in \mathbb{N}$ be the set of sequences of elements of $\mathbb{S}$ of length n, called traces. $\mathbb{S}^0 = \{\ni\}$ where $\ni$ is the empty trace. $\mathbb{S}^n = \{\pi_0 \ldots \pi_{n-1} \mid \forall i \in [0, n[\,.\, \pi_i \in \mathbb{S}\}$ when $n > 0$. If $\pi \in \mathbb{S}^n$, then its length is $|\pi| = n$. We let $\mathbb{S}^* \triangleq \bigcup_{n \in \mathbb{N}} \mathbb{S}^n$ and $\mathbb{S}^+ \triangleq \bigcup_{n \in \mathbb{N}^+} \mathbb{S}^n$ where $\mathbb{N}^+ = \mathbb{N} \setminus \{0\}$. $\mathbb{S}^\infty$ is the set of infinite traces of states in $\mathbb{S}$. If $\pi \in \mathbb{S}^\infty$, then $|\pi| = \infty$.

If $\sigma \in \mathbb{S}$, we let $\vec{\sigma} = \{\pi \in \mathbb{S}^1 \mid \pi_0 = \sigma\}$ be the traces of length 1 consisting of state σ. We let $\vec{\sigma}^* = \{\pi \in \mathbb{S}^n \mid n \in \mathbb{N}^+ \wedge \pi_0 = \sigma\}$ be the finite traces starting with state σ and $\vec{\sigma}^\infty = \{\pi \in \mathbb{S}^\infty \mid \pi_0 = \sigma\}$ be the infinite ones. If $\mathrm{\tau} \in \wp(\mathbb{S} \times \mathbb{S})$, we let $\vec{\mathrm{\tau}} = \{\pi \in \mathbb{S}^2 \mid \langle \pi_0, \pi_1 \rangle \in \mathrm{\tau}\}$ be the traces of length 2 consisting of a transition $\mathrm{\tau}$. If $\sigma \in \mathbb{S}$ and $\pi \in \mathbb{S}^n$, we let $\pi' = \pi\sigma \in \mathbb{S}^{n+1}$ be such that $\forall n \in [0, n[\,.\, \pi'_i = \pi_i \wedge \pi'_n = \sigma$. Similarly for $\sigma\pi$. In particular $\sigma\ni = \ni\sigma = \vec{\sigma}$.

Trace concatenation $\mathbin{\dot\frown}$ is defined such that the empty trace $\ni$ is the neutral element $\pi \mathbin{\dot\frown} \ni = \ni \mathbin{\dot\frown} \pi = \pi$ and $\pi\sigma \mathbin{\dot\frown} \sigma\pi' = \pi\sigma\pi'$ for finite traces. This is extended to sets of traces as $X \mathbin{\dot\frown} Y \triangleq (\!|\, \ni \in X \mathrel{?} Y \mathrel{:} \emptyset \,|\!) \cup (\!|\, \ni \in Y \mathrel{?} X \mathrel{:} \emptyset \,|\!) \cup \{\pi\sigma\pi' \mid \pi\sigma \in X \wedge \sigma\pi' \in Y\}$.

The prefix trace semantics of a transition system $\tau = \langle \mathbb{S}, \iota_0, \mathrm{\tau} \rangle$ is

$$
\begin{aligned}
\boldsymbol{\mathcal{S}}^n_{\mathrm{t}}[\![\mathrm{\tau}]\!] &\triangleq \{\pi \in \mathbb{S}^n \mid \forall i \in [0, n-1[\,.\, \langle \pi_i, \pi_{i+1} \rangle \in \mathrm{\tau}\} \\
\boldsymbol{\mathcal{S}}_{\mathrm{t}}[\![\mathrm{\tau}]\!] &\triangleq \bigcup_{n \in \mathbb{N}^+} \boldsymbol{\mathcal{S}}^n_{\mathrm{t}}[\![\mathrm{\tau}]\!]
\end{aligned}
\tag{44.55}
$$

so that, obviously, $\boldsymbol{\mathcal{S}}^1_{\mathrm{t}}[\![\tau]\!] = \mathbb{S}^1$ and $\boldsymbol{\mathcal{S}}^{n+1}_{\mathrm{t}}[\![\tau]\!] = \boldsymbol{\mathcal{S}}^n_{\mathrm{t}}[\![\mathrm{\tau}]\!] \mathbin{\dot\frown} \vec{\mathrm{\tau}} = \vec{\mathrm{\tau}} \mathbin{\dot\frown} \boldsymbol{\mathcal{S}}^n_{\mathrm{t}}[\![\mathrm{\tau}]\!]$. The prefix trace semantics of a transition system $\tau = \langle \mathbb{S}, \iota_0, \mathrm{\tau} \rangle$ starting with initial state ι_0 are $\boldsymbol{\mathcal{S}}_{\mathrm{t}}[\![\mathrm{\tau}]\!] \cap \vec{\iota_0}^*$.

— Prove that the function $X \mapsto \mathbb{S}^1 \cup \vec{\mathrm{\tau}} \mathbin{\dot\frown} X$ preserves nonempty joins (hence is continuous).

— Prove that the iterates $\langle X^n,\ n \in \mathbb{N}\rangle$ of the function $X \mapsto \mathbb{S}^1 \cup \vec{\tau} \frown X$ from $\varnothing$ are $X^n = \bigcup_{i=1}^{n} \boldsymbol{\mathcal{S}}^i_{\mathrm{t}}[\![\tau]\!]$.

— Prove that $\boldsymbol{\mathcal{S}}_{\mathrm{t}}[\![\tau]\!] = \mathsf{lfp}^{\subseteq}\ X \mapsto \mathbb{S}^1 \cup \vec{\tau} \frown X$.

Exercise 44.56 (temporal specification) Continuing exercise 44.54, a temporal specification is a transition system $\mathcal{T} = \langle \wp(\mathbb{S}),\ I_0,\ T\rangle$ where the transition relation T is total, that is $\forall \Sigma \in \wp(\mathbb{S})\,.\,\exists \Sigma' \in \wp(\mathbb{S})\,.\,\langle \Sigma,\ \Sigma'\rangle \in T$. This can always be obtained by, for example, duplicating final states infinitely often. For example, the temporal specification $\mathbb{S} = \mathbb{Z}$, $I_0 = \{0\}, \tau = \{\langle \{0\},\ \{0, 2\}\rangle\} \cup \{\langle Z,\ \{0\} \cup \{z + 4 \mid z \in Z\}\rangle \mid Z \in \wp(\mathbb{Z})\}$ specifies that the initial state is 0 and then 0 or multiples of 2 or 4 modulo 4.

Invariance is a special case of temporal specification that does not change over time, that is, is the same everywhere along traces. For example $\mathcal{T} = \langle \wp(\mathbb{Z}),\ \mathbb{N},\ \{\langle \mathbb{N},\ \mathbb{N}\rangle\}\rangle$ is an invariance specification that the state is always positive or zero during execution.

The semantics of a specification $\mathcal{T}$ is the set of infinite sequences of sets of states P starting with the initial specification I_0 and then related by the total transition relation T.

$$\boldsymbol{\mathcal{S}}[\![T]\!] \quad \triangleq \quad \{P \in \wp(\mathbb{S})^{\infty} \mid \forall i \in \mathbb{N}\,.\,\langle P_i,\ P_{i+1}\rangle \in T\} \tag{44.57}$$

$\boldsymbol{\mathcal{S}}[\![T]\!] \cap \vec{I_0}^{\infty}$ restricts the temporal specification $\mathcal{T} = \langle \wp(\mathbb{S}),\ I_0,\ T\rangle$ to initial states satisfying I_0.

Let us define the abstraction of a specification semantics $\mathsf{T} \in \wp(\mathbb{S})^{\infty}$ into a set of nonempty prefix traces in $\mathbb{S}^+$.

$$\alpha^{\mathbb{T}}(P) \quad \triangleq \quad \bigcup_{n \in \mathbb{N}^+} \{\pi \in \mathbb{S}^n \mid \forall i \in [0, n[\,.\,\pi_i \in P_i\} \tag{44.58}$$

$$\alpha^{\mathbb{T}}(\mathsf{T}) \quad \triangleq \quad \bigcup \{\alpha^{\mathbb{T}}(P) \mid P \in \mathsf{T}\} \tag{44.59}$$

— Prove that $\langle \wp(\wp(\mathbb{S})^{\infty}),\ \subseteq\rangle \underset{\alpha^{\mathbb{T}}}{\overset{\gamma^{\mathbb{T}}}{\leftrightarrows}} \langle \wp(\mathbb{S}^+),\ \subseteq\rangle$.

Exercise 44.60 (satisfaction) Continuing exercises 44.54 and 44.56, a transition system $\tau = \langle \mathbb{S},\ \iota_0,\ \tau\rangle$ satisfies the specification $\mathcal{T} = \langle \wp(\mathbb{S}),\ I_0,\ T\rangle$ (written $\tau \vdash T$) if the traces of τ are all described by the specification $\mathcal{T}$.

$$\tau \vdash \mathcal{T} \quad \triangleq \quad (\boldsymbol{\mathcal{S}}_{\mathrm{t}}[\![\tau]\!] \cap \vec{\iota_0}^{*}) \subseteq \alpha^{\mathbb{T}}(\boldsymbol{\mathcal{S}}[\![T]\!] \cap \vec{I_0}^{\infty}) \tag{44.61}$$

The satisfaction abstraction is

$$\alpha^{\mathcal{T}}(X) \quad \triangleq \quad \langle \sigma,\ \Sigma\rangle \mapsto \{\pi \in X \mid \pi_0 = \sigma\} \subseteq \alpha^{\mathbb{T}}(\{P \in \boldsymbol{\mathcal{S}}[\![T]\!] \mid P_0 = \Sigma\}) \tag{44.62}$$

so that $\tau \vdash \mathcal{T} = \alpha^{\mathcal{T}}(\boldsymbol{\mathcal{S}}_{\mathrm{t}}[\![\tau]\!])\langle \iota_0,\ I_0\rangle$.

— Prove that $\langle (\mathbb{S} \times \wp(\mathbb{S})) \to \wp(\mathbb{S})^*,\ \subseteq\rangle \underset{\alpha^{\mathcal{T}}}{\overset{\gamma^{\mathcal{T}}}{\leftrightarrows}} \langle (\mathbb{S} \times \wp(\mathbb{S})) \to \mathbb{B},\ \dot{\Leftarrow}\rangle$.

Exercise 44.63 (model checking) Continuing exercises 44.54, 44.56, and 44.60, model checking of a transition τ for a temporal specification $\mathcal{T}$ is an algorithm to check satisfaction $\tau \vdash T$. We define

$$\mathsf{mc}(\sigma, \Sigma) \quad \triangleq \quad \alpha^{\mathcal{T}}(\boldsymbol{\mathcal{S}}_{\mathrm{t}}[\![\tau]\!])\langle \sigma,\ \Sigma\rangle \tag{44.64}$$

so that $\tau \vdash \mathcal{T} = \mathsf{mc}(\iota_0, I_0)$.

— Given a transition system $\tau = \langle \mathbb{S}, \iota_0, \tau \rangle$ and a specification $\mathscr{T} = \langle \wp(\mathbb{S}), I_0, T \rangle$, prove the commutation condition

$$\alpha^{\mathscr{T}}(\mathbb{S}^1 \cup \vec{\tau} \,\bar{;}\, X) = \langle \sigma, \Sigma \rangle \mapsto \sigma \in \Sigma \wedge \bigwedge_{\langle \sigma, \sigma' \rangle \in \tau} \bigvee_{\langle \Sigma, \Sigma' \rangle \in T} \sigma' \in \Sigma' \wedge \alpha^{\mathscr{T}}(X)\langle \sigma', \Sigma' \rangle \quad (44.65)$$

— Given a transition system $\tau = \langle \mathbb{S}, \iota_0, \tau \rangle$ and a specification $\mathscr{T} = \langle \wp(\mathbb{S}), I_0, T \rangle$, prove that

$$\mathsf{mc}(\iota_0, I_0) = \Big(\mathsf{gfp}^{\dot{\Rightarrow}}\, X \mapsto \langle \sigma, \Sigma \rangle \mapsto ((\sigma \in \Sigma) \wedge \bigwedge_{\langle \sigma, \sigma' \rangle \in \tau} \bigvee_{\langle \Sigma, \Sigma' \rangle \in T} X(\sigma', \Sigma'))\Big)\langle \iota_0, I_0 \rangle \quad (44.66)$$

44.9 Conclusion

We have illustrated the idea that model checking is an abstract interpretation, as first introduced in [254]. The consequence is that the model checking algorithm can be formally derived by calculational design and so is sound by construction. The model construction is automatic whereas it might be expensive to design [352]. The model is sound, which is in contrast with approaches in which the model is postulated without justification.

Understanding model checking as an abstract interpretation also yields specification-preserving abstract model checking [280, 817–819] as well as abstraction refinement algorithms [276, 414, 816].

Regular expressions were introduced by Stephen Cole Kleene in [581] to specify "events" (that is, sequences of states) and their temporal order. Specifications by temporal logics are not commonly accepted by programmers (who may be more familiar with regular expressions). For example, in [683], the specifications had to be written by academics. The trace semantics for regular expressions always relates variable environments to the initial environment, so that one cannot express sequences of variable increments. This would be possible with a more complex syntax and semantics of regular expressions allowing for recording intermediate or final values of variables during computations, not only the initial ones.

For more expressive specifications beyond regular expressions with intersection and complement [151], Büchi automata can be used to represent maximal finite and infinite traces [507]. Beyond the regular setting, one can also consider quantifiers, for example, [408].

We do not argue that (extended) regular expressions or path expressions [163] would be more convenient temporal specifications because in practice such specifications are usually very complex and must often be modified whenever the program is modified. However, in the context of program security using runtime safety monitors [644, 862], these monitors could be used as a static specification to be model checked. The redundant parts of the runtime monitoring could be eliminated whereas those corresponding to false alarms in the analysis would be preserved (similar to, for example, the elimination of array bound checks found redundant by a static analysis).

Specifications by regular expressions may also be useful to define a subset of the program executions that have to be model-checked by other means. This would allow for user-guided modular proofs based on partitions of the program execution traces [672, 836].

Regular expressions have also be used to abstract flowcharts [236], graphs [968], strings [50, 51, 747], languages [347, 704], and so forth.

As an alternative approach, a regular expression can be automatically extracted by static analysis of the program trace semantics that recognizes all feasible execution paths and usually more [283, 291, 681]. Then model checking a regular specification becomes a regular language inclusion problem [601].

Scalability is always a problem with model checking and the regular software model checking of this chapter is no exception. The static analyzer Astrée has parameters allowing to unroll loops of which the effect is to refine an invariance analysis into a finite prefix trace analysis with invariants attached along the trace instead of a program point [113]. This definitely increases precision. However, if the parameter is chosen too large, the analysis blows up [268]. The same phenomenon will definitely happen with the abstraction considered in this chapter, so further abstractions are necessary.

Most often, abstract model checking uses the homomorphic/partitioning abstractions of exercise 11.6, for example, [75, 79]. The reason is that the abstraction of a transition system on concrete states is a transition system on abstract states so model checkers are reusable in the abstract. However, excluding edgy abstractions as in [226], state-based finite abstraction is very restrictive [416] and does not guarantee scalability (e.g., SLAM [76, 77, 81, 82], timing out on half of the analyses; it was ultimately abandoned).

Finite overapproximating abstractions of properties such as termination may be both unsound and incomplete. Programs have unbounded finite and infinite execution traces. Hence the infinite traces as well as large enough finite traces are abstracted into a lasso (this execution comes back to an identical previous abstract state) so this makes termination proofs impossible in the abstract. Moreover, nontermination proofs are unsound because there may be lassos in the abstract and none in the concrete.

Convergence of model checking requires expressivity restrictions on both the considered models of computation and the considered temporal logics. For some expressive models of computation and temporal logics, state finiteness is not enough to guarantee termination of model checking [254, 416]. Finite enumeration is limited, even with symbolic encodings, and induction is ultimately required to reason on programs (e.g., [977]).

44.10 Solutions to Selected Exercises

Solution to exercise 44.15 Very informally,

$(\mathsf{R}_1 \mid \mathsf{R}_2)^*$

$\approx (\mathsf{R}_1 \mid \mathsf{R}_2)(\mathsf{R}_1 \mid \mathsf{R}_2)\ldots(\mathsf{R}_1 \mid \mathsf{R}_2)$ ⦃definition of *, 0 or more times⦄

$\approx \mathsf{R}_1^*\mathsf{R}_2^*\mathsf{R}_1^*\mathsf{R}_2^*\ldots\mathsf{R}_1^*\mathsf{R}_2^*$ ⦃grouping/ungrouping consecutive R_1's and R_2's using ε if necessary⦄

$\approx (\mathsf{R}_1{}^*\mathsf{R}_2{}^*)^*$ ⦃definition of *, 0 or more times⦄

Solution to exercise 44.17

Proof of lemma 44.16 The proof is by structural induction. The base cases for ε and $\mathsf{L : B}$ are trivial. The cases $\mathsf{R_1 \mid R_2}$ and $\mathsf{(R)}$ are an easy induction. Otherwise,

– $\boldsymbol{\mathcal{S}}^{\mathsf{r}}[\![\mathsf{dnf(R_1R_2)}]\!]$

$= \boldsymbol{\mathcal{S}}^{\mathsf{r}}[\![\mid_{i=1}^{n_1} \mid_{j=1}^{n_2} \mathsf{R_1^i R_2^j}]\!]$ ⁅definition of dnf⁆

$= \bigcup_{i=1}^{n_1}\bigcup_{j=1}^{n_2} \boldsymbol{\mathcal{S}}^{\mathsf{r}}[\![\mathsf{R_1^i R_2^j}]\!]$ ⁅definition of $\boldsymbol{\mathcal{S}}^{\mathsf{r}}$⁆

$= \bigcup_{i=1}^{n_1}\bigcup_{j=1}^{n_2} \boldsymbol{\mathcal{S}}^{\mathsf{r}}[\![\mathsf{R_1^i}]\!] \odot \boldsymbol{\mathcal{S}}^{\mathsf{r}}[\![\mathsf{R_2^j}]\!]$ ⁅definition of $\boldsymbol{\mathcal{S}}^{\mathsf{r}}$⁆

$= \bigcup_{i=1}^{n_1} \boldsymbol{\mathcal{S}}^{\mathsf{r}}[\![\mathsf{R_1^i}]\!] \bullet \bigcup_{j=1}^{n_2} \boldsymbol{\mathcal{S}}^{\mathsf{r}}[\![\mathsf{R_2^j}]\!]$ ⁅definition of $\bullet$⁆

$= \boldsymbol{\mathcal{S}}^{\mathsf{r}}[\![\mid_{i=1}^{n_1} \mathsf{R_1^i}]\!] \bullet \boldsymbol{\mathcal{S}}^{\mathsf{r}}[\![\mid_{j=1}^{n_2} \mathsf{R_2^j}]\!]$ ⁅definition of $\boldsymbol{\mathcal{S}}^{\mathsf{r}}$⁆

$= \boldsymbol{\mathcal{S}}^{\mathsf{r}}[\![\mathsf{dnf(R_1)}]\!] \bullet \boldsymbol{\mathcal{S}}^{\mathsf{r}}[\![\mathsf{dnf(R_2)}]\!]$ ⁅definition of dnf where $\mathsf{R_k^1 \mid \ldots \mid R_k^{n_k} = dnf(R_k)}, k = 1, 2$⁆

$= \boldsymbol{\mathcal{S}}^{\mathsf{r}}[\![\mathsf{R_1}]\!] \bullet \boldsymbol{\mathcal{S}}^{\mathsf{r}}[\![\mathsf{R_2}]\!]$ ⁅induction hypothesis⁆

$= \boldsymbol{\mathcal{S}}^{\mathsf{r}}[\![\mathsf{R_1R_2}]\!]$ ⁅definition of $\boldsymbol{\mathcal{S}}^{\mathsf{r}}$⁆

– $\boldsymbol{\mathcal{S}}^{\mathsf{r}}[\![\mathsf{dnf(R^*)}]\!]$

$= \boldsymbol{\mathcal{S}}^{\mathsf{r}}[\![\mathsf{((R^1)^* \ldots (R^n)^*)^*}]\!]$ ⁅definition of dnf where ⁆

$= \boldsymbol{\mathcal{S}}^{\mathsf{r}}[\![\mathsf{(R^1 \mid \ldots \mid R^n)^*}]\!]$ ⁅exercise 44.15⁆

$= \bigcup_{k\geqslant 0} \boldsymbol{\mathcal{S}}^{\mathsf{r}}[\![\mathsf{(R^1 \mid \ldots \mid R^n)}]\!]^k$ ⁅definition of $\boldsymbol{\mathcal{S}}^{\mathsf{r}}$⁆

$= \bigcup_{k\geqslant 0} \boldsymbol{\mathcal{S}}^{\mathsf{r}}[\![\mathsf{dnf(R)}]\!]^k$ ⁅$\mathsf{R^1 \mid \ldots \mid R^n = dnf(R)}$⁆

$= \bigcup_{k\geqslant 0} \boldsymbol{\mathcal{S}}^{\mathsf{r}}[\![\mathsf{R}]\!]^k$ ⁅induction hypothesis⁆

$= \boldsymbol{\mathcal{S}}^{\mathsf{r}}[\![\mathsf{R^*}]\!]$ ⁅definition of $\boldsymbol{\mathcal{S}}^{\mathsf{r}}$⁆

The proof for $\mathsf{R^+}$ is similar. □

Solution to exercise 44.56 This is a join homomorphic/partitioning abstraction of exercise 11.8.

45 Flow-Insensitive Static Analysis

The abstract interpreter of chapter 21 is flow-sensitive in that the information provided on execution is specific to each program point. We design by calculus a flow-insensitive abstract interpreter such that the information provided on execution is the same for all program points. The global information can be Cartesian or relational (e.g., [698]).

Contents

45.1 Flow-Sensitive Abstract Interpreter

Let $\mathbb{D}^{\sharp} \triangleq \langle \mathbb{P}^{\sharp}, \sqsubseteq^{\sharp}, \bot^{\sharp}, \sqcup^{\sharp}, \mathsf{assign}^{\sharp}[\![\mathsf{x}, \mathsf{A}]\!], \mathsf{test}^{\sharp}[\![\mathsf{B}]\!], \overline{\mathsf{test}}^{\sharp}[\![\mathsf{B}]\!]\rangle$ be a well defined abstract domain satisfying definition 21.1. The instance $\widehat{\boldsymbol{\mathcal{S}}}^{\sharp}$ of the generic structural abstract interpreter of section 21.2 for that abstract domain $\mathbb{D}^{\sharp}$ returns an abstract property pertinent to each program point because $\widehat{\boldsymbol{\mathcal{S}}}^{\sharp}[\![\mathsf{S}]\!] \in \mathbb{P}^{\sharp} \rightarrow (\mathsf{labx}[\![\mathsf{S}]\!] \rightarrow \mathbb{P}^{\sharp})$ for each program component $\mathsf{S} \in \mathbb{P}\mathbb{c}$.

45.2 Flow-Insensitive Abstraction

The flow-insensitive abstraction joins all local program properties attached to program points into a single global property holding at any program point.

$$
\begin{aligned}
\alpha_{\mathcal{I}} &\in (\mathsf{labx}[\![\mathsf{S}]\!] \rightarrow \mathbb{P}^{\sharp}) \rightarrow \mathbb{P}^{\sharp} \qquad (45.1)\\
\alpha_{\mathcal{I}}(P) &\triangleq \bigsqcup_{\ell \in \mathsf{labx}[\![\mathsf{S}]\!]}^{\sharp} P(\ell)\\
\dot{\alpha}_{\mathcal{I}} &\in (\mathbb{P}^{\sharp} \rightarrow (\mathsf{labx}[\![\mathsf{S}]\!] \rightarrow \mathbb{P}^{\sharp})) \rightarrow (\mathbb{P}^{\sharp} \rightarrow \mathbb{P}^{\sharp})\\
\dot{\alpha}_{\mathcal{I}}(\boldsymbol{\mathcal{S}}) &\triangleq P \mapsto \alpha_{\mathcal{I}}(\boldsymbol{\mathcal{S}}(P))
\end{aligned}
$$

This is a Galois retraction

$$
\langle \mathsf{labx}[\![\mathsf{S}]\!] \rightarrow \mathbb{P}^{\sharp}, \dot{\sqsubseteq}^{\sharp}\rangle \xrightleftharpoons[\alpha_{\mathcal{I}}]{\gamma_{\mathcal{I}}} \langle \mathbb{P}^{\sharp}, \sqsubseteq^{\sharp}\rangle
$$

45.3 Flow-Insensitive Abstract Interpreter

The structural flow-insensitive abstract interpreter $\widehat{\mathcal{S}}^{\natural}_{g}[\![\mathsf{S}]\!] \in \mathbb{P}^{\natural} \to \mathbb{P}^{\natural}$ joins abstract properties at all program points.

Flow-insensitive abstract semantics of a program $\mathsf{P} ::= \mathsf{Sl}\ \ell'$

$$\widehat{\mathcal{S}}^{\natural}_{g}[\![\mathsf{P}]\!]\,\overline{P} \triangleq \widehat{\mathcal{S}}^{\natural}_{g}[\![\mathsf{Sl}]\!]\,\overline{P} \tag{45.2}$$

Flow-insensitive abstract semantics of a statement list $\mathsf{Sl} ::= \mathsf{Sl}'\ \mathsf{S}$

$$\widehat{\mathcal{S}}^{\natural}_{g}[\![\mathsf{Sl}]\!]\,\overline{P} = \widehat{\mathcal{S}}^{\natural}_{g}[\![\mathsf{Sl}']\!]\,\overline{P} \sqcup^{\natural} \widehat{\mathcal{S}}^{\natural}_{g}[\![\mathsf{S}]\!]\,\overline{P} \tag{45.3}$$

Flow-insensitive abstract semantics of an empty statement list $\mathsf{Sl} ::= \epsilon$

$$\widehat{\mathcal{S}}^{\natural}_{g}[\![\mathsf{Sl}]\!]\,\overline{P} = \overline{P} \tag{45.4}$$

Flow-insensitive abstract semantics of an assignment statement $\mathsf{S} ::= \mathsf{x} = \mathsf{A}\,;$

$$\widehat{\mathcal{S}}^{\natural}_{g}[\![\mathsf{S}]\!]\,\overline{P} = \overline{P} \sqcup^{\natural} \mathsf{assign}^{\natural}[\![\mathsf{x}, \mathsf{A}]\!]\,\overline{P} \tag{45.5}$$

Flow-insensitive abstract semantics of a skip statement $\mathsf{S} ::= ;$

$$\widehat{\mathcal{S}}^{\natural}_{g}[\![\mathsf{S}]\!]\,\overline{P} = \overline{P} \tag{45.6}$$

Flow-insensitive abstract semantics of a conditional statement $\mathsf{S} ::= \mathtt{if}\,(\mathsf{B})\,\mathsf{S}_t$

$$\widehat{\mathcal{S}}^{\natural}_{g}[\![\mathsf{S}]\!]\,\overline{P} = \overline{P} \sqcup^{\natural} \widehat{\mathcal{S}}^{\natural}_{g}[\![\mathsf{S}_t]\!]\,(\mathsf{test}^{\natural}[\![\mathsf{B}]\!]\,\overline{P}) \tag{45.7}$$

Flow-insensitive abstract semantics of a conditional statement $\mathsf{S} ::= \mathtt{if}\,(\mathsf{B})\,\mathsf{S}_t$ $\mathtt{else}\ \mathsf{S}_f$

$$\widehat{\mathcal{S}}^{\natural}_{g}[\![\mathsf{S}]\!]\,\overline{P} = \overline{P} \sqcup^{\natural} \widehat{\mathcal{S}}^{\natural}_{g}[\![\mathsf{S}_t]\!]\,(\mathsf{test}^{\natural}[\![\mathsf{B}]\!]\,\overline{P}) \sqcup^{\natural} \widehat{\mathcal{S}}^{\natural}_{g}[\![\mathsf{S}_f]\!]\,\overline{\mathsf{test}}^{\natural}[\![\mathsf{B}]\!]\,\overline{P}) \tag{45.8}$$

Flow-insensitive abstract semantics of an iteration statement $\mathsf{S} ::= \mathtt{while}\ \ell\ (\mathsf{B})\ \mathsf{S}_b$

$$\widehat{\mathcal{S}}^{\natural}_{g}[\![\mathsf{S}]\!]\,\overline{P} = \mathsf{lfp}^{\subseteq}\,(\mathcal{F}^{\natural}_{g}[\![\mathtt{while}\ \ell\ (\mathsf{B})\ \mathsf{S}_b]\!]\,\overline{P}) \tag{45.9}$$

$$\mathcal{F}^{\natural}_{g}[\![\mathtt{while}\ \ell\ (\mathsf{B})\ \mathsf{S}_b]\!]\,\overline{P}\,X = \overline{P} \sqcup^{\natural} \widehat{\mathcal{S}}^{\natural}_{g}[\![\mathsf{S}]\!]\,(\mathsf{test}^{\natural}[\![\mathsf{B}]\!]\,X)$$

Flow-insensitive abstract semantics of a break statement $\mathsf{S} ::= \ell\ \mathtt{break}\,;$

$$\widehat{\mathcal{S}}^{\natural}_{g}[\![\mathsf{S}]\!]\,\overline{P} = \overline{P} \tag{45.10}$$

Flow-insensitive abstract semantics of a compound statement $\mathsf{S} ::= \{\,\mathsf{Sl}\,\}$

$$\widehat{\mathcal{S}}^{\natural}_{g}[\![\mathsf{S}]\!]\,\overline{P} = \widehat{\mathcal{S}}^{\natural}_{g}[\![\mathsf{Sl}]\!]\,\overline{P} \tag{45.11}$$

Exercise 45.12 (Project: flow-insensitive abstract interpreter) Transform the flow-sensitive abstract interpreter of exercise 21.22 into the previously described flow-insensitive abstract interpreter.

45.4 Calculation Design of the Flow-Insensitive Abstract Interpreter

The soundness of the flow-insensitive abstract interpreter is expressed in terms of the flow-sensitive abstract interpreter and consists of joining the abstract properties.

Theorem 45.13 The flow-insensitive abstract semantics $\widehat{\boldsymbol{S}}^{\sharp}_{\mathfrak{g}}$ of section 45.3 on a domain $\mathbb{D}^{\sharp}$ is a sound abstraction of the flow-sensitive abstract semantics $\widehat{\boldsymbol{S}}^{\sharp}$ of section 21.2 on the same domain $\mathbb{D}^{\sharp}$, $\dot{\alpha}_{\mathfrak{g}}(\widehat{\boldsymbol{S}}^{\sharp}[\![\mathsf{S}]\!]) \sqsubseteq^{\sharp} \widehat{\boldsymbol{S}}^{\sharp}_{\mathfrak{g}}[\![\mathsf{S}]\!]$.

Proof of theorem 45.13 The proof is by structural calculational design with the same structure as the proof of theorem 27.4. Let us consider two typical cases.

— For the assignment statement $\mathsf{S} ::= \mathsf{x} = \mathsf{A}\,;$,

$$\begin{aligned}
&\dot{\alpha}_{\mathfrak{g}}(\widehat{\boldsymbol{S}}^{\sharp}[\![\mathsf{x} = \mathsf{A}\,;]\!])\overline{P}\\
=\ &\alpha_{\mathfrak{g}}(\widehat{\boldsymbol{S}}^{\sharp}[\![\mathsf{x} = \mathsf{A}\,;]\!]\overline{P}) && \text{(definition (45.1) of } \dot{\alpha}_{\mathfrak{g}}\text{)}\\
=\ &\alpha_{\mathfrak{g}}((\!|\, \ell = \mathsf{at}[\![\mathsf{S}]\!] \ ?\ \overline{P} \ [\!]\ \ell = \mathsf{after}[\![\mathsf{S}]\!] \ ?\ \mathsf{assign}^{\sharp}[\![\mathsf{x}, \mathsf{A}]\!]\,\overline{P} : \bot^{\sharp} \,|\!)) && \text{(definition (21.7) of } \widehat{\boldsymbol{S}}^{\sharp}\text{)}\\
=\ &\overline{P} \sqcup^{\sharp} \mathsf{assign}^{\sharp}[\![\mathsf{x}, \mathsf{A}]\!]\,\overline{P} && \text{(definition (45.1) of } \alpha_{\mathfrak{g}} \text{ and } \mathsf{labx}[\![\mathsf{S}]\!] = \{\mathsf{at}[\![\mathsf{S}]\!], \mathsf{after}[\![\mathsf{S}]\!]\}\text{)}\\
\triangleq\ &\widehat{\boldsymbol{S}}^{\sharp}_{\mathfrak{g}}[\![\mathsf{S}]\!]\,\overline{P} && \text{(as defined in (45.5))}
\end{aligned}$$

— For the iteration statement $\mathsf{S} ::= \mathtt{while}\ \ell\ (\mathsf{B})\ \mathsf{S}_b$, we apply the fixpoint approximation corollary 18.16 by looking for a transformer $\mathscr{F}^{\sharp}_{\mathfrak{g}}[\![\mathtt{while}\ \ell\ (\mathsf{B})\ \mathsf{S}_b]\!]\,\overline{P}$ semicommuting with $\mathscr{F}^{\sharp}[\![\mathtt{while}\ \ell\ (\mathsf{B})\ \mathsf{S}_b]\!]\,\overline{P}$ in (21.11).

$$\begin{aligned}
&\alpha_{\mathfrak{g}}(\mathscr{F}^{\sharp}[\![\mathtt{while}\ \ell\ (\mathsf{B})\ \mathsf{S}_b]\!]\,\overline{P}\,X)\\
=\ &\bigsqcup^{\sharp}_{\ell \in \mathsf{labx}[\![\mathsf{S}]\!]} \mathscr{F}^{\sharp}[\![\mathtt{while}\ \ell\ (\mathsf{B})\ \mathsf{S}_b]\!]\,\overline{P}\,X\,\ell' && \text{(definition (45.1) of } \alpha_{\mathfrak{g}}\text{)}\\
=\ &\bigsqcup^{\sharp}_{\ell \in \mathsf{labx}[\![\mathsf{S}]\!]} (\!|\, \ell' = \ell \ ?\ \overline{P} \sqcup^{\sharp} \widehat{\boldsymbol{S}}^{\sharp}[\![\mathsf{S}_b]\!]\,(\mathsf{test}^{\sharp}[\![\mathsf{B}]\!]X(\ell))\,\ell\\
&\qquad [\!]\ \ell' \in \mathsf{in}[\![\mathsf{S}_b]\!] \setminus \{\ell\} \ ?\ \widehat{\boldsymbol{S}}^{\sharp}[\![\mathsf{S}_b]\!]\,(\mathsf{test}^{\sharp}[\![\mathsf{B}]\!]X(\ell))\,\ell'\\
&\qquad [\!]\ \ell' = \mathsf{after}[\![\mathsf{S}]\!] \ ?\ \overline{\mathsf{test}}^{\sharp}[\![\mathsf{B}]\!]X(\ell) \sqcup^{\sharp} \bigsqcup^{\sharp}_{\ell'' \in \mathsf{breaks\text{-}of}[\![\mathsf{S}_b]\!]} \widehat{\boldsymbol{S}}^{\sharp}[\![\mathsf{S}_b]\!]\,(\mathsf{test}^{\sharp}[\![\mathsf{B}]\!]X(\ell))\,\ell''\\
&\qquad : \bot^{\sharp} \,|\!) && \text{(definition of (21.11) of } \mathscr{F}^{\sharp}[\![\mathtt{while}\ \ell\ (\mathsf{B})\ \mathsf{S}_b]\!]\text{)}\\
=\ &\overline{P} \sqcup^{\sharp} \widehat{\boldsymbol{S}}^{\sharp}[\![\mathsf{S}_b]\!]\,(\mathsf{test}^{\sharp}[\![\mathsf{B}]\!]X(\ell))\,\ell \sqcup^{\sharp} \bigsqcup^{\sharp}_{\ell' \in \mathsf{in}[\![\mathsf{S}_b]\!]\setminus\{\ell\}} \widehat{\boldsymbol{S}}^{\sharp}[\![\mathsf{S}_b]\!]\,(\mathsf{test}^{\sharp}[\![\mathsf{B}]\!]X(\ell))\,\ell' \sqcup^{\sharp} \overline{\mathsf{test}}^{\sharp}[\![\mathsf{B}]\!]X(\ell) \sqcup^{\sharp}\\
&\qquad \bigsqcup^{\sharp}_{\ell'' \in \mathsf{breaks\text{-}of}[\![\mathsf{S}_b]\!]} \widehat{\boldsymbol{S}}^{\sharp}[\![\mathsf{S}_b]\!]\,(\mathsf{test}^{\sharp}[\![\mathsf{B}]\!]X(\ell))\,\ell''\\
&&& \text{(definition of } \mathsf{labx}[\![\mathsf{S}]\!] = \{\ell\} \cup \mathsf{in}[\![\mathsf{S}_b]\!] \cup \{\mathsf{after}[\![\mathsf{S}]\!]\}\text{)}\\
=\ &\overline{P} \sqcup^{\sharp} \bigsqcup^{\sharp}_{\ell' \in \mathsf{labx}[\![\mathsf{S}_b]\!]} \widehat{\boldsymbol{S}}^{\sharp}[\![\mathsf{S}_b]\!]\,(\mathsf{test}^{\sharp}[\![\mathsf{B}]\!]X(\ell))\,\ell'
\end{aligned}$$

(grouping equal terms under $\sqcup^{\sharp}$,

- $\mathsf{labx}[\![\mathsf{S}_b]\!] = \mathsf{in}[\![\mathsf{S}_b]\!] \cup \{\ell\}$ so the term $\widehat{\boldsymbol{S}}^{\sharp}[\![\mathsf{S}_b]\!]\,(\mathsf{test}^{\sharp}[\![\mathsf{B}]\!]X(\ell))\,\ell$ can be eliminated;
- $\ell \in \mathsf{labx}[\![\mathsf{S}_b]\!]$, $\widehat{\boldsymbol{S}}^{\sharp}[\![\mathsf{S}_b]\!]\,(\mathsf{test}^{\sharp}[\![\mathsf{B}]\!]X(\ell))\,\ell = X(\ell)$, $X(\ell) \sqcup^{\sharp} \overline{\mathsf{test}}^{\sharp}[\![\mathsf{B}]\!]X(\ell) = X(\ell)$ so the term $\overline{\mathsf{test}}^{\sharp}[\![\mathsf{B}]\!]X(\ell)$ can be eliminates, and
- $\mathsf{breaks\text{-}of}[\![\mathsf{S}_b]\!] \subseteq \mathsf{in}[\![\mathsf{S}_b]\!]$ so that the term $\bigsqcup^{\sharp}_{\ell'' \in \mathsf{breaks\text{-}of}[\![\mathsf{S}_b]\!]} \widehat{\boldsymbol{S}}^{\sharp}[\![\mathsf{S}_b]\!]\,(\mathsf{test}^{\sharp}[\![\mathsf{B}]\!]X(\ell))\,\ell''$

can be eliminated)

$$\begin{aligned}
=\ &\overline{P} \sqcup^{\sharp} \alpha_{\mathfrak{g}}(\widehat{\boldsymbol{S}}^{\sharp}[\![\mathsf{S}_b]\!]\,(\mathsf{test}^{\sharp}[\![\mathsf{B}]\!]X(\ell))) && \text{(definition (45.1) of } \alpha_{\mathfrak{g}}\text{)}\\
\sqsubseteq^{\sharp}\ &\overline{P} \sqcup^{\sharp} \alpha_{\mathfrak{g}}(\widehat{\boldsymbol{S}}^{\sharp}[\![\mathsf{S}_b]\!]\,(\mathsf{test}^{\sharp}[\![\mathsf{B}]\!]\,\alpha_{\mathfrak{g}}(X)))
\end{aligned}$$

$$\text{\{by (45.1) so that } X(\ell) \sqsubseteq^{¤} \bigsqcup^{¤}_{\ell' \in \mathsf{labx}[\![\mathsf{S}]\!]} X(\ell') = \alpha_{g}(X) \text{ because } \ell \in \mathsf{labx}[\![\mathsf{S}]\!] \text{ and } \mathsf{test}^{¤}[\![\mathsf{B}]\!]$$
$$\text{as well as } \widehat{\boldsymbol{\mathcal{S}}}^{¤}[\![\mathsf{S}_b]\!] \text{ are increasing\}}$$

$$\sqsubseteq^{¤} \overline{P} \sqcup^{¤} \widehat{\boldsymbol{\mathcal{S}}}^{¤}_{g}[\![\mathsf{S}]\!] (\mathsf{test}^{¤}[\![\mathsf{B}]\!]\, \dot{\alpha}_{g}(X)) \qquad \text{\{structural induction hypothesis and } \sqcup^{¤} \text{ increasing\}}$$
$$= \boldsymbol{\mathcal{F}}^{¤}_{g}[\![\mathtt{while}\ \ell\ (\mathsf{B})\ \mathsf{S}_b]\!]\, \overline{P}\, \dot{\alpha}_{g}(X) \qquad \text{\{as defined in (45.9)\}} \quad \square$$

45.5 Well Definedness of the Flow-Insensitive Abstract Interpreter

Theorem 45.14 The flow-insensitive abstract semantics $\widehat{\boldsymbol{\mathcal{S}}}^{¤}_{g}[\![\mathsf{S}]\!]$ on a well defined abstract domain of definition 21.1 is well defined.

Proof of theorem 45.14 This directly follows by structural induction from definition 21.1. □

45.6 Abstraction of the Flow-Insensitive Abstract Interpreter

Theorem 27.4 shows that the sound approximate abstraction of definition 27.1 of a sound abstract flow-sensitive semantics is a sound abstract flow-sensitive semantics. This obviously extends to flow-insensitive semantics.

Theorem 45.15 (soundness of the flow-insensitive abstract interpreter) Let $\widehat{\boldsymbol{\mathcal{S}}}^{¤}_{g}$ and $\widehat{\boldsymbol{\mathcal{S}}}^{\sharp}_{g}$ be structural flow-insensitive abstract interpreters for well defined concrete $\mathbb{D}^{¤}$ and abstract domains $\mathbb{D}^{\sharp}$ by definition 21.1 such that $\mathbb{P}^{\sharp}$ is an approximate abstraction of $\mathbb{P}^{¤}$ by definition 27.1.I. Then for all $\overline{P} \in \mathbb{P}^{\sharp}$,

$$\widehat{\boldsymbol{\mathcal{S}}}^{¤}_{g}[\![\mathsf{S}]\!](\gamma(\overline{P})) \mathrel{\dot{\sqsubseteq}^{¤}} \dot{\gamma}(\widehat{\boldsymbol{\mathcal{S}}}^{\sharp}_{g}[\![\mathsf{S}]\!](\overline{P}))$$

Proof of theorem 45.15 The proof follows from the definitions 21.1 and 27.1-I by structural induction and essentially mimics that of theorem 27.4. □

Exercise 45.16 (Project: flow [in]sensitive analyses) Design and implement an abstract interpreter to handle both flow-sensitive and flow-insensitive analyses are obtained by a mere change of abstract domain.

45.7 Conclusion

Flow insensitive static analysis is used, for example, for typing in languages like C [569] and Pascal [986] in which the type of variables is stored globally in the symbol table, to globally qualify types [378], in chapter 46 for pointer analysis (based on the empirical assumption that static pointers point to the same address everywhere in the program), and for other purposes. It is rather imprecise and not always more efficient than flow-sensitive static analysis. The program structure has to be traversed iteratively in any case and properties have to be memorized once globally instead of once per loop head, which may not be a significative difference, in particular in using widenings to avoid space and time explosions.

More generally, the abstraction of a sensitive to an insensitive static analysis relies on an abstraction joining cases. *Path insensitivity* joins path properties into flow-sensitive state properties attached to program points, as illustrated in the reachability semantics of chapter 19. *Flow sensitive* static analyses can be abstracted into flow-insensitive analyses by the abstraction (45.1) joining local properties attached to program points into a single global property. A *field-sensitive* static analysis can be abstracted into a field-insensitive static analysis by joining the properties attached to field of a data structure into a single structure property. A *context-sensitive* static analysis can be abstracted into a context-insensitive static analysis by joining the preconditions of procedure calls into a single precondition.

Flow-insensitive static analysis is not always, if ever, a good solution. It is supposed to be a solution to the problem that keeping information at each program point is costly. That is true, but keeping information at each program point is also useless. Using a structural approach, you need to keep the information at the loop heads only to check for fixpoint convergence and do extrapolations and interpolations, and in fact you need only the pertinent information. After convergence, one more iteration will propagate the information at whichever program point where it is needed. This is the approach taken in Astrée [124]. Having global information only usually diminishes the precision and sometimes the convergence speed, because fixpoints have to climb slowly higher in the lattice.

46 Points-To Analysis

Pointer analysis, or points-to analysis, consists of automatically inferring where pointer variables of a program can point. We consider static memory allocation where the memory addresses (where pointer variables of a program can point) are determined statically by the compiler (as opposed to dynamic memory allocation by malloc in C [569] during program execution).

Example 46.1 The points-to analysis of the following C program

```
#include <stdio.h>
#include <stdlib.h>
int main () {
    int x,*p,**q,*r;
    x = &p;
    p = &x;
    q = &p;
    r = *q;
    *r = &r;
}
```

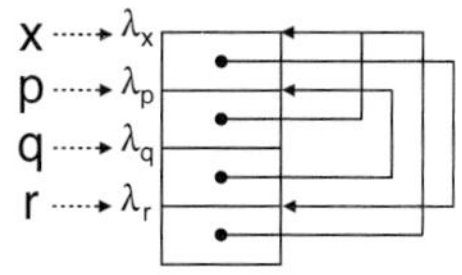

Memory μ:

$\mu(\lambda_x)$	λ_r
$\mu(\lambda_p)$	λ_x
$\mu(\lambda_q)$	λ_p
$\mu(\lambda_r)$	λ_x

should be able to determine that p points to x on program exit. The memory location (or address) λ_y of the program variables y is chosen arbitrarily by the compiler. λ is a fixed but unknown parameter of the program semantics and analysis (so the compiler has a free choice of memory allocation). $\mu(\lambda_y)$ is the value of variable y in memory μ. It may evolve during program computation.

Points-to analysis started at the end of the 1970s [92, 975] and has been extensively studied since then [316, 498, 517, 556]. Our objective is to explore the formal design of classic points-to analyses [43, 908] by abstract interpretation as opposed to the resolution of postulated constraints [43] or a postulated type inference system [908].

Contents

46.1 Syntax of the Pointer Language

The syntax of programs of section 4.1 is extended as follows:

p, q, … , x, y, …	$\in$	$\mathbb{V}$	variables
p, q, …	$\in$	$\vec{\mathbb{V}} \subseteq \mathbb{V}$	pointer variables
A	::=	…	arithmetic expressions
	|	x	variable
	|	NULL	null pointer
	|	&x	address of variable
	|	*p	dereferenced pointer variable
B	::=	…	Boolean expressions
	|	x == y	equality
	|	p == NULL	pointer equality to nil
S	::=	…	statement
	|	x = A ;	assignment
	|	*p = A ;	assignment to a dereferenced pointer

Variables are statically allocated in memory at locations/addresses determined by the compiler, linker, and loader. These locations can be values of pointer variables. The null pointer NULL designates no location at all. The referencing &x returns the location of variable x in memory. Dereferencing *p a pointer variable p returns the value from the location/memory address that is pointed to by the pointer variable x (that is the location contained in the location of variable p). One can test for nullity p == NULL or x == y equality of pointer variables. Assignment is of an integer or pointer value to a variable location x = A ; or an indirect assignment *p = A ; to a dereferenced pointer variable. Some languages like Pascal [986] do not have such primitives but use them implicitly, for example, when passing parameters by reference (also called *by variable*) or for array parameters (to avoid a copy of the array).

46.2 Pointer Semantic Domains

The compiler, linker, and loader of an operating system initially assign a unique memory location/address $\lambda_{\mathsf{x}} \in \mathbb{L}$ to each program variable $\mathsf{x} \in \mathbb{V}$. It is fixed for a given program execution and may vary for different executions. This is formalized by the injective memory allocation function $\boldsymbol{\lambda} \triangleq \mathsf{x} \in \mathbb{V} \mapsto \lambda_{\mathsf{x}} \in \mathbb{L}$. The injection ensures that the locations of variables are unique (no two different variables can be allocated at the same location $\mathsf{x} \neq \mathsf{y} \Rightarrow \lambda_{\mathsf{x}} \neq \lambda_{\mathsf{y}}$).

z	$\in$	$\mathbb{Z}$	integers
l	$\in$	$\mathbb{L}$	enumerable set of locations (addresses) ($\mathbb{L} \cap \mathbb{Z} = \emptyset$)
λ	$\in$	$\mathbb{L}_{\mathbb{V}} \triangleq \mathbb{V} \rightarrowtail \mathbb{L}$	locations of variables (unique location λ_{x} of variable x in memory)

Values returned by expressions are integers, locations, the null pointer `nil` (denoted `NULL`), or an error Ω (e.g., when dereferencing a null pointer, which in practice often stops the execution with a segmentation fault [raised after a memory access violation in hardware with memory protection] or bus error [e.g., in misaligned memory access that hardware cannot access]).

$$\begin{array}{rcll} \mathtt{nil} & \notin & \mathbb{L} & \text{null pointer (value denoted by } \mathtt{NULL}) \\ v \in \mathbb{V} & \triangleq & \mathbb{Z} \cup \mathbb{L} \cup \{\mathtt{nil}\} & \text{values} \\ \Omega & \notin & \mathbb{V} & \text{error} \\ v \in \mathbb{V}_\Omega & \triangleq & \mathbb{V} \cup \{\Omega\} & \text{values or error} \end{array}$$

In case of error, execution is assumed to stop immediately. This is contrary to some programming languages such as C [569] in which the program behavior is undefined in the case of error. In C, errors need not be signaled during execution, so the undefined execution may have unpredictable effects like erasing data or destroying the program code. Of course, it is impossible to define the semantics of such programs. So, more restrictively, we consider such undefined behaviors to be an error that immediately stops execution. The soundness of the analysis of C under this hypothesis (which is *not* satisfied by C) is precisely defined in section 46.12.

A memory $\mu \in \mathbb{M}_\lambda$ maps locations λ_{x} of variables $\mathsf{x} \in \mathbb{V}$ for the given memory allocation λ to their content $\mu(\lambda_{\mathsf{x}})$ which is the value of x. Of course, the $\mu(\lambda_{\mathsf{x}})$ will vary during execution, as opposed to λ_{x}, which is determined at compile time and fixed during execution.

$$\begin{array}{rcccll} l & \in & \mathbb{L}_\lambda & \triangleq & \{\lambda_{\mathsf{x}} \in \mathbb{L} \mid \mathsf{x} \in \mathbb{V}\} & \text{set of locations of all variables} \\ \mu & \in & \mathbb{M}_\lambda & \triangleq & \mathbb{L}_\lambda \to \mathbb{V} & \text{memories (values of locations)} \end{array}$$

So $\mu \in \mathbb{L} \nrightarrow \mathbb{V}$ but need not be defined for addresses not allocated to a program variable. We can therefore refine the notion of values as follows:

$$v \in \mathbb{V}_\lambda \quad \triangleq \quad \mathbb{Z} \cup \mathbb{L}_\lambda \cup \{\mathtt{nil}\} \qquad \text{values}$$

Assertional invariant properties $P \in \mathbb{P}^{\mathbb{M}}_\lambda \triangleq \wp(\mathbb{M}_\lambda)$ are sets of memories for the given memory allocation λ (but not an error because execution stops when the evaluation of a pointer expression is erroneous).

$$P \in \mathbb{P}^{\mathbb{M}}_\lambda \quad \triangleq \quad \wp(\mathbb{M}_\lambda) \qquad \text{memory properties}$$

We use the notation $\wp(\mathbb{M}^{\varrho}_\lambda)$ of section 19.1.3 to handle uniformly assertional and relational properties.

46.3 Prefix and Maximal Trace Semantics of the Pointer Language

Because we postulate the reachability semantics of the pointer language in section 46.4, we do not need to provide a prefix and maximal trace semantics of the pointer language (as in chapters 6 and 7). This is easier in considering a stateful semantics, as in chapter 42, where states are memories $\mu \in \mathbb{M}_\lambda$ for a given memory allocation λ (see exercise 42.8).

We need only to extend the semantics of arithmetic and Boolean expressions in section 3.6 to allow for pointer operations and comparison, as follows.

The evaluation of an expression may return an error Ω. NULL denotes the null pointer nil. &x denotes the location of the variable x. It is assumed that the locations λ allocated by the compiler for different variables cannot overlap and be erroneous, so $\boldsymbol{\lambda} \triangleq \mathsf{x} \in \mathbb{V} \mapsto \lambda_{\mathsf{x}} \in \mathbb{L}$ is injective. The value of a variable x is obtained by looking at its location λ_{x} and then at the value $\mu(\lambda_{\mathsf{x}})$ in the memory at that location. For *p there is one more level of indirection $\mu(\mu(\lambda_{\mathsf{p}}))$ unless p is not a pointer, in which case this is a runtime error. Otherwise the definition is similar to (3.4), except that errors are propagated (as a way of handling exceptions).

Definition of the semantics of arithmetic, pointer, and Boolean expressions

$$\mathscr{A} \in \lambda \in \mathbb{L}\mathbb{v} \to \mathbb{A} \to \mathbb{M}_\lambda \to \mathbb{V}_\lambda \cup \{\Omega\} \tag{46.2}$$

$$\mathscr{A}_\lambda[\![1]\!]\,\mu \triangleq 1$$

$$\mathscr{A}_\lambda[\![\mathsf{x}]\!]\,\mu \triangleq \mu(\lambda_{\mathsf{x}})$$

$$\mathscr{A}_\lambda[\![\mathsf{A}_1 - \mathsf{A}_2]\!]\,\mu \triangleq (\!(\, \mathscr{A}_\lambda[\![\mathsf{A}_1]\!]\,\mu \in \mathbb{Z} \wedge \mathscr{A}_\lambda[\![\mathsf{A}_2]\!]\,\mu \in \mathbb{Z} \;?\; \mathscr{A}_\lambda[\![\mathsf{A}_1]\!]\,\mu - \mathscr{A}_\lambda[\![\mathsf{A}_2]\!]\,\mu \;:\; \Omega \,)\!)$$

$$\mathscr{A}_\lambda[\![\mathsf{NULL}]\!]\,\mu \triangleq \mathtt{nil}$$

$$\mathscr{A}_\lambda[\![\&\mathsf{x}]\!]\,\mu \triangleq \lambda_{\mathsf{x}}$$

$$\mathscr{A}_\lambda[\![*\mathsf{p}]\!]\,\mu \triangleq (\!(\, \mu(\lambda_{\mathsf{p}}) \in \mathbb{Z} \cup \{\mathtt{nil}\} \;?\; \Omega \;:\; \mu(\mu(\lambda_{\mathsf{p}})) \,)\!)$$

$$\mathscr{B} \in \lambda \in \mathbb{L}\mathbb{v} \to \mathbb{B} \to \mathbb{M}_\lambda \to \mathbb{B} \cup \{\Omega\}$$

$$\mathscr{B}_\lambda[\![\mathsf{A}_1 < \mathsf{A}_2]\!]\,\mu \triangleq (\!(\, \mathscr{A}_\lambda[\![\mathsf{A}_1]\!]\,\mu \in \mathbb{Z} \wedge \mathscr{A}_\lambda[\![\mathsf{A}_2]\!]\,\mu \in \mathbb{Z} \;?\; \mathscr{A}_\lambda[\![\mathsf{A}_1]\!]\,\mu < \mathscr{A}_\lambda[\![\mathsf{A}_2]\!]\,\mu \;:\; \Omega \,)\!)$$

$$\mathscr{B}_\lambda[\![\mathsf{p} == \mathsf{NULL}]\!]\,\mu \triangleq (\mu(\lambda_{\mathsf{p}}) = \mathtt{nil})$$

$$\mathscr{B}_\lambda[\![\mathsf{x} == \mathsf{y}]\!]\,\mu \triangleq (\mu(\lambda_{\mathsf{x}}) = \mu(\lambda_{\mathsf{y}}))$$

$$\mathscr{B}_\lambda[\![\mathsf{B}_1 \text{ nand } \mathsf{B}_2]\!]\,\mu \triangleq (\!(\, \mathscr{B}_\lambda[\![\mathsf{B}_1]\!]\,\mu \neq \Omega \wedge \mathscr{B}_\lambda[\![\mathsf{B}_2]\!]\,\mu \neq \Omega \;?\; \mathscr{B}_\lambda[\![\mathsf{B}_1]\!]\,\mu \uparrow \mathscr{B}_\lambda[\![\mathsf{B}_2]\!]\,\mu \;:\; \Omega \,)\!)$$

46.4 Reachability Semantics of the Pointer Language

- We extend chapter 19, "Structural Forward Reachability Semantics," and postulate the reachability semantics $\widehat{\boldsymbol{\mathcal{S}}}^{\vec{e}}_{\lambda}$ of the pointer manipulating statements. Of course, it could be obtained by abstraction as shown in chapter 20, "Calculational Design of the Forward Reachability Semantics," from an extension of the prefix trace semantics of chapter 6, "Structural Deductive Stateless Prefix Trace Semantics," to include pointers.
- The reachability semantics $\widehat{\boldsymbol{\mathcal{S}}}^{\vec{e}}_{\lambda}[\![\mathsf{S}]\!]\,\mathscr{P}_0\,\ell$ defines the memory states reachable at program point $\ell \in \mathsf{labx}[\![\mathsf{S}]\!]$ of program component $\mathsf{S} \in \mathbb{P}\mathbb{c}$ when its execution starts in an initial memory state satisfying precondition $\mathscr{P}_0$. The reachability semantics $\widehat{\boldsymbol{\mathcal{S}}}^{\vec{e}}_{\lambda}[\![\mathsf{S}]\!]$ is relative to a given assignment λ of a location λ_{x} to each variable $\mathsf{x} \in \mathsf{vars}[\![\mathsf{S}]\!]$ of program component S.
- The forward reachability semantics $\widehat{\boldsymbol{\mathcal{S}}}^{\vec{e}}[\![\mathsf{S}]\!]$ of a program component $\mathsf{S} \in \mathbb{P}\mathbb{c}$ in chapter 19 operates on environments mapping variables to values,

$$\widehat{\boldsymbol{\mathcal{S}}}^{\vec{e}}[\![\mathsf{S}]\!] \quad \in \quad \wp(\mathbb{E}\mathbb{v}^{e}) \to \mathsf{labx}[\![\mathsf{S}]\!] \to \wp(\mathbb{E}\mathbb{v}^{e})$$

The forward reachability semantics is here operating on memories (for a fixed allocation λ of locations to variables),

$$\widehat{\boldsymbol{\mathcal{S}}}^{\vec{\varrho}}_{\lambda}[\![\mathsf{S}]\!] \quad \in \quad \wp(\mathbb{M}^{\varrho}_{\lambda}) \to \mathsf{labx}[\![\mathsf{S}]\!] \to \wp(\mathbb{M}^{\varrho}_{\lambda})$$

So, the value of a variable x which was $\rho(\mathsf{x})$ for environments $\rho \in \mathbb{E}\mathbb{v}$ is now $\mu(\lambda_{\mathsf{x}})$ for memories $\mu \in \mathbb{M}_{\lambda}$ with given memory allocation λ. By using symbolic addresses, the analysis is independent of the particular memory allocation λ determined at compile and loading time.

- Let us now define the reachability semantics $\widehat{\boldsymbol{\mathcal{S}}}^{\vec{\varrho}}_{\lambda}[\![\mathsf{S}]\!]$ of a program component S for the pointer semantics (covering both assertional $\widehat{\boldsymbol{\mathcal{S}}}^{\vec{r}}_{\lambda}[\![\mathsf{S}]\!]$ and relational $\widehat{\boldsymbol{\mathcal{S}}}^{\vec{R}}_{\lambda}[\![\mathsf{S}]\!]$ cases).
  - The semantics (19.12) of an assignment statement $\mathsf{S} ::= \mathsf{x} = \mathsf{A}\,;$ maps the initial states to themselves on entry and to themselves as modified by the assignment on exit. An arithmetic or pointer assignment modifies the memory (but not the allocation of locations to variables, which are chosen statically by the compiler and loader hence are immutable during execution). Moreover, if the evaluation of the expression A returns an error Ω then the execution stops, so there is no reachable successor state (hence the postcondition is $\varnothing$).

Definition of the reachability of an arithmetic or pointer assignment statement $\mathsf{S} ::= \mathsf{x} = \mathsf{A}\,;$

$$\widehat{\boldsymbol{\mathcal{S}}}^{\vec{\varrho}} \quad \in \quad \lambda \in \mathbb{L}\mathbb{v} \to \mathbb{S} \to \wp(\mathbb{M}^{\vec{\varrho}}_{\lambda}) \to \mathbb{L} \to \wp(\mathbb{M}^{\vec{\varrho}}_{\lambda}) \tag{46.3}$$

$$\begin{aligned}
\widehat{\boldsymbol{\mathcal{S}}}^{\vec{\varrho}}_{\lambda}[\![\mathsf{S}]\!]\,\mathcal{P}_0\,\ell \quad &= \quad (\!|\ \ell = \mathsf{at}[\![\mathsf{S}]\!]\ ?\ \mathcal{P}_0 \\
&\qquad\ \|\ \ell = \mathsf{after}[\![\mathsf{S}]\!]\ ?\ \mathsf{assign}^{\vec{\varrho}}_{\lambda}[\![\mathsf{x}, \mathsf{A}]\!]\ \mathcal{P}_0 \\
&\qquad\ :\ \varnothing\ |\!) \\
\mathsf{assign}^{\vec{r}}_{\lambda}[\![\mathsf{x}, \mathsf{A}]\!]\ \mathcal{P}_0 \quad &\triangleq \quad \{\mu[\lambda_{\mathsf{x}} \leftarrow \boldsymbol{\mathcal{A}}_{\lambda}[\![\mathsf{A}]\!]\,\mu] \mid \mu \in \mathcal{P}_0 \wedge \boldsymbol{\mathcal{A}}_{\lambda}[\![\mathsf{A}]\!]\,\mu \neq \Omega\} \\
\mathsf{assign}^{\vec{R}}_{\lambda}[\![\mathsf{x}, \mathsf{A}]\!]\ \mathcal{P}_0 \quad &\triangleq \quad \{\langle\mu_0,\ \mu[\lambda_{\mathsf{x}} \leftarrow \boldsymbol{\mathcal{A}}_{\lambda}[\![\mathsf{A}]\!]\,\mu]\rangle \mid \langle\mu_0,\ \mu\rangle \in \mathcal{P}_0 \wedge \boldsymbol{\mathcal{A}}_{\lambda}[\![\mathsf{A}]\!]\,\mu \neq \Omega\}
\end{aligned}$$

The difference with (19.12) is that the value of a variable x has to be fetched in memory μ at the location λ_{x} for that variable so $\rho(\mathsf{x}) = \mu(\lambda_{\mathsf{x}})$ and that execution stops on error (there was no error in (19.12) but, for example, integer overflow could have been handled similarly by stopping execution).

  - The assignment to a dereferenced pointer $\star\mathsf{p} = \mathsf{A}\,;$ assigns the value of expression A to the memory pointed to by pointer p. This is an error if p does not contain a pointer, that is, is an integer or `nil` so that $\mu(\lambda_{\mathsf{p}}) \notin \mathbb{L}$. An error can also be raised by the evaluation of expression A (e.g., by dereferencing a null pointer). In case of error, the execution stops, so there is no reachable successor state.

Reachability of a dereferenced pointer assignment statement S ::= ⋆p = A ;

$$\widehat{\mathcal{S}}^{\vec{\varrho}} \in \lambda \in \mathbb{L}\mathbb{v} \to \mathbb{S} \to \wp(\mathbb{M}^{\vec{\varrho}}_{\lambda}) \to \mathbb{L} \to \wp(\mathbb{M}^{\vec{\varrho}}_{\lambda}) \tag{46.4}$$

$$\begin{aligned}\widehat{\mathcal{S}}^{\vec{\varrho}}_{\lambda}[\![\mathsf{S}]\!]\,\mathcal{P}_0\,\ell \;=\;& (\!|\, \ell = \mathsf{at}[\![\mathsf{S}]\!] \;?\; \mathcal{P}_0 \\ & \;|\!|\; \ell = \mathsf{after}[\![\mathsf{S}]\!] \;?\; \mathsf{assign}^{\vec{\varrho}}_{\lambda}[\![\star\mathsf{p}, \mathsf{A}]\!]\;\mathcal{P}_0 \\ & \;:\; \varnothing \,|\!)\end{aligned}$$

$$\mathsf{assign}^{\vec{r}}_{\lambda}[\![\star\mathsf{p}, \mathsf{A}]\!]\;\mathcal{P}_0 \triangleq \{\mu[\mu(\lambda_{\mathsf{p}}) \leftarrow \mathcal{A}_{\lambda}[\![\mathsf{A}]\!]\,\mu] \mid \mu \in \mathcal{P}_0 \wedge \mu(\lambda_{\mathsf{p}}) \in \mathbb{L} \wedge \mathcal{A}_{\lambda}[\![\mathsf{A}]\!]\,\mu \neq \Omega\}$$

$$\mathsf{assign}^{\vec{R}}_{\lambda}[\![\star\mathsf{p}, \mathsf{A}]\!]\;\mathcal{P}_0 \triangleq \{\langle \mu_0, \mu[\mu(\lambda_{\mathsf{p}}) \leftarrow \mathcal{A}_{\lambda}[\![\mathsf{A}]\!]\,\mu]\rangle \mid \langle \mu_0, \mu\rangle \in \mathcal{P}_0 \wedge \mu(\lambda_{\mathsf{p}}) \in \mathbb{L} \wedge \mathcal{A}_{\lambda}[\![\mathsf{A}]\!]\,\mu \neq \Omega\}$$

Example 46.5 Continuing example 46.1, assume that the precondition $\mathcal{P}_0$ has only one memory state. Then the reachability semantics $\widehat{\mathcal{S}}^{\vec{\varrho}}_{\lambda}[\![\mathsf{S}]\!]\,\mathcal{P}_0$ is similar to one execution and goes as follows:

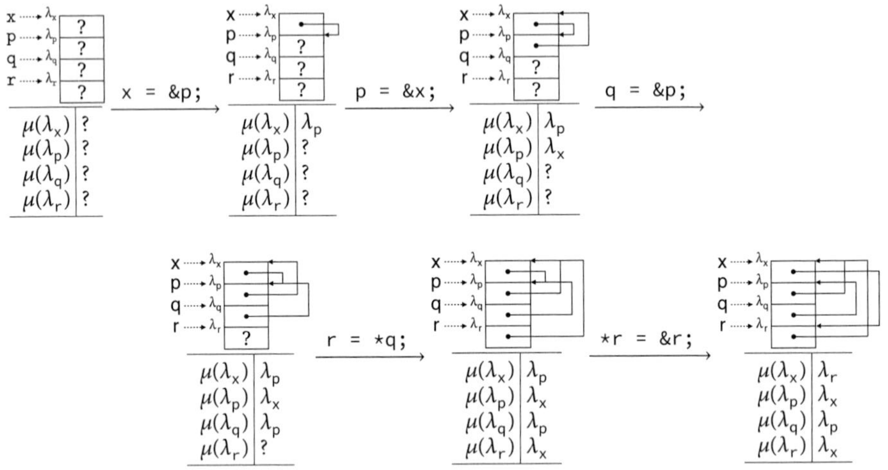

- The reachability test transformers of (19.16) are adapted to the memory model. In case of error $\mathcal{B}_{\lambda}[\![\mathsf{B}]\!]\,\mu = \Omega \notin \{\mathsf{tt}, \mathsf{ff}\}$, the execution stops so there is no reachable successor state.
- Observe that by our definition of reachability, the fatal error Ω is never reachable by our assumption that execution stops in case of error. To improve error reporting, it would also be possible to keep errors in the reachable states (even with, for example, a program point to better track the origin of errors).

Reachability of a test B

$$\begin{aligned}
\mathsf{test}^{\vec{\varrho}}, \overline{\mathsf{test}}^{\vec{\varrho}} &\in \lambda \in \mathbb{L}\mathbb{v} \to \mathbb{B} \to \wp(\mathbb{M}^{\vec{\varrho}}_{\lambda}) \to \wp(\mathbb{M}^{\vec{\varrho}}_{\lambda}) && (46.6)\\
\mathsf{test}^{\vec{r}}_{\lambda}[\![\mathsf{B}]\!]\,\mathscr{P}_0 &\triangleq \{\mu \in \mathscr{P}_0 \mid \mathscr{B}_{\lambda}[\![\mathsf{B}]\!]\,\mu = \mathsf{tt}\}\\
\mathsf{test}^{\vec{R}}_{\lambda}[\![\mathsf{B}]\!]\,\mathscr{P}_0 &\triangleq \{\langle\mu_0, \mu\rangle \in \mathscr{P}_0 \mid \mathscr{B}_{\lambda}[\![\mathsf{B}]\!]\,\mu = \mathsf{tt}\}\\
\overline{\mathsf{test}}^{\vec{r}}_{\lambda}[\![\mathsf{B}]\!]\,\mathscr{P}_0 &\triangleq \{\mu \in \mathscr{P}_0 \mid \mathscr{B}_{\lambda}[\![\mathsf{B}]\!]\,\mu = \mathsf{ff}\}\\
\overline{\mathsf{test}}^{\vec{R}}_{\lambda}[\![\mathsf{B}]\!]\,\mathscr{P}_0 &\triangleq \{\langle\mu_0, \mu\rangle \in \mathscr{P}_0 \mid \mathscr{B}_{\lambda}[\![\mathsf{B}]\!]\,\mu = \mathsf{ff}\}
\end{aligned}$$

- The reachability semantics $\widehat{\boldsymbol{\mathcal{S}}}^{\vec{\varrho}}_{\lambda}[\![\mathsf{S}]\!]\,\mathscr{P}_0\,\ell$ of all other program components in (19.12) is modified in the same way, replacing environments ρ by memories μ for the given location assignment λ and stopping execution when expressions (e.g., Boolean conditions in conditional or iteration statements) return an error.
  For example, the reachability semantics (19.22) and (19.23) of conditional statements and (19.16) of an iteration statement remain the same except for the definition of $\mathsf{test}^{\vec{\varrho}}[\![\mathsf{B}]\!]$ by (46.6) and similarly for $\overline{\mathsf{test}}^{\vec{\varrho}}[\![\mathsf{B}]\!]$. In case the evaluation $\mathscr{B}_{\lambda}[\![\mathsf{B}]\!]\,\mu$ of the Boolean expression B returns an error Ω, $\mathsf{test}^{\vec{\varrho}}[\![\mathsf{B}]\!]X(\ell)$ is the empty set $\emptyset$, and therefore, by theorem 19.36, $\widehat{\boldsymbol{\mathcal{S}}}^{\vec{\varrho}}[\![\mathsf{S}_b]\!]\,(\mathsf{test}^{\vec{\varrho}}[\![\mathsf{B}]\!]X(\ell))\,\ell'$ is also the empty set $\emptyset$, that is, there are no further reachable states because the trace execution is stopped by the error.

Exercise 46.7 Assuming that $\mathbb{M}$ is infinite denumerable, add a memory allocation statement `x = malloc()` allocating a new location not already used in the memory $\mu \in \mathbb{M} \triangleq \mathbb{L} \nrightarrow \mathbb{V}_{\lambda}$ to x.

46.5 Abstract Interpreters for the Pointer Language

We extend the flow-sensitive static analysis of section 21.2 and the flow-insensitive static analysis of section 45.3 to handle the pointer statements of section 46.1.

46.5.1 Flow-Sensitive Abstract Interpreter for the Pointer Language

The abstract interpreter of section 21.2 is designed for an abstract domain

$$\mathbb{D}^{\sharp} \triangleq \langle \mathbb{P}^{\sharp}, \sqsubseteq^{\sharp}, \bot^{\sharp}, \sqcup^{\sharp}, \mathsf{assign}^{\sharp}[\![\mathsf{x}, \mathsf{A}]\!], \mathsf{test}^{\sharp}[\![\mathsf{B}]\!], \overline{\mathsf{test}}^{\sharp}[\![\mathsf{B}]\!]\rangle$$

which is unchanged in definition 27.1 except for the concretization $\gamma_{\lambda} \in \mathbb{P}^{\sharp}_{\lambda} \xrightarrow{\nearrow} \wp(\mathbb{M}^{\varrho}_{\lambda})$ to the memory collecting domain $\wp(\mathbb{M}^{\varrho}_{\lambda})$ instead of the environment collecting domain $\wp(\mathbb{E}\mathbb{v}^{\varrho})$. Equation (21.7) becomes

$$\mathbb{D}^{\sharp}_{\lambda} \triangleq \langle \mathbb{P}^{\sharp}_{\lambda}, \sqsubseteq^{\sharp}_{\lambda}, \bot^{\sharp}_{\lambda}, \sqcup^{\sharp}_{\lambda}, \mathsf{assign}^{\sharp}_{\lambda}[\![\mathsf{x}, \mathsf{A}]\!], \mathsf{assign}^{\sharp}_{\lambda}[\![{*}\mathsf{p}, \mathsf{A}]\!], \mathsf{test}^{\sharp}_{\lambda}[\![\mathsf{B}]\!], \overline{\mathsf{test}}^{\sharp}_{\lambda}[\![\mathsf{B}]\!]\rangle$$

The structural flow-sensitive abstract pointer semantics is defined as follows.

Flow-sensitive abstract semantics of an assignment statement $\mathsf{S} ::= \mathsf{x} = \mathsf{A}\,;$

$$\widehat{\mathcal{S}}^{\natural} \in \lambda \in \mathbb{L}\mathrm{v} \to \mathbb{S} \to \mathbb{P}^{\natural}_{\lambda} \to \mathbb{L} \to \mathbb{P}^{\natural}_{\lambda} \tag{46.8}$$

$$\begin{aligned}\widehat{\mathcal{S}}^{\natural}_{\lambda}[\![\mathsf{S}]\!]\,\overline{P}\,\ell \triangleq\;& (\!\mid \ell = \mathsf{at}[\![\mathsf{S}]\!] \;?\; \overline{P} \\ & \mid\!\mid \ell = \mathsf{after}[\![\mathsf{S}]\!] \;?\; \mathsf{assign}^{\natural}_{\lambda}[\![\mathsf{x}, \mathsf{A}]\!]\,\overline{P} \\ & \mathbin{\S}\, \bot^{\natural}_{\lambda} \,\mid\!)\end{aligned}$$

where $\mathsf{assign}^{\natural}_{\lambda}[\![\mathsf{x}, \mathsf{A}]\!] \in \mathbb{P}^{\natural}_{\lambda} \to \mathbb{P}^{\natural}_{\lambda}$ satisfies $\mathsf{assign}^{\vec{\varrho}}_{\lambda}[\![\mathsf{x}, \mathsf{A}]\!] \circ \gamma_{\lambda} \subseteq \gamma_{\lambda} \circ \mathsf{assign}^{\natural}_{\lambda}[\![\mathsf{x}, \mathsf{A}]\!]$.

Flow-sensitive abstract semantics of a dereferenced pointer assignment statement $\mathsf{S} ::= \star\mathsf{p} = \mathsf{A}\,;$

$$\widehat{\mathcal{S}}^{\natural} \in \lambda \in \mathbb{L}\mathrm{v} \to \mathbb{S} \to \mathbb{P}^{\natural}_{\lambda} \to \mathbb{L} \to \mathbb{P}^{\natural}_{\lambda} \tag{46.9}$$

$$\begin{aligned}\widehat{\mathcal{S}}^{\natural}_{\lambda}[\![\mathsf{S}]\!]\,\overline{P}\,\ell \triangleq\;& (\!\mid \ell = \mathsf{at}[\![\mathsf{S}]\!] \;?\; \overline{P} \\ & \mid\!\mid \ell = \mathsf{after}[\![\mathsf{S}]\!] \;?\; \mathsf{assign}^{\natural}_{\lambda}[\![\star\mathsf{p}, \mathsf{A}]\!]\,\overline{P} \\ & \mathbin{\S}\, \bot^{\natural}_{\lambda} \,\mid\!)\end{aligned}$$

where $\mathsf{assign}^{\natural}_{\lambda}[\![\star\mathsf{p}, \mathsf{A}]\!] \in \mathbb{P}^{\natural}_{\lambda} \to \mathbb{P}^{\natural}_{\lambda}$ satisfies $\mathsf{assign}^{\vec{\varrho}}_{\lambda}[\![\star\mathsf{p}, \mathsf{A}]\!] \circ \gamma_{\lambda} \subseteq \gamma_{\lambda} \circ \mathsf{assign}^{\natural}_{\lambda}[\![\star\mathsf{p}, \mathsf{A}]\!]$.

The abstract semantics of conditional and iteration statements is unchanged except for the use of (46.6) for the collecting semantics of tests.

Theorem 46.10 (well definedness of the reachability semantics and the abstract interpreter) The abstract interpreter $\widehat{\mathcal{S}}^{\natural}_{\lambda}[\![\mathsf{S}]\!]$ for the abstract domain $\mathbb{D}^{\natural}_{\lambda}$ assumed to be well defined by definition 21.1 (and assuming $\mathsf{assign}^{\natural}_{\lambda}[\![\mathsf{x}, \mathsf{A}]\!]$, $\mathsf{test}^{\natural}_{\lambda}[\![\mathsf{B}]\!]$, and $\overline{\mathsf{test}}^{\natural}_{\lambda}[\![\mathsf{B}]\!]$ to be continuous) is well defined and continuous for any program component $\mathsf{S} \in \mathbb{P}\mathbb{c}$.

Proof The language extension of section 46.1 preserves the well definedness of the abstract interpreter in theorem 21.16, that of the reachability semantics (exercise 21.21 and theorem 21.16), as well as the soundness of the abstract interpreter theorem 27.4 (in particular for the collecting semantics (46.3) and (46.4)). □

Exercise 46.11 (Project: flow-sensitive abstract interpreter for the pointer language) Extend the flow-sensitive abstract interpreter of exercise 21.22 to handle the syntax of the pointer language of section 46.1.

46.5.2 Flow-Insensitive Abstract Interpreter for the Pointer Language

As in chapter 45, "Flow-Insensitive Static Analysis," the flow insensitive abstract semantics for the pointer language is obtained by the $\alpha_{\mathcal{G}}$-abstraction of its flow-insensitive semantics, as follows:

Flow-insensitive semantics of an assignment statement S ::= x = A ;

$$\widehat{\mathcal{S}}^{\mathfrak{f}\alpha} \in \lambda \in \mathbb{L}\mathbb{v} \to \mathbb{S} \to \mathbb{P}^{\alpha}_{\lambda} \to \mathbb{P}^{\alpha}_{\lambda} \qquad (46.12)$$
$$\widehat{\mathcal{S}}^{\mathfrak{f}\alpha}_{\lambda}[\![\mathsf{S}]\!]\,\overline{P} \triangleq \overline{P} \sqcup^{\alpha}_{\lambda} \mathsf{assign}^{\alpha}_{\lambda}[\![\mathsf{x}, \mathsf{A}]\!]\,\overline{P}$$

Proof of (46.12)

$$\begin{aligned}
& \dot{\alpha}_{\mathfrak{f}}(\widehat{\mathcal{S}}^{\alpha}[\![\mathsf{x} = \mathsf{A}\,;]\!]\,\lambda)\overline{P} \\
=\;& \alpha_{\mathfrak{f}}(\widehat{\mathcal{S}}^{\alpha}[\![\mathsf{x} = \mathsf{A}\,;]\!]\,\lambda\,\overline{P}) && \text{(definition (45.1) of } \dot{\alpha}_{\mathfrak{f}}\text{)} \\
=\;& \alpha_{\mathfrak{f}}(\llbracket\, \ell = \mathsf{at}[\![\mathsf{S}]\!] \,?\, \overline{P} \,\|\, \ell = \mathsf{after}[\![\mathsf{S}]\!] \,?\, \mathsf{assign}^{\alpha}_{\lambda}[\![\mathsf{x}, \mathsf{A}]\!]\,\overline{P} : \bot^{\alpha} \,\rrbracket) && \text{(definition (46.8) of } \widehat{\mathcal{S}}^{\alpha}\text{)} \\
=\;& \overline{P} \sqcup^{\alpha}_{\lambda} \mathsf{assign}^{\alpha}_{\lambda}[\![\mathsf{x}, \mathsf{A}]\!]\,\overline{P} && \text{(definition (45.1) of } \alpha_{\mathfrak{f}} \text{ and } \mathsf{labx}[\![\mathsf{S}]\!] = \{\mathsf{at}[\![\mathsf{S}]\!], \mathsf{after}[\![\mathsf{S}]\!]\}\text{)} \\
\triangleq\;& \widehat{\mathcal{S}}^{\mathfrak{f}\alpha}_{\lambda}[\![\mathsf{S}]\!]\,\lambda\,\overline{P} && \text{(as defined in (46.12))} \quad \square
\end{aligned}$$

By a similar proof using (45.1) and (46.9), we get (46.13).

Flow-insensitive semantics of a dereferenced pointer assignment statement S ::= ⋆p = A ;

$$\widehat{\mathcal{S}}^{\mathfrak{f}\alpha} \in \lambda \in \mathbb{L}\mathbb{v} \to \mathbb{S} \to \mathbb{P}^{\alpha}_{\lambda} \to \mathbb{P}^{\alpha}_{\lambda} \qquad (46.13)$$
$$\widehat{\mathcal{S}}^{\mathfrak{f}\alpha}_{\lambda}[\![\mathsf{S}]\!]\,\overline{P} = \overline{P} \sqcup^{\alpha}_{\lambda} \mathsf{assign}^{\alpha}_{\lambda}[\![\star\mathsf{p}, \mathsf{A}]\!]\,\overline{P}$$

Theorem 45.14 becomes

Theorem 46.14 The flow-insensitive abstract semantics $\widehat{\mathcal{S}}^{\mathfrak{f}\alpha}_{\lambda}[\![\mathsf{S}]\!]$ on a well defined abstract domain of definition 21.1 is well defined.

Proof of theorem 46.14 This directly follows by structural induction from definition 21.1. □

Exercise 46.15 (Project: flow-insensitive abstract interpreter for the pointer language) Extend the flow-insensitive abstract interpreter of exercise 45.12 to handle the syntax of the pointer language of section 46.1.

46.6 Cartesian Abstract Domain for the Pointer Language

Most existing points-to static analyses are Cartesian. Cartesian points-to analyzers can express that x and y may point only to the locations of u or v but not that x and y may point only to the same location either u or v. For such Cartesian analyses, chapter 28, "Abstract Cartesian Semantics" is most relevant, and the Cartesian semantics can be easily extended to handle the pointer statements of section 46.1.

46.6.1 Cartesian Abstraction for the Pointer Language

Assume we abstract value properties by

$$\langle \wp(\mathbb{V}_{\lambda}), \subseteq \rangle \underset{\alpha^{\alpha}_{\lambda}}{\overset{\gamma^{\alpha}_{\lambda}}{\leftrightarrows}} \langle \mathbb{P}^{\alpha}_{\lambda}, \sqsubseteq^{\alpha}_{\lambda} \rangle. \qquad (46.16)$$

Then the Cartesian abstraction of properties $\wp(\mathbb{M}_{\lambda})$ of memories $\mathbb{M}_{\lambda} \triangleq \mathbb{L}_{\lambda} \to \mathbb{V}_{\lambda}$ is

$$\langle \wp(\mathbb{L}_\lambda \to \mathbb{V}_\lambda),\ \subseteq\rangle \underset{\dot\alpha^{\times}_\lambda}{\overset{\dot\gamma^{\times}_\lambda}{\leftrightarrows}} \langle \mathbb{L}_\lambda \to \wp(\mathbb{V}_\lambda),\ \dot\subseteq\rangle \underset{\dot\alpha^{\text{¤}}_\lambda}{\overset{\dot\gamma^{\text{¤}}_\lambda}{\leftrightarrows}} \langle \mathbb{L}_\lambda \to \mathbb{P}^{\text{¤}}_\lambda,\ \dot\sqsubseteq^{\text{¤}}_\lambda\rangle$$

where

$$\dot\alpha^{\times}_\lambda(P) \quad \triangleq \quad l \in \mathbb{L}_\lambda \mapsto \{\mu(l) \mid \mu \in P\} \tag{46.17}$$

is the Cartesian abstraction (see (28.1)) and $\dot\alpha^{\text{¤}}_\lambda(P) \triangleq l \in \mathbb{L}_\lambda \mapsto \alpha^{\text{¤}}_\lambda(P(l))$ is the pointwise extension of $\alpha^{\text{¤}}_\lambda$. By composition of these Galois connections, we get

$$\langle \wp(\mathbb{L}_\lambda \to \mathbb{V}_\lambda),\ \subseteq\rangle \underset{\dot\alpha^{\times\text{¤}}_\lambda}{\overset{\dot\gamma^{\times\text{¤}}_\lambda}{\leftrightarrows}} \langle \dot{\mathbb{P}}^{\text{¤}}_\lambda,\ \dot\sqsubseteq^{\text{¤}}\rangle \quad \text{where} \quad \dot{\mathbb{P}}^{\text{¤}}_\lambda \triangleq \mathbb{L}_\lambda \to \mathbb{P}^{\text{¤}}_\lambda$$

with $\dot\alpha^{\times\text{¤}}_\lambda(P) \triangleq \dot\alpha^{\text{¤}}_\lambda \circ \dot\alpha^{\times}_\lambda(P) = l \in \mathbb{L}_\lambda \mapsto \alpha^{\text{¤}}_\lambda(\{\mu(l) \mid \mu \in P\})$.

The generic Cartesian abstract domain

$$\mathbb{D}^{\text{¤}} \quad \triangleq \quad \langle \mathbb{P}^{\text{¤}},\ \sqsubseteq^{\text{¤}},\ \bot^{\text{¤}},\ \top^{\text{¤}},\ \sqcup^{\text{¤}},\ \sqcap^{\text{¤}},\ 1^{\text{¤}},\ \ominus^{\text{¤}},\ \overset{\text{¤}}{\ominus}{}^{-1},\ \otimes^{\text{¤}},\ \overline{\otimes}^{\text{¤}}\rangle \tag{28.42}$$

is extended to include abstract pointer operations

$$\mathbb{D}^{\text{¤}}_\lambda \quad \triangleq \quad \langle \mathbb{P}^{\text{¤}}_\lambda,\ \sqsubseteq^{\text{¤}}_\lambda,\ \bot^{\text{¤}}_\lambda,\ \top^{\text{¤}}_\lambda,\ \sqcup^{\text{¤}}_\lambda,\ \sqcap^{\text{¤}}_\lambda,\ 1^{\text{¤}}_\lambda,\ \ominus^{\text{¤}}_\lambda,\ \overset{\text{¤}}{\ominus}{}^{-1}_\lambda,\ \overset{\text{¤}}{\star}{}^{-1}_\lambda,\ \otimes^{\text{¤}},\ \overline{\otimes}^{\text{¤}},\ \mathsf{NULL}^{\text{¤}}_\lambda,\ \boldsymbol{\lambda}^{\text{¤}}_\lambda,\ \star^{\text{¤}}_\lambda\rangle \tag{46.18}$$

with well definedness conditions $\mathsf{NULL}^{\text{¤}}_\lambda \in \mathbb{P}^{\text{¤}}_\lambda$, $\boldsymbol{\lambda}^{\text{¤}}, \star^{\text{¤}} \in \mathbb{L}_{\mathbb{V}} \to \mathcal{V} \to \mathbb{P}^{\text{¤}}_\lambda$.

We must then calculate the structural definitions of $\mathscr{A}^{\text{¤}}_\lambda[\![\mathsf{A}]\!]\,\overline{P}$, $\overset{\text{¤}}{\mathscr{A}}{}^{-1}_\lambda[\![\mathsf{1}]\!]\,\chi\,\overline{P}$, $\mathsf{assign}^{\text{¤}}_\lambda[\![\mathsf{x}, \mathsf{A}]\!]$, $\mathsf{assign}^{\text{¤}}_\lambda[\![\star\mathsf{p}, \mathsf{A}]\!]$, $\mathsf{test}^{\text{¤}}_\lambda[\![\mathsf{B}]\!]$, and $\overline{\mathsf{test}}^{\text{¤}}_\lambda[\![\mathsf{B}]\!]$ to extend (28.17), (28.30), and (28.38) for the pointer language of section 46.1.

46.6.2 Cartesian Forward Reachability of an Arithmetic or Pointer Expressions

The abstract Cartesian reachability of expressions is $\mathscr{A}^{\text{¤}}_\lambda[\![\mathsf{A}]\!] \in (\mathbb{L}_\lambda \to \mathbb{P}^{\text{¤}}_\lambda) \to \mathbb{P}^{\text{¤}}_\lambda$ such that

$$\mathscr{A}^{\text{¤}}_\lambda[\![\mathsf{A}]\!]\,\overline{P} \quad \dot\sqsupseteq^{\text{¤}}_\lambda \quad \alpha^{\text{¤}}_\lambda(\{\mathscr{A}_\lambda[\![\mathsf{A}]\!]\,\mu \mid \mu \in \dot\gamma^{\times\text{¤}}_\lambda(\overline{P})\})$$

To define $\mathscr{A}^{\text{¤}}_\lambda[\![\mathsf{A}]\!]$ structurally, we extend (28.17) as follows (in case $\gamma^{\text{¤}}_\lambda(\bot^{\text{¤}}) = \varnothing$, the smash $\oslash^{\times}_\lambda$ of the Cartesian property $\oslash^{\text{¤}}_\lambda$ ensures strictness by a test for the infimum $\dot\bot^{\text{¤}}_\lambda$)

> *Cartesian abstract forward reachability semantics of an arithmetic or pointer expression* A
>
> The abstract Cartesian semantics of an expression A and an assignment statement $\mathsf{x} = \mathsf{A}\,;$ is defined as follows ($\overline{Q} \in \mathbb{P}^{\text{¤}}_\lambda$, $\overline{P} \in \mathbb{L}_\lambda \to \mathbb{P}^{\text{¤}}_\lambda$)
>
> $$\begin{array}{rcll} \oslash^{\text{¤}}_\lambda(\overline{Q})\,\overline{P} & \triangleq & \overline{Q} & \text{if } \gamma^{\text{¤}}_\lambda(\bot^{\text{¤}}_\lambda) \neq \varnothing \quad \text{smash} \\ & \triangleq & (\!|\,\overline{P} = \dot\bot^{\text{¤}}_\lambda \mathrel{?} \bot^{\text{¤}}_\lambda \mathrel{:} \overline{Q}\,|\!) & \text{if } \gamma^{\text{¤}}_\lambda(\bot^{\text{¤}}_\lambda) = \varnothing \\ \mathscr{A}^{\text{¤}} & \in & \lambda \in \mathbb{A} \to \mathbb{L}_\lambda \to \dot{\mathbb{P}}^{\text{¤}} \to \mathbb{V}_\lambda \quad \text{where} & \dot{\mathbb{P}}^{\text{¤}} \triangleq \mathbb{L}_\lambda \to \mathbb{P}^{\text{¤}}_\lambda \qquad (46.19) \\ \mathscr{A}^{\text{¤}}_\lambda[\![\mathsf{1}]\!]\,\overline{P} & \triangleq & \oslash^{\text{¤}}_\lambda(1^{\text{¤}}_\lambda)\,\overline{P} & 1^{\text{¤}}_\lambda \mathrel{\dot\sqsupseteq^{\text{¤}}_\lambda} \alpha^{\text{¤}}_\lambda(\{1\}) \\ \mathscr{A}^{\text{¤}}_\lambda[\![\mathsf{x}]\!]\,\overline{P} & \triangleq & \overline{P}(\lambda_{\mathsf{x}}) & \\ \mathscr{A}^{\text{¤}}_\lambda[\![\mathsf{A}_1 - \mathsf{A}_2]\!]\,\overline{P} & \triangleq & \oslash^{\text{¤}}_\lambda(\mathscr{A}^{\text{¤}}_\lambda[\![\mathsf{A}_1]\!]\,\overline{P} \ominus^{\text{¤}}_\lambda \mathscr{A}^{\text{¤}}_\lambda[\![\mathsf{A}_2]\!]\,\overline{P})\,\overline{P} & \end{array}$$
>
> where $\overline{P} \ominus^{\text{¤}}_\lambda \overline{Q} \mathrel{\dot\sqsupseteq^{\text{¤}}_\lambda} \alpha^{\text{¤}}_\lambda(\{x - y \mid x \in \gamma^{\text{¤}}_\lambda(\overline{P}) \cap \mathbb{Z} \wedge y \in \gamma^{\text{¤}}_\lambda(\overline{Q}) \cap \mathbb{Z}\})$

$$
\begin{array}{rcl}
\mathscr{A}^{¤}_{\lambda}[\![\mathsf{NULL}]\!]\,\overline{P} & \triangleq & \oslash^{¤}_{\lambda}(\mathsf{NULL}^{¤}_{\lambda})\,\overline{P} \qquad\qquad \mathsf{NULL}^{¤}_{\lambda} \mathrel{\dot{\sqsupseteq}^{¤}_{\lambda}} \alpha^{¤}_{\lambda}(\{\mathtt{nil}\})\\
\mathscr{A}^{¤}_{\lambda}[\![\&\mathsf{x}]\!]\,\overline{P} & \triangleq & \oslash^{¤}_{\lambda}(\boldsymbol{\lambda}^{¤}_{\lambda}[\![\mathsf{x}]\!])\,\overline{P} \qquad\qquad \boldsymbol{\lambda}^{¤}_{\lambda}[\![\mathsf{x}]\!] \mathrel{\dot{\sqsupseteq}^{¤}_{\lambda}} \alpha^{¤}_{\lambda}(\{\lambda_{\mathsf{x}}\})\\
\mathscr{A}^{¤}_{\lambda}[\![\star\mathsf{p}]\!]\,\overline{P} & \triangleq & \oslash^{¤}_{\lambda}(\star^{¤}_{\lambda}[\![\mathsf{p}]\!]\,\overline{P})\,\overline{P}
\end{array}
$$

$$
\text{where } \star^{¤}_{\lambda}[\![\mathsf{p}]\!]\,\overline{P} \mathrel{\dot{\sqsupseteq}^{¤}_{\lambda}} \alpha^{¤}_{\lambda}(\{\mu(\mu(\lambda_{\mathsf{p}})) \mid \mu \in \dot{\gamma}^{\times¤}_{\lambda}(\overline{P}) \wedge \mu(\lambda_{\mathsf{p}}) \in \mathbb{L}\})
$$

Remark 46.20 By choosing $\alpha^{¤}_{\lambda}$ to be the identity, we get the definition of Cartesian reachability $\mathscr{A}^{\times}_{\lambda}[\![\mathsf{A}]\!]\,\overline{P} \triangleq \{\mathscr{A}_{\lambda}[\![\mathsf{A}]\!]\,\mu \mid \mu \in \dot{\gamma}^{\times}_{\lambda}(\overline{P})\} \in (\mathbb{L}_{\lambda} \to \wp(\mathbb{V}_{\lambda}) \to \wp(\mathbb{V}_{\lambda})$ as an instance of $\mathscr{A}^{¤}_{\lambda}[\![\mathsf{A}]\!]\,\overline{P}$ with $\dot{\mathbb{P}}^{¤}_{\lambda} \triangleq \wp(\mathbb{L}_{\lambda} \to \mathbb{V}_{\lambda})$, $\dot{\bot}^{\times}_{\lambda} \triangleq \dot{\varnothing}$, $\not\bot^{\times} \triangleq \varnothing$, $1^{\times}_{\lambda} \triangleq \{1\}$, $P \ominus^{\times}_{\lambda} Q = \{x - y \mid x \in P \cap \mathbb{Z} \wedge y \in Q \cap \mathbb{Z}\}$, $\mathsf{NULL}^{\times}_{\lambda} \triangleq \{\mathtt{nil}\}$, $\boldsymbol{\lambda}^{\times}_{\lambda}[\![\mathsf{x}]\!] \triangleq \{\lambda_{\mathsf{x}}\}$, and $\star^{\times}_{\lambda}[\![\mathsf{p}]\!]\,P \triangleq \{\mu(\mu(\lambda_{\mathsf{p}})) \mid \mu \in P \wedge \mu(\lambda_{\mathsf{p}}) \in \mathbb{L}\}$.

Then lemma 28.14 remains valid, as follows

Lemma 46.21 $\forall P \in \wp(\mathbb{L}_{\lambda} \to \mathbb{V}_{\lambda})\,.\,\{\mathscr{A}_{\lambda}[\![\mathsf{A}]\!]\,\mu \mid \mu \in P \wedge \mathscr{A}_{\lambda}[\![\mathsf{A}]\!]\,\mu \neq \Omega\} \subseteq \mathscr{A}^{\times}_{\lambda}[\![\mathsf{A}]\!](\dot{\alpha}^{\times}_{\lambda}(P))$.

Proof of lemma 46.21

— $\{\mathscr{A}_{\lambda}[\![\mathsf{NULL}]\!]\,\mu \mid \mu \in P \wedge \mathscr{A}_{\lambda}[\![\mathsf{NULL}]\!]\,\mu \neq \Omega\}$

$= (\!| P = \varnothing \mathrel{?} \varnothing \mathrel{:} \{\mathtt{nil}\} |\!)$ $\quad$ ⟅definition (46.2) of $\mathscr{A}_{\lambda}[\![\mathsf{NULL}]\!]\,\mu \triangleq \mathtt{nil} \neq \Omega$⟆

$= (\!| \dot{\alpha}^{\times}_{\lambda}(P) = \dot{\varnothing} \mathrel{?} \varnothing \mathrel{:} \{\mathtt{nil}\} |\!)$

⟅definition $\dot{\alpha}^{\times}_{\lambda}(P) \triangleq l \in \mathbb{L}_{\lambda} \mapsto \{\mu(l) \mid \mu \in P\}$ so $P = \varnothing \Leftrightarrow \dot{\alpha}^{\times}_{\lambda}(P) = \dot{\varnothing}$⟆

$= \oslash^{\times}_{\lambda}(\mathsf{NULL}^{\times}_{\lambda})\,\dot{\alpha}^{\times}_{\lambda}(P)$ $\quad$ ⟅definitions of $\oslash^{\times}_{\lambda}$ and $\mathsf{NULL}^{\times}_{\lambda} \triangleq \{\mathtt{nil}\}$⟆

$= \mathscr{A}^{\times}_{\lambda}[\![\mathsf{NULL}]\!]\,\dot{\alpha}^{\times}_{\lambda}(P)$ $\quad$ ⟅definition (46.19) of $\mathscr{A}^{\times}_{\lambda}$⟆

— $\{\mathscr{A}_{\lambda}[\![\star\mathsf{p}]\!]\,\mu \mid \mu \in P \wedge \mathscr{A}_{\lambda}[\![\star\mathsf{p}]\!]\,\mu \neq \Omega\}$

$= \{(\!| \mu(\lambda_{\mathsf{p}}) \in \mathbb{Z} \cup \{\mathtt{nil}\} \mathrel{?} \Omega \mathrel{:} \mu(\mu(\lambda_{\mathsf{p}})) |\!) \mid \mu \in P \wedge \mu(\lambda_{\mathsf{p}}) \notin \mathbb{Z} \cup \{\mathtt{nil}\}\}$

⟅definition (46.2) of $\mathscr{A}_{\lambda}[\![\star\mathsf{p}]\!]\,\mu$⟆

$= \{\mu(\mu(\lambda_{\mathsf{p}})) \mid \mu \in P \wedge \mu(\lambda_{\mathsf{p}}) \in \mathbb{L}_{\lambda}\}$ $\quad$ ⟅definition of values $\mathbb{V}_{\lambda} \triangleq \mathbb{Z} \cup \mathbb{L}_{\lambda} \cup \{\mathtt{nil}\}$⟆

$\subseteq \{\mu(\mu(\lambda_{\mathsf{p}})) \mid \mu \in \dot{\gamma}^{\times}_{\lambda}(\dot{\alpha}^{\times}_{\lambda}(P)) \wedge \mu(\lambda_{\mathsf{p}}) \in \mathbb{L}_{\lambda}\}$

⟅Cartesian abstraction Galois connection $\langle\wp(\mathbb{L}_{\lambda} \to \mathbb{V}_{\lambda}),\ \subseteq\rangle \xleftarrow[\dot{\alpha}^{\times}_{\lambda}]{\dot{\gamma}^{\times}_{\lambda}} \langle\mathbb{L}_{\lambda} \to \wp(\mathbb{V}_{\lambda}),\ \dot{\subseteq}\rangle$ and exercise 11.35.3⟆

$= (\!| \dot{\alpha}^{\times}_{\lambda}(P) = \dot{\varnothing} \mathrel{?} \varnothing \mathrel{:} \{\mu(\mu(\lambda_{\mathsf{p}})) \mid \mu \in \dot{\gamma}^{\times}_{\lambda}(\dot{\alpha}^{\times}_{\lambda}(P)) \wedge \mu(\lambda_{\mathsf{p}}) \in \mathbb{L}_{\lambda}\} |\!)$ $\quad$ ⟅$\dot{\gamma}^{\times}_{\lambda}(\dot{\alpha}^{\times}_{\lambda}(\dot{\varnothing})) = \varnothing$⟆

$= \oslash^{\times}_{\lambda}(\star^{\times}_{\lambda}[\![\mathsf{p}]\!]\,\dot{\alpha}^{\times}_{\lambda}(P))\,\dot{\alpha}^{\times}_{\lambda}(P)$ $\quad$ ⟅definitions of $\oslash^{\times}_{\lambda}$ and $\star^{\times}_{\lambda}[\![\mathsf{p}]\!]$⟆

$= \mathscr{A}^{\times}_{\lambda}[\![\mathsf{A}]\!]\,(\dot{\alpha}^{\times}_{\lambda}(P))$ $\quad$ ⟅definition (46.19) of $\mathscr{A}^{\times}_{\lambda}[\![\star\mathsf{p}]\!]\,\overline{P} \triangleq \oslash^{\times}_{\lambda}(\star^{\times}_{\lambda}[\![\mathsf{p}]\!]\,\overline{P})\,\overline{P}$⟆

The proof of the other cases is similar. □

46.6.3 Cartesian Forward Reachability of a Pointer Assignment

Cartesian abstract forward reachability semantics of assignment $\mathsf{x} = \mathsf{A}\,;$

$$\mathsf{assign}^{¤}_{\lambda}[\![\mathsf{x}, \mathsf{A}]\!]\,\overline{P} \;\triangleq\; \overline{P}[\lambda_{\mathsf{x}} \leftarrow \mathscr{A}^{¤}_{\lambda}[\![\mathsf{A}]\!]\,\overline{P}] \tag{46.22}$$

Cartesian abstract forward reachability semantics of assignment $\star\mathsf{p} = \mathsf{A}\,;$

$$\begin{aligned}\mathsf{assign}^{¤}_{\lambda}[\![\star\mathsf{p}, \mathsf{A}]\!]\,\overline{P} \;&\triangleq\; \lambda_{\mathsf{y}} \mapsto \overline{P}(\lambda_{\mathsf{y}}) \sqcup^{¤} (\!(\, \boldsymbol{\lambda}^{¤}_{\lambda}[\![y]\!] \sqsubseteq^{¤} \overline{P}(\lambda_{\mathsf{p}})\, ?\, \mathscr{A}^{¤}[\![\mathsf{A}]\!]\,\overline{P} \,:\, \bot^{¤} \,)\!) \\ &\qquad\qquad \text{if } \alpha^{¤}_{\lambda}(\{\lambda_{\mathsf{x}}\}) = \boldsymbol{\lambda}^{¤}_{\lambda}[\![\mathsf{x}]\!] \\ &\triangleq\; \lambda_{\mathsf{y}} \mapsto \overline{P}(\lambda_{\mathsf{y}}) \sqcup^{¤} \mathscr{A}^{¤}[\![\mathsf{A}]\!]\,\overline{P} \qquad \text{otherwise}\end{aligned}$$

The Cartesian assignment theorem 28.15 becomes

Theorem 46.23 (Cartesian assignment for the pointer language)

$$\dot{\alpha}^{\times}_{\lambda} \circ \mathsf{assign}^{\vec{r}}_{\lambda}[\![\mathsf{x}, \mathsf{A}]\!] \;\dot{\subseteq}\; \mathsf{assign}^{\times}_{\lambda}[\![\mathsf{x}, \mathsf{A}]\!] \circ \dot{\alpha}^{\times}_{\lambda}$$

$$\dot{\alpha}^{\times}_{\lambda} \circ \mathsf{assign}^{\vec{r}}_{\lambda}[\![\star\mathsf{p}, \mathsf{A}]\!] \;\dot{\subseteq}\; \mathsf{assign}^{\times}_{\lambda}[\![\star\mathsf{p}, \mathsf{A}]\!] \circ \dot{\alpha}^{\times}_{\lambda}$$

Proof of theorem 46.23

— $\dot{\alpha}^{\times}_{\lambda} \circ \mathsf{assign}^{\vec{r}}_{\lambda}[\![\mathsf{x}, \mathsf{A}]\!]\,\overline{P}$

$= \dot{\alpha}^{\times}_{\lambda}(\{\mu[\lambda_{\mathsf{x}} \leftarrow \mathscr{A}_{\lambda}[\![\mathsf{A}]\!]\,\mu] \mid \mu \in \mathscr{P}_0 \wedge \mathscr{A}_{\lambda}[\![\mathsf{A}]\!]\,\mu \neq \Omega\})$?definition (46.3) of $\mathsf{assign}^{\vec{r}}_{\lambda}[\![_, _]\!]$S

$= l \in \mathbb{L}_{\lambda} \mapsto \{\mu[\lambda_{\mathsf{x}} \leftarrow \mathscr{A}_{\lambda}[\![\mathsf{A}]\!]\,\mu](l) \mid \mu \in \overline{P} \wedge \mathscr{A}_{\lambda}[\![\mathsf{A}]\!]\,\mu \neq \Omega\}$

?definition $\dot{\alpha}^{\times}_{\lambda}(P) \triangleq l \in \mathbb{L}_{\lambda} \mapsto \{\mu(l) \mid \mu \in P\}$S

$= l \in \mathbb{L}_{\lambda} \mapsto (\!(\, \overline{P} = \varnothing\, ?\, \varnothing \parallel l = \lambda_{\mathsf{x}}\, ?\, \{\mathscr{A}_{\lambda}[\![\mathsf{A}]\!]\,\mu \mid \mu \in \overline{P} \wedge \mathscr{A}_{\lambda}[\![\mathsf{A}]\!]\,\mu \neq \Omega\} : \{\mu(l) \mid \mu \in \overline{P} \wedge \mathscr{A}_{\lambda}[\![\mathsf{A}]\!]\,\mu \neq \Omega\}\,)\!)$?definition (19.10) of environment assignmentS

$\subseteq l \in \mathbb{L}_{\lambda} \mapsto (\!(\, \overline{P} = \varnothing\, ?\, \varnothing \parallel l = \lambda_{\mathsf{x}}\, ?\, \{\mathscr{A}_{\lambda}[\![\mathsf{A}]\!]\,\mu \mid \mu \in \overline{P} \wedge \mathscr{A}_{\lambda}[\![\mathsf{A}]\!]\,\mu \neq \Omega\} : \{\mu(l) \mid \mu \in \overline{P}\}\,)\!)$

?not propagating errors to all componentsS

$= l \in \mathbb{L}_{\lambda} \mapsto (\!(\, \overline{P} = \varnothing\, ?\, \varnothing \parallel l = \lambda_{\mathsf{x}}\, ?\, \{\mathscr{A}_{\lambda}[\![\mathsf{A}]\!]\,\mu \mid \mu \in \overline{P} \wedge \mathscr{A}_{\lambda}[\![\mathsf{A}]\!]\,\mu \neq \Omega\} : \dot{\alpha}^{\times}_{\lambda}(\overline{P})(l)\,)\!)$

?definition $\dot{\alpha}^{\times}_{\lambda}(P) \triangleq l \in \mathbb{L}_{\lambda} \mapsto \{\mu(l) \mid \mu \in P\}$S

$\subseteq l \in \mathbb{L}_{\lambda} \mapsto (\!(\, \overline{P} = \varnothing\, ?\, \varnothing \parallel l = \lambda_{\mathsf{x}}\, ?\, \mathscr{A}^{\times}[\![\mathsf{A}]\!](\dot{\alpha}^{\times}_{\lambda}(\overline{P})) : \dot{\alpha}^{\times}_{\lambda}(\overline{P})(l)\,)\!)$?by lemma 46.21S

$= (\!(\, \dot{\alpha}^{\times}_{\lambda}(\overline{P}) = \dot{\varnothing}\, ?\, \dot{\varnothing} : \dot{\alpha}^{\times}_{\lambda}(\overline{P})[\lambda_{\mathsf{x}} \leftarrow \mathscr{A}^{\times}_{\lambda}[\![\mathsf{A}]\!](\dot{\alpha}^{\times}_{\lambda}(\overline{P}))]\,)\!)$

?definition (19.10) of environment assignment and $\overline{P} = \varnothing \Leftrightarrow \dot{\alpha}^{\times}_{\lambda}(\overline{P}) = \dot{\varnothing}$S

$= \dot{\alpha}^{\times}_{\lambda}(\overline{P})[\lambda_{\mathsf{x}} \leftarrow \mathscr{A}^{\times}_{\lambda}[\![\mathsf{A}]\!](\dot{\alpha}^{\times}_{\lambda}(\overline{P}))]$

?because $\mathscr{A}^{\times}_{\lambda}[\![\mathsf{A}]\!](\dot{\varnothing}) = \varnothing$ implies $\dot{\varnothing}[\lambda_{\mathsf{x}} \leftarrow \mathscr{A}^{\times}_{\lambda}[\![\mathsf{A}]\!](\dot{\varnothing})] = \dot{\varnothing}$S

$= \mathsf{assign}^{\times}_{\lambda}[\![\mathsf{x}, \mathsf{A}]\!]\,\lambda\,(\dot{\alpha}^{\times}_{\lambda}(\overline{P}))$?definition (46.22) of $\mathsf{assign}^{\times}_{\lambda}[\![\mathsf{x}, \mathsf{A}]\!]\,\overline{P} \triangleq \overline{P}[\lambda_{\mathsf{x}} \leftarrow \mathscr{A}^{\times}_{\lambda}[\![\mathsf{A}]\!]\overline{P}]$S

— $\dot{\alpha}^{\times}_{\lambda} \circ \mathsf{assign}^{\vec{r}}_{\lambda}[\![\star\mathsf{p}, \mathsf{A}]\!]\,\overline{P}$

$= \dot{\alpha}^{\times}_{\lambda}(\{\mu[\mu(\lambda_{\mathsf{p}}) \leftarrow \mathscr{A}_{\lambda}[\![\mathsf{A}]\!]\,\mu] \mid \mu \in \overline{P} \wedge \mu(\lambda_{\mathsf{p}}) \in \mathbb{L} \wedge \mathscr{A}_{\lambda}[\![\mathsf{A}]\!]\,\mu \neq \Omega\})$

?definition (46.4) of $\mathsf{assign}^{\vec{r}}_{\lambda}[\![\star\mathsf{p}, \mathsf{A}]\!]$S

$= l \mapsto \{\mu'(l) \mid \mu' \in \{\mu[\mu(\lambda_{\mathsf{p}}) \leftarrow \mathscr{A}_{\lambda}[\![\mathsf{A}]\!]\,\mu] \mid \mu \in \overline{P} \wedge \mu(\lambda_{\mathsf{p}}) \in \mathbb{L} \wedge \mathscr{A}_{\lambda}[\![\mathsf{A}]\!]\,\mu \neq \Omega\}\}$

⟨definition $\dot{\alpha}^{\times}_{\lambda}(P) \triangleq l \in \mathbb{L}_{\lambda} \mapsto \{\mu'(l) \mid \mu' \in P\}$⟩

$= \; l \mapsto \{\mu[\mu(\lambda_{\mathsf{p}}) \leftarrow \mathscr{A}_{\lambda}\llbracket \mathsf{A} \rrbracket \mu](l) \mid \mu \in \overline{P} \wedge \mu(\lambda_{\mathsf{p}}) \in \mathbb{L} \wedge \mathscr{A}_{\lambda}\llbracket \mathsf{A} \rrbracket \mu \neq \Omega\}$ ⟨definition of $\in$⟩

$\dot{\subseteq} \; l \mapsto \{\mu(l) \mid \mu \in \overline{P}\} \cup \{\mathscr{A}_{\lambda}\llbracket \mathsf{A} \rrbracket \mu \mid l = \mu(\lambda_{\mathsf{p}}) \wedge \mu \in \overline{P} \wedge \mu(\lambda_{\mathsf{p}}) \in \mathbb{L} \wedge \mathscr{A}_{\lambda}\llbracket \mathsf{A} \rrbracket \mu \neq \Omega\}$

⟨definition (19.10) of environment assignment because either $l = \mu(\lambda_{\mathsf{p}})$ and $\{\mu(\mathscr{A}_{\lambda}\llbracket \mathsf{A} \rrbracket \mu) \mid \mu \in \overline{P} \wedge \mu(\lambda_{\mathsf{p}}) \in \mathbb{L} \wedge \mathscr{A}_{\lambda}\llbracket \mathsf{A} \rrbracket \mu \neq \Omega\}$ is overapproximated by $\{\mu(l) \mid \mu \in \overline{P})\} \cup \{\mathscr{A}_{\lambda}\llbracket \mathsf{A} \rrbracket \mu \mid \mu \in \overline{P} \wedge \mu(\lambda_{\mathsf{p}}) \in \mathbb{L} \wedge \mathscr{A}_{\lambda}\llbracket \mathsf{A} \rrbracket \mu \neq \Omega\}$ or $l \neq \mu(\lambda_{\mathsf{p}})$ and $\{\mu(l) \mid \mu \in \overline{P} \wedge \mu(\lambda_{\mathsf{p}}) \in \mathbb{L} \wedge \mathscr{A}_{\lambda}\llbracket \mathsf{A} \rrbracket \mu \neq \Omega\}$ is overapproximated by $\{\mu(l) \mid \mu \in \overline{P}\}$ ⟩

$= \; l \mapsto \{\mu(l) \mid \mu \in \overline{P}\} \cup \{v \mid l = \mu(\lambda_{\mathsf{p}}) \wedge \mu \in \overline{P} \wedge \mu(\lambda_{\mathsf{p}}) \in \mathbb{L} \wedge v \in \{\mathscr{A}_{\lambda}\llbracket \mathsf{A} \rrbracket \mu \mid \mu \in \overline{P} \wedge \mathscr{A}_{\lambda}\llbracket \mathsf{A} \rrbracket \mu \neq \Omega\}\}$ ⟨definition of $\in$⟩

$\dot{\subseteq} \; l \mapsto \{\mu(l) \mid \mu \in \overline{P}\} \cup \{v \mid l = \mu(\lambda_{\mathsf{p}}) \wedge \mu \in \overline{P} \wedge \mu(\lambda_{\mathsf{p}}) \in \mathbb{L} \wedge v \in \mathscr{A}^{\times}_{\lambda}\llbracket \mathsf{A} \rrbracket(\dot{\alpha}^{\times}_{\lambda}(P))\}$

⟨lemma 46.21⟩

$\dot{\subseteq} \; l \mapsto \{\mu(l) \mid \mu \in \overline{P}\} \cup \{v \mid l \in \{\mu(\lambda_{\mathsf{p}}) \mid \mu \in \overline{P}\} \wedge v \in \mathscr{A}^{\times}_{\lambda}\llbracket \mathsf{A} \rrbracket(\dot{\alpha}^{\times}_{\lambda}(P))\}$ ⟨definition of $\subseteq$⟩

$= \; l \mapsto \dot{\alpha}^{\times}_{\lambda}(P)\, l \cup \{v \mid l \in \dot{\alpha}^{\times}_{\lambda}(P)\, \lambda_{\mathsf{p}} \wedge v \in \mathscr{A}^{\times}_{\lambda}\llbracket \mathsf{A} \rrbracket(\dot{\alpha}^{\times}_{\lambda}(P))\}$

⟨definition $\dot{\alpha}^{\times}_{\lambda}(P) \triangleq l \in \mathbb{L}_{\lambda} \mapsto \{\mu(l) \mid \mu \in P\}$⟩

$= \; l \mapsto \dot{\alpha}^{\times}_{\lambda}(P)\, l \cup (\!| l \in \dot{\alpha}^{\times}_{\lambda}(P)\, \lambda_{\mathsf{p}} \;?\; \mathscr{A}^{\times}_{\lambda}\llbracket \mathsf{A} \rrbracket(\dot{\alpha}^{\times}_{\lambda}(P)) \;:\; \varnothing |\!)$ ⟨definition of the conditional⟩

$= \; \mathsf{assign}^{\times}_{\lambda}\llbracket \star\mathsf{p}, \mathsf{A} \rrbracket\, (\dot{\alpha}^{\times}_{\lambda}(\overline{P}))$

⟨remark 46.20 and definition (46.22) of $\mathsf{assign}^{\times}_{\lambda}\llbracket \star\mathsf{p}, \mathsf{A} \rrbracket \triangleq l \mapsto \overline{P}(l) \cup (\!| l \in \overline{P}(\lambda_{\mathsf{p}}) \;?\; \mathscr{A}^{¤}_{\lambda}\llbracket \mathsf{A} \rrbracket\, \overline{P} \;:\; \varnothing |\!)$⟩ □

Theorem 28.19 becomes

> **Theorem 46.24 (abstract Cartesian assignment)** If $1^{\times} \sqsubseteq^{\times} \gamma^{¤}_{\lambda}(1^{¤})$, $\ominus^{\times}$ is $\sqsubseteq^{\times}$-increasing argumentwise, and $\forall X, Y \in \mathbb{P}^{¤}_{\lambda} \,.\, \gamma^{¤}_{\lambda}(X) \ominus^{\times} \gamma^{¤}_{\lambda}(Y) \sqsubseteq^{\times} \gamma^{¤}_{\lambda}(X \ominus^{¤} Y)$ then
>
> $$\begin{aligned} \mathscr{A}^{\times}_{\lambda}\llbracket \mathsf{A} \rrbracket \circ \dot{\gamma}_{¤} &\sqsubseteq^{\times} \gamma^{¤}_{\lambda} \circ \mathscr{A}^{¤}_{\lambda}\llbracket \mathsf{A} \rrbracket \\ \mathsf{assign}^{\times}_{\lambda}\llbracket \mathsf{x}, \mathsf{A} \rrbracket \circ \dot{\gamma}_{¤} &\dot{\sqsubseteq}^{\times} \dot{\gamma}_{¤} \circ \mathsf{assign}^{¤}_{\lambda}\llbracket \mathsf{x}, \mathsf{A} \rrbracket \\ \mathsf{assign}^{\times}_{\lambda}\llbracket \star\mathsf{p}, \mathsf{A} \rrbracket \circ \dot{\gamma}_{¤} &\dot{\sqsubseteq}^{\times} \dot{\gamma}_{¤} \circ \mathsf{assign}^{¤}_{\lambda}\llbracket \star\mathsf{p}, \mathsf{A} \rrbracket \end{aligned}$$

Proof of theorem 46.24 Let $Q \in \mathbb{P}^{\times}$, $\overline{Q} \in \mathbb{P}^{¤}_{\lambda}$ and $\overline{P} \in \mathbb{V} \to \mathbb{P}^{¤}_{\lambda}$ such that $Q \sqsubseteq^{\times} \gamma^{¤}_{\lambda}(\overline{Q})$. Let us prove that

$$\circledcirc^{\times}_{\lambda}(Q)\, (\dot{\gamma}_{¤}(\overline{P})) \;\sqsubseteq^{\times}\; \gamma^{¤}_{\lambda}(\circledcirc^{¤}_{\lambda}(\overline{Q})\, \overline{P}) \tag{46.24.a}$$

Let us first consider the case when $\gamma^{¤}_{\lambda}(\bot^{¤}) = \varnothing$.

$\circledcirc^{\times}_{\lambda}(Q)\, (\dot{\gamma}_{¤}(\overline{P}))$

$= \; (\!| \dot{\gamma}_{¤}(\overline{P}) = \dot{\bot}^{\times} \;?\; \bot^{\times} \;:\; Q |\!)$

⟨remark 46.20 and definition (46.19) of $\circledcirc^{\times}_{\lambda}(Q)$ because $\gamma^{¤}_{\lambda}(\bot^{¤}) = \varnothing$⟩

$= \; (\!| \dot{\gamma}_{¤}(\overline{P}) = \dot{\varnothing} \;?\; \varnothing \;:\; Q |\!)$ ⟨definitions $\dot{\bot}^{\times} \triangleq \dot{\varnothing}$ and $\bot^{\times} \triangleq \varnothing$ in remark 46.20⟩

$\sqsubseteq^{\times} (\!| \dot{\gamma}_{¤}(\overline{P}) = \dot{\varnothing} \;?\; \gamma^{¤}_{\lambda}(\bot^{¤}) \;:\; \gamma^{¤}_{\lambda}(\overline{Q}) |\!)$ ⟨$\varnothing \subseteq \gamma^{¤}_{\lambda}(\bot^{¤})$ because $\sqsubseteq^{\times} = \subseteq$, $Q \sqsubseteq^{\times} \gamma^{¤}_{\lambda}(\overline{Q})$⟩

$= \; \gamma^{¤}_{\lambda}((\!| \dot{\gamma}_{¤}(\overline{P}) = \dot{\varnothing} \;?\; \bot^{¤} \;:\; \overline{Q} |\!))$ ⟨factoring out of the conditional⟩

$= \; \gamma^{¤}_{\lambda}((\!| \dot{\gamma}_{¤}(\overline{P}) = \dot{\gamma}_{¤}(\dot{\bot}^{¤}) \;?\; \bot^{¤} \;:\; \overline{Q} |\!))$ ⟨because $\gamma^{¤}_{\lambda}(\bot^{¤}) = \varnothing$ implies $\dot{\gamma}_{¤}(\dot{\bot}^{¤}) = \dot{\varnothing}$ pointwise⟩

$\sqsubseteq^{\times}\ \gamma^{\sharp}_{\lambda}(\llparenthesis\,\overline{P} = \dot{\bot}^{\sharp}\ ?\ \bot^{\sharp} : \overline{Q}\,\rrparenthesis)$

$\langle$because either $\overline{P} = \dot{\bot}^{\sharp}$, which implies $\dot{\gamma}_{\sharp}(\overline{P}) = \dot{\gamma}_{\sharp}(\dot{\bot}^{\sharp})$ and $\dot{\bot}^{\sharp} \sqsubseteq^{\times} \dot{\bot}^{\sharp}$ or $\overline{P} \neq \dot{\bot}^{\sharp}$ and $\dot{\gamma}_{\sharp}(\overline{P}) = \dot{\gamma}_{\sharp}(\dot{\bot}^{\sharp})$ and $\bot^{\sharp} \sqsubseteq^{\times} \overline{Q}$ or else $\overline{P} \neq \dot{\bot}^{\sharp}$ and $\dot{\gamma}_{\sharp}(\overline{P}) \neq \dot{\gamma}_{\sharp}(\dot{\bot}^{\sharp})$ and $\overline{Q} \sqsubseteq^{\times} \overline{Q}\rangle$

$= \ \gamma^{\sharp}_{\lambda}(\oslash^{\sharp}_{\lambda}(\overline{Q})\,\overline{P})$ $\langle$definition (46.19) of $\oslash^{\sharp}_{\lambda}$ when $\gamma^{\sharp}_{\lambda}(\bot^{\sharp}) \neq \varnothing\rangle$

Otherwise $\gamma^{\sharp}_{\lambda}(\bot^{\sharp}) \neq \varnothing$.

$\oslash^{\times}_{\lambda}(Q)\,(\dot{\gamma}_{\sharp}(\overline{P}))$

$= \ Q$ $\langle$definition (46.19) of $\oslash^{\times}_{\lambda}(Q)$ because $\gamma^{\sharp}_{\lambda}(\bot^{\sharp}) \neq \varnothing\rangle$

$\sqsubseteq^{\times}\ \gamma^{\sharp}_{\lambda}(\overline{Q})$ $\langle$by hypothesis.$\rangle$

$= \ \gamma^{\sharp}_{\lambda}(\oslash^{\sharp}_{\lambda}(\overline{Q})\,\overline{P})$ $\langle$definition (46.19) of $\oslash^{\sharp}_{\lambda}$ when $\gamma^{\sharp}_{\lambda}(\bot^{\sharp}) \neq \varnothing\rangle$

Observe that $\oslash^{\times}_{\lambda}(Q)\,P$ is $\subseteq$-increasing in Q.

Let us now prove theorem 46.24. Recall that $\sqsubseteq^{\times}$ is $\subseteq$.

— $\mathscr{A}^{\times}_{\lambda}[\![\mathtt{1}]\!](\dot{\gamma}_{\sharp}(\overline{P}))$

$= \ \oslash^{\times}_{\lambda}(1^{\times})\,(\dot{\gamma}_{\sharp}(\overline{P}))$ $\langle$remark 46.20 and definition (46.19) of $\mathscr{A}^{\times}_{\lambda}[\![\mathtt{1}]\!]\ \rangle$

$\sqsubseteq^{\times}\ \gamma^{\sharp}_{\lambda}(\oslash^{\sharp}_{\lambda}(1^{\sharp})\,\overline{P})$

$\langle$(46.24.a) because $1^{\times} \triangleq \{1\}$ by remark 46.20 and $1^{\sharp} \mathrel{\dot{\sqsupseteq}}^{\sharp}_{\lambda} \alpha^{\sharp}_{\lambda}(\{1\})$ by (46.19) so $1^{\times} \sqsubseteq^{\times} \gamma^{\sharp}_{\lambda}(1^{\sharp})$ by (46.16)$\rangle$

$= \ \gamma^{\sharp}_{\lambda}(\mathscr{A}^{\sharp}[\![\mathtt{1}]\!](\overline{P}))$ $\langle$definition (46.19) of $\mathscr{A}^{\sharp}[\![\mathtt{1}]\!]\,\overline{P}\rangle$

— The proof that $\mathscr{A}^{\times}[\![\mathtt{x}]\!] \circ \dot{\gamma}_{\sharp} \sqsubseteq^{\times} \gamma^{\sharp}_{\lambda} \circ \mathscr{A}^{\sharp}[\![\mathtt{x}]\!]$ is similar.

— $\mathscr{A}^{\times}[\![\mathtt{A}_1 - \mathtt{A}_2]\!](\dot{\gamma}_{\sharp}(\overline{P}))$

$= \ \oslash^{\times}_{\lambda}(\mathscr{A}^{\times}_{\lambda}[\![\mathtt{A}_1]\!]\,(\dot{\gamma}_{\sharp}(\overline{P})) \ominus^{\times} \mathscr{A}^{\times}_{\lambda}[\![\mathtt{A}_2]\!]\,(\dot{\gamma}_{\sharp}(\overline{P})))\,(\dot{\gamma}_{\sharp}(\overline{P}))$ $\langle$definition (46.19) of $\mathscr{A}^{\times}[\![\mathtt{A}_1 - \mathtt{A}_2]\!]\rangle$

$\sqsubseteq^{\times}\ \oslash^{\times}_{\lambda}(\dot{\gamma}_{\sharp}(\mathscr{A}^{\sharp}_{\lambda}[\![\mathtt{A}_1]\!]\,\overline{P}) \ominus^{\times}\ \dot{\gamma}_{\sharp}(\mathscr{A}^{\sharp}_{\lambda}[\![\mathtt{A}_2]\!]\,\overline{P}))\,(\dot{\gamma}_{\sharp}(\overline{P}))$

$\langle$structural induction hypothesis of theorem 46.24, $\ominus^{\times}$ is $\sqsubseteq^{\times}$-increasing argumentwise, and $\oslash^{\times}_{\lambda}$ is increasing in its first parameter$\rangle$

$\sqsubseteq^{\times}\ \oslash^{\times}_{\lambda}(\dot{\gamma}_{\sharp}(\mathscr{A}^{\sharp}_{\lambda}[\![\mathtt{A}_1]\!]\,\overline{P} \ominus^{\sharp}_{\lambda} \mathscr{A}^{\sharp}_{\lambda}[\![\mathtt{A}_2]\!]\,\overline{P}))\,(\dot{\gamma}_{\sharp}(\overline{P}))$

$\langle$hypothesis $\overline{P} \ominus^{\sharp}_{\lambda} \overline{Q} \mathrel{\dot{\sqsupseteq}}^{\sharp}_{\lambda} \alpha^{\sharp}_{\lambda}(\gamma^{\sharp}_{\lambda}(\overline{P}) \ominus^{\times} \gamma^{\sharp}_{\lambda}(\overline{Q}))$ of (46.19) where $P \ominus^{\times} Q = \{x - y \mid x \in P \cap \mathbb{Z} \wedge y \in Q \cap \mathbb{Z}\}$ by remark 46.20 so $(\gamma^{\sharp}_{\lambda}(\overline{P}) \ominus^{\times} \gamma^{\sharp}_{\lambda}(\overline{Q})) \subseteq \gamma^{\sharp}_{\lambda}(\overline{P} \ominus^{\sharp}_{\lambda} \overline{Q})$ by the Galois connection (46.16). $\rangle$

$\sqsubseteq^{\times}\ \gamma^{\sharp}_{\lambda}(\oslash^{\sharp}_{\lambda}(\mathscr{A}^{\sharp}_{\lambda}[\![\mathtt{A}_1]\!]\,\overline{P} \ominus^{\sharp}_{\lambda} \mathscr{A}^{\sharp}_{\lambda}[\![\mathtt{A}_2]\!]\,\overline{P})\,\overline{P})$

$\langle$(46.24.a) with reflexivity $\dot{\gamma}_{\sharp}(\mathscr{A}^{\sharp}_{\lambda}[\![\mathtt{A}_1]\!]\,\overline{P} \ominus^{\sharp}_{\lambda} \mathscr{A}^{\sharp}_{\lambda}[\![\mathtt{A}_2]\!]\,\overline{P}) \sqsubseteq^{\times} \gamma^{\sharp}_{\lambda}(\mathscr{A}^{\sharp}_{\lambda}[\![\mathtt{A}_1]\!]\,\overline{P} \ominus^{\sharp}_{\lambda} \mathscr{A}^{\sharp}_{\lambda}[\![\mathtt{A}_2]\!]\,\overline{P})\rangle$

$= \ \gamma^{\sharp}_{\lambda}(\mathscr{A}^{\sharp}[\![\mathtt{A}_1 - \mathtt{A}_2]\!]\,\overline{P})$ $\langle$definition (46.19) of $\mathscr{A}^{\sharp}[\![\mathtt{A}_1 - \mathtt{A}_2]\!]\,\overline{P}\rangle$

— $\mathscr{A}^{\times}_{\lambda}[\![\mathtt{NULL}]\!](\dot{\gamma}_{\sharp}(\overline{P}))$

$= \ \oslash^{\times}_{\lambda}(\mathtt{NULL}^{\times})\,(\dot{\gamma}_{\sharp}(\overline{P}))$ $\langle$definition (46.19) of $\mathscr{A}^{\times}_{\lambda}[\![\mathtt{NULL}]\!]$ and $\mathtt{NULL}^{\times} = \{\mathtt{nil}\}$ by remark 46.20$\rangle$

$\sqsubseteq^{\times}\ \gamma^{\sharp}_{\lambda}(\oslash^{\sharp}_{\lambda}(\mathtt{NULL}^{\sharp}_{\lambda})\,\overline{P})$

$$\langle\text{by (46.24.a) because } \mathsf{NULL}^{\text{¤}}_{\lambda} \mathrel{\dot{\sqsupseteq}}^{\text{¤}}_{\lambda} \alpha^{\text{¤}}_{\lambda}(\{\mathtt{nil}\}) = \alpha^{\text{¤}}_{\lambda}(\mathsf{NULL}^{\times}) \text{ so by } \mathsf{NULL}^{\times} \sqsubseteq^{\times} \gamma^{\text{¤}}_{\lambda}(\mathsf{NULL}^{\text{¤}}_{\lambda}) \text{ by the Galois connection (46.16)}\rangle$$

$$= \gamma^{\text{¤}}_{\lambda}(\mathscr{A}^{\text{¤}}_{\lambda}[\![\mathsf{NULL}]\!]\overline{P}) \qquad \langle\text{definition (46.19) of } \mathscr{A}^{\text{¤}}_{\lambda}[\![\mathsf{NULL}]\!]\rangle$$

— $\mathscr{A}^{\times}_{\lambda}[\![\&\mathsf{x}]\!](\dot{\gamma}_{\text{¤}}(\overline{P}))$

$$= \oslash^{\times}_{\lambda}(\boldsymbol{\lambda}^{\times}_{\lambda}[\![\mathsf{x}]\!])\,(\dot{\gamma}_{\text{¤}}(\overline{P})) \qquad \langle\text{definition (46.19) of } \mathscr{A}^{\times}_{\lambda}[\![\&\mathsf{x}]\!] \text{ and } \boldsymbol{\lambda}^{\times}_{\lambda}[\![\mathsf{x}]\!] = \{\lambda_{\mathsf{x}}\} \text{ by remark 46.20}\rangle$$

$$\sqsubseteq^{\times} \gamma^{\text{¤}}_{\lambda}(\oslash^{\text{¤}}_{\lambda}(\boldsymbol{\lambda}^{\text{¤}}_{\lambda}[\![\mathsf{x}]\!])\,\overline{P})$$

$$\langle\text{by (46.24.a) because } \boldsymbol{\lambda}^{\text{¤}}_{\lambda}[\![\mathsf{x}]\!] \mathrel{\dot{\sqsupseteq}}^{\text{¤}}_{\lambda} \alpha^{\text{¤}}_{\lambda}(\{\lambda_{\mathsf{x}}\}) = \alpha^{\text{¤}}_{\lambda}(\boldsymbol{\lambda}^{\times}_{\lambda}[\![\mathsf{x}]\!]) \text{ so } \boldsymbol{\lambda}^{\times}_{\lambda}[\![\mathsf{x}]\!] \sqsubseteq^{\times} \gamma^{\text{¤}}_{\lambda}(\boldsymbol{\lambda}^{\text{¤}}_{\lambda}[\![\mathsf{x}]\!]) \text{ by the Galois connection (46.16)}\rangle$$

$$= \gamma^{\text{¤}}_{\lambda}(\mathscr{A}^{\text{¤}}_{\lambda}[\![\&\mathsf{x}]\!]\overline{P}) \qquad \langle\text{definition (46.19) of } \mathscr{A}^{\text{¤}}_{\lambda}[\![\&\mathsf{x}]\!]\rangle$$

— $\mathscr{A}^{\times}_{\lambda}[\![\star\mathsf{p}]\!](\dot{\gamma}_{\text{¤}}(\overline{P}))$

$$= \oslash^{\times}_{\lambda}(\star^{\times}_{\lambda}[\![\mathsf{p}]\!])\,(\dot{\gamma}_{\text{¤}}(\overline{P}))$$

$$\langle\text{definition (46.19) of } \mathscr{A}^{\times}_{\lambda}[\![\star\mathsf{p}]\!] \text{ and } \star^{\times}_{\lambda}[\![\mathsf{p}]\!]\,P = \{\mu(\mu(\lambda_{\mathsf{p}})) \mid \mu \in P \wedge \mu(\lambda_{\mathsf{p}}) \in \mathbb{L}\} \text{ by remark 46.20}\rangle$$

$$\sqsubseteq^{\times} \gamma^{\text{¤}}_{\lambda}(\oslash^{\text{¤}}_{\lambda}(\star^{\text{¤}}_{\lambda}[\![\mathsf{p}]\!])\,\overline{P})$$

$$\langle\text{by (46.24.a) because } \star^{\text{¤}}_{\lambda}[\![\mathsf{p}]\!]\,\overline{P} \mathrel{\dot{\sqsupseteq}}^{\text{¤}}_{\lambda} \alpha^{\text{¤}}_{\lambda}(\{\mu(\mu(\lambda_{\mathsf{p}})) \mid \mu \in \dot{\gamma}^{\times\text{¤}}_{\lambda}(\overline{P}) \wedge \mu(\lambda_{\mathsf{p}}) \in \mathbb{L}\}) = \alpha^{\text{¤}}_{\lambda}(\star^{\times}_{\lambda}[\![\mathsf{p}]\!]\,(\dot{\gamma}^{\times\text{¤}}_{\lambda}(\overline{P}))) \text{ so } \star^{\times}_{\lambda}[\![\mathsf{p}]\!]\,(\dot{\gamma}^{\times\text{¤}}_{\lambda}(\overline{P})) \sqsubseteq^{\times} \gamma^{\text{¤}}_{\lambda}(\star^{\text{¤}}_{\lambda}[\![\mathsf{p}]\!]\,\overline{P})) \text{ by the Galois connection (46.16)}\rangle$$

$$= \gamma^{\text{¤}}_{\lambda}(\mathscr{A}^{\text{¤}}_{\lambda}[\![\star\mathsf{p}]\!]\overline{P}) \qquad \langle\text{definition (46.19) of } \mathscr{A}^{\text{¤}}_{\lambda}[\![\star\mathsf{p}]\!]\rangle$$

— $\mathsf{assign}^{\times}_{\lambda}[\![\mathsf{x}, \mathsf{A}]\!](\dot{\gamma}_{\text{¤}}(\overline{P}))$

$$= \dot{\gamma}_{\text{¤}}(\overline{P})[\lambda_{\mathsf{x}} \leftarrow \mathscr{A}^{\text{¤}}_{\lambda}[\![\mathsf{A}]\!]\overline{P}] \qquad \langle\text{definition (46.22) of } \mathsf{assign}^{\text{¤}}_{\lambda}[\![\mathsf{x}, \mathsf{A}]\!]\,\overline{P}\rangle$$

$$\dot{\sqsubseteq}^{\times} \dot{\gamma}_{\text{¤}}(\overline{P})[\lambda_{\mathsf{x}} \leftarrow \gamma^{\text{¤}}_{\lambda}(\mathscr{A}^{\text{¤}}[\![\mathsf{A}]\!](\overline{P}))]$$

$$\langle \mathscr{A}^{\times}[\![\mathsf{A}]\!] \circ \dot{\gamma}_{\text{¤}} \sqsubseteq^{\times} \gamma^{\text{¤}}_{\lambda} \circ \mathscr{A}^{\text{¤}}[\![\mathsf{A}]\!] \text{ and pointwise definition of } \dot{\sqsubseteq}^{\times}\rangle$$

$$= \dot{\gamma}_{\text{¤}}(\overline{P}[\lambda_{\mathsf{x}} \leftarrow \mathscr{A}^{\text{¤}}[\![\mathsf{A}]\!](\overline{P})]) \qquad \langle\text{definition (19.10) of environment assignment}\rangle$$

$$= \dot{\gamma}_{\text{¤}}(\mathsf{assign}^{\text{¤}}_{\lambda}[\![\mathsf{x}, \mathsf{A}]\!]\overline{P}) \qquad \langle\text{(46.22) of } \mathsf{assign}^{\text{¤}}_{\lambda}[\![\mathsf{x}, \mathsf{A}]\!]\overline{P}\rangle$$

— $\mathsf{assign}^{\times}_{\lambda}[\![\star\mathsf{p}, \mathsf{A}]\!](\dot{\gamma}_{\text{¤}}(\overline{P}))$

$$= l \mapsto \dot{\gamma}_{\text{¤}}(\overline{P})(l) \cup \llparenthesis\, l \in \dot{\gamma}_{\text{¤}}(\overline{P})(\lambda_{\mathsf{p}}) \mathrel{?} \mathscr{A}^{\text{¤}}_{\lambda}[\![\mathsf{A}]\!]\,\dot{\gamma}_{\text{¤}}(\overline{P}) \mathrel{\text{⁝}} \varnothing \,\rrparenthesis$$

$$\langle\text{definition (46.22) of } \mathsf{assign}^{\text{¤}}_{\lambda}[\![\star\mathsf{p}, \mathsf{A}]\!]\,\dot{\gamma}_{\text{¤}}(\overline{P})\rangle$$

$$= l \mapsto \gamma^{\text{¤}}_{\lambda}(\overline{P}(l)) \cup \llparenthesis\, l \in \gamma^{\text{¤}}_{\lambda}(\overline{P}(\lambda_{\mathsf{p}})) \mathrel{?} \mathscr{A}^{\text{¤}}_{\lambda}[\![\mathsf{A}]\!]\,\dot{\gamma}_{\text{¤}}(\overline{P}) \mathrel{\text{⁝}} \varnothing \,\rrparenthesis \qquad \langle\text{pointwise definition of } \dot{\gamma}_{\text{¤}}\rangle$$

$$\dot{\sqsubseteq}^{\times} l \mapsto \gamma^{\text{¤}}_{\lambda}(\overline{P}(l)) \cup \llparenthesis\, l \in \gamma^{\text{¤}}_{\lambda}(\overline{P}(\lambda_{\mathsf{p}})) \mathrel{?} \gamma^{\text{¤}}_{\lambda}(\mathscr{A}^{\text{¤}}[\![\mathsf{A}]\!]\,\overline{P}) \mathrel{\text{⁝}} \gamma^{\text{¤}}_{\lambda}(\bot^{\text{¤}}) \,\rrparenthesis$$

$$\langle \mathscr{A}^{\times}[\![\mathsf{A}]\!] \circ \dot{\gamma}_{\text{¤}} \sqsubseteq^{\times} \gamma^{\text{¤}}_{\lambda} \circ \mathscr{A}^{\text{¤}}[\![\mathsf{A}]\!] \text{ and pointwise definition of } \dot{\sqsubseteq}^{\times}\rangle$$

$$= l \mapsto \gamma^{\text{¤}}_{\lambda}(\overline{P}(l)) \cup \gamma^{\text{¤}}_{\lambda}(\llparenthesis\, l \in \gamma^{\text{¤}}_{\lambda}(\overline{P}(\lambda_{\mathsf{p}})) \mathrel{?} \mathscr{A}^{\text{¤}}[\![\mathsf{A}]\!]\,\overline{P} \mathrel{\text{⁝}} \bot^{\text{¤}} \,\rrparenthesis)$$

$$\langle\text{factorizing } \gamma^{\text{¤}}_{\lambda} \text{ over the conditional}\rangle$$

$$\dot{\sqsubseteq}^{\times} l \mapsto \gamma^{\text{¤}}_{\lambda}(\overline{P}(l) \sqcup^{\text{¤}} \llparenthesis\, l \in \gamma^{\text{¤}}_{\lambda}(\overline{P}(\lambda_{\mathsf{p}})) \mathrel{?} \mathscr{A}^{\text{¤}}[\![\mathsf{A}]\!]\,\overline{P} \mathrel{\text{⁝}} \bot^{\text{¤}} \,\rrparenthesis) \qquad \langle \gamma^{\text{¤}}_{\lambda} \text{ increasing}\rangle$$

$$= l \mapsto \gamma^{\text{¤}}_{\lambda}(\overline{P}(l) \sqcup^{\text{¤}} \llparenthesis\, \alpha^{\text{¤}}_{\lambda}(\{l\}) \sqsubseteq^{\text{¤}} \overline{P}(\lambda_{\mathsf{p}}) \mathrel{?} \mathscr{A}^{\text{¤}}[\![\mathsf{A}]\!]\,\overline{P} \mathrel{\text{⁝}} \bot^{\text{¤}} \,\rrparenthesis) \qquad \langle\text{Galois connection}\rangle$$

- if $\alpha^{\text{¤}}_{\lambda}(\{\lambda_{\mathsf{x}}\}) = \boldsymbol{\lambda}^{\text{¤}}_{\lambda}[\![\mathsf{x}]\!]$ then

$= \;\dot{\gamma}_{¤}(l \mapsto \overline{P}(l) \sqcup^{¤} (\!| \boldsymbol{\lambda}^{¤}_{\lambda}[\![\mathsf{x}]\!] \sqsubseteq^{¤} \overline{P}(\lambda_{\mathsf{p}}) \,?\, \mathscr{A}^{¤}[\![\mathsf{A}]\!]\,\overline{P} \,\text{:}\, \bot^{¤} |\!))$ ⟨pointwise definition of $\dot{\gamma}_{¤}$⟩

$= \;\dot{\gamma}_{¤}(\mathsf{assign}^{¤}[\![\star\mathsf{p}, \mathsf{A}]\!]\overline{P})$ ⟨definition (46.22) of $\mathsf{assign}^{¤}[\![\star\mathsf{p}, \mathsf{A}]\!]\overline{P}$⟩

- Otherwise $\alpha^{¤}_{\lambda}(\{\lambda_{\mathsf{x}}\}) \dot{\sqsubset}^{¤} \boldsymbol{\lambda}^{¤}_{\lambda}[\![\mathsf{x}]\!]$ by (46.19), so we have to overapproximate

$\dot{\sqsubseteq}^{\times} \; l \mapsto \gamma^{¤}_{\lambda}(\overline{P}(l) \sqcup^{¤} \mathscr{A}^{¤}[\![\mathsf{A}]\!]\,\overline{P})$ ⟨$\gamma^{¤}_{\lambda}$ increasing⟩

$= \;\dot{\gamma}_{¤}(l \mapsto \overline{P}(l) \sqcup^{¤} \mathscr{A}^{¤}[\![\mathsf{A}]\!]\,\overline{P})$ ⟨pointwise definition of $\dot{\gamma}_{¤}$⟩

$= \;\dot{\gamma}_{¤}(\mathsf{assign}^{¤}[\![\star\mathsf{p}, \mathsf{A}]\!]\overline{P})$ ⟨definition (46.22) of $\mathsf{assign}^{¤}[\![\star\mathsf{p}, \mathsf{A}]\!]\overline{P}$⟩ □

46.6.4 Cartesian Backward Accessibility of an Arithmetic or Pointer Expressions

The Cartesian backward accessibility abstraction of expressions is $\mathscr{A}^{\overset{¤}{-1}}_{\lambda}[\![\mathsf{A}]\!] \in \mathbb{P}^{¤}_{\lambda} \to (\mathbb{L}_{\lambda} \to \mathbb{P}^{¤}_{\lambda}) \to (\mathbb{L}_{\lambda} \to \mathbb{P}^{¤}_{\lambda})$,

$$\mathscr{A}^{\overset{¤}{-1}}_{\lambda}[\![\mathsf{A}]\!]\,\chi\,P \;\triangleq\; \alpha^{¤}_{\lambda}(\{\mu \in \dot{\gamma}^{\times¤}_{\lambda}(\overline{P}) \mid \mathscr{A}_{\lambda}[\![\mathsf{A}]\!]\,\mu \in \dot{\gamma}^{\times¤}_{\lambda}(\chi)\}).$$

so that we extend (28.30)

Cartesian backward accessibility semantics of an arithmetic or pointer expression A

$$\mathscr{A}^{\overset{¤}{-1}}_{\lambda}[\![\mathsf{A}]\!] \;\in\; \mathbb{P}^{¤}_{\lambda} \to (\mathbb{L}_{\lambda} \to \mathbb{P}^{¤}_{\lambda}) \to (\mathbb{L}_{\lambda} \to \mathbb{P}^{¤}_{\lambda}) \tag{46.25}$$

$$\mathscr{A}^{\overset{¤}{-1}}_{\lambda}[\![1]\!]\,\chi\,\overline{P} \;\triangleq\; (\!| 1^{¤} \sqcap^{¤} \chi \neq \bot^{¤} \,?\, \overline{P} \,\text{:}\, \dot{\bot}^{¤} |\!)$$

$$\mathscr{A}^{\overset{¤}{-1}}_{\lambda}[\![\lambda_{\mathsf{x}}]\!]\,\chi\,\overline{P} \;\triangleq\; (\!| \overline{P}(\lambda_{\mathsf{x}}) \sqcap^{¤} \chi \neq \bot^{¤} \,?\, \overline{P}[\lambda_{\mathsf{x}} \leftarrow \overline{P}(\lambda_{\mathsf{x}}) \sqcap^{¤} \chi] \,\text{:}\, \dot{\bot}^{¤} |\!)$$

$$\mathscr{A}^{\overset{¤}{-1}}_{\lambda}[\![\mathsf{A}_1 - \mathsf{A}_2]\!]\,\chi\,\overline{P} \;\triangleq\; \mathsf{let}\ \langle \chi_1, \chi_2 \rangle = \ominus^{\overset{¤}{-1}}_{\lambda}\,\chi\,\langle \mathscr{A}^{¤}_{\lambda}[\![\mathsf{A}_1]\!]\,\overline{P},\, \mathscr{A}^{¤}_{\lambda}[\![\mathsf{A}_2]\!]\,\overline{P} \rangle\ \mathsf{in}$$
$$\mathscr{A}^{\overset{¤}{-1}}_{\lambda}[\![\mathsf{A}_1]\!]\,\chi_1\,\overline{P} \,\dot{\sqcap}^{¤}\, \mathscr{A}^{\overset{¤}{-1}}_{\lambda}[\![\mathsf{A}_2]\!]\,\chi_2\,\overline{P}$$

where $\ominus^{\overset{¤}{-1}}_{\lambda}\,P\,\langle Q_1, Q_2 \rangle \dot{\sqsupseteq}^{¤}_{\lambda} \langle \alpha^{¤}_{\lambda}(\{v_1 \in \gamma^{¤}_{\lambda}(Q_1) \mid \exists v_2 \in \gamma^{¤}_{\lambda}(Q_2) \,.\, (v_1 - v_2) \in \gamma^{¤}_{\lambda}(P)\}),$
$\alpha^{¤}_{\lambda}(\{v_2 \in \gamma^{¤}_{\lambda}(Q_2) \mid \exists v_1 \in \gamma^{¤}_{\lambda}(Q_1) \,.\, (v_1 - v_2) \in \gamma^{¤}_{\lambda}(P)\}) \rangle$

$$\mathscr{A}^{\overset{¤}{-1}}_{\lambda}[\![\mathsf{NULL}]\!]\,\chi\,\overline{P} \;\triangleq\; (\!| \mathsf{NULL}^{¤}_{\lambda} \sqcap^{¤} \chi \neq \bot^{¤} \,?\, \overline{P} \,\text{:}\, \dot{\bot}^{¤} |\!)$$

$$\mathscr{A}^{\overset{¤}{-1}}_{\lambda}[\![\&\mathsf{x}]\!]\,\chi\,\overline{P} \;\triangleq\; (\!| \boldsymbol{\lambda}^{¤}_{\lambda}[\![\lambda_{\mathsf{x}}]\!] \sqcap^{¤} \chi \neq \bot^{¤} \,?\, \overline{P}[\lambda_{\mathsf{x}} \leftarrow \boldsymbol{\lambda}^{¤}_{\lambda}[\![\mathsf{x}]\!] \sqcap^{¤} \chi] \,\text{:}\, \dot{\bot}^{¤} |\!)$$

$$\mathscr{A}^{\overset{¤}{-1}}_{\lambda}[\![\star\mathsf{p}]\!]\,\chi\,\overline{P} \;\triangleq\; \star^{\overset{¤}{-1}}[\![\mathsf{p}]\!]\,\lambda\,\chi\,\overline{P}$$

where $\star^{\overset{¤}{-1}}[\![\mathsf{p}]\!]\,\lambda\,\chi\,\overline{P} \dot{\sqsupseteq}^{¤}_{\lambda} \lambda_{\mathsf{z}} \mapsto \alpha^{¤}_{\lambda}(\{\mu(\lambda_{\mathsf{z}}) \in \gamma^{¤}_{\lambda}(\overline{P}(\lambda_{\mathsf{z}})) \mid \mu(\mu(\lambda_{\mathsf{p}})) \in \chi \cap \gamma^{¤}_{\lambda}(\overline{P}(\mu(\lambda_{\mathsf{p}})))\})$

Remark 46.26 By choosing $\alpha^{¤}_{\lambda}$ to be the identity, we get the definitions of $\dot{\sqsubseteq}^{\times} \triangleq \dot{\subseteq}$, $\ominus^{\overset{¤}{-1}}\,P\,\langle Q_1, Q_2 \rangle \triangleq \langle \{v_1 \in Q_1 \mid \exists v_2 \in Q_2 \,.\, (v_1 - v_2) \in P\},\, \{v_2 \in Q_2 \mid \exists v_1 \in Q_1 \,.\, (v_1 - v_2) \in P\} \rangle$ and $\mathscr{A}^{\overset{\times}{-1}}_{\lambda}[\![\mathsf{A}]\!]$ as instances of $\ominus^{\overset{¤}{-1}}_{\lambda}$ and $\mathscr{A}^{\overset{¤}{-1}}_{\lambda}[\![\mathsf{A}]\!]$. in (46.25).

Then $\mathscr{A}^{\overset{\times}{-1}}_{\lambda}[\![\mathsf{A}]\!]\,\chi$ is a lower closure operator on $\mathcal{V} \to \wp(\mathbb{V}_{\lambda})$ and $\mathscr{A}^{\overset{\times}{-1}}_{\lambda}[\![\mathsf{A}]\!]\,\chi$ is $\dot{\subseteq}$-increasing in χ.
Theorem 28.29 becomes

Theorem 46.27 $\dot{\alpha}^{\times}_{\lambda} \circ \mathscr{A}^{\overset{\times}{-1}}_{\lambda}[\![\mathsf{A}]\!]\,\chi \,\dot{\subseteq}\, \mathscr{A}^{\overset{\times}{-1}}_{\lambda}[\![\mathsf{A}]\!]\,\chi \circ \dot{\alpha}^{\times}_{\lambda}$.

Proof of theorem 46.27 We need to extend definition (28.22) and theorem 28.24 as follows

$$\begin{array}{rcl}\mathscr{A}^{-1}[\![\mathtt{A}]\!]\,\lambda\,\chi\,P & \triangleq & \{\mu \in P \mid \mathscr{A}_\lambda[\![\mathtt{A}]\!]\,\mu \in \chi\} \\ \mathscr{A}^{-1}[\![\mathtt{x}]\!]\,\lambda\,\chi\,P & = & \{\mu \in P \mid \mu(\lambda_\mathtt{x}) \in \chi\} \\ \mathscr{A}^{-1}[\![\mathtt{NULL}]\!]\,\lambda\,\chi\,P & = & (\!|\, \mathtt{nil} \in \chi \mathbin{?} P \mathbin{:} \varnothing \,|\!) \\ \mathscr{A}^{-1}[\![\mathtt{\&x}]\!]\,\lambda\,\chi\,P & = & (\!|\, \lambda_\mathtt{x} \in \chi \mathbin{?} P \mathbin{:} \varnothing \,|\!) \\ \mathscr{A}^{-1}[\![\mathtt{*p}]\!]\,\lambda\,\chi\,P & = & \{\mu \in P \mid \mu(\lambda_\mathtt{p}) \in \mathbb{L} \wedge \mu(\mu(\lambda_\mathtt{p})) \in \chi\}\end{array} \tag{46.28}$$

Let us prove (46.28).

— $\mathscr{A}^{-1}[\![\mathtt{x}]\!]\,\lambda\,\chi\,P$

$= \{\mu \in P \mid \mathscr{A}_\lambda[\![\mathtt{x}]\!]\,\mu \in \chi\}$ ⟅(46.28)⟆

$= \{\mu \in P \mid \mu(\lambda_\mathtt{x}) \in \chi\}$ ⟅(46.2)⟆

— $\mathscr{A}^{-1}[\![\mathtt{NULL}]\!]\,\lambda\,\chi\,P$

$= \{\mu \in P \mid \mathscr{A}_\lambda[\![\mathtt{NULL}]\!]\,\mu \in \chi\}$ ⟅(46.28)⟆

$= \{\mu \in P \mid \mathtt{nil} \in \chi\}$ ⟅(46.2)⟆

$= (\!|\, \mathtt{nil} \in \chi \mathbin{?} P \mathbin{:} \varnothing \,|\!)$ ⟅definition of the conditional⟆

— $\mathscr{A}^{-1}[\![\mathtt{\&x}]\!]\,\lambda\,\chi\,P$

$= \{\mu \in P \mid \mathscr{A}_\lambda[\![\mathtt{\&x}]\!]\,\mu \in \chi\}$ ⟅(46.28)⟆

$= \{\mu \in P \mid \lambda_\mathtt{x} \in \chi\}$ ⟅(46.2)⟆

$= (\!|\, \lambda_\mathtt{x} \in \chi \mathbin{?} P \mathbin{:} \varnothing \,|\!)$ ⟅definition of the conditional⟆

— $\mathscr{A}^{-1}[\![\mathtt{*p}]\!]\,\lambda\,\chi\,P$

$= \{\mu \in P \mid \mathscr{A}_\lambda[\![\mathtt{*p}]\!]\,\mu \in \chi\}$ ⟅(46.28)⟆

$= \{\mu \in P \mid (\!|\, \mu(\lambda_\mathtt{p}) \in \mathbb{L} \mathbin{?} \mu(\mu(\lambda_\mathtt{p})) \mathbin{:} \Omega \,|\!) \in \chi\}$ ⟅(46.2) and section 46.2⟆

$= \{\mu \in P \mid \mu(\lambda_\mathtt{p}) \in \mathbb{L} \wedge \mu(\mu(\lambda_\mathtt{p})) \in \chi\}$ ⟅because $\Omega \notin \mathbb{L}$⟆

The proof of theorem 46.27 is by structural induction on $\mathtt{A}$ for $\mathscr{P}_0 \in \wp(\mathbb{V} \to \mathbb{V}_\lambda)$. The proof for integer expressions $\mathtt{1}$ and $\mathtt{A}_1 - \mathtt{A}_2$ is very similar to that of theorem 28.29. Let us consider pointer expressions.

— $\dot{\alpha}^{\times}_{\lambda} \circ \mathscr{A}^{\overleftarrow{\times}_1}_{\lambda}[\![\mathtt{x}]\!]\,\chi\,(\mathscr{P}_0)$

$= \dot{\alpha}^{\times}_{\lambda}(\{\mu \in \mathscr{P}_0 \mid \mu(\lambda_\mathtt{x}) \in \chi\})$ ⟅(46.28)⟆

$= \lambda_\mathtt{z} \in \mathbb{L} \mapsto \{\mu(\lambda_\mathtt{z}) \mid \mu \in \{\mu \in \mathscr{P}_0 \mid \mu(\lambda_\mathtt{x}) \in \chi\}\}$

⟅definition (46.17) of the Cartesian abstraction $\dot{\alpha}^{\times}_{\lambda}$⟆

$= \lambda_\mathtt{z} \in \mathbb{L} \mapsto \{\mu(\lambda_\mathtt{z}) \mid \mu \in \mathscr{P}_0 \wedge \mu(\lambda_\mathtt{x}) \in \chi\}$ ⟅definition of $\in$⟆

$\dot{\subseteq}\ \lambda_\mathtt{z} \in \mathbb{L} \mapsto \{\mu(\lambda_\mathtt{z}) \in \dot{\alpha}^{\times}_{\lambda}(\mathscr{P}_0)(\lambda_\mathtt{z}) \mid \mu(\lambda_\mathtt{x}) \in \chi \cap \dot{\alpha}^{\times}_{\lambda}(\mathscr{P}_0)(\lambda_\mathtt{x})\}$

⟅definition $\dot{\alpha}^{\times}_{\lambda}(P) \triangleq l \in \mathbb{L}_\lambda \mapsto \{\mu(l) \mid \mu \in P\}$ of the Cartesian abstraction (46.17) so that $\mu \in \mathscr{P}_0$ implies $\mu(\lambda) \in \dot{\alpha}^{\times}_{\lambda}(\mathscr{P}_0)(\lambda)$⟆

$= (\!| \chi \cap \dot{\alpha}^{\times}_{\lambda}(\mathscr{P}_0)(\lambda_x) \neq \emptyset \mathbin{?} \lambda_z \in \mathbb{L} \mapsto (\!| \lambda_z = \lambda_x \mathbin{?} \{\mu(\lambda_x) \in \dot{\alpha}^{\times}_{\lambda}(\mathscr{P}_0)(\lambda_x) \mid \mu(\lambda_x) \in \chi \cap \dot{\alpha}^{\times}_{\lambda}(\mathscr{P}_0)(\lambda_x)\} \mathbin{:} \{\mu(\lambda_z) \in \dot{\alpha}^{\times}_{\lambda}(\mathscr{P}_0)(\lambda_z) \mid \mu(\lambda_x) \in \chi \cap \dot{\alpha}^{\times}_{\lambda}(\mathscr{P}_0)(\lambda_x)\} |\!) \mathbin{:} \lambda_z \in \mathbb{L} \mapsto \emptyset |\!)$

$\wr$definition of the conditional$\wr$

$\dot{\subseteq} (\!| \chi \cap \dot{\alpha}^{\times}_{\lambda}(\mathscr{P}_0)(\lambda_x) \neq \emptyset \mathbin{?} \lambda_z \in \mathbb{L} \mapsto (\!| \lambda_z = \lambda_x \mathbin{?} \chi \cap \dot{\alpha}^{\times}_{\lambda}(\mathscr{P}_0)(\lambda_x) \mathbin{:} \dot{\alpha}^{\times}_{\lambda}(\mathscr{P}_0)(\lambda_z) |\!) \mathbin{:} \lambda_z \in \mathbb{L} \mapsto \emptyset |\!)$ $\wr\{\mu(\lambda_z) \in \dot{\alpha}^{\times}_{\lambda}(\mathscr{P}_0)(\lambda_z) \mid \mu(\lambda_x) \in \chi \cap \dot{\alpha}^{\times}_{\lambda}(\mathscr{P}_0)(\lambda_x)\} \subseteq \dot{\alpha}^{\times}_{\lambda}(\mathscr{P}_0)(\lambda_z)\wr$

$= (\!| \dot{\alpha}^{\times}_{\lambda}(\mathscr{P}_0)(\lambda_x) \cap \chi \neq \emptyset \mathbin{?} \dot{\alpha}^{\times}_{\lambda}(\mathscr{P}_0)[\lambda_x \leftarrow \dot{\alpha}^{\times}_{\lambda}(\mathscr{P}_0)(\lambda_x) \cap \chi] \mathbin{:} \lambda_z \in \mathbb{L} \mapsto \emptyset |\!)$

$\wr$definition (19.10) of environment assignment$\wr$

$= \mathscr{A}^{\dot{\times}_1}_{\lambda}[\![\lambda_x]\!]\, \chi\, \dot{\alpha}^{\times}_{\lambda}(\mathscr{P}_0)$ $\wr$definition (46.25) and remark 46.26 $\wr$

— $\dot{\alpha}^{\times}_{\lambda} \circ \mathscr{A}^{\dot{\times}_1}_{\lambda}[\![\texttt{NULL}]\!]\, \chi\, (\mathscr{P}_0)$

$= \dot{\alpha}^{\times}_{\lambda}((\!| \texttt{nil} \in \chi \mathbin{?} \mathscr{P}_0 \mathbin{:} \emptyset |\!))$ $\wr(46.28)\wr$

$= (\!| \texttt{nil} \in \chi \mathbin{?} \dot{\alpha}^{\times}_{\lambda}(\mathscr{P}_0) \mathbin{:} \dot{\alpha}^{\times}_{\lambda}(\emptyset) |\!)$ $\wr$distributing $\dot{\alpha}^{\times}_{\lambda}$ over the conditional$\wr$

$= (\!| \{\texttt{nil}\} \cap \chi \neq \emptyset \mathbin{?} \dot{\alpha}^{\times}_{\lambda}(\mathscr{P}_0) \mathbin{:} \lambda \in \mathbb{L} \mapsto \emptyset) |\!)$ $\wr$definition of $\in$ and $\dot{\alpha}^{\times}_{\lambda}(P) \triangleq l \in \mathbb{L}_{\lambda} \mapsto \{\mu(l) \mid \mu \in P\}\wr$

$= (\!| \texttt{NULL}^{\times} \sqcap^{\times} \chi \neq \bot^{\times} \mathbin{?} \dot{\alpha}^{\times}_{\lambda}(\mathscr{P}_0) \mathbin{:} \dot{\bot}^{\times} |\!)$ $\wr$remark 46.20$\wr$

$= \mathscr{A}^{\dot{\times}_1}_{\lambda}[\![\texttt{NULL}]\!]\, \chi\, \dot{\alpha}^{\times}_{\lambda}(\mathscr{P}_0)$ $\wr$definition (46.25) of $\mathscr{A}^{\dot{\times}_1}_{\lambda}[\![\texttt{NULL}]\!]\, \chi\wr$

— $\dot{\alpha}^{\times}_{\lambda} \circ \mathscr{A}^{\dot{\times}_1}_{\lambda}[\![\texttt{\&x}]\!]\, \chi\, (\mathscr{P}_0)$

$= \dot{\alpha}^{\times}_{\lambda}((\!| \lambda_x \in \chi \mathbin{?} \mathscr{P}_0 \mathbin{:} \emptyset |\!))$ $\wr(46.28)\wr$

$= (\!| \lambda_x \in \chi \mathbin{?} \dot{\alpha}^{\times}_{\lambda}(\mathscr{P}_0) \mathbin{:} \dot{\alpha}^{\times}_{\lambda}(\emptyset) |\!)$ $\wr$distributing $\dot{\alpha}^{\times}_{\lambda}$ over the conditional$\wr$

$= (\!| \{\lambda_x\} \cap \chi \neq \emptyset \mathbin{?} \dot{\alpha}^{\times}_{\lambda}(\mathscr{P}_0) \mathbin{:} \lambda_z \in \mathbb{L} \mapsto \emptyset) |\!)$ $\wr$definition of $\in$ and $\dot{\alpha}^{\times}_{\lambda}(P) \triangleq l \in \mathbb{L}_{\lambda} \mapsto \{\mu(l) \mid \mu \in P\}\wr$

$= (\!| \boldsymbol{\lambda}^{\times}_{\lambda_z}[\![x]\!] \sqcap^{\times} \chi \neq \bot^{\times} \mathbin{?} \dot{\alpha}^{\times}_{\lambda}(\mathscr{P}_0) \mathbin{:} \dot{\bot}^{\times} |\!)$ $\wr\boldsymbol{\lambda}^{\times}_{\lambda_z}[\![x]\!] \triangleq \{\lambda_x\}$ by remark 46.20$\wr$

$= \mathscr{A}^{\dot{\times}_1}_{\lambda}[\![\texttt{\&x}]\!]\, \chi\, \dot{\alpha}^{\times}_{\lambda}(\mathscr{P}_0)$ $\wr$definition (46.25) of $\mathscr{A}^{\dot{\times}_1}_{\lambda}[\![\texttt{\&x}]\!]\, \chi\wr$

— $\dot{\alpha}^{\times}_{\lambda} \circ \mathscr{A}^{\dot{\times}_1}_{\lambda}[\![\texttt{*p}]\!]\, \chi\, (\mathscr{P}_0)$

$= \dot{\alpha}^{\times}_{\lambda}(\{\mu \in \mathscr{P}_0 \mid \mu(\lambda_p) \in \mathbb{L} \wedge \mu(\mu(\lambda_p)) \in \chi\})$ $\wr(46.28)\wr$

$= \lambda_z \mapsto \{\mu(\lambda_z) \mid \mu \in \{\mu \in \mathscr{P}_0 \mid \mu(\lambda_p) \in \mathbb{L} \wedge \mu(\mu(\lambda_p)) \in \chi\}\}$

$\wr$definition $\dot{\alpha}^{\times}_{\lambda}(P) \triangleq l \in \mathbb{L}_{\lambda} \mapsto \{\mu(l) \mid \mu \in P\}\wr$

$= \lambda_z \mapsto \{\mu(\lambda_z) \mid \mu \in \mathscr{P}_0 \wedge \mu(\lambda_p) \in \mathbb{L} \wedge \mu(\mu(\lambda_p)) \in \chi\}$ $\wr$definition of $\in\wr$

$= \lambda_z \mapsto \{\mu(\lambda_z) \in \dot{\alpha}^{\times}_{\lambda}(\mathscr{P}_0)\, \lambda_z \mid \mu \in \mathscr{P}_0 \wedge \mu(\lambda_p) \in \mathbb{L} \wedge \mu(\mu(\lambda_p)) \in \chi \cap \dot{\alpha}^{\times}_{\lambda}(\mathscr{P}_0)\, \mu(\lambda_p)\}$

$\wr$definition $\dot{\alpha}^{\times}_{\lambda}(P) \triangleq l \in \mathbb{L}_{\lambda} \mapsto \{\mu(l) \mid \mu \in P\}$ so $\mu \in \mathscr{P}_0$ implies $\mu(\lambda_z) \in \dot{\alpha}^{\times}_{\lambda}(\mathscr{P}_0)\, \lambda_z$ and $\mu(\mu(\lambda_p)) \in \dot{\alpha}^{\times}_{\lambda}(\mathscr{P}_0)\, \mu(\lambda_p)\wr$

$\dot{\subseteq} \lambda_z \mapsto \{\mu(\lambda_z) \in \dot{\alpha}^{\times}_{\lambda}(\mathscr{P}_0)\, \lambda_z \mid \mu(\mu(\lambda_p)) \in \chi \cap \dot{\alpha}^{\times}_{\lambda}(\mathscr{P}_0)\, \mu(\lambda_p)\}$ $\wr$definition of $\subseteq\wr$

$= \star^{\dot{\times}_1}[\![\texttt{p}]\!]\, \lambda_z\, \chi\, \dot{\alpha}^{\times}_{\lambda}(\mathscr{P}_0)$ $\wr$definition (46.25) of $\star^{\dot{\times}_1}$ and remark 46.20$\wr$

$= \mathscr{A}^{\dot{\times}_1}_{\lambda}[\![\texttt{*p}]\!]\, \chi\, \dot{\alpha}^{\times}_{\lambda}(\mathscr{P}_0)$ $\wr$definition (46.25) of $\mathscr{A}^{\dot{\times}_1}_{\lambda}[\![\texttt{*p}]\!]\, \chi\wr$ □

For the pointer semantics, theorem 28.31 becomes

> **Theorem 46.29** If $\gamma^{¤}_{\lambda} \in \langle \mathbb{P}^{¤}_{\lambda}, \sqsubseteq^{¤}\rangle \xrightarrow{\sqcap} \langle \mathbb{P}^{\times}, \sqsubseteq^{\times}\rangle$ preserves finite meets, is strict $(\gamma^{¤}_{\lambda}(\bot^{¤}) = \varnothing)$, $1 \in \gamma^{¤}_{\lambda}(1^{¤})$, the hypotheses of theorem 46.24 hold, $(\ominus^{\overset{\times}{-1}}\ \chi)$ is increasing argumentwise, and $\ominus^{\overset{\times}{-1}}\ \gamma^{¤}_{\lambda}(\chi)\ \langle \gamma^{¤}_{\lambda}(P_1), \gamma^{¤}_{\lambda}(P_2)\rangle \mathrel{\dot{\subseteq}} \dot{\gamma}_{¤}(\ominus^{\overset{¤}{-1}}_{\lambda}\ \chi\ \langle P_1, P_2\rangle)$ then
>
> $$\mathscr{A}^{\overset{\times}{-1}}_{\lambda}[\![\mathsf{A}]\!]\ \gamma^{¤}_{\lambda}(\chi) \circ \dot{\gamma}_{¤} \quad \dot{\subseteq} \quad \dot{\gamma}_{¤} \circ \mathscr{A}^{\overset{¤}{-1}}_{\lambda}[\![\mathsf{A}]\!]\ \chi.$$

Proof of theorem 46.29 We proceed by structural induction on A. The proofs in case A is $\mathtt{1}$, x, or $\mathsf{A}_1 - \mathsf{A}_2$ are very similar to those of theorem 28.31. The proof that $\mathtt{nil} \in \gamma^{¤}_{\lambda}(\chi) \Rightarrow \mathsf{NULL}^{¤}_{\lambda} \sqcap^{¤} \chi \neq \bot^{¤}$ is also the same as for $\mathtt{1}$. It remains to consider

– $\mathscr{A}^{\overset{\times}{-1}}_{\lambda}[\![\mathsf{NULL}]\!]\ \gamma^{¤}_{\lambda}(\chi)\ (\dot{\gamma}_{¤}(\overline{P}))$

$= (\!|\ \mathtt{nil} \in \gamma^{¤}_{\lambda}(\chi)\ \text{?}\ \dot{\gamma}_{¤}(\overline{P}) : \dot{\varnothing}\ |\!)$ ⟨definition (46.25) of $\mathscr{A}^{\overset{\times}{-1}}_{\lambda}[\![\mathsf{NULL}]\!]$ and remark 46.20⟩

$\dot{\subseteq} (\!|\ \mathsf{NULL}^{¤} \sqcap^{¤} \chi = \bot^{¤}\ \text{?}\ \dot{\gamma}_{¤}(\overline{P}) : \dot{\gamma}_{¤}(\dot{\bot}^{¤})\ |\!)$

⟨because either $\mathtt{nil} \in \gamma^{¤}_{\lambda}(\chi)$ so $\mathsf{NULL}^{¤} \sqcap^{¤} \chi \neq \bot^{¤}$ and we conclude by reflexivity or $\mathtt{nil} \notin \gamma^{¤}_{\lambda}(\chi)$ and $\dot{\varnothing}$ is the infimum⟩

$= \dot{\gamma}_{¤}((\!|\ \mathsf{NULL}^{¤} \sqcap^{¤} \chi = \bot^{¤}\ \text{?}\ \overline{P} : \dot{\bot}^{¤}\ |\!))$ ⟨factoring $\dot{\gamma}_{¤}$ out of the conditional⟩

$= \dot{\gamma}_{¤}(\mathscr{A}^{\overset{\times}{-1}}_{\lambda}[\![\mathsf{NULL}]\!]\ \chi\ \overline{P})$ ⟨definition (46.25) of $\mathscr{A}^{\overset{\times}{-1}}_{\lambda}[\![\mathsf{NULL}]\!]$⟩

– $\mathscr{A}^{\overset{\times}{-1}}_{\lambda}[\![\&\mathsf{x}]\!]\ \gamma^{¤}_{\lambda}(\chi)\ (\dot{\gamma}_{¤}(\overline{P}))$

$= (\!|\ \dot{\gamma}_{¤}(\overline{P})(\lambda_{\mathsf{x}}) \cap \gamma^{¤}_{\lambda}(\chi) \neq \varnothing\ \text{?}\ \dot{\gamma}_{¤}(\overline{P})[\lambda_{\mathsf{x}} \leftarrow \dot{\gamma}_{¤}(\overline{P})(\lambda_{\mathsf{x}}) \cap \gamma^{¤}_{\lambda}(\chi)] : \dot{\varnothing}\ |\!)$

⟨definition (46.25) of $\mathscr{A}^{\overset{\times}{-1}}_{\lambda}[\![\&\mathsf{x}]\!]$ and remark 46.20⟩

$= (\!|\ \gamma^{¤}_{\lambda}(\overline{P}(\lambda_{\mathsf{x}}) \sqcap^{¤} \chi) \neq \varnothing\ \text{?}\ \gamma^{¤}_{\lambda}(\overline{P}[\lambda_{\mathsf{x}} \leftarrow \overline{P}(\lambda_{\mathsf{x}}) \sqcap^{¤} \chi]) : \dot{\varnothing}\ |\!)$

⟨pointwise definition of $\dot{\gamma}_{¤}$, $\gamma^{¤}_{\lambda}$ preserves meets, and definition (19.10) of functional assignment⟩

$\dot{\subseteq} (\!|\ \overline{P}(\lambda_{\mathsf{x}}) \sqcap^{¤} \chi \neq \bot^{¤}\ \text{?}\ \dot{\gamma}_{¤}(\overline{P}[\lambda_{\mathsf{x}} \leftarrow \overline{P}(\lambda_{\mathsf{x}}) \sqcap^{¤} \chi]) : \dot{\gamma}_{¤}(\dot{\bot}^{¤})\ |\!)$

⟨$\gamma^{¤}_{\lambda}$ is strict so $\overline{P}(\lambda_{\mathsf{x}}) \sqcap^{¤} \chi = \bot^{¤} \Rightarrow \gamma^{¤}_{\lambda}(\overline{P}(\lambda_{\mathsf{x}}) \sqcap^{¤} \chi) = \gamma^{¤}_{\lambda}(\bot^{¤}) = \varnothing$ and, by contraposition, $\gamma^{¤}_{\lambda}(\overline{P}(\lambda_{\mathsf{x}}) \sqcap^{¤} \chi) \neq \varnothing$ implies $\overline{P}(\lambda_{\mathsf{x}}) \sqcap^{¤} \chi \neq \bot^{¤}$ so we conclude by reflexivity or $\dot{\varnothing}$ is the infimum⟩

$= \dot{\gamma}_{¤}((\!|\ \overline{P}(\lambda_{\mathsf{x}}) \sqcap^{¤} \chi \neq \bot^{¤}\ \text{?}\ \overline{P}[\lambda_{\mathsf{x}} \leftarrow \overline{P}(\lambda_{\mathsf{x}}) \sqcap^{¤} \chi] : \dot{\bot}^{¤}\ |\!))$

⟨factoring $\dot{\gamma}_{¤}$ out of the conditional⟩

$= \dot{\gamma}_{¤}(\mathscr{A}^{\overset{\times}{-1}}_{\lambda}[\![\&\mathsf{x}]\!]\ \chi\ \overline{P})$ ⟨definition (46.25) of $\mathscr{A}^{\overset{\times}{-1}}_{\lambda}[\![\&\mathsf{x}]\!]$⟩

– $\mathscr{A}^{\overset{\times}{-1}}_{\lambda}[\![\star\mathsf{p}]\!]\ \gamma^{¤}_{\lambda}(\chi)\ (\dot{\gamma}_{¤}(\overline{P}))$

$= \star^{\overset{¤}{-1}}[\![\mathsf{p}]\!]\ \lambda\ \gamma^{¤}_{\lambda}(\chi)\ (\dot{\gamma}_{¤}(\overline{P}))$ ⟨definition (46.25) of $\mathscr{A}^{\overset{\times}{-1}}_{\lambda}[\![\star\mathsf{p}]\!]$⟩

$= \{\mu(\lambda) \in (\dot{\gamma}_{¤}(\overline{P}))\ \lambda \mid \mu(\mu(\lambda_{\mathsf{p}})) \in \gamma^{¤}_{\lambda}(\chi) \cap (\dot{\gamma}_{¤}(\overline{P}))\ \mu(\lambda_{\mathsf{p}})\}$

⟨definition (46.25) of $\star^{\overset{¤}{-1}}$ and remark 46.20⟩

$= \{\mu(\lambda) \in \gamma^{¤}_{\lambda}(\overline{P}(\lambda)) \mid \mu(\mu(\lambda_{\mathsf{p}})) \in \gamma^{¤}_{\lambda}(\chi) \cap \gamma^{¤}_{\lambda}(\overline{P}(\mu(\lambda_{\mathsf{p}}))\}$ ⟨pointwise definition of $\dot{\gamma}_{¤}$⟩

$= \gamma^{¤}_{\lambda}(\alpha^{¤}_{\lambda}(\{\mu(\lambda) \in \gamma^{¤}_{\lambda}(\overline{P}(\lambda)) \mid \mu(\mu(\lambda_{\mathsf{p}})) \in \gamma^{¤}_{\lambda}(\chi) \cap \gamma^{¤}_{\lambda}(\overline{P}(\mu(\lambda_{\mathsf{p}}))\}))$ ⟨$\gamma^{¤}_{\lambda} \circ \alpha^{¤}_{\lambda}$ is extensive⟩

$= \gamma^{¤}_{\lambda}(\star^{\overset{¤}{-1}}[\![\mathsf{p}]\!]\ \lambda\ \chi\ \overline{P})$ ⟨hypothesis (46.25) on $\star^{\overset{¤}{-1}}[\![\mathsf{p}]\!]\ \lambda\ \chi\ \overline{P}$⟩

$= \dot{\gamma}_{¤}(\mathscr{A}^{\overset{¤}{-1}}_{\lambda}[\![\star\mathsf{p}]\!]\ \chi)\ \overline{P}$ ⟨definition (46.25) of $\mathscr{A}^{\overset{¤}{-1}}_{\lambda}[\![\star\mathsf{p}]\!]$ and pointiwse definition of $\dot{\gamma}_{¤}$⟩ □

46.6.5 Cartesian Forward Reachability of Tests

For tests, (28.38) is completed by pointer equality tests.

Cartesian abstract forward reachability semantics of an arithmetic or pointer Boolean expressions

$$\mathsf{test}^{¤}, \overline{\mathsf{test}}^{¤} \in \lambda \in \mathbb{L} \to \mathbb{B} \to \dot{\mathbb{P}}^{¤}_{\lambda} \to \dot{\mathbb{P}}^{¤}_{\lambda} \quad \text{where} \quad \dot{\mathbb{P}}^{¤}_{\lambda} \triangleq \mathbb{L}_{\lambda} \to \mathbb{P}^{¤}_{\lambda} \tag{46.30}$$

$$\mathsf{test}^{¤}_{\lambda}[\![\mathsf{p} == \mathsf{NULL}]\!]\,\overline{P} \triangleq (\!|\, \overline{P}(\lambda_{\mathsf{p}}) \sqcap^{¤} \mathsf{NULL}^{¤}_{\lambda} \neq \mathsf{NULL}^{¤}_{\lambda} \mathrel{?} \dot{\bot}^{¤} \mathrel{:} \overline{P}[\lambda_{\mathsf{p}} \leftarrow \mathsf{NULL}^{¤}_{\lambda}] \,|\!)$$

$$\overline{\mathsf{test}}^{¤}_{\lambda}[\![\mathsf{p} == \mathsf{NULL}]\!]\,\overline{P} \triangleq (\!|\, \overline{P}(\lambda_{\mathsf{p}}) \sqsubseteq^{¤} \mathsf{NULL}^{¤}_{\lambda} \mathrel{?} \dot{\bot}^{¤} \mathrel{:} \overline{P} \,|\!)$$

$$\mathsf{test}^{¤}_{\lambda}[\![\mathsf{x} == \mathsf{y}]\!]\,\overline{P} \triangleq \mathsf{let}\, a = \overline{P}(\lambda_{\mathsf{x}}) \sqcap^{¤} \overline{P}(\lambda_{\mathsf{y}})\, \mathsf{in}\, (\!|\, a = \bot^{¤} \mathrel{?} \dot{\bot}^{¤} \mathrel{:} \overline{P}[\lambda_{\mathsf{x}} \leftarrow a][\lambda_{\mathsf{y}} \leftarrow a] \,|\!)$$

$$\overline{\mathsf{test}}^{¤}_{\lambda}[\![\mathsf{x} == \mathsf{y}]\!]\,\overline{P} \triangleq (\!|\, \overline{P}(\lambda_{\mathsf{x}}) \sqsubseteq^{¤} \mathsf{NULL}^{¤}_{\lambda} \wedge \overline{P}(l) \sqsubseteq^{¤} \mathsf{NULL}^{¤}_{\lambda} \mathrel{?} \dot{\bot}^{¤} \mathrel{:} \overline{P} \,|\!)$$

Remark 46.31 Following remark 46.20, by choosing $\alpha^{¤}_{\lambda}$ to be the identity, we get $\mathsf{test}^{\times} \in \mathbb{L}_{\lambda} \to \mathbb{B} \to (\mathbb{L}_{\lambda} \to \wp(\mathbb{V}_{\lambda})) \to (\mathbb{L}_{\lambda} \to \wp(\mathbb{V}_{\lambda}))$ where $\mathsf{test}^{\times}_{\lambda}[\![\mathsf{B}]\!]/\overline{\mathsf{test}}^{\times}_{\lambda}[\![\mathsf{B}]\!]$ is the instance of $\mathsf{test}^{¤}_{\lambda}[\![\mathsf{B}]\!]/\overline{\mathsf{test}}^{¤}_{\lambda}[\![\mathsf{B}]\!]$ with $\dot{\mathbb{P}}^{¤} \triangleq \wp(\mathbb{L} \to \mathbb{V}_{\lambda})$, $\bot^{\times} \triangleq \emptyset$, $\dot{\bot}^{\times} \triangleq \dot{\emptyset}$, $\sqcap^{¤} \triangleq \cap$, $\sqsubseteq^{¤} \triangleq \subseteq$, and $\mathsf{NULL}^{\times} \triangleq \{\mathtt{nil}\}$.

Theorem 28.35 is extended as

Theorem 46.32 (Cartesian test) $\dot{\alpha}^{\times}_{\lambda} \circ \mathsf{test}^{\vec{r}}_{\lambda}[\![\mathsf{B}]\!] \mathrel{\dot{\subseteq}} \mathsf{test}^{\times}_{\lambda}[\![\mathsf{B}]\!] \circ \dot{\alpha}^{\times}_{\lambda}$ and $\dot{\alpha}^{\times}_{\lambda} \circ \overline{\mathsf{test}}^{\vec{r}}_{\lambda}[\![\mathsf{B}]\!] \mathrel{\dot{\subseteq}} \overline{\mathsf{test}}^{\times}_{\lambda}[\![\mathsf{B}]\!] \circ \dot{\alpha}^{\times}_{\lambda}$.

Proof of theorem 46.32 Notice that $\dot{\alpha}^{\times}_{\lambda} \in \wp(\mathbb{L}_{\lambda} \to \mathbb{V}_{\lambda}) \to (\mathbb{L}_{\lambda} \to \wp(\mathbb{V}_{\lambda}))$, $\mathsf{test}^{\vec{r}}[\![\mathsf{B}]\!] \in \mathbb{L} \to \wp(\mathbb{L} \to \mathbb{V}_{\lambda}) \to \wp(\mathbb{L} \to \mathbb{V}_{\lambda})$ and $\mathsf{test}^{\times}[\![\mathsf{B}]\!] \in \mathbb{L}_{\lambda} \to (\mathbb{L}_{\lambda} \to \wp(\mathbb{V}_{\lambda})) \to (\mathbb{L}_{\lambda} \to \wp(\mathbb{V}_{\lambda}))$ so that $\dot{\alpha}^{\times}_{\lambda} \circ \mathsf{test}^{\vec{r}}_{\lambda}[\![\mathsf{B}]\!]$, $\mathsf{test}^{\times}_{\lambda}[\![\mathsf{B}]\!] \circ \dot{\alpha}^{\times}_{\lambda} \in \wp(\mathbb{L} \to \mathbb{V}_{\lambda}) \to (\mathbb{L}_{\lambda} \to \wp(\mathbb{V}_{\lambda}))$. The cases not considered in the proof of theorem 28.35 are the following.

— $\dot{\alpha}^{\times}_{\lambda} \circ \mathsf{test}^{\vec{r}}_{\lambda}[\![\mathsf{p} == \mathsf{NULL}]\!]\,(\mathscr{P}_0)$

$= \dot{\alpha}^{\times}_{\lambda}(\mathsf{test}^{\vec{r}}_{\lambda}[\![\mathsf{p} == \mathsf{NULL}]\!]\,\mathscr{P}_0)$ ⟨definition of the function composition $\circ$⟩

$= \dot{\alpha}^{\times}_{\lambda}(\{\mu \in \mathscr{P}_0 \mid \mathscr{B}_{\lambda}[\![\mathsf{p} == \mathsf{NULL}]\!]\,\mu = \mathtt{tt}\})$ ⟨definition (46.6) of $\mathsf{test}^{\vec{r}}_{\lambda}[\![\mathsf{B}]\!]\,\mathscr{P}_0$⟩

$= \dot{\alpha}^{\times}_{\lambda}(\{\mu \in \mathscr{P}_0 \mid \mu(\lambda_{\mathsf{p}}) = \mathtt{nil}\})$ ⟨definition (46.2) of $\mathscr{B}_{\lambda}[\![\mathsf{p} == \mathsf{NULL}]\!]\,\mu$⟩

$= \lambda \mapsto \{\mu(\lambda) \mid \mu \in \{\mu \in \mathscr{P}_0 \mid \mu(\lambda_{\mathsf{p}}) = \mathtt{nil}\}\}$

⟨definition $\dot{\alpha}^{\times}_{\lambda}(P)\,l \triangleq \{\mu(l) \mid \mu \in P\}$ of the Cartesian abstraction⟩

$= \lambda \mapsto \{\mu(\lambda) \mid \mu \in \mathscr{P}_0 \wedge \mu(\lambda_{\mathsf{p}}) = \mathtt{nil}\}$ ⟨definition of $\in$⟩

$= (\!|\, (\dot{\alpha}^{\times}_{\lambda}(\mathscr{P}_0))(\lambda_{\mathsf{p}}) \cap \{\mathtt{nil}\} \neq \{\mathtt{nil}\} \mathrel{?} \dot{\emptyset} \mathrel{:} (\dot{\alpha}^{\times}_{\lambda}(\mathscr{P}_0))[\lambda_{\mathsf{p}} \leftarrow \{\mathtt{nil}\}] \,|\!)$

⟨the Cartesian abstraction is $\dot{\alpha}^{\times}_{\lambda}(P)\,l \triangleq \{\mu(l) \mid \mu \in P\}$ so $\mu \in P$ implies $\mu(\lambda) \in \dot{\alpha}^{\times}_{\lambda}(P)\,\lambda$ and $\mu(\lambda_{\mathsf{p}}) = \mathtt{nil}$ implies $\mu = \mu[\lambda_{\mathsf{p}} \leftarrow \mathtt{nil}]$ ⟩

$= (\!|\, (\dot{\alpha}^{\times}_{\lambda}(\mathscr{P}_0))(\lambda_{\mathsf{p}}) \sqcap^{\times} \mathsf{NULL}^{\times} \neq \mathsf{NULL}^{\times} \mathrel{?} \dot{\bot}^{¤} \mathrel{:} (\dot{\alpha}^{\times}_{\lambda}(\mathscr{P}_0))[\lambda_{\mathsf{p}} \leftarrow \mathsf{NULL}^{\times}] \,|\!)$ ⟨remark 46.31⟩

$= \mathsf{test}^{\times}_{\lambda}[\![\mathsf{p} == \mathsf{NULL}]\!]\,(\dot{\alpha}^{\times}_{\lambda}(\mathscr{P}_0))$

⟨definition (46.30) of $\mathsf{test}^{\times}_{\lambda}[\![\mathsf{p} == \mathsf{NULL}]\!]$ and remark 46.31⟩

— $\dot{\alpha}^{\times}_{\lambda}(\overline{\mathsf{test}}^{\vec{r}}_{\lambda}[\![\mathsf{p} == \mathsf{NULL}]\!]\mathscr{P}_0)$

$$= \quad \dot{\alpha}^{\times}_{\lambda}(\{\mu \in \mathscr{P}_0 \mid \mathscr{B}_{\lambda}[\![\mathtt{p == NULL}]\!]\mu = \mathrm{ff}\}) \qquad \text{\{definition (46.6) of } \overline{\mathsf{test}}^{\vec{r}}[\![\mathsf{B}]\!]\,\lambda\,\mathscr{P}_0\text{\}}$$

$$= \quad \lambda \mapsto \{\mu(\lambda) \mid \mu \in \mathscr{P}_0 \wedge \mu(\lambda_{\mathtt{p}}) \neq \mathtt{nil}\}$$

{definition (46.2) of $\mathscr{B}_{\lambda}[\![\mathtt{p == NULL}]\!]\mu$, definition $\dot{\alpha}^{\times}_{\lambda}(P) \triangleq l \in \mathbb{L}_{\lambda} \mapsto \{\mu(l) \mid \mu \in P\}$ of the Cartesian abstraction, and definition of $\in$}

$$\dot{\subseteq} \quad \lambda \mapsto (\!| \{\mu(\lambda_{\mathtt{p}}) \mid \mu \in \mathscr{P}_0\} \subseteq \{\mathtt{nil}\} \;?\; \varnothing \;\text{:}\; \{\mu(\lambda) \mid \mu \in \mathscr{P}_0\} |\!)$$

{if $\{\mu(\lambda_{\mathtt{p}}) \mid \mu \in \mathscr{P}_0\} \subseteq \{\mathtt{nil}\}$ then $\mu(\lambda_{\mathtt{p}}) = \mathtt{nil}$ so $\{\mu(\lambda) \mid \mu \in \mathscr{P}_0 \wedge \mu(\lambda_{\mathtt{p}}) \neq \mathtt{nil}\}$ is empty. Otherwise, $\{\mu(\lambda) \mid \mu \in \mathscr{P}_0 \wedge \mu(\lambda_{\mathtt{p}}) \neq \mathtt{nil}\} \subseteq \{\mu(\lambda) \mid \mu \in \mathscr{P}_0\}$}

$$= \quad \lambda \mapsto (\!| \dot{\alpha}^{\times}_{\lambda}(\mathscr{P}_0)(\lambda_{\mathtt{p}}) \subseteq \{\mathtt{nil}\} \;?\; \dot{\varnothing}(\lambda) \;\text{:}\; \dot{\alpha}^{\times}_{\lambda}(\mathscr{P}_0)\,\lambda |\!)$$

{pointwise definition of $\dot{\varnothing}$ and definition $\dot{\alpha}^{\times}_{\lambda}(P) \triangleq l \in \mathbb{L}_{\lambda} \mapsto \{\mu(l) \mid \mu \in P\}$}

$$= \quad \lambda \mapsto (\!| \dot{\alpha}^{\times}_{\lambda}(\mathscr{P}_0)(\lambda_{\mathtt{p}}) \sqsubseteq^{\times} \mathtt{NULL}^{\times} \;?\; \dot{\bot}^{\times} \;\text{:}\; \dot{\alpha}^{\times}_{\lambda}(\mathscr{P}_0) |\!)(\lambda)$$

{definition of the conditional and remark 46.31}

$$= \quad (\!| \dot{\alpha}^{\times}_{\lambda}(\mathscr{P}_0)(\lambda_{\mathtt{p}}) \sqsubseteq^{\times} \mathtt{NULL}^{\times} \;?\; \dot{\bot}^{\times} \;\text{:}\; \dot{\alpha}^{\times}_{\lambda}(\mathscr{P}_0) |\!) \qquad \text{\{definition of the application\}}$$

$$= \quad \overline{\mathsf{test}}^{\times}_{\lambda}[\![\mathtt{p == NULL}]\!](\dot{\alpha}^{\times}_{\lambda}(\mathscr{P}_0)) \qquad \text{\{definition (46.30) of } \overline{\mathsf{test}}^{\times}_{\lambda}[\![\mathtt{p == NULL}]\!]\text{\}}$$

$$— \quad \dot{\alpha}^{\times}_{\lambda}(\mathsf{test}^{\vec{r}}_{\lambda}[\![\mathtt{x == y}]\!]\,\mathscr{P}_0)$$

$$= \quad \dot{\alpha}^{\times}_{\lambda}(\{\mu \in \mathscr{P}_0 \mid \mathscr{B}_{\lambda}[\![\mathtt{x == y}]\!]\mu = \mathrm{tt}\}) \qquad \text{\{definition (46.6) of } \mathsf{test}^{\vec{r}}[\![\mathsf{B}]\!]\,\lambda\,\mathscr{P}_0\text{\}}$$

$$= \quad \dot{\alpha}^{\times}_{\lambda}(\{\mu \in \mathscr{P}_0 \mid \mu(\lambda_{\mathtt{x}}) = \mu(l)\}) \qquad \text{\{definition (46.2) of } \mathscr{B}_{\lambda}[\![\mathtt{x == y}]\!]\mu\text{\}}$$

$$= \quad \lambda \mapsto \{\mu(\lambda) \mid \mu \in \{\mu \in \mathscr{P}_0 \mid \mu(\lambda_{\mathtt{x}}) = \mu(l)\}\}$$

{definition (46.16) of $\dot{\alpha}^{\times}_{\lambda}(P) \triangleq l \in \mathbb{L}_{\lambda} \mapsto \{\mu(l) \mid \mu \in P\}$}

$$= \quad \lambda \mapsto \{\mu(\lambda) \mid \mu \in \mathscr{P}_0 \wedge \mu(\lambda_{\mathtt{x}}) = \mu(l)\} \qquad \text{\{definition of } \in\text{\}}$$

$$= \quad \lambda \mapsto \mathsf{let}\; a = \dot{\alpha}^{\times}_{\lambda}(\mathscr{P}_0)\,\lambda_{\mathtt{x}} \cap \dot{\alpha}^{\times}_{\lambda}(\mathscr{P}_0)\,l \;\mathsf{in}\; (\!| a = \varnothing \;?\; \varnothing \;\text{:}\; (\dot{\alpha}^{\times}_{\lambda}(\mathscr{P}_0))[\lambda_{\mathtt{x}} \leftarrow a][l \leftarrow a]\,\lambda |\!)$$

{— If $a = \dot{\alpha}^{\times}_{\lambda}(\mathscr{P}_0)\,\lambda_{\mathtt{x}} \cap \dot{\alpha}^{\times}_{\lambda}(\mathscr{P}_0)\,l = \{\mu(\lambda_{\mathtt{x}}) \mid \mu \in \mathscr{P}_0\} \cap \{\mu(l) \mid \mu \in \mathscr{P}_0\} = \varnothing$ then $\forall \mu \in \mathscr{P}_0 \,.\, \lambda_{\mathtt{x}} \neq l$ which implies that $\{\mu(\lambda) \mid \mu \in \mathscr{P}_0 \wedge \mu(\lambda_{\mathtt{x}}) = \mu(l)\} = \varnothing$ which is the true alternative of the conditional.

— Otherwise, there are two cases.

- if $\lambda \neq \lambda_{\mathtt{x}}$ and $\lambda \neq l$ then $\{\mu(\lambda) \mid \mu \in \mathscr{P}_0 \wedge \mu(\lambda_{\mathtt{x}}) = \mu(l)\} \subseteq \{\mu(\lambda) \mid \mu \in \mathscr{P}_0\} = \dot{\alpha}^{\times}_{\lambda}(\mathscr{P}_0)\,\lambda = (\dot{\alpha}^{\times}_{\lambda}(\mathscr{P}_0))[\lambda_{\mathtt{x}} \leftarrow a][l \leftarrow a]\,\lambda$ by definition (19.10) of assignment because $\lambda \neq \lambda_{\mathtt{x}}$ and $\lambda \neq l$;
- otherwise $\lambda = \lambda_{\mathtt{x}}$ or $\lambda = l$. Consider $\lambda = \lambda_{\mathtt{x}}$, the other case is symmetric. We have $\{\mu(\lambda) \mid \mu \in \mathscr{P}_0 \wedge \mu(\lambda_{\mathtt{x}}) = \mu(l)\} = \{\mu(\lambda_{\mathtt{x}}) \mid \mu \in \mathscr{P}_0 \wedge \mu(\lambda_{\mathtt{x}}) = \mu(l)\} \subseteq \{\mu(\lambda_{\mathtt{x}}) \mid \mu \in \mathscr{P}_0\} \cap \{\mu(l) \mid \mu \in \mathscr{P}_0\} = \dot{\alpha}^{\times}_{\lambda}(\mathscr{P}_0)\,\lambda_{\mathtt{x}} \cap \dot{\alpha}^{\times}_{\lambda}(\mathscr{P}_0)\,l = a = (\dot{\alpha}^{\times}_{\lambda}(\mathscr{P}_0))[\lambda_{\mathtt{x}} \leftarrow a][l \leftarrow a]\,\lambda_{\mathtt{x}} = (\dot{\alpha}^{\times}_{\lambda}(\mathscr{P}_0))[\lambda_{\mathtt{x}} \leftarrow a][l \leftarrow a]\,\lambda$ because $\lambda = \lambda_{\mathtt{x}}$. }

$$= \quad \mathsf{let}\; a = \dot{\alpha}^{\times}_{\lambda}(\mathscr{P}_0)\,\lambda_{\mathtt{x}} \cap \dot{\alpha}^{\times}_{\lambda}(\mathscr{P}_0)\,l \;\mathsf{in}\; (\!| a = \varnothing \;?\; \dot{\varnothing} \;\text{:}\; (\dot{\alpha}^{\times}_{\lambda}(\mathscr{P}_0))[\lambda_{\mathtt{x}} \leftarrow a][l \leftarrow a] |\!)$$

{definition of a function}

$$= \quad \mathsf{test}^{\times}_{\lambda}[\![\mathtt{x == y}]\!](\dot{\alpha}^{\times}_{\lambda}(\mathscr{P}_0)) \qquad \text{\{definition (46.30) of } \mathsf{test}^{\times}_{\lambda}[\![\mathtt{x == y}]\!] \text{ and remark 46.31\}}$$

$$— \quad \dot{\alpha}^{\times}_{\lambda}(\overline{\mathsf{test}}^{\vec{r}}_{\lambda}[\![\mathtt{x == y}]\!]\,\mathscr{P}_0)$$

$$= \quad \lambda \mapsto \{\mu(\lambda) \mid \mu \in \mathscr{P}_0 \wedge \mu(\lambda_{\mathtt{x}}) \neq \mu(l)\}$$

{definition (46.6) of $\overline{\mathsf{test}}^{\vec{r}}[\![\mathsf{B}]\!]\,\lambda\,\mathscr{P}_0$, definition (46.2) of $\mathscr{B}_{\lambda}[\![\mathtt{x == y}]\!]\mu$, definition (46.16) of $\dot{\alpha}^{\times}_{\lambda}(P) \triangleq l \in \mathbb{L}_{\lambda} \mapsto \{\mu(l) \mid \mu \in P\}$, and definition of $\in$}

$$\dot{\subseteq}\quad \lambda \mapsto (\!|\, \ddot{\alpha}_\lambda^\times(\mathscr{P}_0)\,\lambda_\mathsf{x} \subseteq \{\mathtt{nil}\} \wedge \ddot{\alpha}_\lambda^\times(\mathscr{P}_0)\, l \subseteq \{\mathtt{nil}\} \;?\; \varnothing \,\text{:}\, \ddot{\alpha}_\lambda^\times(\mathscr{P}_0)\,\lambda \,|\!)$$

⦃if $\{\mu(\lambda_\mathsf{x}) \mid \mu \in \mathscr{P}_0\} = \ddot{\alpha}_\lambda^\times(\mathscr{P}_0)\,\lambda_\mathsf{x} \subseteq \{\mathtt{nil}\}$ and $\{\mu(l) \mid \mu \in \mathscr{P}_0\} = \ddot{\alpha}_\lambda^\times(\mathscr{P}_0)\, l \subseteq \{\mathtt{nil}\}$ then $\forall \mu \in \mathscr{P}_0 \,.\, \mu(\lambda_\mathsf{x}) = \mu(l) = \mathtt{nil}$ so that $\{\mu(\lambda) \mid \mu \in \mathscr{P}_0 \wedge \mu(\lambda_\mathsf{x}) \neq \mu(l)\} = \varnothing$. Otherwise $\{\mu(\lambda) \mid \mu \in \mathscr{P}_0 \wedge \mu(\lambda_\mathsf{x}) \neq \mu(l)\} \subseteq \{\mu(\lambda) \mid \mu \in \mathscr{P}_0\} = \ddot{\alpha}_\lambda^\times(\mathscr{P}_0)\,\lambda$⦄

$= \overline{\mathsf{test}}_\lambda^\times[\![\mathsf{x == y}]\!](\ddot{\alpha}_\lambda^\times(\mathscr{P}_0))$ ⦃definition (46.30) of $\overline{\mathsf{test}}_\lambda^\times[\![\mathsf{x == y}]\!]$ and remark 46.31⦄ □

With static pointers, theorem 28.39 on abstract Cartesian tests becomes

> **Theorem 46.33 (abstract Cartesian test)** If $\otimes^\times$ is increasing argumentwise, $\otimes^\times \langle \gamma_\lambda^{¤}(X), \gamma_\lambda^{¤}(Y)\rangle \dot{\sqsubseteq}^\times \dot{\gamma}_{¤}(\otimes^{¤} \langle X, Y\rangle))$, and $\gamma_\lambda^{¤}$ preserves finite glbs then $\mathsf{test}_\lambda^\times[\![\mathsf{B}]\!] \circ \dot{\gamma}_{¤} \dot{\sqsubseteq}^\times \dot{\gamma}_{¤} \circ \mathsf{test}_\lambda^{¤}[\![\mathsf{B}]\!]$ and $\overline{\mathsf{test}}_\lambda^\times[\![\mathsf{B}]\!] \circ \dot{\gamma}_{¤} \dot{\sqsubseteq}^\times \dot{\gamma}_{¤} \circ \overline{\mathsf{test}}_\lambda^{¤}[\![\mathsf{B}]\!]$.

Proof of theorem 46.33 The cases not considered in the proof of theorem 28.39 are the following.

— $\mathsf{test}_\lambda^\times[\![\mathsf{x == y}]\!](\dot{\gamma}_{¤}(\overline{P}))$

$= \mathsf{let}\ a' = \gamma_\lambda^{¤}(\overline{P}(\lambda_\mathsf{x}) \sqcap^{¤} \overline{P}(l))\ \mathsf{in}\ (\!|\, a' = \varnothing \;?\; \dot{\varnothing} \,\text{:}\, (\dot{\gamma}_{¤}(\overline{P}))[\lambda_\mathsf{x} \leftarrow a'][l \leftarrow a'] \,|\!)$

⦃definition (46.30) of $\mathsf{test}_\lambda^{¤}$, remark 46.31, pointwise definition of $\dot{\gamma}_{¤}$, and $\gamma_\lambda^{¤}$ preserves meets⦄

$\dot{\subseteq}\ \mathsf{let}\ a = \overline{P}(\lambda_\mathsf{x}) \sqcap^{¤} \overline{P}(l)\ \mathsf{in}\ (\!|\, a = \bot^{¤} \;?\; \dot{\gamma}_{¤}(\dot{\bot}^{¤}) \,\text{:}\, \dot{\gamma}_{¤}(\overline{P}[\lambda_\mathsf{x} \leftarrow a][l \leftarrow a]) \,|\!)$

⦃if $a' = \varnothing$ then $\dot{\varnothing} \dot{\subseteq} \dot{\gamma}_{¤}(\dot{\bot}^{¤})$. Otherwise $a' = \gamma_\lambda^{¤}(a)$ so, by pointwise definition of $\dot{\gamma}_{¤}$, $\dot{\gamma}_{¤}(\overline{P}[\lambda_\mathsf{x} \leftarrow a][l \leftarrow a]) = \dot{\gamma}_{¤}(\overline{P})[\lambda_\mathsf{x} \leftarrow \gamma_\lambda^{¤}(a)][l \leftarrow \gamma_\lambda^{¤}(a)]) = (\dot{\gamma}_{¤}(\overline{P}))[\lambda_\mathsf{x} \leftarrow a'][l \leftarrow a']$⦄

$= \dot{\gamma}_{¤}(\mathsf{test}_\lambda^{¤}[\![\mathsf{x == y}]\!]\,\overline{P})$ ⦃definition (46.30) of $\mathsf{test}_\lambda^{¤}$. Moreover $\dot{\sqsubseteq}^\times \triangleq \dot{\subseteq}$ by remark 46.26⦄

— $\mathsf{test}_\lambda^\times[\![\mathsf{p == NULL}]\!](\dot{\gamma}_{¤}(\overline{P}))$

$= (\!|\, \gamma_\lambda^{¤}(\overline{P}(\lambda_\mathsf{p})) \cap \{\mathtt{nil}\} \neq \{\mathtt{nil}\} \;?\; \dot{\varnothing} \,\text{:}\, (\dot{\gamma}_{¤}(\overline{P}))[\lambda_\mathsf{p} \leftarrow \{\mathtt{nil}\}] \,|\!)$

⦃definition (46.30) of $\mathsf{test}_\lambda^{¤}$, remark 46.31, and pointwise de. $\dot{\gamma}_{¤}$⦄

$= (\!|\, \overline{P}(\lambda_\mathsf{p}) \sqcap^{¤} \mathsf{NULL}_\lambda^{¤} \neq \mathsf{NULL}_\lambda^{¤} \;?\; \dot{\gamma}_{¤}(\dot{\bot}^{¤}) \,\text{:}\, \dot{\gamma}_{¤}(\overline{P}[\lambda_\mathsf{p} \leftarrow \mathsf{NULL}_\lambda^{¤}]) \,|\!)$

⦃— If $\overline{P}(\lambda_\mathsf{p}) \sqcap^{¤} \mathsf{NULL}_\lambda^{¤} = \mathsf{NULL}_\lambda^{¤}$ then $\gamma_\lambda^{¤}(\overline{P}(\lambda_\mathsf{p})) \cap \{\mathtt{nil}\} = \gamma_\lambda^{¤}(\overline{P}(\lambda_\mathsf{p}) \sqcap^{¤} \mathsf{NULL}_\lambda^{¤}) = \gamma_\lambda^{¤}(\mathsf{NULL}_\lambda^{¤}) = \{\mathtt{nil}\}$ in which case $(\dot{\gamma}_{¤}(\overline{P}))[\lambda_\mathsf{p} \leftarrow \{\mathtt{nil}\}] = \overline{P}[\lambda_\mathsf{p} \leftarrow \gamma_\lambda^{¤}(\mathsf{NULL}_\lambda^{¤})] = \dot{\gamma}_{¤}(\overline{P}[\lambda_\mathsf{p} \leftarrow \mathsf{NULL}_\lambda^{¤}])$ by pointwise definition of $\dot{\gamma}_{¤}$.

— Otherwise, $\dot{\varnothing} \dot{\subseteq} \dot{\gamma}_{¤}(\dot{\bot}^{¤})$⦄

$= \dot{\gamma}_{¤}(\mathsf{test}_\lambda^{¤}[\![\mathsf{p == NULL}]\!]\,\overline{P})$

⦃definition (46.30) of $\mathsf{test}_\lambda^{¤}$. Moreover $\dot{\sqsubseteq}^\times \triangleq \dot{\subseteq}$ by remark 46.26⦄

— The remaining cases are similar. □

The soundness theorem 28.44 can be extended as follows:

> **Theorem 46.34 (well definedness and soundness of the Cartesian static analysis)** The Cartesian abstract domain $\mathbb{D}^{\dot{¤}}$, pointwise extension of $\mathbb{D}^{¤}$ in (46.18) for $\mathbb{L}$ is well defined according to definition 21.1 so the flow-sensitive abstract interpreter

$\widehat{\mathcal{S}}^{\dot{\alpha}}[\![\mathsf{S}]\!]$ and the flow-insensitive abstract interpreter $\widehat{\mathcal{S}}_{\lambda}^{gf\alpha}[\![\mathsf{S}]\!]$ are well defined and a sound abstraction of the assertional forward reachability semantics $\mathcal{S}^{\vec{r}}[\![\mathsf{S}]\!]$ of section 46.4 for any program component $\mathsf{S} \in \mathbb{P}\mathbb{c}$.

Proof of theorem 46.34 By Theorems 46.24, 46.29, and 46.33, the Cartesian abstract domain $\mathbb{D}^{\dot{\alpha}}$ is well defined according to definition 21.1 because $\langle \mathbb{P}_{\lambda}^{\alpha}, \sqsubseteq^{\alpha}\rangle$ is assumed to be a poset of properties with the infimum $\bot^{\alpha}$, the lub $\sqcup^{\alpha}$ which is well defined for pairs of properties, and $\sqsubseteq^{\alpha}$-increasing chains (so $\langle \mathbb{P}_{\lambda}^{\alpha}, \sqsubseteq^{\alpha}\rangle$ is a join-semilattice and a CPO), the assignment assign^{α} is well defined in $(\mathbb{V} \times \mathbb{E}) \to \mathbb{P}_{\lambda}^{\alpha} \xrightarrow{\nearrow} \mathbb{P}_{\lambda}^{\alpha}$ (see (46.8)/(46.8) and (46.12)/(46.13)), and the tests $\mathsf{test}^{\alpha}[\![\mathsf{B}]\!]$ and $\overline{\mathsf{test}}^{\alpha}[\![\mathsf{B}]\!]$ are well defined in $\mathbb{B} \to \mathbb{P}_{\lambda}^{\alpha} \xrightarrow{\nearrow} \mathbb{P}_{\lambda}^{\alpha}$.

$\dot{\alpha}_{\lambda}^{\times\alpha}$ is an approximate domain abstraction according to definition 27.1.I. We conclude, by theorem 46.10 and theorem 27.4, that the flow-sensitive abstract interpreter of (21.3)—(21.13) of chapter 21 augmented by (46.8) and (46.10), instantiated with the Cartesian abstract domain $\mathbb{D}^{\dot{\alpha}}$ of abstract reachability properties is a sound abstraction of the assertional forward reachability semantics $\mathcal{S}^{\vec{r}}[\![\mathsf{S}]\!]$ of chapter 19 augmented by (46.3),(46.4), and (46.6).

A similar result holds for the flow-insensitive abstract interpreter $\widehat{\mathcal{S}}_{\lambda}^{gf\alpha}[\![\mathsf{S}]\!]$ of (45.11)—(45.11) augmented by (46.12) and (46.13) using a generalization of theorem 45.13 for the pointer language. □

Exercise 46.35 (Project: Cartesian flow-[in]sensitive abstract interpreter for the pointer language) Extend the Cartesian reachability interpreter of exercise 28.46 into flow-sensitive (exercise 46.11) and flow-insensitive (exercise 46.15) Cartesian abstract interpreters for the pointer language of section 46.1. The Cartesian integer and pointer domain $\mathbb{D}^{\alpha}$ of (46.18) should be a parameter of these Cartesian abstract interpreters.

46.7 Potential Points-to Cartesian Abstract Domain

The flow-insensitive *points-to analysis* of Lars Ole Andersen [43] collects the locations to which the pointer variables may point. Hence it is a Cartesian abstraction, and no relation between pointer variables can be inferred. It is *potential* (it collects *may point to* information as opposed to definite *ust point to* information). In points-to analysis, null pointers and runtime errors are abstracted away. To get flow-sensitive or flow-insensitive pointer analyzers, it remains to instantiate the abstract domain (46.18). So, we do not propose a word-by-word presentation of Lars Ole Andersen 's algorithm but rather a fresh interpretation as an abstract domain.

46.7.1 Points-to Abstraction

Value properties are abstracted by $\langle \wp(\mathbb{V}_{\lambda}), \subseteq\rangle \underset{\alpha^{\equiv}}{\overset{\gamma^{\equiv}}{\leftrightarrows}} \langle \mathbb{P}_{\lambda}^{\equiv}, \subseteq\rangle$ where $\mathbb{P}_{\lambda}^{\equiv} \triangleq \wp(\mathbb{L})$ and

$$\begin{aligned} \alpha^{\equiv}(V) &\triangleq V \cap \mathbb{L} \\ \gamma^{\equiv}(\overline{P}) &\triangleq \overline{P} \cup \{\mathtt{nil}\} \cup \mathbb{Z}. \end{aligned} \tag{46.36}$$

It follows that the Cartesian pointer abstraction following (46.16) is

$$\langle \wp(\mathbb{L}_\lambda \to \mathbb{V}_\lambda),\ \subseteq\rangle \xrightleftharpoons[\dot{\alpha}^{\exists}_{\times}]{\dot{\gamma}^{\exists}_{\times}} \langle \mathbb{L}_\lambda \to \mathbb{P}^{\exists}_{\lambda},\ \dot{\subseteq}\rangle \tag{46.37}$$

with $\dot{\alpha}^{\exists}_{\times}(P)\,l = \alpha^{\exists}(\{\mu(l) \in \mathbb{L} \mid \mu \in P\})$ and $\dot{\gamma}^{\exists}_{\times}(\overline{P}) = \{\mu \mid \forall l \in \mathbb{L}\,.\,\mu(l) \in \gamma^{\exists}(\overline{P}(l))\}$. So if a variable x has a non-null pointer value, it can point only to $\overline{P}(\lambda_{\mathsf{x}})$.

Exercise 46.38 Change the analysis thereafter (up to and including section 46.8) for $\alpha^{\exists}(V) \triangleq V \cap (\mathbb{L} \cup \{\mathtt{nil}\})$ to keep track of null pointers.

46.7.2 Points-to Abstract Domain

The Cartesian pointer abstract domain is

$$\mathbb{D}^{\exists}_{\lambda} \triangleq \langle \mathbb{P}^{\exists}_{\lambda}, \sqsubseteq^{\exists}_{\lambda}, \bot^{\exists}_{\lambda}, \top^{\exists}_{\lambda}, \sqcup^{\exists}_{\lambda}, \sqcap^{\exists}_{\lambda}, 1^{\exists}_{\lambda}, \ominus^{\exists}_{\lambda}, \overleftarrow{\ominus}^{\exists}_{\lambda}, \overleftarrow{\star}^{\exists}_{\lambda}, \oslash^{\exists}_{\lambda}, \overline{\oslash}^{\exists}_{\lambda}, \mathsf{NULL}^{\exists}_{\lambda}, \boldsymbol{\lambda}^{\exists}_{\lambda}, \star^{\exists}_{\lambda}\rangle \tag{46.39}$$

An abstract property $\overline{P} \in \mathbb{P}^{\exists}_{\lambda} \triangleq \wp(\mathbb{L})$ is a set of locations that may be pointed to (or may be null $\mathtt{nil}$ or an integer) so $\sqsubseteq^{\exists}_{\lambda} \triangleq \subseteq$, $\bot^{\exists}_{\lambda} \triangleq \emptyset$, $\top^{\exists}_{\lambda} \triangleq \mathbb{L}$, $\sqcup^{\exists}_{\lambda} \triangleq \cup$, and $\sqcap^{\exists}_{\lambda} \triangleq \cap$.

Remark 46.40 The traditional abstract domain [43] is $\wp(\mathbb{V})$ where variables $\mathsf{x} \in \mathbb{V}$ are confused with their location λ_{x}. This is impractical for languages in which a variable x may not be visible although its location λ_{x} is still accessible.

46.7.3 Calculational Design of the Abstract Reachability Points-to Semantics of an Assignment

The abstract evaluation $\mathcal{A}^{\exists}_{\lambda}[\![\mathsf{A}]\!]$ of an expression A is defined by (46.19) with

Points-to reachability semantics of arithmetic and pointer expressions

$$\begin{array}{rcl@{\qquad}rcl}
1^{\exists}_{\lambda} & \triangleq & \emptyset & \overleftarrow{\ominus}^{\exists}_{\lambda}\, P\, \langle Q_1, Q_2\rangle & \triangleq & \emptyset \\
P \ominus^{\exists}_{\lambda} Q & \triangleq & \emptyset & \mathsf{NULL}^{\exists}_{\lambda} & \triangleq & \emptyset \\
P \oslash^{\exists}_{\lambda} Q & \triangleq & \emptyset & \boldsymbol{\lambda}^{\exists}_{\lambda}[\![x]\!] & \triangleq & \{\lambda_{\mathsf{x}}\} \\
P \overline{\oslash}^{\exists}_{\lambda} Q & \triangleq & \emptyset & \star^{\exists}_{\lambda}[\![\mathsf{p}]\!]\,\overline{P} & \triangleq & \bigcup_{l \in \overline{P}(\lambda_{\mathsf{p}})} \overline{P}(l) \\
 & & & \overleftarrow{\star}^{\exists}_{\lambda}[\![\mathsf{p}]\!]\,\chi\,\overline{P} & \triangleq & (\!|\, \chi \cap \star^{\exists}_{\lambda}[\![\mathsf{p}]\!]\,\overline{P} = \emptyset \mathrel{?} \dot{\emptyset} \mathrel{:} \overline{P} \,|\!)
\end{array} \tag{46.41}$$

Proof of (46.41) We must show that the soundness hypotheses of (46.19) are satisfied, by calculational design.

— For the arithmetic operators $1^{\exists}_{\lambda}$ and $\ominus^{\exists}_{\lambda}$, the abstraction (46.36) is the empty set.

— Similar for $\mathsf{NULL}^{\exists}_{\lambda} = \alpha^{\exists}(\{\mathtt{nil}\}) = \{\mathtt{nil}\} \cap \mathbb{L} = \emptyset$ ⟨soundness condition from (46.19)⟩

— $\boldsymbol{\lambda}^{\exists}_{\lambda}[\![x]\!] = \alpha^{\exists}(\{\lambda_{\mathsf{x}}\}) = \{\lambda_{\mathsf{x}}\} \cap \mathbb{L} = \{\lambda_{\mathsf{x}}\}$ ⟨soundness condition from (46.19)⟩

— $\alpha^{\exists}(\{v_1 \in \gamma^{\exists}(Q_1) \mid \exists v_2 \in \gamma^{\exists}(Q_2)\,.\,(v_1 - v_2) \in \gamma^{\exists}(P)\})$

⟨soundness condition from (46.19)⟩

$= \alpha^{\exists}(\{v_1 \in Q_1 \cup \{\mathtt{nil}\} \cup \mathbb{Z} \mid \exists v_2 \in Q_2 \cup \{\mathtt{nil}\} \cup \mathbb{Z}\,.\,(v_1 - v_2) \in P \cup \{\mathtt{nil}\} \cup \mathbb{Z}\})$

⟨definition of $\gamma^{\exists}$⟩

$= \alpha^{\exists}(\{v_1 \in \mathbb{Z} \mid \exists v_2 \in \mathbb{Z}\,.\,(v_1 - v_2) \in \mathbb{Z}\} \cup \{\Omega\})$

$$\begin{aligned}
&\quad \langle Q_1, Q_2 \in \wp(\mathbb{L}) \text{ so if } v_1 \in Q_1 \cup \{\texttt{nil}\} \text{ or } v_2 \in Q_2 \cup \{\texttt{nil}\} \text{ then } (v_1 - v_2) = \Omega \text{ else} \\
&\quad v_1 \in \mathbb{Z} \text{ and } v_2 \in \mathbb{Z} \text{ so } (v_1 - v_2) \in \mathbb{Z} \text{ because } P \in \wp(\mathbb{L}) \text{ and } \mathbb{L} \cap (\{\texttt{nil}\} \cup \mathbb{Z}) = \emptyset \rangle \\
= \;& \alpha^{\equiv}(\mathbb{Z} \cup \{\Omega\}) && \langle \text{definition of the integer minus } - \rangle \\
= \;& \emptyset && \langle \text{definition (46.36) of } \alpha^{\equiv} \rangle \\
= \;& \overset{\equiv}{\ominus}{}^{-1}_{\lambda} P \,\langle Q_1, Q_2 \rangle && \langle \text{definition (46.41) of } \overset{\equiv}{\ominus}{}^{-1}_{\lambda} \rangle
\end{aligned}$$

$$\begin{aligned}
- \;& \alpha^{\equiv}(\{\mu(\mu(\lambda_{\mathrm{p}})) \mid \mu \in \dot{\gamma}^{\equiv}_{\times}(\overline{P}) \wedge \mu(\lambda_{\mathrm{p}}) \in \mathbb{L}\}) && \langle \text{soundness condition from (46.19)} \rangle \\
= \;& \alpha^{\equiv}(\{\mu(\mu(\lambda_{\mathrm{p}})) \mid \mu \in \mathbb{M} \wedge \forall \lambda_{\mathrm{z}} \in \mathbb{L} \,.\, \mu(\lambda_{\mathrm{z}}) \in \overline{P}(\lambda_{\mathrm{z}}) \cup \mathbb{Z} \cup \{\texttt{nil}\} \wedge \mu(\lambda_{\mathrm{p}}) \in \mathbb{L}\}) \\
& && \langle \text{definition of } \dot{\gamma}^{\equiv}_{\times} \rangle \\
= \;& \alpha^{\equiv}(\{\mu(\mu(\lambda_{\mathrm{p}})) \mid \mu \in \mathbb{M} \wedge \forall \lambda_{\mathrm{z}} \in \mathbb{L} \,.\, \mu(\lambda_{\mathrm{z}}) \in \overline{P}(\lambda_{\mathrm{z}}) \cup \mathbb{Z} \cup \{\texttt{nil}\} \wedge \mu(\lambda_{\mathrm{p}}) \in \overline{P}(\lambda_{\mathrm{p}})\}) \\
& && \langle \text{because } \overline{P} \in \mathbb{L} \to \wp(\mathbb{L}) \rangle \\
\sqsubseteq^{\equiv}_{\lambda} \;& \alpha^{\equiv}(\{\mu(l)) \mid \mu \in \mathbb{M} \wedge \mu(l) \in \overline{P}(l) \cup \mathbb{Z} \cup \{\texttt{nil}\} \wedge l \in \overline{P}(\lambda_{\mathrm{p}})\}) \\
& && \langle \text{letting } l = \mu(\lambda_{\mathrm{p}}) \text{ and } \alpha^{\equiv} \text{ increasing for } \sqsubseteq^{\equiv}_{\lambda} \triangleq \subseteq \rangle \\
= \;& \{\mu(l)) \mid \mu \in \mathbb{M} \wedge \mu(l) \in \overline{P}(l) \wedge l \in \overline{P}(\lambda_{\mathrm{p}})\} && \langle \text{definition of } \alpha^{\equiv} \text{ and } \overline{P} \in \mathbb{L} \to \wp(\mathbb{L}) \rangle \\
\sqsubseteq^{\equiv}_{\lambda} \;& \{\lambda_{\mathrm{z}}) \mid \lambda_{\mathrm{z}} \in \overline{P}(l) \wedge l \in \overline{P}(\lambda_{\mathrm{p}})\} && \langle \text{letting } \lambda_{\mathrm{z}} = \mu(l) \rangle \\
= \;& \bigcup \{\overline{P}(l) \mid l \in \overline{P}(\lambda_{\mathrm{p}})\} && \langle \text{definition of } \cup \rangle \\
\triangleq \;& \star^{\mathtt{¤}}_{\lambda}[\![\mathtt{p}]\!]\,\overline{P} && \langle \text{definition (46.41) of } \star^{\mathtt{¤}}_{\lambda}[\![\mathtt{p}]\!] \rangle
\end{aligned}$$

$$\begin{aligned}
- \;& \lambda_{\mathrm{z}} \mapsto \alpha^{\equiv}(\{\mu(\lambda_{\mathrm{z}}) \in \gamma^{\equiv}(\overline{P}(\lambda_{\mathrm{z}})) \mid \mu(\mu(\lambda_{\mathrm{p}})) \in \chi \cap \gamma^{\equiv}(\overline{P}(\mu(\lambda_{\mathrm{p}}))\}) \\
& && \langle \text{soundness condition (46.25)} \rangle \\
= \;& \lambda_{\mathrm{z}} \mapsto \alpha^{\equiv}(\{\mu(\lambda_{\mathrm{z}}) \in (\overline{P}(\lambda_{\mathrm{z}}) \cup \{\texttt{nil}\} \cup \mathbb{Z}) \mid \mu(\mu(\lambda_{\mathrm{p}})) \in \chi \cap (\overline{P}(\mu(\lambda_{\mathrm{p}})) \cup \{\texttt{nil}\} \cup \mathbb{Z})\}) \\
& && \langle \text{definition (46.36) of } \gamma^{\equiv} \rangle \\
= \;& \lambda_{\mathrm{z}} \mapsto \alpha^{\equiv}(\{\mu(\lambda_{\mathrm{z}}) \in \overline{P}(\lambda_{\mathrm{z}}) \mid \mu(\mu(\lambda_{\mathrm{p}})) \in \chi \cap \overline{P}(\mu(\lambda_{\mathrm{p}}))\}) \\
& && \langle \overline{P} \in \mathbb{L} \to \wp(\mathbb{L}),\ \chi \in \wp(\mathbb{L}), \text{ and } \mathbb{L} \cap (\{\texttt{nil}\} \cup \mathbb{Z}) = \emptyset \rangle \\
= \;& \lambda_{\mathrm{z}} \mapsto \{\mu(\lambda_{\mathrm{z}}) \in \overline{P}(\lambda_{\mathrm{z}}) \mid \mu(\mu(\lambda_{\mathrm{p}})) \in \chi \cap \overline{P}(\mu(\lambda_{\mathrm{p}}))\} \\
& && \langle \text{definition (46.36) of } \alpha^{\equiv} \text{ with } \overline{P}(\lambda_{\mathrm{z}}) \in \wp(\mathbb{L}) \rangle \\
\dot{\sqsubseteq}^{\mathtt{¤}} \;& (\!|\, \chi \cap \star^{\mathtt{¤}}_{\lambda}[\![\mathtt{p}]\!]\,\overline{P} = \emptyset \;?\; \dot{\emptyset} \;\mathbin{\$}\; \overline{P} \,|\!) \\
&\quad \langle \text{if } \chi \cap \star^{\mathtt{¤}}_{\lambda}[\![\mathtt{p}]\!]\,\overline{P} = \emptyset \text{ then } \chi \cap \overline{P}(\mu(\lambda_{\mathrm{p}})) = \emptyset \text{ otherwise } \overline{P} \text{ is an overapproxima-} \\
&\quad \text{tion for } \dot{\sqsubseteq}^{\mathtt{¤}} \triangleq \subseteq \rangle \\
= \;& \star^{\mathtt{¤}\,-1}[\![\mathtt{p}]\!]\,\lambda\,\chi\,\overline{P} && \langle \text{definition (46.41) of } \star^{\mathtt{¤}\,-1}[\![\mathtt{p}]\!] \rangle \quad \square
\end{aligned}$$

46.7.4 Abstract Reachability Points-to Semantics of Tests

The Cartesian abstract forward reachability semantics of an arithmetic or pointer Boolean expressions is provided by (46.30) using the operations of the points-to abstract domain (46.39).

Exercise 46.42 (Project: points-to abstract domain) Instantiate the flow-sensitive abstract interpreter of exercise 46.11 to the points-to abstract domain of section 46.7.2.

46.8 Flow-insensitive Potential Points-to Analysis

To get Lars Ole Andersen's points-to analysis [43], we simply have to instantiate the flow-insensitive abstract interpreter of section 45.3 defined by (45.2)–(45.11) extended by (46.12) and (46.13) for flow-insensitive points-to analysis. $\widehat{\mathcal{S}}^{\overline{\equiv}}_{\mathfrak{g}}[\![\mathsf{S}]\!]\,\lambda \in (\mathbb{L} \to \wp(\mathbb{L})) \to (\mathbb{L} \to \wp(\mathbb{L}))$ where λ is a symbolic encoding of memory addresses λ_{x} assigned statically to pointer variables x by the compiler, loader, and linker. For example, λ_{x} can be x or the variables can be totally ordered and λ_{x} is the rank of x in this total order. If $\overline{P} \in \mathbb{L} \to \wp(\mathbb{L})$ assigns a set of locations $\overline{P}(\lambda_{\mathsf{x}})$ to which location λ_{x} of x may point to before executing the program component $\mathsf{S} \in \mathbb{P}\mathrm{c}$, $\widehat{\mathcal{S}}^{\overline{\equiv}}_{\mathfrak{g}}[\![\mathsf{S}]\!]\,\lambda\,\overline{P}\,\lambda_{\mathsf{x}}$ provides the set of locations $\overline{P}(\lambda_{\mathsf{x}})$ to which location λ_{x} of x may point to after executing S.

We obtain Lars Ole Andersen's [43] points-to flow-insensitive analysis $\widehat{\mathcal{S}}^{\overline{\equiv}}_{\mathfrak{g}}$ as the instantiation of the flow-insensitive abstract interpreter $\widehat{\mathcal{S}}^{\mathfrak{g}\sharp}_{\lambda}(\mathbb{D}^{\times}(\mathbb{D}^{\overline{\equiv}}_{\lambda}))$ parameterized by the Cartesian abstract domain $\mathbb{D}^{\times}(\mathbb{D}^{\overline{\equiv}}_{\lambda})$ itself parameterized by the points-to abstract domain $\mathbb{D}^{\overline{\equiv}}_{\lambda}$. We provide the following expansion (but the modular implementation is much preferable).

Cartesian flow-insensitive points-to abstract semantics

$$\widehat{\mathcal{S}}^{\overline{\equiv}}_{\mathfrak{g}} \in \mathbb{S} \to \mathbb{L}_{\lambda} \to (\mathbb{L}_{\lambda} \to \wp(\mathbb{L}_{\lambda})) \to (\mathbb{L}_{\lambda} \to \wp(\mathbb{L}_{\lambda}))$$

Cartesian flow-insensitive points-to semantics of a program $\mathsf{P} ::= \mathsf{Sl}\ \ell'$

$$\widehat{\mathcal{S}}^{\overline{\equiv}}_{\mathfrak{g}}[\![\mathsf{S}]\!]\,\lambda\,\overline{P} \triangleq \widehat{\mathcal{S}}^{\overline{\equiv}}_{\mathfrak{g}}[\![\mathsf{Sl}]\!]\,\lambda\,\overline{P} \qquad (46.43)\ [\text{from } (45.2)]$$

Cartesian flow-insensitive points-to semantics of a statement list $\mathsf{Sl} ::= \mathsf{Sl}'\ \mathsf{S}$

$$\widehat{\mathcal{S}}^{\overline{\equiv}}_{\mathfrak{g}}[\![\mathsf{Sl}]\!]\,\lambda\,\overline{P} = \widehat{\mathcal{S}}^{\overline{\equiv}}_{\mathfrak{g}}[\![\mathsf{Sl}']\!]\,\lambda\,\overline{P}\,\dot{\cup}\,\widehat{\mathcal{S}}^{\overline{\equiv}}_{\mathfrak{g}}[\![\mathsf{S}]\!]\,\lambda\,\overline{P} \qquad (46.44)\ [\text{from } (45.3)]$$

Cartesian flow-insensitive points-to semantics of an empty statement list $\mathsf{Sl} ::= \epsilon$

$$\widehat{\mathcal{S}}^{\overline{\equiv}}_{\mathfrak{g}}[\![\mathsf{Sl}]\!]\,\lambda\,\overline{P} = \overline{P} \qquad (46.45)\ [\text{from } (45.4)]$$

Cartesian flow-insensitive points-to semantics of an assignment statement $\mathsf{S} ::= \mathsf{x} = \mathsf{A}\,;$

$$\widehat{\mathcal{S}}^{\overline{\equiv}}_{\mathfrak{g}}[\![\mathsf{S}]\!]\,\lambda\,\overline{P} = \overline{P}[\lambda_{\mathsf{x}} \leftarrow \overline{P}(\lambda_{\mathsf{x}}) \cup \mathcal{A}^{\overline{\equiv}}_{\lambda}[\![\mathsf{A}]\!]\,\overline{P}] \qquad (46.46)\ [\text{from } (45.5)]$$

Cartesian flow-insensitive points-to semantics of a dereferenced pointer assignment statement $\mathsf{S} ::= \star\mathsf{p} = \mathsf{A}\,;$

$$\widehat{\mathcal{S}}^{\overline{\equiv}}_{\mathfrak{g}}[\![\mathsf{S}]\!]\,\lambda\,\overline{P} = l \mapsto \overline{P}(l) \cup (\!|\, l \in \overline{P}(\lambda_{\mathsf{p}})\ ?\ \mathcal{A}^{\overline{\equiv}}_{\lambda}[\![\mathsf{A}]\!]\,\overline{P} : \varnothing \,|\!) \qquad (46.47)\ [\text{from } (46.13)]$$

Cartesian flow-insensitive points-to semantics of a skip statement $\mathsf{S} ::= ;$

$$\widehat{\mathcal{S}}^{\overline{\equiv}}_{\mathfrak{g}}[\![\mathsf{S}]\!]\,\lambda\,\overline{P} = \overline{P} \qquad (46.48)\ [\text{from } (45.6)]$$

Cartesian flow-insensitive points-to semantics of a conditional statement $\mathsf{S} ::= \mathtt{if}\,(\mathsf{B})\,\mathsf{S}_t$

$$\widehat{\mathcal{S}}^{\overline{\equiv}}_{\mathfrak{g}}[\![\mathsf{S}]\!]\,\lambda\,\overline{P} = \overline{P}\,\dot{\cup}\,\widehat{\mathcal{S}}^{\overline{\equiv}}_{\mathfrak{g}}[\![\mathsf{S}_t]\!]\,\lambda\,(\mathrm{test}^{\overline{\equiv}}_{\lambda}[\![\mathsf{B}]\!]\,\overline{P}) \qquad (46.49)\ [\text{from } (45.7)]$$

Cartesian flow-insensitive points-to semantics of a conditional statement $\mathsf{S} ::= \mathtt{if}\,(\mathsf{B})\,\mathsf{S}_t\ \mathtt{else}\ \mathsf{S}_f$

$$\widehat{\mathcal{S}}^{\mp}_{g}[\![\mathsf{S}]\!]\,\lambda\,\overline{P} \;=\; \overline{P}\,\dot{\cup}\,\widehat{\mathcal{S}}^{\mp}_{g}[\![\mathsf{S}_t]\!]\,\lambda\,(\mathsf{test}^{\mp}_{\lambda}[\![\mathsf{B}]\!]\,\overline{P})\,\dot{\cup}\;\widehat{\mathcal{S}}^{\mp}_{g}[\![\mathsf{S}_f]\!]\,\lambda\,(\overline{\mathsf{test}}^{\mp}_{\lambda}[\![\mathsf{B}]\!]\,\overline{P}) \qquad (46.50)\ [\text{from } (45.8)]$$

Cartesian flow-insensitive points-to semantics of an iteration statement S ::= **while** ℓ (B) S_b

$$\widehat{\mathcal{S}}^{\mp}_{g}[\![\mathsf{S}]\!]\,\lambda\,\overline{P} \;=\; \mathsf{lfp}^{\dot{\subseteq}}\,(\mathcal{F}^{\mp}_{g}[\![\mathtt{while}\,\ell\,(\mathsf{B})\,\mathsf{S}_b]\!]\,\lambda\,\overline{P}) \qquad (46.51)\ [\text{from } (45.8)]$$

$$\mathcal{F}^{\mp}_{g}[\![\mathtt{while}\,\ell\,(\mathsf{B})\,\mathsf{S}_b]\!]\,\lambda\,\overline{P}\,X \;=\; \overline{P}\,\dot{\cup}\,\widehat{\mathcal{S}}^{\mp}_{g}[\![\mathsf{S}_b]\!]\,\lambda\,(\mathsf{test}^{\mp}_{\lambda}[\![\mathsf{B}]\!]\,X)$$

Cartesian flow-insensitive points-to semantics of a break statement S ::= ℓ **break** ;

$$\widehat{\mathcal{S}}^{\mp}_{g}[\![\mathsf{S}]\!]\,\lambda\,\overline{P} \;=\; \overline{P} \qquad (46.52)\ [\text{from } (45.10)]$$

Cartesian flow-insensitive points-to semantics of a compound statement S ::= { Sl }

$$\widehat{\mathcal{S}}^{\mp}_{g}[\![\mathsf{S}]\!]\,\lambda\,\overline{P} \;=\; \widehat{\mathcal{S}}^{\mp}_{g}[\![\mathsf{Sl}]\!]\,\lambda\,\overline{P} \qquad (46.53)\ [\text{from } (45.11)]$$

Proof

— For (46.46),

$\overline{P}\,\dot{\cup}\,\mathsf{assign}^{\mp}[\![\mathsf{x},\mathsf{A}]\!]\,\lambda\,\overline{P}$ ⦃from (45.5)⦄

$= \overline{P}\,\dot{\cup}\,\overline{P}[\lambda_{\mathsf{x}} \leftarrow \mathcal{A}^{\mp}_{\lambda}[\![\mathsf{A}]\!]\overline{P}]$ ⦃ definition (46.22) of $\mathsf{assign}^{\mp}[\![\mathsf{x},\mathsf{A}]\!]\,\lambda\,\overline{P}$ ⦄

$= \overline{P}[\lambda_{\mathsf{x}} \leftarrow \overline{P}(\lambda_{\mathsf{x}}) \cup \mathcal{A}^{\mp}_{\lambda}[\![\mathsf{A}]\!]\overline{P}]$ ⦃pointwise definitions of $\dot{\cup}$ and (19.10) of assignment ⦄

— For (46.47),

$\overline{P}\,\dot{\sqcup}^{\mp}\,\mathsf{assign}^{\mp}[\![\star\mathsf{p},\mathsf{A}]\!]\,\lambda\,\overline{P}$ ⦃from (46.13)⦄

$= \overline{P}\,\dot{\cup}\,l \mapsto \overline{P}(l) \cup (\!|\, l \in \overline{P}(\lambda_{\mathsf{p}})\ ?\ \mathcal{A}^{\mp}_{\lambda}[\![\mathsf{A}]\!]\,\overline{P} \mathbin{⁏} \varnothing \,|\!)$

⦃definition (46.22) of $\mathsf{assign}^{\mp}[\![\star\mathsf{p},\mathsf{A}]\!]\,\lambda\,\overline{P} \triangleq l \mapsto \overline{P}(l) \sqcup^{\mp}_{\lambda} (\!|\, \boldsymbol{\lambda}^{\mp}_{\lambda}[\![x]\!] \sqsubseteq^{\mp}_{\lambda} \overline{P}(\lambda_{\mathsf{p}})\ ?\ \mathcal{A}^{\mp}_{\lambda}[\![\mathsf{A}]\!]\,\overline{P} \mathbin{⁏} \bot^{\mp} \,|\!)$ because $\alpha^{\mp}(\{\lambda_{\mathsf{x}}\}) = \{\lambda_{\mathsf{x}}\} \cap \mathbb{L} = \{\lambda_{\mathsf{x}}\} \triangleq \boldsymbol{\lambda}^{\mp}_{\lambda}[\![y]\!]$⦄

$= l \mapsto \overline{P}(l) \cup (\!|\, l \in \overline{P}(\lambda_{\mathsf{p}})\ ?\ \mathcal{A}^{\mp}_{\lambda}[\![\mathsf{A}]\!]\,\overline{P} \mathbin{⁏} \varnothing \,|\!)$ ⦃pointwise definition of $\dot{\cup}$⦄ □

Example 46.54 Continuing examples 46.1 and 46.5, the abstract semantics $\widehat{\mathcal{S}}^{\mp}_{g}[\![\mathsf{Sl}]\!]\,\lambda\,\overline{P}_0$ for an abstract precondition $\overline{P}_0 = \dot{\varnothing}$ (e.g. pointer variables are initially `nil`) are

$$\overline{P}_0 \xrightarrow{\texttt{x = \&p;}} \overline{P}_0[\lambda_{\mathsf{x}} \leftarrow \{\lambda_{\mathsf{p}}\}]$$
$$\xrightarrow{\texttt{p = \&x;}} \overline{P}_0[\lambda_{\mathsf{x}} \leftarrow \{\lambda_{\mathsf{p}}\}][\lambda_{\mathsf{p}} \leftarrow \{\lambda_{\mathsf{x}}\}]$$
$$\xrightarrow{\texttt{q = \&p;}} \overline{P}_0[\lambda_{\mathsf{x}} \leftarrow \{\lambda_{\mathsf{p}}\}][\lambda_{\mathsf{p}} \leftarrow \{\lambda_{\mathsf{x}}\}][\lambda_{\mathsf{q}} \leftarrow \{\lambda_{\mathsf{p}}\}]$$
$$\xrightarrow{\texttt{r = *q;}} \overline{P}_0[\lambda_{\mathsf{x}} \leftarrow \{\lambda_{\mathsf{p}}\}][\lambda_{\mathsf{p}} \leftarrow \{\lambda_{\mathsf{x}}\}][\lambda_{\mathsf{q}} \leftarrow \{\lambda_{\mathsf{p}}\}][\lambda_{\mathsf{r}} \leftarrow \{\lambda_{\mathsf{x}}\}]$$
$$\xrightarrow{\texttt{*r = \&r;}} \overline{P}_0[\lambda_{\mathsf{x}} \leftarrow \{\lambda_{\mathsf{p}}, \lambda_{\mathsf{r}}\}][\lambda_{\mathsf{p}} \leftarrow \{\lambda_{\mathsf{x}}\}][\lambda_{\mathsf{q}} \leftarrow \{\lambda_{\mathsf{p}}\}][\lambda_{\mathsf{r}} \leftarrow \{\lambda_{\mathsf{x}}\}]$$

Note the imprecision due to flow-insensitivity, x is found to point either to λ_{p} or λ_{r} on program exit.

Theorem 46.55 (soundness of the flow-insensitive potential points-to analysis) The points-to analysis is sound in that $\forall \mathtt{S} \in \mathbb{S} \,.\, \ddot{\alpha}^{\equiv}_{\lambda}(\widehat{\mathcal{S}}^{\vec{\varrho}}_{\mathfrak{g}}[\![\mathtt{S}]\!]\,\lambda) \mathrel{\dot{\subseteq}} \widehat{\mathcal{S}}^{\equiv}_{\mathfrak{g}}[\![\mathtt{S}]\!]\,\lambda$.

Proof of theorem 46.55 The reachability semantics is an instance of the flow-sensitive abstract interpreter of section 21.2 (and the language extension for assignment and tests (46.3), (46.4), and (46.6) handled in this chapter 46). This flow-sensitive abstract interpreter is overapproximated by the flow-insensitive abstract interpreter of chapter 45 (and the language extension (46.12) and (46.13) for assignment and tests handled in this chapter 46) which, by theorem 45.14 and theorem 45.15 (and the complement in this chapter) are well defined and sound. Both flow-sensitive and flow-insensitive abstract interpreters can be overapproximated by a Cartesian abstraction, which is well defined and sound by theorem 46.34. This theorem 46.34 also shows that a further Cartesian abstraction of a Cartesian abstraction is also well defined and sound. From the approximate abstraction (46.37), the points-to semantics of equations 46.43–46.53 is a sound instance of this flow-insensitive abstract interpreter and so is itself well defined and sound. □

A flow-sensitive version of Lars Ole Andersen's points-to analysis [43] is $\widehat{\mathcal{S}}^{\equiv} = \widehat{\mathcal{S}}^{\alpha}(\mathbb{D}^{\times}(\mathbb{D}^{\equiv}_{\lambda}))$, obtained by instantiation of the flow-sensitive abstract interpreter.

Exercise 46.56 (Project: points-to analysis) Instantiate the flow-sensitive or flow-insensitive abstract interpreter of exercise 46.15 to the points-to abstract domain of section 46.7.2. For example `l1: x = &z; l2: y = &z; l3: z = &x; l4: *y = *x; l5:` yields `l4: x=[\z], y=[\z], z=[\x];` and `l5: x=[\z], y=[\z], z=[\z];` where `\z` is λ_z for the flow-sensitive analysis and `x=[\z], y=[\z], z=[\x,\z];` for the flow-insensitive one.

Exercise 46.57 (Project: type inference for pointers) Implement a flow-insensitive static analysis to infer the type $t ::= \mathtt{int} \mid \star t$ of the program variables.

Exercise 46.58 Continuing exercise 46.7, extend the points-to analysis to the memory allocation statement `x = malloc()`. There will be now potentially infinitely many allocated concrete locations that can be abstracted to one single abstract location, one abstract location per allocation statement, or by using a widening.

46.8.1 Constraint Resolution

Lars Ole Andersen's points-to analysis [43] is presented as a constraint-solving problem. As shown in section 23.3, the least solution to equation $X = F(X)$ is also, by Tarski's fixpoint theorem 15.6, the least solution to constraints $F(X) \subseteq X$, which for powerset abstract domains can be decomposed into more elementary constraints using the transformations $X \cup Y \subseteq Z \mapsto X \subseteq Z \wedge Y \subseteq Z$, $\{\lambda_{\mathsf{x}}\} \subseteq X \mapsto \lambda_{\mathsf{x}} \in X$, and the inference rules $\dfrac{\lambda_{\mathsf{x}} \in X,\ X \subseteq Y}{\lambda_{\mathsf{x}} \in Y}$, $\dfrac{\lambda_{\mathsf{q}} \in \overline{P}(\lambda_{\mathsf{p}}),\ \lambda_{\mathsf{r}} \in \overline{P}(\mathsf{q}),\ \displaystyle\bigcup_{\lambda_{\mathsf{q}} \in \overline{P}(\lambda_{\mathsf{p}})} \overline{P}(\mathsf{q}) \subseteq \overline{P}(\mathsf{r})}{\lambda_{\mathsf{r}} \in \overline{P}(\mathsf{r})}$ etc. [251]. These rewritings make the transformer F not immediately recognizable [250, 251]. Solving the constraints is then a fixpoint iteration propagating constraints [47].

Exercise 46.59 Modify the points-to semantics $\widehat{\mathcal{S}}^{\overline{\equiv}}_{\mathrm{g}}[\![\mathsf{S}]\!]\,\lambda\,\overline{P}$ to directly generate the constraints to be satisfied by the solution, and prove the soundness of the points-to analysis (including the constraint solver).

46.9 Flow-insensitive Potential Extrapolated Points-to Analysis

The points-to analysis of Bjarne Steensgaard [908] is flow-insensitive and potential (may point to). It can be understood as Andersen analysis equipped with a widening to be performed at each step of the flow-insensitive analysis of the program P.

$$\nabla^{\overline{\equiv}}_{\lambda} P \;\triangleq\; \mathsf{lfp}^{\subseteq}\, X \mapsto P \;\dot{\cup} \bigcup_{\substack{\mathsf{p}\,\in\,\mathsf{vars}[\![\mathsf{P}]\!] \\ X(\lambda_{\mathsf{p}}) = \{\lambda_{\mathsf{q}_1},\ldots,\lambda_{\mathsf{q}_n}\} \\ i\in[1,n],\; n>1}}^{\cdot} X[\lambda_{\mathsf{q}_i} \leftarrow \bigcup_{j=1}^{n} X(\lambda_{\mathsf{q}_j})]$$

The widening $\nabla^{\overline{\equiv}}_{\lambda}$ collapses locations of pointers of which the locations appear in the same pointed-to set. This reduces the number of pointed-to sets to be considered and decreases precision.

Example 46.60

$\dot{\varnothing}$

```
p = &x; q = &p; r =&q; p' = &y; q' = &p'; r'
=&q';
```

$\{\mathsf{p}\mapsto\{\lambda_{\mathsf{x}}\},\mathsf{q}\mapsto\{\lambda_{\mathsf{p}}\},\mathsf{r}\mapsto\{\lambda_{\mathsf{q}}\},\mathsf{p}'\mapsto\{\lambda_{\mathsf{y}}\},\mathsf{q}'\mapsto\{\lambda_{\mathsf{p}'}\},\mathsf{r}'\mapsto\{\lambda_{\mathsf{q}'}\}\}$

```
r = r';
```

$\{\mathsf{p}\mapsto\{\lambda_{\mathsf{x}}\},\mathsf{q}\mapsto\{\lambda_{\mathsf{p}}\},\mathsf{r}\mapsto\{\lambda_{\mathsf{q}},\lambda_{\mathsf{q}'}\},\mathsf{p}'\mapsto\{\lambda_{\mathsf{y}}\},\mathsf{q}'\mapsto\{\lambda_{\mathsf{p}'}\},\mathsf{r}'\mapsto\{\lambda_{\mathsf{q}'}\}\}$ Andersen

$\{\mathsf{p}\mapsto\{\lambda_{\mathsf{x}}\},\mathsf{q}\mapsto\{\lambda_{\mathsf{p}},\lambda_{\mathsf{p}'}\},\mathsf{r}\mapsto\{\lambda_{\mathsf{q}},\lambda_{\mathsf{q}'}\},\mathsf{p}'\mapsto\{\lambda_{\mathsf{y}}\},\mathsf{q}'\mapsto\{\lambda_{\mathsf{p}},\lambda_{\mathsf{p}'}\},\mathsf{r}'\mapsto\{\lambda_{\mathsf{q}'}\}\}$ Widening iteration 1

$\{\mathsf{p}\mapsto\{\lambda_{\mathsf{x}},\lambda_{\mathsf{y}}\},\mathsf{q}\mapsto\{\lambda_{\mathsf{p}},\lambda_{\mathsf{p}'}\},\mathsf{r}\mapsto\{\lambda_{\mathsf{q}},\lambda_{\mathsf{q}'}\},\mathsf{p}'\mapsto\{\lambda_{\mathsf{x}},\lambda_{\mathsf{y}}\},\mathsf{q}'\mapsto\{\lambda_{\mathsf{p}},\lambda_{\mathsf{p}'}\},\mathsf{r}'\mapsto\{\lambda_{\mathsf{q}'}\}\}$ Widening iteration 2

Precision is lost because, for example, location λ_{y} is not reachable from r.

This widening does not depend upon the fixpoint iterations, so this abstraction can trivially be understood as a Galois connection. A better cost/precision compromise is to use Steensgaard's widening at loop heads for Andersen abstraction. [294, 461, 881, 882] essentially consider variants of this widening. By varying the widening, one can adjust the analysis to a good compromise between cost and precision.

A further abstraction of Bjarne Steensgaard's points-to analysis [238] is as follows. Given $\overline{P}\in\mathbb{L}\to\wp(\mathbb{L})$, define $\lambda_{\mathsf{x}}\equiv_{\overline{P}}\lambda_{\mathsf{y}}$ if and only if $\overline{P}(\lambda_{\mathsf{x}})\cap\overline{P}(\lambda_{\mathsf{x}})=\varnothing$. Define $\alpha(\overline{P})\triangleq\{[\lambda_{\mathsf{x}}]_{\equiv_{\overline{P}}}\mid\lambda_{\mathsf{x}}\in\mathbb{L}\}$ so that, from locations in different equivalence classes, it is impossible to reach a shared location.

46.10 The Hierarchy of Points-to Static Analyzers

We have described a hierarchy of points-to static analyzers at various levels of abstraction that can be depicted as in figure 46.10.

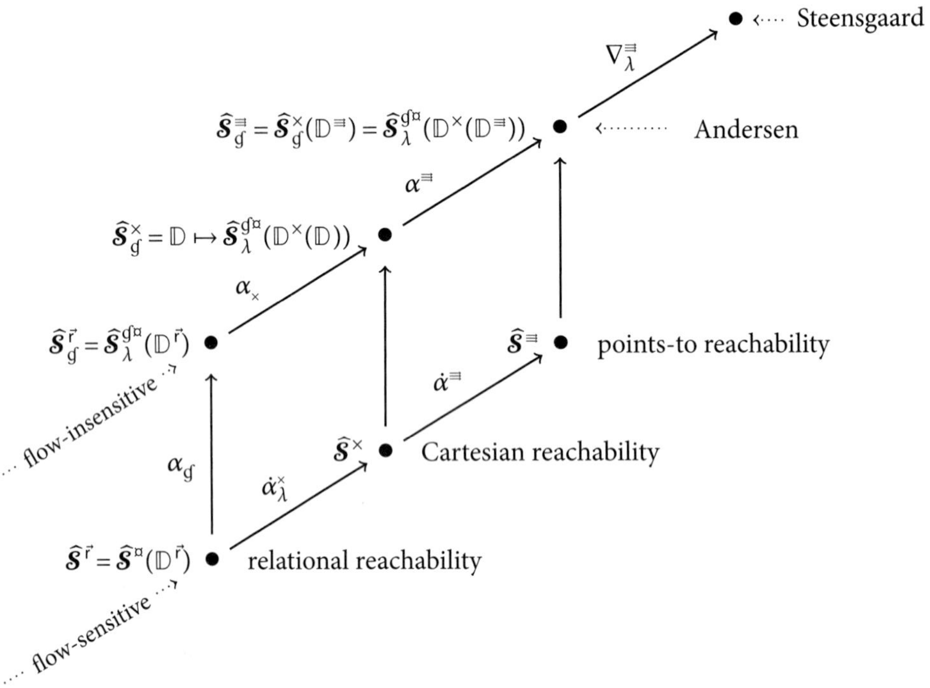

Figure 46.61
Hierarchy of points-to static analyzers

46.11 Memory Models

Points-to analysis is usually performed before all other analyses. Being flow-insensitive, it is imprecise and nevertheless often does not scale up. Moreover the imprecision is forwarded to the later analyses. A flow-sensitive pointer demand-driven analysis performed as a reduced product with the other analyses as, for example, [266] is more precise and efficient, and can be controlled by widenings.

For example, the memory locations of numerical variables will be handled using chapter 33, "Static Interval Analysis," chapter 38, "Linear Equality Analysis," chapter 40, "Zone and Octagon Analysis," or other numerical domains (replacing values of variables in environments by the content of locations in memory).

Languages like C [569] have memory models more complicated than the very simple one considered in this chapter (with pointer arithmetics, fields of structures, `union`, etc.). See [691, 824], and [693, ch. 5] as well as [178, ch. 21] for static analysis using such memory models.

For the analysis of procedures; see [410], which introduces an efficient representation for a memory transformer of a procedure.

Exercise 46.62 (Project: numerical analysis for the pointer language) Modify the flow-sensitive abstract interpreter with a pointer abstraction parameterized by a numerical abstract domain and instantiate with intervals and zones.

46.12 Soundness

By theorem 46.55, the pointer analysis $\widehat{\boldsymbol{\mathcal{S}}}_{\mathfrak{g}}^{\overrightarrow{\equiv}}[\![\mathsf{S}]\!]$ of a program component $\mathsf{S} \in \mathbb{P}\mathbb{c}$ has been proved sound with respect to the reachability semantics $\widehat{\boldsymbol{\mathcal{S}}}_{\mathfrak{g}}^{\vec{\varrho}}$ of section 46.4.

The semantics of languages like C [569] is different. For example, an integer can be used as a pointer, which by (46.2), $\widehat{\boldsymbol{\mathcal{S}}}_{\mathfrak{g}}^{\vec{\varrho}}[\![\mathsf{S}]\!]$ defines as an error. In absence of error, the semantics coincide. This means that the pointer analysis $\widehat{\boldsymbol{\mathcal{S}}}_{\mathfrak{g}}^{\vec{\varrho}}[\![\mathsf{S}]\!]$ is valid for a C program component $\mathsf{S} \in \mathbb{P}\mathbb{c}$ only if, and up to the point of execution where, the reachability semantics $\widehat{\boldsymbol{\mathcal{S}}}^{\vec{\varrho}}[\![\mathsf{S}]\!]$ produces an error. The pointer analysis $\widehat{\boldsymbol{\mathcal{S}}}_{\mathfrak{g}}^{\vec{\varrho}}[\![\mathsf{S}]\!]$ can be easily refined to collect errors. Then the analysis applied to C is sound, in absence of error and for erroneous executions up to the first occurrence of an error.

This is the point of view taken in Astrée [266], that the analysis is guaranteed to be correct in the absence of runtime errors with unpredictable behavior. In the case there is such a runtime error with unpredictable behavior during execution, the analysis is guaranteed to be sound before the occurrence of the first runtime error with unpredictable behavior and no guarantee is provided on the soundness of the analysis after this runtime error with unpredictable behavior. Moreover, all such runtime errors with unpredictable behavior are reported by the static analysis (and should be corrected in priority, if necessary). This provides a precise definition of soundness with respect to the semantics of C.

46.13 Conclusion

Static analyses based on constraint resolution are sometimes understood as different from abstract interpretation. This is often the case for Andersen points-to analysis, which as shown in section 46.7, is a simple abstract interpretation. This is also the case for set-based analysis [475] which is also an abstract interpretation [251].

The advantage of using preexisting constraint solvers like Banshee [595] is to save the programming of the fixpoint iterator.

The inconvenience is that there is neither a universal constraint solver nor a specialized constraint solver for each possible abstract domain, nor a universal way of combining different constraint solvers that use different data formats. Using different solvers for different abstract domains of the abstract interpreter is impractical because it is hard to have different constraint solvers cooperate (e.g., through a reduced product [274]). Moreover their precision/performance ratio can hardly be adapted by adjusting the fixpoint extrapolation and interpolation operations that are often implicitly embedded in the constraint solver. In contrast, our solution is to use abstract domain libraries with standardized interfaces to a reduced product called by the fixpoint iterator and reducer [113, section III.K].

On the basis of the arbitrary claim that "pointer analysis is a prerequisite for many program analyses," many static analyzers start with a points-to analysis and then use it to perform the analysis of value properties of other types. This is an ill-conceived design. It is much preferable to perform the points-to analysis online (e.g. as a component in a reduced product), which is more precise and less costly than an offline analysis (e.g. [113, sections III.F & III.G]). For efficiency, on-demand analysis makes it possible to restrict the pointer variables for which the analysis is performed [476].

[476] introduces a demand-driven [342] version of Andersen's flow-insensitive points-to analysis (to perform only computations that contribute to answering given points-to queries issued by the compiler). [97] is a points-to analysis that handles missing code by combining runtime monitoring and specifications. [556] is a informal comparison of heap static analyses. With the same objective, [316] shows that, more concisely, it is sufficient to compare pointer semantics and their abstractions (and widenings).

By Rice's theorem 9.12, a static analysis problem is never definitely solved. So, starting from classic static analysis algorithms is often an error. This is the case of points-to analysis where classic abstraction choices are not immutable. Understood as arbitrary choices rather than definite solutions, calculational design can be useful for exploring various alternatives just by concentrating on the semantics and varying its abstractions. In particular, the widening can offer a flexible solution to master combinatorial explosion. Of course, automated calculational design would be extremely helpful.

47 Dependency Analysis

Dependency analysis determines how entities (e.g., values of variables) computed in a program execution depend upon other entities (e.g., initial values of variables). For example, (the value of) y depends on (the value of) x in y = x ; (but not if x = y always holds before the assignment). This includes many problems in programming languages. Classic examples are Dorothy Denning and Peter Denning's information flow [312] to check "that a program cannot cause supposedly nonconfidential results to depend on confidential input data"; this was reformulated as a type system by [750, 959] and Mark Weiser's program slicing [976], which consists of determining the set of programs statements that may affect the values at some point of interest (e.g., to locate the source of errors at that point of interest). These definitions are syntactic, that is, based on the program syntax. For example, y depends on x in y=x−x. Another classic example is Ellis Cohen's and Joseph Goguen and José Meseguer noninterference [199, 421, 422], such as public output data that do not depend on private input data. For example, the public y does not depend on the private x in y = x − x ; because y is always 0. The definition of interference is semantic, that is, based on the program semantics. Unfortunately, it applies only to input-output semantics. We introduce abstract dependency semantic generalizing information flow, such as slicing, noninterference, dye, and taint analyses, and so forth. A coarse approximation yields classic syntactic definitions.

Contents

47.1 Syntactic Versus Semantic Dependency

Many definitions of dependency are purely syntactic, that is, depend only on a program control graph and on the syntax of operations performed on data, for example, [618]. This is typical of dataflow analysis considered in chapter 41. Such syntactic definitions are postulated (no justification is given) and imprecise (syntax is a very rough approximation of semantics). This does not mean that it is impossible to provide rigorous soundness proofs (as done in chapter 41 for syntactic liveness/deadness). But such soundness proofs ultimately require a definition of dependency based on the semantics of programs, which is the subject of this chapter.

47.2 Semantic Properties

Recall the maximal trace semantics $\mathcal{S}^{+\infty}[\![\mathsf{S}]\!]$ of chapter 7, which given a finite initialization trace $\pi_0 \in \mathbb{T}^+$ ending on entry $\mathsf{at}[\![\mathsf{S}]\!]$ of statement S, returns the maximal finite or infinite executions $\pi_0 \frown \pi$ of S, extending this initialization trace π_0 by a maximal finite or infinite continuation trace $\pi \in \mathbb{T}^{+\infty}$. There is no such trace if π_0 does not end $\mathsf{at}[\![\mathsf{S}]\!]$. We have considered semantics to be total functions $\mathcal{S} \in \mathbb{T}^+ \to \wp(\mathbb{T}^{+\infty})$ by defining $\mathcal{S}(\pi) \triangleq \emptyset$ when $\pi \notin \mathsf{dom}(\mathcal{S})$. Therefore, up to the isomorphism between a relation and its right image (exercise 11.52), we will consider that $\mathcal{S} \in \wp(\mathbb{T}^+ \times \mathbb{T}^{+\infty})$ (see section 6.10). So up to this isomorphism, semantic properties in $\wp(\mathbb{T}^+ \to \wp(\mathbb{T}^{+\infty}))$ can be considered to be elements of $\wp(\wp(\mathbb{T}^+ \times \mathbb{T}^{+\infty}))$. Our task is to define dependency as such a semantic property. We start with simple cases.

47.3 Functional Dependency

Functional dependency, that is, dependency for pure mathematical functions, is based on the idea that changing a parameter changes the result. For $f(x, y) = x + 1$ on integers, changing the actual parameter x changes $f(x, y)$, while changing the actual parameter y does not change $f(x, y)$. Therefore f depends on x but not on y. The function $f(x, y) = (x - x) + (y - y)$ does not depend on its integer parameters because it is always 0 (a conclusion that a purely syntactic definition cannot reach). The function $f(x, y) = x + y$ depends on both parameters. Let $\mathcal{Fd}^{ni}$ be the property "to be a function of n parameters that depends on its ith parameters, $1 \leqslant i \leqslant n$." Formally,

$$\begin{aligned}\mathcal{Fd}^{ni} &\triangleq \{f \mid \exists x_1, \ldots, x_n, x_i' \,.\, f(x_1, \ldots, x_{i-1}, x_i, x_{i+1}, \ldots, x_n) \neq \\ &\qquad\qquad f(x_1, \ldots, x_{i-1}, x_i', x_{i+1}, \ldots, x_n)\}. \qquad (47.1)\\ \mathcal{Fd} &\triangleq \bigcup_{n\in\mathbb{N}^+} \bigcup_{1\leqslant i\leqslant n} \mathcal{Fd}^{ni} \in \wp(\bigcup_{n\in\mathbb{N}^+} \mathbb{Z}^n \to \mathbb{Z})\end{aligned}$$

$\mathcal{Fd}$ is a property of integer functions of any arity (in $\bigcup_{n\in\mathbb{N}^+} \mathbb{Z}^n \to \mathbb{Z}$). $\mathcal{Fd}$ is the property "to be a function that depends on at least one of its parameters", and its complement $\neg\mathcal{Fd}$ is "to be a constant function." (Here we exclude any "function" that has side effects, for example that uses global variables, that returns a random number, and more generally that is not a pure mathematical function.)

Exercise 47.2 Prove that the semantics $\boldsymbol{S}[\![\mathsf{E}]\!]\rho$ of an expression E in environment ρ in (3.4) does not depend on the values $\rho(\mathsf{x})$ of the variables $\mathsf{x} \notin \mathsf{vars}[\![\mathsf{E}]\!]$ not appearing in the expression E.

47.4 Noninterference

Given low variables L (e.g., *public* respectively *trusted*) and high variables H (e.g., *private/confidential* respectively *untrusted*) noninterference is defined by [421, 422] as "if an execution starts with equal low data then if and when it finishes, the low data are equal (so that changing the high data only cannot change the low ones)." For example, changing the *private* variables H does not change the *public* L variables, so no confidential information in H can flow to the *public* variables L. Similarly, changing the *untrusted* variables H does not change the *trusted* L variables so that spoiled information in H cannot flow to the *trusted* variables L.

The noninterference semantic property is therefore $\mathcal{N}i(L, H) \in \wp(\wp(\mathbb{T}^+ \times \mathbb{T}^\infty))$ such that

$$\begin{aligned}\mathcal{N}i(L, H) \;\triangleq\; & \{\Pi \in \wp(\mathbb{T}^+ \times \mathbb{T}^\infty) \mid \forall \langle \pi_0, \pi\rangle, \langle \pi_0', \pi'\rangle \in \Pi \cap (\mathbb{T}^+ \times \mathbb{T}^+) \,.\\ & \quad (\forall \mathsf{x} \in L \,.\, \varrho(\pi_0)\mathsf{x} = \varrho(\pi_0')\mathsf{x}) \Rightarrow (\forall \mathsf{x} \in L \,.\, \varrho(\pi_0 \frown \pi)\mathsf{x} = \varrho(\pi_0' \frown \pi')\mathsf{x})\}\end{aligned}$$

The maximal trace semantics $\boldsymbol{S}^{+\infty}[\![\mathsf{S}]\!]$ has the noninterference property $\mathcal{N}i(L, H)$ if and only if $\boldsymbol{S}^{+\infty}[\![\mathsf{S}]\!] \in \mathcal{N}i(L, H)$, which considering the collecting semantics, is the property implication $\{\boldsymbol{S}^{+\infty}[\![\mathsf{S}]\!]\} \subseteq \mathcal{N}i(L, H)$.

Notice that following [421, 422], interference in the course of execution or between infinite executions is *not* taken into account by this definition. More generally, the definition of interference has many variants (e.g. [675, 1000]).

Exercise 47.3 [676, section 2.1] Prove that noninterference cannot be defined as a trace property in $\wp(\mathbb{T}^+ \times \mathbb{T}^\infty)$.

47.5 Examples of Dependencies

According to [312], "Information *flows* from object x to object y, denoted $\mathsf{x} \rightsquigarrow \mathsf{y}$, whenever information stored in x is transferred to, or used to derive information transferred to, object y." When $\mathsf{x} \rightsquigarrow \mathsf{y}$ we say that x *flows* to y, and in considering the inverse relation, that y *depends* on x. In this definition [312], $\rightsquigarrow$ is *global* (valid anywhere in the program) and *static* (valid for any execution of the program).

Most definitions of in/dependency ([177, 199, 421, 947, 958]) are of the form "changing (part of) the input (for instance, x) may/should not change (part of) the output (for instance, y)," sometimes including nontermination [947]. Therefore, "information from object x" is the initial value x_0 of variable x and "object y" is the final value of variable y. These definitions are *relational* (that is, relative to a correspondence between initial and final values of variables, or $\perp$ for nontermination) and *static* (valid for any execution of the program).

We would like to consider a *static* definition of dependency that is *local*, that ism relative to a correspondence between the initial values of variables and their values when reaching a program point, if ever, including for nonterminating executions. The following examples illustrate this intuition and show how it may be made more precise.

Example 47.4 (explicit dependency) Consider ℓ_1 y = x ; ℓ_2. Changing the initial value y_0 of y will change the value of y at the entry point ℓ_1. Changing the initial value x_0 of x will not change the value of y at ℓ_1. So at the entry point ℓ_1, y depends on y_0 but not on x_0.

The value of y at exit point ℓ_2 is x_0 so changing the initial value y_0 of y will not change the value of y at ℓ_2. Changing the initial value x_0 of x will change the value of y at ℓ_2. So at ℓ_2, y depends on x_0 but not on y_0. Such a dependency at ℓ_2 is called explicit in [312] because it does not depend on the program control.

We have x $\not\rightsquigarrow$ y at ℓ_1 and x $\rightsquigarrow$ y at ℓ_2. This shows that the program point of interest must be specified and we write x $\not\rightsquigarrow^{\ell_2}$ y and x $\rightsquigarrow^{\ell_2}$ y.

Example 47.5 (implicit dependency) Consider the program $\mathsf{P_a} \triangleq \ell_1$ y = 1 ; **if** ℓ_2 (x == 0) { ; ℓ_4 } ℓ_5. Changing the initial value x_0 of x will change whether program control ℓ_4 is reached or not. If ℓ_4 is reached then the value of y at ℓ_4 will always be 1 so y does not depend on x_0 at ℓ_4 (nor at ℓ_5).

Consider now $\mathsf{P_b} \triangleq \ell_1$ y = 1 ; **if** ℓ_2 (x == 0) { ℓ_3 y = x ; ℓ_4 } ℓ_5. Changing the initial value x_0 of x will change whether program control points ℓ_3 and ℓ_4 are reached or not. If ℓ_4 is reached then the value of y at ℓ_4 will always be 0, so y does not depend on x_0 at ℓ_4. However, depending on the initial value x_0 of x, the value of y at ℓ_5 will be either 1 (when $x_0 \neq 0$) or 0 (when $x_0 = 0$), so y depends on x_0 at ℓ_5. Such a dependency at ℓ_5 is called implicit in [312] because it depends on the program control.

Example 47.6 (timely dependency) Consider the program **while** (0 == 0) ℓ y = x ;. The sequence of values taken by x at ℓ is $x_0, x_0, x_0, \ldots$, and it is $y_0, x_0, x_0, \ldots$ for y. So x depends on x_0 while y depends on x_0 and y_0 at ℓ. For y, considering only one possible value during the iterations would be insufficient to determine the dependency upon initial values, and in general, we have to consider the full sequence of successive values of y at a given program point ℓ.

Example 47.7 (value dependency) Consider the program

while (0 == 0) { ℓ x = x − 1 ; **if** (x == 0) y = y + 1 ; }.

If $x_0 \leqslant 0$, the sequence of values of y at ℓ is the infinite sequence $y_0, y_0, y_0, \ldots$. If $x_0 > 0$, it is $\overbrace{y_0, \ldots, y_0}^{x_0\,\texttt{times}}, y_0 + 1, y_0 + 1, \ldots$. Hence we can find two executions of the program with different initial values x_0 of x such that the sequences of values of y at ℓ have a common prefix but differ at least by one value after that prefix. So, y depends on x_0 at ℓ.

Example 47.8 (timing channel) Consider the program `int x, y;` **while** (x > 0) ℓ x = x − 1; (where we have added a declaration to show that the program involves variable y). If the initial value of x is $x_0 \leqslant 0$ then sequence of values taken by y at ℓ is empty. Otherwise, if $x_0 > 0$, it is $y_0 \cdot y_0 \cdot \ldots \cdot y_0$ repeated x_0 times. Therefore changing x_0 changes this sequence of values. Depending on x_0, we can find sequences of values of y at ℓ that differ in length, but along these sequences we cannot find a point where they differ in value. So, there is no transfer of the information x_0 to y at ℓ and hence no dependency.

In security, this is a covert channel [612, 726] (more precisely a timing channel [847, 849, 990]) that may or may not be considered as observable. Traditionally, in dependency analysis, timing channels are not considered to be at the origin of dependencies [312, 535]. However, timing validation is required, for example, to qualify real-time systems [984].

Example 47.9 (empty observation) Consider the program

$$\texttt{if}\,(\texttt{x} == \texttt{0})\ \{\ \ell_1\ \texttt{y} = \texttt{x}\ ;\ell_2\ \}\ \ell_3.$$

What are the values of y observed at ℓ_2? If x == 0 this is 0, whereas if x != 0 there is no possible observation of the value of y at ℓ_2. So we may consider that an empty observation is a valid observation, in which case y depends on x at ℓ_2. This is certainly a frequent point of view in security. On the contrary, we may exclude empty observations, in which case y does not depend on x at ℓ_2. This is certainly a frequent point of view in compilation, because y is constant at ℓ_2. Notice that in both cases, y depends on x at ℓ_3 because we can observe different values depending on the test.

Exercise 47.10 In ℓ_1 y = 0 ;ℓ_2 y = x ;ℓ_3, does $\texttt{x} \overset{\ell}{\rightsquigarrow} \texttt{y}$ for $\ell \in \{\ell_1, \ell_2, \ell_3\}$?

Exercise 47.11 Consider the programs

$$\texttt{P}_\texttt{a} \triangleq \texttt{if}\ \ell_0\ (\texttt{x} == \texttt{1})\ \{\ \ell_1\ \texttt{y} = \texttt{x}\ ;\ell_2\ \}\ \texttt{else}\ \{\ \ell_3\ \texttt{y} = \texttt{2}\ ;\ell_4\ \};\ell_5\ \texttt{y} = \texttt{3}\ ;\ell_6 \text{ and}$$

$$\texttt{P}_\texttt{b} \triangleq \texttt{if}\ \ell_0\ (\texttt{x} == \texttt{1})\ \{\ \ell_1\ \texttt{y} = \texttt{x}\ ;\ell_2\ \}\ \texttt{else}\ \{\ \ell_3\ \texttt{y} = \texttt{1}\ ;\ell_4\ \};\ell_5\ \texttt{y} = \texttt{3}\ ;\ell_6.$$

What is the dependency of y on the initial value of x at each program points of $\texttt{P}_\texttt{a}$ and $\texttt{P}_\texttt{b}$?

Exercise 47.12 Consider the programs $\texttt{P}_0 \triangleq$ x = 0 ; **while** ℓ (0 == 0) x = x + 1 ; and $\texttt{P}_u \triangleq$ **while** ℓ (0 == 0) x = x + 1 ;. For each of them, does x depends on its initial value x_0 at ℓ ?

Exercise 47.13 Consider the program **while** ℓ_0 (x < 5) { ℓ_1 x = x + 1 ;ℓ_2 y = y + 1 ; }ℓ_3. The relational invariant $\texttt{x} - x_0 = \texttt{y} - y_0$ holds at ℓ_0. Does y depend on x_0 at ℓ_0?

Exercise 47.14 Consider the program

```
ℓ0 i = 0;
ℓ1 y = 0;
while ℓ2 (0 == 0) {
   if ℓ3 (x >= 0 || i % 2 == 0)
      ℓ4 y = y + 1;
   ℓ5 i = i + 1;
}
ℓ6
```

Does y at ℓ_5 depend on the initial value x_0 of x? And at ℓ_4? Answer the same question, changing the iteration condition 0 == 0 to i < 5.

Exercise 47.15 (liveness) Show that liveness/deadness in chapter 41 is a dependency property.

47.6 Semantic Definition of Dependency

Informally, we say that the initial value x_0 of variable x flows to variable y at program point ℓ (or y depends on the initial value of x), written $\mathsf{x} \overset{\ell}{\rightsquigarrow} \mathsf{y}$, if and only if changing the initial value x_0 of x will change the sequence of values taken by y whenever execution reaches program point ℓ.

47.6.1 Sequence of Values of a Variable at a Program Point

Given an initialization trace $\pi_0 \in \mathbb{T}^+$ followed by a nonempty trace $\pi \in \mathbb{T}^{+\infty}$, let us define the sequence $\mathsf{seqval}[\![\mathsf{y}]\!]^\ell(\pi_0, \pi)$ of values of the variable y at program point ℓ along the trace π continuing π_0 as follows:

$$
\begin{aligned}
\mathsf{seqval}[\![\mathsf{y}]\!]^\ell(\pi_0, \ell) &\triangleq \varrho(\pi_0)\mathsf{y} && && (47.16)\\
\mathsf{seqval}[\![\mathsf{y}]\!]^\ell(\pi_0, \ell') &\triangleq \ni && \text{(empty sequence)} && \text{when } \ell' \neq \ell\\
\mathsf{seqval}[\![\mathsf{y}]\!]^\ell(\pi_0, \ell \xrightarrow{\mathsf{a}} \ell''\pi) &\triangleq \varrho(\pi_0)\mathsf{y} \cdot \mathsf{seqval}[\![\mathsf{y}]\!]^\ell(\pi_0 \mathbin{\widehat{\ }} \ell \xrightarrow{\mathsf{a}} \ell'', \ell''\pi)\\
\mathsf{seqval}[\![\mathsf{y}]\!]^\ell(\pi_0, \ell' \xrightarrow{\mathsf{a}} \ell''\pi) &\triangleq \mathsf{seqval}[\![\mathsf{y}]\!]^\ell(\pi_0 \mathbin{\widehat{\ }} \ell' \xrightarrow{\mathsf{a}} \ell'', \ell''\pi) && && \text{when } \ell' \neq \ell
\end{aligned}
$$

$\mathsf{seqval}[\![\mathsf{y}]\!]^\ell(\pi_0, \pi)$ is the empty sequence $\ni$ when ℓ does not appear in π. We rely on intuition that this definition applies to finite and infinite traces (by passing to the limit as in section 7.2).

The sequence of values of variable y at a program point ℓ abstracts away the position in traces where the values are observed. Hence execution time (represented as the number of steps that have been executed to reach a given position in the trace) is abstracted away.

Exercise 47.17 (value observation abstraction) Show that $\mathsf{seqval}[\![\mathsf{y}]\!]^\ell$ defines an abstraction of the program semantics.

47.6.2 Differences Between Sequences of Values of a Variable at a Program Point

The definition of dependency of y on the initial value of x involves the comparison of sequences ω and ω' of the successive values of variable y taken at some program point for two different executions that differ on the initial value of x. By *differ*, we mean that the sequences may have a common prefix but must eventually have a different value at some position in the sequences.

$$\mathsf{diff}(\omega, \omega') \triangleq \exists \omega_0, \omega_1, \omega_1', \nu, \nu' \,.\, \omega = \omega_0 \cdot \nu \cdot \omega_1 \wedge \omega' = \omega_0 \cdot \nu' \cdot \omega_1' \wedge \nu \neq \nu' \qquad (47.18)$$

Observe that $\neg\mathsf{diff}(\omega, \omega')$ implies either that $\omega = \omega'$ (the futures are the same so there is no dependency) or one is a strict prefix of the other (this is a timing channel abstracted away in this definition (47.18) of dependency). Because ω and ω' in (47.18) must contain at least one value; they cannot be empty.

47.6.3 Definition of Value Dependency

Let us define $\mathcal{D}^\ell\langle\mathsf{x}, \mathsf{y}\rangle$ to mean that "the sequence of values of variable y at ℓ depends upon the initial value of x," also written $\mathsf{x} \overset{\ell}{\rightsquigarrow} \mathsf{y}$, to mean that the initial value of x flows to y. So, there are two execution traces whose initial values are the same except for x, for which the sequences of values of y at program point ℓ on these two execution traces do differ.

Definition 47.19 (dependency $\mathcal{D}$)

$$\mathcal{D}\ell\langle \mathsf{x}, \mathsf{y}\rangle \triangleq \{\Pi \in \wp(\mathbb{T}^+ \times \mathbb{T}^{+\infty}) \mid \exists \langle \pi_0, \pi_1\rangle, \langle \pi'_0, \pi'_1\rangle \in \Pi . \quad (\forall \mathsf{z} \in \mathbb{V} \setminus \{\mathsf{x}\} . \varrho(\pi_0)\mathsf{z} = \varrho(\pi'_0)\mathsf{z}) \wedge \mathsf{diff}(\mathsf{seqval}[\![\mathsf{y}]\!]\ell(\pi_0, \pi_1), \mathsf{seqval}[\![\mathsf{y}]\!]\ell(\pi'_0, \pi'_1))\} \tag{47.19}$$

We do not need to require $\varrho(\pi_0)\mathsf{x} \neq \varrho(\pi'_0)\mathsf{x}$ in (47.19), because , by exercises 17.13 and 17.21, the language being deterministic, the computations would be the same for the same initial values of variables.

Exercise 47.20 What would be definition 47.19 for a nondeterministic language (including, for example, inputs or calls to a random number generator)?

Our definition of dependency is a timely (example 47.6) and value dependency (example 47.7) with no empty (example 47.9) and timing channel (example 47.8) observation, and contrary to [312], makes no artificial distinction between explicit (example 47.4) and implicit (example 47.5) flows.

Definition 47.19 is an instance of a general scheme to define dependency by (a) an abstraction of the past (the initial value of variables), (b) an abstraction of the future (seqval[[y]]), (c) the difference between past abstractions (the initial values of variables only differ for x), and (d) the difference between futures (diff). It states that the abstraction of the future depends on the abstraction of the past if and only if there exist two executions with different past abstractions and different future abstractions.

We can now provide a formal definition of the flow notation $\rightsquigarrow^{\ell}_{\mathsf{P}}$ informally introduced in section 47.5.

Definition 47.21 (value dependency flow) At program point ℓ of program P, variable y depends on the initial value of variable x (or $\mathsf{x} \rightsquigarrow^{\ell}_{\mathsf{P}} \mathsf{y}$, that is, the initial value of variable x flows to variable y at program point ℓ) if and only if

$$\mathsf{x} \rightsquigarrow^{\ell}_{\mathsf{P}} \mathsf{y} \triangleq (\boldsymbol{\mathcal{S}}^{+\infty}[\![\mathsf{P}]\!] \in \mathcal{D}\ell\langle \mathsf{x}, \mathsf{y}\rangle). \tag{47.21}$$

$$\mathsf{x} \not\rightsquigarrow^{\ell}_{\mathsf{P}} \mathsf{y} \triangleq \neg(\mathsf{x} \rightsquigarrow^{\ell}_{\mathsf{P}} \mathsf{y})$$

In (47.19) and (47.21) both explicit and implicit flows are comprehended in the same definition.

47.6.4 Prefix Versus Maximal Trace Semantics Based Dependency

Because timing channels are ignored, the use of the prefix trace semantics $\boldsymbol{\mathcal{S}}^*[\![\mathsf{P}]\!]$ is equivalent to that of the maximal trace semantics $\boldsymbol{\mathcal{S}}^{+\infty}[\![\mathsf{P}]\!]$ in the definition (47.21) of dependency. This is formally stated by the following:

Lemma 47.23 (value dependency for finite prefix traces)

$$\mathsf{x} \rightsquigarrow^{\ell}_{\mathsf{P}} \mathsf{y} = (\boldsymbol{\mathcal{S}}^*[\![\mathsf{P}]\!] \in \mathcal{D}\ell\langle \mathsf{x}, \mathsf{y}\rangle).$$

Proof of lemma 47.23 Assume that $\mathsf{x} \overset{\ell}{\rightsquigarrow}_{\mathsf{P}} \mathsf{y}$ that is, by (47.21), $\boldsymbol{\mathcal{S}}^{+\infty}[\![\mathsf{P}]\!] \in \mathcal{D}\ell\langle\mathsf{x}, \mathsf{y}\rangle$. Then by (47.19), $\exists\langle\pi_0, \pi_1\ell\pi_2\rangle, \langle\pi'_0, \pi'_1\ell\pi'_2\rangle \in \boldsymbol{\mathcal{S}}^{+\infty}[\![\mathsf{P}]\!] . (\forall \mathsf{z} \in \mathbb{V} \setminus \{\mathsf{x}\} . \varrho(\pi_0)\mathsf{z} = \varrho(\pi'_0)\mathsf{z}) \wedge \mathsf{diff}(\mathsf{seqval}[\![\mathsf{y}]\!]^{\ell}(\pi_0 \frown \pi_1\ell, \ell\pi_2), \mathsf{seqval}[\![\mathsf{y}]\!]^{\ell}(\pi'_0 \frown \pi'_1\ell, \ell\pi'_2))$. Because ℓ does occur in $\pi_1\ell\pi_2$ and in $\pi'_1\ell\pi'_2$, we can consider the first occurrence, that is, assume that $\ell \notin \pi_1 \wedge \ell \notin \pi'_1$. By definition (47.16) of $\mathsf{seqval}[\![\mathsf{y}]\!]^{\ell}$ and (47.18) of diff, $\ell\pi_2$ has the form $\ell\overline{\pi}_0\ell' \xrightarrow{\mathsf{a}} \ell\overline{\pi}_2$ whereas $\ell\pi'_2$ has the form $\ell\overline{\pi}_0\ell' \xrightarrow{\mathsf{a}'} \ell\overline{\pi}'_2$ with $\varrho(\ell\overline{\pi}_0\ell' \xrightarrow{\mathsf{a}} \ell)\mathsf{y} \neq \varrho(\ell\overline{\pi}_0\ell' \xrightarrow{\mathsf{a}'} \ell)\mathsf{y}$. We have $\langle\pi_0, \ell\overline{\pi}_0\ell' \xrightarrow{\mathsf{a}} \ell\rangle, \langle\pi'_0, \ell\overline{\pi}_0\ell' \xrightarrow{\mathsf{a}'} \ell\rangle \in \boldsymbol{\mathcal{S}}^{*}[\![\mathsf{S}]\!]$, $\mathsf{diff}(\mathsf{seqval}[\![\mathsf{y}]\!]^{\ell}(\pi_0 \frown \pi_1\ell, \ell\overline{\pi}_0\ell' \xrightarrow{\mathsf{a}} \ell), \mathsf{seqval}[\![\mathsf{y}]\!]^{\ell}(\pi'_0 \frown \pi'_1\ell, \ell\overline{\pi}_0\ell' \xrightarrow{\mathsf{a}'} \ell)) = \mathsf{tt}$, and $\boldsymbol{\mathcal{S}}^{*}[\![\mathsf{P}]\!] \in \wp(\mathbb{T}^{+} \times \mathbb{T}^{+}) \subseteq \wp(\mathbb{T}^{+} \times \mathbb{T}^{+\infty})$ so $\boldsymbol{\mathcal{S}}^{*}[\![\mathsf{P}]\!] \in \mathcal{D}\ell\langle\mathsf{x}, \mathsf{y}\rangle$.

Inversely, assume that $\boldsymbol{\mathcal{S}}^{*}[\![\mathsf{P}]\!] \in \mathcal{D}\ell\langle\mathsf{x}, \mathsf{y}\rangle$. Then by (47.19), $\exists\langle\pi_0, \pi_1\ell\pi_2\rangle, \langle\pi'_0, \pi'_1\ell\pi'_2\rangle \in \boldsymbol{\mathcal{S}}^{*}[\![\mathsf{P}]\!] . (\forall \mathsf{z} \in \mathbb{V} \setminus \{\mathsf{x}\} . \varrho(\pi_0)\mathsf{z} = \varrho(\pi'_0)\mathsf{z}) \wedge \ell \notin \pi_1 \wedge \ell \notin \pi'_1 \wedge \mathsf{diff}(\mathsf{seqval}[\![\mathsf{y}]\!]^{\ell}(\pi_0 \frown \pi_1\ell, \ell\pi_2), \mathsf{seqval}[\![\mathsf{y}]\!]^{\ell}(\pi'_0 \frown \pi'_1\ell, \ell\pi'_2))$. $\ell\pi_2$ and $\ell\pi'_2$ can be maximally extended to a maximal finite trace or, by definition of lim, (7.6), (7.7), and $\boldsymbol{\mathcal{S}}^{*}[\![\mathsf{S}]\!]$ into an infinite trace $\ell\pi_2 \frown \pi_3$ and $\ell\pi'_2 \frown \pi'_3$. (The proof that a non-maximal prefix trace can be extended is by structural induction, showing that if a finite trace is defined by a rule of $\boldsymbol{\mathcal{S}}^{*}[\![\mathsf{S}]\!]$ in section 6.6, then there is another rule that can be used to extend this prefix trace.) Then, by definition (47.16) of $\mathsf{seqval}[\![\mathsf{y}]\!]^{\ell}$ and (47.18) of diff, $\mathsf{diff}(\mathsf{seqval}[\![\mathsf{y}]\!]^{\ell}(\pi_0 \frown \pi_1\ell, \ell\pi_2 \frown \pi_3), \mathsf{seqval}[\![\mathsf{y}]\!]^{\ell}(\pi'_0 \frown \pi'_1\ell, \ell\pi'_2 \frown \pi'_3)) = \mathsf{tt}$. Moreover $\langle\pi_0, \pi_1\ell\pi_2 \frown \pi_3\rangle, \langle\pi'_0, \pi'_1\ell\pi'_2 \frown \pi'_3\rangle \in \boldsymbol{\mathcal{S}}^{+\infty}[\![\mathsf{S}]\!]$ and $\boldsymbol{\mathcal{S}}^{+\infty}[\![\mathsf{P}]\!] \in \wp(\mathbb{T}^{+} \times \mathbb{T}^{+\infty})$ so $\boldsymbol{\mathcal{S}}^{+\infty}[\![\mathsf{P}]\!] \in \mathcal{D}\ell\langle\mathsf{x}, \mathsf{y}\rangle$.

We conclude by antisymmetry of inclusion that $\{\mathsf{x} \overset{\ell}{\rightsquigarrow}_{\mathsf{P}} \mathsf{y} \mid \boldsymbol{\mathcal{S}}^{+\infty}[\![\mathsf{P}]\!] \in \mathcal{D}\ell\langle\mathsf{x}, \mathsf{y}\rangle\} = \{\mathsf{x} \overset{\ell}{\rightsquigarrow}_{\mathsf{P}} \mathsf{y} \mid \boldsymbol{\mathcal{S}}^{*}[\![\mathsf{P}]\!] \in \mathcal{D}\ell\langle\mathsf{x}, \mathsf{y}\rangle\}$. □

47.6.5 Value Dependency Abstraction

The value dependency abstraction of a semantic property $\mathcal{S} \in \wp(\wp(\mathbb{T}^{+} \times \mathbb{T}^{+\infty}))$ is

Definition 47.24 (value dependency abstraction α^{d})

$$\alpha^{\mathrm{d}}(\mathcal{S})\ell \triangleq \{\langle\mathsf{x}, \mathsf{y}\rangle \mid \mathcal{S} \subseteq \mathcal{D}\ell\langle\mathsf{x}, \mathsf{y}\rangle\} \in \mathbb{P}^{\mathrm{d}}(\ell) \triangleq \wp(\mathbb{V} \times \mathbb{V}) \quad (47.25)$$

Lemma 47.26 There is a Galois connection $\langle\wp(\wp(\mathbb{T}^{+} \times \mathbb{T}^{+\infty})), \subseteq\rangle \underset{\alpha^{\mathrm{d}}}{\overset{\gamma^{\mathrm{d}}}{\leftrightarrows}} \langle\mathbb{P}^{\mathrm{d}}, \supseteq^{\mathrm{d}}\rangle$ where the concretization of a dependency property $\mathbf{D}$ is

$$\gamma^{\mathrm{d}}(\mathbf{D}) \triangleq \bigcap_{\ell \in \mathbb{L}} \bigcap_{\langle\mathsf{x}, \mathsf{y}\rangle \in \mathbf{D}(\ell)} \mathcal{D}\ell\langle\mathsf{x}, \mathsf{y}\rangle$$

The intuition is that the more semantics $\boldsymbol{S}$ have semantic property $\mathcal{S}$, the less dependencies can be found, that is, $\mathcal{S} \subseteq \mathcal{S}' \Rightarrow \alpha^{\mathrm{d}}(\mathcal{S})\ell \supseteq \alpha^{\mathrm{d}}(\mathcal{S}')\ell$. The reason is that the dependencies must exist for all semantics $\boldsymbol{S}$ having semantic property $\mathcal{S}$. Otherwise stated, the less dependencies you consider, the more semantics will exactly have these dependencies.

This is different from the observation than larger semantics have more dependencies $\boldsymbol{S} \subseteq \boldsymbol{S}' \Rightarrow \alpha^{\mathrm{d}}(\{\boldsymbol{S}\})\ell \subseteq \alpha^{\mathrm{d}}(\{\boldsymbol{S}'\})\ell$ because $\boldsymbol{S} \in \mathcal{D}\ell\langle\mathsf{x}, \mathsf{y}\rangle \Rightarrow \boldsymbol{S}' \in \mathcal{D}\ell\langle\mathsf{x}, \mathsf{y}\rangle$.

Proof of lemma 47.26

$$
\begin{array}{ll}
\alpha^{\mathrm{d}}(\mathcal{S}) \supseteq^{\mathrm{d}} \mathbf{D} & \\
\Leftrightarrow \alpha^{\mathrm{d}}(\mathcal{S}) \dot{\supseteq} \mathbf{D} & \wr\text{definition of } \subseteq^{\mathrm{d}}\wr \\
\Leftrightarrow \mathbf{D} \dot{\subseteq} \alpha^{\mathrm{d}}(\mathcal{S}) & \wr\text{definition of the inverse}\wr \\
\Leftrightarrow \forall \ell \in \mathbb{L} . \mathbf{D}(\ell) \subseteq \alpha^{\mathrm{d}}(\mathcal{S})\ell & \wr\text{pointwise definition of } \dot{\subseteq}\wr \\
\Leftrightarrow \forall \ell \in \mathbb{L} . \mathbf{D}(\ell) \subseteq \{\langle \mathtt{x}, \mathtt{y}\rangle \mid \mathcal{S} \subseteq \mathcal{D}\ell\langle \mathtt{x}, \mathtt{y}\rangle)\} & \wr\text{definition (47.25) of } \alpha^{\mathrm{d}}(\mathcal{S})\ell\wr \\
\Leftrightarrow \forall \ell \in \mathbb{L} . \forall \langle \mathtt{x}, \mathtt{y}\rangle \in \mathbf{D}(\ell) . \mathcal{S} \subseteq \mathcal{D}\ell\langle \mathtt{x}, \mathtt{y}\rangle) & \wr\text{definition of } \subseteq\wr \\
\Leftrightarrow \forall \ell \in \mathbb{L} . \forall \langle \mathtt{x}, \mathtt{y}\rangle \in \mathbf{D}(\ell) . \forall S \in \mathcal{S} . S \in \mathcal{D}\ell\langle \mathtt{x}, \mathtt{y}\rangle) & \wr\text{definition of } \subseteq\wr \\
\Leftrightarrow \forall S \in \mathcal{S} . \forall \ell \in \mathbb{L} . \forall \langle \mathtt{x}, \mathtt{y}\rangle \in \mathbf{D}(\ell) . S \in \mathcal{D}\ell\langle \mathtt{x}, \mathtt{y}\rangle) & \wr\text{definition of } \forall\wr \\
\Leftrightarrow \mathcal{S} \subseteq \{S \mid \forall \ell \in \mathbb{L} . \forall \langle \mathtt{x}, \mathtt{y}\rangle \in \mathbf{D}(\ell) . S \in \mathcal{D}\ell\langle \mathtt{x}, \mathtt{y}\rangle\} & \wr\text{definition of } \subseteq\wr \\
\Leftrightarrow \mathcal{S} \subseteq \bigcap_{\ell \in \mathbb{L}} \bigcap_{\langle \mathtt{x}, \mathtt{y}\rangle \in \mathbf{D}(\ell)} \mathcal{D}\ell\langle \mathtt{x}, \mathtt{y}\rangle & \wr\text{definition of } \bigcap\wr \\
\Leftrightarrow \mathcal{S} \subseteq \gamma^{\mathrm{d}}(\mathbf{D}) & \wr\text{definition of } \gamma^{\mathrm{d}}\wr \quad \square
\end{array}
$$

Value dependency semantics is an abstraction of the collecting trace semantics.

Corollary 47.27 (value dependency for finite prefix traces)

$$
\ell \mapsto \{\langle \mathtt{x}, \mathtt{y}\rangle \mid \mathtt{x} \overset{\ell}{\rightsquigarrow}_{\mathtt{P}} \mathtt{y}\} \quad = \quad \alpha^{\mathrm{d}}(\{\boldsymbol{\mathcal{S}}^{+\infty}[\![\mathtt{P}]\!]\}) \quad = \quad \alpha^{\mathrm{d}}(\{\boldsymbol{\mathcal{S}}^{*}[\![\mathtt{P}]\!]\})
$$

Proof of corollary 47.27

$$
\begin{array}{ll}
\ell \mapsto \{\langle \mathtt{x}, \mathtt{y}\rangle \mid \mathtt{x} \overset{\ell}{\rightsquigarrow}_{\mathtt{P}} \mathtt{y}\} & \\
= \ell \mapsto \{\langle \mathtt{x}, \mathtt{y}\rangle \mid \boldsymbol{\mathcal{S}}^{+\infty}[\![\mathtt{P}]\!] \in \mathcal{D}\ell\langle \mathtt{x}, \mathtt{y}\rangle\} & \wr(47.21)\wr \\
= \ell \mapsto \{\langle \mathtt{x}, \mathtt{y}\rangle \mid \{\boldsymbol{\mathcal{S}}^{+\infty}[\![\mathtt{P}]\!]\} \subseteq \mathcal{D}\ell\langle \mathtt{x}, \mathtt{y}\rangle\} & \wr\text{definition of } \subseteq\wr \\
= \alpha^{\mathrm{d}}(\{\boldsymbol{\mathcal{S}}^{+\infty}[\![\mathtt{P}]\!]\}) & \wr(47.25)\wr \\
= \alpha^{\mathrm{d}}(\{\boldsymbol{\mathcal{S}}^{*}[\![\mathtt{P}]\!]\}) & \wr\text{lemma 47.23}\wr \quad \square
\end{array}
$$

47.7 Exact, Definite, and Potential Value Dependency Semantics

The exact value dependency semantics $\boldsymbol{\mathcal{S}}^{\mathrm{d}}[\![\mathtt{S}]\!]$ abstracts the maximal trace semantics, or equivalently, by lemma 47.23, the prefix trace semantics by the dependency abstraction α^{d}. Dependency is a nontrivial property, so by Rice's theorem 9.12, $\alpha^{\mathrm{d}}(\{\boldsymbol{\mathcal{S}}^{*}[\![\mathtt{S}]\!]\})$ is not computable. Therefore, static analysis must content itself with approximations (or unsoundness that we disapprove of). There are two possibilities. Definite value dependency is an underapproximation of value dependency (so $\varnothing$ is a correct underapproximation). Potential value dependency is an overapproximation of value dependency (so $\mathbb{V} \times \mathbb{V}$ is a correct overapproximation).

47.7.1 Exact Value Dependency Semantics

The exact value dependency semantics $\boldsymbol{S}^{\mathrm{d}}$ abstracts the maximal trace semantics, or equivalently, by lemma 47.23, the prefix trace semantics by the dependency abstraction. Recall that by example 47.37, it cannot be given an equivalent structural definition.

Definition 47.28 (exact dependency semantics $\boldsymbol{S}^{\mathrm{d}}$)

$$\boldsymbol{S}^{\mathrm{d}}[\![\mathtt{S}]\!] \triangleq \alpha^{\mathrm{d}}(\{\boldsymbol{S}^{+\infty}[\![\mathtt{S}]\!]\})$$

Example 47.29 We have seen in example 47.4 that for the assignment $\mathtt{S} \triangleq \ell_1\ \mathtt{y = x\ ;}\ell_2$, we have

ℓ	ℓ_1	ℓ_2
$\boldsymbol{S}^{\mathrm{d}}[\![\mathtt{S}]\!]\ell$	$\{\langle\mathtt{x},\mathtt{x}\rangle, \langle\mathtt{y},\mathtt{y}\rangle\}$	$\{\langle\mathtt{y},\mathtt{x}\rangle, \langle\mathtt{y},\mathtt{y}\rangle\}$

because the initial value of y flows to x and y at ℓ_2. Changing the initial values of y will change x and y at ℓ_2.

Example 47.30 Continuing exercise 47.11, in the statement

$$\mathtt{P_a} \triangleq \mathtt{if}\ \ell_0\ \mathtt{(x == 1)\ \{}\ \ell_1\ \mathtt{y = x\ ;}\ell_2\ \mathtt{\}\ else\ \{}\ \ell_3\ \mathtt{y = 2\ ;}\ell_4\ \mathtt{\};}\ell_5\ \mathtt{y = 3\ ;}\ell_6. \tag{a}$$

y at ℓ_5 depends on x. Therefore

ℓ	ℓ_0	ℓ_1	ℓ_2	ℓ_3	ℓ_4	ℓ_5	ℓ_6
$\boldsymbol{S}^{\mathrm{d}}[\![\mathtt{P_a}]\!]\ell$	$\{\langle\mathtt{x},\mathtt{x}\rangle, \langle\mathtt{y},\mathtt{y}\rangle\}$	$\{\langle\mathtt{y},\mathtt{y}\rangle\}$	$\varnothing$	$\{\langle\mathtt{x},\mathtt{x}\rangle, \langle\mathtt{y},\mathtt{y}\rangle\}$	$\{\langle\mathtt{x},\mathtt{x}\rangle\}$	$\{\langle\mathtt{x},\mathtt{x}\rangle, \langle\mathtt{x},\mathtt{y}\rangle\}$	$\{\langle\mathtt{x},\mathtt{x}\rangle\}$

For example, changing the initial value of variables x and y cannot change their value at ℓ_2 which is always 1. For statement

$$\mathtt{P_b} \triangleq \mathtt{if}\ \ell_0\ \mathtt{(x == 1)\ \{}\ \ell_1\ \mathtt{y = x\ ;}\ell_2\ \mathtt{\}\ else\ \{}\ \ell_3\ \mathtt{y = 1\ ;}\ell_4\ \mathtt{\};}\ell_5\ \mathtt{y = 3\ ;}\ell_6. \tag{b}$$

the dependency is different at ℓ_5

ℓ	ℓ_0	ℓ_1	ℓ_2	ℓ_3	ℓ_4	ℓ_5	ℓ_6
$\boldsymbol{S}^{\mathrm{d}}[\![\mathtt{P_b}]\!]\ell$	$\{\langle\mathtt{x},\mathtt{x}\rangle, \langle\mathtt{y},\mathtt{y}\rangle\}$	$\{\langle\mathtt{y},\mathtt{y}\rangle\}$	$\varnothing$	$\{\langle\mathtt{x},\mathtt{x}\rangle, \langle\mathtt{y},\mathtt{y}\rangle\}$	$\{\langle\mathtt{x},\mathtt{x}\rangle\}$	$\{\langle\mathtt{x},\mathtt{x}\rangle\}$	$\{\langle\mathtt{x},\mathtt{x}\rangle\}$

because y is constant equal to 1.

Exercise 47.31 Consider the program $\mathtt{S} \triangleq \mathtt{if}\ \ell_0\ \mathtt{(x == 1)\ \{}\ \ell_1\ \mathtt{y = x\ ;}\ell_2\ \mathtt{\}}\ \ell_3$ as well as $\mathtt{S'} \triangleq \mathtt{if}\ \ell_0\ \mathtt{(x\ != 1)\ \{}\ \ell_1\ \mathtt{y = x\ ;}\ell_2\ \mathtt{\}}\ \ell_3$. What is the dependency at ℓ_2 for S and S′?

By Rice's theorem 9.12, the exact dependency semantics $\boldsymbol{S}^{\mathrm{d}}$ is not computable (and moreover not in structural form), a static analysis must perform an approximation by underapproximation or overapproximating dependencies.

47.7.2 Definite Value Dependency Semantics

Definite value dependency is an underapproximation of value dependency (so $\varnothing$ is a correct underapproximation).

Definite dependency semantics $\overline{\boldsymbol{S}}^{\forall}_{\text{diff}}$

$$\overline{\boldsymbol{S}}^{\forall}_{\text{diff}}[\![\mathsf{S}]\!] \;\dot{\subseteq}\; \alpha^{\mathrm{d}}(\{\boldsymbol{S}^{+\infty}[\![\mathsf{S}]\!]\})$$

47.7.3 Potential value dependency semantics

Potential value dependency is an overapproximation of value dependency (so $\mathbb{V} \times \mathbb{V}$ is a correct over-approximation).

Potential dependency semantics $\overline{\boldsymbol{S}}^{\exists}_{\text{diff}}$

$$\alpha^{\mathrm{d}}(\{\boldsymbol{S}^{+\infty}[\![\mathsf{S}]\!]\}) \;\dot{\subseteq}\; \overline{\boldsymbol{S}}^{\exists}_{\text{diff}}[\![\mathsf{S}]\!] \tag{47.32}$$

We are interested in potential value dependency, that is, an overapproximation of value dependency, which is the concept needed, for example, in compilation or security, looking for more dependencies than there actually are.

Example 47.33 Continuing example 47.30, the program $\mathsf{P_b}$ in 47.30-(b) has the same potential value dependency analysis than $\mathsf{P_a}$ in 47.30-(a) because the value of y is not tracked. A simple reduction with a Cartesian constancy analysis of section 28.8.3 would make case (a) more precise. Moreover, we look for a structural definition $\widehat{\overline{\boldsymbol{S}}}^{\exists}_{\text{diff}}[\![\mathsf{S}]\!] = \overline{\boldsymbol{S}}^{\exists}_{\text{diff}}[\![\mathsf{S}]\!]$.

Example 47.34 For the test $\mathsf{S} \triangleq \mathtt{if}\ \ell_0\ (\mathsf{x} == 1)\ \{\ \ell_1\ \mathsf{y} = \mathsf{z}\ ;\ell_2\ \};\ell_3$, we have the potential value dependency

ℓ	ℓ_0	ℓ_1	ℓ_2	ℓ_3
$\overline{\boldsymbol{S}}^{\exists}_{\text{diff}}[\![\mathsf{S}]\!]\,\ell$	$\{\langle \mathsf{x}, \mathsf{x}\rangle, \langle \mathsf{y}, \mathsf{y}\rangle, \langle \mathsf{z}, \mathsf{z}\rangle\}$	$\{\langle \mathsf{y}, \mathsf{y}\rangle, \langle \mathsf{z}, \mathsf{z}\rangle\}$	$\{\langle \mathsf{z}, \mathsf{y}\rangle, \langle \mathsf{z}, \mathsf{z}\rangle\}$	$\{\langle \mathsf{x}, \mathsf{x}\rangle, \langle \mathsf{x}, \mathsf{y}\rangle, \langle \mathsf{y}, \mathsf{y}\rangle, \langle \mathsf{z}, \mathsf{y}\rangle, \langle \mathsf{z}, \mathsf{z}\rangle\}$

Observe that this is an overapproximation because, for example, z flows to y at ℓ_3 only when $\mathsf{x} = 1$ at ℓ_0.

47.8 Design of an Abstract Structural Dependency Semantics, Verification Method, and Static Analysis

47.8.1 Syntactic Dependency Property

Following the tradition established by [312], the calculational design of the potential dependency semantics $\widehat{\overline{\boldsymbol{S}}}^{\exists}_{\text{diff}}[\![\mathsf{S}]\!]$ in section 47.9 does not take values of variables into account so, as noticed in remark 47.39, is imprecise.

A syntactic dependency property $\mathbf{D} \in \mathbb{P}^{\mathrm{d}}$ attaches to each program point ℓ a set of pairs $\langle \mathsf{x}, \mathsf{y}\rangle \in \mathbf{D}(\ell)$ of variables x and y meaning that the value of variable y at ℓ depends upon the initial value of x (or the initial value of x may flow to the value of y at ℓ, written $\mathsf{x} \overset{\ell}{\rightsquigarrow} \mathsf{y}$). This is a syntactic property because the values of variables as defined by the program semantics are abstracted away. This point of view on dependency is traditional [312, 976]. The syntactic dependency domain $\mathbb{P}^{\mathrm{d}}$, that is, the set of all such dependency properties $\mathbf{D}$, is therefore

$$\mathbf{D} \in \mathbb{P}^{\mathrm{d}} \;\triangleq\; \mathbb{L} \to \wp(\mathbb{V} \times \mathbb{V}) \tag{47.35}$$

where $\wp(\mathbb{V} \times \mathbb{V})$ is the set of relations on program variables $\mathbb{V}$. It is a finite complete lattice $\langle \mathbb{P}^{\mathrm{d}}, \subseteq^{\mathrm{d}}, \bot^{\mathrm{d}}, \top^{\mathrm{d}}, \cap^{\mathrm{d}}, \cup^{\mathrm{d}} \rangle$ defined pointwise that is partially ordered by pointwise subset inclusion $\subseteq^{\mathrm{d}} \triangleq \dot{\subseteq}$, $\bot^{\mathrm{d}} \triangleq \ell \mapsto \emptyset$, $\top^{\mathrm{d}} \triangleq \ell \mapsto \mathbb{V} \times \mathbb{V}$, and so on.

47.8.2 Design of an Abstract Structural Dependency Semantics

Now that we have chosen a semantics $\boldsymbol{S}^*[\![\mathsf{S}]\!]$ and an abstraction α^{d}, we should follow section 19.6 to get a structural abstract dependency semantics $\widehat{\boldsymbol{S}}^{\mathrm{d}}[\![\mathsf{S}]\!] = \alpha^{\mathrm{d}}(\boldsymbol{S}^*[\![\mathsf{S}]\!])$, from which we would derive a sound and complete dependency verification method, and, by a further abstractions, various sound dependency static analyses.

Example 47.36 This is the methodology that we followed for reachability, with semantics $\boldsymbol{S}^*[\![\mathsf{P}]\!]$, abstraction $\mathrm{post}^{\vec{\varrho}}$ in (20.15), structural reachability semantics in chapter 19, and abstract reachability/invariance verification method in chapter 25 (equivalently expressed as a logic in chapter 26). Various further abstractions in chapters 28, 31, 33, 38, and 40 yield various Cartesian or relational numerical analyses that can be combined by the reduced product of chapter 36.

The same way that invariants in reachability proof methods must be structurally inductive to allow for a proof by induction (see section 25.2), the abstract semantics $\alpha^{\mathrm{d}}(\boldsymbol{S}^*[\![\mathsf{S}]\!])$ must be structurally inductive to allow for a structural specification $\widehat{\boldsymbol{S}}^{\mathrm{d}}[\![\mathsf{S}]\!]$ of the form

$$\begin{cases} \widehat{\boldsymbol{S}}^{\mathrm{d}}[\![\mathsf{S}]\!] & \triangleq \quad f^{\mathrm{d}}[\![\mathsf{S}]\!](\prod_{\mathsf{S}' \lhd \mathsf{S}} \widehat{\boldsymbol{S}}^{\mathrm{d}}[\![\mathsf{S}']\!]) \\ \mathsf{S} \in \mathbb{P}\mathbb{c} \end{cases} \tag{19.40}$$

As shown by exercise 47.11 (for y at $\ell_0, \ell_1, \ell_2, \ell_3, \ell_4$, and ℓ_6 for $\mathsf{P_a}$ an also ℓ_5 for $\mathsf{P_b}$), dependency depends on the value of variables. In particular, variables with constant values are independent of the initial values of the variables (for example, y at ℓ_2 in $\mathsf{P_a}$ of example 47.5). The following example 47.37 shows that the abstract dependency semantics $\alpha^{\mathrm{d}}(\boldsymbol{S}^*[\![\mathsf{S}]\!])$ is not structural (i.e., defined by induction on the program syntax) because it does not take value of variables into account.

Example 47.37 (structural compositionality) In the following statement, x and y at ℓ_1 depend on x at ℓ_0.

```
                               /* x = x0, y = y0 */
ℓ0 y = x ;
ℓ1                             /* x = x0, y = x0 */
```

In the following statement, x and y at ℓ_2 depend on x at ℓ_1.

```
                               /* x = x0', y = y0' */
ℓ1 y = y - x ;
ℓ2                             /* x = x0', y = y0' - x0' */
```

In the sequential composition of the two statements

$$
\begin{array}{ll}
 & /\star\ x = x_0, y = y_0\ \star/ \\
\ell_0\ \mathtt{y = x\,;} & /\star\ x = x_0, y = x_0\ \star/ \\
\ell_1\ \mathtt{y = y - x\,;} & /\star\ x = x_0, y = 0\ \star/ \\
\ell_2 &
\end{array}
$$

y at ℓ_2 depends on x at ℓ_1 which depends on x at ℓ_0 so that, by composition, y at ℓ_2 depends on x at ℓ_0. However, y = 0 at ℓ_2 so y at ℓ_2 does not depend on x at ℓ_0.

Exercise 47.38 Propose a refinement of the analysis of example 47.37, that is, exact by structural induction.

In conclusion $\widehat{\boldsymbol{S}}^{\mathrm{d}}[\![\mathtt{S}]\!] \neq \alpha^{\mathrm{d}}(\boldsymbol{S}^{*}[\![\mathtt{S}]\!])$ because values of variables are not taken into account. To get a structural definition of dependency that is sound and complete, we would need a more refined abstract domain **D** and abstraction α^{d} taking values of variables into account, more precisely, using relational invariants between initial and current values of variables at the program points, as defined in section 19.1.2. We do not pursue this plan but instead go directly to the structural syntactic definition of dependency, which by Rice's theorem 9.12 is necessarily incomplete.

Remark 47.39 (imprecision of syntactic dependency) Notice that values of variables can be taken into account owing to a reduced product (chapter 36) with a static analysis either Cartesian in part IX (such as the value constant propagation static analysis of [572, 969]) or relational in part XI (which is necessary to handle example 47.37, such as the symbolic constant propagation static analysis of [560, 585, 586, 728]).

47.9 Calculational Design of the Structural Abstract Potential Value Dependency Semantics

Our objective is to design a structural abstract potential value dependency semantics that is sound as stated by the following:

> **Theorem 47.40 (soundness of the potential value dependency semantics)** For all program components S, the abstract value dependency semantics $\widehat{\overline{\boldsymbol{S}}}^{\exists}_{\mathsf{diff}}[\![\mathtt{S}]\!]$ defined by (47.41) to (47.63) is sound as specified by the definition (47.32) of potential value dependency.

Proof of theorem 47.40 The proof is by structural induction and proceeds by calculational design for each program component, as shown in the proofs of (47.41) to (47.63) given subsequently. □

This is soundness because the analysis never misses a value dependency as defined semantically by (47.32). This is incomplete because , in general $\dot{\subseteq}$ cannot be =, by Rice's theorem 9.12.

At a statement, variables have their initial value and so depend only on themselves. (For loops (47.63), more dependencies may originate from the iterations.)

> *Abstract potential dependency semantics at a program component* S *that is not an iteration*
>
> $$\widehat{\widehat{\boldsymbol{\mathcal{S}}}}^{\exists}_{\text{diff}}[\![\mathsf{S}]\!]\,\mathsf{at}[\![\mathsf{S}]\!] \;=\; \mathbb{1}_{\mathbb{V}} \qquad (47.41)$$
>
> (where $\mathbb{1}_{\mathbb{V}} \triangleq \{\langle \mathsf{x}, \mathsf{x}\rangle \mid \mathsf{x} \in \mathbb{V}\}$ is the identity on the set $\mathbb{V}$ of program variables).

Proof of (47.41)

$\alpha^{\mathrm{d}}(\{\boldsymbol{\mathcal{S}}^{+\infty}[\![\mathsf{S}]\!]\})\,(\mathsf{at}[\![\mathsf{S}]\!])$

$= \alpha^{\mathrm{d}}(\{\boldsymbol{\mathcal{S}}^{*}[\![\mathsf{S}]\!]\})\,(\mathsf{at}[\![\mathsf{S}]\!])$ ⟅lemma 47.23⟆

$= \{\langle \mathsf{x}', \mathsf{y}\rangle \mid \boldsymbol{\mathcal{S}}^{*}[\![\mathsf{S}]\!] \in \mathcal{D}(\mathsf{at}[\![\mathsf{S}]\!])\langle \mathsf{x}', \mathsf{y}\rangle\}$ ⟅definition (47.25) of α^{d}⟆

$= \{\langle \mathsf{x}', \mathsf{y}\rangle \mid \exists \langle \pi_0, \pi_1\rangle, \langle \pi_0', \pi_1'\rangle \in \boldsymbol{\mathcal{S}}^{*}[\![\mathsf{S}]\!]\,.\,(\forall \mathsf{z} \in \mathbb{V} \setminus \{\mathsf{x}'\}\,.\,\varrho(\pi_0)\mathsf{z} = \varrho(\pi_0')\mathsf{z}) \wedge$
$\mathsf{diff}(\mathsf{seqval}[\![\mathsf{y}]\!](\mathsf{at}[\![\mathsf{S}]\!])(\pi_0, \pi_1), \mathsf{seqval}[\![\mathsf{y}]\!](\mathsf{at}[\![\mathsf{S}]\!])(\pi_0', \pi_1'))\}$
⟅definition (47.19) of $\mathcal{D}(\mathsf{at}[\![\mathsf{S}]\!])\langle \mathsf{x}', \mathsf{y}\rangle$⟆

$= \{\langle \mathsf{x}', \mathsf{y}\rangle \mid \exists \langle \pi_0\mathsf{at}[\![\mathsf{S}]\!], \mathsf{at}[\![\mathsf{S}]\!]\pi_1\rangle, \langle \pi_0'\mathsf{at}[\![\mathsf{S}]\!], \mathsf{at}[\![\mathsf{S}]\!]\pi_1'\rangle \in \boldsymbol{\mathcal{S}}^{*}[\![\mathsf{S}]\!]\,.\,(\forall \mathsf{z} \in \mathbb{V} \setminus \{\mathsf{x}'\}\,.$
$\varrho(\pi_0\mathsf{at}[\![\mathsf{S}]\!])\mathsf{z} = \varrho(\pi_0'\mathsf{at}[\![\mathsf{S}]\!])\mathsf{z}) \wedge \mathsf{diff}(\mathsf{seqval}[\![\mathsf{y}]\!](\mathsf{at}[\![\mathsf{S}]\!])(\pi_0\mathsf{at}[\![\mathsf{S}]\!], \mathsf{at}[\![\mathsf{S}]\!]\pi_1),$
$\mathsf{seqval}[\![\mathsf{y}]\!](\mathsf{at}[\![\mathsf{S}]\!])(\pi_0'\mathsf{at}[\![\mathsf{S}]\!], \mathsf{at}[\![\mathsf{S}]\!]\pi_1'))\}$
⟅definition of $\boldsymbol{\mathcal{S}}^{*}[\![\mathsf{S}]\!]$ so that if $\langle \pi, \pi'\rangle \in \boldsymbol{\mathcal{S}}^{*}[\![\mathsf{S}]\!]$ then π ends $\mathsf{at}[\![\mathsf{S}]\!]$ and π' starts $\mathsf{at}[\![\mathsf{S}]\!]$⟆

If S is the empty statement list ϵ then, by definition (6.15) of $\boldsymbol{\mathcal{S}}^{*}[\![\mathsf{S}]\!]$, we have $\pi_1 = \pi_1' = \ni$. By definition (47.16) of $\mathsf{seqval}[\![\mathsf{y}]\!]$, we have $\mathsf{seqval}[\![\mathsf{y}]\!](\mathsf{at}[\![\mathsf{S}]\!])(\pi_0\mathsf{at}[\![\mathsf{S}]\!], \mathsf{at}[\![\mathsf{S}]\!]) = \varrho(\pi_0\mathsf{at}[\![\mathsf{S}]\!])\mathsf{y}$ as well as $\mathsf{seqval}[\![\mathsf{y}]\!](\mathsf{at}[\![\mathsf{S}]\!])(\pi_0'\mathsf{at}[\![\mathsf{S}]\!], \mathsf{at}[\![\mathsf{S}]\!]) = \varrho(\pi_0'\mathsf{at}[\![\mathsf{S}]\!])\mathsf{y}$. Then, by exercise 6.8 and the cardinality of $\mathbb{V}$ is at least 2, we have $\forall \mathsf{z} \in \mathbb{V} \setminus \{\mathsf{x}'\}\,.\,\varrho(\pi_0\mathsf{at}[\![\mathsf{S}]\!])\mathsf{z} = \varrho(\pi_0'\mathsf{at}[\![\mathsf{S}]\!])\mathsf{z}) \wedge \varrho(\pi_0\mathsf{at}[\![\mathsf{S}]\!])\mathsf{y} \neq \varrho(\pi_0'\mathsf{at}[\![\mathsf{S}]\!])\mathsf{y}$ if and only if $\mathsf{y} = \mathsf{x}'$. We get

$= \{\langle \mathsf{x}', \mathsf{x}'\rangle \mid \mathsf{x}' \in \mathbb{V}\} \;\triangleq\; \mathbb{1}_{\mathbb{V}}$

— For all other program components $\mathsf{S} \in \mathbb{P}\mathbb{c}$, definition of $\boldsymbol{\mathcal{S}}^{*}[\![\mathsf{S}]\!]$ in chapter 7 shows that any execution from $\mathsf{at}[\![\mathsf{S}]\!]$ has at least one transition. Therefore we are in the situation

$= \{\langle \mathsf{x}', \mathsf{y}\rangle \mid \exists \langle \pi_0\mathsf{at}[\![\mathsf{S}]\!], \mathsf{at}[\![\mathsf{S}]\!] \xrightarrow{\mathsf{a}} \ell \mathbin{\overset{\frown}{\cdot}} \pi_2\rangle, \langle \pi_0'\mathsf{at}[\![\mathsf{S}]\!], \mathsf{at}[\![\mathsf{S}]\!] \xrightarrow{\mathsf{a}'} \ell' \mathbin{\overset{\frown}{\cdot}} \pi_2'\rangle \in \boldsymbol{\mathcal{S}}^{*}[\![\mathsf{S}]\!]\,.\,(\forall \mathsf{z} \in$
$\mathbb{V} \setminus \{\mathsf{x}'\}\,.\,\varrho(\pi_0\mathsf{at}[\![\mathsf{S}]\!])\mathsf{z} = \varrho(\pi_0'\mathsf{at}[\![\mathsf{S}]\!])\mathsf{z}) \wedge \mathsf{diff}(\mathsf{seqval}[\![\mathsf{y}]\!](\mathsf{at}[\![\mathsf{S}]\!])(\pi_0\mathsf{at}[\![\mathsf{S}]\!], \mathsf{at}[\![\mathsf{S}]\!] \xrightarrow{\mathsf{a}}$
$\ell \mathbin{\overset{\frown}{\cdot}} \pi_2), \mathsf{seqval}[\![\mathsf{y}]\!](\mathsf{at}[\![\mathsf{S}]\!])(\pi_0'\mathsf{at}[\![\mathsf{S}]\!], \mathsf{at}[\![\mathsf{S}]\!] \xrightarrow{\mathsf{a}'} \ell' \mathbin{\overset{\frown}{\cdot}} \pi_2'))\}$

$= \{\langle \mathsf{x}', \mathsf{y}\rangle \mid \exists \langle \pi_0\mathsf{at}[\![\mathsf{S}]\!], \mathsf{at}[\![\mathsf{S}]\!] \xrightarrow{\mathsf{a}} \ell \mathbin{\overset{\frown}{\cdot}} \pi_2\rangle, \langle \pi_0'\mathsf{at}[\![\mathsf{S}]\!], \mathsf{at}[\![\mathsf{S}]\!] \xrightarrow{\mathsf{a}'} \ell' \mathbin{\overset{\frown}{\cdot}} \pi_2'\rangle \in$
$\boldsymbol{\mathcal{S}}^{*}[\![\mathsf{S}]\!]\,.\,(\forall \mathsf{z} \in \mathbb{V} \setminus \{\mathsf{x}'\}\,.\,\varrho(\pi_0\mathsf{at}[\![\mathsf{S}]\!])\mathsf{z} = \varrho(\pi_0'\mathsf{at}[\![\mathsf{S}]\!])\mathsf{z}) \wedge \mathsf{diff}(\varrho(\pi_0\mathsf{at}[\![\mathsf{S}]\!])\mathsf{y} \cdot$
$\mathsf{seqval}[\![\mathsf{y}]\!](\mathsf{at}[\![\mathsf{S}]\!])(\pi_0\mathsf{at}[\![\mathsf{S}]\!] \xrightarrow{\mathsf{a}} \ell, \pi_2), \varrho(\pi_0'\mathsf{at}[\![\mathsf{S}]\!])\mathsf{y} \cdot \mathsf{seqval}[\![\mathsf{y}]\!](\mathsf{at}[\![\mathsf{S}]\!])(\pi_0'\mathsf{at}[\![\mathsf{S}]\!] \xrightarrow{\mathsf{a}'}$
$\ell', \pi_2'))\}$ ⟅definition (47.16) of $\mathsf{seqval}[\![\mathsf{y}]\!]$⟆

— If the program component $\mathsf{S} \in \mathbb{P}\mathbb{c}$ is not an iteration, then by definition of $\boldsymbol{\mathcal{S}}^{*}[\![\mathsf{S}]\!]$ in chapter 7, execution never comes back to $\mathsf{at}[\![\mathsf{S}]\!]$ so that by definition (47.16) of $\mathsf{seqval}[\![\mathsf{y}]\!]$, we have $\mathsf{seqval}[\![\mathsf{y}]\!](\mathsf{at}[\![\mathsf{S}]\!])(\pi_0\mathsf{at}[\![\mathsf{S}]\!] \xrightarrow{\mathsf{a}} \ell, \pi_2) = \ni$ and similarly $\mathsf{seqval}[\![\mathsf{y}]\!](\mathsf{at}[\![\mathsf{S}]\!])(\pi_0'\mathsf{at}[\![\mathsf{S}]\!] \xrightarrow{\mathsf{a}'} \ell', \pi_2') = \ni$. Again, by exercise 6.8 and the cardinality of $\mathbb{V}$ is at least 2, we have $\forall z \in \mathcal{V} \setminus \{x'\} . \varrho(\pi_0 \text{at}[\![S]\!])z = \varrho(\pi'_0 \text{at}[\![S]\!])z) \wedge \varrho(\pi_0 \text{at}[\![S]\!])y \neq \varrho(\pi'_0 \text{at}[\![S]\!])y$ if and only if $y = x'$. We get

$= \{\langle x', x'\rangle \mid x' \in \mathcal{V}\} \quad \triangleq \quad \mathbb{1}_{\mathcal{V}}$ □

Outside a statement, there is no possible potential dependency because executions never reach that point.

Abstract potential dependency semantics outside a program component S

$$\ell \notin \text{labx}[\![S]\!] \quad \Rightarrow \quad \widehat{\boldsymbol{\mathcal{S}}}^{\exists}_{\text{diff}}[\![S]\!]\,\ell \quad = \quad \varnothing \tag{47.42}$$

Exercise 47.43 Prove (47.42).

After an assignment $x = A$;, the unmodified variables $y \neq x$ depend upon their initial value at the assignment. The assigned-to variable x depends on the variables $\widehat{\boldsymbol{\mathcal{S}}}^{\exists}_{\text{diff}}[\![A]\!]$ on which the assigned expression A does depend. As noticed in remark 47.39, it is empty if A is constant. A trivial syntactic overapproximation of the functional dependency $\widehat{\boldsymbol{\mathcal{S}}}^{\exists}_{\text{diff}}[\![A]\!]$ of expression A is the set of variables $\text{vars}[\![A]\!]$ of this expression.

Abstract potential dependency semantics of an assignment statement $S ::= x = A$;

$$\widehat{\boldsymbol{\mathcal{S}}}^{\exists}_{\text{diff}}[\![S]\!]\,\ell \quad = \quad (\!|\ \ell = \text{at}[\![S]\!] \ \text{?}\ \mathbb{1}_{\mathcal{V}} \tag{47.44}$$
$$\quad \|\ \ell = \text{after}[\![S]\!] \ \text{?}\ \{\langle y, x\rangle \mid y \in \widehat{\boldsymbol{\mathcal{S}}}^{\exists}_{\text{diff}}[\![A]\!]\} \cup \{\langle y, y\rangle \mid y \neq x\}$$
$$\quad \text{:}\ \varnothing\ |\!)$$
$$\widehat{\boldsymbol{\mathcal{S}}}^{\exists}_{\text{diff}}[\![A]\!] \quad \triangleq \quad \{y \mid \exists \rho \in \mathbb{E}\text{v} . \exists \nu \in \mathbb{V} . \mathscr{A}[\![A]\!]\rho \neq \mathscr{A}[\![A]\!]\rho[y \leftarrow \nu]\} \quad \subseteq \quad \text{vars}[\![A]\!]$$

Proof of (47.44) The cases $\ell = \text{at}[\![S]\!]$ was handled in (47.41) and $\ell \notin \text{labx}[\![S]\!]$ in (47.42). It remains the case $\ell = \text{after}[\![S]\!]$.

$\alpha^{\text{d}}(\{\boldsymbol{\mathcal{S}}^{+\infty}[\![S]\!]\})\ \text{after}[\![S]\!]$

$= \alpha^{\text{d}}(\{\boldsymbol{\mathcal{S}}^{*}[\![S]\!]\})\ \text{after}[\![S]\!]$ (lemma 47.23)

$= \{\langle x', y\rangle \mid \boldsymbol{\mathcal{S}}^{*}[\![S]\!] \in \mathcal{D}(\text{after}[\![S]\!])\langle x', y\rangle\}$ (definition (47.25) of α^{d} and definition of $\subseteq$)

$= \{\langle x', y\rangle \mid \exists \langle \pi_0, \pi_1\rangle, \langle \pi'_0, \pi'_1\rangle \in \boldsymbol{\mathcal{S}}^{*}[\![S]\!] . \forall z \in \mathcal{V} \setminus \{x'\} . \varrho(\pi_0)z = \varrho(\pi'_0)z \wedge \text{diff}(\text{seqval}[\![y]\!](\text{after}[\![S]\!])(\pi_0, \pi_1), \text{seqval}[\![y]\!](\text{after}[\![S]\!])(\pi'_0, \pi'_1))\}$

(definition of $\in$ and definition (47.19) of $\mathcal{D}\ell\langle x', y\rangle$)

$= \{\langle x', y\rangle \mid \exists \langle \pi_0, \pi_1\rangle, \langle \pi'_0, \pi'_1\rangle \in \{\langle \pi\text{at}[\![S]\!], \text{at}[\![S]\!] \xrightarrow{x = \mathscr{A}[\![A]\!]\varrho(\pi\text{at}[\![S]\!])} \text{after}[\![S]\!]\rangle \mid \pi\text{at}[\![S]\!] \in \mathbb{T}^{+}\} . \forall z \in \mathcal{V} \setminus \{x'\} . \varrho(\pi_0)z = \varrho(\pi'_0)z \wedge \text{diff}(\text{seqval}[\![y]\!](\text{after}[\![S]\!])(\pi_0, \pi_1), \text{seqval}[\![y]\!](\text{after}[\![S]\!])(\pi'_0, \pi'_1))\}$

(definition (17.2) of the assignment prefix finite trace semantics)

$$= \{\langle \mathsf{x}', \mathsf{y}\rangle \mid \exists \langle \pi_0 \mathsf{at}[\![\mathsf{S}]\!], \mathsf{at}[\![\mathsf{S}]\!] \xrightarrow{\mathsf{x}=\mathcal{A}[\![\mathsf{A}]\!]\varrho(\pi_0\mathsf{at}[\![\mathsf{S}]\!])} \mathsf{after}[\![\mathsf{S}]\!]\rangle, \langle \pi'_0 \mathsf{at}[\![\mathsf{S}]\!], \mathsf{at}[\![\mathsf{S}]\!] \xrightarrow{\mathsf{x}=\mathcal{A}[\![\mathsf{A}]\!]\varrho(\pi'_0\mathsf{at}[\![\mathsf{S}]\!])} \mathsf{after}[\![\mathsf{S}]\!]\rangle . \forall \mathsf{z} \in \mathbb{V} \setminus \{\mathsf{x}'\} . \varrho(\pi_0\mathsf{at}[\![\mathsf{S}]\!])\mathsf{z} = \varrho(\pi'_0\mathsf{at}[\![\mathsf{S}]\!])\mathsf{z} \wedge \mathsf{diff}(\mathsf{seqval}[\![\mathsf{y}]\!](\mathsf{after}[\![\mathsf{S}]\!])(\pi_0\mathsf{at}[\![\mathsf{S}]\!], \mathsf{at}[\![\mathsf{S}]\!] \xrightarrow{\mathsf{x}=\mathcal{A}[\![\mathsf{A}]\!]\varrho(\pi_0\mathsf{at}[\![\mathsf{S}]\!])} \mathsf{after}[\![\mathsf{S}]\!]), \mathsf{seqval}[\![\mathsf{y}]\!](\mathsf{after}[\![\mathsf{S}]\!])(\pi'_0\mathsf{at}[\![\mathsf{S}]\!], \mathsf{at}[\![\mathsf{S}]\!] \xrightarrow{\mathsf{x}=\mathcal{A}[\![\mathsf{A}]\!]\varrho(\pi'_0\mathsf{at}[\![\mathsf{S}]\!])} \mathsf{after}[\![\mathsf{S}]\!]))\}$$

〖definition of $\in$〗

$$= \{\langle \mathsf{x}', \mathsf{y}\rangle \mid \exists \langle \pi_0 \mathsf{at}[\![\mathsf{S}]\!], \mathsf{at}[\![\mathsf{S}]\!] \xrightarrow{\mathsf{x}=\mathcal{A}[\![\mathsf{A}]\!]\varrho(\pi_0\mathsf{at}[\![\mathsf{S}]\!])} \mathsf{after}[\![\mathsf{S}]\!]\rangle, \langle \pi'_0 \mathsf{at}[\![\mathsf{S}]\!], \mathsf{at}[\![\mathsf{S}]\!] \xrightarrow{\mathsf{x}=\mathcal{A}[\![\mathsf{A}]\!]\varrho(\pi'_0\mathsf{at}[\![\mathsf{S}]\!])} \mathsf{after}[\![\mathsf{S}]\!]\rangle . (\forall \mathsf{z} \in \mathbb{V} \setminus \{\mathsf{x}'\} . \varrho(\pi_0\mathsf{at}[\![\mathsf{S}]\!])\mathsf{z} = \varrho(\pi'_0\mathsf{at}[\![\mathsf{S}]\!])\mathsf{z}) \wedge \mathsf{diff}(\varrho(\pi_0\mathsf{at}[\![\mathsf{S}]\!] \xrightarrow{\mathsf{x}=\mathcal{A}[\![\mathsf{A}]\!]\varrho(\pi_0\mathsf{at}[\![\mathsf{S}]\!])} \mathsf{after}[\![\mathsf{S}]\!])\mathsf{y}, \varrho(\pi'_0\mathsf{at}[\![\mathsf{S}]\!] \xrightarrow{\mathsf{x}=\mathcal{A}[\![\mathsf{A}]\!]\varrho(\pi'_0\mathsf{at}[\![\mathsf{S}]\!])} \mathsf{after}[\![\mathsf{S}]\!])\mathsf{y})\}$$

〖definition (47.16) of the future $\mathsf{seqval}[\![\mathsf{y}]\!]$〗

$$= \{\langle \mathsf{x}', \mathsf{y}\rangle \mid \exists \langle \pi_0 \mathsf{at}[\![\mathsf{S}]\!], \mathsf{at}[\![\mathsf{S}]\!] \xrightarrow{\mathsf{x}=\mathcal{A}[\![\mathsf{A}]\!]\varrho(\pi_0\mathsf{at}[\![\mathsf{S}]\!])} \mathsf{after}[\![\mathsf{S}]\!]\rangle, \langle \pi'_0 \mathsf{at}[\![\mathsf{S}]\!], \mathsf{at}[\![\mathsf{S}]\!] \xrightarrow{\mathsf{x}=\mathcal{A}[\![\mathsf{A}]\!]\varrho(\pi'_0\mathsf{at}[\![\mathsf{S}]\!])} \mathsf{after}[\![\mathsf{S}]\!]\rangle . (\forall \mathsf{z} \in \mathbb{V} \setminus \{\mathsf{x}'\} . \varrho(\pi_0\mathsf{at}[\![\mathsf{S}]\!])\mathsf{z} = \varrho(\pi'_0\mathsf{at}[\![\mathsf{S}]\!])\mathsf{z}) \wedge ((\varrho(\pi_0\mathsf{at}[\![\mathsf{S}]\!])\mathsf{y} \neq \varrho(\pi'_0\mathsf{at}[\![\mathsf{S}]\!])\mathsf{y}) \vee (\varrho(\pi_0\mathsf{at}[\![\mathsf{S}]\!])\mathsf{y} = \varrho(\pi'_0\mathsf{at}[\![\mathsf{S}]\!])\mathsf{y} \wedge \varrho(\pi_0\mathsf{at}[\![\mathsf{S}]\!] \xrightarrow{\mathsf{x}=\mathcal{A}[\![\mathsf{A}]\!]\varrho(\pi_0\mathsf{at}[\![\mathsf{S}]\!])} \mathsf{after}[\![\mathsf{S}]\!])\mathsf{y} \neq \varrho(\pi'_0\mathsf{at}[\![\mathsf{S}]\!] \xrightarrow{\mathsf{x}=\mathcal{A}[\![\mathsf{A}]\!]\varrho(\pi'_0\mathsf{at}[\![\mathsf{S}]\!])} \mathsf{after}[\![\mathsf{S}]\!])\mathsf{y})\}$$

〖(47.18) so that $\mathsf{diff}(a \cdot b, c \cdot d)$ if and only if (1) $a \neq c$ or (2) $a = c \wedge b \neq d$.〗

$$= \{\langle \mathsf{x}', \mathsf{y}\rangle \mid \exists \langle \pi_0 \mathsf{at}[\![\mathsf{S}]\!], \mathsf{at}[\![\mathsf{S}]\!] \xrightarrow{\mathsf{x}=\mathcal{A}[\![\mathsf{A}]\!]\varrho(\pi_0\mathsf{at}[\![\mathsf{S}]\!])} \mathsf{after}[\![\mathsf{S}]\!]\rangle, \langle \pi'_0 \mathsf{at}[\![\mathsf{S}]\!], \mathsf{at}[\![\mathsf{S}]\!] \xrightarrow{\mathsf{x}=\mathcal{A}[\![\mathsf{A}]\!]\varrho(\pi'_0\mathsf{at}[\![\mathsf{S}]\!])} \mathsf{after}[\![\mathsf{S}]\!]\rangle . (\forall \mathsf{z} \in \mathbb{V} \setminus \{\mathsf{x}'\} . \varrho(\pi_0\mathsf{at}[\![\mathsf{S}]\!])\mathsf{z} = \varrho(\pi'_0\mathsf{at}[\![\mathsf{S}]\!])\mathsf{z}) \wedge ((\mathsf{y} = \mathsf{x}') \vee (\mathsf{y} = \mathsf{x} \wedge \mathcal{A}[\![\mathsf{A}]\!]\varrho(\pi_0\mathsf{at}[\![\mathsf{S}]\!]) \neq \mathcal{A}[\![\mathsf{A}]\!]\varrho(\pi'_0\mathsf{at}[\![\mathsf{S}]\!])))\}$$

〖definition (6.6) of ϱ〗

$$\subseteq \{\langle \mathsf{x}', \mathsf{y}\rangle \mid ((\mathsf{y} = \mathsf{x}') \vee (\mathsf{y} = \mathsf{x} \wedge \exists \rho, \nu . \mathcal{A}[\![\mathsf{A}]\!]\rho \neq \mathcal{A}[\![\mathsf{A}]\!]\rho[\mathsf{x}' \leftarrow \nu]))\} \tag{47.45}$$

〖letting $\rho = \varrho(\pi_0\mathsf{at}[\![\mathsf{S}]\!])$ and $\nu = \varrho(\pi'_0\mathsf{at}[\![\mathsf{S}]\!])(\mathsf{x}')$ so that $\forall \mathsf{z} \in \mathbb{V} \setminus \{\mathsf{x}'\}$. $\varrho(\pi_0\mathsf{at}[\![\mathsf{S}]\!])\mathsf{z} = \varrho(\pi'_0\mathsf{at}[\![\mathsf{S}]\!])\mathsf{z}$ implies that $\varrho(\pi'_0\mathsf{at}[\![\mathsf{S}]\!]) = \rho[\mathsf{x}' \leftarrow \nu]$.〗

$$= \{\langle \mathsf{x}', \mathsf{x}'\rangle \mid \mathsf{x}' \neq \mathsf{x}\} \cup \{\langle \mathsf{x}', \mathsf{x}\rangle \mid \exists \rho, \nu . \mathcal{A}[\![\mathsf{A}]\!]\rho \neq \mathcal{A}[\![\mathsf{A}]\!]\rho[\mathsf{x}' \leftarrow \nu]\}$$

〖case analysis〗

$$= \{\langle \mathsf{x}', \mathsf{x}'\rangle \mid \mathsf{x}' \neq \mathsf{x}\} \cup \{\langle \mathsf{x}', \mathsf{x}\rangle \mid \mathsf{x}' \in \widehat{\boldsymbol{S}}^{\exists}_{\mathsf{diff}}[\![\mathsf{A}]\!]\}$$

〖by defining the functional dependency of an expression A as $\widehat{\boldsymbol{S}}^{\exists}_{\mathsf{diff}}[\![\mathsf{A}]\!] \triangleq \{\mathsf{x}' \mid \exists \rho, \nu . \mathcal{A}[\![\mathsf{A}]\!]\rho \neq \mathcal{A}[\![\mathsf{A}]\!]\rho[\mathsf{x}' \leftarrow \nu]\}$ in (47.44)〗 □

Equality holds in (47.45) if every environment is computable, that is, $\forall \rho \in \mathbb{E}\mathbb{v} . \exists \pi \in \mathbb{T}^+ . \varrho(\pi) = \rho$. Therefore imprecision comes only from $\widehat{\boldsymbol{S}}^{\exists}_{\mathsf{diff}}[\![\mathsf{A}]\!]$, that is, when it is impossible to evaluate A in all possible environments.

Exercise 47.46 In (47.44), $\widehat{\boldsymbol{S}}^{\exists}_{\mathsf{diff}}[\![\mathsf{A}]\!]$ is overapproximated by $\mathsf{vars}[\![\mathsf{A}]\!]$. Propose a more precise approximation of $\widehat{\boldsymbol{S}}^{\exists}_{\mathsf{diff}}[\![\mathsf{A}]\!]$.

Notice that a reduced product with a reachability analysis providing an invariant $\mathsf{at}[\![\mathsf{S}]\!]$ will limit the possible values of the environments ρ and $\rho[\mathsf{x}' \leftarrow \nu]$ in $\widehat{\boldsymbol{S}}^{\exists}_{\mathsf{diff}}[\![\mathsf{A}]\!]$ and therefore make the dependency analysis more precise.

Abstract potential dependency semantics of a conditional statement $\mathsf{S} ::= \mathtt{if}\,(\mathsf{B})\,\mathsf{S}_t$

$$\widehat{\overline{\boldsymbol{\mathcal{S}}}}^{\exists}_{\text{diff}}[\![\mathsf{S}]\!]\,\ell \;=\; (\![\,\ell = \mathsf{at}[\![\mathsf{S}]\!] \;?\; \mathbb{1}_{\mathbb{V}} \qquad \text{(a)} \quad (47.47)$$

$$[\![\,\ell \in \mathsf{in}[\![\mathsf{S}_t]\!] \;?\; \widehat{\overline{\boldsymbol{\mathcal{S}}}}^{\exists}_{\text{diff}}[\![\mathsf{S}_t]\!]\,\ell \;]\,\mathsf{nondet}(\mathsf{B},\mathsf{B}) \qquad \text{(b)}$$

$$[\![\,\ell = \mathsf{after}[\![\mathsf{S}]\!] \;?\; \widehat{\overline{\boldsymbol{\mathcal{S}}}}^{\exists}_{\text{diff}}[\![\mathsf{S}_t]\!]\,\mathsf{after}[\![\mathsf{S}_t]\!] \;]\,\mathsf{nondet}(\mathsf{B},\mathsf{B}) \qquad \text{(c.1)}$$

$$\cup\; \mathbb{1}_{\mathbb{V}} \;]\,\mathsf{nondet}(\neg\mathsf{B},\neg\mathsf{B}) \qquad \text{(c.2)}$$

$$\cup\; \mathsf{nondet}(\mathsf{B},\neg\mathsf{B}) \times \mathsf{mod}[\![\mathsf{S}_t]\!] \qquad \text{(c.3)}$$

$$\,\text{:}\; \varnothing \,]\!) \qquad \text{(d)}$$

where determinacy ($\mathsf{det}(\mathsf{B}_1, \mathsf{B}_2)$) and nondeterminacy ($\mathsf{nondet}(\mathsf{B}_1, \mathsf{B}_2) \subseteq \mathbb{vars}[\![\mathsf{B}_1]\!] \cup \mathbb{vars}[\![\mathsf{B}_2]\!]$) are

$$\mathsf{det}(\mathsf{B}_1, \mathsf{B}_2) \;\subseteq\; \{\mathsf{x} \mid \forall \rho, \rho' \,.\, (\mathscr{B}[\![\mathsf{B}_1]\!]\rho \wedge \mathscr{B}[\![\mathsf{B}_2]\!]\rho') \Rightarrow (\rho(\mathsf{x}) = \rho'(\mathsf{x}))\} \qquad (47.48)$$

$$\mathsf{nondet}(\mathsf{B}_1, \mathsf{B}_2) \;\supseteq\; \mathbb{V} \setminus \mathsf{det}(\mathsf{B}_1, \mathsf{B}_2)$$

and the variables $\mathsf{mod}[\![\mathsf{S}]\!]$ modified by a statement S are

$$\mathsf{mod}[\![\mathsf{S}]\!] \;\supseteq\; \{\mathsf{x} \mid \exists \pi_0, \pi_1 \,.\, \langle \pi_0\mathsf{at}[\![\mathsf{S}]\!],\, \mathsf{at}[\![\mathsf{S}]\!]\pi_1\mathsf{after}[\![\mathsf{S}]\!]\rangle \in \widehat{\boldsymbol{\mathcal{S}}}^{+\infty}[\![\mathsf{S}]\!] \qquad (47.49)$$
$$\wedge\, \boldsymbol{\varrho}(\pi_0\mathsf{at}[\![\mathsf{S}]\!]\pi_1\mathsf{after}[\![\mathsf{S}]\!])\mathsf{x} \neq \boldsymbol{\varrho}(\pi_0\mathsf{at}[\![\mathsf{S}]\!])\mathsf{x}\}$$

In the style of [312], a purely syntactic and very rough overapproximation would be

$$\mathsf{mod}[\![\mathsf{x} = \mathsf{E}\,;]\!] \;\triangleq\; \{\mathsf{x}\} \qquad (47.50)$$

$$\mathsf{mod}[\![;]\!] \;\triangleq\; \mathsf{mod}[\![\,\epsilon\,]\!] \;\triangleq\; \mathsf{mod}[\![\mathtt{break}\,;]\!] \;\triangleq\; \varnothing$$

$$\mathsf{mod}[\![\mathtt{while}\,(\mathsf{B})\,\mathsf{S}]\!] \;=\; \mathsf{mod}[\![\mathtt{if}\,(\mathsf{B})\,\mathsf{S}]\!] \;\triangleq\; \mathsf{mod}[\![\mathsf{S}]\!]$$

$$\mathsf{mod}[\![\mathtt{if}\,(\mathsf{B})\,\mathsf{S}_t\;\mathtt{else}\;\mathsf{S}_f]\!] \;\triangleq\; \mathsf{mod}[\![\mathsf{S}_t]\!] \cup \mathsf{mod}[\![\mathsf{S}_f]\!]$$

$$\mathsf{mod}[\![\{\,\mathsf{Sl}\,\}]\!] \;\triangleq\; \mathsf{mod}[\![\mathsf{Sl}]\!]$$

$$\mathsf{mod}[\![\mathsf{Sl}\;\mathsf{S}]\!] \;\triangleq\; \mathsf{mod}[\![\mathsf{Sl}]\!] \cup \mathsf{mod}[\![\mathsf{S}]\!]$$

— On entry (47.47.a), which is an instance of (47.41), variables in $\mathbb{V}$ depend only upon themselves as specified by the identity relation $\mathbb{1}_{\mathbb{V}}$.

— The reasoning in (47.47.b) is that if a variable y depends at ℓ on the initial value of a variable x at $\mathsf{at}[\![\mathsf{S}_t]\!]$, it depends in the same way on that initial value of the variable x at $\mathsf{at}[\![\mathsf{S}]\!]$ because the test B has no side effect. However, (47.47.b) also takes into account that if S_t can be reached only for a unique value of the variable x and the branch is not taken for all other values of x then the variable y does not depend on x in S_t because empty observations are disallowed by the abstraction (47.25) using the definition (47.18) of diff.

By definition (47.48) of [non-]determinacy $\mathsf{det}(\mathsf{B}_1, \mathsf{B}_2)$ [$\mathsf{nondet}(\mathsf{B}_1, \mathsf{B}_2)$], if $\mathsf{x} \in \mathsf{det}(\mathsf{B}_1, \mathsf{B}_2)$ in (47.48) then B_1 and B_2 can both be true for at most one value of x (e.g., $\mathsf{det}(\mathsf{x{=}{=}1}, \mathsf{x{=}{=}1}) = \{\mathsf{x}\}$ and $\mathsf{det}(\mathsf{x{=}{=}1}, \mathsf{x\,!{=}1}) = \varnothing$). It is underapproximated by $\varnothing$. Its complement $\mathsf{nondet}(\mathsf{B}_1, \mathsf{B}_2)$ in (47.48) is the set of variables for which B_1 and B_2 may both be true for different values of variable x. It is overapproximated by all variables $\mathbb{V}$. A better solution is to use a reduced product with a value or symbolic constant propagation analysis as in section 47.10.

If $\mathsf{x} \notin \mathsf{nondet}(\mathsf{B}, \mathsf{B})$ in (47.47.b) then $\mathsf{x} \in \mathsf{det}(\mathsf{B}, \mathsf{B})$ so the value of x is constant in S_t so no variable y in S_t can depend on x. For example, dependency at ℓ in **if** (x == 1) { y = x ; ℓ } is the same as x = 1 ;y = x ; ℓ, which is the same as y = 1 ; ℓ so y does not depends on x at ℓ.

— (47.47.c) determines dependencies after S and so compares two possible executions of that statement. In case (47.47.c.1) both executions go through the true branch. In case (47.47.c.2) both executions go through the false branch, while in case (47.47.c.3) the executions take different branches.

– In case (47.47.c.1) when the test is true tt for both executions, the executions of the true branch S_t terminate and control after S_t reaches the program point after S (recall that $\mathsf{after}[\![\mathsf{S}_t]\!] = \mathsf{after}[\![\mathsf{S}]\!]$). The dependencies after S_t propagate after S but only in case of nondeterminism, for example, for variables that are not constant.

– The second case in (47.47.c.2) is for those executions for which the test B is false ff. Variables depend on themselves $\mathsf{at}[\![\mathsf{S}]\!]$ and control moves to $\mathsf{after}[\![\mathsf{S}]\!]$ so that dependencies are the same there, but only for variables that can reach $\mathsf{after}[\![\mathsf{S}]\!]$ with different values on different executions as indicated by the restriction to $\mathsf{nondet}(\neg\mathsf{B}, \neg\mathsf{B})$.

– The third case in (47.47.c.3) is for pairs of executions, one through the true branch and the other through the false branch. In that case, y depends on x only if x does not force execution to always take the same branch, meaning that $\mathsf{x} \in \mathsf{nondet}(\neg\mathsf{B}, \neg\mathsf{B})$. If y is not modified by the execution through S_t then its value after S is always the same as its value $\mathsf{at}[\![\mathsf{S}]\!]$ (because y is not modified on the false branch either). In that case changing y $\mathsf{at}[\![\mathsf{S}]\!]$ would not change y after S, so that in that situation, y does not depend on x. Therefore (47.47.c.3) requires that $\mathsf{y} \in \mathsf{mod}[\![\mathsf{S}_t]\!]$.

The very rough overapproximation (47.50) of the variables $\mathsf{mod}[\![\mathsf{S}]\!]$ modified by a statement S can be improved by a reduced product with a reachability analysis, for example, when the variable is constant on exit of S or by a relational analysis such that linear equalities [560] show that it is equal to its initial value.

— Finally in case (47.47.d) the program point ℓ is not reachable in S so, as stated in (47.42) there is not a dependency at ℓ originating from S.

Example 47.51 Consider $\mathsf{S} ::= \ell\ \mathsf{L} = \mathsf{H}\ ;\ell'$. We have $\hat{\overline{\boldsymbol{S}}}^{\exists}_{\mathsf{diff}}[\![\mathsf{S}]\!]\,\ell = \{\langle \mathsf{x}, \mathsf{x}\rangle \mid \mathsf{x} \in \mathbb{V}\}$ and $\hat{\overline{\boldsymbol{S}}}^{\exists}_{\mathsf{diff}}[\![\mathsf{S}]\!]\,\ell' = \{\langle \mathsf{H}, \mathsf{L}\rangle\} \cup \{\langle \mathsf{x}, \mathsf{x}\rangle \mid \mathsf{x} \in \mathbb{V} \setminus \{\mathsf{L}\}\}$.

We have $\mathsf{nondet}(\mathsf{H}, \mathsf{H}) = \mathsf{nondet}(\neg\mathsf{H}, \neg\mathsf{H}) = \{\mathsf{L}\}$ so that for the statement

$$\mathsf{S}' ::= \{\ \mathbf{if}\ \ell_1\ (\mathsf{H})\ \ell_2\ \mathsf{L} = \mathsf{H}\ ;\ell_3\ \mathbf{else}\ \ell_4\ \mathsf{L} = \mathsf{H}\ ;\ell_5\ \}\ell_6,$$

we have

$$\begin{array}{rcl}
& & \hat{\overline{\boldsymbol{S}}}^{\exists}_{\mathsf{diff}}[\![\mathsf{S}']\!]\,\ell_1 = \{\langle \mathsf{x}, \mathsf{x}\rangle \mid \mathsf{x} \in \mathbb{V}\} \\
\hat{\overline{\boldsymbol{S}}}^{\exists}_{\mathsf{diff}}[\![\mathsf{S}']\!]\,\ell_2 & = & \hat{\overline{\boldsymbol{S}}}^{\exists}_{\mathsf{diff}}[\![\mathsf{S}']\!]\,\ell_4 = \{\langle \mathsf{x}, \mathsf{x}\rangle \mid \mathsf{x} \in \mathbb{V} \setminus \{\mathsf{H}\}\} \\
\hat{\overline{\boldsymbol{S}}}^{\exists}_{\mathsf{diff}}[\![\mathsf{S}']\!]\,\ell_3 & = & \hat{\overline{\boldsymbol{S}}}^{\exists}_{\mathsf{diff}}[\![\mathsf{S}']\!]\,\ell_5 = \{\langle \mathsf{x}, \mathsf{x}\rangle \mid \mathsf{x} \in \mathbb{V} \setminus \{\mathsf{L}\}\} \\
& & \hat{\overline{\boldsymbol{S}}}^{\exists}_{\mathsf{diff}}[\![\mathsf{S}']\!]\,\ell_6 = \{\langle \mathsf{H}, \mathsf{L}\rangle\} \cup \{\langle \mathsf{x}, \mathsf{x}\rangle \mid \mathsf{x} \in \mathbb{V} \setminus \{\mathsf{L}\}\}
\end{array}$$

This is different and more precise than, for example, [42, 312] because L does not depend on H at ℓ_3 and ℓ_5 because , by definition of nondet, L is constant at ℓ_3 and ℓ_5. So this is equivalent to ℓ_2 `L = true ;`ℓ_3 and ℓ_4 `L = false ;`ℓ_5 which obviously would create no dependency. In contrast [42, 312] maintain an imprecise control dependence context, denoting (a superset of) the variables that at least one test surrounding S depends on. section 47.10 provides other examples of increased precision when taking values of variables into account.

Example 47.52

- $\texttt{S}_0 = \{\, \texttt{x} = \texttt{x} + 1\,;\, \texttt{y} = 1\,;\, \texttt{z} = 1\,;\, \}$,
  $\alpha^{\mathrm{d}}(\{\boldsymbol{\mathcal{S}}^{+\infty}[\![\texttt{S}_0]\!]\})\ \mathsf{after}[\![\texttt{S}_0]\!] = \{\langle \texttt{x},\ \texttt{x}\rangle\}$ because changing the initial value of x changes the final value of x but not those of y or z (always 1), whereas changing the initial value of y or z changes neither the final value of x nor those of y or z.
- $\texttt{S}_1 = \texttt{if}\,(\texttt{y} \,!\!= 1)\,\{\, \texttt{x} = \texttt{x} + 1\,;\, \texttt{y} = 1\,;\, \texttt{z} = 1\,;\, \}$,
  $\alpha^{\mathrm{d}}(\{\boldsymbol{\mathcal{S}}^{+\infty}[\![\texttt{S}_1]\!]\})\ \mathsf{after}[\![\texttt{S}_1]\!] = \{\langle \texttt{x},\ \texttt{x}\rangle, \langle \texttt{y},\ \texttt{x}\rangle, \langle \texttt{y},\ \texttt{z}\rangle, \langle \texttt{z},\ \texttt{z}\rangle\}$ because changing the initial value of x changes the final value of x but not that of y or z, changing the initial value of y may change the final value of x or z but not that of y (always equal to 1), but changing the initial value of z may change the final value of z (e.g., when the test is false).
- $\texttt{S}_2 = \texttt{if}\,(\texttt{y} \,!\!= 2)\,\{\, \texttt{x} = \texttt{x} + 1\,;\, \texttt{y} = 1\,;\, \texttt{z} = 1\,;\, \}$,
  $\alpha^{\mathrm{d}}(\{\boldsymbol{\mathcal{S}}^{+\infty}[\![\texttt{S}_2]\!]\})\ \mathsf{after}[\![\texttt{S}_2]\!] = \{\langle \texttt{x},\ \texttt{x}\rangle, \langle \texttt{y},\ \texttt{x}\rangle, \langle \texttt{y},\ \texttt{y}\rangle, \langle \texttt{y},\ \texttt{z}\rangle, \langle \texttt{z},\ \texttt{z}\rangle\}$ because changing the initial value of x changes the final value of x but not that of y or z, changing the initial value of y may change both the final values of x, y and z, whereas changing the initial value of z may change only the final value of z.
- $\texttt{S}_3 = \texttt{if}\,(\texttt{y} == \texttt{y})\,\{\, \texttt{x} = \texttt{x} + 1\,;\, \texttt{y} = 1\,;\, \texttt{z} = 1\,;\, \}$,
  $\alpha^{\mathrm{d}}(\{\boldsymbol{\mathcal{S}}^{+\infty}[\![\texttt{S}_3]\!]\})\ \mathsf{after}[\![\texttt{S}_3]\!] = \{\langle \texttt{x},\ \texttt{x}\rangle\}$ because $\texttt{S}_3$ is semantically equivalent to $\texttt{S}_0$.
- $\hat{\overline{\boldsymbol{\mathcal{S}}}}{}^{\exists}_{\mathsf{diff}}[\![\texttt{S}_1]\!]\ \mathsf{after}[\![\texttt{S}_1]\!] = \hat{\overline{\boldsymbol{\mathcal{S}}}}{}^{\exists}_{\mathsf{diff}}[\![\texttt{S}_2]\!]\ \mathsf{after}[\![\texttt{S}_2]\!] = \hat{\overline{\boldsymbol{\mathcal{S}}}}{}^{\exists}_{\mathsf{diff}}[\![\texttt{S}_3]\!]\ \mathsf{after}[\![\texttt{S}_3]\!] = \{\langle \texttt{x},\ \texttt{x}\rangle, \langle \texttt{y},\ \texttt{x}\rangle, \langle \texttt{y},\ \texttt{y}\rangle, \langle \texttt{y},\ \texttt{z}\rangle, \langle \texttt{z},\ \texttt{z}\rangle\}$ is a sound overapproximation ignoring values of variables.
- $\texttt{S}_4 = \texttt{if}\,(0 == 0)\,\{\, \texttt{x} = \texttt{x} + 1\,;\, \texttt{y} = 1\,;\, \texttt{z} = 1\,;\, \}$,
  $\alpha^{\mathrm{d}}(\{\boldsymbol{\mathcal{S}}^{+\infty}[\![\texttt{S}_4]\!]\})\ \mathsf{after}[\![\texttt{S}_4]\!] = \{\langle \texttt{x},\ \texttt{x}\rangle\}$ because $\texttt{S}_4$ is semantically equivalent to $\texttt{S}_0$ and $\texttt{S}_3$.
  $\hat{\overline{\boldsymbol{\mathcal{S}}}}{}^{\exists}_{\mathsf{diff}}[\![\texttt{S}_4]\!]\ \mathsf{after}[\![\texttt{S}_4]\!] = \{\langle \texttt{x},\ \texttt{x}\rangle, \langle \texttt{y},\ \texttt{y}\rangle, \langle \texttt{z},\ \texttt{z}\rangle\}$ is a sound overapproximation ignoring that the test is always true. Moreover, the test tt has no variable that can flow to variables assigned in the tt branch.
- Notice that (47.47) does not take values of variables into account very precisely, but only the nondeterminacy of expressions.

B	nondet(B, B)	nondet(B, ¬B)	nondet(¬B, ¬B)
y != 1	{y}	{y}	∅
y != 2	{y}	{y}	∅
y == y	{y}	{y}	{y}
0 == 0	∅	∅	∅

- so both examples $\texttt{S}_1$ and $\texttt{S}_2$ above are handled in the same way .

The proof of (47.47) is available at https://github.com/PrAbsInt/.

Abstract potential dependency semantics of a conditional statement $\mathsf{S} ::= \mathtt{if}\,(\mathsf{B})\,\mathsf{S}_t$ $\mathtt{else}\ \mathsf{S}_f$

$$\widehat{\overline{\boldsymbol{S}}}^{\exists}_{\text{diff}}[\![\mathsf{S}]\!]\ell = \Big(\ell = \mathsf{at}[\![\mathsf{S}]\!] \ ?\ \mathbb{1}_{\mathbb{V}} \quad (47.53)$$
$$\Big\| \ \ell \in \mathsf{in}[\![\mathsf{S}_t]\!] \ ?\ \widehat{\overline{\boldsymbol{S}}}^{\exists}_{\text{diff}}[\![\mathsf{S}_t]\!]\ell \ \rceil\, \mathsf{nondet}(\mathsf{B}, \mathsf{B})$$
$$\Big\| \ \ell \in \mathsf{in}[\![\mathsf{S}_f]\!] \ ?\ \widehat{\overline{\boldsymbol{S}}}^{\exists}_{\text{diff}}[\![\mathsf{S}_f]\!]\ell \ \rceil\, \mathsf{nondet}(\neg\mathsf{B}, \neg\mathsf{B})$$
$$\Big\| \ \ell = \mathsf{after}[\![\mathsf{S}]\!] \ ?\ \widehat{\overline{\boldsymbol{S}}}^{\exists}_{\text{diff}}[\![\mathsf{S}_t]\!]\,\mathsf{after}[\![\mathsf{S}_t]\!] \ \rceil\, \mathsf{nondet}(\mathsf{B}, \mathsf{B})$$
$$\cup\, \widehat{\overline{\boldsymbol{S}}}^{\exists}_{\text{diff}}[\![\mathsf{S}_f]\!]\,\mathsf{after}[\![\mathsf{S}_f]\!] \ \rceil\, \mathsf{nondet}(\neg\mathsf{B}, \neg\mathsf{B})$$
$$\cup\, \mathsf{nondet}(\mathsf{B}, \neg\mathsf{B}) \times (\mathsf{mod}[\![\mathsf{S}_t]\!] \cup \mathsf{mod}[\![\mathsf{S}_f]\!])$$
$$:\ \varnothing\Big)$$

Abstract potential dependency semantics of an empty statement list $\mathtt{Sl} ::= \epsilon$

$$\widehat{\overline{\boldsymbol{S}}}^{\exists}_{\text{diff}}[\![\mathtt{Sl}]\!]\ell \ \triangleq\ \Big(\ell = \mathsf{at}[\![\mathtt{Sl}]\!] \ ?\ \mathbb{1}_{\mathbb{V}} : \varnothing\Big) \quad (47.54)$$

Abstract semantics of a skip statement $\mathsf{S} ::= ;$

$$\widehat{\overline{\boldsymbol{S}}}^{\exists}_{\text{diff}}[\![\mathsf{S}]\!]\ell = \Big(\ell \in \{\mathsf{at}[\![\mathsf{S}]\!], \mathsf{after}[\![\mathsf{S}]\!]\} \ ?\ \mathbb{1}_{\mathbb{V}} : \varnothing\Big) \quad (47.55)$$

Abstract semantics of a break statement $\mathsf{S} ::= \ell\ \mathtt{break}\,;$

$$\widehat{\overline{\boldsymbol{S}}}^{\exists}_{\text{diff}}[\![\mathsf{S}]\!]\ell = \Big(\ell \in \{\mathsf{at}[\![\mathsf{S}]\!], \mathsf{break\text{-}to}[\![\mathsf{S}]\!]\} \ ?\ \mathbb{1}_{\mathbb{V}} : \varnothing\Big) \quad (47.56)$$

Abstract semantics of a compound statement $\mathsf{S} ::= \{\ \mathtt{Sl}\ \}$

$$\widehat{\overline{\boldsymbol{S}}}^{\exists}_{\text{diff}}[\![\mathsf{S}]\!] = \widehat{\overline{\boldsymbol{S}}}^{\exists}_{\text{diff}}[\![\mathtt{Sl}]\!] \quad (47.57)$$

Exercise 47.58 Prove the soundness of (47.53), (47.54), (47.55), (47.56), and (47.57).

For the statement list and the iteration statement, we will need the following lemma expressing that dependency for the sequential composition $\ell_0\mathsf{S}\ \ell'\mathsf{S}'$ of statements S and S' is the relational composition $\mathbin{⨟}$ of their dependencies. More precisely, at ℓ in S' of $\ell_0\mathsf{S}\ \ell'\mathsf{S}'$, y depends on the initial value of x at ℓ_0 if and only if y depends on the initial value of some variable z at ℓ' on entry of S' and z at ℓ' after S depends on the initial value of x at ℓ_0 on entry of S.

Lemma 47.59 Let $\boldsymbol{S}$ and $\boldsymbol{S}'$ be semantics of program components where ℓ' is after the first and at the second component and ℓ is in the second components or iterates of the fixpoint semantics of an iteration statement where ℓ' is at the iteration and ℓ in the body. Define

$$\mathbf{D} \triangleq \{\langle \mathsf{x}, \mathsf{y}\rangle \mid \exists \pi_0 \ell_0 \pi_1 \ell' \pi_2 \ell \pi_3,\ \pi_0' \ell_0 \pi_1' \ell' \pi_2' \ell \pi_3' .$$
$$\ell_0 \pi_1 \ell' \in \mathcal{S}(\pi_0 \ell_0) \wedge \ell' \pi_2 \ell \pi_3 \in \mathcal{S}'(\pi_0 \ell_0 \pi_1 \ell') \wedge$$
$$\ell_0 \pi_1' \ell' \in \mathcal{S}(\pi_0' \ell_0) \wedge \ell' \pi_2' \ell \pi_3' \in \mathcal{S}'(\pi_0' \ell_0 \pi_1' \ell') \wedge$$
$$(\forall \mathsf{z} \in \mathbb{V} \setminus \{\mathsf{x}\} . \varrho(\pi_0 \ell_0)\mathsf{z} = \varrho(\pi_0' \ell_0)\mathsf{z}) \wedge$$
$$\mathsf{diff}(\mathsf{seqval}[\![\mathsf{y}]\!]\ell(\pi_0 \ell_0 \pi_1 \ell' \pi_2 \ell,\ \ell \pi_3), \mathsf{seqval}[\![\mathsf{y}]\!]\ell(\pi_0' \ell_0 \pi_1' \ell' \pi_2' \ell,\ \ell \pi_3'))\} \tag{47.59}$$

Then $\mathbf{D} \subseteq \alpha^{\mathrm{d}}(\{\mathcal{S}\})\,\ell' \fatsemi \alpha^{\mathrm{d}}(\{\mathcal{S}'\})\,\ell$.

Proof of lemma 47.59 Assume $\mathbf{D} \neq \varnothing$. Let us show that (47.59) implies that y at ℓ must depend on some variable z at ℓ' such that z at ℓ' depends on x at ℓ_0. The proof is by reductio ad absurdum.

We assume (47.59), that is, that y at ℓ depends on the value of x at ℓ_0 and that either y at ℓ depends on no variable z at ℓ' or else if y at ℓ depends on some variable z at ℓ' then variable z at ℓ' does not depend on x at ℓ_0. In both cases, we derive a contradiction.

— (1) — y at ℓ depends on no variable z at ℓ'

Let us look at the value $\varrho(\pi)\mathsf{y}$ and $\varrho(\pi')\mathsf{y}$ of y at ℓ on traces $\pi = \pi_0 \ell_0 \pi_1 \ell' \pi_2 \ell \pi_3$ and $\pi' = \pi_0' \ell_0 \pi_1' \ell' \pi_2' \ell \pi_3'$ as defined in (6.6) of section 6.2 by an assignment to y previously at ℓ_1 on π and ℓ_1' on π'.

- These assignments cannot be of the same constant on both traces because then the dependency $\mathsf{diff}(\mathsf{seqval}[\![\mathsf{y}]\!]\ell(\pi_0 \ell_0 \pi_1 \ell' \pi_2 \ell,\ \ell \pi_3), \mathsf{seqval}[\![\mathsf{y}]\!]\ell(\pi_0' \ell_0 \pi_1' \ell' \pi_2' \ell,\ \ell \pi_3'))$ is a contradiction.
- These assignments could be of different constants on both traces. However then there must be a test B_1 at ℓ_2 along π and π' that evaluates to different values, hence is not a constant (by the program syntax it cannot be different tests on both traces, because there is a common ancestor). It follows, by definition (47.1) of functional dependency, that it depends on some variable y_1 that evaluates to different Boolean values on π and π'.
- Or, the assignment to y previously at ℓ_1 on π is not a constant (the other case is symmetric) and so is from an expression E_1. By definition (47.1) of functional dependency, E_1 depends on some variable y_1 in order to evaluate to different values of y on π and π'.

In both cases we are back to the same dependency problem for variables $\ldots, \mathsf{y}_2, \mathsf{y}_1, \mathsf{y}$ at a strictly earlier points $\ldots, \ell_2, \ell_1, \ell$ in the traces, so this cannot go on forever. There are two cases.

- If we end up at ℓ', we get a contradiction because y at ℓ depends on no variable z at ℓ'.
- Else, we end up at ℓ_n for some variable y_n before ℓ' in the part $\pi_0 \ell_0 \pi_1 \ell'$ of $\pi = \pi_0 \ell_0 \pi_1 \ell' \pi_2 \ell \pi_3$. In that case, y is unchanged on $\ell' \pi_2 \ell$ and hence depends upon itself, in contradiction to y at ℓ, which depends on no variable z at ℓ', including itself.

— (2) — No variable z at ℓ' depends on x at ℓ_0.

We can also assume that $\mathtt{y}$ at ℓ depends on some variables $\mathtt{z}$ at ℓ' (the contrary has been shown previously to be impossible). However changing $\mathtt{x}$ at ℓ_0 will not change these variables $\mathtt{z}$ at ℓ' so will not change $\mathtt{y}$ at ℓ in contradiction to the dependence of $\mathtt{y}$ at ℓ on variable $\mathtt{x}$ at $\mathsf{at}[\![\mathtt{Sl}']\!]$.

It follows that

$$
\begin{aligned}
&(47.59)\\
\subseteq\ & \{\langle \mathtt{x}, \mathtt{y}\rangle \mid \exists \mathtt{z} \in \mathbb{V} \,.\, \exists \langle \pi_0\ell_0, \ell_0\pi_1\ell'\pi_2\rangle, \langle \pi'_0\ell_0, \ell_0\pi'_1\ell'\pi'_2\rangle \in \boldsymbol{\mathcal{S}} \,.\, (\forall \mathtt{u} \in \mathbb{V} \setminus \{\mathtt{x}\} \,.\, \varrho(\pi_0\ell_0)\mathtt{u} = \\
& \varrho(\pi'_0\ell_0)\mathtt{u}) \wedge (\mathsf{seqval}[\![\mathtt{z}]\!]\,(\ell')\,(\pi_0\ell_0\pi_1\ell', \ell'\pi_2) \neq \mathsf{seqval}[\![\mathtt{z}]\!]\,(\ell')\,(\pi'_0\ell_0\pi'_1\ell', \ell'\pi'_2))\\
& \wedge\\
& \exists \langle \pi_0\ell', \ell'\pi_1\ell\pi_2\rangle, \langle \pi'_0\ell', \ell'\pi'_1\ell\pi'_2\rangle \in \boldsymbol{\mathcal{S}}' \,.\, (\forall \mathtt{u} \in \mathbb{V} \setminus \{\mathtt{z}\} \,.\, \varrho(\pi_0\ell')\mathtt{u} = \varrho(\pi'_0\ell')\mathtt{u}) \wedge\\
& \mathsf{diff}(\mathsf{seqval}[\![\mathtt{y}]\!]\ell(\pi_0\ell'\pi_1\ell, \ell\pi_2), \mathsf{seqval}[\![\mathtt{y}]\!]\ell(\pi'_0\ell'\pi'_1\ell, \ell\pi'_2))\}\\
& \qquad \text{⦃some } \mathtt{z} \text{ at } \ell' \text{ depends upon } \mathtt{x} \text{ at } \ell_0 \text{ and } \mathtt{y} \text{ at } \ell \text{ depends on } \mathtt{z} \text{ at } \ell'\text{⦄}\\
=\ & \{\langle \mathtt{x}, \mathtt{y}\rangle \mid \exists \mathtt{z} \,.\, \boldsymbol{\mathcal{S}} \in \{\Pi \in \wp(\mathbb{T}^{+\infty} \times \mathbb{T}^{+\infty}) \mid \exists \langle \pi_0\ell_0, \ell_0\pi_1\ell'\pi_2\rangle, \langle \pi'_0\ell_0, \ell_0\pi'_1\ell'\pi'_2\rangle \in\\
& \Pi \,.\, (\forall \mathtt{u} \in \mathbb{V} \setminus \{\mathtt{x}\} \,.\, \varrho(\pi_0\ell_0)\mathtt{u} = \varrho(\pi'_0\ell_0)\mathtt{u}) \wedge (\mathsf{seqval}[\![\mathtt{z}]\!]\,(\ell')\,(\pi_0\ell_0\pi_1\ell', \ell'\pi_2) \neq\\
& \mathsf{seqval}[\![\mathtt{z}]\!]\,(\ell')\,(\pi'_0\ell_0\pi'_1\ell', \ell'\pi'_2))\}\\
& \wedge\\
& \boldsymbol{\mathcal{S}}' \in \{\Pi \in \wp(\mathbb{T}^{+\infty} \times \mathbb{T}^{+\infty}) \mid \exists \langle \pi_0\ell', \ell'\pi_1\ell\pi_2\rangle, \langle \pi'_0\ell', \ell'\pi'_1\ell\pi'_2\rangle \in \Pi \,.\, (\forall \mathtt{u} \in \mathbb{V} \setminus \{\mathtt{z}\} \,.\\
& \varrho(\pi_0\ell')\mathtt{u} = \varrho(\pi'_0\ell')\mathtt{u}) \wedge \mathsf{diff}(\mathsf{seqval}[\![\mathtt{y}]\!]\ell(\pi_0\ell'\pi_1\ell, \ell\pi_2), \mathsf{seqval}[\![\mathtt{y}]\!]\ell(\pi'_0\ell'\pi'_1\ell, \ell\pi'_2))\}\}\\
& \qquad \text{⦃definition of } \in\text{⦄}\\
=\ & \{\langle \mathtt{x}, \mathtt{y}\rangle \mid \exists \mathtt{z} \,.\, \boldsymbol{\mathcal{S}} \in \mathcal{D}\ell'\,\langle \mathtt{x}, \mathtt{z}\rangle \wedge \boldsymbol{\mathcal{S}}' \in \mathcal{D}\ell\langle \mathtt{z}, \mathtt{y}\rangle\} \qquad \text{⦃definition (47.19) of } \mathcal{D}\ell\langle \mathtt{x}, \mathtt{y}\rangle\text{⦄}\\
=\ & \{\langle \mathtt{x}, \mathtt{y}\rangle \mid \exists \mathtt{z} \,.\, \langle \mathtt{x}, \mathtt{z}\rangle \in \{\langle \mathtt{x}, \mathtt{z}\rangle \mid \{\boldsymbol{\mathcal{S}}\} \subseteq \mathcal{D}\ell'\,\langle \mathtt{x}, \mathtt{z}\rangle\} \wedge \langle \mathtt{z}, \mathtt{y}\rangle \in \{\langle \mathtt{x}, \mathtt{y}\rangle \mid \{\boldsymbol{\mathcal{S}}'\} \subseteq \mathcal{D}\ell\langle \mathtt{x},\\
& \mathtt{y}\rangle\}\} \qquad \text{⦃definitions of } \in \text{ and of } \subseteq\text{⦄}\\
=\ & \{\langle \mathtt{x}, \mathtt{y}\rangle \mid \exists \mathtt{z} \,.\, \langle \mathtt{x}, \mathtt{z}\rangle \in \alpha^{\mathsf{d}}(\{\boldsymbol{\mathcal{S}}\})\,\ell' \wedge \langle \mathtt{z}, \mathtt{y}\rangle \in \alpha^{\mathsf{d}}(\{\boldsymbol{\mathcal{S}}'\})\,\ell\} \qquad \text{⦃definition (47.25) of } \alpha^{\mathsf{d}}\text{⦄}\\
=\ & \alpha^{\mathsf{d}}(\{\boldsymbol{\mathcal{S}}\})\,\ell' \,\fatsemi\, \alpha^{\mathsf{d}}(\{\boldsymbol{\mathcal{S}}'\})\,\ell\\
& \qquad \text{⦃definition } r_1 \fatsemi r_2 \triangleq \{\langle x, z\rangle \mid \exists y \,.\, \langle x, y\rangle \in r_1 \wedge \langle y, z\rangle \in r_2\} \text{ in section 2.2.2⦄} \qquad \square
\end{aligned}
$$

Abstract potential dependency semantics of a statement list $\mathtt{Sl} ::= \mathtt{Sl}'\ \mathtt{S}$

$$
\begin{aligned}
\widehat{\boldsymbol{\mathcal{S}}}^{\exists}_{\mathsf{diff}}[\![\mathtt{Sl}]\!]\,\ell \ \triangleq\ & (\!|\ \ell \in \mathsf{labx}[\![\mathtt{Sl}']\!] \ ?\ \widehat{\boldsymbol{\mathcal{S}}}^{\exists}_{\mathsf{diff}}[\![\mathtt{Sl}']\!]\,\ell && \text{(a)} \quad (47.60)\\
& \ \ [\!]\ \ell \in \mathsf{labx}[\![\mathtt{S}]\!] \setminus \{\mathsf{at}[\![\mathtt{S}]\!]\} \ ?\ \widehat{\boldsymbol{\mathcal{S}}}^{\exists}_{\mathsf{diff}}[\![\mathtt{Sl}']\!]\ \mathsf{at}[\![\mathtt{S}]\!] \,\fatsemi\, \widehat{\boldsymbol{\mathcal{S}}}^{\exists}_{\mathsf{diff}}[\![\mathtt{S}]\!]\,\ell && \text{(b)}\\
& \ \ :\ \emptyset\ |\!) && \text{(c)}
\end{aligned}
$$

— The first case (47.60.a) looks for dependencies at a program point ℓ inside $\mathtt{Sl}'$, so that by structural induction, this is $\widehat{\boldsymbol{\mathcal{S}}}^{\exists}_{\mathsf{diff}}[\![\mathtt{Sl}']\!]\,\ell$.

— The second case (47.60.b) looks for dependencies at a program point ℓ inside $\mathtt{S}$. We exclude the case $\ell = \mathsf{at}[\![\mathtt{S}]\!]$ because $\mathsf{at}[\![\mathtt{S}]\!] = \mathsf{after}[\![\mathtt{Sl}']\!] \in \mathsf{labx}[\![\mathtt{Sl}']\!]$, so this case has already been handled in the previous case (47.60.a).

— Otherwise in (47.60.b), a variable $\mathtt{y}$ at ℓ in $\mathtt{S}$ depends on the initial value of a variable $\mathtt{x}$ on entry $\mathsf{at}[\![\mathtt{Sl} ::= \mathtt{Sl}'\ \mathtt{S}]\!] = \mathsf{at}[\![\mathtt{Sl}']\!] \in \mathsf{labx}[\![\mathtt{Sl}']\!]$ if and only if $\mathtt{y}$ at ℓ in $\mathtt{S}$ depends on the initial value of some variable $\mathtt{z}$ on entry $\mathsf{at}[\![\mathtt{S}]\!] = \mathsf{after}[\![\mathtt{Sl}']\!]$ of statement $\mathtt{S}$ and the value of z at that point depends on the initial value of a variable x on entry $\mathsf{at}[\![\mathtt{Sl}]\!]$ of the statement list $\mathtt{Sl}$. So there exists z such that $\langle \mathtt{y},\ \mathtt{z}\rangle \in \widehat{\overline{\boldsymbol{\mathcal{S}}}}^{\exists}_{\mathsf{diff}}[\![\mathtt{Sl}']\!]\ \mathsf{after}[\![\mathtt{Sl}']\!] = \widehat{\overline{\boldsymbol{\mathcal{S}}}}^{\exists}_{\mathsf{diff}}[\![\mathtt{Sl}']\!]\ \mathsf{at}[\![\mathtt{S}]\!]$ and $\langle \mathtt{z},\ \mathtt{y}\rangle \in \widehat{\overline{\boldsymbol{\mathcal{S}}}}^{\exists}_{\mathsf{diff}}[\![\mathtt{S}]\!]\ \ell$, meaning that $\langle \mathtt{x},\ \mathtt{y}\rangle \in \widehat{\overline{\boldsymbol{\mathcal{S}}}}^{\exists}_{\mathsf{diff}}[\![\mathtt{Sl}']\!]\ \mathsf{at}[\![\mathtt{S}]\!] \fatsemi \widehat{\overline{\boldsymbol{\mathcal{S}}}}^{\exists}_{\mathsf{diff}}[\![\mathtt{S}]\!]\ \ell$ is in their composition. As shown by example 47.37, the precision can be improved by a reduced product with a value analysis.

Proof of (47.60)

$$\begin{aligned}
& \alpha^{\mathrm{d}}(\{\boldsymbol{\mathcal{S}}^{+\infty}[\![\mathtt{Sl}'\ \mathtt{S}]\!]\})\,\ell \\
={}& \alpha^{\mathrm{d}}(\{\boldsymbol{\mathcal{S}}^{*}[\![\mathtt{S}]\!]\})\,\ell && \text{⟅lemma 47.23⟆} \\
={}& \{\langle \mathtt{x}',\ \mathtt{y}\rangle \mid \boldsymbol{\mathcal{S}}^{*}[\![\mathtt{Sl}'\mathtt{S}]\!] \in \mathcal{D}(\ell)\langle \mathtt{x}',\ \mathtt{y}\rangle\} && \text{⟅definition (47.25) of } \alpha^{\mathrm{d}}\text{⟆} \\
={}& \{\langle \mathtt{x}',\ \mathtt{y}\rangle \mid \exists \langle \pi_0,\ \pi_1\rangle, \langle \pi_0',\ \pi_1'\rangle \in \boldsymbol{\mathcal{S}}^{*}[\![\mathtt{Sl}'\mathtt{S}]\!]\,.\,(\forall \mathtt{z} \in \mathbb{V}\setminus\{\mathtt{x}'\}\,.\,\varrho(\pi_0)\mathtt{z} = \varrho(\pi_0')\mathtt{z}) \wedge \\
& \quad \mathsf{diff}(\mathsf{seqval}[\![\mathtt{y}]\!]\ell(\pi_0, \pi_1), \mathsf{seqval}[\![\mathtt{y}]\!]\ell(\pi_0', \pi_1'))\} && \text{⟅definition (47.19) of } \mathcal{D}\ell\langle \mathtt{x},\ \mathtt{y}\rangle\text{⟆} \\
={}& \{\langle \mathtt{x}',\ \mathtt{y}\rangle \mid \exists \langle \pi_0,\ \pi_1\rangle, \langle \pi_0',\ \pi_1'\rangle \in \widehat{\boldsymbol{\mathcal{S}}}^{*}[\![\mathtt{Sl}']\!] \cup \{\langle \pi,\ \pi' \frown \pi''\rangle \mid \pi' \in \\
& \quad \widehat{\boldsymbol{\mathcal{S}}}^{*}[\![\mathtt{Sl}']\!](\pi) \wedge \pi'' \in \widehat{\boldsymbol{\mathcal{S}}}^{*}[\![\mathtt{S}]\!](\pi \frown \pi')\}\,.\,(\forall \mathtt{z} \in \mathbb{V}\setminus\{\mathtt{x}'\}\,.\,\varrho(\pi_0)\mathtt{z} = \varrho(\pi_0')\mathtt{z}) \wedge \\
& \quad \mathsf{diff}(\mathsf{seqval}[\![\mathtt{y}]\!]\ell(\pi_0, \pi_1), \mathsf{seqval}[\![\mathtt{y}]\!]\ell(\pi_0', \pi_1'))\}
\end{aligned}$$

⟅definition of $\widehat{\boldsymbol{\mathcal{S}}}^{*}[\![\mathtt{Sl}]\!](\pi) = \widehat{\boldsymbol{\mathcal{S}}}^{*}[\![\mathtt{Sl}']\!](\pi) \cup \bigcup\{\widehat{\boldsymbol{\mathcal{S}}}^{*}[\![\mathtt{S}]\!](\pi') \mid \pi' \in \widehat{\boldsymbol{\mathcal{S}}}^{*}[\![\mathtt{Sl}']\!](\pi)\}$ by (7.7), (6.13), (6.14) and (7.6)⟆

$$\begin{aligned}
={}& \llparenthesis\, \ell \in \mathsf{labx}[\![\mathtt{Sl}']\!] \mathrel{?} \{\langle \mathtt{x}',\ \mathtt{y}\rangle \mid \exists \langle \pi_0,\ \pi_1\rangle, \langle \pi_0',\ \pi_1'\rangle \in \widehat{\boldsymbol{\mathcal{S}}}^{*}[\![\mathtt{Sl}']\!] \cup \{\langle \pi,\ \pi' \frown \pi''\rangle \mid \\
& \quad \pi' \in \widehat{\boldsymbol{\mathcal{S}}}^{*}[\![\mathtt{Sl}']\!](\pi) \wedge \pi'' \in \widehat{\boldsymbol{\mathcal{S}}}^{*}[\![\mathtt{S}]\!](\pi \frown \pi')\}\,.\,(\forall \mathtt{z} \in \mathbb{V}\setminus\{\mathtt{x}'\}\,.\,\varrho(\pi_0)\mathtt{z} = \varrho(\pi_0')\mathtt{z}) \wedge \\
& \quad \mathsf{diff}(\mathsf{seqval}[\![\mathtt{y}]\!]\ell(\pi_0, \pi_1), \mathsf{seqval}[\![\mathtt{y}]\!]\ell(\pi_0', \pi_1'))\} \mathrel{:} \varnothing \,\rrparenthesis \\
& \cup \\
& \llparenthesis\, \ell \in \mathsf{labx}[\![\mathtt{S}]\!]\setminus\{\mathsf{at}[\![\mathtt{S}]\!]\} \mathrel{?} \{\langle \mathtt{x}',\ \mathtt{y}\rangle \mid \exists \langle \pi_0,\ \pi_1\rangle, \langle \pi_0',\ \pi_1'\rangle \in \widehat{\boldsymbol{\mathcal{S}}}^{*}[\![\mathtt{Sl}']\!] \cup \{\langle \pi,\ \pi' \frown \pi''\rangle \mid \\
& \quad \pi' \in \widehat{\boldsymbol{\mathcal{S}}}^{*}[\![\mathtt{Sl}']\!](\pi) \wedge \pi'' \in \widehat{\boldsymbol{\mathcal{S}}}^{*}[\![\mathtt{S}]\!](\pi \frown \pi')\}\,.\,(\forall \mathtt{z} \in \mathbb{V}\setminus\{\mathtt{x}'\}\,.\,\varrho(\pi_0)\mathtt{z} = \varrho(\pi_0')\mathtt{z}) \wedge \\
& \quad \mathsf{diff}(\mathsf{seqval}[\![\mathtt{y}]\!]\ell(\pi_0, \pi_1), \mathsf{seqval}[\![\mathtt{y}]\!]\ell(\pi_0', \pi_1'))\} \mathrel{:} \varnothing \,\rrparenthesis
\end{aligned}$$

⟅definition of $\mathsf{labx}[\![\mathtt{Sl}'\ \mathtt{S}]\!]$ so that either $\ell \in \mathsf{labx}[\![\mathtt{Sl}']\!]$ or $\ell \in \mathsf{labx}[\![\mathtt{S}]\!]\setminus\{\mathsf{at}[\![\mathtt{S}]\!]\}$. Definition of $\cup$ to join the two cases.⟆

$$\begin{aligned}
={}& \llparenthesis\, \ell \in \mathsf{labx}[\![\mathtt{Sl}']\!] \mathrel{?} \{\langle \mathtt{x}',\ \mathtt{y}\rangle \mid \exists \langle \pi_0,\ \pi_1\ell\pi_2\rangle, \langle \pi_0',\ \pi_1'\ell\pi_2'\rangle \in \widehat{\boldsymbol{\mathcal{S}}}^{*}[\![\mathtt{Sl}']\!]\,.\,\varrho(\pi_0)\mathtt{z} = \varrho(\pi_0')\mathtt{z}) \wedge \\
& \quad \ell \notin \pi_1 \wedge \ell \notin \pi_1' \wedge \mathsf{diff}(\mathsf{seqval}[\![\mathtt{y}]\!]\ell(\pi_0 \frown \pi_1\ell,\ \ell\pi_2), \mathsf{seqval}[\![\mathtt{y}]\!]\ell(\pi_0' \frown \pi_1'\ell,\ \ell\pi_2'))\} \mathrel{:} \varnothing \,\rrparenthesis \\
& \cup \\
& \llparenthesis\, \ell \in \mathsf{labx}[\![\mathtt{S}]\!]\setminus\{\mathsf{at}[\![\mathtt{S}]\!]\} \mathrel{?} \{\langle \mathtt{x}',\ \mathtt{y}\rangle \mid \exists \langle \pi_0,\ \pi_1\ell\pi_2\rangle, \langle \pi_0',\ \pi_1'\ell\pi_2'\rangle \in \{\langle \pi,\ \pi' \frown \pi''\rangle \mid \pi' \in \\
& \quad \widehat{\boldsymbol{\mathcal{S}}}^{+}[\![\mathtt{Sl}']\!](\pi) \wedge \pi'' \in \widehat{\boldsymbol{\mathcal{S}}}^{*}[\![\mathtt{S}]\!](\pi \frown \pi')\}\,.\,(\forall \mathtt{z} \in \mathbb{V}\setminus\{\mathtt{x}'\}\,.\,\varrho(\pi_0)\mathtt{z} = \varrho(\pi_0')\mathtt{z}) \wedge \ell \notin \pi_1 \wedge \ell \notin \\
& \quad \pi_1' \wedge \mathsf{diff}(\mathsf{seqval}[\![\mathtt{y}]\!]\ell(\pi_0 \frown \pi_1\ell,\ \ell\pi_2), \mathsf{seqval}[\![\mathtt{y}]\!]\ell(\pi_0' \frown \pi_1'\ell,\ \ell\pi_2'))\} \mathrel{:} \varnothing \,\rrparenthesis
\end{aligned}$$

⟅definition (47.18) of diff and (47.16) of $\mathsf{seqval}[\![\mathtt{y}]\!]\ell(\pi, \pi')$ so that necessarily $\ell \in \pi'$ and therefore π' can be decomposed into $\pi' = \pi_0'\ell\pi_1'$ where $\ell \notin \pi_0'$. Definition (47.16) of $\mathsf{seqval}[\![\mathtt{y}]\!]$ so that $\mathsf{seqval}[\![\mathtt{y}]\!]\ell(\pi_0,\ \pi_1\ell\pi_2) = \mathsf{seqval}[\![\mathtt{y}]\!]\ell(\pi_0 \frown \pi_1\ell,\ \ell\pi_2)$ when $\ell \notin \pi_1$. Definition of $\widehat{\boldsymbol{\mathcal{S}}}^{*}[\![\mathtt{S}]\!]$ so that $\pi'\ell\pi'' \in \widehat{\boldsymbol{\mathcal{S}}}^{*}[\![\mathtt{S}]\!]\pi$ only if $\ell \in \mathsf{labx}[\![\mathtt{S}]\!]$⟆

$$\begin{aligned}
\subseteq{}& \llparenthesis\, \ell \in \mathsf{labx}[\![\mathtt{Sl}']\!] \mathrel{?} \widehat{\overline{\boldsymbol{\mathcal{S}}}}^{\exists}_{\mathsf{diff}}[\![\mathtt{Sl}']\!]\,\ell \mathrel{:} \varnothing \,\rrparenthesis \cup \\
& \llparenthesis\, \ell \in \mathsf{labx}[\![\mathtt{S}]\!]\setminus\{\mathsf{at}[\![\mathtt{S}]\!]\} \mathrel{?} \{\langle \mathtt{x}',\ \mathtt{y}\rangle \mid \exists \langle \pi_0,\ \pi_1\ell\pi_2\rangle, \langle \pi_0',\ \pi_1'\ell\pi_2'\rangle \in \{\langle \pi,\ \pi' \frown \pi''\rangle \mid \pi' \in \\
& \quad \widehat{\boldsymbol{\mathcal{S}}}^{*}[\![\mathtt{Sl}']\!](\pi) \wedge \pi'' \in \widehat{\boldsymbol{\mathcal{S}}}^{*}[\![\mathtt{S}]\!](\pi \frown \pi')\}\,.\,(\forall \mathtt{z} \in \mathbb{V}\setminus\{\mathtt{x}'\}\,.\,\varrho(\pi_0)\mathtt{z} = \varrho(\pi_0')\mathtt{z}) \wedge \ell \notin \pi_1 \wedge \ell \notin \\
& \quad \pi_1' \wedge \mathsf{diff}(\mathsf{seqval}[\![\mathtt{y}]\!]\ell(\pi_0 \frown \pi_1\ell,\ \ell\pi_2), \mathsf{seqval}[\![\mathtt{y}]\!]\ell(\pi_0' \frown \pi_1'\ell,\ \ell\pi_2'))\} \mathrel{:} \varnothing \,\rrparenthesis
\end{aligned}$$

$\wr$definition $\alpha^{\mathrm{d}}(\{\boldsymbol{\mathcal{S}}^{*}[\![\mathtt{Sl}']\!]\})\,\ell$ and structural induction hypothesis$\int$

$$\begin{aligned}
\subseteq\ & (\!|\ \ell \in \mathsf{labx}[\![\mathtt{Sl}']\!] \mathbin{?} \widehat{\boldsymbol{\mathcal{S}}}^{\exists}_{\mathrm{diff}}[\![\mathtt{Sl}']\!]\,\ell \mathbin{:} \varnothing\ |\!) \cup \\
& (\!|\ \ell \in \mathsf{labx}[\![\mathtt{S}]\!] \setminus \{\mathsf{at}[\![\mathtt{S}]\!]\} \mathbin{?} \{\langle \mathtt{x}', \mathtt{y}\rangle \mid \exists \pi_0, \pi_1, \pi_2, \pi_3, \pi_0', \pi_1', \pi_2', \pi_3' \,.\, \pi_1 \mathsf{at}[\![\mathtt{S}]\!] \in \\
& \widehat{\boldsymbol{\mathcal{S}}}^{*}[\![\mathtt{Sl}']\!](\pi_0) \wedge \mathsf{at}[\![\mathtt{S}]\!]\pi_2\ell\pi_3 \in \widehat{\boldsymbol{\mathcal{S}}}^{*}[\![\mathtt{S}]\!](\pi_0 \mathbin{\widehat{}} \pi_1\mathsf{at}[\![\mathtt{S}]\!]) \wedge \pi_1'\mathsf{at}[\![\mathtt{S}]\!] \in \widehat{\boldsymbol{\mathcal{S}}}^{*}[\![\mathtt{Sl}']\!](\pi_0') \wedge \\
& \mathsf{at}[\![\mathtt{S}]\!]\pi_2'\ell\pi_3' \in \widehat{\boldsymbol{\mathcal{S}}}^{*}[\![\mathtt{S}]\!](\pi_0' \mathbin{\widehat{}} \pi_1'\mathsf{at}[\![\mathtt{S}]\!]) \wedge (\forall \mathtt{z} \in \mathbb{V} \setminus \{\mathtt{x}'\} \,.\, \boldsymbol{\varrho}(\pi_0)\mathtt{z} = \boldsymbol{\varrho}(\pi_0')\mathtt{z}) \wedge \\
& \mathsf{diff}(\mathsf{seqval}[\![\mathtt{y}]\!]\ell(\pi_0 \mathbin{\widehat{}} \pi_1\mathsf{at}[\![\mathtt{S}]\!]\pi_2\ell,\ \ell\pi_3), \mathsf{seqval}[\![\mathtt{y}]\!]\ell(\pi_0' \mathbin{\widehat{}} \pi_1'\mathsf{at}[\![\mathtt{S}]\!]\pi_2'\ell,\ \ell\pi_3'))\} \mathbin{:} \varnothing\ |\!)
\end{aligned}$$

$\wr$definition of $\in$, $\ell \in \mathsf{labx}[\![\mathtt{S}]\!] \setminus \{\mathsf{at}[\![\mathtt{S}]\!]\}$, $\mathsf{after}[\![\mathtt{Sl}']\!] = \mathsf{at}[\![\mathtt{S}]\!]$, definition of $\widehat{\boldsymbol{\mathcal{S}}}^{*}[\![\mathtt{S}]\!]$ so that $\ell\pi' \in \widehat{\boldsymbol{\mathcal{S}}}^{*}[\![\mathtt{S}]\!](\pi\ell')$ implies $\ell = \ell' = \mathsf{at}[\![\mathtt{S}]\!]$, ignoring that $\ell \notin \pi_1, \pi_2 \wedge \ell \notin \pi_1', \pi_2'$, and $\widehat{\boldsymbol{\mathcal{S}}}^{+}[\![\mathtt{Sl}']\!] \subseteq \widehat{\boldsymbol{\mathcal{S}}}^{*}[\![\mathtt{Sl}']\!]$$\int$

$$\begin{aligned}
\subseteq\ & (\!|\ \ell \in \mathsf{labx}[\![\mathtt{Sl}']\!] \mathbin{?} \widehat{\boldsymbol{\mathcal{S}}}^{\exists}_{\mathrm{diff}}[\![\mathtt{Sl}']\!]\,\ell \mathbin{:} \varnothing\ |\!) \cup \\
& (\!|\ \ell \in \mathsf{labx}[\![\mathtt{S}]\!] \setminus \{\mathsf{at}[\![\mathtt{S}]\!]\} \mathbin{?} \alpha^{\mathrm{d}}(\{\widehat{\boldsymbol{\mathcal{S}}}^{*}[\![\mathtt{Sl}']\!]\})\,\mathsf{at}[\![\mathtt{S}]\!] \mathbin{\mathring{;}} \alpha^{\mathrm{d}}(\{\widehat{\boldsymbol{\mathcal{S}}}^{*}[\![\mathtt{S}]\!]\})\,\ell \mathbin{:} \varnothing\ |\!) \qquad \wr(47.59)\int
\end{aligned}$$

$$\begin{aligned}
\subseteq\ & (\!|\ \ell \in \mathsf{labx}[\![\mathtt{Sl}']\!] \mathbin{?} \widehat{\boldsymbol{\mathcal{S}}}^{\exists}_{\mathrm{diff}}[\![\mathtt{Sl}']\!]\,\ell \mathbin{:} \varnothing\ |\!) \cup \\
& (\!|\ \ell \in \mathsf{labx}[\![\mathtt{S}]\!] \setminus \{\mathsf{at}[\![\mathtt{S}]\!]\} \mathbin{?} \widehat{\boldsymbol{\mathcal{S}}}^{\exists}_{\mathrm{diff}}[\![\mathtt{Sl}']\!]\,\mathsf{at}[\![\mathtt{S}]\!] \mathbin{\mathring{;}} \widehat{\boldsymbol{\mathcal{S}}}^{\exists}_{\mathrm{diff}}[\![\mathtt{S}]\!]\,\ell \mathbin{:} \varnothing\ |\!) \qquad \wr\text{induction hypothesis}\int \quad \square
\end{aligned}$$

Exercise 47.61 Calculate $\widehat{\boldsymbol{\mathcal{S}}}^{\exists}_{\mathrm{diff}}[\![\mathtt{Sl}]\!]\,\ell_2$ for the statement list $\mathtt{Sl} \triangleq \ell_0\ \mathtt{y} = \mathtt{x}\ ;\ell_1\ \mathtt{y} = \mathtt{y} - \mathtt{x}\ ;\ell_2$ of example 47.37.

For the iteration, we will need the following

$\lceil$ **Lemma 47.62** $\alpha^{\mathrm{d}}(\{\{\langle \pi, \pi'\rangle \in \boldsymbol{\mathcal{S}}[\![\mathtt{S}]\!] \mid \boldsymbol{\mathcal{B}}[\![\mathtt{B}]\!]\boldsymbol{\varrho}(\pi)\}\})\,\ell \subseteq \alpha^{\mathrm{d}}(\{\boldsymbol{\mathcal{S}}[\![\mathtt{S}]\!]\})\,\ell \rceil\, \mathsf{nondet}(\mathtt{B}, \mathtt{B})$

Proof of lemma 47.62

$$\alpha^{\mathrm{d}}(\{\{\langle \pi, \pi'\rangle \in \boldsymbol{\mathcal{S}}[\![\mathtt{S}]\!] \mid \boldsymbol{\mathcal{B}}[\![\mathtt{B}]\!]\boldsymbol{\varrho}(\pi)\}\})\,\ell$$

$$= \{\langle \mathtt{x}, \mathtt{y}\rangle \mid \{\langle \pi, \pi'\rangle \in \boldsymbol{\mathcal{S}}[\![\mathtt{S}]\!] \mid \boldsymbol{\mathcal{B}}[\![\mathtt{B}]\!]\boldsymbol{\varrho}(\pi)\} \in \mathcal{D}\ell\langle \mathtt{x}, \mathtt{y}\rangle\}$$

$\wr$definition (47.25) of α^{d} and definition of $\in$$\int$

$$\begin{aligned}
= &\ \{\langle \mathtt{x}, \mathtt{y}\rangle \mid \exists \langle \pi_0, \pi_1\rangle, \langle \pi_0', \pi_1'\rangle \in \{\langle \pi, \pi'\rangle \in \boldsymbol{\mathcal{S}}[\![\mathtt{S}]\!] \mid \boldsymbol{\mathcal{B}}[\![\mathtt{B}]\!]\boldsymbol{\varrho}(\pi)\} \,.\, (\forall \mathtt{z} \in \mathbb{V} \setminus \{\mathtt{x}\} \,.\, \boldsymbol{\varrho}(\pi_0)\mathtt{z} = \\
& \boldsymbol{\varrho}(\pi_0')\mathtt{z}) \wedge \mathsf{diff}(\mathsf{seqval}[\![\mathtt{y}]\!]\ell(\pi_0, \pi_1), \mathsf{seqval}[\![\mathtt{y}]\!]\ell(\pi_0', \pi_1'))\}
\end{aligned}$$

$\wr$definition (47.19) of $\mathcal{D}$ and (47.18) of $\mathsf{diff}$$\int$

$$\begin{aligned}
= &\ \{\langle \mathtt{x}, \mathtt{y}\rangle \mid \exists \langle \pi_0, \pi_1\rangle \in \boldsymbol{\mathcal{S}}[\![\mathtt{S}]\!], \langle \pi_0', \pi_1'\rangle \in \boldsymbol{\mathcal{S}}[\![\mathtt{S}]\!] \,.\, \boldsymbol{\mathcal{B}}[\![\mathtt{B}]\!]\boldsymbol{\varrho}(\pi_0) \wedge \boldsymbol{\mathcal{B}}[\![\mathtt{B}]\!]\boldsymbol{\varrho}(\pi_0') \wedge (\forall \mathtt{z} \in \\
& \mathbb{V} \setminus \{\mathtt{x}\} \,.\, \boldsymbol{\varrho}(\pi_0)\mathtt{z} = \boldsymbol{\varrho}(\pi_0')\mathtt{z}) \wedge \mathsf{diff}(\mathsf{seqval}[\![\mathtt{y}]\!]\ell(\pi_0, \pi_1), \mathsf{seqval}[\![\mathtt{y}]\!]\ell(\pi_0', \pi_1'))\}
\end{aligned}$$

$\wr$definition of $\in$$\int$

$$\begin{aligned}
\subseteq &\ \{\langle \mathtt{x}, \mathtt{y}\rangle \mid \exists \langle \pi_0, \pi_1\rangle \in \boldsymbol{\mathcal{S}}[\![\mathtt{S}]\!], \langle \pi_0', \pi_1'\rangle \in \boldsymbol{\mathcal{S}}[\![\mathtt{S}]\!] \,.\, (\forall \mathtt{z} \in \mathbb{V} \setminus \{\mathtt{x}\} \,.\, \boldsymbol{\varrho}(\pi_0)\mathtt{z} = \boldsymbol{\varrho}(\pi_0')\mathtt{z}) \wedge \\
& \mathsf{diff}(\mathsf{seqval}[\![\mathtt{y}]\!]\ell(\pi_0, \pi_1), \mathsf{seqval}[\![\mathtt{y}]\!]\ell(\pi_0', \pi_1'))\} \rceil\, \mathsf{nondet}(\mathtt{B}, \mathtt{B})
\end{aligned}$$

$\wr$because if $\mathtt{x} \notin \mathsf{nondet}(\mathtt{B}, \mathtt{B})$ then $\mathtt{x} \in \mathsf{det}(\mathtt{B}, \mathtt{B})$ and so $\boldsymbol{\mathcal{B}}[\![\mathtt{B}]\!]\boldsymbol{\varrho}(\pi_0)$ and $\boldsymbol{\mathcal{B}}[\![\mathtt{B}]\!]\boldsymbol{\varrho}(\pi_0')$ imply that $\boldsymbol{\varrho}(\pi_0)\mathtt{x} = \boldsymbol{\varrho}(\pi_0')\mathtt{x}$. By $\forall \mathtt{z} \in \mathbb{V} \setminus \{\mathtt{x}\} \,.\, \boldsymbol{\varrho}(\pi_0)\mathtt{z} = \boldsymbol{\varrho}(\pi_0')\mathtt{z})$, it follows that $\boldsymbol{\varrho}(\pi_0) = \boldsymbol{\varrho}(\pi_0')$. Because all program components $\mathtt{S} \in \mathbb{P}\mathrm{c}$ of the language are deterministic (see exercises 17.13 and 17.21), it follows that $\pi_1 = \pi_1'$. Therefore $\mathsf{seqval}[\![\mathtt{y}]\!]\ell(\pi_0, \pi_1) = \mathsf{seqval}[\![\mathtt{y}]\!]\ell(\pi_0', \pi_1'))$. By (47.18), diff is false and in this case the set is empty.$\int$

$= (\alpha^{\mathrm{d}}(\{\boldsymbol{\mathcal{S}}[\![\mathsf{S}]\!]\})\,\ell)\,\rceil\,\mathsf{nondet}(\mathsf{B},\mathsf{B})$

$\wr$definitions (47.18) of diff, (47.19) of $\mathcal{D}$, and (47.25) of $\alpha^{\mathrm{d}}\wr$ □

Abstract potential dependency semantics of an iteration statement $\mathsf{S} ::= \mathbf{while}\,\ell\,(\mathsf{B})\,\mathsf{S}_b$

$$\widehat{\boldsymbol{\mathcal{S}}}^{\exists}_{\mathsf{diff}}[\![\mathsf{S}]\!]\,\ell' = (\mathsf{lfp}^{\dot{\subseteq}}\,\mathcal{F}^{\mathsf{diff}}[\![\mathbf{while}\,\ell\,(\mathsf{B})\,\mathsf{S}_b]\!])\,\ell' \tag{47.63}$$

$$\mathcal{F}^{\mathsf{diff}}[\![\mathbf{while}\,\ell\,(\mathsf{B})\,\mathsf{S}_b]\!]\,X\,\ell' =$$

$$(\![\,\ell' = \ell \;?\; \mathbb{1}_{\mathbb{V}} \cup \Big(X(\ell)\,{}_{9}^{\circ}\,(\widehat{\boldsymbol{\mathcal{S}}}^{\exists}_{\mathsf{diff}}[\![\mathsf{S}_b]\!]\,\ell\,\rceil\,\mathsf{nondet}(\mathsf{B},\mathsf{B}))\Big) \tag{a}$$

$$[\!]\;\ell' \in \mathsf{in}[\![\mathsf{S}_b]\!] \;?\; X(\ell)\,{}_{9}^{\circ}\,(\widehat{\boldsymbol{\mathcal{S}}}^{\exists}_{\mathsf{diff}}[\![\mathsf{S}_b]\!]\,\ell'\,\rceil\,\mathsf{nondet}(\mathsf{B},\mathsf{B})) \tag{b}$$

$$[\!]\;\ell' = \mathsf{after}[\![\mathsf{S}]\!] \;?\; X(\ell) \cup (X(\ell)\,{}_{9}^{\circ}\,(\mathbb{V} \times \mathsf{mod}[\![\mathsf{S}_b]\!])) \tag{c}$$

$$\cup\Big(X(\ell)\,{}_{9}^{\circ}\,\Big(\Big(\bigcup_{\ell'' \in \mathsf{breaks\text{-}of}[\![\mathsf{S}_b]\!]} \widehat{\boldsymbol{\mathcal{S}}}^{\exists}_{\mathsf{diff}}[\![\mathsf{S}_b]\!]\,\ell''\Big)\,\rceil\,\mathsf{nondet}(\mathsf{B},\mathsf{B})\Big)\Big)$$

$$\,{}_{\circ}^{\circ}\,\varnothing\,]\!) \tag{d}$$

Remark 47.64 By exercise 15.8, the term $\mathbb{1}_{\mathbb{V}} \cup X(\ell) \cup (X(\ell)\,{}_{9}^{\circ}\,\widehat{\boldsymbol{\mathcal{S}}}^{\exists}_{\mathsf{diff}}[\![\mathsf{S}_b]\!]\,\ell)$ can be simplified to $\mathbb{1}_{\mathbb{V}} \cup (X(\ell)\,{}_{9}^{\circ}\,\widehat{\boldsymbol{\mathcal{S}}}^{\exists}_{\mathsf{diff}}[\![\mathsf{S}_b]\!]\,\ell)$.

Because $\mathcal{F}^{\mathsf{diff}}[\![\mathbf{while}\,\ell\,(\mathsf{B})\,\mathsf{S}_b]\!] \in \mathbb{P}^{\mathrm{d}} \xrightarrow{\nearrow} \mathbb{P}^{\mathrm{d}}$ is $\dot{\subseteq}$-monotone and the abstract domain $\langle \mathbb{P}^{\mathrm{d}}, \dot{\subseteq}, \ell \mapsto \varnothing, \ell \mapsto \mathbb{V} \times \mathbb{V}, \dot{\cup}, \dot{\cap} \rangle$ in (47.35) is a complete lattice, the least fixpoint $\mathsf{lfp}^{\dot{\subseteq}}\,\mathcal{F}^{\mathsf{diff}}[\![\mathbf{while}\,\ell\,(\mathsf{B})\,\mathsf{S}_b]\!]$ of $\mathcal{F}^{\mathsf{diff}}[\![\mathbf{while}\,\ell\,(\mathsf{B})\,\mathsf{S}_b]\!]$ exists by Tarski's theorem 15.6. Moreover, because $\mathbb{P}^{\mathrm{d}}$ is finite (at least in considering only the program labels $\mathbb{L}$ and variables $\mathbb{V}$ occurring in a program), the abstract properties of $\mathbb{P}^{\mathrm{d}}$ have a finite computer memory representation and the limit of Tarski–Kantorovich's iterates of theorem 15.21 can be computed in finitely many iterations, which, using chaotic iterations in theorem 22.4, yields an effective static analysis algorithm.

$\mathsf{lfp}^{\dot{\subseteq}}\,\mathcal{F}^{\mathsf{diff}}[\![\mathbf{while}\,\ell\,(\mathsf{B})\,\mathsf{S}_b]\!]$ is the least solution to the system of equations

$$\left\{\begin{array}{lcl} X(\ell') & = & \mathcal{F}^{\mathsf{diff}}[\![\mathbf{while}\,\ell\,(\mathsf{B})\,\mathsf{S}_b]\!]\,X\,\ell' \\ \ell' \in \mathbb{L} & & \end{array}\right.$$

which can be understood as follows:

(a) On loop entry the variables depend on themselves. After several iterations the dependency is the composition of the dependencies of the previous iterations $X(\ell)$ composed with those $\widehat{\boldsymbol{\mathcal{S}}}^{\exists}_{\mathsf{diff}}[\![\mathsf{S}_b]\!]\,\ell$ created by one more iteration. The composition ${}_{9}^{\circ}$ is that used for a list of statements in (47.60). The restriction $\rceil\mathsf{nondet}(\mathsf{B},\mathsf{B})$ eliminates dependencies on variables with a single possible value on loop [re-]entry;

(b) The initial value of a variable x on loop entry flows to a variable z at $\ell' \in \mathsf{in}[\![\mathsf{S}_b]\!]$ in the loop body if and only if the initial value of a variable x on loop entry flows to some variable y at loop entry ℓ after 0 or more iterations (so $\langle \mathsf{x}, \mathsf{y} \rangle \in X(\ell)$) and the value of variable y at the loop body flows to z at ℓ' (so $\langle \mathsf{y}, \mathsf{z} \rangle \in \widehat{\boldsymbol{\mathcal{S}}}^{\exists}_{\mathsf{diff}}[\![\mathsf{S}_b]\!]\,\ell'$). Moreover, the restriction nondet(B, B) eliminates the case when x is constant after the loop test B.

(The case when y is constant after the loop test B has been recursively eliminated by $\hat{\overline{\boldsymbol{\mathcal{S}}}}^{\exists}_{\text{diff}}[\![\mathsf{S}_b]\!]\,\ell'$);

(c) This case determines the dependencies on loop exit $\ell' = \mathsf{after}[\![\mathsf{S}]\!]$, not knowing the values of variables, so the number of iterations and how the loop is exited is unknown. Therefore all cases must be considered.

- The term $X(\ell)$ (where $\ell \triangleq \mathsf{at}[\![\mathsf{S}]\!] = \mathsf{after}[\![\mathsf{S}_b]\!]$) corresponds to the case when the loop is entered and iterated 0 or more times before exiting, so that either the loop is never entered so each variable depends on itself by (47.63.a) or the dependencies on exit of the loop are those after the last iteration of the body;
- The term $(X(\ell) \,\mathring{\varsigma}\, (\mathcal{V} \times \mathsf{mod}[\![\mathsf{S}_b]\!]))$ covers dependencies originating from two executions decided by the initial values of a variable x such that in one case the loop is entered and exited and for another value it is immediately exited. The variables y modified in the loop body depend on x, as was the case in (47.47.c.3) for the conditional;
- The term $\bigcup_{\ell'' \in \mathsf{breaks\text{-}of}[\![\mathsf{S}]\!]} (X(\ell) \,\mathring{\varsigma}\, \hat{\overline{\boldsymbol{\mathcal{S}}}}^{\exists}_{\text{diff}}[\![\mathsf{S}_b]\!]\,\ell'')$ propagates the dependencies at the **break**; statements within the loop body to the break point $\mathsf{after}[\![\mathsf{S}]\!]$ after the loop; and
- The term (47.63.c) can be refined into (47.63.c′) as shown in case (3) of the proof of (47.63). There, nondet checks those cases for which it is sure that no two distinct executions can be found in the definition of dependency;

(d) The iteration statement **while** ℓ (B) S_b introduces no dependency outside its reachable points.

Example 47.65 (fixpoint dependency of an iteration)

$$\mathsf{S} \;=\; \mathbf{while}\;\ell_0\;(\mathsf{tt})\;\{\;\ell_1\;\mathsf{y} = \mathsf{z}\;;\ell_2\;\mathsf{z} = \mathsf{x}\;;\;\}\ell_3.$$

The system of equations $X = f^{\mathrm{d}}[\![\mathsf{S}]\!](X)$ is

$$\begin{cases} X(\ell_0) &= \{\langle \mathsf{v}, \mathsf{v}\rangle \mid \mathsf{v} \in \mathcal{V}\} \cup (X(\ell_2) \,\mathring{\varsigma}\, \{\langle \mathsf{x}, \mathsf{x}\rangle, \langle \mathsf{x}, \mathsf{z}\rangle, \langle \mathsf{y}, \mathsf{y}\rangle\}) \\ X(\ell_1) &= X(\ell_0) \\ X(\ell_2) &= X(\ell_2) \cup (X(\ell_1) \,\mathring{\varsigma}\, \{\langle \mathsf{x}, \mathsf{x}\rangle, \langle \mathsf{z}, \mathsf{y}\rangle, \langle \mathsf{z}, \mathsf{z}\rangle\}) \\ X(\ell_3) &= \varnothing \end{cases}$$

The chaotic iterations are

ℓ	ℓ_0, ℓ_1	ℓ_2	ℓ_3
$X^0(\ell)$	$\varnothing$	$\varnothing$	$\varnothing$
$X^1(\ell)$	$\{\langle x, x\rangle, \langle y, y\rangle, \langle z, z\rangle\}$	$\{\langle x, x\rangle, \langle z, y\rangle, \langle z, z\rangle\}$	$\varnothing$
$X^2(\ell)$	$\{\langle x, x\rangle, \langle x, z\rangle, \langle y, y\rangle, \langle z, y\rangle, \langle z, z\rangle\}$	$\{\langle x, x\rangle, \langle x, y\rangle, \langle x, z\rangle, \langle z, y\rangle, \langle z, z\rangle\}$	$\varnothing$
$X^3(\ell)$	$\{\langle x, x\rangle, \langle x, y\rangle, \langle x, z\rangle, \langle y, y\rangle, \langle z, y\rangle, \langle z, z\rangle\}$	$\{\langle x, x\rangle, \langle x, y\rangle, \langle x, z\rangle, \langle z, y\rangle, \langle z, z\rangle\}$	$\varnothing$
$X^4(\ell)$	$X^3(\ell_0) = X^3(\ell_1)$	$X^3(\ell_2)$	$\varnothing$

For example, the initial value x_0 of x flows to x at ℓ_0 on loop entry, to z after the first iteration and hence to y after the second iteration. The initial value y_0 of y flows only to y at ℓ_0 on loop entry. The initial value z_0 of z flows to z at ℓ_0 on loop entry and then to y after the first iteration.

Exercise 47.66 Continuing example 47.65, prove that for $\mathsf{S}_b = \{\ell_1\ \mathsf{y} = \mathsf{z}\ ;\ell_2\ \mathsf{z} = \mathsf{x}\ ;\}\ \ell_0$, we have $\widehat{\boldsymbol{\mathcal{S}}}^{\exists}_{\text{diff}}[\![\mathsf{S}_b]\!]\ \ell_0 = \{\langle \mathsf{x}, \mathsf{x}\rangle, \langle \mathsf{x}, \mathsf{z}\rangle, \langle \mathsf{z}, \mathsf{y}\rangle\}$.

The proof of (47.63) is available at https://github.com/PrAbsInt/.

Exercise 47.67 Propose a sound and complete structural dependency semantics (i.e., with no $\subseteq$ derivation steps in the calculations) and derive a structural dependency proof method from this dependency semantics using fixpoint induction for loops; see, for example, [94].

47.10 Reduced Product with a Relational Value Analysis

$\widehat{\boldsymbol{\mathcal{S}}}^{\exists}_{\text{diff}}[\![\mathsf{A}]\!]$ in (47.44) handles the case $\widehat{\boldsymbol{\mathcal{S}}}^{\exists}_{\text{diff}}[\![\mathsf{x} - \mathsf{x}]\!] = \varnothing$ while $\mathbb{vars}[\![\mathsf{x} - \mathsf{x}]\!] = \{\mathsf{x}\}$. As shown in example 47.37, even more precision can be achieved by considering reachable environments only. The abstraction $\alpha_{\mathfrak{r}}(\boldsymbol{\mathcal{S}}[\![\mathsf{S}]\!])\ell$ of the trace semantics $\boldsymbol{\mathcal{S}}[\![\mathsf{S}]\!]$ of a program component S by the classic relation abstraction

$$\alpha_{\mathfrak{r}}(\mathcal{S}) \triangleq \ell \in \mathbb{L} \mapsto \{\langle \varrho(\pi_0), \varrho(\pi_0 \frown \pi_1\ell)\rangle \mid \langle \pi_0, \pi_1\ell\rangle \in \mathcal{S}\}$$

provides a relation between the initial value of variables and their value at a program point ℓ of S (the relation is empty if ℓ is not in S).

Then the dependency analysis can be refined using this relational value information.

- For the assignment (47.44), the imprecision is due only to the term $\widehat{\boldsymbol{\mathcal{S}}}^{\exists}_{\text{diff}}[\![\mathsf{A}]\!]$ because it is impossible to evaluate the arithmetic expression A in all reachable environments on entry of the assignment (see step (47.45) of the calculation design). However, a relational value static analysis can provide relevant information.
  For example, using a constant propagation either Cartesian [572, 969] or relational in [560, 586, 728], or a zone/octagon analysis [692], $\mathsf{y} \in \widehat{\boldsymbol{\mathcal{S}}}^{\exists}_{\text{diff}}[\![\mathsf{A}_1 - \mathsf{A}_2]\!]$ only if this analysis cannot prove that $\mathsf{A}_1 - \mathsf{A}_2$ is constant.
- For the conditional (47.47) and the iteration (47.63), the relational value static analysis can provide relevant information on nondeterminacy nondet in (47.48).
- For sequential composition, conditionals, and iteration, it is also possible to refine the calculational design to improve compositionality. For example, in `if (H) L=X; else L=X;`, the above relational value analyses yield $\mathsf{L}{=}\mathsf{X}_0$ on exit of the conditional so L does not depend on H.

This is better implemented by the reduced product of section 36.3 [212, 241, 999] and side conditions in the dependency analysis (such as $\widehat{\boldsymbol{\mathcal{S}}}^{\exists}_{\text{diff}}[\![\mathsf{A}]\!]$ and nondet) refining dependencies using relational value information provided by the other domains in the product. This separation of concerns greatly simplifies the design of the analysis [268].

47.11 Examples of Derived Dependency Semantics and Analyzes

From the definition of dependency that we have given, we can derive other existing definitions or static analysis algorithms.

47.11.1 Independence

Definite independence is the complement of potential dependency. [41, 42, 752] introduced a Hoare-like logic to statically check independences. It also takes nontermination into account and so relies on a different definition of $\neg\mathcal{D}\ell\langle \mathsf{x}, \mathsf{y}\rangle$. It is recognized that definite independence is an abstract interpretation but this is not used to design the logic, which remains empirical.

47.11.2 Abstract Noninterference/Dependency

The abstraction $\alpha^{\mathrm{d}}(\mathcal{S})$ of the semantic property $\mathcal{S}$ in (47.25) is meaningful for any semantic property $\mathcal{S}$, including abstract ones, as considered in abstract noninterference/dependency [413]. Given a structural semantic definition of this abstract property, the principle of design by calculational design of the abstract dependency remains the same.

47.11.3 Forward and Backward Dependency

Dependency information is useful for program slicing [976]. The semantics (6.16)–(17.4) considered in chapter 6 is forward, defining the continuation in a program component of an initialization computation ending on entry of that program component. This forward dependency is adequate for forward slicing [42]. The dependency abstraction may be applied to a backward semantics defining the reachability in a program component of a finalization computation starting at that end of a program component or on a break (see chapter 50). This backward dependency would certainly be more useful as a basis for slicing [976], or abstract slicing [509, 668, 835].

47.11.4 Flow-Insensitive Dependency

The flow-insensitive dependency uses the flow-insensitive abstraction (45.1) to join all local dependencies attached to program points into a single dependency property holding at any program point.

$$\alpha^{\Gamma}(\mathbf{D}) \triangleq \bigcup_{\ell \in \mathbb{L}} \mathbf{D}(\ell)$$

$$\boldsymbol{\mathcal{S}}^{\Gamma}[\![\mathsf{S}]\!] \triangleq \alpha^{\Gamma}(\alpha^{\mathrm{d}}(\{\boldsymbol{\mathcal{S}}^{+\infty}[\![\mathsf{S}]\!]\}))$$

47.11.5 Dye Dependency Analysis

By analogy with dye-tracer tests in hydrology to determine the possible origins of spring discharges or resurgences by water source coloring and flow tracing, it has been suggested to decorate the initial values of variables with labels such as color annotations and to track their diffusion and mixtures to determine dependencies [177]. This postulated definition of dependency can be proved sound by observing that the initial color of variables can be designated by the name of these variables and that the color mix at point ℓ for variable y is $\{\mathsf{x} \mid \boldsymbol{\mathcal{S}}^{+\infty}[\![\mathsf{P}]\!] \in \mathcal{D}\ell\langle \mathsf{x}, \mathsf{y}\rangle\}$. Note that in the postulated instrumented semantics, the definition of diff remains implicit as defined by the arbitrarily selected color mixing rules. In general, instrumented semantics [541] are not a good idea, unless the instrumentation is semantically justified with respect to the noninstrumented semantics, in which case it is useless because one can directly use the noninstrumented semantics.

47.11.6 Dualistic Dependency Analysis

Assume the initial values of variables (more generally inputs) are partitioned into positive $\mathbb{P}$ and negative $\mathbb{N}$ variables, $\mathbb{V} = \mathbb{P} \cup \mathbb{N}$ and $\mathbb{P} \cap \mathbb{N} = \emptyset$. The dualistic abstraction $\alpha^{\delta}(\mathbf{D})$ of a dependency property $\mathbf{D} \in \mathbb{L} \to \wp(\mathbb{V} \times \mathbb{V})$ (47.35) attaches to each program point ℓ the set of variables y that, at that program point ℓ, depend upon the initial value of at least one variable $\mathsf{x} \in \mathbb{P}$ that is positive.

$$\alpha^{\delta}(\mathbf{D})\ell \quad \triangleq \quad \{\mathsf{y} \mid \exists \mathsf{x} \in \mathbb{P} \,.\, \langle \mathsf{x},\, \mathsf{y} \rangle \in \mathbf{D}(\ell)\}$$

A dualistic analysis is an overapproximation of the abstract dualistic semantics

$$\boldsymbol{\mathcal{S}}^{\delta}[\![\mathsf{S}]\!] \quad \triangleq \quad \alpha^{\delta}(\alpha^{\mathrm{d}}(\{\boldsymbol{\mathcal{S}}^{+\infty}[\![\mathsf{S}]\!]\}))$$

assigning the each program point ℓ, a set $\boldsymbol{\mathcal{S}}^{\delta}[\![\mathsf{S}]\!]\ell \in \wp(\mathbb{V})$ of potentially positive variables. It can be used to underapproximate at each program point the set of variables that definitely do not dependent on initially positive variable. Examples are taint analysis in which positive is tainted and negative is untainted as widely used to check privacy/security [633, 645, 823]. Another example is binding time analysis in partial evaluation [465, 542] in which positive is dynamic and negative is static. A final example is absence of interference [199, 421, 422, 750, 959] in which positive is high (private/untrusted) and negative is low (public/trusted).

47.11.7 Tracking Analysis

Assume the initial values of variables (more generally inputs) are partitioned into tracked $\mathbb{T}$ and untracked $\mathbb{U}$ variables, $\mathbb{V} = \mathbb{T} \cup \mathbb{U}$ and $\mathbb{T} \cap \mathbb{U} = \emptyset$. The tracking abstraction $\alpha^{\tau}(\mathbf{D})$ of a dependency property $\mathbf{D} \in \mathbb{L} \to \wp(\mathbb{V} \times \mathbb{V})$ (47.35) attaches to each program point ℓ the set of variables y that, at that program point ℓ, depend upon the initial value of at least one tracked variable $\mathsf{x} \in \mathbb{T}$.

$$\alpha^{\tau}(\mathbf{D})\ell \quad \triangleq \quad \{\mathsf{y} \mid \exists \mathsf{x} \in \mathbb{T} \,.\, \langle \mathsf{x},\, \mathsf{y} \rangle \in \mathbf{D}(\ell)\}$$

A tracking analysis is an overapproximation of the abstract tracking semantics

$$\boldsymbol{\mathcal{S}}^{\tau}[\![\mathsf{S}]\!] \quad \triangleq \quad \alpha^{\tau}(\alpha^{\mathrm{d}}(\{\boldsymbol{\mathcal{S}}^{+\infty}[\![\mathsf{S}]\!]\}))$$

assigning the each program point ℓ, a set $\boldsymbol{\mathcal{S}}^{\tau}[\![\mathsf{S}]\!]\ell \in \wp(\mathbb{V})$ of variables potentially depending on tracked variables. It can be used to underapproximate at each program point the set of variables that definitely do not dependent on initial values of tracked variables. Examples are taint analysis (tracked is tainted, untracked is untainted) in privacy/security checks [633]; binding time analysis in offline partial evaluation [465, 542] (tracked is dynamic, untracked is static) and absence of interference [142, 199, 421, 422, 959] (tracked is high (private/untrusted), untracked is low (public/trusted)).

47.11.8 Taint Analysis

Taint analysis is an abstraction of dependency analysis widely used to check privacy/security [633]. Assume that initial values of variables (more generally inputs) are partitioned into tainted $\mathbb{T}$ and untainted $\mathbb{U}$ variables, $\mathbb{V} = \mathbb{T} \cup \mathbb{U}$ and $\mathbb{T} \cap \mathbb{U} = \emptyset$. The taint abstraction $\alpha^{\tau}(\mathbf{D})$ of a dependency property $\mathbf{D} \in \mathbb{L} \to \wp(\mathbb{V} \times \mathbb{V})$ (47.35) attaches to each program point ℓ the set of variables y that, at that program point ℓ, depend upon the initial value of at least one variable $\mathsf{x} \in \mathbb{T}$ which is tainted.

$$\alpha^{\tau}(\mathbf{D})\ell \triangleq \{\mathsf{y} \mid \exists \mathsf{x} \in \mathbb{T} . \langle \mathsf{x}, \mathsf{y} \rangle \in \mathbf{D}(\ell)\}$$

A taint analysis is an overapproximation of the abstract taint semantics

$$\boldsymbol{\mathcal{S}}^{\tau}[\![\mathsf{S}]\!] \triangleq \alpha^{\tau}(\alpha^{\mathrm{d}}(\{\boldsymbol{\mathcal{S}}^{+\infty}[\![\mathsf{S}]\!]\}))$$

assigning the each program point ℓ, a set $\boldsymbol{\mathcal{S}}^{\tau}[\![\mathsf{S}]\!]\ell \in \wp(\mathbb{V})$ of potentially tainted variables.

Exercise 47.68 Show that α^{τ} defines a Galois connection. Design a structural syntactic taint analysis $\widehat{\boldsymbol{\mathcal{S}}}^{\tau}[\![\mathsf{S}]\!]$ in the style of dataflow analyses of sections 41.3 and 41.4.

47.12 Conclusion

We have designed by calculus a potential value dependency analysis between the initial value of variables and their value when reaching a program point during execution (in particular the final values of variables that can be used for noninterference analysis in section 47.4).

Our formal definition follows and formalizes the intuition provided by [312], "Information flows from object x to object y, denoted $\mathsf{x} \rightsquigarrow \mathsf{y}$, whenever information stored in x is transferred to, or used to derive information transferred to, object y. A program statement specifies a flow $\mathsf{x} \rightsquigarrow \mathsf{y}$ if execution of the statement could result in a flow $\mathsf{x} \rightsquigarrow \mathsf{y}$." However, in a spirit of contradiction, we have excluded counterfactual empty flows (which, alternatively, can be easily incorporated). Following [312], it is then easy to check that these inferred flows satisfy a specification of secure flows. An alternative [622, 759] consists of monitoring an abstraction of the program semantics dynamically.

The analysis is not postulated but derived formally by abstract interpretation of the trace semantics. Therefore our definition is concise and coherent. For example, postulated static analyses offer no guarantee that the same notion of dependency is applied uniformly for all program statements. We found no need for extra notions like (hyper)^{n}properties [56], nonstandard abstract interpretation [947], postulated instrumented semantics [761, section 4], multisemantics [158], and so forth.

The proofs are long because we have to consider all possibilities for the choice of execution traces in definition 47.19.

As shown by [105, 473] and examples 47.30 and 47.34, taking values into account will definitely improve the precision of the dependency analysis. One possible implementation is by a reduced product of a dependency analysis with a reachability analysis [212, 999]. There are several examples of implementations of dependence analyses e.g. [334, 434, 435].

The definition 47.19 of potential dependency (and its complement independency) depends largely on the notion of the value of variables. These notions can definitely be different in other languages like logic (LP) and constraint logic (CLP) programming languages (although the link is clearly established in [488]). A pioneer work for logic programming is [736], describing a multivariant analysis for detecting independence of program variables. Beyond the classic read-write dependencies for imperative programs, more general notions of independence are needed. For traditional logic programming, [491, 735] defines "strict" and "non-strict" independence (and some later studies based on "search-space preservation" [86]); [85, 88] defines notions of independence for constraint-based languages; whereas [658] consider independence for logic languages with concurrency ("dynamic scheduling"), with their analyses [87]. The main application is automatic compile-time parallelization [152, 442, 734]. These static analyses are implemented in the Ciao modular multiparadigm logic/constraint programming environment [489].

Dependency analysis has many applications, for example, to data flow analysis considered in chapter 41 and [947], to program development in [110], to inferring the dependencies of the terms bound to program variables for global analysis of logic programs in [736], to privacy analysis in [646], to slicing in [2, 668, 834], to unused input data in [947], to databases [526], and so forth. A more refined approach for security is responsibility analysis [311]. A dependency analysis of the fixpoint equations in chapter 18, "Fixpoint Abstraction," yields the worklist and other iteration strategies mentioned in section 22.5 (e.g., [396]).

The data-dependence analysis used to detect parallelism in sequential code [702, 773] is also an abstract interpretation; see [935], [936, ch. 5].

More generally, dependences are abstract interpretations. The semantics is a set of executions by pairs $\langle \pi\ell, \ell\pi' \rangle$ where $\pi\ell$ is the past before reaching ℓ and $\ell\pi'$ is the continuation. We define an abstraction of the past $\pi\ell$ (it was the initial value of variables in our case). We define an abstraction of the future (it was the sequence of values of a variable at a program point (47.16) in our case). We define the difference between past abstractions (the initial values of variables only differ for one variable in our case). We define the difference between futures (it was diff in our case). Then the abstraction of the future depends on the abstraction of the past if and only if there are two executions with different past abstractions and different future abstractions. By varying the past/future abstractions and their difference, we can express the dependency abstractions introduced in the literature. Good examples are [413] for (abstract) noninterference and [95] for mitigation against side-channel attacks.

47.13 Solutions to Selected Exercises

Solution to exercise 47.11 In $\mathsf{P_a}$, y does not depend on x at ℓ_0, ℓ_1, ℓ_2, ℓ_3, ℓ_4, and ℓ_6 because, when execution reaches any of these program points, the value of y will always be the same. However, y depends on x at ℓ_5 because , depending on the initial value of x and the value of y at ℓ_5 will be either 1 or 2. For program $\mathsf{P_b}$, y does not depend on x at any of the program points $\ell_0, \ell_1, \ldots, \ell_5, \ell_6$.

Solution to exercise 47.12 The sequence of values of x at ℓ is 0, 1, 2, …17, … 42, … for P_0 (so x does not depend on its initial value x_0 at ℓ) and x_0, x_0+1, x_0+2,…x_0+17…x_0+42 …for P_u (so x depends on its initial value x_0 at ℓ because changing x_0 will change the first element of this sequence).

Solution to exercise 47.13 The sequence of values y at ℓ_0 is a finite prefix of the infinite sequence y_0, y_0+1, y_0+2, … of which the length depends on the initial value x_0 of x. This is a timing channel and there is no dependency, even if y and x_0 are related at ℓ_0.

Solution to exercise 47.15 The input value x_0 of some variable x is dead at some program point ℓ if and only if changing x_0 will not change the value of x if and when reaching ℓ. This compares two program executions starting with different initial values x_0 of x. However, the definition $\alpha^l_{use,mod}[\![\mathsf{S}]\!]\, L_b, L_e\, \langle\pi_0,\, \pi\rangle$ of liveness in (41.2) seems to involve only one trace π continuing a single initialization trace π_0 because the terms $use[\![\mathsf{a}]\!]\varrho(\pi_0)$ and $mod[\![\mathsf{a}]\!]\varrho(\pi_0)$ in (41.2) seem to involve only the trace π_0 to get the value $\varrho(\pi_0)$ of variables. However, the semantic definition (41.10) of $use[\![\mathsf{a}]\!]\varrho(\pi_0)$ involves the comparison of the value of variable y in environment $\rho=\varrho(\pi_0)$ with another environment $\rho[\mathsf{y}\leftarrow\nu]$ in which this value $\rho(\mathsf{y})$ is modified to ν. This environment $\rho[\mathsf{y}\leftarrow\nu]$ is an abstraction of another execution that would reach the program point ℓ with a different value of y. This is an overapproximation because such an execution might not exist. So $\rho[\mathsf{y}\leftarrow\nu]$ implicitly refers to another execution in which y would have a value different from its value on $\pi_0 \mathbin{\hat{\ }} \pi$. In conclusion liveness/deadness in chapter 41 involves two executions but does this implicitly in order to follow an established tradition in dataflow analysis. This fact is also completely eluded in presentations of liveness based of model checking [857] and [191, ch 16].

Solution to exercise 47.17 Let the future of a semantic property $\mathscr{P} \in \wp(\wp(\mathbb{T}^+ \times \mathbb{T}^{+\infty}))$ be $\alpha^{\ell}_{\mathsf{f}}[\![\mathsf{y}]\!](\mathscr{P}) \triangleq \{\mathsf{seqval}[\![\mathsf{y}]\!]^{\ell}(\pi_0,\pi) \mid \langle\pi_0,\, \pi\rangle \in \mathscr{P}\}$. Then $\langle\wp(\wp(\mathbb{T}^+ \times \mathbb{T}^{+\infty})),\, \subseteq\rangle \xleftrightarrow[\alpha^{\ell}_{\mathsf{f}}[\![\mathsf{y}]\!]]{\gamma^{\ell}_{\mathsf{f}}[\![\mathsf{y}]\!]} \langle\wp(\mathbb{V}^{+\infty}),\, \subseteq\rangle$.

Solution to exercise 47.20 There exist initialization traces π_0 and π'_0 with same initial values of variables but for x (i.e., $\forall \mathsf{z} \in \mathbb{V} \setminus \{\mathsf{x}\}\,.\,\varrho(\pi_0)\mathsf{z} = \varrho(\pi'_0)\mathsf{z} \wedge \varrho(\pi_0)\mathsf{x} \neq \varrho(\pi'_0)\mathsf{x}$) and continuations π_1 and π'_1 (i.e., $\langle\pi_0,\, \pi_1\rangle, \langle\pi'_0,\, \pi'_1\rangle \in \Pi$) sharing a common prefix (i.e., ω_0 in (47.18)), *where all nondeterministic choices are identical*, but have different values ($\nu \neq \nu'$ in (47.18)) of y at ℓ after that common prefix (i.e., ω_0). For example, $\mathsf{x} \rightsquigarrow \mathsf{z}$ in `input y; z = y + x ;`.

48 The Herbrand Abstract Domain of Symbolic Terms

In mathematical logic, Jacques Herbrand introduced *ground terms* [487, ch. 1][1] to denote a basic mathematical object (e.g., 0) or an operation on objects (e.g., $+(1, 2)$) as well as *symbolic terms*, that is, *terms with variables* (where the variables x are unknowns standing for any ground term [487, ch. 2]) (e.g., $+(1, x)$). Gordon Plotkin [800, 801] and John Reynolds [827] proved that the set of symbolic terms form a complete lattice in which the partial order $\preceq^{\nu}$ on terms is "*less general*," for example, $+(1, 2) \preceq^{\nu} +(1, y) \preceq^{\nu} +(x, y) \preceq^{\nu} z$.

Following [224], we consider the problem of abstracting ground term properties by terms with variables. A ground term property is a set of ground terms. For example, $P_0 = \{0, -(0, 0), -(1, 1), -(2, 2), \ldots, -(1, -(2, 1)), \ldots, -(-(0, 1), -(0, 1)), \ldots\}$ is the property of evaluating to zero. Such properties can be abstracted by symbolic terms. For example, $+(1, x)$ abstracts $P_1 = \{+(1, 0), +(1, 1), \ldots, +(1, -(-(0, 1), -(0, 1))), \ldots\}$). The abstraction is in general imprecise (e.g., for P_1), and we introduce sufficient conditions for exactness (e.g., for P_0) using a Galois retraction between ground term properties and symbolic terms.

The abstract domain of symbolic terms is used in refutation theorem-proving on the basis of the resolution rule of inference [838, 839], the programming language Prolog [203, 204, 596, 912], deductive databases [697], type inference [500], the analysis of general-purpose logic programming languages [90, 270, 490], pointer analysis in imperative[908] or logic languages [735], uninterpreted functions analysis [390], and so forth.

Contents

1. English translation in [486].

48.1 Ground Terms

The *signature* **F** defines a set of function symbols $f \backslash n$ (f for brevity), each one with an *arity* n, that is, a fixed number of parameters (0 for constants). The round parentheses (,) and comma , do not belong to **F**.

$$f\backslash n,\ g\backslash n,\ h\backslash n \quad \in \quad \mathbf{F}[n] \qquad \text{function symbols of arity } n \geqslant 0$$

$$f, g, h \quad \in \quad \mathbf{F} \quad = \quad \bigcup_{n \in \mathbb{N}} \mathbf{F}[n] \qquad \text{function symbols (signature)}$$

We assume that the signature **F** has at least two different function symbols. Ground terms denote uninterpreted functional expressions.

$$\begin{array}{lll} \mathbf{t} \in \mathbf{T} \ ::= & & \text{ground terms (Herbrand universe)} \\ & f\backslash 0 & \text{constants of arity 0} \\ \mid & f\backslash n(\mathbf{t}_1, \ldots, \mathbf{t}_n) & \text{term of arity } n \in \mathbb{N}^+ \end{array}$$

The set **T** of all ground terms is called the *Herbrand universe* with signature **F**.

48.2 The Complete Lattice of Ground Term Properties

Ground term properties form a complete lattice ordered by inclusion

$$\langle \wp(\mathbf{T}), \subseteq, \emptyset, \mathbf{T}, \cup, \cap \rangle \qquad \text{ground term properties} \tag{48.1}$$

48.3 Terms with Variables

A term with variables (also called *symbolic term*) abstracts a set of terms. For example, the set of ground terms $\{+(0,1), +(0,+(1,1)), +(0,+(1,+(1,1))), +(0,+(1,+(1,+(1,1))))), \ldots\}$ can be abstracted by the term $+(0,\alpha)$ with variable α. The abstraction can be very imprecise. For example, $\{0, +(0,1)\}$ would be abstracted by variable α of which concretization is the set of all ground terms. So, the abstraction is precise enough only for a set of terms with adequate regularity properties.

$$\begin{array}{lll} \alpha, \beta, \gamma \ \in \ \mathbb{V}_{t} & & \text{term variables} \\ \boldsymbol{\tau} \in \mathbf{T}^{\nu} \ ::= & & \text{terms with variables} \\ & f\backslash 0 & \text{constants} \\ \mid & f\backslash n(\boldsymbol{\tau}_1, \ldots, \boldsymbol{\tau}_n) & \text{term of arity } n \in \mathbb{N} \\ \mid & \alpha & \text{term variable} \end{array} \tag{48.2}$$

The round parentheses (,), comma , , and variables $\alpha \in \mathbb{V}_{t}$ do not belong to **F**.

We write $\mathsf{vars}[\![\boldsymbol{\tau}]\!]$ for the free variables of a term $\boldsymbol{\tau}$.

$$\mathsf{vars}[\![\alpha]\!] \triangleq \{\alpha\} \qquad \alpha \in \mathbb{V}_{t} \tag{48.3}$$

$$\mathsf{vars}[\![f(\boldsymbol{\tau}_1, \ldots, \boldsymbol{\tau}_n)]\!] \triangleq \bigcup_{i=1}^{n} \mathsf{vars}[\![\boldsymbol{\tau}_i]\!]$$

By default *term* means *term with or without variables* and we say *ground term* for *term without variable* and *symbolic term* for *term with variables*. The tag $^{\nu}$ means *with variables*.

Example 48.4 Type expressions in OCaml [628] are terms with variables (using a quote '*ident* to stand for the type variable named *ident*, the infix notation `->` for the function type, and the postfix notation `list` for lists)

```
# List.map;;
- : ('a -> 'b) -> 'a list -> 'b list = <fun>
```

In prefix notation that would be $\texttt{->}(\texttt{->}(\alpha, \beta), \texttt{->}(\texttt{list}(\alpha), \texttt{list}(\beta)))$.

The *syntactic replacement* $\boldsymbol{\tau}[\alpha \leftarrow \boldsymbol{\tau}']$ on terms with variables $\boldsymbol{\tau}$ replaces all instances of a variable α in the term $\boldsymbol{\tau}$ by another term with variables $\boldsymbol{\tau}'$.

$$\begin{aligned} \alpha[\alpha \leftarrow \boldsymbol{\tau}'] &\triangleq \boldsymbol{\tau}' && (48.5)\\ \beta[\alpha \leftarrow \boldsymbol{\tau}'] &\triangleq \beta && \text{when } \beta \neq \alpha\\ f(\boldsymbol{\tau}_1, \ldots, \boldsymbol{\tau}_n)[\alpha \leftarrow \boldsymbol{\tau}'] &\triangleq f(\boldsymbol{\tau}_1[\alpha \leftarrow \boldsymbol{\tau}'], \ldots, \boldsymbol{\tau}_n[\alpha \leftarrow \boldsymbol{\tau}']) \end{aligned}$$

48.4 Term Environments/Assignments

An assignment is an environment mapping variables to ground terms.

$$\boldsymbol{\varrho} \in \mathbf{P}^{\mathbb{V}} \triangleq \mathbb{V}_{t} \to \mathbf{T} \qquad \text{assignment (term environment)} \qquad (48.6)$$

An assignment can be homomorphically extended to a term with variables, as follows:

$$\boldsymbol{\varrho}(f(\boldsymbol{\tau}_1, \ldots, \boldsymbol{\tau}_n)) \triangleq f(\boldsymbol{\varrho}(\boldsymbol{\tau}_1), \ldots, \boldsymbol{\varrho}(\boldsymbol{\tau}_n)) \qquad (48.7)$$

The intuition is that $\boldsymbol{\varrho} \in \mathbf{T}^{\mathbb{V}} \to \mathbf{T}$ is the evaluation $\boldsymbol{\varrho}(\boldsymbol{\tau})$ of term $\boldsymbol{\tau}$ by replacing variables α of $\boldsymbol{\tau}$ by their value $\boldsymbol{\varrho}(\alpha)$ which is a ground term. Environment assignment $\boldsymbol{\varrho}[\alpha \leftarrow \mathbf{t}]$ (19.10) can be used to change the value of a variable α to $\mathbf{t}$.

We use the same notation for syntactic replacement (48.5) and environment assignment (19.10) because of the following lemma 48.8 showing that instantiation of syntactic replacement and environment assignment commute.

Lemma 48.8 $\boldsymbol{\varrho}(\boldsymbol{\tau}[\alpha \leftarrow \boldsymbol{\tau}']) = \boldsymbol{\varrho}[\alpha \leftarrow \boldsymbol{\varrho}(\boldsymbol{\tau}')](\boldsymbol{\tau})$.

Proof of lemma 48.8 By structural induction on $\boldsymbol{\tau}$.

- If $\boldsymbol{\tau} = \alpha$ then, by (48.5), $\boldsymbol{\varrho}(\boldsymbol{\tau}[\alpha \leftarrow \boldsymbol{\tau}']) = \boldsymbol{\varrho}(\boldsymbol{\tau}'[\alpha \leftarrow \alpha]) = \boldsymbol{\varrho}(\boldsymbol{\tau}')$ while, by (19.10), $\boldsymbol{\varrho}[\alpha \leftarrow \boldsymbol{\varrho}(\boldsymbol{\tau}')](\boldsymbol{\tau}) = \boldsymbol{\varrho}[\alpha \leftarrow \boldsymbol{\varrho}(\boldsymbol{\tau}')](\alpha) = \boldsymbol{\varrho}(\boldsymbol{\tau}')$, as required;
- If $\boldsymbol{\tau} = \beta \neq \alpha$ then, by (48.5), $\boldsymbol{\varrho}(\boldsymbol{\tau}[\alpha \leftarrow \boldsymbol{\tau}']) = \boldsymbol{\varrho}(\boldsymbol{\tau}'[\alpha \leftarrow \beta]) = \boldsymbol{\varrho}(\beta)$. This is equal to $\boldsymbol{\varrho}[\alpha \leftarrow \boldsymbol{\varrho}(\boldsymbol{\tau}')](\boldsymbol{\tau}) = \boldsymbol{\varrho}[\alpha \leftarrow \boldsymbol{\varrho}(\boldsymbol{\tau}')](\beta) = \boldsymbol{\varrho}(\beta)$, by (19.10);
- Otherwise, $\boldsymbol{\tau} = f(\boldsymbol{\tau}_1, \ldots, \boldsymbol{\tau}_n)$ so that, by (48.5), (48.7), induction hypothesis, and (48.7) again, $\boldsymbol{\varrho}(\boldsymbol{\tau}'[\alpha \leftarrow \boldsymbol{\tau}]) = \boldsymbol{\varrho}(f(\boldsymbol{\tau}_1, \ldots, \boldsymbol{\tau}_n)[\alpha \leftarrow \boldsymbol{\tau}']) = f(\boldsymbol{\varrho}(\boldsymbol{\tau}_1[\alpha \leftarrow \boldsymbol{\tau}']), \ldots, \boldsymbol{\varrho}(\boldsymbol{\tau}_n[\boldsymbol{\tau}' \leftarrow \alpha])) = f(\boldsymbol{\varrho}[\alpha \leftarrow \boldsymbol{\varrho}(\boldsymbol{\tau}')](\boldsymbol{\tau}_1), \ldots, \boldsymbol{\varrho}[\alpha \leftarrow \boldsymbol{\varrho}(\boldsymbol{\tau}')](\boldsymbol{\tau}_n)) = \boldsymbol{\varrho}[\alpha \leftarrow \boldsymbol{\varrho}(\boldsymbol{\tau}')](f(\boldsymbol{\tau}_1, \ldots, \boldsymbol{\tau}_n)) = \boldsymbol{\varrho}[\alpha \leftarrow \boldsymbol{\varrho}(\boldsymbol{\tau}')](\boldsymbol{\tau})$. $\square$

Let us call $\boldsymbol{\varrho}(\boldsymbol{\tau})$ the *ground instance* of $\boldsymbol{\tau}$ for the assignment $\boldsymbol{\varrho}$. Unless it is reduced to a variable, a term with variables cannot have the same instance as any of its variables (this is known as *occurs check*).

Lemma 48.9 For all variables $\alpha \in \text{vars}[\![\boldsymbol{\tau}]\!]$ of a term with variables $\boldsymbol{\tau} \in \mathbf{P}^{\mathbb{V}} \setminus \mathbb{V}_{t}$, there is no assignment $\boldsymbol{\varrho} \in \mathbf{P}^{\mathbb{V}} \triangleq \mathbb{V}_{t} \to \mathbf{T}$ such that $\boldsymbol{\varrho}(\alpha) = \boldsymbol{\varrho}(\boldsymbol{\tau})$.

Proof of lemma 48.9 Because $\boldsymbol{\tau} \in \mathbf{P}^{\mathrm{v}} \setminus \mathbb{V}_{\mathfrak{t}}$, we have $\boldsymbol{\tau} = f^1(\boldsymbol{\tau}^1_1, \ldots, f^n(\boldsymbol{\tau}^n_1, \ldots, \alpha, \ldots, \boldsymbol{\tau}^n_{m^n}), \ldots, \boldsymbol{\tau}^1_{m^1})$ where $f^i \in \mathbf{F}_{m^i}$ with $m^i > 0, i \in [1, n]$. By (48.7), we have $\boldsymbol{\varrho}(\boldsymbol{\tau}) = f^1(\boldsymbol{\tau}^1_1, \ldots, f^n(\boldsymbol{\tau}^n_1, \ldots, \boldsymbol{\varrho}(\alpha), \ldots, \boldsymbol{\tau}^n_{m^n}) = f^1(\boldsymbol{\tau}^1_1, \ldots, f^n(\boldsymbol{\tau}^n_1, \ldots, f^1(\boldsymbol{\tau}^1_1, \ldots, f^n(\boldsymbol{\tau}^n_1, \ldots, \boldsymbol{\varrho}(\alpha), \ldots, \boldsymbol{\tau}^n_{m^n}), \ldots, \boldsymbol{\tau}^n_{m^n})$ $= \ldots$. So $\boldsymbol{\varrho}(\boldsymbol{\tau})$ is an infinite object not in $\mathbf{P}^{\mathrm{v}}$. □

48.5 The Symbolic Abstraction

The symbolic abstraction abstracts a property of ground terms into a term with variables. The symbolic abstraction is easily defined by its concretization.

$$\begin{aligned} \mathit{ground}(\boldsymbol{\tau}) &\triangleq \{\boldsymbol{\varrho}(\boldsymbol{\tau}) \mid \boldsymbol{\varrho} \in \mathbf{P}^{\mathrm{v}}\} \quad \text{ground instances} \\ \mathit{ground}(\overline{\varnothing}^{\mathrm{v}}) &\triangleq \varnothing \end{aligned} \tag{48.10}$$

Because all terms with variables $\boldsymbol{\tau} \in \mathbf{P}^{\mathrm{v}}$ have a nonempty concretization $\mathit{ground}(\boldsymbol{\tau})$, we add the empty term $\overline{\varnothing}^{\mathrm{v}} \notin \mathbf{P}^{\mathrm{v}}$ to denote the empty set $\varnothing$ with $\boldsymbol{\varrho}(\overline{\varnothing}^{\mathrm{v}}) = \overline{\varnothing}^{\mathrm{v}}$.

Remark 48.11 (the symbolic abstraction is relational) An important remark on the definition of *ground* in (48.10) is that all instances $\boldsymbol{\varrho}(\alpha)$ of a variable α in a term with variables $\boldsymbol{\tau}$ are the same in a given instance $\boldsymbol{\varrho}(\boldsymbol{\tau})$ of the term. For example, $f(a, b) \notin \mathit{ground}(f(\alpha, \alpha))$ when $a \neq b$. However, two terms with variables equal up to variable renaming have the same concretization. Therefore, the names attributed to the same instances of variables in terms with variables do not matter. For example, $\mathit{ground}(f(\alpha, \alpha)) = \mathit{ground}(f(\beta, \beta)) = \{f(\mathbf{t}, \mathbf{t}) \mid \mathbf{t} \in \mathbf{T}\}$.

48.6 The Herbrand Symbolic Abstract Domain

The symbolic abstract domain is an abstraction of the complete lattice of ground term properties (48.1).

48.6.1 The Subsumption Partial Order

We define the preorder $\preceq^{\mathrm{v}}$ on terms with variables, called *subsumption*, as the inclusion of sets of their ground instances. (This will be shown to be equivalent to the classic definition in theorem 48.31.)

$$(\boldsymbol{\tau} \preceq^{\mathrm{v}} \boldsymbol{\tau}') \triangleq (\mathit{ground}(\boldsymbol{\tau}) \subseteq \mathit{ground}(\boldsymbol{\tau}')) \tag{48.12}$$

This is a preorder $\langle \mathbf{T}^{\mathrm{v}} \cup \{\overline{\varnothing}^{\mathrm{v}}\}, \preceq^{\mathrm{v}}\rangle$ with infimum $\overline{\varnothing}^{\mathrm{v}}$. For example, $f(a, b) \preceq^{\mathrm{v}} f(\alpha, b) \preceq^{\mathrm{v}} f(\alpha, \beta) \preceq^{\mathrm{v}} \gamma$.

Exercise 48.13 Prove that for all terms with variables $\boldsymbol{\tau}, \boldsymbol{\tau}' \in \mathbf{T}^{\mathrm{v}}$, we have $\boldsymbol{\tau} \preceq^{\mathrm{v}} \boldsymbol{\tau}'$ if and only if $\forall \boldsymbol{\varrho} \in \mathbf{P}^{\mathrm{v}} . \exists \boldsymbol{\varrho}' \in \mathbf{P}^{\mathrm{v}} . \boldsymbol{\varrho}(\boldsymbol{\tau}) = \boldsymbol{\varrho}'(\boldsymbol{\tau}')$.

The corresponding equivalence relation is $\simeq^{\mathrm{v}}$. The quotient is a partial order $\langle \mathbb{P}^{\mathrm{H}}, \preceq_{\simeq^{\mathrm{v}}}\rangle$ where

$$\begin{aligned} \boldsymbol{\tau} \simeq^{\mathrm{v}} \boldsymbol{\tau}' &\triangleq \boldsymbol{\tau} \preceq^{\mathrm{v}} \boldsymbol{\tau}' \wedge \boldsymbol{\tau}' \preceq^{\mathrm{v}} \boldsymbol{\tau} \\ \mathbb{P}^{\mathrm{H}} &\triangleq (\mathbf{T}^{\mathrm{v}} \cup \{\overline{\varnothing}^{\mathrm{v}}\})/_{\simeq^{\mathrm{v}}} \triangleq \{[\boldsymbol{\tau}]_{\simeq^{\mathrm{v}}} \mid \boldsymbol{\tau} \in \mathbf{T}^{\mathrm{v}} \cup \{\overline{\varnothing}^{\mathrm{v}}\}\} \\ [\boldsymbol{\tau}]_{\simeq^{\mathrm{v}}} \preceq_{\simeq^{\mathrm{v}}} [\boldsymbol{\tau}']_{\simeq^{\mathrm{v}}} &\triangleq \exists \overline{\boldsymbol{\tau}} \in [\boldsymbol{\tau}]_{\simeq^{\mathrm{v}}}, \overline{\boldsymbol{\tau}}' \in [\boldsymbol{\tau}']_{\simeq^{\mathrm{v}}} . \overline{\boldsymbol{\tau}} \preceq^{\mathrm{v}} \overline{\boldsymbol{\tau}}' \end{aligned} \tag{48.14}$$

For example $f(\alpha, \alpha) \simeq^{\mathrm{v}} f(\beta, \beta)$ and $[f(\alpha, \alpha)]_{\simeq^{\mathrm{v}}} = \{f(\gamma, \gamma) \mid \gamma \in \mathbb{V}_{\mathfrak{t}}\}$. More generally, equivalent terms are equal up to variable renaming.

Exercise 48.15 A renaming is an assignment $\rho \in \mathbb{V}_t \rightarrowtail\!\!\!\twoheadrightarrow \mathbb{V}_t$ between variables extended to terms with variables by (48.7), that is $\rho(f(\tau_1, \ldots, \tau_n)) = f(\rho(\tau_1), \ldots, \rho(\tau_n))$. Prove that $\forall \tau, \tau' \in \mathbf{T}^{\mathbb{V}} . (\tau \simeq^{\mathbb{V}} \tau') \Leftrightarrow (\exists \rho \in \mathsf{vars}[\![\tau]\!] \rightarrowtail\!\!\!\twoheadrightarrow \mathsf{vars}[\![\tau']\!] . \rho(\tau) = \tau')$, that is, equivalent terms have a bijective renaming of their variables and reciprocally.

The comparison of equivalence classes is equivalent to the comparison of the representatives of these classes.

Exercise 48.16 $[\tau_1]_{\simeq^{\mathbb{V}}} \preceq_{\simeq^{\mathbb{V}}} [\tau_2]_{\simeq^{\mathbb{V}}} \Leftrightarrow \tau_1 \preceq^{\mathbb{V}} \tau_2$.

48.6.2 The Symbolic Abstraction Function

The abstraction of $\{f(a,a), f(b,b), f(c,c)\}$ is $f(\alpha,\alpha)$ because the parameters of f are equal whereas $\{f(a,b), f(b,a), f(a,a)\}$ is $f(\beta,\gamma)$ because the parameters of f are not always related. The abstraction function must select variables in order to identify equal parameters on all instances of f. For this purpose, we encode sets as families, for example, sequences $\langle f(a,a),\ f(b,b),\ f(c,c)\rangle$ and $\langle f(a,b),\ f(b,a),\ f(a,a)\rangle$. In the first case, the subterms all yield $\langle a,\ b,\ c\rangle$, which is abstracted by a variable such as α. In the second case we get $\langle a,\ b,\ a\rangle$ encoded by β and $\langle b,\ a,\ a\rangle$, which is different so is encoded by a different variable such as γ. Notice that the variable name does not matter and that the order in the sequences does not matter either (so sets of ground terms encoded differently as index families will have the same abstraction, up to variable renaming via a bijection between variables; see exercise 48.20).

We arbitrarily define a scheme to name sets of ground terms by a unique variable thanks to an injective function

$$\boldsymbol{\nu} \in (\Delta \to \mathbf{T}) \rightarrowtail \mathbb{V}_t \qquad \text{(naming scheme)} \qquad (48.17)$$

assigning a variable $\boldsymbol{\nu}(\{\mathbf{t}_i \mid i \in \Delta\})$ to any arbitrary family of ground terms $\{\mathbf{t}_i \mid i \in \Delta\}$. Injectivity ensures uniqueness, that is, different families of terms are abstracted by different variables.

The abstraction is called the *least common generalization* (*lcg*).

$$\begin{aligned}
lcg[\boldsymbol{\nu}](\varnothing) &\triangleq \overline{\varnothing}^{\mathbb{V}} \qquad (48.18)\\
lcg[\boldsymbol{\nu}](\{f_i(\mathbf{t}_i^1, \ldots, \mathbf{t}_i^{n_i}) \mid i \in \Delta\}) &\triangleq \text{if } \forall i, j \in \Delta . f_i = f_j = f \wedge n_i = n_j = n \text{ then}\\
&\qquad \text{let } T^k = lcg[\boldsymbol{\nu}](\{\mathbf{t}_i^k \mid i \in \Delta\}), k = 1, \ldots, n \text{ in}\\
&\qquad\quad f(T^1, \ldots, T^n)\\
&\qquad \text{else } \boldsymbol{\nu}(\{f_i(\mathbf{t}^1, \ldots, \mathbf{t}^{n_i}) \mid i \in \Delta\})
\end{aligned}$$

If all the terms in the family have the same structure then the abstraction proceeds recursively, else the family is abstracted by a variable. Equalities between all subterms of the family are preserved by the abstraction because the families of these subterms are abstracted by the same variable when they have different structures.

Example 48.19 Assume that $\boldsymbol{\nu}(\langle a,\ b\rangle) = \alpha$ and $\boldsymbol{\nu}(\langle b,\ a\rangle) = \beta$, then

$$\begin{aligned}
&lcg[\boldsymbol{\nu}](\langle f(g(a,a), h(b,b), a, b),\ f(g(b,b), h(a,a), b, a)\rangle)\\
= \ & f(lcg[\boldsymbol{\nu}](\langle g(a,a),\ g(b,b)\rangle), lcg[\boldsymbol{\nu}](\langle h(b,b),\ h(a,a)\rangle), lcg[\boldsymbol{\nu}](\langle a,\ b\rangle), lcg[\boldsymbol{\nu}](\langle b,\ a\rangle))\\
= \ & f(g(lcg[\boldsymbol{\nu}](\langle a,\ b\rangle), lcg[\boldsymbol{\nu}](\langle a,\ b\rangle)), h(lcg[\boldsymbol{\nu}](\langle b,\ a\rangle), lcg[\boldsymbol{\nu}](\langle b,\ a\rangle)), lcg[\boldsymbol{\nu}](\langle a,\ b\rangle),\\
& lcg[\boldsymbol{\nu}](\langle b,\ a\rangle))
\end{aligned}$$

$$= f(g(\boldsymbol{v}(\langle a, b\rangle), \boldsymbol{v}(\langle a, b\rangle)), h(\boldsymbol{v}(\langle b, a\rangle), \boldsymbol{v}(\langle b, a\rangle)), \boldsymbol{v}(\langle a, b\rangle), \boldsymbol{v}(\langle b, a\rangle))$$
$$= f(g(\alpha, \alpha)), h(\beta, \beta), \alpha, \beta)$$

The definition (48.18) of the symbolic abstraction $lcg[\boldsymbol{v}]$ is independent of the naming scheme $\boldsymbol{v}$.

Exercise 48.20 If $\boldsymbol{v}, \boldsymbol{v}' \in (\Delta \to \mathbf{T}) \rightarrowtail \mathbb{V}_{t}$ then $\forall T \in \wp(\mathbf{T}) \,.\, lcg[\boldsymbol{v}](T) \simeq^{\mathrm{v}} lcg[\boldsymbol{v}'](T)$.

We now want to identify a Galois connection with abstraction $lcg[\boldsymbol{v}]$ and concretization *ground*. Several preliminary results are needed. First, the symbolic abstraction $lcg[\boldsymbol{v}]$ is $\preceq^{\mathrm{v}}$-increasing.

Lemma 48.21 Let $\Delta \subseteq \Delta'$ be index sets and $\mathbf{t} \in \Delta' \to \mathbf{T}$ (and therefore $\{\mathbf{t}_i \mid i \in \Delta\} \subseteq \{\mathbf{t}_i \mid i \in \Delta'\}$). Then $lcg[\boldsymbol{v}](\{\mathbf{t}_i \mid i \in \Delta\}) \preceq^{\mathrm{v}} lcg[\boldsymbol{v}](\{\mathbf{t}_i \mid i \in \Delta'\})$.

Proof of lemma 48.21 By exercise 48.13, we must prove that we have $\forall \boldsymbol{\varrho} \in \mathbf{P}^{\mathrm{v}} \,.\, \exists \boldsymbol{\varrho}' \in \mathbf{P}^{\mathrm{v}} \,.\, \boldsymbol{\varrho}(lcg[\boldsymbol{v}](\{\mathbf{t}_i \mid i \in \Delta\})) = \boldsymbol{\varrho}'(lcg[\boldsymbol{v}](\{\mathbf{t}_i \mid i \in \Delta'\}))$. Given any $\boldsymbol{\varrho}$, let us define $\boldsymbol{\varrho}'$ such that, for any $\{\mathbf{t}_i \mid i \in \Delta'\} \in \Delta' \to \mathbf{T}$, we have (including the case of constants when $n = 0$ or $m = 0$)

- If $\{\mathbf{t}_i \mid i \in \Delta\} = \{f(\mathbf{t}_i^1, \ldots, \mathbf{t}_i^n) \mid i \in \Delta\}$ and $\exists j \in \Delta' \setminus \Delta \,.\, \mathbf{t}_j = g(\mathbf{t}_j'^1, \ldots, \mathbf{t}_j'^m)$ with $g \neq f$ then
$$\boldsymbol{\varrho}'(\boldsymbol{v}(\{\mathbf{t}_i \mid i \in \Delta'\})) \triangleq \boldsymbol{\varrho}(lcg[\boldsymbol{v}](\{\mathbf{t}_i \mid i \in \Delta\}))$$
- Otherwise, $\{\mathbf{t}_i \mid i \in \Delta'\} = \{f(\mathbf{t}_i^1, \ldots, \mathbf{t}_i^n) \mid i \in \Delta'\}$, in which case
$$\boldsymbol{\varrho}'(\boldsymbol{v}(\{\mathbf{t}_i \mid i \in \Delta'\})) \triangleq \boldsymbol{\varrho}(\boldsymbol{v}(\{\mathbf{t}_i \mid i \in \Delta\}))$$

Let us show that $\boldsymbol{\varrho}(lcg[\boldsymbol{v}](\{\mathbf{t}_i \mid i \in \Delta\})) = \boldsymbol{\varrho}'(lcg[\boldsymbol{v}](\{\mathbf{t}_i \mid i \in \Delta'\}))$, by structural induction on $lcg[\boldsymbol{v}](\{\mathbf{t}_i \mid i \in \Delta\})$. There are two cases.

- If $lcg[\boldsymbol{v}](\{\mathbf{t}_i \mid i \in \Delta\})$ is a variable $\alpha = \boldsymbol{v}(\{\mathbf{t}_i \mid i \in \Delta\})$ then $\exists j, k \in \Delta \,.\, \mathbf{t}_j = f(\mathbf{t}_j^1, \ldots, \mathbf{t}_j^{n_j}) \wedge \mathbf{t}_k = g(\mathbf{t}_k^1, \ldots, \mathbf{t}_k^{n_k}) \wedge f \neq g$. By (48.18), because $\mathbf{t}_j, \mathbf{t}_k \in \{\mathbf{t}_i \mid i \in \Delta'\}$, $lcg[\boldsymbol{v}](\{\mathbf{t}_i \mid i \in \Delta'\})$ is a variable $\beta = \boldsymbol{v}(\{\mathbf{t}_i \mid i \in \Delta'\})$. By our definition of $\boldsymbol{\varrho}'$, we have $\boldsymbol{\varrho}(lcg[\boldsymbol{v}](\{\mathbf{t}_i \mid i \in \Delta\})) = \boldsymbol{\varrho}(\alpha) = \boldsymbol{\varrho}'(\beta) = \boldsymbol{\varrho}'(lcg[\boldsymbol{v}](\{\mathbf{t}_i \mid i \in \Delta'\})))$;
- Otherwise $\{\mathbf{t}_i \mid i \in \Delta'\} = \{f(\mathbf{t}_i^1, \ldots, \mathbf{t}_i^n) \mid i \in \Delta'\}$ and then

$$\begin{aligned}
& \boldsymbol{\varrho}(lcg[\boldsymbol{v}](\{\mathbf{t}_i \mid i \in \Delta\})) \\
= {} & \boldsymbol{\varrho}(lcg[\boldsymbol{v}](\{f(\mathbf{t}_i^1, \ldots, \mathbf{t}_i^n) \mid i \in \Delta\})) && \text{(}\Delta \subseteq \Delta'\text{)} \\
= {} & \boldsymbol{\varrho}(f(lcg[\boldsymbol{v}](\{\mathbf{t}_1^i \mid i \in \Delta\}), \ldots, lcg[\boldsymbol{v}](\{\mathbf{t}_n^i \mid i \in \Delta\}))) && \text{(definition (48.18) of } lcg[\boldsymbol{v}]\text{)} \\
= {} & f(\boldsymbol{\varrho}(lcg[\boldsymbol{v}](\{\mathbf{t}_1^i \mid i \in \Delta\})), \ldots, \boldsymbol{\varrho}(lcg[\boldsymbol{v}](\{\mathbf{t}_n^i \mid i \in \Delta\})) && \text{((48.7))} \\
= {} & f(\boldsymbol{\varrho}'(lcg[\boldsymbol{v}](\{\mathbf{t}_1^i \mid i \in \Delta'\})), \ldots, \boldsymbol{\varrho}'(lcg[\boldsymbol{v}](\{\mathbf{t}_n^i \mid i \in \Delta'\})) && \text{(induction hypothesis)} \\
= {} & \boldsymbol{\varrho}'(f(lcg[\boldsymbol{v}](\langle \mathbf{t}_1^i, i \in \Delta'\rangle)), \ldots, lcg[\boldsymbol{v}](\{\mathbf{t}_n^i \mid i \in \Delta'\})) && \text{((48.7))} \\
= {} & \boldsymbol{\varrho}'(lcg[\boldsymbol{v}](\langle \mathbf{t}_i, i \in \Delta'\rangle)) && \text{(definition (48.18) of } lcg[\boldsymbol{v}]\text{)} \quad \square
\end{aligned}$$

Let us prove that the abstraction of a set of terms overapproximates any term of the set.

Lemma 48.22 Let Δ be a nonempty set and $\mathbf{t} \in \Delta \to \mathbf{T}$ be a family of terms. Then $\forall j \in \Delta \,.\, \mathbf{t}_j \preceq^{\mathrm{v}} lcg[\mathbf{v}](\{\mathbf{t}_i \mid i \in \Delta\})$, that is, $\forall j \in \Delta \,.\, \exists \varrho' \in \mathbf{P}^{\mathrm{v}} \,.\, \mathbf{t}_j = \varrho'(lcg[\mathbf{v}](\{\mathbf{t}_i \mid i \in \Delta\}))$.

Proof of lemma 48.22 By exercise 48.13, we must prove that $\forall j \in \Delta \,.\, \forall \varrho \in \mathbf{P}^{\mathrm{v}} \,.\, \exists \varrho' \in \mathbf{P}^{\mathrm{v}} \,.\, \varrho(\mathbf{t}_j) = \varrho'(lcg[\mathbf{v}](\{\mathbf{t}_i \mid i \in \Delta\}))$, By $\mathbf{t}_j \in \mathbf{T}$ has no variable so $\forall \varrho \in \mathbf{P}^{\mathrm{v}} \,.\, \varrho(\mathbf{t}_j) = \mathbf{t}_j$. It follows that we have to prove that $\forall j \in \Delta \,.\, \exists \varrho' \in \mathbf{P}^{\mathrm{v}} \,.\, \mathbf{t}_j = \varrho'(lcg[\mathbf{v}](\{\mathbf{t}_i \mid i \in \Delta\}))$.

For any element of j be of Δ, let us define $\varrho'(\beta) \triangleq \mathbf{t}'_j$ for all $\beta = \mathbf{v}(\{\mathbf{t}'_i \mid i \in \Delta\})$ and $\{\mathbf{t}'_i \mid i \in \Delta\} \in \Delta \to \mathbf{T}$.

The proof that $\mathbf{t}_j = \varrho'(lcg[\mathbf{v}](\{\mathbf{t}_i \mid i \in \Delta\}))$ is by structural induction on $lcg[\mathbf{v}](\{\mathbf{t}_i \mid i \in \Delta\}) \in \mathbf{T}^{\mathrm{v}}$.

- If $lcg[\mathbf{v}](\{\mathbf{t}_i \mid i \in \Delta\})$ is a ground term $a \in \mathbf{F}_0$ then, by (48.18), $\forall i \in \Delta \,.\, \mathbf{t}_i = a$ so $\mathbf{t}_j = a = \varrho'(a) = \varrho'(lcg[\mathbf{v}](\{\mathbf{t}_i \mid i \in \Delta\}))$.
- If $lcg[\mathbf{v}](\{\mathbf{t}_i \mid i \in \Delta\})$ is a variable $\alpha = \mathbf{v}(\{\mathbf{t}_i \mid i \in \Delta\})$ then, by (48.18), there exist $i, k \in \Delta$ such that $\mathbf{t}_i = f(\mathbf{t}_{i,1}, \ldots, \mathbf{t}_{i,n})$ and $\mathbf{t}_k = g(\mathbf{t}_{k,1}, \ldots, \mathbf{t}_{k,m})$ with $f \neq g$. Then, by definition of ϱ', $\varrho'(\alpha) = \mathbf{t}_j$.
- Otherwise, by (48.18), we have $\forall i \in \Delta \,.\, \mathbf{t}_i = f(\mathbf{t}_{i,1}, \ldots, \mathbf{t}_{i,n})$, so by structural induction hypothesis, $\forall \ell \in [1, n] \,.\, \mathbf{t}_{j,\ell} = \varrho'(lcg[\mathbf{v}](\{\mathbf{t}_{i,\ell} \mid i \in \Delta\}))$. Therefore $\mathbf{t}_j = f(\mathbf{t}_{j,1}, \ldots, \mathbf{t}_{j,n}) = f(\varrho'(lcg[\mathbf{v}](\{\mathbf{t}_{i,1} \mid i \in \Delta\})), \ldots, \varrho'(lcg[\mathbf{v}](\{\mathbf{t}_{i,n} \mid i \in \Delta\}))) = \varrho'(f(lcg[\mathbf{v}](\{\mathbf{t}_{i,1} \mid i \in \Delta\})), \ldots, lcg[\mathbf{v}](\{\mathbf{t}_{i,n} \mid i \in \Delta\})) = \varrho'(lcg[\mathbf{v}](\{f(\mathbf{t}_{i,1}, \ldots, \mathbf{t}_{i,n}) \mid i \in \Delta\})) = \varrho'(lcg[\mathbf{v}](\{\mathbf{t}_{i,1} \mid i \in \Delta)\}))$. □

The following corollary shows that the symbolic abstraction is an overapproximation of properties of ground terms, that is, $ground \circ lcg[\mathbf{v}]()$ is extensive.

Corollary 48.23 If Δ is a nonempty set and $\{\mathbf{t}_i \mid i \in \Delta\} \in \Delta \to \mathbf{T}$, then $\{\mathbf{t}_i \mid i \in \Delta\} \subseteq ground(lcg[\mathbf{v}](\{\mathbf{t}_i \mid i \in \Delta\}))$.

Proof of corollary 48.23

$$
\begin{aligned}
& \{\mathbf{t}_i \mid i \in \Delta\} \subseteq ground(lcg[\mathbf{v}](\{\mathbf{t}_i \mid i \in \Delta\})) \\
\Leftrightarrow\ & \{\mathbf{t}_i \mid i \in \Delta\} \subseteq \{\varrho(lcg[\mathbf{v}](\{\mathbf{t}_i \mid i \in \Delta\})) \mid \varrho \in \mathbf{P}^{\mathrm{v}}\} && \text{(definition (48.12) of } ground\text{)} \\
\Leftrightarrow\ & \forall j \in \Delta \,.\, \exists \varrho \in \mathbf{P}^{\mathrm{v}} \,.\, \mathbf{t}_j = \varrho(lcg[\mathbf{v}](\{\mathbf{t}_i \mid i \in \Delta\})) && \text{(definition of } \subseteq\text{)}
\end{aligned}
$$

which is true by lemma 48.22. □

The following corollary shows that the abstraction of the concretization of a term with variables loses no information.

Corollary 48.24 For all $\boldsymbol{\tau} \in \mathbf{T}^{\mathrm{v}} \,.\, ground \circ lcg[\mathbf{v}] \circ ground(\boldsymbol{\tau}) = ground(\boldsymbol{\tau})$.

Proof of corollary 48.24 By corollary 48.23 (where $ground(\boldsymbol{\tau}) = \{\mathbf{t}_i \mid i \in \Delta\}$), $ground(\boldsymbol{\tau}) \subseteq ground \circ lcg[\mathbf{v}] \circ ground(\boldsymbol{\tau})$. It remains to prove that

$$
\begin{aligned}
& ground \circ lcg[\mathbf{v}] \circ ground(\boldsymbol{\tau}) \subseteq ground(\boldsymbol{\tau}) \\
\Leftrightarrow\ & \{\varrho(lcg[\mathbf{v}] \circ ground(\boldsymbol{\tau})) \mid \varrho \in \mathbf{P}^{\mathrm{v}}\} \subseteq \{\varrho'(\boldsymbol{\tau}) \mid \varrho' \in \mathbf{P}^{\mathrm{v}}\} && \text{(definition (48.10) of } ground\text{)} \\
\Leftrightarrow\ & \forall \varrho \in \mathbf{P}^{\mathrm{v}} \,.\, \exists \varrho' \in \mathbf{P}^{\mathrm{v}} \,.\, \varrho(lcg[\mathbf{v}] \circ ground(\boldsymbol{\tau})) = \varrho'(\boldsymbol{\tau}) && \text{(definition of } \subseteq\text{)}
\end{aligned}
$$

which holds by lemma 48.22. □

48.6.3 The Symbolic Term Galois Connection

In order to take into account the equivalence of terms with variables up to variable renaming (see exercise 48.15), we reason on the quotient partial order of terms $\langle \mathbb{P}^{\mathrm{H}}, \preceq_{\simeq^{\mathrm{v}}}\rangle$. We extend the concretization (48.10) and the abstraction (48.18) to equivalence classes as follows:

$$\begin{aligned} lcg_{\simeq^{\mathrm{v}}}[\boldsymbol{v}](\{\mathbf{t}_i \mid i \in \Delta\}) &\triangleq [lcg[\boldsymbol{v}](\{\mathbf{t}_i \mid i \in \Delta\})]_{\simeq^{\mathrm{v}}} \\ ground_{\simeq^{\mathrm{v}}}([\boldsymbol{\tau}]_{\simeq^{\mathrm{v}}}) &\triangleq ground(\boldsymbol{\tau}). \end{aligned} \tag{48.25}$$

Theorem 48.26 For any naming scheme $\boldsymbol{v} \in (\Delta \to \mathbf{T}) \rightarrowtail \mathbb{V}_{\mathfrak{t}}$,

$$\langle \wp(\mathbf{T}), \subseteq\rangle \xleftarrow[lcg_{\simeq^{\mathrm{v}}}[\boldsymbol{v}]]{ground_{\simeq^{\mathrm{v}}}} \twoheadrightarrow \langle \mathbb{P}^{\mathrm{H}}, \preceq_{\simeq^{\mathrm{v}}}\rangle \tag{48.27}$$

This definition of the Galois retraction is independent of the choice of the naming scheme $\boldsymbol{v}$.

Proof of theorem 48.26 By definition of a Galois connection, we must prove that for all families of terms $\{\mathbf{t}_i \mid i \in \Delta\} \in \wp(\mathbf{T})$ and term with variables $\boldsymbol{\tau}' \in \mathbf{T}^{\mathrm{V}} \cup \{\overline{\varnothing}^{\mathrm{V}}\}$,

$$lcg_{\simeq^{\mathrm{v}}}[\boldsymbol{v}](\{\mathbf{t}_i \mid i \in \Delta\}) \preceq_{\simeq^{\mathrm{v}}} [\boldsymbol{\tau}']_{\simeq^{\mathrm{v}}} \quad \Leftrightarrow \quad \{\mathbf{t}_i \mid i \in \Delta\} \subseteq ground_{\simeq^{\mathrm{v}}}([\boldsymbol{\tau}']_{\simeq^{\mathrm{v}}}).$$

This is obvious for $\Delta = \varnothing$ because $lcg_{\simeq^{\mathrm{v}}}[\boldsymbol{v}](\varnothing) = \overline{\varnothing}^{\mathrm{V}}$ which is the infimum. Otherwise, we have

$$\begin{aligned}
& lcg_{\simeq^{\mathrm{v}}}[\boldsymbol{v}](\{\mathbf{t}_i \mid i \in \Delta\}) \preceq_{\simeq^{\mathrm{v}}} [\boldsymbol{\tau}']_{\simeq^{\mathrm{v}}} \\
\Leftrightarrow\ & [lcg[\boldsymbol{v}](\{\mathbf{t}_i \mid i \in \Delta\})]_{\simeq^{\mathrm{v}}} \preceq_{\simeq^{\mathrm{v}}} [\boldsymbol{\tau}']_{\simeq^{\mathrm{v}}} && \text{(definition (48.25) of } lcg_{\simeq^{\mathrm{v}}}[\boldsymbol{v}]\text{)} \\
\Leftrightarrow\ & lcg[\boldsymbol{v}](\{\mathbf{t}_i \mid i \in \Delta\}) \preceq^{\mathrm{V}} \boldsymbol{\tau}' && \text{(definition (48.14) of } \preceq_{\simeq^{\mathrm{v}}} \text{ and } \simeq^{\mathrm{V}}\text{)} \\
\Leftrightarrow\ & ground \circ lcg[\boldsymbol{v}](\{\mathbf{t}_i \mid i \in \Delta\}) \subseteq ground(\boldsymbol{\tau}') && \text{(definition (48.12) of } \preceq^{\mathrm{V}}\text{)} \\
\Leftrightarrow\ & \{\mathbf{t}_i \mid i \in \Delta\} \subseteq ground(\boldsymbol{\tau}')
\end{aligned}$$

(($\Rightarrow$) by corollary 48.23 and transitivity;
($\Leftarrow$) By lemma 48.21, $lcg[\boldsymbol{v}]$ is increasing. By definition (48.12) of $\preceq^{\mathrm{V}}$, $ground$ is increasing. Their composition is increasing so $ground \circ lcg[\boldsymbol{v}](\{\mathbf{t}_i \mid i \in \Delta\}) \subseteq ground \circ lcg[\boldsymbol{v}] \circ ground(\boldsymbol{\tau}') = ground(\boldsymbol{\tau}')$ by corollary 48.24.)

$$\Leftrightarrow\ \{\mathbf{t}_i \mid i \in \Delta\} \subseteq ground_{\simeq^{\mathrm{v}}}([\boldsymbol{\tau}']_{\simeq^{\mathrm{v}}}) \qquad \text{(definition (48.25)) of } ground_{\simeq^{\mathrm{v}}}\text{)}$$

Moreover $ground_{\simeq^{\mathrm{v}}}$ is injective so (48.27) is a Galois retraction.

By exercise 48.20, if $\boldsymbol{v}, \boldsymbol{v}' \in (\Delta \to \mathbf{T}) \rightarrowtail \mathbb{V}_{\mathfrak{t}}$ then $\forall T \in \wp(\mathbf{T}) \,.\, lcg[\boldsymbol{v}](T) \simeq^{\mathrm{V}} lcg[\boldsymbol{v}'](T)$ so that this definition of the Galois retraction (48.27) is independent of the choice of the naming scheme $\boldsymbol{v}$. □

Observe that it follows immediately, by exercise 11.34, that

$$\langle ground_{\simeq^{\mathrm{v}}}(\mathbb{P}^{\mathrm{H}}), \subseteq\rangle \xleftarrow[lcg_{\simeq^{\mathrm{v}}}[\boldsymbol{v}]]{ground_{\simeq^{\mathrm{v}}}} \twoheadrightarrow \langle \mathbb{P}^{\mathrm{H}}, \preceq_{\simeq^{\mathrm{v}}}\rangle$$

is a Galois isomorphism, an essential remark for completeness in typing [224].

48.6.4 The Symbolic Abstract Domain is a Complete Lattice

By exercise 11.64, the image of a complete lattice by a Galois retraction is a complete lattice. Hence the terms with variables form a complete lattice because they are the image of the complete lattice of properties of ground terms by the Galois retraction (48.27).

Corollary 48.28 (symbolic abstract domain) For any naming scheme $\boldsymbol{v} \in (\Delta \rightarrow \mathbf{T}) \rightarrowtail \mathbb{V}_{t}$, $\langle \mathbb{P}^{\mathrm{H}}, \preceq_{\simeq^{\mathrm{v}}}, [\overline{\varnothing}^{\mathrm{v}}]_{\simeq^{\mathrm{v}}}, [\alpha]_{\simeq^{\mathrm{v}}}, LCG_{\simeq^{\mathrm{v}}}, GCI_{\simeq^{\mathrm{v}}}\rangle$ is a complete lattice where $\alpha \in \mathbb{V}_{t}$, the least upper bound is $LCG_{\simeq^{\mathrm{v}}}(S) \triangleq lcg_{\simeq^{\mathrm{v}}}[\boldsymbol{v}](\bigcup ground_{\simeq^{\mathrm{v}}}(S))$ (binary lcg for symbolic terms and $lcg_{\simeq^{\mathrm{v}}}$ for term classes), and the greatest lower bound is $GCI_{\simeq^{\mathrm{v}}}(S) \triangleq lcg_{\simeq^{\mathrm{v}}}[\boldsymbol{v}](\bigcap ground_{\simeq^{\mathrm{v}}}(S))$ (binary gci and $gci_{\simeq^{\mathrm{v}}}$). This characterization of the lattice operations is independent of the naming scheme $\boldsymbol{v}$ that is used.

Proof of corollary 48.28 Because $\langle \wp(\mathbf{T}), \subseteq, \varnothing, \mathbf{T}, \bigcup, \bigcap\rangle$ is a complete lattice and (48.27) is a Galois retraction, it follows that, for any naming scheme $\boldsymbol{v} \in (\Delta \rightarrow \mathbf{T}) \rightarrowtail \mathbb{V}_{t}$, its image $lcg_{\simeq^{\mathrm{v}}}[\boldsymbol{v}](\wp(\mathbf{T})) = \mathbb{P}^{\mathrm{H}}$ by $lcg_{\simeq^{\mathrm{v}}}[\boldsymbol{v}]$ is also a complete lattice $\langle \mathbb{P}^{\mathrm{H}}, \preceq_{\simeq^{\mathrm{v}}}, [\overline{\varnothing}^{\mathrm{v}}]_{\simeq^{\mathrm{v}}}, [\alpha]_{\simeq^{\mathrm{v}}}, LCG_{\simeq^{\mathrm{v}}}, GCI_{\simeq^{\mathrm{v}}}\rangle$ where the infimum is $lcg_{\simeq^{\mathrm{v}}}[\boldsymbol{v}](\varnothing) = [\overline{\varnothing}^{\mathrm{v}}]_{\simeq^{\mathrm{v}}}$, the supremum is $lcg_{\simeq^{\mathrm{v}}}[\boldsymbol{v}](\mathbf{T}) = [\boldsymbol{v}(\langle \mathbf{t}, \mathbf{t} \in \mathbf{T}\rangle)]_{\simeq^{\mathrm{v}}} \simeq^{\mathrm{v}} [\alpha]_{\simeq^{\mathrm{v}}}$, $\alpha \in \mathbb{V}_{t}$, the least upper bound is $LCG_{\simeq^{\mathrm{v}}}(S) \triangleq lcg_{\simeq^{\mathrm{v}}}[\boldsymbol{v}](\bigcup ground_{\simeq^{\mathrm{v}}}(S))$ and the greatest lower bound is $GCI_{\simeq^{\mathrm{v}}}(S) \triangleq lcg_{\simeq^{\mathrm{v}}}[\boldsymbol{v}](\bigcap ground_{\simeq^{\mathrm{v}}}(S))$.

By exercise 48.20, if $\boldsymbol{v}, \boldsymbol{v}' \in (\Delta \rightarrow \mathbf{T}) \rightarrowtail \mathbb{V}_{t}$ then $\forall T \in \wp(\mathbf{T}) \,.\, lcg[\boldsymbol{v}](T) \simeq^{\mathrm{v}} lcg[\boldsymbol{v}'](T)$ so that this characterization of the lattice operations is independent of the choice of the naming scheme $\boldsymbol{v}$. □

We use $lcg_{\simeq^{\mathrm{v}}}$ (respectively $gci_{\simeq^{\mathrm{v}}}$) for the binary version of $LCG_{\simeq^{\mathrm{v}}}$ (respect. $GCI_{\simeq^{\mathrm{v}}}$). In programs, we encode classes $[\boldsymbol{\tau}]_{\simeq^{\mathrm{v}}}$ by the representant $\boldsymbol{\tau}$.

Observe that ground terms $[\mathbf{t}]_{\simeq^{\mathrm{v}}} \in \mathbb{P}^{\mathrm{H}}$ belongs to the abstract domain and abstract the concrete property $\{\mathbf{t}\}$ of being that ground term. Then $lcg_{\simeq^{\mathrm{v}}}[\boldsymbol{v}](\{\mathbf{t}\}) = LCG_{\simeq^{\mathrm{v}}}(\{[\mathbf{t}]_{\simeq^{\mathrm{v}}}\})$ (because, by (48.27), we have

$$
\begin{aligned}
& LCG_{\simeq^{\mathrm{v}}}(\{[\mathbf{t}]_{\simeq^{\mathrm{v}}}\}) \\
\triangleq\ & lcg_{\simeq^{\mathrm{v}}}[\boldsymbol{v}](\bigcup ground_{\simeq^{\mathrm{v}}}(\{[\mathbf{t}]_{\simeq^{\mathrm{v}}}\})) \\
=\ & lcg_{\simeq^{\mathrm{v}}}[\boldsymbol{v}](\bigcup ground_{\simeq^{\mathrm{v}}}(\{\mathbf{t}\})) \\
=\ & lcg_{\simeq^{\mathrm{v}}}[\boldsymbol{v}](\bigcup\{\mathbf{t}\}) \\
=\ & lcg_{\simeq^{\mathrm{v}}}[\boldsymbol{v}](\{\mathbf{t}\})).
\end{aligned}
$$

This explains why the abstraction and the lub in the complete lattice have been given the same name.

48.7 The Classic Definition of the Subsumption Partial Order Using Substitutions

The subsumption preorder $\preceq^{\mathrm{v}}$ is classically defined syntactically, using substitutions [839, pp. 180–188] (instead of (48.12)) [800, 801, 827]. We show that this classic syntactic definition is equivalent to the semantic definition (48.12) based on the interpretation of terms with variables as properties of ground terms.

48.7.1 Substitutions

The same way that assignments (48.6) record ground values of variables, we use substitutions to record symbolic values of some variables, so substitutions are partial functions

$$\vartheta \in \Sigma \triangleq \mathbb{V}_t \nrightarrow \mathbf{T}^\nu \tag{48.29}$$

mapping variables α in its domain $\mathsf{dom}(\vartheta)$ to terms with variables $\vartheta(\alpha)$.

A substitution is extended to a total function $\vartheta \in \mathbb{V}_t \rightarrow \mathbf{T}^\nu$ and homomorphically to terms with variables, as follows

$$\begin{aligned} \vartheta(\alpha) &\triangleq \alpha && \text{when} \quad \alpha \notin \mathsf{dom}(\vartheta) \\ \vartheta(f(\tau_1, \ldots, \tau_n)) &\triangleq f(\vartheta(\tau_1), \ldots, \vartheta(\tau_n)) \end{aligned} \tag{48.30}$$

Observe that the substitution is carried out simultaneously on all variable occurrences.

The empty substitution ε is totally undefined, that is, $\mathsf{dom}(\varepsilon) = \emptyset$. It's total extension is the identity $\forall \alpha \in \mathbb{V}_t \,.\, \varepsilon(\alpha) = \alpha$. By structural induction on terms with variables, we have $\forall \tau \in \mathbf{T}^\nu \,.\, \varepsilon(\tau) = \tau$.

48.7.2 The Classic Characterization of the Subsumption Preorder Using Substitutions

The following theorem 48.31 shows that the syntactic and semantic definitions of subsumption are equivalent. It follows that the subsumption lattice of [519, 800, 801, 827] is the complete lattice considered in corollary 48.28 because the partial order is the same (although defined differently).

Theorem 48.31 $\forall \tau_1, \tau_2 \in \mathbf{T}^\nu \,.\, [\tau_1]_{\simeq^\nu} \preceq_{\simeq^\nu} [\tau_2]_{\simeq^\nu} \Leftrightarrow \exists \vartheta \in \Sigma \,.\, \vartheta(\tau_2) = \tau_1$.

Proof of theorem 48.31 Let us first show that for all $\tau_1, \tau_2 \in \mathbf{T}^\nu$,

$$(\forall \varrho \in \mathbf{P}^\nu \,.\, \exists \varrho' \in \mathbf{P}^\nu \,.\, \varrho(\tau_1) = \varrho'(\tau_2)) \Leftrightarrow (\exists \vartheta \in \Sigma \,.\, \vartheta(\tau_2) = \tau_1) \tag{48.32}$$

($\Leftarrow$) Choose $\varrho' = \alpha \mapsto \varrho(\vartheta(\alpha))$ so that, by structural induction on terms with variables, $\forall \tau \in \mathbf{T}^\nu \,.\, \varrho'(\tau) = \varrho(\vartheta(\tau))$. Then $\vartheta(\tau_2) = \tau_1$ implies $\varrho(\vartheta(\tau_2)) = \varrho(\tau_1)$ and so $\varrho'(\tau_2) = \varrho(\tau_1)$.
($\Rightarrow$) We assume that $\forall \varrho \in \mathbf{P}^\nu \,.\, \exists \varrho' \in \mathbf{P}^\nu \,.\, \varrho(\tau_1) = \varrho'(\tau_2)$. The structural proof is by cases on the pair $\langle \tau_1, \tau_2 \rangle$ ordered lexicographically. It consists of constructing ϑ given ϱ and ϱ'.

- $\tau_1 = a \in \mathbf{F}_0$
  - $\tau_2 = b \in \mathbf{F}_0$. If $b \neq a$, the hypothesis is false and the implication is true. Otherwise $b = a$ and any substitution has $\vartheta(\tau_1) = \vartheta(a) = a = \vartheta(\tau_2)$.
  - $\tau_2 = \beta \in \mathbb{V}_t$. Any substitution s.t. $\vartheta(\beta) = a$ has $\vartheta(\tau_2) = \vartheta(\beta) = a = \tau_1$.
  - $\tau_2 = g(\tau''_1, \ldots, \tau''_m)$, $g \in \mathbf{F}_m$. $\forall \varrho, \varrho' \,.\, \varrho(\tau_1) \neq \varrho'(\tau_2)$ so that the hypothesis is false and the implication is true.
- $\tau_1 = \alpha \in \mathbb{V}_t$
  - $\tau_2 = b \in \mathbf{F}_0$. If $\varrho(\alpha) \neq b$, the hypothesis is false and the implication is true. Otherwise $\varrho(\alpha) = b$.
  - $\tau_2 = \beta \in \mathbb{V}_t$. Any substitution such that $\vartheta(\beta) = \alpha$ will do.

- $\tau_2 = g(\tau_1'', \ldots, \tau_m'')$, $g \in \mathbf{F}_m$. Because we assume that the signature $\mathbf{F}$ has at least two different function symbols, there is a term τ with this different symbol at the root. For ϱ such that $\varrho(\alpha) = \tau$ there is no ϱ' such that $\varrho(\alpha) = \tau = \varrho'(\tau_2) = g(\varrho'(\tau_1''), \ldots, \varrho'(\tau_m''))$. In that case, the hypothesis is false and the implication is true.

- $\tau_1 = f(\tau_1', \ldots, \tau_n')$, $f \in \mathbf{F}_n$
  - $\tau_2 = b \in \mathbf{F}_0$. $\forall \varrho, \varrho' . \varrho(\tau_1) \neq \varrho'(\tau_2)$ so the hypothesis is false and the implication is true.
  - $\tau_2 = \beta \in \mathcal{V}_{\mathfrak{t}}$. Simply choose $\vartheta(\beta) = \tau_1$.
  - $\tau_2 = g(\tau_1'', \ldots, \tau_m'')$, $g \in \mathbf{F}_m$. The hypothesis that $\varrho(\tau_1) = \varrho'(\tau_2)$, that is, $f(\varrho(\tau_1'), \ldots, \varrho(\tau_n')) = g(\varrho'(\tau_1''), \ldots, \varrho'(\tau_m''))$, implies that $f = g$, $m = n$, and $\forall i \in [1, n] . \varrho(\tau_j') = \varrho'(\tau_j'')$. So, by structural induction hypothesis, $\exists \vartheta_j . \vartheta_j(\tau_j'') = \tau_j'$, that is, $\tau_1 = f(\tau_1', \ldots, \tau_n') = f(\vartheta_1(\tau_1''), \ldots, \vartheta_n(\tau_n''))$. We now have to define ϑ such that $\vartheta(\tau_2) = \tau_1$.
    - For the variables α that do not occur in τ_2, we choose $\vartheta(\alpha) = \alpha$;
    - For variables α that occur once or more in τ_2, say in τ_j'' and τ_k'', $j, k \in [1, n]$, we have

$$\begin{array}{rcl} \tau_j'' & = & f_j^1(\ldots f_j^{m_j}(\ldots, \alpha, \ldots) \ldots) \\ \tau_k'' & = & f_k^1(\ldots f_k^{m_k}(\ldots, \alpha, \ldots) \ldots) \end{array}$$

    There are two subcases:
      - If for all occurrences, the substitutions are identical, we choose $\vartheta(\alpha) = \vartheta_j(\alpha) = \vartheta_k(\alpha)$;
      - Otherwise, $\exists \alpha, j \neq k . \vartheta_j(\alpha) \neq \vartheta_k(\alpha)$[2] and so $\vartheta_j(\tau_j'') \neq \vartheta_k(\tau_k'')$. Therefore, $\exists \varrho . \varrho(\vartheta_j(\tau_j'')) \neq \varrho(\vartheta_k(\tau_k''))$. By hypothesis, $\exists \varrho' . \varrho(\tau_1) = \varrho'(\tau_2)$ so $\varrho(\tau_1) = f(\varrho(\tau_1'), \ldots, \varrho(\tau_n')) = f(\varrho(\vartheta_1(\tau_1'')), \ldots, \varrho(\vartheta_n(\tau_n''))) = f(\varrho'(\tau_1''), \ldots, \varrho'(\tau_n'')) = \varrho'(\tau_2)$. If follows that $\varrho(\tau_j') = \varrho(\vartheta_j(\tau_j'')) = \varrho'(\tau_j'')$ and $\varrho(\tau_k') = \varrho(\vartheta_k(\tau_k'')) = \varrho'(\tau_k'')$. For the term τ_j'', we have $\varrho'(\tau_j'') = f_j^1(\ldots f_j^{m_j}(\ldots, \varrho'(\alpha), \ldots) \ldots) = \varrho(\vartheta_j(\tau_j''))$ $= f_j^1(\ldots f_j^{m_j}(\ldots, \varrho(\vartheta_j(\alpha)), \ldots) \ldots)$ and so $\varrho'(\alpha) = \varrho(\vartheta_j(\alpha))$. Similarly, for the term τ_k'', we get $\varrho'(\alpha) = \varrho(\vartheta_k(\alpha))$. It follows that $\varrho(\vartheta_j(\alpha)) = \varrho(\vartheta_k(\alpha))$, a contradiction. This case is therefore impossible. □

In conclusion, in all cases which are possible, the hypothesis $\forall \varrho . \exists \varrho' . \varrho(\tau_1) = \varrho'(\tau_2)$ implies that $\exists \vartheta \in \mathbb{\Sigma} . \vartheta(\tau_2) = \tau_1$. It follows that

$$\begin{array}{lr}
[\tau_1]_{\simeq^\nu} \preceq_{\simeq^\nu} [\tau_2]_{\simeq^\nu} & \\
\Leftrightarrow \tau_1 \preceq^\nu \tau_2 & \text{(exercise 48.16)} \\
\Leftrightarrow ground(\tau_1) \subseteq ground(\tau_2) & \text{(definition (48.12) of } \preceq^\nu\text{)} \\
\Leftrightarrow \{\varrho(\tau_1) \mid \varrho \in \mathbf{P}^\nu\} \subseteq \{\varrho'(\tau_2) \mid \varrho' \in \mathbf{P}^\nu\} & \text{((48.10))} \\
\Leftrightarrow \forall \varrho \in \mathbf{P}^\nu . \exists \varrho' \in \mathbf{P}^\nu . \varrho(\tau_1) = \varrho'(\tau_2) & \text{(definition of } \subseteq\text{)} \\
\Leftrightarrow \exists \vartheta \in \mathbb{\Sigma} . \vartheta(\tau_2) = \tau_1 & \text{((48.32))} \quad \square
\end{array}$$

2. An example is $f(\alpha, \beta) \not\preceq^\nu f(\alpha, \alpha)$ with $\vartheta_1(\alpha) = \alpha$ and $\vartheta_2(\alpha) = \beta$.

48.8 Subsumption Lattice Operations

We now provide effective algorithms for the subsumption lattice operations of comparison, greatest lower bound, and least upper bound.

48.8.1 The Subsumption Partial Order Algorithm

The subsumption algorithm $\mathsf{leq}(\boldsymbol{\tau}_1, \boldsymbol{\tau}_2)$ in figure 48.34 checks that $\boldsymbol{\tau}_1 \preceq^{\mathcal{V}} \boldsymbol{\tau}_2$ (and so that $[\boldsymbol{\tau}_1]_{\simeq^{\mathcal{V}}} \preceq_{\simeq^{\mathcal{V}}} [\boldsymbol{\tau}_2]_{\simeq^{\mathcal{V}}}$ by exercise 48.16). It proceeds by structural induction (48.33.5)–(48.33.9). The base cases (48.33.1) and (48.33.3) check that $\boldsymbol{\tau}_1 \preceq^{\mathcal{V}} \alpha$ where $\alpha \in \mathbb{V}_{t}$ is a variable. A substitution ϑ_0 (initially empty ε) is used as a table recording the correspondence between variables α of $\boldsymbol{\tau}_2$ and subterms of $\boldsymbol{\tau}_1$ so that, by (48.33.3), occurrences of α later found in $\boldsymbol{\tau}_2$ are checked to all have the same subterm instances in $\boldsymbol{\tau}_1$ (see example 48.36).

Subsumption algorithm ($\boldsymbol{\tau}_1, \boldsymbol{\tau}_2 \in \mathbf{T}^{\mathcal{V}}$ *and* $\vartheta_0 \in \Sigma$)

$$
\begin{array}{lr}
\mathsf{let\ rec\ leq}(\boldsymbol{\tau}_1, \boldsymbol{\tau}_2, \vartheta_0) = & (48.33)\\
\quad \mathsf{if}\ \boldsymbol{\tau}_2 = \alpha \in \mathbb{V}_{t} \wedge \alpha \notin \mathsf{dom}(\vartheta_0)\ \mathsf{then} & (1)\\
\qquad \langle \mathsf{tt},\ \vartheta_0[\alpha \leftarrow \boldsymbol{\tau}_1]\rangle & (2)\\
\quad \mathsf{elsif}\ \boldsymbol{\tau}_2 = \alpha \in \mathbb{V}_{t} \wedge \alpha \in \mathsf{dom}(\vartheta_0) \wedge \vartheta_0(\alpha) = \boldsymbol{\tau}_1\ \mathsf{then} & (3)\\
\qquad \langle \mathsf{tt},\ \vartheta_0\rangle & (4)\\
\quad \mathsf{elsif}\ \boldsymbol{\tau}_1 = f(\boldsymbol{\tau}_1^1, \ldots, \boldsymbol{\tau}_1^n) \wedge \boldsymbol{\tau}_2 = f(\boldsymbol{\tau}_2^1, \ldots, \boldsymbol{\tau}_2^n)\ \mathsf{then} & (5)\\
\qquad \mathsf{let}\langle b_1,\ \vartheta_1\rangle = \mathsf{leq}(\boldsymbol{\tau}_1^1, \boldsymbol{\tau}_2^1, \vartheta_0)\ \mathsf{in} & (6)\\
\qquad\quad \mathsf{let}\langle b_2,\ \vartheta_2\rangle = \mathsf{leq}(\boldsymbol{\tau}_1^2, \boldsymbol{\tau}_2^2, \vartheta_1)\ \mathsf{in} & (7)\\
\qquad\quad \ldots & \ldots\\
\qquad\qquad \mathsf{let}\langle b_n,\ \vartheta_n\rangle = \mathsf{leq}(\boldsymbol{\tau}_1^n, \boldsymbol{\tau}_2^n, \vartheta_{n-1})\ \mathsf{in} & (8)\\
\qquad\qquad\quad \langle b_1 \wedge b_2 \wedge \cdots \wedge b_n,\ \vartheta_n\rangle \quad /\star\ \langle \mathsf{tt},\ \vartheta_0\rangle \text{ for } n=0 \star/ & (9)\\
\quad \mathsf{else}\ \langle \mathsf{ff},\ \vartheta_0\rangle & (10)\\
\mathsf{let\ leq}(\boldsymbol{\tau}_1, \boldsymbol{\tau}_2) = & (11)\\
\quad \mathsf{if}\ \boldsymbol{\tau}_1 = \overline{\varnothing}^{\mathcal{V}}\ \mathsf{then\ tt\ elsif}\ \boldsymbol{\tau}_2 = \overline{\varnothing}^{\mathcal{V}}\ \mathsf{then\ ff\ else\ let}\ \langle b,\ \vartheta\rangle = \mathsf{leq}(\boldsymbol{\tau}_1, \boldsymbol{\tau}_2, \varepsilon)\ \mathsf{in}\ b & (12)
\end{array}
$$

Figure 48.34
The subsumption algorithm

Algorithm (48.33) can be easily optimized, for example, to avoid useless recursive calls as soon as one b_i is ff in (48.33.6–8), by eliminating recursion using an iteration and a stack [934, BURRY and UNBURRY instructions, p. 35-36] and [320], or by keeping the substitution ϑ_0 global (see lemma 48.38), and so forth.

Example 48.35 Notice that in $\mathsf{leq}(\boldsymbol{\tau}_1, \boldsymbol{\tau}_2, \vartheta_0)$ the scope of variables of $\boldsymbol{\tau}_1$ and $\boldsymbol{\tau}_2$ is limited to each of these terms and that $\vartheta_0 \in \mathsf{vars}[\![\boldsymbol{\tau}_2]\!] \nrightarrow \mathsf{vars}[\![\boldsymbol{\tau}_1]\!]$ so there is no possible ambiguity between homonym variables in the two terms.

$$
\begin{array}{lr}
\mathsf{leq}(f(g(\alpha), g(\beta)), f(g(\alpha), \beta), \varnothing) & (48.33)\\
\quad \mathsf{leq}(g(\alpha), g(\alpha), \varnothing) & (48.33.6)\\
\quad\quad \mathsf{leq}(\alpha, \alpha, \varnothing) & (48.33.6)\\
\quad\quad = \langle \mathsf{tt}, \{\langle \alpha, \alpha\rangle\}\rangle & (48.33.2)\\
\quad = \langle \mathsf{tt}, \{\langle \alpha, \alpha\rangle\}\rangle & (48.33.9)\\
\quad \mathsf{leq}(g(\beta), \beta, \{\langle \alpha, \alpha\rangle\}) & (48.33.7)\\
\quad = \langle \mathsf{tt}, \{\langle \alpha, \alpha\rangle, \langle \beta, g(\beta)\rangle\}\rangle & (48.33.2)\\
= \langle \mathsf{tt}, \{\langle \alpha, \alpha\rangle, \langle \beta, g(\beta)\rangle\}\rangle & (48.33.9)
\end{array}
$$

Example 48.36 This example demonstrate the use of the substitution ϑ_0 to track dependencies between variables.

$$
\begin{array}{lr}
\mathsf{leq}(f(g(g(\alpha)), g(\beta)), f(\gamma, \gamma), \varnothing) & (48.33)\\
\quad \mathsf{leq}(g(g(\alpha)), \gamma, \varnothing) & (48.33.6)\\
\quad = \langle \mathsf{tt}, \{\langle \gamma, g(g(\alpha))\rangle\}\rangle & (48.33.2)\\
\quad \mathsf{leq}(g(\beta), \gamma, \{\langle \gamma, g(g(\alpha))\rangle\}) & (48.33.7)\\
\quad = \langle \mathsf{ff}, \{\langle \gamma, g(g(\alpha))\rangle\}\rangle & (48.33.10)\\
= \langle \mathsf{ff}, \{\langle \gamma, g(g(\alpha))\rangle\}\rangle & (48.33.9)
\end{array}
$$

Lemma 48.37 The leq algorithm terminates.

Proof of lemma 48.37 The proof of termination of $\mathsf{leq}(\boldsymbol{\tau}_1, \boldsymbol{\tau}_2, \vartheta_0)$ is by structural induction on $\boldsymbol{\tau}_2$ which is either a variable so that there is no recursive call at (48.33.2), (48.33.4), or (48.33.10) or $\boldsymbol{\tau}_2$ has the form $f(\boldsymbol{\tau}_2^1, \dots, \boldsymbol{\tau}_2^n)$ and termination either follows from (48.33.10) when the test (48.33.5) is false or, when it is true, the recursive calls (48.33.6) to (48.33.8) do terminate by induction hypothesis so that the main call terminates at (48.33.9). □

The following lemma 48.38 shows that the substitution table ϑ_0 is successively extended along calls, while keeping its previous values unchanged.

Lemma 48.38 For all $\boldsymbol{\tau}_1^0, \boldsymbol{\tau}_2^0, \boldsymbol{\tau}_1, \boldsymbol{\tau}_2 \in \mathbf{T}^\nu$ and $\vartheta_0, \vartheta' \in \mathbb{V}_{\mathfrak{t}} \nrightarrow \mathbf{T}^\nu$, if $\mathsf{leq}(\boldsymbol{\tau}_1, \boldsymbol{\tau}_2, \vartheta_0)$ is (recursively) called from the main call $\mathsf{leq}(\boldsymbol{\tau}_1^0, \boldsymbol{\tau}_2^0)$ and returns $\langle b, \vartheta'\rangle = \mathsf{leq}(\boldsymbol{\tau}_1, \boldsymbol{\tau}_2, \vartheta_0)$, then

$$
\begin{array}{c}
\vartheta_0, \vartheta' \in \mathbb{V}_{\mathfrak{t}} \nrightarrow \mathbf{T}^\nu \wedge \mathsf{dom}(\vartheta') = \mathsf{dom}(\vartheta_0) \cup \mathsf{vars}[\![\boldsymbol{\tau}_2]\!]\\
\wedge\, \forall \alpha \in \mathsf{dom}(\vartheta_0) \,.\, \vartheta_0(\alpha) = \vartheta'(\alpha)
\end{array}
\qquad (48.39)
$$

Proof of lemma 48.38 By induction on the sequence of calls to leq. For the basis $\vartheta_0 = \boldsymbol{\varepsilon}$ so $\mathsf{dom}(\vartheta_0) = \varnothing$ and (48.39) holds vacuously for ϑ_0. For the induction step, by case analysis.

- In case (48.33.2), $\vartheta_0 \in \mathbb{V}_{\mathfrak{t}} \nrightarrow \mathbf{T}^\nu$ by induction hypothesis, $b = \mathsf{tt}$, $\boldsymbol{\tau}_2 = \alpha$, the test $\alpha \notin \mathsf{dom}(\vartheta_0)$ at (48.33.1), and $\vartheta' = \boldsymbol{\tau}_1[\alpha \leftarrow \vartheta_0]$ at (48.33.2) imply $\vartheta' \in \mathbb{V}_{\mathfrak{t}} \nrightarrow \mathbf{T}^\nu$. Moreover, $\alpha \notin \mathsf{dom}(\vartheta_0)$ so that $\mathsf{dom}(\vartheta') = \mathsf{dom}(\boldsymbol{\tau}_1[\alpha \leftarrow \vartheta_0]) = \mathsf{dom}(\vartheta_0) \cup \{\alpha\} = \mathsf{dom}(\vartheta_0) \cup \mathsf{vars}[\![\boldsymbol{\tau}_2]\!]$. Finally, for all $\beta \in \mathsf{dom}(\vartheta_0)$, $\beta \neq \alpha$ by (48.33.1), so $\vartheta'(\beta) = \boldsymbol{\tau}_1[\alpha \leftarrow \vartheta_0](\beta) = \vartheta_0(\beta)$, proving (48.39);

- In case (48.33.4), (48.33.10), and (48.33.9) when $n=0$, we have $\vartheta' = \vartheta_0$ so (48.39) follows immediately;
- Finally in case (48.33.9) when $n>0$, we proceed by recurrence with recurrence hypothesis $\vartheta_i \in \mathbb{V}_t \not\to \mathbf{T}^{\mathsf{v}} \wedge \mathsf{dom}(\vartheta_i) = \mathsf{dom}(\vartheta_0) \cup \bigcup_{j=1}^{i} \mathsf{vars}[\![\tau_2^j]\!] \wedge \forall \alpha \in \mathsf{dom}(\vartheta_0) \,.\, \vartheta_0(\alpha) = \vartheta_i(\alpha)$.
  - By induction hypothesis (48.39) on the sequence of calls to leq, this is true for $i=1$;
  - Assume that the recurrence hypothesis is true after the ith call. By induction hypothesis (48.39) on the sequence of calls to leq, we have $\vartheta_{i+1} \in \mathbb{V}_t \not\to \mathbf{T}^{\mathsf{v}} \wedge \mathsf{dom}(\vartheta_{i+1}) = \mathsf{dom}(\vartheta_i) \cup \mathsf{vars}[\![\tau_2^{i+1}]\!] \wedge \forall \alpha \in \mathsf{dom}(\vartheta_i) \,.\, \vartheta_i(\alpha) = \vartheta_{i+1}(\alpha)$ after the next call $\langle b_{i+1}, \vartheta_{i+1}\rangle = \mathsf{leq}(\tau_1^{i+1}, \tau_2^{i+1}, \vartheta_i)$. This implies that $\mathsf{dom}(\vartheta_{i+1}) = \mathsf{dom}(\vartheta_i) \cup \mathsf{vars}[\![\tau_2^{i+1}]\!] = \mathsf{dom}(\vartheta_0) \cup \bigcup_{j=1}^{i} \mathsf{vars}[\![\tau_2^j]\!] \cup \mathsf{vars}[\![\tau_2^{i+1}]\!] = \mathsf{dom}(\vartheta_0) \cup \bigcup_{j=1}^{i+1} \mathsf{vars}[\![\tau_2^j]\!]$. Moreover, by the recurrence hypothesis, $\forall \alpha \in \mathsf{dom}(\vartheta_0) \,.\, \vartheta_0(\alpha) = \vartheta_i(\alpha) = \vartheta_{i+1}(\alpha)$, proving the recurrence hypothesis for $i+1$.

  By recurrence for $i=n$, by (48.33.9), we have $\vartheta' = \vartheta_n \in \mathbb{V}_t \not\to \mathbf{T}^{\mathsf{v}}$. We have $\mathsf{dom}(\vartheta') = \mathsf{dom}(\vartheta_n) = \mathsf{dom}(\vartheta_0) \cup \bigcup_{j=1}^{n} \mathsf{vars}[\![\tau_2^j]\!] = \mathsf{dom}(\vartheta_0) \cup \mathsf{vars}[\![f(\tau_2^1, \ldots, \tau_2^n)]\!] = \mathsf{dom}(\vartheta_0) \cup \mathsf{vars}[\![\tau_2]\!]$ by (48.33.5). Moreover, $\forall \alpha \in \mathsf{dom}(\vartheta_0) \,.\, \vartheta_0(\alpha) = \vartheta_n(\alpha) = \vartheta'(\alpha)$, proving (48.39) for the case (48.33.9). □

It remains to observe, by induction on the sequence of calls to leq, that for the main call (48.33.12), $\vartheta' \in \mathbb{V}_t \not\to \mathbf{T}^{\mathsf{v}}$ and $\mathsf{dom}(\vartheta') = \mathsf{dom}(\varepsilon) \cup \mathsf{vars}[\![\tau_2^0]\!] = \mathsf{vars}[\![\tau_2^0]\!]$. □

The following lemma 48.40 shows the soundness of the subsumption algorithm $\mathsf{leq}(\tau_1, \tau_2)$ in figure 48.34. If it returns true (tt) then τ_1 is smaller than or equal to τ_2.

Lemma 48.40 For any substitution $\vartheta \in \mathbb{V}_t \not\to \mathbf{T}^{\mathsf{v}}$ and ground assignment $\varrho \in \mathbb{V}_t \to \mathbf{T}$, define

$$\widetilde{\vartheta}(\varrho) \triangleq \varrho[\alpha \leftarrow \varrho(\vartheta(\alpha)), \alpha \in \mathsf{dom}(\vartheta)] \quad (\text{so that } \forall \alpha \in \mathsf{dom}(\vartheta) \,.\, \widetilde{\vartheta}(\varrho)\alpha = \varrho(\vartheta(\alpha))) \tag{48.41}$$

For all $\tau_1^0, \tau_2^0, \tau_1, \tau_2 \in \mathbf{T}^{\mathsf{v}}$ and $\vartheta_0, \vartheta' \in \mathbb{V}_t \not\to \mathbf{T}^{\mathsf{v}}$, if $\mathsf{leq}(\tau_1, \tau_2, \vartheta_0)$ is (recursively) called from the main call $\mathsf{leq}(\tau_1^0, \tau_2^0)$ and returns $\langle b, \vartheta'\rangle = \mathsf{leq}(\tau_1, \tau_2, \vartheta_0)$, then

$$b \Rightarrow (\forall \varrho \in \mathbb{V}_t \to \mathbf{T} \,.\, \varrho(\tau_1) = \widetilde{\vartheta'}(\varrho)(\tau_2)). \tag{48.42}$$

It follows that $\mathsf{leq}(\tau_1^0, \tau_2^0) \Rightarrow ([\tau_1^0]_{\simeq^{\mathsf{v}}} \preceq_{\simeq^{\mathsf{v}}} [\tau_2^0]_{\simeq^{\mathsf{v}}})$.

Proof of lemma 48.40 The proof of (48.42) is by induction on the sequence of calls to leq and case analysis for the nontrivial case $b = \mathsf{tt}$.

- In case (48.33.2), by (48.33.1), (48.41), and $\alpha \in \mathsf{dom}(\vartheta_0[\alpha \leftarrow \tau_1])$, we have $\widetilde{\vartheta'}(\varrho)(\tau_2) = \widetilde{\vartheta'}(\varrho)(\alpha) = \varrho[\beta \leftarrow \varrho(\vartheta'(\beta)), \beta \in \mathsf{dom}(\vartheta')](\alpha) = \varrho[\beta \leftarrow \varrho(\vartheta_0[\alpha \leftarrow \tau_1](\beta)), \beta \in \mathsf{dom}(\vartheta_0[\alpha \leftarrow \tau_1])](\alpha) = \varrho(\vartheta_0[\alpha \leftarrow \tau_1](\alpha)) = \varrho(\tau_1)$;
- In case (48.33.4), by (48.33.3) and (48.41), we have $\alpha \in \mathsf{dom}(\vartheta_0)$ so that $\widetilde{\vartheta'}(\varrho)(\tau_2) = \widetilde{\vartheta_0}(\varrho)(\tau_2) = \widetilde{\vartheta_0}(\varrho)(\alpha) = \varrho[\beta \leftarrow \varrho(\vartheta_0(\beta)), \beta \in \mathsf{dom}(\vartheta_0)](\alpha) = \varrho(\vartheta_0(\alpha)) = \varrho(\tau_1)$ because $\vartheta_0(\alpha) = \tau_1$ by (48.33.3);
- In case (48.33.9), all b_i, $i \in [1, n]$ must be tt for b to be tt. We proceed by recurrence, with recurrence hypothesis $\forall \varrho \in \mathbb{V}_t \to \mathbf{T} \,.\, \forall j \in [1, i] \,.\, \widetilde{\vartheta_i}(\varrho)(\tau_2^j) = \varrho(\tau_1^j)$.

- For $i = 1$, $\widetilde{\vartheta_1}(\varrho)(\tau_2^1)) = \varrho(\tau_1^1)$ follows from the recursive call (48.33.6), by the induction on the sequence of calls to leq;
- Assume that the recurrence hypothesis holds for $i \in [1, n[$. After the next call a$\langle b_{i+1}, \vartheta_{i+1}\rangle = \mathsf{leq}(\tau_1^{i+1}, \tau_2^{i+1}, \vartheta_i)$, we have, by induction hypothesis (48.42) for this next call, that $\widetilde{\vartheta_{i+1}}(\varrho)(\tau_2^{i+1}) = \varrho(\tau_1^{i+1})$. In the remaining cases, $j \in [1, i+1[$, we have

$$\begin{aligned}
&\varrho(\tau_1^j)\\
=\ &\widetilde{\vartheta_i}(\varrho)(\tau_2^j) && \text{(recurrence hypothesis)}\\
=\ &\varrho[\alpha \leftarrow \varrho(\vartheta_i(\alpha)), \alpha \in \mathsf{dom}(\vartheta_i)](\tau_2^j) && \text{(48.41)}\\
=\ &\varrho[\alpha \leftarrow \varrho(\vartheta_{i+1}(\alpha)), \alpha \in \mathsf{dom}(\vartheta_{i+1})](\tau_2^j)
\end{aligned}$$

(by lemma 48.38, $\mathsf{dom}(\vartheta_i) = \mathsf{dom}(\vartheta_0) \cup \bigcup_{j=1}^{i} \mathsf{vars}[\![\tau_2^j]\!]$ so $\mathsf{vars}[\![\tau_2^j]\!] \subseteq \mathsf{dom}(\vartheta_i)$ and $\forall \alpha \in \mathsf{dom}(\vartheta_i) . \vartheta_i(\alpha) = \vartheta_{i+1}(\alpha)$. The proof is by structural induction on τ_2^j.

In the base case, $\tau_2^j = \beta \in \mathbb{V}_{\mathfrak{t}}$ and $\beta \in \mathsf{dom}(\vartheta_i) \subseteq \mathsf{dom}(\vartheta_{i+1})$ so $\varrho[\alpha \leftarrow \varrho(\vartheta_{i+1}(\alpha)), \alpha \in \mathsf{dom}(\vartheta_{i+1})](\beta) = \varrho(\vartheta_{i+1}(\beta)) = \varrho(\vartheta_i(\beta)) = \varrho[\alpha \leftarrow \varrho(\vartheta_i(\alpha)), \alpha \in \mathsf{dom}(\vartheta_i)](\beta)$.

For the induction step, $\tau_2^j = g(\tau_2^{j,1}, \ldots, \tau_2^{j,m})$. By (48.7) and induction hypothesis, $\varrho[\alpha \leftarrow \varrho(\vartheta_{i+1}(\alpha)), \alpha \in \mathsf{dom}(\vartheta_{i+1})](g(\tau_2^{j,1}, \ldots, \tau_2^{j,m}))$ $=$ $g(\varrho[\alpha \leftarrow \varrho(\vartheta_{i+1}(\alpha)), \alpha \in \mathsf{dom}(\vartheta_{i+1})](\tau_2^{j,1}), \ldots, \varrho[\alpha \leftarrow \varrho(\vartheta_{i+1}(\alpha)), \alpha \in \mathsf{dom}(\vartheta_{i+1})](\tau^{j,m})_2))$ $=$ $g(\varrho[\alpha \leftarrow \varrho(\vartheta_i(\alpha)), \alpha \in \mathsf{dom}(\vartheta_i)](\tau_2^{j,1}), \ldots, \varrho[\alpha \leftarrow \varrho(\vartheta_i(\alpha)), \alpha \in \mathsf{dom}(\vartheta_i)](\tau^{j,m})_2))$ $=$ $\varrho[\alpha \leftarrow \varrho(\vartheta_i(\alpha)), \alpha \in \mathsf{dom}(\vartheta_i)](g(\tau_2^{j,1}, \ldots, \tau_2^{j,m}))$.)

$$= \widetilde{\vartheta_i}(\varrho)(\tau_2^j) \qquad \text{(48.41)}$$

By recurrence, for $i = n$, we have $\forall \varrho \in \mathbb{V}_{\mathfrak{t}} \to \mathbf{T} . \forall j \in [1, n] . \widetilde{\vartheta_n}(\varrho)(\tau_2^j) = \tau_1^j$ hence $\widetilde{\vartheta_n}(\varrho)(f(\tau_2^1, \ldots, \tau_2^n)) = f(\tau_1^1, \ldots, \tau_1^n)$, so by (48.33.5), $\widetilde{\vartheta'}(\varrho) = \widetilde{\vartheta_n}(\varrho)(\tau_2) = \tau_1$, proving (48.42).

By (48.33.12), if $\mathsf{leq}(\tau_1, \tau_2) = \mathtt{tt}$ then there exists ϑ such that $\mathsf{leq}(\tau_1, \tau_2, \varepsilon) = \langle \mathtt{tt}, \vartheta \rangle$. So $\forall \varrho . \exists \varrho' . \varrho'(\tau_2) = \varrho(\tau_1)$ by choosing $\varrho' = \widetilde{\vartheta}(\varrho)$ which satisfies (48.42). By exercise 48.13, this implies $\tau_1 \preceq^{\mathrm{v}} \tau_2$, which, by exercise 48.16, implies $[\tau_1]_{\simeq^{\mathrm{v}}} \preceq_{\simeq^{\mathrm{v}}} [\tau_2]_{\simeq^{\mathrm{v}}}$. □

Let us prove the converse of lemma 48.40. The following lemma 48.43 shows the completeness of the subsumption algorithm $\mathsf{leq}(\tau_1, \tau_2)$ in figure 48.34. If τ_1 is smaller than or equal to τ_2 then it returns true ($\mathtt{tt}$).

Lemma 48.43 For all $\tau_1, \tau_2 \in \mathbf{T}^{\mathrm{v}}$ $([\tau_1]_{\simeq^{\mathrm{v}}} \preceq_{\simeq^{\mathrm{v}}} [\tau_2]_{\simeq^{\mathrm{v}}}) \Rightarrow \mathsf{leq}(\tau_1, \tau_2)$.

Proof of lemma 48.43 Let $\tau_1^0, \tau_2^0 \in \mathbf{T}^{\mathrm{v}}$ such that $[\tau_1^0]_{\simeq^{\mathrm{v}}} \preceq_{\simeq^{\mathrm{v}}} [\tau_2^0]_{\simeq^{\mathrm{v}}}$. By theorem 48.31, there exists a substitution $\vartheta \in \mathbb{\Sigma}$ such that $\vartheta(\tau_2^0) = \tau_1^0$.

Let us show, by induction on the sequence of calls to leq that for all $\tau_1, \tau_2 \in \mathbf{T}^{\mathrm{v}}$ and $\vartheta_0 \in \mathbb{V}_{\mathfrak{t}} \not\rightarrow \mathbf{T}^{\mathrm{v}}$, if $\mathsf{leq}(\tau_1, \tau_2, \vartheta_0)$ is (recursively) called from the main call $\mathsf{leq}(\tau_1^0, \tau_2^0)$ and returns $\langle b, \vartheta' \rangle$ then $\vartheta_0 = \vartheta \lceil \mathsf{dom}(\vartheta_0)$, $b = \mathtt{tt}$, and $\vartheta' = \vartheta \lceil \mathsf{dom}(\vartheta')$.

- For the basis (48.33.12), $\vartheta_0 = \varepsilon$ so $\mathsf{dom}(\vartheta_0) = \varnothing$ and therefore $\vartheta_0 = \vartheta \lceil \mathsf{dom}(\vartheta_0) = \varepsilon$. Moreover $\vartheta(\tau_2^0) = \tau_1^0$, $\tau_1 = \tau_1^0$, and $\tau_2 = \tau_2^0$ imply $\vartheta(\tau_2) = \tau_1$.

- For the induction step assume that $\vartheta_0 = \vartheta \lceil \mathsf{dom}(\vartheta_0)$ and $\vartheta(\boldsymbol{\tau}_2) = \boldsymbol{\tau}_1$. We proceed by cases.
  - In case (48.33.2), b is $\mathtt{tt}$. (48.33.1) implies that $\alpha \notin \mathsf{dom}(\vartheta_0)$ so the hypothesis $\vartheta(\boldsymbol{\tau}_2) = \vartheta(\alpha) = \boldsymbol{\tau}_1$ implies that $\vartheta' = \boldsymbol{\tau}_1[\alpha \leftarrow \vartheta_0] = \vartheta(\alpha)[\alpha \leftarrow (\vartheta \lceil \mathsf{dom}(\vartheta_0))] = \vartheta \lceil (\mathsf{dom}(\vartheta_0) \cup \{\alpha\}) = \vartheta \lceil \mathsf{dom}(\vartheta')$;
  - In case (48.33.4), b is $\mathtt{tt}$, by hypothesis $\vartheta_0 = \vartheta \lceil \mathsf{dom}(\vartheta_0)$ so $\vartheta' = \vartheta \lceil \mathsf{dom}(\vartheta)$ because $\vartheta' = \vartheta_0$;
  - In case (48.33.9), we proceed by recurrence.
    - For the basis, $i = 1$, we have $\vartheta_0 = \vartheta \lceil \mathsf{dom}(\vartheta_0)$ by hypothesis, so the recursive call $\mathsf{leq}(\boldsymbol{\tau}_1^1, \boldsymbol{\tau}_2^1, \vartheta_0)$ returns $\langle b_1, \vartheta_1 \rangle$ such that $b_1 = \mathtt{tt}$ and $\vartheta_1 = \vartheta \lceil \mathsf{dom}(\vartheta_1)$.
    - Assume, by recurrence hypothesis, that $\vartheta_i = \vartheta \lceil \mathsf{dom}(\vartheta_i)$, $i \in [1, n[$. Then, by induction hypothesis on the sequence of calls to leq, the next call $\mathsf{leq}(\boldsymbol{\tau}_1^{i+1}, \boldsymbol{\tau}_2^{i+1}, \vartheta_i)$ returns $\langle b_{i+1}, \vartheta_{i+1} \rangle$ such that $b_{i+1} = \mathtt{tt}$ and $\vartheta_{i+1} = \vartheta \lceil \mathsf{dom}(\vartheta_{i+1})$.

  By recurrence for (48.33.9), we have $b = b_1 \wedge b_2 \wedge \cdots \wedge b_n = \mathtt{tt}$ and $\vartheta_n = \vartheta \lceil \mathsf{dom}(\vartheta_n)$ so $\vartheta' = \vartheta \lceil \mathsf{dom}(\vartheta')$ because $\vartheta' = \vartheta_n$. □

By induction on the sequence of calls to leq, we conclude that $b = \mathtt{tt}$ for the main call at (48.33.12). □

We conclude that the leq algorithm (48.33) is totally correct.

Theorem 48.44 (subsumption ordering) For all $\boldsymbol{\tau}_1, \boldsymbol{\tau}_2 \in \mathbf{T}^{\mathrm{v}}$, the $\mathsf{leq}(\boldsymbol{\tau}_1, \boldsymbol{\tau}_2)$ algorithm (48.33) always terminates and returns $\mathsf{leq}(\boldsymbol{\tau}_1, \boldsymbol{\tau}_2) = ([\boldsymbol{\tau}_1]_{\simeq^{\mathrm{v}}} \preceq_{\simeq^{\mathrm{v}}} [\boldsymbol{\tau}_2]_{\simeq^{\mathrm{v}}})$.

Proof of theorem 48.44 By lemmas 48.37, 48.40, and 48.43, and antisymmetry. □

48.8.2 Greatest Lower Bound: the Unification and Greatest Common Instance Algorithms

48.8.2.1 The greatest common instance algorithm is based on unification. By the Galois retraction (48.27), the greatest lower bound in the lattice corollary 48.28 is $gci([\boldsymbol{\tau}_1]_{\simeq^{\mathrm{v}}}, [\boldsymbol{\tau}_2]_{\simeq^{\mathrm{v}}}) \triangleq lcg_{\simeq^{\mathrm{v}}}[\mathbf{v}](ground_{\simeq^{\mathrm{v}}}([\boldsymbol{\tau}_1]_{\simeq^{\mathrm{v}}}) \cap ground_{\simeq^{\mathrm{v}}}([\boldsymbol{\tau}_1]_{\simeq^{\mathrm{v}}}))$. By the Galois retraction (48.27), this is equivalent to requiring that $ground_{\simeq^{\mathrm{v}}}([\boldsymbol{\tau}_1]_{\simeq^{\mathrm{v}}}, [\boldsymbol{\tau}_2]_{\simeq^{\mathrm{v}}}) = ground_{\simeq^{\mathrm{v}}}([\boldsymbol{\tau}_1]_{\simeq^{\mathrm{v}}}) \cap ground_{\simeq^{\mathrm{v}}}([\boldsymbol{\tau}_1]_{\simeq^{\mathrm{v}}})$. Let us evaluate

$$\begin{aligned}
& ground_{\simeq^{\mathrm{v}}}([\boldsymbol{\tau}_1]_{\simeq^{\mathrm{v}}}) \cap ground_{\simeq^{\mathrm{v}}}([\boldsymbol{\tau}_2]_{\simeq^{\mathrm{v}}}) \\
= \ & ground(\boldsymbol{\tau}_1) \cap ground(\boldsymbol{\tau}_2) && \text{(exercise 48.16)} \\
= \ & \{\varrho(\boldsymbol{\tau}_1) \mid \varrho \in \mathbf{P}^{\mathrm{v}}\} \cap \{\varrho(\boldsymbol{\tau}_2) \mid \varrho \in \mathbf{P}^{\mathrm{v}}\} && \text{((48.10))} \\
= \ & \{\varrho(\boldsymbol{\tau}_1) \mid \varrho, \varrho' \in \mathbf{P}^{\mathrm{v}} \wedge \varrho(\boldsymbol{\tau}_1) = \varrho'(\boldsymbol{\tau}_2)\} && \text{(definition of } \cap) \\
= \ & \{\varrho \rceil \mathsf{vars}[\![\boldsymbol{\tau}_1]\!](\boldsymbol{\tau}_1) \mid \varrho, \varrho' \in \mathbf{P}^{\mathrm{v}} \wedge \varrho \rceil \mathsf{vars}[\![\boldsymbol{\tau}_1]\!](\boldsymbol{\tau}_1) = \varrho' \rceil \mathsf{vars}[\![\boldsymbol{\tau}_2]\!](\boldsymbol{\tau}_2)\}
\end{aligned}$$

(by (48.7), so that the assignment can be restricted to the variables of the term it is applied to)

$$= \ \{\varrho''(\boldsymbol{\tau}_1) \mid \varrho'' \in \mathbf{P}^{\mathrm{v}} \wedge \varrho''(\boldsymbol{\tau}_1) = \varrho''(\boldsymbol{\tau}_2)\}$$

≀assuming $\mathsf{vars}[\![\boldsymbol{\tau}_1]\!] \cap \mathsf{vars}[\![\boldsymbol{\tau}_2]\!] = \varnothing$ and defining $\boldsymbol{\varrho}''(\alpha) \triangleq \boldsymbol{\varrho}(\alpha)$ when $\alpha \in \mathsf{vars}[\![\boldsymbol{\tau}_1]\!]$ and $\boldsymbol{\varrho}''(\alpha) \triangleq \boldsymbol{\varrho}'(\alpha)$ when $\alpha \in \mathsf{vars}[\![\boldsymbol{\tau}_2]\!]$. It follows, by (48.7), that $\boldsymbol{\varrho}''(\boldsymbol{\tau}_1) = \boldsymbol{\varrho}(\boldsymbol{\tau}_1)$ and $\boldsymbol{\varrho}''(\boldsymbol{\tau}_2) = \boldsymbol{\varrho}'(\boldsymbol{\tau}_2)$.∫

Because knowing $\boldsymbol{\varrho}'$, it is easy to compute $\boldsymbol{\varrho}'(\boldsymbol{\tau})$, we look for

$$\{\boldsymbol{\varrho}' \mid \boldsymbol{\varrho}' \in \mathbf{P}^{\mathcal{V}} \wedge \boldsymbol{\varrho}'(\boldsymbol{\tau}_1) = \boldsymbol{\varrho}'(\boldsymbol{\tau}_2)\}$$

$$= \{\boldsymbol{\varrho} \circ \vartheta \mid \boldsymbol{\varrho} \in \mathbf{P}^{\mathcal{V}} \wedge \vartheta \in \mathbb{\Sigma} \wedge \boldsymbol{\varrho} \circ \vartheta(\boldsymbol{\tau}_1) = \boldsymbol{\varrho} \circ \vartheta(\boldsymbol{\tau}_2)\} \tag{48.45}$$

≀($\subseteq$) by choosing $\vartheta = \boldsymbol{\varepsilon}$ to be the identity;

($\supseteq$) by choosing $\boldsymbol{\varrho}' = \boldsymbol{\varrho} \circ \vartheta$, applying (48.30) first and then (48.7) for the composed application to terms with variables∫

Reasoning in the symbolic abstract domain corollary 48.28, given terms with variables $\boldsymbol{\tau}_1$ and $\boldsymbol{\tau}_2$ with no common variable, (48.45) shows that we have to find a substitution ϑ making them equal $\vartheta(\boldsymbol{\tau}_1) = \vartheta(\boldsymbol{\tau}_2)$ (hence $[\vartheta(\boldsymbol{\tau}_1)]_{\doteq^{\mathcal{V}}} = [\vartheta(\boldsymbol{\tau}_2)]_{\doteq^{\mathcal{V}}}$ by exercise 48.16). This may not be possible (as in $f(a, \alpha) = f(b, \beta)$ where $a, b \in$ and $a \neq b$). In that case we choose ϑ to be the error substitution $\Omega^{\mathrm{r}}_{\mathrm{s}}$. Formally, the error substitution $\Omega^{\mathrm{r}}_{\mathrm{s}}$ can be defined as ($f \in \mathbf{F}_1$)

$$\Omega^{\mathrm{r}}_{\mathrm{s}} \quad \triangleq \quad X \in \mathbb{V}_{\mathfrak{t}} \mapsto f(X) \tag{48.46}$$

which concretization $\gamma^{\mathrm{r}}_{\mathrm{s}}(\Omega^{\mathrm{r}}_{\mathrm{s}}) = \varnothing$ is, by lemma 48.9, empty.

48.8.2.2 Herbrand unification algorithm. This equation solving problem was first stated and solved by Jacques Herbrand in his thesis [487, ch. 5, 2.4, pp. 96–97][3] with a recursive procedure, by structural induction on the terms to produce "normal equalities" of a variable to a term (that is, the substitution ϑ where the "normal equalities" are $\alpha = \vartheta(\alpha)$). There may be several solutions (e.g., $\vartheta(\alpha) = \vartheta(\beta) = \gamma$ or $\vartheta(\alpha) = \vartheta(\beta) = \vartheta(\gamma) = a$ for $f(x, y) = f(z, z)$) but Jacques Herbrand's algorithm produces the most general one (of which all others are instances).

Alan Robinson gave a name (*unification*) to the problem and its solution (*most general unifier* also called *principal unifier*) and proposed another algorithm [838], [839, ch. 11, pp. 182-198]. Instead of a recursion by structural induction, Alan Robinson's algorithm [839, p. 191], [800, p. 155], [827, pp. 138-139] looks for common paths in terms that lead to different subterms, the paths being explored in some predefined order, by an algorithm called the "difference" [839, p. 188]. Alberto Martelli and Ugo Montanari rephrased Jacques Herbrand's algorithm in figure 48.48 using deductive rules [660].

In case of failure at (48.47.2), the algorithm returns the error substitution $\Omega^{\mathrm{r}}_{\mathrm{s}}$ but could as well raise an exception. The algorithm can also be optimized, for example, using a stack for derecursivisation [912, p. 90]. Expressed as a fixpoint and translated into deductive rules by theorem 16.16, one gets [660].

Lines (48.47.1–6) are explained by Jacques Herbrand as follows: "If one of the equations to be satisfied equates $f(\boldsymbol{\tau}_1^1, \ldots, \boldsymbol{\tau}_1^n)$ to $g(\boldsymbol{\tau}_2^1, \ldots, \boldsymbol{\tau}_2^m)$, either the elementary functions f and g are different, and the equation cannot be satisfied [(48.47.2)], or

3. Page 148 of the English translation [486].

Herbrand unification algorithm ($\boldsymbol{\tau}_1, \boldsymbol{\tau}_2 \in \mathbf{T}^{\mathsf{v}}$ *and* $T_0 \in \Sigma$)

let rec unify$(\boldsymbol{\tau}_1, \boldsymbol{\tau}_2, \vartheta_0) =$ (48.47)
  if $\boldsymbol{\tau}_1 = f(\boldsymbol{\tau}_1^1, \ldots, \boldsymbol{\tau}_1^n) \wedge \boldsymbol{\tau}_2 = g(\boldsymbol{\tau}_2^1, \ldots, \boldsymbol{\tau}_2^m)$ then (1)
    if $f \neq g$ then Ω_s^r (2)
    else let $\vartheta_1 =$ unify$(\boldsymbol{\tau}_1^1, \boldsymbol{\tau}_2^1, \vartheta_0)$ in /⋆ term decomposition ⋆/ (3)
      let $\vartheta_2 =$ unify$(\boldsymbol{\tau}_1^2, \boldsymbol{\tau}_2^2, \vartheta_1)$ in (4)
      … …
      let $\vartheta_n =$ unify$(\boldsymbol{\tau}_1^n, \boldsymbol{\tau}_2^n, \vartheta_{n-1})$ in (5)
        if $\exists i \in [1, n] \,.\, \vartheta_i = \Omega_s^r$ then Ω_s^r else ϑ_n (6)
  elsif $\boldsymbol{\tau}_1 = \boldsymbol{\tau}_2 = \alpha \in \mathbb{V}_t$ then ϑ_0 /⋆ variable erasure ⋆/ (7)
  elsif $\boldsymbol{\tau}_1 = \alpha \in \mathbb{V}_t \wedge \alpha \in \mathsf{vars}[\![\boldsymbol{\tau}_2]\!]$ then Ω_s^r /⋆ occurs check ⋆/ (8)
  elsif $\boldsymbol{\tau}_1 = \alpha \in \mathbb{V}_t$ then (9)
    if $\alpha \notin$ dom(ϑ_0) then (10)
      $\{\langle \alpha, \boldsymbol{\tau}_2 \rangle\} \circ \vartheta_0$ /⋆ variable elimination ⋆/ (11)
    else unify$(\vartheta_0(\alpha), \boldsymbol{\tau}_2, \vartheta_0)$ (12)
  else unify$(\boldsymbol{\tau}_2, \boldsymbol{\tau}_1, \vartheta_0)$ /⋆ equation orientation, $\boldsymbol{\tau}_2 = \beta \in \mathbb{V}_t$ ⋆/ (13)
let mgu$(\boldsymbol{\tau}_1, \boldsymbol{\tau}_2) =$ unify$(\boldsymbol{\tau}_1, \boldsymbol{\tau}_2, \varepsilon)$ (14)
let gci$(\boldsymbol{\tau}_1, \boldsymbol{\tau}_2) =$ /⋆ precondition: $\mathsf{vars}[\![\boldsymbol{\tau}_1]\!] \cap \mathsf{vars}[\![\boldsymbol{\tau}_2]\!] = \varnothing$ ⋆/ (15)
  if $\boldsymbol{\tau}_1 = \overline{\varnothing}^{\mathsf{v}} \vee \boldsymbol{\tau}_2 = \overline{\varnothing}^{\mathsf{v}}$ then $\overline{\varnothing}^{\mathsf{v}}$ (16)
  else let $\vartheta =$ mgu$(\boldsymbol{\tau}_1, \boldsymbol{\tau}_2)$ in (17)
    if $\vartheta = \Omega_s^r$ then $\overline{\varnothing}^{\mathsf{v}}$ else $\vartheta(\boldsymbol{\tau}_1)$ /⋆ $= \vartheta(\boldsymbol{\tau}_2)$ ⋆/ (18)

Figure 48.48
The greatest lower bound algorithm

they are the same, and then we turn to those equations that equate the $\boldsymbol{\tau}_1^j$ to the $\boldsymbol{\tau}_2^j$ [(48.47.3–6)]."

The explanation of Jacques Herbrand for lines (48.47.8–13) is "If one of the equalities to be satisfied equates a restraint variable α to an individual [$\boldsymbol{\tau}_2$], either this individual contains α and the equation cannot be satisfied [occurs check at (48.47.8)], or else the individual does not contain α, and then the equation will be one of the associated equations [$\alpha = \vartheta(\alpha)$] that we are looking for [(48.47.11)]; and we shall replace α by this function [$\vartheta(\alpha)$] in the other equalities to be satisfied [(48.47.12)]."

Remark 48.49 Jacques Herbrand considers equations as unordered pairs whereas in (48.47), the parameters are ordered as usual in programming languages. Therefore (48.47.13) has to consider the symmetric case, that could also be handled by duplicating lines (48.47.8–12) to avoid the recursive call (as will be considered in the following proofs).

Example 48.50

$\mathsf{unify}(f(\alpha, g(\beta), \beta, \alpha), f(g(\alpha'), \beta', \gamma', g(\delta')), \varepsilon)$

$\quad\mathsf{unify}(\alpha, g(\alpha'), \varepsilon)$ (48.47.3)

$\quad = \{\langle\alpha,\ g(\alpha')\rangle\} \circ \varepsilon$ (48.47.11)

$\quad = \{\langle\alpha,\ g(\alpha')\rangle\}$

$\quad\mathsf{unify}(g(\beta), \beta', \{\langle\alpha,\ g(\alpha')\rangle\})$ (48.47.4)

$\quad\quad\mathsf{unify}(\beta', g(\beta), \{\langle\alpha,\ g(\alpha')\rangle\})$ (48.47.13)

$\quad\quad = \{\langle\alpha,\ g(\alpha')\rangle, \langle\beta',\ g(\beta)\rangle\}$ (48.47.11)

$\quad = \{\langle\alpha,\ g(\alpha')\rangle, \langle\beta',\ g(\beta)\rangle\}$

$\quad\mathsf{unify}(\beta, \gamma', \{\langle\alpha,\ g(\alpha')\rangle, \langle\beta',\ g(\beta)\rangle\})$ (48.47....)

$\quad = \{\langle\beta,\ \gamma'\rangle\} \circ \{\langle\alpha,\ g(\alpha')\rangle, \langle\beta',\ g(\beta)\rangle\}$ (48.47.11)

$\quad = \{\langle\alpha,\ g(\alpha')\rangle, \langle\beta',\ g(\gamma')\rangle, \langle\beta,\ \gamma'\rangle\}$

$\quad\mathsf{unify}(\alpha, g(\delta'), \{\langle\alpha,\ g(\alpha')\rangle, \langle\beta',\ g(\gamma')\rangle, \langle\beta,\ \gamma'\rangle\})$ (48.47.5)

$\quad\quad\mathsf{unify}(g(\alpha'), g(\delta'), \{\langle\alpha,\ g(\alpha')\rangle, \langle\beta',\ g(\gamma')\rangle, \langle\beta,\ \gamma'\rangle\})$ (48.47.12)

$\quad\quad\quad\mathsf{unify}(\alpha', \delta', \{\langle\alpha,\ g(\alpha')\rangle, \langle\beta',\ g(\gamma')\rangle, \langle\beta,\ \gamma'\rangle\})$ (48.47.2)

$\quad\quad\quad = \{\langle\alpha,\ g(\delta')\rangle, \langle\beta',\ g(\gamma')\rangle, \langle\beta,\ \gamma'\rangle, \langle\alpha',\ \delta'\rangle\}$ (48.47.11)

$\quad\quad = \{\langle\alpha,\ g(\delta')\rangle, \langle\beta',\ g(\gamma')\rangle, \langle\beta,\ \gamma'\rangle, \langle\alpha',\ \delta'\rangle\}$ (48.47.6)

$\quad = \{\langle\alpha,\ g(\delta')\rangle, \langle\beta',\ g(\gamma')\rangle, \langle\beta,\ \gamma'\rangle, \langle\alpha',\ \delta'\rangle\}$

$= \{\langle\alpha,\ g(\delta')\rangle, \langle\beta',\ g(\gamma')\rangle, \langle\beta,\ \gamma'\rangle, \langle\alpha',\ \delta'\rangle\}$ (48.47.6)

This example shows that the algorithm (48.47) would be incorrect if the precondition $\mathsf{vars}[\![\boldsymbol{\tau}_1]\!] \cap \mathsf{vars}[\![\boldsymbol{\tau}_2]\!] = \varnothing$ were not satisfied (e.g., if $\alpha = \alpha'$ then $\mathsf{unify}(\alpha, g(\alpha'), \varepsilon)$ is an occurs check). The case $\langle\beta',\ g(\gamma')\rangle$ also shows that the substitution may involve a map of the variables of one term into subterms of this same term.

48.8.2.3 Equations, most general unifier. After Jacques Herbrand, unification is understood as solving equations $\boldsymbol{\tau}_1 \doteq \boldsymbol{\tau}_2$ [4] between terms with variables $\boldsymbol{\tau}_1$ and $\boldsymbol{\tau}_2$ and producing a solution in the form of a substitution ϑ ("normal equations," that is, solved equations $\{\alpha \doteq \vartheta(\alpha) \mid \alpha \in \mathsf{dom}(\vartheta)\}$ in Herbrand's terminology).

Observe that by the occurs check (48.47.8), the substitution returned by unification is always idempotent, that is, $\vartheta \circ \vartheta = \vartheta$. A nonidempotent substitution would contradict the fact that terms are finite (by lemma 48.9).

The meaning $\gamma_{\mathrm{e}}(\boldsymbol{\tau}_1 \doteq \boldsymbol{\tau}_2)$ of the equation $\boldsymbol{\tau}_1 \doteq \boldsymbol{\tau}_2$ is the set of *all* its ground solutions.

$$\gamma_{\mathrm{e}}(\boldsymbol{\tau}_1 \doteq \boldsymbol{\tau}_2) \triangleq \{\boldsymbol{\varrho} \in \mathbf{P}^{\mathbb{V}} \mid \boldsymbol{\varrho}(\boldsymbol{\tau}_1) = \boldsymbol{\varrho}(\boldsymbol{\tau}_2)\}. \tag{48.51}$$

A symbolic solution to the equation $\boldsymbol{\tau}_1 \doteq \boldsymbol{\tau}_2$ (more precisely an underapproximation of the solution to this equation) is a substitution $\vartheta \in \Sigma \triangleq \mathbb{V}_{\mathfrak{t}} \not\rightarrow \mathbf{T}^{\mathbb{V}}$, understood as a solved equation, that is, with relational concretization

$$\begin{aligned} \gamma_{\mathrm{s}}^{\mathrm{r}}(\vartheta) &\triangleq \gamma_{\mathrm{e}}(\{\alpha \doteq \vartheta(\alpha) \mid \alpha \in \mathsf{dom}(\vartheta)\}) \\ &= \{\boldsymbol{\varrho} \in \mathbf{P}^{\mathbb{V}} \mid \forall \alpha \in \mathbb{V}_{\mathfrak{t}}\,.\,\boldsymbol{\varrho}(\alpha) = \boldsymbol{\varrho}(\vartheta(\alpha))\} \qquad (48.52) \\ &= \{\boldsymbol{\varrho} \in \mathbf{P}^{\mathbb{V}} \mid \forall \boldsymbol{\tau} \in \mathbf{T}^{\mathbb{V}}\,.\,\boldsymbol{\varrho}(\boldsymbol{\tau}) = \boldsymbol{\varrho}(\vartheta(\boldsymbol{\tau}))\}^{5} \end{aligned}$$

characterizing *some* of the ground solutions to the equations,

$$\gamma_{\mathrm{s}}^{\mathrm{r}}(\vartheta) \subseteq \gamma_{\mathrm{e}}(\boldsymbol{\tau}_1 \doteq \boldsymbol{\tau}_2). \tag{48.53}$$

4. We use the symbol $\doteq$ and not $=$ because, in general, $\boldsymbol{\tau}_1$ and $\boldsymbol{\tau}_2$ are different.

5. This can be proved by structural induction on $\boldsymbol{\tau} \in \mathbf{T}^{\mathbb{V}}$ using (48.7) and (48.30).

The *most general unifier* $\mathsf{mgu}(\boldsymbol{\tau}_1 \doteq \boldsymbol{\tau}_2)$ is a substitution exactly characterizing all ground solutions to the equations, that is, $\gamma^{\mathrm{r}}_{\mathrm{s}}(\mathsf{mgu}(\boldsymbol{\tau}_1 \doteq \boldsymbol{\tau}_2)) = \gamma_{\mathrm{e}}(\boldsymbol{\tau}_1 \doteq \boldsymbol{\tau}_2)$. It is the most general in that any other solution ϑ' satisfying (48.53) will have $\gamma^{\mathrm{r}}_{\mathrm{s}}(\vartheta') \subseteq \gamma^{\mathrm{r}}_{\mathrm{s}}(\mathsf{mgu}(\boldsymbol{\tau}_1 \doteq \boldsymbol{\tau}_2))$. Of course the most general unifier is not unique and must be understood up to the congruence $\simeq^{\mathrm{v}}$ (48.14) on terms.

48.8.2.4 Proof of total correctness of the unify, mgu, and gci algorithms. The correctness proof involves a number of preliminary definitions and results that involve properties of the successive calls to unify. A *preinvariant* is an assertion which is always true before the call. It depends on the parameters $\boldsymbol{\tau}_1$, $\boldsymbol{\tau}_2$, ϑ_0 of the call but also on the parameters $\boldsymbol{\tau}^0_1$, $\boldsymbol{\tau}^0_2$ of the main call at (48.47.18). The *postinvariant* is an assertion which is always after before the call. It depends on the parameters of the main call $\boldsymbol{\tau}^0_1$, $\boldsymbol{\tau}^0_2$ at (48.47.18), the parameters $\boldsymbol{\tau}_1$, $\boldsymbol{\tau}_2$ of the call, and the results ϑ returned by the function call.

Let us define

$$\mathsf{covars}[\![\vartheta]\!] \quad \triangleq \quad \bigcup\{\mathsf{vars}[\![\vartheta(\alpha)]\!] \mid \alpha \in \mathsf{dom}(\vartheta)\} \tag{48.54}$$

to be the set of variables found in the terms belonging to the image of the domain of the substitution ϑ.

The following lemma 48.55, states that the variables in intermediate terms and substitutions are always those of the parameters of the main call at (48.47.18). Moreover, ϑ_0 is idempotent (owing to the occurs check).

Lemma 48.55 For all $\boldsymbol{\tau}^0_1, \boldsymbol{\tau}^0_2, \boldsymbol{\tau}_1, \boldsymbol{\tau}_2 \in \mathbf{T}^{\mathrm{v}}$ and $\vartheta_0, \vartheta \in \mathbb{V}_{t} \nrightarrow \mathbf{T}^{\mathrm{v}}$, if $\mathsf{unify}(\boldsymbol{\tau}_1, \boldsymbol{\tau}_2, \vartheta_0)$ is (recursively) called from the main call $\mathsf{mgu}(\boldsymbol{\tau}^0_1, \boldsymbol{\tau}^0_2)$ with $\mathsf{vars}[\![\boldsymbol{\tau}^0_1]\!] \cap \mathsf{vars}[\![\boldsymbol{\tau}^0_2]\!] = \varnothing$ and $\mathsf{unify}(\boldsymbol{\tau}_1, \boldsymbol{\tau}_2, \vartheta_0)$ returns $\vartheta \neq \Omega^{\mathrm{r}}_{\mathrm{s}}$, then

$$\begin{aligned} \text{preinvariant} \quad & (\mathsf{vars}[\![\boldsymbol{\tau}_1]\!] \cup \mathsf{vars}[\![\boldsymbol{\tau}_2]\!]) \subseteq (\mathsf{vars}[\![\boldsymbol{\tau}^0_1]\!] \cup \mathsf{vars}[\![\boldsymbol{\tau}^0_2]\!]) \wedge \\ & (\mathsf{dom}(\vartheta_0) \cup \mathsf{covars}[\![\vartheta_0]\!]) \subseteq (\mathsf{vars}[\![\boldsymbol{\tau}^0_1]\!] \cup \mathsf{vars}[\![\boldsymbol{\tau}^0_2]\!]) \wedge \\ & \mathsf{dom}(\vartheta_0) \cap \mathsf{covars}[\![\vartheta_0]\!] = \varnothing \end{aligned} \tag{48.56}$$

$$\begin{aligned} \text{postinvariant} \quad & (\mathsf{dom}(\vartheta) \cup \mathsf{covars}[\![\vartheta]\!]) \subseteq (\mathsf{vars}[\![\boldsymbol{\tau}^0_1]\!] \cup \mathsf{vars}[\![\boldsymbol{\tau}^0_2]\!]) \wedge \\ & \mathsf{dom}(\vartheta) \cap \mathsf{covars}[\![\vartheta]\!] = \varnothing \end{aligned} \tag{48.57}$$

Proof of lemma 48.55 By induction on the sequence of calls to unify and, for any given call, by recurrence to handle the recursive calls at (48.47.3), …, (48.47.5), and by case analysis. The cases where $\vartheta \neq \Omega^{\mathrm{r}}_{\mathrm{s}}$ are as follows:

- In case (48.47.6), $(\mathsf{vars}[\![\boldsymbol{\tau}^i_1]\!] \cup \mathsf{vars}[\![\boldsymbol{\tau}^i_2]\!]) \subseteq (\mathsf{vars}[\![\boldsymbol{\tau}_1]\!] \cup \mathsf{vars}[\![\boldsymbol{\tau}_2]\!]) \subseteq (\mathsf{vars}[\![\boldsymbol{\tau}^0_1]\!] \cup \mathsf{vars}[\![\boldsymbol{\tau}^0_2]\!])$, $i \in [1, n]$ by definition (48.3) of vars and preinvariant (48.56).
  We have $(\mathsf{dom}(\vartheta_i) \cup \mathsf{covars}[\![\vartheta_i]\!]) \subseteq (\mathsf{vars}[\![\boldsymbol{\tau}^0_1]\!] \cup \mathsf{vars}[\![\boldsymbol{\tau}^0_2]\!])$ and $\mathsf{dom}(\vartheta_i) \cap \mathsf{covars}[\![\vartheta_i]\!] = \varnothing$, $i \in [1, n]$ by recurrence. For the basis $i = 0$, by the preinvariant (48.56). For the recurrence step, by the postinvariant (48.57). The result ϑ satisfies the postinvariant (48.57) because $\vartheta = \vartheta_n$;
- In cases (48.47.7) and (48.47.12), $\vartheta = \vartheta_0$ so the postinvariant directly follows from the preinvariant.

- In case (48.47.11), we have $\vartheta = \{\langle\alpha, \boldsymbol{\tau}_2\rangle\} \circ \vartheta_0$ so $\mathsf{dom}(\vartheta) = \mathsf{dom}(\vartheta_0) \cup \{\alpha\}$ and $\mathsf{covars}[\![\vartheta]\!] \subseteq (\mathsf{covars}[\![\vartheta_0]\!] \cup \mathsf{vars}[\![\boldsymbol{\tau}_2]\!])$. It follows that $\mathsf{dom}(\vartheta) \cup \mathsf{covars}[\![\vartheta]\!] \subseteq (\mathsf{vars}[\![\boldsymbol{\tau}_1^0]\!] \cup \mathsf{vars}[\![\boldsymbol{\tau}_2^0]\!])$ by the preinvariant (48.56).
  Moreover, $\mathsf{dom}(\vartheta) \cap \mathsf{covars}[\![\vartheta]\!] \subseteq (\mathsf{dom}(\vartheta_0) \cup \{\alpha\}) \cap (\mathsf{covars}[\![\vartheta_0]\!] \cup \mathsf{vars}[\![\boldsymbol{\tau}_2]\!]) = \varnothing$ by the preinvariant (48.56) and the test (48.47.8) so $\alpha \notin \mathsf{vars}[\![\boldsymbol{\tau}_2]\!]$, and the test (48.47.10) so that $\alpha \notin \mathsf{dom}(\vartheta_0)$.
- The case (48.47.13) is handled by symmetry, as in remark 48.49. □

Lemma 48.58 The program gci is well-typed and commutative, and it terminates.

Proof of lemma 48.58 unify has type $(\mathbf{T}^\mathcal{V} \times \mathbf{T}^\mathcal{V} \times (\mathcal{V}_{\mathfrak{t}} \nrightarrow \mathbf{T}^\mathcal{V})) \rightarrow (\mathcal{V}_{\mathfrak{t}} \nrightarrow \mathbf{T}^\mathcal{V})$, mgu has type $(\mathbf{T}^\mathcal{V} \times \mathbf{T}^\mathcal{V}) \rightarrow (\mathcal{V}_{\mathfrak{t}} \nrightarrow \mathbf{T}^\mathcal{V})$ and gci has type $(\mathbf{T}^\mathcal{V} \times \mathbf{T}^\mathcal{V}) \rightarrow \mathbf{T}^\mathcal{V}$. The code is symmetric in $\boldsymbol{\tau}_1$ and $\boldsymbol{\tau}_2$, proving commutativity $\mathsf{unify}(\boldsymbol{\tau}_1, \boldsymbol{\tau}_2, \vartheta_0) = \mathsf{unify}(\boldsymbol{\tau}_2, \boldsymbol{\tau}_1, \vartheta_0)$, $\mathsf{mgu}(\boldsymbol{\tau}_1, \boldsymbol{\tau}_2) = \mathsf{mgu}(\boldsymbol{\tau}_2, \boldsymbol{\tau}_1)$, and $\mathsf{gci}(\boldsymbol{\tau}_1, \boldsymbol{\tau}_2) = \mathsf{gci}(\boldsymbol{\tau}_2, \boldsymbol{\tau}_1)$ (which we already know from corollary 48.28).

By (48.47.14), we must show that the call of $\mathsf{unify}(\boldsymbol{\tau}_1, \boldsymbol{\tau}_2, \varepsilon)$ does terminate. By reductio ad absurdum, assume there is an infinite sequence of calls $\mathsf{unify}(\boldsymbol{\tau}_1^0, \boldsymbol{\tau}_2^0, \varepsilon)$, …, $\mathsf{unify}(\boldsymbol{\tau}_1^i, \boldsymbol{\tau}_2^i, \vartheta_i)$, $\mathsf{unify}(\boldsymbol{\tau}_1^{i+1}, \boldsymbol{\tau}_2^{i+1}, \vartheta_{i+1})$, ….

If the recursive call at rank i is one of (48.47.3), …, (48.47.5), we have $\boldsymbol{\tau}_1^{i+1} \lhd \boldsymbol{\tau}_1^i$ and $\boldsymbol{\tau}_2^{i+1} \lhd \boldsymbol{\tau}_2^i$. If it is at (48.47.12), we have (by tests (48.47.9) and (48.47.10))

$$\boldsymbol{\tau}_1^i = \alpha^i \in \mathsf{dom}(\vartheta_i) \wedge \boldsymbol{\tau}_1^{i+1} = \vartheta_i(\alpha^i) \tag{48.59}$$

and symmetrically for (48.47.13). Because the strict subcomponent/subterm $\lhd$ is well founded, we can find in the infinite sequence of calls, from some rank k on, an infinite sequence of cases (48.59) (or its symmetric). By (48.47.12) (or symmetrically (48.47.13)), from that rank k on, $\vartheta_i = \vartheta_k$ is unchanged. So we can extract from the infinite sequence of calls an infinite subsequence

$$\boldsymbol{\tau}_1^i = \alpha^i \in \mathsf{dom}(\vartheta_k) \wedge \boldsymbol{\tau}_1^{i+1} = \vartheta_k(\alpha^i), \quad i \geqslant k$$

Because, by (48.56) of lemma 48.55, $(\mathsf{dom}(\vartheta_k) \cup \mathsf{covars}[\![\vartheta_k]\!]) \subseteq (\mathsf{vars}[\![\boldsymbol{\tau}_1^0]\!] \cup \mathsf{vars}[\![\boldsymbol{\tau}_2^0]\!])$ is finite, this implies that there exists a circularity $\alpha^{i+1} \in \vartheta_k(\alpha^i), \alpha^{i+2} \in \vartheta_k(\alpha^{i+1}), \ldots, \alpha^{i+j+1} \in \vartheta_k(\alpha^{i+j})$ with $\alpha^{i+j+1} = \alpha^i$, in contradiction to $\mathsf{dom}(\vartheta_k) \cap \mathsf{covars}[\![\vartheta_k]\!] = \varnothing$ by (48.56) of lemma 48.55. □

Exercise 48.60 Prove that for all variables $\alpha \in \mathcal{V}_{\mathfrak{t}}$ and terms with variables $\boldsymbol{\tau}, \boldsymbol{\tau}' \in \mathbf{T}^\mathcal{V}$, we have $\{\langle\alpha, \boldsymbol{\tau}\rangle\}(\boldsymbol{\tau}') = \boldsymbol{\tau}'[\alpha \leftarrow \boldsymbol{\tau}]$.

Let us write $\boldsymbol{\tau}[\beta \in \mathsf{vars}[\![\boldsymbol{\tau}]\!] \leftarrow \vartheta(\beta)]$ for the simultaneous functional substitution of all variables $\beta \in \mathsf{vars}[\![\boldsymbol{\tau}]\!]$ of $\boldsymbol{\tau}$ by $\vartheta(\beta)$ that is

$$\begin{aligned}
\alpha[\beta \in \mathsf{vars}[\![\boldsymbol{\tau}]\!] \leftarrow \vartheta(\beta)] &\triangleq \vartheta(\alpha) \quad \text{if} \quad \alpha \in \mathsf{vars}[\![\boldsymbol{\tau}]\!] \cap \mathsf{dom}(\vartheta) \\
\alpha[\beta \in \mathsf{vars}[\![\boldsymbol{\tau}]\!] \leftarrow \vartheta(\beta)] &\triangleq \alpha \qquad\;\; \text{if} \quad \alpha \notin \mathsf{dom}(\vartheta) \\
f(\boldsymbol{\tau}_1, \ldots, \boldsymbol{\tau}_n)[\beta \in \mathsf{vars}[\![\boldsymbol{\tau}]\!] \leftarrow \vartheta(\beta)] &\triangleq \\
& f(\boldsymbol{\tau}_1[\beta \in \mathsf{vars}[\![\boldsymbol{\tau}]\!] \leftarrow \vartheta(\beta)], \ldots, \boldsymbol{\tau}_n[\beta \in \mathsf{vars}[\![\boldsymbol{\tau}]\!] \leftarrow \vartheta(\beta)])
\end{aligned} \tag{48.61}$$

Exercise 48.62 Prove that for all variables $\alpha, \beta \in \mathcal{V}_{\mathfrak{t}}$, terms with variables $\boldsymbol{\tau}, \boldsymbol{\tau}' \in \mathbf{T}^\mathcal{V}$, and substitution $\vartheta \in \mathcal{V}_{\mathfrak{t}} \nrightarrow \mathbf{T}^\mathcal{V}$, we have $\boldsymbol{\tau}[\beta \in \mathsf{vars}[\![\boldsymbol{\tau}]\!] \leftarrow \vartheta(\beta)] = \vartheta(\boldsymbol{\tau})$.

The following lemma 48.63 shows that $\mathsf{unify}(\boldsymbol{\tau}_1, \boldsymbol{\tau}_2, \vartheta_0)$ returns a solution to the equations $\boldsymbol{\tau}_1 \doteq \boldsymbol{\tau}_2$ and $\{\alpha \doteq \vartheta(\alpha) \mid \alpha \in \mathsf{dom}(\vartheta)\}$, if any.

Lemma 48.63 $\forall \boldsymbol{\tau}_1, \boldsymbol{\tau}_2 \in \mathbf{T}^{\mathrm{v}} . \forall \vartheta_0 \in \Sigma . \gamma^{\mathrm{r}}_{\mathrm{s}}(\mathsf{unify}(\boldsymbol{\tau}_1, \boldsymbol{\tau}_2, \vartheta_0)) = \gamma_{\mathrm{e}}(\boldsymbol{\tau}_1 \doteq \boldsymbol{\tau}_2) \cap \gamma^{\mathrm{r}}_{\mathrm{s}}(\vartheta_0)$.

The proof of lemma 48.63 is available at https://github.com/PrAbsInt/.

Theorem 48.64 shows that $\mathsf{mgu}(\boldsymbol{\tau}_1, \boldsymbol{\tau}_2)$ is the most general unifier of terms with variables $\boldsymbol{\tau}_1$ and $\boldsymbol{\tau}_2$.

Theorem 48.64 (mgu) $\forall \boldsymbol{\tau}_1, \boldsymbol{\tau}_2 \in \mathbf{T}^{\mathrm{v}} . \gamma^{\mathrm{r}}_{\mathrm{s}}(\mathsf{mgu}(\boldsymbol{\tau}_1, \boldsymbol{\tau}_2)) = \gamma_{\mathrm{e}}(\boldsymbol{\tau}_1 \doteq \boldsymbol{\tau}_2)$.

Proof of theorem 48.64 By (48.47.14) and lemma 48.63,

$$\gamma^{\mathrm{r}}_{\mathrm{s}}(\mathsf{mgu}(\boldsymbol{\tau}_1, \boldsymbol{\tau}_2)) = \gamma^{\mathrm{r}}_{\mathrm{s}}(\mathsf{unify}(\boldsymbol{\tau}_1, \boldsymbol{\tau}_2, \varepsilon)) = \gamma_{\mathrm{e}}(\boldsymbol{\tau}_1 \doteq \boldsymbol{\tau}_2) \cap \gamma^{\mathrm{r}}_{\mathrm{s}}(\varepsilon) = \gamma_{\mathrm{e}}(\boldsymbol{\tau}_1 \doteq \boldsymbol{\tau}_2)$$

because $\gamma^{\mathrm{r}}_{\mathrm{s}}(\varepsilon) = \mathbf{P}^{\mathrm{v}}$ by definition (48.52) of $\gamma^{\mathrm{r}}_{\mathrm{s}}$. □

Finally, theorem 48.66 shows that $\mathsf{gci}(\boldsymbol{\tau}_1, \boldsymbol{\tau}_2)$ is the greatest common instance of terms with variables $\boldsymbol{\tau}_1$ and $\boldsymbol{\tau}_2$. We need the following

Lemma 48.65 $\forall \boldsymbol{\tau} \in \mathbf{T}^{\mathrm{v}} . \forall \vartheta \in \Sigma . (\mathsf{dom}(\vartheta) \cap \mathbb{covars}[\![\vartheta]\!] = \emptyset) \Rightarrow (\mathit{ground}(\vartheta(\boldsymbol{\tau})) = \{\boldsymbol{\varrho}(\boldsymbol{\tau}) \mid \boldsymbol{\varrho} \in \gamma^{\mathrm{r}}_{\mathrm{s}}(\vartheta)\})$.

Proof of lemma 48.65 By (48.52), we have $\{\boldsymbol{\varrho}(\boldsymbol{\tau}) \mid \boldsymbol{\varrho} \in \gamma^{\mathrm{r}}_{\mathrm{s}}(\vartheta)\} = \{\boldsymbol{\varrho}(\boldsymbol{\tau}) \mid \forall \alpha \in \mathbb{V}_{t} . \boldsymbol{\varrho}(\alpha) = \boldsymbol{\varrho}(\vartheta(\alpha))\}$. Therefore, by (48.10), we have to prove that this is equal to $\{\boldsymbol{\varrho}'(\vartheta(\boldsymbol{\tau})) \mid \boldsymbol{\varrho}' \in \mathbf{P}^{\mathrm{v}}\}$.

- For ($\subseteq$), take $\boldsymbol{\varrho}' = \boldsymbol{\varrho}$. By structural induction, from $\forall \alpha \in \mathbb{V}_{t} . \boldsymbol{\varrho}(\alpha) = \boldsymbol{\varrho}(\vartheta(\alpha))$, we infer that $\boldsymbol{\varrho}(\boldsymbol{\tau}) = \boldsymbol{\varrho}(\vartheta(\boldsymbol{\tau})) = \boldsymbol{\varrho}'(\vartheta(\boldsymbol{\tau}))$;
- For ($\supseteq$), take $\forall \alpha \in \mathbb{V}_{t} . \boldsymbol{\varrho}(\alpha) = \boldsymbol{\varrho}'(\vartheta(\alpha))$.
  - By (48.7), (48.30), $\forall \alpha \in \mathbb{V}_{t} . \boldsymbol{\varrho}(\alpha) = \boldsymbol{\varrho}'(\vartheta(\alpha))$, and structural induction on $\boldsymbol{\tau}$, we have $\boldsymbol{\varrho}'(\vartheta(\boldsymbol{\tau})) = \boldsymbol{\varrho}(\boldsymbol{\tau})$;
  - Let $\alpha \in \mathbb{V}_{t}$ be any term variable. Either $\alpha \notin \mathsf{dom}(\vartheta(\alpha))$, and we have $\vartheta(\alpha) = \alpha$ and so $\boldsymbol{\varrho}(\alpha) = \boldsymbol{\varrho}(\vartheta(\alpha))$. Otherwise, $\alpha \in \mathsf{dom}(\vartheta(\alpha))$ in which case, by hypothesis $\mathsf{dom}(\vartheta) \cap \mathbb{covars}[\![\vartheta]\!] = \emptyset$, we have $\forall \beta \in \mathbb{vars}[\![\vartheta(\alpha)]\!] . \beta \notin \mathsf{dom}(\vartheta)$ so that $\boldsymbol{\varrho}(\beta) = \boldsymbol{\varrho}'(\vartheta(\beta)) = \boldsymbol{\varrho}'(\beta)$. By (48.30) and structural induction on $\vartheta(\alpha)$, this implies that $\boldsymbol{\varrho}'(\vartheta(\alpha)) = \boldsymbol{\varrho}(\vartheta(\alpha))$. This allows us to conclude that $\boldsymbol{\varrho}(\vartheta(\alpha)) = \boldsymbol{\varrho}(\alpha)$ because we have defined $\boldsymbol{\varrho}(\alpha) = \boldsymbol{\varrho}'(\vartheta(\alpha))$. □

$\mathsf{gci}(\boldsymbol{\tau}_1, \boldsymbol{\tau}_2)$ returns the $\preceq_{\doteq^{\mathrm{v}}}$-least upper bound of $[\boldsymbol{\tau}_1]_{\doteq^{\mathrm{v}}}$ and $[\boldsymbol{\tau}_2]_{\doteq^{\mathrm{v}}}$.

Theorem 48.66 (lub) $\forall \boldsymbol{\tau}_1, \boldsymbol{\tau}_2 \in \mathbf{T}^{\mathrm{v}} . [\mathsf{gci}(\boldsymbol{\tau}_1, \boldsymbol{\tau}_2)]_{\doteq^{\mathrm{v}}} = [\boldsymbol{\tau}_1]_{\doteq^{\mathrm{v}}} \, \mathit{gci}_{\doteq^{\mathrm{v}}} \, [\boldsymbol{\tau}_2]_{\doteq^{\mathrm{v}}}$.

Proof of theorem 48.66 The case $\mathsf{mgu}(\boldsymbol{\tau}_1, \boldsymbol{\tau}_2) = \Omega^{\mathrm{r}}_{\mathrm{s}}$ at (48.47.18) is trivial because $\overline{\emptyset}^{\mathrm{v}}$ is the infimum of the lattice. Otherwise, we have

$$\begin{aligned} & \mathit{ground}(\boldsymbol{\tau}_1) \cap \mathit{ground}(\boldsymbol{\tau}_2) \\ = \; & \{\boldsymbol{\varrho}(\boldsymbol{\tau}_1) \mid \boldsymbol{\varrho} \in \mathbf{P}^{\mathrm{v}}\} \cap \{\boldsymbol{\varrho}'(\boldsymbol{\tau}_2) \mid \boldsymbol{\varrho}' \in \mathbf{P}^{\mathrm{v}}\} && \text{(definition (48.10) of } \mathit{ground}\text{)} \end{aligned}$$

$= \{\varrho(\boldsymbol{\tau}_j) \mid \varrho(\boldsymbol{\tau}_1) = \varrho(\boldsymbol{\tau}_2)\}, \quad j = 1,2$ ⟨definition of $\cap$⟩

$= \{\varrho(\boldsymbol{\tau}_j) \mid \varrho \in \gamma_{\mathrm{e}}(\boldsymbol{\tau}_1 \doteq \boldsymbol{\tau}_2)\}, \quad j = 1,2$ ⟨definition (48.51) of γ_{e}⟩

$= \{\varrho(\boldsymbol{\tau}_j) \mid \varrho \in \gamma_{\mathrm{s}}^{\mathrm{r}}(\mathsf{mgu}(\boldsymbol{\tau}_1, \boldsymbol{\tau}_2))\}, \quad j = 1,2$ ⟨theorem 48.64⟩

$= \mathit{ground}(\mathsf{mgu}(\boldsymbol{\tau}_1, \boldsymbol{\tau}_2)(\boldsymbol{\tau}_j)) \quad j = 1,2$

⟨by (48.57) and lemma 48.65 where $\boldsymbol{\tau} = \boldsymbol{\tau}_j$ and $\vartheta = \mathsf{mgu}(\boldsymbol{\tau}_1, \boldsymbol{\tau}_2)$⟩

$= \mathit{ground}(\mathsf{gci}(\boldsymbol{\tau}_1, \boldsymbol{\tau}_2))$ ⟨(48.47.18)⟩

By the Galois retraction (48.27), we conclude that

$$\mathsf{gci}(\boldsymbol{\tau}_1, \boldsymbol{\tau}_2) = lcg_{\simeq^{\mathrm{v}}}[\boldsymbol{v}](\mathit{ground}(\boldsymbol{\tau}_1) \cap \mathit{ground}(\boldsymbol{\tau}_2))$$

and therefore $[\mathsf{gci}(\boldsymbol{\tau}_1, \boldsymbol{\tau}_2)]_{\simeq^{\mathrm{v}}} = [\boldsymbol{\tau}_1]_{\simeq^{\mathrm{v}}}\, gci_{\simeq^{\mathrm{v}}}\, [\boldsymbol{\tau}_2]_{\simeq^{\mathrm{v}}}$ in the subsumption lattice corollary 48.28. □

Exercise 48.67 Lemma 48.63 shows how to solve a finite system of equations $E = \{\boldsymbol{\tau}_1^i \doteq \boldsymbol{\tau}_2^i \mid i \in \Delta\}$. Starting from $\vartheta_0 = \boldsymbol{\varepsilon}$, pick any equation $\boldsymbol{\tau}_1^i \doteq \boldsymbol{\tau}_2^i$ left in E and compute $\vartheta_{j+1} = \mathsf{unify}(\boldsymbol{\tau}_1^i, \boldsymbol{\tau}_2^i, \vartheta_j)$. Repeat until $\vartheta_{j+1} = \Omega_{\mathrm{s}}^{\mathrm{r}}$ or no equation is left. Prove that this algorithm is correct.

48.8.3 Least Upper Bound: the Least Common Generalization Algorithm

The concept of *anti-unification* or *most specific generalization* or *least common generalization* was introduced by Gordon Plotkin [800] and John Reynolds [827]. For example, the anti-unification of $+(1,2)$ and $+(1,3)$ is $+(1,x)$. [17] introduced a rule-based presentation of the least common generalization of figure 48.69, which is essentially equivalent to the following (a proof is given in [17] that the rules produce an upper bound, but not that it is the least one).

Example 48.70

$\mathsf{lub}(f(a,b), f(b,a), \varnothing)$
$\quad \mathsf{lub}(a, b, \varnothing)$ (48.68.2)
$\quad = \langle\alpha,\ \{\langle\alpha,\ \langle a,\ b\rangle\rangle\}\rangle$ (48.68.9)
$\quad \mathsf{lub}(b, a, \{\langle\alpha,\ \langle a,\ b\rangle\rangle\})$ (48.68.3)
$\quad = \langle\beta,\ \{\langle\alpha,\ \langle a,\ b\rangle\rangle, \langle\beta,\ \langle b,\ a\rangle\rangle\}\rangle$ (48.68.9)
$= \langle f(\alpha,\beta),\ \{\langle\alpha,\ \langle a,\ b\rangle\rangle, \langle\beta,\ \langle b,\ a\rangle\rangle\}\rangle$ (48.68.5)

Example 48.71

$\mathsf{lub}(f(a,b), f(a,b), \varnothing)$
$\quad \mathsf{lub}(a, b, \varnothing)$ (48.68.2)
$\quad = \langle\alpha,\ \{\langle\alpha,\ \langle a,\ b\rangle\rangle\}\rangle$ (48.68.9)
$\quad \mathsf{lub}(a, b, \{\langle\alpha,\ \langle a,\ b\rangle\rangle\})$ (48.68.3)
$\quad = \langle\alpha,\ \{\langle\alpha,\ \langle a,\ b\rangle\rangle\}\rangle$ (48.68.7)
$= \langle f(\alpha,\alpha),\ \{\langle\alpha,\ \langle a,\ b\rangle\rangle\}\rangle$ (48.68.5)

The proof of various lemmas preparing for the proof of total correctness of the lcg algorithm is available at https://github.com/PrAbsInt/.

least common generalization $(\boldsymbol{\tau}_1, \boldsymbol{\tau}_2 \in \mathbf{T}^{\mathrm{v}}$ *and* $T_0 \in \mathbb{V}_{\mathfrak{t}} \nrightarrow \mathbf{T}^{\mathrm{v}} \times \mathbf{T}^{\mathrm{v}})$

$$
\begin{array}{l}
\mathsf{let\ rec\ lub}(\boldsymbol{\tau}_1, \boldsymbol{\tau}_2, T_0) = \qquad (48.68)\\
\quad \mathsf{if}\ \boldsymbol{\tau}_1 = f(\boldsymbol{\tau}_1^1, \ldots, \boldsymbol{\tau}_1^n) \wedge \boldsymbol{\tau}_2 = f(\boldsymbol{\tau}_2^1, \ldots, \boldsymbol{\tau}_2^n)\ \mathsf{then}\\
\quad\quad \mathsf{let}\ \langle \boldsymbol{\tau}^1, T_1\rangle = \mathsf{lub}(\boldsymbol{\tau}_1^1, \boldsymbol{\tau}_2^1, T_0)\ \mathsf{in}\\
\quad\quad\quad \mathsf{let}\ \langle \boldsymbol{\tau}^2, T_2\rangle = \mathsf{lub}(\boldsymbol{\tau}_1^2, \boldsymbol{\tau}_2^2, T_1)\ \mathsf{in}\\
\quad\quad\quad \ldots\\
\quad\quad\quad\quad \mathsf{let}\ \langle \boldsymbol{\tau}^n, T_n\rangle = \mathsf{lub}(\boldsymbol{\tau}_1^n, \boldsymbol{\tau}_2^n, T_{n-1})\ \mathsf{in}\\
\quad\quad\quad\quad\quad \langle f(\boldsymbol{\tau}^1, \ldots, \boldsymbol{\tau}^n), T_n\rangle\\
\quad \mathsf{elsif}\ \exists \beta \in \mathsf{dom}(T_0)\,.\,T_0(\beta) = \langle \boldsymbol{\tau}_1, \boldsymbol{\tau}_2\rangle\ \mathsf{then}\\
\quad\quad \langle \beta, T_0\rangle\\
\quad \mathsf{else\ let}\ \beta \in \mathbb{V}_{\mathfrak{t}} \setminus \mathsf{dom}(T_0)\ \mathsf{in}\\
\quad\quad \langle \beta, \langle \boldsymbol{\tau}_1, \boldsymbol{\tau}_2\rangle[\beta \leftarrow T_0]\rangle\\
\mathsf{let\ lcg}(\boldsymbol{\tau}_1, \boldsymbol{\tau}_2) =\\
\quad \mathsf{if}\ \boldsymbol{\tau}_1 = \overline{\varnothing}^{\mathrm{v}}\ \mathsf{then}\ \boldsymbol{\tau}_2\ \mathsf{elsif}\ \boldsymbol{\tau}_2 = \overline{\varnothing}^{\mathrm{v}}\ \mathsf{then}\ \boldsymbol{\tau}_1\\
\quad \mathsf{else\ let}\ \langle \boldsymbol{\tau}, T\rangle = \mathsf{lub}(\boldsymbol{\tau}_1, \boldsymbol{\tau}_2, \varnothing)\ \mathsf{in}\ \boldsymbol{\tau}
\end{array}
$$

Figure 48.69
The least upper bound algorithm

Theorem 48.103 proves the total correctness of the lcg algorithm for calculating lubs in Herbrand abstract domain.

Theorem 48.103 (lub) The lcg algorithm terminates and returns

$$[\mathsf{lcg}(\boldsymbol{\tau}_1, \boldsymbol{\tau}_2)]_{\simeq^{\mathrm{v}}} = [\boldsymbol{\tau}_1]_{\simeq^{\mathrm{v}}}\ lcg_{\simeq^{\mathrm{v}}}\ [\boldsymbol{\tau}_2]_{\simeq^{\mathrm{v}}}.$$

The proof of theorem 48.103 is available at https://github.com/PrAbsInt/.

48.9 Conclusion

We have shown that a concrete property represented by a set of ground terms can be overapproximated by a term with variables. Of course a concrete property P represented by a set of terms with variables can be overapproximated by the term with variables $LCG_{\simeq^{\mathrm{v}}}(P)$ (this is Galois connection of exercise 11.8 where h is the identity). These abstractions have been extensively used in type inference (see chapter 49).

$\gamma_{\mathsf{s}}^{\mathsf{c}}(\vartheta) \triangleq \{\alpha \in \mathbb{V}_{\mathfrak{t}} \mapsto \boldsymbol{\varrho}(\vartheta(\alpha)) \mid \boldsymbol{\varrho} \in \mathbf{P}^{\mathrm{v}}\}$ to approximate environment properties and $\gamma_{\mathsf{s}}^{\mathsf{r}}(\vartheta)$ at (48.52) and (48.53) to approximate equations in solved form are two distinct Cartesian and relational interpretations of substitutions. Both are lower adjoints of Galois retractions that yield two different lattice structure for substitutions. [346, 776] merge them into one structure for idempotent substitutions, which may be confusing (because $\gamma_{\mathsf{s}}^{\mathsf{c}}(\vartheta)$ does not require ϑ to be idempotent). γ_{e} at (48.51) is also the lower adjoint of

Galois retraction which yields the complete lattice of systems of equations [619]. All these constructions are based on the fact that the image of a complete lattice by a Galois retraction is a complete lattice (exercise 11.64).

One can avoid variables by representing a term with variables as a rooted directed acyclic graph (DAG), that is, the term syntax tree where the leaf nodes that have the same variable are joined [781] (mathematically represented by a tree and an equivalence relation between leaves that have the same variable). The lattice structure of symbolic terms generalizes to rooted order-sorted feature (OSF) graphs [12, 18].

A deductive synthesis of the unification algorithm was proposed by [653]. Unification has also been extended to higher-order terms (with higher-order variables taking symbols in $\mathbf{F}$ as values) [518, 519, 782], as used in the Isabelle proof assistant [783]. Fuzzy unification [879] is another extension allowing comparison of different symbols of $\mathbf{F}$ with the same arity up to some fuzziness degree of similarity. [16, 17] extend this fuzzy unification to symbols of different arities.

Without the expensive occurs check, Prolog I unification was incorrect. To solve this problem, Alain Colmerauer extended unification to infinite trees in Prolog II [202]. It is supported in the abstract machine [718] by most modern Prolog family languages such as Ciao [489]. It does not cost any more than the unification without the occurs check of the original Prolog systems. The other source of incorrectness of Prolog I was the depth-first search. Nowadays the search method (breadth-first, iterative-deepening, tabling, delays, and others) is a parameter for the programmer.

48.10 Solutions to Selected Exercises

Solution to exercise 48.13 The case of $\varnothing$ is trivial. Otherwise,

$$\begin{aligned}
&\boldsymbol{\tau} \preceq^{\mathrm{v}} \boldsymbol{\tau}' \\
\Leftrightarrow\ & \mathit{ground}(\boldsymbol{\tau}) \subseteq \mathit{ground}(\boldsymbol{\tau}') && \text{(definition (48.12) of } \preceq^{\mathrm{v}}\text{)}\\
\Leftrightarrow\ & \{\varrho(\boldsymbol{\tau}) \mid \varrho \in \mathbf{P}^{\mathrm{v}}\} \subseteq \{\varrho'(\boldsymbol{\tau}') \mid \varrho' \in \mathbf{P}^{\mathrm{v}}\} && \text{(definition (48.10) of } \mathit{ground}\text{)}\\
\Leftrightarrow\ & \forall \varrho \in \mathbf{P}^{\mathrm{v}} \,.\, \exists \varrho' \in \mathbf{P}^{\mathrm{v}} \,.\, \varrho(\boldsymbol{\tau}) = \varrho'(\boldsymbol{\tau}') && \text{(definition of } \subseteq\text{)}
\end{aligned}$$

Solution to exercise 48.15 ($\Rightarrow$) Assume $\boldsymbol{\tau} \simeq^{\mathrm{v}} \boldsymbol{\tau}'$ so that, by definition of $\simeq^{\mathrm{v}}$ and exercise 48.13, $\forall \varrho \,.\, \exists \varrho' \,.\, \varrho(\boldsymbol{\tau}) = \varrho'(\boldsymbol{\tau}')$ and $\forall \varrho' \,.\, \exists \varrho \,.\, \varrho(\boldsymbol{\tau}) = \varrho'(\boldsymbol{\tau}')$. Let us define a relation $\rho \in \wp(\mathsf{vars}[\![\boldsymbol{\tau}]\!] \times \mathsf{vars}[\![\boldsymbol{\tau}']\!])$, starting from $\varnothing$ as follows:

- If $\boldsymbol{\tau}$ is a variable α and $\boldsymbol{\tau}'$ is not then $\boldsymbol{\tau} \not\preceq^{\mathrm{v}} \boldsymbol{\tau}'$ so $\boldsymbol{\tau}'$ must be a variable β and we let $\langle \alpha, \beta \rangle \in \rho$.
- It $\boldsymbol{\tau} = f(\boldsymbol{\tau}_1, \ldots, \boldsymbol{\tau}_{})$ then $\boldsymbol{\tau} \simeq^{\mathrm{v}} \boldsymbol{\tau}'$ implies that $\forall n \in [1, n] \,.\, \boldsymbol{\tau}_k \simeq^{\mathrm{v}} \boldsymbol{\tau}'_k$, so by structural induction, there is a relation $\rho_k \in \wp(\mathsf{vars}[\![\boldsymbol{\tau}_k]\!] \times \mathsf{vars}[\![\boldsymbol{\tau}'_k]\!])$ so we take $\rho = \bigcup_{k=1}^{n} \rho_k$.

We have to prove that ρ is a function. By contradiction, if ρ is not a function then there is a variable α of $\boldsymbol{\tau}$ and two variables β and γ of $\boldsymbol{\tau}'$ with at least one which is not α, say γ. Then instances of γ in $\boldsymbol{\tau}'$ cannot be replicated with α in $\boldsymbol{\tau}$ so the two terms cannot be equivalent. Now if ρ is not injective there there are two variables β and γ of $\boldsymbol{\tau}$ with only one correspondent α of $\boldsymbol{\tau}$, so again the instances of β and γ cannot be matched with α.

Finally, if ρ is not surjective, then there is a variable γ with arbitrary instantiations in $\boldsymbol{\tau}'$ with no correspondent in $\boldsymbol{\tau}$, which again prevents equivalence. In conclusion, ρ is a bijection.

($\Leftarrow$) Conversely let $\rho \in \mathsf{vars}[\![\boldsymbol{\tau}]\!] \rightarrowtail\!\!\!\!\twoheadrightarrow \mathsf{vars}[\![\boldsymbol{\tau}']\!]$ be such $\rho(\boldsymbol{\tau}) = \boldsymbol{\tau}'$. Given any $\boldsymbol{\varrho} \in \mathbf{P}^{\mathrm{v}}$, define $\boldsymbol{\varrho}'(\alpha) \triangleq \boldsymbol{\varrho}(\rho^{-1}(\alpha))$, $\alpha \in \mathbb{V}_{t}$. Let us show, by structural induction on α, that $\boldsymbol{\varrho}(\boldsymbol{\tau}) = \boldsymbol{\varrho}'(\boldsymbol{\tau}')$.

- If $\boldsymbol{\tau} = \alpha$ then $\rho \in \mathsf{vars}[\![\boldsymbol{\tau}]\!] \rightarrowtail\!\!\!\!\twoheadrightarrow \mathsf{vars}[\![\boldsymbol{\tau}']\!]$ and $\rho(\boldsymbol{\tau}) = \boldsymbol{\tau}'$ imply that $\boldsymbol{\tau}' = \beta$ is the variable $\beta = \rho(\boldsymbol{\tau})$. It follows that $\boldsymbol{\varrho}'(\boldsymbol{\tau}') = \boldsymbol{\varrho}'(\beta) \triangleq \boldsymbol{\varrho}(\rho^{-1}(\beta)) = \boldsymbol{\varrho}(\alpha) = \boldsymbol{\varrho}(\boldsymbol{\tau})$;
- Otherwise, $\boldsymbol{\tau} = f(\boldsymbol{\tau}_1, \ldots, \boldsymbol{\tau}_n)$ so that $\rho(\boldsymbol{\tau}) = \rho(f(\boldsymbol{\tau}_1, \ldots, \boldsymbol{\tau}_n)) = f(\rho(\boldsymbol{\tau}_1), \ldots, \rho(\boldsymbol{\tau}_n)) = \boldsymbol{\tau}'$ implies that $\boldsymbol{\tau}' = f(\boldsymbol{\tau}_1', \ldots, \boldsymbol{\tau}_n')$ with $\boldsymbol{\tau}_1' = \rho(\boldsymbol{\tau}_i), i \in [1, n]$. It follows, by (48.7) and induction hypothesis, that $\boldsymbol{\varrho}'(\boldsymbol{\tau}') = \boldsymbol{\varrho}'(f(\boldsymbol{\tau}_1', \ldots, \boldsymbol{\tau}_n')) = f(\boldsymbol{\varrho}'(\boldsymbol{\tau}_1'), \ldots, \boldsymbol{\varrho}'(\boldsymbol{\tau}_n')) = f(\boldsymbol{\varrho}(\boldsymbol{\tau}_1), \ldots, \boldsymbol{\varrho}(\boldsymbol{\tau}_n)) = \boldsymbol{\varrho}(f(\boldsymbol{\tau}_1), \ldots, \boldsymbol{\tau}_n)) = \boldsymbol{\varrho}(\boldsymbol{\tau})$.

Solution to exercise 48.16

$$
\begin{aligned}
& [\boldsymbol{\tau}_1]_{\simeq^{\mathrm{v}}} \preceq_{\simeq^{\mathrm{v}}} [\boldsymbol{\tau}_2]_{\simeq^{\mathrm{v}}} \\
\Leftrightarrow\ & \exists \boldsymbol{\tau}_1' \in [\boldsymbol{\tau}_1]_{\simeq^{\mathrm{v}}}, \boldsymbol{\tau}_2' \in [\boldsymbol{\tau}_2]_{\simeq^{\mathrm{v}}} \,.\, \boldsymbol{\tau}_1' \preceq^{\mathrm{v}} \boldsymbol{\tau}_2' && \text{(definition (48.14) of } \preceq_{\simeq^{\mathrm{v}}}\text{)} \\
\Leftrightarrow\ & \exists \boldsymbol{\tau}_1', \boldsymbol{\tau}_2' \,.\, \boldsymbol{\tau}_1' \simeq^{\mathrm{v}} \boldsymbol{\tau}_1 \wedge \boldsymbol{\tau}_2' \simeq^{\mathrm{v}} \boldsymbol{\tau}_2 \wedge \boldsymbol{\tau}_1' \preceq^{\mathrm{v}} \boldsymbol{\tau}_2' && \text{(definition of } [\boldsymbol{\tau}]_{\simeq^{\mathrm{v}}}\text{)} \\
\Leftrightarrow\ & \exists \boldsymbol{\tau}_1', \boldsymbol{\tau}_2' \,.\, \boldsymbol{\tau}_1 \preceq^{\mathrm{v}} \boldsymbol{\tau}_1' \wedge \boldsymbol{\tau}_1' \preceq^{\mathrm{v}} \boldsymbol{\tau}_2' \wedge \boldsymbol{\tau}_2' \preceq^{\mathrm{v}} \boldsymbol{\tau}_2 \wedge \boldsymbol{\tau}_1' \preceq^{\mathrm{v}} \boldsymbol{\tau}_1 \wedge \boldsymbol{\tau}_2 \preceq^{\mathrm{v}} \boldsymbol{\tau}_2' && \text{(definition of } \simeq^{\mathrm{v}}\text{)} \\
\Leftrightarrow\ & \boldsymbol{\tau}_1 \preceq^{\mathrm{v}} \boldsymbol{\tau}_2 \\
& \quad \text{((}\Rightarrow\text{) transitivity} \\
& \quad \text{(}\Leftarrow\text{) choosing } \boldsymbol{\tau}_1' = \boldsymbol{\tau}_1, \boldsymbol{\tau}_2' = \boldsymbol{\tau}_2, \text{ and reflexivity)}
\end{aligned}
$$

Solution to exercise 48.20 Because $\mathbf{v} \in (\Delta \to \mathbf{T}) \rightarrowtail \mathbb{V}_{t}$ is injective, it has a left inverse (improperly) denoted $\mathbf{v}^{-1}$ such that $\mathbf{v}^{-1} \circ \mathbf{v} = \mathbb{1}_{\wp(\mathbf{T})}$ is the identity on $\wp(\mathbf{T})$ (encoded as families $\Delta \to \mathbf{T}$). Define $\rho \triangleq \mathbf{v}' \circ \mathbf{v}^{-1}$. Given $T \in \wp(\mathbf{T})$, let us show that $\rho(lcg[\mathbf{v}](T)) = lcg[\mathbf{v}'](T)$, by structural induction and case analysis on the definition (48.18) of $lcg[\mathbf{v}]$.

1. If $T = \varnothing$ then $\rho(lcg[\mathbf{v}](T)) = \rho(lcg[\mathbf{v}](\varnothing)) = \overline{\varnothing}^{\mathrm{v}} = lcg[\mathbf{v}'](\varnothing) = lcg[\mathbf{v}'](T)$;
2. Else, if $lcg[\mathbf{v}](T) = \mathbf{v}(T)$ then $lcg[\mathbf{v}'](T) = \mathbf{v}'(T)$ so that $\rho(lcg[\mathbf{v}](T)) = \rho(\mathbf{v}(T)) = \mathbf{v}' \circ \mathbf{v}^{-1} \circ \mathbf{v}(T) = \mathbf{v}' \circ \mathbb{1}_{\wp(\mathbf{T})}(T) = \mathbf{v}'(T) = lcg[\mathbf{v}'](T)$;
3. Otherwise, $T = \{f(\mathbf{t}_i^1, \ldots, \mathbf{t}_i^n) \mid i \in \Delta\}$. It follows, by induction hypothesis, that $\rho(T_{\mathbf{v}}^k) = \rho(lcg[\mathbf{v}](\{\mathbf{t}^k \mid i \in \Delta\})) = lcg[\mathbf{v}'](\{\mathbf{t}^k \mid i \in \Delta\}) = T_{\mathbf{v}'}^k$, $k = 1, \ldots, n$. Therefore, by (48.7), $\rho(lcg[\mathbf{v}](T)) = \rho(f(T_{\mathbf{v}}^1, \ldots, T_{\mathbf{v}}^n)) = f(\rho(T_{\mathbf{v}}^1), \ldots, \rho(T_{\mathbf{v}}^n)) = f(T_{\mathbf{v}'}^1, \ldots, T_{\mathbf{v}'}^n) = lcg[\mathbf{v}'](T)$.

If $\alpha \in \mathsf{vars}[\![lcg[\mathbf{v}'](T)]\!]$ then case 1. of the above proof shows that $\alpha = lcg[\mathbf{v}'](T') = \mathbf{v}'(T')$ for some $T' \in \wp(\mathbf{T})$ and therefore $\rho(lcg[\mathbf{v}](T')) = \rho(\mathbf{v}(T')) = \mathbf{v}' \circ \mathbf{v}^{-1} \circ \mathbf{v}(T') = \mathbf{v}' \circ \mathbb{1}_{\wp(\mathbf{T})}(T') = \mathbf{v}'(T') = \alpha$, proving that $\rho \in \mathsf{vars}[\![lcg[\mathbf{v}](T)]\!] \twoheadrightarrow \mathsf{vars}[\![lcg[\mathbf{v}'](T)]\!]$ is surjective.

If $\alpha_1, \alpha_2 \in \mathsf{vars}[\![lcg[\mathbf{v}](T)]\!]$ then case 1. of the above proof shows that $\alpha_1 = lcg[\mathbf{v}](\alpha_1) = \mathbf{v}(\alpha_1)$ and $\alpha_2 = lcg[\mathbf{v}](\alpha_2) = \mathbf{v}(\alpha_2)$ for some $\alpha_1, \alpha_2 \in \wp(\mathbf{T})$. Assume that $\rho(\alpha_1) = \rho(\alpha_2)$. Then we have $\mathbf{v}' \circ \mathbf{v}^{-1}(\alpha_1) = \mathbf{v}' \circ \mathbf{v}^{-1}(\alpha_2)$ that is $\mathbf{v}' \circ \mathbf{v}^{-1}(\mathbf{v}(\alpha_1)) = \mathbf{v}' \circ \mathbf{v}^{-1}(\mathbf{v}(\alpha_2))$, which implies $\mathbf{v}'(\alpha_1) = \mathbf{v}'(\alpha_2)$ because $\mathbf{v}^{-1} \circ \mathbf{v} = \mathbb{1}_{\wp(\mathbf{T})}$. It follows that $\alpha_1 = \alpha_2$ because $\mathbf{v}'$ is injective. Therefore $\alpha_1 = \mathbf{v}(\alpha_1) = \mathbf{v}(\alpha_2) = \alpha_2$, proving that ρ is injective.

It follows that $\rho \in \mathsf{vars}[\![lcg[\mathbf{v}](T)]\!] \rightarrowtail\!\!\!\!\twoheadrightarrow \mathsf{vars}[\![lcg[\mathbf{v}'](T)]\!]$ is bijective so that $lcg[\mathbf{v}](T) \simeq^{\mathrm{v}} lcg[\mathbf{v}'](T)$ by exercise 48.15.

49 Typing

A type system is a deductive system (see section 16.2) for automatically checking that programs "do not go wrong" [685], that is, to guarantee the absence of certain errors (like using an integer in place of a function, but not dividing by zero). We consider monotypes à la Church-Hindley [185, 500]. Types of programs are defined by a deductive system of rules and implemented by a type inference algorithm based on solving type constraints. Classically, soundness is proved with respect to an operational semantics by *progress*, that is, typing rules are always applicable to well-typed expressions until a type is found and *subject reduction*, that is, preservation of types during evaluation. From a more general point of view, typing has been shown to be an abstract interpretation of a denotational semantics [224].

This chapter 49 introduces monomorphic types for data structures of a simple extension of our imperative language using an abstract interpretation of the collecting semantics of the extended language.

To make the slogan "typable programs cannot go wrong" precise, we distinguish between static and dynamic types. Dynamic types ensure that all operations of the language are well defined (e.g., it is not allowed to create a list of mixed elements such as pairs of integers and single integers or to take the tail of a null list). Dynamic types must be checked at runtime, which is costly. This can be partly avoided by static typing. Statically typable programs cannot produce a subset of the dynamic errors called static errors (e.g., an attempt to create mixed lists) but can still produce some dynamic errors (e.g., taking the tail of a null list).

Exercise 49.1 Prove that automatically checking that a program has no static or dynamic errors is undecidable.

Contents

49.1 Values

We build values $\mathbb{U}$ out of Booleans $\mathbb{B}$, integers $\mathbb{Z}$, the null list `nil`, pairs $\langle x,\ y\rangle$, $x, y \in \mathbb{U}$, and lists $x :: y$, $x, y \in \mathbb{U}$. Moreover, operations on values can be erroneous and return a static error Ω^σ (to be detected at compile time) or a dynamic error Ω^δ (which may have to be detected at runtime).

Such a definition allows to built inhomogeneous lists such as $1 :: \langle 0, \texttt{nil}\rangle :: \texttt{nil} \in \mathbb{U}$. Such inhomogeneous lists are disallowed in some functional programming languages because all elements of the list must have the same type. Moreover, errors are terminals so that $1 :: \langle \Omega^\sigma, \texttt{nil}\rangle :: \Omega^\delta \in \mathbb{U}$ is just Ω^δ. We have to introduce typed values to prevents such inhomogeneous lists with elements of different types.

Exercise 49.2 Provide a formal definition of the set $\mathbb{U}$ of untyped values.

49.2 Dynamic Types

Values manipulated by programs at runtime have dynamic types, represented by ground terms, as follows (assuming $t_1, t_2 \neq \textbf{\textit{err}}$).

$t \in \mathbb{T}^\delta$	$::=$	***bool***	dynamic Boolean type
	$\mid$	***int***	dynamic integer type
	$\mid$	***nil***	dynamic type of a null list
	$\mid$	$t_1 \star t_2$	dynamic type of pairs
	$\mid$	t_1 ***list***	dynamic type of lists
	$\mid$	***err***	dynamic error type

For dynamically typed languages, the type of values is maintained at runtime together with those values on order to check that operations do not go wrong according to their dynamic type (e.g., ***int*** – ***int*** is correct but ***int*** – ***bool*** is wrong). The null list `nil` is a problem, for example, [40].

A list $x :: y :: \texttt{nil}$ is constructed by adding an element x as head of a list $y :: \texttt{nil}$ which is constructed by adding y as head of the null list `nil`. If element x has type t, the list $y :: \texttt{nil}$ should have type t ***list***, so y should have type t while `nil` should have type t ***list***. Because t is arbitrary, `nil` should have all the types $\{t\ \textbf{\textit{list}} \mid t \in \mathbb{T}^\delta\}$.

The type ***nil*** can be thought of as a finite representation of these infinitely many types $\{t\ \textbf{\textit{list}} \mid t \in \mathbb{T}^\delta\}$ of `nil`. This means that when a list of some type t ***list*** is needed, we can use the object `nil` of type ***nil*** instead. So, we have the following equivalence relation on dynamic types:

$$
\begin{aligned}
t_1\ \textbf{\textit{list}} \equiv^\delta \textbf{\textit{nil}} \;=\; \textbf{\textit{nil}} \equiv^\delta t_1\ \textbf{\textit{list}} &\;\triangleq\; \mathtt{tt} && \text{true}\\
t_1\ \textbf{\textit{list}} \equiv^\delta t_2\ \textbf{\textit{list}} &\;\triangleq\; t_1 \equiv^\delta t_2 \\
t_1 \star t_2 \equiv^\delta t_3 \star t_4 &\;\triangleq\; t_1 \equiv^\delta t_3 \wedge t_2 \equiv^\delta t_4 \\
t_1 \equiv^\delta t_2 &\;\triangleq\; t_1 = t_2 && \text{otherwise}
\end{aligned}
$$

meaning that objects of equivalent types can be use in place of one another.

49.3 Typed Values

The values of a given type are defined as follows (where $t_1, t_2 \neq \textit{err}$).

$$
\begin{array}{rcll}
\mathbb{V}[\![\textit{bool}]\!] & \triangleq & \mathbb{B} & \text{Booleans} \quad (49.3) \\
\mathbb{V}[\![\textit{int}]\!] & \triangleq & \mathbb{Z} & \text{integers} \\
\mathbb{V}[\![\textit{nil}]\!] & \triangleq & \{\texttt{nil}\} & \text{null list} \\
\mathbb{V}[\![t_1 \star t_2]\!] & \triangleq & \{\langle x,\ y\rangle \mid x \in \mathbb{V}[\![t_1]\!] \wedge y \in \mathbb{V}[\![t_2]\!]\} & \text{pairs} \\
\mathbb{V}^0[\![t_1\ \textit{list}]\!] & \triangleq & \{\texttt{nil}\} & \text{null list} \\
\mathbb{V}^{n+1}[\![t_1\ \textit{list}]\!] & \triangleq & \{x :: y \mid x \in \mathbb{V}[\![t_1]\!] \wedge y \in \mathbb{V}^n[\![t_1\ \textit{list}]\!]\} & \text{lists} \\
\mathbb{V}[\![t_1\ \textit{list}]\!] & \triangleq & \bigcup_{n\in\mathbb{N}} \mathbb{V}^n[\![t_1\ \textit{list}]\!] & \\
\mathbb{V} & \triangleq & \bigcup\{\mathbb{V}[\![t]\!] \mid t \in \mathbb{T}^\delta \setminus \{\textit{err}\}\} & \text{typed values} \\
\mathbb{V}[\![\textit{err}]\!] & \triangleq & \{\Omega^\sigma, \Omega^\delta\} & \text{static and dynamic errors} \\
\mathbb{V}^{\Omega^\delta} & \triangleq & \mathbb{V} \cup \{\Omega^\delta\} & \text{values or dynamic errors} \\
\mathbb{V}^{\Omega} & \triangleq & \mathbb{V}^{\Omega^\delta} \cup \{\Omega^\sigma\} & \text{values or errors}
\end{array}
$$

49.4 Dynamic Type of Untyped Values

Given an untyped value $x \in \mathbb{U}$, $\tau^\delta(x)$ is its type if $x \in \mathbb{V}$ else $\tau^\delta(x) = \textit{err}$, meaning that the value is untypable or is the result of an erroneous operation. The type of `nil` is ***nil*** (although `nil` is also a list of any type t_1 ***list***).

$$
\begin{array}{rcll}
\tau^\delta(b) & \triangleq & \textit{bool} & b \in \mathbb{B} \quad (49.4) \\
\tau^\delta(z) & \triangleq & \textit{int} & z \in \mathbb{Z} \\
\tau^\delta(\texttt{nil}) & \triangleq & \textit{nil} & \\
\tau^\delta(\langle x,\ y\rangle) & \triangleq & (\!|\ \exists t_1 \neq \textit{err} \,.\, \tau^\delta(x) = t_1 \wedge \exists t_2 \neq \textit{err} \,.\, \tau^\delta(y) = t_2 \ ?\ t_1 \star t_2 \ :\ \textit{err}\ |\!) & \\
\tau^\delta(x :: y) & \triangleq & (\!|\ \exists t \neq \textit{err} \,.\, \tau^\delta(x) = t \wedge \tau^\delta(y) \equiv^\delta t\ \textit{list}\ ?\ t\ \ \textit{list} : \textit{err}\ |\!) & \\
\tau^\delta(x) & \triangleq & \textit{err}, \qquad \text{otherwise} &
\end{array}
$$

Example 49.5

untyped value in $\mathbb{U}$	dynamic type	
`1 :: 2 :: nil`	***int list***	list of integers
`nil :: (nil :: nil)`	***nil list***	list of lists
`(nil :: nil) :: nil`	***nil list list***	
`(1 :: nil) :: nil`	***int list list***	list of integer lists
`nil :: (1 :: nil) :: nil`	***int list list***	
`(1, nil) :: (2, (3 :: nil)) :: nil`	***(int ★ int list) list***	list of pairs of an integer and integer list
`nil :: 1`	***err***	(1 is not a list)
`nil :: (2 :: nil)`	***err***	(2 is not a list)

The semantics of the language implements only well formed values (e.g., all elements of lists must have the same type).

$$\mathbb{V} \quad \triangleq \quad \{v \in \mathbb{U} \mid \tau^\delta(v) \neq err\}.$$

This can be checked either by a dynamic or a static type system.

Exercise 49.6 Write a program, in the language of your choice that infers the dynamic type of a value.

49.5 Homogeneous Lists

$\mathbb{V}[\![t\ list]\!]$ is be the set of all possibly empty homogeneous lists in which the elements all have dynamic type $t \neq err$.

Lemma 49.7 (list values) For all $t \neq err$,

$$\begin{aligned}\mathbb{V}[\![t\ list]\!] &= \{x_1 :: x_2 :: \ldots x_n :: \mathtt{nil} \mid n \in \mathbb{N}^+ \wedge \forall i \in [1,n]\,.\,\tau^\delta(x_i) = t\} \cup \{\mathtt{nil}\} \\ &= \{x \in \mathbb{U} \mid \tau^\delta(x) \equiv^\delta t\ list\} \\ &= \mathsf{lfp}^{\subseteq}\, \mathbb{F}[\![t]\!]\end{aligned}$$

where $\mathbb{F}[\![t]\!] \triangleq \mathcal{X} \mapsto (\{\langle hd,\ tl\rangle \mid \tau^\delta(hd) = t \wedge tl \in \mathcal{X}\} \cup \{\mathtt{nil}\})$.

Notice that the fixpoint definition corresponds to an Ocaml definition like `type intlist = Nil | Cons of (int * intlist)`.

Proof of lemma 49.7

— By recurrence, $\mathbb{V}^0[\![t\ list]\!] = \{\mathtt{nil}\}$.
$\mathbb{V}^1[\![t\ list]\!] = \{x :: y \mid x \in \mathbb{V}[\![t]\!] \wedge y \in \mathbb{V}^0[\![t\ list]\!]\} = \{x :: \mathtt{nil} \mid x \in \mathbb{V}[\![t]\!]\}$.
If, by recurrence hypothesis, $\mathbb{V}^n[\![t\ list]\!] = \{x_1 :: x_2 :: \ldots x_n :: \mathtt{nil} \mid \forall i \in [1,n]\,.\,\tau^\delta(x_i) = t\}$ then $\mathbb{V}^{n+1}[\![t\ list]\!] = \{x :: y \mid x \in \mathbb{V}[\![t]\!] \wedge y \in \mathbb{V}^n[\![t\ list]\!]\} = \{x :: x_1 :: x_2 :: \ldots x_n :: \mathtt{nil} \mid x \in \mathbb{V}[\![t]\!] \wedge \forall i \in [1,n]\,.\,\tau^\delta(x_i) = t\} = \{x'_1 :: x'_2 :: \ldots x'_{n+1} :: \mathtt{nil} \mid n \in \mathbb{N}^+ \wedge \forall i \in [1,n+1]\,.\,\tau^\delta(x'_i) = t\}$ ⦃letting $x'_1 = x,\ x'_2 = x_1, \ldots x'_{n+1} = x_n$⦄.
It follows that $\mathbb{V}[\![t_1\ list]\!] = \bigcup_{n\in\mathbb{N}} \mathbb{V}^n[\![t\ list]\!] = \{x_1 :: x_2 :: \ldots x_n :: \mathtt{nil} \mid n \in \mathbb{N}^+ \wedge \forall i \in [1,n]\,.\,\tau^\delta(x_i) = t\} \cup \{\mathtt{nil}\}$.
— if $x \in \mathbb{V}[\![t\ list]\!]$ then $x = x_1 :: x_2 :: \ldots x_n :: \mathtt{nil} \in \mathbb{U}$ where $\forall i \in [1,n]\,.\,\tau^\delta(x_i) = t$ so that $\tau^\delta(x) = \boldsymbol{nil}$ for $n = 0$ and $\tau^\delta(x) = t\ list$ for $n > 0$, proving $\mathbb{V}[\![t\ list]\!] \subseteq \{x \in \mathbb{U} \mid \tau^\delta(x) \equiv^\delta t\ list\}$. Inversely, if $\tau^\delta(x) \in \{nil, t\ list$ then either $x = \mathtt{nil}$ when $\tau^\delta(x) = \boldsymbol{nil}$ or else $\tau^\delta(x) = t\ list$ in which case, by recurrence, $x = x_1 :: x_2 :: \ldots x_n :: \mathtt{nil} \in \mathbb{V}[\![t\ list]\!]$. We conclude by antisymmetry.
— Let $\mathbb{F}[\![t]\!]^k$, $k \in \mathbb{N}$ be the iterates of $\mathbb{F}[\![t]\!]$ from $\mathbb{F}[\![t]\!]^0 = \emptyset$. $\mathbb{F}[\![t]\!]^1 = \{\mathtt{nil}\}$, $\mathbb{F}[\![t]\!]^2 = \{\mathtt{nil}\} \cup \{\langle x_1,\ \mathtt{nil}\rangle \mid \tau^\delta(x_1) = t\}$. By recurrence, $\mathbb{F}[\![t]\!]^k$ contains all lists of elements of type t of length less than k. Passing to the limit $\mathsf{lfp}^{\subseteq}\, \mathbb{F}[\![t]\!] = \mathbb{F}[\![t]\!]^\omega = \mathbb{F}[\![t]\!]^{\omega+1} = \mathbb{V}[\![t\ list]\!]$. □

49.6 Syntax of Symbolic Expressions

We extend expressions of section 3.4 to handle pairs and lists.

$\mathtt{x}, \mathtt{y}, \ldots$	$\in \mathbb{V}$			variables
$\mathtt{A}$	$\in \mathcal{A}$	$::=$	$\mathtt{1} \mid \mathtt{x} \mid \mathtt{A}_1 - \mathtt{A}_2$	arithmetic expressions
$\mathtt{S}$	$\in \mathcal{Se}$	$::=$	$\mathtt{A} \mid [] \mid \mathtt{S}_1 , \mathtt{S}_2 \mid \mathtt{S}_1 :: \mathtt{S}_2$	symbolic expressions
		$\mid$	$\mathtt{hd(S)} \mid \mathtt{tl(S)}$	
$\mathtt{B}$	$\in \mathcal{B}$	$::=$	$\mathtt{A}_1 < \mathtt{A}_2 \mid \mathtt{S}_1 = [] \mid \mathtt{B}_1 \ \mathsf{nand}\ \mathtt{B}_2$	Boolean expressions
$\mathtt{E}$	$\in \mathcal{E}$	$::=$	$\mathtt{S} \mid \mathtt{B}$	expressions

Empty brackets [] denotes the null list `nil`, a comma , denotes pairs, and :: denotes lists. In many functional languages, $\mathtt{S}_1 :: \ldots :: \mathtt{S}_n :: \mathtt{nil}$ is denoted $[\mathtt{S}_1; \ldots ; \mathtt{S}_n]$. $\mathtt{hd}([\mathtt{S}_1; \ldots ; \mathtt{S}_n]) = \mathtt{S}_1$ and $\mathtt{tl}([\mathtt{S}_1; \mathtt{S}_2; \ldots ; \mathtt{S}_n]) = [\mathtt{S}_2; \ldots ; \mathtt{S}_n]$. hd(S), and tl(S) is a static error Ω^σ if S is not a list and is a dynamic error Ω^δ if S is a `nil` list. Both types of errors, Ω^σ and Ω^δ, must be caught by a dynamic type system at runtime. Only static errors Ω^σ must be caught by a static type system at compile time. (Sophisticated typing systems might also be able to catch some, if not all, dynamic errors Ω^δ; see exercise 49.34).

49.7 Semantics of Symbolic Expressions

Given an environment $\rho \in \mathbb{E}\mathrm{v} \triangleq \mathbb{V} \to \mathbb{V}^{\Omega^\delta}$ mapping variables $\mathtt{x} \in \mathbb{V}$ to their symbolic or dynamic error value $\rho(\mathtt{x}) \in \mathbb{V}^{\Omega^\delta}$, the evaluation of an expression E returns symbolic or erroneous value $\mathscr{E}[\![\mathtt{E}]\!]\rho \in \mathbb{V}^{\Omega}$.

$$\mathscr{E}[\![\mathtt{E}]\!] \quad \in \quad \mathbb{E}\mathrm{v} \to \mathbb{V}^{\Omega}$$

The evaluation of arithmetic expressions in section 3.6 is essentially unchanged except for the check that arithmetic and comparison operators must be applied to integers (not to pairs or lists).

$$\begin{array}{rcl}
\mathscr{A}[\![\mathtt{1}]\!]\rho & \triangleq & 1 \\
\mathscr{A}[\![\mathtt{x}]\!]\rho & \triangleq & \rho(\mathtt{x}) \\
\mathscr{A}[\![\mathtt{A}_1 - \mathtt{A}_2]\!]\rho & \triangleq & \mathsf{let}\ v_1 = \mathscr{A}[\![\mathtt{A}_1]\!]\rho\ \mathsf{and}\ v_2 = \mathscr{A}[\![\mathtt{A}_2]\!]\rho\ \mathsf{in} \\
& & (\!| v_1 \in \mathbb{Z} \wedge v_2 \in \mathbb{Z} \mathrel{?} v_1 - v_2 \mathrel{:} \Omega^\sigma |\!) \\
\mathscr{B}[\![\mathtt{A}_1 < \mathtt{A}_2]\!]\rho & \triangleq & \mathsf{let}\ v_1 = \mathscr{A}[\![\mathtt{A}_1]\!]\rho\ \mathsf{and}\ v_2 = \mathscr{A}[\![\mathtt{A}_2]\!]\rho\ \mathsf{in} \\
& & (\!| v_1 \in \mathbb{Z} \wedge v_2 \in \mathbb{Z} \mathrel{?} v_1 < v_2 \mathrel{:} \Omega^\sigma |\!)
\end{array}$$

The variables might be pairs or lists. In the absence of overflows and division by zero, these errors are the only possible errors and they are static.

The evaluation of symbolic expressions is defined as follows. Types are checked at runtime and return a static error Ω^σ for errors checkable at compile time and a Ω^δ for those checkable at runtime time. Any operation with an erroneous argument is erroneous. The well-formedness of lists is checked when constructed (initial list values of variables are assumed to be well formed).

$$
\begin{aligned}
\mathscr{E}[\![[]]\!]\rho &\triangleq \texttt{nil} \qquad (49.8)\\
\mathscr{E}[\![\mathsf{S}_1\,,\mathsf{S}_2]\!]\rho &\triangleq \text{let } v_1 = \mathscr{E}[\![\mathsf{S}_1]\!]\rho \text{ and } v_2 = \mathscr{E}[\![\mathsf{S}_2]\!]\rho \text{ in}\\
&\qquad (\!\![\, v_1 = \Omega^\sigma \vee v_2 = \Omega^\sigma \mathrel{?} \Omega^\sigma\\
&\qquad \| \, v_1 = \Omega^\delta \vee v_2 = \Omega^\delta \mathrel{?} \Omega^\delta\\
&\qquad \mathrel{\text{⸬}} \langle v_1,\, v_2\rangle \,]\!\!)\\
\mathscr{E}[\![\mathsf{S}_1 :: \mathsf{S}_2]\!]\rho &\triangleq \text{let } v_1 = \mathscr{E}[\![\mathsf{S}_1]\!]\rho \text{ and } v_2 = \mathscr{E}[\![\mathsf{S}_2]\!]\rho \text{ in}\\
&\qquad (\!\![\, v_1 = \Omega^\sigma \vee v_2 = \Omega^\sigma \mathrel{?} \Omega^\sigma\\
&\qquad \| \, v_1 = \Omega^\delta \vee v_2 = \Omega^\delta \mathrel{?} \Omega^\delta\\
&\qquad \| \, \tau^\delta(v_1 :: v_2) \neq \textbf{\textit{err}} \mathrel{?} v_1 :: v_2 \mathrel{\text{⸬}} \Omega^\sigma \,]\!\!)\\
\mathscr{E}[\![\texttt{hd}(\mathsf{S}_1)]\!]\rho &\triangleq \text{let } v_1 = \mathscr{E}[\![\mathsf{S}_1]\!]\rho \text{ in}\\
&\qquad (\!\![\, v_1 = \Omega^\sigma \mathrel{?} \Omega^\sigma \,\|\, v_1 = \Omega^\delta \vee v_1 = \texttt{nil} \mathrel{?} \Omega^\delta\\
&\qquad \| \, \exists t \neq \textbf{\textit{err}}\,.\,\tau^\delta(v_1) = t \textbf{\textit{ list}} \mathrel{?} \text{let } v_h :: v_t = v_1 \text{ in } v_h \mathrel{\text{⸬}} \Omega^\sigma \,]\!\!)\\
\mathscr{E}[\![\texttt{tl}(\mathsf{S}_1)]\!]\rho &\triangleq \text{let } v_1 = \mathscr{E}[\![\mathsf{S}_1]\!]\rho \text{ in}\\
&\qquad (\!\![\, v_1 = \Omega^\sigma \mathrel{?} \Omega^\sigma \,\|\, v_1 = \Omega^\delta \vee v_1 = \texttt{nil} \mathrel{?} \Omega^\delta\\
&\qquad \| \, \exists t \neq \textbf{\textit{err}}\,.\,\tau^\delta(v_1) = t \textbf{\textit{ list}} \mathrel{?} \text{let } v_h :: v_t = v_1 \text{ in } v_t \mathrel{\text{⸬}} \Omega^\sigma \,]\!\!)
\end{aligned}
$$

Exercise 49.9 Change the definition of the semantics in case variables are not assumed to initially contain well formed values.

The evaluation of Boolean expressions in section 3.6 is changed to incorporate the check of lists for nullity. It is a static error Ω^σ if the check for nullity is not on a list. Therefore the value of a Boolean expressions is in $\mathbb{B} \cup \{\Omega^\sigma, \Omega^\delta\}$.

$$
\begin{aligned}
\mathscr{B}[\![\mathsf{S}_1 = []]\!]\rho &\triangleq \text{let } v_1 = \mathscr{E}[\![\mathsf{S}_1]\!]\rho \text{ in}\\
&\qquad (\!\![\, v_1 = \Omega^\sigma \mathrel{?} \Omega^\sigma \,\|\, v_1 = \Omega^\delta \mathrel{?} \Omega^\delta\\
&\qquad \| \, v_1 = \texttt{nil} \mathrel{?} \text{tt} \,\|\, \exists v_h, v_t\,.\,v_1 = v_h :: v_t \mathrel{?} \text{ff} \mathrel{\text{⸬}} \Omega^\sigma \,]\!\!)\\
\mathscr{B}[\![\mathsf{B}_1 \texttt{ nand } \mathsf{B}_2]\!]\rho &\triangleq \text{let } v_1 = \mathscr{B}[\![\mathsf{B}_1]\!]\rho \text{ and } v_2 = \mathscr{B}[\![\mathsf{B}_2]\!]\rho \text{ in}\\
&\qquad (\!\![\, v_1 = \Omega^\sigma \vee v_2 = \Omega^\sigma \mathrel{?} \Omega^\sigma\\
&\qquad \| \, v_1 = \Omega^\delta \vee v_2 = \Omega^\delta \mathrel{?} \Omega^\delta\\
&\qquad (\!\![\, v_1 \in \mathbb{B} \wedge v_2 \in \mathbb{B} \mathrel{?} v_1 \uparrow v_2 \mathrel{\text{⸬}} \Omega^\sigma \,]\!\!)
\end{aligned}
$$

Exercise 49.10 Write a program (in the language of your choice) to implement the evaluation of an expression (given in abstract syntax) in a given environment.

49.8 Monomorphic Static Typing

Monomorphic static typing checks statically, at compile time, that evaluation of an expression E will never return a static error Ω^σ. This does not necessarily prevent dynamic errors Ω^δ that can be detected only at runtime (or by more sophisticated static analyses).

49.8.1 Monomorphic Static Types

Monomorphic static types are ground terms with signature $\mathsf{F}_0 = \{\texttt{bool}, \texttt{int}\}$, $\mathsf{F}_1 = \{\texttt{list}\}$ in postfix notation, $\mathsf{F}_2 = \{*\}$ in infix notation, and otherwise $\mathsf{F}_n = \varnothing, n > 2$.

$$
\begin{array}{llll}
\boldsymbol{\mu} \in \mathbf{M} & ::= & & \text{monotypes} \\
& & \texttt{bool} & \text{Boolean type} \\
& \mid & \texttt{int} & \text{integer type} \\
& \mid & \boldsymbol{\mu}_1 * \boldsymbol{\mu}_2 & \text{pair type} \\
& \mid & \boldsymbol{\mu}_1\ \texttt{list} & \text{list type} \\
\boldsymbol{M} \in \wp(\mathbf{M}) & & & \text{set of monotypes}
\end{array}
$$

For example, $1 :: \texttt{nil}$ has type $\texttt{int list}$ whereas $\texttt{nil} :: 1$ is not typable.

49.8.2 Semantics of Monomorphic Static Types

A monomorphic type denotes a set of well formed values in $\mathbb{V}$ or a dynamic error Ω^δ, so $\boldsymbol{\gamma}[\![\boldsymbol{\mu}]\!] \in \wp(\mathbb{V}^{\Omega^\delta})$.

$$
\begin{aligned}
\boldsymbol{\gamma}[\![\texttt{int}]\!] &\triangleq \mathbb{Z} \cup \{\Omega^\delta\} && (49.11)\\
\boldsymbol{\gamma}[\![\texttt{bool}]\!] &\triangleq \mathbb{B} \cup \{\Omega^\delta\} \\
\boldsymbol{\gamma}[\![\boldsymbol{\mu}_1 * \boldsymbol{\mu}_2]\!] &\triangleq \{\langle v_1, v_2\rangle \mid v_1 \in \boldsymbol{\gamma}[\![\boldsymbol{\mu}_1]\!] \setminus \{\Omega^\delta\} \wedge v_2 \in \boldsymbol{\gamma}[\![\boldsymbol{\mu}_2]\!] \setminus \{\Omega^\delta\}\} \cup \{\Omega^\delta\} \\
\boldsymbol{\gamma}[\![\boldsymbol{\mu}_1\ \texttt{list}]\!] &\triangleq \{v_1 :: \ldots :: v_n :: \texttt{nil} \mid && (49.12)\\
&\qquad n \in \mathbb{N}^+ \wedge \forall i \in [1, n] \,.\, v_i \in \boldsymbol{\gamma}[\![\boldsymbol{\mu}_1]\!] \setminus \{\Omega^\delta\}\} \cup \{\texttt{nil}, \Omega^\delta\}
\end{aligned}
$$

By defining the following correspondence between static and dynamic types:

$$
\begin{aligned}
\delta(\texttt{bool}) &\triangleq \textit{\textbf{bool}} \\
\delta(\texttt{int}) &\triangleq \textit{\textbf{int}} \\
\delta(\boldsymbol{\mu}_1 * \boldsymbol{\mu}_2) &\triangleq \delta(\boldsymbol{\mu}_1) \bigstar \delta(\boldsymbol{\mu}_2) \\
\delta(\boldsymbol{\mu}_1\ \texttt{list}) &\triangleq \delta(\boldsymbol{\mu}_1)\ \textit{\textbf{list}}
\end{aligned}
$$

we have

$$
\boldsymbol{\gamma}[\![\boldsymbol{\mu}]\!] = \{v \in \mathbb{V} \mid \tau^\delta(v) = \delta(\boldsymbol{\mu})\} \cup \{\Omega^\delta\} \cup (\!\![\, \boldsymbol{\mu} = \boldsymbol{\mu}_1\ \texttt{list}\ ?\ \{\texttt{nil}\} \,:\, \varnothing \,]\!\!) \qquad (49.13)
$$

Proof of (49.13) The proof is by structural induction on $\boldsymbol{\mu}$.

— If $\boldsymbol{\mu} = \texttt{bool}$ then $\{v \in \mathbb{V} \mid \tau^\delta(v) = \delta(\texttt{bool})\} \cup \{\Omega^\delta\} = \{v \in \mathbb{V} \mid \tau^\delta(v) = \textit{\textbf{bool}}\} \cup \{\Omega^\delta\} = \mathbb{B} \cup \{\Omega^\delta\}$ ⟨by (49.4)⟩ $= \boldsymbol{\gamma}[\![\texttt{bool}]\!]$ ⟨by (49.11)⟩.

— The proof is similar for $\boldsymbol{\mu} = \texttt{int}$.

— If $\boldsymbol{\mu} = \boldsymbol{\mu}_1 * \boldsymbol{\mu}_2$ then $\{v \in \mathbb{V} \mid \tau^\delta(v) = \delta(\boldsymbol{\mu}_1 * \boldsymbol{\mu}_2)\} \cup \{\Omega^\delta\} = \{v \in \mathbb{V} \mid \tau^\delta(v) = \delta(\boldsymbol{\mu}_1) \bigstar \delta(\boldsymbol{\mu}_2)\} \cup \{\Omega^\delta\}$ ⟨definition of δ⟩ $= \{\langle x, y\rangle \in \mathbb{V} \mid \tau^\delta(\langle x, y\rangle) = \delta(\boldsymbol{\mu}_1) \bigstar \delta(\boldsymbol{\mu}_2)\} \cup \{\Omega^\delta\}$ ⟨definition (49.4) of τ^δ so that only pairs have a pair type $\bigstar$⟩ $= \{\langle x, y\rangle \in \mathbb{V} \mid \tau^\delta(x) = \delta(\boldsymbol{\mu}_1) \wedge \tau^\delta(y) = \delta(\boldsymbol{\mu}_2)\} \cup \{\Omega^\delta\}$ ⟨definition of δ so that $\delta(\boldsymbol{\mu}) \neq \textit{\textbf{err}}$ and (49.4) of $\tau^\delta(\langle x, y\rangle)$⟩ $= \{\langle x, y\rangle \in \mathbb{V} \mid x \in \boldsymbol{\gamma}[\![\boldsymbol{\mu}_1]\!] \wedge y \in \boldsymbol{\gamma}[\![\boldsymbol{\mu}_2]\!]\} \cup \{\Omega^\delta\}$ ⟨induction hypothesis⟩ $= \{\langle x, y\rangle \in \mathbb{V} \mid x \in \boldsymbol{\gamma}[\![\boldsymbol{\mu}_1]\!] \setminus \{\Omega^\delta\} \wedge y \in \boldsymbol{\gamma}[\![\boldsymbol{\mu}_2]\!] \setminus \{\Omega^\delta\}\} \cup \{\Omega^\delta\}$ ⟨definition (49.3) of $\mathbb{V}$ so that $\Omega^\delta \notin \mathbb{V}$⟩ $= \boldsymbol{\gamma}[\![\boldsymbol{\mu}_1 * \boldsymbol{\mu}_2]\!]$ ⟨definition (49.11) of $\boldsymbol{\gamma}$⟩.

— If $\mu = \mu_1$ list then $\gamma[\![\mu_1 \texttt{ list}]\!] = \{v \in \mathbb{V} \mid \tau^\delta(v) = \delta(\mu)\} \cup \{\Omega^\delta\} \cup (\!\mid \mu_1 \texttt{ list} = \mu_1 \texttt{ list} \,?\, \{\texttt{nil}\} \,\text{:}\, \varnothing \mid\!) = \{v \in \mathbb{V} \mid \tau^\delta(v) = \delta(\mu_1 \texttt{ list})\} \cup \{\texttt{nil}, \Omega^\delta\} = \{v \in \mathbb{V} \mid \tau^\delta(v) = \delta(\mu_1)\ \textit{list}\} \cup \{\texttt{nil}, \Omega^\delta\}$ ⟅definition of δ⟆ $= \{x :: y \in \mathbb{V} \mid \tau^\delta(x :: y) = \delta(\mu_1)\ \textit{list}\} \cup \{\texttt{nil}, \Omega^\delta\}$ ⟅definition (49.4) of τ^δ so that only lists have a list type *list*⟆ $= \{x :: y \in \mathbb{V} \mid (\!\mid \exists t \neq \textit{err} \,.\, \tau^\delta(x) = t \wedge \tau^\delta(y) \equiv^\delta t\ \textit{list} \,?\, t\ \ \textit{list} \,\text{:}\, \textit{err} \mid\!) = \delta(\mu_1)\ \textit{list}\} \cup \{\texttt{nil}, \Omega^\delta\}$ ⟅definition (49.4) of $\tau^\delta(x :: y)$⟆ $= \{x :: y \in \mathbb{V} \mid \tau^\delta(x) = \delta(\mu_1) \wedge \tau^\delta(y) \in \{\textit{nil}, \delta(\mu_1)\ \textit{list}\}\} \cup \{\texttt{nil}, \Omega^\delta\}$ ⟅because $\forall \mu_1 \,.\, \delta(\mu_1) \neq \textit{err} \wedge \delta(\mu_1)\ \textit{list} \neq \textit{err}$ and so $t = \delta(\mu_1)$⟆ $= \{v \in \mathbb{U} \mid \tau^\delta(v) \in \{\textit{nil}, \delta(\mu_1)\ \textit{list}\}\} \cup \{\texttt{nil}, \Omega^\delta\}$ ⟅definition (49.4) of τ^δ and $\forall \mu_1 \,.\, \delta(\mu_1) \neq \textit{err}$⟆ $= \mathbb{V}[\![\delta(\mu_1)\ \textit{list}]\!] \cup \{\Omega^\delta\}$ ⟅lemma 49.7⟆ $= \{x_1 :: x_2 :: \ldots x_n :: \texttt{nil} \mid n \in \mathbb{N}^+ \wedge \forall i \in [1, n] \,.\, \tau^\delta(x_i) = \delta(\mu_1)\} \cup \{\texttt{nil}, \Omega^\delta\}$ ⟅lemma 49.7 and $\forall \mu_1 \,.\, \delta(\mu_1) \neq \textit{err}$⟆ $= \{v_1 :: \ldots :: v_n :: \texttt{nil} \mid n \in \mathbb{N}^+ \wedge \forall i \in [1, n] \,.\, v_i \in \gamma[\![\mu_1]\!] \setminus \{\Omega^\delta\}\} \cup \{\texttt{nil}, \Omega^\delta\}$ ⟅structural induction hypothesis⟆ $= \gamma[\![\mu_1 \texttt{ list}]\!]$ ⟅definition (49.11) of γ⟆. □

49.8.3 Type Assignments

The type of an expression (like x − x) can be determined only if we know the type $\Gamma(\mathsf{x})$ of its free variables $\mathsf{x} \in \mathbb{V}$. This is provided by a type assignment Γ.

$$\Gamma \in \mathbf{A} \quad \triangleq \quad \mathbb{V} \to \mathbf{M} \qquad \text{monotype assignment}$$

49.8.4 Semantics of Type Assignments

The type assignment Γ denotes an environment property in $\wp(\mathbb{E}_\mathbb{V})$ that is a set of environments assigning values to program variables.

$$\gamma[\![\Gamma]\!] \quad \triangleq \quad \{\rho \in \mathbb{V} \to \mathbb{V}^{\Omega^\delta} \mid \forall \mathsf{x} \in \mathbb{V} \,.\, \rho(\mathsf{x}) \in \gamma[\![\Gamma(\mathsf{x})]\!]\} \tag{49.14}$$

49.8.5 Typings

A typing $\langle \Gamma, \mu \rangle \in \Theta$ expresses the property that under the hypothesis that the variables x of an expression E have type $\Gamma(\mathsf{x})$ then the expression E has type μ.

$$\theta \in \Theta \quad \triangleq \quad \mathbf{A} \times \mathbf{M} \qquad \text{monotyping} \tag{49.15}$$

49.8.6 Semantics of Typings

A typing $\langle \Gamma, \mu \rangle$ denotes an expression semantic property in $\wp(\mathbb{E}_\mathbb{V} \to \mathbb{V}^\Omega)$.

$$\gamma[\![\langle \Gamma, \mu \rangle]\!] \quad \triangleq \quad \{f \in \mathbb{E}_\mathbb{V} \to \mathbb{V}^\Omega \mid \forall \rho \in \gamma[\![\Gamma]\!] \,.\, f(\rho) \in \gamma[\![\mu]\!]\} \tag{49.16}$$

49.8.7 Type Judgments

We write the type judgment $\Gamma \vdash \mathsf{E} : \mu$ to mean that expression E has monotype μ in monotype assignment Γ and $\Gamma \nvdash \mathsf{E}$ to mean that expression is not typable in type assignment Γ.

Example 49.17 If $\Gamma(\mathsf{x}) = \texttt{int}$ then $\Gamma \vdash \mathsf{x} - 1 : \texttt{int}$. If $\Gamma(\mathsf{x}) = \texttt{int} * \texttt{int}$ then $\Gamma \nvdash \mathsf{x} - 1$. For all Γ, $\Gamma \vdash 1 :: [] : \texttt{int list}$ and $\Gamma \nvdash [] :: 1$.

49.8.8 Semantics of Type Judgments

The type judgment $\Gamma \vdash \mathsf{E} : \mu$ states that the semantics $\mathscr{E}[\![\mathsf{E}]\!]$ of this expression E has the property $\gamma[\![\langle \Gamma, \mu \rangle]\!]$.

$$\Gamma \vdash \mathsf{E} : \mu \quad \triangleq \quad \mathscr{E}[\![\mathsf{E}]\!] \in \gamma[\![\langle \Gamma, \mu \rangle]\!] \tag{49.18}$$

Otherwise stated, $\boldsymbol{\Gamma} \vdash \mathsf{E} : \boldsymbol{\mu}$ means that the expression E (understand its semantics $\mathscr{E}[\![\mathsf{E}]\!]$) has the semantics property $\boldsymbol{\gamma}[\![\langle \boldsymbol{\Gamma}, \boldsymbol{\mu} \rangle]\!]$ denoted by the typing $\langle \boldsymbol{\Gamma}, \boldsymbol{\mu} \rangle$.

49.8.9 Soundness: Typable Expressions Cannot Go Wrong

An expression E is typable if and only if there exists a type judgment $\boldsymbol{\Gamma} \vdash \mathsf{E} : \boldsymbol{\mu}$. Typable expressions cannot go wrong because the execution of a typable expression cannot produce a static error (but it can produce a dynamic one such as hd([])).

A typable expression cannot go wrong in that its evaluation never returns a static error Ω^{σ}.

Lemma 49.19 (typable expressions cannot go wrong) If $\boldsymbol{\Gamma} \vdash \mathsf{E} : \boldsymbol{\mu}$ then $\forall \rho \in \boldsymbol{\gamma}[\![\boldsymbol{\Gamma}]\!] .\ \mathscr{E}[\![\mathsf{E}]\!]\rho \neq \Omega^{\sigma}$.

Notice that *go wrong* means evaluate to a static error Ω^{σ}, which does not exclude evaluating to a dynamic error Ω^{δ}.

Proof of lemma 49.19

$$
\begin{array}{ll}
& \boldsymbol{\Gamma} \vdash \mathsf{E} : \boldsymbol{\mu} \\
\Leftrightarrow & \mathscr{E}[\![\mathsf{E}]\!] \in \boldsymbol{\gamma}[\![\langle \boldsymbol{\Gamma}, \boldsymbol{\mu} \rangle]\!] \qquad \wr(49.18)\wr \\
\Leftrightarrow & \forall \rho \in \boldsymbol{\gamma}[\![\boldsymbol{\Gamma}]\!] .\ \mathscr{E}[\![\mathsf{E}]\!](\rho) \in \boldsymbol{\gamma}[\![\boldsymbol{\mu}]\!] \qquad \wr(49.16)\wr \\
\Leftrightarrow & \forall \rho \in \boldsymbol{\gamma}[\![\boldsymbol{\Gamma}]\!] .\ \mathscr{E}[\![\mathsf{E}]\!](\rho) \in \{v \in \mathbb{V} \mid \tau^{\delta}(v) = \delta(\boldsymbol{\mu})\} \cup \{\Omega^{\delta}\} \qquad \wr(49.13) \text{ and } \Omega^{\sigma} \notin \mathbb{V}\wr \quad \square
\end{array}
$$

Notice that we have proved type soundness in lemma 49.19 by considering the semantics of types only. This is to be contrasted with the classic approach [796, section 8.3], which proceeds by induction on a derivation of the type $\boldsymbol{\mu}$ of E according to type inference rules that have to be guessed a priori. In contrast, we use the semantic interpretation $\boldsymbol{\gamma}$ of types to later design the type inference rules in figure 49.41 by calculus (in section 49.9.3), which are then sound by construction.

49.8.10 Expression Type

An expression has in general many typings. For example, the expression x has all typings $\{\langle \boldsymbol{\Gamma}, \boldsymbol{\Gamma}(\mathsf{x}) \rangle \mid \boldsymbol{\Gamma} \in \mathbf{A}\}$. Therefore the abstract property of the expression x, that is, its type, is the set of all these typings.

$$\boldsymbol{\tau} \in \mathbf{T} \quad \triangleq \quad \wp(\boldsymbol{\Theta}) \qquad \text{type (property)} \tag{49.20}$$

49.8.11 Semantics of an Expression Type

If an expression has many typings $\boldsymbol{\theta} \in \boldsymbol{\tau}$, its properties must be those $\boldsymbol{\gamma}[\![\boldsymbol{\theta}]\!]$ shared by all these typings $\boldsymbol{\theta} \in \boldsymbol{\tau}$.

$$\boldsymbol{\gamma}[\![\boldsymbol{\tau}]\!] \quad \triangleq \quad \bigcap \{\boldsymbol{\gamma}[\![\boldsymbol{\theta}]\!] \mid \boldsymbol{\theta} \in \boldsymbol{\tau}\} \tag{49.21}$$

It follows, by the dual of exercise 11.39 that there is a Galois connection between expression properties and expression types.

$$\langle \wp(\mathbb{E}\mathbb{v} \rightarrow \mathbb{V}^{\Omega}), \subseteq \rangle \underset{\boldsymbol{\alpha}}{\overset{\boldsymbol{\gamma}}{\leftrightarrows}} \langle \mathbf{T}, \supseteq \rangle \tag{49.22}$$

49.8.12 Type Abstract Semantics of Expressions

The type abstract semantics of expressions is defined as an overapproximation of the type abstraction of the collecting semantics.

$$\boldsymbol{\alpha}(\{\mathscr{E}[\![\mathsf{E}]\!]\}) = \{\langle\boldsymbol{\Gamma}, \boldsymbol{\mu}\rangle \mid \boldsymbol{\Gamma} \vdash \mathsf{E} : \boldsymbol{\mu}\} \overset{\triangle}{\supseteq} \boldsymbol{\tau}[\![\mathsf{E}]\!] \tag{49.23}$$

Proof of (49.23)

$$\begin{aligned}
&\boldsymbol{\alpha}(\{\mathscr{E}[\![\mathsf{E}]\!]\})\\
=\ &\bigcup\{\boldsymbol{\tau} \mid \{\mathscr{E}[\![\mathsf{E}]\!]\} \subseteq \boldsymbol{\gamma}[\![\boldsymbol{\tau}]\!]\} && \text{(Galois connection (49.22) and lemma 11.42)}\\
=\ &\bigcup\{\boldsymbol{\tau} \mid \mathscr{E}[\![\mathsf{E}]\!] \in \boldsymbol{\gamma}[\![\boldsymbol{\tau}]\!]\} && \text{(definition of } \subseteq\text{)}\\
=\ &\bigcup\{\boldsymbol{\tau} \mid \mathscr{E}[\![\mathsf{E}]\!] \in \bigcap\{\boldsymbol{\gamma}[\![\boldsymbol{\theta}]\!] \mid \boldsymbol{\theta} \in \boldsymbol{\tau}\}\} && \text{(definition (49.21) of } \boldsymbol{\gamma}[\![\boldsymbol{\tau}]\!]\text{)}\\
=\ &\bigcup\{\boldsymbol{\tau} \mid \forall \boldsymbol{\theta} \in \boldsymbol{\tau} \,.\, \mathscr{E}[\![\mathsf{E}]\!] \in \boldsymbol{\gamma}[\![\boldsymbol{\theta}]\!]\} && \text{(definition of } \bigcap\text{)}\\
=\ &\{\langle\boldsymbol{\Gamma}, \boldsymbol{\mu}\rangle \mid \mathscr{E}[\![\mathsf{E}]\!] \in \boldsymbol{\gamma}[\![\langle\boldsymbol{\Gamma}, \boldsymbol{\mu}\rangle]\!]\} && \text{(definitions of } \bigcup \text{ and (49.15) of } \boldsymbol{\theta} = \langle\boldsymbol{\Gamma}, \boldsymbol{\mu}\rangle\text{)}\\
=\ &\{\langle\boldsymbol{\Gamma}, \boldsymbol{\mu}\rangle \mid \boldsymbol{\Gamma} \vdash \mathsf{E} : \boldsymbol{\mu}\} && \text{(definition (49.18) of } \boldsymbol{\Gamma} \vdash \mathsf{E} : \boldsymbol{\mu}\text{)} \quad \square
\end{aligned}$$

Example 49.24 (type of an expression that always goes wrong) Assume that we have $\forall \rho \in \mathbb{E}\mathbb{v} \,.\, \mathscr{E}[\![\mathsf{E}]\!]\rho = \Omega^{\delta}$. For example, $\mathsf{E} = \mathsf{hd}([])$ or $\mathsf{E} = \mathtt{tl}([])$.

By definition (49.11) of $\boldsymbol{\gamma}[\![\boldsymbol{\mu}]\!]$, $\forall \boldsymbol{\mu} \in \mathbf{M} \,.\, \forall \rho \in \mathbb{E}\mathbb{v} \,.\, \mathscr{E}[\![\mathsf{E}]\!]\rho = \Omega^{\delta} \in \boldsymbol{\gamma}[\![\boldsymbol{\mu}]\!]$ proving $\forall \boldsymbol{\Gamma} \in \mathbf{A} \,.\, \mathscr{E}[\![\mathsf{E}]\!] \in \boldsymbol{\gamma}[\![\langle\boldsymbol{\Gamma}, \boldsymbol{\mu}\rangle]\!]$ by (49.16), hence $\boldsymbol{\Gamma} \vdash \mathsf{E} : \boldsymbol{\mu}$ by (49.18). This implies $\boldsymbol{\tau}[\![\mathsf{E}]\!] = \{\langle\boldsymbol{\Gamma}, \boldsymbol{\mu}\rangle \mid \boldsymbol{\Gamma} \in \mathbf{A} \wedge \boldsymbol{\mu} \in \mathbf{M}\} = \mathbf{A} \times \mathbf{M} = \boldsymbol{\Theta}$ by (49.23) and (49.15).

Reciprocally, if $\boldsymbol{\tau}[\![\mathsf{E}]\!] = \boldsymbol{\Theta}$ then, by (49.21), $\boldsymbol{\gamma}[\![\boldsymbol{\tau}[\![\mathsf{E}]\!]]\!] = \rho \in \mathbb{E}\mathbb{v} \mapsto \Omega^{\delta}$.

In conclusion, $\boldsymbol{\tau}[\![\mathsf{E}]\!] = \boldsymbol{\Theta}$ if and only if E always evaluates to a dynamic error Ω^{δ}.

Example 49.25 (not typable expressions) An expression that is *not typable* has type $\boldsymbol{\tau}[\![\mathsf{E}]\!] = \varnothing$. Then by (49.22),

$$\begin{aligned}
&\boldsymbol{\gamma}[\![\boldsymbol{\tau}[\![\mathsf{E}]\!]]\!]\\
=\ &\boldsymbol{\gamma}[\![\varnothing]\!]\\
=\ &\bigcap\{\boldsymbol{\gamma}[\![\boldsymbol{\theta}]\!] \mid \boldsymbol{\theta} \in \varnothing\} && \text{(by (49.22))}\\
=\ &\bigcap\varnothing\\
=\ &\mathbb{E}\mathbb{v} \to \mathbb{V}^{\Omega}
\end{aligned}$$

This is the true property tt; nothing is known of the evaluations of E that may return any value, a static error Ω^{σ}, or a dynamic one Ω^{δ}.

49.8.13 Calculational Design of the Typing Rules

We get the typing rules by calculating

$$\begin{aligned}
&\boldsymbol{\tau}[\![\mathsf{E}]\!]\\
\overset{\triangle}{\subseteq}\ &\boldsymbol{\alpha}(\{\mathscr{E}[\![\mathsf{E}]\!]\}) && \text{(by definition, } \supseteq\text{-overapproximation of the collecting semantics)}\\
=\ &\{\langle\boldsymbol{\Gamma}, \boldsymbol{\mu}\rangle \mid \mathscr{E}[\![\mathsf{E}]\!] \in \boldsymbol{\gamma}[\![\langle\boldsymbol{\Gamma}, \boldsymbol{\mu}\rangle]\!]\} && \text{(proof of (49.23))}\\
=\ &\{\langle\boldsymbol{\Gamma}, \boldsymbol{\mu}\rangle \mid \boldsymbol{\Gamma} \vdash \mathsf{E} : \boldsymbol{\mu}\} && \text{(definition (49.18) of } \boldsymbol{\Gamma} \vdash \mathsf{E} : \boldsymbol{\mu}\text{)} \quad (49.26)
\end{aligned}$$

$= \{\langle \boldsymbol{\Gamma}, \boldsymbol{\mu}\rangle \mid \forall \rho \in \boldsymbol{\gamma}[\![\boldsymbol{\Gamma}]\!] \,.\, \mathscr{E}[\![\mathsf{E}]\!](\rho) \in \boldsymbol{\gamma}[\![\boldsymbol{\mu}]\!]\}$

$\wr$definition (49.16) of $\boldsymbol{\gamma}[\![\langle \boldsymbol{\Gamma}, \boldsymbol{\mu}\rangle]\!]$ and definition of $\in \wr$

By the Galois connection (49.22), an overapproximation of the concrete properties in $\langle \wp(\mathbb{E}\mathbb{v} \to \mathbb{V}^{\Omega}), \subseteq\rangle$ consists of selecting a subset of the possible types in $\langle \mathbf{T}, \supseteq\rangle$.

Example 49.27 Continuing examples 49.24 and 49.25, $\boldsymbol{\Theta} \supseteq \varnothing$ so the type $\boldsymbol{\tau}[\![\mathsf{E}]\!] = \boldsymbol{\Theta}$ (i.e., the evaluation of E is always a runtime error) can be overapproximated by $\boldsymbol{\tau}[\![\mathsf{E}]\!] = \varnothing$ (i.e., E is *not typable*).

We go on by structural induction on E.

— $\mathsf{E} = 1$

$\{\langle \boldsymbol{\Gamma}, \boldsymbol{\mu}\rangle \mid \forall \rho \in \boldsymbol{\gamma}[\![\boldsymbol{\Gamma}]\!] \,.\, \mathscr{A}[\![1]\!](\rho) \in \boldsymbol{\gamma}[\![\boldsymbol{\mu}]\!]\}$

$= \{\langle \boldsymbol{\Gamma}, \boldsymbol{\mu}\rangle \mid 1 \in \boldsymbol{\gamma}[\![\boldsymbol{\mu}]\!]\}$ $\wr \forall \rho \,.\, \mathscr{A}[\![1]\!](\rho) = 1 \wr$

$= \{\langle \boldsymbol{\Gamma}, \mathtt{int}\rangle \mid \boldsymbol{\Gamma} \in \mathbf{A}\}$

$\wr$by definition (49.11) of $\boldsymbol{\gamma}[\![\boldsymbol{\mu}]\!]$ such that $\boldsymbol{\mu} = \mathtt{int}$ is the only monotype such that $1 \in \boldsymbol{\gamma}[\![\boldsymbol{\mu}]\!] \wr$

— $\mathsf{E} = \mathsf{x}$

$\{\langle \boldsymbol{\Gamma}, \boldsymbol{\mu}\rangle \mid \forall \rho \in \boldsymbol{\gamma}[\![\boldsymbol{\Gamma}]\!] \,.\, \mathscr{A}[\![\mathsf{x}]\!](\rho) \in \boldsymbol{\gamma}[\![\boldsymbol{\mu}]\!]\}$

$= \{\langle \boldsymbol{\Gamma}, \boldsymbol{\mu}\rangle \mid \forall \rho \in \boldsymbol{\gamma}[\![\boldsymbol{\Gamma}]\!] \,.\, \rho(\mathsf{x}) \in \boldsymbol{\gamma}[\![\boldsymbol{\mu}]\!]\}$ $\wr \mathscr{A}[\![\mathsf{x}]\!](\rho) = \rho(\mathsf{x}) \wr$

$= \{\langle \boldsymbol{\Gamma}, \boldsymbol{\mu}\rangle \mid \forall \rho \in \mathcal{V} \to \mathbb{V}^{\Omega^{\delta}} \,.\, \rho(\mathsf{x}) \in \boldsymbol{\gamma}[\![\boldsymbol{\Gamma}(\mathsf{x})]\!] \Rightarrow \rho(\mathsf{x}) \in \boldsymbol{\gamma}[\![\boldsymbol{\mu}]\!]\}$

$\wr$definition (49.14) of $\boldsymbol{\gamma}[\![\boldsymbol{\Gamma}]\!]$ and definition of $\in \wr$

$= \{\langle \boldsymbol{\Gamma}, \boldsymbol{\mu}\rangle \mid \boldsymbol{\gamma}[\![\boldsymbol{\Gamma}(\mathsf{x})]\!] \subseteq \boldsymbol{\gamma}[\![\boldsymbol{\mu}]\!]\}$ $\wr$definition of $\subseteq \wr$

$= \{\langle \boldsymbol{\Gamma}, \boldsymbol{\Gamma}(\mathsf{x})\rangle \mid \boldsymbol{\Gamma} \in \mathbf{A}\}$

$\wr$by definition (49.11) of $\boldsymbol{\gamma}[\![\boldsymbol{\mu}]\!]$ such that $\boldsymbol{\gamma}[\![\boldsymbol{\mu}]\!] \subseteq \boldsymbol{\gamma}[\![\boldsymbol{\mu}']\!]$ if and only if $\boldsymbol{\mu} = \boldsymbol{\mu}' \wr$

— $\mathsf{E} = \mathsf{A}_1 - \mathsf{A}_2$

$\{\langle \boldsymbol{\Gamma}, \boldsymbol{\mu}\rangle \mid \forall \rho \in \boldsymbol{\gamma}[\![\boldsymbol{\Gamma}]\!] \,.\, \mathscr{A}[\![\mathsf{A}_1 - \mathsf{A}_2]\!](\rho) \in \boldsymbol{\gamma}[\![\boldsymbol{\mu}]\!]\}$

$= \{\langle \boldsymbol{\Gamma}, \boldsymbol{\mu}\rangle \mid \forall \rho \in \boldsymbol{\gamma}[\![\boldsymbol{\Gamma}]\!] \,.\, \mathsf{let}\ v_1 = \mathscr{A}[\![\mathsf{A}_1]\!]\rho\ \mathsf{and}\ v_2 = \mathscr{A}[\![\mathsf{A}_2]\!]\rho\ \mathsf{in}\ (\!|\, v_1 \in \mathbb{Z} \wedge v_2 \in \mathbb{Z} \,?\, v_1 - v_2 \,\mathrm{:}\, \Omega^{\sigma} \,|\!) \in \boldsymbol{\gamma}[\![\boldsymbol{\mu}]\!]\}$ $\wr$definition of $\mathscr{A}[\![\mathsf{A}_1 - \mathsf{A}_2]\!]\rho \wr$

$= \{\langle \boldsymbol{\Gamma}, \boldsymbol{\mu}\rangle \mid \forall \rho \in \boldsymbol{\gamma}[\![\boldsymbol{\Gamma}]\!] \,.\, \mathscr{A}[\![\mathsf{A}_1]\!]\rho \in \mathbb{Z} \wedge \mathscr{A}[\![\mathsf{A}_2]\!]\rho \in \mathbb{Z} \wedge \mathscr{A}[\![\mathsf{A}_1]\!]\rho - \mathscr{A}[\![\mathsf{A}_2]\!]\rho \in \boldsymbol{\gamma}[\![\boldsymbol{\mu}]\!]\}$

$\wr$by definition (49.11) of $\boldsymbol{\gamma}[\![\boldsymbol{\mu}]\!]$ such that $\Omega^{\sigma} \notin \boldsymbol{\gamma}[\![\boldsymbol{\mu}]\!] \wr$

$= \{\langle \boldsymbol{\Gamma}, \mathtt{int}\rangle \mid \forall \rho \in \boldsymbol{\gamma}[\![\boldsymbol{\Gamma}]\!] \,.\, \mathscr{A}[\![\mathsf{A}_1]\!]\rho \in \mathbb{Z} \wedge \mathscr{A}[\![\mathsf{A}_2]\!]\rho \in \mathbb{Z}\}$

$\wr$because $\mathscr{A}[\![\mathsf{A}_1]\!]\rho \in \mathbb{Z}$ and $\mathscr{A}[\![\mathsf{A}_2]\!]\rho \in \mathbb{Z}$ imply that $\mathscr{A}[\![\mathsf{A}_1]\!]\rho - \mathscr{A}[\![\mathsf{A}_2]\!]\rho \in \mathbb{Z}$ and definition (49.11) of $\boldsymbol{\gamma}[\![\boldsymbol{\mu}]\!]$ such that $\boldsymbol{\mu} = \mathtt{int}$ is the only monotype such that $\mathscr{A}[\![\mathsf{A}_1]\!]\rho - \mathscr{A}[\![\mathsf{A}_2]\!]\rho \in \boldsymbol{\gamma}[\![\boldsymbol{\mu}]\!] \wr$

$= \{\langle \boldsymbol{\Gamma}, \mathtt{int}\rangle \mid \forall \rho \in \boldsymbol{\gamma}[\![\boldsymbol{\Gamma}]\!] \,.\, \mathscr{A}[\![\mathsf{A}_1]\!]\rho \in \boldsymbol{\gamma}[\![\mathtt{int}]\!] \wedge \mathscr{A}[\![\mathsf{A}_2]\!]\rho \in \boldsymbol{\gamma}[\![\mathtt{int}]\!]\}$

$$\text{\{by definition (49.11) of } \boldsymbol{\gamma}[\![\mathtt{int}]\!] \text{ and } \mathscr{A}[\![\mathsf{A}_1]\!]\rho \neq \Omega^\delta \text{\}}$$

$= \{\langle \boldsymbol{\Gamma}, \mathtt{int}\rangle \mid \mathscr{A}[\![\mathsf{A}_1]\!] \in \boldsymbol{\gamma}[\![\langle \boldsymbol{\Gamma}, \mathtt{int}\rangle]\!] \wedge \mathscr{A}[\![\mathsf{A}_2]\!] \in \boldsymbol{\gamma}[\![\langle \boldsymbol{\Gamma}, \mathtt{int}\rangle]\!]\}$

(definition (49.16) of $\boldsymbol{\gamma}[\![\langle \boldsymbol{\Gamma}, \boldsymbol{\mu}\rangle]\!]$)

$= \{\langle \boldsymbol{\Gamma}, \mathtt{int}\rangle \mid \boldsymbol{\Gamma} \vdash \mathsf{A}_1 : \mathtt{int} \wedge \boldsymbol{\Gamma} \vdash \mathsf{A}_2 : \mathtt{int}\}$ (definition (49.18) of type judgments $\boldsymbol{\Gamma} \vdash \mathsf{E} : \boldsymbol{\mu}$)

— $\mathsf{E} = \mathsf{S}_1 , \mathsf{S}_2$

$\{\langle \boldsymbol{\Gamma}, \boldsymbol{\mu}\rangle \mid \forall \rho \in \boldsymbol{\gamma}[\![\boldsymbol{\Gamma}]\!] \,.\, \mathscr{E}[\![\mathsf{S}_1 , \mathsf{S}_2]\!](\rho) \in \boldsymbol{\gamma}[\![\boldsymbol{\mu}]\!]\}$

$= \{\langle \boldsymbol{\Gamma}, \boldsymbol{\mu}\rangle \mid \forall \rho \in \boldsymbol{\gamma}[\![\boldsymbol{\Gamma}]\!] \,.\, \text{let } v_1 = \mathscr{E}[\![\mathsf{S}_1]\!]\rho \text{ and } v_2 = \mathscr{E}[\![\mathsf{S}_2]\!]\rho \text{ in } (\!| v_1 = \Omega^\sigma \vee v_2 = \Omega^\sigma \mathrel{?} \Omega^\sigma \mathrel{|\!|} v_1 = \Omega^\delta \vee v_2 = \Omega^\delta \mathrel{?} \Omega^\delta \mathrel{:} \langle v_1, v_2\rangle |\!) \in \boldsymbol{\gamma}[\![\boldsymbol{\mu}]\!]\}$ (definition (49.8) of $\mathscr{E}[\![\mathsf{S}_1 , \mathsf{S}_2]\!]\rho$)

$= \{\langle \boldsymbol{\Gamma}, \boldsymbol{\mu}\rangle \mid \forall \rho \in \boldsymbol{\gamma}[\![\boldsymbol{\Gamma}]\!] \,.\, \text{let } v_1 = \mathscr{E}[\![\mathsf{S}_1]\!]\rho \text{ and } v_2 = \mathscr{E}[\![\mathsf{S}_2]\!]\rho \text{ in } (\!| v_1 = \Omega^\delta \vee v_2 = \Omega^\delta \mathrel{?} \Omega^\delta \mathrel{:} \langle v_1, v_2\rangle |\!) \in \boldsymbol{\gamma}[\![\boldsymbol{\mu}]\!]\}$ (by definition (49.11) of $\boldsymbol{\gamma}[\![\boldsymbol{\mu}]\!]$ such that $\Omega^\sigma \notin \boldsymbol{\gamma}[\![\boldsymbol{\mu}]\!]$)

$= \{\langle \boldsymbol{\Gamma}, \boldsymbol{\mu}\rangle \mid \forall \rho \in \boldsymbol{\gamma}[\![\boldsymbol{\Gamma}]\!] \,.\, \text{let } v_1 = \mathscr{E}[\![\mathsf{S}_1]\!]\rho \text{ and } v_2 = \mathscr{E}[\![\mathsf{S}_2]\!]\rho \text{ in } (\!| v_1 = \Omega^\delta \vee v_2 = \Omega^\delta \mathrel{?} \Omega^\delta \mathrel{:} \langle v_1, v_2\rangle |\!) \in (\{\langle v_1, v_2\rangle \mid v_1 \in \boldsymbol{\gamma}[\![\boldsymbol{\mu}_1]\!] \setminus \{\Omega^\delta\} \wedge v_2 \in \boldsymbol{\gamma}[\![\boldsymbol{\mu}_2]\!] \setminus \{\Omega^\delta\}\} \cup \{\Omega^\delta\}) \wedge \boldsymbol{\mu} = \boldsymbol{\mu}_1 * \boldsymbol{\mu}_2\}$

(definition (49.11) of $\boldsymbol{\gamma}[\![\boldsymbol{\mu}]\!]$ so that $\boldsymbol{\mu} = \boldsymbol{\mu}_1 * \boldsymbol{\mu}_2$ is the only type which concretization is a set of pairs or Ω^δ)

$= \{\langle \boldsymbol{\Gamma}, \boldsymbol{\mu}_1 * \boldsymbol{\mu}_2\rangle \mid \forall \rho \in \boldsymbol{\gamma}[\![\boldsymbol{\Gamma}]\!] \,.\, \mathscr{E}[\![\mathsf{S}_1]\!]\rho \in \boldsymbol{\gamma}[\![\boldsymbol{\mu}_1]\!] \wedge \mathscr{E}[\![\mathsf{S}_2]\!]\rho \in \boldsymbol{\gamma}[\![\boldsymbol{\mu}_2]\!]\}$

(definition of $\in$ and of the conditional)

$= \{\langle \boldsymbol{\Gamma}, \boldsymbol{\mu}_1 * \boldsymbol{\mu}_2\rangle \mid \mathscr{E}[\![\mathsf{S}_1]\!] \in \boldsymbol{\gamma}[\![\langle \boldsymbol{\Gamma}, \boldsymbol{\mu}_1\rangle]\!] \wedge \mathscr{E}[\![\mathsf{S}_2]\!] \in \boldsymbol{\gamma}[\![\langle \boldsymbol{\Gamma}, \boldsymbol{\mu}_2\rangle]\!]\}$

(definition (49.16) of typing concretization $\boldsymbol{\gamma}[\![\langle \boldsymbol{\Gamma}, \boldsymbol{\mu}\rangle]\!]$)

$= \{\langle \boldsymbol{\Gamma}, \boldsymbol{\mu}_1 * \boldsymbol{\mu}_2\rangle \mid \boldsymbol{\Gamma} \vdash \mathsf{S}_1 : \boldsymbol{\mu}_1 \wedge \boldsymbol{\Gamma} \vdash \mathsf{S}_2 : \boldsymbol{\mu}_2\}$ (definition (49.18) of $\boldsymbol{\Gamma} \vdash \mathsf{E} : \boldsymbol{\mu}$)

— $\mathsf{E} = []$

$\{\langle \boldsymbol{\Gamma}, \boldsymbol{\mu}\rangle \mid \forall \rho \in \boldsymbol{\gamma}[\![\boldsymbol{\Gamma}]\!] \,.\, \mathscr{E}[\![[]]\!](\rho) \in \boldsymbol{\gamma}[\![\boldsymbol{\mu}]\!]\}$

$= \{\langle \boldsymbol{\Gamma}, \boldsymbol{\mu}\rangle \mid \mathtt{nil} \in \boldsymbol{\gamma}[\![\boldsymbol{\mu}]\!]\}$ (definition (49.8) of $\mathscr{E}[\![[]]\!](\rho) = \mathtt{nil}$)

$= \{\langle \boldsymbol{\Gamma}, \boldsymbol{\mu}' \ \mathtt{list}\rangle \mid \boldsymbol{\mu}' \in \mathbf{M}\}$

(by definition (49.11) of $\boldsymbol{\gamma}[\![\boldsymbol{\mu}]\!]$ such that $\boldsymbol{\mu} = \boldsymbol{\mu}' \ \mathtt{list}$, $\boldsymbol{\mu}' \in \mathbf{M}$ are the only monotypes such that $\mathtt{nil} \in \boldsymbol{\gamma}[\![\boldsymbol{\mu}]\!]$)

— $\mathsf{E} = \mathsf{S}_1 :: \mathsf{S}_2$

$\{\langle \boldsymbol{\Gamma}, \boldsymbol{\mu}\rangle \mid \forall \rho \in \boldsymbol{\gamma}[\![\boldsymbol{\Gamma}]\!] \,.\, \mathscr{E}[\![\mathsf{S}_1 :: \mathsf{S}_2]\!](\rho) \in \boldsymbol{\gamma}[\![\boldsymbol{\mu}]\!]\}$

$= \{\langle \boldsymbol{\Gamma}, \boldsymbol{\mu}\rangle \mid \forall \rho \in \boldsymbol{\gamma}[\![\boldsymbol{\Gamma}]\!] \,.\, \text{let } v_1 = \mathscr{E}[\![\mathsf{S}_1]\!]\rho \text{ and } v_2 = \mathscr{E}[\![\mathsf{S}_2]\!]\rho \text{ in } (\!| v_1 = \Omega^\delta \vee v_2 = \Omega^\delta \mathrel{?} \Omega^\delta \mathrel{|\!|} \tau^\delta(v_1 :: v_2) \neq \mathit{err} \mathrel{?} v_1 :: v_2 \mathrel{:} \Omega^\sigma |\!) \in \boldsymbol{\gamma}[\![\boldsymbol{\mu}]\!]\}$ (definition (49.8) of $\mathscr{E}[\![\mathsf{S}_1 :: \mathsf{S}_2]\!]\rho$)

$= \{\langle \boldsymbol{\Gamma}, \boldsymbol{\mu}\rangle \mid \forall \rho \in \boldsymbol{\gamma}[\![\boldsymbol{\Gamma}]\!] \,.\, \text{let } v_1 = \mathscr{E}[\![\mathsf{S}_1]\!]\rho \text{ and } v_2 = \mathscr{E}[\![\mathsf{S}_2]\!]\rho \text{ in } (\!| v_1 = \Omega^\delta \vee v_2 = \Omega^\delta \mathrel{?} \Omega^\delta \mathrel{|\!|} (\!| \exists t \neq \mathit{err} \,.\, \tau^\delta(v_1) = t \wedge \tau^\delta(v_2)) \equiv^\delta t\ \mathit{list} \mathrel{?} t\ \ \mathit{list} \mathrel{:} \mathit{err} |\!) \neq \mathit{err} \mathrel{?} v_1 :: v_2 \mathrel{:} \Omega^\sigma |\!) \in \boldsymbol{\gamma}[\![\boldsymbol{\mu}]\!]\}$

(definition (49.4) of τ^δ)

$= \{\langle \boldsymbol{\Gamma}, \boldsymbol{\mu}'\rangle \mid \forall \rho \in \boldsymbol{\gamma}[\![\boldsymbol{\Gamma}]\!] \,.\, \text{let } v_1 = \mathscr{E}[\![\mathsf{S}_1]\!]\rho \text{ and } v_2 = \mathscr{E}[\![\mathsf{S}_2]\!]\rho \text{ in } (\!| v_1 = \Omega^\delta \vee v_2 = \Omega^\delta \mathrel{?} \Omega^\delta \in \boldsymbol{\gamma}[\![\boldsymbol{\mu}']\!] \mathrel{:} \exists t \neq \mathit{err} \,.\, \tau^\delta(v_1) = t \wedge \tau^\delta(v_2) \equiv^\delta t\ \mathit{list} \wedge v_1 :: v_2 \in \boldsymbol{\gamma}[\![\boldsymbol{\mu}']\!] |\!)\}$

$\wr$ definition of $\boldsymbol{\gamma}[\![\boldsymbol{\mu}]\!]$ so that $\Omega^\sigma \notin \boldsymbol{\gamma}[\![\boldsymbol{\mu}]\!]$ and renaming $\boldsymbol{\mu}$ into $\boldsymbol{\mu}'$ $\wr$

$= \{\langle \boldsymbol{\Gamma},\ \boldsymbol{\mu}\ \mathtt{list}\rangle \mid \forall \rho \in \boldsymbol{\gamma}[\![\boldsymbol{\Gamma}]\!] \,.\, \mathsf{let}\ v_1 = \mathscr{E}[\![\mathsf{S}_1]\!]\rho\ \mathsf{and}\ v_2 = \mathscr{E}[\![\mathsf{S}_2]\!]\rho\ \mathsf{in}\ (\!|\ v_1 = \Omega^\delta \vee v_2 = \Omega^\delta\ ?$
$\Omega^\delta \in \boldsymbol{\gamma}[\![\boldsymbol{\mu}\ \mathtt{list}]\!] : \exists t \neq \mathit{err} \,.\, \tau^\delta(v_1) = t \wedge \tau^\delta(v_2) \equiv^\delta t\ \mathit{list} \wedge v_1 :: v_2 \in \boldsymbol{\gamma}[\![\boldsymbol{\mu}\ \mathtt{list}]\!]\ |\!)\}$

$\wr$ definition (49.11) of $\boldsymbol{\gamma}[\![\boldsymbol{\mu}']\!]$ so that $\boldsymbol{\mu}' = \boldsymbol{\mu}\ \mathtt{list}$ is the only type which concretization is Ω^δ or a set of lists $v_1 :: v_2$ $\wr$

$= \{\langle \boldsymbol{\Gamma},\ \boldsymbol{\mu}\ \mathtt{list}\rangle \mid \forall \rho \in \boldsymbol{\gamma}[\![\boldsymbol{\Gamma}]\!] \,.\, \mathsf{let}\ v_1 = \mathscr{E}[\![\mathsf{S}_1]\!]\rho\ \mathsf{and}\ v_2 = \mathscr{E}[\![\mathsf{S}_2]\!]\rho\ \mathsf{in}\ (\!|\ v_1 = \Omega^\delta \vee v_2 = \Omega^\delta\ ?$
$\Omega^\delta \in \boldsymbol{\gamma}[\![\boldsymbol{\mu}\ \mathtt{list}]\!] : \exists t \neq \mathit{err} \,.\, \tau^\delta(v_1) = t \wedge \tau^\delta(v_2) \equiv^\delta t\ \mathit{list} \wedge \tau^\delta(v_1 :: v_2) = \delta(\boldsymbol{\mu})\ \mathit{list})\ |\!)\}$

$\wr$ by (49.13) and definition of δ $\wr$

$= \{\langle \boldsymbol{\Gamma},\ \boldsymbol{\mu}\ \mathtt{list}\rangle \mid \forall \rho \in \boldsymbol{\gamma}[\![\boldsymbol{\Gamma}]\!] \,.\, \mathsf{let}\ v_1 = \mathscr{E}[\![\mathsf{S}_1]\!]\rho\ \mathsf{and}\ v_2 = \mathscr{E}[\![\mathsf{S}_2]\!]\rho\ \mathsf{in}\ (\!|\ v_1 = \Omega^\delta \vee v_2 = \Omega^\delta\ ?$
$\Omega^\delta \in \boldsymbol{\gamma}[\![\boldsymbol{\mu}\ \mathtt{list}]\!] : \exists t \neq \mathit{err} \,.\, \tau^\delta(v_1) = t \wedge \tau^\delta(v_2) \equiv^\delta t\ \mathit{list} \wedge \tau^\delta(v_1) = \delta(\boldsymbol{\mu}) \wedge \tau^\delta(v_2) \equiv^\delta$
$\delta(\boldsymbol{\mu})\ \mathit{list}\ |\!)\}$

$\wr$ definition (49.4) of τ^δ $\wr$

$= \{\langle \boldsymbol{\Gamma},\ \boldsymbol{\mu}\ \mathtt{list}\rangle \mid \forall \rho \in \boldsymbol{\gamma}[\![\boldsymbol{\Gamma}]\!] \,.\, \mathsf{let}\ v_1 = \mathscr{E}[\![\mathsf{S}_1]\!]\rho\ \mathsf{and}\ v_2 = \mathscr{E}[\![\mathsf{S}_2]\!]\rho\ \mathsf{in}\ (\!|\ v_1 = \Omega^\delta \vee v_2 = \Omega^\delta\ ?$
$\Omega^\delta \in \boldsymbol{\gamma}[\![\boldsymbol{\mu}\ \mathtt{list}]\!] : \tau^\delta(v_1) = \delta(\boldsymbol{\mu}) \wedge \tau^\delta(v_2) \equiv^\delta \delta(\boldsymbol{\mu}\ \mathit{list})\ |\!)\}$

$\wr$ letting $t = \delta(\boldsymbol{\mu})$ and definition of δ $\wr$

$= \{\langle \boldsymbol{\Gamma},\ \boldsymbol{\mu}\ \mathtt{list}\rangle \mid \forall \rho \in \boldsymbol{\gamma}[\![\boldsymbol{\Gamma}]\!] \,.\, \mathsf{let}\ v_1 = \mathscr{E}[\![\mathsf{S}_1]\!]\rho\ \mathsf{and}\ v_2 = \mathscr{E}[\![\mathsf{S}_2]\!]\rho\ \mathsf{in}\ (\!|\ v_1 = \Omega^\delta \vee v_2 = \Omega^\delta\ ?$
$\Omega^\delta \in \boldsymbol{\gamma}[\![\boldsymbol{\mu}\ \mathtt{list}]\!] : v_1 \in \boldsymbol{\gamma}[\![\boldsymbol{\mu}]\!] \wedge (v_2 = \mathtt{nil} \vee v_2 \in \boldsymbol{\gamma}[\![\boldsymbol{\mu}\ \mathtt{list}]\!]\ |\!)\}$

$\wr$ by (49.13) and definition (49.4) of τ^δ $\wr$

$= \{\langle \boldsymbol{\Gamma},\ \boldsymbol{\mu}\ \mathtt{list}\rangle \mid \forall \rho \in \boldsymbol{\gamma}[\![\boldsymbol{\Gamma}]\!] \,.\, \mathscr{E}[\![\mathsf{S}_1]\!]\rho \in \boldsymbol{\gamma}[\![\boldsymbol{\mu}]\!] \wedge \mathscr{E}[\![\mathsf{S}_2]\!]\rho \in \boldsymbol{\gamma}[\![\boldsymbol{\mu}\ \mathtt{list}]\!]\}$

$\wr$ definition (49.11) of $\boldsymbol{\gamma}$ so that $\forall \boldsymbol{\mu} \,.\, \Omega^\delta \in \boldsymbol{\gamma}[\![\boldsymbol{\mu}]\!]$ and $\mathtt{nil} \in \boldsymbol{\gamma}[\![\boldsymbol{\mu}\ \mathtt{list}]\!]$ $\wr$

$= \{\langle \boldsymbol{\Gamma},\ \boldsymbol{\mu}\ \mathtt{list}\rangle \mid \forall \rho \in \boldsymbol{\gamma}[\![\boldsymbol{\Gamma}]\!] \,.\, \mathscr{E}[\![\mathsf{S}_1]\!]\rho \in \boldsymbol{\gamma}[\![\boldsymbol{\mu}]\!] \wedge \forall \rho \in \boldsymbol{\gamma}[\![\boldsymbol{\Gamma}]\!] \,.\, \mathscr{E}[\![\mathsf{S}_2]\!]\rho \in \boldsymbol{\gamma}[\![\boldsymbol{\mu}\ \mathtt{list}]\!]\}$

$\wr$ definitions of $\forall$ and $\wedge$ $\wr$

$= \{\langle \boldsymbol{\Gamma},\ \boldsymbol{\mu}\ \mathtt{list}\rangle \mid \mathscr{E}[\![\mathsf{S}_1]\!] \in \{f \in \mathbb{E}\mathrm{v} \to \mathbb{V}^\Omega \mid \forall \rho \in \boldsymbol{\gamma}[\![\boldsymbol{\Gamma}]\!] \,.\, f(\rho) \in \boldsymbol{\gamma}[\![\boldsymbol{\mu}]\!]\} \wedge \mathscr{E}[\![\mathsf{S}_2]\!] \in \{f \in \mathbb{E}\mathrm{v} \to$
$\mathbb{V}^\Omega \mid \forall \rho \in \boldsymbol{\gamma}[\![\boldsymbol{\Gamma}]\!] \,.\, f(\rho) \in \boldsymbol{\gamma}[\![\boldsymbol{\mu}\ \mathtt{list}]\!]\}\}$

$\wr$ definition of $\in$ $\wr$

$= \{\langle \boldsymbol{\Gamma},\ \boldsymbol{\mu}\ \mathtt{list}\rangle \mid \mathscr{E}[\![\mathsf{S}_1]\!] \in \boldsymbol{\gamma}[\![\langle \boldsymbol{\Gamma},\ \boldsymbol{\mu}\rangle]\!] \wedge \mathscr{E}[\![\mathsf{S}_2]\!] \in \boldsymbol{\gamma}[\![\langle \boldsymbol{\Gamma},\ \boldsymbol{\mu}\ \mathtt{list}\rangle]\!]\}$

$\wr$ definition (49.16) of typing concretization $\boldsymbol{\gamma}[\![\langle \boldsymbol{\Gamma},\ \boldsymbol{\mu}\rangle]\!]$ $\wr$

$= \{\langle \boldsymbol{\Gamma},\ \boldsymbol{\mu}\ \mathtt{list}\rangle \mid \boldsymbol{\Gamma} \vdash \mathsf{S}_1 : \boldsymbol{\mu} \wedge \boldsymbol{\Gamma} \vdash \mathsf{S}_2 : \boldsymbol{\mu}\ \mathtt{list}\}$ $\wr$ definition (49.18) of $\boldsymbol{\Gamma} \vdash \mathsf{E} : \boldsymbol{\mu}$ $\wr$

— $\mathsf{E} = \mathsf{hd}(\mathsf{S})$

$\{\langle \boldsymbol{\Gamma},\ \boldsymbol{\mu}\rangle \mid \forall \rho \in \boldsymbol{\gamma}[\![\boldsymbol{\Gamma}]\!] \,.\, \mathscr{E}[\![\mathsf{hd}(\mathsf{S})]\!](\rho) \in \boldsymbol{\gamma}[\![\boldsymbol{\mu}]\!]\}$

$= \{\langle \boldsymbol{\Gamma},\ \boldsymbol{\mu}\rangle \mid \forall \rho \in \boldsymbol{\gamma}[\![\boldsymbol{\Gamma}]\!] \,.\, \mathsf{let}\ v_1 = \mathscr{E}[\![\mathsf{S}]\!]\rho\ \mathsf{in}\ (\!|\ v_1 = \Omega^\sigma\ ?\ \Omega^\sigma \parallel v_1 = \Omega^\delta \vee v_1 = \mathtt{nil}\ ?\ \Omega^\delta \parallel \exists t \neq$
$\mathit{err} \,.\, \tau^\delta(v_1) = t\ \mathit{list}\ ?\ \mathsf{let}\ v_h :: v_t = v_1\ \mathsf{in}\ v_h : \Omega^\sigma\ |\!) \in \boldsymbol{\gamma}[\![\boldsymbol{\mu}]\!]\}$

$\wr$ definition (49.8) of $\mathscr{E}[\![\mathsf{hd}(\mathsf{S})]\!]$ $\wr$

$= \{\langle \boldsymbol{\Gamma},\ \boldsymbol{\mu}\rangle \mid \forall \rho \in \boldsymbol{\gamma}[\![\boldsymbol{\Gamma}]\!] \,.\, \mathsf{let}\ v_1 = \mathscr{E}[\![\mathsf{S}]\!]\rho\ \mathsf{in}\ (\!|\ v_1 = \Omega^\sigma\ ?\ \Omega^\sigma \parallel v_1 = \Omega^\delta \vee v_1 = \mathtt{nil}\ ?\ \Omega^\delta \parallel \exists t \neq$
$\mathit{err} \,.\, \exists n > 0 \,.\, v_1 = x_1 :: x_2 :: \ldots x_n :: \mathtt{nil} \wedge \forall i \in [1, n] \,.\, \tau^\delta(x_i) = t\ ?\ x_1 : \Omega^\sigma\ |\!) \in \boldsymbol{\gamma}[\![\boldsymbol{\mu}]\!]\}$

$\wr$ lemma 49.7 and definition (49.4) of $\tau^\delta(v_1) \neq \mathit{nil}$ when $v_1 \neq \mathtt{nil}$ $\wr$

$= \{\langle \boldsymbol{\Gamma},\ \boldsymbol{\mu}\rangle \mid \forall \rho \in \boldsymbol{\gamma}[\![\boldsymbol{\Gamma}]\!] \,.\, \mathsf{let}\ v_1 = \mathscr{E}[\![\mathsf{S}]\!]\rho\ \mathsf{in}\ v_1 = \Omega^\delta \vee v_1 = \mathtt{nil} \vee \exists t \neq \mathit{err} \,.\, \exists n > 0 \,.\, v_1 =$
$x_1 :: x_2 :: \ldots x_n :: \mathtt{nil} \wedge \forall i \in [1, n] \,.\, \tau^\delta(x_i) = t \wedge x_1 \in \boldsymbol{\gamma}[\![\boldsymbol{\mu}]\!]\}$

$\wr$ definition (49.11) of $\boldsymbol{\gamma}[\![\boldsymbol{\mu}]\!]$ so that $\Omega^\sigma \notin \boldsymbol{\gamma}[\![\boldsymbol{\mu}]\!]$ $\wr$

$$= \{\langle\boldsymbol{\Gamma}, \boldsymbol{\mu}\rangle \mid \forall\rho\in\boldsymbol{\gamma}[\![\boldsymbol{\Gamma}]\!] . \text{let } v_1 = \mathscr{E}[\![\mathsf{S}]\!]\rho \text{ in } v_1 = \Omega^\delta \vee v_1 = \texttt{nil} \vee \exists n>0 . v_1 = x_1 :: x_2 :: \ldots x_n :: \texttt{nil} \wedge \forall i\in[1,n] . \tau^\delta(x_i)\in\boldsymbol{\gamma}[\![\boldsymbol{\mu}]\!]\}$$

$\langle$ $\forall i\in[1,n] . \tau^\delta(x_i) = \tau^\delta(x_1) = t \wedge x_1 \in \boldsymbol{\gamma}[\![\boldsymbol{\mu}]\!]$ so $\tau^\delta(x_1) = \delta(\boldsymbol{\mu}) \neq$ *err* by (49.13) and definition of δ $\rangle$

$$= \{\langle\boldsymbol{\Gamma}, \boldsymbol{\mu}\rangle \mid \forall\rho\in\boldsymbol{\gamma}[\![\boldsymbol{\Gamma}]\!] . \mathscr{E}[\![\mathsf{S}]\!](\rho) \in \{v_1 :: \ldots :: v_n :: \texttt{nil} \mid n\in\mathbb{N}^+ \wedge \forall i\in[1,n] . v_i \in \boldsymbol{\gamma}[\![\boldsymbol{\mu}]\!]\setminus\{\Omega^\delta\}\} \cup \{\texttt{nil}, \Omega^\delta\}\} \qquad \langle\text{definition of } \in\rangle$$

$$= \{\langle\boldsymbol{\Gamma}, \boldsymbol{\mu}\rangle \mid \forall\rho\in\boldsymbol{\gamma}[\![\boldsymbol{\Gamma}]\!] . \mathscr{E}[\![\mathsf{S}]\!](\rho) \in \boldsymbol{\gamma}[\![\boldsymbol{\mu}\ \texttt{list}]\!]\} \qquad \langle\text{definition (49.11) of } \boldsymbol{\gamma}[\![\boldsymbol{\mu}\ \texttt{list}]\!]\rangle$$

$$= \{\langle\boldsymbol{\Gamma}, \boldsymbol{\mu}\rangle \mid \mathscr{E}[\![\mathsf{S}]\!] \in \{f\in\mathbb{E}\mathbb{v}\to\mathbb{V}^\Omega \mid \forall\rho\in\boldsymbol{\gamma}[\![\boldsymbol{\Gamma}]\!] . f(\rho)\in\boldsymbol{\gamma}[\![\boldsymbol{\mu}\ \texttt{list}]\!]\}\} \qquad \langle\text{definition of } \in\rangle$$

$$= \{\langle\boldsymbol{\Gamma}, \boldsymbol{\mu}\rangle \mid \mathscr{E}[\![\mathsf{S}]\!] \in \boldsymbol{\gamma}[\![\langle\boldsymbol{\Gamma}, \boldsymbol{\mu}\ \texttt{list}\rangle]\!]\}$$

$\langle$definition (49.16) of typing concretization $\boldsymbol{\gamma}[\![\langle\boldsymbol{\Gamma}, \boldsymbol{\mu}\rangle]\!]\rangle$

$$= \{\langle\boldsymbol{\Gamma}, \boldsymbol{\mu}\rangle \mid \boldsymbol{\Gamma}\vdash\mathsf{S} : \boldsymbol{\mu}\ \texttt{list}\} \qquad \langle\text{definition (49.18) of } \boldsymbol{\Gamma}\vdash\mathsf{E}:\boldsymbol{\mu}\rangle$$

— The proof is similar when $\mathsf{E} = \texttt{tl(S)}$

This is a structural definition 16.24 of sets and each one can be given an equivalent deductive definition 16.10. Using the judgment notation (49.18) this yields the typing rules of figure 49.29.

Monomorphic typing rules

$$\boldsymbol{\Gamma}\vdash\texttt{1}:\texttt{int} \quad [\text{cte}] \qquad \boldsymbol{\Gamma}\vdash\mathsf{x}:\boldsymbol{\Gamma}(\mathsf{x}) \quad [\text{var}] \qquad (49.28)$$

$$\frac{\boldsymbol{\Gamma}\vdash\mathsf{A}_1:\texttt{int},\ \boldsymbol{\Gamma}\vdash\mathsf{A}_2:\texttt{int}}{\boldsymbol{\Gamma}\vdash\mathsf{A}_1-\mathsf{A}_2:\texttt{int}}\ [\text{mns}] \qquad \frac{\boldsymbol{\Gamma}\vdash\mathsf{S}_1:\boldsymbol{\mu}_1,\ \boldsymbol{\Gamma}\vdash\mathsf{S}_2:\boldsymbol{\mu}_2}{\boldsymbol{\Gamma}\vdash\mathsf{S}_1,\mathsf{S}_2:\boldsymbol{\mu}_1 * \boldsymbol{\mu}_2}\ [\text{pair}]$$

$$\boldsymbol{\Gamma}\vdash[\,]:\boldsymbol{\mu}\ \texttt{list}\ [\text{nil}] \qquad \frac{\boldsymbol{\Gamma}\vdash\mathsf{S}_1:\boldsymbol{\mu},\ \boldsymbol{\Gamma}\vdash\mathsf{S}_2:\boldsymbol{\mu}\ \texttt{list}}{\boldsymbol{\Gamma}\vdash\mathsf{S}_1::\mathsf{S}_2:\boldsymbol{\mu}\ \texttt{list}}\ [\text{cons}]$$

$$\frac{\boldsymbol{\Gamma}\vdash\mathsf{S}:\boldsymbol{\mu}\ \texttt{list}}{\boldsymbol{\Gamma}\vdash\mathsf{hd(S)}:\boldsymbol{\mu}}\ [\text{hd}] \qquad \frac{\boldsymbol{\Gamma}\vdash\mathsf{S}:\boldsymbol{\mu}\ \texttt{list}}{\boldsymbol{\Gamma}\vdash\texttt{tl(S)}:\boldsymbol{\mu}\ \texttt{list}}\ [\text{tl}]$$

$$\frac{\boldsymbol{\Gamma}\vdash\mathsf{A}_1:\texttt{int},\ \boldsymbol{\Gamma}\vdash\mathsf{A}_2:\texttt{int}}{\boldsymbol{\Gamma}\vdash\mathsf{A}_1<\mathsf{A}_2:\texttt{bool}}\ [\text{lt}] \qquad \frac{\boldsymbol{\Gamma}\vdash\mathsf{S}:\boldsymbol{\mu}\ \texttt{list}}{\boldsymbol{\Gamma}\vdash\mathsf{S}=[\,]:\texttt{bool}}\ [\text{eq}]$$

$$\frac{\boldsymbol{\Gamma}\vdash\mathsf{B}_1:\texttt{bool},\ \boldsymbol{\Gamma}\vdash\mathsf{B}_2:\texttt{bool}}{\boldsymbol{\Gamma}\vdash\mathsf{B}_1\ \mathsf{nand}\ \mathsf{B}_2:\texttt{bool}}\ [\text{nand}]$$

Figure 49.29
Monomorphic typing rules

Example 49.30 We have $\boldsymbol{\Gamma}[\mathsf{x}\leftarrow\texttt{int list}][\mathsf{y}\leftarrow\texttt{int}]\vdash\mathsf{hd(x)}-\mathsf{y}:\texttt{int}$ by

$$\boldsymbol{\Gamma}[\mathsf{x}\leftarrow\texttt{int list}][\mathsf{y}\leftarrow\texttt{int}]\vdash\mathsf{x}:\texttt{int list} \qquad \langle[\text{var}]\rangle \qquad (49.31)$$

$$\boldsymbol{\Gamma}[\mathsf{x}\leftarrow\texttt{int list}][\mathsf{y}\leftarrow\texttt{int}]\vdash\mathsf{hd(x)}:\texttt{int} \qquad \langle(49.31), [\text{hd}]\rangle \qquad (49.32)$$

$\mathbf{\Gamma}$[x ← int list][y ← int] ⊢ y : int ⦃[var]⦄ (49.33)

$\mathbf{\Gamma}$[x ← int list][y ← int] ⊢ hd(x) − y : int ⦃(49.32), (49.33), [mns]⦄

Typing with any other type environment will fail because [mns] requires that both hd(x) and y have type int. This is obviously not the case if $\mathbf{\Gamma}$[x ← int list][y ← int](y) ≠ int.

This is also not the case if $\mathbf{\Gamma}$[x ← int list][y ← int](x) ≠ int list because the only rule applicable to hd(x) is [hd] which requires for hd(x) to be of type int to have hd(x) be of type int list.

It follows that $\boldsymbol{\tau}$⟦hd(x) − y⟧ = {⟨$\mathbf{\Gamma}$, $\boldsymbol{\mu}$⟩ | $\mathbf{\Gamma}$ ⊢ hd(x) − y : $\boldsymbol{\mu}$} = {⟨$\mathbf{\Gamma}$[x ← int list][y ← int], int⟩ | $\mathbf{\Gamma} \in \mathbf{A}$}.

Exercise 49.34 Propose improvements for the hd and tl rules that would improve the precision of the typing (i.e., increase the number of definitely incorrect expressions that would be rejected by the typing rules).

Exercise 49.35 Design the Boolean typing rules by calculational design.

Exercise 49.36 Note that, for example, the empty list [] has infinitely many types {$\boldsymbol{\mu}$ list | $\boldsymbol{\mu} \in \mathbf{M}$} so it is not possible to compute this set by enumeration. However, it possible to check the typing rules (49.28) for a given expression, given type environment, and given type of the expression. Write such a program (in the language of your choice) to implement the checking of the monotype of an expression (given in abstract syntax) in a given type environment according to the typing rules (49.28).

49.9 Monomorphic Static Type Inference Algorithm

The typing rules section 49.8.13 in Alonso Church style are not algorithmic because types are sets that may be infinite (e.g., $\boldsymbol{\tau}$⟦hd([])⟧ = $\boldsymbol{\Theta}$ in example 49.24). Roger Hindley [499] proposed to finitely represent these infinite sets by type schemes, that is, monomorphic types with variables. This is an abstract interpretation based on the symbolic term abstract domain of chapter 48. This abstraction is in general imprecise. However, due to restrictions on monomorphic types and the considered language, the abstraction turned out to be exact for the sets involved in the typing rules of section 49.8.13 [224]. This exactness is obtained by restrictions on the types (e.g., there is no union of types like Boolean or integer, list elements must all be of the same type, and others) and the programming language (e.g., the type of branches in a conditional must be the same, and all recursive calls of a procedure must have the same type).

49.9.1 Monotypes with Variables

A monotype with variables is a term with variables which abstracts a set of monotypes.

$$\begin{array}{rcll}
\alpha, \beta & \in & \mathfrak{V} & \text{type variables} \\
\boldsymbol{\mu} \in \mathbf{M}^{\mathcal{V}} & ::= & & \text{monotypes with variables} \\
& & \texttt{int} & \quad \text{integer type} \\
& | & \texttt{bool} & \quad \text{Boolean type} \\
& | & \boldsymbol{\mu}_1 * \boldsymbol{\mu}_2 & \quad \text{pair type} \\
& | & \boldsymbol{\mu}_1 \; \texttt{list} & \quad \text{list type} \\
& | & \alpha & \quad \text{type variable} \\
\mathcal{M} \in \wp(\mathbf{M}^{\mathcal{V}}) & & & \text{sets of monotypes with variables} \\
\boldsymbol{\Gamma} \in \boldsymbol{A}^{\mathcal{V}} & \triangleq & \mathbb{V} \to \mathbf{M}^{\mathcal{V}} & \text{monotype with variables assignment}
\end{array}$$

49.9.2 Abstraction of Sets of Monotypes to Monotypes with Variables

We use the following abstraction of sets of monotypes to monotypes with variables studied in chapter 48. By (48.27) and corollary 48.28, we have a Galois retraction onto a complete lattice (where the naming scheme $\boldsymbol{\nu}$ characterizes how type variables are chosen).

$$\langle \wp(\mathbf{M}), \subseteq, \varnothing, \mathbf{M}, \cup, \cap \rangle \xrightleftharpoons[lcg_{\simeq^{\mathcal{V}}}[\boldsymbol{\nu}]]{ground_{\simeq^{\mathcal{V}}}} \langle (\mathbf{M}^{\mathcal{V}} \cup \{\overline{\varnothing}^{\mathcal{V}}\})/_{\simeq^{\mathcal{V}}}, \preceq_{\simeq^{\mathcal{V}}}, [\overline{\varnothing}^{\mathcal{V}}]_{\simeq^{\mathcal{V}}}, [\alpha]_{\simeq^{\mathcal{V}}}, LCG_{\simeq^{\mathcal{V}}}, GCI_{\simeq^{\mathcal{V}}} \rangle \tag{49.37}$$

In general, the abstraction is imprecise, that is, if $\boldsymbol{M} \in \wp(\mathbf{M})$ then $\boldsymbol{M} \subsetneq ground(lcg[\boldsymbol{\nu}](\boldsymbol{M}))$. However, the sets $\boldsymbol{\tau}[\![\mathsf{E}]\!]$ considered in the monomorphic typing rules (49.28) turn out to belong to the ground instances of the monotypes with variables.

$$\mathbf{PM} \triangleq \{ground(\boldsymbol{\mu}) \mid \boldsymbol{\mu} \in \mathbf{M}^{\mathcal{V}}\} \cup \{\varnothing\} \subsetneq \wp(\mathbf{M}). \tag{49.38}$$

In that case, by exercise 11.34, we have the Galois isomorphism

$$\langle \mathbf{PM}, \subseteq, \varnothing, \mathbf{M}, \cup, \cap \rangle \xrightleftharpoons[lcg_{\simeq^{\mathcal{V}}}[\boldsymbol{\nu}]]{ground_{\simeq^{\mathcal{V}}}} \langle (\mathbf{M}^{\mathcal{V}} \cup \{\overline{\varnothing}^{\mathcal{V}}\})/_{\simeq^{\mathcal{V}}}, \preceq_{\simeq^{\mathcal{V}}}, [\overline{\varnothing}^{\mathcal{V}}]_{\simeq^{\mathcal{V}}}, [\alpha]_{\simeq^{\mathcal{V}}}, LCG_{\simeq^{\mathcal{V}}}, GCI_{\simeq^{\mathcal{V}}} \rangle \tag{49.39}$$

The abstraction is therefore exact for the concrete domain **PM**. This is extended pointwise to type environments and componentwise to typings.

49.9.3 Type Inference for Expressions

To infer the expression type, we need a finite representation of the set typings of this expression of section 49.8.12. We represent such a set of monotypes by the monotype with variables which is the least common generalization abstraction of this set of monotypes.

$$\boldsymbol{\tau}^{\mathcal{V}}[\![\mathsf{E}]\!] \triangleq lcg[\boldsymbol{\nu}](\boldsymbol{\tau}[\![\mathsf{E}]\!])$$

In general, there is a loss of information, for example, $lcg[\boldsymbol{\nu}](\{\texttt{int}, \texttt{bool}\}) = [\alpha]_{\simeq^{\mathcal{V}}}$ (where $\alpha = \boldsymbol{\nu}(\langle \texttt{int}, \texttt{bool} \rangle)$). However the expressions and the typing rules have been designed so that there is no expression with type $\{\texttt{int}, \texttt{bool}\}$ (because the union of types is carefully forbidden). Therefore the least common generalization abstraction of the set of monotypes of an expression is a monotype with variables that exactly abstracts this set. Therefore the loss of information is only for the set of monotypes that are the set of types for no expression.

This is the same for type environments. If $\mathbf{\Gamma}_1 \vdash \mathtt{S} : \boldsymbol{\mu}_1$ and $\mathbf{\Gamma}_2 \vdash \mathtt{S} : \boldsymbol{\mu}_2$ and $lcg[\boldsymbol{v}](\{\mathbf{\Gamma}_1(\mathtt{x}), \mathbf{\Gamma}_1(\mathtt{x})\}) = \alpha$ then x is not used in S. Otherwise, x is used in S by some operation of section 49.6 and none of these operations can take arguments or deliver results with types that do not unify. No operation in the language can be generic (i.e., have arguments with types that do not unify[1]) and polymorphic type arguments are not touched by the operations that manipulate these arguments.[2]

We check that the abstraction is exact for sets of monotypes typing expressions by proving that

$$\boldsymbol{\tau}[\![\mathtt{E}]\!] \triangleq ground(\boldsymbol{\tau}^{v}[\![\mathtt{E}]\!])$$

We go on by structural induction on E. We let $\boldsymbol{\Gamma}^{v}$ be the most general environment such that $\boldsymbol{\Gamma}^{v} \in \mathbb{V} \to \mathfrak{W}$ and $\forall \mathtt{x}, \mathtt{y} \in \mathbb{V}\,.\,(\mathtt{x} \neq \mathtt{y}) \Rightarrow (\boldsymbol{\Gamma}^{v}(\mathtt{x}) \neq \boldsymbol{\Gamma}^{v}(\mathtt{y}))$, considered up to renaming of type variables.

— $\mathtt{E} = 1$

$$
\begin{aligned}
- \quad & \boldsymbol{\tau}^{v}[\![\mathtt{E}]\!] = lcg[\boldsymbol{v}](\boldsymbol{\tau}[\![\mathtt{E}]\!]) \\
= \quad & lcg[\boldsymbol{v}](\{\langle \mathbf{\Gamma}, \mathtt{int}\rangle \mid \mathbf{\Gamma} \in \mathbf{A}\}) \\
= \quad & \langle \boldsymbol{\Gamma}^{v}, \mathtt{int}\rangle
\end{aligned}
$$

$$
\begin{aligned}
- \quad & ground(\boldsymbol{\tau}^{v}[\![\mathtt{E}]\!]) \\
= \quad & \{\langle \mathbf{\Gamma}, \mathtt{int}\rangle \mid \mathbf{\Gamma} \in \mathbf{A}\} = \boldsymbol{\tau}[\![\mathtt{E}]\!]
\end{aligned}
$$

— $\mathtt{E} = \mathtt{x}$

$$
\begin{aligned}
- \quad & \boldsymbol{\tau}^{v}[\![\mathtt{E}]\!] = lcg[\boldsymbol{v}](\boldsymbol{\tau}[\![\mathtt{E}]\!]) \\
= \quad & lcg[\boldsymbol{v}](\{\langle \mathbf{\Gamma}, \mathbf{\Gamma}(\mathtt{x})\rangle \mid \mathbf{\Gamma} \in \mathbf{A}\}) \\
= \quad & \langle \boldsymbol{\Gamma}^{v}, \boldsymbol{\Gamma}^{v}(\mathtt{x})\rangle
\end{aligned}
$$

$$
\begin{aligned}
- \quad & ground(\boldsymbol{\tau}^{v}[\![\mathtt{E}]\!]) \\
= \quad & \{\langle \mathbf{\Gamma}, \mathbf{\Gamma}(\mathtt{x})\rangle \mid \mathbf{\Gamma} \in \mathbf{A}\} = \boldsymbol{\tau}[\![\mathtt{E}]\!]
\end{aligned}
$$

— $\mathtt{E} = \mathtt{A}_1 - \mathtt{A}_2$

$$
\begin{aligned}
- \quad & \boldsymbol{\tau}^{v}[\![\mathtt{E}]\!] = lcg[\boldsymbol{v}](\boldsymbol{\tau}[\![\mathtt{E}]\!]) \\
= \quad & lcg[\boldsymbol{v}](\{\langle \mathbf{\Gamma}, \mathtt{int}\rangle \mid \mathbf{\Gamma} \vdash \mathtt{A}_1 : \mathtt{int} \wedge \mathbf{\Gamma} \vdash \mathtt{A}_2 : \mathtt{int}\}) \\
= \quad & lcg[\boldsymbol{v}](\{\langle \mathbf{\Gamma}, \mathtt{int}\rangle \mid \mathscr{A}[\![\mathtt{A}_1]\!] \in \boldsymbol{\gamma}[\![\langle \mathbf{\Gamma}, \mathtt{int}\rangle]\!] \wedge \mathscr{A}[\![\mathtt{A}_2]\!] \in \boldsymbol{\gamma}[\![\langle \mathbf{\Gamma}, \mathtt{int}\rangle]\!]\}) \\
& \hfill \text{(definition (49.18) of } \mathbf{\Gamma} \vdash \mathtt{E} : \boldsymbol{\mu}\text{)}
\end{aligned}
$$

1. For example, in OCaml + is integer addition, `add_big_int` is addition of arbitrary-precision integers, `+.` is float addition, `add` is addition on complex numbers, and so forth.

2. This means that polymorphic arguments must be represented in the same way, whichever monotype their instances may have, for example, by pointers in OCaml.

$= \ lcg[\mathbf{v}](\{\langle \boldsymbol{\Gamma}, \mathtt{int}\rangle \mid \forall \rho \in \boldsymbol{\gamma}[\![\boldsymbol{\Gamma}]\!] \,.\, \mathscr{A}[\![\mathsf{A}_1]\!](\rho) \in \boldsymbol{\gamma}[\![\mathtt{int}]\!] \wedge \forall \rho \in \boldsymbol{\gamma}[\![\boldsymbol{\Gamma}]\!] \,.\, \mathscr{A}[\![\mathsf{A}_2]\!](\rho) \in \boldsymbol{\gamma}[\![\mathtt{int}]\!]\})$

⦃definition (49.16) of $\boldsymbol{\gamma}[\![\langle \boldsymbol{\Gamma}, \boldsymbol{\mu}\rangle]\!]$⦄

$= \ lcg[\mathbf{v}](\{\langle \boldsymbol{\Gamma}, \mathtt{int}\rangle \mid \boldsymbol{\Gamma} \in \mathbf{A}\} \cap \{\langle \boldsymbol{\Gamma}, \boldsymbol{\mu}\rangle \mid \forall \rho \in \boldsymbol{\gamma}[\![\boldsymbol{\Gamma}]\!] \,.\, \mathscr{A}[\![\mathsf{A}_1]\!](\rho) \in \boldsymbol{\gamma}[\![\boldsymbol{\mu}]\!]\} \cap$
$\{\langle \boldsymbol{\Gamma}, \boldsymbol{\mu}'\rangle \mid \forall \rho \in \boldsymbol{\gamma}[\![\boldsymbol{\Gamma}]\!] \,.\, \mathscr{A}[\![\mathsf{A}_2]\!](\rho) \in \boldsymbol{\gamma}[\![\boldsymbol{\mu}']\!]\})$ ⦃definition $\cap$ so that $\boldsymbol{\mu} = \boldsymbol{\mu}' = \mathtt{int}$⦄

$\preceq^{v} gci(\{lcg[\mathbf{v}](\{\langle \boldsymbol{\Gamma}, \mathtt{int}\rangle \mid \boldsymbol{\Gamma} \in \mathbf{A}\}), lcg[\mathbf{v}](\{\langle \boldsymbol{\Gamma}, \boldsymbol{\mu}\rangle \mid \forall \rho \in \boldsymbol{\gamma}[\![\boldsymbol{\Gamma}]\!] \,.\, \mathscr{A}[\![\mathsf{A}_1]\!](\rho) \in$
$\boldsymbol{\gamma}[\![\boldsymbol{\mu}]\!]\}), lcg[\mathbf{v}](\{\langle \boldsymbol{\Gamma}, \boldsymbol{\mu}'\rangle \mid \forall \rho \in \boldsymbol{\gamma}[\![\boldsymbol{\Gamma}]\!] \,.\, \mathscr{A}[\![\mathsf{A}_2]\!](\rho) \in \boldsymbol{\gamma}[\![\boldsymbol{\mu}']\!]\})\})$

⦃Let $\langle \mathcal{M}_i \in \wp(\mathbf{M}),\ i \in \Delta\rangle$ be a family of sets of monotypes. By definition of the glb $\cap$, $\forall j \in \Delta \,.\, \bigcap_{i\in\Delta} \mathcal{M}_i \subseteq \mathcal{M}_j$. By the Galois connection (48.27), $lcg[\mathbf{v}]$ is increasing so $\forall j \in \Delta \,.\, lcg[\mathbf{v}](\bigcap_{i\in\Delta} \mathcal{M}_i) \preceq^{v} lcg[\mathbf{v}](\mathcal{M}_j)$ and by definition of the glb gci in corollary 48.28, $lcg[\mathbf{v}](\bigcap_{i\in\Delta} \mathcal{M}_i) \preceq^{v} \underset{j\in\Delta}{gci}(lcg[\mathbf{v}](\mathcal{M}_j))$⦄

$= \ gci(\{lcg[\mathbf{v}](\{\langle \boldsymbol{\Gamma}, \mathtt{int}\rangle \mid \boldsymbol{\Gamma} \in \mathbf{A}\}), lcg[\mathbf{v}](\{\langle \boldsymbol{\Gamma}, \boldsymbol{\mu}\rangle \mid \boldsymbol{\Gamma} \vdash \mathsf{A}_1 : \boldsymbol{\mu}\}), lcg[\mathbf{v}](\{\langle \boldsymbol{\Gamma}, \boldsymbol{\mu}'\rangle \mid$
$\boldsymbol{\Gamma} \vdash \mathsf{A}_2 : \boldsymbol{\mu}'\})\})$ ⦃definition (49.16) of $\boldsymbol{\gamma}[\![\langle \boldsymbol{\Gamma}, \boldsymbol{\mu}\rangle]\!]$ and (49.18) of $\boldsymbol{\Gamma} \vdash \mathsf{E} : \boldsymbol{\mu}$⦄

å $\preceq^{v} gci(\langle \boldsymbol{\Gamma}^{v}, \mathtt{int}\rangle, lcg[\mathbf{v}](\boldsymbol{\tau}[\![\mathsf{A}_1]\!]), lcg[\mathbf{v}](\boldsymbol{\tau}[\![\mathsf{A}_2]\!]))$ ⦃(49.26)⦄

$= \ gci(\langle \boldsymbol{\Gamma}^{v}, \mathtt{int}\rangle, \boldsymbol{\tau}^{v}[\![\mathsf{A}_1]\!], \boldsymbol{\tau}^{v}[\![\mathsf{A}_2]\!])$ ⦃induction hypothesis⦄

Example 49.40 The difference of a pair of integers and a Boolean is not typable because the monotypes $\mathtt{int}$, $\mathtt{int} * \mathtt{int}$, and $\mathtt{bool}$ are not unifiable so that $gci(\langle \boldsymbol{\Gamma}^{v}, \mathtt{int}\rangle, \langle \boldsymbol{\Gamma}^{v}, \mathtt{int} * \mathtt{int}\rangle, \langle \boldsymbol{\Gamma}^{v}, \mathtt{bool}\rangle) = \overline{\varnothing}^{v}$.

$- \ ground(\boldsymbol{\tau}^{v}[\![\mathsf{E}]\!])$

$= \ ground(gci(\langle \boldsymbol{\Gamma}^{v}, \mathtt{int}\rangle, \boldsymbol{\tau}^{v}[\![\mathsf{A}_1]\!], \boldsymbol{\tau}^{v}[\![\mathsf{A}_2]\!]))$

$= \ ground(\langle \boldsymbol{\Gamma}^{v}, \mathtt{int}\rangle) \cap ground(\boldsymbol{\tau}^{v}[\![\mathsf{A}_1]\!]) \cap ground(\boldsymbol{\tau}^{v}[\![\mathsf{A}_2]\!])$

⦃*ground* preserves glbs by (49.37)⦄

$= \ \{\langle \boldsymbol{\Gamma}, \mathtt{int}\rangle \mid \boldsymbol{\Gamma} \in \mathbf{A}\} \cap \boldsymbol{\tau}[\![\mathsf{A}_1]\!] \cap \boldsymbol{\tau}[\![\mathsf{A}_2]\!]$ ⦃instantiation of $\boldsymbol{\Gamma}^{v}$ and induction hypothesis⦄

$= \ \{\langle \boldsymbol{\Gamma}, \mathtt{int}\rangle \mid \boldsymbol{\Gamma} \in \mathbf{A}\} \cap \{\langle \boldsymbol{\Gamma}, \boldsymbol{\mu}\rangle \mid \boldsymbol{\Gamma} \vdash \mathsf{A}_1 : \boldsymbol{\mu}\} \cap \{\langle \boldsymbol{\Gamma}, \boldsymbol{\mu}'\rangle \mid \boldsymbol{\Gamma} \vdash \mathsf{A}_2 : \boldsymbol{\mu}'\}$ ⦃(49.26)⦄

$= \ \{\langle \boldsymbol{\Gamma}, \mathtt{int}\rangle \mid \boldsymbol{\Gamma} \vdash \mathsf{A}_1 : \mathtt{int} \wedge \boldsymbol{\Gamma} \vdash \mathsf{A}_2 : \mathtt{int}\} = \boldsymbol{\tau}[\![\mathsf{E}]\!]$

— $\mathsf{E} = \mathsf{S}_1\,,\,\mathsf{S}_2$

$\boldsymbol{\tau}^{v}[\![\mathsf{E}]\!] = lcg[\mathbf{v}](\boldsymbol{\tau}[\![\mathsf{E}]\!])$

$= \ lcg[\mathbf{v}](\{\langle \boldsymbol{\Gamma}, \boldsymbol{\mu}_1 * \boldsymbol{\mu}_2\rangle \mid \boldsymbol{\Gamma} \vdash \mathsf{S}_1 : \boldsymbol{\mu}_1 \wedge \boldsymbol{\Gamma} \vdash \mathsf{S}_2 : \boldsymbol{\mu}_2\})$

$= \ lcg[\mathbf{v}](\{\langle \boldsymbol{\Gamma}, \boldsymbol{\mu}_1 * \boldsymbol{\mu}_2\rangle \mid \mathscr{E}[\![\mathsf{S}_1]\!] \in \boldsymbol{\gamma}[\![\langle \boldsymbol{\Gamma}, \boldsymbol{\mu}_1\rangle]\!] \wedge \mathscr{E}[\![\mathsf{S}_2]\!] \in \boldsymbol{\gamma}[\![\langle \boldsymbol{\Gamma}, \boldsymbol{\mu}_2\rangle]\!]\})$

⦃definition (49.18) of $\boldsymbol{\Gamma} \vdash \mathsf{E} : \boldsymbol{\mu}$⦄

$= \ lcg[\mathbf{v}](\{\langle \boldsymbol{\Gamma}, \boldsymbol{\mu}_1 * \boldsymbol{\mu}_2\rangle \mid \forall \rho \in \boldsymbol{\gamma}[\![\boldsymbol{\Gamma}]\!] \,.\, \mathscr{E}[\![\mathsf{S}_1]\!](\rho) \in \boldsymbol{\gamma}[\![\boldsymbol{\mu}_1]\!] \wedge \forall \rho \in \boldsymbol{\gamma}[\![\boldsymbol{\Gamma}]\!] \,.\, \mathscr{E}[\![\mathsf{S}_2]\!](\rho) \in \boldsymbol{\gamma}[\![\boldsymbol{\mu}_2]\!]\})$

⦃definition (49.16) of $\boldsymbol{\gamma}[\![\langle \boldsymbol{\Gamma}, \boldsymbol{\mu}\rangle]\!]$⦄

$= \ lcg[\mathbf{v}](\{\langle \boldsymbol{\Gamma}, \boldsymbol{\mu}_1 * \boldsymbol{\mu}_2'\rangle \mid \forall \rho \in \boldsymbol{\gamma}[\![\boldsymbol{\Gamma}]\!] \,.\, \mathscr{E}[\![\mathsf{S}_1]\!](\rho) \in \boldsymbol{\gamma}[\![\boldsymbol{\mu}_1]\!] \wedge \boldsymbol{\mu}_2' \in \mathbf{M}^{v}\} \cap \{\langle \boldsymbol{\Gamma}, \boldsymbol{\mu}_1' * \boldsymbol{\mu}_2\rangle \mid \forall \rho \in$
$\boldsymbol{\gamma}[\![\boldsymbol{\Gamma}]\!] \,.\, \mathscr{E}[\![\mathsf{S}_2]\!](\rho) \in \boldsymbol{\gamma}[\![\boldsymbol{\mu}_2]\!] \wedge \boldsymbol{\mu}_1' \in \mathbf{M}^{v}\})$ ⦃definition of $\cap$ so that $\boldsymbol{\mu}_1 = \boldsymbol{\mu}_1'$ and $\boldsymbol{\mu}_2 = \boldsymbol{\mu}_2'$⦄

$$\preceq^{\nu} gci(lcg[\mathbf{v}](\{\langle \boldsymbol{\Gamma}, \boldsymbol{\mu}_1 * \boldsymbol{\mu}_2' \rangle \mid \forall \rho \in \boldsymbol{\gamma}[\![\boldsymbol{\Gamma}]\!] . \mathscr{E}[\![\mathsf{S}_1]\!](\rho) \in \boldsymbol{\gamma}[\![\boldsymbol{\mu}_1]\!] \wedge \boldsymbol{\mu}_2' \in \mathbf{M}^{\nu}\}), lcg[\mathbf{v}](\{\langle \boldsymbol{\Gamma}, \boldsymbol{\mu}_1' * \boldsymbol{\mu}_2 \rangle \mid \forall \rho \in \boldsymbol{\gamma}[\![\boldsymbol{\Gamma}]\!] . \mathscr{E}[\![\mathsf{S}_2]\!](\rho) \in \boldsymbol{\gamma}[\![\boldsymbol{\mu}_2]\!] \wedge \boldsymbol{\mu}_1' \in \mathbf{M}^{\nu}\}))$$

$$\text{\wr} lcg[\mathbf{v}](\textstyle\bigcap_{i\in\Delta} \mathcal{M}_i) \preceq^{\nu} \underset{j\in\Delta}{gci}(lcg[\mathbf{v}](\mathcal{M}_j)) \text{\wr}$$

$$= gci(\{lcg[\mathbf{v}](\{\langle \boldsymbol{\Gamma}, \boldsymbol{\mu}_1 * \boldsymbol{\mu}_2' \rangle \mid \boldsymbol{\Gamma} \vdash \mathsf{S}_1 : \boldsymbol{\mu}_1 \wedge \boldsymbol{\mu}_2' \in \mathbf{M}^{\nu}\}), lcg[\mathbf{v}](\{\langle \boldsymbol{\Gamma}, \boldsymbol{\mu}_1' * \boldsymbol{\mu}_2 \rangle \mid \boldsymbol{\Gamma} \vdash \mathsf{S}_2 : \boldsymbol{\mu}_2 \wedge \boldsymbol{\mu}_1' \in \mathbf{M}^{\nu}\})\}) \quad \text{\wr definition (49.16) of } \boldsymbol{\gamma}[\![\langle \boldsymbol{\Gamma}, \boldsymbol{\mu} \rangle]\!] \text{ and (49.18) of } \boldsymbol{\Gamma} \vdash \mathsf{E} : \boldsymbol{\mu} \text{\wr}$$

$$= gci(\{lcg[\mathbf{v}](\{\langle \boldsymbol{\Gamma}, \boldsymbol{\mu}_1 * \boldsymbol{\mu}_2' \rangle \mid \langle \boldsymbol{\Gamma}, \boldsymbol{\mu}_1 \rangle \in \boldsymbol{\tau}[\![\mathsf{S}_1]\!] \wedge \boldsymbol{\mu}_2' \in \mathbf{M}^{\nu}\}), lcg[\mathbf{v}](\{\langle \boldsymbol{\Gamma}, \boldsymbol{\mu}_1' * \boldsymbol{\mu}_2 \rangle \mid \langle \boldsymbol{\Gamma}, \boldsymbol{\mu}_2 \rangle \in \boldsymbol{\tau}[\![\mathsf{S}_2]\!] \wedge \boldsymbol{\mu}_1' \in \mathbf{M}^{\nu}\})\}) \quad \text{\wr(49.26)\wr}$$

$$= \text{let } \langle \boldsymbol{\Gamma}_1, \boldsymbol{\mu}_1 \rangle = lcg[\mathbf{v}](\boldsymbol{\tau}[\![\mathsf{S}_1]\!]) \text{ and } \langle \boldsymbol{\Gamma}_2, \boldsymbol{\mu}_2 \rangle = lcg[\mathbf{v}](\boldsymbol{\tau}[\![\mathsf{S}_2]\!]) \text{ in } gci(\langle \boldsymbol{\Gamma}_1, \boldsymbol{\mu}_1 * \alpha \rangle, \langle \boldsymbol{\Gamma}_2, \beta * \boldsymbol{\mu}_2 \rangle) \quad \text{\wr definition of } lcg[\mathbf{v}] \text{\wr}$$

$$\preceq^{\nu} \text{let } \langle \boldsymbol{\Gamma}_1, \boldsymbol{\mu}_1 \rangle = \boldsymbol{\tau}^{\nu}[\![\mathsf{S}_1]\!] \text{ and } \langle \boldsymbol{\Gamma}_2, \boldsymbol{\mu}_2 \rangle = \boldsymbol{\tau}^{\nu}[\![\mathsf{S}_2]\!] \text{ in } gci(\langle \boldsymbol{\Gamma}_1, \boldsymbol{\mu}_1 * \alpha \rangle, \langle \boldsymbol{\Gamma}_2, \beta * \boldsymbol{\mu}_2 \rangle) \quad \text{\wr induction hypothesis and } gci \text{ increasing \wr}$$

– $ground(\boldsymbol{\tau}^{\nu}[\![\mathsf{E}]\!])$

$$= \text{let } \langle \boldsymbol{\Gamma}_1, \boldsymbol{\mu}_1 \rangle = \boldsymbol{\tau}^{\nu}[\![\mathsf{S}_1]\!] \text{ and } \langle \boldsymbol{\Gamma}_2, \boldsymbol{\mu}_2 \rangle = \boldsymbol{\tau}^{\nu}[\![\mathsf{S}_2]\!] \text{ in } ground(gci(\langle \boldsymbol{\Gamma}_1, \boldsymbol{\mu}_1 * \alpha \rangle, \langle \boldsymbol{\Gamma}_2, \beta * \boldsymbol{\mu}_2 \rangle))$$

$$= \text{let } \langle \boldsymbol{\Gamma}_1, \boldsymbol{\mu}_1 \rangle = \boldsymbol{\tau}^{\nu}[\![\mathsf{S}_1]\!] \text{ and } \langle \boldsymbol{\Gamma}_2, \boldsymbol{\mu}_2 \rangle = \boldsymbol{\tau}^{\nu}[\![\mathsf{S}_2]\!] \text{ in } ground(\langle \boldsymbol{\Gamma}_1, \boldsymbol{\mu}_1 * \alpha \rangle) \cap ground(\langle \boldsymbol{\Gamma}_2, \beta * \boldsymbol{\mu}_2 \rangle) \quad \text{\wr Galois connection (49.37) so } gci \text{ preserves glbs\wr}$$

$$= \{\langle \boldsymbol{\Gamma}, \boldsymbol{\mu}_1 * \boldsymbol{\mu}' \rangle \mid \boldsymbol{\Gamma} \vdash \mathsf{S}_1 : \boldsymbol{\mu}_1 \wedge \boldsymbol{\mu}' \in \mathbf{M}\} \cap \{\langle \boldsymbol{\Gamma}, \boldsymbol{\mu}'' * \boldsymbol{\mu}_2 \rangle \mid \boldsymbol{\mu}'' \in \mathbf{M} \wedge \boldsymbol{\Gamma} \vdash \mathsf{S}_2 : \boldsymbol{\mu}_2\} \quad \text{\wr induction hypothesis and } ground(\alpha) = \mathbf{M} \text{\wr}$$

$$= \{\langle \boldsymbol{\Gamma}, \boldsymbol{\mu}_1 * \boldsymbol{\mu}_2 \rangle \mid \boldsymbol{\Gamma} \vdash \mathsf{S}_1 : \boldsymbol{\mu}_1 \wedge \boldsymbol{\Gamma} \vdash \mathsf{S}_2 : \boldsymbol{\mu}_2\} = \boldsymbol{\tau}[\![\mathsf{E}]\!] \quad \text{\wr definition of } \cap \text{\wr}$$

— $\mathsf{E} = [\,]$

$$\boldsymbol{\tau}^{\nu}[\![\mathsf{E}]\!] = lcg[\mathbf{v}](\boldsymbol{\tau}[\![\mathsf{E}]\!])$$

$$= lcg[\mathbf{v}](\{\langle \boldsymbol{\Gamma}, \boldsymbol{\mu}\ \mathtt{list} \rangle \mid \boldsymbol{\Gamma} \in \mathbf{A} \wedge \boldsymbol{\mu} \in \mathbf{M}\})$$

$$= \langle \boldsymbol{\Gamma}^{\nu}, \alpha\ \mathtt{list} \rangle$$

– $ground(\boldsymbol{\tau}^{\nu}[\![\mathsf{E}]\!])$

$$= ground(\langle \boldsymbol{\Gamma}^{\nu}, \alpha\ \mathtt{list} \rangle)$$

$$= \{\langle \boldsymbol{\Gamma}, \boldsymbol{\mu}\ \mathtt{list} \rangle \mid \boldsymbol{\Gamma} \in \mathbf{A} \wedge \boldsymbol{\mu} \in \mathbf{M}\} = \boldsymbol{\tau}[\![\mathsf{E}]\!]$$

— $\mathsf{E} = \mathsf{S}_1 :: \mathsf{S}_2$

$$\boldsymbol{\tau}^{\nu}[\![\mathsf{E}]\!] = lcg[\mathbf{v}](\boldsymbol{\tau}[\![\mathsf{E}]\!])$$

$$= lcg[\mathbf{v}](\{\langle \boldsymbol{\Gamma}, \boldsymbol{\mu}\ \mathtt{list} \rangle \mid \boldsymbol{\Gamma} \vdash \mathsf{S}_1 : \boldsymbol{\mu} \wedge \boldsymbol{\Gamma} \vdash \mathsf{S}_2 : \boldsymbol{\mu}\ \mathtt{list}\})$$

$$= lcg[\mathbf{v}](\{\langle \boldsymbol{\Gamma}, \boldsymbol{\mu}\ \mathtt{list} \rangle \mid \mathscr{E}[\![\mathsf{S}_1]\!] \in \boldsymbol{\gamma}[\![\langle \boldsymbol{\Gamma}, \boldsymbol{\mu} \rangle]\!] \wedge \mathscr{E}[\![\mathsf{S}_2]\!] \in \boldsymbol{\gamma}[\![\langle \boldsymbol{\Gamma}, \boldsymbol{\mu}\ \mathtt{list} \rangle]\!]\}) \quad \text{\wr definition (49.18) of } \boldsymbol{\Gamma} \vdash \mathsf{E} : \boldsymbol{\mu} \text{\wr}$$

$$= lcg[\mathbf{v}](\{\langle \boldsymbol{\Gamma}, \boldsymbol{\mu}\ \mathtt{list} \rangle \mid \forall \rho \in \boldsymbol{\gamma}[\![\boldsymbol{\Gamma}]\!] . \mathscr{E}[\![\mathsf{S}_1]\!](\rho) \in \boldsymbol{\gamma}[\![\boldsymbol{\mu}]\!] \wedge \forall \rho \in \boldsymbol{\gamma}[\![\boldsymbol{\Gamma}]\!] . \mathscr{E}[\![\mathsf{S}_2]\!](\rho) \in \boldsymbol{\gamma}[\![\boldsymbol{\mu}\ \mathtt{list}]\!]\}) \quad \text{\wr definition (49.16) of } \boldsymbol{\gamma}[\![\langle \boldsymbol{\Gamma}, \boldsymbol{\mu} \rangle]\!] \text{\wr}$$

$= \ lcg[\boldsymbol{v}](\{\langle \boldsymbol{\Gamma}, \boldsymbol{\mu}\ \mathtt{list}\rangle \mid \forall \rho \in \boldsymbol{\gamma}[\![\boldsymbol{\Gamma}]\!] \,.\, \mathscr{E}[\![\mathsf{S}_1]\!](\rho) \in \boldsymbol{\gamma}[\![\boldsymbol{\mu}]\!]\} \cap \{\langle \boldsymbol{\Gamma}', \boldsymbol{\mu}'\ \mathtt{list}\rangle \mid \forall \rho \in \boldsymbol{\gamma}[\![\boldsymbol{\Gamma}']\!] \,.\, \mathscr{E}[\![\mathsf{S}_2]\!](\rho) \in \boldsymbol{\gamma}[\![\boldsymbol{\mu}'\ \mathtt{list}]\!]\})$ ⟨definition of $\cap$ so that $\boldsymbol{\Gamma} = \boldsymbol{\Gamma}'$ and $\boldsymbol{\mu} = \boldsymbol{\mu}'$⟩

$\preceq^{\nu}\ gci(lcg[\boldsymbol{v}](\{\langle \boldsymbol{\Gamma}, \boldsymbol{\mu}\ \mathtt{list}\rangle \mid \forall \rho \in \boldsymbol{\gamma}[\![\boldsymbol{\Gamma}]\!] \,.\, \mathscr{E}[\![\mathsf{S}_1]\!](\rho) \in \boldsymbol{\gamma}[\![\boldsymbol{\mu}]\!]\}), lcg[\boldsymbol{v}](\{\langle \boldsymbol{\Gamma}', \boldsymbol{\mu}'\ \mathtt{list}\rangle \mid \forall \rho \in \boldsymbol{\gamma}[\![\boldsymbol{\Gamma}']\!] \,.\, \mathscr{E}[\![\mathsf{S}_2]\!](\rho) \in \boldsymbol{\gamma}[\![\boldsymbol{\mu}'\ \mathtt{list}]\!]\}))$ ⟨$lcg[\boldsymbol{v}](\bigcap_{i\in\Delta} \mathcal{M}_i) \preceq^{\nu} \underset{j\in\Delta}{gci}(lcg[\boldsymbol{v}](\mathcal{M}_j))$⟩

$= \ gci(\{lcg[\boldsymbol{v}](\{\langle \boldsymbol{\Gamma}, \boldsymbol{\mu}\ \mathtt{list}\rangle \mid \boldsymbol{\Gamma} \vdash \mathsf{S}_1 : \boldsymbol{\mu}\}), lcg[\boldsymbol{v}](\{\langle \boldsymbol{\Gamma}, \boldsymbol{\mu}\ \mathtt{list}\rangle \mid \boldsymbol{\Gamma} \vdash \mathsf{S}_2 : \boldsymbol{\mu}\ \mathtt{list}\})\})$ ⟨definition (49.16) of $\boldsymbol{\gamma}[\![\langle \boldsymbol{\Gamma}, \boldsymbol{\mu}\rangle]\!]$ and (49.18) of $\boldsymbol{\Gamma} \vdash \mathsf{E} : \boldsymbol{\mu}$⟩

$= \ gci(\{lcg[\boldsymbol{v}](\{\langle \boldsymbol{\Gamma}, \boldsymbol{\mu}\ \mathtt{list}\rangle \mid \langle \boldsymbol{\Gamma}, \boldsymbol{\mu}\rangle \in \boldsymbol{\tau}[\![\mathsf{S}_1]\!]\}), lcg[\boldsymbol{v}](\{\langle \boldsymbol{\Gamma}, \boldsymbol{\mu}\ \mathtt{list}\rangle \mid \langle \boldsymbol{\Gamma}, \boldsymbol{\mu}\ \mathtt{list}\rangle \in \boldsymbol{\tau}[\![\mathsf{S}_2]\!]\})\})$ ⟨(49.26)⟩

$= \ \text{let } \langle \boldsymbol{\Gamma}_1, \boldsymbol{\mu}_1\rangle = lcg[\boldsymbol{v}](\boldsymbol{\tau}[\![\mathsf{S}_1]\!]) \text{ and } \langle \boldsymbol{\Gamma}_2, \boldsymbol{\mu}_2\rangle = lcg[\boldsymbol{v}](\boldsymbol{\tau}[\![\mathsf{S}_2]\!]) \text{ in } gci(\langle \boldsymbol{\Gamma}_1, \boldsymbol{\mu}_1\ \mathtt{list}\rangle, \langle \boldsymbol{\Gamma}_2, \boldsymbol{\mu}_2\rangle)$ ⟨definition of $lcg[\boldsymbol{v}]$⟩

$\preceq^{\nu}\ \text{let } \langle \boldsymbol{\Gamma}_1, \boldsymbol{\mu}_1\rangle = \boldsymbol{\tau}^{\nu}[\![\mathsf{S}_1]\!] \text{ and } \langle \boldsymbol{\Gamma}_2, \boldsymbol{\mu}_2\rangle = \boldsymbol{\tau}^{\nu}[\![\mathsf{S}_2]\!] \text{ in } gci(\langle \boldsymbol{\Gamma}_1, \boldsymbol{\mu}_1\ \mathtt{list}\rangle, \langle \boldsymbol{\Gamma}_2, \boldsymbol{\mu}_2\rangle)$ ⟨induction hypothesis and *gci* increasing⟩

– $ground(\boldsymbol{\tau}^{\nu}[\![\mathsf{E}]\!])$

$= \ ground(\text{let } \langle \boldsymbol{\Gamma}_1, \boldsymbol{\mu}_1\rangle = \boldsymbol{\tau}^{\nu}[\![\mathsf{S}_1]\!] \text{ and } \langle \boldsymbol{\Gamma}_2, \boldsymbol{\mu}_2\rangle = \boldsymbol{\tau}^{\nu}[\![\mathsf{S}_2]\!] \text{ in } gci(\langle \boldsymbol{\Gamma}_1, \boldsymbol{\mu}_1\ \mathtt{list}\rangle, \langle \boldsymbol{\Gamma}_2, \boldsymbol{\mu}_2\rangle))$

$= \ \text{let } \langle \boldsymbol{\Gamma}_1, \boldsymbol{\mu}_1\rangle = \boldsymbol{\tau}^{\nu}[\![\mathsf{S}_1]\!] \text{ and } \langle \boldsymbol{\Gamma}_2, \boldsymbol{\mu}_2\rangle = \boldsymbol{\tau}^{\nu}[\![\mathsf{S}_2]\!] \text{ in } ground(\langle \boldsymbol{\Gamma}_1, \boldsymbol{\mu}_1\ \mathtt{list}\rangle) \cap ground(\langle \boldsymbol{\Gamma}_2, \boldsymbol{\mu}_2\rangle)$ ⟨Galois connection (49.37) so *gci* preserves glbs⟩

$= \ \{\langle \boldsymbol{\Gamma}, \boldsymbol{\mu}\ \mathtt{list}\rangle \mid \langle \boldsymbol{\Gamma}, \boldsymbol{\mu}\rangle \in ground(\boldsymbol{\tau}^{\nu}[\![\mathsf{S}_1]\!]) \wedge \langle \boldsymbol{\Gamma}, \boldsymbol{\mu}\ \mathtt{list}\rangle \in ground(\boldsymbol{\tau}^{\nu}[\![\mathsf{S}_2]\!])\}$ ⟨definition of $\cap$⟩

$= \ \{\langle \boldsymbol{\Gamma}, \boldsymbol{\mu}\ \mathtt{list}\rangle \mid \boldsymbol{\Gamma} \vdash \mathsf{S}_1 : \boldsymbol{\mu} \wedge \boldsymbol{\Gamma} \vdash \mathsf{S}_2 : \boldsymbol{\mu}\ \mathtt{list}\} = \boldsymbol{\tau}[\![\mathsf{E}]\!]$ ⟨induction hypothesis⟩

— $\mathsf{E} = \mathtt{tl}(\mathsf{S})$

$\boldsymbol{\tau}^{\nu}[\![\mathsf{E}]\!] = lcg[\boldsymbol{v}](\boldsymbol{\tau}[\![\mathsf{E}]\!])$

$= \ lcg[\boldsymbol{v}](\{\langle \boldsymbol{\Gamma}, \boldsymbol{\mu}\ \mathtt{list}\rangle \mid \boldsymbol{\Gamma} \vdash \mathsf{S} : \boldsymbol{\mu}\ \mathtt{list}\})$

$= \ lcg[\boldsymbol{v}](\{\langle \boldsymbol{\Gamma}, \boldsymbol{\mu}\ \mathtt{list}\rangle \mid \langle \boldsymbol{\Gamma}, \boldsymbol{\mu}\ \mathtt{list}\rangle \in \boldsymbol{\tau}[\![\mathsf{S}]\!]\})$ ⟨(49.26)⟩

$= \ lcg[\boldsymbol{v}](\{\langle \boldsymbol{\Gamma}, \boldsymbol{\mu}\ \mathtt{list}\rangle \mid \boldsymbol{\Gamma} \in \mathbf{A} \wedge \boldsymbol{\mu} \in \mathbf{M}\} \cap \boldsymbol{\tau}[\![\mathsf{S}]\!])$ ⟨definition of $\cap$⟩

$\preceq^{\nu}\ gci(lcg[\boldsymbol{v}](\{\langle \boldsymbol{\Gamma}, \boldsymbol{\mu}\ \mathtt{list}\rangle \mid \boldsymbol{\Gamma} \in \mathbf{A} \wedge \boldsymbol{\mu} \in \mathbf{M}\}), lcg[\boldsymbol{v}](\boldsymbol{\tau}[\![\mathsf{S}]\!]))$ ⟨$lcg[\boldsymbol{v}](\bigcap_{i\in\Delta} \mathcal{M}_i) \preceq^{\nu} \underset{j\in\Delta}{gci}(lcg[\boldsymbol{v}](\mathcal{M}_j))$⟩

$\preceq^{\nu}\ gci(\langle \boldsymbol{\Gamma}^{\nu}, \alpha\ \mathtt{list}\rangle, \boldsymbol{\tau}^{\nu}[\![\mathsf{S}]\!])$ ⟨induction hypothesis, definition of $lcg[\boldsymbol{v}]$, and *gci* increasing⟩

– $ground(\boldsymbol{\tau}^{\nu}[\![\mathsf{E}]\!])$

$= \ ground(gci(\langle \boldsymbol{\Gamma}^{\nu}, \alpha\ \mathtt{list}\rangle, \boldsymbol{\tau}^{\nu}[\![\mathsf{S}]\!]))$

$= \ ground(\langle \boldsymbol{\Gamma}^{\nu}, \alpha\ \mathtt{list}\rangle) \cap ground(\boldsymbol{\tau}^{\nu}[\![\mathsf{S}]\!])$

⟨Galois connection (49.37) so *gci* preserves glbs⟩

$= \{\langle \Gamma, \mu\ \mathtt{list}\rangle \mid \Gamma \in \mathbf{A} \wedge \mu \in \mathbf{M}\} \cap \tau[\![S]\!]$ ⟅definition of *ground* and induction hypothesis⟆

$= \{\langle \Gamma, \mu\ \mathtt{list}\rangle \mid \Gamma \vdash S : \mu\ \mathtt{list}\} = \tau[\![E]\!]$ ⟅definition $\cap$ and (49.26)⟆

— $E = \mathsf{hd}(S)$

$\tau^{\nu}[\![E]\!] = lcg[\nu](\tau[\![E]\!])$

$= lcg[\nu](\{\langle \Gamma, \mu\rangle \mid \Gamma \vdash S : \mu\ \mathtt{list}\})$

$= lcg[\nu](\{\langle \Gamma, \mu\rangle \mid \langle \Gamma, \mu\ \mathtt{list}\rangle \in \tau[\![S]\!]\})$ ⟅(49.26)⟆

$= \mathsf{let}\ \langle \Gamma, \mu\ \mathtt{list}\rangle = lcg[\nu](\{\langle \Gamma, \mu\ \mathtt{list}\rangle \mid \langle \Gamma, \mu\ \mathtt{list}\rangle \in \tau[\![S]\!]\})\ \mathsf{in}\ \langle \Gamma, \mu\rangle$ ⟅definition of $lcg[\nu]$⟆

$= \mathsf{let}\ \langle \Gamma, \mu\ \mathtt{list}\rangle = lcg[\nu](\{\langle \Gamma, \mu\ \mathtt{list}\rangle \mid \Gamma \in \mathbf{A} \wedge \mu \in \mathbf{M}\} \cap \tau[\![S]\!])\ \mathsf{in}\ \langle \Gamma, \mu\rangle$ ⟅definition of $\cap$⟆

$\preceq^{\nu}\ \mathsf{let}\ \langle \Gamma, \mu\ \mathtt{list}\rangle = gci(lcg[\nu](\{\langle \Gamma, \mu\ \mathtt{list}\rangle \mid \Gamma \in \mathbf{A} \wedge \mu \in \mathbf{M}\}), lcg[\nu](\tau[\![S]\!]))\ \mathsf{in}\ \langle \Gamma, \mu\rangle$ ⟅$lcg[\nu](\bigcap_{i\in\Delta} \mathcal{M}_i) \preceq^{\nu} \underset{j\in\Delta}{gci}(lcg[\nu](\mathcal{M}_j))$⟆

$\preceq^{\nu}\ \mathsf{let}\ \langle \Gamma, \mu\ \mathtt{list}\rangle = gci(\langle \Gamma^{\nu}, \alpha\ \mathtt{list}\rangle, \tau^{\nu}[\![S]\!])\ \mathsf{in}\ \langle \Gamma, \mu\rangle$ ⟅induction hypothesis, definition of $lcg[\nu]$, and *gci* increasing⟆

– $ground(\tau^{\nu}[\![E]\!])$

$= ground(\mathsf{let}\ \langle \Gamma, \mu\ \mathtt{list}\rangle = gci(\langle \Gamma^{\nu}, \alpha\ \mathtt{list}\rangle, \tau^{\nu}[\![S]\!])\ \mathsf{in}\ \langle \Gamma, \mu\rangle)$

$= \{\langle \Gamma, \mu\rangle \mid \langle \Gamma, \mu\ \mathtt{list}\rangle \in ground(gci(\langle \Gamma^{\nu}, \alpha\ \mathtt{list}\rangle, \tau^{\nu}[\![S]\!]))\}$

$= \{\langle \Gamma, \mu\rangle \mid \langle \Gamma, \mu\ \mathtt{list}\rangle \in ground(\langle \Gamma^{\nu}, \alpha\ \mathtt{list}\rangle) \cap ground(\tau^{\nu}[\![S]\!])\}$ ⟅Galois connection (49.37) so *gci* preserves glbs⟆

$= \{\langle \Gamma, \mu\rangle \mid \langle \Gamma, \mu\ \mathtt{list}\rangle \in ground(\tau^{\nu}[\![S]\!])\}$ ⟅definition of *ground*⟆

$= \{\langle \Gamma, \mu\rangle \mid \Gamma \vdash S : \mu\ \mathtt{list}\} = \tau[\![E]\!]$ ⟅(49.26)⟆

49.9.4 Type Inference Rules for Expressions

To derive the type inference rules of figure 49.41, we use the definition of *gci* in (48.47.15–18). Assuming that (after proper renaming) no two different pairs of a type assignment and a monotype with variables share a common variable in the set of typings with variables $\mathcal{T}$, we specify $gci(\mathcal{T})$ by the rule of inference $\dfrac{\vartheta = \mathsf{mgu}(\mathcal{T}), \langle \Gamma, \mu\rangle \in \mathcal{T}}{\langle \vartheta(\Gamma), \vartheta(\mu)\rangle}$ (because, by section 49.9.2, $gci(\mathcal{T}) = \langle \vartheta(\Gamma), \vartheta(\mu)\rangle$ is defined componentwise).

Example 49.42 Continuing example 49.30, in order to type $\mathsf{hd(x)} - \mathsf{y}$, the only applicable rule is [mns]. Its premises tell us that we have to type $\mathsf{hd(x)}$ and y.

– To type $\mathsf{hd(x)}$, the only applicable rule is [hd]. Its premise tells us that we have to type x.

– The only applicable rule to type x is [var] and we get $\Gamma^{\nu} \vdash \mathsf{x} : \Gamma^{\nu}(\mathsf{x})$.

Type inference rules

$$\boldsymbol{\Gamma}^{v} \vdash \texttt{1} : \texttt{int} \qquad \text{[cst]}$$

$$\boldsymbol{\Gamma}^{v} \vdash \texttt{x} : \boldsymbol{\Gamma}^{v}(\texttt{x}) \qquad \text{[var]}$$

$$\frac{\boldsymbol{\Gamma}_1 \vdash \mathsf{A}_1 : \boldsymbol{\mu}_1,\ \boldsymbol{\Gamma}_2 \vdash \mathsf{A}_2 : \boldsymbol{\mu}_2, \vartheta = \mathsf{mgu}(\{\langle \boldsymbol{\Gamma}_1, \boldsymbol{\mu}_1\rangle, \langle \boldsymbol{\Gamma}_2, \boldsymbol{\mu}_2\rangle, \langle \boldsymbol{\Gamma}^{v}, \texttt{int}\rangle\})}{\vartheta(\boldsymbol{\Gamma}^{v}) \vdash \mathsf{A}_1 - \mathsf{A}_2 : \texttt{int}} \qquad \text{[mns]}$$

$$\frac{\boldsymbol{\Gamma}_1 \vdash \mathsf{S}_1 : \boldsymbol{\mu}_1,\ \boldsymbol{\Gamma}_2 \vdash \mathsf{S}_2 : \boldsymbol{\mu}_2, \vartheta = \mathsf{mgu}(\{\langle \boldsymbol{\Gamma}_1, \boldsymbol{\mu}_1 * \alpha\rangle, \langle \boldsymbol{\Gamma}_2, \beta * \boldsymbol{\mu}_2\rangle\})}{\vartheta(\boldsymbol{\Gamma}_1) \vdash \mathsf{S}_1 , \mathsf{S}_2 : \vartheta(\boldsymbol{\mu}_1 * \boldsymbol{\mu}_2)} \qquad \text{[pair]}$$

$$\boldsymbol{\Gamma}^{v} \vdash [\,] : \alpha\ \texttt{list} \qquad \text{[nil]}$$

$$\frac{\boldsymbol{\Gamma}_1 \vdash \mathsf{S}_1 : \boldsymbol{\mu}_1,\ \boldsymbol{\Gamma}_2 \vdash \mathsf{S}_2 : \boldsymbol{\mu}_2, \vartheta = \mathsf{mgu}(\{\langle \boldsymbol{\Gamma}_1, \boldsymbol{\mu}_1\ \texttt{list}\rangle, \langle \boldsymbol{\Gamma}_2, \boldsymbol{\mu}_2\rangle\})}{\vartheta(\boldsymbol{\Gamma}_1) \vdash \mathsf{S}_1 :: \mathsf{S}_2 : \vartheta(\boldsymbol{\mu}_1)\ \texttt{list}} \qquad \text{[cons]}$$

$$\frac{\boldsymbol{\Gamma} \vdash \mathsf{S} : \boldsymbol{\mu}, \vartheta = \mathsf{mgu}(\{\langle \boldsymbol{\Gamma}, \boldsymbol{\mu}\rangle, \langle \boldsymbol{\Gamma}^{v}, \alpha\ \texttt{list}\rangle\})}{\vartheta(\boldsymbol{\Gamma}) \vdash \mathsf{hd}(\mathsf{S}) : \vartheta(\alpha)} \qquad \text{[hd]}$$

$$\frac{\boldsymbol{\Gamma} \vdash \mathsf{S} : \boldsymbol{\mu}, \vartheta = \mathsf{mgu}(\{\langle \boldsymbol{\Gamma}, \boldsymbol{\mu}\rangle, \langle \boldsymbol{\Gamma}^{v}, \alpha\ \texttt{list}\rangle\})}{\vartheta(\boldsymbol{\Gamma}) \vdash \texttt{tl}(\mathsf{S}) : \vartheta(\boldsymbol{\mu})} \qquad \text{[tl]}$$

$$\frac{\boldsymbol{\Gamma} \vdash \mathsf{S} : \boldsymbol{\mu}, \vartheta = \mathsf{mgu}(\{\langle \boldsymbol{\Gamma}, \boldsymbol{\mu}\rangle, \langle \boldsymbol{\Gamma}^{v}, \alpha\ \texttt{list}\rangle\})}{\vartheta(\boldsymbol{\Gamma}) \vdash \mathsf{S} = [\,] : \texttt{bool}} \qquad \text{[nil]}$$

$$\frac{\boldsymbol{\Gamma}_1 \vdash \mathsf{A}_1 : \boldsymbol{\mu}_1,\ \boldsymbol{\Gamma}_2 \vdash \mathsf{A}_2 : \boldsymbol{\mu}_2, \vartheta = \mathsf{mgu}(\{\langle \boldsymbol{\Gamma}_1, \boldsymbol{\mu}_1\rangle, \langle \boldsymbol{\Gamma}_2, \boldsymbol{\mu}_2\rangle, \langle \boldsymbol{\Gamma}^{v}, \texttt{int}\rangle\})}{\vartheta(\boldsymbol{\Gamma}_1) \vdash \mathsf{A}_1 < \mathsf{A}_2 : \texttt{bool}} \qquad \text{[lt]}$$

$$\frac{\boldsymbol{\Gamma}_1 \vdash \mathsf{B}_1 : \boldsymbol{\mu}_1,\ \boldsymbol{\Gamma}_2 \vdash \mathsf{B}_2 : \boldsymbol{\mu}_2, \vartheta = \mathsf{mgu}(\{\langle \boldsymbol{\Gamma}_1, \boldsymbol{\mu}_1\rangle, \langle \boldsymbol{\Gamma}_2, \boldsymbol{\mu}_2\rangle, \langle \boldsymbol{\Gamma}^{v}, \texttt{bool}\rangle\})}{\vartheta(\boldsymbol{\Gamma}_1) \vdash \mathsf{B}_1\ \mathsf{nand}\ \mathsf{B}_2 : \texttt{bool}} \qquad \text{[nand]}$$

Figure 49.41
Monotype inference rules for expressions

– Returning to rule [hd], from $\boldsymbol{\Gamma}^{v} \vdash \texttt{x} : \boldsymbol{\Gamma}^{v}(\texttt{x}), \vartheta = \mathsf{mgu}(\{\langle \boldsymbol{\Gamma}^{v}, \boldsymbol{\Gamma}^{v}(\texttt{x})\rangle, \langle \boldsymbol{\Gamma}^{v}, \alpha_1\ \texttt{list}\rangle\})$, we have $\vartheta(\boldsymbol{\Gamma}^{v}) = \boldsymbol{\Gamma}^{v}[\texttt{x} \leftarrow \alpha_1\ \texttt{list}]$ so we infer $\vartheta(\boldsymbol{\Gamma}^{v}) \vdash \mathsf{hd}(\texttt{x}) : \vartheta(\alpha_1)$ that is $\boldsymbol{\Gamma}^{v}[\texttt{x} \leftarrow \alpha_1\ \texttt{list}] \vdash \mathsf{hd}(\texttt{x}) : \alpha_1$.
– The only applicable rule to type y is [var] and we get $\boldsymbol{\Gamma}^{v} \vdash \texttt{y} : \boldsymbol{\Gamma}^{v}(\texttt{y})$.
– Returning to rule [mns], the premise is $\boldsymbol{\Gamma}^{v}[\texttt{x} \leftarrow \alpha_1\ \texttt{list}] \vdash \mathsf{hd}(\texttt{x}) : \alpha_1$, $\boldsymbol{\Gamma}^{v} \vdash \texttt{y} : \boldsymbol{\Gamma}^{v}(\texttt{y})$, $\vartheta' = \mathsf{mgu}(\{\langle \boldsymbol{\Gamma}^{v}[\texttt{x} \leftarrow \alpha_1\ \texttt{list}], \alpha_1\rangle, \langle \boldsymbol{\Gamma}^{v}, \boldsymbol{\Gamma}^{v}(\texttt{y})\rangle, \langle \boldsymbol{\Gamma}^{v}, \texttt{int}\rangle\})$. We get $\vartheta'(\alpha_1) = \texttt{int}$ and $\vartheta'(\boldsymbol{\Gamma}^{v}) = \boldsymbol{\Gamma}^{v}[\texttt{x} \leftarrow \texttt{int list}][\texttt{y} \leftarrow \texttt{int}]$.
– The conclusion is $\vartheta'(\boldsymbol{\Gamma}^{v}) \vdash \mathsf{hd}(\texttt{x}) - \texttt{y} : \texttt{int} = \boldsymbol{\Gamma}^{v}[\texttt{x} \leftarrow \texttt{int list}][\texttt{y} \leftarrow \texttt{int}] \vdash \mathsf{hd}(\texttt{x}) - \texttt{y} : \texttt{int}$.

The concretization is $\mathit{ground}(\langle \boldsymbol{\Gamma}^{v}[\texttt{x} \leftarrow \texttt{int list}][\texttt{y} \leftarrow \texttt{int}], \texttt{int}\rangle) = \{\langle \boldsymbol{\Gamma}[\texttt{x} \leftarrow \texttt{int list}][\texttt{y} \leftarrow \texttt{int}], \texttt{int}\rangle \mid \boldsymbol{\Gamma} \in \mathbf{A}\}$, as found in example 49.30.

Exercise 49.43 Write a program in the language of your choice to compute the most general unifier of a system of equations between monotypes with variables. For example,

```
# mgu [(Tlist (Tpair (Tint, Tvar "a")), Tlist (Tpair (Tvar "b", Tvar "c")))];;
- : substitution = [("b", Tint); ("a", Tvar "c")]
```

Exercise 49.44 Write a program in the language of your choice to infer the type of expressions (given in abstract syntax).

Observe that if instead of deriving the most general typing (traditionally called *principal type*) $\boldsymbol{\tau}^{\mathsf{v}}[\![\mathsf{E}]\!]$ of expressions E, we want to derive its instances $inst(\boldsymbol{\tau}^{\mathsf{v}}[\![\mathsf{E}]\!]) \triangleq \{\vartheta(\boldsymbol{\tau}^{\mathsf{v}}[\![\mathsf{E}]\!]) \mid \vartheta \in \Sigma\}$ and then we get rules similar to the monomorphic typing rules (49.28) (where the monomorphic typings $\langle \boldsymbol{\Gamma}, \boldsymbol{\mu} \rangle$ are replaced by monomorphic typings with variables $\langle \boldsymbol{\mathit{\Gamma}}, \boldsymbol{\mathit{\mu}} \rangle$).

49.10 Conclusion

Pairs and lists date back to John McCarthy's LISP programming language [673]. Roger Hindley [500] showed how to infer Alonzo Church monotypes for the λ-calculus [185] using type variables and unification. This inspired Robin Milner to design a polymorphic type inference algorithm (so-called algorithm W) in which type variables denotes any fixed but unknown type [685]. The type system was later presented by a structural deductive system [286] that was proved equivalent to the type inference algorithm. This initial work gave rise to numerous refinements [796, 797]. Empirical design of types systems leads to errors, for example, [40].

Generalizing monotypes with variables to represent sets of monotypes, [450] introduced *abstract types*, which stand for (infinitely) many monomorphic instances, at different levels of abstraction.

The understanding of type theory as an abstract interpretation is from [224, 237]. This yields language-independent concepts that apply equally well, for example, to denotational semantics or resolution-based languages [270]. The classic type theory [796, 797] has difficulties with modern typing systems taking data values into account that are easily solved by abstract interpretation. For example, [784] proves the soundness of *liquid types* [842] (i.e., types refined by abstract properties in a finite abstract domain) by abstract interpretation. [395] is an example of a *gradual type system* designed by abstract interpretation that has been fully mechanized [292]. Beyond types, types and effects (describing code side effects [641]) can also be formalized by abstract interpretation [385, 961]. [386] shows that this may be useful to catch and fix soundness errors in postulated types and effects systems.

49.11 Solutions to Selected Exercises

Solution to exercise 49.2

$$\begin{array}{rcll}
\mathbb{U}^0 & \triangleq & \mathbb{B} \cup \mathbb{Z} \cup \{\mathtt{nil}\} & \text{Booleans, integers, and null list} \\
\mathbb{U}^{n+1} & \triangleq & \mathbb{U}^n & \\
 & & \cup \{\langle x,\ y\rangle \mid x, y \in \mathbb{U}^n\} & \text{pairs} \\
 & & \cup \{x :: y \mid x, y \in \mathbb{U}^n\} & \text{lists} \\
\mathbb{U} & \triangleq & \bigcup_{n \in \mathbb{N}} \mathbb{U}^n \cup \{\Omega^\sigma, \Omega^\delta\} & \text{values}
\end{array}$$

Solution to exercise 49.34 hd([]) is a definite dynamic error so can be rejected

$$\frac{\Gamma \vdash \mathtt{S} : \boldsymbol{\mu}\ \mathtt{list},\ \mathtt{S} \neq []}{\Gamma \vdash \mathtt{hd(S)} : \boldsymbol{\mu}}.$$

Of course, this refinement is endless because $(\mathtt{hd} \mid \mathtt{tl})^*([])$ also definitely yield a dynamic error. More generally, a static analysis would be useful.

Solution to exercise 49.36

```
type monotype =
    Mbool
  | Mint
  | Mpair of monotype * monotype
  | Mlist of monotype

(* type environment mapping program variables to monotypes *)
type menvironment = (program_variable * monotype) list

(* check g|-e:m i.e. in type environment g, expression e has monotype m *)
let rec mcheck g e m =
  match e with
  | One -> m = Mint
  | Var x ->
     (try (List.assoc x g) = m
      with Not_found -> false)
  | Minus (e1, e2) ->
     (mcheck g e1 Mint) && (mcheck g e2 Mint) && (m = Mint)
  | Nil ->
     (match m with
      | Mlist _ -> true
      | _ -> false)
  | Pair (e1, e2) ->
     (match m with
      | Mpair (m1, m2) -> (mcheck g e1 m1) && (mcheck g e2 m2)
      | _ -> false)
  | Cons (e1, e2) ->
     (match m with
      | Mlist m' -> (mcheck g e1 m') && (mcheck g e2 m)
      | _ -> false)
  | Hd e1 ->
     (match m with
      | Mlist m' -> (mcheck g e1 m')
      | _ -> false)
  | Tl e1 ->
     (match m with
      | Mlist m' -> (mcheck g e1 m)
      | _ -> false)
  | Less (e1, e2) ->
     (mcheck g e1 Mint) && (mcheck g e2 Mint) && (m = Mbool)
  | Isnil e1 ->
     (match m with
      | Mlist m' -> true
      | _ -> false)
  | Nand (e1, e2) ->
     (mcheck g e1 Mbool) && (mcheck g e2 Mbool) && (m = Mbool)

# mcheck [("x", Mint)] (Cons ((Var "x"), Nil)) (Mlist Mint) ;;
- : bool = true
```

XIII BACKWARD STATIC ANALYSIS

A forward/deductive static analysis computes states reachable from specified initial states of the computation (chapters 19 and 20).

Inversely, a backward/abductive static analysis computes initial states from states specified to be reachable by the computation (chapter 50).

A forward/backward static analysis computes states reachable from specified initial states and reaching specified states of the computation. This is called a reduced forward/backward analysis when it involves the iteration of successive forward and backward analyses (chapter 51).

50 Backward Accessibility Semantics

It is often useful to reason on program executions backward, in reverse execution order, as shown by Dijkstra's weakest (liberal) preconditions [327] (see section 12.4). For example, for the property $x = y$ to hold at a program point ℓ_2 of program ℓ_0 y = 1 ;ℓ_1 x = x − y ;ℓ_2, it is necessary and sufficient that property $x = 2 \times y$ did hold at the previous point ℓ_1 in the execution and so that $x = 2$ did hold at the previous point ℓ_0. Such reasonings can be based on backward/abductive accessibility semantics and are also called abductive (infer *a* to explain *b*) [412] as opposed to forward reachability semantics, also called deductive (infer *b* as a consequence of *a*). An application of backward analysis is the inference of conditions on inputs to avoid an error.[1]

The impossible failure accessibility semantics is the adjoint of the reachability semantics. It specifies a condition on the initial states for reaching only states satisfying a given reachability condition. This is similar to $\widetilde{\mathsf{pre}}$ (12.12).

The possible success accessibility semantics is the conjugate of the impossible failure accessibility semantics. It specifies a condition on the initial states for the existence of an execution reaching a state satisfying a given reachability condition. This is similar to pre (12.11).

Contents

1. For example [470], the guided-missile cruiser *USS Yorktown* suffered a major computer crash on September 21, 1997. After a crew member mistakenly entered a blank, treated as zero, in a data field of an application, the computer system proceeded to divide another quantity by that zero. The operation caused a buffer overflow, in which data leaked from a temporary storage space in memory, and the error eventually brought down the ship's propulsion system. The *USS Yorktown* was dead in the water for more than two hours.

50.1 Definition of the Backward/Abductive Impossible Failure Accessibility Semantics

By definition (20.6) of $\boldsymbol{\mathcal{S}}^{\vec{\mathsf{r}}}\llbracket \mathsf{S} \rrbracket$ and definition (20.1) of $\mathsf{post}^{\vec{\mathsf{r}}}$, we have characterized $\boldsymbol{\mathcal{S}}^{\vec{\mathsf{r}}}\llbracket \mathsf{S} \rrbracket$ (and so its structural version $\widehat{\boldsymbol{\mathcal{S}}}^{\vec{\mathsf{r}}}\llbracket \mathsf{S} \rrbracket$) as follows.

Lemma 50.1 $\widehat{\boldsymbol{\mathcal{S}}}^{\vec{\mathsf{r}}}\llbracket \mathsf{S} \rrbracket\, \mathcal{P}_0 = \mathsf{post}^{\vec{\mathsf{r}}}(\boldsymbol{\mathcal{S}}^{*}\llbracket \mathsf{S} \rrbracket)\, \mathcal{P}_0$ where $\mathsf{post}^{\vec{\mathsf{r}}}(\boldsymbol{\mathcal{S}})\, \mathcal{P}_0 = \ell \in \mathsf{labs}\llbracket \mathsf{S} \rrbracket \mapsto \{\boldsymbol{\varrho}(\pi_0 \ell_0 \pi_1 \ell') \mid \boldsymbol{\varrho}(\pi_0 \ell_0) \in \mathcal{P}_0 \wedge \ell_0 \pi_1 \ell' \in \boldsymbol{\mathcal{S}}(\pi_0 \ell_0) \wedge \ell' = \ell\}$.

and similarly for $\widehat{\boldsymbol{\mathcal{S}}}^{\vec{\mathsf{R}}}\llbracket \mathsf{S} \rrbracket$ and $\mathsf{post}^{\vec{\mathsf{R}}}$.

Exercise 11.39 implies that the reachability semantics $\widehat{\boldsymbol{\mathcal{S}}}^{\vec{\varrho}}\llbracket \mathsf{S} \rrbracket$ is the lower adjoint of a Galois connection, because it preserves arbitrary joins, as shown in theorem 19.36.

Corollary 50.2 $\langle \wp(\mathbb{E}\mathbb{v}^{\varrho}), \subseteq \rangle \underset{\widehat{\boldsymbol{\mathcal{S}}}^{\vec{\varrho}}\llbracket \mathsf{S} \rrbracket}{\overset{\widehat{\boldsymbol{\mathcal{S}}}^{\tilde{\vec{\varrho}}}\llbracket \mathsf{S} \rrbracket}{\leftrightarrows}} \langle \mathsf{labs}\llbracket \mathsf{S} \rrbracket \to \wp(\mathbb{E}\mathbb{v}^{\varrho}), \dot{\subseteq} \rangle$ where $\mathbb{E}\mathbb{v}^{\varrho} = \mathbb{E}\mathbb{v}$ when $\vec{\varrho} = \vec{\mathsf{r}}$ and $\tilde{\vec{\varrho}} = \tilde{\vec{\mathsf{r}}}$ and $\mathbb{E}\mathbb{v}^{\varrho} = \mathbb{E}\mathbb{v} \times \mathbb{E}\mathbb{v}$ when $\vec{\varrho} = \vec{\mathsf{R}}$ and $\tilde{\vec{\varrho}} = \tilde{\vec{\mathsf{R}}}$.

Proof By theorem 19.36 and exercise 11.39. □

By the unicity of the adjoints in a Galois connection, this defines a unique backward trace semantics $\widehat{\boldsymbol{\mathcal{S}}}^{\tilde{\vec{\varrho}}}\llbracket \mathsf{S} \rrbracket$. This backward semantics $\widehat{\boldsymbol{\mathcal{S}}}^{\tilde{\vec{\varrho}}}\llbracket \mathsf{S} \rrbracket$ is the inverse of the forward semantics $\widehat{\boldsymbol{\mathcal{S}}}^{\vec{\varrho}}\llbracket \mathsf{S} \rrbracket$, so the knowledge of one is equivalent to the knowledge of the other. However, this is true in the concrete but not necessarily in the abstract. Hence combining forward reachability and backward accessibility analyses may be useful. As shown in chapter 12, there are many possible backward semantics, so we must characterize them precisely.

50.2 Characterization of the Backward Impossible Failure Accessibility Semantics

Theorem 50.3 states that any execution from any state in $\widehat{\boldsymbol{\mathcal{S}}}^{\tilde{\vec{\mathsf{r}}}}\llbracket \mathsf{S} \rrbracket\, \mathcal{P}_f$ can reach only the of S satisfying $\mathcal{P}_f$.

Theorem 50.3 $\widehat{\boldsymbol{\mathcal{S}}}^{\tilde{\vec{\mathsf{r}}}}\llbracket \mathsf{S} \rrbracket\, \mathcal{P}_f = \widetilde{\mathsf{pre}}^{\tilde{\vec{\mathsf{r}}}}(\boldsymbol{\mathcal{S}}^{*}\llbracket \mathsf{S} \rrbracket)\, \mathcal{P}_f$ where $\widetilde{\mathsf{pre}}^{\tilde{\vec{\mathsf{r}}}}(\boldsymbol{\mathcal{S}})\, \mathcal{P}_f \triangleq \{\boldsymbol{\varrho}(\pi_0 \ell_0) \in \mathbb{E}\mathbb{v} \mid \forall \ell \in \mathsf{labs}\llbracket \mathsf{S} \rrbracket \,.\, (\ell_0 \pi_1 \ell \in \boldsymbol{\mathcal{S}}(\pi_0 \ell_0)) \Rightarrow \boldsymbol{\varrho}(\pi_0 \ell_0 \pi_1 \ell) \in \mathcal{P}_f(\ell)\}$.

Proof of theorem 50.3

$$\widehat{\boldsymbol{\mathcal{S}}}^{\vec{\mathsf{r}}}\llbracket \mathsf{S} \rrbracket\, \mathcal{P}_0 \mathrel{\dot{\subseteq}} \mathcal{P}_f$$

$$\Leftrightarrow \forall \ell \in \mathsf{labs}\llbracket \mathsf{S} \rrbracket \,.\, \{\boldsymbol{\varrho}(\pi_0 \ell_0 \pi_1 \ell') \mid \boldsymbol{\varrho}(\pi_0 \ell_0) \in \mathcal{P}_0 \wedge \ell_0 \pi_1 \ell' \in \boldsymbol{\mathcal{S}}^{*}\llbracket \mathsf{S} \rrbracket\, (\pi_0 \ell_0) \wedge \ell' = \ell\} \subseteq \mathcal{P}_f(\ell)$$

$$\text{(pointwise definition of } \dot{\subseteq} \text{ and lemma 50.1)}$$

$$\Leftrightarrow \forall \ell \in \mathsf{labs}\llbracket \mathsf{S} \rrbracket \,.\, \forall \boldsymbol{\varrho}(\pi_0 \ell_0) \in \mathcal{P}_0 \,.\, (\ell_0 \pi_1 \ell \in \boldsymbol{\mathcal{S}}^{*}\llbracket \mathsf{S} \rrbracket\, (\pi_0 \ell_0)) \Rightarrow \boldsymbol{\varrho}(\pi_0 \ell_0 \pi_1 \ell) \in \mathcal{P}_f(\ell)$$

$$\text{(definition of } \subseteq\text{)}$$

$$\Leftrightarrow\ \forall \varrho(\pi_0\ell_0) \in \mathscr{P}_0 \,.\, \forall \ell \in \mathsf{labs}[\![\mathsf{S}]\!] \,.\, (\ell_0\pi_1\ell \in \boldsymbol{S}^*[\![\mathsf{S}]\!]\,(\pi_0\ell_0)) \Rightarrow \varrho(\pi_0\ell_0\pi_1\ell) \in \mathscr{P}_f(\ell)$$

$$\text{(definition of } \forall\text{)}$$

$$\Leftrightarrow\ \mathscr{P}_0 \subseteq \{\varrho(\pi_0\ell_0) \mid \forall \ell \in \mathsf{labs}[\![\mathsf{S}]\!] \,.\, (\ell_0\pi_1\ell \in \boldsymbol{S}^*[\![\mathsf{S}]\!]\,(\pi_0\ell_0)) \Rightarrow \varrho(\pi_0\ell_0\pi_1\ell) \in \mathscr{P}_f(\ell)\}$$

$$\text{(definition of } \subseteq\text{)}$$

$$\Rightarrow\ \widehat{\boldsymbol{S}}^{\tilde{\bar{\mathsf{r}}}}[\![\mathsf{S}]\!]\,\mathscr{P}_f = \{\varrho(\pi_0\ell_0) \mid \forall \ell \in \mathsf{labs}[\![\mathsf{S}]\!] \,.\, (\ell_0\pi_1\ell \in \boldsymbol{S}^*[\![\mathsf{S}]\!]\,(\pi_0\ell_0)) \Rightarrow \varrho(\pi_0\ell_0\pi_1\ell) \in \mathscr{P}_f(\ell)\}$$

$$\text{(by unicity of the adjoint in the Galois connection of corollary 50.2)} \quad \square$$

Example 50.4 Consider a semantics $\boldsymbol{S}$ illustrated below together with a specification $\mathscr{P}_f$ of the reachable states. Then $\widetilde{\mathsf{pre}}^{\tilde{\bar{\mathsf{r}}}}(\boldsymbol{S})\,\mathscr{P}_f$ is the set of initial states from which the traces of $\boldsymbol{S}$ cannot reach a state not in $\mathscr{P}_f$.

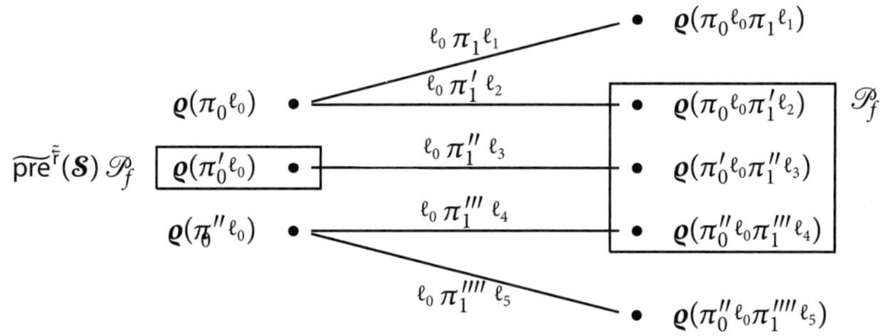

Example 50.5 (absence of errors) By theorem 50.3, we can specify that erroneous states in $\mathscr{E} \in \mathsf{labs}[\![\mathsf{S}]\!] \to \wp(\mathbb{E}\mathrm{v}^{\varrho})$ are unreachable from initial states in $\mathscr{P}_0 \in \wp(\mathbb{E}\mathrm{v}^{\varrho})$ by $\mathscr{P}_0 \subseteq \widehat{\boldsymbol{S}}^{\tilde{\bar{\varrho}}}[\![\mathsf{S}]\!](\ell \mapsto \wp(\mathbb{E}\mathrm{v}^{\varrho}) \setminus \mathscr{E}(\ell))$ using the backward semantics $\widehat{\boldsymbol{S}}^{\tilde{\bar{\varrho}}}[\![\mathsf{S}]\!]$. By corollary 50.2, this is equivalent to $\widehat{\boldsymbol{S}}^{\vec{\varrho}}[\![\mathsf{S}]\!]\,\mathscr{P}_0 \mathrel{\dot{\subseteq}} \ell \mapsto \wp(\mathbb{E}\mathrm{v}^{\varrho}) \setminus \mathscr{E}(\ell)$ using the forward semantics $\widehat{\boldsymbol{S}}^{\vec{\varrho}}[\![\mathsf{S}]\!]$, as also stated by (20.6).

Exercise 50.6 Prove that $\widehat{\boldsymbol{S}}^{\tilde{\bar{\mathsf{R}}}}[\![\mathsf{S}]\!]\,\mathscr{P}_f = \{\langle\rho_0, \varrho(\pi_0\ell_0)\rangle \in \mathbb{E}\mathrm{v} \times \mathbb{E}\mathrm{v} \mid \forall \ell \in \mathsf{labs}[\![\mathsf{S}]\!] \,.\, (\ell_0\pi_1\ell \in \boldsymbol{S}^*[\![\mathsf{S}]\!]\,(\pi_0\ell_0)) \Rightarrow \langle\rho_0, \varrho(\pi_0\ell_0\pi_1\ell)\rangle \in \mathscr{P}_f(\ell)\}$.

50.3 Calculational Design of the Structural Backward Impossible Failure Accessibility Semantics

The structural backward impossible failure accessibility semantics $\widehat{\boldsymbol{S}}^{\tilde{\bar{\varrho}}}[\![\mathsf{S}]\!]$ of a program component S is the following ($\mathscr{P}_f \in \mathbb{L} \to \wp(\mathbb{E}\mathrm{v}^{\vec{\varrho}})$).

- Backward impossible failure accessibility of a program $\mathsf{P} ::= \mathsf{Sl}\ \ell'$

$$\widehat{\boldsymbol{S}}^{\tilde{\bar{\varrho}}}[\![\mathsf{P}]\!]\,\mathscr{P}_f \ \triangleq\ \widehat{\boldsymbol{S}}^{\tilde{\bar{\varrho}}}[\![\mathsf{Sl}]\!]\,\mathscr{P}_f \tag{50.7}$$

- Backward impossible failure accessibility of a statement list $\mathsf{Sl} ::= \mathsf{Sl}'\ \mathsf{S}$

$$\widehat{\boldsymbol{\mathcal{S}}}^{\tilde{\overleftarrow{\varrho}}}[\![\mathtt{Sl}]\!]\,\mathscr{P}_f \;=\; \widehat{\boldsymbol{\mathcal{S}}}^{\tilde{\overleftarrow{\varrho}}}[\![\mathtt{Sl}']\!](\ell \in \mathsf{labs}[\![\mathtt{Sl}']\!] \mapsto \\ (\![\, \ell = \mathsf{after}[\![\mathtt{Sl}']\!] \mathrel{?} \mathscr{P}_f(\ell) \cap \widehat{\boldsymbol{\mathcal{S}}}^{\tilde{\overleftarrow{\varrho}}}[\![\mathtt{S}]\!]\,\mathscr{P}_f \mathrel{:} \mathscr{P}_f(\ell)\,]\!)) \tag{50.8}$$

- Backward impossible failure accessibility of an empty statement list $\mathtt{Sl} ::= \epsilon$

$$\widehat{\boldsymbol{\mathcal{S}}}^{\tilde{\overleftarrow{\varrho}}}[\![\mathtt{Sl}]\!]\,\mathscr{P}_f \;=\; \mathscr{P}_f(\mathsf{at}[\![\mathtt{Sl}]\!]) \tag{50.9}$$

- Backward impossible failure accessibility of an assignment statement $\mathtt{S} ::= \mathtt{x} = \mathtt{A}\,;$

$$\widehat{\boldsymbol{\mathcal{S}}}^{\tilde{\overleftarrow{\varrho}}}[\![\mathtt{S}]\!]\,\mathscr{P}_f \;=\; \mathscr{P}_f(\mathsf{at}[\![\mathtt{S}]\!]) \cap \mathsf{assign}^{\tilde{\overleftarrow{\varrho}}}[\![\mathtt{x}, \mathtt{A}]\!]\,\mathscr{P}_f(\mathsf{after}[\![\mathtt{S}]\!]) \tag{50.10}$$

$$\begin{aligned} \mathsf{assign}^{\tilde{\overleftarrow{\mathsf{r}}}}[\![\mathtt{x}, \mathtt{A}]\!]\,X &\triangleq \{\rho \mid \rho[\mathtt{x} \leftarrow \boldsymbol{\mathscr{A}}[\![\mathtt{A}]\!]\rho] \in X\} \\ \mathsf{assign}^{\tilde{\overleftarrow{\mathsf{R}}}}[\![\mathtt{x}, \mathtt{A}]\!]\,X &\triangleq \{\langle\rho_0,\, \rho\rangle \mid \langle\rho_0,\, \rho[\mathtt{x} \leftarrow \boldsymbol{\mathscr{A}}[\![\mathtt{A}]\!]\rho]\rangle \in X\} \end{aligned} \qquad \text{(see (36.35))}$$

- Backward impossible failure accessibility of a skip statement $\mathtt{S} ::= ;$

$$\widehat{\boldsymbol{\mathcal{S}}}^{\tilde{\overleftarrow{\varrho}}}[\![\mathtt{S}]\!]\,\mathscr{P}_f \;=\; \mathscr{P}_f(\mathsf{at}[\![\mathtt{S}]\!]) \cap \mathscr{P}_f(\mathsf{after}[\![\mathtt{S}]\!]) \tag{50.11}$$

- Backward impossible failure accessibility of a conditional statement $\mathtt{S} ::= \mathtt{if}\,(\mathtt{B})\,\mathtt{S}_t$

$$\widehat{\boldsymbol{\mathcal{S}}}^{\tilde{\overleftarrow{\varrho}}}[\![\mathtt{S}]\!]\,\mathscr{P}_f \;=\; \mathscr{P}_f(\mathsf{at}[\![\mathtt{S}]\!]) \cap \mathsf{test}^{\tilde{\overleftarrow{\varrho}}}[\![\mathtt{B}]\!](\widehat{\boldsymbol{\mathcal{S}}}^{\tilde{\overleftarrow{\varrho}}}[\![\mathtt{S}_t]\!]\,\mathscr{P}_f) \cap \overline{\mathsf{test}}^{\tilde{\overleftarrow{\varrho}}}[\![\mathtt{B}]\!]\,\mathscr{P}_f(\mathsf{after}[\![\mathtt{S}]\!]) \tag{50.12}$$

$$\begin{aligned} \mathsf{test}^{\tilde{\overleftarrow{\mathsf{r}}}}[\![\mathtt{B}]\!]X &\triangleq \{\rho \mid (\boldsymbol{\mathscr{B}}[\![\mathtt{B}]\!]\rho = \mathsf{tt}) \Rightarrow (\rho \in X)\} \\ \mathsf{test}^{\tilde{\overleftarrow{\mathsf{R}}}}[\![\mathtt{B}]\!]X &\triangleq \{\langle\rho_0,\, \rho\rangle \mid (\boldsymbol{\mathscr{B}}[\![\mathtt{B}]\!]\rho = \mathsf{tt}) \Rightarrow (\langle\rho_0,\, \rho\rangle \in X)\} \\ \overline{\mathsf{test}}^{\tilde{\overleftarrow{\mathsf{r}}}}[\![\mathtt{B}]\!]X &\triangleq \{\rho \mid (\boldsymbol{\mathscr{B}}[\![\mathtt{B}]\!]\rho = \mathsf{ff}) \Rightarrow (\rho \in X)\} \\ \overline{\mathsf{test}}^{\tilde{\overleftarrow{\mathsf{R}}}}[\![\mathtt{B}]\!]X &\triangleq \{\langle\rho_0,\, \rho\rangle \mid (\boldsymbol{\mathscr{B}}[\![\mathtt{B}]\!]\rho = \mathsf{ff}) \Rightarrow (\langle\rho_0,\, \rho\rangle \in X)\} \end{aligned}$$

- Backward impossible failure accessibility of a conditional statement $\mathtt{S} ::= \mathtt{if}\,(\mathtt{B})\,\mathtt{S}_t\ \mathtt{else}\ \mathtt{S}_f$

$$\widehat{\boldsymbol{\mathcal{S}}}^{\tilde{\overleftarrow{\varrho}}}[\![\mathtt{S}]\!]\,\mathscr{P}_f \;=\; \mathscr{P}_f(\mathsf{at}[\![\mathtt{S}]\!]) \cap \mathsf{test}^{\tilde{\overleftarrow{\varrho}}}[\![\mathtt{B}]\!](\widehat{\boldsymbol{\mathcal{S}}}^{\tilde{\overleftarrow{\varrho}}}[\![\mathtt{S}_t]\!]\,\mathscr{P}_f) \cap \overline{\mathsf{test}}^{\tilde{\overleftarrow{\varrho}}}[\![\mathtt{B}]\!](\widehat{\boldsymbol{\mathcal{S}}}^{\tilde{\overleftarrow{\varrho}}}[\![\mathtt{S}_f]\!]\,\mathscr{P}_f) \tag{50.13}$$

- Backward impossible failure accessibility of an iteration statement $\mathtt{S} ::= \mathtt{while}\,\ell\,(\mathtt{B})\,\mathtt{S}_b$ (where $\ell = \mathsf{at}[\![\mathtt{S}]\!] = \mathsf{after}[\![\mathtt{S}_b]\!]]$)

$$\widehat{\boldsymbol{\mathcal{S}}}^{\tilde{\overleftarrow{\varrho}}}[\![\mathtt{S}]\!]\,\mathscr{P}_f \;\triangleq\; \mathsf{gfp}^{\dot{\subseteq}}\, \boldsymbol{\mathscr{F}}^{\tilde{\overleftarrow{\varrho}}}[\![\mathtt{while}\,\ell\,(\mathtt{B})\,\mathtt{S}_b]\!]\,\mathscr{P}_f \tag{50.14}$$

$$\boldsymbol{\mathscr{F}}^{\tilde{\overleftarrow{\varrho}}}[\![\mathtt{while}\,\ell\,(\mathtt{B})\,\mathtt{S}_b]\!]\,\mathscr{P}_f \;\in\; \wp(\mathbb{E}\mathbb{v}^{\vec{\varrho}}) \xrightarrow{\nearrow} \wp(\mathbb{E}\mathbb{v}^{\vec{\varrho}})$$

$$\begin{aligned} \boldsymbol{\mathscr{F}}^{\tilde{\overleftarrow{\varrho}}}[\![\mathtt{while}\,\ell\,(\mathtt{B})\,\mathtt{S}_b]\!]\,\mathscr{P}_f\,Y \;\triangleq\; & \mathscr{P}_f(\ell) \cap \overline{\mathsf{test}}^{\tilde{\overleftarrow{\varrho}}}[\![\mathtt{B}]\!](\mathscr{P}_f(\mathsf{after}[\![\mathtt{S}]\!])) \cap \\ & \mathsf{test}^{\tilde{\overleftarrow{\varrho}}}[\![\mathtt{B}]\!](\widehat{\boldsymbol{\mathcal{S}}}^{\tilde{\overleftarrow{\varrho}}}[\![\mathtt{S}_b]\!]\,(\ell' \in \mathsf{labs}[\![\mathtt{S}_b]\!] \mapsto (\![\, \ell' = \ell \mathrel{?} \mathscr{P}_f(\ell) \cap Y \\ & \qquad [\!]\ \ell' \in \mathsf{breaks\text{-}of}[\![\mathtt{S}_b]\!] \mathrel{?} \mathscr{P}_f(\mathsf{after}[\![\mathtt{S}]\!]) \\ & \qquad \mathrel{:} \mathscr{P}_f(\ell')\,]\!))) \end{aligned}$$

- Backward impossible failure accessibility of a break statement $\mathtt{S} ::= \ell\ \mathtt{break}\,;$

$$\widehat{\boldsymbol{\mathcal{S}}}^{\tilde{\tilde{\ell}}}[\![\mathtt{S}]\!]\,\mathscr{P}_f \;=\; \mathscr{P}_f(\mathsf{at}[\![\mathtt{S}]\!]) \cap \mathscr{P}_f(\mathsf{break\text{-}to}[\![\mathtt{S}]\!]) \tag{50.15}$$

- Backward impossible failure accessibility of a compound statement $\mathtt{S} ::= \{\,\mathtt{Sl}\,\}$

$$\widehat{\boldsymbol{\mathcal{S}}}^{\tilde{\tilde{\ell}}}[\![\mathtt{S}]\!] \;=\; \widehat{\boldsymbol{\mathcal{S}}}^{\tilde{\tilde{\ell}}}[\![\mathtt{Sl}]\!] \tag{50.16}$$

Example 50.17 Consider $\mathtt{S} ::= \mathtt{\textbf{while}}\ \ell_1\ (\mathtt{tt})\ \ell_2\ \mathtt{x} = \mathtt{x} + 1\,;\ell_3$ where ℓ_3 is after the iteration. For the iteration body, we have

$$
\begin{aligned}
& \widehat{\boldsymbol{\mathcal{S}}}^{\tilde{\tilde{\ell}}}[\![\ell_2\ \mathtt{x} = \mathtt{x} + 1\,;\ell_1]\!]\,\mathscr{P}' && \text{(where } \ell_1 \text{ is after the assignment)}\\
= {} & \mathscr{P}'(\ell_2) \cap \{\rho \mid \rho[\mathtt{x} \leftarrow \rho(\mathtt{x}) + 1] \in \mathscr{P}'(\ell_1)\} && \text{((50.10) and (3.4))}\\
= {} & \mathscr{P}'(\ell_2) \cap \{x \mid x + 1 \in \mathscr{P}'(\ell_1)\} && \text{(letting } x = \rho(\mathtt{x}) \text{ where } \mathsf{dom}(\rho) = \{\mathtt{x}\}\text{)}\\
= {} & \mathscr{P}'(\ell_2) \cap \{x - 1 \mid x \in \mathscr{P}'(\ell_1)\}
\end{aligned}
$$

$$
\begin{aligned}
- {} & \widehat{\boldsymbol{\mathcal{S}}}^{\tilde{\tilde{\ell}}}[\![\mathtt{\textbf{while}}\ \ell_1\ (\mathtt{tt})\ \ell_2\ \mathtt{x} = \mathtt{x} + 1\,;\ell_3]\!]\,\mathscr{P}_f\\
= {} & \mathsf{gfp}^{\dot\subseteq}\, \boldsymbol{\mathscr{F}}^{\tilde{\tilde{\ell}}}[\![\mathtt{\textbf{while}}\ \ell_1\ (\mathtt{tt})\ \ell_2\ \mathtt{x} = \mathtt{x} + 1\,;\ell_3]\!]\,\mathscr{P}_f && \text{((50.14))}\\
= {} & \mathsf{gfp}^{\dot\subseteq}\, Y \mapsto \mathscr{P}_f(\ell_1) \cap \overline{\mathsf{test}}^{\tilde{\tilde{\ell}}}[\![\mathtt{tt}]\!](\mathscr{P}_f(\ell_3)) \cap \mathsf{test}^{\tilde{\tilde{\ell}}}[\![\mathtt{tt}]\!](\widehat{\boldsymbol{\mathcal{S}}}^{\tilde{\tilde{\ell}}}[\![\ell_2\ \mathtt{x} = \mathtt{x} + 1\,;\ell_1]\!]\ (\ell'' \in \{\ell_1, \ell_2\} \mapsto (\!| \ell'' =\\
& \ell_1\ ?\ \mathscr{P}_f(\ell_1) \cap Y : \mathscr{P}_f(\ell'')\,|\!)))\\
& \qquad \text{((50.14), } \mathsf{breaks\text{-}of}[\![\ell_2\ \mathtt{x} = \mathtt{x} + 1\,;\ell_1]\!] = \varnothing\text{, and and } \mathsf{labs}[\![\ell_2\ \mathtt{x} = \mathtt{x} + 1\,;\ell_1]\!] = \{\ell_1, \ell_2\}\text{)}\\
= {} & \mathsf{gfp}^{\dot\subseteq}\, Y \mapsto \mathscr{P}_f(\ell_1) \cap \widehat{\boldsymbol{\mathcal{S}}}^{\tilde{\tilde{\ell}}}[\![\ell_2\ \mathtt{x} = \mathtt{x} + 1\,;\ell_1]\!]\ (\ell'' \in \{\ell_1, \ell_2\} \mapsto (\!| \ell'' = \ell_1\ ?\ \mathscr{P}_f(\ell_1) \cap Y : \mathscr{P}_f(\ell'')\,|\!))\\
& \qquad \text{(by (50.12) so that } \mathsf{test}^{\tilde{\tilde{\ell}}}[\![\mathtt{tt}]\!]X = X \text{ and } \overline{\mathsf{test}}^{\tilde{\tilde{\ell}}}[\![\mathtt{tt}]\!]X = \mathbb{Z} \text{ by definition of } \boldsymbol{\mathscr{B}} \text{ in}\\
& \text{section 3.6)}\\
= {} & \mathsf{gfp}^{\dot\subseteq}\, Y \mapsto \mathscr{P}_f(\ell_1) \cap \mathscr{P}_f(\ell_2) \cap \{x - 1 \mid x \in \mathscr{P}_f(\ell_1) \cap Y\}\\
& \qquad\qquad \text{(definition (50.10) of } \widehat{\boldsymbol{\mathcal{S}}}^{\tilde{\tilde{\ell}}}[\![\ell_2\ \mathtt{x} = \mathtt{x} + 1\,;\ell_1]\!]\,\mathscr{P}'\text{)}
\end{aligned}
$$

Assume that $\mathscr{P}_f(\ell_2) = \mathbb{N}$ and $\mathscr{P}_f(\ell_1) = \mathscr{P}_f(\ell_3) = \mathbb{Z}$, which requires that x be positive or zero at ℓ_2. We have to calculate $\mathsf{gfp}^{\dot\subseteq}\, Y \mapsto \mathbb{N} \cap \{x - 1 \mid x \in Y\}$, for which the iterates are $Y^0 = \mathbb{Z}^3$, $Y^1 = Y^2 = \mathbb{N}$. This states that x will be positive or zero at ℓ_2 of the iteration $\mathtt{\textbf{while}}\ \ell_1\ (\mathtt{tt})\ \ell_2\ \mathtt{x} = \mathtt{x} + 1\,;\ell_3$ if and only if it is entered with x positive or zero.

Notice that the gfp allows for nontermination while the lfp would not (and would be $\varnothing$, for this example which never terminates).

Proof of (50.7—50.16) By corollary 50.2, we have

$$
\begin{aligned}
& \forall \ell \in \mathbb{L}\,.\,\widehat{\boldsymbol{\mathcal{S}}}^{\vec{\ell}}[\![\mathtt{S}]\!]\,\mathscr{P}_0\,\ell \subseteq \mathscr{P}_f(\ell)\\
\Leftrightarrow {} & \forall \ell \in \mathsf{labs}[\![\mathtt{S}]\!]\,.\,\widehat{\boldsymbol{\mathcal{S}}}^{\vec{\ell}}[\![\mathtt{S}]\!]\,\mathscr{P}_0\,\ell \subseteq \mathscr{P}_f(\ell) && \text{(because } \widehat{\boldsymbol{\mathcal{S}}}^{\vec{\ell}}[\![\mathtt{S}]\!]\,\mathscr{P}_0\,\ell = \varnothing \text{ when } \ell \notin \mathsf{labs}[\![\mathtt{S}]\!]\text{)}\\
\Leftrightarrow {} & \widehat{\boldsymbol{\mathcal{S}}}^{\vec{\ell}}[\![\mathtt{S}]\!]\mathscr{P}_0 \mathrel{\dot\subseteq} \mathscr{P}_f && \text{(pointwise definition of } \dot\subseteq\text{)}\\
\Leftrightarrow {} & \mathscr{P}_0 \subseteq \widehat{\boldsymbol{\mathcal{S}}}^{\tilde{\tilde{\ell}}}[\![\mathtt{S}]\!]\,\mathscr{P}_f && \text{(corollary 50.2.)}
\end{aligned}
$$

By unicity of adjoints in a Galois connection, we can derive $\widehat{\boldsymbol{\mathcal{S}}}^{\tilde{\tilde{\ell}}}[\![\mathtt{S}]\!]$ from $\widehat{\boldsymbol{\mathcal{S}}}^{\vec{\ell}}[\![\mathtt{S}]\!]$ by calculational design.

- For an assignment statement $\mathsf{S} ::= \mathsf{x} = \mathsf{A}\,;$, we have $\langle \wp(\mathbb{E}\mathbb{v}^{\varrho}), \subseteq\rangle \underset{\mathsf{assign}_{\vec{\varrho}}[\![\mathsf{x},\mathsf{A}]\!]}{\overset{\mathsf{assign}^{\tilde{\tilde{\overleftarrow{\varrho}}}}[\![x,\mathsf{A}]\!]}{\leftrightarrows}} \langle \wp(\mathbb{E}\mathbb{v}^{\varrho}), \subseteq\rangle$ (where $\mathbb{E}\mathbb{v}^{\varrho}$ is defined in corollary 50.2).

$$
\begin{aligned}
& \mathsf{assign}_{\vec{\mathsf{r}}}[\![\mathsf{x}, \mathsf{A}]\!]\, \mathscr{P}_0 \subseteq X \\
\Leftrightarrow\ & \{\rho[\mathsf{x} \leftarrow \boldsymbol{\mathscr{A}}[\![\mathsf{A}]\!]\rho] \mid \rho \in \mathscr{P}_0\} \subseteq X && \wr\text{definition (19.12) of } \mathsf{assign}_{\vec{\mathsf{r}}}[\![\mathsf{x}, \mathsf{A}]\!]\, \mathscr{P}_0 \int \\
\Leftrightarrow\ & \forall \rho \in \mathscr{P}_0 \,.\, \rho[\mathsf{x} \leftarrow \boldsymbol{\mathscr{A}}[\![\mathsf{A}]\!]\rho] \in X && \wr\text{definition of } \subseteq \int \\
\Leftrightarrow\ & \mathscr{P}_0 \subseteq \{\rho \mid \rho[\mathsf{x} \leftarrow \boldsymbol{\mathscr{A}}[\![\mathsf{A}]\!]\rho] \in X\} && \wr\text{definition of } \subseteq \int \\
\Leftrightarrow\ & \mathscr{P}_0 \subseteq \mathsf{assign}^{\tilde{\tilde{\overleftarrow{\mathsf{r}}}}}[\![x, \mathsf{A}]\!]\, X && \wr\text{by defining } \mathsf{assign}^{\tilde{\tilde{\overleftarrow{\mathsf{r}}}}}[\![x, \mathsf{A}]\!]\, X \text{ as in (50.10)} \int
\end{aligned}
$$

and similarly for $\mathsf{assign}_{\vec{\mathsf{R}}}[\![\mathsf{x}, \mathsf{A}]\!]\, \mathscr{P}_0 \subseteq X \Leftrightarrow \mathscr{P}_0 \subseteq \mathsf{assign}^{\tilde{\tilde{\overleftarrow{\mathsf{R}}}}}[\![x, \mathsf{A}]\!]\, X$. It follows that

$$
\begin{aligned}
& \widehat{\boldsymbol{\mathcal{S}}}^{\vec{\varrho}}[\![\mathsf{S}]\!]\mathscr{P}_0 \,\dot{\subseteq}\, \mathscr{P}_f \\
\Leftrightarrow\ & \forall \ell \,.\, \llparenthesis\, \ell = \mathsf{at}[\![\mathsf{S}]\!] \;?\; \mathscr{P}_0 \parallel \ell = \mathsf{after}[\![\mathsf{S}]\!] \;?\; \mathsf{assign}_{\vec{\varrho}}[\![\mathsf{x}, \mathsf{A}]\!]\, \mathscr{P}_0 \,:\, \varnothing \,\rrparenthesis \,\dot{\subseteq}\, \mathscr{P}_f(\ell) && \wr\text{by (19.12)} \int \\
\Leftrightarrow\ & \mathscr{P}_0 \subseteq \mathscr{P}_f(\mathsf{at}[\![\mathsf{S}]\!]) \wedge \mathsf{assign}_{\vec{\varrho}}[\![\mathsf{x}, \mathsf{A}]\!]\, \mathscr{P}_0 \subseteq \mathscr{P}_f(\mathsf{after}[\![\mathsf{S}]\!]) \\
& && \wr\text{definition of the conditional and } \forall X \,.\, \varnothing \subseteq X \int \\
\Leftrightarrow\ & \mathscr{P}_0 \subseteq \mathscr{P}_f(\mathsf{at}[\![\mathsf{S}]\!]) \wedge \mathscr{P}_0 \subseteq \mathsf{assign}^{\tilde{\tilde{\overleftarrow{\varrho}}}}[\![x, \mathsf{A}]\!]\, \mathscr{P}_f(\mathsf{after}[\![\mathsf{S}]\!]) && \wr\text{as shown previously} \int \\
\Leftrightarrow\ & \mathscr{P}_0 \subseteq (\mathscr{P}_f(\mathsf{at}[\![\mathsf{S}]\!]) \cap \mathsf{assign}^{\tilde{\tilde{\overleftarrow{\varrho}}}}[\![x, \mathsf{A}]\!]\, \mathscr{P}_f(\mathsf{after}[\![\mathsf{S}]\!])) && \wr\text{definitions of } \subseteq \text{ and } \cap \int \\
\Leftrightarrow\ & \mathscr{P}_0 \subseteq \widehat{\boldsymbol{\mathcal{S}}}^{\tilde{\tilde{\overleftarrow{\varrho}}}}[\![\mathsf{S}]\!]\, \mathscr{P}_f && \wr\text{which yields definition (50.10) of } \widehat{\boldsymbol{\mathcal{S}}}^{\tilde{\tilde{\overleftarrow{\varrho}}}}[\![\mathsf{S}]\!]\, \mathscr{P}_f \int
\end{aligned}
$$

- For a statement list $\mathtt{Sl} ::= \mathtt{Sl}'\ \mathsf{S}$,

$$
\begin{aligned}
& \widehat{\boldsymbol{\mathcal{S}}}^{\vec{\varrho}}[\![\mathtt{Sl}]\!]\mathscr{P}_0 \,\dot{\subseteq}\, \mathscr{P}_f \\
\Leftrightarrow\ & \forall \ell \,.\, \widehat{\boldsymbol{\mathcal{S}}}^{\vec{\varrho}}[\![\mathtt{Sl}]\!]\mathscr{P}_0\, \ell \subseteq \mathscr{P}_f(\ell) && \wr\text{pointwise definition of } \dot{\subseteq} \int \\
\Leftrightarrow\ & \forall \ell \,.\, \llparenthesis\, \ell \in \mathsf{labs}[\![\mathtt{Sl}']\!] \;?\; \widehat{\boldsymbol{\mathcal{S}}}^{\vec{\varrho}}[\![\mathtt{Sl}']\!]\mathscr{P}_0\, \ell \,:\, \varnothing \,\rrparenthesis \cup \llparenthesis\, \ell \in \mathsf{labs}[\![\mathsf{S}]\!] \;?\; \widehat{\boldsymbol{\mathcal{S}}}^{\vec{\varrho}}[\![\mathsf{S}]\!](\widehat{\boldsymbol{\mathcal{S}}}^{\vec{\varrho}}[\![\mathtt{Sl}']\!]\mathscr{P}_0\, \mathsf{at}[\![\mathsf{S}]\!])\, \ell \,: \\
& \quad \varnothing \,\rrparenthesis \subseteq \mathscr{P}_f(\ell) && \wr\text{(19.24)} \int \\
\Leftrightarrow\ & \forall \ell \,.\, \llparenthesis\, \ell \in \mathsf{labs}[\![\mathtt{Sl}']\!] \;?\; \widehat{\boldsymbol{\mathcal{S}}}^{\vec{\varrho}}[\![\mathtt{Sl}']\!]\mathscr{P}_0\, \ell \,:\, \widehat{\boldsymbol{\mathcal{S}}}^{\vec{\varrho}}[\![\mathtt{Sl}']\!]\mathscr{P}_0\, \ell \,\rrparenthesis \cup \llparenthesis\, \ell \in \mathsf{labs}[\![\mathsf{S}]\!] \;? \\
& \quad \widehat{\boldsymbol{\mathcal{S}}}^{\vec{\varrho}}[\![\mathsf{S}]\!](\widehat{\boldsymbol{\mathcal{S}}}^{\vec{\varrho}}[\![\mathtt{Sl}']\!]\mathscr{P}_0\, \mathsf{at}[\![\mathsf{S}]\!])\, \ell \,:\, \widehat{\boldsymbol{\mathcal{S}}}^{\vec{\varrho}}[\![\mathsf{S}]\!](\widehat{\boldsymbol{\mathcal{S}}}^{\vec{\varrho}}[\![\mathtt{Sl}']\!]\mathscr{P}_0\, \mathsf{at}[\![\mathsf{S}]\!])\, \ell \,\rrparenthesis \subseteq \mathscr{P}_f(\ell) && \wr\text{by (19.30)} \int \\
\Leftrightarrow\ & \forall \ell \,.\, (\widehat{\boldsymbol{\mathcal{S}}}^{\vec{\varrho}}[\![\mathtt{Sl}']\!]\mathscr{P}_0\, \ell \cup \widehat{\boldsymbol{\mathcal{S}}}^{\vec{\varrho}}[\![\mathsf{S}]\!](\widehat{\boldsymbol{\mathcal{S}}}^{\vec{\varrho}}[\![\mathtt{Sl}']\!]\mathscr{P}_0\, \mathsf{at}[\![\mathsf{S}]\!])\, \ell) \subseteq \mathscr{P}_f(\ell) \\
& && \wr\text{definition of the conditional} \int \\
\Leftrightarrow\ & (\widehat{\boldsymbol{\mathcal{S}}}^{\vec{\varrho}}[\![\mathtt{Sl}']\!]\mathscr{P}_0 \,\dot{\cup}\, \widehat{\boldsymbol{\mathcal{S}}}^{\vec{\varrho}}[\![\mathsf{S}]\!](\widehat{\boldsymbol{\mathcal{S}}}^{\vec{\varrho}}[\![\mathtt{Sl}']\!]\mathscr{P}_0\, \mathsf{at}[\![\mathsf{S}]\!])) \,\dot{\subseteq}\, \mathscr{P}_f && \wr\text{pointwise definition of } \dot{\subseteq} \int \\
\Leftrightarrow\ & \widehat{\boldsymbol{\mathcal{S}}}^{\vec{\varrho}}[\![\mathtt{Sl}']\!]\mathscr{P}_0 \,\dot{\subseteq}\, \mathscr{P}_f \wedge \widehat{\boldsymbol{\mathcal{S}}}^{\vec{\varrho}}[\![\mathsf{S}]\!](\widehat{\boldsymbol{\mathcal{S}}}^{\vec{\varrho}}[\![\mathtt{Sl}']\!]\mathscr{P}_0\, \mathsf{at}[\![\mathsf{S}]\!]) \,\dot{\subseteq}\, \mathscr{P}_f && \wr X \,\dot{\cup}\, Y \,\dot{\subseteq}\, Z \Leftrightarrow X \,\dot{\subseteq}\, Z \wedge Y \,\dot{\subseteq}\, Z \int \\
\Leftrightarrow\ & \exists \mathscr{P}'_0 \in \wp(\mathbb{E}\mathbb{v}^{\varrho}) \,.\, \widehat{\boldsymbol{\mathcal{S}}}^{\vec{\varrho}}[\![\mathtt{Sl}']\!]\mathscr{P}_0 \,\dot{\subseteq}\, \mathscr{P}_f \wedge \widehat{\boldsymbol{\mathcal{S}}}^{\vec{\varrho}}[\![\mathtt{Sl}']\!]\mathscr{P}_0\, \mathsf{at}[\![\mathsf{S}]\!] \subseteq \mathscr{P}'_0 \wedge \widehat{\boldsymbol{\mathcal{S}}}^{\vec{\varrho}}[\![\mathsf{S}]\!](\mathscr{P}'_0) \,\dot{\subseteq}\, \mathscr{P}_f
\end{aligned}
$$

≀The lower adjoint $\widehat{\boldsymbol{\mathcal{S}}}^{\vec{\varrho}}[\![\mathtt{Sl}']\!]$ of the Galois connection of corollary 50.2 is increasing and for any $\subseteq$-increasing f, $f(g\,x\,a) \,\dot{\sqsubseteq}\, y$ if and only if $\exists z \,.\, (g\,x\,a) \sqsubseteq z \wedge f\,z \,\dot{\sqsubseteq}\, y$. $(\Rightarrow)$ is by choosing $z = (g\,x\,a)$ and reflexivity, $(\Leftarrow)$ is by f increasing and transitivity∫

$$
\Leftrightarrow\ \exists \mathscr{P}'_0 \in \wp(\mathbb{E}\mathbb{v}^{\varrho}) \,.\, \widehat{\boldsymbol{\mathcal{S}}}^{\vec{\varrho}}[\![\mathtt{Sl}']\!]\mathscr{P}_0 \,\dot{\subseteq}\, \ell \mapsto \llparenthesis\, \ell = \mathsf{at}[\![\mathsf{S}]\!] \;?\; \mathscr{P}'_0 \cap \mathscr{P}_f(\ell) \,:\, \mathscr{P}_f(\ell) \,\rrparenthesis \wedge \widehat{\boldsymbol{\mathcal{S}}}^{\vec{\varrho}}[\![\mathsf{S}]\!](\mathscr{P}'_0) \,\dot{\subseteq}\, \mathscr{P}_f
$$

≀pointwise definition of $\dot{\subseteq}$ and definition of the conditional∫

$\Leftrightarrow\ \exists \mathscr{P}'_0 \in \wp(\mathbb{Ev}^{\varrho}) \,.\, \mathscr{P}_0 \subseteq \widehat{\boldsymbol{\mathcal{S}}}^{\tilde{\bar{\varrho}}}[\![\mathtt{Sl}']\!]\ (\ell \mapsto (\![\ \ell = \mathsf{at}[\![\mathsf{S}]\!]\ \text{?}\ \mathscr{P}'_0 \cap \mathscr{P}_f(\ell) \text{ : } \mathscr{P}_f(\ell)\ ]\!)) \wedge \mathscr{P}'_0 \subseteq \widehat{\boldsymbol{\mathcal{S}}}^{\tilde{\bar{\varrho}}}[\![\mathsf{S}]\!]\ \mathscr{P}_f$
(induction hypothesis)

$\Leftrightarrow\ \mathscr{P}_0 \subseteq \widehat{\boldsymbol{\mathcal{S}}}^{\tilde{\bar{\varrho}}}[\![\mathtt{Sl}']\!]\ (\ell \mapsto (\![\ \ell = \mathsf{at}[\![\mathsf{S}]\!]\ \text{?}\ \widehat{\boldsymbol{\mathcal{S}}}^{\tilde{\bar{\varrho}}}[\![\mathsf{S}]\!]\ \mathscr{P}_f \cap \mathscr{P}_f(\ell) \text{ : } \mathscr{P}_f(\ell)\ ]\!))$

($(\Rightarrow)$ $\widehat{\boldsymbol{\mathcal{S}}}^{\tilde{\bar{\varrho}}}[\![\mathtt{Sl}']\!]$ is the upper adjoint of the Galois connection of corollary 50.2 and so is increasing.
$(\Leftarrow)$ choose $\mathscr{P}'_0 = \widehat{\boldsymbol{\mathcal{S}}}^{\tilde{\bar{\varrho}}}[\![\mathsf{S}]\!]\ \mathscr{P}_f$ and reflexivity)

$\Leftrightarrow\ \mathscr{P}_0 \subseteq \widehat{\boldsymbol{\mathcal{S}}}^{\tilde{\bar{\varrho}}}[\![\mathtt{Sl}]\!]\ \mathscr{P}_f$ (by defining $\widehat{\boldsymbol{\mathcal{S}}}^{\tilde{\bar{\varrho}}}$ as (50.8))

- For a test B,

– $\mathsf{test}^{\vec{\mathsf{r}}}[\![\mathsf{B}]\!]\mathscr{P}_0 \subseteq \mathscr{P}'$

$\Leftrightarrow\ \{\rho \in \mathscr{P}_0 \mid \boldsymbol{\mathcal{B}}[\![\mathsf{B}]\!]\rho = \mathsf{tt}\} \subseteq \mathscr{P}'$ (definition (19.22) of $\mathsf{test}^{\vec{\mathsf{r}}}[\![\mathsf{B}]\!]$)

$\Leftrightarrow\ \forall \rho \in \mathscr{P}_0 \,.\, (\boldsymbol{\mathcal{B}}[\![\mathsf{B}]\!]\rho = \mathsf{tt}) \Rightarrow (\rho \in \mathscr{P}')$ (definition of $\subseteq$)

$\Leftrightarrow\ \mathscr{P}_0 \subseteq \{\rho \mid (\boldsymbol{\mathcal{B}}[\![\mathsf{B}]\!]\rho = \mathsf{tt}) \Rightarrow (\rho \in \mathscr{P}')\}$ (definition of $\subseteq$)

$\Leftrightarrow\ \mathscr{P}_0 \subseteq \mathsf{test}^{\tilde{\bar{\mathsf{r}}}}[\![\mathsf{B}]\!]\mathscr{P}$ (by defining as in (50.12))

so $\langle \wp(\mathbb{Ev}), \subseteq\rangle \underset{\mathsf{test}^{\vec{\mathsf{r}}}[\![\mathsf{B}]\!]}{\overset{\mathsf{test}^{\tilde{\bar{\mathsf{r}}}}[\![\mathsf{B}]\!]}{\leftrightarrows}} \langle \wp(\mathbb{Ev}), \subseteq\rangle$. Similarly for $\langle \wp(\mathbb{Ev} \times \mathbb{Ev}), \subseteq\rangle \underset{\mathsf{test}^{\vec{\mathsf{R}}}[\![\mathsf{B}]\!]}{\overset{\mathsf{test}^{\tilde{\bar{\mathsf{R}}}}[\![\mathsf{B}]\!]}{\leftrightarrows}}$ (50.18)
$\langle \wp(\mathbb{Ev} \times \mathbb{Ev}), \subseteq\rangle$. By the convention in section 19.1.3, we summarize both cases as $\langle \wp(\mathbb{Ev}^{\varrho}), \subseteq\rangle \underset{\mathsf{test}^{\vec{\varrho}}[\![\mathsf{B}]\!]}{\overset{\mathsf{test}^{\tilde{\bar{\varrho}}}[\![\mathsf{B}]\!]}{\leftrightarrows}} \langle \wp(\mathbb{Ev}^{\varrho}), \subseteq\rangle$ where $\varrho \in \{\tilde{\bar{\mathsf{r}}}, \vec{\varrho}\}$.

- For a conditional statement $\mathsf{S} ::= \mathbf{if}\,(\mathsf{B})\ \mathsf{S}_t$,

$\widehat{\boldsymbol{\mathcal{S}}}^{\vec{\varrho}}[\![\mathsf{S}]\!]\mathscr{P}_0 \mathrel{\dot{\subseteq}} \mathscr{P}_f$

$\Leftrightarrow\ \forall \ell \,.\, \widehat{\boldsymbol{\mathcal{S}}}^{\vec{\varrho}}[\![\mathsf{S}]\!]\mathscr{P}_0\ \ell \subseteq \mathscr{P}_f(\ell)$ (pointwise definition of $\dot{\subseteq}$)

$\Leftrightarrow\ \forall \ell \,.\, ((\![\ \ell = \mathsf{at}[\![\mathsf{S}]\!]\ \text{?}\ \mathscr{P}_0 \mathrel{\|} \ell \in \mathsf{in}[\![\mathsf{S}_t]\!] \cup (\![\ \mathsf{escape}[\![\mathsf{S}_t]\!]\ \text{?}\ \{\mathsf{break\text{-}to}[\![\mathsf{S}_t]\!]\} \text{ : } \varnothing\ ]\!)\ \text{?}$
$\widehat{\boldsymbol{\mathcal{S}}}^{\vec{\varrho}}[\![\mathsf{S}_t]\!]\ (\mathsf{test}^{\vec{\varrho}}[\![\mathsf{B}]\!]\mathscr{P}_0)\ \ell \mathrel{\|} \ell = \mathsf{after}[\![\mathsf{S}]\!]\ \text{?}\ \widehat{\boldsymbol{\mathcal{S}}}^{\vec{\varrho}}[\![\mathsf{S}_t]\!]\ (\mathsf{test}^{\vec{\varrho}}[\![\mathsf{B}]\!]\mathscr{P}_0)\ \ell \cup (\overline{\mathsf{test}}^{\vec{\varrho}}[\![\mathsf{B}]\!]\mathscr{P}_0) \text{ : } \varnothing\ ]\!)) \subseteq$
$\mathscr{P}_f(\ell)$ (definition (19.22) of $\widehat{\boldsymbol{\mathcal{S}}}^{\vec{\varrho}}[\![\mathsf{S}]\!]\ \mathscr{P}_0\ \ell$)

$\Leftrightarrow\ \mathscr{P}_0 \subseteq \mathscr{P}_f(\mathsf{at}[\![\mathsf{S}]\!]) \wedge \forall \ell \in \mathsf{in}[\![\mathsf{S}_t]\!] \cup (\![\ \mathsf{escape}[\![\mathsf{S}_t]\!]\ \text{?}\ \{\mathsf{break\text{-}to}[\![\mathsf{S}_t]\!]\} \text{ : } \varnothing\ ]\!)\ .$
$\widehat{\boldsymbol{\mathcal{S}}}^{\vec{\varrho}}[\![\mathsf{S}_t]\!]\ (\mathsf{test}^{\vec{\varrho}}[\![\mathsf{B}]\!]\mathscr{P}_0)\ \ell \subseteq \mathscr{P}_f(\ell) \wedge \widehat{\boldsymbol{\mathcal{S}}}^{\vec{\varrho}}[\![\mathsf{S}_t]\!]\ (\mathsf{test}^{\vec{\varrho}}[\![\mathsf{B}]\!]\mathscr{P}_0)\ \mathsf{after}[\![\mathsf{S}]\!] \subseteq \mathscr{P}_f(\mathsf{after}[\![\mathsf{S}]\!]) \wedge$
$\overline{\mathsf{test}}^{\vec{\varrho}}[\![\mathsf{B}]\!]\mathscr{P}_0 \subseteq \mathscr{P}_f(\mathsf{after}[\![\mathsf{S}]\!])$
(definition of the conditional, $\forall X \,.\, \varnothing \subseteq X$, and definition of the lub)

$\Leftrightarrow\ \mathscr{P}_0 \subseteq \mathscr{P}_f(\mathsf{at}[\![\mathsf{S}]\!]) \wedge \forall \ell \in \mathsf{in}[\![\mathsf{S}_t]\!] \cup \{\mathsf{after}[\![\mathsf{S}_t]\!]\} \cup (\![\ \mathsf{escape}[\![\mathsf{S}_t]\!]\ \text{?}\ \{\mathsf{break\text{-}to}[\![\mathsf{S}_t]\!]\} \text{ : } \varnothing\ ]\!)\ .$
$\widehat{\boldsymbol{\mathcal{S}}}^{\vec{\varrho}}[\![\mathsf{S}_t]\!]\ (\mathsf{test}^{\vec{\varrho}}[\![\mathsf{B}]\!]\mathscr{P}_0)\ \ell \subseteq \mathscr{P}_f(\ell) \wedge \overline{\mathsf{test}}^{\vec{\varrho}}[\![\mathsf{B}]\!]\mathscr{P}_0 \subseteq \mathscr{P}_f(\mathsf{after}[\![\mathsf{S}]\!])$
(grouping identical cases and $\mathsf{after}[\![\mathsf{S}]\!] = \mathsf{after}[\![\mathsf{S}_t]\!]$ in section 4.2)

$\Leftrightarrow\ \mathscr{P}_0 \subseteq \mathscr{P}_f(\mathsf{at}[\![\mathsf{S}]\!]) \wedge \forall \ell \in \mathsf{labs}[\![\mathsf{S}_t]\!] \,.\, \widehat{\boldsymbol{\mathcal{S}}}^{\vec{\varrho}}[\![\mathsf{S}_t]\!]\ (\mathsf{test}^{\vec{\varrho}}[\![\mathsf{B}]\!]\mathscr{P}_0)\ \ell \subseteq \mathscr{P}_f(\ell) \wedge \overline{\mathsf{test}}^{\vec{\varrho}}[\![\mathsf{B}]\!]\mathscr{P}_0 \subseteq$
$\mathscr{P}_f(\mathsf{after}[\![\mathsf{S}]\!])$ (definition of $\mathsf{labs}[\![\mathsf{S}_t]\!]$ in section 4.2)

$\Leftrightarrow\ \mathscr{P}_0 \subseteq \mathscr{P}_f(\mathsf{at}[\![\mathsf{S}]\!]) \wedge \exists \mathscr{P} \in \wp(\mathbb{Ev}^{\varrho}) \,.\, \mathsf{test}^{\vec{\varrho}}[\![\mathsf{B}]\!]\mathscr{P}_0 \subseteq \mathscr{P} \wedge \forall \ell \in \mathsf{labs}[\![\mathsf{S}_t]\!] \,.\, \widehat{\boldsymbol{\mathcal{S}}}^{\vec{\varrho}}[\![\mathsf{S}_t]\!]\ \mathscr{P}\ \ell \subseteq$
$\mathscr{P}_f(\ell) \wedge \overline{\mathsf{test}}^{\vec{\varrho}}[\![\mathsf{B}]\!]\mathscr{P}_0 \subseteq \mathscr{P}_f(\mathsf{after}[\![\mathsf{S}]\!])$ ($\widehat{\boldsymbol{\mathcal{S}}}^{\vec{\varrho}}[\![\mathsf{S}_t]\!]$ increasing)

$$\Leftrightarrow \mathscr{P}_0 \subseteq \mathscr{P}_f(\mathsf{at}[\![\mathsf{S}]\!]) \wedge \exists \mathscr{P} \in \wp(\mathbb{E}\mathbb{v}^{\varrho}) \,.\, \mathscr{P}_0 \subseteq \mathsf{test}^{\tilde{\overleftarrow{\varrho}}}[\![\mathsf{B}]\!]\mathscr{P} \wedge \mathscr{P} \subseteq \widehat{\boldsymbol{\mathcal{S}}}^{\tilde{\overleftarrow{\varrho}}}[\![\mathsf{S}_t]\!]\,\mathscr{P}_f \wedge \mathscr{P}_0 \subseteq \overline{\mathsf{test}}^{\tilde{\overleftarrow{\varrho}}}[\![\mathsf{B}]\!]\,\mathscr{P}_f(\mathsf{after}[\![\mathsf{S}]\!])$$

$$\text{\{Galois connections in corollary 50.2 and for } \mathsf{test}^{\vec{\varrho}}[\![\mathsf{B}]\!] \text{ and } \overline{\mathsf{test}}^{\vec{\varrho}}[\![\mathsf{B}]\!]\text{\}}$$

$$\Leftrightarrow \mathscr{P}_0 \subseteq \mathscr{P}_f(\mathsf{at}[\![\mathsf{S}]\!]) \wedge \mathscr{P}_0 \subseteq \mathsf{test}^{\tilde{\overleftarrow{\varrho}}}[\![\mathsf{B}]\!](\widehat{\boldsymbol{\mathcal{S}}}^{\tilde{\overleftarrow{\varrho}}}[\![\mathsf{S}_t]\!]\,\mathscr{P}_f) \wedge \mathscr{P}_0 \subseteq \overline{\mathsf{test}}^{\tilde{\overleftarrow{\varrho}}}[\![\mathsf{B}]\!]\,\mathscr{P}_f(\mathsf{after}[\![\mathsf{S}]\!])$$

$$\text{\{}\mathsf{test}^{\tilde{\overleftarrow{\varrho}}}[\![\mathsf{B}]\!] \text{ is increasing\}}$$

$$\Leftrightarrow \mathscr{P}_0 \subseteq (\mathscr{P}_f(\mathsf{at}[\![\mathsf{S}]\!]) \cap \mathsf{test}^{\tilde{\overleftarrow{\varrho}}}[\![\mathsf{B}]\!](\widehat{\boldsymbol{\mathcal{S}}}^{\tilde{\overleftarrow{\varrho}}}[\![\mathsf{S}_t]\!]\,\mathscr{P}_f) \cap \overline{\mathsf{test}}^{\tilde{\overleftarrow{\varrho}}}[\![\mathsf{B}]\!]\,\mathscr{P}_f(\mathsf{after}[\![\mathsf{S}]\!]))$$

$$\text{\{definition of the lub\}}$$

$$\Leftrightarrow \mathscr{P}_0 \subseteq \widehat{\boldsymbol{\mathcal{S}}}^{\tilde{\overleftarrow{\varrho}}}[\![\mathsf{S}]\!]\,\mathscr{P}_f \qquad \text{\{by defining } \widehat{\boldsymbol{\mathcal{S}}}^{\tilde{\overleftarrow{\varrho}}}[\![\mathsf{S}]\!] \text{ as in (50.12)\}}$$

- For an iteration statement $\mathsf{S} ::= \mathtt{while}\ \ell\ (\mathsf{B})\ \mathsf{S}_b$, we first inverse $\mathscr{F}^{\vec{\varrho}}[\![\mathtt{while}\ \ell\ (\mathsf{B})\ \mathsf{S}_b]\!]$ of (19.16) as shown by the following lemma 50.19.

Lemma 50.19 If $X, Y \in \mathbb{L} \to \wp(\mathbb{E}\mathbb{v}^{\vec{\varrho}})$ then

$$\mathscr{F}^{\vec{\varrho}}[\![\mathtt{while}\ \ell\ (\mathsf{B})\ \mathsf{S}_b]\!]\,\mathscr{P}_0\,X \mathrel{\dot{\subseteq}} (\mathscr{P}_f \cap Y)$$
$$\Leftrightarrow$$
$$\mathscr{P}_0 \subseteq (\mathscr{P}_f(\ell) \cap Y(\ell)) \ \wedge\ X \mathrel{\dot{\subseteq}} \overleftarrow{\mathscr{F}}^{\tilde{\overleftarrow{\varrho}}}[\![\mathtt{while}\ \ell\ (\mathsf{B})\ \mathsf{S}_b]\!]\,\mathscr{P}_f\,Y$$

where $\mathscr{F}^{\vec{\varrho}}[\![\mathtt{while}\ \ell\ (\mathsf{B})\ \mathsf{S}_b]\!]\,\mathscr{P}_0$ is defined by (19.16) and $\overleftarrow{\mathscr{F}}^{\tilde{\overleftarrow{\varrho}}}[\![\mathtt{while}\ \ell\ (\mathsf{B})\ \mathsf{S}_b]\!]\,\mathscr{P}_f \in (\mathbb{L} \to \wp(\mathbb{E}\mathbb{v}^{\vec{\varrho}})) \xrightarrow{\nearrow} (\mathbb{L} \to \wp(\mathbb{E}\mathbb{v}^{\vec{\varrho}}))$ is

$$\overleftarrow{\mathscr{F}}^{\tilde{\overleftarrow{\varrho}}}[\![\mathtt{while}\ \ell\ (\mathsf{B})\ \mathsf{S}_b]\!]\,\mathscr{P}_f\,Y\,\ell' \triangleq \tag{50.20}$$
$$\llparenthesis\, \ell' = \ell \mathrel{?} \overline{\mathsf{test}}^{\tilde{\overleftarrow{\varrho}}}[\![\mathsf{B}]\!](\mathscr{P}_f(\mathsf{after}[\![\mathsf{S}]\!]) \cap Y(\mathsf{after}[\![\mathsf{S}]\!])) \cap$$
$$\mathsf{test}^{\tilde{\overleftarrow{\varrho}}}[\![\mathsf{B}]\!](\widehat{\boldsymbol{\mathcal{S}}}^{\tilde{\overleftarrow{\varrho}}}[\![\mathsf{S}_b]\!]\ (\ell'' \in \mathsf{labs}[\![\mathsf{S}_b]\!] \mapsto \llparenthesis\, \ell'' \in \mathsf{breaks\text{-}of}[\![\mathsf{S}_b]\!] \mathrel{?}$$
$$(\mathscr{P}_f(\mathsf{after}[\![\mathsf{S}]\!]) \cap Y(\mathsf{after}[\![\mathsf{S}]\!])) \fatsemi (\mathscr{P}_f(\ell'') \cap Y(\ell''))\,\rrparenthesis))$$
$$\fatsemi\ \mathbb{E}\mathbb{v}^{\vec{\varrho}}\,\rrparenthesis$$

Proof of lemma 50.19

$$\mathscr{F}^{\vec{\varrho}}[\![\mathtt{while}\ \ell\ (\mathsf{B})\ \mathsf{S}_b]\!]\,\mathscr{P}_0\,X \mathrel{\dot{\subseteq}} (\mathscr{P}_f \mathrel{\dot{\cap}} Y)$$

$$\Leftrightarrow \forall \ell' \in \mathbb{L} \,.\, \mathscr{F}^{\vec{\varrho}}[\![\mathtt{while}\ \ell\ (\mathsf{B})\ \mathsf{S}_b]\!]\,\mathscr{P}_0\,X\,\ell' \subseteq (\mathscr{P}_f(\ell') \cap Y(\ell'))$$

$$\text{\{pointwise definitions of } \dot{\subseteq} \text{ and } \dot{\cap}\text{\}}$$

$$\Leftrightarrow \forall \ell' \in \mathbb{L} \,.\, \llparenthesis\, \ell' = \ell \mathrel{?} \mathscr{P}_0 \cup \widehat{\boldsymbol{\mathcal{S}}}^{\vec{\varrho}}[\![\mathsf{S}_b]\!]\ (\mathsf{test}^{\vec{\varrho}}[\![\mathsf{B}]\!]X(\ell))\,\ell \,[\!]\, \ell' \in \mathsf{in}[\![\mathsf{S}_b]\!] \setminus$$
$$\{\ell\} \mathrel{?} \widehat{\boldsymbol{\mathcal{S}}}^{\vec{\varrho}}[\![\mathsf{S}_b]\!]\ (\mathsf{test}^{\vec{\varrho}}[\![\mathsf{B}]\!]X(\ell))\,\ell' \,[\!]\, \ell' = \mathsf{after}[\![\mathsf{S}]\!] \mathrel{?} \overline{\mathsf{test}}^{\vec{\varrho}}[\![\mathsf{B}]\!](X(\ell)) \cup$$
$$\bigcup_{\ell'' \in \mathsf{breaks\text{-}of}[\![\mathsf{S}_b]\!]} \widehat{\boldsymbol{\mathcal{S}}}^{\vec{\varrho}}[\![\mathsf{S}_b]\!]\ (\mathsf{test}^{\vec{\varrho}}[\![\mathsf{B}]\!]X(\ell))\,\ell'' \fatsemi \varnothing\,\rrparenthesis \subseteq (\mathscr{P}_f(\ell') \cap Y(\ell')) \qquad \text{\{(19.16)\}}$$

$$\Leftrightarrow \mathscr{P}_0 \subseteq (\mathscr{P}_f(\ell) \cap Y(\ell)) \wedge \widehat{\boldsymbol{\mathcal{S}}}^{\vec{\varrho}}[\![\mathsf{S}_b]\!]\ (\mathsf{test}^{\vec{\varrho}}[\![\mathsf{B}]\!]X(\ell))\,\ell \subseteq (\mathscr{P}_f(\ell) \cap Y(\ell)) \wedge \forall \ell' \in \mathsf{in}[\![\mathsf{S}_b]\!] \setminus \{\ell\} \,.$$
$$\widehat{\boldsymbol{\mathcal{S}}}^{\vec{\varrho}}[\![\mathsf{S}_b]\!]\ (\mathsf{test}^{\vec{\varrho}}[\![\mathsf{B}]\!]X(\ell))\,\ell' \subseteq (\mathscr{P}_f(\ell') \cap Y(\ell')) \wedge \overline{\mathsf{test}}^{\vec{\varrho}}[\![\mathsf{B}]\!](X(\ell)) \subseteq (\mathscr{P}_f(\mathsf{after}[\![\mathsf{S}]\!]) \cap$$
$$Y(\mathsf{after}[\![\mathsf{S}]\!])) \wedge \forall \ell'' \in \mathsf{breaks\text{-}of}[\![\mathsf{S}_b]\!] \,.\, \widehat{\boldsymbol{\mathcal{S}}}^{\vec{\varrho}}[\![\mathsf{S}_b]\!]\ (\mathsf{test}^{\vec{\varrho}}[\![\mathsf{B}]\!]X(\ell))\,\ell'' \subseteq (\mathscr{P}_f(\mathsf{after}[\![\mathsf{S}]\!]) \cap$$
$$Y(\mathsf{after}[\![\mathsf{S}]\!])) \qquad \text{\{definitions of the conditional and the lub\}}$$

$\Leftrightarrow \mathscr{P}_0 \subseteq (\mathscr{P}_f(\ell) \cap Y(\ell)) \wedge \forall \ell' \in \mathsf{labs}[\![\mathsf{S}_b]\!] \,.\, \widehat{\boldsymbol{\mathcal{S}}}^{\vec{e}}[\![\mathsf{S}_b]\!]\, (\mathsf{test}^{\vec{e}}[\![\mathsf{B}]\!] X(\ell))\, \ell' \subseteq (\mathscr{P}_f(\ell') \cap Y(\ell')) \wedge \overline{\mathsf{test}}^{\vec{e}}[\![\mathsf{B}]\!](X(\ell)) \subseteq (\mathscr{P}_f(\mathsf{after}[\![\mathsf{S}]\!]) \cap Y(\mathsf{after}[\![\mathsf{S}]\!])) \wedge \forall \ell'' \in \mathsf{breaks\text{-}of}[\![\mathsf{S}_b]\!] \,.\, \widehat{\boldsymbol{\mathcal{S}}}^{\vec{e}}[\![\mathsf{S}_b]\!]\, (\mathsf{test}^{\vec{e}}[\![\mathsf{B}]\!] X(\ell))\, \ell'' \subseteq (\mathscr{P}_f(\mathsf{after}[\![\mathsf{S}]\!]) \cap Y(\mathsf{after}[\![\mathsf{S}]\!]))$

(grouping cases $\ell' = \ell = \mathsf{after}[\![\mathsf{S}_b]\!]$ and $\ell' \in \mathsf{in}[\![\mathsf{S}_b]\!] \setminus \{\ell\}$, where $\mathsf{labs}[\![\mathsf{S}_b]\!] \triangleq \mathsf{in}[\![\mathsf{S}_b]\!] \cup \{\mathsf{after}[\![\mathsf{S}_b]\!]\}$)

$\Leftrightarrow \mathscr{P}_0 \subseteq (\mathscr{P}_f(\ell) \cap Y(\ell)) \wedge \forall \ell' \in \mathsf{labs}[\![\mathsf{S}_b]\!] \,.\, \widehat{\boldsymbol{\mathcal{S}}}^{\vec{e}}[\![\mathsf{S}_b]\!]\, (\mathsf{test}^{\vec{e}}[\![\mathsf{B}]\!] X(\ell))\, \ell' \subseteq (\!|\, \ell' \in \mathsf{breaks\text{-}of}[\![\mathsf{S}_b]\!] \,?\, (\mathscr{P}_f(\mathsf{after}[\![\mathsf{S}]\!]) \cap Y(\mathsf{after}[\![\mathsf{S}]\!])) \,{\text{\textperiodcentered}}\, (\mathscr{P}_f(\ell') \cap Y(\ell')) \,|\!) \wedge \overline{\mathsf{test}}^{\vec{e}}[\![\mathsf{B}]\!](X(\ell)) \subseteq (\mathscr{P}_f(\mathsf{after}[\![\mathsf{S}]\!]) \cap Y(\mathsf{after}[\![\mathsf{S}]\!]))$

(grouping cases $\ell' \in \mathsf{labs}[\![\mathsf{S}_b]\!]$ and $\ell'' \in \mathsf{breaks\text{-}of}[\![\mathsf{S}_b]\!]$ because $\mathsf{breaks\text{-}of}[\![\mathsf{S}_b]\!] \subseteq \mathsf{labs}[\![\mathsf{S}_b]\!]$)

$\Leftrightarrow \mathscr{P}_0 \subseteq (\mathscr{P}_f(\ell) \cap Y(\ell)) \wedge \widehat{\boldsymbol{\mathcal{S}}}^{\vec{e}}[\![\mathsf{S}_b]\!]\, (\mathsf{test}^{\vec{e}}[\![\mathsf{B}]\!] X(\ell)) \mathrel{\dot{\subseteq}} \ell' \in \mathsf{labs}[\![\mathsf{S}_b]\!] \mapsto (\!|\, \ell' \in \mathsf{breaks\text{-}of}[\![\mathsf{S}_b]\!] \,?\, (\mathscr{P}_f(\mathsf{after}[\![\mathsf{S}]\!]) \cap Y(\mathsf{after}[\![\mathsf{S}]\!])) \,:\, (\mathscr{P}_f(\ell') \cap Y(\ell')) \,|\!) \wedge \overline{\mathsf{test}}^{\vec{e}}[\![\mathsf{B}]\!](X(\ell)) \subseteq (\mathscr{P}_f(\mathsf{after}[\![\mathsf{S}]\!]) \cap Y(\mathsf{after}[\![\mathsf{S}]\!]))$ (pointwise definition of $\dot{\subseteq}$)

$\Leftrightarrow \mathscr{P}_0 \subseteq (\mathscr{P}_f(\ell) \cap Y(\ell)) \wedge \mathsf{test}^{\vec{e}}[\![\mathsf{B}]\!] X(\ell) \subseteq \widehat{\boldsymbol{\mathcal{S}}}^{\tilde{\bar{e}}}[\![\mathsf{S}_b]\!]\, (\ell' \in \mathsf{labs}[\![\mathsf{S}_b]\!] \mapsto (\!|\, \ell' \in \mathsf{breaks\text{-}of}[\![\mathsf{S}_b]\!] \,?\, (\mathscr{P}_f(\mathsf{after}[\![\mathsf{S}]\!]) \cap Y(\mathsf{after}[\![\mathsf{S}]\!])) \,:\, (\mathscr{P}_f(\ell') \cap Y(\ell')) \,|\!)) \wedge \overline{\mathsf{test}}^{\vec{e}}[\![\mathsf{B}]\!](X(\ell)) \subseteq (\mathscr{P}_f(\mathsf{after}[\![\mathsf{S}]\!]) \cap Y(\mathsf{after}[\![\mathsf{S}]\!]))$ (Galois connection of corollary 50.2)

$\Leftrightarrow \mathscr{P}_0 \subseteq (\mathscr{P}_f(\ell) \cap Y(\ell)) \wedge X(\ell) \subseteq \mathsf{test}^{\tilde{\bar{e}}}[\![\mathsf{B}]\!](\widehat{\boldsymbol{\mathcal{S}}}^{\tilde{\bar{e}}}[\![\mathsf{S}_b]\!]\, (\ell' \in \mathsf{labs}[\![\mathsf{S}_b]\!] \mapsto (\!|\, \ell' \in \mathsf{breaks\text{-}of}[\![\mathsf{S}_b]\!] \,?\, (\mathscr{P}_f(\mathsf{after}[\![\mathsf{S}]\!]) \cap Y(\mathsf{after}[\![\mathsf{S}]\!])) \,:\, (\mathscr{P}_f(\ell') \cap Y(\ell')) \,|\!))) \wedge X(\ell) \subseteq \overline{\mathsf{test}}^{\tilde{\bar{e}}}[\![\mathsf{B}]\!](\mathscr{P}_f(\mathsf{after}[\![\mathsf{S}]\!]) \cap Y(\mathsf{after}[\![\mathsf{S}]\!]))$ (Galois connection of (50.18) for $\mathsf{test}^{\vec{e}}[\![\mathsf{B}]\!]$ and $\overline{\mathsf{test}}^{\vec{e}}[\![\mathsf{B}]\!]$)

$\Leftrightarrow \mathscr{P}_0 \subseteq (\mathscr{P}_f(\ell) \cap Y(\ell)) \wedge X(\ell) \subseteq \mathsf{test}^{\tilde{\bar{e}}}[\![\mathsf{B}]\!](\widehat{\boldsymbol{\mathcal{S}}}^{\tilde{\bar{e}}}[\![\mathsf{S}_b]\!]\, (\ell'' \in \mathsf{labs}[\![\mathsf{S}_b]\!] \mapsto (\!|\, \ell'' \in \mathsf{breaks\text{-}of}[\![\mathsf{S}_b]\!] \,?\, (\mathscr{P}_f(\mathsf{after}[\![\mathsf{S}]\!]) \cap Y(\mathsf{after}[\![\mathsf{S}]\!])) \,:\, (\mathscr{P}_f(\ell'') \cap Y(\ell'')) \,|\!))) \cap \overline{\mathsf{test}}^{\tilde{\bar{e}}}[\![\mathsf{B}]\!](\mathscr{P}_f(\mathsf{after}[\![\mathsf{S}]\!]) \cap Y(\mathsf{after}[\![\mathsf{S}]\!]))$ (definition of the glb and renaming)

$\Leftrightarrow \mathscr{P}_0 \subseteq (\mathscr{P}_f(\ell) \cap Y(\ell)) \wedge X \mathrel{\dot{\subseteq}} \ell' \mapsto (\!|\, \ell' = \ell \,?\, \mathsf{test}^{\tilde{\bar{e}}}[\![\mathsf{B}]\!](\widehat{\boldsymbol{\mathcal{S}}}^{\tilde{\bar{e}}}[\![\mathsf{S}_b]\!]\, (\ell'' \in \mathsf{labs}[\![\mathsf{S}_b]\!] \mapsto (\!|\, \ell'' \in \mathsf{breaks\text{-}of}[\![\mathsf{S}_b]\!] \,?\, (\mathscr{P}_f(\mathsf{after}[\![\mathsf{S}]\!]) \cap Y(\mathsf{after}[\![\mathsf{S}]\!])) \,:\, (\mathscr{P}_f(\ell'') \cap Y(\ell'')) \,|\!))) \cap \overline{\mathsf{test}}^{\tilde{\bar{e}}}[\![\mathsf{B}]\!](\mathscr{P}_f(\mathsf{after}[\![\mathsf{S}]\!]) \cap Y(\mathsf{after}[\![\mathsf{S}]\!])) \,:\, \mathbb{E}\mathbb{v}^{\vec{e}} \,|\!)$ (pointwise definition of $\dot{\subseteq}$)

$\Leftrightarrow \mathscr{P}_0 \subseteq (\mathscr{P}_f(\ell) \cap Y(\ell)) \wedge X \mathrel{\dot{\subseteq}} \overleftarrow{\boldsymbol{\mathscr{F}}}^{\tilde{\bar{e}}}[\![\texttt{while}\ \ell\ (\mathsf{B})\ \mathsf{S}_b]\!]\ \mathscr{P}_f\ Y$

(by defining $\overleftarrow{\boldsymbol{\mathscr{F}}}^{\tilde{\bar{e}}}[\![\texttt{while}\ \ell\ (\mathsf{B})\ \mathsf{S}_b]\!]\ \mathscr{P}_f$ as in (50.20)) □

We can now derive the backward accessibility fixpoint semantics $\mathsf{gfp}^{\dot{\subseteq}}\ \overleftarrow{\boldsymbol{\mathscr{F}}}^{\tilde{\bar{e}}}[\![\texttt{while}\ \ell\ (\mathsf{B})\ \mathsf{S}_b]\!]\ \mathscr{P}_f$ of the iteration statement $\mathsf{S} ::= \texttt{while}\ \ell\ (\mathsf{B})\ \mathsf{S}_b$ by calculational design.

Lemma 50.21 $\widehat{\boldsymbol{\mathcal{S}}}^{\vec{e}}[\![\mathsf{S}]\!]\ \mathscr{P}_0 \mathrel{\dot{\subseteq}} \mathscr{P}_f \Leftrightarrow \mathscr{P}_0 \subseteq (\mathsf{gfp}^{\dot{\subseteq}}\ \overleftarrow{\boldsymbol{\mathscr{F}}}^{\tilde{\bar{e}}}[\![\texttt{while}\ \ell\ (\mathsf{B})\ \mathsf{S}_b]\!]\ \mathscr{P}_f)\ \ell$

$\widehat{\boldsymbol{\mathcal{S}}}^{\vec{e}}[\![\mathsf{S}]\!]\ \mathscr{P}_0 \mathrel{\dot{\subseteq}} \mathscr{P}_f$

$\Leftrightarrow \mathsf{lfp}^{\dot{\subseteq}}\ (\boldsymbol{\mathscr{F}}^{\vec{e}}[\![\texttt{while}\ \ell\ (\mathsf{B})\ \mathsf{S}_b]\!]\ \mathscr{P}_0) \mathrel{\dot{\subseteq}} \mathscr{P}_f$ ((19.16))

$\Leftrightarrow \exists \mathscr{P} \in \mathsf{labs}[\![\mathsf{S}]\!] \to \wp(\mathbb{E}\mathbb{v}^{e}) \,.\, \boldsymbol{\mathscr{F}}^{\vec{e}}[\![\texttt{while}\ \ell\ (\mathsf{B})\ \mathsf{S}_b]\!]\ \mathscr{P}_0\ \mathscr{P} \mathrel{\dot{\subseteq}} \mathscr{P} \wedge \mathscr{P} \mathrel{\dot{\subseteq}} \mathscr{P}_f$

(fixpoint induction theorem 24.1)

$\Leftrightarrow \exists \mathscr{P} \in \mathsf{labs}[\![\mathsf{S}]\!] \to \wp(\mathbb{E}\mathbb{v}^{e}) \,.\, \boldsymbol{\mathscr{F}}^{\vec{e}}[\![\texttt{while}\ \ell\ (\mathsf{B})\ \mathsf{S}_b]\!]\ \mathscr{P}_0\ \mathscr{P} \mathrel{\dot{\subseteq}} (\mathscr{P}_f \mathbin{\dot{\cap}} \mathscr{P}) \wedge \mathscr{P} \mathrel{\dot{\subseteq}} \mathscr{P}_f$

$$
\begin{aligned}
& \qquad \wr \text{definition of } \dot{\cap} \text{ because } \mathscr{P} \mathrel{\dot{\subseteq}} \mathscr{P}_f \int \\
\Leftrightarrow\ & \exists \mathscr{P} \in \mathsf{labs}[\![\mathsf{S}]\!] \to \wp(\mathbb{E}\mathbb{v}^{\varrho}) \,.\, \mathscr{P}_0 \subseteq (\mathscr{P}_f(\ell) \cap \mathscr{P}(\ell)) \wedge \mathscr{P} \mathrel{\dot{\subseteq}} \overleftarrow{\mathscr{F}}^{\tilde{\bar{\varrho}}}[\![\texttt{while}\ \ell\ (\mathsf{B})\ \mathsf{S}_b]\!]\, \mathscr{P}_f\, \mathscr{P} \wedge \\
& \mathscr{P} \mathrel{\dot{\subseteq}} \mathscr{P}_f \qquad \wr \text{lemma 50.19 where } X = Y = \mathscr{P} \int \\
\Leftrightarrow\ & \exists \mathscr{P} \in \mathsf{labs}[\![\mathsf{S}]\!] \to \wp(\mathbb{E}\mathbb{v}^{\varrho}) \,.\, \mathscr{P}_0 \subseteq \mathscr{P}(\ell) \wedge \mathscr{P} \mathrel{\dot{\subseteq}} \overleftarrow{\mathscr{F}}^{\tilde{\bar{\varrho}}}[\![\texttt{while}\ \ell\ (\mathsf{B})\ \mathsf{S}_b]\!]\, \mathscr{P}_f\, \mathscr{P} \wedge \mathscr{P} \mathrel{\dot{\subseteq}} \mathscr{P}_f \\
& \qquad \wr \text{because } \mathscr{P} \mathrel{\dot{\subseteq}} \mathscr{P}_f \text{ so, pointwise, } (\mathscr{P}_f(\ell) \cap \mathscr{P}(\ell)) = \mathscr{P}(\ell \int \\
\Leftrightarrow\ & \exists \mathscr{P} \in \mathsf{labs}[\![\mathsf{S}]\!] \to \wp(\mathbb{E}\mathbb{v}^{\varrho}) \,.\, \mathscr{P}_0 \subseteq \mathscr{P}(\ell) \wedge \mathscr{P} \mathrel{\dot{\subseteq}} \mathscr{P}_f \mathbin{\dot{\cap}} \overleftarrow{\mathscr{F}}^{\tilde{\bar{\varrho}}}[\![\texttt{while}\ \ell\ (\mathsf{B})\ \mathsf{S}_b]\!]\, \mathscr{P}_f\, \mathscr{P} \\
& \qquad \wr \text{definition of the glb } \dot{\cap} \int \\
\Leftrightarrow\ & \exists \mathscr{P} \in \mathsf{labs}[\![\mathsf{S}]\!] \to \wp(\mathbb{E}\mathbb{v}^{\varrho}) \,.\, \mathscr{P}_0 \subseteq \mathscr{P}(\ell) \wedge \mathscr{P} \mathrel{\dot{\subseteq}} (\mathsf{gfp}^{\dot{\subseteq}}\, Y \mapsto \mathscr{P}_f \mathbin{\dot{\cap}} \overleftarrow{\mathscr{F}}^{\tilde{\bar{\varrho}}}[\![\texttt{while}\ \ell\ (\mathsf{B}) \\
& \mathsf{S}_b]\!]\, \mathscr{P}_f\, Y) \qquad \wr \text{dual Tarski's fixpoint theorem 15.6} \int \\
\Leftrightarrow\ & \mathscr{P}_0 \subseteq (\mathsf{gfp}^{\dot{\subseteq}}\, Y \mapsto \mathscr{P}_f \mathbin{\dot{\cap}} \overleftarrow{\mathscr{F}}^{\tilde{\bar{\varrho}}}[\![\texttt{while}\ \ell\ (\mathsf{B})\ \mathsf{S}_b]\!]\, \mathscr{P}_f\, Y)\, \ell \\
& \qquad \wr (\Rightarrow) \text{ pointwise definition of } \dot{\subseteq} \text{ and transitivity} \\
& \qquad (\Leftarrow) \text{ choosing } \mathscr{P} \triangleq \mathsf{gfp}^{\dot{\subseteq}}\, Y \mapsto \mathscr{P}_f \mathbin{\dot{\cap}} \overleftarrow{\mathscr{F}}^{\tilde{\bar{\varrho}}}[\![\texttt{while}\ \ell\ (\mathsf{B})\ \mathsf{S}_b]\!]\, \mathscr{P}_f\, Y \text{ and} \\
& \qquad \text{reflexivity} \int
\end{aligned}
$$

□

Now by applying the dual of Scott–Kleene's iterative fixpoint theorem 15.26, we observe that the iterates for the fixpoint $\mathsf{gfp}^{\dot{\subseteq}}\, Y \mapsto \mathscr{P}_f \mathbin{\dot{\cap}} \overleftarrow{\mathscr{F}}^{\tilde{\bar{\varrho}}}[\![\texttt{while}\ \ell\ (\mathsf{B})\ \mathsf{S}_b]\!]\, \mathscr{P}_f\, Y$ start from $\ell' \in \mathsf{labs}[\![\mathsf{S}]\!] \mapsto \mathbb{E}\mathbb{v}^{\vec{\varrho}}$ and later iterates have, by definition (50.20) of $\overleftarrow{\mathscr{F}}^{\tilde{\bar{\varrho}}}[\![\mathsf{S}]\!]\, \mathscr{P}_f\, Y\, \ell'$, all components ℓ' but ℓ always equal to $\mathscr{P}_f(\ell')$. It is therefore equivalent to compute iteratively the ℓ component and let the other be constant equal to $\mathscr{P}_f(\ell')$. Therefore $\mathscr{F}^{\tilde{\bar{\varrho}}}[\![\mathsf{S}]\!]$ derives from $\overleftarrow{\mathscr{F}}^{\tilde{\bar{\varrho}}}[\![\mathsf{S}]\!]$ in (50.20) where Y in (50.14) is the component $Y(\ell)$ in (50.20) and all other components are $\mathscr{P}_f(\ell')$ allowing for the simplification $Z \cap \mathbb{E}\mathbb{v}^{\vec{\varrho}} = Z$. □

Exercise 50.22 For a program component S, consider the abstraction α of $\mathscr{P}_f$ into $\ell \mapsto (\!| \ell = \mathsf{after}[\![\mathsf{S}]\!] \mathrel{?} \mathscr{P}_f(\ell) \mathrel{\text{\textonehalf}} \mathbb{E}\mathbb{v}^{\varrho} |\!)$ so that $\widehat{\mathscr{S}}^{\tilde{\bar{\varrho}}}[\![\mathsf{S}]\!]\, \alpha(\mathscr{P}_f)$ is the set of initial states from which executions will reach a final state in $\mathscr{P}_f(\mathsf{after}[\![\mathsf{S}]\!])$ or will not terminate. Design $\widehat{\mathscr{S}}^{\tilde{\bar{\varrho}}}[\![\mathsf{S}]\!]\, \alpha(\mathscr{P}_f)$ by calculation.

Observe that any subset of $\widehat{\mathscr{S}}^{\tilde{\bar{\varrho}}}[\![\mathsf{S}]\!]$ guarantees impossible failure. So a sound impossible failure static analysis should use underapproximations [265, 271, 694].

50.4 Inversion

Corollary 50.2 provides a relationship between forward reachability and backward accessibility semantics. Another symbolic one is the following [222, theorem 10-13] (also called magic transformation in deductive databases [84, 103], logic programming [197, 300, 338, 753], history or prophecy variables [3, 649, 832], and other names).

> **Theorem 50.23** $\widehat{\mathscr{S}}^{\vec{r}}[\![\mathsf{S}]\!]\, \mathscr{P}_0\, \ell \;=\; \{\rho_f \mid \exists \rho_0 \in \mathscr{P}_0 \,.\, \langle \rho_f, \rho_0 \rangle \in \widehat{\mathscr{S}}^{\tilde{\bar{R}}}[\![\mathsf{S}]\!]\, (\ell'' \mapsto (\!| \ell'' = \ell' \mathrel{?} \{\langle \rho_f, \rho_f' \rangle \mid \rho_f = \rho_f'\} \mathrel{:} \mathbb{E}\mathbb{v} \times \mathbb{E}\mathbb{v} |\!))\}$

Theorem 50.23 allows forward assertional reachability $\widehat{\mathscr{S}}^{\vec{r}}[\![\mathsf{S}]\!]$ to be recovered from a backward relational accessibility $\widehat{\mathscr{S}}^{\tilde{\bar{R}}}[\![\mathsf{S}]\!]$.

Proof of theorem 50.23

$$\widehat{\boldsymbol{\mathcal{S}}}^{\vec{r}}[\![\mathsf{S}]\!]\,\mathcal{P}_0\,\ell$$

$$= \{\varrho(\pi_0{}^{\ell_0}\pi_1{}^{\ell'}) \mid \varrho(\pi_0{}^{\ell_0}) \in \mathcal{P}_0 \land {}^{\ell_0}\pi_1{}^{\ell'} \in \boldsymbol{\mathcal{S}}^*[\![\mathsf{S}]\!]\,(\pi_0{}^{\ell_0}) \land \ell' = \ell\}$$

⟅definition (20.6) of $\widehat{\boldsymbol{\mathcal{S}}}^{\vec{r}}[\![\mathsf{S}]\!]$ and (20.1) of $\mathsf{post}^{\vec{r}}$⟆

$$= \{\rho_f \mid \exists \rho_0 \in \mathcal{P}_0\,.\,\langle\rho_f,\ \rho_0\rangle \in \{\langle\varrho(\pi_0{}^{\ell_0}\pi_1{}^{\ell'}),\ \varrho(\pi_0{}^{\ell_0})\rangle \mid {}^{\ell_0}\pi_1{}^{\ell'} \in \boldsymbol{\mathcal{S}}^*[\![\mathsf{S}]\!]\,(\pi_0{}^{\ell_0}) \land \ell' = \ell\}\}$$

⟅definition of $\in$⟆

$$= \{\rho_f \mid \exists \rho_0 \in \mathcal{P}_0\,.\,\langle\rho_f,\ \rho_0\rangle \in \{\langle\rho_f',\ \varrho(\pi_0{}^{\ell_0})\rangle \mid \exists \ell' \in \mathsf{labs}[\![\mathsf{S}]\!]\,.\,{}^{\ell_0}\pi_1{}^{\ell'} \in \boldsymbol{\mathcal{S}}^*[\![\mathsf{S}]\!]\,(\pi_0{}^{\ell_0}) \land \rho_f' = \varrho(\pi_0{}^{\ell_0}\pi_1{}^{\ell'}) \land \ell' = \ell\}\}$$

⟅letting $\rho_f' = \varrho(\pi_0{}^{\ell_0}\pi_1{}^{\ell'})$⟆

$$= \{\rho_f \mid \exists \rho_0 \in \mathcal{P}_0\,.\,\langle\rho_f,\ \rho_0\rangle \in \{\langle\rho_f',\ \varrho(\pi_0{}^{\ell_0})\rangle \mid \exists \ell' \in \mathsf{labs}[\![\mathsf{S}]\!]\,.\,{}^{\ell_0}\pi_1{}^{\ell'} \in \boldsymbol{\mathcal{S}}^*[\![\mathsf{S}]\!]\,(\pi_0{}^{\ell_0}) \land \langle\rho_f',\ \varrho(\pi_0{}^{\ell_0}\pi_1{}^{\ell'})\rangle \in \{\langle\rho_f,\ \rho_f\rangle \mid \rho_f \in \mathbb{E}\mathbb{v}\} \land \ell' = \ell\}\}$$

⟅definition of $\in$⟆

$$= \{\rho_f \mid \exists \rho_0 \in \mathcal{P}_0\,.\,\langle\rho_f,\ \rho_0\rangle \in \{\langle\rho_f',\ \varrho(\pi_0{}^{\ell_0})\rangle \mid \exists \ell' \in \mathsf{labs}[\![\mathsf{S}]\!]\,.\,{}^{\ell_0}\pi_1{}^{\ell'} \in \boldsymbol{\mathcal{S}}^*[\![\mathsf{S}]\!]\,(\pi_0{}^{\ell_0}) \land \langle\rho_f',\ \varrho(\pi_0{}^{\ell_0}\pi_1{}^{\ell'})\rangle \in (\ell'' \mapsto \llparenthesis\, \ell'' = \ell' \;?\; \{\langle\rho_f,\ \rho_f\rangle \mid \rho_f \in \mathbb{E}\mathbb{v}\} \fatsemi \varnothing \,\rrparenthesis(\ell')) \land \ell' = \ell\}\}$$

⟅definition of function application⟆

$$= \{\rho_f \mid \exists \rho_0 \in \mathcal{P}_0\,.\,\langle\rho_f,\ \rho_0\rangle \in \{\langle\rho_f',\ \varrho(\pi_0{}^{\ell_0})\rangle \mid \exists \ell' \in \mathsf{labs}[\![\mathsf{S}]\!]\,.\,{}^{\ell_0}\pi_1{}^{\ell'} \in \boldsymbol{\mathcal{S}}^*[\![\mathsf{S}]\!]\,(\pi_0{}^{\ell_0}) \land \langle\rho_f',\ \varrho(\pi_0{}^{\ell_0}\pi_1{}^{\ell'})\rangle \in (\ell'' \mapsto \llparenthesis\, \ell'' = \ell' \;?\; \{\langle\rho_f,\ \rho_f\rangle \mid \rho_f \in \mathbb{E}\mathbb{v}\} \fatsemi \varnothing \,\rrparenthesis(\ell))\}\}$$

⟅because when $\ell' = \ell$ we have $\langle\rho_f',\ \varrho(\pi_0{}^{\ell_0}\pi_1{}^{\ell'})\rangle \in \varnothing$ which is false.⟆

$$= \{\rho_f \mid \exists \rho_0 \in \mathcal{P}_0\,.\,\langle\rho_f,\ \rho_0\rangle \in \{\langle\rho_f',\ \varrho(\pi_0{}^{\ell_0})\rangle \mid \forall \ell' \in \mathsf{labs}[\![\mathsf{S}]\!]\,.\,{}^{\ell_0}\pi_1{}^{\ell'} \in \boldsymbol{\mathcal{S}}^*[\![\mathsf{S}]\!]\,(\pi_0{}^{\ell_0}) \Rightarrow \langle\rho_f',\ \varrho(\pi_0{}^{\ell_0}\pi_1{}^{\ell'})\rangle \in (\ell'' \mapsto \llparenthesis\, \ell'' = \ell' \;?\; \{\langle\rho_f,\ \rho_f'\rangle \mid \rho_f = \rho_f'\} \fatsemi \mathbb{E}\mathbb{v} \times \mathbb{E}\mathbb{v} \,\rrparenthesis(\ell))\}\}$$

⟅where $\dot\neg X \triangleq (\mathbb{E}\mathbb{v} \times \mathbb{E}\mathbb{v}) \setminus X$ and $\{x \mid P(x)\} = \neg\{x \mid \neg P(x)\}$⟆

$$= \{\rho_f \mid \exists \rho_0 \in \mathcal{P}_0\,.\,\langle\rho_f,\ \rho_0\rangle \in \widehat{\boldsymbol{\mathcal{S}}}^{\tilde{\bar{\mathsf{R}}}}[\![\mathsf{S}]\!]\,(\ell'' \mapsto \llparenthesis\, \ell'' = \ell' \;?\; \{\langle\rho_f,\ \rho_f'\rangle \mid \rho_f = \rho_f'\} \fatsemi \mathbb{E}\mathbb{v} \times \mathbb{E}\mathbb{v} \,\rrparenthesis)\}$$

⟅$\widehat{\boldsymbol{\mathcal{S}}}^{\tilde{\bar{\mathsf{R}}}}[\![\mathsf{S}]\!]\,\mathcal{P}_f = \{\langle\rho_0,\ \varrho(\pi_0{}^{\ell_0})\rangle \mid \forall \ell \in \mathsf{labs}[\![\mathsf{S}]\!]\,.\,({}^{\ell_0}\pi_1{}^{\ell} \in \boldsymbol{\mathcal{S}}^*[\![\mathsf{S}]\!]\,(\pi_0{}^{\ell_0})) \Rightarrow \langle\rho_0,\ \varrho(\pi_0{}^{\ell_0}\pi_1{}^{\ell})\rangle \in \mathcal{P}_f(\ell)\}$ by exercise 50.6⟆ □

Inversely, backward assertional accessibility can be recovered from a forward relational accessibility analysis.

Example 50.24 Consider the forward relational reachability analysis ${}^{\ell_0}\{x = x_0 \land y = y_0\}\ \mathsf{y = 1\,;}{}^{\ell_1}\{x = x_0 \land y = 1\}\ \mathsf{x = x - y\,;}{}^{\ell_2}\{x = x_0 - 1 \land y = 1\}$. Given the reachability condition ${}^{\ell_0}\{\mathsf{tt}\}\ \mathsf{y = 1\,;}{}^{\ell_1}\{\mathsf{tt}\}\ \mathsf{x = x - y\,;}{}^{\ell_2}\{x = y\}$, we get the precondition $(\exists x, y\,.\,x = x_0 - 1 \land y = 1 \land x = y) = (x_0 = 2)$.

Exercise 50.25 Prove that backward assertional accessibility can be recovered from a forward relational accessibility analysis.

50.5 Complement Dual Abstraction

Consider the complement abstraction

$$\langle \wp(\mathbb{E}\mathbb{v}^{\varrho}),\ \subseteq\rangle \xrightleftharpoons[\bar\alpha]{\bar\gamma} \langle \wp(\mathbb{E}\mathbb{v}^{\varrho}),\ \supseteq\rangle \tag{50.26}$$

where $\bar\alpha(X) = \bar\gamma(X) = \neg X \triangleq \mathbb{E}\mathbb{v}^{\varrho} \setminus X$.

Proof of (50.26)

$$
\begin{array}{lll}
& \bar{\alpha}(X) \subseteq Y & \\
\Leftrightarrow & \neg X \subseteq Y & \wr\text{definition of } \bar{\alpha}\wr \\
\Leftrightarrow & X \supseteq \neg Y & \wr\text{definition of } \neg\wr \\
\Leftrightarrow & X \supseteq \bar{\gamma}(Y) & \wr\text{definition of } \bar{\gamma}\wr \quad \square
\end{array}
$$

The complement dual abstraction is

$$
\langle(\mathsf{labs}[\![\mathsf{S}]\!] \rightarrow \wp(\mathbb{E}\mathbb{v}^{\varrho})) \rightarrow \wp(\mathbb{E}\mathbb{v}^{\varrho}),\ \dot{\subseteq}\rangle \underset{\dot{\alpha}[\![\mathsf{S}]\!]}{\overset{\dot{\gamma}[\![\mathsf{S}]\!]}{\leftrightarrows}} \langle(\mathsf{labs}[\![\mathsf{S}]\!] \rightarrow \wp(\mathbb{E}\mathbb{v}^{\varrho})) \rightarrow \wp(\mathbb{E}\mathbb{v}^{\varrho}),\ \dot{\supseteq}\rangle \tag{50.27}
$$

where $\dot{\alpha}[\![\mathsf{S}]\!](\boldsymbol{\mathcal{S}})\,\mathscr{P} = \dot{\gamma}[\![\mathsf{S}]\!](\boldsymbol{\mathcal{S}})\,\mathscr{P} \triangleq \neg\boldsymbol{\mathcal{S}}\,(\dot{\neg}\mathscr{P})$ and $\dot{\neg}\mathscr{P} \triangleq \ell \in \mathsf{labs}[\![\mathsf{S}]\!] \mapsto \neg\mathscr{P}(\ell)$.

Proof of (50.27)

$$
\begin{array}{lll}
& \dot{\alpha}[\![\mathsf{S}]\!](\boldsymbol{\mathcal{S}}) \,\dot{\subseteq}\, \boldsymbol{\mathcal{S}}' & \\
\Leftrightarrow & \forall \mathscr{P} \in \wp(\mathbb{E}\mathbb{v}^{\varrho}) \,.\, \dot{\alpha}[\![\mathsf{S}]\!](\boldsymbol{\mathcal{S}})\,\mathscr{P} \subseteq \boldsymbol{\mathcal{S}}'\,\mathscr{P} & \wr\text{pointwise definiton of } \dot{\subseteq}\wr \\
\Leftrightarrow & \forall \mathscr{P} \in \wp(\mathbb{E}\mathbb{v}^{\varrho}) \,.\, \neg\boldsymbol{\mathcal{S}}\,(\dot{\neg}\mathscr{P}) \subseteq \boldsymbol{\mathcal{S}}'\,\mathscr{P} & \wr\text{definition of } \dot{\alpha}[\![\mathsf{S}]\!](\boldsymbol{\mathcal{S}})\wr \\
\Leftrightarrow & \forall \mathscr{P}' \in \wp(\mathbb{E}\mathbb{v}^{\varrho}) \,.\, \neg\boldsymbol{\mathcal{S}}\,(\mathscr{P}') \subseteq \boldsymbol{\mathcal{S}}'\,(\dot{\neg}\mathscr{P}') & \wr\text{letting } \mathscr{P}' \triangleq \dot{\neg}\mathscr{P}\wr \\
\Leftrightarrow & \forall \mathscr{P}' \in \wp(\mathbb{E}\mathbb{v}^{\varrho}) \,.\, \boldsymbol{\mathcal{S}}\,(\mathscr{P}') \supseteq \neg\boldsymbol{\mathcal{S}}'\,(\dot{\neg}\mathscr{P}') & \wr\text{definition of } \neg\wr \\
\Leftrightarrow & \forall \mathscr{P}' \in \wp(\mathbb{E}\mathbb{v}^{\varrho}) \,.\, \boldsymbol{\mathcal{S}}\,(\mathscr{P}') \supseteq \dot{\gamma}[\![\mathsf{S}]\!](\boldsymbol{\mathcal{S}}')\,\mathscr{P}' & \wr\text{definition of } \dot{\gamma}\wr \\
\Leftrightarrow & \boldsymbol{\mathcal{S}} \,\dot{\supseteq}\, \dot{\gamma}[\![\mathsf{S}]\!](\boldsymbol{\mathcal{S}}') & \wr\text{pointwise definition of } \dot{\supseteq}\wr \quad \square
\end{array}
$$

50.6 Definition of the Backward/Abductive Possible Success Accessibility Semantics

Definition 50.28 $\widehat{\boldsymbol{\mathcal{S}}}^{\overleftarrow{\varrho}}[\![\mathsf{S}]\!] \triangleq \dot{\alpha}[\![\mathsf{S}]\!](\widehat{\boldsymbol{\mathcal{S}}}^{\tilde{\overleftarrow{\varrho}}}[\![\mathsf{S}]\!])$.

It follows that we have the Galois connection

$$
\langle\mathsf{labs}[\![\mathsf{S}]\!] \rightarrow \wp(\mathbb{E}\mathbb{v}^{\varrho}),\ \dot{\subseteq}\rangle \underset{\widehat{\boldsymbol{\mathcal{S}}}^{\overleftarrow{\varrho}}[\![\mathsf{S}]\!]}{\overset{\widehat{\boldsymbol{\mathcal{S}}}^{\tilde{\vec{\varrho}}}[\![\mathsf{S}]\!]}{\leftrightarrows}} \langle\wp(\mathbb{E}\mathbb{v}^{\varrho}),\ \subseteq\rangle \tag{50.29}
$$

Proof of (50.29)

$$
\begin{array}{lll}
& \widehat{\boldsymbol{\mathcal{S}}}^{\overleftarrow{\varrho}}[\![\mathsf{S}]\!](\mathscr{P}_f) \subseteq \mathscr{P}_0 & \\
\Leftrightarrow & \neg\widehat{\boldsymbol{\mathcal{S}}}^{\tilde{\overleftarrow{\varrho}}}[\![\mathsf{S}]\!](\dot{\neg}\mathscr{P}_f) \subseteq \mathscr{P}_0 & \wr\text{definition 50.28 and definition of } \dot{\alpha}[\![\mathsf{S}]\!]\wr \\
\Leftrightarrow & \neg\mathscr{P}_0 \subseteq \widehat{\boldsymbol{\mathcal{S}}}^{\tilde{\overleftarrow{\varrho}}}[\![\mathsf{S}]\!](\dot{\neg}\mathscr{P}_f) & \wr\text{definition of } \neg\wr \\
\Leftrightarrow & \widehat{\boldsymbol{\mathcal{S}}}^{\vec{\varrho}}[\![\mathsf{S}]\!](\neg\mathscr{P}_0) \,\dot{\subseteq}\, \dot{\neg}\mathscr{P}_f & \wr\text{corollary 50.2}\wr \\
\Leftrightarrow & \mathscr{P}_f \,\dot{\subseteq}\, \dot{\neg}\widehat{\boldsymbol{\mathcal{S}}}^{\vec{\varrho}}[\![\mathsf{S}]\!](\neg\mathscr{P}_0) & \wr\text{definition of } \neg\wr \\
\Leftrightarrow & \mathscr{P}_f \,\dot{\subseteq}\, \widehat{\boldsymbol{\mathcal{S}}}^{\tilde{\vec{\varrho}}}[\![\mathsf{S}]\!](\mathscr{P}_0) & \wr\text{by defining } \widehat{\boldsymbol{\mathcal{S}}}^{\tilde{\vec{\varrho}}}[\![\mathsf{S}]\!] \triangleq \dot{\neg}\widehat{\boldsymbol{\mathcal{S}}}^{\vec{\varrho}}[\![\mathsf{S}]\!](\neg\mathscr{P}_0)\wr \quad \square
\end{array}
$$

$\widehat{\boldsymbol{\mathcal{S}}}^{\tilde{\vec{\varrho}}}[\![\mathsf{S}]\!](\mathscr{P}_0)$ is the set of states that can be reached only from $\mathscr{P}_0$ and no other initial states.

Theorem 50.30 below states that there exists an execution from any state in $\widehat{\boldsymbol{\mathcal{S}}}^{\tilde{\overleftarrow{\mathsf{r}}}}[\![\mathsf{S}]\!]\,\mathcal{P}_f$ reaching a state of S satisfying $\mathcal{P}_f$.

Theorem 50.30 $\widehat{\boldsymbol{\mathcal{S}}}^{\overleftarrow{\mathsf{r}}}[\![\mathsf{S}]\!]\,\mathcal{P}_f = \mathsf{pre}^{\overleftarrow{\mathsf{r}}}(\boldsymbol{\mathcal{S}}^*[\![\mathsf{S}]\!])\,\mathcal{P}_f$ where $\mathsf{pre}^{\overleftarrow{\mathsf{r}}}(\boldsymbol{\mathcal{S}})\,\mathcal{P}_f \triangleq \{\boldsymbol{\varrho}(\pi_0\ell_0) \in \mathbb{E}\mathbb{v} \mid \exists \ell \in \mathsf{labs}[\![\mathsf{S}]\!] \,.\, \ell_0\pi_1\ell \in \boldsymbol{\mathcal{S}}(\pi_0\ell_0) \wedge \boldsymbol{\varrho}(\pi_0\ell_0\pi_1\ell) \in \mathcal{P}_f(\ell)\}$.

Proof of theorem 50.30

$$
\begin{aligned}
& \widehat{\boldsymbol{\mathcal{S}}}^{\overleftarrow{\mathsf{r}}}[\![\mathsf{S}]\!]\,\mathcal{P}_f \\
= {} & \dot{\alpha}[\![\mathsf{S}]\!](\widehat{\boldsymbol{\mathcal{S}}}^{\tilde{\overleftarrow{\mathsf{r}}}}[\![\mathsf{S}]\!])\,\mathcal{P}_f && \text{(definition 50.28)} \\
= {} & \neg\widehat{\boldsymbol{\mathcal{S}}}^{\tilde{\overleftarrow{\mathsf{r}}}}[\![\mathsf{S}]\!]\,(\dot{\neg}\mathcal{P}_f) && \text{(definition of } \dot{\alpha}[\![\mathsf{S}]\!]\text{)} \\
= {} & \neg\{\boldsymbol{\varrho}(\pi_0\ell_0) \in \mathbb{E}\mathbb{v} \mid \forall \ell \in \mathsf{labs}[\![\mathsf{S}]\!] \,.\, (\ell_0\pi_1\ell \in \boldsymbol{\mathcal{S}}^*[\![\mathsf{S}]\!]\,(\pi_0\ell_0)) \Rightarrow \boldsymbol{\varrho}(\pi_0\ell_0\pi_1\ell) \notin \mathcal{P}_f(\ell)\} \\
& && \text{(theorem 50.3)} \\
= {} & \{\boldsymbol{\varrho}(\pi_0\ell_0) \in \mathbb{E}\mathbb{v} \mid \exists \ell \in \mathsf{labs}[\![\mathsf{S}]\!] \,.\, \ell_0\pi_1\ell \in \boldsymbol{\mathcal{S}}^*[\![\mathsf{S}]\!]\,(\pi_0\ell_0) \wedge \boldsymbol{\varrho}(\pi_0\ell_0\pi_1\ell) \in \mathcal{P}_f(\ell)\} \\
& && \text{(definition of } \neg\text{)} \quad \square
\end{aligned}
$$

Example 50.31 Consider the semantics $\boldsymbol{\mathcal{S}}$ and the specification $\mathcal{P}_f$ of the reachable states of example 50.4, as illustrated below. Then $\mathsf{pre}^{\overleftarrow{\mathsf{r}}}(\boldsymbol{\mathcal{S}})\,\mathcal{P}_f$ is the set of initial states from which the traces of $\boldsymbol{\mathcal{S}}$ may potentially/occasionally/sometimes reach a state in $\mathcal{P}_f$. So traces from initial states in $\neg\mathsf{pre}^{\overleftarrow{\mathsf{r}}}(\boldsymbol{\mathcal{S}})\,\mathcal{P}_f$ definitely/for sure cannot ever reach a state in $\mathcal{P}_f$.

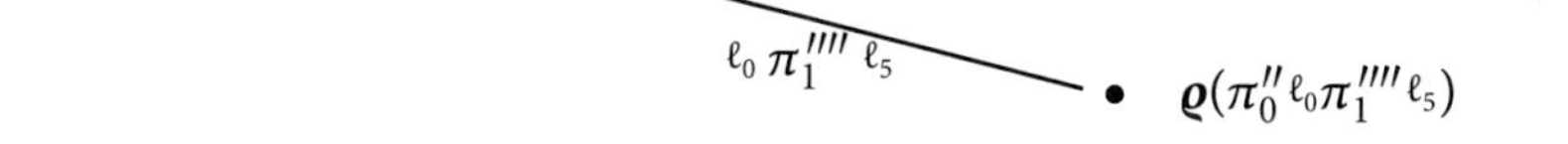

Example 50.32 (absence of errors) Continuing example 50.5, erroneous states in $\mathcal{E} \in \mathsf{labs}[\![\mathsf{S}]\!] \to \wp(\mathbb{E}\mathbb{v}^{\varrho})$ are unreachable from initial states in $\mathcal{P}_0 \in \wp(\mathbb{E}\mathbb{v}^{\varrho})$ if and only if $\widehat{\boldsymbol{\mathcal{S}}}^{\vec{\varrho}}[\![\mathsf{S}]\!]\,\mathcal{P}_0 \mathrel{\dot{\subseteq}} \dot{\neg}\mathcal{E}$ if and only if $\mathcal{P}_0 \subseteq \widehat{\boldsymbol{\mathcal{S}}}^{\tilde{\overleftarrow{\varrho}}}[\![\mathsf{S}]\!]\dot{\neg}\mathcal{E}$ where $\mathcal{E} \triangleq \ell \mapsto \wp(\mathbb{E}\mathbb{v}^{\varrho}) \setminus \mathcal{E}(\ell)$. By complement duality, we get two more equivalent definitions.

$\mathcal{P}_0 \subseteq \widehat{\boldsymbol{\mathcal{S}}}^{\tilde{\overleftarrow{\varrho}}}[\![\mathsf{S}]\!]\dot{\neg}\mathcal{E}$ is equivalent to $\neg\mathcal{P}_0 \supseteq \neg\widehat{\boldsymbol{\mathcal{S}}}^{\tilde{\overleftarrow{\varrho}}}[\![\mathsf{S}]\!]\dot{\neg}\mathcal{E}$, that is $\dot{\alpha}[\![\mathsf{S}]\!](\widehat{\boldsymbol{\mathcal{S}}}^{\tilde{\overleftarrow{\varrho}}}[\![\mathsf{S}]\!])\mathcal{E} \subseteq \neg\mathcal{P}_0$ hence $\widehat{\boldsymbol{\mathcal{S}}}^{\overleftarrow{\varrho}}[\![\mathsf{S}]\!]\mathcal{E} \subseteq \neg\mathcal{P}_0$ by definition 50.28.

Similarly $\widehat{\boldsymbol{\mathcal{S}}}^{\vec{\varrho}}[\![\mathsf{S}]\!]\,\mathcal{P}_0 \mathrel{\dot{\subseteq}} \dot{\neg}\mathcal{E}$ is equivalent to $\dot{\neg}\widehat{\boldsymbol{\mathcal{S}}}^{\vec{\varrho}}[\![\mathsf{S}]\!]\neg(\neg\mathcal{P}_0) \mathrel{\dot{\supseteq}} \mathcal{E}$ that is $\mathcal{E} \mathrel{\dot{\subseteq}} \widehat{\boldsymbol{\mathcal{S}}}^{\tilde{\vec{\varrho}}}[\![\mathsf{S}]\!]\,(\neg\mathcal{P}_0)$ where the complement dual forward reachability semantics $\widehat{\boldsymbol{\mathcal{S}}}^{\tilde{\vec{\varrho}}}[\![\mathsf{S}]\!]$ was defined in exercise 20.19.

Example 50.33 Consider the program

```
if l1: (x < 1)
   l2: y = 0;
else
   l3: y = 1;
l4:
```

with the following interval backward possible success accessibility specification

```
l4: { y:[1,oo] }
```

which is a shorthand for the following $\mathcal{P}_f$

```
l1: {x:_|_; y:_|_}
l2: {x:_|_; y:_|_}
l3: {x:_|_; y:_|_}
l4: {x:T; y:[1, oo]}
```

which holds for execution traces reaching a program point ℓ in a state satisfying $\mathcal{P}_f(\ell)$, and so the only possibility is to have the program terminate at ℓ_4 with y $\geqslant 1$.

The interval backward possible success accessibility static analysis returns

```
if l1: (x < 1) {x:[1, oo]; y:T}
   l2: {x:_|_; y:_|_} y = 0;
else
   l3: {x:T; y:T} y = 1;
l4: {x:T; y:[1, oo]}
```

because y may be strictly positive on program exit only if the alternative `y = 0;` is not entered, which requires initially that x $\geqslant 1$.

Exercise 50.34 Consider the following labeled program and the interval (termination) specification

```
l1: n = (0 - 10);               l1: {n:_|_}
while l2: (n != 0)              l2: {n:_|_}
   l3: n = (n - 1);             l3: {n:_|_}
l4:                             l4: {n:T}
```

What will be the result of the backward possible success accessibility analysis?

Exercise 50.35 Consider the following labeled program (skeleton of an iterative factorial) and the interval (termination) specification:

```
l1: i = n;                      l1: {i:_|_; n:_|_}
while l2: (i != 1)              l2: {i:_|_; n:_|_}
   l3: i = (i - 1);             l3: {i:_|_; n:_|_}
l4:                             l4: {i:T; n:T}
```

What will be the result of the backward possible success accessibility analysis?

Exercise 50.36 Consider the following labeled program and the interval specification

```
while l1: (x < 1)               l1: {x:_|_}
   {                            l2: {x:[-1, -1]}
   }
l2:
```

What will be the result of the backward possible success accessibility analysis?

Exercise 50.37 Consider the following labeled program and invariance specification $\mathscr{P}_f$

```
while l1: (n > 0)                    l1: {n:_|_}
   l2: n = (n - 1);                  l2: {n:_|_}
l3:                                  l3: {n:_|_}
```

What will be the result of the backward possible success accessibility analysis?

50.7 Calculational Design of the Structural Backward Possible Success Accessibility Semantics

The structural backward possible success accessibility semantics $\widehat{\boldsymbol{\mathcal{S}}}^{\overleftarrow{\varrho}}[\![\mathsf{S}]\!]$ of a program component S is the following.

- Backward possible success accessibility of a program $\mathsf{P} ::= \mathsf{Sl}\ \ell'$

$$\widehat{\boldsymbol{\mathcal{S}}}^{\overleftarrow{\varrho}}[\![\mathsf{P}]\!] \triangleq \widehat{\boldsymbol{\mathcal{S}}}^{\overleftarrow{\varrho}}[\![\mathsf{Sl}]\!] \tag{50.38}$$

- Backward possible success accessibility of a statement list $\mathsf{Sl} ::= \mathsf{Sl}'\ \mathsf{S}$

$$\begin{aligned}\widehat{\boldsymbol{\mathcal{S}}}^{\overleftarrow{\varrho}}[\![\mathsf{Sl}]\!]\,\mathscr{P}_f &= \widehat{\boldsymbol{\mathcal{S}}}^{\overleftarrow{\varrho}}[\![\mathsf{Sl}']\!](\ell \in \mathsf{labs}[\![\mathsf{Sl}']\!] \mapsto \\ &\qquad (\!|\, \ell = \mathsf{after}[\![\mathsf{Sl}']\!] \;?\; \mathscr{P}_f(\ell) \cup \widehat{\boldsymbol{\mathcal{S}}}^{\overleftarrow{\varrho}}[\![\mathsf{S}]\!]\,\mathscr{P}_f \;:\; \mathscr{P}_f(\ell)\,|\!))\end{aligned} \tag{50.39}$$

- Backward possible success accessibility of an empty statement list $\mathsf{Sl} ::= \epsilon$

$$\widehat{\boldsymbol{\mathcal{S}}}^{\overleftarrow{\varrho}}[\![\mathsf{Sl}]\!]\,\mathscr{P}_f = \mathscr{P}_f(\mathsf{at}[\![\mathsf{Sl}]\!]) \tag{50.40}$$

- Backward possible success accessibility of an assignment statement $\mathsf{S} ::= \mathsf{x} = \mathsf{A}\,;$

$$\widehat{\boldsymbol{\mathcal{S}}}^{\overleftarrow{\varrho}}[\![\mathsf{S}]\!]\,\mathscr{P}_f = \mathscr{P}_f(\mathsf{at}[\![\mathsf{S}]\!]) \cup \mathsf{assign}_{\overleftarrow{\varrho}}[\![\mathsf{x}, \mathsf{A}]\!]\,\mathscr{P}_f(\mathsf{after}[\![\mathsf{S}]\!]) \tag{50.41}$$

$$\mathsf{assign}_{\overleftarrow{\tau}}[\![\mathsf{x}, \mathsf{A}]\!]\,X \triangleq \{\rho \mid \rho[\mathsf{x} \leftarrow \boldsymbol{\mathcal{A}}[\![\mathsf{A}]\!]\rho] \in X\}$$

$$\mathsf{assign}_{\overleftarrow{R}}[\![\mathsf{x}, \mathsf{A}]\!]\,X \triangleq \{\langle \rho_0, \rho\rangle \mid \langle \rho_0, \rho[\mathsf{x} \leftarrow \boldsymbol{\mathcal{A}}[\![\mathsf{A}]\!]\rho]\rangle \in X\}$$

- Backward possible success accessibility of a skip statement $\mathsf{S} ::= ;$

$$\widehat{\boldsymbol{\mathcal{S}}}^{\overleftarrow{\varrho}}[\![\mathsf{S}]\!]\,\mathscr{P}_f = \mathscr{P}_f(\mathsf{at}[\![\mathsf{S}]\!]) \cup \mathscr{P}_f(\mathsf{after}[\![\mathsf{S}]\!]) \tag{50.42}$$

- Backward possible success accessibility of a conditional statement $\mathsf{S} ::= \mathtt{if}\,(\mathsf{B})\;\mathsf{S}_t$

$$\begin{aligned}\widehat{\boldsymbol{\mathcal{S}}}^{\overleftarrow{\varrho}}[\![\mathsf{S}]\!]\,\mathscr{P}_f &= \mathscr{P}_f(\mathsf{at}[\![\mathsf{S}]\!]) \cup \mathsf{test}^{\overleftarrow{\varrho}}[\![\mathsf{B}]\!](\widehat{\boldsymbol{\mathcal{S}}}^{\overleftarrow{\varrho}}[\![\mathsf{S}_t]\!]\,\mathscr{P}_f) \cup \\ &\qquad \overline{\mathsf{test}}^{\overleftarrow{\varrho}}[\![\mathsf{B}]\!]\,\mathscr{P}_f(\mathsf{after}[\![\mathsf{S}]\!])\end{aligned} \tag{50.43}$$

$$\mathsf{test}^{\overleftarrow{\tau}}[\![\mathsf{B}]\!]X \triangleq X \cap \{\rho \mid \boldsymbol{\mathcal{B}}[\![\mathsf{B}]\!]\rho = \mathsf{tt}\}$$

$$\mathsf{test}^{\overleftarrow{R}}[\![\mathsf{B}]\!]X \triangleq X \cap \{\langle \rho_0, \rho\rangle \mid \boldsymbol{\mathcal{B}}[\![\mathsf{B}]\!]\rho = \mathsf{tt}\}$$

$$\overline{\mathsf{test}}^{\overleftarrow{\tau}}[\![\mathsf{B}]\!]X \triangleq X \cap \{\rho \mid \boldsymbol{\mathcal{B}}[\![\mathsf{B}]\!]\rho = \mathsf{ff}\}$$

$$\overline{\mathsf{test}}^{\overleftarrow{R}}[\![\mathsf{B}]\!]X \triangleq X \cap \{\langle \rho_0, \rho\rangle \mid \boldsymbol{\mathcal{B}}[\![\mathsf{B}]\!]\rho = \mathsf{ff}\}$$

- Backward possible success accessibility of a conditional statement $\mathsf{S} ::= \mathtt{if}\,(\mathsf{B})\;\mathsf{S}_t\;\mathtt{else}\;\mathsf{S}_f$

$$\widehat{\boldsymbol{\mathcal{S}}}^{\overleftarrow{\varrho}}[\![\mathsf{S}]\!]\,\mathscr{P}_f \;=\; \mathscr{P}_f(\mathsf{at}[\![\mathsf{S}]\!]) \cup \mathsf{test}^{\overleftarrow{\varrho}}[\![\mathsf{B}]\!](\widehat{\boldsymbol{\mathcal{S}}}^{\overleftarrow{\varrho}}[\![\mathsf{S}_t]\!]\,\mathscr{P}_f) \cup \overline{\mathsf{test}}^{\overleftarrow{\varrho}}[\![\mathsf{B}]\!](\widehat{\boldsymbol{\mathcal{S}}}^{\overleftarrow{\varrho}}[\![\mathsf{S}_f]\!]\,\mathscr{P}_f) \tag{50.44}$$

- Backward possible success accessibility of an iteration statement $\mathsf{S} ::= \mathtt{while}\,\ell\,(\mathsf{B})\,\mathsf{S}_b$

$$\widehat{\boldsymbol{\mathcal{S}}}^{\overleftarrow{\varrho}}[\![\mathsf{S}]\!]\,\mathscr{P}_f \;\triangleq\; \mathsf{lfp}^{\dot{\subseteq}}\,\boldsymbol{\mathscr{F}}^{\overleftarrow{\varrho}}[\![\mathtt{while}\,\ell\,(\mathsf{B})\,\mathsf{S}_b]\!]\,\mathscr{P}_f \tag{50.45}$$

$$\boldsymbol{\mathscr{F}}^{\overleftarrow{\varrho}}[\![\mathtt{while}\,\ell\,(\mathsf{B})\,\mathsf{S}_b]\!]\,\mathscr{P}_f\,Y \;\triangleq\; \mathscr{P}_f(\ell) \cup \overline{\mathsf{test}}^{\overleftarrow{\varrho}}[\![\mathsf{B}]\!](\mathscr{P}_f(\mathsf{after}[\![\mathsf{S}]\!])) \cup \tag{50.46}$$
$$\mathsf{test}^{\overleftarrow{\varrho}}[\![\mathsf{B}]\!](\widehat{\boldsymbol{\mathcal{S}}}^{\overleftarrow{\varrho}}[\![\mathsf{S}_b]\!]\,(\ell'' \in \mathsf{labs}[\![\mathsf{S}_b]\!] \mapsto \llparenthesis\, \ell'' = \ell \mathrel{?} \mathscr{P}_f(\ell) \cup Y \parallel \ell'' \in \mathsf{breaks\text{-}of}[\![\mathsf{S}_b]\!] \mathrel{?} \mathscr{P}_f(\mathsf{after}[\![\mathsf{S}]\!]) \mathrel{:} \mathscr{P}_f(\ell'')\,\rrparenthesis))$$

- Backward possible success accessibility of a break statement $\mathsf{S} ::= \ell\ \mathtt{break}\,;$

$$\widehat{\boldsymbol{\mathcal{S}}}^{\overleftarrow{\varrho}}[\![\mathsf{S}]\!]\,\mathscr{P}_f \;=\; \mathscr{P}_f(\mathsf{at}[\![\mathsf{S}]\!]) \cup \mathscr{P}_f(\mathsf{break\text{-}to}[\![\mathsf{S}]\!]) \tag{50.47}$$

- Backward possible success accessibility of a compound statement $\mathsf{S} ::= \{\,\mathsf{Sl}\,\}$

$$\widehat{\boldsymbol{\mathcal{S}}}^{\overleftarrow{\varrho}}[\![\mathsf{S}]\!] \;=\; \widehat{\boldsymbol{\mathcal{S}}}^{\overleftarrow{\varrho}}[\![\mathsf{Sl}]\!] \tag{50.48}$$

Proof of (50.45)

$\widehat{\boldsymbol{\mathcal{S}}}^{\overleftarrow{\varrho}}[\![\mathsf{S}]\!]\,\mathscr{P}_f$

$\triangleq \dot{\alpha}[\![\mathsf{S}]\!](\widehat{\boldsymbol{\mathcal{S}}}^{\overleftarrow{\tilde{\varrho}}}[\![\mathsf{S}]\!])$ ⦃definition 50.28⦄

$= \neg\widehat{\boldsymbol{\mathcal{S}}}^{\overleftarrow{\tilde{\varrho}}}[\![\mathsf{S}]\!]\,(\dot{\neg}\mathscr{P}_f)$ ⦃definition of $\dot{\alpha}[\![\mathsf{S}]\!]$⦄

$= \neg(\mathsf{gfp}^{\dot{\subseteq}}\,Y \mapsto \boldsymbol{\mathscr{F}}^{\overleftarrow{\tilde{\varrho}}}[\![\mathtt{while}\,\ell\,(\mathsf{B})\,\mathsf{S}_b]\!]\,(\dot{\neg}\mathscr{P}_f)\,Y)$ ⦃(50.14)⦄

$= (\mathsf{lfp}^{\dot{\subseteq}}\,Y \mapsto \neg\boldsymbol{\mathscr{F}}^{\overleftarrow{\tilde{\varrho}}}[\![\mathtt{while}\,\ell\,(\mathsf{B})\,\mathsf{S}_b]\!]\,(\dot{\neg}\mathscr{P}_f)\,(\neg Y))$

⦃David Park's conjugate fixpoint theorem 15.33⦄

$= (\mathsf{lfp}^{\dot{\subseteq}}\,\boldsymbol{\mathscr{F}}^{\overleftarrow{\varrho}}[\![\mathtt{while}\,\ell\,(\mathsf{B})\,\mathsf{S}_b]\!]\,\mathscr{P}_f)$

by defining

$\boldsymbol{\mathscr{F}}^{\overleftarrow{\varrho}}[\![\mathtt{while}\,\ell\,(\mathsf{B})\,\mathsf{S}_b]\!]\,\mathscr{P}_f\,Y$

$\triangleq \neg\boldsymbol{\mathscr{F}}^{\overleftarrow{\tilde{\varrho}}}[\![\mathtt{while}\,\ell\,(\mathsf{B})\,\mathsf{S}_b]\!]\,(\dot{\neg}\mathscr{P}_f)\,(\neg Y)$

$= \neg((\dot{\neg}\mathscr{P}_f)(\ell) \cap \overline{\mathsf{test}}^{\overleftarrow{\tilde{\varrho}}}[\![\mathsf{B}]\!]((\dot{\neg}\mathscr{P}_f)(\mathsf{after}[\![\mathsf{S}]\!])) \cap \mathsf{test}^{\overleftarrow{\tilde{\varrho}}}[\![\mathsf{B}]\!](\widehat{\boldsymbol{\mathcal{S}}}^{\overleftarrow{\tilde{\varrho}}}[\![\mathsf{S}_b]\!]\,(\ell'' \in \mathsf{labs}[\![\mathsf{S}_b]\!] \mapsto \llparenthesis\, \ell'' = \ell \mathrel{?}$
$(\dot{\neg}\mathscr{P}_f)(\ell) \cap (\neg Y) \parallel \ell'' \in \mathsf{breaks\text{-}of}[\![\mathsf{S}_b]\!] \mathrel{?} (\dot{\neg}\mathscr{P}_f)(\mathsf{after}[\![\mathsf{S}]\!]) \mathrel{:} (\dot{\neg}\mathscr{P}_f)(\ell'')\,\rrparenthesis)))$ ⦃(50.14)⦄

$= \neg((\neg\mathscr{P}_f(\ell)) \cap \overline{\mathsf{test}}^{\overleftarrow{\tilde{\varrho}}}[\![\mathsf{B}]\!](\neg\mathscr{P}_f(\mathsf{after}[\![\mathsf{S}]\!])) \cap \mathsf{test}^{\overleftarrow{\tilde{\varrho}}}[\![\mathsf{B}]\!](\widehat{\boldsymbol{\mathcal{S}}}^{\overleftarrow{\tilde{\varrho}}}[\![\mathsf{S}_b]\!]\,(\ell'' \in \mathsf{labs}[\![\mathsf{S}_b]\!] \mapsto \llparenthesis\, \ell'' = \ell \mathrel{?}$
$(\neg\mathscr{P}_f(\ell)) \cap (\neg Y) \parallel \ell'' \in \mathsf{breaks\text{-}of}[\![\mathsf{S}_b]\!] \mathrel{?} \neg\mathscr{P}_f(\mathsf{after}[\![\mathsf{S}]\!]) \mathrel{:} \neg\mathscr{P}_f(\ell''))\,\rrparenthesis))$

⦃pointwise definition of $\dot{\neg}$⦄

$= \mathscr{P}_f(\ell) \cup \neg\overline{\mathsf{test}}^{\overleftarrow{\tilde{\varrho}}}[\![\mathsf{B}]\!](\neg\mathscr{P}_f(\mathsf{after}[\![\mathsf{S}]\!])) \cup \neg\mathsf{test}^{\overleftarrow{\tilde{\varrho}}}[\![\mathsf{B}]\!]\neg(\neg\widehat{\boldsymbol{\mathcal{S}}}^{\overleftarrow{\tilde{\varrho}}}[\![\mathsf{S}_b]\!]\,\neg(\ell'' \in \mathsf{labs}[\![\mathsf{S}_b]\!] \mapsto \llparenthesis\, \ell'' = \ell \mathrel{?}$
$\mathscr{P}_f(\ell)) \cup Y \parallel \ell'' \in \mathsf{breaks\text{-}of}[\![\mathsf{S}_b]\!] \mathrel{?} \mathscr{P}_f(\mathsf{after}[\![\mathsf{S}]\!]) \mathrel{:} \mathscr{P}_f(\ell''))\,\rrparenthesis)$

⦃De Morgan's laws of section 2.1.3⦄

$= \mathscr{P}_f(\ell) \cup \overline{\mathsf{test}}^{\overleftarrow{\varrho}}[\![\mathsf{B}]\!](\mathscr{P}_f(\mathsf{after}[\![\mathsf{S}]\!])) \cup \mathsf{test}^{\overleftarrow{\varrho}}[\![\mathsf{B}]\!](\neg\widehat{\boldsymbol{\mathcal{S}}}^{\overleftarrow{\tilde{\varrho}}}[\![\mathsf{S}_b]\!]\,\neg(\ell'' \in \mathsf{labs}[\![\mathsf{S}_b]\!] \mapsto \llparenthesis\, \ell'' = \ell \mathrel{?}$
$\mathscr{P}_f(\ell)) \cup Y \parallel \ell'' \in \mathsf{breaks\text{-}of}[\![\mathsf{S}_b]\!] \mathrel{?} \mathscr{P}_f(\mathsf{after}[\![\mathsf{S}]\!]) \mathrel{:} \mathscr{P}_f(\ell''))\,\rrparenthesis)$

⦃by defining $\mathsf{test}^{\overleftarrow{\varrho}}[\![\mathsf{B}]\!]\,X \triangleq \neg\mathsf{test}^{\overleftarrow{\tilde{\varrho}}}[\![\mathsf{B}]\!](\dot{\neg}X)$ and $\overline{\mathsf{test}}^{\overleftarrow{\varrho}}[\![\mathsf{B}]\!]\,X \triangleq \neg\overline{\mathsf{test}}^{\overleftarrow{\tilde{\varrho}}}[\![\mathsf{B}]\!](\dot{\neg}X)$⦄

$$= \mathscr{P}_f(\ell) \cup \overline{\mathsf{test}}^{\overleftarrow{\varrho}}[\![\mathsf{B}]\!](\mathscr{P}_f(\mathsf{after}[\![\mathsf{S}]\!])) \cup \mathsf{test}^{\overleftarrow{\varrho}}[\![\mathsf{B}]\!](\widehat{\boldsymbol{\mathcal{S}}}^{\overleftarrow{\varrho}}[\![\mathsf{S}_b]\!]\ (\ell'' \in \mathsf{labs}[\![\mathsf{S}_b]\!] \mapsto (\!|\ \ell'' = \ell\ ?\ \mathscr{P}_f(\ell)) \cup Y \parallel \ell'' \in \mathsf{breaks\text{-}of}[\![\mathsf{S}_b]\!]\ ?\ \mathscr{P}_f(\mathsf{after}[\![\mathsf{S}]\!]) : \mathscr{P}_f(\ell''))\ |\!)) \qquad \text{(by structural induction hypothesis)}$$

For the tests in (50.43), we have

$$\begin{aligned}
& \mathsf{test}^{\overleftarrow{r}}[\![\mathsf{B}]\!]X \\
\triangleq\ & \neg\mathsf{test}^{\widetilde{\overleftarrow{r}}}[\![\mathsf{B}]\!](\dot{\neg}X) \\
=\ & \neg\{\rho \mid (\mathscr{B}[\![\mathsf{B}]\!]\rho = \mathsf{tt}) \Rightarrow (\rho \in (\dot{\neg}X))\} && \text{((50.12))} \\
=\ & \{\rho \mid (\mathscr{B}[\![\mathsf{B}]\!]\rho = \mathsf{tt}) \wedge \rho \in X\} && \text{(}\neg\{x \mid P(x)\} = \{x \mid \neg P(x)\}\text{)} \\
=\ & X \cap \{\rho \mid \mathscr{B}[\![\mathsf{B}]\!]\rho = \mathsf{tt}\} && \text{(definition of } \cap\text{)}
\end{aligned}$$

The other cases are similar. □

Observe that any superset of $\widehat{\boldsymbol{\mathcal{S}}}^{\widetilde{\overleftarrow{\varrho}}}[\![\mathsf{S}]\!]$ guarantees a possible success. So a sound possible success static analysis should use overapproximations (similar to the reachability semantics $\widehat{\boldsymbol{\mathcal{S}}}^{\overleftarrow{\varrho}}[\![\mathsf{S}]\!]$).

Exercise 50.49 Prove (50.38), (50.39), (50.40), (50.41), (50.42), (50.43), (50.44), (50.47), and (50.48).

Exercise 50.50 Relate $\mathsf{post}^{\vec{\varrho}}$, $\widetilde{\mathsf{pre}}^{\vec{\varrho}}$, $\mathsf{pre}^{\vec{\varrho}}$, and $\widetilde{\mathsf{post}}^{\vec{\varrho}}$ by Galois connections.

50.8 Impossible Failure Versus Possible Success Backward Accessibility Semantics

Let $\mathscr{P}_f$ be a condition on accessible states. Any execution trace starting with a state of the impossible failure backward accessibility semantics $\widehat{\boldsymbol{\mathcal{S}}}^{\widetilde{\overleftarrow{\varrho}}}[\![\mathsf{S}]\!]\ \mathscr{P}_f$ must reach the accessible states in $\mathscr{P}_f$. Therefore $\widehat{\boldsymbol{\mathcal{S}}}^{\widetilde{\overleftarrow{\varrho}}}[\![\mathsf{S}]\!]\ \mathscr{P}_f$ is a sufficient condition for initial states to reach the accessible states (but it may not be necessary as shown by trace π_2 in figure 50.51).

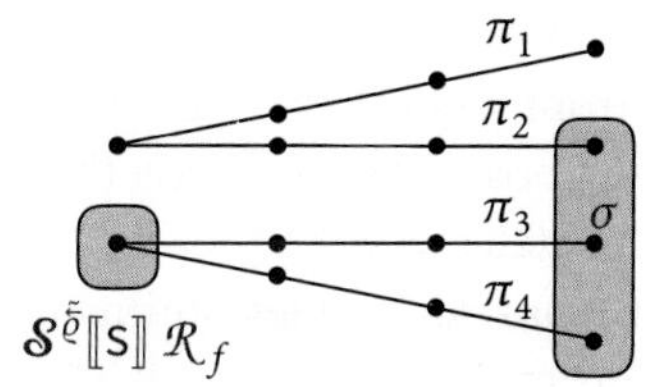

Figure 50.51
Impossible failure

On the contrary, some execution traces starting with a state of the possible success backward accessibility semantics $\widehat{\boldsymbol{\mathcal{S}}}^{\overleftarrow{\varrho}}[\![\mathsf{S}]\!]\ \mathscr{P}_f$ may reach the accessible states, not necessarily all of these traces. So $\widehat{\boldsymbol{\mathcal{S}}}^{\overleftarrow{\varrho}}[\![\mathsf{S}]\!]\ \mathscr{P}_f$ is a necessary condition for initial states to reach the accessible states, but maybe not sufficient (as shown by trace π_1 in figure 50.52).

Notice, however, that in both cases, there is no guarantee that a specific state satisfying $\mathscr{P}_f$ will ever be reached (as shown by state σ in figures 50.51 and 50.52, and except if $\mathscr{P}_f$ holds only for that state).

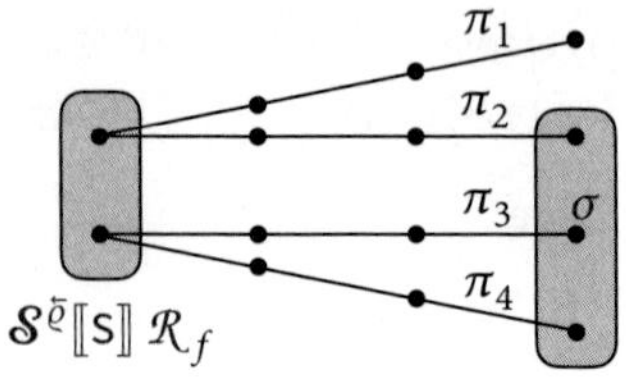

Figure 50.52
Possible success

Exercise 50.53 Are there semantics $\boldsymbol{\mathcal{S}}$ such that $\boldsymbol{\mathcal{S}}^{\tilde{\overleftarrow{\varrho}}}(\mathscr{P}_f) \nsubseteq \boldsymbol{\mathcal{S}}^{\overleftarrow{\varrho}}(\mathscr{P}_f)$?

Example 50.54 Consider an iteration S with reachability specification that $Q \in \wp(\mathbb{E}\mathrm{v})$ should hold on program exit $\mathsf{after}[\![\mathsf{S}]\!]$. Assume that S has the following possible executions (where the possible initial states are $\{a, b, c, d, e, f, g\}$ and the final states are marked $\mathsf{after}[\![\mathsf{S}]\!]$).

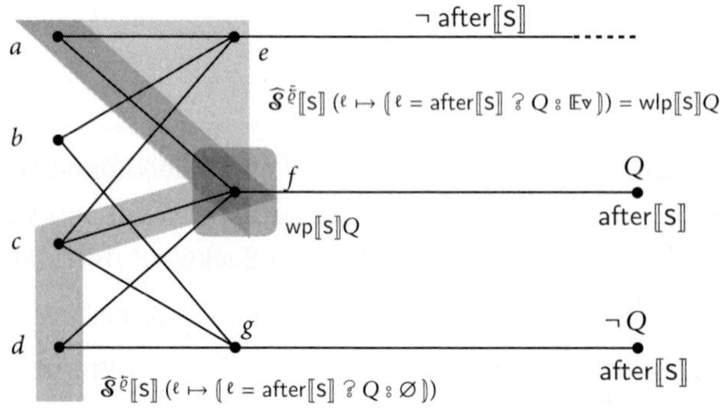

The executions starting from states in $\widehat{\boldsymbol{\mathcal{S}}}^{\tilde{\overleftarrow{\varrho}}}[\![\mathsf{S}]\!]\,(\ell \mapsto (\!|\, \ell = \mathsf{after}[\![\mathsf{S}]\!] \;?\; Q \;:\; \mathbb{E}\mathrm{v} \,|\!)) = \mathsf{wlp}[\![\mathsf{S}]\!]Q = \{a, e, f\}$ will either nonterminate or terminate with Q true. This includes the state e from which it is only possible to never terminate.

The executions starting from states in $\neg(\widehat{\boldsymbol{\mathcal{S}}}^{\tilde{\overleftarrow{\varrho}}}[\![\mathsf{S}]\!]\,(\ell \mapsto (\!|\, \ell = \mathsf{after}[\![\mathsf{S}]\!] \;?\; Q \;:\; \mathbb{E}\mathrm{v} \,|\!))) = \neg\mathsf{wlp}[\![\mathsf{S}]\!]Q = \{b, c, d, g\}$ all have a possibility of failing (but without guarantee).

The executions starting from states in $\widehat{\boldsymbol{\mathcal{S}}}^{\overleftarrow{\varrho}}[\![\mathsf{S}]\!]\,(\ell \mapsto (\!|\, \ell = \mathsf{after}[\![\mathsf{S}]\!] \;?\; Q \;:\; \varnothing \,|\!)) = \{a, f, c, d\}$ have a possibility of success. Because of nondeterminism, this is not a certainty.

The executions starting from initial states in $\neg(\widehat{\boldsymbol{\mathcal{S}}}^{\overleftarrow{\varrho}}[\![\mathsf{S}]\!]\,(\ell \mapsto (\!|\, \ell = \mathsf{after}[\![\mathsf{S}]\!] \;?\; Q \;:\; \varnothing \,|\!))) = \{e, b, g\}$ have a guarantee of either non terminating or failing [265, 271, 694].

The executions starting from states in $(\widehat{\boldsymbol{\mathcal{S}}}^{\tilde{\overleftarrow{\varrho}}}[\![\mathsf{S}]\!] \,\dot{\cap}\, \widehat{\boldsymbol{\mathcal{S}}}^{\overleftarrow{\varrho}}[\![\mathsf{S}]\!])\,(\ell \mapsto (\!|\, \ell = \mathsf{after}[\![\mathsf{S}]\!] \;?\; Q \;:\; \varnothing \,|\!)) = \{a, f\}$ all have a possibility of success but none of failure, without excluding the possibility of nontermination (so that is not $\mathsf{wp}[\![\mathsf{S}]\!]Q = \{f\}$).

Exercise 50.55 Show that if $\boldsymbol{\mathcal{S}}[\![\mathsf{S}]\!](\pi\mathsf{at}[\![\mathsf{S}]\!])$ is a singleton (i.e., the semantics is not empty and deterministic) then $\boldsymbol{\mathcal{S}}^{\tilde{\overleftarrow{\varrho}}}[\![\mathsf{S}]\!]\,(\mathscr{P}_f) = \boldsymbol{\mathcal{S}}^{\overleftarrow{\varrho}}[\![\mathsf{S}]\!]\,(\mathscr{P}_f)$.

50.9 Conclusion

The backward possible success accessibility semantics $\widehat{\boldsymbol{\mathcal{S}}}^{\overleftarrow{\varrho}}[\![\mathsf{S}]\!]$ of a program component S is from [221, 222, 241] in which it was derived from the forward reachability semantics by inversion of the transition system i.e. $\widehat{\boldsymbol{\mathcal{S}}}^{\overleftarrow{\varrho}}[\![\mathsf{S}]\!]\,\mathcal{P}_f = \mathsf{pre}^{\overleftarrow{\varrho}}(\widehat{\boldsymbol{\mathcal{S}}}^{\tau}[\![\mathsf{S}]\!]^*)\,\mathcal{P}_f = \mathsf{post}^{\vec{\varrho}}(((\widehat{\boldsymbol{\mathcal{S}}}^{\tau}[\![\mathsf{S}]\!])^{-1})^*)\,\mathcal{P}_f$. In this chapter, the calculational design of the structural backward impossible failure accessibility semantics $\widehat{\boldsymbol{\mathcal{S}}}^{\tilde{\overleftarrow{\varrho}}}[\![\mathsf{S}]\!]\,\mathcal{P}_f$ was by direct inversion of the structural forward reachability semantics $\widehat{\boldsymbol{\mathcal{S}}}^{\vec{\varrho}}[\![\mathsf{S}]\!]$, and the backward possible success accessibility semantics $\widehat{\boldsymbol{\mathcal{S}}}^{\overleftarrow{\varrho}}[\![\mathsf{S}]\!]\,\mathcal{P}_f$ followed by complement duality. The two backward semantics can be applied to other languages (e.g., [513]) and conjuncted (e.g., [71]). [577] compares the expressive power of forward and backward static analyses for logic programs.

The same way that the forward reachability semantics of chapter 19 was the basis for a forward abstract interpreter (chapter 21) leading to forward analyses and forward invariance proof methods (chapter 25), the backward accessibility semantics of this chapter 50 can be further abstracted to get a backward abstract interpreter ([dual] adjoint to section 21.2), which is the basis for invariance proof methods (such as subgoal induction [721]; see [243, 427]) and automatic static program analysis methods (such as tracing imprecise analysis values back to origin of the imprecision [623]).

Although corollary 50.2 and theorem 50.23 express equivalences between semantics, this may no longer be true in the abstract. Forward and backward analyses are in general different (e.g., [700]) and can be combined as shown in the next chapter 51. These analyses can be used to check temporal properties by abstract interpretation; see [256, 665–667] and the next chapter 51.

[550] is an example of backward analysis in which verification conditions are expressed using Horn clauses [511]. Backward analyses are essential for inferring and proving necessary conditions for program termination, absence of errors, and sufficient correctness conditions [230, 263, 940, 943]; existential temporal properties [217, 942, 944, 948]; tracing alarms to their origin [311, 369]; and responsibility analysis [310, 311].

Exercise 50.56 (Project: backward abstract interpreter) Implement a backward abstract interpreter for possible success. Instantiate it for the intervals of chapter 33. Optionally, parameterize the abstract interpreter by a choice among impossible failure, possible success, or their meet.

50.10 Solutions to Selected Exercises

Solution to exercise 50.6

$$\widehat{\boldsymbol{\mathcal{S}}}^{\vec{\mathsf{R}}}[\![\mathsf{S}]\!]\,\mathcal{P}_0 \mathrel{\dot{\subseteq}} \mathcal{P}_f$$

$$\Leftrightarrow \mathsf{post}^{\vec{\varrho}}[\![\boldsymbol{\mathcal{S}}^*[\![\mathsf{S}]\!]]\!]\,\mathcal{P}_0 \mathrel{\dot{\subseteq}} \mathcal{P}_f \qquad \text{(definition (20.14) of } \widehat{\boldsymbol{\mathcal{S}}}^{\vec{\mathsf{R}}}[\![\mathsf{S}]\!]\text{)}$$

$$\Leftrightarrow \forall \ell \in \mathsf{labs}[\![\mathsf{S}]\!]\,.\,\mathsf{post}^{\vec{\varrho}}[\![\boldsymbol{\mathcal{S}}^*[\![\mathsf{S}]\!]]\!]\,\mathcal{P}_0\,\ell \subseteq \mathcal{P}_f(\ell) \qquad \text{(pointwise definition of } \dot{\subseteq}\text{)}$$

$$\Leftrightarrow \forall \ell \in \mathsf{labs}[\![\mathsf{S}]\!]\,.\,\{\langle \rho_0,\, \boldsymbol{\varrho}(\pi_0\ell_0\pi_1\ell')\rangle \mid \langle \rho_0,\, \boldsymbol{\varrho}(\pi_0\ell_0)\rangle \in \mathcal{P}_0 \wedge \ell_0\pi_1\ell' \in \boldsymbol{\mathcal{S}}^*[\![\mathsf{S}]\!]\,(\pi_0\ell_0) \wedge \ell' = \ell\} \subseteq \mathcal{P}_f(\ell) \qquad \text{(definition (20.9) of } \mathsf{post}^{\vec{\varrho}}\text{)}$$

$\Leftrightarrow \forall \ell \in \mathsf{labs}[\![\mathsf{S}]\!] \,.\, \forall \langle \rho_0, \varrho(\pi_0 \ell_0)\rangle \in \mathscr{P}_0 \,.\, (\ell_0 \pi_1{}^{\ell} \in \boldsymbol{\mathcal{S}}^* [\![\mathsf{S}]\!]\, (\pi_0 \ell_0)) \Rightarrow \langle \rho_0, \varrho(\pi_0 \ell_0 \pi_1 \ell)\rangle \in \mathscr{P}_f(\ell)$
(definition of $\subseteq$)

$\Leftrightarrow \forall \langle \rho_0, \varrho(\pi_0 \ell_0)\rangle \in \mathscr{P}_0 \,.\, \forall \ell \in \mathsf{labs}[\![\mathsf{S}]\!] \,.\, (\ell_0 \pi_1{}^{\ell} \in \boldsymbol{\mathcal{S}}^* [\![\mathsf{S}]\!]\, (\pi_0 \ell_0)) \Rightarrow \langle \rho_0, \varrho(\pi_0 \ell_0 \pi_1 \ell)\rangle \in \mathscr{P}_f(\ell)$
(definition of $\forall$)

$\Leftrightarrow \mathscr{P}_0 \subseteq \{\langle \rho_0, \varrho(\pi_0 \ell_0)\rangle \mid \forall \ell \in \mathsf{labs}[\![\mathsf{S}]\!] \,.\, (\ell_0 \pi_1{}^{\ell} \in \boldsymbol{\mathcal{S}}^* [\![\mathsf{S}]\!]\, (\pi_0 \ell_0)) \Rightarrow \langle \rho_0, \varrho(\pi_0 \ell_0 \pi_1 \ell)\rangle \in \mathscr{P}_f(\ell)\}$
(definition of $\subseteq$)

$\Rightarrow \widehat{\boldsymbol{\mathcal{S}}}^{\widetilde{\widetilde{\mathsf{R}}}} [\![\mathsf{S}]\!]\, \mathscr{P}_f = \{\langle \rho_0, \varrho(\pi_0 \ell_0)\rangle \mid \forall \ell \in \mathsf{labs}[\![\mathsf{S}]\!] \,.\, (\ell_0 \pi_1{}^{\ell} \in \boldsymbol{\mathcal{S}}^* [\![\mathsf{S}]\!]\, (\pi_0 \ell_0)) \Rightarrow \langle \rho_0, \varrho(\pi_0 \ell_0 \pi_1 \ell)\rangle \in \mathscr{P}_f(\ell)\}$ (by unicity of the adjoint in the Galois connection of corollary 50.2)

Solution to exercise 50.34

```
l1: {n:_|_} n = (0 - 10);
while l2: (n != 0) {n:[0, oo]}
   l3: {n:[1, oo]} n = (n - 1);
l4: {n:[0, oo]}
```

The specification only requires the program to terminate without any constraint on the final value of n. This is only possible if $\mathsf{n} \geqslant 0$ within and on entrance of the loop. But n is strictly negative at ℓ_2. So the program will never terminate. This is shown by $\bot$ at ℓ_1 i.e. the only way for the program to terminate is to never execute it!

Solution to exercise 50.35

```
l1: {i:T; n:[1, oo]} i = n;
while l2: (i != 1) {i:[1, oo]; n:T}
   l3: {i:[2, oo]; n:T} i = (i - 1);
l4: {i:T; n:T}
```

The specification requires only that the program terminate without any constraint on the final values of variables. A necessary condition is discovered stating that, initially, n is positive (see exercise 33.10 showing that $\mathsf{n} \leqslant 0$ is sufficient for nontermination, which is not equivalent because the backward analysis infers the precondition whereas the forward analysis has only checked it [277]).

Such backward analyses are useful in termination proofs, e.g. [230] and the analysis of existential temporal properties [938, 945, 948].

Solution to exercise 50.36 The program cannot terminate with $\mathsf{x} = -1$, as shown by the backward analysis

```
while l1: (x < 1) {x:_|_}
   {
   }
l2: {x:[-1, -1]}
```

51 Reduced Forward—Backward Analysis

We study various combinations of forward/deductive/reachability and backward/abductive/accessibility static analyses. The forward reachability semantics $\widehat{\boldsymbol{\mathcal{S}}}^{\vec{\varrho}}[\![\mathsf{S}]\!]$ that we consider in this chapter 51 has been postulated in chapter 19, “Structural Forward Reachability Semantics,” and proved sound and complete in chapter 20, “Calculational Design of the Forward Reachability Semantics.” The backward accessibility semantics $\widehat{\boldsymbol{\mathcal{S}}}^{\overleftarrow{\varrho}}[\![\mathsf{S}]\!]$ that we consider in this chapter is the possible success accessibility semantics of section 50.6.

Contents

51.1 Backward/Abductive—Forward/Deductive Extremal Static Analysis

Take as given an initial precondition $\mathscr{P}_0 \in \wp(\mathbb{E}\mathbb{v}^{\varrho})$ and an invariance specification $\mathscr{P}_f \in \mathsf{labs}[\![\mathsf{S}]\!] \rightarrow \wp(\mathbb{E}\mathbb{v}^{\varrho})$. For example, $\mathscr{P}_f = \ell \mapsto (\![\ \ell = \mathsf{after}[\![\mathsf{S}]\!]\ \text{?}\ \mathbb{E}\mathbb{v}^{\varrho} \mathbin{\text{:}} \varnothing\ ]\!)$ requires termination of program component S.

Recall that $\widehat{\boldsymbol{\mathcal{S}}}^{\vec{\varrho}}[\![\mathsf{S}]\!]\,\mathscr{P}_0$ is the set of local states potentially reachable from the initial states $\mathscr{P}_0$ ((20.6)-lemma 50.1/(20.14)) and that $\widehat{\boldsymbol{\mathcal{S}}}^{\overleftarrow{\varrho}}[\![\mathsf{S}]\!]\mathscr{P}_f$ is the set of initial states that may reach final states $\mathscr{P}_f$ (theorem 50.30).

Assume that we want to restrict the precondition $\mathscr{P}_0$ to those states $\mathscr{P}_0 \cap \widehat{\boldsymbol{\mathcal{S}}}^{\overleftarrow{\varrho}}[\![\mathsf{S}]\!]\mathscr{P}_f$ that may reach states $\mathscr{P}_f$ and to restrict the postcondition $\mathscr{P}_f$ to those states $\widehat{\boldsymbol{\mathcal{S}}}^{\vec{\varrho}}[\![\mathsf{S}]\!]\mathscr{P}_0 \mathbin{\dot{\cap}} \mathscr{P}_f$ that are reachable from the initial precondition $\mathscr{P}_0$.

Example 51.1 Consider the following labeled program and the interval specification $\mathscr{P}_f$ stating that the program execution must reach l4 i.e. terminate.

```
l1: i = n;                      l1: {i:_|_; n:_|_}
while l2: (i != 1)              l2: {i:_|_; n:_|_}
   l3: i = (i - 1);             l3: {i:_|_; n:_|_}
l4:                             l4: {i:T; n:T}
```

The interval backward/forward static analysis (where $\mathscr{P}_0$ is true)

```
<{i:T; n:[1, oo]}, [l1: {i:_|_; n:_|_}; l2: {i:_|_; n:_|_};
 l3: {i:_|_; n:_|_}; l4: {i:[1, 1]; n:[1, oo]}]>
```

infers that the initial condition $\mathsf{n} \geqslant 1$ is necessary for the program to terminate, in which case the invariance specification $\mathscr{P}_f$ can be refined with $\mathsf{n} \geqslant 1$ and $\mathsf{i} = 1$ upon termination, if ever, at ℓ_4. Of course, the initial condition may not be sufficient for termination because of overapproximations.

Hence we are interested in $\langle \mathscr{P}_0, \mathscr{P}_f \rangle \mapsto \langle \mathscr{P}_0 \cap \widehat{\boldsymbol{\mathcal{S}}}^{\overleftarrow{\varrho}}[\![\mathsf{S}]\!]\mathscr{P}_f, \widehat{\boldsymbol{\mathcal{S}}}^{\vec{\varrho}}[\![\mathsf{S}]\!]\mathscr{P}_0 \mathbin{\dot{\cap}} \mathscr{P}_f \rangle$ which is a lower closure operator for the pairwise inclusion order $\subseteq_2$.

The corresponding abstract operator is increasing and reductive but not necessarily idempotent. By the dual of corollary 29.3, a more precise static analysis is obtained by iterated reduction. We say that this reduction is *extremal* because it involves a reduction only at initial and final states (as opposed to the more precise *intermediate* reduction of section 51.4 involving a reduction at all initial, intermediate, and final states).

51.1.1 Which Initial States Can Potentially Reach Which Final States Satisfying an Invariance Specification?

Consider the transformer $\mathfrak{R}^{\varrho}_{0f}$ that restricts the initial states to those that may reach final states and the final states to those that are reachable from these initial states.

$$\mathfrak{R}^{\varrho}_{0f} \in (\wp(\mathbb{E}\mathrm{v}^{\varrho}) \times \mathsf{labs}[\![\mathsf{S}]\!] \to \wp(\mathbb{E}\mathrm{v}^{\varrho})) \to (\wp(\mathbb{E}\mathrm{v}^{\varrho}) \times \mathsf{labs}[\![\mathsf{S}]\!] \to \wp(\mathbb{E}\mathrm{v}^{\varrho})) \quad (51.2)$$

$$\mathfrak{R}^{\varrho}_{0f}\langle \mathscr{P}_0, \mathscr{P}_f \rangle \triangleq \langle \mathscr{P}_0 \cap \widehat{\boldsymbol{\mathcal{S}}}^{\overleftarrow{\varrho}}[\![\mathsf{S}]\!]\mathscr{P}_f, \widehat{\boldsymbol{\mathcal{S}}}^{\vec{\varrho}}[\![\mathsf{S}]\!]\mathscr{P}_0 \mathbin{\dot{\cap}} \mathscr{P}_f \rangle$$

where $\langle \wp(\mathbb{E}\mathrm{v}^{\varrho}), \subseteq \rangle$ is ordered by inclusion, $\langle \mathsf{labs}[\![\mathsf{S}]\!] \to \wp(\mathbb{E}\mathrm{v}^{\varrho}), \dot{\subseteq} \rangle$ is ordered by pointwise inclusion, and $\langle \wp(\mathbb{E}\mathrm{v}^{\varrho}) \times \mathsf{labs}[\![\mathsf{S}]\!] \to \wp(\mathbb{E}\mathrm{v}^{\varrho}), \subseteq_2 \rangle$ is ordered componentwise.

Theorem 51.3 $\mathfrak{R}^{\varrho}_{0f}$ is a lower closure operator on $\langle \wp(\mathbb{E}\mathrm{v}^{\varrho}) \times \mathsf{labs}[\![\mathsf{S}]\!] \to \wp(\mathbb{E}\mathrm{v}^{\varrho}), \subseteq_2 \rangle$.

Proof of theorem 51.3 $\mathfrak{R}^{\vec{r}}_{0f}$ is increasing because the identity and the intersection are increasing and by corollary 50.2 with lemma 11.33.

$\mathfrak{R}^{\vec{r}}_{0f}$ is reductive by definition of intersection.

For idempotence, let us consider $\langle \mathscr{P}'_0, \mathscr{P}'_f \rangle = \mathfrak{R}^{\vec{r}}_{0f}\langle \mathscr{P}_0, \mathscr{P}_f \rangle$ and $\langle \mathscr{P}''_0, \mathscr{P}''_f \rangle = \mathfrak{R}^{\vec{r}}_{0f}\langle \mathscr{P}'_0, \mathscr{P}'_f \rangle$, and prove that $\langle \mathscr{P}''_0, \mathscr{P}''_f \rangle = \langle \mathscr{P}'_0, \mathscr{P}'_f \rangle$. For the first component, we have

$\mathscr{P}''_0$

$= \mathscr{P}_0 \cap \widehat{\boldsymbol{\mathcal{S}}}^{\overleftarrow{r}}[\![\mathsf{S}]\!](\widehat{\boldsymbol{\mathcal{S}}}^{\vec{r}}[\![\mathsf{S}]\!]\,\mathscr{P}_0 \mathbin{\dot{\cap}} \mathscr{P}_f)$ (definition of $\mathscr{P}''_0$)

$= \mathscr{P}_0 \cap \{\varrho(\pi_0\ell_0) \in \mathbb{E}\mathrm{v} \mid \exists \ell \in \mathsf{labs}[\![\mathsf{S}]\!] \,.\, \ell_0\pi_1\ell \in \boldsymbol{\mathcal{S}}^*[\![\mathsf{S}]\!]\,(\pi_0\ell_0) \wedge \varrho(\pi_0\ell_0\pi_1\ell) \in (\widehat{\boldsymbol{\mathcal{S}}}^{\vec{r}}[\![\mathsf{S}]\!]\,\mathscr{P}_0 \mathbin{\dot{\cap}}$ $\mathscr{P}_f)(\ell)\}$ (theorem 50.30)

$= \mathscr{P}_0 \cap \{\varrho(\pi_0\ell_0) \in \mathbb{E}\mathrm{v} \mid \exists \ell \in \mathsf{labs}[\![\mathsf{S}]\!] \,.\, \ell_0\pi_1\ell \in \boldsymbol{\mathcal{S}}^*[\![\mathsf{S}]\!]\,(\pi_0\ell_0) \wedge \varrho(\pi_0\ell_0\pi_1\ell) \in (\{\varrho(\pi_0\ell_0\pi_1\ell') \mid$ $\varrho(\pi_0\ell_0) \in \mathscr{P}_0 \wedge \ell_0\pi_1\ell' \in \boldsymbol{\mathcal{S}}^*[\![\mathsf{S}]\!](\pi_0\ell_0) \wedge \ell' = \ell\} \cap \mathscr{P}_f(\ell))\}$
(lemma 50.1 and pointwise definition of $\dot{\cap}$)

$= \mathscr{P}_0 \cap \{\varrho(\pi_0\ell_0) \in \mathbb{E}\mathrm{v} \mid \exists \ell \in \mathsf{labs}[\![\mathsf{S}]\!] \,.\, \ell_0\pi_1\ell \in \boldsymbol{\mathcal{S}}^*[\![\mathsf{S}]\!]\,(\pi_0\ell_0) \wedge \varrho(\pi_0\ell_0) \in \mathscr{P}_0 \wedge \ell_0\pi_1\ell \in$ $\boldsymbol{\mathcal{S}}^*[\![\mathsf{S}]\!](\pi_0\ell_0) \wedge \varrho(\pi_0\ell_0\pi_1\ell) \in \mathscr{P}_f(\ell))\}$ (definition of $\cap$ and $\in$)

$= \mathscr{P}_0 \cap \{\boldsymbol{\varrho}(\pi_0\ell_0) \in \mathbb{E}\mathbb{v} \mid \exists \ell \in \mathsf{labs}[\![\mathsf{S}]\!] \,.\, \ell_0\pi_1\ell \in \boldsymbol{\mathcal{S}}^*[\![\mathsf{S}]\!](\pi_0\ell_0) \wedge \boldsymbol{\varrho}(\pi_0\ell_0\pi_1\ell) \in \mathscr{P}_f(\ell)\}$ ⦃definition of $\wedge$⦄

$= \mathscr{P}_0 \cap \widehat{\boldsymbol{\mathcal{S}}}^{\overleftarrow{\mathsf{r}}}[\![\mathsf{S}]\!](\mathscr{P}_f)$ ⦃theorem 50.30⦄

$= \mathscr{P}_0'$ ⦃definition of $\mathscr{P}_0'$⦄

For the second component, we have for all $\ell \in \mathsf{labs}[\![\mathsf{S}]\!]$,

$\mathscr{P}_f''(\ell)$

$= ((\widehat{\boldsymbol{\mathcal{S}}}^{\vec{\mathsf{r}}}[\![\mathsf{S}]\!]\,(\mathscr{P}_0 \cap \widehat{\boldsymbol{\mathcal{S}}}^{\overleftarrow{\mathsf{r}}}[\![\mathsf{S}]\!]\mathscr{P}_f)) \mathbin{\dot\cap} \mathscr{P}_f)\,\ell$ ⦃definition of $\mathscr{P}_f''$⦄

$= (\widehat{\boldsymbol{\mathcal{S}}}^{\vec{\mathsf{r}}}[\![\mathsf{S}]\!]\,(\mathscr{P}_0 \cap \widehat{\boldsymbol{\mathcal{S}}}^{\overleftarrow{\mathsf{r}}}[\![\mathsf{S}]\!]\mathscr{P}_f))(\ell) \cap \mathscr{P}_f(\ell)$ ⦃pointwise definition of $\dot\cap$⦄

$= \{\boldsymbol{\varrho}(\pi_0\ell_0\pi_1\ell) \mid \boldsymbol{\varrho}(\pi_0\ell_0) \in (\mathscr{P}_0 \cap \widehat{\boldsymbol{\mathcal{S}}}^{\overleftarrow{\mathsf{r}}}[\![\mathsf{S}]\!]\mathscr{P}_f) \wedge \ell_0\pi_1\ell \in \boldsymbol{\mathcal{S}}(\pi_0\ell_0)\} \cap \mathscr{P}_f(\ell)$ ⦃by lemma 50.1⦄

$= \{\boldsymbol{\varrho}(\pi_0\ell_0\pi_1\ell) \mid \boldsymbol{\varrho}(\pi_0\ell_0) \in (\mathscr{P}_0 \cap \{\boldsymbol{\varrho}(\pi_0\ell_0) \in \mathbb{E}\mathbb{v} \mid \exists \ell' \in \mathsf{labs}[\![\mathsf{S}]\!] \,.\, \ell_0\pi_1\ell' \in \boldsymbol{\mathcal{S}}^*[\![\mathsf{S}]\!](\pi_0\ell_0) \wedge$
$\boldsymbol{\varrho}(\pi_0\ell_0\pi_1\ell') \in \mathscr{P}_f(\ell')\}) \wedge \ell_0\pi_1\ell \in \boldsymbol{\mathcal{S}}(\pi_0\ell_0)\} \cap \mathscr{P}_f(\ell)$ ⦃theorem 50.30⦄

$= \{\boldsymbol{\varrho}(\pi_0\ell_0\pi_1\ell) \mid \boldsymbol{\varrho}(\pi_0\ell_0) \in \mathscr{P}_0 \wedge \exists \ell' \in \mathsf{labs}[\![\mathsf{S}]\!] \,.\, \ell_0\pi_1\ell' \in \boldsymbol{\mathcal{S}}^*[\![\mathsf{S}]\!](\pi_0\ell_0) \wedge \boldsymbol{\varrho}(\pi_0\ell_0\pi_1\ell') \in$
$\mathscr{P}_f(\ell') \wedge \ell_0\pi_1\ell \in \boldsymbol{\mathcal{S}}(\pi_0\ell_0) \wedge \boldsymbol{\varrho}(\pi_0\ell_0\pi_1\ell) \in \mathscr{P}_f(\ell)\}$ ⦃definitions of $\in$ and $\cap$⦄

$= \{\boldsymbol{\varrho}(\pi_0\ell_0\pi_1\ell) \mid \boldsymbol{\varrho}(\pi_0\ell_0) \in \mathscr{P}_0 \wedge \ell_0\pi_1\ell \in \boldsymbol{\mathcal{S}}(\pi_0\ell_0)\} \cap \mathscr{P}_f(\ell)$
⦃definition of $\exists$ with $\ell' = \ell$ and definition of $\cap$⦄

$= \widehat{\boldsymbol{\mathcal{S}}}^{\vec{\mathsf{r}}}[\![\mathsf{S}]\!]\,\mathscr{P}_0\,(\ell) \cap \mathscr{P}_f\ell$ ⦃lemma 50.1⦄

$= \mathscr{P}_f'(\ell)$ ⦃definition of $\mathscr{P}_f'$⦄

The relational case $\mathfrak{R}^{\vec{\mathsf{R}}}_{0f}$ is quite similar. □

51.1.2 Abstraction

Let us consider an abstraction $\langle \wp(\mathbb{E}\mathbb{v}^{\varrho}), \subseteq\rangle \xleftrightarrows[\alpha]{\gamma} \langle \mathscr{A}^{\varrho}, \sqsubseteq\rangle$ into the complete lattice $\langle \mathscr{A}^{\varrho}, \sqsubseteq, \bot, \top, \sqcap, \sqcup\rangle$, which is

- extended pointwise to $\langle \mathsf{labs}[\![\mathsf{S}]\!] \to \wp(\mathbb{E}\mathbb{v}^{\varrho}), \dot\subseteq\rangle \xleftrightarrows[\dot\alpha]{\dot\gamma} \langle \mathsf{labs}[\![\mathsf{S}]\!] \to \mathscr{A}^{\varrho}, \dot\sqsubseteq\rangle$,
- extended component wise to $\langle \wp(\mathbb{E}\mathbb{v}^{\varrho}) \times \mathsf{labs}[\![\mathsf{S}]\!] \to \wp(\mathbb{E}\mathbb{v}^{\varrho}), \subseteq_2\rangle \xleftrightarrows[\alpha_2]{\gamma_2} \langle \mathscr{A}^{\varrho} \times \mathsf{labs}[\![\mathsf{S}]\!] \to \mathscr{A}^{\varrho}, \sqsubseteq_2\rangle$,
- extended functionally to $\langle \wp(\mathbb{E}\mathbb{v}^{\varrho}) \to (\mathsf{labs}[\![\mathsf{S}]\!] \to \wp(\mathbb{E}\mathbb{v}^{\varrho})), \dot{\vec\subseteq}\rangle \xleftrightarrows[\vec\alpha]{\vec\gamma} \langle \mathscr{A}^{\varrho} \to (\mathsf{labs}[\![\mathsf{S}]\!] \to \mathscr{A}^{\varrho}), \dot{\vec\sqsubseteq}\rangle$ with $\vec\alpha(\vec{\boldsymbol{\mathcal{S}}}) \triangleq \dot\alpha \circ \vec{\boldsymbol{\mathcal{S}}} \circ \gamma$ and to $\langle (\mathsf{labs}[\![\mathsf{S}]\!] \to \wp(\mathbb{E}\mathbb{v}^{\varrho})) \to \wp(\mathbb{E}\mathbb{v}^{\varrho}), \dot{\overleftarrow\subseteq}\rangle \xleftrightarrows[\overleftarrow\alpha]{\overleftarrow\gamma} \langle (\mathsf{labs}[\![\mathsf{S}]\!] \to \mathscr{A}^{\varrho}) \to \mathscr{A}^{\varrho}, \dot{\overleftarrow\sqsubseteq}\rangle$ with $\overleftarrow\alpha(\overleftarrow{\boldsymbol{\mathcal{S}}}) \triangleq \alpha \circ \overleftarrow{\boldsymbol{\mathcal{S}}} \circ \dot\gamma$, and
- pairwise to $\langle (\mathsf{labs}[\![\mathsf{S}]\!] \to \wp(\mathbb{E}\mathbb{v}^{\varrho})) \to \wp(\mathbb{E}\mathbb{v}^{\varrho}) \times \wp(\mathbb{E}\mathbb{v}^{\varrho}) \to (\mathsf{labs}[\![\mathsf{S}]\!] \to \wp(\mathbb{E}\mathbb{v}^{\varrho})), \dot{\overleftarrow\subseteq} \times \dot{\vec\subseteq}\rangle \xleftrightarrows[\vec{\overleftarrow\alpha}]{\vec{\overleftarrow\gamma}} \langle (\mathsf{labs}[\![\mathsf{S}]\!] \to \mathscr{A}^{\varrho}) \to \mathscr{A}^{\varrho} \times \mathscr{A}^{\varrho} \to (\mathsf{labs}[\![\mathsf{S}]\!] \to \mathscr{A}^{\varrho}), \dot{\overleftarrow\sqsubseteq} \times \dot{\vec\sqsubseteq}\rangle$.

Define

$$\langle \overleftarrow{\widehat{\boldsymbol{\mathcal{S}}}}^{\varrho}[\![\mathsf{S}]\!], \vec{\widehat{\boldsymbol{\mathcal{S}}}}^{\varrho}[\![\mathsf{S}]\!]\rangle \;\; \dot{\overleftarrow{=}} \times \dot{\vec{=}} \;\; \vec{\overleftarrow\alpha}(\langle \widehat{\boldsymbol{\mathcal{S}}}^{\overleftarrow\varrho}[\![\mathsf{S}]\!], \widehat{\boldsymbol{\mathcal{S}}}^{\vec\varrho}[\![\mathsf{S}]\!]\rangle) \;\triangleq\; \langle \overleftarrow\alpha(\widehat{\boldsymbol{\mathcal{S}}}^{\overleftarrow\varrho}[\![\mathsf{S}]\!]), \vec\alpha(\widehat{\boldsymbol{\mathcal{S}}}^{\vec\varrho}[\![\mathsf{S}]\!])\rangle \tag{51.4}$$

so that we are interested in

$$\overline{\mathfrak{R}}^{\varrho}_{0f} \in (\mathscr{A}^{\varrho} \times \mathsf{labs}[\![\mathsf{S}]\!] \to \mathscr{A}^{\varrho}) \to (\mathscr{A}^{\varrho} \times \mathsf{labs}[\![\mathsf{S}]\!] \to \mathscr{A}^{\varrho}) \tag{51.5}$$

$$\overline{\mathfrak{R}}^{\varrho}_{0f}\langle\overline{\mathscr{P}_0}, \overline{\mathscr{P}_f}\rangle \triangleq \langle\overline{\mathscr{P}_0} \sqcap \widehat{\overline{\boldsymbol{\mathcal{S}}}}^{\overleftarrow{\varrho}}[\![\mathsf{S}]\!]\overline{\mathscr{P}_f}, (\widehat{\overline{\boldsymbol{\mathcal{S}}}}^{\overrightarrow{\varrho}}[\![\mathsf{S}]\!]\overline{\mathscr{P}_0}) \dot{\sqcap} \overline{\mathscr{P}_f}\rangle$$

This is sound because if $\langle\mathscr{P}_0', \mathscr{P}_f'\rangle = \mathfrak{R}^{\varrho}_{0f}\langle\mathscr{P}_0, \mathscr{P}_f\rangle$ and $\langle\overline{\mathscr{P}_0}', \overline{\mathscr{P}_f}'\rangle = \overline{\mathfrak{R}}^{\varrho}_{0f}\langle\overline{\mathscr{P}_0}, \overline{\mathscr{P}_f}\rangle$ then $\langle\mathscr{P}_0', \mathscr{P}_f'\rangle \subseteq_2 \gamma_2(\langle\overline{\mathscr{P}_0}', \overline{\mathscr{P}_f}'\rangle)$ that is $\mathscr{P}_0' \subseteq \gamma(\overline{\mathscr{P}_0}')$ and $\mathscr{P}_f' \dot{\subseteq} \dot{\gamma}(\overline{\mathscr{P}_f}')$.

However, this result may not be optimal. Because of the abstraction, $\overline{\mathfrak{R}}^{\varrho}_{0f}$ may in general be increasing and reductive but not idempotent. In that case, by the dual of corollary 29.3, iterated reduction provides a more precise sound static analysis.

51.2 Backward—Forward Versus Forward—Backward Extremal Static Analysis

Instead of performing independent analyses $\widehat{\overline{\boldsymbol{\mathcal{S}}}}^{\overleftarrow{\varrho}}[\![\mathsf{S}]\!]\overline{\mathscr{P}_f}$ and $\widehat{\overline{\boldsymbol{\mathcal{S}}}}^{\overrightarrow{\varrho}}[\![\mathsf{S}]\!]\overline{\mathscr{P}_0}$ to return $\overline{\mathfrak{R}}^{\varrho}_{0f}\langle\overline{\mathscr{P}_0}, \overline{\mathscr{P}_f}\rangle = \langle\overline{\mathscr{P}_0} \sqcap \widehat{\overline{\boldsymbol{\mathcal{S}}}}^{\overleftarrow{\varrho}}[\![\mathsf{S}]\!]\overline{\mathscr{P}_f}, (\widehat{\overline{\boldsymbol{\mathcal{S}}}}^{\overrightarrow{\varrho}}[\![\mathsf{S}]\!]\overline{\mathscr{P}_0}) \dot{\sqcap} \overline{\mathscr{P}_f}\rangle$ in (51.5), it is possible to do one analysis first and to reuse its result for the second. Depending on the needed information, one can first perform a forward analysis and then a backward one to calculate

$$\langle\overline{\mathscr{P}_0} \sqcap \widehat{\overline{\boldsymbol{\mathcal{S}}}}^{\overleftarrow{\varrho}}[\![\mathsf{S}]\!]\overline{\mathscr{P}_f}, (\widehat{\overline{\boldsymbol{\mathcal{S}}}}^{\overrightarrow{\varrho}}[\![\mathsf{S}]\!]\,(\overline{\mathscr{P}_0} \sqcap \widehat{\overline{\boldsymbol{\mathcal{S}}}}^{\overleftarrow{\varrho}}[\![\mathsf{S}]\!]\overline{\mathscr{P}_f})) \dot{\sqcap} \overline{\mathscr{P}_f}\rangle$$

or to first perform a backward analysis and then a forward one to overapproximate

$$\langle\overline{\mathscr{P}_0} \sqcap \widehat{\overline{\boldsymbol{\mathcal{S}}}}^{\overleftarrow{\varrho}}[\![\mathsf{S}]\!]((\widehat{\overline{\boldsymbol{\mathcal{S}}}}^{\overrightarrow{\varrho}}[\![\mathsf{S}]\!]\overline{\mathscr{P}_0}) \dot{\sqcap} \overline{\mathscr{P}_f}), (\widehat{\overline{\boldsymbol{\mathcal{S}}}}^{\overrightarrow{\varrho}}[\![\mathsf{S}]\!]\overline{\mathscr{P}_0}) \dot{\sqcap} \overline{\mathscr{P}_f}\rangle$$

By reductivity, this is more precise than $\overline{\mathfrak{R}}^{\varrho}_{0f}\langle\overline{\mathscr{P}_0}, \overline{\mathscr{P}_f}\rangle$.

Exercise 51.6 Show that the backward—forward and forward—backward extremal static analyses provide in general different results.

In general the forward/deductive and backward/abductive analyses look for different information and are based on different abstract domains. Examples are Clousot [359, 360] and Zoncolan [334].

51.3 Backward—Forward Static Analysis with Iterated Extremal Reduction

The hypotheses of the reductive iterations theorem 29.2 are satisfied for $f = \mathfrak{R}^{\varrho}_{0f}$, $g = \overline{\mathfrak{R}}^{\varrho}_{0f}$, $\dot{\sqsubseteq} = \dot{\sqsubseteq}_2$, $\alpha = \alpha_2$, and $\gamma = \gamma_2$. Therefore the reduction $\rho(\langle\overline{\mathscr{P}_0}, \overline{\mathscr{P}_f}\rangle) \triangleq \mathsf{gfp}^{\sqsubseteq_2}_{\langle\overline{\mathscr{P}_0}, \overline{\mathscr{P}_f}\rangle} \overline{\mathfrak{R}}^{\varrho}_{0f}$ satisfies $\alpha_2 \circ \mathfrak{R}^{\varrho}_{0f} \circ \gamma_2 \mathrel{\dot{\sqsubseteq}} \rho \mathrel{\dot{\sqsubseteq}} \overline{\mathfrak{R}}^{\varrho}_{0f}$. By the dual of the chaotic iteration theorem 22.4, we can choose to compute the iterates alternatively as follows.[1]

1. Also, by a generalization to asynchronous iterations [220], the computation of the two components of the pair can be done in parallel.

$$
\begin{array}{rcl}
X^0 & \triangleq & \overline{\mathscr{P}_0} \\
Y^0 & \triangleq & \overline{\mathscr{P}_f} \\
\ldots & \ldots & \ldots \\
X^{n+1} & \triangleq & X^n \sqcap \overset{\widehat{\vec{\varrho}}}{\boldsymbol{\mathcal{S}}}[\![\mathtt{S}]\!]Y^n \\
Y^{n+1} & \triangleq & (\overset{\widehat{\overleftarrow{\varrho}}}{\boldsymbol{\mathcal{S}}}[\![\mathtt{S}]\!]\,X^{n+1})\,\dot{\sqcap}\,Y^n \\
\ldots & \ldots & \ldots
\end{array}
$$

If the abstract domain $\langle \mathscr{A}^{\varrho}, \sqsubseteq \rangle$ does not satisfy the descending chain condition, these iterates may not converge. However, corollary 29.3 shows that it is sound to stop the iterates at any rank n. A narrowing (definition 34.14) can be used to enforce convergence.

Example 51.7 The following program (the skeleton of bubble sort [592, section 5.2.2]) was used in [221, section 5.7.5, pp. (5)-53— (5)-56] to illustrate the reduced forward—backward static analysis. The specification is that the program must terminate.

```
l1: i = n;
while l2: (i != 0)
   {
      l3: j = 0;
      while l4: (j != i)
         {
            l5: j = (j + 1);
         }
      l6: i = (i - 1);
   }
l7:
```

```
*** Specification:
l1: {i:_|_; j:_|_; n:_|_}
l2: {i:_|_; j:_|_; n:_|_}
l3: {i:_|_; j:_|_; n:_|_}
l4: {i:_|_; j:_|_; n:_|_}
l5: {i:_|_; j:_|_; n:_|_}
l6: {i:_|_; j:_|_; n:_|_}
l7: {i:T; j:T; n:T}
```

The first backward analysis determines that `n` must necessarily be initially positive for the program to terminate.

```
*** Backward analysis:
l1: {i:T; j:T; n:[0, oo]} i = n;
while l2: (i != 0) {i:[0, oo]; j:T; n:T}
   {
    l3: {i:[1, oo]; j:T; n:T} j = 0;
   while l4: (j != i) {i:[1, oo]; j:T; n:T}
         {
        l5: {i:[1, oo]; j:[-oo, oo-1]; n:T} j = (j + 1);
         }
    l6: {i:[1, oo]; j:T; n:T} i = (i - 1);
   }
l7: {i:T; j:T; n:T}
```

Then the forward analysis determines that if `n` is initially positive then `i` is positive or zero at `l2`. Therefore `i` ,which is decremented, and `n` which is unmodified in the loop body remain positive or zero.

```
*** Forward analysis from r0 = {i:T; j:T; n:[0, oo]}:
l1: {i:T; j:T; n:[0, oo]} i = n;
while l2: (i != 0) {i:[0, oo]; j:T; n:[0, oo]}
   {
    l3: {i:[1, oo]; j:T; n:[0, oo]} j = 0;
    while l4: (j != i) {i:[1, oo]; j:[0, oo]; n:[0, oo]}
         {
         l5: {i:[1, oo]; j:[0, oo]; n:[0, oo]} j = (j + 1);
         }
    l6: {i:[1, oo]; j:[1, oo]; n:[0, oo]} i = (i - 1);
   }
l7: {i:[0, 0]; j:T; n:[0, oo]}
```

Further backward and forward analyses leave this result unchanged. The result of the iterated forward—backward analysis is therefore the pair of the precondition and refined specification

```
<{i:T; j:T; n:[0, oo]},
 [l1: {i:_|_; j:_|_; n:_|_}; l2: {i:_|_; j:_|_; n:_|_};
  ...
  l7: {i:[0, 0]; j:T; n:[0, oo]}]>
```

The precondition $\mathscr{P}_0$ shows that n must necessarily be positive or zero for the program to terminate (but this might not be sufficient), in which case i is zero and n is positive or zero on program exit. Moreover, the local abstract invariants of the last forward analysis overapproximate the reachable states from the precondition $\mathscr{P}_0$.

Exercise 51.8 (Project: forward—backward abstract interpreter) Continuing exercises 21.22, 33.8, and 50.56 implement an iterated forward—backward abstract interpreter, and instantiate it for the intervals of chapter 33.

51.4 Backward—Forward Static Analysis With Iterated Intermediate Reduction

The iterated reduction of section 51.3 differs from the original proposal [221, section 5.7.5] in that local invariants are not reduced locally by those of the inverse analysis but only as the descendants or ascendants of reduced extremally initial or final states, which is less precise. Let us introduce intermediate reduction for structural semantics.

Given a structural abstract semantics of the form

$$\begin{cases} \hat{\mathcal{S}}^{¤}[\![\mathsf{S}]\!] & \triangleq & f^{¤}[\![\mathsf{S}]\!](\prod_{\mathsf{S}' \triangleleft \mathsf{S}} \hat{\mathcal{S}}^{¤}[\![\mathsf{S}']\!]) \\ \mathsf{S} \in \mathbb{P}\mathbb{c} \end{cases} \qquad (17.1) \quad (19.40)$$

on the abstract domain $\mathbb{D}^{¤}[\![\mathsf{S}]\!] \triangleq \langle \mathbb{P}^{¤}[\![\mathsf{S}]\!], \sqsubseteq^{¤}, \bot^{¤}, \sqcup^{¤}, \sqcap^{¤}\rangle$ for the transformer $f^{¤}[\![\mathsf{S}]\!] \in \prod_{\mathsf{S}' \triangleleft \mathsf{S}} \mathbb{P}^{¤}[\![\mathsf{S}']\!] \rightarrow \mathbb{P}^{¤}[\![\mathsf{S}]\!]$ and $X \in \prod_{\mathsf{S} \in \mathbb{P}\mathbb{c}} \mathbb{P}^{¤}[\![\mathsf{S}]\!]$ (standing for the result of the previous inverse analysis), we define

$$\begin{cases} \hat{\dot{\mathcal{S}}}^{¤}[\![\mathsf{S}]\!]\, X & \triangleq & X[\![\mathsf{S}]\!] \sqcap^{¤} f^{¤}[\![\mathsf{S}]\!](\prod_{\mathsf{S}' \triangleleft \mathsf{S}} \hat{\dot{\mathcal{S}}}^{¤}[\![\mathsf{S}']\!]\, X) \\ \mathsf{S} \in \mathbb{P}\mathbb{c} \end{cases} \qquad (51.9)$$

The iterated reduction becomes

$$
\begin{array}{rcl}
X^0 & \triangleq & \overline{\mathcal{P}_0} \\
Y^0 & \triangleq & \overline{\mathcal{P}_f} \\
\ldots & \ldots & \ldots \\
X^{n+1} & \triangleq & \widehat{\mathbf{S}}^{\vec{\varrho}}[\![\mathsf{S}]\!]Y^n \\
Y^{n+1} & \triangleq & \widehat{\mathbf{S}}^{\overleftarrow{\varrho}}[\![\mathsf{S}]\!]\,X^{n+1} \\
\ldots & \ldots & \ldots
\end{array}
$$

Example 51.10 Let us consider the followed example (contrived for illustrational purposes)

```
l1: x = (x - x); l2: y = (y - y); if l3: (x == y) l4: ; l5:
```

(where `(x - x)` and `(y - y)` are intended to create imprecision with interval analysis as shown in section 32.7).

Let the specification $\overline{\mathcal{P}_0}$ be true (i.e., `{x:T; y:T}`) and $\overline{\mathcal{P}_f}$ be

```
l1: {x:_|_; y:_|_} l2: {x:_|_; y:_|_} l3: {x:_|_; y:_|_}
l4: {x:[0, 20]; y:[10, 30]} l5: {x:_|_; y:_|_}
```

stating that the program point `l4` must be reached with $0 \leqslant \mathsf{x} \leqslant 20$ and $10 \leqslant \mathsf{y} \leqslant 30$.

The backward—forward static analysis with iterated extremal reduction of section 51.3 yields the following refined specification.

```
<{x:[-oo+10, oo-10]; y:[-oo+10, oo-10]},                    (51.11)
 [l1: {x:_|_; y:_|_}; l2: {x:_|_; y:_|_}; l3: {x:_|_; y:_|_};
  l4: {x:[0, 20]; y:[10, 30]}; l5: {x:_|_; y:_|_}]>
```

while the backward—forward static analysis with iterated intermediate reduction of section 51.4 yields the more precise (although not exact) result

```
<{x:[-oo+10, oo-10]; y:[-oo+10, oo-10]},                    (51.12)
 [l1: {x:_|_; y:_|_}; l2: {x:_|_; y:_|_}; l3: {x:_|_; y:_|_};
  l4: {x:[10, 20]; y:[10, 20]}; l5: {x:_|_; y:_|_}]>
```

Exercise 51.13 Explain the different results (51.11) and (51.12) for the iterated extremal and intermediate reduction in example 51.10.

Exercise 51.14 (Project: forward—backward abstract interpreter with iterated intermediate reduction) Continuing exercises 21.22, 33.8, 50.56, and 51.8 implement an iterated forward—backward abstract interpreter, and instantiate it for the intervals of chapter 33.

51.5 Conclusion

The combination of forward (reachability/deductive/top-down for logic/constraint languages) and backward (accessibility/abductive/bottom-up) static analyses in sections 51.1 and 51.2, iterated extremal reduction in section 51.3, and intermediate reduction in section 51.4 have increasing precisions and costs. A reduction is more precise but may be too costly (a compromise is to reduce at loop head only). The reduction may also produce slight improvements only with very precise extremal analyses. [930] is an example of forward/backward interval analysis of floating-point numbers augmented by a precision to determine the minimal accuracy on the inputs and on the intermediary results of a program containing floating-point computations in order to ensure the desired accuracy of the outputs.

[332] combines a forward analysis with both impossible failure and possible success backward analyses.

Iterated forward—backward reduction analysis was introduced by [221, section 5.7.5, pp. (5)-53— (5)-56] and [557]; see [253, section 4] and [256]. They apply to languages such as Prolog [244]. [549] is an example of backward—forward static analysis with iterated extremal reduction where verification conditions are expressed with Horn clauses [511]. [72] noticed that this is less precise than the original iterated intermediate reduction and proposed a formulation of the intermediate reduction using Horn clauses. Generalizations to the abstraction of temporal trace properties are proposed by [665–667].

The iterated forward—backward analysis is the basis for combining program analysis with model checking [252, 253] and, associated with other test generation techniques, for providing counterexamples to failed specifications [994, 995].

51.6 Solutions to Selected Exercises

Solution to exercise 51.13 The backward—forward static analysis with iterated extremal reduction of section 51.3 starts with a backward analysis from the specification `l4: { x:[0,20]; y:[10, 30]}` and yields

```
l1: {x:[-oo+10, oo-10]; y:[-oo+10, oo-10]} x = (x - x);
l2: {x:[10, 20]; y:[-oo+10, oo-10]} y = (y - y);
if l3: (x == y){x:[10, 20]; y:[10, 20]}
   l4: {x:[0, 20]; y:[10, 30]} ;
l5: {x:_|_; y:_|_}
```

which is imprecise. Then the forward analysis from `r0 = {x:[-oo+10, oo-10]; y:[-oo+10, oo-10]}` returns

```
l1: {x:[-oo+10, oo-10]; y:[-oo+10, oo-10]} x = (x - x);
l2: {x:T; y:[-oo+10, oo-10]} y = (y - y);
if l3: (x == y){x:T; y:T}
   l4: {x:T; y:T} ;
```

The next backward analysis shows that the extremal reduction has converged and the intersection with the specifications yields (51.11).

The backward—forward static analysis with iterated intermediate reduction starts with the same backward analysis. However, the next forward analysis is

```
l1: {x:[-oo+10, oo-10]; y:[-oo+10, oo-10]} x = (x - x);
l2: {x:[10, 20]; y:[-oo+10, oo-10]} y = (y - y);
if l3: (x == y){x:[10, 20]; y:[10, 20]}
   l4: {x:[10, 20]; y:[10, 20]} ;
l5: {x:_|_; y:_|_}
```

because of the intersection with the backward analysis at `l2` (so that we get `{x:[10, 20]; y:[10, 20]}` instead of `{x:T; y:T}`). The next backward analysis shows that the intermediate reduction has converged and the intersection with the specifications yields (51.12).

XIV SOUND STATIC ANALYSIS TOOLS

We discuss the applications of static analysis in academic and industrial environments (chapter 53), with an emphasis on soundness (chapter 52).

52 Semantic Soundness, Completeness, and Definedness

We sum up the calculational design of verification and static analysis methods by discussing semantic soundness, completeness, and well definedness.

Contents

52.1 Soundness by Calculational Design

Soundness of verification or static analysis methods states that proved or inferred program properties cannot be contradicted by any program execution as defined by the program semantics. Abstract interpretation does not keep this notion in the vagueness of informal explanations [638] but provides a rigorous definition, which is recalled and discussed below.

Formal reasonings on programs P of a language $\mathbb{P}$ are reasonings on their semantics $\boldsymbol{\mathcal{S}}[\![\mathsf{P}]\!] \in \mathcal{D}$ providing a formal model of all possible program executions as an element of the semantic domain $\mathcal{D}$. The semantic domain $\mathcal{D}$ can be chosen at various levels of abstraction as shown in chapters 6, 7, 19, 23, 25, 26, 28, and 42. Program properties are sets of possible semantics in $\mathbb{P} \triangleq \wp(\mathcal{D})$ (chapter 8). The collecting semantics $\{\boldsymbol{\mathcal{S}}[\![\mathsf{P}]\!]\}$ is the strongest semantic property of program P (see section 3.10 and chapter 8).

In general, one is interested in a subset of the program properties encoded in a domain $\overline{\mathbb{P}}$ whose meaning is given by a concretization function $\gamma \in \overline{\mathbb{P}} \rightarrow \wp(\mathcal{D})$. The implication $\subseteq$ on $\wp(\mathcal{D})$ is $\sqsubseteq$ on $\overline{\mathbb{P}}$ so that γ is increasing. If any property in $\wp(\mathcal{D})$ has a best/$\sqsubseteq$-most precise approximation in $\overline{\mathbb{P}}$ then there is a Galois connection $\langle \wp(\mathcal{D}), \subseteq\rangle \xrightleftharpoons[\alpha]{\gamma} \langle \overline{\mathbb{P}}, \sqsubseteq\rangle$.

The program properties of interest are given by the abstract semantics $\overline{\boldsymbol{\mathcal{S}}}[\![\mathsf{P}]\!] \in \overline{\mathbb{P}}$ which is sound if $\{\boldsymbol{\mathcal{S}}[\![\mathsf{P}]\!]\} \subseteq \gamma(\overline{\boldsymbol{\mathcal{S}}}[\![\mathsf{P}]\!])$ i.e. $\boldsymbol{\mathcal{S}}[\![\mathsf{P}]\!] \in \gamma(\overline{\boldsymbol{\mathcal{S}}}[\![\mathsf{P}]\!])$ and, in case of a Galois connection, the most best/precise abstract semantics is $\alpha(\{\boldsymbol{\mathcal{S}}[\![\mathsf{P}]\!]\})$ so the soundness condition is then $\alpha(\{\boldsymbol{\mathcal{S}}[\![\mathsf{P}]\!]\}) \sqsubseteq \overline{\boldsymbol{\mathcal{S}}}[\![\mathsf{P}]\!]$ i.e. $\boldsymbol{\mathcal{S}}[\![\mathsf{P}]\!] \in \gamma(\overline{\boldsymbol{\mathcal{S}}}[\![\mathsf{P}]\!])$.

In general, the semantics $\boldsymbol{S}[\![\mathsf{P}]\!]$ is defined by structural induction on the grammatical structure of the program P, using fixpoints for iterative or recursive constructs (e.g. chapters 17 and 19), so that the abstract semantics $\overline{\boldsymbol{S}}[\![\mathsf{P}]\!]$ can be designed by calculus using induction and fixpoint abstraction theorems (chapters 16 and 24). Another essential idea is the decomposition of the abstract properties of interest $\overline{\mathbb{P}}$ into abstract domains, which can be verified independently, and assembled inductively via functors (like the reduced product of chapter 36). This modularity is essential to scalability [268] and verifiability [563, 564].

52.2 Verification Versus Static Analysis

For either verification or static analysis automation, further abstractions may be needed to get computer-representable properties $\overline{\mathbb{P}}$ (chapter 27). In both cases, a property has to be discovered and proved to be inductive (using either structural and/or fixpoint induction). By Rice's theorem 9.12, the collecting, as well as the abstract semantics, hence the necessary inductive property are in general not effectively computable. The problem can be solved in two ways:

- In verification, the end user must provide the inductive property (chapters 25 and 26). Unless the properties $\overline{\mathbb{P}}$ are decidable, the verification that the given property is inductive cannot be fully automated and requires either user assistance for theorem provers or even providing the proof whose validity is checked by proof assistants.
- For static analysis, the strongest inductive property will be approximated by abstraction (chapter 27) and extrapolation/interpolation (chapter 34) techniques, which yield an effective but incomplete solution.

Because static analyzers must infer an inductive property that is provided to verifiers, and this inductive property might need to be the exact abstraction of the collecting semantics, sound static analysis is more difficult than sound verification, at least for infinite abstract domains (not reducible to finite ones) [277].

52.3 Completeness by Calculational Design

Completeness of formal verification or static analysis methods states that if a program property holds for all program executions as defined by the program semantics, then the formal method can always be used to prove it. Completeness comes in addition to a soundness requirement. In that case, the calculational design will ensure soundness and completeness whenever $\overline{\boldsymbol{S}}[\![\mathsf{P}]\!] = \alpha(\{\boldsymbol{S}[\![\mathsf{P}]\!]\})$.

Completeness is different from computability, and abstraction is not always a synonym for incompleteness. For example, verification methods are often complete although, by Rice's theorem 9.12, not fully automatizable (i.e. either the program property of interest cannot be expressed in $\overline{\mathbb{P}}$, or it can but the inductive proof is not fully mechanizable).

It is always possible to make an abstraction complete either by refining to a more precise one (at the detriment of cost or computability) or by abstracting to a coarser one (at the detriment of precision) [417]. The intuition is that the identity abstraction and the everywhere imprecise abstraction to $\mathtt{tt}$ are always complete. For example, [416] shows that for the past and future temporal logic of [254], any abstraction is incomplete (except the identity and the everywhere imprecise ones). Otherwise stated, any abstraction in that case must be imprecise unless it is a trivial one.

52.4 Improving the Precision in Case of Incompleteness

By Rice theorem 9.12, incompleteness is unavoidable, for all programs. This does not mean that for one program or one family of similar programs one cannot reach zero false alarms, as shown by Astrée for synchronous control command programs [113]. We can indefinitely refine the abstraction, for example, by introducing new abstract domains in a reduced product as in chapter 36 to cope with the origin of imprecision or by refining the widening; see section 34.5. However, if the precision increases the computation cost often increases disproportionately as the class of considered program widens greatly. It is then necessary to ask for user help or resort to other verification methods. Ultimately, the proof must be done by hand, documented in the program, and redone after each program or specification modification; see chapters 25 and 26 where the expressiveness restrictions do not apply to manual proofs.

One possible user help is to select the abstract domains to be included in the analyzer among a wide variety of abstract domains, each one adapted to specific analysis objectives. This is possible for numerical domains but the choice of symbolic domains is still limited at present.

Another possible form of user help is to add invariants to the program. This often does not help when the invariant is overapproximated in the analyzer's abstract domains and the useful information provided by the invariant is lost in its abstraction (e.g., a nonlinear invariant will be linearized). A solution is to express the invariants in a very expressive abstract domain known to the abstract interpreter and used only to reduce the other abstract domains by user-provided invariants [397].

52.5 Handling Alarms

Soundness implies that all alarms be reported whether true or false. If the analysis is too imprecise, the number of alarms may be very large. Because true alarms require program correction, and the user does not know which alarms are true of false, it may be important to propose an order to deal with alarms. For example, in Astrée alarms resulting from the undefinedness of the C semantics should be handled first (see the next section 52.6). Some alarms may result from imprecisions already causing previous alarms, so that a dependency analysis (chapter 47, e.g., [733]), or more precisely a responsibility analysis [311] may be useful to handle the original dominant alarms first [624]. True alarms may be identified by abstract testing [995]. If only false alarms remain, the analysis must ultimately be refined.

52.6 Handling Semantic Undefinedness

The entire issues of soundness and completeness rely on a formal definition of the language semantics. Many semantics leave some erroneous program behaviors undefined. This does not prevent building sound verifiers and analyzers provided that the undefinedness is adequately reflected in the formal program semantics and explicitly reported to the user.

A first issue is that the language semantics is not the one that is implemented (e.g. compilers have errors). A scientific approach is to prove compilers correct [627] or to check the equivalence of the source and object code for each compiled program [833, 1003].

A second issue is that there might be no definition of the language semantics, either formal or informal. In that case, reasoning on program executions is reasoning on undefined objects, which is mathematical nonsense. In that case, there is no way out, except by first defining the language semantics. An example is [381] for Python.

Of course, the verification method or static analysis can be postulated, which is therefore a definition of the semantics at an abstract level [376]. All mathematical theorems would be easier to prove if postulated in this way. However, a correctness proof requires a program (a static analyzer modeled by $\overline{\mathcal{S}}[\![\mathsf{P}]\!]$), a specification $\alpha(\{\mathcal{S}[\![\mathsf{P}]\!]\})$, and a proof that the program semantics satisfies the specification that is $\alpha(\{\mathcal{S}[\![\mathsf{P}]\!]\}) \sqsubseteq \overline{\mathcal{S}}[\![\mathsf{P}]\!]$.

The third and most common issue is that the language has an informal semantics from which a formal semantics can be derived but with imprecisions (e.g. some computations are undefined in the semantics when something is going to happen anyway at runtime) [349, 466, 598, 757]. The formal method is then sound if the expected result can be defined precisely whenever it may differ from the formal semantics. For example, the formally undefined behaviors in C can be classified as follows [63]:

- An *unspecified behavior* occurs when several different behaviors are allowed. If all possible behaviors can be taken into account, this is a sound overapproximation. In practice this may be too imprecise, so all possible behaviors cannot be taken into account. An underapproximation warning should be issued when some possible behaviors may not be taken into account.
  An example is the order of execution of expressions, which may vary for each use. A solution is, for example, a determinacy analysis ensuring that all evaluation orders are equivalent, and issuing a warning otherwise. Another solution is to use a parameter specifying which compiler is to be used and analyzing the expression in the order of the code generated by that compiler, if available.
- An *implementation defined behavior*[1] is an unspecified behavior, but the compiler has to document its choice.
  For example, the size and endianness of integers. This can be handled by parameterizing the analysis. Another example is signed integer overflow, which the compiler may handle as modulo arithmetic or as an error (a sound analysis will signal the potential error and, depending on what might happen at execution [319], e.g. go on with any integer value).
- For *undefined behaviors*, the C standard imposes no requirements at all.
  For example, the program can crash after dereferencing a NULL or dangling pointer or after sequent point violation (modifying a memory location more than once between two sequence points).

1. There are also *locale-specific behaviors* that can be handled as the implementation-defined behaviors.

A sound approach is to signal all undefined behaviors, go on with correct behaviors, with the explicit warning that the analysis is no longer valid after such an error if it ever happens at runtime. Otherwise stated, the semantics defines the maximal executions up to the first undefined behavior, if any, and the analysis is sound for that semantics. Any execution that possibly differs from this semantics must be statically reported.

Handling semantic undefinedness is always possible in a sound way by defining exactly when undefined behaviors may occur in the semantics $\mathcal{S}[\![\mathsf{P}]\!] \in \mathcal{D} \cup$ {*undefined behaviors*}. In case of overapproximation, an error may not always occur at a program point. A solution is to continue the analysis from that program point in the cases when the error with undefined behavior does not occur. This is an underapproximation in $\mathcal{D}$ but all subsequent warnings by the analyzer are still potentially possible (unless it is a false alarm). If the error with undefined behavior does not occur at runtime (e.g. it was a false alarm), the analysis based on the formal semantics coincides with the runtime execution. Otherwise, if the error with undefined behavior does occur at runtime (it was a true alarm), the results of the analysis coincide with the runtime executions up to the point where the first undefined behavior does appear at runtime. No conclusion of the analysis on the runtime executions is valid after the occurrence of the error. It follows that static analysis warnings relative to semantic undefinedness should be considered first. When eliminated by program rewriting or correction, or refinement of the static analyzer, the semantics and runtime executions coincide exactly.

This is the point of view taken by typing in chapter 49 where the semantics makes clear which runtime errors are captured by types (which we called static errors) and those that are not (which we called dynamic errors). Similarly, the soundness of Astrée is precisely defined and reported to the user. Another example is [183] for Clousot [360].

52.7 Handling Missing Code

In many applications, the code calls libraries, for which the source, often written in another language, is not available. If the library or operating system is well documented, it is possible to write stubs that given an abstract precondition return a abstract postcondition that is valid after a library/system call [765]. In absence of documentation, the only sound solution is to return "top/everywhere undefined," which typically will propagate everywhere and produce many false alarms. An intermediate solution is to ask the programmer for a hypothesis after the call. In that case, soundness is obviously relative to the veracity of this hypothesis. This is the solution adopted in Astrée where, for example, end users have to provide intervals of values for inputs. A good practice is to provide larger intervals than necessary to prove the robustness of the code.

52.8 Unsoundness

Commercial sound static analyzers are an order of magnitude harder and costly to design [148] and hence much more expensive than unsound ones. Soundness is hard to sell because it is not certain that end users and their managers and even scientists [638] understand perfectly what soundness means and its benefit when they choose a static analysis tool. So, many commercial analyzers adopt easy and inexpensive unsound solutions.

52.8.1 Omitting and Hiding Alarms

Unsound analyzers often produce many more alarms than sound ones (while missing many other ones). Nevertheless they are often popular with unscrupulous programmers because unsound analyzers hide most alarms, showing only those considered the most probable according to some heuristic.

A sound solution, implemented in Polyspace Code Prover [317, 350], is to distinguish between no error (green in Polyspace Code Prover), definite error (red), or potential error (gray in Polyspace Code Prover). Eliminating definite errors first usually also eliminates false alarms in potential errors. More generally, rather than hiding potential but not definite errors, while omitting some definite errors, it is better to be sound and to help the end user in dealing with alarms. Solutions include exhibiting an order in which definite/potential errors should be handled, for example, [624], and refining the analysis, for example, chapter 51 to ultimately prove correctness or propose a counterexample, for example, [994, 995].

52.8.2 Bounding Execution

Some programs have a bounded execution. It is then sound to take this bound on executions into account. For example, in control-command of an airplane in which a loop is executed every 5 ms and the plane has, because of physical limitations, a maximum flight time of 10 hours (for instance 20 hours by precaution), it is safe to limit the number of iterations of the loop to 20 h/5 ms = 14,400,000. However, if the program is not reinitialized at each stop of the plane, the static analysis with this hypothesis would be unsound beyond 20 hours. For the same reason, unrolling the loop n times less than 14,400,000 times is unsound [119]. So, comparing the precision and performances of n times loop unrolling with bounded execution with widening is unfair [725] without considering soundness and scalaility, because the soundly checked specifications are different, unless $n \geqslant 14{,}400{,}000$, which is unlikely for a large nontrivial control-command program.

52.8.3 Unsoundiness

Many commercial static analyzers are unsound, in a way that is nowhere defined, and their creators remain silent about unsoundness [182]. Of course, truthiness would be easier for charlatans who would not feel the need to tell the truth, including by omission, a situation not so uncommon, even in some scientific circles.

However, this approach is rejected, with very few exceptions, by most scientists who are not ready to abandon mathematical rigor whenever applicable. Abstract interpretation obviously shows that mathematical rigor is, of course, applicable in program verification and analysis.

Unsound methods provide no guarantee and so are essentially debugging tools [114]. Because the debugging is not carried on executions, it is imprecise. Unsound analyses may go on forever, always reporting new bugs, real or imaginary. Unsound analyzers may change the reporting diagnostics, including regressions that no longer report some bugs, after any release of a new version of the tool [114]. In summary, there is no notion of scientific progress without soundness.

Verifiers and static analyzers are programs that can have bugs. Nevertheless, they can and should be proved to satisfy the specification provided by the abstract semantics. Notice that for static analyzers the correctness of inductive property inference need not be checked. Only the inductiveness check must be proved sound (and could be a separate tool extracted from the soundness proof of the abstract semantics).

Of course, even if everything is proved correct, the compiler, operating system, machine, verifier, and static analyzer may have, and will have, bugs. This is not a serious enough pretext to abandon rigor. Mathematicians also produce wrong proofs. This does not prevent them from trying to correct the errors in their proofs or redesign correct ones. At least the formal methods offer a systematic approach to converge toward correct verification or inference of trustworthy program properties.

52.9 Formal Development of Static Analyzers

The advantage of a formal soundness basis for a static analyzer is that it can be constructed by calculational design. Presently this is mainly a guideline for informal design, or an posteriori check of correctness [161, 546]. This paves the way for the design of tools based on algebraic manipulations that could perform the symbolic calculations automatically, as is already the case in computer algebra systems, for example, maple for integration [655].

52.10 Conclusion

Abstract interpretation provides a formal approach to the design of semantics, and program verification and analysis methods. The soundness condition is the basis for the calculational design of these formal methods. Already, metatools can be used to check this calculational design [93, 126, 127, 546]. Beyond post-design verification, *calculational design assistants* will hopefully be built to assist, and hopefully automatize, this formal calculational design.

53 Static Analysis Tools

We quickly review and categorize the broad spectrum of static software analysis tools and the resources available or needed to build these tools. We consider only automatic tools that require only lightweight specifications (such as machine parameterization or assertions to be checked), do not run the code, and need no user intervention to support the analysis.

Contents

53.1 A Broad Spectrum of Static Analysis Tools

Compilers (such as CLang [620]) are widely used in the software development cycle. They provide selected elementary definite errors and potential warnings. Compilers with advanced type checking or inference such as Ocaml are useful to eliminate careless mistakes such as nonmatching formal parameters and actual arguments. Every programmer benefits from such tools.

A second level of static analysis tools is linters. The Unix Lint command [534] was one of the first widely available static analyzers to report errors such as unused variables and functions, use before initialization, unreachable code, unused returned values of functions, type checking (more precise than the C compiler at the time), nonportable features, pointer misalignments, and other errors. Bugs are found mostly by pattern matching but may use more sophisticated data flow analyses. Some code analyzers (such as Microsoft's Roslyn or LGTM from Semmle) are extensible and allow for defining programming style rules for code quality. This is useful mainly for junior programmers to help them think about basic potential problems in their code.

A third level is that of wide-scope, almost always unsound and commercial, static analysis tools. They target a wide range of programming languages, development environments, and deployment platforms. It is therefore difficult to take semantic specificities into account. For example, Coverity [350] "supports 21 languages and over 70 frameworks and templates" in 2021. One emphasis is speed, often based on unsound techniques. For example, [182] mentioned that Coverity analyze loops by considering the very first few iterations only. Software models checkers such as CBMC unroll loops a bounded number of times, so belong to this category of unsound static analyzers.

Another emphasis is the user interface, filtering out less frequent errors at first, providing error explanations, and correction suggestions. This explains why professional programmers enjoy these tools. Empirically based, such bug-finders help but provide no guarantee at all, and so are inadequate for certification.

At the fourth level, static analysis tools are semantics based, sound, and precise. Based on a sound abstraction of the program semantics properties, they need to automatically do mathematical inference. They never miss any error in the well defined category of bugs that they consider (e.g. runtime errors) under well defined and checked hypotheses on runtime behavior hypotheses (e.g. undefined behaviors in C). They require parameterization to specify semantic parameters (such as word size, endianness, etc.). For industrial codes of millions of lines, the analysis time can be minutes to hours. So the objective is not immediate interactive support. They are useful for highly qualified engineers looking for the highest level of robustness, safety, and security. An example is Astrée, originally designed for embedded control-command. When applied to domain-specific applications, they provide no false alarm ([307, 696, 904]). When extended outside their original domain of specialization (e.g. large ARINC 653 and POSIX multithreaded applications for Astrée [696]), they remain precise enough not to be overwhelming.

Of course, this categorization only points out a few tool characteristics in a very large spectrum of formal or informal foundations, scope of analyzed properties, analysis cost and precision, ease of use, intended application domain, and targeted academic or industrial public [269]. Most static analysis tools have characteristics of all categories and so are very hard to evaluate objectively and precisely, in particular for soundness. Examples are C Global Surveyor [955] and IKOS [145] developed by Guillaume Brat and Arnaud Venet for the NASA Mars missions; Kestrel Technology CodeHawk [196]; CodeSonar from GrammaTech; Polyspace Code Prover originally developed by Alain Deutsch; Sparrow originally developed at the Seoul National University [485]; and others.

53.2 On the Design of Static Analyzers

53.2.1 Specifications

Formal language semantics specifications and software developing guidelines should be the reference for building static analyzers. They are mostly missing. Exceptions are MISRA C for the automotive industry [63] and DO-178C for the avionics industry, among others (ISO 26262, IEC 61508, EN 50128, etc.). If syntactic rules are trivial to check, precise and scalable static analysis is indispensable to check for undecidable rules (e.g., rulechecker of AbsInt).

53.2.2 Certification

Sound static analyzers provide a guarantee on the final result of a software development. They are also very useful during the development phase for early bug detection. Unfortunately, software certification is often optional and more an obligation of means during development rather than an obligation of results on the product. The use of static analyzers, even incorrect ones, is therefore a facility rather than an obligation.

53.2.3 Benchmarks

Benchmarks are indispensable during the design of static analyzers to adapt the analysis to the application domain. Most available benchmarks are for academic tools and do not reflect at all the difficulties encountered in industrial code. This is made clear, for example, by [977], showing the limits of academic tools working very well on tailored academic benchmarks but failing on industrial code, even very small ones. Obviously, industrial benchmarks should be made available publicly, for example, [821] or, at least, to tool design and evaluation teams.

53.2.4 Control Architecture

There are essentially two types of static analyzers. The vast majority are based on an intermediate language targeting multiple languages (e.g. Clousot [360] for the .NET framework languages compiling to the common intermediate language (CIL)). Very few use the structural approach propounded in this book (e.g. Astrée and Zoncolan, developed by Francesco Logozzo and Manuel Fähndrich at Facebook for scalable information-flow analysis with application to security and privacy [334]).

The frontend, similar to that of a compiler, is the easiest and most well-understood part of a static analyzer. Hence the effort saved by considering an intermediate language is insignificant compared with the many difficulties they generate (e.g. to report at the source level, to take the program structure and the language specificities into account, etc.). Moreover, if information in the intermediate language is needed, it can be easily added to the program representation.

The control architecture is essential for determining inductive invariant for mathematical induction (by widening and narrowing and their duals). In more complex situations, the structure on which induction is performed may have to be determined during the analysis itself (see, for example, the segmentation concept in [263]).

53.2.5 Property Architecture

Another basic early decision to be taken in the design of static analyzers is how abstract properties will be represented and implemented. We have chosen to use abstract domains combined with functors like the reduced product, as in Astrée. This decomposes a difficult problem into subproblems that are easier to master. A counterexample is Infer in which properties are represented by logical formulas and operations like widening are left implicit and distributed all over the code. A restructuration is apparent in the publicly available code but the kernel is left inextricable.

53.2.6 Extensibility

Because it is very hard to predict which specific inductive properties will be ultimately necessary to cope with specified errors, it is absolutely essential that the property architecture be extensible in a structured way. This is the whole purpose of the abstract domains and the domain functors/combinators. Plug-ins such as Eva plug-in [155] can be used to implement new abstract domains in the static analyzer of the Frama-C [282, 579] source-code analyzer and verifier. It is harder to influence their combinations. Astrée was designed right at the beginning for extensibility [267].

53.2.7 Scalability

Analyzers that do not scale may still be useful to analyze small program parts. They may be used on purely functional modular software. However, most industrial software is not decomposable into small modular functional units. Those programs usually have millions of lines sharing a complex global state. Then scalability becomes a central problem [70]. Scalability must be thought about right at the beginning of the static analyzer design. For example, Astrée does not use a reduced product but a complex rather efficient overapproximation of the reduced product [268]. It's industrial version by AbsInt scales to tens of millions of lines of C with high precision.

53.2.8 Libraries

Abstract domain libraries such as APRON [529], ELINA [897], or PPL (Parma Polyhedra Library) [69, 98, 99] for numerical properties are extremely useful both to spread know-how and to be used in different analyzers. Unfortunately, symbolic properties are much harder to formalize independently of particular languages, and no largely distributed library is available.

53.2.9 Generators

Static analyzer generators, similar to compiler-compilers or compiler generators and going beyond parser generators, might be useful at least for rapid prototyping [980]. An early example is PAG, a generic abstract interpreter covering chapters 21, 23, 28, and 45 [661, 662]. However the technology has not matured enough to be used in most compiler or static analyzer projects. Crab is an analyzer of a programming language independent intermediate language [744]. The SPARTA library provides the basic blocks for building static analyzers based on Abstract Interpretation [905].

53.2.10 Qualification

There are very few requirements on the static analysis tools used in industry. An exception is the DO-333 [377] supplement to DO-178C, providing guidance on the processes necessary to qualify a tool that will be used to claim DO-178C objectives. One must show that if the program has a bug in the category considered by the tool, it must be signaled. So, an unsound static analyzer cannot be qualified (but an imprecise one can). Moreover, signaled bugs must be handled according to the procedures defined by the DO-178C (so an imprecise analyzer will require much more extra work). Astrée is DO-333 qualified, thus saving a lot of redundant testing work.

53.2.11 Library Verification

Some libraries are known to be erroneous (such as APRON's not reporting float overflows or nonterminating on the polyhedra domain because of rounding errors. Rationals should be used but are expensive because the denominators tend to be very large. This is the reason why polyhedra are not used in Astrée). A few abstract domain libraries have been formally verified, such as octagons [544]. It is also possible to check the result of each analysis, a posteriori, after the program analysis [101, 379].

53.2.12 Static Analyzer Verification

The computer-aided verification of static analyzers would go much beyond qualification and ideally would help in the calculational design propounded in this book. Verasco [544, 546] is an example of academic static analyzer verified using the proof assistant COQ [520]. An analyzer like Astrée is about 265,000 lines of OCaml (not counting the much larger user interface) in 2021, so that a formal verification is not in the reach of proof checkers or theorem provers at that time.

53.2.13 Evaluation and Benchmarking

Evaluations of the soundness, precision, and scalability of static analyzers by independent authorities are extremely rare, leaving place for academic fantasies or grandiloquent commercial arguments. An exception is [122], in which only Astrée and Frama-C Eva plug-in were found sound (and even detected unexpected errors in the test set) and precise enough to fully handle all alarms. Analyzers that failed are not reported, although this would be important information. The license of some commercial analyzers even forbids making evaluations public. More honest peer reviews like [122, 493, 977, 982] are necessary to guide research in fruitful directions and orient credulous users in their difficult tool choice. Such work is useful but often difficult to publish.

53.2.14 Industrialization

The industrialization of a static analyzer, most often with an academic origin, is a difficult task going much beyond what the academic world can produce. The adaptation to special needs and standards, the documentation, the user interface, and other essentials most often requires more work that the analyzer itself. This can hardly be done outside an industrial context. This too often means compromises with rigor. But there are exceptions, such as AbsInt (www.absint.com).

53.2.15 Human Resources

The design of a static analyzer is an order of magnitude more complicated than that of a compiler. If elementary static analyzers (such as the programming project in the **Project** index) take a few days or weeks, academic projects a few months or years, an advanced static analyzer such as Astrée requires decades of research and development.

Very few experienced highly qualified engineers can complete this task with success. The successes often come after a strong academic background in the theory of program analysis and a large practical experience. This hugely demanding combination is extremely rare, which might explain why very few scalable, sound, precise, perennial, and open sourced or industrialized analyzers do exist.

53.2.16 Education

The adoption of static analysis in an industrial context requires that end users understand the tool they are using. Most software engineers have a fuzzy understanding of static analysis and could read [261, 837] as an introduction. A larger adoption would require a better education, for example, the use of static analyzers at the very beginning of programmers' education, which beyond typing, which is a rather coarse analysis, is almost never the case. Educational static analyzers are still missing, and educational licenses of professional static analyzers are underused.

53.2.17 Responsibility

Finally, the main obstacle to the wide adoption of static analyzers is that programmers are never held responsible for their errors, even when the human and economic consequences are huge. Contrary to other disciplines, software engineers are guaranteed qualified immunity under the pretext that verification is beyond best practice. Rapid development with user debugging and distant updates is the rule. If best practice included the mandatory use of standards and qualified tools, programmers and their hierarchy could be held accountable for bugs automatically discoverable by such tools, at least the definite ones. This would certainly encourage the community to use reliable tools and boost the static analysis market [981].

53.3 Beyond Program Static Analysis

Beyond application to program analysis and verification, abstract interpretation has numerous other applications such as program development [490]; controller synthesis [516]; program synthesis [962]; test generation [548]; fuzz testing [851]; cellular signaling systems in biology [139, 140, 164, 367]; validation of the local robustness [44, 339, 407, 699, 893, 894] and the absence of gender bias [939] of neural networks; machine learning [946]; verification of datadase query languages [527]; reactive systems [290]; SMT solvers [284]; hybrid systems [458, 481]; Simulink [171] and Lustre [454] models ; games [483, 814]; energy consumption [356]; climate models [826]; quantum computing [793, 997]; and others.

53.4 Conclusion and Perspectives

Although academic tools have severe limitations, some have scaled to an industrial success, for example, for safety critical applications (too often underadvertised due to severe confidentiality constraints and lack of independent comparative evaluations).

As programming languages and computer systems become more complex and software as well as hardware much larger, a lot of problems remain to be solved. For example, the analysis of complex data structures; of hardware and their memory models; of concurrency; of operating systems [178]; of distributed, real-time, and hybrid systems are still in their infancy.

Moreover, program analysis must be broaden to systems analysis of a program together with a model of the environment with which the program interacts. A challenging example is closing the loop in cyber physical systems (instead of analyzing the program under input-output hypotheses), for example, to prove stability and robustness [229]. Another challenging example would be to analyze the code of a social network to prove that it correctly implements the officially published privacy policy.

Finally, by undecidability, soundness, precision, and scalability will remain an interesting challenge, forever.

XV CONCLUSION

54 Conclusion

We briefly revisit the principles that have guided the ideas and work presented in this book.

Contents

54.1 On the Scope of Abstract Interpretation

Verification methods (including model checking) and static analysis methods are the subject of tens of thousands of publications (e.g., internet search engines of "static program analysis" produce millions of results). Abstract interpretation provides a unified understanding of this literature as well as intellectual tools to go far beyond the present-day state of the art. Following [713, p. 1],

> *"We lay down a fundamental principle of generalization by abstraction: The existence of analogies between central features of various theories implies the existence of a general theory which underlies the particular theories and unifies them with respect to those central features."*

More than a disparate collection of unrelated techniques and algorithms (e.g., [191]), abstract interpretation offers a uniform abstract mathematical framework to understand formal methods and to design the semantics, specifications, proof methods, data flow analysis, model checking, typing, static analysis, and other components of computer systems described by a formal language. The abstract interpretation theory is generic, general, uniform, and practically applicable to any system evolving in time (see a few nonconventional examples in section 53.3). It is an effective method for constructing sound-by-construction formal methods and tools.

54.2 Principles of Abstract Interpretation

Abstract interpretation is a mathematical theory of exact or approximate abstraction. We have introduced the theoretical foundations and provided several examples of application to the semantics, verification and static analysis of programming languages.

Hopefully, this book will broaden the understanding of what it means to reason on programs and to reason on methods to reason on and manipulate programs.

The principle of abstraction that we have used recurrently in all chapters is the following.

- Let $\langle \wp(\mathfrak{E}), \subseteq \rangle$ be properties of entities (the so-called *concrete domain*) that we want to prove for a given entity of $\mathfrak{E}$;
- Consider a subset $\mathscr{A} \subseteq \wp(\mathfrak{E})$ of *properties of interest* $\langle \mathscr{A}, \subseteq \rangle$;
- Encode these properties of interest in an *abstract domain* $\langle A, \sqsubseteq \rangle$;
- The decoding function $\gamma \in A \to \mathscr{A}$ is called the *concretization function*;
- Make proofs and analyses using abstract properties only;
- So, because the expressivity is limited to $\mathscr{A} = \gamma(A)$, any concrete property in $\wp(\mathfrak{E})$ must be overapproximated (or dually underapproximated) by an abstract property in A;
- If the abstract proof/analysis succeeds, it is valid in the concrete (*soundness*);
- If all such proofs can always be done in the abstract, the approach is *complete* because no approximation is ever necessary; and
- If the abstract proof/exact analysis may sometime fail, some properties in $\wp(\mathfrak{E}) \setminus \mathscr{A}$ that are essential in the concrete proof are missed in the abstract (*incompleteness*).

In case every concrete property has a best abstraction, we have a Galois connection $\langle \wp(\mathfrak{E}), \subseteq \rangle \underset{\alpha}{\overset{\gamma}{\leftrightarrows}} \langle A, \sqsubseteq \rangle$ (chapter 11).

We have rigorously followed this approach for program semantics, verification, and static analysis. The same abstraction principle can be applied to $\langle A, \sqsubseteq \rangle$, which yields a hierarchy of abstractions.

54.3 Semantics

For semantics, we postulated a prefix stateless trace semantics in chapter 6 that is equivalent to the maximal trace semantics of chapter 7 (equivalent in the sense that each one is an abstraction of the other). This level of abstraction was appropriate for all the applications we considered (including recovering by abstraction of a classic stateful trace semantics in chapter 42).

Were semantics not precise enough, we would have had to propose a refinement and show that the prefix trace semantics is an abstraction of this refinement. This would ensure, by composition of abstractions, that the rest of the book would be consistent with this refined semantics.

For some applications, the prefix trace semantics is too refined. So, we considered several more abstract semantics for invariance/reachability in chapter 19, invariance verification conditions in chapter 25, Hoare logic in chapter 26, Cartesian invariants in chapter 28, stateful prefix traces in chapter 42, and transition systems in chapter 43.

All semantics are presented inductively, in structural form (by induction on the program syntax of chapter 4), and using equivalently either deductive definitions or fixpoint definitions (chapter 15). The advantage is that we can use inductive proofs everywhere, which is indispensable to reason on infinite objects (chapter 16). Moreover, by observing that abstractions are relating abstract domains we were able to show in chapter 21 that all semantics have a similar structure formalized by an abstract interpreter and differ only by the abstract domain taken in parameter. This greatly simplifies soundness proofs that must be done for each particular abstract domain but only once for the abstract interpreter (chapter 27). Moreover, it paves the way for calculational design and hopefully, in the future, of computer-aided design of formal methods. This provides strong design guidelines and ensures correctness by construction [322, 324, 327] as opposed to posteriori verification [501, 504]. The abstract interpreter can itself be the subject of abstract interpretation [278], for example, to perform induction during static analysis.

54.4 Verification

Verification methods in chapters 25 and 26 are derived from abstract semantics by applying structural, fixpoint, or iteration induction studied in chapter 24. They can themselves, when parameterized by an abstract domain, be expressed as abstract semantics.

54.5 Static Analysis

Static analysis aims at automation of verification methods. Because of Rice's theorem in chapter 9, a static analysis method is always in one of the following categories.

1. unsound;
2. sound and potentially nonterminating (or getting out of resources) as illustrated in chapter 44;
3. sound, terminating, and potentially producing false alarms (in all other static analysis chapters).

Abstract interpretation theorizes the last two possibilities (the first empirical and too frequent alternative needing no theory at all because nothing is ever guaranteed).

The main criticisms are that abstract interpretation does not provide a definite method to design useful abstractions and that abstract interpretation-based static analyzers have false alarms.

— The same way that mathematics does not provide the proof of your theorem but tools to help you make the proof, the theory of abstract interpretation does not provide the abstraction to design your static analyzer but tools to help you design your abstraction (such as reduction and the reduced product in chapters 29 and 36). Coming up with the right abstraction providing the best cost/precision ratio is almost never easy. It is presently an art supported by a purely experimental science. This was made clear by the evolutionary design of Astrée or, for example, [974].

— By Rice's theorem, any completely automatic analysis method is doomed to fail on infinitely many programs. For example, decidable methods apply only to restricted classes of programs that do not fit well to practice. When not applicable, the only sound conclusion is that everything is an alarm. One can also argue that computers are finite, so undecidability can be bypassed. But then combinatorial explosion strikes on infinitely many programs. Running out of resources is even worse that false alarms.

Static analysis has been defined by abstraction (chapter 27) presented structurally, and, in absence of a general method to compute direct solutions of fixpoint equations, by iterative fixpoint computation (chapter 15) or approximation (chapter 34) and checking (chapter 35), reducing various iteration strategies to chaotic iterations (chapter 22).

So, the main challenge is to study abstractions well suited for specific program properties and to combine them astutely to achieve enough precision at reasonable cost. It is a matter of both theoretical studies and experimentation. As we say in French, « À l'impossible nul n'est tenu ».

The examples we have considered are both Cartesian and relational, history sensitive and flow sensitive and insensitive, numerical and symbolic, complete and incomplete, monotrace based and multitrace based, all of them being sound.

Exercise 54.1 Categorize the various verification methods and static analyses provided in the book according to the preceding classification.

54.6 The Future of Abstract Interpretation

Abstract interpretation has shown itself to be a transversal idea applicable to a large variety of formal methods and to scale to industrial practice. Much more is certainly to come!

The future of abstract interpretation is in the hand of talented generations of researchers and engineers who master all necessary intricacies of the theory and practice. This book is a first step in that direction to be complemented by a study of the literature.

Hopefully, the readers will take up abstract interpretation and successfully design new abstractions, methods, and tools, and extend its application domains.

54.7 Solutions to Selected Exercises

Solution to exercise 54.1

Chapter	31	33	38	40	41	44	46	47	49
forward	✓	✓	✓	✓		✓	✓	✓	✓
backward					✓				
monotrace based	✓	✓	✓	✓	✓	✓	✓		✓
multitrace based								✓	
Cartesian	✓	✓			✓		✓		
relational			✓	✓		✓		✓	
history sensitive						✓			
flow-sensitive	✓	✓	✓	✓	✓	✓		✓	
flow-insensitive							✓		✓
numerical	✓	✓	✓	✓					
symbolic					✓	✓	✓	✓	✓
complete									
incomplete	✓	✓	✓	✓	✓	✓	✓	✓	✓
sound	✓	✓	✓	✓	✓	✓	✓	✓	✓
unsound									

Bibliography

[1] Martín Abadi, Bowen Alpern, Krzysztof R. Apt, Nissim Francez, Shmuel Katz, Leslie Lamport, and Fred B. Schneider. "Preserving Liveness: Comments on 'Safety and Liveness From a Methodological Point of View'". *Inf. Process. Lett.* 40.3 (1991), pp. 141–142 (p. 161).

[2] Martín Abadi, Anindya Banerjee, Nevin Heintze, and Jon G. Riecke. "A Core Calculus of Dependency." In *POPL*. ACM, 1999, pp. 147–160 (p. 621).

[3] Martín Abadi and Leslie Lamport. "The Existence of Refinement Mappings." *Theor. Comput. Sci.* 82.2 (1991), pp. 253–284 (p. 684).

[4] Parosh Aziz Abdulla, Bengt Jonsson, Marcus Nilsson, and Mayank Saksena. "A Survey of Regular Model Checking." In *CONCUR*. Vol. 3170. Lecture Notes in Computer Science. Springer, 2004, pp. 35–48 (p. 531).

[5] Smbat Abian and Arthur B. Brown. "A Theorem on Partially Ordered Sets with Applications to Fixed Point Theorems." *Canadian Journal of Mathematics.* 13 (1961), pp. 78–82 (p. 172).

[6] Peter Aczel. "An Introduction to Inductive Definitions." In John Barwise, ed. *Handbook of Mathematical Logic.* Amsterdam: North-Holland, 1977. Chap. 7, pp. 739–782 (p. 184).

[7] Peter Aczel. "Non-Well-Founded Sets." In *CSLI Lecture Notes.* Vol. 14. Center for the Study of Language and Information, Stanford University, 1988 (p. 184).

[8] Peter Aczel and Nax Paul Mendler. "A Final Coalgebra Theorem." In *Category Theory and Computer Science.* Vol. 389. Lecture Notes in Computer Science. Springer, 1989, 357–365 (p. 184).

[9] Assalé Adjé, Stéphane Gaubert, and Éric Goubault. "Coupling Policy Iteration with Semi-Definite Relaxation to Compute Accurate Numerical Invariants in Static Analysis." *Logical Methods in Computer Science.* 8.1 (2012) (p. 412).

[10] Wolfgang Ahrendt, Bernhard Beckert, Richard Bubel, Reiner Hähnle, Peter H. Schmitt, and Mattias Ulbrich, eds. *Deductive Software Verification—The KeY Book—From Theory to Practice.* Vol. 10001. Lecture Notes in Computer Science. Springer, 2016 (p. 304).

[11] Wolfgang Ahrendt, Laura Kovács, and Simon Robillard. "Reasoning About Loops Using Vampire in KeY." In *LPAR*. Vol. 9450. Lecture Notes in Computer Science. Springer, 2015, pp. 434–443 (p. 410).

[12] Hassan Aït-Kaci. "A Lattice Theoretic Approach to Computation Based on a Calculus of Partially Ordered Type Structures (Property Inheritance, Nets, Graph Unification)." PhD Thesis. University of Pennsylvania: Computer and Information Science Dept., 1984 (p. 647).

[13] Hassan Aït-Kaci. "Jacc: (Much More Than) Just Another Compiler Compiler." https://hassan-ait-kaci.net/hlt/doc/hlt/jaccdoc/000_START_HERE.html. 2014 (p. 61).

[14] Hassan Aït-Kaci and Samir Amir. "Classifying and Querying Very Large Taxonomies with Bit-Vector Encoding." *J. Intell. Inf. Syst.* 48.1 (2017), pp. 1–25 (p. 480).

[15] Hassan Aït-Kaci, Robert S. Boyer, Patrick Lincoln, and Roger Nasr. "Efficient Implementation of Lattice Operations." *ACM Trans. Program. Lang. Syst.* 11.1 (1989), pp. 115–146 (p. 109, 480).

[16] Hassan Aït-Kaci and Gabriella Pasi. "Fuzzy Unification and Generalization of First-Order Terms Over Similar Signatures." In *LOPSTR*. Vol. 10855. Lecture Notes in Computer Science. Springer, 2017, pp. 218–234 (p. 647).

[17] Hassan Aït-Kaci and Gabriella Pasi. "Fuzzy Lattice Operations on First-Order Terms Over Signatures with Similar Constructors: A Constraint-Based Approach." *Fuzzy Sets Syst.* 391 (2020), pp. 1–46 (p. 645, 647).

[18] Hassan Aït-Kaci, Andreas Podelski, and Seth Copen Goldstein. "Order Sorted Feature Theory Unification." *J. Log. Program.* 30.2 (1997), pp. 99–124 (p. 647).

[19] Marianne Akian, Stéphane Gaubert, and Vassili Kolokoltsov. "Invertibility of Functional Galois Connections." *CRAS, Série I.* 335 (2002), pp. 1–6 (p. 131).

[20] Jade Alglave. "A Shared Memory Poetics." PhD thesis. Université Paris 7—Denis Diderot, Nov. 20, 2010 (p. 192).

[21] Jade Alglave. "A Formal Hierarchy of Weak Memory Models." *Formal Methods in System Design.* 41.2 (2012), pp. 178–210 (p. 79, 190, 526).

[22] Jade Alglave. "Modeling of Architectures." In *SFM*. Vol. 9104. Lecture Notes in Computer Science. Springer, 2015, pp. 97–145 (p. 190, 192).

[23] Jade Alglave and Patrick Cousot. "Syntax and Analytic Semantics of LISA." *CoRR.* abs/1608.06583 (2016) (p. 79, 515, 526).

[24] Jade Alglave and Patrick Cousot. "Ogre and Pythia: An Invariance Proof Method for Weak Consistency Models." In *POPL*. ACM, 2017, pp. 3–18 (p. 304).

[25] Jade Alglave, Luc Maranget, and Michael Tautschnig. "Herding Cats: Modelling, Simulation, Testing, and Data-Mining for Weak Memory." In *PLDI*. ACM, 2014, p. 40 (p. 190).

[26] Frances E. Allen. "Control Flow Analysis." *ACM SIGPLAN Not.* 5.7 (1970), pp. 1–19 (p. 497, 514).

[27] Frances E. Allen. "A Basis for Program Optimization." In *IFIP Congress (1)*. 1971, 385–390 (p. 109, 270, 412, 497).

[28] Frances E. Allen. "Interprocedural Data Flow Analysis." In Jack L. Rosenfeld, ed. *Information Processing 74*. North-Holland, 1974, pp. 398–402 (p. 497).

[29] Frances E. Allen and John Cocke. "A Program Data Flow Analysis Procedure." *Commun. ACM.* 19.3 (1976), pp. 137–147 (p. 109, 497, 506, 514).

[30] Nabil Almashfi and Lunjin Lu. "Analyzing JavaScript Programs Using Octagon Domain." In *CATA 2019 (EPiC Series in Computing)*. Vol. 58. EasyChair, 2019, pp. 155–167 (p. 493).

[31] José Bacelar Almeida, Manuel Barbosa, Gilles Barthe, Benjamin Grégoire, Adrien Koutsos, Vincent Laporte, Tiago Oliveira, and Pierre-Yves Strub. "The Last Mile: High-Assurance and High-Speed Cryptographic Implementations." In *IEEE Symposium on Security and Privacy*. IEEE, 2020, pp. 965–982 (p. 493).

[32] Bowen Alpern and Fred B. Schneider. "Defining Liveness." *Inf. Process. Lett.* 21.4 (1985), pp. 181–185 (p. 151, 156, 159–161).

[33] Bowen Alpern and Fred B. Schneider. "Recognizing Safety and Liveness." *Distributed Computing.* 2.3 (1987), pp. 117–126 (p. 161).

[34] Steven C. Althoen and Renate McLaughlin. "Gauss-Jordan Reduction: A Brief History." *The American Mathematical Monthly.* 94.2 (Feb. 1987), pp. 130–142 (p. 443).

[35] Rajeev Alur, Konstantinos Mamouras, and Dogan Ulus. “Derivatives of Quantitative Regular Expressions.” In *Models, Algorithms, Logics and Tools.* Vol. 10460. Lecture Notes in Computer Science. Springer, 2017, pp. 75–95 (p. 532).

[36] Gianluca Amato, Simone Di Nardo Di Maio, and Francesca Scozzari. “Sum of Abstract Domains.” In *NFM.* Vol. 9058. Lecture Notes in Computer Science. Springer, 2015, pp. 35–49 (p. 433).

[37] Gianluca Amato and Marco Rubino. “Experimental Evaluation of Numerical Domains for Inferring Ranges.” *Electron. Notes Theor. Comput. Sci.* 334 (2018), pp. 3–16 (p. 398).

[38] Gianluca Amato, Marco Rubino, and Francesca Scozzari. “Inferring Linear Invariants with Parallelotopes.” *Sci. Comput. Program.* 148 (2017), pp. 161–188 (p. 492).

[39] Gianluca Amato, Francesca Scozzari, Helmut Seidl, Kalmer Apinis, and Vesal Vojdani. “Efficiently Intertwining Widening and Narrowing.” *Sci. Comput. Program.* 120 (2016), pp. 1–24 (p. 412).

[40] Nada Amin and Ross Tate. “Java and Scala’s Type Systems Are Unsound: The Existential Crisis of Null Pointers.” In *OOPSLA.* ACM, 2016, pp. 838–848 (p. 650, 671).

[41] Torben Amtoft, Sruthi Bandhakavi, and Anindya Banerjee. “A Logic for Information Flow in Object-Oriented Programs.” In *POPL.* ACM, 2006, pp. 91–102 (p. 618).

[42] Torben Amtoft and Anindya Banerjee. “A Logic for Information Flow Analysis with an Application to Forward Slicing of Simple Imperative Programs.” *Sci. Comput. Program.* 64.1 (2007), pp. 3–28 (p. 79, 609, 618).

[43] Lars Ole Andersen. “Program Analysis and Specialization for the C Programming Language.” Ph.D. Thesis. DIKU, University of Copenhagen, June 1994 (p. 559, 581, 582, 584, 586).

[44] Greg Anderson, Shankara Pailoor, Isil Dillig, and Swarat Chaudhuri. “Optimization and Abstraction: A Synergistic Approach for Analyzing Neural Network Robustness.” In *PLDI.* ACM, 2019, pp. 731–744 (p. 718).

[45] Kalmer Apinis, Helmut Seidl, and Vesal Vojdani. “How to Combine Widening and Narrowing for Non-Monotonic Systems of Equations.” In *PLDI.* ACM, 2013, pp. 377–386 (p. 412).

[46] Andrew W. Appel. “Verified Software Toolchain (Invited Talk).” In *ESOP.* Vol. 6602. Lecture Notes in Computer Science. Springer, 2011, pp. 1–17 (p. 304).

[47] Krzysztof R. Apt. “From Chaotic Iteration to Constraint Propagation.” In *ICALP.* Vol. 1256. Lecture Notes in Computer Science. Springer, 1997, pp. 36–55 (p. 257, 270, 291, 586).

[48] Krzysztof R. Apt and Ernst-Rüdiger Olderog. “Fifty Years of Hoare’s Logic.” *Formal Aspects Comput.* 31.6 (2019), pp. 751–807 (p. 293).

[49] Krzysztof R. Apt and Gordon D. Plotkin. “Countable Nondeterminism and Random Assignment.” *J. ACM.* 33.4 (1986), pp. 724–767 (p. 94).

[50] Vincenzo Arceri, Isabella Mastroeni, and Sunyi Xu. “Static Analysis for ECMAScript String Manipulation Programs.” *Appl. Sci.* 10 (2020), p. 3525 (p. 551).

[51] Vincenzo Arceri, Martina Olliaro, Agostino Cortesi, and Isabella Mastroeni. “Completeness of Abstract Domains for String Analysis of JavaScript Programs.” In *ICTAC.* Vol. 11884. Lecture Notes in Computer Science. Springer, 2019, pp. 255–272 (p. 551).

[52] Dean N. Arden. “Delayed-Logic and Finite-State Machines.” In *SWCT (FOCS).* IEEE Computer Society, 1961, pp. 133–151 (p. 170, 531).

[53] Mark A. Armstrong. *Introduction to Topology and Modern Analysis.* McGraw-Hill, 1983 (p. 147).

[54] Emil Artin. *Galois Theory.* 2nd ed. Notre Dame Mathematical Lectures, Number 2. University of Notre Dame Press, 1942 (p. 437).

[55] Andrea Asperti. "The Intensional Content of Rice's Theorem." In *POPL*. ACM, 2008, pp. 113–119 (p. 104).

[56] Mounir Assaf, David A. Naumann, Julien Signoles, Éric Totel, and Frédéric Tronel. "Hypercollecting Semantics and Its Application to Static Analysis of Information Flow." In *POPL*. ACM, 2017, pp. 874–887 (p. 620).

[57] Egidio Astesiano, Michel Bidoit, Hélène Kirchner, Bernd Krieg–Brückner, Peter D. Mosses, Donald Sannella, and Andrzej Tarlecki. "CASL: the Common Algebraic Specification Language." *Theor. Comput. Sci.* 286.2 (2002), pp. 153–196 (p. 349).

[58] Maximilian Baader, Matthew Mirman, and Martin T. Vechev. "Universal Approximation with Certified Networks." In *ICLR*. OpenReview.net, 2020 (p. 398).

[59] Claude–Gaspard Bachet de Méziriac. *Problèmes plaisans et délectables, qui se font par les nombres.* 2nd ed. Pierre Rigaud & Associez, 1624. http://gallica.bnf.fr/ark:/12148/bpt6k5818046p (p. 353).

[60] Kevin Backhouse and Roland Carl Backhouse. "Safety of Abstract Interpretations for Free, via Logical Relations and Galois Connections." *Sci. Comput. Program.* 51.1–2 (2004), pp. 153–196 (p. 131).

[61] Roland Carl Backhouse and Paul F. Hoogendijk. "Elements of a Relational Theory of Datatypes." In *Formal Program Development.* Vol. 755. Lecture Notes in Computer Science. Springer, 1993, pp. 7–42 (p. 117).

[62] John W. Backus et al. "Revised Report on the Algorithmic Language ALGOL 60." *Commun. ACM.* 6.1 (1963), pp. 1–17 (p. 61).

[63] Roberto Bagnara, Abramo Bagnara, and Patricia M. Hill. "The MISRA C Coding Standard and Its Role in the Development and Analysis of Safety– and Security–Critical Embedded Software." In *SAS.* Vol. 11002. Lecture Notes in Computer Science. Springer, 2018, pp. 5–23 (p. 708, 714).

[64] Roberto Bagnara, Michele Chiari, Roberta Gori, and Abramo Bagnara. "A Practical Approach to Verification of Floating-Point C/C++ Programs with math.h/cmath Functions." *ACM Trans. Softw. Eng. Methodol.* 30.1, Art. 9 (2020), 53 p. (P. 386).

[65] Roberto Bagnara, Patricia M. Hill, Elena Mazzi, and Enea Zaffanella. "Widening Operators for Weakly–Relational Numeric Abstractions." In *SAS.* Vol. 3672. Lecture Notes in Computer Science. Springer, 2005, pp. 3–18 (p. 483, 490, 492).

[66] Roberto Bagnara, Patricia M. Hill, and Enea Zaffanella. "Widening Operators for Powerset Domains." *STTT.* 8.4–5 (2006), pp. 449–466 (p. 433).

[67] Roberto Bagnara, Patricia M. Hill, and Enea Zaffanella. "Widening Operators for Powerset Domains." *STTT.* 9.3–4 (2007), pp. 413–414 (p. 433).

[68] Roberto Bagnara, Patricia M. Hill, and Enea Zaffanella. "An Improved Tight Closure Algorithm for Integer Octagonal Constraints." In *VMCAI.* Vol. 4905. Lecture Notes in Computer Science. Springer, 2008, pp. 8–21 (p. 493).

[69] Roberto Bagnara, Patricia M. Hill, and Enea Zaffanella. "The Parma Polyhedra Library: Toward a Complete Set of Numerical Abstractions for the Analysis and Verification of Hardware and Software Systems." *Sci. Comput. Program.* 72.1–2 (2008), pp. 3–21 (p. 250, 492, 716).

[70] Christel Baier and Joost–Pieter Katoen. *Principles of Model Checking.* MIT Press, 2008 (p. 527, 539, 548, 716).

[71] Alexey Bakhirkin, Josh Berdine, and Nir Piterman. "Backward Analysis via Over–Approximate Abstraction and Under–Approximate Subtraction." In *SAS.* Vol. 8723. Lecture Notes in Computer Science. Springer, 2014, pp. 34–50 (p. 693).

[72] Alexey Bakhirkin and David Monniaux. "Combining Forward and Backward Abstract Interpretation of Horn Clauses." In *SAS.* Vol. 10422. Lecture Notes in Computer Science. Springer, 2017, pp. 23–45 (p. 702).

[73] Vasanth Balasundaram and Ken Kennedy. "A Technique for Summarizing Data Access and Its Use in Parallelism Enhancing Transformations." In *PLDI*. ACM, 1989, pp. 41–53 (p. 481).

[74] Paolo Baldan, Francesco Ranzato, and Linpeng Zhang. "A Rice's Theorem for Abstract Semantics." In *ICALP*. Vol. 112. Leibniz International Proceedings in Informatics. Dagstuhl Publishing, Germany, 2021, 112:1–112:19 (p. 104).

[75] Thomas Ball, Ella Bounimova, Rahul Kumar, and Vladimir Levin. "SLAM2: Static Driver Verification with Under 4% False Alarms." In *FMCAD*. IEEE, 2010, pp. 35–42 (p. 551).

[76] Thomas Ball, Byron Cook, Vladimir Levin, and Sriram K. Rajamani. "SLAM and Static Driver Verifier: Technology Transfer of Formal Methods Inside Microsoft." In *IFM*. Vol. 2999. Lecture Notes in Computer Science. Springer, 2004, pp. 1–20 (p. 551).

[77] Thomas Ball, Vladimir Levin, and Sriram K. Rajamani. "A Decade of Software Model Checking with SLAM." *Commun. ACM*. 54.7 (2011), pp. 68–76 (p. 551).

[78] Thomas Ball, Rupak Majumdar, Todd D. Millstein, and Sriram K. Rajamani. "Automatic Predicate Abstraction of C Programs." In *PLDI*. ACM, 2001, pp. 203–213 (p. 319).

[79] Thomas Ball, Andreas Podelski, and Sriram K. Rajamani. "Boolean and Cartesian Abstraction for Model Checking C Programs." In *TACAS*. Vol. 2031. Lecture Notes in Computer Science. Springer, 2001, pp. 268–283 (p. 551).

[80] Thomas Ball, Andreas Podelski, and Sriram K. Rajamani. "Boolean and Cartesian Abstraction for Model Checking C Programs." *STTT*. 5.1 (2003), pp. 49–58 (p. 318, 319, 344).

[81] Thomas Ball and Sriram K. Rajamani. "The SLAM Toolkit." In *CAV*. Vol. 2102. Lecture Notes in Computer Science. Springer, 2001, pp. 260–264 (p. 551).

[82] Thomas Ball and Sriram K. Rajamani. "The SLAM Project: Debugging System Software via Static Analysis." In *POPL*. ACM, 2002, pp. 1–3 (p. 551).

[83] Stefan Banach. "Sur les opérations dans les ensembles abstraits et leur application aux équations intégrales." *Fund. Math. III*. (1922), pp. 133–181 (p. 172).

[84] François Bancilhon, David Maier, Yehoshua Sagiv, and Jeffrey D. Ullman. "Magic Sets and Other Strange Ways to Implement Logic Programs." In *PODS*. ACM, 1986, pp. 1–15 (p. 684).

[85] Maria J. García de la Banda, Manuel V. Hermenegildo, and Kim Marriott. "Independence in Constraint Logic Programs." In *ILPS*. MIT Press, 1993, pp. 130–146 (p. 621).

[86] Maria J. García de la Banda, Manuel V. Hermenegildo, and Kim Marriott. *Independence and Search Space Preservation in Dynamically Scheduled Constraint Logic Languages*. Monografía (Informe Técnico). Facultad De Informática (UPM), 1995. http://oa.upm.es/15085/ (p. 621).

[87] Maria J. García de la Banda, Manuel V. Hermenegildo, and Kim Marriott. "Independence in Dynamically Scheduled Logic Languages." In *ALP*. Vol. 1139. Lecture Notes in Computer Science. Springer, 1996, pp. 47–61 (p. 621).

[88] Maria J. García de la Banda, Manuel V. Hermenegildo, and Kim Marriott. "Independence in CLP Languages." *ACM Trans. Program. Lang. Syst.* 22.2 (2000), pp. 296–339 (p. 621).

[89] Francesco Banterle and Roberto Giacobazzi. "A Fast Implementation of the Octagon Abstract Domain on Graphics Hardware." In *SAS*. Vol. 4634. Lecture Notes in Computer Science. Springer, 2007, pp. 315–332 (p. 493).

[90] Roberto Barbuti, Roberto Giacobazzi, and Giorgio Levi. "A General Framework for Semantics-Based Bottom-Up Abstract Interpretation of Logic Programs." *ACM Trans. Program. Lang. Syst.* 15.1 (1993), pp. 133–181 (p. 623).

[91] Clark W. Barrett and Cesare Tinelli. "Satisfiability Modulo Theories." In *Handbook of Model Checking*. Springer, 2018, pp. 305–343 (p. 289).

[92] Jeffrey M. Barth. "Shifting Garbage Collection Overhead to Compile Time." *Commun. ACM*. 20.7 (1977), pp. 513–518 (p. 559).

[93] Gilles Barthe, Sandrine Blazy, Vincent Laporte, David Pichardie, and Alix Trieu. "Verified Translation Validation of Static Analyses." In *CSF.* IEEE Computer Society, 2017, 405–419 (p. 711).

[94] Gilles Barthe, Juan Manuel Crespo, and César Kunz. "Product Programs and Relational Program Logics." *J. Log. Algebr. Meth. Program.* 85.5 (2016), pp. 847–859 (p. 617).

[95] Gilles Barthe, Benjamin Grégoire, and Vincent Laporte. "Provably Secure Compilation of Side-Channel Countermeasures." *IACR Cryptology ePrint Archive.* 2017 (2017), p. 1233 (p. 621).

[96] Ezio Bartocci, Yliès Falcone, Adrian Francalanza, and Giles Reger. "Introduction to Runtime Verification." In *Lectures on Runtime Verification.* Vol. 10457. Lecture Notes in Computer Science. Springer, 2018, pp. 1–33 (p. 387).

[97] Osbert Bastani, Rahul Sharma, Lazaro Clapp, Saswat Anand, and Alexander Aiken. "Eventually Sound Points-to Analysis with Specifications." In *ECOOP.* Vol. 134. LIPIcs. Schloss Dagstuhl—Leibniz—Zentrum für Informatik, 2019, 11:1–11:28 (p. 590).

[98] Anna Becchi and Enea Zaffanella. "An Efficient Abstract Domain for Not Necessarily Closed Polyhedra." In *SAS.* Vol. 11002. Lecture Notes in Computer Science. Springer, 2018, pp. 146–165 (p. 716).

[99] Anna Becchi and Enea Zaffanella. "Revisiting Polyhedral Analysis for Hybrid Systems." In *SAS.* Vol. 11822. Lecture Notes in Computer Science. Springer, 2019, pp. 183–202 (p. 716).

[100] Matthias Beck and Sinai Robins. *Computing the Continuous Discretely: Integer-Point Enumeration in Polyhedra.* 2nd ed. Undergraduate Texts in Mathematics. Springer, 2015 (p. 387).

[101] Heiko Becker, Nikita Zyuzin, Raphaël Monat, Eva Darulova, Magnus O. Myreen, and Anthony C. J. Fox. "A Verified Certificate Checker for Finite-Precision Error Bounds in Coq and HOL4." In *FMCAD.* IEEE, 2018, pp. 1–10 (p. 716).

[102] Eva Beckschulze, Stefan Kowalewski, and Jörg Brauer. "Access-Based Localization for Octagons." *Electr. Notes Theor. Comput. Sci.* 287 (2012), pp. 29–40 (p. 493).

[103] Catriel Beeri and Raghu Ramakrishnan. "On the Power of Magic." *J. Log. Program.* 10.3&4 (1991), pp. 255–299 (p. 684).

[104] Hans Bekić. "Definable Operation in General Algebras, and the Theory of Automata and Flowcharts." In *Programming Languages and Their Definition.* Vol. 177. Lecture Notes in Computer Science. Springer, 1984, pp. 30–55 (p. 168).

[105] Luciano Bello, Daniel Hedin, and Andrei Sabelfeld. "Value Sensitivity and Observable Abstract Values for Information Flow Control." In *LPAR.* Vol. 9450. Lecture Notes in Computer Science. Springer, 2015, pp. 63–78 (p. 620).

[106] Michaël Benado. "Ensembles ordonnés et fonctions monotones." *Rev. Fac. Ci. Univ. Lisboa.* 14.(II–A) (1973), pp. 165–194 (p. 131).

[107] Florian Benz, Andreas Hildebrandt, and Sebastian Hack. "A Dynamic Program Analysis to Find Floating-Point Accuracy Problems." In *PLDI.* ACM, 2012, pp. 453–462 (p. 387).

[108] Claude Berge. *Théorie des graphes et ses applications.* Dunod, 1958 (p. 462, 480).

[109] Claude Berge. *Graphs and Hypergraphs.* North-Holland, 1973 (p. 459).

[110] Jean-Francois Bergeretti and Bernard Carré. "Information-Flow and Data-Flow Analysis of While-Programs." *ACM Trans. Program. Lang. Syst.* 7.1 (1985), pp. 37–61 (p. 621).

[111] Bernard Berthomieu and Miguel Menasche. "An Enumerative Approach for Analyzing Time Petri Nets." In *IFIP Congress.* 1983, pp. 41–46 (p. 481).

[112] Yves Bertot, Benjamin Grégoire, and Xavier Leroy. "A Structured Approach to Proving Compiler Optimizations Based on Dataflow Analysis." In *TYPES.* Vol. 3839. Lecture Notes in Computer Science. Springer, 2004, pp. 66–81 (p. 514).

[113] Julien Bertrane, Patrick Cousot, Radhia Cousot, Jérôme Feret, Laurent Mauborgne, Antoine Miné, and Xavier Rival. "Static Analysis and Verification of Aerospace Software by Abstract Interpretation." *Foundations and Trends in Programming Languages.* 2.2–3 (2015), pp. 71–190 (p. 304, 344, 349, 365, 415, 492, 551, 589, 590, 707).

[114] Al Bessey, Ken Block, Benjamin Chelf, Andy Chou, Bryan Fulton, Seth Hallem, Charles-Henri Gros, Asya Kamsky, Scott McPeak, and Dawson R. Engler. "A Few Billion Lines of Code Later: Using Static Analysis to Find Bugs in the Real World." *Commun. ACM.* 53.2 (2010), pp. 66–75 (p. 710, 711).

[115] Dirk Beyer, Sumit Gulwani, and David A. Schmidt. "Combining Model Checking and Data-Flow Analysis." In *Handbook of Model Checking.* Springer, 2018, pp. 493–540 (p. 512).

[116] Étienne Bézout. *Recherches sur le degré des équations résultantes de l'évanouissement des inconnues et sur les moyens qu'on doit employer pour trouver ces équations.* Mémoires De lL'Académie Royale des Sciences, 1764, pp. 288–338 (p. 353).

[117] Pavol Bielik, Veselin Raychev, and Martin T. Vechev. "Learning a Static Analyzer From Data." In *CAV (1).* Vol. 10426. Lecture Notes in Computer Science. Springer, 2017, pp. 233–253 (p. 319).

[118] Armin Biere, Cyrille Artho, and Viktor Schuppan. "Liveness Checking as Safety Checking." *Electr. Notes Theor. Comput. Sci.* 66.2 (2002), pp. 160–177 (p. 161).

[119] Armin Biere, Alessandro Cimatti, Edmund M. Clarke Jr., Ofer Strichman, and Yunshan Zhu. "Bounded Model Checking." *Advances in Computers.* 58 (2003), pp. 117–148 (p. 548, 710).

[120] Garrett Birkhoff. *Lattice Theory.* 1st ed. American Mathematical Society, Colloquium Publications, Volume XXV, 1940 (p. 105, 112, 113).

[121] Garrett Birkhoff. *Lattice Theory.* 3rd ed. American Mathematical Society, Colloquium Publications, Volume XXV, 1973 (p. 23, 105, 106).

[122] Paul E. Black and Kanwardeep Singh Walia. *SATE VI Ockham Sound Analysis Criteria.* Tech. rep. Intern. Rep. 8304. National Institute of Standards and Technology, May 2020. https://doi.org/10.6028/NIST.IR.8304 (p. 717).

[123] Bruno Blanchet, Patrick Cousot, Radhia Cousot, Jérôme Feret, Laurent Mauborgne, Antoine Miné, David Monniaux, and Xavier Rival. "Design and Implementation of a Special-Purpose Static Program Analyzer for Safety-Critical Real-Time Embedded Software." In *The Essence of Computation.* Vol. 2566. Lecture Notes in Computer Science. Springer, 2002, pp. 85–108 (p. 406).

[124] Bruno Blanchet, Patrick Cousot, Radhia Cousot, Jérôme Feret, Laurent Mauborgne, Antoine Miné, David Monniaux, and Xavier Rival. "A Static Analyzer for Large Safety-Critical Software." In *PLDI.* ACM, 2003, pp. 196–207 (p. 397, 406, 557).

[125] Sandrine Blazy, David Bühler, and Boris Yakobowski. "Structuring Abstract Interpreters Through State and Value Abstractions." In *VMCAI.* Vol. 10145. Lecture Notes in Computer Science. Springer, 2017, pp. 112–130 (p. 432).

[126] Sandrine Blazy, Vincent Laporte, and David Pichardie. "An Abstract Memory Functor for Verified C Static Analyzers." In *ICFP.* ACM, 2016, pp. 325–337 (p. 711).

[127] Sandrine Blazy, Vincent Laporte, and David Pichardie. "Verified Abstract Interpretation Techniques for Disassembling Low-Level Self-Modifying Code." *J. Autom. Reasoning.* 56.3 (2016), pp. 283–308 (p. 711).

[128] Thomas S. Blyth. *Lattices and Ordered Algebraic Structures.* Universitext. Springer, 2005 (p. 115).

[129] François Bobot, Jean-Christophe Filliâtre, Claude Marché, Guillaume Melquiond, and Andrei Paskevich. *The Why3 Platform, Version 1.3.* University Paris–Sud, CNRS, Inria, Sept. 2020. http://why3.lri.fr (p. 291).

[130] Benoit Boissinot, Florian Brandner, Alain Darte, Benoît Dupont De Dinechin, and Fabrice Rastello. "A Non–Iterative Data–Flow Algorithm for Computing Liveness Sets in Strict SSA Programs." In *APLAS.* Vol. 7078. Lecture Notes in Computer Science. Springer, 2011, pp. 137–154 (p. 506).

[131] George Boole. *An Investigation of the Laws of Thought on Which Are Founded the Mathematical Theories of Logic and Probabilities.* Walton and Maberly, 1884 (p. 9, 10).

[132] Dragan Bošnački. "On Some Galois Connection Based Abstractions for the Mu-Calculus." In *FM.* Vol. 3582. Lecture Notes in Computer Science. Springer, 2005, pp. 366–381 (p. 131).

[133] Amed Bouabdallah. "Propriétés des programmes parallèles." Thèse de doctorat en informatique. Université De Besançon, 1993. http://refdoc-Info.inist.fr/c4/refdoc.html?Cpsidt=158587 (p. 161).

[134] Dominique Boucher and Marc Feeley. "Abstract Compilation: A New Implementation Paradigm for Static Analysis." In *CC.* Vol. 1060. Lecture Notes in Computer Science. Springer, 1996, pp. 192–207 (p. 250).

[135] Olivier Bouissou, Yassamine Seladji, and Alexandre Chapoutot. "Acceleration of the Abstract Fixpoint Computation in Numerical Program Analysis." *J. Symb. Comput.* 47.12 (2012), pp. 1479–1511 (p. 406).

[136] Nicolas Bourbaki. *Théorie des ensembles.* Hermann, Paris, 1956 (p. 172).

[137] Nicolas Bourbaki. *Elements of Mathematics: General Topology.* Addison–Wesley, 1966 (p. 147).

[138] François Bourdoncle. "Efficient Chaotic Iteration Strategies with Widenings." In *Formal Methods in Programming and Their Applications.* Vol. 735. Lecture Notes in Computer Science. Springer, 1993, pp. 128–141 (p. 257, 409).

[139] Pierre Boutillier, Ferdinanda Camporesi, Jean Coquet, Jérôme Feret, Kim Quyên Lý, Nathalie Théret, and Pierre Vignet. "KaSa: A Static Analyzer for Kappa." In *CMSB.* Vol. 11095. Lecture Notes in Computer Science. Springer, 2018, pp. 285–291 (p. 718).

[140] Pierre Boutillier, Ioana Cristescu, and Jérôme Feret. "Counters in Kappa: Semantics, Simulation, and Static Analysis." In *ESOP.* Vol. 11423. Lecture Notes in Computer Science. Springer, 2019, pp. 176–204 (p. 5, 718).

[141] Rémy Boutonnet and Nicolas Halbwachs. "Improving the Results of Program Analysis by Abstract Interpretation Beyond the Decreasing Sequence." *Formal Methods in System Design.* 53.3 (2018), pp. 384–406 (p. 412).

[142] William J. Bowman and Amal Ahmed. "Noninterference for Free." In *ICFP.* ACM, 2015, pp. 101–113 (p. 619).

[143] Stephen P. Boyd and Lieven Vandenberghe. *Convex Optimization.* Cambridge University Press, 2004 (p. 167).

[144] Marius Bozga, Codruta Gîrlea, and Radu Iosif. "Iterating Octagons." In *TACAS.* Vol. 5505. Lecture Notes in Computer Science. Springer, 2009, pp. 337–351 (p. 493).

[145] Guillaume P. Brat, Jorge A. Navas, Nija Shi, and Arnaud J. Venet. "IKOS: A Framework for Static Analysis Based on Abstract Interpretation." In *SEFM.* Vol. 8702. Lecture Notes in Computer Science. Springer, 2014, pp. 271–277 (p. 714).

[146] Vasco Brattka. "A Galois Connection Between Turing Jumps and Limits." *Logical Methods in Computer Science.* 14.3 (2018) (p. 131).

[147] Jörg Brauer, Andy King, and Stefan Kowalewski. "Abstract Interpretation of Microcontroller Code: Intervals Meet Congruences." *Sci. Comput. Program.* 78.7 (2013), pp. 862–883 (p. 427).

[148] Fraser Brown, Andres Nötzli, and Dawson R. Engler. "How to Build Static Checking Systems Using Orders of Magnitude Less Code." In *ASPLOS*. ACM, 2016, pp. 143–157 (p. 709).

[149] Fraser Brown, John Renner, Andres Nôtzli, Sorin Lerner, Hovav Shacham, and Deian Stefan. "Towards a verified range analysis for JavaScript JITs." In *PLDI*. ACM, 2020, pp. 135–150 (p. 387).

[150] Randal E. Bryant. "Graph-Based Algorithms for Boolean Function Manipulation." *IEEE Trans. Computers.* 35.8 (1986), pp. 677–691 (p. 398, 548).

[151] Janusz A. Brzozowski. "Derivatives of Regular Expressions." *J. ACM.* 11.4 (1964), pp. 481–494 (p. 532, 550).

[152] Francisco Bueno, Maria J. García de la Banda, and Manuel V. Hermenegildo. "Effectivness of Abstract Interpretation in Automatic Parallelization: A Case Study in Logic Programming." *ACM Trans. Program. Lang. Syst.* 21.2 (1999), pp. 189–239 (p. 621).

[153] Alexandra Bugariu, Valentin Wüstholz, Maria Christakis, and Peter Müller. "Automatically Testing Implementations of Numerical Abstract Domains." In *ASE*. ACM, 2018, pp. 768–778 (p. 492).

[154] Sándor Buglyó and Árpád Száz. "A More Important Galois Connection Between Distance Functions and Inequality Relations." *Scientia, Series a: Mathematical Sciences.* 18 (2019), pp. 17–38 (p. 131).

[155] David Bühler. "Structuring an Abstract Interpreter Through Value and State Abstractions: EVA, an Evolved Value Analysis for Frama-C. (Structurer un interpréteur abstrait au moyen d'abstractions de valeurs et d'états : EVA, une analyse de valeur évoluée pour Frama-C)." PhD thesis. IRISA, University of Rennes 1, 2017. https://hal.archives-ouvertes.fr/tel-01664726/Document (p. 715).

[156] Jo Van Bulck, Marina Minkin, Ofir Weisse, Daniel Genkin, Baris Kasikci, Frank Piessens, Mark Silberstein, Thomas F. Wenisch, Yuval Yarom, and Raoul Strackx. "Foreshadow: Extracting the Keys to the Intel SGX Kingdom with Transient Out-of-Order Execution." In *USENIX Security Symposium*. USENIX Association, 2018, pp. 991–1008 (p. 79).

[157] Rod M. Burstall. "Proving Properties of Programs by Structural Induction." *Computer Journal.* 12.1 (1969), pp. 41–48 (p. 22, 182).

[158] Gurvan Cabon and Alan Schmitt. "Annotated Multisemantics to Prove Non-Interference Analyses." In *PLAS@CCS*. ACM, 2017, pp. 49–62 (p. 620).

[159] Inma P. Cabrera, Pablo Cordero, Francisca García-Pardo, Manuel Ojeda-Aciego, and Bernard De Baets. "Galois Connections Between a Fuzzy Preordered Structure and a General Fuzzy Structure." *IEEE Trans. Fuzzy Systems.* 26.3 (2018), pp. 1274–1287 (p. 131).

[160] David Cachera, Thomas P. Jensen, Arnaud Jobin, and Florent Kirchner. "Inference of Polynomial Invariants for Imperative Programs: A Farewell to Gröbner Bases." *Sci. Comput. Program.* 93 (2014), pp. 89–109 (p. 457).

[161] David Cachera and David Pichardie. "Programmation d'un interpréteur abstrait certifié en logique constructive." *Technique Et Science Informatiques.* 30.4 (2011), pp. 381–408 (p. 35, 711).

[162] Cristiano Calcagno, Dino Distefano, Jérémy Dubreil, Dominik Gabi, Pieter Hooimeijer, Martino Luca, Peter W. O'Hearn, Irene Papakonstantinou, Jim Purbrick, and Dulma Rodriguez. "Moving Fast with Software Verification." In *NFM*. Vol. 9058. Lecture Notes in Computer Science. Springer, 2015, pp. 3–11 (p. 432).

[163] Roy H. Campbell and A. Nico Habermann. "The Specification of Process Synchronization by Path Expressions." In *Symposium on Operating Systems.* Vol. 16. Lecture Notes in Computer Science. Springer, 1974, pp. 89–102 (p. 550).

[164] Ferdinanda Camporesi, Jérôme Feret, and Jonathan Hayman. "Context-Sensitive Flow Analyses: A Hierarchy of Model Reductions." In *CMSB.* Vol. 8130. Lecture Notes in Computer Science. Springer, 2013, pp. 220–233 (p. 718).

[165] Bernard Carré. *Graphs and Networks.* Clarendon Press, 1979. https://archive.org/details/GraphsAndNetworks (p. 459, 479, 480).

[166] Simon Castellan, Pierre Clairambault, and Peter Dybjer. "Categories with Families: Unityped, Simply Typed, and Dependently Typed." *CoRR.* abs/1904.00827 (2019) (p. 14).

[167] Gabriele Castellini. *Categorical Closure Operators.* Birkhäuser, 2003 (p. 131).

[168] Arthur Cayley. "A Memoir on the Theory of Matrices." *Proceedings of the Royal Society of London, Philosophical Transactions of the Royal Society.* 148 (Jan. 1858), pp. 17–37 (p. 441).

[169] Sagar Chaki, Arie Gurfinkel, and Ofer Strichman. "Decision Diagrams for Linear Arithmetic." In *FMCAD.* IEEE, 2009, pp. 53–60 (p. 398).

[170] Edward Chang, Zohar Manna, and Amir Pnueli. "The Safety-Progress Classification." In Friedrich L. Bauer, Wilfried Brauer, and Helmut Schwichtenberg, eds. *Logic and Algebra of Specification.* Vol. 94. NATO ASI Series (Series F: Computer & Systems Sciences), Springer, 1993, pp. 143–202 (p. 161).

[171] Alexandre Chapoutot and Matthieu Martel. "Abstract Simulation: A Static Analysis of Simulink Models." In *ICESS.* IEEE Computer Society, 2009, pp. 83–92 (p. 718).

[172] Aziem Chawdhary and Andy King. "Compact Difference Bound Matrices." In *APLAS.* Vol. 10695. Lecture Notes in Computer Science. Springer, 2017, pp. 471–490 (p. 492).

[173] Aziem Chawdhary, Edward Robbins, and Andy King. "Simple and Efficient Algorithms for Octagons." In *APLAS.* Vol. 8858. Lecture Notes in Computer Science. Springer, 2014, pp. 296–313 (p. 492).

[174] Junjie Chen and Patrick Cousot. "A Binary Decision Tree Abstract Domain Functor." In *SAS.* Vol. 9291. Lecture Notes in Computer Science. Springer, 2015, pp. 36–53 (p. 433).

[175] Liqian Chen, Jiangchao Liu, Antoine Miné, Deepak Kapur, and Ji Wang. "An Abstract Domain to Infer Octagonal Constraints with Absolute Value." In *SAS.* Vol. 8723. Lecture Notes in Computer Science. Springer, 2014, pp. 101–117 (p. 493).

[176] Liqian Chen, Antoine Miné, Ji Wang, and Patrick Cousot. "An Abstract Domain to Discover Interval Linear Equalities." In *VMCAI.* Vol. 5944. Lecture Notes in Computer Science. Springer, 2010, pp. 112–128 (p. 457).

[177] James Cheney, Amal Ahmed, and Umut A. Acar. "Provenance as Dependency Analysis." *Mathematical Structures in Computer Science.* 21.6 (2011), pp. 1301–1337 (p. 593, 618).

[178] Marc Chevalier. "Proving the Security of Software—Intensive Embedded Systems by Abstract Interpretation. (Analyse de la sécurité de systèmes critiques embarqués à forte composante logicielle par interprétation abstraite)." PhD thesis. Université PSL—Paris, Nov. 2020 (p. 79, 322, 431, 432, 588, 718).

[179] Marc Chevalier and Jérôme Feret. "Sharing Ghost Variables in a Collection of Abstract Domains." In *VMCAI.* Vol. 11990. Lecture Notes in Computer Science. Springer, 2020, pp. 158–179 (p. 432, 433).

[180] Noam Chomsky. "Three Models for the Description of Language." *IRE Transactions on Information Theory.* 2.3 (1956), pp. 113–124. http://dx.doi.org/10.1109/TIT.1956.1056813 (p. 20, 39).

[181] Noam Chomsky and Marcel Paul Schützenberger. "The Algebraic Theory of Context-Free Languages." In P. Braffort and D. Hirshberg, eds. *Computer Programming and Formal Systems.* North-Holland, 1963, pp. 118–161 (p. 184).

[182] Andy Chou. "Static Analysis in Industry (Talk)." In *POPL*. ACM, 2014. https://popl.mpi-sws.org/2014/andy.pdf (p. 393, 710, 713).

[183] Maria Christakis, Peter Müller, and Valentin Wüstholz. "An Experimental Evaluation of Deliberate Unsoundness in a Static Program Analyzer." In *VMCAI*. Vol. 8931. Lecture Notes in Computer Science. Springer, 2015, pp. 336–354 (p. 709).

[184] Alonzo Church. "An Unsolvable Problem of Elementary Number Theory." *American Journal of Mathematics*. 58.2 (Apr. 1936), pp. 345–363 (p. 98).

[185] Alonzo Church. "A Formulation of the Simple Theory of Types." *J. Symb. Log.* 5.2 (1940), pp. 56–68 (p. 649, 671).

[186] Robert Clarisó and Jordi Cortadella. "The Octahedron Abstract Domain." *Sci. Comput. Program.* 64.1 (2007), pp. 115–139 (p. 492).

[187] Edmund M. Clarke Jr. "Programming Language Constructs for Which It Is Impossible to Obtain Good Hoare Axiom Systems." *J. ACM*. 26.1 (1979), pp. 129–147 (p. 295).

[188] Edmund M. Clarke Jr. and E. Allen Emerson. "Design and Synthesis of Synchronization Skeletons Using Branching-Time Temporal Logic." In *Logic of Programs*. Vol. 131. Lecture Notes in Computer Science. Springer, 1981, pp. 52–71 (p. 527).

[189] Edmund M. Clarke Jr., Orna Grumberg, Daniel Kroening, Doron A. Peled, and Helmut Veith. *Model Checking*. 2nd ed. MIT Press, 2018 (p. 527, 539).

[190] Edmund M. Clarke Jr., Orna Grumberg, and David E. Long. "Model Checking and Abstraction." *ACM Trans. Program. Lang. Syst.* 16.5 (1994), pp. 1512–1542 (p. 116).

[191] Edmund M. Clarke Jr., Thomas A. Henzinger, Helmut Veith, and Roderick Bloem, eds. *Handbook of Model Checking*. Springer, 2018 (p. 527, 539, 622, 721).

[192] Edmund M. Clarke Jr., William Klieber, Milos Novácek, and Paolo Zuliani. "Model Checking and the State Explosion Problem." In *LASER Summer School*. Vol. 7682. Lecture Notes in Computer Science. Springer, 2011, pp. 1–30 (p. 33, 548).

[193] Edmund M. Clarke Jr. and Qinsi Wang. "2^5 Years of Model Checking." In *Ershov Memorial Conference*. Vol. 8974. Lecture Notes in Computer Science. Springer, 2014, pp. 26–40 (p. 225).

[194] Michael R. Clarkson and Fred B. Schneider. "Hyperproperties." *Journal of Computer Security*. 18.6 (2010), pp. 1157–1210 (p. 91, 92).

[195] John Cocke and Jacob T. Schwartz. *Programming Languages and Their Compilers : Preliminary Notes*. Courant Institute of Mathematical Sciences, New York University, New York, USA, 1969, Second Revised Version, April 1970 (p. 259, 270).

[196] *CodeHawk*. Kestrel Technology, 2020. https://github.com/static-analysis-engineering/codehawk (p. 714).

[197] Michael Codish and Bart Demoen. "Analyzing Logic Programs Using "PROP"-ositional Logic Programs and a Magic Wand." *J. Log. Program.* 25.3 (1995), pp. 249–274 (p. 684).

[198] Michael Codish, Anne Mulkers, Maurice Bruynooghe, Maria J. García de la Banda, and Manuel V. Hermenegildo. "Improving Abstract Interpretations by Combining Domains." *ACM Trans. Program. Lang. Syst.* 17.1 (1995), pp. 28–44 (p. 432).

[199] Ellis S. Cohen. "Information Transmission in Computational Systems." In *SOSP*. ACM, 1977, pp. 133–139 (p. 591, 593, 619).

[200] Christopher Colby and Peter Lee. "Trace-Based Program Analysis." In *POPL*. ACM Press, 1996, pp. 195–207 (p. 192).

[201] George E. Collins. "Hauptvortrag: Quantifier Elimination for Real Closed Fields by Cylindrical Algebraic Decomposition." In *Automata Theory and Formal Languages*. Vol. 33. Lecture Notes in Computer Science. Springer, 1975, pp. 134–183 (p. 290, 402).

[202] Alain Colmerauer. "Prolog and Infinite Trees." In Keith L. Clark and Sten-Åke Tärnlund. *Logic Programming*. Academic Press, 1982, pp. 231–251 (p. 647).

[203] Alain Colmerauer. "Prolog in 10 Figures." *Commun. ACM.* 28.12 (1985), pp. 1296–1310 (p. 623).

[204] Alain Colmerauer and Philippe Roussel. "The Birth of Prolog." In *HOPL Preprints.* ACM, 1993, pp. 37–52 (p. 623).

[205] Livio Colussi and Elena Marchiori. "A Predicate Transformer for Unification." In *JICSLP.* MIT Press, 1992, pp. 67–81 (p. 144).

[206] João Luiz Dihl Comba and Jorge Stolfi. "Affine Arithmetic and Its Applications to Computer Graphics." *IEEE SIBGRAPI.* (1993), pp. 9–18 (p. 387).

[207] Hubert Comon and Yan Jurski. "Multiple Counters Automata, Safety Analysis and Presburger Arithmetic." In *CAV.* Vol. 1427. Lecture Notes in Computer Science. Springer, 1998, pp. 268–279 (p. 411).

[208] Stephen A. Cook. "Soundness and Completeness of an Axiom System for Program Verification." *SIAM J. Comput.* 7.1 (1978), pp. 70–90 (p. 303).

[209] Stephen A. Cook. "Corrigendum: Soundness and Completeness of an Axiom System for Program Verification." *SIAM J. Comput.* 10.3 (1981), p. 612 (p. 303).

[210] The Coq Development Team. *The Coq Reference Manual, Release 8.12.1.* INRIA, Nov. 13, 2020. https://coq.inria.fr/documentation (p. 225).

[211] Agostino Cortesi, Giulia Costantini, and Pietro Ferrara. "A Survey on Product Operators in Abstract Interpretation." In *Festschrift for Dave Schmidt.* Vol. 129. EPTCS. 2013, pp. 325–336 (p. 432).

[212] Agostino Cortesi, Pietro Ferrara, Raju Halder, and Matteo Zanioli. "Combining Symbolic and Numerical Domains for Information Leakage Analysis." *Trans. Computational Science.* 31 (2018), pp. 98–135 (p. 617, 620).

[213] Agostino Cortesi, Gilberto Filé, Roberto Giacobazzi, Catuscia Palamidessi, and Francesco Ranzato. "Complementation in Abstract Interpretation." *ACM Trans. Program. Lang. Syst.* 19.1 (1997), pp. 7–47 (p. 433).

[214] Agostino Cortesi, Baudouin Le Charlier, and Pascal Van Hentenryck. "Combinations of Abstract Domains for Logic Programming: Open Product and Generic Pattern Construction." *Sci. Comput. Program.* 38.1–3 (2000), pp. 27–71 (p. 432).

[215] Agostino Cortesi and Matteo Zanioli. "Widening and Narrowing Operators for Abstract Interpretation." *Computer Languages, Systems & Structures.* 37.1 (2011), pp. 24–42 (p. 431).

[216] Alexandru Costan, Stéphane Gaubert, Éric Goubault, Matthieu Martel, and Sylvie Putot. "A Policy Iteration Algorithm for Computing Fixed Points in Static Analysis of Programs." In *CAV.* Vol. 3576. Lecture Notes in Computer Science. Springer, 2005, pp. 462–475 (p. 412).

[217] Nathanaël Courant and Caterina Urban. "Precise Widening Operators for Proving Termination by Abstract Interpretation." In *TACAS (1).* Vol. 10205. Lecture Notes in Computer Science. 2017, pp. 136–152 (p. 693).

[218] Richard Courant and Herbert Robbins. *What Is Mathematics? An Elementary Approach to Ideas and Methods (Revised by Ian Stewart).* 2nd ed. Oxford University Press, 1941, 1969, 1996 (p. 9).

[219] Bruno Courcelle. "Fundamental Properties of Infinite Trees." *Theor. Comput. Sci.* 25 (1983), pp. 95–169 (p. 184).

[220] Patrick Cousot. *Asynchronous Iterative Methods for Solving a Fixed Point System of Monotone Equations in a Complete Lattice.* Tech. rep. R.R. 88. 15 p. Laboratoire IMAG, Université de Grenoble Alpes, Sept. 1977 (p. 257, 269, 409, 698).

[221] Patrick Cousot. "Méthodes itératives de construction et d'approximation de points fixes d'opérateurs monotones sur un treillis, analyse sémantique de programmes." Thèse

d'État ès Sciences Mathématiques. Université de Grenoble Alpes, Mar. 21, 1978 (p. 130, 209, 257, 269, 278, 347, 409, 412, 527, 693, 699, 700, 702).

[222] Patrick Cousot. "Semantic Foundations of Program Analysis." In Steven S. Muchnick and Neil D. Jones. *Program Flow Analysis—Theory and Applications.* Prentice Hall Software Series. Prentice Hall, 1981. Chap. 10, pp. 303–342 (p. 141, 144, 684, 693).

[223] Patrick Cousot. "Methods and Logics for Proving Programs." In Jan van Leeuwen, ed. *Handbook of Theoretical Computer Science, Volume B: Formal Models and Semantics (B).* MIT Press, 1990, pp. 841–994 (p. 79).

[224] Patrick Cousot. "Types as Abstract Interpretations." In *POPL.* ACM Press, 1997, pp. 316–331 (p. 410, 432, 623, 630, 649, 663, 671).

[225] Patrick Cousot. "The Calculational Design of a Generic Abstract Interpreter." In M. Broy and R. Steinbrüggen, eds. *Calculational System Design.* NATO ASI Series F. IOS Press, 1999 (p. 334, 344, 374, 431, 506).

[226] Patrick Cousot. "Partial Completeness of Abstract Fixpoint Checking." In *SARA.* Vol. 1864. Lecture Notes in Computer Science. Springer, 2000, pp. 1–25 (p. 319, 414, 415, 551).

[227] Patrick Cousot. "Constructive Design of a Hierarchy of Semantics of a Transition System by Abstract Interpretation." *Theor. Comput. Sci.* 277.1–2 (2002), pp. 47–103 (p. 85, 94, 144, 192, 242).

[228] Patrick Cousot. "Verification by Abstract Interpretation." In *Verification: Theory and Practice.* Vol. 2772. Lecture Notes in Computer Science. Springer, 2003, pp. 243–268 (p. 318).

[229] Patrick Cousot. "Integrating Physical Systems in the Static Analysis of Embedded Control Software." In *APLAS.* Vol. 3780. Lecture Notes in Computer Science. Springer, 2005, pp. 135–138 (p. 718).

[230] Patrick Cousot. "Proving Program Invariance and Termination by Parametric Abstraction, Lagrangian Relaxation and Semidefinite Programming." In *VMCAI.* Vol. 3385. Lecture Notes in Computer Science. Springer, 2005, pp. 1–24 (p. 167, 693, 694).

[231] Patrick Cousot. "Abstracting Induction by Extrapolation and Interpolation." In *VMCAI.* Vol. 8931. Lecture Notes in Computer Science. Springer, 2015, pp. 19–42 (p. 409, 412).

[232] Patrick Cousot. "On Fixpoint/Iteration/Variant Induction Principles for Proving Total Correctness of Programs with Denotational Semantics." In *LOPSTR.* Vol. 12042. Lecture Notes in Computer Science. Springer, 2019, pp. 3–18 (p. 278, 415).

[233] Patrick Cousot. "Calculational design of a regular model checker by abstract interpretation." *Theor. Comput. Sci.* 869 (2021), pp. 62–84 (p. 387, 388).

[234] Patrick Cousot and Radhia Cousot. "Static Determination of Dynamic Properties of Programs." In *Proceedings of the Second International Symposium on Programming.* Dunod, 1976, pp. 106–130 (p. 109, 250, 344, 398, 412).

[235] Patrick Cousot and Radhia Cousot. "Abstract Interpretation: A Unified Lattice Model for Static Analysis of Programs by Construction or Approximation of Fixpoints." In *POPL.* ACM, 1977, pp. 238–252 (p. 4, 23, 109, 144, 210, 398, 412, 431, 527).

[236] Patrick Cousot and Radhia Cousot. "Automatic Synthesis of Optimal Invariant Assertions: Mathematical Foundations." *SIGART Newsl.* 64 (1977), pp. 1–12 (p. 255, 409, 551).

[237] Patrick Cousot and Radhia Cousot. "Static Determination of Dynamic Properties of Generalized Type Unions." In *Language Design for Reliable Software.* 1977, pp. 77–94 (p. 671).

[238] Patrick Cousot and Radhia Cousot. "Static Determination of Dynamic Properties of Recursive Procedures." In *IFIP Conf. on Formal Description of Programming Concepts.* Edited by E.J. Neuhold. North-Holland, 1977, pp. 237–277 (p. 161, 587).

[239] Patrick Cousot and Radhia Cousot. "A Constructive Characterization of the Lattices of All Retractions, Pre-closure, Quasi-closure and Closure Operators on a Complete Lattice." *Portugaliæ Mathematica.* 38.2 (1979), pp. 185–198 (p. 347).

[240] Patrick Cousot and Radhia Cousot. "Constructive Versions of Tarski's Fixed Point Theorems." *Pacific Journal of Mathematics.* 82.1 (1979), pp. 43–57 (p. 172, 207, 208, 249, 278).

[241] Patrick Cousot and Radhia Cousot. "Systematic Design of Program Analysis Frameworks." In *POPL.* ACM Press, 1979, pp. 269–282 (p. 4, 94, 130, 131, 144, 209, 210, 347, 417, 419, 422, 433, 514, 617, 693).

[242] Patrick Cousot and Radhia Cousot. *Reasoning About Program Invariance Proof Methods.* Res. Rep. CRIN-80-P050. Centre de Recherche en Informatique de Nancy (CRIN), Institut National Polytechnique de Lorraine, July 1980, p. 22 (p. 217).

[243] Patrick Cousot and Radhia Cousot. "Induction Principles for Proving Invariance Properties of Programs." In *Tools & Notions for Program Construction: An Advanced Course.* Edited by D. Néel. Cambridge University Press, June 1982, pp. 75–119 (p. 278, 291, 693).

[244] Patrick Cousot and Radhia Cousot. "Abstract Interpretation and Application to Logic Programs." *J. Log. Program.* 13.2&3 (1992), pp. 103–179 (p. 702).

[245] Patrick Cousot and Radhia Cousot. "Abstract Interpretation Frameworks." *J. Log. Comput.* 2.4 (1992), pp. 511–547 (p. 131).

[246] Patrick Cousot and Radhia Cousot. "Comparing the Galois Connection and Widening/Narrowing Approaches to Abstract Interpretation." In *PLILP.* Vol. 631. Lecture Notes in Computer Science. Springer, 1992, pp. 269–295 (p. 409, 412).

[247] Patrick Cousot and Radhia Cousot. "Inductive Definitions, Semantics and Abstract Interpretation." In *POPL.* ACM Press, 1992, pp. 83–94 (p. 85, 184).

[248] Patrick Cousot and Radhia Cousot. "Galois Connection Based Abstract Interpretations for Strictness Analysis (Invited Paper)." In *Formal Methods in Programming and Their Applications.* Vol. 735. Lecture Notes in Computer Science. Springer, 1993, pp. 98–127 (p. 199, 387).

[249] Patrick Cousot and Radhia Cousot. "Higher Order Abstract Interpretation (and Application to Comportment Analysis Generalizing Strictness, Termination, Projection, and PER Analysis)." In *ICCL.* IEEE Computer Society, 1994, pp. 95–112 (p. 113, 131).

[250] Patrick Cousot and Radhia Cousot. "Compositional and Inductive Semantic Definitions in Fixpoint, Equational, Constraint, Closure-Condition, Rule-Based and Game-Theoretic Form." In *CAV.* Vol. 939. Lecture Notes in Computer Science. Springer, 1995, pp. 293–308 (p. 184, 210, 586).

[251] Patrick Cousot and Radhia Cousot. "Formal Language, Grammar and Set-Constraint-Based Program Analysis by Abstract Interpretation." In *FPCA.* ACM, 1995, pp. 170–181 (p. 257, 269, 397, 410, 586, 589).

[252] Patrick Cousot and Radhia Cousot. "Parallel Combination of Abstract Interpretation and Model-Based Automatic Analysis of Software." In *Proceedings of the First ACM SIGPLAN Workshop on Automatic Analysis of Software, AAS'97.* Edited by R. Cleaveland and D. Jackson. ACM Press, Jan. 1997, pp. 91–98 (p. 702).

[253] Patrick Cousot and Radhia Cousot. "Refining Model Checking by Abstract Interpretation." *Autom. Softw. Eng.* 6.1 (1999), pp. 69–95 (p. 702).

[254] Patrick Cousot and Radhia Cousot. "Temporal Abstract Interpretation." In *POPL.* ACM, 2000, pp. 12–25 (p. 497, 501, 550, 551, 707).

[255] Patrick Cousot and Radhia Cousot. "Modular Static Program Analysis." In *CC.* Vol. 2304. Lecture Notes in Computer Science. Springer, 2002, pp. 159–178 (p. 141).

[256] Patrick Cousot and Radhia Cousot. "On Abstraction in Software Verification." In *CAV.* Vol. 2404. Lecture Notes in Computer Science. Springer, 2002, pp. 37–56 (p. 693, 702).

[257] Patrick Cousot and Radhia Cousot. "Systematic Design of Program Transformation Frameworks by Abstract Interpretation." In *POPL*. ACM, 2002, pp. 178–190 (p. 250, 514).

[258] Patrick Cousot and Radhia Cousot. "Parsing as Abstract Interpretation of Grammar Semantics." *Theor. Comput. Sci.* 290.1 (2003), pp. 531–544 (p. 61).

[259] Patrick Cousot and Radhia Cousot. "Basic Concepts of Abstract Interpretation." In *Building the Information Society*. Edited by René Jacquard. Springer, 2004, pp. 359–366 (p. 460, 463, 480).

[260] Patrick Cousot and Radhia Cousot. "Bi-Inductive Structural Semantics." *Inf. Comput.* 207.2 (2009), pp. 258–283 (p. 85, 184, 499, 505).

[261] Patrick Cousot and Radhia Cousot. "A Gentle Introduction to Formal Verification of Computer Systems by Abstract Interpretation." In *Logics and Languages for Reliability and Security*. Edited by J. Esparza, O. Grumberg, and M. Broy. NATO Science Series III: Computer and Systems Sciences. IOS Press, 2010, pp. 1–29 (p. 4, 717).

[262] Patrick Cousot and Radhia Cousot. "Grammar Semantics, Analysis and Parsing by Abstract Interpretation." *Theor. Comput. Sci.* 412.44 (2011), pp. 6135–6192 (p. 61).

[263] Patrick Cousot and Radhia Cousot. "An Abstract Interpretation Framework for Termination." In *POPL*. ACM, 2012, pp. 245–258 (p. 79, 192, 280, 281, 693, 715).

[264] Patrick Cousot and Radhia Cousot. "A Galois Connection Calculus for Abstract Interpretation." In *POPL*. ACM, 2014, pp. 3–4 (p. 131).

[265] Patrick Cousot, Radhia Cousot, Manuel Fähndrich, and Francesco Logozzo. "Automatic Inference of Necessary Preconditions." In *VMCAI*. Vol. 7737. Lecture Notes in Computer Science. Springer, 2013, pp. 128–148 (p. 141, 684, 692).

[266] Patrick Cousot, Radhia Cousot, Jérôme Feret, Laurent Mauborgne, Antoine Miné, David Monniaux, and Xavier Rival. "The Astrée Analyzer." In *ESOP*. Vol. 3444. Lecture Notes in Computer Science. Springer, 2005, pp. 21–30 (p. 588, 589).

[267] Patrick Cousot, Radhia Cousot, Jérôme Feret, Laurent Mauborgne, Antoine Miné, David Monniaux, and Xavier Rival. "Combination of Abstractions in the Astrée Static Analyzer." In *ASIAN*. Vol. 4435. Lecture Notes in Computer Science. Springer, 2006, pp. 272–300 (p. 406, 431, 432, 715).

[268] Patrick Cousot, Radhia Cousot, Jérôme Feret, Laurent Mauborgne, Antoine Miné, and Xavier Rival. "Why Does Astrée Scale Up?." *Formal Methods in System Design*. 35.3 (2009), pp. 229–264 (p. 397, 406, 492, 493, 551, 617, 706, 716).

[269] Patrick Cousot, Radhia Cousot, Jérôme Feret, Antoine Miné, Laurent Mauborgne, David Monniaux, and Xavier Rival. "Varieties of Static Analyzers: A Comparison with Astrée." In *TASE*. IEEE Computer Society, 2007, pp. 3–20 (p. 714).

[270] Patrick Cousot, Radhia Cousot, and Roberto Giacobazzi. "Abstract Interpretation of Resolution-Based Semantics." *Theor. Comput. Sci.* 410.46 (2009), pp. 4724–4746 (p. 623, 671).

[271] Patrick Cousot, Radhia Cousot, and Francesco Logozzo. "Precondition Inference From Intermittent Assertions and Application to Contracts on Collections." In *VMCAI*. Vol. 6538. Lecture Notes in Computer Science. Springer, 2011, pp. 150–168 (p. 141, 684, 692).

[272] Patrick Cousot, Radhia Cousot, Francesco Logozzo, and Michael Barnett. "An Abstract Interpretation Framework for Refactoring with Application to Extract Methods with Contracts." In *OOPSLA*. ACM, 2012, pp. 213–232 (p. 303).

[273] Patrick Cousot, Radhia Cousot, and Laurent Mauborgne. "A Scalable Segmented Decision Tree Abstract Domain." In *Essays in Memory of Amir Pnueli*. Vol. 6200. Lecture Notes in Computer Science. Springer, 2010, pp. 72–95 (p. 433).

[274] Patrick Cousot, Radhia Cousot, and Laurent Mauborgne. "The Reduced Product of Abstract Domains and the Combination of Decision Procedures." In *FOSSACS*. Vol. 6604. Lecture Notes in Computer Science. Springer, 2011, pp. 456–472 (p. 5, 432, 589).

[275] Patrick Cousot, Radhia Cousot, and Laurent Mauborgne. "Theories, Solvers and Static Analysis by Abstract Interpretation." *J. ACM*. 59.6 (2012), 31:1–31:56 (p. 428).

[276] Patrick Cousot, Pierre Ganty, and Jean-François Raskin. "Fixpoint-Guided Abstraction Refinements." In *SAS*. Vol. 4634. Lecture Notes in Computer Science. Springer, 2007, pp. 333–348 (p. 550).

[277] Patrick Cousot, Roberto Giacobazzi, and Francesco Ranzato. "Program Analysis Is Harder Than Verification: A Computability Perspective." In *CAV (2)*. Vol. 10982. Lecture Notes in Computer Science. Springer, 2018, pp. 75–95 (p. 104, 694, 706).

[278] Patrick Cousot, Roberto Giacobazzi, and Francesco Ranzato. "A^2I: abstract^2 interpretation." *Proc. ACM Program. Lang.* 3.POPL (2019), 42:1–42:31 (p. 189, 412, 493, 723).

[279] Patrick Cousot and Nicolas Halbwachs. "Automatic Discovery of Linear Restraints Among Variables of a Program." In *POPL*. ACM Press, 1978, pp. 84–96 (p. 457, 492).

[280] Silvia Crafa, Francesco Ranzato, and Francesco Tapparo. "Saving Space in a Time Efficient Simulation Algorithm." *Fundam. Inform.* 108.1–2 (2011), pp. 23–42 (p. 550).

[281] Robert Croisot. "Applications résiduées." *Annales scientifiques de l'École normale supérieure, Série III.* 73.4 (1956), pp. 453–474 (p. 131).

[282] Pascal Cuoq, Florent Kirchner, Nikolai Kosmatov, Virgile Prevosto, Julien Signoles, and Boris Yakobowski. "Frama-C - A Software Analysis Perspective." In *SEFM*. Vol. 7504. Lecture Notes in Computer Science. Springer, 2012, pp. 233–247 (p. 715).

[283] John Cyphert, Jason Breck, Zachary Kincaid, and Thomas W. Reps. "Refinement of Path Expressions for Static Analysis." *Proc. ACM Program. Lang.* 3.POPL (2019), 45:1–45:29 (p. 551).

[284] Vijay D'Silva, Leopold Haller, and Daniel Kroening. "Abstract Satisfaction." In *POPL*. ACM, 2014, pp. 139–150 (p. 5, 718).

[285] Francesco Dagnino. "Foundations of Regular Coinduction." *CoRR*. abs/2006.02887 (2020) (p. 184).

[286] Luís Damas and Robin Milner. "Principal Type-Schemes for Functional Programs." In *POPL*. ACM Press, 1982, pp. 207–212 (p. 671).

[287] Werner Damm and Bernhard Josko. "A Sound and Relatively* Compete Axiomatization of Clarke's Language L_4." In *Logic of Programs*. Vol. 164. Lecture Notes in Computer Science. Springer, 1983, pp. 161–175 (p. 295).

[288] Werner Damm and Bernhard Josko. "A Sound and Relatively* Complete Hoare Logic for a Language with Higher Type Procedures." *Acta Inf.* 20 (1983), pp. 59–101 (p. 295).

[289] Nasrine Damouche, Matthieu Martel, and Alexandre Chapoutot. "Numerical Program Optimisation by Automatic Improvement of the Accuracy of Computations." *IJIEI*. 6.1/2 (2018), pp. 115–145 (p. 386).

[290] Dennis Dams, Rob Gerth, and Orna Grumberg. "Abstract Interpretation of Reactive Systems." *ACM Trans. Program. Lang. Syst.* 19.2 (1997), pp. 253–291 (p. 718).

[291] Dennis Dams and Kedar S. Namjoshi. "Automata as Abstractions." In *VMCAI*. Vol. 3385. Lecture Notes in Computer Science. Springer, 2005, pp. 216–232 (p. 551).

[292] David Darais and David Van Horn. "Constructive Galois Connections: Taming the Galois Connection Framework for Mechanized Metatheory." In *ICFP*. ACM, 2016, pp. 311–324 (p. 113, 671).

[293] Eva Darulova and Viktor Kuncak. "Towards a Compiler for Reals." *ACM Trans. Program. Lang. Syst.* 39.2 (2017), 8:1–8:28 (p. 370).

[294] Manuvir Das. "Unification-Based Pointer Analysis with Directional Assignments." In *PLDI*. ACM, 2000, pp. 35–46 (p. 587).

[295] Brian A. Davey and Hilary A. Priestley. *Introduction to Lattices and Order.* 2nd ed. Cambridge University Press, 1990, 2002 (p. 105, 109).

[296] Luiz Henrique De Figueiredo and Jorge Stolfi. "Affine Arithmetic: Concepts and Applications." *Numerical Algorithms.* 37.1–4 (2004), pp. 147–158 (p. 370).

[297] Giuseppe De Giacomo and Moshe Y. Vardi. "Linear Temporal Logic and Linear Dynamic Logic on Finite Traces." In *IJCAI*. IJCAI/AAAI, 2013, pp. 854–860 (p. 527).

[298] Giuseppe De Giacomo and Moshe Y. Vardi. "Synthesis for LTL and LDL on Finite Traces." In *IJCAI*. AAAI Press, 2015, pp. 1558–1564 (p. 527).

[299] Leonardo Mendonça De Moura and Nikolaj Bjørner. "Z3: an Efficient SMT Solver." In *TACAS*. Vol. 4963. Lecture Notes in Computer Science. Springer, 2008, pp. 337–340 (p. 290).

[300] Saumya K. Debray and Raghu Ramakrishnan. "Abstract Interpretation of Logic Programs Using Magic Transformations." *J. Log. Program.* 18.2 (1994), pp. 149–176 (p. 684).

[301] Richard Dedekind. *Stetigkeit und irrationale Zahlen.* Braunschweig : F. Vieweg, 1892.

[302] Richard Dedekind. *Was sind und was sollen die Zahlen?* 2nd ed. Friedrich Bieweg Und Sohn, 1893 (p. 9, 171).

[303] Richard Dedekind and Heinrich Weber. "Theorie der algebraischen Functionen einer Verânderlichen." *Journal Für Die Reine Und Angewandte Mathematik.* 92 (1882), pp. 181–290 (p. 438).

[304] Elisabeth Degrand, François Fages, and Sylvain Soliman. "Graphical Conditions for Rate Independence in Chemical Reaction Networks." In *CMSB*. Vol. 12314. Lecture Notes in Computer Science. Springer, 2020, pp. 61–78 (p. 319).

[305] David Delmas, Éric Goubault, Sylvie Putot, Jean Souyris, Karim Tekkal, and Franck Védrine. "Towards an Industrial Use of FLUCTUAT on Safety-Critical Avionics Software." In *FMICS*. Vol. 5825. Lecture Notes in Computer Science. Springer, 2009, 53–69 (p. 386).

[306] David Delmas and Antoine Miné. "Analysis of Program Differences with Numerical Abstract Interpretation." In *PERR 2019, 3rd Workshop on Program Equivalence and Relational Reasoning, ETAPS 2019, Prague, Czech Republic.* Apr. 2019 (p. 493).

[307] David Delmas and Jean Souyris. "Astrée: From Research to Industry." In *SAS*. Vol. 4634. Lecture Notes in Computer Science. Springer, 2007, pp. 437–451 (p. 714).

[308] Erik L. Demeulemeester and Willy S. Herroelen. *Project Scheduling, a Research Handbook.* Springer, 2002 (p. 493).

[309] Klaus Denecke, Marcel Erné, and Shelly L. Wismath. *Galois Connections and Applications.* Springer, 2004 (p. 130).

[310] Chaoqiang Deng. "Responsibility Analysis by Abstract Interpretation." PhD thesis. New York University, Courant Institute of Mathematical STudies, Aug. 2020 (p. 693).

[311] Chaoqiang Deng and Patrick Cousot. "Responsibility Analysis by Abstract Interpretation." In *SAS*. Vol. 11822. Lecture Notes in Computer Science. Springer, 2019, pp. 368–388 (p. 621, 693, 707).

[312] Dorothy E. Denning and Peter J. Denning. "Certification of Programs for Secure Information Flow." *Commun. ACM.* 20.7 (1977), pp. 504–513 (p. 591, 593–595, 597, 601, 607, 609, 620).

[313] Frank DeRemer and Thomas J. Pennello. "Efficient Computation of LALR(1) Look-Ahead Sets." *ACM Trans. Program. Lang. Syst.* 4.4 (1982), pp. 615–649 (p. 61).

[314] Jean-Claude Derniame and Claude Pair. *Problèmes de cheminement dans les graphes.* Dunod, 1971 (p. 462, 479, 480).

[315] René Descartes. *la Géométrie.* Leyde, 1637. https : / / gallica . bnf . fr / ark : /12148 / bpt6k29040s (p. 321).

[316] Alain Deutsch. "Semantic Models and Abstract Interpretation Techniques for Inductive Data Structures and Pointers." In *PEPM.* ACM Press, 1995, pp. 226–229 (p. 559, 590).

[317] Alain Deutsch and Klaus Wissing. "New Test Approach for Embedded Applications." In *Software Engineering (Workshops).* Vol. 106. LNI. GI, 2007, pp. 127–136 (p. 398, 710).

[318] Reinhard Diestel. *Graph Theory.* 5th ed. Graduate Texts in Mathematics. Springer, 2017 (p. 459).

[319] Will Dietz, Peng Li, John Regehr, and Vikram S. Adve. "Understanding Integer Overflow in C/C++." *ACM Trans. Softw. Eng. Methodol.* 25.1 (2015), 2:1–2:29 (p. 291, 708).

[320] Edsger W. Dijkstra. "Recursive Programming." *Numerische Mathematik.* 2.1 (1960), pp. 312–318 (p. 15, 634).

[321] Edsger W. Dijkstra. "Guarded Commands, Nondeterminacy and Formal Derivation of Programs." *Commun. ACM.* 18.8 (1975), pp. 453–457 (p. 141, 142, 144, 281).

[322] Edsger W. Dijkstra. *A Discipline of Programming.* Prentice Hall, 1976 (p. 142, 723).

[323] Edsger W. Dijkstra. "On Subgoal Induction." In *Selected Writings on Computing: A Personal Perspective.* Texts and Monographs in Computer Science. Springer, 1982, pp. 223–224. https://www.cs.utexas.edu/users/EWD/ewd05xx/EWD576.PDF (p. 291).

[324] Edsger W. Dijkstra. *Selected Writings on Computing: A Personal Perspective.* Texts and Monographs in Computer Science. Springer, 1982 (p. 723).

[325] Edsger W. Dijkstra, W. H. J. Feijen, and A. J. M. Van Gasteren. "Derivation of a Termination Detection Algorithm for Distributed Computations." *Inf. Process. Lett.* 16.5 (1983), pp. 217–219 (p. 257).

[326] Edsger W. Dijkstra and Carel S. Scholten. "Termination Detection for Diffusing Computations." *Inf. Process. Lett.* 11.1 (1980), pp. 1–4 (p. 257).

[327] Edsger W. Dijkstra and Carel S. Scholten. *Predicate Calculus and Program Semantics.* Texts and Monographs in Computer Science. Springer, 1990 (p. 141, 143, 675, 723).

[328] Edsger W. Dijkstra and Carel S. Scholten. "The Strongest Postcondition." In *Predicate Calculus and Program Semantics.* Texts and Monographs in Computer Science. Springer, 1990, pp. 209–215 (p. 12, 144).

[329] David L. Dill. "Timing Assumptions and Verification of Finite-State Concurrent Systems." In *Automatic Verification Methods for Finite State Systems.* Vol. 407. Lecture Notes in Computer Science. Springer, 1989, pp. 197–212 (p. 481, 483).

[330] Aleksandar S. Dimovski. "Lifted Static Analysis Using a Binary Decision Diagram Abstract Domain." In *GPCE.* ACM, 2019, pp. 102–114 (p. 433).

[331] Aleksandar S. Dimovski, Claus Brabrand, and Andrzej Wasowski. "Finding Suitable Variability Abstractions for Lifted Analysis." *Formal Asp. Comput.* 31.2 (2019), pp. 231–259 (p. 433).

[332] Aleksandar S. Dimovski and Axel Legay. "Computing Program Reliability Using Forward—Backward Precondition Analysis and Model Counting." In *FASE.* Vol. 12076. Lecture Notes in Computer Science. Springer, 2020, pp. 182–202 (p. 702).

[333] Stefano Dissegna, Francesco Logozzo, and Francesco Ranzato. "An Abstract Interpretation-Based Model of Tracing Just-in-Time Compilation." *ACM Trans. Program. Lang. Syst.* 38.2 (2016), 7:1–7:50 (p. 42).

[334] Dino Distefano, Manuel Fähndrich, Francesco Logozzo, and Peter W. O'Hearn. "Scaling Static Analyses at Facebook." *Commun. ACM.* 62.8 (2019), pp. 62–70 (p. 620, 698, 715).

[335] Florent Adrien Domenach. "Structures latticielles, correspondances de Galois, contraintes et classification symbolique." PhD thesis. Université Panthéon—Sorbonne—Paris I, 2002 (p. 113).

[336] Alastair F. Donaldson, Daniel Kroening, and Philipp Rümmer. "Automatic analysis of DMA races using model checking and k-induction." *Formal Methods Syst. Des.* 39.1 (2011), pp. 83–113 (p. 276).

[337] Gilles Dowek and Jean-Jacques Lévy. *Introduction to the Theory of Programming Languages.* Undergraduate Topics in Computer Science. Springer, 2011 (p. 276).

[338] Włodzimierz Drabent. "A Simple Correctness Proof for Magic Transformation." *TPLP.* 12.6 (2012), pp. 929–936 (p. 684).

[339] Samuel Drews, Aws Albarghouthi, and Loris D'Antoni. "Proving Data—Poisoning Robustness in Decision Trees." In *PLDI.* ACM, 2020, pp. 1083–1097 (p. 718).

[340] Doron Drusinsky. "The Temporal Rover and the ATG Rover." In *SPIN 2000.* Vol. 1885. Lecture Notes in Computer Science. Springer, 2000, pp. 323–330 (p. 388).

[341] Paul Dubreil and Robert Croisot. "Propriétés générales de la résiduation en liaison avec les correspondances de Galois." *Collectanea Mathematica.* 7.2–3 (1954), pp. 193–204 (p. 131).

[342] Evelyn Duesterwald, Rajiv Gupta, and Mary Lou Soffa. "A Practical Framework for Demand-Driven Interprocedural Data Flow Analysis." *ACM Trans. Program. Lang. Syst.* 19.6 (1997), pp. 992–1030 (p. 590).

[343] Berke Durak and Nicolas Pouillard. "The `Ocamlbuild` Users Manual." Institut National de Recherche en Informatique et en Automatique. Sept. 2008. https://ocaml.org/learn/tutorials/ocamlbuild/ (p. 60).

[344] Wojciech Dzik, Jouni Järvinen, and Michiro Kondo. "Intuitionistic Modal Logic with a Galois Connection Has the Finite Model Property." *Logic Journal of the IGPL.* 21.2 (2013), pp. 199–204 (p. 131).

[345] Jay Earley. "An Efficient Context-Free Parsing Algorithm (Reprint)." *Commun. ACM.* 26.1 (1983), pp. 57–61 (p. 61).

[346] Elmar Eder. "Properties of Substitutions and Unifications." *J. Symb. Comput.* 1.1 (1985), pp. 31–46 (p. 646).

[347] Ömer Eğecioğlu. "Strongly Regular Grammars and Regular Approximation of Context-Free Languages." In *Developments in Language Theory.* Vol. 5583. Lecture Notes in Computer Science. Springer, 2009, pp. 207–220 (p. 551).

[348] Matt Elder, Junghee Lim, Tushar Sharma, Tycho Andersen, and Thomas W. Reps. "Abstract Domains of Affine Relations." *ACM Trans. Program. Lang. Syst.* 36.4 (2014), 11:1–11:73 (p. 457).

[349] Chucky Ellison and Grigore Rosu. "An Executable Formal Semantics of C with Applications." In *POPL.* ACM, 2012, pp. 533–544 (p. 708).

[350] Pär Emanuelsson and Ulf Nilsson. "A Comparative Study of Industrial Static Analysis Tools." *Electr. Notes Theor. Comput. Sci.* 217 (2008), pp. 5–21 (p. 710, 713).

[351] E. Allen Emerson and Joseph Y. Halpern. ""Sometimes" and "Not Never" Revisited: On Branching Versus Linear Time Temporal Logic." *J. ACM.* 33.1 (1986), pp. 151–178 (p. 527).

[352] Dawson R. Engler and Madanlal Musuvathi. "Static Analysis Versus Software Model Checking for Bug Finding." In *VMCAI.* Vol. 2937. Lecture Notes in Computer Science. Springer, 2004, pp. 191–210 (p. 550).

[353] Marcel Erné, Jurgen Koslowski, Austin Melton, and George E. Strecker. "A Primer on Galois Connections." *Annals of the New York Academy of Sciences.* (Aug. 1997) (p. 119).

[354] Martín Hötzel Escardó. "Joins in the Frame of Nuclei." *Applied Categorical Structures.* 11.2 (Apr. 2003), pp. 117–124 (p. 167).

[355] Zoltán Ésik and Dexter Kozen. "On Free Ω-Continuous and Regular Ordered Algebras." *Logical Methods in Computer Science.* 15.4 (2019) (p. 527).

[356] Enrico Eugenio and Agostino Cortesi. "WiFi–Related Energy Consumption Analysis of Mobile Devices in a Walkable Area by Abstract Interpretation." In *ICDCIT.* Vol. 10109. Lecture Notes in Computer Science. Springer, 2017, pp. 27–39 (p. 718).

[357] Leonhard Euler. "Solutio Problematis Ad Geometriam Situs Pertinentis." *Commentarii Academiae Scientiarum Petropolitanae.* 8 (1736, Published 1741), pp. 128–140. (P. 480).

[358] François Fages and Sylvain Soliman. "Abstract Interpretation and Types for Systems Biology." *Theor. Comput. Sci.* 403.1 (2008), pp. 52–70 (p. 319).

[359] Manuel Fähndrich. "Static Verification for Code Contracts." In *SAS.* Vol. 6337. Lecture Notes in Computer Science. Springer, 2010, pp. 2–5 (p. 698).

[360] Manuel Fähndrich and Francesco Logozzo. "Static Contract Checking with Abstract Interpretation." In *FoVeOOS.* Vol. 6528. Lecture Notes in Computer Science. Springer, 2010, pp. 10–30 (p. 698, 709, 715).

[361] Yliès Falcone, Klaus Havelund, and Giles Reger. "A Tutorial on Runtime Verification." In M. Broy, D. Peled, and G. Kalus, eds. *Engineering Dependable Software Systems.* Vol. 34. NATO Science for Peace and Security Series, D: Information and Communication Security. IOS Press, 2013, pp. 141–175 (p. 387).

[362] Azadeh Farzan and Zachary Kincaid. "DUET: Static Analysis for Unbounded Parallelism." In *CAV.* Vol. 8044. Lecture Notes in Computer Science. Springer, 2013, pp. 191–196 (p. 493).

[363] Stuart I. Feldman. "Make—A Program for Maintaining Computer Programs." *Softw., Pract. Exper.* 9.4 (1979), pp. 255–65 (p. 59).

[364] Walter Felscher. "Beziehungen zwischen verbands-ähnlichen Algebren und geregelten Menge." *Mathematische Annalen.* 135.5 (Oct. 1958), pp. 369–387 (p. 131).

[365] Jérôme Feret. "Static Analysis of Digital Filters." In *ESOP.* Vol. 2986. Lecture Notes in Computer Science. Springer, 2004, pp. 33–48 (p. 492).

[366] Jérôme Feret. "The Arithmetic–Geometric Progression Abstract Domain." In *VMCAI.* Vol. 3385. Lecture Notes in Computer Science. Springer, 2005, pp. 42–58 (p. 492).

[367] Jérôme Feret and Kim Quyên Lý. "Local Traces: An Over–Approximation of the Behaviour of the Proteins in Rule–Based Models." In *CMSB.* Vol. 9859. Lecture Notes in Computer Science. Springer, 2016, pp. 116–131 (p. 718).

[368] Pierre De Fermat. "Relation des nouvelles découvertes en la science des nombres." Lettre à Pierre De Carcavi. Aug. 1659 (p. 16).

[369] Pietro Ferrara, Luca Olivieri, and Fausto Spoto. "BackFlow: Backward Context–Sensitive Flow Reconstruction of Taint Analysis Results." In *VMCAI.* Vol. 11990. Lecture Notes in Computer Science. Springer, 2020, pp. 23–43 (p. 693).

[370] Gilberto Filé and Francesco Ranzato. "The Powerset Operator on Abstract Interpretations." *Theor. Comput. Sci.* 222.1–2 (1999), pp. 77–111 (p. 433, 514).

[371] Jean–Christophe Filliâtre and Andrei Paskevich. "Why3 - Where Programs Meet Provers." In *ESOP.* Vol. 7792. Lecture Notes in Computer Science. Springer, 2013, pp. 125–128 (p. 291).

[372] Jeffrey Fischer, Ranjit Jhala, and Rupak Majumdar. "Joining Dataflow with Predicates." In *ESEC/SIGSOFT FSE.* ACM, 2005, pp. 227–236 (p. 432).

[373] Michael J. Fischer and Michael O. Rabin. "Super–Exponential Complexity of Presburger Arithmetic." *Proceedings of the SIAM–AMS Symposium in Applied Mathematics.* 7 (1974), pp. 27–41 (p. 290).

[374] Dana Fisman and Amir Pnueli. "Beyond Regular Model Checking." In *FSTTCS*. Vol. 2245. Lecture Notes in Computer Science. Springer, 2001, pp. 156–170 (p. 531).

[375] Robert W. Floyd. "Algorithm 97: Shortest Path." *Commun. ACM*. 5.6 (1962), p. 345 (p. 459, 475).

[376] Robert W. Floyd. "Assigning Meaning to Programs." In *Proc. Symp. in Applied Math.*. Edited by J.T. Schwartz. Vol. 19. Amer. Math. Soc., 1967, pp. 19–32 (p. 3, 94, 216, 281, 291, 304, 708).

[377] Forum for Aeronautical Software. *RTCA DO-333 / EUROCAE ED-216 (Formal Methods Supplement)*. Tech. rep. http://www.rtca.org, https://eshop.eurocae.net/eurocae-documents-and-reports/ed-216/. RTCA, Inc., Radio Technical Commission for Aeronautics, Dec. 13, 2011 (p. 716).

[378] Jeffrey S. Foster, Robert Johnson, John Kodumal, and Alexander Aiken. "Flow-Insensitive Type Qualifiers." *ACM Trans. Program. Lang. Syst.* 28.6 (2006), pp. 1035–1087 (p. 556).

[379] Alexis Fouilhé, David Monniaux, and Michaël Périn. "Efficient Generation of Correctness Certificates for the Abstract Domain of Polyhedra." In *SAS*. Vol. 7935. Lecture Notes in Computer Science. Springer, 2013, pp. 345–365 (p. 716).

[380] Maria João Frade, Ando Saabas, and Tarmo Uustalu. "Foundational Certification of Data-Flow Analyses." In *TASE*. IEEE Computer Society, 2007, pp. 107–116 (p. 514).

[381] Aymeric Fromherz, Abdelraouf Ouadjaout, and Antoine Miné. "Static Value Analysis of Python Programs by Abstract Interpretation." In *NFM*. Vol. 10811. Lecture Notes in Computer Science. Springer, 2018, pp. 185–202. https://hal.sorbonne-universite.fr/hal-01782390 (p. 708).

[382] Peter Fulla and Stanislav Živný. "A Galois Connection for Valued Constraint Languages of Infinite Size." In *ICALP (1)*. Vol. 9134. Lecture Notes in Computer Science. Springer, 2015, pp. 517–528 (p. 131).

[383] Peter Fulla and Stanislav Živný. "A Galois Connection for Weighted (Relational) Clones of Infinite Size." *TOCT*. 8.3 (2016), 9:1–9:21 (p. 131).

[384] Yoshihiko Futamura. "Partial Evaluation of Computation Process, Revisited." *Higher-Order and Symbolic Computation*. 12.4 (1999), pp. 377–380 (p. 250).

[385] Letterio Galletta. "An Abstract Interpretation Framework for Type and Effect Systems." *Fundam. Inform.* 134.3–4 (2014), pp. 355–393 (p. 671).

[386] Letterio Galletta and Giorgio Levi. "An Abstract Semantics for Inference of Types and Effects in a Multi-Tier Web Language." In *WWV*. Vol. 61. EPTCS. 2011, pp. 81–95 (p. 671).

[387] Diane Gallois-Wong, Sylvie Boldo, and Pascal Cuoq. "Optimal Inverse Projection of Floating-Point Addition." *Numer. Algorithms*. 83.3 (2020), pp. 957–986 (p. 397).

[388] Évariste Galois. "Mémoire sur les conditions de résolubilité des équations par radicaux." *Journal de Mathématiques Pures et Appliquées*. XI (1846), pp. 417–433. http://www.bibnum.education.fr/sites/default/files/galois_memoire_sur_la_resolubiblite.pdf (p. 112, 130).

[389] Graeme Gange, Jorge A. Navas, Peter Schachte, Harald Søndergaard, and Peter J. Stuckey. "Interval Analysis and Machine Arithmetic: Why Signedness Ignorance Is Bliss." *ACM Trans. Program. Lang. Syst.* 37.1 (2014), 1:1–1:35 (p. 367, 397).

[390] Graeme Gange, Jorge A. Navas, Peter Schachte, Harald Søndergaard, and Peter J. Stuckey. "An Abstract Domain of Uninterpreted Functions." In *VMCAI*. Vol. 9583. Lecture Notes in Computer Science. Springer, 2016, pp. 85–103 (p. 623).

[391] Graeme Gange, Jorge A. Navas, Peter Schachte, Harald Søndergaard, and Peter J. Stuckey. "Exploiting Sparsity in Difference-Bound Matrices." In *SAS*. Vol. 9837. Lecture Notes in Computer Science. Springer, 2016, pp. 189–211 (p. 492).

[392] Graeme Gange, Jorge A. Navas, Peter Schachte, Harald Søndergaard, and Peter J. Stuckey. "Dissecting Widening: Separating Termination From Information." In *APLAS*. Vol. 11893. Lecture Notes in Computer Science. Springer, 2019, pp. 95–114 (p. 412).

[393] Bernhard Ganter, Sebastian Rudolph, and Gerd Stumme. "Explaining Data with Formal Concept Analysis." In *Reasoning Web*. Vol. 11810. Lecture Notes in Computer Science. Springer, 2019, pp. 153–195 (p. 113).

[394] Bernhard Ganter and Rudolf Wille. *Formal Concept Analysis—Mathematical Foundations*. Springer, 1999 (p. 113, 131).

[395] Ronald Garcia, Alison M. Clark, and Éric Tanter. "Abstracting Gradual Typing." In *POPL*. ACM, 2016, pp. 429–442 (p. 671).

[396] Isabel Garcia-Contreras, José F. Morales, and Manuel V. Hermenegildo. "Multivariant Assertion-Based Guidance in Abstract Interpretation." In *LOPSTR*. Vol. 11408. Lecture Notes in Computer Science. Springer, 2018, pp. 184–201 (p. 621).

[397] Isabel Garcia-Contreras, José F. Morales, and Manuel V. Hermenegildo. "Multivariant Assertion-Based Guidance in Abstract Interpretation." In *LOPSTR*. Vol. 11408. Lecture Notes in Computer Science. Springer, 2018, pp. 184–201 (p. 707).

[398] Carolo Friderico Gavss. *Disqvisitiones Arithmeticae*. English Translation Arthur A. Clarke: Disquisitiones Arithmeticae, Yale University Press, 1966. Gerh. Fleischer, 1801 (p. 352, 355).

[399] Carolo Friderico Gavss. *Disquisitio de elementis ellipticis palladis ex opposiotionibus annorum 1803, 1804, 1805, 1807, 1808, 1809*. Societati Regiae Tradita Nov. 25, 1810, H. Dieterich, 1874, 1811 (p. 442).

[400] Thomas Martin Gawlitza, Jérôme Leroux, Jan Reineke, Helmut Seidl, Grégoire Sutre, and Reinhard Wilhelm. "Polynomial Precise Interval Analysis Revisited." In *Efficient Algorithms*. Vol. 5760. Lecture Notes in Computer Science. Springer, 2009, pp. 422–437 (p. 412).

[401] Thomas Martin Gawlitza and Helmut Seidl. "Precise Fixpoint Computation Through Strategy Iteration." In *ESOP*. Vol. 4421. Lecture Notes in Computer Science. Springer, 2007, pp. 300–315 (p. 412).

[402] Thomas Martin Gawlitza and Helmut Seidl. "Abstract Interpretation Over Zones Without Widening." In *WING@ETAPS/IJCAR*. Vol. 1. EPiC Series in Computing. EasyChair, 2010, pp. 12–43 (p. 493).

[403] Thomas Martin Gawlitza and Helmut Seidl. "Computing Relaxed Abstract Semantics w.r.t. Quadratic Zones Precisely." In *SAS*. Vol. 6337. Lecture Notes in Computer Science. Springer, 2010, pp. 271–286 (p. 493).

[404] Thomas Martin Gawlitza and Helmut Seidl. "Precise Program Analysis Through Strategy Iteration and Optimization." In T. Nipkow, O. Grumberg, and B. Hauptmann, eds. *Software Safety and Security*. Vol. 33. NATO Science for Peace and Security Series - D: Information and Communication Security. IOS Press, 2012, pp. 348–384 (p. 412).

[405] Thomas Martin Gawlitza and Helmut Seidl. "Numerical Invariants Through Convex Relaxation and Max-Strategy Iteration." *Formal Methods in System Design*. 44.2 (2014), pp. 101–148 (p. 412).

[406] Thomas Martin Gawlitza, Helmut Seidl, Assalé Adjé, Stéphane Gaubert, and Éric Goubault. "Abstract Interpretation Meets Convex Optimization." *J. Symb. Comput.* 47.12 (2012), pp. 1416–1446 (p. 412).

[407] Timon Gehr, Matthew Mirman, Dana Drachsler-Cohen, Petar Tsankov, Swarat Chaudhuri, and Martin T. Vechev. "AI^2: Safety and Robustness Certification of Neural Networks with Abstract Interpretation." In *IEEE Symposium on Security and Privacy*. IEEE Computer Society, 2018, pp. 3–18 (p. 718).

[408] Mai Gehrke, Daniela Petrisan, and Luca Reggio. "Quantifiers on Languages and Codensity Monads." In *LICS*. IEEE Computer Society, 2017, pp. 1–12 (p. 550).

[409] Elazar Gershuni, Nadav Amit, Arie Gurfinkel, Nina Narodytska, Jorge A. Navas, Noam Rinetzky, Leonid Ryzhyk, and Mooly Sagiv. "Simple and Precise Static Analysis of Untrusted Linux Kernel Extensions." In *PLDI*. ACM, 2019, pp. 1069–1084 (p. 493).

[410] Pritam M. Gharat, Uday P. Khedker, and Alan Mycroft. "Generalized Points-to Graphs: A Precise and Scalable Abstraction for Points-to Analysis." *ACM Trans. Program. Lang. Syst.* 42.2 (2020), 8:1–8:78 (p. 589).

[411] Khalil Ghorbal, Éric Goubault, and Sylvie Putot. "The Zonotope Abstract Domain Taylor1+." In *CAV*. Vol. 5643. Lecture Notes in Computer Science. Springer, 2009, pp. 627–633 (p. 386, 398).

[412] Roberto Giacobazzi. "Abductive Analysis of Modular Logic Programs." *J. Log. Comput.* 8.4 (1998), pp. 457–483 (p. 675).

[413] Roberto Giacobazzi and Isabella Mastroeni. "Abstract Non-Interference: A Unifying Framework for Weakening Information-Flow." *ACM Trans. Priv. Secur.* 21.2 (2018), 9:1–9:31 (p. 618, 621).

[414] Roberto Giacobazzi and Elisa Quintarelli. "Incompleteness, Counterexamples, and Refinements in Abstract Model Checking." In *SAS*. Vol. 2126. Lecture Notes in Computer Science. Springer, 2001, pp. 356–373 (p. 209, 550).

[415] Roberto Giacobazzi and Francesco Ranzato. "The Reduced Relative Power Operation on Abstract Domains." *Theor. Comput. Sci.* 216.1–2 (1999), pp. 159–211 (p. 433).

[416] Roberto Giacobazzi and Francesco Ranzato. "Incompleteness of States w.r.t. Traces in Model Checking." *Inf. Comput.* 204.3 (2006), pp. 376–407 (p. 551, 706).

[417] Roberto Giacobazzi, Francesco Ranzato, and Francesca Scozzari. "Making Abstract Interpretations Complete." *J. ACM.* 47.2 (2000), pp. 361–416 (p. 209, 706).

[418] Roberto Giacobazzi and Francesca Scozzari. "Intuitionistic Implication in Abstract Interpretation." In *APPIA-GULP-PRODE*. 1997, pp. 33–44 (p. 433).

[419] Seymour Ginsburg and Henry Gordon Rice. "Two Families of Languages Related to ALGOL." *J. ACM.* 9.3 (1962), pp. 350–371 (p. 178, 184).

[420] Kurt Gödel. "Über formal unentscheidbare Sätze der Principia Mathematica und verwandter Systeme, I." *Monatshefte Für Mathematik Und Physik.* 38.1 (1931), pp. 173–198 (p. 98, 290).

[421] Joseph A. Goguen and José Meseguer. "Security Policies and Security Models." In *IEEE Symposium on Security and Privacy*. IEEE Computer Society, 1982, pp. 11–20 (p. 591, 593, 619).

[422] Joseph A. Goguen and José Meseguer. "Unwinding and Inference Control." In *IEEE Symposium on Security and Privacy*. IEEE Computer Society, 1984, pp. 75–87 (p. 591, 593, 619).

[423] David Goldberg. "What Every Computer Scientist Should Know About Floating-Point Arithmetic." *ACM Comput. Surv.* 23.1 (1991), pp. 5–48 (p. 367, 372).

[424] Michel Gondran. "Algèbre linéaire et cheminement dans un graphe." *Revue française d'Automatique, Informatique, Recherche Opérationnelle (R.A.I.R.O.).* Recherche opérationnelle, Tome 9.V1 (1975), pp. 77–99 (p. 462, 480).

[425] Laure Gonnord and Nicolas Halbwachs. "Combining Widening and Acceleration in Linear Relation Analysis." In *SAS*. Vol. 4134. Lecture Notes in Computer Science. Springer, 2006, pp. 144–160 (p. 412).

[426] Laure Gonnord and Peter Schrammel. "Abstract Acceleration in Linear Relation Analysis." *Sci. Comput. Program.* 93 (2014), pp. 125–153 (p. 412).

[427] Mike Gordon and Hélène Collavizza. "Forward with Hoare." In C.B. Jones, A.W. Roscoe, and K.R. Wood, eds. *Reflections on the Work of Charles Antony Richard Hoare.* Springer, 2010, pp. 101–121 (p. 693).

[428] Éric Goubault and Sylvie Putot. "Static Analysis of Numerical Algorithms." In *SAS.* Vol. 4134. Lecture Notes in Computer Science. Springer, 2006, pp. 18–34 (p. 386, 398).

[429] Éric Goubault and Sylvie Putot. "A Zonotopic Framework for Functional Abstractions." *Formal Methods in System Design.* 47.3 (2015), pp. 302–360 (p. 386, 398).

[430] Éric Goubault and Sylvie Putot. "Inner and Outer Reachability for the Verification of Control Systems." In *HSCC.* ACM, 2019, pp. 11–22 (p. 386).

[431] Éric Goubault, Sylvie Putot, Philippe Baufreton, and Jean Gassino. "Static Analysis of the Accuracy in Control Systems: Principles and Experiments." In *FMICS.* Vol. 4916. Lecture Notes in Computer Science. Springer, 2007, pp. 3–20 (p. 386).

[432] Éric Goubault, Sylvie Putot, and Lorenz Sahlmann. "Inner and Outer Approximating Flowpipes for Delay Differential Equations." In *CAV (2).* Vol. 10982. Lecture Notes in Computer Science. Springer, 2018, pp. 523–541 (p. 386, 398).

[433] Éric Goubault, Sylvie Putot, and Franck Védrine. "Modular Static Analysis with Zonotopes." In *SAS.* Vol. 7460. Lecture Notes in Computer Science. Springer, 2012, pp. 24–40 (p. 386, 398).

[434] Jürgen Graf, Martin Hecker, and Martin Mohr. "Using JOANA for Information Flow Control in Java Programs—A Practical Guide." In *Software Engineering (Workshops).* Vol. P-215. LNI. GI, 2013, pp. 123–138 (p. 620).

[435] Jürgen Graf, Martin Hecker, Martin Mohr, and Gregor Snelting. "Tool Demonstration: JOANA." In *POST.* Vol. 9635. Lecture Notes in Computer Science. Springer, 2016, pp. 89–93 (p. 620).

[436] Susanne Graf and Hassen Saïdi. "Verifying Invariants Using Theorem Proving." In *CAV.* Vol. 1102. Lecture Notes in Computer Science. Springer, 1996, pp. 196–207 (p. 35).

[437] Susanne Graf and Hassen Saïdi. "Construction of Abstract State Graphs with PVS." In *CAV.* Vol. 1254. Lecture Notes in Computer Science. Springer, 1997, pp. 72–83 (p. 318).

[438] Philippe Granger. "Static Analysis of Arithmetical Congruences." *International Journal of Computer Mathematics.* 30.3 & 4 (1989), pp. 165–190 (p. 364, 365, 434).

[439] Philippe Granger. "Static Analysis of Linear Congruence Equalities Among Variables of a Program." In *TAPSOFT, Vol.1.* Vol. 493. Lecture Notes in Computer Science. Springer, 1991, pp. 169–192 (p. 365, 457, 492).

[440] Philippe Granger. "Improving the Results of Static Analyses Programs by Local Decreasing Iteration." In *FSTTCS.* Vol. 652. Lecture Notes in Computer Science. Springer, 1992, pp. 68–79 (p. 347).

[441] Philippe Granger. "Static Analyses of Congruence Properties on Rational Numbers (Extended Abstract)." In *SAS.* Vol. 1302. Lecture Notes in Computer Science. Springer, 1997, pp. 278–292 (p. 365, 395).

[442] Daniel Cabeza Gras and Manuel V. Hermenegildo. "Extracting Non–Strict Independent And–Parallelism Using Sharing and Freeness Information." In *SAS.* Vol. 864. Lecture Notes in Computer Science. Springer, 1994, pp. 297–313 (p. 621).

[443] Hermann Grassmann. *Die Lineale Ausdehnungslehre, Ein Neuer Zweig Der Mathematik.* Verlag Von Otto Wigand, 1844 (p. 437).

[444] George Grätzer. *Universal Algebra.* 2nd ed. Springer, 2008 (p. 342).

[445] George Grätzer. *Lattice Theory: Foundation.* Birkhäuser, 2011 (p. 105).

[446] Jeremy Gray. "The History of the Concept of a Finite–Dimensional Vector Space." *Historia Mathematica.* 7 (1980), pp. 65–70 (p. 438).

[447] Raymond Greenlaw, H. James Hoover, and Walter L. Ruzzo. *Limits to Parallel Computation: P-Completeness Theory.* Oxford University Press, 1995 (p. 257).

[448] Sumit Gulwani and George C. Necula. “Discovering Affine Equalities Using Random Interpretation.” In *POPL.* ACM, 2003, pp. 74–84 (p. 319).

[449] Sumit Gulwani and Ashish Tiwari. “Combining Abstract Interpreters.” In *PLDI.* ACM, 2006, pp. 376–386 (p. 432).

[450] Zheng Guo, Michael James, David Justo, Jiaxiao Zhou, Ziteng Wang, Ranjit Jhala, and Nadia Polikarpova. “Program Synthesis by Type-Guided Abstraction Refinement.” *Proc. ACM Program. Lang.* 4.POPL (2020), 12:1–12:28 (p. 671).

[451] Arie Gurfinkel and Sagar Chaki. “Boxes: A Symbolic Abstract Domain of Boxes.” In *SAS.* Vol. 6337. Lecture Notes in Computer Science. Springer, 2010, pp. 287–303 (p. 398, 433).

[452] Mohammad R. Haghighat. *Symbolic Analysis for Parallelizing Compilers.* Kluwer, 1995 (p. 397).

[453] Mohammad R. Haghighat and Constantine D. Polychronopoulos. “Symbolic Analysis for Parallelizing Compilers.” *ACM Trans. Program. Lang. Syst.* 18.4 (1996), pp. 477–518 (p. 397, 513).

[454] Nicolas Halbwachs. “About Synchronous Programming and Abstract Interpretation.” In *SAS.* Vol. 864. Lecture Notes in Computer Science. Springer, 1994, pp. 179–192 (p. 718).

[455] Nicolas Halbwachs and Julien Henry. “When the Decreasing Sequence Fails.” In *SAS.* Vol. 7460. Lecture Notes in Computer Science. Springer, 2012, pp. 198–213 (p. 409).

[456] Nicolas Halbwachs, David Merchat, and Laure Gonnord. “Some Ways to Reduce the Space Dimension in Polyhedra Computations.” *Formal Methods in System Design.* 29.1 (2006), pp. 79–95 (p. 493).

[457] Nicolas Halbwachs, David Merchat, and Catherine Parent-Vigouroux. “Cartesian Factoring of Polyhedra in Linear Relation Analysis.” In *SAS.* Vol. 2694. Lecture Notes in Computer Science. Springer, 2003, pp. 355–365 (p. 493).

[458] Nicolas Halbwachs, Yann-Erick Proy, and Pascal Raymond. “Verification of Linear Hybrid Systems by Means of Convex Approximations.” In *SAS.* Vol. 864. Lecture Notes in Computer Science. Springer, 1994, pp. 223–237 (p. 718).

[459] Nicolas Halbwachs, Yann-Erick Proy, and Patrick Roumanoff. “Verification of Real-Time Systems Using Linear Relation Analysis.” *Formal Methods in System Design.* 11.2 (1997), pp. 157–185 (p. 406).

[460] Pierre Hansen and Dominique De Werra. “Connectivity, Transitivity and Chromaticity: The Pioneering Work of Bernard Roy in Graph Theory.” In *Aiding Decisions with Multiple Criteria: Essays in Honor of Bernard Roy.* Springer, 2002, pp. 23–44 (p. 480).

[461] Ben Hardekopf and Calvin Lin. “The Ant and the Grasshopper: Fast and Accurate Pointer Analysis for Millions of Lines of Code.” In *PLDI.* ACM, 2007, pp. 290–299 (p. 587).

[462] Godfrey H. Hardy and Edward M. Wright. *An Introduction to Theory of Numbers.* 6th ed. Oxford University Press, 2008 (p. 351).

[463] Godfrey H. Hardy, Edward M. Wright, and Andrew J. Wiles. *An Introduction to the Theory of Numbers.* 6th ed. Clarendon Press, 2007 (p. 114).

[464] David Harel, Dexter Kozen, and Rohit Parikh. “Process Logic: Expressiveness, Decidability, Completeness.” *J. Comput. Syst. Sci.* 25.2 (1982), pp. 144–170 (p. 527).

[465] John Hatcliff. “An Introduction to Online and Offline Partial Evaluation Using a Simple Flowchart Language.” In *Partial Evaluation.* Vol. 1706. Lecture Notes in Computer Science. Springer, 1998, pp. 20–82 (p. 619).

[466] Chris Hathhorn, Chucky Ellison, and Grigore Rosu. “Defining the Undefinedness of C.” In *PLDI.* ACM, 2015, pp. 336–345 (p. 708).

[467] Melvin Hausner. *A Vector Space Approach to Geometry.* Dover Books on Mathematics. Dover, 2018 (p. 437).

[468] Klaus Havelund and Allen Goldberg. “Verify Your Runs.” In *VSTTE.* Vol. 4171. Lecture Notes in Computer Science. Springer, 2005, pp. 374–383 (p. 387).

[469] Klaus Havelund, Giles Reger, and Grigore Rosu. “Runtime Verification Past Experiences and Future Projections.” In *Computing and Software Science.* Vol. 10000. Lecture Notes in Computer Science. Springer, 2019, pp. 532–562 (p. 387).

[470] Alden M. Hayashi. “ROUGH SAILING FOR SMART SHIPS, Does Commercial Software Such as Windows NT Compromise Naval Ship Performance?.” *Scientific American.* 279.5 (Nov. 1998), p. 46 (p. 675).

[471] Jingxuan He, Gagandeep Singh, Markus Püschel, and Martin T. Vechev. “Learning Fast and Precise Numerical Analysis.” In *PLDI.* ACM, 2020, pp. 1112–1127 (p. 493).

[472] Matthew S. Hecht. *Flow Analysis of Computer Programs.* Vol. 5. Programming Languages Series. Elsevier North-Holland, 1973 (p. 109, 259, 497, 514).

[473] Daniel Hedin, Luciano Bello, and Andrei Sabelfeld. “Value-Sensitive Hybrid Information Flow Control for a JavaScript-Like Language.” In *CSF.* IEEE Computer Society, 2015, pp. 351–365 (p. 620).

[474] Nevin Heintze. “Practical Aspects of Set Based Analysis.” In *JICSLP.* MIT Press, 1992, pp. 765–779 (p. 397).

[475] Nevin Heintze and Joxan Jaffar. “Set Constraints and Set-Based Analysis.” In *PPCP.* Vol. 874. Lecture Notes in Computer Science. Springer, 1994, pp. 281–298 (p. 269, 589).

[476] Nevin Heintze and Olivier Tardieu. “Demand-Driven Pointer Analysis.” In *PLDI.* ACM, 2001, pp. 24–34 (p. 590).

[477] Matthias Heizmann, Jochen Hoenicke, and Andreas Podelski. “Software Model Checking for People Who Love Automata.” In *CAV.* Vol. 8044. Lecture Notes in Computer Science. Springer, 2013, pp. 36–52 (p. 548).

[478] Peter Henderson. “Finite State Modelling in Program Development.” In *Proceedings of the International Conference on Reliable Software.* ACM, Apr. 1975, pp. 221–227 (p. 528).

[479] Peter Henderson and Peter Quarendon. “Finite State Testing of Structured Programs.” In *Symposium on Programming.* Vol. 19. Lecture Notes in Computer Science. Springer, 1974, pp. 72–80 (p. 528).

[480] Wolfgang Henhapl and Cliff B. Jones. “A Formal Definition of Algol 60 as Described in the 1975 Modified Report.” In *The Vienna Development Method: The Meta-Language.* Vol. 61. Lecture Notes in Computer Science. Springer, 1978, pp. 305–336 (p. 526).

[481] Thomas A. Henzinger and Pei-Hsin Ho. “A Note on Abstract Interpretation Strategies for Hybrid Automata.” In *Hybrid Systems.* Vol. 999. Lecture Notes in Computer Science. Springer, 1994, pp. 252–264 (p. 718).

[482] Thomas A. Henzinger, Anna Lukina, and Christian Schilling. “Outside the Box: Abstraction-Based Monitoring of Neural Networks.” In G. De Giacomo, A. Catalá, B. Dilkina, M. Milano, S. Barro, A. Bugarín, and J. Lang, eds. *ECAI.* Vol. 325. Frontiers in Artificial Intelligence and Applications. IOS Press, 2020, pp. 2433–2440 (p. 398).

[483] Thomas A. Henzinger, Rupak Majumdar, Freddy Y. C. Mang, and Jean-François Raskin. “Abstract Interpretation of Game Properties.” In *SAS.* Vol. 1824. Lecture Notes in Computer Science. Springer, 2000, pp. 220–239 (p. 718).

[484] Kihong Heo, Hakjoo Oh, and Hongseok Yang. “Learning a Variable-Clustering Strategy for Octagon From Labeled Data Generated by a Static Analysis.” In *SAS.* Vol. 9837. Lecture Notes in Computer Science. Springer, 2016, pp. 237–256 (p. 493).

[485] Kihong Heo, Hakjoo Oh, and Kwangkeun Yi. “Selective Conjunction of Context-Sensitivity and Octagon Domain Toward Scalable and Precise Global Static Analysis.” *Softw., Pract. Exper.* 47.11 (2017), pp. 1677–1705 (p. 493, 714).

[486] Jacques Herbrand. "Investigations in Proof Theory." Ch. V of "Logical Writings," edited by Warren D. Goldfarb, Springer Netherlands, 1971, 44–202, English Translation of [487]. Thesis. Université de Paris, June 11, 1930 (p. 623, 639).

[487] Jacques Herbrand. "Recherches sur la théorie de la démonstration." Ch. V of "Écrits logiques," edited by Jean Van Heijenoort, Presses Universitaires de France, 1968, 35–143. Thèse. Université de Paris, June 11, 1930 (p. 10, 623, 639, 751).

[488] Manuel V. Hermenegildo. "Parallelizing Irregular and Pointer-Based Computations Automatically: Perspectives From Logic and Constraint Programming." *Parallel Computing.* 26.13–14 (2000), pp. 1685–1708 (p. 621).

[489] Manuel V. Hermenegildo, Francisco Bueno, Manuel Carro, Pedro López-García, Edison Mera, José F. Morales, and Germán Puebla. "An Overview of Ciao and Its Design Philosophy." *TPLP.* 12.1–2 (2012), pp. 219–252 (p. 621, 647).

[490] Manuel V. Hermenegildo, Germán Puebla, Francisco Bueno, and Pedro López-García. "Program Development Using Abstract Interpretation (And the Ciao System Preprocessor)." In *SAS.* Vol. 2694. Lecture Notes in Computer Science. Springer, 2003, pp. 127–152 (p. 623, 718).

[491] Manuel V. Hermenegildo and Francesca Rossi. "Strict and Nonstrict Independent And-Parallelism in Logic Programs: Correctness, Efficiency, and Compile-Time Conditions." *J. Log. Program.* 22.1 (1995), pp. 1–45 (p. 621).

[492] Horst Herrlich and Miroslav Hušek. "Galois Connections Categorically." *J. Pure Appl. Algebra.* 68 (1990), pp. 165–180 (p. 131).

[493] Jörg Herter, Daniel Kästner, Christoph Mallon, and Reinhard Wilhelm. "Benchmarking static code analyzers." *Reliab. Eng. Syst. Saf.* 188 (2019), pp. 336–346 (p. 717).

[494] Gerhard Hessenberg. "Kettentheorie und Wohlordnung." *Journal Für Die Reine Und Angewandte Mathematik.* 135 (1910), pp. 81–133, 318 (p. 172).

[495] Arend Heyting. *Intuitionism. An Introduction.* 3rd ed. Elsevier North-Holland, 1956, 1971 (p. 123).

[496] Timothy J. Hickey, Qun Ju, and Maarten H. Van Emden. "Interval Arithmetic: From Principles to Implementation." *J. ACM.* 48.5 (2001), pp. 1038–1068 (p. 370).

[497] David Hilbert and Wilhelm Ackermann. *Grundzüge der Theoretischen Logik.* 6th ed. Engl. Trans. "Principles of Mathematical Logic," Lewis M. Hammond, George G. Leckie, F. Steinhardt, AMS Chelsea, 1958, reprinted 2008. Springer, 1928, 1949, reprinted 1959 (p. 9, 10, 12, 65, 532).

[498] Michael Hind. "Pointer Analysis: Haven't We Solved This Problem Yet?" In *PASTE.* ACM, 2001, pp. 54–61 (p. 559).

[499] J. Roger Hindley. "The Principal Type-Scheme of an Object in Combinatory Logic." *Transactions of the American Mathematical Society.* 146 (1969), pp. 29–60 (p. 663).

[500] J. Roger Hindley. *Basic Simple Type Theory.* Cambridge Tracts in Theoretical Computer Science. Cambridge University Press, Jan. 21, 2008 (p. 623, 649, 671).

[501] Charles Antony Richard Hoare. "An Axiomatic Basis for Computer Programming." *Commun. ACM.* 12.10 (1969), pp. 576–580. http://doi.acm.org/10.1145/363235.363259 (p. 3, 94, 216, 281, 293, 294, 301, 303, 304, 723).

[502] Charles Antony Richard Hoare. "Some Properties of Predicate Transformers." *J. ACM.* 25.3 (1978), pp. 461–480 (p. 144).

[503] Charles Antony Richard Hoare. "An Axiomatic Basis for Computer Programming (Reprint)." *Commun. ACM.* 26.1 (1983), pp. 53–56 (p. 304).

[504] Charles Antony Richard Hoare. "Viewpoint—Retrospective: An Axiomatic Basis for Computer Programming." *Commun. ACM.* 52.10 (2009), pp. 30–32 (p. 94, 304, 723).

[505] Charles Antony Richard Hoare and Peter E. Lauer. "Consistent and Complementary Formal Theories of the Semantics of Programming Languages." *Acta Inf.* 3 (1974), pp. 135–153 (p. 94).

[506] Jochen Hoenicke, Rupak Majumdar, and Andreas Podelski. "Thread Modularity at Many Levels: A Pearl in Compositional Verification." In *POPL*. ACM, 2017, pp. 473–485 (p. 344).

[507] Martin Hofmann and Wei Chen. "Abstract Interpretation From Büchi Automata." In *CSL–LICS*. ACM, 2014, 51:1–51:10 (p. 550).

[508] Gerard J. Holzmann. "Does Not Compute." *IEEE Software*. 36.3 (2019), pp. 14–16 (p. 12).

[509] Hyoung Seok Hong, Insup Lee, and Oleg Sokolsky. "Abstract Slicing: A New Approach to Program Slicing Based on Abstract Interpretation and Model Checking." In *SCAM*. IEEE Computer Society, 2005, pp. 25–34 (p. 618).

[510] John E. Hopcroft, Rajeev Motwani, and Jeffrey D. Ullman. *Introduction to Automata Theory, Languages, and Computation*. 2nd ed. Addison–Wesley, 2003 (p. 532).

[511] Alfred Horn. "On Sentences Which Are True of Direct Unions of Algebras." *Journal of Symbolic Logic*. 16.1 (Mar. 1951), pp. 14–21 (p. 693, 702).

[512] Jacob M. Howe and Andy King. "Logahedra: A New Weakly Relational Domain." In *ATVA*. Vol. 5799. Lecture Notes in Computer Science. Springer, 2009, pp. 306–320 (p. 492).

[513] Jacob M. Howe, Andy King, and Lunjin Lu. "Analysing Logic Programs by Reasoning Backwards." In *Program Development in Computational Logic*. Vol. 3049. Lecture Notes in Computer Science. Springer, 2004, pp. 152–188 (p. 693).

[514] Jacob M. Howe, Andy King, and Axel Simon. "Incremental Closure for Systems of Two Variables Per Inequality." *Theor. Comput. Sci.* 768 (2019), pp. 1–42 (p. 492).

[515] Ehud Hrushovski, Joël Ouaknine, Amaury Pouly, and James Worrell. "Polynomial Invariants for Affine Programs." In *LICS*. ACM, 2018, pp. 530–539 (p. 457).

[516] Kyle Hsu, Rupak Majumdar, Kaushik Mallik, and Anne–Kathrin Schmuck. "Multi-Layered Abstraction–Based Controller Synthesis for Continuous–Time Systems." In *HSCC*. ACM, 2018, pp. 120–129 (p. 718).

[517] Paul Hudak. "A Semantic Model of Reference Counting and Its Abstraction (Detailed Summary)." In *LISP and Functional Programming*. 1986, pp. 351–363 (p. 559).

[518] Gérard P. Huet. "Résolution d'équations dans les langages d'ordre 1, 2, …, ω." Thèse d'État ès Sciences Mathématiques. Paris, France: Université Paris VII, Sept. 28, 1976 (p. 647).

[519] Gérard P. Huet. "Confluent Reductions: Abstract Properties and Applications to Term Rewriting Systems." *J. ACM*. 27.4 (1980), pp. 797–821 (p. 632, 647).

[520] Gérard P. Huet and Hugo Herbelin. "30 Years of Research and Development Around Coq." In *POPL*. ACM, 2014, pp. 249–250 (p. 290, 717).

[521] Jean D. Ichbiah, Bernd Krieg–Brückner, Brian A. Wichmann, John G. P. Barnes, Olivier Roubine, and Jean–Claude Héliard. "Rationale for the Design of the Ada Programming Language." *SIGPLAN Notices*. 14.6b (1979), pp. 1–261 (p. 349).

[522] IEEE. *IEEE Standard for Binary Floating–Point Arithmetic*. American National Standards Institute, Institute of Electrical, and Electronic Engineers, ANSI/IEEE Standard 754-1985, 1985 (p. 367, 372).

[523] International Organization for Standardization. "ISO/IEC 19761: Software Engineering—COSMIC: A Functional Size Measurement Method." Mar. 2011. https://www.iso.org/standard/54849.html (p. 23).

[524] Eugene Isaacson and Herbert Bishop Keller. *Analysis of Numerical Methods*. Dover Books on Mathematics, 1994 (p. 254, 386, 393, 443).

[525] Maxime Jacquemin, Sylvie Putot, and Franck Védrine. "A Reduced Product of Absolute and Relative Error Bounds for Floating-Point Analysis." In *SAS*. Vol. 11002. Lecture Notes in Computer Science. Springer, 2018, pp. 223–242 (p. 398).

[526] Angshuman Jana, Raju Halder, Kalahasti Venkata Abhishekh, Sanjeevini Devi Ganni, and Agostino Cortesi. "Extending Abstract Interpretation to Dependency Analysis of Database Applications." *IEEE Trans. Software Eng.* 46.5 (2020), pp. 463–494 (p. 621).

[527] Angshuman Jana, Raju Halder, and Agostino Cortesi. "Verification of Hibernate Query Language by Abstract Interpretation." In *ICIDE (2)*. Vol. 9243. Lecture Notes in Computer Science. Springer, 2015, pp. 116–128 (p. 718).

[528] Bertrand Jeannet. "A Web Interface to the INTERPROC Static Analyzer." 2019. http://pop-Art.inrialpes.fr/interproc/interprocweb.cgi (p. 250).

[529] Bertrand Jeannet and Antoine Miné. "Apron: A Library of Numerical Abstract Domains for Static Analysis." In *CAV*. Vol. 5643. Lecture Notes in Computer Science. Springer, 2009, pp. 661–667 (p. 250, 492, 716).

[530] Thomas P. Jensen. "Strictness Analysis in Logical Form." In *FPCA*. Vol. 523. Lecture Notes in Computer Science. Springer, 1991, pp. 352–366 (p. 121).

[531] Arnaud Jobin. "Dioïdes et idéaux de polynômes en analyse statique. (Static Analysis with Dioids and Polynomial Ideals)." PhD thesis. École normale supérieure De Cachan, France, 2012 (p. 457).

[532] Donald B. Johnson. "Efficient Algorithms for Shortest Paths in Sparse Networks." *J. ACM.* 24.1 (1977), pp. 1–13 (p. 480).

[533] Stephen C. Johnson. "Yacc: Yet Another Compiler-Compiler." *AT&T Bell Laboratories Computer Science Technical Reports.* 32 (1975) (p. 61).

[534] Stephen C. Johnson. "Lint, a C Program Checker." In *Comp. Sci. Tech. Rep.* Bell Laboratories, July 26, 1978, p. 11 (p. 713).

[535] Anita K. Jones and Richard J. Lipton. "The Enforcement of Security Policies for Computation." In *SOSP.* ACM, 1975, pp. 197–206 (p. 595).

[536] Cliff B. Jones. "The META-Language: A Reference Manual." In *The Vienna Development Method: The Meta-Language.* Vol. 61. Lecture Notes in Computer Science. Springer, 1978, pp. 218–277 (p. 526).

[537] Neil D. Jones. *Computability and Complexity—From a Programming Perspective.* Foundations of Computing. MIT Press, 1997. https://www.itu.dk/people/sestoft/pebook/ (p. 97).

[538] Neil D. Jones, Carsten K. Gomard, and Peter Sestoft. *Partial Evaluation and Automatic Program Generation.* Prentice Hall International Series in Computer Science. Prentice Hall, 1993 (p. 250).

[539] Neil D. Jones and Steven S. Muchnick. "The Complexity of Finite Memory Programs with Recursion." *J. ACM.* 25.2 (1978), pp. 312–321 (p. 295).

[540] Neil D. Jones and Steven S. Muchnick. "Complexity of Flow Analysis, Inductive Assertion Synthesis and a Language Due to Dijkstra." In Steven S. Muchnick and Neil D. Jones. *Program Flow Analysis—Theory and Applications.* Prentice Hall Software Series. Prentice Hall, 1981. Chap. 12, pp. 380–393 (p. 321).

[541] Neil D. Jones and Flemming Nielson. "Abstract Interpretation: A Semantics-Based Tool for Program Analysis." In Samson Abramsky and Dov M. Gabbay, eds. *Handbook of Logic in Computer Science.* Vol. 4, Semantic Modelling. Oxford University Press, 1995, pp. 527–636 (p. 619).

[542] Neil D. Jones, Peter Sestoft, and Harald Søndergaard. "Mix: A Self-Applicable Partial Evaluator for Experiments in Compiler Generation." *Lisp and Symbolic Computation.* 2.1 (1989), pp. 9–50 (p. 619).

[543] Wilhelm Jordan. *Handbuch Der Vermessungskunde.* Verlag Der J. B. Metzler'schen Buchhandlung, 1878 (p. 443).

[544] Jacques-Henri Jourdan. "Verasco: A Formally Verified C Static Analyzer. (Verasco: un analyseur statique pour C formellement vérifié)." PhD thesis. Université Paris Diderot, May 26, 2016 (p. 492, 716, 717).

[545] Jacques-Henri Jourdan. "Sparsity Preserving Algorithms for Octagons." *Electr. Notes Theor. Comput. Sci.* 331 (2017), pp. 57–70 (p. 492).

[546] Jacques-Henri Jourdan, Vincent Laporte, Sandrine Blazy, Xavier Leroy, and David Pichardie. "A Formally-Verified C Static Analyzer." In *POPL*. ACM, 2015, pp. 247–259 (p. 35, 131, 304, 432, 492, 711, 717).

[547] Matthieu Journault, Antoine Miné, Raphaël Monat, and Abdelraouf Ouadjaout. "Combinations of Reusable Abstract Domains for a Multilingual Static Analyzer." In *VSTTE*. Vol. 12031. Lecture Notes in Computer Science. Springer, 2019, pp. 1–18. https://www-apr.lip6.fr/~mine/publi/article-mine-al-vstte19.pdf (p. 433).

[548] Jacques Julliand, Olga Kouchnarenko, Pierre-Alain Masson, and Guillaume Voiron. "Test Generation From Event System Abstractions to Cover Their States and Transitions." *Programming and Computer Software.* 44.1 (2018), pp. 1–14 (p. 718).

[549] Bishoksan Kafle and John P. Gallagher. "Constraint Specialisation in Horn Clause Verification." In *PEPM*. ACM, 2015, pp. 85–90 (p. 702).

[550] Bishoksan Kafle, John P. Gallagher, Graeme Gange, Peter Schachte, Harald Søndergaard, and Peter J. Stuckey. "An Iterative Approach to Precondition Inference Using Constrained Horn Clauses." *TPLP.* 18.3-4 (2018), pp. 553–570 (p. 693).

[551] Gilles Kahn. "Natural Semantics." In *STACS*. Vol. 247. Lecture Notes in Computer Science. Springer, 1987, pp. 22–39 (p. 143).

[552] John B. Kam and Jeffrey D. Ullman. "Global Data Flow Analysis and Iterative Algorithms." *J. ACM.* 23.1 (1976), pp. 158–171 (p. 497).

[553] John B. Kam and Jeffrey D. Ullman. "Monotone Data Flow Analysis Frameworks." *Acta Inf.* 7 (1977), pp. 305–317 (p. 412).

[554] Daniel M. Kan. "Adjoint Functors." *Transactions of the American Mathematical Society.* 87.2 (1958), pp. 294–329 (p. 131).

[555] Leonid Kantorovich. "The Method of Successive Approximations for Functional Equations." *Acta Mathematica.* 71 (1939), pp. 63–97 (p. 172).

[556] Vini Kanvar and Uday P. Khedker. "Heap Abstractions for Static Analysis." *ACM Comput. Surv.* 49.2 (2016), 29:1–29:47 (p. 559, 590).

[557] Marc A. Kaplan and Jeffrey D. Ullman. "A Scheme for the Automatic Inference of Variable Types." *J. ACM.* 27.1 (1980), pp. 128–145 (p. 702).

[558] Irving Kaplansky. *Set Theory and Metric Spaces.* 2nd ed. American Mathematical Society, 1977 (p. 9).

[559] Egor George Karpenkov, David Monniaux, and Philipp Wendler. "Program Analysis with Local Policy Iteration." In *VMCAI*. Vol. 9583. Lecture Notes in Computer Science. Springer, 2016, pp. 127–146 (p. 412).

[560] Michael Karr. "Affine Relationships Among Variables of a Program." *Acta Inf.* 6 (1976), pp. 133–151 (p. 453, 456, 603, 608, 617).

[561] Matt Kaufmann and J. Strother Moore. "An ACL2 Tutorial." In *TPHOLs*. Vol. 5170. Lecture Notes in Computer Science. Springer, 2008, pp. 17–21 (p. 290).

[562] Matt Kaufmann and J. Strother Moore. *ACL2, Version 8.3.* University of Texas at Austin, Apr. 14, 2020. https://www.cs.utexas.edu/users/moore/acl2/ (p. 225).

[563] Sven Keidel and Sebastian Erdweg. "Sound and Reusable Components for Abstract Interpretation." *Proc. ACM Program. Lang.* 3.OOPSLA (2019), 176:1–176:28 (p. 706).

[564] Sven Keidel, Casper Bach Poulsen, and Sebastian Erdweg. "Compositional Soundness Proofs of Abstract Interpreters." *Proc. ACM Program. Lang.* 2.ICFP (2018), 72:1–72:26 (p. 706).

[565] Ken Kennedy. "A Global Flow Analysis Algorithm." *Int. J. of Comp. Math.* 3.1–4 (1972), pp. 5–15 (p. 498).

[566] Ken Kennedy. "Node Listings Applied to Data Flow Analysis." In *POPL*. ACM Press, 1975, pp. 10–21 (p. 498).

[567] Ken Kennedy. "A Comparison of Two Algorithms for Global Data Flow Analysis." *SIAM J. Comput.* 5.1 (Mar. 1976), pp. 158–180 (p. 498).

[568] Ken Kennedy. "A Survey of Data Flow Analysis Techniques." In Steven S. Muchnick and Neil D. Jones. *Program Flow Analysis—Theory and Applications.* Prentice Hall Software Series. Prentice Hall, 1981. Chap. 1, pp. 5–54 (p. 497, 514).

[569] Brian W. Kernighan and Dennis Ritchie. *The C Programming Language.* 2nd ed. Prentice-Hall, 1988 (p. 5, 556, 559, 561, 588, 589).

[570] Yonit Kesten, Oded Maler, Monica Marcus, Amir Pnueli, and Elad Shahar. "Symbolic Model Checking with Rich Assertional Languages." In *CAV.* Vol. 1254. Lecture Notes in Computer Science. Springer, 1997, pp. 424–435 (p. 531).

[571] Uday P. Khedker, Amitabha Sanyal, and Bageshri Sathe. *Data Flow Analysis—Theory and Practice.* CRC Press, 2009 (p. 497).

[572] Gary A. Kildall. "A Unified Approach to Global Program Optimization." In *POPL*. ACM Press, 1973, pp. 194–206 (p. 32, 109, 250, 257, 321, 344, 412, 497, 603, 617).

[573] Sol Kim, Kihong Heo, Hakjoo Oh, and Kwangkeun Yi. "Widening with Thresholds via Binary Search." *Softw., Pract. Exper.* 46.10 (2016), pp. 1317–1328 (p. 406).

[574] Sung Kook Kim, Arnaud J. Venet, and Aditya V. Thakur. "Deterministic parallel fixpoint computation." *Proc. ACM Program. Lang.* 4.POPL (2020), 14:1–14:33 (p. 257).

[575] Sung Kook Kim, Arnaud J. Venet, and Aditya V. Thakur. "Memory-Efficient Fixpoint Computation." In *SAS*. Vol. 12389. Lecture Notes in Computer Science. Springer, 2020, pp. 35–64 (p. 257).

[576] Zachary Kincaid, John Cyphert, Jason Breck, and Thomas W. Reps. "Non-Linear Reasoning for Invariant Synthesis." *Proc. ACM Program. Lang.* 2.POPL (2018), 54:1–54:33 (p. 457).

[577] Andy King and Lunjin Lu. "Forward Versus Backward Verification of Logic Programs." In *ICLP.* Vol. 2916. Lecture Notes in Computer Science. Springer, 2003, pp. 315–330 (p. 693).

[578] James C. King. "Symbolic Execution and Program Testing." *Commun. ACM.* 19.7 (1976), pp. 385–394 (p. 548).

[579] Florent Kirchner, Nikolai Kosmatov, Virgile Prevosto, Julien Signoles, and Boris Yakobowski. "Frama-C: A Software Analysis Perspective." *Formal Asp. Comput.* 27.3 (2015), pp. 573–609 (p. 304, 715).

[580] Oleg Kiselyov. "Effects Without Monads: Non-Determinism—Back to the Meta Language." In *ML/OCaml.* Vol. 294. EPTCS. 2017, pp. 15–40.

[581] Stephen Cole Kleene. "Representation of Events in Nerve Nets and Finite Automata." In *Automata Studies.* Princeton University Press, 1951, pp. 3–42 (p. 527, 528, 532, 550).

[582] Stephen Cole Kleene. *Introduction to Meta-Mathematics.* Elsevier North-Holland, 1952 (p. 9–11, 172).

[583] Stephen Cole Kleene. "The Theory of Recursive Functions Approaching Its Centennial." *Bulletin of the American Mathematical Society.* 5.1 (July 1981), pp. 43–61 (p. 172).

[584] Bronisław Knaster. "Un théorème sur les fonctions d´ensembles." *Ann. Soc. Polon. Math.* 5 (1928), pp. 133–134 (p. 167, 172).

[585] Jens Knoop, Dirk Koschützki, and Bernhard Steffen. "Basic–Block Graphs: Living Dinosaurs?" In *CC.* Vol. 1383. Lecture Notes in Computer Science. Springer, 1998, pp. 65–79 (p. 603).

[586] Jens Knoop and Oliver Rüthing. "Constant Propagation on the Value Graph: Simple Constants and Beyond." In *CC.* Vol. 1781. Lecture Notes in Computer Science. Springer, 2000, pp. 94–109 (p. 603, 617).

[587] Donald Ervin Knuth. "On the Translation of Languages From Left to Right." *Information and Control.* 8.6 (1965), pp. 607–639 (p. 61).

[588] Donald Ervin Knuth. "The Remaining Trouble Spots in ALGOL 60." *Commun. ACM.* 10.10 (1967), pp. 611–618 (p. 61).

[589] Donald Ervin Knuth. "Semantics of Context–Free Languages." *Mathematical Systems Theory.* 2.2 (1968), pp. 127–145 (p. 41).

[590] Donald Ervin Knuth. "Correction: Semantics of Context–Free Languages." *Mathematical Systems Theory.* 5.1 (1971), pp. 95–96 (p. 41).

[591] Donald Ervin Knuth. "The Genesis of Attribute Grammars." In *Attribute Grammars and Their Applications.* Vol. 461. Lecture Notes in Computer Science. Springer, 1990, pp. 1–12 (p. 41).

[592] Donald Ervin Knuth. *The Art of Computer Programming, Volume III.* 2nd ed. Addison–Wesley, 1998 (p. 699).

[593] Niklas Kochdumper and Matthias Althoff. "Sparse Polynomial Zonotopes: A Novel Set Representation for Reachability Analysis." *CoRR.* abs/1901.01780 (2019) (p. 398).

[594] Paul Kocher, Jann Horn, Anders Fogh, Daniel Genkin, Daniel Gruss, Werner Haas, Mike Hamburg, Moritz Lipp, Stefan Mangard, Thomas Prescher, Michael Schwarz, and Yuval Yarom. "Spectre Attacks: Exploiting Speculative Execution." *Commun. ACM.* 63.7 (2020), pp. 93–101 (p. 79).

[595] John Kodumal and Alexander Aiken. "Banshee: A Scalable Constraint–Based Analysis Toolkit." In *SAS.* Vol. 3672. Lecture Notes in Computer Science. Springer, 2005, pp. 218–234 (p. 589).

[596] Robert A. Kowalski. "The Early Years of Logic Programming." *Commun. ACM.* 31.1 (1988), pp. 38–43 (p. 623).

[597] Dexter Kozen. "Results on the Propositional Mu–Calculus." *Theor. Comput. Sci.* 27 (1983), pp. 333–354 (p. 527).

[598] Robbert Krebbers, Xavier Leroy, and Freek Wiedijk. "Formal C Semantics: CompCert and the C Standard." In *ITP.* Vol. 8558. Lecture Notes in Computer Science. Springer, 2014, pp. 543–548 (p. 708).

[599] Saul A. Kripke. "Semantical Considerations on Modal Logic." *Proceedings of a Colloquium on Modal and Many–Valued Logics, Helsinki, 23–26 August, 1962, Acta Philosophica Fennica.* 16 (1963), pp. 83–94 (p. 539).

[600] Markus Krötzsch and Grit Malik. "The Tensor Product as a Lattice of Regular Galois Connections." In *ICFCA.* Vol. 3874. Lecture Notes in Computer Science. Springer, 2006, pp. 89–104 (p. 131).

[601] Orna Kupferman. "Automata Theory and Model Checking." In *Handbook of Model Checking.* Springer, 2018, pp. 107–151 (p. 551).

[602] Kazimierz Kuratowski. "Sur la notion de l'ordre dans la théorie des ensembles." *Fundamenta Mathematicæ.* 2.1 (1921), pp. 161–171 (p. 12).

[603] Kazimierz Kuratowski. "Une méthode d'élimination des nombres transfinis des raisonnements mathématiques." *Fundamenta Mathematicæ.* 3 (1922), pp. 76–108 (p. 172).

[604] Kazimierz Kuratowski. *Topologie, Tome I.* 4th ed. Éditions Jacques Gabay, 1958 (p. 148).

[605] Kazimierz Kuratowski. *Topologie, Tome II.* 3rd ed. Éditions Jacques Gabay, 1961 (p. 148).

[606] Kazimierz Kuratowski and Andrzej Mostowski. *Set Theory.* North–Holland, Jan. 1968 (p. 10).

[607] Radosław A. Kycia. "Landauer's Principle as a Special Case of Galois Connection." *Entropy.* 20.12 (2018), p. 971 (p. 131).

[608] Lies Lakhdar–Chaouch, Bertrand Jeannet, and Alain Girault. "Widening with Thresholds for Programs with Complex Control Graphs." In *ATVA.* Vol. 6996. Lecture Notes in Computer Science. Springer, 2011, pp. 492–502 (p. 406).

[609] Joachim Lambek. "Some Galois Connections in Elementary Number Theory." *Journal of Number Theory.* 47 (1994), pp. 371–377 (p. 131).

[610] Leslie Lamport. "How to Make a Correct Multiprocess Program Execute Correctly on a Multiprocessor." *IEEE Trans. Computers.* 46.7 (1997), pp. 779–782 (p. 192).

[611] Leslie Lamport. "Safety, Liveness, and Fairness." May 2019. https://lamport.azurewebsites.net/nobreak tlaa/nobreak safety-liveness.pdf (p. 161).

[612] Butler W. Lampson. "A Note on the Confinement Problem." *Commun. ACM.* 16.10 (1973), pp. 613–615 (p. 595).

[613] William Landi. "Undecidability of Static Analysis." *LOPLAS.* 1.4 (1992), pp. 323–337 (p. 104).

[614] Saunders Mac Lane. *Category for the Working Mathematician.* Graduate Texts in Mathematics. Springer, 1971 (p. 119).

[615] Serge Lang. *Linear Algebra.* 3rd ed. Undergraduate Texts in Mathematics. Springer, 1991 (p. 437).

[616] Serge Lang. *Algebra.* 3rd ed. Graduate Texts in Mathematics. Springer, 2002 (p. 112).

[617] Kim Guldstrand Larsen, Paul Pettersson, and Wang Yi. "Compositional and Symbolic Model Checking of Real–Time Systems." In *RTSS.* IEEE Computer Society, 1995, pp. 76–87 (p. 483).

[618] Janusz Laski and William Stanley. "Program Dependencies (Ch. 6)." In *Software Verification and Analysis: An Integrated, Hands–On Approach.* Springer, 2009, pp. 125–142 (p. 592).

[619] Jean–Louis Lassez, Michael J. Maher, and Kim Marriott. "Unification Revisited." In J. Minker, ed. *Foundations of Deductive Databases and Logic Programming.* Morgan Kaufmann, 1988, pp. 587–625 (p. 647).

[620] Chris Lattner and Vikram S. Adve. "LLVM: A Compilation Framework for Lifelong Program Analysis & Transformation." In *CGO.* IEEE Computer Society, 2004, pp. 75–88 (p. 713).

[621] Baudouin Le Charlier and Pascal Van Hentenryck. "Reexecution in Abstract Interpretation of Prolog." In *JICSLP.* MIT Press, 1992, pp. 750–764 (p. 347).

[622] Gurvan Le Guernic. "Confidentiality Enforcement Using Dynamic Information Flow Analyses." PhD thesis. Kansas State University, 2007 (p. 620).

[623] Hongki Lee, Changhee Park, and Sukyoung Ryu. "Automatically Tracing Imprecision Causes in JavaScript Static Analysis." *The Art, Science, and Engineering of Programming.* 4.2 (2020), Article 2, 27 Pages (p. 693).

[624] Woosuk Lee, Wonchan Lee, Dongok Kang, Kihong Heo, Hakjoo Oh, and Kwangkeun Yi. "Sound Non–Statistical Clustering of Static Analysis Alarms." *ACM Trans. Program. Lang. Syst.* 39.4 (2017), 16:1–16:35 (p. 707, 710).

[625] Daniel J. Lehmann, Amir Pnueli, and Jonathan Stavi. "Impartiality, Justice and Fairness: The Ethics of Concurrent Termination." In *ICALP.* Vol. 115. Lecture Notes in Computer Science. Springer, 1981, pp. 264–277 (p. 12).

[626] Sorin Lerner, David Grove, and Craig Chambers. "Composing Dataflow Analyses and Transformations." In *POPL.* ACM, 2002, pp. 270–282 (p. 432).

[627] Xavier Leroy. "Formal Verification of a Realistic Compiler." *Commun. ACM.* 52.7 (2009), pp. 107–115 (p. 513, 708).

[628] Xavier Leroy, Damien Doligez, Alain Frisch, Jacques Garrigue, Didier Rémy, and Jérôme Vouillon. "The OCaml System, Release 4.12, Documentation and User's Manual." Institut National de Recherche en Informatique et en Automatique. Apr. 7, 2021. http://caml.inria.fr/pub/docs/manual-ocaml/ (p. 5, 33, 34, 57, 61, 342, 625).

[629] Michael E. Lesk. "Lex—a Lexical Analyzer Generator." *AT&T Bell Laboratories Computer Science Technical Reports.* 39 (Oct. 1975) (p. 61).

[630] Jacek Leszczyłowski. "A Theorem on Resolving Equations in the Space of Languages." *Bull. Acad. Polon. Sci. Sgr. ScL Math. Astronom. Phys..* 19 (1971), pp. 967–970 (p. 168).

[631] Martin Leucker, César Sánchez, Torben Scheffel, Malte Schmitz, and Daniel Thoma. "Runtime Verification for Timed Event Streams with Partial Information." In *ArXiv:1907.07761.* July 2019, p. 29 (p. 388).

[632] Tal Lev-Ami and Mooly Sagiv. "TVLA: A System for Implementing Static Analyses." In *SAS.* Vol. 1824. Lecture Notes in Computer Science. Springer, 2000, pp. 280–301 (p. 195).

[633] Li Li, Tegawendé F. Bissyandé, Mike Papadakis, Siegfried Rasthofer, Alexandre Bartel, Damien Octeau, Jacques Klein, and Yves Le Traon. "Static Analysis of Android Apps: A Systematic Literature Review." *Information & Software Technology.* 88 (2017), pp. 67–95 (p. 619, 620).

[634] Leo Liberti. "Undecidability and Hardness in Mixed-Integer Nonlinear Programming." *RAIRO - Operations Research.* 53.1 (2019), pp. 81–109 (p. 167).

[635] Leo Liberti and Fabrizio Marinelli. "Mathematical Programming: Turing Completeness and Applications to Software Analysis." *J. Comb. Optim.* 28.1 (2014), pp. 82–104 (p. 167).

[636] Orna Lichtenstein and Amir Pnueli. "Checking That Finite State Concurrent Programs Satisfy Their Linear Specification." In *POPL.* ACM Press, 1985, pp. 97–107 (p. 548).

[637] Moritz Lipp, Michael Schwarz, Daniel Gruss, Thomas Prescher, Werner Haas, Anders Fogh, Jann Horn, Stefan Mangard, Paul Kocher, Daniel Genkin, Yuval Yarom, and Mike Hamburg. "Meltdown: Reading Kernel Memory From User Space." In *USENIX Security Symposium.* USENIX Association, 2018, pp. 973–990 (p. 79).

[638] Benjamin Livshits, Manu Sridharan, Yannis Smaragdakis, Ondrej Lhoták, José Nelson Amaral, Bor-Yuh Evan Chang, Samuel Z. Guyer, Uday P. Khedker, Anders Møller, and Dimitrios Vardoulakis. "in Defense of Soundiness: A Manifesto." *Commun. ACM.* 58.2 (2015), pp. 44–46 (p. 705, 709).

[639] Francesco Logozzo and Manuel Fähndrich. "Pentagons: A Weakly Relational Abstract Domain for the Efficient Validation of Array Accesses." *Sci. Comput. Program.* 75.9 (2010), pp. 796–807 (p. 492).

[640] Pedro López-García, Francisco Bueno, and Manuel V. Hermenegildo. "Automatic Inference of Determinacy and Mutual Exclusion for Logic Programs Using Mode and Type Analyses." *New Generation Comput.* 28.2 (2010), pp. 177–206.

[641] John M. Lucassen and David K. Gifford. "Polymorphic Effect Systems." In *POPL.* ACM Press, 1988, pp. 47–57 (p. 671).

[642] Michael Lvov and Yulia Tarasich. "The Static Analysis of Linear Loops." In *ICTERI.* Vol. 1356. CEUR Workshop Proceedings. CEUR-WS.org, 2015, pp. 366–381 (p. 457).

[643] Holbrook M. MacNeille. "Partially Ordered Sets." *Trans. Amer. Math. Soc.* 42.3 (1937), pp. 416–460.

[644] Yannis Mallios, Lujo Bauer, Dilsun Kirli Kaynar, and Jay Ligatti. "Enforcing More with Less: Formalizing Target-Aware Run-Time Monitors." In *STM.* Vol. 7783. Lecture Notes in Computer Science. Springer, 2012, pp. 17–32 (p. 528, 550).

[645] Amit Kr Mandal, Pietro Ferrara, Yuliy Khlyebnikov, Agostino Cortesi, and Fausto Spoto. "Cross-Program Taint Analysis for IoT Systems." In *SAC*. ACM, 2020, pp. 1944–1952 (p. 619).

[646] Christopher Mann and Artem Starostin. "A Framework for Static Detection of Privacy Leaks in Android Applications." In *SAC*. ACM, 2012, pp. 1457–1462 (p. 621).

[647] Zohar Manna. "Mathematical Theory of Partial Correctness." *J. Comput. Syst. Sci.* 5.3 (1971), pp. 239–253 (p. 216, 281, 291).

[648] Zohar Manna, Stephen Ness, and Jean Vuillemin. "Inductive Methods for Proving Properties of Programs." *Commun. ACM.* 16.8 (1973), pp. 491–502 (p. 278).

[649] Zohar Manna and Amir Pnueli. "Axiomatic Approach to Total Correctness of Programs." *Acta Inf.* 3 (1974), pp. 243–263 (p. 684).

[650] Zohar Manna and Amir Pnueli. "A Hierarchy of Temporal Properties." In *PODC*. ACM, 1990, pp. 377–410 (p. 161).

[651] Zohar Manna and Amir Pnueli. *The temporal logic of reactive and concurrent systems—specification*. Springer, 1992 (p. 160, 161).

[652] Zohar Manna and Jean Vuillemin. "Fix Point Approach to the Theory of Computation." *Commun. ACM.* 15.7 (1972), pp. 528–536 (p. 278).

[653] Zohar Manna and Richard J. Waldinger. "Deductive Synthesis of the Unification Algorithm." *Sci. Comput. Program.* 1.1–2 (1981), pp. 5–48 (p. 647).

[654] Panagiotis Manolios and Richard J. Trefler. "A Lattice-Theoretic Characterization of Safety and Liveness." In *PODC*. ACM, 2003, pp. 325–333 (p. 161).

[655] Maplesoft. "Maple User Manual." 2020. www.maplesoft.com (p. 711).

[656] George Markowsky. "Chain-Complete Posets and Directed Sets with Applications." *Algebra Universalis.* 6.1 (Nov. 1976), pp. 53–68 (p. 121, 167, 172).

[657] George Markowsky and Linda Markowsky. "Lattice Data Analytics: The Poset of Irreducibles and the MacNeille Completion." In *IDAACS*. IEEE, 2019, pp. 263–268 (p. 121).

[658] Kim Marriott, Maria J. García de la Banda, and Manuel V. Hermenegildo. "Analyzing Logic Programs with Dynamic Scheduling." In *POPL*. ACM Press, 1994, pp. 240–253 (p. 621).

[659] Matthieu Martel. "RangeLab: A Static-Analyzer to Bound the Accuracy of Finite-Precision Computations." In *SYNASC*. IEEE Computer Society, 2011, pp. 118–122 (p. 386, 398).

[660] Alberto Martelli and Ugo Montanari. "An Efficient Unification Algorithm." *ACM Trans. Program. Lang. Syst.* 4.2 (1982), pp. 258–282 (p. 639).

[661] Florian Martin. "PAG - an Efficient Program Analyzer Generator." *STTT.* 2.1 (1998), pp. 46–67 (p. 716).

[662] Florian Martin. "Generating Program Analyzers." PhD thesis. Saarland University, 1999 (p. 716).

[663] François Masdupuy. "Array Abstractions Using Semantic Analysis of Trapezoid Congruences." In *ICS*. ACM, 1992, pp. 226–235 (p. 365).

[664] François Masdupuy. "Semantic Analysis of Interval Congruences." In *Formal Methods in Programming and Their Applications*. Vol. 735. Lecture Notes in Computer Science. Springer, 1993, pp. 142–155 (p. 365, 398).

[665] Damien Massé. "Combining Forward and Backward Analyses of Temporal Properties." In *PADO*. Vol. 2053. Lecture Notes in Computer Science. Springer, 2001, pp. 103–116 (p. 693, 702).

[666] Damien Massé. "Property Checking Driven Abstract Interpretation-Based Static Analysis." In *VMCAI*. Vol. 2575. Lecture Notes in Computer Science. Springer, 2003, pp. 56–69 (p. 693, 702).

[667] Damien Massé. "Abstract Domains for Property Checking Driven Analysis of Temporal Properties." In *AMAST*. Vol. 3116. Lecture Notes in Computer Science. Springer, 2004, pp. 349–363 (p. 693, 702).

[668] Isabella Mastroeni and Damiano Zanardini. "Abstract Program Slicing: An Abstract Interpretation-Based Approach to Program Slicing." *ACM Trans. Comput. Log.* 18.1 (2017), 7:1–7:58 (p. 618, 621).

[669] Laurent Mauborgne. "Abstract Interpretation Using TDGs." In *SAS*. 1994, pp. 363–379 (p. 548).

[670] Laurent Mauborgne. "Abstract Interpretation Using Typed Decision Graphs." *Sci. Comput. Program.* 31.1 (1998), pp. 91–112 (p. 548).

[671] Laurent Mauborgne. "Binary Decision Graphs." In *SAS*. Vol. 1694. Lecture Notes in Computer Science. Springer, 1999, pp. 101–116 (p. 548).

[672] Laurent Mauborgne and Xavier Rival. "Trace Partitioning in Abstract Interpretation Based Static Analyzers." In *ESOP*. Vol. 3444. Lecture Notes in Computer Science. Springer, 2005, pp. 5–20 (p. 551).

[673] John McCarthy. "Recursive Functions of Symbolic Expressions and Their Computation by Machine, Part I." *Commun. ACM*. 3.4 (1960), pp. 184–195 (p. 15, 526, 671).

[674] John McCarthy, Paul W. Abrahams, Daniel J. Edwards, Timothy P. Hart, and Michael I. Levin. *LISP 1.5 Programmer's Manual*. MIT Press, 1966 (p. 526).

[675] Daryl McCullough. "Noninterference and the Composability of Security Properties." In *IEEE Symposium on Security and Privacy*. IEEE Computer Society, 1988, pp. 177–186 (p. 593).

[676] John McLean. "A General Theory of Composition for Trace Sets Closed Under Selective Interleaving Functions." In *IEEE Symposium on Security and Privacy*. IEEE Computer Society, 1994, pp. 79–93 (p. 593).

[677] Kenneth L. McMillan. "Quantified Invariant Generation Using an Interpolating Saturation Prover." In *TACAS*. Vol. 4963. Lecture Notes in Computer Science. Springer, 2008, pp. 413–427 (p. 412).

[678] Kenneth L. McMillan. "Widening and Interpolation." In *SAS*. Vol. 6887. Lecture Notes in Computer Science. Springer, 2011, p. 1 (p. 412).

[679] Peter McMullen. "On Zonotopes." *Trans. Amer. Math. Soc.* 159 (1971), pp. 91–110 (p. 387).

[680] Austin Melton, Bernd S. W. Schröder, and George E. Strecker. "Lagois Connections—a Counterpart to Galois Connections." *Theor. Comput. Sci.* 136.1 (1994), pp. 79–107 (p. 131).

[681] Jan Midtgaard, Flemming Nielson, and Hanne Riis Nielson. "A Parametric Abstract Domain for Lattice-Valued Regular Expressions." In *SAS*. Vol. 9837. Lecture Notes in Computer Science. Springer, 2016, pp. 338–360 (p. 551).

[682] Bogdan Mihaila, Alexander Sepp, and Axel Simon. "Widening as Abstract Domain." In *NASA Formal Methods*. Vol. 7871. Lecture Notes in Computer Science. Springer, 2013, pp. 170–184 (p. 406).

[683] Steven P. Miller, Michael W. Whalen, and Darren D. Cofer. "Software Model Checking Takes Off." *Commun. ACM*. 53.2 (2010), pp. 58–64 (p. 319, 550).

[684] Robert Milne and Christopher Strachey. *A Theory of Programming Language Semantics (2 Vol)*. Springer, 1976 (p. 526).

[685] Robin Milner. "A Theory of Type Polymorphism in Programming." *J. Comput. Syst. Sci.* 17.3 (1978), pp. 348–375 (p. 410, 649, 671).

[686] Robin Milner and Mads Tofte. "Co-Induction in Relational Semantics." *Theor. Comput. Sci.* 87.1 (1991), pp. 209–220 (p. 143, 184).

[687] Antoine Miné. "A New Numerical Abstract Domain Based on Difference-Bound Matrices." In *PADO*. Vol. 2053. Lecture Notes in Computer Science. Springer, 2001, pp. 155–172 (p. 481, 492).

[688] Antoine Miné. "The Octagon Abstract Domain." In *Proceedings of the Eighth Working Conference on Reverse Engineering, WCRE'01, Stuttgart, Germany, October 2-5, 2001*. IEEE Computer Society, 2001, pp. 310–319 (p. 481).

[689] Antoine Miné. "A Few Graph-Based Relational Numerical Abstract Domains." In *SAS*. Vol. 2477. Lecture Notes in Computer Science. Springer, 2002, pp. 117–132 (p. 365).

[690] Antoine Miné. "Weakly Relational Numerical Abstract Domains." Thèse de doctorat. Palaiseau, France: École polytechnique, Dec. 6, 2004 (p. 491, 492).

[691] Antoine Miné. "Field-Sensitive Value Analysis of Embedded C Programs with Union Types and Pointer Arithmetics." In *LCTES*. ACM, 2006, pp. 54–63 (p. 322, 588).

[692] Antoine Miné. "The Octagon Abstract Domain." *Higher-Order and Symbolic Computation*. 19.1 (2006), pp. 31–100 (p. 431, 481, 490–492, 617).

[693] Antoine Miné. "Static Analysis by Abstract Interpretation of Concurrent Programs." Mémoire d'habilitation à diriger les recherches. Paris, France: École normale supérieure, May 28, 2013 (p. 322, 588).

[694] Antoine Miné. "Backward Underapproximations in Numeric Abstract Domains to Automatically Infer Sufficient Program Conditions." *Sci. Comput. Program*. 93 (2014), pp. 154–182 (p. 684, 692).

[695] Antoine Miné. "Tutorial on Static Inference of Numeric Invariants by Abstract Interpretation." *Foundations and Trends in Programming Languages*. 4.3–4 (2017), pp. 120–372 (p. 456, 492).

[696] Antoine Miné and David Delmas. "Towards an Industrial Use of Sound Static Analysis for the Verification of Concurrent Embedded Avionics Software." In *EMSOFT*. IEEE, 2015, pp. 65–74 (p. 714).

[697] Jack Minker, Dietmar Seipel, and Carlo Zaniolo. "Logic and Databases: A History of Deductive Databases." In D.M. Gabbay, J.H. Siekmann, and J. Woods, eds. *Computational Logic*. Vol. 9. Handbook of the History of Logic. Elsevier, 2014, pp. 571–627 (p. 623).

[698] Solène Mirliaz and David Pichardie. "Flow Insensitive Relational Static Analysis." In *NSAD*, 8th International Workshop on Numerical and Symbolic Abstract Domains, Porto, Portugal. 2019. https://staticanalysis.org/nsad2019/papers/NSAD2019_paper_6_WM.pdf (p. 553).

[699] Matthew Mirman, Timon Gehr, and Martin T. Vechev. "Differentiable Abstract Interpretation for Provably Robust Neural Networks." In *ICML*. Vol. 80. Proceedings of Machine Learning Research. PMLR, 2018, pp. 3575–3583 (p. 718).

[700] Ian M. Mitchell. "Comparing Forward and Backward Reachability as Tools for Safety Analysis." In *HSCC*. Vol. 4416. Lecture Notes in Computer Science. Springer, 2007, pp. 428–443 (p. 693).

[701] Eugenio Moggi, Walid Tahab, and Johan Thunberg. "Sound Over-Approximation of Probabilities." *Acta Cybern*. 24.3 (2020), pp. 269–285 (p. 526).

[702] Mahdi Soltan Mohammadi, Kazem Cheshmi, Ganesh Gopalakrishnan, Mary W. Hall, Maryam Mehri Dehnavi, Anand Venkat, Tomofumi Yuki, and Michelle Mills Strout. "Sparse Matrix Code Dependence Analysis Simplification at Compile Time." *CoRR*. abs/1807.10852 (2018) (p. 621).

[703] Markus Mohnen. "A Graph-Free Approach to Data-Flow Analysis." In *CC*. Vol. 2304. Lecture Notes in Computer Science. Springer, 2002, pp. 46–61 (p. 514).

[704] Mehryar Mohri and Mark-Jan Nederhof. "Regular Approximation of Context-Free Grammars Through Transformation." In Jean-Claude Junqua and Gertjan Van Noord.

Robustness in Language and Speech Technology. Vol. 17. Texts, Speech and Language Technology. Springer, 2001, pp. 153–163 (p. 551).

[705] Anders Møller and Michael I. Schwartzbach. "Static Program Analysis." Nov. 2020. https://cs.au.dk/~amoeller/spa/spa.pdf (p. 412).

[706] James Donald Monk. *Introduction to Set Theory.* McGraw–Hill, 1969. http://euclid.colorado.edu/~monkd/monk11.pdf (p. 16).

[707] James Donald Monk and Robert Bonnet. *Handbook of Boolean Algebras (Volumes 1 to 3).* North–Holland, 1983. http://euclid.colorado.edu/~monkd/monk[47,48,49].pdf (p. 171).

[708] David Monniaux. "The Parallel Implementation of the Astrée Static Analyzer." In *APLAS.* Vol. 3780. Lecture Notes in Computer Science. Springer, 2005, pp. 86–96 (p. 257).

[709] David Monniaux. "The Pitfalls of Verifying Floating–Point Computations." *ACM Trans. Program. Lang. Syst.* 30.3 (2008), 12:1–12:41 (p. 367, 372).

[710] David Monniaux. "Automatic Modular Abstractions for Template Numerical Constraints." *Logical Methods in Computer Science.* 6.3 (2010) (p. 402).

[711] David Monniaux. "On the Decidability of the Existence of Polyhedral Invariants in Transition Systems." *Acta Inf.* 56.4 (2019), pp. 385–389 (p. 319).

[712] António Monteiro and Hugo Ribeiro. "L'opération de fermeture et ses invariants dans les systèmes partiellement ordonnés." *Portugaliæ Mathematica.* 3.3 (1942), pp. 171–184 (p. 127, 130).

[713] Eliakim Hastings Moore. *Introduction to a Form of General Analysis.* Yale University Press, 1910 (p. 10, 129, 721).

[714] J. Strother Moore and Claus–Peter Wirth. "Automation of Mathematical Induction as Part of the History of Logic." *FLAP.* 4.5 (2017) (p. 290).

[715] Ramon E. Moore. *Interval Analysis.* Prentice Hall, 1966 (p. 367, 371, 387, 398).

[716] Ramon E. Moore. *Methods and Applications of Interval Analysis.* SIAM Studies in Applied Mathematics. SIAM, 1995 (p. 367, 371, 387).

[717] Ramon E. Moore, R. Baker Kearfott, and Michael J. Cloud. *Introduction to Interval Analysis.* Society for Industrial and Applied Mathematics, Mar. 2009 (p. 367, 371, 387).

[718] José F. Morales, Manuel Carro, and Manuel V. Hermenegildo. "Description and Optimization of Abstract Machines in a Dialect of Prolog." *Theory Pract. Log. Program.* 16.1 (2016), pp. 1–58 (p. 647).

[719] José Morgado. "A Characterization of the Closure Operators by Means of a Single Axiom." *Portugaliæ Mathematica.* 21.3 (1962), pp. 155–156 (p. 121).

[720] Augustus De Morgan. "Mathematical Induction." *The Penny Cyclopedia of the Society for the Diffusion of Useful Knowledge.* 12 (1838). http://education.lms.ac.uk/wp-content/uploads/2011/10/De_Morgan_Mathematical_Induction.pdf (p. 16).

[721] James H. Morris Jr. and Ben Wegbreit. "Subgoal Induction." *Commun. ACM.* 20.4 (1977), pp. 209–222 (p. 291, 693).

[722] Shin–Cheng Mu and José Nuno Oliveira. "Programming From Galois Connections." *J. Log. Algebr. Program.* 81.6 (2012), pp. 680–704 (p. 112).

[723] Steven S. Muchnick. *Advanced Compiler Design and Implementation.* Morgan Kaufmann, 1997 (p. 514).

[724] Steven S. Muchnick and Neil D. Jones. *Program Flow Analysis—Theory and Applications.* Prentice Hall Software Series. Prentice Hall, 1981.

[725] Rajdeep Mukherjee, Peter Schrammel, Leopold Haller, Daniel Kroening, and Tom Melham. "Lifting CDCL to Template–Based Abstract Domains for Program Verification."

In *ATVA*. Vol. 10482. Lecture Notes in Computer Science. Springer, 2017, pp. 307–326 (p. 710).

[726] Elke De Mulder, Thomas Eisenbarth, and Patrick Schaumont. "Identifying and Eliminating Side-Channel Leaks in Programmable Systems." *IEEE Design & Test*. 35.1 (2018), pp. 74–89 (p. 595).

[727] Markus Müller-Olm, Michael Petter, and Helmut Seidl. "Interprocedurally Analyzing Polynomial Identities." In *STACS*. Vol. 3884. Lecture Notes in Computer Science. Springer, 2006, pp. 50–67 (p. 457).

[728] Markus Müller-Olm and Oliver Rüthing. "On the Complexity of Constant Propagation." In *ESOP*. Vol. 2028. Lecture Notes in Computer Science. Springer, 2001, pp. 190–205 (p. 603, 617).

[729] Markus Müller-Olm and Helmut Seidl. "A Note on Karr's Algorithm." In *ICALP*. Vol. 3142. Lecture Notes in Computer Science. Springer, 2004, pp. 1016–1028 (p. 457).

[730] Markus Müller-Olm and Helmut Seidl. "Precise Interprocedural Analysis Through Linear Algebra." In *POPL*. ACM, 2004, pp. 330–341 (p. 457).

[731] Markus Müller-Olm and Helmut Seidl. "Analysis of Modular Arithmetic." In *ESOP*. Vol. 3444. Lecture Notes in Computer Science. Springer, 2005, pp. 46–60 (p. 367).

[732] Markus Müller-Olm and Helmut Seidl. "Analysis of Modular Arithmetic." *ACM Trans. Program. Lang. Syst.* 29.5 (2007), p. 29 (p. 397).

[733] Tukaram Muske, Rohith Talluri, and Alexander Serebrenik. "Reducing Static Analysis Alarms Based on Non-Impacting Control Dependencies." In *APLAS*. Vol. 11893. Lecture Notes in Computer Science. Springer, 2019, pp. 115–135 (p. 707).

[734] Kalyan Muthukumar, Francisco Bueno, Maria J. García de la Banda, and Manuel V. Hermenegildo. "Automatic Compile-Time Parallelization of Logic Programs for Restricted, Goal Level, Independent and Parallelism." *J. Log. Program.* 38.2 (1999), pp. 165–218 (p. 621).

[735] Kalyan Muthukumar and Manuel V. Hermenegildo. "Determination of Variable Dependence Information Through Abstract Interpretation." In *NACLP*. MIT Press, 1989, pp. 166–185 (p. 621, 623).

[736] Kalyan Muthukumar and Manuel V. Hermenegildo. "Compile-Time Derivation of Variable Dependency Using Abstract Interpretation." *J. Log. Program.* 13.2&3 (1992), pp. 315–347 (p. 621).

[737] Alan Mycroft. "The Theory and Practice of Transforming Call-by-Need Into Call-by-Value." In *Symposium on Programming*. Vol. 83. Lecture Notes in Computer Science. Springer, 1980, pp. 269–281 (p. 387).

[738] Alan Mycroft and Neil D. Jones. "A Relational Framework for Abstract Interpretation." In *Programs as Data Objects*. Vol. 217. Lecture Notes in Computer Science. Springer, 1985, pp. 156–171 (p. 131).

[739] Peter Naur. "Generation of Machine Code in ALGOL Compilers." *BIT*. 5.4 (1960), pp. 235–245 (p. 33).

[740] Peter Naur. "The Design of the GIER ALGOL Compiler." *BIT Numerical Mathematics*. 3 (June 1963), 124–140 and 145–166 (p. 18).

[741] Peter Naur. "Checking of Operand Types in ALGOL Compilers." *BIT Numerical Mathematics*. 5 (Sept. 1965), pp. 151–163 (p. 18, 250, 321, 344).

[742] Peter Naur. "Proofs of Algorithms by General Snapshots." *BIT*. 6 (1966), pp. 310–316 (p. 3, 94, 216, 281, 291).

[743] Peter Naur. "Proof Versus Formalization." *BIT Numerical Mathematics*. 34.1 (1994), pp. 148–164 (p. 480).

[744] Jorge A. Navas. "Crab." 2021. https://github.com/seahorn/crab (p. 716).

[745] Henrique Nazaré, Izabela Maffra, Willer Santos, Leonardo Barbosa E Oliveira, Laure Gonnord, and Fernando Magno Quintão Pereira. "Validation of Memory Accesses Through Symbolic Analyses." In *OOPSLA*. ACM, 2014, pp. 791–809 (p. 398).

[746] Nedialko S. Nedialkov, Vladik Kreinovich, and Scott A. Starks. "Interval Arithmetic, Affine Arithmetic, Taylor Series Methods: Why, What Next?." *Numerical Algorithms*. 37.1–4 (2004), pp. 325–336 (p. 387).

[747] Luca Negrini, Vincenzo Arceri, Pietro Ferrara, and Agostino Cortesi. "Twinning Automata and Regular Expressions for String Static Analysis." *CoRR*. abs/2006.02715 (2020) (p. 551).

[748] Greg Nelson and Derek C. Oppen. "Simplification by Cooperating Decision Procedures." *ACM Trans. Program. Lang. Syst.* 1.2 (1979), pp. 245–257 (p. 428, 432).

[749] John Von Neumann. "Zur Einführung der transfiniten Zahlen." *Acta Scientiarum Mathematicarum (University of Szeged)*. 1.4 (1923), pp. 199–208 (p. 11).

[750] Max S. New, William J. Bowman, and Amal Ahmed. "Fully Abstract Compilation via Universal Embedding." In *ICFP*. ACM, 2016, pp. 103–116 (p. 591, 619).

[751] Isaac Newton. "De Analysi per æquationes numero terminorum infinitas." 1669. http://www.newtonproject.ox.ac.uk/view/texts/diplomatic/NATP00204 (p. 171).

[752] Minh Ngo, David A. Naumann, and Tamara Rezk. "Typed–Based Relaxed Noninterference for Free." *CoRR*. abs/1905.00922 (2019) (p. 618).

[753] Ulf Nilsson. "Abstract Interpretation: A Kind of Magic." *Theor. Comput. Sci.* 142.1 (1995), pp. 125–139 (p. 684).

[754] Tobias Nipkow, Lawrence C. Paulson, and Makarius Wenzel. *Isabelle/HOL—a Proof Assistant for Higher-Order Logic*. University of Cambridge and Technische Universität München, Dec. 2016. https://isabelle.in.tum.de (p. 225).

[755] Emmy Noether. "Idealtheorie in Ringbereichen." *Mathematische Annalen*. 83 (1–2 1921), pp. 24–66 (p. 108).

[756] Emmy Noether. "Abstrakter Aufbau der Idealtheorie in algebraischen Zahl- Und Funktionenkörpern." *Mathematische Annalen*. 96 (1–2 1927), pp. 26–61 (p. 108).

[757] Michael Norrish. "C Formalised in HOL." PhD. University of Cambridge (UCAM–CL–TR-453), Dec. 1998 (p. 708).

[758] Michael J. O'Donnell. "A Critique of the Foundations of Hoare-Style Programming Logics." In *Logic of Programs*. Vol. 131. Lecture Notes in Computer Science. Springer, 1981, pp. 349–374 (p. 304).

[759] Rasha Omar, Mostafa M. Abbas, Ahmed El–Mahdy, and Erven Rohou. "Binary–Level Data Dependence Analysis of Hot Execution Regions Using Abstract Interpretation at Runtime." *PLoS ONE*. 15.4 (2020), 23p. (P. 388, 620).

[760] Derek C. Oppen. "A $2^{2^{2^{pn}}}$ Upper Bound on the Complexity of Presburger Arithmetic." *J. Comput. Syst. Sci.* 16.3 (1978), pp. 323–332 (p. 290).

[761] Peter Ørbæk. "Can You Trust Your Data?" ' In *TAPSOFT*. Vol. 915. Lecture Notes in Computer Science. Springer, 1995, pp. 575–589 (p. 620).

[762] Oystein Ore. "Galois Connexions." *Trans. Amer. Math. Soc.* 55.3 (May 1944), pp. 493–513 (p. 112, 119, 130).

[763] Oystein Ore. *Number Theory and Its History*. Dover Books in Mathematics. Dover, 1948 (p. 351, 355).

[764] Oystein Ore. "Review: Jacques Riguet, Relations binaires, fermetures, correspondances De Galois." *J. Symbolic Logic*. 16.1 (Mar. 1951), p. 61. https://projecteuclid.org:443/euclid.jsl/1183731055 (p. 768).

[765] Abdelraouf Ouadjaout and Antoine Miné. "A Library Modeling Language for the Static Analysis of C Programs." In *SAS*. Lecture Notes in Computer Science. Springer, 2020 (p. 709).

[766] Rachid Oucheikh, Ismail Berrada, and Outman El Hichami. "The 4-Octahedron Abstract Domain." In *NETYS*. Vol. 9944. Lecture Notes in Computer Science. Springer, 2016, pp. 311–317 (p. 492).

[767] Rachid Oucheikh, Ismail Berrada, and Outman El Hichami. "A Hypergraph Based Approach for the 4-Constraint Satisfaction Problem Tractability." In *ArXiv:1905.09083*. May 2019, p. 23 (p. 492).

[768] Mendes Oulamara and Arnaud J. Venet. "Abstract Interpretation with Higher-Dimensional Ellipsoids and Conic Extrapolation." In *CAV (1)*. Vol. 9206. Lecture Notes in Computer Science. Springer, 2015, pp. 415–430 (p. 492).

[769] Michael L. Overton. *Numerical Computing with IEEE Floating Point Arithmetic—Including One Theorem, One Rule of Thumb, and One Hundred and One Exercices*. SIAM, 2001 (p. 367).

[770] Scott Owens, John H. Reppy, and Aaron Turon. "Regular-Expression Derivatives Re-Examined." *J. Funct. Program.* 19.2 (2009), pp. 173–190 (p. 532).

[771] Sam Owre, John M. Rushby, and Natarajan Shankar. "PVS: A Prototype Verification System." In *CADE*. Vol. 607. Lecture Notes in Computer Science. Springer, 1992, pp. 748–752 (p. 289).

[772] Oded Padon, Jochen Hoenicke, Giuliano Losa, Andreas Podelski, Mooly Sagiv, and Sharon Shoham. "Reducing Liveness to Safety in First-Order Logic." *Proc. ACM Program. Lang.* 2.POPL (2018), 26:1–26:33 (p. 161).

[773] David A. Padua and Michael Wolfe. "Advanced Compiler Optimizations for Supercomputers." *Commun. ACM*. 29.12 (1986), pp. 1184–1201 (p. 621).

[774] Claude Pair. "Sur des algorithmes pour les problèmes de cheminement dans les graphes finis." In *Théorie des graphes. Journées internationales d'étude. Centre international de calcul, Rome*. Dunod, Gordon, and Breach, July 1966, pp. 271–300 (p. 462, 480).

[775] Claude Pair. "Mille et un algorithmes pour les problèmes de cheminement dans les graphes." *Revue française d'Informatique et de Recherche Opérationnelle (R.I.R.O.)*. B-3 (1970), pp. 125–143 (p. 462, 480).

[776] Catuscia Palamidessi. "Algebraic Properties of Idempotent Substitutions." In *ICALP*. Vol. 443. Lecture Notes in Computer Science. Springer, 1990, pp. 386–399 (p. 646).

[777] David Michael Ritchie Park. "Fixpoint Induction and Proofs of Program Properties." In *Machine Intelligence Volume 5*. Edited by Donald Mitchie and Bernard Meltzer. Edinburgh Univ. Press, 1969. Chap. 3, pp. 59–78. https://www.doc.ic.ac.uk/~shm/MI/mi.html (p. 167, 171, 278).

[778] David Michael Ritchie Park. "On the Semantics of Fair Parallelism." In *Abstract Software Specifications*. Vol. 86. Lecture Notes in Computer Science. Springer, 1979, pp. 504–526 (p. 12, 171, 278).

[779] Joseph C. H. Park, Kwang-Moo Choe, and C. H. Chang. "A New Analysis of LALR Formalisms." *ACM Trans. Program. Lang. Syst.* 7.1 (1985), pp. 159–175 (p. 61).

[780] Dito Pataraia. "A Constructive Proof of Tarski's Fixed-Point Theorem for Dcpo's." Presented at the 65th Peripatetic Seminar on Sheaves and Logic, in Aarhus, Denmark. Nov. 1997 (p. 167).

[781] Mike Paterson and Mark N. Wegman. "Linear Unification." *J. Comput. Syst. Sci.* 16.2 (1978), pp. 158–167 (p. 647).

[782] Lawrence C. Paulson. "Natural Deduction as Higher-Order Resolution." *J. Log. Program.* 3.3 (1986), pp. 237–258 (p. 647).

[783] Lawrence C. Paulson, Tobias Nipkow, and Makarius Wenzel. "From LCF to Isabelle/HOL." *Formal Asp. Comput.* 31.6 (2019), pp. 675–698 (p. 647).

[784] Zvonimir Pavlinovic. "Leveraging Program Analysis for Type Inference." PhD Thesis. New York University: CIMS, CS, Apr. 22, 2019 (p. 671).

[785] Étienne Payet and Fausto Spoto. "Checking Array Bounds by Abstract Interpretation and Symbolic Expressions." In *IJCAR*. Vol. 10900. Lecture Notes in Computer Science. Springer, 2018, pp. 706–722 (p. 493).

[786] Giuseppe Peano. "Intégration par séries des équations différentielles." *Mathematische Annalen.* XXXII (1888), pp. 450–456 (p. 438).

[787] Giuseppe Peano. *Arithmetices Principia, Nova Methodo Exposita.* Fratres Bocca, 1889. https://archive.org/details/arithmeticespri00peangoog/ (p. 9–11, 290).

[788] Giuseppe Peano. *Formulaire de mathématiques.* Bocca Frères, 1895. https://archive.org/details/formulairedemat03peangoog/Page/n9 (p. 9).

[789] Yan Pei, Swarnendu Biswas, Donald S. Fussell, and Keshav Pingali. "An Elementary Introduction to Kalman Filtering." *Commun. ACM.* 62.11 (2019), pp. 122–133 (p. 492).

[790] Charles Sanders Peirce. "On the Algebra of Logic: A Contribution to the Philosophy of Notation." *American Journal of Mathematics.* 7.2 (Jan. 1885), pp. 180–202 (p. 10).

[791] Marie Pelleau, Antoine Miné, Charlotte Truchet, and Frédéric Benhamou. "A Constraint Solver Based on Abstract Domains." In *VMCAI*. Vol. 7737. Lecture Notes in Computer Science. Springer, 2013, pp. 434–454 (p. 250, 493).

[792] Marie Pelleau, Charlotte Truchet, and Frédéric Benhamou. "The Octagon Abstract Domain for Continuous Constraints." *Constraints.* 19.3 (2014), pp. 309–337 (p. 493).

[793] Simon Perdrix. "Quantum Entanglement Analysis Based on Abstract Interpretation." In *SAS*. Vol. 5079. Lecture Notes in Computer Science. Springer, 2008, pp. 270–282 (p. 718).

[794] Mathias Péron and Nicolas Halbwachs. "An Abstract Domain Extending Difference-Bound Matrices with Disequality Constraints." In *VMCAI*. Vol. 4349. Lecture Notes in Computer Science. Springer, 2007, pp. 268–282 (p. 492).

[795] Günter Pickert. "Bemerkungen über Galois-Verbindungen." *Archiv Der Mathematik.* 3 (1952), pp. 285–289 (p. 117).

[796] Benjamin C. Pierce. *Types and Programming Languages.* MIT Press, 2002 (p. 657, 671).

[797] Benjamin C. Pierce. *Advanced Topics in Types and Programming Languages.* MIT Press, 2005 (p. 671).

[798] Andrew M. Pitts. "Nominal Techniques." *SIGLOG News.* 3.1 (2016), pp. 57–72 (p. 22, 78).

[799] Kim Plofker. *Mathematics in India.* Princeton University Press, 2007 (p. 17, 19).

[800] Gordon D. Plotkin. "A Note on Inductive Generalization." In *Machine Intelligence.* Edited by D. Meltzer B.; Michie. Vol. 5. Edinburgh University Press, 1970, pp. 153–163 (p. 623, 631, 632, 639, 645).

[801] Gordon D. Plotkin. "A Further Note on Inductive Generalization." In *Machine Intelligence.* Edited by D. Meltzer B.; Michie. Vol. 6. Edinburgh University Press, 1971, 101–124 (p. 623, 631, 632).

[802] Gordon D. Plotkin. "A Structural Approach to Operational Semantics." *J. Log. Algebr. Program.* 1972-01 (2004), pp. 17–139 (p. 524, 526).

[803] Amir Pnueli. "The Temporal Logic of Programs." In *FOCS*. IEEE Computer Society, 1977, pp. 46–57 (p. 527).

[804] Olivier Ponsini, Claude Michel, and Michel Rueher. "Verifying Floating-Point Programs with Constraint Programming and Abstract Interpretation Techniques." *Autom. Softw. Eng.* 23.2 (2016), pp. 191–217 (p. 192).

[805] Emil Post. "Recursively Enumerable Sets of Positive Integers and Their Decision Problems." *Bull. Amer. Math. Soc.* 50 (1944), pp. 284–316 (p. 100).

[806] François Pottier and Yann Régis-Gianas. "Towards Efficient, Typed LR Parsers." *Electr. Notes Theor. Comput. Sci.* 148.2 (2006), pp. 155–180 (p. 61).

[807] François Pottier and Yann Régis-Gianas. "Menhir Reference Manual." Institut National de Recherche en Informatique et en Automatique. Nov. 22, 2020. http://gallium.inria.fr/~fpottier/menhir/manual.pdf (p. 61).

[808] Vaughan R. Pratt. "Two Easy Theories Whose Combination Is Hard." MIT. Sept. 1977. http://boole.stanford.edu/pub/sefnp.pdf (p. 481).

[809] Mila Dalla Preda, Roberto Giacobazzi, and Niccolò Marastoni. "Formal framework for reasoning about the precision of dynamic analysis." In *SAS*. Lecture Notes in Computer Science. Springer, 2020 (p. 388).

[810] Reese T. Prosser. "Applications of Boolean Matrices to the Analysis of Flow Diagrams." In *1959 Proceedings of the Eastern Joint Computer Conference*. ACM, Dec. 1959, pp. 133–139 (p. 514).

[811] William Pugh. "The Omega Test: A Fast and Practical Integer Programming Algorithm for Dependence Analysis." In *SC*. ACM, 1991, pp. 4–13 (p. 290).

[812] Peng Qi and Christopher D. Manning. "*Arc-Swift*: A Novel Transition System for Dependency Parsing." In *ACL (2)*. Association for Computational Linguistics, 2017, 110–117 (p. 61).

[813] Jean-Pierre Queille and Joseph Sifakis. "Specification and Verification of Concurrent Systems in CESAR." In *Symposium on Programming*. Vol. 137. Lecture Notes in Computer Science. Springer, 1982, pp. 337–351 (p. 527).

[814] Francesco Ranzato. "Abstracting Nash Equilibria of Supermodular Games." *Formal Methods in System Design*. 53.2 (2018), pp. 259–285 (p. 718).

[815] Francesco Ranzato. "Decidability and Synthesis of Abstract Inductive Invariants." In *CONCUR 2020*. Vol. 171. LIPIcs. 2020. Chap. 48 (p. 277).

[816] Francesco Ranzato, Olivia Rossi-Doria, and Francesco Tapparo. "A Forward-Backward Abstraction Refinement Algorithm." In *VMCAI*. Vol. 4905. Lecture Notes in Computer Science. Springer, 2008, pp. 248–262 (p. 550).

[817] Francesco Ranzato and Francesco Tapparo. "Generalized Strong Preservation by Abstract Interpretation." *J. Log. Comput.* 17.1 (2007), pp. 157–197 (p. 550).

[818] Francesco Ranzato and Francesco Tapparo. "Generalizing the Paige-Tarjan Algorithm by Abstract Interpretation." *Inf. Comput.* 206.5 (2008), pp. 620–651 (p. 550).

[819] Francesco Ranzato and Francesco Tapparo. "An Efficient Simulation Algorithm Based on Abstract Interpretation." *Inf. Comput.* 208.1 (2010), pp. 1–22 (p. 550).

[820] Joseph Raphson. *Analysis Æquationum Universalis, seu ad Æquationes Algebraicas resolvendas Methodus generalis, & expedita, ex nova infinitarum serierum methodo, deducta ac demonstrata. Editio secunda cui accessit Appendix de Infinito Infinitarum Serierum progressu ad Equationum Algebraicarum Radices eliciendas. Cui etiam Annexum est; De Spatio reali, seu Ente Infinito Conamen Mathematico-Metaphysicum.* 2nd ed. London: A. & I. Churchill et al., 1690, 1697, 1702 (p. 171).

[821] John Regehr. *Specification of Test Bench*. TOYOTA InfoTechnology Center, U.S.A., 2015. https://github.com/regehr/itc-benchmarks/blob/master/docs/spec.pdf (p. 715).

[822] Yann Régis-Gianas and François Pottier. "A Hoare Logic for Call-by-Value Functional Programs." In *MPC*. Vol. 5133. Lecture Notes in Computer Science. Springer, 2008, pp. 305–335 (p. 295).

[823] Steven P. Reiss. "Continuous Flow Analysis to Detect Security Problems." *CoRR*. abs/1909.13683 (2019) (p. 619).

[824] Thomas W. Reps and Gogul Balakrishnan. "Improved Memory-Access Analysis for x86 Executables." In *CC*. Vol. 4959. Lecture Notes in Computer Science. Springer, 2008, pp. 16–35 (p. 588).

[825] Bernhard Reus. *Limits of Computation—From a Programming Perspective*. Undergraduate Topics in Computer Science. Springer, 2016 (p. 97).

[826] Peter Z. Revesz and Robert J. Woodward. "Estimating the Maximum Rise in Temperature According to Climate Models Using Abstract Interpretation." *Acta Univ. Sapientiæ, Informatica*. 11.1 (2019), pp. 5–23 (p. 718).

[827] John C. Reynolds. "Transformational Systems and the Algebraic Structure of Atomic Formulas." In *Machine Intelligence*. Edited by D. Meltzer B.; Michie. Vol. 5. Edinburgh University Press, 1970, pp. 135–151 (p. 623, 631, 632, 639, 645).

[828] John C. Reynolds. "Types, Abstraction and Parametric Polymorphism." In R.E.A. Mason, ed. *Information Processing 83*. Elsevier, 1983, pp. 513–523 (p. 131).

[829] John C. Reynolds. "Definitional Interpreters Revisited." *Higher-Order and Symbolic Computation*. 11.4 (1998), pp. 355–361 (p. 526).

[830] Henry Gordon Rice. "Classes of Recursively Enumerable Sets and Their Decision Problems." *Trans. Amer. Math. Soc.* 74.1 (1953), pp. 358–366 (p. 100).

[831] Jacques Riguet. "Relations binaires, fermetures, correspondances de Galois." *Bulletin de la S.M.F., Tome 76*. (1948). Review in [764], pp. 114–155. http://www.numdam.org/article/BSMF_1948__76__114_0.pdf (p. 119, 144).

[832] Martin Rinard and Austin Gadient. "Dataflow Analysis with Prophecy and History Variables." *CoRR*. abs/2007.12015 (2020) (p. 684).

[833] Xavier Rival. "Certification of Compiled Assembly Code by Invariant Translation." *STTT*. 6.1 (2004), pp. 15–37 (p. 708).

[834] Xavier Rival. "Abstract Dependences for Alarm Diagnosis." In *APLAS*. Vol. 3780. Lecture Notes in Computer Science. Springer, 2005, pp. 347–363 (p. 621).

[835] Xavier Rival. "Understanding the Origin of Alarms in Astrée." In *SAS*. Vol. 3672. Lecture Notes in Computer Science. Springer, 2005, pp. 303–319 (p. 618).

[836] Xavier Rival and Laurent Mauborgne. "The Trace Partitioning Abstract Domain." *ACM Trans. Program. Lang. Syst.* 29.5 (2007), p. 26 (p. 551).

[837] Xavier Rival and Kwangkeun Yi. *Introduction to Static Analysis*. MIT Press, 2020 (p. 5, 717).

[838] John Alan Robinson. "A Machine-Oriented Logic Based on the Resolution Principle." *J. ACM*. 12.1 (1965), pp. 23–41 (p. 623, 639).

[839] John Alan Robinson. *Logic: Form and Function—the Mechanization of Deductive Reasoning*. Artificial Intelligence. Elsevier North-Holland, 1979 (p. 623, 631, 639).

[840] Enric Rodríguez-Carbonell and Deepak Kapur. "Automatic Generation of Polynomial Invariants of Bounded Degree Using Abstract Interpretation." *Sci. Comput. Program.* 64.1 (2007), pp. 54–75 (p. 457).

[841] Hartley Rogers. *Theory of Recursive Functions and Effective Computability*. MIT Press, 1987 (p. 97).

[842] Patrick Maxim Rondon, Ming Kawaguchi, and Ranjit Jhala. "Liquid Types." In *PLDI*. ACM, 2008, pp. 159–169 (p. 32, 671).

[843] Pierre Roux, Romain Jobredeaux, Pierre-Loïc Garoche, and Éric Feron. "A Generic Ellipsoid Abstract Domain for Linear Time Invariant Systems." In *HSCC*. ACM, 2012, pp. 105–114 (p. 492).

[844] Bernard Roy. "Transitivité et connexité." *C. R. Acad. Sci. Paris*. 249 (1959), pp. 216–218 (p. 459, 475, 479).

[845] Bernard Roy. *Cheminement et connexité dans les graphes, application aux problèmes d'ordonnancement.* 2nd ed. Metra, Jan. 1965, p. 137 (p. 475).

[846] Cindy Rubio-González, Cuong Nguyen, Benjamin Mehne, Koushik Sen, James Demmel, William Kahan, Costin Iancu, Wim Lavrijsen, David H. Bailey, and David Hough. "Floating-Point Precision Tuning Using Blame Analysis." In *ICSE.* ACM, 2016, pp. 1074–1085 (p. 387).

[847] Alejandro Russo, John Hughes, David A. Naumann, and Andrei Sabelfeld. "Closing Internal Timing Channels by Transformation." In *ASIAN.* Vol. 4435. Lecture Notes in Computer Science. Springer, 2006, pp. 120–135 (p. 595).

[848] Barbara G. Ryder and Marvin C. Paull. "Elimination Algorithms for Data Flow Analysis." *ACM Comput. Surv.* 18.3 (1986), pp. 277–316 (p. 506, 514).

[849] Andrei Sabelfeld and Andrew C. Myers. "Language-Based Information-Flow Security." *IEEE Journal on Selected Areas in Communications.* 21.1 (2003), pp. 5–19 (p. 595).

[850] Jacques Sakarovitch. *Elements of Automata Theory.* Cambridge University Press, 2009 (p. 167, 528, 548).

[851] Christopher Salls, Aravind Machiry, Adam Doupe, Yan Shoshitaishvili, Christopher Kruegel, and Giovanni Vigna. "Exploring Abstraction Functions in Fuzzing." In *Proceedings of the IEEE Conference on Communications and Network Security (CNS).* Avignon, France, June 2020 (p. 5, 718).

[852] Sriram Sankaranarayanan. "Reachability Analysis Using Message Passing Over Tree Decompositions." In *CAV (1).* Vol. 12224. Lecture Notes in Computer Science. Springer, 2020, pp. 604–628 (p. 433).

[853] Sriram Sankaranarayanan, Franjo Ivančić, and Aarti Gupta. "Program Analysis Using Symbolic Ranges." In *SAS.* Vol. 4634. Lecture Notes in Computer Science. Springer, 2007, pp. 366–383 (p. 398).

[854] Sriram Sankaranarayanan and Mohamed Amin Ben Sassi. "Template Polyhedra with a Twist." In *SAS.* Vol. 10422. Lecture Notes in Computer Science. Springer, 2017, pp. 321–341 (p. 492).

[855] Sriram Sankaranarayanan, Henny B. Sipma, and Zohar Manna. "Constraint-Based Linear-Relations Analysis." In *SAS.* Vol. 3148. Lecture Notes in Computer Science. Springer, 2004, pp. 53–68 (p. 492).

[856] Sriram Sankaranarayanan, Henny B. Sipma, and Zohar Manna. "Non-Linear Loop Invariant Generation Using Gröbner Bases." In *POPL.* ACM, 2004, pp. 318–329 (p. 457).

[857] David A. Schmidt. "Data Flow Analysis Is Model Checking of Abstract Interpretations." In *POPL.* ACM, 1998, pp. 38–48 (p. 497, 501, 512, 622).

[858] David A. Schmidt. "Abstract Interpretation From a Topological Perspective." In *SAS.* Vol. 5673. Lecture Notes in Computer Science. Springer, 2009, pp. 293–308 (p. 161).

[859] David A. Schmidt. "Inverse-Limit and Topological Aspects of Abstract Interpretation." *Theor. Comput. Sci.* 430 (2012), pp. 23–42 (p. 161).

[860] Gunther Schmidt and Thomas Ströhlein. *Relations and Graphs.* Discrete Mathematics for Computer Scientists. Springer, 2012 (p. 459, 460).

[861] Jürgen Schmidt. "Beiträge zur Filtertheorie. II." *Mathematische Nachrichten.* 10.3–4 (1953), pp. 197–232 (p. 112).

[862] Fred B. Schneider. "Enforceable Security Policies." *ACM Trans. Inf. Syst. Secur.* 3.1 (2000), pp. 30–50 (p. 528, 535, 550).

[863] Fred B. Schneider. "History and Context for Defining Liveness: Winner 2018 Edsger W. Dijkstra Prize." *SIGACT News.* 49.4 (2018), pp. 60–63 (p. 161).

[864] Bernhard Scholz and Johann Blieberger. "A New Elimination-Based Data Flow Analysis Framework Using Annotated Decomposition Trees." In *CC*. Vol. 4420. Lecture Notes in Computer Science. Springer, 2007, pp. 202–217 (p. 514).

[865] Alexander Schrijver. *Combinatorial Optimization, Polyhedra and Efficiency*. Vol. 24. Algorithms and Combinatorics. Springer, 2003 (p. 480).

[866] Ernst Schröder. *Vorlesungen Über die Algebra der Logik (Exakte Logik)*. Vol. 1. B.G. Teubner, 1890 (p. 11).

[867] Viktor Schuppan and Armin Biere. "Liveness Checking as Safety Checking for Infinite State Spaces." *Electr. Notes Theor. Comput. Sci.* 149.1 (2006), pp. 79–96 (p. 161).

[868] Marcel Paul Schützenberger. "On Context-Free Languages and Push-Down Automata." *Information and Control*. 6.3 (1963), pp. 246–264 (p. 178, 184).

[869] Dana S. Scott. "Outline of a Mathematical Theory of Computation." In *Proceedings of the Fourth Annual Princeton Conference on Information Sciences and Systems*. Princeton University Press, Mar. 1970, pp. 169–176 (p. 278).

[870] Dana S. Scott. "The Lattice of Flow Diagrams." In E. Engeler, ed. *Symposium on Semantics of Algorithmic Languages*. Vol. 188. Lecture Notes in Mathematics. Springer, 1971, pp. 311–366 (p. 172).

[871] Dana S. Scott. "Mathematical Concepts in Programming Language Semantics." In *AFIPS Spring Joint Computing Conference*. Vol. 40. AFIPS Conference Proceedings. AFIPS, 1972, pp. 225–234 (p. 192).

[872] Dana S. Scott and Christopher Strachey. *Towards a Mathematical Semantics for Computer Languages*. Technical Report PRG-6. Oxford University Computer Laboratory, Aug. 1971, p. 49. https://www.cs.ox.ac.uk/files/3228/PRG06.pdf (p. 20, 135, 143, 192, 526).

[873] Abraham Seidenberg. "A New Decision Method for Elementary Algebra." *Ann. Math., Ser. 2*. 60 (1954), pp. 356–374 (p. 402).

[874] Helmut Seidl, Thomas Martin Gawlitza, and Martin D. Schwarz. "Parametric Strategy Iteration." In *SCSS*. Vol. 30. EPiC Series in Computing. EasyChair, 2014, pp. 62–76 (p. 412).

[875] Helmut Seidl and Ralf Vogler. "Three Improvements to the Top-Down Solver." In *PPDP*. ACM, 2018, 21:1–21:14 (p. 412).

[876] Koushik Sen and Grigore Rosu. "Generating Optimal Monitors for Extended Regular Expressions." *Electr. Notes Theor. Comput. Sci.* 89.2 (2003), pp. 226–245 (p. 527).

[877] Ilya Sergey, Jan Midtgaard, and Dave Clarke. "Calculating Graph Algorithms for Dominance and Shortest Path." In *MPC*. Vol. 7342. Lecture Notes in Computer Science. Springer, 2012, pp. 132–156 (p. 480).

[878] Ilya Sergey, Vaivaswatha Nagaraj, Jacob Johannsen, Amrit Kumar, Anton Trunov, and Ken Chan Guan Hao. "Safer Smart Contract Programming with Scilla." *Proc. ACM Program. Lang.* 3.OOPSLA (2019), 185:1–185:30 (p. 32).

[879] Maria I. Sessa. "Approximate Reasoning by Similarity-Based SLD Resolution." *Theor. Comput. Sci.* 275.1–2 (2002), pp. 389–426 (p. 647).

[880] Glenn Shafer. *A Mathematical Theory of Evidence*. Princeton University Press, 1976 (p. 95).

[881] Marc Shapiro and Susan Horwitz. "Fast and Accurate Flow-Insensitive Points-to Analysis." In *POPL*. ACM Press, 1997, pp. 1–14 (p. 587).

[882] Marc Shapiro and Susan Horwitz. "The Effects of the Precision of Pointer Analysis." In *SAS*. Vol. 1302. Lecture Notes in Computer Science. Springer, 1997, pp. 16–34 (p. 587).

[883] Micha Sharir. "Structural Analysis: A New Approach to Flow Analysis in Optimizing Compilers." *Comput. Lang.* 5.3 (1980), pp. 141–153 (p. 514).

[884] Tushar Sharma and Thomas W. Reps. "Sound Bit-Precise Numerical Domains." In *VMCAI*. Vol. 10145. Lecture Notes in Computer Science. Springer, 2017, pp. 500–520 (p. 397).

[885] Tushar Sharma and Thomas W. Reps. "A new abstraction framework for affine transformers." *Formal Methods Syst. Des.* 54.1 (2019), pp. 110–143 (p. 457).

[886] Dennis E. Shasha and Marc Snir. "Efficient and Correct Execution of Parallel Programs That Share Memory." *ACM Trans. Program. Lang. Syst.* 10.2 (1988), pp. 282–312 (p. 190).

[887] Zahava Shmuely. "The Structure of Galois Connections." *Pacific Journal of Mathematics.* 54.2 (1974), pp. 209–225 (p. 126).

[888] Victor Shoup. *A Computational Introduction to Number Theory and Algebra.* 2nd ed. Cambridge University Press, 2008. https://www.shoup.net/ntb/ntb-v2.pdf (p. 351).

[889] Paulo F. Silva and José Nuno Oliveira. "'Galculator': Functional Prototype of a Galois-Connection Based Proof Assistant." In *PPDP.* ACM, 2008, pp. 44–55 (p. 112).

[890] George S. Simmons. *Basic Topology.* Springer, 1963 (p. 147).

[891] Axel Simon and Andy King. "Taming the Wrapping of Integer Arithmetic." In *SAS.* Vol. 4634. Lecture Notes in Computer Science. Springer, 2007, pp. 121–136 (p. 367, 397).

[892] Axel Simon and Andy King. "The Two Variable Per Inequality Abstract Domain." *Higher-Order and Symbolic Computation.* 23.1 (2010), pp. 87–143 (p. 492).

[893] Gagandeep Singh, Timon Gehr, Markus Püschel, and Martin T. Vechev. "An Abstract Domain for Certifying Neural Networks." *Proc. ACM Program. Lang.* 3.POPL (2019), 41:1–41:30 (p. 5, 718).

[894] Gagandeep Singh, Timon Gehr, Markus Püschel, and Martin T. Vechev. "Boosting Robustness Certification of Neural Networks." In *ICLR (Poster).* OpenReview.net, 2019 (p. 718).

[895] Gagandeep Singh, Markus Püschel, and Martin T. Vechev. "Making Numerical Program Analysis Fast." In *PLDI.* ACM, 2015, pp. 303–313 (p. 492, 493).

[896] Gagandeep Singh, Markus Püschel, and Martin T. Vechev. "Fast Polyhedra Abstract Domain." In *POPL.* ACM, 2017, pp. 46–59 (p. 407, 493).

[897] Gagandeep Singh, Markus Püschel, and Martin T. Vechev. "A Practical Construction for Decomposing Numerical Abstract Domains." *Proc. ACM Program. Lang.* 2.POPL (2018), 55:1–55:28 (p. 250, 407, 493, 716).

[898] Moritz Sinn, Florian Zuleger, and Helmut Veith. "Complexity and Resource Bound Analysis of Imperative Programs Using Difference Constraints." *J. Autom. Reasoning.* 59.1 (2017), pp. 3–45 (p. 493).

[899] Michel Sintzoff. "Calculating Properties of Programs by Valuations on Specific Models." In *Proceedings of ACM Conference on Proving Assertions About Programs.* ACM, 1972, pp. 203–207 (p. 18, 250, 316, 321, 344).

[900] Seppo Sippu and Eljas Soisalon-Soininen. *Parsing Theory—Volume I: Languages and Parsing.* Vol. 15. EATCS Monographs on Theoretical Computer Science. Springer, 1988. https://doi.org/10.1007/978-3-642-61345-6 (p. 61).

[901] Seppo Sippu and Eljas Soisalon-Soininen. *Parsing Theory—Volume II: LR(k) and LL(k) Parsing.* Vol. 20. EATCS Monographs on Theoretical Computer Science. Springer, 1990 (p. 61).

[902] Michael Sipser. *Introduction to the Theory of Computation.* PWS, 1997 (p. 97).

[903] Raymond Smullyan. *A Beginner's Guide to Mathematical Logic.* Dover Books on Mathematics, 2014 (p. 9).

[904] Jean Souyris and David Delmas. "Experimental Assessment of Astrée on Safety–Critical Avionics Software." In *SAFECOMP.* Vol. 4680. Lecture Notes in Computer Science. Springer, 2007, pp. 479–490 (p. 714).

[905] *SPARTA.* Facebook Incubator, 2020. https://github.com/facebookincubator/SPARTA (p. 716).

[906] Fausto Spoto and Thomas P. Jensen. "Class Analyses as Abstract Interpretations of Trace Semantics." *ACM Trans. Program. Lang. Syst.* 25.5 (2003), pp. 578–630 (p. 192).

[907] Meera Sridhar and Kevin W. Hamlen. "Model Checking In–Lined Reference Monitors." In *VMCAI.* Vol. 5944. Lecture Notes in Computer Science. Springer, 2010, pp. 312–327 (p. 528).

[908] Bjarne Steensgaard. "Points–to Analysis in Almost Linear Time." In *POPL.* ACM Press, 1996, pp. 32–41 (p. 559, 587, 623).

[909] Bernhard Steffen. "Data Flow Analysis as Model Checking." In *TACS.* Vol. 526. Lecture Notes in Computer Science. Springer, 1991, pp. 346–365 (p. 512).

[910] Bernhard Steffen. "Generating Data Flow Analysis Algorithms From Modal Specifications." *Sci. Comput. Program.* 21.2 (1993), pp. 115–139 (p. 512).

[911] Dominic Steinhöfel and Nathan Wasser. "A New Invariant Rule for the Analysis of Loops with Non–Standard Control Flows." In *IFM.* Vol. 10510. Lecture Notes in Computer Science. Springer, 2017, pp. 279–294 (p. 301).

[912] Leon Sterling and Ehud Shapiro. *The Art of Prolog—Advanced Programming Techniques, 2nd Ed.* MIT Press, 1994 (p. 623, 639).

[913] Jorge Stolfi and Luiz Henrique De Figueiredo. "An Introduction to Affine Arithmetic." *Tend. Mat. Apl. Comput., SBMAC.* 4.3 (2003), pp. 297–312 (p. 387).

[914] Joseph E. Stoy. *Denotational Semantics: The Scott–Strachey Approach to Programming Language Theory.* Computer Science Series. MIT Press, 1981 (p. 526).

[915] Christopher Strachey and Christopher P. Wadsworth. "Continuations: A Mathematical Semantics for Handling Full Jumps." *Higher–Order and Symbolic Computation.* 13.1/2 (2000), pp. 135–152 (p. 68).

[916] Martin Sulzmann and Kenny Zhuo Ming Lu. "Solving of Regular Equations Revisited." In *ICTAC.* Vol. 11884. Lecture Notes in Computer Science. Springer, 2019, pp. 392–409 (p. 531).

[917] Patrick Suppes. *Axiomatic Set Theory.* Dover, 1952 (p. 9, 11).

[918] Árpád Száz. "A Particular Galois Connection Between Relations and Set Functions." *Acta Univ. Sapientiæ, Mathematica.* 6.1 (2014), pp. 73–91 (p. 144).

[919] S. Tucker Taft. "AdaCore ParaScope: Advanced Static Analysis for ParaSail, a Parallel Specification and Implementation Language." *IBM Programming Languages Day, T. J. Watson Research Center.* (Nov. 2015). https://researcher.watson.ibm.com/researcher/files/us-pietroferrara/taft.pdf (p. 344).

[920] S. Tucker Taft. "ParaSail: A Pointer–Free Pervasively–Parallel Language for Irregular Computations." *Programming Journal.* 3.3 (2019), p. 7 (p. 344).

[921] Pierre Talbot, David Cachera, Éric Monfroy, and Charlotte Truchet. "Combining Constraint Languages via Abstract Interpretation." In *ICTAI, 31st International Conference on Tools with Artificial Intelligence.* IEEE, 2019 (p. 250).

[922] Pierre Talbot, David Cachera, Éric Monfroy, and Charlotte Truchet. "Octogones entiers pour le problème RCPSP." In *Actes JFPC 2019.* IMT Mines Albi, June 2019, pp. 33–42 (p. 493).

[923] Robert Endre Tarjan. "Depth–First Search and Linear Graph Algorithms." *SIAM J. Comput.* 1.2 (1972), pp. 146–160 (p. 259).

[924] Alfred Tarski. "A Decision Method for Elementary Algebra and Geometry." R-109, RAND Corporation, Santa Monica, CA. Aug. 1948. https://www.rand.org/content/dam/rand/pubs/reports/2008/R109.pdf (p. 402).

[925] Alfred Tarski. "A Lattice Theoretical Fixpoint Theorem and Its Applications." *Pacific J. of Math.* 5 (1955), pp. 285–310 (p. 165, 172).

[926] Alfred Tarski. *Introduction to Logic and to the Methodology of Deductive Sciences.* 4th ed. Oxford University Press, Mar. 1994 (p. 9, 10, 12).

[927] Alfred Tarski. "A Decision Method for Elementary Algebra and Geometry." In *Quantifier Elimination and Cylindrical Algebraic Decomposition.* Texts and Monographs in Symbolic Computation. Springer, 1998, pp. 24–84 (p. 290).

[928] Robert D. Tennent. "The Denotational Semantics of Programming Languages." *Commun. ACM.* 19.8 (1976), pp. 437–453 (p. 276).

[929] Robert D. Tennent. *Semantics of Programming Languages.* Prentice Hall, 1981 (p. 526).

[930] Mg Thushara and Kanagasabapathi Somasundaram. "Forward and Backward Static Analysis for Critical Accuracy in Floating-Point Programs." *Computer Science.* 21.2 (2020) (p. 701).

[931] Alan Turing. "On Computable Numbers, with an Application to the Entscheidungsproblem." *Proceedings of the London Mathematical Society.* 2nd ser. 42 (1937), pp. 230–265 (p. 98).

[932] Alan Turing. "On Computable Numbers, with an Application to the Entscheidungsproblem. A Correction." *Proceedings of the London Mathematical Society.* 2nd ser. 43 (1938), pp. 544–546 (p. 98).

[933] Alan Turing. "Checking a Large Routine." In *Report of a Conference on High Speed Automatic Calculating Machines, University of Cambridge Mathematical Laboratory, Cambridge, England.* 1949, pp. 67–69. http://www.turingarchive.org/browse.php/b/8 (p. 3, 94, 216, 281, 291).

[934] Alan Turing. "Proposal for Development in the Mathematics Division of an Automatic Computing Machine (ACE), (1945)." In *Collected Works of A.M. Turing, Mechanical Intelligence.* Edited by D.C. Ince. North-Holland, 1992, pp. 20–86. http://www.turingarchive.orge/browse.php/C/32 (p. 634).

[935] Stanislav Tzolovski. "Data Dependence as Abstract Interpretations." In *SAS.* Vol. 1302. Lecture Notes in Computer Science. Springer, 1997, p. 366 (p. 621).

[936] Stanislav Tzolovski. "Raffinement d'analyses par interprétation abstraite." Thèse de doctorat. Palaiseau, France: École polytechnique, June 15, 2002 (p. 621).

[937] Ramakrishna Upadrasta and Albert Cohen. "Sub-Polyhedral Scheduling Using (Unit-)Two-Variable-Per-Inequality Polyhedra." In *POPL.* ACM, 2013, pp. 483–496 (p. 493).

[938] Caterina Urban. "Static Analysis by Abstract Interpretation of Functional Temporal Properties of Programs. (Analyse statique par interprétation abstraite de propriétés temporelles fonctionnelles des programmes)." PhD thesis. École normale supérieure, Paris, July 9, 2015 (p. 694).

[939] Caterina Urban. "Static Analysis of Data Science Software." In *SAS.* Vol. 11822. Lecture Notes in Computer Science. Springer, 2019, pp. 17–23 (p. 718).

[940] Caterina Urban. "What Programs Want: Automatic Inference of Input Data Specifications." *CoRR.* abs/2007.10688 (2020) (p. 693).

[941] Caterina Urban, Maria Christakis, Valentin Wüstholz, and Fuyuan Zhang. "Perfectly Parallel Fairness Certification of Neural Networks." *Proc. ACM Program. Lang.* 4.OOPSLA (Nov. 2020), 185:1–185:30 (p. 5).

[942] Caterina Urban, Arie Gurfinkel, and Temesghen Kahsai. "Synthesizing Ranking Functions From Bits and Pieces." In *TACAS*. Vol. 9636. Lecture Notes in Computer Science. Springer, 2016, pp. 54–70 (p. 693).

[943] Caterina Urban and Antoine Miné. "An Abstract Domain to Infer Ordinal-Valued Ranking Functions." In *ESOP*. Vol. 8410. Lecture Notes in Computer Science. Springer, 2014, pp. 412–431 (p. 693).

[944] Caterina Urban and Antoine Miné. "Proving Guarantee and Recurrence Temporal Properties by Abstract Interpretation." In *VMCAI*. Vol. 8931. Lecture Notes in Computer Science. Springer, 2015, pp. 190–208 (p. 693).

[945] Caterina Urban and Antoine Miné. "Inference of Ranking Functions for Proving Temporal Properties by Abstract Interpretation." *Computer Languages, Systems & Structures*. 47 (2017), pp. 77–103 (p. 694).

[946] Caterina Urban and Antoine Miné. "A Review of Formal Methods applied to Machine Learning." *CoRR*. abs/2104.02466 (2021) (p. 718).

[947] Caterina Urban and Peter Müller. "An Abstract Interpretation Framework for Input Data Usage." In *ESOP*. Vol. 10801. Lecture Notes in Computer Science. Springer, 2018, pp. 683–710 (p. 514, 593, 620, 621).

[948] Caterina Urban, Samuel Ueltschi, and Peter Müller. "Abstract Interpretation of CTL Properties." In *SAS*. Vol. 11002. Lecture Notes in Computer Science. Springer, 2018, pp. 402–422 (p. 693, 694).

[949] Francisco J. Valverde-Albacete, Carmen Peláez-Moreno, and Cristina Del Campo. "Activating Generalized Fuzzy Implications From Galois Connections." In *A Passion for Fuzzy Sets*. Vol. 322. Studies in Fuzziness and Soft Computing. Springer, 2015, pp. 201–212 (p. 131).

[950] Joris Van Der Hoeven. "Ball Arithmetic." In A. Beckmann, Ch. Gaßner, and B. Löwe, eds. *International Workshop on Logical Approaches to Barriers in Computing and Complexity*. Preprint-Reihe Mathematik 6. Ernst-Moritz-Arndt-Universität Greifswald, 2010, pp. 179–208 (p. 387).

[951] Joris Van Der Hoeven and Grégoire Lecerf. "Evaluating Straight-Line Programs Over Balls." In *ARITH*. IEEE Computer Society, 2016, pp. 142–149 (p. 387).

[952] Franck Védrine, Maxime Jacquemin, Nikolai Kosmatov, and Julien Signoles. "Runtime Abstract Interpretation for Numerical Accuracy and Robustness." In *VMCAI*. Lecture Notes in Computer Science. Springer, 2021 (p. 387).

[953] Arnaud J. Venet. "Automatic Analysis of Pointer Aliasing for Untyped Programs." *Sci. Comput. Program.* 35.2 (1999), pp. 223–248 (p. 204).

[954] Arnaud J. Venet. "The Gauge Domain: Scalable Analysis of Linear Inequality Invariants." In *CAV*. Vol. 7358. Lecture Notes in Computer Science. Springer, 2012, pp. 139–154 (p. 492).

[955] Arnaud J. Venet and Guillaume P. Brat. "Precise and Efficient Static Array Bound Checking for Large Embedded C Programs." In *PLDI*. ACM, 2004, pp. 231–242 (p. 714).

[956] Ivan Matveevich Vinogradov. *Elements of Number Theory*. Dover, 2016 (p. 351).

[957] Harishankar Vishwanathan, Matan Shachnai, Srinivas Narayana, and Santosh Nagarakatte. "Semantics, Verification, and Efficient Implementations for Tristate Numbers." *CoRR*. abs/2105.05398 (2021) (p. 387).

[958] Dennis M. Volpano. "Safety Versus Secrecy." In *SAS*. Vol. 1694. Lecture Notes in Computer Science. Springer, 1999, pp. 303–311 (p. 593).

[959] Dennis M. Volpano, Cynthia E. Irvine, and Geoffrey Smith. "A Sound Type System for Secure Flow Analysis." *Journal of Computer Security*. 4.2/3 (1996), pp. 167–188 (p. 591, 619).

[960] Hagen Völzer, Daniele Varacca, and Ekkart Kindler. "Defining Fairness." In *CONCUR*. Vol. 3653. Lecture Notes in Computer Science. Springer, 2005, pp. 458–472 (p. 161).

[961] Jérôme Vouillon and Pierre Jouvelot. "Type and Effect Systems via Abstract Interpretation." 1995. http : / / www . cri . ensmp . fr / classement / doc / A - 273 . pdf (p. 671).

[962] Xinyu Wang, Isil Dillig, and Rishabh Singh. "Program Synthesis Using Abstraction Refinement." *Proc. ACM Program. Lang.* 2.POPL (2018), 63:1–63:30 (p. 718).

[963] Zi Wang, Aws Albarghouthi, and Somesh Jha. "Abstract Universal Approximation for Neural Networks." *CoRR*. abs/2007.06093 (July 14, 2020) (p. 398).

[964] Morgan Ward. "The Closure Operators of a Lattice." *Annals of Mathematics*. 43.2 (Apr. 1942), pp. 191–196 (p. 121, 130).

[965] Stephen Warshall. "A Theorem on Boolean Matrices." *J. ACM*. 9.1 (1962), pp. 11–12 (p. 459, 475, 479).

[966] Heinrich Weber. "Die allgemeinen Grundlagen der Galois'schen Gleichungstheorie." *Mathematische Annalen*. XLIII (1893), pp. 521–549 (p. 437).

[967] Ben Wegbreit. "Property Extraction in Well-Founded Property Sets." *IEEE Trans. Software Eng.* 1.3 (1975), pp. 270–285 (p. 109, 250, 321, 344, 412).

[968] Mark N. Wegman. "Summarizing Graphs by Regular Expressions." In *POPL*. ACM Press, 1983, pp. 203–216 (p. 551).

[969] Mark N. Wegman and F. Kenneth Zadeck. "Constant Propagation with Conditional Branches." *ACM Trans. Program. Lang. Syst.* 13.2 (1991), pp. 181–210 (p. 397, 513, 603, 617).

[970] Peter Wegner. "Operational Semantics of Programming Languages." *ACM SIGPLAN Not.* 7.1 (Jan. 1972), pp. 128–141 (p. 79, 519, 526).

[971] Peter Wegner. "The Vienna Definition Language." *ACM Comput. Surv.* 4.1 (1972), pp. 5–63 (p. 526).

[972] Guannan Wei, Yuxuan Chen, and Tiark Rompf. "Staged Abstract Interpreters: Fast and Modular Whole-Program Analysis via Meta-Programming." *Proc. ACM Program. Lang.* 3.OOPSLA (2019), 126:1–126:32 (p. 250).

[973] Jiawang Wei. "Parallel Asynchronous Iterations of Least Fixed Points." *Parallel Computing*. 19.8 (1993), pp. 887–895 (p. 257).

[974] Shiyi Wei, Piotr Mardziel, Andrew Ruef, Jeffrey S. Foster, and Michael Hicks. "Evaluating Design Tradeoffs in Numeric Static Analysis for Java." In *ESOP*. Vol. 10801. Lecture Notes in Computer Science. Springer, 2018, pp. 653–682 (p. 723).

[975] William E. Weihl. "Interprocedural Data Flow Analysis in the Presence of Pointers, Procedure Variables and Label Variables." In *POPL*. ACM Press, 1980, pp. 83–94 (p. 559).

[976] Mark Weiser. "Program Slicing." *IEEE Trans. Software Eng.* 10.4 (1984), pp. 352–357 (p. 591, 601, 618).

[977] Lukas Westhofen, Philipp Berger, and Joost-Pieter Katoen. "Benchmarking Software Model Checkers on Automotive Code." *CoRR*. abs/2003.11689 (2020) (p. 551, 715, 717).

[978] Alfred North Whitehead and Bertrand Russell. *Principia Mathematica, Volume I, II, III*. 2nd ed. Cambridge University Press, 1927 (p. 10).

[979] Adriaan Van Wijngaarden, B. J. Mailloux, J. E. L. Peck, Cornelis H. A. Koster, Michel Sintzoff, C. H. Lindsey, Lambert G. L. T. Meertens, and R. G. Fisker. "Revised Report on the Algorithmic Language ALGOL 68." *Acta Inf.* 5 (1975), pp. 1–236 (p. 349).

[980] Reinhard Wilhelm. "Program Analysis: A Toolmaker's Perspective." *ACM SIGPLAN Notices*. 32.1 (1997), pp. 120–121 (p. 716).

[981] Reinhard Wilhelm. "Software Quality Assurance by Static Program Analysis." In *SWQD*. Vol. 166. Lecture Notes in Business Information Processing. Springer, 2014, pp. 1–11 (p. 718).

[982] Reinhard Wilhelm, Markus Pister, Gernot Gebhard, and Daniel Kästner. "Testing Implementation Soundness of a WCET Analysis Tool." In *A Journey of Embedded and Cyber-Physical Systems*. Springer, 2021, pp. 5–17 (p. 717).

[983] Reinhard Wilhelm, Helmut Seidl, and Sebastian Hack. *Compiler Design - Syntactic and Semantic Analysis*. Springer, 2013 (p. 39).

[984] Reinhard Wilhelm and Björn Wachter. "Abstract Interpretation with Applications to Timing Validation." In *CAV*. Vol. 5123. Lecture Notes in Computer Science. Springer, 2008, pp. 22–36 (p. 595).

[985] Glynn Winskel. "A Note on Powerdomains and Modality." In *FCT*. Vol. 158. Lecture Notes in Computer Science. Springer, 1983, pp. 505–514 (p. 377).

[986] Niklaus Wirth. "The Programming Language Pascal." *Acta Inf.* 1 (1971), pp. 35–63 (p. 398, 556, 560).

[987] Stephen Wolfram. *Mathematica—A System for Doing Mathematics by Computer*. Addison-Wesley, 1988 (p. 290).

[988] Pierre Wolper. "Temporal Logic Can Be More Expressive." *Information and Control*. 56.1/2 (1983), pp. 72–99 (p. 527).

[989] Pierre Wolper and Bernard Boigelot. "Verifying Systems with Infinite but Regular State Spaces." In *CAV*. Vol. 1427. Lecture Notes in Computer Science. Springer, 1998, pp. 88–97 (p. 411).

[990] Meng Wu, Shengjian Guo, Patrick Schaumont, and Chao Wang. "Eliminating Timing Side-Channel Leaks Using Program Repair." In *ISSTA*. ACM, 2018, pp. 15–26 (p. 595).

[991] Meng Wu and Chao Wang. "Abstract Interpretation Under Speculative Execution." In *PLDI*. ACM, 2019, pp. 802–815 (p. 79).

[992] Xueguang Wu, Liqian Chen, and Ji Wang. "An Abstract Domain to Infer Symbolic Ranges Over Nonnegative Parameters." *Electr. Notes Theor. Comput. Sci.* 307 (2014), pp. 33–45 (p. 398).

[993] Wei Yao and Ling-Xia Lu. "Fuzzy Galois Connections on Fuzzy Posets." *Math. Log. Q.* 55.1 (2009), pp. 105–112 (p. 131).

[994] Banghu Yin. "Property Oriented Verification via Iterative Abstract Interpretation." In *ICSE (Companion Volume)*. IEEE, 2019, pp. 162–164 (p. 702, 710).

[995] Banghu Yin, Liqian Chen, Jiangchao Liu, Ji Wang, and Patrick Cousot. "Verifying Numerical Programs via Iterative Abstract Testing." In *SAS*. Vol. 11822. Lecture Notes in Computer Science. Springer, 2019, pp. 247–267 (p. 702, 707, 710).

[996] Greta Yorsh, Thomas W. Reps, and Mooly Sagiv. "Symbolically Computing Most-Precise Abstract Operations for Shape Analysis." In *TACAS*. Vol. 2988. Lecture Notes in Computer Science. Springer, 2004, pp. 530–545 (p. 195).

[997] Nengkun Yu and Jens Palsberg. "Quantum Abstract Interpretation." In *PLDI*. ACM, 2021 (p. 718).

[998] Yannick Zakowski. "Generation of Polynomial Inequalities as Invariants." Symbolic Computation [cs.SC]. 2013. dumas-00854834. 2013. https://dumas.ccsd.cnrs.fr/dumas-00854834/ (p. 457).

[999] Matteo Zanioli and Agostino Cortesi. "Information Leakage Analysis by Abstract Interpretation." In *SOFSEM*. Vol. 6543. Lecture Notes in Computer Science. Springer, 2011, pp. 545–557 (p. 617, 620).

[1000] Steve Zdancewic and Andrew C. Myers. "Observational Determinism for Concurrent Program Security." In *CSFW*. IEEE Computer Society, 2003, p. 29 (p. 593).

[1001] Ernst Zermelo. "Untersuchungen über die Grundlagen der Mengenlehre I." *Mathematische Annalen.* 65.2 (1908), pp. 261–281 (p. 172).

[1002] Ghiles Ziat, Alexandre Maréchal, Marie Pelleau, Antoine Miné, and Charlotte Truchet. "Combination of Boxes and Polyhedra Abstractions for Constraint Solving." In *FM Workshops (2).* Vol. 12233. Lecture Notes in Computer Science. Springer, 2019, pp. 119–135 (p. 432).

[1003] Lenore D. Zuck, Amir Pnueli, Benjamin Goldberg, Clark W. Barrett, Yi Fang, and Ying Hu. "Translation and Run-Time Validation of Loop Transformations." *Formal Methods in System Design.* 27.3 (2005), pp. 335–360 (p. 708).

Author Index

Index

Symbol Index

Project Index